Contents

3   Introduction to Graphing   147

Elementary and Intermediate Algebra

CONCEPTS AND APPLICATIONS

MARVIN L. BITTINGER
Indiana University Purdue University Indianapolis

DAVID J. ELLENBOGEN
Community College of Vermont

BARBARA L. JOHNSON
Indiana University Purdue University Indianapolis

ADDISON-WESLEY

Boston San Francisco New York
London Toronto Sydney Tokyo Singapore Madrid
Mexico City Munich Paris Cape Town Hong Kong Montreal

Editorial Director	Christine Hoag
Editor in Chief	Maureen O'Connor
Executive Project Manager	Kari Heen
Associate Editor	Joanna Doxey
Editorial Assistant	Jonathan Wooding
Production Manager	Ron Hampton
Composition and Production Services	Pre-Press PMG
Art Editor and Photo Researcher	The Davis Group, Inc.
Senior Media Producer	Ceci Fleming
Associate Producer	Jennifer Thomas
Software Development	Eileen Moore and Marty Wright
Marketing Manager	Marlana Voerster
Marketing Coordinator	Nathaniel Koven
Prepress Supervisor	Caroline Fell
Manufacturing Manager	Evelyn Beaton
Senior Manufacturing Buyer	Carol Melville
Senior Media Buyer	Ginny Michaud
Text Designer	The Davis Group, Inc.
Cover Designer	Beth Paquin
Cover Photograph	© Georgette Douwma, Photographer's Choice/Getty Images

Photo Credits
Photo credits appear on page 990.

Library of Congress Cataloging-in-Publication Data
Bittinger, Marvin L.
  Elementary and intermediate algebra : concepts and applications.
  — Fifth ed. / Marvin L. Bittinger, David J. Ellenbogen, Barbara L. Johnson.
    p. cm.
    Includes indexes.

1. Algebra—Textbooks.   I. Ellenbogen, David.   II. Johnson, Barbara L.
    QA152.3.B546 2010
    512. 9—dc22   2008024305

4 5 6 7 8 9 10—RRDJC—12 11

© 2010, 2006, 2002, 1998, 1994. Pearson Education, Inc.

Addison-Wesley
is an imprint of

PEARSON

www.pearsonhighered.com

ISBN-13: 978-0-321-55944-9
ISBN-10: 0-321-55944-4

4 Polynomials 227

7   Functions and Graphs   443

8   Systems of Linear Equations and Problem Solving   505

9   Inequalities and Problem Solving   577

12 Exponential and Logarithmic Functions 783

R   Elementary Algebra Review   939

Appendixes   991

Tables   1003

Preface

It is with great pleasure that we introduce you to the fifth edition of *Elementary and Intermediate Algebra: Concepts and Applications*. Our goal, as always, is to present content that is easy to understand and has the depth required for success in this and future courses. In this edition, faculty will recognize features, applications, and explanations that they have come to rely on and expect. Students and faculty will also find many changes resulting from our own ideas for improvement as well as insights from faculty and students throughout North America. Thus this new edition contains exciting new features and applications, along with updates and refinements to those from previous editions.

Appropriate for a course, or courses, combining the study of elementary and intermediate algebra, this text covers both elementary and intermediate algebra topics without the repetition of instruction necessary in two separate texts. It is one of three texts in an algebra series that also includes *Elementary Algebra: Concepts and Applications*, Eighth Edition, by Bittinger/Ellenbogen, and *Intermediate Algebra: Concepts and Applications*, Eighth Edition, by Bittinger/Ellenbogen.

Approach

Our goal, quite simply, is to help today's students both learn and retain mathematical concepts. To achieve this goal, we feel that we must prepare developmental-mathematics students for the transition from "skills-oriented" elementary and intermediate algebra courses to more "concept-oriented" college-level mathematics courses. This requires that we teach these same students critical thinking skills: to reason mathematically, to communicate mathematically, and to identify and solve mathematical problems. Following are three aspects of our approach that we use to help meet the challenges we all face when teaching developmental mathematics.

Problem Solving

One distinguishing feature of our approach is our treatment of and emphasis on problem solving. We use problem solving and applications to motivate the material wherever possible, and we include real-life applications and problem-solving techniques throughout the text. Problem solving not only encourages students to think about how mathematics can be used, it helps to prepare them for more advanced material in future courses.

In Chapter 2, we introduce our five-step process for solving problems: (1) Familiarize, (2) Translate, (3) Carry out, (4) Check, and (5) State the answer. These steps are then used consistently throughout the text when encountering a problem-solving situation. Repeated use of this problem-solving strategy helps provide students with a starting point for any type of problem they encounter, and frees them to focus on the unique aspects of the particular problem situation. We often use estimation and carefully checked guesses to help with the *Familiarize* and *Check* steps (see pp. 110 and 422–423).

Applications

Interesting applications of mathematics help motivate both students and instructors. Solving applied problems gives students the opportunity to see their conceptual understanding put to use in a real way. In the fifth edition of *Elementary and Intermediate Algebra: Concepts and Applications*, we have increased the number of applications, the number of real-data problems, and the number of reference lines that specify the sources of the real-world data. As in the past, art is integrated into the applications and exercises to aid the student in visualizing the mathematics. (See pp. 111, 190, 260, 364.)

Pedagogy

New!

TRY EXERCISES

Try Exercises. This icon concludes nearly every example by pointing students to one or more parallel exercises from the corresponding exercise set so that they can immediately reinforce the concepts and skills presented in the examples. For easy identification in the exercise sets, the "Try" exercises have a shaded block on the exercise number. (See pp. 56, 256, 415.)

New!

Translating for Success and **Visualizing for Success.** These matching exercises help students learn to associate word problems (through translation) and graphs (through visualization) with their appropriate mathematical equations. (See pp. 134, 361 (Translating); pp. 212, 753 (Visualizing).) Each feature contains a corresponding activity in MyMathLab.

Revised!

Connecting the Concepts. Revised and expanded to include new Mixed Review exercises, this midchapter review helps students understand the big picture and prepare for chapter tests and cumulative reviews by relating the concept at hand to previously learned and upcoming concepts. (See pp. 206, 279, 739.)

Revised!

Study Summary. Found at the end of each chapter and now presented in a two-column format organized by section, this synopsis gives students a fast and effective review of key chapter terms and concepts paired with accompanying examples. (See pp. 139, 218, 367.)

Revised!

Cumulative Review. This review now appears after every chapter to help students retain and apply their knowledge from previous chapters. (See pp. 145, 300, 574.)

Algebraic–Graphical Connections. This feature provides students with a way to visualize concepts that might otherwise prove elusive. (See pp. 350, 416, 704.)

Study Skills. This feature in the margin provides tips for successful study habits that even experienced students will appreciate. Ranging from time management to test preparation, these study skills can be applied in any college course. (See pp. 86, 229, 597.)

Student Notes. These notes in the margin give students extra explanation of the mathematics appearing on that page. These comments are more casual in format than the typical exposition and range from suggestions for avoiding common mistakes to how to best read new notation. (See pp. 79, 312, 728.)

Technology Connection. These optional boxes in each chapter help students use a graphing calculator to better visualize a concept that they have just learned. To connect this optional instruction to the exercise sets, certain exercises are marked with a graphing calculator icon ⊡ to indicate the optional use of technology. (See pp. 164, 351, 791.)

Revised!

Concept Reinforcement Exercises. Now with all answers listed in the answer section at the back of the book, these section and review exercises build students' confidence and comprehension through true/false, matching, and fill-in-the-blank exercises at the start of most exercise sets. To help further student understanding, emphasis is given to new vocabulary and notation developed in the section. (See pp. 10, 165, 760.)

Aha!

Aha! Exercises. These exercises are not more difficult than their neighboring exercises and can be solved quickly, without going through a lengthy computation, if the student has the proper insight. Designed to reward students who "look before they leap," the icon indicates the first time a new insight applies, and then it is up to the student to determine when to use the Aha! method on subsequent exercises. (See pp. 213, 285, 730.)

Revised!

Skill Review Exercises. These exercises, included in Section 1.2 and every section thereafter, review skills and concepts from preceding sections of the text. In most cases, these exercises prepare students for the next section. An introduction to each set directs students to the appropriate sections to review if necessary. On occasion, Skill Review exercises focus on a single topic in greater depth and from multiple perspectives. (See pp. 166, 243, 594.)

Synthesis Exercises.
Synthesis exercises follow the Skill Review exercises at the end of each exercise set. Generally more challenging, these exercises synthesize skills and concepts from earlier sections with the present material, often providing students with deeper insight into the current topic. Aha! exercises are sometimes included as Synthesis exercises. (See pp. 99, 365, 714.)

 ### Writing Exercises.
These appear just before the Skill Review exercises (two basic writing exercises) and also in the Synthesis exercises (at least two more challenging exercises). Writing exercises aid student comprehension by requiring students to use critical thinking to provide explanations of concepts in one or more complete sentences. Because some instructors may collect answers to writing exercises and because more than one answer can be correct, only answers to writing exercises in the review section are included at the back of the text. (See pp. 58, 473, 686.)

Collaborative Corner.
These optional activities for students to explore together usually appear two to three times per chapter at the end of an exercise set. Studies show that students who study in groups generally outperform those who do not, so these exercises are for students who want to solve mathematical problems together. Additional collaborative activities and suggestions for directing collaborative learning appear in the *Instructor and Adjunct Support Manual*. (See pp. 158, 537, 766.)

What's New in the Fifth Edition?

We have rewritten many key topics in response to user and reviewer feedback and have made significant improvements in design, art, pedagogy, and an expanded supplements package. Detailed information about the content changes is available in the form of a conversion guide. Please ask your local Pearson sales consultant for more information. Following is a list of the major changes in this edition.

NEW DESIGN

While incorporating a new layout, a fresh palette of colors, and new features, we have a larger page dimension for an open look and a typeface that is easy to read. As always, it is our goal to make the text look mature without being intimidating. In addition, we continue to pay close attention to the pedagogical use of color to make sure that it is used to present concepts in the clearest possible manner.

CONTENT CHANGES

A variety of content changes have been made throughout the text. Some of the more significant changes are listed below.

What's New in Combined

- Examples and exercises that use real data are updated or replaced with current applications.
- Over 35% of the exercises are new or updated.
- Quick-glance reminders for multistep process are included next to examples. These appear by one multistep example of each type. (See pp. 197, 333, 519.)
- Chapter 2 now includes increased practice of solving for y in a formula.
- Interval notation is introduced when students first solve inequalities in Section 2.6.
- Inequalities are now graphed on number lines using brackets and parentheses. Interval notation can thus be read directly from the graph of an inequality.
- Chapter 3 now gives increased emphasis to units when finding a rate of change.

- Discussion of negative exponents (Section 4.2) now immediately follows the introduction to the rules for manipulating exponents.
- Chapter 5 now makes greater use of prime factorizations as a tool for finding the largest common factor.
- Domains of radical functions are now discussed in Section 9.1, separately from domains of rational functions in Section 9.2.
- The distance formula is now presented in Section 10.7 as one application of the Pythagorean theorem.
- In Chapter 11, the discussion of the discriminant now directly follows the quadratic formula.

ANCILLARIES

The following ancillaries are available to help both instructors and students use this text more effectively.

STUDENT SUPPLEMENTS

New! Chapter Test Prep Video CD

- Watch instructors work through step-by-step solutions to all the chapter test exercises from the textbook. The Chapter Test Prep Video CD is included with each new student text.

New! Worksheets for Classroom or Lab Practice

by Carrie Green
These lab- and classroom-friendly workbooks offer the following resources for every section of the text:

- A list of learning objectives;
- Vocabulary practice problems;
- Extra practice exercises with ample work space.

ISBNs: 0-321-59933-0 and 978-0-321-59933-9

Student's Solutions Manual

by Christine S. Verity

- Contains completely worked-out solutions with step-by-step annotations for all the odd-numbered exercises in the text, with the exception of the writing exercises.
- New! Now contains all solutions to Chapter Review, Chapter Test, and Connecting the Concepts exercises.

ISBNs: 0-321-58623-9 and 978-0-321-58623-0

INSTRUCTOR SUPPLEMENTS

Annotated Instructor's Edition

- Provides answers to all text exercises in color next to the corresponding problems.
- Includes Teaching Tips.
- Icons identify writing 🖋 and graphing calculator 📈 exercises.

ISBNs: 0-321-56726-9 and 978-0-321-56726-0

Instructor's Solutions Manual

by Christine S. Verity

- Contains fully worked-out solutions to the odd-numbered exercises and brief solutions to the even-numbered exercises in the exercise sets.
- Available for download at www.pearsonhighered.com

ISBNs: 0-321-58620-4 and 978-0-321-58620-9

Instructor and Adjunct Support Manual

- Includes resources designed to help both new and adjunct faculty with course preparation and classroom management.
- Offers helpful teaching tips correlated to the sections of the text.

ISBNs: 0-321-58624-7 and 978-0-321-58624-7

Videos on DVD

- A complete set of digitized videos on DVD for use at home or on campus.
- Includes a full lecture for each section of the text, many presented by author team members David J. Ellenbogen and Barbara Johnson.
- Optional subtitles in English are available.

ISBNs: 0-321-59935-7 and 978-0-321-59935-3

InterAct Math® Tutorial Website

www.interactmath.com

- Online practice and tutorial help.
- Retry an exercise with new values each time for unlimited practice and mastery.
- Every exercise is accompanied by an interactive guided solution that gives helpful feedback when an incorrect answer is entered.
- View the steps of a worked-out sample problem similar to those in the text.

Printable Test Bank

by Laurie Hurley

- Contains two multiple-choice tests per chapter, six free-response tests per chapter, and eight final exams.
- Available for download at www.pearsonhighered.com

PowerPoint® Lecture Slides

- Present key concepts and definitions from the text.
- Available for download at www.pearsonhighered.com

TestGen

www.pearsonhighered.com/testgen

- Enables instructors to build, edit, print, and administer tests using a computerized bank of questions developed to cover all text objectives.
- Algorithmically based, TestGen allows instructors to create multiple but equivalent versions of the same question or test with the click of a button.
- Instructors can also modify test bank questions or add new questions.
- Tests can be printed or administered online.

Pearson Math Adjunct Support Center

http://www.pearsontutorservices.com/math-adjunct.html

Staffed by qualified instructors with more than 50 years of combined experience at both the community college and university levels, this center provides assistance for faculty in the following areas:

- Suggested syllabus consultation;
- Tips on using materials packed with the text;
- Book-specific content assistance;
- Teaching suggestions, including advice on classroom strategies.

AVAILABLE FOR STUDENTS AND INSTRUCTORS

MyMathLab® Online Course (access code required)

MyMathLab is a series of text-specific, easily customizable online courses for Pearson Education's textbooks in mathematics and statistics. Powered by CourseCompass™ (our online teaching and learning environment) and MathXL® (our online homework, tutorial, and assessment system), MyMathLab gives you the tools you need to deliver all or a portion of your course online, whether your students are in a lab setting or working from home. MyMathLab provides a rich and flexible set of course materials, featuring free-response exercises that are algorithmically generated for unlimited practice and mastery. Students can also use online tools, such as video lectures, animations, and a multimedia textbook, to independently improve their understanding and performance. Instructors can use MyMathLab's homework and test managers to select and assign online exercises correlated directly to the textbook, and they can also create and assign their own online exercises and import TestGen tests for added flexibility. MyMathLab's online gradebook—designed specifically for mathematics and statistics—automatically tracks students' homework and test results and gives the instructor control over how to calculate final grades. Instructors can also add offline (paper-and-pencil) grades to the gradebook. MyMathLab also includes access to the **Pearson Tutor Center** (www.pearsontutorservices.com). The Tutor Center is staffed by qualified mathematics instructors who provide textbook-specific tutoring for students via toll-free phone, fax, e-mail, and interactive Web sessions. MyMathLab is available to qualified adopters. For more information, visit our website at www.mymathlab.com or contact your sales representative.

MathXL® Online Course (access code required)

MathXL® is a powerful online homework, tutorial, and assessment system that accompanies Pearson Education's textbooks in mathematics or statistics. With MathXL, instructors can create, edit, and assign online homework and tests using algorithmically generated exercises correlated at the objective level to the textbook. They can also create and assign their own online exercises and import TestGen tests for added flexibility. All student work is tracked in MathXL's online gradebook. Students can take chapter tests in MathXL and receive personalized study plans based on their test results. The study plan diagnoses weaknesses and links students directly to tutorial exercises for the objectives they need to study and retest. Students can also access supplemental animations and video clips directly from selected exercises. MathXL is available to qualified adopters. For more information, visit our website at www.mathxl.com, or contact your Pearson sales representative.

MathXL® Tutorials on CD

This interactive tutorial CD-ROM provides algorithmically generated practice exercises that are correlated at the objective level to the exercises in the textbook. Every practice exercise is accompanied by an example and a guided solution designed to involve students in the solution process. Selected exercises may also include a video clip to help students visualize concepts. The software provides helpful feedback for incorrect answers and can generate printed summaries of students' progress.

Acknowledgments

No book can be produced without a team of professionals who take pride in their work and are willing to put in long hours. Laurie Hurley, in particular, deserves extra thanks for her work as developmental editor. Rebecca Hubiak, Laurie Hurley, Holly Martinez, Ann Ostberg, and Christine Verity also deserve special thanks for their careful accuracy checks, well-thought-out suggestions, and uncanny eye for detail. Thanks to Carrie Green, Laurie Hurley, and Christine Verity for their outstanding work in preparing supplements.

We are also indebted to Chris Burditt and Jann MacInnes for their many fine ideas that appear in our Collaborative Corners and Vince McGarry and Janet Wyatt for their recommendations for Teaching Tips featured in the Annotated Instructor's Edition.

Geri Davis, of the Davis Group, Inc., performed superb work as designer, art editor, and photo researcher, and is always a pleasure to work with. Tracy Duff and her colleagues at Pre-Press PMG provided excellent composition and editorial support throughout the production process. Network Graphics generated the graphs, charts, and many of the illustrations. Not only are the people at Network reliable, but they clearly take pride in their work. The many illustrations appear thanks to Bill Melvin—an artist with insight and creativity.

Our team at Pearson deserves special thanks. Acquisitions Editor Randy Welch provided many fine suggestions, remaining involved and accessible throughout the project. Executive Project Manager Kari Heen carefully coordinated tasks and schedules, keeping a widely spread team working together. Associate Editor Joanna Doxey coordinated reviews and assisted in a variety of tasks with patience and creativity. Editorial Assistant Jonathan Wooding responded quickly to all requests, always in a pleasant manner. Production Manager Ron Hampton's attention to detail, willingness to listen, and creative responses helped result in a book that is beautiful to look at. Marketing Manager Marlana Voerster and Marketing Assistant Nathaniel Koven skillfully kept us in touch with the needs of faculty. Our Editor in Chief, Maureen O'Connor, and Editorial Director, Chris Hoag, deserve credit for assembling this fine team.

We also thank the students at Indiana University Purdue University Indianapolis and the Community College of Vermont and the following professors for their thoughtful reviews and insightful comments.

Elementary Algebra: Concepts and Applications, Eighth Edition

Roberta Abarca, *Centralia College*
Darla J. Aguilar, *Pima Community College, Desert Vista Campus*
Bonnie Alcorn, *Waubonsee College*
Eugene Alderman, *South University*
Joseph Berland, *Chabot College*
Paul Blankenship, *Lexington Community College*
Susan Caldiero, *Cosumnes River College*
David Casey, *Citrus College*
Emmett Dennis, *Southern Connecticut State University*
Henri Feiner, *Coastline Community College*
Gary Glaze, *Spokane Falls Community College*
Janet Hansen, *Dixie State College*
Elizabeth Hodes, *Santa Barbara City College*
Weilin Jang, *Austin Community College*
Paulette Kirkpatrick, *Wharton County Junior College*
Susan Knights, *Boise State University*
Jeff Koleno, *Lorain County Community College*
Julianne Labbiento, *Lehigh Carbon Community College*
Kathryn Lavelle, *Westchester Community College*
Amy Marolt, *Northeastern Mississippi Community College*
Rogers Martin, *Louisiana State University, Shreveport*
Ben Mayo, *Yakima Valley Community College*
Laurie McManus, *St. Louis Community College–Meramac*
Carol Metz, *Westchester Community College*
Anne Marie Mosher, *St. Louis Community College–Florissant Valley*
Pedro Mota, *Austin Community College, South Austin Campus*
Brenda M. Norman, *Tidewater Community College*
Kim Nunn, *Northeast State Technical College*
Michael Oppedisano, *Morrisville College SUNY*
Zaddock B. Reid, *San Bernardino Valley College*
Terry Reeves, *Red Rocks Community College*
Terri Seiver, *San Jacinto College–Central*
Timothy Thompson, *Oregon Institute of Technology*
Diane Trimble, *Tulsa Community College, West Campus*
Jennifer Vanden Eynden, *Grossmont College*
Beverly Vredevelt, *Spokane Falls Community College*
Michael Yarbrough, *Cosumnes River College*

Intermediate Algebra: Concepts and Applications, Eighth Edition

Marie Aratari, *Oakland Community College–Orange Ridge Campus*
Barbara Armenta, *Pima Community College*
Douglas Brozovic, *University of North Texas*
Barbara Burke, *Hawaii Pacific University*
Laura Burris, *Sam Houston State University*
Lisa Carnell, *High Point University*
Sharon Edgmon, *Bakersfield College*
Karen Ernst, *Hawkeye College*
Kathy Garrison, *Clayton College and State University*
Cynthia Harrison, *Baton Rouge Community College*
Tracey L. Johnson, *University of Georgia*
Joanne Kawczenski, *Luzerne County Community College*
Rachel Lamp, *North Iowa Area Community College*
Kevin J. Leith, *Central New Mexico Community College*
Stephanie Lochbaum, *Austin Community College*

Debi McCandrew, *Florence-Darlington Technical College*
Bob McCarthy, *Community College of Allegheny County—South Campus*
Doug Mace, *Kirtland Community College*
Timothy McKenna, *University of Michigan–Dearborn*
Rhea Meyerholtz, *Indiana State University*
Bronte Miller, *Patrick Henry Community College*
Kausha Miller, *Lexington Community College*
Rebecca Parrish, *Ohio University*
Kay Petrash, *Sam Houston State University*
Debra Pharo, *Northwestern Michigan College*
Terry Reeves, *Red Rocks Community College*
Kathy Rod, *Wharton County Junior College*
Nicole Saporito, *Luzerne Community College*
Elgin Schilhab, *Austin Community College*
M. Terry Simon, *University of Toledo*
Fran Smith, *Oakland Community College*
Donald Soloman, *University of Wisconsin–Milwaukee*

***Elementary and Intermediate Algebra: Concepts and Applications,* Fifth Edition**

Michael Anzzolin, *Waubonsee Community College*
Jan Archibald, *Ventura College*
Don Brown, *Macon State College*
Gary Carpenter, *Pima Community College, Northwest Campus*
Tim Chappell, *Penn Valley Community College*
Ola Disu, *Tarrant County College*
Anissa Florence, *University of Louisville*
Sandy Gordon, *Central Carolina Technical College*
Sharon Hamsa, *Longview Community College*
Geoffrey Hirsch, *Ohlone College*
Pat Horacek, *Pensacola Junior College*
Sally Keely, *Clark College*
Ana Leon, *Louisville Community College*
Linda Lohman, *Jefferson Community College*
Bob Martin, *Tarrant County College*
Amy Petty, *South Suburban College*
Thomas Pulver, *Waubonsee Community College*
Angela Redmon, *Wenatchee Valley College*
Richard Rupp, *Del Mar College*
Mehdi Sadatmousavi, *Pima Community College*
Ann Thrower, *Kilgore College*

Finally, a special thank-you to all those who so generously agreed to discuss their professional use of mathematics in our chapter openers. These dedicated people all share a desire to make math more meaningful to students. We cannot imagine a finer set of role models.

M.L.B.
D.J.E.
B.L.J.

Introduction to Algebraic Expressions

BRIAN BUSBY
CHIEF METEOROLOGIST
Kansas City, Missouri

All weather measurements are a series of numbers and values. Temperature, relative humidity, wind speed and direction, precipitation amount, and air pressure are all expressed in various numbers and percentages. Because weather systems move north and south, east and west, up and down, *and* over time, high-level math like calculus is the only way to represent that movement. But before you study calculus, you must begin with algebra.

AN APPLICATION

On December 10, Jenna notes that the temperature is −3°F at 6:00 A.M. She predicts that the temperature will rise at a rate of 2° per hour for 3 hr, and then rise at a rate of 3° per hour for 6 hr. She also predicts that the temperature will then fall at a rate of 2° per hour for 3 hr, and then fall at a rate of 5° per hour for 2 hr. What is Jenna's temperature forecast for 8:00 P.M.?

This problem appears as Exercise 135 in Section 1.7.

P roblem solving is the focus of this text. Chapter 1 presents important preliminaries that are needed for the problem-solving approach that is developed in Chapter 2 and used throughout the rest of the book. These preliminaries include a review of arithmetic, a discussion of real numbers and their properties, and an examination of how real numbers are added, subtracted, multiplied, divided, and raised to powers.

1.1   Introduction to Algebra

Algebraic Expressions   ▪   Translating to Algebraic Expressions   ▪   Translating to Equations

This section introduces some basic concepts and expressions used in algebra. Solving real-world problems is an important part of algebra, so we will focus on the wordings and mathematical expressions that often arise in applications.

Algebraic Expressions

Probably the greatest difference between arithmetic and algebra is the use of *variables* in algebra. When a letter can be any one of a set of numbers, that letter is a **variable**. For example, if n represents the number of tickets purchased for a Maroon 5 concert, then n will vary, depending on factors like price and day of the week. This makes n a variable. If each ticket costs \$40, then 3 tickets cost $40 \cdot 3$ dollars, 4 tickets cost $40 \cdot 4$ dollars, and n tickets cost $40 \cdot n$, or $40n$ dollars. Note that both $40 \cdot n$ and $40n$ mean 40 *times* n. The number 40 is an example of a **constant** because it does not change.

Price per Ticket (in dollars)	Number of Tickets Purchased	Total Paid (in dollars)
40	n	$40n$

The expression $40n$ is a **variable expression** because its value varies with the replacement for n. In this case, the total amount paid, $40n$, will change with the number of tickets purchased. In the following chart, we replace n with a variety of values and compute the total amount paid. In doing so, we are **evaluating the expression** $40n$.

Price per Ticket (in dollars), 40	Number of Tickets Purchased, n	Total Paid (in dollars), $40n$
40	400	$16,000
40	500	20,000
40	600	24,000

Variable expressions are examples of *algebraic expressions*. An **algebraic expression** consists of variables and/or numerals, often with operation signs and grouping symbols. Examples are

$$t + 97, \quad 5 \cdot x, \quad 3a - b, \quad 18 \div y, \quad \frac{9}{7}, \quad \text{and} \quad 4r(s + t).$$

Recall that a fraction bar is a division symbol: $\frac{9}{7}$, or 9/7, means $9 \div 7$. Similarly, multiplication can be written in several ways. For example, "5 times x" can be written as $5 \cdot x, 5 \times x, 5(x)$, or simply $5x$. On many calculators, this appears as $5 * x$.

To **evaluate** an algebraic expression, we substitute a number for each variable in the expression. We then calculate the result.

EXAMPLE **1**

Evaluate each expression for the given values.

a) $x + y$ for $x = 37$ and $y = 28$

b) $5ab$ for $a = 2$ and $b = 3$

SOLUTION

a) We substitute 37 for x and 28 for y and carry out the addition:

$$x + y = 37 + 28 = 65.$$

The number 65 is called the **value** of the expression.

b) We substitute 2 for a and 3 for b and multiply:

$$5ab = 5 \cdot 2 \cdot 3 = 10 \cdot 3 = 30. \qquad 5ab \text{ means 5 times } a \text{ times } b.$$

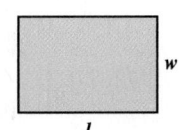 TRY EXERCISE 17

STUDENT NOTES

As we will see later, it is sometimes necessary to use parentheses when substituting a number for a variable. You may wish to use parentheses whenever you substitute. In Example 1, we could write

$$x + y = (37) + (28) = 65$$

and

$$5ab = 5(2)(3) = 30.$$

EXAMPLE **2**

The area A of a rectangle of length l and width w is given by the formula $A = lw$. Find the area when l is 17 in. and w is 10 in.

SOLUTION   We evaluate, using 17 in. for l and 10 in. for w, and carry out the multiplication:

$$A = lw$$
$$A = (17 \text{ in.})(10 \text{ in.})$$
$$A = (17)(10)(\text{in.})(\text{in.})$$
$$A = 170 \text{ in}^2, \text{ or } 170 \text{ square inches.}$$

Note that we always use square units for area and $(\text{in.})(\text{in.}) = \text{in}^2$. Exponents like the 2 within the expression in^2 are discussed further in Section 1.8.

TRY EXERCISE 25

EXAMPLE 3   The area of a triangle with a base of length b and a height of length h is given by the formula $A = \frac{1}{2}bh$. Find the area when b is 8 m (meters) and h is 6.4 m.

SOLUTION   We substitute 8 m for b and 6.4 m for h and then multiply:

$$A = \frac{1}{2}bh$$
$$A = \frac{1}{2}(8\,\text{m})(6.4\,\text{m})$$
$$A = \frac{1}{2}(8)(6.4)(\text{m})(\text{m})$$
$$A = 4(6.4)\,\text{m}^2$$
$$A = 25.6\,\text{m}^2, \text{ or } 25.6 \text{ square meters.}$$

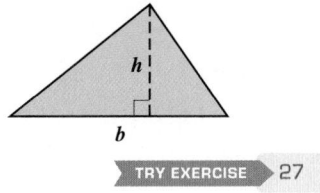

TRY EXERCISE 27

Translating to Algebraic Expressions

Before attempting to translate problems to equations, we need to be able to translate certain phrases to algebraic expressions.

Important Words	Sample Phrase or Sentence	Translation
Addition (+)		
added to	700 pounds was added to the car's weight.	$w + 700$
sum of	The sum of a number and 12	$n + 12$
plus	53 plus some number	$53 + x$
more than	800 more than Biloxi's population	$p + 800$
increased by	Ty's original estimate, increased by 4	$n + 4$
Subtraction (−)		
subtracted from	2 ounces was subtracted from the bag's weight.	$w - 2$
difference of	The difference of two scores	$m - n$
minus	A team of size s, minus 2 injured players	$s - 2$
less than	9 less than the number of volunteers last month	$v - 9$
decreased by	The car's speed, decreased by 8 mph	$s - 8$
Multiplication (·)		
multiplied by	The number of reservations, multiplied by 3	$r \cdot 3$
product of	The product of two numbers	$m \cdot n$
times	5 times the dog's weight	$5w$
twice	Twice the wholesale cost	$2c$
of	$\frac{1}{2}$ of Amelia's salary	$\frac{1}{2}s$
Division (÷)		
divided by	A 2-pound coffee cake, divided by 3	$2 \div 3$
quotient of	The quotient of 14 and 7	$14 \div 7$
divided into	4 divided into the delivery fee	$f \div 4$
ratio of	The ratio of $500 to the price of a new car	$500/p$
per	There were 18 computers per class of size s.	$18/s$

Any variable can be used to represent an unknown quantity; however, it is helpful to choose a descriptive letter. For example, w suggests weight and p suggests population or price. It is important to write down what the chosen variable represents.

EXAMPLE **4**   Translate each phrase to an algebraic expression.

a) Four less than Ava's height, in inches

b) Eighteen more than a number

c) A day's pay, in dollars, divided by eight

SOLUTION   To help think through a translation, we sometimes begin with a specific number in place of a variable.

a) If the height were 60, then 4 less than 60 would mean $60 - 4$. If the height were 65, the translation would be $65 - 4$. If we use h to represent "Ava's height, in inches," the translation of "Four less than Ava's height, in inches" is $h - 4$.

b) If we knew the number to be 10, the translation would be $10 + 18$, or $18 + 10$. If we use t to represent "a number," the translation of "Eighteen more than a number" is

$$t + 18, \quad \text{or} \quad 18 + t.$$

c) We let d represent "a day's pay, in dollars." If the pay were $78, the translation would be $78 \div 8$, or $\frac{78}{8}$. Thus our translation of "A day's pay, in dollars, divided by eight" is

$$d \div 8, \quad \text{or} \quad \frac{d}{8}.$$

 TRY EXERCISE 31

CAUTION!   The order in which we subtract and divide affects the answer! Answering $4 - h$ or $8 \div d$ in Examples 4(a) and 4(c) is incorrect.

EXAMPLE **5**   Translate each phrase to an algebraic expression.

a) Half of some number

b) Seven more than twice the weight

c) Six less than the product of two numbers

d) Nine times the difference of a number and 10

e) Eighty-two percent of last year's enrollment

SOLUTION

Phrase	Variable(s)	Algebraic Expression
a) Half of some number	Let n represent the number.	$\frac{1}{2}n$, or $\frac{n}{2}$, or $n \div 2$
b) Seven more than twice the weight	Let w represent the weight.	$2w + 7$, or $7 + 2w$
c) Six less than the product of two numbers	Let m and n represent the numbers.	$mn - 6$
d) Nine times the difference of a number and 10	Let a represent the number.	$9(a - 10)$
e) Eighty-two percent of last year's enrollment	Let r represent last year's enrollment.	82% of r, or $0.82r$

TRY EXERCISE 45

Translating to Equations

The symbol = ("equals") indicates that the expressions on either side of the equals sign represent the same number. An **equation** is a number sentence with the verb =. Equations may be true, false, or neither true nor false.

EXAMPLE 6   Determine whether each equation is true, false, or neither.

a) $8 \cdot 4 = 32$ **b)** $7 - 2 = 4$ **c)** $x + 6 = 13$

SOLUTION

a) $8 \cdot 4 = 32$ The equation is *true*.

b) $7 - 2 = 4$ The equation is *false*.

c) $x + 6 = 13$ The equation is *neither* true nor false, because we do not know what number x represents.

> ### Solution
>
> A replacement or substitution that makes an equation true is called a *solution*. Some equations have more than one solution, and some have no solution. When all solutions have been found, we have *solved* the equation.

To see if a number is a solution, we evaluate all expressions in the equation. If the values on both sides of the equation are the same, the number is a solution.

EXAMPLE 7   Determine whether 7 is a solution of $x + 6 = 13$.

SOLUTION   We evaluate $x + 6$ and compare both sides of the equation.

$$\frac{x + 6 = 13}{7 + 6 \ \bigl|\ 13}$$
$$13 \stackrel{?}{=} 13$$

Writing the equation

Substituting 7 for x

$13 = 13$ is TRUE.

Since the left-hand side and the right-hand side are the same, 7 is a solution.

> TRY EXERCISE ▶ 57

Although we do not study solving equations until Chapter 2, we can translate certain problem situations to equations now. The words "is the same as," "equal," "is," and "are" often translate to "=."

> **Words indicating equality, = :** "is the same as," "equal," "is," "are"

When translating a problem to an equation, we translate phrases to algebraic expressions, and the entire statement to an equation containing those expressions.

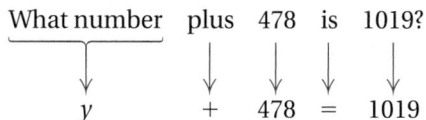   **EXAMPLE   8**   Translate the following problem to an equation.

What number plus 478 is 1019?

SOLUTION   We let y represent the unknown number. The translation then comes almost directly from the English sentence.

$$
\begin{array}{ccccc}
\underbrace{\text{What number}} & \text{plus} & 478 & \text{is} & 1019? \\
\downarrow & \downarrow & \downarrow & \downarrow & \downarrow \\
y & + & 478 & = & 1019
\end{array}
$$

Note that "what number plus 478" translates to "$y + 478$" and "is" translates to "=."

> TRY EXERCISE 63

Sometimes it helps to reword a problem before translating.

EXAMPLE   9   Translate the following problem to an equation.

The Taipei Financial Center, or Taipei 101, in Taiwan is the world's tallest building. At 1666 ft, it is 183 ft taller than the Petronas Twin Towers in Kuala Lumpur. How tall are the Petronas Twin Towers?

Source: *Guinness World Records* 2007

SOLUTION   We let h represent the height, in feet, of the Petronas Towers. A rewording and translation follow:

$$
\begin{array}{ccc}
\textit{Rewording:} & \underbrace{\begin{array}{c}\text{The height of} \\ \text{Taipei 101}\end{array}} \quad \text{is} & \underbrace{\begin{array}{c}\text{183 ft more than the height} \\ \text{of the Petronas Towers}\end{array}} \\
& \downarrow \qquad\qquad \downarrow & \downarrow \\
\textit{Translating:} & 1666 \qquad = & h + 183
\end{array}
$$

> TRY EXERCISE 69

TECHNOLOGY CONNECTION

Technology Connections are activities that make use of features that are common to most graphing calculators. In some cases, students may find the user's manual for their particular calculator helpful for exact keystrokes.

Although all graphing calculators are not the same, most share the following characteristics.

Screen.   The large screen can show graphs and tables as well as the expressions entered. The screen has a different layout for different functions. Computations are performed in the **home screen**. On many calculators, the home screen is accessed by pressing **2ND** (QUIT). The **cursor** shows location on the screen, and the **contrast** (set by **2ND** ⌃ or **2ND** ⌄) determines how dark the characters appear.

Keypad.   There are options written above the keys as well as on them. To access those above the keys, we press **2ND** or **ALPHA** and then the key. Expressions are usually entered as they would appear in print. For example, to evaluate $3xy + x$ for $x = 65$ and $y = 92$, we press 3 (×) 65 (×) 92 (+) 65 and then **ENTER**. The value of the expression, 18005, will appear at the right of the screen.

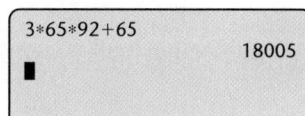

Evaluate each of the following.

1. $27a - 18b$, for $a = 136$ and $b = 13$
2. $19xy - 9x + 13y$, for $x = 87$ and $y = 29$

STUDY SKILLS

Get the Facts

Throughout this textbook, you will find a feature called Study Skills. These tips are intended to help improve your math study skills. On the first day of class, you should complete this chart.

Instructor: Name _____

Office hours and location _____

Phone number _____

Fax number _____

E-mail address _____

Find the names of two students whom you could contact for information or study questions:

1. Name _____

   Phone number _____

   E-mail address _____

2. Name _____

   Phone number _____

   E-mail address _____

Math lab on campus:

Location _____

Hours _____

Phone _____

Tutoring:

Campus location _____

Hours _____

Important supplements:

(See the preface for a complete list of available supplements.)

Supplements recommended by the instructor.

_____

_____

Translating for Success

1. Twice the difference of a number and 11

2. The product of a number and 11 is 2.

3. Twice the difference of two numbers is 11.

4. The quotient of twice a number and 11

5. The quotient of 11 and the product of two numbers

Translate to an expression or an equation and match that translation with one of the choices A–O below. Do not solve.

A. $x = 0.2(11)$

B. $\dfrac{2x}{11}$

C. $2x + 2 = 11$

D. $2(11x + 2)$

E. $11x = 2$

F. $0.2x = 11$

G. $11(2x - y)$

H. $2(x - 11)$

I. $11 + 2x = 2$

J. $2x + y = 11$

K. $2(x - y) = 11$

L. $11(x + 2x)$

M. $2(x + y) = 11$

N. $2 + \dfrac{x}{11}$

O. $\dfrac{11}{xy}$

Answers on page A-1

An additional, animated version of this activity appears in MyMathLab. To use MyMathLab, you need a course ID and a student access code. Contact your instructor for more information.

6. Eleven times the sum of a number and twice the number

7. Twice the sum of two numbers is 11.

8. Two more than twice a number is 11.

9. Twice the sum of 11 times a number and 2

10. Twenty percent of some number is 11.

1.1   EXERCISE SET

 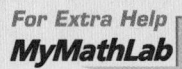

👈 *Concept Reinforcement* *Classify each of the following as either an expression or an equation.*

1. $10n - 1$

2. $3x = 21$

3. $2x - 5 = 9$

4. $5(x + 2)$

5. $38 = 2t$

6. $45 = a - 1$

7. $4a - 5b$

8. $3s + 4t = 19$

9. $2x - 3y = 8$

10. $12 - 4xy$

11. $r(t + 7) + 5$

12. $9a + b$

To the student and the instructor: *The* TRY EXERCISES *for examples are indicated by a shaded block* ▢ *on the exercise number. Complete step-by-step solutions for these exercises appear online at www.pearsonhighered.com/ bittingerellenbogen.*

Evaluate.

13. $5a$, for $a = 9$

14. $11y$, for $y = 7$

15. $12 - r$, for $r = 4$

16. $t + 8$, for $t = 2$

17. $\dfrac{a}{b}$, for $a = 45$ and $b = 9$

18. $\dfrac{c + d}{3}$, for $c = 14$ and $d = 13$

19. $\dfrac{x + y}{4}$, for $x = 2$ and $y = 14$

20. $\dfrac{m}{n}$, for $m = 54$ and $n = 9$

21. $\dfrac{p - q}{7}$, for $p = 55$ and $q = 20$

22. $\dfrac{9m}{q}$, for $m = 6$ and $q = 18$

23. $\dfrac{5z}{y}$, for $z = 9$ and $y = 15$

24. $\dfrac{m - n}{2}$, for $m = 20$ and $n = 8$

Substitute to find the value of each expression.

25. *Hockey.*   The area of a rectangle with base b and height h is bh. A regulation hockey goal is 6 ft wide and 4 ft high. Find the area of the opening.

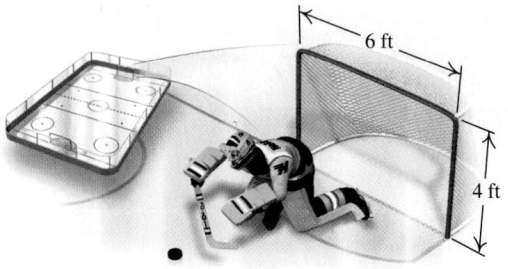

26. *Orbit time.*   A communications satellite orbiting 300 mi above the earth travels about 27,000 mi in one orbit. The time, in hours, for an orbit is

$$\frac{27{,}000}{v},$$

where v is the velocity, in miles per hour. How long will an orbit take at a velocity of 1125 mph?

27. *Zoology.*   A great white shark has triangular teeth. Each tooth measures about 5 cm across the base and has a height of 6 cm. Find the surface area of the front side of one such tooth. (See Example 3.)

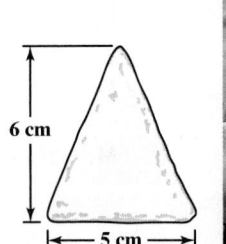 

28. *Work time.*   Javier takes three times as long to do a job as Luis does. Suppose t represents the time it takes Luis to do the job. Then $3t$ represents the time it takes Javier. How long does it take Javier if Luis takes **(a)** 30 sec? **(b)** 90 sec? **(c)** 2 min?

29. *Women's softball.*   A softball player's batting average is h/a, where h is the number of hits and a is the number of "at bats." In the 2007 Women's College World Series, Caitlin Lowe of the Arizona Wildcats had 10 hits in 29 at bats. What was her batting average? Round to the nearest thousandth.

30. *Area of a parallelogram.*   The area of a parallelogram with base *b* and height *h* is *bh*. Find the area of the parallelogram when the height is 6 cm (centimeters) and the base is 7.5 cm.

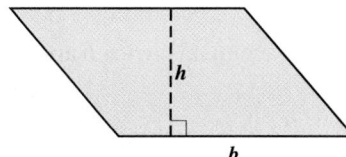

Translate to an algebraic expression.

31. 5 more than Ron's age

32. The product of 4 and *a*

33. 6 times *b*

34. 7 more than Lori's weight

35. 9 less than *c*

36. 4 less than *d*

37. 6 increased by *q*

38. 11 increased by *z*

39. 8 times Mai's speed

40. *m* subtracted from *n*

41. *x* less than *y*

42. 2 less than Than's age

43. *x* divided by *w*

44. The quotient of two numbers

45. The sum of the box's length and height

46. The sum of *d* and *f*

47. The product of 9 and twice *m*

48. Pemba's speed minus twice the wind speed

49. Thirteen less than one quarter of some number

50. Four less than ten times a number

51. Five times the difference of two numbers

52. One third of the sum of two numbers

53. 64% of the women attending

54. 38% of a number

Determine whether the given number is a solution of the given equation.

55. 25; $x + 17 = 42$

56. 75; $93 - y = 28$

57. 93; $a - 28 = 75$

58. 12; $8t = 96$

59. 63; $\dfrac{t}{7} = 9$

60. 52; $\dfrac{x}{8} = 6$

61. 3; $\dfrac{108}{x} = 36$

62. 7; $\dfrac{94}{y} = 12$

Translate each problem to an equation. Do not solve.

63. What number added to 73 is 201?

64. Seven times what number is 1596?

65. When 42 is multiplied by a number, the result is 2352. Find the number.

66. When 345 is added to a number, the result is 987. Find the number.

67. *Chess.*   A chess board has 64 squares. If pieces occupy 19 squares, how many squares are unoccupied?

68. *Hours worked.*   A carpenter charges $35 an hour. How many hours did she work if she billed a total of $3640?

69. *Recycling.*   Currently, Americans recycle or compost 32% of all municipal solid waste. This is the same as recycling or composting 79 million tons. What is the total amount of waste generated?
Source: U.S. EPA, Municipal Solid Waste Department

70. *Travel to work.*   In 2005, the average commuting time to work in New York was 31.2 min. The average commuting time in North Dakota was 14.9 min shorter. How long was the average commute in North Dakota?
Source: American Community Survey

In each of Exercises 71–78, match the phrase or sentence with the appropriate expression or equation from the column on the right.

71. ____ Twice the sum of two numbers

a) $\dfrac{x}{y} + 6$

72. ____ Five less than a number is twelve.

b) $2(x + y) = 48$

73. ____ Twelve more than a number is five.

c) $\dfrac{1}{2} \cdot a \cdot b$

74. ____ Half of the product of two numbers

d) $t + 12 = 5$

75. ____ Three times the sum of a number and five

e) $ab - 1 = 48$

76. ____ Twice the sum of two numbers is 48.

f) $2(m + n)$

77. ____ One less than the product of two numbers is 48.

g) $3(t + 5)$

78. ____ Six more than the quotient of two numbers

h) $x - 5 = 12$

To the student and the instructor: Writing exercises, denoted by ✍, *should be answered using one or more English sentences. Because answers to many writing exercises will vary, solutions are not listed in the answers at the back of the book.*

✍ **79.** What is the difference between a variable, a variable expression, and an equation?

✍ **80.** What does it mean to evaluate an algebraic expression?

Synthesis

To the student and the instructor: Synthesis exercises *are designed to challenge students to extend the concepts or skills studied in each section. Many synthesis exercises will require the assimilation of skills and concepts from several sections.*

✍ **81.** If the lengths of the sides of a square are doubled, is the area doubled? Why or why not?

✍ **82.** Write a problem that translates to $1998 + t = 2006$.

83. Signs of Distinction charges \$120 per square foot for handpainted signs. The town of Belmar commissioned a triangular sign with a base of 3 ft and a height of 2.5 ft. How much will the sign cost?

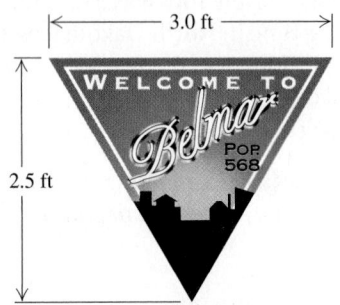

84. Find the area that is shaded.

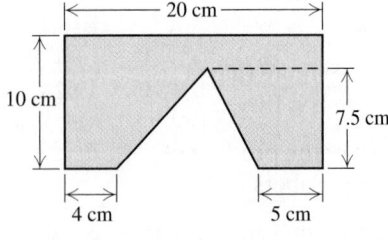

85. Evaluate $\dfrac{x - y}{3}$ when x is twice y and $x = 12$.

86. Evaluate $\dfrac{x + y}{2}$ when y is twice x and $x = 6$.

87. Evaluate $\dfrac{a + b}{4}$ when a is twice b and $a = 16$.

88. Evaluate $\dfrac{a - b}{3}$ when a is three times b and $a = 18$.

Answer each question with an algebraic expression.

89. If $w + 3$ is a whole number, what is the next whole number after it?

90. If $d + 2$ is an odd number, what is the preceding odd number?

Translate to an algebraic expression.

91. The perimeter of a rectangle with length l and width w (perimeter means distance around)

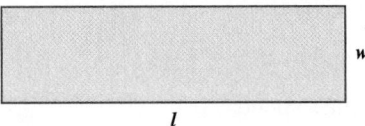

92. The perimeter of a square with side s (perimeter means distance around)

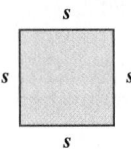

93. Ellie's race time, assuming she took 5 sec longer than Joe and Joe took 3 sec longer than Molly. Assume that Molly's time was t seconds.

94. Ray's age 7 yr from now if he is 2 yr older than Monique and Monique is a years old

✍ **95.** If the length of the height of a triangle is doubled, is its area also doubled? Why or why not?

COLLABORATIVE CORNER

Teamwork

Focus: Group problem solving; working collaboratively

Time: 15 minutes

Group size: 2

Working and studying as a team often enables students to solve problems that are difficult to solve alone.

ACTIVITY

1. The left-hand column below contains the names of 12 colleges. A scrambled list of the names of their sports teams is on the right. As a group, match the names of the colleges to the teams.

1. University of Texas		**a)**	Antelopes
2. Western State College of Colorado		**b)**	Banana Slugs
3. University of North Carolina		**c)**	Sea Warriors
4. University of Massachusetts		**d)**	Gators
5. Hawaii Pacific University		**e)**	Mountaineers
6. University of Nebraska		**f)**	Sailfish
7. University of California, Santa Cruz		**g)**	Longhorns
8. University of Louisiana at Lafayette		**h)**	Tar Heels
9. Grand Canyon University		**i)**	Seawolves
10. Palm Beach Atlantic University		**j)**	Ragin' Cajuns
11. University of Alaska, Anchorage		**k)**	Cornhuskers
12. University of Florida		**l)**	Minutemen

2. After working for 5 min, confer with another group and reach mutual agreement.

3. Does the class agree on all 12 pairs?

4. Do you agree that group collaboration enhances our ability to solve problems?

1.2   The Commutative, Associative, and Distributive Laws

Equivalent Expressions ▪ The Commutative Laws ▪ The Associative Laws ▪ The Distributive Law ▪ The Distributive Law and Factoring

In order to solve equations, we must be able to manipulate algebraic expressions. The commutative, associative, and distributive laws discussed in this section enable us to write *equivalent expressions* that will simplify our work. Indeed, much of this text is devoted to finding equivalent expressions.

Equivalent Expressions

The expressions $4 + 4 + 4$, $3 \cdot 4$, and $4 \cdot 3$ all represent the same number, 12. Expressions that represent the same number are said to be **equivalent**. The equivalent expressions $t + 18$ and $18 + t$ were used on p. 5 when we translated "eighteen more than a number." These expressions are equivalent because they represent the same number for any value of t. We can illustrate this by making some choices for t.

When $t = 3,$ $\quad t + 18 = 3 + 18 = 21$
$\qquad$ and $\quad 18 + t = 18 + 3 = 21.$
When $t = 40,$ $\quad t + 18 = 40 + 18 = 58$
$\qquad$ and $\quad 18 + t = 18 + 40 = 58.$

The Commutative Laws

Recall that changing the order in addition or multiplication does not change the result. Equations like $3 + 78 = 78 + 3$ and $5 \cdot 14 = 14 \cdot 5$ illustrate this idea and show that addition and multiplication are **commutative**.

> **The Commutative Laws**
>
> **For Addition.** For any numbers a and b,
>
> $$a + b = b + a.$$
>
> (Changing the order of addition does not affect the answer.)
>
> **For Multiplication.** For any numbers a and b,
>
> $$ab = ba.$$
>
> (Changing the order of multiplication does not affect the answer.)

EXAMPLE **1** Use the commutative laws to write an expression equivalent to each of the following: **(a)** $y + 5$; **(b)** $9x$; **(c)** $7 + ab$.

SOLUTION

a) $y + 5$ is equivalent to $5 + y$ by the commutative law of addition.

b) $9x$ is equivalent to $x \cdot 9$ by the commutative law of multiplication.

c) $7 + ab$ is equivalent to $ab + 7$ by the commutative law of *addition*.

$\quad$ $7 + ab$ is also equivalent to $7 + ba$ by the commutative law of *multiplication*.

$\quad$ $7 + ab$ is also equivalent to $ba + 7$ by the two commutative laws, used together.

> TRY EXERCISE 11

The Associative Laws

Parentheses are used to indicate groupings. We generally simplify within the parentheses first. For example,

$$3 + (8 + 4) = 3 + 12 = 15$$

and

$$(3 + 8) + 4 = 11 + 4 = 15.$$

Similarly,

$$4 \cdot (2 \cdot 3) = 4 \cdot 6 = 24$$

and

$$(4 \cdot 2) \cdot 3 = 8 \cdot 3 = 24.$$

Note that, so long as only addition or only multiplication appears in an expression, changing the grouping does not change the result. Equations such as $3 + (7 + 5) = (3 + 7) + 5$ and $4(5 \cdot 3) = (4 \cdot 5)3$ illustrate that addition and multiplication are **associative**.

The Associative Laws

For Addition.   For any numbers a, b, and c,

$$a + (b + c) = (a + b) + c.$$

(Numbers can be grouped in any manner for addition.)

For Multiplication.   For any numbers a, b, and c,

$$a \cdot (b \cdot c) = (a \cdot b) \cdot c.$$

(Numbers can be grouped in any manner for multiplication.)

EXAMPLE   2   Use an associative law to write an expression equivalent to each of the following:
(a) $y + (z + 3)$; **(b)** $(8x)y$.

SOLUTION

a) $y + (z + 3)$ is equivalent to $(y + z) + 3$ by the associative law of addition.

b) $(8x)y$ is equivalent to $8(xy)$ by the associative law of multiplication.

TRY EXERCISE 27

When only addition or only multiplication is involved, parentheses do not change the result. For that reason, we sometimes omit them altogether. Thus,

$$x + (y + 7) = x + y + 7 \quad \text{and} \quad l(wh) = lwh.$$

A sum such as $(5 + 1) + (3 + 5) + 9$ can be simplified by pairing numbers that add to 10. The associative and commutative laws allow us to do this:

$$(5 + 1) + (3 + 5) + 9 = 5 + 5 + 9 + 1 + 3$$
$$= 10 + 10 + 3 = 23.$$

EXAMPLE   3   Use the commutative and/or associative laws of addition to write two expressions equivalent to $(7 + x) + 3$. Then simplify.

SOLUTION

$(7 + x) + 3 = (x + 7) + 3$ Using the commutative law; $(x + 7) + 3$ is one equivalent expression.

$= x + (7 + 3)$ Using the associative law; $x + (7 + 3)$ is another equivalent expression.

$= x + 10$ Simplifying   TRY EXERCISE 39

EXAMPLE 4

Use the commutative and/or associative laws of multiplication to write two expressions equivalent to $2(x \cdot 3)$.

SOLUTION

$$2(x \cdot 3) = 2(3x) \qquad \text{Using the commutative law; } 2(3x) \text{ is one equivalent expression.}$$

$$= (2 \cdot 3)x \qquad \text{Using the associative law; } (2 \cdot 3)x \text{ is another equivalent expression.}$$

$$= 6x \qquad \text{Simplifying} \qquad \boxed{\text{TRY EXERCISE} \; 41}$$

The Distributive Law

The *distributive law* is probably the single most important law for manipulating algebraic expressions. Unlike the commutative and associative laws, the distributive law uses multiplication together with addition.

You have already used the distributive law although you may not have realized it at the time. To illustrate, try to multiply $3 \cdot 21$ mentally. Many people find the product, 63, by thinking of 21 as $20 + 1$ and then multiplying 20 by 3 and 1 by 3. The sum of the two products, $60 + 3$, is 63. Note that if the 3 does not multiply *both* 20 and 1, the result will not be correct.

EXAMPLE 5

Compute in two ways: $4(7 + 2)$.

SOLUTION

a) As in the discussion of $3(20 + 1)$ above, to compute $4(7 + 2)$, we can multiply both 7 and 2 by 4 and add the results:

$$4(7 + 2) = 4 \cdot 7 + 4 \cdot 2 \qquad \text{Multiplying both 7 and 2 by 4}$$

$$= 28 + 8 = 36. \qquad \text{Adding}$$

b) By first adding inside the parentheses, we get the same result in a different way:

$$4(7 + 2) = 4(9) \qquad \text{Adding; } 7 + 2 = 9$$

$$= 36. \qquad \text{Multiplying}$$

> **The Distributive Law**
>
> For any numbers a, b, and c,
>
> $$a(b + c) = ab + ac.$$
>
> (The product of a number and a sum can be written as the sum of two products.)

EXAMPLE 6

Multiply: $3(x + 2)$.

SOLUTION   Since $x + 2$ cannot be simplified unless a value for x is given, we use the distributive law:

$$3(x + 2) = 3 \cdot x + 3 \cdot 2 \qquad \text{Using the distributive law}$$

$$= 3x + 6. \qquad \text{Note that } 3 \cdot x \text{ is the same as } 3x.$$

$$\boxed{\text{TRY EXERCISE} \; 47}$$

The expression $3x + 6$ has two *terms*, $3x$ and 6. In general, a **term** is a number, a variable, or a product or a quotient of numbers and/or variables. Thus, t, 29, $5ab$, and $2x/y$ are terms in $t + 29 + 5ab + 2x/y$. Note that terms are separated by plus signs.

EXAMPLE **7**   List the terms in the expression $7s + st + \dfrac{3}{t}$.

SOLUTION   Terms are separated by plus signs, so the terms in $7s + st + \dfrac{3}{t}$ are

$7s$, st, and $\dfrac{3}{t}$.

> TRY EXERCISE ▸ 61

The distributive law can also be used when more than two terms are inside the parentheses.

EXAMPLE **8**   Multiply: $6(s + 2 + 5w)$.

SOLUTION

$$6(s + 2 + 5w) = 6 \cdot s + 6 \cdot 2 + 6 \cdot 5w \quad \text{Using the distributive law}$$
$$= 6s + 12 + (6 \cdot 5)w \quad \text{Using the associative law for multiplication}$$
$$= 6s + 12 + 30w$$

> TRY EXERCISE ▸ 55

Because of the commutative law of multiplication, the distributive law can be used on the "right": $(b + c)a = ba + ca$.

EXAMPLE **9**   Multiply: $(c + 4)5$.

SOLUTION

$$(c + 4)5 = c \cdot 5 + 4 \cdot 5 \quad \text{Using the distributive law on the right}$$
$$= 5c + 20 \quad \text{Using the commutative law; } c \cdot 5 = 5c$$

> TRY EXERCISE ▸ 57

CAUTION! To use the distributive law for removing parentheses, be sure to multiply *each* term inside the parentheses by the multiplier outside. Thus,

$\overline{a(b + c)} = \overline{ab + c}$ but $a(b + c) = ab + ac$.

The Distributive Law and Factoring

If we use the distributive law in reverse, we have the basis of a process called **factoring**: $ab + ac = a(b + c)$. To **factor** an expression means to write an equivalent expression that is a product. The parts of the product are called **factors**. Note that "factor" can be used as either a verb or a noun. Thus in the expression $5t$, the factors are 5 and t. In the expression $4(m + n)$, the factors are 4 and $(m + n)$. A **common factor** is a factor that appears in every term in an expression.

EXAMPLE **10**   Use the distributive law to factor each of the following.

a) $3x + 3y$ **b)** $7x + 21y + 7$

SOLUTION

a) By the distributive law,

$$3x + 3y = 3(x + y). \qquad \text{The common factor for } 3x \text{ and } 3y \text{ is } 3.$$

b) $7x + 21y + 7 = 7 \cdot x + 7 \cdot 3y + 7 \cdot 1$ The common factor is 7.

$$= 7(x + 3y + 1) \qquad \begin{array}{l} \text{Using the distributive law.} \\ \text{Be sure to include both the 1 and} \\ \text{the common factor, 7.} \end{array}$$

> TRY EXERCISE   69

To check our factoring, we multiply to see if the original expression is obtained. For example, to check the **factorization** in Example 10(b), note that

$$7(x + 3y + 1) = 7 \cdot x + 7 \cdot 3y + 7 \cdot 1$$
$$= 7x + 21y + 7.$$

Since $7x + 21y + 7$ is what we started with in Example 10(b), we have a check.

CAUTION!   Do not confuse **terms** with **factors**. Terms are separated by plus signs, and factors are parts of products. The distributive law is used when there are two or more terms inside parentheses. For example, in the expression $a(b \cdot c)$, b and c are factors, not terms. We can use the commutative and associative laws to reorder and regroup the factors, but the distributive law does not apply here. Thus,

$$\cancel{a(b \cdot c) = a \cdot b \cdot a \cdot c} \quad \text{but} \quad a(b \cdot c) = (a \cdot b) \cdot c.$$

1.2   EXERCISE SET

↰ *Concept Reinforcement   Complete each sentence using one of these terms:* commutative, associative, *or* distributive.

1. $8 + t$ is equivalent to $t + 8$ by the _____ law for addition.

2. $3(xy)$ is equivalent to $(3x)y$ by the _____ law for multiplication.

3. $(5b)c$ is equivalent to $5(bc)$ by the _____ law for multiplication.

4. mn is equivalent to nm by the _____ law for multiplication.

5. $x(y + z)$ is equivalent to $xy + xz$ by the _____ law.

6. $(9 + a) + b$ is equivalent to $9 + (a + b)$ by the _____ law for addition.

7. $a + (6 + d)$ is equivalent to $(a + 6) + d$ by the _____ law for addition.

8. $3(t + 4)$ is equivalent to $3(4 + t)$ by the _____ law for addition.

9. $5(x + 2)$ is equivalent to $(x + 2)5$ by the _____ law for multiplication.

10. $2(a + b)$ is equivalent to $2 \cdot a + 2 \cdot b$ by the _____ law.

Use the commutative law of addition to write an equivalent expression.

11. $11 + t$

12. $a + 2$

13. $4 + 8x$

14. $ab + c$

15. $9x + 3y$

16. $3a + 7b$

17. $5(a + 1)$

18. $9(x + 5)$

Use the commutative law of multiplication to write an equivalent expression.

19. $7x$

20. xy

21. st

22. $13m$

23. $5 + ab$

24. $x + 3y$

25. $5(a + 1)$

26. $9(x + 5)$

Use the associative law of addition to write an equivalent expression.

27. $(x + 8) + y$

28. $(5 + m) + r$

29. $u + (v + 7)$

30. $x + (2 + y)$

31. $(ab + c) + d$

32. $(m + np) + r$

Use the associative law of multiplication to write an equivalent expression.

33. $(8x)y$

34. $(4u)v$

35. $2(ab)$

36. $9(7r)$

37. $3[2(a + b)]$

38. $5[x(2 + y)]$

Use the commutative and/or associative laws to write two equivalent expressions. Answers may vary.

39. $s + (t + 6)$

40. $7 + (v + w)$

41. $(17a)b$

42. $x(3y)$

Use the commutative and/or associative laws to show why the expression on the left is equivalent to the expression on the right. Write a series of steps with labels, as in Example 4.

43. $(1 + x) + 2$ is equivalent to $x + 3$

44. $(2a)4$ is equivalent to $8a$

45. $(m \cdot 3)7$ is equivalent to $21m$

46. $4 + (9 + x)$ is equivalent to $x + 13$

Multiply.

47. $2(x + 15)$

48. $3(x + 5)$

49. $4(1 + a)$

50. $6(v + 4)$

51. $8(3 + y)$

52. $7(s + 1)$

53. $10(9x + 6)$

54. $9(6m + 7)$

55. $5(r + 2 + 3t)$

56. $4(5x + 8 + 3p)$

57. $(a + b)2$

58. $(x + 2)7$

59. $(x + y + 2)5$

60. $(2 + a + b)6$

List the terms in each expression.

61. $x + xyz + 1$

62. $9 + 17a + abc$

63. $2a + \dfrac{a}{3b} + 5b$

64. $3xy + 20 + \dfrac{4a}{b}$

65. $4(x + y)$

66. $(7 + y)2$

67. $4x + 4y$

68. $14 + 2y$

Use the distributive law to factor each of the following. Check by multiplying.

69. $2a + 2b$

70. $5y + 5z$

71. $7 + 7y$

72. $13 + 13x$

73. $32x + 4$

74. $20a + 5$

75. $5x + 10 + 15y$

76. $3 + 27b + 6c$

77. $7a + 35b$

78. $8x + 24y$

79. $44x + 11y + 22z$

80. $14a + 56b + 7$

List the factors in each expression.

81. $5n$

82. uv

83. $3(x + y)$

84. $(a + b)12$

85. $7 \cdot a \cdot b$

86. $m \cdot n \cdot 2$

87. $(a - b)(x - y)$

88. $(3 - a)(b + c)$

89. Is subtraction commutative? Why or why not?

90. Is division associative? Why or why not?

Skill Review

To the student and the instructor: Exercises included for Skill Review include skills previously studied in the text. Often these exercises provide preparation for the next section of the text. The numbers in brackets immediately following the directions or exercise indicate the section in which the skill was introduced. The answers to all Skill Review exercises appear at the back of the book. If a Skill Review exercise gives you difficulty, review the material in the indicated section of the text.

Translate to an algebraic expression. [1.1]

91. Half of Kara's salary

92. Twice the sum of m and 3

Synthesis

93. Give an example illustrating the distributive law, and identify the terms and the factors in your example. Explain how you can determine terms and factors in an expression.

94. Explain how the distributive, commutative, and associative laws can be used to show that $2(3x + 4y)$ is equivalent to $6x + 8y$.

Tell whether the expressions in each pairing are equivalent. Then explain why or why not.

95. $8 + 4(a + b)$ and $4(2 + a + b)$

96. $5(a \cdot b)$ and $5 \cdot a \cdot 5 \cdot b$

97. $7 \div 3m$ and $m \cdot 3 \div 7$

98. $(rt + st)5$ and $5t(r + s)$

99. $30y + x \cdot 15$ and $5[2(x + 3y)]$

100. $[c(2 + 3b)]5$ and $10c + 15bc$

101. Evaluate the expressions $3(2 + x)$ and $6 + x$ for $x = 0$. Do your results indicate that $3(2 + x)$ and $6 + x$ are equivalent? Why or why not?

102. Factor $15x + 40$. Then evaluate both $15x + 40$ and the factorization for $x = 4$. Do your results *guarantee* that the factorization is correct? Why or why not? (*Hint:* See Exercise 101.)

COLLABORATIVE CORNER

Mental Addition

Focus: Application of commutative and associative laws

Time: 10 minutes

Group size: 2–3

Legend has it that while still in grade school, the mathematician Carl Friedrich Gauss (1777–1855) was able to add the numbers from 1 to 100 mentally. Gauss did not add them sequentially, but rather paired 1 with 99, 2 with 98, and so on.

ACTIVITY

1. Use a method similar to Gauss's to simplify the following:

   $$1 + 2 + 3 + 4 + 5 + 6 + 7 + 8 + 9 + 10.$$

   One group member should add from left to right as a check.

2. Use Gauss's method to find the sum of the first 25 counting numbers:

   $$1 + 2 + 3 + \cdots + 23 + 24 + 25.$$

   Again, one student should add from left to right as a check.

3. How were the associative and commutative laws applied in parts (1) and (2) above?

4. Now use a similar approach involving both addition and division to find the sum of the first 10 counting numbers:

   $$1 + 2 + 3 + \cdots + 10$$
   $$\underline{+ 10 + 9 + 8 + \cdots + 1}$$

5. Use the approach in step (4) to find the sum of the first 100 counting numbers. Are the associative and commutative laws applied in this method, too? How is the distributive law used in this approach?

1.3   Fraction Notation

Factors and Prime Factorizations   ▪   Fraction Notation   ▪   Multiplication, Division, and Simplification   ▪
More Simplifying   ▪   Addition and Subtraction

This section covers multiplication, addition, subtraction, and division with fractions. Although much of this may be review, note that fraction expressions that contain variables are also included.

Factors and Prime Factorizations

In preparation for work with fraction notation, we first review how *natural numbers* are factored. **Natural numbers** can be thought of as the counting numbers:

$$1, 2, 3, 4, 5, \ldots .^*$$

(The dots indicate that the established pattern continues without ending.)

Since factors are parts of products, to factor a number, we express it as a product of two or more numbers.

Several factorizations of 12 are

$$1 \cdot 12, \quad 2 \cdot 6, \quad 3 \cdot 4, \quad 2 \cdot 2 \cdot 3.$$

It is easy to miss a factor of a number if the factorizations are not written methodically.

EXAMPLE 1   List all factors of 18.

SOLUTION   Beginning at 1, we check all natural numbers to see if they are factors of 18. If they are, we write the factorization. We stop when we have already included the next natural number in a factorization.

1 is a factor of every number.	$1 \cdot 18$
2 is a factor of 18.	$2 \cdot 9$
3 is a factor of 18.	$3 \cdot 6$

4 is *not* a factor of 18.

5 is *not* a factor of 18.

6 is the next natural number, but we have already listed 6 as a factor in the product $3 \cdot 6$.

We need check no additional numbers, because any natural number greater than 6 must be paired with a factor less than 6.

We now write the factors of 18 beginning with 1, going down the list of factorizations writing the first factor, then up the list of factorizations writing the second factor:

$$1, \quad 2, \quad 3, \quad 6, \quad 9, \quad 18.$$

> TRY EXERCISE 15

Some numbers have only two different factors, the number itself and 1. Such numbers are called **prime**.

---

*A similar collection of numbers, the **whole numbers,** includes 0: 0, 1, 2, 3, ...

> ### Prime Number
>
> A *prime number* is a natural number that has exactly two different factors: the number itself and 1. The first several primes are 2, 3, 5, 7, 11, 13, 17, 19, and 23.

If a natural number other than 1 is not prime, we call it **composite**.

EXAMPLE **2**   Label each number as prime, composite, or neither: 29, 4, 1.

SOLUTION

29 is prime. It has exactly two different factors, 29 and 1.

4 is not prime. It has three different factors, 1, 2, and 4. It is composite.

1 is not prime. It does not have two *different* factors. The number 1 is not considered composite. It is neither prime nor composite.

> TRY EXERCISE 5

Every composite number can be factored into a product of prime numbers. Such a factorization is called the **prime factorization** of that composite number.

EXAMPLE **3**   Find the prime factorization of 36.

SOLUTION   We first factor 36 in any way that we can. One way is like this:

$$36 = 4 \cdot 9.$$

The factors 4 and 9 are not prime, so we factor them:

$$36 = 4 \cdot 9$$
$$= 2 \cdot 2 \cdot 3 \cdot 3. \quad \text{2 and 3 are both prime.}$$

The prime factorization of 36 is $2 \cdot 2 \cdot 3 \cdot 3$.

> TRY EXERCISE 25

STUDENT NOTES ————

When writing a factorization, you are writing an equivalent expression for the original number. Some students do this with a tree diagram:

$$36 = 4 \quad \cdot \quad 9$$
$$36 = 2 \cdot 2 \cdot 3 \cdot 3$$

All prime

Fraction Notation

An example of **fraction notation** for a number is

$$\frac{2}{3}. \quad \begin{array}{l} \leftarrow \text{Numerator} \\ \leftarrow \text{Denominator} \end{array}$$

The top number is called the **numerator**, and the bottom number is called the **denominator**. When the numerator and the denominator are the same nonzero number, we have fraction notation for the number 1.

> ### Fraction Notation for 1
>
> For any number *a*, except 0,
>
> $$\frac{a}{a} = 1.$$
>
> (Any nonzero number divided by itself is 1.)

Note that in the definition for fraction notation for the number 1, we have excluded 0. In fact, 0 cannot be the denominator of *any* fraction. In this section, we limit our discussion to natural numbers, so this situation does not arise. Later in this chapter, we will discuss why denominators cannot be 0.

Multiplication, Division, and Simplification

Recall from arithmetic that fractions are multiplied as follows.

> ### Multiplication of Fractions
>
> For any two fractions a/b and c/d,
>
> $$\frac{a}{b} \cdot \frac{c}{d} = \frac{ac}{bd}.$$
>
> (The numerator of the product is the product of the two numerators. The denominator of the product is the product of the two denominators.)

EXAMPLE **4**   Multiply: **(a)** $\dfrac{2}{3} \cdot \dfrac{5}{7}$; **(b)** $\dfrac{4}{x} \cdot \dfrac{8}{y}$.

SOLUTION   We multiply numerators as well as denominators.

a) $\dfrac{2}{3} \cdot \dfrac{5}{7} = \dfrac{2 \cdot 5}{3 \cdot 7} = \dfrac{10}{21}$

b) $\dfrac{4}{x} \cdot \dfrac{8}{y} = \dfrac{4 \cdot 8}{x \cdot y} = \dfrac{32}{xy}$

TRY EXERCISE 53

Two numbers whose product is 1 are **reciprocals**, or **multiplicative inverses**, of each other. All numbers, except zero, have reciprocals. For example,

the reciprocal of $\dfrac{2}{3}$ is $\dfrac{3}{2}$ because $\dfrac{2}{3} \cdot \dfrac{3}{2} = \dfrac{6}{6} = 1$;

the reciprocal of 9 is $\dfrac{1}{9}$ because $9 \cdot \dfrac{1}{9} = \dfrac{9}{9} = 1$; and

the reciprocal of $\dfrac{1}{4}$ is 4 because $\dfrac{1}{4} \cdot 4 = 1$.

Reciprocals are used to rewrite division in an equivalent form that uses multiplication.

> ### Division of Fractions
>
> To divide two fractions, multiply by the reciprocal of the divisor:
>
> $$\frac{a}{b} \div \frac{c}{d} = \frac{a}{b} \cdot \frac{d}{c}.$$

EXAMPLE **5**   Divide: $\dfrac{1}{2} \div \dfrac{3}{5}$.

SOLUTION

$$\dfrac{1}{2} \div \dfrac{3}{5} = \dfrac{1}{2} \cdot \dfrac{5}{3} \qquad \dfrac{5}{3} \text{ is the reciprocal of } \dfrac{3}{5}.$$

$$= \dfrac{5}{6}$$

TRY EXERCISE ▶ 73

When one of the fractions being multiplied is 1, multiplying yields an equivalent expression because of the *identity property of* 1. A similar property could be stated for division, but there is no need to do so here.

> ### The Identity Property of 1
>
> For any number a,
>
> $$a \cdot 1 = 1 \cdot a = a.$$
>
> (Multiplying a number by 1 gives that same number.) The number 1 is called the *multiplicative identity*.

EXAMPLE **6**   Multiply $\dfrac{4}{5} \cdot \dfrac{6}{6}$ to find an expression equivalent to $\dfrac{4}{5}$.

SOLUTION   Since $\frac{6}{6} = 1$, the expression $\frac{4}{5} \cdot \frac{6}{6}$ is equivalent to $\frac{4}{5} \cdot 1$, or simply $\frac{4}{5}$. We have

$$\dfrac{4}{5} \cdot \dfrac{6}{6} = \dfrac{4 \cdot 6}{5 \cdot 6} = \dfrac{24}{30}.$$

Thus, $\frac{24}{30}$ is equivalent to $\frac{4}{5}$.

The steps of Example 6 are reversed by "removing a factor equal to 1"—in this case, $\frac{6}{6}$. By removing a factor that equals 1, we can *simplify* an expression like $\frac{24}{30}$ to an equivalent expression like $\frac{4}{5}$.

To simplify, we factor the numerator and the denominator, looking for the largest factor common to both. This is sometimes made easier by writing prime factorizations. After identifying common factors, we can express the fraction as a product of two fractions, one of which is in the form a/a.

EXAMPLE **7**   Simplify: **(a)** $\dfrac{15}{40}$; **(b)** $\dfrac{36}{24}$.

SOLUTION

a) Note that 5 is a factor of both 15 and 40:

$$\dfrac{15}{40} = \dfrac{3 \cdot 5}{8 \cdot 5} \qquad \text{Factoring the numerator and the denominator, using the common factor, 5}$$

$$= \dfrac{3}{8} \cdot \dfrac{5}{5} \qquad \text{Rewriting as a product of two fractions; } \dfrac{5}{5} = 1$$

$$= \dfrac{3}{8} \cdot 1 = \dfrac{3}{8}. \qquad \text{Using the identity property of 1 (removing a factor equal to 1)}$$

STUDENT NOTES

The following rules can help you quickly determine whether 2, 3, or 5 is a factor of a number.

2 is a factor of a number if the number is even (the ones digit is 0, 2, 4, 6, or 8).

3 is a factor of a number if the sum of its digits is divisible by 3.

5 is a factor of a number if its ones digit is 0 or 5.

b) $\dfrac{36}{24} = \dfrac{2 \cdot 2 \cdot 3 \cdot 3}{2 \cdot 2 \cdot 2 \cdot 3}$ Writing the prime factorizations and identifying common factors; 12/12 could also be used.

$= \dfrac{3}{2} \cdot \dfrac{2 \cdot 2 \cdot 3}{2 \cdot 2 \cdot 3}$ Rewriting as a product of two fractions; $\dfrac{2 \cdot 2 \cdot 3}{2 \cdot 2 \cdot 3} = 1$

$= \dfrac{3}{2} \cdot 1 = \dfrac{3}{2}$ Using the identity property of 1   **TRY EXERCISE 35**

It is always wise to check your result to see if any common factors of the numerator and the denominator remain. (This will never happen if prime factorizations are used correctly.) If common factors remain, repeat the process by removing another factor equal to 1 to simplify your result.

More Simplifying

"Canceling" is a shortcut that you may have used for removing a factor equal to 1 when working with fraction notation. With *great* concern, we mention it as a possible way to speed up your work. Canceling can be used only when removing common factors in numerators and denominators. Canceling *cannot* be used in sums or differences. Our concern is that "canceling" be used with understanding. Example 7(b) might have been done faster as follows:

$$\frac{36}{24} = \frac{\cancel{2} \cdot \cancel{2} \cdot 3 \cdot \cancel{3}}{\cancel{2} \cdot \cancel{2} \cdot 2 \cdot \cancel{3}} = \frac{3}{2}, \quad \text{or} \quad \frac{36}{24} = \frac{3 \cdot \cancel{12}}{2 \cdot \cancel{12}} = \frac{3}{2}, \quad \text{or} \quad \frac{\overset{3}{\cancel{\overset{18}{\cancel{36}}}}}{\underset{2}{\cancel{\underset{12}{\cancel{24}}}}} = \frac{3}{2}.$$

CAUTION!   Unfortunately, canceling is often performed incorrectly:

$$\frac{\cancel{2} + 3}{\cancel{2}} = 3, \qquad \frac{\cancel{4} - 1}{\cancel{4} - 2} = \frac{1}{2}, \qquad \frac{1\cancel{5}}{5\cancel{4}} = \frac{1}{4}.$$

The above cancellations are incorrect because the expressions canceled are *not* factors. For example, in $2 + 3$, the 2 and the 3 are not factors. Correct simplifications are as follows:

$$\frac{2 + 3}{2} = \frac{5}{2}, \qquad \frac{4 - 1}{4 - 2} = \frac{3}{2}, \qquad \frac{15}{54} = \frac{5 \cdot \cancel{3}}{18 \cdot \cancel{3}} = \frac{5}{18}.$$

Remember: **If you can't factor, you can't cancel! If in doubt, don't cancel!**

Sometimes it is helpful to use 1 as a factor in the numerator or the denominator when simplifying.

EXAMPLE 8   Simplify: $\dfrac{9}{72}$.

SOLUTION

$\dfrac{9}{72} = \dfrac{1 \cdot 9}{8 \cdot 9}$ Factoring and using the identity property of 1 to write 9 as $1 \cdot 9$

$= \dfrac{1 \cdot \cancel{9}}{8 \cdot \cancel{9}} = \dfrac{1}{8}$ Simplifying by removing a factor equal to 1: $\dfrac{9}{9} = 1$ **TRY EXERCISE 39**

Addition and Subtraction

When denominators are the same, fractions are added or subtracted by adding or subtracting numerators and keeping the same denominator.

> ### Addition and Subtraction of Fractions
> For any two fractions a/d and b/d,
> $$\frac{a}{d} + \frac{b}{d} = \frac{a+b}{d} \quad \text{and} \quad \frac{a}{d} - \frac{b}{d} = \frac{a-b}{d}.$$

EXAMPLE **9** Add and simplify: $\dfrac{4}{8} + \dfrac{5}{8}$.

SOLUTION The common denominator is 8. We add the numerators and keep the common denominator:

$$\frac{4}{8} + \frac{5}{8} = \frac{4+5}{8} = \frac{9}{8}.$$

You can think of this as
$$4 \cdot \frac{1}{8} + 5 \cdot \frac{1}{8} = 9 \cdot \frac{1}{8}, \text{ or } \frac{9}{8}.$$

TRY EXERCISE 63

In arithmetic, we often write $1\frac{1}{8}$ rather than the "improper" fraction $\frac{9}{8}$. In algebra, $\frac{9}{8}$ is generally more useful and is quite "proper" for our purposes.

When denominators are different, we use the identity property of 1 and multiply to find a common denominator. Then we add, as in Example 9.

EXAMPLE **10** Add or subtract as indicated: **(a)** $\dfrac{7}{8} + \dfrac{5}{12}$; **(b)** $\dfrac{9}{8} - \dfrac{4}{5}$.

SOLUTION

a) The number 24 is divisible by both 8 and 12. We multiply both $\frac{7}{8}$ and $\frac{5}{12}$ by suitable forms of 1 to obtain two fractions with denominators of 24:

$$\frac{7}{8} + \frac{5}{12} = \frac{7}{8} \cdot \frac{3}{3} + \frac{5}{12} \cdot \frac{2}{2}$$

Multiplying by 1. Since $8 \cdot 3 = 24$, we multiply $\frac{7}{8}$ by $\frac{3}{3}$. Since $12 \cdot 2 = 24$, we multiply $\frac{5}{12}$ by $\frac{2}{2}$.

$$= \frac{21}{24} + \frac{10}{24}$$

Performing the multiplication

$$= \frac{31}{24}.$$

Adding fractions

b) $\dfrac{9}{8} - \dfrac{4}{5} = \dfrac{9}{8} \cdot \dfrac{5}{5} - \dfrac{4}{5} \cdot \dfrac{8}{8}$

Using 40 as a common denominator

$$= \frac{45}{40} - \frac{32}{40} = \frac{13}{40}$$

Subtracting fractions

TRY EXERCISE 69

After adding, subtracting, multiplying, or dividing, we may still need to simplify the answer.

EXAMPLE **11** Perform the indicated operation and, if possible, simplify.

a) $\dfrac{7}{10} - \dfrac{1}{5}$ **b)** $8 \cdot \dfrac{5}{12}$ **c)** $\dfrac{\frac{5}{6}}{\frac{25}{9}}$

SOLUTION

a)
$$\dfrac{7}{10} - \dfrac{1}{5} = \dfrac{7}{10} - \dfrac{1}{5} \cdot \dfrac{2}{2} \qquad \text{Using 10 as the common denominator}$$

$$= \dfrac{7}{10} - \dfrac{2}{10}$$

$$= \dfrac{5}{10} = \dfrac{1 \cdot \cancel{5}}{2 \cdot \cancel{5}} = \dfrac{1}{2} \qquad \text{Removing a factor equal to 1: } \dfrac{5}{5} = 1$$

b)
$$8 \cdot \dfrac{5}{12} = \dfrac{8 \cdot 5}{12} \qquad \text{Multiplying numerators and denominators. Think of 8 as } \tfrac{8}{1}.$$

$$= \dfrac{2 \cdot 2 \cdot 2 \cdot 5}{2 \cdot 2 \cdot 3} \qquad \text{Factoring; } \dfrac{4 \cdot 2 \cdot 5}{4 \cdot 3} \text{ can also be used.}$$

$$= \dfrac{\cancel{2} \cdot \cancel{2} \cdot 2 \cdot 5}{\cancel{2} \cdot \cancel{2} \cdot 3} \qquad \text{Removing a factor equal to 1: } \dfrac{2 \cdot 2}{2 \cdot 2} = 1$$

$$= \dfrac{10}{3} \qquad \text{Simplifying}$$

c)
$$\dfrac{\frac{5}{6}}{\frac{25}{9}} = \dfrac{5}{6} \div \dfrac{25}{9} \qquad \text{Rewriting horizontally. Remember that a fraction bar indicates division.}$$

$$= \dfrac{5}{6} \cdot \dfrac{9}{25} \qquad \text{Multiplying by the reciprocal of } \tfrac{25}{9}$$

$$= \dfrac{5 \cdot 3 \cdot 3}{2 \cdot 3 \cdot 5 \cdot 5} \qquad \text{Writing as one fraction and factoring}$$

$$= \dfrac{\cancel{5} \cdot \cancel{3} \cdot 3}{2 \cdot \cancel{3} \cdot \cancel{5} \cdot 5} \qquad \text{Removing a factor equal to 1: } \dfrac{5 \cdot 3}{3 \cdot 5} = 1$$

$$= \dfrac{3}{10} \qquad \text{Simplifying} \qquad \boxed{\text{TRY EXERCISE} \blacktriangleright 65}$$

TECHNOLOGY CONNECTION

Some graphing calculators can perform operations using fraction notation. Others can convert answers given in decimal notation to fraction notation. Often this conversion is done using a command found in a **menu** of options that appears when a key is pressed. To select an item from a menu, we highlight its number and press **ENTER** or simply press the number of the item.

For example, to find fraction notation for $\frac{2}{15} + \frac{7}{12}$, we enter the expression as $2/15 + 7/12$. The answer is given in decimal notation. To convert this to fraction notation, we press **MATH** and select the Frac option. In this case, the notation Ans ▶ Frac shows that the graphing calculator will convert .7166666667 to fraction notation.

```
2/15+7/12
            .7166666667
Ans▶Frac
                  43/60
```

We see that $\frac{2}{15} + \frac{7}{12} = \frac{43}{60}$.

1.3   **EXERCISE SET**

To the student and the instructor: *Beginning in this section, selected exercises are marked with the symbol* Aha!. *Students who pause to inspect an Aha! exercise should find the answer more readily than those who proceed mechanically. This is done to discourage rote memorization. Some later "Aha!" exercises in this exercise set are unmarked, to encourage students to always pause before working a problem.*

↪ *Concept Reinforcement*   *In each of Exercises 1–4, match the description with a number from the list on the right.*

1. ____ A factor of 35   **a)** 2

2. ____ A number that has 3 as a factor   **b)** 7

3. ____ An odd composite number   **c)** 60

4. ____ The only even prime number   **d)** 65

Label each of the following numbers as prime, composite, or neither.

5. 9 **6.** 15 **7.** 41 **8.** 49

9. 77 **10.** 37 **11.** 2 **12.** 1

13. 0 **14.** 16

Write all two-factor factorizations of each number. Then list all the factors of the number.

15. 50 **16.** 70 **17.** 42 **18.** 60

Find the prime factorization of each number. If the number is prime, state this.

19. 39 **20.** 34 **21.** 30

22. 55 **23.** 27 **24.** 98

25. 150 **26.** 54 **27.** 40

28. 56 **29.** 31 **30.** 180

31. 210 **32.** 79 **33.** 115

34. 143

Simplify.

35. $\dfrac{21}{35}$ **36.** $\dfrac{20}{26}$ **37.** $\dfrac{16}{56}$

38. $\dfrac{72}{27}$ **39.** $\dfrac{12}{48}$ **40.** $\dfrac{18}{84}$

41. $\dfrac{52}{13}$ **42.** $\dfrac{132}{11}$ **43.** $\dfrac{19}{76}$

44. $\dfrac{17}{51}$ **45.** $\dfrac{150}{25}$ **46.** $\dfrac{180}{36}$

47. $\dfrac{42}{50}$ **48.** $\dfrac{75}{80}$ **49.** $\dfrac{120}{82}$

50. $\dfrac{75}{45}$ **51.** $\dfrac{210}{98}$ **52.** $\dfrac{140}{350}$

Perform the indicated operation and, if possible, simplify.

53. $\dfrac{1}{2} \cdot \dfrac{3}{5}$ **54.** $\dfrac{11}{10} \cdot \dfrac{8}{5}$ **55.** $\dfrac{9}{2} \cdot \dfrac{4}{3}$

Aha! **56.** $\dfrac{11}{12} \cdot \dfrac{12}{11}$ **57.** $\dfrac{1}{8} + \dfrac{3}{8}$ **58.** $\dfrac{1}{2} + \dfrac{1}{8}$

59. $\dfrac{4}{9} + \dfrac{13}{18}$ **60.** $\dfrac{4}{5} + \dfrac{8}{15}$ **61.** $\dfrac{3}{a} \cdot \dfrac{b}{7}$

62. $\dfrac{x}{5} \cdot \dfrac{y}{z}$ **63.** $\dfrac{4}{n} + \dfrac{6}{n}$ **64.** $\dfrac{9}{x} - \dfrac{5}{x}$

65. $\dfrac{3}{10} + \dfrac{8}{15}$ **66.** $\dfrac{7}{8} + \dfrac{5}{12}$ **67.** $\dfrac{11}{7} - \dfrac{4}{7}$

68. $\dfrac{12}{5} - \dfrac{2}{5}$ **69.** $\dfrac{13}{18} - \dfrac{4}{9}$ **70.** $\dfrac{13}{15} - \dfrac{11}{45}$

Aha! **71.** $\dfrac{20}{30} - \dfrac{2}{3}$ **72.** $\dfrac{5}{7} - \dfrac{5}{21}$ **73.** $\dfrac{7}{6} \div \dfrac{3}{5}$

74. $\dfrac{7}{5} \div \dfrac{10}{3}$ **75.** $\dfrac{8}{9} \div \dfrac{4}{15}$ **76.** $\dfrac{9}{4} \div 9$

77. $12 \div \dfrac{4}{9}$ **78.** $\dfrac{1}{10} \div \dfrac{1}{5}$ Aha! **79.** $\dfrac{7}{13} \div \dfrac{7}{13}$

80. $\dfrac{17}{8} \div \dfrac{5}{6}$ **81.** $\dfrac{\frac{2}{7}}{\frac{5}{3}}$ **82.** $\dfrac{\frac{3}{8}}{\frac{1}{5}}$

83. $\dfrac{\frac{9}{1}}{\frac{1}{2}}$ **84.** $\dfrac{\frac{3}{7}}{6}$

85. Under what circumstances would the sum of two fractions be easier to compute than the product of the same two fractions?

86. Under what circumstances would the product of two fractions be easier to compute than the sum of the same two fractions?

Skill Review

Use a commutative law to write an equivalent expression. There can be more than one correct answer. [1.2]

87. $5(x + 3)$ **88.** $7 + (a + b)$

Synthesis

89. Bryce insists that $(2 + x)/8$ is equivalent to $(1 + x)/4$. What mistake do you think is being made and how could you demonstrate to Bryce that the two expressions are not equivalent?

90. Why are 0 and 1 considered neither prime nor composite?

91. In the following table, the top number can be factored in such a way that the sum of the factors is the bottom number. For example, in the first column, 56 is factored as $7 \cdot 8$, since $7 + 8 = 15$, the bottom number. Find the missing numbers in each column.

Product	56	63	36	72	140	96	168
Factor	7						
Factor	8						
Sum	15	16	20	38	24	20	29

92. *Packaging.* Tritan Candies uses two sizes of boxes, 6 in. long and 8 in. long. These are packed end to end in bigger cartons to be shipped. What is the shortest-length carton that will accommodate boxes of either size without any room left over? (Each carton must contain boxes of only one size; no mixing is allowed.)

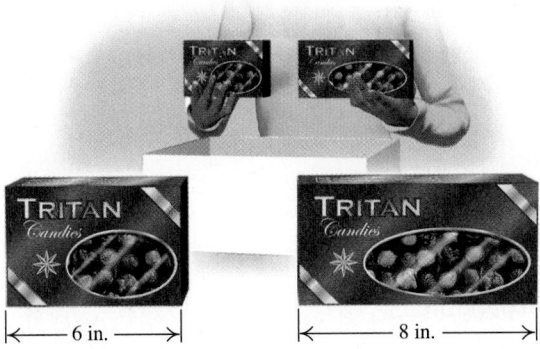

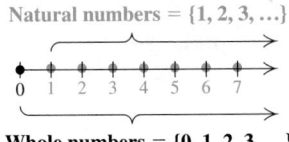

Simplify.

93. $\dfrac{16 \cdot 9 \cdot 4}{15 \cdot 8 \cdot 12}$

94. $\dfrac{9 \cdot 8xy}{2xy \cdot 36}$

95. $\dfrac{45pqrs}{9prst}$

96. $\dfrac{247}{323}$

97. $\dfrac{15 \cdot 4xy \cdot 9}{6 \cdot 25x \cdot 15y}$

98. $\dfrac{10x \cdot 12 \cdot 25y}{2z \cdot 30x \cdot 20y}$

99. $\dfrac{\frac{27ab}{15mn}}{\frac{18bc}{25np}}$

100. $\dfrac{\frac{45xyz}{24ab}}{\frac{30xz}{32ac}}$

101. $\dfrac{5\frac{3}{4}rs}{4\frac{1}{2}st}$

102. $\dfrac{3\frac{5}{7}mn}{2\frac{4}{5}np}$

Find the area of each figure.

103.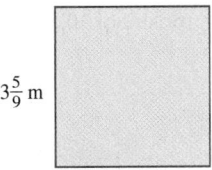

104.

105. Find the perimeter of a square with sides of length $3\frac{5}{9}$ m.

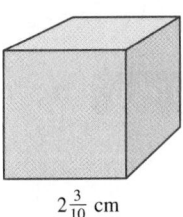

106. Find the perimeter of the rectangle in Exercise 103.

107. Find the total length of the edges of a cube with sides of length $2\frac{3}{10}$ cm.

1.4 **Positive and Negative Real Numbers**

The Integers ▪ The Rational Numbers ▪ Real Numbers and Order ▪ Absolute Value

A **set** is a collection of objects. The set containing 1, 3, and 7 is usually written $\{1, 3, 7\}$. In this section, we examine some important sets of numbers. More on sets can be found in Appendix B.

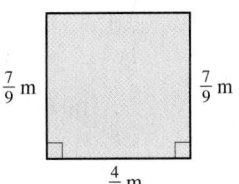

Natural numbers = $\{1, 2, 3, \ldots\}$

Whole numbers = $\{0, 1, 2, 3, \ldots\}$

The Integers

Two sets of numbers were mentioned in Section 1.3. We represent these sets using dots on a number line, as shown at left.

To create the set of *integers,* we include all whole numbers, along with their *opposites.* To find the opposite of a number, we locate the number that is the same distance from 0 but on the other side of the number line. For example,

the opposite of 1 is negative 1, written −1;

and

the opposite of 3 is negative 3, written −3.

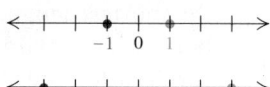

The **integers** consist of all whole numbers and their opposites.

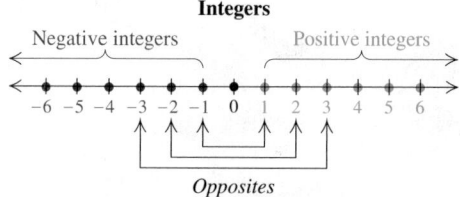

STUDENT NOTES

It is not uncommon in mathematics for a symbol to have more than one meaning in different contexts. The symbol "−" in 5 − 3 indicates subtraction. The same symbol in −10 indicates the opposite of 10, or negative 10.

Opposites are discussed in more detail in Section 1.6. Note that, except for 0, opposites occur in pairs. Thus, 5 is the opposite of −5, just as −5 is the opposite of 5. Note that 0 acts as its own opposite.

> ### Set of Integers
> The set of integers $= \{ \ldots, -4, -3, -2, -1, 0, 1, 2, 3, 4, \ldots \}$.

Integers are associated with many real-world problems and situations.

EXAMPLE **1** State which integer(s) corresponds to each situation.

a) In 2006, there was $13 trillion in outstanding mortgage debt in the United States.
Source: Board of Governors of the Federal Reserve System

b) Part of Death Valley is 200 ft below sea level.

c) To lose one pound of fat, it is necessary for most people to create a 3500-calorie deficit.
Source: World Health Organization

SOLUTION

a) The integer −13,000,000,000,000 corresponds to a debt of $13 trillion.

b) The integer −200 corresponds to 200 ft below sea level.

c) The integer −3500 corresponds to a deficit of 3500 calories.

> TRY EXERCISE 9

The Rational Numbers

A number like $\frac{5}{9}$, although built out of integers, is not itself an integer. Another set of numbers, the **rational numbers**, contains integers, fractions, and decimals. Some examples of rational numbers are

$$\frac{5}{9}, \quad -\frac{4}{7}, \quad 95, \quad -16, \quad 0, \quad \frac{-35}{8}, \quad 2.4, \quad -0.31.$$

In Section 1.7, we show that $-\frac{4}{7}$ can be written as $\frac{-4}{7}$ or $\frac{4}{-7}$. Indeed, every number listed above can be written as an integer over an integer. For example, 95 can be written as $\frac{95}{1}$ and 2.4 can be written as $\frac{24}{10}$. In this manner, any *ratio*nal number can be expressed as the *ratio* of two integers. Rather than attempt to list all rational numbers, we use this idea of ratio to describe the set as follows.

> ### Set of Rational Numbers
>
> The set of rational numbers $= \left\{ \dfrac{a}{b} \,\middle|\, a \text{ and } b \text{ are integers and } b \neq 0 \right\}$.
>
> This is read "the set of all numbers a over b, where a and b are integers and b does not equal zero."

In Section 1.7, we explain why b cannot equal 0.

To *graph* a number is to mark its location on a number line.

EXAMPLE 2   Graph each of the following rational numbers: **(a)** $\frac{5}{2}$; **(b)** -3.2; **(c)** $\frac{11}{8}$.

SOLUTION

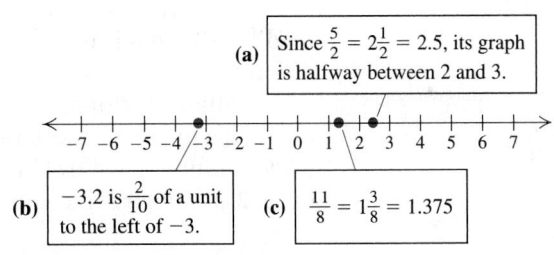

(a) Since $\frac{5}{2} = 2\frac{1}{2} = 2.5$, its graph is halfway between 2 and 3.

(b) -3.2 is $\frac{2}{10}$ of a unit to the left of -3.

(c) $\frac{11}{8} = 1\frac{3}{8} = 1.375$

> TRY EXERCISE   19

It is important to remember that every rational number can be written using fraction notation or decimal notation.

EXAMPLE 3   Convert to decimal notation: $-\frac{5}{8}$.

SOLUTION   We first find decimal notation for $\frac{5}{8}$. Since $\frac{5}{8}$ means $5 \div 8$, we divide.

$$
\begin{array}{r}
0.6\,2\,5 \\
8\overline{)5.0\,0\,0} \\
\underline{4\,8\,0\,0} \\
2\,0\,0 \\
\underline{1\,6\,0} \\
4\,0 \\
\underline{4\,0} \\
0 \quad \leftarrow \text{The remainder is 0.}
\end{array}
$$

Thus, $\frac{5}{8} = 0.625$, so $-\frac{5}{8} = -0.625$.

> TRY EXERCISE   25

Because the division in Example 3 ends with the remainder 0, we consider -0.625 a **terminating decimal**. If we are "bringing down" zeros and a remainder reappears, we have a **repeating decimal**, as shown in the next example.

EXAMPLE 4

Convert to decimal notation: $\frac{7}{11}$.

SOLUTION We divide:

$$
\begin{array}{r}
0.6\,3\,6\,3\ldots \\
11{\overline{)7.0\,0\,0\,0}} \\
6\,6 \\
\hline
4\,0 \\
3\,3 \\
\hline
7\,0 \\
6\,6 \\
\hline
4\,0 \\
\end{array}
$$

4 reappears as a remainder, so the pattern of 6's and 3's in the quotient will continue.

We abbreviate repeating decimals by writing a bar over the repeating part—in this case, $0.\overline{63}$. Thus, $\frac{7}{11} = 0.\overline{63}$.

TRY EXERCISE 29

Although we do not prove it here, every rational number can be expressed as either a terminating or repeating decimal, and every terminating or repeating decimal can be expressed as a ratio of two integers.

Real Numbers and Order

Some numbers, when written in decimal form, neither terminate nor repeat. Such numbers are called **irrational numbers**.

What sort of numbers are irrational? One example is π (the Greek letter *pi*, read "pie"), which is used to find the area and the circumference of a circle: $A = \pi r^2$ and $C = 2\pi r$.

Another irrational number, $\sqrt{2}$ (read "the square root of 2"), is the length of the diagonal of a square with sides of length 1. It is also the number that, when multiplied by itself, gives 2. No rational number can be multiplied by itself to get 2, although some approximations come close:

1.4 is an *approximation* of $\sqrt{2}$ because $(1.4)(1.4) = 1.96$;

1.41 is a better approximation because $(1.41)(1.41) = 1.9881$;

1.4142 is an even better approximation because $(1.4142)(1.4142) = 1.99996164$.

To approximate $\sqrt{2}$ on some calculators, we simply press ② and then ⎷. With other calculators, we press ⎷, ②, and **ENTER**, or consult a manual.

EXAMPLE 5

Graph the real number $\sqrt{3}$ on the number line.

SOLUTION We use a calculator and approximate: $\sqrt{3} \approx 1.732$ ("$\approx$" means "approximately equals"). Then we locate this number on the number line.

TRY EXERCISE 37

The rational numbers and the irrational numbers together correspond to all the points on the number line and make up what is called the **real-number system**.

> **Set of Real Numbers**
>
> The set of real numbers = The set of all numbers corresponding to points on the number line.

The following figure shows the relationships among various kinds of numbers.

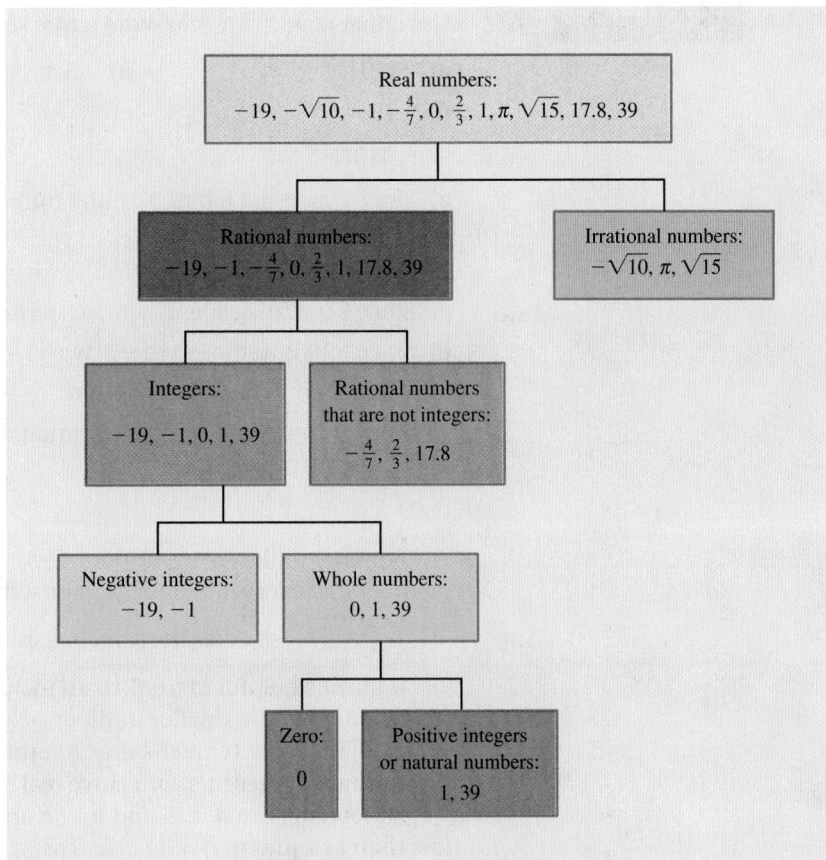

EXAMPLE 6   Which numbers in the following list are **(a)** whole numbers? **(b)** integers? **(c)** rational numbers? **(d)** irrational numbers? **(e)** real numbers?

$$-38, \quad -\frac{8}{5}, \quad 0, \quad 0.\overline{3}, \quad 4.5, \quad \sqrt{30}, \quad 52$$

SOLUTION

a) 0 and 52 are whole numbers.

b) -38, 0, and 52 are integers.

c) $-38, -\frac{8}{5}, 0, 0.\overline{3}, 4.5$, and 52 are rational numbers.

d) $\sqrt{30}$ is an irrational number.

e) $-38, -\frac{8}{5}, 0, 0.\overline{3}, 4.5, \sqrt{30}$, and 52 are real numbers.

▶ TRY EXERCISE 75

Real numbers are named in order on the number line, with larger numbers further to the right. For any two numbers, the one to the left is less than the one to the right. We use the symbol **<** to mean "**is less than**." The sentence $-8 < 6$ means "-8 is less than 6." The symbol **>** means "**is greater than**." The sentence $-3 > -7$ means "-3 is greater than -7."

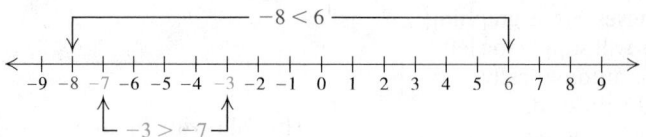

EXAMPLE 7

Use either $<$ or $>$ for ▪ to write a true sentence.

a) 2 ▪ 9   **b)** -3.45 ▪ 1.32 **c)** 6 ▪ -12

d) -18 ▪ -5 **e)** $\frac{7}{11}$ ▪ $\frac{5}{8}$

SOLUTION

a) Since 2 is to the left of 9 on the number line, we know that 2 is less than 9, so $2 < 9$.

b) Since -3.45 is to the left of 1.32, we have $-3.45 < 1.32$.

c) Since 6 is to the right of -12, we have $6 > -12$.

d) Since -18 is to the left of -5, we have $-18 < -5$.

e) We convert to decimal notation: $\frac{7}{11} = 0.\overline{63}$ and $\frac{5}{8} = 0.625$. Thus, $\frac{7}{11} > \frac{5}{8}$.

We also could have used a common denominator: $\frac{7}{11} = \frac{56}{88} > \frac{55}{88} = \frac{5}{8}$.

> **TRY EXERCISES** 41 and 45

Sentences like "$a < -5$" and "$-3 > -8$" are **inequalities**. It is useful to remember that every inequality can be written in two ways. For example,

$$-3 > -8 \quad \text{has the same meaning as} \quad -8 < -3.$$

It may be helpful to think of an inequality sign as an "arrow" with the smaller side pointing to the smaller number.

Note that $a > 0$ means that a represents a positive real number and $a < 0$ means that a represents a negative real number.

Statements like $a \le b$ and $b \ge a$ are also inequalities. We read $a \le b$ as "a **is less than or equal to** b" and $a \ge b$ as "a **is greater than or equal to** b."

EXAMPLE 8

Classify each inequality as true or false.

a) $-3 \le 5$ **b)** $-3 \le -3$ **c)** $-5 \ge 4$

STUDENT NOTES

It is important to remember that just because an equation or inequality is written or printed, it is not necessarily *true*. For instance, $6 = 7$ is an equation and $2 > 5$ is an inequality. Of course, both statements are *false*.

SOLUTION

a) $-3 \le 5$ is *true* because $-3 < 5$ is true.

b) $-3 \le -3$ is *true* because $-3 = -3$ is true.

c) $-5 \ge 4$ is *false* since neither $-5 > 4$ nor $-5 = 4$ is true.

> **TRY EXERCISE** 57

Absolute Value

There is a convenient terminology and notation for the distance a number is from 0 on the number line. It is called the **absolute value** of the number.

> ### Absolute Value
> We write $|a|$, read "the absolute value of a," to represent the number of units that a is from zero.

EXAMPLE   9   Find each absolute value: **(a)** $|-3|$; **(b)** $|7.2|$; **(c)** $|0|$.

SOLUTION

a) $|-3| = 3$ since -3 is 3 units from 0.

b) $|7.2| = 7.2$ since 7.2 is 7.2 units from 0.

c) $|0| = 0$ since 0 is 0 units from itself.

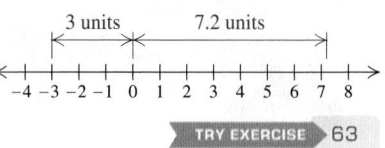

TRY EXERCISE   63

Distance is never negative, so numbers that are opposites have the same absolute value. If a number is nonnegative, its absolute value is the number itself. If a number is negative, its absolute value is its opposite.

TECHNOLOGY CONNECTION

Most graphing calculators use the notation abs (2) to indicate the absolute value of 2. This is often accessed using the NUM option of the MATH key. When using a graphing calculator for this, be sure to distinguish between the ⊖ and ⊙ keys (see p. 46 in Section 1.6).

1.4   ## EXERCISE SET

↪ *Concept Reinforcement*   *In each of Exercises 1–8, fill in the blank using one of the following words:* natural number, whole number, integer, rational number, terminating, repeating, irrational number, absolute value.

1. Division can be used to show that $\frac{4}{7}$ can be written as a(n) _____ decimal.

2. Division can be used to show that $\frac{3}{20}$ can be written as a(n) _____ decimal.

3. If a number is a(n) _____, it is either a whole number or the opposite of a whole number.

4. 0 is the only _____ that is not a natural number.

5. Any number of the form a/b, where a and b are integers, with $b \neq 0$, is an example of a(n) _____.

6. A number like $\sqrt{5}$, which cannot be written precisely in fraction notation or decimal notation, is an example of a(n) _____.

7. If a number is a(n) _____, then it can be thought of as a counting number.

8. When two numbers are opposites, they have the same _____.

State which real number(s) correspond to each situation.

9. *Student loans and grants.*   The maximum amount that a student may borrow each year with a Stafford Loan is $10,500. The maximum annual award for the Nurse Educator Scholarship Program is $27,482.
   Sources: www.studentaid.ed.gov and www.collegezone.com

10. Using a NordicTrack exercise machine, LaToya burned 150 calories. She then drank an isotonic drink containing 65 calories.

11. The highest temperature ever recorded in a desert was 136 degrees Fahrenheit (°F) at Al-Aziziyah in the Sahara, Libya. The coldest temperature recorded in a desert was 4°F below zero in the McMurdo Dry Valleys, Antarctica.
Source: *Guinness World Records* 2007

12. The Dead Sea is 1312 ft below sea level, whereas Mt. Everest is 29,035 ft above sea level.
Source: *Guinness World Records* 2007

13. *Stock market.* The Dow Jones Industrial Average is an indicator of the stock market. On October 12, 1997, the Dow Jones fell a record 554 points. On March 16, 2002, the Dow Jones gained a record 499.19 points.
Source: www.finfacts.ie

14. Ignition occurs 10 sec before liftoff. A spent fuel tank is detached 235 sec after liftoff.

15. Kim deposited $650 in a savings account. Two weeks later, she withdrew $180.

16. *Birth and death rates.* Recently, the world birth rate was 20.09 per thousand. The death rate was 8.37 per thousand.
Source: Central Intelligence Agency, 2007

17. The halfback gained 8 yd on the first play. The quarterback was tackled for a 5-yd loss on the second play.

18. In the 2007 Masters Tournament, golfer Tiger Woods finished 3 over par. In the World Golf Championship, he finished 10 under par.
Source: PGA Tour Inc.

Graph each rational number on a number line.

19. $\frac{10}{3}$ **20.** $-\frac{17}{5}$

21. -4.3 **22.** 3.87

23. -2 **24.** 5

Write decimal notation for each number.

25. $\frac{7}{8}$ **26.** $-\frac{1}{8}$ **27.** $-\frac{3}{4}$

28. $\frac{11}{6}$ **29.** $-\frac{7}{6}$ **30.** $-\frac{5}{12}$

31. $\frac{2}{3}$ **32.** $\frac{1}{4}$ **33.** $-\frac{1}{2}$

34. $-\frac{1}{9}$ Aha! **35.** $\frac{13}{100}$ **36.** $-\frac{9}{20}$

Graph each irrational number on a number line.

37. $\sqrt{5}$ **38.** $\sqrt{92}$

39. $-\sqrt{22}$ **40.** $-\sqrt{54}$

Write a true sentence using either $<$ or $>$.

41. $5 \;\blacksquare\; 0$ **42.** $8 \;\blacksquare\; -8$

43. $-9 \;\blacksquare\; 9$ **44.** $0 \;\blacksquare\; -7$

45. $-8 \;\blacksquare\; -5$ **46.** $-4 \;\blacksquare\; -3$

47. $-5 \;\blacksquare\; -11$ **48.** $-3 \;\blacksquare\; -4$

49. $-12.5 \;\blacksquare\; -10.2$ **50.** $-10.3 \;\blacksquare\; -14.5$

51. $\frac{5}{12} \;\blacksquare\; \frac{11}{25}$ **52.** $-\frac{14}{17} \;\blacksquare\; -\frac{27}{35}$

For each of the following, write a second inequality with the same meaning.

53. $-2 > x$ **54.** $a > 9$

55. $10 \le y$ **56.** $-12 \ge t$

Classify each inequality as either true or false.

57. $-3 \ge -11$ **58.** $5 \le -5$

59. $0 \ge 8$ **60.** $-5 \le 7$

61. $-8 \le -8$ **62.** $10 \ge 10$

Find each absolute value.

63. $|-58|$ **64.** $|-47|$

65. $|-12.2|$ **66.** $|4.3|$

67. $|\sqrt{2}|$ **68.** $|-456|$

69. $\left|-\frac{9}{7}\right|$ **70.** $|-\sqrt{3}|$

71. $|0|$ **72.** $\left|-\frac{3}{4}\right|$

73. $|x|$, for $x = -8$ **74.** $|a|$, for $a = -5$

For Exercises 75–80, consider the following list:
$$-83, \quad -4.7, \quad 0, \quad \frac{5}{9}, \quad 2.\overline{16}, \quad \pi, \quad \sqrt{17}, \quad 62.$$

75. List all rational numbers.

76. List all natural numbers.

77. List all integers.

78. List all irrational numbers.

79. List all real numbers.

80. List all nonnegative integers.

81. Is every integer a rational number? Why or why not?

82. Is every integer a natural number? Why or why not?

Skill Review

83. Evaluate $3xy$ for $x = 2$ and $y = 7$. [1.1]

84. Use a commutative law to write an expression equivalent to $ab + 5$. [1.2]

Synthesis

85. Is the absolute value of a number always positive? Why or why not?

86. How many rational numbers are there between 0 and 1? Justify your answer.

87. Does "nonnegative" mean the same thing as "positive"? Why or why not?

List in order from least to greatest.

88. $13, -12, 5, -17$

89. $-23, 4, 0, -17$

90. $-\frac{2}{3}, \frac{1}{2}, -\frac{3}{4}, -\frac{5}{6}, \frac{3}{8}, \frac{1}{6}$

91. $\frac{4}{5}, \frac{4}{3}, \frac{4}{8}, \frac{4}{6}, \frac{4}{9}, \frac{4}{2}, -\frac{4}{3}$

Write a true sentence using either $<$, $>$, or $=$.

92. $|-5| \;\blacksquare\; |-2|$

93. $|4| \;\blacksquare\; |-7|$

94. $|-8| \;\blacksquare\; |8|$

95. $|23| \;\blacksquare\; |-23|$

96. $|-11| \;\blacksquare\; |5|$

Solve. Consider only integer replacements.

Aha! **97.** $|x| = 19$

98. $|x| < 3$

99. $2 < |x| < 5$

Given that $0.3\overline{3} = \frac{1}{3}$ and $0.6\overline{6} = \frac{2}{3}$, express each of the following as a ratio of two integers.

100. $0.1\overline{1}$

101. $0.9\overline{9}$

102. $5.5\overline{5}$

103. $7.7\overline{7}$

Translate to an inequality.

104. A number a is negative.

105. A number x is nonpositive.

106. The distance from x to 0 is no more than 10.

107. The distance from t to 0 is at least 20.

To the student and the instructor: *The calculator icon, ▧, is used to indicate those exercises designed to be solved with a calculator.*

108. When Helga's calculator gives a decimal value for $\sqrt{2}$ and that value is promptly squared, the result is 2. Yet when that same decimal approximation is entered by hand and then squared, the result is not exactly 2. Why do you suppose this is?

109. Is the following statement true? Why or why not?
$$\sqrt{a^2} = |a| \quad \text{for any real number } a.$$

1.5   Addition of Real Numbers

Adding with the Number Line ▪ Adding Without the Number Line ▪ Problem Solving ▪ Combining Like Terms

We now consider addition of real numbers. To gain understanding, we will use the number line first. After observing the principles involved, we will develop rules that allow us to work more quickly without the number line.

Adding with the Number Line

To add $a + b$ on the number line, we start at a and move according to b.

a) If b is positive, we move to the right (the positive direction).

b) If b is negative, we move to the left (the negative direction).

c) If b is 0, we stay at a.

EXAMPLE **1**

Add: $-4 + 9$.

SOLUTION   To add on the number line, we locate the first number, -4, and then move 9 units to the right. Note that it requires 4 units to reach 0. The difference between 9 and 4 is where we finish.

$$-4 + 9 = 5$$

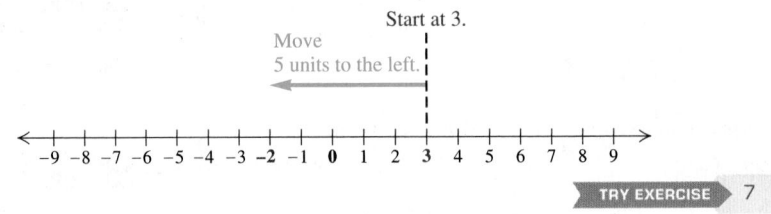

TRY EXERCISE 9

EXAMPLE **2**

Add: $3 + (-5)$.

STUDENT NOTES

Parentheses are essential when a negative sign follows an operation. Just as we would never write $8 \div \times 2$, it is improper to write $3 + -5$.

SOLUTION   We locate the first number, 3, and then move 5 units to the left. Note that it requires 3 units to reach 0. The difference between 5 and 3 is 2, so we finish 2 units to the left of 0.

$$3 + (-5) = -2$$

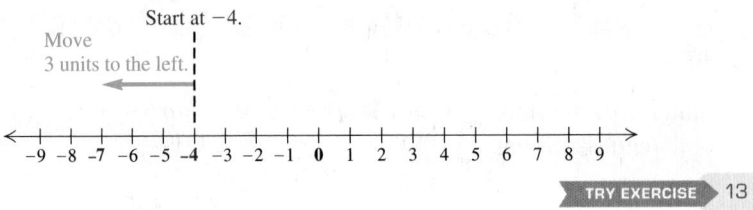

TRY EXERCISE 7

EXAMPLE **3**

Add: $-4 + (-3)$.

SOLUTION   After locating -4, we move 3 units to the left. We finish a total of 7 units to the left of 0.

$$-4 + (-3) = -7$$

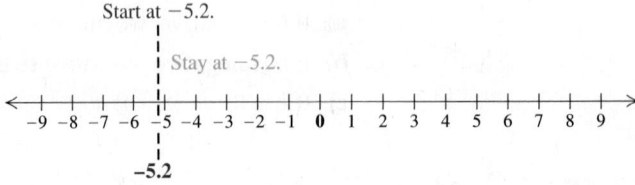

TRY EXERCISE 13

EXAMPLE **4**

Add: $-5.2 + 0$.

SOLUTION   We locate -5.2 and move 0 units. Thus we finish where we started, at -5.2.

$$-5.2 + 0 = -5.2$$

Start at -5.2.

Stay at -5.2.

−9 −8 −7 −6 −5 −4 −3 −2 −1 0 1 2 3 4 5 6 7 8 9

−5.2

TRY EXERCISE 11

From Examples 1–4, the following rules emerge.

Rules for Addition of Real Numbers

1. *Positive numbers*: Add as usual. The answer is positive.
2. *Negative numbers*: Add absolute values and make the answer negative (see Example 3).
3. *A positive number and a negative number*: Subtract the smaller absolute value from the greater absolute value. Then:

   a) If the positive number has the greater absolute value, the answer is positive (see Example 1).
   b) If the negative number has the greater absolute value, the answer is negative (see Example 2).
   c) If the numbers have the same absolute value, the answer is 0.

4. *One number is zero*: The sum is the other number (see Example 4).

Rule 4 is known as the **identity property of 0**.

Identity Property of 0

For any real number a,

$$a + 0 = 0 + a = a.$$

(Adding 0 to a number gives that same number.) The number 0 is called the *additive identity*.

Adding Without the Number Line

The rules listed above can be used without drawing the number line.

EXAMPLE **5**   Add without using the number line.

a) $-12 + (-7)$ b) $-1.4 + 8.5$
c) $-36 + 21$ d) $1.5 + (-1.5)$
e) $-\frac{7}{8} + 0$ f) $\frac{2}{3} + \left(-\frac{5}{8}\right)$

SOLUTION

a) $-12 + (-7) = -19$ Two negatives. *Think:* Add the absolute values, 12 and 7, to get 19. Make the answer *negative*, -19.

b) $-1.4 + 8.5 = 7.1$ A negative and a positive. *Think:* The difference of absolute values is $8.5 - 1.4$, or 7.1. The positive number has the greater absolute value, so the answer is *positive*, 7.1.

c) $-36 + 21 = -15$ A negative and a positive. *Think:* The difference of absolute values is $36 - 21$, or 15. The negative number has the greater absolute value, so the answer is *negative*, -15.

d) $1.5 + (-1.5) = 0$ A negative and a positive. *Think:* Since the numbers are opposites, they have the same absolute value and the answer is 0.

STUDY SKILLS ————————

Two (or a Few More) Heads Are Better Than One

Consider forming a study group with some of your classmates. Exchange telephone numbers, schedules, and any e-mail addresses so that you can coordinate study time for homework and tests.

e) $-\dfrac{7}{8} + 0 = -\dfrac{7}{8}$ **One number is zero.** The sum is the other number, $-\frac{7}{8}$.

f) $\dfrac{2}{3} + \left(-\dfrac{5}{8}\right) = \dfrac{16}{24} + \left(-\dfrac{15}{24}\right)$ This is similar to part (b) above. We find a common denominator and then add.

$\qquad\qquad = \dfrac{1}{24}$

> TRY EXERCISES   15 and 21

If we are adding several numbers, some positive and some negative, the commutative and associative laws allow us to add all the positives, then add all the negatives, and then add the results. Of course, we can also add from left to right, if we prefer.

EXAMPLE   **6**   Add: $15 + (-2) + 7 + 14 + (-5) + (-12)$.

SOLUTION

$15 + (-2) + 7 + 14 + (-5) + (-12)$

$= 15 + 7 + 14 + (-2) + (-5) + (-12)$ Using the commutative law of addition

$= (15 + 7 + 14) + [(-2) + (-5) + (-12)]$ Using the associative law of addition

$= 36 + (-19)$ Adding the positives; adding the negatives

$= 17$ Adding a positive and a negative

> TRY EXERCISE   55

Problem Solving

EXAMPLE   **7**

Interest rates.   Between 1994 and 2007, the average interest rate for a 30-yr fixed-rate mortgage dropped 2.5 percent, rose 1.75 percent, dropped 3.25 percent, and rose 1 percent. By how much did the average interest rate change?

Source: Mortgage-X.com

SOLUTION   The problem translates to a sum:

Rewording: The 1st change   plus   the 2nd change   plus   the 3rd change   plus   the 4th change   is   the total change.

Translating: -2.5 $+$ 1.75 $+$ (-3.25) $+$ 1 $=$ Total change

Adding from left to right, we have

$$-2.5 + 1.75 + (-3.25) + 1 = -0.75 + (-3.25) + 1 = -4 + 1 = -3.$$

The average interest rate dropped 3 percent between 1994 and 2007.

> TRY EXERCISE   59

Combining Like Terms

When two terms have variable factors that are exactly the same, like $5a$ and $-7a$, the terms are called **like**, or **similar**, **terms.*** The distributive law enables us to **combine**, or **collect**, **like terms**. The above rules for addition will again apply.

EXAMPLE **8**   Combine like terms.

a) $-7x + 9x$

b) $2a + (-3b) + (-5a) + 9b$

c) $6 + y + (-3.5y) + 2$

SOLUTION

a) $-7x + 9x = (-7 + 9)x$ Using the distributive law

$\quad\quad\quad\quad\quad = 2x$ Adding -7 and 9

b) $2a + (-3b) + (-5a) + 9b$

$\quad = 2a + (-5a) + (-3b) + 9b$ Using the commutative law of addition

$\quad = (2 + (-5))a + (-3 + 9)b$ Using the distributive law

$\quad = -3a + 6b$ Adding

c) $6 + y + (-3.5y) + 2 = y + (-3.5y) + 6 + 2$ Using the commutative law of addition

$\quad\quad\quad\quad\quad\quad = (1 + (-3.5))y + 6 + 2$ Using the distributive law

$\quad\quad\quad\quad\quad\quad = -2.5y + 8$ Adding

TRY EXERCISE ▶ 69

With practice we can omit some steps, combining like terms mentally. Note that numbers like 6 and 2 in the expression $6 + y + (-3.5y) + 2$ are constants and are also considered to be like terms.

1.5   EXERCISE SET

For Extra Help   *MyMathLab*

↷ *Concept Reinforcement*   *In each of Exercises 1–6, match the term with a like term from the column on the right.*

1. ___ $8n$

2. ___ $7m$

3. ___ 43

4. ___ $28z$

5. ___ $-2x$

6. ___ $-9t$

a) $-3z$

b) $5x$

c) $2t$

d) $-4m$

e) 9

f) $-3n$

Add using the number line.

7. $5 + (-8)$

8. $2 + (-5)$

9. $-6 + 10$

10. $-3 + 8$

11. $-7 + 0$

12. $-6 + 0$

13. $-3 + (-5)$

14. $-4 + (-6)$

Add. Do not use a number line except as a check.

15. $-35 + 0$

16. $-68 + 0$

17. $0 + (-8)$

18. $0 + (-2)$

19. $12 + (-12)$

20. $17 + (-17)$

---

*Like terms are discussed in greater detail in Section 1.8.

21. $-24 + (-17)$

22. $-17 + (-25)$

23. $-13 + 13$

24. $-31 + 31$

25. $20 + (-11)$

26. $8 + (-5)$

27. $10 + (-12)$

28. $9 + (-13)$

29. $-3 + 14$

30. $25 + (-6)$

31. $-24 + (-19)$

32. $11 + (-9)$

33. $19 + (-19)$

34. $-20 + (-6)$

35. $23 + (-5)$

36. $-15 + (-7)$

37. $-31 + (-14)$

38. $40 + (-8)$

39. $40 + (-40)$

40. $-25 + 25$

41. $85 + (-69)$

42. $63 + (-13)$

43. $-3.6 + 2.8$

44. $-6.5 + 4.7$

45. $-5.4 + (-3.7)$

46. $-3.8 + (-9.4)$

47. $\frac{4}{5} + \left(\frac{-1}{5}\right)$

48. $\frac{-2}{7} + \frac{3}{7}$

49. $\frac{-4}{7} + \frac{-2}{7}$

50. $\frac{-5}{9} + \frac{-2}{9}$

51. $-\frac{2}{5} + \frac{1}{3}$

52. $-\frac{4}{13} + \frac{1}{2}$

53. $\frac{-4}{9} + \frac{2}{3}$

54. $\frac{1}{9} + \left(\frac{-1}{3}\right)$

55. $35 + (-14) + (-19) + (-5)$

56. $-28 + (-44) + 17 + 31 + (-94)$

Aha! **57.** $-4.9 + 8.5 + 4.9 + (-8.5)$

58. $24 + 3.1 + (-44) + (-8.2) + 63$

Solve. Write your answer as a complete sentence.

59. *Gasoline prices.* In a recent year, the price of a gallon of 87-octane gasoline was $2.89. The price rose 15¢, then dropped 3¢, and then rose 17¢. By how much did the price change during that period?

60. *Natural gas prices.* In a recent year, the price of a gallon of natural gas was $1.88. The price dropped 2¢, then rose 25¢, and then dropped 43¢. By how much did the price change during that period?

61. *Telephone bills.* Chloe's cell-phone bill for July was $82. She sent a check for $50 and then ran up $63 in charges for August. What was her new balance?

62. *Profits and losses.* The following table lists the profits and losses of Premium Sales over a 3-yr period. Find the profit or loss after this period of time.

Year	Profit or loss
2006	−$26,500
2007	−$10,200
2008	+$32,400

63. *Yardage gained.* In an intramural football game, the quarterback attempted passes with the following results.

First try	13-yd loss
Second try	12-yd gain
Third try	21-yd gain

Find the total gain (or loss).

64. *Account balance.* Aiden has $450 in a checking account. He writes a check for $530, makes a deposit of $75, and then writes a check for $90. What is the balance in the account?

65. *Lake level.* Between October 2003 and February 2005, the south end of the Great Salt Lake dropped $\frac{2}{5}$ ft, rose $1\frac{1}{5}$ ft, and dropped $\frac{1}{2}$ ft. By how much did the level change?
Source: U.S. Geological Survey

66. *Peak elevation.* The tallest mountain in the world, as measured from base to peak, is Mauna Kea in Hawaii. From a base 19,684 ft below sea level, it rises 33,480 ft. What is the elevation of its peak?
Source: *Guinness World Records* 2007

67. *Credit-card bills.* Logan's credit-card bill indicates that he owes $470. He sends a check to the credit-card company for $45, charges another $160 in merchandise, and then pays off another $500 of his bill. What is Logan's new balance?

68. *Class size.* During the first two weeks of the semester, 5 students withdrew from Hailey's algebra class, 8 students were added to the class, and 4 students were dropped as "no-shows." By how many students did the original class size change?

Combine like terms.

69. $7a + 10a$

70. $3x + 8x$

71. $-3x + 12x$

72. $-2m + (-7m)$

73. $4t + 21t$

74. $5a + 8a$

75. $7m + (-9m)$

76. $-4x + 4x$

77. $-8y + (-2y)$

78. $10n + (-17n)$

79. $-3 + 8x + 4 + (-10x)$

80. $8a + 5 + (-a) + (-3)$

Find the perimeter of each figure.

81.
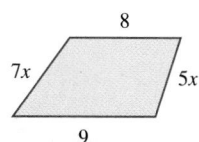

82.
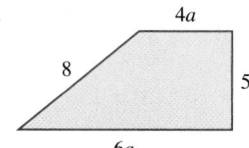

83.
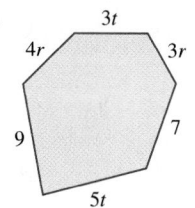

84.
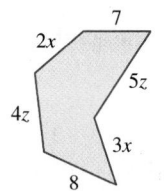

85.
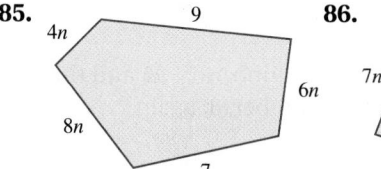

86.
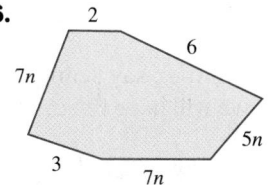

87. Explain in your own words why the sum of two negative numbers is negative.

88. Without performing the actual addition, explain why the sum of all integers from -10 to 10 is 0.

Skill Review

89. Multiply: $7(3z + 2y + 1)$. [1.2]

90. Divide and simplify: $\frac{7}{2} \div \frac{3}{8}$. [1.3]

Synthesis

91. Under what circumstances will the sum of one positive number and several negative numbers be positive?

92. Is it possible to add real numbers without knowing how to calculate $a - b$ with a and b both nonnegative and $a \geq b$? Why or why not?

93. *Banking.* Travis had $257.33 in his checking account. After depositing $152 in the account and writing a check, his account was overdrawn by $42.37. What was the amount of the check?

94. *Sports-card values.* The value of a sports card dropped $12 and then rose $17.50 before settling at $61. What was the original value of the card?

Find the missing term or terms.

95. $4x + \underline{\quad} + (-9x) + (-2y) = -5x - 7y$

96. $-3a + 9b + \underline{\quad} + 5a = 2a - 6b$

97. $3m + 2n + \underline{\quad} + (-2m) = 2n + (-6m)$

98. $\underline{\quad} + 9x + (-4y) + x = 10x - 7y$

Aha! **99.** $7t + 23 + \underline{\quad} + \underline{\quad} = 0$

100. *Geometry.* The perimeter of a rectangle is $7x + 10$. If the length of the rectangle is 5, express the width in terms of x.

101. *Golfing.* After five rounds of golf, a golf pro was 3 under par twice, 2 over par once, 2 under par once, and 1 over par once. On average, how far above or below par was the golfer?

1.6   Subtraction of Real Numbers

Opposites and Additive Inverses   ▪   Subtraction   ▪   Problem Solving

In arithmetic, when a number b is subtracted from another number a, the difference, $a - b$, is the number that when added to b gives a. For example, $45 - 17 = 28$ because $28 + 17 = 45$. We will use this approach to develop an efficient way of finding the value of $a - b$ for any real numbers a and b. Before doing so, however, we must develop some terminology.

Opposites and Additive Inverses

Numbers such as 6 and -6 are *opposites*, or *additive inverses*, of each other. Whenever opposites are added, the result is 0; and whenever two numbers add to 0, those numbers are opposites.

EXAMPLE **1**   Find the opposite of each number: **(a)** 34; **(b)** −8.3; **(c)** 0.

SOLUTION

a) The opposite of 34 is −34: $34 + (−34) = 0$.

b) The opposite of −8.3 is 8.3: $−8.3 + 8.3 = 0$.

c) The opposite of 0 is 0: $0 + 0 = 0$.   TRY EXERCISE 19

To write the opposite, we use the symbol −, as follows.

> **Opposite**
>
> The *opposite*, or *additive inverse*, of a number a is written $−a$ (read "the opposite of a" or "the additive inverse of a").

Note that if we take a number, say 8, and find its opposite, −8, and then find the opposite of the result, we will have the original number, 8, again.

EXAMPLE **2**   Find $−x$ and $−(−x)$ when $x = 16$.

SOLUTION

If $x = 16$, then $−x = −16$.   The opposite of 16 is −16.

If $x = 16$, then $−(−x) = −(−16) = 16$.   The opposite of the opposite of 16 is 16.

TRY EXERCISE 25

> **The Opposite of an Opposite**
>
> For any real number a,
>
> $$−(−a) = a.$$
>
> (The opposite of the opposite of a is a.)

EXAMPLE **3**   Find $−x$ and $−(−x)$ when $x = −3$.

SOLUTION

If $x = −3$, then $−x = −(−3) = 3$.   The opposite of −3 is 3.

Since $−(−x) = x$, it follows that $−(−(−3)) = −3$.   Finding the opposite of an opposite

TRY EXERCISE 31

Note in Example 3 that an extra set of parentheses is used to show that we are substituting the negative number −3 for x. The notation $− −x$ is not used.

A symbol such as −8 is usually read "negative 8." It could be read "the additive inverse of 8," because the additive inverse of 8 is negative 8. It could also be read "the opposite of 8," because the opposite of 8 is −8.

A symbol like $-x$, which has a variable, should be read "the opposite of x" or "the additive inverse of x" and *not* "negative x," since to do so suggests that $-x$ represents a negative number.

The symbol "$-$" is read differently depending on where it appears. For example, $-5 - (-x)$ should be read "negative five minus the opposite of x."

EXAMPLE 4

Write each of the following in words.

a) $2 - 8$ **b)** $5 - (-4)$ **c)** $-6 - (-x)$

SOLUTION

a) $2 - 8$ is read "two minus eight."

b) $5 - (-4)$ is read "five minus negative four."

c) $-6 - (-x)$ is read "negative six minus the opposite of x."

> TRY EXERCISE 11

STUDENT NOTES ⎯⎯⎯⎯

As you read mathematics, it is important to verbalize correctly the words and symbols to yourself. Consistently reading the expression $-x$ as "the opposite of x" is a good step in this direction.

As we saw in Example 3, $-x$ can represent a positive number. This notation can be used to restate a result from Section 1.5 as *the law of opposites*.

> **The Law of Opposites**
>
> For any two numbers a and $-a$,
>
> $$a + (-a) = 0.$$
>
> (When opposites are added, their sum is 0.)

A negative number is said to have a "negative *sign*." A positive number is said to have a "positive *sign*." If we change a number to its opposite, or additive inverse, we say that we have "changed or reversed its sign."

EXAMPLE 5

Change the sign (find the opposite) of each number: **(a)** -3; **(b)** -10; **(c)** 14.

SOLUTION

a) When we change the sign of -3, we obtain 3.

b) When we change the sign of -10, we obtain 10.

c) When we change the sign of 14, we obtain -14.

> TRY EXERCISE 35

Subtraction

Opposites are helpful when subtraction involves negative numbers. To see why, look for a pattern in the following:

Subtracting		*Adding the Opposite*
$9 - 5 = 4$	since $4 + 5 = 9$	$9 + (-5) = 4$
$5 - 8 = -3$	since $-3 + 8 = 5$	$5 + (-8) = -3$
$-6 - 4 = -10$	since $-10 + 4 = -6$	$-6 + (-4) = -10$
$-7 - (-10) = 3$	since $3 + (-10) = -7$	$-7 + 10 = 3$
$-7 - (-2) = -5$	since $-5 + (-2) = -7$	$-7 + 2 = -5$

The matching results suggest that we can subtract by adding the opposite of the number being subtracted. This can always be done and often provides the easiest way to subtract real numbers.

> ## Subtraction of Real Numbers
>
> For any real numbers a and b,
>
> $$a - b = a + (-b).$$
>
> (To subtract, add the opposite, or additive inverse, of the number being subtracted.)

EXAMPLE **6**

Subtract each of the following and then check with addition.

a) $2 - 6$ **b)** $4 - (-9)$ **c)** $-4.2 - (-3.6)$

d) $-1.8 - (-7.5)$ **e)** $\frac{1}{5} - \left(-\frac{3}{5}\right)$

SOLUTION

a) $2 - 6 = 2 + (-6) = -4$ The opposite of 6 is -6. We change the subtraction to addition and add the opposite. *Check:* $-4 + 6 = 2$.

b) $4 - (-9) = 4 + 9 = 13$ The opposite of -9 is 9. We change the subtraction to addition and add the opposite. *Check:* $13 + (-9) = 4$.

c) $-4.2 - (-3.6) = -4.2 + 3.6$ Adding the opposite of -3.6
$= -0.6$ *Check:* $-0.6 + (-3.6) = -4.2$.

d) $-1.8 - (-7.5) = -1.8 + 7.5$ Adding the opposite
$= 5.7$ *Check:* $5.7 + (-7.5) = -1.8$.

e) $\dfrac{1}{5} - \left(-\dfrac{3}{5}\right) = \dfrac{1}{5} + \dfrac{3}{5}$ Adding the opposite

$= \dfrac{1 + 3}{5}$ A common denominator exists so we add in the numerator.

$= \dfrac{4}{5}$

Check: $\dfrac{4}{5} + \left(-\dfrac{3}{5}\right) = \dfrac{4}{5} + \dfrac{-3}{5} = \dfrac{4 + (-3)}{5} = \dfrac{1}{5}$.

TRY EXERCISES 39 and 47

EXAMPLE **7**

Simplify: $8 - (-4) - 2 - (-5) + 3$.

SOLUTION

$8 - (-4) - 2 - (-5) + 3 = 8 + 4 + (-2) + 5 + 3$ To subtract, we add the opposite.

$= 18$

TRY EXERCISE 109

Recall from Section 1.2 that the terms of an algebraic expression are separated by plus signs. This means that the terms of $5x - 7y - 9$ are $5x$, $-7y$, and -9, since $5x - 7y - 9 = 5x + (-7y) + (-9)$.

EXAMPLE **8**

Identify the terms of $4 - 2ab + 7a - 9$.

SOLUTION   We have

$$4 - 2ab + 7a - 9 = 4 + (-2ab) + 7a + (-9),$$ Rewriting as addition

so the terms are 4, $-2ab$, $7a$, and -9.

> **TRY EXERCISE** 117

EXAMPLE **9**

Combine like terms.

a) $1 + 3x - 7x$

b) $-5a - 7b - 4a + 10b$

c) $4 - 3m - 9 + 2m$

SOLUTION

a) $1 + 3x - 7x = 1 + 3x + (-7x)$ Adding the opposite

$\qquad\qquad = 1 + (3 + (-7))x$ Using the distributive law.

$\qquad\qquad = 1 + (-4)x$ Try to do this mentally.

$\qquad\qquad = 1 - 4x$ Rewriting as subtraction to be more concise

b) $-5a - 7b - 4a + 10b = -5a + (-7b) + (-4a) + 10b$ Adding the opposite

$\qquad\qquad\qquad\qquad = -5a + (-4a) + (-7b) + 10b$ Using the commutative law of addition

$\qquad\qquad\qquad\qquad = -9a + 3b$ Combining like terms mentally

c) $4 - 3m - 9 + 2m = 4 + (-3m) + (-9) + 2m$ Rewriting as addition

$\qquad\qquad\qquad = 4 + (-9) + (-3m) + 2m$ Using the commutative law of addition

$\qquad\qquad\qquad = -5 + (-1m)$ We can write $-1m$ as $-m$.

$\qquad\qquad\qquad = -5 - m$

> **TRY EXERCISE** 121

Problem Solving

We use subtraction to solve problems involving differences. These include problems that ask "How much more?" or "How much higher?"

EXAMPLE **10**

Record elevations.   The current world records for the highest parachute jump and the lowest manned vessel ocean dive were both set in 1960. On August 16 of that year, Captain Joseph Kittinger jumped from a height of 102,800 ft above sea level. Earlier, on January 23, Jacques Piccard and Navy Lieutenant Donald Walsh descended in a bathyscaphe 35,797 ft below sea level. What was the difference in elevation between the highest parachute jump and the lowest ocean dive?

Sources: www.firstflight.org and www.seasky.org

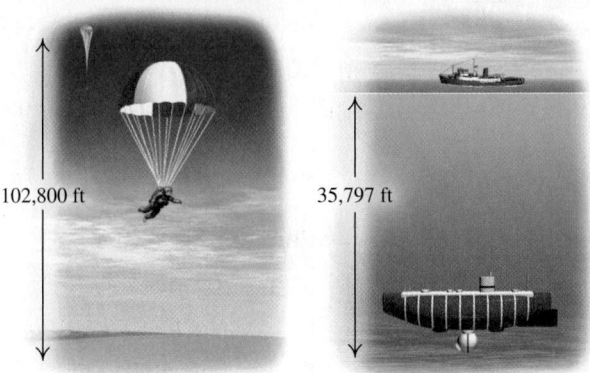

SOLUTION To find the difference between two elevations, we always subtract the lower elevation from the higher elevation:

Higher elevation − Lower elevation

$$102,800 - (-35,797)$$
$$= 102,800 + 35,797$$
$$= 138,597.$$

The parachute jump began 138,597 ft higher than the ocean dive ended.

> **TRY EXERCISE** ▶ 135

1.6 EXERCISE SET

 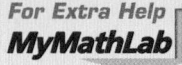

✎ *Concept Reinforcement* *In each of Exercises 1–8, match the expression with the appropriate wording from the column on the right.*

1. _____ $-x$

2. _____ $12 - x$

3. _____ $12 - (-x)$

4. _____ $x - 12$

5. _____ $x - (-12)$

6. _____ $-x - 12$

7. _____ $-x - x$

8. _____ $-x - (-12)$

a) x minus negative twelve

b) The opposite of x minus x

c) The opposite of x minus twelve

d) The opposite of x

e) The opposite of x minus negative twelve

f) Twelve minus the opposite of x

g) Twelve minus x

h) x minus twelve

Write each of the following in words.

9. $6 - 10$

10. $5 - 13$

11. $2 - (-12)$

12. $4 - (-1)$

13. $9 - (-t)$

14. $8 - (-m)$

15. $-x - y$

16. $-a - b$

17. $-3 - (-n)$

18. $-7 - (-m)$

Find the opposite, or additive inverse.

19. 51

20. -17

21. $-\frac{11}{3}$

22. $\frac{7}{2}$

23. -3.14

24. 48.2

Find $-x$ when x is each of the following.

25. -45

26. 26

27. $-\frac{14}{3}$

28. $\frac{1}{328}$

29. 0.101

30. 0

Find $-(-x)$ *when x is each of the following.*

31. 37

32. 29

33. $-\frac{2}{5}$

34. -9.1

Change the sign. (Find the opposite.)

35. -1

36. -7

37. 15

38. 10

Subtract.

39. $7 - 10$

40. $4 - 13$

41. $0 - 6$

42. $0 - 8$

43. $2 - 5$

44. $3 - 13$

45. $-4 - 3$

46. $-5 - 6$

47. $-9 - (-3)$

48. $-9 - (-5)$

Aha! **49.** $-8 - (-8)$

50. $-10 - (-10)$

51. $14 - 19$

52. $12 - 16$

53. $30 - 40$

54. $20 - 27$

55. $0 - 11$

56. $0 - 31$

57. $-9 - (-9)$

58. $-40 - (-40)$

59. $5 - 5$

60. $7 - 7$

61. $4 - (-4)$

62. $6 - (-6)$

63. $-7 - 4$

64. $-6 - 8$

65. $6 - (-10)$

66. $3 - (-12)$

67. $-4 - 15$

68. $-14 - 2$

69. $-6 - (-7)$

70. $-4 - (-7)$

71. $5 - (-12)$

72. $5 - (-6)$

73. $0 - (-3)$

74. $0 - (-5)$

75. $-5 - (-2)$

76. $-3 - (-1)$

77. $-7 - 14$

78. $-9 - 16$

79. $0 - (-10)$

80. $0 - (-1)$

81. $-8 - 0$

82. $-9 - 0$

83. $-52 - 8$

84. $-63 - 11$

85. $2 - 25$

86. $18 - 63$

87. $-4.2 - 3.1$

88. $-10.1 - 2.6$

89. $-1.3 - (-2.4)$

90. $-5.8 - (-7.3)$

91. $3.2 - 8.7$

92. $1.5 - 9.4$

93. $0.072 - 1$

94. $0.825 - 1$

95. $\frac{2}{11} - \frac{9}{11}$

96. $\frac{3}{7} - \frac{5}{7}$

97. $\frac{-1}{5} - \frac{3}{5}$

98. $\frac{-2}{9} - \frac{5}{9}$

99. $-\frac{4}{17} - \left(-\frac{9}{17}\right)$

100. $-\frac{2}{13} - \left(-\frac{5}{13}\right)$

In each of Exercises 101–104, translate the phrase to mathematical language and simplify. See the solution to Example 10.

101. The difference between 3.8 and -5.2

102. The difference between -2.1 and -5.9

103. The difference between 114 and -79

104. The difference between 23 and -17

105. Subtract 32 from -8.

106. Subtract 19 from -7.

107. Subtract -25 from 18.

108. Subtract -31 from -5.

Simplify.

109. $16 - (-12) - 1 - (-2) + 3$

110. $22 - (-18) + 7 + (-42) - 27$

111. $-31 + (-28) - (-14) - 17$

112. $-43 - (-19) - (-21) + 25$

113. $-34 - 28 + (-33) - 44$

114. $39 + (-88) - 29 - (-83)$

Aha! **115.** $-93 + (-84) - (-93) - (-84)$

116. $84 + (-99) + 44 - (-18) - 43$

Identify the terms in each expression.

117. $-3y - 8x$

118. $7a - 9b$

119. $9 - 5t - 3st$

120. $-4 - 3x + 2xy$

Combine like terms.

121. $10x - 13x$

122. $3a - 14a$

123. $7a - 12a + 4$

124. $-9x - 13x + 7$

125. $-8n - 9 + 7n$

126. $-7 + 9n - 8n$

127. $5 - 3x - 11$

128. $2 + 3a - 7$

129. $2 - 6t - 9 - 2t$

130. $-5 + 4b - 7 - 5b$

131. $5y + (-3x) - 9x + 1 - 2y + 8$

132. $14 - (-5x) + 2z - (-32) + 4z - 2x$

133. $13x - (-2x) + 45 - (-21) - 7x$

134. $8t - (-2t) - 14 - (-5t) + 53 - 9t$

Solve.

135. *Temperature extremes.*   The highest temperature ever recorded in the United States is 134°F in Greenland Ranch, California, on July 10, 1913. The lowest temperature ever recorded is -80°F in Prospect Creek, Alaska, on January 23, 1971. How much higher was the temperature in Greenland Ranch than that in Prospect Creek?
Source: Information Please Database 2007, Pearson Education, Inc.

Prospect Creek, AK
−80° F

AK

CA

Greenland Ranch, CA
134° F

136. *Temperature change.*   In just 12 hr on February 21, 1918, the temperature in Granville, North Dakota, rose from -33°F to 50°F. By how much did the temperature change?
Source: Information Please Database 2007, Pearson Education, Inc.

137. *Elevation extremes.*   The lowest elevation in Asia, the Dead Sea, is 1312 ft below sea level. The highest elevation in Asia, Mount Everest, is 29,035 ft. Find the difference in elevation.
Source: Guinness World Records 2007

138. *Elevation extremes.*   The elevation of Mount Whitney, the highest peak in California, is 14,776 ft more than the elevation of Death Valley, California.

If Death Valley is 282 ft below sea level, find the elevation of Mount Whitney.
Source: The Columbia Electronic Encyclopedia, 6th ed., 2007 (New York: Columbia University Press)

139. *Changes in elevation.*   The lowest point in Africa is Lake Assal, which is 156 m below sea level. The lowest point in South America is the Valdes Peninsula, which is 40 m below sea level. How much lower is Lake Assal than the Valdes Peninsula?
Source: Information Please Database 2007, Pearson Education, Inc.

140. *Underwater elevation.*   The deepest point in the Pacific Ocean is the Marianas Trench, with a depth of 10,911 m. The deepest point in the Atlantic Ocean is the Puerto Rico Trench, with a depth of 8648 m. What is the difference in elevation of the two trenches?
Source: Guinness World Records 2007

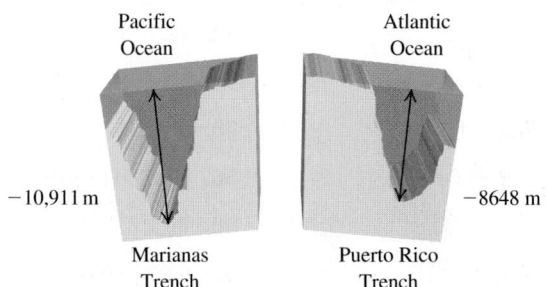

Pacific Ocean

Atlantic Ocean

$-10,911$ m

-8648 m

Marianas Trench

Puerto Rico Trench

141. Jeremy insists that if you can *add* real numbers, then you can also *subtract* real numbers. Do you agree? Why or why not?

142. Are the expressions $-a + b$ and $a + (-b)$ opposites of each other? Why or why not?

Skill Review

143. Find the area of a rectangle when the length is 36 ft and the width is 12 ft. [1.1]

144. Find the prime factorization of 864. [1.3]

Synthesis

145. Explain the different uses of the symbol "$-$". Give examples of each and how they should be read.

146. If a and b are both negative, under what circumstances will $a - b$ be negative?

147. *Power outages.*   During the Northeast's electrical blackout of August 14, 2003, residents of Bloomfield, New Jersey, lost power at 4:00 P.M. One resident returned from vacation at 3:00 P.M. the following day to find the clocks in her apartment reading 8:00 A.M. At what time, and on what day, was power restored?

Tell whether each statement is true or false for all real numbers m and n. Use various replacements for m and n to support your answer.

148. If $m > n$, then $m - n > 0$.

149. If $m > n$, then $m + n > 0$.

150. If m and n are opposites, then $m - n = 0$.

151. If $m = -n$, then $m + n = 0$.

152. A gambler loses a wager and then loses "double or nothing" (meaning the gambler owes twice as much) twice more. After the three losses, the gambler's assets are −$20. Explain how much the gambler originally bet and how the $20 debt occurred.

153. List the keystrokes needed to compute $-9 - (-7)$.

154. If n is positive and m is negative, what is the sign of $n + (-m)$? Why?

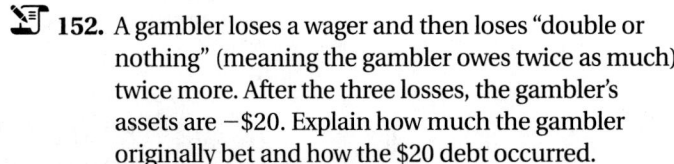

| 1.7 | Multiplication and Division of Real Numbers |

Multiplication ■ Division

We now develop rules for multiplication and division of real numbers. Because multiplication and division are closely related, the rules are quite similar.

Multiplication

We already know how to multiply two nonnegative numbers. To see how to multiply a positive number and a negative number, consider the following pattern in which multiplication is regarded as repeated addition:

This number → decreases by 1 each time.
$$4(-5) = (-5) + (-5) + (-5) + (-5) = -20$$
$$3(-5) = (-5) + (-5) + (-5) = -15$$
$$2(-5) = (-5) + (-5) = -10$$
$$1(-5) = (-5) = -5$$
$$0(-5) = 0 = 0$$
← This number increases by 5 each time.

This pattern illustrates that the product of a negative number and a positive number is negative.

The Product of a Negative Number and a Positive Number

To multiply a positive number and a negative number, multiply their absolute values. The answer is negative.

EXAMPLE 1   Multiply: **(a)** $8(-5)$; **(b)** $-\frac{1}{3} \cdot \frac{5}{7}$.

SOLUTION

a) $8(-5) = -40$ *Think:* $8 \cdot 5 = 40$; make the answer negative.

b) $-\frac{1}{3} \cdot \frac{5}{7} = -\frac{5}{21}$ *Think:* $\frac{1}{3} \cdot \frac{5}{7} = \frac{5}{21}$; make the answer negative.

TRY EXERCISE 11

The pattern developed above includes not just products of positive numbers and negative numbers, but a product involving zero as well.

> ## The Multiplicative Property of Zero
> For any real number a,
> $$0 \cdot a = a \cdot 0 = 0.$$
> (The product of 0 and any real number is 0.)

EXAMPLE 2   Multiply: $173(-452)0$.

SOLUTION   We have

$$173(-452)0 = 173[(-452)0]$$ Because of the associative law of multiplication, we can multiply the last two factors first.

$$= 173[0]$$ Using the multiplicative property of zero

$$= 0.$$ Using the multiplicative property of zero again

Note that whenever 0 appears as a factor, the product is 0.

TRY EXERCISE 33

We can extend the above pattern still further to examine the product of two negative numbers.

This number → $\quad 2(-5) = \quad (-5) + (-5) = -10 \leftarrow$ This number increases
decreases by $\qquad 1(-5) = \qquad\qquad (-5) = \;-5 \qquad$ by 5 each time.
1 each time. $\qquad 0(-5) = \qquad\qquad\quad 0 = \quad 0$
$\qquad\qquad -1(-5) = \qquad\quad -(-5) = \quad 5$
$\qquad\qquad -2(-5) = -(-5) - (-5) = \quad 10$

According to the pattern, the product of two negative numbers is positive.

> ## The Product of Two Negative Numbers
> To multiply two negative numbers, multiply their absolute values. The answer is positive.

EXAMPLE 3   Multiply: **(a)** $(-6)(-8)$; **(b)** $(-1.2)(-3)$.

SOLUTION

a) The absolute value of -6 is 6 and the absolute value of -8 is 8. Thus,

$$(-6)(-8) = 6 \cdot 8 \qquad \text{Multiplying absolute values. The answer is positive.}$$
$$= 48.$$

b) $(-1.2)(-3) = (1.2)(3) \qquad \text{Multiplying absolute values. The answer is positive.}$
$$= 3.6 \qquad \text{Try to go directly to this step.}$$

> TRY EXERCISE 17

When three or more numbers are multiplied, we can order and group the numbers as we please, because of the commutative and associative laws.

EXAMPLE 4   Multiply: **(a)** $-3(-2)(-5)$; **(b)** $-4(-6)(-1)(-2)$.

SOLUTION

a) $-3(-2)(-5) = 6(-5) \qquad \text{Multiplying the first two numbers. The product of two negatives is positive.}$

$$= -30 \qquad \text{The product of a positive and a negative is negative.}$$

b) $-4(-6)(-1)(-2) = 24 \cdot 2 \qquad \text{Multiplying the first two numbers and the last two numbers}$

$$= 48$$

> TRY EXERCISE 43

We can see the following pattern in the results of Example 4.

The product of an even number of negative numbers is positive.

The product of an odd number of negative numbers is negative.

Division

Recall that $a \div b$, or $\dfrac{a}{b}$, is the number, if one exists, that when multiplied by b gives a. For example, to show that $10 \div 2$ is 5, we need only note that $5 \cdot 2 = 10$. Thus division can always be checked with multiplication.

EXAMPLE 5   Divide, if possible, and check your answer.

a) $14 \div (-7)$

b) $\dfrac{-32}{-4}$

c) $\dfrac{-10}{9}$

d) $\dfrac{-17}{0}$

SOLUTION

a) $14 \div (-7) = -2 \qquad$ We look for a number that when multiplied by -7 gives 14. That number is -2. *Check:* $(-2)(-7) = 14$.

b) $\dfrac{-32}{-4} = 8 \qquad$ We look for a number that when multiplied by -4 gives -32. That number is 8. *Check:* $8(-4) = -32$.

c) $\dfrac{-10}{9} = -\dfrac{10}{9} \qquad$ We look for a number that when multiplied by 9 gives -10. That number is $-\frac{10}{9}$. *Check:* $-\frac{10}{9} \cdot 9 = -10$.

d) $\dfrac{-17}{0}$ is **undefined**. We look for a number that when multiplied by 0 gives -17. There is no such number because if 0 is a factor, the product is 0, not -17.

> TRY EXERCISE 57

STUDENT NOTES

Try to regard "undefined" as a mathematical way of saying "we do not give any meaning to this expression."

The rules for signs for division are the same as those for multiplication: The quotient of a positive number and a negative number is negative; the quotient of two negative numbers is positive.

> ## Rules for Multiplication and Division
>
> To multiply or divide two nonzero real numbers:
>
> 1. Using the absolute values, multiply or divide, as indicated.
> 2. If the signs are the same, the answer is positive.
> 3. If the signs are different, the answer is negative.

Had Example 5(a) been written as $-14 \div 7$ or $-\frac{14}{7}$, rather than $14 \div (-7)$, the result would still have been -2. Thus from Examples 5(a)–5(c), we have the following:

$$\frac{-a}{b} = \frac{a}{-b} = -\frac{a}{b} \quad \text{and} \quad \frac{-a}{-b} = \frac{a}{b}.$$

EXAMPLE 6 Rewrite each of the following in two equivalent forms: **(a)** $\frac{5}{-2}$; **(b)** $-\frac{3}{10}$.

SOLUTION We use one of the properties just listed.

a) $\dfrac{5}{-2} = \dfrac{-5}{2}$ and $\dfrac{5}{-2} = -\dfrac{5}{2}$

b) $-\dfrac{3}{10} = \dfrac{-3}{10}$ and $-\dfrac{3}{10} = \dfrac{3}{-10}$

Since $\dfrac{-a}{b} = \dfrac{a}{-b} = -\dfrac{a}{b}$

TRY EXERCISE 81

When a fraction contains a negative sign, it can be helpful to rewrite (or simply visualize) the fraction in an equivalent form.

EXAMPLE 7 Perform the indicated operation: **(a)** $\left(-\frac{4}{5}\right)\left(\frac{-7}{3}\right)$; **(b)** $-\frac{2}{7} + \frac{9}{-7}$.

SOLUTION

a) $\left(-\dfrac{4}{5}\right)\left(\dfrac{-7}{3}\right) = \left(-\dfrac{4}{5}\right)\left(-\dfrac{7}{3}\right)$ Rewriting $\dfrac{-7}{3}$ as $-\dfrac{7}{3}$

$\qquad = \dfrac{28}{15}$ Try to go directly to this step.

b) Given a choice, we generally choose a positive denominator:

$-\dfrac{2}{7} + \dfrac{9}{-7} = \dfrac{-2}{7} + \dfrac{-9}{7}$ Rewriting both fractions with a common denominator of 7

$\qquad = \dfrac{-11}{7}, \text{ or } -\dfrac{11}{7}.$

TRY EXERCISE 101

To divide with fraction notation, it is usually easiest to find a reciprocal and then multiply.

EXAMPLE 8   Find the reciprocal of each number, if it exists.

a) -27 **b)** $\frac{-3}{4}$

c) $-\frac{1}{5}$ **d)** 0

SOLUTION   Recall from Section 1.3 that we can check that two numbers are reciprocals of each other by confirming that their product is 1.

a) The reciprocal of -27 is $\frac{1}{-27}$. More often, this number is written as $-\frac{1}{27}$.
   Check: $(-27)\left(-\frac{1}{27}\right) = \frac{27}{27} = 1.$

b) The reciprocal of $\frac{-3}{4}$ is $\frac{4}{-3}$, or, equivalently, $-\frac{4}{3}$.   *Check:* $\frac{-3}{4} \cdot \frac{4}{-3} = \frac{-12}{-12} = 1.$

c) The reciprocal of $-\frac{1}{5}$ is -5.   *Check:* $-\frac{1}{5}(-5) = \frac{5}{5} = 1.$

d) The reciprocal of 0 does not exist. To see this, recall that there is no number r for which $0 \cdot r = 1$.

TRY EXERCISE 89

EXAMPLE 9   Divide: **(a)** $-\frac{2}{3} \div \left(-\frac{5}{4}\right)$; **(b)** $-\frac{3}{4} \div \frac{3}{10}$.

SOLUTION   We divide by multiplying by the reciprocal of the divisor.

a) $-\dfrac{2}{3} \div \left(-\dfrac{5}{4}\right) = -\dfrac{2}{3} \cdot \left(-\dfrac{4}{5}\right) = \dfrac{8}{15}$ Multiplying by the reciprocal

> Be careful not to change the sign when taking a reciprocal!

b) $-\dfrac{3}{4} \div \dfrac{3}{10} = -\dfrac{3}{4} \cdot \left(\dfrac{10}{3}\right) = -\dfrac{30}{12} = -\dfrac{5}{2} \cdot \dfrac{6}{6} = -\dfrac{5}{2}$ Removing a factor equal to 1: $\frac{6}{6} = 1$

TRY EXERCISE 109

To divide with decimal notation, it is usually easiest to carry out the division.

EXAMPLE 10   Divide: $27.9 \div (-3)$.

SOLUTION

$$27.9 \div (-3) = \frac{27.9}{-3} = -9.3 \qquad \text{Dividing: } 3\overline{)27.9}\ . $$
The answer is negative.

TRY EXERCISE 67

In Example 5(d), we explained why we cannot divide -17 by 0. To see why *no* nonzero number b can be divided by 0, remember that $b \div 0$ would have to be the number that when multiplied by 0 gives b. But since the product of 0 and any number is 0, not b, we say that $b \div 0$ is **undefined** for $b \neq 0$. In the special case of $0 \div 0$, we look for a number r such that $0 \div 0 = r$ and $r \cdot 0 = 0$. But, $r \cdot 0 = 0$ for *any* number r. For this reason, we say that $b \div 0$ is undefined for any choice of b.*

Finally, note that $0 \div 7 = 0$ since $0 \cdot 7 = 0$. This can be written $0/7 = 0$. It is important not to confuse division *by* 0 with division *into* 0.

---

*Sometimes $0 \div 0$ is said to be *indeterminate*.

EXAMPLE **11** Divide, if possible: **(a)** $\frac{0}{-2}$; **(b)** $\frac{5}{0}$.

SOLUTION

a) $\frac{0}{-2} = 0$ We can divide 0 by a nonzero number.
Check: $0(-2) = 0$.

b) $\frac{5}{0}$ is undefined. We cannot divide by 0. **TRY EXERCISE** 73

Division Involving Zero

For any real number a,

$$\frac{a}{0} \text{ is undefined,}$$

and for $a \neq 0$,

$$\frac{0}{a} = 0.$$

It is important *not* to confuse *opposite* with *reciprocal*. Keep in mind that the opposite, or additive inverse, of a number is what we add to the number to get 0. The reciprocal, or multiplicative inverse, is what we multiply the number by to get 1.

Compare the following.

Number	Opposite (Change the sign.)	Reciprocal (Invert but do not change the sign.)
$-\frac{3}{8}$	$\frac{3}{8}$	$-\frac{8}{3}$ $\left(-\frac{3}{8}\right)\left(-\frac{8}{3}\right) = 1$
19	-19	$\frac{1}{19}$ $\quad -\frac{3}{8} + \frac{3}{8} = 0$
$\frac{18}{7}$	$-\frac{18}{7}$	$\frac{7}{18}$
-7.9	7.9	$-\frac{1}{7.9}$, or $-\frac{10}{79}$
0	0	Undefined

1.7 **EXERCISE SET**

↪ *Concept Reinforcement* *In each of Exercises 1–10, replace the blank with either* 0 *or* 1 *to match the description given.*

1. The product of two reciprocals ____

2. The sum of a pair of opposites ____

3. The sum of a pair of additive inverses ____

4. The product of two multiplicative inverses ____

5. This number has no reciprocal. ____

6. This number is its own reciprocal. ____

7. This number is the multiplicative identity. ____

8. This number is the additive identity. ____

9. A nonzero number divided by itself ____

10. Division by this number is undefined. ____

Multiply.

11. $-4 \cdot 10$

12. $-5 \cdot 6$

13. $-8 \cdot 7$

14. $-9 \cdot 2$

15. $4 \cdot (-10)$

16. $9 \cdot (-5)$

17. $-9 \cdot (-8)$

18. $-10 \cdot (-11)$

19. $-6 \cdot 7$

20. $-2 \cdot 5$

21. $-5 \cdot (-9)$

22. $-9 \cdot (-2)$

23. $-19 \cdot (-10)$

24. $-12 \cdot (-10)$

25. $11 \cdot (-12)$

26. $-13 \cdot (-15)$

27. $-25 \cdot (-48)$

28. $15 \cdot (-43)$

29. $4.5 \cdot (-28)$

30. $-49 \cdot (-2.1)$

31. $-5 \cdot (-2.3)$

32. $-6 \cdot 4.8$

33. $(-25) \cdot 0$

34. $0 \cdot (-4.7)$

35. $\frac{2}{5} \cdot \left(-\frac{5}{7}\right)$

36. $\frac{5}{7} \cdot \left(-\frac{2}{3}\right)$

37. $-\frac{3}{8} \cdot \left(-\frac{2}{9}\right)$

38. $-\frac{5}{8} \cdot \left(-\frac{2}{5}\right)$

39. $(-5.3)(2.1)$

40. $(9.5)(-3.7)$

41. $-\frac{5}{9} \cdot \frac{3}{4}$

42. $-\frac{8}{3} \cdot \frac{9}{4}$

43. $3 \cdot (-7) \cdot (-2) \cdot 6$

44. $9 \cdot (-2) \cdot (-6) \cdot 7$

Aha! **45.** $27 \cdot (-34) \cdot 0$

46. $-43 \cdot (-74) \cdot 0$

47. $-\frac{1}{3} \cdot \frac{1}{4} \cdot \left(-\frac{3}{7}\right)$

48. $-\frac{1}{2} \cdot \frac{3}{5} \cdot \left(-\frac{2}{7}\right)$

49. $-2 \cdot (-5) \cdot (-3) \cdot (-5)$

50. $-3 \cdot (-5) \cdot (-2) \cdot (-1)$

51. $(-31) \cdot (-27) \cdot 0 \cdot (-13)$

52. $7 \cdot (-6) \cdot 5 \cdot (-4) \cdot 3 \cdot (-2) \cdot 1 \cdot 0$

53. $(-8)(-9)(-10)$

54. $(-7)(-8)(-9)(-10)$

55. $(-6)(-7)(-8)(-9)(-10)$

56. $(-5)(-6)(-7)(-8)(-9)(-10)$

Divide, if possible, and check. If a quotient is undefined, state this.

57. $18 \div (-2)$

58. $\frac{24}{-3}$

59. $\frac{36}{-9}$

60. $26 \div (-13)$

61. $\frac{-56}{8}$

62. $\frac{-35}{-7}$

63. $\frac{-48}{-12}$

64. $-63 \div (-9)$

65. $-72 \div 8$

66. $\frac{-50}{25}$

67. $-10.2 \div (-2)$

68. $-2 \div 0.8$

69. $-100 \div (-11)$

70. $\frac{-64}{-7}$

71. $\frac{400}{-50}$

72. $-300 \div (-13)$

73. $\frac{48}{0}$

74. $\frac{0}{-5}$

75. $-4.8 \div 1.2$

76. $-3.9 \div 1.3$

77. $\frac{0}{-9}$

78. $0 \div 18$

Aha! **79.** $\frac{9.7(-2.8)0}{4.3}$

80. $\frac{(-4.9)(7.2)}{0}$

Write each expression in two equivalent forms, as in Example 6.

81. $\frac{-8}{3}$

82. $\frac{18}{-7}$

83. $\frac{29}{-35}$

84. $\frac{-10}{3}$

85. $-\frac{7}{3}$

86. $-\frac{4}{15}$

87. $\frac{-x}{2}$

88. $\frac{9}{-a}$

Find the reciprocal of each number, if it exists.

89. $-\frac{4}{5}$

90. $-\frac{13}{11}$

91. $\frac{51}{-10}$

92. $\frac{43}{-24}$

93. -10

94. 34

95. 4.3

96. -1.7

97. $\frac{-9}{4}$

98. $\frac{-6}{11}$

99. 0

100. -1

Perform the indicated operation and, if possible, simplify. If a quotient is undefined, state this.

101. $\left(\frac{-7}{4}\right)\left(-\frac{3}{5}\right)$

102. $\left(-\frac{5}{6}\right)\left(\frac{-1}{3}\right)$

103. $\frac{-3}{8} + \frac{-5}{8}$

104. $\frac{-4}{5} + \frac{7}{5}$

Aha! **105.** $\left(\frac{-9}{5}\right)\left(\frac{5}{-9}\right)$

106. $\left(-\frac{2}{7}\right)\left(\frac{5}{-8}\right)$

107. $\left(-\frac{3}{11}\right) - \left(-\frac{6}{11}\right)$

108. $\left(-\frac{4}{7}\right) - \left(-\frac{2}{7}\right)$

109. $\frac{7}{8} \div \left(-\frac{1}{2}\right)$

110. $\frac{3}{4} \div \left(-\frac{2}{3}\right)$

Aha! **111.** $-\frac{5}{9} \div \left(-\frac{5}{9}\right)$

112. $\frac{-5}{12} \div \frac{15}{7}$

113. $\frac{-3}{10} + \frac{2}{5}$

114. $\frac{-5}{9} + \frac{2}{3}$

115. $\frac{7}{10} \div \left(\frac{-3}{5}\right)$

116. $\left(\frac{-3}{5}\right) \div \frac{6}{15}$

117. $\frac{14}{-9} \div \frac{0}{3}$

118. $\frac{0}{-10} \div \frac{-3}{8}$

119. $\frac{-4}{15} + \frac{2}{-3}$

120. $\frac{3}{-10} + \frac{-1}{5}$

121. Most calculators have a key, often appearing as ①/ⓧ, for finding reciprocals. To use this key, we enter a number and then press ①/ⓧ to find its reciprocal. What should happen if we enter a number and then press the reciprocal key twice? Why?

122. Multiplication can be regarded as repeated addition. Using this idea and a number line, explain why $3 \cdot (-5) = -15$.

Skill Review

123. Simplify: $\frac{264}{468}$. [1.3]

124. Combine like terms: $x + 12y + 11x - 13y - 9$. [1.5]

Synthesis

125. If two nonzero numbers are opposites of each other, are their reciprocals opposites of each other? Why or why not?

126. If two numbers are reciprocals of each other, are their opposites reciprocals of each other? Why or why not?

Translate to an algebraic expression or equation.

127. The reciprocal of a sum

128. The sum of two reciprocals

129. The opposite of a sum

130. The sum of two opposites

131. A real number is its own opposite.

132. A real number is its own reciprocal.

133. Show that the reciprocal of a sum is *not* the sum of the two reciprocals.

134. Which real numbers are their own reciprocals?

135. Jenna is a meteorologist. On December 10, she notes that the temperature is $-3°F$ at 6:00 A.M. She predicts that the temperature will rise at a rate of 2° per hour for 3 hr, and then rise at a rate of 3° per hour for 6 hr. She also predicts that the temperature will then fall at a rate of 2° per hour for 3 hr, and then fall at a rate of 5° per hour for 2 hr. What is Jenna's temperature forecast for 8:00 P.M?

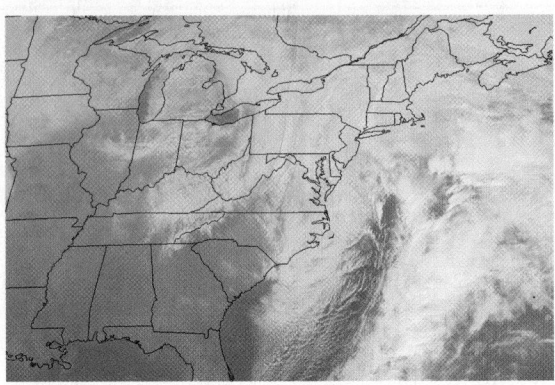

Tell whether each expression represents a positive number or a negative number when m and n are negative.

136. $\frac{m}{-n}$

137. $\frac{-n}{-m}$

138. $-m \cdot \left(\frac{-n}{m}\right)$

139. $-\left(\frac{n}{-m}\right)$

140. $(m + n) \cdot \frac{m}{n}$

141. $(-n - m)\frac{n}{m}$

142. What must be true of m and n if $-mn$ is to be **(a)** positive? **(b)** zero? **(c)** negative?

143. The following is a proof that a positive number times a negative number is negative. Provide a reason for each step. Assume that $a > 0$ and $b > 0$.

$$a(-b) + ab = a[-b + b]$$
$$= a(0)$$
$$= 0$$

Therefore, $a(-b)$ is the opposite of ab.

144. Is it true that for any numbers a and b, if a is larger than b, then the reciprocal of a is smaller than the reciprocal of b? Why or why not?

CONNECTING the CONCEPTS

The rules for multiplication and division of real numbers differ significantly from the rules for addition and subtraction. When simplifying an expression, look at the operation first to determine which set of rules to follow.

Addition

If the signs are the same, add absolute values.

- *Both numbers are positive:* Add as usual. The answer is positive.
- *Both numbers are negative:* Add absolute values and make the answer negative.

If the signs are different, subtract absolute values.

- *The positive number has the greater absolute value:* The answer is positive.
- *The negative number has the greater absolute value:* The answer is negative.
- *The numbers have the same absolute value:* The answer is 0.

If one number is zero, the sum is the other number.

Subtraction

Add the opposite of the number being subtracted.

Multiplication

If the signs are the same, multiply absolute values. The answer is positive.

If the signs are different, multiply absolute values. The answer is negative.

If one number is zero, the product is 0.

Division

Multiply by the reciprocal of the divisor.

MIXED REVIEW

Perform the indicated operation and, if possible, simplify.

1. $-8 + (-2)$ **2.** $-8 \cdot (-2)$

3. $-8 \div (-2)$ **4.** $-8 - (-2)$

5. $12 \cdot (-10)$ **6.** $13 - 20$

7. $-5 - 18$ **8.** $-12 \div 4$

9. $\dfrac{3}{5} - \dfrac{8}{5}$ **10.** $\dfrac{-12}{5} + \left(\dfrac{-3}{5}\right)$

11. $-5.6 + 4.8$ **12.** $1.3 \cdot (-2.9)$

13. $-44.1 \div 6.3$ **14.** $6.6 + (-10.7)$

15. $\dfrac{9}{5} \cdot \left(-\dfrac{20}{3}\right)$ **16.** $-\dfrac{5}{4} \div \left(-\dfrac{3}{4}\right)$

17. $38 - (-62)$ **18.** $-17 + 94$

19. $(-15) \cdot (-12)$ **20.** $-26 - 26$

1.8 Exponential Notation and Order of Operations

Exponential Notation ▪ Order of Operations ▪ Simplifying and the Distributive Law ▪
The Opposite of a Sum

Algebraic expressions often contain *exponential notation*. In this section, we learn how to use exponential notation as well as rules for the *order of operations* in performing certain algebraic manipulations.

STUDY SKILLS

A Journey of 1000 Miles Starts with a Single Step

It is extremely important to include steps when working problems. Doing so allows you and others to follow your thought process. It also helps you to avoid careless errors and to identify specific areas in which you may have made mistakes.

Exponential Notation

A product like $3 \cdot 3 \cdot 3 \cdot 3$, in which the factors are the same, is called a **power**. Powers occur often enough that a simpler notation called **exponential notation** is used. For

$$\underbrace{3 \cdot 3 \cdot 3 \cdot 3}_{\text{4 factors}}, \quad \text{we write} \quad 3^4. \qquad \text{Because } 3^4 = 81, \text{ we can say that } 81 \text{ "is a power of 3."}$$

This is read "three to the fourth power," or simply, "three to the fourth." The number 4 is called an **exponent** and the number 3 a **base**.

Expressions like s^2 and s^3 are usually read "s squared" and "s cubed," respectively. This comes from the fact that a square with sides of length s has an area A given by $A = s^2$ and a cube with sides of length s has a volume V given by $V = s^3$.

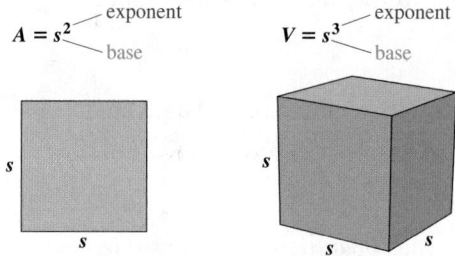

EXAMPLE 1 Write exponential notation for $10 \cdot 10 \cdot 10 \cdot 10 \cdot 10$.

SOLUTION

Exponential notation is 10^5.   5 is the exponent.
10 is the base.

TRY EXERCISE 3

EXAMPLE 2 Simplify: **(a)** 5^2; **(b)** $(-5)^3$; **(c)** $(2n)^3$.

SOLUTION

a) $5^2 = 5 \cdot 5 = 25$ The exponent 2 indicates two factors of 5.

b) $(-5)^3 = (-5)(-5)(-5)$ The exponent 3 indicates three factors of -5.

     $= 25(-5)$ Using the associative law of multiplication

     $= -125$

c) $(2n)^3 = (2n)(2n)(2n)$ The exponent 3 indicates three factors of $2n$.

     $= 2 \cdot 2 \cdot 2 \cdot n \cdot n \cdot n$ Using the associative and commutative laws of multiplication

     $= 8n^3$

TRY EXERCISE 13

On most graphing calculators, grouping symbols, such as a fraction bar, must be replaced with parentheses. For example, to calculate

$$\frac{12(9-7)+4 \cdot 5}{2^4 + 3^2},$$

we enter $(12(9-7)+4 \cdot 5) \div (2^4 + 3^2)$. To enter an exponential expression, we enter the base, press ⌃ and then enter the exponent. The x^2 key can be used to enter an exponent of 2. We can also convert to fraction notation if we wish.

```
(12(9−7)+4*5)/(2^4+3²)
                    1.76
Ans▶Frac
                   44/25
■
```

To determine what the exponent 1 will mean, look for a pattern in the following:

$$7 \cdot 7 \cdot 7 \cdot 7 = 7^4$$
$$7 \cdot 7 \cdot 7 = 7^3$$
$$7 \cdot 7 = 7^2$$
$$? = 7^1$$

The exponent decreases by 1 each time.

The number of factors decreases by 1 each time. To extend the pattern, we say that

$$7 = 7^1.$$

> **Exponential Notation**
>
> For any natural number n,
>
> $$b^n \quad \text{means} \quad \overbrace{b \cdot b \cdot b \cdot b \cdots b}^{n \text{ factors}}.$$

Order of Operations

How should $4 + 2 \times 5$ be computed? If we multiply 2 by 5 and then add 4, the result is 14. If we add 2 and 4 first and then multiply by 5, the result is 30. Since these results differ, the order in which we perform operations matters. If grouping symbols such as parentheses ( ), brackets [ ], braces { }, absolute-value symbols | |, or fraction bars appear, they tell us what to do first. For example,

$$(4 + 2) \times 5 \quad \text{indicates} \quad 6 \times 5, \quad \text{resulting in 30,}$$

and

$$4 + (2 \times 5) \quad \text{indicates} \quad 4 + 10, \quad \text{resulting in 14.}$$

Besides grouping symbols, the following conventions exist for determining the order in which operations should be performed.

Although most scientific and graphing calculators follow the rules for order of operations when evaluating expressions, some calculators do not. Try calculating $4 + 2 \times 5$ on your calculator. If the result shown is 30, your calculator does not follow the rules for order of operations. In this case, you will have to multiply 2×5 first and then add 4.

> **Rules for Order of Operations**
> 1. Calculate within the innermost grouping symbols, ( ), [ ], { }, | |, and above or below fraction bars.
> 2. Simplify all exponential expressions.
> 3. Perform all multiplications and divisions, working from left to right.
> 4. Perform all additions and subtractions, working from left to right.

Thus the correct way to compute $4 + 2 \times 5$ is to first multiply 2 by 5 and then add 4. The result is 14.

EXAMPLE 3

Simplify: $15 - 2 \cdot 5 + 3$.

SOLUTION   When no groupings or exponents appear, we *always* multiply or divide before adding or subtracting:

$$\begin{aligned} 15 - 2 \cdot 5 + 3 &= 15 - 10 + 3 \qquad \text{Multiplying} \\ &= 5 + 3 \\ &= 8. \end{aligned} \quad \text{Subtracting and adding from left to right}$$

▶ TRY EXERCISE 31

Always calculate within parentheses first. When there are exponents and no parentheses, we simplify powers before multiplying or dividing.

EXAMPLE **4**

Simplify: **(a)** $(3 \cdot 4)^2$; **(b)** $3 \cdot 4^2$.

SOLUTION

a) $(3 \cdot 4)^2 = (12)^2$ Working within parentheses first

           $= 144$

b) $3 \cdot 4^2 = 3 \cdot 16$ Simplifying the power

       $= 48$ Multiplying

Note that $(3 \cdot 4)^2 \neq 3 \cdot 4^2$.

> **TRY EXERCISE** 37

> *CAUTION!*   Example 4 illustrates that, in general, $(ab)^2 \neq ab^2$.

EXAMPLE **5**

Evaluate when $x = 5$: **(a)** $(-x)^2$; **(b)** $-x^2$.

SOLUTION

a) $(-x)^2 = (-5)^2 = (-5)(-5) = 25$ We square the opposite of 5.

b) $-x^2 = -5^2 = -25$ We square 5 and then find the opposite.

> **TRY EXERCISE** 15

> *CAUTION!*   Example 5 illustrates that, in general, $(-x)^2 \neq -x^2$.

To simplify $-x^2$, it may help to write

$$-x^2 = (-1)x^2.$$

EXAMPLE **6**

Evaluate $-15 \div 3(6 - a)^3$ when $a = 4$.

SOLUTION

$$-15 \div 3(6 - a)^3 = -15 \div 3(6 - 4)^3 \quad \text{Substituting 4 for } a$$
$$= -15 \div 3(2)^3 \quad \text{Working within parentheses first}$$
$$= -15 \div 3 \cdot 8 \quad \text{Simplifying the exponential expression}$$
$$\left.\begin{array}{l} = -5 \cdot 8 \\ = -40 \end{array}\right\} \quad \text{Dividing and multiplying from left to right}$$

> **TRY EXERCISE** 67

STUDENT NOTES ——————

When simplifying an expression, it is important to copy the entire expression on each line, not just the parts that have been simplified in a given step. As shown in Examples 6 and 7, each line should be equivalent to the line above it.

The symbols ( ), [ ], and { } are all used in the same way. Used inside or next to each other, they make it easier to locate the left and right sides of a grouping. When combinations of grouping symbols are used, we begin with the innermost grouping symbols and work to the outside.

EXAMPLE **7**

Simplify: $8 \div 4 + 3[9 + 2(3 - 5)^3]$.

SOLUTION

$8 \div 4 + 3[9 + 2(3 - 5)^3] = 8 \div 4 + 3[9 + 2(-2)^3]$ Doing the calculations in the innermost grouping symbols first

$= 8 \div 4 + 3[9 + 2(-8)]$ $(-2)^3 = (-2)(-2)(-2)$
$= -8$

$= 8 \div 4 + 3[9 + (-16)]$

$= 8 \div 4 + 3[-7]$ Completing the calculations within the brackets

$= 2 + (-21)$ Multiplying and dividing from left to right

$= -19$

TRY EXERCISE 47

EXAMPLE **8**

Calculate: $\dfrac{12(9 - 7) + 4 \cdot 5}{3^4 + 2^3}$.

SOLUTION   An equivalent expression with brackets is

$[12(9 - 7) + 4 \cdot 5] \div [3^4 + 2^3]$.   Here the grouping symbols are necessary.

In effect, we need to simplify the numerator, simplify the denominator, and then divide the results:

$$\frac{12(9 - 7) + 4 \cdot 5}{3^4 + 2^3} = \frac{12(2) + 4 \cdot 5}{81 + 8}$$

$$= \frac{24 + 20}{89} = \frac{44}{89}.$$

TRY EXERCISE 55

Simplifying and the Distributive Law

Sometimes we cannot simplify within grouping symbols. When a sum or a difference is being grouped, the distributive law provides a method for removing the grouping symbols.

EXAMPLE **9**

Simplify: $5x - 9 + 2(4x + 5)$.

SOLUTION

$5x - 9 + 2(4x + 5) = 5x - 9 + 8x + 10$ Using the distributive law

$= 13x + 1$ Combining like terms

TRY EXERCISE 85

Now that exponents have been introduced, we can make our definition of *like* or *similar terms* more precise. **Like,** or **similar, terms** are either constant terms or terms containing the same variable(s) raised to the same power(s). Thus, 5 and -7, $19xy$ and $2yx$, and $4a^3b$ and a^3b are all pairs of like terms.

EXAMPLE 10 Simplify: $7x^2 + 3[x^2 + 2x] - 5x$.

SOLUTION

$$7x^2 + 3[x^2 + 2x] - 5x = 7x^2 + 3x^2 + 6x - 5x \quad \text{Using the distributive law}$$

$$= 10x^2 + x \quad \text{Combining like terms}$$

> **TRY EXERCISE** 91

The Opposite of a Sum

When a number is multiplied by -1, the result is the opposite of that number. For example, $-1(7) = -7$ and $-1(-5) = 5$.

> **The Property of -1**
>
> For any real number a,
>
> $$-1 \cdot a = -a.$$
>
> (Negative one times a is the opposite of a.)

An expression such as $-(x + y)$ indicates the *opposite*, or *additive inverse*, of the sum of x and y. When a sum within grouping symbols is preceded by a "$-$" symbol, we can multiply the sum by -1 and use the distributive law. In this manner, we can find an equivalent expression for the opposite of a sum.

EXAMPLE 11 Write an expression equivalent to $-(3x + 2y + 4)$ without using parentheses.

SOLUTION

$$-(3x + 2y + 4) = -1(3x + 2y + 4) \quad \text{Using the property of } -1$$

$$= -1(3x) + (-1)(2y) + (-1)4 \quad \text{Using the distributive law}$$

$$= -3x - 2y - 4 \quad \text{Using the associative law and the property of } -1$$

> **TRY EXERCISE** 73

Example 11 illustrates an important property of real numbers.

> **The Opposite of a Sum**
>
> For any real numbers a and b,
>
> $$-(a + b) = -a + (-b) = -a - b.$$
>
> (The opposite of a sum is the sum of the opposites.)

To remove parentheses from an expression like $-(x - 7y + 5)$, we can first rewrite the subtraction as addition:

$$-(x - 7y + 5) = -(x + (-7y) + 5) \quad \text{Rewriting as addition}$$

$$= -x + 7y - 5. \quad \text{Taking the opposite of a sum}$$

This procedure is normally streamlined to one step in which we find the opposite by "removing parentheses and changing the sign of every term":

$$-(x - 7y + 5) = -x + 7y - 5.$$

EXAMPLE 12 Simplify: $3x - (4x + 2)$.

SOLUTION

$$
\begin{aligned}
3x - (4x + 2) &= 3x + [-(4x + 2)] && \text{Adding the opposite of } 4x + 2 \\
&= 3x + [-4x - 2] && \text{Taking the opposite of } 4x + 2 \\
&= 3x + (-4x) + (-2) && \\
&= 3x - 4x - 2 && \text{Try to go directly to this step.} \\
&= -x - 2 && \text{Combining like terms}
\end{aligned}
$$

> **TRY EXERCISE** 81

In practice, the first three steps of Example 12 are generally skipped.

EXAMPLE 13 Simplify: $5t^2 - 2t - (-4t^2 + 9t)$.

SOLUTION

$$
\begin{aligned}
5t^2 - 2t - (-4t^2 + 9t) &= 5t^2 - 2t + 4t^2 - 9t && \text{Removing parentheses} \\
& && \text{and changing the sign} \\
& && \text{of each term inside} \\
&= 9t^2 - 11t && \text{Combining like terms}
\end{aligned}
$$

> **TRY EXERCISE** 89

Expressions such as $7 - 3(x + 2)$ can be simplified as follows:

$$
\begin{aligned}
7 - 3(x + 2) &= 7 + [-3(x + 2)] && \text{Adding the opposite of } 3(x + 2) \\
&= 7 + [-3x - 6] && \text{Multiplying } x + 2 \text{ by } -3 \\
&= 7 - 3x - 6 && \text{Try to go directly to this step.} \\
&= 1 - 3x. && \text{Combining like terms}
\end{aligned}
$$

EXAMPLE 14 Simplify: **(a)** $3n - 2(4n - 5)$; **(b)** $7x^3 + 2 - [5(x^3 - 1) + 8]$.

SOLUTION

a)
$$
\begin{aligned}
3n - 2(4n - 5) &= 3n - 8n + 10 && \text{Multiplying each term inside the} \\
& && \text{parentheses by } -2 \\
&= -5n + 10 && \text{Combining like terms}
\end{aligned}
$$

b)
$$
\begin{aligned}
7x^3 + 2 - [5(x^3 - 1) + 8] &= 7x^3 + 2 - [5x^3 - 5 + 8] && \text{Removing} \\
& && \text{parentheses} \\
&= 7x^3 + 2 - [5x^3 + 3] && \\
&= 7x^3 + 2 - 5x^3 - 3 && \text{Removing brackets} \\
&= 2x^3 - 1 && \text{Combining like terms}
\end{aligned}
$$

> **TRY EXERCISE** 93

As we progress through our study of algebra, it is important that we be able to distinguish between the two tasks of **simplifying an expression** and **solving an equation**. In Chapter 1, we have not solved equations, but we have simplified expressions. This enabled us to write *equivalent expressions* that were simpler than the given expression. In Chapter 2, we will continue to simplify expressions, but we will also begin to solve equations.

1.8 EXERCISE SET

Concept Reinforcement   *In each part of Exercises 1 and 2, name the operation that should be performed first. Do not perform the calculations.*

1. a) $4 + 8 \div 2 \cdot 2$

   b) $7 - 9 + 15$

   c) $5 - 2(3 + 4)$

   d) $6 + 7 \cdot 3$

   e) $18 - 2[4 + (3 - 2)]$

   f) $\dfrac{5 - 6 \cdot 7}{2}$

2. a) $9 - 3 \cdot 4 \div 2$

   b) $8 + 7(6 - 5)$

   c) $5 \cdot [2 - 3(4 + 1)]$

   d) $8 - 7 + 2$

   e) $4 + 6 \div 2 \cdot 3$

   f) $\dfrac{37}{8 - 2 \cdot 2}$

Write exponential notation.

3. $x \cdot x \cdot x \cdot x \cdot x \cdot x$

4. $y \cdot y \cdot y \cdot y \cdot y \cdot y$

5. $(-5)(-5)(-5)$

6. $(-7)(-7)(-7)(-7)$

7. $3t \cdot 3t \cdot 3t \cdot 3t \cdot 3t$

8. $5m \cdot 5m \cdot 5m \cdot 5m \cdot 5m$

9. $2 \cdot n \cdot n \cdot n \cdot n$

10. $8 \cdot a \cdot a \cdot a$

Simplify.

11. 4^2

12. 5^3

13. $(-3)^2$

14. $(-7)^2$

15. -3^2

16. -7^2

17. 4^3

18. 9^1

19. $(-5)^4$

20. 5^4

21. 7^1

22. $(-1)^7$

23. $(-2)^5$

24. -2^5

25. $(3t)^4$

26. $(5t)^2$

27. $(-7x)^3$

28. $(-5x)^4$

29. $5 + 3 \cdot 7$

30. $3 - 4 \cdot 2$

31. $10 \cdot 5 + 1 \cdot 1$

32. $19 - 5 \cdot 4 + 3$

33. $6 - 70 \div 7 - 2$

34. $12 \div 3 + 18 \div 2$

Aha! 35. $14 \cdot 19 \div (19 \cdot 14)$

36. $18 - 6 \div 3 \cdot 2 + 7$

37. $3(-10)^2 - 8 \div 2^2$

38. $9 - 3^2 \div 9(-1)$

39. $8 - (2 \cdot 3 - 9)$

40. $(8 - 2 \cdot 3) - 9$

41. $(8 - 2)(3 - 9)$

42. $32 \div (-2)^2 \cdot 4$

43. $13(-10)^2 + 45 \div (-5)$

44. $2^4 + 2^3 - 10 \div (-1)^4$

45. $5 + 3(2 - 9)^2$

46. $9 - (3 - 5)^3 - 4$

47. $[2 \cdot (5 - 8)]^2$

48. $3(5 - 7)^3 \div 4$

49. $\dfrac{7 + 2}{5^2 - 4^2}$

50. $\dfrac{(5^2 - 3^2)^2}{2 \cdot 6 - 4}$

51. $8(-7) + |3(-4)|$

52. $|10(-5)| + 1(-1)$

53. $36 \div (-2)^2 + 4[5 - 3(8 - 9)^5]$

54. $-48 \div (7 - 9)^3 - 2[1 - 5(2 - 6) + 3^2]$

55. $\dfrac{7^2 - (-1)^7}{5 \cdot 7 - 4 \cdot 3^2 - 2^2}$

56. $\dfrac{(-2)^3 + 4^2}{2 \cdot 3 - 5^2 + 3 \cdot 7}$

57. $\dfrac{-3^3 - 2 \cdot 3^2}{8 \div 2^2 - (6 - |2 - 15|)}$

58. $\dfrac{(-5)^2 - 3 \cdot 5}{3^2 + 4 \cdot |6 - 7| \cdot (-1)^5}$

Evaluate.

59. $9 - 4x$, for $x = 7$

60. $1 + x^3$, for $x = -2$

61. $24 \div t^3$, for $t = -2$

62. $-100 \div a^2$, for $a = -5$

63. $45 \div a \cdot 5$, for $a = -3$

64. $50 \div 2 \cdot t$, for $t = 5$

65. $5x \div 15x^2$, for $x = 3$

66. $6a \div 12a^3$, for $a = 2$

67. $45 \div 3^2 x(x - 1)$, for $x = 3$

68. $-30 \div t(t + 4)^2$, for $t = -6$

69. $-x^2 - 5x$, for $x = -3$

70. $(-x)^2 - 5x$, for $x = -3$

71. $\dfrac{3a - 4a^2}{a^2 - 20}$, for $a = 5$

72. $\dfrac{a^3 - 4a}{a(a - 3)}$, for $a = -2$

Write an equivalent expression without using grouping symbols.

73. $-(9x + 1)$

74. $-(3x + 5)$

75. $-[-7n + 8]$

76. $-(6x - 7)$

77. $-(4a - 3b + 7c)$

78. $-[5n - m - 2p]$

79. $-(3x^2 + 5x - 1)$

80. $-(-9x^3 + 8x + 10)$

Simplify.

81. $8x - (6x + 7)$

82. $2a - (5a - 9)$

83. $2x - 7x - (4x - 6)$

84. $2a + 5a - (6a + 8)$

85. $9t - 7r + 2(3r + 6t)$

86. $4m - 9n + 3(2m - n)$

87. $15x - y - 5(3x - 2y + 5z)$

88. $4a - b - 4(5a - 7b + 8c)$

89. $3x^2 + 11 - (2x^2 + 5)$

90. $5x^4 + 3x - (5x^4 + 3x)$

91. $5t^3 + t + 3(t - 2t^3)$

92. $8n^2 - 3n + 2(n - 4n^2)$

93. $12a^2 - 3ab + 5b^2 - 5(-5a^2 + 4ab - 6b^2)$

94. $-8a^2 + 5ab - 12b^2 - 6(2a^2 - 4ab - 10b^2)$

95. $-7t^3 - t^2 - 3(5t^3 - 3t)$

96. $9t^4 + 7t - 5(9t^3 - 2t)$

97. $5(2x - 7) - [4(2x - 3) + 2]$

98. $3(6x - 5) - [3(1 - 8x) + 5]$

99. Some students use the mnemonic device PEMDAS to help remember the rules for the order of operations. Explain how this can be done and how the order of the letters in PEMDAS could lead a student to a wrong conclusion about the order of some operations.

100. Jake keys $18/2 \cdot 3$ into his calculator and expects the result to be 3. What mistake is he probably making?

Skill Review

Translate to an algebraic expression. [1.1]

101. Nine less than twice a number

102. Half of the sum of two numbers

Synthesis

103. Write the sentence $(-x)^2 \neq -x^2$ in words. Explain why $(-x)^2$ and $-x^2$ are not equivalent.

104. Write the sentence $-|x| \neq -x$ in words. Explain why $-|x|$ and $-x$ are not equivalent.

Simplify.

105. $5t - \{7t - [4r - 3(t - 7)] + 6r\} - 4r$

106. $z - \{2z - [3z - (4z - 5z) - 6z] - 7z\} - 8z$

107. $\{x - [f - (f - x)] + [x - f]\} - 3x$

108. Is it true that for all real numbers a and b,
$$ab = (-a)(-b)?$$
Why or why not?

109. Is it true that for all real numbers a, b, and c,
$$a|b - c| = ab - ac?$$
Why or why not?

If $n > 0$, $m > 0$, and $n \neq m$, classify each of the following as either true or false.

110. $-n + m = -(n + m)$

111. $m - n = -(n - m)$

112. $n(-n - m) = -n^2 + nm$

113. $-m(n - m) = -(mn + m^2)$

114. $-n(-n - m) = n(n + m)$

Evaluate.

Aha! **115.** $[x + 3(2 - 5x) \div 7 + x](x - 3)$, for $x = 3$

Aha! **116.** $[x + 2 \div 3x] \div [x + 2 \div 3x]$, for $x = -7$

117. $\dfrac{x^2 + 2^x}{x^2 - 2^x}$, for $x = 3$

118. $\dfrac{x^2 + 2^x}{x^2 - 2^x}$, for $x = 2$

119. In Mexico, between 500 B.C. and 600 A.D., the Mayans represented numbers using powers of 20 and certain symbols. For example, the symbols

represent $4 \cdot 20^3 + 17 \cdot 20^2 + 10 \cdot 20^1 + 0 \cdot 20^0$. Evaluate this number.
Source: National Council of Teachers of Mathematics, 1906 Association Drive, Reston, VA 22091

120. Examine the Mayan symbols and the numbers in Exercise 119. What numbers do

    and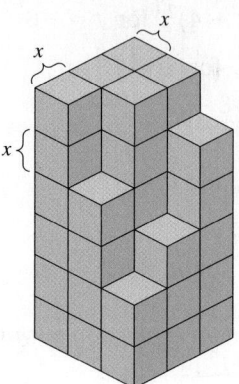

each represent?

121. Calculate the volume of the tower shown below.

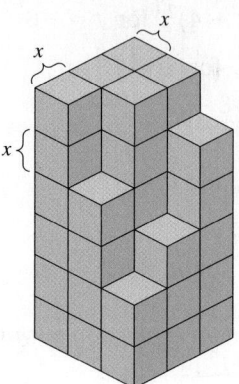

COLLABORATIVE CORNER

Select the Symbols

Focus: Order of operations

Time: 15 minutes

Group size: 2

One way to master the rules for the order of operations is to insert symbols within a display of numbers in order to obtain a predetermined result. For example, the display

　　1　　2　　3　　4　　5

can be used to obtain the result 21 as follows:

　　$(1 + 2) \div 3 + 4 \cdot 5.$

Note that without an understanding of the rules for the order of operations, solving a problem of this sort is impossible.

ACTIVITY

1. Each group should prepare an exercise similar to the example shown above. (Exponents are not allowed.) To do so, first select five single-digit numbers for display. Then insert operations and grouping symbols and calculate the result.

2. Pair with another group. Each group should give the other its result along with its five-number display, and challenge the other group to insert symbols that will make the display equal the result given.

3. Share with the entire class the various mathematical statements developed by each group.

Study Summary

KEY TERMS AND CONCEPTS	EXAMPLES

SECTION 1.1: INTRODUCTION TO ALGEBRA

An **algebraic expression** is a collection of **variables** and **constants** on which operations are performed.

$5ab^3$ is an algebraic expression; 5 is a constant; and a and b are variables.

To **evaluate** an algebraic expression, substitute a number for each variable and carry out the operations. The result is a **value** of that expression.

Evaluate $\dfrac{x+y}{8}$ *for* $x = 15$ *and* $y = 9$.

$$\frac{x+y}{8} = \frac{15+9}{8} = \frac{24}{8} = 3$$

To find the area of a rectangle, a triangle, or a parallelogram, evaluate the appropriate formula for the given values.

Find the area of a triangle with base 3.1 m *and height* 6 m.

$$A = \tfrac{1}{2}bh = \tfrac{1}{2}(3.1\text{ m})(6\text{ m}) = \tfrac{1}{2}(3.1)(6)(\text{m}\cdot\text{m}) = 9.3\text{ m}^2$$

Many problems can be solved by **translating** phrases to algebraic expressions and then forming an equation. The table on p. 4 shows translations of many words that occur in problems.

Translate to an equation. Do not solve.

When 34 is subtracted from a number, the result is 13. What is the number?

Let n represent the number.

Rewording: 34 subtracted from a number   is   13

Translating: $n - 34$ $=$ 13

An **equation** is a number sentence with the verb $=$. A substitution for the variable in an equation that makes the equation true is a **solution** of the equation.

Determine whether 9 *is a solution of* $47 - n = 38$.

$$47 - n = 38$$

$$\frac{47 - 9 \ \big|\ 38}{38 \overset{?}{=} 38 \quad \text{TRUE}}$$

Since $38 = 38$ is true, 9 is a solution.

SECTION 1.2: THE COMMUTATIVE, ASSOCIATIVE, AND DISTRIBUTIVE LAWS

Equivalent expressions represent the same value for any replacement of the variable.

$x + 10$ and $3 + 7 + x$ are equivalent expressions.

The Commutative Laws

$$a + b = b + a;$$
$$ab = ba$$

$3 + (-5) = -5 + 3;$
$8(10) = 10(8)$

The Associative Laws

$$a + (b + c) = (a + b) + c;$$
$$a \cdot (b \cdot c) = (a \cdot b) \cdot c$$

$-5 + (5 + 6) = (-5 + 5) + 6;$
$2 \cdot (5 \cdot 9) = (2 \cdot 5) \cdot 9$

The Distributive Law
$$a(b + c) = ab + ac$$

$$4(x + 2) = 4 \cdot x + 4 \cdot 2 = 4x + 8$$

The distributive law is used to multiply and to **factor** expressions.

Multiply: $3(2x + 5y)$.
$$3(2x + 5y) = 3 \cdot 2x + 3 \cdot 5y = 6x + 15y$$
Factor: $16x + 24y + 8$.
$$16x + 24y + 8 = 8(2x + 3y + 1)$$

SECTION 1.3: FRACTION NOTATION

Natural numbers: $\{1, 2, 3, \ldots\}$
Whole numbers: $\{0, 1, 2, 3, \ldots\}$

15, 39, and 1567 are some natural numbers.
0, 5, 16, and 2890 are some whole numbers.

A **prime** number has only two different factors, the number itself and 1. Natural numbers that have factors other than 1 and the number itself are **composite** numbers.

2, 3, 5, 7, 11, and 13 are the first six prime numbers.
4, 6, 8, 24, and 100 are examples of composite numbers.

The **prime factorization** of a composite number expresses that number as a product of prime numbers.

The prime factorization of 136 is $2 \cdot 2 \cdot 2 \cdot 17$.

For any nonzero number a,
$$\frac{a}{a} = 1.$$

$$\frac{15}{15} = 1 \quad \text{and} \quad \frac{2x}{2x} = 1$$

The Identity Property of 1
$$a \cdot 1 = 1 \cdot a = a$$

The number 1 is called the **multiplicative identity**.

$$\frac{2}{3} = \frac{2}{3} \cdot \frac{5}{5} \quad \text{since} \quad \frac{5}{5} = 1.$$

$$\frac{a}{d} + \frac{b}{d} = \frac{a + b}{d}$$

$$\frac{a}{d} - \frac{b}{d} = \frac{a - b}{d}$$

$$\frac{a}{b} \cdot \frac{c}{d} = \frac{a \cdot c}{b \cdot d}$$

$$\frac{a}{b} \div \frac{c}{d} = \frac{a}{b} \cdot \frac{d}{c}$$

$$\frac{1}{6} + \frac{3}{8} = \frac{4}{24} + \frac{9}{24} = \frac{13}{24}$$

$$\frac{5}{12} - \frac{1}{6} = \frac{5}{12} - \frac{2}{12} = \frac{3}{12} = \frac{1 \cdot 3}{4 \cdot 3} = \frac{1}{4} \cdot \frac{3}{3} = \frac{1}{4} \cdot 1 = \frac{1}{4}$$

$$\frac{2}{5} \cdot \frac{7}{8} = \frac{2 \cdot 7}{5 \cdot 2 \cdot 4} = \frac{7}{20} \qquad \text{Removing a factor equal to 1: } \frac{2}{2} = 1$$

$$\frac{10}{9} \div \frac{4}{15} = \frac{10}{9} \cdot \frac{15}{4} = \frac{2 \cdot 5 \cdot 3 \cdot 5}{3 \cdot 3 \cdot 2 \cdot 2} = \frac{25}{6} \qquad \text{Removing a factor}$$
$$\text{equal to 1: } \frac{2 \cdot 3}{2 \cdot 3} = 1$$

SECTION 1.4: POSITIVE AND NEGATIVE REAL NUMBERS

Integers:
$\{\ldots, -3, -2, -1, 0, 1, 2, 3, \ldots\}$

Rational numbers:
$$\left\{ \frac{a}{b} \,\middle|\, a \text{ and } b \text{ are integers and } b \neq 0 \right\}$$

The rational numbers and the **irrational numbers** make up the set of **real numbers**.

$-25, -2, 0, 1,$ and 2000 are some integers.

$\frac{1}{6}, \frac{-3}{7}, 0, 17, 0.758,$ and $9.\overline{608}$ are some rational numbers.

$\sqrt{7}$ and π are some irrational numbers.

Every rational number can be written using fraction notation or decimal notation. When written in decimal notation, a rational number either **repeats** or **terminates**.

$-\dfrac{1}{16} = -0.0625$ This is a terminating decimal.

$\dfrac{5}{6} = 0.8333\ldots = 0.8\overline{3}$ This is a repeating decimal.

Every real number corresponds to a point on the number line. For any two numbers, the one to the left is less than the one to the right. The symbol $<$ means "**is less than**" and the symbol $>$ means "**is greater than**."

$$4 > -3.1 \qquad -\frac{1}{2} < \sqrt{2}$$

The **absolute value** of a number is the number of units that number is from zero on the number line.

$|3| = 3$ since 3 is 3 units from 0.
$|-3| = 3$ since -3 is 3 units from 0.

SECTION 1.5: ADDITION OF REAL NUMBERS

To **add** two real numbers, use the rules on p. 39.

$-8 + (-3) = -11;$
$-8 + 3 = -5;$
$8 + (-3) = 5;$
$-8 + 8 = 0$

The Identity Property of 0
$$a + 0 = 0 + a = a$$
The number 0 is called the **additive identity**.

$-35 + 0 = -35;$
$0 + \dfrac{2}{9} = \dfrac{2}{9}$

SECTION 1.6: SUBTRACTION OF REAL NUMBERS

The **opposite,** or **additive inverse**, of a number a is written $-a$. The opposite of the opposite of a is a.
$$-(-a) = a$$

Find $-x$ *and* $-(-x)$ *when* $x = -11.$
$$-x = -(-11) = 11;$$
$$-(-x) = -(-(-11)) = -11 \qquad -(-x) = x$$

To **subtract** two real numbers, add the opposite of the number being subtracted.

$-10 - 12 = -10 + (-12) = -22$;
$-10 - (-12) = -10 + 12 = 2$

The **terms** of an expression are separated by plus signs. **Like terms** either are constants or have the same variable factors raised to the same power. Like terms can be **combined** using the distributive law.

In the expression $-2x + 3y + 5x - 7y$:

The terms are $-2x, 3y, 5x$, and $-7y$.

The like terms are $-2x$ and $5x$, and $3y$ and $-7y$.

Combining like terms gives

$$-2x + 3y + 5x - 7y = -2x + 5x + 3y - 7y$$
$$= (-2 + 5)x + (3 - 7)y = 3x - 4y.$$

SECTION 1.7: MULTIPLICATION AND DIVISION OF REAL NUMBERS

To **multiply** or **divide** two real numbers, use the rules on p. 54.

Division by 0 is undefined.

$(-5)(-2) = 10$;
$30 \div (-6) = -5$;
$0 \div (-3) = 0$;
$-3 \div 0$ is undefined.

SECTION 1.8: EXPONENTIAL NOTATION AND ORDER OF OPERATIONS

Exponential notation:

Exponent

$$\overset{}{b^n} = \overbrace{b \cdot b \cdot b \cdots b}^{n \text{ factors}}$$

Base

$6^2 = 6 \cdot 6 = 36$;
$(-6)^2 = (-6) \cdot (-6) = 36$;
$-6^2 = -6 \cdot 6 = -36$;
$(6x)^2 = (6x) \cdot (6x) = 36x^2$

To perform multiple operations, use the rules for **order of operations** on p. 61.

$$-3 + (3 - 5)^3 \div 4(-1) = -3 + (-2)^3 \div 4(-1)$$
$$= -3 + (-8) \div 4(-1)$$
$$= -3 + (-2)(-1)$$
$$= -3 + 2$$
$$= -1$$

The Property of -1

For any real number a,

$$-1 \cdot a = -a.$$

$-1 \cdot 5x = -5x$ and $-5x = -1(5x)$

The Opposite of a Sum

For any real numbers a and b,

$$-(a + b) = -a - b.$$

$-(2x - 3y) = -(2x) - (-3y) = -2x + 3y$

Expressions containing parentheses can be simplified by removing parentheses using the distributive law.

Simplify: $3x^2 - 5(x^2 - 4xy + 2y^2) - 7y^2$.

$$3x^2 - 5(x^2 - 4xy + 2y^2) - 7y^2 = 3x^2 - 5x^2 + 20xy - 10y^2 - 7y^2$$
$$= -2x^2 + 20xy - 17y^2$$

Review Exercises: Chapter 1

Concept Reinforcement *In each of Exercises 1–10, classify the statement as either true or false.*

1. $4x - 5y$ and $12 - 7a$ are both algebraic expressions containing two terms. [1.2]

2. $3t + 1 = 7$ and $8 - 2 = 9$ are both equations. [1.1]

3. The fact that $2 + x$ is equivalent to $x + 2$ is an illustration of the associative law for addition. [1.2]

4. The statement $4(a + 3) = 4 \cdot a + 4 \cdot 3$ illustrates the distributive law. [1.2]

5. The number 2 is neither prime nor composite. [1.3]

6. Every irrational number can be written as a repeating decimal or a terminating decimal. [1.4]

7. Every natural number is a whole number and every whole number is an integer. [1.4]

8. The expressions $9r^2s$ and $5rs^2$ are like terms. [1.8]

9. The opposite of x, written $-x$, never represents a positive number. [1.6]

10. The number 0 has no reciprocal. [1.7]

Evaluate.

11. $8t$, for $t = 3$ [1.1]

12. $\dfrac{x - y}{3}$, for $x = 17$ and $y = 5$ [1.1]

13. $9 - y^2$, for $y = -5$ [1.8]

14. $-10 + a^2 \div (b + 1)$, for $a = 5$ and $b = -6$ [1.8]

Translate to an algebraic expression. [1.1]

15. 7 less than y

16. 10 more than the product of x and z

17. 15 times the difference of Brandt's speed and the wind speed

18. Determine whether 35 is a solution of $x/5 = 8$. [1.1]

19. Translate to an equation. Do not solve. [1.1]

   According to Photo Marketing Association International, in 2006, 14.1 billion prints were made from film. This number is 3.2 billion more than the number of digital prints made. How many digital prints were made in 2006?

20. Use the commutative law of multiplication to write an expression equivalent to $3t + 5$. [1.2]

21. Use the associative law of addition to write an expression equivalent to $(2x + y) + z$. [1.2]

22. Use the commutative and associative laws to write three expressions equivalent to $4(xy)$. [1.2]

Multiply. [1.2]

23. $6(3x + 5y)$

24. $8(5x + 3y + 2)$

Factor. [1.2]

25. $21x + 15y$

26. $22a + 99b + 11$

27. Find the prime factorization of 56. [1.3]

Simplify. [1.3]

28. $\dfrac{20}{48}$

29. $\dfrac{18}{8}$

Perform the indicated operation and, if possible, simplify. [1.3]

30. $\dfrac{5}{12} + \dfrac{3}{8}$

31. $\dfrac{9}{16} \div 3$

32. $\dfrac{2}{3} - \dfrac{1}{15}$

33. $\dfrac{9}{10} \cdot \dfrac{6}{5}$

34. Tell which integers correspond to this situation. [1.4]

   Natalie borrowed \$3600 for an entertainment center. Sean has \$1350 in his savings account.

35. Graph on a number line: $\frac{-1}{3}$. [1.4]

36. Write an inequality with the same meaning as $-3 < x$. [1.4]

37. Classify as true or false: $2 \geq -8$. [1.4]

38. Classify as true or false: $0 \leq -1$. [1.4]

39. Find decimal notation: $-\dfrac{4}{9}$. [1.4]

40. Find the absolute value: $|-1|$. [1.4]

41. Find $-(-x)$ when x is -12. [1.6]

Simplify.

42. $-3 + (-7)$ [1.5]

43. $-\frac{2}{3} + \frac{1}{12}$ [1.5]

44. $10 + (-9) + (-8) + 7$ [1.5]

45. $-3.8 + 5.1 + (-12) + (-4.3) + 10$ [1.5]

46. $-2 - (-10)$ [1.6]

47. $-\frac{9}{10} - \frac{1}{2}$ [1.6]

48. $-3.8 - 4.1$ [1.6]

49. $-9 \cdot (-7)$ [1.7]

50. $-2.7(3.4)$ [1.7]

51. $\frac{2}{3} \cdot \left(-\frac{3}{7}\right)$ [1.7]

52. $2 \cdot (-7) \cdot (-2) \cdot (-5)$ [1.7]

53. $35 \div (-5)$ [1.7]

54. $-5.1 \div 1.7$ [1.7]

55. $-\frac{3}{5} \div \left(-\frac{4}{15}\right)$ [1.7]

56. $120 - 6^2 \div 4 \cdot 8$ [1.8]

57. $(120 - 6^2) \div 4 \cdot 8$ [1.8]

58. $(120 - 6^2) \div (4 \cdot 8)$ [1.8]

59. $16 \div (-2)^3 - 5[3 - 1 + 2(4 - 7)]$ [1.8]

60. $|-3 \cdot 5 - 4 \cdot 8| - 3(-2)$ [1.8]

61. $\dfrac{4(18 - 8) + 7 \cdot 9}{9^2 - 8^2}$ [1.8]

Combine like terms.

62. $11a + 2b + (-4a) + (-3b)$ [1.5]

63. $7x - 3y - 11x + 8y$ [1.6]

64. Find the opposite of -7. [1.6]

65. Find the reciprocal of -7. [1.7]

66. Write exponential notation for $2x \cdot 2x \cdot 2x \cdot 2x$. [1.8]

67. Simplify: $(-5x)^3$. [1.8]

Remove parentheses and simplify. [1.8]

68. $2a - (5a - 9)$

69. $5b + 3(2b - 9)$

70. $11x^4 + 2x + 8(x - x^4)$

71. $2n^2 - 5(-3n^2 + m^2 - 4mn) + 6m^2$

72. $8(x + 4) - 6 - [3(x - 2) + 4]$

Synthesis

73. Explain the difference between a constant and a variable. [1.1]

74. Explain the difference between a term and a factor. [1.2]

75. Describe at least three ways in which the distributive law was used in this chapter. [1.2]

76. Devise a rule for determining the sign of a negative number raised to a power. [1.8]

77. Evaluate $a^{50} - 20a^{25}b^4 + 100b^8$ for $a = 1$ and $b = 2$. [1.8]

78. If $0.090909\ldots = \frac{1}{11}$ and $0.181818\ldots = \frac{2}{11}$, what rational number is named by each of the following?

    a) $0.272727\ldots$ [1.4]   **b)** $0.909090\ldots$ [1.4]

Simplify. [1.8]

79. $-\left|\frac{7}{8} - \left(-\frac{1}{2}\right) - \frac{3}{4}\right|$

80. $(|2.7 - 3| + 3^2 - |-3|) \div (-3)$

Match each phrase in the left column with the most appropriate choice from the right column.

81. _____ A number is nonnegative. [1.4]

82. _____ The product of a number and its reciprocal is 1. [1.7]

83. _____ A number squared [1.8]

84. _____ A sum of squares [1.8]

85. _____ The opposite of an opposite is the original number. [1.6]

86. _____ The order in which numbers are added does not change the result. [1.2]

87. _____ A number is positive. [1.4]

88. _____ The absolute value of a product [1.4]

89. _____ A sum of a number and its reciprocal [1.7]

90. _____ The square of a sum [1.8]

91. _____ The absolute value of one number is less than the absolute value of another number. [1.4]

    a) a^2

    b) $a + b = b + a$

    c) $a > 0$

    d) $a + \dfrac{1}{a}$

    e) $|ab|$

    f) $(a + b)^2$

    g) $|a| < |b|$

    h) $a^2 + b^2$

    i) $a \geq 0$

    j) $a \cdot \dfrac{1}{a} = 1$

    k) $-(-a) = a$

Test: Chapter 1

1. Evaluate $\dfrac{2x}{y}$ for $x = 10$ and $y = 5$.

2. Write an algebraic expression: Nine less than the product of two numbers.

3. Find the area of a triangle when the height h is 30 ft and the base b is 16 ft.

4. Use the commutative law of addition to write an expression equivalent to $3p + q$.

5. Use the associative law of multiplication to write an expression equivalent to $x \cdot (4 \cdot y)$.

6. Determine whether 7 is a solution of $65 - x = 69$.

7. Translate to an equation. Do not solve.

   In the summer of 2005, member utilities of the Florida Reliability Coordinating Council had a demand of 45,950 megawatts. This is only 4250 megawatts less than its maximum production capability. What is the maximum capability of production?
   Source: Energy Information Administration

Multiply.

8. $7(5 + x)$

9. $-5(y - 2)$

Factor.

10. $11 + 44x$

11. $7x + 7 + 49y$

12. Find the prime factorization of 300.

13. Simplify: $\dfrac{24}{56}$.

Write a true sentence using either $<$ or $>$.

14. $-4 \;\blacksquare\; 0$

15. $-3 \;\blacksquare\; -8$

Find the absolute value.

16. $\left|\dfrac{9}{4}\right|$

17. $|-3.8|$

18. Find the opposite of $-\dfrac{2}{3}$.

19. Find the reciprocal of $-\dfrac{4}{7}$.

20. Find $-x$ when x is -10.

21. Write an inequality with the same meaning as $x \le -5$.

Perform the indicated operations and, if possible, simplify.

22. $3.1 - (-4.7)$

23. $-8 + 4 + (-7) + 3$

24. $3.2 - 5.7$

25. $-\dfrac{1}{8} - \dfrac{3}{4}$

26. $4 \cdot (-12)$

27. $-\dfrac{1}{2} \cdot \left(-\dfrac{4}{9}\right)$

28. $-66 \div 11$

29. $-\dfrac{3}{5} \div \left(-\dfrac{4}{5}\right)$

30. $4.864 \div (-0.5)$

31. $-2(16) - |2(-8) - 5^3|$

32. $9 + 7 - 4 - (-3)$

33. $256 \div (-16) \cdot 4$

34. $2^3 - 10[4 - (-2 + 18)3]$

35. Combine like terms: $18y + 30a - 9a + 4y$.

36. Simplify: $(-2x)^4$.

Remove parentheses and simplify.

37. $4x - (3x - 7)$

38. $4(2a - 3b) + a - 7$

39. $3[5(y - 3) + 9] - 2(8y - 1)$

Synthesis

40. Evaluate $\dfrac{5y - x}{2}$ when $x = 20$ and y is 4 less than half of x.

41. Insert one pair of parentheses to make the following a true statement:
    $$9 - 3 - 4 + 5 = 15.$$

Simplify.

42. $|-27 - 3(4)| - |-36| + |-12|$

43. $a - \{3a - [4a - (2a - 4a)]\}$

44. Classify the following as either true or false:
    $$a|b - c| = |ab| - |ac|.$$

Equations, Inequalities, and Problem Solving

DEBORAH ELIAS
EVENT COORDINATOR
Houston, Texas

As an event planner, I am constantly using math. Calculations range from figuring the tax and gratuity percentage to add to the final bill to finding dimensions of linens to fit a table properly. For every client, I also determine a budget with income and expenses.

AN APPLICATION

Event promoters use the formula

$$p = \frac{1.2x}{s}$$

to determine a ticket price p for an event with x dollars of expenses and s anticipated ticket sales. Grand Events expects expenses for an upcoming concert to be $80,000 and anticipates selling 4000 tickets. What should the ticket price be?

Source: *The Indianapolis Star, 2/27/03*

This problem appears as Example 1 in Section 2.3.

S olving equations and inequalities is a recurring theme in much of mathematics. In this chapter, we will study some of the principles used to solve equations and inequalities. We will then use equations and inequalities to solve applied problems.

2.1   Solving Equations

Equations and Solutions   •   The Addition Principle   •   The Multiplication Principle   •
Selecting the Correct Approach

Solving equations is essential for problem solving in algebra. In this section, we study two of the most important principles used for this task.

Equations and Solutions

We have already seen that an equation is a number sentence stating that the expressions on either side of the equals sign represent the same number. Some equations, like $3 + 2 = 5$ or $2x + 6 = 2(x + 3)$, are *always* true and some, like $3 + 2 = 6$ or $x + 2 = x + 3$, are *never* true. In this text, we will concentrate on equations like $x + 6 = 13$ or $7x = 141$ that are *sometimes* true, depending on the replacement value for the variable.

> **Solution of an Equation**
>
> Any replacement for the variable that makes an equation true is called a *solution* of the equation. To *solve* an equation means to find all of its solutions.

To determine whether a number is a solution, we substitute that number for the variable throughout the equation. If the values on both sides of the equals sign are the same, then the number that was substituted is a solution.

EXAMPLE   **1**   Determine whether 7 is a solution of $x + 6 = 13$.

SOLUTION   We have

$$
\begin{array}{ll}
x + 6 = 13 & \text{Writing the equation} \\
\overline{7 + 6 \mid 13} & \text{Substituting 7 for } x \\
13 \overset{?}{=} 13 \quad \text{TRUE} & 13 = 13 \text{ is a true statement.}
\end{array}
$$

Since the left-hand side and the right-hand side are the same, 7 is a solution.

> *CAUTION!*   Note that in Example 1, the solution is 7, not 13.

EXAMPLE   2   Determine whether $\frac{2}{3}$ is a solution of $156x = 117$.

SOLUTION   We have

$$156x = 117 \qquad \text{Writing the equation}$$

$$156\left(\frac{2}{3}\right) \;\Big|\; 117 \qquad \text{Substituting } \tfrac{2}{3} \text{ for } x$$

$$104 \overset{?}{=} 117 \quad \text{FALSE} \qquad \text{The statement } 104 = 117 \text{ is false.}$$

Since the left-hand side and the right-hand side differ, $\frac{2}{3}$ is not a solution.

The Addition Principle

Consider the equation

$$x = 7.$$

We can easily see that the solution of this equation is 7. Replacing x with 7, we get

$$7 = 7, \quad \text{which is true.}$$

Now consider the equation

$$x + 6 = 13.$$

In Example 1, we found that the solution of $x + 6 = 13$ is also 7. Although the solution of $x = 7$ may seem more obvious, because $x + 6 = 13$ and $x = 7$ have identical solutions, the equations are said to be **equivalent**.

> ### Equivalent Equations
> Equations with the same solutions are called *equivalent equations.*

There are principles that enable us to begin with one equation and end up with an equivalent equation, like $x = 7$, for which the solution is obvious. One such principle concerns addition. The equation $a = b$ says that a and b stand for the same number. Suppose this is true, and some number c is added to a. We get the same result if we add c to b, because a and b are the same number.

> ### The Addition Principle
> For any real numbers a, b, and c,
> $$a = b \quad \text{is equivalent to} \quad a + c = b + c.$$

To visualize the addition principle, consider a balance similar to one a jeweler might use. When the two sides of a balance hold equal weight, the balance is level. If weight is then added or removed, equally, on both sides, the balance will remain level.

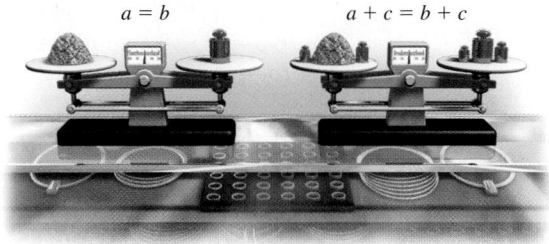

$a = b$ $\qquad\qquad$ $a + c = b + c$

When using the addition principle, we often say that we "add the same number to both sides of an equation." We can also "subtract the same number from both sides," since subtraction can be regarded as the addition of an opposite.

EXAMPLE **3** Solve: $x + 5 = -7$.

SOLUTION We can add any number we like to both sides. Since -5 is the opposite, or additive inverse, of 5, we add -5 to each side:

$$x + 5 = -7$$
$$x + 5 - 5 = -7 - 5 \quad \text{Using the addition principle: adding } -5 \text{ to both sides or subtracting 5 from both sides}$$
$$x + 0 = -12 \quad \text{Simplifying; } x + 5 - 5 = x + 5 + (-5) = x + 0$$
$$x = -12. \quad \text{Using the identity property of 0}$$

The equation $x = -12$ is equivalent to the equation $x + 5 = -7$ by the addition principle, so the solution of $x = -12$ is the solution of $x + 5 = -7$.

It is obvious that the solution of $x = -12$ is the number -12. To check the answer in the original equation, we substitute.

Check:
$$\frac{x + 5 = -7}{-12 + 5 \mid -7}$$
$$-7 \overset{?}{=} -7 \quad \text{TRUE} \qquad -7 = -7 \text{ is true.}$$

The solution of the original equation is -12. **TRY EXERCISE 11**

In Example 3, note that because we added the *opposite*, or *additive inverse*, of 5, the left side of the equation simplified to x plus the *additive identity*, 0, or simply x. These steps effectively replaced the 5 on the left with a 0. To solve $x + a = b$ for x, we add $-a$ to (or subtract a from) both sides.

EXAMPLE **4** Solve: $-6.5 = y - 8.4$.

SOLUTION The variable is on the right side this time. We can isolate y by adding 8.4 to each side:

$$-6.5 = y - 8.4 \quad y - 8.4 \text{ can be regarded as } y + (-8.4).$$
$$-6.5 + 8.4 = y - 8.4 + 8.4 \quad \text{Using the addition principle: Adding 8.4 to both sides "eliminates" } -8.4 \text{ on the right side.}$$
$$1.9 = y. \quad y - 8.4 + 8.4 = y + (-8.4) + 8.4 = y + 0 = y$$

Check:
$$\frac{-6.5 = y - 8.4}{-6.5 \mid 1.9 - 8.4}$$
$$-6.5 \overset{?}{=} -6.5 \quad \text{TRUE} \qquad -6.5 = -6.5 \text{ is true.}$$

The solution is 1.9. **TRY EXERCISE 15**

Note that the equations $a = b$ and $b = a$ have the same meaning. Thus, $-6.5 = y - 8.4$ could have been rewritten as $y - 8.4 = -6.5$.

STUDENT NOTES

We can also think of "undoing" operations in order to isolate a variable. In Example 4, we began with $y - 8.4$ on the right side. To undo the subtraction, we *add* 8.4.

The Multiplication Principle

A second principle for solving equations concerns multiplying. Suppose a and b are equal. If a and b are multiplied by some number c, then ac and bc will also be equal.

> **The Multiplication Principle**
>
> For any real numbers a, b, and c, with $c \neq 0$,
>
> $$a = b \quad \text{is equivalent to} \quad a \cdot c = b \cdot c.$$

EXAMPLE 5   Solve: $\frac{5}{4}x = 10$.

SOLUTION   We can multiply both sides by any nonzero number we like. Since $\frac{4}{5}$ is the reciprocal of $\frac{5}{4}$, we decide to multiply both sides by $\frac{4}{5}$:

$$\frac{5}{4}x = 10$$

$$\frac{4}{5} \cdot \frac{5}{4}x = \frac{4}{5} \cdot 10 \qquad \text{Using the multiplication principle: Multiplying both sides by } \frac{4}{5} \text{ "eliminates" the } \frac{5}{4} \text{ on the left.}$$

$$1 \cdot x = 8 \qquad \text{Simplifying}$$

$$x = 8. \qquad \text{Using the identity property of 1}$$

Check:

$$\frac{5}{4}x = 10$$

$$\begin{array}{c|c} \frac{5}{4} \cdot 8 & 10 \\ \hline \frac{40}{4} & \\ 10 \overset{?}{=} 10 & \text{TRUE} \end{array} \qquad \text{Think of 8 as } \frac{8}{1}.$$

$10 = 10$ is true.

The solution is 8.

TRY EXERCISE 49

In Example 5, to get x alone, we multiplied by the *reciprocal*, or *multiplicative inverse* of $\frac{5}{4}$. We then simplified the left-hand side to x times the *multiplicative identity*, 1, or simply x. These steps effectively replaced the $\frac{5}{4}$ on the left with 1.

Because division is the same as multiplying by a reciprocal, the multiplication principle also tells us that we can "divide both sides by the same nonzero number." That is,

$$\text{if } a = b, \text{ then } \quad \frac{1}{c} \cdot a = \frac{1}{c} \cdot b \quad \text{and} \quad \frac{a}{c} = \frac{b}{c} \qquad (\text{provided } c \neq 0).$$

In a product like $3x$, the multiplier 3 is called the **coefficient**. *When the coefficient of the variable is an integer or a decimal, it is usually easiest to solve an equation by dividing on both sides. When the coefficient is in fraction notation, it is usually easiest to multiply by the reciprocal.*

EXAMPLE 6   Solve: **(a)** $-4x = 9$; **(b)** $-x = 5$; **(c)** $\dfrac{2y}{9} = \dfrac{8}{3}$.

SOLUTION

STUDENT NOTES

In Example 6(a), we can think of undoing the multiplication $-4 \cdot x$ by *dividing* by -4.

a) In $-4x = 9$, the coefficient of x is an integer, so we *divide* on both sides:

$$\frac{-4x}{-4} = \frac{9}{-4} \qquad \text{Using the multiplication principle: Dividing both sides by } -4 \text{ is the same as multiplying by } -\frac{1}{4}.$$

$$1 \cdot x = -\frac{9}{4} \qquad \text{Simplifying}$$

$$x = -\frac{9}{4}. \qquad \text{Using the identity property of 1}$$

Check:

$$\frac{-4x = 9}{-4\left(-\frac{9}{4}\right) \mid 9}$$

$$9 \overset{?}{=} 9 \quad \text{TRUE} \qquad 9 = 9 \text{ is true.}$$

The solution is $-\frac{9}{4}$.

b) To solve an equation like $-x = 5$, remember that when an expression is multiplied or divided by -1, its sign is changed. Here we divide both sides by -1 to change the sign of $-x$:

$$-x = 5 \qquad \text{Note that } -x = -1 \cdot x.$$

$$\frac{-x}{-1} = \frac{5}{-1} \qquad \text{Dividing both sides by } -1. \text{ (Multiplying by } -1 \text{ would also}$$
$$\text{work. Note that the reciprocal of } -1 \text{ is } -1.)$$

$$x = -5. \qquad \text{Note that } \frac{-x}{-1} \text{ is the same as } \frac{x}{1}.$$

Check:

$$\frac{-x = 5}{-(-5) \mid 5}$$

$$5 \overset{?}{=} 5 \quad \text{TRUE} \qquad 5 = 5 \text{ is true.}$$

The solution is -5.

c) To solve an equation like $\dfrac{2y}{9} = \dfrac{8}{3}$, we rewrite the left-hand side as $\dfrac{2}{9} \cdot y$ and then use the multiplication principle, multiplying by the reciprocal of $\dfrac{2}{9}$:

$$\frac{2y}{9} = \frac{8}{3}$$

$$\frac{2}{9} \cdot y = \frac{8}{3} \qquad \text{Rewriting } \frac{2y}{9} \text{ as } \frac{2}{9} \cdot y$$

$$\frac{9}{2} \cdot \frac{2}{9} \cdot y = \frac{9}{2} \cdot \frac{8}{3} \qquad \text{Multiplying both sides by } \frac{9}{2}$$

$$1y = \frac{3 \cdot \cancel{3} \cdot \cancel{2} \cdot 4}{\cancel{2} \cdot \cancel{3}} \qquad \text{Removing a factor equal to 1: } \frac{3 \cdot 2}{2 \cdot 3} = 1$$

$$y = 12.$$

Check:

$$\frac{2y}{9} = \frac{8}{3}$$

$$\frac{\dfrac{2 \cdot 12}{9}}{\qquad} \Bigg| \frac{8}{3}$$

$$\frac{24}{9}$$

$$\frac{8}{3} \overset{?}{=} \frac{8}{3} \quad \text{TRUE} \qquad \frac{8}{3} = \frac{8}{3} \text{ is true.}$$

The solution is 12.   **TRY EXERCISE** 35

Selecting the Correct Approach

It is important that you be able to determine which principle should be used to solve a particular equation.

EXAMPLE 7

Solve: **(a)** $-\frac{2}{3} + x = \frac{5}{2}$; **(b)** $12.6 = 3t$.

SOLUTION

a) To undo addition of $-\frac{2}{3}$, we subtract $-\frac{2}{3}$ from both sides. Subtracting $-\frac{2}{3}$ is the same as adding $\frac{2}{3}$.

$$-\frac{2}{3} + x = \frac{5}{2}$$

$$-\frac{2}{3} + x + \frac{2}{3} = \frac{5}{2} + \frac{2}{3} \qquad \text{Using the addition principle}$$

$$x = \frac{5}{2} + \frac{2}{3}$$

$$x = \frac{5}{2} \cdot \frac{3}{3} + \frac{2}{3} \cdot \frac{2}{2} \qquad \text{Finding a common denominator}$$

$$x = \frac{15}{6} + \frac{4}{6}$$

$$x = \frac{19}{6}$$

Check:
$$-\frac{2}{3} + x = \frac{5}{2}$$

$$\begin{array}{c|c} -\frac{2}{3} + \frac{19}{6} & \frac{5}{2} \\ -\frac{4}{6} + \frac{19}{6} & \\ \frac{15}{6} & \\ \frac{5 \cdot 3}{2 \cdot 3} & \\ \frac{5}{2} \overset{?}{=} \frac{5}{2} & \text{TRUE} \end{array}$$

$-\frac{2}{3} \cdot \frac{2}{2} = -\frac{4}{6}$

Removing a factor equal to 1: $\frac{3}{3} = 1$

$\frac{5}{2} = \frac{5}{2}$ is true.

The solution is $\frac{19}{6}$.

b) To undo multiplication by 3, we either divide both sides by 3 or multiply both sides by $\frac{1}{3}$:

$$12.6 = 3t$$

$$\frac{12.6}{3} = \frac{3t}{3} \qquad \text{Using the multiplication principle}$$

$$4.2 = t. \qquad \text{Simplifying}$$

Check:
$$12.6 = 3t$$

$$\begin{array}{c|c} 12.6 & 3(4.2) \\ 12.6 \overset{?}{=} 12.6 & \text{TRUE} \end{array}$$

$12.6 = 12.6$ is true.

The solution is 4.2.

TRY EXERCISES 59 and 67

2.1 EXERCISE SET

For Extra Help

MyMathLab  | Math PRACTICE | WATCH | DOWNLOAD

Concept Reinforcement *For each of Exercises 1–6, match the statement with the most appropriate choice from the column on the right.*

1. ____ The equations $x + 3 = 7$ and $6x = 24$

2. ____ The expressions $3(x - 2)$ and $3x - 6$

3. ____ A replacement that makes an equation true

4. ____ The role of 9 in $9ab$

5. ____ The principle used to solve $\frac{2}{3} \cdot x = -4$

6. ____ The principle used to solve $\frac{2}{3} + x = -4$

a) Coefficient

b) Equivalent expressions

c) Equivalent equations

d) The multiplication principle

e) The addition principle

f) Solution

For each of Exercises 7–10, match the equation with the step, from the column on the right, that would be used to solve the equation.

7. $6x = 30$ a) Add 6 to both sides.

8. $x + 6 = 30$ b) Subtract 6 from both sides.

9. $\frac{1}{6}x = 30$ c) Multiply both sides by 6.

10. $x - 6 = 30$ d) Divide both sides by 6.

To the student and the instructor: The [TRY EXERCISES▶] for examples are indicated by a shaded block ▢ on the exercise number. Complete step-by-step solutions for these exercises appear online at www.pearsonhighered.com/bittingerellenbogen.

Solve using the addition principle. Don't forget to check!

11. $x + 10 = 21$ **12.** $t + 9 = 47$

13. $y + 7 = -18$ **14.** $x + 12 = -7$

15. $-6 = y + 25$ **16.** $-5 = x + 8$

17. $x - 18 = 23$ **18.** $x - 19 = 16$

19. $12 = -7 + y$ **20.** $15 = -8 + z$

21. $-5 + t = -11$ **22.** $-6 + y = -21$

23. $r + \frac{1}{3} = \frac{8}{3}$ **24.** $t + \frac{3}{8} = \frac{5}{8}$

25. $x - \frac{3}{5} = -\frac{7}{10}$ **26.** $x - \frac{2}{3} = -\frac{5}{6}$

27. $x - \frac{5}{6} = \frac{7}{8}$ **28.** $y - \frac{3}{4} = \frac{5}{6}$

29. $-\frac{1}{5} + z = -\frac{1}{4}$ **30.** $-\frac{2}{3} + y = -\frac{3}{4}$

31. $m - 2.8 = 6.3$ **32.** $y - 5.3 = 8.7$

33. $-9.7 = -4.7 + y$ **34.** $-7.8 = 2.8 + x$

Solve using the multiplication principle. Don't forget to check!

35. $8a = 56$ **36.** $6x = 72$

37. $84 = 7x$ **38.** $45 = 9t$

39. $-x = 38$ **40.** $100 = -x$

Aha! **41.** $-t = -8$ **42.** $-68 = -r$

43. $-7x = 49$ **44.** $-4x = 36$

45. $-1.3a = -10.4$ **46.** $-3.4t = -20.4$

47. $\frac{y}{8} = 11$ **48.** $\frac{a}{4} = 13$

49. $\frac{4}{5}x = 16$ **50.** $\frac{3}{4}x = 27$

51. $\frac{-x}{6} = 9$ **52.** $\frac{-t}{4} = 8$

53. $\frac{1}{9} = \frac{z}{-5}$ **54.** $\frac{2}{7} = \frac{x}{-3}$

Aha! **55.** $-\frac{3}{5}r = -\frac{3}{5}$ **56.** $-\frac{2}{5}y = -\frac{4}{15}$

57. $\frac{-3r}{2} = -\frac{27}{4}$ **58.** $\frac{5x}{7} = -\frac{10}{14}$

Solve. The icon ▤ indicates an exercise designed to give practice using a calculator.

59. $4.5 + t = -3.1$ **60.** $\frac{3}{4}x = 18$

61. $-8.2x = 20.5$ **62.** $t - 7.4 = -12.9$

63. $x - 4 = -19$ **64.** $y - 6 = -14$

65. $t - 3 = -8$ **66.** $t - 9 = -8$

67. $-12x = 14$ **68.** $-15x = 20$

69. $48 = -\frac{3}{8}y$ **70.** $14 = t + 27$

71. $a - \frac{1}{6} = -\frac{2}{3}$ **72.** $-\frac{x}{6} = \frac{2}{9}$

73. $-24 = \frac{8x}{5}$ **74.** $\frac{1}{5} + y = -\frac{3}{10}$

75. $-\frac{4}{3}t = -12$ **76.** $\frac{17}{35} = -x$

▤ **77.** $-483.297 = -794.053 + t$

▤ **78.** $-0.2344x = 2028.732$

📜 **79.** When solving an equation, how do you determine what number to add, subtract, multiply, or divide by on both sides of that equation?

📜 **80.** What is the difference between equivalent expressions and equivalent equations?

Skill Review

To prepare for Section 2.2, review the rules for order of operations (Section 1.8).

Simplify. [1.8]

81. $3 \cdot 4 - 18$

82. $14 - 2(7 - 1)$

83. $16 \div (2 - 3 \cdot 2) + 5$

84. $12 - 5 \cdot 2^3 + 4 \cdot 3$

Synthesis

📜 **85.** To solve $-3.5 = 14t$, Anita adds 3.5 to both sides. Will this form an equivalent equation? Will it help solve the equation? Explain.

📜 **86.** Explain why it is not necessary to state a subtraction principle: For any real numbers a, b, and c, $a = b$ is equivalent to $a - c = b - c$.

Solve for x. Assume a, c, m ≠ 0.

87. $mx = 11.6m$

88. $x - 4 + a = a$

89. $cx + 5c = 7c$

90. $c \cdot \dfrac{21}{a} = \dfrac{7cx}{2a}$

91. $7 + |x| = 30$

92. $ax - 3a = 5a$

93. If $t - 3590 = 1820$, find $t + 3590$.

94. If $n + 268 = 124$, find $n - 268$.

95. Lydia makes a calculation and gets an answer of 22.5. On the last step, she multiplies by 0.3 when she should have divided by 0.3. What should the correct answer be?

96. Are the equations $x = 5$ and $x^2 = 25$ equivalent? Why or why not?

2.2   Using the Principles Together

Applying Both Principles ▪ Combining Like Terms ▪ Clearing Fractions and Decimals ▪ Contradictions and Identities

An important strategy for solving new problems is to find a way to make a new problem look like a problem that we already know how to solve. This is precisely the approach taken in this section. You will find that the last steps of the examples in this section are nearly identical to the steps used for solving the equations of Section 2.1. What is new in this section appears in the early steps of each example.

Applying Both Principles

The addition and multiplication principles, along with the laws discussed in Chapter 1, are our tools for solving equations. In this section, we will find that the sequence and manner in which these tools are used is especially important.

EXAMPLE 1   Solve: $5 + 3x = 17$.

SOLUTION   Were we to evaluate $5 + 3x$, the rules for the order of operations direct us to *first* multiply by 3 and *then* add 5. Because of this, we can isolate $3x$ and then x by reversing these operations: We first subtract 5 from both sides and then divide both sides by 3. Our goal is an equivalent equation of the form $x = a$.

$$5 + 3x = 17$$

$$5 + 3x - 5 = 17 - 5 \qquad \text{Using the addition principle: subtracting 5 from both sides (adding } -5)$$

$$5 + (-5) + 3x = 12 \qquad \text{Using a commutative law. Try to perform this step mentally.}$$

Isolate the *x*-term. $\qquad 3x = 12 \qquad$ Simplifying

$$\frac{3x}{3} = \frac{12}{3} \qquad \text{Using the multiplication principle: dividing both sides by 3 (multiplying by } \tfrac{1}{3})$$

Isolate *x*. $\qquad x = 4 \qquad$ Simplifying

Check:
$$\begin{array}{c|c} 5 + 3x = 17 \\ \hline 5 + 3 \cdot 4 & 17 \\ 5 + 12 & \\ 17 \stackrel{?}{=} 17 & \text{TRUE} \end{array}$$

We use the rules for order of operations: Find the product, $3 \cdot 4$, and then add.

The solution is 4.

TRY EXERCISE 7

EXAMPLE 2 Solve: $\frac{4}{3}x - 7 = 1$.

SOLUTION In $\frac{4}{3}x - 7$, we multiply first and then subtract. To reverse these steps, we first add 7 and then either divide by $\frac{4}{3}$ or multiply by $\frac{3}{4}$.

$$\frac{4}{3}x - 7 = 1$$

$$\frac{4}{3}x - 7 + 7 = 1 + 7 \qquad \text{Adding 7 to both sides}$$

$$\frac{4}{3}x = 8$$

$$\frac{3}{4} \cdot \frac{4}{3}x = \frac{3}{4} \cdot 8 \qquad \text{Multiplying both sides by } \frac{3}{4}$$

$$\left.\begin{array}{l} 1 \cdot x = \dfrac{3 \cdot \cancel{4} \cdot 2}{\cancel{4}} \\[2mm] x = 6 \end{array}\right\} \quad \text{Simplifying}$$

Check:
$$\begin{array}{r|l} \frac{4}{3}x - 7 = 1 \\ \hline \frac{4}{3} \cdot 6 - 7 & 1 \\ 8 - 7 & \\ 1 \stackrel{?}{=} 1 & \text{TRUE} \end{array}$$

The solution is 6. ▸ **TRY EXERCISE** 27

STUDY SKILLS ─────

Use the Answer Section Carefully

When using the answers listed at the back of this book, try not to "work backward" from the answer. If you frequently require two or more attempts to answer an exercise correctly, you probably need to work more carefully and/or reread the section preceding the exercise set. Remember that on quizzes and tests you have only one attempt per problem and no answer section to consult.

EXAMPLE 3 Solve: $45 - t = 13$.

SOLUTION We have

$$45 - t = 13$$

$$45 - t - 45 = 13 - 45 \qquad \text{Subtracting 45 from both sides}$$

$$\left.\begin{array}{l} 45 + (-t) + (-45) = 13 - 45 \\ 45 + (-45) + (-t) = 13 - 45 \end{array}\right\} \quad \text{Try to do these steps mentally.}$$

$$-t = -32 \qquad \text{Try to go directly to this step.}$$

$$(-1)(-t) = (-1)(-32) \qquad \begin{array}{l}\text{Multiplying both sides by } -1 \\ \text{(Dividing by } -1 \text{ would also} \\ \text{work.)}\end{array}$$

$$t = 32.$$

Check:
$$\begin{array}{r|l} 45 - t = 13 \\ \hline 45 - 32 & 13 \\ 13 \stackrel{?}{=} 13 & \text{TRUE} \end{array}$$

The solution is 32. ▸ **TRY EXERCISE** 19

As our skills improve, certain steps can be streamlined.

EXAMPLE 4   Solve: $16.3 - 7.2y = -8.18$.

SOLUTION   We have

$$16.3 - 7.2y = -8.18$$

$$16.3 - 7.2y - 16.3 = -8.18 - 16.3 \qquad \text{Subtracting 16.3 from both sides}$$

$$-7.2y = -24.48 \qquad \text{Simplifying}$$

$$\frac{-7.2y}{-7.2} = \frac{-24.48}{-7.2} \qquad \text{Dividing both sides by } -7.2$$

$$y = 3.4. \qquad \text{Simplifying}$$

Check:

$$\begin{array}{c|c} \hline 16.3 - 7.2y = -8.18 \\ \hline 16.3 - 7.2(3.4) & -8.18 \\ 16.3 - 24.48 & \\ -8.18 \overset{?}{=} -8.18 & \text{TRUE} \end{array}$$

The solution is 3.4.

> TRY EXERCISE ▸ 23

Combining Like Terms

If like terms appear on the same side of an equation, we combine them and then solve. Should like terms appear on both sides of an equation, we can use the addition principle to rewrite all like terms on one side.

EXAMPLE 5   Solve.

a) $3x + 4x = -14$ **b)** $-x + 5 = -8x + 6$
c) $6x + 5 - 7x = 10 - 4x + 7$ **d)** $2 - 5(x + 5) = 3(x - 2) - 1$

SOLUTION

a) $3x + 4x = -14$

$$7x = -14 \qquad \text{Combining like terms}$$

$$\frac{7x}{7} = \frac{-14}{7} \qquad \text{Dividing both sides by 7}$$

$$x = -2 \qquad \text{Simplifying}$$

The check is left to the student. The solution is -2.

b) To solve $-x + 5 = -8x + 6$, we must first write only variable terms on one side and only constant terms on the other. This can be done by subtracting 5 from both sides, to get all constant terms on the right, and adding $8x$ to both sides, to get all variable terms on the left.

$$-x + 5 = -8x + 6$$

Isolate variable terms on one side and constant terms on the other side.

$$-x + 8x + 5 = -8x + 8x + 6 \qquad \text{Adding } 8x \text{ to both sides}$$

$$7x + 5 = 6 \qquad \text{Simplifying}$$

$$7x + 5 - 5 = 6 - 5 \qquad \text{Subtracting 5 from both sides}$$

$$7x = 1 \qquad \text{Combining like terms}$$

$$\frac{7x}{7} = \frac{1}{7} \qquad \text{Dividing both sides by 7}$$

$$x = \frac{1}{7}$$

The check is left to the student. The solution is $\frac{1}{7}$.

Most graphing calculators have a TABLE feature that lists the value of a variable expression for different choices of x. For example, to evaluate $6x + 5 - 7x$ for $x = 0, 1, 2, \ldots$, we first use $\boxed{\text{Y=}}$ to enter $6x + 5 - 7x$ as y_1. We then use $\boxed{\text{2ND}}$ $\boxed{\text{TBLSET}}$ to specify TblStart $= 0$, ΔTbl $= 1$, and select AUTO twice. By pressing $\boxed{\text{2ND}}$ $\boxed{\text{TABLE}}$, we can generate a table in which the value of $6x + 5 - 7x$ is listed for values of x starting at 0 and increasing by ones.

X	Y₁	
0	5	
1	4	
2	3	
3	2	
4	1	
5	0	
6	−1	
X = 0		

1. Create the above table on your graphing calculator. Scroll up and down to extend the table.
2. Enter $10 - 4x + 7$ as y_2. Your table should now have three columns.
3. For what x-value is y_1 the same as y_2? Compare this with the solution of Example 5(c). Is this a reliable way to solve equations? Why or why not?

c) $6x + 5 - 7x = 10 - 4x + 7$

$$\begin{aligned}
-x + 5 &= 17 - 4x & &\text{Combining like terms within each side} \\
-x + 5 + 4x &= 17 - 4x + 4x & &\text{Adding } 4x \text{ to both sides} \\
5 + 3x &= 17 & &\text{Simplifying. This is identical to Example 1.} \\
3x &= 12 & &\text{Subtracting 5 from both sides} \\
\frac{3x}{3} &= \frac{12}{3} & &\text{Dividing both sides by 3} \\
x &= 4
\end{aligned}$$

Check:

$$\begin{array}{c|c}
\multicolumn{2}{c}{6x + 5 - 7x = 10 - 4x + 7} \\
\hline
6 \cdot 4 + 5 - 7 \cdot 4 & 10 - 4 \cdot 4 + 7 \\
24 + 5 - 28 & 10 - 16 + 7 \\
1 \overset{?}{=} 1 & \text{TRUE}
\end{array}$$

The student can confirm that 4 checks and is the solution.

d) $2 - 5(x + 5) = 3(x - 2) - 1$

$$\begin{aligned}
2 - 5x - 25 &= 3x - 6 - 1 & &\text{Using the distributive law. This is now similar to part (c) above.} \\
-5x - 23 &= 3x - 7 & &\text{Combining like terms on each side} \\
\left.\begin{aligned} -5x - 23 + 7 &= 3x \\ -23 + 7 &= 3x + 5x \end{aligned}\right\} & & &\text{Adding 7 and } 5x \text{ to both sides. This isolates the } x\text{-terms on one side and the constant terms on the other.} \\
-16 &= 8x & &\text{Simplifying} \\
\frac{-16}{8} &= \frac{8x}{8} & &\text{Dividing both sides by 8} \\
-2 &= x & &\text{This is equivalent to } x = -2.
\end{aligned}$$

The student can confirm that -2 checks and is the solution.

 **TRY EXERCISE**   39

Clearing Fractions and Decimals

Equations are generally easier to solve when they do not contain fractions or decimals. The multiplication principle can be used to "clear" fractions or decimals, as shown here.

Clearing Fractions	Clearing Decimals
$\frac{1}{2}x + 5 = \frac{3}{4}$	$2.3x + 7 = 5.4$
$4\left(\frac{1}{2}x + 5\right) = 4 \cdot \frac{3}{4}$	$10(2.3x + 7) = 10 \cdot 5.4$
$2x + 20 = 3$	$23x + 70 = 54$

In each case, the resulting equation is equivalent to the original equation, but easier to solve.

The easiest way to clear an equation of fractions is to multiply *both sides* of the equation by the smallest, or *least*, common denominator of the fractions in the equation.

EXAMPLE **6**   Solve: **(a)** $\frac{2}{3}x - \frac{1}{6} = 2x$; **(b)** $\frac{2}{5}(3x + 2) = 8$.

SOLUTION

a) We multiply both sides by 6, the least common denominator of $\frac{2}{3}$ and $\frac{1}{6}$.

$$6\left(\frac{2}{3}x - \frac{1}{6}\right) = 6 \cdot 2x \qquad \text{Multiplying both sides by 6}$$

$$6 \cdot \frac{2}{3}x - 6 \cdot \frac{1}{6} = 6 \cdot 2x \longleftarrow$$

> **CAUTION!**   Be sure the distributive law is used to multiply *all* the terms by 6.

$$4x - 1 = 12x \qquad \text{Simplifying. Note that the fractions are cleared: } 6 \cdot \frac{2}{3} = 4, 6 \cdot \frac{1}{6} = 1, \text{ and } 6 \cdot 2 = 12.$$

$$-1 = 8x \qquad \text{Subtracting } 4x \text{ from both sides}$$

$$\frac{-1}{8} = \frac{8x}{8} \qquad \text{Dividing both sides by 8}$$

$$-\frac{1}{8} = x$$

The student can confirm that $-\frac{1}{8}$ checks and is the solution.

b) To solve $\frac{2}{5}(3x + 2) = 8$, we can multiply both sides by $\frac{5}{2}$ (or divide by $\frac{2}{5}$) to "undo" the multiplication by $\frac{2}{5}$ on the left side.

$$\frac{5}{2} \cdot \frac{2}{5}(3x + 2) = \frac{5}{2} \cdot 8 \qquad \text{Multiplying both sides by } \frac{5}{2}$$

$$3x + 2 = 20 \qquad \text{Simplifying; } \frac{5}{2} \cdot \frac{2}{5} = 1 \text{ and } \frac{5}{2} \cdot \frac{8}{1} = 20$$

$$3x = 18 \qquad \text{Subtracting 2 from both sides}$$

$$x = 6 \qquad \text{Dividing both sides by 3}$$

The student can confirm that 6 checks and is the solution.

> **TRY EXERCISE** ▶ 69

To clear an equation of decimals, we count the greatest number of decimal places in any one number. If the greatest number of decimal places is 1, we multiply both sides by 10; if it is 2, we multiply by 100; and so on. This procedure is the same as multiplying by the least common denominator after converting the decimals to fractions.

EXAMPLE **7**   Solve: $16.3 - 7.2y = -8.18$.

SOLUTION   The greatest number of decimal places in any one number is *two*. Multiplying by 100 will clear all decimals.

$$100(16.3 - 7.2y) = 100(-8.18) \qquad \text{Multiplying both sides by 100}$$

$$100(16.3) - 100(7.2y) = 100(-8.18) \qquad \text{Using the distributive law}$$

$$1630 - 720y = -818 \qquad \text{Simplifying}$$

$$-720y = -818 - 1630 \qquad \text{Subtracting 1630 from both sides}$$

$$-720y = -2448 \qquad \text{Combining like terms}$$

$$y = \frac{-2448}{-720} \qquad \text{Dividing both sides by } -720$$

$$y = 3.4$$

In Example 4, the same solution was found without clearing decimals. Finding the same answer in two ways is a good check. The solution is 3.4.

> **TRY EXERCISE** ▶ 75

STUDENT NOTES ———

Compare the steps of Examples 4 and 7. Note that although the two approaches differ, they yield the same solution. Whenever you can use two approaches to solve a problem, try to do so, both as a check and as a valuable learning experience.

> **An Equation-Solving Procedure**
>
> 1. Use the multiplication principle to clear any fractions or decimals. (This is optional, but can ease computations. See Examples 6 and 7.)
> 2. If necessary, use the distributive law to remove parentheses. Then combine like terms on each side. (See Example 5.)
> 3. Use the addition principle, as needed, to isolate all variable terms on one side. Then combine like terms. (See Examples 1–7.)
> 4. Multiply or divide to solve for the variable, using the multiplication principle. (See Examples 1–7.)
> 5. Check all possible solutions in the original equation. (See Examples 1–4.)

Contradictions and Identities

All of the equations we have examined so far had a solution. Equations that are true for some values (solutions), but not for others, are called **conditional equations**. Equations that have no solution, such as $x + 1 = x + 2$, are called **contradictions**. If, when solving an equation, we obtain an equation that is false for any value of x, the equation has no solution.

EXAMPLE 8   Solve: $3x - 5 = 3(x - 2) + 4$.

SOLUTION

$$3x - 5 = 3(x - 2) + 4$$
$$3x - 5 = 3x - 6 + 4 \qquad \text{Using the distributive law}$$
$$3x - 5 = 3x - 2 \qquad \text{Combining like terms}$$
$$-3x + 3x - 5 = -3x + 3x - 2 \qquad \text{Using the addition principle}$$
$$-5 = -2$$

Since the original equation is equivalent to $-5 = -2$, which is false regardless of the choice of x, the original equation has no solution. There is no solution of $3x - 5 = 3(x - 2) + 4$. The equation is a contradiction. It is *never* true.

> **TRY EXERCISE** 45

Some equations, like $x + 1 = x + 1$, are true for all replacements. Such an equation is called an **identity**.

EXAMPLE 9   Solve: $2x + 7 = 7(x + 1) - 5x$.

SOLUTION

$$2x + 7 = 7(x + 1) - 5x$$
$$2x + 7 = 7x + 7 - 5x \qquad \text{Using the distributive law}$$
$$2x + 7 = 2x + 7 \qquad \text{Combining like terms}$$

The equation $2x + 7 = 2x + 7$ is true regardless of the replacement for x, so all real numbers are solutions. Note that $2x + 7 = 2x + 7$ is equivalent to $2x = 2x$, $7 = 7$, or $0 = 0$. All real numbers are solutions and the equation is an identity.

> **TRY EXERCISE** 33

2.2   EXERCISE SET

Concept Reinforcement   *In each of Exercises 1–6, match the equation with an equivalent equation from the column on the right that could be the next step in finding a solution.*

1. ____ $3x - 1 = 7$

2. ____ $4x + 5x = 12$

3. ____ $6(x - 1) = 2$

4. ____ $7x = 9$

5. ____ $4x = 3 - 2x$

6. ____ $8x - 5 = 6 - 2x$

a) $6x - 6 = 2$

b) $4x + 2x = 3$

c) $3x = 7 + 1$

d) $8x + 2x = 6 + 5$

e) $9x = 12$

f) $x = \frac{9}{7}$

Solve and check. Label any contradictions or identities.

7. $2x + 9 = 25$

8. $3x - 11 = 13$

9. $6z + 5 = 47$

10. $5z + 2 = 57$

11. $7t - 8 = 27$

12. $6x - 5 = 2$

13. $3x - 9 = 1$

14. $5x - 9 = 41$

15. $8z + 2 = -54$

16. $4x + 3 = -21$

17. $-37 = 9t + 8$

18. $-39 = 1 + 5t$

19. $12 - t = 16$

20. $9 - t = 21$

21. $-6z - 18 = -132$

22. $-7x - 24 = -129$

23. $5.3 + 1.2n = 1.94$

24. $6.4 - 2.5n = 2.2$

25. $32 - 7x = 11$

26. $27 - 6x = 99$

27. $\frac{3}{5}t - 1 = 8$

28. $\frac{2}{3}t - 1 = 5$

29. $6 + \frac{7}{2}x = -15$

30. $6 + \frac{5}{4}x = -4$

31. $-\dfrac{4a}{5} - 8 = 2$

32. $-\dfrac{8a}{7} - 2 = 4$

33. $4x = x + 3x$

34. $-3z + 8z = 45$

35. $4x - 6 = 6x$

36. $5x - x = x + 3x$

37. $2 - 5y = 26 - y$

38. $6x - 5 = 7 + 2x$

39. $7(2a - 1) = 21$

40. $5(3 - 3t) = 30$

Aha! 41. $11 = 11(x + 1)$

42. $9 = 3(5x - 2)$

43. $2(3 + 4m) - 6 = 48$

44. $3(5 + 3m) - 8 = 7$

45. $3(x + 4) = 3(x - 1)$

46. $5(x - 7) = 3(x - 2) + 2x$

47. $2r + 8 = 6r + 10$

48. $3b - 2 = 7b + 4$

49. $6x + 3 = 2x + 3$

50. $5y + 3 = 2y + 15$

51. $5 - 2x = 3x - 7x + 25$

52. $10 - 3x = x - 2x + 40$

53. $7 + 3x - 6 = 3x + 5 - x$

54. $5 + 4x - 7 = 4x - 2 - x$

55. $4y - 4 + y + 24 = 6y + 20 - 4y$

56. $5y - 10 + y = 7y + 18 - 5y$

57. $4 + 7x = 7(x + 1)$

58. $3(t + 2) + t = 2(3 + 2t)$

59. $19 - 3(2x - 1) = 7$

60. $5(d + 4) = 7(d - 2)$

61. $7(5x - 2) = 6(6x - 1)$

62. $5(t + 1) + 8 = 3(t - 2) + 6$

63. $2(3t + 1) - 5 = t - (t + 2)$

64. $4x - (x + 6) = 5(3x - 1) + 8$

65. $2(7 - x) - 20 = 7x - 3(2 + 3x)$

66. $5(x - 7) = 3(x - 2) + 2x$

67. $19 - (2x + 3) = 2(x + 3) + x$

68. $13 - (2c + 2) = 2(c + 2) + 3c$

Clear fractions or decimals, solve, and check.

69. $\frac{2}{3} + \frac{1}{4}t = 2$

70. $-\frac{5}{6} + x = -\frac{1}{2} - \frac{2}{3}$

71. $\frac{2}{3} + 4t = 6t - \frac{2}{15}$

72. $\frac{1}{2} + 4m = 3m - \frac{5}{2}$

73. $\frac{1}{3}x + \frac{2}{5} = \frac{4}{5} + \frac{3}{5}x - \frac{2}{3}$

74. $1 - \frac{2}{3}y = \frac{9}{5} - \frac{1}{5}y + \frac{3}{5}$

75. $2.1x + 45.2 = 3.2 - 8.4x$

76. $0.91 - 0.2z = 1.23 - 0.6z$

77. $0.76 + 0.21t = 0.96t - 0.49$

78. $1.7t + 8 - 1.62t = 0.4t - 0.32 + 8$

79. $\frac{2}{5}x - \frac{3}{2}x = \frac{3}{4}x + 3$

80. $\frac{5}{16}y + \frac{3}{8}y = 2 + \frac{1}{4}y$

81. $\frac{1}{3}(2x - 1) = 7$

82. $\frac{1}{5}(4x - 1) = 7$

83. $\frac{3}{4}(3t - 4) = 15$

84. $\frac{3}{2}(2x + 5) = -\frac{15}{2}$

85. $\frac{1}{6}\left(\frac{3}{4}x - 2\right) = -\frac{1}{5}$

86. $\frac{2}{3}\left(\frac{7}{8} - 4x\right) - \frac{5}{8} = \frac{3}{8}$

87. $0.7(3x + 6) = 1.1 - (x - 3)$

88. $0.9(2x - 8) = 4 - (x + 5)$

89. $a + (a - 3) = (a + 2) - (a + 1)$

90. $0.8 - 4(b - 1) = 0.2 + 3(4 - b)$

91. Tyla solves $45 - t = 13$ (Example 3) by adding $t - 13$ to both sides. Is this approach preferable to the one used in Example 3? Why or why not?

92. Why must the rules for the order of operations be understood before solving the equations in this section?

Skill Review

To prepare for Section 2.3, review evaluating algebraic expressions (Section 1.8).

Evaluate. [1.8]

93. $3 - 5a$, for $a = 2$

94. $12 \div 4 \cdot t$, for $t = 5$

95. $7x - 2x$, for $x = -3$

96. $t(8 - 3t)$, for $t = -2$

Synthesis

97. What procedure would you use to solve an equation like $0.23x + \frac{17}{3} = -0.8 + \frac{3}{4}x$? Could your procedure be streamlined? If so, how?

98. Dave is determined to solve $3x + 4 = -11$ by first using the multiplication principle to "eliminate" the 3. How should he proceed and why?

Solve. Label any contradictions or identities.

99. $8.43x - 2.5(3.2 - 0.7x) = -3.455x + 9.04$

100. $0.008 + 9.62x - 42.8 = 0.944x + 0.0083 - x$

101. $-2[3(x - 2) + 4] = 4(5 - x) - 2x$

102. $0 = t - (-6) - (-7t)$

103. $2x(x + 5) - 3(x^2 + 2x - 1) = 9 - 5x - x^2$

104. $x(x - 4) = 3x(x + 1) - 2(x^2 + x - 5)$

105. $9 - 3x = 2(5 - 2x) - (1 - 5x)$

Aha! **106.** $[7 - 2(8 \div (-2))]x = 0$

107. $\dfrac{x}{14} - \dfrac{5x + 2}{49} = \dfrac{3x - 4}{7}$

108. $\dfrac{5x + 3}{4} + \dfrac{25}{12} = \dfrac{5 + 2x}{3}$

109. $2\{9 - 3[-2x - 4]\} = 12x + 42$

110. $-9t + 2 = 2 - 9t - 5(8 \div 4(1 + 3^4))$

CORNER

Step-by-Step Solutions

Focus: Solving linear equations

Time: 20 minutes

Group size: 3

In general, there is more than one correct sequence of steps for solving an equation. This makes it important that you write your steps clearly and logically so that others can follow your approach.

ACTIVITY

1. Each group member should select a different one of the following equations and, on a fresh sheet of paper, perform the first step of the solution.

$$4 - 3(x - 3) = 7x + 6(2 - x)$$
$$5 - 7[x - 2(x - 6)] = 3x + 4(2x - 7) + 9$$
$$4x - 7[2 + 3(x - 5) + x] = 4 - 9(-3x - 19)$$

2. Pass the papers around so that the second and third steps of each solution are performed by the other two group members. Before writing, make sure that the previous step is correct. If a mistake is discovered, return the problem to the person who made the mistake for repairs. Continue passing the problems around until all equations have been solved.

3. Each group should reach a consensus on what the three solutions are and then compare their answers to those of other groups.

2.3   Formulas

Evaluating Formulas ▪ Solving for a Variable

Many applications of mathematics involve relationships among two or more quantities. An equation that represents such a relationship will use two or more letters and is known as a **formula**. Most of the letters in this book are variables, but some are constants. For example, c in $E = mc^2$ represents the speed of light.

Evaluating Formulas

EXAMPLE 1   *Event promotion.*   Event promoters use the formula

$$p = \frac{1.2x}{s}$$

to determine a ticket price p for an event with x dollars of expenses and s anticipated ticket sales. Grand Events expects expenses for an upcoming concert to be $80,000 and anticipates selling 4000 tickets. What should the ticket price be?

Source: *The Indianapolis Star, 2/27/03*

SOLUTION   We substitute 80,000 for x and 4000 for s in the formula and calculate p:

$$p = \frac{1.2x}{s} = \frac{1.2(80,000)}{4000} = 24.$$

The ticket price should be $24.

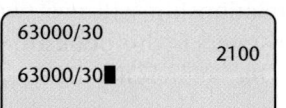 TRY EXERCISE ▶ 1

Solving for a Variable

In the Northeast, the formula $B = 30a$ is used to determine the minimum furnace output B, in British thermal units (Btu's), for a well-insulated home with a square feet of flooring. Suppose that a contractor has an extra furnace and wants to determine the size of the largest (well-insulated) house in which it can be used. The contractor can substitute the amount of the furnace's output in Btu's—say, 63,000—for B, and then solve for a:

$$63,000 = 30a \qquad \text{Replacing } B \text{ with } 63,000$$
$$2100 = a. \qquad \text{Dividing both sides by } 30$$

The home should have no more than 2100 ft^2 of flooring.

Were these calculations to be performed for a variety of furnaces, the contractor would find it easier to first solve $B = 30a$ for a, and *then* substitute values for B. Solving for a variable can be done in much the same way that we solved equations in Sections 2.1 and 2.2.

EXAMPLE **2**

Solve for a: $B = 30a$.

SOLUTION   We have

$$B = 30\overset{\downarrow}{a} \qquad \text{We want this letter alone.}$$
$$\frac{B}{30} = a. \qquad \text{Dividing both sides by } 30$$

The equation $a = B/30$ gives a quick, easy way to determine the floor area of the largest (well-insulated) house that a furnace supplying B Btu's could heat.

TRY EXERCISE ▶ 9

To see how solving a formula is just like solving an equation, compare the following. In (A), we solve as usual; in (B), we show steps but do not simplify; and in (C), we *cannot* simplify because a, b, and c are unknown.

A. $5x + 2 = 12$
$$5x = 12 - 2$$
$$5x = 10$$
$$x = \frac{10}{5} = 2$$

B. $5x + 2 = 12$
$$5x = 12 - 2$$
$$x = \frac{12 - 2}{5}$$

C. $ax + b = c$
$$ax = c - b$$
$$x = \frac{c - b}{a}$$

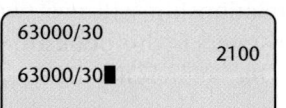

TECHNOLOGY CONNECTION

Suppose that after calculating $63,000 \div 30$, we wish to find $72,000 \div 30$. Pressing **2ND** (ENTRY) gives the following.

```
63000/30
              2100
63000/30▮
```

Moving the cursor left, we can change 63,000 to 72,000 and press **ENTER**.

```
63000/30
              2100
72000/30
              2400
```

1. Verify the work above and then use **2ND** (ENTRY) to find $72,000 \div 90$.

EXAMPLE 3

Circumference of a circle. The formula $C = 2\pi r$ gives the *circumference C* of a circle with radius r. Solve for r.

SOLUTION   The **circumference** is the distance around a circle.

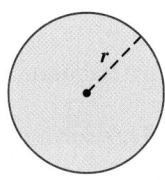

Given a radius r, we can use this equation to find a circle's circumference C.

$$C = 2\pi r \quad \text{We want this variable alone.}$$

$$\frac{C}{2\pi} = \frac{2\pi r}{2\pi} \quad \text{Dividing both sides by } 2\pi$$

Given a circle's circumference C, we can use this equation to find the radius r.

$$\frac{C}{2\pi} = r$$

TRY EXERCISE 13

EXAMPLE 4

Solve for y: $3x - 4y = 10$.

SOLUTION   There is one term that contains y, so we begin by isolating that term on one side of the equation.

$$3x - 4y = 10 \quad \text{We want this variable alone.}$$
$$-4y = 10 - 3x \quad \text{Subtracting } 3x \text{ from both sides}$$
$$-\tfrac{1}{4}(-4y) = -\tfrac{1}{4}(10 - 3x) \quad \text{Multiplying both sides by } -\tfrac{1}{4}$$
$$y = -\tfrac{10}{4} + \tfrac{3}{4}x \quad \text{Multiplying using the distributive law}$$
$$y = -\tfrac{5}{2} + \tfrac{3}{4}x \quad \text{Simplifying the fraction}$$

TRY EXERCISE 33

EXAMPLE 5

Nutrition.   The number of calories K needed each day by a moderately active woman who weighs w pounds, is h inches tall, and is a years old, can be estimated using the formula

$$K = 917 + 6(w + h - a).*$$

Solve for w.

SOLUTION   We reverse the order in which the operations occur on the right side:

We want w alone.

$$K = 917 + 6(w + h - a)$$
$$K - 917 = 6(w + h - a) \quad \text{Subtracting 917 from both sides}$$
$$\frac{K - 917}{6} = w + h - a \quad \text{Dividing both sides by 6}$$
$$\frac{K - 917}{6} + a - h = w. \quad \text{Adding } a \text{ and subtracting } h \text{ on both sides}$$

This formula can be used to estimate a woman's weight, if we know her age, height, and caloric needs.

TRY EXERCISE 43

*Based on information from M. Parker (ed.), *She Does Math!* (Washington, D.C.: Mathematical Association of America, 1995), p. 96.

STUDY SKILLS

Pace Yourself

Most instructors agree that it is better for a student to study for one hour four days in a week, than to study once a week for four hours. Of course, the total weekly study time will vary from student to student. It is common to expect an average of two hours of homework for each hour of class time.

The above steps are similar to those used in Section 2.2 to solve equations. We use the addition and multiplication principles just as before. An important difference that we will see in the next example is that we will sometimes need to factor.

To Solve a Formula for a Given Variable

1. If the variable for which you are solving appears in a fraction, use the multiplication principle to clear fractions.
2. Isolate the term(s), with the variable for which you are solving on one side of the equation.
3. If two or more terms contain the variable for which you are solving, factor the variable out.
4. Multiply or divide to solve for the variable in question.

We can also solve for a letter that represents a constant.

EXAMPLE **6**   *Surface area of a right circular cylinder.*   The formula $A = 2\pi rh + 2\pi r^2$ gives the surface area A of a right circular cylinder of height h and radius r. Solve for π.

SOLUTION   We have

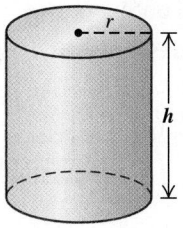

$$A = 2\pi rh + 2\pi r^2 \qquad \text{We want this letter alone.}$$

$$A = \pi(2rh + 2r^2) \qquad \text{Factoring}$$

$$\frac{A}{2rh + 2r^2} = \pi. \qquad \begin{array}{l}\text{Dividing both sides by } 2rh + 2r^2, \\ \text{or multiplying both sides by} \\ 1/(2rh + 2r^2)\end{array}$$

We can also write this as

$$\pi = \frac{A}{2rh + 2r^2}.$$

TRY EXERCISE   47

CAUTION!   Had we performed the following steps in Example 6, we would *not* have solved for π:

$$A = 2\pi rh + 2\pi r^2 \qquad \text{We want } \pi \text{ alone.}$$

$$A - 2\pi r^2 = 2\pi rh \qquad \text{Subtracting } 2\pi r^2 \text{ from both sides}$$

Two occurrences of π

$$\frac{A - 2\pi r^2}{2rh} = \pi. \qquad \text{Dividing both sides by } 2rh$$

The mathematics of each step is correct, but because π occurs on both sides of the formula, *we have not solved the formula for π.* Remember that the letter being solved for should be alone on one side of the equation, with no occurrence of that letter on the other side!

1. *Outdoor concerts.* The formula $d = 344t$ can be used to determine how far d, in meters, sound travels through room-temperature air in t seconds. At a large concert, fans near the back of the crowd experienced a 0.9-sec time lag between the time each word was pronounced on stage (as shown on large video monitors) and the time the sound reached their ears. How far were these fans from the stage?

2. *Furnace output.* Contractors in the Northeast use the formula $B = 30a$ to determine the minimum furnace output B, in British thermal units (Btu's), for a well-insulated house with a square feet of flooring. Determine the minimum furnace output for an 1800-ft^2 house that is well insulated.
Source: U.S. Department of Energy

3. *College enrollment.* At many colleges, the number of "full-time-equivalent" students f is given by
$$f = \frac{n}{15},$$
where n is the total number of credits for which students have enrolled in a given semester. Determine the number of full-time-equivalent students on a campus in which students registered for a total of 21,345 credits.

4. *Distance from a storm.* The formula $M = \frac{1}{5}t$ can be used to determine how far M, in miles, you are from lightning when its thunder takes t seconds to reach your ears. If it takes 10 sec for the sound of thunder to reach you after you have seen the lightning, how far away is the storm?

5. *Federal funds rate.* The Federal Reserve Board sets a target f for the federal funds rate, that is, the interest rate that banks charge each other for overnight borrowing of Federal funds. This target rate can be estimated by
$$f = 8.5 + 1.4(I - U),$$
where I is the core inflation rate over the previous 12 months and U is the seasonally adjusted unemployment rate. If core inflation is 0.025 and unemployment is 0.044, what should the federal funds rate be?
Source: Greg Mankiw, Harvard University, www.gregmankiw .blogspot.com/2006/06/what-would-alan-do.html

6. *Calorie density.* The calorie density D, in calories per ounce, of a food that contains c calories and weighs w ounces is given by
$$D = \frac{c}{w}.^*$$
Eight ounces of fat-free milk contains 84 calories. Find the calorie density of fat-free milk.

7. *Absorption of ibuprofen.* When 400 mg of the painkiller ibuprofen is swallowed, the number of milligrams n in the bloodstream t hours later (for $0 \leq t \leq 6$) is estimated by
$$n = 0.5t^4 + 3.45t^3 - 96.65t^2 + 347.7t.$$
How many milligrams of ibuprofen remain in the blood 1 hr after 400 mg has been swallowed?

8. *Size of a league schedule.* When all n teams in a league play every other team twice, a total of N games are played, where
$$N = n^2 - n.$$
If a soccer league has 7 teams and all teams play each other twice, how many games are played?

In Exercises 9–48, solve each formula for the indicated letter.

9. $A = bh$, for b
(Area of parallelogram with base b and height h)

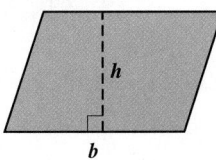

10. $A = bh$, for h

11. $d = rt$, for r
(A distance formula, where d is distance, r is speed, and t is time)

---

*Source: Nutrition Action Healthletter, March 2000, p. 9. Center for Science in the Public Interest, Suite 300; 1875 Connecticut Ave NW, Washington, D.C. 20008.

12. $d = rt$, for t

13. $I = Prt$, for P
(Simple-interest formula, where I is interest, P is principal, r is interest rate, and t is time)

14. $I = Prt$, for t

15. $H = 65 - m$, for m
(To determine the number of heating degree days H for a day with m degrees Fahrenheit as the average temperature)

16. $d = h - 64$, for h
(To determine how many inches d above average an h-inch-tall woman is)

17. $P = 2l + 2w$, for l
(Perimeter of a rectangle of length l and width w)

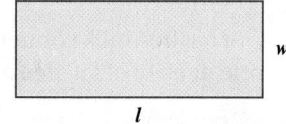

18. $P = 2l + 2w$, for w

19. $A = \pi r^2$, for π
(Area of a circle with radius r)

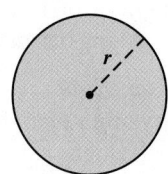

20. $A = \pi r^2$, for r^2

21. $A = \frac{1}{2}bh$, for h
(Area of a triangle with base b and height h)

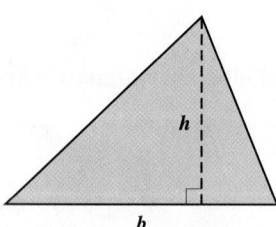

22. $A = \frac{1}{2}bh$, for b

23. $E = mc^2$, for c^2
(A relativity formula from physics)

24. $E = mc^2$, for m

25. $Q = \dfrac{c + d}{2}$, for d **26.** $Q = \dfrac{p - q}{2}$, for p

27. $A = \dfrac{a + b + c}{3}$, for b **28.** $A = \dfrac{a + b + c}{3}$, for c

29. $w = \dfrac{r}{f}$, for r
(To compute the wavelength w of a musical note with frequency f and speed of sound r)

30. $M = \dfrac{A}{s}$, for A
(To compute the Mach number M for speed A and speed of sound s)

31. $F = \dfrac{9}{5}C + 32$, for C
(To convert the Celsius temperature C to the Fahrenheit temperature F)

32. $M = \dfrac{5}{9}n + 18$, for n

33. $2x - y = 1$, for y

34. $3x - y = 7$, for y

35. $2x + 5y = 10$, for y

36. $3x + 2y = 12$, for y

37. $4x - 3y = 6$, for y

38. $5x - 4y = 8$, for y

39. $9x + 8y = 4$, for y

40. $x + 10y = 2$, for y

41. $3x - 5y = 8$, for y

42. $7x - 6y = 7$, for y

43. $z = 13 + 2(x + y)$, for x

44. $A = 115 + \dfrac{1}{2}(p + s)$, for s

45. $t = 27 - \dfrac{1}{4}(w - l)$, for l

46. $m = 19 - 5(x - n)$, for n

47. $A = at + bt$, for t

48. $S = rx + sx$, for x

49. *Area of a trapezoid.* The formula
$$A = \tfrac{1}{2}ah + \tfrac{1}{2}bh$$
can be used to find the area A of a trapezoid with bases a and b and height h. Solve for h. (*Hint*: First clear fractions.)

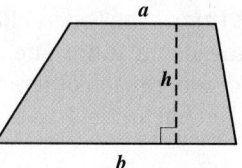

50. *Compounding interest.* The formula

$$A = P + Prt$$

is used to find the amount A in an account when simple interest is added to an investment of P dollars (see Exercise 13). Solve for P.

51. *Chess rating.* The formula

$$R = r + \frac{400(W - L)}{N}$$

is used to establish a chess player's rating R after that player has played N games, won W of them, and lost L of them. Here r is the average rating of the opponents. Solve for L.
Source: The U.S. Chess Federation

52. *Angle measure.* The angle measure S of a sector of a circle is given by

$$S = \frac{360A}{\pi r^2},$$

where r is the radius, A is the area of the sector, and S is in degrees. Solve for r^2.

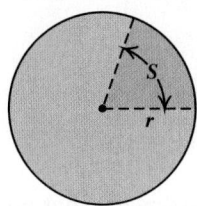

53. Naomi has a formula that allows her to convert Celsius temperatures to Fahrenheit temperatures. She needs a formula for converting Fahrenheit temperatures to Celsius temperatures. What advice can you give her?

54. Under what circumstances would it be useful to solve $d = rt$ for r? (See Exercise 11.)

Skill Review

Review simplifying expressions (Sections 1.6, 1.7, and 1.8).

Perform the indicated operations.

55. $-2 + 5 - (-4) - 17$ [1.6]

56. $-98 \div \frac{1}{2}$ [1.7]

Aha! **57.** $4.2(-11.75)(0)$ [1.7]

58. $(-2)^5$ [1.8]

Simplify. [1.8]

59. $20 \div (-4) \cdot 2 - 3$

60. $5|8 - (2 - 7)|$

Synthesis

61. The equations

$$P = 2l + 2w \quad \text{and} \quad w = \frac{P}{2} - l$$

are equivalent formulas involving the perimeter P, length l, and width w of a rectangle. Devise a problem for which the second of the two formulas would be more useful.

62. While solving $2A = ah + bh$ for h, Lea writes $\frac{2A - ah}{b} = h$. What is her mistake?

63. The Harris–Benedict formula gives the number of calories K needed each day by a moderately active man who weighs w kilograms, is h centimeters tall, and is a years old as

$$K = 21.235w + 7.75h - 10.54a + 102.3.$$

If Janos is moderately active, weighs 80 kg, is 190 cm tall, and needs to consume 2852 calories a day, how old is he?

64. *Altitude and temperature.* Air temperature drops about 1° Celsius (C) for each 100-m rise above ground level, up to 12 km. If the ground level temperature is t°C, find a formula for the temperature T at an elevation of h meters.
Source: *A Sourcebook of School Mathematics*, Mathematical Association of America, 1980

65. *Surface area of a cube.* The surface area A of a cube with side s is given by

$$A = 6s^2.$$

If a cube's surface area is 54 in^2, find the volume of the cube.

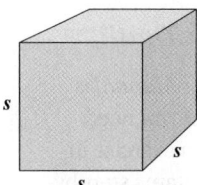

66. *Weight of a fish.* An ancient fisherman's formula for estimating the weight of a fish is

$$w = \frac{lg^2}{800},$$

where w is the weight, in pounds, l is the length, in inches, and g is the girth (distance around the midsection), in inches. Estimate the girth of a 700-lb yellow tuna that is 8 ft long.

67. *Dosage size.* Clark's rule for determining the size of a particular child's medicine dosage c is

$$c = \frac{w}{a} \cdot d,$$

where w is the child's weight, in pounds, and d is the usual adult dosage for an adult weighing a pounds. Solve for a.

Source: Olsen, June Looby, et al., *Medical Dosage Calculations.* Redwood City, CA: Addison-Wesley, 1995

Solve each formula for the given letter.

68. $\frac{y}{z} \div \frac{z}{t} = 1$, for y

69. $ac = bc + d$, for c

70. $qt = r(s + t)$, for t

71. $3a = c - a(b + d)$, for a

72. *Furnace output.* The formula

$$B = 50a$$

is used in New England to estimate the minimum furnace output B, in Btu's, for an old, poorly insulated house with a square feet of flooring. Find an equation for determining the number of Btu's saved by insulating an old house. (*Hint*: See Exercise 2.)

73. Revise the formula in Exercise 63 so that a man's weight in pounds (2.2046 lb = 1 kg) and his height in inches (0.3937 in. = 1 cm) are used.

74. Revise the formula in Example 5 so that a woman's weight in kilograms (2.2046 lb = 1 kg) and her height in centimeters (0.3937 in. = 1 cm) are used.

2.4 Applications with Percent

Converting Between Percent Notation and Decimal Notation ▪ Solving Percent Problems

Percent problems arise so frequently in everyday life that most often we are not even aware of them. In this section, we will solve some real-world percent problems. Before doing so, however, we need to review a few basics.

Converting Between Percent Notation and Decimal Notation

STUDY SKILLS

How Did They Get That!?

The *Student's Solutions Manual* is an excellent resource if you need additional help with an exercise in the exercise sets. It contains step-by-step solutions to the odd-numbered exercises in each exercise set.

Nutritionists recommend that no more than 30% of the calories in a person's diet come from fat. This means that of every 100 calories consumed, no more than 30 should come from fat. Thus, 30% is a ratio of 30 to 100.

Calories consumed

Calories from fat 30%

The percent symbol % means "per hundred." We can regard the percent symbol as part of a name for a number. For example,

30% is defined to mean $\frac{30}{100}$, or $30 \times \frac{1}{100}$, or 30×0.01.

> **Percent Notation**
>
> $n\%$ means $\frac{n}{100}$, or $n \times \frac{1}{100}$, or $n \times 0.01$.

EXAMPLE 1   Convert to decimal notation: **(a)** 78%; **(b)** 1.3%.

SOLUTION

a) 78% = 78 × 0.01   Replacing % with × 0.01
        = 0.78

b) 1.3% = 1.3 × 0.01   Replacing % with × 0.01
         = 0.013

TRY EXERCISE 19

As shown above, multiplication by 0.01 simply moves the decimal point two places to the left.

To convert from percent notation to decimal notation, move the decimal point two places to the left and drop the percent symbol.

EXAMPLE 2   Convert the percent notation in the following sentence to decimal notation: Only 20% of teenagers get 8 hr of sleep a night.

Source: National Sleep Foundation

SOLUTION

20% = 20.0%   0.20.0   20% = 0.20, or simply 0.2

Move the decimal point two places to the left.

TRY EXERCISE 11

The procedure used in Examples 1 and 2 can be reversed:

0.38 = 38 × 0.01
     = 38%.   Replacing × 0.01 with %

To convert from decimal notation to percent notation, move the decimal point two places to the right and write a percent symbol.

EXAMPLE 3   Convert to percent notation: **(a)** 1.27; **(b)** $\frac{1}{4}$; **(c)** 0.3.

SOLUTION

a) We first move the decimal point two places to the right: 1.27.
and then write a % symbol: 127%   This is the same as multiplying 1.27 by 100 and writing %.

b) Note that $\frac{1}{4}$ = 0.25. We move the decimal point two places to the right: 0.25.
and then write a % symbol: 25%   Multiplying by 100 and writing %

c) We first move the decimal point two places to the right (recall that 0.3 = 0.30):   0.30.
and then write a % symbol: 30%   Multiplying by 100 and writing %

TRY EXERCISE 33

Solving Percent Problems

In solving percent problems, we first *translate* the problem to an equation. Then we *solve* the equation using the techniques discussed in Sections 2.1–2.3. The key words in the translation are as follows.

> ## Key Words in Percent Translations
> "**Of**" translates to " · " or " × ".   "**Is**" or "**Was**" translates to " = ".
> "**What**" translates to a variable.   "**%**" translates to "$\times \frac{1}{100}$" or "× 0.01".

EXAMPLE **4**

STUDENT NOTES ————

A way of checking answers is by estimating as follows:

$$11\% \times 49 \approx 10\% \times 50$$
$$= 0.10 \times 50 = 5.$$

Since 5 is close to 5.39, our answer is reasonable.

What is 11% of 49?

SOLUTION

Translate: What is 11% of 49?
              ↓   ↓   ↓   ↓   ↓
              a = 0.11 ·   49   "of" means multiply;
                                        11% = 0.11

$$a = 5.39$$

Thus, 5.39 is 11% of 49. The answer is 5.39.

TRY EXERCISE ▶ 51

EXAMPLE **5**

3 is 16 percent of what?

SOLUTION

Translate: 3   is   16 percent   of   what?
              ↓   ↓   ↓   ↓   ↓
              3   =   0.16   · y

$$\frac{3}{0.16} = y \qquad \text{Dividing both sides by 0.16}$$

$$18.75 = y$$

Thus, 3 is 16 percent of 18.75. The answer is 18.75.

TRY EXERCISE ▶ 47

EXAMPLE **6**

What percent of $50 is $34?

SOLUTION

Translate: What percent   of $50   is $34?
                   ↓   ↓   ↓   ↓   ↓
                   n ·   50   =   34

$$n = \frac{34}{50} \qquad \text{Dividing both sides by 50}$$

$$n = 0.68 = 68\% \qquad \text{Converting to percent notation}$$

Thus, $34 is 68% of $50. The answer is 68%.

TRY EXERCISE ▶ 43

Examples 4–6 represent the three basic types of percent problems. Note that in all the problems, the following quantities are present:

- a percent, expressed in decimal notation in the translation,
- a base amount, indicated by "of" in the problem, and
- a percentage of the base, found by multiplying the base times the percent.

EXAMPLE 7

Discount stores.   In 2006, there were 300 million people in the United States, and 62.2% of them lived within 5 mi of a Wal-Mart store. How many lived within 5 mi of a Wal-Mart store?

Source: *The Wall Street Journal, 9/25/06*

SOLUTION   We first reword and then translate. We let a = the number of people in the United States, in millions, who live within 5 mi of a Wal-Mart store.

Rewording: What   is   62.2%   of   300?

Translating: a =   0.622   ×   300

The letter is by itself. To solve the equation, we need only multiply:

$$a = 0.622 \times 300 = 186.6.$$

Since 186.6 million is 62.2% of 300 million, we have found that in 2006 about 186.6 million people in the United States lived within 5 mi of a Wal-Mart store.

TRY EXERCISE 65

EXAMPLE 8

College enrollment.   About 1.6 million students who graduated from high school in 2006 were attending college in the fall of 2006. This was 66% of all 2006 high school graduates. How many students graduated from high school in 2006?

Source: U.S. Bureau of Labor Statistics

SOLUTION   Before translating the problem to mathematics, we reword and let S represent the total number of students, in millions, who graduated from high school in 2006.

Rewording: 1.6   is   66%   of S.

Translating: 1.6   =   0.66   · S

$$\frac{1.6}{0.66} = S \qquad \text{Dividing both sides by 0.66}$$

$$2.4 \approx S \qquad \begin{array}{l}\text{The symbol } \approx \text{ means}\\ \textit{is approximately equal to.}\end{array}$$

About 2.4 million students graduated from high school in 2006.

TRY EXERCISE 67

EXAMPLE 9

Automobile prices.   Recently, Harken Motors reduced the price of a Flex Fuel 2007 Chevy Impala from the manufacturer's suggested retail price (MSRP) of $20,830 to $18,955.

a) What percent of the MSRP does the sale price represent?
b) What is the percent of discount?

SOLUTION

a) We reword and translate, using n for the unknown percent.

Rewording: What percent   of   20,830   is   18,955?

Translating: n ·   20,830   =   18,955

$$n = \frac{18,955}{20,830} \qquad \begin{array}{l}\text{Dividing both}\\ \text{sides by 20,830}\end{array}$$

$$n \approx 0.91 = 91\% \qquad \begin{array}{l}\text{Converting to}\\ \text{percent notation}\end{array}$$

The sale price is about 91% of the MSRP.

b) Since the original price of $20,830 represents 100% of the MSRP, the sale price represents a discount of $(100 - 91)\%$, or 9%.

Alternatively, we could find the amount of discount and then calculate the percent of discount:

Amount of discount: $20,830 - $18,955 = $1875.

Rewording: <u>What percent</u> of 20,830 is 1875?

Translating: n $\cdot$ 20,830 = 1875

$$n = \frac{1875}{20,830}$$ Dividing both sides by 20,830

$$n \approx 0.09 = 9\%$$ Converting to percent notation

Again we find that the percent of discount is 9%.   **TRY EXERCISE** 69

2.4   EXERCISE SET

 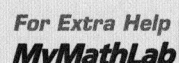

🔖 **Concept Reinforcement**   *In each of Exercises 1–10, match the question with the most appropriate translation from the column on the right. Some choices are used more than once.*

1. ___ What percent of 57 is 23?

2. ___ What percent of 23 is 57?

3. ___ 23 is 57% of what number?

4. ___ 57 is 23% of what number?

5. ___ 57 is what percent of 23?

6. ___ 23 is what percent of 57?

7. ___ What is 23% of 57?

8. ___ What is 57% of 23?

9. ___ 23% of what number is 57?

10. ___ 57% of what number is 23?

a) $a = (0.57)23$

b) $57 = 0.23y$

c) $n \cdot 23 = 57$

d) $n \cdot 57 = 23$

e) $23 = 0.57y$

f) $a = (0.23)57$

Convert the percent notation in each sentence to decimal notation.

11. *Energy use.* Heating accounts for 49% of all household energy use.
Source: Chevron

12. *Energy use.* Water heating accounts for 15% of all household energy use.
Source: Chevron

13. *Drinking water.* Only 1% of the water on earth is suitable for drinking.
Source: www.drinktap.org

Drinking water, 1%

14. *Dehydration.* A 2% drop in water content of the body can affect one's ability to study mathematics.
Source: High Performance Nutrition

15. *College tuition.* Tuition and fees at two-year public colleges increased 4.1 percent in 2006.
Source: College Board 2006 tuition survey

16. *Plant species.* Trees make up about 3.5% of all plant species found in the United States.
Source: South Dakota Project Learning Tree

17. *Women in the workforce.* Women comprise 20% of all database administrators.
Source: U.S. Census Bureau

18. *Women in the workforce.* Women comprise 60% of all accountants and auditors.
Source: U.S. Census Bureau

Convert to decimal notation.

19. 6.25%   **20.** 8.375%

21. 0.2%   **22.** 0.8%

23. 175%   **24.** 250%

Convert the decimal notation in each sentence to percent notation.

25. *NASCAR fans.* Auto racing is the seventh most popular sport in the United States, with 0.38 of the adult population saying they are NASCAR fans.
Source: ESPN Sports poll

26. *Baseball fans.* Baseball is the second most popular sport in the United States, with 0.61 of the adult population saying they are baseball fans.
Source: ESPN Sports poll

27. *Food security.* The USDA defines food security as access to enough nutritious food for a healthy life. In 2005, 0.039 of U.S. households had very low food security.
Source: USDA

28. *Poverty rate.* In 2005, 0.199 of Americans age 65 and older were under the poverty level.
Source: www.census.gov

29. *Music downloads.* In 2006, 0.45 of Americans downloaded music.
Source: Solutions Research Group

30. *Music downloads.* In 2006, 0.23 of Americans paid to download a song.
Source: Solutions Research Group

31. *Composition of the sun.* The sun is 0.7 hydrogen.

32. *Jupiter's atmosphere.* The atmosphere of Jupiter is 0.1 helium.

Convert to percent notation.

33. 0.0009   **34.** 0.0056

35. 1.06   **36.** 1.08

37. 1.8   **38.** 2.4

39. $\dfrac{3}{5}$ **40.** $\dfrac{3}{4}$

41. $\dfrac{8}{25}$ **42.** $\dfrac{5}{8}$

Solve.

43. What percent of 76 is 19?

44. What percent of 125 is 30?

45. What percent of 150 is 39?

46. What percent of 360 is 270?

47. 14 is 30% of what number?

48. 54 is 24% of what number?

49. 0.3 is 12% of what number?

50. 7 is 175% of what number?

51. What number is 1% of one million?

52. What number is 35% of 240?

53. What percent of 60 is 75?

Aha! **54.** What percent of 70 is 70?

55. What is 2% of 40?

56. What is 40% of 2?

Aha! **57.** 25 is what percent of 50?

58. 0.8 is 2% of what number?

59. What percent of 69 is 23?

60. What percent of 40 is 9?

Riding bicycles.   *There are 57 million Americans who ride a bicycle at least occasionally. The following circle graph shows the reasons people ride. In each of Exercises 61–64, determine the number of Americans who ride a bicycle for the given reason.*

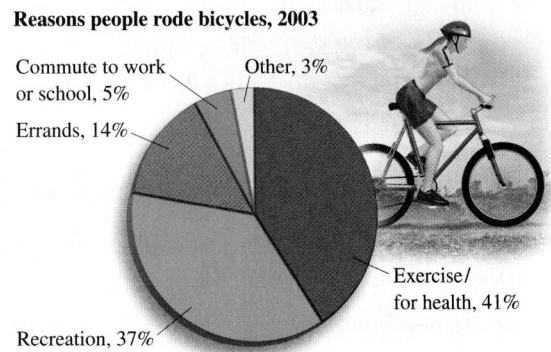

Reasons people rode bicycles, 2003

Commute to work or school, 5%

Errands, 14%

Other, 3%

Exercise/ for health, 41%

Recreation, 37%

Sources: U.S. Census Bureau; Bureau of Transportation Statistics

61. Commute to school or work

62. Run errands

63. Exercise for health

64. Recreation

65. *College graduation.*   To obtain his bachelor's degree in nursing, Cody must complete 125 credit hours of instruction. If he has completed 60% of his requirement, how many credits did Cody complete?

66. *College graduation.*   To obtain her bachelor's degree in journalism, Addy must complete 125 credit hours of instruction. If 20% of Addy's credit hours remain to be completed, how many credits does she still need to take?

67. *Batting average.*   In the 2007 season, Magglio Ordonez of the Detroit Tigers had 216 hits. His batting average was 0.363, the highest in major league baseball for that season. This means that of the total number of at bats, 36.3% were hits. How many at bats did he have?
Source: ESPN

68. *Pass completions.*   At one point in a recent season, Peyton Manning of the Indianapolis Colts had completed 357 passes. This was 62.5% of his attempts. How many attempts did he make?
Source: National Football League

69. *Tipping.*   Trent left a $4 tip for a meal that cost $25.
  a) What percent of the cost of the meal was the tip?
  b) What was the total cost of the meal including the tip?

70. *Tipping.*   Selena left a $12.76 tip for a meal that cost $58.
  a) What percent of the cost of the meal was the tip?
  b) What was the total cost of the meal including the tip?

71. *Crude oil imports.*   In April 2007, crude oil imports to the United States averaged 10.2 million barrels per day. Of this total, 3.4 million came from Canada and Mexico. What percent of crude oil imports came from Canada and Mexico? What percent came from the rest of the world?
Source: Energy Information Administration

72. *Alternative-fuel vehicles.*   Of the 550,000 alternative-fuel vehicles produced in the United States in 2004, 150,000 were E85 flexible-fuel vehicles. What percent of alternative-fuel vehicles used E85? What percent used other alternative fuels?
Source: Energy Information Administration

73. *Student loans.*   Glenn takes out a subsidized federal Stafford loan for $2400. After a year, Glenn decides to pay off the interest, which is 7% of $2400. How much will he pay?

74. *Student loans.*   To finance her community college education, LaTonya takes out a Stafford loan for $3500. After a year, LaTonya decides to pay off the interest, which is 8% of $3500. How much will she pay?

75. *Infant health.*   In a study of 300 pregnant women with "good-to-excellent" diets, 95% had babies in good or excellent health. How many women in this group had babies in good or excellent health?

76. *Infant health.*   In a study of 300 pregnant women with "poor" diets, 8% had babies in good or excellent health. How many women in this group had babies in good or excellent health?

77. *Cost of self-employment.*   Because of additional taxes and fewer benefits, it has been estimated that a self-employed person must earn 20% more than a non–self-employed person performing the same task(s). If Tia earns $16 an hour working for Village

Copy, how much would she need to earn on her own for a comparable income?

78. Refer to Exercise 77. Rik earns $18 an hour working for Round Edge stairbuilders. How much would Rik need to earn on his own for a comparable income?

79. *Budget overruns.* The Indianapolis Central Library expansion, begun in 2002, was expected to cost $103 million. By 2006, library officials estimated the cost would be $45 million over budget. By what percent did the actual cost exceed the initial estimate?
Source: *The Indianapolis Star*, 5/23/06

80. *Fastest swimmer.* In 1990, Tom Jager of the United States set a world record by swimming 50 m at a rate of 2.29 m/s. Previously, the fastest swimming rate on record was 2.26 m/s, set in 1975 by David Holmes Edgar, also of the United States. Calculate the percentage by which the rate increased.
Source: *Guinness Book of World Records* 1975 and 1998

81. A bill at Officeland totaled $47.70. How much did the merchandise cost if the sales tax is 6%?

82. Marta's checkbook shows that she wrote a check for $987 for building materials. What was the price of the materials if the sales tax is 5%?

83. *Deducting sales tax.* A tax-exempt school group received a bill of $157.41 for educational software. The bill incorrectly included sales tax of 6%. How much should the school group pay?

84. *Deducting sales tax.* A tax-exempt charity received a bill of $145.90 for a sump pump. The bill incorrectly included sales tax of 5%. How much does the charity owe?

85. *Body fat.* One author of this text exercises regularly at a local YMCA that recently offered a body-fat percentage test to its members. The device used measures the passage of a very low voltage of electricity through the body. The author's body-fat percentage was found to be 16.5% and he weighs 191 lb. What part, in pounds, of his body weight is fat?

86. *Areas of Alaska and Arizona.* The area of Arizona is 19% of the area of Alaska. The area of Alaska is 586,400 mi^2. What is the area of Arizona?

87. *Direct mail.* Only 2.15% of mailed ads lead to a sale or a response from customers. In 2006, businesses sent out 114 billion pieces of direct mail (catalogs, coupons, and so on). How many pieces of mail led to a response from customers?
Sources: Direct Marketing Association; U.S. Postal Service

88. *Kissing and colds.* In a medical study, it was determined that if 800 people kiss someone else who has a cold, only 56 will actually catch the cold. What percent is this?

89. *Calorie content.* Pepperidge Farm Light Style 7 Grain Bread® has 140 calories in a 3-slice serving. This is 15% less than the number of calories in a serving of regular bread. How many calories are in a serving of regular bread?

90. *Fat content.* Peek Freans Shortbread Reduced Fat Cookies® contain 35 calories of fat in each serving. This is 40% less than the fat content in the leading imported shortbread cookie. How many calories of fat are in a serving of the leading shortbread cookie?

91. Campus Bookbuyers pays $30 for a book and sells it for $60. Is this a 100% markup or a 50% markup? Explain.

92. If Julian leaves a $12 tip for a $90 dinner, is he being generous, stingy, or neither? Explain.

Skill Review

To prepare for Section 2.5, review translating to algebraic expressions and equations (Section 1.1).

Translate to an algebraic expression or equation. [1.1]

93. Twice the length plus twice the width

94. 5% of $180

95. 5 fewer than the number of points Tino scored

96. 15 plus the product of 1.5 and *x*

97. The product of 10 and half of *a*

98. 10 more than three times a number

99. The width is 2 in. less than the length.

100. A number is four times as large as a second number.

Synthesis

101. How is the use of statistics in the following misleading?

   a) A business explaining new restrictions on sick leave cited a recent survey indicating that 40% of all sick days were taken on Monday or Friday.

   b) An advertisement urging summer installation of a security system quoted FBI statistics stating that over 26% of home burglaries occur between Memorial Day and Labor Day.

102. Erin is returning a tent that she bought during a 25%-off storewide sale that has ended. She is offered store credit for 125% of what she paid (not to be used on sale items). Is this fair to Erin? Why or why not?

103. The community of Bardville has 1332 left-handed females. If 48% of the community is female and 15% of all females are left-handed, how many people are in the community?

104. It has been determined that at the age of 10, a girl has reached 84.4% of her final adult height. Dana is 4 ft 8 in. at the age of 10. What will her final adult height be?

105. It has been determined that at the age of 15, a boy has reached 96.1% of his final adult height. Jaraan is 6 ft 4 in. at the age of 15. What will his final adult height be?

106. *Dropout rate.* Between 2002 and 2004, the high school dropout rate in the United States decreased from 105 to 103 per thousand. Calculate the percent by which the dropout rate decreased and use that percentage to estimate dropout rates for the United States in 2005 and in 2006.
Source: www.childrendsdatabank.org

107. *Photography.* A 6-in. by 8-in. photo is framed using a mat meant for a 5-in. by 7-in. photo. What percentage of the photo will be hidden by the mat?

108. Would it be better to receive a 5% raise and then, a year later, an 8% raise or the other way around? Why?

109. Jorge is in the 30% tax bracket. This means that 30¢ of each dollar earned goes to taxes. Which would cost him the least: contributing $50 that is tax-deductible or contributing $40 that is not tax-deductible? Explain.

COLLABORATIVE

CORNER

Sales and Discounts

Focus: Applications and models using percent

Time: 15 minutes

Group size: 3

Materials: Calculators are optional.

Often a store will reduce the price of an item by a fixed percentage. When the sale ends, the items are returned to their original prices. Suppose a department store reduces all sporting goods 20%, all clothing 25%, and all electronics 10%.

ACTIVITY

1. Each group member should select one of the following items: a $50 basketball, an $80 jacket, or a $200 MP3 player. Fill in the first three columns of the first three rows of the chart below.

2. Apply the appropriate discount and determine the sale price of your item. Fill in the fourth column of the chart.

3. Next, find a multiplier that can be used to convert the sale price back to the original price and fill in the remaining column of the chart. Does this multiplier depend on the price of the item?

4. Working as a group, compare the results of part (3) for all three items. Then develop a formula for a multiplier that will restore a sale price to its original price, p, after a discount r has been applied. Complete the fourth row of the table and check that your formula will duplicate the results of part (3).

5. Use the formula from part (4) to find the multiplier that a store would use to return an item to its original price after a "30% off" sale expires. Fill in the last line on the chart.

6. Inspect the last column of your chart. How can these multipliers be used to determine the percentage by which a sale price is increased when a sale ends?

Original Price, p	Discount, r	$1 - r$	Sale Price	Multiplier to convert back to p
p	r	$1 - r$		
	0.30			

2.5 Problem Solving

Five Steps for Problem Solving ▪ Applying the Five Steps

Probably the most important use of algebra is as a tool for problem solving. In this section, we develop a problem-solving approach that is used throughout the remainder of the text.

Five Steps for Problem Solving

In Section 2.4, we solved several real-world problems. To solve them, we first *familiarized* ourselves with percent notation. We then *translated* each problem into an equation, *solved* the equation, *checked* the solution, and *stated* the answer.

Five Steps for Problem Solving in Algebra

1. *Familiarize* yourself with the problem.
2. *Translate* to mathematical language. (This often means writing an equation.)
3. *Carry out* some mathematical manipulation. (This often means *solving* an equation.)
4. *Check* your possible answer in the original problem.
5. *State* the answer clearly, using a complete English sentence.

Of the five steps, the most important is probably the first one: becoming familiar with the problem. Here are some hints for familiarization.

To Become Familiar with a Problem

1. Read the problem carefully. Try to visualize the problem.
2. Reread the problem, perhaps aloud. Make sure you understand all important words and any symbols or abbreviations.
3. List the information given and the question(s) to be answered. Choose a variable (or variables) to represent the unknown and specify exactly what the variable represents. For example, let L = length in centimeters, d = distance in miles, and so on.
4. Look for similarities between the problem and other problems you have already solved. Ask yourself what type of problem this is.
5. Find more information. Look up a formula in a book, at a library, or online. Consult a reference librarian or an expert in the field.
6. Make a table that uses all the information you have available. Look for patterns that may help in the translation.
7. Make a drawing and label it with known and unknown information, using specific units if given.
8. Think of a possible answer and check the guess. Note the manner in which the guess is checked.

Applying the Five Steps

EXAMPLE 1

Bicycling.   After finishing college, Nico spent a week touring Tuscany, Italy, by bicycle. He biked 260 km from Pisa through Siena to Florence. At Siena, he had biked three times as far from Pisa as he would then bike to Florence. How far had he biked, and how far did he have left to go?

SOLUTION

1. **Familiarize.** It is often helpful to make a drawing. In this case, we can use a map of Nico's trip.

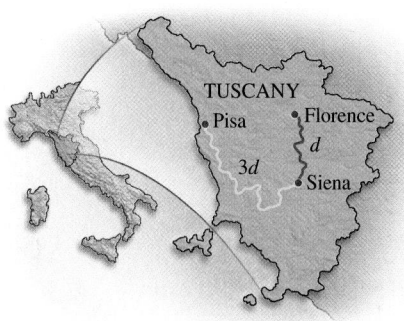

   To gain familiarity, let's suppose that Nico has 50 km to go. Then he would have traveled three times 50 km, or 150 km, already. Since 50 km + 150 km = 200 km and 200 km < 260 km, we see that our guess is too small. Rather than guess again, we let

   d = the distance, in kilometers, from Siena to Florence

   and

   $3d$ = the distance, in kilometers, from Siena to Pisa.

   (We could also let x = the distance to Pisa; then the distance to Florence would be $\frac{1}{3}x$.)

2. **Translate.** The lengths of the two parts of the trip must add up to 260 km. This leads to our translation.

   Rewording: Distance to Florence   plus   distance to Pisa   is   260 km

   Translating: d $+$ $3d$ $=$ 260

3. **Carry out.** We solve the equation:

   $d + 3d = 260$

   $4d = 260$ Combining like terms

   $d = 65.$ Dividing both sides by 4

4. **Check.** As predicted in the *Familiarize* step, d is greater than 50 km. If $d =$ 65 km, then $3d = 195$ km. Since 65 km + 195 km = 260 km, we have a check.

5. **State.** At Siena, Nico had biked 195 km and had 65 km left to go to arrive in Florence.

TRY EXERCISE 9

   Before we solve the next problem, we need to learn some additional terminology regarding integers.

   The following are examples of **consecutive integers:** 16, 17, 18, 19, 20; and −31, −30, −29, −28. Note that consecutive integers can be represented in the form $x, x + 1, x + 2$, and so on.

The following are examples of **consecutive even integers:** 16, 18, 20, 22, 24; and −52, −50, −48, −46. Note that consecutive even integers can be represented in the form $x, x + 2, x + 4$, and so on.

The following are examples of **consecutive odd integers:** 21, 23, 25, 27, 29; and −71, −69, −67, −65. Note that consecutive odd integers can be also represented in the form $x, x + 2, x + 4$, and so on.

EXAMPLE **2**

Interstate mile markers. U.S. interstate highways post numbered markers at every mile to indicate location in case of an emergency. The sum of two consecutive mile markers on I-70 in Kansas is 559. Find the numbers on the markers.

Source: Federal Highway Administration, Ed Rotalewski

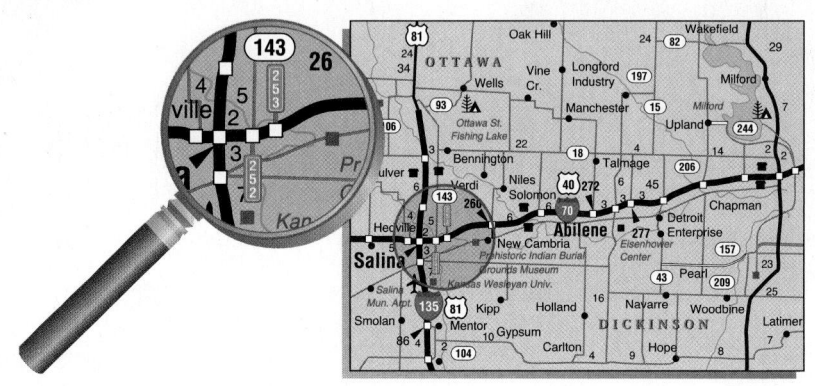

x	$x + 1$	Sum of x and $x + 1$
114	115	229
252	253	505
302	303	605

SOLUTION

1. **Familiarize.** The numbers on the mile markers are consecutive positive integers. Thus if we let $x =$ the smaller number, then $x + 1 =$ the larger number.

   To become familiar with the problem, we can make a table, as shown at left. First, we guess a value for x; then we find $x + 1$. Finally, we add the two numbers and check the sum.

   From the table, we see that the first marker will be between 252 and 302. We could continue guessing and solve the problem this way, but let's work on developing our algebra skills.

2. **Translate.** We reword the problem and translate as follows.

   Rewording: First integer plus second integer is 559.

   Translating: $x \quad + \quad (x + 1) \quad = \quad 559$

3. **Carry out.** We solve the equation:

   $$x + (x + 1) = 559$$
   $$2x + 1 = 559 \qquad \text{Using an associative law and combining like terms}$$
   $$2x = 558 \qquad \text{Subtracting 1 from both sides}$$
   $$x = 279. \qquad \text{Dividing both sides by 2}$$

   If x is 279, then $x + 1$ is 280.

4. **Check.** Our possible answers are 279 and 280. These are consecutive positive integers and $279 + 280 = 559$, so the answers check.

5. **State.** The mile markers are 279 and 280.

 **TRY EXERCISE** 13

EXAMPLE 3

Color printers.   Egads Computer Corporation rents a Xerox Phaser 8400 Color Laser Printer for $300 a month. A new art gallery is leasing a printer for a 2-month advertising campaign. The ink and paper for the brochures will cost an additional 21.5¢ per copy. If the gallery allots a budget of $3000, how many brochures can they print?

Source: egadscomputer.com

SOLUTION

1. **Familiarize.** Suppose that the art gallery prints 20,000 brochures. Then the cost is the monthly charges plus ink and paper cost, or

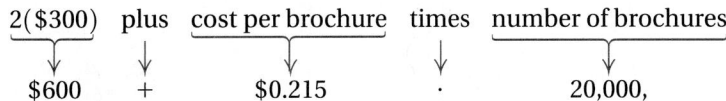

   which is $4900. Our guess of 20,000 is too large, but we have familiarized ourselves with the way in which a calculation is made. Note that we convert 21.5¢ to $0.215 so that all information is in the same unit, dollars. We let c = the number of brochures that can be printed for $3000.

2. **Translate.** We reword the problem and translate as follows.

   Rewording: Monthly cost   plus   ink and paper cost   is $3000.

   Translating: 2($300)   +   ($0.215)c = $3000

3. **Carry out.** We solve the equation:

$$2(300) + 0.215c = 3000$$
$$600 + 0.215c = 3000$$
$$0.215c = 2400 \qquad \text{Subtracting 600 from both sides}$$
$$c = \frac{2400}{0.215} \qquad \text{Dividing both sides by 0.215}$$
$$c \approx 11{,}162. \qquad \text{We round } \textit{down} \text{ to avoid going over the budget.}$$

4. **Check.** We check in the original problem. The cost for 11,162 brochures is 11,162($0.215) = $2399.83. The rental for 2 months is 2($300) = $600. The total cost is then $2399.83 + $600 = $2999.83, which is just under the amount that was allotted. Our answer is less than 20,000, as we expected from the *Familiarize* step.

5. **State.** The art gallery can make 11,162 brochures with the rental allotment of $3000.

> TRY EXERCISE ▶ 37

STUDENT NOTES

For most students, the most challenging step is step (2), "Translate." The table on p. 4 (Section 1.1) can be helpful in this regard.

EXAMPLE 4

Perimeter of NBA court.   The perimeter of an NBA basketball court is 288 ft. The length is 44 ft longer than the width. Find the dimensions of the court.

Source: National Basketball Association

SOLUTION

1. **Familiarize.** Recall that the perimeter of a rectangle is twice the length plus twice the width. Suppose the court were 30 ft wide. The length would then be 30 + 44, or 74 ft, and the perimeter would be 2 · 30 ft + 2 · 74 ft, or 208 ft. This shows that in order for the perimeter to be 288 ft, the width must exceed 30 ft. Instead of guessing again, we let w = the width of the court, in feet.

Since the court is "44 ft longer than it is wide," we let $w + 44 =$ the length of the court, in feet.

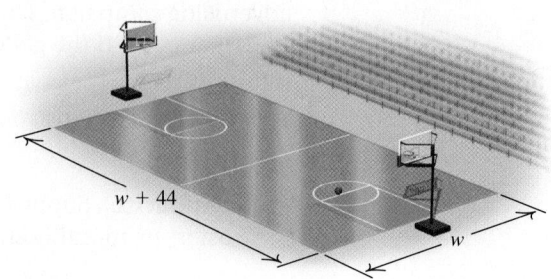

2. **Translate.** To translate, we use $w + 44$ as the length and 288 as the perimeter. To double the length, $w + 44$, parentheses are essential.

Rewording: Twice the length   plus   twice the width   is   288 ft.

Translating: $2(w + 44)$ $+$ $2w$ $=$ 288

3. **Carry out.** We solve the equation:

$$2(w + 44) + 2w = 288$$
$$2w + 88 + 2w = 288 \qquad \text{Using the distributive law}$$
$$4w + 88 = 288 \qquad \text{Combining like terms}$$
$$4w = 200$$
$$w = 50.$$

The dimensions appear to be $w = 50$ ft, and $l = w + 44 = 94$ ft.

4. **Check.** If the width is 50 ft and the length is 94 ft, then the court is 44 ft longer than it is wide. The perimeter is $2(50 \text{ ft}) + 2(94 \text{ ft}) = 100 \text{ ft} + 188 \text{ ft}$, or 288 ft, as specified. We have a check.

5. **State.** An NBA court is 50 ft wide and 94 ft long.   TRY EXERCISE ▶ 25

STUDENT NOTES

Get in the habit of writing what each variable represents before writing an equation. In Example 4, you might write

width $= w$,

length $= w + 44$

before translating the problem to an equation. This step becomes more important as problems become more complex.

> *CAUTION!* Always be sure to answer the original problem completely. For instance, in Example 1 we needed to find *two* numbers: the distances from *each* city to Siena. Similarly, in Example 4 we needed to find two dimensions, not just the width. Be sure to label each answer with the proper unit.

EXAMPLE 5

Selling at an auction. Jared is selling his collection of Transformers at an auction. He wants to be left with $1150 after paying a seller's premium of 8% on the final bid (hammer price) for the collection. What must the hammer price be in order for him to clear $1150?

SOLUTION

1. **Familiarize.** Suppose the collection sells for $1200. The 8% seller's premium can be determined by finding 8% of $1200:

$$8\% \text{ of } \$1200 = 0.08(\$1200) = \$96.$$

Subtracting this premium from $1200 would leave Jared with

$$\$1200 - \$96 = \$1104.$$

This shows that in order for Jared to clear $1150, the collection must sell for more than $1200. We let $x =$ the hammer price, in dollars. Jared then must pay a seller's premium of $0.08x$.

2. **Translate.** We reword the problem and translate as follows.

Rewording: $\underbrace{\text{Hammer price}}$ $\underbrace{\text{less}}$ $\underbrace{\text{seller's premium}}$ $\underbrace{\text{is}}$ $\underbrace{\text{amount remaining.}}$

Translating: x $-$ $0.08x$ $=$ $\$1150$

3. **Carry out.** We solve the equation:

$$x - 0.08x = 1150$$
$$1x - 0.08x = 1150$$
$$0.92x = 1150 \qquad \text{Combining like terms. Had we noted that after the premium has been paid, 92\% remains, we could have begun with this equation.}$$

$$x = \frac{1150}{0.92} \qquad \text{Dividing both sides by 0.92}$$

$$x = 1250.$$

4. **Check.** To check, we first find 8% of \$1250:

$$8\% \text{ of } \$1250 = 0.08(\$1250) = \$100. \qquad \text{This is the premium.}$$

Next, we subtract the premium to find the remaining amount:

$$\$1250 - \$100 = \$1150.$$

Since, after Jared pays the seller's premium, he is left with \$1150, our answer checks. Note that the \$1250 hammer price is greater than \$1200, as predicted in the *Familiarize* step.

5. **State.** Jared's collection must sell for \$1250 in order for him to be left with \$1150.

TRY EXERCISE ▸ 7

EXAMPLE 6

Cross section of a roof. In a triangular gable end of a roof, the angle of the peak is twice as large as the angle on the back side of the house. The measure of the angle on the front side is 20° greater than the angle on the back side. How large are the angles?

SOLUTION

1. **Familiarize.** We make a drawing. In this case, the measure of the back angle is x, the measure of the front angle is $x + 20$, and the measure of the peak angle is $2x$.

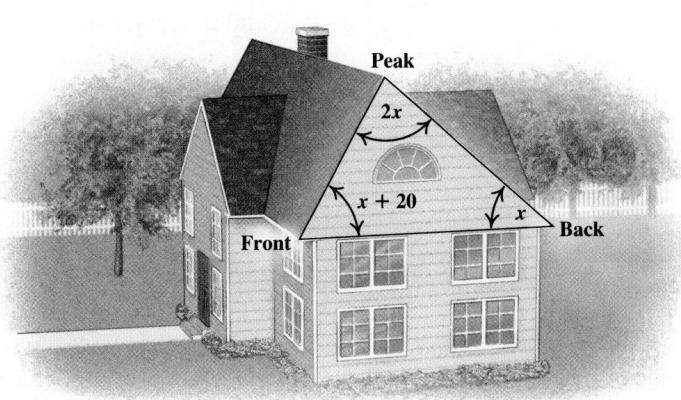

2. Translate. To translate, we need to recall that the sum of the measures of the angles in a triangle is 180°.

Rewording: Measure of   measure of   measure of
                 back angle   + front angle + peak angle   is   180°.

Translating: x + $(x + 20)$ + $2x$ =   180

3. Carry out. We solve:

$$x + (x + 20) + 2x = 180$$
$$4x + 20 = 180$$
$$4x = 160$$
$$x = 40.$$

The measures for the angles appear to be:

Back angle: $x = 40°$,

Front angle: $x + 20 = 40 + 20 = 60°$,

Peak angle: $2x = 2(40) = 80°$.

4. Check. Consider 40°, 60°, and 80°, as listed above. The measure of the front angle is 20° greater than the measure of the back angle, the measure of the peak angle is twice the measure of the back angle, and the sum is 180°. These numbers check.

5. State. The measures of the angles are 40°, 60°, and 80°.   **TRY EXERCISE** 31

We close this section with some tips to aid you in problem solving.

Problem-Solving Tips

1. The more problems you solve, the more your skills will improve.
2. Look for patterns when solving problems. Each time you study an example or solve an exercise, you may observe a pattern for problems found later.
3. Clearly define variables before translating to an equation.
4. Consider the dimensions of the variables and constants in the equation. The variables that represent length should all be in the same unit, those that represent money should all be in dollars or all in cents, and so on.
5. Make sure that units appear in the answer whenever appropriate and that you completely answer the original problem.

2.5   EXERCISE SET

 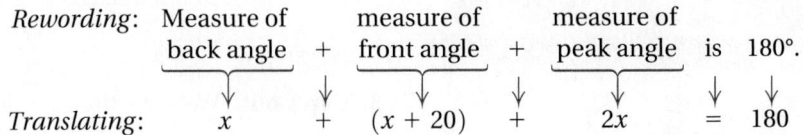

Solve. Even though you might find the answer quickly in some other way, practice using the five-step problem-solving process in order to build the skill of problem solving.

1. Three less than twice a number is 19. What is the number?

2. Two fewer than ten times a number is 78. What is the number?

3. Five times the sum of 3 and twice some number is 70. What is the number?

4. Twice the sum of 4 and three times some number is 34. What is the number?

5. *Price of an iPod.*   Kyle paid $120 for an iPod nano during a 20%-off sale. What was the regular price?

6. *Price of sneakers.*   Amy paid $102 for a pair of New Balance 1122 running shoes during a 15%-off sale. What was the regular price?

7. *Price of a calculator.*   Kayla paid $137.80, including 6% tax, for her graphing calculator. How much did the calculator itself cost?

8. *Price of a printer.*   Laura paid $219.45, including 5% tax, for an all-in-one color printer. How much did the printer itself cost?

9. *Unicycling.*   In 2005, Ken Looi of New Zealand set a record by covering 235.3 mi in 24 hr on his unicycle. After 8 hr, he was approximately twice as far from the finish line as he was from the start. How far had he traveled?
Source: *Guinness World Records* 2007

10. *Sled-dog racing.*   The Iditarod sled-dog race extends for 1049 mi from Anchorage to Nome. If a musher is twice as far from Anchorage as from Nome, how many miles has the musher traveled?

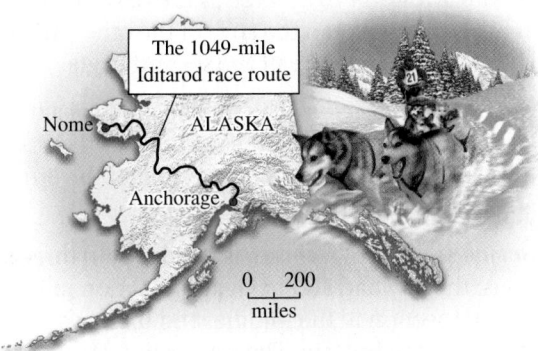

11. *Indy Car racing.*   In April 2008, Danica Patrick won the Indy Japan 300 with a time of 01:51:02.6739 for the 300-mi race. At one point, Patrick was 20 mi closer to the finish than to the start. How far had Patrick traveled at that point?

12. *NASCAR racing.*   In June 2007, Carl Edwards won the Michigan 400 with a time of 2:42:5 for the 400-mi race. At one point, Edwards was 80 mi closer to the finish than to the start. How far had Edwards traveled at that point?

13. *Apartment numbers.*   The apartments in Erica's apartment house are consecutively numbered on each floor. The sum of her number and her next-door neighbor's number is 2409. What are the two numbers?

14. *Apartment numbers.*   The apartments in Brian's apartment house are numbered consecutively on each floor. The sum of his number and his next-door neighbor's number is 1419. What are the two numbers?

15. *Street addresses.*   The houses on the west side of Lincoln Avenue are consecutive odd numbers. Sam and Colleen are next-door neighbors and the sum of their house numbers is 572. Find their house numbers.

16. *Street addresses.*   The houses on the south side of Elm Street are consecutive even numbers. Wanda and Larry are next-door neighbors and the sum of their house numbers is 794. Find their house numbers.

17. The sum of three consecutive page numbers is 99. Find the numbers.

18. The sum of three consecutive page numbers is 60. Find the numbers.

19. *Longest marriage.*   As half of the world's longest-married couple, the woman was 2 yr younger than her husband. Together, their ages totaled 204 yr. How old were the man and the woman?
Source: *Guinness World Records* 2007

20. *Oldest bride.*   The world's oldest bride was 19 yr older than her groom. Together, their ages totaled 185 yr. How old were the bride and the groom?
Source: *Guinness World Records* 2007

21. *e-mail.*   In 2006, approximately 125 billion e-mail messages were sent each day. The number of spam messages was about four times the number of non-spam messages. How many of each type of message were sent each day in 2006?
Source: Ferris Research

22. *Home remodeling.*   In 2005, Americans spent a total of $26 billion to remodel bathrooms and kitchens. They spent $5 billion more on kitchens than on bathrooms. How much was spent on each?
Source: Joint Center for Housing Studies, Harvard University

23. *Page numbers.*   The sum of the page numbers on the facing pages of a book is 281. What are the page numbers?

24. *Perimeter of a triangle.*   The perimeter of a triangle is 195 mm. If the lengths of the sides are consecutive odd integers, find the length of each side.

25. *Hancock Building dimensions.* The top of the John Hancock Building in Chicago is a rectangle whose length is 60 ft more than the width. The perimeter is 520 ft. Find the width and the length of the rectangle. Find the area of the rectangle.

26. *Dimensions of a state.* The perimeter of the state of Wyoming is 1280 mi. The width is 90 mi less than the length. Find the width and the length.

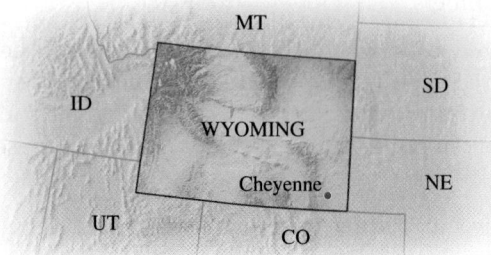

27. A rectangular community garden is to be enclosed with 92 m of fencing. In order to allow for compost storage, the garden must be 4 m longer than it is wide. Determine the dimensions of the garden.

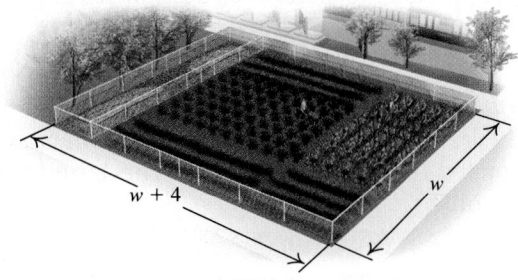

28. *Perimeter of a high school basketball court.* The perimeter of a standard high school basketball court is 268 ft. The length is 34 ft longer than the width. Find the dimensions of the court.
Source: Indiana High School Athletic Association

29. *Two-by-four.* The perimeter of a cross section of a "two-by-four" piece of lumber is $10\frac{1}{2}$ in. The length is twice the width. Find the actual dimensions of the cross section of a two-by-four.

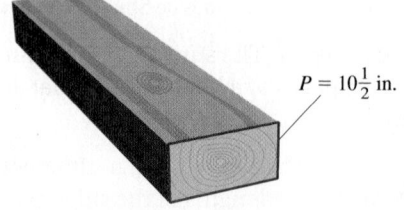

$P = 10\frac{1}{2}$ in.

30. *Standard billboard sign.* A standard rectangular highway billboard sign has a perimeter of 124 ft. The length is 6 ft more than three times the width. Find the dimensions of the sign.

$3w + 6$

w

31. *Angles of a triangle.* The second angle of an architect's triangle is three times as large as the first. The third angle is 30° more than the first. Find the measure of each angle.

32. *Angles of a triangle.* The second angle of a triangular garden is four times as large as the first. The third angle is 45° less than the sum of the other two angles. Find the measure of each angle.

33. *Angles of a triangle.* The second angle of a triangular kite is four times as large as the first. The third angle is 5° more than the sum of the other two angles. Find the measure of the second angle.

34. *Angles of a triangle.* The second angle of a triangular building lot is three times as large as the first. The third angle is 10° more than the sum of the other two angles. Find the measure of the third angle.

35. *Rocket sections.* A rocket is divided into three sections: the payload and navigation section in the top, the fuel section in the middle, and the rocket engine section in the bottom. The top section is one-sixth the length of the bottom section. The middle section is one-half the length of the bottom section. The total length is 240 ft. Find the length of each section.

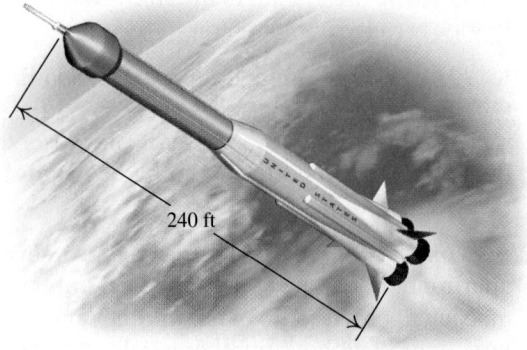

240 ft

36. *Gourmet sandwiches.* Jenny, Demi, and Drew buy an 18-in. long gourmet sandwich and take it back to their apartment. Since they have different appetites, Jenny cuts the sandwich so that Demi gets half of what Jenny gets and Drew gets three-fourths of what Jenny gets. Find the length of each person's sandwich.

37. *Taxi rates.* In Chicago, a taxi ride costs $2.25 plus $1.80 for each mile traveled. Debbie has budgeted $18 for a taxi ride (excluding tip). How far can she travel on her $18 budget?
Source: City of Chicago

38. *Taxi fares.* In New York City, taxis charge $2.50 plus $2.00 per mile for off-peak fares. How far can Ralph travel for $17.50 (assuming an off-peak fare)?
Source: New York City Taxi and Limousine Commission

39. *Truck rentals.* Truck-Rite Rentals rents trucks at a daily rate of $49.95 plus 39¢ per mile. Concert Productions has budgeted $100 for renting a truck to haul equipment to an upcoming concert. How far can they travel in one day and stay within their budget?

40. *Truck rentals.* Fine Line Trucks rents an 18-ft truck for $42 plus 35¢ per mile. Judy needs a truck for one day to deliver a shipment of plants. How far can she drive and stay within a budget of $70?

41. *Complementary angles.* The sum of the measures of two *complementary* angles is 90°. If one angle measures 15° more than twice the measure of its complement, find the measure of each angle.

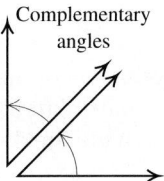

Complementary angles

42. *Complementary angles.* Two angles are complementary. (See Exercise 41.) The measure of one angle is $1\frac{1}{2}$ times the measure of the other. Find the measure of each angle.

43. *Supplementary angles.* The sum of the measures of two *supplementary* angles is 180°. If the measure of one angle is $3\frac{1}{2}$ times the measure of the other, find the measure of each angle.

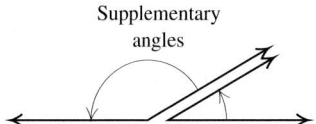

Supplementary angles

44. *Supplementary angles.* Two angles are supplementary. (See Exercise 43.) If one angle measures 45° less than twice the measure of its supplement, find the measure of each angle.

45. *Copier paper.* The perimeter of standard-size copier paper is 99 cm. The width is 6.3 cm less than the length. Find the length and the width.

46. *Stock prices.* Sarah's investment in Jet Blue stock grew 28% to $448. How much did she originally invest?

47. *Savings interest.* Janeka invested money in a savings account at a rate of 6% simple interest. After 1 yr, she has $6996 in the account. How much did Janeka originally invest?

48. *Credit cards.* The balance in Will's Mastercard® account grew 2%, to $870, in one month. What was his balance at the beginning of the month?

49. *Scrabble®.* In a single game on October 12, 2006, Michael Cresta and Wayne Yorra set three North American Scrabble records: the most points in one game by one player, the most total points in the game, and the most points on a single turn. Cresta scored 340 points more than Yorra, and together they scored 1320 points. What was the winning score?
Source: www.slate.com

50. *Color printers.*   The art gallery in Example 3 decides to raise its budget to $5000 for the 2-month period. How many brochures can they print for $5000?

51. *Selling a home.*   The Brannons are planning to sell their home. If they want to be left with $117,500 after paying 6% of the selling price to a realtor as a commission, for how much must they sell the house?

52. *Budget overruns.*   The massive roadworks project in Boston known as The Big Dig cost approximately $14.6 billion. This cost was 484% more than the original estimate. What was the original estimate of the cost of The Big Dig?
Sources: Taxpayers for Common Sense; www.msnbc.cmsn.com

53. *Cricket chirps and temperature.*   The equation $T = \frac{1}{4}N + 40$ can be used to determine the temperature T, in degrees Fahrenheit, given the number of times N a cricket chirps per minute. Determine the number of chirps per minute for a temperature of 80°F.

54. *Race time.*   The equation $R = -0.028t + 20.8$ can be used to predict the world record in the 200-m dash, where R is the record in seconds and t is the number of years since 1920. In what year will the record be 18.0 sec?

55. Sean claims he can solve most of the problems in this section by guessing. Is there anything wrong with this approach? Why or why not?

56. When solving Exercise 20, Beth used a to represent the bride's age and Ben used a to represent the groom's age. Is one of these approaches preferable to the other? Why or why not?

Skill Review

To prepare for Section 2.6, review inequalities (Section 1.4).

Write a true sentence using either $<$ or $>$. [1.4]

57. $-8 \ \blacksquare \ 1$

58. $-2 \ \blacksquare \ -5$

59. $\frac{1}{2} \ \blacksquare \ 0$

60. $-3 \ \blacksquare \ -1$

Write a second inequality with the same meaning. [1.4]

61. $x \geq -4$

62. $x < 5$

63. $5 > y$

64. $-10 \leq t$

Synthesis

65. Write a problem for a classmate to solve. Devise it so that the problem can be translated to the equation $x + (x + 2) + (x + 4) = 375$.

66. Write a problem for a classmate to solve. Devise it so that the solution is "Audrey can drive the rental truck for 50 mi without exceeding her budget."

67. *Discounted dinners.*   Kate's "Dining Card" entitles her to $10 off the price of a meal after a 15% tip has been added to the cost of the meal. If, after the discount, the bill is $32.55, how much did the meal originally cost?

68. *Test scores.*   Pam scored 78 on a test that had 4 fill-in questions worth 7 points each and 24 multiple-choice questions worth 3 points each. She had one fill-in question wrong. How many multiple-choice questions did Pam get right?

69. *Gettysburg Address.*   Abraham Lincoln's 1863 Gettysburg Address refers to the year 1776 as "four *score* and seven years ago." Determine what a score is.

70. One number is 25% of another. The larger number is 12 more than the smaller. What are the numbers?

71. A storekeeper goes to the bank to get $10 worth of change. She requests twice as many quarters as half dollars, twice as many dimes as quarters, three times as many nickels as dimes, and no pennies or dollars. How many of each coin did the storekeeper get?

72. *Perimeter of a rectangle.*   The width of a rectangle is three fourths of the length. The perimeter of the rectangle becomes 50 cm when the length and the width are each increased by 2 cm. Find the length and the width.

73. *Discounts.*   In exchange for opening a new credit account, Macy's Department Stores® subtracts 10% from all purchases made the day the account is established. Julio is opening an account and has a coupon for which he receives 10% off the first day's reduced price of a camera. If Julio's final price is $77.75, what was the price of the camera before the two discounts?

74. *Sharing fruit.*   Apples are collected in a basket for six people. One third, one fourth, one eighth, and one fifth of the apples are given to four people, respectively. The fifth person gets ten apples, and one apple remains for the sixth person. Find the original number of apples in the basket.

75. *eBay purchases.*   An eBay seller charges $9.99 for the first DVD purchased and $6.99 for all others. For shipping and handling, he charges the full shipping fee of $3 for the first DVD, one half of the shipping charge for the second item, and one third of the shipping charge per item for all remaining items. The total cost of a shipment (excluding tax) was $45.45. How many DVDs were in the shipment?

76. *Winning percentage.*   In a basketball league, the Falcons won 15 of their first 20 games. In order to win 60% of the total number of games, how many more games will they have to play, assuming they win only half of the remaining games?

77. *Taxi fares.*   In New York City, a taxi ride costs $2.50 plus 40¢ per $\frac{1}{5}$ mile and 40¢ per minute stopped in traffic. Due to traffic, Glenda's taxi took 20 min to complete what is usually a 10-min drive. If she is charged $18.50 for the ride, how far did Glenda travel?
Source: New York City Taxi and Limousine Commission

78. *Test scores.*   Ella has an average score of 82 on three tests. Her average score on the first two tests is 85. What was the score on the third test?

79. A school purchases a piano and must choose between paying $2000 at the time of purchase or $2150 at the end of one year. Which option should the school select and why?

80. Annette claims the following problem has no solution: "The sum of the page numbers on facing pages is 191. Find the page numbers." Is she correct? Why or why not?
Aha!

81. The perimeter of a rectangle is 101.74 cm. If the length is 4.25 cm longer than the width, find the dimensions of the rectangle.

82. The second side of a triangle is 3.25 cm longer than the first side. The third side is 4.35 cm longer than the second side. If the perimeter of the triangle is 26.87 cm, find the length of each side.

2.6   Solving Inequalities

Solutions of Inequalities ▪ Graphs of Inequalities ▪ Set-Builder and Interval Notation ▪
Solving Inequalities Using the Addition Principle ▪ Solving Inequalities Using the Multiplication Principle ▪
Using the Principles Together

Many real-world situations translate to *inequalities*. For example, a student might need to register for *at least* 12 credits; an elevator might be designed to hold *at most* 2000 pounds; a tax credit might be allowable for families with incomes of *less than* $25,000; and so on. Before solving applications of this type, we must adapt our equation-solving principles to the solving of inequalities.

Solutions of Inequalities

Recall from Section 1.4 that an inequality is a number sentence containing $>$ (is greater than), $<$ (is less than), $\geq$ (is greater than or equal to), or $\leq$ (is less than or equal to). Inequalities like

$$-7 > x, \quad t < 5, \quad 5x - 2 \geq 9, \quad \text{and} \quad -3y + 8 \leq -7$$

are true for some replacements of the variable and false for others.

Any value for the variable that makes an inequality true is called a **solution**. The set of all solutions is called the **solution set**. When all solutions of an inequality are found, we say that we have **solved** the inequality.

EXAMPLE   1   Determine whether the given number is a solution of $x < 2$: **(a)** -3; **(b)** 2.

SOLUTION

a) Since $-3 < 2$ is true, -3 is a solution.

b) Since $2 < 2$ is false, 2 is not a solution.

 TRY EXERCISE   9

EXAMPLE   2   Determine whether the given number is a solution of $y \geq 6$: **(a)** 6; **(b)** -4.

SOLUTION

a) Since $6 \geq 6$ is true, 6 is a solution.

b) Since $-4 \geq 6$ is false, -4 is not a solution.

TRY EXERCISE   11

Graphs of Inequalities

Because the solutions of inequalities like $x < 2$ are too numerous to list, it is helpful to make a drawing that represents all the solutions. The **graph** of an inequality is such a drawing. Graphs of inequalities in one variable can be drawn on the number line by shading all points that are solutions. Parentheses are used to indicate endpoints that are *not* solutions and brackets to indicate endpoints that *are* solutions.*

EXAMPLE   3   Graph each inequality: **(a)** $x < 2$; **(b)** $y \geq -3$; **(c)** $-2 < x \leq 3$.

SOLUTION

a) The solutions of $x < 2$ are those numbers less than 2. They are shown on the graph by shading all points to the left of 2. The parenthesis at 2 and the shading to its left indicate that 2 is *not* part of the graph, but numbers like 1.2 and 1.99 are.

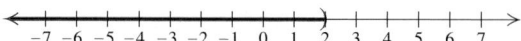

b) The solutions of $y \geq -3$ are shown on the number line by shading the point for -3 and all points to the right of -3. The bracket at -3 indicates that -3 *is* part of the graph.

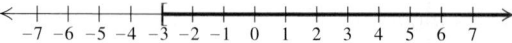

STUDENT NOTES

Note that $-2 < x < 3$ means $-2 < x$ and $x < 3$. Because of this, statements like $2 < x < 1$ make no sense—no number is both greater than 2 and less than 1.

c) The inequality $-2 < x \leq 3$ is read "-2 is less than x *and* x is less than or equal to 3," or "x is greater than -2 *and* less than or equal to 3." To be a solution of $-2 < x \leq 3$, a number must be a solution of both $-2 < x$ *and* $x \leq 3$. The number 1 is a solution, as are -0.5, 1.9, and 3. The parenthesis indicates that -2 is *not* a solution, whereas the bracket indicates that 3 *is* a solution. The other solutions are shaded.

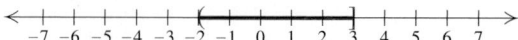

TRY EXERCISE   17

*An alternative notation uses open dots to indicate endpoints that are not solutions and closed dots to indicate endpoints that are solutions. Using this notation, the solutions of $x < 2$ are graphed as ←++++++o++→ and the solutions of $y \geq -3$ are graphed as ←+●++++++→

Set–Builder and Interval Notation

To write the solution set of $x < 3$, we can use **set-builder notation:**

$$\{x\,|\,x < 3\}.$$

This is read "The set of all x such that x is less than 3."

Another way to write solutions of an inequality in one variable is to use **interval notation**. Interval notation uses parentheses, ( ), and brackets, [ ].

If a and b are real numbers with $a < b$, we define the **open interval (a, b)** as the set of all numbers x for which $a < x < b$. Using set-builder notation, we write

$$(a, b) = \{x\,|\,a < x < b\}. \qquad \text{Parentheses are used to exclude endpoints.}$$

Its graph excludes the endpoints:

$$(a, b)$$

$\longleftarrow \quad \underset{a}{(} \rule{3cm}{0.4pt} \underset{b}{)} \quad \longrightarrow \qquad \{x\,|\,a < x < b\}$

The **closed interval $[a, b]$** is defined as the set of all numbers x for which $a \le x \le b$. Thus,

$$[a, b] = \{x\,|\,a \le x \le b\}. \qquad \text{Brackets are used to include endpoints.}$$

Its graph includes the endpoints:

$$[a, b]$$

$\longleftarrow \quad \underset{a}{[} \rule{3cm}{0.4pt} \underset{b}{]} \quad \longrightarrow \qquad \{x\,|\,a \le x \le b\}$

There are two kinds of **half-open intervals**, defined as follows:

1. $(a, b] = \{x\,|\,a < x \le b\}$. This is open on the left. Its graph is as follows:

$$(a, b]$$

$\longleftarrow \quad \underset{a}{(} \rule{3cm}{0.4pt} \underset{b}{]} \quad \longrightarrow \qquad \{x\,|\,a < x \le b\}$

2. $[a, b) = \{x\,|\,a \le x < b\}$. This is open on the right. Its graph is as follows:

$$[a, b)$$

$\longleftarrow \quad \underset{a}{[} \rule{3cm}{0.4pt} \underset{b}{)} \quad \longrightarrow \qquad \{x\,|\,a \le x < b\}$

We use the symbols ∞ and $-\infty$ to represent positive infinity and negative infinity, respectively. Thus the notation (a, ∞) represents the set of all real numbers greater than a, and $(-\infty, a)$ represents the set of all real numbers less than a.

$$(a, \infty)$$

$\longleftarrow \quad \underset{a}{(} \rule{3cm}{0.4pt} \quad \longrightarrow \qquad \{x\,|\,x > a\}$

$$(-\infty, a)$$

$\longleftarrow \quad \rule{3cm}{0.4pt} \underset{a}{)} \quad \longrightarrow \qquad \{x\,|\,x < a\}$

The notation $[a, \infty)$ or $(-\infty, a]$ is used when we want to include the endpoint a.

CAUTION! Do not confuse the *interval* (a, b) with the *ordered pair* (a, b). The context in which the notation appears should make the meaning clear.

STUDENT NOTES

You may have noticed which inequality signs in set-builder notation correspond to brackets and which correspond to parentheses. The relationship could be written informally as

$$\le \quad \ge \qquad [\,]$$
$$< \quad > \qquad (\,).$$

EXAMPLE 4   Graph $y \ge -2$ on a number line and write the solution set using both set-builder and interval notations.

SOLUTION   Using set-builder notation, we write the solution set as $\{y\,|\,y \ge -2\}$.

Using interval notation, we write $[-2, \infty)$.

To graph the solution, we shade all numbers to the right of -2 and use a bracket to indicate that -2 is also a solution.

$\longleftarrow \!\!\! \underset{-7}{|}\ \underset{-6}{|}\ \underset{-5}{|}\ \underset{-4}{|}\ \underset{-3}{|}\ \underset{-2}{[}\ \underset{-1}{|}\ \underset{0}{|}\ \underset{1}{|}\ \underset{2}{|}\ \underset{3}{|}\ \underset{4}{|}\ \underset{5}{|}\ \underset{6}{|}\ \underset{7}{|}\!\!\! \longrightarrow$

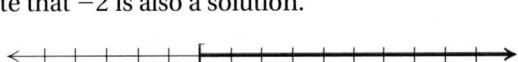
TRY EXERCISE ▶ 27

Solving Inequalities Using the Addition Principle

Consider a balance similar to one that appears in Section 2.1. When one side of the balance holds more weight than the other, the balance tips in that direction. If equal amounts of weight are then added to or subtracted from both sides of the balance, the balance remains tipped in the same direction.

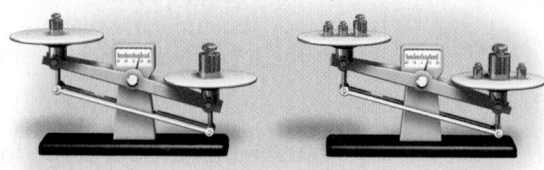

The balance illustrates the idea that when a number, such as 2, is added to (or subtracted from) both sides of a true inequality, such as $3 < 7$, we get another true inequality:

$$3 + 2 < 7 + 2, \quad \text{or} \quad 5 < 9.$$

Similarly, if we add -4 to both sides of $x + 4 < 10$, we get an *equivalent* inequality:

$$x + 4 + (-4) < 10 + (-4), \quad \text{or} \quad x < 6.$$

We say that $x + 4 < 10$ and $x < 6$ are **equivalent**, which means that both inequalities have the same solution set.

The Addition Principle for Inequalities

For any real numbers a, b, and c:

$a < b$ is equivalent to $a + c < b + c$;

$a \leq b$ is equivalent to $a + c \leq b + c$;

$a > b$ is equivalent to $a + c > b + c$;

$a \geq b$ is equivalent to $a + c \geq b + c$.

As with equations, our goal is to isolate the variable on one side.

EXAMPLE **5**

Solve $x + 2 > 8$ and then graph the solution.

SOLUTION   We use the addition principle, subtracting 2 from both sides:

$$x + 2 - 2 > 8 - 2 \qquad \text{Subtracting 2 from, or adding } -2 \text{ to, both sides}$$
$$x > 6.$$

From the inequality $x > 6$, we can determine the solutions easily. Any number greater than 6 makes $x > 6$ true and is a solution of that inequality as well as the inequality $x + 2 > 8$. Using set-builder notation, the solution set is $\{x \mid x > 6\}$. Using interval notation, the solution set is $(6, \infty)$. The graph is as follows:

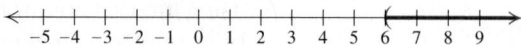

Because most inequalities have an infinite number of solutions, we cannot possibly check them all. A partial check can be made using one of the possible solutions. For this example, we can substitute any number greater than 6—say, 6.1—into the original inequality:

$$\frac{x + 2 > 8}{6.1 + 2 \mid 8}$$
$$8.1 \overset{?}{>} 8 \quad \text{TRUE} \quad 8.1 > 8 \text{ is a true statement.}$$

Since $8.1 > 8$ is true, 6.1 is a solution. Any number greater than 6 is a solution.

> **TRY EXERCISE** 43

EXAMPLE   6

Solve $3x - 1 \le 2x - 5$ and then graph the solution.

SOLUTION   We have

$$3x - 1 \le 2x - 5$$
$$3x - 1 + 1 \le 2x - 5 + 1 \qquad \text{Adding 1 to both sides}$$
$$3x \le 2x - 4 \qquad \text{Simplifying}$$
$$3x - 2x \le 2x - 4 - 2x \qquad \text{Subtracting } 2x \text{ from both sides}$$
$$x \le -4. \qquad \text{Simplifying}$$

The graph is as follows:

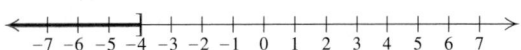

The student should check that any number less than or equal to -4 is a solution. The solution set is $\{x \mid x \le -4\}$, or $(-\infty, -4]$.

> **TRY EXERCISE** 47

TECHNOLOGY CONNECTION

As a partial check of Example 5, we can let $y_1 = 3x - 1$ and $y_2 = 2x - 5$. We set TblStart $= -5$ and ΔTbl $= 1$ in the TBLSET menu to get the following table. By scrolling up or down, you can note that for $x \le -4$, we have $y_1 \le y_2$.

X	Y₁	Y₂
−5	−16	−15
−4	−13	−13
−3	−10	−11
−2	−7	−9
−1	−4	−7
0	−1	−5
1	2	−3

X = −5

Solving Inequalities Using the Multiplication Principle

There is a multiplication principle for inequalities similar to that for equations, but it must be modified when multiplying both sides by a negative number. Consider the true inequality

$$3 < 7.$$

If we multiply both sides by a *positive* number—say, 2—we get another true inequality:

$$3 \cdot 2 < 7 \cdot 2, \quad \text{or} \quad 6 < 14. \qquad \text{TRUE}$$

If we multiply both sides by a negative number—say, -2—we get a *false* inequality:

$$3 \cdot (-2) < 7 \cdot (-2), \quad \text{or} \quad -6 < -14. \qquad \text{FALSE}$$

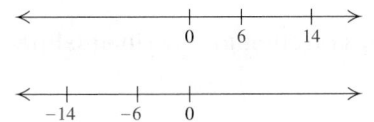

The fact that $6 < 14$ is true, but $-6 < -14$ is false, stems from the fact that the negative numbers, in a sense, *mirror* the positive numbers. Whereas 14 is to the *right* of 6, the number -14 is to the *left* of -6. Thus if we reverse the inequality symbol in $-6 < -14$, we get a true inequality:

$$-6 > -14. \qquad \text{TRUE}$$

> ### The Multiplication Principle for Inequalities
>
> For any real numbers a and b, and for any *positive* number c:
>
> $$a < b \quad \text{is equivalent to} \quad ac < bc, \quad \text{and}$$
> $$a > b \quad \text{is equivalent to} \quad ac > bc.$$
>
> For any real numbers a and b, and for any *negative* number c:
>
> $$a < b \quad \text{is equivalent to} \quad ac > bc, \quad \text{and}$$
> $$a > b \quad \text{is equivalent to} \quad ac < bc.$$
>
> Similar statements hold for $\leq$ and $\geq$.

> *CAUTION!* When multiplying or dividing both sides of an inequality by a negative number, don't forget to reverse the inequality symbol!

EXAMPLE **7** Solve and graph each inequality: **(a)** $\frac{1}{4}x < 7$; **(b)** $-2y \leq 18$.

SOLUTION

a) $\qquad \frac{1}{4}x < 7$

$\qquad 4 \cdot \frac{1}{4}x < 4 \cdot 7 \qquad$ Multiplying both sides by 4, the reciprocal of $\frac{1}{4}$

$\qquad\qquad\qquad\qquad$ The symbol stays the same, since 4 is positive.

$\qquad\qquad x < 28 \qquad$ Simplifying

The solution set is $\{x \mid x < 28\}$, or $(-\infty, 28)$. The graph is shown at left.

b) $\quad -2y \leq 18$

$\qquad \dfrac{-2y}{-2} \geq \dfrac{18}{-2} \qquad$ Multiplying both sides by $-\frac{1}{2}$, or dividing both sides by -2

$\qquad\qquad\qquad\qquad$ *At this step*, we reverse the inequality, because $-\frac{1}{2}$ is negative.

$\qquad\quad y \geq -9 \qquad$ Simplifying

As a partial check, we substitute a number greater than -9, say -8, into the original inequality:

$$\frac{-2y \leq 18}{-2(-8) \mid 18}$$
$$16 \overset{?}{\leq} 18 \quad \text{TRUE} \qquad 16 \leq 18 \text{ is a true statement.}$$

The solution set is $\{y \mid y \geq -9\}$, or $[-9, \infty)$. The graph is shown at left.

TRY EXERCISE 59

Using the Principles Together

We use the addition and multiplication principles together to solve inequalities much as we did when solving equations.

EXAMPLE **8** Solve: **(a)** $6 - 5y > 7$; **(b)** $2x - 9 < 7x + 1$.

SOLUTION

a) $\qquad\quad 6 - 5y > 7$

$\qquad -6 + 6 - 5y > -6 + 7 \qquad$ Adding -6 to both sides

$\qquad\qquad\quad -5y > 1 \qquad\qquad$ Simplifying

$$-\tfrac{1}{5} \cdot (-5y) < -\tfrac{1}{5} \cdot 1$$ Multiplying both sides by $-\tfrac{1}{5}$, or dividing both sides by -5

Remember to reverse the inequality symbol!

$$y < -\tfrac{1}{5}$$ Simplifying

As a partial check, we substitute a number smaller than $-\tfrac{1}{5}$, say -1, into the original inequality:

$$\frac{6 - 5y > 7}{\begin{array}{c|c} 6 - 5(-1) & 7 \\ 6 - (-5) & \end{array}}$$

$$11 \overset{?}{>} 7 \quad \text{TRUE} \qquad 11 > 7 \text{ is a true statement.}$$

The solution set is $\left\{y \mid y < -\tfrac{1}{5}\right\}$, or $\left(-\infty, -\tfrac{1}{5}\right)$. We show the graph in the margin for reference.

b)
$$2x - 9 < 7x + 1$$
$$2x - 9 - 1 < 7x + 1 - 1 \qquad \text{Subtracting 1 from both sides}$$
$$2x - 10 < 7x \qquad \text{Simplifying}$$
$$2x - 10 - 2x < 7x - 2x \qquad \text{Subtracting } 2x \text{ from both sides}$$
$$-10 < 5x \qquad \text{Simplifying}$$
$$\frac{-10}{5} < \frac{5x}{5} \qquad \text{Dividing both sides by 5}$$
$$-2 < x \qquad \text{Simplifying}$$

The solution set is $\{x \mid -2 < x\}$, or $\{x \mid x > -2\}$, or $(-2, \infty)$.

TRY EXERCISE ▶ 69

All of the equation-solving techniques used in Sections 2.1 and 2.2 can be used with inequalities provided we remember to reverse the inequality symbol when multiplying or dividing both sides by a negative number.

EXAMPLE 9   Solve: **(a)** $16.3 - 7.2p \le -8.18$; **(b)** $3(x - 9) - 1 \le 2 - 5(x + 6)$.

SOLUTION

a) The greatest number of decimal places in any one number is *two*. Multiplying both sides by 100 will clear decimals. Then we proceed as before.

$$16.3 - 7.2p \le -8.18$$
$$100(16.3 - 7.2p) \le 100(-8.18) \qquad \text{Multiplying both sides by 100}$$
$$100(16.3) - 100(7.2p) \le 100(-8.18) \qquad \text{Using the distributive law}$$
$$1630 - 720p \le -818 \qquad \text{Simplifying}$$
$$-720p \le -818 - 1630 \qquad \text{Subtracting 1630 from both sides}$$
$$-720p \le -2448 \qquad \text{Simplifying;} \\ -818 - 1630 = -2448$$
$$p \ge \frac{-2448}{-720} \qquad \text{Dividing both sides by } -720$$

Remember to reverse the symbol!

$$p \ge 3.4$$

The solution set is $\{p \mid p \ge 3.4\}$, or $[3.4, \infty)$.

b) $3(x - 9) - 1 \leq 2 - 5(x + 6)$

$$3x - 27 - 1 \leq 2 - 5x - 30 \qquad \text{Using the distributive law to remove parentheses}$$

$$3x - 28 \leq -5x - 28 \qquad \text{Simplifying}$$

$$3x - 28 + 28 \leq -5x - 28 + 28 \qquad \text{Adding 28 to both sides}$$

$$3x \leq -5x$$

$$3x + 5x \leq -5x + 5x \qquad \text{Adding } 5x \text{ to both sides}$$

$$8x \leq 0$$

$$x \leq 0 \qquad \text{Dividing both sides by 8}$$

The solution set is $\{x \mid x \leq 0\}$, or $(-\infty, 0]$.

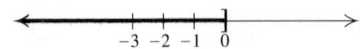

▶ **TRY EXERCISE** 83

2.6 EXERCISE SET

✎ *Concept Reinforcement* *Insert the symbol* $<, >, \leq, or \geq$ *to make each pair of inequalities equivalent.*

1. $-5x \leq 30; \quad x \,\blacksquare\, -6$

2. $-7t \geq 56; \quad t \,\blacksquare\, -8$

3. $-2t > -14; \quad t \,\blacksquare\, 7$

4. $-3x < -15; \quad x \,\blacksquare\, 5$

Classify each pair of inequalities as "equivalent" or "not equivalent."

5. $x < -2; \quad -2 > x$

6. $t > -1; \quad -1 < t$

7. $-4x - 1 \leq 15;$
$\quad -4x \leq 16$

8. $-2t + 3 \geq 11;$
$\quad -2t \geq 14$

Determine whether each number is a solution of the given inequality.

9. $x > -4$

   a) 4   **b)** -6 **c)** -4

10. $t < 3$

   a) -3 **b)** 3   **c)** $2\frac{19}{20}$

11. $y \leq 19$

   a) 18.99   **b)** 19.01   **c)** 19

12. $n \geq -4$

   a) 0   **b)** -4.1 **c)** -3.9

13. $c \geq -7$

   a) 0   **b)** -5.4 **c)** 7.1

14. $a > 6$

   a) 6   **b)** -6.7 **c)** 0

15. $z < -3$

   a) 0   **b)** $-3\frac{1}{3}$ **c)** 1

16. $m \leq -2$

   a) $-1\frac{9}{10}$ **b)** 0   **c)** $-2\frac{1}{3}$

Graph on a number line.

17. $y < 2$ **18.** $x \leq 7$

19. $x \geq -1$ **20.** $t > -2$

21. $0 \leq t$ **22.** $1 \leq m$

23. $-5 \leq x < 2$

24. $-3 < x \leq 5$

25. $-4 < x < 0$

26. $0 \leq x \leq 5$

Graph each inequality, and write the solution set using both set-builder notation and interval notation.

27. $y < 6$ **28.** $x > 4$

29. $x \geq -4$ **30.** $t \leq 6$

31. $t > -3$ **32.** $y < -3$

33. $x \leq -7$ **34.** $x \geq -6$

Describe each graph using set-builder notation and interval notation.

35.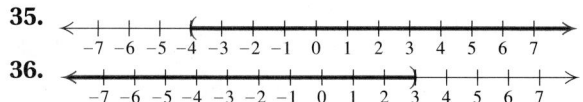

36.

37.

38.

39.

40.

41.

42.

Solve using the addition principle. Graph and write set-builder notation and interval notation for each answer.

43. $y + 6 > 9$

44. $x + 8 \leq -10$

45. $n - 6 < 11$

46. $n - 4 > -3$

47. $2x \leq x - 9$

48. $3x \leq 2x + 7$

49. $y + \frac{1}{3} \leq \frac{5}{6}$

50. $x + \frac{1}{4} \leq \frac{1}{2}$

51. $t - \frac{1}{8} > \frac{1}{2}$

52. $y - \frac{1}{3} > \frac{1}{4}$

53. $-9x + 17 > 17 - 8x$

54. $-8n + 12 > 12 - 7n$

Aha! **55.** $-23 < -t$

56. $19 < -x$

57. $10 - y \leq -12$

58. $3 - y \geq -6$

Solve using the multiplication principle. Graph and write set-builder notation and interval notation for each answer.

59. $4x < 28$

60. $3x \geq 24$

61. $-24 > 8t$

62. $-16x < -64$

63. $1.8 \geq -1.2n$

64. $9 \leq -2.5a$

65. $-2y \leq \frac{1}{5}$

66. $-2x \geq \frac{1}{5}$

67. $-\frac{8}{5} > 2x$

68. $-\frac{5}{8} < -10y$

Solve using the addition and multiplication principles.

69. $2 + 3x < 20$

70. $7 + 4y < 31$

71. $4t - 5 \leq 23$

72. $15x - 7 \leq -7$

73. $39 > 3 - 9x$

74. $5 > 5 - 7y$

75. $5 - 6y > 25$

76. $8 - 2y > 9$

77. $-3 < 8x + 7 - 7x$

78. $-5 < 9x + 8 - 8x$

79. $6 - 4y > 6 - 3y$

80. $7 - 8y > 5 - 7y$

81. $7 - 9y \leq 4 - 7y$

82. $6 - 13y \leq 4 - 12y$

83. $2.1x + 43.2 > 1.2 - 8.4x$

84. $0.96y - 0.79 \leq 0.21y + 0.46$

85. $1.7t + 8 - 1.62t < 0.4t - 0.32 + 8$

86. $0.7n - 15 + n \geq 2n - 8 - 0.4n$

87. $\frac{x}{3} + 4 \leq 1$

88. $\frac{2}{3} - \frac{x}{5} < \frac{4}{15}$

89. $3 < 5 - \frac{t}{7}$

90. $2 > 9 - \frac{x}{5}$

91. $4(2y - 3) \leq -44$

92. $3(2y - 3) > 21$

93. $8(2t + 1) > 4(7t + 7)$

94. $3(t - 2) \geq 9(t + 2)$

95. $3(r - 6) + 2 < 4(r + 2) - 21$

96. $5(t + 3) + 9 \geq 3(t - 2) - 10$

97. $\frac{4}{5}(3x + 4) \leq 20$

98. $\frac{2}{3}(2x - 1) \geq 10$

99. $\frac{2}{3}\left(\frac{7}{8} - 4x\right) - \frac{5}{8} < \frac{3}{8}$

100. $\frac{3}{4}\left(3x - \frac{1}{2}\right) - \frac{2}{3} < \frac{1}{3}$

101. Are the inequalities $x > -3$ and $x \geq -2$ equivalent? Why or why not?

102. Are the inequalities $t < -7$ and $t \leq -8$ equivalent? Why or why not?

Skill Review

Review simplifying expressions (Section 1.8).

Simplify. [1.8]

103. $5x - 2(3 - 6x)$

104. $8m - n - 3(2m + 5n)$

105. $x - 2[4y + 3(8 - x) - 1]$

106. $5 - 3t - 4[6 + 5(2t - 1) + t]$

107. $3[5(2a - b) + 1] - 5[4 - (a - b)]$

108. $9x - 2\{4 - 5[6 - 2(x + 1) - x]\}$

Synthesis

109. Explain how it is possible for the graph of an inequality to consist of just one number. (*Hint*: See Example 3c.)

110. The statements of the addition and multiplication principles begin with *conditions* set for the variables. Explain the conditions given for each principle.

Solve.

Aha! **111.** $x < x + 1$

112. $6[4 - 2(6 + 3t)] > 5[3(7 - t) - 4(8 + 2t)] - 20$

113. $27 - 4[2(4x - 3) + 7] \geq 2[4 - 2(3 - x)] - 3$

Solve for x.

114. $\frac{1}{2}(2x + 2b) > \frac{1}{3}(21 + 3b)$

115. $-(x + 5) \geq 4a - 5$

116. $y < ax + b$ (Assume $a < 0$.)

117. $y < ax + b$ (Assume $a > 0$.)

118. Graph the solutions of $|x| < 3$ on a number line.

Aha! **119.** Determine the solution set of $|x| > -3$.

120. Determine the solution set of $|x| < 0$.

CONNECTING the CONCEPTS

The procedure for solving inequalities is very similar to that used to solve equations. There are, however, two important differences.

- The multiplication principle for inequalities differs from the multiplication principle for equations: When we multiply or divide on both sides of an inequality by a *negative* number, we must *reverse* the direction of the inequality.

- The solution set of an equation like those we solved in this chapter typically consists of one number. The solution set of an inequality typically consists of a set of numbers and is written using set-builder notation.

Compare the following solutions.

Solve: $2 - 3x = x + 10$.

SOLUTION

$$2 - 3x = x + 10$$
$$-3x = x + 8 \qquad \text{Subtracting 2 from both sides}$$
$$-4x = 8 \qquad \text{Subtracting } x \text{ from both sides}$$
$$x = -2 \qquad \text{Dividing both sides by } -4$$

The solution is -2.

Solve: $2 - 3x > x + 10$.

SOLUTION

$$2 - 3x > x + 10$$
$$-3x > x + 8 \qquad \text{Subtracting 2 from both sides}$$
$$-4x > 8 \qquad \text{Subtracting } x \text{ from both sides}$$
$$x < -2 \qquad \text{Dividing both sides by } -4 \text{ and reversing the direction of the inequality symbol}$$

The solution is $\{x | x < -2\}$, or $(-\infty, -2)$.

MIXED REVIEW

Solve.

1. $x - 6 = 15$

2. $x - 6 \leq 15$

3. $3x = -18$

4. $3x > -18$

5. $-3x > -18$

6. $5x + 2 = 17$

7. $7 - 3x = 8$

8. $4y - 7 < 5$

9. $3 - t \geq 19$

10. $2 + 3n = 5n - 9$

11. $3 - 5a > a + 9$

12. $1.2x - 3.4 < 0.4x + 5.2$

13. $\frac{2}{3}(x + 5) \geq -4$

14. $\frac{n}{5} - 6 = 15$

15. $0.5x - 2.7 = 3x + 7.9$

16. $5(6 - t) = -45$

17. $8 - \frac{y}{3} \leq 7$

18. $\frac{1}{3}x - \frac{5}{6} = \frac{3}{2} - \frac{1}{6}x$

19. $-15 > 7 - 5x$

20. $10 \geq -2(a - 5)$

2.7	Solving Applications with Inequalities

Translating to Inequalities ■ Solving Problems

The five steps for problem solving can be used for problems involving inequalities.

Translating to Inequalities

Before solving problems that involve inequalities, we list some important phrases to look for. Sample translations are listed as well.

Important Words	Sample Sentence	Translation
is at least	Ming walks at least 2 mi a day.	$m \geq 2$
is at most	At most 5 students dropped the course.	$n \leq 5$
cannot exceed	The width cannot exceed 40 ft.	$w \leq 40$
must exceed	The speed must exceed 15 mph.	$s > 15$
is less than	Kamal's weight is less than 120 lb.	$w < 120$
is more than	Boston is more than 200 mi away.	$d > 200$
is between	The film was between 90 and 100 min long.	$90 < t < 100$
minimum	Ned drank a minimum of 5 glasses of water a day.	$w \geq 5$
maximum	The maximum penalty is $100.	$p \leq 100$
no more than	Alan weighs no more than 90 lb.	$w \leq 90$
no less than	Mallory scored no less than 8.3.	$s \geq 8.3$

The following phrases deserve special attention.

> **Translating "at least" and "at most"**
>
> The quantity x is at least some amount q: $x \geq q$.
> (If x is *at least* q, it cannot be less than q.)
>
> The quantity x is at most some amount q: $x \leq q$.
> (If x is *at most* q, it cannot be more than q.)

Solving Problems

EXAMPLE **1**

Catering costs. To cater a party, Curtis' Barbeque charges a $50 setup fee plus $15 per person. The cost of Hotel Pharmacy's end-of-season softball party cannot exceed $450. How many people can attend the party?

SOLUTION

1. **Familiarize.** Suppose that 20 people were to attend the party. The cost would then be $50 + \$15 \cdot 20$, or $350. This shows that more than 20 people could attend without exceeding $450. Instead of making another guess, we let $n =$ the number of people in attendance.

2. **Translate.** The cost of the party will be $50 for the setup fee plus $15 times the number of people attending. We can reword as follows:

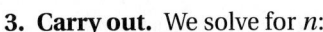

Rewording:	The setup fee	plus	the cost of the meals	cannot exceed	$450.
	↓	↓	↓	↓	↓
Translating:	50	+	$15 \cdot n$	$\leq$	450

3. **Carry out.** We solve for n:

$$50 + 15n \leq 450$$
$$15n \leq 400 \qquad \text{Subtracting 50 from both sides}$$
$$n \leq \frac{400}{15} \qquad \text{Dividing both sides by 15}$$
$$n \leq 26\frac{2}{3}. \qquad \text{Simplifying}$$

4. **Check.** Although the solution set of the inequality is all numbers less than or equal to $26\frac{2}{3}$, since n represents the number of people in attendance, we round *down* to 26. If 26 people attend, the cost will be $50 + \$15 \cdot 26$, or $440, and if 27 attend, the cost will exceed $450.

5. **State.** At most 26 people can attend the party.

> TRY EXERCISE ▶ 23

> *CAUTION!* Solutions of problems should always be checked using the original wording of the problem. In some cases, answers might need to be whole numbers or integers or rounded off in a particular direction.

Some applications with inequalities involve *averages*, or *means*. You are already familiar with the concept of averages from grades in courses that you have taken.

> **Average, or Mean**
>
> To find the **average**, or **mean**, of a set of numbers, add the numbers and then divide by the number of addends.

EXAMPLE 2

Financial aid.   Full-time students in a health-care education program can receive financial aid and employee benefits from Covenant Health System by working at Covenant while attending school and also agreeing to work there after graduation. Students who work an average of at least 16 hr per week receive extra pay and part-time employee benefits. For the first three weeks of September, Dina worked 20 hr, 12 hr, and 14 hr. How many hours must she work during the fourth week in order to average at least 16 hr per week for the month?

Source: Covenant Health Systems

SOLUTION

1. **Familiarize.** Suppose Dina works 10 hr during the fourth week. Her average for the month would be

$$\frac{20\,\text{hr} + 12\,\text{hr} + 14\,\text{hr} + 10\,\text{hr}}{4} = 14\,\text{hr}.$$ There are 4 addends, so we divide by 4.

This shows that Dina must work more than 10 hr during the fourth week, if she is to average at least 16 hr of work per week. We let x represent the number of hours Dina works during the fourth week.

2. **Translate.** We reword the problem and translate as follows:

Rewording: The average number of hours worked   should be   at least   16 hr.

Translating: $\dfrac{20 + 12 + 14 + x}{4}$ $\geq$ 16

3. **Carry out.** Because of the fraction, it is convenient to use the multiplication principle first:

$$\frac{20 + 12 + 14 + x}{4} \geq 16$$

$$4\left(\frac{20 + 12 + 14 + x}{4}\right) \geq 4 \cdot 16$$ Multiplying both sides by 4

$$20 + 12 + 14 + x \geq 64$$

$$46 + x \geq 64$$ Simplifying

$$x \geq 18.$$ Subtracting 46 from both sides

4. **Check.** As a partial check, we show that if Dina works 18 hr, she will average at least 16 hr per week:

$$\frac{20 + 12 + 14 + 18}{4} = \frac{64}{4} = 16.$$ Note that 16 is at least 16.

5. **State.** Dina will average at least 16 hr of work per week for September if she works at least 18 hr during the fourth week.

TRY EXERCISE 27

Translating for Success

1. *Consecutive integers.* The sum of two consecutive even integers is 102. Find the integers.

2. *Salary increase.* After Susanna earned a 5% raise, her new salary was $25,750. What was her former salary?

3. *Dimensions of a rectangle.* The length of a rectangle is 6 in. more than the width. The perimeter of the rectangle is 102 in. Find the length and the width.

4. *Population.* The population of Kelling Point is decreasing at a rate of 5% per year. The current population is 25,750. What was the population the previous year?

5. *Reading assignment.* Quinn has 6 days to complete a 150-page reading assignment. How many pages must he read the first day so that he has no more than 102 pages left to read on the 5 remaining days?

Translate each word problem to an equation or an inequality and select a correct translation from A–O.

A. $0.05(25{,}750) = x$

B. $x + 2x = 102$

C. $2x + 2(x + 6) = 102$

D. $150 - x \le 102$

E. $x - 0.05x = 25{,}750$

F. $x + (x + 2) = 102$

G. $x + (x + 6) > 102$

H. $x + 5x = 150$

I. $x + 0.05x = 25{,}750$

J. $x + (2x + 6) = 102$

K. $x + (x + 1) = 102$

L. $102 + x > 150$

M. $0.05x = 25{,}750$

N. $102 + 5x > 150$

O. $x + (x + 6) = 102$

Answers on page A-5

An additional, animated version of this activity appears in MyMathLab. To use MyMathLab, you need a course ID and a student access code. Contact your instructor for more information.

6. *Numerical relationship.* One number is 6 more than twice another. The sum of the numbers is 102. Find the numbers.

7. *DVD collections.* Together Mindy and Ken have 102 DVDs. If Ken has 6 more DVDs than Mindy, how many does each have?

8. *Sales commissions.* Kirk earns a commission of 5% on his sales. One year he earned commissions totaling $25,750. What were his total sales for the year?

9. *Fencing.* Jess has 102 ft of fencing that he plans to use to enclose two dog runs. The perimeter of one run is to be twice the perimeter of the other. Into what lengths should the fencing be cut?

10. *Quiz scores.* Lupe has a total of 102 points on the first 6 quizzes in her sociology class. How many total points must she earn on the 5 remaining quizzes in order to have more than 150 points for the semester?

2.7	EXERCISE SET

Concept Reinforcement *In each of Exercises 1–8, match the sentence with one of the following:*

$$a < b; \quad a \le b; \quad b < a; \quad b \le a.$$

1. *a* is at least *b*

2. *a* exceeds *b*.

3. *a* is at most *b*.

4. *a* is exceeded by *b*.

5. *b* is no more than *a*.

6. *b* is no less than *a*.

7. *b* is less than *a*.

8. *b* is more than *a*.

Translate to an inequality.

9. A number is less than 10.

10. A number is greater than or equal to 4.

11. The temperature is at most $-3°C$.

12. The average credit-card debt is more than $8000.

13. To rent a car, a driver must have a minimum of 5 yr driving experience.

14. The Barringdean Shopping Center is no more than 20 mi away.

15. The age of the Mayan altar exceeds 1200 yr.

16. The maximum safe exposure limit of formaldehyde is 2 parts per million.

17. Tania earns between $12 and $15 an hour.

18. Leslie's test score was at least 85.

19. Wind speeds were greater than 50 mph.

20. The costs of production of that software cannot exceed $12,500.

21. A room at Pine Tree Bed and Breakfast costs no more than $120 a night.

22. The cost of gasoline was at most $4 per gallon.

Use an inequality and the five-step process to solve each problem.

23. *Furnace repairs.* RJ's Plumbing and Heating charges $55 plus $40 per hour for emergency service. Gary remembers being billed over $150 for an emergency call. How long was RJ's there?

24. *College tuition.* Karen's financial aid stipulates that her tuition not exceed $1000. If her local community college charges a $35 registration fee plus $375 per course, what is the greatest number of courses for which Karen can register?

25. *Graduate school.* An unconditional acceptance into the Master of Business Administration (MBA) program at Arkansas State University will be given to students whose GMAT score plus 200 times the undergraduate grade point average is at least 950. Robbin's GMAT score was 500. What must her grade point average be in order to be unconditionally accepted into the program?
Source: graduateschool.astate.edu

26. *Car payments.* As a rule of thumb, debt payments (other than mortgages) should be less than 8% of a consumer's monthly gross income. Oliver makes $54,000 a year and has a $100 student-loan payment every month. What size car payment can he afford?
Source: money.cnn.com

27. *Quiz average.* Rod's quiz grades are 73, 75, 89, and 91. What scores on a fifth quiz will make his average quiz grade at least 85?

28. *Nutrition.* Following the guidelines of the U.S. Department of Agriculture, Dale tries to eat at least 5 half-cup servings of vegetables each day. For the first six days of one week, she had 4, 6, 7, 4, 6, and 4 servings. How many servings of vegetables should Dale eat on Saturday, in order to average at least 5 servings per day for the week?

29. *College course load.* To remain on financial aid, Millie needs to complete an average of at least 7 credits per quarter each year. In the first three quarters of 2008, Millie completed 5, 7, and 8 credits. How many credits of course work must Millie complete in the fourth quarter if she is to remain on financial aid?

30. *Music lessons.* Band members at Colchester Middle School are expected to average at least 20 min of practice time per day. One week Monroe practiced 15 min, 28 min, 30 min, 0 min, 15 min, and 25 min. How long must he practice on the seventh day if he is to meet expectations?

31. *Baseball.* In order to qualify for a batting title, a major league baseball player must average at least 3.1 plate appearances per game. For the first nine games of the season, a player had 5, 1, 4, 2, 3, 4, 4, 3, and 2 plate appearances. How many plate appearances must the player have in the tenth game in order to average at least 3.1 per game?
Source: Major League Baseball

32. *Education.* The Mecklenberg County Public Schools stipulate that a standard school day will average at least $5\frac{1}{2}$ hr, excluding meal breaks. For the first four days of one school week, bad weather resulted in school days of 4 hr, $6\frac{1}{2}$ hr, $3\frac{1}{2}$ hr, and $6\frac{1}{2}$ hr. How long must the Friday school day be in order to average at least $5\frac{1}{2}$ hr for the week?
Source: www.meck.k12.va.us

33. *Perimeter of a triangle.* One side of a triangle is 2 cm shorter than the base. The other side is 3 cm longer than the base. What lengths of the base will allow the perimeter to be greater than 19 cm?

34. *Perimeter of a sign.* The perimeter of a rectangular sign is not to exceed 50 ft. The length is to be twice the width. What widths will meet these conditions?

35. *Well drilling.* All Seasons Well Drilling offers two plans. Under the "pay-as-you-go" plan, they charge $500 plus $8 a foot for a well of any depth. Under their "guaranteed-water" plan, they charge a flat fee of $4000 for a well that is guaranteed to provide adequate water for a household. For what depths would it save a customer money to use the pay-as-you-go plan?

36. *Cost of road service.* Rick's Automotive charges $50 plus $15 for each (15-min) unit of time when making a road call. Twin City Repair charges $70 plus $10 for each unit of time. Under what circumstances would it be more economical for a motorist to call Rick's?

37. *Insurance-covered repairs.* Most insurance companies will replace a vehicle if an estimated repair exceeds 80% of the "blue-book" value of the vehicle. Michele's insurance company paid $8500 for repairs to her Subaru after an accident. What can be concluded about the blue-book value of the car?

38. *Insurance-covered repairs.* Following an accident, Jeff's Ford pickup was replaced by his insurance company because the damage was so extensive. Before the damage, the blue-book value of the truck was $21,000. How much would it have cost to repair the truck? (See Exercise 37.)

39. *Sizes of packages.* The U.S. Postal Service defines a "package" as a parcel for which the sum of the length and the girth is less than 84 in. (Length is the longest side of a package and girth is the distance around the other two sides of the package.) A box has a fixed girth of 29 in. Determine (in terms of an inequality) those lengths for which the box is considered a "package."

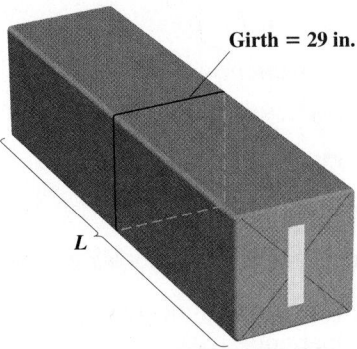

Girth = 29 in.

L

40. *Sizes of envelopes.* Rhetoric Advertising is a direct-mail company. It determines that for a particular campaign, it can use any envelope with a fixed width of $3\frac{1}{2}$ in. and an area of at least $17\frac{1}{2}$ in². Determine (in terms of an inequality) those lengths that will satisfy the company constraints.

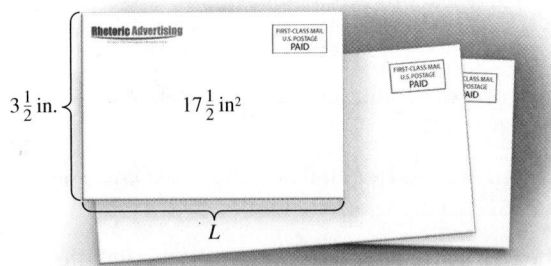

$3\frac{1}{2}$ in. $17\frac{1}{2}$ in²

L

41. *Body temperature.* A person is considered to be feverish when his or her temperature is higher than 98.6°F. The formula $F = \frac{9}{5}C + 32$ can be used to convert Celsius temperatures C to Fahrenheit temperatures F. For which Celsius temperatures is a person considered feverish?

42. *Gold temperatures.*   Gold stays solid at Fahrenheit temperatures below 1945.4°. Determine (in terms of an inequality) those Celsius temperatures for which gold stays solid. Use the formula given in Exercise 41.

43. *Area of a triangular sign.*   Zoning laws in Harrington prohibit displaying signs with areas exceeding 12 ft². If Flo's Marina is ordering a triangular sign with an 8-ft base, how tall can the sign be?

44. *Area of a triangular flag.*   As part of an outdoor education course, Trisha needs to make a bright-colored triangular flag with an area of at least 3 ft². What heights can the triangle be if the base is $1\frac{1}{2}$ ft?

45. *Fat content in foods.*   Reduced Fat Skippy® peanut butter contains 12 g of fat per serving. In order for a food to be labeled "reduced fat," it must have at least 25% less fat than the regular item. What can you conclude about the number of grams of fat in a serving of the regular Skippy peanut butter?
Source: Best Foods

46. *Fat content in foods.*   Reduced Fat Chips Ahoy!® cookies contain 5 g of fat per serving. What can you conclude about the number of grams of fat in regular Chips Ahoy! cookies (see Exercise 45)?
Source: Nabisco Brands, Inc.

47. *Weight gain.*   In the last weeks before the yearly Topsfield Weigh In, heavyweight pumpkins gain about 26 lb per day. Charlotte's heaviest pumpkin weighs 532 lb on September 5. For what dates will its weight exceed 818 lb?
Source: Based on a story in the *Burlington Free Press*

48. *Pond depth.*   On July 1, Garrett's Pond was 25 ft deep. Since that date, the water level has dropped $\frac{2}{3}$ ft per week. For what dates will the water level not exceed 21 ft?

49. *Cell-phone budget.*   Liam has budgeted $60 a month for his cell phone. For his service, he pays a monthly fee of $39.95, plus taxes of $6.65, plus 10¢ for each text message sent or received. How many text messages can he send or receive and not exceed his budget?

50. *Banquet costs.*   The women's volleyball team can spend at most $700 for its awards banquet at a local restaurant. If the restaurant charges a $100 setup fee plus $24 per person, at most how many can attend?

51. *World records in the mile run.*   The formula
$$R = -0.0065t + 4.3259$$
can be used to predict the world record, in minutes, for the 1-mi run t years after 1900. Determine (in terms of an inequality) those years for which the world record will be less than 3.6 min.
Source: Based on information from Information Please Database 2007, Pearson Education, Inc.

52. *Women's records in the women's 1500-m run.*   The formula
$$R = -0.0026t + 4.0807$$
can be used to predict the world record, in minutes, for the 1500-m run t years after 1900. Determine (in terms of an inequality) those years for which the world record will be less than 3.8 min.
Source: Based on information from *Track and Field*

53. *Toll charges.*   The equation

$$y = 0.06x + 0.50$$

can be used to determine the approximate cost y, in dollars, of driving x miles on the Pennsylvania Turnpike. For what mileages x will the cost be at most $14?

54. *Price of a movie ticket.*   The average price of a movie ticket can be estimated by the equation

$$P = 0.169Y - 333.04,$$

where Y is the year and P is the average price, in dollars. For what years will the average price of a movie ticket be at least $7? (Include the year in which the $7 ticket first occurs.)
Source: National Association of Theatre Owners

55. If f represents Fran's age and t represents Todd's age, write a sentence that would translate to $t + 3 < f$.

56. Explain how the meanings of "Five more than a number" and "Five is more than a number" differ.

Skill Review

Review operations with real numbers (Sections 1.5–1.8).

Simplify.

57. $-2 + (-5) - 7$ [1.6]

58. $\dfrac{1}{2} \div \left(-\dfrac{3}{4}\right)$ [1.7]

59. $3 \cdot (-10) \cdot (-1) \cdot (-2)$ [1.7]

60. $-6.3 + (-4.8)$ [1.5]

61. $(3 - 7) - (4 - 8)$ [1.8]

62. $3 - 2 + 5 \cdot 10 \div 5^2 \cdot 2$ [1.8]

63. $\dfrac{-2 - (-6)}{8 - 10}$ [1.8]

64. $\dfrac{1 - (-7)}{-3 - 5}$ [1.8]

Synthesis

65. Write a problem for a classmate to solve. Devise the problem so the answer is "At most 18 passengers can go on the boat." Design the problem so that at least one number in the solution must be rounded down.

66. Write a problem for a classmate to solve. Devise the problem so the answer is "The Rothmans can drive 90 mi without exceeding their truck rental budget."

67. *Ski wax.*   Green ski wax works best between 5° and 15° Fahrenheit. Determine those Celsius temperatures for which green ski wax works best. (See Exercise 41.)

68. *Parking fees.*   Mack's Parking Garage charges $4.00 for the first hour and $2.50 for each additional hour. For how long has a car been parked when the charge exceeds $16.50?

Aha! **69.** The area of a square can be no more than 64 cm². What lengths of a side will allow this?

Aha! **70.** The sum of two consecutive odd integers is less than 100. What is the largest pair of such integers?

71. *Nutritional standards.*   In order for a food to be labeled "lowfat," it must have fewer than 3 g of fat per serving. Reduced-fat tortilla chips contain 60% less fat than regular nacho cheese tortilla chips, but still cannot be labeled lowfat. What can you conclude about the fat content of a serving of nacho cheese tortilla chips?

72. *Parking fees.*   When asked how much the parking charge is for a certain car (see Exercise 68), Mack replies, "between 14 and 24 dollars." For how long has the car been parked?

73. *Frequent buyer bonus.*   Alice's Books allows customers to select one free book for every 10 books purchased. The price of that book cannot exceed the average cost of the 10 books. Neoma has bought 9 books whose average cost is $12 per book. How much should her tenth book cost if she wants to select a $15 book for free?

74. *Grading.*   After 9 quizzes, Blythe's average is 84. Is it possible for Blythe to improve her average by two points with the next quiz? Why or why not?

75. *Discount card.*   Barnes & Noble offers a member card for $25 a year. This card entitles a customer to a 40% discount off list price on hardcover bestsellers, a 20% discount on adult hardcovers, and a 10% discount on other purchases. Describe two sets of circumstances for which an individual would save money by becoming a member.
Source: Barnes & Noble

Study Summary

KEY TERMS AND CONCEPTS	EXAMPLES

SECTION 2.1: SOLVING EQUATIONS

Equivalent equations share the same solution.

$3x - 1 = 10$, $3x = 11$, and $x = \dfrac{11}{3}$ are equivalent equations.

The Addition Principle for Equations

$a = b$ is equivalent to
$a + c = b + c$.

$x + 5 = -2$ is equivalent to
$x + 5 + (-5) = -2 + (-5)$ and to
$x = -7$.

The Multiplication Principle for Equations

$a = b$ is equivalent to $ac = bc$,
for $c \neq 0$.

$-\frac{1}{3}x = 7$ is equivalent to
$(-3)(-\frac{1}{3}x) = (-3)(7)$ and to
$x = -21$.

SECTION 2.2: USING THE PRINCIPLES TOGETHER

We can **clear fractions** by multiplying both sides of an equation by the least common multiple of the denominators in the equation.

Solve: $\frac{1}{2}x - \frac{1}{3} = \frac{1}{6}x + \frac{2}{3}$.

$6\left(\frac{1}{2}x - \frac{1}{3}\right) = 6\left(\frac{1}{6}x + \frac{2}{3}\right)$ Multiplying by 6, the least common denominator

$6 \cdot \frac{1}{2}x - 6 \cdot \frac{1}{3} = 6 \cdot \frac{1}{6}x + 6 \cdot \frac{2}{3}$ Using the distributive law

$3x - 2 = x + 4$ Simplifying

$2x = 6$ Subtracting x from and adding 2 to both sides

$x = 3$

We can **clear decimals** by multiplying both sides by a power of 10. If there is at most one decimal place in any one number, multiply by 10. If there are at most two decimal places, multiply by 100, and so on.

Solve: $3.6t - 1.5 = 2 - 0.8t$.

$10(3.6t - 1.5) = 10(2 - 0.8t)$ Multiplying both sides by 10 because the greatest number of decimal places is 1.

$36t - 15 = 20 - 8t$ Using the distributive law

$44t = 35$ Adding $8t$ and 15 to both sides

$t = \dfrac{35}{44}$ Dividing both sides by 44

SECTION 2.3: FORMULAS

A **formula** uses letters to show a relationship among two or more quantities. Formulas can be solved for a given letter using the addition and multiplication principles.

Solve: $x = \frac{2}{5}y + 7$ *for* y.

$x = \frac{2}{5}y + 7$ We are solving for y.

$x - 7 = \frac{2}{5}y$ Isolating the term containing y

$\frac{5}{2}(x - 7) = \frac{5}{2} \cdot \frac{2}{5}y$ Multiplying both sides by $\frac{5}{2}$

$\frac{5}{2}x - \frac{5}{2} \cdot 7 = 1 \cdot y$ Using the distributive law

$\frac{5}{2}x - \frac{35}{2} = y$ We have solved for y.

SECTION 2.4: APPLICATIONS WITH PERCENT

Percent Notation

$n\%$ means $\dfrac{n}{100}$, or $n \times \dfrac{1}{100}$, or $n \times 0.01$

$31\% = 0.31$; $\dfrac{1}{8} = 0.125 = 12.5\%$;
$2.9\% = 0.029$; $2.94 = 294\%$

Key Words in Percent Translations

"Of" translates to " $\cdot$ " or "$\times$"

"What" translates to a variable

"Is" or "Was" translates to "$=$"

"%" translates to "$\times \frac{1}{100}$" or "$\times 0.01$"

$$\underbrace{\text{What percent}}_{n} \quad \underbrace{\text{of}}_{\cdot} \quad \underbrace{60}_{60} \quad \underbrace{\text{is}}_{=} \quad \underbrace{7.2?}_{7.2}$$

$$n = \frac{7.2}{60}$$

$$n = 0.12$$

Thus, 7.2 is 12% of 60.

SECTION 2.5: PROBLEM SOLVING

Five Steps for Problem Solving in Algebra

1. *Familiarize* yourself with the problem.
2. *Translate* to mathematical language. (This often means writing an equation.)
3. *Carry out* some mathematical manipulation. (This often means *solving* an equation.)
4. *Check* your possible answer in the original problem.
5. *State* the answer clearly.

The perimeter of a rectangle is 70 cm. The width is 5 cm longer than half the length. Find the length and the width.

1. **Familiarize.** Look up, if necessary, the formula for the perimeter of a rectangle:

   $$P = 2l + 2w.$$

   We are looking for two values, the length and the width. We can describe the width in terms of the length:

   $$w = \tfrac{1}{2}l + 5.$$

2. **Translate.**

   Rewording: Twice the length plus twice the width is the perimeter.

   Translating: $2l$ $+$ $2\left(\tfrac{1}{2}l + 5\right)$ $=$ 70

3. **Carry out.** Solve the equation:

$$
\begin{aligned}
2l + 2\left(\tfrac{1}{2}l + 5\right) &= 70 \\
2l + l + 10 &= 70 &&\text{Using the distributive law} \\
3l + 10 &= 70 &&\text{Combining like terms} \\
3l &= 60 &&\text{Subtracting 10 from both sides} \\
l &= 20. &&\text{Dividing both sides by 3}
\end{aligned}
$$

   If $l = 20$, then $w = \tfrac{1}{2}l + 5 = \tfrac{1}{2} \cdot 20 + 5 = 10 + 5 = 15$.

4. **Check.** The width should be 5 cm longer than half the length. Since half the length is 10 cm, and 15 cm is 5 cm longer, this statement checks. The perimeter should be 70 cm. Since $2l + 2w = 2(20) + 2(15) = 40 + 30 = 70$, this statement also checks.

5. **State.** The length is 20 cm and the width is 15 cm.

SECTION 2.6: SOLVING INEQUALITIES

An **inequality** is any sentence containing $<$, $>$, $\leq$, $\geq$, or $\neq$. Solution sets of inequalities can be **graphed** and written in **set-builder notation** or **interval notation**.

Interval Notation	Set-builder Notation	Graph
(a, b)	$\{x \mid a < x < b\}$	
$[a, b]$	$\{x \mid a \leq x \leq b\}$	
$[a, b)$	$\{x \mid a \leq x < b\}$	
$(a, b]$	$\{x \mid a < x \leq b\}$	
(a, ∞)	$\{x \mid a < x\}$	
$(-\infty, a)$	$\{x \mid x < a\}$	

The Addition Principle for Inequalities

For any real numbers a, b, and c,

$\quad a < b$ is equivalent to $a + c < b + c$;
$\quad a > b$ is equivalent to $a + c > b + c$.

Similar statements hold for $\leq$ and $\geq$.

$x + 3 \leq 5$ is equivalent to
$x + 3 - 3 \leq 5 - 3$ and to
$\quad x \leq 2$.

The Multiplication Principle for Inequalities

For any real numbers a and b, and for any *positive* number c,

$\quad a < b$ is equivalent to $ac < bc$;
$\quad a > b$ is equivalent to $ac > bc$.

For any real numbers a and b, and for any *negative* number c,

$\quad a < b$ is equivalent to $ac > bc$;
$\quad a > b$ is equivalent to $ac < bc$.

Similar statements hold for $\leq$ and $\geq$.

$3x > 9$ is equivalent to
$\frac{1}{3} \cdot 3x > \frac{1}{3} \cdot 9$ The inequality symbol does not change because $\frac{1}{3}$ is positive.
$\quad x > 3$.

$-3x > 9$ is equivalent to
$-\frac{1}{3} \cdot -3x < -\frac{1}{3} \cdot 9$ The inequality symbol is reversed because $-\frac{1}{3}$ is negative.
$\quad x < -3$.

SECTION 2.7: SOLVING APPLICATIONS WITH INEQUALITIES

Many real-world problems can be solved by translating the problem to an inequality and applying the five-step problem-solving strategy.

Translate to an inequality.

The test score must exceed 85.	$s > 85$
At most 15 volunteers greeted visitors.	$v \leq 15$
Ona makes no more than $100 a week.	$w \leq 100$
Herbs need at least 4 hr of sun a day.	$h \geq 4$

Review Exercises: Chapter 2

➥ *Concept Reinforcement* *Classify each statement as either true or false.*

1. $5x - 4 = 2x$ and $3x = 4$ are equivalent equations. [2.1]

2. $5 - 2t < 9$ and $t > 6$ are equivalent inequalities. [2.6]

3. Some equations have no solution. [2.1]

4. Consecutive odd integers are 2 units apart. [2.5]

5. For any number a, $a \le a$. [2.6]

6. The addition principle is always used before the multiplication principle. [2.2]

7. A 10% discount results in a sale price that is 90% of the original price. [2.4]

8. Often it is impossible to list all solutions of an inequality number by number. [2.6]

Solve. Label any contradictions or identities.

9. $x + 9 = -16$ [2.1]

10. $-8x = -56$ [2.1]

11. $-\dfrac{x}{5} = 13$ [2.1]

12. $-8 = n - 11$ [2.1]

13. $\frac{2}{5}t = -8$ [2.1]

14. $x - 0.1 = 1.01$ [2.1]

15. $-\frac{2}{3} + x = -\frac{1}{6}$ [2.1]

16. $4y + 11 = 5$ [2.2]

17. $5 - x = 13$ [2.2]

18. $3t + 7 = t - 1$ [2.2]

19. $7x - 6 = 25x$ [2.2]

20. $\frac{1}{4}x - \frac{5}{8} = \frac{3}{8}$ [2.2]

21. $14y = 23y - 17 - 9y$ [2.2]

22. $0.22y - 0.6 = 0.12y + 3 - 0.8y$ [2.2]

23. $\frac{1}{4}x - \frac{1}{8}x = 3 - \frac{1}{16}x$ [2.2]

24. $6(4 - n) = 18$ [2.2]

25. $4(5x - 7) = -56$ [2.2]

26. $8(x - 2) = 4(x - 4)$ [2.2]

27. $3(x - 4) + 2 = x + 2(x - 5)$ [2.2]

Solve each formula for the given letter. [2.3]

28. $C = \pi d$, for d

29. $V = \dfrac{1}{3}Bh$, for B

30. $5x - 2y = 10$, for y

31. $tx = ax + b$, for x

32. Find decimal notation: 1.2%. [2.4]

33. Find percent notation: $\frac{11}{25}$. [2.4]

34. What percent of 60 is 42? [2.4]

35. 49 is 35% of what number? [2.4]

Determine whether each number is a solution of $x \le -5$. [2.6]

36. -3

37. -7

38. 4

Graph on a number line. [2.6]

39. $5x - 6 < 2x + 3$

40. $-2 < x \le 5$

41. $t > 0$

Solve. Write the answers in set-builder notation and interval notation. [2.6]

42. $t + \frac{2}{3} \ge \frac{1}{6}$

43. $9x \ge 63$

44. $2 + 6y > 20$

45. $7 - 3y \ge 27 + 2y$

46. $3x + 5 < 2x - 6$

47. $-4y < 28$

48. $3 - 4x < 27$

49. $4 - 8x < 13 + 3x$

50. $13 \le -\frac{2}{3}t + 5$

51. $7 \le 1 - \frac{3}{4}x$

Solve.

52. In 2006, U.S. retailers lost a record $41.6 billion due to theft and fraud. Of this amount, $20 billion was due to employee theft. What percent of the total loss was employee theft? [2.4]
Source: www.wwaytv3.com

53. An 18-ft beam is cut into two pieces. One piece is 2 ft longer than the other. How long are the pieces? [2.5]

54. In 2004, a total of 103,000 students from China and Japan enrolled in U.S. colleges and universities. The number of Japanese students was 10,000 more than half the number of Chinese students. How many Chinese students and how many Japanese students enrolled in the United States? [2.5]
Source: Institute of International Education

55. The sum of two consecutive odd integers is 116. Find the integers. [2.5]

56. The perimeter of a rectangle is 56 cm. The width is 6 cm less than the length. Find the width and the length. [2.5]

57. After a 25% reduction, a picnic table is on sale for $120. What was the regular price? [2.4]

58. From 2000 to 2006, the number of U.S. wireless-phone subscribers increased by 114 percent to 233 million. How many subscribers were there in 2000? [2.4]
Source: Cellular Telecommunications and Internet Association

59. The measure of the second angle of a triangle is 50° more than that of the first. The measure of the third angle is 10° less than twice the first. Find the measures of the angles. [2.5]

60. The U.S. Centers for Disease Control recommends that for a typical 2000-calorie daily diet, no more than 65 g of fat be consumed. In the first three days of a four-day vacation, Teresa consumed 55 g, 80 g, and 70 g of fat. Determine how many grams of fat Teresa can consume on the fourth day if she is to average no more than 65 g of fat per day. [2.7]

61. *Blueprints.* To make copies of blueprints, Vantage Reprographics charges a $6 setup fee plus $4 per copy. Myra can spend no more than $65 for the copying. What number of copies will allow her to stay within budget? [2.7]

Synthesis

62. How does the multiplication principle for equations differ from the multiplication principle for inequalities? [2.1], [2.6]

63. Explain how checking the solutions of an equation differs from checking the solutions of an inequality. [2.1], [2.6]

64. A study of sixth- and seventh-graders in Boston revealed that, on average, the students spent 3 hr 20 min per day watching TV or playing video and computer games. This represents 108% more than the average time spent reading or doing homework. How much time each day was spent, on average, reading or doing homework? [2.4]
Source: Harvard School of Public Health

65. In June 2007, a team of Brazilian scientists exploring the Amazon measured its length as 65 mi longer than the Nile. If the combined length of both rivers is 8385 mi, how long is each river? [2.5]
Source: news.nationalgeographic.com

66. Kent purchased a book online at 25% off the retail price. The shipping charges were $4.95. If the amount due was $16.95, what was the retail price of the book? [2.4], [2.5]

Solve.

67. $2|n| + 4 = 50$ [1.4], [2.2]

68. $|3n| = 60$ [1.4], [2.1]

69. $y = 2a - ab + 3$, for a [2.3]

70. The Maryland Heart Center gives the following steps to calculate the number of fat grams needed daily by a moderately active woman. Write the steps as one formula relating the number of fat grams F to a woman's weight w, in pounds. [2.3]

1. Calculate the total number of calories per day.
   ____ pounds × 12 calories = ____ total calories per day
2. Take the total number of calories and multiply by 30 percent.
   ____ calories per day × 0.30 = ____ calories from fat per day.
3. Take the number of calories from fat per day and divide by 9 (there are 9 calories per gram of fat).
   ____ calories from fat per day divided by 9 = ____ fat grams per day

Test: Chapter 2

Solve. Label any contradictions or identities.

1. $t + 7 = 16$

2. $t - 3 = 12$

3. $6x = -18$

4. $-\frac{4}{7}x = -28$

5. $3t + 7 = 2t - 5$

6. $\frac{1}{2}x - \frac{3}{5} = \frac{2}{5}$

7. $8 - y = 16$

8. $4.2x + 3.5 = 1.2 - 2.5x$

9. $4(x + 2) = 36$

10. $\frac{5}{6}(3x + 1) = 20$

11. $13t - (5 - 2t) = 5(3t - 1)$

Solve. Write the answers in set-builder notation and interval notation.

12. $x + 6 > 1$

13. $14x + 9 > 13x - 4$

14. $-5y \geq 65$

15. $4y \leq -30$

16. $4n + 3 < -17$

17. $3 - 5x > 38$

18. $\frac{1}{2}t - \frac{1}{4} \leq \frac{3}{4}t$

19. $5 - 9x \geq 19 + 5x$

Solve each formula for the given letter.

20. $A = 2\pi rh$, for r

21. $w = \dfrac{P + l}{2}$, for l

22. Find decimal notation: 230%.

23. Find percent notation: 0.003.

24. What number is 18.5% of 80?

25. What percent of 75 is 33?

Graph on a number line.

26. $y < 4$

27. $-2 \leq x \leq 2$

Solve.

28. The perimeter of a rectangular calculator is 36 cm. The length is 4 cm greater than the width. Find the width and the length.

29. In 1948, Earl Shaffer became the first person to hike all 2100 mi of the Appalachian trail—from Springer Mountain, Georgia, to Mt. Katahdin, Maine. Shaffer repeated the feat 50 years later, and at age 79 became the oldest person to hike the entire trail. When Shaffer stood atop Big Walker Mountain, Virginia, he was three times as far from the northern end of the trail as from the southern end. At that point, how far was he from each end of the trail?

30. The perimeter of a triangle is 249 mm. If the sides are consecutive odd integers, find the length of each side.

31. By lowering the temperature of their electric hot-water heater from 140°F to 120°F, the Kellys' average electric bill dropped by 7% to $60.45. What was their electric bill before they lowered the temperature of their hot water?

32. *Mass transit.* Local light rail service in Denver, Colorado, costs $1.50 per trip (one way). A monthly pass costs $54. Gail is a student at Community College of Denver. Express as an inequality the number of trips per month that Gail should make if the pass is to save her money.
Source: rtd-denver.com

Synthesis

Solve.

33. $c = \dfrac{2cd}{a - d}$, for d

34. $3|w| - 8 = 37$

35. Translate to an inequality.

A plant marked "partial sun" needs at least 4 hr but no more than 6 hr of sun each day.
Source: www.yardsmarts.com

36. A concert promoter had a certain number of tickets to give away. Five people got the tickets. The first got one third of the tickets, the second got one fourth of the tickets, and the third got one fifth of the tickets. The fourth person got eight tickets, and there were five tickets left for the fifth person. Find the total number of tickets given away.

Cumulative Review: Chapters 1–2

Simplify.

1. $18 + (-30)$ [1.5]
2. $\frac{1}{2} - \left(-\frac{1}{4}\right)$ [1.6]
3. $-1.2(3.5)$ [1.7]
4. $-5 \div \left(-\frac{1}{2}\right)$ [1.7]
5. $150 - 10^2 \div 25 \cdot 4$ [1.8]
6. $(150 - 10^2) \div (25 \cdot 4)$ [1.8]

Remove parentheses and simplify. [1.8]

7. $5x - (3x - 1)$
8. $2(t + 6) - 12t$
9. $3[4n - 5(2n - 1)] - 3(n - 7)$
10. Graph on a number line: $-\frac{5}{2}$. [1.4]
11. Find the absolute value: $|27|$. [1.4]
12. Factor: $12x + 18y + 30z$. [1.8]

Solve.

13. $12 = -2x$ [2.1]
14. $4x - 7 = 3x + 9$ [2.1]
15. $\frac{2}{3}t + 7 = 13$ [2.2]
16. $9(2a - 1) = 4$ [2.2]
17. $12 - 3(5x - 1) = x - 1$ [2.2]
18. $3(x + 1) - 2 = 8 - 5(x + 7)$ [2.2]

Solve each formula for the given letter. [2.3]

19. $\frac{1}{2}x = 2yz$, for z
20. $4x - 9y = 1$, for y
21. $an = p - rn$, for n
22. Find decimal notation: 183%. [2.4]
23. Find percent notation: $\frac{3}{8}$. [2.4]
24. Graph on a number line: $t > -\frac{5}{2}$. [2.6]

Solve. Write the answer in set-builder notation and interval notation. [2.6]

25. $4t + 10 \le 2$
26. $8 - t > 5$
27. $4 < 10 - \dfrac{x}{5}$
28. $4(2n - 3) \le 2(5n - 8)$

Solve.

29. The total attendance at NCAA basketball games during the 2006–2007 school year was 33 million. This was 31.25% less than the total attendance at NCAA

football games during that year. What was the total attendance at NCAA football games during the 2006–2007 school year? [2.4]
Source: NCAA

30. On an average weekday, a full-time college student spends a total of 7.1 hr in educational activities and in leisure activities. The average student spends 0.7 hr more in leisure activities than in educational activities. On an average weekday, how many hours does a full-time college student spend on educational activities? [2.5]
Source: U.S. Bureau of Labor Statistics

31. The wavelength w, in meters per cycle, of a musical note is given by

$$w = \frac{r}{f},$$

where r is the speed of the sound, in meters per second, and f is the frequency, in cycles per second. The speed of sound in air is 344 m/sec. What is the wavelength of a note whose frequency in air is 24 cycles per second? [2.3]

32. A 24-ft ribbon is cut into two pieces. One piece is 6 ft longer than the other. How long are the pieces? [2.5]

33. Juanita has budgeted an average of $65 a month for entertainment. For the first five months of the year, she has spent $88, $15, $125, $50, and $60. How much can Juanita spend in the sixth month without exceeding her average budget? [2.7]

34. In 2006, about 17 million Americans had diabetes. The U.S. Centers for Disease Control predicts that by 2050, 50 million Americans may have the disease. By what percent would the number of Americans with diabetes increase? [2.4]

35. The second angle of a triangle is twice as large as the first. The third angle is 5° more than four times the first. Find the measure of the largest angle. [2.5]

36. The length of a rectangular frame is 53 cm. For what widths would the perimeter be greater than 160 cm? [2.7]

Synthesis

37. Simplify: $t - \{t - [3t - (2t - t) - t] - 4t\} - t$. [1.8]

38. Solve: $3|n| + 10 = 25$. [2.2]

39. Lindy sold her Fender acoustic guitar on eBay using i-soldit.com. The i-soldit location she used charges 35% of the first $500 of the selling price and 20% of the amount over $500. After these charges were deducted from the selling price, she received $745. For how much did her guitar sell? [2.4], [2.5]
Source: Based on information in *The Wall Street Journal*, 9/11/07

Introduction to Graphing

HEATHER HUTH
DIRECTOR OF VOLUNTEERS
New Orleans, Louisiana

We average about 750 volunteers a month who come to Beacon of Hope to help us clean New Orleans' neighborhoods. We service 22 neighborhoods, so I use math to figure out how many volunteers to send to each location, the quantity of supplies each will need, and how much time is required to complete the task.

AN APPLICATION

An increasing number of college students are donating time and energy in volunteer service. The number of college students volunteering grew from 2.7 million in 2002 to 3.3 million in 2005. Graph the given data and estimate the number of college students who volunteered in 2004, and then predict the number of college students who will volunteer in 2010.

Source: Corporation for National and Community Service

This problem appears as Example 7 in Section 3.7.

We now begin our study of graphing. First we will examine graphs as they commonly appear in newspapers or magazines and develop some terminology. Following that, we will graph certain equations and study the connection between rate and slope. We will also learn how graphs can be used as a problem-solving tool in many applications. Our work in this chapter centers on equations that contain two variables.

3.1   Reading Graphs, Plotting Points, and Scaling Graphs

Problem Solving with Bar, Circle, and Line Graphs   ▪   Points and Ordered Pairs   ▪   Numbering the Axes Appropriately

Today's print and electronic media make almost constant use of graphs. In this section, we consider problem solving with bar graphs, line graphs, and circle graphs. Then we examine graphs that use a coordinate system.

Problem Solving with Bar, Circle, and Line Graphs

A *bar graph* is a convenient way of showing comparisons. In every bar graph, certain categories, such as levels of education in the example below, are paired with certain numbers.

EXAMPLE   **1**

Lifetime earnings.   Getting a college degree usually means delaying the start of a career. As the bar graph below shows, this loss in earnings is more than made up over a worker's lifetime.

Source: U.S. Census Bureau

Studying Can Pay

Lifetime earnings estimates, full-time workers ages 25 to 64 (in millions)

$5, 4, 3, 2, 1

Less than high school | High school | Associate's degree | Bachelor's degree | Master's degree | Doctoral degree | Professional degree

Levels of education

a) Keagan plans to get an associate's degree. How much can he expect to make in his lifetime?

b) Isabella would like to make at least $2 million in her lifetime. What level of education should she pursue?

SOLUTION

a) Since level of education is shown on the horizontal scale, we go to the top of the bar above the label "associate's degree." Then we move horizontally from the top of the bar to the vertical scale, which shows earnings. We read there that Keagan can expect to make about $1.6 million in his lifetime.

b) By moving up the vertical scale to $2 million and then moving horizontally, we see that the first bar to reach a height of $2 million or higher corresponds to a bachelor's degree. Thus Isabella should pursue a bachelor's, master's, doctoral, or professional degree in order to make at least $2 million in her lifetime.

TRY EXERCISE ▶ 5

Circle graphs, or *pie charts,* are often used to show what percent of the whole each particular item in a group represents.

EXAMPLE 2

Student aid. The circle graph below shows the sources for student aid in 2006 and the percentage of aid students received from each source. In that year, the total amount of aid distributed was $134.8 billion. About 5,387,000 students received a federal Pell grant. What was the average amount of the aid per recipient?

Source: Trends in Student Aid 2006, www.collegeboard.com

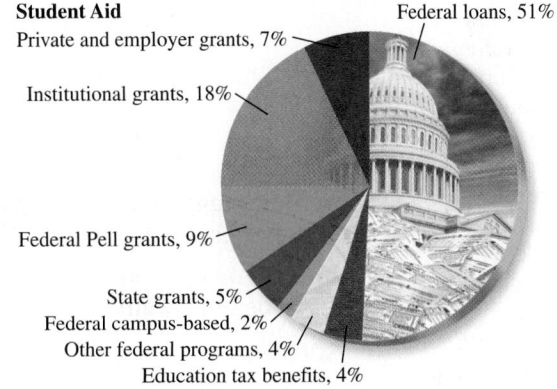

Student Aid
Private and employer grants, 7%
Federal loans, 51%
Institutional grants, 18%
Federal Pell grants, 9%
State grants, 5%
Federal campus-based, 2%
Other federal programs, 4%
Education tax benefits, 4%

SOLUTION

1. Familiarize. The problem involves percents, so if we were unsure of how to solve percent problems, we might review Section 2.4.

The solution of this problem will involve two steps. We are told the total amount of student aid distributed. In order to find the average amount of a Pell grant, we must first calculate the total of all Pell grants and then divide by the number of students.

We let g = the average amount of a Pell grant in 2006.

2. Translate. From the circle graph, we see that federal Pell grants were 9% of the total amount of aid. The total amount distributed was $134.8 billion, or $134,800,000,000, so we have

Find the value of all Pell grants.

$$\text{the value of all Pell grants} = 0.09(134,800,000,000)$$
$$= 12,132,000,000.$$

Then we reword the problem and translate as follows:

Calculate the average amount of a Pell grant.

Rewording: The average amount of a Pell grant is the value of all Pell grants divided by the number of recipients.

Translating: g = 12,132,000,000 ÷ 5,387,000

3. Carry out. We solve the equation:

$$g = 12,132,000,000 \div 5,387,000$$
$$\approx 2252. \quad \text{Rounding to the nearest dollar}$$

4. **Check.** If each student received $2252, the total amount of aid distributed through Pell grants would be $2252 · 5,387,000, or $12,131,524,000. Since this is approximately 9% of the total student aid for 2006, our answer checks.

5. **State.** In 2006, the average Pell grant was $2252. ▶ TRY EXERCISE ▶ 9

EXAMPLE **3**

Exercise and pulse rate. The following *line graph* shows the relationship between a person's resting pulse rate and months of regular exercise.* Note that the symbol ⌇ is used to indicate that counting on the vertical scale begins at 50.

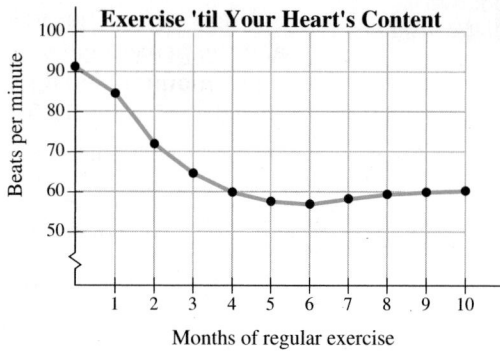

a) How many months of regular exercise are required to lower the pulse rate as much as possible?

b) How many months of regular exercise are needed to achieve a pulse rate of 65 beats per minute?

SOLUTION

a) The lowest point on the graph occurs above the number 6. Thus, after 6 months of regular exercise, the pulse rate is lowered as much as possible.

b) To determine how many months of exercise are needed to lower a person's resting pulse rate to 65, we locate 65 midway between 60 and 70 on the vertical scale. From that location, we move right until the line is reached. At that point, we move down to the horizontal scale and read the number of months required, as shown.

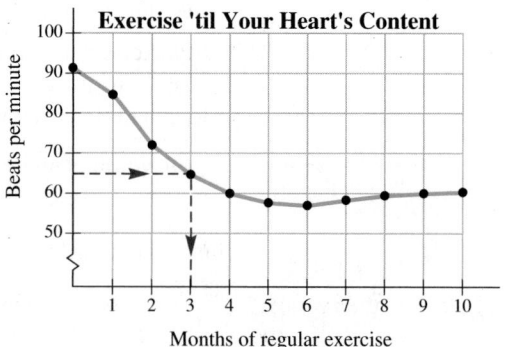

The pulse rate is 65 beats per minute after 3 months of regular exercise.

▶ TRY EXERCISE ▶ 17

---

*Data from *Body Clock* by Dr. Martin Hughes (New York: Facts on File, Inc.), p. 60.

TECHNOLOGY CONNECTION

The portion of the coordinate plane shown by a graphing calculator is called a **viewing window**. We indicate the **dimensions** of the window by setting a minimum *x*-value, a maximum *x*-value, a minimum *y*-value, and a maximum *y*-value. The **scale** by which we count must also be chosen. Window settings are often abbreviated in the form [L, R, B, T], with the letters representing **L**eft, **R**ight, **B**ottom, and **T**op endpoints. The window [−10, 10, −10, 10] is called the **standard viewing window**. On most graphing calculators, a standard viewing window can be set up using an option in the ZOOM menu.

To set up a [−100, 100, −5, 5] window, we press ⌞WINDOW⌟ and use the following settings. A scale of 10 for the *x*-axis is large enough for the marks on the *x*-axis to be distinct.

```
WINDOW
  Xmin=-100
  Xmax=100
  Xscl=10
  Ymin=-5
  Ymax=5
  Yscl=1
  Xres=1
```

When ⌞GRAPH⌟ is pressed, a graph extending from −100 to 100 along the *x*-axis (counted by 10's) and from −5 to 5 along the *y*-axis appears. When the arrow keys are pressed, a cursor can be

(continued)

moved, its coordinates appearing at the bottom of the window.

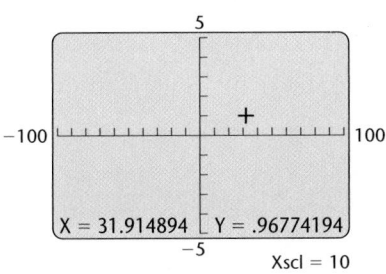

Xscl = 10

Set up the following viewing windows, choosing an appropriate scale for each axis. Then move the cursor and practice reading coordinates.

1. $[-10, 10, -10, 10]$
2. $[-5, 5, 0, 100]$
3. $[-1, 1, -0.1, 0.1]$

Points and Ordered Pairs

The line graph in Example 3 contains a collection of points. Each point pairs up a number of months of exercise with a pulse rate. To create such a graph, we **graph**, or **plot**, pairs of numbers on a plane. This is done using two perpendicular number lines called **axes** (pronounced "ak-sēz"; singular, **axis**). The point at which the axes cross is called the **origin**. Arrows on the axes indicate the positive directions.

Consider the pair $(3, 4)$. The numbers in such a pair are called **coordinates**. The **first coordinate** in this case is 3 and the **second coordinate** is 4.* To plot, or graph, $(3, 4)$, we start at the origin, move horizontally to the 3, move up vertically 4 units, and then make a "dot." Thus, $(3, 4)$ is located above 3 on the first axis and to the right of 4 on the second axis.

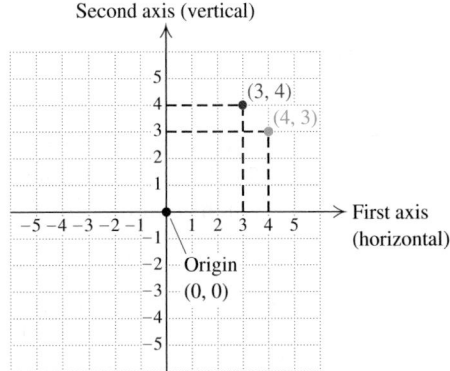

The point $(4, 3)$ is also plotted in the figure above. Note that $(3, 4)$ and $(4, 3)$ are different points. For this reason, coordinate pairs are called **ordered pairs**—the order in which the numbers appear is important.

EXAMPLE   4   Plot the point $(-3, 4)$.

SOLUTION   The first number, -3, is negative. Starting at the origin, we move 3 units in the negative horizontal direction (3 units to the left). The second number, 4, is positive, so we move 4 units in the positive vertical direction (up). The point $(-3, 4)$ is above -3 on the first axis and to the left of 4 on the second axis.

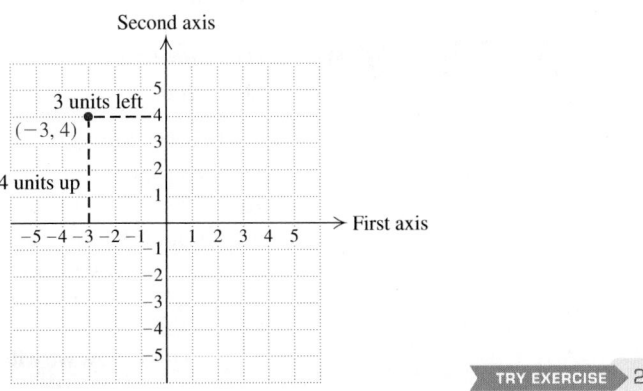

> TRY EXERCISE   21

---

*The first coordinate is called the *abscissa* and the second coordinate is called the *ordinate*. The plane is called the *Cartesian coordinate plane* after the French mathematician René Descartes (1595–1650).

To find the coordinates of a point, we see how far to the right or left of the origin the point is and how far above or below the origin it is. Note that the coordinates of the origin itself are $(0, 0)$.

EXAMPLE **5**

STUDENT NOTES ───────

It is important that you remember that the first coordinate of an ordered pair is always represented on the horizontal axis. It is better to go slowly and plot points correctly than to go quickly and require more than one attempt.

Find the coordinates of points *A*, *B*, *C*, *D*, *E*, *F*, and *G*.

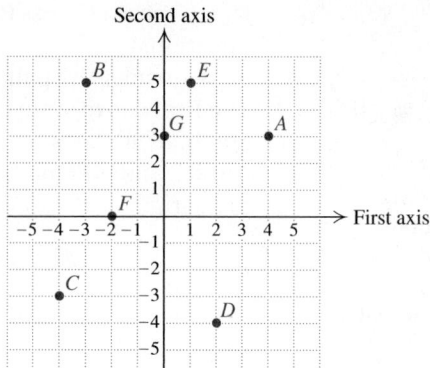

SOLUTION   Point *A* is 4 units to the right of the origin and 3 units above the origin. Its coordinates are $(4, 3)$. The coordinates of the other points are as follows:

B: $(-3, 5)$;   *C*: $(-4, -3)$;   *D*: $(2, -4)$;

E: $(1, 5)$;   *F*: $(-2, 0)$;   *G*: $(0, 3)$.

▶ **TRY EXERCISE** 27

The variables *x* and *y* are commonly used when graphing on a plane. Coordinates of ordered pairs are often labeled

(*x*-coordinate, *y*-coordinate).

The first, or horizontal, axis is labeled the *x*-axis, and the second, or vertical, axis is labeled the *y*-axis.

Numbering the Axes Appropriately

In Examples 4 and 5, each square on the grid shown is 1 unit long and 1 unit high: The **scale** of both the *x*-axis and the *y*-axis is 1. Often it is necessary to use a different scale on one or both of the axes.

EXAMPLE **6**

Use a grid 10 squares wide and 10 squares high to plot $(-34, 450)$, $(48, 95)$, and $(10, -200)$.

SOLUTION   Since *x*-coordinates vary from a low of -34 to a high of 48, the 10 horizontal squares must span $48 - (-34)$, or 82 units. Because 82 is not a multiple of 10, we round *up* to the next multiple of 10, which is 90. Dividing 90 by 10, we find that if each square is 9 units wide (has a scale of 9), we could represent all the *x*-values. However, since it is more convenient to count by 10's, we will instead use a scale of 10. Starting at 0, we count backward to -40 and forward to 60.

This is how we will arrange the *x*-axis.

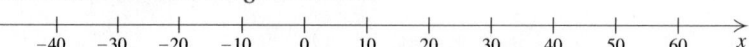

There is more than one correct way to cover the values from -34 to 48 using 10 increments. For instance, we could have counted from -60 to 90, using a scale

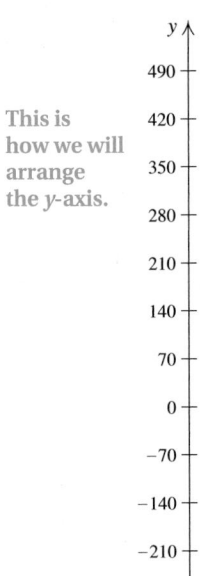

This is how we will arrange the y-axis.

of 15. In general, we try to use the smallest range and scale that will cover the given coordinates. Scales that are multiples of 2, 5, or 10 are especially convenient. It is essential that the numbering always begin at the origin.

Since we must be able to show y-values from -200 to 450, the 10 vertical squares must span $450 - (-200)$, or 650 units. For convenience, we round 650 *up* to 700 and then divide by 10: $700 \div 10 = 70$. Using 70 as the scale, we count *down* from 0 until we pass -200 and *up* from 0 until we pass 450, as shown at left.

Next, we combine our work with the x-values and the y-values to draw a graph in which the x-axis extends from -40 to 60 with a scale of 10 and the y-axis extends from -210 to 490 with a scale of 70. To correctly locate the axes on the grid, the two 0's must coincide where the axes cross. Finally, once the graph has been numbered, we plot the points as shown below.

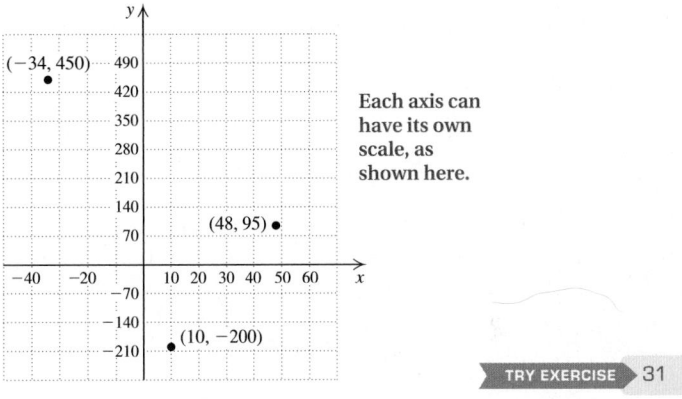

Each axis can have its own scale, as shown here.

TRY EXERCISE 31

The horizontal and vertical axes divide the plane into four regions, or **quadrants**, as indicated by Roman numerals in the following figure. Note that the point $(-4, 5)$ is in the second quadrant and the point $(5, -5)$ is in the fourth quadrant. The points $(3, 0)$ and $(0, 1)$ are on the axes and are not considered to be in any quadrant.

Second quadrant:
First coordinate negative, second coordinate positive:
$(-, +)$

First quadrant:
Both coordinates positive:
$(+, +)$

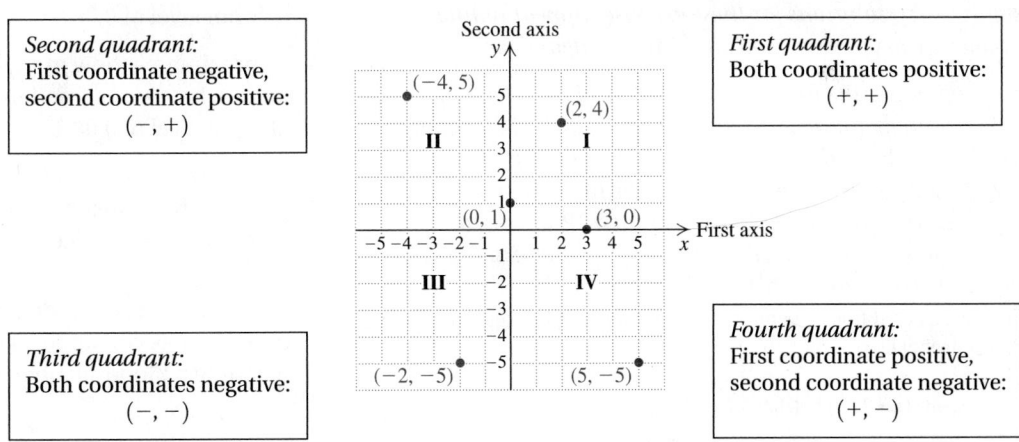

Third quadrant:
Both coordinates negative:
$(-, -)$

Fourth quadrant:
First coordinate positive, second coordinate negative:
$(+, -)$

3.1   EXERCISE SET

🖋 *Concept Reinforcement* *In each of Exercises 1–4, match the set of coordinates with the graph on the right that would be the best for plotting the points.*

1. ____ $(-9, 3), (-2, -1), (4, 5)$

2. ____ $(-2, -1), (1, 5), (7, 3)$

3. ____ $(-2, -9), (2, 1), (4, -6)$

4. ____ $(-2, -1), (-9, 3), (-4, -6)$

a)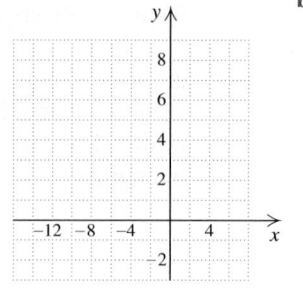

b)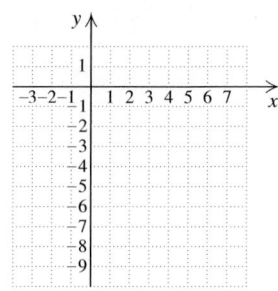

c)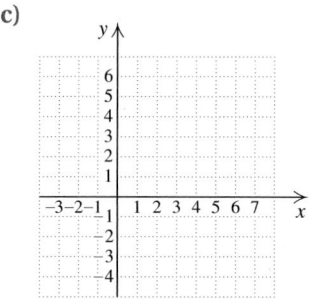

d)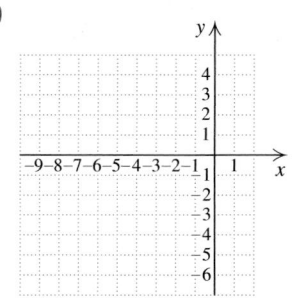

The ▸ TRY EXERCISES ▸ *for examples are indicated by a shaded block* ⬜ *on the exercise number. Complete step-by-step solutions for these exercises appear online at www.pearsonhighered.com/bittingerellenbogen.*

Driving under the influence. *A blood-alcohol level of 0.08% or higher makes driving illegal in the United States. This bar graph shows how many drinks a person of a certain weight would need to consume in 1 hr to achieve a blood-alcohol level of 0.08%. Note that a 12-oz beer, a 5-oz glass of wine, or a cocktail containing $1\frac{1}{2}$ oz of distilled liquor all count as one drink.*
Source: Adapted from soberup.com and vsa.vassar.edu/~source/drugs/alcohol.html

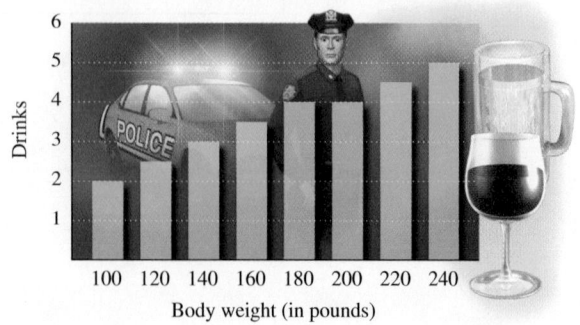

Friends Don't Let Friends Drive Drunk

5. Approximately how many drinks would a 100-lb person have consumed in 1 hr to reach a blood-alcohol level of 0.08%?

6. Approximately how many drinks would a 160-lb person have consumed in 1 hr to reach a blood-alcohol level of 0.08%?

7. What can you conclude about the weight of someone who has consumed 3 drinks in 1 hr without reaching a blood-alcohol level of 0.08%?

8. What can you conclude about the weight of someone who has consumed 4 drinks in 1 hr without reaching a blood-alcohol level of 0.08%?

Student aid. *Use the information in Example 2 to answer Exercises 9–12.*

9. In 2006, there were 13,334,170 full-time equivalent students in U.S. colleges and universities. What was the average federal loan per full-time equivalent student?

10. In 2006, there were 13,334,170 full-time equivalent students in U.S. colleges and universities. What was the average education tax benefit received per full-time equivalent student?

11. Approximately 8.6% of campus-based federal student aid is given to students in two-year public institutions. How much campus-based aid did students at two-year public institutions receive in 2006?

12. Approximately 17.7% of Pell grant dollars is given to students in for-profit institutions. How much did students in for-profit institutions receive in Pell grants in 2006?

Sorting solid waste.   Use the following pie chart to answer Exercises 13–16.

Sorting Solid Waste

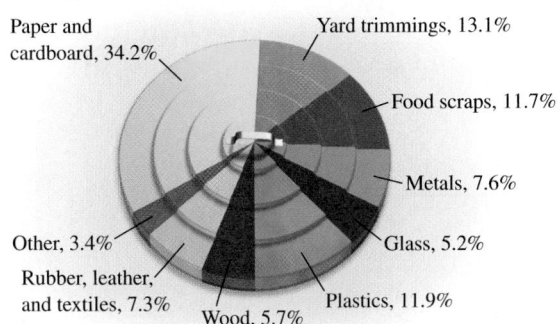

Source: Environmental Protection Agency

13. In 2005, Americans generated 245 million tons of waste. How much of the waste was plastic?

14. In 2005, the average American generated 4.5 lb of waste per day. How much of that was paper and cardboard?

15. Americans are recycling about 25.3% of all glass that is in the waste stream. How much glass did Americans recycle in 2005? (See Exercise 13.)

16. Americans are recycling about 61.9% of all yard trimmings. What amount of yard trimmings did the average American recycle per day in 2005? (Use the information in Exercise 14.)

Home video spending.   The line graph below shows U.S. consumer spending on home-video movies.

Home Video Spending

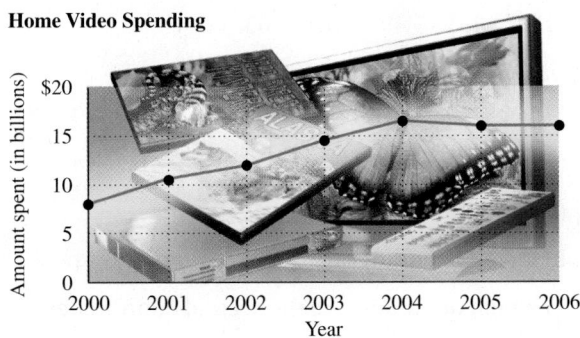

Source: Adams Media Research

17. Approximately how much was spent on home videos in 2002?

18. Approximately how much was spent on home videos in 2006?

19. In what year was approximately $10.5 billion spent on home videos?

20. In what year was approximately $14.5 billion spent on home videos?

Plot each group of points.

21. $(1, 2), (-2, 3), (4, -1), (-5, -3), (4, 0), (0, -2)$

22. $(-2, -4), (4, -3), (5, 4), (-1, 0), (-4, 4), (0, 5)$

23. $(4, 4), (-2, 4), (5, -3), (-5, -5), (0, 4), (0, -4),$ $(-4, 0), (0, 0)$

24. $(2, 5), (-1, 3), (3, -2), (-2, -4), (0, 0), (0, -5),$ $(5, 0), (-5, 0)$

25. *Text messaging.*   Listed below are estimates of the number of text messages sent in the United States. Make a line graph of the data.

Year	Monthly Text Messages (in millions)
2000	12
2001	34
2002	931
2003	1221
2004	2862
2005	7253
2006	8000 (estimated)

Source: CSCA

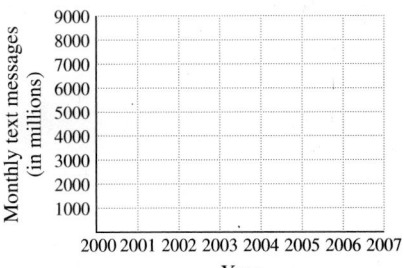

26. *Ozone layer.* Listed below are estimates of the ozone level. Make a line graph of the data, listing years on the horizontal scale.

Year	Ozone Level (in Dobson Units)
2000	287.1
2001	288.2
2002	285.8
2003	285.0
2004	281.2
2005	283.5

Source: johnstonsarchive.net

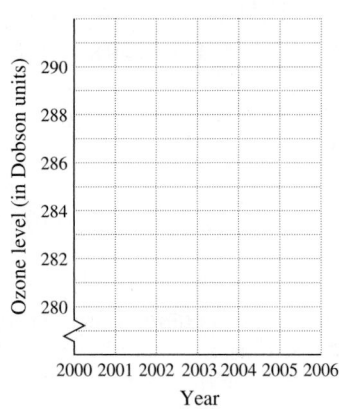

In Exercises 27–30, find the coordinates of points A, B, C, D, and E.

27.

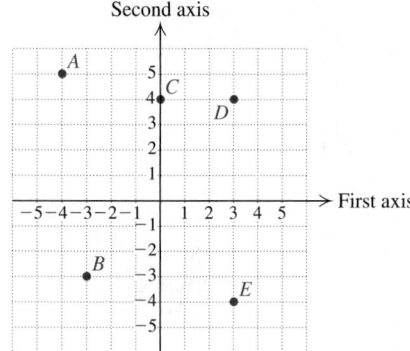

28.

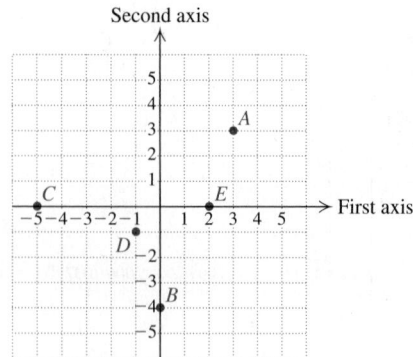

29.

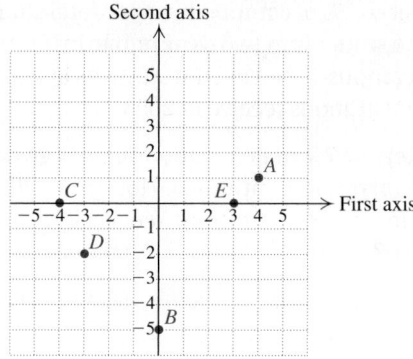

30.

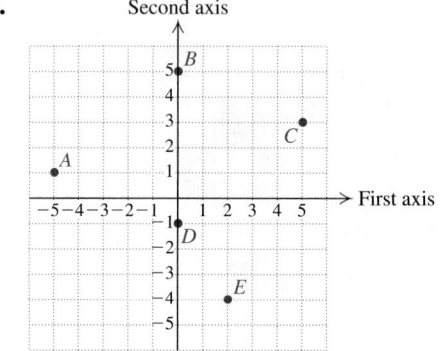

In Exercises 31–40, use a grid 10 squares wide and 10 squares high to plot the given coordinates. Choose your scale carefully. Scales may vary.

31. $(-75, 5), (-18, -2), (9, -4)$

32. $(-13, 3), (48, -1), (62, -4)$

33. $(-1, 83), (-5, -14), (5, 37)$

34. $(2, -79), (4, -25), (-4, 12)$

35. $(-10, -4), (-16, 7), (3, 15)$

36. $(5, -16), (-7, -4), (12, 3)$

37. $(-100, -5), (350, 20), (800, 37)$

38. $(750, -8), (-150, 17), (400, 32)$

39. $(-83, 491), (-124, -95), (54, -238)$

40. $(738, -89), (-49, -6), (-165, 53)$

In which quadrant is each point located?

41. $(7, -2)$ **42.** $(-1, -4)$ **43.** $(-4, -3)$

44. $(1, -5)$ **45.** $(2, 1)$ **46.** $(-4, 6)$

47. $(-4.9, 8.3)$ **48.** $(7.5, 2.9)$

49. In which quadrants are the first coordinates positive?

50. In which quadrants are the second coordinates negative?

51. In which quadrants do both coordinates have the same sign?

52. In which quadrants do the first and second coordinates have opposite signs?

53. The following graph was included in a mailing sent by Agway® to their oil customers in 2000. What information is missing from the graph and why is the graph misleading?

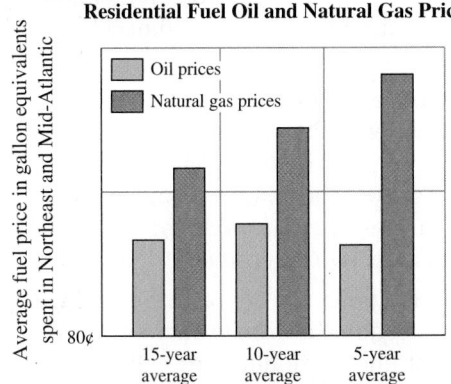

Residential Fuel Oil and Natural Gas Prices

Source: Energy Research Center, Inc. *3/1/99–2/29/00

54. What do all points plotted on the vertical axis of a graph have in common?

Skill Review

To prepare for Section 3.2, review solving for a variable (Section 2.3).

Solve for y. [2.3]

55. $5y = 2x$

56. $2y = -3x$

57. $x - y = 8$

58. $2x + 5y = 10$

59. $2x + 3y = 5$

60. $5x - 8y = 1$

Synthesis

61. In an article about consumer spending on home videos (see the graph used for Exercise 17), the *Wall Street Journal* (9/2/06) stated that "the movie industry has hit a wall." To what were they referring?

62. Describe what the result would be if the first and second coordinates of every point in the following graph of an arrow were interchanged.

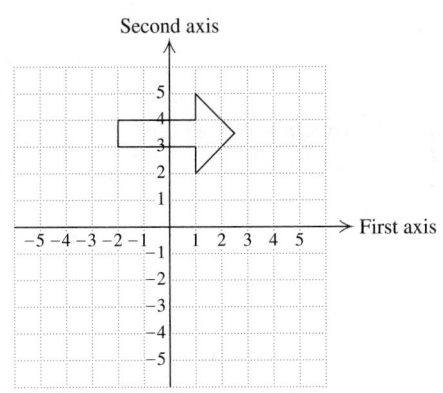

63. In which quadrant(s) could a point be located if its coordinates are opposites of each other?

64. In which quadrant(s) could a point be located if its coordinates are reciprocals of each other?

65. The points $(-1, 1)$, $(4, 1)$, and $(4, -5)$ are three vertices of a rectangle. Find the coordinates of the fourth vertex.

66. The pairs $(-2, -3)$, $(-1, 2)$, and $(4, -3)$ can serve as three (of four) vertices for three different parallelograms. Find the fourth vertex of each parallelogram.

67. Graph eight points such that the sum of the coordinates in each pair is 7. Answers may vary.

68. Find the perimeter of a rectangle if three of its vertices are $(5, -2)$, $(-3, -2)$, and $(-3, 3)$.

69. Find the area of a triangle whose vertices have coordinates $(0, 9)$, $(0, -4)$, and $(5, -4)$.

Coordinates on the globe.   Coordinates can also be used to describe the location on a sphere: 0° latitude is the equator and 0° longitude is a line from the North Pole to the South Pole through France and Algeria. In the figure shown here, hurricane Clara is at a point about 260 mi northwest of Bermuda near latitude 36.0° North, longitude 69.0° West.

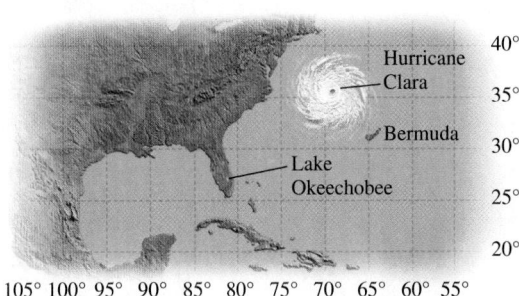

70. Approximate the latitude and the longitude of Bermuda.

71. Approximate the latitude and the longitude of Lake Okeechobee.

72. The graph accompanying Example 3 flattens out. Why do you think this occurs?

73. In the *Star Trek* science-fiction series, a three-dimensional coordinate system is used to locate objects in space. If the center of a planet is used as the origin, how many "quadrants" will exist? Why? If possible, sketch a three-dimensional coordinate system and label each "quadrant."

COLLABORATIVE

CORNER

You Sank My Battleship!

Focus: Graphing points; logical questioning

Time: 15–25 minutes

Group size: 3–5

Materials: Graph paper

In the game Battleship®, a player places a miniature ship on a grid that only that player can see. An opponent guesses at coordinates that might "hit" the "hidden" ship. The following activity is similar to this game.

ACTIVITY

1. Using only integers from −10 to 10 (inclusive), one group member should secretly record the coordinates of a point on a slip of paper. (This point is the hidden "battleship.")

2. The other group members can then ask up to 10 "yes/no" questions in an effort to determine the coordinates of the secret point. Be sure to phrase each question mathematically (for example, "Is the *x*-coordinate negative?")

3. The group member who selected the point should answer each question. On the basis of the answer given, another group member should cross out the points no longer under consideration. All group members should check that this is done correctly.

4. If the hidden point has not been determined after 10 questions have been answered, the secret coordinates should be revealed to all group members.

5. Repeat parts (1)–(4) until each group member has had the opportunity to select the hidden point and answer questions.

3.2 Graphing Linear Equations

Solutions of Equations • Graphing Linear Equations • Applications

We have seen how bar, line, and circle graphs can represent information. Now we begin to learn how graphs can be used to represent solutions of equations.

Solutions of Equations

When an equation contains two variables, solutions are ordered pairs in which each number in the pair replaces a letter in the equation. Unless stated otherwise, the first number in each pair replaces the variable that occurs first alphabetically.

EXAMPLE **1** Determine whether each of the following pairs is a solution of $4b - 3a = 22$: **(a)** $(2, 7)$; **(b)** $(1, 6)$.

SOLUTION

a) We substitute 2 for a and 7 for b (alphabetical order of variables):

$$\begin{array}{c|c} 4b - 3a = 22 \\ \hline 4(7) - 3(2) & 22 \\ 28 - 6 & \\ & 22 \overset{?}{=} 22 \quad \text{TRUE} \end{array}$$

Since $22 = 22$ is *true*, the pair $(2, 7)$ *is* a solution.

b) In this case, we replace a with 1 and b with 6:

$$\begin{array}{c|c} 4b - 3a = 22 \\ \hline 4(6) - 3(1) & 22 \\ 24 - 3 & \\ & 21 \overset{?}{=} 22 \quad \text{FALSE} \qquad 21 \neq 22 \end{array}$$

Since $21 = 22$ is *false*, the pair $(1, 6)$ is *not* a solution.   **TRY EXERCISE** ▶ 7

EXAMPLE ▪ **2**   Show that the pairs $(3, 7)$, $(0, 1)$, and $(-3, -5)$ are solutions of $y = 2x + 1$. Then graph the three points to determine another pair that is a solution.

SOLUTION   To show that a pair is a solution, we substitute, replacing x with the first coordinate and y with the second coordinate of each pair:

$$\begin{array}{c|c} y = 2x + 1 \\ \hline 7 & 2 \cdot 3 + 1 \\ & 6 + 1 \\ 7 \overset{?}{=} 7 & \text{TRUE} \end{array} \qquad \begin{array}{c|c} y = 2x + 1 \\ \hline 1 & 2 \cdot 0 + 1 \\ & 0 + 1 \\ 1 \overset{?}{=} 1 & \text{TRUE} \end{array} \qquad \begin{array}{c|c} y = 2x + 1 \\ \hline -5 & 2(-3) + 1 \\ & -6 + 1 \\ -5 \overset{?}{=} -5 & \text{TRUE} \end{array}$$

In each of the three cases, the substitution results in a true equation. Thus the pairs $(3, 7)$, $(0, 1)$, and $(-3, -5)$ are all solutions. We graph them as shown at left.

Note that the three points appear to "line up." Will other points that line up with these points also represent solutions of $y = 2x + 1$? To find out, we use a ruler and draw a line passing through $(-3, -5)$, $(0, 1)$, and $(3, 7)$.

The line appears to pass through $(2, 5)$. Let's check to see if this pair is a solution of $y = 2x + 1$:

$$\begin{array}{c|c} y = 2x + 1 \\ \hline 5 & 2 \cdot 2 + 1 \\ & 4 + 1 \\ 5 \overset{?}{=} 5 & \text{TRUE} \end{array}$$

We see that $(2, 5)$ *is* a solution. You should perform a similar check for at least one other point that appears to be on the line.   **TRY EXERCISE** ▶ 13

Example 2 leads us to suspect that *any* point on the line passing through $(3, 7)$, $(0, 1)$, and $(-3, -5)$ represents a solution of $y = 2x + 1$. In fact, every solution of $y = 2x + 1$ is represented by a point on this line and every point on this line represents a solution. The line is called the **graph** of the equation.

Graphing Linear Equations

Equations like $y = 2x + 1$ or $4b - 3a = 22$ are said to be **linear** because the graph of each equation is a line. In general, any equation that can be written in the form $y = mx + b$ or $Ax + By = C$ (where m, b, A, B, and C are constants and A and B are not both 0) is linear.

To *graph* an equation is to make a drawing that represents its solutions. Linear equations can be graphed as follows.

To Graph a Linear Equation

1. Select a value for one coordinate and calculate the corresponding value of the other coordinate. Form an ordered pair. This pair is one solution of the equation.
2. Repeat step (1) to find a second ordered pair. A third ordered pair should be found to use as a check.
3. Plot the ordered pairs and draw a straight line passing through the points. The line represents all solutions of the equation.

EXAMPLE **3**

Graph: $y = -3x + 1$.

SOLUTION Since $y = -3x + 1$ is in the form $y = mx + b$, the equation is linear and the graph is a straight line. We select a convenient value for x, compute y, and form an ordered pair. Then we repeat the process for other choices of x.

If $x = 2$, then $y = -3 \cdot 2 + 1 = -5$, and $(2, -5)$ is a solution.
If $x = 0$, then $y = -3 \cdot 0 + 1 = 1$, and $(0, 1)$ is a solution.
If $x = -1$, then $y = -3(-1) + 1 = 4$, and $(-1, 4)$ is a solution.

Results are often listed in a table, as shown below. The points corresponding to each pair are then plotted.

Calculate ordered pairs.

$$y = -3x + 1$$

x	y	(x, y)
2	−5	$(2, -5)$
0	1	$(0, 1)$
−1	4	$(-1, 4)$

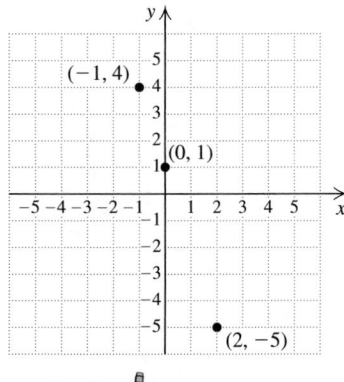

(1) Choose x.
(2) Compute y.
(3) Form the pair (x, y).
(4) Plot the points.

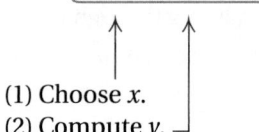

Plot the points.

Note that all three points line up. If they didn't, we would know that we had made a mistake, because the equation is linear. When only two points are plotted, an error is more difficult to detect.

Draw the graph.

Finally, we use a ruler or other straight-edge to draw a line. We add arrowheads to the ends of the line to indicate that it extends indefinitely beyond the edge of the grid drawn. Every point on the line represents a solution of $y = -3x + 1$.

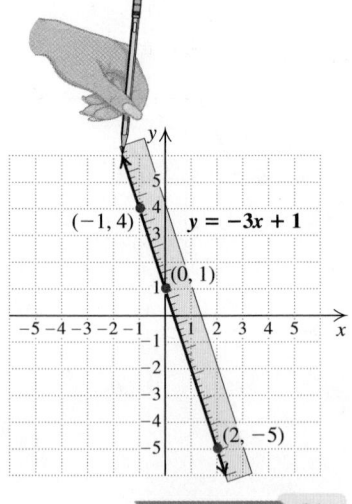

TRY EXERCISE 21

EXAMPLE 4   Graph: $y = 2x - 3$.

SOLUTION   We select some convenient x-values and compute y-values.

If $x = 0$,   then $y = 2 \cdot 0 - 3 = -3$,   and $(0, -3)$ is a solution.

If $x = 1$,   then $y = 2 \cdot 1 - 3 = -1$,   and $(1, -1)$ is a solution.

If $x = 4$,   then $y = 2 \cdot 4 - 3 = 5$,   and $(4, 5)$ is a solution.

$y = 2x - 3$

x	y	(x, y)
0	−3	$(0, -3)$
1	−1	$(1, -1)$
4	5	$(4, 5)$

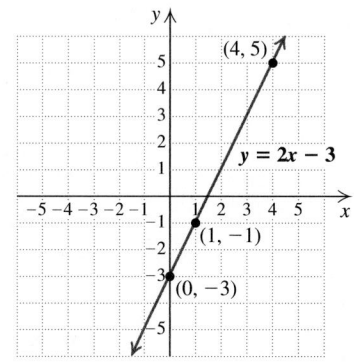

TRY EXERCISE 27

EXAMPLE 5   Graph: $4x + 2y = 12$.

SOLUTION   To form ordered pairs, we can replace either variable with a number and then calculate the other coordinate:

If $y = 0$, we have $4x + 2 \cdot 0 = 12$
$$4x = 12$$
$$x = 3,$$

so $(3, 0)$ is a solution.

If $x = 0$, we have $4 \cdot 0 + 2y = 12$
$$2y = 12$$
$$y = 6,$$

so $(0, 6)$ is a solution.

If $y = 2$, we have $4x + 2 \cdot 2 = 12$
$$4x + 4 = 12$$
$$4x = 8$$
$$x = 2,$$

so $(2, 2)$ is a solution.

$4x + 2y = 12$

x	y	(x, y)
3	0	$(3, 0)$
0	6	$(0, 6)$
2	2	$(2, 2)$

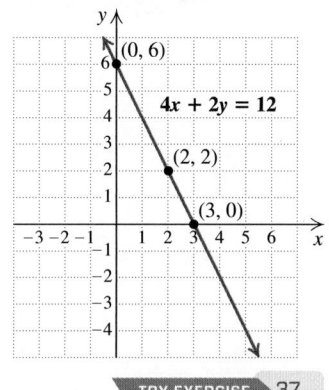

TRY EXERCISE 37

Note that in Examples 3 and 4 the variable y is isolated on one side of the equation. This generally simplifies calculations, so it is important to be able to solve for y before graphing.

EXAMPLE **6**

Graph $3y = 2x$ by first solving for y.

SOLUTION To isolate y, we divide both sides by 3, or multiply both sides by $\frac{1}{3}$:

$$3y = 2x$$
$$\frac{1}{3} \cdot 3y = \frac{1}{3} \cdot 2x \qquad \text{Using the multiplication principle to}$$
$$\text{multiply both sides by } \tfrac{1}{3}$$
$$\left.\begin{array}{l} 1y = \frac{2}{3} \cdot x \\ y = \frac{2}{3}x. \end{array}\right\} \quad \text{Simplifying}$$

Because all the equations above are equivalent, we can use $y = \frac{2}{3}x$ to draw the graph of $3y = 2x$.

To graph $y = \frac{2}{3}x$, we can select x-values that are multiples of 3. This will allow us to avoid fractions when the corresponding y-values are computed.

$$\left.\begin{array}{ll} \text{If } x = 3, & \text{then } y = \frac{2}{3} \cdot 3 = 2. \\ \text{If } x = -3, & \text{then } y = \frac{2}{3}(-3) = -2. \\ \text{If } x = 6, & \text{then } y = \frac{2}{3} \cdot 6 = 4. \end{array}\right\} \quad \begin{array}{l} \text{Note that when multiples of 3 are} \\ \text{substituted for } x, \text{ the } y\text{-coordinates} \\ \text{are not fractions.} \end{array}$$

The following table lists these solutions. Next, we plot the points and see that they form a line. Finally, we draw and label the line.

$$3y = 2x, \ \text{ or } \ y = \frac{2}{3}x$$

x	y	(x, y)
3	2	$(3, 2)$
-3	-2	$(-3, -2)$
6	4	$(6, 4)$

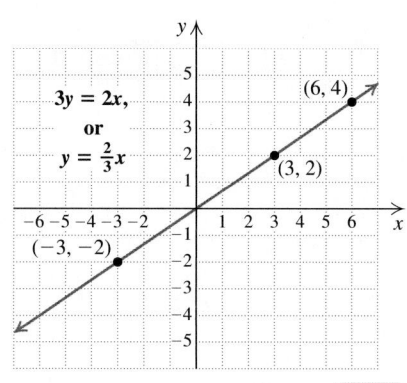

TRY EXERCISE 41

EXAMPLE **7**

Graph $x + 5y = -10$ by first solving for y.

SOLUTION We have

$$x + 5y = -10$$
$$5y = -x - 10 \qquad \text{Adding } -x \text{ to both sides}$$
$$y = \frac{1}{5}(-x - 10) \qquad \text{Multiplying both sides by } \tfrac{1}{5}$$
$$y = -\frac{1}{5}x - 2. \qquad \text{Using the distributive law}$$

> *CAUTION!* It is very important to multiply *both* $-x$ and -10 by $\frac{1}{5}$.

Thus, $x + 5y = -10$ is equivalent to $y = -\frac{1}{5}x - 2$. It is important to note that if we now choose x-values that are multiples of 5, we can avoid fractions when calculating the corresponding y-values.

$$\text{If } x = 5, \quad \text{then } y = -\frac{1}{5} \cdot 5 - 2 = -1 - 2 = -3.$$
$$\text{If } x = 0, \quad \text{then } y = -\frac{1}{5} \cdot 0 - 2 = 0 - 2 = -2.$$
$$\text{If } x = -5, \quad \text{then } y = -\frac{1}{5}(-5) - 2 = 1 - 2 = -1.$$

$$x + 5y = -10, \text{ or } y = -\frac{1}{5}x - 2$$

x	y	(x, y)
5	-3	$(5, -3)$
0	-2	$(0, -2)$
-5	-1	$(-5, -1)$

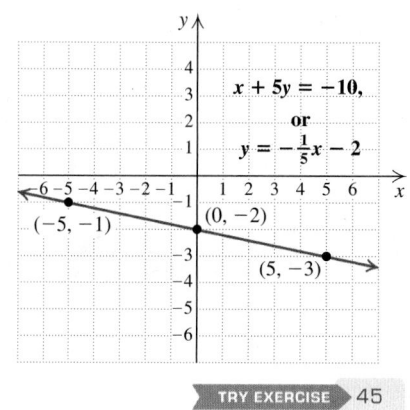

TRY EXERCISE ▶ 45

Applications

Linear equations appear in many real-life situations.

EXAMPLE 8

Fuel efficiency.   A typical tractor-trailer will move 18 tons of air per mile at 55 mph. Air resistance increases with speed, causing fuel efficiency to decrease at higher speeds. At highway speeds, a certain truck's fuel efficiency t, in miles per gallon (mpg), can be given by

$$t = -0.1s + 13.1,$$

where s is the speed of the truck, in miles per hour (mph). Graph the equation and then use the graph to estimate the fuel efficiency at 66 mph.

Source: Based on data from Kenworth Truck Co.

SOLUTION   We graph $t = -0.1s + 13.1$ by first selecting values for s and then calculating the associated values t. Since the equation is true for highway speeds, we use $s \geq 50$.

s	t
50	8.1
60	7.1
70	6.1

If $s = 50$,   then $t = -0.1(50) + 13.1 = 8.1$.
If $s = 60$,   then $t = -0.1(60) + 13.1 = 7.1$.
If $s = 70$,   then $t = -0.1(70) + 13.1 = 6.1$.

Because we are *selecting* values for s and *calculating* values for t, we represent s on the horizontal axis and t on the vertical axis. Counting by 5's horizontally, beginning at 50, and by 0.5 vertically, beginning at 4, will allow us to plot all three pairs, as shown below.

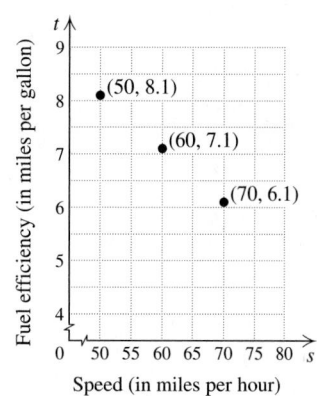

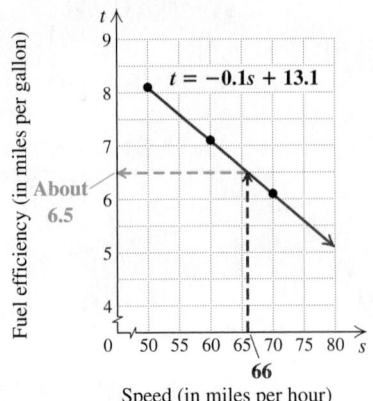

Fuel efficiency (in miles per gallon)

$t = -0.1s + 13.1$

About 6.5

Speed (in miles per hour)

66

Since the three points line up, our calculations are probably correct. We draw a line, beginning at (50, 8.1). To estimate the fuel efficiency at 66 mph, we locate the point on the line that is above 66 and then find the value on the *t*-axis that corresponds to that point, as shown at left. The fuel efficiency at 66 mph is about 6.5 mpg.

▶ TRY EXERCISE 49

CAUTION! When the coordinates of a point are read from a graph, as in Example 8, values should not be considered exact.

Many equations in two variables have graphs that are not straight lines. Three such graphs are shown below. As before, each graph represents the solutions of the given equation. Graphing calculators are especially helpful when drawing these *nonlinear* graphs. Nonlinear graphs are studied later in this text and in more advanced courses.

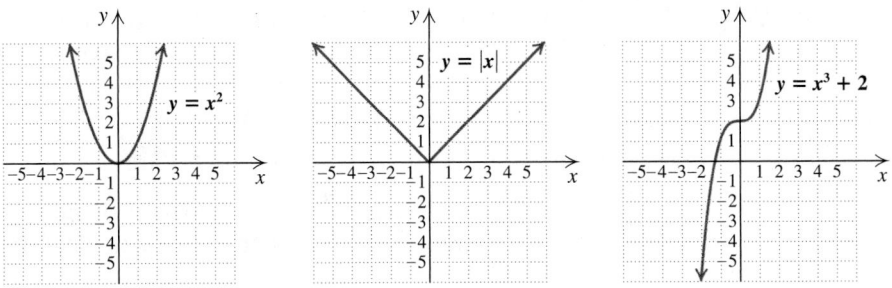

$y = x^2$ $y = |x|$ $y = x^3 + 2$

TECHNOLOGY CONNECTION

Most graphing calculators require that *y* be alone on one side before the equation is entered. For example, to graph $5y + 4x = 13$, we would first solve for *y*. The student can check that solving for *y* yields the equation $y = -\frac{4}{5}x + \frac{13}{5}$.

We press Y= , enter $-\frac{4}{5}x + \frac{13}{5}$ as Y1, and press GRAPH . The graph is shown here in the standard viewing window $[-10, 10, -10, 10]$.

Using a graphing calculator, graph each of the following. Select the "standard" $[-10, 10, -10, 10]$ window.

1. $y = -5x + 6.5$ **2.** $y = 3x + 4.5$
3. $7y - 4x = 22$ **4.** $5y + 11x = -20$
5. $2y - x^2 = 0$ **6.** $y + x^2 = 8$

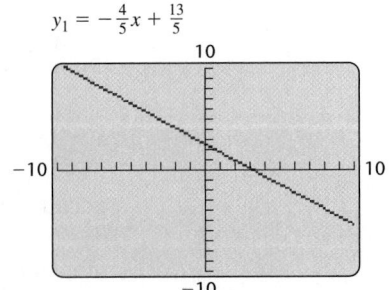

$y_1 = -\frac{4}{5}x + \frac{13}{5}$

🖑 **Concept Reinforcement**   *Classify each statement as either true or false.*

1. A linear equation in two variables has at most one solution.

2. Every solution of $y = 3x - 7$ is an ordered pair.

3. The graph of $y = 3x - 7$ represents all solutions of the equation.

4. If a point is on the graph of $y = 3x - 7$, the corresponding ordered pair is a solution of the equation.

5. To find a solution of $y = 3x - 7$, we can choose any value for x and calculate the corresponding value for y.

6. The graph of every equation is a straight line.

Determine whether each equation has the given ordered pair as a solution.

7. $y = 4x - 7$; $(2, 1)$

8. $y = 5x + 8$; $(0, 8)$

9. $3y + 4x = 19$; $(5, 1)$

10. $5x - 3y = 15$; $(0, 5)$

11. $4m - 5n = 7$; $(3, -1)$

12. $3q - 2p = -8$; $(1, -2)$

In Exercises 13–20, an equation and two ordered pairs are given. Show that each pair is a solution of the equation. Then graph the two pairs to determine another solution. Answers may vary.

13. $y = x + 3$; $(-1, 2), (4, 7)$

14. $y = x - 2$; $(3, 1), (-2, -4)$

15. $y = \frac{1}{2}x + 3$; $(4, 5), (-2, 2)$

16. $y = \frac{1}{2}x - 1$; $(6, 2), (0, -1)$

17. $y + 3x = 7$; $(2, 1), (4, -5)$

18. $2y + x = 5$; $(-1, 3), (7, -1)$

19. $4x - 2y = 10$; $(0, -5), (4, 3)$

20. $6x - 3y = 3$; $(1, 1), (-1, -3)$

Graph each equation.

21. $y = x + 1$

22. $y = x - 1$

23. $y = -x$

24. $y = x$

25. $y = 2x$

26. $y = -3x$

27. $y = 2x + 2$

28. $y = 3x - 2$

29. $y = -\frac{1}{2}x$

30. $y = \frac{1}{4}x$

31. $y = \frac{1}{3}x - 4$

32. $y = \frac{1}{2}x + 1$

33. $x + y = 4$

34. $x + y = -5$

35. $x - y = -2$

36. $y - x = 3$

37. $x + 2y = -6$

38. $x + 2y = 8$

39. $y = -\frac{2}{3}x + 4$

40. $y = \frac{3}{2}x + 1$

41. $4x = 3y$

42. $2x = 5y$

43. $5x - y = 0$

44. $3x - 5y = 0$

45. $6x - 3y = 9$

46. $8x - 4y = 12$

47. $6y + 2x = 8$

48. $8y + 2x = -4$

49. **Student aid.**   The average award a of federal student financial assistance per student is approximated by
$$a = 0.08t + 2.5,$$
where a is in thousands of dollars and t is the number of years since 1994. Graph the equation and use the graph to estimate the average amount of federal student aid per student in 2010.
Source: Based on data from U.S. Department of Education, Office of Postsecondary Education

50. **Value of a color copier.**   The value of Dupliographic's color copier is given by
$$v = -0.68t + 3.4,$$
where v is the value, in thousands of dollars, t years from the date of purchase. Graph the equation and use the graph to estimate the value of the copier after $2\frac{1}{2}$ yr.

51. **FedEx mailing costs.**   Recently, the cost c, in dollars, of shipping a FedEx Priority Overnight package weighing 1 lb or more a distance of 1001 to 1400 mi was given by
$$c = 3.1w + 29.07,$$
where w is the package's weight, in pounds. Graph the equation and use the graph to estimate the cost of shipping a $6\frac{1}{2}$-lb package.
Source: Based on data from FedEx.com

52. **Increasing life expectancy.**   A smoker is 15 times more likely to die of lung cancer than a nonsmoker. An ex-smoker who stopped smoking t years ago is

w times more likely to die of lung cancer than a nonsmoker, where

$$w = 15 - t.$$

Graph the equation and use the graph to estimate how much more likely it is for Sandy to die of lung cancer than Polly, if Polly never smoked and Sandy quit $2\frac{1}{2}$ yr ago.
Source: Data from *Body Clock* by Dr. Martin Hughes, p. 60. New York: Facts on File, Inc.

53. *Scrapbook pricing.* The price *p*, in dollars, of an 8-in. by 8-in. assembled scrapbook is given by

$$p = 3.5n + 9,$$

where *n* is the number of pages in the scrapbook. Graph the equation and use the graph to estimate the price of a scrapbook containing 25 pages.
Source: www.scrapbooksplease.com

54. *Value of computer software.* The value *v* of a shop-keeper's inventory software program, in hundreds of dollars, is given by

$$v = -\tfrac{3}{4}t + 6,$$

where *t* is the number of years since the shopkeeper first bought the program. Graph the equation and use the graph to estimate what the program is worth 4 yr after it was first purchased.

55. *Bottled water.* The number of gallons of bottled water *w* consumed by the average American in one year is given by

$$w = 1.6t + 16.7,$$

where *t* is the number of years since 2000. Graph the equation and use the graph to predict the number of gallons consumed by the average American in 2010.
Source: Based on data from Beverage Marketing Corporation

56. *Record temperature drop.* On January 22, 1943, the temperature *T*, in degrees Fahrenheit, in Spearfish, South Dakota, could be approximated by

$$T = -2m + 54,$$

where *m* is the number of minutes since 9:00 A.M. that morning. Graph the equation and use the graph to estimate the temperature at 9:15 A.M.
Source: Based on information from the National Oceanic Atmospheric Administration

57. *Cost of college.* The cost *T*, in hundreds of dollars, of tuition and fees at many community colleges can be approximated by

$$T = \tfrac{5}{4}c + 2,$$

where *c* is the number of credits for which a student registers. Graph the equation and use the graph to estimate the cost of tuition and fees when a student registers for 4 three-credit courses.

Bartonville Community College
Ferndell Hall

Tuition $125 / credit
Fees
Registration $85
Transcripts $55
Computer $45
Activities $15

Activities include: Math team, fencing, drama club

58. *Cost of college.* The cost *C*, in thousands of dollars, of a year at a private four-year college (all expenses) can be approximated by

$$C = \tfrac{13}{10}t + 21,$$

where *t* is the number of years since 1995. Graph the equation and use the graph to predict the cost of a year at a private four-year college in 2012.
Source: Based on information in *Statistical Abstract of the United States,* 2007

59. The equations $3x + 4y = 8$ and $y = -\tfrac{3}{4}x + 2$ are equivalent. Which equation would be easier to graph and why?

60. Suppose that a linear equation is graphed by plotting three points and that the three points line up with each other. Does this *guarantee* that the equation is being correctly graphed? Why or why not?

Skill Review

Review solving equations and formulas (Sections 2.2 and 2.3).

Solve and check. [2.2]

61. $5x + 3 \cdot 0 = 12$

62. $3 \cdot 0 - 8y = 6$

63. $5x + 3(2 - x) = 12$

64. $3(y - 5) - 8y = 6$

Solve. [2.3]

65. $A = \dfrac{T + Q}{2}$, for *Q*

66. $pq + p = w$, for *p*

67. $Ax + By = C$, for *y*

68. $\dfrac{y - k}{m} = x - h$, for *y*

Synthesis

69. Janice consistently makes the mistake of plotting the x-coordinate of an ordered pair using the y-axis, and the y-coordinate using the x-axis. How will Janice's incorrect graph compare with the appropriate graph?

70. Explain how the graph in Example 8 can be used to determine the speed for which the fuel efficiency is 6 mpg.

71. *Bicycling.* Long Beach Island in New Jersey is a long, narrow, flat island. For exercise, Laura routinely bikes to the northern tip of the island and back. Because of the steady wind, she uses one gear going north and another for her return. Laura's bike has 21 gears and the sum of the two gears used on her ride is always 24. Write and graph an equation that represents the different pairings of gears that Laura uses. Note that there are no fraction gears on a bicycle.

In Exercises 72–75, try to find an equation for the graph shown.

72.

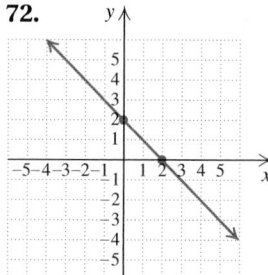

73.

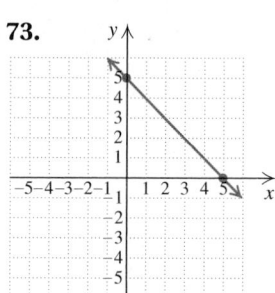

74.

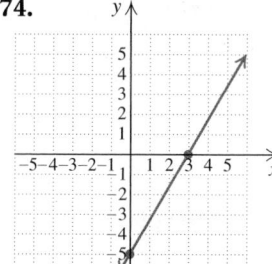

75.

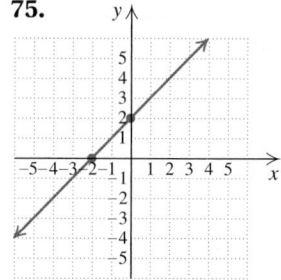

76. Translate to an equation:

d dimes and n nickels total $1.75.

Then graph the equation and use the graph to determine three different combinations of dimes and nickels that total $1.75 (see also Exercise 90).

77. Translate to an equation:

d $25 dinners and l $5 lunches total $225.

Then graph the equation and use the graph to determine three different combinations of lunches and dinners that total $225 (see also Exercise 90).

Use the suggested x-values $-3, -2, -1, 0, 1, 2,$ and 3 to graph each equation.

78. $y = |x|$

Aha! **79.** $y = -|x|$

Aha! **80.** $y = |x| - 2$

81. $y = x^2$

82. $y = x^2 + 1$

For Exercises 83–88, use a graphing calculator to graph the equation. Use a $[-10, 10, -10, 10]$ window.

83. $y = -2.8x + 3.5$

84. $y = 4.5x + 2.1$

85. $y = 2.8x - 3.5$

86. $y = -4.5x - 2.1$

87. $y = x^2 + 4x + 1$

88. $y = -x^2 + 4x - 7$

89. Example 8 discusses fuel efficiency. If fuel costs $3.50 a gallon, how much money will a truck driver save on a 500-mi trip by driving at 55 mph instead of 70 mph? How many gallons of fuel will be saved?

90. Study the graph of Exercises 76 and 77. Does *every* point on the graph represent a solution of the associated problem? Why or why not?

3.3 Graphing and Intercepts

Intercepts ▪ Using Intercepts to Graph ▪ Graphing Horizontal or Vertical Lines

Unless a line is horizontal or vertical, it will cross both axes. Often, finding the points where the axes are crossed gives us a quick way of graphing linear equations.

Intercepts

In Example 5 of Section 3.2, we graphed $4x + 2y = 12$ by plotting the points $(3, 0)$, $(0, 6)$, and $(2, 2)$ and then drawing the line.

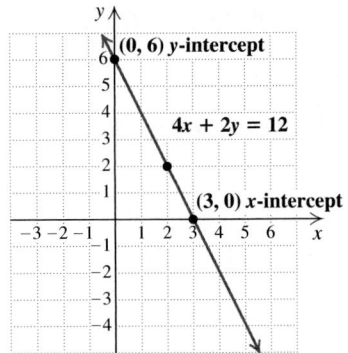

- The point at which a graph crosses the y-axis is called the **y-intercept**. In the figure above, the y-intercept is $(0, 6)$. The x-coordinate of a y-intercept is always 0.

- The point at which a graph crosses the x-axis is called the **x-intercept**. In the figure above, the x-intercept is $(3, 0)$. The y-coordinate of an x-intercept is always 0.

It is possible for the graph of a curve to have more than one y-intercept or more than one x-intercept.

EXAMPLE 1 For the graph shown below, **(a)** give the coordinates of any x-intercepts and **(b)** give the coordinates of any y-intercepts.

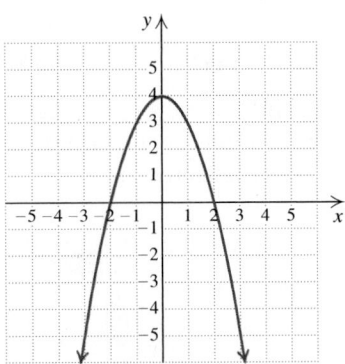

SOLUTION

a) The x-intercepts are the points at which the graph crosses the x-axis. For the graph shown, the x-intercepts are $(-2, 0)$ and $(2, 0)$.

b) The y-intercept is the point at which the graph crosses the y-axis. For the graph shown, the y-intercept is $(0, 4)$.

> TRY EXERCISE 7

Using Intercepts to Graph

It is important to know how to locate a graph's intercepts from the equation being graphed.

> **To Find Intercepts**
>
> To find the y-intercept(s) of an equation's graph, replace x with 0 and solve for y.
>
> To find the x-intercept(s) of an equation's graph, replace y with 0 and solve for x.

EXAMPLE **2**   Find the y-intercept and the x-intercept of the graph of $2x + 4y = 20$.

SOLUTION   To find the y-intercept, we let $x = 0$ and solve for y:

$$2 \cdot 0 + 4y = 20 \qquad \text{Replacing } x \text{ with } 0$$
$$4y = 20$$
$$y = 5.$$

Thus the y-intercept is $(0, 5)$.

To find the x-intercept, we let $y = 0$ and solve for x:

$$2x + 4 \cdot 0 = 20 \qquad \text{Replacing } y \text{ with } 0$$
$$2x = 20$$
$$x = 10.$$

Thus the x-intercept is $(10, 0)$.

> TRY EXERCISE 15

Since two points are sufficient to graph a line, intercepts can be used to graph linear equations.

EXAMPLE **3**   Graph $2x + 4y = 20$ using intercepts.

SOLUTION   In Example 2, we showed that the y-intercept is $(0, 5)$ and the x-intercept is $(10, 0)$. Before drawing a line, we plot a third point as a check. We substitute any convenient value for x and solve for y.

If we let $x = 5$, then

$$2 \cdot 5 + 4y = 20 \qquad \text{Substituting 5 for } x$$

$$10 + 4y = 20$$

$$4y = 10 \qquad \text{Subtracting 10 from both sides}$$

$$y = \tfrac{10}{4}, \text{ or } 2\tfrac{1}{2}. \qquad \text{Solving for } y$$

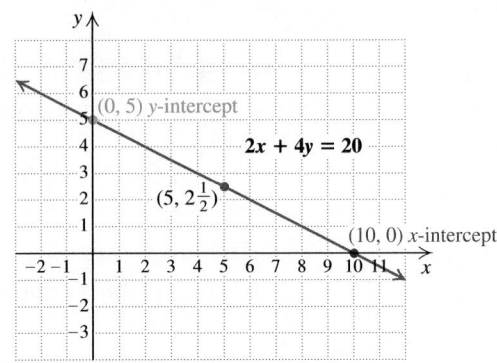

The point $\left(5, 2\tfrac{1}{2}\right)$ appears to line up with the intercepts, so our work is probably correct. To finish, we draw and label the line.

TRY EXERCISE ▶ 25

Note that when we solved for the y-intercept, we replaced x with 0 and simplified $2x + 4y = 20$ to $4y = 20$. Thus, to find the y-intercept, we can momentarily ignore the x-term and solve the remaining equation.

In a similar manner, when we solved for the x-intercept, we simplified $2x + 4y = 20$ to $2x = 20$. Thus, to find the x-intercept, we can momentarily ignore the y-term and then solve this remaining equation.

EXAMPLE **4**   Graph $3x - 2y = 60$ using intercepts.

SOLUTION   To find the y-intercept, we let $x = 0$. This amounts to temporarily ignoring the x-term and then solving:

$$-2y = 60 \qquad \text{For } x = 0, \text{ we have } 3 \cdot 0 - 2y, \text{ or simply } -2y.$$

$$y = -30.$$

The y-intercept is $(0, -30)$.

To find the x-intercept, we let $y = 0$. This amounts to temporarily disregarding the y-term and then solving:

$$3x = 60 \qquad \text{For } y = 0, \text{ we have } 3x - 2 \cdot 0, \text{ or simply } 3x.$$

$$x = 20.$$

The x-intercept is $(20, 0)$.

To find a third point, we can replace x with 4 and solve for y:

$$3 \cdot 4 - 2y = 60 \qquad \text{Numbers other than 4 can be used for } x.$$

$$12 - 2y = 60$$

$$-2y = 48$$

$$y = -24. \qquad \text{This means that } (4, -24) \text{ is on the graph.}$$

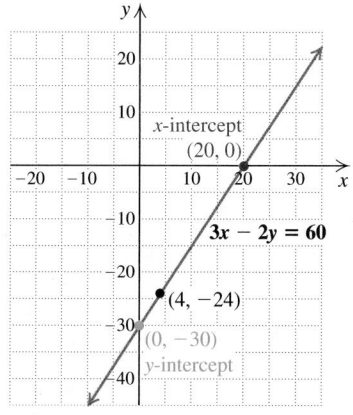

In order for us to graph all three points, the y-axis of our graph must go down to at least -30 and the x-axis must go up to at least 20. Using a scale of 5 units per square allows us to display both intercepts and $(4, -24)$, as well as the origin.

The point $(4, -24)$ appears to line up with the intercepts, so we draw and label the line, as shown at left.

TRY EXERCISE ▶ 45

TECHNOLOGY CONNECTION

When an equation has been entered into a graphing calculator, we may not be able to see both intercepts. For example, if $y = -0.8x + 17$ is graphed in the window $[-10, 10, -10, 10]$, neither intercept is visible.

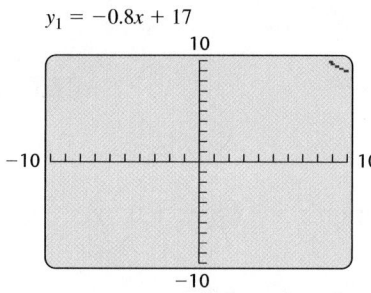

To better view the intercepts, we can change the window dimensions or we can zoom out. The ZOOM feature allows us to reduce or magnify a graph or a portion

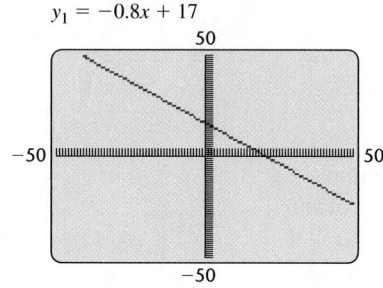

of a graph. Before zooming, the ZOOM *factors* must be set in the memory of the ZOOM key. If we zoom out with factors set at 5, both intercepts are visible but the axes are heavily drawn, as shown in the preceding figure.

This suggests that the *scales* of the axes should be changed. To do this, we use the WINDOW menu and set Xscl to 5 and Yscl to 5. The resulting graph has tick marks 5 units apart and clearly shows both intercepts. Other choices for Xscl and Yscl can also be made.

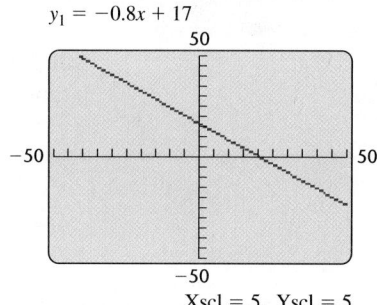

Xscl = 5, Yscl = 5

Graph each equation so that both intercepts can be easily viewed. Zoom or adjust the window settings so that tick marks can be clearly seen on both axes.

1. $y = -0.72x - 15$ 2. $y - 2.13x = 27$
3. $5x + 6y = 84$ 4. $2x - 7y = 150$
5. $19x - 17y = 200$ 6. $6x + 5y = 159$

Graphing Horizontal or Vertical Lines

The equations graphed in Examples 3 and 4 are both in the form $Ax + By = C$. We have already stated that any equation in the form $Ax + By = C$ is linear, provided A and B are not both zero. What if A or B (but not both) is zero? We will find that when A is zero, there is no x-term and the graph is a horizontal line. We will also find that when B is zero, there is no y-term and the graph is a vertical line.

EXAMPLE **5**

Graph: $y = 3$.

SOLUTION We can regard the equation $y = 3$ as $0 \cdot x + y = 3$. No matter what number we choose for x, we find that y must be 3 if the equation is to be solved. Consider the following table.

STUDENT NOTES ————

Many students draw horizontal lines when they should be drawing vertical lines and vice versa. To avoid this mistake, first locate the correct number on the axis whose label is given. Thus, to graph $x = 2$, we locate 2 on the x-axis and then draw a line perpendicular to that axis at that point. Note that the graph of $x = 2$ on a plane is a line, whereas the graph of $x = 2$ on a number line is a point.

$y = 3$

Choose any number for x. ⟶

y must be 3.

x	y	(x, y)
-2	3	$(-2, 3)$
0	3	$(0, 3)$
4	3	$(4, 3)$

All pairs will have 3 as the y-coordinate.

When we plot the ordered pairs $(-2, 3)$, $(0, 3)$, and $(4, 3)$ and connect the points, we obtain a horizontal line. Any ordered pair of the form $(x, 3)$ is a solution, so the line is parallel to the x-axis with y-intercept $(0, 3)$. Note that the graph of $y = 3$ has no x-intercept.

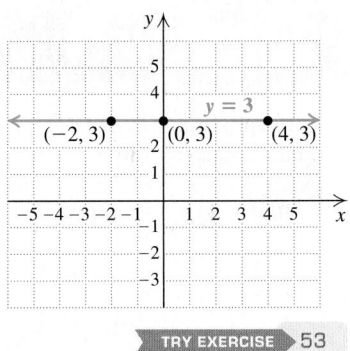

TRY EXERCISE ▶ 53

EXAMPLE **6**

Graph: $x = -4$.

SOLUTION We can regard the equation $x = -4$ as $x + 0 \cdot y = -4$. We make up a table with all -4's in the x-column.

$x = -4$

x must be -4. ⟶

Any number can be used for y.

x	y	(x, y)
-4	-5	$(-4, -5)$
-4	1	$(-4, 1)$
-4	3	$(-4, 3)$

All pairs will have -4 as the x-coordinate.

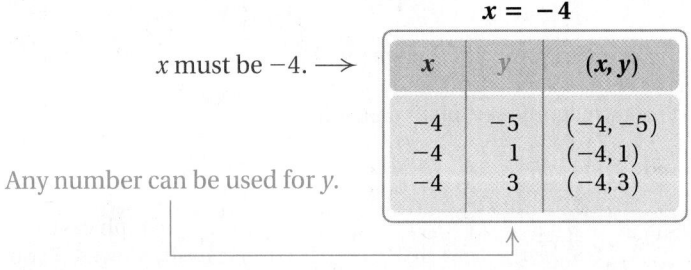

When we plot the ordered pairs $(-4, -5)$, $(-4, 1)$, and $(-4, 3)$ and connect them, we obtain a vertical line. Any ordered pair of the form $(-4, y)$ is a solution. The line is parallel to the y-axis with x-intercept $(-4, 0)$. Note that the graph of $x = -4$ has no y-intercept.

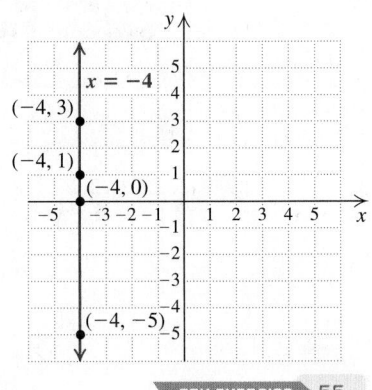

TRY EXERCISE ▶ 55

Linear Equations in One Variable

The graph of $y = b$ is a horizontal line, with y-intercept $(0, b)$.

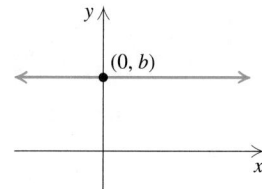

The graph of $x = a$ is a vertical line, with x-intercept $(a, 0)$.

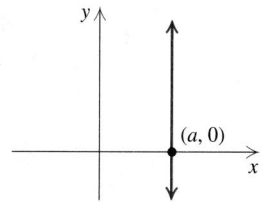

EXAMPLE **7**   Write an equation for each graph.

a)
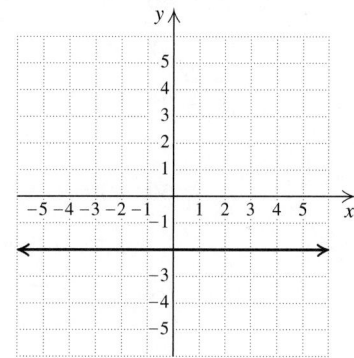

b)
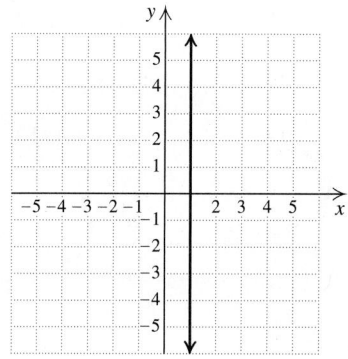

SOLUTION

a) Note that every point on the horizontal line passing through $(0, -2)$ has -2 as the y-coordinate. Thus the equation of the line is $y = -2$.

b) Note that every point on the vertical line passing through $(1, 0)$ has 1 as the x-coordinate. Thus the equation of the line is $x = 1$.   **TRY EXERCISE** 71

3.3   EXERCISE SET

For Extra Help   **MyMathLab**   Math**XL** PRACTICE   WATCH   DOWNLOAD

Concept Reinforcement   *In each of Exercises 1–6, match the phrase with the most appropriate choice from the column on the right.*

1. ____ A vertical line

2. ____ A horizontal line

3. ____ A y-intercept

4. ____ An x-intercept

5. ____ A third point as a check

6. ____ Use a scale of 10 units per square.

a) $2x + 5y = 100$

b) $(3, -2)$

c) $(1, 0)$

d) $(0, 2)$

e) $y = 3$

f) $x = -4$

For Exercises 7–14, list **(a)** *the coordinates of the y-intercept and* **(b)** *the coordinates of all x-intercepts.*

7.
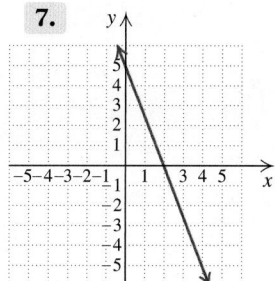

8.
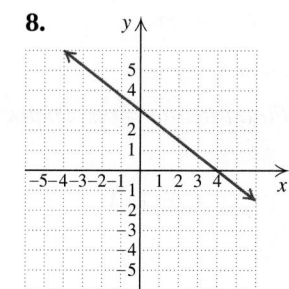

9.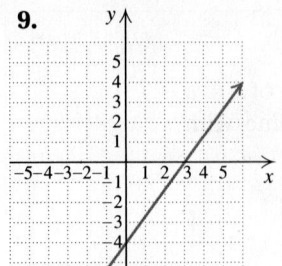

10.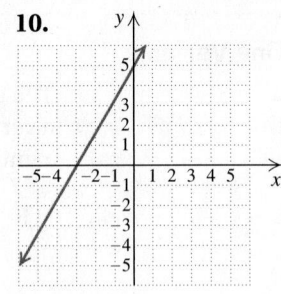

11.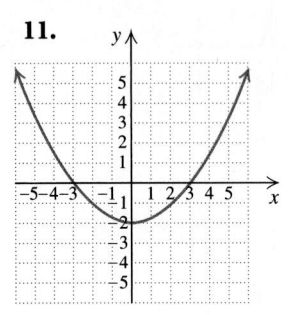

12.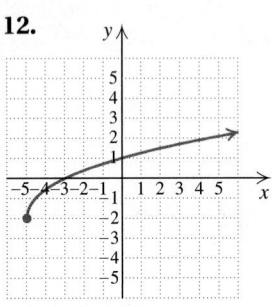

13.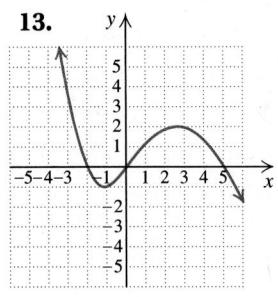

14.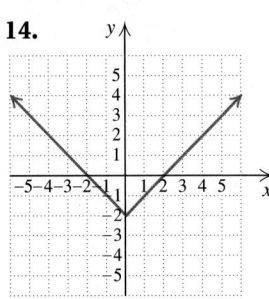

For Exercises 15–24, list **(a)** *the coordinates of any y-intercept and* **(b)** *the coordinates of any x-intercept. Do not graph.*

15. $3x + 5y = 15$

16. $2x + 7y = 14$

17. $9x - 2y = 36$

18. $10x - 3y = 60$

19. $-4x + 5y = 80$

20. $-5x + 6y = 100$

Aha! **21.** $x = 12$

22. $y = 10$

23. $y = -9$

24. $x = -5$

Find the intercepts. Then graph.

25. $3x + 5y = 15$ **26.** $2x + y = 6$

27. $x + 2y = 4$ **28.** $2x + 5y = 10$

29. $-x + 2y = 8$ **30.** $-x + 3y = 9$

31. $3x + y = 9$ **32.** $2x - y = 8$

33. $y = 2x - 6$ **34.** $y = -3x + 6$

35. $5x - 10 = 5y$ **36.** $3x - 9 = 3y$

37. $2x - 5y = 10$ **38.** $2x - 3y = 6$

39. $6x + 2y = 12$

40. $4x + 5y = 20$

41. $4x + 3y = 16$

42. $3x + 2y = 8$

43. $2x + 4y = 1$

44. $3x - 6y = 1$

45. $5x - 3y = 180$

46. $10x + 7y = 210$

47. $y = -30 + 3x$

48. $y = -40 + 5x$

49. $-4x = 20y + 80$

50. $60 = 20x - 3y$

51. $y - 3x = 0$

52. $x + 2y = 0$

Graph.

53. $y = 1$ **54.** $y = 4$

55. $x = 3$ **56.** $x = 6$

57. $y = -2$ **58.** $y = -4$

59. $x = -1$ **60.** $x = -6$

61. $y = -15$ **62.** $x = 20$

63. $y = 0$ **64.** $y = \frac{3}{2}$

65. $x = -\frac{5}{2}$ **66.** $x = 0$

67. $-4x = -100$

68. $12y = -360$

69. $35 + 7y = 0$

70. $-3x - 24 = 0$

Write an equation for each graph.

71.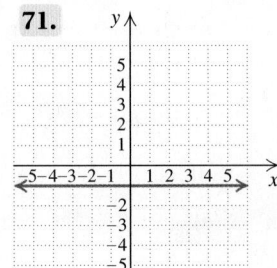

72.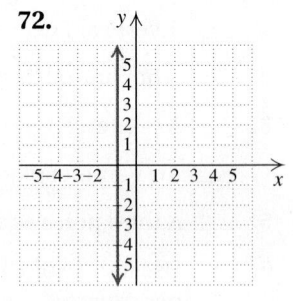

73.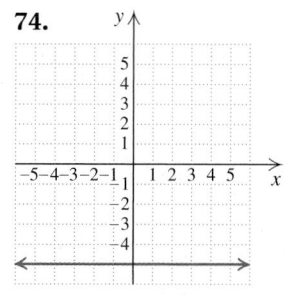

74.

75.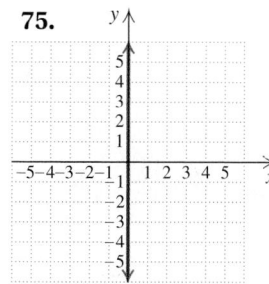

76.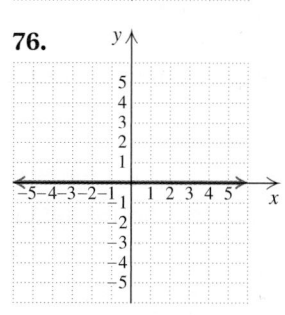

77. Explain in your own words why the graph of $y = 8$ is a horizontal line.

78. Explain in your own words why the graph of $x = -4$ is a vertical line.

Skill Review

Review translating to algebraic expressions (Section 1.1).

Translate to an algebraic expression. [1.1]

79. 7 less than d

80. 5 more than w

81. The sum of 7 and four times a number

82. The product of 3 and a number

83. Twice the sum of two numbers

84. Half of the sum of two numbers

Synthesis

85. Describe what the graph of $x + y = C$ will look like for any choice of C.

86. If the graph of a linear equation has one point that is both the x- and the y-intercepts, what is that point? Why?

87. Write an equation for the x-axis.

88. Write an equation of the line parallel to the x-axis and passing through $(3, 5)$.

89. Write an equation of the line parallel to the y-axis and passing through $(-2, 7)$.

90. Find the coordinates of the point of intersection of the graphs of $y = x$ and $y = 6$.

91. Find the coordinates of the point of intersection of the graphs of the equations $x = -3$ and $y = 4$.

92. Write an equation of the line shown in Exercise 7.

93. Write an equation of the line shown in Exercise 10.

94. Find the value of C such that the graph of $3x + C = 5y$ has an x-intercept of $(-4, 0)$.

95. Find the value of C such that the graph of $4x = C - 3y$ has a y-intercept of $(0, -8)$.

96. For A and B nonzero, the graphs of $Ax + D = C$ and $By + D = C$ will be parallel to an axis. Explain why.

97. Find the x-intercept of the graph of $Ax + D = C$.

In Exercises 98–103, find the intercepts of each equation algebraically. Then adjust the window and scale so that the intercepts can be checked graphically with no further window adjustments.

98. $3x + 2y = 50$

99. $2x - 7y = 80$

100. $y = 1.3x - 15$

101. $y = 0.2x - 9$

102. $25x - 20y = 1$

103. $50x + 25y = 1$

3.4 Rates

Rates of Change • Visualizing Rates

Rates of Change

Because graphs make use of two axes, they allow us to visualize how two quantities change with respect to each other. A number accompanied by units is used to represent this type of change and is referred to as a *rate*.

> ### Rate
>
> A *rate* is a ratio that indicates how two quantities change with respect to each other.

Rates occur often in everyday life:

A business whose customer base grows by 1500 customers over a period of 2 yr has an average *growth rate* of $\frac{1500}{2}$, or 750, customers per year.

A vehicle traveling 260 mi in 4 hr is moving at a *rate* of $\frac{260}{4}$, or 65, mph (miles per hour).

A class of 25 students pays a total of $93.75 to visit a museum. The *rate* is $\frac{\$93.75}{25}$, or $3.75, per student.

> *CAUTION!* To calculate a rate, it is important to keep track of the units being used.

EXAMPLE 1

On January 3, Alisha rented a Ford Focus with a full tank of gas and 9312 mi on the odometer. On January 7, she returned the car with 9630 mi on the odometer.* If the rental agency charged Alisha $108 for the rental and needed 12 gal of gas to fill up the gas tank, find the following rates.

a) The car's rate of gas consumption, in miles per gallon

b) The average cost of the rental, in dollars per day

c) The car's rate of travel, in miles per day

SOLUTION

a) The rate of gas consumption, in miles per gallon, is found by dividing the number of miles traveled by the number of gallons used for that amount of driving:

$$\text{Rate, in miles per gallon} = \frac{9630\,\text{mi} - 9312\,\text{mi}}{12\,\text{gal}}$$

The word "per" indicates division.

$$= \frac{318\,\text{mi}}{12\,\text{gal}}$$

$$= 26.5\,\text{mi/gal} \quad \text{Dividing}$$

$$= 26.5\ \text{miles per gallon.}$$

─────────────

*For all problems concerning rentals, assume that the pickup time was later in the day than the return time so that no late fees were applied.

b) The average cost of the rental, in dollars per day, is found by dividing the cost of the rental by the number of days:

$$\text{Rate, in dollars per day} = \frac{108 \text{ dollars}}{4 \text{ days}} \qquad \begin{array}{l}\text{From January 3 to}\\ \text{January 7 is}\\ 7 - 3 = 4 \text{ days.}\end{array}$$

$$= 27 \text{ dollars/day}$$

$$= \$27 \text{ per day.}$$

c) The car's rate of travel, in miles per day, is found by dividing the number of miles traveled by the number of days:

$$\text{Rate, in miles per day} = \frac{318 \text{ mi}}{4 \text{ days}} \qquad \begin{array}{l}9630 \text{ mi} - 9312 \text{ mi} = 318 \text{ mi;}\\ \text{From January 3 to January 7}\\ \text{is } 7 - 3 = 4 \text{ days.}\end{array}$$

$$= 79.5 \text{ mi/day}$$

$$= 79.5 \text{ mi per day.} \qquad \boxed{\text{TRY EXERCISE} \blacktriangleright 7}$$

CAUTION!   Units are a vital part of real-world problems. They must be considered in the translation of a problem and included in the answer to a problem.

Many problems involve a rate of travel, or *speed*. The **speed** of an object is found by dividing the distance traveled by the time required to travel that distance.

EXAMPLE 2

Transportation.   An Atlantic City Express bus makes regular trips between Paramus and Atlantic City, New Jersey. At 6:00 P.M., the bus is at mileage marker 40 on the Garden State Parkway, and at 8:00 P.M. it is at marker 170. Find the average speed of the bus.

SOLUTION   Speed is the distance traveled divided by the time spent traveling:

$$\text{Bus speed} = \frac{\text{Distance traveled}}{\text{Time spent traveling}}$$

$$= \frac{\text{Change in mileage}}{\text{Change in time}}$$

$$= \frac{130 \text{ mi}}{2 \text{ hr}} \qquad \begin{array}{l}170 \text{ mi} - 40 \text{ mi} = 130 \text{ mi;}\\ 8:00 \text{ P.M.} - 6:00 \text{ P.M.} = 2 \text{ hr}\end{array}$$

$$= 65 \frac{\text{mi}}{\text{hr}}$$

$$= 65 \text{ miles per hour.} \qquad \begin{array}{l}\text{This } average \text{ speed does not}\\ \text{indicate by how much the bus speed}\\ \text{may vary along the route.}\end{array}$$

$$\boxed{\text{TRY EXERCISE} \blacktriangleright 13}$$

Visualizing Rates

Graphs allow us to visualize a rate of change. As a rule, the quantity listed in the numerator appears on the vertical axis and the quantity listed in the denominator appears on the horizontal axis.

EXAMPLE **3**

Recycling. Between 1991 and 2006, the amount of paper recycled in the United States increased at a rate of approximately 1.5 million tons per year. In 1991, approximately 31 million tons of paper was recycled. Draw a graph to represent this information.

Source: Based on information from American Forest and Paper Association

SOLUTION   To label the axes, note that the rate is given as 1.5 million tons per year, or

$$1.5 \text{ million } \frac{\text{tons}}{\text{yr}}. \qquad \begin{matrix} \longleftarrow \text{Numerator: vertical axis} \\ \longleftarrow \text{Denominator: horizontal axis} \end{matrix}$$

We list *Amount of paper recycled (in millions of tons)* on the vertical axis and *Year* on the horizontal axis. (See the figure on the left below.)

Next, we select a scale for each axis that allows us to plot the given information. If we count by increments of 10 million on the vertical axis, we can show 31 million tons for 1991 and increasing amounts for later years. On the horizontal axis, we count by increments of 2 years to make certain that both 1991 and 2006 are included. (See the figure in the middle below.)

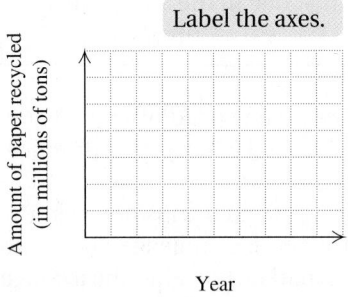
Label the axes.

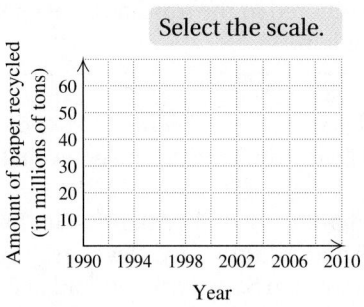
Select the scale.

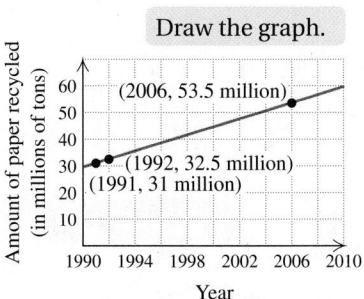
Draw the graph.

We now plot the point corresponding to (1991, 31 million). Then, to display the rate of growth, we move from that point to a second point that represents 1.5 million more tons 1 year later.

(1991, 31 million)	Beginning point
(1991 + 1, 31 million + 1.5 million)	1.5 million more tons, 1 year later
(1992, 32.5 million)	A second point on the graph

Similarly, we can find the coordinates for 2006. Since 2006 is 15 years after 1991, we add 15 to the year and 15(1.5 million) = 22.5 million to the amount.

(1991, 31 million)	Beginning point
(1991 + 15, 31 million + 22.5 million)	15(1.5) million more tons, 15 years later
(2006, 53.5 million)	A third point on the graph

After plotting the three points, we draw a line through them, as shown in the figure on the right above. This gives us the graph.   TRY EXERCISE ▶ 19

EXAMPLE **4**

Banking. Nadia prepared the following graph from data collected on a recent day at a branch bank.

a) What rate can be determined from the graph?

b) What is that rate?

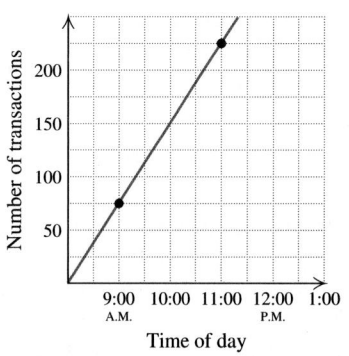

SOLUTION

a) Because the vertical axis shows the number of transactions and the horizontal axis lists the time in hour-long increments, we can find the rate *Number of transactions per hour.*

b) The points (9:00, 75) and (11:00, 225) are both on the graph. This tells us that in the 2 hours between 9:00 and 11:00, there were $225 - 75 = 150$ transactions. Thus the rate is

$$\frac{225 \text{ transactions} - 75 \text{ transactions}}{11:00 - 9:00} = \frac{150 \text{ transactions}}{2 \text{ hours}}$$

$$= 75 \text{ transactions per hour.}$$

Note that this is an *average* rate.

> **TRY EXERCISE** 29

3.4   EXERCISE SET

For Extra Help
MyMathLab
Math XL PRACTICE
WATCH
DOWNLOAD

🖐 ***Concept Reinforcement***   *For Exercises 1–6, fill in the missing units for each rate.*

1. If Eva biked 100 miles in 5 hours, her average rate was 20 _____.

2. If it took Lauren 18 hours to read 6 chapters, her average rate was 3 _____.

3. If Denny's ticket cost $300 for a 150-mile flight, his average rate was 2 _____.

4. If Geoff planted 36 petunias along a 12-ft sidewalk, his average rate was 3 _____.

5. If Christi ran 8 errands in 40 minutes, her average rate was 5 _____.

6. If Ben made 8 cakes using 20 cups of flour, his average rate was $2\frac{1}{2}$ _____.

Solve. For Exercises 7–14, round answers to the nearest cent.

7. *Car rentals.*   Late on June 5, Gaya rented a Ford Focus with a full tank of gas and 13,741 mi on the odometer. On June 8, she returned the car with 14,131 mi on the odometer. The rental agency charged Gaya $118 for the rental and needed 13 gal of gas to fill up the tank.

   a) Find the car's rate of gas consumption, in miles per gallon.
   b) Find the average cost of the rental, in dollars per day.
   c) Find the average rate of travel, in miles per day.
   🖩 d) Find the rental rate, in cents per mile.

8. *SUV rentals.*   On February 10, Oscar rented a Chevy Trailblazer with a full tank of gas and 13,091 mi on the odometer. On February 12, he returned the vehicle with 13,322 mi on the odometer. The rental agency charged $92 for the rental and needed 14 gal of gas to fill the tank.

   a) Find the SUV's rate of gas consumption, in miles per gallon.

   b) Find the average cost of the rental, in dollars per day.

   c) Find the average rate of travel, in miles per day.

   d) Find the rental rate, in cents per mile.

9. *Bicycle rentals.*   At 9:00, Jodi rented a mountain bike from The Bike Rack. She returned the bicycle at 11:00, after cycling 14 mi. Jodi paid $15 for the rental.

   a) Find Jodi's average speed, in miles per hour.

   b) Find the rental rate, in dollars per hour.

   c) Find the rental rate, in dollars per mile.

10. *Bicycle rentals.*   At 2:00, Braden rented a mountain bike from The Slick Rock Cyclery. He returned the bike at 5:00, after cycling 18 mi. Braden paid $12 for the rental.

    a) Find Braden's average speed, in miles per hour.

    b) Find the rental rate, in dollars per hour.

    c) Find the rental rate, in dollars per mile.

11. *Proofreading.*   Sergei began proofreading at 9:00 A.M., starting at the top of page 93. He worked until 2:00 P.M. that day and finished page 195. He billed the publishers $110 for the day's work.

    a) Find the rate of pay, in dollars per hour.

    b) Find the average proofreading rate, in number of pages per hour.

    c) Find the rate of pay, in dollars per page.

12. *Temporary help.*   A typist for Kelly Services reports to 3E's Properties for work at 10:00 A.M. and leaves at 6:00 P.M. after having typed from the end of page 8 to the end of page 50 of a proposal. 3E's pays $120 for the typist's services.

    a) Find the rate of pay, in dollars per hour.

    b) Find the average typing rate, in number of pages per hour.

    c) Find the rate of pay, in dollars per page.

13. *National debt.*   The U.S. federal budget debt was $5770 billion in 2001 and $8612 billion in 2006. Find the rate at which the debt was increasing.
    Source: U.S. Office of Management and Budget

14. *Four-year-college tuition.*   The average tuition at a public four-year college was $3983 in 2001 and $5948 in 2005. Find the rate at which tuition was increasing.
    Source: U.S. National Center for Education Statistics

15. *Elevators.*   At 2:38, Lara entered an elevator on the 34th floor of the Regency Hotel. At 2:40, she stepped off at the 5th floor.

    a) Find the elevator's average rate of travel, in number of floors per minute.

    b) Find the elevator's average rate of travel, in seconds per floor.

16. *Snow removal.*   By 1:00 P.M., Olivia had already shoveled 2 driveways, and by 6:00 P.M. that day, the number was up to 7.

    a) Find Olivia's average shoveling rate, in number of driveways per hour.

    b) Find Olivia's average shoveling rate, in hours per driveway.

17. *Mountaineering.*   The fastest ascent of Mt. Everest was accomplished by the Sherpa guide Pemba Dorje of Nepal in 2004. Pemba Dorje climbed from base camp, elevation 17,552 ft, to the summit, elevation 29,028 ft, in 8 hr 10 min.
    Source: *Guinness Book of World Records* 2006 Edition

    a) Find Pemba Dorje's average rate of ascent, in feet per minute.

    b) Find Pemba Dorje's average rate of ascent, in minutes per foot.

18. *Mountaineering.*   As part of an ill-fated expedition to climb Mt. Everest in 1996, author Jon Krakauer departed "The Balcony," elevation 27,600 ft, at 7:00 A.M. and reached the summit, elevation 29,028 ft, at 1:25 P.M.
Source: Krakauer, Jon, *Into Thin Air, the Illustrated Edition.* New York: Random House, 1998

  a) Find Krakauer's average rate of ascent, in feet per minute.
  b) Find Krakauer's average rate of ascent, in minutes per foot.

In Exercises 19–28, draw a linear graph to represent the given information. Be sure to label and number the axes appropriately (see Example 3).

19. *Landfills.*   In 2006, 35,700,000 tons of paper was deposited in landfills in the United States, and this figure was decreasing by 700,000 tons per year.
Source: Based on data from American Forest and Paper Association

20. *Health insurance.*   In 2005, the average cost for health insurance for a family was about $11,000 and the figure was rising at a rate of about $1100 per year.
Source: Based on data from Kaiser/HRET Survey of Health Benefits

21. *Prescription drug sales.*   In 2006, there were sales of approximately $11 billion of asthma drug products in the United States, and the figure was increasing at a rate of about $1.2 billion per year.
Source: *The Wall Street Journal,* 6/28/2007

22. *Violent crimes.*   In 2004, there were approximately 21.1 violent crimes per 1000 population in the United States, and the figure was dropping at a rate of about 1.2 crimes per 1000 per year.
Source: U.S. Bureau of Justice Statistics

23. *Train travel.*   At 3:00 P.M., the Boston–Washington Metroliner had traveled 230 mi and was cruising at a rate of 90 miles per hour.

24. *Plane travel.*   At 4:00 P.M., the Seattle–Los Angeles shuttle had traveled 400 mi and was cruising at a rate of 300 miles per hour.

25. *Wages.*   By 2:00 P.M., Diane had earned $50. She continued earning money at a rate of $15 per hour.

26. *Wages.*   By 3:00 P.M., Arnie had earned $70. He continued earning money at a rate of $12 per hour.

27. *Telephone bills.*   Roberta's phone bill was already $7.50 when she made a call for which she was charged at a rate of $0.10 per minute.

28. *Telephone bills.*   At 3:00 P.M., Larry's phone bill was $6.50 and increasing at a rate of 7¢ per minute.

In Exercises 29–38, use the graph provided to calculate a rate of change in which the units of the horizontal axis are used in the denominator.

29. *Call center.*   The following graph shows data from a technical assistance call center. At what rate are calls being handled?

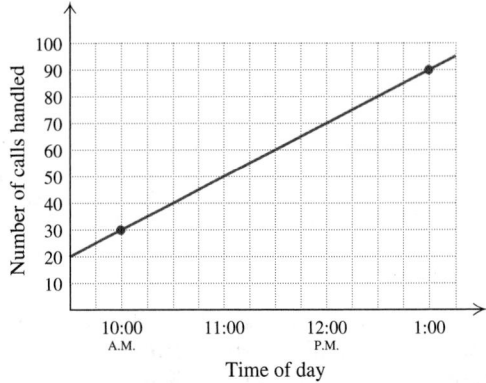

30. *Hairdresser.*   Eve's Custom Cuts has a graph displaying data from a recent day of work. At what rate does Eve work?

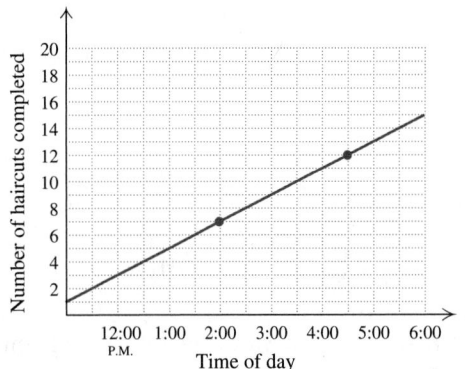

31. *Train travel.*   The following graph shows data from a recent train ride from Chicago to St. Louis. At what rate did the train travel?

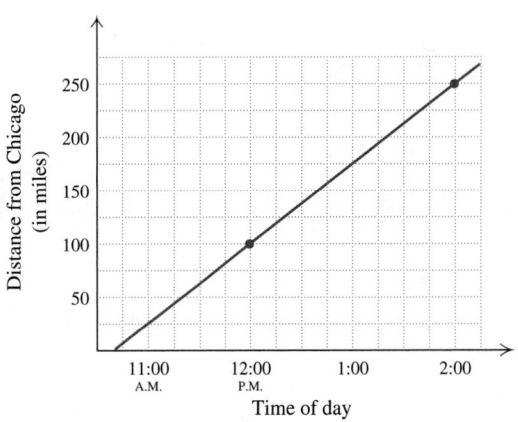

32. *Train travel.* The following graph shows data from a recent train ride from Denver to Kansas City. At what rate did the train travel?

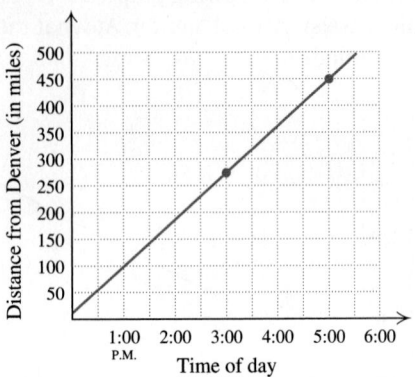

33. *Cost of a telephone call.* The following graph shows data from a recent phone call between the United States and the Netherlands. At what rate was the customer being billed?

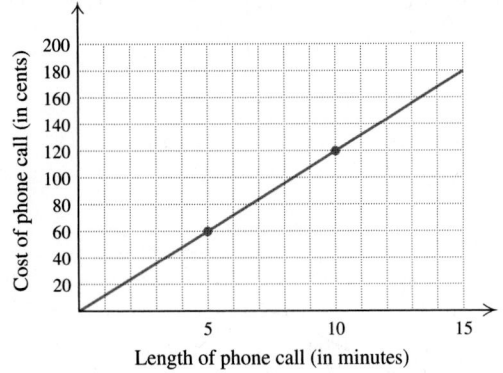

34. *Cost of a telephone call.* The following graph shows data from a recent phone call between the United States and South Korea. At what rate was the customer being billed?

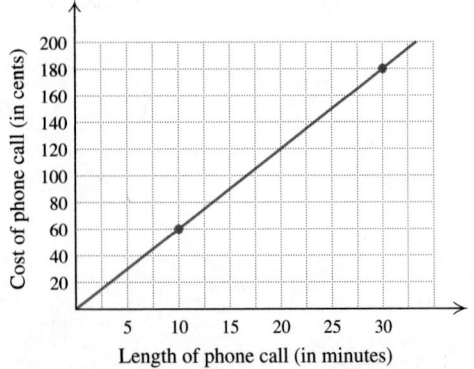

35. *Population.* The following graph shows data regarding the population of Youngstown,

Ohio. At what rate was the population changing?

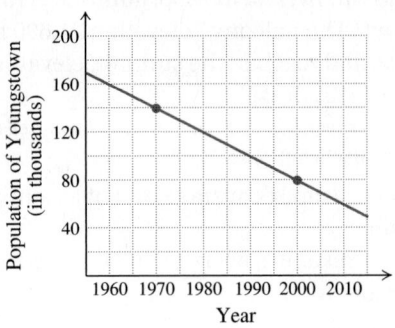

36. *Depreciation of an office machine.* Data regarding the value of a particular color copier is represented in the following graph. At what rate is the value changing?

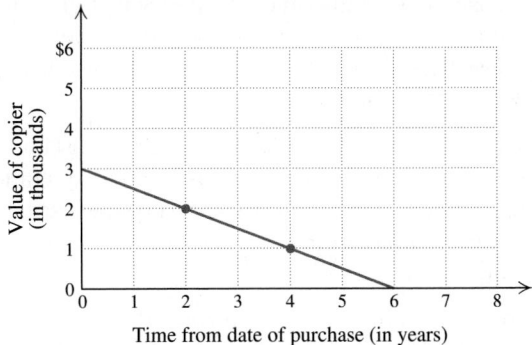

37. *Gas mileage.* The following graph shows data for a 2008 Toyota Prius driven on interstate highways. At what rate was the vehicle consuming gas?
Source: www.fueleconomy.gov

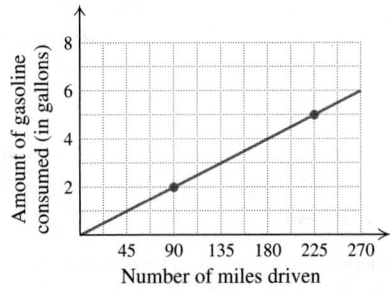

38. *Gas mileage.* The following graph shows data for a 2008 Chevy Malibu driven on city streets. At what rate was the vehicle consuming gas?
Source: Chevrolet

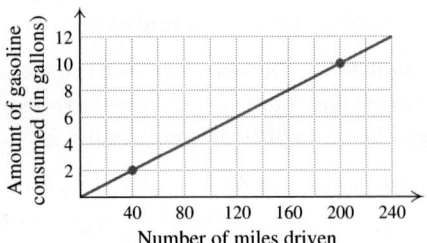

In each of Exercises 39–44, match the description with the most appropriate graph from the choices below. Scales are intentionally omitted. Assume that of the three sports listed, swimming is the slowest and biking is the fastest.

39. ____ Robin trains for triathlons by running, biking, and then swimming every Saturday.

40. ____ Gene trains for triathlons by biking, running, and then swimming every Sunday.

41. ____ Shirley trains for triathlons by swimming, biking, and then running every Sunday.

42. ____ Evan trains for triathlons by swimming, running, and then biking every Saturday.

43. ____ Angie trains for triathlons by biking, swimming, and then running every Sunday.

44. ____ Mick trains for triathlons by running, swimming, and then biking every Saturday.

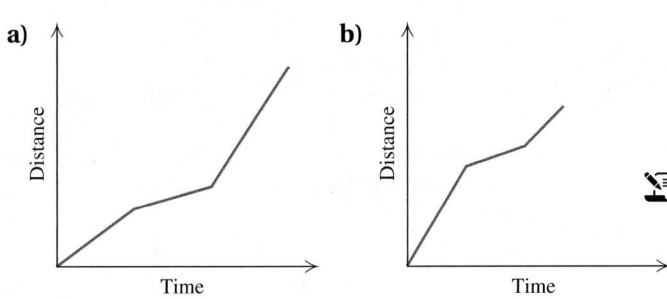

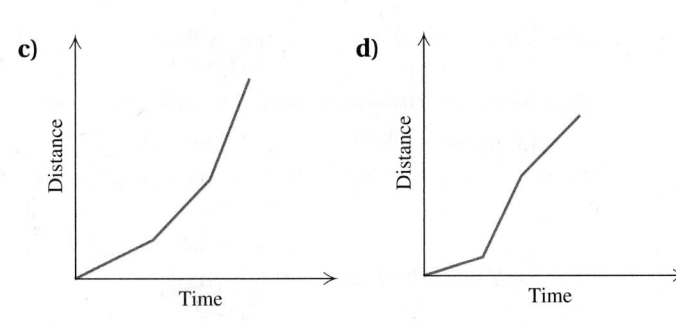

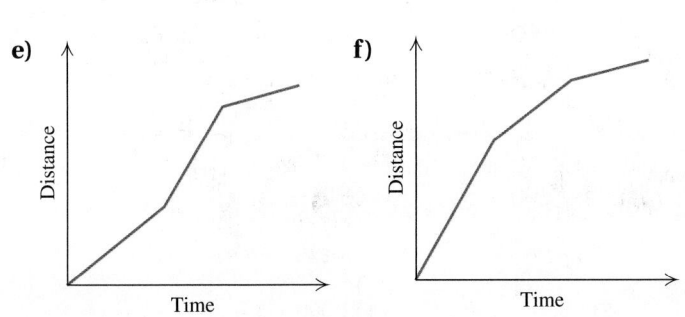

45. What does a negative rate of travel indicate? Explain.

46. Explain how to convert from kilometers per hour to meters per second.

Skill Review

To prepare for Section 3.5, review subtraction and order of operations (Sections 1.6 and 1.8).

Simplify.

47. $-2 - (-7)$ [1.6]

48. $-9 - (-3)$ [1.6]

49. $\dfrac{5 - (-4)}{-2 - 7}$ [1.8]

50. $\dfrac{8 - (-4)}{2 - 11}$ [1.8]

51. $\dfrac{-4 - 8}{11 - 2}$ [1.8]

52. $\dfrac{-5 - (-3)}{4 - 6}$ [1.8]

53. $\dfrac{-6 - (-6)}{-2 - 7}$ [1.8]

54. $\dfrac{-3 - 5}{-1 - (-1)}$ [1.8]

Synthesis

55. How would the graphs of Jon's and Jenny's total earnings compare in each of the following situations?
  a) Jon earns twice as much per hour as Jenny.
  b) Jon and Jenny earn the same hourly rate, but Jenny received a bonus for a cost-saving suggestion.
  c) Jon is paid by the hour, and Jenny is paid a weekly salary.

56. Write an exercise similar to those in Exercises 7–18 for a classmate to solve. Design the problem so that the solution is "The motorcycle's rate of gas consumption was 65 miles per gallon."

57. *Aviation.* A Boeing 737 climbs from sea level to a cruising altitude of 31,500 ft at a rate of 6300 ft/min. After cruising for 3 min, the jet is forced to land, descending at a rate of 3500 ft/min. Represent the flight with a graph in which altitude is measured on the vertical axis and time on the horizontal axis.

58. *Wages with commissions.* Each salesperson at Mike's Bikes is paid $140 a week plus 13% of all sales up to $2000, and then 20% on any sales in excess of $2000. Draw a graph in which sales are measured on the horizontal axis and wages on the vertical axis. Then use the graph to estimate the wages paid when a salesperson sells $2700 in merchandise in one week.

59. *Taxi fares.* The driver of a New York City Yellow Cab recently charged $2 plus 50¢ for each fifth of a mile traveled. Draw a graph that could be used to determine the cost of a fare.

60. *Gas mileage.* Suppose that a Kawasaki motorcycle travels three times as far as a Chevy Malibu on the same amount of gas (see Exercise 38). Draw a graph that reflects this information.

61. *Aviation.* Tim's F-16 jet is moving forward at a deck speed of 95 mph aboard an aircraft carrier that is traveling 39 mph in the same direction. How fast is the jet traveling, in minutes per mile, with respect to the sea?

62. *Navigation.* In 3 sec, Penny walks 24 ft, to the bow (front) of a tugboat. The boat is cruising at a rate of 5 ft/sec. What is Penny's rate of travel with respect to land?

63. *Running.* Anne ran from the 4-km mark to the 7-km mark of a 10-km race in 15.5 min. At this rate, how long would it take Anne to run a 5-mi race?

64. *Running.* Jerod ran from the 2-mi marker to the finish line of a 5-mi race in 25 min. At this rate, how long would it take Jerod to run a 10-km race?

65. Alex picks apples twice as fast as Ryan. By 4:30, Ryan had already picked 4 bushels of apples. Fifty minutes later, his total reached $5\frac{1}{2}$ bushels. Find Alex's picking rate. Give your answer in number of bushels per hour.

66. At 3:00 P.M., Catanya and Chad had already made 46 candles. By 5:00 P.M., the total reached 100 candles. Assuming a constant production rate, at what time did they make their 82nd candle?

CORNER

COLLABORATIVE

Determining Depreciation Rates

Focus: Modeling, graphing, and rates

Time: 30 minutes

Group size: 3

Materials: Graph paper and straightedges

From the minute a new car is driven out of the dealership, it *depreciates*, or drops in value with the passing of time. The N.A.D.A. Official Used Car Guide is a periodic listing of the trade-in values of used cars. The data below are taken from two such reports from 2007.

ACTIVITY

1. Each group member should select a different one of the cars listed in the table below as his or her own. Assuming that the values are dropping linearly, each student should draw a line representing the trade-in value of his or her car. Draw all three lines on the same graph. Let the horizontal axis represent the time, in months, since January 2007, and let the vertical axis represent the trade-in value of each car. Decide as a group how many months or dollars each square should represent. Make the drawings as neat as possible.

2. At what *rate* is each car depreciating and how are the different rates illustrated in the graph of part (1)?

3. If one of the three cars had to be sold in January 2009, which one would your group sell and why? Compare answers with other groups.

Car	Trade-in Value in January 2007	Trade-in Value in June 2007
2005 Mustang V6 Coupe	$13,625	$13,125
2005 Nissan Sentra SE-R	$11,825	$11,000
2005 Volkswagen Jetta Sedan GL	$12,600	$11,425

3.5 Slope

Rate and Slope ▪ Horizontal and Vertical Lines ▪ Applications

In Section 3.4, we introduced *rate* as a method of measuring how two quantities change with respect to each other. In this section, we will discuss how rate can be related to the slope of a line.

Rate and Slope

Automated digitization machines use robotic arms to carefully turn pages of books so that they can take pictures of each page. Suppose that a large university library purchased a DL-1500 and an APT 1200. The DL-1500 digitizes 3 volumes of an encyclopedia every 2 hr. The APT 1200 digitizes 6 volumes of an encyclopedia every 5 hr. The following tables list the number of books digitized after various amounts of time for each machine.

Source: Based on information from Kirtas and 4DigitalBooks

DL-1500	
Hours Elapsed	**Books Digitized**
0	0
2	3
4	6
6	9
8	12

APT 1200	
Hours Elapsed	**Books Digitized**
0	0
5	6
10	12
15	18
20	24

We now graph the pairs of numbers listed in the tables, using the horizontal axis for the number of hours elapsed and the vertical axis for the number of books digitized.

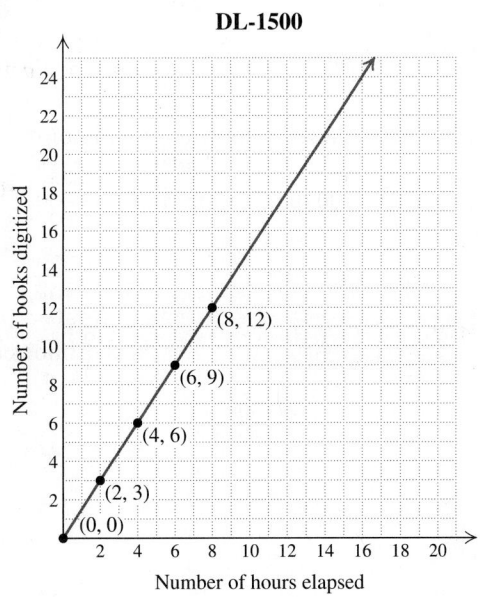

DL-1500

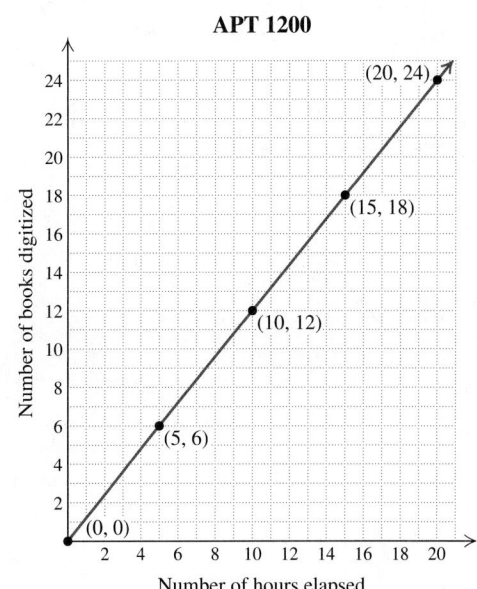

APT 1200

By comparing the number of books digitized by each machine over a specified period of time, we can compare the two rates. For example, the DL-1500 digitizes 3 books every 2 hr, so its *rate* is $3 \div 2 = \frac{3}{2}$ books per hour. Since the APT 1200 digitizes 6 books every 5 hr, its rate is $6 \div 5 = \frac{6}{5}$ books per hour. Note that the rate of the DL-1500 is greater so its graph is steeper.

The rates $\frac{3}{2}$ and $\frac{6}{5}$ can also be found using the coordinates of any two points that are on the line. For example, we can use the points $(6, 9)$ and $(8, 12)$ to find the digitization rate for the DL-1500. To do so, remember that these coordinates tell us that after 6 hr, 9 books have been digitized, and after 8 hr, 12 books have been digitized. In the 2 hr between the 6-hr and 8-hr points, $12 - 9$, or 3, books were digitized. Thus we have

$$\text{DL-1500 digitization rate} = \frac{\text{change in number of books digitized}}{\text{corresponding change in time}}$$

$$= \frac{12 - 9 \text{ books}}{8 - 6 \text{ hr}}$$

$$= \frac{3 \text{ books}}{2 \text{ hr}} = \frac{3}{2} \text{ books per hour.}$$

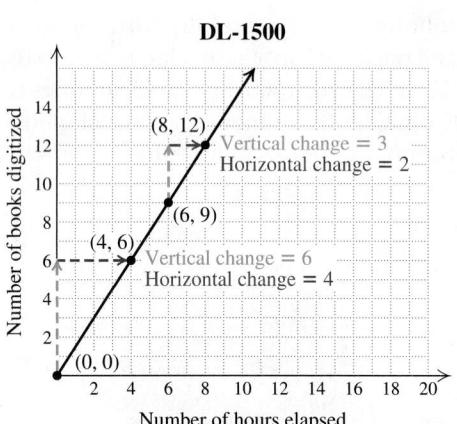

DL-1500

Because the line is straight, the same rate is found using *any* pair of points on the line. For example, using $(0, 0)$ and $(4, 6)$, we have

$$\text{DL-1500 digitization rate} = \frac{6 - 0 \text{ books}}{4 - 0 \text{ hr}} = \frac{6 \text{ books}}{4 \text{ hr}} = \frac{3}{2} \text{ books per hour.}$$

Note that the rate is always the vertical change divided by the corresponding horizontal change.

EXAMPLE **1** Use the graph of book digitization by the APT 1200 to find the rate at which books are digitized.

SOLUTION We can use any two points on the line, such as $(15, 18)$ and $(20, 24)$:

$$\text{APT 1200 digitization rate} = \frac{\text{change in number of books digitized}}{\text{corresponding change in time}}$$

$$= \frac{24 - 18 \text{ books}}{20 - 15 \text{ hr}}$$

$$= \frac{6 \text{ books}}{5 \text{ hr}}$$

$$= \frac{6}{5} \text{ books per hour.}$$

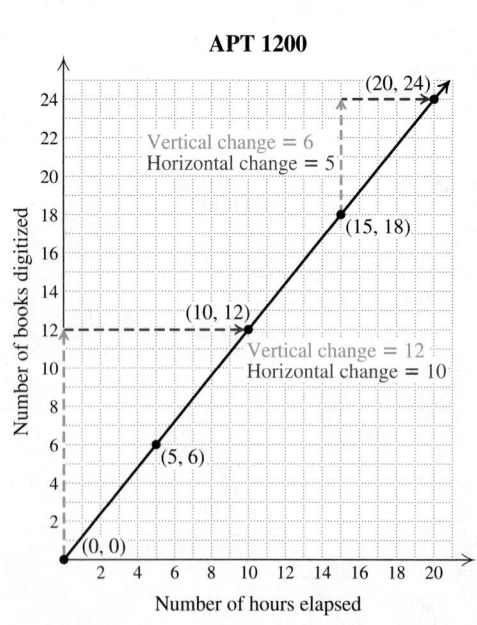

APT 1200

As a check, we can use another pair of points, like $(0, 0)$ and $(10, 12)$:

$$\text{APT 1200 digitization rate} = \frac{12 - 0 \text{ books}}{10 - 0 \text{ hr}}$$

$$= \frac{12 \text{ books}}{10 \text{ hr}}$$

$$= \frac{6}{5} \text{ books per hour.}$$

TRY EXERCISE 11

When the axes of a graph are simply labeled x and y, the ratio of vertical change to horizontal change is the rate at which y is changing with respect to x. This ratio is a measure of a line's slant, or **slope**.

Consider a line passing through $(2, 3)$ and $(6, 5)$, as shown below. We find the ratio of vertical change, or *rise*, to horizontal change, or *run*, as follows:

$$\text{Ratio of vertical change to horizontal change} = \frac{\text{change in } y}{\text{change in } x} = \frac{\text{rise}}{\text{run}}$$

$$= \frac{5 - 3}{6 - 2} \left. \vphantom{\frac{5-3}{6-2}} \right\} \quad \textbf{Note that these}$$
$$= \frac{2}{4}, \text{ or } \frac{1}{2}. \left. \vphantom{\frac{2}{4}} \right\} \quad \begin{array}{l} \textbf{calculations can be} \\ \textbf{performed without} \\ \textbf{viewing a graph.} \end{array}$$

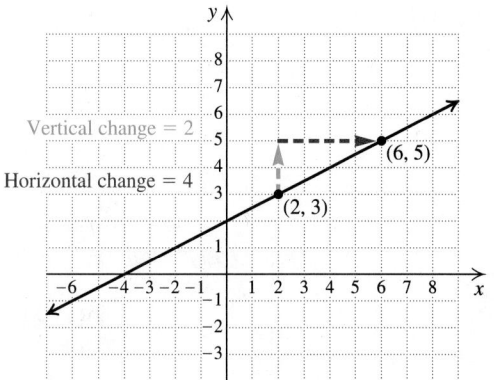

Thus the y-coordinates of points on this line increase at a rate of 2 units for every 4-unit increase in x, which is 1 unit for every 2-unit increase in x, or $\frac{1}{2}$ unit for every 1-unit increase in x. The slope of the line is $\frac{1}{2}$.

In the box below, the *subscripts* 1 and 2 are used to distinguish two arbitrary points, point 1 and point 2, from each other. The slightly lowered 1's and 2's are not exponents but are used to denote x-values (or y-values) that may differ from each other.

Slope

The *slope* of the line containing points (x_1, y_1) and (x_2, y_2) is given by

$$m = \frac{\text{change in } y}{\text{change in } x} = \frac{\text{rise}}{\text{run}} = \frac{y_2 - y_1}{x_2 - x_1}.$$

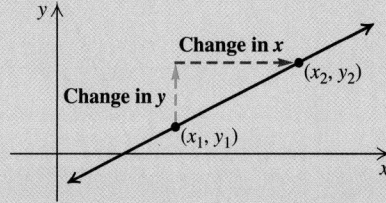

EXAMPLE **2** Graph the line containing the points $(-4, 3)$ and $(2, -6)$ and find the slope.

SOLUTION The graph is shown below. From $(-4, 3)$ to $(2, -6)$, the change in y, or rise, is $-6 - 3$, or -9. The change in x, or run, is $2 - (-4)$, or 6. Thus,

$$\text{Slope} = \frac{\text{change in } y}{\text{change in } x}$$

$$= \frac{\text{rise}}{\text{run}}$$

$$= \frac{-6 - 3}{2 - (-4)}$$

$$= \frac{-9}{6}$$

$$= -\frac{9}{6}, \text{ or } -\frac{3}{2}.$$

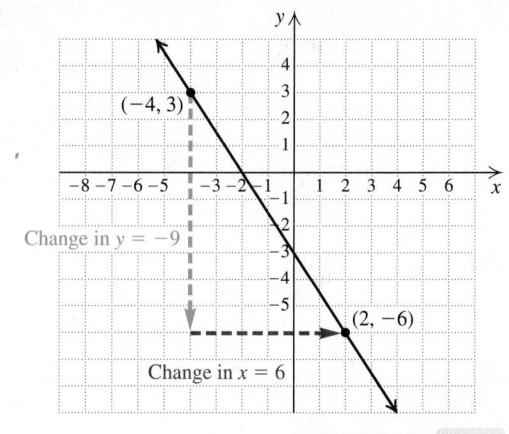

Change in $y = -9$

Change in $x = 6$

TRY EXERCISE 39

STUDENT NOTES

You may wonder which point should be regarded as (x_1, y_1) and which should be (x_2, y_2). To see that the math works out the same either way, perform both calculations on your own.

CAUTION! When we use the formula

$$m = \frac{y_2 - y_1}{x_2 - x_1},$$

it makes no difference which point is considered (x_1, y_1). What matters is that we subtract the y-coordinates in the same order that we subtract the x-coordinates.

To illustrate, we reverse *both* of the subtractions in Example 2. The slope is still $-\frac{3}{2}$:

$$\text{Slope} = \frac{\text{change in } y}{\text{change in } x} = \frac{3 - (-6)}{-4 - 2} = \frac{9}{-6} = -\frac{3}{2}.$$

As shown in the graphs below, a line with positive slope slants up from left to right, and a line with negative slope slants down from left to right. The larger the absolute value of the slope, the steeper the line.

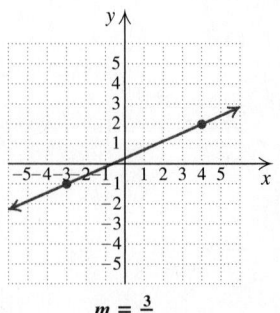

$m = \frac{3}{7}$

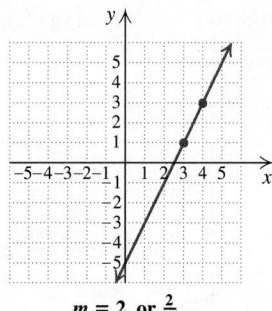

$m = 2$, or $\frac{2}{1}$

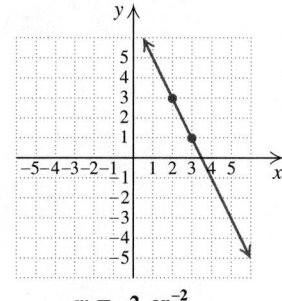

$m = -2$, or $\frac{-2}{1}$

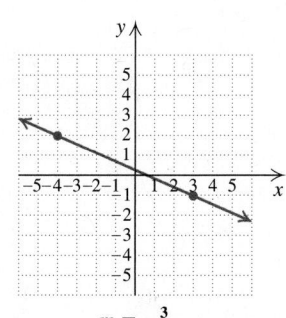

$m = -\frac{3}{7}$

Horizontal and Vertical Lines

What about the slope of a horizontal line or a vertical line?

EXAMPLE 3   Find the slope of the line $y = 4$.

SOLUTION   Consider the points $(2, 4)$ and $(-3, 4)$, which are on the line. The change in y, or the rise, is $4 - 4$, or 0. The change in x, or the run, is $-3 - 2$, or -5. Thus,

$$m = \frac{4 - 4}{-3 - 2}$$

$$= \frac{0}{-5}$$

$$= 0.$$

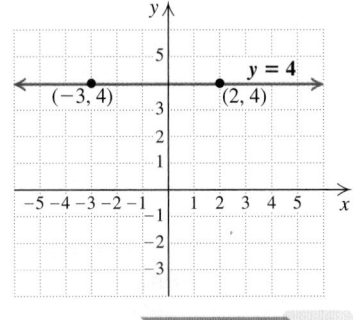

Any two points on a horizontal line have the same y-coordinate. Thus the change in y is 0, so the slope is 0.

TRY EXERCISE 55

A horizontal line has slope 0.

EXAMPLE 4   Find the slope of the line $x = -3$.

SOLUTION   Consider the points $(-3, 4)$ and $(-3, -2)$, which are on the line. The change in y, or the rise, is $-2 - 4$, or -6. The change in x, or the run, is $-3 - (-3)$, or 0. Thus,

$$m = \frac{-2 - 4}{-3 - (-3)}$$

$$= \frac{-6}{0}. \quad \text{(undefined)}$$

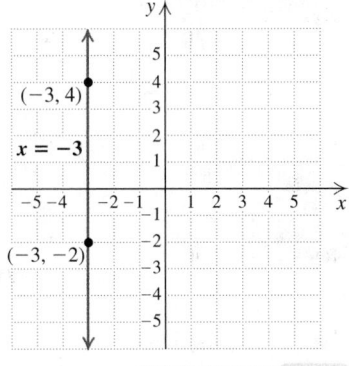

Since division by 0 is not defined, the slope of this line is not defined. The answer to a problem of this type is "The slope of this line is undefined."

TRY EXERCISE 57

The slope of a vertical line is undefined.

Applications

We have seen that slope has many real-world applications, ranging from car speed to production rate. Some applications use slope to measure steepness. For example, numbers like 2%, 3%, and 6% are often used to represent the **grade** of a road, a measure of a road's steepness. That is, since $3\% = \frac{3}{100}$, a 3% grade means that for every horizontal distance of 100 ft, the road rises or drops 3 ft. The concept of grade also occurs in skiing or snowboarding, where a 7% grade is considered very tame, but a 70% grade is considered steep.

EXAMPLE 5   *Skiing.*   Among the steepest skiable terrain in North America, the Headwall on Mount Washington, in New Hampshire, drops 720 ft over a horizontal distance of 900 ft. Find the grade of the Headwall.

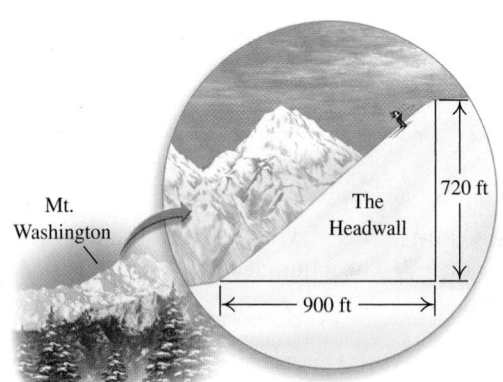

SOLUTION   The grade of the Headwall is its slope, expressed as a percent:

$$
\left.\begin{aligned}
m &= \frac{720}{900} \\
&= \frac{8}{10} \\
&= 80\%.
\end{aligned}\right\} \quad \text{Grade is slope expressed as a percent.}
$$

▶ TRY EXERCISE 63

Carpenters use slope when designing stairs, ramps, or roof pitches. Another application occurs in the engineering of a dam—the force or strength of a river depends on how much the river drops over a specified distance.

3.5   EXERCISE SET

For Extra Help   MyMathLab      Math XL PRACTICE   WATCH   DOWNLOAD

🖐 *Concept Reinforcement*   *State whether each of the following rates is positive, negative, or zero.*

1. The rate at which a teenager's height changes

2. The rate at which an elderly person's height changes

3. The rate at which a pond's water level changes during a drought

4. The rate at which a pond's water level changes during the rainy season

5. The rate at which a runner's distance from the starting point changes during a race

6. The rate at which a runner's distance from the finish line changes during a race

7. The rate at which the number of U.S. senators changes

8. The rate at which the number of people in attendance at a basketball game changes in the moments before the opening tipoff

9. The rate at which the number of people in attendance at a basketball game changes in the moments after the final buzzer sounds

10. The rate at which a person's I.Q. changes during his or her sleep

11. *Blogging.*   Find the rate at which a career blogger is paid.

Source: Based on information from Wired

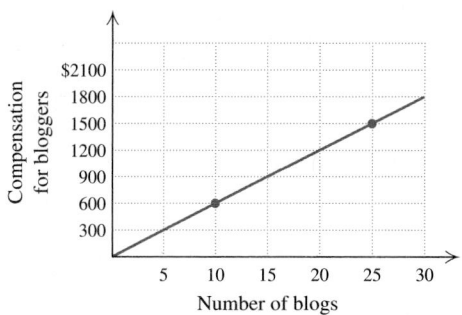

12. *Fitness.*   Find the rate at which a runner burns calories.

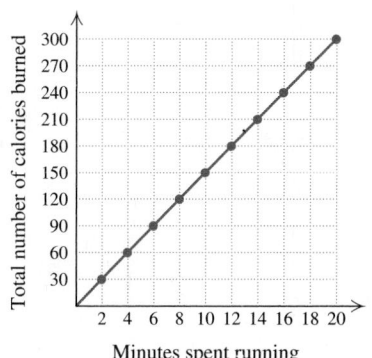

13. *Cell-phone prices.*   Find the rate of change in the average price of a new cell phone.

Source: Based on information from Market Reporter at PriceGrabber.com 2006

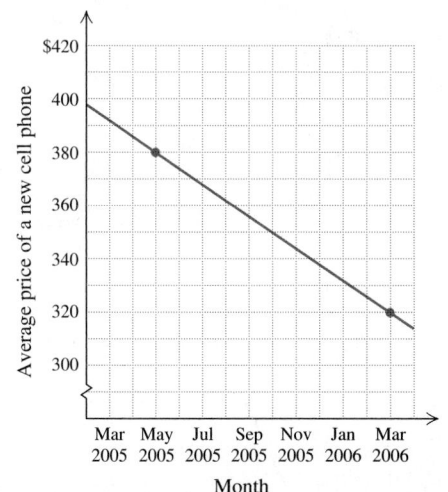

14. *Retail sales.*   Find the rate of change in the percentage of department stores' share of total retail sales.

Source: Based on information from the National Retail Federation

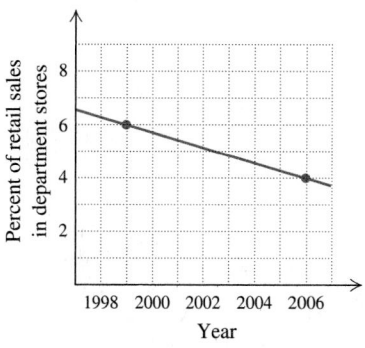

15. *College admission tests.*   Find the rate of change in SAT verbal scores with respect to family income.

Source: Based on 2004–2005 data from the National Center for Education Statistics

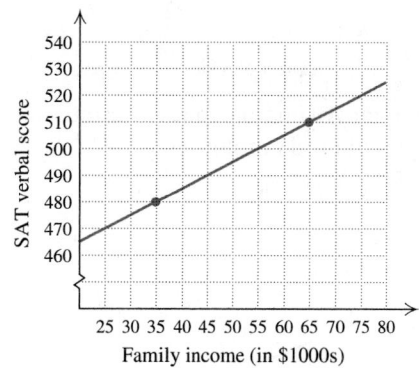

16. *Long-term care.*   Find the rate of change in Medicaid spending on long-term care.

Source: Based on data from Thomson Medstat, prepared by AARP Public Policy Institute

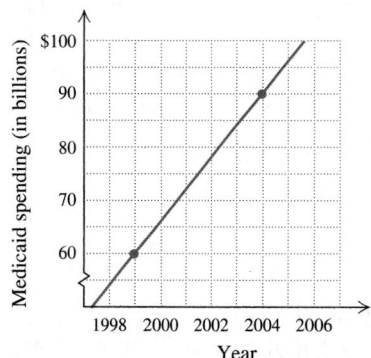

17. *Meteorology.*   Find the rate of change in the temperature in Spearfish, Montana, on January 22, 1943, as shown below.

Source: National Oceanic Atmospheric Administration

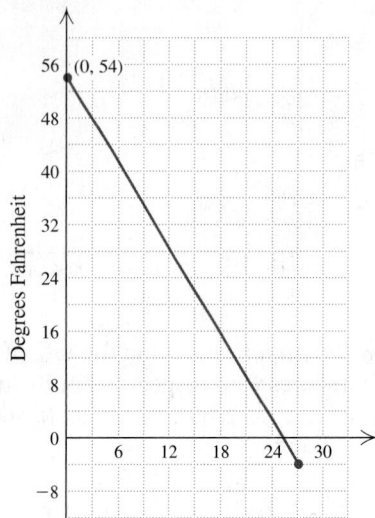

Number of minutes after 9 A.M.

18. Find the rate of change in the number of union-represented Ford employees.

Source: Ford Motor Co.

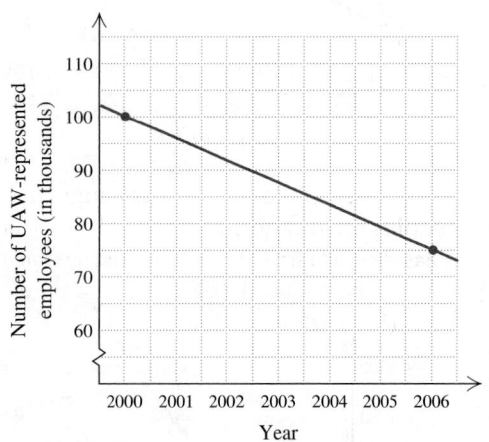

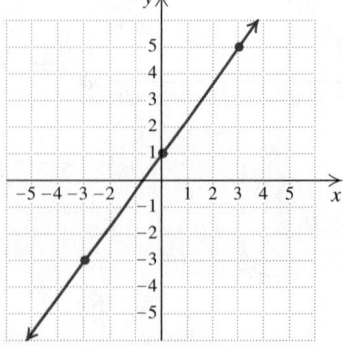

Find the slope, if it is defined, of each line. If the slope is undefined, state this.

19.

20.

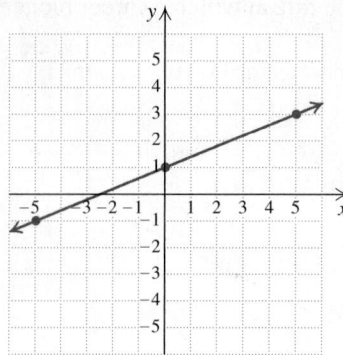

21.

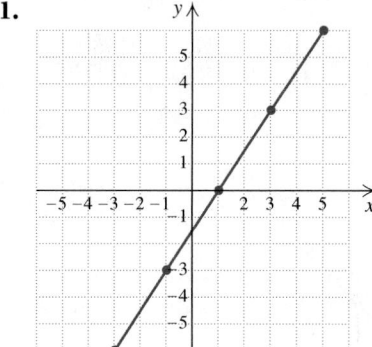

22.

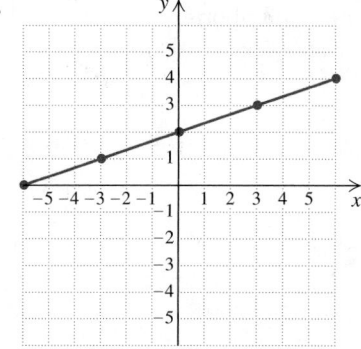

23.

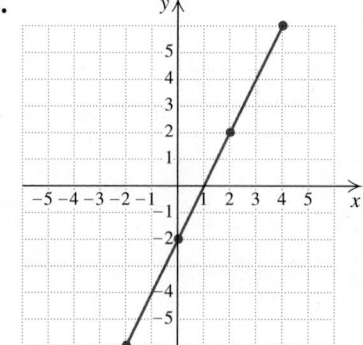

24.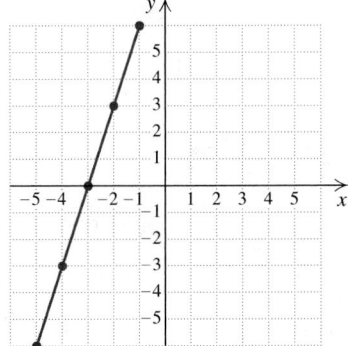

25.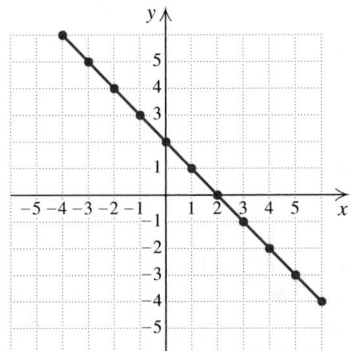

26.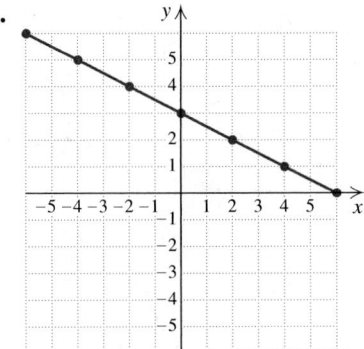

27.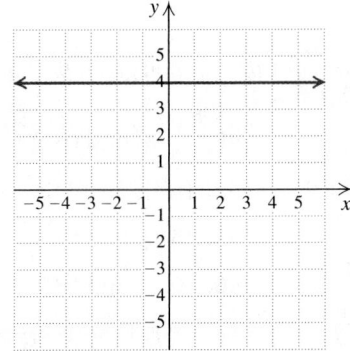

28.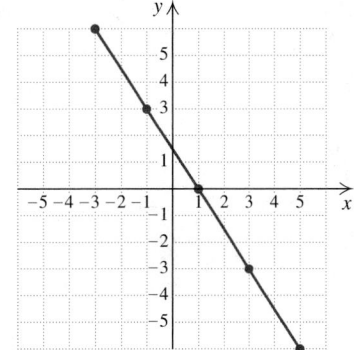

29.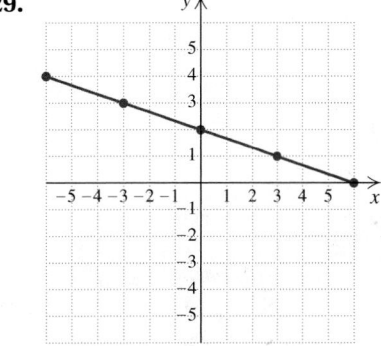

30.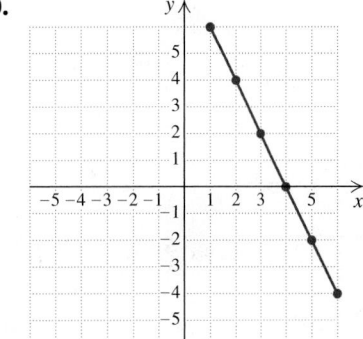

31.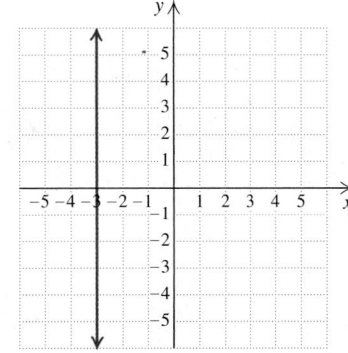

32.

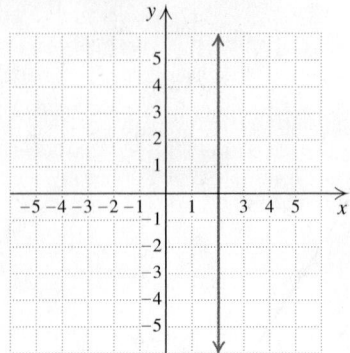

33.

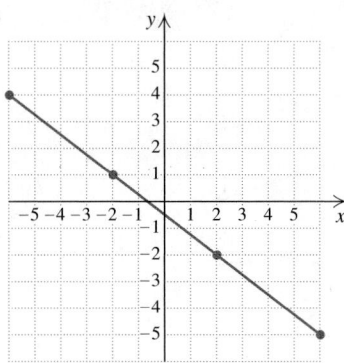

34.

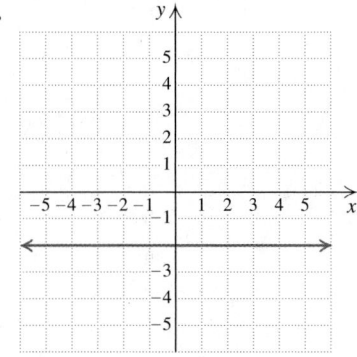

35.

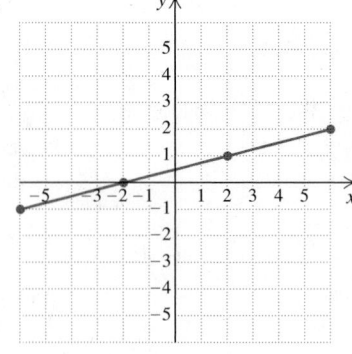

36.

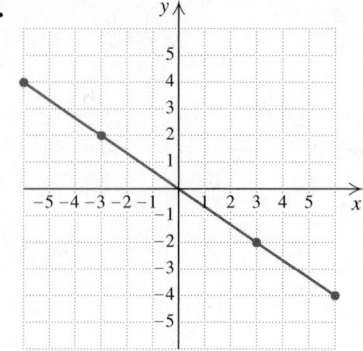

37.

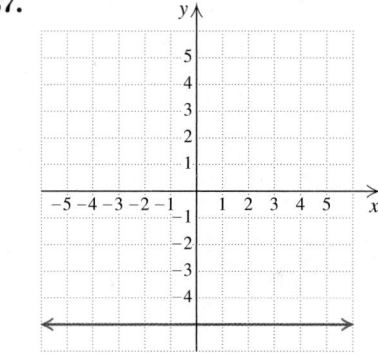

38.

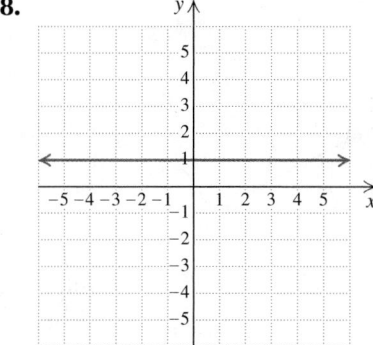

Find the slope of the line containing each given pair of points. If the slope is undefined, state this.

39. $(1, 3)$ and $(5, 8)$ **40.** $(1, 8)$ and $(6, 9)$

41. $(-2, 4)$ and $(3, 0)$ **42.** $(-4, 2)$ and $(2, -3)$

43. $(-4, 0)$ and $(5, 6)$ **44.** $(3, 0)$ and $(6, 9)$

45. $(0, 7)$ and $(-3, 10)$ **46.** $(0, 9)$ and $(-5, 0)$

47. $(-2, 3)$ and $(-6, 5)$ **48.** $(-1, 4)$ and $(5, -8)$

Aha! **49.** $\left(-2, \frac{1}{2}\right)$ and $\left(-5, \frac{1}{2}\right)$

50. $(-5, -1)$ and $(2, -3)$

51. $(5, -4)$ and $(2, -7)$

52. $(-10, 3)$ and $(-10, 4)$

53. $(6, -4)$ and $(6, 5)$

54. $(5, -2)$ and $(-4, -2)$

Find the slope of each line whose equation is given. If the slope is undefined, state this.

55. $y = 5$ **56.** $y = 13$

57. $x = -8$ **58.** $x = 18$

59. $x = 9$ **60.** $x = -7$

61. $y = -10$ **62.** $y = -4$

63. *Surveying.* Lick Skillet Road, near Boulder, Colorado, climbs 792 ft over a horizontal distance of 5280 ft. What is the grade of the road?

64. *Navigation.* Capital Rapids drops 54 ft vertically over a horizontal distance of 1080 ft. What is the slope of the rapids?

65. *Construction.* Part of New Valley rises 28 ft over a horizontal distance of 80 ft, and is too steep to build on. What is the slope of the land?

66. *Engineering.* At one point, Yellowstone's Beartooth Highway rises 315 ft over a horizontal distance of 4500 ft. Find the grade of the road.

67. *Carpentry.* Find the slope (or pitch) of the roof.

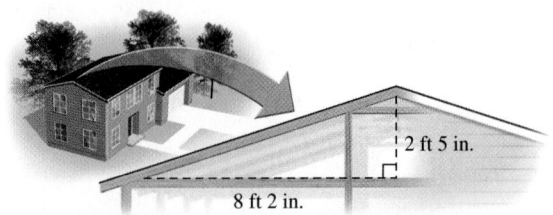

2 ft 5 in.

8 ft 2 in.

68. *Exercise.* Find the slope (or grade) of the treadmill.

0.4 ft

5 ft

69. *Bicycling.* To qualify as a rated climb on the Tour de France, a grade must average at least 4%. The ascent of Dooley Mountain, Oregon, part of the Elkhorn Classic, begins at 3500 ft and climbs to 5400 ft over a horizontal distance of 37,000 ft. What is the grade of the road? Would it qualify as a rated climb if it were part of the Tour de France?

Source: barkercityherald.com

70. *Construction.* Public buildings regularly include steps with 7-in. risers and 11-in. treads. Find the grade of such a stairway.

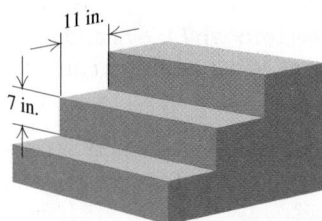

11 in.

7 in.

71. Explain why the order in which coordinates are subtracted to find slope does not matter so long as *y*-coordinates and *x*-coordinates are subtracted in the same order.

72. If one line has a slope of -3 and another has a slope of 2, which line is steeper? Why?

Skill Review

To prepare for Section 3.6, review solving a formula for a variable and graphing linear equations (Sections 2.3 and 3.2).

Solve. [2.3]

73. $ax + by = c$, for y **74.** $rx - mn = p$, for r

75. $ax - by = c$, for y **76.** $rs + nt = q$, for t

Graph. [3.2]

77. $8x + 6y = 24$ **78.** $3y = 4$

Synthesis

79. The points $(-4, -3)$, $(1, 4)$, $(4, 2)$, and $(-1, -5)$ are vertices of a quadrilateral. Use slopes to explain why the quadrilateral is a parallelogram.

80. Which is steeper and why: a ski slope that is 50° or one with a grade of 100%?

81. The plans below are for a skateboard "Fun Box". For the ramps labeled A, find the slope or grade.
Source: www.heckler.com

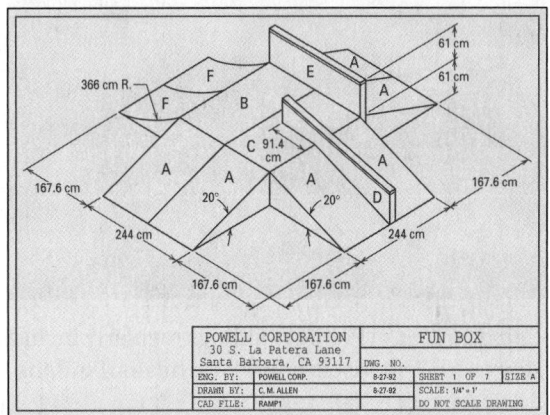

82. A line passes through $(4, -7)$ and never enters the first quadrant. What numbers could the line have for its slope?

83. A line passes through $(2, 5)$ and never enters the second quadrant. What numbers could the line have for its slope?

84. *Architecture.* Architects often use the equation $x + y = 18$ to determine the height y, in inches, of the riser of a step when the tread is x inches wide. Express the slope of stairs designed with this equation without using the variable y.

In Exercises 85 and 86, the slope of the line is $-\frac{2}{3}$, but the numbering on one axis is missing. How many units should each tick mark on that unnumbered axis represent?

85.

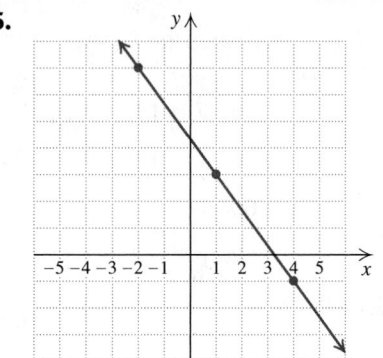

86.

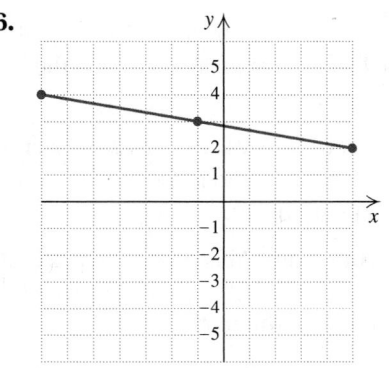

3.6 Slope–Intercept Form

Using the *y*-intercept and the Slope to Graph a Line ■ Equations in Slope–Intercept Form ■ Graphing and Slope–Intercept Form ■ Parallel and Perpendicular Lines

If we know the slope and the *y*-intercept of a line, it is possible to graph the line. In this section, we will discover that a line's slope and *y*-intercept can be determined directly from the line's equation, provided the equation is written in a certain form.

Using the *y*-intercept and the Slope to Graph a Line

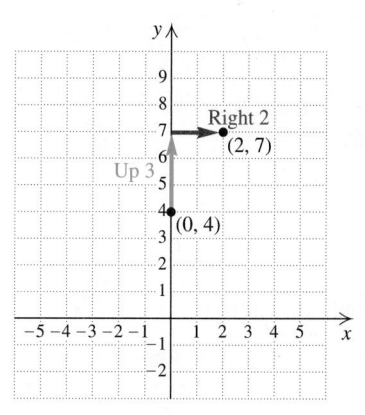

DL -1500

Let's modify the book-digitization situation that first appeared in Section 3.5. Suppose that as the information technologist arrives, 4 books had already been digitized by the DL-1500. If the rate of $\frac{3}{2}$ books per hour remains in effect, the table and graph shown here can be made.

DL-1500	
Hours Elapsed	**Books Digitized**
0	4
2	7
4	10
6	13
8	16

To confirm that the digitization rate is still $\frac{3}{2}$, we calculate the slope. Recall that

$$\text{Slope} = \frac{\text{change in } y}{\text{change in } x} = \frac{\text{rise}}{\text{run}} = \frac{y_2 - y_1}{x_2 - x_1},$$

where (x_1, y_1) and (x_2, y_2) are any two points on the graphed line. Here we select $(0, 4)$ and $(2, 7)$:

$$\text{Slope} = \frac{\text{change in } y}{\text{change in } x} = \frac{7 - 4}{2 - 0} = \frac{3}{2}.$$

Knowing that the slope is $\frac{3}{2}$, we could have drawn the graph by plotting $(0, 4)$ and from there moving *up* 3 units and *to the right* 2 units. This would have located the point $(2, 7)$. Using $(0, 4)$ and $(2, 7)$, we can then draw the line. This is the method used in the next example.

EXAMPLE 1

Draw a line that has slope $\frac{1}{4}$ and *y*-intercept $(0, 2)$.

SOLUTION We plot $(0, 2)$ and from there move *up* 1 unit and *to the right* 4 units. This locates the point $(4, 3)$. We plot $(4, 3)$ and draw a line passing through $(0, 2)$ and $(4, 3)$, as shown on the right below.

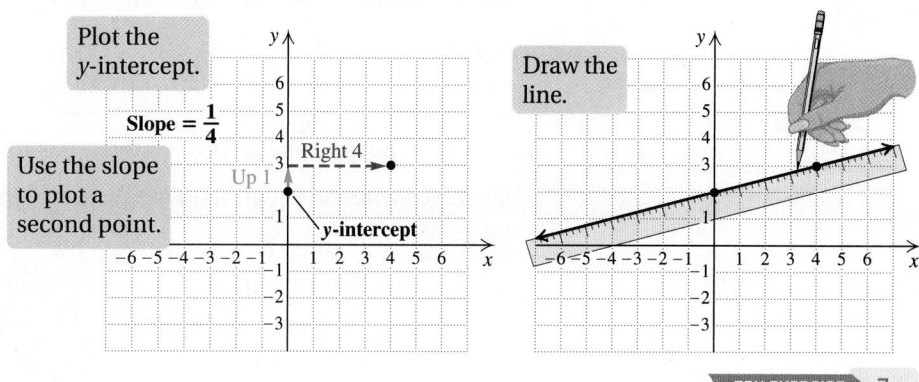

Plot the *y*-intercept.

Slope = $\frac{1}{4}$

Use the slope to plot a second point.

Up 1   Right 4

y-intercept

Draw the line.

TRY EXERCISE 7

Equations in Slope–Intercept Form

It is possible to read the slope and the y-intercept of a line directly from its equation. Recall from Section 3.3 that to find the y-intercept of an equation's graph, we replace x with 0 and solve the resulting equation for y. For example, to find the y-intercept of the graph of $y = 2x + 3$, we replace x with 0 and solve as follows:

$$y = 2x + 3$$
$$= 2 \cdot 0 + 3 = 0 + 3 = 3. \qquad \text{The } y\text{-intercept is } (0, 3).$$

The y-intercept of the graph of $y = 2x + 3$ is $(0, 3)$. It can be similarly shown that the graph of $y = mx + b$ has the y-intercept $(0, b)$.

To calculate the slope of the graph of $y = 2x + 3$, we need two ordered pairs that are solutions of the equation. The y-intercept $(0, 3)$ is one pair; a second pair, $(1, 5)$, can be found by substituting 1 for x. We then have

$$\text{Slope} = \frac{\text{change in } y}{\text{change in } x} = \frac{5 - 3}{1 - 0} = \frac{2}{1} = 2.$$

Note that the slope, 2, is also the x-coefficient in $y = 2x + 3$. It can be similarly shown that the graph of any equation of the form $y = mx + b$ has slope m (see Exercise 79).

STUDENT NOTES

An equation for a given line can be written in many different forms. Note that in the slope–intercept form, the equation is solved for y.

The Slope–Intercept Equation

The equation $y = mx + b$ is called the *slope–intercept equation*. The equation represents a line of slope m with y-intercept $(0, b)$.

The equation of any nonvertical line can be written in this form.

EXAMPLE 2 Find the slope and the y-intercept of each line whose equation is given.

a) $y = \frac{4}{5}x - 8$ **b)** $2x + y = 5$ **c)** $3x - 4y = 7$

SOLUTION

a) We rewrite $y = \frac{4}{5}x - 8$ as $y = \frac{4}{5}x + (-8)$. Now we simply read the slope and the y-intercept from the equation:

$$y = \frac{4}{5}x + (-8).$$

The slope is $\frac{4}{5}$. The y-intercept is $(0, -8)$.

b) We first solve for y to find an equivalent equation in the form $y = mx + b$:

$$2x + y = 5$$
$$y = -2x + 5. \qquad \text{Adding } -2x \text{ to both sides}$$

The slope is -2. The y-intercept is $(0, 5)$.

c) We rewrite the equation in the form $y = mx + b$:

$$3x - 4y = 7$$
$$-4y = -3x + 7 \qquad \text{Adding } -3x \text{ to both sides}$$
$$y = -\frac{1}{4}(-3x + 7) \qquad \text{Multiplying both sides by } -\frac{1}{4}$$
$$y = \frac{3}{4}x - \frac{7}{4}. \qquad \text{Using the distributive law}$$

The slope is $\frac{3}{4}$. The y-intercept is $\left(0, -\frac{7}{4}\right)$.

TRY EXERCISE 19

EXAMPLE 3   A line has slope $-\frac{12}{5}$ and y-intercept $(0, 11)$. Find an equation of the line.

SOLUTION   We use the slope–intercept equation, substituting $-\frac{12}{5}$ for m and 11 for b:

$$y = mx + b = -\tfrac{12}{5}x + 11.$$

The desired equation is $y = -\frac{12}{5}x + 11$.

TRY EXERCISE 35

EXAMPLE 4   *Threatened species.*   A threatened species is a species that is likely to become endangered in the future. Determine an equation for the graph of threatened species in the Blue Mountains.

Source: Based on information from Sustainable Blue Mountains

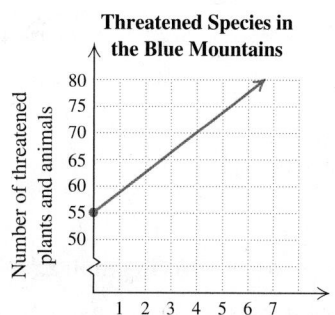

SOLUTION   To write an equation for a line, we can use slope–intercept form, provided the slope and the y-intercept are known. From the graph, we see that $(0, 55)$ is the y-intercept. Looking closely, we see that the line passes through $(4, 70)$. We can either count squares on the graph or use the formula to calculate the slope:

$$m = \frac{\text{change in } y}{\text{change in } x} = \frac{70 - 55}{4 - 0} = \frac{15}{4}.$$

The desired equation is

$$y = \frac{15}{4}x + 55, \qquad \text{Using } \tfrac{15}{4} \text{ for } m \text{ and } 55 \text{ for } b$$

where y is the number of threatened species in the Blue Mountains x years after 2000.

TRY EXERCISE 43

Graphing and Slope–Intercept Form

In Example 1, we drew a graph, knowing only the slope and the y-intercept. In Example 2, we determined the slope and the y-intercept of a line by examining its equation. We now combine the two procedures to develop a quick way to graph a linear equation.

EXAMPLE 5   Graph: **(a)** $y = \frac{3}{4}x + 5$;   **(b)** $2x + 3y = 3$.

SOLUTION

a)   From the equation $y = \frac{3}{4}x + 5$, we see that the slope of the graph is $\frac{3}{4}$ and the y-intercept is $(0, 5)$. We plot $(0, 5)$ and then consider the slope, $\frac{3}{4}$. Starting at $(0, 5)$, we plot a second point by moving *up* 3 units (since the numerator is *positive* and corresponds to the change in y) and *to the right* 4 units (since the denominator is *positive* and corresponds to the change in x). We reach a new point, $(4, 8)$.

Since $\frac{3}{4} = \frac{-3}{-4}$, we can again start at the *y*-intercept, $(0, 5)$, but move *down* 3 units (since the numerator is *negative* and corresponds to the change in *y*) and *to the left* 4 units (since the denominator is *negative* and corresponds to the change in *x*). We reach another point, $(-4, 2)$. Once two or three points have been plotted, the line representing all solutions of $y = \frac{3}{4}x + 5$ can be drawn.

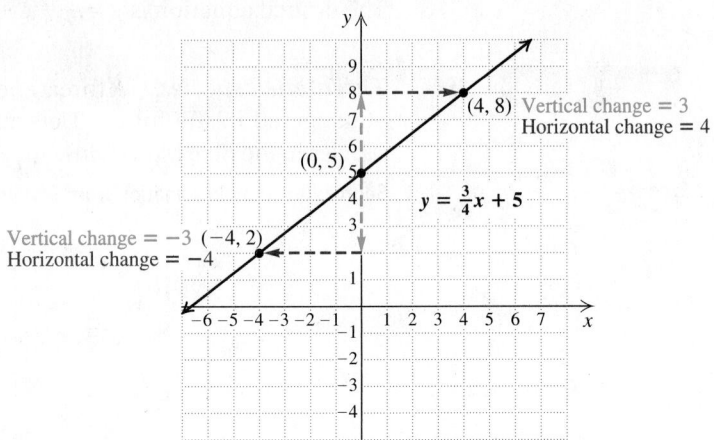

b) To graph $2x + 3y = 3$, we first rewrite it in slope–intercept form:

$$2x + 3y = 3$$

$$3y = -2x + 3 \qquad \text{Adding } -2x \text{ to both sides}$$

$$y = \tfrac{1}{3}(-2x + 3) \qquad \text{Multiplying both sides by } \tfrac{1}{3}$$

$$y = -\tfrac{2}{3}x + 1. \qquad \text{Using the distributive law}$$

To graph $y = -\frac{2}{3}x + 1$, we first plot the *y*-intercept, $(0, 1)$. We can think of the slope as $\frac{-2}{3}$. Starting at $(0, 1)$ and using the slope, we find a second point by moving *down* 2 units (since the numerator is *negative*) and *to the right* 3 units (since the denominator is *positive*). We plot the new point, $(3, -1)$. In a similar manner, we can move from the point $(3, -1)$ to locate a third point, $(6, -3)$. The line can then be drawn.

Since $-\frac{2}{3} = \frac{2}{-3}$, an alternative approach is to again plot $(0, 1)$, but this time move *up* 2 units (since the numerator is *positive*) and *to the left* 3 units (since the denominator is *negative*). This leads to another point on the graph, $(-3, 3)$.

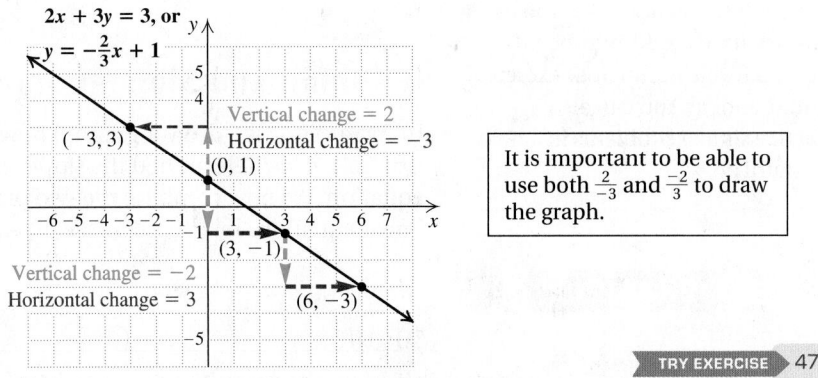

It is important to be able to use both $\frac{2}{-3}$ and $\frac{-2}{3}$ to draw the graph.

TRY EXERCISE 47

Slope–intercept form allows us to quickly determine the slope of a line by simply inspecting its equation. This can be especially helpful when attempting to decide whether two lines are parallel or perpendicular.

TECHNOLOGY CONNECTION

Using a standard $[-10, 10, -10, 10]$ window, graph the equations $y_1 = \frac{2}{3}x + 1$, $y_2 = \frac{3}{8}x + 1$, $y_3 = \frac{2}{3}x + 5$, and $y_4 = \frac{3}{8}x + 5$. If you can, use your graphing calculator in the MODE that graphs equations *simultaneously*. Once all lines have been drawn, try to decide which equation corresponds to each line. After matching equations with lines, you can check your matches by using TRACE and the up and down arrow keys to move from one line to the next. The number of the equation will appear in a corner of the screen.

1. Graph $y_1 = -\frac{3}{4}x - 2$, $y_2 = -\frac{1}{5}x - 2$, $y_3 = -\frac{3}{4}x - 5$, and $y_4 = -\frac{1}{5}x - 5$ using the SIMULTANEOUS mode. Then match each line with the corresponding equation. Check using TRACE.

Parallel and Perpendicular Lines

Two lines are parallel if they lie in the same plane and do not intersect no matter how far they are extended. If two lines are vertical, they are parallel. How can we tell if nonvertical lines are parallel? The answer is simple: We look at their slopes.

> **Slope and Parallel Lines**
>
> Two lines are parallel if they have the same slope or if both lines are vertical.

EXAMPLE 6   Determine whether the graphs of $y = -3x + 4$ and $6x + 2y = -10$ are parallel.

SOLUTION   We compare the slopes of the two lines to determine whether the graphs are parallel.

One of the two equations given is in slope–intercept form:

$y = -3x + 4.$ The slope is -3 and the y-intercept is $(0, 4)$.

To find the slope of the other line, we need to rewrite the other equation in slope–intercept form:

$$6x + 2y = -10$$
$$2y = -6x - 10 \qquad \text{Adding } -6x \text{ to both sides}$$
$$y = -3x - 5. \qquad \text{The slope is } -3 \text{ and the } y\text{-intercept is } (0, -5).$$

Since both lines have slope -3 but different y-intercepts, the graphs are parallel. There is no need for us to actually graph either equation.   **TRY EXERCISE** 63

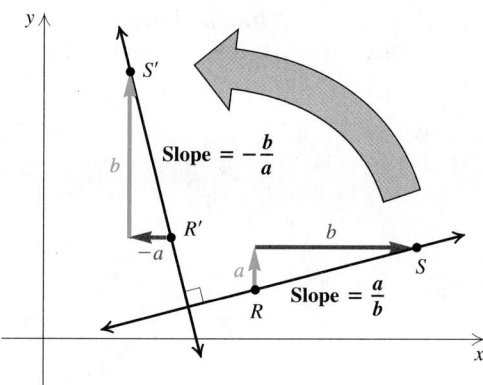

Two lines are perpendicular if they intersect at a right angle. If one line is vertical and another is horizontal, they are perpendicular. There are other instances in which two lines are perpendicular.

Consider a line $\overleftrightarrow{RS}$ as shown at left, with slope a/b. Then think of rotating the figure 90° to get a line $\overleftrightarrow{R'S'}$ perpendicular to $\overleftrightarrow{RS}$. For the new line, the rise and the run are interchanged, but the run is now negative. Thus the slope of the new line is $-b/a$. Let's multiply the slopes:

$$\frac{a}{b}\left(-\frac{b}{a}\right) = -1.$$

This can help us determine which lines are perpendicular.

> **Slope and Perpendicular Lines**
>
> Two lines are perpendicular if the product of their slopes is -1 or if one line is vertical and the other line is horizontal.

Thus, if one line has slope m ($m \neq 0$), the slope of a line perpendicular to it is $-1/m$. That is, we take the reciprocal of m ($m \neq 0$) and change the sign.

EXAMPLE 7 Determine whether the graphs of $2x + y = 8$ and $y = \frac{1}{2}x + 7$ are perpendicular.

SOLUTION One of the two equations given is in slope–intercept form:

$$y = \tfrac{1}{2}x + 7. \quad \text{The slope is } \tfrac{1}{2}.$$

To find the slope of the other line, we rewrite the other equation in slope–intercept form:

$$2x + y = 8$$
$$y = -2x + 8. \quad \text{Adding } -2x \text{ to both sides}$$

The slope of the line is -2.

The lines are perpendicular if the product of their slopes is -1. Since

$$\frac{1}{2}(-2) = -1,$$

the graphs are perpendicular.

TRY EXERCISE ▶ 69

EXAMPLE 8 Write a slope–intercept equation for the line whose graph is described.

a) Parallel to the graph of $2x - 3y = 7$, with y-intercept $(0, -1)$

b) Perpendicular to the graph of $2x - 3y = 7$, with y-intercept $(0, -1)$

SOLUTION We begin by determining the slope of the line represented by $2x - 3y = 7$:

$$2x - 3y = 7$$
$$-3y = -2x + 7 \quad \text{Adding } -2x \text{ to both sides}$$
$$y = \tfrac{2}{3}x - \tfrac{7}{3}. \quad \text{Dividing both sides by } -3$$

The slope is $\frac{2}{3}$.

a) A line parallel to the graph of $2x - 3y = 7$ has a slope of $\frac{2}{3}$. Since the y-intercept is $(0, -1)$, the slope–intercept equation is

$$y = \tfrac{2}{3}x - 1. \quad \text{Substituting in } y = mx + b$$

b) A line perpendicular to the graph of $2x - 3y = 7$ has a slope that is the opposite of the reciprocal of $\frac{2}{3}$, or $-\frac{3}{2}$. Since the y-intercept is $(0, -1)$, the slope–intercept equation is

$$y = -\tfrac{3}{2}x - 1. \quad \text{Substituting in } y = mx + b$$

TRY EXERCISE ▶ 75

3.6   EXERCISE SET

For Extra Help

↪ *Concept Reinforcement*   *In each of Exercises 1–6, match the phrase with the most appropriate choice from the column on the right.*

1. ____ The slope of the graph of $y = 3x - 2$ a) $\left(0, \frac{3}{4}\right)$

2. ____ The slope of the graph of $y = 2x - 3$ b) 2

3. ____ The slope of the graph of $y = \frac{2}{3}x + 3$ c) $(0, -3)$

4. ____ The y-intercept of the graph of $y = 2x - 3$ d) $\frac{2}{3}$

5. ____ The y-intercept of the graph of $y = 3x - 2$ e) $(0, -2)$

6. ____ The y-intercept of the graph of $y = \frac{2}{3}x + \frac{3}{4}$ f) 3

Draw a line that has the given slope and y-intercept.

7. Slope $\frac{2}{3}$; y-intercept $(0, 1)$

8. Slope $\frac{3}{5}$; y-intercept $(0, -1)$

9. Slope $\frac{5}{3}$; y-intercept $(0, -2)$

10. Slope $\frac{1}{2}$; y-intercept $(0, 0)$

11. Slope $-\frac{1}{3}$; y-intercept $(0, 5)$

12. Slope $-\frac{4}{5}$; y-intercept $(0, 6)$

13. Slope 2; y-intercept $(0, 0)$

14. Slope -2; y-intercept $(0, -3)$

15. Slope -3; y-intercept $(0, 2)$

16. Slope 3; y-intercept $(0, 4)$

Aha!   17. Slope 0; y-intercept $(0, -5)$

18. Slope 0; y-intercept $(0, 1)$

Find the slope and the y-intercept of each line whose equation is given.

19. $y = -\frac{2}{7}x + 5$

20. $y = -\frac{3}{8}x + 4$

21. $y = \frac{1}{3}x + 7$

22. $y = \frac{4}{5}x + 1$

23. $y = \frac{9}{5}x - 4$

24. $y = -\frac{9}{10}x - 5$

25. $-3x + y = 7$

26. $-4x + y = 7$

27. $4x + 2y = 8$

28. $3x + 4y = 12$

Aha!   29. $y = 3$

30. $y - 3 = 5$

31. $2x - 5y = -8$

32. $12x - 6y = 9$

33. $9x - 8y = 0$

34. $7x = 5y$

Find the slope–intercept equation for the line with the indicated slope and y-intercept.

35. Slope 5; y-intercept $(0, 7)$

36. Slope -4; y-intercept $\left(0, -\frac{3}{5}\right)$

37. Slope $\frac{7}{8}$; y-intercept $(0, -1)$

38. Slope $\frac{5}{7}$; y-intercept $(0, 4)$

39. Slope $-\frac{5}{3}$; y-intercept $(0, -8)$

40. Slope $\frac{3}{4}$; y-intercept $(0, -35)$

Aha!   41. Slope 0; y-intercept $\left(0, \frac{1}{3}\right)$

42. Slope 7; y-intercept $(0, 0)$

Determine an equation for each graph shown.

43.

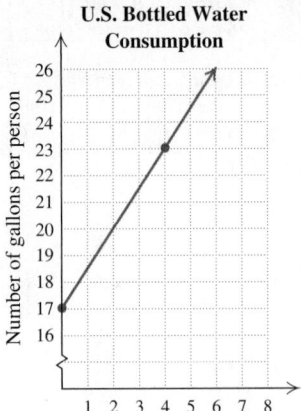

**U.S. Bottled Water
Consumption**

Number of gallons per person

Number of years since 2000

Based on information from the International Bottled
Water Association

44.

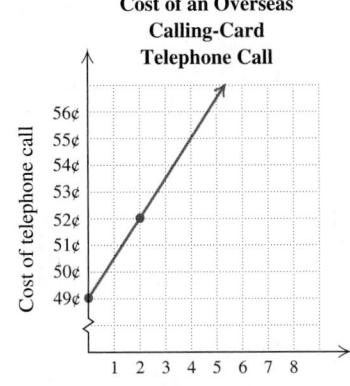

**Cost of an Overseas
Calling-Card
Telephone Call**

Cost of telephone call

Number of minutes

Source: www.pennytalk.com

45.

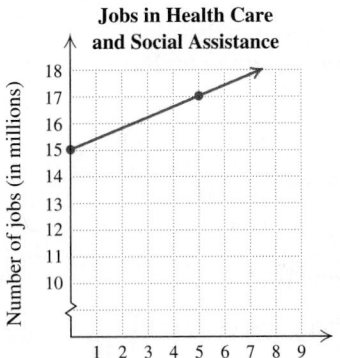

**Jobs in Health Care
and Social Assistance**

Number of jobs (in millions)

Number of years since 2000

Source: U.S. Bureau of Labor Statistics

46.

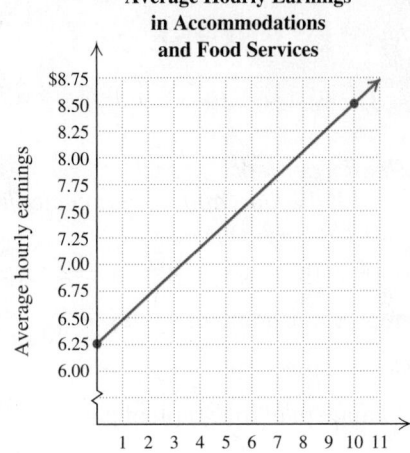

**Average Hourly Earnings
in Accommodations
and Food Services**

Average hourly earnings

Number of years since 1995

Source: U.S. Bureau of Labor Statistics

Graph.

47. $y = \frac{2}{3}x + 2$ **48.** $y = -\frac{2}{3}x - 3$

49. $y = -\frac{2}{3}x + 3$ **50.** $y = \frac{2}{3}x - 2$

51. $y = \frac{3}{2}x + 3$ **52.** $y = \frac{3}{2}x - 2$

53. $y = -\frac{4}{3}x + 3$ **54.** $y = -\frac{3}{2}x - 2$

55. $2x + y = 1$ **56.** $3x + y = 2$

57. $3x + y = 0$ **58.** $2x + y = 0$

59. $4x + 5y = 15$ **60.** $2x + 3y = 9$

61. $x - 4y = 12$ **62.** $x + 5y = 20$

*Determine whether each pair of equations represents
parallel lines.*

63. $y = \frac{3}{4}x + 6,$ **64.** $y = \frac{1}{3}x - 2,$
    $y = \frac{3}{4}x - 2$ $y = -\frac{1}{3}x + 1$

65. $y = 2x - 5,$ **66.** $y = -3x + 1,$
    $4x + 2y = 9$ $6x + 2y = 8$

67. $3x + 4y = 8,$ **68.** $3x = 5y - 2,$
    $7 - 12y = 9x$ $10y = 4 - 6x$

*Determine whether each pair of equations represents
perpendicular lines.*

69. $y = 4x - 5,$ **70.** $2x - 5y = -3,$
    $4y = 8 - x$ $2x + 5y = 4$

71. $x - 2y = 5,$ **72.** $y = -x + 7,$
    $2x + 4y = 8$ $y - x = 3$

73. $2x + 3y = 1,$
    $3x - 2y = 1$

74. $y = 5 - 3x,$
    $3x - y = 8$

Write a slope–intercept equation of the line whose graph is described.

75. Parallel to the graph of $y = 5x - 7$; y-intercept $(0, 11)$

76. Parallel to the graph of $2x - y = 1$; y-intercept $(0, -3)$

77. Perpendicular to the graph of $2x + y = 0$; y-intercept $(0, 0)$

78. Perpendicular to the graph of $y = \frac{1}{3}x + 7$; y-intercept $(0, 5)$

Aha! **79.** Parallel to the graph of $y = x$; y-intercept $(0, 3)$

Aha! **80.** Perpendicular to the graph of $y = x$; y-intercept $(0, 0)$

81. Perpendicular to the graph of $x + y = 3$; y-intercept $(0, -4)$

82. Parallel to the graph of $3x + 2y = 5$; y-intercept $(0, -1)$

83. Can a horizontal line be graphed using the method of Example 5? Why or why not?

84. If two lines are perpendicular, does it follow that the lines have slopes that are negative reciprocals of each other? Why or why not?

Skill Review

To prepare for Section 3.7, review solving a formula for a variable and subtracting real numbers (Sections 1.6 and 2.3).

Solve. [2.3]

85. $y - k = m(x - h)$, for y

86. $y - 9 = -2(x + 4)$, for y

Simplify. [1.6]

87. $-10 - (-3)$

88. $8 - (-5)$

89. $-4 - 5$

90. $-6 - 5$

Synthesis

91. Explain how it is possible for an incorrect graph to be drawn, even after plotting three points that line up.

92. Which would you prefer, and why: graphing an equation of the form $y = mx + b$ or graphing an equation of the form $Ax + By = C$?

93. Show that the slope of the line given by $y = mx + b$ is m. (*Hint*: Substitute both 0 and 1 for x to find two pairs of coordinates. Then use the formula, Slope $=$ change in y/change in x.)

94. Write an equation of the line with the same slope as the line given by $5x + 2y = 8$ and the same y-intercept as the line given by $3x - 7y = 10$.

95. Write an equation of the line parallel to the line given by $2x - 6y = 10$ and having the same y-intercept as the line given by $9x + 6y = 18$.

96. Write an equation of the line parallel to the line given by $3x - 2y = 8$ and having the same y-intercept as the line given by $2y + 3x = -4$.

97. Write an equation of the line perpendicular to the line given by $3x - 5y = 8$ and having the same y-intercept as the line given by $2x + 4y = 12$.

98. Write an equation of the line perpendicular to the line given by $2x + 3y = 7$ and having the same y-intercept as the line given by $5x + 2y = 10$.

99. Write an equation of the line perpendicular to the line given by $3x - 2y = 9$ and having the same y-intercept as the line given by $2x + 5y = 0$.

100. Write an equation of the line perpendicular to the line given by $2x + 5y = 6$ that passes through $(2, 6)$. (*Hint*: Draw a graph.)

CONNECTING the CONCEPTS

If any two points are plotted on a plane, there is only one line that will go through both points. Thus, if we know that the graph of an equation is a straight line, we need only find two points that are on that line. Then we can plot those points and draw the line that goes through them.

The different graphing methods discussed in this chapter present efficient ways of finding two points that are on the graph of the equation. Following is a general strategy for graphing linear equations.

1. Make sure that the equation is linear. Linear equations can always be written in the form $Ax + By = C$. (A or B may be 0.)
2. Graph the line. Use substitution to find two points on the line, or use a more efficient method based on the form of the equation.
3. Check the graph by finding another point that should be on the line and determining whether it does actually fall on the line.

Form of Equation	Graph
$x = a$	Draw a vertical line through $(a, 0)$.
$y = b$	Draw a horizontal line through $(0, b)$.
$y = mx + b$	Plot the y-intercept $(0, b)$. Start at the y-intercept and count off the rise and run using the slope m to find another point. Draw a line through the two points.
$Ax + By = C$	Determine the x- and y-intercepts $(a, 0)$ and $(0, b)$. Draw a line through the two points.

MIXED REVIEW

For each equation, (a) determine whether it is linear and (b) if it is linear, graph the line.

1. $x = 3$
2. $y = -2$
3. $y = \frac{1}{2}x + 3$
4. $4x + 3y = 12$
5. $y - 5 = x$
6. $x + y = -2$
7. $3xy = 6$
8. $2x = 3y$
9. $3 - y = 4$
10. $y = x^2 - 4$
11. $2y = 9x - 10$
12. $x + 8 = 7$
13. $2x - 6 = 3y$
14. $y - 2x = 4$
15. $2y - x = 4$
16. $y = \frac{1}{x}$
17. $x - 2y = 0$
18. $y = 4 - x$
19. $y = 4 + x$
20. $4x - 5y = 20$

3.7   Point–Slope Form

Writing Equations in Point–Slope Form   •   Graphing and Point–Slope Form   •
Estimations and Predictions Using Two Points

There are many applications in which a slope—or a rate of change—and an ordered pair are known. When the ordered pair is the y-intercept, an equation in slope–intercept form can be easily produced. When the ordered pair represents a point other than the y-intercept, a different form, known as *point–slope form*, is more convenient.

Writing Equations in Point–Slope Form

Consider a line with slope 2 passing through the point $(4, 1)$, as shown in the figure. In order for a point (x, y) to be on the line, the coordinates x and y must be solutions of the slope equation

$$\frac{y - 1}{x - 4} = 2.$$ If (x, y) is not on the line, this equation will not be true.

Take a moment to examine this equation. Pairs like $(5, 3)$ and $(3, -1)$ are on the line and are solutions, since

$$\frac{3 - 1}{5 - 4} = 2 \quad \text{and} \quad \frac{-1 - 1}{3 - 4} = 2.$$

When $x \neq 4$, then $x - 4 \neq 0$, and we can multiply on both sides of the slope equation by $x - 4$:

$$(x - 4) \cdot \frac{y - 1}{x - 4} = 2(x - 4)$$

$$y - 1 = 2(x - 4).$$ Removing a factor equal to 1: $\frac{x - 4}{x - 4} = 1$

Every point on the line is a solution of this equation.

This is considered **point–slope form** for the line shown in the figure at left. A point–slope equation can be written any time a line's slope and a point on the line are known.

> ### The Point–Slope Equation
>
> The equation $y - y_1 = m(x - x_1)$ is called the *point–slope equation* for the line with slope m that contains the point (x_1, y_1).

Point–slope form is especially useful in more advanced mathematics courses, where problems similar to the following often arise.

EXAMPLE   1   Write a point–slope equation for the line with slope $\frac{1}{5}$ that contains the point $(7, 2)$.

SOLUTION   We substitute $\frac{1}{5}$ for m, 7 for x_1, and 2 for y_1:

$$y - y_1 = m(x - x_1)$$ Using the point–slope equation
$$y - 2 = \frac{1}{5}(x - 7).$$ Substituting

TRY EXERCISE 13

EXAMPLE **2** Write a point–slope equation for the line with slope $-\frac{4}{3}$ that contains the point $(1, -6)$.

SOLUTION We substitute $-\frac{4}{3}$ for m, 1 for x_1, and -6 for y_1:

$$y - y_1 = m(x - x_1) \qquad \text{Using the point–slope equation}$$
$$y - (-6) = -\frac{4}{3}(x - 1). \qquad \text{Substituting} \qquad \boxed{\text{TRY EXERCISE} \; 19}$$

EXAMPLE **3** Write the slope–intercept equation for the line with slope 2 that contains the point $(3, 1)$.

SOLUTION There are two parts to this solution. First, we write an equation in point–slope form:

$$y - y_1 = m(x - x_1)$$
$$y - 1 = 2(x - 3). \qquad \text{Substituting} \qquad \boxed{\text{Write in point–slope form.}}$$

Next, we find an equivalent equation of the form $y = mx + b$:

$$y - 1 = 2(x - 3)$$
$$y - 1 = 2x - 6 \qquad \text{Using the distributive law} \qquad \boxed{\text{Write in slope–intercept form.}}$$
$$y = 2x - 5. \qquad \begin{array}{l}\text{Adding 1 to both sides to get} \\ \text{slope–intercept form}\end{array} \qquad \boxed{\text{TRY EXERCISE} \; 27}$$

STUDENT NOTES

There are several forms in which a line's equation can be written. For instance, as shown in Example 3, $y - 1 = 2(x - 3), y - 1 = 2x - 6$, and $y = 2x - 5$ all are equations for the same line.

EXAMPLE **4** Consider the line given by the equation $8y = 7x - 24$.

a) Write the slope–intercept equation for a parallel line passing through $(-1, 2)$.

b) Write the slope–intercept equation for a perpendicular line passing through $(-1, 2)$.

SOLUTION Both parts (a) and (b) require us to find the slope of the line given by $8y = 7x - 24$. To do so, we solve for y to find slope–intercept form:

$$8y = 7x - 24$$
$$y = \frac{7}{8}x - 3. \qquad \text{Multiplying both sides by } \frac{1}{8}$$
$$\underuparrow{\qquad\qquad} \text{The slope is } \frac{7}{8}.$$

a) The slope of any parallel line will be $\frac{7}{8}$. The point–slope equation yields

$$y - 2 = \frac{7}{8}[x - (-1)] \qquad \begin{array}{l}\text{Substituting } \frac{7}{8} \text{ for the slope and} \\ (-1, 2) \text{ for the point}\end{array}$$
$$y - 2 = \frac{7}{8}[x + 1]$$
$$y = \frac{7}{8}x + \frac{7}{8} + 2 \qquad \begin{array}{l}\text{Using the distributive law and} \\ \text{adding 2 to both sides}\end{array}$$
$$y = \frac{7}{8}x + \frac{23}{8}.$$

b) The slope of a perpendicular line is given by the opposite of the reciprocal of $\frac{7}{8}$, or $-\frac{8}{7}$. The point–slope equation yields

$$y - 2 = -\frac{8}{7}[x - (-1)] \qquad \begin{array}{l}\text{Substituting } -\frac{8}{7} \text{ for the slope and} \\ (-1, 2) \text{ for the point}\end{array}$$
$$y - 2 = -\frac{8}{7}[x + 1]$$
$$y = -\frac{8}{7}x - \frac{8}{7} + 2 \qquad \begin{array}{l}\text{Using the distributive law and} \\ \text{adding 2 to both sides}\end{array}$$
$$y = -\frac{8}{7}x + \frac{6}{7}. \qquad \boxed{\text{TRY EXERCISE} \; 39}$$

TECHNOLOGY CONNECTION

To check that the graphs of $y = \frac{7}{8}x - 3$ and $y = -\frac{8}{7}x + \frac{6}{7}$ are perpendicular, we use the ZSQUARE option of the ZOOM menu to create a "squared" window. This corrects for distortion that would result from the units on the axes being of different lengths.

1. Show that the graphs of $y = \frac{3}{4}x + 2$ and $y = -\frac{4}{3}x - 1$ appear to be perpendicular.
2. Show that the graphs of $y = -\frac{2}{5}x - 4$ and $y = \frac{5}{2}x + 3$ appear to be perpendicular.
3. To see that this type of check is not foolproof, graph

   $$y = \frac{31}{40}x + 2$$

   and $y = -\frac{40}{30}x - 1$.

   Are the lines perpendicular? Why or why not?

Graphing and Point–Slope Form

When we know a line's slope and a point that is on the line, we can draw the graph, much as we did in Section 3.6. For example, the information given in the statement of Example 3 is sufficient for drawing a graph.

EXAMPLE 5

Graph the line with slope 2 that passes through $(3, 1)$.

SOLUTION   We plot $(3, 1)$, move *up* 2 and *to the right* 1 $\left(\text{since } 2 = \frac{2}{1}\right)$, and draw the line.

Plot the point.

Use the slope to plot a second point.

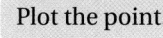
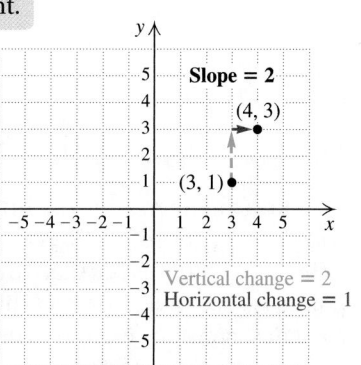

Draw the line.

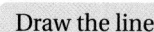
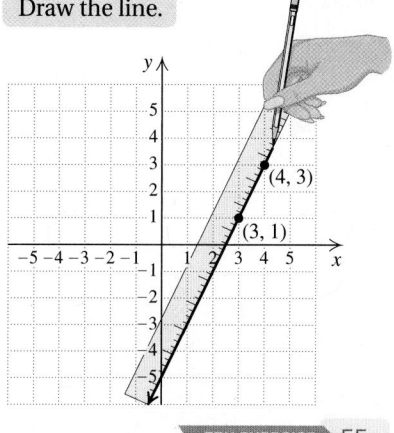

TRY EXERCISE 55

EXAMPLE 6

Graph: $y - 2 = 3(x - 4)$.

SOLUTION   Since $y - 2 = 3(x - 4)$ is in point–slope form, we know that the line has slope 3, or $\frac{3}{1}$, and passes through the point $(4, 2)$. We plot $(4, 2)$ and then find a second point by moving *up* 3 units and *to the right* 1 unit. The line can then be drawn, as shown below.

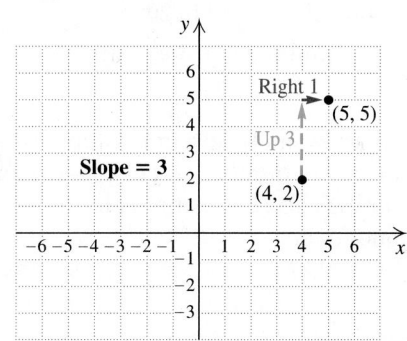
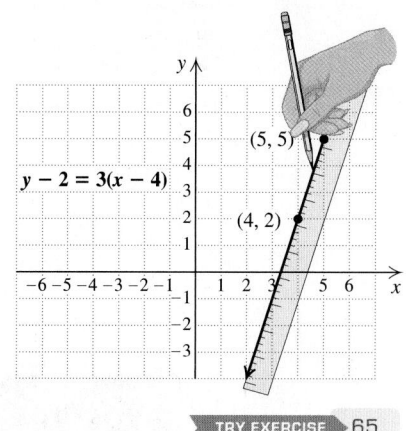

TRY EXERCISE 65

EXAMPLE 7

Graph: $y + 4 = -\frac{5}{2}(x + 3)$.

SOLUTION   Once we have written the equation in point–slope form, $y - y_1 = m(x - x_1)$, we can proceed much as we did in Example 6. To find an equivalent equation in point–slope form, we subtract opposites instead of adding:

$$y + 4 = -\frac{5}{2}(x + 3)$$
$$y - (-4) = -\frac{5}{2}(x - (-3)).$$ Subtracting a negative instead of adding a positive. This is now in point–slope form.

From this last equation, $y - (-4) = -\frac{5}{2}(x - (-3))$, we see that the line passes through $(-3, -4)$ and has slope $-\frac{5}{2}$, or $\frac{5}{-2}$.

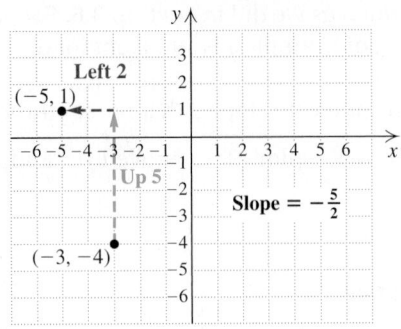

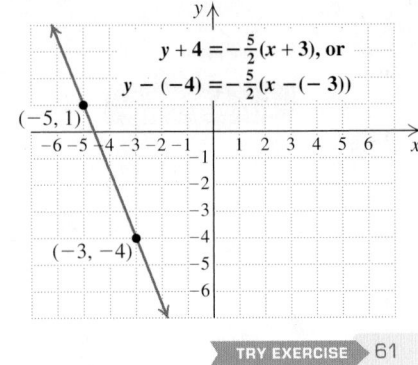

TRY EXERCISE ▸ 61

Estimations and Predictions Using Two Points

We can estimate real-life quantities that are not already known by using two points with known coordinates. When the unknown point is located *between* the two points, this process is called **interpolation**. If a graph passing through the known points is *extended* to predict future values, the process is called **extrapolation**. In statistics, methods exist for using a set of several points to interpolate or extrapolate values using curves other than lines.

EXAMPLE **8**

Student volunteers. An increasing number of college students are donating time and energy in volunteer service. The number of college-student volunteers grew from 2.7 million in 2002 to 3.3 million in 2005.

Source: Corporation for National and Community Service

a) Graph the line passing through the given data points, letting $x =$ the number of years since 2000.

b) Determine an equation for the line and estimate the number of college students who volunteered in 2004.

c) Predict the number of college students who will volunteer in 2010.

SOLUTION

a) We first draw and label a horizontal axis to display the year and a vertical axis to display the number of college-student volunteers, in millions. Next, we number the axes, choosing scales that will include both the given values and the values to be estimated.

Since $x =$ the number of years since 2000, we plot the points $(2, 2.7)$ and $(5, 3.3)$ and draw a line passing through both points.

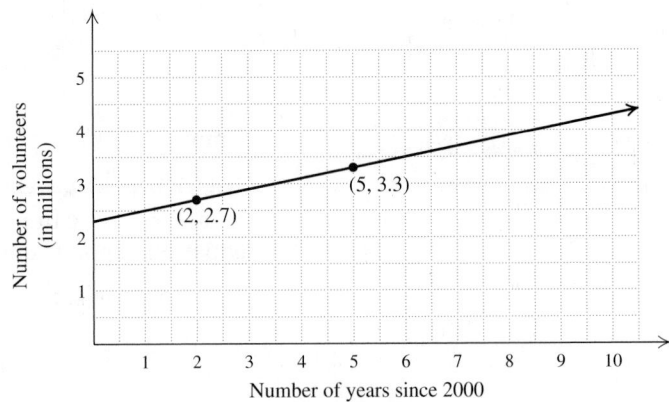

b) To find an equation for the line, we first calculate its slope:

$$m = \frac{\text{change in } y}{\text{change in } x} = \frac{3.3 - 2.7}{5 - 2} = \frac{0.6}{3} = 0.2.$$

The number of college-student volunteers increased at a rate of 0.2 million students per year. We can use either of the given points to write a point–slope equation for the line. Let's use $(2, 2.7)$ and then write an equivalent equation in slope–intercept form:

$y - 2.7 = 0.2(x - 2)$ This is a point–slope equation.

$y - 2.7 = 0.2x - 0.4$ Using the distributive law

$y = 0.2x + 2.3.$ Adding 2.7 to both sides. This is slope–intercept form.

To estimate the number of college-student volunteers in 2004, we substitute 4 for x in the slope–intercept equation:

$y = 0.2 \cdot 4 + 2.3 = 3.1.$

As the graph confirms, in 2004 there were about 3.1 million college-student volunteers.

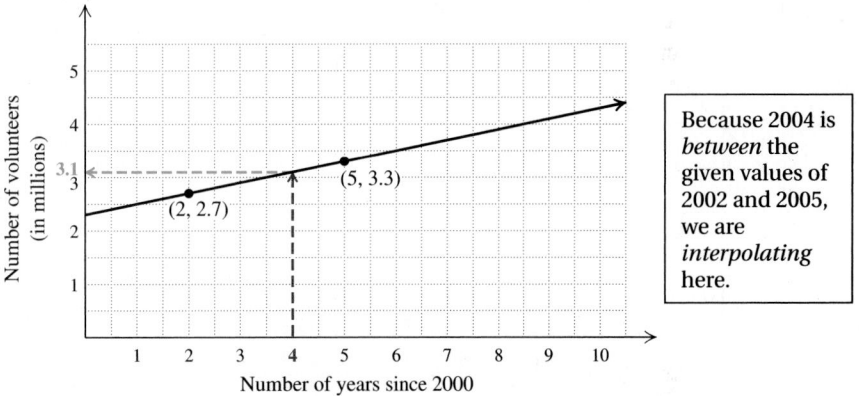

Because 2004 is *between* the given values of 2002 and 2005, we are *interpolating* here.

c) To predict the number of college-student volunteers in 2010, we again substitute for x in the slope–intercept equation:

$y = 0.2 \cdot 10 + 2.3 = 4.3.$ 2010 is 10 yr after 2000.

As we can see from the graph, if the trend continues, there will be about 4.3 million college-student volunteers in 2010.

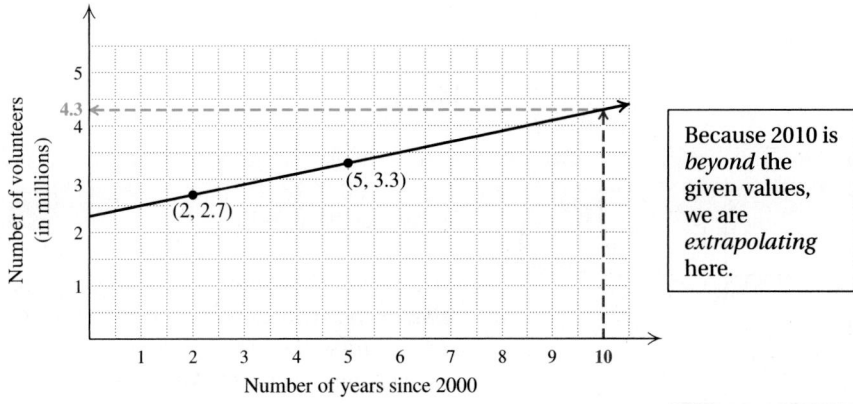

Because 2010 is *beyond* the given values, we are *extrapolating* here.

TRY EXERCISE 71

Visualizing for Success

A
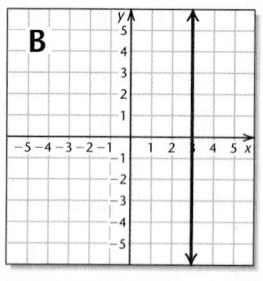

B
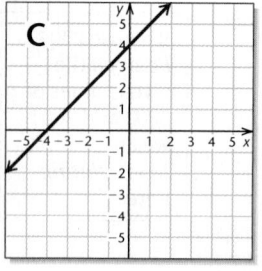

C
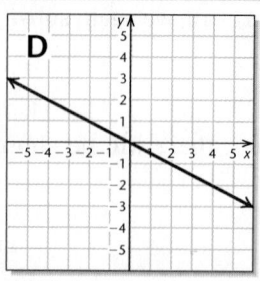

D
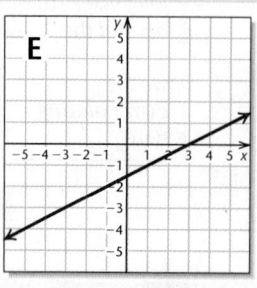

E

F
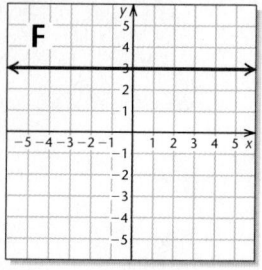

G
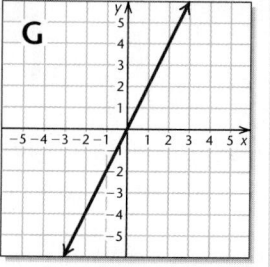

H
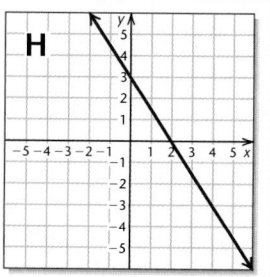

I
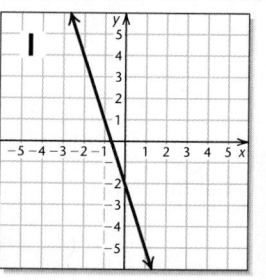

J
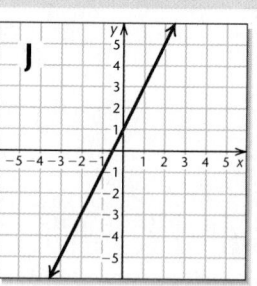

Match each equation with its graph.

1. $y = x + 4$

2. $y = 2x$

3. $y = 3$

4. $x = 3$

5. $y = -\frac{1}{2}x$

6. $2x - 3y = 6$

7. $y = -3x - 2$

8. $3x + 2y = 6$

9. $y - 3 = 2(x - 1)$

10. $y + 2 = \frac{1}{2}(x + 1)$

Answers on page A-13

An additional, animated version of this activity appears in MyMathLab. To use MyMathLab, you need a course ID and a student access code. Contact your instructor for more information.

🔖 **Concept Reinforcement**   *In each of Exercises 1–8, match the given information about a line with the appropriate equation from the column on the right.*

1. ____ Slope 5; includes (2, 3)

2. ____ Slope 5; includes (3, 2)

3. ____ Slope −5; includes (2, 3)

4. ____ Slope −5; includes (3, 2)

5. ____ Slope −5; includes (−2, −3)

6. ____ Slope 5; includes (−2, −3)

7. ____ Slope −5; includes (−3, −2)

8. ____ Slope 5; includes (−3, −2)

a) $y + 3 = 5(x + 2)$

b) $y - 2 = 5(x - 3)$

c) $y + 2 = 5(x + 3)$

d) $y - 3 = -5(x - 2)$

e) $y + 3 = -5(x + 2)$

f) $y + 2 = -5(x + 3)$

g) $y - 3 = 5(x - 2)$

h) $y - 2 = -5(x - 3)$

In each of Exercises 9–12, match the graph with the appropriate equation from the column on the right.

9.

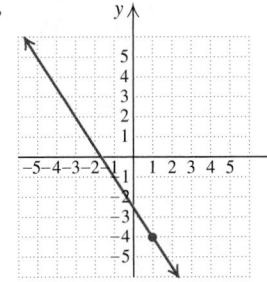

a) $y - 4 = -\frac{3}{2}(x + 1)$

b) $y - 4 = \frac{3}{2}(x + 1)$

c) $y + 4 = -\frac{3}{2}(x - 1)$

d) $y + 4 = \frac{3}{2}(x - 1)$

10.

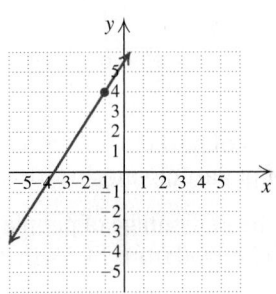

11.

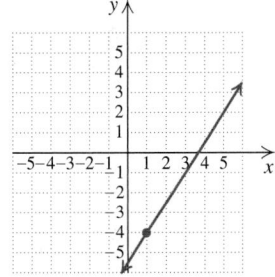

12.

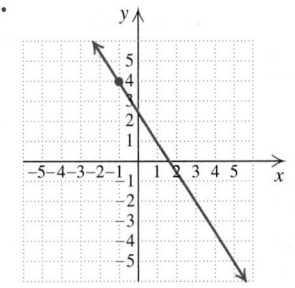

Write a point–slope equation for the line with the given slope that contains the given point.

13. $m = 3;\ (1, 6)$

14. $m = 2;\ (3, 7)$

15. $m = \frac{3}{5};\ (2, 8)$

16. $m = \frac{2}{3};\ (4, 1)$

17. $m = -4;\ (3, 1)$

18. $m = -5;\ (6, 2)$

19. $m = \frac{3}{2};\ (5, -4)$

20. $m = -\frac{4}{3};\ (7, -1)$

21. $m = -\frac{5}{4};\ (-2, 6)$

22. $m = \frac{7}{2};\ (-3, 4)$

23. $m = -2;\ (-4, -1)$

24. $m = -3;\ (-2, -5)$

25. $m = 1;\ (-2, 8)$

26. $m = -1;\ (-3, 6)$

Write the slope–intercept equation for the line with the given slope that contains the given point.

27. $m = 4;\ (3, 5)$

28. $m = 3;\ (6, 2)$

29. $m = \frac{7}{4};\ (4, -2)$

30. $m = \frac{8}{3};\ (3, -4)$

31. $m = -2;\ (-3, 7)$

32. $m = -3;\ (-2, 1)$

33. $m = -4;\ (-2, -1)$

34. $m = -5;\ (-1, -4)$

35. $m = \frac{2}{3};\ (5, 6)$

36. $m = \frac{3}{2};\ (7, 4)$

Aha! **37.** $m = -\frac{5}{6};\ (0, 4)$

38. $m = -\frac{3}{4};\ (0, 5)$

Write an equation of the line containing the specified point and parallel to the indicated line.

39. $(2, 5)$, $x - 2y = 3$

40. $(1, 4)$, $3x + y = 5$

Aha! **41.** $(0, -5)$, $y = 4x + 3$

42. $(0, 2)$, $y = x - 11$

43. $(-2, -3)$, $2x + 3y = -7$

44. $(-7, 0)$, $5x + 2y = 6$

45. $(5, -4)$, $x = 2$

46. $(-3, 6)$, $y = 7$

Write an equation of the line containing the specified point and perpendicular to the indicated line.

47. $(3, 1)$, $2x - 3y = 4$

48. $(6, 0)$, $5x + 4y = 1$

49. $(-4, 2)$, $x + y = 6$

50. $(-2, -5)$, $x - 2y = 3$

Aha! **51.** $(0, 6)$, $2x - 5 = y$

52. $(0, -7)$, $4x + 3 = y$

53. $(-3, 7)$, $y = 5$

54. $(4, -2)$, $x = 1$

55. Graph the line with slope $\frac{4}{3}$ that passes through the point $(1, 2)$.

56. Graph the line with slope $\frac{2}{5}$ that passes through the point $(3, 4)$.

57. Graph the line with slope $-\frac{3}{4}$ that passes through the point $(2, 5)$.

58. Graph the line with slope $-\frac{3}{2}$ that passes through the point $(1, 4)$.

Graph.

59. $y - 5 = \frac{1}{3}(x - 2)$ **60.** $y - 2 = \frac{1}{2}(x - 1)$

61. $y - 1 = -\frac{1}{4}(x - 3)$ **62.** $y - 1 = -\frac{1}{2}(x - 3)$

63. $y + 2 = \frac{2}{3}(x - 1)$ **64.** $y - 1 = \frac{3}{4}(x + 5)$

65. $y + 4 = 3(x + 1)$

66. $y + 3 = 2(x + 1)$

67. $y - 4 = -2(x + 1)$

68. $y + 3 = -1(x - 4)$

69. $y + 1 = -\frac{3}{5}(x - 2)$

70. $y - 2 = -\frac{2}{3}(x + 1)$

In Exercises 71–78, graph the given data and determine an equation for the related line (as in Example 8). Then use the equation to answer parts (a) and (b).

71. *Birth rate among teenagers.*   The birth rate among teenagers, measured in births per 1000 females age 15–19, fell steadily from 62.1 in 1991 to 41.1 in 2007.
Source: National Center for Health Statistics

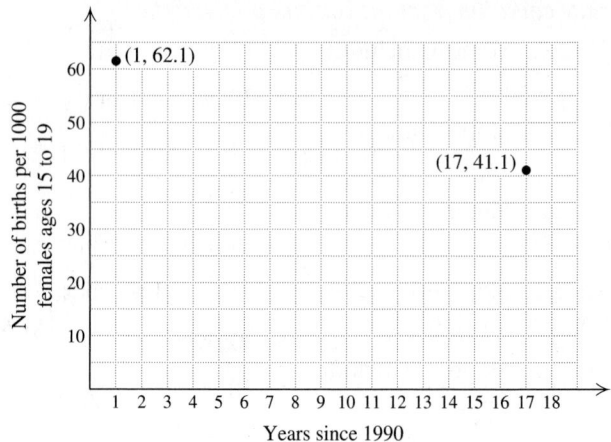

a) Calculate the birth rate among teenagers in 1999.
b) Predict the birth rate among teenagers in 2008.

72. *Food-stamp program participation.*   Participation in the U.S. food-stamp program grew from approximately 17.1 million people in 2000 to approximately 23 million in 2004.
Source: U.S. Department of Agriculture

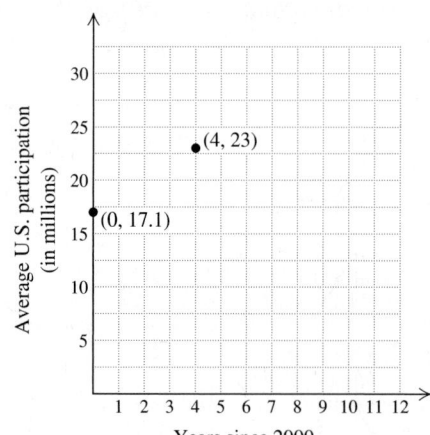

a) Calculate the number of participants in 2001.
b) Predict the number of participants in 2010.

73. *Cigarette smoking.*   The percentage of people age 25–44 who smoke has changed from 14.2% in 2001 to 10.8% in 2004.
Source: Office on Smoking and Health, National Center for Chronic Disease Prevention and Health Promotion
a) Calculate the percentage of people age 25–44 who smoked in 2002.
b) Predict the percentage of people age 25–44 who will smoke in 2010.

74. *Cigarette smoking.* The percentage of people age 18–24 who smoke has dropped from 5.2% in 2001 to 3.4% in 2004.

Source: Office on Smoking and Health, National Center for Chronic Disease Prevention and Health Promotion

a) Calculate the percentage of people age 18–24 who smoked in 2003.

b) Estimate the percentage of people age 18–24 who smoked in 2008.

75. *College enrollment.* U.S. college enrollment has grown from approximately 14.3 million in 1995 to 17.4 million in 2005.

Source: National Center for Education Statistics

a) Calculate the U.S. college enrollment for 2002.

b) Predict the U.S. college enrollment for 2010.

76. *High school enrollment.* U.S. high school enrollment has changed from approximately 13.7 million in 1995 to 16.3 million in 2005.

Source: National Center for Education Statistics

a) Calculate the U.S. high school enrollment for 2002.

b) Predict the U.S. high school enrollment for 2010.

77. *Aging population.* The number of U.S. residents over the age of 65 was approximately 31 million in 1990 and 36.3 million in 2004.

Source: U.S. Census Bureau

Aha!   **a)** Calculate the number of U.S. residents over the age of 65 in 1997.

b) Predict the number of U.S. residents over the age of 65 in 2010.

78. *Urban population.* The percentage of the U.S. population that resides in metropolitan areas increased from about 78% in 1980 to about 83% in 2006.

a) Calculate the percentage of the U.S. population residing in metropolitan areas in 1992.

b) Predict the percentage of the U.S. population residing in metropolitan areas in 2012.

Write the slope–intercept equation for the line containing the given pair of points.

79. $(2, 3)$ and $(4, 1)$

80. $(6, 8)$ and $(3, 5)$

81. $(-3, 1)$ and $(3, 5)$

82. $(-3, 4)$ and $(3, 1)$

83. $(5, 0)$ and $(0, -2)$

84. $(-2, 0)$ and $(0, 3)$

85. $(-4, -1)$ and $(1, 9)$

86. $(-3, 5)$ and $(-1, -3)$

87. Can equations for horizontal or vertical lines be written in point–slope form? Why or why not?

88. Describe a situation in which it is easier to graph the equation of a line in point–slope form than in slope–intercept form.

Skill Review

To prepare for Chapter 4, review exponential notation and order of operations (Section 1.8).

Simplify. [1.8]

89. $(-5)^3$

90. $(-2)^6$

91. -2^6

92. $3 \cdot 2^4 - 5 \cdot 2^3$

93. $2 - (3 - 2^2) + 10 \div 2 \cdot 5$

94. $(5 - 7)^2(3 - 2 \cdot 2)$

Synthesis

95. Describe a procedure that can be used to write the slope–intercept equation for any nonvertical line passing through two given points.

96. Any nonvertical line has many equations in point–slope form, but only one in slope–intercept form. Why is this?

Graph.

Aha!   **97.** $y - 3 = 0(x - 52)$

98. $y + 4 = 0(x + 93)$

Write the slope–intercept equation for each line shown.

99.

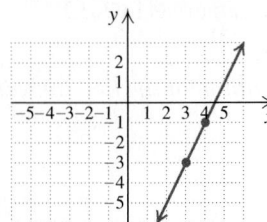

100.

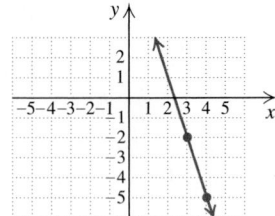

101.

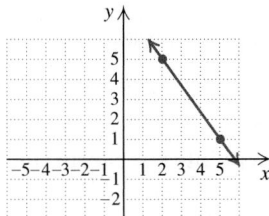

102.

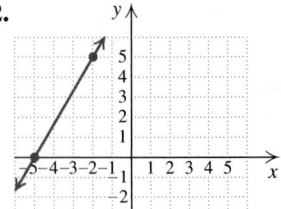

Aha! **103.** Write an equation of the line parallel to the line given by $y = 3 - 4x$ that passes through $(0, 7)$.

104. Write the slope–intercept equation of the line that has the same y-intercept as the line $x - 3y = 6$ and contains the point $(5, -1)$.

105. Write the slope–intercept equation of the line that contains the point $(-1, 5)$ and is parallel to the line passing through $(2, 7)$ and $(-1, -3)$.

106. Write the slope–intercept equation of the line that has x-intercept $(-2, 0)$ and is parallel to $4x - 8y = 12$.

Another form of a linear equation is the double-intercept *form:* $\dfrac{x}{a} + \dfrac{y}{b} = 1$. *From this form, we can read the* x-intercept $(a, 0)$ *and the* y-intercept $(0, b)$ *directly.*

107. Find the x-intercept and the y-intercept of the graph of the line given by $\dfrac{x}{2} + \dfrac{y}{5} = 1$.

108. Find the x-intercept and the y-intercept of the graph of the line given by $\dfrac{x}{10} - \dfrac{y}{3} = 1$.

109. Write the equation $4y - 3x = 12$ in double-intercept form and find the intercepts.

110. Write the equation $6x + 5y = 30$ in double-intercept form and find the intercepts.

111. Why is slope–intercept form more useful than point–slope form when using a graphing calculator? How can point–slope form be modified so that it is more easily used with graphing calculators?

CONNECTING the CONCEPTS

Any line can be described by a number of equivalent equations. We write the equation in the form that is most useful for us. For example, all four of the equations below describe the given line.

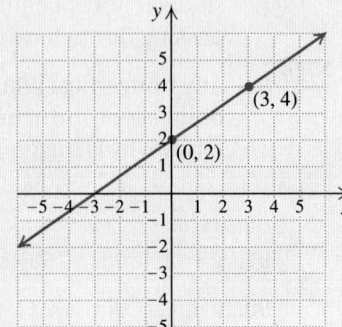

$2x - 3y = -6;$

$y = \tfrac{2}{3}x + 2;$

$y - 4 = \tfrac{2}{3}(x - 3);$

$2x + 6 = 3y$

Form of a Linear Equation	Example	Uses
Standard form: $Ax + By = C$	$2x - 3y = -6$	Finding x- and y-intercepts; Graphing using intercepts
Slope–intercept form: $y = mx + b$	$y = \frac{2}{3}x + 2$	Finding slope and y-intercept; Graphing using slope and y-intercept; Writing an equation given slope and y-intercept
Point–slope form: $y - y_1 = m(x - x_1)$	$y - 4 = \frac{2}{3}(x - 3)$	Finding slope and a point on the line; Graphing using slope and a point on the line; Writing an equation given slope and a point on the line

MIXED REVIEW

Tell whether each equation is in standard form, slope–intercept form, point–slope form, or none of these.

1. $y = -\frac{1}{2}x - 7$

2. $5x - 8y = 10$

3. $x = y + 2$

4. $\frac{1}{2}x + \frac{1}{3}y = 5$

5. $y - 2 = 5(x + 1)$

6. $3y + 7 = x$

Write each equation in standard form.

7. $2x = 5y + 10$

8. $x = y + 2$

9. $y = 2x + 7$

10. $y = -\frac{1}{2}x + 3$

11. $y - 2 = 3(x + 7)$

12. $x - 7 = 11$

Write each equation in slope–intercept form.

13. $2x - 7y = 8$

14. $y + 5 = -(x + 3)$

15. $8x = y + 3$

16. $6x + 10y = 30$

17. $9y = 5 - 8x$

18. $x - y = 3x + y$

19. $2 - 3y = 5y + 6$

20. $3(x - 4) = 6(y - 2)$

Study Summary

KEY TERMS AND CONCEPTS	EXAMPLES

SECTION 3.1: READING GRAPHS, PLOTTING POINTS, AND SCALING GRAPHS

Ordered pairs, like $(-3, 2)$ or $(4, 3)$, can be **plotted** or **graphed** using a **coordinate system** that uses two **axes**, which are most often labeled x and y. The axes intersect at the **origin**, $(0, 0)$, and divide a plane into four **quadrants**.

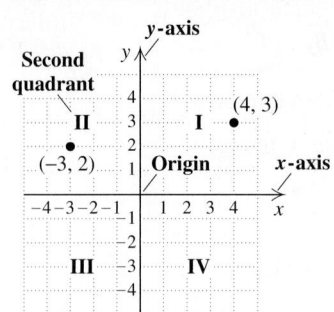

SECTION 3.2: GRAPHING LINEAR EQUATIONS

To **graph** an equation means to make a drawing that represents all of its solutions.

A **linear equation**, such as $y = 2x - 7$ or $2x + 3y = 12$, has a graph that is a straight line.

Any linear equation can be graphed by finding two ordered pairs that are solutions of the equation, plotting the corresponding points, and drawing a line through those points. Plotting a third point serves as a check.

x	y	(x, y)
1	2	$(1, 2)$
0	−1	$(0, -1)$
−1	−4	$(-1, -4)$

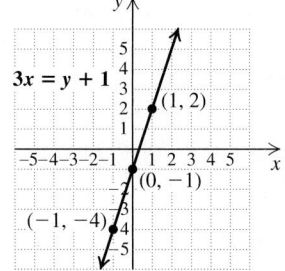

SECTION 3.3: GRAPHING AND INTERCEPTS

Intercepts
To find a y-intercept $(0, b)$, let $x = 0$ and solve for y.
To find an x-intercept $(a, 0)$, let $y = 0$ and solve for x.

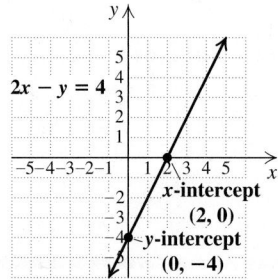

Horizontal Lines
The slope of a horizontal line is 0.
The graph of $y = b$ is a horizontal line, with y-intercept $(0, b)$.

Vertical Lines
The slope of a vertical line is undefined.
The graph of $x = a$ is a vertical line, with x-intercept $(a, 0)$.

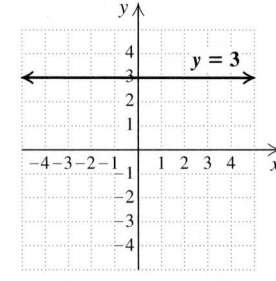

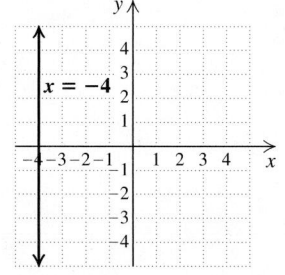

SECTION 3.4: RATES

A **rate** is a ratio that indicates how two quantities change with respect to each other.

Lara had $1500 in her savings account at the beginning of February, and $2400 at the beginning of May. Find the rate at which Lara is saving.

$$\text{Savings rate} = \frac{\text{Amount saved}}{\text{Number of months}}$$
$$= \frac{\$2400 - \$1500}{3 \text{ months}}$$
$$= \frac{\$900}{3 \text{ months}} = \$300 \text{ per month}$$

SECTION 3.5: SLOPE

Slope

$$\text{Slope} = m = \frac{\text{change in } y}{\text{change in } x}$$
$$= \frac{\text{rise}}{\text{run}} = \frac{y_2 - y_1}{x_2 - x_1}$$

The slope of the line containing the points $(-1, -4)$ and $(2, -6)$ is

$$m = \frac{-6 - (-4)}{2 - (-1)} = \frac{-2}{3} = -\frac{2}{3}.$$

A line with positive slope slants up from left to right.

A line with negative slope slants down from left to right.

The slope of a horizontal line is 0.

The slope of a vertical line is undefined.

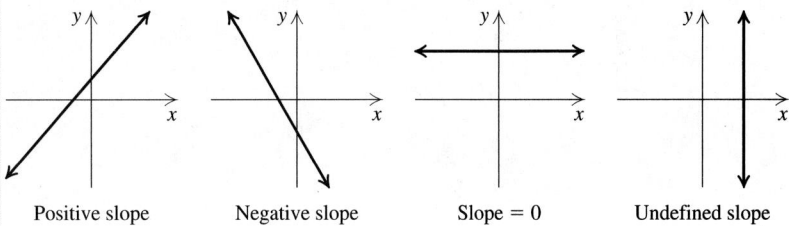

| Positive slope | Negative slope | Slope = 0 | Undefined slope |

SECTION 3.6: SLOPE–INTERCEPT FORM

Slope–Intercept Form

$$y = mx + b$$

The slope of the line is m.
The y-intercept of the line is $(0, b)$.

For the line given by $y = \frac{2}{3}x - 8$:

The slope is $\frac{2}{3}$ and the y-intercept is $(0, -8)$.

To graph a line written in slope–intercept form, plot the y-intercept and count off the slope.

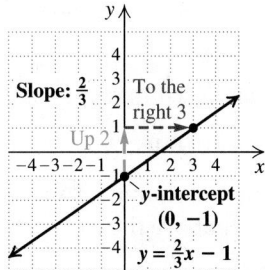

Parallel and Perpendicular Lines

Two lines are parallel if they have the same slope or if both are vertical.

Determine whether the graphs of $y = \frac{2}{3}x - 5$ and $3y - 2x = 7$ are parallel.

$$y = \frac{2}{3}x - 5 \qquad 3y - 2x = 7$$

The slope is $\frac{2}{3}$.

$$3y = 2x + 7$$
$$y = \frac{2}{3}x + \frac{7}{3}$$

The slope is $\frac{2}{3}$.

Since the slopes are the same, the graphs are parallel.

Two lines are perpendicular if the product of their slopes is -1 or if one line is vertical and the other line is horizontal.	*Determine whether the graphs of $y = \frac{2}{3}x - 5$ and $2y + 3x = 1$ are perpendicular.* $\qquad y = \frac{2}{3}x - 5 \qquad\qquad 2y + 3x = 1$ $\qquad$ The slope is $\frac{2}{3}$. $\qquad\qquad 2y = -3x + 1$ $\qquad\qquad\qquad\qquad\qquad\qquad y = -\frac{3}{2}x + \frac{1}{2}$ $\qquad\qquad\qquad\qquad\qquad$ The slope is $-\frac{3}{2}$. Since $\frac{2}{3}\left(-\frac{3}{2}\right) = -1$, the graphs are perpendicular.

SECTION 3.7: POINT–SLOPE FORM

Point–Slope Form $\qquad y - y_1 = m(x - x_1)$ The slope of the line is m. The line passes through (x_1, y_1).	*Write a point–slope equation for the line with slope -2 that contains the point $(3, -5)$.* $\qquad\quad y - y_1 = m(x - x_1)$ $\qquad\quad y - (-5) = -2(x - 3)$

Review Exercises: Chapter 3

🪝 **Concept Reinforcement** *Classify each statement as either true or false.*

1. Not every ordered pair lies in one of the four quadrants. [3.1]

2. The equation of a vertical line cannot be written in slope–intercept form. [3.6]

3. Equations for lines written in slope–intercept form appear in the form $Ax + By = C$. [3.6]

4. Every horizontal line has an x-intercept. [3.3]

5. A line's slope is a measure of rate. [3.5]

6. A positive rate of ascent means that an airplane is flying increasingly higher above the earth. [3.4]

7. Any two points on a line can be used to determine the line's slope. [3.5]

8. Knowing a line's slope is enough to write the equation of the line. [3.6]

9. Knowing two points on a line is enough to write the equation of the line. [3.7]

10. Parallel lines that are not vertical have the same slope. [3.6]

The following circle graph shows the percentage of online searches done in July 2006 that were performed by a particular search engine. [3.1]
Source: NetRatings for SearchEngineWatch.com

Online Searches

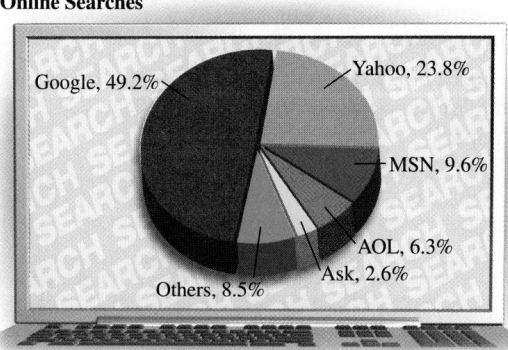

11. There were 5.6 billion searches done by home and business Internet users in July 2006. How many searches were done using Yahoo?

12. About 55% of the online searches done by Waterworks Graphics are image searches. In July 2006, Waterworks employees did 4200 online searches. If their search engine use is typical, how many image searches did they do using Google?

Plot each point. [3.1]

13. $(5, -1)$ **14.** $(2, 3)$ **15.** $(-4, 0)$

In which quadrant is each point located? [3.1]

16. $(-8, -7)$ **17.** $(15.3, -13.8)$ **18.** $\left(-\frac{1}{2}, \frac{1}{10}\right)$

Find the coordinates of each point in the figure. [3.1]

19. A **20.** B **21.** C

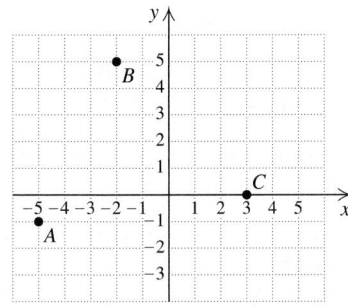

22. Use a grid 10 squares wide and 10 squares high to plot $(-65, -2)$, $(-10, 6)$, and $(25, 7)$. Choose the scale carefully. [3.1]

23. Determine whether the equation $y = 2x + 7$ has *each* ordered pair as a solution: **(a)** $(3, 1)$; **(b)** $(-3, 1)$. [3.2]

24. Show that the ordered pairs $(0, -3)$ and $(2, 1)$ are solutions of the equation $2x - y = 3$. Then use the graph of the two points to determine another solution. Answers may vary. [3.2]

Graph.

25. $y = x - 5$ [3.2]   **26.** $y = -\frac{1}{4}x$ [3.2]

27. $y = -x + 4$ [3.2]   **28.** $4x + y = 3$ [3.2]

29. $4x + 5 = 3$ [3.3]   **30.** $5x - 2y = 10$ [3.3]

31. *TV viewing.* The average number of daily viewers v, in millions, of ABC's soap opera "General Hospital" is given by $v = -\frac{1}{4}t + 9$, where t is the number of years since 2000. Graph the equation and use the graph to predict the average number of daily viewers of "General Hospital" in 2008. [3.2]
Source: Nielsen Media Research

32. At 4:00 P.M., Jesse's Honda Civic was at mile marker 17 of Interstate 290 in Chicago. At 4:45 P.M., Jesse was at mile marker 23. [3.4]

  a) Find Jesse's driving rate, in number of miles per minute.

  b) Find Jesse's driving rate, in number of minutes per mile.

33. *Gas mileage.* The following graph shows data for a Ford Explorer driven on city streets. At what rate was the vehicle consuming gas? [3.4]

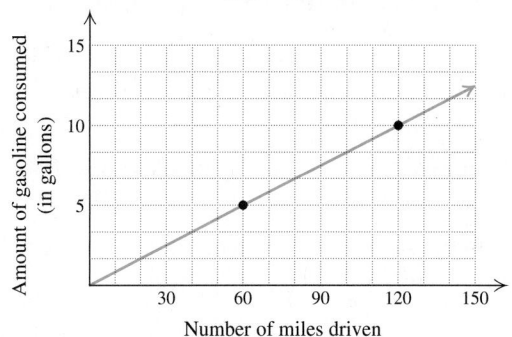

Find the slope of each line. [3.5]

34.

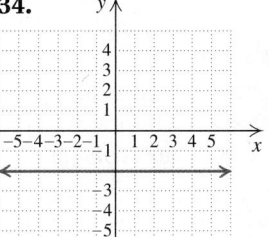

35.

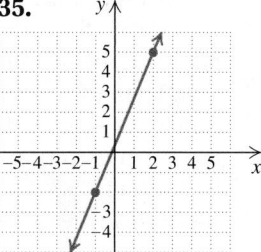

36.

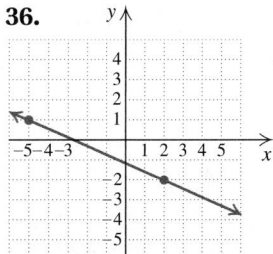

Find the slope of the line containing the given pair of points. If it is undefined, state this. [3.5]

37. $(-2, 5)$ and $(3, -1)$

38. $(6, 5)$ and $(-2, 5)$

39. $(-3, 0)$ and $(-3, 5)$

40. $(-8.3, 4.6)$ and $(-9.9, 1.4)$

41. *Architecture.* To meet federal standards, a wheelchair ramp cannot rise more than 1 ft over a horizontal distance of 12 ft. Express this slope as a grade. [3.5]

42. Find the *x*-intercept and the *y*-intercept of the line given by $5x - 8y = 80$. [3.3]

43. Find the slope and the *y*-intercept of the line given by $3x + 5y = 45$. [3.6]

Determine whether each pair of lines is parallel, perpendicular, or neither. [3.6]

44. $y + 5 = -x$,
$x - y = 2$

45. $3x - 5 = 7y$,
$7y - 3x = 7$

46. Write the slope–intercept equation of the line with slope $\frac{3}{8}$ and *y*-intercept $(0, 7)$. [3.6]

47. Write a point–slope equation for the line with slope $-\frac{1}{3}$ that contains the point $(-2, 9)$. [3.7]

48. The average tuition at a public two-year college was $1359 in 2001 and $1847 in 2005. Graph the data and determine an equation for the related line. Then **(a)** calculate the average tuition at a public two-year college in 2004 and **(b)** predict the average tuition at a public two-year college in 2012. [3.7]
Source: U.S. National Center for Education Statistics

49. Write the slope–intercept equation for the line with slope 5 that contains the point $(3, -10)$. [3.7]

50. Write the slope–intercept equation for the line that is perpendicular to the line $3x - 5y = 9$ and that contains the point $(2, -5)$ [3.7]

Graph.

51. $y = \frac{2}{3}x - 5$ [3.6]

52. $2x + y = 4$ [3.3]

53. $y = 6$ [3.3]

54. $x = -2$ [3.3]

55. $y + 2 = -\frac{1}{2}(x - 3)$ [3.7]

Synthesis

56. Can two perpendicular lines share the same *y*-intercept? Why or why not? [3.3]

57. Is it possible for a graph to have only one intercept? Why or why not? [3.3]

58. Find the value of *m* in $y = mx + 3$ such that $(-2, 5)$ is on the graph. [3.2]

59. Find the value of *b* in $y = -5x + b$ such that $(3, 4)$ is on the graph. [3.2]

60. Find the area and the perimeter of a rectangle for which $(-2, 2)$, $(7, 2)$, and $(7, -3)$ are three of the vertices. [3.1]

61. Find three solutions of $y = 4 - |x|$. [3.2]

Test: Chapter 3

 CHAPTER **Test Prep** VIDEO CD *Step-by-step test solutions are found on the video CD in the front of this book.*

Volunteering. *The following pie chart shows the types of organizations in which college students volunteer.*
Source: Corporation for National and Community Service

Volunteering by College Students, 2005

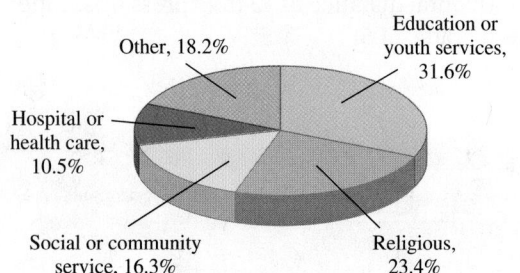

Other, 18.2%

Education or youth services, 31.6%

Hospital or health care, 10.5%

Social or community service, 16.3%

Religious, 23.4%

1. At Rolling Hills College, 25% of the 1200 students volunteer. If their choice of organizations is typical, how many students will volunteer in education or youth services?

2. At Valley University, $\frac{1}{3}$ of the 3900 students volunteer. If their choice of organizations is typical, how many students will volunteer in hospital or health-care services?

In which quadrant is each point located?

3. $(-2, -10)$

4. $(-1.6, 2.3)$

Find the coordinates of each point in the figure.

5. *A*

6. *B*

7. *C*

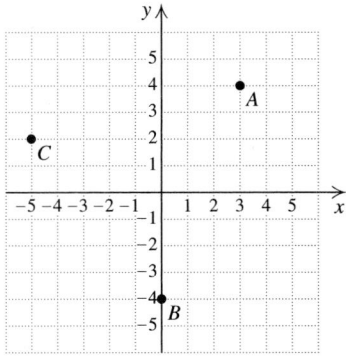

Graph.

8. $y = 2x - 1$

9. $2x - 4y = -8$

10. $y + 1 = 6$

11. $y = \frac{3}{4}x$

12. $2x - y = 3$

13. $x = -1$

Find the slope of the line containing each pair of points. If it is undefined, state this.

14. $(3, -2)$ and $(4, 3)$

15. $(-5, 6)$ and $(-1, -3)$

16. $(4, 7)$ and $(4, -8)$

17. *Running.* Jon reached the 3-km mark of a race at 2:15 P.M. and the 6-km mark at 2:24 P.M. What is his running rate?

18. At one point Filbert Street, the steepest street in San Francisco, drops 63 ft over a horizontal distance of 200 ft. Find the road grade.

19. Find the *x*-intercept and the *y*-intercept of the line given by $5x - y = 30$.

20. Find the slope and the *y*-intercept of the line given by $y - 8x = 10$.

21. Write the slope–intercept equation of the line with slope $-\frac{1}{3}$ and *y*-intercept $(0, -11)$.

Determine without graphing whether each pair of lines is parallel, perpendicular, or neither.

22. $4y + 2 = 3x,$
$-3x + 4y = -12$

23. $y = -2x + 5,$
$2y - x = 6$

24. Write the slope–intercept equation of the line that is perpendicular to the line $2x - 5y = 8$ and that contains the point $(-3, 2)$.

25. *Aerobic exercise.* A person's target heart rate is the number of beats per minute that brings the most aerobic benefit to his or her heart. The target heart rate for a 20-year-old is 150 beats per minute; for a 60-year-old, it is 120 beats per minute.

a) Graph the data and determine an equation for the related line. Let a = age and r = target heart rate, in number of beats per minute.

b) Calculate the target heart rate for a 36-year-old.

Graph.

26. $y = \frac{1}{4}x - 2$

27. $y + 4 = -\frac{1}{2}(x - 3)$

Synthesis

28. Write an equation of the line that is parallel to the graph of $2x - 5y = 6$ and has the same *y*-intercept as the graph of $3x + y = 9$.

29. A diagonal of a square connects the points $(-3, -1)$ and $(2, 4)$. Find the area and the perimeter of the square.

30. List the coordinates of three other points that are on the same line as $(-2, 14)$ and $(17, -5)$. Answers may vary.

Cumulative Review: Chapters 1–3

1. Evaluate $\dfrac{x}{5y}$ for $x = 70$ and $y = 2$. [1.1]

2. Multiply: $6(2a - b + 3)$. [1.2]

3. Factor: $8x - 4y + 4$. [1.2]

4. Find the prime factorization of 54. [1.3]

5. Find decimal notation: $-\frac{3}{20}$. [1.4]

6. Find the absolute value: $|-37|$. [1.4]

7. Find the opposite of $-\frac{1}{10}$. [1.6]

8. Find the reciprocal of $-\frac{1}{10}$. [1.7]

9. Find decimal notation: 36.7%. [2.4]

Simplify.

10. $\frac{3}{5} - \frac{5}{12}$ [1.3]

11. $3.4 + (-0.8)$ [1.5]

12. $(-2)(-1.4)(2.6)$ [1.7]

13. $\frac{3}{8} \div \left(-\frac{9}{10}\right)$ [1.7]

14. $1 - [32 \div (4 + 2^2)]$ [1.8]

15. $-5 + 16 \div 2 \cdot 4$ [1.8]

16. $y - (3y + 7)$ [1.8]

17. $3(x - 1) - 2[x - (2x + 7)]$ [1.8]

Solve.

18. $2.7 = 5.3 + x$ [2.1]

19. $\frac{5}{3}x = -45$ [2.1]

20. $3x - 7 = 41$ [2.2]

21. $\dfrac{3}{4} = \dfrac{-n}{8}$ [2.1]

22. $14 - 5x = 2x$ [2.2]

23. $3(5 - x) = 2(3x + 4)$ [2.2]

24. $\frac{1}{4}x - \frac{2}{3} = \frac{3}{4} + \frac{1}{3}x$ [2.2]

25. $y + 5 - 3y = 5y - 9$ [2.2]

26. $x - 28 < 20 - 2x$ [2.6]

27. $2(x + 2) \geq 5(2x + 3)$ [2.6]

28. Solve $A = 2\pi rh + \pi r^2$ for h. [2.3]

29. In which quadrant is the point $(3, -1)$ located? [3.1]

30. Graph on a number line: $-1 < x \leq 2$. [2.6]

31. Use a grid 10 squares wide and 10 squares high to plot $(-150, -40)$, $(40, -7)$, and $(0, 6)$. Choose the scale carefully. [3.1]

Graph.

32. $x = 3$ [3.3]

33. $2x - 5y = 10$ [3.3]

34. $y = -2x + 1$ [3.2]

35. $y = \frac{2}{3}x$ [3.2]

36. $y = -\frac{3}{4}x + 2$ [3.6]

37. $2y - 5 = 3$ [3.3]

Find the coordinates of the x- and y-intercepts. Do not graph.

38. $2x - 7y = 21$ [3.3]

39. $y = 4x + 5$ [3.3]

40. Find the slope and the y-intercept of the line given by $3x - y = 2$. [3.6]

41. Find the slope of the line containing the points $(-4, 1)$ and $(2, -1)$. [3.5]

42. Write an equation of the line with slope $\frac{2}{7}$ and y-intercept $(0, -4)$. [3.6]

43. Write a point–slope equation of the line with slope $-\frac{3}{8}$ that contains the point $(-6, 4)$. [3.7]

44. Write the slope–intercept form of the equation in Exercise 43. [3.6]

45. Determine an equation for the following graph. [3.6], [3.7]

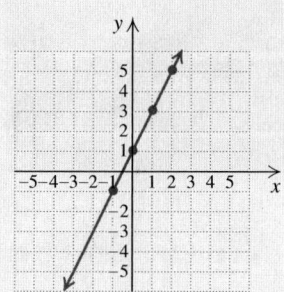

Solve.

46. U.S. bicycle sales rose from 15 million in 1995 to 20 million in 2005. Find the rate of change of bicycle sales. [3.4]
Sources: National Bicycle Dealers Association; U.S. Department of Transportation

47. A 150-lb person will burn 240 calories per hour when riding a bicycle at 6 mph. The same person will burn 410 calories per hour when cycling at 12 mph. [3.7]
Source: American Heart Association

   a) Graph the data and determine an equation for the related line. Let r = the rate at which the person is cycling and c = the number of calories burned per hour.
   b) Use the equation of part (a) to estimate the number of calories burned per hour by a 150-lb person cycling at 10 mph.

48. Americans spent an estimated $238 billion on home remodeling in 2006. This was $\frac{17}{15}$ of the amount spent on remodeling in 2005. How much was spent on remodeling in 2005? [2.5]
Source: National Association of Home Builders' Remodelers Council

49. In 2005, the mean earnings of individuals with a high school diploma was $29,448. This was about 54% of the mean earnings of those with a bachelor's degree. What were the mean earnings of individuals with a bachelor's degree in 2005? [2.4]
Source: U.S. Census Bureau

50. Recently there were 132 million Americans with either O-positive or O-negative blood. Those with O-positive blood outnumbered those with O-negative blood by 90 million. How many Americans had O-negative blood? [2.5]
Source: American Red Cross

52.

53.

Synthes

54. Anya's s _____ year is $26,780. This reflects _____ y increase in February and then a 3% cost-of-living adjustment in June. What was her salary at the beginning of the year? [2.4]

Solve. If no solution exists, state this.

55. $4|x| - 13 = 3$ [1.4], [2.2]

56. $4(x + 2) = 9(x - 2) + 16$ [2.2]

57. $2(x + 3) + 4 = 0$ [2.2]

58. $\dfrac{2 + 5x}{4} = \dfrac{11}{28} + \dfrac{8x + 3}{7}$ [2.2]

59. $5(7 + x) = (x + 6)5$ [2.2]

60. Solve $p = \dfrac{2}{m + Q}$ for Q. [2.3]

61. The points $(-3, 0)$, $(0, 7)$, $(3, 0)$, and $(0, -7)$ are vertices of a parallelogram. Find four equations of lines that intersect to form the parallelogram. [3.6]

Polynomials

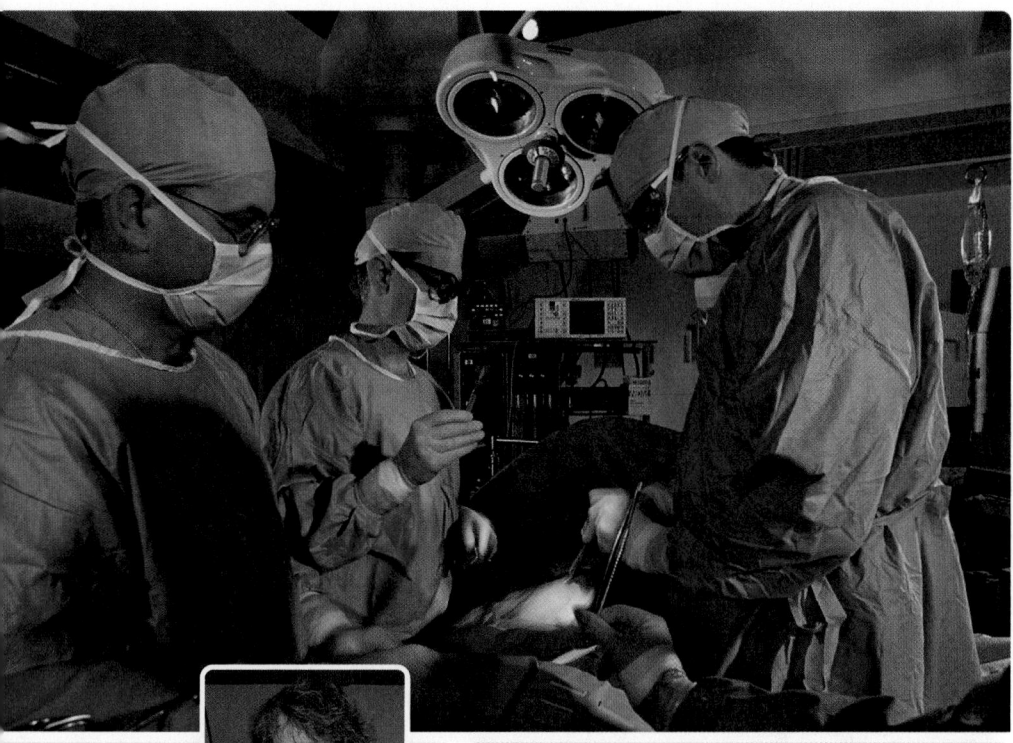

SHELLEY ZOMAK
TRANSPLANT COORDINATOR
(NURSE)
Pittsburgh, Pennsylvania

It is very important that a transplant recipient take the medications prescribed to maintain a determined blood concentration of the drug. As a transplant coordinator, I use math to educate patients on the correct dosing of medications, since the doctor's orders do not always match the pills available from the pharmacy. Some medications need to be adjusted more frequently on the basis of the patient's weight.

AN APPLICATION

Often a patient needing a kidney transplant has a willing kidney donor who does not match the patient medically. A kidney-paired donation matches donor–recipient pairs. Two kidney transplants are performed simultaneously, with each patient receiving the kidney of a stranger. The number k of such "kidney swaps" t years after 2003 can be approximated by

$$k = 14.3t^3 - 56t^2 + 57.7t + 19.$$

Estimate the number of kidney-paired donations in 2006.

Source: Based on data from United Network for Organ Sharing

This problem appears as Exercise 69 in Section 4.3.

lgebraic expressions such as $16t^2$, $5a^2 - 3ab$, and $3x^2 - 7x + 5$ are called polynomials. Polynomials occur frequently in applications and appear in most branches of mathematics. Thus learning to add, subtract, multiply, and divide polynomials is an important part of nearly every course in elementary algebra. The focus of this chapter is finding equivalent expressions, not solving equations.

4.1   Exponents and Their Properties

Multiplying Powers with Like Bases ▪ Dividing Powers with Like Bases ▪ Zero as an Exponent ▪
Raising a Power to a Power ▪ Raising a Product or a Quotient to a Power

In Section 4.3, we begin our study of polynomials. Before doing so, however, we must develop some rules for working with exponents.

Multiplying Powers with Like Bases

Recall from Section 1.8 that an expression like a^3 means $a \cdot a \cdot a$. We can use this fact to find the product of two expressions that have the same base:

$$a^3 \cdot a^2 = (a \cdot a \cdot a)(a \cdot a) \qquad \text{There are three factors in } a^3 \text{ and two factors in } a^2.$$

$$a^3 \cdot a^2 = a \cdot a \cdot a \cdot a \cdot a \qquad \text{Using an associative law}$$

$$a^3 \cdot a^2 = a^5.$$

Note that the exponent in a^5 is the sum of the exponents in $a^3 \cdot a^2$. That is, $3 + 2 = 5$. Similarly,

$$b^4 \cdot b^3 = (b \cdot b \cdot b \cdot b)(b \cdot b \cdot b)$$

$$b^4 \cdot b^3 = b^7, \quad \text{where } 4 + 3 = 7.$$

Adding the exponents gives the correct result.

STUDENT NOTES

There are several rules for manipulating exponents in this section. One way to remember them all is to replace variables with numbers (other than 1) and see what the results suggest. For example, multiplying $2^2 \cdot 2^3$ and examining the result is a fine way of reminding yourself that $a^m \cdot a^n = a^{m+n}$.

> **The Product Rule**
>
> For any number a and any positive integers m and n,
>
> $$a^m \cdot a^n = a^{m+n}.$$
>
> (To multiply powers with the same base, keep the base and add the exponents.)

EXAMPLE   1   Multiply and simplify each of the following. (Here "simplify" means express the product as one base to a power whenever possible.)

a) $2^3 \cdot 2^8$

b) $5^3 \cdot 5^8 \cdot 5^1$

c) $(r + s)^7(r + s)^6$

d) $(a^3b^2)(a^3b^5)$

STUDY SKILLS

*Helping Yourself by
Helping Others*

When you feel confident in your
command of a topic, don't hesitate
to help classmates experiencing
trouble. Your understanding and
retention of a concept will deepen
when you explain it to someone
else and your classmate will
appreciate your help.

SOLUTION

a) $2^3 \cdot 2^8 = 2^{3+8}$ Adding exponents: $a^m \cdot a^n = a^{m+n}$

$\qquad = 2^{11}$ | ***CAUTION!*** The base is unchanged: $2^3 \cdot 2^8 \neq 4^{11}$. |

b) $5^3 \cdot 5^8 \cdot 5^1 = 5^{3+8+1}$ Adding exponents

$\qquad = 5^{12}$ | ***CAUTION!*** $5^{12} \neq 5 \cdot 12$. |

c) $(r+s)^7(r+s)^6 = (r+s)^{7+6}$ The base here is $r+s$.

$\qquad = (r+s)^{13}$ | ***CAUTION!*** $(r+s)^{13} \neq r^{13} + s^{13}$. |

d) $(a^3b^2)(a^3b^5) = a^3b^2a^3b^5$ Using an associative law

$\qquad = a^3a^3b^2b^5$ Using a commutative law

$\qquad = a^6b^7$ Adding exponents | TRY EXERCISE 15 |

Dividing Powers with Like Bases

Recall that any expression that is divided or multiplied by 1 is unchanged. This,
together with the fact that anything (besides 0) divided by itself is 1, can lead to a
rule for division:

$$\frac{a^5}{a^2} = \frac{a \cdot a \cdot a \cdot a \cdot a}{a \cdot a}$$

$$\frac{a^5}{a^2} = \frac{a \cdot a \cdot a}{1} \cdot \frac{a \cdot a}{a \cdot a}$$

$$\frac{a^5}{a^2} = \frac{a \cdot a \cdot a}{1} \cdot 1$$

$$\frac{a^5}{a^2} = a \cdot a \cdot a = a^3.$$

Note that the exponent in a^3 is the difference of the exponents in a^5/a^2. Similarly,

$$\frac{x^4}{x^3} = \frac{x \cdot x \cdot x \cdot x}{x \cdot x \cdot x} = \frac{x}{1} \cdot \frac{x \cdot x \cdot x}{x \cdot x \cdot x} = \frac{x}{1} \cdot 1 = x = x^1.$$

Subtracting the exponents gives the correct result.

> ### The Quotient Rule
>
> For any nonzero number a and any positive integers m and n for which
> $m > n$,
>
> $$\frac{a^m}{a^n} = a^{m-n}.$$
>
> (To divide powers with the same base, subtract the exponent of the
> denominator from the exponent of the numerator.)

EXAMPLE **2**   Divide and simplify. (Here "simplify" means express the quotient as one base to
a power whenever possible.)

a) $\dfrac{x^8}{x^2}$ **b)** $\dfrac{7^9}{7^4}$ **c)** $\dfrac{(5a)^{12}}{(5a)^4}$ **d)** $\dfrac{4p^5q^7}{6p^2q}$

SOLUTION

a) $\dfrac{x^8}{x^2} = x^{8-2}$ Subtracting exponents: $\dfrac{a^m}{a^n} = a^{m-n}$

$\quad\quad = x^6$

b) $\dfrac{7^9}{7^4} = 7^{9-4}$

> **CAUTION!** The base is unchanged:
>
> $$\dfrac{7^9}{7^4} \neq 1^5.$$

$\quad\quad = 7^5$

c) $\dfrac{(5a)^{12}}{(5a)^4} = (5a)^{12-4} = (5a)^8$ The base here is $5a$.

d) $\dfrac{4p^5q^7}{6p^2q} = \dfrac{4}{6} \cdot \dfrac{p^5}{p^2} \cdot \dfrac{q^7}{q^1}$ Note that the 4 and 6 are factors, not exponents!

$\quad\quad = \dfrac{2}{3} \cdot p^{5-2} \cdot q^{7-1} = \dfrac{2}{3}p^3q^6$ Using the quotient rule twice; simplifying   **TRY EXERCISE** ▶ 33

Zero as an Exponent

The quotient rule can be used to help determine what 0 should mean when it appears as an exponent. Consider a^4/a^4, where a is nonzero. Since the numerator and the denominator are the same,

$$\frac{a^4}{a^4} = 1.$$

On the other hand, using the quotient rule would give us

$$\frac{a^4}{a^4} = a^{4-4} = a^0. \quad \text{Subtracting exponents}$$

Since $a^0 = a^4/a^4 = 1$, this suggests that $a^0 = 1$ for any nonzero value of a.

> ### The Exponent Zero
>
> For any real number a, with $a \neq 0$,
>
> $$a^0 = 1.$$
>
> (Any nonzero number raised to the 0 power is 1.)

Note that in the above box, 0^0 is not defined. For this text, we will assume that expressions like a^m do not represent 0^0.

EXAMPLE   **3**   Simplify: **(a)** 1948^0; **(b)** $(-9)^0$; **(c)** $(3x)^0$; **(d)** $3x^0$; **(e)** $(-1)9^0$; **(f)** -9^0.

SOLUTION

a) $1948^0 = 1$ Any nonzero number raised to the 0 power is 1.

b) $(-9)^0 = 1$ Any nonzero number raised to the 0 power is 1. The base here is -9.

c) $(3x)^0 = 1$, for any $x \neq 0$.   The parentheses indicate that the base is $3x$.

d) Since $3x^0$ means $3 \cdot x^0$, the base is x. Recall that simplifying exponential expressions is done before multiplication in the rules for order of operations:

$$3x^0 = 3 \cdot 1 = 3, \quad \text{for any } x \neq 0.$$

e) $(-1)9^0 = (-1)1 = -1$ The base here is 9.

f) -9^0 is read "the opposite of 9^0" and is equivalent to $(-1)9^0$:

$$-9^0 = (-1)9^0 = (-1)1 = -1.$$

Note from parts (b), (e), and (f) that $-9^0 = (-1)9^0$ and $-9^0 \neq (-9)^0$.

> **TRY EXERCISE** 49

CAUTION! $-9^0 \neq (-9)^0$, and, in general, $-a^n \neq (-a)^n$.

Raising a Power to a Power

Consider an expression like $(7^2)^4$:

$$(7^2)^4 = (7^2)(7^2)(7^2)(7^2)$$ There are four factors of 7^2.

$$(7^2)^4 = (7 \cdot 7)(7 \cdot 7)(7 \cdot 7)(7 \cdot 7)$$ We could also use the product rule.

$$(7^2)^4 = 7 \cdot 7 \cdot 7 \cdot 7 \cdot 7 \cdot 7 \cdot 7 \cdot 7$$ Using an associative law

$$(7^2)^4 = 7^8.$$

Note that the exponent in 7^8 is the product of the exponents in $(7^2)^4$. Similarly,

$$(y^5)^3 = y^5 \cdot y^5 \cdot y^5$$ There are three factors of y^5.

$$(y^5)^3 = (y \cdot y \cdot y \cdot y \cdot y)(y \cdot y \cdot y \cdot y \cdot y)(y \cdot y \cdot y \cdot y \cdot y)$$

$$(y^5)^3 = y^{15}.$$

Once again, we get the same result if we multiply exponents:

$$(y^5)^3 = y^{5 \cdot 3} = y^{15}.$$

> ### The Power Rule
>
> For any number a and any whole numbers m and n,
>
> $$(a^m)^n = a^{mn}.$$
>
> (To raise a power to a power, multiply the exponents and leave the base unchanged.)

Remember that for this text we assume that 0^0 is not considered.

EXAMPLE 4   Simplify: **(a)** $(3^5)^4$; **(b)** $(m^2)^5$.

SOLUTION

a) $(3^5)^4 = 3^{5 \cdot 4}$ Multiplying exponents: $(a^m)^n = a^{mn}$

$ = 3^{20}$

b) $(m^2)^5 = m^{2 \cdot 5}$

$ = m^{10}$

> **TRY EXERCISE** 57

Raising a Product or a Quotient to a Power

When an expression inside parentheses is raised to a power, the inside expression is the base. Let's compare $2a^3$ and $(2a)^3$:

$2a^3 = 2 \cdot a \cdot a \cdot a;$ The base is a.

$(2a)^3 = (2a)(2a)(2a)$ The base is $2a$.

$(2a)^3 = (2 \cdot 2 \cdot 2)(a \cdot a \cdot a)$

$(2a)^3 = 2^3a^3$

$(2a)^3 = 8a^3.$

We see that $2a^3$ and $(2a)^3$ are *not* equivalent. Note too that $(2a)^3$ can be simplified by cubing each factor in $2a$. This leads to the following rule for raising a product to a power.

> ### Raising a Product to a Power
>
> For any numbers a and b and any whole number n,
>
> $$(ab)^n = a^nb^n.$$
>
> (To raise a product to a power, raise each factor to that power.)

EXAMPLE **5**

Simplify: **(a)** $(4a)^3$; **(b)** $(-5x^4)^2$; **(c)** $(a^7b)^2(a^3b^4)$.

SOLUTION

a) $(4a)^3 = 4^3a^3 = 64a^3$ Raising each factor to the third power and simplifying

b) $(-5x^4)^2 = (-5)^2(x^4)^2$ Raising each factor to the second power Parentheses are important here.

$\qquad = 25x^8$ Simplifying $(-5)^2$ and using the power rule

c) $(a^7b)^2(a^3b^4) = (a^7)^2b^2a^3b^4$ Raising a product to a power

$\qquad = a^{14}b^2a^3b^4$ Multiplying exponents

$\qquad = a^{17}b^6$ Adding exponents   **TRY EXERCISE** 63

CAUTION!   The rule $(ab)^n = a^nb^n$ applies only to *products* raised to a power, not to sums or differences. For example, $(3 + 4)^2 \neq 3^2 + 4^2$ since $49 \neq 9 + 16$. Similarly, $(5x)^2 = 5^2 \cdot x^2$, but $(5 + x)^2 \neq 5^2 + x^2$.

There is a similar rule for raising a quotient to a power.

> ## Raising a Quotient to a Power
>
> For any numbers a and b, $b \neq 0$, and any whole number n,
>
> $$\left(\frac{a}{b}\right)^n = \frac{a^n}{b^n}.$$
>
> (To raise a quotient to a power, raise the numerator to the power and divide by the denominator to the power.)

EXAMPLE 6   Simplify: **(a)** $\left(\dfrac{x}{5}\right)^2$; **(b)** $\left(\dfrac{5}{a^4}\right)^3$; **(c)** $\left(\dfrac{3a^4}{b^3}\right)^2$.

SOLUTION

a) $\left(\dfrac{x}{5}\right)^2 = \dfrac{x^2}{5^2} = \dfrac{x^2}{25}$ Squaring the numerator and the denominator

b) $\left(\dfrac{5}{a^4}\right)^3 = \dfrac{5^3}{(a^4)^3}$ Raising a quotient to a power

$= \dfrac{125}{a^{4 \cdot 3}} = \dfrac{125}{a^{12}}$ Using the power rule and simplifying

c) $\left(\dfrac{3a^4}{b^3}\right)^2 = \dfrac{(3a^4)^2}{(b^3)^2}$ Raising a quotient to a power

$= \dfrac{3^2(a^4)^2}{b^{3 \cdot 2}} = \dfrac{9a^8}{b^6}$ Raising a product to a power and using the power rule

TRY EXERCISE 75

In the following summary of definitions and rules, we assume that no denominators are 0 and that 0^0 is not considered.

> ## Definitions and Properties of Exponents
>
> For any whole numbers m and n,
>
> | | |
> |---|---|
> | 1 as an exponent: | $a^1 = a$ |
> | 0 as an exponent: | $a^0 = 1$ |
> | The Product Rule: | $a^m \cdot a^n = a^{m+n}$ |
> | The Quotient Rule: | $\dfrac{a^m}{a^n} = a^{m-n}$ |
> | The Power Rule: | $(a^m)^n = a^{mn}$ |
> | Raising a product to a power: | $(ab)^n = a^n b^n$ |
> | Raising a quotient to a power: | $\left(\dfrac{a}{b}\right)^n = \dfrac{a^n}{b^n}$ |

4.1   EXERCISE SET

 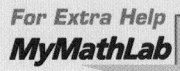 

↪ *Concept Reinforcement*   *In each of Exercises 1–8, complete the sentence using the most appropriate phrase from the column on the right.*

1. To raise a product to a power, ____

2. To raise a quotient to a power, ____

3. To raise a power to a power, ____

4. To divide powers with the same base, ____

5. Any nonzero number raised to the 0 power ____

6. To multiply powers with the same base, ____

7. To square a fraction, ____

8. To square a product, ____

a) keep the base and add the exponents.

b) multiply the exponents and leave the base unchanged.

c) square the numerator and square the denominator.

d) square each factor.

e) raise each factor to that power.

f) raise the numerator to the power and divide by the denominator to the power.

g) is one.

h) subtract the exponent of the denominator from the exponent of the numerator.

Identify the base and the exponent in each expression.

9. $(2x)^5$

10. $(x + 1)^0$

11. $2x^3$

12. $-y^6$

13. $\left(\dfrac{4}{y}\right)^7$

14. $(-5x)^4$

Simplify. Assume that no denominator is 0 and that 0^0 is not considered.

15. $d^3 \cdot d^{10}$

16. $8^4 \cdot 8^3$

17. $a^6 \cdot a$

18. $y^7 \cdot y^9$

19. $6^5 \cdot 6^{10}$

20. $t^0 \cdot t^{16}$

21. $(3y)^4(3y)^8$

22. $(2t)^8(2t)^{17}$

23. $(8n)(8n)^9$

24. $(5p)^0(5p)^1$

25. $(a^2b^7)(a^3b^2)$

26. $(m - 3)^4(m - 3)^5$

27. $(x + 3)^5(x + 3)^8$

28. $(a^8b^3)(a^4b)$

29. $r^3 \cdot r^7 \cdot r^0$

30. $s^4 \cdot s^5 \cdot s^2$

31. $(mn^5)(m^3n^4)$

32. $(a^3b)(ab)^4$

33. $\dfrac{7^5}{7^2}$

34. $\dfrac{4^7}{4^3}$

35. $\dfrac{t^8}{t}$

36. $\dfrac{x^7}{x}$

37. $\dfrac{(5a)^7}{(5a)^6}$

38. $\dfrac{(3m)^9}{(3m)^8}$

Aha! 39. $\dfrac{(x + y)^8}{(x + y)^8}$

40. $\dfrac{(9x)^{10}}{(9x)^2}$

41. $\dfrac{(r + s)^{12}}{(r + s)^4}$

42. $\dfrac{(a - b)^4}{(a - b)^3}$

43. $\dfrac{12d^9}{15d^2}$

44. $\dfrac{10n^7}{15n^3}$

45. $\dfrac{8a^9b^7}{2a^2b}$

46. $\dfrac{12r^{10}s^7}{4r^2s}$

47. $\dfrac{x^{12}y^9}{x^0y^2}$

48. $\dfrac{a^{10}b^{12}}{a^2b^0}$

Simplify.

49. t^0 when $t = 15$

50. y^0 when $y = 38$

51. $5x^0$ when $x = -22$

52. $7m^0$ when $m = 1.7$

53. $7^0 + 4^0$

54. $(8 + 5)^0$

55. $(-3)^1 - (-3)^0$

56. $(-4)^0 - (-4)^1$

Simplify. Assume that no denominator is 0 and that 0^0 is not considered.

57. $(x^3)^{11}$

58. $(a^5)^8$

59. $(5^8)^4$

60. $(2^5)^2$

61. $(t^{20})^4$

62. $(x^{25})^6$

63. $(10x)^2$

64. $(5a)^2$

65. $(-2a)^3$

66. $(-3x)^3$

67. $(-5n^7)^2$

68. $(-4m^4)^2$

69. $(a^2b)^7$

70. $(xy^4)^9$

71. $(r^5t)^3(r^2t^8)$

72. $(a^4b^6)(a^2b)^5$

73. $(2x^5)^3(3x^4)$

74. $(5x^3)^2(2x^7)$

75. $\left(\dfrac{x}{5}\right)^3$

76. $\left(\dfrac{2}{a}\right)^4$

77. $\left(\dfrac{7}{6n}\right)^2$

78. $\left(\dfrac{4x}{3}\right)^3$

79. $\left(\dfrac{a^3}{b^8}\right)^6$

80. $\left(\dfrac{x^5}{y^2}\right)^7$

81. $\left(\dfrac{x^2y}{z^3}\right)^4$

82. $\left(\dfrac{a^4}{b^2c}\right)^5$

83. $\left(\dfrac{a^3}{-2b^5}\right)^4$

84. $\left(\dfrac{x^5}{-3y^3}\right)^4$

85. $\left(\dfrac{5x^7y}{-2z^4}\right)^3$

86. $\left(\dfrac{-4p^5}{3m^2n^3}\right)^3$

Aha! 87. $\left(\dfrac{4x^3y^5}{3z^7}\right)^0$

88. $\left(\dfrac{5a^7}{2b^5c}\right)^0$

89. Explain in your own words why $-5^2 \neq (-5)^2$.

90. Under what circumstances should exponents be added?

Skill Review

To prepare for Section 4.2, review operations with integers (Sections 1.5–1.7).

Perform the indicated operations.

91. $-10 - 14$ [1.6]

92. $-3(5)$ [1.7]

93. $-16 + 5$ [1.5]

94. $12 - (-4)$ [1.6]

95. $-3 + (-11)$ [1.5]

96. $-8 - (-12)$ [1.6]

Synthesis

97. Under what conditions does a^n represent a negative number? Why?

98. Using the quotient rule, explain why 9^0 is 1.

99. Suppose that the width of a square is three times the width of a second square. How do the areas of the squares compare? Why?

100. Suppose that the width of a cube is twice the width of a second cube. How do the volumes of the cubes compare? Why?

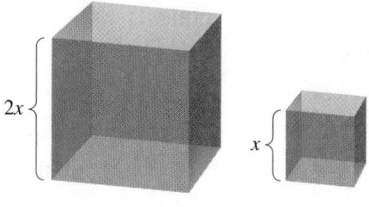

Find a value of the variable that shows that the two expressions are not equivalent. Answers may vary.

101. $3x^2$; $(3x)^2$

102. $(a + 5)^2$; $a^2 + 5^2$

103. $\dfrac{t^6}{t^2}$; t^3

104. $\dfrac{a + 7}{7}$; a

Simplify.

105. $y^{4x} \cdot y^{2x}$

106. $a^{10k} \div a^{2k}$

107. $\dfrac{x^{5t}(x^t)^2}{(x^{3t})^2}$

108. $\dfrac{\left(\frac{1}{2}\right)^3\left(\frac{2}{3}\right)^4}{\left(\frac{5}{6}\right)^3}$

109. Solve for x:
$$\dfrac{t^{26}}{t^x} = t^x.$$

Replace ▓ with $>$, $<$, or $=$ to write a true sentence.

110. 3^5 ▓ 3^4

111. 4^2 ▓ 4^3

112. 4^3 ▓ 5^3

113. 4^3 ▓ 3^4

114. 9^7 ▓ 3^{13}

115. 25^8 ▓ 125^5

Use the fact that $10^3 \approx 2^{10}$ to estimate each of the following powers of 2. Then compute the power of 2 with a calculator and find the difference between the exact value and the approximation.

116. 2^{14}

117. 2^{22}

118. 2^{26}

119. 2^{31}

In computer science, 1 KB of memory refers to 1 kilobyte, or 1×10^3 bytes, of memory. This is really an approximation of 1×2^{10} bytes (since computer memory actually uses powers of 2).

120. The TI-84 Plus graphing calculator has 480 KB of "FLASH ROM." How many bytes is this?

121. The TI-84 Plus Silver Edition graphing calculator has 1.5 MB (megabytes) of FLASH ROM, where 1 MB is 1000 KB (see Exercise 120). How many bytes of FLASH ROM does this calculator have?

Negative Integers as Exponents ▪ Scientific Notation ▪
Multiplying and Dividing Using Scientific Notation

We now attach a meaning to negative exponents. Once we understand both positive exponents and negative exponents, we can study a method for writing numbers known as *scientific notation.*

Negative Integers as Exponents

Let's define negative exponents so that the rules that apply to whole-number exponents will hold for all integer exponents. To do so, consider a^{-5} and the rule for adding exponents:

$$a^{-5} = a^{-5} \cdot 1 \qquad \text{Using the identity property of 1}$$

$$= \frac{a^{-5}}{1} \cdot \frac{a^5}{a^5} \qquad \text{Writing 1 as } \frac{a^5}{a^5} \text{ and } a^{-5} \text{ as } \frac{a^{-5}}{1}$$

$$= \frac{a^{-5+5}}{a^5} \qquad \text{Adding exponents}$$

$$= \frac{1}{a^5}. \qquad -5 + 5 = 0 \text{ and } a^0 = 1$$

This leads to our definition of negative exponents.

> ### Negative Exponents
> For any real number a that is nonzero and any integer n,
> $$a^{-n} = \frac{1}{a^n}.$$
> (The numbers a^{-n} and a^n are reciprocals of each other.)

EXAMPLE **1** Express using positive exponents and, if possible, simplify.

a) m^{-3} **b)** 4^{-2} **c)** $(-3)^{-2}$ **d)** ab^{-1}

STUDY SKILLS

Indicate the Highlights

Most students find it helpful to draw a star or use a color, felt-tipped highlighter to indicate important concepts or trouble spots that require further study. Most campus bookstores carry a variety of highlighters that permit you to brightly color written material while keeping it easy to read.

SOLUTION

a) $m^{-3} = \dfrac{1}{m^3}$ m^{-3} is the reciprocal of m^3.

b) $4^{-2} = \dfrac{1}{4^2} = \dfrac{1}{16}$ 4^{-2} is the reciprocal of 4^2. Note that $4^{-2} \neq 4(-2)$.

c) $(-3)^{-2} = \dfrac{1}{(-3)^2} = \dfrac{1}{(-3)(-3)} = \dfrac{1}{9}$ $(-3)^{-2}$ is the reciprocal of $(-3)^2$. Note that $(-3)^{-2} \neq -\dfrac{1}{3^2}$.

d) $ab^{-1} = a\left(\dfrac{1}{b^1}\right) = a\left(\dfrac{1}{b}\right) = \dfrac{a}{b}$ b^{-1} is the reciprocal of b^1. Note that the base is b, not ab.

TRY EXERCISE ▸ 5

> *CAUTION!*   A negative exponent does not, in itself, indicate that an expression is negative. As shown in Example 1,
>
> $$4^{-2} \neq 4(-2) \quad \text{and} \quad (-3)^{-2} \neq -\frac{1}{3^2}.$$

The following is another way to illustrate why negative exponents are defined as they are.

On this side, we divide by 5 at each step.	$125 = 5^3$	On this side, the exponents decrease by 1.
	$25 = 5^2$	
	$5 = 5^1$	
	$1 = 5^0$	
	$\frac{1}{5} = 5^?$	
	$\frac{1}{25} = 5^?$	

To continue the pattern, it follows that

$$\frac{1}{5} = \frac{1}{5^1} = 5^{-1}, \qquad \frac{1}{25} = \frac{1}{5^2} = 5^{-2}, \quad \text{and, in general,} \quad \frac{1}{a^n} = a^{-n}.$$

EXAMPLE 2   Express $\dfrac{1}{x^7}$ using negative exponents.

SOLUTION   We know that $\dfrac{1}{a^n} = a^{-n}$. Thus,

$$\frac{1}{x^7} = x^{-7}.$$

> TRY EXERCISE ▶ 25

The rules for powers still hold when exponents are negative.

EXAMPLE 3   Simplify. Do not use negative exponents in the answer.

a) $t^5 \cdot t^{-2}$ **b)** $(5x^2y^{-3})^4$ **c)** $\dfrac{x^{-4}}{x^{-5}}$

d) $\dfrac{1}{t^{-5}}$ **e)** $\dfrac{s^{-3}}{t^{-5}}$ **f)** $\dfrac{-10x^{-3}y}{5x^2y^5}$

SOLUTION

a) $t^5 \cdot t^{-2} = t^{5+(-2)} = t^3$ Adding exponents

b) $(5x^2y^{-3})^4 = 5^4(x^2)^4(y^{-3})^4$ Raising each factor to the fourth power

$$= 625x^8y^{-12} = \frac{625x^8}{y^{12}}$$

c) $\dfrac{x^{-4}}{x^{-5}} = x^{-4-(-5)} = x^1 = x$ We subtract exponents even if the exponent in the denominator is negative.

d) Since $\dfrac{1}{a^n} = a^{-n}$, we have $\dfrac{1}{t^{-5}} = t^{-(-5)} = t^5$.

e) $\dfrac{s^{-3}}{t^{-5}} = s^{-3} \cdot \dfrac{1}{t^{-5}}$

$$= \frac{1}{s^3} \cdot t^5 = \frac{t^5}{s^3}$$ Using the result from part (d) above

f) $\dfrac{-10x^{-3}y}{5x^2y^5} = \dfrac{-10}{5} \cdot \dfrac{x^{-3}}{x^2} \cdot \dfrac{y^1}{y^5}$ Note that the -10 and 5 are factors.

$\qquad\qquad = -2 \cdot x^{-3-2} \cdot y^{1-5}$ Using the quotient rule twice; simplifying

$\qquad\qquad = -2x^{-5}y^{-4} = \dfrac{-2}{x^5y^4}$ ▸ TRY EXERCISE 33

The result from Example 3(e) can be generalized.

Factors and Negative Exponents

For any nonzero real numbers a and b and any integers m and n,

$$\frac{a^{-n}}{b^{-m}} = \frac{b^m}{a^n}.$$

(A factor can be moved to the other side of the fraction bar if the sign of the exponent is changed.)

EXAMPLE **4**   Simplify: $\dfrac{-15x^{-7}}{5y^2z^{-4}}$.

SOLUTION   We can move the factors x^{-7} and z^{-4} to the other side of the fraction bar if we change the sign of each exponent:

$\dfrac{-15x^{-7}}{5y^2z^{-4}} = \dfrac{-15}{5} \cdot \dfrac{x^{-7}}{y^2z^{-4}}$ We can simply divide the constant factors.

$\qquad\qquad = -3 \cdot \dfrac{z^4}{y^2x^7}$

$\qquad\qquad = \dfrac{-3z^4}{x^7y^2}.$ ▸ TRY EXERCISE 55

Another way to change the sign of the exponent is to take the reciprocal of the base. To understand why this is true, note that

$$\left(\frac{s}{t}\right)^{-5} = \frac{s^{-5}}{t^{-5}} = \frac{t^5}{s^5} = \left(\frac{t}{s}\right)^5.$$

This often provides the easiest way to simplify an expression containing a negative exponent.

Reciprocals and Negative Exponents

For any nonzero real numbers a and b and any integer n,

$$\left(\frac{a}{b}\right)^{-n} = \left(\frac{b}{a}\right)^n.$$

(Any base to a power is equal to the reciprocal of the base raised to the opposite power.)

EXAMPLE 5   Simplify: $\left(\dfrac{x^4}{2y}\right)^{-3}$.

SOLUTION

$$\left(\dfrac{x^4}{2y}\right)^{-3} = \left(\dfrac{2y}{x^4}\right)^3 \qquad \text{Taking the reciprocal of the base and changing the sign of the exponent}$$

$$= \dfrac{(2y)^3}{(x^4)^3} \qquad \text{Raising a quotient to a power by raising both the numerator and the denominator to the power}$$

$$= \dfrac{2^3 y^3}{x^{12}} \qquad \text{Raising a product to a power; using the power rule in the denominator}$$

$$= \dfrac{8y^3}{x^{12}} \qquad \text{Cubing 2}$$

TRY EXERCISE 73

Scientific Notation

When we are working with the very large numbers or very small numbers that frequently occur in science, **scientific notation** provides a useful way of writing them. The following are examples of scientific notation.

The mass of the earth:

6.0×10^{24} kilograms (kg) = 6,000,000,000,000,000,000,000,000 kg

The mass of a hydrogen atom:

1.7×10^{-24} g = 0.0000000000000000000000017 g

Scientific Notation

Scientific notation for a number is an expression of the type

$$N \times 10^m,$$

where N is at least 1 but less than 10 (that is, $1 \le N < 10$), N is expressed in decimal notation, and m is an integer.

Converting from scientific notation to decimal notation involves multiplying by a power of 10. Consider the following.

Scientific Notation	Multiplication	Decimal Notation
4.52×10^2	4.52×100	452.
4.52×10^1	4.52×10	45.2
4.52×10^0	4.52×1	4.52
4.52×10^{-1}	4.52×0.1	0.452
4.52×10^{-2}	4.52×0.01	0.0452

Note that when m, the power of 10, is positive, the decimal point moves right m places in decimal notation. When m is negative, the decimal point moves left $|m|$ places. We generally try to perform this multiplication mentally.

EXAMPLE 6 Convert to decimal notation: **(a)** 7.893×10^5; **(b)** 4.7×10^{-8}.

SOLUTION

a) Since the exponent is positive, the decimal point moves to the right:

7.89300.⤴ $7.893 \times 10^5 = 789,300$ The decimal point moves to the right 5 places.

5 places

b) Since the exponent is negative, the decimal point moves to the left:

0.00000004.7 $4.7 \times 10^{-8} = 0.000000047$ The decimal point moves to the left 8 places.

8 places

TRY EXERCISE 85

To convert from decimal notation to scientific notation, this procedure is reversed.

EXAMPLE 7 Write in scientific notation: **(a)** 83,000; **(b)** 0.0327.

SOLUTION

a) We need to find m such that $83,000 = 8.3 \times 10^m$. To change 8.3 to 83,000 requires moving the decimal point 4 places to the right. This can be accomplished by multiplying by 10^4. Thus,

$$83,000 = 8.3 \times 10^4. \quad \text{This is scientific notation.}$$

b) We need to find m such that $0.0327 = 3.27 \times 10^m$. To change 3.27 to 0.0327 requires moving the decimal point 2 places to the left. This can be accomplished by multiplying by 10^{-2}. Thus,

$$0.0327 = 3.27 \times 10^{-2}. \quad \text{This is scientific notation.}$$

TRY EXERCISE 93

Conversions to and from scientific notation are often made mentally. Remember that positive exponents are used to represent large numbers and negative exponents are used to represent small numbers between 0 and 1.

Multiplying and Dividing Using Scientific Notation

Products and quotients of numbers written in scientific notation are found using the rules for exponents.

EXAMPLE 8 Simplify.

a) $(1.8 \times 10^9) \cdot (2.3 \times 10^{-4})$

b) $(3.41 \times 10^5) \div (1.1 \times 10^{-3})$

SOLUTION

a) $(1.8 \times 10^9) \cdot (2.3 \times 10^{-4})$

$\quad = 1.8 \times 2.3 \times 10^9 \times 10^{-4}$ Using the associative and commutative laws

$\quad = 4.14 \times 10^{9+(-4)}$ Adding exponents

$\quad = 4.14 \times 10^5$

b) $(3.41 \times 10^5) \div (1.1 \times 10^{-3})$

$$= \frac{3.41 \times 10^5}{1.1 \times 10^{-3}}$$

$$= \frac{3.41}{1.1} \times \frac{10^5}{10^{-3}}$$

$$= 3.1 \times 10^{5-(-3)} \qquad \text{Subtracting exponents}$$

$$= 3.1 \times 10^8$$

> TRY EXERCISE 103

When a problem is stated using scientific notation, we generally use scientific notation for the answer. This often requires an additional conversion.

EXAMPLE 9   Simplify.

a) $(3.1 \times 10^5) \cdot (4.5 \times 10^{-3})$ 　　　　　　　**b)** $(7.2 \times 10^{-7}) \div (8.0 \times 10^6)$

SOLUTION

a) We have

$$(3.1 \times 10^5) \cdot (4.5 \times 10^{-3}) = 3.1 \times 4.5 \times 10^5 \times 10^{-3}$$
$$= 13.95 \times 10^2.$$

Our answer is not yet in scientific notation because 13.95 is not between 1 and 10. We convert to scientific notation as follows:

$$13.95 \times 10^2 = 1.395 \times 10^1 \times 10^2 \qquad \text{Substituting } 1.395 \times 10^1 \text{ for } 13.95$$
$$= 1.395 \times 10^3. \qquad \text{Adding exponents}$$

b) $(7.2 \times 10^{-7}) \div (8.0 \times 10^6) = \dfrac{7.2 \times 10^{-7}}{8.0 \times 10^6} = \dfrac{7.2}{8.0} \times \dfrac{10^{-7}}{10^6}$

$$= 0.9 \times 10^{-13}$$

$$= 9.0 \times 10^{-1} \times 10^{-13} \qquad \text{Substituting } 9.0 \times 10^{-1} \text{ for } 0.9$$

$$= 9.0 \times 10^{-14} \qquad \text{Adding exponents}$$

> TRY EXERCISE 111

TECHNOLOGY CONNECTION

A key labeled $\boxed{10^x}$, $\boxed{\wedge}$, or $\boxed{\text{EE}}$ is used to enter scientific notation into a calculator. Sometimes this is a secondary function, meaning that another key—often labeled SHIFT or **2ND** —must be pressed first.

To check Example 8(a), we press

　　1.8 $\boxed{\text{EE}}$ 9 $\boxed{\times}$ 2.3 $\boxed{\text{EE}}$ $\boxed{(-)}$ 4.

When we then press $\boxed{=}$ or **ENTER**, the result 4.14E5 appears. This represents 4.14×10^5. On many calculators, the MODE Sci must be selected in order to display scientific notation.

    1.8E9*2.3E−4
                        4.14E5

On some calculators, this appears as

    4.14 05

or

    4.14e+05

Calculate each of the following.

1. $(3.8 \times 10^9) \cdot (4.5 \times 10^7)$
2. $(2.9 \times 10^{-8}) \div (5.4 \times 10^6)$
3. $(9.2 \times 10^7) \div (2.5 \times 10^{-9})$

4.2   EXERCISE SET

For Extra Help
MyMathLab   Math XL PRACTICE   WATCH   DOWNLOAD

↪ *Concept Reinforcement*   *Match each expression with an equivalent expression from the column on the right.*

1. ___ $\left(\dfrac{x^3}{y^2}\right)^{-2}$ **a)** $\dfrac{y^6}{x^9}$

2. ___ $\left(\dfrac{y^2}{x^3}\right)^{-2}$ **b)** $\dfrac{x^9}{y^6}$

3. ___ $\left(\dfrac{y^{-2}}{x^{-3}}\right)^{-3}$ **c)** $\dfrac{y^4}{x^6}$

4. ___ $\left(\dfrac{x^{-3}}{y^{-2}}\right)^{-3}$ **d)** $\dfrac{x^6}{y^4}$

Express using positive exponents. Then, if possible, simplify.

5. 2^{-3} **6.** 10^{-5} **7.** $(-2)^{-6}$

8. $(-3)^{-4}$ **9.** t^{-9} **10.** x^{-7}

11. xy^{-2} **12.** $a^{-3}b$ **13.** $r^{-5}t$

14. xy^{-9} **15.** $\dfrac{1}{a^{-8}}$ **16.** $\dfrac{1}{z^{-6}}$

17. 7^{-1} **18.** 3^{-1} **19.** $\left(\dfrac{3}{5}\right)^{-2}$

20. $\left(\dfrac{3}{4}\right)^{-2}$ **21.** $\left(\dfrac{x}{2}\right)^{-5}$ **22.** $\left(\dfrac{a}{2}\right)^{-4}$

23. $\left(\dfrac{s}{t}\right)^{-7}$ **24.** $\left(\dfrac{r}{v}\right)^{-5}$

Express using negative exponents.

25. $\dfrac{1}{9^2}$ **26.** $\dfrac{1}{5^2}$ **27.** $\dfrac{1}{y^3}$

28. $\dfrac{1}{t^4}$ **29.** $\dfrac{1}{5}$ **30.** $\dfrac{1}{8}$

31. $\dfrac{1}{t}$ **32.** $\dfrac{1}{m}$

Simplify. Do not use negative exponents in the answer.

33. $2^{-5} \cdot 2^8$ **34.** $5^{-8} \cdot 5^{10}$

35. $x^{-3} \cdot x^{-9}$ **36.** $x^{-4} \cdot x^{-7}$

37. $t^{-3} \cdot t$ **38.** $y^{-5} \cdot y$

39. $(n^{-5})^3$ **40.** $(m^{-5})^{10}$

41. $(t^{-3})^{-6}$ **42.** $(a^{-4})^{-7}$

43. $(t^4)^{-3}$ **44.** $(x^4)^{-5}$

45. $(mn)^{-7}$ **46.** $(ab)^{-9}$

47. $(3x^{-4})^2$ **48.** $(2a^{-5})^3$

49. $(5r^{-4}t^3)^2$ **50.** $(4x^5y^{-6})^3$

51. $\dfrac{t^{12}}{t^{-2}}$ **52.** $\dfrac{x^7}{x^{-2}}$

53. $\dfrac{y^{-7}}{y^{-3}}$ **54.** $\dfrac{z^{-6}}{z^{-2}}$

55. $\dfrac{15y^{-7}}{3y^{-10}}$ **56.** $\dfrac{-12a^{-5}}{2a^{-8}}$

57. $\dfrac{2x^6}{x}$ **58.** $\dfrac{3x}{x^{-1}}$

59. $\dfrac{-15a^{-7}}{10b^{-9}}$ **60.** $\dfrac{12x^{-6}}{8y^{-10}}$

Aha! **61.** $\dfrac{t^{-7}}{t^{-7}}$ **62.** $\dfrac{a^{-5}}{b^{-7}}$

63. $\dfrac{8x^{-3}}{y^{-7}z^{-1}}$ **64.** $\dfrac{10a^{-1}}{b^{-7}c^{-3}}$

65. $\dfrac{3t^4}{s^{-2}u^{-4}}$ **66.** $\dfrac{5x^{-8}}{y^{-3}z^2}$

67. $(x^4y^5)^{-3}$ **68.** $(t^5x^3)^{-4}$

69. $(3m^{-5}n^{-3})^{-2}$ **70.** $(2y^{-4}z^{-2})^{-3}$

71. $(a^{-5}b^7c^{-2})(a^{-3}b^{-2}c^6)$

72. $(x^3y^{-4}z^{-5})(x^{-4}y^{-2}z^9)$

73. $\left(\dfrac{a^4}{3}\right)^{-2}$ **74.** $\left(\dfrac{y^2}{2}\right)^{-2}$

75. $\left(\dfrac{m^{-1}}{n^{-4}}\right)^3$ **76.** $\left(\dfrac{x^2y}{z^{-5}}\right)^3$

77. $\left(\dfrac{2a^2}{3b^4}\right)^{-3}$ **78.** $\left(\dfrac{a^2b}{2d^3}\right)^{-5}$

Aha! **79.** $\left(\dfrac{5x^{-2}}{3y^{-2}z}\right)^0$ **80.** $\left(\dfrac{4a^3b^{-2}}{5c^{-3}}\right)^1$

81. $\dfrac{-6a^3b^{-5}}{-3a^7b^{-8}}$ **82.** $\dfrac{12x^{-2}y^4}{-3xy^{-7}}$

83. $\dfrac{10x^{-4}yz^7}{8x^7y^{-3}z^{-3}}$ **84.** $\dfrac{9a^6b^{-4}c^7}{27a^{-4}b^5c^9}$

Convert to decimal notation.

85. 4.92×10^3

86. 8.13×10^4

87. 8.92×10^{-3}

88. 7.26×10^{-4}

89. 9.04×10^8

90. 1.35×10^7

91. 3.497×10^{-6}

92. 9.043×10^{-3}

Convert to scientific notation.

93. 36,000,000

94. 27,400

95. 0.00583

96. 0.0814

97. 78,000,000,000

98. 3,700,000,000,000

99. 0.000000527

100. 0.0000506

101. 0.000001032

102. 0.00000008

Multiply or divide, and write scientific notation for the result.

103. $(3 \times 10^5)(2 \times 10^8)$

104. $(3.1 \times 10^7)(2.1 \times 10^{-4})$

105. $(3.8 \times 10^9)(6.5 \times 10^{-2})$

106. $(7.1 \times 10^{-7})(8.6 \times 10^{-5})$

107. $(8.7 \times 10^{-12})(4.5 \times 10^{-5})$

108. $(4.7 \times 10^5)(6.2 \times 10^{-12})$

109. $\dfrac{8.5 \times 10^8}{3.4 \times 10^{-5}}$

110. $\dfrac{5.6 \times 10^{-2}}{2.5 \times 10^5}$

111. $(4.0 \times 10^3) \div (8.0 \times 10^8)$

112. $(1.5 \times 10^{-3}) \div (1.6 \times 10^{-6})$

113. $\dfrac{7.5 \times 10^{-9}}{2.5 \times 10^{12}}$

114. $\dfrac{3.0 \times 10^{-2}}{6.0 \times 10^{10}}$

115. Without performing actual computations, explain why 3^{-29} is smaller than 2^{-29}.

116. Explain why each of the following is not scientific notation:

$$12.6 \times 10^8;$$
$$4.8 \times 10^{1.7};$$
$$0.207 \times 10^{-5}.$$

Skill Review

To prepare for Section 4.3, review combining like terms and evaluating expressions (Sections 1.6 and 1.8).

Combine like terms. [1.6]

117. $9x + 2y - x - 2y$

118. $5a - 7b - 8a + b$

119. $-3x + (-2) - 5 - (-x)$

120. $2 - t - 3t - r - 7$

Evaluate. [1.8]

121. $4 + x^3$, for $x = 10$

122. $-x^2 - 5x + 3$, for $x = -2$

Synthesis

123. Explain what requirements must be met in order for x^{-n} to represent a negative integer.

124. Explain why scientific notation cannot be used without an understanding of the rules for exponents.

125. Write the reciprocal of 1.25×10^{-6} in scientific notation.

126. Write the reciprocal of 2.5×10^9 in scientific notation.

127. Write $8^{-3} \cdot 32 \div 16^2$ as a power of 2.

128. Write $81^3 \div 27 \cdot 9^2$ as a power of 3.

Simplify each of the following. Use a calculator only where indicated.

Aha! **129.** $\dfrac{125^{-4}(25^2)^4}{125}$

130. $(13^{-12})^2 \cdot 13^{25}$

131. $[(5^{-3})^2]^{-1}$

132. $5^0 - 5^{-1}$

133. $3^{-1} + 4^{-1}$

134. $\dfrac{4.2 \times 10^8[(2.5 \times 10^{-5}) \div (5.0 \times 10^{-9})]}{3.0 \times 10^{-12}}$

135. $\dfrac{27^{-2}(81^2)^3}{9^8}$

136. $\dfrac{7.4 \times 10^{29}}{(5.4 \times 10^{-6})(2.8 \times 10^8)}$

137. $\dfrac{5.8 \times 10^{17}}{(4.0 \times 10^{-13})(2.3 \times 10^4)}$

138. $\dfrac{(7.8 \times 10^7)(8.4 \times 10^{23})}{2.1 \times 10^{-12}}$

139. $\dfrac{(2.5 \times 10^{-8})(6.1 \times 10^{-11})}{1.28 \times 10^{-3}}$

140. Determine whether each of the following is true for all pairs of integers m and n and all positive numbers x and y.

a) $x^m \cdot y^n = (xy)^{mn}$

b) $x^m \cdot y^m = (xy)^{2m}$

c) $(x - y)^m = x^m - y^m$

Solve. Write scientific notation for each answer.

141. *Ecology.* In one year, a large tree can remove from the air the same amount of carbon dioxide produced by a car traveling 500 mi. If New York City contains approximately 600,000 trees, how many miles of car traffic can those trees clean in a year?

Sources: Colorado Tree Coalition; New York City Department of Parks and Recreations

142. *Computer technology.* One gigabit is about 1 billion bits of information. In 2007, Intel Corp. began making silicon modulators that can encode data onto a beam of light at a rate of 40 gigabits per second. If 25 of these communication lasers are packed on a single chip, how many bits per second could that chip encode?

Source: *The Wall Street Journal,* 7/25/2007

143. *Hotel management.* The new Four Seasons Hotel in Seattle contains 110,000 ft^2 of condominium space. If these condos sold for about \$2100 per ft^2, how much money did the hotel make selling the condominiums?

Source: seattletimes.nwsource.com

144. *Coral reefs.* There are 10 million bacteria per square centimeter of coral in a coral reef. The coral reefs near the Hawaiian Islands cover 14,000 km^2. How many bacteria are there in Hawaii's coral reefs?

Sources: livescience.com; U.S. Geological Survey

145. *Hospital care.* In 2005, 115 million patients visited emergency rooms in the United States. If the average visit lasted 3.3 hr, how many minutes in all did people spend in emergency rooms in 2005?

Source: *The Indianapolis Star,* 7/25/07

CONNECTING the CONCEPTS

The following properties of exponents hold for all integers m and n, assuming that no denominator is 0 and that 0^0 is not considered.

Definitions and Properties of Exponents

The following summary assumes that no denominators are 0 and that 0^0 is not considered. For any integers m and n,

1 as an exponent:	$a^1 = a$
0 as an exponent:	$a^0 = 1$
Negative exponents:	$a^{-n} = \dfrac{1}{a^n},$
	$\dfrac{a^{-n}}{b^{-m}} = \dfrac{b^m}{a^n},$
	$\left(\dfrac{a}{b}\right)^{-n} = \left(\dfrac{b}{a}\right)^n$
The Product Rule:	$a^m \cdot a^n = a^{m+n}$
The Quotient Rule:	$\dfrac{a^m}{a^n} = a^{m-n}$
The Power Rule:	$(a^m)^n = a^{mn}$
Raising a product to a power:	$(ab)^n = a^n b^n$
Raising a quotient to a power:	$\left(\dfrac{a}{b}\right)^n = \dfrac{a^n}{b^n}$

MIXED REVIEW

Simplify. Do not use negative exponents in the answer.

1. $x^4 x^{10}$

2. $x^{-4} x^{-10}$

3. $\dfrac{x^{-4}}{x^{10}}$

4. $\dfrac{x^4}{x^{-10}}$

5. $(x^{-4})^{-10}$

6. $(x^4)^{10}$

7. $\dfrac{1}{c^{-8}}$

8. c^{-8}

9. $(2x^3 y)^4$

10. $(2x^3 y)^{-4}$

11. $(3xy^{-1}z^5)^0$

12. $(a^2 b)(a^3 b^{-1})$

13. $\left(\dfrac{a^3}{b^4}\right)^5$

14. $\left(\dfrac{a^3}{b^4}\right)^{-5}$

15. $\dfrac{30x^4 y^3}{12xy^7}$

16. $\dfrac{12ab^{-8}}{14a^{-1}b^{-3}}$

17. $\dfrac{7p^{-5}}{xt^{-6}}$

18. $\left(\dfrac{3a^{-1}}{4b^{-3}}\right)^{-2}$

19. $(2p^2 q^4)(3pq^5)^2$

20. $(2xy^{-1})^{-1}(3x^2 y^{-3})^2$

4.3   Polynomials

Terms   ■   Types of Polynomials   ■   Degree and Coefficients   ■   Combining Like Terms   ■
Evaluating Polynomials and Applications

We now examine an important algebraic expression known as a *polynomial*. Certain polynomials have appeared earlier in this text so you already have some experience working with them.

Terms

At this point, we have seen a variety of algebraic expressions like

$$3a^2b^4, \quad 2l + 2w, \quad \text{and} \quad 5x^2 + x - 2.$$

Within these expressions, $3a^2b^4$, $2l$, $2w$, $5x^2$, x, and -2 are examples of *terms*. A **term** (see p. 17) can be a number (like -2), a variable (like x), a product of numbers and/or variables (like $3a^2b^4$, $2l$, $2w$, or $5x^2$), or a quotient of numbers and/or variables (like $7/t$ or $(a^2b^3)/(4c)$).*

Types of Polynomials

If a term is a product of constants and/or variables, it is called a **monomial**. Note that a term, but not a monomial, can include division by a variable. A **polynomial** is a monomial or a sum of monomials.

Examples of monomials: $7, \quad t, \quad 2l, \quad 2w, \quad 5x^3y, \quad \frac{3}{7}a^5$

Examples of polynomials: $4x + 7, \quad \frac{2}{3}t^2, \quad 6a + 7, \quad -5n^2 + m - 1, \quad 42r^5,$
$\quad x, \quad 0$

When a polynomial is written as a sum of monomials, each monomial is called a *term of the polynomial.*

EXAMPLE   **1**   Identify the terms of the polynomial $3t^4 - 5t^6 - 4t + 2$.

SOLUTION   The terms are $3t^4$, $-5t^6$, $-4t$, and 2. We can see this by rewriting all subtractions as additions of opposites:

$$3t^4 - 5t^6 - 4t + 2 = 3t^4 + (-5t^6) + (-4t) + 2.$$

These are the terms of the polynomial.

▸ **TRY EXERCISE** 9

A polynomial that is composed of two terms is called a **binomial**, whereas those composed of three terms are called **trinomials**. Polynomials with four or more terms have no special name.

---

*Later in this text, expressions like $5x^{3/2}$ and $2a^{-7}b$ will be discussed. Such expressions are also considered terms.

Monomials	Binomials	Trinomials	No Special Name
$4x^2$	$2x + 4$	$3t^3 + 4t + 7$	$4x^3 - 5x^2 + xy - 8$
9	$3a^5 + 6bc$	$6x^7 - 8z^2 + 4$	$z^5 + 2z^4 - z^3 + 7z + 3$
$-7a^{19}b^5$	$-9x^7 - 6$	$4x^2 - 6x - \frac{1}{2}$	$4x^6 - 3x^5 + x^4 - x^3 + 2x - 1$

The following algebraic expressions are *not* polynomials:

(1) $\dfrac{x+3}{x-4}$,   **(2)** $5x^3 - 2x^2 + \dfrac{1}{x}$,   **(3)** $\dfrac{1}{x^3 - 2}$.

Expressions (1) and (3) are not polynomials because they represent quotients, not sums. Expression (2) is not a polynomial because $1/x$ is not a monomial.

Degree and Coefficients

The **degree of a term of a polynomial** is the number of variable factors in that term. Thus the degree of $7t^2$ is 2 because $7t^2$ has two variable factors: $7t^2 = 7 \cdot t \cdot t$. We will revisit the meaning of degree in Section 4.7 when polynomials in several variables are examined.

EXAMPLE **2**   Determine the degree of each term: **(a)** $8x^4$; **(b)** $3x$; **(c)** 7.

SOLUTION

a) The degree of $8x^4$ is 4. x^4 represents 4 variable factors: $x \cdot x \cdot x \cdot x$.

b) The degree of $3x$ is 1.   There is 1 variable factor.

c) The degree of 7 is 0.   There is no variable factor.

The degree of a constant polynomial, as in Example 2(c), is 0 since there are no variable factors. There is an exception to this statement, however. Since $0 = 0x = 0x^2 = 0x^3$ and so on, we say that the polynomial 0 has *no* degree.

The part of a term that is a constant factor is the **coefficient** of that term. Thus the coefficient of $3x$ is 3, and the coefficient for the term 7 is simply 7.

EXAMPLE **3**   Identify the coefficient of each term in the polynomial

$$4x^3 - 7x^2y + x - 8.$$

SOLUTION

The coefficient of $4x^3$ is 4.

The coefficient of $-7x^2y$ is -7.

The coefficient of the third term is 1, since $x = 1x$.

The coefficient of -8 is simply -8.   **TRY EXERCISE** 13

The **leading term** of a polynomial is the term of highest degree. Its coefficient is called the **leading coefficient** and its degree is referred to as the **degree of the polynomial**. To see how this terminology is used, consider the polynomial

$$3x^2 - 8x^3 + 5x^4 + 7x - 6.$$

The *terms* are $3x^2$, $-8x^3$, $5x^4$, $7x$,   and -6.

The *coefficients* are   3, -8,   5,   7,   and -6.

The *degree of each term* is   2,   3,   4,   1,   and   0.

The *leading term* is $5x^4$ and the *leading coefficient* is 5.

The *degree of the polynomial* is 4.

Combining Like Terms

Recall from Section 1.8 that *like*, or *similar*, *terms* are either constant terms or terms containing the same variable(s) raised to the same power(s). To simplify certain polynomials, we can often *combine*, or *collect*, like terms.

EXAMPLE **4**   Identify the like terms in $4x^3 + 5x - 7x^2 + 2x^3 + x^2$.

SOLUTION

Like terms:	$4x^3$ and	$2x^3$	Same variable and exponent
Like terms:	$-7x^2$ and	x^2	Same variable and exponent

EXAMPLE **5**   Write an equivalent expression by combining like terms.

a) $2x^3 + 6x^3$ **b)** $5x^2 + 7 + 2x^4 - 6x^2 - 11 - 2x^4$

c) $7a^3 - 5a^2 + 9a^3 + a^2$ **d)** $\frac{2}{3}x^4 - x^3 - \frac{1}{6}x^4 + \frac{2}{5}x^3 - \frac{3}{10}x^3$

STUDENT NOTES

Remember that when we combine like terms, we are not solving equations, but are forming equivalent expressions.

SOLUTION

a) $2x^3 + 6x^3 = (2 + 6)x^3$ Using the distributive law
$ = 8x^3$

b) $5x^2 + 7 + 2x^4 - 6x^2 - 11 - 2x^4$
$ = 5x^2 - 6x^2 + 2x^4 - 2x^4 + 7 - 11$
$ = (5 - 6)x^2 + (2 - 2)x^4 + (7 - 11)$
$ = -1x^2 + 0x^4 + (-4)$ These steps are often done mentally.
$ = -x^2 - 4$

c) $7a^3 - 5a^2 + 9a^3 + a^2 = 7a^3 - 5a^2 + 9a^3 + 1a^2$ $a^2 = 1 \cdot a^2 = 1a^2$
$ = 16a^3 - 4a^2$

d) $\frac{2}{3}x^4 - x^3 - \frac{1}{6}x^4 + \frac{2}{5}x^3 - \frac{3}{10}x^3 = \left(\frac{2}{3} - \frac{1}{6}\right)x^4 + \left(-1 + \frac{2}{5} - \frac{3}{10}\right)x^3$
$\phantom{\frac{2}{3}x^4 - x^3 - \frac{1}{6}x^4 + \frac{2}{5}} = \left(\frac{4}{6} - \frac{1}{6}\right)x^4 + \left(-\frac{10}{10} + \frac{4}{10} - \frac{3}{10}\right)x^3$
$\phantom{\frac{2}{3}x^4 - x^3 - \frac{1}{6}x^4 + \frac{2}{5}} = \frac{3}{6}x^4 - \frac{9}{10}x^3$
$\phantom{\frac{2}{3}x^4 - x^3 - \frac{1}{6}x^4 + \frac{2}{5}} = \frac{1}{2}x^4 - \frac{9}{10}x^3$ There are no similar terms, so we are done.

> **TRY EXERCISE** 41

Note in Example 5 that the solutions are written so that the term of highest degree appears first, followed by the term of next highest degree, and so on. This is known as **descending order** and is the form in which answers will normally appear.

Evaluating Polynomials and Applications

When each variable in a polynomial is replaced with a number, the polynomial then represents a number, or *value*, that can be calculated using the rules for order of operations.

EXAMPLE **6**   Evaluate $-x^2 + 3x + 9$ for $x = -2$.

SOLUTION   For $x = -2$, we have

Substitute.
$-x^2 + 3x + 9 = -(-2)^2 + 3(-2) + 9$ The negative sign in front of x^2 remains.
$ = -(4) + 3(-2) + 9$

Simplify.
$ = -4 + (-6) + 9$
$ = -10 + 9 = -1.$

> **TRY EXERCISE** 51

EXAMPLE 7

Games in a sports league. In a sports league of n teams in which each team plays every other team twice, the total number of games to be played is given by the polynomial

$$n^2 - n.$$

A girls' soccer league has 10 teams. How many games are played if each team plays every other team twice?

SOLUTION We evaluate the polynomial for $n = 10$:

$$n^2 - n = 10^2 - 10$$
$$= 100 - 10 = 90.$$

The league plays 90 games.

TRY EXERCISE 61

EXAMPLE 8

Vehicle miles traveled. The average annual number of vehicle miles traveled per vehicle (VMT), in thousands, for a driver of age a can be approximated by the polynomial

$$-0.003a^2 + 0.2a + 8.6.$$

Find the VMT per vehicle for a 20-year-old driver.

Source: Based on information from the Energy Information Administration

SOLUTION To find the VMT per vehicle for a 20-year-old driver, we evaluate the polynomial for $a = 20$:

$$-0.003a^2 + 0.2a + 8.6 = -0.003(20)^2 + 0.2(20) + 8.6$$
$$= -0.003 \cdot 400 + 4 + 8.6$$
$$= -1.2 + 4 + 8.6$$
$$= 11.4.$$

The average annual number of VMT per vehicle by a 20-year-old driver is 11.4 thousand, or 11,400.

TRY EXERCISE 65

Sometimes, a graph can be used to estimate the value of a polynomial visually.

EXAMPLE 9

Vehicle miles traveled. In the following graph, the polynomial from Example 8 has been graphed by evaluating it for several choices of a. Use the graph to estimate the number of vehicle miles traveled each year, per vehicle, by a 30-year-old driver.

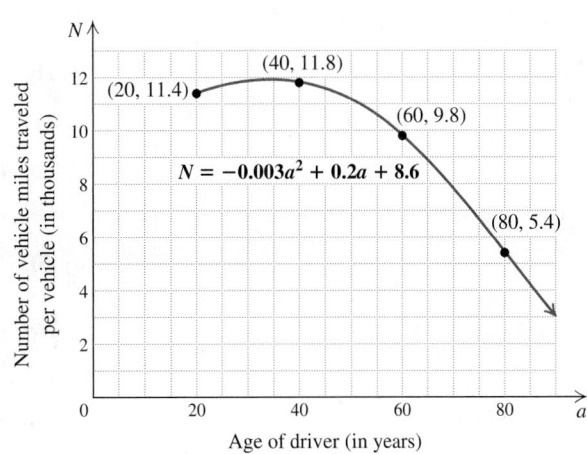

TECHNOLOGY CONNECTION

One way to evaluate a polynomial is to use the TRACE key. For example, to evaluate $-0.003a^2 + 0.2a + 8.6$ in Example 9 for $a = 30$, we can enter the polynomial as $y = -0.003x^2 + 0.2x + 8.6$. We then use TRACE and enter an x-value of 30.

(*continued*)

The value of the polynomial appears as *y*, and the cursor automatically appears at (30, 11.9). The Value option of the CALC menu works in a similar way.

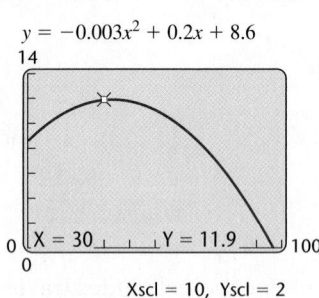

$$y = -0.003x^2 + 0.2x + 8.6$$

Xscl = 10, Yscl = 2

1. Use TRACE or CALC Value to find the value of $-0.003a^2 + 0.2a + 8.6$ for $a = 60$.

SOLUTION   To estimate the number of vehicle miles traveled by a 30-year-old driver, we locate 30 on the horizontal axis. From there, we move vertically until we meet the curve at some point. From that point, we move horizontally to the *N*-axis.

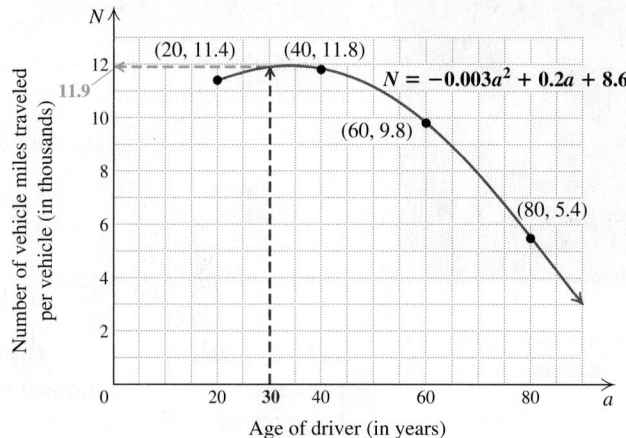

The average number of vehicle miles traveled each year, per vehicle, by a 30-year-old driver is 11.9 thousand, or 11,900. (For $a = 30$, the value of $-0.003a^2 + 0.2a + 8.6$ is approximately 11.9.)

TRY EXERCISE 71

4.3   EXERCISE SET

For Extra Help
MyMathLab   Math XL PRACTICE    WATCH   DOWNLOAD

Concept Reinforcement   *In each of Exercises 1–8, match the description with the most appropriate algebraic expression from the column on the right.*

1. _____ A polynomial with four terms

2. _____ A polynomial with 7 as its leading coefficient

3. _____ A trinomial written in descending order

4. _____ A polynomial with degree 5

5. _____ A binomial with degree 7

6. _____ A monomial of degree 0

7. _____ An expression with two terms that is not a binomial

8. _____ An expression with three terms that is not a trinomial

a) $8x^3 + \dfrac{2}{x^2}$

b) $5x^4 + 3x^3 - 4x + 7$

c) $\dfrac{3}{x} - 6x^2 + 9$

d) $8t - 4t^5$

e) 5

f) $6x^2 + 7x^4 - 2x^3$

g) $4t - 2t^7$

h) $3t^2 + 4t + 7$

Identify the terms of each polynomial.

9. $8x^3 - 11x^2 + 6x + 1$

10. $5a^3 + 4a^2 - a - 7$

11. $-t^6 - 3t^3 + 9t - 4$

12. $n^5 - 4n^3 + 2n - 8$

Determine the coefficient and the degree of each term in each polynomial.

13. $8x^4 + 2x$

14. $9a^3 - 4a^2$

15. $9t^2 - 3t + 4$

16. $7x^4 + 5x - 3$

17. $6a^5 + 9a + a^3$

18. $4t^8 - t + 6t^5$

19. $x^4 - x^3 + 4x - 3$

20. $2a^5 + a^2 + 8a + 10$

For each of the following polynomials, (a) list the degree of each term; (b) determine the leading term and the leading coefficient; and (c) determine the degree of the polynomial.

21. $5t + t^3 + 8t^4$

22. $1 + 6n + 4n^2$

23. $3a^2 - 7 + 2a^4$

24. $9x^4 + x^2 + x^7 - 12$

25. $8 + 6x^2 - 3x - x^5$

26. $9a - a^4 + 3 + 2a^3$

27. Complete the following table for the polynomial
$7x^2 + 8x^5 - 4x^3 + 6 - \frac{1}{2}x^4$.

Term	Coefficient	Degree of the Term	Degree of the Polynomial
		5	
$-\frac{1}{2}x^4$			
	-4		
		2	
	6		

28. Complete the following table for the polynomial
$-3x^4 + 6x^3 - 2x^2 + 8x + 7$.

Term	Coefficient	Degree of the Term	Degree of the Polynomial
	-3		
$6x^3$			
		2	
		1	
	7		

Classify each polynomial as a monomial, a binomial, a trinomial, or a polynomial with no special name.

29. $x^2 - 23x + 17$

30. $-9x^2$

31. $x^3 - 7x + 2x^2 - 4$

32. $t^3 + 4$

33. $y + 8$

34. $3x^8 + 12x^3 - 9$

35. 17

36. $2x^4 - 7x^3 + x^2 + x - 6$

Combine like terms. Write all answers in descending order.

37. $5n^2 + n + 6n^2$

38. $5a + 7a^2 + 3a$

39. $3a^4 - 2a + 2a + a^4$

40. $9b^5 + 3b^2 - 2b^5 - 3b^2$

41. $7x^3 - 11x + 5x + x^2$

42. $3x^4 - 7x + x^4 - 2x$

43. $4b^3 + 5b + 7b^3 + b^2 - 6b$

44. $6x^2 + 2x^4 - 2x^2 - x^4 - 4x^2 + x$

45. $10x^2 + 2x^3 - 3x^3 - 4x^2 - 6x^2 - x^4$

46. $12t^6 - t^3 + 8t^6 + 4t^3 - t^7 - 3t^3$

47. $\frac{1}{5}x^4 + 7 - 2x^2 + 3 - \frac{2}{15}x^4 + 2x^2$

48. $\frac{1}{6}x^3 + 3x^2 - \frac{1}{3}x^3 + 7 + x^2 - 10$

49. $8.3a^2 + 3.7a - 8 - 9.4a^2 + 1.6a + 0.5$

50. $1.4y^3 + 2.9 - 7.7y - 1.3y - 4.1 + 9.6y^3$

Evaluate each polynomial for $x = 3$ and for $x = -3$.

51. $-4x + 9$

52. $-6x + 5$

53. $2x^2 - 3x + 7$

54. $4x^2 - 6x + 9$

55. $-3x^3 + 7x^2 - 4x - 8$

56. $-2x^3 - 3x^2 + 4x + 2$

57. $2x^4 - \frac{1}{9}x^3$

58. $\frac{1}{3}x^4 - 2x^3$

59. $-x - x^2 - x^3$

60. $-x^2 - 3x^3 - x^4$

Back-to-college expenses. The amount of money, in billions of dollars, spent on shoes for college can be estimated by the polynomial
$$0.4t + 1.13,$$
where t is the number of years since 2004.
Source: Based on data from the National Retail Federation

61. Estimate the amount spent on shoes for college in 2006.

62. Estimate the amount spent on shoes for college in 2010.

63. *Skydiving.* During the first 13 sec of a jump, the number of feet that a skydiver falls in t seconds is approximated by the polynomial
$$11.12t^2.$$
Approximately how far has a skydiver fallen 10 sec after having jumped from a plane?

11.12 t^2

64. *Skydiving.* For jumps that exceed 13 sec, the polynomial $173t - 369$ can be used to approximate the distance, in feet, that a skydiver has fallen in t seconds. Approximately how far has a skydiver fallen 20 sec after having jumped from a plane?

Circumference. *The circumference of a circle of radius r is given by the polynomial $2\pi r$, where π is an irrational number. For an approximation of π, use 3.14.*

65. Find the circumference of a circle with radius 10 cm.

66. Find the circumference of a circle with radius 5 ft.

Area of a circle. *The area of a circle of radius r is given by the polynomial πr^2. Use 3.14 as an approximation for π.*

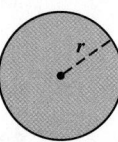

67. Find the area of a circle with radius 7 m.

68. Find the area of a circle with radius 6 ft.

Kidney transplants. *Often a patient needing a kidney transplant has a willing kidney donor who does not match the patient medically. A kidney-paired donation matches donor–recipient pairs. Two kidney transplants are performed simultaneously, with each patient receiving the kidney of a stranger. The number k of such "kidney swaps" t years after 2003 can be approximated by*

$$k = 14.3t^3 - 56t^2 + 57.7t + 19.$$

Use the following graph for Exercises 69 and 70.
Source: Based on data from United Network for Organ Sharing

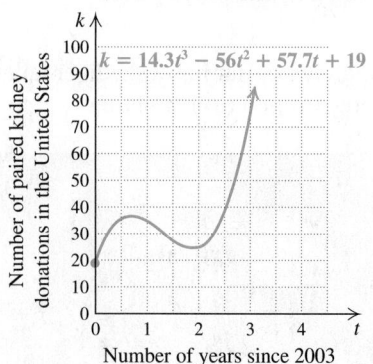

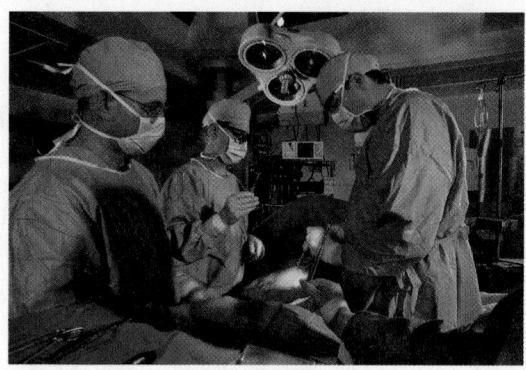

69. Estimate the number of kidney-paired donations in 2006.

70. Estimate the number of kidney-paired donations in 2004.

Memorizing words. *Participants in a psychology experiment were able to memorize an average of M words in t minutes, where $M = -0.001t^3 + 0.1t^2$. Use the following graph for Exercises 71–74.*

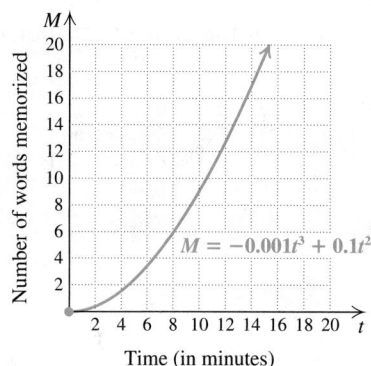

71. Estimate the number of words memorized after 10 min.

72. Estimate the number of words memorized after 14 min.

73. Find the approximate value of M for $t = 8$.

74. Find the approximate value of M for $t = 12$.

Body mass index. *The body mass index, or BMI, is one measure of a person's health. The average BMI B for males of age x, where x is between 2 and 20, is approximated by*

$$B = -0.003x^3 + 0.13x^2 - 1.2x + 18.6.$$

Use the following graph for Exercises 75 and 76.
Source: Based on information from the National Center for Health Statistics

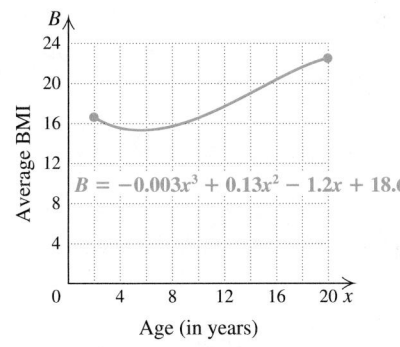

$$B = -0.003x^3 + 0.13x^2 - 1.2x + 18.6$$

Average BMI vs. Age (in years)

75. Approximate the average BMI for 4-year-old males; for 14-year-old males.

76. Approximate the average BMI for 10-year-old males; for 16-year-old males.

77. Explain how it is possible for a term to not be a monomial.

78. Is it possible to evaluate polynomials without understanding the rules for order of operations? Why or why not?

Skill Review

To prepare for Section 4.4, review simplifying expressions containing parentheses (Section 1.8).

Simplify. [1.8]

79. $2x + 5 - (x + 8)$ **80.** $3x - 7 - (5x - 1)$

81. $4a + 3 - (-2a + 6)$ **82.** $\frac{1}{2}t - \frac{1}{4} - \left(\frac{3}{2}t + \frac{3}{4}\right)$

83. $4t^4 + 8t - (5t^4 - 9t)$

84. $0.1a^2 + 5 - (-0.3a^2 + a - 6)$

Synthesis

85. Suppose that the coefficients of a polynomial are all integers and the polynomial is evaluated for some integer. Must the value of the polynomial then also be an integer? Why or why not?

86. Is it easier to evaluate a polynomial before or after like terms have been combined? Why?

87. Construct a polynomial in x (meaning that x is the variable) of degree 5 with four terms, with coefficients that are consecutive even integers. Write in descending order.

Revenue, cost, and profit. *Gigabytes Electronics is selling a new type of computer monitor.* Total revenue *is the total amount of money taken in and* total cost *is the total amount paid for producing the items. The firm estimates that for the monitor's first year, revenue from the sale of x monitors is*

$$250x - 0.5x^2 \text{ dollars},$$

and the total cost is given by

$$4000 + 0.6x^2 \text{ dollars}.$$

Profit *is the difference between revenue and cost.*

88. Find the profit when 20 monitors are produced and sold.

89. Find the profit when 30 monitors are produced and sold.

Simplify.

90. $\frac{9}{2}x^8 + \frac{1}{9}x^2 + \frac{1}{2}x^9 + \frac{9}{2}x + \frac{9}{2}x^9 + \frac{8}{9}x^2 + \frac{1}{2}x - \frac{1}{2}x^8$

91. $(3x^2)^3 + 4x^2 \cdot 4x^4 - x^4(2x)^2 + ((2x)^2)^3 - 100x^2(x^2)^2$

92. A polynomial in x has degree 3. The coefficient of x^2 is 3 less than the coefficient of x^3. The coefficient of x is three times the coefficient of x^2. The remaining constant is 2 more than the coefficient of x^3. The sum of the coefficients is -4. Find the polynomial.

93. Use the graph for Exercises 75 and 76 to determine the ages for which the average BMI is 16.

94. *Path of the Olympic arrow.* The Olympic flame at the 1992 Summer Olympics was lit by a flaming arrow. As the arrow moved d meters horizontally from the archer, its height h, in meters, was approximated by the polynomial

$$-0.0064d^2 + 0.8d + 2.$$

Complete the table for the choices of d given. Then plot the points and draw a graph representing the path of the arrow.

d	$-0.0064d^2 + 0.8d + 2$
0	
30	
60	
90	
120	

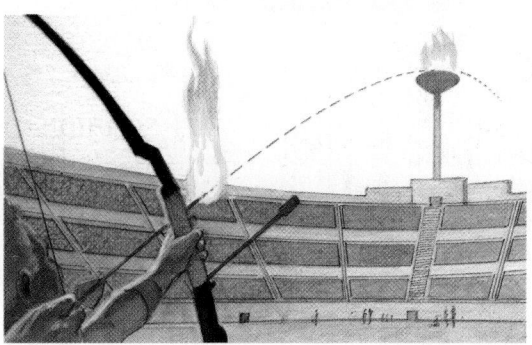

▨ *Semester averages.*   *Professor Sakima calculates a student's average for her course using*

$$A = 0.3q + 0.4t + 0.2f + 0.1h,$$

with q, t, f, and h representing a student's quiz average, test average, final exam score, and homework average, respectively. In Exercises 95 and 96, find the given student's course average rounded to the nearest tenth.

95. Galina: quizzes: 60, 85, 72, 91; final exam: 84; tests: 89, 93, 90; homework: 88

96. Nigel: quizzes: 95, 99, 72, 79; final exam: 91; tests: 68, 76, 92; homework: 86

In Exercises 97 and 98, complete the table for the given choices of t. Then plot the points and connect them with a smooth curve representing the graph of the polynomial.

97.

t	$-t^2 + 10t - 18$
3	
4	
5	
6	
7	

98.

t	$-t^2 + 6t - 4$
1	
2	
3	
4	
5	

4.4   Addition and Subtraction of Polynomials

Addition of Polynomials ▪ Opposites of Polynomials ▪ Subtraction of Polynomials ▪ Problem Solving

Addition of Polynomials

To add two polynomials, we write a plus sign between them and combine like terms.

EXAMPLE 1   Write an equivalent expression by adding.

a) $(-5x^3 + 6x - 1) + (4x^3 + 3x^2 + 2)$

b) $\left(\frac{2}{3}x^4 + 3x^2 - 7x + \frac{1}{2}\right) + \left(-\frac{1}{3}x^4 + 5x^3 - 3x^2 + 3x - \frac{1}{2}\right)$

SOLUTION

a) $(-5x^3 + 6x - 1) + (4x^3 + 3x^2 + 2)$

$= -5x^3 + 6x - 1 + 4x^3 + 3x^2 + 2$ Writing without parentheses

$= -5x^3 + 4x^3 + 3x^2 + 6x - 1 + 2$ Using the commutative and associative laws to write like terms together

$= (-5 + 4)x^3 + 3x^2 + 6x + (-1 + 2)$ Combining like terms; using the distributive law

$= -x^3 + 3x^2 + 6x + 1$ Note that $-1x^3 = -x^3$.

b) $\left(\frac{2}{3}x^4 + 3x^2 - 7x + \frac{1}{2}\right) + \left(-\frac{1}{3}x^4 + 5x^3 - 3x^2 + 3x - \frac{1}{2}\right)$

$\quad = \left(\frac{2}{3} - \frac{1}{3}\right)x^4 + 5x^3 + (3 - 3)x^2 + (-7 + 3)x + \left(\frac{1}{2} - \frac{1}{2}\right)$ Combining like terms

$\quad = \frac{1}{3}x^4 + 5x^3 - 4x$ **TRY EXERCISE** 9

After some practice, polynomial addition is often performed mentally.

EXAMPLE 2

Add: $(2 - 3x + x^2) + (-5 + 7x - 3x^2 + x^3)$.

SOLUTION   We have

$(2 - 3x + x^2) + (-5 + 7x - 3x^2 + x^3)$

$\quad = (2 - 5) + (-3 + 7)x + (1 - 3)x^2 + x^3$ You might do this step mentally.

$\quad = -3 + 4x - 2x^2 + x^3.$ Then you would write only this.

TRY EXERCISE 17

In the polynomials of the last example, the terms are arranged according to degree, from least to greatest. Such an arrangement is called *ascending order*. As a rule, answers are written in ascending order when the polynomials in the original problem are given in ascending order. If the polynomials in the original problem are given in descending order, the answer is usually written in descending order.

We can also add polynomials by writing like terms in columns. Sometimes this makes like terms easier to see.

EXAMPLE 3

Add: $9x^5 - 2x^3 + 6x^2 + 3$ and $5x^4 - 7x^2 + 6$ and $3x^6 - 5x^5 + x^2 + 5$.

SOLUTION   We arrange the polynomials with like terms in columns.

$$
\begin{array}{l}
9x^5 \qquad\quad - 2x^3 + 6x^2 + \ 3 \\
\qquad\quad 5x^4 \qquad\quad\ - 7x^2 + \ 6 \qquad \text{We leave spaces for missing terms.} \\
\underline{3x^6 - 5x^5 \qquad\qquad\quad + 1x^2 + \ 5} \qquad \text{Writing } x^2 \text{ as } 1x^2 \\
3x^6 + 4x^5 + 5x^4 - 2x^3 \qquad\quad + 14 \qquad \text{Adding}
\end{array}
$$

The answer is $3x^6 + 4x^5 + 5x^4 - 2x^3 + 14$.   **TRY EXERCISE** 23

Opposites of Polynomials

In Section 1.8, we used the property of -1 to show that the opposite of a sum is the sum of the opposites. This idea can be extended.

The Opposite of a Polynomial

To find an equivalent polynomial for the *opposite*, or *additive inverse*, of a polynomial, change the sign of every term. This is the same as multiplying the polynomial by -1.

EXAMPLE 4

Write two equivalent expressions for the opposite of $4x^5 - 7x^3 - 8x + \frac{5}{6}$.

SOLUTION

i) $-\left(4x^5 - 7x^3 - 8x + \frac{5}{6}\right)$ This is one representation of the opposite of $4x^5 - 7x^3 - 8x + \frac{5}{6}$.

ii) $-4x^5 + 7x^3 + 8x - \frac{5}{6}$ Changing the sign of every term

Thus, $-\left(4x^5 - 7x^3 - 8x + \frac{5}{6}\right)$ and $-4x^5 + 7x^3 + 8x - \frac{5}{6}$ are equivalent. Both expressions represent the opposite of $4x^5 - 7x^3 - 8x + \frac{5}{6}$.   **TRY EXERCISE** 27

EXAMPLE 5   Simplify: $-\left(-7x^4 - \frac{5}{9}x^3 + 8x^2 - x + 67\right)$.

SOLUTION   We have

$$-\left(-7x^4 - \frac{5}{9}x^3 + 8x^2 - x + 67\right) = 7x^4 + \frac{5}{9}x^3 - 8x^2 + x - 67.$$

The same result can be found by multiplying by -1:

$$-\left(-7x^4 - \frac{5}{9}x^3 + 8x^2 - x + 67\right)$$
$$= -1(-7x^4) + (-1)\left(-\frac{5}{9}x^3\right) + (-1)(8x^2) + (-1)(-x) + (-1)67$$
$$= 7x^4 + \frac{5}{9}x^3 - 8x^2 + x - 67.$$ **TRY EXERCISE** 31

Subtraction of Polynomials

We can now subtract one polynomial from another by adding the opposite of the polynomial being subtracted.

EXAMPLE 6   Write an equivalent expression by subtracting.

a) $(9x^5 + x^3 - 2x^2 + 4) - (-2x^5 + x^4 - 4x^3 - 3x^2)$

b) $(7x^5 + x^3 - 9x) - (3x^5 - 4x^3 + 5)$

SOLUTION

a) $(9x^5 + x^3 - 2x^2 + 4) - (-2x^5 + x^4 - 4x^3 - 3x^2)$
$$= 9x^5 + x^3 - 2x^2 + 4 + 2x^5 - x^4 + 4x^3 + 3x^2   \text{Adding the opposite}$$
$$= 11x^5 - x^4 + 5x^3 + x^2 + 4   \text{Combining like terms}$$

b) $(7x^5 + x^3 - 9x) - (3x^5 - 4x^3 + 5)$
$$= 7x^5 + x^3 - 9x + (-3x^5) + 4x^3 - 5   \text{Adding the opposite}$$
$$= 7x^5 + x^3 - 9x - 3x^5 + 4x^3 - 5   \text{Try to go directly to this step.}$$
$$= 4x^5 + 5x^3 - 9x - 5   \text{Combining like terms}$$

TRY EXERCISE 39

To subtract using columns, we first replace the coefficients in the polynomial being subtracted with their opposites. We then add as before.

EXAMPLE 7   Write in columns and subtract: $(5x^2 - 3x + 6) - (9x^2 - 5x - 3)$.

SOLUTION

i) $5x^2 - 3x + 6$ Writing similar terms in columns
      $-(9x^2 - 5x - 3)$

ii) $5x^2 - 3x + 6$ Changing signs and removing parentheses
       $-9x^2 + 5x + 3$

iii) $5x^2 - 3x + 6$
        $-9x^2 + 5x + 3$
        $\overline{-4x^2 + 2x + 9}$ Adding   **TRY EXERCISE** 53

If you can do so without error, you can arrange the polynomials in columns, mentally find the opposite of each term being subtracted, and write the answer. Lining up like terms is important and may require leaving some blank space.

EXAMPLE **8**   Write in columns and subtract: $(x^3 + x^2 - 12) - (-2x^3 + x^2 - 3x + 6)$.

SOLUTION   We have

$$
\begin{array}{r}
x^3 + x^2 \qquad\quad - 12 \qquad \text{Leaving a blank space for the missing term}\\
-(-2x^3 + x^2 - 3x + \ 6)\\
\hline
3x^3 \qquad\quad + 3x - 18
\end{array}
$$

▸ TRY EXERCISE ▸ 55

CAUTION!   Be sure to subtract every term of the polynomial being subtracted when using columns.

Problem Solving

EXAMPLE **9**   Find a polynomial for the sum of the areas of rectangles A, B, C, and D.

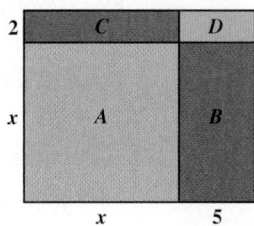

SOLUTION

1. **Familiarize.** Recall that the area of a rectangle is the product of its length and width.

2. **Translate.** We translate the problem to mathematical language. The sum of the areas is a sum of products. We find each product and then add:

Area of A	plus	area of B	plus	area of C	plus	area of D
↓	↓	↓	↓	↓	↓	↓
$x \cdot x$	$+$	$5x$	$+$	$2x$	$+$	$2 \cdot 5.$

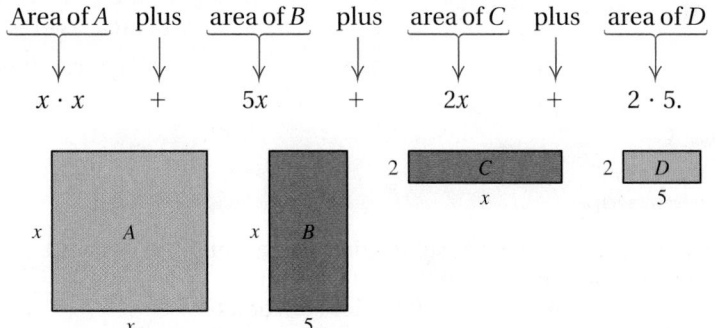

3. **Carry out.** We simplify $x \cdot x$ and $2 \cdot 5$ and combine like terms:

$$x^2 + 5x + 2x + 10 = x^2 + 7x + 10.$$

4. **Check.** A partial check is to replace x with a number, say 3. Then we evaluate $x^2 + 7x + 10$ and compare that result with an alternative calculation:

$$3^2 + 7 \cdot 3 + 10 = 9 + 21 + 10 = 40.$$

When we substitute 3 for x and calculate the total area by regarding the figure as one large rectangle, we should also get 40:

Total area $= (x + 5)(x + 2) = (3 + 5)(3 + 2) = 8 \cdot 5 = 40.$

Our check is only partial, since it is possible for an incorrect answer to equal 40 when evaluated for $x = 3$. This would be unlikely, especially if a second choice of x, say $x = 5$, also checks. We leave that check to the student.

5. **State.** A polynomial for the sum of the areas is $x^2 + 7x + 10$.

TRY EXERCISE ▶ 61

EXAMPLE 10   A 16-ft wide round fountain is built in a square city park that measures x ft by x ft. Find a polynomial for the remaining area of the park.

SOLUTION

1. **Familiarize.** We make a drawing of the square park and the circular fountain, and let x represent the length of a side of the park.

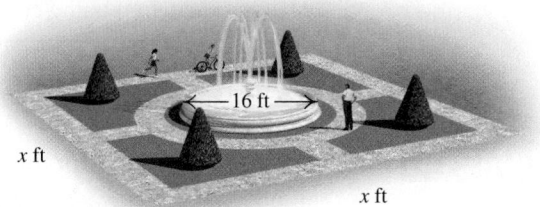

The area of a square is given by $A = s^2$, and the area of a circle is given by $A = \pi r^2$. Note that a circle with a diameter of 16 ft has a radius of 8 ft.

2. **Translate.** We reword the problem and translate as follows.

Rewording: <u>Area of park</u> <u>minus</u> <u>area of fountain</u> <u>is</u> <u>area left over.</u>

Translating: $x \text{ ft} \cdot x \text{ ft}$ $-$ $\pi \cdot 8 \text{ ft} \cdot 8 \text{ ft}$ $=$ Area left over

3. **Carry out.** We carry out the multiplication:

$$x^2 \text{ ft}^2 - 64\pi \text{ ft}^2 = \text{Area left over.}$$

4. **Check.** As a partial check, note that the units in the answer are square feet (ft^2), a measure of area, as expected.

5. **State.** The remaining area of the park is $(x^2 - 64\pi) \text{ ft}^2$.   TRY EXERCISE ▶ 69

TECHNOLOGY CONNECTION

To check polynomial addition or subtraction, we can let y_1 = the expression before the addition or subtraction has been performed and y_2 = the simplified sum or difference. If the addition or subtraction is correct, y_1 will equal y_2 and $y_2 - y_1$ will be 0. We enter $y_2 - y_1$ as y_3, using **VARS**. Below is a check of Example 6(b) in which

$$y_1 = (7x^5 + x^3 - 9x) - (3x^5 - 4x^3 + 5),$$
$$y_2 = 4x^5 + 5x^3 - 9x - 5,$$

and

$$y_3 = y_2 - y_1.$$

We graph only y_3. If indeed y_1 and y_2 are equivalent, then y_3 should equal 0. This means its graph should coincide with the x-axis. The TRACE or TABLE features can confirm

that y_3 is always 0, or we can select y_3 to be drawn bold at the **Y=** window.

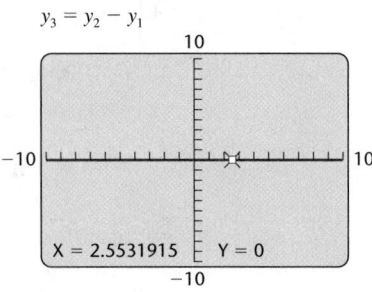

$y_3 = y_2 - y_1$

1. Use a graphing calculator to check Examples 1, 2, and 6.

4.4   EXERCISE SET

For Extra Help
MyMathLab   Math XL
PRACTICE WATCH DOWNLOAD

☞ **Concept Reinforcement**   *For Exercises 1–4, replace ▨ with the correct expression or operation sign.*

1. $(3x^2 + 2) + (6x^2 + 7) = (3 + 6) \ ▨ + (2 + 7)$

2. $(5t - 6) + (4t + 3) = (5 + 4)t + (\ ▨\ + 3)$

3. $(9x^3 - x^2) - (3x^3 + x^2) = 9x^3 - x^2 - 3x^3 \ ▨\ x^2$

4. $(-2n^3 + 5) - (n^2 - 2) = -2n^3 + 5 - n^2 \ ▨\ 2$

Add.

5. $(3x + 2) + (x + 7)$

6. $(x + 1) + (12x + 10)$

7. $(2t + 7) + (-8t + 1)$

8. $(4t - 3) + (-11t + 2)$

9. $(x^2 + 6x + 3) + (-4x^2 - 5)$

10. $(x^2 - 5x + 4) + (8x - 9)$

11. $(7t^2 - 3t - 6) + (2t^2 + 4t + 9)$

12. $(8a^2 + 4a - 7) + (6a^2 - 3a - 1)$

13. $(4m^3 - 7m^2 + m - 5) + (4m^3 + 7m^2 - 4m - 2)$

14. $(5n^3 - n^2 + 4n + 11) + (2n^3 - 4n^2 + n - 11)$

15. $(3 + 6a + 7a^2 + a^3) + (4 + 7a - 8a^2 + 6a^3)$

16. $(7 + 4t - 5t^2 + 6t^3) + (2 + t + 6t^2 - 4t^3)$

17. $(3x^6 + 2x^4 - x^3 + 5x) + (-x^6 + 3x^3 - 4x^2 + 7x^4)$

18. $(4x^5 - 6x^3 - 9x + 1) + (3x^4 + 6x^3 + 9x^2 + x)$

19. $\left(\frac{3}{5}x^4 + \frac{1}{2}x^3 - \frac{2}{3}x + 3\right) + \left(\frac{2}{5}x^4 - \frac{1}{4}x^3 - \frac{3}{4}x^2 - \frac{1}{6}x\right)$

20. $\left(\frac{1}{3}x^9 + \frac{1}{5}x^5 - \frac{1}{2}x^2 + 7\right) + \left(-\frac{1}{5}x^9 + \frac{1}{4}x^4 - \frac{3}{5}x^5\right)$

21. $(5.3t^2 - 6.4t - 9.1) + (4.2t^3 - 1.8t^2 + 7.3)$

22. $(4.9a^3 + 3.2a^2 - 5.1a) + (2.1a^2 - 3.7a + 4.6)$

23. $\begin{array}{r} -4x^3 + 8x^2 + 3x - 2 \\ -\ 4x^2 + 3x + 2 \\ \hline \end{array}$

24. $\begin{array}{r} -3x^4 + 6x^2 + 2x - 4 \\ -\ 3x^2 + 2x + 4 \\ \hline \end{array}$

25. $\begin{array}{r} 0.05x^4 + 0.12x^3 - \ 0.5x^2 \\ -\ 0.02x^3 + 0.02x^2 + 2x \\ 1.5x^4 \qquad\quad + 0.01x^2 \qquad + 0.15 \\ 0.25x^3 \qquad\qquad\qquad + 0.85 \\ -0.25x^4 \qquad\quad + \quad 10x^2 \qquad - 0.04 \\ \hline \end{array}$

26. $\begin{array}{r} 0.15x^4 + 0.10x^3 - \ 0.9x^2 \\ -\ 0.01x^3 + 0.01x^2 + x \\ 1.25x^4 \qquad\quad + 0.11x^2 \qquad + 0.01 \\ 0.27x^3 \qquad\qquad\qquad + 0.99 \\ -0.35x^4 \qquad\quad + \quad 15x^2 \qquad - 0.03 \\ \hline \end{array}$

Write two equivalent expressions for the opposite of each polynomial, as in Example 4.

27. $-3t^3 + 4t^2 - 7$

28. $-x^3 - 5x^2 + 2x$

29. $x^4 - 8x^3 + 6x$

30. $5a^3 + 2a - 17$

Simplify.

31. $-(9x - 10)$

32. $-(-5x + 8)$

33. $-(3a^4 - 5a^2 + 1.2)$

34. $-(-6a^3 + 0.2a^2 - 7)$

35. $-\left(-4x^4 + 6x^2 + \frac{3}{4}x - 8\right)$

36. $-\left(3x^5 - 2x^3 - \frac{3}{5}x^2 + 16\right)$

Subtract.

37. $(3x + 1) - (5x + 8)$

38. $(7x + 3) - (3x + 2)$

39. $(-9t + 12) - (t^2 + 3t - 1)$

40. $(a^2 - 3a - 2) - (2a^2 - 6a - 2)$

41. $(4a^2 + a - 7) - (3 - 8a^3 - 4a^2)$

42. $(-4x^2 + 2x) - (-5x^2 + 2x^3 + 3)$

43. $(1.2x^3 + 4.5x^2 - 3.8x) - (-3.4x^3 - 4.7x^2 + 23)$

44. $(0.5x^4 - 0.6x^2 + 0.7) - (2.3x^4 + 1.8x - 3.9)$

Aha! **45.** $(7x^3 - 2x^2 + 6) - (6 - 2x^2 + 7x^3)$

46. $(8x^5 + 3x^4 + x - 1) - (8x^5 + 3x^4 - 1)$

47. $(3 + 5a + 3a^2 - a^3) - (2 + 4a - 9a^2 + 2a^3)$

48. $(7 + t - 5t^2 + 2t^3) - (1 + 2t - 4t^2 + 5t^3)$

49. $\left(\frac{5}{8}x^3 - \frac{1}{4}x - \frac{1}{3}\right) - \left(-\frac{1}{2}x^3 + \frac{1}{4}x - \frac{1}{3}\right)$

50. $\left(\frac{1}{5}x^3 + 2x^2 - \frac{3}{10}\right) - \left(-\frac{2}{5}x^3 + 2x^2 + \frac{7}{1000}\right)$

51. $(0.07t^3 - 0.03t^2 + 0.01t) - (0.02t^3 + 0.04t^2 - 1)$

52. $(0.9a^3 + 0.2a - 5) - (0.7a^4 - 0.3a - 0.1)$

53. $\quad x^3 + 3x^2 + 1$
$\quad \underline{-(x^3 + x^2 - 5)}$

54. $\quad x^2 + 5x + 6$
$\quad \underline{-(x^2 + 2x + 1)}$

55. $\quad 4x^4 - 2x^3$
$\quad \underline{-(7x^4 + 6x^3 + 7x^2)}$

56. $\quad 5x^4 + 6x^3 - 9x^2$
$\quad \underline{-(-6x^4 + x^2)}$

57. Solve.

  a) Find a polynomial for the sum of the areas of the rectangles shown in the figure.

  b) Find the sum of the areas when $x = 5$ and $x = 7$.

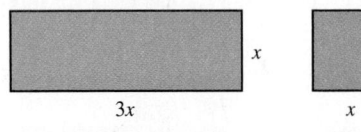
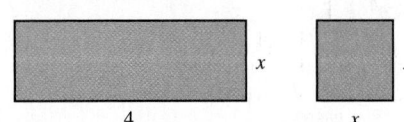

58. Solve. Leave the answers in terms of π.

  a) Find a polynomial for the sum of the areas of the circles shown in the figure.

  b) Find the sum of the areas when $r = 5$ and $r = 11.3$.

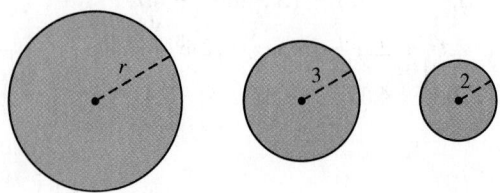

Find a polynomial for the perimeter of each figure in Exercises 59 and 60.

59.

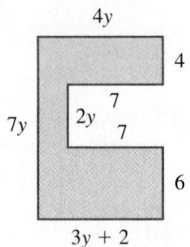

60.

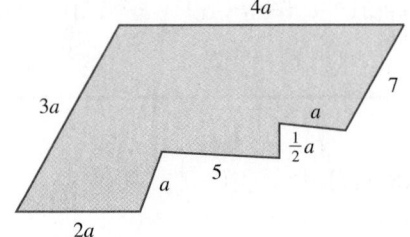

Find two algebraic expressions for the area of each figure. First, regard the figure as one large rectangle, and then regard the figure as a sum of four smaller rectangles.

61.

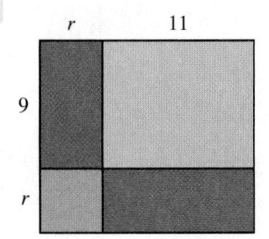

62.

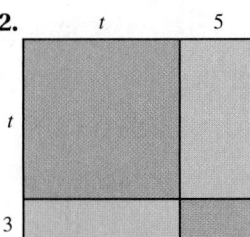

63.

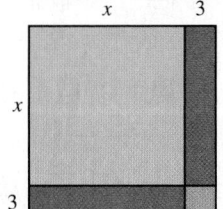

64.

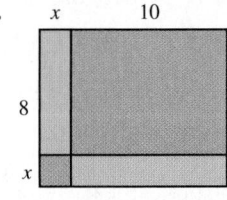

Find a polynomial for the shaded area of each figure.

65.

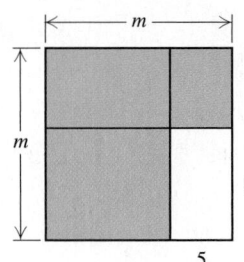

66.

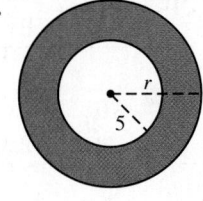

67.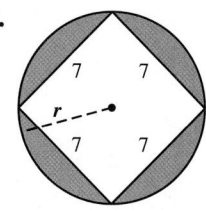

68.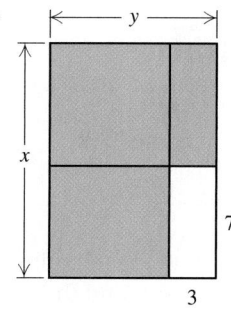

Simplify.

85. $(6t^2 - 7t) + (3t^2 - 4t + 5) - (9t - 6)$

86. $(3x^2 - 4x + 6) - (-2x^2 + 4) + (-5x - 3)$

87. $4(x^2 - x + 3) - 2(2x^2 + x - 1)$

88. $3(2y^2 - y - 1) - (6y^2 - 3y - 3)$

89. $(345.099x^3 - 6.178x) - (94.508x^3 - 8.99x)$

69. A 2-ft by 6-ft bath enclosure is installed in a new bathroom measuring x ft by x ft. Find a polynomial for the remaining floor area.

70. A 5-ft by 7-ft Jacuzzi™ is installed on an outdoor deck measuring y ft by y ft. Find a polynomial for the remaining area of the deck.

71. A 12-ft wide round patio is laid in a garden measuring z ft by z ft. Find a polynomial for the remaining area of the garden.

72. A 10-ft wide round water trampoline is floating in a pool measuring x ft by x ft. Find a polynomial for the remaining surface area of the pool.

73. A 12-m by 12-m mat includes a circle of diameter d meters for wrestling. Find a polynomial for the area of the mat outside the wrestling circle.

74. A 2-m by 3-m rug is spread inside a tepee that has a diameter of x meters. Find a polynomial for the area of the tepee's floor that is not covered.

75. Explain why parentheses are used in the statement of the solution of Example 10: $(x^2 - 64\pi)$ ft².

76. Is the sum of two trinomials always a trinomial? Why or why not?

Skill Review

To prepare for Section 4.5, review multiplying using the distributive law and multiplying with exponential notation (Sections 1.8 and 4.1).

Simplify.

77. $2(x^2 - x + 3)$ [1.8]

78. $-5(3x^2 - 2x - 7)$ [1.8]

79. $x^2 \cdot x^6$ [4.1]

80. $y^6 \cdot y$ [4.1]

81. $2n \cdot n^2$ [4.1]

82. $-6n^4 \cdot n^8$ [4.1]

Synthesis

83. What can be concluded about two polynomials whose sum is zero?

84. Which, if any, of the commutative, associative, and distributive laws are needed for adding polynomials? Why?

Find a polynomial for the surface area of the right rectangular solid.

90.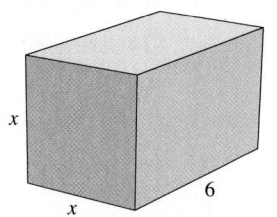

91.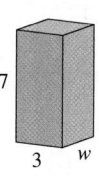

92.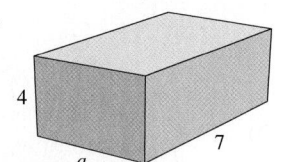

93.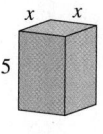

94. Find a polynomial for the total length of all edges in the figure appearing in Exercise 93.

95. Find a polynomial for the total length of all edges in the figure appearing in Exercise 90.

96. *Total profit.* Hadley Electronics is marketing a new digital camera. Total revenue is the total amount of money taken in. The firm determines that when it sells x cameras, its total revenue is given by

$$R = 175x - 0.4x^2.$$

Total cost is the total cost of producing x cameras. Hadley Electronics determines that the total cost of producing x cameras is given by

$$C = 5000 + 0.6x^2.$$

The total profit P is

(Total Revenue) − (Total Cost) = $R - C$.

a) Find a polynomial for total profit.

b) What is the total profit on the production and sale of 75 cameras?

c) What is the total profit on the production and sale of 120 cameras?

97. Does replacing each occurrence of the variable x in $4x^7 - 6x^3 + 2x$ with its opposite result in the opposite of the polynomial? Why or why not?

4.5 Multiplication of Polynomials

Multiplying Monomials ▪ Multiplying a Monomial and a Polynomial ▪
Multiplying Any Two Polynomials ▪ Checking by Evaluating

We now multiply polynomials using techniques based largely on the distributive, associative, and commutative laws and the rules for exponents.

Multiplying Monomials

Consider $(3x)(4x)$. We multiply as follows:

$$
\begin{aligned}
(3x)(4x) &= 3 \cdot x \cdot 4 \cdot x && \text{Using an associative law} \\
&= 3 \cdot 4 \cdot x \cdot x && \text{Using a commutative law} \\
&= (3 \cdot 4) \cdot x \cdot x && \text{Using an associative law} \\
&= 12x^2.
\end{aligned}
$$

To Multiply Monomials

To find an equivalent expression for the product of two monomials, multiply the coefficients and then multiply the variables using the product rule for exponents.

EXAMPLE **1**

Multiply to form an equivalent expression.

a) $(5x)(6x)$

b) $(-a)(3a)$

c) $(7x^5)(-4x^3)$

SOLUTION

a) $\begin{aligned}(5x)(6x) &= (5 \cdot 6)(x \cdot x) && \text{Multiplying the coefficients; multiplying} \\ & && \text{the variables} \\ &= 30x^2 && \text{Simplifying}\end{aligned}$

b) $\begin{aligned}(-a)(3a) &= (-1a)(3a) && \text{Writing } -a \text{ as } -1a \text{ can ease calculations.} \\ &= (-1 \cdot 3)(a \cdot a) && \text{Using an associative law and a} \\ & && \text{commutative law} \\ &= -3a^2\end{aligned}$

c) $\begin{aligned}(7x^5)(-4x^3) &= 7(-4)(x^5 \cdot x^3) \\ &= -28x^{5+3} \\ &= -28x^8\end{aligned}$ $\left.\begin{array}{l}\\\\\end{array}\right\}$ Using the product rule for exponents

TRY EXERCISE ▸ 13

After some practice, you can try writing only the answer.

STUDENT NOTES

Remember that when we compute $(3 \cdot 5)(2 \cdot 4)$, each factor is used only once, even if we change the order:

$$
\begin{aligned}
(3 \cdot 5)(2 \cdot 4) &= (5 \cdot 2)(3 \cdot 4) \\
&= 10 \cdot 12 \\
&= 120.
\end{aligned}
$$

In the same way,

$$
\begin{aligned}
(3 \cdot x)(2 \cdot x) &= (3 \cdot 2)(x \cdot x) \\
&= 6x^2.
\end{aligned}
$$

Some students mistakenly "reuse" a factor.

Multiplying a Monomial and a Polynomial

To find an equivalent expression for the product of a monomial, such as $5x$, and a polynomial, such as $2x^2 - 3x + 4$, we use the distributive law.

EXAMPLE 2

Multiply: **(a)** $x(x + 3)$; **(b)** $5x(2x^2 - 3x + 4)$.

SOLUTION

a) $x(x + 3) = x \cdot x + x \cdot 3$ Using the distributive law

$\quad\quad\quad\quad = x^2 + 3x$

b) $5x(2x^2 - 3x + 4) = (5x)(2x^2) - (5x)(3x) + (5x)(4)$ Using the distributive law

$\quad\quad\quad\quad\quad\quad\quad = 10x^3 - 15x^2 + 20x$ Performing the three multiplications

> TRY EXERCISE 29

The product in Example 2(a) can be visualized as the area of a rectangle with width x and length $x + 3$.

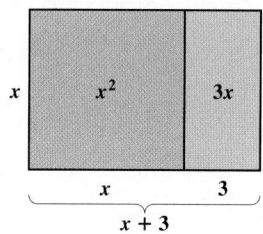

Note that the total area can be expressed as $x(x + 3)$ or, by adding the two smaller areas, $x^2 + 3x$.

> ### The Product of a Monomial and a Polynomial
> To multiply a monomial and a polynomial, multiply each term of the polynomial by the monomial.

Try to do this mentally, when possible. Remember that we multiply coefficients and, when the bases match, add exponents.

EXAMPLE 3

Multiply: $2x^2(x^3 - 7x^2 + 10x - 4)$.

SOLUTION

$\quad\quad\quad\quad\quad$ *Think:* $\underbrace{2x^2 \cdot x^3} - \underbrace{2x^2 \cdot 7x^2} + \underbrace{2x^2 \cdot 10x} - \underbrace{2x^2 \cdot 4}$

$2x^2(x^3 - 7x^2 + 10x - 4) = 2x^5 \quad - \quad 14x^4 \quad + \quad 20x^3 \quad - \quad 8x^2$

> TRY EXERCISE 31

Multiplying Any Two Polynomials

Before considering the product of *any* two polynomials, let's look at products when both polynomials are binomials.

To find an equivalent expression for the product of two binomials, we again begin by using the distributive law. This time, however, it is a *binomial* rather than a monomial that is being distributed.

EXAMPLE **4**

Multiply each pair of binomials.

a) $x + 5$ and $x + 4$ **b)** $4x - 3$ and $x - 2$

SOLUTION

a) $(x + 5)\,(x + 4) = (x + 5)\,x + (x + 5)\,4$ Using the distributive law

$\qquad\qquad\qquad = x(x + 5) + 4(x + 5)$ Using the commutative law for multiplication

$\qquad\qquad\qquad = x \cdot x + x \cdot 5 + 4 \cdot x + 4 \cdot 5$ Using the distributive law (twice)

$\qquad\qquad\qquad = x^2 + 5x + 4x + 20$ Multiplying the monomials

$\qquad\qquad\qquad = x^2 + 9x + 20$ Combining like terms

b) $(4x - 3)\,(x - 2) = (4x - 3)\,x - (4x - 3)\,2$ Using the distributive law

$\qquad\qquad\qquad = x(4x - 3) - 2(4x - 3)$ Using the commutative law for multiplication. This step is often omitted.

$\qquad\qquad\qquad = x \cdot 4x - x \cdot 3 - 2 \cdot 4x - 2(-3)$ Using the distributive law (twice)

$\qquad\qquad\qquad = 4x^2 - 3x - 8x + 6$ Multiplying the monomials

$\qquad\qquad\qquad = 4x^2 - 11x + 6$ Combining like terms

TRY EXERCISE 37

To visualize the product in Example 4(a), consider a rectangle of length $x + 5$ and width $x + 4$, as shown here.

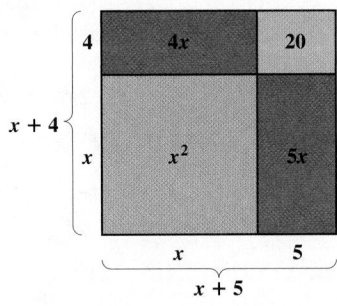

The total area can be expressed as $(x + 5)(x + 4)$ or, by adding the four smaller areas, $x^2 + 5x + 4x + 20$.

Let's consider the product of a binomial and a trinomial. Again we make repeated use of the distributive law.

EXAMPLE 5   Multiply: $(x^2 + 2x - 3)(x + 4)$.

SOLUTION

$$(x^2 + 2x - 3)\ (x + 4)$$

$$= (x^2 + 2x - 3)\ x + (x^2 + 2x - 3)\ 4 \qquad \text{Using the distributive law}$$

$$= x(x^2 + 2x - 3) + 4(x^2 + 2x - 3) \qquad \text{Using the commutative law}$$

$$= x \cdot x^2 + x \cdot 2x - x \cdot 3 + 4 \cdot x^2 + 4 \cdot 2x - 4 \cdot 3 \qquad \text{Using the distributive law (twice)}$$

$$= x^3 + 2x^2 - 3x + 4x^2 + 8x - 12 \qquad \text{Multiplying the monomials}$$

$$= x^3 + 6x^2 + 5x - 12 \qquad \text{Combining like terms}$$

> **TRY EXERCISE** 57

Perhaps you have discovered the following in the preceding examples.

The Product of Two Polynomials

To multiply two polynomials P and Q, select one of the polynomials, say P. Then multiply each term of P by every term of Q and combine like terms.

To use columns for long multiplication, multiply each term in the top row by every term in the bottom row. We write like terms in columns, and then add the results. Such multiplication is like multiplying with whole numbers.

$$
\begin{array}{r}
321 \\
\times\ 12 \\
\hline
642 \\
321 \\
\hline
3852
\end{array}
\qquad
\begin{array}{r}
300 + 20 + 1 \\
\times \qquad\quad 10 + 2 \\
\hline
600 + 40 + 2 \\
3000 + 200 + 10 \\
\hline
3000 + 800 + 50 + 2
\end{array}
\qquad
\begin{array}{l}
\\
\\
\text{Multiplying the top row by 2} \\
\text{Multiplying the top row by 10} \\
\text{Adding}
\end{array}
$$

EXAMPLE 6   Multiply: $(5x^4 - 2x^2 + 3x)(x^2 + 2x)$.

SOLUTION

$$
\begin{array}{r}
5x^4 \qquad\quad - 2x^2 + 3x \\
x^2 + 2x \\
\hline
10x^5 \qquad\quad - 4x^3 + 6x^2 \\
5x^6 \qquad - 2x^4 + 3x^3 \\
\hline
5x^6 + 10x^5 - 2x^4 - x^3 + 6x^2
\end{array}
$$

Note that each polynomial is written in descending order, and space is left for missing terms.

Multiplying the top row by $2x$

Multiplying the top row by x^2

Combining like terms

Line up like terms in columns.

> **TRY EXERCISE** 61

With practice, you will be able to skip some steps. Sometimes we multiply horizontally, while still aligning like terms as we write the product.

EXAMPLE 7 Multiply: $(2x^3 + 3x^2 - 4x + 6)(3x + 5)$.

SOLUTION

$$\overbrace{\hphantom{(2x^3 + 3x^2 - 4x + 6)(3x + 5) = 6x^4 + 9x^3 - 12x^2 + 18x}}^{\text{Multiplying by } 3x}$$

$$(2x^3 + 3x^2 - 4x + 6)(3x + 5) = \begin{array}{r} 6x^4 + 9x^3 - 12x^2 + 18x \\ + \underline{10x^3 + 15x^2 - 20x + 30} \end{array}$$

Multiplying by 5

$$= 6x^4 + 19x^3 + 3x^2 - 2x + 30$$

TRY EXERCISE 65

Checking by Evaluating

How can we be certain that our multiplication (or addition or subtraction) of polynomials is correct? One check is to simply review our calculations. A different type of check, used in Example 9 of Section 4.4, makes use of the fact that equivalent expressions have the same value when evaluated for the same replacement. Thus a quick, partial, check of Example 7 can be made by selecting a convenient replacement for x (say, 1) and comparing the values of the expressions $(2x^3 + 3x^2 - 4x + 6)(3x + 5)$ and $6x^4 + 19x^3 + 3x^2 - 2x + 30$:

$$(2x^3 + 3x^2 - 4x + 6)(3x + 5) = (2 \cdot 1^3 + 3 \cdot 1^2 - 4 \cdot 1 + 6)(3 \cdot 1 + 5)$$
$$= (2 + 3 - 4 + 6)(3 + 5)$$
$$= 7 \cdot 8 = 56;$$

$$6x^4 + 19x^3 + 3x^2 - 2x + 30 = 6 \cdot 1^4 + 19 \cdot 1^3 + 3 \cdot 1^2 - 2 \cdot 1 + 30$$
$$= 6 + 19 + 3 - 2 + 30$$
$$= 28 - 2 + 30 = 56.$$

Since the value of both expressions is 56, the multiplication in Example 7 is very likely correct.

It is possible, by chance, for two expressions that are not equivalent to share the same value when evaluated. For this reason, checking by evaluating is only a partial check. Consult your instructor for the checking approach that he or she prefers.

TECHNOLOGY CONNECTION

Tables can also be used to check polynomial multiplication. To illustrate, we can check Example 7 by entering $y_1 = (2x^3 + 3x^2 - 4x + 6)(3x + 5)$ and $y_2 = 6x^4 + 19x^3 + 3x^2 - 2x + 30$.

When TABLE is then pressed, we are shown two columns of values—one for y_1 and one for y_2. If our multiplication was correct, the columns of values will match.

X	Y1	Y2
−3	36	36
−2	−10	−10
−1	22	22
0	30	30
1	56	56
2	286	286
3	1050	1050

X = −3

1. Form a table and scroll up and down to check Example 6.
2. Check Example 7 using the method discussed in Section 4.4: Let

$$y_1 = (2x^3 + 3x^2 - 4x + 6)(3x + 5),$$
$$y_2 = 6x^4 + 19x^3 + 3x^2 - 2x + 30,$$

and

$$y_3 = y_2 - y_1.$$

Then check that y_3 is always 0.

4.5	**EXERCISE SET**

Concept Reinforcement *In each of Exercises 1–6, match the expression with the correct result from the column on the right. Choices may be used more than once.*

1. _____ $3x^2 \cdot 2x^4$

2. _____ $3x^8 + 5x^8$

3. _____ $4x^3 \cdot 2x^5$

4. _____ $3x^5 \cdot 2x^3$

5. _____ $4x^6 + 2x^6$

6. _____ $4x^4 \cdot 2x^2$

a) $6x^8$

b) $8x^6$

c) $6x^6$

d) $8x^8$

Multiply.

7. $(3x^5)7$

8. $2x^3 \cdot 11$

9. $(-x^3)(x^4)$

10. $(-x^2)(-x)$

11. $(-x^6)(-x^2)$

12. $(-x^5)(x^3)$

13. $4t^2(9t^2)$

14. $(6a^8)(3a^2)$

15. $(0.3x^3)(-0.4x^6)$

16. $(-0.1x^6)(0.2x^4)$

17. $\left(-\frac{1}{4}x^4\right)\left(\frac{1}{5}x^8\right)$

18. $\left(-\frac{1}{5}x^3\right)\left(-\frac{1}{3}x\right)$

19. $(-5n^3)(-1)$

20. $19t^2 \cdot 0$

21. $11x^5(-4x^5)$

22. $12x^3(-5x^3)$

23. $(-4y^5)(6y^2)(-3y^3)$

24. $7x^2(-2x^3)(2x^6)$

25. $5x(4x + 1)$

26. $3x(2x - 7)$

27. $(a - 9)3a$

28. $(a - 7)4a$

29. $x^2(x^3 + 1)$

30. $-2x^3(x^2 - 1)$

31. $-3n(2n^2 - 8n + 1)$

32. $4n(3n^3 - 4n^2 - 5n + 10)$

33. $-5t^2(3t + 6)$

34. $7t^2(2t + 1)$

35. $\frac{2}{3}a^4\left(6a^5 - 12a^3 - \frac{5}{8}\right)$

36. $\frac{3}{4}t^5\left(8t^6 - 12t^4 + \frac{12}{7}\right)$

37. $(x + 3)(x + 4)$

38. $(x + 7)(x + 3)$

39. $(t + 7)(t - 3)$

40. $(t - 4)(t + 3)$

41. $(a - 0.6)(a - 0.7)$

42. $(a - 0.4)(a - 0.8)$

43. $(x + 3)(x - 3)$

44. $(x + 6)(x - 6)$

45. $(4 - x)(7 - 2x)$

46. $(5 + x)(5 + 2x)$

47. $\left(t + \frac{3}{2}\right)\left(t + \frac{4}{3}\right)$

48. $\left(a - \frac{2}{5}\right)\left(a + \frac{5}{2}\right)$

49. $\left(\frac{1}{4}a + 2\right)\left(\frac{3}{4}a - 1\right)$

50. $\left(\frac{2}{5}t - 1\right)\left(\frac{3}{5}t + 1\right)$

Draw and label rectangles similar to those following Examples 2 and 4 to illustrate each product.

51. $x(x + 5)$

52. $x(x + 2)$

53. $(x + 1)(x + 2)$

54. $(x + 3)(x + 1)$

55. $(x + 5)(x + 3)$

56. $(x + 4)(x + 6)$

Multiply and check.

57. $(x^2 - x + 3)(x + 1)$

58. $(x^2 + x - 7)(x + 2)$

59. $(2a + 5)(a^2 - 3a + 2)$

60. $(3t - 4)(t^2 - 5t + 1)$

61. $(y^2 - 7)(3y^4 + y + 2)$

62. $(a^2 + 4)(5a^3 - 3a - 1)$

Aha! 63. $(3x + 2)(7x + 4x + 1)$

64. $(4x - 5x - 3)(1 + 2x^2)$

65. $(x^2 + 5x - 1)(x^2 - x + 3)$

66. $(x^2 - 3x + 2)(x^2 + x + 1)$

67. $\left(5t^2 - t + \frac{1}{2}\right)(2t^2 + t - 4)$

68. $(2t^2 - 5t - 4)\left(3t^2 - t + \frac{1}{2}\right)$

69. $(x + 1)(x^3 + 7x^2 + 5x + 4)$

70. $(x + 2)(x^3 + 5x^2 + 9x + 3)$

71. Is it possible to understand polynomial multiplication without understanding the distributive law? Why or why not?

72. The polynomials

$$(a + b + c + d) \quad \text{and} \quad (r + s + m + p)$$

are multiplied. Without performing the multiplication, determine how many terms the product will contain. Provide a justification for your answer.

Skill Review

Review simplifying expressions using the rules for order of operations (Section 1.8).

Simplify. [1.8]

73. $(9 - 3)(9 + 3) + 3^2 - 9^2$

74. $(7 + 2)(7 - 2) + 2^2 - 7^2$

75. $5 + \dfrac{7 + 4 + 2 \cdot 5}{7}$

76. $11 - \dfrac{2 + 6 \cdot 3 + 4}{6}$

77. $(4 + 3 \cdot 5 + 5) \div 3 \cdot 4$

78. $(2 + 2 \cdot 7 + 4) \div 2 \cdot 5$

Synthesis

79. Under what conditions will the product of two binomials be a trinomial?

80. How can the following figure be used to show that $(x + 3)^2 \neq x^2 + 9$?

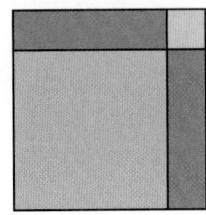

Find a polynomial for the shaded area of each figure.

81.

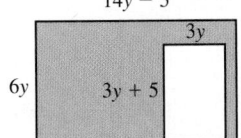

82.

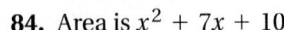

For each figure, determine what the missing number must be in order for the figure to have the given area.

83. Area is $x^2 + 8x + 15$

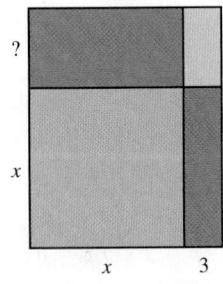

84. Area is $x^2 + 7x + 10$

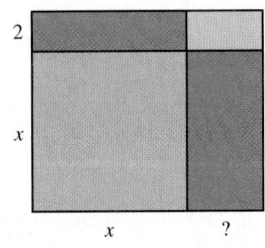

85. A box with a square bottom and no top is to be made from a 12-in.–square piece of cardboard. Squares with side x are cut out of the corners and the sides are folded up. Find the polynomials for the volume and the outside surface area of the box.

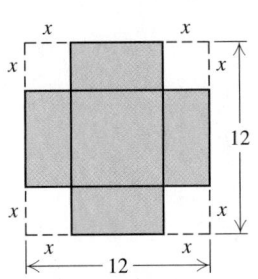

86. Find a polynomial for the volume of the solid shown below.

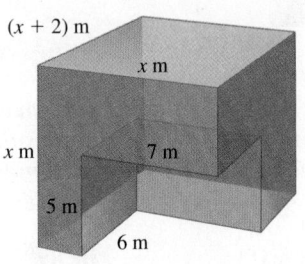

87. An open wooden box is a cube with side x cm. The box, including its bottom, is made of wood that is 1 cm thick. Find a polynomial for the interior volume of the cube.

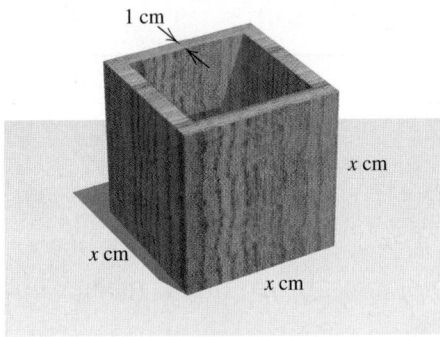

88. A side of a cube is $(x + 2)$ cm long. Find a polynomial for the volume of the cube.

89. A rectangular garden is twice as long as it is wide and is surrounded by a sidewalk that is 4 ft wide (see the figure below). The area of the sidewalk is 256 ft². Find the dimensions of the garden.

Compute and simplify.

90. $(x + 3)(x + 6) + (x + 3)(x + 6)$

Aha! **91.** $(x - 2)(x - 7) - (x - 7)(x - 2)$

92. $(x + 5)^2 - (x - 3)^2$

93. $(x + 2)(x + 4)(x - 5)$

94. $(x - 3)^3$

Aha! **95.** Extend the pattern and simplify

$$(x - a)(x - b)(x - c)(x - d) \cdots (x - z).$$

96. Use a graphing calculator to check your answers to Exercises 25, 45, and 57. Use graphs, tables, or both, as directed by your instructor.

COLLABORATIVE CORNER

Slick Tricks with Algebra

Focus: Polynomial multiplication
Time: 15 minutes
Group size: 2

Consider the following dialogue.

Jinny: Cal, let me do a number trick with you. Think of a number between 1 and 7. I'll have you perform some manipulations to this number, you'll tell me the result, and I'll tell you your number.

Cal: OK. I've thought of a number.

Jinny: Good. Write it down so I can't see it. Now double it, and then subtract x from the result.

Cal: Hey, this is algebra!

Jinny: I know. Now square your binomial. After you're through squaring, subtract x^2.

Cal: How did you know I had an x^2? I *thought* this was rigged!

Jinny: It is. Now divide each of the remaining terms by 4 and tell me either your constant term

or your x-term. I'll tell you the other term and the number you chose.

Cal: OK. The constant term is 16.

Jinny: Then the other term is $-4x$ and the number you chose was 4.

Cal: You're right! How did you do it?

ACTIVITY

1. Each group member should follow Jinny's instructions. Then determine how Jinny determined Cal's number and the other term.

2. Suppose that, at the end, Cal told Jinny the x-term. How would Jinny have determined Cal's number and the other term?

3. Would Jinny's "trick" work with *any* real number? Why do you think she specified numbers between 1 and 7?

4.6   Special Products

Products of Two Binomials   •   Multiplying Sums and Differences of Two Terms   •
Squaring Binomials   •   Multiplications of Various Types

We can observe patterns in the products of two binomials. These patterns allow us to compute such products quickly.

Products of Two Binomials

In Section 4.5, we found the product $(x + 5)(x + 4)$ by using the distributive law a total of three times (see p. 264). Note that each term in $x + 5$ is multiplied by each term in $x + 4$. To shorten our work, we can go right to this step:

$$(x + 5)(x + 4) = x \cdot x + x \cdot 4 + 5 \cdot x + 5 \cdot 4$$
$$= x^2 + 4x + 5x + 20$$
$$= x^2 + 9x + 20.$$

Note that $x \cdot x$ is found by multiplying the *First* terms of each binomial, $x \cdot 4$ is found by multiplying the *Outer* terms of the two binomials, $5 \cdot x$ is the product of the *Inner* terms of the two binomials, and $5 \cdot 4$ is the product of the *Last* terms of each binomial:

$$\underbrace{\text{First}}_{\text{terms}} \quad \underbrace{\text{Outer}}_{\text{terms}} \quad \underbrace{\text{Inner}}_{\text{terms}} \quad \underbrace{\text{Last}}_{\text{terms}}$$

$$(x + 5)(x + 4) = x \cdot x + 4 \cdot x + 5 \cdot x + 5 \cdot 4.$$

To remember this shortcut for multiplying, we use the initials **FOIL**.

The FOIL Method

To multiply two binomials, $A + B$ and $C + D$, multiply the First terms AC, the Outer terms AD, the Inner terms BC, and then the Last terms BD. Then combine like terms, if possible.

$$(A + B)(C + D) = AC + AD + BC + BD$$

1. Multiply First terms: AC.
2. Multiply Outer terms: AD.
3. Multiply Inner terms: BC.
4. Multiply Last terms: BD.

$$\downarrow$$
FOIL

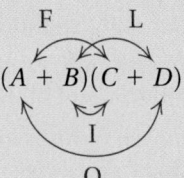

Because addition is commutative, the individual multiplications can be performed in any order. Both FLOI and FIOL yield the same result as FOIL, but FOIL is most easily remembered and most widely used.

EXAMPLE **1** Form an equivalent expression by multiplying: $(x + 8)(x^2 + 5)$.

SOLUTION

$$(x + 8)(x^2 + 5) = x^3 + 5x + 8x^2 + 40 \qquad \text{There are no like terms.}$$
$$= x^3 + 8x^2 + 5x + 40 \qquad \text{Writing in descending order}$$

> TRY EXERCISE 5

After multiplying, remember to combine any like terms.

EXAMPLE **2** Multiply to form an equivalent expression.

a) $(x + 7)(x + 4)$ b) $(y + 3)(y - 2)$
c) $(4t^3 + 5t)(3t^2 - 2)$ d) $(3 - 4x)(7 - 5x^3)$

SOLUTION

a) $(x + 7)(x + 4) = x^2 + 4x + 7x + 28 \qquad$ Using FOIL
$ = x^2 + 11x + 28 \qquad$ Combining like terms

b) $(y + 3)(y - 2) = y^2 - 2y + 3y - 6$
$ = y^2 + y - 6$

c) $(4t^3 + 5t)(3t^2 - 2) = 12t^5 - 8t^3 + 15t^3 - 10t$ Remember to add
exponents when
multiplying terms
with the same base.

$$= 12t^5 + 7t^3 - 10t$$

d) $(3 - 4x)(7 - 5x^3) = 21 - 15x^3 - 28x + 20x^4$

$$= 21 - 28x - 15x^3 + 20x^4$$ In general, if the original
binomials are written
in *ascending* order, the
answer is also written
that way.

> **TRY EXERCISE** 9

Multiplying Sums and Differences of Two Terms

Consider the product of the sum and the difference of the same two terms, such as

$$(x + 5)(x - 5).$$

Since this is the product of two binomials, we can use FOIL. In doing so, we find
that the "outer" and "inner" products are opposites:

a) $(x + 5)(x - 5) = x^2 - 5x + 5x - 25$
$$= x^2 - 25;$$

b) $(3a - 2)(3a + 2) = 9a^2 + 6a - 6a - 4$ The "outer" and "inner" terms
$$= 9a^2 - 4;$$ "drop out." Their sum is zero.

c) $\left(x^3 + \frac{2}{7}\right)\left(x^3 - \frac{2}{7}\right) = x^6 - \frac{2}{7}x^3 + \frac{2}{7}x^3 - \frac{4}{49}$
$$= x^6 - \frac{4}{49}.$$

Because opposites always add to zero, for products like $(x + 5)(x - 5)$ we can
use a shortcut that is faster than FOIL.

> ### The Product of a Sum and a Difference
>
> The product of the sum and the difference of the same two terms is the
> square of the first term minus the square of the second term:
>
> $$(A + B)(A - B) = \underbrace{A^2 - B^2}.$$
>
> This is called a *difference of squares*.

EXAMPLE 3   Multiply.

a) $(x + 4)(x - 4)$

b) $(5 + 2w)(5 - 2w)$

c) $(3a^4 - 5)(3a^4 + 5)$

SOLUTION

$$(A + B)(A - B) = A^2 - B^2$$
$$\downarrow \quad \downarrow \quad \downarrow \quad \downarrow \quad \quad \downarrow \quad \downarrow$$
a) $(x + 4)(x - 4) = x^2 - 4^2$ Saying the words can help: "The square of
the first term, x^2, minus the square of the
second, 4^2"

$$= x^2 - 16$$ Simplifying

b) $(5 + 2w)(5 - 2w) = 5^2 - (2w)^2$

$$= 25 - 4w^2 \qquad \text{Squaring both 5 and } 2w$$

c) $(3a^4 - 5)(3a^4 + 5) = (3a^4)^2 - 5^2$

$$= 9a^8 - 25 \qquad \text{Remember to multiply exponents when raising a power to a power.}$$

> **TRY EXERCISE** ▶ 41

Squaring Binomials

Consider the square of a binomial, such as $(x + 3)^2$. This can be expressed as $(x + 3)(x + 3)$. Since this is the product of two binomials, we can use FOIL. But again, this product occurs so often that a faster method has been developed. Look for a pattern in the following:

a) $(x + 3)^2 = (x + 3)(x + 3)$

$$= x^2 + 3x + 3x + 9$$

$$= x^2 + 6x + 9;$$

b) $(5 - 3p)^2 = (5 - 3p)(5 - 3p)$

$$= 25 - 15p - 15p + 9p^2$$

$$= 25 - 30p + 9p^2;$$

c) $(a^3 - 7)^2 = (a^3 - 7)(a^3 - 7)$

$$= a^6 - 7a^3 - 7a^3 + 49$$

$$= a^6 - 14a^3 + 49.$$

Perhaps you noticed that in each product the "outer" product and the "inner" product are identical. The other two terms, the "first" product and the "last" product, are squares.

The Square of a Binomial

The square of a binomial is the square of the first term, plus twice the product of the two terms, plus the square of the last term:

$$(A + B)^2 = A^2 + 2AB + B^2;$$
$$(A - B)^2 = A^2 - 2AB + B^2.$$

These are called *perfect-square trinomials.**

EXAMPLE 4   Write an equivalent expression for each square of a binomial.

a) $(x + 7)^2$

b) $(t - 5)^2$

c) $(3a + 0.4)^2$

d) $(5x - 3x^4)^2$

SOLUTION

$$(A + B)^2 = A^2 + 2 \cdot A \cdot B + B^2$$

a) $(x + 7)^2 = x^2 + 2 \cdot x \cdot 7 + 7^2$

> Saying the words can help: "The square of the first term, x^2, plus twice the product of the terms, $2 \cdot 7x$, plus the square of the second term, 7^2"

$$= x^2 + 14x + 49$$

---

*Another name for these is *trinomial squares*.

b) $(t - 5)^2 = t^2 - 2 \cdot t \cdot 5 + 5^2$
$$= t^2 - 10t + 25$$

c) $(3a + 0.4)^2 = (3a)^2 + 2 \cdot 3a \cdot 0.4 + 0.4^2$
$$= 9a^2 + 2.4a + 0.16$$

d) $(5x - 3x^4)^2 = (5x)^2 - 2 \cdot 5x \cdot 3x^4 + (3x^4)^2$
$$= 25x^2 - 30x^5 + 9x^8 \qquad \text{Using the rules for exponents}$$

> TRY EXERCISE ▶ 57

CAUTION!   Although the square of a product is the product of the squares, the square of a sum is *not* the sum of the squares. That is, $(AB)^2 = A^2 B^2$, but

The term $2AB$ is missing.

$$(A + B)^2 \neq A^2 + B^2.$$

To confirm this inequality, note that

$$(7 + 5)^2 = 12^2 = 144,$$

whereas

$$7^2 + 5^2 = 49 + 25 = 74, \quad \text{and} \quad 74 \neq 144.$$

Geometrically, $(A + B)^2$ can be viewed as the area of a square with sides of length $A + B$:

$$(A + B)(A + B) = (A + B)^2.$$

This is equal to the sum of the areas of the four smaller regions:

$$A^2 + AB + AB + B^2 = A^2 + 2AB + B^2.$$

Thus,

$$(A + B)^2 = A^2 + 2AB + B^2.$$

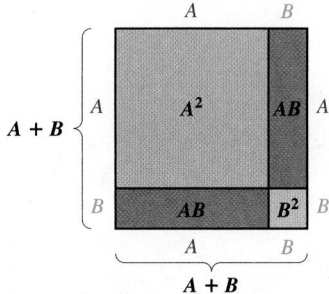

Note that the areas A^2 and B^2 do not fill the area $(A + B)^2$. Two additional areas, AB and AB, are needed.

Multiplications of Various Types

Recognizing patterns often helps when new problems are encountered. To simplify a new multiplication problem, always examine what type of product it is so that the best method for finding that product can be used. To do this, ask yourself questions similar to the following.

Multiplying Two Polynomials

1. Is the multiplication the product of a monomial and a polynomial? If so, multiply each term of the polynomial by the monomial.
2. Is the multiplication the product of two binomials? If so:

   a) Is it the product of the sum and the difference of the *same* two terms? If so, use the pattern
   $$(A + B)(A - B) = A^2 - B^2.$$

   b) Is the product the square of a binomial? If so, use the pattern
   $$(A + B)(A + B) = (A + B)^2 = A^2 + 2AB + B^2,$$
   or
   $$(A - B)(A - B) = (A - B)^2 = A^2 - 2AB + B^2.$$

   c) If neither (a) nor (b) applies, use FOIL.

3. Is the multiplication the product of two polynomials other than those above? If so, multiply each term of one by every term of the other. Use columns if you wish.

EXAMPLE 5   Multiply.

a) $(x + 3)(x - 3)$

b) $(t + 7)(t - 5)$

c) $(x + 7)(x + 7)$

d) $2x^3(9x^2 + x - 7)$

e) $(p + 3)(p^2 + 2p - 1)$

f) $\left(3x - \frac{1}{4}\right)^2$

SOLUTION

a) $(x + 3)(x - 3) = x^2 - 9$ This is the product of the sum and the difference of the same two terms.

b) $(t + 7)(t - 5) = t^2 - 5t + 7t - 35$ Using FOIL
$$= t^2 + 2t - 35$$

c) $(x + 7)(x + 7) = x^2 + 14x + 49$ This is the square of a binomial, $(x + 7)^2$.

d) $2x^3(9x^2 + x - 7) = 18x^5 + 2x^4 - 14x^3$ Multiplying each term of the trinomial by the monomial

e) We multiply each term of $p^2 + 2p - 1$ by every term of $p + 3$:
$$(p + 3)(p^2 + 2p - 1) = p^3 + 2p^2 - p \qquad \text{Multiplying by } p$$
$$+ 3p^2 + 6p - 3 \qquad \text{Multiplying by 3}$$
$$= p^3 + 5p^2 + 5p - 3.$$

f) $\left(3x - \frac{1}{4}\right)^2 = 9x^2 - 2(3x)\left(\frac{1}{4}\right) + \frac{1}{16}$ Squaring a binomial
$$= 9x^2 - \frac{3}{2}x + \frac{1}{16}$$

TRY EXERCISE ▶ 69

Visualizing for Success

1

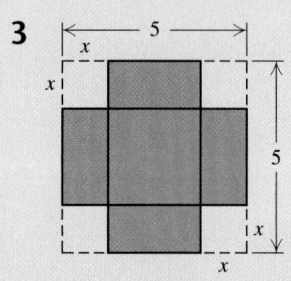

6

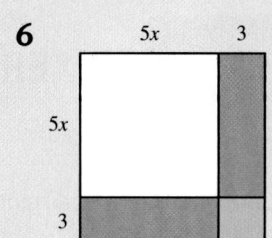

In each of Exercises 1–10, find two algebraic expressions for the shaded area of the figure from the list below.

A. $9 - 4x^2$

B. $x^2 - (x - 6)^2$

C. $(x + 3)(x - 3)$

D. $10^2 + 2^2$

E. $8x + 15$

F. $(x + 5)(x + 3) - x^2$

G. $x^2 - 6x + 9$

H. $(3 - 2x)^2 + 4x(3 - 2x)$

I. $(x + 3)^2$

J. $(5x + 3)^2 - 25x^2$

K. $(5 - 2x)^2 + 4x(5 - 2x)$

L. $x^2 - 9$

M. 104

N. $x^2 - 15$

O. $12x - 36$

P. $30x + 9$

Q. $(x - 5)(x - 3) + 3(x - 5) + 5(x - 3)$

R. $(x - 3)^2$

S. $25 - 4x^2$

T. $x^2 + 6x + 9$

Answers on page A-18

An additional, animated version of this activity appears in MyMathLab. To use MyMathLab, you need a course ID and a student access code. Contact your instructor for more information.

2

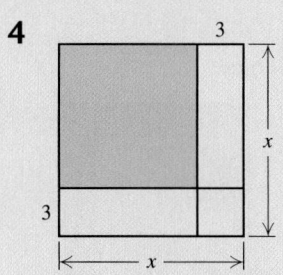

3

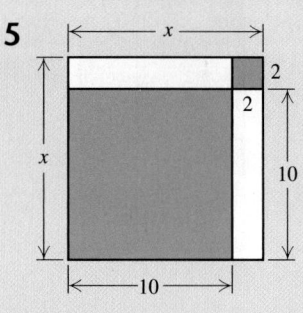

4

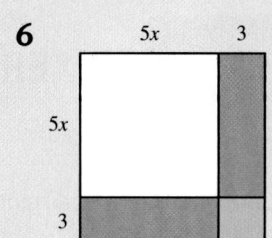

5

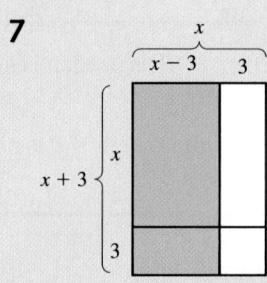

7

8

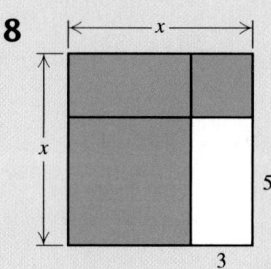

9

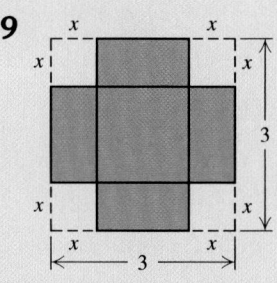

10

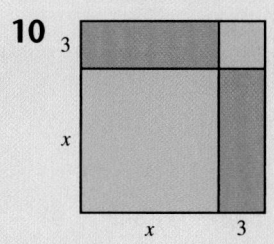

🖐 *Concept Reinforcement* *Classify each statement as either true or false.*

1. FOIL is simply a memory device for finding the product of two binomials.

2. The square of a binomial cannot be found using FOIL.

3. Once FOIL is used, it is always possible to combine like terms.

4. The square of $A + B$ is not the sum of the squares of A and B.

Multiply.

5. $(x^2 + 2)(x + 3)$ **6.** $(x - 5)(x^2 - 6)$

7. $(t^4 - 2)(t + 7)$ **8.** $(n^3 + 8)(n - 4)$

9. $(y + 2)(y - 3)$ **10.** $(a + 2)(a + 2)$

11. $(3x + 2)(3x + 5)$ **12.** $(4x + 1)(2x + 7)$

13. $(5x - 3)(x + 4)$ **14.** $(4x - 5)(4x + 5)$

15. $(3 - 2t)(5 - t)$ **16.** $(7 - a)(4 - 3a)$

17. $(x^2 + 3)(x^2 - 7)$ **18.** $(x^2 + 2)(x^2 - 8)$

19. $\left(p - \frac{1}{4}\right)\left(p + \frac{1}{4}\right)$ **20.** $\left(q + \frac{3}{4}\right)\left(q + \frac{3}{4}\right)$

21. $(x - 0.3)(x - 0.3)$ **22.** $(x - 0.1)(x + 0.1)$

23. $(-3n + 2)(n + 7)$ **24.** $(-m + 5)(2m - 9)$

25. $(x + 10)(x + 10)$ **26.** $(x + 12)(x + 12)$

27. $(1 - 3t)(1 + 5t^2)$ **28.** $(1 + 2t)(1 - 3t^2)$

29. $(x^2 + 3)(x^3 - 1)$ **30.** $(x^4 - 3)(2x + 1)$

31. $(3x^2 - 2)(x^4 - 2)$ **32.** $(x^{10} + 3)(x^{10} - 3)$

33. $(2t^3 + 5)(2t^3 + 5)$ **34.** $(5t^2 + 1)(2t^2 + 3)$

35. $(8x^3 + 5)(x^2 + 2)$ **36.** $(5 - 4x^5)(5 + 4x^5)$

37. $(10x^2 + 3)(10x^2 - 3)$ **38.** $(7x - 2)(2x - 7)$

Multiply. Try to recognize the type of product before multiplying.

39. $(x + 8)(x - 8)$ **40.** $(x + 1)(x - 1)$

41. $(2x + 1)(2x - 1)$ **42.** $(4n + 7)(4n - 7)$

43. $(5m^2 + 4)(5m^2 - 4)$ **44.** $(3x^4 + 2)(3x^4 - 2)$

45. $(9a^3 + 1)(9a^3 - 1)$ **46.** $(t^2 - 0.2)(t^2 + 0.2)$

47. $(x^4 + 0.1)(x^4 - 0.1)$ **48.** $(a^3 + 5)(a^3 - 5)$

49. $\left(t - \frac{3}{4}\right)\left(t + \frac{3}{4}\right)$ **50.** $\left(m - \frac{2}{3}\right)\left(m + \frac{2}{3}\right)$

51. $(x + 3)^2$ **52.** $(2x - 1)^2$

53. $(7x^3 - 1)^2$ **54.** $(5x^3 + 2)^2$

55. $\left(a - \frac{2}{5}\right)^2$ **56.** $\left(t - \frac{1}{5}\right)^2$

57. $(t^4 + 3)^2$ **58.** $(a^3 + 6)^2$

59. $(2 - 3x^4)^2$ **60.** $(5 - 2t^3)^2$

61. $(5 + 6t^2)^2$ **62.** $(3p^2 - p)^2$

63. $(7x - 0.3)^2$ **64.** $(4a - 0.6)^2$

65. $7n^3(2n^2 - 1)$ **66.** $5m^3(4 - 3m^2)$

67. $(a - 3)(a^2 + 2a - 4)$ **68.** $(x^2 - 5)(x^2 + x - 1)$

69. $(7 - 3x^4)(7 - 3x^4)$ **70.** $(x - 4x^3)^2$

71. $5x(x^2 + 6x - 2)$ **72.** $6x(-x^5 + 6x^2 + 9)$

73. $(q^5 + 1)(q^5 - 1)$ **74.** $(p^4 + 2)(p^4 - 2)$

75. $3t^2(5t^3 - t^2 + t)$ **76.** $-5x^3(x^2 + 8x - 9)$

77. $(6x^4 - 3x)^2$ **78.** $(8a^3 + 5)(8a^3 - 5)$

79. $(9a + 0.4)(2a^3 + 0.5)$ **80.** $(2a - 0.7)(8a^3 - 0.5)$

81. $\left(\frac{1}{5} - 6x^4\right)\left(\frac{1}{5} + 6x^4\right)$ **82.** $\left(3 + \frac{1}{2}t^5\right)\left(3 + \frac{1}{2}t^5\right)$

83. $(a + 1)(a^2 - a + 1)$

84. $(x - 5)(x^2 + 5x + 25)$

Find the total area of all shaded rectangles.

85.
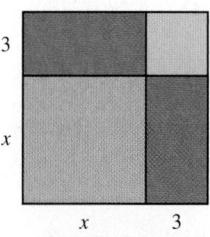

86.
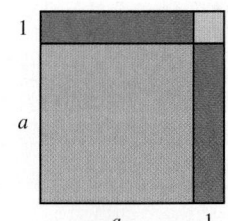

87.
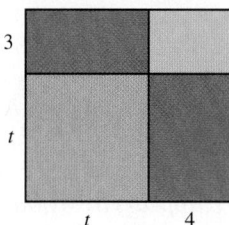

88.
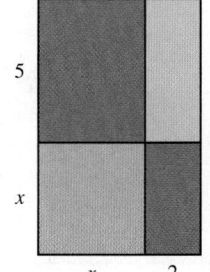

89.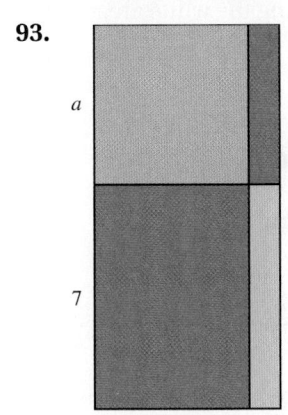

90.

91.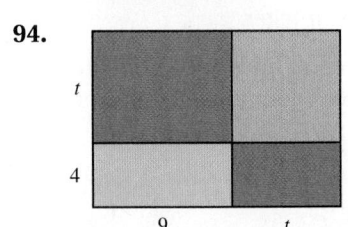

92.

93.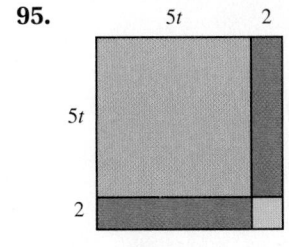

94.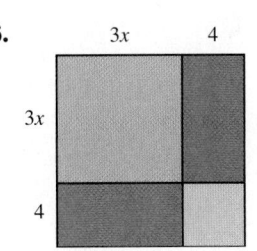

95.

96.

Draw and label rectangles similar to those in Exercises 85–96 to illustrate each of the following.

97. $(x + 5)^2$

98. $(x + 8)^2$

99. $(t + 9)^2$

100. $(a + 12)^2$

101. $(3 + x)^2$

102. $(7 + t)^2$

103. Kristi feels that since she can find the product of any two binomials using FOIL, she needn't study the other special products. What advice would you give her?

104. Under what conditions is the product of two binomials a binomial?

Skill Review

Review problem solving and solving a formula for a variable (Sections 2.3 and 2.5).

Solve. [2.5]

105. *Energy use.* Under typical use, a refrigerator, a freezer, and a washing machine together use 297 kilowatt-hours per month (kWh/mo). A refrigerator uses 21 times as much energy as a washing machine, and a freezer uses 11 times as much energy as a washing machine. How much energy is used by each appliance?

106. *Advertising.* North American advertisers spent $9.4 billion on search-engine marketing in 2006. This was a 62% increase over the amount spent in 2005. How much was spent in 2005?
Source: Search Engine Marketing Professional Organization

Solve. [2.3]

107. $5xy = 8$, for y

108. $3ab = c$, for a

109. $ax - by = c$, for x

110. $ax - by = c$, for y

Synthesis

111. By writing $19 \cdot 21$ as $(20 - 1)(20 + 1)$, Justin can find the product mentally. How do you think he does this?

112. The product $(A + B)^2$ can be regarded as the sum of the areas of four regions (as shown following Example 4). How might one visually represent $(A + B)^3$? Why?

Multiply.

Aha! **113.** $(4x^2 + 9)(2x + 3)(2x - 3)$

114. $(9a^2 + 1)(3a - 1)(3a + 1)$

Aha! **115.** $(3t - 2)^2(3t + 2)^2$

116. $(5a + 1)^2(5a - 1)^2$

117. $(t^3 - 1)^4(t^3 + 1)^4$

 118. $(32.41x + 5.37)^2$

Calculate as the difference of squares.

119. 18×22 [*Hint*: $(20 - 2)(20 + 2)$.]

120. 93×107

Solve.

121. $(x + 2)(x - 5) = (x + 1)(x - 3)$

122. $(2x + 5)(x - 4) = (x + 5)(2x - 4)$

Find a polynomial for the total shaded area in each figure.

123.

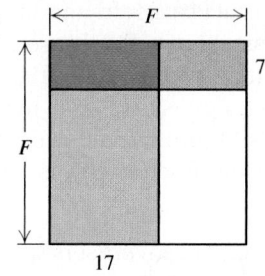

124.

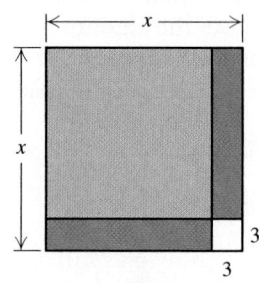

125.

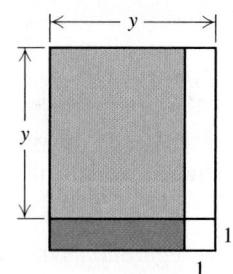

126. Find $(10 - 2x)^2$ by subtracting the white areas from 10^2.

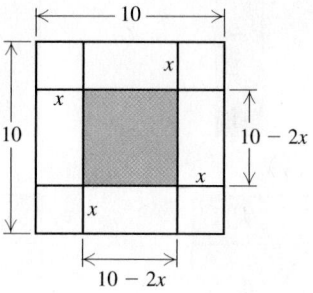

127. Find $(y - 2)^2$ by subtracting the white areas from y^2.

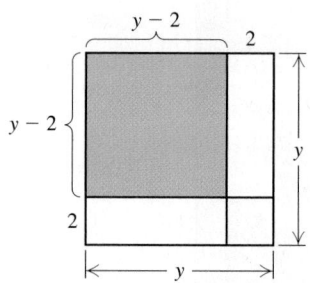

128. Find three consecutive integers for which the sum of the squares is 65 more than three times the square of the smallest integer.

129. Use a graphing calculator and the method developed on p. 258 to check your answers to Exercises 22, 47, and 83.

CONNECTING the CONCEPTS

When writing equivalent polynomial expressions, look first at the operation that you are asked to perform.

Operation	Procedure	Examples
Addition	Combine like terms.	$(2x^3 + 3x^2 - 5x - 7) + (9x^3 - 11x + 8)$ $= \underbrace{2x^3 + 9x^3} + 3x^2 \underbrace{- 5x - 11x} \underbrace{- 7 + 8}$ $= \quad 11x^3 \quad + 3x^2 \quad -16x \quad + 1$
Subtraction	Add the opposite of the polynomial being subtracted.	$(9x^4 - 3x^2 + x - 7) - (4x^3 - 8x^2 - 9x + 11)$ $= 9x^4 - 3x^2 + x - 7 + (-4x^3 + 8x^2 + 9x - 11)$ $= 9x^4 - 3x^2 + x - 7 - 4x^3 + 8x^2 + 9x - 11$ $= 9x^4 - 4x^3 + 5x^2 + 10x - 18$
Multiplication	Multiply each term of one polynomial by every term of the other.	$(x^2 - 5x)(3x^4 - 7x^3 + 1)$ $= x^2(3x^4 - 7x^3 + 1) - 5x(3x^4 - 7x^3 + 1)$ $= 3x^6 - 7x^5 + x^2 - 15x^5 + 35x^4 - 5x$ $= 3x^6 - 22x^5 + 35x^4 + x^2 - 5x$
	Special products: $(A + B)(A - B) = A^2 - B^2;$ $(A + B)^2 = A^2 + 2AB + B^2;$ $(A - B)^2 = A^2 - 2AB + B^2;$ $(A + B)(C + D)$ $\quad = AC + AD + BC + BD$	$(x + 5)(x - 5) = x^2 - 25;$ $(2x + 3)^2 = (2x)^2 + 2(2x)(3) + (3)^2 = 4x^2 + 12x + 9;$ $(x^2 - 1)^2 = (x^2)^2 - 2(x^2)(1) + (1)^2 = x^4 - 2x^2 + 1;$ $(x^2 + 3)(x - 2) = x^2(x) + x^2(-2) + 3(x) + 3(-2)$ $\quad = x^3 - 2x^2 + 3x - 6$

MIXED REVIEW

Identify the operation to be performed. Then simplify to form an equivalent expression.

1. $(3x^2 - 2x + 6) + (5x - 3)$

2. $(9x + 6) - (3x - 7)$

3. $6x^3(8x^2 - 7)$

4. $(3x + 2)(2x - 1)$

5. $(9x^3 - 7x + 3) - (5x^2 - 10)$

6. $(2x + 1)(x^2 + x - 3)$

7. $(9x + 1)(9x - 1)$

8. $(8x^3 + 5x) + (9x^4 - 6x^3 - 10x)$

Perform the indicated operation to form an equivalent expression.

9. $(4x^2 - x - 7) - (10x^2 - 3x + 5)$

10. $(3x + 8)(3x + 7)$

11. $8x^5(5x^4 - 6x^3 + 2)$

12. $(t^9 + 3t^6 - 8t^2) + (5t^7 - 3t^6 + 8t^2)$

13. $(2m - 1)^2$

14. $(x - 1)(x^2 + x + 1)$

15. $(5x^3 - 6x^2 - 2x) + (6x^2 + 2x + 3)$

16. $(c + 3)(c - 3)$

17. $(4y^3 + 7)^2$

18. $(3a^4 - 9a^3 - 7) - (4a^3 + 13a^2 - 3)$

19. $(4t^2 - 5)(4t^2 + 5)$

20. $(a^4 + 3)(a^4 - 8)$

4.7   Polynomials in Several Variables

Evaluating Polynomials • Like Terms and Degree • Addition and Subtraction • Multiplication

Thus far, the polynomials that we have studied have had only one variable. Polynomials such as

$$5x + x^2y - 3y + 7, \quad 9ab^2c - 2a^3b^2 + 8a^2b^3, \quad \text{and} \quad 4m^2 - 9n^2$$

contain two or more variables. In this section, we will add, subtract, multiply, and evaluate such **polynomials in several variables**.

Evaluating Polynomials

To evaluate a polynomial in two or more variables, we substitute numbers for the variables. Then we compute, using the rules for order of operations.

EXAMPLE 1   Evaluate the polynomial $4 + 3x + xy^2 + 8x^3y^3$ for $x = -2$ and $y = 5$.

SOLUTION   We substitute -2 for x and 5 for y:

$$4 + 3x + xy^2 + 8x^3y^3 = 4 + 3(-2) + (-2) \cdot 5^2 + 8(-2)^3 \cdot 5^3$$
$$= 4 - 6 - 50 - 8000 = -8052.$$

> TRY EXERCISE   9

EXAMPLE 2   *Surface area of a right circular cylinder.*   The surface area of a right circular cylinder is given by the polynomial

$$2\pi rh + 2\pi r^2,$$

where h is the height and r is the radius of the base. A 12-oz can has a height of 4.7 in. and a radius of 1.2 in. Approximate its surface area to the nearest tenth of a square inch.

SOLUTION   We evaluate the polynomial for $h = 4.7$ in. and $r = 1.2$ in. If 3.14 is used to approximate π, we have

$$2\pi rh + 2\pi r^2 \approx 2(3.14)(1.2 \text{ in.})(4.7 \text{ in.}) + 2(3.14)(1.2 \text{ in.})^2$$
$$\approx 2(3.14)(1.2 \text{ in.})(4.7 \text{ in.}) + 2(3.14)(1.44 \text{ in}^2)$$
$$\approx 35.4192 \text{ in}^2 + 9.0432 \text{ in}^2 \approx 44.4624 \text{ in}^2.$$

If the π key of a calculator is used, we have

$$2\pi rh + 2\pi r^2 \approx 2(3.141592654)(1.2 \text{ in.})(4.7 \text{ in.})$$
$$+ 2(3.141592654)(1.2 \text{ in.})^2$$
$$\approx 44.48495197 \text{ in}^2.$$

Note that the unit in the answer (square inches) is a unit of area. The surface area is about 44.5 in^2 (square inches).

> TRY EXERCISE   13

Like Terms and Degree

Recall that the degree of a monomial is the number of variable factors in the term. For example, the degree of $5x^2$ is 2 because there are two variable factors in $5 \cdot x \cdot x$. Similarly, the degree of $5a^2b^4$ is 6 because there are 6 variable factors in $5 \cdot a \cdot a \cdot b \cdot b \cdot b \cdot b$. Note that 6 can be found by adding the exponents 2 and 4.

As we learned in Section 4.3, the degree of a polynomial is the degree of the term of highest degree.

EXAMPLE 3 Identify the coefficient and the degree of each term and the degree of the polynomial

$$9x^2y^3 - 14xy^2z^3 + xy + 4y + 5x^2 + 7.$$

SOLUTION

Term	Coefficient	Degree	Degree of the Polynomial
$9x^2y^3$	9	5	
$-14xy^2z^3$	-14	6	6
xy	1	2	
$4y$	4	1	
$5x^2$	5	2	
7	7	0	

TRY EXERCISE 21

STUDY SKILLS

Use Your Voice

Don't be hesitant to ask questions in class at appropriate times. Most instructors welcome questions and encourage students to ask them. Other students probably have the same questions you do, so asking questions can help others as well as you.

Note in Example 3 that although both xy and $5x^2$ have degree 2, they are *not* like terms. *Like,* or *similar, terms* either have exactly the same variables with exactly the same exponents or are constants. For example,

$$8a^4b^7 \text{ and } 5b^7a^4 \text{ are like terms}$$

and

$$-17 \text{ and } 3 \text{ are like terms,}$$

but

$$-2x^2y \text{ and } 9xy^2 \text{ are } not \text{ like terms.}$$

As always, combining like terms is based on the distributive law.

EXAMPLE 4 Combine like terms to form equivalent expressions.

a) $9x^2y + 3xy^2 - 5x^2y - xy^2$

b) $7ab - 5ab^2 + 3ab^2 + 6a^3 + 9ab - 11a^3 + b - 1$

SOLUTION

a) $9x^2y + 3xy^2 - 5x^2y - xy^2 = (9 - 5)x^2y + (3 - 1)xy^2$
$$= 4x^2y + 2xy^2 \quad \text{Try to go directly to this step.}$$

b) $7ab - 5ab^2 + 3ab^2 + 6a^3 + 9ab - 11a^3 + b - 1$
$$= -5a^3 - 2ab^2 + 16ab + b - 1 \quad \text{We choose to write descending powers of } a. \text{ Other, equivalent, forms can also be used.}$$

TRY EXERCISE 25

Addition and Subtraction

The procedure used for adding polynomials in one variable is used to add polynomials in several variables.

Add.

a) $(-5x^3 + 3y - 5y^2) + (8x^3 + 4x^2 + 7y^2)$

b) $(5ab^2 - 4a^2b + 5a^3 + 2) + (3ab^2 - 2a^2b + 3a^3b - 5)$

STUDENT NOTES ———

Always read the problem carefully. The difference between

$$(-5x^3 - 3y) + (8x^3 + 4x^2)$$

and

$$(-5x^3 - 3y)(8x^3 + 4x^2)$$

is enormous. To avoid wasting time working on an incorrectly copied exercise, be sure to double-check that you have written the correct problem in your notebook.

SOLUTION

a) $(-5x^3 + 3y - 5y^2) + (8x^3 + 4x^2 + 7y^2)$

$\qquad = (-5 + 8)x^3 + 4x^2 + 3y + (-5 + 7)y^2$ Try to do this step mentally.

$\qquad = 3x^3 + 4x^2 + 3y + 2y^2$

b) $(5ab^2 - 4a^2b + 5a^3 + 2) + (3ab^2 - 2a^2b + 3a^3b - 5)$

$\qquad = 8ab^2 - 6a^2b + 5a^3 + 3a^3b - 3$ ▶ **TRY EXERCISE** 33

When subtracting a polynomial, remember to find the opposite of each term in that polynomial and then add.

Subtract: $(4x^2y + x^3y^2 + 3x^2y^3 + 6y) - (4x^2y - 6x^3y^2 + x^2y^2 - 5y)$.

SOLUTION

$$(4x^2y + x^3y^2 + 3x^2y^3 + 6y) - (4x^2y - 6x^3y^2 + x^2y^2 - 5y)$$

$\qquad = 4x^2y + x^3y^2 + 3x^2y^3 + 6y - 4x^2y + 6x^3y^2 - x^2y^2 + 5y$

$\qquad = 7x^3y^2 + 3x^2y^3 - x^2y^2 + 11y$ Combining like terms

▶ **TRY EXERCISE** 35

Multiplication

To multiply polynomials in several variables, multiply each term of one polynomial by every term of the other, just as we did in Sections 4.5 and 4.6.

Multiply: $(3x^2y - 2xy + 3y)(xy + 2y)$.

SOLUTION

$$3x^2y - 2xy + 3y$$
$$\underline{\qquad\qquad xy + 2y}$$
$$6x^2y^2 - 4xy^2 + 6y^2 \qquad \text{Multiplying by } 2y$$
$$\underline{3x^3y^2 - 2x^2y^2 + 3xy^2 \qquad\qquad \text{Multiplying by } xy}$$
$$3x^3y^2 + 4x^2y^2 - \ xy^2 + 6y^2 \qquad \text{Adding}$$

▶ **TRY EXERCISE** 51

The special products discussed in Section 4.6 can speed up our work.

Multiply.

a) $(p + 5q)(2p - 3q)$

b) $(3x + 2y)^2$

c) $(a^3 - 7a^2b)^2$

d) $(3x^2y + 2y)(3x^2y - 2y)$

e) $(-2x^3y^2 + 5t)(2x^3y^2 + 5t)$

f) $(2x + 3 - 2y)(2x + 3 + 2y)$

SOLUTION

$$\begin{array}{cccc} & \text{F} & \text{O} & \text{I} & \text{L} \end{array}$$

a) $(p + 5q)(2p - 3q) = 2p^2 - 3pq + 10pq - 15q^2$
$$= 2p^2 + 7pq - 15q^2 \quad \text{Combining like terms}$$

$$(A + B)^2 = A^2 + 2 \cdot A \cdot B + B^2$$

b) $(3x + 2y)^2 = (3x)^2 + 2(3x)(2y) + (2y)^2$ Using the pattern for squaring a binomial
$$= 9x^2 + 12xy + 4y^2$$

$$(A - B)^2 = A^2 - 2 \cdot A \cdot B + B^2$$

c) $(a^3 - 7a^2b)^2 = (a^3)^2 - 2(a^3)(7a^2b) + (7a^2b)^2$ Squaring a binomial
$$= a^6 - 14a^5b + 49a^4b^2 \quad \text{Using the rules for exponents}$$

$$(A + B)(A - B) = A^2 - B^2$$

d) $(3x^2y + 2y)(3x^2y - 2y) = (3x^2y)^2 - (2y)^2$ Using the pattern for multiplying the sum and the difference of two terms
$$= 9x^4y^2 - 4y^2 \quad \text{Using the rules for exponents}$$

e) $(-2x^3y^2 + 5t)(2x^3y^2 + 5t) = (5t - 2x^3y^2)(5t + 2x^3y^2)$ Using the commutative law for addition twice
$$= (5t)^2 - (2x^3y^2)^2 \quad \text{Multiplying the sum and the difference of the same two terms}$$
$$= 25t^2 - 4x^6y^4$$

$$(A - B)(A + B) = A^2 - B^2$$

f) $(2x + 3 - 2y)(2x + 3 + 2y) = (2x + 3)^2 - (2y)^2$ Multiplying a sum and a difference
$$= 4x^2 + 12x + 9 - 4y^2 \quad \text{Squaring a binomial}$$

TRY EXERCISE ▸ 57

In Example 8, we recognized patterns that might elude some students, particularly in parts (e) and (f). In part (e), we *can* use FOIL, and in part (f), we *can* use long multiplication, but doing so is much slower. By carefully inspecting a problem before "jumping in," we can save ourselves considerable work. At least one instructor refers to this as "working smart" instead of "working hard."*

*Thanks to Pauline Kirkpatrick of Wharton County Junior College for this language.

TECHNOLOGY CONNECTION

One way to evaluate the polynomial in Example 1 for $x = -2$ and $y = 5$ is to store -2 to X and 5 to Y and enter the polynomial.

```
-2 → X
                      -2
5 → Y
                       5
4+3X+XY²+8X^3Y^3
                   -8052
■
```

Evaluate.

1. $3x^2 - 2y^2 + 4xy + x$, for $x = -6$ and $y = 2.3$
2. $a^2b^2 - 8c^2 + 4abc + 9a$, for $a = 11$, $b = 15$, and $c = -7$

👈 **Concept Reinforcement** *Each of the expressions in Exercises 1–8 can be regarded as either* (**a**) *the square of a binomial,* (**b**) *the product of the sum and the difference of the same two terms, or* (**c**) *neither of the above. Select the appropriate choice for each expression.*

1. $(3x + 5y)^2$

2. $(4x - 9y)(4x + 9y)$

3. $(5a + 6b)(-6b + 5a)$

4. $(4a - 3b)(4a - 3b)$

5. $(r - 3s)(5r + 3s)$

6. $(2x - 7y)(7y - 2x)$

7. $(4x - 9y)(4x - 9y)$

8. $(2r + 9t)^2$

Evaluate each polynomial for $x = 5$ *and* $y = -2$.

9. $x^2 - 2y^2 + 3xy$

10. $x^2 + 5y^2 - 4xy$

Evaluate each polynomial for $x = 2$, $y = -3$, *and* $z = -4$.

11. $xy^2z - z$

12. $xy - x^2z + yz^2$

Lung capacity. *The polynomial*

$$0.041h - 0.018A - 2.69$$

can be used to estimate the lung capacity, in liters, of a female with height h, in centimeters, and age A, in years.

13. Find the lung capacity of a 20-year-old woman who is 160 cm tall.

14. Find the lung capacity of a 40-year-old woman who is 165 cm tall.

15. *Female caloric needs.* The number of calories needed each day by a moderately active woman who weighs w pounds, is h inches tall, and is a years old can be estimated by the polynomial

$$917 + 6w + 6h - 6a.$$

Rachel is moderately active, weighs 125 lb, is 64 in. tall, and is 27 yr old. What are her daily caloric needs?
Source: Parker, M., *She Does Math.* Mathematical Association of America

16. *Male caloric needs.* The number of calories needed each day by a moderately active man who weighs w kilograms, is h centimeters tall, and is a years old can be estimated by the polynomial

$$19.18w + 7h - 9.52a + 92.4.$$

One of the authors of this text is moderately active, weighs 87 kg, is 185 cm tall, and is 59 yr old. What are his daily caloric needs?
Source: Parker, M., *She Does Math.* Mathematical Association of America

Surface area of a silo. *A silo is a structure that is shaped like a right circular cylinder with a half sphere on top. The surface area of a silo of height h and radius r (including the area of the base) is given by the polynomial* $2\pi rh + \pi r^2$.

17. A coffee grinder is shaped like a silo, with a height of 7 in. and a radius of $1\frac{1}{2}$ in. Find the surface area of the coffee grinder. Use 3.14 for π.

18. A $1\frac{1}{2}$-oz bottle of roll-on deodorant has a height of 4 in. and a radius of $\frac{3}{4}$ in. Find the surface area of the bottle if the bottle is shaped like a silo. Use 3.14 for π.

Altitude of a launched object. *The altitude of an object, in meters, is given by the polynomial*

$$h + vt - 4.9t^2,$$

where h is the height, in meters, at which the launch occurs, v is the initial upward speed (or velocity), in meters per second, and t is the number of seconds for which the object is airborne.

19. A bocce ball is thrown upward with an initial speed of 18 m/sec by a person atop the Leaning Tower of Pisa, which is 50 m above the ground. How high will the ball be 2 sec after it has been thrown?

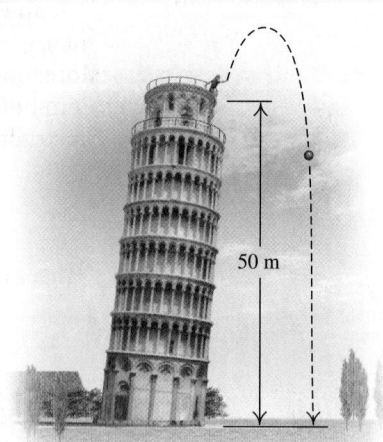

50 m

20. A golf ball is launched upward with an initial speed of 30 m/sec by a golfer atop the Washington Monument, which is 160 m above the ground. How high above the ground will the ball be after 3 sec?

Identify the coefficient and the degree of each term of each polynomial. Then find the degree of each polynomial.

21. $3x^2y - 5xy + 2y^2 - 11$

22. $xy^3 + 7x^3y^2 - 6xy^4 + 2$

23. $7 - abc + a^2b + 9ab^2$

24. $3p - pq - 7p^2q^3 - 8pq^6$

Combine like terms.

25. $3r + s - r - 7s$

26. $9a + b - 8a - 5b$

27. $5xy^2 - 2x^2y + x + 3x^2$

28. $m^3 + 2m^2n - 3m^2 + 3mn^2$

29. $6u^2v - 9uv^2 + 3vu^2 - 2v^2u + 11u^2$

30. $3x^2 + 6xy + 3y^2 - 5x^2 - 10xy$

31. $5a^2c - 2ab^2 + a^2b - 3ab^2 + a^2c - 2ab^2$

32. $3s^2t + r^2t - 9ts^2 - st^2 + 5t^2s - 7tr^2$

Add or subtract, as indicated.

33. $(6x^2 - 2xy + y^2) + (5x^2 - 8xy - 2y^2)$

34. $(7r^3 + rs - 5r^2) - (2r^3 - 3rs + r^2)$

35. $(3a^4 - 5ab + 6ab^2) - (9a^4 + 3ab - ab^2)$

36. $(2r^2t - 5rt + rt^2) - (7r^2t + rt - 5rt^2)$

Aha! **37.** $(5r^2 - 4rt + t^2) + (-6r^2 - 5rt - t^2) + (-5r^2 + 4rt - t^2)$

38. $(2x^2 - 3xy + y^2) + (-4x^2 - 6xy - y^2) + (4x^2 + 6xy + y^2)$

39. $(x^3 - y^3) - (-2x^3 + x^2y - xy^2 + 2y^3)$

40. $(a^3 + b^3) - (-5a^3 + 2a^2b - ab^2 + 3b^3)$

41. $(2y^4x^3 - 3y^3x) + (5y^4x^3 - y^3x) - (9y^4x^3 - y^3x)$

42. $(5a^2b - 7ab^2) - (3a^2b + ab^2) + (a^2b - 2ab^2)$

43. Subtract $7x + 3y$ from the sum of $4x + 5y$ and $-5x + 6y$.

44. Subtract $5a + 2b$ from the sum of $2a + b$ and $3a - 4b$.

Multiply.

45. $(4c - d)(3c + 2d)$

46. $(5x + y)(2x - 3y)$

47. $(xy - 1)(xy + 5)$

48. $(ab + 3)(ab - 5)$

49. $(2a - b)(2a + b)$

50. $(a - 3b)(a + 3b)$

51. $(5rt - 2)(4rt - 3)$

52. $(3xy - 1)(4xy + 2)$

53. $(m^3n + 8)(m^3n - 6)$

54. $(9 - u^2v^2)(2 - u^2v^2)$

55. $(6x - 2y)(5x - 3y)$

56. $(7a - 6b)(5a + 4b)$

57. $(pq + 0.1)(-pq + 0.1)$

58. $(rt + 0.2)(-rt + 0.2)$

59. $(x + h)^2$

60. $(a - r)^2$

61. $(4a - 5b)^2$

62. $(2x + 5y)^2$

63. $(ab + cd^2)(ab - cd^2)$

64. $(p^3 - 5q)(p^3 + 5q)$

65. $(2xy + x^2y + 3)(xy + y^2)$

66. $(5cd - c^2 - d^2)(2c - c^2d)$

Aha! **67.** $(a + b - c)(a + b + c)$

68. $(x + y + 2z)(x + y - 2z)$

69. $[a + b + c][a - (b + c)]$

70. $(a + b + c)(a - b - c)$

Find the total area of each shaded area.

71.

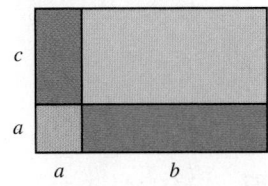

72.

73.

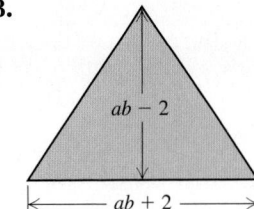

74.

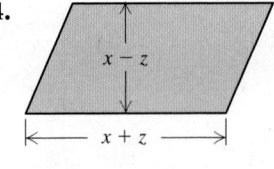

75.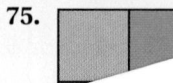

76.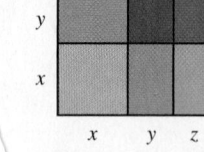

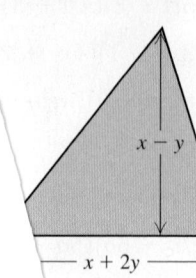

 Drau...ose in Exercises 71, 72, 75...

79. (...

80. (*m* ...

81. (*a* ... *)*

82. (*r* + ... *)*

83. Is it possible for a polynomial in 4 variables to have a degree less than 4? Why or why not?

84. A fourth-degree monomial is multiplied by a third-degree monomial. What is the degree of the product? Explain your reasoning.

Skill Review

To prepare for Section 4.8, review subtraction of polynomials using columns (Section 4.4).

Subtract. [4.4]

85.
$$\begin{array}{r} x^2 - 3x - 7 \\ -(\quad 5x - 3) \\ \hline \end{array}$$

86.
$$\begin{array}{r} 2x^3 \quad\ - x + 3 \\ -(\quad x^2 \quad\ - 1) \\ \hline \end{array}$$

87.
$$\begin{array}{r} 3x^2 + \ x + 5 \\ -(3x^2 + 3x) \\ \hline \end{array}$$

88.
$$\begin{array}{r} 4x^3 - 3x^2 + x \\ -(4x^3 - 8x^2) \\ \hline \end{array}$$

89.
$$\begin{array}{r} 5x^3 - \ 2x^2 + 1 \\ -(5x^3 - 15x^2) \\ \hline \end{array}$$

90.
$$\begin{array}{r} 2x^2 + 5x - 3 \\ -(2x^2 + 6x) \\ \hline \end{array}$$

Synthesis

91. The concept of "leading term" was intentionally not discussed in this section. Why?

92. Explain how it is possible for the sum of two trinomials in several variables to be a binomial in one variable.

Find a polynomial for the shaded area. (Leave results in terms of π where appropriate.)

93.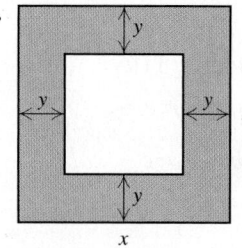

94.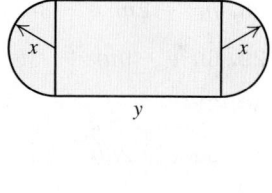

95.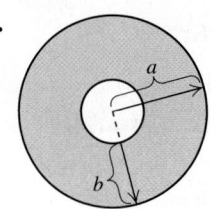

96.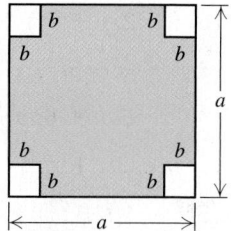

97. Find a polynomial for the total volume of the figure shown.

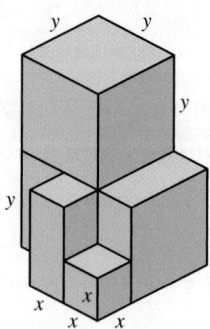

98. Find the shaded area in this figure using each of the approaches given below. Then check that both answers match.

a) Find the shaded area by subtracting the area of the unshaded square from the total area of the figure.

b) Find the shaded area by adding the areas of the three shaded rectangles.

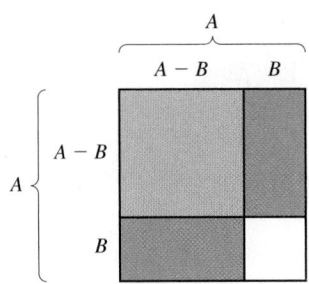

Find a polynomial for the surface area of each solid object shown. (Leave results in terms of π.)

99.

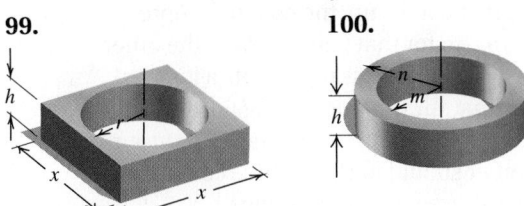

100.

101. The observatory at Danville University is shaped like a silo that is 40 ft high and 30 ft wide (see Exercise 17). The Heavenly Bodies Astronomy Club is to paint the exterior of the observatory using paint that covers 250 ft² per gallon. How many gallons should they purchase? Explain your reasoning.

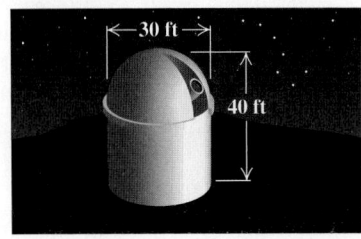

102. Multiply: $(x + a)(x - b)(x - a)(x + b)$.

The computer application Excel allows values for cells in a spreadsheet to be calculated from values in other cells. For example, if the cell C1 contains the formula

$$= A1 + 2*B1,$$

the value in C1 will be the sum of the value in A1 and twice the value in B1. This formula is a polynomial in the two variables A1 and B1.

103. The cell D4 contains the formula

$$= 2*A4 + 3*B4.$$

What is the value in D4 if the value in A4 is 5 and the value in B4 is 10?

104. The cell D6 contains the formula

$$= A1 - 0.2*B1 + 0.3*C1.$$

What is the value in D6 if the value in A1 is 10, the value in B1 is -3, and the value in C1 is 30?

105. *Interest compounded annually.* An amount of money P that is invested at the yearly interest rate r grows to the amount $P(1 + r)^t$ after t years. Find a polynomial that can be used to determine the amount to which P will grow after 2 yr.

106. *Yearly depreciation.* An investment P that drops in value at the yearly rate r drops in value to

$$P(1 - r)^t$$

after t years. Find a polynomial that can be used to determine the value to which P has dropped after 2 yr.

107. Suppose that $10,400 is invested at 8.5% compounded annually. How much is in the account at the end of 5 yr? (See Exercise 105.)

108. A $90,000 investment in computer hardware is depreciating at a yearly rate of 12.5%. How much is the investment worth after 4 yr? (See Exercise 106.)

COLLABORATIVE

CORNER

Finding the Magic Number

Focus: Evaluating polynomials in several variables

Time: 15–25 minutes

Group size: 3

Materials: A coin for each person

When a team nears the end of its schedule in first place, fans begin to discuss the team's "magic number." A team's magic number is the combined number of wins by that team and losses by the second-place team that guarantee the leading team a first-place finish. For example, if the Cubs' magic number is 3 over the Reds, any combination of Cubs wins and Reds losses that totals 3 will guarantee a first-place finish for the Cubs. A team's magic number is computed using the polynomial

$$G - P - L + 1,$$

where G is the length of the season, in games, P is the number of games that the leading team has played, and L is the total number of games that the second-place team has lost minus the total number of games that the leading team has lost.

ACTIVITY

1. The standings below are from a fictitious league. Each group should calculate the Jaguars' magic number with respect to the Catamounts as well as the Jaguars' magic number with respect to the Wildcats.

(Assume that the schedule is 162 games long.)

	W	L
Jaguars	92	64
Catamounts	90	66
Wildcats	89	66

2. Each group member should play the role of one of the teams, using coin tosses to simulate the remaining games. If a group member correctly predicts the side (heads or tails) that comes up, the coin toss represents a win for that team. Should the other side appear, the toss represents a loss. Assume that these games are against other (unlisted) teams in the league. Each group member should perform three coin tosses and then update the standings.

3. Recalculate the two magic numbers, using the updated standings from part (2).

4. Slowly—one coin toss at a time—play out the remainder of the season. Record all wins and losses, update the standings, and recalculate the magic numbers each time all three group members have completed a round of coin tosses.

5. Examine the work in part (4) and explain why a magic number of 0 indicates that a team has been eliminated from contention.

4.8   Division of Polynomials

Dividing by a Monomial   •   Dividing by a Binomial

In this section, we study division of polynomials. We will find that polynomial division is similar to division in arithmetic.

Dividing by a Monomial

We first consider division by a monomial. When dividing a monomial by a monomial, we use the quotient rule of Section 4.1 to subtract exponents when bases are the same. For example,

$$\frac{15x^{10}}{3x^4} = 5x^{10-4}$$
$$= 5x^6$$

> **CAUTION!** The coefficients are divided but the exponents are subtracted.

and

$$\frac{42a^2b^5}{-3ab^2} = \frac{42}{-3}a^{2-1}b^{5-2} \qquad \text{Recall that } a^m/a^n = a^{m-n}.$$
$$= -14ab^3.$$

To divide a polynomial by a monomial, we note that since

$$\frac{A}{C} + \frac{B}{C} = \frac{A + B}{C},$$

it follows that

$$\frac{A + B}{C} = \frac{A}{C} + \frac{B}{C}. \qquad \begin{array}{l}\text{Switching the left side and the right side} \\ \text{of the equation}\end{array}$$

This is actually how we perform divisions like $86 \div 2$. Although we might simply write

$$\frac{86}{2} = 43,$$

we are really saying

$$\frac{80 + 6}{2} = \frac{80}{2} + \frac{6}{2} = 40 + 3.$$

Similarly, to divide a polynomial by a monomial, we divide each term by the monomial:

$$\frac{80x^5 + 6x^7}{2x^3} = \frac{80x^5}{2x^3} + \frac{6x^7}{2x^3}$$
$$= \frac{80}{2}x^{5-3} + \frac{6}{2}x^{7-3} \qquad \begin{array}{l}\text{Dividing coefficients and} \\ \text{subtracting exponents}\end{array}$$
$$= 40x^2 + 3x^4.$$

EXAMPLE 1   Divide $x^4 + 15x^3 - 6x^2$ by $3x$.

SOLUTION   We divide each term of $x^4 + 15x^3 - 6x^2$ by $3x$:

$$\frac{x^4 + 15x^3 - 6x^2}{3x} = \frac{x^4}{3x} + \frac{15x^3}{3x} - \frac{6x^2}{3x}$$
$$= \frac{1}{3}x^{4-1} + \frac{15}{3}x^{3-1} - \frac{6}{3}x^{2-1} \qquad \begin{array}{l}\text{Dividing coefficients} \\ \text{and subtracting} \\ \text{exponents}\end{array}$$
$$= \frac{1}{3}x^3 + 5x^2 - 2x. \qquad \text{This is the quotient.}$$

To check, we multiply our answer by $3x$, using the distributive law:

$$3x\left(\frac{1}{3}x^3 + 5x^2 - 2x\right) = 3x \cdot \frac{1}{3}x^3 + 3x \cdot 5x^2 - 3x \cdot 2x$$
$$= x^4 + 15x^3 - 6x^2.$$

This is the polynomial that was being divided, so our answer, $\frac{1}{3}x^3 + 5x^2 - 2x$, checks.

TRY EXERCISE 5

EXAMPLE 2   Divide and check: $(10a^5b^4 - 2a^3b^2 + 6a^2b) \div (-2a^2b)$.

SOLUTION   We have

$$\frac{10a^5b^4 - 2a^3b^2 + 6a^2b}{-2a^2b} = \frac{10a^5b^4}{-2a^2b} - \frac{2a^3b^2}{-2a^2b} + \frac{6a^2b}{-2a^2b}$$

We divide coefficients and subtract exponents.

$$= -\frac{10}{2}a^{5-2}b^{4-1} - \left(-\frac{2}{2}\right)a^{3-2}b^{2-1} + \left(-\frac{6}{2}\right)$$
$$= -5a^3b^3 + ab - 3.$$

Check: $-2a^2b(-5a^3b^3 + ab - 3)$
$$= -2a^2b(-5a^3b^3) + (-2a^2b)(ab) + (-2a^2b)(-3)$$
$$= 10a^5b^4 - 2a^3b^2 + 6a^2b$$

Our answer, $-5a^3b^3 + ab - 3$, checks.

TRY EXERCISE 7

Dividing by a Binomial

The divisors in Examples 1 and 2 have just one term. For divisors with more than one term, we use long division, much as we do in arithmetic. Polynomials are written in descending order and any missing terms in the dividend are written in, using 0 for the coefficients.

EXAMPLE 3   Divide $x^2 + 5x + 6$ by $x + 3$.

SOLUTION   We begin by dividing x^2 by x:

Divide the first term, x^2, by the first term in the divisor: $x^2/x = x$. Ignore the term 3 for the moment.

$$x + 3 \overline{)x^2 + 5x + 6}$$
$$-(x^2 + 3x)$$ **Multiply** $x + 3$ by x, using the distributive law.
$$2x.$$ **Subtract** both x^2 and $3x$:
$$x^2 + 5x - (x^2 + 3x) = 2x.$$

Now we "bring down" the next term—in this case, 6. The current remainder, $2x + 6$, now becomes the focus of our division. We divide $2x$ by x.

$$\begin{array}{r} x + 2 \\ x + 3\overline{)x^2 + 5x + 6} \\ -(x^2 + 3x) \\ \hline 2x + 6 \end{array}$$

Divide $2x$ by x: $2x/x = 2$.

$-(2x + 6)$ ← Multiply 2 by the divisor, $x + 3$, using the distributive law.

0 ← Subtract $(2x + 6) - (2x + 6) = 0$.

The quotient is $x + 2$. The notation R 0 indicates a remainder of 0, although a remainder of 0 is generally not listed in an answer.

Check: To check, we multiply the quotient by the divisor and add any remainder to see if we get the dividend:

$$\underbrace{(x + 3)}_{\text{Divisor}} \quad \underbrace{(x + 2)}_{\text{Quotient}} + \underbrace{0}_{\text{Remainder}} = \underbrace{x^2 + 5x + 6}_{\text{Dividend}}.$$

Our answer, $x + 2$, checks.

> TRY EXERCISE 17

EXAMPLE   4   Divide: $(2x^2 + 5x - 1) \div (2x - 1)$.

SOLUTION   We begin by dividing $2x^2$ by $2x$:

Divide the first term by the first term: $2x^2/(2x) = x$.

$$\begin{array}{r} x \\ 2x - 1\overline{)2x^2 + 5x - 1} \\ -(2x^2 - x) \\ \hline 6x \end{array}$$

Multiply $2x - 1$ by x.

Subtract by changing signs and adding: $(2x^2 + 5x) - (2x^2 - x) = 6x$.

> **CAUTION!** Write the parentheses around the polynomial being subtracted to remind you to subtract all its terms.

Now, we bring down the -1 and divide $6x - 1$ by $2x - 1$.

$$\begin{array}{r} x + 3 \\ 2x - 1\overline{)2x^2 + 5x - 1} \\ -(2x^2 - x) \\ \hline 6x - 1 \\ -(6x - 3) \\ \hline 2 \end{array}$$

Divide $6x$ by $2x$: $6x/(2x) = 3$.

Multiply 3 by the divisor, $2x - 1$.

Subtract. Note that $-1 - (-3) = -1 + 3 = 2$.

The answer is $x + 3$ with R 2.

Another way to write $x + 3$ R 2 is as

$$\underbrace{x + 3}_{\text{Quotient}} + \frac{\overset{\text{Remainder}}{2}}{\underset{\text{Divisor}}{2x - 1}}.$$

(This is the way answers will be given at the back of the book.)

Check: To check, we multiply the divisor by the quotient and add the remainder:

$$(2x - 1)(x + 3) + 2 = 2x^2 + 5x - 3 + 2$$
$$= 2x^2 + 5x - 1. \quad \text{Our answer checks.}$$

> TRY EXERCISE 29

Our division procedure ends when the degree of the remainder is less than that of the divisor. Check that this was indeed the case in Example 4.

EXAMPLE **5** Divide each of the following.

a) $(x^3 + 1) \div (x + 1)$ **b)** $(x^4 - 3x^2 + 4x - 3) \div (x^2 - 5)$

SOLUTION

a)
$$
\begin{array}{r}
x^2 - x + 1 \\
x + 1\overline{\smash{)}x^3 + 0x^2 + 0x + 1} \longleftarrow \text{Writing in the missing terms}\\
\underline{-(x^3 + x^2)}\\
-x^2 + 0x \longleftarrow \text{Subtracting } x^3 + x^2 \text{ from } x^3 + 0x^2 \text{ and}\\
\underline{-(-x^2 - x)} \text{bringing down the } 0x\\
x + 1 \longleftarrow \text{Subtracting } -x^2 - x \text{ from } -x^2 + 0x \text{ and}\\
\underline{-(x + 1)} \text{bringing down the } 1\\
0
\end{array}
$$

The answer is $x^2 - x + 1$.

Check: $(x + 1)(x^2 - x + 1) = x^3 - x^2 + x + x^2 - x + 1$
$$= x^3 + 1.$$

b)
$$
\begin{array}{r}
x^2 + 2 \\
x^2 - 5\overline{\smash{)}x^4 + 0x^3 - 3x^2 + 4x - 3} \text{Writing in the missing term}\\
\underline{-(x^4 - 5x^2)}\\
2x^2 + 4x - 3 \longleftarrow \text{Subtracting } x^4 - 5x^2 \text{ from}\\
\underline{-(2x^2 - 10)} x^4 - 3x^2 \text{ and bringing down } 4x - 3\\
4x + 7 \longleftarrow \text{Subtracting } 2x^2 - 10 \text{ from}\\
2x^2 + 4x - 3
\end{array}
$$

Since the remainder, $4x + 7$, is of lower degree than the divisor, the division process stops. The answer is

$$x^2 + 2 + \frac{4x + 7}{x^2 - 5}.$$

Check: $(x^2 - 5)(x^2 + 2) + 4x + 7 = x^4 + 2x^2 - 5x^2 - 10 + 4x + 7$
$$= x^4 - 3x^2 + 4x - 3.$$

TRY EXERCISE 35

4.8 **EXERCISE SET**

Divide and check.

1. $\dfrac{40x^6 - 25x^3}{5}$

2. $\dfrac{16a^5 - 24a^2}{8}$

3. $\dfrac{u - 2u^2 + u^7}{u}$

4. $\dfrac{50x^5 - 7x^4 + 2x}{x}$

5. $(18t^3 - 24t^2 + 6t) \div (3t)$

6. $(20t^3 - 15t^2 + 30t) \div (5t)$

7. $(42x^5 - 36x^3 + 9x^2) \div (6x^2)$

8. $(24x^6 + 18x^4 + 8x^3) \div (4x^3)$

9. $(32t^5 + 16t^4 - 8t^3) \div (-8t^3)$

10. $(36t^6 - 27t^5 - 9t^2) \div (-9t^2)$

11. $\dfrac{8x^2 - 10x + 1}{2x}$

12. $\dfrac{9x^2 + 3x - 2}{3x}$

13. $\dfrac{5x^3y + 10x^5y^2 + 15x^2y}{5x^2y}$

14. $\dfrac{12a^3b^2 + 4a^4b^5 + 16ab^2}{4ab^2}$

15. $\dfrac{9r^2s^2 + 3r^2s - 6rs^2}{-3rs}$

16. $\dfrac{4x^4y - 8x^6y^2 + 12x^8y^6}{4x^4y}$

17. $(x^2 - 8x + 12) \div (x - 2)$

18. $(x^2 + 2x - 15) \div (x + 5)$

19. $(t^2 - 10t - 20) \div (t - 5)$

20. $(t^2 + 8t - 15) \div (t + 4)$

21. $(2x^2 + 11x - 5) \div (x + 6)$

22. $(3x^2 - 2x - 13) \div (x - 2)$

23. $\dfrac{t^3 + 27}{t + 3}$ **24.** $\dfrac{a^3 + 8}{a + 2}$

25. $\dfrac{a^2 - 21}{a - 5}$ **26.** $\dfrac{t^2 - 13}{t - 4}$

27. $(5x^2 - 16x) \div (5x - 1)$

28. $(3x^2 - 7x + 1) \div (3x - 1)$

29. $(6a^2 + 17a + 8) \div (2a + 5)$

30. $(10a^2 + 19a + 9) \div (2a + 3)$

31. $\dfrac{2t^3 - 9t^2 + 11t - 3}{2t - 3}$

32. $\dfrac{8t^3 - 22t^2 - 5t + 12}{4t + 3}$

33. $(x^3 - x^2 + x - 1) \div (x - 1)$

34. $(t^3 - t^2 + t - 1) \div (t + 1)$

35. $(t^4 + 4t^2 + 3t - 6) \div (t^2 + 5)$

36. $(t^4 - 2t^2 + 4t - 5) \div (t^2 - 3)$

37. $(6x^4 - 3x^2 + x - 4) \div (2x^2 + 1)$

38. $(4x^4 - 4x^2 - 3) \div (2x^2 - 3)$

39. How is the distributive law used when dividing a polynomial by a binomial?

40. On an assignment, Emmy Lou *incorrectly* writes
$$\frac{12x^3 - 6x}{3x} = 4x^2 - 6x.$$
What mistake do you think she is making and how might you convince her that a mistake has been made?

Skill Review

Review graphing linear equations (Chapter 3).

Graph.

41. $3x - 4y = 12$ [3.3]   **42.** $y = -\frac{2}{3}x + 4$ [3.6]

43. $3y - 2 = 7$ [3.3]   **44.** $8x = 4y$ [3.2]

45. Find the slope of the line containing the points $(3, 2)$ and $(-7, 5)$. [3.5]

46. Find the slope and the y-intercept of the line given by $2y = 8x + 7$. [3.6]

47. Find the slope–intercept form of the line with slope -5 and y-intercept $(0, -10)$. [3.6]

48. Find the slope–intercept form of the line containing the points $(6, 3)$ and $(-2, -7)$. [3.7]

Synthesis

49. Explain how to form trinomials for which division by $x - 5$ results in a remainder of 3.

50. Under what circumstances will the quotient of two binomials have more than two terms?

Divide.

51. $(10x^{9k} - 32x^{6k} + 28x^{3k}) \div (2x^{3k})$

52. $(45a^{8k} + 30a^{6k} - 60a^{4k}) \div (3a^{2k})$

53. $(6t^{3h} + 13t^{2h} - 4t^h - 15) \div (2t^h + 3)$

54. $(x^4 + a^2) \div (x + a)$

55. $(5a^3 + 8a^2 - 23a - 1) \div (5a^2 - 7a - 2)$

56. $(15y^3 - 30y + 7 - 19y^2) \div (3y^2 - 2 - 5y)$

57. Divide the sum of $4x^5 - 14x^3 - x^2 + 3$ and $2x^5 + 3x^4 + x^3 - 3x^2 + 5x$ by $3x^3 - 2x - 1$.

58. Divide $5x^7 - 3x^4 + 2x^2 - 10x + 2$ by the sum of $(x - 3)^2$ and $5x - 8$.

If the remainder is 0 when one polynomial is divided by another, the divisor is a factor *of the dividend. Find the value(s) of c for which $x - 1$ is a factor of each polynomial.*

59. $x^2 - 4x + c$

60. $2x^2 - 3cx - 8$

61. $c^2x^2 + 2cx + 1$

Study Summary

KEY TERMS AND CONCEPTS	EXAMPLES

SECTION 4.1: EXPONENTS AND THEIR PROPERTIES

(Assume that no denominators are 0 and that 0^0 is not considered.)

For any integers m and n:

1 as an exponent:	$a^1 = a$	$3^1 = 3$
0 as an exponent:	$a^0 = 1$	$3^0 = 1$
The Product Rule:	$a^m \cdot a^n = a^{m+n}$	$3^5 \cdot 3^9 = 3^{5+9} = 3^{14}$
The Quotient Rule:	$\dfrac{a^m}{a^n} = a^{m-n}$	$\dfrac{3^7}{3} = 3^{7-1} = 3^6$
The Power Rule:	$(a^m)^n = a^{mn}$	$(3^4)^2 = 3^{4\cdot2} = 3^8$
Raising a product to a power:	$(ab)^n = a^n b^n$	$(3x^5)^4 = 3^4(x^5)^4 = 81x^{20}$
Raising a quotient to a power:	$\left(\dfrac{a}{b}\right)^n = \dfrac{a^n}{b^n}$	$\left(\dfrac{3}{x}\right)^6 = \dfrac{3^6}{x^6}$

SECTION 4.2: NEGATIVE EXPONENTS AND SCIENTIFIC NOTATION

$a^{-n} = \dfrac{1}{a^n}$;

$\dfrac{a^{-n}}{b^{-m}} = \dfrac{b^m}{a^n}$;

$\left(\dfrac{a}{b}\right)^{-n} = \left(\dfrac{b}{a}\right)^n$

$3^{-2} = \dfrac{1}{3^2} = \dfrac{1}{9}$;

$\dfrac{3^{-7}}{x^{-5}} = \dfrac{x^5}{3^7}$;

$\left(\dfrac{3}{x}\right)^{-2} = \left(\dfrac{x}{3}\right)^2$

Scientific notation is given by $N \times 10^m$, where m is an integer, N is in decimal notation, and $1 \le N < 10$.

$4100 = 4.1 \times 10^3$;

$5 \times 10^{-3} = 0.005$

SECTION 4.3: POLYNOMIALS

A **polynomial** is a monomial or a sum of monomials.

When a polynomial is written as a sum of monomials, each monomial is a **term** of the polynomial.

The **degree of a term** of a polynomial is the number of variable factors in that term.

The **coefficient** of a term is the part of the term that is a constant factor.

The **leading term** of a polynomial is the term of highest degree.

The **leading coefficient** is the coefficient of the leading term.

The **degree of the polynomial** is the degree of the leading term.

Polynomial: $10x - x^3 - \frac{1}{2}x^2 + 3x^5 + 7$

Term	$10x$	$-x^3$	$-\frac{1}{2}x^2$	$3x^5$	7
Degree of term	1	3	2	5	0
Coefficient of term	10	-1	$-\frac{1}{2}$	3	7
Leading term			$3x^5$		
Leading coefficient			3		
Degree of polynomial			5		

A **monomial** has one term.
A **binomial** has two terms.
A **trinomial** has three terms.

Monomial (one term): $4x^3$
Binomial (two terms): $x^2 - 5$
Trinomial (three terms): $3t^3 + 2t - 10$

Like terms, or **similar terms**, are either constant terms or terms containing the same variable(s) raised to the same power(s). These can be **combined** within a polynomial.

Combine like terms: $3y^4 + 6y^2 - 7 - y^4 - 6y^2 + 8$.

$$3y^4 + 6y^2 - 7 - y^4 - 6y^2 + 8 = \underbrace{3y^4 - y^4}_{} + \underbrace{6y^2 - 6y^2}_{} \underbrace{- 7 + 8}_{}$$
$$= \quad 2y^4 \quad + \quad 0 \quad + \quad 1$$
$$= 2y^4 + 1$$

To **evaluate** a polynomial, replace the variable with a number. The **value** is calculated using the rules for order of operations.

Evaluate $t^3 - 2t^2 - 5t + 1$ *for* $t = -2$.
$$t^3 - 2t^2 - 5t + 1 = (-2)^3 - 2(-2)^2 - 5(-2) + 1$$
$$= -8 - 2(4) - (-10) + 1$$
$$= -8 - 8 + 10 + 1$$
$$= -5$$

SECTION 4.4: ADDITION AND SUBTRACTION OF POLYNOMIALS

Add polynomials by combining like terms.

$$(2x^2 - 3x + 7) + (5x^3 + 3x - 9)$$
$$= 2x^2 + (-3x) + 7 + 5x^3 + 3x + (-9)$$
$$= 5x^3 + 2x^2 - 2$$

Subtract polynomials by adding the opposite of the polynomial being subtracted.

$$(2x^2 - 3x + 7) - (5x^3 + 3x - 9)$$
$$= 2x^2 - 3x + 7 + (-5x^3 - 3x + 9)$$
$$= 2x^2 - 3x + 7 - 5x^3 - 3x + 9$$
$$= -5x^3 + 2x^2 - 6x + 16$$

SECTION 4.5: MULTIPLICATION OF POLYNOMIALS

Multiply polynomials by multiplying each term of one polynomial by each term of the other.

$$(x + 2)(x^2 - x - 1)$$
$$= x \cdot x^2 - x \cdot x - x \cdot 1 + 2 \cdot x^2 - 2 \cdot x - 2 \cdot 1$$
$$= x^3 - x^2 - x + 2x^2 - 2x - 2$$
$$= x^3 + x^2 - 3x - 2$$

SECTION 4.6: SPECIAL PRODUCTS

FOIL (First, Outer, Inner, Last):

$$(A + B)(C + D) = AC + AD + BC + BD$$

$$(x + 3)(x - 2) = x^2 - 2x + 3x - 6$$
$$= x^2 + x - 6$$

The product of a sum and a difference:

$$(A + B)(A - B) = A^2 - B^2$$

$A^2 - B^2$ is called a **difference of squares**.

$$(t^3 + 5)(t^3 - 5) = (t^3)^2 - 5^2$$
$$= t^6 - 25$$

The square of a binomial:
$$(A + B)^2 = A^2 + 2AB + B^2;$$
$$(A - B)^2 = A^2 - 2AB + B^2$$
$A^2 + 2AB + B^2$ and $A^2 - 2AB + B^2$ are called **perfect-square trinomials**.

$$(5x + 3)^2 = (5x)^2 + 2(5x)(3) + 3^2 = 25x^2 + 30x + 9;$$
$$(5x - 3)^2 = (5x)^2 - 2(5x)(3) + 3^2 = 25x^2 - 30x + 9$$

SECTION 4.7: POLYNOMIALS IN SEVERAL VARIABLES

To **evaluate** a polynomial, replace each variable with a number and simplify.

Evaluate $4 - 3xy + x^2y$ for $x = 5$ and $y = -1$.
$$4 - 3xy + x^2y = 4 - 3(5)(-1) + (5)^2(-1)$$
$$= 4 - (-15) + (-25)$$
$$= -6$$

The **degree** of a term is the number of variables in the term or the sum of the exponents of the variables.

The degree of $-19x^3yz^2$ is 6.

Add, subtract, and multiply polynomials in several variables in the same way as polynomials in one variable.

$$(3xy^2 - 4x^2y + 5xy) + (xy - 6x^2y) = 3xy^2 - 10x^2y + 6xy;$$
$$(3xy^2 - 4x^2y + 5xy) - (xy - 6x^2y) = 3xy^2 + 2x^2y + 4xy;$$
$$(2a^2b + 3a)(5a^2b - a) = 10a^4b^2 + 13a^3b - 3a^2$$

SECTION 4.8: DIVISION OF POLYNOMIALS

To divide a polynomial by a monomial, divide each term by the monomial. Divide coefficients and subtract exponents.

$$\frac{3t^5 - 6t^4 + 4t^2 + 9t}{3t} = \frac{3t^5}{3t} - \frac{6t^4}{3t} + \frac{4t^2}{3t} + \frac{9t}{3t}$$
$$= t^4 - 2t^3 + \frac{4}{3}t + 3$$

To divide a polynomial by a binomial, use long division.

Divide: $(x^2 + 5x - 2) \div (x - 3)$.

$$
\begin{array}{r}
x + 8 \\
x - 3 \overline{)x^2 + 5x - 2} \\
\underline{-(x^2 - 3x)} \\
8x - 2 \\
\underline{-(8x - 24)} \\
22
\end{array}
$$

$$(x^2 + 5x - 2) \div (x - 3) = x + 8 + \frac{22}{x - 3}$$

Review Exercises: Chapter 4

🐋 **Concept Reinforcement** *Classify each statement as either true or false.*

1. When two polynomials that are written in descending order are added, the result is generally written in descending order. [4.4]

2. The product of the sum and the difference of the same two terms is a difference of squares. [4.6]

3. When a binomial is squared, the result is a perfect-square trinomial. [4.6]

4. FOIL can be used whenever two polynomials are being multiplied. [4.6]

5. The degree of a polynomial cannot exceed the value of the polynomial's leading coefficient. [4.3]

6. Scientific notation is used only for extremely large numbers. [4.2]

7. FOIL can be used with polynomials in several variables. [4.7]

8. A positive number raised to a negative exponent can never represent a negative number. [4.2]

Simplify. [4.1]

9. $n^3 \cdot n^8 \cdot n$

10. $(7x)^8 \cdot (7x)^2$

11. $t^6 \cdot t^0$

12. $\dfrac{4^5}{4^2}$

13. $\dfrac{(a+b)^4}{(a+b)^4}$

14. $\dfrac{-18c^9 d^3}{2c^5 d}$

15. $(-2xy^2)^3$

16. $(2x^3)(-3x)^2$

17. $(a^2 b)(ab)^5$

18. $\left(\dfrac{2t^5}{3s^4}\right)^2$

19. Express using a positive exponent: 8^{-6}. [4.2]

20. Express using a negative exponent: $\dfrac{1}{a^9}$. [4.2]

Simplify. Do not use negative exponents in the answer. [4.2]

21. $4^5 \cdot 4^{-7}$

22. $\dfrac{6a^{-5}b}{3a^8 b^{-8}}$

23. $(w^3)^{-5}$

24. $(2x^{-3}y)^{-2}$

25. $\left(\dfrac{2x}{y}\right)^{-3}$

26. Convert to decimal notation: 4.7×10^8. [4.2]

27. Convert to scientific notation: 0.0000109. [4.2]

Multiply or divide and write scientific notation for the result. [4.2]

28. $(3.8 \times 10^4)(5.5 \times 10^{-1})$

29. $\dfrac{1.28 \times 10^{-8}}{2.5 \times 10^{-4}}$

Identify the terms of each polynomial. [4.3]

30. $8x^2 - x + \frac{2}{3}$

31. $-4y^5 + 7y^2 - 3y - 2$

List the coefficients of the terms in each polynomial. [4.3]

32. $9x^2 - x + 7$

33. $7n^4 - \frac{5}{6}n^2 - 4n + 10$

For each polynomial, (a) list the degree of each term; (b) determine the leading term and the leading coefficient; and (c) determine the degree of the polynomial. [4.3]

34. $4t^2 + 6 + 15t^5$

35. $-2x^5 + 7 - 3x^2 + x$

Classify each polynomial as a monomial, a binomial, a trinomial, or a polynomial with no special name. [4.3]

36. $4x^3 - 5x + 3$

37. $4 - 9t^3 - 7t^4 + 10t^2$

38. $7y^2$

Combine like terms and write in descending order. [4.3]

39. $3x - x^2 + 4x$

40. $\frac{3}{4}x^3 + 4x^2 - x^3 + 7$

41. $-4t^3 + 2t + 4t^3 + 8 - t - 9$

42. $-a + \frac{1}{3} + 20a^5 - 1 - 6a^5 - 2a^2$

Evaluate each polynomial for $x = -2$. [4.3]

43. $9x - 6$

44. $x^2 - 3x + 6$

Add or subtract. [4.4]

45. $(8x^4 - x^3 + x - 4) + (x^5 + 7x^3 - 3x - 5)$

46. $(5a^5 - 2a^3 - 9a^2) + (2a^5 + a^3) + (-a^5 - 3a^2)$

47. $(y^2 + 8y - 7) - (4y^2 - 10)$

48. $(3x^5 - 4x^4 + 2x^2 + 3) - (2x^5 - 4x^4 + 3x^3 + 4x^2 - 5)$

49.
$$\begin{array}{r} -\frac{3}{4}x^4 + \frac{1}{2}x^3 \qquad\qquad + \frac{7}{8} \\ -\frac{1}{4}x^3 - x^2 - \frac{7}{4}x \\ +\frac{3}{2}x^4 \qquad\qquad + \frac{2}{3}x^2 \qquad - \frac{1}{2} \end{array}$$

50.
$$\begin{array}{r} 2x^5 \qquad - x^3 \qquad + x + 3 \\ -(3x^5 - x^4 + 4x^3 + 2x^2 - x + 3) \end{array}$$

51. The length of a rectangle is 3 m greater than its width.

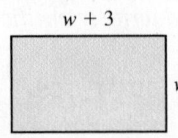

a) Find a polynomial for the perimeter. [4.4]
b) Find a polynomial for the area. [4.5]

Multiply.

52. $5x^2(-6x^3)$ [4.5]

53. $(7x + 1)^2$ [4.6]

54. $(a - 7)(a + 4)$ [4.6]

55. $(d - 8)(d + 8)$ [4.6]

56. $(4x^2 - 5x + 1)(3x - 2)$ [4.5]

57. $(x - 8)^2$ [4.6]

58. $3t^2(5t^3 - 2t^2 + 4t)$ [4.5]

59. $(2a + 9)(2a - 9)$ [4.6]

60. $(x - 0.8)(x - 0.5)$ [4.6]

61. $(x^4 - 2x + 3)(x^3 + x - 1)$ [4.5]

62. $(4y^3 - 5)^2$ [4.6]

63. $(2t^2 + 3)(t^2 - 7)$ [4.6]

64. $\left(a - \frac{1}{2}\right)\left(a + \frac{2}{3}\right)$ [4.6]

65. $(-7 + 2n)(7 + 2n)$ [4.6]

66. Evaluate $2 - 5xy + y^2 - 4xy^3 + x^6$ for $x = -1$ and $y = 2$. [4.7]

Identify the coefficient and the degree of each term of each polynomial. Then find the degree of each polynomial. [4.7]

67. $x^5y - 7xy + 9x^2 - 8$

68. $a^3b^8c^2 - c^{22} + a^5c^{10}$

Combine like terms. [4.7]

69. $u + 3v - 5u + v - 7$

70. $6m^3 + 3m^2n + 4mn^2 + m^2n - 5mn^2$

Add or subtract. [4.7]

71. $(4a^2 - 10ab - b^2) + (-2a^2 - 6ab + b^2)$

72. $(6x^3y^2 - 4x^2y - 6x) - (-5x^3y^2 + 4x^2y + 6x^2 - 6)$

Multiply. [4.7]

73. $(2x + 5y)(x - 3y)$

74. $(5ab - cd^2)^2$

75. Find a polynomial for the shaded area. [4.7]

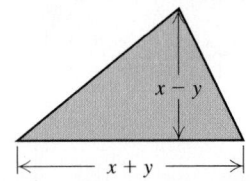

Divide. [4.8]

76. $(3y^5 - y^2 + 12y) \div (3y)$

77. $(6x^3 - 5x^2 - 13x + 13) \div (2x + 3)$

78. $\dfrac{t^4 + t^3 + 2t^2 - t - 3}{t + 1}$

Synthesis

79. Explain why $5x^3$ and $(5x)^3$ are not equivalent expressions. [4.1]

80. A binomial is squared and the result, written in descending order, is $x^2 - 6x + 9$. Is it possible to determine what binomial was squared? Why or why not? [4.6]

81. Determine, without performing the multiplications, the degree of each product. [4.5]
a) $(x^5 - 6x^2 + 3)(x^4 + 3x^3 + 7)$
b) $(x^7 - 4)^4$

82. Simplify:
$$(-3x^5 \cdot 3x^3 - x^6(2x)^2 + (3x^4)^2 + (2x^2)^4 - 20x^2(x^3)^2)^2. \quad [4.1], [4.3]$$

83. A polynomial has degree 4. The x^2-term is missing. The coefficient of x^4 is two times the coefficient of x^3. The coefficient of x is 3 less than the coefficient of x^4. The remaining coefficient is 7 less than the coefficient of x. The sum of the coefficients is 15. Find the polynomial. [4.3]

Aha! **84.** Multiply: $[(x - 5) - 4x^3][(x - 5) + 4x^3]$. [4.6]

85. Solve: $(x - 7)(x + 10) = (x - 4)(x - 6)$. [2.2], [4.6]

86. *Blood donors.* Every 4–6 weeks, Jordan donates 1.14×10^6 cubic millimeters (two pints) of whole blood, from which platelets are removed and the blood returned to the body. In one cubic millimeter of blood, there are about 2×10^5 platelets. Approximate the number of platelets in Jordan's typical donation. [4.2]

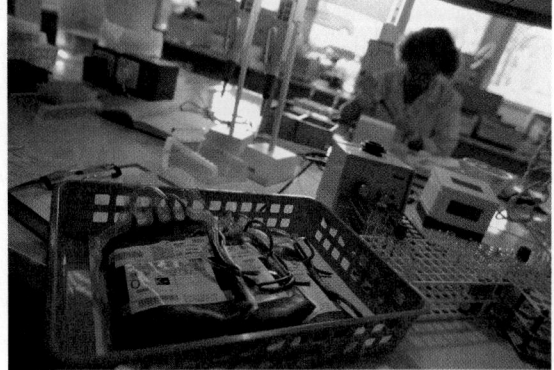

CHAPTER Test Prep VIDEO CD
Step-by-step test solutions are found on the video CD in the front of this book.

Simplify.

1. $x^7 \cdot x \cdot x^5$

2. $\dfrac{3^8}{3^7}$

3. $\dfrac{(3m)^4}{(3m)^4}$

4. $(t^5)^9$

5. $(5x^4y)(-2x^5y)^3$

6. $\dfrac{24a^7b^4}{20a^2b}$

7. Express using a positive exponent: y^{-7}.

8. Express using a negative exponent: $\dfrac{1}{5^6}$.

Simplify.

9. $t^{-4} \cdot t^{-5}$

10. $\dfrac{9x^3y^2}{3x^8y^{-3}}$

11. $(2a^3b^{-1})^{-4}$

12. $\left(\dfrac{ab}{c}\right)^{-3}$

13. Convert to scientific notation: 3,060,000,000.

14. Convert to decimal notation: 5×10^{-8}.

Multiply or divide and write scientific notation for the result.

15. $\dfrac{5.6 \times 10^6}{3.2 \times 10^{-11}}$

16. $(2.4 \times 10^5)(5.4 \times 10^{16})$

17. Classify $4x^2y - 7y^3$ as a monomial, a binomial, a trinomial, or a polynomial with no special name.

18. Identify the coefficient of each term of the polynomial:
$$3x^5 - x + \tfrac{1}{9}.$$

19. Determine the degree of each term, the leading term and the leading coefficient, and the degree of the polynomial:
$$2t^3 - t + 7t^5 + 4.$$

20. Evaluate $x^2 + 5x - 1$ for $x = -3$.

Combine like terms and write in descending order.

21. $4a^2 - 6 + a^2$

22. $y^2 - 3y - y + \tfrac{3}{4}y^2$

23. $3 - x^2 + 8x + 5x^2 - 6x - 2x + 4x^3$

Add or subtract.

24. $(3x^5 + 5x^3 - 5x^2 - 3) + (x^5 + x^4 - 3x^2 + 2x - 4)$

25. $\left(x^4 + \tfrac{2}{3}x + 5\right) + \left(4x^4 + 5x^2 + \tfrac{1}{3}x\right)$

26. $(5a^4 + 3a^3 - a^2 - 2a - 1) - (7a^4 - a^2 - a + 6)$

27. $(t^3 - 0.3t^2 - 20) - (t^4 - 1.5t^3 + 0.3t^2 - 11)$

Multiply.

28. $-2x^2(3x^2 - 3x - 5)$

29. $\left(x - \tfrac{1}{3}\right)^2$

30. $(5t - 7)(5t + 7)$

31. $(3b + 5)(2b - 1)$

32. $(x^6 - 4)(x^8 + 4)$

33. $(8 - y)(6 + 5y)$

34. $(2x + 1)(3x^2 - 5x - 3)$

35. $(8a^3 + 3)^2$

36. Evaluate $2x^2y - 3y^2$ for $x = -3$ and $y = 2$.

37. Combine like terms:
$$2x^3y - y^3 + xy^3 + 8 - 6x^3y - x^2y^2 + 11.$$

38. Subtract:
$$(8a^2b^2 - ab + b^3) - (-6ab^2 - 7ab - ab^3 + 5b^3).$$

39. Multiply: $(3x^5 - y)(3x^5 + y)$.

Divide.

40. $(12x^4 + 9x^3 - 15x^2) \div (3x^2)$

41. $(6x^3 - 8x^2 - 14x + 13) \div (3x + 2)$

Synthesis

42. The height of a box is 1 less than its length, and the length is 2 more than its width. Express the volume in terms of the length.

43. Solve: $x^2 + (x - 7)(x + 4) = 2(x - 6)^2$.

44. Simplify: $2^{-1} - 4^{-1}$.

45. Every day about 12.4 billion spam e-mails are sent. If each spam e-mail wastes 4 sec, how many hours are wasted each day due to spam?
Source: spam-filter-review.toptenreviews.com

Cumulative Review: Chapters 1–4

1. Evaluate $\dfrac{2x + y}{5}$ for $x = 12$ and $y = 6$. [1.1]

2. Evaluate $5x^2y - xy + y^2$ for $x = -1$ and $y = -2$. [4.7]

Simplify.

3. $\frac{1}{15} - \frac{2}{9}$ [1.3]

4. $2 - [10 - (5 + 12 \div 2^2 \cdot 3)]$ [1.8]

5. $2y - (y - 7) + 3$ [1.8]

6. $t^4 \cdot t^7 \cdot t$ [4.1]

7. $\dfrac{-100x^6y^8}{25xy^5}$ [4.1]

8. $(2a^2b)(5ab^3)^2$ [4.1]

9. Factor: $10a - 6b + 12$. [1.2]

10. Find the absolute value: $\left|\dfrac{11}{16}\right|$. [1.4]

11. Determine the degree of the polynomial
$-x^4 + 5x^3 + 3x^6 - 1$. [4.3]

12. Combine like terms and write in descending order:
$-\frac{1}{2}t^3 + 3t^2 + \frac{3}{4}t^3 + 0.1 - 8t^2 - 0.45$. [4.3]

13. In which quadrant is $(-2, 5)$ located? [3.1]

14. Graph on a number line: $x > -1$. [2.6]

Graph.

15. $3y + 2x = 0$ [3.2]

16. $3y - 2x = 12$ [3.3]

17. $3y = 2$ [3.3]

18. $3y = 2x + 9$ [3.6]

19. Find the slope and the y-intercept of the line given by $y = \frac{1}{10}x + \frac{3}{8}$. [3.6]

20. Find the slope of the line containing the points $(2, 3)$ and $(-6, 8)$. [3.5]

21. Write an equation of the line with slope $-\frac{2}{3}$ and y-intercept $(0, -10)$. [3.6]

22. Find the coordinates of the x- and y-intercepts of the graph of $2x + 5y = 8$. Do not graph. [3.3]

Solve.

23. $\frac{1}{6}n = -\frac{2}{3}$ [2.1]

24. $3 - 5x = 0$ [2.2]

25. $5y + 7 = 8y - 1$ [2.2]

26. $0.4t - 0.5 = 8.3$ [2.2]

27. $5(3 - x) = 2 + 7(x - 1)$ [2.2]

28. $2 - (x - 7) = 8 - 4(x + 5)$ [2.2]

29. $-\frac{1}{2}t \le 4$ [2.6]

30. $3x - 5 > 9x - 8$ [2.6]

31. Solve $c = \dfrac{5pq}{2t}$ for t. [2.3]

Add or subtract.

32. $(2u^2v - uv^2 + uv) + (3u^2 - v^2u + 5vu^2)$ [4.7]

33. $(3x^3 - 2x^2 + 6x) - (x^2 - 6x + 7)$ [4.4]

34. $(2x^5 - x^4 - x) - (x^5 - x^4 + x)$ [4.4]

Multiply.

35. $10(2a - 3b + 7)$ [1.2]

36. $8x^3(-2x^2 - 6x + 7)$ [4.5]

37. $(3a + 7)(2a - 1)$ [4.6]

38. $(x - 2)(x^2 + x - 5)$ [4.5]

39. $(4t^2 + 3)^2$ [4.6]

40. $\left(\frac{1}{2}x + 1\right)\left(\frac{1}{2}x - 1\right)$ [4.6]

41. $(2r^2 + s)(3r^2 - 4s)$ [4.7]

42. Divide: $(x^2 - x + 3) \div (x - 1)$. [4.8]

Simplify. Do not use negative exponents in the answer. [4.2]

43. 7^{-10}

44. $x^{-8} \cdot x$

45. $\left(\dfrac{4s}{3t^{-5}}\right)^{-2}$

46. $(3x^{-7}y^{-2})^{-1}$

Solve.

47. In 2007, Europe and the United States together had installed wind turbines capable of producing about 60 thousand megawatts of electricity. Europe's wind-turbine capacity was four times that of the United States. What was Europe's wind-turbine capacity in 2007? [2.5]
Source: BP Statistical Review of World Energy, 2007

48. In the first four months of 2007, U.S. electric utilities used coal to generate 644 megawatt hours of electricity. This was 49.6% of the total amount of electricity generated. How much electricity was generated in the first four months of 2007? [2.4]
Source: U.S. Energy Information Administration

49. Antonio's energy-efficient washer and dryer use a total of 70 kilowatt hours (kWh) of electricity each month. The dryer uses 10 more than twice as many kilowatt hours as the washer. How many kilowatt hours does each appliance use in a month? [2.5]

50. A typical two-person household will use 195 kWh of electricity each month to heat water. The usage increases to 315 kWh per month for a four-person household. [3.7]
Source: Lee County Electric Cooperative
a) Graph the data and determine an equation for the related line. Let w represent the number of kilowatt hours used each month and n represent the number of people in a household.
b) Use the equation of part (a) to estimate the number of kilowatt hours used each month by a five-person household.

51. Abrianna's contract specifies that she cannot work more than 40 hr per week. For the first 4 days of one week, she worked 8, 10, 7, and 9 hr. How many hours can she work on the fifth day without violating her contract? [2.7]

52. In 2006, Brazil owed about \$240 billion in external debt. This amount was $\frac{3}{2}$ of Russia's debt. How much did Russia owe in 2006? [2.5]

53. In 2006, the average GPA of incoming freshmen at the University of South Carolina was 3.73. This was a 3.6% increase over the average GPA in 2001. What was the average GPA in 2001? [2.4]
Source: *The Wall Street Journal*, 11/10/06

54. U.S. retail losses due to crime have increased from \$26 billion in 1998 to \$38 billion in 2005. Find the rate of change of retail losses due to crime. [3.4]
Source: University of Florida

Synthesis

Solve. If no solution exists, state this.

55. $3x - 2(x + 6) = 4(x - 3)$ [2.2]

56. $x - (2x - 1) = 3x - 4(x + 1) + 10$ [2.2]

57. $(x - 2)(x + 3) = (4 - x)^2$ [2.2], [4.6]

58. Find the equation of a line with the same slope as $y = \frac{1}{2}x - 7$ and the same y-intercept as $2y = 5x + 8$. [3.6]

Simplify.

59. $7^{-1} + 8^0$ [4.2]

60. $-2x^5(x^7) + (x^3)^4 - (4x^5)^2(-x^2)$ [4.1], [4.3]

Polynomials and Factoring

CHRIS GJERSVIK
PRINCIPAL NETWORK ENGINEER
Upper Saddle River, New Jersey

The math skills we use in networking–telecommunications help us in so many ways. For instance, we use math when planning the budgets for a computer network or phone system and when determining capacity of a network. We need to accurately size network circuits so they have enough bandwidth to carry phone calls, Internet access, streaming video, e-mail, and other such applications. Finally, we use math when we need to analyze network traffic and break it down into its most basic form, which uses the binary numeral system.

AN APPLICATION

The number N, in millions, of broadband cable and DSL subscribers in the United States t years after 1998 can be approximated by

$$N = 0.3t^2 + 0.6t.$$

When will there be 36 million broadband cable and DSL subscribers?

Source: Based on information from Leichtman Research Group

This problem appears as Exercise 19 in Section 5.8.

1n Chapter 1, we learned that *factoring* is multiplying reversed. Thus factoring polynomials requires a solid command of the multiplication methods learned in Chapter 4. Factoring is an important skill that will be used to solve equations and simplify other types of expressions found later in the study of algebra.

In Sections 5.1–5.6, we factor polynomials to find equivalent expressions that are products. In Sections 5.7 and 5.8, we use factoring to solve equations, including those that arise from real-world problems.

5.1 Introduction to Factoring

Factoring Monomials ▪ Factoring When Terms Have a Common Factor ▪
Factoring by Grouping ▪ Checking by Evaluating

Just as a number like 15 can be factored as $3 \cdot 5$, a polynomial like $x^2 + 7x$ can be factored as $x(x + 7)$. In both cases, we ask ourselves, "What was multiplied to obtain the given result?" The situation is much like a popular television game show in which an "answer" is given and participants must find the "question" to which the answer corresponds.

> **Factoring**
>
> To *factor* a polynomial is to find an equivalent expression that is a product. An equivalent expression of this type is called a *factorization* of the polynomial.

Factoring Monomials

To factor a monomial, we find two monomials whose product is equivalent to the original monomial. For example, $20x^2$ can be factored as $2 \cdot 10x^2$, $4x \cdot 5x$, or $10x \cdot 2x$, as well as several other ways. To check, we multiply.

EXAMPLE 1 Find three factorizations of $15x^3$.

SOLUTION

a) $15x^3 = (3 \cdot 5)(x \cdot x^2)$ Thinking of how 15 and x^3 can be factored

$\quad\quad = (3x)(5x^2)$ The factors are $3x$ and $5x^2$. *Check*: $3x \cdot 5x^2 = 15x^3$.

b) $15x^3 = (3 \cdot 5)(x^2 \cdot x)$

$\quad\quad = (3x^2)(5x)$ The factors are $3x^2$ and $5x$. *Check*: $3x^2 \cdot 5x = 15x^3$.

c) $15x^3 = ((-5)(-3))x^3$

$\quad\quad = (-5)(-3x^3)$ The factors are -5 and $-3x^3$.
                            Check: $(-5)(-3x^3) = 15x^3$.

$(3x)(5x^2)$, $(3x^2)(5x)$, and $(-5)(-3x^3)$ are all factorizations of $15x^3$. Other factorizations exist as well.

TRY EXERCISE 9

Recall from Section 1.2 that the word "factor" can be a verb or a noun, depending on the context in which it appears.

Factoring When Terms Have a Common Factor

To multiply a polynomial of two or more terms by a monomial, we use the distributive law: $a(b + c) = ab + ac$. To factor a polynomial with two or more terms of the form $ab + ac$, we use the distributive law with the sides of the equation switched: $ab + ac = a(b + c)$.

Multiply   *Factor*

$3(x + 2y - z)$ $3x + 6y - 3z$
$= 3 \cdot x + 3 \cdot 2y - 3 \cdot z$ $= 3 \cdot x + 3 \cdot 2y - 3 \cdot z$
$= 3x + 6y - 3z$ $= 3(x + 2y - z)$

In the factorization on the right, note that since 3 appears as a factor of $3x$, $6y$, and $-3z$, it is a *common factor* for all the terms of the trinomial $3x + 6y - 3z$.

When we factor, we are forming an equivalent expression that is a product.

EXAMPLE **2**   Factor to form an equivalent expression: $10y + 15$.

SOLUTION   We write the prime factorization of both terms to determine any common factors:

The prime factorization of $10y$ is $2 \cdot 5 \cdot y$;
The prime factorization of 15 is $3 \cdot 5$.   } 5 is a common factor.

We "factor out" the common factor 5 using the distributive law:

> We can always check a factorization by multiplying.

$10y + 15 = 5 \cdot 2y + 5 \cdot 3$ Try to do this step mentally.
$= 5(2y + 3)$.   Using the distributive law

Check: $5(2y + 3) = 5 \cdot 2y + 5 \cdot 3 = 10y + 15$.

The factorization of $10y + 15$ is $5(2y + 3)$.   **TRY EXERCISE** 15

We generally factor out the *largest* common factor.

EXAMPLE **3**   Factor: $8a - 12$.

SOLUTION   Lining up common factors in columns can help us determine the largest common factor:

The prime factorization of $8a$ is $2 \cdot 2 \cdot 2 \cdot \quad a$;
The prime factorization of 12 is $2 \cdot 2 \cdot \quad 3$.

Since both factorizations include two factors of 2, the largest common factor is $2 \cdot 2$, or 4:

$8a - 12 = 4 \cdot 2a - 4 \cdot 3$ 4 is a factor of $8a$ and of 12.
$8a - 12 = 4(2a - 3)$.   Try to go directly to this step.

Check: $4(2a - 3) = 4 \cdot 2a - 4 \cdot 3 = 8a - 12$, as expected.

The factorization of $8a - 12$ is $4(2a - 3)$.   **TRY EXERCISE** 17

CAUTION! $2 \cdot 2 \cdot 2a - 2 \cdot 2 \cdot 3$ is a factorization of the *terms* of $8a - 12$ but not of the polynomial itself. The factorization of $8a - 12$ is $4(2a - 3)$.

A common factor may contain a variable.

EXAMPLE **4**

Factor: $24x^5 + 30x^2$.

SOLUTION

The prime factorization of $24x^5$ is $2 \cdot 2 \cdot 2 \cdot 3 \cdot \quad x \cdot x \cdot x \cdot x \cdot x.$
The prime factorization of $30x^2$ is $2 \cdot \qquad 3 \cdot 5 \cdot x \cdot x.$

The largest common factor is $2 \cdot 3 \cdot x \cdot x$, or $6x^2$.

$$24x^5 + 30x^2 = 6x^2 \cdot 4x^3 + 6x^2 \cdot 5 \qquad \text{Factoring each term}$$
$$= 6x^2(4x^3 + 5) \qquad \text{Factoring out } 6x^2$$

Check: $6x^2(4x^3 + 5) = 6x^2 \cdot 4x^3 + 6x^2 \cdot 5 = 24x^5 + 30x^2$, as expected.

The factorization of $24x^5 + 30x^2$ is $6x^2(4x^3 + 5)$. 〉**TRY EXERCISE** 〉27

The largest common factor of a polynomial is the largest common factor of the coefficients times the largest common factor of the variable(s) in all the terms. Suppose in Example 4 that you did not recognize the *largest* common factor, and removed only part of it, as follows:

$$24x^5 + 30x^2 = 2x^2 \cdot 12x^3 + 2x^2 \cdot 15 \qquad 2x^2 \text{ is a common factor.}$$
$$= 2x^2(12x^3 + 15). \qquad 12x^3 + 15 \text{ itself contains a common factor.}$$

Note that $12x^3 + 15$ still has a common factor, 3. To find the largest common factor, continue factoring out common factors, as follows, until no more exist:

$$24x^5 + 30x^2 = 2x^2[3(4x^3 + 5)] \qquad \text{Factoring } 12x^3 + 15. \text{ Remember to rewrite the first common factor, } 2x^2.$$
$$= 6x^2(4x^3 + 5). \qquad \text{Using an associative law; } 2x^2 \cdot 3 = 6x^2$$

Since $4x^3 + 5$ cannot be factored any further, we say that we have factored *completely*. When we are directed simply to factor, it is understood that we should always factor completely.

EXAMPLE **5**

Factor: $12x^5 - 15x^4 + 27x^3$.

SOLUTION

The prime factorization of $12x^5$ is $2 \cdot 2 \cdot 3 \cdot \qquad x \cdot x \cdot x \cdot x \cdot x.$
The prime factorization of $15x^4$ is $\qquad 3 \cdot \qquad 5 \cdot x \cdot x \cdot x \cdot x.$
The prime factorization of $27x^3$ is $\qquad 3 \cdot 3 \cdot 3 \cdot \qquad x \cdot x \cdot x.$

The largest common factor is $3 \cdot x \cdot x \cdot x$, or $3x^3$.

$$12x^5 - 15x^4 + 27x^3 = 3x^3 \cdot 4x^2 - 3x^3 \cdot 5x + 3x^3 \cdot 9$$
$$= 3x^3(4x^2 - 5x + 9)$$

Since $4x^2 - 5x + 9$ has no common factor, we are done, except for a check:

$$3x^3(4x^2 - 5x + 9) = 3x^3 \cdot 4x^2 - 3x^3 \cdot 5x + 3x^3 \cdot 9$$
$$= 12x^5 - 15x^4 + 27x^3,$$

as expected. The factorization of $12x^5 - 15x^4 + 27x^3$ is $3x^3(4x^2 - 5x + 9)$.

〉**TRY EXERCISE** 〉31

Note in Examples 4 and 5 that the *largest* common variable factor is the *smallest* power of x in the original polynomial.

With practice, we can determine the largest common factor without writing the prime factorization of each term. Then, to factor, we write the largest common factor and parentheses and then fill in the parentheses. It is customary for the leading coefficient of the polynomial inside the parentheses to be positive.

EXAMPLE 6

Factor: **(a)** $8r^3s^2 + 16rs^3$; **(b)** $-3xy + 6xz - 3x$.

SOLUTION

a) $8r^3s^2 + 16rs^3 = 8rs^2(r^2 + 2s)$ Try to go directly to this step.

The largest common factor is $8rs^2$. $\longleftarrow$ $\begin{cases} 8r^3s^2 = 2 \cdot 2 \cdot 2 \cdot \quad r \cdot r^2 \cdot s^2 \\ 16rs^3 = 2 \cdot 2 \cdot 2 \cdot 2 \cdot r \cdot \quad s^2 \cdot s \end{cases}$

Check: $8rs^2(r^2 + 2s) = 8r^3s^2 + 16rs^3$.

STUDENT NOTES

The 1 in $(y - 2z + 1)$ plays an important role in Example 6(b). If left out of the factorization, the term $-3x$ would not appear in the check.

b) $-3xy + 6xz - 3x = -3x(y - 2z + 1)$ Note that either $-3x$ or $3x$ can be the largest common factor.

We generally factor out a negative when the first coefficient is negative. The way we factor can depend on the situation in which we are working. We might also factor as follows:

$$-3xy + 6xz - 3x = 3x(-y + 2z - 1).$$

The checks are left to the student.   **TRY EXERCISE** 35

In some texts, the largest common factor is referred to as the *greatest* common factor. We have avoided this language because, as shown in Example 6(b), the largest common factor may represent a negative value that is actually *less* than other common factors.

Tips for Factoring

1. Factor out the largest common factor, if one exists.
2. The common factor multiplies a polynomial with the same number of terms as the original polynomial.
3. Factoring can always be checked by multiplying. Multiplication should yield the original polynomial.

Factoring by Grouping

Sometimes algebraic expressions contain a common factor with two or more terms.

EXAMPLE 7

Factor: $x^2(x + 1) + 2(x + 1)$.

SOLUTION   The binomial $x + 1$ is a factor of both $x^2(x + 1)$ and $2(x + 1)$. Thus, $x + 1$ is a common factor:

$$x^2(x + 1) + 2(x + 1) = (x + 1)x^2 + (x + 1)2$$ Using a commutative law twice. Try to do this step mentally.

$$= (x + 1)(x^2 + 2).$$ Factoring out the common factor, $x + 1$

To check, we could simply reverse the above steps.
The factorization is $(x + 1)(x^2 + 2)$.   **TRY EXERCISE** 37

In Example 7, the common binomial factor was clearly visible. How do we find such a factor in a polynomial like $5x^3 - x^2 + 15x - 3$? Although there is no factor, other than 1, common to all four terms, $5x^3 - x^2$ and $15x - 3$ can be grouped and factored separately:

$$5x^3 - x^2 = x^2(5x - 1) \quad \text{and} \quad 15x - 3 = 3(5x - 1).$$

Note that $5x^3 - x^2$ and $15x - 3$ share a common factor, $5x - 1$. This means that the original polynomial, $5x^3 - x^2 + 15x - 3$, can be factored:

$$5x^3 - x^2 + 15x - 3 = (5x^3 - x^2) + (15x - 3) \qquad \text{Each binomial has a common factor.}$$
$$= x^2(5x - 1) + 3(5x - 1) \qquad \text{Factoring each binomial}$$
$$= (5x - 1)(x^2 + 3). \qquad \text{Factoring out the common factor, } 5x - 1$$

Check:
$$(5x - 1)(x^2 + 3) = 5x \cdot x^2 + 5x \cdot 3 - 1 \cdot x^2 - 1 \cdot 3$$
$$= 5x^3 - x^2 + 15x - 3.$$

If a polynomial can be split into groups of terms and the groups share a common factor, then the original polynomial can be factored. This method, known as **factoring by grouping**, can be tried on any polynomial with four or more terms.

EXAMPLE 8 Factor by grouping.

a) $2x^3 + 8x^2 + x + 4$ **b)** $8x^4 + 6x - 28x^3 - 21$

SOLUTION

a)
$$2x^3 + 8x^2 + x + 4 = (2x^3 + 8x^2) + (x + 4)$$
$$= 2x^2(x + 4) + 1(x + 4) \qquad \text{Factoring } 2x^3 + 8x^2 \text{ to find a common binomial factor. Writing the 1 helps with the next step.}$$
$$= (x + 4)(2x^2 + 1) \qquad \text{Factoring out the common factor, } x + 4. \text{ The 1 is essential in the factor } 2x^2 + 1.$$

> *CAUTION!* Be sure to include the term 1. The check below shows why it is essential.

Check:
$$(x + 4)(2x^2 + 1) = x \cdot 2x^2 + x \cdot 1 + 4 \cdot 2x^2 + 4 \cdot 1 \qquad \text{Using FOIL}$$
$$= 2x^3 + x + 8x^2 + 4$$
$$= 2x^3 + 8x^2 + x + 4. \qquad \text{Using a commutative law}$$

The factorization is $(x + 4)(2x^2 + 1)$.

b) We have a choice of either

$$8x^4 + 6x - 28x^3 - 21 = (8x^4 + 6x) + (-28x^3 - 21)$$
$$= 2x(4x^3 + 3) + 7(-4x^3 - 3) \longleftarrow \text{No common factor}$$

or

$$8x^4 + 6x - 28x^3 - 21 = (8x^4 + 6x) + (-28x^3 - 21)$$
$$= 2x(4x^3 + 3) + (-7)(4x^3 + 3). \longleftarrow \text{Common factor}$$

Because of the common factor $4x^3 + 3$, we choose the latter:

$$8x^4 + 6x - 28x^3 - 21 = 2x(4x^3 + 3) + (-7)(4x^3 + 3)$$
$$= (4x^3 + 3)(2x + (-7)) \qquad \text{Try to do this step mentally.}$$
$$= (4x^3 + 3)(2x - 7). \qquad \text{The common factor } 4x^3 + 3 \text{ was factored out.}$$

Check: $(4x^3 + 3)(2x - 7) = 8x^4 - 28x^3 + 6x - 21$
$$= 8x^4 + 6x - 28x^3 - 21. \qquad \text{This is the original polynomial.}$$

The factorization is $(4x^3 + 3)(2x - 7)$.

> TRY EXERCISE ▸ 43

Although factoring by grouping can be useful, some polynomials, like $x^3 + x^2 + 2x - 2$, cannot be factored this way. Factoring polynomials of this type is beyond the scope of this text.

Checking by Evaluating

One way to check a factorization is to multiply. A second type of check, discussed toward the end of Section 4.5, uses the fact that equivalent expressions have the same value when evaluated for the same replacement. Thus a quick, partial check of Example 8(a) can be made by using a convenient replacement for x (say, 1) and evaluating both $2x^3 + 8x^2 + x + 4$ and $(x + 4)(2x^2 + 1)$:

$$2 \cdot 1^3 + 8 \cdot 1^2 + 1 + 4 = 2 + 8 + 1 + 4$$
$$= 15;$$
$$(1 + 4)(2 \cdot 1^2 + 1) = 5 \cdot 3$$
$$= 15.$$

Since the value of both expressions is the same, the factorization is probably correct.

Keep in mind the possibility that two expressions that are not equivalent may share the same value when evaluated at a certain value. Because of this, unless several values are used (at least one more than the degree of the polynomial, it turns out), evaluating offers only a partial check. Consult with your instructor before making extensive use of this type of check.

TECHNOLOGY CONNECTION

A partial check of a factorization can be performed using a table or a graph. To check Example 8(a), we let

$$y_1 = 2x^3 + 8x^2 + x + 4 \quad \text{and} \quad y_2 = (x + 4)(2x^2 + 1).$$

Then we set up a table in AUTO mode (see p. 88). If the factorization is correct, the values of y_1 and y_2 will be the same regardless of the table settings used.

ΔTBL = 1

X	Y₁	Y₂
0	4	4
1	15	15
2	54	54
3	133	133
4	264	264
5	459	459
6	730	730

X = 0

We can also graph $y_1 = 2x^3 + 8x^2 + x + 4$ and $y_2 = (x + 4)(2x^2 + 1)$. If the graphs appear to coincide, the factorization is probably correct. The TRACE feature can be used to confirm this.

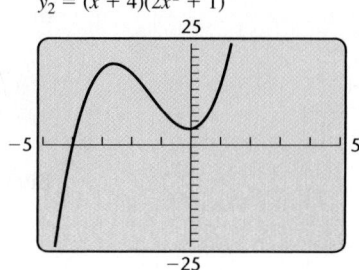

$y_1 = 2x^3 + 8x^2 + x + 4$,
$y_2 = (x + 4)(2x^2 + 1)$

Yscl = 2

Use a table or a graph to determine whether each factorization is correct.

1. $x^2 - 7x - 8 = (x - 8)(x + 1)$
2. $4x^2 - 5x - 6 = (4x + 3)(x - 2)$
3. $5x^2 + 17x - 12 = (5x + 3)(x - 4)$
4. $10x^2 + 37x + 7 = (5x - 1)(2x + 7)$
5. $12x^2 - 17x - 5 = (6x + 1)(2x - 5)$
6. $12x^2 - 17x - 5 = (4x + 1)(3x - 5)$
7. $x^2 - 4 = (x - 2)(x - 2)$
8. $x^2 - 4 = (x + 2)(x - 2)$

5.1 EXERCISE SET

⤴ Concept Reinforcement *In each of Exercises 1–8, match the phrase with the most appropriate choice from the column on the right.*

1. ____ A factorization of $35a^2b$
2. ____ A factor of $35a^2b$
3. ____ A common factor of $5x + 10$ and $4x + 8$
4. ____ A factorization of $3x^4 - 9x^2$
5. ____ A factorization of $9x^4 - 3x^2$
6. ____ A common factor of $2x + 10$ and $4x + 8$
7. ____ A factor of $3a + 6a^2$
8. ____ A factorization of $3a + 6a^2$

a) $3a(1 + 2a)$
b) $x + 2$
c) $3x^2(3x^2 - 1)$
d) $1 + 2a$
e) $3x^2(x^2 - 3)$
f) $5a^2$
g) 2
h) $7a \cdot 5ab$

Find three factorizations for each monomial. Answers may vary.

9. $14x^3$
10. $22x^3$
11. $-15a^4$
12. $-8t^5$
13. $25t^5$
14. $9a^4$

Factor. Remember to use the largest common factor and to check by multiplying. Factor out a negative factor if the first coefficient is negative.

15. $8x + 24$
16. $10x + 50$
17. $6x - 30$
18. $7x - 21$
19. $2x^2 + 2x - 8$
20. $6x^2 + 3x - 15$
21. $3t^2 + t$
22. $2t^2 + t$
23. $-5y^2 - 10y$
24. $-4y^2 - 12y$
25. $x^3 + 6x^2$
26. $5x^4 - x^2$

27. $16a^4 - 24a^2$

28. $25a^5 + 10a^3$

29. $-6t^6 + 9t^4 - 4t^2$

30. $-10t^5 + 15t^4 + 9t^3$

31. $6x^8 + 12x^6 - 24x^4 + 30x^2$

32. $10x^4 - 30x^3 - 50x - 20$

33. $x^5y^5 + x^4y^3 + x^3y^3 - x^2y^2$

34. $x^9y^6 - x^7y^5 + x^4y^4 + x^3y^3$

35. $-35a^3b^4 + 10a^2b^3 - 15a^3b^2$

36. $-21r^5t^4 - 14r^4t^6 + 21r^3t^6$

Factor.

37. $n(n - 6) + 3(n - 6)$

38. $b(b + 5) + 3(b + 5)$

39. $x^2(x + 3) - 7(x + 3)$

40. $3z^2(2z + 9) + (2z + 9)$

41. $y^2(2y - 9) + (2y - 9)$

42. $x^2(x - 7) - 3(x - 7)$

Factor by grouping, if possible, and check.

43. $x^3 + 2x^2 + 5x + 10$

44. $z^3 + 3z^2 + 7z + 21$

45. $5a^3 + 15a^2 + 2a + 6$

46. $3a^3 + 2a^2 + 6a + 4$

47. $9n^3 - 6n^2 + 3n - 2$

48. $10x^3 - 25x^2 + 2x - 5$

49. $4t^3 - 20t^2 + 3t - 15$

50. $8a^3 - 2a^2 + 12a - 3$

51. $7x^3 + 5x^2 - 21x - 15$

52. $5x^3 + 4x^2 - 10x - 8$

53. $6a^3 + 7a^2 + 6a + 7$

54. $7t^3 - 5t^2 + 7t - 5$

55. $2x^3 + 12x^2 - 5x - 30$

56. $x^3 - x^2 - 2x + 5$

57. $p^3 + p^2 - 3p + 10$

58. $w^3 + 7w^2 + 4w + 28$

59. $y^3 + 8y^2 - 2y - 16$

60. $3x^3 + 18x^2 - 5x - 25$

61. $2x^3 - 8x^2 - 9x + 36$

62. $20g^3 - 4g^2 - 25g + 5$

63. In answering a factoring problem, Taylor says the largest common factor is $-5x^2$ and Madison says the largest common factor is $5x^2$. Can they both be correct? Why or why not?

64. Write a two-sentence paragraph in which the word "factor" is used at least once as a noun and once as a verb.

Skill Review

To prepare for Section 5.2, review multiplying binomials using FOIL (Section 4.6).

Multiply. [4.6]

65. $(x + 2)(x + 7)$

66. $(x - 2)(x - 7)$

67. $(x + 2)(x - 7)$

68. $(x - 2)(x + 7)$

69. $(a - 1)(a - 3)$

70. $(t + 3)(t + 5)$

71. $(t - 5)(t + 10)$

72. $(a + 4)(a - 6)$

Synthesis

73. Azrah recognizes that evaluating provides only a partial check of her factoring. Because of this, she often performs a second check with a different replacement value. Is this a good idea? Why or why not?

74. Holly factors $12x^2y - 18xy^2$ as $6xy \cdot 2x - 6xy \cdot 3y$. Is this the factorization of the polynomial? Why or why not?

Factor, if possible.

75. $4x^5 + 6x^2 + 6x^3 + 9$

76. $x^6 + x^2 + x^4 + 1$

77. $2x^4 + 2x^3 - 4x^2 - 4x$

78. $x^3 + x^2 - 2x + 2$

Aha! **79.** $5x^5 - 5x^4 + x^3 - x^2 + 3x - 3$

Aha! **80.** $ax^2 + 2ax + 3a + x^2 + 2x + 3$

81. Write a trinomial of degree 7 for which $8x^2y^3$ is the largest common factor. Answers may vary.

5.2 Factoring Trinomials of the Type $x^2 + bx + c$

When the Constant Term Is Positive ● When the Constant Term Is Negative ●
Prime Polynomials

We now learn how to factor trinomials like

$$x^2 + 5x + 4 \quad \text{or} \quad x^2 + 3x - 10,$$

for which no common factor exists and the leading coefficient is 1. Recall that when factoring, we are writing an equivalent expression that is a product. For these trinomials, the factors will be binomials.

As preparation for the factoring that follows, compare the following multiplications:

$$
\begin{array}{cccc}
\text{F} & \text{O} & \text{I} & \text{L} \\
\downarrow & \downarrow & \downarrow & \downarrow
\end{array}
$$

$$
\begin{aligned}
(x + 2)(x + 5) &= x^2 + 5x + 2x + 2 \cdot 5 \\
&= x^2 + 7x + 10; \\
(x - 2)(x - 5) &= x^2 - 5x - 2x + (-2)(-5) \\
&= x^2 - 7x + 10; \\
(x + 3)(x - 7) &= x^2 - 7x + 3x + 3(-7) \\
&= x^2 - 4x - 21; \\
(x - 3)(x + 7) &= x^2 + 7x - 3x + (-3)7 \\
&= x^2 + 4x - 21.
\end{aligned}
$$

Note that for all four products:

- The product of the two binomials is a trinomial.
- The coefficient of x in the trinomial is the sum of the constant terms in the binomials.
- The constant term in the trinomial is the product of the constant terms in the binomials.

These observations lead to a method for factoring certain trinomials. We first consider trinomials that have a positive constant term, just as in the first two multiplications above.

When the Constant Term Is Positive

To factor a polynomial like $x^2 + 7x + 10$, we think of FOIL in reverse. The x^2 resulted from x times x, which suggests that the first term of each binomial factor is x. Next, we look for numbers p and q such that

$$x^2 + 7x + 10 = (x + p)(x + q).$$

To get the middle term and the last term of the trinomial, we need two numbers, p and q, whose product is 10 and whose sum is 7. Those numbers are 2 and 5. Thus the factorization is

$$(x + 2)(x + 5).$$

Check: $(x + 2)(x + 5) = x^2 + 5x + 2x + 10$
$$= x^2 + 7x + 10.$$

EXAMPLE 1

A GEOMETRIC APPROACH TO EXAMPLE 1

In Section 4.5, we saw that the product of two binomials can be regarded as the sum of the areas of four rectangles (see p. 264). Thus we can regard the factoring of $x^2 + 5x + 6$ as a search for p and q so that the sum of areas A, B, C, and D is $x^2 + 5x + 6$.

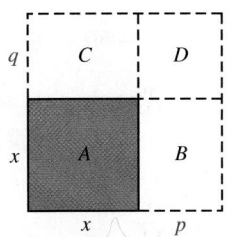

Note that area D is simply the product of p and q. In order for area D to be 6, p and q must be either 1 and 6 or 2 and 3. We illustrate both below.

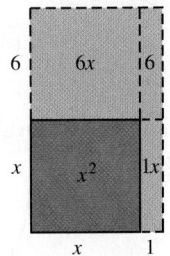 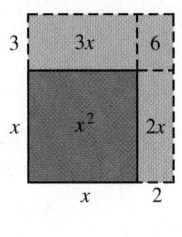

When p and q are 1 and 6, the total area is $x^2 + 7x + 6$, but when p and q are 2 and 3, as shown on the right, the total area is $x^2 + 5x + 6$, as desired. Thus the factorization of $x^2 + 5x + 6$ is $(x + 2)(x + 3)$.

Factor to form an equivalent expression: $x^2 + 5x + 6$.

SOLUTION Think of FOIL in reverse. The first term of each factor is x:

$$(x + \quad)(x + \quad).$$

To complete the factorization, we need a constant term for each binomial. The constants must have a product of 6 and a sum of 5. We list some pairs of numbers that multiply to 6 and then check the sum of each pair of factors.

Pairs of Factors of 6	Sums of Factors
1, 6	7
2, 3	5 ← The numbers we seek are 2 and 3.
−1, −6	−7
−2, −3	−5

Every pair has a product of 6.   *One* pair has a sum of 5.

Since

$$2 \cdot 3 = 6 \quad \text{and} \quad 2 + 3 = 5,$$

the factorization of $x^2 + 5x + 6$ is $(x + 2)(x + 3)$.

Check: $(x + 2)(x + 3) = x^2 + 3x + 2x + 6$
$$= x^2 + 5x + 6.$$

Thus, $(x + 2)(x + 3)$ is a product that is equivalent to $x^2 + 5x + 6$.

Note that since 5 and 6 are both positive, when factoring $x^2 + 5x + 6$ we need not consider negative factors of 6. Note too that changing the signs of the factors changes only the sign of the sum (see the table above).

TRY EXERCISE 7

At the beginning of this section, we considered the multiplication $(x - 2)(x - 5)$. For this product, the resulting trinomial, $x^2 - 7x + 10$, has a positive constant term but a negative coefficient of x. This is because the *product* of two negative numbers is always positive, whereas the *sum* of two negative numbers is always negative.

To Factor $x^2 + bx + c$ When c Is Positive

When the constant term c of a trinomial is positive, look for two numbers with the same sign. Select pairs of numbers with the sign of b, the coefficient of the middle term.

$$x^2 - 7x + 10 = (x - 2)(x - 5);$$

$$x^2 + 7x + 10 = (x + 2)(x + 5)$$

EXAMPLE **2**

Factor: $y^2 - 8y + 12$.

SOLUTION Since the constant term is positive and the coefficient of the middle term is negative, we look for a factorization of 12 in which both factors are negative. Their sum must be -8.

Pairs of Factors of 12	Sums of Factors
$-1, -12$	-13
$-2,\ -6$	-8 ← We need a sum of -8.
$-3,\ -4$	-7

The numbers we need are -2 and -6.

STUDENT NOTES

It is important to be able to list *all* the pairs of factors of a number. See Example 1 on p. 21 for an organized approach for listing pairs of factors.

The factorization of $y^2 - 8y + 12$ is $(y - 2)(y - 6)$. The check is left to the student.

TRY EXERCISE 13

When the Constant Term Is Negative

As we saw in two of the multiplications at the start of this section, the product of two binomials can have a negative constant term:

$$(x + 3)(x - 7) = x^2 - 4x - 21$$

and

$$(x - 3)(x + 7) = x^2 + 4x - 21.$$

It is important to note that when the signs of the constants in the binomials are reversed, only the sign of the middle term of the trinomial changes.

EXAMPLE **3**

Factor: $x^2 - 8x - 20$.

SOLUTION The constant term, -20, must be expressed as the product of a negative number and a positive number. Since the sum of these two numbers must be negative (specifically, -8), the negative number must have the greater absolute value.

Pairs of Factors of -20	Sums of Factors
$1, -20$	-19
$2, -10$	-8 ← The numbers we need are 2 and -10.
$4,\ -5$	-1
$5,\ -4$	1
$10,\ -2$	8
$20,\ -1$	19

Because in these three pairs, the positive number has the greater absolute value, these sums are all positive. For this problem, these pairs can be eliminated even before calculating the sum.

The numbers that we are looking for are 2 and -10.

Check: $(x + 2)(x - 10) = x^2 - 10x + 2x - 20$
$$= x^2 - 8x - 20.$$

The factorization of $x^2 - 8x - 20$ is $(x + 2)(x - 10)$.

TRY EXERCISE 21

> **To Factor $x^2 + bx + c$ When c Is Negative**
>
> When the constant term c of a trinomial is negative, look for a positive number and a negative number that multiply to c. Select pairs of numbers for which the number with the larger absolute value has the same sign as b, the coefficient of the middle term.
>
> $$x^2 - 4x - 21 = (x + 3)(x - 7);$$
>
> $$x^2 + 4x - 21 = (x - 3)(x + 7)$$

EXAMPLE   4

Factor: $t^2 - 24 + 5t$.

SOLUTION   It helps to first write the trinomial in descending order: $t^2 + 5t - 24$. The factorization of the constant term, -24, must have one factor positive and one factor negative. The sum must be 5, so the positive factor must have the larger absolute value. Thus we consider only pairs of factors in which the positive factor has the larger absolute value.

Pairs of Factors of -24	Sums of Factors
$-1, 24$	23
$-2, 12$	10
$-3,\ 8$	5 ←
$-4,\ 6$	2

The numbers we need are -3 and 8.

The factorization is $(t - 3)(t + 8)$. The check is left to the student.

TRY EXERCISE 23

Polynomials in two or more variables, such as $a^2 + 4ab - 21b^2$, are factored in a similar manner.

EXAMPLE   5

Factor: $a^2 + 4ab - 21b^2$.

SOLUTION   It may help to write the trinomial in the equivalent form

$$a^2 + 4ba - 21b^2.$$

We now regard $-21b^2$ as the "constant" term and $4b$ as the "coefficient" of a. Then we try to express $-21b^2$ as a product of two factors whose sum is $4b$. Those factors are $-3b$ and $7b$.

Check: $(a - 3b)(a + 7b) = a^2 + 7ab - 3ba - 21b^2$
$$= a^2 + 4ab - 21b^2.$$

The factorization of $a^2 + 4ab - 21b^2$ is $(a - 3b)(a + 7b)$.

TRY EXERCISE 55

Prime Polynomials

EXAMPLE **6**

Factor: $x^2 - x + 5$.

SOLUTION Since 5 has very few factors, we can easily check all possibilities.

Pairs of Factors of 5	Sums of Factors
5, 1	6
−5, −1	−6

Since there are no factors whose sum is −1, the polynomial is *not* factorable into binomials.

> **TRY EXERCISE** 37

In this text, a polynomial like $x^2 - x + 5$ that cannot be factored further is said to be **prime**. In more advanced courses, other types of numbers are considered. There, polynomials like $x^2 - x + 5$ can be factored and are not considered prime.

Often factoring requires two or more steps. Remember, when told to factor, we should *factor completely*. This means that the final factorization should contain only prime polynomials.

EXAMPLE **7**

Factor: $-2x^3 + 20x^2 - 50x$.

SOLUTION *Always* look first for a common factor. Since the leading coefficient is negative, we begin by factoring out $-2x$:

$$-2x^3 + 20x^2 - 50x = -2x(x^2 - 10x + 25).$$

Now consider $x^2 - 10x + 25$. Since the constant term is positive and the coefficient of the middle term is negative, we look for a factorization of 25 in which both factors are negative. Their sum must be −10.

Pairs of Factors of 25	Sums of Factors	
−25, −1	−26	
−5, −5	−10 ←	The numbers we need are −5 and −5.

The factorization of $x^2 - 10x + 25$ is $(x - 5)(x - 5)$, or $(x - 5)^2$.

> **CAUTION!** When factoring involves more than one step, be careful to write out the *entire* factorization.

Check: $-2x(x - 5)(x - 5) = -2x[x^2 - 10x + 25]$ Multiplying binomials
$$= -2x^3 + 20x^2 - 50x. \quad \text{Using the distributive law}$$

The factorization of $-2x^3 + 20x^2 - 50x$ is $-2x(x - 5)(x - 5)$, or $-2x(x - 5)^2$.

> **TRY EXERCISE** 27

Once any common factors have been factored out, the following summary can be used to factor $x^2 + bx + c$.

STUDENT NOTES ────────

Whenever a new set of parentheses is created while factoring, check the expression inside the parentheses to see if it can be factored further.

> **To Factor $x^2 + bx + c$**
> 1. Find a pair of factors that have c as their product and b as their sum.
>   a) If c is positive, both factors will have the same sign as b.
>   b) If c is negative, one factor will be positive and the other will be negative. Select the factors such that the factor with the larger absolute value has the same sign as b.
> 2. Check by multiplying.

Note that each polynomial has a unique factorization (except for the order in which the factors are written).

5.2 EXERCISE SET

🐦 *Concept Reinforcement   For Exercises 1–6, assume that a polynomial of the form $x^2 + bx + c$ can be factored as $(x + p)(x + q)$. Complete each sentence by replacing each blank with either "positive" or "negative."*

1. If b is positive and c is positive, then p will be _____ and q will be _____.

2. If b is negative and c is positive, then p will be _____ and q will be _____.

3. If p is negative and q is negative, then b must be _____ and c must be _____.

4. If p is positive and q is positive, then b must be _____ and c must be _____.

5. If b, c, and p are all negative, then q must be _____.

6. If b and c are negative and p is positive, then q must be _____.

Factor completely. Remember to look first for a common factor. Check by multiplying. If a polynomial is prime, state this.

7. $x^2 + 8x + 16$
8. $x^2 + 9x + 20$
9. $x^2 + 11x + 10$
10. $y^2 + 8y + 7$
11. $x^2 + 10x + 21$
12. $x^2 + 6x + 9$
13. $t^2 - 9t + 14$
14. $a^2 - 9a + 20$
15. $b^2 - 5b + 4$
16. $x^2 - 10x + 25$
17. $a^2 - 7a + 12$
18. $z^2 - 8z + 7$
19. $d^2 - 7d + 10$
20. $x^2 - 8x + 15$
21. $x^2 - 2x - 15$
22. $x^2 - x - 42$

23. $x^2 + 2x - 15$
24. $x^2 + x - 42$
25. $2x^2 - 14x - 36$
26. $3y^2 - 9y - 84$
27. $-x^3 + 6x^2 + 16x$
28. $-x^3 + x^2 + 42x$
29. $4y - 45 + y^2$
30. $7x - 60 + x^2$
31. $x^2 - 72 + 6x$
32. $-2x - 99 + x^2$
33. $-5b^2 - 35b + 150$
34. $-c^4 - c^3 + 56c^2$
35. $x^5 - x^4 - 2x^3$
36. $2a^2 - 4a - 70$
37. $x^2 + 5x + 10$
38. $x^2 + 11x + 18$
39. $32 + 12t + t^2$
40. $y^2 - y + 1$
41. $x^2 + 20x + 99$
42. $x^2 + 20x + 100$
43. $3x^3 - 63x^2 - 300x$
44. $2x^3 - 40x^2 + 192x$
45. $-2x^2 + 42x + 144$
46. $-4x^2 - 40x - 100$
47. $y^2 - 20y + 96$
48. $144 - 25t + t^2$
49. $-a^6 - 9a^5 + 90a^4$
50. $-a^4 - a^3 + 132a^2$
51. $t^2 + \frac{2}{3}t + \frac{1}{9}$
52. $x^2 - \frac{2}{5}x + \frac{1}{25}$
53. $11 + w^2 - 4w$
54. $6 + p^2 + 2p$
55. $p^2 - 7pq + 10q^2$
56. $a^2 - 2ab - 3b^2$
57. $m^2 + 5mn + 5n^2$
58. $x^2 - 11xy + 24y^2$
59. $s^2 - 4st - 12t^2$
60. $b^2 + 8bc - 20c^2$
61. $6a^{10} + 30a^9 - 84a^8$
62. $5a^8 - 20a^7 - 25a^6$

📝 63. Without multiplying $(x - 17)(x - 18)$, explain why it cannot possibly be a factorization of $x^2 + 35x + 306$.

📝 64. Shari factors $x^3 - 8x^2 + 15x$ as $(x^2 - 5x)(x - 3)$. Is she wrong? Why or why not? What advice would you offer?

Skill Review

To prepare for Section 5.3, review multiplying binomials using FOIL (Section 4.6).

Multiply. [4.6]

65. $(2x + 3)(3x + 4)$ **66.** $(2x + 3)(3x - 4)$

67. $(2x - 3)(3x + 4)$ **68.** $(2x - 3)(3x - 4)$

69. $(5x - 1)(x - 7)$ **70.** $(x + 6)(3x - 5)$

Synthesis

71. When searching for a factorization, why do we list pairs of numbers with the correct *product* instead of pairs of numbers with the correct *sum*?

72. When factoring $x^2 + bx + c$ with a large value of c, Riley begins by writing out the prime factorization of c. What is the advantage of doing this?

73. Find all integers b for which $a^2 + ba - 50$ can be factored.

74. Find all integers m for which $y^2 + my + 50$ can be factored.

Factor completely.

75. $y^2 - 0.2y - 0.08$ **76.** $x^2 + \frac{1}{2}x - \frac{3}{16}$

77. $-\frac{1}{3}a^3 + \frac{1}{3}a^2 + 2a$ **78.** $-a^7 + \frac{25}{7}a^5 + \frac{30}{7}a^6$

79. $x^{2m} + 11x^m + 28$ **80.** $-t^{2n} + 7t^n - 10$

Aha! **81.** $(a + 1)x^2 + (a + 1)3x + (a + 1)2$

82. $ax^2 - 5x^2 + 8ax - 40x - (a - 5)9$
(*Hint*: See Exercise 81.)

83. Find the volume of a cube if its surface area is $6x^2 + 36x + 54$ square meters.

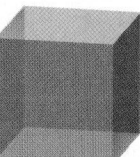

Find a polynomial in factored form for the shaded area in each figure. (Use π in your answers where appropriate.)

84.

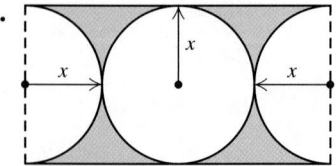

85.

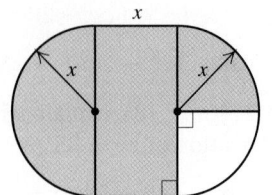

86.

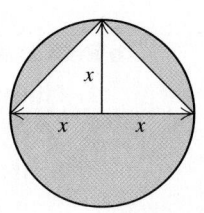

87.

88.

89.

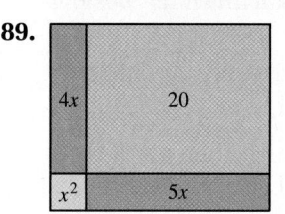

90. A census taker asks a woman, "How many children do you have?"

"Three," she answers.

"What are their ages?"

She responds, "The product of their ages is 36. The sum of their ages is the house number next door."

The math-savvy census taker walks next door, reads the house number, appears puzzled, and returns to the woman, asking, "Is there something you forgot to tell me?"

"Oh yes," says the woman. "I'm sorry. The oldest child is at the park."

The census taker records the three ages, thanks the woman for her time, and leaves.

How old is each child? Explain how you reached this conclusion. (*Hint*: Consider factorizations.)
Source: Adapted from Anita Harnadek, *Classroom Quickies*. Pacific Grove, CA: Critical Thinking Press and Software

CORNER

Visualizing Factoring

Focus: Visualizing factoring

Time: 20–30 minutes

Group size: 3

Materials: Graph paper and scissors

The product $(x + 2)(x + 3)$ can be regarded as the area of a rectangle with width $x + 2$ and length $x + 3$. Similarly, factoring a polynomial like $x^2 + 5x + 6$ can be thought of as determining the length and the width of a rectangle that has area $x^2 + 5x + 6$. This is the approach used below.

ACTIVITY

1. **a)** To factor $x^2 + 11x + 10$ geometrically, the group needs to cut out shapes like those below to represent x^2, $11x$, and 10. This can be done by either tracing the figures below or by selecting a value for x, say 4, and using the squares on the graph paper to cut out the following:

    x^2: Using the value selected for x, cut out a square that is x units on each side.

    $11x$: Using the value selected for x, cut out a rectangle that is 1 unit wide and x units long. Repeat this to form 11 such strips.

    10: Cut out two rectangles with whole-number dimensions and an area of 10. One should be 2 units by 5 units and the other 1 unit by 10 units.

    b) The group, working together, should then attempt to use one of the two rectangles with area 10, along with all of the other shapes, to piece together one large rectangle. Only one of the rectangles with area 10 will work.

    c) From the large rectangle formed in part (b), use the length and the width to determine the factorization of $x^2 + 11x + 10$. Where do the dimensions of the rectangle representing 10 appear in the factorization?

2. Repeat step (1) above, but this time use the other rectangle with area 10, and use only 7 of the 11 strips, along with the x^2-shape. Piece together the shapes to form one large rectangle. What factorization do the dimensions of this rectangle suggest?

3. Cut out rectangles with area 12 and use the above approach to factor $x^2 + 8x + 12$. What dimensions should be used for the rectangle with area 12?

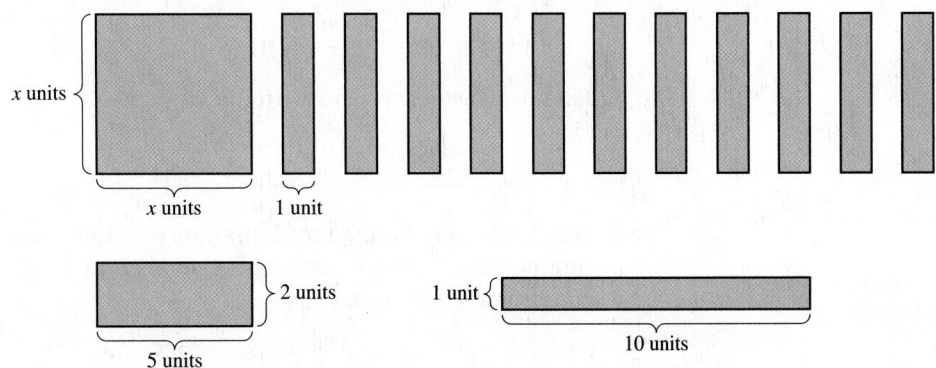

5.3 Factoring Trinomials of the Type $ax^2 + bx + c$

Factoring with FOIL ■ The Grouping Method

In Section 5.2, we learned a FOIL-based method for factoring trinomials of the type $x^2 + bx + c$. Now we learn to factor trinomials in which the leading, or x^2, coefficient is not 1. First we will use another FOIL-based method and then we will use an alternative method that involves factoring by grouping. Use the method that you prefer or the one recommended by your instructor.

Factoring with FOIL

Before factoring trinomials of the type $ax^2 + bx + c$, consider the following:

$$\overset{\text{F}\qquad\ \text{O}\qquad\ \ \text{I}\qquad\ \text{L}}{(2x + 5)(3x + 4) = 6x^2 + 8x + 15x + 20}$$
$$= 6x^2 +\quad 23x\quad + 20.$$

To factor $6x^2 + 23x + 20$, we could reverse the multiplication and look for two binomials whose product is this trinomial. We see from the multiplication above that:

- the product of the First terms must be $6x^2$;
- the product of the Outer terms plus the product of the Inner terms must be $23x$; and
- the product of the Last terms must be 20.

How can such a factorization be found without first seeing the corresponding multiplication? Our first approach relies on trial and error and FOIL.

To Factor $ax^2 + bx + c$ Using FOIL

1. Make certain that all common factors have been removed. If any remain, factor out the largest common factor.
2. Find two First terms whose product is ax^2:

$$(\blacksquare x +\quad)(\blacksquare x +\quad) = ax^2 + bx + c.$$
$$\underset{\text{FOIL}}{\underline{\qquad\qquad\qquad}}$$

3. Find two Last terms whose product is c:

$$(\ x + \blacksquare)(\ x + \blacksquare) = ax^2 + bx + c.$$
$$\underset{\text{FOIL}}{\underline{\qquad\qquad\qquad}}$$

4. Check by multiplying to see if the sum of the Outer and Inner products is bx. If necessary, repeat steps 2 and 3 until the correct combination is found.

$$(\blacksquare x + \blacksquare)(\blacksquare x + \blacksquare) = ax^2 + bx + c.$$
$$\underset{\text{O}}{\underset{\text{I}}{\underline{\qquad\qquad\qquad}}} \qquad\qquad \underset{\text{FOIL}}{\underline{\quad}}$$

If no correct combination exists, state that the polynomial is prime.

EXAMPLE **1**   Factor: $3x^2 - 10x - 8$.

SOLUTION

1. First, check for a common factor. In this case, there is none (other than 1 or -1).

2. Find two **First** terms whose product is $3x^2$.

The only possibilities for the **First** terms are $3x$ and x:

$(3x +\ )(x +\ )$.

3. Find two **Last** terms whose product is -8. There are four pairs of factors of -8 and each can be listed in two ways:

$$
\begin{array}{ccc}
-1,\ \ 8 & & 8, -1 \\
1, -8 & \text{and} & -8,\ \ 1 \\
-2,\ \ 4 & & 4, -2 \\
2, -4 & & -4,\ \ 2.
\end{array}
$$

Important! Since the First terms are not identical, we must consider the pairs of factors in both orders.

4. Knowing that all **First** and **Last** products will check, systematically inspect the **O**uter and **I**nner products resulting from steps (2) and (3). Look for the combination in which the sum of the products is the middle term, $-10x$. Our search ends as soon as the correct combination is found. If none exists, we state that the polynomial is prime.

Pair of Factors	*Corresponding Trial*	*Product*	
$-1,\ \ 8$	$(3x - 1)(x + 8)$	$3x^2 + 24x - x - 8$ $= 3x^2 + 23x - 8$	Wrong middle term
$1, -8$	$(3x + 1)(x - 8)$	$3x^2 - 24x + x - 8$ $= 3x^2 - 23x - 8$	Wrong middle term
$-2,\ \ 4$	$(3x - 2)(x + 4)$	$3x^2 + 12x - 2x - 8$ $= 3x^2 + 10x - 8$	Wrong middle term
$2, -4$	$(3x + 2)(x - 4)$	$3x^2 - 12x + 2x - 8$ $= 3x^2 - 10x - 8$	Correct middle term!
$8, -1$	$(3x + 8)(x - 1)$	$3x^2 - 3x + 8x - 8$ $= 3x^2 + 5x - 8$	Wrong middle term
$-8,\ \ 1$	$(3x - 8)(x + 1)$	$3x^2 + 3x - 8x - 8$ $= 3x^2 - 5x - 8$	Wrong middle term
$4, -2$	$(3x + 4)(x - 2)$	$3x^2 - 6x + 4x - 8$ $= 3x^2 - 2x - 8$	Wrong middle term
$-4,\ \ 2$	$(3x - 4)(x + 2)$	$3x^2 + 6x - 4x - 8$ $= 3x^2 + 2x - 8$	Wrong middle term

The correct factorization is $(3x + 2)(x - 4)$. ⬅

TRY EXERCISE 5

Two observations can be made from Example 1. First, we listed all possible trials even though we generally stop after finding the correct factorization. We did this to show that **each trial differs only in the middle term of the product**. Second, note that as in Section 5.2, **only the sign of the middle term changes when the signs in the binomials are reversed**.

EXAMPLE 2   Factor: $10x^2 + 37x + 7$.

SOLUTION

1. There is no factor (other than 1 or -1) common to all three terms.

2. Because $10x^2$ factors as $10x \cdot x$ or $5x \cdot 2x$, we have two possibilities:

   $$(10x +)(x +) \quad \text{or} \quad (5x +)(2x +).$$

3. There are two pairs of factors of 7 and each can be listed in two ways:

$$\begin{matrix} 1, & 7 \\ -1, & -7 \end{matrix} \quad \text{and} \quad \begin{matrix} 7, & 1 \\ -7, & -1. \end{matrix}$$

4. Look for **O**uter and **I**nner products for which the sum is the middle term. Because all coefficients in $10x^2 + 37x + 7$ are positive, we need consider only those combinations involving positive factors of 7.

Trial	*Product*	
$(10x + 1)(x + 7)$	$10x^2 + 70x + 1x + 7$ $= 10x^2 + 71x + 7$	Wrong middle term
$(10x + 7)(x + 1)$	$10x^2 + 10x + 7x + 7$ $= 10x^2 + 17x + 7$	Wrong middle term
$(5x + 7)(2x + 1)$	$10x^2 + 5x + 14x + 7$ $= 10x^2 + 19x + 7$	Wrong middle term
$(5x + 1)(2x + 7)$	$10x^2 + 35x + 2x + 7$ $= 10x^2 + 37x + 7$	Correct middle term!

The correct factorization is $(5x + 1)(2x + 7)$.

> TRY EXERCISE   9

A GEOMETRIC APPROACH TO EXAMPLE 2

The factoring of $10x^2 + 37x + 7$ can be regarded as a search for *r* and *s* so that the sum of areas *A*, *B*, *C*, and *D* is $10x^2 + 37x + 7$.

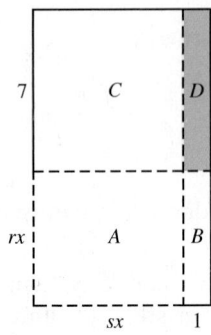

Because *A* must be $10x^2$, the product *rs* must be 10. Only when *r* is 2 and *s* is 5 will the sum of areas *B* and *C* be $37x$ (see below).

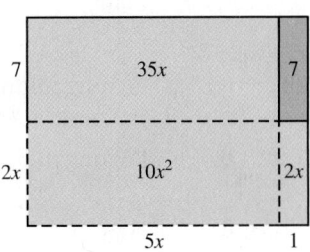

EXAMPLE 3   Factor: $24x^3 - 76x^2 + 40x$.

SOLUTION

1. First, we factor out the largest common factor, $4x$:

   $$4x(6x^2 - 19x + 10).$$

2. Next, we factor $6x^2 - 19x + 10$. Since $6x^2$ can be factored as $3x \cdot 2x$ or $6x \cdot x$, we have two possibilities:

   $$(3x +)(2x +) \quad \text{or} \quad (6x +)(x +).$$

3. There are four pairs of factors of 10 and each can be listed in two ways:

$$\begin{matrix} 1, & 10 \\ -1, & -10 \\ 2, & 5 \\ -2, & -5 \end{matrix} \quad \text{and} \quad \begin{matrix} 10, & 1 \\ -10, & -1 \\ 5, & 2 \\ -5, & -2. \end{matrix}$$

4. The two possibilities from step (2) and the eight possibilities from step (3) give $2 \cdot 8$, or 16 possibilities for factorizations. With careful consideration,

we can eliminate some possibilities without multiplying. Since the sign of the middle term, $-19x$, is negative, but the sign of the last term, 10, is positive, the two factors of 10 must both be negative. This means only four pairings from step (3) need be considered. We first try these factors with $(3x + \quad)(2x + \quad)$, looking for **O**uter and **I**nner products for which the sum is $-19x$. If none gives the correct factorization, then we will consider $(6x + \quad)(x + \quad)$.

Trial	*Product*	
$(3x - 1)(2x - 10)$	$6x^2 - 30x - 2x + 10$	
	$= 6x^2 - 32x + 10$	Wrong middle term
$(3x - 10)(2x - 1)$	$6x^2 - 3x - 20x + 10$	
	$= 6x^2 - 23x + 10$	Wrong middle term
$(3x - 2)(2x - 5)$	$6x^2 - 15x - 4x + 10$	
	$= 6x^2 - 19x + 10$	Correct middle term!
$(3x - 5)(2x - 2)$	$6x^2 - 6x - 10x + 10$	
	$= 6x^2 - 16x + 10$	Wrong middle term

Since we have a correct factorization, we need not consider

$$(6x + \quad)(x + \quad).$$

Look again at the possibility $(3x - 5)(2x - 2)$. Without multiplying, we can reject such a possibility. To see why, note that

$$(3x - 5)(2x - 2) = (3x - 5)2(x - 1).$$

The expression $2x - 2$ has a common factor, 2. But we removed the *largest* common factor in step (1). If $2x - 2$ were one of the factors, then 2 would be *another* common factor in addition to the original, $4x$. Thus, $(2x - 2)$ cannot be part of the factorization of $6x^2 - 19x + 10$. Similar reasoning can be used to reject $(3x - 1)(2x - 10)$ as a possible factorization.

Once the largest common factor is factored out, none of the remaining factors can have a common factor.

The factorization of $6x^2 - 19x + 10$ is $(3x - 2)(2x - 5)$, but do not forget the common factor! The factorization of $24x^3 - 76x^2 + 40x$ is

$$4x(3x - 2)(2x - 5).$$

▸ **TRY EXERCISE** 15

STUDENT NOTES

Keep your work organized so that you can see what you have already considered. For example, when factoring $6x^2 - 19x + 10$, we can list all possibilities and cross out those in which a common factor appears:

$\cancel{(3x - 1)(2x - 10)}$
$(3x - 10)(2x - 1)$
$(3x - 2)(2x - 5)$
$\cancel{(3x - 5)(2x - 2)}$
$(6x - 1)(x - 10)$
$\cancel{(6x - 10)(x - 1)}$
$\cancel{(6x - 2)(x - 5)}$
$(6x - 5)(x - 2)$

By being organized and not erasing, we can see that there are only four possible factorizations.

Tips for Factoring $ax^2 + bx + c$

To factor $ax^2 + bx + c$ $(a > 0)$:

- Make sure that any common factor has been factored out.
- Once the largest common factor has been factored out of the original trinomial, no binomial factor can contain a common factor (other than 1 or -1).
- If c is positive, then the signs in both binomial factors must match the sign of b.
- Reversing the signs in the binomials reverses the sign of the middle term of their product.
- Organize your work so that you can keep track of which possibilities you have checked.
- Remember to include the largest common factor—if there is one—in the final factorization.
- Always check by multiplying.

EXAMPLE **4**

Factor: $10x + 8 - 3x^2$.

SOLUTION   An important problem-solving strategy is to find a way to make new problems look like problems we already know how to solve. The factoring tips above apply only to trinomials of the form $ax^2 + bx + c$, with $a > 0$. This leads us to rewrite $10x + 8 - 3x^2$ in descending order:

$$10x + 8 - 3x^2 = -3x^2 + 10x + 8.$$ Using the commutative law to write descending order

Although $-3x^2 + 10x + 8$ looks similar to the trinomials we have factored, the tips above require a positive leading coefficient. This can be found by factoring out -1:

$$-3x^2 + 10x + 8 = -1(3x^2 - 10x - 8)$$ Factoring out -1 changes the signs of the coefficients.

$$= -1(3x + 2)(x - 4).$$ Using the result from Example 1

The factorization of $10x + 8 - 3x^2$ is $-1(3x + 2)(x - 4)$.   **TRY EXERCISE** 31

EXAMPLE **5**

Factor: $6r^2 - 13rs - 28s^2$.

SOLUTION   In order for the product of the first terms to be $6r^2$ and the product of the last terms to be $-28s^2$, the binomial factors will be of the form

$$(\blacksquare r + \blacksquare s)(\blacksquare r + \blacksquare s).$$

We verify that no common factor exists and then examine the first term, $6r^2$. There are two possibilities:

$$(2r + \ )(3r + \ ) \quad \text{or} \quad (6r + \ )(r + \ ).$$

The last term, $-28s^2$, has the following pairs of factors:

$s, -28s$		$-28s, \quad s$
$-s, \quad 28s$		$28s, \quad -s$
$2s, -14s$	and	$-14s, \quad 2s$
$-2s, \quad 14s$		$14s, -2s$
$4s, \quad -7s$		$-7s, \quad 4s$
$-4s, \quad 7s$		$7s, -4s.$

Note that listing the pairs of factors of $-28s^2$ is just like listing the pairs of factors of -28, except that each factor also contains a factor of s.

Some trials, like $(2r + 28s)(3r - s)$ and $(2r + 14s)(3r - 2s)$, cannot be correct because both $(2r + 28s)$ and $(2r + 14s)$ contain a common factor, 2. We try $(2r + 7s)(3r - 4s)$:

$$(2r + 7s)(3r - 4s) = 6r^2 - 8rs + 21rs - 28s^2$$
$$= 6r^2 + 13rs - 28s^2.$$

Our trial is incorrect, but only because of the sign of the middle term. To correctly factor $6r^2 - 13rs - 28s^2$, we simply change the signs in the binomials:

$$(2r - 7s)(3r + 4s) = 6r^2 + 8rs - 21rs - 28s^2$$
$$= 6r^2 - 13rs - 28s^2.$$

The correct factorization of $6r^2 - 13rs - 28s^2$ is $(2r - 7s)(3r + 4s)$.

TRY EXERCISE 67

The Grouping Method

Another method of factoring trinomials of the type $ax^2 + bx + c$ is known as the *grouping method*. The grouping method relies on rewriting $ax^2 + bx + c$ in the form $ax^2 + px + qx + c$ and then factoring by grouping. To develop this method, consider the following*:

$$(2x + 5)(3x + 4) = 2x \cdot 3x + 2x \cdot 4 + 5 \cdot 3x + 5 \cdot 4 \qquad \text{Using FOIL}$$
$$= 2 \cdot 3 \cdot x^2 + 2 \cdot 4x + 5 \cdot 3x + 5 \cdot 4$$
$$= 2 \cdot 3 \cdot x^2 + (2 \cdot 4 + 5 \cdot 3)x + 5 \cdot 4$$

$$\qquad\qquad\quad a \qquad\qquad\qquad b \qquad\qquad\quad c$$

$$= 6x^2 \quad + \quad 23x \quad + \quad 20.$$

Note that reversing these steps shows that $6x^2 + 23x + 20$ can be rewritten as $6x^2 + 8x + 15x + 20$ and then factored by grouping. Note that the numbers that add to b (in this case, $2 \cdot 4$ and $5 \cdot 3$) also multiply to ac (in this case, $2 \cdot 3 \cdot 5 \cdot 4$).

To Factor $ax^2 + bx + c$, Using the Grouping Method

1. Factor out the largest common factor, if one exists.
2. Multiply the leading coefficient a and the constant c.
3. Find a pair of factors of ac whose sum is b.
4. Rewrite the middle term, bx, as a sum or a difference using the factors found in step (3).
5. Factor by grouping.
6. Include any common factor from step (1) and check by multiplying.

EXAMPLE **6**   Factor: $3x^2 - 10x - 8$.

SOLUTION

1. First, we note that there is no common factor (other than 1 or -1).

2. We multiply the leading coefficient, 3, and the constant, -8:

   $$3(-8) = -24.$$

3. We next look for a factorization of -24 in which the sum of the factors is the coefficient of the middle term, -10.

Pairs of Factors of -24	Sums of Factors
1, -24	-23
-1, 24	23
2, -12	-10 ← $\qquad$ 2 + (−12) = −10
-2, 12	10
3, -8	-5
-3, 8	5
4, -6	-2
-4, 6	2

We normally stop listing pairs of factors once we have found the one we are after.

---

*This discussion was inspired by a lecture given by Irene Doo at Austin Community College.

4. Next, we express the middle term as a sum or a difference using the factors found in step (3):

$$-10x = 2x - 12x.$$

5. We now factor by grouping as follows:

$$3x^2 - 10x - 8 = 3x^2 + 2x - 12x - 8$$

Substituting $2x - 12x$ for $-10x$. We could also use $-12x + 2x$.

$$= x(3x + 2) - 4(3x + 2)$$

Factoring by grouping; see Section 5.1

$$= (3x + 2)(x - 4).$$

Factoring out the common factor, $3x + 2$

6. *Check:* $(3x + 2)(x - 4) = 3x^2 - 12x + 2x - 8 = 3x^2 - 10x - 8.$

The factorization of $3x^2 - 10x - 8$ is $(3x + 2)(x - 4)$. **TRY EXERCISE** 51

EXAMPLE **7**   Factor: $8x^3 + 22x^2 - 6x$.

SOLUTION

1. We factor out the largest common factor, $2x$:

$$8x^3 + 22x^2 - 6x = 2x(4x^2 + 11x - 3).$$

2. To factor $4x^2 + 11x - 3$ by grouping, we multiply the leading coefficient, 4, and the constant term, -3:

$$4(-3) = -12.$$

3. We next look for factors of -12 that add to 11.

Pairs of Factors of -12	Sums of Factors
1, -12	-11
-1, 12	11 ←
.	.
.	.
.	.

Since $-1 + 12 = 11$, there is no need to list other pairs of factors.

4. We then rewrite the $11x$ in $4x^2 + 11x - 3$ using

$$11x = -1x + 12x, \quad \text{or} \quad 11x = 12x - 1x.$$

5. Next, we factor by grouping:

$$4x^2 + 11x - 3 = 4x^2 - 1x + 12x - 3$$

Rewriting the middle term; $12x - 1x$ could also be used.

$$= x(4x - 1) + 3(4x - 1)$$

Factoring by grouping. Note the common factor, $4x - 1$.

$$= (4x - 1)(x + 3).$$

Factoring out the common factor

6. The factorization of $4x^2 + 11x - 3$ is $(4x - 1)(x + 3)$. But don't forget the common factor, $2x$. The factorization of the original trinomial is

$$2x(4x - 1)(x + 3).$$ **TRY EXERCISE** 57

5.3   EXERCISE SET

Concept Reinforcement   *In each of Exercises 1–4, match the polynomial with the correct factorization from the column on the right.*

1. ___ $12x^2 + 16x - 3$ a) $(7x - 1)(2x + 3)$

2. ___ $14x^2 + 19x - 3$ b) $(6x + 1)(2x - 3)$

3. ___ $14x^2 - 19x - 3$ c) $(6x - 1)(2x + 3)$

4. ___ $12x^2 - 16x - 3$ d) $(7x + 1)(2x - 3)$

Factor completely. If a polynomial is prime, state this.

5. $2x^2 + 7x - 4$ 6. $3x^2 + x - 4$

7. $3x^2 - 17x - 6$ 8. $5x^2 - 19x - 4$

9. $4t^2 + 12t + 5$ 10. $6t^2 + 17t + 7$

11. $15a^2 - 14a + 3$ 12. $10a^2 - 11a + 3$

13. $6x^2 + 17x + 12$ 14. $6x^2 + 19x + 10$

15. $6x^2 - 10x - 4$ 16. $5t^3 - 21t^2 + 18t$

17. $7t^3 + 15t^2 + 2t$ 18. $15t^2 + 20t - 75$

19. $10 - 23x + 12x^2$ 20. $-20 + 31x - 12x^2$

21. $-35x^2 - 34x - 8$ 22. $28x^2 + 38x - 6$

23. $4 + 6t^2 - 13t$ 24. $9 + 8t^2 - 18t$

25. $25x^2 + 40x + 16$ 26. $49t^2 + 42t + 9$

27. $20y^2 + 59y - 3$ 28. $25a^2 - 23a - 2$

29. $14x^2 + 73x + 45$ 30. $35x^2 - 57x - 44$

31. $-2x^2 + 15 + x$ 32. $2t^2 - 19 - 6t$

33. $-6x^2 - 33x - 15$ 34. $-12x^2 - 28x + 24$

35. $10a^2 - 8a - 18$ 36. $20y^2 - 25y + 5$

37. $12x^2 + 68x - 24$ 38. $6x^2 + 21x + 15$

39. $4x + 1 + 3x^2$ 40. $-9 + 18x^2 + 21x$

Factor. Use factoring by grouping even though it would seem reasonable to first combine like terms.

41. $x^2 + 3x - 2x - 6$ 42. $x^2 + 4x - 2x - 8$

43. $8t^2 - 6t - 28t + 21$

44. $35t^2 - 40t + 21t - 24$

45. $6x^2 + 4x + 15x + 10$ 46. $3x^2 - 2x + 3x - 2$

47. $2y^2 + 8y - y - 4$ 48. $7n^2 + 35n - n - 5$

49. $6a^2 - 8a - 3a + 4$ 50. $10a^2 - 4a - 5a + 2$

Factor completely. If a polynomial is prime, state this.

51. $16t^2 + 23t + 7$ 52. $9t^2 + 14t + 5$

53. $-9x^2 - 18x - 5$ 54. $-16x^2 - 32x - 7$

55. $10x^2 + 30x - 70$ 56. $10a^2 + 25a - 15$

57. $18x^3 + 21x^2 - 9x$ 58. $6x^3 - 4x^2 - 10x$

59. $89x + 64 + 25x^2$ 60. $47 - 42y + 9y^2$

61. $168x^3 + 45x^2 + 3x$

62. $144x^5 - 168x^4 + 48x^3$

63. $-14t^4 + 19t^3 + 3t^2$

64. $-70a^4 + 68a^3 - 16a^2$

65. $132y + 32y^2 - 54$ 66. $220y + 60y^2 - 225$

67. $2a^2 - 5ab + 2b^2$ 68. $3p^2 - 16pq - 12q^2$

69. $8s^2 + 22st + 14t^2$ 70. $10s^2 + 4st - 6t^2$

71. $27x^2 - 72xy + 48y^2$

72. $-30a^2 - 87ab - 30b^2$

73. $-24a^2 + 34ab - 12b^2$ 74. $15a^2 - 5ab - 20b^2$

75. $19x^3 - 3x^2 + 14x^4$ 76. $10x^5 - 2x^4 + 22x^3$

77. $18a^7 + 8a^6 + 9a^8$ 78. $40a^8 + 16a^7 + 25a^9$

79. Asked to factor $2x^2 - 18x + 36$, Kay *incorrectly* answers

$$2x^2 - 18x + 36 = 2(x^2 + 9x + 18)$$
$$= 2(x + 3)(x + 6).$$

If this were a 10-point quiz question, how many points would you take off? Why?

80. Asked to factor $4x^2 + 28x + 48$, Herb *incorrectly* answers

$$4x^2 + 28x + 48 = (2x + 6)(2x + 8)$$
$$= 2(x + 3)(x + 4).$$

If this were a 10-point quiz question, how many points would you take off? Why?

Skill Review

To prepare for Section 5.4, review the special products in Section 4.6.

Multiply. [4.6]

81. $(x - 2)^2$ 82. $(x + 2)^2$

83. $(x + 2)(x - 2)$ 84. $(5t - 3)^2$

85. $(4a + 1)^2$

86. $(2n + 7)(2n - 7)$

87. $(3c - 10)^2$

88. $(1 - 5a)^2$

89. $(8n + 3)(8n - 3)$

90. $(9 - y)(9 + y)$

Synthesis

91. Explain how you would prove to a fellow student that a given trinomial is prime.

92. For the trinomial $ax^2 + bx + c$, suppose that a is the product of three different prime factors and c is the product of another two prime factors. How many possible factorizations (like those in Example 1) exist? Explain how you determined your answer.

Factor. If a polynomial is prime, state this.

93. $18x^2y^2 - 3xy - 10$

94. $8x^2y^3 + 10xy^2 + 2y$

95. $9a^2b^3 + 25ab^2 + 16$

96. $-9t^{10} - 12t^5 - 4$

97. $16t^{10} - 8t^5 + 1$

98. $9a^2b^2 - 15ab - 2$

99. $-15x^{2m} + 26x^m - 8$

100. $-20x^{2n} - 16x^n - 3$

101. $3a^{6n} - 2a^{3n} - 1$

102. $a^{2n+1} - 2a^{n+1} + a$

103. $7(t - 3)^{2n} + 5(t - 3)^n - 2$

104. $3(a + 1)^{n+1}(a + 3)^2 - 5(a + 1)^n(a + 3)^3$

CONNECTING the CONCEPTS

In Sections 5.1–5.3, we have considered factoring out a common factor, factoring by grouping, and factoring with FOIL. The following is a good strategy to follow when you encounter a mixed set of factoring problems.

1. Factor out any common factor.
2. Try factoring by grouping for polynomials with four terms.
3. Try factoring with FOIL for trinomials. If the leading coefficient of the trinomial is not 1, you may instead try factoring by grouping.

Polynomial	Number of Terms	Factorization
$12y^5 - 6y^4 + 30y^2$	3	There is a common factor: $6y^2$. $12y^5 - 6y^4 + 30y^2 = 6y^2(2y^3 - y^2 + 5)$. The trinomial in the parentheses cannot be factored further.
$t^4 - 5t^3 - 3t + 15$	4	There is no common factor. We factor by grouping. $t^4 - 5t^3 - 3t + 15 = t^3(t - 5) - 3(t - 5)$ $= (t - 5)(t^3 - 3)$
$-4x^4 + 4x^3 + 80x^2$	3	There is a common factor: $-4x^2$. $-4x^4 + 4x^3 + 80x^2 = -4x^2(x^2 - x - 20)$ The trinomial in the parentheses can be factored: $-4x^4 + 4x^3 + 80x^2 = -4x^2(x^2 - x - 20)$ $= -4x^2(x + 4)(x - 5)$.
$10n^2 - 17n + 3$	3	There is no common factor. We factor with FOIL or by grouping. $10n^2 - 17n + 3 = (2n - 3)(5n - 1)$

MIXED REVIEW

Factor completely. If a polynomial is prime, state this.

1. $6x^5 - 18x^2$

2. $x^2 + 10x + 16$

3. $2x^2 + 13x - 7$

4. $x^3 + 3x^2 + 2x + 6$

5. $5x^2 + 40x - 100$

6. $x^2 - 2x - 5$

7. $7x^2y - 21xy - 28y$

8. $15a^4 - 27a^2b^2 + 21a^2b$

9. $b^2 - 14b + 49$

10. $12x^2 - x - 1$

11. $c^3 + c^2 - 4c - 4$

12. $2x^2 + 30x - 200$

13. $t^2 + t - 10$

14. $15d^2 - 30d + 75$

15. $15p^2 + 16pq + 4q^2$

16. $-2t^3 - 10t^2 - 12t$

17. $x^2 + 4x - 77$

18. $10c^2 + 20c + 10$

19. $5 + 3x - 2x^2$

20. $2m^3n - 10m^2n - 6mn + 30n$

5.4   Factoring Perfect-Square Trinomials and Differences of Squares

Recognizing Perfect-Square Trinomials ▪ Factoring Perfect-Square Trinomials ▪
Recognizing Differences of Squares ▪ Factoring Differences of Squares ▪ Factoring Completely

In Section 4.6, we studied special products of certain binomials. Reversing these rules provides shortcuts for factoring certain polynomials.

Recognizing Perfect-Square Trinomials

Some trinomials are squares of binomials. For example, $x^2 + 10x + 25$ is the square of the binomial $x + 5$, because

$$(x + 5)^2 = x^2 + 2 \cdot x \cdot 5 + 5^2 = x^2 + 10x + 25.$$

A trinomial that is the square of a binomial is called a **perfect-square trinomial**.

In Section 4.6, we considered squaring binomials as a special-product rule:

$$(A + B)^2 = A^2 + 2AB + B^2;$$
$$(A - B)^2 = A^2 - 2AB + B^2.$$

Reading the right-hand sides first, we can use these equations to factor perfect-square trinomials. Note that in order for a trinomial to be the square of a binomial, it must have the following:

1. Two terms, A^2 and B^2, must be squares, such as

   $$4, \quad x^2, \quad 81m^2, \quad 16t^2.$$

2. Neither A^2 nor B^2 is being subtracted.

3. The remaining term is either $2 \cdot A \cdot B$ or $-2 \cdot A \cdot B$, where A and B are the square roots of A^2 and B^2.

EXAMPLE 1

Determine whether each of the following is a perfect-square trinomial.

a) $x^2 + 6x + 9$ **b)** $t^2 - 8t - 9$ **c)** $16x^2 + 49 - 56x$

SOLUTION

a) To see if $x^2 + 6x + 9$ is a perfect-square trinomial, note that:

1. Two terms, x^2 and 9, are squares.
2. Neither x^2 nor 9 is being subtracted.
3. The remaining term, $6x$, is $2 \cdot x \cdot 3$, where x and 3 are the square roots of x^2 and 9.

Thus, $x^2 + 6x + 9$ *is* a perfect-square trinomial.

b) To see if $t^2 - 8t - 9$ is a perfect-square trinomial, note that:

1. Both t^2 and 9, are squares. But:
2. Since 9 is being subtracted, $t^2 - 8t - 9$ *is not* a perfect-square trinomial.

c) To see if $16x^2 + 49 - 56x$ is a perfect-square trinomial, it helps to first write it in descending order:

$$16x^2 - 56x + 49.$$

Next, note that:

1. Two terms, $16x^2$ and 49, are squares.
2. There is no minus sign before $16x^2$ or 49.
3. Twice the product of the square roots, $2 \cdot 4x \cdot 7$, is $56x$. The remaining term, $-56x$, is the opposite of this product.

Thus, $16x^2 + 49 - 56x$ *is* a perfect-square trinomial.   **TRY EXERCISE** 11

STUDENT NOTES ———

If you're not already quick to recognize the squares that represent $1^2, 2^2, 3^2, \ldots, 12^2$, this would be a good time to memorize these numbers.

STUDY SKILLS ———

Fill in Your Blanks

Don't hesitate to write out any missing steps that you'd like to see included. For instance, in Example 1(c), we state (in red) that $16x^2$ is a square. To solidify your understanding, you may want to write $4x \cdot 4x = 16x^2$ in the margin of your text.

Factoring Perfect–Square Trinomials

Either of the factoring methods discussed in Section 5.3 can be used to factor perfect-square trinomials, but a faster method is to recognize the following patterns.

> ### Factoring a Perfect-Square Trinomial
> $$A^2 + 2AB + B^2 = (A + B)^2; \qquad A^2 - 2AB + B^2 = (A - B)^2$$

Each factorization uses the square roots of the squared terms and the sign of the remaining term. To verify these equations, you should compute $(A + B)(A + B)$ and $(A - B)(A - B)$.

EXAMPLE 2

Factor: **(a)** $x^2 + 6x + 9$; **(b)** $t^2 + 49 - 14t$; **(c)** $16x^2 - 40x + 25$.

SOLUTION

a) $x^2 + 6x + 9 = x^2 + 2 \cdot x \cdot 3 + 3^2 = (x + 3)^2$ The sign of the middle term is positive.

$$A^2 + 2 \quad A \quad B + B^2 = (A + B)^2$$

b) $t^2 + 49 - 14t = t^2 - 14t + 49$ Using a commutative law to write in descending order

$$= t^2 - 2 \cdot t \cdot 7 + 7^2 = (t - 7)^2$$

$$A^2 - 2 \quad A \quad B + B^2 = (A - B)^2$$

c) $16x^2 - 40x + 25 = (4x)^2 - 2 \cdot 4x \cdot 5 + 5^2 = (4x - 5)^2$ Recall that $(4x)^2 = 16x^2$.

$A^2 \quad - 2 \quad A \quad B + B^2 = (A - B)^2$ **TRY EXERCISE** 19

Polynomials in more than one variable can also be perfect-square trinomials.

EXAMPLE **3**

Factor: $4p^2 - 12pq + 9q^2$.

SOLUTION   We have

$$4p^2 - 12pq + 9q^2 = (2p)^2 - 2(2p)(3q) + (3q)^2$$ Recognizing the perfect-square trinomial

$$= (2p - 3q)^2.$$ The sign of the middle term is negative.

Check: $(2p - 3q)(2p - 3q) = 4p^2 - 12pq + 9q^2$.

The factorization is $(2p - 3q)^2$.   **TRY EXERCISE** 43

EXAMPLE **4**

Factor: $-75m^3 - 60m^2 - 12m$.

SOLUTION   *Always* look first for a common factor. This time there is one. We factor out $-3m$ so that the leading coefficient of the polynomial inside the parentheses is positive:

Factor out the common factor.

$$-75m^3 - 60m^2 - 12m = -3m[25m^2 + 20m + 4]$$

$$= -3m[(5m)^2 + 2(5m)(2) + 2^2]$$ Recognizing the perfect-square trinomial. Try to do this mentally.

Factor the perfect-square trinomial.

$$= -3m(5m + 2)^2.$$

Check: $-3m(5m + 2)^2 = -3m(5m + 2)(5m + 2)$

$$= -3m(25m^2 + 20m + 4)$$

$$= -75m^3 - 60m^2 - 12m.$$

The factorization is $-3m(5m + 2)^2$.   **TRY EXERCISE** 31

Recognizing Differences of Squares

Some binomials represent the difference of two squares. For example, the binomial $16x^2 - 9$ is a difference of two expressions, $16x^2$ and 9, that are squares. To see this, note that $16x^2 = (4x)^2$ and $9 = 3^2$.

Any expression, like $16x^2 - 9$, that can be written in the form $A^2 - B^2$ is called a **difference of squares**. Note that in order for a binomial to be a difference of squares, it must have the following.

1. There must be two expressions, both squares, such as

$$25, \quad t^2, \quad 4x^2, \quad 1, \quad x^6, \quad 49y^8, \quad 100x^2y^2.$$

2. The terms in the binomial must have different signs.

Note that in order for an expression to be a square, its coefficient must be a perfect square and the power(s) of the variable(s) must be even.

EXAMPLE **5** Determine whether each of the following is a difference of squares.

a) $9x^2 - 64$ **b)** $25 - t^3$ **c)** $-4x^{10} + 36$

SOLUTION

a) To see if $9x^2 - 64$ is a difference of squares, note that:

  1. The first expression is a square: $9x^2 = (3x)^2$.
     The second expression is a square: $64 = 8^2$.
  2. The terms have different signs.

  Thus, $9x^2 - 64$ is a difference of squares, $(3x)^2 - 8^2$.

b) To see if $25 - t^3$ is a difference of squares, note that:

  1. The expression t^3 is not a square.

  Thus, $25 - t^3$ is not a difference of squares.

c) To see if $-4x^{10} + 36$ is a difference of squares, note that:

  1. The expressions $4x^{10}$ and 36 are squares: $4x^{10} = (2x^5)^2$ and $36 = 6^2$.
  2. The terms have different signs.

  Thus, $-4x^{10} + 36$ is a difference of squares, $6^2 - (2x^5)^2$. It is often useful to
  rewrite $-4x^{10} + 36$ in the equivalent form $36 - 4x^{10}$. **TRY EXERCISE** 51

Factoring Differences of Squares

To factor a difference of squares, we reverse a pattern from Section 4.6.

Factoring a Difference of Squares
  $A^2 - B^2 = (A + B)(A - B)$

Once we have identified the expressions that are playing the roles of A and B,
the factorization can be written directly. To verify this equation, simply multiply
$(A + B)(A - B)$.

EXAMPLE **6** Factor: **(a)** $x^2 - 4$; **(b)** $1 - 9p^2$; **(c)** $s^6 - 16t^{10}$; **(d)** $50x^2 - 8x^8$.

SOLUTION

a) $x^2 - 4 = x^2 - 2^2 = (x + 2)(x - 2)$
$$A^2 - B^2 = (A + B)(A - B)$$

b) $1 - 9p^2 = 1^2 - (3p)^2 = (1 + 3p)(1 - 3p)$
$$A^2 - B^2 = (A + B)(A - B)$$

c) $s^6 - 16t^{10} = (s^3)^2 - (4t^5)^2$ Using the rules for powers
$$A^2 - B^2$$

$= (s^3 + 4t^5)(s^3 - 4t^5)$ Try to go directly to this step.
$$(A + B)(A - B)$$

d) *Always* look first for a common factor. This time there is one, $2x^2$:

Factor out the common factor.

$$50x^2 - 8x^8 = 2x^2(25 - 4x^6)$$
$$= 2x^2[5^2 - (2x^3)^2] \quad \text{Recognizing } A^2 - B^2.$$
$$\text{Try to do this mentally.}$$

Factor the difference of squares.

$$= 2x^2(5 + 2x^3)(5 - 2x^3).$$

Check: $2x^2(5 + 2x^3)(5 - 2x^3) = 2x^2(25 - 4x^6)$
$$= 50x^2 - 8x^8.$$

The factorization of $50x^2 - 8x^8$ is $2x^2(5 + 2x^3)(5 - 2x^3)$.

> **TRY EXERCISE** 57

CAUTION!   Note in Example 6 that a difference of squares is *not* the square of the difference; that is,

$$A^2 - B^2 \neq (A - B)^2. \quad \text{To see this, note that}$$
$$(A - B)^2 = A^2 - 2AB + B^2.$$

Factoring Completely

Sometimes, as in Examples 4 and 6(d), a *complete* factorization requires two or more steps. Factoring is complete when no factor can be factored further.

EXAMPLE 7   Factor: $y^4 - 16$.

SOLUTION   We have

Factor a difference of squares.

$$y^4 - 16 = (y^2)^2 - 4^2 \quad \text{Recognizing } A^2 - B^2$$
$$= (y^2 + 4)(y^2 - 4) \quad \text{Note that } y^2 - 4 \text{ is not prime.}$$

Factor another difference of squares.

$$= (y^2 + 4)(y + 2)(y - 2). \quad \text{Note that } y^2 - 4 \text{ is itself a}$$
$$\text{difference of squares.}$$

Check: $(y^2 + 4)(y + 2)(y - 2) = (y^2 + 4)(y^2 - 4)$
$$= y^4 - 16.$$

The factorization is $(y^2 + 4)(y + 2)(y - 2)$.

> **TRY EXERCISE** 79

Note in Example 7 that the factor $y^2 + 4$ is a *sum* of squares that cannot be factored further.

CAUTION!   There is no general formula for factoring a sum of squares. In particular,

$$A^2 + B^2 \neq (A + B)^2.$$

As you proceed through the exercises, these suggestions may prove helpful.

Tips for Factoring

1. Always look first for a common factor! If there is one, factor it out.
2. Be alert for perfect-square trinomials and for binomials that are differences of squares. Once recognized, they can be factored without trial and error.
3. Always factor completely.
4. Check by multiplying.

5.4 EXERCISE SET

🖐 *Concept Reinforcement Identify each of the following as a perfect-square trinomial, a difference of squares, a prime polynomial, or none of these.*

1. $4x^2 + 49$ **2.** $x^2 - 64$

3. $t^2 - 100$ **4.** $x^2 - 5x + 4$

5. $9x^2 + 6x + 1$ **6.** $a^2 - 8a + 16$

7. $2t^2 + 10t + 6$ **8.** $-25x^2 - 9$

9. $16t^2 - 25$ **10.** $4r^2 + 20r + 25$

Determine whether each of the following is a perfect-square trinomial.

11. $x^2 + 18x + 81$ **12.** $x^2 - 16x + 64$

13. $x^2 - 10x - 25$ **14.** $x^2 - 14x - 49$

15. $x^2 - 3x + 9$ **16.** $x^2 + 4x + 4$

17. $9x^2 + 25 - 30x$ **18.** $36x^2 + 16 - 24x$

Factor completely. Remember to look first for a common factor and to check by multiplying. If a polynomial is prime, state this.

19. $x^2 + 16x + 64$ **20.** $x^2 + 10x + 25$

21. $x^2 - 10x + 25$ **22.** $x^2 - 16x + 64$

23. $5p^2 + 20p + 20$ **24.** $3p^2 - 12p + 12$

25. $1 - 2t + t^2$ **26.** $1 + t^2 + 2t$

27. $18x^2 + 12x + 2$ **28.** $25x^2 + 10x + 1$

29. $49 - 56y + 16y^2$ **30.** $75 - 60m + 12m^2$

31. $-x^5 + 18x^4 - 81x^3$ **32.** $-2x^2 + 40x - 200$

33. $2n^3 + 40n^2 + 200n$ **34.** $x^3 + 24x^2 + 144x$

35. $20x^2 + 100x + 125$ **36.** $27m^2 - 36m + 12$

37. $49 - 42x + 9x^2$ **38.** $64 - 112x + 49x^2$

39. $16x^2 + 24x + 9$ **40.** $2a^2 + 28a + 98$

41. $2 + 20x + 50x^2$ **42.** $9x^2 + 30x + 25$

43. $9p^2 + 12pq + 4q^2$ **44.** $x^2 - 3xy + 9y^2$

45. $a^2 - 12ab + 49b^2$

46. $25m^2 - 20mn + 4n^2$

47. $-64m^2 - 16mn - n^2$

48. $-81p^2 + 18pq - q^2$

49. $-32s^2 + 80st - 50t^2$

50. $-36a^2 - 96ab - 64b^2$

Determine whether each of the following is a difference of squares.

51. $x^2 - 100$ **52.** $x^2 + 49$

53. $n^4 + 1$ **54.** $n^4 - 81$

55. $-1 + 64t^2$ **56.** $-12 + 25t^2$

Factor completely. Remember to look first for a common factor. If a polynomial is prime, state this.

57. $x^2 - 25$ **58.** $x^2 - 36$

59. $p^2 - 9$ **60.** $q^2 + 1$

61. $-49 + t^2$ **62.** $-64 + m^2$

63. $6a^2 - 24$ **64.** $x^2 - 8x + 16$

65. $49x^2 - 14x + 1$ **66.** $3t^2 - 3$

67. $200 - 2t^2$ **68.** $98 - 8w^2$

69. $-80a^2 + 45$

70. $25x^2 - 4$

71. $5t^2 - 80$

72. $-4t^2 + 64$

73. $8x^2 - 162$

74. $24x^2 - 54$

75. $36x - 49x^3$

76. $16x - 81x^3$

77. $49a^4 - 20$

78. $25a^4 - 9$

79. $t^4 - 1$

80. $x^4 - 16$

81. $-3x^3 + 24x^2 - 48x$

82. $-2a^4 + 36a^3 - 162a^2$

83. $75t^3 - 27t$

84. $80s^4 - 45s^2$

85. $a^8 - 2a^7 + a^6$

86. $x^8 - 8x^7 + 16x^6$

87. $10a^2 - 10b^2$

88. $6p^2 - 6q^2$

89. $16x^4 - y^4$

90. $98x^2 - 32y^2$

91. $18t^2 - 8s^2$

92. $a^4 - 81b^4$

93. Explain in your own words how to determine whether a polynomial is a perfect-square trinomial.

94. Explain in your own words how to determine whether a polynomial is a difference of squares.

Skill Review

To prepare for Section 5.5, review the product and power rules for exponents and multiplication of polynomials (Sections 4.1 and 4.5).

Simplify. [4.1]

95. $(2x^2y^4)^3$

96. $(-5x^2y)^3$

Multiply. [4.5]

97. $(x + 1)(x + 1)(x + 1)$

98. $(x - 1)^3$

99. $(p + q)^3$

100. $(p - q)^3$

Synthesis

101. Leon concludes that since $x^2 - 9 = (x - 3)(x + 3)$, it must follow that $x^2 + 9 = (x + 3)(x - 3)$. What mistake(s) is he making?

102. Write directions that would enable someone to construct a polynomial that contains a perfect-square trinomial, a difference of squares, and a common factor.

Factor completely. If a polynomial is prime, state this.

103. $x^8 - 2^8$

104. $3x^2 - \frac{1}{3}$

105. $18x^3 - \frac{8}{25}x$

106. $0.81t - t^3$

107. $(y - 5)^4 - z^8$

108. $x^2 - \left(\frac{1}{x}\right)^2$

109. $-x^4 + 8x^2 + 9$

110. $-16x^4 + 96x^2 - 144$

Aha! **111.** $(y + 3)^2 + 2(y + 3) + 1$

112. $49(x + 1)^2 - 42(x + 1) + 9$

113. $27p^3 - 45p^2 - 75p + 125$

114. $a^{2n} - 49b^{2n}$

115. $81 - b^{4k}$

116. $9b^{2n} + 12b^n + 4$

117. Subtract $(x^2 + 1)^2$ from $x^2(x + 1)^2$.

Factor by grouping. Look for a grouping of three terms that is a perfect-square trinomial.

118. $t^2 + 4t + 4 - 25$

119. $y^2 + 6y + 9 - x^2 - 8x - 16$

Find c such that each polynomial is the square of a binomial.

120. $cy^2 + 6y + 1$

121. $cy^2 - 24y + 9$

122. Find the value of a if $x^2 + a^2x + a^2$ factors into $(x + a)^2$.

123. Show that the difference of the squares of two consecutive integers is the sum of the integers. (*Hint*: Use x for the smaller number.)

5.5 Factoring Sums or Differences of Cubes

Formulas for Factoring Sums or Differences of Cubes • Using the Formulas

Formulas for Factoring Sums or Differences of Cubes

We have seen that a difference of two squares can always be factored, but a *sum* of two squares is usually prime. The situation is different with cubes: The difference *or sum* of two cubes can always be factored. To see this, consider the following products:

$$(A + B)(A^2 - AB + B^2) = A(A^2 - AB + B^2) + B(A^2 - AB + B^2)$$
$$= A^3 - A^2B + AB^2 + A^2B - AB^2 + B^3$$
$$= A^3 + B^3 \quad \text{Combining like terms}$$

and

$$(A - B)(A^2 + AB + B^2) = A(A^2 + AB + B^2) - B(A^2 + AB + B^2)$$
$$= A^3 + A^2B + AB^2 - A^2B - AB^2 - B^3$$
$$= A^3 - B^3. \quad \text{Combining like terms}$$

These products allow us to factor a sum or a difference of two cubes. Observe how the location of the + and − signs changes.

> **Factoring a Sum or a Difference of Two Cubes**
> $$A^3 + B^3 = (A + B)(A^2 - AB + B^2);$$
> $$A^3 - B^3 = (A - B)(A^2 + AB + B^2)$$

Using the Formulas

Remembering this list of cubes may prove helpful when factoring.

N	0.2	0.1	0	1	2	3	4	5	6
N^3	0.008	0.001	0	1	8	27	64	125	216

We say that 2 is the *cube root* of 8, that 3 is the cube root of 27, and so on.

EXAMPLE **1** Write an equivalent expression by factoring: $x^3 + 27$.

SOLUTION We first observe that

$$x^3 + 27 = x^3 + 3^3. \quad \text{This is a sum of cubes.}$$

Next, in one set of parentheses, we write the first cube root, x, plus the second cube root, 3:

$$(x + 3)(\qquad).$$

To get the other factor, we think of $x + 3$ and do the following:

Square the first term: x^2.

Multiply the terms and then change the sign: $-3x$.

Square the second term: 3^2, or 9.

$$(x + 3)(x^2 - 3x + 9).$$

Check: $(x + 3)(x^2 - 3x + 9) = x^3 - 3x^2 + 9x + 3x^2 - 9x + 27$

$$= x^3 + 27.   \text{Combining like terms}$$

Thus, $x^3 + 27 = (x + 3)(x^2 - 3x + 9)$.

> TRY EXERCISE   13

In Example 2, you will see that the pattern used to write the trinomial factor in Example 1 can be used when factoring a *difference* of two cubes as well.

EXAMPLE   2   Factor.

a) $125x^3 - y^3$

b) $m^6 + 64$

c) $128y^7 - 250x^6y$

d) $r^6 - s^6$

SOLUTION

a) We have

$$125x^3 - y^3 = (5x)^3 - y^3.   \text{This is a difference of cubes.}$$

In one set of parentheses, we write the cube root of the first term, $5x$, minus the cube root of the second term, y:

$$(5x - y)(\quad).   \text{This can be regarded as } 5x \text{ plus the cube root of } (-y)^3, \text{ since } -y^3 = (-y)^3.$$

To get the other factor, we think of $5x + y$ and do the following:

Square the first term: $(5x)^2$, or $25x^2$.

Multiply the terms and then change the sign: $5xy$.

Square the second term: $(-y)^2 = y^2$.

$$(5x - y)(25x^2 + 5xy + y^2).$$

STUDENT NOTES

If you think of $A^3 - B^3$ as $A^3 + (-B)^3$, it is then sufficient to remember only the pattern for factoring a sum of two cubes. Be sure to simplify your result if you do this.

Check:

$$(5x - y)(25x^2 + 5xy + y^2) = 125x^3 + 25x^2y + 5xy^2 - 25x^2y - 5xy^2 - y^3$$

$$= 125x^3 - y^3.   \text{Combining like terms}$$

Thus, $125x^3 - y^3 = (5x - y)(25x^2 + 5xy + y^2)$.

b) We have

$$m^6 + 64 = (m^2)^3 + 4^3.   \text{Rewriting as a sum of quantities cubed}$$

Next, we reuse the pattern used in Example 1:

$$A^3 + B^3 = (A + B)(A^2 - A \cdot B + B^2)$$

$$(m^2)^3 + 4^3 = (m^2 + 4)((m^2)^2 - m^2 \cdot 4 + 4^2)$$

$$= (m^2 + 4)(m^4 - 4m^2 + 16).   \text{The check is left to the student.}$$

c) We have

$$128y^7 - 250x^6y = 2y(64y^6 - 125x^6)$$ Remember: *Always* look for a common factor.

$$= 2y[(4y^2)^3 - (5x^2)^3].$$ Rewriting as a difference of quantities cubed

To factor $(4y^2)^3 - (5x^2)^3$, we reuse the pattern in part (a) above:

$$A^3 \quad - \quad B^3 \quad = (A \quad - \quad B)(\quad A^2 \ + \ A \cdot B \ + \ B^2)$$

$$(4y^2)^3 - (5x^2)^3 = (4y^2 - 5x^2)((4y^2)^2 + 4y^2 \cdot 5x^2 + (5x^2)^2)$$

$$= (4y^2 - 5x^2)(16y^4 + 20x^2y^2 + 25x^4).$$

The check is left to the student. We have

$$128y^7 - 250x^6y = 2y(4y^2 - 5x^2)(16y^4 + 20x^2y^2 + 25x^4).$$

d) We have

$$r^6 - s^6 = (r^3)^2 - (s^3)^2$$

$$= (r^3 + s^3)(r^3 - s^3)$$ Factoring a difference of two *squares*

$$= (r + s)(r^2 - rs + s^2)(r - s)(r^2 + rs + s^2).$$ Factoring the sum and the difference of two cubes

To check, read the steps in reverse order and inspect the multiplication.

TRY EXERCISES ▶ 31 and 41

In Example 2(d), suppose we first factored $r^6 - s^6$ as a difference of two cubes:

$$(r^2)^3 - (s^2)^3 = (r^2 - s^2)(r^4 + r^2s^2 + s^4)$$

$$= (r + s)(r - s)(r^4 + r^2s^2 + s^4).$$

In this case, we might have missed some factors; $r^4 + r^2s^2 + s^4$ can be factored as $(r^2 - rs + s^2)(r^2 + rs + s^2)$, but we probably would never have suspected that such a factorization exists. Given a choice, it is generally better to factor as a difference of squares before factoring as a sum or a difference of cubes.

Useful Factoring Facts

Sum of cubes: $A^3 + B^3 = (A + B)(A^2 - AB + B^2)$

Difference of cubes: $A^3 - B^3 = (A - B)(A^2 + AB + B^2)$

Difference of squares: $A^2 - B^2 = (A + B)(A - B)$

There is no formula for factoring a sum of two squares.

5.5 EXERCISE SET

Concept Reinforcement *Classify each binomial as either a sum of cubes, a difference of cubes, a difference of squares, or none of these.*

1. $x^3 - 1$

2. $8 + t^3$

3. $9x^4 - 25$

4. $9x^2 + 25$

5. $1000t^3 + 1$

6. $x^3y^3 - 27z^3$

7. $25x^2 + 8x$

8. $100y^8 - 25x^4$

9. $s^{21} - t^{15}$

10. $14x^3 - 2x$

Factor completely.

11. $x^3 - 64$

12. $t^3 - 27$

13. $z^3 + 1$

14. $x^3 + 8$

15. $t^3 - 1000$

16. $m^3 + 125$

17. $27x^3 + 1$

18. $8a^3 + 1$

19. $64 - 125x^3$

20. $27 - 8t^3$

21. $x^3 - y^3$

22. $y^3 - z^3$

23. $a^3 + \frac{1}{8}$

24. $x^3 + \frac{1}{27}$

25. $8t^3 - 8$

26. $2y^3 - 128$

27. $54x^3 + 2$

28. $8a^3 + 1000$

29. $rs^4 + 64rs$

30. $ab^5 + 1000ab^2$

31. $5x^3 - 40z^3$

32. $2y^3 - 54z^3$

33. $y^3 - \frac{1}{1000}$

34. $x^3 - \frac{1}{8}$

35. $x^3 + 0.001$

36. $y^3 + 0.125$

37. $64x^6 - 8t^6$

38. $125c^6 - 8d^6$

39. $54y^4 - 128y$

40. $3z^5 - 3z^2$

41. $z^6 - 1$

42. $t^6 + 1$

43. $t^6 + 64y^6$

44. $p^6 - q^6$

45. $x^{12} - y^3z^{12}$

46. $a^9 + b^{12}c^{15}$

47. How could you use factoring to convince someone that $x^3 + y^3 \neq (x + y)^3$?

48. Is the following statement true or false and why? If A^3 and B^3 have a common factor, then A and B have a common factor.

Skill Review

Review graphing linear equations (Chapter 3).

49. Find the slope of the line containing the points $(-2, -5)$ and $(3, -6)$. [3.5]

50. Find the slope of the line given by $y - 3 = \frac{1}{4}x$. [3.6]

Graph.

51. $2x - 5y = 10$ [3.3]

52. $-5x = 10$ [3.3]

53. $y = \frac{2}{3}x - 1$ [3.6]

54. $y - 2 = -2(x + 4)$ [3.7]

Synthesis

55. Explain how the geometric model below can be used to verify the formula for factoring $a^3 - b^3$.

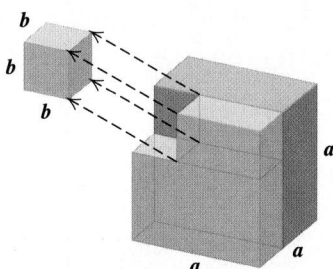

56. Explain how someone could construct a binomial that is both a difference of two cubes and a difference of two squares.

Factor.

57. $x^{6a} - y^{3b}$

58. $2x^{3a} + 16y^{3b}$

Aha! **59.** $(x + 5)^3 + (x - 5)^3$

60. $\frac{1}{16}x^{3a} + \frac{1}{2}y^{6a}z^{9b}$

61. $5x^3y^6 - \frac{5}{8}$

62. $x^3 - (x + y)^3$

63. $x^{6a} - (x^{2a} + 1)^3$

64. $(x^{2a} - 1)^3 - x^{6a}$

65. $t^4 - 8t^3 - t + 8$

5.6 Factoring: A General Strategy

Choosing the Right Method

Thus far, each section in this chapter has examined one or two different methods for factoring polynomials. In practice, when the need for factoring a polynomial arises, we must decide on our own which method to use. Regardless of the polynomial with which we are faced, the guidelines listed below can always be used.

To Factor a Polynomial

A. Always look for a common factor first. If there is one, factor out the largest common factor. Be sure to include it in your final answer.

B. Then look at the number of terms.

Two terms: Try factoring as a difference of squares first: $A^2 - B^2 = (A + B)(A - B)$. Next, try factoring as a sum or a difference of cubes: $A^3 + B^3 = (A + B)(A^2 - AB + B^2)$ and $A^3 - B^3 = (A - B)(A^2 + AB + B^2)$.

Three terms: If the trinomial is a perfect-square trinomial, factor accordingly: $A^2 + 2AB + B^2 = (A + B)^2$ or $A^2 - 2AB + B^2 = (A - B)^2$. If it is not a perfect-square trinomial, try using FOIL or grouping.

Four terms: Try factoring by grouping.

C. Always *factor completely*. When a factor can itself be factored, be sure to factor it. Remember that some polynomials, like $x^2 + 9$, are prime.

D. Check by multiplying.

Choosing the Right Method

EXAMPLE **1** Factor: $5t^4 - 80$.

SOLUTION

A. We look for a common factor:

$$5t^4 - 80 = 5(t^4 - 16). \qquad \text{5 is the largest common factor.}$$

B. The factor $t^4 - 16$ is a difference of squares: $(t^2)^2 - 4^2$. We factor it, being careful to rewrite the 5 from step (A):

$$5t^4 - 80 = 5(t^2 + 4)(t^2 - 4). \qquad t^4 - 16 = (t^2 + 4)(t^2 - 4)$$

C. Since $t^2 - 4$ is a difference of squares, we continue factoring:

$$5t^4 - 80 = 5(t^2 + 4)(t^2 - 4) = 5(t^2 + 4)(t - 2)(t + 2).$$

This is a sum of squares, which cannot be factored.

D. *Check:* $5(t^2 + 4)(t - 2)(t + 2) = 5(t^2 + 4)(t^2 - 4)$
$$= 5(t^4 - 16) = 5t^4 - 80.$$

The factorization is $5(t^2 + 4)(t - 2)(t + 2)$.

TRY EXERCISE 5

Factor: $2x^3 + 10x^2 + x + 5$.

SOLUTION

A. We look for a common factor. There is none.

B. Because there are four terms, we try factoring by grouping:

$2x^3 + 10x^2 + x + 5$

$= (2x^3 + 10x^2) + (x + 5)$ Separating into two binomials

$= 2x^2(x + 5) + 1(x + 5)$ Factoring out the largest common factor from each binomial. The 1 serves as an aid.

$= (x + 5)(2x^2 + 1)$. Factoring out the common factor, $x + 5$

C. Nothing can be factored further, so we have factored completely.

D. *Check:* $(x + 5)(2x^2 + 1) = 2x^3 + x + 10x^2 + 5$

$= 2x^3 + 10x^2 + x + 5$.

The factorization is $(x + 5)(2x^2 + 1)$. **TRY EXERCISE** 13

EXAMPLE 3

Factor: $-n^5 + 2n^4 + 35n^3$.

SOLUTION

A. We note that there is a common factor, $-n^3$:

$-n^5 + 2n^4 + 35n^3 = -n^3(n^2 - 2n - 35)$.

B. The factor $n^2 - 2n - 35$ is not a perfect-square trinomial. We factor it using trial and error:

$-n^5 + 2n^4 + 35n^3 = -n^3(n^2 - 2n - 35)$

$= -n^3(n - 7)(n + 5)$.

C. Nothing can be factored further, so we have factored completely.

D. *Check:* $-n^3(n - 7)(n + 5) = -n^3(n^2 - 2n - 35)$

$= -n^5 + 2n^4 + 35n^3$.

The factorization is $-n^3(n - 7)(n + 5)$. **TRY EXERCISE** 21

EXAMPLE 4

Factor: $x^2 - 20x + 100$.

SOLUTION

A. We look first for a common factor. There is none.

B. This polynomial is a perfect-square trinomial. We factor it accordingly:

$x^2 - 20x + 100 = x^2 - 2 \cdot x \cdot 10 + 10^2$ Try to do this step mentally.

$= (x - 10)^2$.

C. Nothing can be factored further, so we have factored completely.

D. *Check:* $(x - 10)(x - 10) = x^2 - 20x + 100$.

The factorization is $(x - 10)(x - 10)$, or $(x - 10)^2$. **TRY EXERCISE** 7

EXAMPLE   **5**

Factor: $6x^2y^4 - 21x^3y^5 + 3x^2y^6$.

SOLUTION

A. We first factor out the largest common factor, $3x^2y^4$:

$$6x^2y^4 - 21x^3y^5 + 3x^2y^6 = 3x^2y^4(2 - 7xy + y^2).$$

B. The constant term in $2 - 7xy + y^2$ is not a square, so we do not have a perfect-square trinomial. Note that x appears only in $-7xy$. The product of a form like $(1 - y)(2 - y)$ has no x in the middle term. Thus, $2 - 7xy + y^2$ cannot be factored.

C. Nothing can be factored further, so we have factored completely.

D. *Check:* $3x^2y^4(2 - 7xy + y^2) = 6x^2y^4 - 21x^3y^5 + 3x^2y^6$.

The factorization is $3x^2y^4(2 - 7xy + y^2)$.   **TRY EXERCISE** 33

EXAMPLE   **6**

Factor: $x^6 - 64$.

SOLUTION

A. We look first for a common factor. There is none (other than 1 or -1).

B. There are two terms, a difference of squares: $(x^3)^2 - (8)^2$. We factor:

$$x^6 - 64 = (x^3 + 8)(x^3 - 8). \qquad \text{Note that } x^6 = (x^3)^2.$$

C. One factor is a sum of two cubes, and the other factor is a difference of two cubes. We factor both:

$$x^6 - 64 = (x + 2)(x^2 - 2x + 4)(x - 2)(x^2 + 2x + 4).$$

The factorization is complete because no factor can be factored further.

D. *Check:* $(x + 2)(x^2 - 2x + 4)(x - 2)(x^2 + 2x + 4) = (x^3 + 8)(x^3 - 8)$
$$= x^6 - 64.$$

The factorization is $(x + 2)(x^2 - 2x + 4)(x - 2)(x^2 + 2x + 4)$.

   TRY EXERCISE 31

EXAMPLE   **7**

Factor: $-25m^2 - 20mn - 4n^2$.

SOLUTION

A. We look first for a common factor. Since all the terms are negative, we factor out a -1:

$$-25m^2 - 20mn - 4n^2 = -1(25m^2 + 20mn + 4n^2).$$

B. There are three terms in the parentheses. Note that the first term and the last term are squares: $25m^2 = (5m)^2$ and $4n^2 = (2n)^2$. We see that twice the product of $5m$ and $2n$ is the middle term,

$$2 \cdot 5m \cdot 2n = 20mn,$$

so the trinomial is a perfect square. To factor, we write a binomial squared:

$$-25m^2 - 20mn - 4n^2 = -1(25m^2 + 20mn + 4n^2)$$
$$= -1(5m + 2n)^2.$$

C. Nothing can be factored further, so we have factored completely.

D. *Check:* $-1(5m + 2n)^2 = -1(25m^2 + 20mn + 4n^2)$
$$= -25m^2 - 20mn - 4n^2.$$

The factorization is $-1(5m + 2n)^2$, or $-(5m + 2n)^2$.   **TRY EXERCISE** 59

EXAMPLE **8**   Factor: $x^2y^2 + 7xy + 12$.

SOLUTION

A. We look first for a common factor. There is none.

B. Since only one term is a square, we do not have a perfect-square trinomial. We use trial and error, thinking of the product xy as a single variable:

$$(xy + \quad)(xy + \quad).$$

We factor the last term, 12. All the signs are positive, so we consider only positive factors. Possibilities are 1, 12 and 2, 6 and 3, 4. The pair 3, 4 gives a sum of 7 for the coefficient of the middle term. Thus,

$$x^2y^2 + 7xy + 12 = (xy + 3)(xy + 4).$$

C. Nothing can be factored further, so we have factored completely.

D. *Check:* $(xy + 3)(xy + 4) = x^2y^2 + 7xy + 12.$

The factorization is $(xy + 3)(xy + 4)$.   ▶ TRY EXERCISE ▶ 61

Compare the variables appearing in Example 7 with those in Example 8. Note that if the leading term contains one variable and a different variable is in the last term, as in Example 7, each binomial contains two variable terms. When two variables appear in the leading term and no variables appear in the last term, as in Example 8, each binomial contains one term that has two variables and one term that is a constant.

EXAMPLE **9**   Factor: $a^4 - 16b^4$.

SOLUTION

A. We look first for a common factor. There is none.

B. There are two terms. Since $a^4 = (a^2)^2$ and $16b^4 = (4b^2)^2$, we see that we have a difference of squares. Thus,

$$a^4 - 16b^4 = (a^2 + 4b^2)(a^2 - 4b^2).$$

C. The factor $(a^2 - 4b^2)$ is itself a difference of squares. Thus,

$$a^4 - 16b^4 = (a^2 + 4b^2)(a + 2b)(a - 2b). \qquad \text{Factoring } a^2 - 4b^2$$

D. *Check:* $(a^2 + 4b^2)(a + 2b)(a - 2b) = (a^2 + 4b^2)(a^2 - 4b^2)$
$$= a^4 - 16b^4.$$

The factorization is $(a^2 + 4b^2)(a + 2b)(a - 2b)$.   ▶ TRY EXERCISE ▶ 53

5.6   EXERCISE SET

〰 *Concept Reinforcement*   *In each of Exercises 1–4, complete the sentence.*

1. As a first step when factoring polynomials, always check for a _____ .

2. When factoring a trinomial, if two terms are not squares, it cannot be a _____ .

3. If a polynomial has four terms and no common factor, it may be possible to factor by _____ .

4. It is always possible to check a factorization by _____ .

Factor completely. If a polynomial is prime, state this.

5. $5a^2 - 125$

6. $10c^2 - 810$

7. $y^2 + 49 - 14y$

8. $a^2 + 25 + 10a$

9. $3t^2 + 16t + 21$

10. $8t^2 + 31t - 4$

11. $x^3 + 18x^2 + 81x$

12. $x^3 - 24x^2 + 144x$

13. $x^3 - 5x^2 - 25x + 125$

14. $x^3 + 3x^2 - 4x - 12$

15. $27t^3 - 3t$

16. $98t^2 - 18$

17. $9x^3 + 12x^2 - 45x$

18. $20x^3 - 4x^2 - 72x$

19. $t^2 + 25$

20. $4x^2 + 20x - 144$

21. $6y^2 + 18y - 240$

22. $4n^2 + 81$

23. $-2a^6 + 8a^5 - 8a^4$

24. $-x^5 - 14x^4 - 49x^3$

25. $5x^5 - 80x$

26. $4x^4 - 64$

27. $t^4 - 9$

28. $9 + t^8$

29. $-x^6 + 2x^5 - 7x^4$

30. $-x^5 + 4x^4 - 3x^3$

31. $x^3 - y^3$

32. $8t^3 + 1$

33. $ax^2 + ay^2$

34. $12n^2 + 24n^3$

35. $2\pi rh + 2\pi r^2$

36. $4\pi r^2 + 2\pi r$

Aha! **37.** $(a + b)5a + (a + b)3b$

38. $5c(a^3 + b) - (a^3 + b)$

39. $x^2 + x + xy + y$

40. $n^2 + 2n + np + 2p$

41. $a^2 - 2a - ay + 2y$

42. $2x^2 - 4x + xz - 2z$

43. $3x^2 + 13xy - 10y^2$

44. $-x^2 - y^2 - 2xy$

45. $8m^3n - 32m^2n^2 + 24mn$ **46.** $a^2 - 7a - 6$

47. $4b^2 + a^2 - 4ab$

48. $7p^4 - 7q^4$

49. $16x^2 + 24xy + 9y^2$

50. $6a^2b^3 + 12a^3b^2 - 3a^4b^2$

51. $m^2 - 5m + 8$

52. $25z^2 + 10zy + y^2$

53. $a^4b^4 - 16$

54. $a^5 - 4a^4b - 5a^3b^2$

55. $80cd^2 - 36c^2d + 4c^3$

56. $2p^2 + pq + q^2$

57. $64t^6 - 1$

58. $m^6 - 1$

59. $-12 - x^2y^2 - 8xy$

60. $m^2n^2 - 4mn - 32$

61. $5p^2q^2 + 25pq - 30$

62. $a^4b^3 + 2a^3b^2 - 15a^2b$

63. $54a^4 + 16ab^3$

64. $54x^3y - 250y^4$

65. $x^6 + x^5y - 2x^4y^2$

66. $2s^6t^2 + 10s^3t^3 + 12t^4$

67. $36a^2 - 15a + \frac{25}{16}$

68. $a^2 + 2a^2bc + a^2b^2c^2$

69. $\frac{1}{81}x^2 - \frac{8}{27}x + \frac{16}{9}$ **70.** $\frac{1}{4}a^2 + \frac{1}{3}ab + \frac{1}{9}b^2$

71. $1 - 16x^{12}y^{12}$ **72.** $b^4a - 81a^5$

73. $4a^2b^2 + 12ab + 9$ **74.** $9c^2 + 6cd + d^2$

75. $z^4 + 6z^3 - 6z^2 - 36z$

76. $t^5 - 2t^4 + 5t^3 - 10t^2$

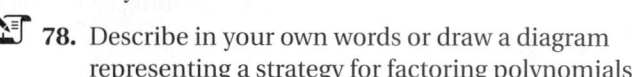 **77.** Kelly factored $16 - 8x + x^2$ as $(x - 4)^2$, while Tony factored it as $(4 - x)^2$. Are they both correct? Why or why not?

78. Describe in your own words or draw a diagram representing a strategy for factoring polynomials.

Skill Review

To prepare for Section 5.7, review solving equations (Section 2.2).

Solve. [2.2]

79. $8x - 9 = 0$

80. $3x + 5 = 0$

81. $2x + 7 = 0$

82. $4x - 1 = 0$

83. $3 - x = 0$

84. $22 - 2x = 0$

85. $2x - 5 = 8x + 1$

86. $3(x - 1) = 9 - x$

Synthesis

87. There are third-degree polynomials in x that we are not yet able to factor, despite the fact that they are not prime. Explain how such a polynomial could be created.

88. Describe a method that could be used to find a binomial of degree 16 that can be expressed as the product of prime binomial factors.

Factor.

89. $-(x^5 + 7x^3 - 18x)$ **90.** $18 + a^3 - 9a - 2a^2$

91. $-x^4 + 7x^2 + 18$ **92.** $-3a^4 + 15a^2 - 12$

Aha! **93.** $y^2(y + 1) - 4y(y + 1) - 21(y + 1)$

94. $y^2(y - 1) - 2y(y - 1) + (y - 1)$

95. $(y + 4)^2 + 2x(y + 4) + x^2$

96. $6(x - 1)^2 + 7y(x - 1) - 3y^2$

97. $2(a + 3)^4 - (a + 3)^3(b - 2) - (a + 3)^2(b - 2)^2$

98. $5(t - 1)^5 - 6(t - 1)^4(s - 1) + (t - 1)^3(s - 1)^2$

99. $49x^4 + 14x^2 + 1 - 25x^6$

COLLABORATIVE CORNER

Matching Factorizations*

Focus: Factoring

Time: 20 minutes

Group size: Begin with the entire class. If there is an odd number of students, the instructor should participate.

Materials: Prepared sheets of paper, pins or tape. On half of the sheets, the instructor writes a polynomial. On the remaining sheets, the instructor writes the factorization of those polynomials. The polynomials and factorizations should be similar; for example,

$$x^2 - 2x - 8, \quad (x - 2)(x - 4),$$
$$x^2 - 6x + 8, \quad (x - 1)(x - 8),$$
$$x^2 - 9x + 8, \quad (x + 2)(x - 4).$$

ACTIVITY

1. As class members enter the room, the instructor pins or tapes either a polynomial or a factorization to the back of each student. Class members are told only whether their sheet of paper contains a polynomial or a factorization.

2. After all students are wearing a sheet of paper, they should mingle with one another, attempting to match up their factorization with the appropriate polynomial or vice versa. They may ask questions of one another that relate to factoring and polynomials. Answers to the questions should be yes or no. For example, a legitimate question might be "Is my last term negative?" or "Do my factors have opposite signs?"

3. The game is over when all factorization/polynomial pairs have "found" one another.

*Thanks to Jann MacInnes of Florida Community College at Jacksonville–Kent Campus for suggesting this activity.

5.7   Solving Polynomial Equations by Factoring

The Principle of Zero Products   ■   Factoring to Solve Equations

When we factor a polynomial, we are forming an *equivalent expression*. We now use our factoring skills to *solve equations*. We already know how to solve linear equations like $x + 2 = 7$ and $2x = 9$. The equations we will learn to solve in this section contain a variable raised to a power greater than 1 and will usually have more than one solution.

Whenever two polynomials are set equal to each other, we have a *polynomial equation*. The degree of a polynomial equation is the same as the highest degree of any term in the equation. Second-degree equations like $4t^2 - 9 = 0$ and $x^2 + 6x + 5 = 0$ are called **quadratic equations**.

> ### Quadratic Equation
> A *quadratic equation* is an equation equivalent to one of the form
> $$ax^2 + bx + c = 0,$$
> where a, b, and c are constants, with $a \neq 0$.

In order to solve quadratic equations, we need to develop a new principle.

The Principle of Zero Products

Suppose we are told that the product of two numbers is 6. On the basis of this information, it is impossible to know the value of either number—the product could be $2 \cdot 3, 6 \cdot 1, 12 \cdot \frac{1}{2}$, and so on. However, if we are told that the product of two numbers is 0, we know that at least one of the two factors must itself be 0. For example, if $(x + 3)(x - 2) = 0$, we can conclude that either $x + 3$ is 0 or $x - 2$ is 0.

> ### The Principle of Zero Products
>
> An equation $AB = 0$ is true if and only if $A = 0$ or $B = 0$, or both.
> (A product is 0 if and only if at least one factor is 0.)

EXAMPLE 1

STUDY SKILLS

Identify the Highlights

If you haven't already tried one, consider using a highlighter as you read. By highlighting sentences or phrases that you find especially important, you will make it easier to review important material in the future. Be sure to keep your highlighter available when you study.

Solve: $(x + 3)(x - 2) = 0$.

SOLUTION We are looking for all values of x that will make the equation true. The equation tells us that the product of $x + 3$ and $x - 2$ is 0. In order for the product to be 0, at least one factor must be 0. Thus we look for any value of x for which $x + 3 = 0$, as well as any value of x for which $x - 2 = 0$, that is, either

$$x + 3 = 0 \quad or \quad x - 2 = 0. \qquad \text{Using the principle of zero products. There are two equations to solve.}$$

We solve each equation:

$$\begin{array}{ccc} x + 3 = 0 & or & x - 2 = 0 \\ x = -3 & or & x = 2. \end{array}$$

Both -3 and 2 should be checked in the original equation.

Check: For -3:

$$\frac{(x + 3)(x - 2) = 0}{(-3 + 3)(-3 - 2) \mid 0}$$

The factor $x + 3$ is 0 when $x = -3$. $\rightarrow 0(-5)$

$$0 \overset{?}{=} 0 \quad \text{TRUE}$$

For 2:

$$\frac{(x + 3)(x - 2) = 0}{(2 + 3)(2 - 2) \mid 0}$$

The factor $x - 2$ is 0 when $x = 2$. $5(0)$

$$0 \overset{?}{=} 0 \quad \text{TRUE}$$

The solutions are -3 and 2.

TRY EXERCISE 5

When we are using the principle of zero products, the word "or" is meant to emphasize that any one of the factors could be the one that represents 0.

EXAMPLE 2

Solve: $3(5x + 1)(x - 7) = 0$.

SOLUTION The factors in this equation are 3, $5x + 1$, and $x - 7$. Since the factor 3 is constant, the only way in which $3(5x + 1)(x - 7)$ can be 0 is for one of the other factors to be 0, that is,

$$\begin{array}{ccc} 5x + 1 = 0 & or & x - 7 = 0 \\ 5x = -1 & or & x = 7 \\ x = -\frac{1}{5} & or & x = 7. \end{array}$$

Using the principle of zero products

Solving the two equations separately

$5x + 1 = 0$ when $x = -\frac{1}{5}$; $x - 7 = 0$ when $x = 7$

Check: For $-\frac{1}{5}$:

$$\frac{3(5x + 1)(x - 7) = 0}{3\left(5\left(-\frac{1}{5}\right) + 1\right)\left(-\frac{1}{5} - 7\right)\ \bigg|\ 0}$$

$$3(-1 + 1)\left(-7\tfrac{1}{5}\right)$$

$$3(0)\left(-7\tfrac{1}{5}\right)$$

$$0 \overset{?}{=} 0 \quad \text{TRUE}$$

For 7:

$$\frac{3(5x + 1)(x - 7) = 0}{3(5(7) + 1)(7 - 7)\ \bigg|\ 0}$$

$$3(35 + 1)0$$

$$0 \overset{?}{=} 0 \quad \text{TRUE}$$

The solutions are $-\frac{1}{5}$ and 7.

TRY EXERCISE ▶ 9

The constant factor 3 in Example 2 is never 0 and is not a solution of the equation. However, a variable factor such as x or t *can* equal 0, and must be considered when using the principle of zero products.

EXAMPLE 3   Solve: $7t(t - 5) = 0$.

SOLUTION   We have

$$7 \cdot t(t - 5) = 0 \qquad \text{The factors are 7, } t, \text{ and } t - 5.$$

$$t = 0 \quad or \quad t - 5 = 0 \qquad \text{Using the principle of zero products}$$

$$t = 0 \quad or \qquad t = 5. \qquad \text{Solving. Note that the constant factor, 7, is never 0.}$$

The solutions are 0 and 5. The check is left to the student.

TRY EXERCISE ▶ 15

Factoring to Solve Equations

By factoring and using the principle of zero products, we can now solve a variety of quadratic equations.

EXAMPLE 4   Solve: $x^2 + 5x + 6 = 0$.

SOLUTION   This equation differs from those solved in Chapter 2. There are no like terms to combine, and there is a squared term. We first factor the polynomial. Then we use the principle of zero products:

$$x^2 + 5x + 6 = 0$$

$$(x + 2)(x + 3) = 0 \qquad \text{Factoring}$$

$$x + 2 = 0 \quad or \quad x + 3 = 0 \qquad \text{Using the principle of zero products}$$

$$x = -2 \quad or \qquad x = -3.$$

Check: For -2:

$$\frac{x^2 + 5x + 6 = 0}{(-2)^2 + 5(-2) + 6\ \bigg|\ 0}$$

$$4 - 10 + 6$$

$$-6 + 6$$

$$0 \overset{?}{=} 0 \quad \text{TRUE}$$

For -3:

$$\frac{x^2 + 5x + 6 = 0}{(-3)^2 + 5(-3) + 6\ \bigg|\ 0}$$

$$9 - 15 + 6$$

$$-6 + 6$$

$$0 \overset{?}{=} 0 \quad \text{TRUE}$$

The solutions are -2 and -3.

TRY EXERCISE ▶ 21

The principle of zero products applies even when there is a common factor.

EXAMPLE **5**

Solve: $x^2 + 7x = 0$.

SOLUTION Although there is no constant term, because of the x^2-term, the equation is still quadratic. The methods of Chapter 2 are not sufficient, so we try factoring:

$$x^2 + 7x = 0$$
$$x(x + 7) = 0 \qquad \text{Factoring out the largest common factor, } x$$
$$x = 0 \quad or \quad x + 7 = 0 \qquad \text{Using the principle of zero products}$$
$$x = 0 \quad or \qquad x = -7.$$

The solutions are 0 and -7. The check is left to the student.

> TRY EXERCISE 27

STUDENT NOTES

Checking for a common factor is an important step that is often overlooked. In Example 5, the equation must be factored. If we "divide both sides by x," we will not find the solution 0.

> **CAUTION!** We *must* have 0 on one side of the equation before the principle of zero products can be used. Get all nonzero terms on one side and 0 on the other.

EXAMPLE **6**

Solve: **(a)** $x^2 - 8x = -16$; **(b)** $4t^2 = 25$.

SOLUTION

a) We first add 16 to get 0 on one side:

$$x^2 - 8x = -16$$
$$x^2 - 8x + 16 = 0 \qquad \text{Adding 16 to both sides to get 0 on one side}$$
$$(x - 4)(x - 4) = 0 \qquad \text{Factoring}$$
$$x - 4 = 0 \quad or \quad x - 4 = 0 \qquad \text{Using the principle of zero products}$$
$$x = 4 \quad or \qquad x = 4.$$

There is only one solution, 4. The check is left to the student.

b) We have

$$4t^2 = 25$$
$$4t^2 - 25 = 0 \qquad \text{Subtracting 25 from both sides to get 0 on one side}$$
$$(2t - 5)(2t + 5) = 0 \qquad \text{Factoring a difference of squares}$$

$$\left.\begin{array}{l} 2t - 5 = 0 \quad or \quad 2t + 5 = 0 \\ 2t = 5 \quad or \qquad 2t = -5 \\ t = \frac{5}{2} \quad or \qquad t = -\frac{5}{2}. \end{array}\right\} \quad \text{Solving the two equations separately}$$

The solutions are $\frac{5}{2}$ and $-\frac{5}{2}$. The check is left to the student.

> TRY EXERCISE 41

When solving quadratic equations by factoring, remember that a factorization is not useful unless 0 is on the other side of the equation.

EXAMPLE **7**

Solve: $(x + 3)(2x - 1) = 9$.

SOLUTION Be careful with an equation like this! Since we need 0 on one side, we first multiply out the product on the left and then subtract 9 from both sides.

$$(x + 3)(2x - 1) = 9 \qquad \text{This is not a product equal to 0.}$$
$$2x^2 + 5x - 3 = 9 \qquad \text{Multiplying on the left}$$
$$2x^2 + 5x - 3 - 9 = 9 - 9 \qquad \text{Subtracting 9 from both sides to}$$
$$\text{get 0 on one side}$$
$$2x^2 + 5x - 12 = 0 \qquad \text{Combining like terms}$$
$$(2x - 3)(x + 4) = 0 \qquad \text{Factoring. Now we have a product}$$
$$\text{equal to 0.}$$
$$2x - 3 = 0 \quad or \quad x + 4 = 0 \qquad \text{Using the principle of zero}$$
$$\text{products}$$
$$2x = 3 \quad or \qquad x = -4$$
$$x = \tfrac{3}{2} \quad or \qquad x = -4$$

Check: For $\tfrac{3}{2}$:   For -4:

$$\begin{array}{c|c} (x + 3)(2x - 1) = 9 \\ \hline \left(\tfrac{3}{2} + 3\right)\left(2 \cdot \tfrac{3}{2} - 1\right) & 9 \\ \left(\tfrac{9}{2}\right)(2) & \\ & 9 \overset{?}{=} 9 \quad \text{TRUE} \end{array} \qquad \begin{array}{c|c} (x + 3)(2x - 1) = 9 \\ \hline (-4 + 3)(2(-4)-1) & 9 \\ (-1)(-9) & \\ & 9 \overset{?}{=} 9 \quad \text{TRUE} \end{array}$$

The solutions are $\tfrac{3}{2}$ and -4.   **TRY EXERCISE** 43

We can use the principle of zero products to solve polynomials with degree greater than 2, if they can be factored.

EXAMPLE 8   Solve: $3x^3 - 30x = 9x^2$.

SOLUTION   We have

$$3x^3 - 30x = 9x^2$$
$$3x^3 - 9x^2 - 30x = 0 \qquad \text{Getting 0 on one side and writing}$$
$$\text{in descending order}$$
$$3x(x^2 - 3x - 10) = 0 \qquad \text{Factoring out a common factor}$$
$$3x(x + 2)(x - 5) = 0 \qquad \text{Factoring the trinomial}$$
$$3x = 0 \quad or \quad x + 2 = 0 \quad or \quad x - 5 = 0 \qquad \text{Using the principle of zero}$$
$$\text{products}$$
$$x = 0 \quad or \qquad x = -2 \quad or \qquad x = 5.$$

Check:

$$\begin{array}{c|c} 3x^3 - 30x = 9x^2 \\ \hline 3 \cdot 0^3 - 30 \cdot 0 & 9 \cdot 0^2 \\ 0 - 0 & 9 \cdot 0 \\ 0 \overset{?}{=} 0 & \text{TRUE} \end{array} \qquad \begin{array}{c|c} 3x^3 - 30x = 9x^2 \\ \hline 3(-2)^3 - 30(-2) & 9(-2)^2 \\ 3(-8) + 60 & 9 \cdot 4 \\ -24 + 60 & 36 \\ 36 \overset{?}{=} 36 & \text{TRUE} \end{array}$$

$$\begin{array}{c|c} 3x^3 - 30x = 9x^2 \\ \hline 3 \cdot 5^3 - 30 \cdot 5 & 9 \cdot 5^2 \\ 3 \cdot 125 - 150 & 9 \cdot 25 \\ 375 - 150 & 225 \\ 225 \overset{?}{=} 225 & \text{TRUE} \end{array}$$

The solutions are 0, -2, and 5.   **TRY EXERCISE** 53

ALGEBRAIC–GRAPHICAL CONNECTION

When graphing equations in Chapter 3, we found the x-intercept by replacing y with 0 and solving for x. This procedure is also used to find the x-intercepts when graphing equations of the form $y = ax^2 + bx + c$. Although the details of creating such graphs is left for Chapter 11, we consider them briefly here from the standpoint of finding x-intercepts. The graphs are shaped as shown. Note that each x-intercept represents a solution of $ax^2 + bx + c = 0$.

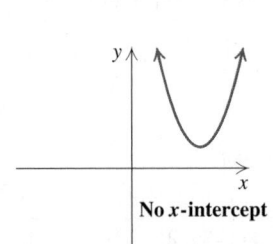

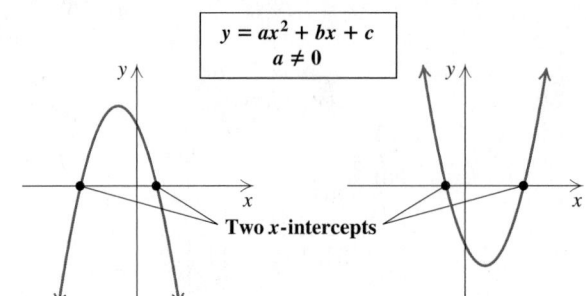

$$y = ax^2 + bx + c$$
$$a \neq 0$$

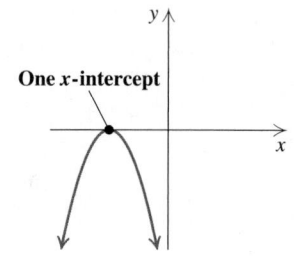

EXAMPLE **9** Find the x-intercepts for the graph of the equation shown. (The grid is intentionally not included.)

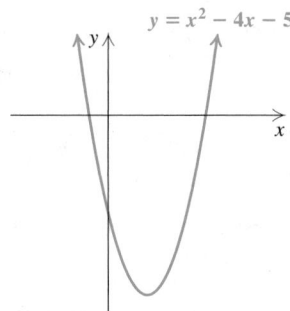

$$y = x^2 - 4x - 5$$

SOLUTION To find the x-intercepts, we let $y = 0$ and solve for x:

$$0 = x^2 - 4x - 5 \qquad \text{Substituting 0 for } y$$
$$0 = (x - 5)(x + 1) \qquad \text{Factoring}$$
$$x - 5 = 0 \quad or \quad x + 1 = 0 \qquad \text{Using the principle of zero products}$$
$$x = 5 \quad or \qquad x = -1. \qquad \text{Solving for } x$$

The x-intercepts are $(5, 0)$ and $(-1, 0)$.

▶ TRY EXERCISE 61

TECHNOLOGY CONNECTION

A graphing calculator allows us to solve polynomial equations even when an equation cannot be solved by factoring. For example, to solve $x^2 - 3x - 5 = 0$, we can let $y_1 = x^2 - 3x - 5$ and $y_2 = 0$. Selecting a bold line type to the left of y_2 in the (Y=) window makes the line easier to see. Using the INTERSECT option of the CALC menu, we select the two graphs in which we are interested, along with a guess. The graphing calculator displays the point of intersection.

An alternative method uses only y_1 and the ZERO option of the CALC menu. This option requires you to enter an x-value to the left of each x-intercept as a LEFT BOUND. An x-value to the right of the x-intercept is then entered as a RIGHT BOUND. Finally, a GUESS value between the two bounds is entered and the x-intercept, or ZERO, is displayed.

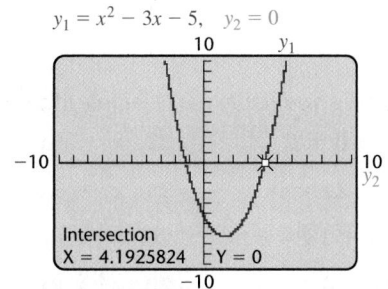

Use a graphing calculator to find the solutions, if they exist, accurate to two decimal places.

1. $x^2 + 4x - 3 = 0$
2. $x^2 - 5x - 2 = 0$
3. $x^2 + 13.54x + 40.95 = 0$
4. $x^2 - 4.43x + 6.32 = 0$
5. $1.235x^2 - 3.409x = 0$

5.7   EXERCISE SET

 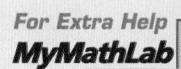

🔁 *Concept Reinforcement*   *For each of Exercises 1–4, match the phrase with the most appropriate choice from the column on the right.*

1. ___ The name of equations of the type $ax^2 + bx + c = 0$, with $a \neq 0$

2. ___ The maximum number of solutions of quadratic equations

3. ___ The idea that $A \cdot B = 0$ if and only if $A = 0$ or $B = 0$

4. ___ The number that a product must equal before the principle of zero products is used

a) 2

b) 0

c) Quadratic

d) The principle of zero products

Solve using the principle of zero products.

5. $(x + 2)(x + 9) = 0$

6. $(x + 3)(x + 10) = 0$

7. $(2t - 3)(t + 6) = 0$

8. $(5t - 8)(t - 1) = 0$

9. $4(7x - 1)(10x - 3) = 0$

10. $6(4x - 3)(2x + 9) = 0$

11. $x(x - 7) = 0$

12. $x(x + 2) = 0$

13. $\left(\frac{2}{3}x - \frac{12}{11}\right)\left(\frac{7}{4}x - \frac{1}{12}\right) = 0$

14. $\left(\frac{1}{9} - 3x\right)\left(\frac{1}{5} + 2x\right) = 0$

15. $6n(3n + 8) = 0$

16. $10n(4n - 5) = 0$

17. $(20 - 0.4x)(7 - 0.1x) = 0$

18. $(1 - 0.05x)(1 - 0.3x) = 0$

19. $(3x - 2)(x + 5)(x - 1) = 0$

20. $(2x + 1)(x + 3)(x - 5) = 0$

Solve by factoring and using the principle of zero products.

21. $x^2 - 7x + 6 = 0$ **22.** $x^2 - 6x + 5 = 0$

23. $x^2 + 4x - 21 = 0$ **24.** $x^2 - 7x - 18 = 0$

25. $n^2 + 11n + 18 = 0$ **26.** $n^2 + 8n + 15 = 0$

27. $x^2 - 10x = 0$ **28.** $x^2 + 8x = 0$

29. $6t + t^2 = 0$ **30.** $3t - t^2 = 0$

31. $x^2 - 36 = 0$ **32.** $x^2 - 100 = 0$

33. $4t^2 = 49$ **34.** $9t^2 = 25$

35. $0 = 25 + x^2 + 10x$ **36.** $0 = 6x + x^2 + 9$

37. $64 + x^2 = 16x$ **38.** $x^2 + 1 = 2x$

39. $4t^2 = 8t$ **40.** $12t = 3t^2$

41. $4y^2 = 7y + 15$ **42.** $12y^2 - 5y = 2$

43. $(x - 7)(x + 1) = -16$

44. $(x + 2)(x - 7) = -18$

45. $15z^2 + 7 = 20z + 7$

46. $14z^2 - 3 = 21z - 3$

47. $36m^2 - 9 = 40$

48. $81x^2 - 5 = 20$

49. $(x + 3)(3x + 5) = 7$

50. $(x - 1)(5x + 4) = 2$

51. $3x^2 - 2x = 9 - 8x$

52. $x^2 - 2x = 18 + 5x$

53. $x^2(2x - 1) = 3x$

54. $x^2 = x(10 - 3x^2)$

55. $(2x - 5)(3x^2 + 29x + 56) = 0$

56. $(4x + 9)(15x^2 - 7x - 2) = 0$

57. Use this graph to solve $x^2 - 3x - 4 = 0$.

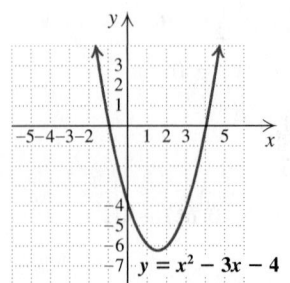

58. Use this graph to solve $x^2 + x - 6 = 0$.

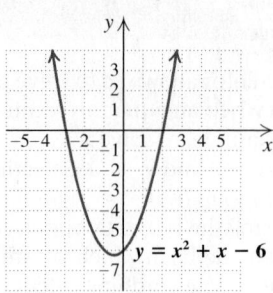

59. Use this graph to solve $-x^2 - x + 6 = 0$.

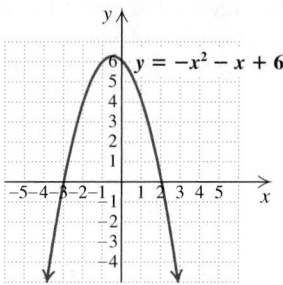

60. Use this graph to solve $-x^2 + 2x + 3 = 0$.

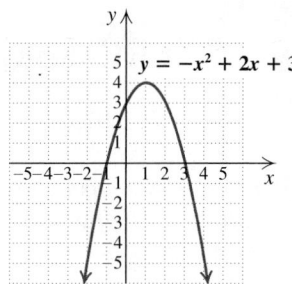

Find the x-intercepts for the graph of each equation. Grids are intentionally not included.

61. $y = x^2 - x - 6$ **62.** $y = x^2 + 3x - 4$

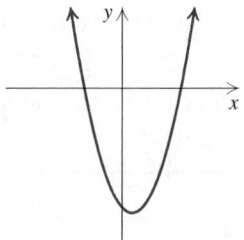   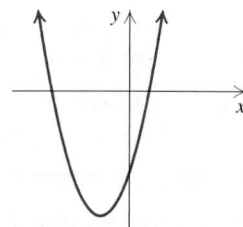

63. $y = x^2 + 2x - 8$ **64.** $y = x^2 - 2x - 15$

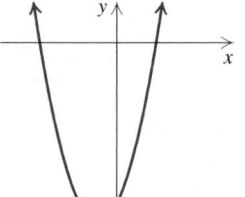   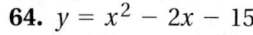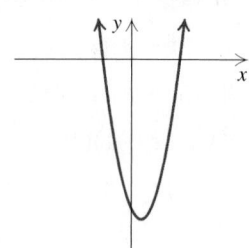

65. $y = 2x^2 + 3x - 9$ **66.** $y = 2x^2 + x - 10$

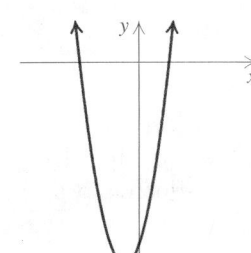   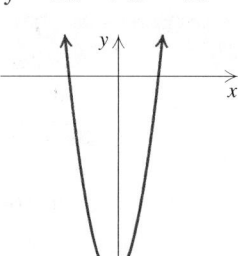

67. The equation $x^2 + 1 = 0$ has no real-number solutions. What implications does this have for the graph of $y = x^2 + 1$?

68. What is the difference between a quadratic polynomial and a quadratic equation?

Skill Review

To prepare for Section 5.8, review solving problems using the five-step strategy (Section 2.5).

Translate to an algebraic expression. [1.1]

69. The square of the sum of two numbers

70. The sum of the squares of two numbers

71. The product of two consecutive integers

Solve. [2.5]

72. In 2005, shoppers spent $22.8 billion on gifts for Mother's Day and for Father's Day combined. They spent $4.8 billion more for Mother's Day than for Father's Day. How much did shoppers spend for each holiday?
Source: National Retail Federation

73. The first angle of a triangle is four times as large as the second. The measure of the third angle is 30° less than that of the second. How large are the angles?

74. A rectangular table top is twice as long as it is wide. The perimeter of the table is 192 in. What are the dimensions of the table?

Synthesis

75. What is wrong with solving $x^2 = 3x$ by dividing both sides of the equation by x?

76. When the principle of zero products is used to solve a quadratic equation, will there always be two different solutions? Why or why not?

77. Find an equation with integer coefficients that has the given numbers as solutions. For example, 3 and -2 are solutions to $x^2 - x - 6 = 0$.
a) $-4, 5$ **b)** $-1, 7$ **c)** $\frac{1}{4}, 3$
d) $\frac{1}{2}, \frac{1}{3}$ **e)** $\frac{2}{3}, \frac{3}{4}$ **f)** $-1, 2, 3$

Solve.

78. $16(x - 1) = x(x + 8)$

79. $a(9 + a) = 4(2a + 5)$

80. $(t - 5)^2 = 2(5 - t)$

81. $-x^2 + \frac{9}{25} = 0$

82. $a^2 = \frac{49}{100}$

Aha! **83.** $(t + 1)^2 = 9$

84. $\frac{27}{25} x^2 = \frac{1}{3}$

85. For each equation on the left, find an equivalent equation on the right.
a) $x^2 + 10x - 2 = 0$ $4x^2 + 8x + 36 = 0$
b) $(x - 6)(x + 3) = 0$ $(2x + 8)(2x - 5) = 0$
c) $5x^2 - 5 = 0$ $9x^2 - 12x + 24 = 0$
d) $(2x - 5)(x + 4) = 0$ $(x + 1)(5x - 5) = 0$
e) $x^2 + 2x + 9 = 0$ $x^2 - 3x - 18 = 0$
f) $3x^2 - 4x + 8 = 0$ $2x^2 + 20x - 4 = 0$

86. Explain how to construct an equation that has seven solutions.

87. Explain how the graph in Exercise 59 can be used to visualize the solutions of
$$-x^2 - x + 6 = 4.$$

Use a graphing calculator to find the solutions of each equation. Round solutions to the nearest hundredth.

88. $-x^2 + 0.63x + 0.22 = 0$

89. $x^2 - 9.10x + 15.77 = 0$

90. $6.4x^2 - 8.45x - 94.06 = 0$

91. $x^2 + 13.74x + 42.00 = 0$

92. $0.84x^2 - 2.30x = 0$

93. $1.23x^2 + 4.63x = 0$

94. $x^2 + 1.80x - 5.69 = 0$

CONNECTING the CONCEPTS

Recall that an *equation* is a statement that two *expressions* are equal. When we simplify expressions, combine expressions, and form equivalent expressions, each result is an expression. When we are asked to solve an equation, the result is one or more numbers. Remember to read the directions to an exercise carefully so you do not attempt to "solve" an expression.

MIXED REVIEW

For Exercises 1–6, tell whether each is an example of an expression or an equation.

1. $x^2 - 25$

2. $x^2 - 25 = 0$

3. $x^2 + 2x = 5$

4. $(x + 3)(2x - 1)$

5. $x(x + 3) - 2(2x - 7) - (x - 5)$

6. $x = 10$

7. Add the expressions:

$(2x^3 - 5x + 1) + (x^2 - 3x - 1).$

8. Subtract the expressions:

$(x^2 - x - 5) - (3x^2 - x + 6).$

9. Solve the equation: $t^2 - 100 = 0.$

10. Multiply the expressions: $(3a - 2)(2a - 5).$

11. Factor the expression: $n^2 - 10n + 9.$

12. Solve the equation: $x^2 + 16 = 10x.$

13. Solve: $4t^2 + 20t + 25 = 0.$

14. Add: $(3x^3 - 5x + 1) + (4x^3 + 7x - 8).$

15. Factor: $16x^2 - 81.$

16. Solve: $y^2 - 5y - 24 = 0.$

17. Subtract: $(a^2 - 2) - (5a^2 + a + 9).$

18. Factor: $18x^4 - 24x^3 + 20x^2.$

19. Solve: $3x^2 + 5x + 2 = 0.$

20. Multiply: $4x^2(2x^3 - 5x^2 + 3).$

5.8 Solving Applications

Applications • The Pythagorean Theorem

Applications

We can use the five-step problem-solving process and our new methods of solving quadratic equations to solve new types of problems.

EXAMPLE **1**

Race numbers. Terry and Jody each entered a boat in the Lakeport Race. The racing numbers of their boats were consecutive numbers, the product of which was 156. Find the numbers.

SOLUTION

STUDY SKILLS ————

Finishing a Chapter

Reaching the end of a chapter often signals the arrival of a quiz or test. Almost always, it indicates the end of a particular area of study. Success in future work will depend on your mastery of previous topics studied. Use the chapter review and test to solidify your understanding of a chapter's material.

1. **Familiarize.** Consecutive numbers are one apart, like 49 and 50. Let $x =$ the first boat number; then $x + 1 =$ the next boat number.

2. **Translate.** We reword the problem before translating:

Rewording: The first boat number   times   the next boat number   is   156.

Translating: x $\cdot$ $(x + 1)$ $=$ 156

3. **Carry out.** We solve the equation as follows:

$$x(x + 1) = 156$$
$$x^2 + x = 156 \quad \text{Multiplying}$$
$$x^2 + x - 156 = 0 \quad \text{Subtracting 156 to get 0 on one side}$$
$$(x - 12)(x + 13) = 0 \quad \text{Factoring}$$
$$x - 12 = 0 \quad or \quad x + 13 = 0 \quad \text{Using the principle of zero products}$$
$$x = 12 \quad or \quad x = -13. \quad \text{Solving each equation}$$

4. **Check.** The solutions of the equation are 12 and -13. Since race numbers are not negative, -13 must be rejected. On the other hand, if x is 12, then $x + 1$ is 13 and $12 \cdot 13 = 156$. Thus the solution 12 checks.

5. **State.** The boat numbers for Terry and Jody were 12 and 13.

TRY EXERCISE 5

EXAMPLE **2**

Manufacturing. Wooden Work, Ltd., builds cutting boards that are twice as long as they are wide. The most popular board that Wooden Work makes has an area of 800 cm². What are the dimensions of the board?

SOLUTION

1. **Familiarize.** We first make a drawing. Recall that the area of any rectangle is Length · Width. We let x = the width of the board, in centimeters. The length is then $2x$, since the board is twice as long as it is wide.

2. **Translate.** We reword and translate as follows:

Rewording: The area of the rectangle   is   800 cm².

Translating: $2x \cdot x$ = 800

3. **Carry out.** We solve the equation as follows:

$$2x \cdot x = 800$$
$$2x^2 = 800$$
$$2x^2 - 800 = 0 \qquad \text{Subtracting 800 to get 0 on one side of the equation}$$
$$2(x^2 - 400) = 0 \qquad \text{Factoring out a common factor of 2}$$
$$2(x - 20)(x + 20) = 0 \qquad \text{Factoring a difference of squares}$$
$$(x - 20)(x + 20) = 0 \qquad \text{Dividing both sides by 2}$$
$$x - 20 = 0 \quad or \quad x + 20 = 0 \qquad \text{Using the principle of zero products}$$
$$x = 20 \quad or \qquad x = -20. \qquad \text{Solving each equation}$$

4. **Check.** The solutions of the equation are 20 and −20. Since the width must be positive, −20 cannot be a solution. To check 20 cm, we note that if the width is 20 cm, then the length is 2 · 20 cm = 40 cm and the area is 20 cm · 40 cm = 800 cm². Thus the solution 20 checks.

5. **State.** The cutting board is 20 cm wide and 40 cm long.   **TRY EXERCISE** ▸ 9

EXAMPLE ▸ **3**   ***Dimensions of a leaf.***   Each leaf of one particular *Philodendron* species is approximately a triangle. A typical leaf has an area of 320 in². If the leaf is 12 in. longer than it is wide, find the length and the width of the leaf.

SOLUTION

1. **Familiarize.** The formula for the area of a triangle is Area = $\frac{1}{2}$ · (base) · (height). We let b = the width, in inches, of the triangle's base and $b + 12$ = the height, in inches.

2. **Translate.** We reword and translate as follows:

Rewording: The area of the leaf   is   320 in².

Translating: $\frac{1}{2} \cdot b(b + 12)$ = 320

3. Carry out. We solve the equation as follows:

$$\tfrac{1}{2} \cdot b \cdot (b + 12) = 320$$

$\tfrac{1}{2}(b^2 + 12b) = 320$	Multiplying
$b^2 + 12b = 640$	Multiplying by 2 to clear fractions
$b^2 + 12b - 640 = 0$	Subtracting 640 to get 0 on one side
$(b + 32)(b - 20) = 0$	Factoring
$b + 32 = 0 \quad or \quad b - 20 = 0$	Using the principle of zero products
$b = -32 \quad or \qquad b = 20.$	

4. Check. The width must be positive, so -32 cannot be a solution. Suppose the base is 20 in. The height would be $20 + 12$, or 32 in., and the area $\tfrac{1}{2}(20)(32)$, or 320 in². These numbers check in the original problem.

5. State. The leaf is 32 in. long and 20 in. wide.

▶ TRY EXERCISE ▶ 13

EXAMPLE 4

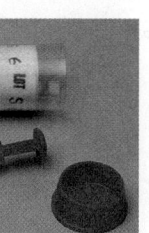

Medicine. For certain people suffering an extreme allergic reaction, the drug epinephrine (adrenaline) is sometimes prescribed. The number of micrograms N of epinephrine in an adult's bloodstream t minutes after 250 micrograms have been injected can be approximated by

$$-10t^2 + 100t = N.$$

How long after an injection will there be about 210 micrograms of epinephrine in the bloodstream?

Source: Based on information in Chohan, Naina, Rita M. Doyle, and Patricia Nayle (eds.), *Nursing Handbook*, 21st ed. Springhouse, PA: Springhouse Corporation, 2001

SOLUTION

1. Familiarize. To familiarize ourselves with this problem, we could calculate N for different choices of t. We leave this for the student. Note that there may be two solutions, one on each side of the time at which the drug's effect peaks.

2. Translate. To find the length of time after injection when 210 micrograms are in the bloodstream, we replace N with 210 in the formula above:

$$-10t^2 + 100t = 210. \qquad \text{Substituting 210 for } N. \text{ This is now an equation in one variable.}$$

3. Carry out. We solve the equation as follows:

$-10t^2 + 100t = 210$	
$-10t^2 + 100t - 210 = 0$	Subtracting 210 from both sides to get 0 on one side
$-10(t^2 - 10t + 21) = 0$	Factoring out the largest common factor, -10
$-10(t - 3)(t - 7) = 0$	Factoring
$t - 3 = 0 \quad or \quad t - 7 = 0$	Using the principle of zero products
$t = 3 \quad or \qquad t = 7.$	

As a visual check for Example 4, we can either let $y_1 = -10x^2 + 100x$ and $y_2 = 210$, or let $y_1 = -10x^2 + 100x - 210$ and $y_2 = 0$. In either case, the points of intersection occur at $x = 3$ and $x = 7$, as shown.

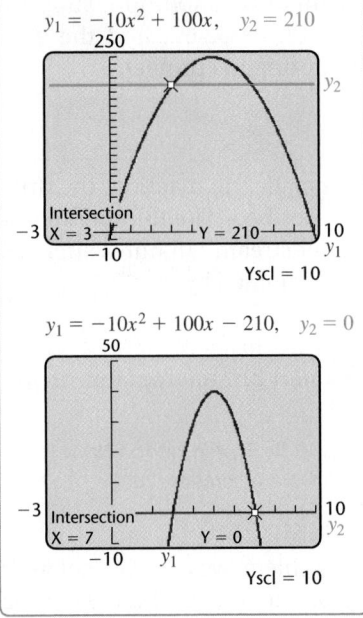

4. **Check.** Since $-10 \cdot 3^2 + 100 \cdot 3 = -90 + 300 = 210$, the number 3 checks. Since $-10 \cdot 7^2 + 100 \cdot 7 = -490 + 700 = 210$, the number 7 also checks.

5. **State.** There will be 210 micrograms of epinephrine in the bloodstream approximately 3 minutes and 7 minutes after injection.

TRY EXERCISE 17

The Pythagorean Theorem

The following problems involve the Pythagorean theorem, which relates the lengths of the sides of a *right* triangle. A triangle is a **right triangle** if it has a 90°, or *right*, angle. The side opposite the 90° angle is called the **hypotenuse**. The other sides are called **legs**.

The Pythagorean Theorem

In any right triangle, if a and b are the lengths of the legs and c is the length of the hypotenuse, then

$$a^2 + b^2 = c^2, \quad \text{or}$$

$$(\text{Leg})^2 + (\text{Other leg})^2 = (\text{Hypotenuse})^2.$$

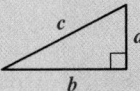

The equation $a^2 + b^2 = c^2$ is called the **Pythagorean equation**.*

The Pythagorean theorem is named for the Greek mathematician Pythagoras (569?–500? B.C.). We can think of this relationship as adding areas.

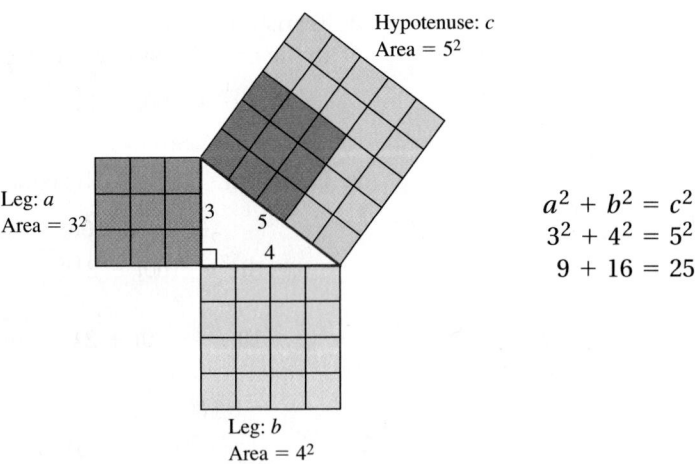

If we know the lengths of any two sides of a right triangle, we can use the Pythagorean equation to determine the length of the third side.

---

*The *converse* of the Pythagorean theorem is also true. That is, if $a^2 + b^2 = c^2$, then the triangle is a right triangle.

EXAMPLE   5

Travel.   A zipline canopy tour in Alaska includes a cable that slopes downward from a height of 135 ft to a height of 100 ft. The trees that the cable connects are 120 ft apart. Find the minimum length of the cable.

SOLUTION

1. **Familiarize.** We first make a drawing or visualize the situation. The difference in height between the platforms is 35 ft. Note that the cable must have some extra length to allow for the rider's movement, but we will approximate its length as the hypotenuse of a right triangle, as shown.

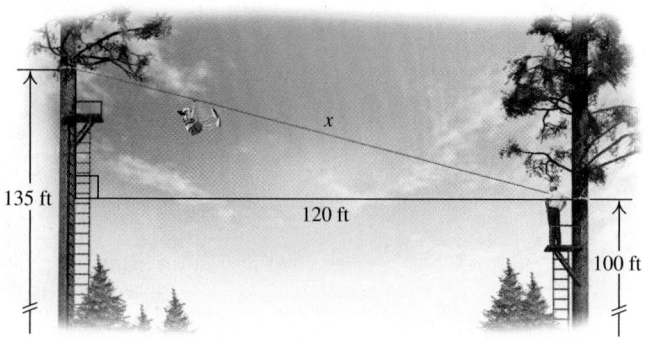

2. **Translate.** Since a right triangle is formed, we can use the Pythagorean theorem:

$$a^2 + b^2 = c^2$$
$$35^2 + 120^2 = x^2. \qquad \text{Substituting}$$

3. **Carry out.** We solve the equation as follows:

$1225 + 14{,}400 = x^2$	Squaring 35 and 120
$15{,}625 = x^2$	Adding
$0 = x^2 - 15{,}625$	Subtracting 15,625 from both sides
$0 = (x + 125)(x - 125)$	Note that $15{,}625 = 125^2$. A calculator would be helpful here.
$x + 125 = 0 \quad or \quad x - 125 = 0$	Using the principle of zero products
$x = -125 \quad or \qquad x = 125.$	

4. **Check.** Since the length of the cable must be positive, -125 is not a solution. If the length is 125 ft, we have $35^2 + 120^2 = 1225 + 14{,}400 = 15{,}625$, which is 125^2. Thus the solution 125 checks.

5. **State.** The minimum length of the cable is 125 ft.

> **TRY EXERCISE** 27

EXAMPLE **6** *Bridge design.* A 50-ft diagonal brace on a bridge connects a support at the center of the bridge to a side support on the bridge. The horizontal distance that it spans is 10 ft longer than the height that it reaches on the side of the bridge. Find both distances.

SOLUTION

1. Familiarize. We first make a drawing. The diagonal brace and the missing distances form the hypotenuse and the legs of a right triangle. We let x = the length of the vertical leg. Then $x + 10$ = the length of the horizontal leg. The hypotenuse has length 50 ft.

2. Translate. Since the triangle is a right triangle, we can use the Pythagorean theorem:

$$a^2 + b^2 = c^2$$
$$x^2 + (x + 10)^2 = 50^2. \quad \text{Substituting}$$

3. Carry out. We solve the equation as follows:

$x^2 + (x^2 + 20x + 100) = 2500$	Squaring
$2x^2 + 20x + 100 = 2500$	Combining like terms
$2x^2 + 20x - 2400 = 0$	Subtracting 2500 to get 0 on one side
$2(x^2 + 10x - 1200) = 0$	Factoring out a common factor
$2(x + 40)(x - 30) = 0$	Factoring. A calculator would be helpful here.
$x + 40 = 0 \quad or \quad x - 30 = 0$	Using the principle of zero products
$x = -40 \quad or \quad x = 30.$	

4. Check. The integer -40 cannot be a length of a side because it is negative. If the length is 30 ft, $x + 10 = 40$, and $30^2 + 40^2 = 900 + 1600 = 2500$, which is 50^2. So the solution 30 checks.

5. State. The height that the brace reaches on the side of the bridge is 30 ft, and the distance that it reaches to the middle of the bridge is 40 ft.

 **TRY EXERCISE** 31

Translating for Success

1. Angle measures. The measures of the angles of a triangle are three consecutive integers. Find the measures of the angles.

2. Rectangle dimensions. The area of a rectangle is 3604 ft². The length is 15 ft longer than the width. Find the dimensions of the rectangle.

3. Sales tax. Claire paid $3604 for a used pickup truck. This included 6% for sales tax. How much did the truck cost before tax?

4. Wire cutting. A 180-m wire is cut into three pieces. The third piece is 2 m longer than the first. The second is two-thirds as long as the first. How long is each piece?

5. Perimeter. The perimeter of a rectangle is 240 ft. The length is 2 ft greater than the width. Find the length and the width.

Translate each word problem to an equation and select a correct translation from equations A–O.

A. $2x \cdot x = 288$

B. $x(x + 60) = 7021$

C. $59 = x \cdot 60$

D. $x^2 + (x + 15)^2 = 3604$

E. $x^2 + (x + 70)^2 = 130^2$

F. $0.06x = 3604$

G. $2(x + 2) + 2x = 240$

H. $\dfrac{1}{2}x(x - 1) = 1770$

I. $x + \dfrac{2}{3}x + (x + 2) = 180$

J. $0.59x = 60$

K. $x + 0.06x = 3604$

L. $2x^2 + x = 288$

M. $x(x + 15) = 3604$

N. $x^2 + 60 = 7021$

O. $x + (x + 1) + (x + 2) = 180$

Answers on page A-22

An additional, animated version of this activity appears in MyMathLab. To use MyMathLab, you need a course ID and a student access code. Contact your instructor for more information.

6. Cell-phone tower. A guy wire on a cell-phone tower is 130 ft long and is attached to the top of the tower. The height of the tower is 70 ft longer than the distance from the point on the ground where the wire is attached to the bottom of the tower. Find the height of the tower.

7. Sales meeting attendance. PTQ Corporation holds a sales meeting in Tucson. Of the 60 employees, 59 of them attend the meeting. What percent attend the meeting?

8. Dimensions of a pool. A rectangular swimming pool is twice as long as it is wide. The area of the surface is 288 ft². Find the dimensions of the pool.

9. Dimensions of a triangle. The height of a triangle is 1 cm less than the length of the base. The area of the triangle is 1770 cm². Find the height and the length of the base.

10. Width of a rectangle. The length of a rectangle is 60 ft longer than the width. Find the width if the area of the rectangle is 7021 ft².

5.8 EXERCISE SET

Solve. Use the five-step problem-solving approach.

1. A number is 6 less than its square. Find all such numbers.

2. A number is 30 less than its square. Find all such numbers.

3. One leg of a right triangle is 2 m longer than the other leg. The length of the hypotenuse is 10 m. Find the length of each side.

4. One leg of a right triangle is 7 cm shorter than the other leg. The length of the hypotenuse is 13 cm. Find the length of each side.

5. *Parking-space numbers.* The product of two consecutive parking spaces is 132. Find the parking-space numbers.

6. *Page numbers.* The product of the page numbers on two facing pages of a book is 420. Find the page numbers.

7. The product of two consecutive even integers is 168. Find the integers.

8. The product of two consecutive odd integers is 195. Find the integers.

9. *Construction.* The front porch on Trent's new home is five times as long as it is wide. If the area of the porch is 180 ft^2, find the dimensions.

10. *Furnishings.* The work surface of Anita's desk is a rectangle that is twice as long as it is wide. If the area

of the desktop is 18 ft^2, find the length and the width of the desk.

11. *Design.* The screen of the TI-84 Plus graphing calculator is nearly rectangular. The length of the rectangle is 2 cm more than the width. If the area of the rectangle is 24 cm^2, find the length and the width.

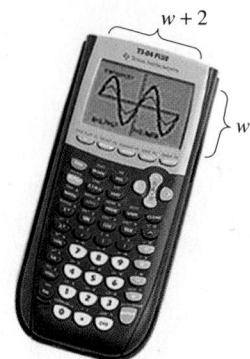

12. *Area of a garden.* The length of a rectangular garden is 4 m greater than the width. The area of the garden is 96 m^2. Find the length and the width.

13. *Dimensions of a triangle.* The height of a triangle is 3 in. less than the length of the base. If the area of the triangle is 54 in^2, find the height and the length of the base.

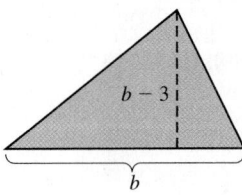

14. *Dimensions of a triangle.* A triangle is 10 cm wider than it is tall. The area is 48 cm^2. Find the height and the base.

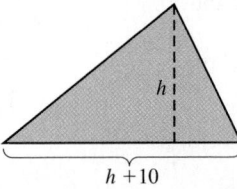

15. *Dimensions of a sail.* The height of the jib sail on a Lightning sailboat is 5 ft greater than the length of its "foot." If the area of the sail is 42 ft^2, find the length of the foot and the height of the sail.

16. *Road design.* A triangular traffic island has a base half as long as its height. Find the base and the height if the island has an area of 64 m^2.

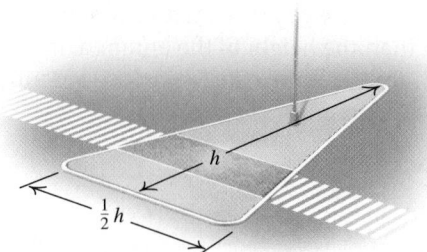

17. *Medicine.* For many people suffering from constricted bronchial muscles, the drug Albuterol is prescribed. The number of micrograms A of Albuterol in a person's bloodstream t minutes after 200 micrograms have been inhaled can be approximated by

$$A = -50t^2 + 200t.$$

How long after an inhalation will there be about 150 micrograms of Albuterol in the bloodstream?
Source: Based on information in Chohan, Naina, Rita M. Doyle, and Patricia Nayle (eds.), *Nursing Handbook*, 21st ed. Springhouse, PA: Springhouse Corporation, 2001

18. *Medicine.* For adults with certain heart conditions, the drug Primacor (milrinone lactate) is prescribed. The number of milligrams M of Primacor in the bloodstream of a 132-lb patient t hours after a 3-mg dose has been injected can be approximated by

$$M = -\frac{1}{2}t^2 + \frac{5}{2}t.$$

How long after an injection will there be about 2 mg in the bloodstream?
Source: Based on information in Chohan, Naina, Rita M. Doyle, and Patricia Nayle (eds.), *Nursing Handbook*, 21st ed. Springhouse, PA: Springhouse Corporation, 2001

19. *High-speed Internet.* The number N, in millions, of broadband cable and DSL subscribers in the United States t years after 1998 can be approximated by

$$N = 0.3t^2 + 0.6t.$$

When will there be 36 million broadband cable and DSL subscribers?
Source: Based on information from Leichtman Research Group

20. *Wave height.* The height of waves in a storm depends on the speed of the wind. Assuming the wind has no obstructions for a long distance, the maximum wave height H for a wind speed x can be approximated by

$$H = 0.006x^2 + 0.6x.$$

Here H is in feet and x is in knots (nautical miles per hour). For what wind speed would the maximum wave height be 6.6 ft?
Source: Based on information from cimss.ssec.wisc.edu

Games in a league's schedule. *In a sports league of x teams in which all teams play each other twice, the total number N of games played is given by*

$$x^2 - x = N.$$

Use this formula for Exercises 21 and 22.

21. The Colchester Youth Soccer League plays a total of 240 games, with all teams playing each other twice. How many teams are in the league?

22. The teams in a women's softball league play each other twice, for a total of 132 games. How many teams are in the league?

Number of handshakes. *The number of possible handshakes H within a group of n people is given by $H = \frac{1}{2}(n^2 - n)$. Use this formula for Exercises 23–26.*

23. At a meeting, there are 12 people. How many handshakes are possible?

24. At a party, there are 25 people. How many handshakes are possible?

25. *High-fives.* After winning the championship, all San Antonio Spurs teammates exchanged "high-fives." Altogether there were 66 high-fives. How many players were there?

26. *Toasting.* During a toast at a party, there were 105 "clicks" of glasses. How many people took part in the toast?

27. *Construction.* The diagonal braces in a lookout tower are 15 ft long and span a horizontal distance of 12 ft. How high does each brace reach vertically?

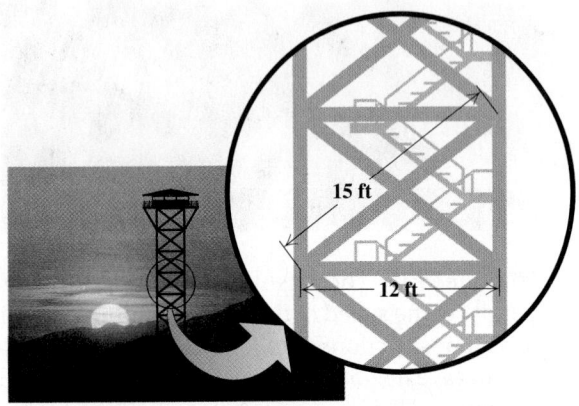

28. *Reach of a ladder.* Twyla has a 26-ft ladder leaning against her house. If the bottom of the ladder is 10 ft from the base of the house, how high does the ladder reach?

29. *Roadway design.* Elliott Street is 24 ft wide when it ends at Main Street in Brattleboro, Vermont. A 40-ft long diagonal crosswalk allows pedestrians to cross Main Street to or from either corner of Elliott Street (see the figure). Determine the width of Main Street.

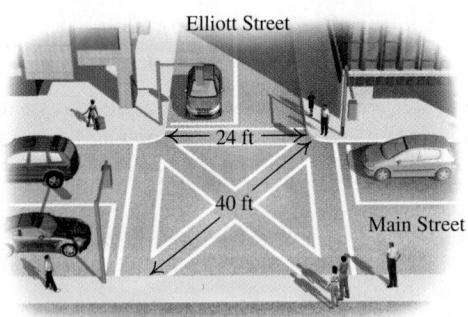

Elliott Street

24 ft

40 ft

Main Street

30. *Aviation.* Engine failure forced Robbin to pilot her Cessna 150 to an emergency landing. To land, Robbin's plane glided 17,000 ft over a 15,000-ft stretch of deserted highway. From what altitude did the descent begin?

31. *Archaeology.* Archaeologists have discovered that the 18th-century garden of the Charles Carroll House in Annapolis, Maryland, was a right triangle. One leg of the triangle was formed by a 400-ft long sea wall. The hypotenuse of the triangle was 200 ft longer than the other leg. What were the dimensions of the garden?
Source: www.bsos.umd.edu

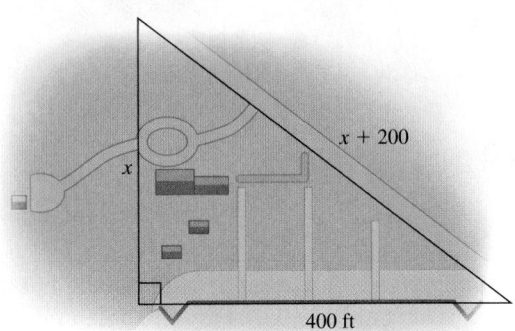

$x + 200$

x

400 ft

32. *Guy wire.* The guy wire on a TV antenna is 1 m longer than the height of the antenna. If the guy wire is anchored 3 m from the foot of the antenna, how tall is the antenna?

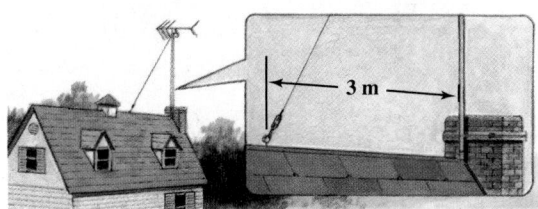

3 m

33. *Architecture.* An architect has allocated a rectangular space of 264 ft^2 for a square dining room and a 10-ft wide kitchen, as shown in the figure. Find the dimensions of each room.

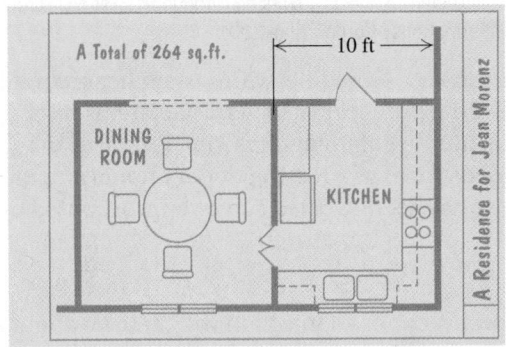

A Total of 264 sq.ft.

10 ft

DINING ROOM

KITCHEN

A Residence for Jean Morenz

34. *Design.* A window panel for a sun porch consists of a 7-ft high rectangular window stacked above a square window. The windows have the same width. If the total area of the window panel is 18 ft², find the dimensions of each window.

7 ft

Height of a rocket. *For Exercises 35–38, assume that a water rocket is launched upward with an initial velocity of 48 ft/sec. Its height h, in feet, after t seconds, is given by* $h = 48t - 16t^2$.

35. Determine the height of the rocket $\frac{1}{2}$ sec after it has been launched.

36. Determine the height of the rocket 2.5 sec after it has been launched.

37. When will the rocket be exactly 32 ft above the ground?

38. When will the rocket crash into the ground?

39. Do we now have the ability to solve *any* problem that translates to a quadratic equation? Why or why not?

40. Write a problem for a classmate to solve such that only one of two solutions of a quadratic equation can be used as an answer.

Skill Review

To prepare for Chapter 6, review addition, subtraction, multiplication, and division using fraction notation (Sections 1.3, 1.5, 1.6, and 1.7).

Simplify.

41. $-\frac{3}{5} \cdot \frac{4}{7}$ [1.7]

42. $-\frac{3}{5} \div \frac{4}{7}$ [1.7]

43. $-\frac{5}{6} - \frac{1}{6}$ [1.6]

44. $\frac{3}{4} + \left(-\frac{5}{2}\right)$ [1.5]

45. $-\frac{3}{8} \cdot \left(-\frac{10}{15}\right)$ [1.7]

46. $\dfrac{-\dfrac{8}{15}}{-\dfrac{2}{3}}$ [1.7]

47. $\frac{5}{24} + \frac{3}{28}$ [1.3]

48. $\frac{5}{6} - \left(-\frac{2}{9}\right)$ [1.6]

Synthesis

The converse of the Pythagorean theorem is also true. That is, if $a^2 + b^2 = c^2$, then the triangle is a right triangle (where a and b are the lengths of the legs and c is the length of the hypotenuse). Use this result to answer Exercises 49 and 50.

49. An archaeologist has measuring sticks of 3 ft, 4 ft, and 5 ft. Explain how she could draw a 7-ft by 9-ft rectangle on a piece of land being excavated.

50. Explain how measuring sticks of 5 cm, 12 cm, and 13 cm can be used to draw a right triangle that has two 45° angles.

51. *Sailing.* The mainsail of a Lightning sailboat is a right triangle in which the hypotenuse is called the leech. If a 24-ft tall mainsail has a leech length of 26 ft and if Dacron® sailcloth costs $1.50 per square foot, find the cost of the fabric for a new mainsail.

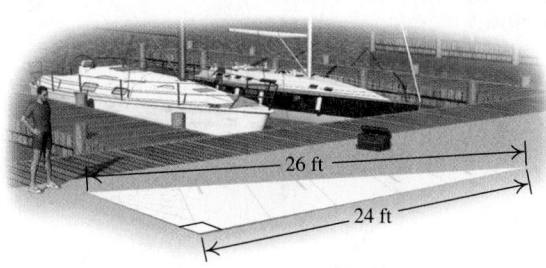

26 ft
24 ft

52. *Roofing.* A *square* of shingles covers 100 ft² of surface area. How many squares will be needed to re-shingle the house shown?

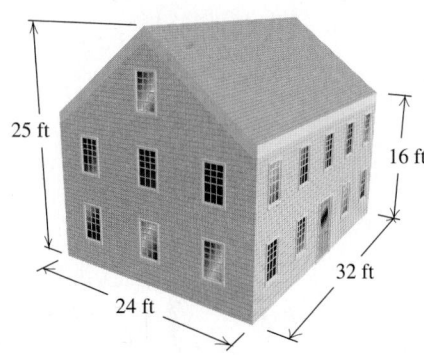

25 ft
16 ft
24 ft
32 ft

53. Solve for *x*.

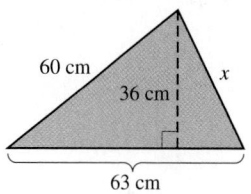

60 cm
36 cm
x
63 cm

54. *Pool sidewalk.* A cement walk of uniform width is built around a 20-ft by 40-ft rectangular pool. The total area of the pool and the walk is 1500 ft². Find the width of the walk.

55. *Folding sheet metal.* An open rectangular gutter is made by turning up the sides of a piece of metal 20 in. wide, as shown. The area of the cross-section of the gutter is 48 in². Find the possible depths of the gutter.

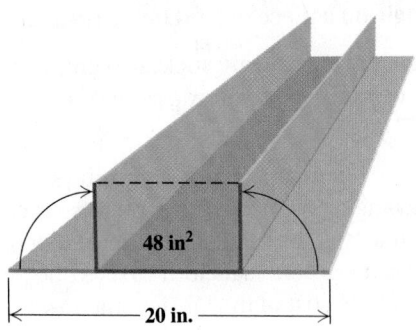

48 in²

20 in.

56. Find a polynomial for the shaded area in the figure below.

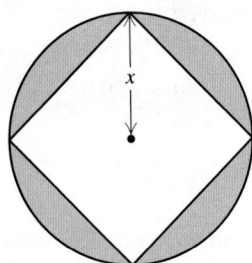

x

57. *Telephone service.* Use the information in the figure below to determine the height of the telephone pole.

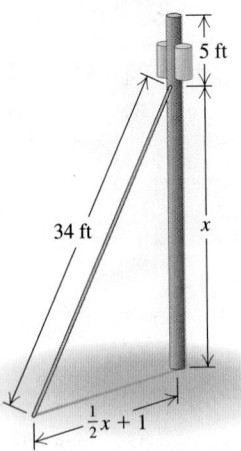

5 ft

34 ft

x

$\frac{1}{2}x + 1$

58. *Dimensions of a closed box.* The total surface area of a closed box is 350 m². The box is 9 m high and has a square base and lid. Find the length of a side of the base.

Medicine. For certain people with acid reflux, the drug Pepcid (famotidine) is used. The number of milligrams N of Pepcid in an adult's bloodstream t hours after a 20-mg tablet has been swallowed can be approximated by

$$N = -0.009t\,(t-12)^3.$$

Use a graphing calculator with the window $[-1, 13, -1, 25]$ *and the* TRACE *feature to answer Exercises 59–61.*

Source: Based on information in Chohan, Naina, Rita M. Doyle, and Patricia Nayle (eds.), *Nursing Handbook*, 21st ed. Springhouse, PA: Springhouse Corporation, 2001

59. Approximately how long after a tablet has been swallowed will there be 18 mg in the bloodstream?

60. Approximately how long after a tablet has been swallowed will there be 10 mg in the bloodstream?

61. Approximately how long after a tablet has been swallowed will the peak dosage in the bloodstream occur?

Study Summary

KEY TERMS AND CONCEPTS	EXAMPLES

SECTION 5.1: INTRODUCTION TO FACTORING

To **factor** a polynomial means to write it as a product. Always begin by factoring out the **largest common factor**.

Factor: $12x^4 - 30x^3$.
$$12x^4 - 30x^3 = 6x^3(2x - 5)$$

Some polynomials with four terms can be **factored by grouping**.

Factor: $3x^3 - x^2 - 6x + 2$.
$$3x^3 - x^2 - 6x + 2 = x^2(3x - 1) - 2(3x - 1)$$
$$= (3x - 1)(x^2 - 2)$$

SECTION 5.2: FACTORING TRINOMIALS OF THE TYPE $x^2 + bx + c$

Some trinomials of the type $x^2 + bx + c$ can be factored by reversing the steps of FOIL.

Factor: $x^2 - 11x + 18$.

Pairs of Factors of 18	Sums of Factors
$-1, -18$	-19
$-2, \; -9$	-11

18 is positive and -11 is negative, so both factors will be negative.

← The numbers we need are -2 and -9.

The factorization is $(x - 2)(x - 9)$.

SECTION 5.3: FACTORING TRINOMIALS OF THE TYPE $ax^2 + bx + c$

One method for factoring trinomials of the type $ax^2 + bx + c$ is a FOIL-based method.

Factor: $6x^2 - 5x - 6$.

The factors will be in the form $(3x + \;)(2x + \;)$ or $(6x + \;)(x + \;)$. We list all pairs of factors of -6, and check possible products by multiplying any possibilities that do not contain a common factor.

$$(3x - 2)(2x + 3) = 6x^2 + 5x - 6,$$
$$(3x + 2)(2x - 3) = 6x^2 - 5x - 6. \leftarrow \text{This is the correct product, so we stop here.}$$

The factorization is $(3x + 2)(2x - 3)$.

Another method for factoring trinomials of the type $ax^2 + bx + c$ involves factoring by grouping.

Factor: $6x^2 - 5x - 6$.

Multiply the leading coefficient and the constant term: $6(-6) = -36$. Look for factors of -36 that add to -5.

Pairs of Factors of -36	Sums of Factors
$1, -36$	-35
$2, -18$	-16
$3, -12$	-9
$4, \; -9$	-5

-5 is negative, so the negative factor must have the greater absolute value.

← The numbers we want are 4 and -9.

Rewrite $-5x$ as $4x - 9x$ and factor by grouping:
$$6x^2 - 5x - 6 = 6x^2 + 4x - 9x - 6$$
$$= 2x(3x + 2) - 3(3x + 2)$$
$$= (3x + 2)(2x - 3).$$

SECTION 5.4: FACTORING PERFECT-SQUARE TRINOMIALS AND DIFFERENCES OF SQUARES

Factoring a perfect-square trinomial

$$A^2 + 2AB + B^2 = (A + B)^2;$$
$$A^2 - 2AB + B^2 = (A - B)^2$$

Factor: $y^2 + 100 - 20y$.

$$A^2 - 2AB + B^2 = (A - B)^2$$

$$y^2 + 100 - 20y = y^2 - 20y + 100 = (y - 10)^2$$

Factoring a difference of squares

$$A^2 - B^2 = (A + B)(A - B)$$

Factor: $9t^2 - 1$.

$$A^2 - B^2 = (A + B)(A - B)$$

$$9t^2 - 1 = (3t + 1)(3t - 1)$$

SECTION 5.5: FACTORING SUMS OR DIFFERENCES OF CUBES

Factoring a sum or a difference of cubes

$$A^3 + B^3 = (A + B)(A^2 - AB + B^2)$$
$$A^3 - B^3 = (A - B)(A^2 + AB + B^2)$$

$$x^3 + 1000 = (x + 10)(x^2 - 10x + 100)$$
$$z^6 - 8w^3 = (z^2 - 2w)(z^4 + 2wz^2 + 4w^2)$$

SECTION 5.6: FACTORING: A GENERAL STRATEGY

To factor a polynomial:

A. Factor out the largest common factor.

B. Look at the number of terms.

Two terms: If a difference of squares, use $A^2 - B^2 = (A + B)(A - B)$.
If a sum of cubes, use
$A^3 + B^3 = (A + B)(A^2 - AB + B^2)$.
If a difference of cubes, use
$A^3 - B^3 = (A - B)(A^2 + AB + B^2)$.

Three terms: If a trinomial square, use
$A^2 + 2AB + B^2 = (A + B)^2$ or
$A^2 - 2AB + B^2 = (A - B)^2$.
Otherwise, try FOIL or grouping.

Four terms: Try factoring by grouping.

C. Factor completely.

D. Check by multiplying.

Factor: $5x^5 - 80x$.

$$5x^5 - 80x = 5x(x^4 - 16)$$ $5x$ is the largest common factor.

$$= 5x(x^2 + 4)(x^2 - 4)$$ $x^4 - 16$ is a difference of squares.

$$= 5x(x^2 + 4)(x + 2)(x - 2)$$ $x^2 - 4$ is also a difference of squares.

Check: $5x(x^2 + 4)(x + 2)(x - 2) = 5x(x^2 + 4)(x^2 - 4)$
$$= 5x(x^4 - 16) = 5x^5 - 80x.$$

Factor: $-x^2y^2 - 3xy + 10$.

$$-x^2y^2 - 3xy + 10 = -(x^2y^2 + 3xy - 10)$$ Factor out -1 to make the leading coefficient positive.

$$= -(xy + 5)(xy - 2)$$

Check: $-(xy + 5)(xy - 2) = -(x^2y^2 + 3xy - 10)$
$$= -x^2y^2 - 3xy + 10.$$

SECTION 5.7: SOLVING POLYNOMIAL EQUATIONS BY FACTORING

The Principle of Zero Products

An equation $AB = 0$ is true if and only if $A = 0$ or $B = 0$, or both.

Solve: $x^2 + 7x = 30$.

$$x^2 + 7x = 30$$
$$x^2 + 7x - 30 = 0$$ Getting 0 on one side
$$(x + 10)(x - 3) = 0$$ Factoring
$$x + 10 = 0 \quad or \quad x - 3 = 0$$ Using the principle of zero products
$$x = -10 \quad or \quad x = 3$$

The solutions are -10 and 3.

SECTION 5.8: SOLVING APPLICATIONS

The Pythagorean Theorem

In any right triangle, if a and b are the lengths of the legs and c is the length of the hypotenuse, then

$a^2 + b^2 = c^2$, or

$(\text{Leg})^2 + (\text{Other leg})^2 = (\text{Hypotenuse})^2.$

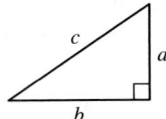

Aaron has a 25-ft ladder leaning against his house. The height that the ladder reaches on the house is 17 ft more than the distance that the bottom of the ladder is from the house. Find both distances.

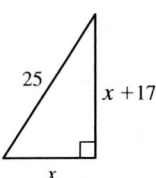

$$x^2 + (x + 17)^2 = 25^2$$
$$x^2 + x^2 + 34x + 289 = 625 \qquad \text{Squaring}$$
$$2x^2 + 34x - 336 = 0$$
$$2(x + 24)(x - 7) = 0 \qquad \text{Factoring}$$
$$x + 24 = 0 \quad or \quad x - 7 = 0 \qquad \text{Using the principle of zero products}$$
$$x = -24 \quad or \qquad x = 7.$$

Since -24 cannot be a length, we check 7. When $x = 7$,
$x + 17 = 24$, and $7^2 + 24^2 = 49 + 576 = 625 = 25^2$.

The ladder reaches 24 ft up the side of the house, and the bottom of the ladder is 7 ft from the house.

Review Exercises: Chapter 5

🍂 **Concept Reinforcement** *Classify each statement as either true or false.*

1. The largest common variable factor is the largest power of the variable in the polynomial. [5.1]

2. A prime polynomial has no common factor other than 1 or −1. [5.2]

3. Every perfect-square trinomial can be expressed as a binomial squared. [5.4]

4. Every binomial can be regarded as a difference of squares. [5.4]

5. Every quadratic equation has two different solutions. [5.7]

6. The principle of zero products can be applied whenever a product equals 0. [5.7]

7. In a right triangle, the hypotenuse is always longer than either leg. [5.8]

8. The Pythagorean theorem can be applied to any triangle that has an angle measuring at least 90°. [5.8]

Find three factorizations of each monomial. [5.1]

9. $20x^3$

10. $-18x^5$

Factor completely. If a polynomial is prime, state this.

11. $12x^4 - 18x^3$ [5.1]

12. $8a^2 - 12a$ [5.1]

13. $100t^2 - 1$ [5.4]

14. $x^2 + x - 12$ [5.2]

15. $x^2 + 14x + 49$ [5.4]

16. $12x^3 + 12x^2 + 3x$ [5.4]

17. $6x^3 + 9x^2 + 2x + 3$ [5.1]

18. $6a^2 + a - 5$ [5.3]

19. $25t^2 + 9 - 30t$ [5.4]

20. $48t^2 - 28t + 6$ [5.1]

21. $81a^4 - 1$ [5.4]

22. $9x^3 + 12x^2 - 45x$ [5.3]

23. $2x^3 - 250$ [5.5]

24. $x^4 + 4x^3 - 2x - 8$ [5.1]

25. $a^2b^4 - 64$ [5.4]

26. $-8x^6 + 32x^5 - 4x^4$ [5.1]

27. $75 + 12x^2 - 60x$ [5.4]

28. $y^2 + 9$ [5.4]

29. $-t^3 + t^2 + 42t$ [5.2]

30. $4x^2 - 25$ [5.4]

31. $n^2 - 60 - 4n$ [5.2]

32. $5z^2 - 30z + 10$ [5.1]

33. $4t^2 + 13t + 10$ [5.3]

34. $2t^2 - 7t - 4$ [5.3]

35. $7x^3 + 35x^2 + 28x$ [5.2]

36. $8y^3 + 27x^6$ [5.5]

37. $20x^2 - 20x + 5$ [5.4]

38. $-6x^3 + 150x$ [5.4]

39. $15 - 8x + x^2$ [5.2]

40. $3x + x^2 + 5$ [5.2]

41. $x^2y^2 + 6xy - 16$ [5.2]

42. $12a^2 + 84ab + 147b^2$ [5.4]

43. $m^2 + 5m + mt + 5t$ [5.1]

44. $32x^4 - 128y^4z^4$ [5.4]

45. $6m^2 + 2mn + n^2 + 3mn$ [5.1], [5.3]

46. $6r^2 + rs - 15s^2$ [5.3]

Solve. [5.7]

47. $(x - 9)(x + 11) = 0$

48. $x^2 + 2x - 35 = 0$

49. $16x^2 = 9$

50. $3x^2 + 2 = 5x$

51. $2x^2 - 7x = 30$

52. $(x + 1)(x - 2) = 4$

53. $9t - 15t^2 = 0$

54. $3x^2 + 3 = 6x$

55. The square of a number is 12 more than the number. Find all such numbers. [5.8]

56. The formula $x^2 - x = N$ can be used to determine the total number of games played, N, in a league of x teams in which all teams play each other twice. Serena referees for a soccer league in which all teams play each other twice and a total of 90 games is played. How many teams are in the league? [5.8]

57. Find the x-intercepts for the graph of $y = 2x^2 - 3x - 5$. [5.7]

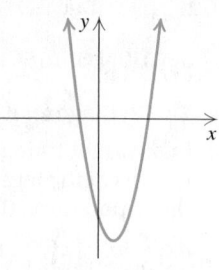

58. The front of a house is a triangle that is as wide as it is tall. Its area is 98 ft^2. Find the height and the base. [5.8]

59. Josh needs to add a diagonal brace to his LEGO® robot. The brace must span a height of 8 holes and a width of 6 holes. How long should the brace be? [5.8]

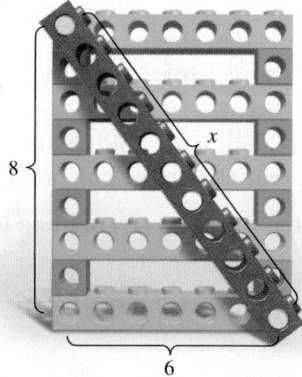

Synthesis

60. On a quiz, Celia writes the factorization of $4x^2 - 100$ as $(2x - 10)(2x + 10)$. If this were a 10-point question, how many points would you give Celia? Why? [5.4]

61. How do the equations solved in this chapter differ from those solved in previous chapters? [5.7]

Solve.

62. The pages of a book measure 15 cm by 20 cm. Margins of equal width surround the printing on each page and constitute one half of the area of the page. Find the width of the margins. [5.8]

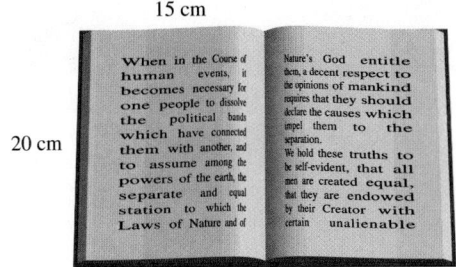

15 cm

20 cm

When in the Course of human events, it becomes necessary for one people to dissolve the political bands which have connected them with another, and to assume among the powers of the earth, the separate and equal station to which the Laws of Nature and of

Nature's God entitle them, a decent respect to the opinions of mankind requires that they should declare the causes which impel them to the separation. We hold these truths to be self-evident, that all men are created equal, that they are endowed by their Creator with certain unalienable

63. The cube of a number is the same as twice the square of the number. Find the number. [5.8]

64. The length of a rectangle is two times its width. When the length is increased by 20 cm and the width is decreased by 1 cm, the area is 160 cm². Find the original length and width. [5.8]

65. The length of each side of a square is increased by 5 cm to form a new square. The area of the new square is $2\frac{1}{4}$ times the area of the original square. Find the area of each square. [5.8]

Solve. [5.7]

66. $(x - 2)2x^2 + x(x - 2) - (x - 2)15 = 0$

Aha! **67.** $x^2 + 25 = 0$

Test: Chapter 5

CHAPTER
Test Prep
VIDEO CD

Step-by-step test solutions are found on the video CD in the front of this book.

1. Find three factorizations of $12x^4$.

Factor completely. If a polynomial is prime, state this.

2. $x^2 - 13x + 36$

3. $x^2 + 25 - 10x$

4. $6y^2 - 8y^3 + 4y^4$

5. $x^3 + x^2 + 2x + 2$

6. $t^7 - 3t^5$

7. $a^3 + 3a^2 - 4a$

8. $28x - 48 + 10x^2$

9. $4t^2 - 25$

10. $x^2 - x - 6$

11. $-6m^3 - 9m^2 - 3m$

12. $3r^3 - 3$

13. $45r^2 + 60r + 20$

14. $3x^4 - 48$

15. $49t^2 + 36 + 84t$

16. $x^4 + 2x^3 - 3x - 6$

17. $x^2 + 3x + 6$

18. $4x^2 - 4x - 15$

19. $6t^3 + 9t^2 - 15t$

20. $3m^2 - 9mn - 30n^2$

Solve.

21. $x^2 - 6x + 5 = 0$

22. $2x^2 - 7x = 15$

23. $4t - 10t^2 = 0$

24. $25t^2 = 1$

25. $x(x - 1) = 20$

26. Find the x-intercepts for the graph of $y = 3x^2 - 5x - 8$.

27. The length of a rectangle is 6 m more than the width. The area of the rectangle is 40 m². Find the length and the width.

28. The number of possible handshakes H within a group of n people is given by $H = \frac{1}{2}(n^2 - n)$. At a meeting, everyone shook hands once with everyone else. If there were 45 handshakes, how many people were at the meeting?

29. A mason wants to be sure she has a right corner in a building's foundation. She marks a point 3 ft from the corner along one wall and another point 4 ft from the corner along the other wall. If the corner is a right angle, what should the distance be between the two marked points?

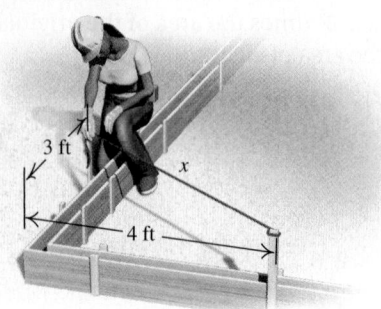

3 ft
x
4 ft

Synthesis

30. *Dimensions of an open box.* A rectangular piece of cardboard is twice as long as it is wide. A 4-cm square is cut out of each corner, and the sides

are turned up to make a box with an open top. The volume of the box is 616 cm^3. Find the original dimensions of the cardboard.

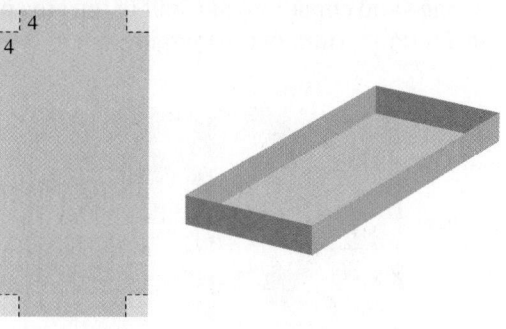

31. Factor: $(a + 3)^2 - 2(a + 3) - 35$.

32. Solve: $20x(x + 2)(x - 1) = 5x^3 - 24x - 14x^2$.

Cumulative Review: Chapters 1–5

Simplify. Do not use negative exponents in the answer.

1. $\frac{3}{8} \div \frac{3}{4}$ [1.3]

2. $\frac{3}{8} \cdot \frac{3}{4}$ [1.3]

3. $\frac{3}{8} + \frac{3}{4}$ [1.3]

4. $-2 + (20 \div 4)^2 - 6 \cdot (-1)^3$ [1.8]

5. $(3x^2y^3)^{-2}$ [4.2]

6. $(t^2)^3 \cdot t^4$ [4.1]

7. $(3x^4 - 2x^2 + x - 7) + (5x^3 + 2x^2 - 3)$ [4.4]

8. $(a^2b - 2ab^2 + 3b^3) - (4a^2b - ab^2 + b^3)$ [4.7]

9. $\dfrac{3t^3s^{-1}}{12t^{-5}s}$ [4.2]

10. $\left(\dfrac{-2x^2y}{3z^4}\right)^3$ [4.1]

11. Evaluate $-x$ for $x = -8$. [1.6]

12. Evaluate $-(-x)$ for $x = -8$. [1.6]

13. Determine the leading term of the polynomial
$4x^3 - 6x^2 - x^4 + 7$. [4.3]

14. Divide: $(8x^4 - 20x^3 + 2x^2 - 4x) \div (4x)$. [4.8]

Multiply.

15. $-4t^8(t^3 - 2t - 5)$ [4.5]

16. $(3x - 5)^2$ [4.6]

17. $(10x^5 + y)(10x^5 - y)$ [4.7]

18. $(x - 1)(x^2 - x - 1)$ [4.5]

Factor completely.

19. $c^2 - 1$ [5.4]

20. $5x + 5y + 10x^2 + 10xy$ [5.1]

21. $4r^2 - 4rt + t^2$ [5.4]

22. $6x^2 - 19x + 10$ [5.3]

23. $10y^2 + 40$ [5.1]

24. $x^2y - 3xy + 2y$ [5.2]

25. $12x^2 - 5xy - 2y^2$ [5.3]

26. $125a^3 + 64b^3$ [5.5]

Solve.

27. $\frac{1}{3} + 2x = \frac{1}{2}$ [2.2]

28. $3(t - 1) = 2 - (t + 1)$ [2.2]

29. $8y - 6(y - 2) = 3(2y + 7)$ [2.2]

30. $3x - 7 \geq 4 - 8x$ [2.6]

31. $(x - 1)(x + 3) = 0$ [5.7]

32. $x^2 + x = 12$ [5.7]

33. $3x^2 = 12$ [5.7]

34. $3x^2 = 12x$ [5.7]

35. Solve $a = bc + dc$ for c. [2.3]

36. Find the slope of the line containing the points $(6, 7)$ and $(-2, 7)$. [3.5]

37. Find the slope and the y-intercept of the line given by $2x + y = 5$. [3.6]

38. Write the slope–intercept equation for the line with slope 5 and y-intercept $\left(0, -\frac{1}{3}\right)$. [3.6]

39. Write the slope–intercept equation for the line with slope 5 that contains the point $\left(-\frac{1}{3}, 0\right)$. [3.7]

Graph.

40. $4(x + 1) = 8$ [3.3]

41. $x + y = 5$ [3.3]

42. $y = \frac{3}{2}x - 2$ [3.6]

43. $3x + 5y = 10$ [3.6]

44. Use a grid 10 squares wide and 10 squares high to plot $(5, 40)$, $(18, -60)$, and $(30, -22)$. Choose the scale carefully. [3.1]

Solve.

45. On average, men talk 97 min more per month on cell phones than do women. The sum of men's average minutes and women's average minutes is 647 min. What is the average number of minutes per month that men talk on cell phones? [2.5]
Source: *International Communications Research for Cingular Wireless*

46. The number of cell-phone subscribers increased from 680,000 in 1986 to 233,000,000 in 2006. What was the average rate of increase? [3.4]
Source: CTIA – The Wireless Association

47. In 2007, there were 1.2 billion Internet users worldwide. Of these, 5% spoke French. How many Internet users spoke French? [2.4]
Source: Internetworldstats.com

48. The number of people in the United States, in thousands, who are on a waiting list for an organ transplant can be approximated by the polynomial $2.38t + 77.38$, where t is the number of years since 2000. Estimate the number of people on a waiting list for an organ transplant in 2010. [4.3]
Source: Based on information from The Organ Procurement and Transplantation Network

49. A 13-ft ladder is placed against a building in such a way that the distance from the top of the ladder to the ground is 7 ft more than the distance from the bottom of the ladder to the building. Find both distances. [5.8]

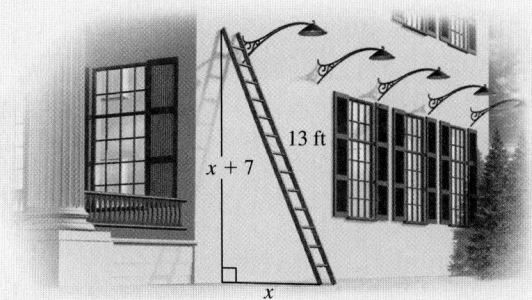

50. A rectangular table in Arlo's House of Tunes is six times as long as it is wide. If the area of the table is 24 ft^2, find the length and the width of the table. [5.8]

51. Donna's quiz grades are 8, 3, 5, and 10. What scores on the fifth quiz will make her average quiz grade at least 7? [2.7]

52. The average amount of sodium in a serving of Chef Boyardee foods dropped from 1100 mg in 2003 to 900 mg in 2007. [3.7]
Source: *The Indianapolis Star*, 11/25/07

a) Graph the data and determine an equation for the related line. Let s represent the average amount of sodium per serving and t the number of years after 2000.

b) Use the equation of part (a) to estimate the average amount of sodium in a serving of Chef Boyardee foods in 2006.

Synthesis

53. Solve $x = \dfrac{abx}{2 - b}$ for b. [2.3]

54. Write an equation of the line parallel to the x-axis and passing through $(-6, -8)$. [3.3]

55. a) Multiply: $(3y + 2 + x)(3y + 2 - x)$. [4.6]
b) Factor: $9y^2 + 12y + 4 - x^2$. [5.4]

56. Solve: $6x^3 + 4x^2 = 2x$. [5.7]

Rational Expressions and Equations

FRED JENKINS
TECHNICAL SERVICES
MANAGER
Goose Creek, South Carolina

As a chemist involved in the
quality control of all chemicals
used and manufactured where
I work, I use mathematics to
calculate purities, density, refrac-
tive index, acidity, and other
parameters that determine
whether a material meets
intended specifications. Even
though many calculations are
performed by instrument
software, it is important for me
as a chemist and manager to
know how to verify results and
troubleshoot with math
when problems occur.

AN APPLICATION

As one alternative to gasoline-powered
vehicles, flex fuel vehicles can use
both regular gasoline and E85, a fuel
containing 85% ethanol. Because using
E85 results in a lower fuel economy, its
price per gallon must be lower than
the price of gasoline in order to make
it an economical fuel option. A 2007
Chevrolet Tahoe gets 21 mpg on the
highway using gasoline and only
15 mpg on the highway using E85.
If the price of gasoline is $3.36 a
gallon, what must the price of E85 be
in order for the fuel cost per mile to be
the same?

Source: www.caranddriver.com

This problem appears as Example 7 in
Section 6.7.

R ational expressions are similar to fractions in arithmetic, in that both are ratios of two expressions. We now learn how to simplify, add, subtract, multiply, and divide rational expressions. These skills will then be used to solve the equations that arise from real-life problems like the one on the preceding page.

6.1 Rational Expressions

Simplifying Rational Expressions • Factors That Are Opposites

Just as a rational number is any number that can be written as a quotient of two integers, a **rational expression** is any expression that can be written as a quotient of two polynomials. The following are examples of rational expressions:

$$\frac{7}{3}, \quad \frac{5}{x+6}, \quad \frac{t^2 - 5t + 6}{4t^2 - 7}.$$

Rational expressions are examples of *algebraic fractions*. They are also examples of *fraction expressions*.

Because rational expressions indicate division, we must be careful to avoid denominators that are 0. For example, in the expression

$$\frac{x+3}{x-7},$$

when x is replaced with 7, the denominator is 0, and the expression is undefined:

$$\frac{x+3}{x-7} = \frac{7+3}{7-7} = \frac{10}{0}. \longleftarrow \text{As explained in Chapter 1, division by 0 is undefined.}$$

When x is replaced with a number other than 7—say, 6—the expression *is* defined because the denominator is not 0:

$$\frac{x+3}{x-7} = \frac{6+3}{6-7} = \frac{9}{-1} = -9.$$

The expression is also defined when $x = -3$:

$$\frac{x+3}{x-7} = \frac{-3+3}{-3-7} = \frac{0}{-10} = 0. \qquad \text{0 divided by a nonzero number is 0.}$$

Any replacement for the variable that makes the *denominator* 0 will cause an expression to be undefined.

EXAMPLE **1** Find all numbers for which the rational expression

$$\frac{x+4}{x^2 - 3x - 10}$$

is undefined.

SOLUTION   The value of the numerator has no bearing on whether or not a rational expression is defined. To determine which numbers make the rational expression undefined, we set the *denominator* equal to 0 and solve:

$$x^2 - 3x - 10 = 0 \qquad \text{We set the denominator equal to 0.}$$
$$(x - 5)(x + 2) = 0 \qquad \text{Factoring}$$
$$x - 5 = 0 \quad or \quad x + 2 = 0 \qquad \text{Using the principle of zero products}$$
$$x = 5 \quad or \qquad x = -2. \qquad \text{Solving each equation}$$

Check:

For $x = 5$:

$$\frac{x + 4}{x^2 - 3x - 10} = \frac{5 + 4}{5^2 - 3 \cdot 5 - 10} \qquad \text{There are no restrictions on the numerator.}$$

$$= \frac{9}{25 - 15 - 10} = \frac{9}{0}. \qquad \text{This expression is undefined, as expected.}$$

For $x = -2$:

$$\frac{x + 4}{x^2 - 3x - 10} = \frac{-2 + 4}{(-2)^2 - 3(-2) - 10}$$

$$= \frac{2}{4 + 6 - 10} = \frac{2}{0}. \qquad \text{This expression is undefined, as expected.}$$

Thus, $\dfrac{x + 4}{x^2 - 3x - 10}$ is undefined for $x = 5$ and $x = -2$. ▶ **TRY EXERCISE** ▶ 7

Simplifying Rational Expressions

A rational expression is said to be *simplified* when the numerator and the denominator have no factors (other than 1) in common. To simplify a rational expression, we first factor the numerator and the denominator. We then identify factors common to the numerator and the denominator, rewrite the expression as a product of two rational expressions (one of which is equal to 1), and then remove the factor equal to 1. The process is identical to that used in Section 1.3 to simplify $\frac{15}{40}$:

$$\frac{15}{40} = \frac{3 \cdot 5}{8 \cdot 5} \qquad \text{Factoring the numerator and the denominator. Note the common factor, 5.}$$

$$= \frac{3}{8} \cdot \frac{5}{5} \qquad \text{Rewriting as a product of two fractions}$$

$$= \frac{3}{8} \cdot 1 \qquad \frac{5}{5} = 1$$

$$= \frac{3}{8}. \qquad \text{Using the identity property of 1 to remove the factor 1}$$

Similar steps are followed when simplifying rational expressions: We factor and remove a factor equal to 1, using the fact that

$$\frac{ab}{cb} = \frac{a}{c} \cdot \frac{b}{b}.$$

EXAMPLE 2 Simplify: $\dfrac{8x^2}{24x}$.

SOLUTION

$$\dfrac{8x^2}{24x} = \dfrac{8 \cdot x \cdot x}{3 \cdot 8 \cdot x} \qquad \text{Factoring the numerator and the denominator.}$$
$$\text{Note the common factor of } 8 \cdot x.$$

$$= \dfrac{x}{3} \cdot \dfrac{8x}{8x} \qquad \text{Rewriting as a product of two rational expressions}$$

$$= \dfrac{x}{3} \cdot 1 \qquad \dfrac{8x}{8x} = 1$$

$$= \dfrac{x}{3} \qquad \text{Removing the factor 1} \qquad \boxed{\text{TRY EXERCISE} \; 17}$$

We say that $\dfrac{8x^2}{24x}$ *simplifies to* $\dfrac{x}{3}$.* In the work that follows, we assume that all denominators are nonzero.

EXAMPLE 3 Simplify: $\dfrac{5a + 15}{10}$.

SOLUTION

$$\dfrac{5a + 15}{10} = \dfrac{5(a + 3)}{5 \cdot 2} \qquad \text{Factoring the numerator and the denominator.}$$
$$\text{Note the common factor of 5.}$$

$$= \dfrac{5}{5} \cdot \dfrac{a + 3}{2} \qquad \text{Rewriting as a product of two rational expressions}$$

$$= 1 \cdot \dfrac{a + 3}{2} \qquad \dfrac{5}{5} = 1$$

$$= \dfrac{a + 3}{2} \qquad \text{Removing the factor 1} \qquad \boxed{\text{TRY EXERCISE} \; 19}$$

The result in Example 3 can be partially checked using a replacement for a—say, $a = 2$.

Original expression:

$$\dfrac{5a + 15}{10} = \dfrac{5 \cdot 2 + 15}{10}$$

$$= \dfrac{25}{10} = \dfrac{5}{2} \qquad \text{The results are the same.}$$

Simplified expression:

$$\dfrac{a + 3}{2} = \dfrac{2 + 3}{2}$$

$$= \dfrac{5}{2}$$

> To see why this check is not foolproof, see Exercise 65.

If we do not get the same result when evaluating both expressions, we know that a mistake has been made. For example, if $(5a + 15)/10$ is *incorrectly* simplified as $(a + 15)/2$ and we evaluate using $a = 2$, we have the following.

Original expression:

$$\dfrac{5a + 15}{10} = \dfrac{5 \cdot 2 + 15}{10}$$

$$= \dfrac{5}{2} \qquad \text{The results are different.}$$

Incorrectly simplified expression:

$$\dfrac{a + 15}{2} = \dfrac{2 + 15}{2}$$

$$= \dfrac{17}{2}$$

This demonstrates that a mistake has been made.

---

*In more advanced courses, we would *not* say that $8x^2/(24x)$ simplifies to $x/3$, but would instead say that $8x^2/(24x)$ simplifies to $x/3$ *with the restriction that $x \neq 0$.*

Sometimes the common factor has two or more terms.

EXAMPLE 4 Simplify.

a) $\dfrac{6x - 12}{7x - 14}$ b) $\dfrac{18t^2 + 6t}{6t^2 + 15t}$ c) $\dfrac{x^2 + 3x + 2}{x^2 - 1}$

SOLUTION

a) $\dfrac{6x - 12}{7x - 14} = \dfrac{6(x - 2)}{7(x - 2)}$ Factoring the numerator and the denominator. Note the common factor of $x - 2$.

$= \dfrac{6}{7} \cdot \dfrac{x - 2}{x - 2}$ Rewriting as a product of two rational expressions

$= \dfrac{6}{7} \cdot 1$ $\dfrac{x - 2}{x - 2} = 1$

$= \dfrac{6}{7}$ Removing the factor 1

b) $\dfrac{18t^2 + 6t}{6t^2 + 15t} = \dfrac{3t \cdot 2(3t + 1)}{3t(2t + 5)}$ Factoring the numerator and the denominator. Note the common factor of $3t$.

$= \dfrac{3t}{3t} \cdot \dfrac{2(3t + 1)}{2t + 5}$ Rewriting as a product of two rational expressions

$= 1 \cdot \dfrac{2(3t + 1)}{2t + 5}$ $\dfrac{3t}{3t} = 1$

$= \dfrac{2(3t + 1)}{2t + 5}$ Removing the factor 1. The numerator and the denominator have no common factor so the simplification is complete.

c) $\dfrac{x^2 + 3x + 2}{x^2 - 1} = \dfrac{(x + 1)(x + 2)}{(x + 1)(x - 1)}$ Factoring; $x + 1$ is the common factor.

$= \dfrac{x + 1}{x + 1} \cdot \dfrac{x + 2}{x - 1}$ Rewriting as a product of two rational expressions

$= 1 \cdot \dfrac{x + 2}{x - 1}$ $\dfrac{x + 1}{x + 1} = 1$

$= \dfrac{x + 2}{x - 1}$ Removing the factor 1

TRY EXERCISE 23

Canceling is a shortcut that can be used—and easily *misused*—to simplify rational expressions. As stated in Section 1.3, canceling must be done with care and understanding. Essentially, canceling streamlines the process of removing a factor equal to 1. Example 4(c) could have been streamlined as follows:

$\dfrac{x^2 + 3x + 2}{x^2 - 1} = \dfrac{\cancel{(x + 1)}(x + 2)}{\cancel{(x + 1)}(x - 1)}$ When a factor equal to 1 is noted, it is "canceled": $\dfrac{x + 1}{x + 1} = 1$.

$= \dfrac{x + 2}{x - 1}.$ Simplifying

TECHNOLOGY CONNECTION

We can use the TABLE feature as a partial check that rational expressions have been simplified correctly. To check the simplification in Example 4(c),

$$\frac{x^2 + 3x + 2}{x^2 - 1} = \frac{x + 2}{x - 1},$$

we enter $y_1 = (x^2 + 3x + 2)/(x^2 - 1)$ and $y_2 = (x + 2)/(x - 1)$ and select the mode AUTO to look at a table of values of y_1 and y_2. The values should match for all allowable replacements.

X	Y₁	Y₂
−4	.4	.4
−3	.25	.25
−2	0	0
−1	ERROR	−.5
0	−2	−2
1	ERROR	ERROR
2	4	4
X = −4		

The ERROR messages indicate that −1 and 1 are not allowable replacements in y_1 and 1 is not an allowable replacement in y_2. For all other numbers, y_1 and y_2 are the same, so the simplification appears to be correct.

Use the TABLE feature to determine whether each of the following appears to be correct.

1. $\dfrac{8x^2}{24x} = \dfrac{x}{3}$

2. $\dfrac{5x + 15}{10} = \dfrac{x + 3}{2}$

3. $\dfrac{x + 3}{x} = 3$

4. $\dfrac{x^2 + 3x - 4}{x^2 - 16} = \dfrac{x - 1}{x + 4}$

CAUTION!   Canceling is often used incorrectly. The following cancellations are *incorrect*:

$$\frac{\cancel{x} + 7}{\cancel{x} + 3}, \qquad \frac{a^2 - \cancel{5}}{\cancel{5}}, \qquad \frac{6x^2 + 5\cancel{x} + 1}{4x^2 - 3\cancel{x}}.$$

Wrong!   Wrong!   Wrong!

None of the above cancellations removes a factor equal to 1. Factors are parts of products. For example, in $x \cdot 7$, x and 7 are factors, but in $x + 7$, x and 7 are terms, *not* factors. Only factors can be canceled.

EXAMPLE **5**   Simplify: $\dfrac{3x^2 - 2x - 1}{x^2 - 3x + 2}$.

SOLUTION   We factor the numerator and the denominator and look for common factors:

$$\frac{3x^2 - 2x - 1}{x^2 - 3x + 2} = \frac{(3x + 1)\cancel{(x - 1)}}{(x - 2)\cancel{(x - 1)}} \qquad \begin{array}{l} \text{Try to visualize this as} \\[4pt] \dfrac{3x + 1}{x - 2} \cdot \dfrac{x - 1}{x - 1}. \end{array}$$

$$= \frac{3x + 1}{x - 2}. \qquad \begin{array}{l} \text{Removing a factor equal to 1:} \\[4pt] \dfrac{x - 1}{x - 1} = 1 \end{array}$$

> **TRY EXERCISE** ❯ 31

Factors That Are Opposites

Consider

$$\frac{x - 4}{8 - 2x}, \quad \text{or, equivalently,} \quad \frac{x - 4}{2(4 - x)}.$$

At first glance, the numerator and the denominator do not appear to have any common factors. But $x - 4$ and $4 - x$ are opposites, or additive inverses, of each other. Thus we can find a common factor by factoring out -1 in one expression.

EXAMPLE **6**   Simplify $\dfrac{x - 4}{8 - 2x}$ and check by evaluating.

SOLUTION   We have

$$\frac{x - 4}{8 - 2x} = \frac{x - 4}{2(4 - x)} \qquad \text{Factoring}$$

$$= \frac{x - 4}{2(-1)(x - 4)} \qquad \text{Note that } 4 - x = -x + 4 = -1(x - 4).$$

$$= \frac{x - 4}{-2(x - 4)} \qquad \begin{array}{l} \text{Had we originally factored out } -2, \text{ we} \\ \text{could have gone directly to this step.} \end{array}$$

$$= \frac{1}{-2} \cdot \frac{x - 4}{x - 4} \qquad \begin{array}{l} \text{Rewriting as a product. It is important} \\ \text{to write the 1 in the numerator.} \end{array}$$

$$= -\frac{1}{2}. \qquad \begin{array}{l} \text{Removing a factor equal to 1:} \\ (x - 4)/(x - 4) = 1 \end{array}$$

As a partial check, note that for any choice of x other than 4, the value of the rational expression is $-\frac{1}{2}$. For example, if $x = 5$, then

$$\frac{x - 4}{8 - 2x} = \frac{5 - 4}{8 - 2 \cdot 5}$$

$$= \frac{1}{-2} = -\frac{1}{2}.$$

TRY EXERCISE ▸ 47

6.1 EXERCISE SET

 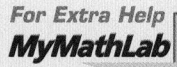 

🖐 **Concept Reinforcement** *In each of Exercises 1–6, match the rational expression with the list of numbers in the column on the right for which the rational expression is undefined.*

1. ___ $\dfrac{x - 5}{(x - 2)(x + 3)}$

    a) $-1, 4$

2. ___ $\dfrac{3t}{(t + 1)(t - 4)}$

    b) $-3, 5$

    c) $-\dfrac{4}{3}, \dfrac{1}{2}$

3. ___ $\dfrac{a + 7}{a^2 - a - 12}$

    d) $-3, 4$

4. ___ $\dfrac{m - 3}{m^2 - 2m - 15}$

    e) $-3, 2$

    f) $\dfrac{1}{3}, \dfrac{2}{3}$

5. ___ $\dfrac{2t + 7}{(2t - 1)(3t + 4)}$

6. ___ $\dfrac{4x - 1}{(3x - 1)(3x - 2)}$

List all numbers for which each rational expression is undefined.

7. $\dfrac{18}{-11x}$

8. $\dfrac{13}{-5t}$

9. $\dfrac{y - 3}{y + 5}$

10. $\dfrac{a + 6}{a - 10}$

11. $\dfrac{t - 5}{3t - 15}$

12. $\dfrac{x^2 - 4}{5x + 10}$

13. $\dfrac{x^2 - 25}{x^2 - 3x - 28}$

14. $\dfrac{p^2 - 9}{p^2 - 7p + 10}$

15. $\dfrac{t^2 + t - 20}{2t^2 + 11t - 6}$

16. $\dfrac{x^2 + 2x + 1}{3x^2 - x - 14}$

Simplify by removing a factor equal to 1. Show all steps.

17. $\dfrac{50a^2b}{40ab^3}$

18. $\dfrac{-24x^4y^3}{6x^7y}$

19. $\dfrac{6t + 12}{6t - 18}$

20. $\dfrac{5n - 30}{5n + 5}$

21. $\dfrac{21t - 7}{24t - 8}$

22. $\dfrac{10n + 25}{8n + 20}$

23. $\dfrac{a^2 - 9}{a^2 + 4a + 3}$

24. $\dfrac{a^2 + 5a + 6}{a^2 - 9}$

Simplify, if possible. Then check by evaluating, as in Example 6.

25. $\dfrac{-36x^8}{54x^5}$

26. $\dfrac{45a^4}{30a^6}$

27. $\dfrac{-2y + 6}{-8y}$

28. $\dfrac{-4m^2 + 4m}{-8m^2 + 12m}$

29. $\dfrac{t^2 - 16}{t^2 - t - 20}$

30. $\dfrac{a^2 - 4}{a^2 + 5a + 6}$

31. $\dfrac{3a^2 + 9a - 12}{6a^2 - 30a + 24}$

32. $\dfrac{2t^2 - 6t + 4}{4t^2 + 12t - 16}$

33. $\dfrac{x^2 - 8x + 16}{x^2 - 16}$

34. $\dfrac{x^2 - 25}{x^2 + 10x + 25}$

35. $\dfrac{n - 2}{n^3 - 8}$

36. $\dfrac{n^6 + 27}{n^2 + 3}$

37. $\dfrac{t^2 - 1}{t + 1}$

38. $\dfrac{a^2 - 1}{a - 1}$

39. $\dfrac{y^2 + 4}{y + 2}$

40. $\dfrac{m^2 + 9}{m + 3}$

41. $\dfrac{5x^2 + 20}{10x^2 + 40}$

42. $\dfrac{6x^2 + 54}{4x^2 + 36}$

43. $\dfrac{y^2 + 6y}{2y^2 + 13y + 6}$

44. $\dfrac{t^2 + 2t}{2t^2 + t - 6}$

45. $\dfrac{4x^2 - 12x + 9}{10x^2 - 11x - 6}$

46. $\dfrac{4x^2 - 4x + 1}{6x^2 + 5x - 4}$

47. $\dfrac{10 - x}{x - 10}$ **48.** $\dfrac{x - 8}{8 - x}$

49. $\dfrac{7t - 14}{2 - t}$ **50.** $\dfrac{3 - n}{5n - 15}$

51. $\dfrac{a - b}{4b - 4a}$ **52.** $\dfrac{2p - 2q}{q - p}$

53. $\dfrac{3x^2 - 3y^2}{2y^2 - 2x^2}$ **54.** $\dfrac{7a^2 - 7b^2}{3b^2 - 3a^2}$

Aha! **55.** $\dfrac{7s^2 - 28t^2}{28t^2 - 7s^2}$ **56.** $\dfrac{9m^2 - 4n^2}{4n^2 - 9m^2}$

57. Explain how simplifying is related to the identity property of 1.

58. If a rational expression is undefined for $x = 5$ and $x = -3$, what is the degree of the denominator? Why?

Skill Review

To prepare for Section 6.2, review multiplication and division using fraction notation (Section 1.7).

Simplify.

59. $-\dfrac{2}{15} \cdot \dfrac{10}{7}$ [1.7]   **60.** $\left(\dfrac{3}{4}\right)\left(\dfrac{-20}{9}\right)$ [1.7]

61. $\dfrac{5}{8} \div \left(-\dfrac{1}{6}\right)$ [1.7]

62. $\dfrac{7}{10} \div \left(-\dfrac{8}{15}\right)$ [1.7]

63. $\dfrac{7}{9} - \dfrac{2}{3} \cdot \dfrac{6}{7}$ [1.8]

64. $\dfrac{2}{3} - \left(\dfrac{3}{4}\right)^2$ [1.8]

Synthesis

65. Keith *incorrectly* simplifies

$$\frac{x^2 + x - 2}{x^2 + 3x + 2} \quad \text{as} \quad \frac{x - 1}{x + 2}.$$

He then checks his simplification by evaluating both expressions for $x = 1$. Use this situation to explain why evaluating is not a foolproof check.

66. How could you convince someone that $a - b$ and $b - a$ are opposites of each other?

Simplify.

67. $\dfrac{16y^4 - x^4}{(x^2 + 4y^2)(x - 2y)}$

68. $\dfrac{(x - 1)(x^4 - 1)(x^2 - 1)}{(x^2 + 1)(x - 1)^2(x^4 - 2x^2 + 1)}$

69. $\dfrac{x^5 - 2x^3 + 4x^2 - 8}{x^7 + 2x^4 - 4x^3 - 8}$

70. $\dfrac{10t^4 - 8t^3 + 15t - 12}{8 - 10t + 12t^2 - 15t^3}$

71. $\dfrac{(t^4 - 1)(t^2 - 9)(t - 9)^2}{(t^4 - 81)(t^2 + 1)(t + 1)^2}$

72. $\dfrac{(t + 2)^3(t^2 + 2t + 1)(t + 1)}{(t + 1)^3(t^2 + 4t + 4)(t + 2)}$

73. $\dfrac{(x^2 - y^2)(x^2 - 2xy + y^2)}{(x + y)^2(x^2 - 4xy - 5y^2)}$

74. $\dfrac{x^4 - y^4}{(y - x)^4}$

75. Select any number x, multiply by 2, add 5, multiply by 5, subtract 25, and divide by 10. What do you get? Explain how this procedure can be used for a number trick.

6.2 Multiplication and Division

Multiplication • Division

Multiplication and division of rational expressions are similar to multiplication and division with fractions. In this section, we again assume that all denominators are nonzero.

Multiplication

Recall that to multiply fractions, we multiply numerator times numerator and denominator times denominator. Rational expressions are multiplied in a similar way.

> **The Product of Two Rational Expressions**
>
> To multiply rational expressions, multiply numerators and multiply denominators:
>
> $$\frac{A}{B} \cdot \frac{C}{D} = \frac{AC}{BD}.$$
>
> Then factor and, if possible, simplify the result.

For example,

$$\frac{3}{5} \cdot \frac{8}{11} = \frac{3 \cdot 8}{5 \cdot 11} \quad \text{and} \quad \frac{x}{3} \cdot \frac{x+2}{y} = \frac{x(x+2)}{3y}.$$

Fraction bars are grouping symbols, so parentheses are needed when writing some products. Because we generally simplify, we often leave products involving variables in factored form. There is no need to multiply further.

EXAMPLE 1   Multiply and, if possible, simplify.

a) $\dfrac{5a^3}{4} \cdot \dfrac{2}{5a}$

b) $(x^2 - 3x - 10) \cdot \dfrac{x+4}{x^2 - 10x + 25}$

c) $\dfrac{10x + 20}{2x^2 - 3x + 1} \cdot \dfrac{x^2 - 1}{5x + 10}$

SOLUTION

a) $\dfrac{5a^3}{4} \cdot \dfrac{2}{5a} = \dfrac{5a^3(2)}{4(5a)}$ 　　Forming the product of the numerators and the product of the denominators

$\phantom{\dfrac{5a^3}{4} \cdot \dfrac{2}{5a}} = \dfrac{5 \cdot a \cdot a \cdot a \cdot 2}{2 \cdot 2 \cdot 5 \cdot a}$ 　　Factoring the numerator and the denominator

$\phantom{\dfrac{5a^3}{4} \cdot \dfrac{2}{5a}} = \dfrac{\cancel{5} \cdot \cancel{a} \cdot a \cdot a \cdot \cancel{2}}{\cancel{2} \cdot 2 \cdot \cancel{5} \cdot \cancel{a}}$ 　　Removing a factor equal to 1: $\dfrac{2 \cdot 5 \cdot a}{2 \cdot 5 \cdot a} = 1$

$\phantom{\dfrac{5a^3}{4} \cdot \dfrac{2}{5a}} = \dfrac{a^2}{2}$

b) $(x^2 - 3x - 10) \cdot \dfrac{x + 4}{x^2 - 10x + 25}$

$= \dfrac{x^2 - 3x - 10}{1} \cdot \dfrac{x + 4}{x^2 - 10x + 25}$ Writing $x^2 - 3x - 10$ as a rational expression

$= \dfrac{(x^2 - 3x - 10)(x + 4)}{1(x^2 - 10x + 25)}$ Multiplying the numerators and the denominators

$= \dfrac{(x - 5)(x + 2)(x + 4)}{(x - 5)(x - 5)}$ Factoring the numerator and the denominator

$= \dfrac{(\cancel{x - 5})(x + 2)(x + 4)}{(\cancel{x - 5})(x - 5)}$ Removing a factor equal to 1: $\dfrac{x - 5}{x - 5} = 1$

$= \dfrac{(x + 2)(x + 4)}{x - 5}$

c) $\dfrac{10x + 20}{2x^2 - 3x + 1} \cdot \dfrac{x^2 - 1}{5x + 10}$

$= \dfrac{(10x + 20)(x^2 - 1)}{(2x^2 - 3x + 1)(5x + 10)}$ Multiply.

$= \dfrac{5(2)(x + 2)(x + 1)(x - 1)}{(x - 1)(2x - 1)5(x + 2)}$ Factor.   Try to go directly to this step.

$= \dfrac{\cancel{5}(2)(\cancel{x + 2})(x + 1)(\cancel{x - 1})}{(\cancel{x - 1})(2x - 1)\cancel{5}(\cancel{x + 2})}$ Simplify. $\dfrac{5(x + 2)(x - 1)}{5(x + 2)(x - 1)} = 1$

$= \dfrac{2(x + 1)}{2x - 1}$ **TRY EXERCISE** 19

Because our results are often used in problems that require factored form, there is no need to multiply out the numerator or the denominator.

Division

As with fractions, reciprocals of rational expressions are found by interchanging the numerator and the denominator. For example,

the reciprocal of $\dfrac{2}{7}$ is $\dfrac{7}{2}$,   and   the reciprocal of $\dfrac{3x}{x + 5}$ is $\dfrac{x + 5}{3x}$.

> ### The Quotient of Two Rational Expressions
>
> To divide by a rational expression, multiply by its reciprocal:
>
> $$\dfrac{A}{B} \div \dfrac{C}{D} = \dfrac{A}{B} \cdot \dfrac{D}{C} = \dfrac{AD}{BC}.$$
>
> Then factor and, if possible, simplify.

For an explanation of why we divide this way, see Exercise 55 in Section 6.5.

EXAMPLE **2**   Divide: **(a)** $\dfrac{x}{5} \div \dfrac{7}{y}$; **(b)** $(x + 2) \div \dfrac{x - 1}{x + 3}$.

SOLUTION

a) $\dfrac{x}{5} \div \dfrac{7}{y} = \dfrac{x}{5} \cdot \dfrac{y}{7}$ Multiplying by the reciprocal of the divisor

$\qquad = \dfrac{xy}{35}$ Multiplying rational expressions

b) $(x + 2) \div \dfrac{x - 1}{x + 3} = \dfrac{x + 2}{1} \cdot \dfrac{x + 3}{x - 1}$ Multiplying by the reciprocal of the divisor. Writing $x + 2$ as $\dfrac{x + 2}{1}$ can be helpful.

$\qquad = \dfrac{(x + 2)(x + 3)}{x - 1}$ **TRY EXERCISE** 41

As usual, we should simplify when possible. Often that requires us to factor one or more polynomials, hoping to discover a common factor that appears in both the numerator and the denominator.

EXAMPLE **3**   Divide and, if possible, simplify: $\dfrac{x + 1}{x^2 - 1} \div \dfrac{x + 1}{x^2 - 2x + 1}$.

SOLUTION

$\dfrac{x + 1}{x^2 - 1} \div \dfrac{x + 1}{x^2 - 2x + 1} = \dfrac{x + 1}{x^2 - 1} \cdot \dfrac{x^2 - 2x + 1}{x + 1}$ Rewrite as multiplication.

$\qquad = \dfrac{(x + 1)(x - 1)(x - 1)}{(x + 1)(x - 1)(x + 1)}$ Multiply.   Factor.

$\qquad = \dfrac{\cancel{(x + 1)}\cancel{(x - 1)}(x - 1)}{\cancel{(x + 1)}\cancel{(x - 1)}(x + 1)}$ Simplify.

$\qquad = \dfrac{x - 1}{x + 1}$ $\dfrac{(x + 1)(x - 1)}{(x + 1)(x - 1)} = 1$

TRY EXERCISE 47

EXAMPLE **4**   Divide and, if possible, simplify.

a) $\dfrac{a^2 + 3a + 2}{a^2 + 4} \div (5a^2 + 10a)$ **b)** $\dfrac{x^2 - 2x - 3}{x^2 - 4} \div \dfrac{x + 1}{x + 5}$

SOLUTION

a) $\dfrac{a^2 + 3a + 2}{a^2 + 4} \div (5a^2 + 10a)$

$\qquad = \dfrac{a^2 + 3a + 2}{a^2 + 4} \cdot \dfrac{1}{5a^2 + 10a}$ Multiplying by the reciprocal of the divisor

$\qquad = \dfrac{(a + 2)(a + 1)}{(a^2 + 4)5a(a + 2)}$ Multiplying rational expressions and factoring

$\qquad = \dfrac{\cancel{(a + 2)}(a + 1)}{(a^2 + 4)5a\cancel{(a + 2)}}$

$\qquad = \dfrac{a + 1}{(a^2 + 4)5a}$ Removing a factor equal to 1: $\dfrac{a + 2}{a + 2} = 1$

TECHNOLOGY CONNECTION

In performing a partial check of Example 4(b), we must be careful placing parentheses. We enter the original expression as $y_1 = ((x^2 - 2x - 3)/(x^2 - 4))/((x + 1)/(x + 5))$ and the simplified expression as $y_2 = ((x - 3)(x + 5))/((x - 2)(x + 2))$. Comparing values of y_1 and y_2, we see that

(*continued*)

the simplification is probably correct.

X	Y₁	Y₂
−5	ERROR	0
−4	−.5833	−.5833
−3	−2.4	−2.4
−2	ERROR	ERROR
−1	ERROR	5.3333
0	3.75	3.75
1	4	4

X = −5

1. Check Example 4(a).
2. Why are there 3 ERROR messages shown for y_1 on the screen above, and only 1 for y_2?

b) $\dfrac{x^2 - 2x - 3}{x^2 - 4} \div \dfrac{x + 1}{x + 5}$

$= \dfrac{x^2 - 2x - 3}{x^2 - 4} \cdot \dfrac{x + 5}{x + 1}$ Multiplying by the reciprocal of the divisor

$= \dfrac{(x - 3)(x + 1)(x + 5)}{(x - 2)(x + 2)(x + 1)}$ Multiplying rational expressions and factoring

$= \dfrac{(x - 3)\cancel{(x + 1)}(x + 5)}{(x - 2)(x + 2)\cancel{(x + 1)}}$

$= \dfrac{(x - 3)(x + 5)}{(x - 2)(x + 2)}$ Removing a factor equal to 1: $\dfrac{x + 1}{x + 1} = 1$

TRY EXERCISE ▶ 51

| 6.2 | **EXERCISE SET** | 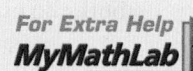  |

Multiply. Leave each answer in factored form.

1. $\dfrac{3x}{8} \cdot \dfrac{x + 2}{5x - 1}$

2. $\dfrac{2x}{7} \cdot \dfrac{3x + 5}{x - 1}$

3. $\dfrac{a - 4}{a + 6} \cdot \dfrac{a + 2}{a + 6}$

4. $\dfrac{a + 3}{a + 6} \cdot \dfrac{a + 3}{a - 1}$

5. $\dfrac{2x + 3}{4} \cdot \dfrac{x + 1}{x - 5}$

6. $\dfrac{x + 2}{3x - 4} \cdot \dfrac{4}{5x + 6}$

7. $\dfrac{n - 4}{n^2 + 4} \cdot \dfrac{n + 4}{n^2 - 4}$

8. $\dfrac{t + 3}{t^2 - 2} \cdot \dfrac{t + 3}{t^2 - 4}$

9. $\dfrac{y + 6}{1 + y} \cdot \dfrac{y - 3}{y + 3}$

10. $\dfrac{m + 4}{m + 8} \cdot \dfrac{2 + m}{m + 5}$

Multiply and, if possible, simplify.

11. $\dfrac{8t^3}{5t} \cdot \dfrac{3}{4t}$

12. $\dfrac{18}{a^5} \cdot \dfrac{2a^2}{3a}$

13. $\dfrac{3c}{d^2} \cdot \dfrac{8d}{6c^3}$

14. $\dfrac{3x^2y}{2} \cdot \dfrac{4}{xy^3}$

15. $\dfrac{x^2 - 3x - 10}{(x - 2)^2} \cdot (x - 2)$

16. $(t + 2) \cdot \dfrac{t^2 - 5t + 6}{(t + 2)^2}$

17. $\dfrac{n^2 - 6n + 5}{n + 6} \cdot \dfrac{n - 6}{n^2 + 36}$

18. $\dfrac{a + 2}{a - 2} \cdot \dfrac{a^2 + 4}{a^2 + 5a + 4}$

19. $\dfrac{a^2 - 9}{a^2} \cdot \dfrac{7a}{a^2 + a - 12}$

20. $\dfrac{x^2 + 10x - 11}{9x} \cdot \dfrac{x^3}{x + 11}$

21. $\dfrac{4v - 8}{5v} \cdot \dfrac{15v^2}{4v^2 - 16v + 16}$

22. $\dfrac{m - 2}{3m + 9} \cdot \dfrac{m^2 + 6m + 9}{2m^2 - 8}$

23. $\dfrac{t^2 + 2t - 3}{t^2 + 4t - 5} \cdot \dfrac{t^2 - 3t - 10}{t^2 + 5t + 6}$

24. $\dfrac{x^2 + 5x + 4}{x^2 - 6x + 8} \cdot \dfrac{x^2 + 5x - 14}{x^2 + 8x + 7}$

25. $\dfrac{12y + 12}{5y + 25} \cdot \dfrac{3y^2 - 75}{8y^2 - 8}$

26. $\dfrac{9t^2 - 900}{5t^2 - 20} \cdot \dfrac{5t + 10}{3t - 30}$

Aha! **27.** $\dfrac{x^2 + 4x + 4}{(x - 1)^2} \cdot \dfrac{x^2 - 2x + 1}{(x + 2)^2}$

28. $\dfrac{x^2 + 7x + 12}{x^2 + 6x + 8} \cdot \dfrac{4 - x^2}{x^2 + x - 6}$

29. $\dfrac{t^2 - 4t + 4}{2t^2 - 7t + 6} \cdot \dfrac{2t^2 + 7t - 15}{t^2 - 10t + 25}$

30. $\dfrac{5y^2 - 4y - 1}{3y^2 + 5y - 12} \cdot \dfrac{y^2 + 6y + 9}{y^2 - 2y + 1}$

31. $(10x^2 - x - 2) \cdot \dfrac{4x^2 - 8x + 3}{10x^2 - 11x - 6}$

32. $\dfrac{2x^2 - 5x + 3}{6x^2 - 5x - 1} \cdot (6x^2 + 13x + 2)$

33. $\dfrac{c^3 + 8}{c^5 - 4c^3} \cdot \dfrac{c^6 - 4c^5 + 4c^4}{c^2 - 2c + 4}$

34. $\dfrac{t^3 - 27}{t^4 - 9t^2} \cdot \dfrac{t^5 - 6t^4 + 9t^3}{t^2 + 3t + 9}$

Find the reciprocal of each expression.

35. $\dfrac{2x}{9}$

36. $\dfrac{3 - x}{x^2 + 4}$

37. $a^4 + 3a$

38. $\dfrac{1}{a^2 - b^2}$

Divide and, if possible, simplify.

39. $\dfrac{5}{9} \div \dfrac{3}{4}$

40. $\dfrac{3}{8} \div \dfrac{4}{7}$

41. $\dfrac{x}{4} \div \dfrac{5}{x}$

42. $\dfrac{5}{x} \div \dfrac{x}{12}$

43. $\dfrac{a^5}{b^4} \div \dfrac{a^2}{b}$

44. $\dfrac{x^5}{y^2} \div \dfrac{x^2}{y}$

45. $\dfrac{t - 3}{6} \div \dfrac{t + 1}{8}$

46. $\dfrac{10}{a + 3} \div \dfrac{15}{a}$

47. $\dfrac{4y - 8}{y + 2} \div \dfrac{y - 2}{y^2 - 4}$

48. $\dfrac{x^2 - 1}{x} \div \dfrac{x + 1}{2x - 2}$

49. $\dfrac{a}{a - b} \div \dfrac{b}{b - a}$

50. $\dfrac{x - y}{6} \div \dfrac{y - x}{3}$

51. $(n^2 + 5n + 6) \div \dfrac{n^2 - 4}{n + 3}$

52. $(v^2 - 1) \div \dfrac{(v + 1)(v - 3)}{v^2 + 9}$

53. $\dfrac{-3 + 3x}{16} \div \dfrac{x - 1}{5}$

54. $\dfrac{-4 + 2x}{15} \div \dfrac{x - 2}{3}$

55. $\dfrac{x - 1}{x + 2} \div \dfrac{1 - x}{4 + x^2}$

56. $\dfrac{-12 + 4x}{12} \div \dfrac{6 - 2x}{6}$

57. $\dfrac{a + 2}{a - 1} \div \dfrac{3a + 6}{a - 5}$

58. $\dfrac{t - 3}{t + 2} \div \dfrac{4t - 12}{t + 1}$

59. $(2x - 1) \div \dfrac{2x^2 - 11x + 5}{4x^2 - 1}$

60. $(a + 7) \div \dfrac{3a^2 + 14a - 49}{a^2 + 8a + 7}$

61. $\dfrac{w^2 - 14w + 49}{2w^2 - 3w - 14} \div \dfrac{3w^2 - 20w - 7}{w^2 - 6w - 16}$

62. $\dfrac{2m^2 + 59m - 30}{m^2 - 10m + 25} \div \dfrac{2m^2 - 21m + 10}{m^2 + m - 30}$

63. $\dfrac{c^2 + 10c + 21}{c^2 - 2c - 15} \div (5c^2 + 32c - 21)$

64. $\dfrac{z^2 - 2z + 1}{z^2 - 1} \div (4z^2 - z - 3)$

65. $\dfrac{x - y}{x^2 + 2xy + y^2} \div \dfrac{x^2 - y^2}{x^2 - 5xy + 4y^2}$

66. $\dfrac{a^2 - b^2}{a^2 - 4ab + 4b^2} \div \dfrac{a^2 - 3ab + 2b^2}{a - 2b}$

67. $\dfrac{x^3 - 64}{x^3 + 64} \div \dfrac{x^2 - 16}{x^2 - 4x + 16}$

68. $\dfrac{8y^3 - 27}{64y^3 - 1} \div \dfrac{4y^2 - 9}{16y^2 + 4y + 1}$

69. $\dfrac{8a^3 + b^3}{2a^2 + 3ab + b^2} \div \dfrac{8a^2 - 4ab + 2b^2}{4a^2 + 4ab + b^2}$

70. $\dfrac{x^3 + 8y^3}{2x^2 + 5xy + 2y^2} \div \dfrac{x^3 - 2x^2y + 4xy^2}{8x^2 - 2y^2}$

71. Why is it important to insert parentheses when multiplying rational expressions such as

$$\dfrac{x + 2}{5x - 7} \cdot \dfrac{3x - 1}{x + 4}?$$

72. As a first step in dividing $\dfrac{x}{3}$ by $\dfrac{7}{x}$, Jan canceled the x's.

Explain why this was incorrect, and show the correct division.

Skill Review

To prepare for Section 6.3, review addition and subtraction with fraction notation (Sections 1.3 and 1.6) and subtraction of polynomials (Section 4.4).

Simplify.

73. $\dfrac{3}{4} + \dfrac{5}{6}$ [1.3]

74. $\dfrac{7}{8} + \dfrac{5}{6}$ [1.3]

75. $\dfrac{2}{9} - \dfrac{1}{6}$ [1.3]

76. $\dfrac{3}{10} - \dfrac{7}{15}$ [1.6]

77. $2x^2 - x + 1 - (x^2 - x - 2)$ [4.4]

78. $3x^2 + x - 7 - (5x^2 + 5x - 8)$ [4.4]

Synthesis

79. Is the reciprocal of a product the product of the two reciprocals? Why or why not?

80. Explain why the quotient

$$\dfrac{x + 3}{x - 5} \div \dfrac{x - 7}{x + 1}$$

is undefined for $x = 5$, $x = -1$, and $x = 7$, but *is* defined for $x = -3$.

81. Find the reciprocal of $2\frac{1}{3}x$.

82. Find the reciprocal of $7.25x$.

Simplify.

83. $(x - 2a) \div \dfrac{a^2x^2 - 4a^4}{a^2x + 2a^3}$

84. $\dfrac{2a^2 - 5ab}{c - 3d} \div (4a^2 - 25b^2)$

85. $\dfrac{3x^2 - 2xy - y^2}{x^2 - y^2} \div (3x^2 + 4xy + y^2)^2$

86. $\dfrac{3a^2 - 5ab - 12b^2}{3ab + 4b^2} \div (3b^2 - ab)^2$

Aha! **87.** $\dfrac{a^2 - 3b}{a^2 + 2b} \cdot \dfrac{a^2 - 2b}{a^2 + 3b} \cdot \dfrac{a^2 + 2b}{a^2 - 3b}$

88. $\dfrac{y^2 - 4xy}{y - x} \div \dfrac{16x^2y^2 - y^4}{4x^2 - 3xy - y^2} \div \dfrac{4}{x^3y^3}$

89. $\dfrac{z^2 - 8z + 16}{z^2 + 8z + 16} \div \dfrac{(z - 4)^5}{(z + 4)^5} \div \dfrac{3z + 12}{z^2 - 16}$

90. $\dfrac{(t + 2)^3}{(t + 1)^3} \div \dfrac{t^2 + 4t + 4}{t^2 + 2t + 1} \cdot \dfrac{t + 1}{t + 2}$

91. $\dfrac{a^4 - 81b^4}{a^2c - 6abc + 9b^2c} \cdot \dfrac{a + 3b}{a^2 + 9b^2} \div \dfrac{a^2 + 6ab + 9b^2}{(a - 3b)^2}$

92. $\dfrac{3y^3 + 6y^2}{y^2 - y - 12} \div \dfrac{y^2 - y}{y^2 - 2y - 8} \cdot \dfrac{y^2 + 5y + 6}{y^2}$

93. Use a graphing calculator to check that

$$\dfrac{x - 1}{x^2 + 2x + 1} \div \dfrac{x^2 - 1}{x^2 - 5x + 4}$$

is equivalent to

$$\dfrac{x^2 - 5x + 4}{(x + 1)^3}.$$

CORNER

Currency Exchange

Focus: Least common multiples and proportions

Time: 20 minutes

Group size: 2

International travelers usually exchange currencies. Recently one New Zealand dollar was worth 76 U.S. cents. Use this exchange rate for the activity that follows.

ACTIVITY

1. Within each group of two students, one student should play the role of a U.S. citizen visiting New Zealand. The other student should play the role of a New Zealander visiting the United States. Use the exchange rate of one New Zealand dollar for 76 U.S. cents.

2. The "U.S." student should exchange $76 U.S. and $100 U.S. for New Zealand money. The "New Zealand" student should exchange $76 New Zealand and $100 New Zealand for U.S. money.

3. The exchanges in part (2) should indicate that coins smaller than a dollar are needed to exchange $76 New Zealand for U.S. funds, or

to exchange $100 U.S. for New Zealand money. What is the smallest amount of New Zealand dollars that can be exchanged for a whole-number amount of U.S. dollars? What is the smallest amount of U.S. dollars that can be exchanged for a whole-number amount of New Zealand dollars? (*Hint*: See part 2.)

4. Use the results from part (3) to find two other amounts of U.S. dollars that can be exchanged for a whole-number amount of New Zealand dollars. Answers may vary.

5. Find the smallest number a for which both conversions—from a New Zealand dollars to U.S. funds and from a U.S. dollars to New Zealand funds—use only whole numbers. (*Hint*: Use LCMs and the results of part 2 above.)

6. At one time in 2008, one Polish zloty was worth about 40 U.S. cents. Find the smallest number a for which both conversions—from a Polish zlotys to U.S. funds and from a U.S. dollars to Polish funds—use only whole numbers. (*Hint*: See part 5.)

<table>
<tr><td>**6.3**</td><td>## Addition, Subtraction, and Least Common Denominators</td></tr>
</table>

Addition When Denominators Are the Same ▪ Subtraction When Denominators Are the Same ▪ Least Common Multiples and Denominators

Addition When Denominators Are the Same

Recall that to add fractions having the same denominator, like $\frac{2}{7}$ and $\frac{3}{7}$, we add the numerators and keep the common denominator: $\frac{2}{7} + \frac{3}{7} = \frac{5}{7}$. The same procedure is used when rational expressions share a common denominator.

> ### The Sum of Two Rational Expressions
>
> To add when the denominators are the same, add the numerators and keep the common denominator:
> $$\frac{A}{B} + \frac{C}{B} = \frac{A + C}{B}.$$

In this section, we again assume that all denominators are nonzero.

EXAMPLE **1** Add. Simplify the result, if possible.

a) $\dfrac{4}{a} + \dfrac{3 + a}{a}$

b) $\dfrac{3x}{x - 5} + \dfrac{2x + 1}{x - 5}$

c) $\dfrac{2x^2 + 3x - 7}{2x + 1} + \dfrac{x^2 + x - 8}{2x + 1}$

d) $\dfrac{x - 5}{x^2 - 9} + \dfrac{2}{x^2 - 9}$

SOLUTION

a) $\dfrac{4}{a} + \dfrac{3 + a}{a} = \dfrac{7 + a}{a}$ When the denominators are alike, add the numerators and keep the common denominator.

b) $\dfrac{3x}{x - 5} + \dfrac{2x + 1}{x - 5} = \dfrac{5x + 1}{x - 5}$ The denominators are alike, so we add the numerators.

c) $\dfrac{2x^2 + 3x - 7}{2x + 1} + \dfrac{x^2 + x - 8}{2x + 1} = \dfrac{(2x^2 + 3x - 7) + (x^2 + x - 8)}{2x + 1}$

$= \dfrac{3x^2 + 4x - 15}{2x + 1}$ Combining like terms

$= \dfrac{(3x - 5)(x + 3)}{2x + 1}$ Factoring. There are no common factors, so we cannot simplify further.

d) $\dfrac{x - 5}{x^2 - 9} + \dfrac{2}{x^2 - 9} = \dfrac{x - 3}{x^2 - 9}$ Combining like terms in the numerator: $x - 5 + 2 = x - 3$

$= \dfrac{x - 3}{(x - 3)(x + 3)}$ Factoring

$= \dfrac{1 \cdot (x - 3)}{(x - 3)(x + 3)}$ Removing a factor equal to 1: $\dfrac{x - 3}{x - 3} = 1$

$= \dfrac{1}{x + 3}$

TRY EXERCISE 7

STUDY SKILLS

Visualize the Steps

If you have completed all assignments and are studying for a quiz or test, don't feel that you need to redo every assigned problem. A more productive use of your time would be to work through one problem of each type. Then read through the other problems, visualizing the steps that lead to a solution. When you are unsure of how to solve a problem, work that problem in its entirety, seeking outside help as needed.

Subtraction When Denominators Are the Same

When two fractions have the same denominator, we subtract one numerator from the other and keep the common denominator: $\frac{5}{7} - \frac{2}{7} = \frac{3}{7}$. The same procedure is used with rational expressions.

The Difference of Two Rational Expressions

To subtract when the denominators are the same, subtract the second numerator from the first numerator and keep the common denominator:

$$\frac{A}{B} - \frac{C}{B} = \frac{A - C}{B}.$$

CAUTION! The fraction bar under a numerator is a grouping symbol, just like parentheses. Thus, when a numerator is subtracted, it is important to subtract *every* term in that numerator.

EXAMPLE 2 Subtract and, if possible, simplify: **(a)** $\dfrac{3x}{x + 2} - \dfrac{x - 5}{x + 2}$; **(b)** $\dfrac{x^2}{x - 4} - \dfrac{x + 12}{x - 4}$.

SOLUTION

a) $\dfrac{3x}{x + 2} - \dfrac{x - 5}{x + 2} = \dfrac{3x - (x - 5)}{x + 2}$ The parentheses are needed to make sure that we subtract both terms.

$= \dfrac{3x - x + 5}{x + 2}$ Removing the parentheses and changing signs (using the distributive law)

$= \dfrac{2x + 5}{x + 2}$ Combining like terms

b) $\dfrac{x^2}{x - 4} - \dfrac{x + 12}{x - 4} = \dfrac{x^2 - (x + 12)}{x - 4}$ Remember the parentheses!

$= \dfrac{x^2 - x - 12}{x - 4}$ Removing parentheses (using the distributive law)

$= \dfrac{(x - 4)(x + 3)}{x - 4}$ Factoring, in hopes of simplifying

$= \dfrac{(x - 4)(x + 3)}{x - 4}$ Removing a factor equal to 1: $\dfrac{x - 4}{x - 4} = 1$

$= x + 3$

TRY EXERCISE 21

Least Common Multiples and Denominators

Thus far, every pair of rational expressions that we have added or subtracted shared a common denominator. To add or subtract rational expressions that have different denominators, we must first find equivalent rational expressions that *do* have a common denominator.

In algebra, we find a common denominator much as we do in arithmetic. Recall that to add $\frac{1}{12}$ and $\frac{7}{30}$, we first identify the smallest number that contains both 12 and 30 as factors. Such a number, the **least common multiple (LCM)** of the denominators, is then used as the **least common denominator (LCD)**.

Let's find the LCM of 12 and 30 using a method that can also be used with polynomials. We begin by writing the prime factorizations of 12 and 30:

$$12 = 2 \cdot 2 \cdot 3;$$
$$30 = 2 \cdot 3 \cdot 5.$$

The LCM must include the factors of each number, so it must include each prime factor the greatest number of times that it appears in either of the factorizations. To find the LCM for 12 and 30, we select one factorization, say

$$2 \cdot 2 \cdot 3,$$

and note that because it lacks a factor of 5, it does not contain the entire factorization of 30. If we multiply $2 \cdot 2 \cdot 3$ by 5, every prime factor occurs just often enough to contain both 12 and 30 as factors.

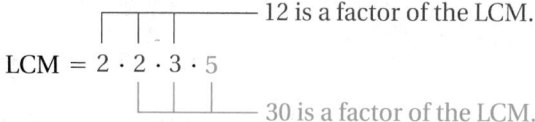

LCM $= 2 \cdot 2 \cdot 3 \cdot 5$

12 is a factor of the LCM.

30 is a factor of the LCM.

Note that each prime factor—2, 3, and 5—is used the greatest number of times that it appears in either of the individual factorizations. The factor 2 occurs twice and the factors 3 and 5 once each.

To Find the Least Common Denominator (LCD)

1. Write the prime factorization of each denominator.
2. Select one of the factorizations and inspect it to see if it completely contains the other factorization.

   a) If it does, it represents the LCM of the denominators.
   b) If it does not, multiply that factorization by any factors of the other denominator that it lacks. The final product is the LCM of the denominators.

The LCD is the LCM of the denominators. It should contain each factor the greatest number of times that it occurs in any of the individual factorizations.

EXAMPLE 3   Find the LCD of $\dfrac{5}{36x^2}$ and $\dfrac{7}{24x}$.

SOLUTION

1. We begin by writing the prime factorizations of $36x^2$ and $24x$:

$$36x^2 = 2 \cdot 2 \cdot 3 \cdot 3 \cdot x \cdot x;$$
$$24x = 2 \cdot 2 \cdot 2 \cdot 3 \cdot x.$$

2. We select the factorization of $36x^2$. Except for a third factor of 2, this factorization contains the entire factorization of $24x$. Thus we multiply $36x^2$ by a third factor of 2.

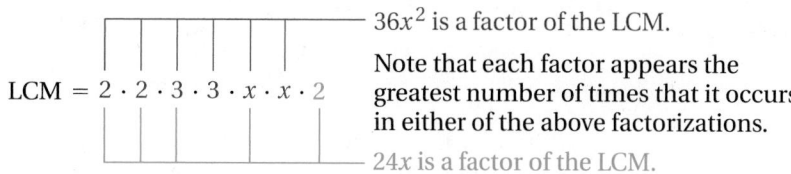

$36x^2$ is a factor of the LCM.

LCM $= 2 \cdot 2 \cdot 3 \cdot 3 \cdot x \cdot x \cdot 2$

Note that each factor appears the greatest number of times that it occurs in either of the above factorizations.

$24x$ is a factor of the LCM.

Since $2^3 \cdot 3^2 \cdot x^2$, or $72x^2$, is the smallest multiple of both $36x^2$ and $24x$, the LCM of the denominators is $72x^2$. The LCD of the expressions is $72x^2$.

Let's add $\dfrac{1}{12}$ and $\dfrac{7}{30}$:

$$\frac{1}{12} + \frac{7}{30} = \frac{1}{2 \cdot 2 \cdot 3} + \frac{7}{2 \cdot 3 \cdot 5}.$$ The least common denominator (LCD) is $2 \cdot 2 \cdot 3 \cdot 5$.

We found above that the LCD is $2 \cdot 2 \cdot 3 \cdot 5$, or 60. To get the LCD, we see that the first denominator needs a factor of 5, and the second denominator needs another factor of 2. Therefore, we multiply $\frac{1}{12}$ by 1, using $\frac{5}{5}$, and we multiply $\frac{7}{30}$ by 1, using $\frac{2}{2}$. Since $a \cdot 1 = a$, for any number a, the values of the fractions are not changed.

$$\frac{1}{12} + \frac{7}{30} = \frac{1}{2 \cdot 2 \cdot 3} \cdot \frac{5}{5} + \frac{7}{2 \cdot 3 \cdot 5} \cdot \frac{2}{2} \qquad \frac{5}{5} = 1 \text{ and } \frac{2}{2} = 1$$

$$= \frac{5}{60} + \frac{14}{60} \qquad \qquad \text{Both denominators are now the LCD.}$$

$$= \frac{19}{60} \qquad \qquad \text{Adding the numerators and keeping the LCD}$$

Expressions like $\dfrac{5}{36x^2}$ and $\dfrac{7}{24x}$ are added in much the same manner. In Example 3, we found that the LCD is $2 \cdot 2 \cdot 2 \cdot 3 \cdot 3 \cdot x \cdot x$, or $72x^2$. To obtain equivalent expressions with this LCD, we multiply each expression by 1, using the missing factors of the LCD to write 1:

$$\frac{5}{36x^2} + \frac{7}{24x} = \frac{5}{2 \cdot 2 \cdot 3 \cdot 3 \cdot x \cdot x} + \frac{7}{2 \cdot 2 \cdot 2 \cdot 3 \cdot x}$$

$$= \frac{5}{2 \cdot 2 \cdot 3 \cdot 3 \cdot x \cdot x} \cdot \frac{2}{2} + \frac{7}{2 \cdot 2 \cdot 2 \cdot 3 \cdot x} \cdot \frac{3 \cdot x}{3 \cdot x}$$

↑   ↑

The LCD requires another factor of 2.   The LCD requires additional factors of 3 and x.

$$= \frac{10}{72x^2} + \frac{21x}{72x^2} \qquad \text{Both denominators are now the LCD.}$$

$$= \frac{21x + 10}{72x^2}.$$

You now have the "big picture" of why LCMs are needed when adding rational expressions. For the remainder of this section, we will practice finding LCMs and rewriting rational expressions so that they have the LCD as the denominator. In Section 6.4, we will return to the addition and subtraction of rational expressions.

EXAMPLE **4**   For each pair of polynomials, find the least common multiple.

a) $15a$ and $35b$

b) $21x^3y^6$ and $7x^5y^2$

c) $x^2 + 5x - 6$ and $x^2 - 1$

SOLUTION

a) We write the prime factorizations and then construct the LCM, starting with the factorization of $15a$.

$$15a = 3 \cdot 5 \cdot a$$
$$35b = 5 \cdot 7 \cdot b$$

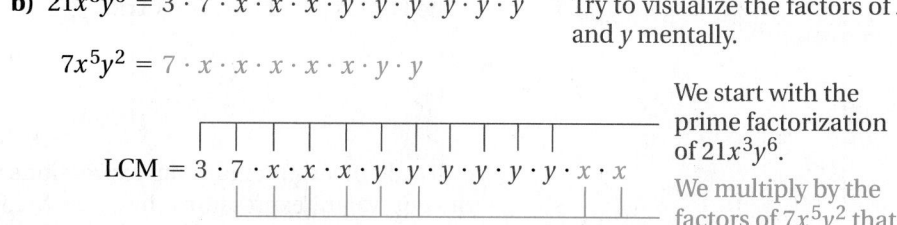

15a is a factor of the LCM.

$$\text{LCM} = 3 \cdot 5 \cdot a \cdot 7 \cdot b$$

Each factor appears the greatest number of times that it occurs in either of the above factorizations.

35b is a factor of the LCM.

The LCM is $3 \cdot 5 \cdot a \cdot 7 \cdot b$, or $105ab$.

b) $21x^3y^6 = 3 \cdot 7 \cdot x \cdot x \cdot x \cdot y \cdot y \cdot y \cdot y \cdot y \cdot y$ Try to visualize the factors of x and y mentally.

$$7x^5y^2 = 7 \cdot x \cdot x \cdot x \cdot x \cdot x \cdot y \cdot y$$

$$\text{LCM} = 3 \cdot 7 \cdot x \cdot x \cdot x \cdot y \cdot y \cdot y \cdot y \cdot y \cdot y \cdot x \cdot x$$

We start with the prime factorization of $21x^3y^6$.

We multiply by the factors of $7x^5y^2$ that are lacking.

Note that we used the highest power of each factor in $21x^3y^6$ and $7x^5y^2$. The LCM is $21x^5y^6$.

c) $x^2 + 5x - 6 = (x - 1)(x + 6)$
$\quad\quad\; x^2 - 1 = (x - 1)(x + 1)$

$$\text{LCM} = (x - 1)(x + 6)(x + 1)$$

We start with the factorization of $x^2 + 5x - 6$.

We multiply by the factor of $x^2 - 1$ that is missing.

The LCM is $(x - 1)(x + 6)(x + 1)$. There is no need to multiply this out.

> **TRY EXERCISE** 43

The procedure above can be used to find the LCM of three or more polynomials as well. We factor each polynomial and then construct the LCM using each factor the greatest number of times that it appears in any one factorization.

EXAMPLE 5   For each group of polynomials, find the LCM.

a) $12x$, $16y$, and $8xyz$

b) $x^2 + 4$, $x + 1$, and 5

STUDENT NOTES

If you prefer, the LCM for a group of three polynomials can be found by finding the LCM of two of them and then finding the LCM of that result and the remaining polynomial.

SOLUTION

a) $12x = 2 \cdot 2 \cdot 3 \cdot x$
$16y = 2 \cdot 2 \cdot 2 \cdot 2 \cdot y$
$8xyz = 2 \cdot 2 \cdot 2 \cdot x \cdot y \cdot z$

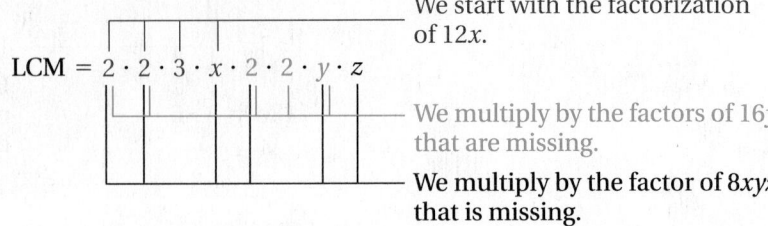

LCM $= 2 \cdot 2 \cdot 3 \cdot x \cdot 2 \cdot 2 \cdot y \cdot z$

We start with the factorization of $12x$.

We multiply by the factors of $16y$ that are missing.

We multiply by the factor of $8xyz$ that is missing.

The LCM is $2^4 \cdot 3 \cdot xyz$, or $48xyz$.

b) Since $x^2 + 4, x + 1$, and 5 are not factorable, the LCM is their product: $5(x^2 + 4)(x + 1)$.

TRY EXERCISE ▶ 51

To add or subtract rational expressions with different denominators, we first write equivalent expressions that have the LCD. To do this, we multiply each rational expression by a carefully constructed form of 1.

EXAMPLE 6

Find equivalent expressions that have the LCD:

$$\frac{x + 3}{x^2 + 5x - 6}, \quad \frac{x + 7}{x^2 - 1}.$$

SOLUTION From Example 4(c), we know that the LCD is

$$(x + 6)(x - 1)(x + 1).$$

Since

$$x^2 + 5x - 6 = (x + 6)(x - 1),$$

the factor of the LCD that is missing from the first denominator is $x + 1$. We multiply by 1 using $(x + 1)/(x + 1)$:

$$\frac{x + 3}{x^2 + 5x - 6} = \frac{x + 3}{(x + 6)(x - 1)} \cdot \frac{x + 1}{x + 1}$$
$$= \frac{(x + 3)(x + 1)}{(x + 6)(x - 1)(x + 1)}.$$

Finding an equivalent expression that has the least common denominator

For the second expression, we have

$$x^2 - 1 = (x + 1)(x - 1).$$

The factor of the LCD that is missing is $x + 6$. We multiply by 1 using $(x + 6)/(x + 6)$:

$$\frac{x + 7}{x^2 - 1} = \frac{x + 7}{(x + 1)(x - 1)} \cdot \frac{x + 6}{x + 6}$$
$$= \frac{(x + 7)(x + 6)}{(x + 1)(x - 1)(x + 6)}.$$

Finding an equivalent expression that has the least common denominator

We leave the results in factored form. In Section 6.4, we will carry out the actual addition and subtraction of such rational expressions.

TRY EXERCISE ▶ 61

6.3 EXERCISE SET

🐦 *Concept Reinforcement* *Use one or more words to complete each of the following sentences.*

1. To add two rational expressions when the denominators are the same, add _____ and keep the common _____.

2. When a numerator is being subtracted, use parentheses to make sure to subtract every _____ in that numerator.

3. The least common multiple of two denominators is usually referred to as the _____ and is abbreviated _____.

4. The least common denominator of two fractions must contain the prime _____ of both _____.

Perform the indicated operation. Simplify, if possible.

5. $\dfrac{3}{t} + \dfrac{5}{t}$

6. $\dfrac{8}{y^2} + \dfrac{2}{y^2}$

7. $\dfrac{x}{12} + \dfrac{2x+5}{12}$

8. $\dfrac{a}{7} + \dfrac{3a-4}{7}$

9. $\dfrac{4}{a+3} + \dfrac{5}{a+3}$

10. $\dfrac{5}{x+2} + \dfrac{8}{x+2}$

11. $\dfrac{11}{4x-7} - \dfrac{3}{4x-7}$

12. $\dfrac{9}{2x+3} - \dfrac{5}{2x+3}$

13. $\dfrac{3y+8}{2y} - \dfrac{y+1}{2y}$

14. $\dfrac{5+3t}{4t} - \dfrac{2t+1}{4t}$

15. $\dfrac{5x+7}{x+3} + \dfrac{x+11}{x+3}$

16. $\dfrac{3x+4}{x-1} + \dfrac{2x-9}{x-1}$

17. $\dfrac{5x+7}{x+3} - \dfrac{x+11}{x+3}$

18. $\dfrac{3x+4}{x-1} - \dfrac{2x-9}{x-1}$

19. $\dfrac{a^2}{a-4} + \dfrac{a-20}{a-4}$

20. $\dfrac{x^2}{x+5} + \dfrac{7x+10}{x+5}$

21. $\dfrac{y^2}{y+2} - \dfrac{5y+14}{y+2}$

22. $\dfrac{t^2}{t-3} - \dfrac{8t-15}{t-3}$

Aha! **23.** $\dfrac{t^2-5t}{t-1} + \dfrac{5t-t^2}{t-1}$

24. $\dfrac{y^2+6y}{y+2} + \dfrac{2y+12}{y+2}$

25. $\dfrac{x-6}{x^2+5x+6} + \dfrac{9}{x^2+5x+6}$

26. $\dfrac{x-5}{x^2-4x+3} + \dfrac{2}{x^2-4x+3}$

27. $\dfrac{t^2-5t}{t^2+6t+9} + \dfrac{4t-12}{t^2+6t+9}$

28. $\dfrac{y^2-7y}{y^2+8y+16} + \dfrac{6y-20}{y^2+8y+16}$

29. $\dfrac{2y^2+3y}{y^2-7y+12} - \dfrac{y^2+4y+6}{y^2-7y+12}$

30. $\dfrac{3a^2+7}{a^2-2a-8} - \dfrac{7+3a^2}{a^2-2a-8}$

31. $\dfrac{3-2x}{x^2-6x+8} + \dfrac{7-3x}{x^2-6x+8}$

32. $\dfrac{1-2t}{t^2-5t+4} + \dfrac{4-3t}{t^2-5t+4}$

33. $\dfrac{x-9}{x^2+3x-4} - \dfrac{2x-5}{x^2+3x-4}$

34. $\dfrac{5-3x}{x^2-2x+1} - \dfrac{x+1}{x^2-2x+1}$

Find the LCM.

35. 15, 36

36. 18, 30

37. 8, 9

38. 12, 15

39. 6, 12, 15

40. 8, 32, 50

Find the LCM.

41. $18t^2,\ 6t^5$

42. $8x^5,\ 24x^2$

43. $15a^4b^7,\ 10a^2b^8$

44. $6a^2b^7,\ 9a^5b^2$

45. $2(y-3),\ 6(y-3)$

46. $4(x-1),\ 8(x-1)$

47. $x^2-2x-15,\ x^2-9$

48. $t^2-4,\ t^2+7t+10$

49. $t^3+4t^2+4t,\ t^2-4t$

50. $y^3-y^2,\ y^4-y^2$

51. $6xz^2,\ 8x^2y,\ 15y^3z$

52. $12s^3t,\ 15sv^2,\ 6t^4v$

53. $a+1,\ (a-1)^2,\ a^2-1$

54. $x-2,\ (x+2)^2,\ x^2-4$

55. $2n^2+n-1,\ 2n^2+3n-2$

56. $m^2-2m-3,\ 2m^2+3m+1$

57. $6x^3-24x^2+18x,\ 4x^5-24x^4+20x^3$

58. $9x^3-9x^2-18x,\ 6x^5-24x^4+24x^3$

59. $2x^3-2,\ x^2-1$

60. $3a^3+24,\ a^2-4$

Find equivalent expressions that have the LCD.

61. $\dfrac{5}{6t^4}, \dfrac{s}{18t^2}$

62. $\dfrac{7}{10y^2}, \dfrac{x}{5y^6}$

63. $\dfrac{7}{3x^4y^2}, \dfrac{4}{9xy^3}$

64. $\dfrac{3}{2a^2b}, \dfrac{7}{8ab^2}$

65. $\dfrac{2x}{x^2 - 4}, \dfrac{4x}{x^2 + 5x + 6}$

66. $\dfrac{5x}{x^2 - 9}, \dfrac{2x}{x^2 + 11x + 24}$

67. Explain why the product of two numbers is not always their least common multiple.

68. If the LCM of two numbers is their product, what can you conclude about the two numbers?

Skill Review

To prepare for Section 6.4, review opposites (Sections 1.7 and 1.8).

Write each number in two equivalent forms. [1.7]

69. $-\dfrac{5}{8}$

70. $\dfrac{4}{-11}$

Write an equivalent expression without parentheses. [1.8]

71. $-(x - y)$

72. $-(3 - a)$

Multiply and simplify. [1.8]

73. $-1(2x - 7)$

74. $-1(a - b)$

Synthesis

75. If the LCM of two third-degree polynomials is a sixth-degree polynomial, what can be concluded about the two polynomials?

76. If the LCM of a binomial and a trinomial is the trinomial, what relationship exists between the two expressions?

Perform the indicated operations. Simplify, if possible.

77. $\dfrac{6x - 1}{x - 1} + \dfrac{3(2x + 5)}{x - 1} + \dfrac{3(2x - 3)}{x - 1}$

78. $\dfrac{2x + 11}{x - 3} \cdot \dfrac{3}{x + 4} + \dfrac{-1}{4 + x} \cdot \dfrac{6x + 3}{x - 3}$

79. $\dfrac{x^2}{3x^2 - 5x - 2} - \dfrac{2x}{3x + 1} \cdot \dfrac{1}{x - 2}$

80. $\dfrac{x + y}{x^2 - y^2} + \dfrac{x - y}{x^2 - y^2} - \dfrac{2x}{x^2 - y^2}$

African artistry. In Southeast Mozambique, the design of every woven handbag, or gipatsi *(plural,* sipatsi*) is created by repeating two or more geometric patterns.*

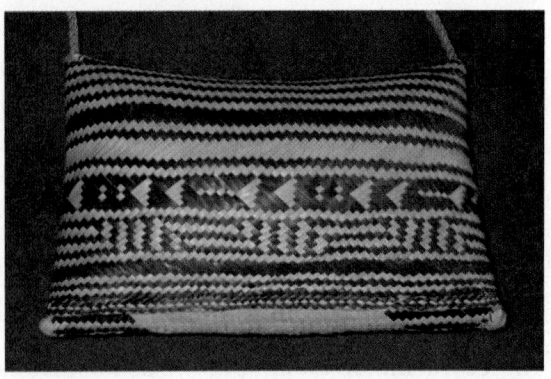

Each pattern encircles the bag, sharing the strands of fabric with any pattern above or below. The length, or period, of each pattern is the number of strands required to construct the pattern. For a gipatsi to be considered beautiful, each individual pattern must fit a whole number of times around the bag.
Source: Gerdes, Paulus, *Women, Art and Geometry in Southern Africa.* Asmara, Eritrea: Africa World Press, Inc., p. 5

81. A weaver is using two patterns to create a gipatsi. Pattern A is 10 strands long, and pattern B is 3 strands long. What is the smallest number of strands that can be used to complete the gipatsi?

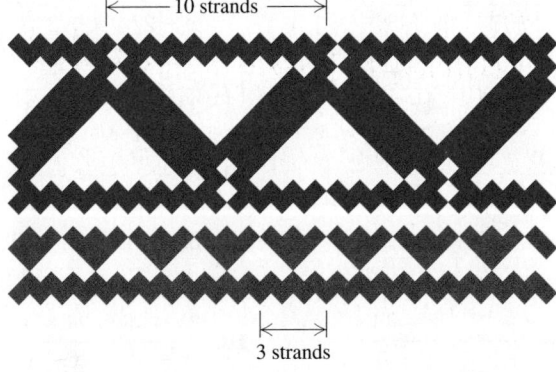

82. A weaver is using a four-strand pattern, a six-strand pattern, and an eight-strand pattern. What is the smallest number of strands that can be used to complete the gipatsi?

83. For technical reasons, the number of strands is generally a multiple of 4. Answer Exercise 79 with this additional requirement in mind.

Find the LCM.

84. 80, 96, 108

85. $4x^2 - 25$, $6x^2 - 7x - 20$, $(9x^2 + 24x + 16)^2$

86. $9n^2 - 9$, $(5n^2 - 10n + 5)^2$, $15n - 15$

87. *Copiers.*   The Brother® MFC240C copier can print 20 color pages per minute. The Lexmark X5470 copier can print 18 color pages per minute. If both machines begin printing at the same instant, how long will it be until they again begin printing a page at exactly the same time?

88. *Running.*   Kim and Trey leave the starting point of a fitness loop at the same time. Kim jogs a lap in 6 min and Trey jogs one in 8 min. Assuming they continue to run at the same pace, when will they next meet at the starting place?

89. *Bus schedules.*   Beginning at 5:00 A.M., a hotel shuttle bus leaves Salton Airport every 25 min, and the downtown shuttle bus leaves the airport every 35 min. What time will it be when both shuttles again leave at the same time?

90. *Appliances.*   Smoke detectors last an average of 10 yr, water heaters an average of 12 yr, and refrigerators an average of 15 yr. If an apartment house is equipped with new smoke detectors, water heaters, and refrigerators in 2010, in what year will all three appliances need to be replaced at once?
Source: Demesne.info

91. Explain how evaluating can be used to perform a partial check on the result of Example 1(d):

$$\frac{x - 5}{x^2 - 9} + \frac{2}{x^2 - 9} = \frac{1}{x + 3}.$$

92. On p. 391, the second step in finding an LCD is to select one of the factorizations of the denominators. Does it matter which one is selected? Why or why not?

<table>
<tr><td>**6.4**</td><td># Addition and Subtraction with Unlike Denominators</td></tr>
</table>

Adding and Subtracting with LCDs   ■   When Factors Are Opposites

Adding and Subtracting with LCDs

We now know how to rewrite two rational expressions in equivalent forms that use the LCD. Once rational expressions share a common denominator, they can be added or subtracted just as in Section 6.3.

> **To Add or Subtract Rational Expressions Having Different Denominators**
>
> 1. Find the LCD.
> 2. Multiply each rational expression by a form of 1 made up of the factors of the LCD missing from that expression's denominator.
> 3. Add or subtract the numerators, as indicated. Write the sum or the difference over the LCD.
> 4. Simplify, if possible.

EXAMPLE **1**

Add: $\dfrac{5x^2}{8} + \dfrac{7x}{12}$.

SOLUTION

1. First, we find the LCD:

$$\left.\begin{array}{l} 8 = 2 \cdot 2 \cdot 2 \\ 12 = 2 \cdot 2 \cdot 3 \end{array}\right\} \quad \text{LCD} = 2 \cdot 2 \cdot 2 \cdot 3, \text{ or } 24.$$

2. The denominator 8 must be multiplied by 3 in order to obtain the LCD. The denominator 12 must be multiplied by 2 in order to obtain the LCD. Thus we multiply the first expression by $\frac{3}{3}$ and the second expression by $\frac{2}{2}$ to get the LCD:

$$\dfrac{5x^2}{8} + \dfrac{7x}{12} = \dfrac{5x^2}{2 \cdot 2 \cdot 2} + \dfrac{7x}{2 \cdot 2 \cdot 3}$$

$$= \dfrac{5x^2}{2 \cdot 2 \cdot 2} \cdot \dfrac{3}{3} + \dfrac{7x}{2 \cdot 2 \cdot 3} \cdot \dfrac{2}{2} \qquad \begin{array}{l}\text{Multiplying each}\\ \text{expression by a form of 1}\\ \text{to get the LCD}\end{array}$$

$$= \dfrac{15x^2}{24} + \dfrac{14x}{24}.$$

3. Next, we add the numerators:

$$\dfrac{15x^2}{24} + \dfrac{14x}{24} = \dfrac{15x^2 + 14x}{24}.$$

4. Since $15x^2 + 14x$ and 24 have no common factor,

$$\dfrac{15x^2 + 14x}{24}$$

cannot be simplified any further. **TRY EXERCISE** 13

Subtraction is performed in much the same way.

EXAMPLE **2**

Subtract: $\dfrac{7}{8x} - \dfrac{5}{12x^2}$.

SOLUTION We follow the four steps shown above. First, we find the LCD:

$$\left.\begin{array}{l} 8x = 2 \cdot 2 \cdot 2 \cdot x \\ 12x^2 = 2 \cdot 2 \cdot 3 \cdot x \cdot x \end{array}\right\} \quad \text{LCD} = 2 \cdot 2 \cdot 3 \cdot x \cdot x \cdot 2, \text{ or } 24x^2.$$

The denominator $8x$ must be multiplied by $3x$ in order to obtain the LCD. The denominator $12x^2$ must be multiplied by 2 in order to obtain the LCD. Thus we multiply by $\dfrac{3x}{3x}$ and $\dfrac{2}{2}$ to get the LCD. Then we subtract and, if possible, simplify.

$$\dfrac{7}{8x} - \dfrac{5}{12x^2} = \dfrac{7}{8x} \cdot \dfrac{3x}{3x} - \dfrac{5}{12x^2} \cdot \dfrac{2}{2}$$

$$= \dfrac{21x}{24x^2} - \dfrac{10}{24x^2} \longleftarrow \boxed{\begin{array}{l}\textit{CAUTION!}\;\; \text{Do not simplify}\\ \textit{these}\; \text{rational expressions or you}\\ \text{will lose the LCD.}\end{array}}$$

$$= \dfrac{21x - 10}{24x^2} \qquad \begin{array}{l}\text{This cannot be simplified,}\\ \text{so we are done.}\end{array}$$ **TRY EXERCISE** 5

When denominators contain polynomials with two or more terms, the same steps are used.

EXAMPLE 3   Add: $\dfrac{2a}{a^2 - 1} + \dfrac{1}{a^2 + a}$.

SOLUTION   First, we find the LCD:

Find the LCD.
$$\left. \begin{array}{l} a^2 - 1 = (a - 1)(a + 1) \\ a^2 + a = a(a + 1). \end{array} \right\} \quad \text{LCD} = (a - 1)(a + 1)a$$

We multiply by a form of 1 to get the LCD in each expression:

Write each expression with the LCD.

$$\dfrac{2a}{a^2 - 1} + \dfrac{1}{a^2 + a} = \dfrac{2a}{(a - 1)(a + 1)} \cdot \dfrac{a}{a} + \dfrac{1}{a(a + 1)} \cdot \dfrac{a - 1}{a - 1}$$

Multiplying by $\dfrac{a}{a}$ and $\dfrac{a - 1}{a - 1}$ to get the LCD

$$= \dfrac{2a^2}{(a - 1)(a + 1)a} + \dfrac{a - 1}{a(a + 1)(a - 1)}$$

Add numerators.
$$= \dfrac{2a^2 + a - 1}{a(a - 1)(a + 1)}$$ Adding numerators

Simplify.
$$= \dfrac{(2a - 1)\cancel{(a + 1)}}{a(a - 1)\cancel{(a + 1)}}$$

$$= \dfrac{2a - 1}{a(a - 1)}.$$

Simplifying by factoring and removing a factor equal to 1:
$$\dfrac{a + 1}{a + 1} = 1$$

▶ TRY EXERCISE 33

EXAMPLE 4   Perform the indicated operations.

a) $\dfrac{x + 4}{x - 2} - \dfrac{x - 7}{x + 5}$

b) $\dfrac{t}{t^2 + 11t + 30} + \dfrac{-5}{t^2 + 9t + 20}$

c) $\dfrac{x}{x^2 + 5x + 6} - \dfrac{2}{x^2 + 3x + 2}$

STUDENT NOTES

As you can see, adding or subtracting rational expressions can involve many steps. Therefore, it is important to double-check each step of your work as you work through each problem. Waiting to inspect your work at the end of each problem is usually a less efficient use of your time.

SOLUTION

a) First, we find the LCD. It is just the product of the denominators:

$$\text{LCD} = (x - 2)(x + 5).$$

We multiply by a form of 1 to get the LCD in each expression. Then we subtract and try to simplify.

$$\dfrac{x + 4}{x - 2} - \dfrac{x - 7}{x + 5} = \dfrac{x + 4}{x - 2} \cdot \dfrac{x + 5}{x + 5} - \dfrac{x - 7}{x + 5} \cdot \dfrac{x - 2}{x - 2}$$

$$= \dfrac{x^2 + 9x + 20}{(x - 2)(x + 5)} - \dfrac{x^2 - 9x + 14}{(x - 2)(x + 5)}$$

Multiplying out numerators (but not denominators)

$$= \dfrac{x^2 + 9x + 20 - (x^2 - 9x + 14)}{(x - 2)(x + 5)}$$

When subtracting a numerator with more than one term, parentheses are important.

$$= \dfrac{x^2 + 9x + 20 - x^2 + 9x - 14}{(x - 2)(x + 5)}$$

Removing parentheses and subtracting every term

$$= \dfrac{18x + 6}{(x - 2)(x + 5)}$$

$$= \dfrac{6(3x + 1)}{(x - 2)(x + 5)}$$ We cannot simplify.

b) $\dfrac{t}{t^2 + 11t + 30} + \dfrac{-5}{t^2 + 9t + 20}$

$= \dfrac{t}{(t + 5)(t + 6)} + \dfrac{-5}{(t + 5)(t + 4)}$ Factoring the denominators in order to find the LCD. The LCD is $(t + 5)(t + 6)(t + 4)$.

$= \dfrac{t}{(t + 5)(t + 6)} \cdot \dfrac{t + 4}{t + 4} + \dfrac{-5}{(t + 5)(t + 4)} \cdot \dfrac{t + 6}{t + 6}$ Multiplying to get the LCD

$= \dfrac{t^2 + 4t}{(t + 5)(t + 6)(t + 4)} + \dfrac{-5t - 30}{(t + 5)(t + 6)(t + 4)}$ Multiplying in each numerator

$= \dfrac{t^2 + 4t - 5t - 30}{(t + 5)(t + 6)(t + 4)}$ Adding numerators

$= \dfrac{t^2 - t - 30}{(t + 5)(t + 6)(t + 4)}$ Combining like terms in the numerator

$= \dfrac{\cancel{(t + 5)}(t - 6)}{\cancel{(t + 5)}(t + 6)(t + 4)}$

$= \dfrac{t - 6}{(t + 6)(t + 4)}$ Always simplify the result, if possible, by removing a factor equal to 1; here $\dfrac{t + 5}{t + 5} = 1$.

c) $\dfrac{x}{x^2 + 5x + 6} - \dfrac{2}{x^2 + 3x + 2}$

> Find the LCD.

$= \dfrac{x}{(x + 2)(x + 3)} - \dfrac{2}{(x + 2)(x + 1)}$ Factoring denominators. The LCD is $(x + 2)(x + 3)(x + 1)$.

> Write each expression with the LCD.

$= \dfrac{x}{(x + 2)(x + 3)} \cdot \dfrac{x + 1}{x + 1} - \dfrac{2}{(x + 2)(x + 1)} \cdot \dfrac{x + 3}{x + 3}$

$= \dfrac{x^2 + x}{(x + 2)(x + 3)(x + 1)} - \dfrac{2x + 6}{(x + 2)(x + 3)(x + 1)}$

> Subtract numerators.

$= \dfrac{x^2 + x - (2x + 6)}{(x + 2)(x + 3)(x + 1)}$ Don't forget the parentheses!

$= \dfrac{x^2 + x - 2x - 6}{(x + 2)(x + 3)(x + 1)}$ Remember to subtract each term in $2x + 6$.

$= \dfrac{x^2 - x - 6}{(x + 2)(x + 3)(x + 1)}$ Combining like terms in the numerator

> Simplify.

$= \dfrac{\cancel{(x + 2)}(x - 3)}{\cancel{(x + 2)}(x + 3)(x + 1)}$

$= \dfrac{x - 3}{(x + 3)(x + 1)}$ Factoring and simplifying; $\dfrac{x + 2}{x + 2} = 1$

TRY EXERCISE ▸ 45

When Factors Are Opposites

When one denominator is the opposite of the other, we can first multiply either expression by 1 using $-1/-1$.

EXAMPLE 5   Add: **(a)** $\dfrac{t}{2} + \dfrac{3}{-2}$; **(b)** $\dfrac{x}{x-5} + \dfrac{7}{5-x}$.

SOLUTION

a) $\dfrac{t}{2} + \dfrac{3}{-2} = \dfrac{t}{2} + \dfrac{3}{-2} \cdot \dfrac{-1}{-1}$ Multiplying by 1 using $\dfrac{-1}{-1}$

$\qquad = \dfrac{t}{2} + \dfrac{-3}{2}$ The denominators are now the same.

$\qquad = \dfrac{t + (-3)}{2}$

$\qquad = \dfrac{t - 3}{2}$

b) Recall that when an expression of the form $a - b$ is multiplied by -1, the subtraction is reversed: $-1(a - b) = -a + b = b + (-a) = b - a$. Since $x - 5$ and $5 - x$ are opposites, we can find a common denominator by multiplying one of the rational expressions by $-1/-1$. Because polynomials are usually written in descending order, we choose to reverse the subtraction in the second denominator:

$\dfrac{x}{x-5} + \dfrac{7}{5-x} = \dfrac{x}{x-5} + \dfrac{7}{5-x} \cdot \dfrac{-1}{-1}$ Multiplying by 1, where $1 = \dfrac{-1}{-1}$

$\qquad = \dfrac{x}{x-5} + \dfrac{-7}{-5+x}$

$\qquad = \dfrac{x}{x-5} + \dfrac{-7}{x-5}$ Note that $-5 + x = x + (-5) = x - 5$.

$\qquad = \dfrac{x-7}{x-5}$.

> **TRY EXERCISE** 53

Sometimes, after factoring to find the LCD, we find a factor in one denominator that is the opposite of a factor in the other denominator. When this happens, multiplication by $-1/-1$ can again be helpful.

EXAMPLE 6   Perform the indicated operations and simplify.

a) $\dfrac{x}{x^2 - 25} + \dfrac{3}{5 - x}$

b) $\dfrac{x+9}{x^2-4} + \dfrac{6-x}{4-x^2} - \dfrac{1+x}{x^2-4}$

SOLUTION

a) $\dfrac{x}{x^2 - 25} + \dfrac{3}{5 - x} = \dfrac{x}{(x-5)(x+5)} + \dfrac{3}{5-x}$ Factoring

$\qquad = \dfrac{x}{(x-5)(x+5)} + \dfrac{3}{5-x} \cdot \dfrac{-1}{-1}$ Multiplication by $-1/-1$ changes $5 - x$ to $x - 5$.

$\qquad = \dfrac{x}{(x-5)(x+5)} + \dfrac{-3}{x-5}$ $(5 - x)(-1) = x - 5$

$\qquad = \dfrac{x}{(x-5)(x+5)} + \dfrac{-3}{(x-5)} \cdot \dfrac{x+5}{x+5}$ The LCD is $(x-5)(x+5)$.

$\qquad = \dfrac{x}{(x-5)(x+5)} + \dfrac{-3x-15}{(x-5)(x+5)}$

$\qquad = \dfrac{-2x-15}{(x-5)(x+5)}$

The TABLE feature can be used to check addition or subtraction of rational expressions. Below we check Example 6(a), using $y_1 = x/(x^2 - 25) + 3/(5 - x)$ and $y_2 = (-2x - 15)/((x - 5)(x + 5))$.

ΔTBL = 1

X	Y₁	Y₂
1	.70833	.70833
2	.90476	.90476
3	1.3125	1.3125
4	2.5556	2.5556
5	ERROR	ERROR
6	−2.455	−2.455
7	−1.208	−1.208
X = 1		

Because the values for y_1 and y_2 match, we have a check.

b) Since $4 - x^2$ is the opposite of $x^2 - 4$, multiplying the second rational expression by $-1/-1$ will lead to a common denominator:

$$\frac{x + 9}{x^2 - 4} + \frac{6 - x}{4 - x^2} - \frac{1 + x}{x^2 - 4} = \frac{x + 9}{x^2 - 4} + \frac{6 - x}{4 - x^2} \cdot \frac{-1}{-1} - \frac{1 + x}{x^2 - 4}$$

$$= \frac{x + 9}{x^2 - 4} + \frac{x - 6}{x^2 - 4} - \frac{1 + x}{x^2 - 4}$$

$$= \frac{x + 9 + x - 6 - 1 - x}{x^2 - 4} \quad \text{Adding and subtracting numerators}$$

$$= \frac{x + 2}{x^2 - 4}$$

$$= \frac{(x + 2) \cdot 1}{(x + 2)(x - 2)} \quad \left.\begin{matrix} \\ \\ \end{matrix}\right\} \text{Simplifying}$$

$$= \frac{1}{x - 2}.$$

TRY EXERCISE 59

6.4 EXERCISE SET

Concept Reinforcement In Exercises 1–4, the four steps for adding rational expressions with different denominators are listed. Fill in the missing word or words for each step.

1. To add or subtract when the denominators are different, first find the _____.

2. Multiply each rational expression by a form of 1 made up of the factors of the LCD that are _____ from that expression's _____.

3. Add or subtract the _____, as indicated. Write the sum or the difference over the _____.

4. _____, if possible.

Perform the indicated operation. Simplify, if possible.

5. $\dfrac{3}{x^2} + \dfrac{5}{x}$

6. $\dfrac{6}{x} + \dfrac{7}{x^2}$

7. $\dfrac{1}{6r} - \dfrac{3}{8r}$

8. $\dfrac{4}{9t} - \dfrac{7}{6t}$

9. $\dfrac{3}{uv^2} + \dfrac{4}{u^3v}$

10. $\dfrac{8}{cd^2} + \dfrac{1}{c^2d}$

11. $\dfrac{-2}{3xy^2} - \dfrac{6}{x^2y^3}$

12. $\dfrac{8}{9t^3} - \dfrac{5}{6t^2}$

13. $\dfrac{x + 3}{8} + \dfrac{x - 2}{6}$

14. $\dfrac{x - 4}{9} + \dfrac{x + 5}{12}$

15. $\dfrac{x - 2}{6} - \dfrac{x + 1}{3}$

16. $\dfrac{a + 2}{2} - \dfrac{a - 4}{4}$

17. $\dfrac{a + 3}{15a} + \dfrac{2a - 1}{3a^2}$

18. $\dfrac{5a + 1}{2a^2} + \dfrac{a + 2}{6a}$

19. $\dfrac{4z - 9}{3z} - \dfrac{3z - 8}{4z}$

20. $\dfrac{x - 1}{4x} - \dfrac{2x + 3}{x}$

21. $\dfrac{3c + d}{cd^2} + \dfrac{c - d}{c^2d}$

22. $\dfrac{u + v}{u^2v} + \dfrac{2u + v}{uv^2}$

23. $\dfrac{4x + 2t}{3xt^2} - \dfrac{5x - 3t}{x^2t}$

24. $\dfrac{5x + 3y}{2x^2y} - \dfrac{3x + 4y}{xy^2}$

25. $\dfrac{3}{x - 2} + \dfrac{3}{x + 2}$

26. $\dfrac{5}{x - 1} + \dfrac{5}{x + 1}$

27. $\dfrac{t}{t + 3} - \dfrac{1}{t - 1}$

28. $\dfrac{y}{y - 3} + \dfrac{12}{y + 4}$

29. $\dfrac{3}{x + 1} + \dfrac{2}{3x}$

30. $\dfrac{2}{x + 5} + \dfrac{3}{4x}$

31. $\dfrac{3}{2t^2 - 2t} - \dfrac{5}{2t - 2}$

32. $\dfrac{8}{3t^2 - 15t} - \dfrac{3}{2t - 10}$

33. $\dfrac{3a}{a^2 - 9} + \dfrac{a}{a + 3}$

34. $\dfrac{5p}{p^2 - 16} + \dfrac{p}{p - 4}$

35. $\dfrac{6}{z + 4} - \dfrac{2}{3z + 12}$

36. $\dfrac{t}{t - 3} - \dfrac{5}{4t - 12}$

37. $\dfrac{5}{q - 1} + \dfrac{2}{(q - 1)^2}$

38. $\dfrac{3}{w + 2} + \dfrac{7}{(w + 2)^2}$

39. $\dfrac{t - 3}{t^3 - 1} - \dfrac{2}{1 - t^3}$

40. $\dfrac{1 - 6m}{1 - m^3} - \dfrac{5}{m^3 - 1}$

41. $\dfrac{3a}{4a - 20} + \dfrac{9a}{6a - 30}$

42. $\dfrac{4a}{5a - 10} + \dfrac{3a}{10a - 20}$

Aha! **43.** $\dfrac{x}{x - 5} + \dfrac{x}{5 - x}$

44. $\dfrac{x + 4}{x} + \dfrac{x}{x + 4}$

45. $\dfrac{6}{a^2 + a - 2} + \dfrac{4}{a^2 - 4a + 3}$

46. $\dfrac{x}{x^2 + 2x + 1} + \dfrac{1}{x^2 + 5x + 4}$

47. $\dfrac{x}{x^2 + 9x + 20} - \dfrac{4}{x^2 + 7x + 12}$

48. $\dfrac{x}{x^2 + 5x + 6} - \dfrac{2}{x^2 + 3x + 2}$

49. $\dfrac{3z}{z^2 - 4z + 4} + \dfrac{10}{z^2 + z - 6}$

50. $\dfrac{3}{x^2 - 9} + \dfrac{2}{x^2 - x - 6}$

Aha! **51.** $\dfrac{-7}{x^2 + 25x + 24} - \dfrac{0}{x^2 + 11x + 10}$

52. $\dfrac{x}{x^2 + 17x + 72} - \dfrac{1}{x^2 + 15x + 56}$

53. $\dfrac{5x}{4} - \dfrac{x - 2}{-4}$

54. $\dfrac{x}{6} - \dfrac{2x - 3}{-6}$

55. $\dfrac{y^2}{y - 3} + \dfrac{9}{3 - y}$

56. $\dfrac{t^2}{t - 2} + \dfrac{4}{2 - t}$

57. $\dfrac{c - 5}{c^2 - 64} - \dfrac{5 - c}{64 - c^2}$

58. $\dfrac{b - 4}{b^2 - 49} + \dfrac{b - 4}{49 - b^2}$

59. $\dfrac{4 - p}{25 - p^2} + \dfrac{p + 1}{p - 5}$

60. $\dfrac{y + 2}{y - 7} + \dfrac{3 - y}{49 - y^2}$

61. $\dfrac{x}{x - 4} - \dfrac{3}{16 - x^2}$

62. $\dfrac{x}{3 - x} - \dfrac{2}{x^2 - 9}$

63. $\dfrac{a}{a^2 - 1} + \dfrac{2a}{a - a^2}$

64. $\dfrac{3x + 2}{3x + 6} + \dfrac{x}{4 - x^2}$

65. $\dfrac{4x}{x^2 - y^2} - \dfrac{6}{y - x}$

66. $\dfrac{4 - a^2}{a^2 - 9} - \dfrac{a - 2}{3 - a}$

Perform the indicated operations. Simplify, if possible.

67. $\dfrac{x - 3}{2 - x} - \dfrac{x + 3}{x + 2} + \dfrac{x + 6}{4 - x^2}$

68. $\dfrac{t - 5}{1 - t} - \dfrac{t + 4}{t + 1} + \dfrac{t + 2}{t^2 - 1}$

69. $\dfrac{2x + 5}{x + 1} + \dfrac{x + 7}{x + 5} - \dfrac{5x + 17}{(x + 1)(x + 5)}$

70. $\dfrac{x + 5}{x + 3} + \dfrac{x + 7}{x + 2} - \dfrac{7x + 19}{(x + 3)(x + 2)}$

71. $\dfrac{1}{x + y} + \dfrac{1}{x - y} - \dfrac{2x}{x^2 - y^2}$

72. $\dfrac{2r}{r^2 - s^2} + \dfrac{1}{r + s} - \dfrac{1}{r - s}$

73. What is the advantage of using the *least* common denominator—rather than just *any* common denominator—when adding or subtracting rational expressions?

74. Describe a procedure that can be used to add any two rational expressions.

Skill Review

To prepare for Section 6.5, review division of fractions and rational expressions (Sections 1.3, 1.7, and 6.2).

Simplify.

75. $-\dfrac{3}{8} \div \dfrac{11}{4}$ [1.7]

76. $-\dfrac{7}{12} \div \left(-\dfrac{3}{4}\right)$ [1.7]

77. $\dfrac{\frac{3}{4}}{\frac{5}{6}}$ [1.3]

78. $\dfrac{\frac{8}{15}}{\frac{9}{10}}$ [1.3]

79. $\dfrac{2x + 6}{x - 1} \div \dfrac{3x + 9}{x - 1}$ [6.2]

80. $\dfrac{x^2 - 9}{x^2 - 4} \div \dfrac{x^2 + 6x + 9}{x^2 + 4x + 4}$ [6.2]

Synthesis

81. How could you convince someone that

$$\dfrac{1}{3 - x} \quad \text{and} \quad \dfrac{1}{x - 3}$$

are opposites of each other?

82. Are parentheses as important for adding rational expressions as they are for subtracting rational expressions? Why or why not?

Write expressions for the perimeter and the area of each rectangle.

83.

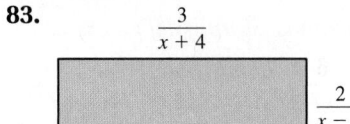

$$\frac{3}{x+4}$$

$$\frac{2}{x-5}$$

84.

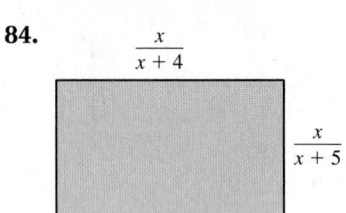

$$\frac{x}{x+4}$$

$$\frac{x}{x+5}$$

Perform the indicated operations.

85. $\dfrac{x^2}{3x^2 - 5x - 2} - \dfrac{2x}{3x+1} \cdot \dfrac{1}{x-2}$

86. $\dfrac{2x+11}{x-3} \cdot \dfrac{3}{x+4} + \dfrac{2x+1}{4+x} \cdot \dfrac{3}{3-x}$

Aha! **87.** $\left(\dfrac{x}{x+7} - \dfrac{3}{x+2}\right)\left(\dfrac{x}{x+7} + \dfrac{3}{x+2}\right)$

88. $\dfrac{1}{ay - 3a + 2xy - 6x} - \dfrac{xy + ay}{a^2 - 4x^2}\left(\dfrac{1}{y-3}\right)^2$

89. $\left(\dfrac{a}{a-b} + \dfrac{b}{a+b}\right)\left(\dfrac{1}{3a+b} + \dfrac{2a+6b}{9a^2 - b^2}\right)$

90. $\dfrac{2x^2 + 5x - 3}{2x^2 - 9x + 9} + \dfrac{x+1}{3-2x} + \dfrac{4x^2 + 8x + 3}{x-3} \cdot \dfrac{x+3}{9-4x^2}$

91. Express

$$\frac{a - 3b}{a - b}$$

as a sum of two rational expressions with denominators that are opposites of each other. Answers may vary.

92. Use a graphing calculator to check the answer to Exercise 29.

93. Why does the word ERROR appear in the table displayed on p. 402?

CONNECTING the CONCEPTS

The process of adding and subtracting rational expressions is significantly different from multiplying and dividing them. The first thing you should take note of when combining rational expressions is the operation sign.

Operation	Need Common Denominator?	Procedure	Tips and Cautions
Addition	Yes	Write with a common denominator. Add numerators. Keep denominator.	Do not simplify after writing with the LCD. Instead, simplify after adding the numerators.
Subtraction	Yes	Write with a common denominator. Subtract numerators. Keep denominator.	Use parentheses around the numerator being subtracted. Simplify after subtracting the numerators.
Multiplication	No	Multiply numerators. Multiply denominators.	Do not carry out the multiplications. Instead, factor and try to simplify.
Division	No	Multiply by the reciprocal of the divisor.	Begin by rewriting as a multiplication using the reciprocal of the divisor.

MIXED REVIEW

Tell what operation is being performed. Then perform the operation and, if possible, simplify.

1. $\dfrac{3}{5x} + \dfrac{2}{x^2}$

2. $\dfrac{3}{5x} \cdot \dfrac{2}{x^2}$

3. $\dfrac{3}{5x} \div \dfrac{2}{x^2}$

4. $\dfrac{3}{5x} - \dfrac{2}{x^2}$

5. $\dfrac{2x - 6}{5x + 10} \cdot \dfrac{x + 2}{6x - 12}$

6. $\dfrac{2}{x + 3} \cdot \dfrac{3}{x + 4}$

7. $\dfrac{2}{x - 5} \div \dfrac{6}{x - 5}$

8. $\dfrac{x}{x + 2} - \dfrac{1}{x - 1}$

9. $\dfrac{2}{x + 3} + \dfrac{3}{x + 4}$

10. $\dfrac{5}{2x - 1} + \dfrac{10x}{1 - 2x}$

11. $\dfrac{3}{x - 4} - \dfrac{2}{4 - x}$

12. $\dfrac{(x - 2)(2x + 3)}{(x + 1)(x - 5)} \div \dfrac{(x - 2)(x + 1)}{(x - 5)(x + 3)}$

13. $\dfrac{a}{6a - 9b} - \dfrac{b}{4a - 6b}$

14. $\dfrac{x^2 - 16}{x^2 - x} \cdot \dfrac{x^2}{x^2 - 5x + 4}$

15. $\dfrac{x + 1}{x^2 - 7x + 10} + \dfrac{3}{x^2 - x - 2}$

16. $\dfrac{3u^2 - 3}{4} \div \dfrac{4u + 4}{3}$

17. $\dfrac{t + 2}{10} + \dfrac{2t + 1}{15}$

18. $(t^2 + t - 20) \cdot \dfrac{t + 5}{t - 4}$

19. $\dfrac{a^2 - 2a + 1}{a^2 - 4} \div (a^2 - 3a + 2)$

20. $\dfrac{2x - 7}{x} - \dfrac{3x - 5}{2}$

6.5   Complex Rational Expressions

Using Division to Simplify   ■   Multiplying by the LCD

A **complex rational expression** is a rational expression that has one or more rational expressions within its numerator or denominator. Here are some examples:

$$\dfrac{1 + \dfrac{2}{x}}{3}, \quad \dfrac{\dfrac{x + y}{7}}{\dfrac{2x}{x + 1}}, \quad \dfrac{\dfrac{4}{3} + \dfrac{1}{5}}{\dfrac{2}{x} - \dfrac{x}{y}}$$

These are rational expressions within the complex rational expression.

When we simplify a complex rational expression, we rewrite it so that it is no longer complex. We will consider two methods for simplifying complex rational expressions. Each method offers certain advantages.

Using Division to Simplify (Method 1)

Our first method for simplifying complex rational expressions involves rewriting the expression as a quotient of two rational expressions.

To Simplify a Complex Rational Expression by Dividing

1. Add or subtract, as needed, to get a single rational expression in the numerator.
2. Add or subtract, as needed, to get a single rational expression in the denominator.
3. Divide the numerator by the denominator (invert and multiply).
4. If possible, simplify by removing a factor equal to 1.

The key here is to express a complex rational expression as one rational expression divided by another. We can then proceed as in Section 6.2.

EXAMPLE **1** Simplify: $\dfrac{\dfrac{x}{x-3}}{\dfrac{4}{5x-15}}$.

SOLUTION   Here the numerator and the denominator are already single rational expressions. This allows us to start by dividing (step 3), as in Section 6.2:

$$\frac{\dfrac{x}{x-3}}{\dfrac{4}{5x-15}} = \frac{x}{x-3} \div \frac{4}{5x-15} \qquad \text{Rewriting with a division symbol}$$

$$= \frac{x}{x-3} \cdot \frac{5x-15}{4} \qquad \begin{array}{l}\text{Multiplying by the reciprocal of the}\\ \text{divisor (inverting and multiplying)}\end{array}$$

$$= \frac{x}{x-3} \cdot \frac{5(x-3)}{4} \qquad \begin{array}{l}\text{Factoring and removing a factor}\\ \text{equal to 1: } \dfrac{x-3}{x-3}=1\end{array}$$

$$= \frac{5x}{4}.$$

> **TRY EXERCISE** 21

All four steps of the division method are used in Example 2(a).

EXAMPLE **2** Simplify.

a) $\dfrac{\dfrac{5}{2a}+\dfrac{1}{a}}{\dfrac{1}{4a}-\dfrac{5}{6}}$

b) $\dfrac{\dfrac{x^2}{y}-\dfrac{5}{x}}{xz}$

SOLUTION

a) $\dfrac{\dfrac{5}{2a}+\dfrac{1}{a}}{\dfrac{1}{4a}-\dfrac{5}{6}} = \dfrac{\dfrac{5}{2a}+\dfrac{1}{a}\cdot\dfrac{2}{2}}{\dfrac{1}{4a}\cdot\dfrac{3}{3}-\dfrac{5}{6}\cdot\dfrac{2a}{2a}}$ $\left.\begin{array}{l}\ \\ \ \end{array}\right\}$ ← Multiplying by 1 to get the LCD, $2a$, for the numerator of the complex rational expression

$\left.\begin{array}{l}\ \\ \ \end{array}\right\}$ ← Multiplying by 1 to get the LCD, $12a$, for the denominator of the complex rational expression

1. Add to get a single rational expression in the numerator.

2. Subtract to get a single rational expression in the denominator.

3. Divide the numerator by the denominator.

4. Simplify.

$$= \frac{\dfrac{5}{2a} + \dfrac{2}{2a}}{\dfrac{3}{12a} - \dfrac{10a}{12a}} = \frac{\dfrac{7}{2a}}{\dfrac{3-10a}{12a}}$$
$\longleftarrow$ Adding
$\swarrow$ Subtracting

$$= \frac{7}{2a} \div \frac{3-10a}{12a}$$ Rewriting with a division symbol. This is often done mentally.

$$= \frac{7}{2a} \cdot \frac{12a}{3-10a}$$ Multiplying by the reciprocal of the divisor (inverting and multiplying)

$$= \frac{7}{2a} \cdot \frac{2a \cdot 6}{3-10a}$$ Removing a factor equal to 1: $\dfrac{2a}{2a} = 1$

$$= \frac{42}{3-10a}$$

b) $$\frac{\dfrac{x^2}{y} - \dfrac{5}{x}}{xz} = \frac{\dfrac{x^2}{y} \cdot \dfrac{x}{x} - \dfrac{5}{x} \cdot \dfrac{y}{y}}{xz}$$ $\longleftarrow$ Multiplying by 1 to get the LCD, xy, for the numerator of the complex rational expression

$$= \frac{\dfrac{x^3}{xy} - \dfrac{5y}{xy}}{xz}$$

$$= \frac{\dfrac{x^3 - 5y}{xy}}{xz}$$ $\longleftarrow$ Subtracting
$\longleftarrow$ If you prefer, write xz as $\dfrac{xz}{1}$.

$$= \frac{x^3 - 5y}{xy} \div (xz)$$ Rewriting with a division symbol

$$= \frac{x^3 - 5y}{xy} \cdot \frac{1}{xz}$$ Multiplying by the reciprocal of the divisor (inverting and multiplying)

$$= \frac{x^3 - 5y}{x^2 yz}$$

TRY EXERCISE ▶ 11

Multiplying by the LCD (Method 2)

A second method for simplifying complex rational expressions relies on multiplying by a carefully chosen expression that is equal to 1. This multiplication by 1 will result in an expression that is no longer complex.

> **To Simplify a Complex Rational Expression by Multiplying by the LCD**
>
> **1.** Find the LCD of *all* rational expressions within the complex rational expression.
> **2.** Multiply the complex rational expression by an expression equal to 1. Write 1 as the LCD over itself (LCD/LCD).
> **3.** Simplify. No fraction expressions should remain within the complex rational expression.
> **4.** Factor and, if possible, simplify.

EXAMPLE 3   Simplify: $\dfrac{\dfrac{1}{2} + \dfrac{3}{4}}{\dfrac{5}{6} - \dfrac{3}{8}}$.

SOLUTION

1. The LCD of $\frac{1}{2}$, $\frac{3}{4}$, $\frac{5}{6}$, and $\frac{3}{8}$ is 24.

2. We multiply by an expression equal to 1:

$$\dfrac{\dfrac{1}{2}+\dfrac{3}{4}}{\dfrac{5}{6}-\dfrac{3}{8}} = \dfrac{\dfrac{1}{2}+\dfrac{3}{4}}{\dfrac{5}{6}-\dfrac{3}{8}} \cdot \dfrac{24}{24}.$$ Multiplying by an expression equal to 1, using the LCD: $\dfrac{24}{24}=1$

3. Using the distributive law, we perform the multiplication:

$$\dfrac{\dfrac{1}{2}+\dfrac{3}{4}}{\dfrac{5}{6}-\dfrac{3}{8}} \cdot \dfrac{24}{24} = \dfrac{\left(\dfrac{1}{2}+\dfrac{3}{4}\right)24}{\left(\dfrac{5}{6}-\dfrac{3}{8}\right)24}$$ ← Multiplying the numerator by 24 Don't forget the parentheses! ← Multiplying the denominator by 24

$$= \dfrac{\dfrac{1}{2}(24)+\dfrac{3}{4}(24)}{\dfrac{5}{6}(24)-\dfrac{3}{8}(24)}$$ Using the distributive law

$$= \dfrac{12+18}{20-9}, \quad \text{or} \quad \dfrac{30}{11}.$$ Simplifying

4. The result, $\frac{30}{11}$, cannot be factored or simplified, so we are done.

> **TRY EXERCISE** ▶ **5**

Multiplying like this effectively clears fractions in both the top and the bottom of the complex rational expression. Compare the steps in Example 4(a) with those in Example 2(a).

EXAMPLE **4** Simplify.

a) $\dfrac{\dfrac{5}{2a}+\dfrac{1}{a}}{\dfrac{1}{4a}-\dfrac{5}{6}}$

b) $\dfrac{1-\dfrac{1}{x}}{1-\dfrac{1}{x^2}}$

SOLUTION

1. Find the LCD.

a) The denominators within the complex expression are $2a$, a, $4a$, and 6, so the LCD is $12a$. We multiply by 1 using $(12a)/(12a)$:

2. Multiply by LCD/LCD.

$$\dfrac{\dfrac{5}{2a}+\dfrac{1}{a}}{\dfrac{1}{4a}-\dfrac{5}{6}} = \dfrac{\dfrac{5}{2a}+\dfrac{1}{a}}{\dfrac{1}{4a}-\dfrac{5}{6}} \cdot \dfrac{12a}{12a} = \dfrac{\dfrac{5}{2a}(12a)+\dfrac{1}{a}(12a)}{\dfrac{1}{4a}(12a)-\dfrac{5}{6}(12a)}.$$ Using the distributive law

When we multiply by $12a$, all fractions in the numerator and the denominator of the complex rational expression are cleared:

3., 4. Simplify.

$$\dfrac{\dfrac{5}{2a}(12a)+\dfrac{1}{a}(12a)}{\dfrac{1}{4a}(12a)-\dfrac{5}{6}(12a)} = \dfrac{30+12}{3-10a} = \dfrac{42}{3-10a}.$$ All fractions have been cleared.

Be careful to place parentheses properly when entering complex rational expressions into a graphing calculator. Remember to enclose the entire numerator of the complex rational expression in one set of parentheses and the entire denominator in another. For example, we enter the expression in Example 4(a) as

$$y_1 = (5/(2x) + 1/x) / (1/(4x) - 5/6).$$

1. Write Example 4(b) as you would to enter it into a graphing calculator.
2. When must the numerator of a rational expression be enclosed in parentheses? the denominator?

b) $\dfrac{1 - \dfrac{1}{x}}{1 - \dfrac{1}{x^2}} = \dfrac{1 - \dfrac{1}{x}}{1 - \dfrac{1}{x^2}} \cdot \dfrac{x^2}{x^2}$ The LCD is x^2 so we multiply by 1 using x^2/x^2.

$$= \dfrac{1 \cdot x^2 - \dfrac{1}{x} \cdot x^2}{1 \cdot x^2 - \dfrac{1}{x^2} \cdot x^2}$$ Using the distributive law

$$= \dfrac{x^2 - x}{x^2 - 1}$$ All fractions have been cleared within the complex rational expression.

$$= \dfrac{x(x - 1)}{(x + 1)(x - 1)}$$ Factoring and simplifying: $\dfrac{x - 1}{x - 1} = 1$

$$= \dfrac{x}{x + 1}$$

> TRY EXERCISE 15

It is important to understand both of the methods studied in this section. Sometimes, as in Example 1, the complex rational expression is either given as—or easily written as—a quotient of two rational expressions. In these cases, Method 1 (using division) is probably the easier method to use. Other times, as in Example 4, it is not difficult to find the LCD of all denominators in the complex rational expression. When this occurs, it is usually easier to use Method 2 (multiplying by the LCD). The more practice you get using both methods, the better you will be at selecting the easier method for any given problem.

6.5 EXERCISE SET

🍮 *Concept Reinforcement* *In each of Exercises 1–4, use the method listed to match the accompanying expression with the expression in the column on the right that arises when that method is used.*

1. $\dfrac{\dfrac{5}{x^2} + \dfrac{1}{x}}{\dfrac{7}{2} - \dfrac{3}{4x}}$;

   Multiplying by the LCD (Method 2)

2. $\dfrac{\dfrac{5}{x^2} + \dfrac{1}{x}}{\dfrac{7}{2} - \dfrac{3}{4x}}$;

   Using division to simplify (Method 1)

3. $\dfrac{\dfrac{4}{5x} - \dfrac{1}{10}}{\dfrac{8}{x^2} + \dfrac{7}{2}}$;

   Using division to simplify (Method 1)

4. $\dfrac{\dfrac{4}{5x} - \dfrac{1}{10}}{\dfrac{8}{x^2} + \dfrac{7}{2}}$;

   Multiplying by the LCD (Method 2)

a) $\dfrac{\dfrac{5 + x}{x^2}}{\dfrac{14x - 3}{4x}}$

b) $\dfrac{\dfrac{8 - x}{10x}}{\dfrac{16 + 7x^2}{2x^2}}$

c) $\dfrac{\dfrac{4}{5x} \cdot 10x^2 - \dfrac{1}{10} \cdot 10x^2}{\dfrac{8}{x^2} \cdot 10x^2 + \dfrac{7}{2} \cdot 10x^2}$

d) $\dfrac{\dfrac{5}{x^2} \cdot 4x^2 + \dfrac{1}{x} \cdot 4x^2}{\dfrac{7}{2} \cdot 4x^2 - \dfrac{3}{4x} \cdot 4x^2}$

Simplify. Use either method or the method specified by your instructor.

5. $\dfrac{1 + \dfrac{1}{4}}{2 + \dfrac{3}{4}}$

6. $\dfrac{3 + \dfrac{1}{4}}{1 + \dfrac{1}{2}}$

7. $\dfrac{\dfrac{1}{2} + \dfrac{1}{3}}{\dfrac{1}{4} - \dfrac{1}{6}}$

8. $\dfrac{\dfrac{2}{5} - \dfrac{1}{10}}{\dfrac{7}{20} - \dfrac{4}{15}}$

9. $\dfrac{\dfrac{x}{4} + x}{\dfrac{4}{x} + x}$

10. $\dfrac{\dfrac{1}{c} + 2}{\dfrac{1}{c} - 5}$

11. $\dfrac{\dfrac{10}{t}}{\dfrac{2}{t^2} - \dfrac{5}{t}}$

12. $\dfrac{\dfrac{5}{x} - \dfrac{2}{x^2}}{\dfrac{2}{x^2}}$

13. $\dfrac{\dfrac{2a - 5}{3a}}{\dfrac{a - 7}{6a}}$

14. $\dfrac{\dfrac{a + 5}{a^2}}{\dfrac{a - 2}{3a}}$

15. $\dfrac{\dfrac{x}{6} - \dfrac{3}{x}}{\dfrac{1}{3} + \dfrac{1}{x}}$

16. $\dfrac{\dfrac{2}{x} + \dfrac{x}{4}}{\dfrac{3}{4} - \dfrac{2}{x}}$

17. $\dfrac{\dfrac{1}{s} - \dfrac{1}{5}}{\dfrac{s - 5}{s}}$

18. $\dfrac{\dfrac{1}{9} - \dfrac{1}{n}}{\dfrac{n + 9}{9}}$

19. $\dfrac{\dfrac{1}{t^2} + 1}{\dfrac{1}{t} - 1}$

20. $\dfrac{2 + \dfrac{1}{x}}{2 - \dfrac{1}{x^2}}$

21. $\dfrac{\dfrac{x^2}{x^2 - y^2}}{\dfrac{x}{x + y}}$

22. $\dfrac{\dfrac{a^2}{a - 2}}{\dfrac{3a}{a^2 - 4}}$

23. $\dfrac{\dfrac{7}{c^2} + \dfrac{4}{c}}{\dfrac{6}{c} - \dfrac{3}{c^3}}$

24. $\dfrac{\dfrac{4}{t^3} - \dfrac{1}{t^2}}{\dfrac{3}{t} + \dfrac{5}{t^2}}$

25. $\dfrac{\dfrac{2}{7a^4} - \dfrac{1}{14a}}{\dfrac{3}{5a^2} + \dfrac{2}{15a}}$

26. $\dfrac{\dfrac{5}{4x^3} - \dfrac{3}{8x}}{\dfrac{3}{2x} + \dfrac{3}{4x^3}}$

Aha! **27.** $\dfrac{\dfrac{x}{5y^3} + \dfrac{3}{10y}}{\dfrac{3}{10y} + \dfrac{x}{5y^3}}$

28. $\dfrac{\dfrac{a}{6b^3} + \dfrac{4}{9b^2}}{\dfrac{5}{6b} - \dfrac{1}{9b^3}}$

29. $\dfrac{\dfrac{3}{ab^4} + \dfrac{4}{a^3b}}{\dfrac{5}{a^3b} - \dfrac{3}{ab}}$

30. $\dfrac{\dfrac{2}{x^2y} + \dfrac{3}{xy^2}}{\dfrac{3}{xy^2} + \dfrac{2}{x^2y}}$

31. $\dfrac{t - \dfrac{9}{t}}{t + \dfrac{4}{t}}$

32. $\dfrac{s + \dfrac{2}{s}}{s - \dfrac{3}{s}}$

33. $\dfrac{\dfrac{1}{a} + \dfrac{1}{b}}{\dfrac{1}{a^3} + \dfrac{1}{b^3}}$

34. $\dfrac{x - y}{\dfrac{1}{x^3} - \dfrac{1}{y^3}}$

35. $\dfrac{3 + \dfrac{4}{ab^3}}{\dfrac{3 + a}{a^2b}}$

36. $\dfrac{5 + \dfrac{3}{x^2y}}{\dfrac{3 + x}{x^3y}}$

37. $\dfrac{t + 5 + \dfrac{3}{t}}{t + 2 + \dfrac{1}{t}}$

38. $\dfrac{a + 3 + \dfrac{2}{a}}{a + 2 + \dfrac{5}{a}}$

39. $\dfrac{x - 2 - \dfrac{1}{x}}{x - 5 - \dfrac{4}{x}}$

40. $\dfrac{x - 3 - \dfrac{2}{x}}{x - 4 - \dfrac{3}{x}}$

41. Is it possible to simplify complex rational expressions without knowing how to divide rational expressions? Why or why not?

42. Why is the distributive law important when simplifying complex rational expressions?

Skill Review

To prepare for Section 6.6, review solving linear and quadratic equations (Sections 2.2 and 5.7).

Solve.

43. $3x - 5 + 2(4x - 1) = 12x - 3$ [2.2]

44. $(x - 1)7 - (x + 1)9 = 4(x + 2)$ [2.2]

45. $\dfrac{3}{4}x - \dfrac{5}{8} = \dfrac{3}{8}x + \dfrac{7}{4}$ [2.2]

46. $\dfrac{5}{9} - \dfrac{2x}{3} = \dfrac{5x}{6} + \dfrac{4}{3}$ [2.2]

47. $x^2 - 7x + 12 = 0$ [5.7]

48. $x^2 + 13x - 30 = 0$ [5.7]

Synthesis

49. Which of the two methods presented would you use to simplify Exercise 30? Why?

50. Which of the two methods presented would you use to simplify Exercise 22? Why?

In Exercises 51–54, find all x-values for which the given expression is undefined.

51. $\dfrac{\dfrac{x - 5}{x - 6}}{\dfrac{x - 7}{x - 8}}$

52. $\dfrac{\dfrac{x + 1}{x + 2}}{\dfrac{x + 3}{x + 4}}$

53. $\dfrac{\dfrac{2x + 3}{5x + 4}}{\dfrac{3}{7} - \dfrac{x^2}{21}}$

54. $\dfrac{\dfrac{3x - 5}{2x - 7}}{\dfrac{4x}{5} - \dfrac{8}{15}}$

55. Use multiplication by the LCD (Method 2) to show that

$$\dfrac{A}{B} \div \dfrac{C}{D} = \dfrac{A}{B} \cdot \dfrac{D}{C}.$$

(*Hint*: Begin by forming a complex rational expression.)

56. The formula

$$\frac{P\left(1 + \dfrac{i}{12}\right)^2}{\dfrac{\left(1 + \dfrac{i}{12}\right)^2 - 1}{\dfrac{i}{12}}},$$

where P is a loan amount and i is an interest rate, arises in certain business situations. Simplify this expression. (*Hint*: Expand the binomials.)

Simplify.

57. $\dfrac{\dfrac{x}{x + 5} + \dfrac{3}{x + 2}}{\dfrac{2}{x + 2} - \dfrac{x}{x + 5}}$

58. $\dfrac{\dfrac{5}{x + 2} - \dfrac{3}{x - 2}}{\dfrac{x}{x - 1} + \dfrac{x}{x + 1}}$

Aha! **59.** $\left[\dfrac{\dfrac{x - 1}{x - 1} - 1}{\dfrac{x + 1}{x - 1} + 1}\right]^5$

60. $1 + \dfrac{1}{1 + \dfrac{1}{1 + \dfrac{1}{x}}}$

61. $\dfrac{\dfrac{z}{1 - \dfrac{z}{2 + 2z}} - 2z}{\dfrac{2z}{5z - 2} - 3}$

62. Find the simplified form for the reciprocal of
$$\frac{2}{x - 1} - \frac{1}{3x - 2}.$$

63. Under what circumstance(s) will there be no restrictions on the variable appearing in a complex rational expression?

64. Use a graphing calculator to check Example 2(a).

6.6   Solving Rational Equations

Solving a New Type of Equation   •   A Visual Interpretation

Our study of rational expressions allows us to solve a type of equation that we could not have solved prior to this chapter.

Solving a New Type of Equation

A **rational**, or **fraction**, **equation** is an equation containing one or more rational expressions, often with the variable in a denominator. Here are some examples:

$$\frac{2}{3} + \frac{5}{6} = \frac{x}{9}, \qquad t + \frac{7}{t} = -5, \qquad \frac{x^2}{x - 1} = \frac{1}{x - 1}.$$

> **To Solve a Rational Equation**
> 1. List any restrictions that exist. Numbers that make a denominator equal 0 can never be solutions.
> 2. Clear the equation of fractions by multiplying both sides by the LCM of the denominators.
> 3. Solve the resulting equation using the addition principle, the multiplication principle, and the principle of zero products, as needed.
> 4. Check the possible solution(s) in the original equation.

When clearing an equation of fractions, we use the terminology LCM instead of LCD because we are *not* adding or subtracting rational expressions.

EXAMPLE **1**	Solve: $\dfrac{x}{6} - \dfrac{x}{8} = \dfrac{1}{12}$.

STUDY SKILLS ─────────

Does More Than One Solution Exist?

Keep in mind that many problems—in math and elsewhere—have more than one solution. When asked to solve an equation, we are expected to find any and all solutions of the equation.

SOLUTION Because no variable appears in a denominator, no restrictions exist. The LCM of 6, 8, and 12 is 24, so we multiply both sides by 24:

$$24\left(\frac{x}{6} - \frac{x}{8}\right) = 24 \cdot \frac{1}{12}$$
Using the multiplication principle to multiply both sides by the LCM. Parentheses are important!

$$24 \cdot \frac{x}{6} - 24 \cdot \frac{x}{8} = 24 \cdot \frac{1}{12}$$
Using the distributive law

Be sure to multiply *each* term by the LCM.

$$\left.\begin{array}{c} \dfrac{24x}{6} - \dfrac{24x}{8} = \dfrac{24}{12} \\ 4x - 3x = 2 \end{array}\right\}$$
Simplifying. Note that all fractions have been cleared. If fractions remain, we have either made a mistake or have not used the LCM of the denominators.

$$x = 2.$$

Check:

$$\begin{array}{c|c} \dfrac{x}{6} - \dfrac{x}{8} = \dfrac{1}{12} \\ \hline \dfrac{2}{6} - \dfrac{2}{8} & \dfrac{1}{12} \\ \dfrac{1}{3} - \dfrac{1}{4} & \\ \dfrac{4}{12} - \dfrac{3}{12} & \\ \dfrac{1}{12} \stackrel{?}{=} \dfrac{1}{12} & \text{TRUE} \end{array}$$

This checks, so the solution is 2.

> **TRY EXERCISE** 5

Recall that the multiplication principle states that $a = b$ is equivalent to $a \cdot c = b \cdot c$, *provided c is not zero.* Because rational equations often have variables in a denominator, clearing fractions will now require us to multiply both sides by a variable expression. Since a variable expression could represent 0, *multiplying both sides of an equation by a variable expression does not always produce an equivalent equation.* Thus checking each solution in the original equation is essential.

EXAMPLE **2**	Solve.

a) $\dfrac{2}{3x} + \dfrac{1}{x} = 10$ **b)** $x + \dfrac{6}{x} = -5$

c) $1 + \dfrac{3x}{x + 2} = \dfrac{-6}{x + 2}$ **d)** $\dfrac{3}{x - 5} + \dfrac{1}{x + 5} = \dfrac{2}{x^2 - 25}$

e) $\dfrac{x^2}{x - 1} = \dfrac{1}{x - 1}$

SOLUTION

List restrictions.

a) If $x = 0$, both denominators are 0. We list this restriction:

$$x \neq 0.$$

We now clear the equation of fractions and solve:

$$\frac{2}{3x} + \frac{1}{x} = 10 \qquad \text{We cannot have } x = 0. \text{ The LCM is } 3x.$$

Clear fractions.

$$3x\left(\frac{2}{3x} + \frac{1}{x}\right) = 3x \cdot 10 \qquad \begin{array}{l}\text{Using the multiplication principle to} \\ \text{multiply both sides by the LCM.} \\ \textit{Don't forget the parentheses!}\end{array}$$

$$\cancel{3x} \cdot \frac{2}{\cancel{3x}} + 3\cancel{x} \cdot \frac{1}{\cancel{x}} = 3x \cdot 10 \qquad \text{Using the distributive law}$$

$$2 + 3 = 30x \qquad \begin{array}{l}\text{Removing factors equal to 1:} \\ (3x)/(3x) = 1 \text{ and } x/x = 1. \text{ This} \\ \text{clears all fractions.}\end{array}$$

Solve.

$$5 = 30x$$

$$\frac{5}{30} = x, \quad \text{so } x = \frac{1}{6}. \qquad \begin{array}{l}\text{Dividing both sides by 30, or} \\ \text{multiplying both sides by } 1/30\end{array}$$

Since $\frac{1}{6} \neq 0$, and 0 is the only restricted value, $\frac{1}{6}$ *should* check.

Check.   *Check:*

$$\frac{2}{3x} + \frac{1}{x} = 10$$

$$\begin{array}{c|c} \dfrac{2}{3 \cdot \frac{1}{6}} + \dfrac{1}{\frac{1}{6}} & 10 \\[2ex] \dfrac{2}{\frac{1}{2}} + \dfrac{1}{\frac{1}{6}} & \\[2ex] 2 \cdot \frac{2}{1} + 1 \cdot \frac{6}{1} & \\[1ex] 4 + 6 & \\[1ex] 10 \overset{?}{=} 10 & \text{TRUE} \end{array}$$

The solution is $\frac{1}{6}$.

b) Again, note that if $x = 0$, the expression $6/x$ is undefined. We list the restriction:

$$x \neq 0.$$

We now clear the equation of fractions and solve:

$$x + \frac{6}{x} = -5 \qquad \text{We cannot have } x = 0. \text{ The LCM is } x.$$

$$x\left(x + \frac{6}{x}\right) = x(-5) \qquad \begin{array}{l}\text{Multiplying both sides by the LCM.} \\ \textit{Don't forget the parentheses!}\end{array}$$

$$x \cdot x + \cancel{x} \cdot \frac{6}{\cancel{x}} = -5x \qquad \text{Using the distributive law}$$

$$x^2 + 6 = -5x \qquad \begin{array}{l}\text{Removing a factor equal to 1: } x/x = 1. \\ \text{We are left with a quadratic equation.}\end{array}$$

$$x^2 + 5x + 6 = 0 \qquad \begin{array}{l}\text{Using the addition principle to add } 5x \text{ to} \\ \text{both sides}\end{array}$$

$$(x + 3)(x + 2) = 0 \qquad \text{Factoring}$$

$$x + 3 = 0 \quad or \quad x + 2 = 0 \qquad \text{Using the principle of zero products}$$

$$x = -3 \quad or \qquad x = -2. \qquad \begin{array}{l}\text{The only restricted value is 0, so} \\ \text{both answers should check.}\end{array}$$

Check: For -3:   For -2:

$$x + \frac{6}{x} = -5 \qquad\qquad x + \frac{6}{x} = -5$$

$$\begin{array}{c|c} -3 + \dfrac{6}{-3} & -5 \\[2mm] -3 - 2 & \\ & -5 \overset{?}{=} -5 \quad \text{TRUE} \end{array} \qquad \begin{array}{c|c} -2 + \dfrac{6}{-2} & -5 \\[2mm] -2 - 3 & \\ & -5 \overset{?}{=} -5 \quad \text{TRUE} \end{array}$$

Both of these check, so there are two solutions, -3 and -2.

c) The only denominator is $x + 2$. We set this equal to 0 and solve:

$$x + 2 = 0$$
$$x = -2.$$

If $x = -2$, the rational expressions are undefined. We list the restriction:

$$x \neq -2.$$

We clear fractions and solve:

$$1 + \frac{3x}{x + 2} = \frac{-6}{x + 2}$$ We cannot have $x = -2$. The LCM is $x + 2$.

$$(x + 2)\left(1 + \frac{3x}{x + 2}\right) = (x + 2)\frac{-6}{x + 2}$$ Multiplying both sides by the LCM. *Don't forget the parentheses!*

$$(x + 2) \cdot 1 + \cancel{(x + 2)}\frac{3x}{\cancel{x + 2}} = \cancel{(x + 2)}\frac{-6}{\cancel{x + 2}}$$ Using the distributive law; removing a factor equal to 1: $(x + 2)/(x + 2) = 1$

$$x + 2 + 3x = -6$$

$$4x + 2 = -6$$

$$4x = -8$$

$$x = -2.$$ Above, we stated that $x \neq -2$.

Because of the above restriction, -2 must be rejected as a solution. The check below simply confirms this.

Check:
$$1 + \frac{3x}{x + 2} = \frac{-6}{x + 2}$$

$$\begin{array}{c|c} 1 + \dfrac{3(-2)}{-2 + 2} & \dfrac{-6}{-2 + 2} \\[3mm] 1 + \dfrac{-6}{0} \overset{?}{=} \dfrac{-6}{0} & \text{FALSE} \end{array}$$

The equation has no solution.

d) The denominators are $x - 5$, $x + 5$, and $x^2 - 25$. Setting them equal to 0 and solving, we find that the rational expressions are undefined when $x = 5$ or $x = -5$. We list the restrictions:

$$x \neq 5, \quad x \neq -5.$$

We clear fractions and solve:

$$\frac{3}{x-5} + \frac{1}{x+5} = \frac{2}{x^2-25}$$

We cannot have $x = 5$ or $x = -5$. The LCM is $(x-5)(x+5)$.

$$(x-5)(x+5)\left(\frac{3}{x-5} + \frac{1}{x+5}\right) = (x-5)(x+5)\frac{2}{(x-5)(x+5)}$$

$$\frac{\cancel{(x-5)}(x+5)3}{\cancel{x-5}} + \frac{(x-5)\cancel{(x+5)}}{\cancel{x+5}} = \frac{2\cancel{(x-5)(x+5)}}{\cancel{(x-5)(x+5)}}$$

Using the distributive law

$$(x+5)3 + (x-5) = 2$$

Removing factors equal to 1: $(x-5)/(x-5) = 1$, $(x+5)/(x+5) = 1$, and $\dfrac{(x-5)(x+5)}{(x-5)(x+5)} = 1$

$$3x + 15 + x - 5 = 2$$ Using the distributive law

$$4x + 10 = 2$$

$$4x = -8$$

$$x = -2.$$ $-2 \neq 5$ and $-2 \neq -5$, so -2 *should* check.

The student can check to confirm that -2 is the solution.

e) If $x = 1$, the denominators are 0. We list the restriction:

$$x \neq 1.$$

We clear fractions and solve:

$$\frac{x^2}{x-1} = \frac{1}{x-1}$$ We cannot have $x = 1$.

$$\cancel{(x-1)} \cdot \frac{x^2}{\cancel{x-1}} = \cancel{(x-1)} \cdot \frac{1}{\cancel{x-1}}$$

Multiplying both sides by $x - 1$, the LCM

$$x^2 = 1$$ Removing a factor equal to 1: $(x-1)/(x-1) = 1$

$$x^2 - 1 = 0$$ Subtracting 1 from both sides

$$(x-1)(x+1) = 0$$ Factoring

$$x - 1 = 0 \quad or \quad x + 1 = 0$$ Using the principle of zero products

$$x = 1 \quad or \quad x = -1.$$ Above, we stated that $x \neq 1$.

Because of the above restriction, 1 must be rejected as a solution. The student should check in the original equation that -1 *does* check. The solution is -1.

TRY EXERCISES 21 and 29

A Visual Interpretation

ALGEBRAIC – GRAPHICAL CONNECTION

We can obtain a visual check of the solutions of a rational equation by graphing. For example, consider the equation

$$\frac{x}{4} + \frac{x}{2} = 6.$$

We can examine the solution by graphing the equations

$$y = \frac{x}{4} + \frac{x}{2} \quad \text{and} \quad y = 6$$

using the same set of axes, as shown below.

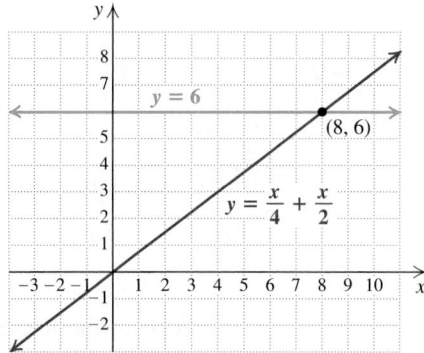

The y-values for each equation will be the same where the graphs intersect. The x-value of that point will yield that value, so it will be a solution of the equation. It appears from the graph that when $x = 8$, the value of $x/4 + x/2$ is 6. We can check by substitution:

$$\frac{x}{4} + \frac{x}{2} = \frac{8}{4} + \frac{8}{2} = 2 + 4 = 6.$$

Thus the solution is 8.

TECHNOLOGY CONNECTION

We can use a table to check possible solutions of rational equations. Consider the equation in Example 2(e),

$$\frac{x^2}{x-1} = \frac{1}{x-1},$$

and the possible solutions that were found, 1 and -1. To check these solutions, we enter $y_1 = x^2/(x-1)$ and $y_2 = 1/(x-1)$. After setting Indpnt to Ask and Depend to Auto in the TBLSET menu, we display the table and enter $x = 1$. The ERROR messages indicate that 1 is not a solution because it is not an allowable replacement for x in the equation. Next, we enter $x = -1$. Since y_1 and y_2 have the same value, we know that the equation is true when $x = -1$, and thus -1 is a solution.

X	Y₁	Y₂
1	ERROR	ERROR
−1	−.5	−.5

X =

Use a graphing calculator to check the possible solutions of Example 1 and Example 2, parts (a)–(d).

> **Concept Reinforcement** *Classify each statement as either true or false.*

1. Every rational equation has at least one solution.

2. It is possible for a rational equation to have more than one solution.

3. When both sides of an equation are multiplied by a variable expression, the result is not always an equivalent equation.

4. All the equation-solving principles studied thus far may be needed when solving a rational equation.

Solve. If no solution exists, state this.

5. $\dfrac{3}{5} - \dfrac{2}{3} = \dfrac{x}{6}$

6. $\dfrac{5}{8} - \dfrac{3}{5} = \dfrac{x}{10}$

7. $\dfrac{1}{3} + \dfrac{5}{6} = \dfrac{1}{x}$

8. $\dfrac{3}{5} + \dfrac{1}{8} = \dfrac{1}{x}$

9. $\dfrac{1}{8} + \dfrac{1}{12} = \dfrac{1}{t}$

10. $\dfrac{1}{6} + \dfrac{1}{10} = \dfrac{1}{t}$

11. $y + \dfrac{4}{y} = -5$

12. $n + \dfrac{3}{n} = -4$

13. $\dfrac{x}{6} - \dfrac{6}{x} = 0$

14. $\dfrac{x}{7} - \dfrac{7}{x} = 0$

15. $\dfrac{2}{x} = \dfrac{5}{x} - \dfrac{1}{4}$

16. $\dfrac{3}{t} = \dfrac{4}{t} - \dfrac{1}{5}$

17. $\dfrac{5}{3t} + \dfrac{3}{t} = 1$

18. $\dfrac{3}{4x} + \dfrac{5}{x} = 1$

19. $\dfrac{n+2}{n-6} = \dfrac{1}{2}$

20. $\dfrac{a-4}{a+6} = \dfrac{1}{3}$

21. $x + \dfrac{12}{x} = -7$

22. $x + \dfrac{8}{x} = -9$

23. $\dfrac{3}{x-4} = \dfrac{5}{x+1}$

24. $\dfrac{1}{x+3} = \dfrac{4}{x-1}$

25. $\dfrac{a}{6} - \dfrac{a}{10} = \dfrac{1}{6}$

26. $\dfrac{t}{8} - \dfrac{t}{12} = \dfrac{1}{8}$

27. $\dfrac{x+1}{3} - 1 = \dfrac{x-1}{2}$

28. $\dfrac{x+2}{5} - 1 = \dfrac{x-2}{4}$

29. $\dfrac{y+3}{y-3} = \dfrac{6}{y-3}$

30. $\dfrac{3}{a+7} = \dfrac{a+10}{a+7}$

31. $\dfrac{3}{x+4} = \dfrac{5}{x}$

32. $\dfrac{2}{x+3} = \dfrac{7}{x}$

33. $\dfrac{n+1}{n+2} = \dfrac{n-3}{n+1}$

34. $\dfrac{n+2}{n-3} = \dfrac{n+1}{n-2}$

35. $\dfrac{5}{t-2} + \dfrac{3t}{t-2} = \dfrac{4}{t^2-4t+4}$

36. $\dfrac{4}{t-3} + \dfrac{2t}{t-3} = \dfrac{12}{t^2-6t+9}$

37. $\dfrac{x}{x+5} - \dfrac{5}{x-5} = \dfrac{14}{x^2-25}$

38. $\dfrac{5}{x+1} + \dfrac{2x}{x^2-1} = \dfrac{1}{x+1}$

39. $\dfrac{5}{t-3} - \dfrac{30}{t^2-9} = 1$

40. $\dfrac{1}{y+3} + \dfrac{1}{y-3} = \dfrac{1}{y^2-9}$

41. $\dfrac{7}{6-a} = \dfrac{a+1}{a-6}$

42. $\dfrac{t-12}{t-10} = \dfrac{1}{10-t}$

Aha! **43.** $\dfrac{-2}{x+2} = \dfrac{x}{x+2}$

44. $\dfrac{3}{2x-6} = \dfrac{x}{2x-6}$

45. $\dfrac{12}{x} = \dfrac{x}{3}$

46. $\dfrac{x}{2} = \dfrac{18}{x}$

47. When solving rational equations, why do we multiply each side by the LCM of the denominators?

48. Explain the difference between adding rational expressions and solving rational equations.

Skill Review

To prepare for Section 6.7, review solving applications and rates of change (Sections 2.5, 3.4, and 5.8).

49. The sum of two consecutive odd numbers is 276. Find the numbers. [2.5]

50. The length of a rectangular picture window is 3 yd greater than the width. The area of the rectangle is 10 yd². Find the perimeter. [5.8]

51. The height of a triangle is 3 cm longer than its base. If the area of the triangle is 54 cm², find the measurements of the base and the height. [5.8]

52. The product of two consecutive even integers is 48. Find the numbers. [5.8]

53. *Human physiology.*   Between June 9 and June 24, Seth's beard grew 0.9 cm. Find the rate at which Seth's beard grows. [3.4]

54. *Gardening.*   Between July 7 and July 12, Carla's string beans grew 1.4 in. Find the growth rate of the string beans. [3.4]

Synthesis

55. Describe a method that can be used to create rational equations that have no solution.

56. How can a graph be used to determine how many solutions an equation has?

Solve.

57. $1 + \dfrac{x-1}{x-3} = \dfrac{2}{x-3} - x$

58. $\dfrac{4}{y-2} + \dfrac{3}{y^2-4} = \dfrac{5}{y+2} + \dfrac{2y}{y^2-4}$

59. $\dfrac{12-6x}{x^2-4} = \dfrac{3x}{x+2} - \dfrac{3-2x}{2-x}$

60. $\dfrac{x}{x^2+3x-4} + \dfrac{x+1}{x^2+6x+8} = \dfrac{2x}{x^2+x-2}$

61. $7 - \dfrac{a-2}{a+3} = \dfrac{a^2-4}{a+3} + 5$

62. $\dfrac{x^2}{x^2-4} = \dfrac{x}{x+2} - \dfrac{2x}{2-x}$

63. $\dfrac{1}{x-1} + x - 5 = \dfrac{5x-4}{x-1} - 6$

64. $\dfrac{5-3a}{a^2+4a+3} - \dfrac{2a+2}{a+3} = \dfrac{3-a}{a+1}$

65. $\dfrac{\dfrac{1}{x}+1}{x} = \dfrac{\dfrac{1}{x}}{2}$

66. $\dfrac{\dfrac{1}{3}}{x} = \dfrac{1-\dfrac{1}{x}}{x}$

67. Use a graphing calculator to check your answers to Exercises 13, 21, 31, and 57.

CONNECTING ⬆ the CONCEPTS

An equation contains an equals sign; an expression does not. Be careful not to confuse simplifying an expression with solving an equation. When expressions are simplified, the result is an equivalent expression. When equations are solved, the result is a solution. Compare the following.

Simplify: $\dfrac{x-1}{6x} + \dfrac{4}{9}$.

> The equals signs indicate that all the expressions are equivalent.

SOLUTION

$\dfrac{x-1}{6x} + \dfrac{4}{9} = \dfrac{x-1}{6x} \cdot \dfrac{3}{3} + \dfrac{4}{9} \cdot \dfrac{2x}{2x}$

$= \dfrac{3x-3}{18x} + \dfrac{8x}{18x}$ Writing with the LCD, $18x$

$= \dfrac{11x-3}{18x}$ The result is an expression equivalent to $\dfrac{x-1}{6x} + \dfrac{4}{9}$.

Solve: $\dfrac{x-1}{6x} = \dfrac{4}{9}$.

SOLUTION

$\dfrac{x-1}{6x} = \dfrac{4}{9}$

> Each line is an equivalent equation.

$18x \cdot \dfrac{x-1}{6x} = 18x \cdot \dfrac{4}{9}$ Multiplying by the LCM, $18x$

$3 \cdot \cancel{6x} \cdot \dfrac{x-1}{\cancel{6x}} = 2 \cdot \cancel{9} \cdot x \cdot \dfrac{4}{\cancel{9}}$

$3(x-1) = 2x \cdot 4$

$3x - 3 = 8x$

$-3 = 5x$

$-\dfrac{3}{5} = x$ The result is a solution; $-\dfrac{3}{5}$ is the solution of $\dfrac{x-1}{6x} = \dfrac{4}{9}$.

MIXED REVIEW

Tell whether each of the following is an expression or an equation. Then simplify the expression or solve the equation.

1. Simplify: $\dfrac{4x^2 - 8x}{4x^2 + 4x}$.

2. Add and, if possible, simplify: $\dfrac{2}{5n} + \dfrac{3}{2n - 1}$.

3. Solve: $\dfrac{3}{y} - \dfrac{1}{4} = \dfrac{1}{y}$.

4. Simplify: $\dfrac{\dfrac{1}{z} + 1}{\dfrac{1}{z^2 - 1}}$.

5. Solve: $\dfrac{5}{x + 3} = \dfrac{3}{x + 2}$.

6. Multiply and, if possible, simplify:

$$\frac{8t + 8}{2t^2 + t - 1} \cdot \frac{t^2 - 1}{t^2 - 2t + 1}.$$

7. Subtract and, if possible, simplify:

$$\frac{2a}{a + 1} - \frac{4a}{1 - a^2}.$$

8. Solve: $\dfrac{15}{x} - \dfrac{15}{x + 2} = 2$.

9. Divide and, if possible, simplify:

$$\frac{18x^2}{25} \div \frac{12x}{5}.$$

10. Solve: $\dfrac{20}{x} = \dfrac{x}{5}$.

6.7 Applications Using Rational Equations and Proportions

Problems Involving Work ▪ Problems Involving Motion ▪ Problems Involving Proportions

In many areas of study, applications involving rates, proportions, or reciprocals translate to rational equations. By using the five steps for problem solving and the lessons of Section 6.6, we can now solve such problems.

Problems Involving Work

EXAMPLE 1

The roof of Finn and Paige's townhouse needs to be reshingled. Finn can do the job alone in 8 hr and Paige can do the job alone in 10 hr. How long will it take the two of them, working together, to reshingle the roof?

SOLUTION

1. Familiarize. This *work problem* is a type of problem we have not yet encountered. Work problems are often *incorrectly* translated to mathematical language in several ways.

a) Add the times together: $8\,\text{hr} + 10\,\text{hr} = 18\,\text{hr}.$ ← Incorrect

This cannot be the correct approach since Finn and Paige working together should not take longer than either of them working alone.

b) Average the times: $(8\,\text{hr} + 10\,\text{hr})/2 = 9\,\text{hr}.$ ← Incorrect

Again, this is longer than it would take Finn to do the job alone.

c) Assume that each person does half the job. ← Incorrect

Finn would reshingle $\frac{1}{2}$ the roof in $\frac{1}{2}(8 \text{ hr})$, or 4 hr, and Paige would reshingle $\frac{1}{2}$ the roof in $\frac{1}{2}(10 \text{ hr})$, or 5 hr, so Finn would finish an hour before Paige. The problem assumes that the two are working together, so Finn will help Paige after completing his half. This tells us that the job will take between 4 and 5 hr.

Each incorrect approach started with the time it took each worker to do the job. The correct approach instead focuses on the *rate* of work, or the amount of the job that each person completes in 1 hr.

Since Finn takes 8 hr to reshingle the entire roof, in 1 hr he reshingles $\frac{1}{8}$ of the roof. Since Paige takes 10 hr to reshingle the entire roof, in 1 hr she reshingles $\frac{1}{10}$ of the roof. Thus Finn works at a rate of $\frac{1}{8}$ roof per hour, and Paige works at a rate of $\frac{1}{10}$ roof per hour.

Working together, Finn and Paige reshingle $\frac{1}{8} + \frac{1}{10}$ roof in 1 hr, so their rate, as a team, is $\frac{1}{8} + \frac{1}{10} = \frac{5}{40} + \frac{4}{40} = \frac{9}{40}$ roof per hour.

We are looking for the time required to reshingle 1 entire roof, not just a part of it. Setting up a table helps us organize the information.

Time	Fraction of the Roof Reshingled		
	By Finn	**By Paige**	**Together**
1 hr	$\frac{1}{8}$	$\frac{1}{10}$	$\frac{1}{8} + \frac{1}{10}$, or $\frac{9}{40}$
2 hr	$\frac{1}{8} \cdot 2$	$\frac{1}{10} \cdot 2$	$\left(\frac{1}{8} + \frac{1}{10}\right)2$, or $\frac{9}{40} \cdot 2$, or $\frac{9}{20}$
3 hr	$\frac{1}{8} \cdot 3$	$\frac{1}{10} \cdot 3$	$\left(\frac{1}{8} + \frac{1}{10}\right)3$, or $\frac{9}{40} \cdot 3$, or $\frac{27}{40}$
t hr	$\frac{1}{8} \cdot t$	$\frac{1}{10} \cdot t$	$\left(\frac{1}{8} + \frac{1}{10}\right)t$, or $\frac{9}{40} \cdot t$

STUDY SKILLS

*Take Advantage
of Free Checking*

It is always wise to check an answer, if it is possible to do so. When an applied problem is being solved, it is usually possible to check an answer in the equation from which it came. While such a check can be helpful, it is even more important to check the answer with the words of the original problem.

2. **Translate.** From the table, we see that t must be some number for which

Fraction of roof done by Finn in t hr ⟶ $\dfrac{1}{8} \cdot t + \dfrac{1}{10} \cdot t = 1,$ ⟵ Fraction of roof done by Paige in t hr

or

$$\frac{t}{8} + \frac{t}{10} = 1.$$

3. **Carry out.** We solve the equation:

$$\frac{t}{8} + \frac{t}{10} = 1 \qquad \text{The LCD is 40.}$$

$$40\left(\frac{t}{8} + \frac{t}{10}\right) = 40 \cdot 1 \qquad \text{Multiplying to clear fractions}$$

$$\frac{40t}{8} + \frac{40t}{10} = 40$$

$$5t + 4t = 40 \qquad \text{Simplifying}$$

$$9t = 40$$

$$t = \frac{40}{9}, \text{ or } 4\frac{4}{9}.$$

4. **Check.** In $\frac{40}{9}$ hr, Finn reshingles $\frac{1}{8} \cdot \frac{40}{9}$, or $\frac{5}{9}$, of the roof and Paige reshingles $\frac{1}{10} \cdot \frac{40}{9}$, or $\frac{4}{9}$, of the roof. Together, they reshingle $\frac{5}{9} + \frac{4}{9}$, or 1 roof. The fact that our solution is between 4 and 5 hr (see step 1 above) is also a check.

5. **State.** It will take $4\frac{4}{9}$ hr for Finn and Paige, working together, to reshingle the roof.

> **TRY EXERCISE** 7

EXAMPLE 2

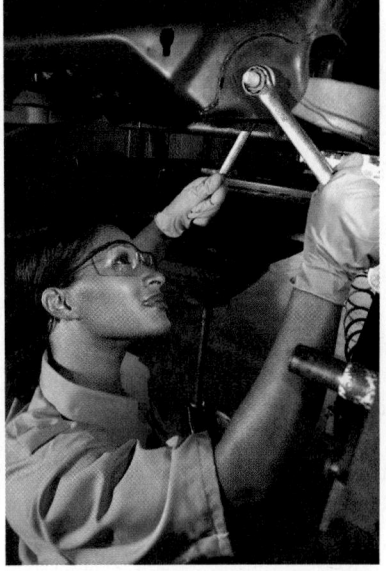

It takes Manuel 9 hr longer than Zoe to rebuild an engine. Working together, they can do the job in 20 hr. How long would it take each, working alone, to rebuild an engine?

SOLUTION

1. **Familiarize.** Unlike Example 1, this problem does not provide us with the times required by the individuals to do the job alone. Let's have $z =$ the number of hours it would take Zoe working alone and $z + 9 =$ the number of hours it would take Manuel working alone.

2. **Translate.** Using the same reasoning as in Example 1, we see that Zoe completes $\frac{1}{z}$ of the job in 1 hr and Manuel completes $\frac{1}{z + 9}$ of the job in 1 hr. Thus, in 20 hr, Zoe completes $\frac{1}{z} \cdot 20$ of the job and Manuel completes $\frac{1}{z + 9} \cdot 20$ of the job.

We know that, together, Zoe and Manuel can complete the entire job in 20 hr. This gives the following:

Fraction of job done by Zoe in 20 hr ⎯ $\frac{1}{z} \cdot 20 + \frac{1}{z + 9} \cdot 20 = 1,$ ⎯ Fraction of job done by Manuel in 20 hr

or $\dfrac{20}{z} + \dfrac{20}{z + 9} = 1.$

3. **Carry out.** We solve the equation:

$$\frac{20}{z} + \frac{20}{z + 9} = 1 \qquad \text{The LCD is } z(z + 9).$$

$$z(z + 9)\left(\frac{20}{z} + \frac{20}{z + 9}\right) = z(z + 9)1 \qquad \text{Multiplying to clear fractions}$$

$$(z + 9)20 + z \cdot 20 = z(z + 9) \qquad \text{Distributing and simplifying}$$

$$40z + 180 = z^2 + 9z$$

$$0 = z^2 - 31z - 180 \qquad \text{Getting 0 on one side}$$

$$0 = (z - 36)(z + 5) \qquad \text{Factoring}$$

$$z - 36 = 0 \quad or \quad z + 5 = 0 \qquad \text{Principle of zero products}$$

$$z = 36 \quad or \qquad z = -5.$$

4. **Check.** Since negative time has no meaning in the problem, -5 is not a solution to the original problem. The number 36 checks since, if Zoe takes 36 hr alone and Manuel takes $36 + 9 = 45$ hr alone, in 20 hr they would have finished

$$\frac{20}{36} + \frac{20}{45} = \frac{5}{9} + \frac{4}{9} = 1 \text{ complete rebuild.}$$

5. **State.** It would take Zoe 36 hr to rebuild an engine alone, and Manuel 45 hr.

> **TRY EXERCISE** 13

The equations used in Examples 1 and 2 can be generalized as follows.

> ## The Work Principle
>
> If
>
> a = the time needed for A to complete the work alone,
>
> b = the time needed for B to complete the work alone, and
>
> t = the time needed for A and B to complete the work together,
>
> then
>
> $$\frac{t}{a} + \frac{t}{b} = 1.$$
>
> The following are equivalent equations that can also be used:
>
> $$\frac{1}{a} \cdot t + \frac{1}{b} \cdot t = 1 \quad \text{and} \quad \frac{1}{a} + \frac{1}{b} = \frac{1}{t}.$$

Problems Involving Motion

Problems dealing with distance, rate (or speed), and time are called **motion problems**. To translate them, we use either the basic motion formula, $d = rt$, or the formulas $r = d/t$ or $t = d/r$, which can be derived from $d = rt$.

EXAMPLE 3

On her road bike, Olivia bikes 15 km/h faster than Jason does on his mountain bike. In the time it takes Olivia to travel 80 km, Jason travels 50 km. Find the speed of each bicyclist.

SOLUTION

1. **Familiarize.** Let's guess that Jason is going 10 km/h. Then the following would be true.

   Olivia's speed would be $10 + 15$, or 25 km/h.

   Jason would cover 50 km in $50/10 = 5$ hr.

   Olivia would cover 80 km in $80/25 = 3.2$ hr.

   Since the times are not the same, our guess is wrong, but we can make some observations.

   If r = the rate, in kilometers per hour, of Jason's bike, then the rate of Olivia's bike $= r + 15$.

   Jason's travel time is the same as Olivia's travel time.

   We make a drawing and construct a table, listing the information we know.

STUDENT NOTES

You need remember only the motion formula $d = rt$. Then you can divide both sides by t to get $r = d/t$, or you can divide both sides by r to get $t = d/r$.

r km/h
50 km

(*r* + 15) km/h
80 km

	Distance	Speed	Time
Jason's Mountain Bike	50	r	
Olivia's Road Bike	80	$r + 15$	

2. **Translate.** By looking at how we checked our guess, we see that we can fill in the **Time** column of the table using the formula *Time = Distance/Rate*, as follows.

	Distance	Speed	Time
Jason's Mountain Bike	50	r	$50/r$
Olivia's Road Bike	80	$r + 15$	$80/(r + 15)$

Since we know that the times are the same, we can write an equation:

$$\frac{50}{r} = \frac{80}{r + 15}.$$

3. **Carry out.** We solve the equation:

$$\frac{50}{r} = \frac{80}{r + 15} \qquad \text{The LCD is } r(r + 15).$$

$$r(r + 15)\frac{50}{r} = r(r + 15)\frac{80}{r + 15} \qquad \text{Multiplying to clear fractions}$$

$$50r + 750 = 80r \qquad \text{Simplifying}$$

$$750 = 30r$$

$$25 = r.$$

4. **Check.** If our answer checks, Jason's mountain bike is going 25 km/h and Olivia's road bike is going $25 + 15 = 40$ km/h.

Traveling 80 km at 40 km/h, Olivia is riding for $\frac{80}{40} = 2$ hr. Traveling 50 km at 25 km/h, Jason is riding for $\frac{50}{25} = 2$ hr. Our answer checks since the two times are the same.

5. **State.** Olivia's speed is 40 km/h, and Jason's speed is 25 km/h.

> **TRY EXERCISE** 19

In the next example, although distance is the same in both directions, the key to the translation lies in an additional piece of given information.

EXAMPLE **4** A Hudson River tugboat goes 10 mph in still water. It travels 24 mi upstream and 24 mi back in a total time of 5 hr. What is the speed of the current?

Sources: Based on information from the Department of the Interior, U.S. Geological Survey, and *The Tugboat Captain*, Montgomery County Community College

SOLUTION

1. **Familiarize.** Let's guess that the speed of the current is 4 mph. Then the following would be true.

The tugboat would move $10 - 4 = 6$ mph upstream.

The tugboat would move $10 + 4 = 14$ mph downstream.

To travel 24 mi upstream would require $\frac{24}{6} = 4$ hr.

To travel 24 mi downstream would require $\frac{24}{14} = 1\frac{5}{7}$ hr.

Since the total time, $4 + 1\frac{5}{7} = 5\frac{5}{7}$ hr, is not the 5 hr mentioned in the problem, our guess is wrong, but we can make some observations.

If c = the current's rate, in miles per hour, we have the following.

The tugboat's speed upstream is $(10 - c)$ mph.

The tugboat's speed downstream is $(10 + c)$ mph.

The total travel time is 5 hr.

We make a sketch and construct a table, listing the information we know.

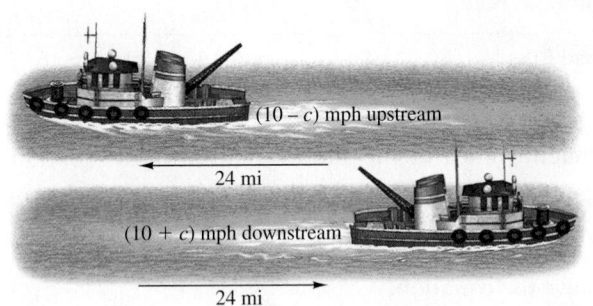

	Distance	Speed	Time
Upstream	24	$10 - c$	
Downstream	24	$10 + c$	

2. **Translate.** From examining our guess, we see that the time traveled can be represented using the formula *Time = Distance/Rate*:

	Distance	Speed	Time
Upstream	24	$10 - c$	$24/(10 - c)$
Downstream	24	$10 + c$	$24/(10 + c)$

Since the total time upstream and back is 5 hr, we use the last column of the table to form an equation:

$$\frac{24}{10 - c} + \frac{24}{10 + c} = 5.$$

3. **Carry out.** We solve the equation:

$$\frac{24}{10 - c} + \frac{24}{10 + c} = 5 \qquad \text{The LCD is } (10 - c)(10 + c).$$

$$(10 - c)(10 + c)\left[\frac{24}{10 - c} + \frac{24}{10 + c}\right] = (10 - c)(10 + c)5 \qquad \text{Multiplying to clear fractions}$$

$$24(10 + c) + 24(10 - c) = (100 - c^2)5$$

$$480 = 500 - 5c^2 \qquad \text{Simplifying}$$

$$5c^2 - 20 = 0$$

$$5(c^2 - 4) = 0$$

$$5(c - 2)(c + 2) = 0$$

$$c = 2 \quad or \quad c = -2.$$

4. **Check.** Since speed cannot be negative in this problem, -2 cannot be a solution. You should confirm that 2 checks in the original problem.

5. **State.** The speed of the current is 2 mph.

TRY EXERCISE ▶ 21

Problems Involving Proportions

A **ratio** of two quantities is their quotient. For example, 37% is the ratio of 37 to 100, or $\frac{37}{100}$. A **proportion** is an equation stating that two ratios are equal.

> ### Proportion
>
> An equality of ratios,
>
> $$\frac{A}{B} = \frac{C}{D},$$
>
> is called a *proportion*. The numbers within a proportion are said to be *proportional* to each other.

Proportions arise in geometry when we are studying *similar triangles*. If two triangles are **similar**, then their corresponding angles have the same measure and their corresponding sides are proportional. To illustrate, if triangle *ABC* is similar to triangle *RST*, then angles *A* and *R* have the same measure, angles *B* and *S* have the same measure, angles *C* and *T* have the same measure, and

$$\frac{a}{r} = \frac{b}{s} = \frac{c}{t}.$$

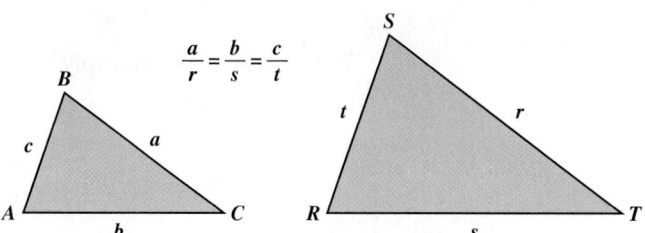

EXAMPLE 5

Similar triangles.   Triangles *ABC* and *XYZ* are similar. Solve for *z* if $x = 10$, $a = 8$, and $c = 5$.

SOLUTION   We make a drawing, write a proportion, and then solve. Note that side *a* is always opposite angle *A*, side *x* is always opposite angle *X*, and so on.

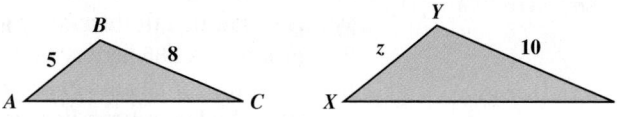

We have

$$\frac{z}{5} = \frac{10}{8}$$ The proportions $\frac{5}{z} = \frac{8}{10}$, $\frac{5}{8} = \frac{z}{10}$, or $\frac{8}{5} = \frac{10}{z}$ could also be used.

$$40 \cdot \frac{z}{5} = 40 \cdot \frac{10}{8}$$ Multiplying both sides by the LCM, 40

$$8z = 50$$ Simplifying

$$z = \frac{50}{8}, \text{ or } 6.25.$$

TRY EXERCISE ▶ 33

EXAMPLE 6

Architecture. A *blueprint* is a scale drawing of a building representing an architect's plans. Ellia is adding 12 ft to the length of an apartment and needs to indicate the addition on an existing blueprint. If a 10-ft long bedroom is represented by $2\frac{1}{2}$ in. on the blueprint, how much longer should Ellia make the drawing in order to represent the addition?

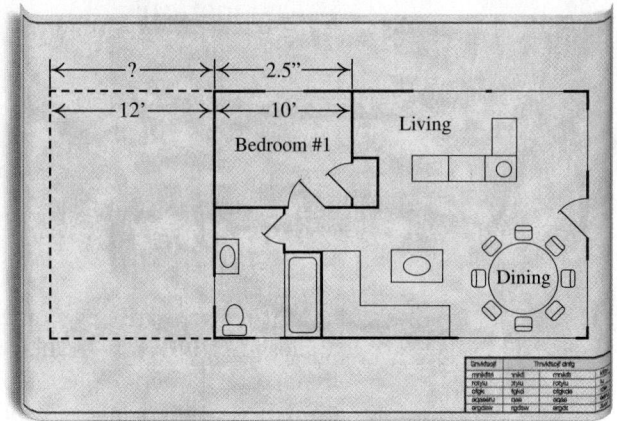

SOLUTION   We let w represent the width, in inches, of the addition that Ellia is drawing. Because the drawing must be to scale, we have

$$\text{Inches on drawing} \rightarrow \frac{w}{12} = \frac{2.5}{10}. \leftarrow \text{Inches on drawing}$$
$$\text{Feet in real life} \rightarrow \qquad\qquad \leftarrow \text{Feet in real life}$$

To solve for w, we multiply both sides by the LCM of the denominators, 60:

$$60 \cdot \frac{w}{12} = 60 \cdot \frac{2.5}{10}$$

$$5w = 6 \cdot 2.5 \qquad \text{Simplifying}$$

$$w = \frac{15}{5}, \text{ or } 3.$$

Ellia should make the blueprint 3 in. longer.   ▶ **TRY EXERCISE** 37

Proportions can be used to solve a variety of applied problems.

EXAMPLE 7

Alternative fuels.   As one alternative to gasoline-powered vehicles, flex fuel vehicles can use both regular gasoline and E85, a fuel containing 85% ethanol. The ethanol in E85 is derived from corn and other plants, a renewable resource, and burning E85 results in less pollution than gasoline. Because using E85 results in a lower fuel economy, its price per gallon must be lower than the price of gasoline in order to make it an economical fuel option. A 2007 Chevrolet Tahoe gets 21 mpg on the highway using gasoline and only 15 mpg on the highway using E85. If the price of gasoline is $3.36 a gallon, what must the price of E85 be in order for the fuel cost per mile to be the same?

Source: www.caranddriver.com

Alternative fuels

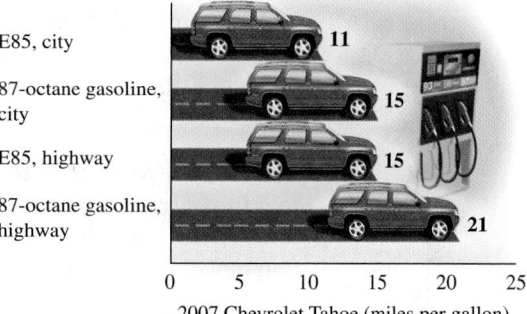

E85, city 11

87-octane gasoline, city 15

E85, highway 15

87-octane gasoline, highway .. 21

2007 Chevrolet Tahoe (miles per gallon)

Source: www.caranddriver.com

SOLUTION

1. Familiarize. We let x = the price per gallon of E85, and organize the given information:

Gasoline miles per gallon: 21,

E85 miles per gallon: 15,

Gasoline price per gallon: $3.36,

E85 price per gallon: x.

We need to use the given information to find the fuel cost per mile. If we divide the number of dollars per gallon by the number of miles per gallon, we can find the number of dollars per mile:

$$\frac{\text{dollars}}{\text{gal}} \div \frac{\text{mi}}{\text{gal}} = \frac{\text{dollars}}{\cancel{\text{gal}}} \cdot \frac{\cancel{\text{gal}}}{\text{mi}} = \frac{\text{dollars}}{\text{mi}}.$$

Thus we form ratios of the form

$$\frac{\text{dollars per gallon}}{\text{miles per gallon}}.$$

2. Translate. We form a proportion in which the ratio of price per gallon and number of miles per gallon is expressed in two ways:

$$\begin{array}{c} \text{Gasoline price} \to \\ \text{Gasoline mpg} \to \end{array} \frac{\$3.36}{21} = \frac{x}{15}. \begin{array}{c} \leftarrow \text{E85 price} \\ \leftarrow \text{E85 mpg} \end{array}$$

3. Carry out. To solve for x, we multiply both sides of the equation by the LCM, 105:

$$105 \cdot \frac{3.36}{21} = 105 \cdot \frac{x}{15}$$

$$5 \cdot \cancel{21} \cdot \frac{3.36}{\cancel{21}} = 7 \cdot \cancel{15} \cdot \frac{x}{\cancel{15}} \qquad \text{Removing factors equal to 1: } 21/21 = 1 \text{ and } 15/15 = 1$$

$$16.8 = 7x$$

$$2.4 = x. \qquad \text{Dividing both sides by 7 and simplifying}$$

4. Check. If E85 costs $2.40 per gallon, the fuel cost per mile will be $2.40/15 = $0.16 per mile. The fuel cost per mile for gasoline is $3.36/21 = $0.16 per mile. The fuel costs are the same.

5. State. The price of E85 must be $2.40 per gallon.

TRY EXERCISE ▶ 45

EXAMPLE **8** *Wildlife population.* To determine the number of brook trout in River Denys, Cape Breton, Nova Scotia, a team of volunteers and professionals caught and marked 1190 brook trout. Later, they captured 915 brook trout, of which 24 were marked. Estimate the number of brook trout in River Denys.

Source: www.gov.ns.ca

SOLUTION We let T = the brook trout population in River Denys. If we assume that the percentage of marked trout in the second group of trout captured is the same as the percentage of marked trout in the entire river, we can form a proportion in which this percentage is expressed in two ways:

Trout originally marked $\rightarrow$ $\dfrac{1190}{T} = \dfrac{24}{915}$ $\leftarrow$ Marked trout in second group
Entire population $\rightarrow$ $\qquad\qquad\qquad$ $\leftarrow$ Total trout in second group

To solve for T, we multiply by the LCM, $915T$:

$$915T \cdot \frac{1190}{T} = 915T \cdot \frac{24}{915} \qquad \text{Multiplying both sides by } 915T$$

$$915 \cdot 1190 = 24T \qquad\qquad \text{Removing factors equal to 1:}$$
$$ \qquad\qquad T/T = 1 \text{ and } 915/915 = 1$$

$$\frac{915 \cdot 1190}{24} = T \text{ or } T \approx 45{,}369. \qquad \text{Dividing both sides by 24}$$

There are about 45,369 brook trout in the river. **TRY EXERCISE** 57

1. *Search-engine ads.* In 2006, North American advertisers spent \$9.4 billion in marketing through Internet search engines such as Google®. This was a 62% increase over the amount spent in 2005. How much was spent in 2005?
Source: Search Engine Marketing Professional Organization

2. *Bicycling.* The speed of one bicyclist is 2 km/h faster than the speed of another bicyclist. The first bicyclist travels 60 km in the same amount of time that it takes the second to travel 50 km. Find the speed of each bicyclist.

3. *Filling time.* A swimming pool can be filled in 5 hr by hose A alone and in 6 hr by hose B alone. How long would it take to fill the tank if both hoses were working?

4. *Payroll.* In 2007, the total payroll for Kraftside Productions was \$9.4 million. Of this amount, 62% was paid to employees working on an assembly line. How much money was paid to assembly-line workers?

Translating for Success

Translate each word problem to an equation and select a correct translation from equations A–O.

A. $2x + 2(x + 1) = 613$

B. $x^2 + (x + 1)^2 = 613$

C. $\dfrac{60}{x + 2} = \dfrac{50}{x}$

D. $x = 62\% \cdot 9.4$

E. $\dfrac{197}{7} = \dfrac{x}{30}$

F. $x + (x + 1) = 613$

G. $\dfrac{7}{197} = \dfrac{x}{30}$

H. $x^2 + (x + 2)^2 = 612$

I. $x^2 + (x + 1)^2 = 612$

J. $\dfrac{50}{x + 2} = \dfrac{60}{x}$

K. $x + 62\% \cdot x = 9.4$

L. $\dfrac{5 + 6}{2} = t$

M. $x^2 + (x + 1)^2 = 452$

N. $\dfrac{1}{5} + \dfrac{1}{6} = \dfrac{1}{t}$

O. $x^2 + (x + 2)^2 = 452$

Answers on page A-26

An additional, animated version of this activity appears in MyMathLab. To use MyMathLab, you need a course ID and a student access code. Contact your instructor for more information.

5. *Cycling distance.* A bicyclist traveled 197 mi in 7 days. At this rate, how many miles could the cyclist travel in 30 days?

6. *Sides of a square.* If the sides of a square are increased by 2 ft, the area of the original square plus the area of the enlarged square is 452 ft². Find the length of a side of the original square.

7. *Consecutive integers.* The sum of two consecutive integers is 613. Find the integers.

8. *Sums of squares.* The sum of the squares of two consecutive odd integers is 612. Find the integers.

9. *Sums of squares.* The sum of the squares of two consecutive integers is 613. Find the integers.

10. *Rectangle dimensions.* The length of a rectangle is 1 ft longer than its width. Find the dimensions of the rectangle such that the perimeter of the rectangle is 613 ft.

Concept Reinforcement *Find each rate.*

1. If Sandy can decorate a cake in 2 hr, what is her rate?

2. If Eric can decorate a cake in 3 hr, what is his rate?

3. If Sandy can decorate a cake in 2 hr and Eric can decorate the same cake in 3 hr, what is their rate, working together?

4. If Lisa and Mark can mow a lawn together in 1 hr, what is their rate?

5. If Lisa can mow a lawn by herself in 3 hr, what is her rate?

6. If Lisa and Mark can mow a lawn together in 1 hr, and Lisa can mow the same lawn by herself in 3 hr, what is Mark's rate, working alone?

7. *Home restoration.* Bryan can refinish the floor of an apartment in 8 hr. Caroline can refinish the floor in 6 hr. How long will it take them, working together, to refinish the floor?

8. *Custom embroidery.* Chandra can embroider logos on a team's sweatshirts in 6 hr. Traci, a new employee, needs 9 hr to complete the same job. Working together, how long will it take them to do the job?

9. *Multifunction copiers.* The Aficio SP C210SF can copy Mousa's dissertation in 7 min. The MX-3501N can copy the same document in 6 min. If the two machines work together, how long would they take to copy the dissertation?
   Source: Manufacturers' marketing brochures

10. *Fax machines.* The FAXPHONE L80 can fax a year-end report in 7 min while the HP 1050 Fax Series can fax the same report in 14 min. How long would it take the two machines, working together, to fax the

report? (Assume that the recipient has at least two machines for incoming faxes.)
Source: Manufacturers' marketing brochures

11. *Pumping water.* A $\frac{1}{2}$ HP Wayne stainless-steel sump pump can remove water from Carmen's flooded basement in 42 min. The $\frac{1}{2}$ HP Craftsman Professional sump pump can complete the same job in 35 min. How long would it take the two pumps together to pump out the basement?
   Source: Based on data from manufacturers

12. *Hotel management.* The Honeywell Enviracaire Silent Comfort air cleaner can clean the air in an 11-ft by 17-ft conference room in 10 min. The Blueair 201 Air Purifier can clean the air in a room of the same size in 12 min. How long would it take the two machines together to clean the air in such a room?
   Source: Based on information from manufacturers' and retailers' websites

13. *Photocopiers.* The HP Officejet H470 Mobile Printer takes twice the time required by the HP Officejet Pro K5400 color printer to photocopy brochures for the New Bretton Arts Council annual arts fair. If, working together, the two machines can complete the job in 45 min, how long would it take each machine, working alone, to copy the brochures?
   Source: www.shoppinghp.com

14. *Forest fires.* The Erickson Air-Crane helicopter can scoop water and douse a certain forest fire four times as fast as an S-58T helicopter. Working together, the two helicopters can douse the fire in 8 hr. How long would it take each helicopter, working alone, to douse the fire?
   Sources: Based on information from www.emergency.com and www.arishelicopters.com

15. *Hotel management.* The Austin Healthmate 400 can purify the air in a conference hall in 15 fewer minutes than it takes the Airgle 750 Air Purifier to do the same job. Together the two machines can purify the air in the conference hall in 10 min. How long would it take each machine, working alone, to purify the air in the room?
   Source: Based on information from manufacturers' and retailers' websites

16. *Photo printing.* It takes the Canon PIXMA iP6310D 15 min longer to print a set of photo proofs than it takes the HP Officejet H470b Mobile Printer. Together it would take them $\frac{180}{7}$, or $25\frac{5}{7}$ min to print the photos. How long would it take each machine, working alone, to print the photos?
Sources: www.shoppinghp.com; www.staples.com

17. *Sorting recyclables.* Together, it takes Kim and Chris 2 hr 55 min to sort recyclables. Alone, Kim would require 2 hr more than Chris. How long would it take Chris to do the job alone? (*Hint*: Convert minutes to hours or hours to minutes.)

18. *Paving.* Together, Steve and Bill require 4 hr 48 min to pave a driveway. Alone, Steve would require 4 hr more than Bill. How long would it take Bill to do the job alone? (*Hint*: Convert minutes to hours.)

19. *Train speeds.* A B & M freight train is traveling 14 km/h slower than an AMTRAK passenger train. The B & M train travels 330 km in the same time that it takes the AMTRAK train to travel 400 km. Find their speeds. Complete the following table as part of the familiarization.

Distance = Rate · Time

	Distance (in km)	Speed (in km/h)	Time (in hours)
B & M	330		
AMTRAK	400	r	$\frac{400}{r}$

20. *Speed of travel.* A loaded Roadway truck is moving 40 mph faster than a New York Railways freight train. In the time that it takes the train to travel 150 mi, the truck travels 350 mi. Find their speeds. Complete the following table as part of the familiarization.

Distance = Rate · Time

	Distance (in miles)	Speed (in miles per hour)	Time (in hours)
Truck	350	r	$\frac{350}{r}$
Train	150		

21. *Kayaking.* The speed of the current in Catamount Creek is 3 mph. Sean can kayak 4 mi upstream in the same time it takes him to kayak 10 mi downstream. What is the speed of Sean's kayak in still water?

22. *Boating.* The current in the Lazy River moves at a rate of 4 mph. Nicole's dinghy motors 6 mi upstream in the same time it takes to motor 12 mi downstream. What is the speed of the dinghy in still water?

23. *Moving sidewalks.* The moving sidewalk at O'Hare Airport in Chicago moves 1.8 ft/sec. Walking on the moving sidewalk, Roslyn travels 105 ft forward in the time it takes to travel 51 ft in the opposite direction. How fast does Roslyn walk on a nonmoving sidewalk?

24. *Moving sidewalks.* Newark Airport's moving sidewalk moves at a speed of 1.7 ft/sec. Walking on the moving sidewalk, Drew can travel 120 ft forward in the same time it takes to travel 52 ft in the opposite direction. What is Drew's walking speed on a nonmoving sidewalk?

Aha! **25.** *Tractor speed.* Manley's tractor is just as fast as Caledonia's. It takes Manley 1 hr more than it takes Caledonia to drive to town. If Manley is 20 mi from town and Caledonia is 15 mi from town, how long does it take Caledonia to drive to town?

26. *Boat speed.* Tory and Emilio's motorboats travel at the same speed. Tory pilots her boat 40 km before docking. Emilio continues for another 2 hr, traveling a total of 100 km before docking. How long did it take Tory to navigate the 40 km?

27. *Boating.* LeBron's Mercruiser travels 15 km/h in still water. He motors 140 km downstream in the same time it takes to travel 35 km upstream. What is the speed of the river?

28. *Boating.*   Annette's paddleboat travels 2 km/h in still water. The boat is paddled 4 km downstream in the same time it takes to go 1 km upstream. What is the speed of the river?

29. *Shipping.*   A barge moves 7 km/h in still water. It travels 45 km upriver and 45 km downriver in a total time of 14 hr. What is the speed of the current?

30. *Aviation.*   A Citation CV jet travels 460 mph in still air and flies 525 mi into the wind and 525 mi with the wind in a total of 2.3 hr. Find the wind speed.
Source: Blue Star Jets, Inc.

31. *Train travel.*   A freight train covered 120 mi at a certain speed. Had the train been able to travel 10 mph faster, the trip would have been 2 hr shorter. How fast did the train go?

32. *Moped speed.*   Cameron's moped travels 8 km/h faster than Ellia's. Cameron travels 69 km in the same time that Ellia travels 45 km. Find the speed of each person's moped.

Geometry.   *For each pair of similar triangles, find the value of the indicated letter.*

33. *b*

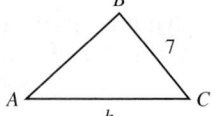 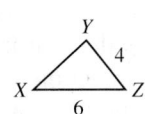

34. *a*

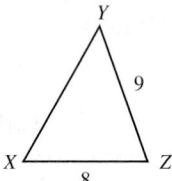 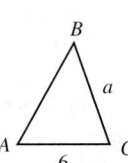

35. *f*

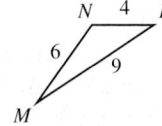 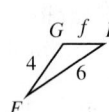

36. *r*

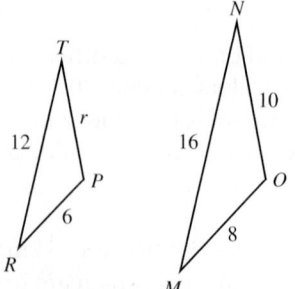

Architecture.   *Use the blueprint below to find the indicated length.*

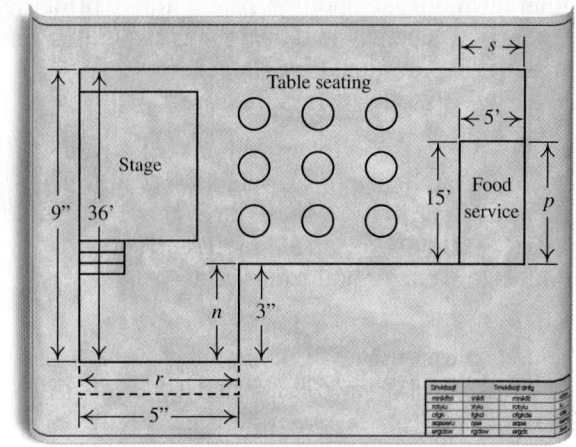

37. *p*, in inches on blueprint

38. *s*, in inches on blueprint

39. *r*, in feet on actual building

40. *n*, in feet on actual building

Find the indicated length.

41. *l*

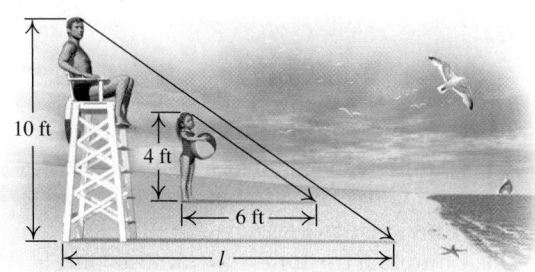

42. *h*

Graphing. Find the indicated length.

43. *r*

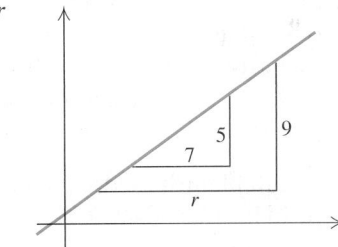

44. *s*

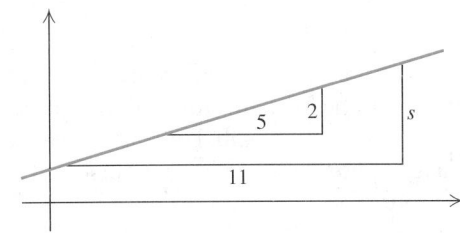

45. *Text messaging.* Brett sent or received 384 text messages in 8 days. At this rate, how many text messages would he send or receive in 30 days?

46. *Burning calories.* The average 140-lb adult burns about 380 calories bicycling 10 mi at a moderate rate. How far should the average 140-lb adult ride in order to burn 100 calories?
Source: *ACE Fitness Matters*, Volume 1, Number 4, 1997

47. *Illegal immigration.* Between 2001 and 2006, the Border Patrol caught 12,334 people trying to cross illegally from Canada to the United States along a 295-mi stretch of the border. If this rate were the same for the entire 5525-mi border between the two countries, how many would the Border Patrol catch in those same years along the entire border?
Source: *The Wall Street Journal*, July 10, 2007

Aha! **48.** *Photography.* Aziza snapped 234 photos over a period of 14 days. At this rate, how many would she take in 42 days?

49. *Mileage.* The Honda Civic Hybrid is a gasoline–electric car that travels approximately 180 mi on 4 gal of gas. Find the amount of gas required for an 810-mi trip.
Source: www.greenhybrid.com

50. *Baking.* In a potato bread recipe, the ratio of milk to flour is $\frac{3}{13}$. If 5 cups of milk are used, how many cups of flour are used?

51. *Wing aspect ratio.* The wing aspect ratio for a bird or an airplane is the ratio of the wing span to the wing width. Generally, higher aspect ratios are more efficient during low speed flying. Herons and storks, both waders, have comparable wing aspect ratios. A grey heron has a wing span of 180 cm and a wing width of 24 cm. A white stork has a wing span of 200 cm. What is the wing width of a stork?
Source: birds.ecoport.org

Aha! **52.** *Money.* The ratio of the weight of copper to the weight of zinc in a U.S. penny is $\frac{1}{39}$. If 50 kg of zinc is being turned into pennies, how much copper is needed?
Source: United States Mint

53. *Light bulbs.* A sample of 220 compact fluorescent light bulbs contained 8 defective bulbs. How many defective bulbs would you expect in a batch of 1430 bulbs?

54. *Flash drives.* A sample of 150 flash drives contained 7 defective drives. How many defective flash drives would you expect in a batch of 2700 flash drives?

55. *Veterinary science.* The amount of water needed by a small dog depends on its weight. A moderately active 8-lb Shih Tzu needs approximately 12 oz of water per day. How much water does a moderately active 5-lb Bolognese require each day?
Source: www.smalldogsparadise.com

56. *Miles driven.* Emmanuel is allowed to drive his leased car for 45,000 mi in 4 yr without penalty. In the first $1\frac{1}{2}$ yr, Emmanuel has driven 16,000 mi. At this rate will he exceed the mileage allowed for 4 yr?

57. *Environmental science.* To determine the number of humpback whales in a pod, a marine biologist, using tail markings, identifies 27 members of the pod. Several weeks later, 40 whales from the pod are randomly sighted. Of the 40 sighted, 12 are from the 27 originally identified. Estimate the number of whales in the pod.

58. *Fox population.* To determine the number of foxes in King County, a naturalist catches, tags, and then releases 25 foxes. Later, 36 foxes are caught; 4 of them have tags. Estimate the fox population of the county.

59. *Weight on the moon.* The ratio of the weight of an object on the moon to the weight of that object on Earth is 0.16 to 1.

a) How much would a 12-ton rocket weigh on the moon?

b) How much would a 180-lb astronaut weigh on the moon?

60. *Weight on Mars.* The ratio of the weight of an object on Mars to the weight of that object on Earth is 0.4 to 1.

a) How much would a 12-ton rocket weigh on Mars?

b) How much would a 120-lb astronaut weigh on Mars?

61. Is it correct to assume that two workers will complete a task twice as quickly as one person working alone? Why or why not?

62. If two triangles are exactly the same shape and size, are they similar? Why or why not?

Skill Review

To prepare for Chapter 7, review graphing linear equations (Section 3.2).

Graph. [3.2]

63. $y = 2x - 6$

64. $y = -2x + 6$

65. $3x + 2y = 12$

66. $x - 3y = 6$

67. $y = -\dfrac{3}{4}x + 2$

68. $y = \dfrac{2}{5}x - 4$

Synthesis

69. Two steamrollers are paving a parking lot. Working together, will the two steamrollers take less than half as long as the slower steamroller would working alone? Why or why not?

70. Two fuel lines are filling a freighter with oil. Will the faster fuel line take more or less than twice as long to fill the freighter by itself? Why?

71. *Filling a bog.* The Norwich cranberry bog can be filled in 9 hr and drained in 11 hr. How long will it take to fill the bog if the drainage gate is left open?

72. *Filling a tub.* Jillian's hot tub can be filled in 10 min and drained in 8 min. How long will it take to empty a full tub if the water is left on?

73. *Escalators.* Together, a 100-cm wide escalator and a 60-cm wide escalator can empty a 1575-person auditorium in 14 min. The wider escalator moves twice as many people as the narrower one. How many people per hour does the 60-cm wide escalator move?
Source: *McGraw-Hill Encyclopedia of Science and Technology*

74. *Aviation.* A Coast Guard plane has enough fuel to fly for 6 hr, and its speed in still air is 240 mph. The plane departs with a 40-mph tailwind and returns to the same airport flying into the same wind. How far can the plane travel under these conditions?

75. *Boating.* Shoreline Travel operates a 3-hr paddleboat cruise on the Missouri River. If the speed of the boat in still water is 12 mph, how far upriver can the pilot travel against a 5-mph current before it is time to turn around?

76. *Travel by car.* Angenita drives to work at 50 mph and arrives 1 min late. She drives to work at 60 mph and arrives 5 min early. How far does Angenita live from work?

77. *Grading.* Alma can grade a batch of placement exams in 3 hr. Kevin can grade a batch in 4 hr. If they work together to grade a batch of exams, what percentage of the exams will have been graded by Alma?

78. According to the U.S. Census Bureau, Population Division, in July 2007, there was one birth every 7 sec, one death every 13 sec, and one new international migrant every 27 sec. How many seconds does it take for a net gain of one person?

79. *Photocopying.* The printer in an admissions office can print a 500-page document in 50 min, while the printer in the business office can print the same document in 40 min. If the two printers work together to print the document, with the faster machine starting on page 1 and the slower machine working backwards from page 500, at what page will the two machines meet to complete the job?

80. *Distances.* The shadow from a 40-ft cliff just reaches across a water-filled quarry at the same time that a 6-ft tall diver casts a 10-ft shadow. How wide is the quarry?

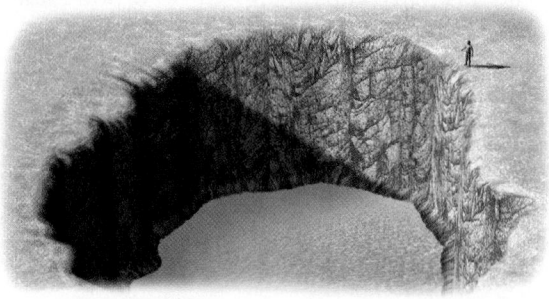

81. Given that

$$\frac{A}{B} = \frac{C}{D},$$

write three other proportions using A, B, C, and D.

Average speed is defined as total distance divided by total time.

82. Ferdaws drove 200 km. For the first 100 km of the trip, she drove at a speed of 40 km/h. For the second half of the trip, she traveled at a speed of 60 km/h. What was the average speed of the entire trip? (It was *not* 50 km/h.)

83. For the first 50 mi of a 100-mi trip, Garry drove 40 mph. What speed would he have to travel for the last half of the trip so that the average speed for the entire trip would be 45 mph?

84. If two triangles are similar, are their areas and perimeters proportional? Why or why not?

85. Are the equations

$$\frac{A + B}{B} = \frac{C + D}{D} \quad \text{and} \quad \frac{A}{B} = \frac{C}{D}$$

equivalent? Why or why not?

COLLABORATIVE CORNER

Sharing the Workload

Focus: Modeling, estimation, and work problems

Time: 15–20 minutes

Group size: 3

Materials: Paper, pencils, textbooks, and a watch

Many tasks can be done by two people working together. If both people work at the same rate, each does half the task, and the project is completed in half the time. However, when the work rates differ, the faster worker performs more than half of the task.

ACTIVITY

1. The project is to write down (but not answer) Review Exercises 27–36 from Chapter 6 (p. 439) on a sheet of paper. The problems should be spaced apart and written clearly so that they can be used for studying in the future. Two of the members in each group should write down the exercises, one working slowly and one working quickly. The third group member should record the time required for each to write down all 10 exercises.

2. Using the times from step (1), calculate how long it will take the two workers, working together, to complete the task.

3. Next, have the same workers as in step (1)— working at the same speeds as in step (1)— perform the task together. To do this, one person should begin writing with Exercise 27, while the other worker begins with Exercise 36 and lists the problems counting backward. The third member is again the timekeeper and should observe when the two workers have written all the exercises. To avoid collision, each of the two writers should use a separate sheet of paper.

4. Compare the actual experimental time from part (3) with the time predicted by the model in part (2). List reasons that might account for any discrepancy.

5. Let t_1, t_2, and t_3 represent the times required for the first worker, the second worker, and the two workers together, respectively, to complete a task. Then develop a model that can be used to find t_3 when t_1 and t_2 are known.

Time Required for Worker A, Working Alone	Time Required for Worker B, Working Alone	Estimated Time for the Two Workers, Working Together	Actual Time Required for the Two Workers, Working Together

Study Summary

KEY TERMS AND CONCEPTS	EXAMPLES

SECTION 6.1: RATIONAL EXPRESSIONS

A **rational expression** can be written as a quotient of two polynomials and is undefined when the denominator is 0. We simplify rational expressions by removing a factor equal to 1.

$$\frac{x^2 - 3x - 4}{x^2 - 1} = \frac{(x + 1)(x - 4)}{(x + 1)(x - 1)}$$

Factoring the numerator and the denominator

$$= \frac{x - 4}{x - 1}$$

$$\frac{x + 1}{x + 1} = 1$$

SECTION 6.2: MULTIPLICATION AND DIVISION

The Product of Two Rational Expressions

$$\frac{A}{B} \cdot \frac{C}{D} = \frac{AC}{BD}$$

$$\frac{5v + 5}{v - 2} \cdot \frac{2v^2 - 8v + 8}{v^2 - 1}$$

$$= \frac{5(v + 1) \cdot 2(v - 2)(v - 2)}{(v - 2)(v + 1)(v - 1)}$$

Multiplying numerators, multiplying denominators, and factoring

$$= \frac{10(v - 2)}{v - 1}$$

$$\frac{(v + 1)(v - 2)}{(v + 1)(v - 2)} = 1$$

The Quotient of Two Rational Expressions

$$\frac{A}{B} \div \frac{C}{D} = \frac{A}{B} \cdot \frac{D}{C} = \frac{AD}{BC}$$

$$(x^2 - 5x - 6) \div \frac{x^2 - 1}{x + 6}$$

$$= \frac{x^2 - 5x - 6}{1} \cdot \frac{x + 6}{x^2 - 1}$$

Multiplying by the reciprocal of the divisor

$$= \frac{(x - 6)(x + 1)(x + 6)}{(x + 1)(x - 1)}$$

Multiplying numerators, multiplying denominators, and factoring

$$= \frac{(x - 6)(x + 6)}{x - 1}$$

$$\frac{x + 1}{x + 1} = 1$$

SECTION 6.3: ADDITION, SUBTRACTION, AND LEAST COMMON DENOMINATORS

The Sum of Two Rational Expressions

$$\frac{A}{B} + \frac{C}{B} = \frac{A + C}{B}$$

$$\frac{7x + 8}{x + 1} + \frac{4x + 3}{x + 1} = \frac{7x + 8 + 4x + 3}{x + 1}$$

Adding numerators and keeping the denominator

$$= \frac{11x + 11}{x + 1}$$

$$= \frac{11(x + 1)}{x + 1}$$

Factoring

$$= 11$$

$$\frac{x + 1}{x + 1} = 1$$

The Difference of Two Rational Expressions

$$\frac{A}{B} - \frac{C}{B} = \frac{A - C}{B}$$

$$\frac{7x + 8}{x + 1} - \frac{4x + 3}{x + 1} = \frac{7x + 8 - (4x + 3)}{x + 1}$$

Subtracting numerators and keeping the denominator. The parentheses are necessary.

$$= \frac{7x + 8 - 4x - 3}{x + 1}$$

Removing parentheses

$$= \frac{3x + 5}{x + 1}$$

The **least common denominator (LCD)** of rational expressions is the **least common multiple (LCM)** of the denominators. To find the LCM, write the prime factorizations of the denominators. The LCM contains each factor the greatest number of times that it occurs in any of the individual factorizations.

Find the LCM of $m^2 - 5m + 6$ and $m^2 - 4m + 4$.

$$\left.\begin{array}{l} m^2 - 5m + 6 = (m - 2)(m - 3) \\ m^2 - 4m + 4 = (m - 2)(m - 2) \end{array}\right\} \quad \text{Factoring each expression}$$

LCM $= (m - 2)(m - 2)(m - 3)$ The LCM contains 2 factors of $(m - 2)$ and 1 factor of $(m - 3)$.

SECTION 6.4: ADDITION AND SUBTRACTION WITH UNLIKE DENOMINATORS

To add or subtract rational expressions with different denominators, first rewrite the expressions as equivalent expressions with a common denominator.

$$\frac{2x}{x^2 - 16} + \frac{x}{x - 4}$$

$$= \frac{2x}{(x + 4)(x - 4)} + \frac{x}{x - 4}$$
Factoring denominators: The LCD is $(x + 4)(x - 4)$.

$$= \frac{2x}{(x + 4)(x - 4)} + \frac{x}{x - 4} \cdot \frac{x + 4}{x + 4}$$
Multiplying by 1 to get the LCD in the second expression

$$= \frac{2x}{(x + 4)(x - 4)} + \frac{x^2 + 4x}{(x + 4)(x - 4)}$$

$$= \frac{x^2 + 6x}{(x + 4)(x - 4)}$$
Adding numerators and keeping the denominator. This cannot be simplified.

SECTION 6.5: COMPLEX RATIONAL EXPRESSIONS

Complex rational expressions contain one or more rational expressions within the numerator and/or the denominator. They can be simplified either by using division or by multiplying by a form of 1 to clear the fractions.

Using division to simplify:

$$\frac{\dfrac{1}{6} - \dfrac{1}{x}}{\dfrac{6 - x}{6}} = \frac{\dfrac{1}{6} \cdot \dfrac{x}{x} - \dfrac{1}{x} \cdot \dfrac{6}{6}}{\dfrac{6 - x}{6}} = \frac{\dfrac{x - 6}{6x}}{\dfrac{6 - x}{6}}$$
Subtracting to get a single rational expression in the numerator

$$= \frac{x - 6}{6x} \div \frac{6 - x}{6} = \frac{x - 6}{6x} \cdot \frac{6}{6 - x}$$
Dividing the numerator by the denominator

$$= \frac{6(x - 6)}{6x(-1)(x - 6)} = \frac{1}{-x} = -\frac{1}{x}$$
Factoring and simplifying; $\dfrac{6(x - 6)}{6(x - 6)} = 1$

Multiplying by 1:

$$\frac{\dfrac{4}{x}}{\dfrac{3}{x} + \dfrac{2}{x^2}} = \frac{\dfrac{4}{x}}{\dfrac{3}{x} + \dfrac{2}{x^2}} \cdot \frac{x^2}{x^2}$$

$$= \frac{\dfrac{4}{x} \cdot \dfrac{x^2}{1}}{\left(\dfrac{3}{x} + \dfrac{2}{x^2}\right) \cdot \dfrac{x^2}{1}}$$
The LCD of all the denominators is x^2; multiplying by $\dfrac{x^2}{x^2}$

$$= \frac{\dfrac{4 \cdot x \cdot x}{x}}{\dfrac{3 \cdot x \cdot x}{x} + \dfrac{2 \cdot x^2}{x^2}} = \frac{4x}{3x + 2}$$
The fractions are cleared.

SECTION 6.6: SOLVING RATIONAL EQUATIONS

To Solve a Rational Equation

1. List any restrictions.
2. Clear the equation of fractions.
3. Solve the resulting equation.
4. Check the possible solution(s) in the original equation.

Solve: $\dfrac{2}{x+1} = \dfrac{1}{x-2}$.   The restrictions are $x \neq -1, x \neq 2$.

$$\frac{2}{x+1} = \frac{1}{x-2}$$

$$(x+1)(x-2) \cdot \frac{2}{x+1} = (x+1)(x-2) \cdot \frac{1}{x-2}$$

$$2(x-2) = x+1$$

$$2x-4 = x+1$$

$$x = 5$$

Check: Since $\dfrac{2}{5+1} = \dfrac{1}{5-2}$, the solution is 5.

SECTION 6.7: APPLICATIONS USING RATIONAL EQUATIONS AND PROPORTIONS

The Work Principle

If a = the time needed for A to complete the work alone,

b = the time needed for B to complete the work alone,

and

t = the time needed for A and B to complete the work together, then:

$$\frac{t}{a} + \frac{t}{b} = 1; \quad \frac{1}{a} \cdot t + \frac{1}{b} \cdot t = 1;$$

$$\frac{1}{a} + \frac{1}{b} = \frac{1}{t}.$$

It takes Manuel 9 hr longer than Zoe to rebuild an engine. Working together, they can do the job in 20 hr. How long would it take each, working alone, to rebuild an engine?

We model the situation using the work principle, with $a = z$, $b = z + 9$, and $t = 20$:

$$\frac{20}{z} + \frac{20}{z+9} = 1 \qquad \text{Using the work principle}$$

$$z = 36. \qquad \text{Solving the equation}$$

It would take Zoe 36 hr to rebuild an engine alone, and Manuel 45 hr.

See Example 2 in Section 6.7 for a complete solution of this problem.

The Motion Formula

$$d = r \cdot t, \quad r = \frac{d}{t}, \quad \text{or} \quad t = \frac{d}{r}$$

On her road bike, Olivia bikes 15 km/h faster than Jason does on his mountain bike. In the time it takes Olivia to travel 80 km, Jason travels 50 km. Find the speed of each bicyclist.

Jason's speed: r km/h   Jason's time: $50/r$ hr
Olivia's speed: $(r + 15)$ km/h   Olivia's time: $80/(r + 15)$ hr

The times are equal:

$$\frac{50}{r} = \frac{80}{r+15}$$

$$r = 25 \qquad \text{Solving the equation}$$

Olivia's speed is 40 km/h, and Jason's speed is 25 km/h.

See Example 3 in Section 6.7 for a complete solution of this problem.

In geometry, proportions arise in the study of **similar triangles**.

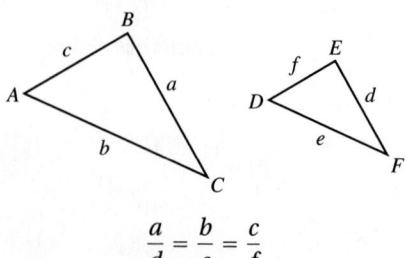

$$\frac{a}{d} = \frac{b}{e} = \frac{c}{f}$$

Triangles DEF and UVW are similar. Solve for u.

$$\frac{6}{8} = \frac{4}{u}$$

$$u = \frac{32}{6} = \frac{16}{3}$$

Review Exercises: Chapter 6

👤 *Concept Reinforcement* *Classify each statement as either true or false.*

1. Every rational expression can be simplified. [6.1]

2. The expression $(t - 3)/(t^2 - 4)$ is undefined for $t = 2$. [6.1]

3. The expression $(t - 3)/(t^2 - 4)$ is undefined for $t = 3$. [6.1]

4. To multiply rational expressions, a common denominator is never required. [6.2]

5. To divide rational expressions, a common denominator is never required. [6.2]

6. To add rational expressions, a common denominator is never required. [6.3]

7. To subtract rational expressions, a common denominator is never required. [6.3]

8. The number 0 can never be a solution of a rational equation. [6.6]

List all numbers for which each expression is undefined. [6.1]

9. $\dfrac{17}{-x^2}$

10. $\dfrac{9}{2a + 10}$

11. $\dfrac{x - 5}{x^2 - 36}$

12. $\dfrac{x^2 + 3x + 2}{x^2 + x - 30}$

13. $\dfrac{-6}{(t + 2)^2}$

Simplify. [6.1]

14. $\dfrac{3x^2 - 9x}{3x^2 + 15x}$

15. $\dfrac{14x^2 - x - 3}{2x^2 - 7x + 3}$

16. $\dfrac{6y^2 - 36y + 54}{4y^2 - 36}$

17. $\dfrac{5x^2 - 20y^2}{2y - x}$

Multiply or divide and, if possible, simplify. [6.2]

18. $\dfrac{a^2 - 36}{10a} \cdot \dfrac{2a}{a + 6}$

19. $\dfrac{6y - 12}{2y^2 + 3y - 2} \cdot \dfrac{y^2 - 4}{8y - 8}$

20. $\dfrac{16 - 8t}{3} \div \dfrac{t - 2}{12t}$

21. $\dfrac{4x^4}{x^2 - 1} \div \dfrac{2x^3}{x^2 - 2x + 1}$

22. $\dfrac{x^2 + 1}{x - 2} \cdot \dfrac{2x + 1}{x + 1}$

23. $(t^2 + 3t - 4) \div \dfrac{t^2 - 1}{t + 4}$

Find the LCM. [6.3]

24. $10a^3b^8$, $12a^5b$

25. $x^2 - x$, $x^5 - x^3$, x^4

26. $y^2 - y - 2$, $y^2 - 4$

Add or subtract and, if possible, simplify.

27. $\dfrac{x + 6}{x + 3} + \dfrac{9 - 4x}{x + 3}$ [6.3]

28. $\dfrac{6x - 3}{x^2 - x - 12} - \dfrac{2x - 15}{x^2 - x - 12}$ [6.3]

29. $\dfrac{3x - 1}{2x} - \dfrac{x - 3}{x}$ [6.4]

30. $\dfrac{2a + 4b}{5ab^2} - \dfrac{5a - 3b}{a^2b}$ [6.4]

31. $\dfrac{y^2}{y - 2} + \dfrac{6y - 8}{2 - y}$ [6.4]

32. $\dfrac{t}{t + 1} + \dfrac{t}{1 - t^2}$ [6.4]

33. $\dfrac{d^2}{d - 2} + \dfrac{4}{2 - d}$ [6.4]

34. $\dfrac{1}{x^2 - 25} - \dfrac{x - 5}{x^2 - 4x - 5}$ [6.4]

35. $\dfrac{3x}{x + 2} - \dfrac{x}{x - 2} + \dfrac{8}{x^2 - 4}$ [6.4]

36. $\dfrac{3}{4t} + \dfrac{3}{3t + 2}$ [6.4]

Simplify. [6.5]

37. $\dfrac{\dfrac{1}{z} + 1}{\dfrac{1}{z^2} - 1}$

38. $\dfrac{\dfrac{5}{2x^2}}{\dfrac{3}{4x} + \dfrac{4}{x^3}}$

39. $\dfrac{\dfrac{c}{d} - \dfrac{d}{c}}{\dfrac{1}{c} + \dfrac{1}{d}}$

Solve. [6.6]

40. $\dfrac{3}{x} - \dfrac{1}{4} = \dfrac{1}{2}$

41. $\dfrac{3}{x + 4} = \dfrac{1}{x - 1}$

42. $x + \dfrac{6}{x} = -7$

Solve. [6.7]

43. Jackson can sand the oak floors and stairs in a two-story home in 12 hr. Charis can do the same job in 9 hr. How long would it take if they worked together? (Assume that two sanders are available.)

44. A research company uses employees' computers to process data while the employee is not using the computer. An Intel Core 2 Quad processor can process a data file in 15 sec less time than an Intel Core 2 Duo processor. Working together, the computers can process the file in 18 sec. How long does it take each computer to process the file?

45. The distance by highway between Richmond and Waterbury is 70 km, and the distance by rail is 60 km. A car and a train leave Richmond at the same time and arrive in Waterbury at the same time, the car having traveled 15 km/h faster than the train. Find the speed of the car and the speed of the train.

46. The Black River's current is 6 mph. A boat travels 50 mi downstream in the same time that it takes to travel 30 mi upstream. What is the speed of the boat in still water?

47. To estimate the harbor seal population in Bristol Bay, scientists radio-tagged 33 seals. Several days later, they collected a sample of 40 seals, and 24 of them were tagged. Estimate the seal population of the bay.

48. Triangles *ABC* and *XYZ* are similar. Find the value of *x*.

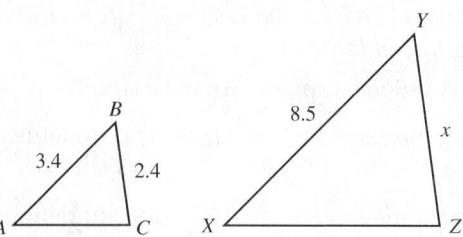

Synthesis

49. For what procedures in this chapter is the LCM of denominators used to clear fractions? [6.5], [6.6]

50. A student always uses the common denominator found by multiplying the denominators of the expressions being added. How could this approach be improved? [6.3]

Simplify.

51. $\dfrac{2a^2 + 5a - 3}{a^2} \cdot \dfrac{5a^3 + 30a^2}{2a^2 + 7a - 4} \div \dfrac{a^2 + 6a}{a^2 + 7a + 12}$ [6.2]

52. $\dfrac{12a}{(a - b)(b - c)} - \dfrac{2a}{(b - a)(c - b)}$ [6.4]

Aha! **53.** $\dfrac{5(x - y)}{(x - y)(x + 2y)} - \dfrac{5(x - 3y)}{(x + 2y)(x - 3y)}$ [6.3]

Test: Chapter 6

Step-by-step test solutions are found on the video CD in the front of this book.

List all numbers for which each expression is undefined.

1. $\dfrac{2 - x}{5x}$

2. $\dfrac{5}{x + 8}$

3. $\dfrac{x - 7}{x^2 - 1}$

4. $\dfrac{x^2 + x - 30}{x^2 - 3x + 2}$

5. Simplify: $\dfrac{6x^2 + 17x + 7}{2x^2 + 7x + 3}$.

Multiply or divide and, if possible, simplify.

6. $\dfrac{t^2 - 9}{12t} \cdot \dfrac{8t^2}{t^2 - 4t + 3}$

7. $\dfrac{25y^2 - 1}{9y^2 - 6y} \div \dfrac{5y^2 + 9y - 2}{3y^2 + y - 2}$

8. $\dfrac{4a^2 + 1}{4a^2 - 1} \div \dfrac{4a^2}{4a^2 + 4a + 1}$

9. $(x^2 + 6x + 9) \cdot \dfrac{(x - 3)^2}{x^2 - 9}$

10. Find the LCM:

$$y^2 - 9, \ y^2 + 10y + 21, \ y^2 + 4y - 21.$$

Add or subtract, and, if possible, simplify.

11. $\dfrac{2 + x}{x^3} + \dfrac{7 - 4x}{x^3}$

12. $\dfrac{5 - t}{t^2 + 1} - \dfrac{t - 3}{t^2 + 1}$

13. $\dfrac{2x - 4}{x - 3} + \dfrac{x - 1}{3 - x}$

14. $\dfrac{2x - 4}{x - 3} - \dfrac{x - 1}{3 - x}$

15. $\dfrac{7}{t-2} + \dfrac{4}{t}$

16. $\dfrac{y}{y^2+6y+9} + \dfrac{1}{y^2+2y-3}$

17. $\dfrac{1}{x-1} + \dfrac{4}{x^2-1} - \dfrac{2}{x^2-2x+1}$

Simplify.

18. $\dfrac{9-\dfrac{1}{y^2}}{3-\dfrac{1}{y}}$

19. $\dfrac{\dfrac{x}{8}-\dfrac{8}{x}}{\dfrac{1}{8}+\dfrac{1}{x}}$

Solve.

20. $\dfrac{1}{t} + \dfrac{1}{3t} = \dfrac{1}{2}$

21. $\dfrac{15}{x} - \dfrac{15}{x-2} = -2$

22. Kopy Kwik has 2 copiers. One can copy a year-end report in 20 min. The other can copy the same document in 30 min. How long would it take both machines, working together, to copy the report?

23. The average 140-lb adult burns about 320 calories walking 4 mi at a moderate speed. How far should the average 140-lb adult walk in order to burn 100 calories?
Source: www.walking.about.com

24. Ryan drives 20 km/h faster than Alicia. In the same time that Alicia drives 225 km, Ryan drives 325 km. Find the speed of each car.

25. Pe'rez and Rema work together to mulch the flower beds around an office complex in $2\frac{6}{7}$ hr. Working alone, it would take Pe'rez 6 hr more than it would take Rema. How long would it take each of them to complete the landscaping working alone?

Synthesis

26. Simplify: $1 - \dfrac{1}{1-\dfrac{1}{1-\dfrac{1}{a}}}$.

27. The square of a number is the opposite of the number's reciprocal. Find the number.

Cumulative Review: Chapters 1–6

1. Use the commutative law of multiplication to write an expression equivalent to $a + bc$. [1.2]

2. Evaluate $-x^2$ for $x = 5$. [1.8]

3. Evaluate $(-x)^2$ for $x = 5$. [1.8]

4. Simplify: $-3[2(x-3) - (x+5)]$. [1.8]

Solve.

5. $5(x-2) = 40$ [2.2]

6. $49 = x^2$ [5.7]

7. $-18n = 30$ [2.1]

8. $4x - 3 = 9x - 11$ [2.2]

9. $4(y-5) = -2(y-2)$ [2.2]

10. $x^2 + 11x + 10 = 0$ [5.7]

11. $\dfrac{4}{9}t + \dfrac{2}{3} = \dfrac{1}{3}t - \dfrac{2}{9}$ [2.2]

12. $\dfrac{4}{x} + x = 5$ [6.6]

13. $6 - y \geq 2y + 8$ [2.6]

14. $\dfrac{2}{x-3} = \dfrac{5}{3x+1}$ [6.6]

15. $2x^2 + 7x = 4$ [5.7]

16. $4(x+7) < 5(x-3)$ [2.6]

17. $\dfrac{t^2}{t+5} = \dfrac{25}{t+5}$ [6.6]

18. $(2x+7)(x-5) = 0$ [5.7]

19. $\dfrac{2}{x^2-9} + \dfrac{5}{x-3} = \dfrac{3}{x+3}$ [6.6]

Solve each formula. [2.3]

20. $3a - b + 9 = c$, for b

21. $\frac{3}{4}(x+2y) = z$, for y

Graph. [3.2], [3.3], [3.6]

22. $y = \frac{3}{4}x + 5$

23. $x = -3$

24. $4x + 5y = 20$

25. $y = 6$

26. Find the slope of the line containing the points $(1, 5)$ and $(2, 3)$. [3.5]

27. Find the slope and the y-intercept of the line given by $2x - 4y = 1$. [3.6]

28. Write the slope–intercept equation of the line with slope $-\frac{5}{8}$ and y-intercept $(0, -4)$. [3.6]

Simplify.

29. $\dfrac{x^{-5}}{x^{-3}}$ [4.2]

30. $y \cdot y^{-8}$ [4.2]

31. $-(2a^2b^7)^2$ [4.1]

32. Subtract: [4.4]

$$(-8y^2 - y + 2) - (y^3 - 6y^2 + y - 5).$$

Multiply.

33. $-5(3a - 2b + c)$ [1.2]

34. $(2x^2 - 1)(x^3 + x - 3)$ [4.5]

35. $(6x - 5y)^2$ [4.6]

36. $(3n + 2)(n - 5)$ [4.6]

37. $(2x^3 + 1)(2x^3 - 1)$ [4.6]

Factor.

38. $6x - 2x^2 - 24x^4$ [5.1]

39. $16x^2 - 81$ [5.4]

40. $10t^3 + 10$ [5.5]

41. $8x^2 + 10x + 3$ [5.3]

42. $6x^2 - 28x + 16$ [5.3]

43. $25t^2 + 40t + 16$ [5.4]

44. $x^2y^2 - xy - 20$ [5.2]

45. $x^4 + 2x^3 - 3x - 6$ [5.1]

Simplify.

46. $\dfrac{4t - 20}{t^2 - 16} \cdot \dfrac{t - 4}{t - 5}$ [6.2]

47. $\dfrac{x^2 - 1}{x^2 - x - 2} \div \dfrac{x - 1}{x - 2}$ [6.2]

48. $\dfrac{5ab}{a^2 - b^2} + \dfrac{a + b}{a - b}$ [6.4]

49. $\dfrac{x + 2}{4 - x} - \dfrac{x + 3}{x - 4}$ [6.4]

50. $\dfrac{1 + \dfrac{2}{x}}{1 - \dfrac{4}{x^2}}$ [6.5]

51. $\dfrac{3y + \dfrac{2}{y}}{y - \dfrac{3}{y^2}}$ [6.5]

Divide. [4.8]

52. $\dfrac{18x^4 - 15x^3 + 6x^2 + 12x + 3}{3x^2}$

53. $(15x^4 - 12x^3 + 6x^2 + 2x + 18) \div (x + 3)$

54. For each order, HanBooks.com charges a shipping fee of $5.50 plus $1.99 per book. The shipping cost for Dae's book order was $35.35. How many books did she order? [2.5]
Source: hanbooks.com

55. In 2006, the attendance at movie theaters was 1448.5 billion. This was 425% more than the number of admissions to theme parks. How many theme park admissions were there in 2006? [2.4]
Source: MPAA, PricewaterhouseCoopers

56. A pair of 1976 two-dollar bills with consecutive serial numbers is being sold in an auction. The sum of the serial numbers is 66,679,015. What are the serial numbers of the bills? [2.5]

57. Nikki is laying out two square flower gardens in a client's lawn. Each side of one garden is 2 ft longer than each side of the smaller garden. Together, the area of the gardens is 340 ft². Find the length of a side of the smaller garden. [5.8]

58. It takes Wes 25 min to file a week's worth of receipts. Corey, a new employee, takes 75 min to do the same job. How long would it take if they worked together? [6.7]

59. A game warden catches, tags, and then releases 18 antelope. A month later, a sample of 30 antelope is caught and released and 6 of them have tags. Use this information to estimate the size of the antelope population in that area. [6.7]

60. Rachel burned 450 calories in a workout. She burned twice as many in her aerobics session as she did doing calisthenics. How many calories did she burn doing calisthenics? [2.5]

Synthesis

61. Solve: $\frac{1}{3}|n| + 8 = 56$. [1.4], [2.2]

Aha! **62.** Multiply: $[4y^3 - (y^2 - 3)][4y^3 + (y^2 - 3)]$. [4.6]

63. Solve: $x(x^2 + 3x - 28) - 12(x^2 + 3x - 28) = 0$. [5.7]

64. Solve: $\dfrac{2}{x - 3} \cdot \dfrac{3}{x + 3} - \dfrac{4}{x^2 - 7x + 12} = 0$. [6.6]

Functions and Graphs

MELANIE CHAMBERS
MATH TEACHER/RECIPIENT OF
THE MILKEN FAMILY
FOUNDATION'S NATIONAL
EDUCATOR AWARD
Cedar Hill, Texas

Being proficient in math not only has multiplied my career options but also opens many doors for those students who choose to sharpen their math skills. As a math instructor, I use math to calculate students' performance on assessments, disaggregate data to differentiate instruction to better meet student learning needs, and demonstrate daily how real-world problem solving often entails mathematical analysis.

AN APPLICATION

According to the National Assessment of Educational Progress (NAEP), the percentage of fourth-graders who are proficient in math has grown from 21% in 1996 to 24% in 2000 and 39% in 2007.

Estimate the percentage of fourth-graders who showed proficiency in 2004 and predict the percentage who will demonstrate proficiency in 2011.

Source: nationsreportcard.gov

This problem appears as Example 6 in Section 7.1.

1 n this chapter we introduce the concept of a *function*. As we will see, many functions are described using linear, polynomial, and rational expressions. Functions can also often be visualized graphically, as well as added, subtracted, multiplied, and divided. Near the end of the chapter, we solve formulas using equation-solving techniques studied earlier, and we use function notation when describing *direct* and *inverse variation*.

7.1 Introduction to Functions

Correspondences and Functions and Equations • Applications • Functions and Graphs • Function Notation

We now develop the idea of a *function*—one of the most important concepts in mathematics.

Correspondences and Functions

When forming ordered pairs to graph equations, we often say that the first coordinate of each ordered pair *corresponds* to the second coordinate. In much the same way, a function is a special kind of correspondence between two sets. For example,

To each person in a class	there corresponds	a date of birth.
To each bar code in a store	there corresponds	a price.
To each real number	there corresponds	the cube of that number.

In each example, the first set is called the **domain**. The second set is called the **range**. For any member of the domain, there is *exactly one* member of the range to which it corresponds. This kind of correspondence is called a **function**.

EXAMPLE **1** Determine whether each correspondence is a function.

STUDENT NOTES

Note that not all correspondences are functions.

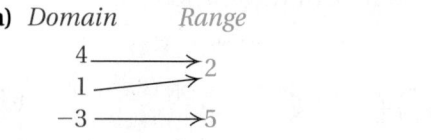

a) *Domain*   *Range*

b) *Domain*   *Range*

SOLUTION

a) The correspondence *is* a function because each member of the domain corresponds to *exactly one* member of the range.

b) The correspondence *is not* a function because a member of the domain (General Motors) corresponds to more than one member of the range.

TRY EXERCISE 9

> **Function**
>
> A *function* is a correspondence between a first set, called the *domain*, and a second set, called the *range*, such that each member of the domain corresponds to *exactly one* member of the range.

EXAMPLE   2   Determine whether each correspondence is a function.

Domain	*Correspondence*	*Range*
a) People in a doctor's waiting room	Each person's weight	A set of positive numbers
b) $\{-2, 0, 1, 2\}$	Each number's square	$\{0, 1, 4\}$
c) Authors of best-selling books	The titles of books written by each author	A set of book titles

SOLUTION

a) The correspondence *is* a function, because each person has *only one* weight.

b) The correspondence *is* a function, because every number has *only one* square.

c) The correspondence *is not* a function, because some authors have written *more than one* book.

▶ TRY EXERCISE   17

Although the correspondence in Example 2(c) is not a function, it is a *relation*.

> **Relation**
>
> A *relation* is a correspondence between a first set, called the *domain*, and a second set, called the *range*, such that each member of the domain corresponds to *at least one* member of the range.

Functions and Graphs

The functions in Examples 1(a) and 2(b) can be expressed as sets of ordered pairs. Example 1(a) can be written $\{(-3, 5), (1, 2), (4, 2)\}$ and Example 2(b) can be written $\{(-2, 4), (0, 0), (1, 1), (2, 4)\}$. We can graph these functions as follows.

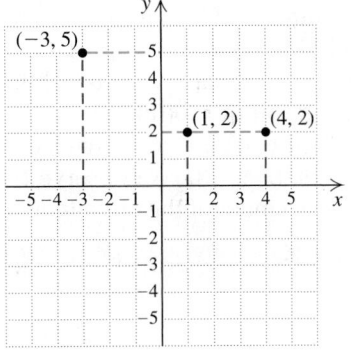

The function $\{(-3, 5), (1, 2), (4, 2)\}$
Domain is $\{-3, 1, 4\}$
Range is $\{5, 2\}$

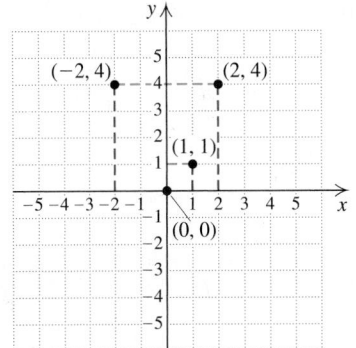

The function $\{(-2, 4), (0, 0), (1, 1), (2, 4)\}$
Domain is $\{-2, 0, 1, 2\}$
Range is $\{4, 0, 1\}$

We can find the domain and the range of a function directly from its graph. The domain is read from the horizontal axis and the range is read from the vertical axis. Note in the graphs above that if we move along the red dashed lines from the points to the horizontal axis, we find the members, or elements, of the domain. Similarly, if we move along the blue dashed lines from the points to the vertical axis, we find the elements of the range.

Functions are generally named using lowercase or uppercase letters. The function in the following example is named *f*.

EXAMPLE **3** For the function *f* represented below, determine each of the following.

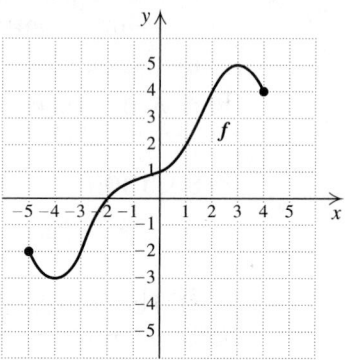

a) The member of the range that is paired with 2

b) The member of the domain that is paired with −3

SOLUTION

a) To determine what member of the range is paired with 2, first note that we are considering 2 in the domain. Thus we locate 2 on the horizontal axis. Next, we find the point directly above 2 on the graph of *f*. From that point, we can look to the vertical axis to find the corresponding *y*-coordinate, 4. Thus, 4 is the member of the range that is paired with 2.

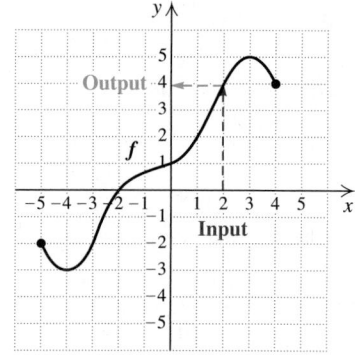

b) To determine what member of the domain is paired with −3, we note that we are considering −3 in the range. Thus we locate −3 on the vertical axis. From there we look at the graph of *f* to find any points for which −3 is the second coordinate. One such point exists, $(-4, -3)$. We observe that −4 is the only element of the domain paired with −3.

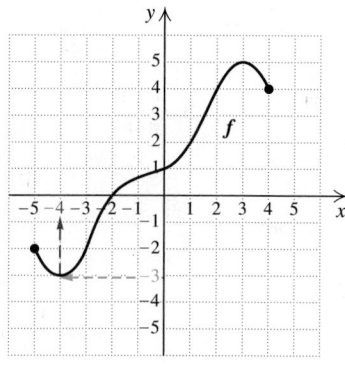

A closed dot on a graph, such as in Example 3, indicates that the point is part of the function. An open dot indicates that the point is *not* part of the function. (See Exercises 33 and 34 on p. 452.)

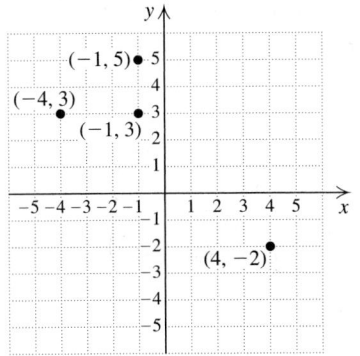

Recall that a function is a correspondence in which each member of the domain corresponds to *exactly* one member of the range. Thus the correspondence

$$\{(-1, 3), (4, -2), (-4, 3), (-1, 5)\}$$

is not a function because the member -1 of the domain corresponds to the members 3 and 5 of the range. Note on the graph at left that the point $(-1, 5)$ is directly above the point $(-1, 3)$.

Any time two points, such as $(-1, 3)$ and $(-1, 5)$, lie on the same vertical line, the graph containing those points cannot represent a function. This observation is the basis of the *vertical-line test*.

The Vertical-Line Test

If it is possible for a vertical line to cross a graph more than once, then the graph is not the graph of a function.

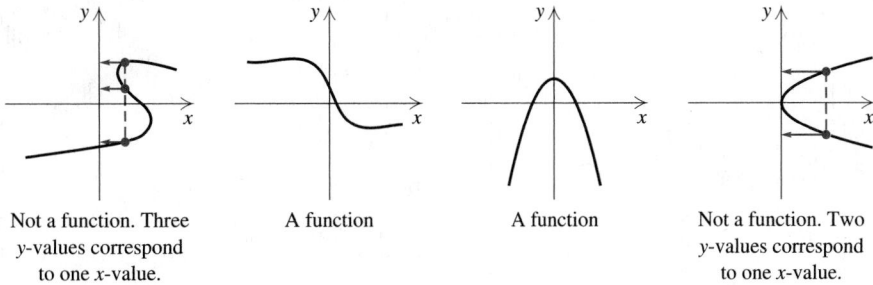

| Not a function. Three y-values correspond to one x-value. | A function | A function | Not a function. Two y-values correspond to one x-value. |

Function Notation and Equations

To understand function notation, it helps to imagine a "function machine." Think of putting a member of the domain (an *input*) into the machine. The machine is programmed to produce the appropriate member of the range (the *output*).

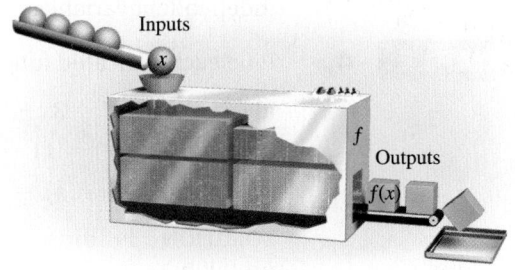

The function pictured has been named f. Here x represents an input, and $f(x)$ represents the corresponding output. In function notation, "$f(x)$" is read "f of x," or "f at x," or "the value of f at x." In Example 3(a), we showed that $f(2) = 4$, read "f of 2 equals 4."

CAUTION! $f(x)$ ***does not*** mean f times x.

Most functions are described by equations. For example, $f(x) = 2x + 3$ describes the function that takes an input x, multiplies it by 2, and then adds 3.

$$f(x) = \underset{\text{Double}}{2x} \quad \underset{\text{Add 3}}{+\ 3}$$
$\overset{\text{Input}}{}$

To calculate the output $f(4)$, we take the input 4, double it, and add 3 to get 11. That is, we substitute 4 into the formula for $f(x)$:

$$f(4) = 2 \cdot 4 + 3$$
$$= 11.$$

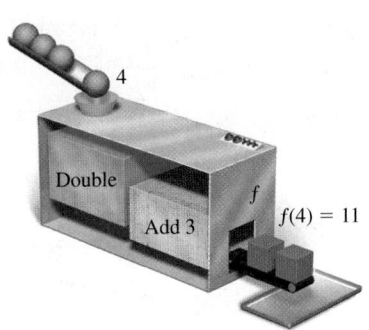

Sometimes, in place of $f(x) = 2x + 3$, we write $y = 2x + 3$, where it is understood that the value of y, the *dependent variable*, depends on our choice of x, the *independent variable*. To understand why $f(x)$ notation is so useful, consider two equivalent statements:

a) If $f(x) = 2x + 3$, then $f(4) = 11$.

b) If $\ y\ = 2x + 3$, then the value of y is 11 when x is 4.

The notation used in part (a) is far more concise and emphasizes that x is the independent variable.

EXAMPLE **4**   Find each indicated function value.

a) $f(5)$, for $f(x) = 3x + 2$ **b)** $g(-2)$, for $g(r) = 5r^2 + 3r$

c) $h(4)$, for $h(x) = 11$ **d)** $F(a) + 1$, for $F(x) = 2x - 7$

e) $F(a + 1)$, for $F(x) = 2x - 7$

SOLUTION   Finding function values is much like evaluating an algebraic expression.

a) $f(5) = 3(5) + 2 = 17$

b) $g(-2) = 5(-2)^2 + 3(-2)$
$$= 5 \cdot 4 - 6 = 14$$

c) For the function given by $h(x) = 11$, all inputs share the same output, 11. Therefore, $h(4) = 11$. The function h is an example of a *constant function*.

d) $F(a) + 1 = 2(a) - 7 + 1$ The input is a; $F(a) = 2a - 7$
$$= 2a - 6$$

e) $F(a + 1) = 2(a + 1) - 7$ The input is $a + 1$.
$$= 2a + 2 - 7 = 2a - 5$$

TRY EXERCISE 41

STUDENT NOTES

In Example 4(e), it is important to note that the parentheses on the left are for function notation, whereas those on the right indicate multiplication.

Note that whether we write $f(x) = 3x + 2$, or $f(t) = 3t + 2$, or $f(\blacksquare) = 3\blacksquare + 2$, we still have $f(5) = 17$. The variable in the parentheses (the independent variable) is the variable used in the algebraic expression. The letter chosen for the independent variable is not as important as the algebraic manipulations to which it is subjected.

Applications

Function notation is often used in formulas. For example, to emphasize that the area A of a circle is a function of its radius r, instead of

$$A = \pi r^2,$$

we can write

$$A(r) = \pi r^2.$$

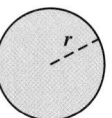

EXAMPLE 5

A typical adult dosage of an antihistamine is 24 mg. Young's rule for determining the dosage size $c(a)$ for a typical child of age a is

$$c(a) = \frac{24a}{a + 12}.^*$$

What should the dosage be for a typical 8-yr-old child?

SOLUTION   We find $c(8)$:

$$c(8) = \frac{24(8)}{8 + 12} = \frac{192}{20} = 9.6.$$

The dosage for a typical 8-yr-old child is 9.6 mg.

▸ **TRY EXERCISE** ▸ 47

When a function is given as a graph, we can use the graph to estimate an unknown function value using known values.

EXAMPLE 6

Elementary school math proficiency.   According to the National Assessment of Educational Progress (NAEP), the percentage of fourth-graders who are proficient in math has grown from 21% in 1996 to 24% in 2000 and 39% in 2007. Estimate the percentage of fourth-graders who showed proficiency in 2004 and predict the percentage who will demonstrate proficiency in 2011.

Source: nationsreportcard.gov

SOLUTION

1., 2. Familiarize., Translate.   The given information enables us to plot and connect three points. We let the horizontal axis represent the year and the vertical axis the percentage of fourth-graders demonstrating mathematical proficiency. We label the function itself P.

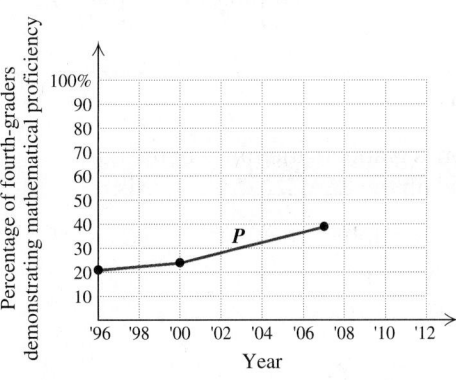

---

*Source: Olsen, June Looby, Leon J. Ablon, and Anthony Patrick Giangrasso, *Medical Dosage Calculations*, 6th ed.

3. Carry out. To estimate the percentage of fourth-graders showing mathematical proficiency in 2004, we locate the point directly above the year 2004. We then estimate its second coordinate by moving horizontally from that point to the *y*-axis. Although our result is not exact, we see that $P(2004) \approx 33$.

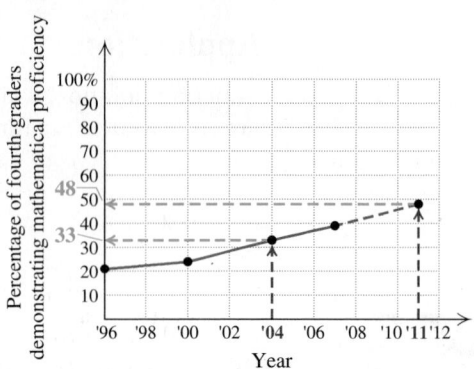

To predict the percentage of fourth-graders showing proficiency in 2011, we extend the graph and extrapolate. It appears that $P(2011) \approx 48$.

4. Check. A precise check requires consulting an outside information source. Since 33% is between 24% and 39% and 48% is greater than 39%, our estimates seem plausible.

5. State. In 2004, about 33% of all fourth-graders showed proficiency in math. By 2011, that figure is predicted to grow to 48%.

> TRY EXERCISE 59

7.1 EXERCISE SET

Concept Reinforcement *Complete each of the following sentences.*

1. A function is a special kind of _____ between two sets.

2. In any function, each member of the domain is paired with _____ one member of the range.

3. For any function, the set of all inputs, or first values, is called the _____.

4. For any function, the set of all outputs, or second values, is called the _____.

5. When a function is graphed, members of the domain are located on the _____ axis.

6. When a function is graphed, members of the range are located on the _____ axis.

7. The notation $f(3)$ is read _____.

8. The _____-line test can be used to determine whether or not a graph represents a function.

Determine whether each correspondence is a function.

9.    **10.**

11. *Girl's age*   *Average daily*
        (in months)   *weight gain (in grams)*

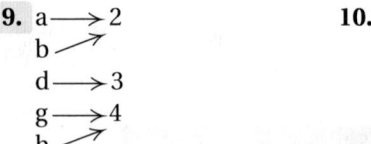

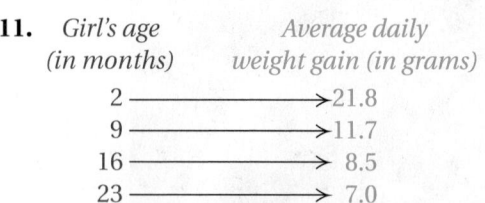

Source: *American Family Physician,* December 1993, p. 1435

12.

Boy's age (in months)	Average daily weight gain (in grams)
2	24.3
9	11.7
16	8.2
23	7.0

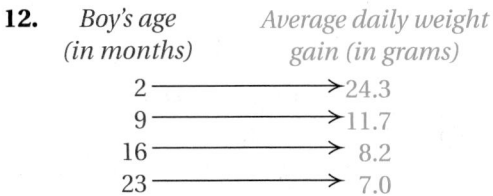

Source: *American Family Physician*, December 1993, p. 1435

13. *Birthday*   *Celebrity*

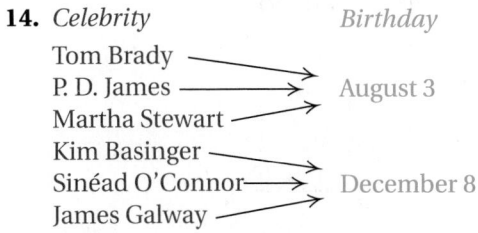

Johnny Depp
June 9   Michael J. Fox
Amanda Lassiter

Michael Andretti
October 5   Chester A. Arthur
Kate Winslet

Source: www.leannesbirthdays.com

14. *Celebrity*   *Birthday*

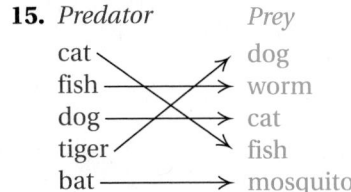

Tom Brady
P. D. James   August 3
Martha Stewart

Kim Basinger
Sinéad O'Connor   December 8
James Galway

Source: www.leannesbirthdays.com

15. *Predator*   *Prey*

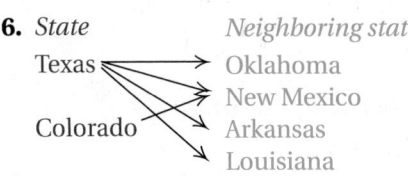

cat   dog
fish   worm
dog   cat
tiger   fish
bat   mosquito

16. *State*   *Neighboring state*

Texas   Oklahoma
     New Mexico
Colorado   Arkansas
     Louisiana

Determine whether each of the following is a function. Identify any relations that are not functions.

	Domain	Correspondence	Range
17.	A pile of USB flash drives	The storage capacity of each flash drive	A set of storage capacities
18.	The members of a rock band	An instrument the person can play	A set of instruments
19.	The players on a team	The uniform number of each player	A set of numbers
20.	A set of triangles	The area of each triangle	A set of numbers

*For each graph of a function, determine **(a)** $f(1)$ and **(b)** any x-values for which $f(x) = 2$.*

21.
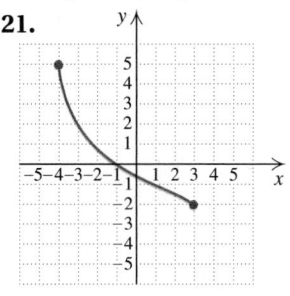

22.
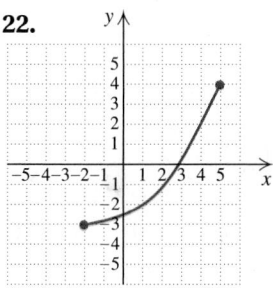

23.
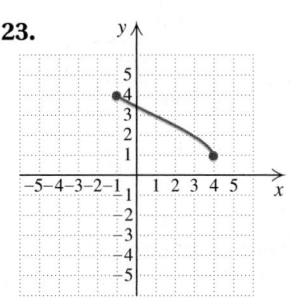

24.
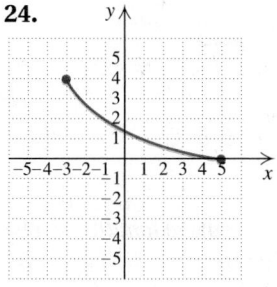

25.
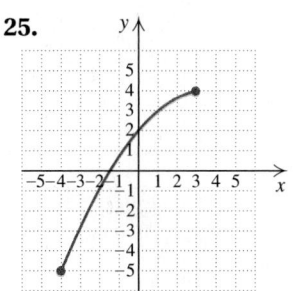

26.
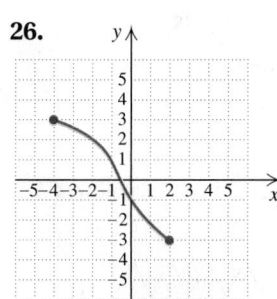

27.
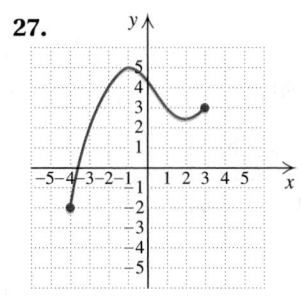

28.
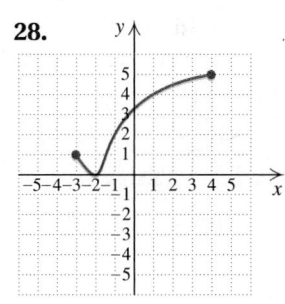

29.
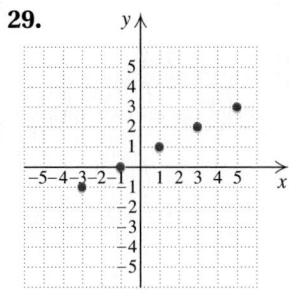

30.
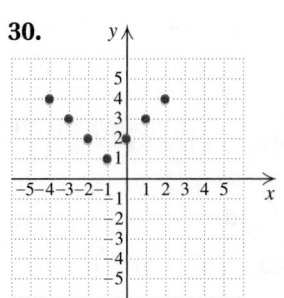

31.

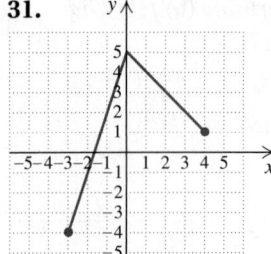

32.

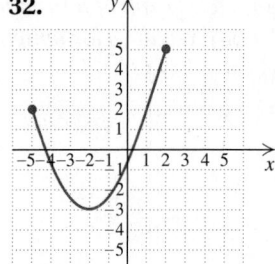

33.

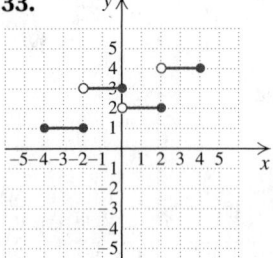

34.

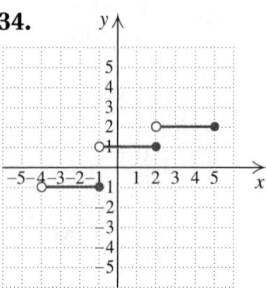

Determine whether each of the following is the graph of a function.

35.

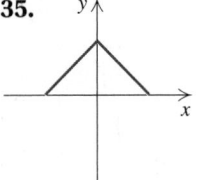

36.

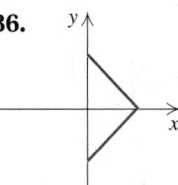

37.

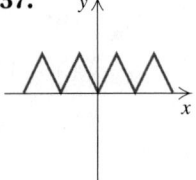

38.

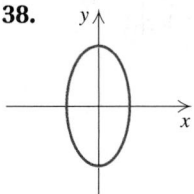

39.

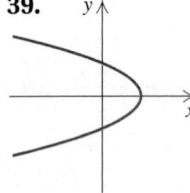

40.

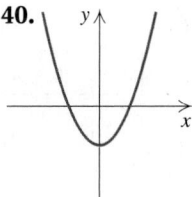

Find the function values.

41. $g(x) = 2x + 5$

  a) $g(0)$ **b)** $g(-4)$ **c)** $g(-7)$
  d) $g(8)$ **e)** $g(a + 2)$ **f)** $g(a) + 2$

42. $h(x) = 5x - 1$

  a) $h(4)$ **b)** $h(8)$ **c)** $h(-3)$
  d) $h(-4)$ **e)** $h(a - 1)$ **f)** $h(a) + 3$

43. $f(n) = 5n^2 + 4n$

  a) $f(0)$ **b)** $f(-1)$ **c)** $f(3)$
  d) $f(t)$ **e)** $f(2a)$ **f)** $f(3) - 9$

44. $g(n) = 3n^2 - 2n$

  a) $g(0)$ **b)** $g(-1)$ **c)** $g(3)$
  d) $g(t)$ **e)** $g(2a)$ **f)** $g(3) - 4$

45. $f(x) = \dfrac{x - 3}{2x - 5}$

  a) $f(0)$ **b)** $f(4)$ **c)** $f(-1)$
  d) $f(3)$ **e)** $f(x + 2)$ **f)** $f(a + h)$

46. $r(x) = \dfrac{3x - 4}{2x + 5}$

  a) $r(0)$ **b)** $r(2)$ **c)** $r\left(\frac{4}{3}\right)$
  d) $r(-1)$ **e)** $r(x + 3)$ **f)** $r(a + h)$

The function A described by $A(s) = s^2 \dfrac{\sqrt{3}}{4}$ gives the area of an equilateral triangle with side s.

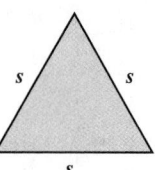

47. Find the area when a side measures 4 cm.

48. Find the area when a side measures 6 in.

The function V described by $V(r) = 4\pi r^2$ gives the surface area of a sphere with radius r.

49. Find the surface area when the radius is 3 in.

50. Find the surface area when the radius is 5 cm.

Archaeology. The function H described by

$$H(x) = 2.75x + 71.48$$

can be used to predict the height, in centimeters, of a woman whose humerus (the bone from the elbow to the shoulder) is x cm long. Predict the height of a woman whose humerus is the length given.

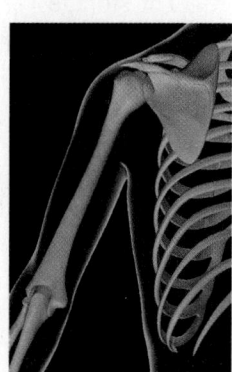

51. 34 cm   **52.** 31 cm

Chemistry.   *The function F described by*

$$F(C) = \tfrac{9}{5}C + 32$$

gives the Fahrenheit temperature corresponding to the Celsius temperature C.

53. Find the Fahrenheit temperature equivalent to −5° Celsius.

54. Find the Fahrenheit temperature equivalent to 10° Celsius.

Heart attacks and cholesterol.   *For Exercises 55 and 56, use the following graph, which shows the annual heart attack rate per 10,000 men as a function of blood cholesterol level.**

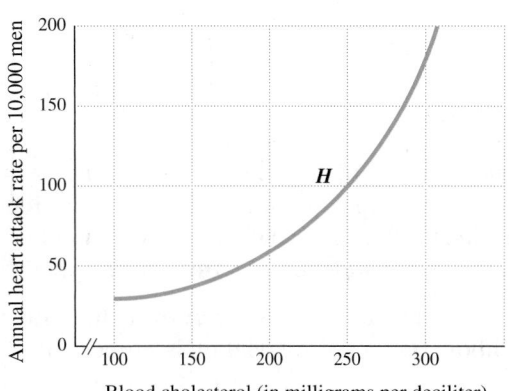

55. Approximate the annual heart attack rate for those men whose blood cholesterol level is 225 mg/dl. That is, find H(225).

56. Approximate the annual heart attack rate for those men whose blood cholesterol level is 275 mg/dl. That is, find H(275).

Films.   *For Exercises 57 and 58, use the following graph, which shows the number of movies released in the United States.*
Source: Nash Information Services

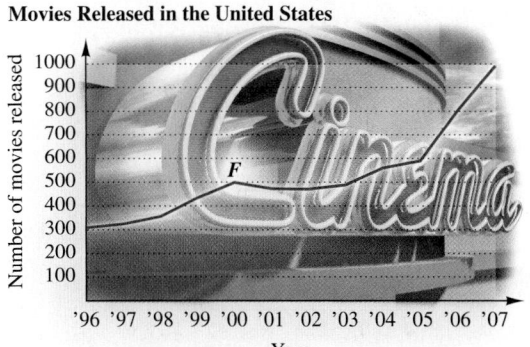

*Copyright 1989, CSPI. Adapted from *Nutrition Action Health-letter* (1875 Connecticut Avenue, N.W., Suite 300, Washington, DC 20009-5728. $24 for 10 issues).

57. Approximate the number of movies released in 2000. That is, find F(2000).

58. Approximate the number of movies released in 2007. That is, find F(2007).

Energy-saving lightbulbs.   *An energy bill signed into law in 2007 requires the United States to phase out standard incandescent lightbulbs. A more efficient replacement is the compact fluorescent (CFL) bulb. The table below lists incandescent wattage and the CFL wattage required to create the same amount of light.*
Source: U.S. Department of Energy

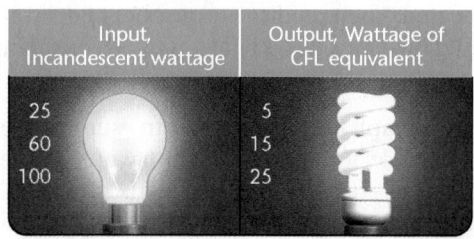

59. Use the data in the figure above to draw a graph. Estimate the wattage of a CFL bulb that creates light equivalent to a 75-watt incandescent bulb. Then predict the wattage of a CFL bulb that creates light equivalent to a 120-watt incandescent bulb.

60. Use the graph from Exercise 59 to estimate the wattage of a CFL bulb that creates light equivalent to a 40-watt incandescent bulb. Then predict the wattage of a CFL bulb that creates light equivalent to a 150-watt incandescent bulb.

Blood alcohol level.   *The following table can be used to predict the number of drinks required for a person of a specified weight to be considered legally intoxicated (blood alcohol level of 0.08 or above). One 12-oz glass of beer, a 5-oz glass of wine, or a cocktail containing 1 oz of a distilled liquor all count as one drink. Assume that all drinks are consumed within one hour.*

Input, Body Weight (in pounds)	Output, Number of Drinks
100	2.5
160	4
180	4.5
200	5

61. Use the data in the table above to draw a graph and to estimate the number of drinks that a 140-lb person would have to drink to be considered intoxicated. Then predict the number of drinks it would take for a 230-lb person to be considered intoxicated.

62. Use the graph from Exercise 61 to estimate the number of drinks that a 120-lb person would have to drink to be considered intoxicated. Then predict the number of drinks it would take for a 250-lb person to be considered intoxicated.

63. *Retailing.* Mountain View Gifts is experiencing constant growth. They recorded a total of $250,000 in sales in 2003 and $285,000 in 2008. Use a graph that displays the store's total sales as a function of time to estimate sales for 2004 and for 2011.

64. Use the graph in Exercise 63 to estimate sales for 2006 and for 2012.

Researchers at Yale University have suggested that the following graphs* may represent three different aspects of love.

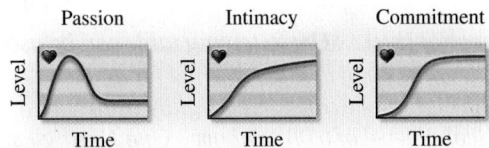

Passion   Intimacy   Commitment

65. In what unit would you measure time if the horizontal length of each graph were ten units? Why?

66. Do you agree with the researchers that these graphs should be shaped as they are? Why or why not?

---

*From "A Triangular Theory of Love," by R. J. Sternberg, 1986, *Psychological Review,* **93**(2), 119–135. Copyright 1986 by the American Psychological Association, Inc. Reprinted by permission.

Skill Review

Review solving equations (Sections 2.2, 5.7, and 6.6).

Solve.

67. $2(x - 5) - 3 = 4 - (x - 1)$ [2.2]

68. $x^2 = 36$ [5.7]

69. $\dfrac{1}{x} = -2$ [6.6]

70. $\dfrac{1}{x} = x$ [6.6]

71. $(x - 2)(x + 3) = 6$ [5.7]

72. $\dfrac{1}{3}x + 2 = \dfrac{5}{4} + 3x$ [2.2]

73. $\dfrac{x + 1}{x} = 8$ [6.6]

74. $(x - 2)^2 = 36$ [5.7]

Synthesis

75. Jaylan is asked to write a function relating the number of fish in an aquarium to the amount of food needed for the fish. Which quantity should he choose as the independent variable? Why?

76. Explain in your own words why every function is a relation, but not every relation is a function.

For Exercises 77 and 78, let $f(x) = 3x^2 - 1$ and $g(x) = 2x + 5$.

77. Find $f(g(-4))$ and $g(f(-4))$.

78. Find $f(g(-1))$ and $g(f(-1))$.

79. If f represents the function in Exercise 15, find $f(f(f(f(\text{tiger}))))$.

Pregnancy. *For Exercises 80–83, use the following graph of a woman's "stress test." This graph shows the size of a pregnant woman's contractions as a function of time.*

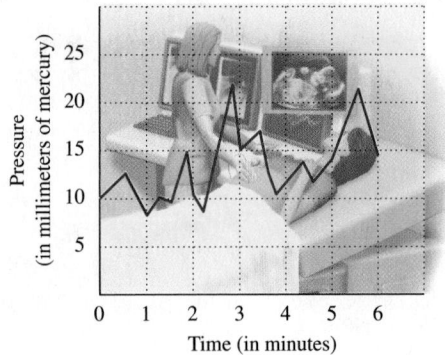

Stress test

80. How large is the largest contraction that occurred during the test?

81. At what time during the test did the largest contraction occur?

82. On the basis of the information provided, how large a contraction would you expect 60 seconds after the end of the test? Why?

83. What is the frequency of the largest contraction?

84. Suppose that a function g is such that $g(-1) = -7$ and $g(3) = 8$. Find a formula for g if $g(x)$ is of the form $g(x) = mx + b$, where m and b are constants.

85. The *greatest integer function* $f(x) = [\![x]\!]$ is defined as follows: $[\![x]\!]$ is the greatest integer that is less than or equal to x. For example, if $x = 3.74$, then $[\![x]\!] = 3$; and if $x = -0.98$, then $[\![x]\!] = -1$. Graph the greatest integer function for $-5 \le x \le 5$. (The notation $f(x) = \text{INT}(x)$ is used in many graphing calculators and computer programs.)

86. *Energy expenditure.*   On the basis of the information given below, what burns more energy: walking $4\frac{1}{2}$ mph for two hours or bicycling 14 mph for one hour?

Approximate Energy Expenditure by a 150-Pound Person in Various Activities

Activity	Calories per Hour
Walking, $2\frac{1}{2}$ mph	210
Bicycling, $5\frac{1}{2}$ mph	210
Walking, $3\frac{3}{4}$ mph	300
Bicycling, 13 mph	660

Source: Based on material prepared by Robert E. Johnson, M.D., Ph.D., and colleagues, University of Illinois.

7.2   Domain and Range

Determining the Domain and the Range   ▪   Restrictions on Domain   ▪   Functions Defined Piecewise

In Section 7.1, we saw that a function is a correspondence from a set called the *domain* to a set called the *range*. In this section, we look more closely at the concepts of domain and range.

Determining the Domain and the Range

When a function is given as a set of ordered pairs, the domain is the set of all first coordinates and the range is the set of all second coordinates.

EXAMPLE 1   Find the domain and the range for the function f given by

$$f = \{(2, 0), (-1, 5), (8, 0), (-3, 2)\}.$$

SOLUTION   The first coordinates are $2, -1, 8$, and -3. The second coordinates are $0, 5$, and 2. Thus we have

Domain of $f = \{2, -1, 8, -3\}$ and

Range of $f = \{0, 5, 2\}$.

> TRY EXERCISE   7

We can also determine the domain and the range of a function from its graph.

EXAMPLE **2**

Find the domain and the range of the function *f* below.

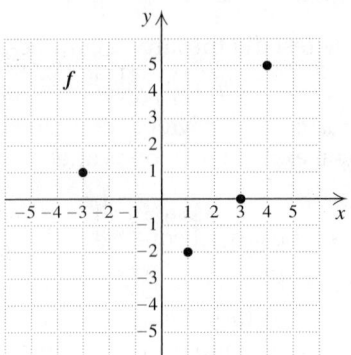

SOLUTION Here *f* can be written $\{(-3, 1), (1, -2), (3, 0), (4, 5)\}$. The domain is the set of all first coordinates, $\{-3, 1, 3, 4\}$, and the range is the set of all second coordinates, $\{1, -2, 0, 5\}$.

▶ **TRY EXERCISE** 11

In Example 2, we could also have found the domain and the range directly, without first writing *f*, by observing the *x*- and *y*-values used in the graph.

EXAMPLE **3**

Find the domain and the range of the function *f* shown here.

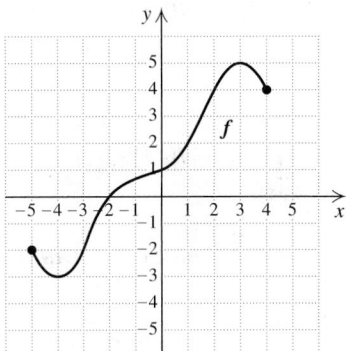

SOLUTION The domain of the function is the set of all *x*-values that are used in the points on the curve. Because there are no breaks in the graph of *f*, these extend continuously from −5 to 4 and can be viewed as the curve's shadow, or *projection*, on the *x*-axis. Thus the domain is $\{x \mid -5 \le x \le 4\}$, or $[-5, 4]$, shown below left.

The range of the function is the set of all *y*-values that are used in the points on the curve. These extend continuously from −3 to 5, and can be viewed as the curve's projection on the *y*-axis. Thus the range is $\{y \mid -3 \le y \le 5\}$, or $[-3, 5]$, shown below right.

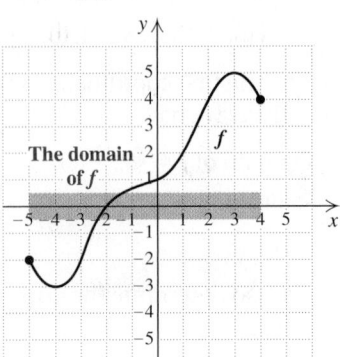

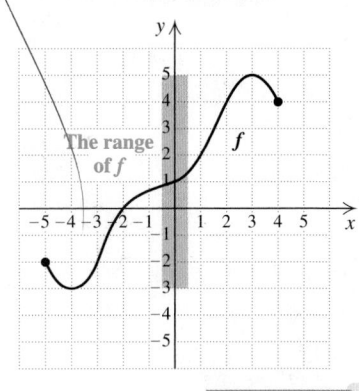

▶ **TRY EXERCISE** 15

In Example 3, the *endpoints* $(-5, -2)$ and $(4, 4)$ emphasize that the function is not defined for values of x less than -5 or greater than 4.

The graphs of some functions have no endpoints. Thus a function may have a domain and/or a range that extends without bound toward positive infinity or negative infinity.

EXAMPLE 4   For the function g represented below, determine **(a)** the domain of g and **(b)** the range of g.

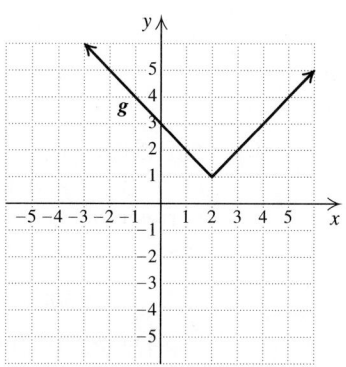

SOLUTION

a) The domain of g is the set of all x-values that are used in the points on the curve. The arrows on the ends of the graph indicate that it extends both left and right without end. Thus the shadow, or projection, of the graph on the x-axis is the entire x-axis. (See the graph on the left below.) The domain is $\{x \,|\, x \text{ is a real number}\}$, or $(-\infty, \infty)$.

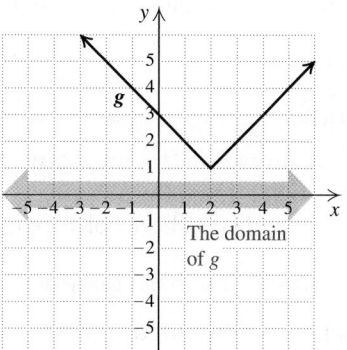

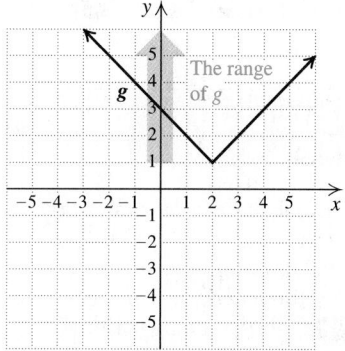

b) The range of g is the set of all y-values that are used in the points on the curve. The arrows on the ends of the graph indicate that it extends up without end. Thus the projection of the graph on the y-axis is the portion of the y-axis greater than or equal to 1. (See the graph on the right above.) The range is $\{y \,|\, y \geq 1\}$, or $[1, \infty)$.

▶ TRY EXERCISE ▶ 25

The set of all real numbers is often abbreviated R. Thus, in Example 4, we could write

Domain of g = R.

EXAMPLE   5   Find the domain and the range of the function f shown here.

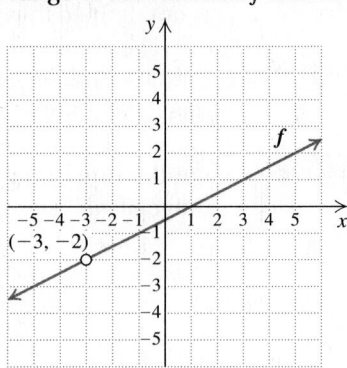

SOLUTION   The domain of f is the set of all x-values that are used in points on the curve. The open dot in the graph at $(-3, -2)$ indicates that there is no y-value that corresponds to $x = -3$; that is, the function is not defined for $x = -3$. Thus, -3 is not in the domain of the function, and

> Domain of $f = \{x \mid x$ is a real number *and* $x \neq -3\}$.

There is no function value at $(-3, -2)$, so -2 is not in the range of the function. Thus we have

> Range of $f = \{y \mid y$ is a real number *and* $y \neq -2\}$.   **TRY EXERCISE** ▸ 27

When a function is described by an equation, we assume that the domain is the set of all real numbers for which function values can be calculated. If an x-value is not in the domain of a function, the graph of the function will not include any point above or below that x-value.

EXAMPLE   6   For each equation, determine the domain of f.

a) $f(x) = |x|$ **b)** $f(x) = \dfrac{7}{2x - 6}$ **c)** $f(t) = \dfrac{t + 1}{t^2 - 4}$

SOLUTION

a) We ask ourselves, "Is there any number x for which we cannot compute $|x|$?" Since we can find the absolute value of *any* number, the answer is no. Thus the domain of f is $\mathbb{R}$, the set of all real numbers.

b) Is there any number x for which $\dfrac{7}{2x - 6}$ cannot be computed? Since $\dfrac{7}{2x - 6}$ cannot be computed when $2x - 6$ is 0, the answer is yes. To determine what x-value causes the denominator to be 0, we solve an equation:

$$2x - 6 = 0 \qquad \text{Setting the denominator equal to 0}$$
$$2x = 6 \qquad \text{Adding 6 to both sides}$$
$$x = 3. \qquad \text{Dividing both sides by 2}$$

Thus, 3 is *not* in the domain of f, whereas all other real numbers are. The domain of f is $\{x \mid x$ is a real number *and* $x \neq 3\}$.

c) The expression $\dfrac{t + 1}{t^2 - 4}$ is undefined when $t^2 - 4 = 0$:

$$t^2 - 4 = 0 \qquad \text{Setting the denominator equal to 0}$$
$$(t + 2)(t - 2) = 0 \qquad \text{Factoring}$$
$$t + 2 = 0 \quad \text{or} \quad t - 2 = 0 \qquad \begin{array}{l}\text{Using the principle of}\\\text{zero products}\end{array}$$
$$t = -2 \quad \text{or} \quad t = 2. \qquad \begin{array}{l}\text{Solving; these are the values for}\\\text{which } (t + 1)/(t^2 - 4) \text{ is undefined.}\end{array}$$

TECHNOLOGY CONNECTION

To visualize Example 6(a), note that the graph of $y_1 = |x|$ (which is entered $y_1 = \text{abs}(x)$, using the NUM option of the MATH menu) appears without interruption for any piece of the x-axis that we examine.

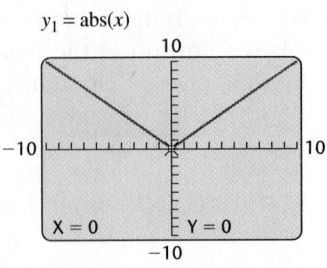

In contrast, the graph of $y_2 = \dfrac{7}{2x - 6}$ in Example 6(b) has a break at $x = 3$.

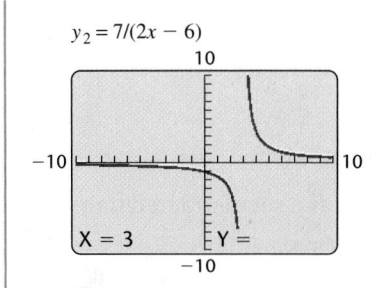

$y_2 = 7/(2x - 6)$

Thus we have

> Domain of $f = \{t \,|\, t$ is a real number *and* $t \neq -2$ *and* $t \neq 2\}$.

Note that when the numerator, $t + 1$, is zero, the function value is 0 and *is* defined.

TRY EXERCISE ▶ 31

Restrictions on Domain

If a function is used as a model for an application, the problem situation may require restrictions on the domain; for example, length and time are generally nonnegative, and a person's age does not increase indefinitely.

EXAMPLE 7   *Prize tee shirts.*   During intermission at sporting events, it has become common for team mascots to use a powerful slingshot to launch tightly rolled tee shirts into the stands. The height $h(t)$, in feet, of an airborne tee shirt t seconds after being launched can be approximated by

$$h(t) = -15t^2 + 70t + 25.$$

What is the domain of the function?

SOLUTION   The expression $-15t^2 + 70t + 25$ can be evaluated for any number t, so any restrictions on the domain will come from the problem situation.

First, we note that t cannot be negative, since it represents time from launch, so we have $t \geq 0$. If we make a drawing, we also note that the function will not be defined for values of t that make the height negative.

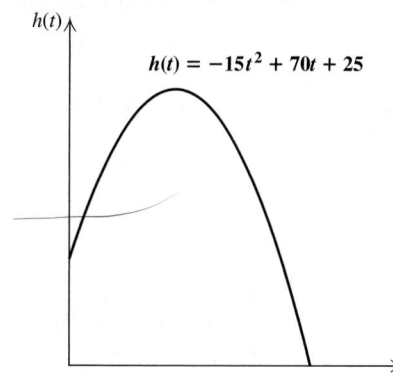

$h(t) = -15t^2 + 70t + 25$

Thus an upper limit for t will be the positive value of t for which $h(t) = 0$. Solving, we obtain

$$
\begin{aligned}
h(t) &= 0 \\
-15t^2 + 70t + 25 &= 0 && \text{Substituting} \\
\left.\begin{array}{l} -5(3t^2 - 14t - 5) = 0 \\ -5(3t + 1)(t - 5) = 0 \end{array}\right\} && \text{Factoring} \\
3t + 1 = 0 \quad or \quad t - 5 &= 0 && \text{Using the principle of zero products} \\
\left.\begin{array}{lcl} 3t = -1 & or & t = 5 \\ t = -\dfrac{1}{3} & or & t = 5. \end{array}\right\} && \text{Solving for } t
\end{aligned}
$$

We already know that $-\frac{1}{3}$ is not in the domain of the function because of the restriction $t \geq 0$ above.

The tee shirt will hit the ground after 5 sec, so we have $t \leq 5$. Putting the two restrictions together, we have $t \geq 0$ *and* $t \leq 5$, so the

> Domain of $h = \{t \,|\, t$ is a real number *and* $0 \leq t \leq 5\}$, or $[0, 5]$.

TRY EXERCISE ▶ 45

If the domain of a function is not specifically listed, it can be determined from a table, a graph, an equation, or an application.

Domain of a Function

The domain of a function $f(x)$ is the set of all inputs x.

- If the correspondence is listed in a table or as a set of ordered pairs, the domain is the set of all first coordinates.
- If the function is described by a graph, the domain is the set of all x-coordinates of the points on the graph.
- If the function is described by an equation, the domain is the set of all numbers for which the value can be calculated.
- If the function is used in an application, the domain is the set of all numbers that make sense in the problem.

Functions Defined Piecewise

Piecewise-defined functions are described by different equations for various parts of their domains. For example, the function $f(x) = |x|$ is described by

$$f(x) = \begin{cases} x, & \text{if } x \geq 0, \\ -x, & \text{if } x < 0. \end{cases}$$

To evaluate a piecewise-defined function for an input a, we first determine what part of the domain a belongs to. Then we use the formula corresponding to that part of the domain.

EXAMPLE **8** Find each function value for the function f given by

$$f(x) = |x| = \begin{cases} x, & \text{if } x \geq 0, \\ -x, & \text{if } x < 0. \end{cases}$$

a) $f(4)$ **b)** $f(-10)$

SOLUTION

a) Since $4 \geq 0$, we use the equation $f(x) = x$. Thus, $f(4) = 4$.

b) Since $-10 < 0$, we use the equation $f(x) = -x$. Thus, $f(-10) = -(-10) = 10$.

> **TRY EXERCISE** 53

EXAMPLE **9** Find each function value for the function g given by

$$g(x) = \begin{cases} x + 2, & \text{if } x \leq -2, \\ x^2, & \text{if } -2 < x \leq 5, \\ 3x, & \text{if } x > 5. \end{cases}$$

a) $g(-2)$ **b)** $g(3)$ **c)** $g(10)$

SOLUTION

a) Since $-2 \leq -2$, we use the first equation, $g(x) = x + 2$:

$$g(-2) = -2 + 2 = 0.$$

b) Since $-2 < 3 \leq 5$, we use the second equation, $g(x) = x^2$:

$$g(3) = 3^2 = 9.$$

c) Since $10 > 5$, we use the last equation, $g(x) = 3x$:

$$g(10) = 3 \cdot 10 = 30.$$

> **TRY EXERCISE** 55

7.2   **EXERCISE SET**

🦢 *Concept Reinforcement*   *For Exercises 1–6, use the function f given by*

$$f(x) = \begin{cases} x - 5, & \text{if } x < -6, \\ 2x^2, & \text{if } -6 \le x < -1, \\ |x|, & \text{if } -1 \le x < 10, \\ 3x + 1, & \text{if } x \ge 10. \end{cases}$$

Write the letter of the equation that should be used to find each function value. Letters may be used more than once or not at all.

1. ____ $f(0)$
2. ____ $f(15)$
3. ____ $f(10)$
4. ____ $f(-6)$
5. ____ $f(-1)$
6. ____ $f(-3)$

a) $f(x) = x - 5$
b) $f(x) = 2x^2$
c) $f(x) = |x|$
d) $f(x) = 3x + 1$

Find the domain and the range for each function given.

7. $f = \{(2, 8), (9, 3), (-2, 10), (-4, 4)\}$

8. $g = \{(1, 2), (2, 3), (3, 4), (4, 5)\}$

9. $g = \{(0, 0), (4, -2), (-5, 0), (-1, -2)\}$

10. $f = \{(3, 7), (2, 7), (1, 7), (0, 7)\}$

For each graph of a function f, determine the domain and the range of f.

11.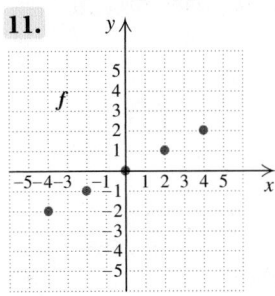

12.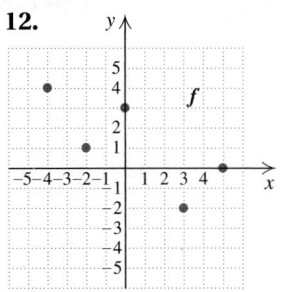

13.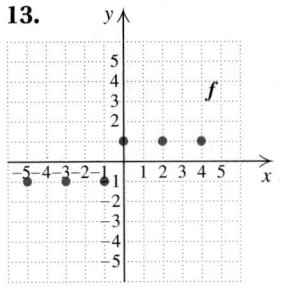

14.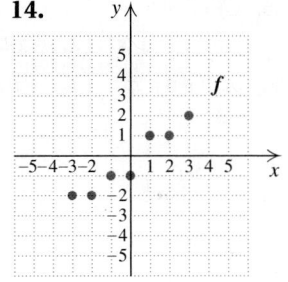

15.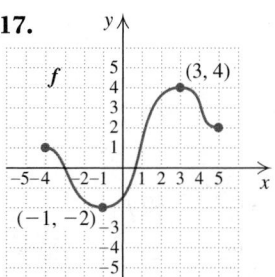

16.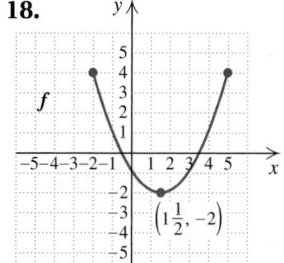

17.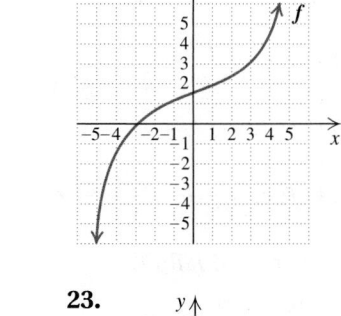

18.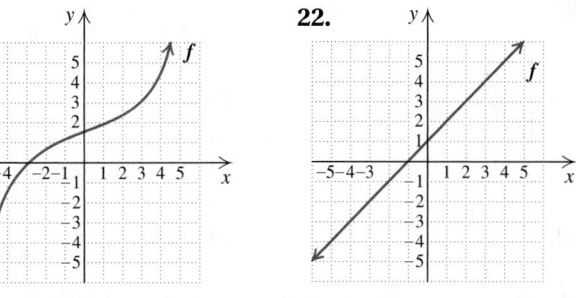

19.

20.

21.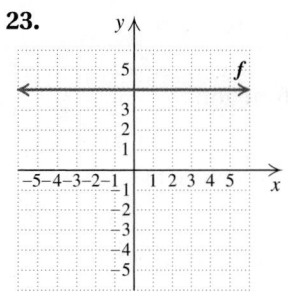

22.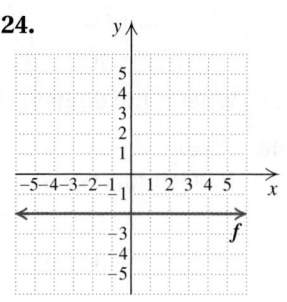

23.

24.

25.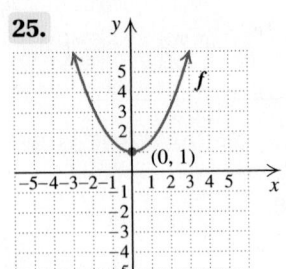

26.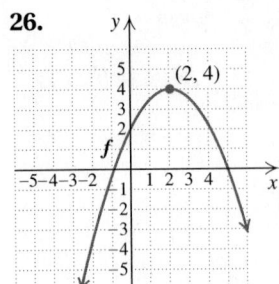

27.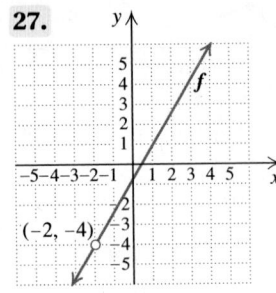

28.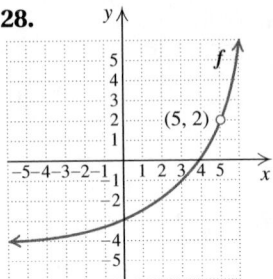

29.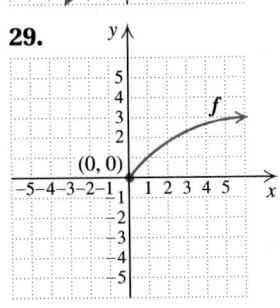

30.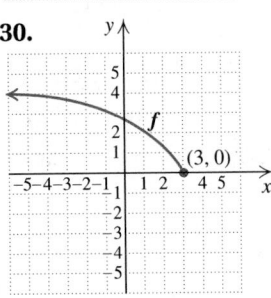

Find the domain of f.

31. $f(x) = \dfrac{5}{x-3}$

32. $f(x) = \dfrac{7}{6-x}$

33. $f(x) = \dfrac{x}{2x-1}$

34. $f(x) = \dfrac{2x}{4x+3}$

35. $f(x) = 2x + 1$

36. $f(x) = x^2 + 3$

37. $f(x) = |5 - x|$

38. $f(x) = |3x - 4|$

39. $f(x) = \dfrac{5}{x^2 - 9}$

40. $f(x) = \dfrac{x}{x^2 - 2x + 1}$

41. $f(x) = x^2 - 9$

42. $f(x) = x^2 - 2x + 1$

43. $f(x) = \dfrac{2x - 7}{x^2 + 8x + 7}$

44. $f(x) = \dfrac{x + 5}{2x^2 - x - 3}$

45. *Records in the 400-m run.* The record R for the 400-m run t years after 1930 is given by

$$R(t) = 46.8 - 0.075t.$$

What is the domain of the function?

46. *Records in the 1500-m run.* The record R for the 1500-m run t years after 1930 is given by

$$R(t) = 3.85 - 0.0075t.$$

What is the domain of the function?

47. *Consumer demand.* The amount A of coffee that consumers are willing to buy at price p is given by

$$A(p) = -2.5p + 26.5.$$

What is the domain of the function?

48. *Seller's supply.* The amount A of coffee that suppliers are willing to supply at price p is given by

$$A(p) = 2p - 11.$$

What is the domain of the function?

49. *Pressure at sea depth.* The pressure P, in atmospheres, at a depth d feet beneath the surface of the ocean is given by

$$P(d) = 0.03d + 1.$$

What is the domain of the function?

50. *Perimeter.* The perimeter P of an equilateral triangle with sides of length s is given by

$$P(s) = 3s.$$

What is the domain of the function?

51. *Fireworks displays.* The height h, in feet, of a "weeping willow" fireworks display, t seconds after having been launched from an 80-ft high rooftop, is given by

$$h(t) = -16t^2 + 64t + 80.$$

What is the domain of the function?

52. *Safety flares.* The height h, in feet, of a safety flare, t seconds after having been launched from a height of 224 ft, is given by

$$h(t) = -16t^2 + 80t + 224.$$

What is the domain of the function?

Find the indicated function values for each function.

53. $f(x) = \begin{cases} x, & \text{if } x < 0, \\ 2x + 1, & \text{if } x \geq 0 \end{cases}$

   a) $f(-5)$ **b)** $f(0)$ **c)** $f(10)$

54. $g(x) = \begin{cases} x - 5, & \text{if } x \leq 5, \\ 3x, & \text{if } x > 5 \end{cases}$

   a) $g(0)$ **b)** $g(5)$ **c)** $g(6)$

55. $G(x) = \begin{cases} x - 5, & \text{if } x < -1, \\ x, & \text{if } -1 \leq x \leq 2, \\ x + 2, & \text{if } x > 2 \end{cases}$

   a) $G(0)$ **b)** $G(2)$ **c)** $G(5)$

56. $F(x) = \begin{cases} 2x, & \text{if } x \leq 0, \\ x, & \text{if } 0 < x \leq 3, \\ -5x, & \text{if } x > 3 \end{cases}$

   a) $F(-1)$ **b)** $F(3)$ **c)** $F(10)$

57. $f(x) = \begin{cases} x^2 - 10, & \text{if } x < -10, \\ x^2, & \text{if } -10 \le x \le 10, \\ x^2 + 10, & \text{if } x > 10 \end{cases}$

   a) $f(-10)$ **b)** $f(10)$ **c)** $f(11)$

58. $f(x) = \begin{cases} 2x^2 - 3, & \text{if } x < 2, \\ x^2, & \text{if } 2 \le x \le 4, \\ 5x - 7, & \text{if } x > 4 \end{cases}$

   a) $f(0)$ **b)** $f(3)$ **c)** $f(6)$

59. Explain why the domain of the function given by
$f(x) = \dfrac{x + 3}{2}$ is $\mathbb{R}$, but the domain of the function
given by $g(x) = \dfrac{2}{x + 3}$ is not $\mathbb{R}$.

60. Chloe asserts that for a function described by a set of ordered pairs, the range of the function will always have the same number of elements as there are ordered pairs. Is she correct? Why or why not?

Skill Review

To prepare for Section 7.3, review graphs of linear equations (Section 3.6).

Graph. [3.6]

61. $y = 2x - 3$ **62.** $y = x + 5$

Find the slope and the y-intercept of each line. [3.6]

63. $y = \dfrac{2}{3}x - 4$ **64.** $y = -\dfrac{1}{4}x + 6$

65. $y = \dfrac{4}{3}x$ **66.** $y = -5x$

Synthesis

67. Ramiro states that $f(x) = \dfrac{x^2}{x}$ and $g(x) = x$ represent the same function. Is he correct? Why or why not?

68. Explain why the domain of a function can be viewed as the projection of its graph on the x-axis.

Sketch the graph of a function for which the domain and range are as given. Graphs may vary.

69. Domain: $\mathbb{R}$; range: $\mathbb{R}$

70. Domain: $\{3, 1, 4\}$; range: $\{0, 5\}$

71. Domain: $\{x \mid 1 \le x \le 5\}$; range: $\{y \mid 0 \le y \le 2\}$

72. Domain: $\{x \mid x$ is a real number *and* $x \ne 1\}$; range: $\{y \mid y$ is a real number *and* $y \ne -2\}$

For each graph of a function f, determine the domain and the range of f.

73.
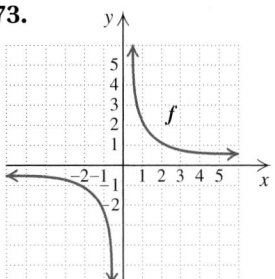

74.
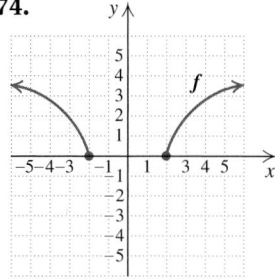

75.
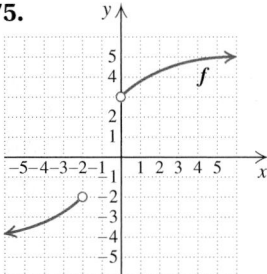

76.
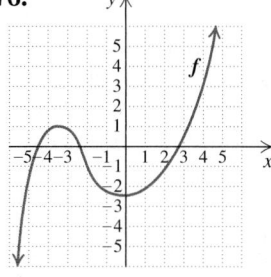

Graph each function on a graphing calculator and estimate its domain and range from the graph.

77. $f(x) = |x - 3|$ **78.** $f(x) = |x| - 3$

79. $f(x) = \dfrac{3}{x - 2}$ **80.** $f(x) = \dfrac{-1}{x + 3}$

81. Use a graphing calculator to estimate the range of the function in Exercise 51.

82. Use a graphing calculator to estimate the range of the function in Exercise 52.

83.–88. *For Exercises 83–88, graph the functions given in each of Exercises 53–58, respectively.*

89. A graphing calculator will interpret an expression like $x \ge 1$ as true or false, depending on the value of x. If the expression is true, the graphing calculator assigns a value of 1 to the expression. If the expression is false, the graphing calculator assigns a value of 0. To graph a piecewise-defined function using a graphing calculator, multiply each part of the definition by its domain, using the TEST menu to enter the inequality symbol. Thus the function in Exercise 53 is entered as $y_1 = x(x < 0) + (2x + 1)(x \ge 0)$. Use a graphing calculator in DOT mode to check your answers to Exercises 83–88.

7.3   Graphs of Functions

Linear Functions   •   Nonlinear Functions

A function can be classified both by the type of equation that is used and by the type of graph it represents. In this section, we will graph a variety of functions that are described by different equations.

Linear Functions

In Chapter 3, we graphed *linear equations*. Here we review such graphs and determine which types of linear graphs represent functions.

Any linear equation can be written in *standard form* $Ax + By = C$. If $B \neq 0$, the equation can also be written in *slope–intercept form*. The *point–slope form* is often used to write equations.

Equations of Lines

Standard form:	$Ax + By = C$
Slope–intercept form:	$y = mx + b$
Point–slope form:	$y - y_1 = m(x - x_1)$

Two points determine a line. If we know that an equation is linear, we can graph the equation by plotting two points that are on the line and drawing the line that goes through those points.

When an equation is written in slope–intercept form $y = mx + b$, the slope of the line is m and the y-intercept is $(0, b)$. Knowing the y-intercept gives us one point on the line, and we can use the slope to determine another point.

EXAMPLE **1**   Graph: $4y = -3x + 8$.

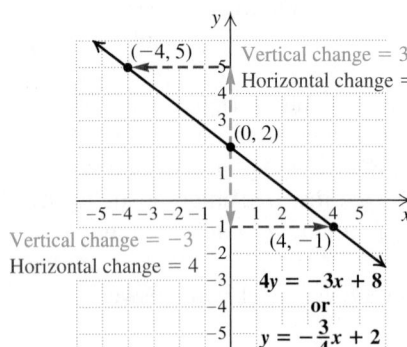

SOLUTION   To graph $4y = -3x + 8$, we first rewrite it in slope–intercept form:

$$4y = -3x + 8$$
$$y = \tfrac{1}{4}(-3x + 8) \qquad \text{Multiplying both sides by } \tfrac{1}{4}$$
$$y = -\tfrac{3}{4}x + 2. \qquad \text{Using the distributive law}$$

The slope is $-\tfrac{3}{4}$ and the y-intercept is $(0, 2)$. We plot $(0, 2)$ and think of the slope as either $\tfrac{-3}{4}$ or $\tfrac{3}{-4}$. Using the form $\tfrac{-3}{4}$, we start at $(0, 2)$ and move *down* 3 units (since the numerator is *negative*) and *to the right* 4 units (since the denominator is *positive*). We plot the new point, $(4, -1)$.

Alternatively, we can think of the slope as $\tfrac{3}{-4}$. Starting at $(0, 2)$, we move *up* 3 units (since the numerator is *positive*) and *to the left* 4 units (since the denominator is *negative*). This leads to another point on the graph, $(-4, 5)$. Using the points found, we draw and label the graph at left.

TRY EXERCISE   7

The graphs of equations of the form $y = b$ are horizontal lines with a slope of 0, and the graphs of equations of the form $x = a$ are vertical lines. The slope of a vertical line is undefined.

We can use the vertical-line test to determine which types of linear graphs represent functions. Consider the following graphs.

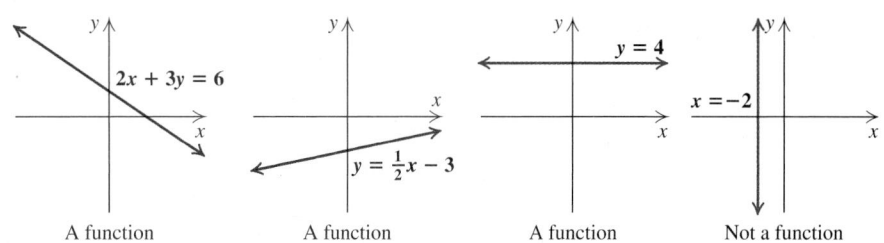

A function   A function   A function   Not a function

Any vertical line that passes through the graphs of $2x + 3y = 6$, $y = \frac{1}{2}x - 3$, and $y = 4$ will cross the graph only once. However, the vertical line through the point $(-2, 0)$ will cross the graph of $x = -2$ at *every* point. In general, *any* straight line that is not vertical is the graph of a function. A **linear function** is a function described by any linear equation whose graph is not vertical. A horizontal line represents a **constant function**.

> ## Linear Function
>
> A function described by an equation of the form $f(x) = mx + b$ is a *linear function*. Its graph is a straight line with slope m and y-intercept $(0, b)$.
>
> When $m = 0$, the function described by $f(x) = b$ is called a *constant function*. Its graph is a horizontal line through $(0, b)$.

EXAMPLE 2

Graph: $f(x) = 3x + 2$.

SOLUTION The notations

$$f(x) = 3x + 2 \quad \text{and} \quad y = 3x + 2$$

are often used interchangeably. The function notation emphasizes that the second coordinate in each ordered pair is determined by the first coordinate of that pair.

We graph $f(x) = 3x + 2$ in the same way that we would graph $y = 3x + 2$. The vertical axis can be labeled y or $f(x)$. We could use a table of values or, since this is a linear function, use the slope and the y-intercept to graph the function.

Since $f(x) = 3x + 2$ is in the form $f(x) = mx + b$, we can tell from the equation that the slope is 3, or $\frac{3}{1}$, and the y-intercept is $(0, 2)$. We plot $(0, 2)$ and go *up* 3 units and *to the right* 1 unit to determine another point on the line, $(1, 5)$. After we have sketched the line, a third point can be calculated as a check.

> **TRY EXERCISE** 19

EXAMPLE 3

Graph: $f(x) = -3$.

SOLUTION This is a constant function. For every input x, the output is -3. The graph is a horizontal line.

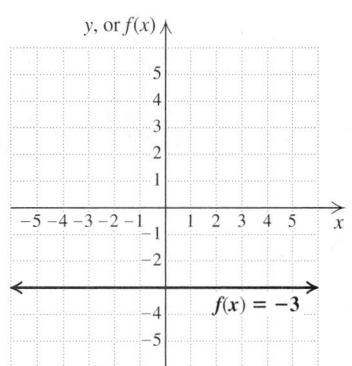

> **TRY EXERCISE** 23

Linear functions are common in today's world.

EXAMPLE **4**   *Cell-phone costs.*   In 2008, an Apple iPhone cost $400. AT&T offered a plan including 450 daytime minutes a month for $60 per month. Formulate a mathematical model for the cost. Then use the model to estimate the number of months required for the total cost to reach $700.

Source: www.apple.com

SOLUTION

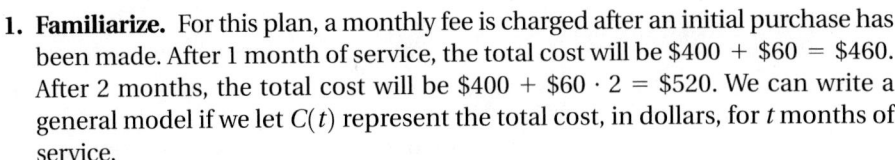

1. **Familiarize.** For this plan, a monthly fee is charged after an initial purchase has been made. After 1 month of service, the total cost will be $400 + $60 = $460. After 2 months, the total cost will be $400 + $60 · 2 = $520. We can write a general model if we let $C(t)$ represent the total cost, in dollars, for t months of service.

2. **Translate.** We reword and translate as follows:

Rewording: The total cost   is   the cost of the phone   plus $60 per month.

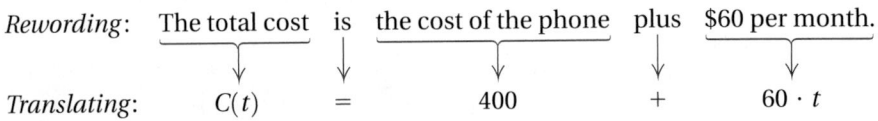

Translating: $C(t)$ $=$ 400 $+$ $60 \cdot t$

where $t \geq 0$ (since there cannot be a negative number of months).

3. **Carry out.** To determine the time required for the total cost to reach $700, we substitute 700 for $C(t)$ and solve for t:

$C(t) = 400 + 60t$

$700 = 400 + 60t$ Substituting

$300 = 60t$ Subtracting 400 from both sides

$5 = t$.   Dividing both sides by 60

4. **Check.** We evaluate:

$C(5) = 60 \cdot 5 + 400 = 300 + 400 = 700.$

5. **State.** It takes 5 months for the total cost to reach $700.   **TRY EXERCISE** 25

Since $f(x) = mx + b$ can be evaluated for any choice of x, the domain of all linear functions is $\mathbb{R}$, the set of all real numbers.

The second coordinate of every ordered pair in a constant function $f(x) = b$ is the number b. The range of a constant function thus consists of one number, b. For a nonconstant linear function, the graph extends indefinitely both up and down, so the range is the set of all real numbers, or $\mathbb{R}$.

Domain and Range of a Linear Function

The domain of any linear function $f(x) = mx + b$ is

$\{x \mid x \text{ is a real number}\}$, or $\mathbb{R}$.

The range of any linear function $f(x) = mx + b, m \neq 0$, is

$\{y \mid y \text{ is a real number}\}$, or $\mathbb{R}$.

The range of any constant function $f(x) = b$ is $\{b\}$.

EXAMPLE **5**   Determine the domain and the range of each of the following functions.

a) f, where $f(x) = 2x - 10$ **b)** g, where $g(x) = 4$

SOLUTION

a) Since $f(x) = 2x - 10$ describes a linear function, but not a constant function,

Domain of $f = \mathbb{R}$ and

Range of $f = \mathbb{R}$.

b) The function described by $g(x) = 4$ is a constant function. Thus,

Domain of $g = \mathbb{R}$ and

Range of $g = \{4\}$.

TRY EXERCISE 41

The graphs of nonlinear functions can get quite complex. We will now define several types of nonlinear functions and discuss some of their characteristics. The detailed study of their graphs appears later in this text or in other courses.

Nonlinear Functions

A function for which the graph is not a straight line is a **nonlinear function**. Some important types of nonlinear functions are listed below.

Type of function	*Example*
Absolute-value function	$f(x) = |x|$
Polynomial function	$p(x) = x^3 - 4x^2 + 1$
Quadratic function	$q(x) = x^2 + 5x + 2$
Rational function	$r(x) = \dfrac{x + 1}{x - 2}$

Note that linear and quadratic functions are special kinds of polynomial functions.

EXAMPLE 6   State whether each equation describes a linear function, an absolute-value function, a general polynomial function, a quadratic function, or a rational function.

a) $f(x) = x^2 - 9$

b) $g(x) = \dfrac{3}{x}$

c) $h(x) = \dfrac{1}{4}x - 16$

d) $v(x) = 4x^4 - 13$

SOLUTION

a) Since f is described by a polynomial equation of degree 2, f is a *quadratic function.*

b) Since g is described by a rational equation, g is a *rational function.*

c) The function h is described by a linear equation, so h is a *linear function.* Note that although $\frac{1}{4}$ is a fraction, there are no variables in a denominator.

d) Since v is described by a polynomial equation, v is a *polynomial function.*

Since the graphs of nonlinear functions are not straight lines, we usually need to calculate more than two or three points to determine the shape of the graph.

EXAMPLE 7   Graph the function given by $f(x) = |x|$, and determine the domain and the range of f.

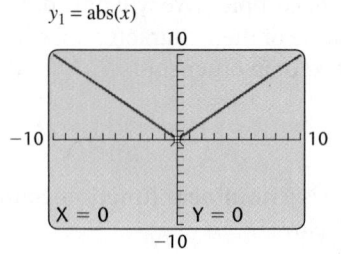
SOLUTION   We calculate function values for several choices of x and list the results in a table.

$$f(0) = |0| = 0,$$
$$f(1) = |1| = 1,$$
$$f(2) = |2| = 2,$$
$$f(-1) = |-1| = 1,$$
$$f(-2) = |-2| = 2$$

x	$f(x) = |x|$	$(x, f(x))$
0	0	$(0, 0)$
1	1	$(1, 1)$
2	2	$(2, 2)$
-1	1	$(-1, 1)$
-2	2	$(-2, 2)$

When we plot these points, we observe a pattern. The value of the function is 0 when x is 0. Function values increase both as x increases from 0 and as x decreases from 0. The graph of f is V-shaped, with the "point" of the V at the origin.

Because we can find the absolute value of any real number, we have

Domain of $f = \mathbb{R}$, or $(-\infty, \infty)$.

Because the absolute value of a number is never negative, we have

Range of $f = \{y \mid y \geq 0\}$, or $[0, \infty)$.

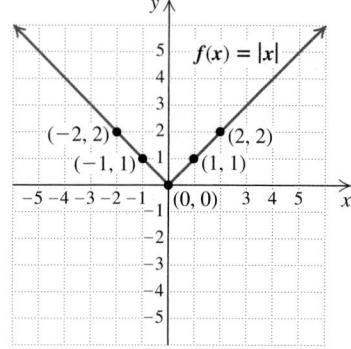

▶ **TRY EXERCISE** ▶ 65

Graphs of *polynomial functions* generally become more complex as the degree of the polynomial increases. In Chapter 11, we will study in greater detail the graphs of *quadratic functions*, or functions of the form

$$q(x) = ax^2 + bx + c, a \neq 0.$$

Since a polynomial can be evaluated for any real number, the domain of a polynomial is the set of all real numbers.

A *rational function* contains a variable in a denominator; thus its domain may be restricted. Division by zero is undefined, so any values of the variable that make a denominator 0 are not in the domain of the function.

EXAMPLE **8**   Determine the domain of f.

a) $f(x) = x^3 + 5x^2 - 4x + 1$ **b)** $f(x) = \dfrac{x^2 - 4}{x + 2}$

SOLUTION

a) $f(x) = x^3 + 5x^2 - 4x + 1$ describes a polynomial function. The domain of any polynomial function is $\mathbb{R}$, so the domain of f is $\mathbb{R}$.

b) $f(x) = \dfrac{x^2 - 4}{x + 2}$ describes a rational function. Note that $f(x)$ is undefined for $x + 2 = 0$, or, equivalently, for $x = -2$. Thus the domain of $f = \{x \mid x \text{ is a real number and } x \neq -2\}$.

▶ **TRY EXERCISE** ▶ 43

Visualizing for Success

A
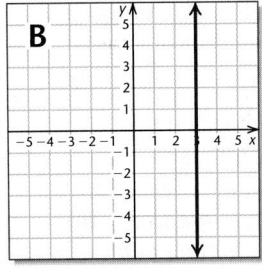

B
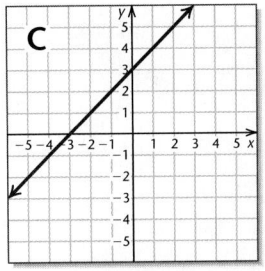

C
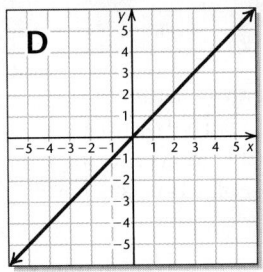

D
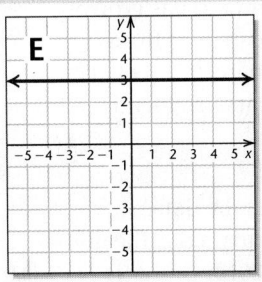

E

F
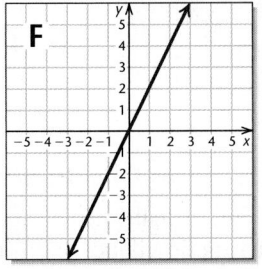

G
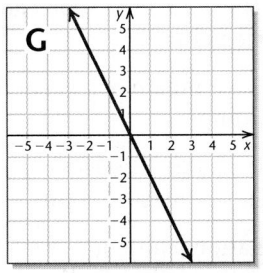

H
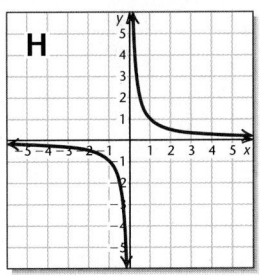

I
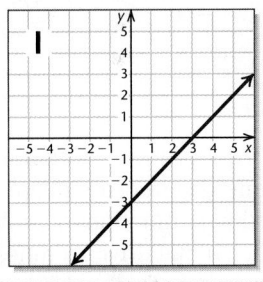

J
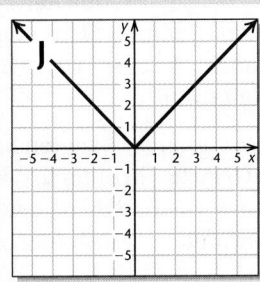

Match each equation or function with its graph.

1. $f(x) = x$

2. $f(x) = |x|$

3. $f(x) = x^2$

4. $f(x) = 3$

5. $x = 3$

6. $f(x) = x + 3$

7. $f(x) = x - 3$

8. $f(x) = 2x$

9. $f(x) = -2x$

10. $f(x) = \dfrac{1}{x}$

Answers on page A-28

An alternate, animated version of this activity appears in MyMathLab. To use MyMathLab, you need a course ID and a student access code. Contact your instructor for more information.

7.3 **EXERCISE SET**

🐦 *Concept Reinforcement*   *Answer true or false.*

1. The vertical-line test states that a graph is not that of a function if it contains a vertical line.

2. The graph of a constant function is a horizontal line.

3. The domain of a constant function consists of a single element.

4. The domain of a linear function is the set of all real numbers.

5. Linear functions are typically written in slope–intercept form.

6. Rational functions may have some restrictions on their domains.

Graph.

7. $y = 2x - 1$ **8.** $y = \frac{1}{3}x + 2$

9. $y = -\frac{2}{3}x + 3$ **10.** $y = -4x - 2$

11. $3y = 6 - 4x$ **12.** $5y = 2x - 15$

13. $x - y = 4$ **14.** $x + y = 3$

15. $y = -2$ **16.** $y = 3$

17. $x = 4$ **18.** $x = -1$

19. $f(x) = x + 3$ **20.** $f(x) = 2 - x$

21. $f(x) = \frac{3}{4}x + 1$ **22.** $f(x) = 3x + 2$

23. $g(x) = 4$ **24.** $g(x) = -5$

25. *Truck rentals.*   Titanium Trucks charges $30 for a one-day truck rental, plus $0.75 per mile. Formulate a linear function to model the cost $C(d)$ of a one-day rental driven d miles, and determine the number of miles driven if the total cost is $75.

26. *Taxis.*   A taxi ride in New York City costs $2.50 plus $2.00 per mile.* Formulate a linear function to model the cost $C(d)$ of a d-mile taxi ride, and determine the length of a ride that cost $23.50.

27. *Hair growth.*   Lauren had her hair cut to a length of 5 inches in order to donate the hair to Locks of Love. Her hair then grew at a rate of $\frac{1}{2}$ inch per month. Formulate a linear function to model the length $L(t)$

---

*Rates are higher between 4 P.M. and 8 P.M. (Source: Based on data from New York City Taxi and Limousine Commission, 2007)

of Lauren's hair t months after she had the haircut, and determine when her hair will be 15 inches long.

28. *Landscaping.*   On Saturday, Shelby Lawncare cut the lawn at Great Harrington Community College to a height of 2 in. Since then, the grass has grown at a rate of $\frac{1}{8}$ in. per day. Formulate a linear function to model the length $L(t)$ of the lawn t days after having been cut, and determine when the grass will be $3\frac{1}{2}$ in. high.

29. *Organic cotton.*   In 2006, 5960 acres in the U.S. were planted with organic cotton. This number is increasing by 849 acres each year. Formulate a linear function to model the number of acres of organic cotton $A(t)$ that is planted t years after 2006, and determine when 10,205 acres of organic cotton will be planted.
Source: Based on data from the Organic Trade Association

30. *Catering.*   Chrissie's Catering charges a setup fee of $75 plus $25 a person for catering a party. Formulate a linear function to model the cost $C(x)$ for a party for x people, and determine the number of people at a party if the cost was $775.

In Exercises 31–40, assume that a constant rate of change exists for each model formed.

31. *Automobile production.*   As demand has grown, worldwide production of small cars rose from 14.5 million in 2002 to 19 million in 2007. Let $a(t)$ represent the number of small cars produced t years after 2000.
Source: *The Wall Street Journal*, 10/22/07

a) Find a linear function that fits the data.
b) Use the function from part (a) to predict the number of small cars produced in 2013.
c) In what year will 25 million small cars be produced?

32. *Convention attendees.*   In recent years, Las Vegas has become a popular location for conventions. The number of convention attendees in Las Vegas rose from 4.6 million in 2002 to 6.1 million in 2006. Let $v(t)$ represent the number of convention attendees in Las Vegas t years after 2000.
Source: Las Vegas Convention and Visitors Authority

a) Find a linear function that fits the data.
b) Use the function from part (a) to predict the number of convention attendees in Las Vegas in 2011.
c) In what year will there be 8 million convention attendees in Las Vegas?

33. *Life expectancy of females in the United States.*
In 1994, the life expectancy of females was 79.0 yr. In 2004, it was 80.4 yr. Let $E(t)$ represent life expectancy and t the number of years since 1990.
Source: Statistical Abstract of the United States, 2007

a) Find a linear function that fits the data.
Aha! b) Use the function of part (a) to predict the life expectancy of females in 2012.

34. *Life expectancy of males in the United States.*   In 1994, the life expectancy of males was 72.4 yr. In 2004, it was 75.2 yr. Let $E(t)$ represent life expectancy and t the number of years since 1990.
Source: Statistical Abstract of the United States, 2007

a) Find a linear function that fits the data.
b) Use the function of part (a) to predict the life expectancy of males in 2012.

35. *PAC contributions.*   In 2002, Political Action Committees (PACs) contributed $282 million to federal candidates. In 2006, the figure rose to $372.1 million. Let $A(t)$ represent the amount of PAC contributions, in millions, and t the number of years since 2000.
Source: Federal Election Commission

PAC contributions to federal candidates

2002 $282 million

2006 $372.1 million

a) Find a linear function that fits the data.
b) Use the function of part (a) to predict the amount of PAC contributions in 2010.

36. *Recycling.*   In 2000, Americans recycled 52.7 million tons of solid waste. In 2005, the figure grew to 58.4 million tons. Let $N(t)$ represent the number of tons recycled, in millions, and t the number of years since 2000.
Sources: U.S. EPA; Franklin Associates, Ltd.

a) Find a linear function that fits the data.
b) Use the function of part (a) to predict the amount recycled in 2012.

37. *Online banking.*   In 2000, about 16 million Americans conducted at least some of their banking online. By 2005, that number had risen to about 63 million. Let $N(t)$ represent the number of Americans using online banking, in millions, t years after 2000.

a) Find a linear function that fits the data.
Aha! b) Use the function of part (a) to predict the number of Americans who will use online banking in 2010.
c) In what year will 157 million Americans use online banking?

38. *Records in the 100-meter run.*   In 1999, the record for the 100-m run was 9.79 sec. In 2007, it was 9.77 sec. Let $R(t)$ represent the record in the 100-m run and t the number of years since 1999.
Sources: International Association of Athletics Federation; Guinness World Records

a) Find a linear function that fits the data.
b) Use the function of part (a) to predict the record in 2015 and in 2030.
c) When will the record be 9.6 sec?

39. *National Park land.* In 1994, the National Park system consisted of about 74.9 million acres. By 2005, the figure had grown to 79 million acres. Let $A(t)$ represent the amount of land in the National Park system, in millions of acres, t years after 1990.

Source: *Statistical Abstract of the United States*, 2007

a) Find a linear function that fits the data.

b) Use the function of part (a) to predict the amount of land in the National Park system in 2010.

40. *Pressure at sea depth.* The pressure 100 ft beneath the ocean's surface is approximately 4 atm (atmospheres), whereas at a depth of 200 ft, the pressure is about 7 atm.

a) Find a linear function that expresses pressure as a function of depth.

b) Use the function of part (a) to determine the pressure at a depth of 690 ft.

Classify each function as a linear function, an absolute-value function, a quadratic function, another polynomial function, or a rational function, and determine the domain of the function.

41. $f(x) = \dfrac{1}{3}x - 7$

42. $g(x) = \dfrac{x}{x + 1}$

43. $p(x) = x^2 + x + 1$

44. $t(x) = |x - 7|$

45. $f(t) = \dfrac{12}{3t + 4}$

46. $g(n) = 15 - 10n$

47. $f(x) = 0.02x^4 - 0.1x + 1.7$

48. $f(a) = 2|a + 3|$

49. $f(x) = \dfrac{x}{2x - 5}$

50. $g(x) = \dfrac{2x}{3x - 4}$

51. $f(n) = \dfrac{4n - 7}{n^2 + 3n + 2}$

52. $h(x) = \dfrac{x - 5}{2x^2 - 2}$

53. $f(n) = 200 - 0.1n$

54. $g(t) = \dfrac{t^2 - 3t + 7}{8}$

Given the graph of each function, determine the range of f.

55.

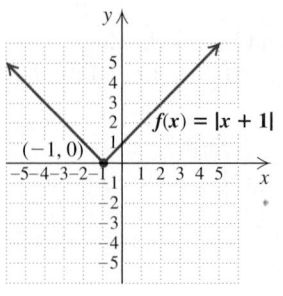

56.

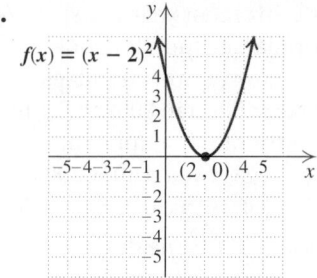

57.

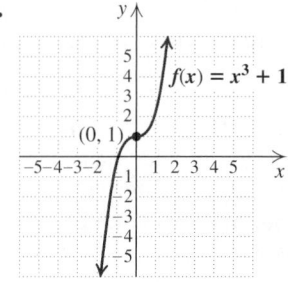

58.

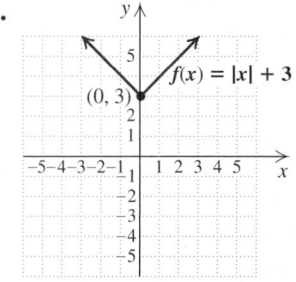

59.

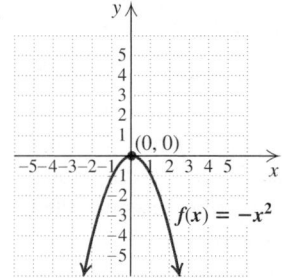

60.

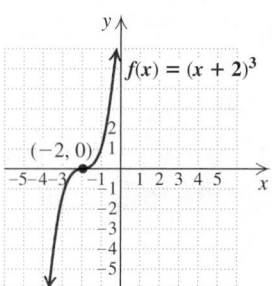

Graph each function and determine its domain and range.

61. $f(x) = x + 3$

62. $f(x) = 2x - 1$

63. $f(x) = -1$

64. $g(x) = 2$

65. $f(x) = |x| + 1$

66. $g(x) = |x - 3|$

67. $g(x) = x^2$

68. $f(x) = x^2 + 2$

69. Bob believes that the domain and range of all polynomial functions is $\mathbb{R}$. How could you convince him that he is mistaken?

70. Explain why the range of a constant function consists of only one number.

Skill Review

To prepare for Section 7.4, review polynomial operations (Sections 4.4 and 4.6).

Perform the indicated operations.

71. $(x^2 + 2x + 7) + (3x^2 - 8)$ [4.4]

72. $(3x^3 - x^2 + x) - (x^3 + 2x - 7)$ [4.4]

73. $(2x + 1)(x - 7)$ [4.6]

74. $(x - 3)(x + 4)$ [4.6]

75. $(x^3 + x^2 - 4x + 7) - (3x^2 - x + 2)$ [4.4]

76. $(2x^2 + x - 3) + (x^3 + 7)$ [4.4]

Synthesis

77. In 2004, Political Action Committees contributed $310.5 million to federal candidates. Does this information make your answer to Exercise 35(b) seem too low or too high? Why?

78. On the basis of your answers to Exercises 33 and 34, would you predict that at some point in the future the life expectancy of males will exceed that of females? Why or why not?

Given that $f(x) = mx + b$, classify each of the following as true or false.

79. $f(c + d) = f(c) + f(d)$

80. $f(cd) = f(c)f(d)$

81. $f(kx) = kf(x)$

82. $f(c - d) = f(c) - f(d)$

For Exercises 83–86, assume that a linear equation models each situation.

83. *Temperature conversion.* Water freezes at 32° Fahrenheit and at 0° Celsius. Water boils at 212°F and at 100°C. What Celsius temperature corresponds to a room temperature of 70°F?

84. *Depreciation of a computer.* After 6 mos of use, the value of Don's computer had dropped to $900. After 8 mos, the value had gone down to $750. How much did the computer cost originally?

85. *Cell-phone charges.* The total cost of Tam's cell phone was $410 after 5 mos of service and $690 after 9 mos. What costs had Tam already incurred when her service just began? Assume that Tam's monthly charge is constant.

86. *Operating expenses.* The total cost for operating Ming's Wings was $7500 after 4 mos and $9250 after 7 mos. Predict the total cost after 10 mos.

87. For a linear function g, $g(3) = -5$ and $g(7) = -1$.
   a) Find an equation for g.
   b) Find $g(-2)$.
   c) Find a such that $g(a) = 75$.

88. When several data points are available and they appear to be nearly collinear, a procedure known as *linear regression* can be used to find an equation for the line that most closely fits the data.

   a) Use a graphing calculator with a LINEAR REGRESSION option and the table that follows to find a linear function that predicts the wattage of a CFL (compact fluorescent) lightbulb as a function of the wattage of a standard incandescent bulb of equivalent brightness. Round coefficients to the nearest thousandth.

Energy Conservation

Incandescent Wattage	CFL Equivalent
25 W	5 W
50 W	9 W
60 W	15 W
100 W	25 W
120 W	28 W

Source: U.S. Department of Energy

   b) Use the function from part (a) to estimate the CFL wattage that is equivalent to a 75-watt incandescent bulb. Then compare your answer with the corresponding answer to Exercise 59 in Section 7.1. Which answer seems more reliable? Why?

CONNECTING the CONCEPTS

A function is a correspondence. This correspondence can be listed as a set of ordered pairs or described by an equation. The correspondence is between two sets, the domain and the range, and each member of the domain corresponds to exactly one member of the range.

For the function f: $\{(2, 3), (0, -4), (-8, 7)\}$:

The domain is $\{-8, 0, 2\}$.

The range is $\{-4, 3, 7\}$.

The input 2 corresponds to the output 3.

$f(2) = 3$

For the function given by $f(x) = x^2$:

The domain is $\mathbb{R}$.

The range is $[0, \infty)$.

The input -3 corresponds to the output 9.

$f(-3) = 9$

The independent variable is x.

MIXED REVIEW

Let $f = \{(3, 6), (4, 8), (-1, -2), (0, 0)\}$.

1. Find the domain of f.

2. Find the range of f.

3. Find $f(-1)$.

4. Graph f.

Let $g(x) = x - 1$ and $h(x) = \dfrac{2}{x}$.

5. Find the domain of g.

6. Find the domain of h.

7. Find $h(10)$.

8. Find $g(1) + h(1)$.

9. Find $(g \cdot h)(-2)$.

10. Find the domain of h/g.

11. Determine whether g is a linear function, a quadratic function, or a rational function.

12. Determine whether h is a linear function, a quadratic function, or a rational function.

13. Graph g and determine its range.

Use the following graph of F for Exercises 14–17.

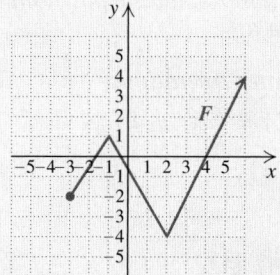

14. Determine from the graph whether F is a function.

15. Find $F(2)$.

16. Find the domain of F.

17. Find the range of F.

Let $G(x) = \begin{cases} 1 - x, & \text{if } x < 1, \\ 10, & \text{if } x = 1, \\ x + 1, & \text{if } x > 1. \end{cases}$

18. Find $G(3)$.

19. Find $G(1)$.

20. Find $G(-12)$.

7.4   The Algebra of Functions

The Sum, Difference, Product, or Quotient of Two Functions   ▪   Domains and Graphs

We now examine four ways in which functions can be combined.

The Sum, Difference, Product, or Quotient of Two Functions

Suppose that a is in the domain of two functions, f and g. The input a is paired with $f(a)$ by f and with $g(a)$ by g. The outputs can then be added to get $f(a) + g(a)$.

Let $f(x) = x + 4$ and $g(x) = x^2 + 1$. Find $f(2) + g(2)$.

SOLUTION   We visualize two function machines. Because 2 is in the domain of each function, we can compute $f(2)$ and $g(2)$.

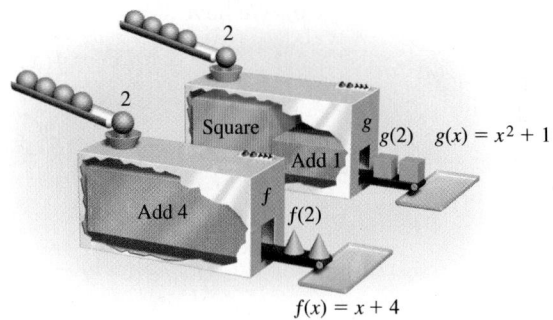

Since

$$f(2) = 2 + 4 = 6 \quad \text{and} \quad g(2) = 2^2 + 1 = 5,$$

we have

$$f(2) + g(2) = 6 + 5 = 11.$$

▶ **TRY EXERCISE**   7

In Example 1, suppose that we were to write $f(x) + g(x)$ as $(x + 4) + (x^2 + 1)$, or $f(x) + g(x) = x^2 + x + 5$. This could then be regarded as a "new" function. The notation $(f + g)(x)$ is generally used to denote a function formed in this manner. Similar notations exist for subtraction, multiplication, and division of functions.

The Algebra of Functions

If f and g are functions and x is in the domain of both functions, then:

1. $(f + g)(x) = f(x) + g(x)$;
2. $(f - g)(x) = f(x) - g(x)$;
3. $(f \cdot g)(x) = f(x) \cdot g(x)$;
4. $(f/g)(x) = f(x)/g(x)$, provided $g(x) \neq 0$.

EXAMPLE **2** For $f(x) = x^2 - x$ and $g(x) = x + 2$, find the following.

a) $(f + g)(4)$ **b)** $(f - g)(x)$ and $(f - g)(-1)$

c) $(f/g)(x)$ and $(f/g)(-4)$ **d)** $(f \cdot g)(4)$

SOLUTION

a) Since $f(4) = 4^2 - 4 = 12$ and $g(4) = 4 + 2 = 6$, we have

$$(f + g)(4) = f(4) + g(4)$$
$$= 12 + 6 \qquad \text{Substituting}$$
$$= 18.$$

Alternatively, we could first find $(f + g)(x)$:

$$(f + g)(x) = f(x) + g(x)$$
$$= x^2 - x + x + 2$$
$$= x^2 + 2. \qquad \text{Combining like terms}$$

Thus,

$$(f + g)(4) = 4^2 + 2 = 18. \qquad \text{Our results match.}$$

b) We have

$$(f - g)(x) = f(x) - g(x)$$
$$= x^2 - x - (x + 2) \qquad \text{Substituting}$$
$$= x^2 - 2x - 2. \qquad \begin{array}{l}\text{Removing parentheses and} \\ \text{combining like terms}\end{array}$$

Thus,

$$(f - g)(-1) = (-1)^2 - 2(-1) - 2 \qquad \begin{array}{l}\text{Using } (f - g)(x) \text{ is faster than} \\ \text{using } f(x) - g(x).\end{array}$$
$$= 1. \qquad \text{Simplifying}$$

c) We have

$$(f/g)(x) = f(x)/g(x)$$
$$= \frac{x^2 - x}{x + 2}. \qquad \text{We assume that } x \neq -2.$$

Thus,

$$(f/g)(-4) = \frac{(-4)^2 - (-4)}{-4 + 2} \qquad \text{Substituting}$$
$$= \frac{20}{-2} = -10.$$

d) Using our work in part (a), we have

$$(f \cdot g)(4) = f(4) \cdot g(4)$$
$$= 12 \cdot 6$$
$$= 72.$$

Alternatively, we could first find $(f \cdot g)(x)$:

$$
\begin{aligned}
(f \cdot g)(x) &= f(x) \cdot g(x) \\
&= (x^2 - x)(x + 2) \\
&= x^3 + x^2 - 2x. \qquad \text{Multiplying and combining like terms.}
\end{aligned}
$$

Then

$$
\begin{aligned}
(f \cdot g)(4) &= 4^3 + 4^2 - 2 \cdot 4 \\
&= 64 + 16 - 8 \\
&= 72.
\end{aligned}
$$

▶ TRY EXERCISE 17

Domains and Graphs

Although applications involving products and quotients of functions rarely appear in newspapers, situations involving sums or differences of functions often do appear in print. For example, the following graphs are similar to those published by the California Department of Education to promote breakfast programs in which students eat a balanced meal of fruit or juice, toast or cereal, and 2% or whole milk. The combination of carbohydrate, protein, and fat gives a sustained release of energy, delaying the onset of hunger for several hours.

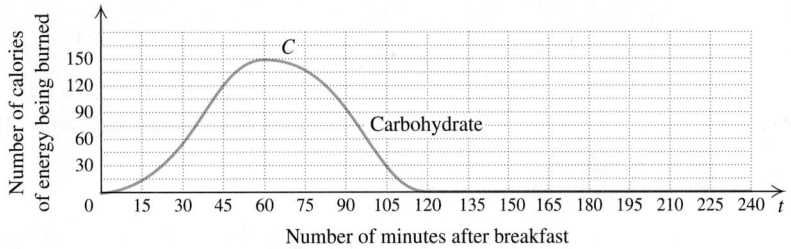

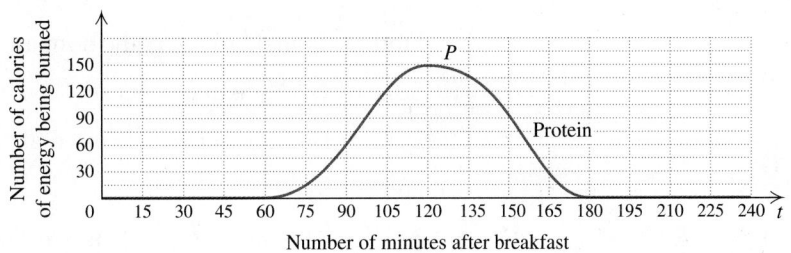

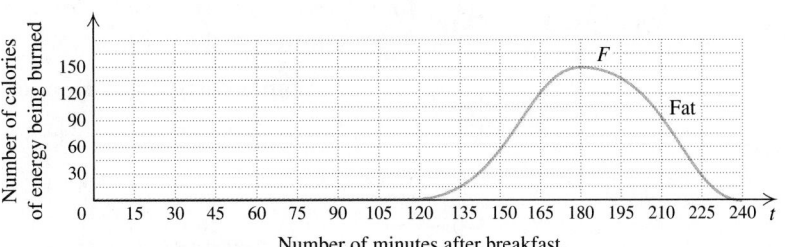

When the three graphs are superimposed, and the calorie expenditures added, it becomes clear that a balanced meal results in a steady, sustained supply of energy.

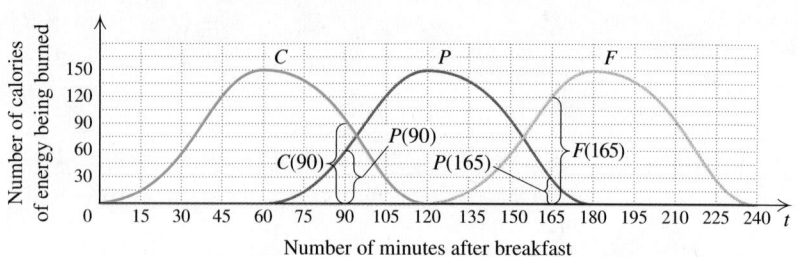

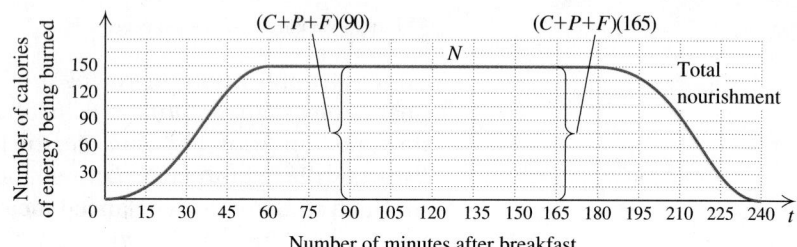

For any point $(t, N(t))$, we have

$$N(t) = (C + P + F)(t) = C(t) + P(t) + F(t).$$

To find $(f + g)(a)$, $(f - g)(a)$, $(f \cdot g)(a)$, or $(f/g)(a)$, we must know that $f(a)$ and $g(a)$ exist. This means a must be in the domain of both f and g.

EXAMPLE **3** Let

$$f(x) = \frac{5}{x} \quad \text{and} \quad g(x) = \frac{2x - 6}{x + 1}.$$

Find the domain of $f + g$, the domain of $f - g$, and the domain of $f \cdot g$.

SOLUTION Note that because division by 0 is undefined, we have

Domain of $f = \{x \mid x$ is a real number *and* $x \neq 0\}$

and

Domain of $g = \{x \mid x$ is a real number *and* $x \neq -1\}$.

In order to find $f(a) + g(a)$, $f(a) - g(a)$, or $f(a) \cdot g(a)$, we must know that a is in *both* of the above domains. Thus,

Domain of $f + g =$ Domain of $f - g =$ Domain of $f \cdot g$

$= \{x \mid x$ is a real number *and* $x \neq 0$ *and* $x \neq -1\}$.

> **TRY EXERCISE** 43

Suppose that for $f(x) = x^2 - x$ and $g(x) = x + 2$, we want to find $(f/g)(-2)$. Finding $f(-2)$ and $g(-2)$ poses no problem:

$$f(-2) = 6 \quad \text{and} \quad g(-2) = 0;$$

but then

$$(f/g)(-2) = f(-2)/g(-2)$$

$$= 6/0. \quad \text{Division by 0 is undefined.}$$

Thus, although -2 is in the domain of both f and g, it is not in the domain of f/g.

We can also see this by writing $(f/g)(x)$:

$$(f/g)(x) = \frac{f(x)}{g(x)} = \frac{x^2 - x}{x + 2}.$$

Since $x + 2 = 0$ when $x = -2$, the domain of f/g must exclude -2.

Determining the Domain

The domain of $f + g$, $f - g$, or $f \cdot g$ is the set of all values common to the domains of f and g.

The domain of f/g is the set of all values common to the domains of f and g, excluding any values for which $g(x)$ is 0.

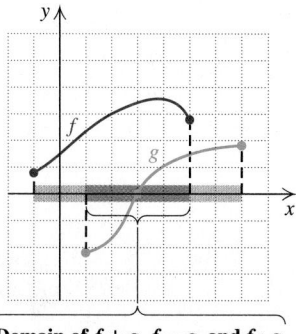

Domain of $f + g$, $f - g$, and $f \cdot g$

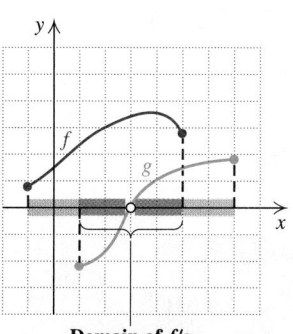

Domain of f/g

EXAMPLE 4

Given $f(x) = 1/x$ and $g(x) = 2x - 7$, find the domains of $f + g$, $f - g$, $f \cdot g$, and f/g.

SOLUTION   We first find the domain of f and the domain of g:

The domain of f is $\{x \mid x \text{ is a real number } and\ x \neq 0\}$.

The domain of g is $\mathbb{R}$.

The domains of $f + g$, $f - g$, and $f \cdot g$ are the set of all elements common to the domains of f and g. This consists of all real numbers except 0.

$$\text{The domain of } f + g = \text{the domain of } f - g = \text{the domain of } f \cdot g$$
$$= \{x \mid x \text{ is a real number } and\ x \neq 0\}.$$

Because we cannot divide by 0, the domain of f/g must also exclude any values of x for which $g(x)$ is 0. We determine those values by solving $g(x) = 0$:

$$g(x) = 0$$
$$2x - 7 = 0 \qquad \text{Replacing } g(x) \text{ with } 2x - 7$$
$$2x = 7$$
$$x = \tfrac{7}{2}.$$

The domain of f/g is the domain of the sum, difference, and product of f and g, found above, excluding $\tfrac{7}{2}$.

$$\text{The domain of } f/g = \left\{ x \mid x \text{ is a real number } and\ x \neq 0 \ and\ x \neq \tfrac{7}{2} \right\}.$$

TRY EXERCISE ▶ 55

STUDENT NOTES

The concern over a denominator being 0 arises throughout this course. Try to develop the habit of checking for any possible input values that would create a denominator of 0 whenever you work with functions.

TECHNOLOGY CONNECTION

A partial check of Example 4 can be performed by setting up a table so the TBLSTART is 0 and the increment of change (ΔTbl) is 0.7. (Other choices, like 0.1, will also work.) Next, we let $y_1 = 1/x$ and $y_2 = 2x - 7$. Using Y-VARS to write $y_3 = y_1 + y_2$ and $y_4 = y_1/y_2$, we can create the table of values shown here. Note that when x is 3.5, a value for y_3 can be found, but y_4 is undefined. If we "de-select" y_1 and y_2 as we enter them, the columns for y_3 and y_4 appear without scrolling through the table.

X	Y₃	Y₄
0	ERROR	ERROR
.7	−4.171	−.2551
1.4	−3.486	−.1701
2.1	−2.324	−.1701
2.8	−1.043	−.2551
3.5	.28571	ERROR
4.2	1.6381	.17007

X = 0

Use a similar approach to partially check Example 3.

Division by 0 is not the only condition that can force restrictions on the domain of a function. In Chapter 10, we will examine functions similar to that given by $f(x) = \sqrt{x}$, for which the concern is taking the square root of a negative number.

7.4 EXERCISE SET

🖐 **Concept Reinforcement** *Make each of the following sentences true by selecting the correct word for each blank.*

1. If f and g are functions, then $(f + g)(x)$ is the _____ of the functions.
   sum/difference

2. One way to compute $(f - g)(2)$ is to _____ $g(2)$ from $f(2)$.
   erase/subtract

3. One way to compute $(f - g)(2)$ is to simplify $f(x) - g(x)$ and then _____ the result
   evaluate/substitute
   for $x = 2$.

4. The domain of $f + g, f - g$, and $f \cdot g$ is the set of all values common to the _____ of f and g.
   domains/ranges

5. The domain of f/g is the set of all values common to the domains of f and g, _____ any
   including/excluding
   values for which $g(x)$ is 0.

6. The height of $(f + g)(a)$ on a graph is the _____ of the heights of $f(a)$ and $g(a)$.
   product/sum

Let $f(x) = -2x + 3$ and $g(x) = x^2 - 5$. Find each of the following.

7. $f(3) + g(3)$ **8.** $f(4) + g(4)$

9. $f(1) - g(1)$ **10.** $f(2) - g(2)$

11. $f(-2) \cdot g(-2)$ **12.** $f(-1) \cdot g(-1)$

13. $f(-4)/g(-4)$ **14.** $f(3)/g(3)$

15. $g(1) - f(1)$ **16.** $g(-3)/f(-3)$

17. $(f + g)(x)$ **18.** $(g - f)(x)$

Let $F(x) = x^2 - 2$ and $G(x) = 5 - x$. Find each of the following.

19. $(F + G)(x)$ **20.** $(F + G)(a)$

21. $(F - G)(3)$ **22.** $(F - G)(2)$

23. $(F \cdot G)(a)$ **24.** $(G \cdot F)(x)$

25. $(F/G)(x)$ **26.** $(G - F)(x)$

27. $(G/F)(-2)$ **28.** $(F/G)(-1)$

29. $(F + F)(1)$ **30.** $(G \cdot G)(6)$

The following graph shows the number of births in the United States, in millions, from 1970–2004. Here $C(t)$ represents the number of Caesarean section births, $B(t)$ the number of non-Caesarean section births, and $N(t)$ the total number of births in year t.

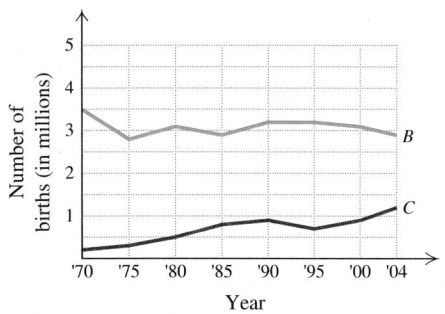

Source: National Center for Health Statistics

31. Use estimates of $C(2004)$ and $B(2004)$ to estimate $N(2004)$.

32. Use estimates of $C(1985)$ and $B(1985)$ to estimate $N(1985)$.

In 2004, a study comparing high doses of the cholesterol-lowering drugs Lipitor and Pravachol indicated that patients taking Lipitor were significantly less likely to have heart attacks or require angioplasty or surgery.

   In the graph below, $L(t)$ is the percentage of patients on Lipitor (80 mg) and $P(t)$ is the percentage of patients on Pravachol (40 mg) who suffered heart problems or death t years after beginning to take the medication.
Source: *New York Times,* March 9, 2004

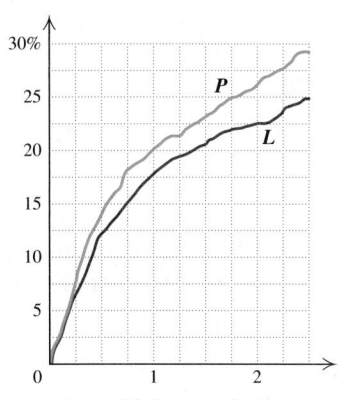

Years of follow-up of patients
Source: New England Journal of Medicine

33. Use estimates of $P(2)$ and $L(2)$ to estimate $(P - L)(2)$.

34. Use estimates of $P(1)$ and $L(1)$ to estimate $(P - L)(1)$.

Often function addition is represented by stacking the individual functions directly on top of each other. The graph below indicates how U.S. municipal solid waste has been managed. The braces indicate the values of the individual functions.

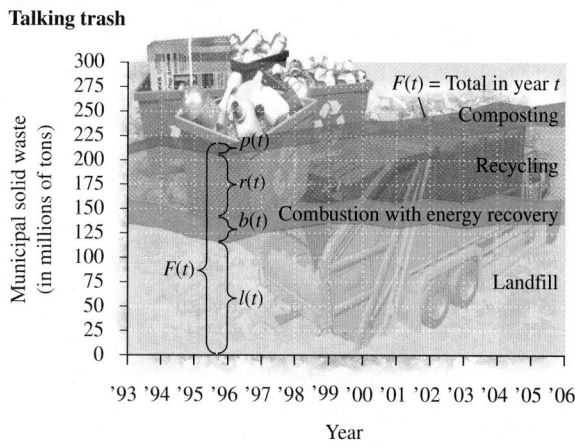

Source: Environmental Protection Agency

35. Estimate $(p + r)('05)$. What does it represent?

36. Estimate $(p + r + b)('05)$. What does it represent?

37. Estimate $F('96)$. What does it represent?

38. Estimate $F('06)$. What does it represent?

39. Estimate $(F - p)('04)$. What does it represent?

40. Estimate $(F - l)('03)$. What does it represent?

For each pair of functions f and g, determine the domain of the sum, difference, and product of the two functions.

41. $f(x) = x^2,$
    $g(x) = 7x - 4$

42. $f(x) = 5x - 1,$
    $g(x) = 2x^2$

43. $f(x) = \dfrac{1}{x + 5},$
    $g(x) = 4x^3$

44. $f(x) = 3x^2,$
    $g(x) = \dfrac{1}{x - 9}$

45. $f(x) = \dfrac{2}{x},$
    $g(x) = x^2 - 4$

46. $f(x) = x^3 + 1,$
    $g(x) = \dfrac{5}{x}$

47. $f(x) = x + \dfrac{2}{x - 1},$
    $g(x) = 3x^3$

48. $f(x) = 9 - x^2,$
    $g(x) = \dfrac{3}{x + 6} + 2x$

49. $f(x) = \dfrac{3}{2x + 9}$,

$g(x) = \dfrac{5}{1 - x}$

50. $f(x) = \dfrac{5}{3 - x}$,

$g(x) = \dfrac{1}{4x - 1}$

For each pair of functions f and g, determine the domain of f/g.

51. $f(x) = x^4$,
$g(x) = x - 3$

52. $f(x) = 2x^3$,
$g(x) = 5 - x$

53. $f(x) = 3x - 2$,
$g(x) = 2x + 8$

54. $f(x) = 5 + x$,
$g(x) = 6 - 2x$

55. $f(x) = \dfrac{3}{x - 4}$,

$g(x) = 5 - x$

56. $f(x) = \dfrac{1}{2 - x}$,

$g(x) = 7 + x$

57. $f(x) = \dfrac{2x}{x + 1}$,

$g(x) = 2x + 5$

58. $f(x) = \dfrac{7x}{x - 2}$,

$g(x) = 3x + 7$

For Exercises 59–66, consider the functions F and G as shown.

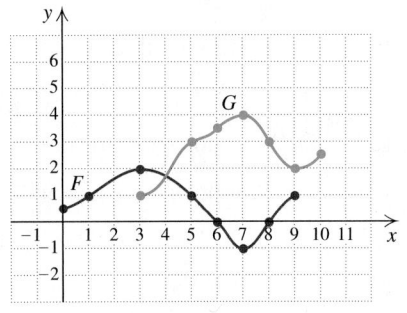

59. Determine $(F + G)(5)$ and $(F + G)(7)$.

60. Determine $(F \cdot G)(6)$ and $(F \cdot G)(9)$.

61. Determine $(G - F)(7)$ and $(G - F)(3)$.

62. Determine $(F/G)(3)$ and $(F/G)(7)$.

63. Find the domains of F, G, $F + G$, and F/G.

64. Find the domains of $F - G$, $F \cdot G$, and G/F.

65. Graph $F + G$.

66. Graph $G - F$.

In the following graph, S(t) represents the number of gallons of carbonated soft drinks consumed by the average American in year t, M(t) the number of gallons of milk, J(t) the number of gallons of fruit juice, and W(t) the number of gallons of bottled water.

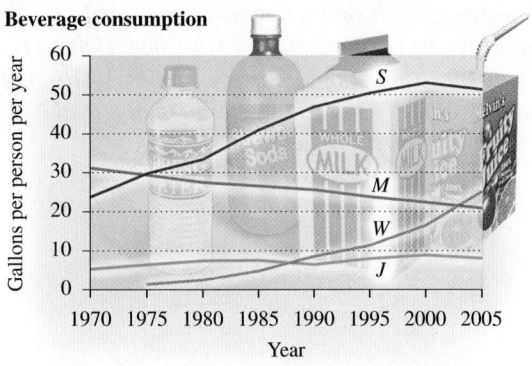

Beverage consumption

Source: Economic Research Service, U. S. Department of Agriculture

67. Between what years did the average American drink more soft drinks than juice, bottled water, and milk combined? Explain how you determined this.

68. Examine the graphs before Exercises 31 and 32. Did the total number of births increase or decrease from 1970 to 2004? Did the percent of births by Caesarean section increase or decrease from 1970 to 2004? Explain how you determined your answers.

Skill Review

To prepare for Section 7.5, review solving a formula for a variable (Section 2.3).

Solve. [2.3]

69. $ac = b$, for c

70. $x - wz = y$, for w

71. $pq - rq = st$, for q

72. $ab = d - cb$, for b

73. $ab - cd = 3b + d$, for b

74. $ab - cd = 3b + d$, for d

Synthesis

75. Examine the graphs following Example 2 and explain how they might be modified to represent the absorption of 200 mg of Advil® taken four times a day.

76. If $f(x) = c$, where c is some positive constant, describe how the graphs of $y = g(x)$ and $y = (f + g)(x)$ will differ.

77. Find the domain of F/G, if

$$F(x) = \dfrac{1}{x - 4} \quad \text{and} \quad G(x) = \dfrac{x^2 - 4}{x - 3}.$$

78. Find the domain of f/g, if

$$f(x) = \dfrac{3x}{2x + 5} \quad \text{and} \quad g(x) = \dfrac{x^4 - 1}{3x + 9}.$$

79. Sketch the graph of two functions f and g such that the domain of f/g is

$$\{x \mid -2 \le x \le 3 \ and \ x \ne 1\}.$$

80. Find the domains of $f + g, f - g, f \cdot g$, and f/g, if
$f = \{(-2, 1), (-1, 2), (0, 3), (1, 4), (2, 5)\}$
and
$g = \{(-4, 4), (-3, 3), (-2, 4), (-1, 0), (0, 5), (1, 6)\}.$

81. Find the domain of m/n, if
$$m(x) = 3x \quad for \ -1 < x < 5$$
and
$$n(x) = 2x - 3.$$

82. For f and g as defined in Exercise 80, find
$(f + g)(-2), (f \cdot g)(0)$, and $(f/g)(1)$.

83. Write equations for two functions f and g such that the domain of $f + g$ is
$\{x \mid x$ is a real number $and \ x \ne -2 \ and \ x \ne 5\}.$

84. Let $y_1 = 2.5x + 1.5$, $y_2 = x - 3$, and $y_3 = y_1/y_2$. Depending on whether the CONNECTED or DOT mode is used, the graph of y_3 appears as follows. Use algebra to determine which graph more accurately represents y_3.

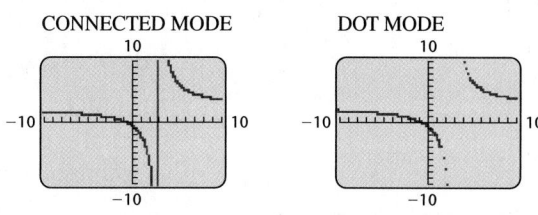

85. Using the window $[-5, 5, -1, 9]$, graph $y_1 = 5$, $y_2 = x + 2$, and $y_3 = \sqrt{x}$. Then predict what shape the graphs of $y_1 + y_2, y_1 + y_3$, and $y_2 + y_3$ will take. Use a graphing calculator to check each prediction.

86. Use the TABLE feature on a graphing calculator to check your answers to Exercises 45, 47, 55, and 57. (See the Technology Connection on p. 480.)

COLLABORATIVE CORNER

Time On Your Hands

Focus: The algebra of functions

Time: 10–15 minutes

Group size: 2–3

The graph and the data at right chart the average retirement age $R(x)$ and life expectancy $E(x)$ of U.S. citizens in year x.

ACTIVITY

1. Working as a team, perform the appropriate calculations and then graph $E - R$.

2. What does $(E - R)(x)$ represent? In what fields of study or business might the function $E - R$ prove useful?

3. Should E and R really be calculated separately for men and women? Why or why not?

4. What advice would you give to someone considering early retirement?

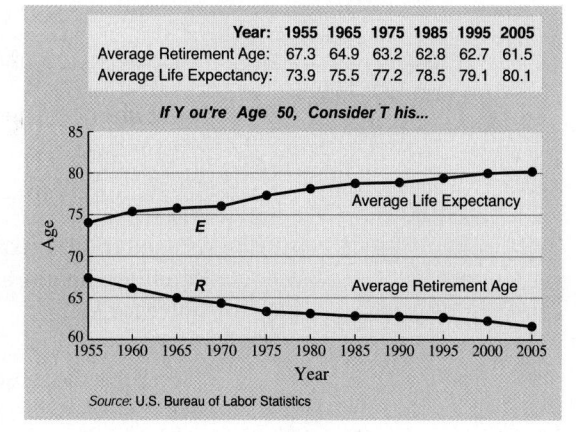

Year:	1955	1965	1975	1985	1995	2005
Average Retirement Age:	67.3	64.9	63.2	62.8	62.7	61.5
Average Life Expectancy:	73.9	75.5	77.2	78.5	79.1	80.1

Source: U.S. Bureau of Labor Statistics

7.5 Formulas, Applications, and Variation

Formulas ● Direct Variation ● Inverse Variation ● Joint Variation and Combined Variation

Formulas

Formulas occur frequently as mathematical models. Many formulas contain rational expressions, and to solve such formulas for a specified letter, we proceed as when solving rational equations.

EXAMPLE 1

Electronics. The formula

$$\frac{1}{R} = \frac{1}{r_1} + \frac{1}{r_2}$$

is used by electricians to determine the resistance R of two resistors r_1 and r_2 connected in parallel.* Solve for r_1.

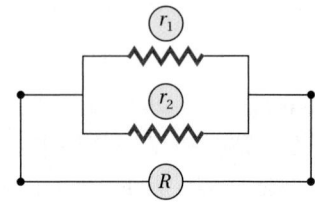

SOLUTION We use the same approach as in Section 6.6:

$$Rr_1r_2 \cdot \frac{1}{R} = Rr_1r_2 \cdot \left(\frac{1}{r_1} + \frac{1}{r_2}\right) \quad \text{Multiplying both sides by the LCD to clear fractions}$$

$$Rr_1r_2 \cdot \frac{1}{R} = Rr_1r_2 \cdot \frac{1}{r_1} + Rr_1r_2 \cdot \frac{1}{r_2} \quad \text{Multiplying to remove parentheses}$$

$$r_1r_2 = Rr_2 + Rr_1. \quad \text{Simplifying by removing factors equal to 1: } \frac{R}{R} = 1; \frac{r_1}{r_1} = 1; \frac{r_2}{r_2} = 1$$

At this point it is tempting to multiply by $1/r_2$ to get r_1 alone on the left, *but* note that there is an r_1 on the right. We must get all the terms involving r_1 on the *same side* of the equation.

$$r_1r_2 - Rr_1 = Rr_2 \quad \text{Subtracting } Rr_1 \text{ from both sides}$$

$$r_1(r_2 - R) = Rr_2 \quad \text{Factoring out } r_1 \text{ in order to combine like terms}$$

$$r_1 = \frac{Rr_2}{r_2 - R} \quad \text{Dividing both sides by } r_2 - R \text{ to get } r_1 \text{ alone}$$

This formula can be used to calculate r_1 whenever R and r_2 are known.

TRY EXERCISE ▸ 17

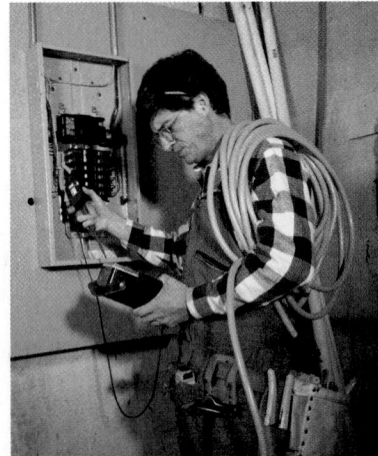

EXAMPLE 2

Astronomy. The formula

$$\frac{V^2}{R^2} = \frac{2g}{R + h}$$

is used to find a satellite's *escape velocity* V, where R is a planet's radius, h is the satellite's height above the planet, and g is the planet's gravitational constant. Solve for h.

_____

*Recall that the subscripts 1 and 2 merely indicate that r_1 and r_2 are different variables representing similar quantities.

SOLUTION   We first multiply by the LCD, $R^2(R + h)$, to clear fractions:

$$\frac{V^2}{R^2} = \frac{2g}{R + h}$$

$$R^2(R + h)\frac{V^2}{R^2} = R^2(R + h)\frac{2g}{R + h} \qquad \text{Multiplying to clear fractions}$$

$$\frac{R^2(R + h)V^2}{R^2} = \frac{R^2(R + h)2g}{R + h}$$

$$(R + h)V^2 = R^2 \cdot 2g. \qquad \text{Removing factors equal to 1:}$$
$$\frac{R^2}{R^2} = 1 \text{ and } \frac{R + h}{R + h} = 1$$

Remember: We are solving for h. Although we *could* distribute V^2, since h appears only within the factor $R + h$, it is easier to divide both sides by V^2:

$$\frac{(R + h)V^2}{V^2} = \frac{2R^2g}{V^2} \qquad \text{Dividing both sides by } V^2$$

$$R + h = \frac{2R^2g}{V^2} \qquad \text{Removing a factor equal to 1: } \frac{V^2}{V^2} = 1$$

$$h = \frac{2R^2g}{V^2} - R. \qquad \text{Subtracting } R \text{ from both sides}$$

The last equation can be used to determine the height of a satellite above a planet when the planet's radius and gravitational constant, along with the satellite's escape velocity, are known.

TRY EXERCISE ▶ 29

EXAMPLE   **3**   *Acoustics (the Doppler Effect).*   The formula

$$f = \frac{sg}{s + v}$$

is used to determine the frequency f of a sound that is moving at velocity v toward a listener who hears the sound as frequency g. Here s is the speed of sound in a particular medium. Solve for s.

STUDENT NOTES

The steps used to solve equations are precisely the same steps used to solve formulas. If you feel "rusty" in this regard, study the earlier section in which this type of equation first appeared. Then make sure that you can consistently solve those equations before returning to the work with formulas.

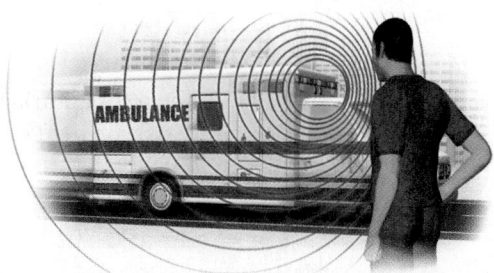

SOLUTION   We first clear fractions by multiplying by the LCD, $s + v$:

$$f \cdot (s + v) = \frac{sg}{s + v}(s + v)$$

$$fs + fv = sg. \qquad \text{The variable for which we are solving, } s,$$
$$\text{appears on both sides, forcing us to}$$
$$\text{distribute on the left side.}$$

Next, we must get all terms containing *s* on one side:

$$fv = sg - fs \qquad \text{Subtracting } fs \text{ from both sides}$$

$$fv = s(g - f) \qquad \text{Factoring out } s. \text{ This is like combining like terms.}$$

$$\frac{fv}{g - f} = s. \qquad \text{Dividing both sides by } g - f$$

Since *s* is isolated on one side, we have solved for *s*. This last equation can be used to determine the speed of sound whenever *f*, *v*, and *g* are known.

> **TRY EXERCISE** 19

STUDY SKILLS ———

Putting Math to Use

One excellent way to study math is to use it in your everyday life. The concepts of many sections can be reinforced if you look for real-life situations in which the math can be used.

> **To Solve a Rational Equation for a Specified Variable**
> 1. Multiply both sides by the LCD to clear fractions, if necessary.
> 2. Multiply to remove parentheses, if necessary.
> 3. Get all terms with the specified variable alone on one side.
> 4. Factor out the specified variable if it is in more than one term.
> 5. Multiply or divide on both sides to isolate the specified variable.

Variation

To extend our study of formulas and functions, we now examine three real-world situations: direct variation, inverse variation, and combined variation.

DIRECT VARIATION

A computer technician earns $22 per hour. In 1 hr, $22 is earned. In 2 hr, $44 is earned. In 3 hr, $66 is earned, and so on. This gives rise to a set of ordered pairs:

$$(1, 22), (2, 44), (3, 66), (4, 88), \quad \text{and so on.}$$

Note that the ratio of earnings *E* to time *t* is $\frac{22}{1}$ in every case.

If a situation is modeled by pairs for which the ratio is constant, we say there is **direct variation**. Here earnings *vary directly* as the time:

We have $\dfrac{E}{t} = 22$, so $E = 22t$ or, using function notation, $E(t) = 22t$.

> **Direct Variation**
> When a situation is modeled by a linear function of the form $f(x) = kx$, or $y = kx$, where *k* is a nonzero constant, we say that there is *direct variation*, that *y varies directly* as *x*, or that *y is proportional to x*. The number *k* is called the *variation constant*, or *constant of proportionality*.

Note that for $k > 0$, any equation of the form $y = kx$ indicates that as *x* increases, *y* increases as well.

EXAMPLE 4   Find the variation constant and an equation of variation if y varies directly as x, and $y = 32$ when $x = 2$.

SOLUTION   We know that $(2, 32)$ is a solution of $y = kx$. Therefore,

$$32 = k \cdot 2 \qquad \text{Substituting}$$

$$\frac{32}{2} = k, \quad \text{or} \quad k = 16. \qquad \text{Solving for } k$$

The variation constant is 16. The equation of variation is $y = 16x$. The notation $y(x) = 16x$ or $f(x) = 16x$ is also used.   **TRY EXERCISE** 43

EXAMPLE 5

Ocean waves.   The speed v of a train of ocean waves varies directly as the swell period t, or time between successive waves. Waves with a swell period of 12 sec are traveling 21 mph. How fast are waves traveling that have a swell period of 20 sec?

Source: www.rodntube.com

SOLUTION

1. **Familiarize.** Because of the phrase "v . . . varies directly as . . . t," we express the speed of the wave v, in miles per hour, as a function of the swell period t, in seconds. Thus, $v(t) = kt$, where k is the variation constant. Because we are using ratios, we can use the units "seconds" and "miles per hour" without converting sec to hr or hr to sec. Knowing that waves with a swell period of 12 sec are traveling 21 mph, we have $v(12) = 21$.

2. **Translate.** We find the variation constant using the data and then use it to write the equation of variation:

$$v(t) = kt$$
$$v(12) = k \cdot 12 \qquad \text{Replacing } t \text{ with } 12$$
$$21 = k \cdot 12 \qquad \text{Replacing } v(12) \text{ with } 21$$
$$\frac{21}{12} = k \qquad \text{Solving for } k$$
$$1.75 = k. \qquad \text{This is the variation constant.}$$

The equation of variation is $v(t) = 1.75t$. This is the translation.

3. **Carry out.** To find the speed of waves with a swell period of 20 sec, we compute $v(20)$:

$$v(t) = 1.75t$$
$$v(20) = 1.75(20) \qquad \text{Substituting } 20 \text{ for } t$$
$$= 35.$$

4. **Check.** To check, we could reexamine all our calculations. Note that our answer seems reasonable since the ratios 21/12 and 35/20 are both 1.75.

5. **State.** Waves with a swell period of 20 sec are traveling 35 mph.

TRY EXERCISE 55

INVERSE VARIATION

Suppose a bus travels 20 mi. At 20 mph, the trip takes 1 hr. At 40 mph, it takes $\frac{1}{2}$ hr. At 60 mph, it takes $\frac{1}{3}$ hr, and so on. This gives pairs of numbers, all having the same product:

$$(20, 1), \left(40, \tfrac{1}{2}\right), \left(60, \tfrac{1}{3}\right), \left(80, \tfrac{1}{4}\right), \quad \text{and so on.}$$

Note that the product of each pair is 20. When a situation is modeled by pairs for which the product is constant, we say that there is **inverse variation**. Since $r \cdot t = 20$, we have

$$t = \frac{20}{r} \quad \text{or, using function notation,} \quad t(r) = \frac{20}{r}.$$

Inverse Variation

When a situation is modeled by a rational function of the form $f(x) = k/x$, or $y = k/x$, where k is a nonzero constant, we say that there is *inverse variation*, that *y varies inversely as x*, or that *y is inversely proportional to x*. The number k is called the *variation constant*, or *constant of proportionality*.

Note that for $k > 0$, any equation of the form $y = k/x$ indicates that as x increases, y decreases.

EXAMPLE 6   Find the variation constant and an equation of variation if y varies inversely as x, and $y = 32$ when $x = 0.2$.

SOLUTION   We know that $(0.2, 32)$ is a solution of

$$y = \frac{k}{x}.$$

Therefore,

$$32 = \frac{k}{0.2} \qquad \text{Substituting}$$

$$(0.2)32 = k$$

$$6.4 = k. \qquad \text{Solving for } k$$

The variation constant is 6.4. The equation of variation is

$$y = \frac{6.4}{x}.$$

> **TRY EXERCISE** 49

There are many real-life quantities that vary inversely.

EXAMPLE 7   *Movie downloads.*   The time t that it takes to download a movie file varies inversely as the transfer speed s of the Internet connection. A typical full-length movie file will transfer in 48 min at a transfer speed of 256 KB/s (kilobytes per second). How long will it take to transfer the same movie file at a transfer speed of 32 KB/s?

Source: www.xsvidmovies.com

SOLUTION

1. **Familiarize.** Because of the phrase ". . . varies inversely as the transfer speed," we express the download time t, in minutes, as a function of the transfer speed s, in kilobytes per second. Thus, $t(s) = k/s$.

2. **Translate.** We use the given information to solve for k. We will then use that result to write the equation of variation.

$$t(s) = \frac{k}{s}$$

$$t(256) = \frac{k}{256} \qquad \text{Replacing } s \text{ with } 256$$

$$48 = \frac{k}{256} \qquad \text{Replacing } t(256) \text{ with } 48$$

$$12{,}288 = k.$$

The equation of variation is $t(s) = 12{,}288/s$. This is the translation.

3. **Carry out.** To find the download time at a transfer speed of 32 KB/s, we calculate $t(32)$:

$$t(32) = \frac{12{,}288}{32} = 384.$$

4. **Check.** Note that, as expected, as the transfer speed goes *down*, the download time goes *up*. Also, the products $48 \cdot 256$ and $32 \cdot 384$ are both 12,288.

5. **State.** At a transfer speed of 32 KB/s, it will take 384 min, or 6 hr 24 min, to download the movie file.

▸ TRY EXERCISE ▸ 57

JOINT VARIATION AND COMBINED VARIATION

When a variable varies directly with more than one other variable, we say that there is *joint variation*. For example, in the formula for the volume of a right circular cylinder, $V = \pi r^2 h$, we say that V varies *jointly* as h and the square of r.

> **Joint Variation**
>
> y varies *jointly* as x and z if, for some nonzero constant k, $y = kxz$.

EXAMPLE **8**   Find an equation of variation if y varies jointly as x and z, and $y = 30$ when $x = 2$ and $z = 3$.

SOLUTION   We have

$$y = kxz,$$

so

$$30 = k \cdot 2 \cdot 3$$

$$k = 5. \qquad \text{The variation constant is 5.}$$

The equation of variation is $y = 5xz$.

▸ TRY EXERCISE ▸ 73

Joint variation is one form of *combined variation*. In general, when a variable varies directly and/or inversely, at the same time, with more than one other variable, there is **combined variation**. Examples 8 and 9 are both examples of combined variation.

EXAMPLE 9   Find an equation of variation if y varies jointly as x and z and inversely as the square of w, and $y = 105$ when $x = 3$, $z = 20$, and $w = 2$.

SOLUTION   The equation of variation is of the form

$$y = k \cdot \frac{xz}{w^2},$$

so, substituting, we have

$$105 = k \cdot \frac{3 \cdot 20}{2^2}$$

$$105 = k \cdot 15$$

$$k = 7.$$

Thus,

$$y = 7 \cdot \frac{xz}{w^2}.$$

▶ TRY EXERCISE 75

7.5   **EXERCISE SET**

↩ *Concept Reinforcement*   *Match each statement with the correct term that completes it from the list on the right.*

1. To clear fractions, we can multiply both sides of an equation by the ____.

2. With direct variation, pairs of numbers have a constant ____.

3. With inverse variation, pairs of numbers have a constant ____.

4. If $y = k/x$, then y varies ____ as x.

5. If $y = kx$, then y varies ____ as x.

6. If $y = kxz$, then y varies ____ as x and z.

a) Directly

b) Inversely

c) Jointly

d) LCD

e) Product

f) Ratio

Determine whether each situation represents direct variation or inverse variation.

7. Two painters can scrape a house in 9 hr, whereas three painters can scrape the house in 6 hr.

8. Andres planted 5 bulbs in 20 min and 7 bulbs in 28 min.

9. Salma swam 2 laps in 7 min and 6 laps in 21 min.

10. It took 2 band members 80 min to set up for a show; with 4 members working, it took 40 min.

11. It took 3 hr for 4 volunteers to wrap the campus' collection of Toys for Tots, but only 1.5 hr with 8 volunteers working.

12. Ayana's air conditioner cooled off 1000 ft^3 in 10 min and 3000 ft^3 in 30 min.

Solve each formula for the specified variable.

13. $f = \dfrac{L}{d}$; d

14. $\dfrac{W_1}{W_2} = \dfrac{d_1}{d_2}$; W_1

15. $s = \dfrac{(v_1 + v_2)t}{2}$; v_1

16. $s = \dfrac{(v_1 + v_2)t}{2}$; t

17. $\dfrac{t}{a} + \dfrac{t}{b} = 1$; b

18. $\dfrac{1}{R} = \dfrac{1}{r_1} + \dfrac{1}{r_2}$; R

19. $R = \dfrac{gs}{g + s}$; g

20. $K = \dfrac{rt}{r - t}$; t

21. $I = \dfrac{nE}{R + nr}$; n

22. $I = \dfrac{nE}{R + nr}$; r

23. $\dfrac{1}{p} + \dfrac{1}{q} = \dfrac{1}{f}$; q

24. $\dfrac{1}{p} + \dfrac{1}{q} = \dfrac{1}{f}$; p

25. $S = \dfrac{H}{m(t_1 - t_2)}$; t_1

26. $S = \dfrac{H}{m(t_1 - t_2)}$; H

27. $\dfrac{E}{e} = \dfrac{R + r}{r}$; r

28. $\dfrac{E}{e} = \dfrac{R + r}{R}$; R

29. $S = \dfrac{a}{1 - r}$; r

30. $S = \dfrac{a - ar^n}{1 - r}$; a

Aha! **31.** $c = \dfrac{f}{(a + b)c}$; $a + b$

32. $d = \dfrac{g}{d(c + f)}$; $c + f$

33. *Interest.*   The formula

$$P = \dfrac{A}{1 + r}$$

is used to determine what principal P should be invested for one year at $(100 \cdot r)\%$ simple interest in order to have A dollars after a year. Solve for r.

34. *Taxable interest.*   The formula

$$I_t = \dfrac{I_f}{1 - T}$$

gives the *taxable interest rate* I_t equivalent to the *tax-free interest rate* I_f for a person in the $(100 \cdot T)\%$ tax bracket. Solve for T.

35. *Average speed.*   The formula

$$v = \dfrac{d_2 - d_1}{t_2 - t_1}$$

gives an object's average speed v when that object has traveled d_1 miles in t_1 hours and d_2 miles in t_2 hours. Solve for t_1.

36. *Average acceleration.*   The formula

$$a = \dfrac{v_2 - v_1}{t_2 - t_1}$$

gives a vehicle's *average acceleration* when its velocity changes from v_1 at time t_1 to v_2 at time t_2. Solve for t_2.

37. *Work rate.*   The formula

$$\dfrac{1}{t} = \dfrac{1}{a} + \dfrac{1}{b}$$

gives the total time t required for two workers to complete a job, if the workers' individual times are a and b. Solve for t.

38. *Planetary orbits.*   The formula

$$\dfrac{x^2}{a^2} + \dfrac{y^2}{b^2} = 1$$

can be used to plot a planet's elliptical orbit of width $2a$ and length $2b$ (see p. 869 in Section 10.2). Solve for b^2.

39. *Semester average.*   The formula

$$A = \dfrac{2Tt + Qq}{2T + Q}$$

gives a student's average A after T tests and Q quizzes, where each test counts as 2 quizzes, t is the test average, and q is the quiz average. Solve for Q.

40. *Astronomy.*   The formula

$$L = \dfrac{dR}{D - d},$$

where D is the diameter of the sun, d is the diameter of the earth, R is the earth's distance from the sun, and L is some fixed distance, is used in calculating when lunar eclipses occur. Solve for D.

41. *Body-fat percentage.*   The YMCA calculates men's body-fat percentage p using the formula

$$p = \dfrac{-98.42 + 4.15c - 0.082w}{w},$$

where c is the waist measurement, in inches, and w is the weight, in pounds. Solve for w.
Source: YMCA guide to Physical Fitness Assessment

42. *Preferred viewing distance.* Researchers model the distance D from which an observer prefers to watch television in "picture heights"—that is, multiples of the height of the viewing screen. The preferred viewing distance is given by

$$D = \frac{3.55H + 0.9}{H},$$

where D is in picture heights and H is in meters. Solve for H.

Source: www.tid.es, Telefonica Investigación y Desarrollo, S.A. Unipersonal

Find the variation constant and an equation of variation if y varies directly as x and the following conditions apply.

43. $y = 30$ when $x = 5$

44. $y = 80$ when $x = 16$

45. $y = 3.4$ when $x = 2$

46. $y = 2$ when $x = 5$

47. $y = 2$ when $x = \frac{1}{5}$

48. $y = 0.9$ when $x = 0.5$

Find the variation constant and an equation of variation in which y varies inversely as x, and the following conditions exist.

49. $y = 5$ when $x = 20$

50. $y = 40$ when $x = 8$

51. $y = 11$ when $x = 4$

52. $y = 9$ when $x = 10$

53. $y = 27$ when $x = \frac{1}{3}$

54. $y = 81$ when $x = \frac{1}{9}$

55. *Hooke's law.* Hooke's law states that the distance d that a spring is stretched by a hanging object varies directly as the mass m of the object. If the distance is 20 cm when the mass is 3 kg, what is the distance when the mass is 5 kg?

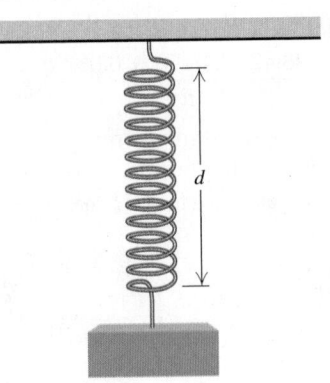

56. *Ohm's law.* The electric current I, in amperes, in a circuit varies directly as the voltage V. When 15 volts are applied, the current is 5 amperes. What is the current when 18 volts are applied?

57. *Work rate.* The time T required to do a job varies inversely as the number of people P working. It takes 5 hr for 7 volunteers to pick up rubbish from 1 mi of roadway. How long would it take 10 volunteers to complete the job?

58. *Pumping rate.* The time t required to empty a tank varies inversely as the rate r of pumping. If a Briggs and Stratton pump can empty a tank in 45 min at the rate of 600 kL/min, how long will it take the pump to empty the tank at 1000 kL/min?

59. *Water from melting snow.* The number of centimeters W of water produced from melting snow varies directly as the number of centimeters S of snow. Meteorologists know that under certain conditions, 150 cm of snow will melt to 16.8 cm of water. The average annual snowfall in Alta, Utah, is 500 in. Assuming the above conditions, how much water will replace the 500 in. of snow?

60. *Gardening.* The number of calories burned by a gardener is directly proportional to the time spent gardening. It takes 30 min to burn 180 calories. How long would it take to burn 240 calories when gardening?

Source: www.healthstatus.com

Aha! **61.** *Mass of water in a human.* The number of kilograms W of water in a human body varies directly as the mass of the body. A 96-kg person contains 64 kg of water. How many kilograms of water are in a 48-kg person?

62. *Weight on Mars.* The weight M of an object on Mars varies directly as its weight E on Earth. A person who weighs 95 lb on Earth weighs 38 lb on Mars. How much would a 100-lb person weigh on Mars?

63. *String length and frequency.*   The frequency of a string is inversely proportional to its length. A violin string that is 33 cm long vibrates with a frequency of 260 Hz. What is the frequency when the string is shortened to 30 cm?

64. *Wavelength and frequency.*   The wavelength W of a radio wave varies inversely as its frequency F. A wave with a frequency of 1200 kilohertz has a length of 300 meters. What is the length of a wave with a frequency of 800 kilohertz?

65. *Ultraviolet index.*   At an ultraviolet, or UV, rating of 4, those people who are less sensitive to the sun will burn in 75 min. Given that the number of minutes it takes to burn, t, varies inversely with the UV rating, u, how long will it take less sensitive people to burn when the UV rating is 14?
Source: *The Electronic Textbook of Dermatology* at www.telemedicine.org

66. *Current and resistance.*   The current I in an electrical conductor varies inversely as the resistance R of the conductor. If the current is $\frac{1}{2}$ ampere when the resistance is 240 ohms, what is the current when the resistance is 540 ohms?

67. *Air pollution.*   The average U.S. household of 2.6 people released 0.94 ton of carbon monoxide into the environment in a recent year. How many tons were released nationally? Use 305,000,000 as the U.S. population.
Sources: Based on data from the U.S. Environmental Protection Agency and the U.S. Census Bureau

68. *Relative aperture.*   The relative aperture, or f-stop, of a 23.5-mm lens is directly proportional to the focal length F of the lens. If a lens with a 150-mm focal length has an f-stop of 6.3, find the f-stop of a 23.5-mm lens with a focal length of 80 mm.

Find an equation of variation in which:

69. y varies directly as the square of x, and $y = 50$ when $x = 10$.

70. y varies directly as the square of x, and $y = 0.15$ when $x = 0.1$.

71. y varies inversely as the square of x, and $y = 50$ when $x = 10$.

72. y varies inversely as the square of x, and $y = 0.15$ when $x = 0.1$.

73. y varies jointly as x and z, and $y = 105$ when $x = 14$ and $z = 5$.

74. y varies jointly as x and z, and $y = \frac{3}{2}$ when $x = 2$ and $z = 10$.

75. y varies jointly as w and the square of x and inversely as z, and $y = 49$ when $w = 3$, $x = 7$, and $z = 12$.

76. y varies directly as x and inversely as w and the square of z, and $y = 4.5$ when $x = 15$, $w = 5$, and $z = 2$.

77. *Stopping distance of a car.*   The stopping distance d of a car after the brakes have been applied varies directly as the square of the speed r. Once the brakes are applied, a car traveling 60 mph can stop in 138 ft. What stopping distance corresponds to a speed of 40 mph?
Source: Based on data from Edmunds.com

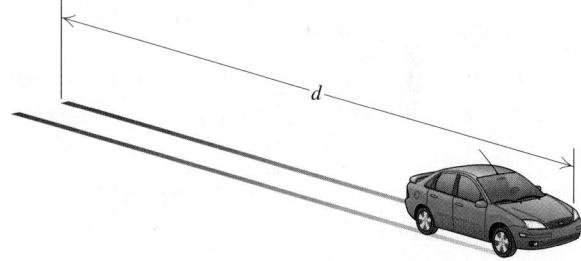

78. *Reverberation time.*   A sound's reverberation time T is the time it takes for the sound level to decrease by 60 dB (decibels) after the sound has been turned off. Reverberation time varies directly as the volume V of a room and inversely as the sound absorption A of the room. A given sound has a reverberation time of 1.5 sec in a room with a volume of 90 m^3 and a sound absorption of 9.6. What is the reverberation time of the same sound in a room with a volume of 84 m^3 and a sound absorption of 10.5?
Source: www.isover.co.uk

79. *Volume of a gas.* The volume V of a given mass of a gas varies directly as the temperature T and inversely as the pressure P. If $V = 231 \text{ cm}^3$ when $T = 300°\text{K}$ (Kelvin) and $P = 20 \text{ lb/cm}^2$, what is the volume when $T = 320°\text{K}$ and $P = 16 \text{ lb/cm}^2$?

80. *Intensity of a signal.* The intensity I of a television signal varies inversely as the square of the distance d from the transmitter. If the intensity is 25 W/m^2 at a distance of 2 km, what is the intensity 6.25 km from the transmitter?

81. *Atmospheric drag.* Wind resistance, or atmospheric drag, tends to slow down moving objects. Atmospheric drag W varies jointly as an object's surface area A and velocity v. If a car traveling at a speed of 40 mph with a surface area of 37.8 ft^2 experiences a drag of 222 N (Newtons), how fast must a car with 51 ft^2 of surface area travel in order to experience a drag force of 430 N?

82. *Drag force.* The drag force F on a boat varies jointly as the wetted surface area A and the square of the velocity of the boat. If a boat traveling 6.5 mph experiences a drag force of 86 N when the wetted surface area is 41.2 ft^2, find the wetted surface area of a boat traveling 8.2 mph with a drag force of 94 N.

83. If y varies directly as x, does doubling x cause y to be doubled as well? Why or why not?

84. Which exercise did you find easier to work: Exercise 15 or Exercise 19? Why?

Skill Review

To prepare for Chapter 8, review solving an equation for y and translating phrases to algebraic expressions (Sections 1.1 and 2.3).

Solve. [2.3]

85. $x - 6y = 3$, for y

86. $3x - 8y = 5$, for y

87. $5x + 2y = -3$, for y

88. $x + 8y = 4$, for y

Translate each of the following. Do not solve. [1.1]

89. Five more than twice a number is 49.

90. Three less than half of some number is 57.

91. The sum of two consecutive integers is 145.

92. The difference between a number and its opposite is 20.

Synthesis

93. Suppose that the number of customer complaints is inversely proportional to the number of employees hired. Will a firm reduce the number of complaints more by expanding from 5 to 10 employees, or from 20 to 25? Explain. Consider using a graph to help justify your answer.

94. Why do you think subscripts are used in Exercises 15 and 25 but not in Exercises 27 and 28?

95. *Escape velocity.* A satellite's escape velocity is 6.5 mi/sec, the radius of the earth is 3960 mi, and the earth's gravitational constant is 32.2 ft/sec^2. How far is the satellite from the surface of the earth? (See Example 2.)

96. The *harmonic mean* of two numbers a and b is a number M such that the reciprocal of M is the average of the reciprocals of a and b. Find a formula for the harmonic mean.

97. *Health-care.* Young's rule for determining the size of a particular child's medicine dosage c is

$$c = \frac{a}{a + 12} \cdot d,$$

where a is the child's age and d is the typical adult dosage. If a child's age is doubled, the dosage increases. Find the ratio of the larger dosage to the smaller dosage. By what percent does the dosage increase?

Source: Olsen, June Looby, Leon J. Ablon, and Anthony Patrick Giangrasso, *Medical Dosage Calculations*, 6th ed.

98. Solve for x:

$$x^2\left(1 - \frac{2pq}{x}\right) = \frac{2p^2q^3 - pq^2x}{-q}.$$

99. *Average acceleration.* The formula

$$a = \frac{\dfrac{d_4 - d_3}{t_4 - t_3} - \dfrac{d_2 - d_1}{t_2 - t_1}}{t_4 - t_2}$$

can be used to approximate average acceleration, where the d's are distances and the t's are the corresponding times. Solve for t_1.

100. If y varies inversely as the cube of x and x is multiplied by 0.5, what is the effect on y?

101. *Intensity of light.*   The intensity I of light from a bulb varies directly as the wattage of the bulb and inversely as the square of the distance d from the bulb. If the wattage of a light source and its distance from reading matter are both doubled, how does the intensity change?

102. Describe in words the variation represented by $W = \dfrac{km_1 M_1}{d^2}$. Assume k is a constant.

103. *Tension of a musical string.*   The tension T on a string in a musical instrument varies jointly as the string's mass per unit length m, the square of its length l, and the square of its fundamental frequency f. A 2-m long string of mass 5 gm/m with a fundamental frequency of 80 has a tension of 100 N (Newtons). How long should the same string be if its tension is going to be changed to 72 N?

104. *Volume and cost.*   A peanut butter jar in the shape of a right circular cylinder is 4 in. high and 3 in. in diameter and sells for $1.20. If we assume that cost is proportional to volume, how much should a jar 6 in. high and 6 in. in diameter cost?

105. *Golf distance finder.*   A device used in golf to estimate the distance d to a hole measures the size s that the 7-ft pin *appears* to be in a viewfinder. The viewfinder uses the principle, diagrammed here, that s gets bigger when d gets smaller. If $s = 0.56$ in. when $d = 50$ yd, find an equation of variation that expresses d as a function of s. What is d when $s = 0.40$ in.?

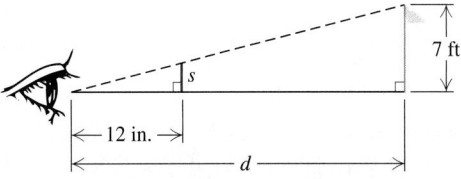

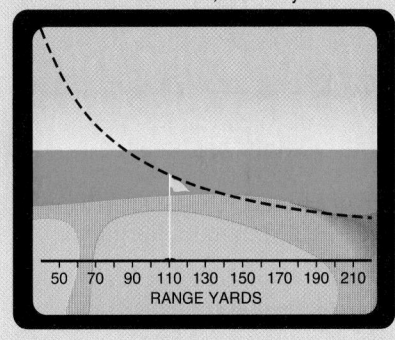

HOW IT WORKS:
Just sight the flagstick through the viewfinder…
fit flag between top dashed line and the solid line below…
…read the distance, 50 – 220 yards.

| 50 | 70 | 90 | 110 | 130 | 150 | 170 | 190 | 210 |
RANGE YARDS

Nothing to focus.
·
Gives you exact distance that your ball
lies from the flagstick.
·
Choose proper club on every approach shot.
·
Figure new pin placement instantly.
·
Train your naked eye for formal and tournament play.
·
Eliminate the need to remember every stake,
tree, and bush on the course.

Study Summary

KEY TERMS AND CONCEPTS	EXAMPLES

SECTION 7.1: INTRODUCTION TO FUNCTIONS

A **function** is a correspondence between a first set, called the **domain**, and a second set, called the **range**, such that each member of the domain corresponds to *exactly one* member of the range.

The correspondence $f: \left\{\left(-1, \frac{1}{2}\right), (0, 1), (1, 2), (2, 4), (3, 8)\right\}$ is a function.

The domain of $f = \{-1, 0, 1, 2, 3\}$.

The range of $f = \left\{\frac{1}{2}, 1, 2, 4, 8\right\}$.

$f(-1) = \frac{1}{2}$

The input -1 corresponds to the output $\frac{1}{2}$.

The Vertical-Line Test

If it is possible for a vertical line to cross a graph more than once, then the graph is not the graph of a function.

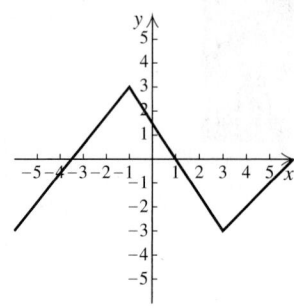

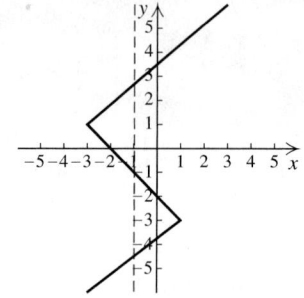

This *is* the graph of a function.   This *is not* the graph of a function.

SECTION 7.2: DOMAIN AND RANGE

The domain of a function is the set of all x-coordinates of the points on the graph.

The range of a function is the set of all y-coordinates of the points on the graph.

Consider the function given by
$f(x) = |x| - 3$.

    The domain of the function is $\mathbb{R}$.

    The range of the function
    is $\{y \mid y \geq -3\}$, or $[-3, \infty)$.

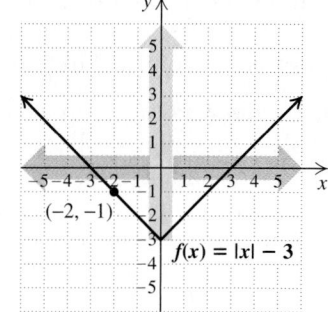

Unless otherwise stated, the domain of a function is the set of all numbers for which function values can be calculated.

Consider the function given by $f(x) = \dfrac{x + 2}{x - 7}$.

Function values cannot be calculated when the denominator is 0. Since $x - 7 = 0$ when $x = 7$, the domain of f is

$$\{x \mid x \text{ is a real number } and\ x \neq 7\}.$$

SECTION 7.3: GRAPHS OF FUNCTIONS

Linear Function

$$f(x) = mx + b$$

Constant Function

$$f(x) = b$$

Graph: $f(x) = \frac{1}{2}x - 3$.

We plot the y-intercept, $(0, -3)$. From there, we count off a slope of $\frac{1}{2}$: We go up 1 unit and to the right 2 units to the point $(2, -2)$. We then draw the graph.

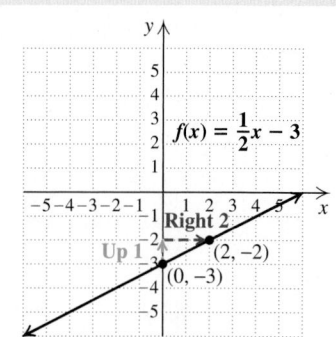

Graphs of nonlinear functions are not straight lines.

x	$f(x) = x^2 - 3$	$(x, f(x))$
0	-3	$(0, -3)$
-1	-2	$(-1, -2)$
1	-2	$(1, -2)$
-2	1	$(-2, 1)$
2	1	$(2, 1)$

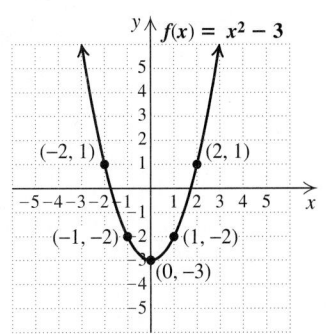

f is a quadratic function.

Domain of $f = \mathbb{R}$, or $(-\infty, \infty)$.

Range of $f = \{y | y \geq -3\}$, or $[-3, \infty)$.

SECTION 7.4: THE ALGEBRA OF FUNCTIONS

For $f(x) = x^2 + 3x$ and $g(x) = x - 5$:

$(f + g)(x) = f(x) + g(x)$

$$(f + g)(x) = f(x) + g(x)$$
$$= x^2 + 3x + x - 5 = x^2 + 4x - 5;$$

$(f - g)(x) = f(x) - g(x)$

$$(f - g)(x) = f(x) - g(x)$$
$$= x^2 + 3x - (x - 5) = x^2 + 2x + 5;$$

$(f \cdot g)(x) = f(x) \cdot g(x)$

$$(f \cdot g)(x) = f(x) \cdot g(x)$$
$$= (x^2 + 3x)(x - 5) = x^3 - 2x^2 - 15x;$$

$(f/g)(x) = f(x)/g(x)$, provided $g(x) \neq 0$

$$(f/g)(x) = f(x)/g(x), \text{ provided } g(x) \neq 0$$
$$= \frac{x^2 + 3x}{x - 5}, \text{ provided } x \neq 5. \qquad g(5) = 0$$

SECTION 7.5: FORMULAS, APPLICATIONS, AND VARIATION

Direct Variation

$$y = kx$$

If y varies directly as x and y = 45 when x = 0.15, find the equation of variation.

$$y = kx$$
$$45 = k(0.15)$$
$$300 = k$$

The equation of variation is $y = 300x$.

Inverse Variation

$$y = \frac{k}{x}$$

If y varies inversely as x and y = 45 when x = 0.15, find the equation of variation.

$$y = \frac{k}{x}$$
$$45 = \frac{k}{0.15}$$
$$6.75 = k$$

The equation of variation is $y = \dfrac{6.75}{x}$.

Joint Variation	If y varies jointly as x and z and $y = 40$ when $x = 5$ and $z = 4$, find the equation of variation.
$y = kxz$	$$y = kxz$$ $$40 = k \cdot 5 \cdot 4$$ $$2 = k$$ The equation of variation is $y = 2xz$.

Review Exercises: Chapter 7

🔖 *Concept Reinforcement* *Classify each of the following as either true or false.*

1. Every function is a relation. [7.1]

2. When we are discussing functions, the notation $f(3)$ does not mean $f \cdot 3$. [7.1]

3. If a graph includes both $(9, 5)$ and $(7, 5)$, it cannot represent a function. [7.1]

4. The domain and the range of a function can be the same set of numbers. [7.2]

5. The horizontal-line test is a quick way to determine whether a graph represents a function. [7.1]

6. In a piecewise-defined function, the function values are determined using more than one rule. [7.2]

7. $(f + g)(x) = f(x) + g(x)$ is not an example of the distributive law when f and g are functions. [7.4]

8. In order for $(f/g)(a)$ to exist, we must have $g(a) \neq 0$. [7.4]

9. If x varies inversely as y, then there exists some constant k for which $x = k/y$. [7.5]

10. If 2 people can decorate for a party in 5 hr, and 10 people can decorate for the same party in 4 hr, the situation represents inverse variation. [7.5]

11. For the following graph of f, determine **(a)** $f(2)$; **(b)** the domain of f; **(c)** any x-values for which $f(x) = 2$; and **(d)** the range of f. [7.1], [7.2]

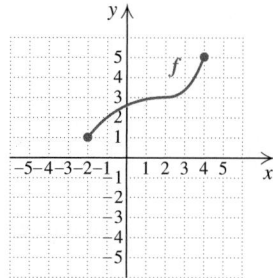

12. Find $g(-3)$ for $g(x) = \dfrac{x}{x + 1}$. [7.1]

13. Find $f(2a)$ for $f(x) = x^2 + 2x - 3$. [7.1]

14. The function $A(t) = 0.233t + 5.87$ can be used to estimate the median age of cars in the United States t years after 1990. (In this context, a median age of 3 yr means that half the cars are more than 3 yr old and half are less.) Predict the median age of cars in 2010; that is, find $A(20)$. [7.1]

Source: The Polk Co.

15. The following table shows the U.S. minimum hourly wage. Use the data in the table to draw a graph and to estimate the U.S. minimum hourly wage in 2012. [7.1]

Input, Year	Output, U.S. Minimum Hourly Wage
1997	$5.15
2007	5.85
2008	6.55
2009	7.25

For each of the graphs in Exercises 16–19, **(a)** *determine whether the graph represents a function and* **(b)** *if so, determine the domain and the range of the function.* [7.1], [7.2]

16.

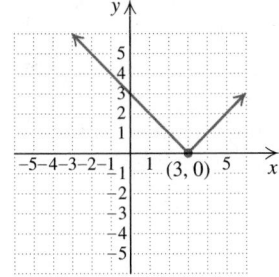

17.

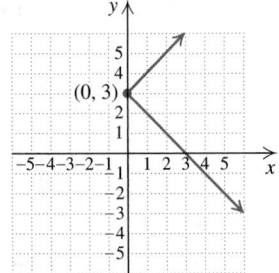

18.

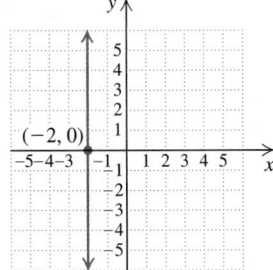

19.

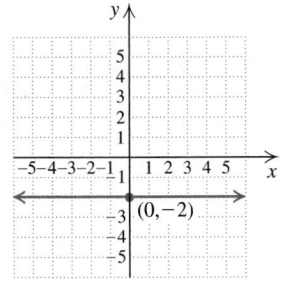

Find the domain of each function.

20. $f(x) = 3x^2 - 7$ [7.2]

21. $g(x) = \dfrac{x^2}{x - 1}$ [7.2]

22. $f(t) = \dfrac{1}{t^2 + 5t + 4}$ [7.2]

23. If a service agreement is cancelled, the amount that Vale Appliances will refund on the agreement is given by the function

$$r(t) = 900 - 15t,$$

where t is the number of weeks since the date of purchase. What is the domain of the function? [7.2]

24. For the function given by

$$f(x) = \begin{cases} 2 - x, & \text{if } x \le -2, \\ x^2, & \text{if } -2 < x \le 5, \\ x + 10, & \text{if } x > 5, \end{cases}$$

find **(a)** $f(-3)$; **(b)** $f(-2)$; **(c)** $f(4)$; and **(d)** $f(25)$. [7.2]

25. It costs $90 plus $30 a month to join the Family Fitness Center. Formulate a linear function to model the cost $C(t)$ for t months of membership, and determine the time required for the cost to reach $300. [7.3]

26. *Records in the 200-meter run.* In 1983, the record for the 200-m run was 19.75 sec. In 2007, it was 19.32 sec. Let $R(t)$ represent the record in the 200-m run and t the number of years since 1980. [7.3]

Source: International Association of Athletics Federation

a) Find a linear function that fits the data.

b) Use the function of part (a) to predict the record in 2013 and in 2020.

Classify each function as a linear function, an absolute-value function, a quadratic function, another polynomial function, or a rational function. [7.3]

27. $f(x) = |3x - 7|$

28. $g(x) = 4x^5 - 8x^3 + 7$

29. $p(x) = x^2 + x - 10$

30. $h(n) = 4n - 17$

31. $s(t) = \dfrac{t + 1}{t + 2}$

Graph each function and determine its domain and range. [7.3]

32. $f(x) = 3$

33. $f(x) = 2x + 1$

34. $g(x) = |x + 1|$

Let $g(x) = 3x - 6$ and $h(x) = x^2 + 1$. Find the following.

35. $(g \cdot h)(4)$ [7.4]

36. $(g - h)(-2)$ [7.4]

37. $(g/h)(-1)$ [7.4]

38. The domains of $g + h$ and $g \cdot h$ [7.4]

39. The domain of h/g [7.4]

Solve. [7.5]

40. $I = \dfrac{2V}{R + 2r}$, for r

41. $S = \dfrac{H}{m(t_1 - t_2)}$, for m

42. $\dfrac{1}{ac} = \dfrac{2}{ab} - \dfrac{3}{bc}$, for c

43. $T = \dfrac{A}{v(t_2 - t_1)}$, for t_1

44. Find an equation of variation in which y varies directly as x, and $y = 30$ when $x = 4$. [7.5]

45. Find an equation of variation in which y varies inversely as x, and $y = 3$ when $x = \frac{1}{4}$. [7.5]

46. Find an equation of variation in which y varies jointly as x and the square of w and inversely as z, and $y = 150$ when $x = 6$, $w = 10$, and $z = 2$. [7.5]

Solve. [7.5]

47. For those people with highly sensitive skin, an ultra-violet, or UV, rating of 6 will cause sunburn after 10 min. Given that the number of minutes it takes to burn t varies inversely as the UV rating u, how long will it take a highly sensitive person to burn on a day with a UV rating of 4?
Source: *The Electronic Textbook of Dermatology* found at www.telemedicine.org

48. The amount of waste generated by a family varies directly as the number of people in the family. The average U.S. family has 3.2 people and generates 14.4 lb of waste daily. How many pounds of waste would be generated daily by a family of 5?
Sources: Based on data from the U.S. Census Bureau and the U.S. Statistical Abstract 2007

49. *Electrical safety.* The amount of time t needed for an electrical shock to stop a 150-lb person's heart varies inversely as the square of the current flowing through the body. It is known that a 0.089-amp current is deadly to a 150-lb person after 3.4 sec. How long would it take a 0.096-amp current to be deadly?
Source: Safety Consulting Services

Synthesis

50. If two functions have the same domain and range, are the functions identical? Why or why not? [7.2]

51. Jenna believes that 0 is never in the domain of a rational function. Is she correct? Why or why not? [7.2]

52. Treasure Tea charges $7.99 for each package of loose tea. Shipping charges are $2.95 per package plus $20 per order for overnight delivery. Find a linear function for determining the cost of one order of x packages of tea, including shipping and overnight delivery. [7.3]

53. Determine the domain and the range of the function graphed below. [7.2]

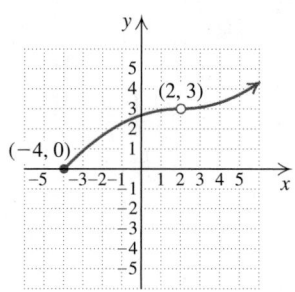

1. For the following graph of f, determine **(a)** $f(-2)$; **(b)** the domain of f; **(c)** any x-value for which $f(x) = \frac{1}{2}$; and **(d)** the range of f.

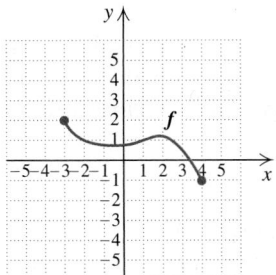

2. There were 41.9 million international visitors to the United States in 2002 and 51.0 million visitors in 2006. Draw a graph and estimate the number of international visitors in 2005.
   Sources: U.S. Department of Commerce, ITA, Office of Travel and Tourism Industries; Global Insight, Inc.

*For each of the following graphs, **(a)** determine whether the graph represents a function and **(b)** if so, determine the domain and the range of the function.*

3.

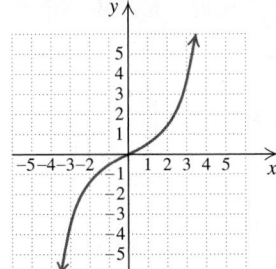

4.

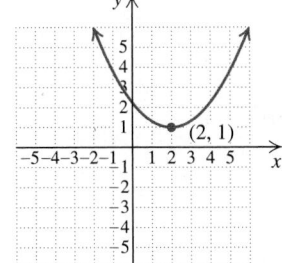

5.

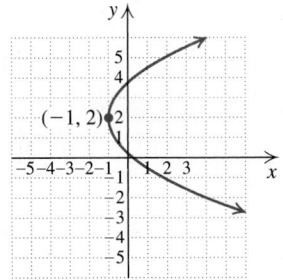

6. The distance d, in miles, that Kerry is from Chicago is given by the function $d(t) = 240 - 60t$, where t is the number of hours since he left Indianapolis. What is the domain of the function?

7. For the function given by

$$f(x) = \begin{cases} x^2, & \text{if } x < 0, \\ 3x - 5, & \text{if } 0 \le x \le 2, \\ x + 7, & \text{if } x > 2, \end{cases}$$

find **(a)** $f(0)$; **(b)** $f(3)$.

8. Porter paid \$180 for his phone. His monthly service fee is \$55. Formulate a linear function to model the cost $C(t)$ for t months of service, and determine the amount of time required for the total cost to reach \$840.

9. If you rent a truck for one day and drive it 250 mi, the cost is \$100. If you rent it for one day and drive it 300 mi, the cost is \$115. Let $C(m)$ represent the cost, in dollars, of driving m miles.
   a) Find a linear function that fits the data.
   b) Use the function to determine how much it will cost to rent the truck for one day and drive it 500 mi.

Classify each function as a linear function, a quadratic function, another polynomial function, an absolute-value function, or a rational function. Then find the domain of each function.

10. $f(x) = \dfrac{1}{4}x + 7$

11. $g(x) = \dfrac{3}{x^2 - 16}$

12. $p(x) = 4x^2 + 7$

Graph each function and determine its domain and range.

13. $f(x) = \dfrac{1}{3}x - 2$ 14. $g(x) = x^2 - 1$

15. $h(x) = -\dfrac{1}{2}$

Find the following, given that $g(x) = \dfrac{1}{x}$ and $h(x) = 2x + 1$.

16. $g(-1)$ 17. $h(5a)$

18. $(g + h)(x)$

19. The domain of g

20. The domain of $g + h$

21. The domain of g/h

22. Solve $R = \dfrac{gs}{g + s}$ for s.

23. Find an equation of variation in which y varies directly as x, and $y = 10$ when $x = 20$.

24. The number of workers n needed to clean a stadium after a game varies inversely as the amount of time t allowed for the cleanup. If it takes 25 workers to clean the stadium when there are 6 hr allowed for the job, how many workers are needed if the stadium must be cleaned in 5 hr?

25. The surface area of a balloon varies directly as the square of its radius. The area is 325 in² when the radius is 5 in. What is the area when the radius is 7 in.?

Synthesis

26. The function $f(t) = 5 + 15t$ can be used to determine a bicycle racer's location, in miles from the starting line, measured t hours after passing the 5-mi mark.

   a) How far from the start will the racer be 1 hr and 40 min after passing the 5-mi mark?

   b) Assuming a constant rate, how fast is the racer traveling?

27. Given that $f(x) = 5x^2 + 1$ and $g(x) = 4x - 3$, find an expression for $h(x)$ so that the domain of $f/g/h$ is $\left\{ x \mid x \text{ is a real number } and \ x \neq \frac{3}{4} \ and \ x \neq \frac{2}{7} \right\}$.
Answers may vary.

Cumulative Review: Chapters 1–7

1. Evaluate
$$\frac{2x - y^2}{x + y}$$
for $x = 3$ and $y = -4$. [1.8]

2. Convert to scientific notation: 391,000,000. [4.2]

3. Determine the slope and the y-intercept for the line given by $7x - 4y = 12$. [3.6]

4. Find an equation for the line that passes through the points $(-1, 7)$ and $(4, -3)$. [3.7]

5. If
$$f(x) = \frac{x - 3}{x^2 - 11x + 30},$$
find **(a)** $f(3)$ and **(b)** the domain of f. [7.1], [7.2]

Graph on a plane.

6. $5x = y$ [3.2]   **7.** $8y + 2x = 16$ [3.3]

8. $f(x) = -4$ [7.2]   **9.** $y = \frac{1}{3}x - 2$ [3.6]

Perform the indicated operations and simplify.

10. $(8x^3y^2)(-3xy^2)$ [4.1]

11. $(5x^2 - 2x + 1)(3x^2 + x - 2)$ [4.5]

12. $(3x^2 + y)^2$ [4.6]

13. $(2x^2 - 9)(2x^2 + 9)$ [4.6]

14. $(-5m^3n^2 - 3mn^3) + (-4m^2n^2 + 4m^3n^2) - (2mn^3 - 3m^2n^2)$ [4.4]

15. $\dfrac{y^2 - 36}{2y + 8} \cdot \dfrac{y + 4}{y + 6}$ [6.2]

16. $\dfrac{x^4 - 1}{x^2 - x - 2} \div \dfrac{x^2 + 1}{x - 2}$ [6.2]

17. $\dfrac{5ab}{a^2 - b^2} + \dfrac{a + b}{a - b}$ [6.4]

18. $\dfrac{2}{m + 1} + \dfrac{3}{m - 5} - \dfrac{m^2 - 1}{m^2 - 4m - 5}$ [6.4]

19. $y - \dfrac{2}{3y}$ [6.4]

20. Simplify: $\dfrac{\dfrac{1}{x} - \dfrac{1}{y}}{x + y}$. [6.5]

Factor.

21. $4x^3 + 400x$ [5.1]

22. $x^2 + 8x - 84$ [5.2]

23. $16y^2 - 25$ [5.4]

24. $64x^3 + 8$ [5.5]

25. $t^2 - 16t + 64$ [5.4]

26. $x^6 - x^2$ [5.4]

27. $\frac{1}{8}b^3 - c^3$ [5.5]

28. $3t^2 + 17t - 28$ [5.3]

29. $x^5 - x^3y + x^2y - y^2$ [5.1]

Solve.

30. $8x = 1 + 16x^2$ [5.7]

31. $288 = 2y^2$ [5.7]

32. $\frac{1}{3}x - \frac{1}{5} \geq \frac{1}{5}x - \frac{1}{3}$ [2.6]

33. $5(x - 2) - (x - 3) = 7x - 2(5 - x)$ [2.2]

34. $\dfrac{6}{x - 5} = \dfrac{2}{2x}$ [6.6]

35. $\dfrac{3x}{x - 2} - \dfrac{6}{x + 2} = \dfrac{24}{x^2 - 4}$ [6.6]

36. $P = \dfrac{4a}{a + b}$, for a [7.5]

37. Find the slope of the line containing $(2, 5)$ and $(1, 10)$. [3.5]

38. Find the slope of the line given by $f(x) = 8x + 3$. [7.3]

39. Find the slope of the line given by $y + 6 = -4$. [3.5]

40. Find an equation of the line containing $(5, -2)$ and perpendicular to the line given by $x - y = 5$. [3.7]

Find the following, given that $f(x) = x + 5$ and $g(x) = x^2 - 1$.

41. $g(-10)$ [7.1]   **42.** $(g/f)(x)$ [7.4]

43. Find the domain of f if $f(x) = \dfrac{x}{x + 6}$. [7.2]

44. Determine the domain and the range of the function f represented below. [7.2]

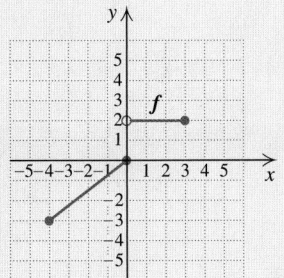

45. *Broadway revenue.* Gross revenue from Broadway shows has grown from \$20 million in 1986–1987 to \$939 million in 2006–2007. Let $r(t)$ represent gross revenue, in millions of dollars, from Broadway shows t seasons after the 1986–1987 season. [7.3]
Source: The League of American Theatres and Producers

a) Find a linear function that fits the data.

b) Use the function from part (a) to predict the gross revenue from Broadway shows in 2009–2010.

c) In what season will the gross revenue from Broadway shows reach \$1.4 billion?

46. *Broadway performances.* In January 2006, *The Phantom of the Opera* became the longest-running Broadway show with 7486 performances. By January 2008, the show had played 8302 times. Calculate the rate at which the number of performances was rising. [3.4]

47. *Quilting.* A rectangular quilted wall hanging is 4 in. longer than it is wide. The area of the quilt is 320 in². Find the perimeter of the quilt. [5.8]

48. *Hotel management.* The IQAir HealthPro Plus air purifier can clean the air in a 20-ft by 25-ft meeting room in 5 fewer minutes than it takes the Austin Healthmate HM400 to do the same job. Together the two machines can purify the air in the room in 6 min. How long would it take each machine, working alone, to purify the air in the room? [6.7]
Source: Manufacturers' and retailers' websites

49. *Driving delays.* According to the National Surface Transportation Policy and Revenue Study Commission, the best-case scenario for driving delays due to road work in 2055 will be 250% of the delays in 2005. If the commission predicts 30 billion hr of driving delays in 2055, how many hours of driving delays were there in 2005? [2.5]

50. *Driving time.* The time t that it takes for Johann to drive to work varies inversely as his speed. On a day when Johann averages 45 mph, it takes him 20 min to drive to work. How long will it take him to drive to work when he averages only 40 mph? [7.5]

51. *Disaster relief.* Six months after Hurricane Katrina struck the Gulf Coast in 2005, $2.18 billion of relief money had been distributed to disaster victims. This was $\frac{2}{3}$ of the amount raised by charity for disaster relief. How much money was still to be distributed? [2.5]
Source: www.washingtonpost.com

Synthesis

52. Multiply: $(x - 4)^3$. [4.5]

53. Find all roots for $f(x) = x^4 - 34x^2 + 225$. [5.6], [7.3]

Solve.

54. $\dfrac{18}{x - 9} + \dfrac{10}{x + 5} = \dfrac{28x}{x^2 - 4x - 45}$ [6.6]

55. $16x^3 = x$ [5.7]

56. *Photo books.* An Everyday Photo Book costs $12 for the first 20 pages plus $0.75 for each additional page. Formulate a piecewise-defined function to model the cost $C(x)$ for a book with x pages. [7.2]
Source: snapfish.com

Systems of Linear Equations and Problem Solving

JUDITH L. BRONSTEIN
NATURALIST/ECOLOGIST
Tucson, Arizona

As an ecologist, I often use equations both to estimate numbers of organisms and to make predictions about how those numbers will change under different conditions. In this way, math helps us to answer questions like these: How quickly is the human population growing, and will food production be able to keep up? How many polar bears are left in the wild, how will their numbers be affected as the climate continues to change, and what approaches should be most successful for preserving them?

AN APPLICATION

The number of plant species listed as threatened or endangered has more than tripled in the past 20 years. In 2008, there were 746 species of plants in the United States that were considered threatened or endangered. The number of species considered threatened was 4 less than one-fourth of the number considered endangered. How many U.S. plant species were considered endangered and how many were considered threatened in 2008?

Source: U.S. Fish and Wildlife Service

This problem appears as Example 1 in Section 8.1 and as Example 1 in Section 8.3.

505

The most difficult part of problem solving is almost always translating the problem situation to mathematical language. In this chapter, we study *systems of equations* and how to solve them using graphing, substitution, elimination, and matrices. Systems of equations often provide the easiest way to model real-world situations in fields such as psychology, sociology, business, education, engineering, and science.

8.1 Systems of Equations in Two Variables

Translating ▪ Identifying Solutions ▪ Solving Systems Graphically

Translating

Problems involving two unknown quantities are often translated most easily using two equations in two unknowns. Together these equations form a **system of equations**. We look for a solution to the problem by attempting to find a pair of numbers for which *both* equations are true.

EXAMPLE 1

Endangered species.   The number of plant species listed as threatened (likely to become endangered) or endangered (in danger of becoming extinct) has more than tripled in the past 20 years. In 2008, there were 746 species of plants in the United States that were considered threatened or endangered. The number of species considered threatened was 4 less than one-fourth of the number considered endangered. How many U.S. plant species were considered endangered and how many were considered threatened in 2008?

Source: U.S. Fish and Wildlife Service

SOLUTION

1. **Familiarize.** Often statements of problems contain information that has no bearing on the question asked. In this case, the fact that the number of threatened or endangered species has tripled in the past 20 years does not help us solve the problem. Instead, we focus on the number of endangered species and the number of threatened species in 2008. We let t represent the number of threatened plant species and d represent the number of endangered plant species in 2008.

2. **Translate.** There are two statements to translate. First, we look at the total number of endangered or threatened species of plants:

   Rewording: The number of threatened species   plus   the number of endangered species   was   746.

   Translating: t $+$ d $=$ 746

   The second statement compares the two amounts, d and t:

   Rewording: The number of threatened species   was   4 less than one-fourth of the number of endangered species.

   Translating: t $=$ $\frac{1}{4}d - 4$

STUDY SKILLS

Speak Up

Don't be hesitant to ask questions in class at appropriate times. Most instructors welcome questions and encourage students to ask them. Other students in your class probably have the same questions you do.

We have now translated the problem to a pair, or **system**, **of equations**:

$$t + d = 746,$$

$$t = \frac{1}{4}d - 4.$$

We complete the solution of this problem in Section 8.3.

TRY EXERCISE ▶ 41

System of Equations

A *system of equations* is a set of two or more equations, in two or more variables, for which a common solution is sought.

Problems like Example 1 *can* be solved using one variable; however, as problems become complicated, you will find that using more than one variable (and more than one equation) is often the preferable approach.

EXAMPLE 2

Jewelry design.   A jewelry designer purchased 80 beads for a total of $39 (excluding tax) to make a necklace. Some of the beads were sterling silver beads that cost 40¢ each and the rest were gemstone beads that cost 65¢ each. How many of each type did the designer buy?

SOLUTION

1. **Familiarize.** To familiarize ourselves with this problem, let's guess that the designer bought 20 beads at 40¢ each and 60 beads at 65¢ each. The total cost would then be

$$20 \cdot 40¢ + 60 \cdot 65¢ = 800¢ + 3900¢, \quad \text{or} \quad 4700¢.$$

Since 4700¢ = $47 and $47 ≠ $39, our guess is incorrect. Rather than guess again, let's see how algebra can be used to translate the problem.

2. **Translate.** We let s = the number of silver beads and g = the number of gemstone beads. Since the cost of each bead is given in cents and the total cost is in dollars, we must choose one of the units to use throughout the problem. We choose to work in cents, so the total cost is 3900¢. The information can be organized in a table, which will help with the translating.

Type of Bead	Silver	Gemstone	Total	
Number Bought	s	g	80	→ $s + g = 80$
Price	40¢	65¢		
Amount	$40s$¢	$65g$¢	3900¢	→ $40s + 65g = 3900$

The first row of the table and the first sentence of the problem indicate that a total of 80 beads were bought:

$$s + g = 80.$$

Since each silver bead cost 40¢ and s beads were bought, $40s$ represents the amount paid, in cents, for the silver beads. Similarly, $65g$ represents the amount paid, in cents, for the gemstone beads. This leads to a second equation:

$$40s + 65g = 3900.$$

We now have the following system of equations as the translation:

$$s + g = 80,$$
$$40s + 65g = 3900.$$

We will complete the solution of this problem in Section 8.3.

TRY EXERCISE ▶ 49

Identifying Solutions

A *solution* of a system of two equations in two variables is an ordered pair of numbers that makes *both* equations true.

EXAMPLE **3**   Determine whether $(-4, 7)$ is a solution of the system

$$x + y = 3,$$
$$5x - y = -27.$$

SOLUTION   As discussed in Chapter 3, unless stated otherwise, we use alphabetical order of the variables. Thus we replace x with -4 and y with 7:

$x + y = 3$	$5x - y = -27$
$-4 + 7 \mid 3$	$5(-4) - 7 \mid -27$
$3 \overset{?}{=} 3$ TRUE	$-20 - 7$
	$-27 \overset{?}{=} -27$ TRUE

> *CAUTION!*   Be sure to check the ordered pair in *both* equations.

The pair $(-4, 7)$ makes both equations true, so it is a solution of the system. We can also describe the solution by writing $x = -4$ and $y = 7$. Set notation can also be used to list the solution set $\{(-4, 7)\}$.

TRY EXERCISE ▶ 9

Solving Systems Graphically

Recall that the graph of an equation is a drawing that represents its solution set. If we graph the equations in Example 3, we find that $(-4, 7)$ is the only point common to both lines. Thus one way to solve a system of two equations is to graph both equations and identify any points of intersection. **The coordinates of each point of intersection represent a solution of that system.**

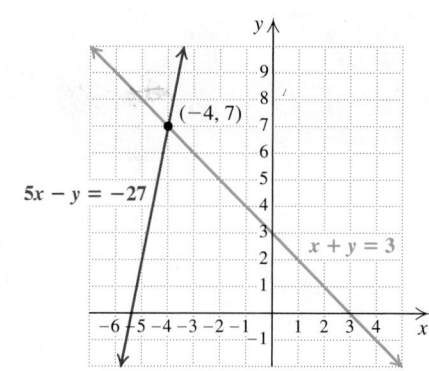

$$x + y = 3,$$
$$5x - y = -27$$

The point of intersection of the graphs is $(-4, 7)$.

The solution of the system is $(-4, 7)$.

Most pairs of lines have exactly one point in common. We will soon see, however, that this is not always the case.

EXAMPLE **4**   Solve each system graphically.

a) $y - x = 1,$ **b)** $y = -3x + 5,$ **c)** $3y - 2x = 6,$
     $y + x = 3$ $y = -3x - 2$ $-12y + 8x = -24$

SOLUTION

a) We graph each equation using any method studied in Chapter 3. All ordered pairs from line L_1 are solutions of the first equation. All ordered pairs from line L_2 are solutions of the second equation. The point of intersection has coordinates that make *both* equations true. Apparently, $(1, 2)$ is the solution. Graphs are not always accurate, so solving by graphing may yield approximate answers. Our check below shows that $(1, 2)$ is indeed the solution.

Graph both equations.

Look for any points in common.

$$y - x = 1,$$
$$y + x = 3$$

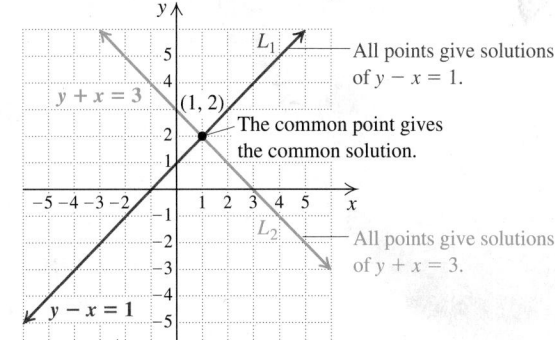

Check.

Check:

$y - x = 1$
$2 - 1 \mid 1$
$1 \overset{?}{=} 1$ TRUE

$y + x = 3$
$2 + 1 \mid 3$
$3 \overset{?}{=} 3$ TRUE

b) We graph the equations. The lines have the same slope, -3, and different y-intercepts, so they are parallel. There is no point at which they cross, so the system has no solution.

$$y = -3x + 5,$$
$$y = -3x - 2$$

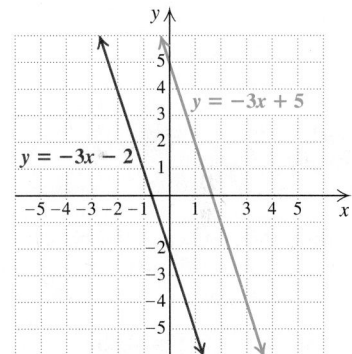

STUDENT NOTES

Although the system in Example 4(c) is true for an infinite number of ordered pairs, those pairs must be of a certain form. Only pairs that are solutions of $3y - 2x = 6$ or $-12y + 8x = -24$ are solutions of the system. It is incorrect to think that *all* ordered pairs are solutions.

c) We graph the equations and find that the same line is drawn twice. Thus any solution of one equation is a solution of the other. Each equation has an infinite number of solutions, so the system itself has an infinite number of solutions. We check one solution, $(0, 2)$, which is the y-intercept of each equation.

$$3y - 2x = 6,$$
$$-12y + 8x = -24$$

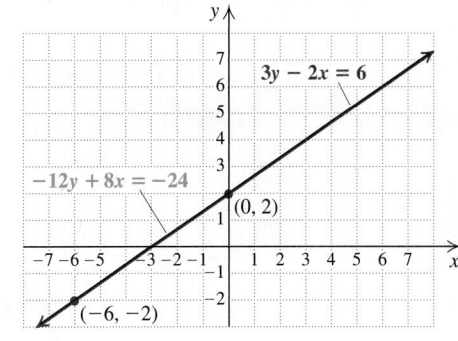

Check:

$$\begin{array}{c|c} 3y - 2x = 6 \\ \hline 3(2) - 2(0) & 6 \\ 6 - 0 & \\ & 6 \overset{?}{=} 6 \quad \text{TRUE} \end{array}$$

$$\begin{array}{c|c} -12y + 8x = -24 \\ \hline -12(2) + 8(0) & -24 \\ -24 + 0 & \\ & -24 \overset{?}{=} -24 \quad \text{TRUE} \end{array}$$

You can check that $(-6, -2)$ is another solution of both equations. In fact, any pair that is a solution of one equation is a solution of the other equation as well. Thus the solution set is

$$\{(x, y) \mid 3y - 2x = 6\}$$

or, in words, "the set of all pairs (x, y) for which $3y - 2x = 6$." Since the two equations are equivalent, we could have written instead $\{(x, y) \mid -12y + 8x = -24\}$.

TRY EXERCISE ▸ 17

When we graph a system of two linear equations in two variables, one of the following three outcomes will occur.

1. The lines have one point in common, and that point is the only solution of the system (see Example 4a). Any system that has *at least one solution* is said to be **consistent**.

2. The lines are parallel, with no point in common, and the system has *no solution* (see Example 4b). This type of system is called **inconsistent**.

3. The lines coincide, sharing the same graph. Because every solution of one equation is a solution of the other, the system has an infinite number of solutions (see Example 4c). Since it has at least one solution, this type of system is also consistent.

TECHNOLOGY CONNECTION

On most graphing calculators, an INTERSECT option allows us to find the coordinates of the intersection directly.

To illustrate, consider the following system:

$$3.45x + 4.21y = 8.39,$$
$$7.12x - 5.43y = 6.18.$$

After solving for y in each equation, we obtain the graph below. Using INTERSECT, we see that, to the nearest hundredth, the coordinates of the intersection are $(1.47, 0.79)$.

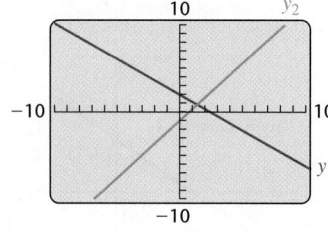

$y_1 = (8.39 - 3.45x)/4.21,$
$y_2 = (6.18 - 7.12x)/(-5.43)$

Use a graphing calculator to solve each of the following systems. Round all x- and y-coordinates to the nearest hundredth.

1. $y = -5.43x + 10.89,$
   $y = 6.29x - 7.04$
2. $y = 123.52x + 89.32,$
   $y = -89.22x + 33.76$
3. $2.18x + 7.81y = 13.78,$
   $5.79x - 3.45y = 8.94$
4. $-9.25x - 12.94y = -3.88,$
   $21.83x + 16.33y = 13.69$

When one equation in a system can be obtained by multiplying both sides of another equation by a constant, the two equations are said to be **dependent**. Thus the equations in Example 4(c) are dependent, but those in Examples 4(a) and 4(b) are **independent**. For systems of three or more equations, the definitions of dependent and independent will be slightly modified.

ALGEBRAIC–GRAPHICAL CONNECTION

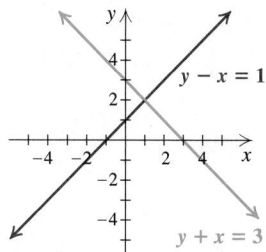

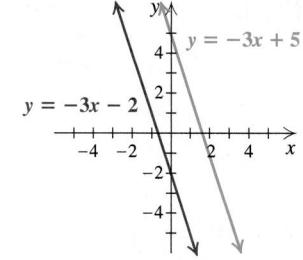

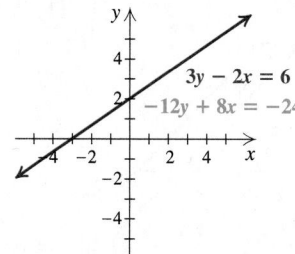

Graphs intersect at one point.

The system

$$y - x = 1,$$
$$y + x = 3$$

is *consistent* and has one solution.

Since neither equation is a multiple of the other, the equations are *independent*.

Graphs are parallel.

The system

$$y = -3x - 2,$$
$$y = -3x + 5$$

is *inconsistent* because there is no solution.

Since neither equation is a multiple of the other, the equations are *independent*.

Equations have the same graph.

The system

$$3y - 2x = 6,$$
$$-12y + 8x = -24$$

is *consistent* and has an infinite number of solutions.

Since one equation is a multiple of the other, the equations are *dependent*.

Graphing is helpful when solving systems because it allows us to "see" the solution. It can also be used on systems of nonlinear equations, and in many applications, it provides a satisfactory answer. However, graphing often lacks precision, especially when fraction or decimal solutions are involved. In Section 8.2, we will develop two algebraic methods of solving systems. Both methods produce exact answers.

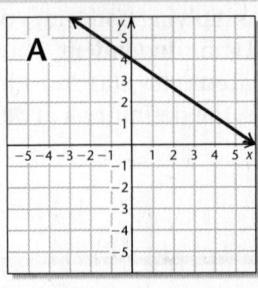

A

Visualizing for Success

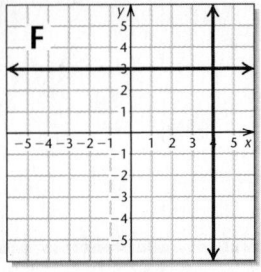

F

Match each equation or system of equations with its graph.

1. $x + y = 2$,
   $x - y = 2$

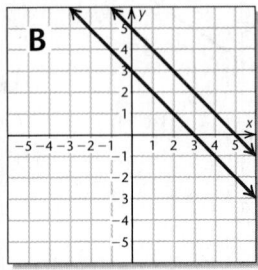

B

2. $y = \frac{1}{3}x - 5$

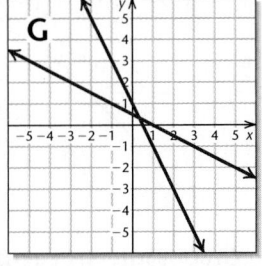

G

3. $4x - 2y = -8$

4. $2x + y = 1$,
   $x + 2y = 1$

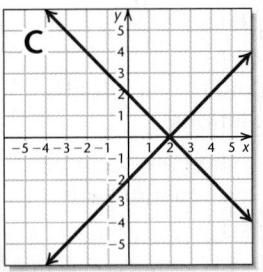

C

5. $8y + 32 = 0$

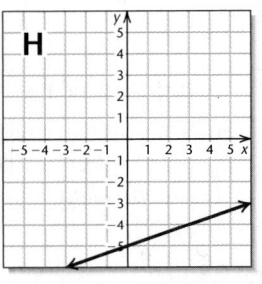

H

6. $f(x) = -x + 4$

7. $\frac{2}{3}x + y = 4$

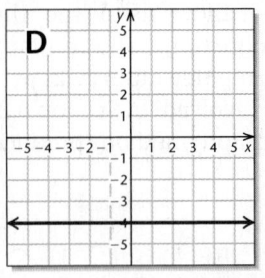

D

8. $x = 4$,
   $y = 3$

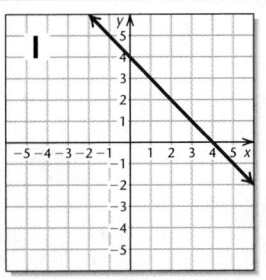

I

9. $y = \frac{1}{2}x + 3$,
   $2y - x = 6$

10. $y = -x + 5$,
    $y = 3 - x$

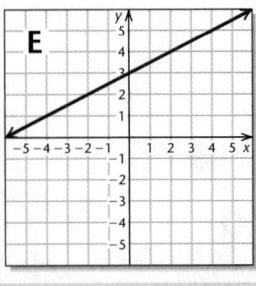

E

Answers on page A-32

An additional, animated version of this activity appears in MyMathLab. To use MyMathLab, you need a course ID and a student access code. Contact your instructor for more information.

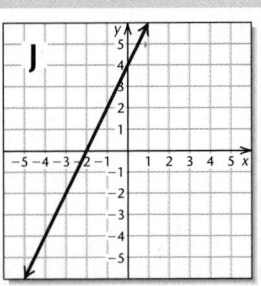

J

| **8.1** | **EXERCISE SET** |  |

Concept Reinforcement   *Classify each statement as either true or false.*

1. Every system of equations has at least one solution.

2. It is possible for a system of equations to have an infinite number of solutions.

3. Every point of intersection of the graphs of the equations in a system corresponds to a solution of the system.

4. The graphs of the equations in a system of two equations may coincide.

5. The graphs of the equations in a system of two equations could be parallel lines.

6. Any system of equations that has at most one solution is said to be consistent.

7. Any system of equations that has more than one solution is said to be inconsistent.

8. The equations $x + y = 5$ and $2(x + y) = 2(5)$ are dependent.

Determine whether the ordered pair is a solution of the given system of equations. Remember to use alphabetical order of variables.

9. $(2, 3)$; $2x - y = 1,$
$5x - 3y = 1$

10. $(4, 0)$; $2x + 7y = 8,$
$x - 9y = 4$

11. $(-5, 1)$; $x + 5y = 0,$
$y = 2x + 9$

12. $(-1, -2)$; $x + 3y = -7,$
$3x - 2y = 12$

13. $(0, -5)$; $x - y = 5,$
$y = 3x - 5$

14. $(5, 2)$; $a + b = 7,$
$2a - 8 = b$

Aha! **15.** $(3, -1)$; $3x - 4y = 13,$
$6x - 8y = 26$

16. $(4, -2)$; $-3x - 2y = -8,$
$8 = 3x + 2y$

Solve each system graphically. Be sure to check your solution. If a system has an infinite number of solutions, use set-builder notation to write the solution set. If a system has no solution, state this.

17. $x - y = 1,$
$x + y = 5$

18. $x + y = 6,$
$x - y = 4$

19. $3x + y = 5,$
$x - 2y = 4$

20. $2x - y = 4,$
$5x - y = 13$

21. $2y = 3x + 5,$
$x = y - 3$

22. $4x - y = 9,$
$x - 3y = 16$

23. $x = y - 1,$
$2x = 3y$

24. $a = 1 + b,$
$b = 5 - 2a$

25. $y = -1,$
$x = 3$

26. $y = 2,$
$x = -4$

27. $t + 2s = -1,$
$s = t + 10$

28. $b + 2a = 2,$
$a = -3 - b$

29. $2b + a = 11,$
$a - b = 5$

30. $y = -\frac{1}{3}x - 1,$
$4x - 3y = 18$

31. $y = -\frac{1}{4}x + 1,$
$2y = x - 4$

32. $6x - 2y = 2,$
$9x - 3y = 1$

33. $y - x = 5,$
$2x - 2y = 10$

34. $y = x + 2,$
$3y - 2x = 4$

35. $y = 3 - x,$
$2x + 2y = 6$

36. $2x - 3y = 6,$
$3y - 2x = -6$

37. For the systems in the odd-numbered exercises 17–35, which are consistent?

38. For the systems in the even-numbered exercises 18–36, which are consistent?

39. For the systems in the odd-numbered exercises 17–35, which contain dependent equations?

40. For the systems in the even-numbered exercises 18–36, which contain dependent equations?

Translate each problem situation to a system of equations. Do not attempt to solve, but save for later use.

41. The sum of two numbers is 10. The first number is $\frac{2}{3}$ of the second number. What are the numbers?

42. The sum of two numbers is 30. The first number is twice the second number. What are the numbers?

43. *e-mail usage.* In 2007, the average e-mail user sent 578 personal and business e-mails each week. The number of business e-mails was 30 more than the number of personal e-mails. How many of each type were sent each week?
Source: *JupiterResearch*

44. *Nontoxic furniture polish.* A nontoxic wood furniture polish can be made by mixing mineral (or olive) oil with vinegar. To make a 16-oz batch for a squirt bottle, Jazmun uses an amount of mineral oil that is 4 oz more than twice the amount of vinegar. How much of each ingredient is required?
Sources: Based on information from Chittenden Solid Waste District and *Clean House, Clean Planet* by Karen Logan

45. *Geometry.* Two angles are supplementary.* One angle is 3° less than twice the other. Find the measures of the angles.

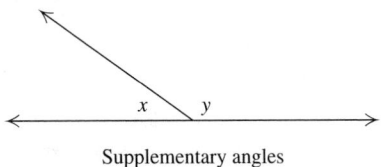

Supplementary angles

46. *Geometry.* Two angles are complementary.† The sum of the measures of the first angle and half the second angle is 64°. Find the measures of the angles.

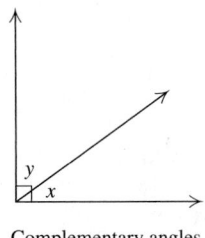

Complementary angles

47. *Basketball scoring.* Wilt Chamberlain once scored 100 points, setting a record for points scored in an NBA game. Chamberlain took only two-point shots and (one-point) foul shots and made a total of 64 shots. How many shots of each type did he make?

48. *Basketball scoring.* The Fenton College Cougars made 40 field goals in a recent basketball game, some 2-pointers and the rest 3-pointers. Altogether the 40 baskets counted for 89 points. How many of each type of field goal was made?

---

*The sum of the measures of two supplementary angles is 180°.
†The sum of the measures of two complementary angles is 90°.

49. *Retail sales.* Simply Souvenirs sold 45 hats and tee shirts. The hats sold for $14.50 each and the tee shirts for $19.50 each. In all, $697.50 was taken in for the souvenirs. How many of each type of souvenir were sold?

50. *Retail sales.* Cool Treats sold 60 ice cream cones. Single-dip cones sold for $2.50 each and double-dip cones for $4.15 each. In all, $179.70 was taken in for the cones. How many of each size cone were sold?

51. *Sales of pharmaceuticals.* In 2008, the Diabetic Express charged $83.29 for a 10-mL vial of Humalog insulin and $76.76 for a 10-mL vial of Lantus insulin. If a total of $3981.66 was collected for 50 vials of insulin, how many vials of each type were sold?

52. *Fundraising.* The Buck Creek Fire Department served 250 dinners. A child's plate cost $5.50 and an adult's plate cost $9.00. A total of $1935 was collected. How many of each type of plate was served?

53. *Lacrosse.* The perimeter of an NCAA men's lacrosse field is 340 yd. The length is 50 yd longer than the width. Find the dimensions.

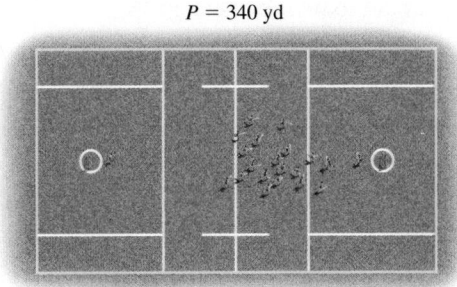

$P = 340$ yd

54. *Tennis.* The perimeter of a standard tennis court used for doubles is 228 ft. The width is 42 ft less than the length. Find the dimensions.

55. Write a problem for a classmate to solve that requires writing a system of two equations. Devise the problem so that the solution is "The Fever made 6 three-point baskets and 31 two-point baskets."

56. Write a problem for a classmate to solve that can be translated into a system of two equations. Devise the problem so that the solution is "In 2009, Diana took five 3-credit classes and two 4-credit classes."

Skill Review

To prepare for Section 8.2, review solving equations and formulas (Sections 2.2 and 2.3).

Solve. [2.2]

57. $3x + 2(5x - 1) = 6$

58. $4(3y + 2) - 7y = 3$

59. $9y = 5 - (y + 6)$

60. $2x - (x - 7) = 18$

Solve. [2.3]

61. $3x - y = 4$, for y

62. $5y - 2x = 7$, for x

Synthesis

Advertising media. *For Exercises 63 and 64, consider the following graph showing the U.S. market share for various advertising media.*

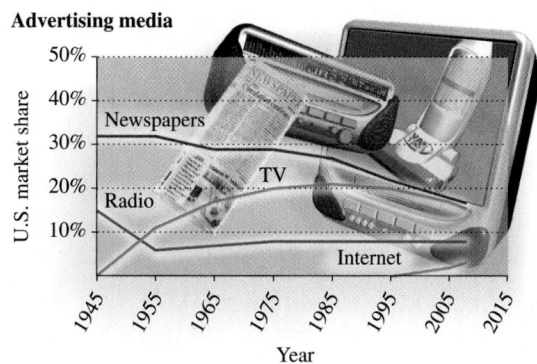

Advertising media

Source: *The Wall Street Journal,* 12/30/07

63. In what year did no one medium have a higher advertising market share than the others? Explain.

64. Will the Internet advertising market share ever exceed that of radio? TV? newspapers? If so, when? Explain your answers.

65. For each of the following conditions, write a system of equations.
   a) $(5, 1)$ is a solution.
   b) There is no solution.
   c) There is an infinite number of solutions.

66. A system of linear equations has $(1, -1)$ and $(-2, 3)$ as solutions. Determine:
   a) a third point that is a solution, and
   b) how many solutions there are.

67. The solution of the following system is $(4, -5)$. Find A and B.
$$Ax - 6y = 13,$$
$$x - By = -8.$$

Translate to a system of equations. Do not solve.

68. *Ages.* Tyler is twice as old as his son. Ten years ago, Tyler was three times as old as his son. How old are they now?

69. *Work experience.* Dell and Juanita are mathematics professors at a state university. Together, they have 46 years of service. Two years ago, Dell had taught 2.5 times as many years as Juanita. How long has each taught at the university?

70. *Design.* A piece of posterboard has a perimeter of 156 in. If you cut 6 in. off the width, the length becomes four times the width. What are the dimensions of the original piece of posterboard?

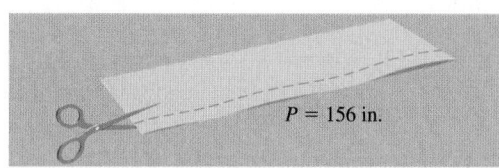

$P = 156$ in.

71. *Nontoxic scouring powder.* A nontoxic scouring powder is made up of 4 parts baking soda and 1 part vinegar. How much of each ingredient is needed for a 16-oz mixture?

72. Solve Exercise 41 graphically.

73. Solve Exercise 44 graphically.

Solve graphically.

74. $y = |x|$,
$3y - x = 8$

75. $x - y = 0$,
$y = x^2$

In Exercises 76–79, use a graphing calculator to solve each system of linear equations for x and y. Round all coordinates to the nearest hundredth.

76. $y = 8.23x + 2.11$,
$y = -9.11x - 4.66$

77. $y = -3.44x - 7.72$,
$y = 4.19x - 8.22$

78. $14.12x + 7.32y = 2.98$,
$21.88x - 6.45y = -7.22$

79. $5.22x - 8.21y = -10.21$,
$-12.67x + 10.34y = 12.84$

8.2 Solving by Substitution or Elimination

The Substitution Method ▪ The Elimination Method

The Substitution Method

Algebraic (nongraphical) methods for solving systems are often superior to graphing, especially when fractions are involved. One algebraic method, the *substitution method*, relies on having a variable isolated.

EXAMPLE 1

Solve the system

$$x + y = 4, \quad (1)$$
$$x = y + 1. \quad (2)$$

For easy reference, we have numbered the equations.

SOLUTION Equation (2) says that x and $y + 1$ name the same number. Thus we can substitute $y + 1$ for x in equation (1):

$$x + y = 4 \qquad \text{Equation (1)}$$
$$(y + 1) + y = 4. \qquad \text{Substituting } y + 1 \text{ for } x$$

We solve this last equation, using methods learned earlier:

$$(y + 1) + y = 4$$
$$2y + 1 = 4 \qquad \text{Removing parentheses and combining like terms}$$
$$2y = 3 \qquad \text{Subtracting 1 from both sides}$$
$$y = \tfrac{3}{2}. \qquad \text{Dividing both sides by 2}$$

We now return to the original pair of equations and substitute $\tfrac{3}{2}$ for y in either equation so that we can solve for x. For this problem, calculations are slightly easier if we use equation (2):

$$x = y + 1 \qquad \text{Equation (2)}$$
$$= \tfrac{3}{2} + 1 \qquad \text{Substituting } \tfrac{3}{2} \text{ for } y$$
$$= \tfrac{3}{2} + \tfrac{2}{2} = \tfrac{5}{2}.$$

We obtain the ordered pair $\left(\tfrac{5}{2}, \tfrac{3}{2}\right)$. A check ensures that it is a solution.

Check:

$$\begin{array}{c|c} x + y = 4 \\ \hline \tfrac{5}{2} + \tfrac{3}{2} & 4 \\ \tfrac{8}{2} & \\ 4 \stackrel{?}{=} 4 & \text{TRUE} \end{array} \qquad \begin{array}{c|c} x = y + 1 \\ \hline \tfrac{5}{2} & \tfrac{3}{2} + 1 \\ & \tfrac{3}{2} + \tfrac{2}{2} \\ \tfrac{5}{2} \stackrel{?}{=} \tfrac{5}{2} & \text{TRUE} \end{array}$$

Since $\left(\tfrac{5}{2}, \tfrac{3}{2}\right)$ checks, it is the solution.

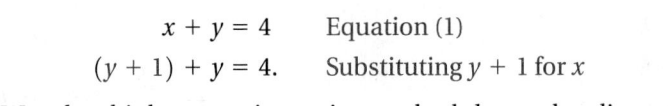 TRY EXERCISE 7

The exact solution to Example 1 is difficult to find graphically because it involves fractions. The graph shown serves as a partial check and provides a visualization of the problem.

If neither equation in a system has a variable alone on one side, we first isolate a variable in one equation and then substitute.

EXAMPLE 2

Solve the system

$$2x + y = 6, \quad (1)$$
$$3x + 4y = 4. \quad (2)$$

60 km/h
d kilometers
t hours

Trains meet here

A visualization of Example 1. Note that the coordinates of the intersection are not obvious.

SOLUTION   First, we select an equation and solve for one variable. We can isolate y by subtracting $2x$ from both sides of equation (1):

$$2x + y = 6 \qquad (1)$$
$$y = 6 - 2x. \qquad (3) \qquad \text{Subtracting } 2x \text{ from both sides}$$

Next, we proceed as in Example 1, by substituting:

$$3x + 4(6 - 2x) = 4 \qquad \text{Substituting } 6 - 2x \text{ for } y \text{ in equation (2).}$$
$$\text{Use parentheses!}$$
$$3x + 24 - 8x = 4 \qquad \text{Distributing to remove parentheses}$$
$$3x - 8x = 4 - 24 \qquad \text{Subtracting 24 from both sides}$$
$$-5x = -20$$
$$x = 4. \qquad \text{Dividing both sides by } -5$$

Next, we substitute 4 for x in either equation (1), (2), or (3). It is easiest to use equation (3) because it has already been solved for y:

$$y = 6 - 2x$$
$$= 6 - 2(4)$$
$$= 6 - 8 = -2.$$

The pair $(4, -2)$ appears to be the solution. We check in equations (1) and (2).

Check:

$2x + y = 6$		$3x + 4y = 4$	
$2(4) + (-2)$	6	$3(4) + 4(-2)$	4
$8 - 2$		$12 - 8$	
	$6 \overset{?}{=} 6$ TRUE		$4 \overset{?}{=} 4$ TRUE

Since $(4, -2)$ checks, it is the solution.    **TRY EXERCISE** 11

A visualization of Example 2

Some systems have no solution, as we saw graphically in Section 8.1. How do we recognize such systems if we are solving by an algebraic method?

EXAMPLE   3   Solve the system

$$y = -3x + 5, \qquad (1)$$
$$y = -3x - 2. \qquad (2)$$

SOLUTION   We solved this system graphically in Example 4(b) of Section 8.1, and found that the lines are parallel and the system has no solution. Let's now try to solve the system by substitution. Proceeding as in Example 1, we substitute $-3x - 2$ for y in the first equation:

$$-3x - 2 = -3x + 5 \qquad \text{Substituting } -3x - 2 \text{ for } y \text{ in equation (1)}$$
$$-2 = 5. \qquad \text{Adding } 3x \text{ to both sides; } -2 = 5 \text{ is a contradiction. The equation is always false.}$$

Since there is no solution of $-2 = 5$, there is no solution of the system. We state that there is no solution.   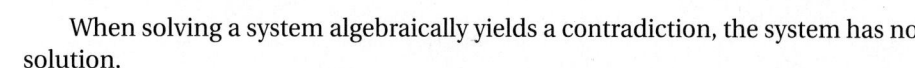 **TRY EXERCISE** 21

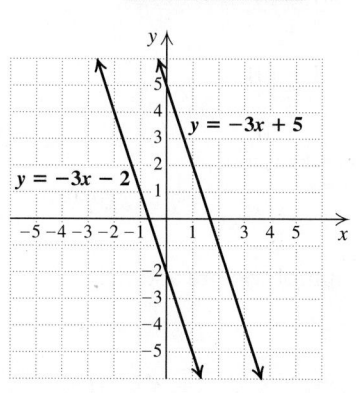

A visualization of Example 3

When solving a system algebraically yields a contradiction, the system has no solution.

As we will see in Example 7, when solving a system of two equations algebraically yields an identity, the system has an infinite number of solutions.

The Elimination Method

The *elimination method* for solving systems of equations makes use of the *addition principle*: If $a = b$, then $a + c = b + c$. Consider the following system:

$$2x - 3y = 0,$$
$$-4x + 3y = -1.$$

Note that the $-3y$ in one equation and the $3y$ in the other are opposites. If we add all terms on the left side of the equations, the sum of $-3y$ and $3y$ is 0, so in effect, the variable y is "eliminated."

EXAMPLE **4**   Solve the system

$$2x - 3y = 0,   \quad (1)$$
$$-4x + 3y = -1.   \quad (2)$$

SOLUTION   Note that according to equation (2), $-4x + 3y$ and -1 are the same number. Thus we can use the addition principle to work vertically and add $-4x + 3y$ to the left side of equation (1) and -1 to the right side:

$$
\begin{array}{ll}
2x - 3y = 0 & (1) \\
\underline{-4x + 3y = -1} & (2) \\
-2x + 0y = -1. & \text{Adding}
\end{array}
$$

This eliminates the variable y, and leaves an equation with just one variable, x, for which we solve:

$$-2x = -1$$
$$x = \tfrac{1}{2}.$$

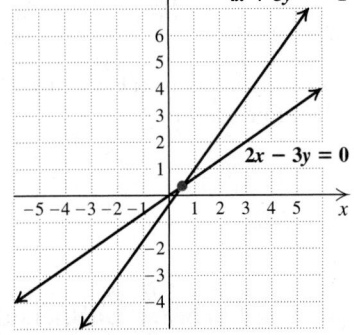

Next, we substitute $\tfrac{1}{2}$ for x in equation (1) and solve for y:

$$2 \cdot \tfrac{1}{2} - 3y = 0 \qquad \text{Substituting. We also could have used equation (2).}$$
$$1 - 3y = 0$$
$$-3y = -1, \text{ so } y = \tfrac{1}{3}.$$

Check:

$$
\begin{array}{c|c}
2x - 3y = 0 \\
\hline
2\left(\tfrac{1}{2}\right) - 3\left(\tfrac{1}{3}\right) & 0 \\
1 - 1 \\
0 \overset{?}{=} 0 & \text{TRUE}
\end{array}
\qquad
\begin{array}{c|c}
-4x + 3y = -1 \\
\hline
-4\left(\tfrac{1}{2}\right) + 3\left(\tfrac{1}{3}\right) & -1 \\
-2 + 1 \\
-1 \overset{?}{=} -1 & \text{TRUE}
\end{array}
$$

A visualization of Example 4

Since $\left(\tfrac{1}{2}, \tfrac{1}{3}\right)$ checks, it is the solution. See also the graph at left.

TRY EXERCISE 23

To eliminate a variable, we must sometimes multiply before adding.

EXAMPLE **5**   Solve the system

$$5x + 4y = 22,   \quad (1)$$
$$-3x + 8y = 18.   \quad (2)$$

STUDENT NOTES

It is wise to double-check each step of your work as you go, rather than checking all steps at the end of a problem. Finding and correcting an error as it occurs saves you time in the long run. One common error is to forget to multiply *both* sides of the equation when you use the multiplication principle.

SOLUTION If we add the left sides of the two equations, we will not eliminate a variable. However, if the $4y$ in equation (1) were changed to $-8y$, we would. To accomplish this change, we multiply both sides of equation (1) by -2:

$$
\begin{array}{ll}
-10x - 8y = -44 & \text{Multiplying both sides of equation (1) by } -2 \\
\underline{-3x + 8y = 18} & \\
-13x + 0 = -26 & \text{Adding} \\
x = 2. & \text{Solving for } x
\end{array}
$$

Then

$$
\begin{array}{ll}
-3 \cdot 2 + 8y = 18 & \text{Substituting 2 for } x \text{ in equation (2)} \\
-6 + 8y = 18 & \\
\left. \begin{array}{l} 8y = 24 \\ y = 3. \end{array} \right\} & \text{Solving for } y
\end{array}
$$

We obtain $(2, 3)$, or $x = 2$, $y = 3$. We leave it to the student to confirm that this checks and is the solution.

TRY EXERCISE 29

Sometimes we must multiply twice in order to make two terms become opposites.

EXAMPLE 6

Solve the system

$$
\begin{array}{ll}
2x + 3y = 17, & (1) \\
5x + 7y = 29. & (2)
\end{array}
$$

SOLUTION We multiply so that the x-terms will be eliminated when we add.

Eliminate x.

$$
2x + 3y = 17, \xrightarrow{\text{Multiplying both sides by 5}} 10x + 15y = 85
$$

Solve for y.

$$
5x + 7y = 29 \xrightarrow{\text{Multiplying both sides by } -2} \underline{-10x - 14y = -58}
$$

$$
\begin{array}{ll}
0 + y = 27 & \text{Adding} \\
y = 27 &
\end{array}
$$

Next, we substitute to find x:

$$
\begin{array}{ll}
2x + 3 \cdot 27 = 17 & \text{Substituting 27 for } y \text{ in equation (1)} \\
2x + 81 = 17 & \\
\left. \begin{array}{l} 2x = -64 \\ x = -32. \end{array} \right\} & \text{Solving for } x
\end{array}
$$

Solve for x.

Check.

Check:

$$
\begin{array}{c|c}
2x + 3y = 17 & \\
\hline
2(-32) + 3(27) & 17 \\
-64 + 81 & \\
17 \stackrel{?}{=} 17 & \text{TRUE}
\end{array}
\qquad
\begin{array}{c|c}
5x + 7y = 29 & \\
\hline
5(-32) + 7(27) & 29 \\
-160 + 189 & \\
29 \stackrel{?}{=} 29 & \text{TRUE}
\end{array}
$$

We obtain $(-32, 27)$, or $x = -32$, $y = 27$, as the solution.

TRY EXERCISE 31

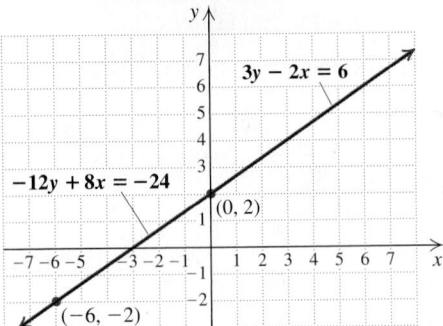

A visualization of Example 7

Solve the system

$$3y - 2x = 6, \qquad (1)$$
$$-12y + 8x = -24. \qquad (2)$$

SOLUTION   We graphed this system in Example 4(c) of Section 8.1, and found that the lines coincide and the system has an infinite number of solutions. Suppose we were to solve this system using the elimination method:

$$
\begin{array}{ll}
12y - 8x = 24 & \text{Multiplying both sides of equation (1) by 4} \\
\underline{-12y + 8x = -24} & \\
0 = 0. & \text{We obtain an identity; } 0 = 0 \text{ is always true.}
\end{array}
$$

Note that both variables have been eliminated and what remains is an identity—that is, an equation that is always true. Any pair that is a solution of equation (1) is also a solution of equation (2). The equations are dependent and the solution set is infinite:

$$\{(x, y) \mid 3y - 2x = 6\}, \text{ or equivalently, } \{(x, y) \mid -12y + 8x = -24\}.$$

> **TRY EXERCISE** 47

Example 3 and Example 7 illustrate how to tell algebraically whether a system of two equations is inconsistent or whether the equations are dependent.

> ### Rules for Special Cases
>
> When solving a system of two linear equations in two variables:
>
> 1. If we obtain an identity such as $0 = 0$, then the system has an infinite number of solutions. The equations are dependent and, since a solution exists, the system is consistent.*
> 2. If we obtain a contradiction such as $0 = 7$, then the system has no solution. The system is inconsistent.

Should decimals or fractions appear, it often helps to *clear* before solving.

EXAMPLE 8   Solve the system

$$0.2x + 0.3y = 1.7,$$
$$\tfrac{1}{7}x + \tfrac{1}{5}y = \tfrac{29}{35}.$$

SOLUTION   We have

$$
\begin{array}{ll}
0.2x + 0.3y = 1.7, & \longrightarrow \text{ Multiplying both sides by 10} \longrightarrow \quad 2x + 3y = 17 \\
\tfrac{1}{7}x + \tfrac{1}{5}y = \tfrac{29}{35} & \longrightarrow \text{ Multiplying both sides by 35} \longrightarrow \quad 5x + 7y = 29.
\end{array}
$$

We multiplied both sides of the first equation by 10 to clear the decimals. Multiplication by 35, the least common denominator, clears the fractions in the second equation. The problem now happens to be identical to Example 6. The solution is $(-32, 27)$, or $x = -32, y = 27$.

> **TRY EXERCISE** 35

The steps for each algebraic method for solving systems of two equations are given below. Note that in both methods, we find the value of one variable and then substitute to find the corresponding value of the other variable.

---

*Consistent systems and dependent equations are discussed in greater detail in Section 8.4.

> **To Solve a System Using Substitution**
> 1. Isolate a variable in one of the equations (unless one is already isolated).
> 2. Substitute for that variable in the other equation, using parentheses.
> 3. Solve for the remaining variable.
> 4. Substitute the value of the second variable in any of the equations, and solve for the first variable.
> 5. Form an ordered pair and check in the original equations.

> **To Solve a System Using Elimination**
> 1. Write both equations in standard form.
> 2. Multiply both sides of one or both equations by a constant, if necessary, so that the coefficients of one of the variables are opposites.
> 3. Add the left sides and the right sides of the resulting equations. One variable should be eliminated in the sum.
> 4. Solve for the remaining variable.
> 5. Substitute the value of the second variable in any of the equations, and solve for the first variable.
> 6. Form an ordered pair and check in the original equations.

8.2   EXERCISE SET

For Extra Help
MyMathLab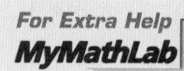

Concept Reinforcement   *In each of Exercises 1–6, match the system listed with the choice from the column on the right that would be a subsequent step in solving the system.*

1. ___ $3x - 4y = 6,$
$5x + 4y = 1$

2. ___ $2x - y = 8,$
$y = 5x + 3$

3. ___ $x - 2y = 3,$
$5x + 3y = 4$

4. ___ $8x + 6y = -15,$
$5x - 3y = 8$

5. ___ $y = 4x - 7,$
$6x + 3y = 19$

6. ___ $y = 4x - 1,$
$y = -\frac{2}{3}x - 1$

a) $-5x + 10y = -15,$
$5x + 3y = 4$

b) The lines intersect at $(0, -1).$

c) $6x + 3(4x - 7) = 19$

d) $8x = 7$

e) $2x - (5x + 3) = 8$

f) $8x + 6y = -15,$
$10x - 6y = 16$

For Exercises 7–54, if a system has an infinite number of solutions, use set-builder notation to write the solution set. If a system has no solution, state this.

Solve using the substitution method.

7. $y = 3 - 2x,$
$3x + y = 5$

8. $3y + x = 4,$
$x = 2y - 1$

9. $3x + 5y = 3,$
$x = 8 - 4y$

10. $9x - 2y = 3,$
$3x - 6 = y$

11. $3s - 4t = 14,$
$5s + t = 8$

12. $m - 2n = 16,$
$4m + n = 1$

13. $4x - 2y = 6,$
$2x - 3 = y$

14. $t = 4 - 2s,$
$t + 2s = 6$

15. $-5s + t = 11,$
$4s + 12t = 4$

16. $5x + 6y = 14,$
$-3y + x = 7$

17. $2x + 2y = 2,$
$3x - y = 1$

18. $4p - 2q = 16,$
$5p + 7q = 1$

19. $2a + 6b = 4,$
$3a - b = 6$

20. $3x - 4y = 5,$
$2x - y = 1$

21. $2x - 3 = y,$
$y - 2x = 1$

22. $a - 2b = 3,$
$3a = 6b + 9$

Solve using the elimination method.

23. $x + 3y = 7,$
$-x + 4y = 7$

24. $2x + y = 6,$
$x - y = 3$

25. $x - 2y = 11,$
$3x + 2y = 17$

26. $5x - 3y = 8,$
$-5x + y = 4$

27. $9x + 3y = -3,$
$2x - 3y = -8$

28. $6x - 3y = 18,$
$6x + 3y = -12$

29. $5x + 3y = 19,$
$x - 6y = 11$

30. $3x + 2y = 3,$
$9x - 8y = -2$

31. $5r - 3s = 24,$
$3r + 5s = 28$

32. $5x - 7y = -16,$
$2x + 8y = 26$

33. $6s + 9t = 12,$
$4s + 6t = 5$

34. $10a + 6b = 8,$
$5a + 3b = 2$

35. $\frac{1}{2}x - \frac{1}{6}y = 10,$
$\frac{2}{5}x + \frac{1}{2}y = 8$

36. $\frac{1}{3}x + \frac{1}{5}y = 7,$
$\frac{1}{6}x - \frac{2}{5}y = -4$

37. $\frac{x}{2} + \frac{y}{3} = \frac{7}{6},$
$\frac{2x}{3} + \frac{3y}{4} = \frac{5}{4}$

38. $\frac{2x}{3} + \frac{3y}{4} = \frac{11}{12},$
$\frac{x}{3} + \frac{7y}{18} = \frac{1}{2}$

Aha! **39.** $12x - 6y = -15,$
$-4x + 2y = 5$

40. $8s + 12t = 16,$
$6s + 9t = 12$

41. $0.3x + 0.2y = 0.3,$
$0.5x + 0.4y = 0.4$

42. $0.3x + 0.2y = 5,$
$0.5x + 0.4y = 11$

Solve using any appropriate method.

43. $a - 2b = 16,$
$b + 3 = 3a$

44. $5x - 9y = 7,$
$7y - 3x = -5$

45. $10x + y = 306,$
$10y + x = 90$

46. $3(a - b) = 15,$
$4a = b + 1$

47. $6x - 3y = 3,$
$4x - 2y = 2$

48. $x + 2y = 8,$
$x = 4 - 2y$

49. $3s - 7t = 5,$
$7t - 3s = 8$

50. $2s - 13t = 120,$
$-14s + 91t = -840$

51. $0.05x + 0.25y = 22,$
$0.15x + 0.05y = 24$

52. $2.1x - 0.9y = 15,$
$-1.4x + 0.6y = 10$

53. $13a - 7b = 9,$
$2a - 8b = 6$

54. $3a - 12b = 9,$
$4a - 5b = 3$

55. Describe a procedure that can be used to write an inconsistent system of equations.

56. Describe a procedure that can be used to write a system that has an infinite number of solutions.

Skill Review

To prepare for Section 8.3, review solving problems using the five-step problem-solving strategy (Section 2.5).

Solve. [2.5]

57. *Energy consumption.*   With average use, a toaster oven and a convection oven together consume 15 kilowatt hours (kWh) of electricity each month. A convection oven uses four times as much electricity as a toaster oven. How much does each use per month?
Source: Lee County Electric Cooperative

58. *Test scores.*   Ellia needs to average 80 on her tests in order to earn a B in her math class. Her average after 4 tests is 77.5. What score is needed on the fifth test in order to raise the average to 80?

59. *Real estate.*   After her house had been on the market for 6 months, Gina reduced the price to $94,500. This was $\frac{9}{10}$ of the original asking price. How much did Gina originally ask for her house?

60. *Car rentals.*   National Car Rental rents minivans to a university for $69 a day plus 30¢ per mile. An English professor rented a minivan for 2 days to take a group of students to a seminar. The bill was $225. How far did the professor drive the van?
Source: www.nationalcar.com

61. *Carpentry.*   Anazi cuts a 96-in. piece of wood trim into three pieces. The second piece is twice as long as the first. The third piece is one-tenth as long as the second. How long is each piece?

62. *Telephone calls.*   Terri's voice over the Internet (VoIP) phone service charges $0.36 for the first minute of each call and $0.06 for each additional $\frac{1}{2}$ minute. One month she was charged $28.20 for 35 calls. How many minutes did she use?

Synthesis

63. Some systems are more easily solved by substitution and some are more easily solved by elimination. What guidelines could be used to help someone determine which method to use?

64. Explain how it is possible to solve Exercise 39 mentally.

65. If $(1, 2)$ and $(-3, 4)$ are two solutions of $f(x) = mx + b$, find m and b.

66. If $(0, -3)$ and $\left(-\frac{3}{2}, 6\right)$ are two solutions of $px - qy = -1$, find p and q.

67. Determine a and b for which $(-4, -3)$ is a solution of the system

$$ax + by = -26,$$
$$bx - ay = 7.$$

68. Solve for x and y in terms of a and b:

$$5x + 2y = a,$$
$$x - y = b.$$

Solve.

69. $\dfrac{x + y}{2} - \dfrac{x - y}{5} = 1,$

$\dfrac{x - y}{2} + \dfrac{x + y}{6} = -2$

70. $3.5x - 2.1y = 106.2,$
$4.1x + 16.7y = -106.28$

Each of the following is a system of nonlinear equations. However, each is reducible to linear, since an appropriate substitution (say, u for $1/x$ and v for $1/y$) yields a linear system. Make such a substitution, solve for the new variables, and then solve for the original variables.

71. $\dfrac{2}{x} + \dfrac{1}{y} = 0,$

$\dfrac{5}{x} + \dfrac{2}{y} = -5$

72. $\dfrac{1}{x} - \dfrac{3}{y} = 2,$

$\dfrac{6}{x} + \dfrac{5}{y} = -34$

73. A student solving the system

$$17x + 19y = 102,$$
$$136x + 152y = 826$$

graphs both equations on a graphing calculator and gets the following screen. The student then (incorrectly) concludes that the equations are dependent and the solution set is infinite. How can algebra be used to convince the student that a mistake has been made?

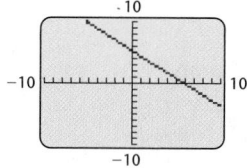

CONNECTING the CONCEPTS

We now have three different methods for solving systems of equations. Each method has certain strengths and weaknesses, as outlined below.

Method	Strengths	Weaknesses
Graphical	Solutions are displayed graphically. Can be used with any system that can be graphed.	For some systems, only approximate solutions can be found graphically. The graph drawn may not be large enough to show the solution.
Substitution	Yields exact solutions. Easy to use when a variable has a coefficient of 1.	Introduces extensive computations with fractions when solving more complicated systems. Solutions are not displayed graphically.
Elimination	Yields exact solutions. Easy to use when fractions or decimals appear in the system. The preferred method for systems of 3 or more equations in 3 or more variables (see Section 8.4).	Solutions are not displayed graphically.

(continued)

When selecting a method to use, consider the strengths and weaknesses listed above. If possible, begin solving the system mentally to help discover the method that seems best suited for that particular system.

MIXED REVIEW

Solve using the best method.

1. $x = y$,
   $x + y = 2$

2. $x + y = 10$,
   $x - y = 8$

3. $y = \frac{1}{2}x + 1$,
   $y = 2x - 5$

4. $y = 2x - 3$,
   $x + y = 12$

5. $x = 5$,
   $y = 10$

6. $3x + 5y = 8$,
   $3x - 5y = 4$

7. $2x - y = 1$,
   $2y - 4x = 3$

8. $x = 2 - y$,
   $3x + 3y = 6$

9. $x + 2y = 3$,
   $3x = 4 - y$

10. $9x + 8y = 0$,
    $11x - 7y = 0$

11. $10x + 20y = 40$,
    $x - y = 7$

12. $y = \frac{5}{3}x + 7$,
    $y = \frac{5}{3}x - 8$

13. $2x - 5y = 1$,
    $3x + 2y = 11$

14. $\dfrac{x}{2} + \dfrac{y}{3} = \dfrac{2}{3}$,
    $\dfrac{x}{5} + \dfrac{5y}{2} = \dfrac{1}{4}$

15. $1.1x - 0.3y = 0.8$,
    $2.3x + 0.3y = 2.6$

16. $y = -3$,
    $x = 11$

17. $x - 2y = 5$,
    $3x - 15 = 6y$

18. $12x - 19y = 13$,
    $8x + 19y = 7$

19. $0.2x + 0.7y = 1.2$,
    $0.3x - 0.1y = 2.7$

20. $\frac{1}{4}x = \frac{1}{3}y$,
    $\frac{1}{2}x - \frac{1}{15}y = 2$

8.3   Solving Applications: Systems of Two Equations

Total-Value and Mixture Problems ▪ Motion Problems

You are in a much better position to solve problems now that you know how systems of equations can be used. Using systems often makes the translating step easier.

EXAMPLE **1**

Patch of Brighamia Insignis with flower

Endangered species. The number of plant species listed as threatened (likely to become endangered) or endangered (in danger of becoming extinct) has more than tripled in the past 20 years. In 2008, there were 746 species of plants in the United States that were considered threatened or endangered. The number considered threatened was 4 less than one-fourth of the number considered endangered. How many U.S. plant species were considered endangered and how many were considered threatened in 2008?

Source: U.S. Fish and Wildlife Service

SOLUTION   The *Familiarize* and *Translate* steps were completed in Example 1 of Section 8.1. The resulting system of equations is

$$t + d = 746,$$
$$t = \tfrac{1}{4}d - 4,$$

where d is the number of endangered plant species and t is the number of threatened plant species in the United States in 2008.

3. Carry out. We solve the system of equations. Since one equation already has a variable isolated, let's use the substitution method:

$$t + d = 746$$
$$\tfrac{1}{4}d - 4 + d = 746 \qquad \text{Substituting } \tfrac{1}{4}d - 4 \text{ for } t$$
$$\tfrac{5}{4}d - 4 = 746 \qquad \text{Combining like terms}$$
$$\tfrac{5}{4}d = 750 \qquad \text{Adding 4 to both sides}$$
$$d = \tfrac{4}{5} \cdot 750 \qquad \text{Multiplying both sides by } \tfrac{4}{5}\text{: } \tfrac{4}{5} \cdot \tfrac{5}{4} = 1$$
$$d = 600. \qquad \text{Simplifying}$$

Next, using either of the original equations, we substitute and solve for t:

$$t = \tfrac{1}{4} \cdot 600 - 4 = 150 - 4 = 146.$$

4. Check. The sum of 600 and 146 is 746, so the total number of species is correct. Since 4 less than one-fourth of 600 is $150 - 4$, or 146, the numbers check.

5. State. In 2008, there were 600 endangered plant species and 146 threatened plant species in the United States.

> TRY EXERCISE ▶ 45

Total-Value and Mixture Problems

Jewelry design. In order to make a necklace, a jewelry designer purchased 80 beads for a total of $39 (excluding tax). Some of the beads were sterling silver beads that cost 40¢ each and the rest were gemstone beads that cost 65¢ each. How many of each type did the designer buy?

SOLUTION The *Familiarize* and *Translate* steps were completed in Example 2 of Section 8.1.

3. Carry out. We are to solve the system of equations

$$s + g = 80, \qquad (1)$$
$$40s + 65g = 3900, \qquad (2) \qquad \text{Working in cents rather than dollars}$$

where s is the number of silver beads bought and g is the number of gemstone beads bought. Because both equations are in the form $Ax + By = C$, let's use the elimination method to solve the system. We can eliminate s by multiplying both sides of equation (1) by -40 and adding them to the corresponding sides of equation (2):

$$-40s - 40g = -3200 \qquad \text{Multiplying both sides of equation (1) by } -40$$
$$\underline{40s + 65g = 3900}$$
$$25g = 700 \qquad \text{Adding}$$
$$g = 28. \qquad \text{Solving for } g$$

To find s, we substitute 28 for g in equation (1) and then solve for s:

$$s + g = 80 \qquad \text{Equation (1)}$$
$$s + 28 = 80 \qquad \text{Substituting 28 for } g$$
$$s = 52. \qquad \text{Solving for } s$$

We obtain $(28, 52)$, or $g = 28$ and $s = 52$.

4. **Check.** We check in the original problem. Recall that g is the number of gemstone beads and s the number of silver beads.

Number of beads:	$g + s = 28 + 52 = 80$
Cost of gemstone beads:	$65g = 65 \times 28 = 1820¢$
Cost of silver beads:	$40s = 40 \times 52 = \underline{2080¢}$
	Total $= 3900¢$

The numbers check.

5. **State.** The designer bought 28 gemstone beads and 52 silver beads.

> **TRY EXERCISE** 15

Example 2 involved two types of items (silver beads and gemstone beads), the quantity of each type bought, and the total value of the items. We refer to this type of problem as a *total-value problem*.

EXAMPLE 3

Blending teas. Tea Pots n Treasures sells loose Oolong tea for $2.15 an ounce. Donna mixed Oolong tea with shaved almonds that sell for $0.95 an ounce to create the Market Street Oolong blend that sells for $1.85 an ounce. One week, she made 300 oz of Market Street Oolong. How much tea and how much shaved almonds did Donna use?

SOLUTION

1. **Familiarize.** This problem is similar to Example 2. Rather than silver beads and gemstone beads, we have ounces of tea and ounces of almonds. Instead of a different price for each type of bead, we have a different price per ounce for each ingredient. Finally, rather than knowing the total cost of the beads, we know the weight and the price per ounce of the mixture. Thus we can find the total value of the blend by multiplying 300 ounces times $1.85 per ounce. We let l = the number of ounces of Oolong tea and a = the number of ounces of shaved almonds.

2. **Translate.** Since a 300-oz batch was made, we must have

$$l + a = 300.$$

To find a second equation, note that the total value of the 300-oz blend must match the combined value of the separate ingredients:

Rewording: The value of the Oolong tea   plus   the value of the almonds   is   the value of the Market Street blend.

Translating: $l \cdot \$2.15 \quad + \quad a \cdot \$0.95 \quad = \quad 300 \cdot \$1.85$

These equations can also be obtained from a table.

	Oolong Tea	Almonds	Market Street Blend	
Number of Ounces	l	a	300	$\longrightarrow l + a = 300$
Price per Ounce	$2.15	$0.95	$1.85	
Value of Tea	$2.15l	$0.95a	$300 \cdot \$1.85$, or $555	$\longrightarrow 2.15l + 0.95a = 555$

Clearing decimals in the second equation, we have $215l + 95a = 55{,}500$. We have translated to a system of equations:

$$l + \quad a = 300, \qquad (1)$$
$$215l + 95a = 55{,}500. \qquad (2)$$

3. **Carry out.** We can solve using substitution. When equation (1) is solved for l, we have $l = 300 - a$. Substituting $300 - a$ for l in equation (2), we find a:

$215(300 - a) + 95a = 55{,}500$	Substituting
$64{,}500 - 215a + 95a = 55{,}500$	Using the distributive law
$-120a = -9000$	Combining like terms; subtracting $64{,}500$ from both sides
$a = 75.$	Dividing both sides by -120

We have $a = 75$ and, from equation (1) above, $l + a = 300$. Thus, $l = 225$.

4. **Check.** Combining 225 oz of Oolong tea and 75 oz of almonds will give a 300-oz blend. The value of 225 oz of Oolong is $225(\$2.15)$, or $\$483.75$. The value of 75 oz of almonds is $75(\$0.95)$, or $\$71.25$. Thus the combined value of the blend is $\$483.75 + \$71.25$, or $\$555$. A 300-oz blend priced at $\$1.85$ an ounce would also be worth $\$555$, so our answer checks.

5. **State.** The Market Street blend was made by combining 225 oz of Oolong tea and 75 oz of almonds.

TRY EXERCISE ▶ 23

EXAMPLE 4

Student loans. Rani's student loans totaled $\$9600$. Part was a PLUS loan made at 8.5% interest and the rest was a Stafford loan made at 6.8% interest. After one year, Rani's loans accumulated $\$729.30$ in interest. What was the original amount of each loan?

SOLUTION

1. **Familiarize.** We begin with a guess. If $\$3000$ was borrowed at 8.5% and $\$6600$ was borrowed at 6.8%, the two loans would total $\$9600$. The interest would then be $0.085(\$3000)$, or $\$255$, and $0.068(\$6600)$, or $\$448.80$, for a total of only $\$703.80$ in interest. Our guess was wrong, but checking the guess familiarized us with the problem. More than $\$3000$ was borrowed at the higher rate.

2. **Translate.** We let $p =$ the amount of the PLUS loan and $s =$ the amount of the Stafford loan. Next, we organize a table in which the entries in each column come from the formula for simple interest:

Principal · Rate · Time = Interest.

	PLUS Loan	Stafford Loan	Total	
Principal	p	s	$\$9600$	$\longrightarrow p + s = 9600$
Rate of Interest	8.5%	6.8%		
Time	1 yr	1 yr		
Interest	$0.085p$	$0.068s$	$\$729.30$	$\longrightarrow 0.085p + 0.068s = 729.30$

The total amount borrowed is found in the first row of the table:

$$p + s = 9600.$$

A second equation, representing the accumulated interest, can be found in the last row:

$$0.085p + 0.068s = 729.30, \quad \text{or} \quad 85p + 68s = 729,300. \qquad \text{Clearing decimals}$$

3. **Carry out.** The system can be solved by elimination:

$$
\begin{array}{l}
p + s = 9600, \\
85p + 68s = 729,300.
\end{array}
\quad \xrightarrow[\text{sides by } -85]{\text{Multiplying both}} \quad
\begin{array}{r}
-85p - 85s = -816,000 \\
\underline{85p + 68s = 729,300} \\
-17s = -86,700
\end{array}
$$

$$p + s = 9600 \xleftarrow{\hspace{2cm}} s = 5100$$

$$p + 5100 = 9600$$

$$p = 4500.$$

We find that $p = 4500$ and $s = 5100$.

4. **Check.** The total amount borrowed is $4500 + $5100, or $9600. The interest on $4500 at 8.5% for 1 yr is 0.085($4500), or $382.50. The interest on $5100 at 6.8% for 1 yr is 0.068($5100), or $346.80. The total amount of interest is $382.50 + $346.80, or $729.30, so the numbers check.

5. **State.** The PLUS loan was for $4500 and the Stafford loan was for $5100.

 **TRY EXERCISE** 29

Before proceeding to Example 5, briefly scan Examples 2–4 for similarities. Note that in each case, one of the equations in the system is a simple sum while the other equation represents a sum of products. Example 5 continues this pattern with what is commonly called a *mixture problem.*

Problem-Solving Tip

When solving a problem, see if it is patterned or modeled after a problem that you have already solved.

EXAMPLE 5 *Mixing fertilizers.* Nature's Green Gardening, Inc., carries two brands of fertilizer containing nitrogen and water. "Gentle Grow" is 3% nitrogen and "Sun Saver" is 8% nitrogen. Nature's Green needs to combine the two types of solutions into a 90-L mixture that is 6% nitrogen. How much of each brand should be used?

SOLUTION

1. **Familiarize.** We make a drawing and note that we must consider not only the size of the mixture, but also its strength. Let's make a guess to gain familiarity with the problem.

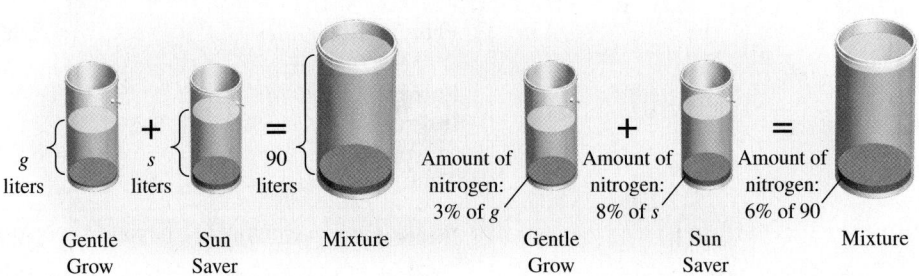

g liters	s liters	90 liters	Amount of nitrogen: 3% of g	Amount of nitrogen: 8% of s	Amount of nitrogen: 6% of 90
Gentle Grow	Sun Saver	Mixture	Gentle Grow	Sun Saver	Mixture

The total amount of the mixture must be 90 L. The total amount of the nitrogen must be 6% of 90 L, or 5.4 L.

Suppose that 40 L of Gentle Grow and 50 L of Sun Saver are mixed. The resulting mixture will be the right size, 90 L, but will it be the right strength? To find out, note that 40 L of Gentle Grow would contribute $0.03(40) = 1.2$ L of nitrogen to the mixture while 50 L of Sun Saver would contribute $0.08(50) = 4$ L of nitrogen to the mixture. The total amount of nitrogen in the mixture would then be $1.2 + 4$, or 5.2 L. But we want 6% of 90, or 5.4 L, to be nitrogen. Our guess of 40 L and 50 L is close but incorrect. Checking our guess has familiarized us with the problem.

2. **Translate.** Let $g =$ the number of liters of Gentle Grow and $s =$ the number of liters of Sun Saver. The information can be organized in a table.

	Gentle Grow	Sun Saver	Mixture	
Number of Liters	g	s	90	→ $g + s = 90$
Percent of Nitrogen	3%	8%	6%	
Amount of Nitrogen	$0.03g$	$0.08s$	0.06×90, or 5.4 liters	→ $0.03g + 0.08s = 5.4$

If we add g and s in the first row, we get one equation. It represents the total amount of mixture: $g + s = 90$.

If we add the amounts of nitrogen listed in the third row, we get a second equation. This equation represents the amount of nitrogen in the mixture: $0.03g + 0.08s = 5.4$.

After clearing decimals, we have translated the problem to the system

$$g + s = 90, \qquad (1)$$
$$3g + 8s = 540. \qquad (2)$$

3. **Carry out.** We use the elimination method to solve the system:

$$-3g - 3s = -270 \qquad \text{Multiplying both sides of equation (1) by } -3$$

$$\underline{3g + 8s = 540}$$

$$5s = 270 \qquad \text{Adding}$$

$$s = 54; \qquad \text{Solving for } s$$

$$g + 54 = 90 \qquad \text{Substituting into equation (1)}$$

$$g = 36. \qquad \text{Solving for } g$$

4. **Check.** Remember, g is the number of liters of Gentle Grow and s is the number of liters of Sun Saver.

Total amount of mixture:	$g + s = 36 + 54 = 90$
Total amount of nitrogen:	3% of 36 + 8% of 54 = 1.08 + 4.32 = 5.4
Percentage of nitrogen in mixture:	$\dfrac{\text{Total amount of nitrogen}}{\text{Total amount of mixture}} = \dfrac{5.4}{90} = 6\%$

The numbers check in the original problem.

5. **State.** Nature's Green Gardening should mix 36 L of Gentle Grow with 54 L of Sun Saver.

TRY EXERCISE 25

Motion Problems

When a problem deals with distance, speed (rate), and time, recall the following.

> ### Distance, Rate, and Time Equations
>
> If r represents rate, t represents time, and d represents distance, then:
>
> $$d = rt, \qquad r = \frac{d}{t}, \quad \text{and} \quad t = \frac{d}{r}.$$

Be sure to remember at least one of these equations. The others can be obtained by multiplying or dividing on both sides as needed.

EXAMPLE 6   *Train travel.*   A Vermont Railways freight train, loaded with logs, leaves Boston, heading to Washington, D.C., at a speed of 60 km/h. Two hours later, an Amtrak® Metroliner leaves Boston, bound for Washington, D.C., on a parallel track at 90 km/h. At what point will the Metroliner catch up to the freight train?

SOLUTION

1. **Familiarize.** Let's make a guess and check to see if it is correct. Suppose the trains meet after traveling 180 km. We can then calculate the time for each train.

	Distance	*Rate*	*Time*
Freight Train	180 km	60 km/h	$\frac{180}{60} = 3$ hr
Metroliner	180 km	90 km/h	$\frac{180}{90} = 2$ hr

We see that the distance cannot be 180 km, since the difference in travel times for the trains is *not* 2 hr. Although our guess is wrong, we can use a similar chart to organize the information in this problem.

The distance at which the trains meet is unknown, but we do know that the trains will have traveled the same distance when they meet. We let $d =$ this distance.

The time that the trains are running is also unknown, but we do know that the freight train has a 2-hr head start. Thus if we let $t =$ the number of hours that the freight train is running before they meet, then $t - 2$ is the number of hours that the Metroliner runs before catching up to the freight train.

60 km/h
d kilometers
t hours

90 km/h
d kilometers
$t - 2$ hours

Trains meet here – – –

2. **Translate.** We can organize the information in a chart. Each row is determined by the formula *Distance* = *Rate* · *Time*.

	Distance	Rate	Time	
Freight Train	d	60	t	$\longrightarrow d = 60t$
Metroliner	d	90	$t - 2$	$\longrightarrow d = 90(t - 2)$

Using *Distance = Rate · Time* twice, we get two equations:

$$d = 60t, \qquad (1)$$
$$d = 90(t - 2). \qquad (2)$$

3. **Carry out.** We solve the system using substitution:

$$60t = 90(t - 2) \qquad \text{Substituting } 60t \text{ for } d \text{ in equation (2)}$$
$$60t = 90t - 180$$
$$-30t = -180$$
$$t = 6.$$

The time for the freight train is 6 hr, which means that the time for the Metroliner is 6 − 2, or 4 hr. Remember that it is distance, not time, that the problem asked for. Thus for $t = 6$, we have $d = 60 \cdot 6 = 360$ km.

4. **Check.** At 60 km/h, the freight train will travel $60 \cdot 6$, or 360 km, in 6 hr. At 90 km/h, the Metroliner will travel $90 \cdot (6 - 2) = 360$ km in 4 hr. The numbers check.

5. **State.** The freight train will catch up to the Metroliner at a point 360 km from Boston.

▶ TRY EXERCISE 37

STUDENT NOTES

Always be careful to answer the question asked in the problem. In Example 6, the problem asks for distance, not time. Answering "6 hr" would be incorrect.

EXAMPLE 7

Jet travel. A Boeing 747-400 jet flies 4 hr west with a 60-mph tailwind. Returning *against* the wind takes 5 hr. Find the speed of the jet with no wind.

SOLUTION

1. **Familiarize.** We imagine the situation and make a drawing. Note that the wind *speeds up* the jet on the outbound flight but *slows down* the jet on the return flight.

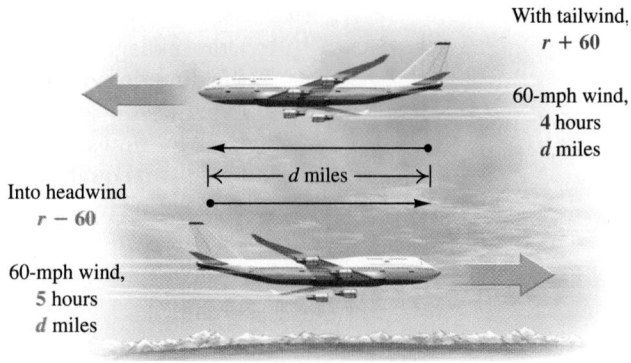

Let's make a guess of the jet's speed if there were no wind. Note that the distances traveled each way must be the same.

Speed with no wind:	400 mph
Speed with the wind:	400 + 60 = 460 mph
Speed against the wind:	400 − 60 = 340 mph
Distance with the wind:	460 · 4 = 1840 mi
Distance against the wind:	340 · 5 = 1700 mi

Since the distances are not the same, our guess of 400 mph is incorrect.

We let r = the speed, in miles per hour, of the jet in still air. Then $r + 60$ = the jet's speed with the wind and $r - 60$ = the jet's speed against the wind. We also let d = the distance traveled, in miles.

2. **Translate.** The information can be organized in a chart. The distances traveled are the same, so we use *Distance* = *Rate* (or *Speed*) · *Time*. Each row of the chart gives an equation.

	Distance	Rate	Time	
With Wind	d	$r + 60$	4	→ $d = (r + 60)4$
Against Wind	d	$r - 60$	5	→ $d = (r - 60)5$

The two equations constitute a system:

$$d = (r + 60)4, \quad (1)$$
$$d = (r - 60)5. \quad (2)$$

3. **Carry out.** We solve the system using substitution:

$$(r - 60)5 = (r + 60)4 \qquad \text{Substituting } (r - 60)5 \text{ for } d \text{ in equation (1)}$$
$$5r - 300 = 4r + 240 \qquad \text{Using the distributive law}$$
$$r = 540. \qquad \text{Solving for } r$$

4. **Check.** When $r = 540$, the speed with the wind is $540 + 60 = 600$ mph, and the speed against the wind is $540 - 60 = 480$ mph. The distance with the wind, $600 \cdot 4 = 2400$ mi, matches the distance into the wind, $480 \cdot 5 = 2400$ mi, so we have a check.

5. **State.** The speed of the jet with no wind is 540 mph. ▸ **TRY EXERCISE** ▸ 39

Tips for Solving Motion Problems

1. Draw a diagram using an arrow or arrows to represent distance and the direction of each object in motion.
2. Organize the information in a chart.
3. Look for times, distances, or rates that are the same. These often can lead to an equation.
4. Translating to a system of equations allows for the use of two variables.
5. Always make sure that you have answered the question asked.

8.3   EXERCISE SET

For Extra Help **MyMathLab** **Math XL**   WATCH   DOWNLOAD
PRACTICE

1.–14. For Exercises 1–14, solve Exercises 41–54 from pp. 513–514.

15. *Recycled paper.*   Staples® recently charged $3.79 per ream (package of 500 sheets) of regular paper and $5.49 per ream of paper made of recycled fibers. Last semester, Valley College spent $582.44 for 116 reams of paper. How many of each type were purchased?

16. *Photocopying.*   Quick Copy recently charged 49¢ per page for color copies and 7¢ per page for black-and-white copies. If Shirlee's bill for 90 copies was $11.34, how many copies of each type were made?

17. *Lighting.*   Lowe's Home Improvement recently sold 13-watt Feit Electric Ecobulbs® for $5 each and 18-watt Ecobulbs® for $6 each. If River County Hospital purchased 200 such bulbs for a total of $1140, how many of each type did they purchase?

18. *Office supplies.*   Staples® recently charged $17.99 per box of Pilot Precise® rollerball pens and $7.49 per box for Bic® Matic Grip mechanical pencils. If Kelling Community College purchased 120 such boxes for a total of $1234.80, how many boxes of each type did they purchase?

19. *Sales.*   Recently, officedepot.com sold a black HP C7115A Laser Jet print cartridge for $64.99 and a color Apple computer M3908GA ink cartridge for $58.99. During a promotion offering free shipping, a total of 450 of these cartridges was purchased for a total of $27,625.50. How many of each type were purchased?

20. *Sales.*   Office Max® recently advertised a three-subject notebook for $2.49 and a five-subject notebook for $3.79. At the start of a recent spring semester, a combination of 50 of these notebooks was sold for a total of $166.10. How many of each type were sold?

Aha! **21.** *Blending coffees.*   The Roasted Bean charges $13.00 per pound for Fair Trade Organic Mexican coffee and $11.00 per pound for Fair Trade Organic Peruvian coffee. How much of each type should be used to make a 28-lb blend that sells for $12.00 per pound?

22. *Mixed nuts.*   Oh Nuts! sells pistachio kernels for $6.50 per pound and almonds for $8.00 per pound. How much of each type should be used to make a 50-lb mixture that sells for $7.40 per pound?

23. *Event planning.*   As part of the refreshments for Yvette's 25th birthday party, Kim plans to provide a bowl of M&M candies. She wants to mix custom-printed M&Ms costing 60¢ per ounce with bulk M&Ms costing 25¢ per ounce to create 20 lb of a mixture costing 32¢ per ounce. How much of each type of M&M should she use?
Source: www.mymms.com

24. *Blending spices.*   Spice of Life sells ground sumac for $1.35 an ounce and ground thyme for $1.85 an ounce. Aman wants to make a 20-oz Zahtar seasoning blend using the two spices that sells for $1.65 an ounce. How much of each spice should Aman use?

25. *Catering.*   Cati's Catering is planning an office reception. The office administrator has requested a candy mixture that is 25% chocolate. Cati has available mixtures that are either 50% chocolate or 10% chocolate. How much of each type should be mixed to get a 20-lb mixture that is 25% chocolate?

26. *Ink remover.*   Etch Clean Graphics uses one cleanser that is 25% acid and a second that is 50% acid. How many liters of each should be mixed to get 30 L of a solution that is 40% acid?

27. *Blending granola.*   Deep Thought Granola is 25% nuts and dried fruit. Oat Dream Granola is 10% nuts and dried fruit. How much of Deep Thought and how much of Oat Dream should be mixed to form a 20-lb batch of granola that is 19% nuts and dried fruit?

28. *Livestock feed.*   Soybean meal is 16% protein and corn meal is 9% protein. How many pounds of each should be mixed to get a 350-lb mixture that is 12% protein?

29. *Student loans.*   Stacey's two student loans totaled $12,000. One of her loans was at 6.5% simple interest and the other at 7.2%. After one year, Stacey owed $811.50 in interest. What was the amount of each loan?

30. *Investments.*   A self-employed contractor nearing retirement made two investments totaling $15,000. In one year, these investments yielded $1023 in simple interest. Part of the money was invested at 6% and the rest at 7.5%. How much was invested at each rate?

31. *Automotive maintenance.* "Steady State" antifreeze is 18% alcohol and "Even Flow" is 10% alcohol. How many liters of each should be mixed to get 20 L of a mixture that is 15% alcohol?

32. *Chemistry.* E-Chem Testing has a solution that is 80% base and another that is 30% base. A technician needs 150 L of a solution that is 62% base. The 150 L will be prepared by mixing the two solutions on hand. How much of each should be used?

33. *Octane ratings.* The octane rating of a gasoline is a measure of the amount of isooctane in the gas. Manufacturers recommend using 93-octane gasoline on retuned motors. How much 87-octane gas and 95-octane gas should Yousef mix in order to make 10 gal of 93-octane gas for his retuned Ford F-150?
Source: Champlain Electric and Petroleum Equipment

34. *Octane ratings.* The octane rating of a gasoline is a measure of the amount of isooctane in the gas. Subaru recommends 91-octane gasoline for the 2008 Legacy 3.0 R. How much 87-octane gas and 93-octane gas should Kelsey mix in order to make 12 gal of 91-octane gas for her Legacy?
Sources: Champlain Electric and Petroleum Equipment: Dean Team Ballwin

35. *Food science.* The following bar graph shows the milk fat percentages in three dairy products. How many pounds each of whole milk and cream should be mixed to form 200 lb of milk for cream cheese?

Milk fat

Percent milk fat	32 28 24 20 16 12 8 4 0

Whole milk   Milk for cream cheese   Cream

36. *Food science.* How much lowfat milk (1% fat) and how much whole milk (4% fat) should be mixed to make 5 gal of reduced fat milk (2% fat)?

37. *Train travel.* A train leaves Danville Union and travels north at a speed of 75 km/h. Two hours later, an express train leaves on a parallel track and travels north at 125 km/h. How far from the station will they meet?

38. *Car travel.* Two cars leave Salt Lake City, traveling in opposite directions. One car travels at a speed of 80 km/h and the other at 96 km/h. In how many hours will they be 528 km apart?

39. *Canoeing.* Kahla paddled for 4 hr with a 6-km/h current to reach a campsite. The return trip against the same current took 10 hr. Find the speed of Kahla's canoe in still water.

40. *Boating.* Cody's motorboat took 3 hr to make a trip downstream with a 6-mph current. The return trip against the same current took 5 hr. Find the speed of the boat in still water.

41. *Point of no return.* A plane flying the 3458-mi trip from New York City to London has a 50-mph tailwind. The flight's *point of no return* is the point at which the flight time required to return to New York is the same as the time required to continue to London. If the speed of the plane in still air is 360 mph, how far is New York from the point of no return?

42. *Point of no return.* A plane is flying the 2553-mi trip from Los Angeles to Honolulu into a 60-mph headwind. If the speed of the plane in still air is 310 mph, how far from Los Angeles is the plane's point of no return? (See Exercise 41.)

43. *Architecture.*   The rectangular ground floor of the John Hancock building has a perimeter of 860 ft. The length is 100 ft more than the width. Find the length and the width.

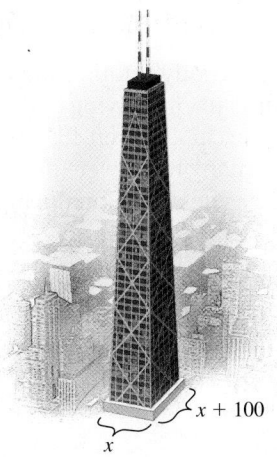

$x + 100$

x

44. *Real estate.*   The perimeter of a rectangular ocean-front lot is 190 m. The width is one-fourth of the length. Find the dimensions.

45. In 2007, Nintendo Co. sold three times as many Wii game machines in Japan as Sony Corp. sold PlayStation 3 consoles. Together, they sold 4.84 million game machines in Japan. How many of each were sold?
Source: Bloomberg.com

46. *Hockey rankings.*   Hockey teams receive 2 points for a win and 1 point for a tie. The Wildcats once won a championship with 60 points. They won 9 more games than they tied. How many wins and how many ties did the Wildcats have?

47. *Video rentals.*   At one time, Netflix offered an unlimited 1 DVD at-a-time rental plan for $8.99 per month and a rental plan with a limit of 2 DVDs per month for $4.99. During one week, 250 new subscribers paid $1975.50 for these plans. How many of each type of plan were purchased?

48. *Radio airplay.*   Roscoe must play 12 commercials during his 1-hr radio show. Each commercial is either 30 sec or 60 sec long. If the total commercial time during that hour is 10 min, how many commercials of each type does Roscoe play?

49. *Making change.*   Monica makes a $9.25 purchase at the bookstore with a $20 bill. The store has no bills and gives her the change in quarters and fifty-cent pieces. There are 30 coins in all. How many of each kind are there?

50. *Teller work.*   Sabina goes to a bank and gets change for a $50 bill consisting of all $5 bills and $1 bills. There are 22 bills in all. How many of each kind are there?

51. In what ways are Examples 3 and 4 similar? In what sense are their systems of equations similar?

52. Write at least three study tips of your own for someone beginning this exercise set.

Skill Review

To prepare for Section 8.4, review evaluating expressions with three variables (Sections 1.1, 1.3, and 1.8).

Evaluate.

53. $2x - 3y - z$, for $x = 5$, $y = 2$, and $z = 3$ [1.1]

54. $4x + y - 6z$, for $x = \frac{1}{2}$, $y = \frac{1}{2}$, and $z = \frac{1}{3}$ [1.3]

55. $x + y + 2z$, for $x = 1$, $y = -4$, and $z = -5$ [1.8]

56. $3a - b + 2c$, for $a = 1$, $b = -6$, and $c = 4$ [1.8]

57. $a - 2b - 3c$, for $a = -2$, $b = 3$, and $c = -5$ [1.8]

58. $2a - 5b - c$, for $a = \frac{1}{4}$, $b = -\frac{1}{4}$, and $c = -\frac{3}{2}$ [1.8]

Synthesis

59. Suppose that in Example 3 you are asked only for the amount of almonds needed for the Market Street blend. Would the method of solving the problem change? Why or why not?

60. Write a problem similar to Example 2 for a classmate to solve. Design the problem so that the solution is "The bakery sold 24 loaves of bread and 18 packages of sandwich rolls."

61. *Recycled paper.*   Unable to purchase 60 reams of paper that contains 20% post-consumer fiber, the Naylor School bought paper that was either 0% post-consumer fiber or 30% post-consumer fiber. How many reams of each should be purchased in order to use the same amount of post-consumer fiber as if the 20% post-consumer fiber paper were available?

62. *Automotive maintenance.*   The radiator in Natalie's car contains 6.3 L of antifreeze and water. This mixture is 30% antifreeze. How much of this mixture should she drain and replace with pure antifreeze so that there will be a mixture of 50% antifreeze?

63. *Metal alloys.*   In order for a metal to be labeled "sterling silver," the silver alloy must contain at least 92.5% pure silver. Nicole has 32 oz of coin silver, which is 90% pure silver. How much pure silver must she add to the coin silver in order to have a sterling-silver alloy?
Source: *The Jewelry Repair Manual*, R. Allen Hardy, Courier Dover Publications, 1996, p. 271.

64. *Exercise.*   Elyse jogs and walks to school each day. She averages 4 km/h walking and 8 km/h jogging. From home to school is 6 km and Elyse makes the trip in 1 hr. How far does she jog in a trip?

65. *Book sales.*   *American Economic History* can be purchased as a three-volume set for $88 or each volume can be purchased separately for $39. An economics class spent $1641 for 51 volumes. How many three-volume sets were ordered?
Source: National History Day, www.nhd.org

66. The tens digit of a two-digit positive integer is 2 more than three times the units digit. If the digits are interchanged, the new number is 13 less than half the given number. Find the given integer. (*Hint*: Let x = the tens-place digit and y = the units-place digit; then $10x + y$ is the number.)

67. *Wood stains.*   Williams' Custom Flooring has 0.5 gal of stain that is 20% brown and 80% neutral. A customer orders 1.5 gal of a stain that is 60% brown and 40% neutral. How much pure brown stain and how much neutral stain should be added to the original 0.5 gal in order to make up the order?*

68. *Train travel.*   A train leaves Union Station for Central Station, 216 km away, at 9 A.M. One hour later, a train leaves Central Station for Union Station. They meet at noon. If the second train had started at 9 A.M. and the first train at 10:30 A.M., they would still have met at noon. Find the speed of each train.

69. *Fuel economy.*   Grady's station wagon gets 18 miles per gallon (mpg) in city driving and 24 mpg in highway driving. The car is driven 465 mi on 23 gal of gasoline. How many miles were driven in the city and how many were driven on the highway?

70. *Biochemistry.*   Industrial biochemists routinely use a machine to mix a buffer of 10% acetone by adding 100% acetone to water. One day, instead of adding 5 L of acetone to create a vat of buffer, a machine added 10 L. How much additional water was needed to bring the concentration down to 10%?

71. See Exercise 67 above. Let x = the amount of pure brown stain added to the original 0.5 gal. Find a function $P(x)$ that can be used to determine the percentage of brown stain in the 1.5-gal mixture. On a graphing calculator, draw the graph of P and use INTERSECT to confirm the answer to Exercise 67.

72. *Gender.*   Phil and Phyllis are twins. Phyllis has twice as many brothers as she has sisters. Phil has the same number of brothers as sisters. How many girls and how many boys are in the family?

---

*This problem was suggested by Professor Chris Burditt of Yountville, California.

COLLABORATIVE

CORNER

How Many Two's? How Many Three's?

Focus: Systems of linear equations

Time: 20 minutes

Group size: 3

The box score at right, from the 2008 NBA All-Star game, contains information on how many field goals (worth either 2 or 3 points) and free throws (worth 1 point) each player attempted and made. For example, the line "Allen 10-14 3-5 28" means that the East's Ray Allen made 10 field goals out of 14 attempts and 3 free throws out of 5 attempts, for a total of 28 points.

ACTIVITY

1. Work as a group to develop a system of two equations in two unknowns that can be used to determine how many 2-pointers and how many 3-pointers were made by the West.

2. Each group member should solve the system from part (1) in a different way: one person algebraically, one person by making a table and methodically checking all combinations of 2- and 3-pointers, and one person by

guesswork. Compare answers when this has been completed.

3. Determine, as a group, how many 2- and 3-pointers the East made.

East (134)

James 12–22 1–1 27, Bosh 7–15 0–2 14, Howard 7–7 2–3 16, Wade 7–12 0–2 14, Kidd 1–2 0–0 2, Hamilton 4–9 0–0 9, Wallace 1–5 0–0 3, Billups 3–10 0–1 6, Jamison 1–3 0–0 2, Pierce 5–9 0–0 10, Johnson 1–2 0–0 3, Allen 10–14 3–5 28
Totals 59–110 6–14 134

West (128)

Anthony 8–17 2–3 18, Duncan 2–7 0–0 4, Ming 2–5 2–2 6, Bryant 0–0 0–0 0, Iverson 3–7 1–2 7, Nash 4–8 0–0 8, Stoudemire 8–11 1–3 18, Nowitzki 5–14 2–2 13, Paul 7–14 0–0 16, West 3–6 0–0 6, Roy 8–10 0–0 18, Boozer 7–15 0–2 14
Totals 57–114 8–14 128

| East | 34 | 40 | 32 | 28 — 134 |
| West | 28 | 37 | 28 | 35 — 128 |

8.4 Systems of Equations in Three Variables

Identifying Solutions ▪ Solving Systems in Three Variables ▪ Dependency, Inconsistency, and Geometric Considerations

Some problems translate directly to two equations. Others more naturally call for a translation to three or more equations. In this section, we learn how to solve systems of three linear equations. Later, we will use such systems in problem-solving situations.

Identifying Solutions

A **linear equation in three variables** is an equation equivalent to one in the form $Ax + By + Cz = D$, where A, B, C, and D are real numbers. We refer to the form $Ax + By + Cz = D$ as *standard form* for a linear equation in three variables.

A solution of a system of three equations in three variables is an ordered triple (x, y, z) that makes *all three* equations true. The numbers in an ordered triple correspond to the variables in alphabetical order unless otherwise indicated.

EXAMPLE   1   Determine whether $\left(\frac{3}{2}, -4, 3\right)$ is a solution of the system

$$4x - 2y - 3z = 5,$$
$$-8x - y + z = -5,$$
$$2x + y + 2z = 5.$$

SOLUTION   We substitute $\left(\frac{3}{2}, -4, 3\right)$ into the three equations, using alphabetical order:

$$
\begin{array}{c|c}
\multicolumn{2}{c}{4x - 2y - 3z = 5} \\
\hline
4 \cdot \frac{3}{2} - 2(-4) - 3 \cdot 3 & 5 \\
6 + 8 - 9 & \\
5 \overset{?}{=} 5 & \text{TRUE}
\end{array}
\qquad
\begin{array}{c|c}
\multicolumn{2}{c}{-8x - y + z = -5} \\
\hline
-8 \cdot \frac{3}{2} - (-4) + 3 & -5 \\
-12 + 4 + 3 & \\
-5 \overset{?}{=} -5 & \text{TRUE}
\end{array}
$$

$$
\begin{array}{c|c}
\multicolumn{2}{c}{2x + y + 2z = 5} \\
\hline
2 \cdot \frac{3}{2} + (-4) + 2 \cdot 3 & 5 \\
3 - 4 + 6 & \\
5 \overset{?}{=} 5 & \text{TRUE}
\end{array}
$$

The triple makes all three equations true, so it is a solution.

> **TRY EXERCISE   7**

Solving Systems in Three Variables

The graph of a linear equation in three variables is a plane. Because a three-dimensional coordinate system is required, solving systems in three variables graphically is difficult. The substitution method *can* be used but becomes cumbersome unless one or more of the equations has only two variables. Fortunately, the elimination method works well for a system of three equations in three variables. We first eliminate one variable to form a system of two equations in two variables. Once that simpler system has been solved, we substitute into one of the three original equations and solve for the third variable.

EXAMPLE   2   Solve the following system of equations:

$$
\begin{aligned}
x + y + z &= 4, & (1) \\
x - 2y - z &= 1, & (2) \\
2x - y - 2z &= -1. & (3)
\end{aligned}
$$

SOLUTION   We select *any* two of the three equations and work to get an equation in two variables. Let's add equations (1) and (2):

$$
\begin{array}{rll}
x + y + z = 4 & (1) \\
x - 2y - z = 1 & (2) \\
\hline
2x - y = 5. & (4) & \text{Adding to eliminate } z
\end{array}
$$

> **CAUTION!** Be sure to eliminate the same variable in both pairs of equations.

Next, we select a different pair of equations and eliminate the *same variable* that we did above. Let's use equations (1) and (3) to again eliminate z. Be careful! A common error is to eliminate a different variable in this step.

$$
\begin{array}{l}
x + y + z = 4, \\
2x - y - 2z = -1
\end{array}
\xrightarrow[\text{of equation (1) by 2}]{\text{Multiplying both sides}}
\begin{array}{rl}
2x + 2y + 2z = 8 \\
2x - y - 2z = -1 \\
\hline
4x + y = 7 & (5)
\end{array}
$$

Now we solve the resulting system of equations (4) and (5). That solution will give us two of the numbers in the solution of the original system.

$$2x - y = 5 \quad (4)$$
$$\underline{4x + y = 7} \quad (5)$$
$$6x \quad\quad = 12 \quad \text{Adding}$$
$$x = 2$$

Note that we now have two equations in two variables. Had we not eliminated the *same* variable in both of the above steps, this would not be the case.

We can use either equation (4) or (5) to find y. We choose equation (5):

$$4x + y = 7 \quad (5)$$
$$4 \cdot 2 + y = 7 \quad \text{Substituting 2 for } x \text{ in equation (5)}$$
$$8 + y = 7$$
$$y = -1.$$

We now have $x = 2$ and $y = -1$. To find the value for z, we use any of the original three equations and substitute to find the third number, z. Let's use equation (1) and substitute our two numbers in it:

$$x + y + z = 4 \quad (1)$$
$$2 + (-1) + z = 4 \quad \text{Substituting 2 for } x \text{ and } -1 \text{ for } y$$
$$1 + z = 4$$
$$z = 3.$$

We have obtained the triple $(2, -1, 3)$. It should check in *all three* equations:

$$\frac{x + y + z = 4}{2 + (-1) + 3 \,\mid\, 4}$$
$$4 \overset{?}{=} 4 \quad \text{TRUE}$$

$$\frac{x - 2y - z = 1}{2 - 2(-1) - 3 \,\mid\, 1}$$
$$1 \overset{?}{=} 1 \quad \text{TRUE}$$

$$\frac{2x - y - 2z = -1}{2 \cdot 2 - (-1) - 2 \cdot 3 \,\mid\, -1}$$
$$-1 \overset{?}{=} -1 \quad \text{TRUE}$$

The solution is $(2, -1, 3)$.

> **TRY EXERCISE** 9

Solving Systems of Three Linear Equations

To use the elimination method to solve systems of three linear equations:

1. Write all equations in the standard form $Ax + By + Cz = D$.
2. Clear any decimals or fractions.
3. Choose a variable to eliminate. Then select two of the three equations and work to get one equation in which the selected variable is eliminated.
4. Next, use a different pair of equations and eliminate the same variable that you did in step (3).
5. Solve the system of equations that resulted from steps (3) and (4).
6. Substitute the solution from step (5) into one of the original three equations and solve for the third variable. Then check.

EXAMPLE 3   Solve the system

$$4x - 2y - 3z = 5, \quad (1)$$
$$-8x - y + z = -5, \quad (2)$$
$$2x + y + 2z = 5. \quad (3)$$

SOLUTION

Write in standard form.

1., 2. The equations are already in standard form with no fractions or decimals.

3. Next, select a variable to eliminate. We decide on y because the y-terms are opposites of each other in equations (2) and (3). We add:

Eliminate a variable. (We choose y.)

$$-8x - y + z = -5 \quad (2)$$
$$\underline{2x + y + 2z = 5} \quad (3)$$
$$-6x + 3z = 0. \quad (4) \qquad \text{Adding}$$

4. We use another pair of equations to create a second equation in x and z. That is, we eliminate the same variable, y, as in step (3). We use equations (1) and (3):

Eliminate the same variable using a different pair of equations.

$$4x - 2y - 3z = 5,$$
$$2x + y + 2z = 5 \quad \xrightarrow[\text{of equation (3) by 2}]{\text{Multiplying both sides}} \quad \begin{aligned} 4x - 2y - 3z &= 5 \\ \underline{4x + 2y + 4z} &= 10 \\ 8x + z &= 15. \end{aligned} \quad (5)$$

5. Now we solve the resulting system of equations (4) and (5). That allows us to find two parts of the ordered triple.

Solve the system of two equations in two variables.

$$-6x + 3z = 0,$$
$$8x + z = 15 \quad \xrightarrow[\text{of equation (5) by } -3]{\text{Multiplying both sides}} \quad \begin{aligned} -6x + 3z &= 0 \\ \underline{-24x - 3z} &= -45 \\ -30x &= -45 \\ x &= \tfrac{-45}{-30} = \tfrac{3}{2} \end{aligned}$$

We use equation (5) to find z:

$$8x + z = 15$$
$$8 \cdot \tfrac{3}{2} + z = 15 \qquad \text{Substituting } \tfrac{3}{2} \text{ for } x$$
$$12 + z = 15$$
$$z = 3.$$

6. Finally, we use any of the original equations and substitute to find the third number, y. We choose equation (3):

Solve for the remaining variable.

$$2x + y + 2z = 5 \quad (3)$$
$$2 \cdot \tfrac{3}{2} + y + 2 \cdot 3 = 5 \qquad \text{Substituting } \tfrac{3}{2} \text{ for } x \text{ and } 3 \text{ for } z$$
$$3 + y + 6 = 5$$
$$y + 9 = 5$$
$$y = -4.$$

Check.

The solution is $\left(\tfrac{3}{2}, -4, 3\right)$. The check was performed as Example 1.

TRY EXERCISE ▸ 23

Sometimes, certain variables are missing at the outset.

EXAMPLE 4   Solve the system

$$x + y + z = 180, \quad (1)$$
$$x - z = -70, \quad (2)$$
$$2y - z = 0. \quad (3)$$

SOLUTION

1., 2. The equations appear in standard form with no fractions or decimals.

3., 4. Note that there is no y in equation (2). Thus, at the outset, we already have y eliminated from one equation. We need another equation with y eliminated, so we work with equations (1) and (3):

$$x + y + z = 180, \quad \xrightarrow[\text{of equation (1) by } -2]{\text{Multiplying both sides}} \quad \begin{array}{rcl} -2x - 2y - 2z &=& -360 \\ 2y - z &=& 0 \end{array}$$

$$2y - z = 0 \qquad \qquad \qquad \overline{\hphantom{-2x}\ -2x \hphantom{xxx} -\ 3z\ =\ -360.} \quad (4)$$

5., 6. Now we solve the resulting system of equations (2) and (4):

$$\begin{array}{rcl} x - z &=& -70, \\ -2x - 3z &=& -360 \end{array} \quad \xrightarrow[\text{of equation (2) by } 2]{\text{Multiplying both sides}} \quad \begin{array}{rcl} 2x - 2z &=& -140 \\ -2x - 3z &=& -360 \end{array}$$

$$\overline{\hphantom{xxxxxxxx} -5z\ =\ -500}$$

$$z = 100.$$

Continuing as in Examples 2 and 3, we get the solution $(30, 50, 100)$. The check is left to the student.

> **TRY EXERCISE** 27

Dependency, Inconsistency, and Geometric Considerations

Each equation in Examples 2, 3, and 4 has a graph that is a plane in three dimensions. The solutions are points common to the planes of each system. Since three planes can have an infinite number of points in common or no points at all in common, we need to generalize the concept of *consistency*.

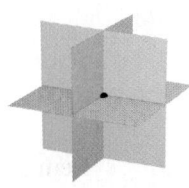   

Planes intersect at one point. System is *consistent* and has one solution.

Planes intersect along a common line. System is *consistent* and has an infinite number of solutions.

Three parallel planes. System is *inconsistent;* it has no solution.

Planes intersect two at a time, with no point common to all three. System is *inconsistent;* it has no solution.

> **Consistency**
>
> A system of equations that has at least one solution is said to be **consistent**.
>
> A system of equations that has no solution is said to be **inconsistent**.

EXAMPLE 5   Solve:

$$\begin{array}{rcl} y + 3z &=& 4, \quad (1) \\ -x - y + 2z &=& 0, \quad (2) \\ x + 2y + z &=& 1. \quad (3) \end{array}$$

SOLUTION   The variable x is missing in equation (1). By adding equations (2) and (3), we can find a second equation in which x is missing:

$$\begin{array}{rcl} -x - y + 2z &=& 0 \quad (2) \\ x + 2y + z &=& 1 \quad (3) \end{array}$$

$$\overline{\hphantom{xxxxxx} y + 3z\ =\ 1. \quad (4)} \qquad \text{Adding}$$

Equations (1) and (4) form a system in y and z. We solve as before:

$$y + 3z = 4, \quad \xrightarrow[\text{of equation (1) by } -1]{\text{Multiplying both sides}} \quad -y - 3z = -4$$
$$y + 3z = 1 \qquad\qquad\qquad\qquad\qquad\qquad\quad y + 3z = 1$$

This is a contradiction. $\xrightarrow{} 0 = -3.$ Adding

Since we end up with a *false* equation, or contradiction, we state that the system has no solution. It is *inconsistent*.   **TRY EXERCISE** 15

The notion of *dependency* from Section 8.1 can also be extended.

EXAMPLE 6   Solve:

$$2x + y + z = 3, \quad (1)$$
$$x - 2y - z = 1, \quad (2)$$
$$3x + 4y + 3z = 5. \quad (3)$$

SOLUTION   Our plan is to first use equations (1) and (2) to eliminate z. Then we will select another pair of equations and again eliminate z:

$$2x + y + z = 3$$
$$x - 2y - z = 1$$
$$\overline{3x - y = 4.} \quad (4)$$

Next, we use equations (2) and (3) to eliminate z again:

$$x - 2y - z = 1, \quad \xrightarrow[\text{of equation (2) by 3}]{\text{Multiplying both sides}} \quad 3x - 6y - 3z = 3$$
$$3x + 4y + 3z = 5 \qquad\qquad\qquad\qquad\qquad\quad 3x + 4y + 3z = 5$$
$$\overline{6x - 2y = 8.} \quad (5)$$

We now try to solve the resulting system of equations (4) and (5):

$$3x - y = 4, \quad \xrightarrow[\text{of equation (4) by } -2]{\text{Multiplying both sides}} \quad -6x + 2y = -8$$
$$6x - 2y = 8 \qquad\qquad\qquad\qquad\qquad\qquad\quad 6x - 2y = 8$$
$$\overline{ 0 = 0.} \quad (6)$$

Equation (6), which is an identity, indicates that equations (1), (2), and (3) are *dependent*. This means that the original system of three equations is equivalent to a system of two equations. One way to see this is to observe that two times equation (1), minus equation (2), is equation (3). Thus removing equation (3) from the system does not affect the solution of the system.* In writing an answer to this problem, we simply state that "the equations are dependent."

 **TRY EXERCISE** 21

Recall that when dependent equations appeared in Section 8.1, the solution sets were always infinite in size and were written in set-builder notation. There, all systems of dependent equations were *consistent*. This is not always the case for

---

*A set of equations is dependent if at least one equation can be expressed as a sum of multiples of other equations in that set.

systems of three or more equations. The following figures illustrate some possibilities geometrically.

The planes intersect along a common line. The equations are *dependent* and the system is *consistent*. There is an infinite number of solutions.

The planes coincide. The equations are *dependent* and the system is *consistent*. There is an infinite number of solutions.

Two planes coincide. The third plane is parallel. The equations are *dependent* and the system is *inconsistent*. There is no solution.

8.4 EXERCISE SET

☙ *Concept Reinforcement* Classify each statement as either true or false.

1. $3x + 5y + 4z = 7$ is a linear equation in three variables.

2. Every system of three equations in three unknowns has at least one solution.

3. It is not difficult to solve a system of three equations in three unknowns by graphing.

4. If, when we are solving a system of three equations, a false equation results from adding a multiple of one equation to another, the system is inconsistent.

5. If, when we are solving a system of three equations, an identity results from adding a multiple of one equation to another, the equations are dependent.

6. Whenever a system of three equations contains dependent equations, there is an infinite number of solutions.

---

7. Determine whether $(2, -1, -2)$ is a solution of the system
$$x + y - 2z = 5,$$
$$2x - y - z = 7,$$
$$-x - 2y - 3z = 6.$$

8. Determine whether $(-1, -3, 2)$ is a solution of the system
$$x - y + z = 4,$$
$$x - 2y - z = 3,$$
$$3x + 2y - z = 1.$$

Solve each system. If a system's equations are dependent or if there is no solution, state this.

9. $x - y - z = 0,$
   $2x - 3y + 2z = 7,$
   $-x + 2y + z = 1$

10. $x + y - z = 0,$
    $2x - y + z = 3,$
    $-x + 5y - 3z = 2$

11. $x - y - z = 1,$
    $2x + y + 2z = 4,$
    $x + y + 3z = 5$

12. $x + y - 3z = 4,$
    $2x + 3y + z = 6,$
    $2x - y + z = -14$

13. $3x + 4y - 3z = 4,$
    $5x - y + 2z = 3,$
    $x + 2y - z = -2$

14. $2x - 3y + z = 5,$
    $x + 3y + 8z = 22,$
    $3x - y + 2z = 12$

15. $x + y + z = 0,$
    $2x + 3y + 2z = -3,$
    $-x - 2y - z = 1$

16. $3a - 2b + 7c = 13,$
    $a + 8b - 6c = -47,$
    $7a - 9b - 9c = -3$

17. $2x - 3y - z = -9,$
    $2x + 5y + z = 1,$
    $x - y + z = 3$

18. $4x + y + z = 17,$
    $x - 3y + 2z = -8,$
    $5x - 2y + 3z = 5$

Aha! 19. $a + b + c = 5,$
    $2a + 3b - c = 2,$
    $2a + 3b - 2c = 4$

20. $u - v + 6w = 8,$
    $3u - v + 6w = 14,$
    $-u - 2v - 3w = 7$

21. $-2x + 8y + 2z = 4,$
    $x + 6y + 3z = 4,$
    $3x - 2y + z = 0$

22. $x - y + z = 4,$
    $5x + 2y - 3z = 2,$
    $4x + 3y - 4z = -2$

23. $2u - 4v - w = 8,$
    $3u + 2v + w = 6,$
    $5u - 2v + 3w = 2$

24. $4p + q + r = 3,$
    $2p - q + r = 6,$
    $2p + 2q - r = -9$

25. $r + \frac{3}{2}s + 6t = 2,$
$2r - 3s + 3t = 0.5,$
$r + s + t = 1$

26. $5x + 3y + \frac{1}{2}z = \frac{7}{2},$
$0.5x - 0.9y - 0.2z = 0.3,$
$3x - 2.4y + 0.4z = -1$

27. $4a + 9b = 8,$
$8a + 6c = -1,$
$6b + 6c = -1$

28. $3p + 2r = 11,$
$q - 7r = 4,$
$p - 6q = 1$

29. $x + y + z = 57,$
$-2x + y = 3,$
$x - z = 6$

30. $x + y + z = 105,$
$10y - z = 11,$
$2x - 3y = 7$

31. $a - 3c = 6,$
$b + 2c = 2,$
$7a - 3b - 5c = 14$

32. $2a - 3b = 2,$
$7a + 4c = \frac{3}{4},$
$2c - 3b = 1$

Aha! **33.** $x + y + z = 83,$
$y = 2x + 3,$
$z = 40 + x$

34. $l + m = 7,$
$3m + 2n = 9,$
$4l + n = 5$

35. $x + z = 0,$
$x + y + 2z = 3,$
$y + z = 2$

36. $x + y = 0,$
$x + z = 1,$
$2x + y + z = 2$

37. $x + y + z = 1,$
$-x + 2y + z = 2,$
$2x - y = -1$

38. $y + z = 1,$
$x + y + z = 1,$
$x + 2y + 2z = 2$

39. Rondel always begins solving systems of three equations in three variables by using the first two equations to eliminate x. Is this a good approach? Why or why not?

40. Describe a method for writing an inconsistent system of three equations in three variables.

Skill Review

To prepare for Section 8.5, review translating sentences to equations (Section 1.1).

Translate each sentence to an equation. [1.1]

41. One number is half another.

42. The difference of two numbers is twice the first number.

43. The sum of three consecutive numbers is 100.

44. The sum of three numbers is 100.

45. The product of two numbers is five times a third number.

46. The product of two numbers is twice their sum.

Synthesis

47. Is it possible for a system of three linear equations to have exactly two ordered triples in its solution set? Why or why not?

48. Describe a procedure that could be used to solve a system of four equations in four variables.

Solve.

49. $\dfrac{x + 2}{3} - \dfrac{y + 4}{2} + \dfrac{z + 1}{6} = 0,$
$\dfrac{x - 4}{3} + \dfrac{y + 1}{4} - \dfrac{z - 2}{2} = -1,$
$\dfrac{x + 1}{2} + \dfrac{y}{2} + \dfrac{z - 1}{4} = \dfrac{3}{4}$

50. $w + x - y + z = 0,$
$w - 2x - 2y - z = -5,$
$w - 3x - y + z = 4,$
$2w - x - y + 3z = 7$

51. $w + x + y + z = 2,$
$w + 2x + 2y + 4z = 1,$
$w - x + y + z = 6,$
$w - 3x - y + z = 2$

For Exercises 52 and 53, let u represent $1/x$, v represent $1/y$, and w represent $1/z$. Solve for u, v, and w, and then solve for x, y, and z.

52. $\dfrac{2}{x} + \dfrac{2}{y} - \dfrac{3}{z} = 3,$
$\dfrac{1}{x} - \dfrac{2}{y} - \dfrac{3}{z} = 9,$
$\dfrac{7}{x} - \dfrac{2}{y} + \dfrac{9}{z} = -39$

53. $\dfrac{2}{x} - \dfrac{1}{y} - \dfrac{3}{z} = -1,$
$\dfrac{2}{x} - \dfrac{1}{y} + \dfrac{1}{z} = -9,$
$\dfrac{1}{x} + \dfrac{2}{y} - \dfrac{4}{z} = 17$

Determine k so that each system is dependent.

54. $x - 3y + 2z = 1,$
$2x + y - z = 3,$
$9x - 6y + 3z = k$

55. $5x - 6y + kz = -5,$
$x + 3y - 2z = 2,$
$2x - y + 4z = -1$

In each case, three solutions of an equation in x, y, and z are given. Find the equation.

56. $Ax + By + Cz = 12;$
$\left(1, \frac{3}{4}, 3\right), \left(\frac{4}{3}, 1, 2\right),$ and $(2, 1, 1)$

57. $z = b - mx - ny;$
$(1, 1, 2), (3, 2, -6),$ and $\left(\frac{3}{2}, 1, 1\right)$

58. Write an inconsistent system of equations that contains dependent equations.

59. Kadi and Ahmed both correctly solve the system

$x + 2y - z = 1,$
$-x - 2y + z = 3,$
$2x + 4y - 2z = 2.$

Kadi states "the equations are dependent" while Ahmed states "there is no solution." How did each person reach the conclusion?

COLLABORATIVE CORNER

Finding the Preferred Approach

Focus: Systems of three linear equations

Time: 10–15 minutes

Group size: 3

Consider the six steps outlined on p. 539 along with the following system:

$$2x + 4y = 3 - 5z,$$
$$0.3x = 0.2y + 0.7z + 1.4,$$
$$0.04x + 0.03y = 0.07 + 0.04z.$$

ACTIVITY

1. Working independently, each group member should solve the system above. One person should begin by eliminating x, one should first eliminate y, and one should first eliminate z. Write neatly so that others can follow your steps.

2. Once all group members have solved the system, compare your answers. If the answers do not check, exchange notebooks and check each other's work. If a mistake is detected, allow the person who made the mistake to make the repair.

3. Decide as a group which of the three approaches above (if any) ranks as easiest and which (if any) ranks as most difficult. Then compare your rankings with the other groups in the class.

8.5   Solving Applications: Systems of Three Equations

Applications of Three Equations in Three Unknowns

Solving systems of three or more equations is important in many applications. Such systems arise in the natural and social sciences, business, and engineering. To begin, let's first look at a purely numerical application.

EXAMPLE 1

The sum of three numbers is 4. The first number minus twice the second, minus the third is 1. Twice the first number minus the second, minus twice the third is -1. Find the numbers.

SOLUTION

1. **Familiarize.** There are three statements involving the same three numbers. Let's label these numbers x, y, and z.

2. **Translate.** We can translate directly as follows.

The sum of the three numbers is 4.
$$x + y + z = 4$$

The first number minus twice the second minus the third is 1.
$$x - 2y - z = 1$$

Twice the first number minus the second minus twice the third is -1.
$$2x - y - 2z = -1$$

We now have a system of three equations:

$$x + y + z = 4,$$
$$x - 2y - z = 1,$$
$$2x - y - 2z = -1.$$

3. **Carry out.** We need to solve the system of equations. Note that we found the solution, $(2, -1, 3)$, in Example 2 of Section 8.4.

4. **Check.** The first statement of the problem says that the sum of the three numbers is 4. That checks, because $2 + (-1) + 3 = 4$. The second statement says that the first number minus twice the second, minus the third is 1: $2 - 2(-1) - 3 = 1$. That checks. The check of the third statement is left to the student.

5. **State.** The three numbers are 2, -1, and 3.

▶ **TRY EXERCISE** 1

EXAMPLE 2   *Architecture.* In a triangular cross section of a roof, the largest angle is 70° greater than the smallest angle. The largest angle is twice as large as the remaining angle. Find the measure of each angle.

SOLUTION

1. **Familiarize.** The first thing we do is make a drawing, or a sketch.

STUDENT NOTES

It is quite likely that you are expected to remember that the sum of the measures of the angles in any triangle is 180°. You may want to ask your instructor which other formulas from geometry and elsewhere you are expected to know.

Since we don't know the size of any angle, we use x, y, and z to represent the three measures, from smallest to largest. Recall that the measures of the angles in any triangle add up to 180°.

2. **Translate.** This geometric fact about triangles gives us one equation:

$$x + y + z = 180.$$

Two of the statements can be translated almost directly.

The largest angle	is	70° greater than the smallest angle.
↓	↓	↓
z	$=$	$x + 70$

The largest angle	is	twice as large as the remaining angle.
↓	↓	↓
z	$=$	$2y$

We now have a system of three equations:

$$x + y + z = 180, \qquad x + y + z = 180,$$
$$x + 70 = z, \quad \text{or} \quad x \qquad - z = -70, \qquad \text{Rewriting in}$$
$$2y = z; \qquad\qquad 2y - z = 0. \qquad \text{standard form}$$

3. **Carry out.** The system was solved in Example 4 of Section 8.4. The solution is $(30, 50, 100)$.

4. **Check.** The sum of the numbers is 180, so that checks. The measure of the largest angle, 100°, is 70° greater than the measure of the smallest angle, 30°, so that checks. The measure of the largest angle is also twice the measure of the remaining angle, 50°. Thus we have a check.

5. **State.** The angles in the triangle measure 30°, 50°, and 100°.

> TRY EXERCISE ▶ 5

EXAMPLE 3 *Downloads.* Kaya frequently downloads music, TV shows, and iPod games. In January, she downloaded 5 songs, 10 TV shows, and 3 games for a total of $40. In February, she spent a total of $135 for 25 songs, 25 TV shows, and 12 games. In March, she spent a total of $56 for 15 songs, 8 TV shows, and 5 games. Assuming each song is the same price, each TV show is the same price, and each iPod game is the same price, how much does each cost?

Source: www.iTunes.com

SOLUTION

1. **Familiarize.** We let s = the cost, in dollars, per song, t = the cost, in dollars, per TV show, and g = the cost, in dollars, per game. Then in January, Kaya spent $5 \cdot s$ for songs, $10 \cdot t$ for TV shows, and $3 \cdot g$ for iPod games. The sum of these amounts was $40. Each month's downloads will translate to an equation.

2. **Translate.** We can organize the information in a table.

	Cost of Songs	Cost of TV Shows	Cost of iPod Games	Total Cost	
January	$5s$	$10t$	$3g$	40	⟶ $5s + 10t + 3g = 40$
February	$25s$	$25t$	$12g$	135	⟶ $25s + 25t + 12g = 135$
March	$15s$	$8t$	$5g$	56	⟶ $15s + 8t + 5g = 56$

We now have a system of three equations:

$$5s + 10t + 3g = 40, \quad (1)$$
$$25s + 25t + 12g = 135, \quad (2)$$
$$15s + 8t + 5g = 56. \quad (3)$$

3. **Carry out.** We begin by using equations (1) and (2) to eliminate s.

$$\begin{array}{l} 5s + 10t + 3g = 40, \\ 25s + 25t + 12g = 135 \end{array} \xrightarrow[\text{of equation (1) by } -5]{\text{Multiplying both sides}} \begin{array}{r} -25s - 50t - 15g = -200 \\ 25s + 25t + 12g = 135 \\ \hline -25t - 3g = -65 \quad (4) \end{array}$$

We then use equations (1) and (3) to again eliminate s.

$$\begin{array}{l} 5s + 10t + 3g = 40, \\ 15s + 8t + 5g = 56 \end{array} \xrightarrow[\text{of equation (1) by } -3]{\text{Multiplying both sides}} \begin{array}{r} -15s - 30t - 9g = -120 \\ 15s + 8t + 5g = 56 \\ \hline -22t - 4g = -64 \quad (5) \end{array}$$

Now we solve the resulting system of equations (4) and (5).

$-25t - 3g = -65$ Multiplying both sides of equation (4) by -4 → $100t + 12g = 260$

$-22t - 4g = -64$ Multiplying both sides of equation (5) by 3   → $-66t - 12g = -192$

$$34t = 68$$
$$t = 2$$

To find g, we use equation (4):

$$-25t - 3g = -65$$
$$-25 \cdot 2 - 3g = -65 \qquad \text{Substituting 2 for } t$$
$$-50 - 3g = -65$$
$$-3g = -15$$
$$g = 5.$$

Finally, we use equation (1) to find s:

$$5s + 10t + 3g = 40$$
$$5s + 10 \cdot 2 + 3 \cdot 5 = 40 \qquad \text{Substituting 2 for } t \text{ and 5 for } g$$
$$5s + 20 + 15 = 40$$
$$5s + 35 = 40$$
$$5s = 5$$
$$s = 1.$$

4. Check. If a song costs \$1, a TV show costs \$2, and an iPod game costs \$5, then the total cost for each month's downloads is as follows:

January: $5 \cdot \$1 + 10 \cdot \$2 + 3 \cdot \$5 = \$5 + \$20 + \$15 = \$40$;

February: $25 \cdot \$1 + 25 \cdot \$2 + 12 \cdot \$5 = \$25 + \$50 + \$60 = \$135$;

March: $15 \cdot \$1 + 8 \cdot \$2 + 5 \cdot \$5 = \$15 + \$16 + \$25 = \$56.$

This checks with the information given in the problem.

5. State. A song costs \$1, a TV show costs \$2, and an iPod game costs \$5.

TRY EXERCISE 19

8.5   EXERCISE SET

Solve.

1. The sum of three numbers is 85. The second is 7 more than the first. The third is 2 more than four times the second. Find the numbers.

2. The sum of three numbers is 5. The first number minus the second plus the third is 1. The first minus the third is 3 more than the second. Find the numbers.

3. The sum of three numbers is 26. Twice the first minus the second is 2 less than the third. The third is the second minus three times the first. Find the numbers.

4. The sum of three numbers is 105. The third is 11 less than ten times the second. Twice the first is 7 more than three times the second. Find the numbers.

5. *Geometry.*   In triangle *ABC*, the measure of angle *B* is three times that of angle *A*. The measure of angle *C* is 20° more than that of angle *A*. Find the angle measures.

6. *Geometry.*   In triangle *ABC*, the measure of angle *B* is twice the measure of angle *A*. The measure of angle *C* is 80° more than that of angle *A*. Find the angle measures.

7. *Scholastic Aptitude Test.*   Many high-school students take the Scholastic Aptitude Test (SAT). Beginning in March 2005, students taking the SAT received three scores: a critical reading score, a mathematics score, and a writing score. The average total score of 2007 high-school seniors who took the SAT was 1511. The average mathematics score exceeded the reading score by 13 points and the average writing score was 8 points less than the reading score. What was the average score for each category?
Source: College Entrance Examination Board

8. *Advertising.*   In 2006, U.S. companies spent a total of $123.4 billion on newspaper, television, and magazine ads. The total amount spent on television ads was $7.4 billion more than the amount spent on newspaper and magazine ads together. The amount spent on magazine ads was $2 billion more than the amount spent on newspaper ads. How much was spent on each form of advertising?
Source: TNS Media Intelligence

9. *Nutrition.*   Most nutritionists now agree that a healthy adult diet includes 25–35 g of fiber each day. A breakfast of 2 bran muffins, 1 banana, and a 1-cup serving of Wheaties® contains 9 g of fiber; a breakfast of 1 bran muffin, 2 bananas, and a 1-cup serving of Wheaties® contains 10.5 g of fiber; and a breakfast of 2 bran muffins and a 1-cup serving of Wheaties® contains 6 g of fiber. How much fiber is in each of these foods?
Sources: usda.gov and InteliHealth.com

10. *Nutrition.*   Refer to Exercise 9. A breakfast consisting of 2 pancakes and a 1-cup serving of strawberries contains 4.5 g of fiber, whereas a breakfast of 2 pancakes and a 1-cup serving of Cheerios® contains 4 g of fiber. When a meal consists of 1 pancake, a 1-cup serving of Cheerios®, and a 1-cup serving of strawberries, it contains 7 g of fiber. How much fiber is in each of these foods?
Source: InteliHealth.com

Aha! **11.** *Automobile pricing.*   The basic model of a 2008 Jeep Grand Cherokee Rocky Mountain (2WD) with a tow package costs $30,815. When equipped with a tow package and a rear backup camera, the vehicle's price rose to $31,565. The cost of the basic model with a rear camera was $31,360. Find the basic price, the cost of a tow package, and the cost of a rear camera.
Source: www.jeep.com

12. *Telemarketing.*   Sven, Tina, and Laurie can process 740 telephone orders per day. Sven and Tina together can process 470 orders, while Tina and Laurie together can process 520 orders per day. How many orders can each person process alone?

13. *Coffee prices.*   Reba works at a Starbucks® coffee shop where a 12-oz cup of coffee costs $1.65, a 16-oz cup costs $1.85, and a 20-oz cup costs $1.95. During one busy period, Reba served 55 cups of coffee, emptying six 144-oz "brewers" while collecting a total of $99.65. How many cups of each size did Reba fill?

| 12 oz | 16 oz | 20 oz |
| $1.65 | $1.85 | $1.95 |

14. *Restaurant management.*   Chick-fil-A® recently sold small lemonades for $1.29, medium lemonades for $1.49, and large lemonades for $1.85. During a lunch-time rush, Chris sold 40 lemonades for a total of $59.40. The number of small and large drinks, combined, was 10 fewer than the number of medium drinks. How many drinks of each size were sold?

15. *Small-business loans.* Chelsea took out three loans for a total of $120,000 to start an organic orchard. Her bank loan was at an interest rate of 8%, the small-business loan was at an interest rate of 5%, and the mortgage on her house was at an interest rate of 4%. The total simple interest due on the loans in one year was $5750. The annual simple interest on the mortgage was $1600 more than the interest on the bank loan. How much did she borrow from each source?

16. *Investments.* A business class divided an imaginary investment of $80,000 among three mutual funds. The first fund grew by 10%, the second by 6%, and the third by 15%. Total earnings were $8850. The earnings from the first fund were $750 more than the earnings from the third. How much was invested in each fund?

17. *Gold alloys.* Gold used to make jewelry is often a blend of gold, silver, and copper. The relative amounts of the metals determine the color of the alloy. Red gold is 75% gold, 5% silver, and 20% copper. Yellow gold is 75% gold, 12.5% silver, and 12.5% copper. White gold is 37.5% gold and 62.5% silver. If 100 g of red gold costs $2265.40, 100 g of yellow gold costs $2287.75, and 100 g of white gold costs $1312.50, how much do gold, silver, and copper cost?
Source: World Gold Council

18. *Blending teas.* Verity has recently created three custom tea blends. A 5-oz package of Southern Sandalwood sells for $13.15 and contains 2 oz of Keemun tea, 2 oz of Assam tea, and 1 oz of a berry blend. A 4-oz package of Golden Sunshine sells for $12.50 and contains 3 oz of Assam tea and 1 oz of the berry blend. A 6-oz package of Mountain Morning sells for $12.50 and contains 2 oz of the berry blend, 3 oz of Keemun tea, and 1 oz of Assam tea. What is the price per ounce of Keemun tea, Assam tea, and the berry blend?

19. *Nutrition.* A dietician in a hospital prepares meals under the guidance of a physician. Suppose that for a particular patient a physician prescribes a meal to have 800 calories, 55 g of protein, and 220 mg of vitamin C. The dietician prepares a meal of roast beef, baked potatoes, and broccoli according to the data in the following table.

Serving Size	Calories	Protein (in grams)	Vitamin C (in milligrams)
Roast Beef, 3 oz	300	20	0
Baked Potato, 1	100	5	20
Broccoli, 156 g	50	5	100

How many servings of each food are needed in order to satisfy the doctor's orders?

20. *Nutrition.* Repeat Exercise 19 but replace the broccoli with asparagus, for which a 180-g serving contains 50 calories, 5 g of protein, and 44 mg of vitamin C. Which meal would you prefer eating?

21. Students in a Listening Responses class bought 40 tickets for a piano concert. The number of tickets purchased for seats in either the first mezzanine or the main floor was the same as the number purchased for seats in the second mezzanine. First mezzanine seats cost $52, main floor seats cost $38, and second mezzanine seats cost $28. The total cost of the tickets was $1432. How many of each type of ticket were purchased?

22. *Basketball scoring.*   The New York Knicks recently scored a total of 92 points on a combination of 2-point field goals, 3-point field goals, and 1-point foul shots. Altogether, the Knicks made 50 baskets and 19 more 2-pointers than foul shots. How many shots of each kind were made?

23. *World population growth.*   The world population is projected to be 9.4 billion in 2050. At that time, there is expected to be approximately 3.5 billion more people in Asia than in Africa. The population for the rest of the world will be approximately 0.3 billion less than two-fifths the population of Asia. Find the projected populations of Asia, Africa, and the rest of the world in 2050.
Source: U.S. Census Bureau

24. *History.*   Find the year in which the first U.S. transcontinental railroad was completed. The following are some facts about the number. The sum of the digits in the year is 24. The ones digit is 1 more than the hundreds digit. Both the tens and the ones digits are multiples of 3.

25. Problems like Exercises 13 and 14 could be classified as total-value problems. How do these problems differ from the total-value problems of Section 8.3?

26. Write a problem for a classmate to solve. Design the problem so that it translates to a system of three equations in three variables.

Skill Review

To prepare for Section 8.6, review simplifying expressions (Section 1.8).

Simplify. [1.8]

27. $-2(2x - 3y)$

28. $-(x - 6y)$

29. $-6(x - 2y) + (6x - 5y)$

30. $3(2a + 4b) + (5a - 12b)$

31. $-(2a - b - 6c)$

32. $-10(5a + 3b - c)$

33. $-2(3x - y + z) + 3(-2x + y - 2z)$

34. $(8x - 10y + 7z) + 5(3x + 2y - 4z)$

Synthesis

35. Consider Exercise 22. Suppose there were no foul shots made. Would there still be a solution? Why or why not?

36. Consider Exercise 13. Suppose Reba collected $50. Could the problem still be solved? Why or why not?

37. *Health insurance.*   In 2008, UNICARE® health insurance for a 35-year-old and his or her spouse cost $174/month. That rate increased to $221/month if a child were included and $263/month if two children were included. The rate dropped to $134/month for just the applicant and one child. Find the separate costs for insuring the applicant, the spouse, the first child, and the second child.
Source: UNICARE Life and Health Insurance Company® through www.ehealth.com

38. Find a three-digit positive integer such that the sum of all three digits is 14, the tens digit is 2 more than the ones digit, and if the digits are reversed, the number is unchanged.

39. *Ages.*   Tammy's age is the sum of the ages of Carmen and Dennis. Carmen's age is 2 more than the sum of the ages of Dennis and Mark. Dennis's age is four times Mark's age. The sum of all four ages is 42. How old is Tammy?

40. *Ticket revenue.*   A magic show's audience of 100 people consists of adults, students, and children. The ticket prices are $10 for adults, $3 for students, and 50¢ for children. The total amount of money taken in is $100. How many adults, students, and children are in attendance? Does there seem to be some information missing? Do some more careful reasoning.

41. *Sharing raffle tickets.*   Hal gives Tom as many raffle tickets as Tom first had and Gary as many as Gary first had. In like manner, Tom then gives Hal and Gary as many tickets as each then has. Similarly, Gary gives Hal and Tom as many tickets as each then has. If each finally has 40 tickets, with how many tickets does Tom begin?

42. Find the sum of the angle measures at the tips of the star in this figure.

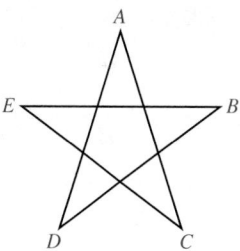

8.6 | Elimination Using Matrices

Matrices and Systems • Row-Equivalent Operations

In solving systems of equations, we perform computations with the constants. If we agree to keep all like terms in the same column, we can simplify writing a system by omitting the variables. For example, the system

$$3x + 4y = 5,$$
$$x - 2y = 1$$

simplifies to

$$\begin{array}{rrr} 3 & 4 & 5 \\ 1 & -2 & 1 \end{array}$$

if we do not write the variables, the operation of addition, and the equals signs.

Matrices and Systems

In the example above, we have written a rectangular array of numbers. Such an array is called a **matrix** (plural, **matrices**). We ordinarily write brackets around matrices. The following are matrices:

$$\begin{bmatrix} -3 & 1 \\ 0 & 5 \end{bmatrix}, \quad \begin{bmatrix} 2 & 0 & -1 & 3 \\ -5 & 2 & 7 & -1 \\ 4 & 5 & 3 & 0 \end{bmatrix}, \quad \begin{bmatrix} 2 & 3 \\ 7 & 15 \\ -2 & 23 \\ 4 & 1 \end{bmatrix}$$

The individual numbers are called *elements* or *entries.*

The **rows** of a matrix are horizontal, and the **columns** are vertical.

$$\begin{bmatrix} 5 & -2 & 2 \\ 1 & 0 & 1 \\ 0 & 1 & 2 \end{bmatrix} \begin{array}{l} \longrightarrow \text{row 1} \\ \longrightarrow \text{row 2} \\ \longrightarrow \text{row 3} \end{array}$$

column 1 column 2 column 3

Let's see how matrices can be used to solve a system.

EXAMPLE **1** Solve the system

$$5x - 4y = -1,$$
$$-2x + 3y = 2.$$

As an aid for understanding, we list the corresponding system in the margin.

$$5x - 4y = -1,$$
$$-2x + 3y = 2$$

SOLUTION We write a matrix using only coefficients and constants, listing *x*-coefficients in the first column and *y*-coefficients in the second. A dashed line separates the coefficients from the constants:

$$\begin{bmatrix} 5 & -4 & \vdots & -1 \\ -2 & 3 & \vdots & 2 \end{bmatrix}.$$

Consult the notes in the margin for further information.

Our goal is to transform

$$\begin{bmatrix} 5 & -4 & \vdots & -1 \\ -2 & 3 & \vdots & 2 \end{bmatrix} \quad \text{into the form} \quad \begin{bmatrix} a & b & \vdots & c \\ 0 & d & \vdots & e \end{bmatrix}.$$

We can then reinsert the variables *x* and *y*, form equations, and complete the solution.

Our calculations are similar to those that we would do if we wrote the entire equations. The first step is to multiply and/or interchange the rows so that each number in the first column below the first number is a multiple of that number. Here that means multiplying Row 2 by 5. This corresponds to multiplying both sides of the second equation by 5.

$$5x - 4y = -1,$$
$$-10x + 15y = 10$$

$$\begin{bmatrix} 5 & -4 & \vdots & -1 \\ -10 & 15 & \vdots & 10 \end{bmatrix}$$

New Row 2 = 5(Row 2 from the step above)
$$= 5(-2 \quad 3 \quad \vdots \quad 2) = (-10 \quad 15 \quad \vdots \quad 10)$$

Next, we multiply the first row by 2, add this to Row 2, and write that result as the "new" Row 2. This corresponds to multiplying the first equation by 2 and adding the result to the second equation in order to eliminate a variable. Write out these computations as necessary.

$$5x - 4y = -1,$$
$$7y = 8$$

$$\begin{bmatrix} 5 & -4 & \vdots & -1 \\ 0 & 7 & \vdots & 8 \end{bmatrix}$$

2(Row 1) $= 2(5 \quad -4 \quad \vdots \quad -1) = (10 \quad -8 \quad \vdots \quad -2)$
New Row 2 = $(10 \quad -8 \quad \vdots \quad -2) + (-10 \quad 15 \quad \vdots \quad 10)$
$= (0 \quad 7 \quad \vdots \quad 8)$

If we now reinsert the variables, we have

$$5x - 4y = -1, \qquad (1) \qquad \text{From Row 1}$$
$$7y = 8. \qquad (2) \qquad \text{From Row 2}$$

Solving equation (2) for y gives us

$$7y = 8 \qquad (2)$$
$$y = \tfrac{8}{7}.$$

Next, we substitute $\tfrac{8}{7}$ for y in equation (1):

$$5x - 4y = -1 \qquad (1)$$
$$5x - 4 \cdot \tfrac{8}{7} = -1 \qquad \text{Substituting } \tfrac{8}{7} \text{ for } y \text{ in equation (1)}$$
$$x = \tfrac{5}{7}. \qquad \text{Solving for } x$$

The solution is $\left(\tfrac{5}{7}, \tfrac{8}{7}\right)$. The check is left to the student.

TRY EXERCISE 7

EXAMPLE 2   Solve the system

$$2x - y + 4z = -3,$$
$$x \qquad - 4z = 5,$$
$$6x - y + 2z = 10.$$

SOLUTION   We first write a matrix, using only the constants. Where there are missing terms, we must write 0's:

$$2x - y + 4z = -3,$$
$$x \qquad - 4z = 5,$$
$$6x - y + 2z = 10$$

$$\begin{bmatrix} 2 & -1 & 4 & \vdots & -3 \\ 1 & 0 & -4 & \vdots & 5 \\ 6 & -1 & 2 & \vdots & 10 \end{bmatrix}.$$

Our goal is to transform the matrix to one of the form

$$ax + by + cz = d,$$
$$ey + fz = g,$$
$$hz = i$$

$$\begin{bmatrix} a & b & c & \vdots & d \\ 0 & e & f & \vdots & g \\ 0 & 0 & h & \vdots & i \end{bmatrix}. \qquad \text{This matrix is in row-echelon form.}$$

A matrix of this form can be rewritten as a system of equations that is equivalent to the original system, and from which a solution can be easily found.

The first step is to multiply and/or interchange the rows so that each number in the first column is a multiple of the first number in the first row. In this case, we begin by interchanging Rows 1 and 2:

$$x \qquad - 4z = 5,$$
$$2x - y + 4z = -3,$$
$$6x - y + 2z = 10$$

$$\begin{bmatrix} 1 & 0 & -4 & \vdots & 5 \\ 2 & -1 & 4 & \vdots & -3 \\ 6 & -1 & 2 & \vdots & 10 \end{bmatrix}.$$

This corresponds to interchanging the first two equations.

Next, we multiply the first row by -2, add it to the second row, and replace Row 2 with the result:

$$x \qquad - 4z = 5,$$
$$-y + 12z = -13,$$
$$6x - y + 2z = 10$$

$$\begin{bmatrix} 1 & 0 & -4 & \vdots & 5 \\ 0 & -1 & 12 & \vdots & -13 \\ 6 & -1 & 2 & \vdots & 10 \end{bmatrix}.$$

$-2(1 \quad 0 \quad -4 \quad \vdots \quad 5) = (-2 \quad 0 \quad 8 \quad \vdots \quad -10)$ and
$(-2 \quad 0 \quad 8 \quad \vdots \quad -10) + (2 \quad -1 \quad 4 \quad \vdots \quad -3) =$
$(0 \quad -1 \quad 12 \quad \vdots \quad -13)$

Now we multiply the first row by -6, add it to the third row, and replace Row 3 with the result:

$$x \qquad - 4z = 5,$$
$$-y + 12z = -13,$$
$$-y + 26z = -20$$

$$\begin{bmatrix} 1 & 0 & -4 & \vdots & 5 \\ 0 & -1 & 12 & \vdots & -13 \\ 0 & -1 & 26 & \vdots & -20 \end{bmatrix}.$$

$-6(1 \quad 0 \quad -4 \quad \vdots \quad 5) = (-6 \quad 0 \quad 24 \quad \vdots \quad -30)$ and
$(-6 \quad 0 \quad 24 \quad \vdots \quad -30) + (6 \quad -1 \quad 2 \quad \vdots \quad 10) =$
$(0 \quad -1 \quad 26 \quad \vdots \quad -20)$

Next, we multiply Row 2 by -1, add it to the third row, and replace Row 3 with the result:

$$x \qquad - 4z = 5,$$
$$-y + 12z = -13,$$
$$14z = -7$$

$$\begin{bmatrix} 1 & 0 & -4 & \vdots & 5 \\ 0 & -1 & 12 & \vdots & -13 \\ 0 & 0 & 14 & \vdots & -7 \end{bmatrix}.$$

$-1(0 \quad -1 \quad 12 \quad \vdots \quad -13) = (0 \quad 1 \quad -12 \quad \vdots \quad 13)$ and
$(0 \quad 1 \quad -12 \quad \vdots \quad 13) + (0 \quad -1 \quad 26 \quad \vdots \quad -20) =$
$(0 \quad 0 \quad 14 \quad \vdots \quad -7)$

Reinserting the variables gives us

$$x \qquad -4z \quad = \quad 5,$$
$$-y + 12z \quad = \quad -13,$$
$$14z \quad = \quad -7.$$

We now solve this last equation for z and get $z = -\frac{1}{2}$. Next, we substitute $-\frac{1}{2}$ for z in the preceding equation and solve for y: $-y + 12\left(-\frac{1}{2}\right) = -13$, so $y = 7$. Since there is no y-term in the first equation of this last system, we need only substitute $-\frac{1}{2}$ for z to solve for x: $x - 4\left(-\frac{1}{2}\right) = 5$, so $x = 3$. The solution is $\left(3, 7, -\frac{1}{2}\right)$. The check is left to the student.

TRY EXERCISE 13

The operations used in the preceding example correspond to those used to produce equivalent systems of equations, that is, systems of equations that have the same solution. We call the matrices **row-equivalent** and the operations that produce them **row-equivalent operations.**

Row-Equivalent Operations

Row-Equivalent Operations

Each of the following row-equivalent operations produces a row-equivalent matrix:

a) Interchanging any two rows.
b) Multiplying all elements of a row by a nonzero constant.
c) Replacing a row with the sum of that row and a multiple of another row.

STUDENT NOTES

Note that row-equivalent matrices are not *equal*. It is the solutions of the corresponding systems that are the same.

The best overall method for solving systems of equations is by row-equivalent matrices; even computers are programmed to use them. Matrices are part of a branch of mathematics known as linear algebra. They are also studied in many courses in finite mathematics.

TECHNOLOGY CONNECTION

Row-equivalent operations can be performed on a graphing calculator. For example, to interchange the first and second rows of the matrix, as in step (1) of Example 2 above, we enter the matrix as matrix **A** and select "rowSwap" from the MATRIX MATH menu. Some graphing calculators will not automatically store the matrix produced using a row-equivalent operation, so when several operations are to be performed in succession, it is helpful to store the result of each operation as it is produced. In the window at right, we see both the matrix produced by the rowSwap operation and the indication that this matrix is stored, using **STO▸**, as matrix **B**.

```
rowSwap([A],1,2)→[B]
[[1   0  -4   5]
 [2  -1   4  -3]
 [6  -1   2  10]]
```

1. Use a graphing calculator to proceed through all the steps in Example 2.

8.6   EXERCISE SET

For Extra Help   *MyMathLab*   Math
PRACTICE   WATCH   DOWNLOAD

Concept Reinforcement   *Complete each of the following statements.*

1. A(n) _____ is a rectangular array of numbers.

2. The rows of a matrix are _____ and the _____ are vertical.

3. Each number in a matrix is called a(n) _____ or element.

4. The plural of the word matrix is _____.

5. As part of solving a system using matrices, we can interchange any two _____.

6. Before we reinsert the variables, the leftmost column in the matrix has zeros in all rows except the _____ one.

Solve using matrices.

7. $x + 2y = 11,$
   $3x - y = 5$

8. $x + 3y = 16,$
   $6x + y = 11$

9. $3x + y = -1,$
   $6x + 5y = 13$

10. $2x - y = 6,$
    $8x + 2y = 0$

11. $6x - 2y = 4,$
    $7x + y = 13$

12. $3x + 4y = 7,$
    $-5x + 2y = 10$

13. $3x + 2y + 2z = 3,$
    $x + 2y - z = 5,$
    $2x - 4y + z = 0$

14. $4x - y - 3z = 19,$
    $8x + y - z = 11,$
    $2x + y + 2z = -7$

15. $p - 2q - 3r = 3,$
    $2p - q - 2r = 4,$
    $4p + 5q + 6r = 4$

16. $x + 2y - 3z = 9,$
    $2x - y + 2z = -8,$
    $3x - y - 4z = 3$

17. $3p + 2r = 11,$
    $q - 7r = 4,$
    $p - 6q = 1$

18. $4a + 9b = 8,$
    $8a + 6c = -1,$
    $6b + 6c = -1$

19. $2x + 2y - 2z - 2w = -10,$
    $w + y + z + x = -5,$
    $x - y + 4z + 3w = -2,$
    $w - 2y + 2z + 3x = -6$

20. $-w - 3y + z + 2x = -8,$
$x + y - z - w = -4,$
$w + y + z + x = 22,$
$x - y - z - w = -14$

Solve using matrices.

21. *Coin value.*   A collection of 42 coins consists of dimes and nickels. The total value is $3.00. How many dimes and how many nickels are there?

22. *Coin value.*   A collection of 43 coins consists of dimes and quarters. The total value is $7.60. How many dimes and how many quarters are there?

23. *Snack mix.*   Bree sells a dried-fruit mixture for $5.80 per pound and Hawaiian macadamia nuts for $14.75 per pound. She wants to blend the two to get a 15-lb mixture that she will sell for $9.38 per pound. How much of each should she use?

24. *Mixing paint.*   Higher quality paint typically contains more solids. Alex has available paint that contains 45% solids and paint that contains 25% solids. How much of each should he use to create 20 gal of paint that contains 39% solids?

25. *Investments.*   Elena receives $212 per year in simple interest from three investments totaling $2500. Part is invested at 7%, part at 8%, and part at 9%. There is $1100 more invested at 9% than at 8%. Find the amount invested at each rate.

26. *Investments.*   Miguel receives $306 per year in simple interest from three investments totaling $3200. Part is invested at 8%, part at 9%, and part at 10%. There is $1900 more invested at 10% than at 9%. Find the amount invested at each rate.

27. Explain how you can recognize dependent equations when solving with matrices.

28. Explain how you can recognize an inconsistent system when solving with matrices.

Skill Review

To prepare for Section 8.7, review order of operations (Section 1.8).

Simplify. [1.8]

29. $3(-1) - (-4)(5)$

30. $7(-5) - 2(-8)$

31. $-2(5 \cdot 3 - 4 \cdot 6) - 3(2 \cdot 7 - 15) + 4(3 \cdot 8 - 5 \cdot 4)$

32. $6(2 \cdot 7 - 3(-4)) - 4(3(-8) - 10) + 5(4 \cdot 3 - (-2)7)$

Synthesis

33. If the matrices

$$\begin{bmatrix} a_1 & b_1 & \vdots & c_1 \\ d_1 & e_1 & \vdots & f_1 \end{bmatrix} \quad \text{and} \quad \begin{bmatrix} a_2 & b_2 & \vdots & c_2 \\ d_2 & e_2 & \vdots & f_2 \end{bmatrix}$$

share the same solution, does it follow that the corresponding entries are all equal to each other ($a_1 = a_2$, $b_1 = b_2$, etc.)? Why or why not?

34. Explain how the row-equivalent operations make use of the addition, multiplication, and distributive properties.

35. The sum of the digits in a four-digit number is 10. Twice the sum of the thousands digit and the tens digit is 1 less than the sum of the other two digits. The tens digit is twice the thousands digit. The ones digit equals the sum of the thousands digit and the hundreds digit. Find the four-digit number.

36. Solve for x and y:
$$ax + by = c,$$
$$dx + ey = f.$$

8.7   Determinants and Cramer's Rule

Determinants of 2 × 2 Matrices ▪ Cramer's Rule: 2 × 2 Systems ▪ Cramer's Rule: 3 × 3 Systems

Determinants of 2 × 2 Matrices

When a matrix has m rows and n columns, it is called an "m by n" matrix. Thus its *dimensions* are denoted by $m \times n$. If a matrix has the same number of rows and columns, it is called a **square matrix**. Associated with every square matrix is a number called its **determinant**, defined as follows for 2×2 matrices.

2 × 2 Determinants

The determinant of a two-by-two matrix $\begin{bmatrix} a & c \\ b & d \end{bmatrix}$ is denoted $\begin{vmatrix} a & c \\ b & d \end{vmatrix}$ and is defined as follows:

$$\begin{vmatrix} a & c \\ b & d \end{vmatrix} = ad - bc.$$

EXAMPLE 1

Evaluate: $\begin{vmatrix} 2 & -5 \\ 6 & 7 \end{vmatrix}$.

SOLUTION   We multiply and subtract as follows:

$$\begin{vmatrix} 2 & -5 \\ 6 & 7 \end{vmatrix} = 2 \cdot 7 - 6 \cdot (-5) = 14 + 30 = 44.$$

▶ TRY EXERCISE 7

STUDY SKILLS

Find the Highlights

If you do not already own one, consider purchasing a highlighter to use as you read this text and work on the exercises. Often the best time to highlight an important sentence or step in an example is after you have read through the section the first time.

Cramer's Rule: 2 × 2 Systems

One of the many uses for determinants is in solving systems of linear equations in which the number of variables is the same as the number of equations and the constants are not all 0. Let's consider a system of two equations:

$$a_1x + b_1y = c_1,$$
$$a_2x + b_2y = c_2.$$

If we use the elimination method, a series of steps can show that

$$x = \frac{c_1b_2 - c_2b_1}{a_1b_2 - a_2b_1} \quad \text{and} \quad y = \frac{a_1c_2 - a_2c_1}{a_1b_2 - a_2b_1}.$$

These fractions can be rewritten using determinants.

Cramer's Rule: 2 × 2 Systems

The solution of the system

$$a_1x + b_1y = c_1,$$
$$a_2x + b_2y = c_2,$$

if it is unique, is given by

$$x = \frac{\begin{vmatrix} c_1 & b_1 \\ c_2 & b_2 \end{vmatrix}}{\begin{vmatrix} a_1 & b_1 \\ a_2 & b_2 \end{vmatrix}}, \quad y = \frac{\begin{vmatrix} a_1 & c_1 \\ a_2 & c_2 \end{vmatrix}}{\begin{vmatrix} a_1 & b_1 \\ a_2 & b_2 \end{vmatrix}}.$$

These formulas apply only if the denominator is not 0. If the denominator *is* 0, then one of two things happens:

1. If the denominator is 0 and the numerators are also 0, then the equations in the system are dependent.
2. If the denominator is 0 and at least one numerator is not 0, then the system is inconsistent.

To use Cramer's rule, we find the determinants and compute x and y as shown above. Note that the denominators are identical and the coefficients of x and y appear in the same position as in the original equations. In the numerator of x, the constants c_1 and c_2 replace a_1 and a_2. In the numerator of y, the constants c_1 and c_2 replace b_1 and b_2.

EXAMPLE 2   Solve using Cramer's rule:

$$2x + 5y = 7,$$
$$5x - 2y = -3.$$

SOLUTION   We have

$$x = \dfrac{\begin{vmatrix} 7 & 5 \\ -3 & -2 \end{vmatrix}}{\begin{vmatrix} 2 & 5 \\ 5 & -2 \end{vmatrix}}$$

← The constants $\begin{smallmatrix}7\\-3\end{smallmatrix}$ form the first column.

← The columns are the coefficients of the variables.

$$= \frac{7(-2) - (-3)5}{2(-2) - 5 \cdot 5} = \frac{1}{-29} = -\frac{1}{29}$$

and

$$y = \dfrac{\begin{vmatrix} 2 & 7 \\ 5 & -3 \end{vmatrix}}{\begin{vmatrix} 2 & 5 \\ 5 & -2 \end{vmatrix}}$$

← The constants $\begin{smallmatrix}7\\-3\end{smallmatrix}$ form the second column.

← The denominator is the same as in the expression for x.

$$= \frac{2(-3) - 5 \cdot 7}{-29} = \frac{-41}{-29} = \frac{41}{29}.$$

The solution is $\left(-\frac{1}{29}, \frac{41}{29}\right)$. The check is left to the student.   **TRY EXERCISE** 17

Cramer's Rule: 3 × 3 Systems

Cramer's rule can be extended for systems of three linear equations. However, before doing so, we must define what a 3 × 3 determinant is.

---

3 × 3 Determinants

The determinant of a three-by-three matrix can be defined as follows:

Subtract.   Add.

$$\begin{vmatrix} a_1 & b_1 & c_1 \\ a_2 & b_2 & c_2 \\ a_3 & b_3 & c_3 \end{vmatrix} = a_1 \begin{vmatrix} b_2 & c_2 \\ b_3 & c_3 \end{vmatrix} - a_2 \begin{vmatrix} b_1 & c_1 \\ b_3 & c_3 \end{vmatrix} + a_3 \begin{vmatrix} b_1 & c_1 \\ b_2 & c_2 \end{vmatrix}$$

STUDENT NOTES ―――――

Cramer's rule and the evaluation of determinants rely on patterns. Recognizing and remembering the patterns will help you understand and use the definitions.

Note that the a's come from the first column. Note too that the 2 × 2 determinants above can be obtained by crossing out the row and the column in which the a occurs.

For a_1:   For a_2:   For a_3:

$$\begin{vmatrix} a_1 & b_1 & c_1 \\ a_2 & b_2 & c_2 \\ a_3 & b_3 & c_3 \end{vmatrix} \qquad \begin{vmatrix} a_1 & b_1 & c_1 \\ a_2 & b_2 & c_2 \\ a_3 & b_3 & c_3 \end{vmatrix} \qquad \begin{vmatrix} a_1 & b_1 & c_1 \\ a_2 & b_2 & c_2 \\ a_3 & b_3 & c_3 \end{vmatrix}$$

EXAMPLE **3**   Evaluate:

$$\begin{vmatrix} -1 & 0 & 1 \\ -5 & 1 & -1 \\ 4 & 8 & 1 \end{vmatrix}.$$

SOLUTION   We have

Subtract.   Add.

$$\begin{vmatrix} -1 & 0 & 1 \\ -5 & 1 & -1 \\ 4 & 8 & 1 \end{vmatrix} = -1 \begin{vmatrix} 1 & -1 \\ 8 & 1 \end{vmatrix} - (-5) \begin{vmatrix} 0 & 1 \\ 8 & 1 \end{vmatrix} + 4 \begin{vmatrix} 0 & 1 \\ 1 & -1 \end{vmatrix}$$

$$= -1(1 + 8) + 5(0 - 8) + 4(0 - 1)$$ Evaluating the three determinants

$$= -9 - 40 - 4 = -53.$$

TRY EXERCISE 11

TECHNOLOGY CONNECTION

Determinants can be evaluated on most graphing calculators using **2ND** **MATRIX**. After entering a matrix, we select the determinant operation from the MATRIX MATH menu and enter the name of the matrix. The graphing calculator will return the value of the determinant of the matrix. For example, if

$$\mathbf{A} = \begin{bmatrix} 1 & 6 & -1 \\ -3 & -5 & 3 \\ 0 & 4 & 2 \end{bmatrix},$$

we have

```
det ([A])
              26
```

1. Confirm the calculations in Example 4.

Cramer's Rule: 3 × 3 Systems

The solution of the system

$$a_1 x + b_1 y + c_1 z = d_1,$$
$$a_2 x + b_2 y + c_2 z = d_2,$$
$$a_3 x + b_3 y + c_3 z = d_3$$

can be found using the following determinants:

$$D = \begin{vmatrix} a_1 & b_1 & c_1 \\ a_2 & b_2 & c_2 \\ a_3 & b_3 & c_3 \end{vmatrix}, \quad D_x = \begin{vmatrix} d_1 & b_1 & c_1 \\ d_2 & b_2 & c_2 \\ d_3 & b_3 & c_3 \end{vmatrix},$$

D contains only coefficients.
In D_x the *d*'s replace the *a*'s.

$$D_y = \begin{vmatrix} a_1 & d_1 & c_1 \\ a_2 & d_2 & c_2 \\ a_3 & d_3 & c_3 \end{vmatrix}, \quad D_z = \begin{vmatrix} a_1 & b_1 & d_1 \\ a_2 & b_2 & d_2 \\ a_3 & b_3 & d_3 \end{vmatrix}.$$

In D_y, the *d*'s replace the *b*'s.
In D_z, the *d*'s replace the *c*'s.

If a unique solution exists, it is given by

$$x = \frac{D_x}{D}, \qquad y = \frac{D_y}{D}, \qquad z = \frac{D_z}{D}.$$

EXAMPLE **4**   Solve using Cramer's rule:

$$x - 3y + 7z = 13,$$
$$x + y + z = 1,$$
$$x - 2y + 3z = 4.$$

SOLUTION   We compute D, D_x, D_y, and D_z:

$$D = \begin{vmatrix} 1 & -3 & 7 \\ 1 & 1 & 1 \\ 1 & -2 & 3 \end{vmatrix} = -10; \qquad D_x = \begin{vmatrix} 13 & -3 & 7 \\ 1 & 1 & 1 \\ 4 & -2 & 3 \end{vmatrix} = 20;$$

$$D_y = \begin{vmatrix} 1 & 13 & 7 \\ 1 & 1 & 1 \\ 1 & 4 & 3 \end{vmatrix} = -6; \qquad D_z = \begin{vmatrix} 1 & -3 & 13 \\ 1 & 1 & 1 \\ 1 & -2 & 4 \end{vmatrix} = -24.$$

Then

$$x = \frac{D_x}{D} = \frac{20}{-10} = -2;$$

$$y = \frac{D_y}{D} = \frac{-6}{-10} = \frac{3}{5};$$

$$z = \frac{D_z}{D} = \frac{-24}{-10} = \frac{12}{5}.$$

The solution is $\left(-2, \frac{3}{5}, \frac{12}{5}\right)$. The check is left to the student.   **TRY EXERCISE** 21

In Example 4, we need not have evaluated D_z. Once x and y were found, we could have substituted them into one of the equations to find z.

To use Cramer's rule, we divide by D, provided $D \neq 0$. If $D = 0$ and at least one of the other determinants is not 0, then the system is inconsistent. If *all* the determinants are 0, then the equations in the system are dependent.

8.7   EXERCISE SET

Concept Reinforcement   *Classify each statement as either true or false.*

1. A square matrix has the same number of rows and columns.

2. A 3×4 matrix has 3 rows and 4 columns.

3. A determinant is a number.

4. Cramer's rule exists only for 2×2 systems.

5. Whenever Cramer's rule yields a denominator that is 0, the system has no solution.

6. Whenever Cramer's rule yields a numerator that is 0, the equations are dependent.

Evaluate.

7. $\begin{vmatrix} 3 & 5 \\ 4 & 8 \end{vmatrix}$

8. $\begin{vmatrix} 3 & 2 \\ 2 & -3 \end{vmatrix}$

9. $\begin{vmatrix} 10 & 8 \\ -5 & -9 \end{vmatrix}$

10. $\begin{vmatrix} 3 & 2 \\ -7 & 11 \end{vmatrix}$

11. $\begin{vmatrix} 1 & 4 & 0 \\ 0 & -1 & 2 \\ 3 & -2 & 1 \end{vmatrix}$

12. $\begin{vmatrix} 2 & 4 & -2 \\ 1 & 0 & 2 \\ 0 & 1 & 3 \end{vmatrix}$

13. $\begin{vmatrix} -1 & -2 & -3 \\ 3 & 4 & 2 \\ 0 & 1 & 2 \end{vmatrix}$

14. $\begin{vmatrix} 5 & 2 & 2 \\ 0 & 1 & -1 \\ 3 & 3 & 1 \end{vmatrix}$

15. $\begin{vmatrix} -4 & -2 & 3 \\ -3 & 1 & 2 \\ 3 & 4 & -2 \end{vmatrix}$

16. $\begin{vmatrix} 2 & -1 & 1 \\ 1 & 2 & -1 \\ 3 & 4 & -3 \end{vmatrix}$

Solve using Cramer's rule.

17. $5x + 8y = 1,$
    $3x + 7y = 5$

18. $3x - 4y = 6,$
    $5x + 9y = 10$

19. $5x - 4y = -3,$
    $7x + 2y = 6$

20. $-2x + 4y = 3,$
    $3x - 7y = 1$

21. $3x - y + 2z = 1,$
$\quad\ x - y + 2z = 3,$
$\ -2x + 3y + z = 1$

22. $3x + 2y - z = 4,$
$\quad 3x - 2y + z = 5,$
$\quad 4x - 5y - z = -1$

23. $2x - 3y + 5z = 27,$
$\quad\ x + 2y - \ z = -4,$
$\quad 5x - \ y + 4z = 27$

24. $\ x - \ y + 2z = -3,$
$\quad x + 2y + 3z = 4,$
$\ 2x + \ y + \ z = -3$

25. $\ r - 2s + 3t = 6,$
$\ 2r - \ s - \ t = -3,$
$\quad r + \ s + \ t = 6$

26. $a \qquad\ - 3c = 6,$
$\qquad b + 2c = 2,$
$\ 7a - 3b - 5c = 14$

27. Describe at least one of the patterns that you see in Cramer's rule.

28. Which version of Cramer's rule do you find more useful: the version for 2×2 systems or the version for 3×3 systems? Why?

Skill Review

To prepare for Section 8.8, review functions (Sections 7.1 and 7.4).

Find each of the following, given $f(x) = 80x + 2500$ and $g(x) = 150x$.

29. $f(90)$ [7.1]

30. $(g - f)(x)$ [7.4]

31. $(g - f)(10)$ [7.4]

32. $(g - f)(100)$ [7.4]

33. All values of x for which $f(x) = g(x)$ [7.1]

34. All values of x for which $(g - f)(x) = 0$ [7.4]

Synthesis

35. Cramer's rule states that if $a_1x + b_1y = c_1$ and $a_2x + b_2y = c_2$ are dependent, then
$$\begin{vmatrix} a_1 & b_1 \\ a_2 & b_2 \end{vmatrix} = 0.$$
Explain why this will always happen.

36. Under what conditions can a 3×3 system of linear equations be consistent but unable to be solved using Cramer's rule?

Solve.

37. $\begin{vmatrix} y & -2 \\ 4 & 3 \end{vmatrix} = 44$

38. $\begin{vmatrix} 2 & x & -1 \\ -1 & 3 & 2 \\ -2 & 1 & 1 \end{vmatrix} = -12$

39. $\begin{vmatrix} m + 1 & -2 \\ m - 2 & 1 \end{vmatrix} = 27$

40. Show that an equation of the line through (x_1, y_1) and (x_2, y_2) can be written
$$\begin{vmatrix} x & y & 1 \\ x_1 & y_1 & 1 \\ x_2 & y_2 & 1 \end{vmatrix} = 0.$$

8.8 Business and Economics Applications

Break-Even Analysis • Supply and Demand

Break-Even Analysis

The money that a business spends to manufacture a product is its *cost*. The **total cost** of production can be thought of as a function C, where $C(x)$ is the cost of producing x units. When the company sells the product, it takes in money. This is *revenue* and can be thought of as a function R, where $R(x)$ is the **total revenue** from the sale of x units. **Total profit** is the money taken in less the money spent, or total revenue minus total cost. Total profit from the production and sale of x units is a function P given by

$$\textbf{Profit = Revenue − Cost,} \quad \text{or} \quad P(x) = R(x) - C(x).$$

If $R(x)$ is greater than $C(x)$, there is a gain and $P(x)$ is positive. If $C(x)$ is greater than $R(x)$, there is a loss and $P(x)$ is negative. When $R(x) = C(x)$, the company breaks even.

There are two kinds of costs. First, there are costs like rent, insurance, machinery, and so on. These costs, which must be paid regardless of how many items are produced, are called *fixed costs*. Second, costs for labor, materials, marketing, and so on are called *variable costs*, because they vary according to the amount being produced. The sum of the fixed cost and the variable cost gives the **total cost**.

> **CAUTION!**   Do not confuse "cost" with "price." When we discuss the *cost* of an item, we are referring to what it costs to produce the item. The *price* of an item is what a consumer pays to purchase the item and is used when calculating revenue.

EXAMPLE **1**   *Manufacturing chairs.*   Renewable Designs is planning to make a new chair. Fixed costs will be $90,000, and it will cost $25 to produce each chair (variable costs). Each chair sells for $48.

a) Find the total cost $C(x)$ of producing x chairs.

b) Find the total revenue $R(x)$ from the sale of x chairs.

c) Find the total profit $P(x)$ from the production and sale of x chairs.

d) What profit will the company realize from the production and sale of 3000 chairs? of 8000 chairs?

e) Graph the total-cost, total-revenue, and total-profit functions using the same set of axes. Determine the break-even point.

SOLUTION

a) Total cost, in dollars, is given by

$$C(x) = \text{(Fixed costs) plus (Variable costs)},$$

or $C(x) = \quad 90{,}000 \quad + \quad 25x$

where x is the number of chairs produced.

b) Total revenue, in dollars, is given by

$R(x) = 48x.$ $48 times the number of chairs sold. We assume that every chair produced is sold.

c) Total profit, in dollars, is given by

$$P(x) = R(x) - C(x) \qquad \text{Profit is revenue minus cost.}$$
$$= 48x - (90{,}000 + 25x)$$
$$= 23x - 90{,}000.$$

d) Profits will be

$$P(3000) = 23 \cdot 3000 - 90{,}000 = -\$21{,}000$$

when 3000 chairs are produced and sold, and

$$P(8000) = 23 \cdot 8000 - 90,000 = \$94,000$$

when 8000 chairs are produced and sold. Thus the company loses money if only 3000 chairs are sold, but makes money if 8000 are sold.

e) The graphs of each of the three functions are shown below:

$C(x) = 90,000 + 25x,$ This represents the cost function.

$R(x) = 48x,$ This represents the revenue function.

$P(x) = 23x - 90,000.$ This represents the profit function.

$C(x)$, $R(x)$, and $P(x)$ are all in dollars.

The revenue function has a graph that goes through the origin and has a slope of 48. The cost function has an intercept on the \$-axis of 90,000 and has a slope of 25. The profit function has an intercept on the \$-axis of $-90,000$ and has a slope of 23. It is shown by the red and black dashed line. The red portion of the dashed line shows a "negative" profit, which is a loss. (That is what is known as "being in the red.") The black portion of the dashed line shows a "positive" profit, or gain. (That is what is known as "being in the black.")

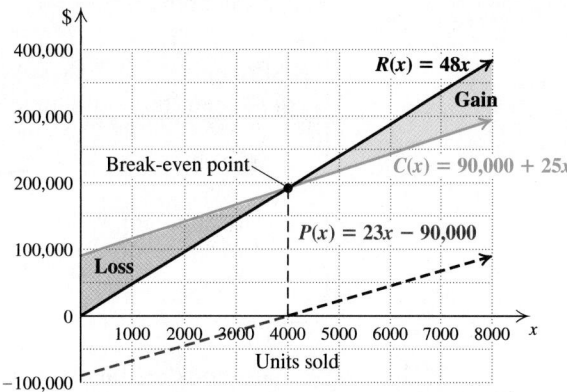

STUDENT NOTES

If you plan to study business or economics, you may want to consult the material in this section when these topics arise in your other courses.

Gains occur where revenue exceeds cost. Losses occur where revenue is less than cost. The **break-even point** occurs where the graphs of R and C cross. Thus to find the break-even point, we solve a system:

$$R(x) = 48x,$$
$$C(x) = 90,000 + 25x.$$

Since both revenue and cost are in *dollars* and they are equal at the break-even point, the system can be rewritten as

$$d = 48x, \qquad (1)$$
$$d = 90,000 + 25x \qquad (2)$$

and solved using substitution:

$48x = 90,000 + 25x$ Substituting $48x$ for d in equation (2)

$23x = 90,000$

$x \approx 3913.04.$

The firm will break even if it produces and sells about 3913 chairs (3913 will yield a tiny loss and 3914 a tiny gain), and takes in a total of $R(3913) = 48 \cdot 3913 = \$187,824$ in revenue. Note that the x-coordinate of the break-even point can also be found by solving $P(x) = 0$. The break-even point is (3913 chairs, \$187,824).

TRY EXERCISE 9

Supply and Demand

As the price of coffee varies, so too does the amount sold. The table and graph below show that *consumers will demand less as the price goes up.*

Demand Function, *D*

Price, *p*, per Kilogram	Quantity, *D(p)* (in millions of kilograms)
$ 8.00	25
9.00	20
10.00	15
11.00	10
12.00	5

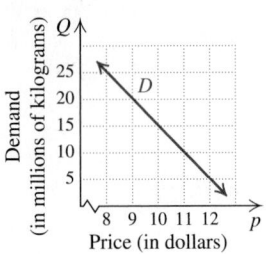

As the price of coffee varies, the amount made available varies as well. The table and graph below show that *sellers will supply more as the price goes up.*

Supply Function, *S*

Price, *p*, per Kilogram	Quantity, *S(p)* (in millions of kilograms)
$ 9.00	5
9.50	10
10.00	15
10.50	20
11.00	25

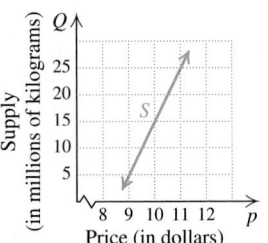

Let's look at the above graphs together. We see that as price increases, demand decreases. As price increases, supply increases. The point of intersection is called the **equilibrium point**. At that price, the amount that the seller will supply is the same amount that the consumer will buy. The situation is similar to a buyer and a seller negotiating the price of an item. The equilibrium point is the price and quantity that they finally agree on.

Any ordered pair of coordinates from the graph is (price, quantity), because the horizontal axis is the price axis and the vertical axis is the quantity axis. If *D* is a demand function and *S* is a supply function, then the equilibrium point is where demand equals supply:

$$D(p) = S(p).$$

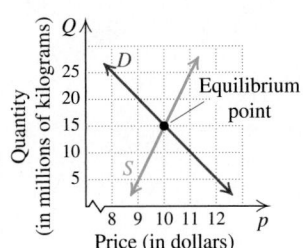

EXAMPLE 2   Find the equilibrium point for the demand and supply functions given:

$$D(p) = 1000 - 60p, \quad (1)$$
$$S(p) = 200 + 4p. \quad (2)$$

SOLUTION Since both demand and supply are *quantities* and they are equal at the equilibrium point, we rewrite the system as

$$q = 1000 - 60p, \quad (1)$$
$$q = 200 + 4p. \quad (2)$$

We substitute $200 + 4p$ for q in equation (1) and solve:

$200 + 4p = 1000 - 60p$	Substituting $200 + 4p$ for q in equation (1)
$200 + 64p = 1000$	Adding $60p$ to both sides
$64p = 800$	Adding -200 to both sides
$p = \frac{800}{64} = 12.5.$	

Thus the equilibrium price is $12.50 per unit.

To find the equilibrium quantity, we substitute $12.50 into either $D(p)$ or $S(p)$. We use $S(p)$:

$$S(12.5) = 200 + 4(12.5) = 200 + 50 = 250.$$

Therefore, the equilibrium quantity is 250 units, and the equilibrium point is ($12.50, 250).

▶ TRY EXERCISE 19

8.8 EXERCISE SET

 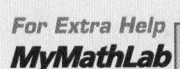 

✎ *Concept Reinforcement* *In each of Exercises 1–8, match the word or phrase with the most appropriate choice from the column on the right.*

1. ____ Total cost

2. ____ Fixed costs

3. ____ Variable costs

4. ____ Total revenue

5. ____ Total profit

6. ____ Price

7. ____ Break-even point

8. ____ Equilibrium point

a) The amount of money that a company takes in

b) The sum of fixed costs and variable costs

c) The point at which total revenue equals total cost

d) What consumers pay per item

e) The difference between total revenue and total cost

f) What companies spend whether or not a product is produced

g) The point at which supply equals demand

h) The costs that vary according to the number of items produced

---

For each of the following pairs of total-cost and total-revenue functions, find **(a)** *the total-profit function and* **(b)** *the break-even point.*

9. $C(x) = 35x + 200{,}000,$
$R(x) = 55x$

10. $C(x) = 20x + 500{,}000,$
$R(x) = 70x$

11. $C(x) = 15x + 3100,$
$R(x) = 40x$

12. $C(x) = 30x + 49{,}500,$
$R(x) = 85x$

13. $C(x) = 40x + 22{,}500,$
$R(x) = 85x$

14. $C(x) = 20x + 10{,}000$,
$R(x) = 100x$

15. $C(x) = 24x + 50{,}000$,
$R(x) = 40x$

16. $C(x) = 40x + 8010$,
$R(x) = 58x$

Aha! **17.** $C(x) = 75x + 100{,}000$,
$R(x) = 125x$

18. $C(x) = 20x + 120{,}000$,
$R(x) = 50x$

Find the equilibrium point for each of the following pairs of demand and supply functions.

19. $D(p) = 2000 - 15p$,
$S(p) = 740 + 6p$

20. $D(p) = 1000 - 8p$,
$S(p) = 350 + 5p$

21. $D(p) = 760 - 13p$,
$S(p) = 430 + 2p$

22. $D(p) = 800 - 43p$,
$S(p) = 210 + 16p$

23. $D(p) = 7500 - 25p$,
$S(p) = 6000 + 5p$

24. $D(p) = 8800 - 30p$,
$S(p) = 7000 + 15p$

25. $D(p) = 1600 - 53p$,
$S(p) = 320 + 75p$

26. $D(p) = 5500 - 40p$,
$S(p) = 1000 + 85p$

Solve.

27. *Manufacturing MP3 players.*   SoundGen, Inc., is planning to manufacture a new type of MP3 player/cell phone. The fixed costs for production are $45,000. The variable costs for producing each unit are estimated to be $40. The revenue from each unit is to be $130. Find the following.

a) The total cost $C(x)$ of producing x MP3/ cell phones

b) The total revenue $R(x)$ from the sale of x MP3/cell phones

c) The total profit $P(x)$ from the production and sale of x MP3/cell phones

d) The profit or loss from the production and sale of 3000 MP3/cell phones; of 400 MP3/cell phones

e) The break-even point

28. *Computer manufacturing.*   Current Electronics is planning to introduce a new laptop computer. The fixed costs for production are $125,300. The variable costs for producing each computer are $450. The revenue from each computer is $800. Find the following.

a) The total cost $C(x)$ of producing x computers

b) The total revenue $R(x)$ from the sale of x computers

c) The total profit $P(x)$ from the production and sale of x computers

d) The profit or loss from the production and sale of 100 computers; of 400 computers

e) The break-even point

29. *Pet safety.*   Ava designed and is now producing a pet car seat. The fixed costs for setting up production are $10,000. The variable costs for producing each seat are $30. The revenue from each seat is to be $80. Find the following.

a) The total cost $C(x)$ of producing x seats

b) The total revenue $R(x)$ from the sale of x seats

c) The total profit $P(x)$ from the production and sale of x seats

d) The profit or loss from the production and sale of 2000 seats; of 50 seats

e) The break-even point

30. *Manufacturing caps.*   Martina's Custom Printing is planning on adding painter's caps to its product line. For the first year, the fixed costs for setting up production are $16,404. The variable costs for producing a dozen caps are $6.00. The revenue on each dozen caps will be $18.00. Find the following.

a) The total cost $C(x)$ of producing x dozen caps

b) The total revenue $R(x)$ from the sale of x dozen caps

c) The total profit $P(x)$ from the production and sale of x dozen caps

d) The profit or loss from the production and sale of 3000 dozen caps; of 1000 dozen caps

e) The break-even point

31. In Example 1, the slope of the line representing Revenue is the sum of the slopes of the other two lines. This is not a coincidence. Explain why.

32. Variable costs and fixed costs are often compared to the slope and the *y*-intercept, respectively, of an equation for a line. Explain why you feel this analogy is or is not valid.

Skill Review

To prepare for Chapter 9, review solving equations using the addition and multiplication principles (Section 2.2).

Solve. [2.2]

33. $4x - 3 = 21$

34. $5 - x = 7$

35. $3x - 5 = 12x + 6$

36. $x - 4 = 9x - 10$

37. $3 - (x + 2) = 7$

38. $1 - 3(2x + 1) = 3 - 5x$

Synthesis

39. Rosie claims that since her fixed costs are $3000, she need sell only 10 custom birdbaths at $300 each in order to break even. Does this sound plausible? Why or why not?

40. In this section, we examined supply and demand functions for coffee. Does it seem realistic to you for the graph of *D* to have a constant slope? Why or why not?

41. *Yo-yo production.* Bing Boing Hobbies is willing to produce 100 yo-yo's at $2.00 each and 500 yo-yo's at $8.00 each. Research indicates that the public will buy 500 yo-yo's at $1.00 each and 100 yo-yo's at $9.00 each. Find the equilibrium point.

42. *Loudspeaker production.* Sonority Speakers, Inc., has fixed costs of $15,400 and variable costs of $100 for each pair of speakers produced. If the speakers sell for $250 per pair, how many pairs of speakers must be produced (and sold) in order to have enough profit to cover the fixed costs of two additional facilities? Assume that all fixed costs are identical.

Use a graphing calculator to solve.

43. *Dog food production.* Puppy Love, Inc., will soon begin producing a new line of puppy food. The marketing department predicts that the demand function will be $D(p) = -14.97p + 987.35$ and the supply function will be $S(p) = 98.55p - 5.13$.

   a) To the nearest cent, what price per unit should be charged in order to have equilibrium between supply and demand?

   b) The production of the puppy food involves $87,985 in fixed costs and $5.15 per unit in variable costs. If the price per unit is the value you found in part (a), how many units must be sold in order to break even?

44. *Computer production.* Brushstroke Computers, Inc., is planning a new line of computers, each of which will sell for $970. The fixed costs in setting up production are $1,235,580 and the variable costs for each computer are $697.

   a) What is the break-even point?

   b) The marketing department at Brushstroke is not sure that $970 is the best price. Their demand function for the new computers is given by $D(p) = -304.5p + 374,580$ and their supply function is given by $S(p) = 788.7p - 576,504$. To the nearest dollar, what price *p* would result in equilibrium between supply and demand?

   c) If the computers are sold for the equilibrium price found in part (b), what is the break-even point?

Study Summary

KEY TERMS AND CONCEPTS	EXAMPLES

SECTION 8.1: SYSTEMS OF EQUATIONS IN TWO VARIABLES

A **system of equations** is a set of two or more equations, in two or more variables. A solution of a system of equations must make all the equations true.

A system is **consistent** if it has at least one solution. Otherwise it is **inconsistent**.

The equations in a system are **dependent** if one of them can be written as a multiple and/or a sum of the other equation(s). Otherwise, they are **independent**.

Systems of two equations in two unknowns can be solved graphically.

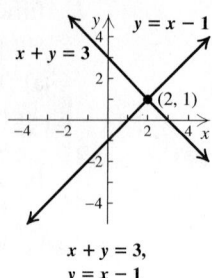

$$x + y = 3,$$
$$y = x - 1$$

The graphs intersect at (2, 1).
The solution is (2, 1).
The system is consistent.
The equations are independent.

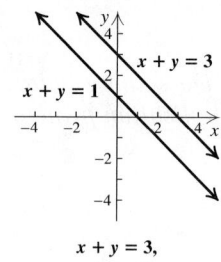

$$x + y = 3,$$
$$x + y = 1$$

The graphs do not intersect.
There is no solution.
The system is inconsistent.
The equations are independent.

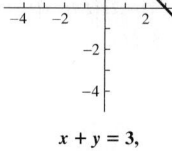

$$x + y = 3,$$
$$2x + 2y = 6$$

The graphs are the same.
The solution set is
$\{(x, y) \mid x + y = 3\}$.
The system is consistent.
The equations are dependent.

SECTION 8.2: SOLVING BY SUBSTITUTION OR ELIMINATION

Systems of equations can be solved using substitution.

Solve:

$$2x + 3y = 8,$$
$$x = y + 1.$$

The solution is $\left(\frac{11}{5}, \frac{6}{5}\right)$.

Substitute and solve for y:

$$2(y + 1) + 3y = 8$$
$$2y + 2 + 3y = 8$$
$$y = \frac{6}{5}.$$

Substitute and solve for x:

$$x = y + 1$$
$$x = \frac{6}{5} + 1$$
$$x = \frac{11}{5}.$$

Systems of equations can be solved using elimination.

Solve:

$$4x - 2y = 6,$$
$$3x + y = 7.$$

The solution is (2, 1).

Eliminate y and solve for x:

$$4x - 2y = 6$$
$$\underline{6x + 2y = 14}$$
$$10x \quad\;\;\; = 20$$
$$x = 2.$$

Substitute and solve for y:

$$3x + y = 7$$
$$3 \cdot 2 + y = 7$$
$$y = 1.$$

SECTION 8.3: SOLVING APPLICATIONS: SYSTEMS OF TWO EQUATIONS

Total-value, **mixture**, and **motion problems** often translate directly to systems of equations.

Total Value

A jewelry designer purchased 80 beads for a total of $39 to make a necklace. Some of the beads were sterling silver beads that cost 40¢ each and the rest were gemstone beads that cost 65¢ each. How many of each type were bought? (See Example 2 on pp. 525–526 for a solution.)

Mixture

Nature's Green Gardening, Inc., carries two brands of fertilizer containing nitrogen and water. "Gentle Grow" is 3% nitrogen and "Sun Saver" is 8% nitrogen. Nature's Green needs to combine the two types of solutions into a 90-L mixture that is 6% nitrogen. How much of each brand should be used? (See Example 5 on pp. 528–529 for a solution.)

Motion

A Boeing 747-400 jet flies 4 hr west with a 60-mph tailwind. Returning against the wind takes 5 hr. Find the speed of the jet with no wind. (See Example 7 on pp. 531–532 for a solution.)

SECTION 8.4: SYSTEMS OF EQUATIONS IN THREE VARIABLES

Systems of three equations in three variables are usually easiest to solve using elimination.

Solve:

$$x + y - z = 3, \quad (1)$$
$$-x + y + 2z = -5, \quad (2)$$
$$2x - y - 3z = 9 \quad (3)$$

Eliminate x using two equations:

$$x + y - z = 3 \quad (1)$$
$$\underline{-x + y + 2z = -5} \quad (2)$$
$$2y + z = -2.$$

Eliminate x again using two different equations:

$$-2x - 2y + 2z = -6 \quad (1)$$
$$\underline{2x - y - 3z = 9} \quad (3)$$
$$-3y - z = 3.$$

Solve the system of two equations for y and z:

$$2y + z = -2$$
$$\underline{-3y - z = 3}$$
$$-y = 1$$
$$y = -1$$

$$2(-1) + z = -2$$
$$z = 0.$$

Substitute and solve for x:

$$x + y - z = 3$$
$$x + (-1) - 0 = 3$$
$$x = 4.$$

The solution is $(4, -1, 0)$.

SECTION 8.5: SOLVING APPLICATIONS: SYSTEMS OF THREE EQUATIONS

Many problems with three unknowns can be solved after translating to a system of three equations.

In a triangular cross section of a roof, the largest angle is 70° greater than the smallest angle. The largest angle is twice as large as the remaining angle. Find the measure of each angle. (See Example 2 on pp. 546–547 for a solution.)

SECTION 8.6: ELIMINATION USING MATRICES

A **matrix** (plural, **matrices**) is a rectangular array of numbers. The individual numbers are called **entries** or **elements**.

$$\begin{bmatrix} 1 & 3 & -4 \\ -2 & 5 & 11 \end{bmatrix} \begin{matrix} \longrightarrow \text{row 1} \\ \longrightarrow \text{row 2} \end{matrix}$$

column 1 column 2 column 3

The *dimensions* of this matrix are 2×3, read "two by three."

By using **row-equivalent** operations, we can solve systems of equations using matrices.

Solve: $x + 4y = 1,$
$$2x - y = 3.$$

Write as a matrix in row-echelon form:

$$\begin{bmatrix} 1 & 4 & | & 1 \\ 2 & -1 & | & 3 \end{bmatrix} \longrightarrow \begin{bmatrix} 1 & 4 & | & 1 \\ 0 & -9 & | & 1 \end{bmatrix}.$$

Rewrite as equations and solve:

$$-9y = 1 \longrightarrow x + 4\left(-\tfrac{1}{9}\right) = 1$$
$$y = -\tfrac{1}{9} \qquad x = \tfrac{13}{9}.$$

The solution is $\left(\tfrac{13}{9}, -\tfrac{1}{9}\right)$.

SECTION 8.7: DETERMINANTS AND CRAMER'S RULE

A **determinant** is a number associated with a square matrix.

Determinant of a 2 × 2 Matrix

$$\begin{vmatrix} a & c \\ b & d \end{vmatrix} = ad - bc$$

Determinant of a 3 × 3 Matrix

$$\begin{vmatrix} a_1 & b_1 & c_1 \\ a_2 & b_2 & c_2 \\ a_3 & b_3 & c_3 \end{vmatrix} =$$

$$a_1 \begin{vmatrix} b_2 & c_2 \\ b_3 & c_3 \end{vmatrix} - a_2 \begin{vmatrix} b_1 & c_1 \\ b_3 & c_3 \end{vmatrix} + a_3 \begin{vmatrix} b_1 & c_1 \\ b_2 & c_2 \end{vmatrix}$$

$$\begin{vmatrix} 2 & 3 \\ -1 & 5 \end{vmatrix} = 2 \cdot 5 - (-1)(3) = 13$$

$$\begin{vmatrix} 2 & 3 & 2 \\ 0 & 1 & 0 \\ -1 & 5 & -4 \end{vmatrix} = 2 \begin{vmatrix} 1 & 0 \\ 5 & -4 \end{vmatrix} - 0 \begin{vmatrix} 3 & 2 \\ 5 & -4 \end{vmatrix} + (-1) \begin{vmatrix} 3 & 2 \\ 1 & 0 \end{vmatrix}$$

$$= 2(-4 - 0) - 0 - 1(0 - 2)$$

$$= -8 + 2 = -6$$

We can use matrices and **Cramer's rule** to solve systems of equations.

Cramer's rule for 2 × 2 matrices is given on p. 557.

Cramer's rule for 3 × 3 matrices is given on p. 559.

Solve:

$x - 3y = 7,$
$2x + 5y = 4.$

$$x = \frac{\begin{vmatrix} 7 & -3 \\ 4 & 5 \end{vmatrix}}{\begin{vmatrix} 1 & -3 \\ 2 & 5 \end{vmatrix}}; \qquad y = \frac{\begin{vmatrix} 1 & 7 \\ 2 & 4 \end{vmatrix}}{\begin{vmatrix} 1 & -3 \\ 2 & 5 \end{vmatrix}}$$

$$x = \frac{47}{11} \qquad\qquad y = \frac{-10}{11}$$

The solution is $\left(\frac{47}{11}, -\frac{10}{11}\right)$.

SECTION 8.8: BUSINESS AND ECONOMICS APPLICATIONS

The **break-even point** occurs where the **revenue** equals the **cost**, or where **profit** is 0.

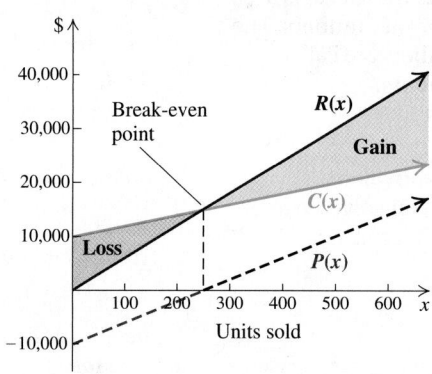

An **equilibrium point** occurs where the **supply** equals the **demand.**

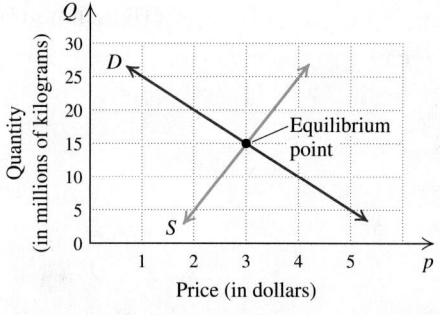

Review Exercises: Chapter 8

🖎 *Concept Reinforcement*   *Complete each of the following sentences.*

1. The system

$$5x + 3y = 7,$$
$$y = 2x + 1$$

is most easily solved using the _____ method. [8.2]

2. The system

$$-2x + 3y = 8,$$
$$2x + 2y = 7$$

is most easily solved using the _____ method. [8.2]

3. Of the methods used to solve systems of equations, the _____ method may yield only approximate solutions. [8.1], [8.2]

4. When one equation in a system is a multiple of another equation in that system, the equations are said to be _____. [8.1]

5. A system for which there is no solution is said to be _____. [8.1]

6. When we are using an algebraic method to solve a system of equations, obtaining a _____ tells us that the system is inconsistent. [8.2]

7. When we are graphing to solve a system of two equations, if there is no solution, the lines will be _____. [8.1]

8. When a matrix has the same number of rows and columns, it is said to be _____. [8.7]

9. Cramer's rule is a formula in which the numerator and the denominator of each fraction is a(n) _____. [8.7]

10. At the break-even point, the value of the profit function is _____. [8.8]

For Exercises 11–19, if a system has an infinite number of solutions, use set-builder notation to write the solution set. If a system has no solution, state this.

Solve graphically. [8.1]

11. $y = x - 3,$
    $y = \frac{1}{4}x$

12. $2x - 3y = 12,$
    $4x + y = 10$

Solve using the substitution method. [8.2]

13. $5x - 2y = 4,$
    $x = y - 2$

14. $y = x + 2,$
    $y - x = 8$

15. $x - 3y = -2,$
    $7y - 4x = 6$

Solve using the elimination method. [8.2]

16. $2x + 5y = 8,$
    $6x - 5y = 10$

17. $4x - 7y = 18,$
    $9x + 14y = 40$

18. $3x - 5y = 9,$
    $5x - 3y = -1$

19. $1.5x - 3 = -2y,$
    $3x + 4y = 6$

Solve. [8.3]

20. Ana bought two melons and one pineapple for $8.96. If she had purchased one melon and two pineapples, she would have spent $1.49 more. What is the price of a melon? of a pineapple?

21. A freight train leaves Houston at midnight traveling north at a speed of 44 mph. One hour later, a passenger train, going 55 mph, travels north from Houston on a parallel track. How many hours will the passenger train travel before it overtakes the freight train?

22. D'Andre wants 14 L of fruit punch that is 10% juice. At the store, he finds only punch that is 15% juice or punch that is 8% juice. How much of each should he purchase?

Solve. If a system's equations are dependent or if there is no solution, state this. [8.4]

23. $\ \ x + 4y + 3z = 2,$
    $2x + \ \ y + \ \ z = 10,$
    $-x + \ \ y + 2z = 8$

24. $4x + 2y - 6z = 34,$
    $2x + \ \ y + 3z = 3,$
    $6x + 3y - 3z = 37$

25. $\ \ 2x - 5y - 2z = -4,$
    $\ \ 7x + 2y - 5z = -6,$
    $-2x + 3y + 2z = 4$

26. $3x + \ \ y \ \ \ \ \ \ = 2,$
    $\ \ x + 3y + z = 0,$
    $\ \ x \ \ \ \ \ \ + z = 2$

Solve.

27. In triangle *ABC*, the measure of angle *A* is four times the measure of angle *C*, and the measure of angle *B* is 45° more than the measure of angle *C*. What are the measures of the angles of the triangle? [8.5]

28. A nontoxic floor wax can be made from lemon juice and food-grade linseed oil. The amount of oil should be twice the amount of lemon juice. How much of each ingredient is needed to make 32 oz of floor wax? (The mix should be spread with a rag and buffed when dry.) [8.3]

29. The sum of the average number of times a man, a woman, and a one-year-old child cry each month is 56.7. A woman cries 3.9 more times than a man. The average number of times a one-year-old cries per month is 43.3 more than the average number of times combined that a man and a woman cry. What is the average number of times per month that each cries? [8.5]

Solve using matrices. Show your work. [8.6]

30. $3x + 4y = -13,$
$5x + 6y = 8$

31. $3x - y + z = -1,$
$2x + 3y + z = 4,$
$5x + 4y + 2z = 5$

Evaluate. [8.7]

32. $\begin{vmatrix} -2 & -5 \\ 3 & 10 \end{vmatrix}$

33. $\begin{vmatrix} 2 & 3 & 0 \\ 1 & 4 & -2 \\ 2 & -1 & 5 \end{vmatrix}$

Solve using Cramer's rule. Show your work. [8.7]

34. $2x + 3y = 6,$
$x - 4y = 14$

35. $2x + y + z = -2,$
$2x - y + 3z = 6,$
$3x - 5y + 4z = 7$

36. Find the equilibrium point for the demand and supply functions
$$S(p) = 60 + 7p$$
and
$$D(p) = 120 - 13p. \quad [8.8]$$

37. Danae is beginning to produce organic honey. For the first year, the fixed costs for setting up production are $54,000. The variable costs for producing each pint of honey are $4.75. The revenue from each pint of honey is $9.25. Find the following. [8.8]

a) The total cost $C(x)$ of producing x pints of honey

b) The total revenue $R(x)$ from the sale of x pints of honey

c) The total profit $P(x)$ from the production and sale of x pints of honey

d) The profit or loss from the production and sale of 5000 pints of honey; of 15,000 pints of honey

e) The break-even point

Synthesis

38. How would you go about solving a problem that involves four variables? [8.5]

39. Explain how a system of equations can be both dependent and inconsistent. [8.4]

40. Danae is leaving a job that pays $36,000 a year to make honey (see Exercise 37). How many pints of honey must she produce and sell in order to make as much money as she earned at her previous job? [8.8]

41. Recently, Staples® charged $5.99 for a 2-count pack of Bic® Round Stic Grip mechanical pencils and $7.49 for a 12-count pack of Bic® Matic Grip mechanical pencils. Wiese Accounting purchased 138 of these two types of mechanical pencils for a total of $157.26. How many packs of each did they buy? [8.3]

42. Solve graphically:
$$y = x + 2,$$
$$y = x^2 + 2. \quad [8.1]$$

43. The graph of $f(x) = ax^2 + bx + c$ contains the points $(-2, 3)$, $(1, 1)$, and $(0, 3)$. Find a, b, and c and give a formula for the function. [8.5]

Test: Chapter 8

 CHAPTER **Test Prep** VIDEO CD

Step-by-step test solutions are found on the video CD in the front of this book.

For Exercises 1–6, if a system has an infinite number of solutions, use set-builder notation to write the solution set. If a system has no solution, state this.

1. Solve graphically:
$$2x + y = 8,$$
$$y - x = 2.$$

Solve using the substitution method.

2. $x + 3y = -8,$
$4x - 3y = 23$

3. $2x - 4y = -6,$
$x = 2y - 3$

Solve using the elimination method.

4. $3x - y = 7,$
$x + y = 1$

5. $4y + 2x = 18,$
$3x + 6y = 26$

6. $4x - 6y = 3,$
$6x - 4y = -3$

7. The perimeter of a standard basketball court is 288 ft. The length is 44 ft longer than the width. Find the dimensions.

P = 288 ft

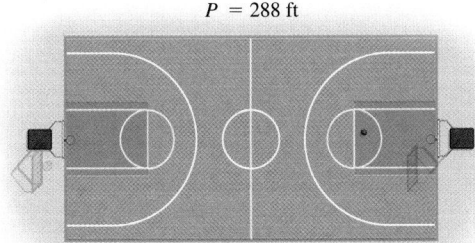

8. Pepperidge Farm® Goldfish is a snack food for which 40% of its calories come from fat. Rold Gold® Pretzels receive 9% of their calories from fat. How many grams of each would be needed to make 620 g of a snack mix for which 15% of the calories are from fat?

9. A truck leaves Gaston at noon traveling 55 mph. An hour later, a car leaves Gaston following the same route as the truck but traveling 65 mph. In how many hours will the car catch up to the truck?

Solve. If a system's equations are dependent or if there is no solution, state this.

10. $-3x + y - 2z = 8,$
$-x + 2y - z = 5,$
$2x + y + z = -3$

11. $6x + 2y - 4z = 15,$
$-3x - 4y + 2z = -6,$
$4x - 6y + 3z = 8$

12. $2x + 2y = 0,$
$4x + 4z = 4,$
$2x + y + z = 2$

13. $3x + 3z = 0,$
$2x + 2y = 2,$
$3y + 3z = 3$

Solve using matrices.

14. $4x + y = 12,$
$3x + 2y = 2$

15. $x + 3y - 3z = 12,$
$3x - y + 4z = 0,$
$-x + 2y - z = 1$

Evaluate.

16. $\begin{vmatrix} 4 & -2 \\ 3 & -5 \end{vmatrix}$

17. $\begin{vmatrix} 3 & 4 & 2 \\ -2 & -5 & 4 \\ 0 & 5 & -3 \end{vmatrix}$

18. Solve using Cramer's rule:
$3x + 4y = -1,$
$5x - 2y = 4.$

19. An electrician, a carpenter, and a plumber are hired to work on a house. The electrician earns $30 per hour, the carpenter $28.50 per hour, and the plumber $34 per hour. The first day on the job, they worked a total of 21.5 hr and earned a total of $673.00. If the plumber worked 2 more hours than the carpenter did, how many hours did each work?

20. Find the equilibrium point for the demand and supply functions
$$D(p) = 79 - 8p \quad \text{and} \quad S(p) = 37 + 6p,$$
where *p* is the price, in dollars, $D(p)$ is the number of units demanded, and $S(p)$ is the number of units supplied.

21. Kick Back, Inc., is producing a new hammock. For the first year, the fixed costs for setting up production are $44,000. The variable costs for producing each hammock are $25. The revenue from each hammock is $80. Find the following.

a) The total cost $C(x)$ of producing *x* hammocks

b) The total revenue $R(x)$ from the sale of *x* hammocks

c) The total profit $P(x)$ from the production and sale of *x* hammocks

d) The profit or loss from the production and sale of 300 hammocks; of 900 hammocks

e) The break-even point

Synthesis

22. The graph of the function $f(x) = mx + b$ contains the points $(-1, 3)$ and $(-2, -4)$. Find *m* and *b*.

23. Some of the world's best and most expensive coffee is Hawaii's Kona coffee. In order for coffee to be labeled "Kona Blend," it must contain at least 30% Kona beans. Bean Town Roasters has 40 lb of Mexican coffee. How much Kona coffee must they add if they wish to market it as Kona Blend?

1. Evaluate $9t \div 6t^3$ for $t = -2$. [1.8]

2. Find the opposite of $-\frac{1}{10}$. [1.6]

3. Find the reciprocal of $-\frac{1}{10}$. [1.7]

4. Remove parentheses and simplify:
   $$3x^2 - 2(-5x^2 + y) + y. \quad [1.8]$$

Simplify. Do not leave negative exponents in your answers.

5. $40 - 8^2 \div 4 \cdot 4$ [1.8]

6. $\dfrac{|-2 \cdot 5 - 3 \cdot 4|}{5^2 - 2 \cdot 7}$ [1.8]

7. $x^4 \cdot x^{-6} \cdot x^{13}$ [4.2]

8. $(6x^2y^3)^2(-2x^0y^4)^{-3}$ [4.2]

9. $\dfrac{-10a^7b^{-11}}{25a^{-4}b^{22}}$ [4.2]

10. $\left(\dfrac{3x^4y^{-2}}{4x^{-5}}\right)^4$ [4.2]

11. $(5.5 \times 10^{-3})(3.4 \times 10^8)$ [4.2]

12. $\dfrac{2.42 \times 10^5}{6.05 \times 10^{-2}}$ [4.2]

Solve.

13. $2x + 1 = 5(2 - x)$ [2.2]

14. $x^2 + 5x + 6 = 0$ [5.7]

15. $t + \dfrac{6}{t} = 5$ [6.6]

16. $\frac{1}{2}t + \frac{1}{6} = \frac{1}{3} - t$ [2.2]

17. $2y + 9 \leq 5y + 11$ [2.6]

18. $n^2 = 100$ [5.7]

19. $3x + y = 5,$
    $y = x + 1$ [8.2]

20. $\dfrac{4}{x - 1} = \dfrac{3}{x + 2}$ [6.6]

21. $6x^2 = x + 2$ [5.7]

22. $x + y = 10,$
    $x - y = 4$ [8.2]

23. $x + y + z = -5,$
    $2x + 3y - 2z = 8,$
    $x - y + 4z = -21$ [8.4]

24. $2x + 5y - 3z = -11,$
    $-5x + 3y - 2z = -7,$
    $3x - 2y + 5z = 12$ [8.4]

Solve each formula. [2.3]

25. $t = \frac{1}{3}pq$, for p

26. $A = \dfrac{r + s}{2}$, for s

Add and, if possible, simplify.

27. $(8a^2 - 6a - 7) + (3a^3 + 6a - 7)$ [4.4]

28. $\dfrac{m + n}{2m + n} + \dfrac{n}{m - n}$ [6.4]

Subtract.

29. $(8a^2 - 6a - 7) - (3a^3 + 6a - 7)$ [4.4]

30. $\dfrac{x + 5}{x - 2} - \dfrac{x - 1}{2 - x}$ [6.4]

Multiply.

31. $(5a^2 + b)(5a^2 - b)$ [4.7]

32. $-3x^2(5x^3 - 6x^2 - 2x + 1)$ [4.5]

33. $(2n + 5)^2$ [4.6]

34. $(8t^2 + 5)(t^3 + 4)$ [4.6]

35. $\dfrac{x^2 - 2x + 1}{x^2 - 4} \cdot \dfrac{x^2 + 4x + 4}{x^2 - 3x + 2}$ [6.2]

Divide.

36. $(2x^2 - 5x - 3) \div (x - 3)$ [4.8]

37. $\dfrac{x^2 - x}{4x^2 + 8x} \div \dfrac{x^2 - 1}{2x}$ [6.2]

Factor completely.

38. $x^3 + x^2 + 2x + 2$ [5.1]

39. $m^4 - 1$ [5.4]

40. $2x^3 + 18x^2 + 40x$ [5.2]

41. $x^4 - 6x^2 + 9$ [5.4]

42. $4x^2 - 2x - 10$ [5.1]

43. $10x^2 - 29x + 10$ [5.3]

Graph.

44. $2x + y = 6$ [3.3]

45. $f(x) = -2x + 8$ [7.3]

46. $2x = 10$ [3.3]

47. $x = 2y$ [3.2]

48. Find the x-intercept and the y-intercept of the line given by $10x - 15y = 60$. [3.3]

49. Write the slope–intercept equation for the line with slope -2 and y-intercept $\left(0, \frac{4}{7}\right)$. [3.6]

50. Find an equation in slope–intercept form of the line containing the points $(-6, 3)$ and $(4, 2)$. [3.7]

51. Determine whether the lines given by the following equations are parallel, perpendicular, or neither:
    $$2x = 4y + 7,$$
    $$x - 2y = 5. \quad [3.6]$$

52. Find an equation of the line containing the point $(2, 1)$ and perpendicular to the line $x - 2y = 5$. [3.7]

53. For the graph of f shown, determine the domain, the range, $f(-3)$, and any value of x for which $f(x) = 5$. [7.2]

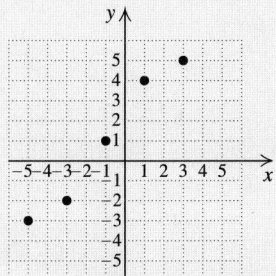

54. Determine the domain of the function given by

$$f(x) = \frac{7}{x + 10}. \quad [7.2]$$

Given $g(x) = 4x - 3$ and $h(x) = -2x^2 + 1$, find the following function values.

55. $h(4)$ [7.1]

56. $-g(0)$ [7.1]

57. $(g - h)(a)$ [7.4]

58. Evaluate: $\begin{vmatrix} 2 & -3 \\ 4 & 1 \end{vmatrix}.$ [8.7]

Solve.

59. A library was mistakenly charged sales tax on an order of children's books. The invoice, including 5% sales tax, was for $1323. How much should the library have been charged? [2.4]

60. A snowmobile is traveling 40 mph faster than a dog sled. In the same time that the dog sled travels 24 mi, the snowmobile travels 104 mi. Find the speeds of the sled and the snowmobile. [6.7]

61. Each nurse practitioner at the Midway Clinic is required to see an average of at least 40 patients per day. During the first 4 workdays of one week, Michael saw 50, 35, 42, and 38 patients. How many patients must he see on the fifth day in order to meet his requirements? [2.7]

62. Tia and Avery live next door to each other on Meachin Street. Their house numbers are consecutive odd numbers, and the product of their house numbers is 143. Find the house numbers. [5.8]

63. Long-distance calls made using the "Ruby" prepaid calling card cost 1.4¢ per minute plus a maintenance fee of 99¢ per week. The "Sapphire" plan has a maintenance fee of only 69¢ per week, but calls cost 1.7¢ per minute. For what number of minutes per week will the two cards cost the same? [8.3]
Source: www.enjoyprepaid.com

64. The second angle of a triangle is twice as large as the first. The third angle is 15° less than the sum of the first two angles. Find the measure of each angle. [2.5]

65. A newspaper uses self-employed copywriters to write its advertising copy. If they use 12 copywriters, each person works 35 hr per week. How many hours per week would each person work if the newspaper uses 10 copywriters? [7.5]

Synthesis

66. Solve $t = px - qx$ for x. [2.3]

67. Write the slope–intercept equation of the line that contains the point $(-2, 3)$ and is parallel to the line $2x - y = 7$. [3.7]

68. Simplify: $\dfrac{t^2 - 9}{2t + 1} \div \dfrac{t^2 - 4t + 3}{6t^2 + 3t} \cdot \dfrac{2t^3 - t^2 + t}{4t + 12}.$ [6.2]

69. Given that $f(x) = mx + b$ and that $f(5) = -3$ when $f(-4) = 2$, find m and b. [7.1], [8.3]

70. The maximum length of a postcard that can be mailed with one postcard stamp is $\frac{7}{4}$ in. longer than the maximum width. The maximum area is $\frac{51}{2}$ in². Find the maximum length and width of a postcard. [5.8]
Source: USPS

Inequalities and Problem Solving

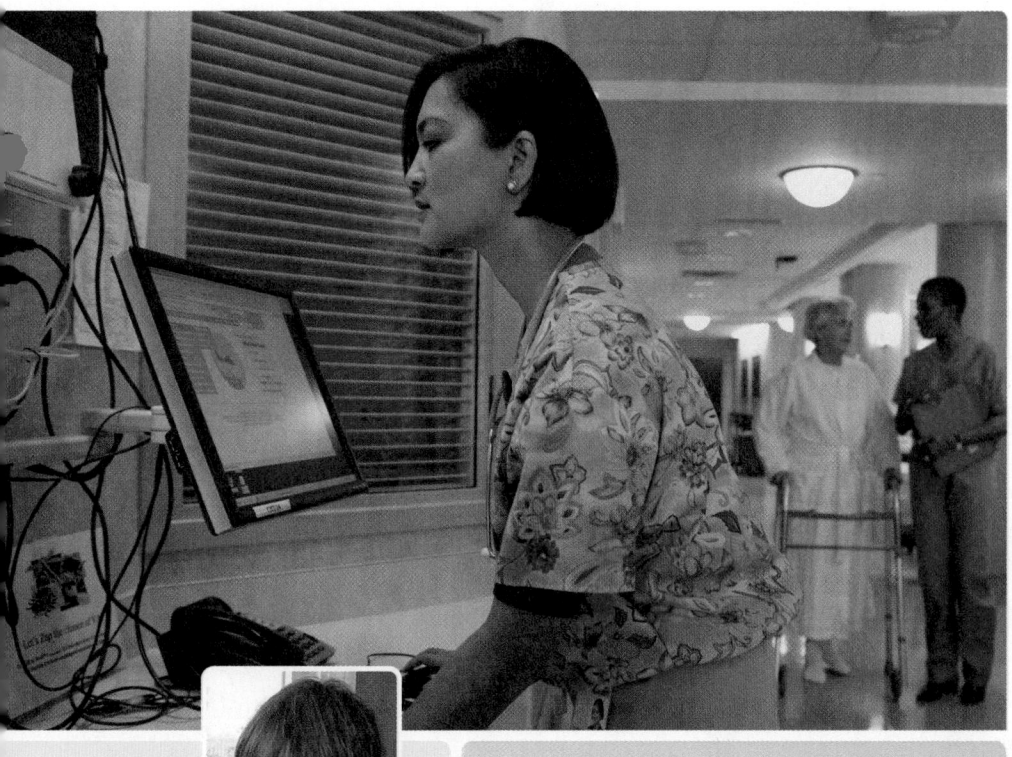

ANNE SAMUELS
REGISTERED NURSE/
NURSE MANAGEMENT
Springfield, Massachusetts

As a nurse, I use math every
day to calculate dosages of
medication. If an order prescribes
a dosage of 2.5 mg/kg twice a
day, I need to know the patient's
weight, change that weight from
pounds to kilograms, and then
calculate the dosage. As a nurse
manager, I use math to determine
the number of nurses needed
to care for patients and to work
out a budget for salaries
and equipment.

AN APPLICATION

The number of registered nurses $R(t)$
employed in the United States, in
millions, t years after 2000, can be
approximated by

$$R(t) = 0.05t + 2.2.$$

Determine (using an inequality) those
years for which more than 3 million
registered nurses will be employed in
the United States.

Source: Based on data from the U.S. Department
of Health and Human Services and the Bureau of
Labor Statistics

This problem appears as Example 5 in
Section 9.1.

nequalities are mathematical sentences containing symbols such as $<$ (is less than). In this chapter, we use the principles for solving inequalities developed in Chapter 2 to solve compound inequalities. We also combine our knowledge of inequalities and systems of equations to solve systems of inequalities.

| 9.1 | **Inequalities and Domain** |

Solving Inequalities Graphically ▪ Domain ▪ Problem Solving

Solving Inequalities Graphically

Recall from Chapter 1 that an **inequality** is any sentence containing $<, >, \leq, \geq$, or $\neq$ (see Section 1.4)—for example,

$$-2 < a, \quad x > 4, \quad x + 3 \leq 6, \quad 6 - 7y \geq 10y - 4, \quad \text{and} \quad 5x \neq 10.$$

Any replacement for the variable that makes an inequality true is called a **solution**. The set of all solutions is called the **solution set**. When all solutions of an inequality are found, we say that we have **solved** the inequality.

We can use two principles, developed in Chapter 2, to solve inequalities.

The Addition Principle for Inequalities

For any real numbers a, b, and c:

$$a < b \text{ is equivalent to } a + c < b + c;$$
$$a > b \text{ is equivalent to } a + c > b + c.$$

Similar statements hold for $\leq$ and $\geq$.

The Multiplication Principle for Inequalities

For any real numbers a and b, and for any *positive* number c,

$$a < b \text{ is equivalent to } ac < bc;$$
$$a > b \text{ is equivalent to } ac > bc.$$

For any real numbers a and b, and for any *negative* number c,

$$a < b \text{ is equivalent to } ac > bc;$$
$$a > b \text{ is equivalent to } ac < bc.$$

Similar statements hold for $\leq$ and $\geq$.

EXAMPLE **1**   Solve: $2x + 4 < -x + 1$.

SOLUTION

$$2x + 4 < -x + 1$$
$$2x + 4 - 4 < -x + 1 - 4 \qquad \text{Subtracting 4 from both sides}$$
$$2x < -x - 3$$
$$2x + x < -x - 3 + x \qquad \text{Adding } x \text{ to both sides}$$
$$3x < -3$$
$$\frac{3x}{3} < \frac{-3}{3} \qquad \begin{array}{l}\text{Dividing both sides by 3. The } < \text{ symbol} \\ \text{stays the same since 3 is positive.}\end{array}$$
$$x < -1$$

The solution set is $\{x | x < -1\}$, or $(-\infty, -1)$.   **TRY EXERCISE** 19

We now look at a graphical method for solving inequalities.

To solve the inequality in Example 1, $2x + 4 < -x + 1$, we let $f(x) = 2x + 4$ and $g(x) = -x + 1$. Consider the graphs of the functions $f(x) = 2x + 4$ and $g(x) = -x + 1$.

Graphs of $f(x)$ and $g(x)$

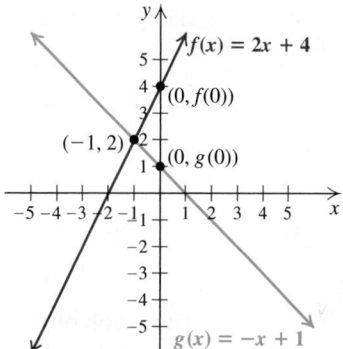

The graphs intersect at the point $(-1, 2)$. Thus, when $x = -1$, $f(x) = g(x)$. At all x-values except -1, either $f(x) > g(x)$ or $f(x) < g(x)$. Note from the graphs that $f(x) > g(x)$ when the graph of f lies above the graph of g. Also, $f(x) < g(x)$ when the graph of f lies below the graph of g.

Compare $f(0)$ and $g(0)$. Note from the graphs that $f(0)$ lies above $g(0)$. In fact, for all values of x greater than -1, $f(x) > g(x)$. For all values of x less than -1, $f(x) < g(x)$. In this way, the point of intersection of the graphs marks the endpoint of the solution set of an inequality.

Note that using the graphs of $f(x)$ and $g(x)$ to solve an inequality is not the same as graphing the solutions of the inequality.

For $f(x) = 2x + 4$ and $g(x) = -x + 1$, compare the following.

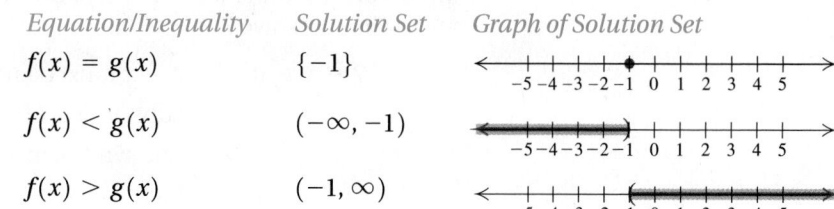

EXAMPLE **2**

TECHNOLOGY CONNECTION

On most calculators, $4x - 1 \geq 5x - 2$ can be solved by graphing $y_1 = 4x - 1 \geq 5x - 2$ ($\geq$ is often found by pressing **2ND** **MATH**). The solution set is then displayed as an interval (shown by a horizontal line 1 unit above the x-axis).

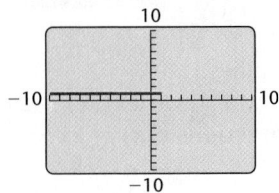

A check can also be made by graphing $y_1 = 4x - 1$ and $y_2 = 5x - 2$ and identifying those x-values for which $y_1 \geq y_2$.

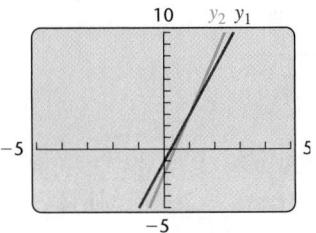

The INTERSECT option helps us find that $y_1 = y_2$ when $x = 1$. Note that $y_1 \geq y_2$ for x-values in the interval $(-\infty, 1]$.

Solve graphically: $-3x + 1 \geq x - 7$.

SOLUTION   We let $f(x) = -3x + 1$ and $g(x) = x - 7$, and graph both functions. The solution set will consist of the interval for which the graph of f lies on or above the graph of g.

To find the point of intersection, we solve the system of equations

$$y = -3x + 1,$$
$$y = x - 7.$$

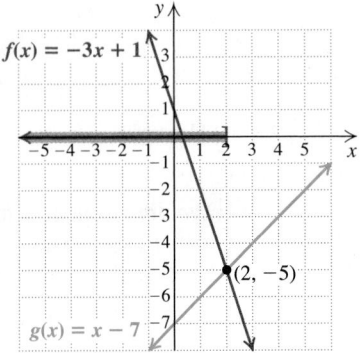

Using substitution, we have

$$-3x + 1 = x - 7$$
$$-4x = -8$$
$$x = 2.$$

Then $y = -3(2) + 1 = -5$.

Thus the point of intersection is

$$(2, -5).$$

The graph of f lies *on* the graph of g when $x = 2$. It lies *above* the graph of g when $x < 2$. Thus the solution of $-3x + 1 \geq x - 7$ is

$$\{x \mid x \leq 2\}, \quad \text{or} \quad (-\infty, 2].$$

This set is indicated by the purple shading on the x-axis.   ▶ **TRY EXERCISE** ▷ 33

Domain

Although radical notation is not discussed in detail until Chapter 10, we know that only nonnegative numbers have square roots that are real numbers. Thus finding the domain of a radical function often involves solving an inequality.

EXAMPLE **3**

Find the domain of f if $f(x) = \sqrt{7 - x}$.

SOLUTION   In order for $\sqrt{7 - x}$ to exist as a real number, $7 - x$ must be nonnegative. Thus we solve $7 - x \geq 0$:

$$7 - x \geq 0 \qquad \text{$7 - x$ must be nonnegative.}$$
$$-x \geq -7 \qquad \text{Subtracting 7 from both sides}$$
$$\qquad\qquad\qquad \text{The symbol must be reversed.}$$
$$x \leq 7. \qquad \text{Multiplying both sides by -1}$$

When $x \leq 7$, the expression $7 - x$ is nonnegative. Thus the domain of f is $\{x \mid x \leq 7\}$, or $(-\infty, 7]$.   ▶ **TRY EXERCISE** ▷ 43

EXAMPLE 4

Find the domain of g if $g(x) = \sqrt{\frac{1}{2}x - 10}$.

SOLUTION   In order for $\sqrt{\frac{1}{2}x - 10}$ to exist as a real number, $\frac{1}{2}x - 10$ must be nonnegative. Thus we solve $\frac{1}{2}x - 10 \geq 0$:

$$\frac{1}{2}x - 10 \geq 0 \qquad \frac{1}{2}x - 10 \text{ must be nonnegative.}$$

$$\frac{1}{2}x \geq 10 \qquad \text{Adding 10 to both sides}$$

$$x \geq 20. \qquad \text{Multiplying both sides by 2.}$$
$$\text{The symbol} \geq \text{remains the same.}$$

Thus the domain of g is $\{x \mid x \geq 20\}$, or $[20, \infty)$.   TRY EXERCISE 49

Problem Solving

Many problem-solving situations translate to inequalities.

EXAMPLE 5

Registered nurses.   The number of registered nurses $R(t)$ employed in the United States, in millions, t years after 2000 can be approximated by

$$R(t) = 0.05t + 2.2.$$

Determine (using an inequality) those years for which more than 3 million registered nurses will be employed in the United States.

Source: Based on data from the U.S. Department of Health and Human Services and the Bureau of Labor Statistics

SOLUTION

1. **Familiarize.** We already have a formula. The number 0.05 tells us that employment of registered nurses is growing at a rate of 0.05 million (or 50,000) per year. The number 2.2 tells us that in 2000, there were approximately 2.2 million registered nurses employed in the United States.

2. **Translate.** We are asked to find the years for which *more than* 3 million registered nurses will be employed in the United States. Thus we have

$$R(t) > 3$$
$$0.05t + 2.2 > 3. \qquad \text{Substituting } 0.05t + 2.2 \text{ for } R(t)$$

3. **Carry out.** We solve the inequality:

$$0.05t + 2.2 > 3$$
$$0.05t > 0.8 \qquad \text{Subtracting 2.2 from both sides}$$
$$t > 16. \qquad \text{Dividing both sides by 0.05}$$

Note that this corresponds to years after 2016.

4. **Check.** We can partially check our answer by finding $R(t)$ for a value of t greater than 16. For example,

$$R(20) = 0.05 \cdot 20 + 2.2 = 3.2, \text{ and } 3.2 > 3.$$

5. **State.** More than 3 million registered nurses will be employed in the United States for years after 2016.   TRY EXERCISE 57

EXAMPLE **6**   *Job offers.*   After graduation, Rose had two job offers in sales:

Uptown Fashions: A salary of $600 per month, plus a commission of 4% of sales;

Ergo Designs: A salary of $800 per month, plus a commission of 6% of sales in excess of $10,000.

If sales always exceed $10,000, for what amount of sales would Uptown Fashions provide higher pay?

SOLUTION

1. **Familiarize.** Listing the given information in a table will be helpful.

Uptown Fashions Monthly Income	Ergo Designs Monthly Income
$600 salary 4% of sales *Total*: $600 + 4% of sales	$800 salary 6% of sales over $10,000 *Total*: $800 + 6% of sales over $10,000

Next, suppose that Rose sold a certain amount—say, $12,000—in one month. Which offer would be better? Working for Uptown, she would earn $600 plus 4% of $12,000, or $600 + 0.04(12,000) = \$1080$. Since with Ergo Designs commissions are paid only on sales in excess of $10,000, Rose would earn $800 plus 6% of ($12,000 − $10,000$), or $800 + 0.06(2000) = \$920$.

For monthly sales of $12,000, Uptown pays better. Similar calculations will show that for sales of $30,000 a month, Ergo pays better. To determine *all* values for which Uptown pays more money, we must solve an inequality that is based on the calculations above.

2. **Translate.** We let S = the amount of monthly sales, in dollars, and will assume $S > 10,000$ so that both plans will pay a commission. Examining the calculations in the *Familiarize* step, we see that monthly income from Uptown is $600 + 0.04S$ and from Ergo is $800 + 0.06(S − 10,000)$. We want to find all values of S for which

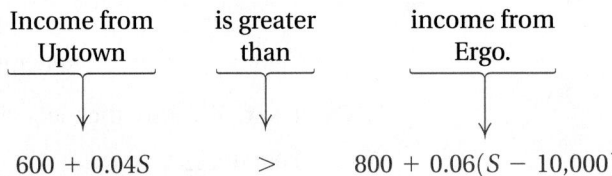

Income from Uptown	is greater than	income from Ergo.
$600 + 0.04S$	$>$	$800 + 0.06(S − 10,000)$

3. **Carry out.** We solve the inequality:

$600 + 0.04S > 800 + 0.06(S − 10,000)$

$600 + 0.04S > 800 + 0.06S − 600$ Using the distributive law

$600 + 0.04S > 200 + 0.06S$ Combining like terms

$400 > 0.02S$ Subtracting 200 and $0.04S$ from both sides

$20,000 > S,$ or $S < 20,000.$ Dividing both sides by 0.02

4. **Check.** The above steps indicate that income from Uptown Fashions is higher than income from Ergo Designs for sales less than $20,000. In the *Familiarize* step, we saw that for sales of $12,000, Uptown pays more. Since $12,000 < 20,000$, this is a partial check.

5. **State.** When monthly sales are less than $20,000, Uptown Fashions provides the higher pay.

TRY EXERCISE 59

9.1 EXERCISE SET

Concept Reinforcement *Classify each of the following as equivalent inequalities, equivalent equations, equivalent expressions, or not equivalent.*

1. $5x + 7 = 6 - 3x$, $8x + 7 = 6$

2. $2(4x + 1)$, $8x + 2$

3. $x - 7 > -2$, $x > 5$

4. $t + 3 < 1$, $t < 2$

5. $-4t \le 12$, $t \le -3$

6. $\frac{3}{5}a + \frac{1}{5} = 2$, $3a + 1 = 10$

7. $6a + 9$, $3(2a + 3)$

8. $-4x \ge -8$, $x \ge 2$

9. $-\frac{1}{2}x < 7$, $x > 14$

10. $-\frac{1}{3}t \le -5$, $t \ge 15$

Solve algebraically.

11. $3x + 1 < 7$

12. $2x - 5 \ge 9$

13. $3 - x \ge 12$

14. $8 - x < 15$

15. $\dfrac{2x + 7}{5} < -9$

16. $\dfrac{5y + 13}{4} > -2$

17. $\dfrac{3t - 7}{-4} \le 5$

18. $\dfrac{2t - 9}{-3} \ge 7$

19. $3 - 8y \ge 9 - 4y$

20. $4m + 7 \ge 9m - 3$

21. $5(t - 3) + 4t < 2(7 + 2t)$

22. $2(4 + 2x) > 2x + 3(2 - 5x)$

23. $5[3m - (m + 4)] > -2(m - 4)$

24. $8x - 3(3x + 2) - 5 \ge 3(x + 4) - 2x$

Solve each inequality using the given graph.

25. $f(x) \ge g(x)$

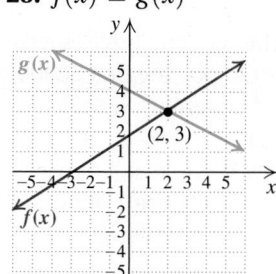

26. $f(x) < g(x)$

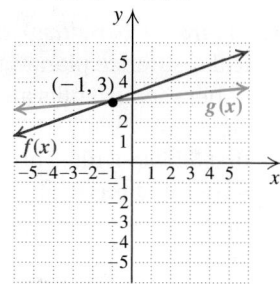

27. $f(x) < g(x)$

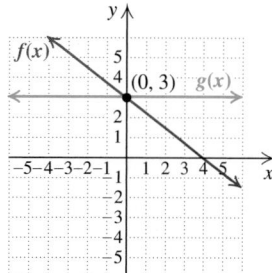

28. $f(x) \ge g(x)$

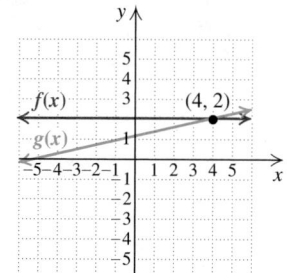

Solve graphically.

29. $x - 3 < 4$

30. $x + 4 \ge 6$

31. $2x - 3 \ge 1$

32. $3x + 1 < 1$

33. $x + 3 > 2x - 5$

34. $3x - 5 \le 3 - x$

35. $\frac{1}{2}x - 2 \le 1 - x$

36. $x + 5 > \frac{1}{3}x - 1$

37. Let $f(x) = 7 - 3x$ and $g(x) = 2x - 3$. Find all values of x for which $f(x) \le g(x)$.

38. Let $f(x) = 8x - 9$ and $g(x) = 3x - 11$. Find all values of x for which $f(x) \le g(x)$.

39. Let $y_1 = 2x - 7$ and $y_2 = 5x - 9$. Find all values of x for which $y_1 < y_2$.

40. Let $y_1 = 2x + 1$ and $y_2 = -\frac{1}{2}x + 6$. Find all values of x for which $y_1 < y_2$.

Find the domain of each function.

41. $f(x) = \sqrt{x - 10}$

42. $f(x) = \sqrt{x + 2}$

43. $f(x) = \sqrt{3 - x}$

44. $f(x) = \sqrt{11 - x}$

45. $f(x) = \sqrt{2x + 7}$

46. $f(x) = \sqrt{8 - 5x}$

47. $f(x) = \sqrt{8 - 2x}$

48. $f(x) = \sqrt{2x - 10}$

49. $f(x) = \sqrt{\dfrac{1}{3}x + 5}$

50. $f(x) = \sqrt{2x + \dfrac{1}{2}}$

51. $f(x) = \sqrt{\dfrac{x - 5}{4}}$

52. $f(x) = \sqrt{\dfrac{x + 7}{12}}$

Solve.

53. *Photography.* Eli will photograph a wedding for a flat fee of $900 or for an hourly rate of $120. For what lengths of time would the hourly rate be less expensive?

54. *Truck rentals.* Jenn can rent a moving truck for either $99 with unlimited mileage or $49 plus 80¢ per mile. For what mileages would the unlimited mileage plan save money?

55. *Exam scores.* There are 80 questions on a college entrance examination. Two points are awarded for each correct answer, and one half point is deducted for each incorrect answer. How many questions does Tami need to answer correctly in order to score at least 100 on the test? Assume that Tami answers every question.

56. *Insurance claims.* After a serious automobile accident, most insurance companies will replace the damaged car with a new one if repair costs exceed 80% of the NADA, or "blue-book," value of the car. Lorenzo's car recently sustained $9200 worth of damage but was not replaced. What was the blue-book value of his car?

57. *Crude-oil production.* The yearly U.S. production of crude oil $C(t)$, in millions of barrels, t years after 2000 can be approximated by

$$C(t) = -40.5t + 2159.$$

Determine (using an inequality) those years for which domestic production will drop below 1750 million barrels.

Source: U.S. Energy Information Administration

58. *HDTVs.* The percentage of U.S. households $p(t)$ with an HDTV t years after 2005 can be approximated by

$$p(t) = 8t + 12.5.$$

Determine (using an inequality) those years for which more than half of all U.S. households will have an HDTV.

Source: Based on data from Consumer Electronics Association

59. *Wages.* Toni can be paid in one of two ways:

    Plan A: A salary of $400 per month, plus a commission of 8% of gross sales;

    Plan B: A salary of $610 per month, plus a commission of 5% of gross sales.

For what amount of gross sales should Toni select plan A?

60. *Wages.* Eric can be paid for his masonry work in one of two ways:

    Plan A: $300 plus $9.00 per hour;

    Plan B: Straight $12.50 per hour.

Suppose that the job takes n hours. For what values of n is plan B better for Eric?

61. *Insurance benefits.* Under the "Green Badge" medical insurance plan, Carlee would pay the first $2000 of her medical bills and 30% of all remaining bills. Under the "Blue Seal" plan, Carlee would pay the first $2500 of bills, but only 20% of the rest. For what amount of medical bills will the "Blue Seal" plan save Carlee money? (Assume that her bills will exceed $2500.)

62. *Checking accounts.*   North Bank charges $10 per month for a student checking account. The first 8 checks are free, and each additional check costs $0.75. South Bank offers a student checking account with no monthly charge. The first 8 checks are free, and each additional check costs $3. For what numbers of checks is the South Bank plan more expensive? (Assume that the student will always write more than 8 checks.)

63. *Body fat percentage.*   The function given by

$$F(d) = (4.95/d - 4.50) \times 100$$

can be used to estimate the body fat percentage $F(d)$ of a person with an average body density d, in kilograms per liter.

a) A man is considered obese if his body fat percentage is at least 25%. Find the body densities of an obese man.

b) A woman is considered obese if her body fat percentage is at least 32%. Find the body densities of an obese woman.

64. *Temperature conversion.*   The function

$$C(F) = \frac{5}{9}(F - 32)$$

can be used to find the Celsius temperature $C(F)$ that corresponds to $F°$ Fahrenheit.

a) Gold is solid at Celsius temperatures less than 1063°C. Find the Fahrenheit temperatures for which gold is solid.

b) Silver is solid at Celsius temperatures less than 960.8°C. Find the Fahrenheit temperatures for which silver is solid.

65. *Manufacturing.*   Bright Ideas is planning to make a new kind of lamp. Fixed costs will be $90,000, and variable costs will be $25 for the production of each lamp. The total-cost function for x lamps is

$$C(x) = 90,000 + 25x.$$

The company makes $48 in revenue for each lamp sold. The total-revenue function for x lamps is

$$R(x) = 48x.$$

(See Section 8.8.)

a) When $R(x) < C(x)$, the company loses money. Find the values of x for which the company loses money.

b) When $R(x) > C(x)$, the company makes a profit. Find the values of x for which the company makes a profit.

66. *Publishing.*   The demand and supply functions for a locally produced poetry book are approximated by

$$D(p) = 2000 - 60p \quad \text{and}$$
$$S(p) = 460 + 94p,$$

where p is the price, in dollars (see Section 8.8).

a) Find those values of p for which demand exceeds supply.

b) Find those values of p for which demand is less than supply.

67. How is the solution of $x + 3 = 8$ related to the solution sets of

$$x + 3 > 8 \quad \text{and} \quad x + 3 < 8?$$

68. Can a negative number be in the domain of a radical function? Why or why not?

Skill Review

To prepare for Section 9.2, review finding domains of rational functions (Section 7.2).

Find the domain of f. [7.2]

69. $f(x) = \dfrac{5}{x}$

70. $f(x) = \dfrac{3}{x - 6}$

71. $f(x) = \dfrac{x - 2}{2x + 1}$

72. $f(x) = \dfrac{x + 3}{5x - 7}$

73. $f(x) = \dfrac{x + 10}{8}$

74. $f(x) = \dfrac{3}{x} + 5$

Synthesis

75. The percentage of the U.S. population that owns an HDTV cannot exceed 100%. How does this affect the answer to Exercise 58?

76. Explain how the addition principle can be used to avoid ever needing to multiply or divide both sides of an inequality by a negative number.

Solve for x and y. Assume that a, b, c, d, and m are positive constants.

77. $3ax + 2x \geq 5ax - 4$; assume $a > 1$

78. $6by - 4y \leq 7by + 10$

79. $a(by - 2) \geq b(2y + 5)$; assume $a > 2$

80. $c(6x - 4) < d(3 + 2x)$; assume $3c > d$

81. $c(2 - 5x) + dx > m(4 + 2x)$;
assume $5c + 2m < d$

82. $a(3 - 4x) + cx < d(5x + 2)$;
assume $c > 4a + 5d$

Determine whether each statement is true or false. If false, give an example that shows this.

83. For any real numbers a, b, c, and d, if $a < b$ and $c < d$, then $a - c < b - d$.

84. For all real numbers x and y, if $x < y$, then $x^2 < y^2$.

85. Are the inequalities

$$x < 3 \quad \text{and} \quad x + \frac{1}{x} < 3 + \frac{1}{x}$$

equivalent? Why or why not?

86. Are the inequalities

$$x < 3 \quad \text{and} \quad 0 \cdot x < 0 \cdot 3$$

equivalent? Why or why not?

Solve. Then graph.

87. $x + 5 \leq 5 + x$

88. $x + 8 < 3 + x$

89. $x^2 > 0$

90. $y_1 < y_2$

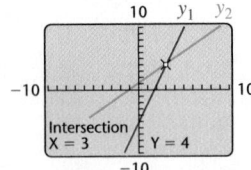

91. $y_1 \geq y_2$

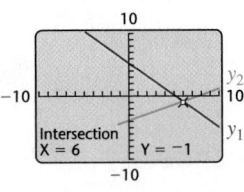

92. The graphs of $f(x) = 2x + 1$, $g(x) = -\frac{1}{2}x + 3$, and $h(x) = x - 1$ are as shown below. Solve each inequality, referring only to the figure.

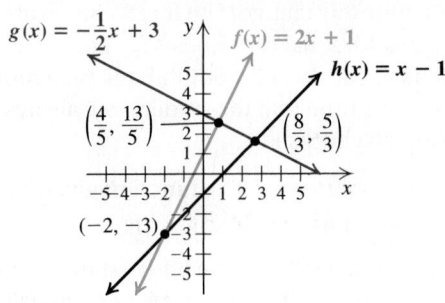

a) $2x + 1 \leq x - 1$
b) $x - 1 > -\frac{1}{2}x + 3$
c) $-\frac{1}{2}x + 3 < 2x + 1$

93. Assume that the graphs of $y_1 = -\frac{1}{2}x + 5$, $y_2 = x - 1$, and $y_3 = 2x - 3$ are as shown below. Solve each inequality, referring only to the figure.

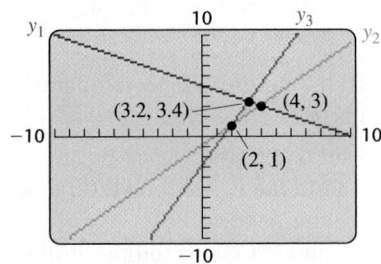

a) $-\frac{1}{2}x + 5 > x - 1$
b) $x - 1 \leq 2x - 3$
c) $2x - 3 \geq -\frac{1}{2}x + 5$

94. Using an approach similar to that in the Technology Connection on p. 580, use a graphing calculator to check your answers to Exercises 11, 37, and 40.

95. Use a graphing calculator to confirm the domains of the functions in Exercises 41, 43, and 47.

CORNER

Saving on Shipping Costs

Focus: Inequalities and problem solving
Time: 20–30 minutes
Group size: 2–3

For overnight delivery packages weighing up to 10 lb sent by Express Mail, the United States Postal Service charges (as of May 2008) $19.00 for up to one pound delivered to Zone 3 and, on average, $2.12 for each pound or part of a pound after the first. UPS Next Day Air charges $22.05 for a one-pound delivery to Zone 3 and each additional pound or part of a pound costs $1.45.*

ACTIVITY

1. One group member should determine the function p, where $p(x)$ represents the

cost, in dollars, of mailing x pounds using Express Mail.

2. One member should determine the function r, where $r(x)$ represents the cost, in dollars, of shipping x pounds using UPS Next Day Air.

3. A third member should graph p and r on the same set of axes.

4. Finally, working together, use the graph to determine those weights for which Express Mail is less expensive than UPS Next Day Air shipping. Express your answer in both set-builder notation and interval notation.

*This activity is based on an article by Michael Contino in *Mathematics Teacher*, May 1995.

9.2   Intersections, Unions, and Compound Inequalities

Intersections of Sets and Conjunctions of Sentences   ●   Unions of Sets and Disjunctions of Sentences   ●
Interval Notation and Domains

Two inequalities joined by the word "and" or the word "or" are called **compound inequalities**. Thus, "$x < -3 \; or \; x > 0$" and "$x < 5 \; and \; x > 3$" are two examples of compound inequalities. To discuss how to solve compound inequalities, we must first study ways in which sets can be combined.

Intersections of Sets and Conjunctions of Sentences

The **intersection** of two sets A and B is the set of all elements that are common to both A and B. We denote the intersection of sets A and B as

$$A \cap B.$$

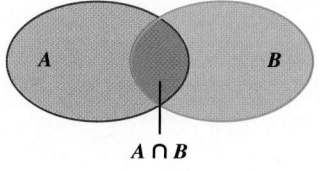

The intersection of two sets is represented by the purple region shown in the figure at left. For example, if $A = \{$all students who are taking a math class$\}$ and $B = \{$all students who are taking a history class$\}$, then $A \cap B = \{$all students who are taking a math class *and* a history class$\}$.

EXAMPLE **1** Find the intersection: $\{1, 2, 3, 4, 5\} \cap \{-2, -1, 0, 1, 2, 3\}$.

SOLUTION The numbers 1, 2, and 3 are common to both sets, so the intersection is $\{1, 2, 3\}$.

TRY EXERCISE 11

When two or more sentences are joined by the word *and* to make a compound sentence, the new sentence is called a **conjunction** of the sentences. The following is a conjunction of inequalities:

$$-2 < x \quad and \quad x < 1.$$

A number is a solution of a conjunction if it is a solution of *both* of the separate parts. For example, -1 is a solution because it is a solution of $-2 < x$ as well as $x < 1$; that is, -1 is *both* greater than -2 *and* less than 1.

The solution set of a conjunction is the intersection of the solution sets of the individual sentences.

EXAMPLE **2** Graph and write interval notation for the conjunction

$$-2 < x \quad and \quad x < 1.$$

SOLUTION We first graph $-2 < x$, then $x < 1$, and finally the conjunction $-2 < x$ *and* $x < 1$.

$\{x \mid -2 < x\}$

$(-2, \infty)$

$\{x \mid x < 1\}$

$(-\infty, 1)$

$\{x \mid -2 < x\} \cap \{x \mid x < 1\}$
$= \{x \mid -2 < x \text{ and } x < 1\}$

$(-2, 1)$

Because there are numbers that are both greater than -2 and less than 1, the solution set of the conjunction $-2 < x$ *and* $x < 1$ is the interval $(-2, 1)$. In set-builder notation, this is written $\{x \mid -2 < x < 1\}$, the set of all numbers that are *simultaneously* greater than -2 *and* less than 1.

TRY EXERCISE 33

For $a < b$,

$$a < x \quad and \quad x < b \quad \text{can be abbreviated} \quad a < x < b;$$

and, equivalently,

$$b > x \quad and \quad x > a \quad \text{can be abbreviated} \quad b > x > a.$$

Mathematical Use of the Word "and"

The word "and" corresponds to "intersection" and to the symbol "$\cap$". Any solution of a conjunction must make each part of the conjunction true.

EXAMPLE **3** Solve and graph: $-1 \le 2x + 5 < 13$.

SOLUTION This inequality is an abbreviation for the conjunction

$$-1 \le 2x + 5 \quad and \quad 2x + 5 < 13.$$

The word *and* corresponds to set *intersection*. To solve the conjunction, we solve each of the two inequalities separately and then find the intersection of the solution sets:

$$-1 \leq 2x + 5 \quad and \quad 2x + 5 < 13$$
$$-6 \leq 2x \qquad and \qquad 2x < 8 \qquad \text{Subtracting 5 from both sides of each inequality}$$
$$-3 \leq x \qquad and \qquad x < 4. \qquad \text{Dividing both sides of each inequality by 2}$$

The solution of the conjunction is the intersection of the two separate solution sets.

$\{x \mid -3 \leq x\}$ $[-3, \infty)$

$\{x \mid x < 4\}$ $(-\infty, 4)$

$\{x \mid -3 \leq x\} \cap \{x \mid x < 4\}$
$= \{x \mid -3 \leq x < 4\}$ $[-3, 4)$

We can abbreviate the answer as $-3 \leq x < 4$. The solution set is $\{x \mid -3 \leq x < 4\}$, or, in interval notation, $[-3, 4)$.

TRY EXERCISE 45

The steps in Example 3 are often combined as follows:

$$-1 \leq 2x + 5 < 13$$
$$-1 - 5 \leq 2x + 5 - 5 < 13 - 5 \qquad \text{Subtracting 5 from all three regions}$$
$$-6 \leq 2x < 8$$
$$-3 \leq x < 4. \qquad \text{Dividing by 2 in all three regions}$$

Such an approach saves some writing and will prove useful in Section 9.3.

CAUTION! The abbreviated form of a conjunction, like $-3 \leq x < 4$, can be written only if both inequality symbols point in the same direction. It is *not acceptable* to write a sentence like $-1 > x < 5$ since doing so does not indicate if *both* $-1 > x$ and $x < 5$ must be true or if it is enough for one of the separate inequalities to be true.

EXAMPLE 4

Solve and graph: $2x - 5 \geq -3$ *and* $5x + 2 \geq 17$.

SOLUTION We first solve each inequality, retaining the word *and*:

$$2x - 5 \geq -3 \quad and \quad 5x + 2 \geq 17$$
$$2x \geq 2 \quad and \quad 5x \geq 15$$
$$x \geq 1 \quad and \quad x \geq 3.$$

Keep the word "and."

Next, we find the intersection of the two separate solution sets.

$\{x \mid x \geq 1\}$ $[1, \infty)$

$\{x \mid x \geq 3\}$ $[3, \infty)$

$\{x \mid x \geq 1\} \cap \{x \mid x \geq 3\}$
$= \{x \mid x \geq 3\}$ $[3, \infty)$

The numbers common to both sets are those greater than or equal to 3. Thus the solution set is $\{x \mid x \geq 3\}$, or, in interval notation, $[3, \infty)$. You should check that any number in $[3, \infty)$ satisfies the conjunction whereas numbers outside $[3, \infty)$ do not.

TRY EXERCISE ▶ 67

Sometimes there is no way to solve both parts of a conjunction at once.

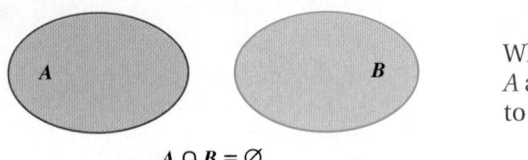

When $A \cap B = \emptyset$, A and B are said to be *disjoint*.

$$A \cap B = \emptyset$$

EXAMPLE 5

Solve and graph: $2x - 3 > 1 \; and \; 3x - 1 < 2$.

SOLUTION We solve each inequality separately:

$$2x - 3 > 1 \quad and \quad 3x - 1 < 2$$
$$2x > 4 \quad and \quad 3x < 3$$
$$x > 2 \quad and \quad x < 1.$$

The solution set is the intersection of the individual inequalities.

$\{x \mid x > 2\}$ $(2, \infty)$

$\{x \mid x < 1\}$ $(-\infty, 1)$

$\{x \mid x > 2\} \cap \{x \mid x < 1\}$
$= \{x \mid x > 2 \; and \; x < 1\} = \emptyset$ $\emptyset$

Since no number is both greater than 2 and less than 1, the solution set is the empty set, $\emptyset$.

TRY EXERCISE ▶ 69

Unions of Sets and Disjunctions of Sentences

The **union** of two sets A and B is the collection of elements belonging to A and/or B. We denote the union of A and B by

$$A \cup B.$$

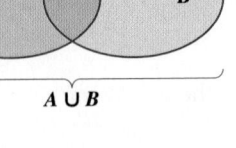

$A \cup B$

The union of two sets is often pictured as shown at left. For example, if $A = \{$all students who are taking a math class$\}$ and $B = \{$all students who are taking a history class$\}$, then $A \cup B = \{$all students who are taking a math class *or* a history class$\}$. Note that this set includes students who are taking a math class *and* a history class.

EXAMPLE 6

Find the union: $\{2, 3, 4\} \cup \{3, 5, 7\}$.

SOLUTION The numbers in either or both sets are 2, 3, 4, 5, and 7, so the union is $\{2, 3, 4, 5, 7\}$.

TRY EXERCISE ▶ 13

STUDENT NOTES ———

Remember that the union or the intersection of two sets is itself a set and should be written with set braces.

When two or more sentences are joined by the word *or* to make a compound sentence, the new sentence is called a **disjunction** of the sentences. Here is an example:

$$x < -3 \quad or \quad x > 3.$$

A number is a solution of a disjunction if it is a solution of at least one of the separate parts. For example, -5 is a solution of this disjunction since -5 is a solution of $x < -3$.

> *The solution set of a disjunction is the union of the solution sets of the individual sentences.*

EXAMPLE 7 Graph and write interval notation for the disjunction

$$x < -3 \quad or \quad x > 3.$$

SOLUTION We first graph $x < -3$, then $x > 3$, and finally the disjunction $x < -3$ *or* $x > 3$.

$\{x \mid x < -3\}$ $(-\infty, -3)$

$\{x \mid x > 3\}$ $(3, \infty)$

$\{x \mid x < -3\} \cup \{x \mid x > 3\}$
$= \{x \mid x < -3 \text{ or } x > 3\}$ $(-\infty, -3) \cup (3, \infty)$

The solution set of $x < -3$ *or* $x > 3$ is $\{x \mid x < -3 \text{ or } x > 3\}$, or, in interval notation, $(-\infty, -3) \cup (3, \infty)$. There is no simpler way to write the solution.

> TRY EXERCISE 27

Mathematical Use of the Word "or"

The word "or" corresponds to "union" and to the symbol "$\cup$". For a number to be a solution of a disjunction, it must be in *at least one* of the solution sets of the individual sentences.

EXAMPLE 8 Solve and graph: $7 + 2x < -1 \text{ or } 13 - 5x \le 3$.

SOLUTION We solve each inequality separately, retaining the word *or*:

$$7 + 2x < -1 \quad or \quad 13 - 5x \le 3$$
$$2x < -8 \quad or \quad -5x \le -10$$

Dividing by a negative and reversing the symbol

$$x < -4 \quad or \quad x \ge 2.$$

To find the solution set of the disjunction, we consider the individual graphs. We graph $x < -4$ and then $x \ge 2$. Then we take the union of the graphs.

$\{x \mid x < -4\}$ $(-\infty, -4)$

$\{x \mid x \ge 2\}$ $[2, \infty)$

$\{x \mid x < -4\} \cup \{x \mid x \ge 2\}$
$= \{x \mid x < -4 \text{ or } x \ge 2\}$ $(-\infty, -4) \cup [2, \infty)$

The solution set is $\{x \mid x < -4 \text{ or } x \ge 2\}$, or $(-\infty, -4) \cup [2, \infty)$.

> TRY EXERCISE 63

> *CAUTION!* A compound inequality like
>
> $$x < -4 \quad or \quad x \geq 2,$$
>
> as in Example 8, *cannot* be expressed as $2 \leq x < -4$ because to do so would be to say that x is *simultaneously* less than -4 and greater than or equal to 2. No number is both less than -4 *and* greater than 2, but many are less than -4 *or* greater than 2.

EXAMPLE **9**   Solve: $-2x - 5 < -2 \text{ or } x - 3 < -10$.

SOLUTION   We solve the individual inequalities, retaining the word *or*:

$$-2x - 5 < -2 \quad or \quad x - 3 < -10$$
$$-2x < 3 \quad or \quad x < -7$$

Dividing by a negative and reversing the symbol ———— Keep the word "or."

$$x > -\tfrac{3}{2} \quad or \quad x < -7.$$

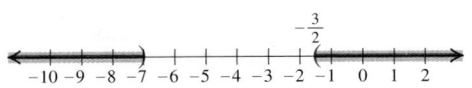

The solution set is $\left\{x \mid x < -7 \text{ or } x > -\tfrac{3}{2}\right\}$, or $(-\infty, -7) \cup \left(-\tfrac{3}{2}, \infty\right)$.

TRY EXERCISE 65

EXAMPLE **10**   Solve: $3x - 11 < 4 \text{ or } 4x + 9 \geq 1$.

SOLUTION   We solve the individual inequalities separately, retaining the word *or*:

$$3x - 11 < 4 \quad or \quad 4x + 9 \geq 1$$
$$3x < 15 \quad or \quad 4x \geq -8$$
$$x < 5 \quad or \quad x \geq -2.$$

Keep the word "or."

To find the solution set, we first look at the individual graphs.

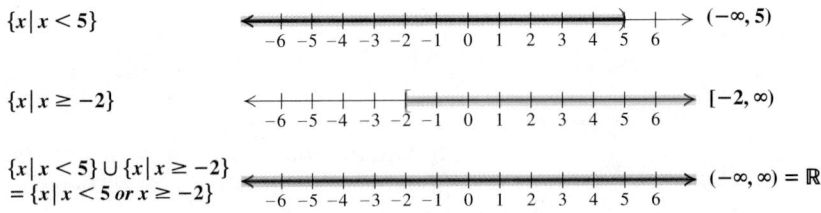

$\{x \mid x < 5\}$ $(-\infty, 5)$

$\{x \mid x \geq -2\}$ $[-2, \infty)$

$\{x \mid x < 5\} \cup \{x \mid x \geq -2\}$
$= \{x \mid x < 5 \text{ or } x \geq -2\}$ $(-\infty, \infty) = \mathbb{R}$

Since *all* numbers are less than 5 or greater than or equal to -2, the two sets fill the entire number line. Thus the solution set is $\mathbb{R}$, the set of all real numbers.

TRY EXERCISE 51

Interval Notation and Domains

In Section 7.2, we saw that if $g(x) = \dfrac{5x - 2}{x - 3}$, then the number 3 is not in the domain of g. We can represent the domain of g using set-builder notation or interval notation.

EXAMPLE 11 Use interval notation to write the domain of g if $g(x) = \dfrac{5x - 2}{x - 3}$.

SOLUTION The expression $\dfrac{5x - 2}{x - 3}$ is not defined when the denominator is 0. We set $x - 3$ equal to 0 and solve:

$$x - 3 = 0$$
$$x = 3. \qquad \text{The number 3 is } not \text{ in the domain.}$$

We have the domain of $g = \{x \,|\, x \text{ is a real number } and \ x \neq 3\}$. If we graph this set, we see that the domain can be written as a union of two intervals.

$$(-\infty, 3) \cup (3, \infty)$$

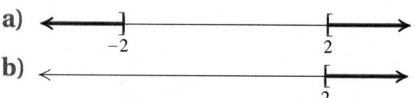

Thus the domain of $g = (-\infty, 3) \cup (3, \infty)$.

TRY EXERCISE 73

9.2 EXERCISE SET

For Extra Help
MyMathLab   MathXL   WATCH   DOWNLOAD
PRACTICE

🖐 **Concept Reinforcement** *In each of Exercises 1–10, match the set with the most appropriate choice from the column on the right.*

1. ____ $\{x \,|\, x < -2 \text{ or } x > 2\}$

2. ____ $\{x \,|\, x < -2 \text{ and } x > 2\}$

3. ____ $\{x \,|\, x > -2\} \cap \{x \,|\, x < 2\}$

4. ____ $\{x \,|\, x \leq -2\} \cup \{x \,|\, x \geq 2\}$

5. ____ $\{x \,|\, x \leq -2\} \cup \{x \,|\, x \leq 2\}$

6. ____ $\{x \,|\, x \leq -2\} \cap \{x \,|\, x \leq 2\}$

7. ____ $\{x \,|\, x \geq -2\} \cap \{x \,|\, x \geq 2\}$

8. ____ $\{x \,|\, x \geq -2\} \cup \{x \,|\, x \geq 2\}$

9. ____ $\{x \,|\, x \leq 2\} \text{ and } \{x \,|\, x \geq -2\}$

10. ____ $\{x \,|\, x \leq 2\} \text{ or } \{x \,|\, x \geq -2\}$

a) ◄────┤──────────┤────►
        −2   2

b) ◄──────────────┤────►
                 2

c) ◄───────┤──────────┤────►
       −2   2

d) ◄───────┤──────────────►
       −2

e) ◄───────────────┤────►
                 2

f) ◄───────)──────────)────►
       −2   2

g) ◄───────┤──────────────►
       −2

h) ◄───────)──────────(────►
       −2   2

i) $\mathbb{R}$

j) $\varnothing$

Find each indicated intersection or union.

11. $\{2, 4, 16\} \cap \{4, 16, 256\}$

12. $\{1, 2, 4\} \cup \{4, 6, 8\}$

13. $\{0, 5, 10, 15\} \cup \{5, 15, 20\}$

14. $\{2, 5, 9, 13\} \cap \{5, 8, 10\}$

15. $\{a, b, c, d, e, f\} \cap \{b, d, f\}$

16. $\{u, v, w\} \cup \{u, w\}$

17. $\{x, y, z\} \cup \{u, v, x, y, z\}$

18. $\{m, n, o, p\} \cap \{m, o, p\}$

19. $\{3, 6, 9, 12\} \cap \{5, 10, 15\}$

20. $\{1, 5, 9\} \cup \{4, 6, 8\}$

21. $\{1, 3, 5\} \cup \varnothing$

22. $\{1, 3, 5\} \cap \varnothing$

Graph and write interval notation for each compound inequality.

23. $1 < x < 3$

24. $0 \le y \le 5$

25. $-6 \le y \le 0$

26. $-8 < x \le -2$

27. $x \le -1 \, or \, x > 4$

28. $x < -5 \, or \, x > 1$

29. $x \le -2 \, or \, x > 1$

30. $x \le -5 \, or \, x > 2$

31. $-4 \le -x < 2$

32. $x > -7 \, and \, x < -2$

33. $x > -2 \, and \, x < 4$

34. $3 > -x \ge -1$

35. $5 > a \, or \, a > 7$

36. $t \ge 2 \, or \, -3 > t$

37. $x \ge 5 \, or \, -x \ge 4$

38. $-x < 3 \, or \, x < -6$

39. $7 > y \, and \, y \ge -3$

40. $6 > -x \ge 0$

41. $-x < 7 \, and \, -x \ge 0$

42. $x \ge -3 \, and \, x < 3$

Aha! **43.** $t < 2 \, or \, t < 5$

44. $t > 4 \, or \, t > -1$

Solve and graph each solution set.

45. $-3 \le x + 2 < 9$

46. $-1 < x - 3 < 5$

47. $0 < t - 4 \, and \, t - 1 \le 7$

48. $-6 \le t + 1 \, and \, t + 8 < 2$

49. $-7 \le 2a - 3 \, and \, 3a + 1 < 7$

50. $-4 \le 3n + 5 \, and \, 2n - 3 \le 7$

Aha! **51.** $x + 3 \le -1 \, or \, x + 3 > -2$

52. $x + 5 < -3 \, or \, x + 5 \ge 4$

53. $-10 \le 3x - 1 \le 5$

54. $-18 \le 4x + 2 \le 30$

55. $5 > \dfrac{x - 3}{4} > 1$

56. $3 \ge \dfrac{x - 1}{2} \ge -4$

57. $-2 \le \dfrac{x + 2}{-5} \le 6$

58. $-10 \le \dfrac{x + 6}{-3} \le -8$

59. $2 \le f(x) \le 8$, where $f(x) = 3x - 1$

60. $7 \ge g(x) \ge -2$, where $g(x) = 3x - 5$

61. $-21 \le f(x) < 0$, where $f(x) = -2x - 7$

62. $4 > g(t) \ge 2$, where $g(t) = -3t - 8$

63. $f(t) < 3 \, or \, f(t) > 8$, where $f(t) = 5t + 3$

64. $g(x) \le -2 \, or \, g(x) \ge 10$, where $g(x) = 3x - 5$

65. $6 > 2a - 1 \, or \, -4 \le -3a + 2$

66. $3a - 7 > -10 \, or \, 5a + 2 \le 22$

67. $a + 3 < -2 \, and \, 3a - 4 < 8$

68. $1 - a < -2 \, and \, 2a + 1 > 9$

69. $3x + 2 < 2 \, and \, 3 - x < 1$

70. $2x - 1 > 5 \, and \, 2 - 3x > 11$

71. $2t - 7 \le 5 \, or \, 5 - 2t > 3$

72. $5 - 3a \le 8 \, or \, 2a + 1 > 7$

For $f(x)$ as given, use interval notation to write the domain of f.

73. $f(x) = \dfrac{9}{x + 6}$

74. $f(x) = \dfrac{2}{x - 5}$

75. $f(x) = \dfrac{1}{x}$

76. $f(x) = -\dfrac{6}{x}$

77. $f(x) = \dfrac{x + 3}{2x - 8}$

78. $f(x) = \dfrac{x - 1}{3x + 6}$

79. Why can the conjunction $2 < x \, and \, x < 5$ be rewritten as $2 < x < 5$, but the disjunction $2 < x \, or \, x < 5$ cannot be rewritten as $2 < x < 5$?

80. Can the solution set of a disjunction be empty? Why or why not?

Skill Review

To prepare for Section 9.3, review graphing and solving equations by graphing (Sections 7.3 and 8.1).

Graph. [7.3]

81. $g(x) = 2x$

82. $f(x) = 4$

83. $g(x) = -3$

84. $f(x) = |x|$

Solve by graphing. [8.1]

85. $x + 4 = 3$

86. $x - 1 = -5$

Synthesis

87. What can you conclude about a, b, c, and d, if $[a, b] \cup [c, d] = [a, d]$? Why?

88. What can you conclude about a, b, c, and d, if $[a, b] \cap [c, d] = [a, b]$? Why?

89. Use the accompanying graph of $f(x) = 2x - 5$ to solve $-7 < 2x - 5 < 7$.

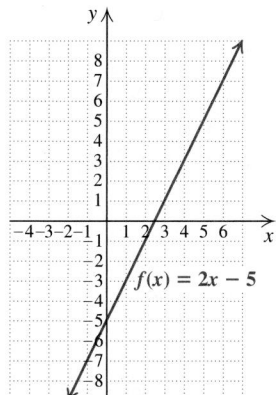

90. Use the accompanying graph of $g(x) = 4 - x$ to solve $4 - x < -2$ or $4 - x > 7$.

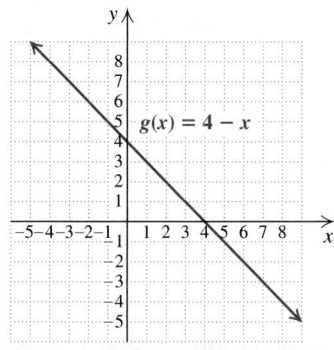

91. *Counseling.* The function given by

$$s(t) = 500t + 16,500$$

can be used to estimate the number of student visits to Cornell University's counseling center t years after 2000. For what years is the number of student visits between 18,000 and 21,000?
Source: Based on data from Cornell University

92. *Pressure at sea depth.* The function given by

$$P(d) = 1 + \frac{d}{33}$$

gives the pressure, in atmospheres (atm), at a depth of d feet in the sea. For what depths d is the pressure at least 1 atm and at most 7 atm?

93. *Converting dress sizes.* The function given by

$$f(x) = 2(x + 10)$$

can be used to convert dress sizes x in the United States to dress sizes $f(x)$ in Italy. For what dress sizes in the United States will dress sizes in Italy be between 32 and 46?

94. *Solid-waste generation.* The function given by

$$w(t) = 0.0125t + 4.525$$

can be used to estimate the number of pounds of solid waste, $w(t)$, produced daily, on average, by each person in the United States, t years after 2000. For what years will waste production range from 4.6 to 4.8 lb per person per day?

95. *Body fat percentage.* The function given by

$$F(d) = (4.95/d - 4.50) \times 100$$

can be used to estimate the body fat percentage $F(d)$ of a person with an average body density d, in kilograms per liter. A woman's body fat percentage is considered acceptable if $25 \le F(d) \le 31$. What body densities are considered acceptable for a woman?

96. *Temperatures of liquids.* The formula

$$C = \tfrac{5}{9}(F - 32)$$

can be used to convert Fahrenheit temperatures F to Celsius temperatures C.

a) Gold is liquid for Celsius temperatures C such that $1063° \le C < 2660°$. Find a comparable inequality for Fahrenheit temperatures.

b) Silver is liquid for Celsius temperatures C such that $960.8° \le C < 2180°$. Find a comparable inequality for Fahrenheit temperatures.

97. *Minimizing tolls.* A $6.00 toll is charged to cross the bridge to Sanibel Island from mainland Florida. A six-month reduced-fare pass costs $50 and reduces the toll to $2.00. A six-month unlimited-trip pass costs $300 and allows for free crossings. How many crossings in six months does it take for the reduced-fare pass to be the more economical choice?
Source: www.leewayinfo.com

Solve and graph.

98. $4a - 2 \le a + 1 \le 3a + 4$

99. $4m - 8 > 6m + 5$ *or* $5m - 8 < -2$

100. $x - 10 < 5x + 6 \leq x + 10$

101. $3x < 4 - 5x < 5 + 3x$

Determine whether each sentence is true or false for all real numbers a, b, and c.

102. If $-b < -a$, then $a < b$.

103. If $a \leq c$ and $c \leq b$, then $b > a$.

104. If $a < c$ and $b < c$, then $a < b$.

105. If $-a < c$ and $-c > b$, then $a > b$.

For f(x) as given, use interval notation to write the domain of f.

106. $f(x) = \dfrac{\sqrt{5 + 2x}}{x - 1}$ **107.** $f(x) = \dfrac{\sqrt{3 - 4x}}{x + 7}$

108. For $f(x) = \sqrt{x - 5}$ and $g(x) = \sqrt{9 - x}$, use interval notation to write the domain of $f + g$.

109. Let $y_1 = -1$, $y_2 = 2x + 5$, and $y_3 = 13$. Then use the graphs of y_1, y_2, and y_3 to check the solution to Example 3.

110. Let $y_1 = -2x - 5$, $y_2 = -2$, $y_3 = x - 3$, and $y_4 = -10$. Then use the graphs of y_1, y_2, y_3, and y_4 to check the solution to Example 9.

111. Use a graphing calculator to check your answers to Exercises 45–48 and Exercises 63–66.

112. On many graphing calculators, the TEST key provides access to inequality symbols, while the LOGIC option of that same key accesses the conjunction *and* and the disjunction *or*. Thus, if $y_1 = x > -2$ and $y_2 = x < 4$, Exercise 33 can be checked by forming the expression $y_3 = y_1$ *and* y_2. The interval(s) in the solution set appears as a horizontal line 1 unit above the x-axis. (Be careful to "deselect" y_1 and y_2 so that only y_3 is drawn.) Use the TEST key to check Exercises 35, 39, 41, and 43.

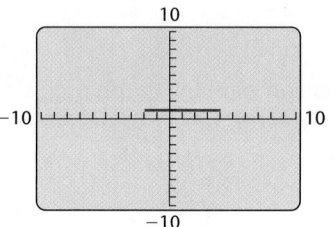

CORNER

Reduce, Reuse, and Recycle

Focus: Compound inequalities
Time: 15–20 minutes
Group size: 2

In the United States, the amount of solid waste (rubbish) being recovered is slowly catching up to the amount being generated. In 2002, each person generated, on average, 4.55 lb of solid waste every day, of which 1.34 lb was recovered. In 2006, each person generated, on average, 4.60 lb of solid waste, of which 1.50 lb was recovered.

Source: U.S. Environmental Protection Agency

ACTIVITY

Assume that the amount of solid waste being generated and the amount being recovered are both increasing linearly. One group member should find a linear function w for which $w(t)$ represents the number of pounds of waste generated per person per day t years after 2000. The other group member should find a linear function r for which $r(t)$ represents the number of pounds recovered per person per day t years after 2000. Finally, working together, the group should determine those years for which the amount recovered will be more than $\frac{1}{3}$ of but less than $\frac{1}{2}$ of the amount generated.

COLLABORATIVE

| 9.3 | **Absolute-Value Equations and Inequalities** |

Equations with Absolute Value   ▪   Inequalities with Absolute Value

Equations with Absolute Value

Recall from Section 1.4 the definition of absolute value.

> **Absolute Value**
>
> The absolute value of x, denoted $|x|$, is defined as
>
> $$|x| = \begin{cases} x, & \text{if } x \geq 0, \\ -x, & \text{if } x < 0. \end{cases}$$
>
> (When x is nonnegative, the absolute value of x is x. When x is negative, the absolute value of x is the opposite of x.)

To better understand this definition, suppose x is -5. Then $|x| = |-5| = 5$, and 5 is the opposite of -5. This shows that when x represents a negative number, the absolute value of x is the opposite of x (which is positive).

Since distance is always nonnegative, we can think of a number's absolute value as its distance from zero on the number line.

EXAMPLE 1   Find the solution set: **(a)** $|x| = 4$; **(b)** $|x| = 0$; **(c)** $|x| = -7$.

SOLUTION

a) We interpret $|x| = 4$ to mean that the number x is 4 units from zero on the number line. There are two such numbers, 4 and -4. Thus the solution set is $\{-4, 4\}$.

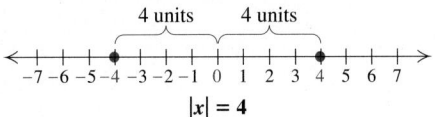

b) We interpret $|x| = 0$ to mean that x is 0 units from zero on the number line. The only number that satisfies this is 0 itself. Thus the solution set is $\{0\}$.

c) Since distance is always nonnegative, it doesn't make sense to talk about a number that is -7 units from zero. Remember: The absolute value of a number is nonnegative. Thus, $|x| = -7$ has no solution; the solution set is $\varnothing$.

> TRY EXERCISE 15

Example 1 leads us to the following principle for solving equations.

> **The Absolute-Value Principle for Equations**
>
> For any positive number p and any algebraic expression X:
>
> **a)** The solutions of $|X| = p$ are those numbers that satisfy
>
> $$X = -p \quad or \quad X = p.$$
>
> **b)** The equation $|X| = 0$ is equivalent to the equation $X = 0$.
>
> **c)** The equation $|X| = -p$ has no solution.

EXAMPLE **2** Find the solution set: **(a)** $|2x + 5| = 13$; **(b)** $|4 - 7x| = -8$.

SOLUTION

a) We use the absolute-value principle, knowing that $2x + 5$ must be either 13 or -13:

$$|X| = p$$
$$|2x + 5| = 13 \quad \text{Substituting}$$
$$2x + 5 = -13 \quad or \quad 2x + 5 = 13$$
$$2x = -18 \quad or \quad 2x = 8$$
$$x = -9 \quad or \quad x = 4.$$

Check: For -9:

$$\frac{|2x + 5| = 13}{\begin{array}{c|c} |2(-9) + 5| & 13 \\ |-18 + 5| & \\ |-13| & \\ & 13 \stackrel{?}{=} 13 \quad \text{TRUE} \end{array}}$$

For 4:

$$\frac{|2x + 5| = 13}{\begin{array}{c|c} |2 \cdot 4 + 5| & 13 \\ |8 + 5| & \\ |13| & \\ & 13 \stackrel{?}{=} 13 \quad \text{TRUE} \end{array}}$$

The number $2x + 5$ is 13 units from zero if x is replaced with -9 or 4. The solution set is $\{-9, 4\}$.

b) The absolute-value principle reminds us that absolute value is always nonnegative. The equation $|4 - 7x| = -8$ has no solution. The solution set is $\varnothing$.

> **TRY EXERCISE** 21

To use the absolute-value principle, we must be sure that the absolute-value expression is alone on one side of the equation.

EXAMPLE **3** Given that $f(x) = 2|x + 3| + 1$, find all x for which $f(x) = 15$.

SOLUTION Since we are looking for $f(x) = 15$, we substitute:

$$f(x) = 15$$
$$2|x + 3| + 1 = 15 \quad \text{Replacing } f(x) \text{ with } 2|x + 3| + 1$$
$$2|x + 3| = 14 \quad \text{Subtracting 1 from both sides}$$
$$|x + 3| = 7 \quad \text{Dividing both sides by 2}$$
$$x + 3 = -7 \quad or \quad x + 3 = 7 \quad \text{Using the absolute-value principle for equations}$$
$$x = -10 \quad or \quad x = 4.$$

The student should check that $f(-10) = f(4) = 15$. The solution set is $\{-10, 4\}$.

> **TRY EXERCISE** 43

EXAMPLE **4** Solve: $|x - 2| = 3$.

SOLUTION Because this equation is of the form $|a - b| = c$, it can be solved in two different ways.

CAUTION!   There are two solutions of $|x - 2| = 3$. Simply solving $x - 2 = 3$ will yield only one of those solutions.

Method 1.   We interpret $|x - 2| = 3$ as stating that the number $x - 2$ is 3 units from zero. Using the absolute-value principle, we replace X with $x - 2$ and p with 3:

$$|X| = p$$
$$|x - 2| = 3$$
$$x - 2 = -3 \quad or \quad x - 2 = 3 \qquad \text{Using the absolute-value principle}$$
$$x = -1 \quad or \qquad x = 5.$$

Method 2.   This approach is helpful in calculus. The expressions $|a - b|$ and $|b - a|$ can be used to represent the *distance between a and b* on the number line. For example, the distance between 7 and 8 is given by $|8 - 7|$ or $|7 - 8|$. From this viewpoint, the equation $|x - 2| = 3$ states that the distance between x and 2 is 3 units. We draw a number line and locate all numbers that are 3 units from 2.

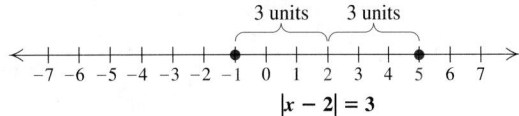

The solutions of $|x - 2| = 3$ are -1 and 5.

Check: The check consists of observing that both methods give the same solutions. The solution set is $\{-1, 5\}$.

TRY EXERCISE 25

Sometimes an equation has two absolute-value expressions. Consider $|a| = |b|$. This means that a and b are the same distance from zero.

If a and b are the same distance from zero, then either they are the same number or they are opposites.

EXAMPLE   **5**

Solve: $|2x - 3| = |x + 5|$.

SOLUTION   The given equation tells us that $2x - 3$ and $x + 5$ are the same distance from zero. This means that they are either the same number or opposites:

This assumes the two numbers are the same.   This assumes the two numbers are opposites.

$$2x - 3 = x + 5 \quad or \quad 2x - 3 = -(x + 5)$$
$$x - 3 = 5 \qquad or \quad 2x - 3 = -x - 5$$
$$x = 8 \qquad or \quad 3x - 3 = -5$$
$$3x = -2$$
$$x = -\tfrac{2}{3}.$$

The check is left to the student. The solutions are 8 and $-\frac{2}{3}$ and the solution set is $\left\{-\frac{2}{3}, 8\right\}$.

TRY EXERCISE 47

Inequalities with Absolute Value

Our methods for solving equations with absolute value can be adapted for solving inequalities. Inequalities of this sort arise regularly in more advanced courses.

EXAMPLE   **6**

Solve $|x| < 4$. Then graph.

SOLUTION   The solutions of $|x| < 4$ are all numbers whose *distance from zero is less than* 4. By substituting or by looking at the number line, we can see that numbers like $-3, -2, -1, -\frac{1}{2}, -\frac{1}{4}, 0, \frac{1}{4}, \frac{1}{2}, 1, 2,$ and 3 are all solutions. In fact, the solutions are all the numbers between -4 and 4. The solution set is $\{x | -4 < x < 4\}$, or, in interval notation, $(-4, 4)$. The graph is as follows:

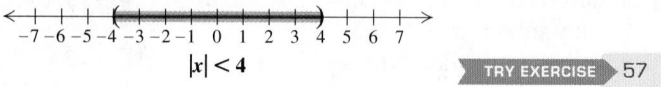

$|x| < 4$

> TRY EXERCISE 57

EXAMPLE 7

Solve $|x| \geq 4$. Then graph.

SOLUTION The solutions of $|x| \geq 4$ are all numbers that are at least 4 units from zero—in other words, those numbers x for which $x \leq -4$ or $4 \leq x$. The solution set is $\{x | x \leq -4 \text{ or } x \geq 4\}$. In interval notation, the solution set is $(-\infty, -4] \cup [4, \infty)$. We can check mentally with numbers like $-4.1, -5, 4.1,$ and 5. The graph is as follows:

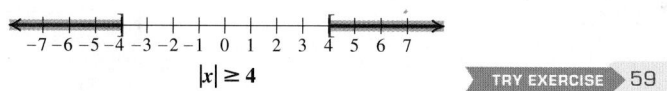

$|x| \geq 4$

> TRY EXERCISE 59

Examples 1, 6, and 7 illustrate three types of problems in which absolute-value symbols appear. The general principle for solving such problems follows.

Principles for Solving Absolute-Value Problems

For any positive number p and any expression X:

a) The solutions of $|X| = p$ are those numbers that satisfy

$$X = -p \quad \text{or} \quad X = p.$$

b) The solutions of $|X| < p$ are those numbers that satisfy

$$-p < X < p.$$

c) The solutions of $|X| > p$ are those numbers that satisfy

$$X < -p \quad \text{or} \quad p < X.$$

STUDENT NOTES

Simply ignoring the absolute-value symbol and solving the resulting equation or inequality will lead to only *part* of a solution.

$|X| = p$ corresponds to two equations.

$|X| < p$ corresponds to a conjunction.

$|X| > p$ corresponds to a disjunction.

The above principles are true for any positive number p.

If p is negative, any value of X will satisfy the inequality $|X| > p$ because absolute value is never negative. Thus, $|2x - 7| > -3$ is true for any real number x, and the solution set is $\mathbb{R}$.

If p is not positive, the inequality $|X| < p$ has no solution. Thus, $|2x - 7| < -3$ has no solution, and the solution set is $\varnothing$.

EXAMPLE **8**   Solve $|3x - 2| < 4$. Then graph.

SOLUTION   The number $3x - 2$ must be less than 4 units from zero. This is of the form $|X| < p$, so part (b) of the principles listed above applies:

$|X| < p$	This corresponds to $-p < X < p$.
$|3x - 2| < 4$	Replacing X with $3x - 2$ and p with 4
$-4 < 3x - 2 < 4$	The number $3x - 2$ must be within 4 units of zero.
$-2 < 3x < 6$	Adding 2
$-\frac{2}{3} < x < 2.$	Multiplying by $\frac{1}{3}$

The solution set is $\left\{ x \mid -\frac{2}{3} < x < 2 \right\}$. In interval notation, the solution is $\left(-\frac{2}{3}, 2 \right)$. The graph is as follows:

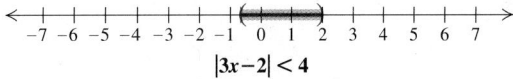

$|3x-2| < 4$

TRY EXERCISE   61

EXAMPLE **9**   Given that $f(x) = |4x + 2|$, find all x for which $f(x) \geq 6$.

SOLUTION   We have

$$f(x) \geq 6,$$

or $|4x + 2| \geq 6.$ Substituting

To solve, we use part (c) of the principles listed above. In this case, X is $4x + 2$ and p is 6:

$|X| \geq p$	This corresponds to $X < -p \text{ or } p < X$.
$|4x + 2| \geq 6$	Replacing X with $4x + 2$ and p with 6
$4x + 2 \leq -6 \;\; or \;\; 6 \leq 4x + 2$	The number $4x + 2$ must be at least 6 units from zero.
$4x \leq -8 \;\; or \;\; 4 \leq 4x$	Adding -2
$x \leq -2 \;\; or \;\; 1 \leq x.$	Multiplying by $\frac{1}{4}$

The solution set is $\{ x \mid x \leq -2 \text{ or } x \geq 1 \}$. In interval notation, the solution is $(-\infty, -2] \cup [1, \infty)$. The graph is as follows:

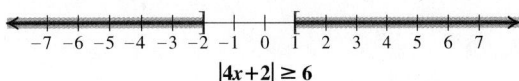

$|4x+2| \geq 6$

TRY EXERCISE   87

ALGEBRAIC–GRAPHICAL CONNECTION

We can visualize Examples 1(a), 6, and 7 by graphing $f(x) = |x|$ and $g(x) = 4$.

Solve: $|x| = 4$.

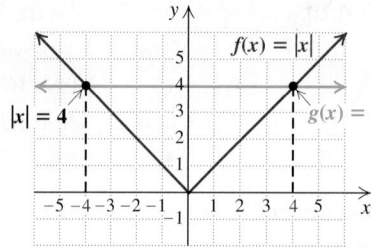

The graphs intersect at $(-4, 4)$ and $(4, 4)$.

$|x| = 4$ when $x = -4$ or $x = 4$.

The solution set of $|x| = 4$ is $\{-4, 4\}$.

Solve: $|x| < 4$.

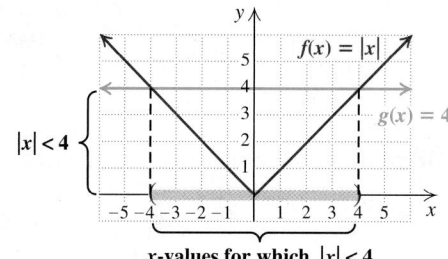

The graphs intersect at $(-4, 4)$ and $(4, 4)$.

$|x| < 4$ when $-4 < x < 4$.

The solution set of $|x| < 4$ is $(-4, 4)$.

Solve: $|x| \geq 4$.

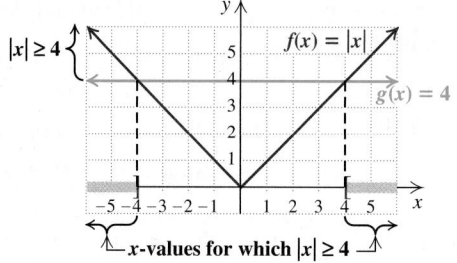

The graphs intersect at $(-4, 4)$ and $(4, 4)$.

$|x| \geq 4$ when $x \leq -4$ or $x \geq 4$.

The solution set of $|x| \geq 4$ is $(-\infty, -4] \cup [4, \infty)$.

TECHNOLOGY CONNECTION

To enter an absolute-value function on a graphing calculator, we press **MATH** and use the abs(option in the NUM menu. To solve $|4x + 2| = 6$, we graph $y_1 = \text{abs}(4x + 2)$ and $y_2 = 6$.

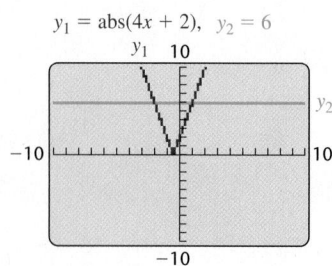

$y_1 = \text{abs}(4x + 2), \quad y_2 = 6$

Using the INTERSECT option of the CALC menu, we find that the graphs intersect at $(-2, 6)$ and $(1, 6)$. The x-coordinates -2 and 1 are the solutions. To solve $|4x + 2| \geq 6$, note where the graph of y_1 is *on or above* the line $y = 6$. The corresponding x-values are the solutions of the inequality.

1. How can the same graph be used to solve $|4x + 2| < 6$?
2. Solve Example 8.
3. Use a graphing calculator to show that $|4x + 2| = -6$ has no solution.

9.3 EXERCISE SET

Concept Reinforcement *Classify each statement as either true or false.*

1. $|x|$ is never negative.

2. $|x|$ is always positive.

3. If x is negative, then $|x| = -x$.

4. The number a is $|a|$ units from 0.

5. The distance between a and b can be expressed as $|a - b|$.

6. There are two solutions of $|3x - 8| = 17$.

7. There is no solution of $|4x + 9| > -5$.

8. All real numbers are solutions of $|2x - 7| < -3$.

Match each equation or inequality with an equivalent statement from the column on the right. Letters may be used more than once or not at all.

9. $|x - 3| = 5$ a) The solution set is $\varnothing$.

10. $|x - 3| < 5$ b) The solution set is $\mathbb{R}$.

11. $|x - 3| > 5$ c) $x - 3 > 5$

12. $|x - 3| < -5$ d) $x - 3 < -5 \text{ or } x - 3 > 5$

13. $|x - 3| = -5$ e) $x - 3 = 5$

14. $|x - 3| > -5$ f) $x - 3 < 5$

                             g) $x - 3 = -5 \text{ or } x - 3 = 5$

                             h) $-5 < x - 3 < 5$

Solve.

15. $|x| = 10$

16. $|x| = 5$

Aha! 17. $|x| = -1$

18. $|x| = -8$

19. $|p| = 0$

20. $|y| = 7.3$

21. $|2x - 3| = 4$

22. $|5x + 2| = 7$

23. $|3x + 5| = -8$

24. $|7x - 2| = -9$

25. $|x - 2| = 6$

26. $|x - 3| = 11$

27. $|x - 7| = 1$

28. $|x - 4| = 5$

29. $|t| + 1.1 = 6.6$

30. $|m| + 3 = 3$

31. $|5x| - 3 = 37$

32. $|2y| - 5 = 13$

33. $7|q| + 2 = 9$

34. $5|z| + 2 = 17$

35. $\left|\dfrac{2x - 1}{3}\right| = 4$

36. $\left|\dfrac{4 - 5x}{6}\right| = 3$

37. $|5 - m| + 9 = 16$

38. $|t - 7| + 1 = 4$

39. $5 - 2|3x - 4| = -5$

40. $3|2x - 5| - 7 = -1$

41. Let $f(x) = |2x + 6|$. Find all x for which $f(x) = 8$.

42. Let $f(x) = |2x - 4|$. Find all x for which $f(x) = 10$.

43. Let $f(x) = |x| - 3$. Find all x for which $f(x) = 5.7$.

44. Let $f(x) = |x| + 7$. Find all x for which $f(x) = 18$.

45. Let $f(x) = \left|\dfrac{1 - 2x}{5}\right|$. Find all x for which $f(x) = 2$.

46. Let $f(x) = \left|\dfrac{3x + 4}{3}\right|$. Find all x for which $f(x) = 1$.

Solve.

47. $|x - 7| = |2x + 1|$

48. $|3x + 2| = |x - 6|$

49. $|x + 4| = |x - 3|$

50. $|x - 9| = |x + 6|$

51. $|3a - 1| = |2a + 4|$

52. $|5t + 7| = |4t + 3|$

Aha! 53. $|n - 3| = |3 - n|$

54. $|y - 2| = |2 - y|$

55. $|7 - 4a| = |4a + 5|$

56. $|6 - 5t| = |5t + 8|$

Solve and graph.

57. $|a| \le 3$

58. $|x| < 5$

59. $|t| > 0$

60. $|t| \ge 1$

61. $|x - 1| < 4$

62. $|x - 1| < 3$

63. $|n + 2| \le 6$

64. $|a + 4| \le 0$

65. $|x - 3| + 2 > 7$

66. $|x - 4| + 5 > 10$

Aha! 67. $|2y - 9| > -5$

68. $|3y - 4| > -8$

69. $|3a + 4| + 2 \ge 8$

70. $|2a + 5| + 1 \ge 9$

71. $|y - 3| < 12$

72. $|p - 2| < 3$

73. $9 - |x + 4| \leq 5$

74. $12 - |x - 5| \leq 9$

75. $6 + |3 - 2x| > 10$

76. $|7 - 2y| < -8$

Aha! **77.** $|5 - 4x| < -6$

78. $7 + |4a - 5| \leq 26$

79. $\left|\dfrac{1 + 3x}{5}\right| > \dfrac{7}{8}$

80. $\left|\dfrac{2 - 5x}{4}\right| \geq \dfrac{2}{3}$

81. $|m + 3| + 8 \leq 14$

82. $|t - 7| + 3 \geq 4$

83. $25 - 2|a + 3| > 19$

84. $30 - 4|a + 2| > 12$

85. Let $f(x) = |2x - 3|$. Find all x for which $f(x) \leq 4$.

86. Let $f(x) = |5x + 2|$. Find all x for which $f(x) \leq 3$.

87. Let $f(x) = 5 + |3x - 4|$. Find all x for which $f(x) \geq 16$.

88. Let $f(x) = |2 - 9x|$. Find all x for which $f(x) \geq 25$.

89. Let $f(x) = 7 + |2x - 1|$. Find all x for which $f(x) < 16$.

90. Let $f(x) = 5 + |3x + 2|$. Find all x for which $f(x) < 19$.

91. Explain in your own words why -7 is not a solution of $|x| < 5$.

92. Explain in your own words why $[6, \infty)$ is only part of the solution of $|x| \geq 6$.

Skill Review

To prepare for Section 9.4, review graphing equations and solving systems of equations (Sections 3.3, 3.6, and 8.2).

Graph.

93. $3x - y = 6$ [3.3]

94. $y = \frac{1}{2}x - 1$ [3.6]

95. $x = -2$ [3.3]

96. $y = 4$ [3.3]

Solve using substitution or elimination. [8.2]

97. $x - 3y = 8,$
$2x + 3y = 4$

98. $x - 2y = 3,$
$x = y + 4$

99. $y = 1 - 5x,$
$2x - y = 4$

100. $3x - 2y = 4,$
$5x - 3y = 5$

Synthesis

101. Describe a procedure that could be used to solve any equation of the form $g(x) < c$ graphically.

102. Explain why the inequality $|x + 5| \geq 2$ can be interpreted as "the number x is at least 2 units from -5."

103. From the definition of absolute value, $|x| = x$ only when $x \geq 0$. Solve $|3t - 5| = 3t - 5$ using this same reasoning.

Solve.

104. $|3x - 5| = x$

105. $|x + 2| > x$

106. $2 \leq |x - 1| \leq 5$

107. $|5t - 3| = 2t + 4$

108. $t - 2 \leq |t - 3|$

Find an equivalent inequality with absolute value.

109. $-3 < x < 3$

110. $-5 \leq y \leq 5$

111. $x \leq -6 \, or \, 6 \leq x$

112. $x < -4 \, or \, 4 < x$

113. $x < -8 \, or \, 2 < x$

114. $-5 < x < 1$

115. x is less than 2 units from 7.

116. x is less than 1 unit from 5.

Write an absolute-value inequality for which the interval shown is the solution.

117.
$$\longleftarrow \overset{}{\underset{-7\,-6\,-5\,-4\,-3\,-2\,-1\ \ 0\ \ 1\ \ 2\ \ 3\ \ 4\ \ 5\ \ 6\ \ 7}{\vert\ \vert\ \vert\ \vert\ \vert\ \vert\ \vert\ \blacksquare\!\!=\!\!=\!\!=\!\!\blacksquare}} \longrightarrow$$

118.
$$\longleftarrow \overset{}{\underset{-5\,-4\,-3\,-2\,-1\ \ 0\ \ 1\ \ 2\ \ 3\ \ 4\ \ 5\ \ 6\ \ 7\ \ 8\ \ 9}{\vert\ \blacksquare\!\!=\!\!=\!\!=\!\!\blacksquare\ \vert\ \vert}} \longrightarrow$$

119.
$$\longleftarrow \overset{}{\underset{-7\,-6\,-5\,-4\,-3\,-2\,-1\ \ 0\ \ 1\ \ 2\ \ 3\ \ 4\ \ 5\ \ 6\ \ 7}{(\!=\!=\!=\!\blacksquare\ \vert\ \vert\ \vert}} \longrightarrow$$

120.
$$\longleftarrow \overset{}{\underset{0\ \ 1\ \ 2\ \ 3\ \ 4\ \ 5\ \ 6\ \ 7\ \ 8\ \ 9\ 10\,11\,12\,13\,14}{\vert\ \vert\ \blacksquare\!\!=\!\!=\!\!=\!\!\blacksquare\ \vert\ \vert}} \longrightarrow$$

121. *Bungee jumping.* A bungee jumper is bouncing up and down so that her distance d above a river satisfies the inequality $|d - 60 \text{ ft}| \leq 10 \text{ ft}$ (see the figure below). If the bridge from which she jumped is 150 ft above the river, how far is the bungee jumper from the bridge at any given time?

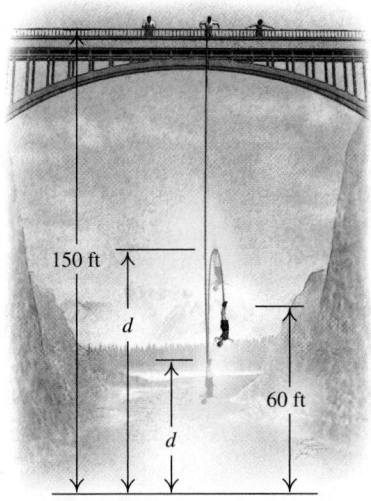

122. *Water level.* Depending on how dry or wet the weather has been, water in a well will rise and fall. The distance d, in feet, that a well's water level is below the ground satisfies the inequality $|d - 15| \leq 2.5$ (see the figure below).

a) Solve for d.

b) How tall a column of water is in the well at any given time?

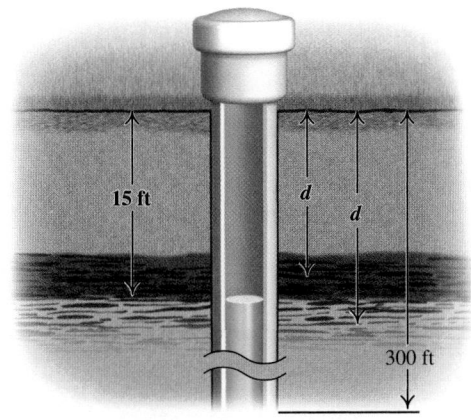

15 ft

d

d

300 ft

123. Use this graph of $f(x) = |2x - 6|$ to solve $|2x - 6| \leq 4$.

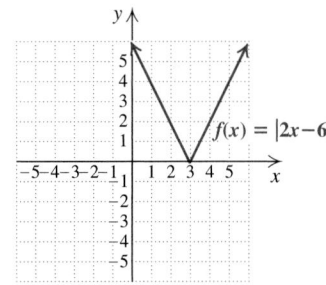

$f(x) = |2x-6|$

124. Is it possible for an equation in x of the form $|ax + b| = c$ to have exactly one solution? Why or why not?

125. Isabel is using the following graph to solve $|x - 3| < 4$. How can you tell that a mistake has been made in entering $y = \text{abs}(x - 3)$?

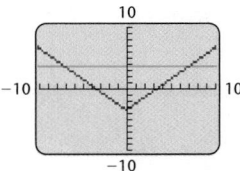

CONNECTING the CONCEPTS

In Chapters 1 and 4, we have learned to solve a variety of equations and inequalities. As we continue our study of algebra, we will learn to solve additional types of equations and inequalities, some of which will require new principles for solving. Following is a list of the principles we have used so far to solve equations and inequalities. Unless otherwise stated, a, b, and c can represent any real number.

The Addition and Multiplication Principles for Equations

$a = b$ is equivalent to $a + c = b + c$.

$a = b$ is equivalent to $ac = bc$, provided $c \neq 0$.

The Addition Principle for Inequalities

$a < b$ is equivalent to $a + c < b + c$.

$a > b$ is equivalent to $a + c > b + c$.

The Multiplication Principle for Inequalities

For any *positive* number c,

$a < b$ is equivalent to $ac < bc$;

$a > b$ is equivalent to $ac > bc$.

For any *negative* number c,

$a < b$ is equivalent to $ac > bc$;

$a > b$ is equivalent to $ac < bc$.

The Absolute-Value Principles for Equations and Inequalities

For any positive number p and any algebraic expression X:

The solutions of $|X| = p$ are those numbers that satisfy

$$X = -p \ \text{ or } \ X = p.$$

The solutions of $|X| < p$ are those numbers that satisfy

$$-p < X < p.$$

The solutions of $|X| > p$ are those numbers that satisfy

$$X < -p \ \text{ or } \ p < X.$$

MIXED REVIEW

Solve.

1. $2x + 3 = 7$

2. $3x - 1 > 8$

3. $3(t - 5) = 4 - (t + 1)$

4. $|2x + 1| = 7$

5. $-x \leq 6$

6. $5|t| < 20$

7. $2(3n + 6) - n = 4 - 3(n + 1)$

8. $3(2a + 9) = 5(3a - 7) - 6a$

9. $2 + |3x| = 10$

10. $|x - 3| \leq 10$

11. $\frac{1}{2}x - 7 = \frac{3}{4} + \frac{1}{4}x$

12. $|t| < 0$

13. $|2x + 5| + 1 \geq 13$

14. $2(x - 3) - x = 5x + 7 - 4x$

15. $|m + 6| - 8 < 10$

16. $\left|\dfrac{x + 2}{5}\right| = 8$

17. $4 - |7 - t| \leq 1$

18. $0.3x + 0.7 = 0.5x$

19. $8 - 5|a + 6| > 3$

20. $|5x + 7| + 9 \geq 4$

9.4   Inequalities in Two Variables

Graphs of Linear Inequalities ■ Systems of Linear Inequalities

In Section 2.6, we graphed inequalities in one variable on a number line. Now we graph inequalities in two variables on a plane.

Graphs of Linear Inequalities

When the equals sign in a linear equation is replaced with an inequality sign, a **linear inequality** is formed. Solutions of linear inequalities are ordered pairs.

EXAMPLE   **1**

Determine whether $(-3, 2)$ and $(6, -7)$ are solutions of $5x - 4y > 13$.

SOLUTION   Below, on the left, we replace x with -3 and y with 2. On the right, we replace x with 6 and y with -7.

$$\begin{array}{r|l} \multicolumn{2}{l}{5x - 4y > 13} \\ \hline 5(-3) - 4 \cdot 2 & 13 \\ -15 - 8 & \\ -23 \overset{?}{>} 13 & \text{FALSE} \end{array} \qquad \begin{array}{r|l} \multicolumn{2}{l}{5x - 4y > 13} \\ \hline 5(6) - 4(-7) & 13 \\ 30 + 28 & \\ 58 \overset{?}{>} 13 & \text{TRUE} \end{array}$$

Since $-23 > 13$ is false, $(-3, 2)$ *is not* a solution.

Since $58 > 13$ is true, $(6, -7)$ *is* a solution.

TRY EXERCISE   7

STUDENT NOTES

Pay careful attention to the inequality symbol when determining whether an ordered pair is a solution of an inequality. Writing the symbol at the end of the check, as in Example 1, will help you compare the numbers correctly.

The graph of a linear equation is a straight line. The graph of a linear inequality is a **half-plane**, with a **boundary** that is a straight line. To find the equation of the boundary, we replace the inequality sign with an equals sign.

EXAMPLE 2

Graph: $y \leq x$.

SOLUTION   We first graph the equation of the boundary, $y = x$. Every solution of $y = x$ is an ordered pair, like $(3, 3)$, in which both coordinates are the same. The graph of $y = x$ is shown on the left below. Since the inequality symbol is $\leq$, the line is drawn solid and is part of the graph of $y \leq x$.

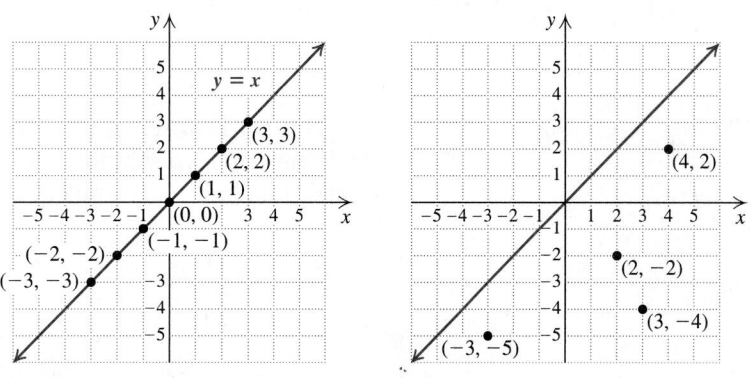

Note that in the graph on the right each ordered pair on the half-plane below $y = x$ contains a y-coordinate that is less than the x-coordinate. All these pairs represent solutions of $y \leq x$. We check one pair, $(4, 2)$, as follows:

$$\frac{y \leq x}{2 \overset{?}{\leq} 4} \quad \text{TRUE}$$

It turns out that *any* point on the same side of $y = x$ as $(4, 2)$ is also a solution. Thus, if one point in a half-plane is a solution, then *all* points in that half-plane are solutions.

We finish drawing the solution set by shading the half-plane below $y = x$. The complete solution set consists of the shaded half-plane as well as the boundary line itself.

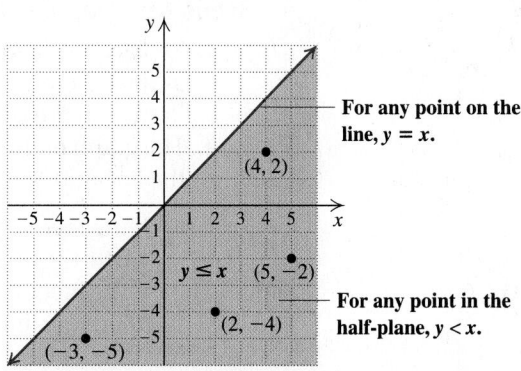

For any point on the line, $y = x$.

For any point in the half-plane, $y < x$.

TRY EXERCISE ▸ 11

From Example 2, we see that for any inequality of the form $y \leq f(x)$ or $y < f(x)$, we shade *below* the graph of $y = f(x)$.

EXAMPLE **3**   Graph: $8x + 3y > 24$.

SOLUTION   First, we sketch the graph of $8x + 3y = 24$. Since the inequality sign is $>$, points on this line do not represent solutions of the inequality, and the line is drawn dashed. Points representing solutions of $8x + 3y > 24$ are in either the half-plane above the line or the half-plane below the line. To determine which, we select a point that is not on the line and check whether it is a solution of $8x + 3y > 24$. Let's use $(1, 1)$ as this *test point*:

$$
\begin{array}{c|c}
\multicolumn{2}{c}{8x + 3y > 24} \\
\hline
8(1) + 3(1) & 24 \\
8 + 3 & \\
11 \overset{?}{>} 24 & \text{FALSE}
\end{array}
$$

Since $11 > 24$ is *false*, $(1, 1)$ is not a solution. Thus no point in the half-plane containing $(1, 1)$ is a solution. The points in the other half-plane *are* solutions, so we shade that half-plane and obtain the graph shown below.

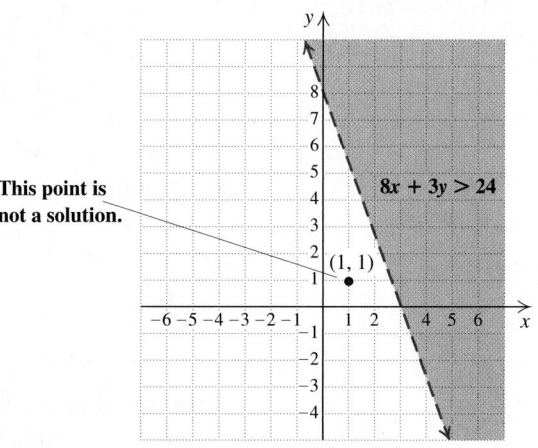

This point is not a solution. $(1, 1)$ $8x + 3y > 24$

> **TRY EXERCISE**   19

Steps for Graphing Linear Inequalities

1. Replace the inequality sign with an equals sign and graph this line as the boundary. If the inequality symbol is $<$ or $>$, draw the line dashed. If the symbol is $\le$ or $\ge$, draw the line solid.

2. The graph of the inequality consists of a half-plane on one side of the line and, if the line is solid, the line as well.

   a) If the inequality is of the form $y < mx + b$ or $y \le mx + b$, shade *below* the line.
      If the inequality is of the form $y > mx + b$ or $y \ge mx + b$, shade *above* the line.

   b) If y is not isolated, either solve for y and graph as in part (a) or simply graph the boundary and use a test point not on the line (as in Example 3). If the test point *is* a solution, shade the half-plane containing the point. If it is not a solution, shade the other half-plane.

EXAMPLE **4**   Graph: $6x - 2y < 12$.

SOLUTION   We could graph $6x - 2y = 12$ and use a test point, as in Example 3. Instead, let's solve $6x - 2y < 12$ for y:

$$6x - 2y < 12$$
$$-2y < -6x + 12 \qquad \text{Adding } -6x \text{ to both sides}$$
$$y > 3x - 6. \qquad \text{Dividing both sides by } -2 \text{ and reversing the } < \text{ symbol}$$

The graph consists of the half plane above the dashed boundary line $y = 3x - 6$ (see the graph below).

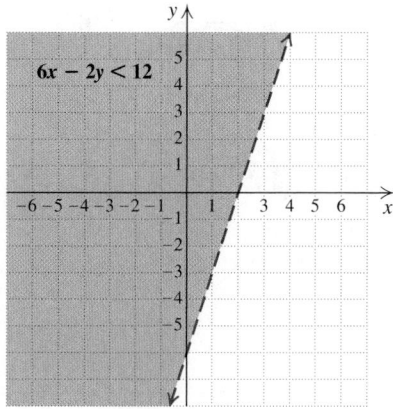

> **TRY EXERCISE** 21

EXAMPLE **5**   Graph $x > -3$ on a plane.

SOLUTION   There is only one variable in this inequality. If we graph the inequality on a line, its graph is as follows:

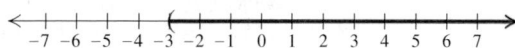

However, we can also write this inequality as $x + 0y > -3$ and graph it on a plane. We can use the same technique as in the examples above. First, we graph the boundary $x = -3$ in the plane, using a dashed line. Then we test some point, say, $(2, 5)$:

$$\frac{x + 0y > -3}{2 + 0 \cdot 5 \mid -3}$$
$$2 \overset{?}{>} -3 \quad \text{TRUE}$$

Since $(2, 5)$ is a solution, all points in the half-plane containing $(2, 5)$ are solutions. We shade that half-plane. Another approach is to simply note that the solutions of $x > -3$ are all pairs with first coordinates greater than -3.

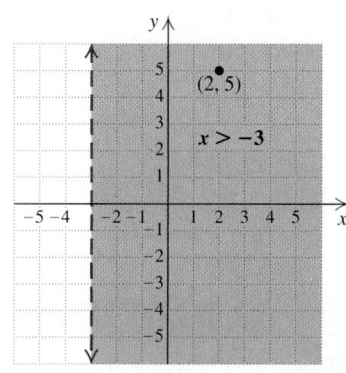

> **TRY EXERCISE** 25

TECHNOLOGY CONNECTION

On most graphing calculators, an inequality like $y < \frac{6}{5}x + 3.49$ can be drawn by entering $(6/5)x + 3.49$ as y_1, moving the cursor to the GraphStyle icon just to the left of y_1, pressing **ENTER** until ◤ appears, and then pressing **GRAPH**.

Many calculators have an INEQUALZ program that is accessed using the **APPS** key. Running this program allows us to write inequalities at the **Y=** screen by pressing **ALPHA** and then one of the five keys just below the screen.

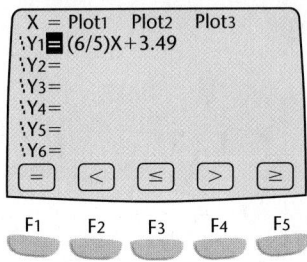

Although the graphs should be identical regardless of the method used, when we are using INEQUALZ, the boundary line appears dashed when < or > is selected.

$y_1 < (6/5)x + 3.49$, or
◤ $y_1 = (6/5)x + 3.49$

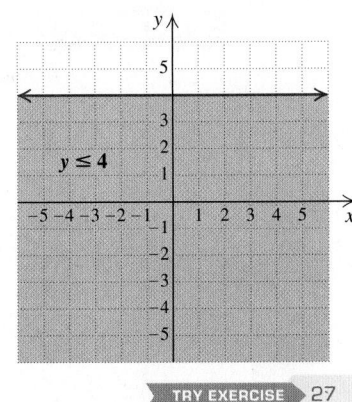

Graph each of the following. Solve for y first if necessary.

1. $y > x + 3.5$ **2.** $7y \le 2x + 5$
3. $8x - 2y < 11$ **4.** $11x + 13y + 4 \ge 0$

EXAMPLE **6**

Graph $y \le 4$ on a plane.

SOLUTION   The inequality is of the form $y \le mx + b$ (with $m = 0$), so we shade below the solid horizontal line representing $y = 4$.

This inequality can also be graphed by drawing $y = 4$ and testing a point above or below the line. The student should check that this results in a graph identical to the one at right.

TRY EXERCISE 27

Systems of Linear Inequalities

To graph a system of equations, we graph the individual equations and then find the intersection of the graphs. We do the same thing for a system of inequalities: We graph each inequality and find the intersection of the graphs.

EXAMPLE **7**

Graph the system

$$x + y \le 4,$$
$$x - y < 4.$$

SOLUTION   To graph $x + y \leq 4$, we graph $x + y = 4$ using a solid line. Since the test point $(0, 0)$ *is* a solution and $(0, 0)$ is below the line, we shade the half-plane below the graph red. The arrows near the ends of the line are another way of indicating the half-plane containing solutions.

Next, we graph $x - y < 4$. We graph $x - y = 4$ using a dashed line and consider $(0, 0)$ as a test point. Again, $(0, 0)$ is a solution, so we shade that side of the line blue. The solution set of the system is the region that is shaded purple (both red and blue) and part of the line $x + y = 4$.

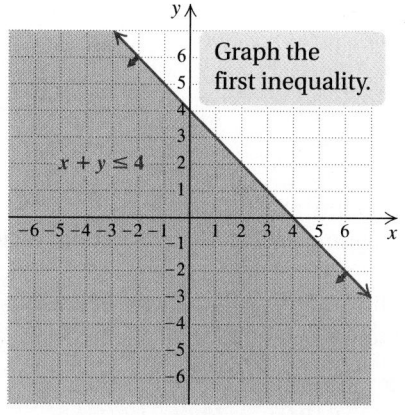
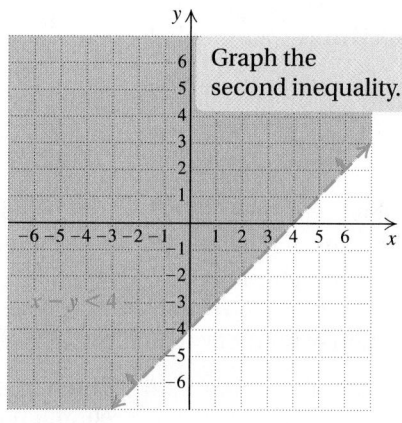
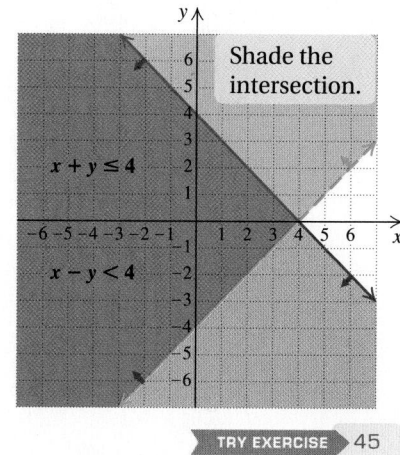

TRY EXERCISE 45

EXAMPLE   8

Graph: $-2 < x \leq 3$.

STUDENT NOTES ——————

If you don't use differently colored pencils or pens to shade different regions, consider using a pencil to make slashes that tilt in different directions in each region. You may also find it useful to attach arrows to the lines, as in the graphs shown.

SOLUTION   This is a system of inequalities:

$$-2 < x,$$
$$x \leq 3.$$

We graph the equation $-2 = x$, and see that the graph of the first inequality is the half-plane to the right of the boundary $-2 = x$. It is shaded red.

We graph the second inequality, starting with the boundary line $x = 3$. The inequality's graph is the line and the half-plane to its left. It is shaded blue.

The solution set of the system is the region that is the intersection of the individual graphs. Since it is shaded both blue and red, it appears to be purple. All points in this region have x-coordinates that are greater than -2 but do not exceed 3.

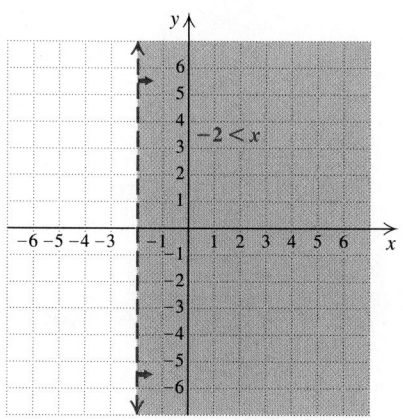
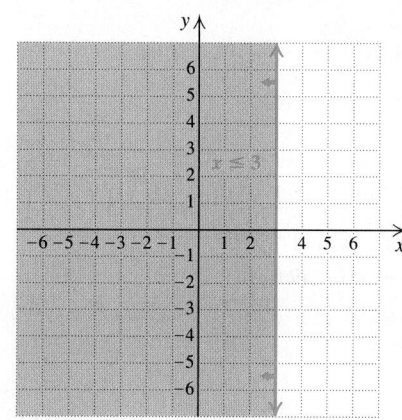
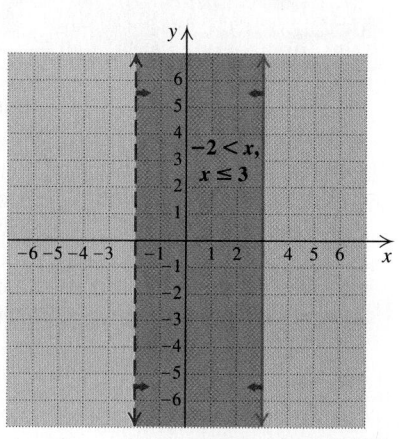

TRY EXERCISE 31

A system of inequalities may have a graph that consists of a polygon and its interior. In Section 9.5, we will have use for the corners, or *vertices* (singular, *vertex*), of such a graph.

EXAMPLE 9 Graph the system of inequalities. Find the coordinates of any vertices formed.

$$6x - 2y \leq 12, \quad (1)$$
$$y - 3 \leq 0, \quad (2)$$
$$x + y \geq 0 \quad (3)$$

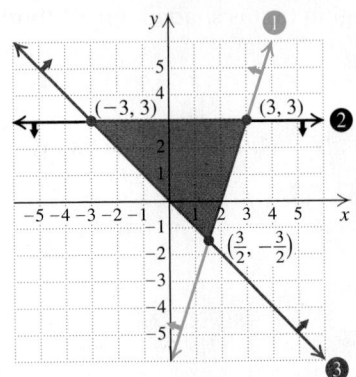

SOLUTION We graph the boundaries

$$6x - 2y = 12,$$
$$y - 3 = 0,$$
and $$x + y = 0$$

using solid lines. The regions for each inequality are indicated by the arrows near the ends of the lines. We note where the regions overlap and shade the region of solutions purple.

To find the vertices, we solve three different systems of two equations. The system of boundary equations from inequalities (1) and (2) is

$$6x - 2y = 12,$$ The student can use graphing, substitution, or
$$y - 3 = 0.$$ elimination to solve these systems.

Solving, we obtain the vertex $(3, 3)$.

The system of boundary equations from inequalities (1) and (3) is

$$6x - 2y = 12,$$
$$x + y = 0.$$

Solving, we obtain the vertex $\left(\frac{3}{2}, -\frac{3}{2}\right)$.

The system of boundary equations from inequalities (2) and (3) is

$$y - 3 = 0,$$
$$x + y = 0.$$

Solving, we obtain the vertex $(-3, 3)$.   **TRY EXERCISE** ▶ 49

TECHNOLOGY CONNECTION

Systems of inequalities can be graphed by solving for y and then graphing each inequality as in the Technology Connection on p. 610. To graph systems directly using the INEQUALZ application, enter the correct inequalities, press GRAPH, and then press ALPHA and Shades (F1 or F2). At the SHADES menu, select Ineq Intersection to see the final graph. To find the vertices, or points of intersection, select PoI-Trace from the graph menu.

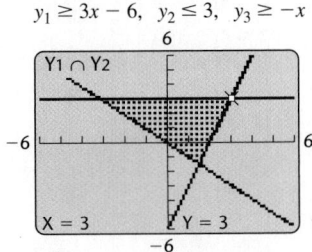

$y_1 \geq 3x - 6, \quad y_2 \leq 3, \quad y_3 \geq -x$

1. Use a graphing calculator to check the solution of Example 7.

Visualizing for Success

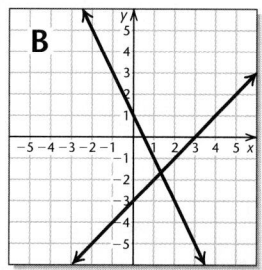

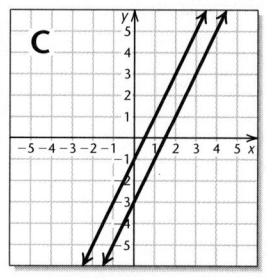

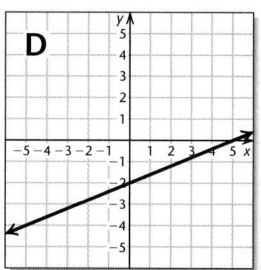

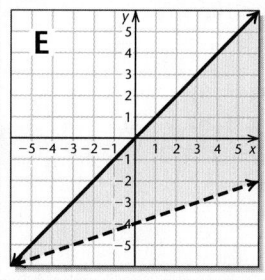

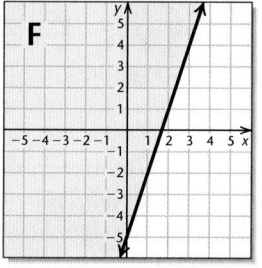

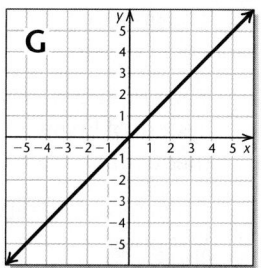

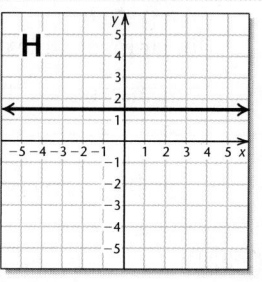

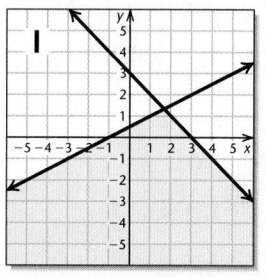

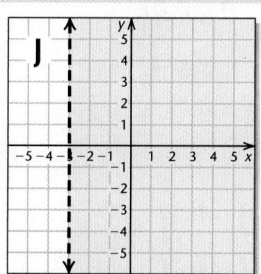

Match each equation, inequality, or system of equations or inequalities with its graph.

1. $x - y = 3,$
    $2x + y = 1$

2. $3x - y \leq 5$

3. $x > -3$

4. $y = \dfrac{1}{3}x - 4$

5. $y > \dfrac{1}{3}x - 4,$
    $y \leq x$

6. $x = y$

7. $y = 2x - 1,$
    $y = 2x - 3$

8. $2x - 5y = 10$

9. $x + y \leq 3,$
    $2y \leq x + 1$

10. $y = \dfrac{3}{2}$

Answers on page A-37

9.4 EXERCISE SET

Concept Reinforcement *In each of Exercises 1–6, match the phrase with the most appropriate choice from the column on the right.*

1. _____ A solution of a linear inequality

2. _____ The graph of a linear inequality

3. _____ The graph of a system of linear inequalities

4. _____ Often a convenient test point

5. _____ The name for the corners of a graph of a system of linear inequalities

6. _____ A dashed line

a) $(0, 0)$

b) Vertices

c) A half-plane

d) The intersection of two or more half-planes

e) An ordered pair that satisfies the inequality

f) Indicates the line is not part of the solution

Determine whether each ordered pair is a solution of the given inequality.

7. $(-2, 3)$; $2x - y > -4$

8. $(1, -6)$; $3x + y \geq -3$

9. $(5, 8)$; $3y - 5x \leq 0$

10. $(6, 20)$; $5y - 8x < 40$

Graph on a plane.

11. $y \geq \frac{1}{2}x$

12. $y \leq 3x$

13. $y > x - 3$

14. $y < x + 3$

15. $y \leq x + 2$

16. $y \geq x - 5$

17. $x - y \leq 4$

18. $x + y < 4$

19. $2x + 3y < 6$

20. $3x + 4y \leq 12$

21. $2y - x \leq 4$

22. $2y - 3x > 6$

23. $2x - 2y \geq 8 + 2y$

24. $3x - 2 \leq 5x + y$

25. $x > -2$

26. $x \geq 3$

27. $y \leq 6$

28. $y < -1$

29. $-2 < y < 7$

30. $-4 < y < -1$

31. $-5 \leq x < 4$

32. $-2 < y \leq 1$

33. $0 \leq y \leq 3$

34. $0 \leq x \leq 6$

Graph each system.

35. $y > x$,
$y < -x + 3$

36. $y < x$,
$y > -x + 1$

37. $y \leq x$,
$y \leq 2x - 5$

38. $y \geq x$,
$y \leq -x + 4$

39. $y \leq -3$,
$x \geq -1$

40. $y \geq -3$,
$x \geq 1$

41. $x > -4$,
$y < -2x + 3$

42. $x < 3$,
$y > -3x + 2$

43. $y \leq 5$,
$y \geq -x + 4$

44. $y \geq -2$,
$y \geq x + 3$

45. $x + y \leq 6$,
$x - y \leq 4$

46. $x + y < 1$,
$x - y < 2$

47. $y + 3x > 0$,
$y + 3x < 2$

48. $y - 2x \geq 1$,
$y - 2x \leq 3$

Graph each system of inequalities. Find the coordinates of any vertices formed.

49. $y \leq 2x - 3$,
$y \geq -2x + 1$,
$x \leq 5$

50. $2y - x \leq 2$,
$y - 3x \geq -4$,
$y \geq -1$

51. $x + 2y \leq 12$,
$2x + y \leq 12$,
$x \geq 0$,
$y \geq 0$

52. $x - y \leq 2$,
$x + 2y \geq 8$,
$y \leq 4$

53. $8x + 5y \leq 40$,
$x + 2y \leq 8$,
$x \geq 0$,
$y \geq 0$

54. $4y - 3x \geq -12$,
$4y + 3x \geq -36$,
$y \leq 0$,
$x \leq 0$

55. $y - x \geq 2$,
$y - x \leq 4$,
$2 \leq x \leq 5$

56. $3x + 4y \geq 12$,
$5x + 6y \leq 30$,
$1 \leq x \leq 3$

57. Explain in your own words why the boundary line is drawn dashed for the symbols < and > and why it is drawn solid for the symbols ≤ and ≥.

58. When graphing linear inequalities, Ron makes a habit of always shading above the line when the symbol ≥ is used. Is this wise? Why or why not?

Skill Review

To prepare for Section 9.5, review solving applications using the five-step problem-solving strategy (Sections 2.5 and 8.3).

Solve.

59. *Interest rate.* What rate of interest is required in order for a principal of $1560 to earn $25.35 in half a year? [2.5]

60. *Interest.* Luke invested $5000 in two accounts. He put $2200 in an account paying 4% simple interest and the rest in an account paying 5% simple interest. How much interest did he earn in one year from both accounts? [2.5]

61. *Investments.* Gina invested $10,000 in two accounts, one paying 3% simple interest and one paying 5% simple interest. After one year, she had earned $428 from both accounts. How much did she invest in each? [8.3]

62. *Catering.* Janice provided 20 lb of fresh vegetables for a reception. Carrots were $1.50 per pound and broccoli was $2.50 per pound. If she spent $38, how much of each vegetable did she buy? [8.3]

63. *Admissions.* There were 170 tickets sold for a high school basketball game. Tickets were $1 each for students and $3 each for adults. The total amount of money collected was $386. How many of each type of ticket were sold? [8.3]

64. *Agriculture.* Josh planted 400 acres in corn and soybeans. He planted 80 more acres in corn than he did in soybeans. How many acres of each did he plant? [8.3]

Synthesis

65. Explain how a system of linear inequalities could have a solution set containing exactly one pair.

66. In Example 7 on pp. 610–611, is the point (4, 0) part of the solution set? Why or why not?

Graph.

67. $x + y > 8,$
$x + y \leq -2$

68. $x + y \geq 1,$
$-x + y \geq 2,$
$x \geq -2,$
$y \geq 2,$
$y \leq 4,$
$x \leq 2$

69. $x - 2y \leq 0,$
$-2x + y \leq 2,$
$x \leq 2,$
$y \leq 2,$
$x + y \leq 4$

70. Write four systems of four inequalities that describe a 2-unit by 2-unit square that has (0, 0) as one of the vertices.

71. *Luggage size.* Unless an additional fee is paid, most major airlines will not check any luggage for which the sum of the item's length, width, and height exceeds 62 in. The U.S. Postal Service will ship a package only if the sum of the package's length and girth (distance around its midsection) does not exceed 130 in. Video Promotions is ordering several 30-in. long cases that will be both mailed and checked as luggage. Using w and h for width and height (in inches), respectively, write and graph an inequality that represents all acceptable combinations of width and height.

Sources: U.S. Postal Service; www.case2go.com

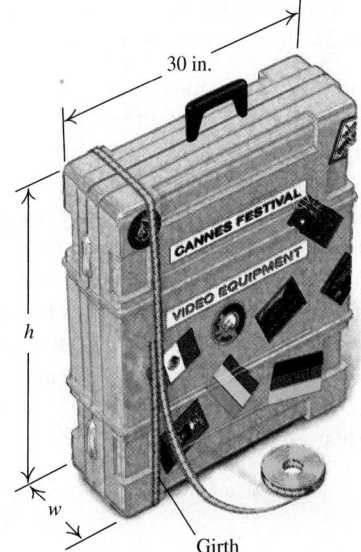

72. *Hockey wins and losses.* The Skating Stars believe they need at least 60 points for the season in order to make the playoffs. A win is worth 2 points and a tie is worth 1 point. Graph a system of inequalities that describes the situation. (*Hint*: Let w = the number of wins and t = the number of ties.)

73. *Graduate-school admissions.* Students entering the Master of Science program in Computer Science and Engineering at University of Texas Arlington must meet minimum score requirements on the Graduate Records Examination (GRE). The GRE Quantitative score must be at least 700 and the GRE Verbal score must be at least 400. The sum of the GRE Quantitative and Verbal scores must be at least 1150. Both scores have a maximum of 800. Using q for the quantitative score and v for the verbal score, write and graph a system of inequalities that represents all combinations that meet the requirements for entrance into the program.
Source: University of Texas Arlington

74. *Widths of a basketball floor.* Sizes of basketball floors vary due to building sizes and other constraints such as cost. The length L is to be at most 94 ft and the width W is to be at most 50 ft. Graph a system of inequalities that describes the possible dimensions of a basketball floor.

75. *Elevators.* Many elevators have a capacity of 1 metric ton (1000 kg). Suppose that c children, each weighing 35 kg, and a adults, each 75 kg, are on an elevator. Graph a system of inequalities that indicates when the elevator is overloaded.

76. *Age of marriage.* The following rule of thumb for determining an appropriate difference in age between a bride and a groom appears in many Internet blogs: *The younger spouse's age should be at least seven more than half the age of the older spouse.* Let b = the age of the bride, in years, and g = the age of the groom, in years. Write and graph a system of inequalities that represents all combinations of ages that follow this rule of thumb. Should a minimum or maximum age for marriage exist? How would the graph of the system of inequalities change with such a requirement?

77. *Waterfalls.* In order for a waterfall to be classified as a classical waterfall, its height must be less than twice its crest width, and its crest width cannot exceed one-and-a-half times its height. The tallest waterfall in the world is about 3200 ft high. Let h represent a waterfall's height, in feet, and w the crest width, in feet. Write and graph a system of inequalities that represents all possible combinations of heights and crest widths of classical waterfalls.

78. Use a graphing calculator to check your answers to Exercises 35–48. Then use INTERSECT to determine any point(s) of intersection.

79. Use a graphing calculator to graph each inequality.
a) $3x + 6y > 2$
b) $x - 5y \leq 10$
c) $13x - 25y + 10 \leq 0$
d) $2x + 5y > 0$

CONNECTING ↕ the CONCEPTS

We have now solved a variety of equations, inequalities, systems of equations, and systems of inequalities. Below is a list of the different types of problems we have solved, illustrations of each type, and descriptions of the solutions. Note that a solution set may be empty.

Type	Example	Solution	Graph
Linear equations in one variable	$2x - 8 = 3(x + 5)$	A number	
Linear inequalities in one variable	$-3x + 5 > 2$	A set of numbers; an interval	
Linear equations in two variables	$2x + y = 7$	A set of ordered pairs; a line	
Linear inequalities in two variables	$x + y \geq 4$	A set of ordered pairs; a half-plane	
System of equations in two variables	$x + y = 3,$ $5x - y = -27$	An ordered pair or a set of ordered pairs	
System of inequalities in two variables	$6x - 2y \leq 12,$ $y - 3 \leq 0,$ $x + y \geq 0$	A set of ordered pairs; a region of a plane	

MIXED REVIEW

Graph each solution on a number line.

1. $x + 2 = 7$

2. $x + 2 > 7$

3. $x + 2 \leq 7$

4. $3(x - 7) - 2 = 5 - (2 - x)$

5. $6 - 2x \geq 8$

6. $7 > 5 - x$

Graph on a plane.

7. $x + y = 2$

8. $x + y < 2$

9. $x + y \geq 2$

10. $y = 3x - 3$

11. $x = 4$

12. $2x - 5y = -10$

13. $x + y = 1,$
    $x - y = 1$

14. $y \geq 1 - x,$
    $y \leq x - 3,$
    $y \leq 2$

15. $2x + y < 6$

16. $x > 6y - 6$

17. $4x = 3y$

18. $y = 2x - 3,$
    $y = -\frac{1}{2}x + 1$

19. $x - y \leq 3,$
    $y \geq 2x,$
    $2y - x \leq 2$

20. $3y = 8$

9.5 Applications Using Linear Programming

Objective Functions and Constraints • Linear Programming

There are many real-world situations in which we need to find a greatest value (a maximum) or a least value (a minimum). For example, most businesses like to know how to make the *most* profit with the *least* expense possible. Some such problems can be solved using systems of inequalities.

Objective Functions and Constraints

Often a quantity we wish to maximize depends on two or more other quantities. For example, a gardener's profits P might depend on the number of shrubs s and the number of trees t that are planted. If the gardener makes a $10 profit from each shrub and an $18 profit from each tree, the total profit, in dollars, is given by the **objective function**

$$P = 10s + 18t.$$

Thus the gardener might be tempted to simply plant lots of trees since they yield the greater profit. This would be a good idea were it not for the fact that the number of trees and shrubs planted—and thus the total profit—is subject to the demands, or **constraints**, of the situation. For example, to improve drainage, the gardener might be required to plant at least 3 shrubs. Thus the objective function would be subject to the *constraint*

$$s \geq 3.$$

Because of limited space, the gardener might also be required to plant no more than 10 plants. This would subject the objective function to a *second* constraint:

$$s + t \leq 10.$$

Finally, the gardener might be told to spend no more than $700 on the plants. If the shrubs cost $40 each and the trees cost $100 each, the objective function is subject to a *third* constraint:

The cost of the shrubs   plus   the cost of the trees   cannot exceed $700.

$$40s \qquad + \qquad 100t \qquad \leq \qquad 700$$

In short, the gardener wishes to maximize the objective function

$$P = 10s + 18t,$$

subject to the constraints

$$s \geq 3,$$
$$s + t \leq 10,$$
$$40s + 100t \leq 700,$$
$$s \geq 0,$$
$$t \geq 0.$$

Because the number of trees and shrubs cannot be negative

These constraints form a system of linear inequalities that can be graphed.

Linear Programming

The gardener's problem is "How many shrubs and trees should be planted, subject to the constraints listed, in order to maximize profit?" To solve such a problem, we use a result from a branch of mathematics known as **linear programming**.

The Corner Principle

Suppose an objective function $F = ax + by + c$ depends on x and y (with a, b, and c constant). Suppose also that F is subject to constraints on x and y, which form a system of linear inequalities. If F has a minimum or a maximum value, then it can be found as follows:

1. Graph the system of inequalities and find the vertices.
2. Find the value of the objective function at each vertex. The greatest and the least of those values are the maximum and the minimum of the function, respectively.
3. The ordered pair at which the maximum or minimum occurs indicates the choice of (x, y) for which that maximum or minimum occurs.

This result was proven during World War II, when linear programming was developed to help allocate troops and supplies bound for Europe.

EXAMPLE 1   Solve the gardener's problem discussed above.

SOLUTION   We are asked to maximize $P = 10s + 18t$, subject to the constraints

$$s \geq 3,$$
$$s + t \leq 10,$$
$$40s + 100t \leq 700,$$
$$s \geq 0,$$
$$t \geq 0.$$

We graph the system, using the techniques of Section 9.4. The portion of the graph that is shaded represents all pairs that satisfy the constraints. It is sometimes called the *feasible region*.

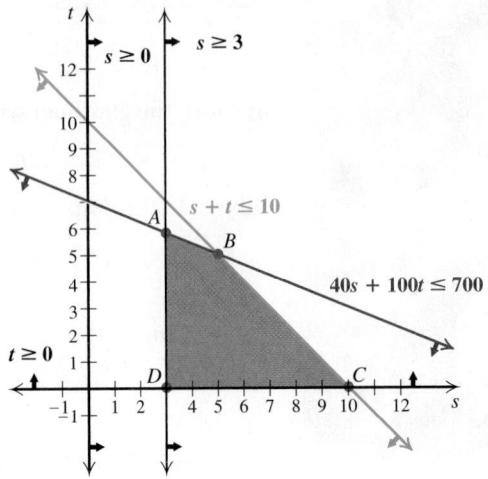

According to the corner principle, P is maximized at one of the vertices of the shaded region. To determine the coordinates of the vertices, we solve the following systems:

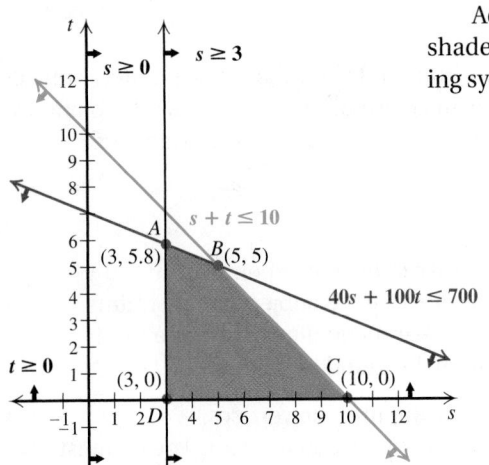

A: $\left.\begin{array}{r} 40s + 100t = 700, \\ s = 3; \end{array}\right\}$
The student can verify that the solution of this system is $(3, 5.8)$. The coordinates of point A are $(3, 5.8)$.

B: $\left.\begin{array}{r} s + t = 10, \\ 40s + 100t = 700; \end{array}\right\}$
The student can verify that the solution of this system is $(5, 5)$. The coordinates of point B are $(5, 5)$.

C: $\left.\begin{array}{r} s + t = 10, \\ t = 0; \end{array}\right\}$
The solution of this system is $(10, 0)$. The coordinates of point C are $(10, 0)$.

D: $\left.\begin{array}{r} t = 0, \\ s = 3. \end{array}\right\}$
The solution of this system is $(3, 0)$. The coordinates of point D are $(3, 0)$.

We now find the value of P at each vertex.

Vertex (s, t)	Profit $P = 10s + 18t$	
A $(3, 5.8)$	$10(3) + 18(5.8) = 134.4$	
B $(5, 5)$	$10(5) + 18(5) = 140$	⟵ Maximum
C $(10, 0)$	$10(10) + 18(0) = 100$	
D $(3, 0)$	$10(3) + 18(0) = 30$	⟵ Minimum

The greatest value of P occurs at $(5, 5)$. Thus profit is maximized at $140 if the gardener plants 5 shrubs and 5 trees. Incidentally, we have also shown that profit is minimized at $30 if 3 shrubs and 0 trees are planted.

▶ TRY EXERCISE ▶ 13

EXAMPLE 2   *Grading.*   For his history grade, Cy can write book summaries for 70 points each or research papers for 80 points each. He estimates that each book summary will take 9 hr and each research paper will take 15 hr and that he will have at most 120 hr to spend. He may turn in a total of no more than 12 summaries or papers. How many of each should he write in order to receive the highest score?

SOLUTION

1. **Familiarize.** Since we are looking for the number of book summaries and the number of research papers, we let b = the number of book summaries and r = the number of research papers. Cy is limited by the number of hours he can spend and by the number of summaries and papers he can turn in. These two limits are the constraints.

2. **Translate.** We organize the information in a table.

Type	Number of Points for Each	Time Required for Each	Number Written	Total Time for Each Type	Total Points for Each Type
Book summary	70	9 hr	b	$9b$	$70b$
Research paper	80	15 hr	r	$15r$	$80r$
Total			$b + r \leq 12$	$9b + 15r \leq 120$	

↑ Because no more than 12 may be turned in

↑ Because the time cannot exceed 120 hr

↑ We wish to maximize the total score.

STUDENT NOTES

It is very important that you clearly label what each variable represents. It is also important to clearly label what the function is that is being maximized or minimized and how that function is evaluated.

Let T represent the total score. We see from the table that

$$T = 70b + 80r.$$

We wish to maximize T subject to the number and time constraints:

$$b + r \leq 12,$$
$$9b + 15r \leq 120,$$
$$b \geq 0, \Big\}$$
$$r \geq 0. \Big\}$$ We include this because the number of summaries and papers cannot be negative.

3. **Carry out.** We graph the system and evaluate T at each vertex. The graph is as follows:

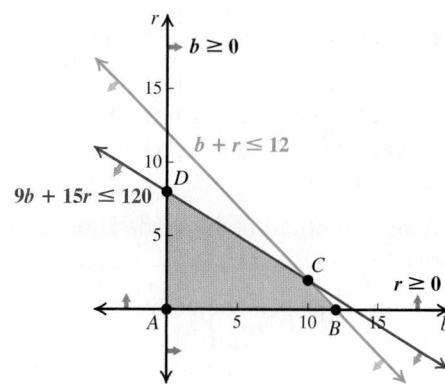

We find the coordinates of each vertex by solving a system of two linear equations. The coordinates of point A are obviously $(0, 0)$. To find the coordinates of point C, we solve the system

$$b + r = 12, \qquad (1)$$
$$9b + 15r = 120. \qquad (2)$$

We multiply both sides of equation (1) by -9 and add:

$$
\begin{array}{r}
-9b - 9r = -108 \\
9b + 15r = 120 \\
\hline
6r = 12 \\
r = 2.
\end{array}
$$

Substituting, we find that $b = 10$. Thus the coordinates of C are $(10, 2)$. Point B is the intersection of $b + r = 12$ and $r = 0$, so B is $(12, 0)$. Point D is the intersection of $9b + 15r = 120$ and $b = 0$, so D is $(0, 8)$. Computing the score for each ordered pair, we obtain the table at left.

The greatest score in the table is 860, obtained when Cy writes 10 book summaries and 2 research papers.

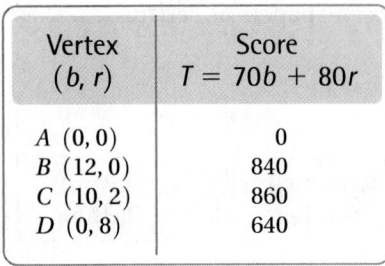

Vertex (b, r)	Score $T = 70b + 80r$
A $(0, 0)$	0
B $(12, 0)$	840
C $(10, 2)$	860
D $(0, 8)$	640

4. **Check.** We can check that $T \le 860$ for any other pair in the shaded region. This is left to the student.

5. **State.** In order to maximize his score, Cy should write 10 book summaries and 2 research papers.

▶ **TRY EXERCISE** 19

9.5 EXERCISE SET

↪ *Concept Reinforcement Complete each of the following sentences.*

1. In linear programming, the quantity we wish to maximize or minimize is given by the _____ function.

2. In linear programming, the demands arising from the given situation are known as _____.

3. To solve a linear programming problem, we make use of the _____ principle.

4. The shaded portion of a graph that represents all points that satisfy a problem's constraints is known as the _____ region.

5. In linear programming, the corners of the shaded portion of the graph are referred to as _____.

6. If it exists, the maximum value of an objective function occurs at a _____ of the feasible region.

Find the maximum and the minimum values of each objective function and the values of x and y at which they occur.

7. $F = 2x + 14y$,
   subject to
   $5x + 3y \le 34$,
   $3x + 5y \le 30$,
   $x \ge 0$,
   $y \ge 0$

8. $G = 7x + 8y$,
   subject to
   $3x + 2y \le 12$,
   $2y - x \le 4$,
   $x \ge 0$,
   $y \ge 0$

9. $P = 8x - y + 20$,
   subject to
   $6x + 8y \le 48$,
   $0 \le y \le 4$,
   $0 \le x \le 7$

10. $Q = 24x - 3y + 52$,
    subject to
    $5x + 4y \le 20$,
    $0 \le y \le 4$,
    $0 \le x \le 3$

11. $F = 2y - 3x$,
    subject to
    $y \le 2x + 1$,
    $y \ge -2x + 3$,
    $x \le 3$

12. $G = 5x + 2y + 4$,
    subject to
    $y \le 2x + 1$,
    $y \ge -x + 3$,
    $x \le 5$

13. *Lunch-time profits.* Art sells gumbo and sandwiches. To stay in business, Art must sell at least 10 orders of gumbo and 30 sandwiches each day. Because of limited space, no more than 40 orders of gumbo or 70 sandwiches can be made. The total number of orders cannot exceed 90. If profit is $1.65 per gumbo order and $1.05 per sandwich, how many of each item should Art sell in order to maximize profit?

14. *Gas mileage.* Caroline owns a car and a moped. She has at most 12 gal of gasoline to be used between the car and the moped. The car's tank holds at most 18 gal and the moped's 3 gal. The mileage for the car is 20 mpg and for the moped is 100 mpg. How many gallons of gasoline should each vehicle use if Caroline wants to travel as far as possible? What is the maximum number of miles?

15. *Photo albums.* Photo Perfect prints pages of photographs for albums. A page containing 4 photos costs $3 and a page containing 6 photos costs $5. Ann can spend no more than $90 for photo pages of her recent vacation, and can use no more than 20 pages in her album. What combination of 4-photo pages and 6-photo pages will maximize the number of photos she can display? What is the maximum number of photos that she can display?

16. *Milling.* Picture Rocks Lumber can convert logs into either lumber or plywood. In a given week, the mill can turn out 400 units of production, of which 100 units of lumber and 150 units of plywood are required by regular customers. The profit on a unit of lumber is $20 and on a unit of plywood is $30. How many units of each type should the mill produce in order to maximize profit?

Aha! **17.** *Investing.* Rosa is planning to invest up to $40,000 in corporate or municipal bonds, or both. She must invest from $6000 to $22,000 in corporate bonds, and she refuses to invest more than $30,000 in municipal bonds. The interest on corporate bonds is 8% and on municipal bonds is $7\frac{1}{2}$%. This is simple interest for one year. How much should Rosa invest in each type of bond in order to earn the most interest? What is the maximum interest?

18. *Investing.* Jamaal is planning to invest up to $22,000 in City Bank or the Southwick Credit Union, or both. He wants to invest at least $2000 but no more than $14,000 in City Bank. Because of insurance limitations, he will invest no more than $15,000 in the Southwick Credit Union. The interest in City Bank is 6% and in the credit union is $6\frac{1}{2}$%. This is simple interest for one year. How much should Jamaal invest in each bank in order to earn the most interest? What is the maximum interest?

19. *Test scores.* Corinna is taking a test in which short-answer questions are worth 10 points each and essay questions are worth 15 points each. She estimates that it will take 3 min to answer each short-answer question and 6 min to answer each essay question. The total time allowed is 60 min, and no more than 16 questions can be answered. Assuming that all her answers are correct, how many questions of each type should Corinna answer to get the best score?

20. *Test scores.* Edy is about to take a test that contains short-answer questions worth 4 points each and word problems worth 7 points each. Edy must do at least 5 short-answer questions, but time restricts doing more than 10. She must do at least 3 word problems, but time restricts doing more than 10. Edy can do no more than 18 questions in total. How many of each type of question must Edy do in order to maximize her score? What is this maximum score?

21. *Grape growing.* Auggie's vineyard consists of 240 acres upon which he wishes to plant Merlot and Cabernet grapes. Profit per acre of Merlot is $400 and profit per acre of Cabernet is $300. Furthermore, the total number of hours of labor available during the harvest season is 3200. Each acre of Merlot requires 20 hr of labor and each acre of Cabernet requires 10 hr of labor. Determine how the land should be divided between Merlot and Cabernet in order to maximize profit.

22. *Coffee blending.* The Coffee Peddler has 1440 lb of Sumatran coffee and 700 lb of Kona coffee. A batch of Hawaiian Blend requires 8 lb of Kona and 12 lb of Sumatran, and yields a profit of $90. A batch of Classic Blend requires 4 lb of Kona and 16 lb of Sumatran, and yields a $55 profit. How many batches of each kind should be made in order to maximize profit? What is the maximum profit? (*Hint*: Organize the information in a table.)

23. *Nutrition.* Becca is supposed to have at least 15 mg but no more than 45 mg of iron each day. She should also have at least 1500 mg but no more than 2500 mg of calcium per day. One serving of goat cheese contains 1 mg of iron, 500 mg of calcium, and 264 calories. One serving of hazelnuts contains 5 mg of iron, 100 mg of calcium, and 628 calories. How many servings of goat cheese and how many servings of hazelnuts should Becca eat in order to meet the daily requirements of iron and calcium but minimize the total number of calories?

24. *Textile production.* It takes Cosmic Stitching 2 hr of cutting and 4 hr of sewing to make a knit suit. To make a worsted suit, it takes 4 hr of cutting and 2 hr of sewing. At most 20 hr per day are available for cutting and at most 16 hr per day are available for sewing. The profit on a knit suit is $68 and on a worsted suit is $62. How many of each kind of suit should be made in order to maximize profit?

25. Before a student begins work in this section, what three sections of the text would you suggest he or she study? Why?

26. What does the use of the word "constraint" in this section have in common with the use of the word in everyday speech?

Skill Review

Review function notation and domains of functions (Sections 7.1 and 7.2).

27. If $f(x) = 4x - 7$, find $f(a) + h$. [7.1]

28. If $f(x) = 4x - 7$, find $f(a + h)$. [7.1]

Find the domain of f.

29. $f(x) = \dfrac{x - 5}{2x + 1}$ [7.2], [9.2]

30. $f(x) = \dfrac{3x}{x^2 + 1}$ [7.2]

31. $f(x) = \sqrt{2x + 8}$ [9.1]

32. $f(x) = \dfrac{3x}{x^2 - 1}$ [7.2], [9.2]

Synthesis

33. Explain how Exercises 17 and 18 can be answered by logical reasoning without linear programming.

34. Write a linear programming problem for a classmate to solve. Devise the problem so that profit must be maximized subject to at least two (nontrivial) constraints.

35. *Airplane production.* Alpha Tours has two types of airplanes, the T3 and the S5, and contracts requiring accommodations for a minimum of 2000 first-class, 1500 tourist-class, and 2400 economy-class passengers. The T3 costs $30 per mile to operate and can accommodate 40 first-class, 40 tourist-class, and 120 economy-class passengers, whereas the S5 costs $25 per mile to operate and can accommodate 80 first-class, 30 tourist-class, and 40 economy-class passengers. How many of each type of airplane should be used in order to minimize the operating cost?

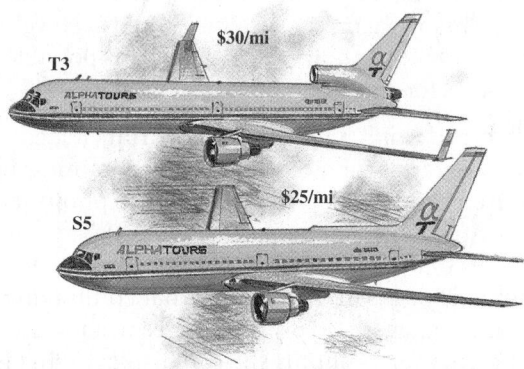

36. *Airplane production.* A new airplane, the T4, is now available, having an operating cost of $37.50 per mile and accommodating 40 first-class, 40 tourist-class, and 80 economy-class passengers. If the T3 of Exercise 35 were replaced with the T4, how many S5's and how many T4's would be needed in order to minimize the operating cost?

37. *Furniture production.* P. J. Edward Furniture Design produces chairs and sofas. The chairs require 20 ft of wood, 1 lb of foam rubber, and 2 sq yd of fabric. The sofas require 100 ft of wood, 50 lb of foam rubber, and 20 sq yd of fabric. The company has 1900 ft of wood, 500 lb of foam rubber, and 240 sq yd of fabric. The chairs can be sold for $80 each and the sofas for $1200 each. How many of each should be produced in order to maximize income?

Study Summary

KEY TERMS AND CONCEPTS	EXAMPLES

SECTION 9.1: INEQUALITIES AND DOMAIN

Inequalities can be solved graphically by determining the x-values for which one graph lies above or below another.

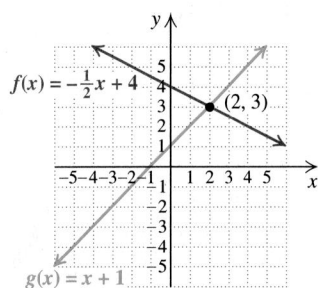

From the graph above, we can read the solution sets of the following inequalities.

Inequality	*Solution Set*
$f(x) < g(x)$	$\{x \mid x > 2\}$, or $(2, \infty)$
$f(x) \leq g(x)$	$\{x \mid x \geq 2\}$, or $[2, \infty)$
$f(x) > g(x)$	$\{x \mid x < 2\}$, or $(-\infty, 2)$
$f(x) \geq g(x)$	$\{x \mid x \leq 2\}$, or $(-\infty, 2]$

Only nonnegative numbers have square roots that are real numbers.

Find the domain of f if $f(x) = \sqrt{2x - 3}$.

$$2x - 3 \geq 0 \qquad 2x - 3 \text{ must be nonnegative}$$
$$2x \geq 3 \qquad \text{Adding 3 to both sides}$$
$$x \geq \frac{3}{2} \qquad \text{Dividing both sides by 2}$$

The domain of f is $\left\{x \mid x \geq \frac{3}{2}\right\}$, or $\left[\frac{3}{2}, \infty\right)$.

SECTION 9.2: INTERSECTIONS, UNIONS, AND COMPOUND INEQUALITIES

A **conjunction** consists of two or more sentences joined by the word *and*. The solution set of the conjunction is the **intersection** of the solution sets of the individual sentences.

$$-4 \leq x - 1 \leq 5$$
$$-4 \leq x - 1 \quad and \quad x - 1 \leq 5$$
$$-3 \leq x \quad\quad and \quad\quad x \leq 6$$

The solution set is $\{x \mid -3 \leq x \leq 6\}$, or $[-3, 6]$.

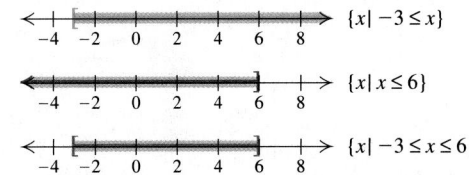

A **disjunction** consists of two or more sentences joined by the word *or*. The solution set of the disjunction is the **union** of the solution sets of the individual sentences.

$$2x + 9 < 1 \quad or \quad 5x - 2 \geq 3$$
$$2x < -8 \quad or \quad 5x \geq 5$$
$$x < -4 \quad or \quad x \geq 1$$

The solution set is $\{x \mid x < -4 \ or \ x \geq 1\}$, or $(-\infty, -4) \cup [1, \infty)$.

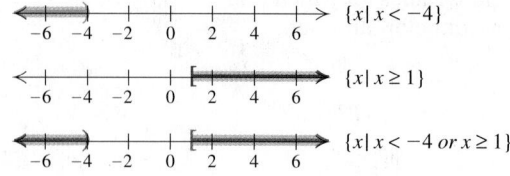

SECTION 9.3: ABSOLUTE-VALUE EQUATIONS AND INEQUALITIES

The Absolute-Value Principles for Equations and Inequalities

For any positive number p and any algebraic expression X:

a) The solutions of $|X| = p$ are those numbers that satisfy
$$X = -p \quad or \quad X = p.$$

b) The solutions of $|X| < p$ are those numbers that satisfy
$$-p < X < p.$$

c) The solutions of $|X| > p$ are those numbers that satisfy
$$X < -p \quad or \quad p < X.$$

If $|X| = 0$, then $X = 0$. If p is negative, then $|X| = p$ and $|X| < p$ have no solution, and any value of X will satisfy $|X| > p$.

$$|x + 3| = 4$$
$$x + 3 = 4 \quad or \quad x + 3 = -4 \qquad \text{Using part (a)}$$
$$x = 1 \quad or \qquad x = -7$$

The solution set is $\{-7, 1\}$.

$$|x + 3| < 4$$
$$-4 < x + 3 < 4 \qquad \text{Using part (b)}$$
$$-7 < x < 1$$

The solution set is $\{x \mid -7 < x < 1\}$, or $(-7, 1)$.

$$|x + 3| \geq 4$$
$$x + 3 \leq -4 \quad or \quad 4 \leq x + 3 \qquad \text{Using part (c)}$$
$$x \leq -7 \quad or \quad 1 \leq x$$

The solution set is $\{x \mid x \leq -7 \ or \ x \geq 1\}$, or $(-\infty, -7] \cup [1, \infty)$.

SECTION 9.4: INEQUALITIES IN TWO VARIABLES

To graph a linear inequality:

1. Graph the **boundary line**. Draw a dashed line if the inequality symbol is $<$ or $>$, and draw a solid line if the inequality symbol is $\leq$ or $\geq$.

2. Determine which side of the boundary line contains the solution set, and shade that **half-plane**.

Graph: $x + y < -1$.

1. Graph $x + y = -1$ using a dashed line.

2. Choose a test point not on the line: $(0, 0)$.

$$\frac{x + y < -1}{0 + 0 \ | \ -1}$$
$$0 \overset{?}{<} -1 \quad \text{FALSE}$$

Since $0 < -1$ is false, shade the half-plane that does *not* contain $(0, 0)$.

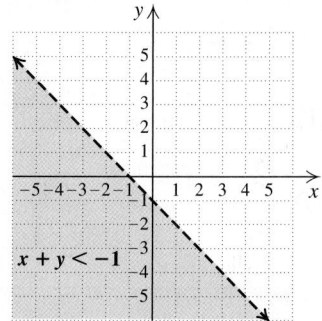

SECTION 9.5: APPLICATIONS USING LINEAR PROGRAMMING

The Corner Principle

The maximum or minimum value of an **objective function** over a *feasible region* is the maximum or minimum value of the function at a **vertex** of that region.

Maximize $F = x + 2y$ subject to

$$x + y \le 5,$$
$$x \ge 0,$$
$$y \ge 1.$$

1. Graph the feasible region.
2. Find the value of F at the vertices.

Vertex	$F = x + 2y$
$(0, 1)$	2
$(0, 5)$	10
$(4, 1)$	6

⟵ The maximum value of F is 10.

Review Exercises: Chapter 9

↪ *Concept Reinforcement*   *Classify each statement as either true or false.*

1. If x cannot exceed 10, then $x \le 10$. [9.1]

2. It is always true that if $a > b$, then $ac > bc$. [9.1]

3. The solution of $|3x - 5| \le 8$ is a closed interval. [9.3]

4. The inequality $2 < 5x + 1 < 9$ is equivalent to $2 < 5x + 1$ *or* $5x + 1 < 9$. [9.2]

5. The solution set of a disjunction is the union of two solution sets. [9.2]

6. The equation $|x| = r$ has no solution when r is negative. [9.3]

7. $|f(x)| > 3$ is equivalent to $f(x) < -3$ *or* $f(x) > 3$. [9.3]

8. A test point is used to determine whether the line in a linear inequality is drawn solid or dashed. [9.4]

9. The graph of a system of linear inequalities is always a half-plane. [9.4]

10. The corner principle states that every objective function has a maximum or minimum value. [9.5]

Solve algebraically. [9.1]

11. $-6x - 5 < 4$

12. $-\dfrac{1}{2}x - \dfrac{1}{4} > \dfrac{1}{2} - \dfrac{1}{4}x$

13. $0.3y - 7 < 2.6y + 15$

14. $-2(x - 5) \ge 6(x + 7) - 12$

Solve graphically. [9.1]

15. $x - 2 < 3$

16. $4 - 3x > 1$

17. $x - 1 \le 2x + 3$

18. $\dfrac{1}{2}x \ge \dfrac{1}{3}x + 1$

19. Let $f(x) = 3x + 2$ and $g(x) = 10 - x$. Find all values of x for which $f(x) \le g(x)$. [9.1]

Solve. [9.1]

20. Mariah has two offers for a summer job. She can work in a sandwich shop for \$8.40 an hour, or she can do carpentry work for \$16 an hour. In order to do the carpentry work, she must spend \$950 for tools. For how many hours must Mariah work in order for carpentry to be more profitable than the sandwich shop?

21. Clay is going to invest \$9000, part at 3% and the rest at 3.5%. What is the most he can invest at 3% and still be guaranteed \$300 in interest each year?

22. Find the intersection:

$$\{a, b, c, d\} \cap \{a, c, e, f, g\}. \quad [9.2]$$

23. Find the union:

$$\{a, b, c, d\} \cup \{a, c, e, f, g\}. \quad [9.2]$$

Graph and write interval notation. [9.2]

24. $x \leq 2$ *and* $x > -3$

25. $x \leq 3$ *or* $x > -5$

Solve and graph each solution set. [9.2]

26. $-3 < x + 5 \leq 5$

27. $-15 < -4x - 5 < 0$

28. $3x < -9$ *or* $-5x < -5$

29. $2x + 5 < -17$ *or* $-4x + 10 \leq 34$

30. $2x + 7 \leq -5$ *or* $x + 7 \geq 15$

31. $f(x) < -5$ *or* $f(x) > 5$, where $f(x) = 3 - 5x$

For $f(x)$ as given, use interval notation to write the domain of f.

32. $f(x) = \dfrac{2x}{x + 3}$ [9.2]

33. $f(x) = \sqrt{5x - 10}$ [9.1]

34. $f(x) = \sqrt{1 - 4x}$ [9.1]

Solve. [9.3]

35. $|x| = 11$

36. $|t| \geq 21$

37. $|x - 8| = 3$

38. $|4a + 3| < 11$

39. $|3x - 4| \geq 15$

40. $|2x + 5| = |x - 9|$

41. $|5n + 6| = -11$

42. $\left| \dfrac{x + 4}{6} \right| \leq 2$

43. $2|x - 5| - 7 > 3$

44. $19 - 3|x + 1| \geq 4$

45. Let $f(x) = |8x - 3|$. Find all x for which $f(x) < 0$. [9.3]

46. Graph $x - 2y \geq 6$ on a plane. [9.4]

Graph each system of inequalities. Find the coordinates of any vertices formed. [9.4]

47. $x + 3y > -1$,
$x + 3y < 4$

48. $x - 3y \leq 3$,
$x + 3y \geq 9$,
$y \leq 6$

49. Find the maximum and the minimum values of

$$F = 3x + y + 4$$

subject to

$$y \leq 2x + 1,$$
$$x \leq 7,$$
$$y \geq 3. \quad [9.5]$$

50. Custom Computers has two manufacturing plants. The Oregon plant cannot produce more than 60 computers per week, while the Ohio plant cannot produce more than 120 computers per week. The Electronics Outpost sells at least 160 Custom computers each week. It costs \$40 to ship a computer to The Electronics Outpost from the Oregon plant and \$25 to ship from the Ohio plant. How many computers should be shipped from each plant in order to minimize cost? [9.5]

Synthesis

51. Explain in your own words why $|X| = p$ has two solutions when p is positive and no solution when p is negative. [9.3]

52. Explain why the graph of the solution of a system of linear inequalities is the intersection, not the union, of the individual graphs. [9.4]

53. Solve: $|2x + 5| \leq |x + 3|$. [9.3]

54. Classify as true or false: If $x < 3$, then $x^2 < 9$. If false, give an example showing why. [9.1]

55. Super Lock manufactures brass doorknobs with a 2.5-in. diameter and a ± 0.003-in. manufacturing tolerance, or allowable variation in diameter. Write the tolerance as an inequality with absolute value. [9.3]

Test: Chapter 9

CHAPTER Test Prep VIDEO CD

Step-by-step test solutions are found on the video CD in the front of this book.

Solve algebraically.

1. $-4y - 3 \geq 5$

2. $3(7 - x) < 2x + 5$

3. $-2(3x - 1) - 5 \geq 6x - 4(3 - x)$

Solve graphically.

4. $3 - x < 2$

5. $2x - 3 \geq x + 1$

6. Let $f(x) = -5x - 1$ and $g(x) = -9x + 3$. Find all values of x for which $f(x) > g(x)$.

7. Dani can rent a van for either \$80 with unlimited mileage or \$45 with 100 free miles and an extra charge of 40¢ for each mile over 100. For what numbers of miles traveled would the unlimited mileage plan save Dani money?

8. A refrigeration repair company charges \$80 for the first half-hour of work and \$60 for each additional hour. Blue Mountain Camp has budgeted \$200 to repair its walk-in cooler. For what lengths of a service call will the budget not be exceeded?

9. Find the intersection:

$$\{a, e, i, o, u\} \cap \{a, b, c, d, e\}.$$

10. Find the union:

$$\{a, e, i, o, u\} \cup \{a, b, c, d, e\}.$$

For $f(x)$ as given, use interval notation to write the domain of f.

11. $f(x) = \sqrt{6 - 3x}$

12. $f(x) = \dfrac{x}{x - 7}$

Solve and graph each solution set.

13. $-5 < 4x + 1 \leq 3$

14. $3x - 2 < 7$ *or* $x - 2 > 4$

15. $-3x > 12$ *or* $4x \geq -10$

16. $1 \leq 3 - 2x \leq 9$

17. $|n| = 15$

18. $|a| > 5$

19. $|3x - 1| < 7$

20. $|-5t - 3| \geq 10$

21. $|2 - 5x| = -12$

22. $g(x) < -3$ *or* $g(x) > 3$, where $g(x) = 4 - 2x$

23. Let $f(x) = |2x - 1|$ and $g(x) = |2x + 7|$. Find all values of x for which $f(x) = g(x)$.

24. Graph $y \leq 2x + 1$ on a plane.

Graph the system of inequalities. Find the coordinates of any vertices formed.

25. $x + y \geq 3,$
$x - y \geq 5$

26. $2y - x \geq -7,$
$2y + 3x \leq 15,$
$y \leq 0,$
$x \leq 0$

27. Find the maximum and the minimum values of

$F = 5x + 3y$

subject to

$x + y \leq 15,$
$1 \leq x \leq 6,$
$0 \leq y \leq 12.$

28. Swift Cuts makes \$12 on each manicure and \$18 on each haircut. A manicure takes 30 min and a haircut takes 50 min, and there are 5 stylists who each work 6 hr a day. If the salon can schedule 50 appointments a day, how many should be manicures and how many haircuts in order to maximize profit? What is the maximum profit?

Synthesis

Solve. Write the solution set using interval notation.

29. $|2x - 5| \leq 7$ *and* $|x - 2| \geq 2$

30. $7x < 8 - 3x < 6 + 7x$

31. Write an absolute-value inequality for which the interval shown is the solution.

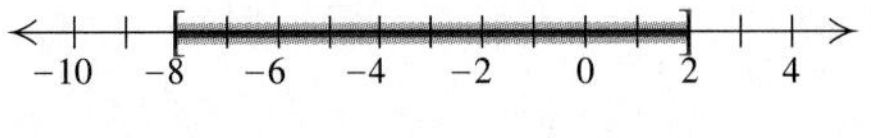

Simplify. Do not leave negative exponents in your answers.

1. $3 + 24 \div 2^2 \cdot 3 - (6 - 7)$ [4.2]

2. $3c - [8 - 2(1 - c)]$ [4.2]

3. -10^{-2} [4.2]

4. $(3xy^{-4})(-2x^3y)$ [4.2]

5. $\left(\dfrac{18a^2b^{-1}}{12a^{-1}b}\right)^2$ [4.2]

6. $\dfrac{2x - 10}{x^3 - 125}$ [6.1]

Perform the indicated operations.

7. $(x - 5)(x + 5)$ [4.6]

8. $(3n - 2)(5n + 7)$ [4.6]

9. $\dfrac{1}{x + 5} - \dfrac{1}{x - 5}$ [6.4]

10. $\dfrac{x^2 - 3x}{2x^2 - x - 3} \div \dfrac{x^3}{x^2 - 2x - 3}$ [6.2]

Factor.

11. $25x^2 - 50x + 25$ [5.4]

12. $8mn + 14n - 12m - 21$ [5.1]

13. $8t^2 + 800$ [5.1]

Solve.

14. $3(x - 2) = 14 - x$ [2.2]

15. $x - 2 < 6$ *or* $2x + 1 > 5$ [9.2]

16. $x^2 - 2x - 3 = 5$ [5.7]

17. $\dfrac{3}{x + 1} = \dfrac{x}{4}$ [6.6]

18. $y = \frac{1}{2}x - 7,$
    $2x - 4y = 3$ [8.2]

19. $x + 3y = 8,$
    $2x - 3y = 7$ [8.2]

20. $|2x - 1| = 8$ [9.3]

21. $9(x - 3) - 4x < 2 - (3 - x)$ [2.6], [9.1]

22. $|4t| > 12$ [9.3]

23. $|3x - 2| \le 8$ [9.3]

Graph on a plane.

24. $y = \frac{2}{3}x - 4$ [3.6]

25. $x = -3$ [3.3]

26. $3x - y = 3$ [3.3]

27. $x + y \ge -2$ [9.4]

28. $f(x) = -x + 1$ [7.3]

29. $x - 2y > 4,$
    $x + 2y \ge -2$ [9.4]

30. Find the slope and the y-intercept of the line given by $4x - 9y = 18$. [3.6]

31. Write a slope–intercept equation for the line with slope -7 and containing the point $(-3, -4)$. [3.7]

32. Find an equation of the line with y-intercept $(0, 4)$ and perpendicular to the line given by $3x + 2y = 1$. [3.6]

33. For the graph of f shown, determine the domain and the range of f. [7.2]

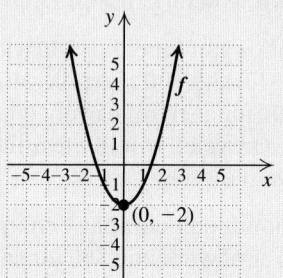

34. Determine the domain of the function given by $f(x) = \dfrac{3}{2x + 5}$. [7.2], [9.2]

35. Find $g(-2)$ if $g(x) = 3x^2 - 5x$. [7.1]

36. Find $(f - g)(x)$ if $f(x) = x^2 + 3x$ and $g(x) = 9 - 3x$. [7.4]

37. Graph the solution set of $-3 \le f(x) \le 2$, where $f(x) = 1 - x$. [9.2]

38. Find the domain of h/g if $h(x) = \dfrac{1}{x}$ and $g(x) = 3x - 1$. [7.4]

39. Solve for t: $at - dt = c$. [2.3]

40. The Baqueira, a resort in Spain, uses 549 snow cannons to make snow for its 4344 acres of ski runs. How many snow cannons should a resort containing 1448 acres of runs use in order to make a comparable amount of snow? [6.7]
    Sources: www.bluebookski.com

41. *Water usage.*   In dry climates, it takes about 11,600 gal of water to produce a pound of beef and a pound of wheat. The pound of beef requires 7000 more gallons of water than the pound of wheat. How much water does it take to produce each? [8.3]

Source: *The Wall Street Journal*, 1/28/08

42. *Book sales.*   U.S. sales of books and maps were $34.6 billion in 2001 and $41.4 billion in 2004. Let $b(t)$ represent U.S. book sales t years after 2001. [7.3]

Source: Bureau of Economic Analysis

a) Find a linear function that fits the data.

b) Use the function from part (a) to predict U.S. book sales in 2010.

c) In what year will U.S. book sales be $50 billion?

43. *Fundraising.*   Michelle is planning a fundraising dinner for Happy Hollow Children's Camp. The banquet facility charges a rental fee of $1500, but will waive the rental fee if more than $6000 is spent for catering. Michelle knows that 150 people will attend the dinner. [9.1]

a) How much should each dinner cost in order for the rental fee to be waived?

b) For what costs per person will the total cost (including the rental fee) exceed $6000?

44. *Perimeter of a rectangle.*   The perimeter of a rectangle is 32 cm. If five times the width equals three times the length, what are the dimensions of the rectangle? [8.3]

45. *Utility bills.*   One month Lori and Tony spent $920 for electricity, rent, and cell phone. The electric bill was $\frac{1}{4}$ of the rent, and the phone bill was $40 less than the electric bill. How much was the rent? [8.5]

46. *Banking.*   Banks charge a fee to a customer whose checking account does not contain enough money to pay for a debit-card purchase or a written check. These insufficient-funds fees totaled $35 billion in the United States in a recent year. This was 70% of the total fee income of banks. What was the total fee income of banks in that year? [2.5]

47. *Catering.*   Dan charges $35 per person for a vegetarian meal and $40 per person for a steak dinner. For one event, he served 28 dinners for a total cost of $1060. How many dinners were vegetarian and how many were steak? [8.3]

Synthesis

48. If $(2, 6)$ and $(-1, 5)$ are two solutions of $f(x) = mx + b$, find m and b. [8.2]

49. Find k such that the line containing $(-2, k)$ and $(3, 8)$ is parallel to the line containing $(1, 6)$ and $(4, -2)$. [3.6]

50. Use interval notation to write the domain of the function given by

$$f(x) = \frac{\sqrt{x + 4}}{x}. \quad [9.2]$$

51. Simplify: $\dfrac{2^{a-1} \cdot 2^{4a}}{2^{3(-2a+5)}}$. [4.2]

Exponents and Radicals

MICHAEL MANOLAKIS
FIREFIGHTER
Boston, Massachusetts

Firefighters use math every day. Hosing down a fire may seem simple, but there are considerations to take into account when getting water from its source to the end of the hose where the "pipeman" is attacking the fire. For example, the safest force or psi (pounds per square inch) needed at the tip of a hose is determined by its size. The safest force needed at the tip of a $1\frac{3}{4}$-in. line is 120 psi. To achieve this pressure, we must consider the distance and height of the hose line from the source of water. These two factors are necessary to accurately calculate and deliver the appropriate psi at the tip.

AN APPLICATION

The velocity of water flow, in feet per second, from a nozzle is given by

$$v(p) = 12.1\sqrt{p},$$

where p is the nozzle pressure, in pounds per square inch. Find the nozzle pressure if the water flow is 100 feet per second.

Source: Houston Fire Department Continuing Education

This problem appears as Exercise 65 in Section 10.6.

I n this chapter, we learn about square roots, cube roots, fourth roots, and so on. These roots can be expressed in radical notation and appear in both radical expressions and radical equations. Exponents that are fractions are also studied and will ease some of our work with radicals. The chapter closes with an introduction to the complex-number system.

10.1   Radical Expressions and Functions

Square Roots and Square-Root Functions • Expressions of the Form $\sqrt{a^2}$ • Cube Roots • Odd and Even *n*th Roots

In this section, we consider roots, such as square roots and cube roots. We look at the symbolism that is used and ways in which symbols can be manipulated to get equivalent expressions. All of this will be important in problem solving.

Square Roots and Square-Root Functions

When a number is multiplied by itself, we say that the number is squared. Often we need to know what number was squared in order to produce some value *a*. If such a number can be found, we call that number a *square root* of *a*.

> **Square Root**
>
> The number *c* is a *square root* of *a* if $c^2 = a$.

For example,

    9 has −3 and 3 as square roots because $(-3)^2 = 9$ and $3^2 = 9$.

    25 has −5 and 5 as square roots because $(-5)^2 = 25$ and $5^2 = 25$.

    −4 does not have a real-number square root because there is no real number *c* for which $c^2 = -4$.

    Note that every positive number has two square roots, whereas 0 has only itself as a square root. Negative numbers do not have real-number square roots, although later in this chapter we introduce the *complex-number* system in which such square roots do exist.

EXAMPLE **1**   Find the two square roots of 36.

SOLUTION   The square roots are 6 and −6, because $6^2 = 36$ and $(-6)^2 = 36$.

> ▶ TRY EXERCISE ▶ 9

STUDENT NOTES ─────

It is important to remember the difference between *the* square root of 9 and *a* square root of 9. *A* square root of 9 means either 3 or −3, whereas *the* square root of 9, denoted $\sqrt{9}$, means the principal square root of 9, or 3.

    Whenever we refer to *the* square root of a number, we mean the nonnegative square root of that number. This is often referred to as the *principal square root* of the number.

> **Principal Square Root**
>
> The *principal square root* of a nonnegative number is its nonnegative square root. The symbol $\sqrt{}$ is called a *radical sign* and is used to indicate the principal square root of the number over which it appears.

EXAMPLE **2**   Simplify each of the following.

a) $\sqrt{25}$ **b)** $\sqrt{\dfrac{25}{64}}$ **c)** $-\sqrt{64}$ **d)** $\sqrt{0.0049}$

SOLUTION

a) $\sqrt{25} = 5$ $\sqrt{}$ indicates the principal square root. Note that $\sqrt{25} \neq -5$.

b) $\sqrt{\dfrac{25}{64}} = \dfrac{5}{8}$ Since $\left(\dfrac{5}{8}\right)^2 = \dfrac{25}{64}$

c) $-\sqrt{64} = -8$ Since $\sqrt{64} = 8, -\sqrt{64} = -8.$

d) $\sqrt{0.0049} = 0.07$ $(0.07)(0.07) = 0.0049.$ Note too that
$$\sqrt{0.0049} = \sqrt{\dfrac{49}{10,000}} = \dfrac{7}{100}.$$

> **TRY EXERCISE** 19

In addition to being read as "the principal square root of a," $\sqrt{a}$ is also read as "the square root of a," "root a," or "radical a." Any expression in which a radical sign appears is called a *radical expression*. The following are radical expressions:
$$\sqrt{5}, \quad \sqrt{a}, \quad -\sqrt{3x}, \quad \sqrt{\dfrac{y^2 + 7}{y}}, \quad \sqrt{x} + 8.$$

The expression under the radical sign is called the **radicand**. In the expressions above, the radicands are 5, a, $3x$, $(y^2 + 7)/y$, and x, respectively.

Values for square roots found on calculators are, for the most part, approximations. For example, a calculator will show a number like

    2.23606798

for $\sqrt{5}$. The exact value of $\sqrt{5}$ is not given by any repeating or terminating decimal. In general, for any whole number a that is not a perfect square, $\sqrt{a}$ is a nonterminating, nonrepeating decimal or an *irrational number*.

The square-root function, given by
$$f(x) = \sqrt{x},$$

has $[0, \infty)$ as its domain and $[0, \infty)$ as its range. We can draw its graph by selecting convenient values for x and calculating the corresponding outputs. Once these ordered pairs have been graphed, a smooth curve can be drawn.

$f(x) = \sqrt{x}$

x	$\sqrt{x}$	$(x, f(x))$
0	0	$(0, 0)$
1	1	$(1, 1)$
4	2	$(4, 2)$
9	3	$(9, 3)$

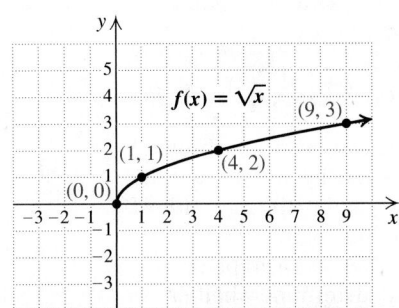

EXAMPLE **3**   For each function, find the indicated function value.

a) $f(x) = \sqrt{3x - 2}; \; f(1)$ **b)** $g(z) = -\sqrt{6z + 4}; \; g(3)$

SOLUTION

a) $f(1) = \sqrt{3 \cdot 1 - 2}$ Substituting
$ = \sqrt{1} = 1$ Simplifying

b) $g(3) = -\sqrt{6 \cdot 3 + 4}$ Substituting

$= -\sqrt{22}$ Simplifying. This answer is exact.

≈ -4.69041576 Using a calculator to write an approximation

TRY EXERCISE ⟩ 35

Expressions of the Form $\sqrt{a^2}$

As the next example shows, $\sqrt{a^2}$ does not always simplify to a.

EXAMPLE **4**   Evaluate $\sqrt{a^2}$ for the following values: **(a)** 5; **(b)** 0; **(c)** −5.

SOLUTION

a) $\sqrt{5^2} = \sqrt{25} = 5$

⎿————————⏌ Same

b) $\sqrt{0^2} = \sqrt{0} = 0$

⎿————————⏌ Same

c) $\sqrt{(-5)^2} = \sqrt{25} = 5$

⎿————————⏌ Opposites   Note that $\sqrt{(-5)^2} \neq -5$.

You may have noticed that evaluating $\sqrt{a^2}$ is just like evaluating $|a|$.

> ## Simplifying $\sqrt{a^2}$
> For any real number a,
> $$\sqrt{a^2} = |a|.$$
> (The principal square root of a^2 is the absolute value of a.)

When a radicand is the square of a variable expression, like $(x + 5)^2$ or $36t^2$, absolute-value signs are needed when simplifying. We use absolute-value signs unless we know that the expression being squared is nonnegative. This ensures that our result is never negative.

EXAMPLE **5**   Simplify each expression. Assume that the variable can represent any real number.

a) $\sqrt{(x + 1)^2}$ **b)** $\sqrt{x^2 - 8x + 16}$

c) $\sqrt{a^8}$ **d)** $\sqrt{t^6}$

TECHNOLOGY CONNECTION

To see the necessity of absolute-value signs, let y_1 represent the left side and y_2 the right side of each of the following equations. Then use a graph or table to determine whether these equations are true.

1. $\sqrt{x^2} \overset{?}{=} x$

2. $\sqrt{x^2} \overset{?}{=} |x|$

3. $x \overset{?}{=} |x|$

SOLUTION

a) $\sqrt{(x + 1)^2} = |x + 1|$ Since $x + 1$ can be negative (for example, if $x = -3$), absolute-value notation is required.

b) $\sqrt{x^2 - 8x + 16} = \sqrt{(x - 4)^2} = |x - 4|$ Since $x - 4$ can be negative, absolute-value notation is required.

c) Note that $(a^4)^2 = a^8$ and that a^4 is never negative. Thus,

$\sqrt{a^8} = a^4$.   Absolute-value notation is unnecessary here.

d) Note that $(t^3)^2 = t^6$. Thus,

$$\sqrt{t^6} = |t^3|.$$ Since t^3 can be negative, absolute-value notation is required.

TRY EXERCISE 43

If we assume that the expression being squared is nonnegative, then absolute-value notation is not necessary.

EXAMPLE 6 Simplify each expression. Assume that no radicands were formed by squaring negative quantities.

a) $\sqrt{y^2}$ **b)** $\sqrt{a^{10}}$ **c)** $\sqrt{9x^2 - 6x + 1}$

SOLUTION

a) $\sqrt{y^2} = y$ We assume that y is nonnegative, so no absolute-value notation is necessary. When y *is* negative, $\sqrt{y^2} \neq y$.

b) $\sqrt{a^{10}} = a^5$ Assuming that a^5 is nonnegative. Note that $(a^5)^2 = a^{10}$.

c) $\sqrt{9x^2 - 6x + 1} = \sqrt{(3x-1)^2} = 3x - 1$ Assuming that $3x - 1$ is nonnegative

TRY EXERCISE 69

Cube Roots

We often need to know what number cubed produces a certain value. When such a number is found, we say that we have found a *cube root*. For example,

2 is the cube root of 8 because $2^3 = 2 \cdot 2 \cdot 2 = 8$;

-4 is the cube root of -64 because $(-4)^3 = (-4)(-4)(-4) = -64$.

> **Cube Root**
>
> The number c is the *cube root* of a if $c^3 = a$. In symbols, we write $\sqrt[3]{a}$ to denote the cube root of a.

Each real number has only one real-number cube root. The cube-root function, given by

$$f(x) = \sqrt[3]{x},$$

has $\mathbb{R}$ as its domain and $\mathbb{R}$ as its range. To draw its graph, we select convenient values for x and calculate the corresponding outputs. Once these ordered pairs have been graphed, a smooth curve is drawn. Note that the cube root of a positive number is positive, and the cube root of a negative number is negative.

$$f(x) = \sqrt[3]{x}$$

x	$\sqrt[3]{x}$	$(x, f(x))$
0	0	$(0, 0)$
1	1	$(1, 1)$
8	2	$(8, 2)$
-1	-1	$(-1, -1)$
-8	-2	$(-8, -2)$

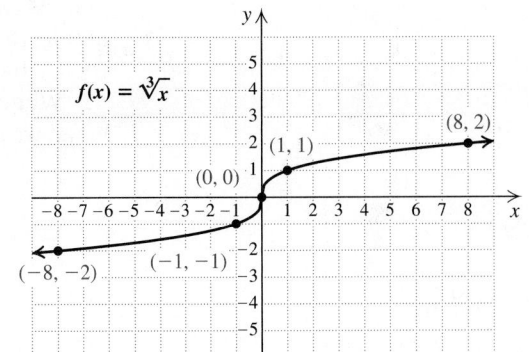

EXAMPLE **7**   For each function, find the indicated function value.

a) $f(y) = \sqrt[3]{y}$; $f(125)$

b) $g(x) = \sqrt[3]{x - 1}$; $g(-26)$

SOLUTION

a) $f(125) = \sqrt[3]{125} = 5$ Since $5 \cdot 5 \cdot 5 = 125$

b) $g(-26) = \sqrt[3]{-26 - 1}$

$\qquad = \sqrt[3]{-27}$

$\qquad\quad = -3$ Since $(-3)(-3)(-3) = -27$

> **TRY EXERCISE** 89

EXAMPLE **8**   Simplify: $\sqrt[3]{-8y^3}$.

SOLUTION

$$\sqrt[3]{-8y^3} = -2y \qquad \text{Since } (-2y)(-2y)(-2y) = -8y^3$$

> **TRY EXERCISE** 83

Odd and Even *n*th Roots

The 4th root of a number a is the number c for which $c^4 = a$. There are also 5th roots, 6th roots, and so on. We write $\sqrt[n]{a}$ for the principal nth root. The number n is called the *index* (plural, *indices*). When the index is 2, we do not write it.

When the index n is odd, we are taking an *odd root*. Note that every number has exactly one real root when n is odd. Odd roots of positive numbers are positive and odd roots of negative numbers are negative. Absolute-value signs are not used when finding odd roots.

EXAMPLE **9**   Simplify each expression.

a) $\sqrt[5]{32}$ **b)** $\sqrt[5]{-32}$ **c)** $-\sqrt[5]{32}$

d) $-\sqrt[5]{-32}$ **e)** $\sqrt[7]{x^7}$ **f)** $\sqrt[9]{(t - 1)^9}$

SOLUTION

a) $\sqrt[5]{32} = 2$ Since $2^5 = 32$

b) $\sqrt[5]{-32} = -2$ Since $(-2)^5 = -32$

c) $-\sqrt[5]{32} = -2$ Taking the opposite of $\sqrt[5]{32}$

d) $-\sqrt[5]{-32} = -(-2) = 2$ Taking the opposite of $\sqrt[5]{-32}$

e) $\sqrt[7]{x^7} = x$ No absolute-value signs are needed.

f) $\sqrt[9]{(t - 1)^9} = t - 1$

> **TRY EXERCISE** 81

When the index n is even, we are taking an *even root*. Every positive real number has two real nth roots when n is even—one positive and one negative. For example, the fourth roots of 16 are -2 and 2. Negative numbers do not have real nth roots when n is even.

When n is even, the notation $\sqrt[n]{a}$ indicates the nonnegative nth root. Thus, when we simplify even nth roots, absolute-value signs are often required.

Compare the following.

Odd Root	Even Root
$\sqrt[3]{8} = 2$	$\sqrt[4]{16} = 2$
$\sqrt[3]{-8} = -2$	$\sqrt[4]{-16}$ is not a real number.
$\sqrt[3]{x^3} = x$	$\sqrt[4]{x^4} = |x|$

EXAMPLE **10** Simplify each expression, if possible. Assume that variables can represent any real number.

a) $\sqrt[4]{81}$ b) $-\sqrt[4]{81}$ c) $\sqrt[4]{-81}$
d) $\sqrt[4]{81x^4}$ e) $\sqrt[6]{(y + 7)^6}$

SOLUTION

a) $\sqrt[4]{81} = 3$ Since $3^4 = 81$

b) $-\sqrt[4]{81} = -3$ Taking the opposite of $\sqrt[4]{81}$

c) $\sqrt[4]{-81}$ cannot be simplified. $\sqrt[4]{-81}$ is not a real number.

d) $\sqrt[4]{81x^4} = |3x|$, or $3|x|$ Use absolute-value notation since x could represent a negative number.

e) $\sqrt[6]{(y + 7)^6} = |y + 7|$ Use absolute-value notation since $y + 7$ is negative for $y < -7$.

▶ **TRY EXERCISE** ▶ 59

We summarize as follows.

Simplifying nth Roots

n	a	$\sqrt[n]{a}$	$\sqrt[n]{a^n}$
Even	Positive	Positive	$|a|$
	Negative	Not a real number	
Odd	Positive	Positive	a
	Negative	Negative	

EXAMPLE **11** Determine the domain of g if $g(x) = \sqrt[6]{7 - 3x}$.

SOLUTION Since the index is even, the radicand, $7 - 3x$, must be nonnegative. We solve the inequality:

$7 - 3x \geq 0$ We cannot find the 6th root of a negative number.

$-3x \geq -7$

$x \leq \frac{7}{3}.$ Multiplying both sides by $-\frac{1}{3}$ and reversing the inequality

Thus,

Domain of $g = \left\{ x \mid x \leq \frac{7}{3} \right\}$

$= \left(-\infty, \frac{7}{3} \right].$

▶ **TRY EXERCISE** ▶ 95

TECHNOLOGY CONNECTION

To enter cube or higher roots on a graphing calculator, select options 4 or 5 of the **MATH** menu. The characters $6\sqrt[x]{}$ indicate the sixth root.

1. Use a **TABLE** or **GRAPH** and **TRACE** to check the solution of Example 11.

Visualizing for Success

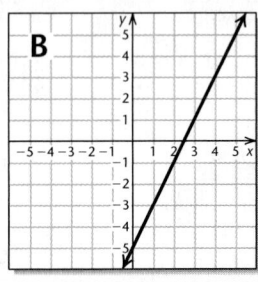

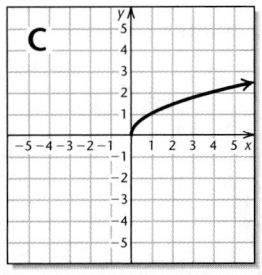

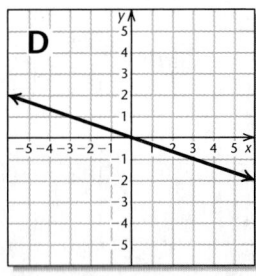

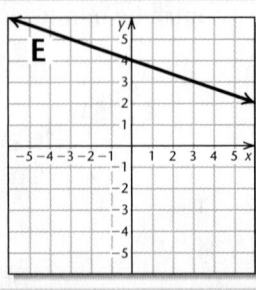

Match each function with its graph.

1. $f(x) = 2x - 5$

2. $f(x) = x^2 - 1$

3. $f(x) = \sqrt{x}$

4. $f(x) = x - 2$

5. $f(x) = -\frac{1}{3}x$

6. $f(x) = 2x$

7. $f(x) = 4 - x$

8. $f(x) = |2x - 5|$

9. $f(x) = -2$

10. $f(x) = -\frac{1}{3}x + 4$

Answers on page A-41

An additional, animated version of this activity appears in MyMathLab. To use MyMathLab, you need a course ID and a student access code. Consult your instructor for more information.

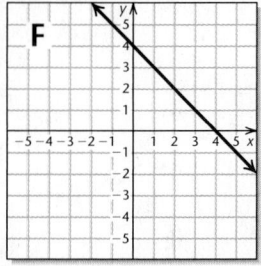

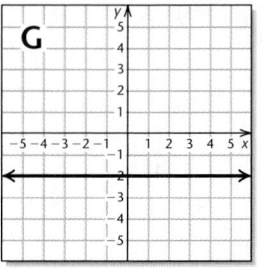

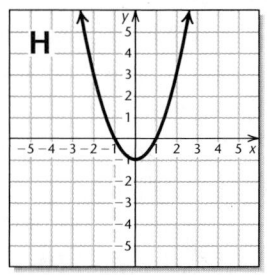

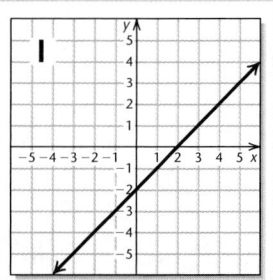

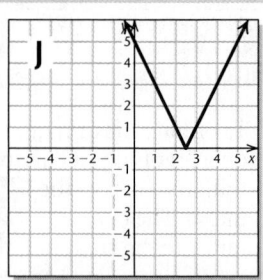

10.1   EXERCISE SET

For Extra Help
MyMathLab
Math XL
PRACTICE   WATCH   DOWNLOAD

👉 **Concept Reinforcement**   *Select the appropriate word to complete each of the following.*

1. Every positive number has _____ square root(s).   one/two

2. The principal square root is never _____.   negative/positive

3. For any _____ number *a*, we have   negative/positive $\sqrt{a^2} = a.$

4. For any _____ number *a*, we have   negative/positive $\sqrt{a^2} = -a.$

5. If *a* is a whole number that is not a perfect square, then $\sqrt{a}$ is a(n) _____ number.   irrational/rational

6. The domain of the function *f* given by $f(x) = \sqrt[3]{x}$ is all _____ numbers.   whole/real/positive

7. If $\sqrt[4]{x}$ is a real number, then *x* must be _____.   negative/positive/nonnegative.

8. If $\sqrt[3]{x}$ is negative, then *x* must be _____.   negative/positive

---

For each number, find all of its square roots.

9. 64   **10.** 81

11. 100   **12.** 121

13. 400   **14.** 2500

15. 625   **16.** 225

Simplify.

17. $\sqrt{49}$ **18.** $\sqrt{144}$

19. $-\sqrt{16}$ **20.** $-\sqrt{100}$

21. $\sqrt{\dfrac{36}{49}}$ **22.** $\sqrt{\dfrac{4}{9}}$

23. $-\sqrt{169}$ **24.** $-\sqrt{196}$

25. $-\sqrt{\dfrac{16}{81}}$ **26.** $-\sqrt{\dfrac{81}{144}}$

27. $\sqrt{0.04}$ **28.** $\sqrt{0.36}$

29. $\sqrt{0.0081}$ **30.** $\sqrt{0.0016}$

Identify the radicand and the index for each expression.

31. $5\sqrt{p^2 + 4}$ **32.** $-7\sqrt{y^2 - 8}$

33. $x^2 y^3 \sqrt[5]{\dfrac{x}{y + 4}}$ **34.** $\dfrac{a^2}{b} \sqrt[6]{a(a + b)}$

For each function, find the specified function value, if it exists.

35. $f(t) = \sqrt{5t - 10}$; $f(3), f(2), f(1), f(-1)$

36. $g(x) = \sqrt{x^2 - 25}$; $g(-6), g(3), g(6), g(13)$

37. $t(x) = -\sqrt{2x^2 - 1}$; $t(5), t(0), t(-1), t\left(-\dfrac{1}{2}\right)$

38. $p(z) = \sqrt{2z - 20}$; $p(4), p(10), p(12), p(0)$

39. $f(t) = \sqrt{t^2 + 1}$; $f(0), f(-1), f(-10)$

40. $g(x) = -\sqrt{(x + 1)^2}$; $g(-3), g(4), g(-5)$

Simplify. Remember to use absolute-value notation when necessary. If a root cannot be simplified, state this.

41. $\sqrt{100x^2}$ **42.** $\sqrt{16t^2}$

43. $\sqrt{(8 - t)^2}$ **44.** $\sqrt{(a + 3)^2}$

45. $\sqrt{y^2 + 16y + 64}$ **46.** $\sqrt{x^2 - 4x + 4}$

47. $\sqrt{4x^2 + 28x + 49}$ **48.** $\sqrt{9x^2 - 30x + 25}$

49. $-\sqrt[4]{256}$ **50.** $-\sqrt[4]{625}$

51. $\sqrt[3]{-1}$ **52.** $-\sqrt[3]{-1000}$

53. $-\sqrt[5]{-\dfrac{32}{243}}$ **54.** $\sqrt[5]{-\dfrac{1}{32}}$

55. $\sqrt[6]{x^6}$ **56.** $\sqrt[8]{y^8}$

57. $\sqrt[9]{t^9}$ **58.** $\sqrt[5]{a^5}$

59. $\sqrt[4]{(6a)^4}$ **60.** $\sqrt[4]{(7b)^4}$

61. $\sqrt[10]{(-6)^{10}}$ **62.** $\sqrt[12]{(-10)^{12}}$

63. $\sqrt[414]{(a + b)^{414}}$ **64.** $\sqrt[1976]{(2a + b)^{1976}}$

65. $\sqrt{a^{22}}$ **66.** $\sqrt{x^{10}}$

67. $\sqrt{-25}$ **68.** $\sqrt{-16}$

Simplify. Assume that no radicands were formed by raising negative quantities to even powers.

69. $\sqrt{16x^2}$ **70.** $\sqrt{25t^2}$

71. $-\sqrt{(3t)^2}$ **72.** $-\sqrt{(7c)^2}$

73. $\sqrt{(-5b)^2}$ **74.** $\sqrt{(-10a)^2}$

75. $\sqrt{a^2 + 2a + 1}$ **76.** $\sqrt{9 - 6y + y^2}$

77. $\sqrt[3]{27}$ **78.** $-\sqrt[3]{64}$

79. $\sqrt[4]{16x^4}$ **80.** $\sqrt[4]{81x^4}$

81. $\sqrt[5]{(x-1)^5}$

82. $-\sqrt[5]{(7y)^5}$

83. $-\sqrt[3]{-125y^3}$

84. $\sqrt[3]{-64x^3}$

85. $\sqrt{t^{18}}$

86. $\sqrt{a^{14}}$

87. $\sqrt{(x-2)^8}$

88. $\sqrt{(x+3)^{10}}$

For each function, find the specified function value, if it exists.

89. $f(x) = \sqrt[3]{x+1}$; $f(7), f(26), f(-9), f(-65)$

90. $g(x) = -\sqrt[3]{2x-1}$; $g(0), g(-62), g(-13), g(63)$

91. $g(t) = \sqrt[4]{t-3}$; $g(19), g(-13), g(1), g(84)$

92. $f(t) = \sqrt[4]{t+1}$; $f(0), f(15), f(-82), f(80)$

Determine the domain of each function described.

93. $f(x) = \sqrt{x-6}$

94. $g(x) = \sqrt{x+8}$

95. $g(t) = \sqrt[4]{t+8}$

96. $f(x) = \sqrt[4]{x-9}$

97. $g(x) = \sqrt[4]{10-2x}$

98. $g(t) = \sqrt[3]{2t-6}$

99. $f(t) = \sqrt[5]{2t+7}$

100. $f(t) = \sqrt[6]{4+3t}$

101. $h(z) = -\sqrt[6]{5z+2}$

102. $d(x) = -\sqrt[4]{5-7x}$

Aha! **103.** $f(t) = 7 + \sqrt[8]{t^8}$

104. $g(t) = 9 + \sqrt[6]{t^6}$

105. Explain how to write the negative square root of a number using radical notation.

106. Does the square root of a number's absolute value always exist? Why or why not?

Skill Review

To prepare for Section 10.2, review exponents (Sections 4.1 and 4.2).

Simplify. Do not use negative exponents in your answer. [4.1], [4.2]

107. $(a^2b)(a^4b)$

108. $(3xy^8)(5x^2y)$

109. $(5x^2y^{-3})^3$

110. $(2a^{-1}b^2c)^{-3}$

111. $\left(\dfrac{10x^{-1}y^5}{5x^2y^{-1}}\right)^{-1}$

112. $\left(\dfrac{8x^3y^{-2}}{2xz^4}\right)^{-2}$

Synthesis

113. Under what conditions does the *n*th root of x^3 exist? Explain your reasoning.

114. Under what conditions does the *n*th root of x^2 exist? Explain your reasoning.

115. *Biology.* The number of species *S* of plants in Guyana in an area of *A* hectares can be estimated using the formula
$$S = 88.63\sqrt[4]{A}.$$
The Kaieteur National Park in Guyana has an area of 63,000 hectares. How many species of plants are in the park?
Source: Hans ter Steege, "A Perspective on Guyana and its Plant Richness," as found on www.bio.uu.nl

116. *Spaces in a parking lot.* A parking lot has attendants to park the cars. The number *N* of stalls needed for waiting cars before attendants can get to them is given by the formula $N = 2.5\sqrt{A}$, where *A* is the number of arrivals in peak hours. Find the number of spaces needed for the given number of arrivals in peak hours: **(a)** 25; **(b)** 36; **(c)** 49; **(d)** 64.

Determine the domain of each function described. Then draw the graph of each function.

117. $f(x) = \sqrt{x+5}$

118. $g(x) = \sqrt{x}+5$

119. $g(x) = \sqrt{x}-2$

120. $f(x) = \sqrt{x-2}$

121. Find the domain of *f* if
$$f(x) = \frac{\sqrt{x+3}}{\sqrt[4]{2-x}}.$$

122. Find the domain of *g* if
$$g(x) = \frac{\sqrt[4]{5-x}}{\sqrt[6]{x+4}}.$$

123. Find the domain of *F* if $F(x) = \dfrac{x}{\sqrt{x^2-5x-6}}$.

124. Use a graphing calculator to check your answers to Exercises 41, 45, and 59. On some graphing calculators, a MATH key is needed to enter higher roots.

125. Use a graphing calculator to check your answers to Exercises 117 and 118. (See Exercise 124.)

10.2 Rational Numbers as Exponents

Rational Exponents ▪ Negative Rational Exponents ▪ Laws of Exponents ▪ Simplifying Radical Expressions

We have already considered natural-number exponents and integer exponents. We now expand the study of exponents further to include all rational numbers. This will give meaning to expressions like $7^{1/3}$ and $(2x)^{-4/5}$. Such notation will help us simplify certain radical expressions.

Rational Exponents

When defining rational exponents, we want the rules for exponents to hold for rational exponents just as they do for integer exponents. In particular, we still want to add exponents when multiplying.

If $a^{1/2} \cdot a^{1/2} = a^{1/2+1/2} = a^1$,   then $a^{1/2}$ should mean $\sqrt{a}$.

If $a^{1/3} \cdot a^{1/3} \cdot a^{1/3} = a^{1/3+1/3+1/3} = a^1$, then $a^{1/3}$ should mean $\sqrt[3]{a}$.

$$a^{1/n} = \sqrt[n]{a}$$

$a^{1/n}$ means $\sqrt[n]{a}$. When a is nonnegative, n can be any natural number greater than 1. When a is negative, n can be any odd natural number greater than 1.

Thus, $a^{1/5} = \sqrt[5]{a}$ and $a^{1/10} = \sqrt[10]{a}$. Note that the denominator of the exponent becomes the index and the base becomes the radicand.

EXAMPLE **1**   Write an equivalent expression using radical notation and, if possible, simplify.

a) $16^{1/2}$ **b)** $(-8)^{1/3}$ **c)** $(abc)^{1/5}$ **d)** $(25x^{16})^{1/2}$

SOLUTION

a) $16^{1/2} = \sqrt{16} = 4$

b) $(-8)^{1/3} = \sqrt[3]{-8} = -2$ The denominator of the exponent becomes the index. The base becomes the radicand. Recall that for square roots, the index 2 is understood without being written.

c) $(abc)^{1/5} = \sqrt[5]{abc}$

d) $(25x^{16})^{1/2} = 25^{1/2}x^8 = \sqrt{25} \cdot x^8 = 5x^8$ ▸ TRY EXERCISE 11

EXAMPLE **2**   Write an equivalent expression using exponential notation.

a) $\sqrt[5]{9ab}$ **b)** $\sqrt[7]{\dfrac{x^3y}{4}}$ **c)** $\sqrt{5x}$

SOLUTION   Parentheses are required to indicate the base.

a) $\sqrt[5]{9ab} = (9ab)^{1/5}$

b) $\sqrt[7]{\dfrac{x^3y}{4}} = \left(\dfrac{x^3y}{4}\right)^{1/7}$ The index becomes the denominator of the exponent. The radicand becomes the base.

c) $\sqrt{5x} = (5x)^{1/2}$ The index 2 is understood without being written. We assume $x \geq 0$.   ▸ TRY EXERCISE 31

How shall we define $a^{2/3}$? If the property for multiplying exponents is to hold, we must have $a^{2/3} = (a^{1/3})^2$ and $a^{2/3} = (a^2)^{1/3}$. This would suggest that $a^{2/3} = (\sqrt[3]{a})^2$ and $a^{2/3} = \sqrt[3]{a^2}$. We make our definition accordingly.

Positive Rational Exponents

For any natural numbers m and n ($n \neq 1$) and any real number a for which $\sqrt[n]{a}$ exists,

$$a^{m/n} \quad \text{means} \quad (\sqrt[n]{a})^m, \quad \text{or} \quad \sqrt[n]{a^m}.$$

EXAMPLE 3

Write an equivalent expression using radical notation and simplify.

a) $27^{2/3}$　　　　　　　　　　　　　**b)** $25^{3/2}$

SOLUTION

a) $27^{2/3}$ means $(\sqrt[3]{27})^2$ or, equivalently, $\sqrt[3]{27^2}$. Let's see which is easier to simplify:

$$(\sqrt[3]{27})^2 = 3^2 \qquad \sqrt[3]{27^2} = \sqrt[3]{729}$$
$$= 9; \qquad\qquad\quad = 9.$$

The simplification on the left is probably easier for most people.

b) $25^{3/2}$ means $(\sqrt[2]{25})^3$ or, equivalently, $\sqrt[2]{25^3}$ (the index 2 is normally omitted). Since $\sqrt{25}$ is more commonly known than $\sqrt{25^3}$, we use that form:

$$25^{3/2} = (\sqrt{25})^3 = 5^3 = 125.$$

> **TRY EXERCISE** 23

STUDENT NOTES

It is important to remember both meanings of $a^{m/n}$. When the root of the base a is known, $(\sqrt[n]{a})^m$ is generally easier to work with. When it is not known, $\sqrt[n]{a^m}$ is often more convenient.

EXAMPLE 4

Write an equivalent expression using exponential notation.

a) $\sqrt[3]{9^4}$　　　　　　　　　　　　　**b)** $(\sqrt[4]{7xy})^5$

SOLUTION

a) $\sqrt[3]{9^4} = 9^{4/3}$

b) $(\sqrt[4]{7xy})^5 = (7xy)^{5/4}$

The index becomes the denominator of the fraction that is the exponent.

> **TRY EXERCISE** 37

Negative Rational Exponents

Recall from Section 4.2 that $x^{-2} = 1/x^2$. Negative rational exponents behave similarly.

Negative Rational Exponents

For any rational number m/n and any nonzero real number a for which $a^{m/n}$ exists,

$$a^{-m/n} \quad \text{means} \quad \frac{1}{a^{m/n}}.$$

TECHNOLOGY CONNECTION

To approximate $7^{2/3}$, we enter
7 ⌢ (2/3).

1. Why are the parentheses needed above?
2. Compare the graphs of $y_1 = x^{1/2}$, $y_2 = x$, and $y_3 = x^{3/2}$ and determine those x-values for which $y_1 > y_3$.

CAUTION! A negative exponent does not indicate that the expression in which it appears is negative: $a^{-1} \neq -a$.

EXAMPLE 5   Write an equivalent expression with positive exponents and, if possible, simplify.

a) $9^{-1/2}$ b) $(5xy)^{-4/5}$ c) $64^{-2/3}$

d) $4x^{-2/3}y^{1/5}$ e) $\left(\dfrac{3r}{7s}\right)^{-5/2}$

SOLUTION

a) $9^{-1/2} = \dfrac{1}{9^{1/2}}$ $9^{-1/2}$ is the reciprocal of $9^{1/2}$.

   Since $9^{1/2} = \sqrt{9} = 3$, the answer simplifies to $\dfrac{1}{3}$.

b) $(5xy)^{-4/5} = \dfrac{1}{(5xy)^{4/5}}$ $(5xy)^{-4/5}$ is the reciprocal of $(5xy)^{4/5}$.

c) $64^{-2/3} = \dfrac{1}{64^{2/3}}$ $64^{-2/3}$ is the reciprocal of $64^{2/3}$.

   Since $64^{2/3} = \left(\sqrt[3]{64}\right)^2 = 4^2 = 16$, the answer simplifies to $\dfrac{1}{16}$.

d) $4x^{-2/3}y^{1/5} = 4 \cdot \dfrac{1}{x^{2/3}} \cdot y^{1/5} = \dfrac{4y^{1/5}}{x^{2/3}}$

e) In Section 4.2, we found that $(a/b)^{-n} = (b/a)^n$. This property holds for *any* negative exponent:

   $$\left(\dfrac{3r}{7s}\right)^{-5/2} = \left(\dfrac{7s}{3r}\right)^{5/2}.$$ Writing the reciprocal of the base and changing the sign of the exponent

> TRY EXERCISE 53

Laws of Exponents

The same laws hold for rational exponents as for integer exponents.

> ### Laws of Exponents
>
> For any real numbers a and b and any rational exponents m and n for which a^m, a^n, and b^m are defined:
>
> **1.** $a^m \cdot a^n = a^{m+n}$ When multiplying, add exponents if the bases are the same.
>
> **2.** $\dfrac{a^m}{a^n} = a^{m-n}$ When dividing, subtract exponents if the bases are the same. (Assume $a \neq 0$.)
>
> **3.** $(a^m)^n = a^{m \cdot n}$ To raise a power to a power, multiply the exponents.
>
> **4.** $(ab)^m = a^m b^m$ To raise a product to a power, raise each factor to the power and multiply.

EXAMPLE 6   Use the laws of exponents to simplify.

a) $3^{1/5} \cdot 3^{3/5}$ b) $\dfrac{a^{1/4}}{a^{1/2}}$

c) $(7.2^{2/3})^{3/4}$ d) $(a^{-1/3}b^{2/5})^{1/2}$

SOLUTION

a) $3^{1/5} \cdot 3^{3/5} = 3^{1/5+3/5} = 3^{4/5}$ Adding exponents

b) $\dfrac{a^{1/4}}{a^{1/2}} = a^{1/4-1/2} = a^{1/4-2/4}$ Subtracting exponents after finding a common denominator

$$= a^{-1/4}, \text{ or } \frac{1}{a^{1/4}} \qquad a^{-1/4} \text{ is the reciprocal of } a^{1/4}.$$

c) $(7.2^{2/3})^{3/4} = 7.2^{(2/3)(3/4)} = 7.2^{6/12}$ Multiplying exponents

$$= 7.2^{1/2} \qquad \begin{array}{l}\text{Using arithmetic to simplify}\\ \text{the exponent}\end{array}$$

d) $(a^{-1/3}b^{2/5})^{1/2} = a^{(-1/3)(1/2)} \cdot b^{(2/5)(1/2)}$ Raising a product to a power and multiplying exponents

$$= a^{-1/6}b^{1/5}, \text{ or } \frac{b^{1/5}}{a^{1/6}}$$

> **TRY EXERCISE** ▶ 69

Simplifying Radical Expressions

Many radical expressions contain radicands or factors of radicands that are powers. When these powers and the index share a common factor, rational exponents can be used to simplify the expression.

> **To Simplify Radical Expressions**
> 1. Convert radical expressions to exponential expressions.
> 2. Use arithmetic and the laws of exponents to simplify.
> 3. Convert back to radical notation as needed.

EXAMPLE **7**   Use rational exponents to simplify. Do not use exponents that are fractions in the final answer.

a) $\sqrt[6]{(5x)^3}$ **b)** $\sqrt[5]{t^{20}}$

c) $\left(\sqrt[3]{ab^2c}\right)^{12}$ **d)** $\sqrt{\sqrt[3]{x}}$

SOLUTION

a) $\sqrt[6]{(5x)^3} = (5x)^{3/6}$ Converting to exponential notation

$$= (5x)^{1/2} \qquad \text{Simplifying the exponent}$$

$$= \sqrt{5x} \qquad \text{Returning to radical notation}$$

b) $\sqrt[5]{t^{20}} = t^{20/5}$ Converting to exponential notation

$$= t^4 \qquad \text{Simplifying the exponent}$$

c) $\left(\sqrt[3]{ab^2c}\right)^{12} = (ab^2c)^{12/3}$ Converting to exponential notation

$$= (ab^2c)^4 \qquad \text{Simplifying the exponent}$$

$$= a^4b^8c^4 \qquad \text{Using the laws of exponents}$$

d) $\sqrt{\sqrt[3]{x}} = \sqrt{x^{1/3}}$ Converting the radicand to exponential notation

$$= (x^{1/3})^{1/2} \qquad \text{Try to go directly to this step.}$$

$$= x^{1/6} \qquad \text{Using the laws of exponents}$$

$$= \sqrt[6]{x} \qquad \text{Returning to radical notation}$$

> **TRY EXERCISE** ▶ 87

TECHNOLOGY CONNECTION

One way to check Example 7(a) is to let $y_1 = (5x)^{3/6}$ and $y_2 = \sqrt{5x}$. Then use **GRAPH** or **TABLE** to see if $y_1 = y_2$. An alternative is to let $y_3 = y_2 - y_1$ and see if $y_3 = 0$. Check Example 7(a) using one of these two methods.

1. Why are rational exponents especially useful when working on a graphing calculator?

10.2 **EXERCISE SET**

Concept Reinforcement *In each of Exercises 1–8, match the expression with the equivalent expression from the column on the right.*

1. ___ $x^{2/5}$ a) $x^{3/5}$

2. ___ $x^{5/2}$ b) $\left(\sqrt[5]{x}\right)^4$

3. ___ $x^{-5/2}$ c) $\sqrt{x^5}$

4. ___ $x^{-2/5}$ d) $x^{1/2}$

5. ___ $x^{1/5} \cdot x^{2/5}$ e) $\dfrac{1}{\left(\sqrt{x}\right)^5}$

6. ___ $(x^{1/5})^{5/2}$

7. ___ $\sqrt[5]{x^4}$ f) $\sqrt[4]{x^5}$

8. ___ $\left(\sqrt[4]{x}\right)^5$ g) $\sqrt[5]{x^2}$

h) $\dfrac{1}{\left(\sqrt[5]{x}\right)^2}$

Note: Assume for all exercises that all variables are nonnegative and that all denominators are nonzero.

Write an equivalent expression using radical notation and, if possible, simplify.

9. $y^{1/3}$ 10. $t^{1/4}$

11. $36^{1/2}$ 12. $125^{1/3}$

13. $32^{1/5}$ 14. $81^{1/4}$

15. $64^{1/2}$ 16. $100^{1/2}$

17. $(xyz)^{1/2}$ 18. $(ab)^{1/4}$

19. $(a^2b^2)^{1/5}$ 20. $(x^3y^3)^{1/4}$

21. $t^{5/6}$ 22. $a^{3/2}$

23. $16^{3/4}$ 24. $4^{7/2}$

25. $125^{4/3}$ 26. $9^{5/2}$

27. $(81x)^{3/4}$ 28. $(125a)^{2/3}$

29. $(25x^4)^{3/2}$ 30. $(9y^6)^{3/2}$

Write an equivalent expression using exponential notation.

31. $\sqrt[3]{18}$ 32. $\sqrt[4]{10}$

33. $\sqrt{30}$ 34. $\sqrt{22}$

35. $\sqrt{x^7}$ 36. $\sqrt{a^3}$

37. $\sqrt[5]{m^2}$ 38. $\sqrt[5]{n^4}$

39. $\sqrt[4]{pq}$ 40. $\sqrt[3]{cd}$

41. $\sqrt[5]{xy^2z}$ 42. $\sqrt[7]{x^3y^2z^2}$

43. $\left(\sqrt{3mn}\right)^3$ 44. $\left(\sqrt[3]{7xy}\right)^4$

45. $\left(\sqrt[5]{8x^2y}\right)^5$ 46. $\left(\sqrt[6]{2a^5b}\right)^7$

47. $\dfrac{2x}{\sqrt[3]{z^2}}$ 48. $\dfrac{3a}{\sqrt[5]{c^2}}$

Write an equivalent expression with positive exponents and, if possible, simplify.

49. $a^{-1/4}$ 50. $m^{-1/3}$

51. $(2rs)^{-3/4}$ 52. $(5xy)^{-5/6}$

53. $\left(\dfrac{1}{16}\right)^{-3/4}$ 54. $\left(\dfrac{1}{8}\right)^{-2/3}$

55. $\dfrac{8c}{a^{-3/5}}$ 56. $\dfrac{3b}{a^{-5/7}}$

57. $2a^{3/4}b^{-1/2}c^{2/3}$ 58. $5x^{-2/3}y^{4/5}z$

59. $3^{-5/2}a^3b^{-7/3}$ 60. $2^{-1/3}x^4y^{-2/7}$

61. $\left(\dfrac{2ab}{3c}\right)^{-5/6}$ 62. $\left(\dfrac{7x}{8yz}\right)^{-3/5}$

63. $\dfrac{6a}{\sqrt[4]{b}}$ 64. $\dfrac{5y}{\sqrt[3]{z}}$

Use the laws of exponents to simplify. Do not use negative exponents in any answers.

65. $11^{1/2} \cdot 11^{1/3}$ 66. $5^{1/4} \cdot 5^{1/8}$

67. $\dfrac{3^{5/8}}{3^{-1/8}}$ 68. $\dfrac{8^{7/11}}{8^{-2/11}}$

69. $\dfrac{4.3^{-1/5}}{4.3^{-7/10}}$ 70. $\dfrac{2.7^{-11/12}}{2.7^{-1/6}}$

71. $(10^{3/5})^{2/5}$ 72. $(5^{5/4})^{3/7}$

73. $a^{2/3} \cdot a^{5/4}$ 74. $x^{3/4} \cdot x^{1/3}$

Aha! 75. $(64^{3/4})^{4/3}$ 76. $(27^{-2/3})^{3/2}$

77. $(m^{2/3}n^{-1/4})^{1/2}$ 78. $(x^{-1/3}y^{2/5})^{1/4}$

Use rational exponents to simplify. Do not use fraction exponents in the final answer.

79. $\sqrt[9]{x^3}$ 80. $\sqrt[12]{a^3}$

81. $\sqrt[3]{y^{15}}$ 82. $\sqrt[4]{y^{40}}$

83. $\sqrt[12]{a^6}$ 84. $\sqrt[30]{x^5}$

85. $\left(\sqrt[7]{xy}\right)^{14}$ 86. $\left(\sqrt[3]{ab}\right)^{15}$

87. $\sqrt[4]{(7a)^2}$

88. $\sqrt[8]{(3x)^2}$

89. $\left(\sqrt[8]{2x}\right)^6$

90. $\left(\sqrt[10]{3a}\right)^5$

91. $\sqrt{\sqrt[5]{m}}$

92. $\sqrt[6]{\sqrt{n}}$

93. $\sqrt[4]{(xy)^{12}}$

94. $\sqrt{(ab)^6}$

95. $\left(\sqrt[5]{a^2b^4}\right)^{15}$

96. $\left(\sqrt[3]{x^2y^5}\right)^{12}$

97. $\sqrt[3]{\sqrt[4]{xy}}$

98. $\sqrt[5]{\sqrt[3]{2a}}$

99. If $f(x) = (x + 5)^{1/2}(x + 7)^{-1/2}$, find the domain of f. Explain how you found your answer.

100. Let $f(x) = 5x^{-1/3}$. Under what condition will we have $f(x) > 0$? Why?

Skill Review

To prepare for Section 10.3, review multiplying and factoring polynomials (Sections 4.5 and 5.4).

Multiply. [4.5]

101. $(x + 5)(x - 5)$

102. $(x - 2)(x^2 + 2x + 4)$

Factor. [5.4]

103. $4x^2 + 20x + 25$

104. $9a^2 - 24a + 16$

105. $5t^2 - 10t + 5$

106. $3n^2 + 12n + 12$

Synthesis

107. Explain why $\sqrt[3]{x^6} = x^2$ for any value of x, whereas $\sqrt{x^6} = x^3$ only when $x \geq 0$.

108. If $g(x) = x^{3/n}$, in what way does the domain of g depend on whether n is odd or even?

Use rational exponents to simplify.

109. $\sqrt{x\sqrt[3]{x^2}}$

110. $\sqrt[4]{\sqrt[3]{8x^3y^6}}$

111. $\sqrt[14]{c^2 - 2cd + d^2}$

Music.   The function given by $f(x) = k2^{x/12}$ can be used to determine the frequency, in cycles per second, of a musical note that is x half-steps above a note with frequency k. *

112. The frequency of concert A for a trumpet is 440 cycles per second. Find the frequency of the A that is two octaves (24 half-steps) above concert A (few trumpeters can reach this note.)

---

*This application was inspired by information provided by Dr. Homer B. Tilton of Pima Community College East.

113. Show that the G that is 7 half-steps (a "perfect fifth") above middle C (262 cycles per second) has a frequency that is about 1.5 times that of middle C.

114. Show that the C sharp that is 4 half-steps (a "major third") above concert A (see Exercise 112) has a frequency that is about 25% greater than that of concert A.

115. *Road pavement messages.*   In a psychological study, it was determined that the proper length L of the letters of a word printed on pavement is given by

$$L = \frac{0.000169d^{2.27}}{h},$$

where d is the distance of a car from the lettering and h is the height of the eye above the surface of the road. All units are in meters. This formula says that from a vantage point h meters above the surface of the road, if a driver is to be able to recognize a message d meters away, that message will be the most recognizable if the length of the letters is L. Find L to the nearest tenth of a meter, given d and h.

a) $h = 1\,\text{m}, d = 60\,\text{m}$
b) $h = 0.9906\,\text{m}, d = 75\,\text{m}$
c) $h = 2.4\,\text{m}, d = 80\,\text{m}$
d) $h = 1.1\,\text{m}, d = 100\,\text{m}$

116. *Baseball.*   The statistician Bill James has found that a baseball team's winning percentage P can be approximated by

$$P = \frac{r^{1.83}}{r^{1.83} + \sigma^{1.83}},$$

where r is the total number of runs scored by that team and σ (sigma) is the total number of runs scored by their opponents. During a recent season, the San Francisco Giants scored 799 runs and their opponents scored 749 runs. Use James's formula to predict the Giants' winning percentage (the team actually won 55.6% of their games).

Source: M. Bittinger, *One Man's Journey Through Mathematics.* Boston: Addison–Wesley, 2004

117. *Forestry.*   The total wood volume T, in cubic feet, in a California black oak can be estimated using the formula

$$T = 0.936\, d^{1.97} h^{0.85},$$

where d is the diameter of the tree at breast height and h is the total height of the tree. How much wood is in a California black oak that is 3 ft in diameter at breast height and 80 ft high?

Source: Norman H. Pillsbury and Michael L. Kirkley, 1984. Equations for total, wood, and saw-log volume for thirteen California hardwoods, USDA Forest Service PNW Research Note No. 414: 52 p.

118. *Physics.*   The equation $m = m_0(1 - v^2 c^{-2})^{-1/2}$, developed by Albert Einstein, is used to determine the mass m of an object that is moving v meters per second and has mass m_0 before the motion begins. The constant c is the speed of light, approximately 3×10^8 m/sec. Suppose that a particle with mass 8 mg is accelerated to a speed of $\frac{9}{5} \times 10^8$ m/sec. Without using a calculator, find the new mass of the particle.

119. Using a graphing calculator, select **MODE** SIMUL and the FORMAT EXPROFF. Then graph

$$y_1 = x^{1/2}, \qquad y_2 = 3x^{2/5},$$
$$y_3 = x^{4/7}, \quad \text{and} \quad y_4 = \tfrac{1}{5}x^{3/4}.$$

Looking only at coordinates, match each graph with its equation.

COLLABORATIVE CORNER

Are Equivalent Fractions Equivalent Exponents?

Focus: Functions and rational exponents

Time: 10–20 minutes

Group size: 3

Materials: Graph paper

In arithmetic, we have seen that $\frac{1}{3}, \frac{1}{6} \cdot 2$, and $2 \cdot \frac{1}{6}$ all represent the same number. Interestingly,

$$f(x) = x^{1/3},$$
$$g(x) = (x^{1/6})^2, \quad \text{and}$$
$$h(x) = (x^2)^{1/6}$$

represent three *different* functions.

ACTIVITY

1. Selecting a variety of values for x and using the definition of positive rational exponents, one group member should graph f, a second group member should graph g, and a third group member should graph h. Be sure to check whether negative x-values are in the domain of the function.

2. Compare the three graphs and check each other's work. How and why do the graphs differ?

3. Decide as a group which graph, if any, would best represent the graph of $k(x) = x^{2/6}$. Then be prepared to explain your reasoning to the entire class. (*Hint:* Study the definition of $a^{m/n}$ on p. 644 carefully.)

10.3 Multiplying Radical Expressions

Multiplying Radical Expressions ▪ Simplifying by Factoring ▪ Multiplying and Simplifying

Multiplying Radical Expressions

Note that $\sqrt{4}\sqrt{25} = 2 \cdot 5 = 10$. Also $\sqrt{4 \cdot 25} = \sqrt{100} = 10$. Likewise,

$$\sqrt[3]{27}\sqrt[3]{8} = 3 \cdot 2 = 6 \quad \text{and} \quad \sqrt[3]{27 \cdot 8} = \sqrt[3]{216} = 6.$$

These examples suggest the following.

> **The Product Rule for Radicals**
>
> For any real numbers $\sqrt[n]{a}$ and $\sqrt[n]{b}$,
>
> $$\sqrt[n]{a} \cdot \sqrt[n]{b} = \sqrt[n]{a \cdot b}.$$
>
> (The product of two nth roots is the nth root of the product of the two radicands.)

Rational exponents can be used to derive this rule:

$$\sqrt[n]{a} \cdot \sqrt[n]{b} = a^{1/n} \cdot b^{1/n} = (a \cdot b)^{1/n} = \sqrt[n]{a \cdot b}.$$

EXAMPLE **1** Multiply.

a) $\sqrt{2} \cdot \sqrt{7}$ **b)** $\sqrt{x + 3}\sqrt{x - 3}$

c) $\sqrt[3]{4} \cdot \sqrt[3]{5}$ **d)** $\sqrt[4]{\dfrac{y}{5}} \cdot \sqrt[4]{\dfrac{7}{x}}$

SOLUTION

a) When no index is written, roots are understood to be square roots with an unwritten index of two. We apply the product rule:

$$\sqrt{2} \cdot \sqrt{7} = \sqrt{2 \cdot 7}$$
$$= \sqrt{14}.$$

b) $\sqrt{x + 3}\sqrt{x - 3} = \sqrt{(x + 3)(x - 3)}$ The product of two square roots is
$$= \sqrt{x^2 - 9}$$ the square root of the product.

CAUTION! $\sqrt{x^2 - 9} \neq \sqrt{x^2} - \sqrt{9}.$

c) Both $\sqrt[3]{4}$ and $\sqrt[3]{5}$ have indices of three, so to multiply we can use the product rule:

$$\sqrt[3]{4} \cdot \sqrt[3]{5} = \sqrt[3]{4 \cdot 5} = \sqrt[3]{20}.$$

d) $\sqrt[4]{\dfrac{y}{5}} \cdot \sqrt[4]{\dfrac{7}{x}} = \sqrt[4]{\dfrac{y}{5} \cdot \dfrac{7}{x}} = \sqrt[4]{\dfrac{7y}{5x}}$ In Section 10.4, we discuss other ways to write answers like this.

TRY EXERCISE 7

To check Example 1(b), let $y_1 = \sqrt{x + 3}\sqrt{x - 3}$ and $y_2 = \sqrt{x^2 - 9}$ and compare:

$y_1 = \sqrt{(x + 3)}\sqrt{(x - 3)}$

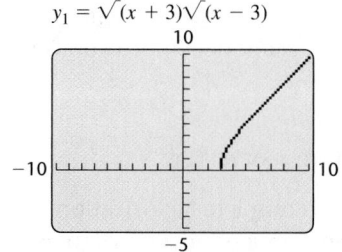

$y_2 = \sqrt{(x^2 - 9)}$

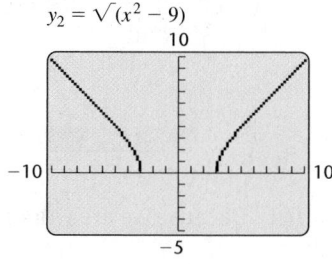

Because $y_1 = y_2$ for all *x*-values that can be used in *both* y_1 and y_2, Example 1(b) *is* correct.

1. Why do the graphs above differ in appearance? (*Hint*: What are the domains of the two related functions?)

CAUTION!   The product rule for radicals applies only when radicals have the same index:

$$\sqrt[n]{a} \cdot \sqrt[m]{b} \ne \sqrt[nm]{a \cdot b}.$$

Simplifying by Factoring

The number *p* is a *perfect square* if there exists a rational number *q* for which $q^2 = p$. We say that *p* is a *perfect cube* if $q^3 = p$ for some rational number *q*. In general, *p* is a *perfect nth power* if $q^n = p$ for some rational number *q*. Thus, 16 and $\frac{1}{10,000}$ are both perfect 4th powers since $2^4 = 16$ and $\left(\frac{1}{10}\right)^4 = \frac{1}{10,000}$.

The product rule allows us to simplify $\sqrt[n]{ab}$ whenever *ab* contains a factor that is a perfect *n*th power.

> **Using the Product Rule to Simplify**
>
> $$\sqrt[n]{ab} = \sqrt[n]{a} \cdot \sqrt[n]{b}.$$
>
> $\left(\sqrt[n]{a} \text{ and } \sqrt[n]{b} \text{ must both be real numbers.}\right)$

To illustrate, suppose we wish to simplify $\sqrt{20}$. Since this is a *square* root, we check to see if there is a factor of 20 that is a perfect square. There is one, 4, so we express 20 as $4 \cdot 5$ and use the product rule:

$$\sqrt{20} = \sqrt{4 \cdot 5} \qquad \text{Factoring the radicand (4 is a perfect square)}$$
$$= \sqrt{4} \cdot \sqrt{5} \qquad \text{Factoring into two radicals}$$
$$= 2\sqrt{5}. \qquad \text{Finding the square root of 4}$$

> **To Simplify a Radical Expression with Index *n* by Factoring**
>
> 1. Express the radicand as a product in which one factor is the largest perfect *n*th power possible.
> 2. Rewrite the expression as the *n*th root of each factor.
> 3. Simplify the expression containing the perfect *n*th power.
> 4. Simplification is complete when no radicand has a factor that is a perfect *n*th power.

It is often safe to assume that a radicand does not represent a negative number raised to an even power. We will henceforth make this assumption—unless functions are involved—and discontinue use of absolute-value notation when taking even roots.

EXAMPLE   2   Simplify by factoring: **(a)** $\sqrt{200}$; **(b)** $\sqrt{18x^2y}$; **(c)** $\sqrt[3]{-72}$; **(d)** $\sqrt[4]{162x^6}$.

SOLUTION

a) $\sqrt{200} = \sqrt{100 \cdot 2}$ 100 is the largest perfect-square factor of 200.
$\qquad\qquad = \sqrt{100} \cdot \sqrt{2} = 10\sqrt{2}$

Express the radicand as a product. **b)** $\sqrt{18x^2y} = \sqrt{9 \cdot 2 \cdot x^2 \cdot y}$ $9x^2$ is the largest perfect-square factor of $18x^2y$.

Rewrite as the *n*th root of each factor. $= \sqrt{9x^2} \cdot \sqrt{2y}$ Factoring into two radicals

Simplify. $= 3x\sqrt{2y}$ Taking the square root of $9x^2$

c) $\sqrt[3]{-72} = \sqrt[3]{-8 \cdot 9}$ -8 is a perfect-cube (third-power) factor of -72.

$= \sqrt[3]{-8} \cdot \sqrt[3]{9} = -2\sqrt[3]{9}$

d) $\sqrt[4]{162x^6} = \sqrt[4]{81 \cdot 2 \cdot x^4 \cdot x^2}$ $81 \cdot x^4$ is the largest perfect fourth-power factor of $162x^6$.

$= \sqrt[4]{81x^4} \cdot \sqrt[4]{2x^2}$ Factoring into two radicals

$= 3x\sqrt[4]{2x^2}$ Taking the fourth root of $81x^4$

Let's look at this example another way. We write a complete factorization and look for quadruples of factors. Each quadruple makes a perfect fourth power:

$$\sqrt[4]{162x^6} = \sqrt[4]{\boxed{3 \cdot 3 \cdot 3 \cdot 3} \cdot 2 \cdot \boxed{x \cdot x \cdot x \cdot x} \cdot x \cdot x} \quad \begin{array}{l} 3 \cdot 3 \cdot 3 \cdot 3 = 3^4 \\ \text{and} \\ x \cdot x \cdot x \cdot x = x^4 \end{array}$$

$$= 3 \cdot x \cdot \sqrt[4]{2 \cdot x \cdot x}$$

$$= 3x\sqrt[4]{2x^2}.$$

> TRY EXERCISE 31

EXAMPLE 3   If $f(x) = \sqrt{3x^2 - 6x + 3}$, find a simplified form for $f(x)$. Because we are working with a function, assume that x can be any real number.

SOLUTION

$$f(x) = \sqrt{3x^2 - 6x + 3}$$

$$= \sqrt{3(x^2 - 2x + 1)} \quad \left. \right\} \quad \text{Factoring the radicand; } x^2 - 2x + 1$$

$$= \sqrt{(x-1)^2 \cdot 3} \quad \text{is a perfect square.}$$

$$= \sqrt{(x-1)^2} \cdot \sqrt{3} \quad \text{Factoring into two radicals}$$

$$= |x - 1|\sqrt{3} \quad \text{Finding the square root of } (x-1)^2$$

> TRY EXERCISE 43

TECHNOLOGY CONNECTION

To check Example 3, let $y_1 = \sqrt{}(3x^2 - 6x + 3)$, $y_2 = \text{abs}(x - 1)\sqrt{}(3)$, and $y_3 = (x - 1)\sqrt{}(3)$. Do the graphs all coincide? Why or why not?

EXAMPLE 4   Simplify: **(a)** $\sqrt{x^7y^{11}z^9}$; **(b)** $\sqrt[3]{16a^7b^{14}}$.

SOLUTION

a) There are many ways to factor $x^7y^{11}z^9$. Because of the square root (index of 2), we identify the largest exponents that are multiples of 2:

$$\sqrt{x^7y^{11}z^9} = \sqrt{x^6 \cdot x \cdot y^{10} \cdot y \cdot z^8 \cdot z} \quad \begin{array}{l} \text{The largest perfect-square} \\ \text{factor is } x^6y^{10}z^8. \end{array}$$

$$= \sqrt{x^6y^{10}z^8}\sqrt{xyz} \quad \text{Factoring into two radicals}$$

$$= x^{6/2}y^{10/2}z^{8/2}\sqrt{xyz} \quad \begin{array}{l} \text{Converting to rational expo-} \\ \text{nents. Try to do this mentally.} \end{array}$$

$$= x^3y^5z^4\sqrt{xyz}. \quad \text{Simplifying}$$

Check: $(x^3y^5z^4\sqrt{xyz})^2 = (x^3)^2(y^5)^2(z^4)^2(\sqrt{xyz})^2$

$$= x^6 \cdot y^{10} \cdot z^8 \cdot xyz = x^7y^{11}z^9.$$

Our check shows that $x^3y^5z^4\sqrt{xyz}$ is the square root of $x^7y^{11}z^9$.

b) There are many ways to factor $16a^7b^{14}$. Because of the cube root (index of 3), we identify factors with the largest exponents that are multiples of 3:

$$\sqrt[3]{16a^7b^{14}} = \sqrt[3]{8 \cdot 2 \cdot a^6 \cdot a \cdot b^{12} \cdot b^2} \qquad \text{The largest perfect-cube} \\ \text{factor is } 8a^6b^{12}.$$

$$= \sqrt[3]{8a^6b^{12}}\sqrt[3]{2ab^2} \qquad \text{Rewriting as a product of} \\ \text{cube roots}$$

$$= 2a^2b^4\sqrt[3]{2ab^2}. \qquad \text{Simplifying the expression} \\ \text{containing the perfect cube}$$

As a check, let's redo the problem using a complete factorization of the radicand:

$$\sqrt[3]{16a^7b^{14}} = \sqrt[3]{\boxed{2 \cdot 2 \cdot 2} \cdot 2 \cdot \boxed{a \cdot a \cdot a} \cdot \boxed{a \cdot a \cdot a} \cdot a \cdot \boxed{b \cdot b \cdot b} \cdot \boxed{b \cdot b \cdot b} \cdot \boxed{b \cdot b \cdot b} \cdot \boxed{b \cdot b \cdot b} \cdot b \cdot b}$$

Each triple of factors makes a cube.

$$= 2 \cdot a \cdot a \cdot b \cdot b \cdot b \cdot b \cdot \sqrt[3]{2 \cdot a \cdot b \cdot b}$$

$$= 2a^2b^4\sqrt[3]{2ab^2}. \qquad \text{Our answer checks.}$$

> **TRY EXERCISE** 51

> *Remember*: To simplify an nth root, identify factors in the radicand with exponents that are multiples of n.

Multiplying and Simplifying

We have used the product rule for radicals to find products and also to simplify radical expressions. For some radical expressions, it is possible to do both: First find a product and then simplify.

EXAMPLE **5** Multiply and simplify.

a) $\sqrt{15}\sqrt{6}$ 　　　　　　**b)** $3\sqrt[3]{25} \cdot 2\sqrt[3]{5}$ 　　　　　　**c)** $\sqrt[4]{8x^3y^5}\sqrt[4]{4x^2y^3}$

SOLUTION

a) $\sqrt{15}\sqrt{6} = \sqrt{15 \cdot 6} \qquad \text{Multiplying radicands}$

$$= \sqrt{90} = \sqrt{9}\sqrt{10} \qquad \text{9 is a perfect square.}$$

$$= 3\sqrt{10}$$

b) $3\sqrt[3]{25} \cdot 2\sqrt[3]{5} = 3 \cdot 2 \cdot \sqrt[3]{25 \cdot 5} \qquad \text{Using a commutative law;} \\ \text{multiplying radicands}$

$$= 6 \cdot \sqrt[3]{125} \qquad \text{125 is a perfect cube.}$$

$$= 6 \cdot 5, \text{ or } 30$$

c) $\sqrt[4]{8x^3y^5}\sqrt[4]{4x^2y^3} = \sqrt[4]{32x^5y^8} \qquad \text{Multiplying radicands}$

$$= \sqrt[4]{16x^4y^8 \cdot 2x} \qquad \text{Identifying the largest perfect} \\ \text{fourth-power factor}$$

$$= \sqrt[4]{16x^4y^8}\sqrt[4]{2x} \qquad \text{Factoring into radicals}$$

$$= 2xy^2\sqrt[4]{2x} \qquad \text{Finding the fourth root;} \\ \text{assume } x \geq 0.$$

The checks are left to the student.

> **TRY EXERCISE** 65

10.3 EXERCISE SET

🐦 **Concept Reinforcement** *Classify each statement as either true or false.*

1. For any real numbers $\sqrt[n]{a}$ and $\sqrt[n]{b}$,
$\sqrt[n]{a} \cdot \sqrt[n]{b} = \sqrt[n]{ab}$.

2. For any real numbers $\sqrt[n]{a}$ and $\sqrt[n]{b}$,
$\sqrt[n]{a} + \sqrt[n]{b} = \sqrt[n]{a+b}$.

3. For any real numbers $\sqrt[n]{a}$ and $\sqrt[m]{b}$,
$\sqrt[n]{a} \cdot \sqrt[m]{b} = \sqrt[nm]{ab}$.

4. For $x > 0$, $\sqrt{x^2 - 9} = x - 3$.

5. The expression $\sqrt[3]{X}$ is not simplified if X contains a factor that is a perfect cube.

6. It is often possible to simplify $\sqrt{A \cdot B}$ even though $\sqrt{A}$ and $\sqrt{B}$ cannot be simplified.

Multiply.

7. $\sqrt{3}\,\sqrt{10}$

8. $\sqrt{6}\,\sqrt{5}$

9. $\sqrt[3]{7}\,\sqrt[3]{5}$

10. $\sqrt[3]{2}\,\sqrt[3]{3}$

11. $\sqrt[4]{6}\,\sqrt[4]{9}$

12. $\sqrt[4]{4}\,\sqrt[4]{10}$

13. $\sqrt{2x}\,\sqrt{13y}$

14. $\sqrt{5a}\,\sqrt{6b}$

15. $\sqrt[5]{8y^3}\,\sqrt[5]{10y}$

16. $\sqrt[5]{9t^2}\,\sqrt[5]{2t}$

17. $\sqrt{y-b}\,\sqrt{y+b}$

18. $\sqrt{x-a}\,\sqrt{x+a}$

19. $\sqrt[3]{0.7y}\,\sqrt[3]{0.3y}$

20. $\sqrt[3]{0.5x}\,\sqrt[3]{0.2x}$

21. $\sqrt[5]{x-2}\,\sqrt[5]{(x-2)^2}$

22. $\sqrt[4]{x-1}\,\sqrt[4]{x^2+x+1}$

23. $\sqrt{\dfrac{2}{t}}\,\sqrt{\dfrac{3s}{11}}$

24. $\sqrt{\dfrac{7p}{6}}\,\sqrt{\dfrac{5}{q}}$

25. $\sqrt[7]{\dfrac{x-3}{4}}\,\sqrt[7]{\dfrac{5}{x+2}}$

26. $\sqrt[6]{\dfrac{a}{b-2}}\,\sqrt[6]{\dfrac{3}{b+2}}$

Simplify by factoring.

27. $\sqrt{12}$

28. $\sqrt{300}$

29. $\sqrt{45}$

30. $\sqrt{27}$

31. $\sqrt{8x^9}$

32. $\sqrt{75y^5}$

33. $\sqrt{120}$

34. $\sqrt{350}$

35. $\sqrt{36a^4b}$

36. $\sqrt{175y^8}$

37. $\sqrt[3]{8x^3y^2}$

38. $\sqrt[3]{27ab^6}$

39. $\sqrt[3]{-16x^6}$

40. $\sqrt[3]{-32a^6}$

Find a simplified form of $f(x)$. Assume that x can be any real number.

41. $f(x) = \sqrt[3]{40x^6}$

42. $f(x) = \sqrt[3]{27x^5}$

43. $f(x) = \sqrt{49(x-3)^2}$

44. $f(x) = \sqrt{81(x-1)^2}$

45. $f(x) = \sqrt{5x^2 - 10x + 5} \cdot$

46. $f(x) = \sqrt{2x^2 + 8x + 8}$

Simplify. Assume that no radicands were formed by raising negative numbers to even powers.

47. $\sqrt{a^{10}b^{11}}$

48. $\sqrt{x^8y^7}$

49. $\sqrt[3]{x^5y^6z^{10}}$

50. $\sqrt[3]{a^6b^7c^{13}}$

51. $\sqrt[4]{16x^5y^{11}}$

52. $\sqrt[5]{-32a^7b^{11}}$

53. $\sqrt[5]{x^{13}y^8z^{17}}$

54. $\sqrt[5]{a^6b^8c^9}$

55. $\sqrt[3]{-80a^{14}}$

56. $\sqrt[4]{810x^9}$

Multiply and simplify. Assume that no radicands were formed by raising negative numbers to even powers.

57. $\sqrt{5}\,\sqrt{10}$

58. $\sqrt{2}\,\sqrt{6}$

59. $\sqrt{6}\,\sqrt{33}$

60. $\sqrt{10}\,\sqrt{35}$

61. $\sqrt[3]{9}\,\sqrt[3]{3}$

62. $\sqrt[3]{2}\,\sqrt[3]{4}$

Aha! 63. $\sqrt{24y^5}\,\sqrt{24y^5}$

64. $\sqrt{120t^9}\,\sqrt{120t^9}$

65. $\sqrt[3]{5a^2}\,\sqrt[3]{2a}$

66. $\sqrt[3]{7x}\,\sqrt[3]{3x^2}$

67. $\sqrt{2x^5}\,\sqrt{10x^2}$

68. $\sqrt{5a^7}\,\sqrt{15a^3}$

69. $\sqrt[3]{s^2t^4}\,\sqrt[3]{s^4t^6}$

70. $\sqrt[3]{x^2y^4}\,\sqrt[3]{x^2y^6}$

71. $\sqrt[3]{(x-y)^2}\,\sqrt[3]{(x-y)^{10}}$

72. $\sqrt[3]{(t+4)^5}\,\sqrt[3]{(t+4)}$

73. $\sqrt[4]{20a^3b^7}\,\sqrt[4]{4a^2b^5}$

74. $\sqrt[4]{9x^7y^2}\,\sqrt[4]{9x^2y^9}$

75. $\sqrt[5]{x^3(y+z)^6}\,\sqrt[5]{x^3(y+z)^4}$

76. $\sqrt[5]{a^3(b-c)^4}\,\sqrt[5]{a^7(b-c)^4}$

📝 77. Explain how you could convince a friend that
$\sqrt{x^2 - 16} \neq \sqrt{x^2} - \sqrt{16}$.

📝 78. Why is it incorrect to say that, in general, $\sqrt{x^2} = x$?

Skill Review

Review simplifying rational expressions (Sections 6.1–6.5).

Perform the indicated operation and, if possible, simplify.

79. $\dfrac{15a^2x}{8b} \cdot \dfrac{24b^2x}{5a}$ [6.2]

80. $\dfrac{x^2 - 1}{x^2 - 4} \div \dfrac{x^2 - x - 2}{x^2 + x - 2}$ [6.2]

81. $\dfrac{x - 3}{2x - 10} - \dfrac{3x - 5}{x^2 - 25}$ [6.4]

82. $\dfrac{6x}{25y^2} + \dfrac{3y}{10x}$ [6.4]

83. $\dfrac{a^{-1} + b^{-1}}{ab}$ [6.5]

84. $\dfrac{\dfrac{1}{x + 1} - \dfrac{2}{x}}{\dfrac{3}{x} + \dfrac{1}{x + 1}}$ [6.5]

Synthesis

85. Explain why it is true that $\sqrt[n]{ab} = \sqrt[n]{a} \cdot \sqrt[n]{b}$ for any real numbers $\sqrt[n]{a}$ and $\sqrt[n]{b}$.

86. Is the equation $\sqrt{(2x + 3)^8} = (2x + 3)^4$ always, sometimes, or never true? Why?

87. *Radar range.*   The function given by

$$R(x) = \frac{1}{2}\sqrt[4]{\frac{x \cdot 3.0 \times 10^6}{\pi^2}}$$

can be used to determine the maximum range $R(x)$, in miles, of an ARSR-3 surveillance radar with a peak power of x watts. Determine the maximum radar range when the peak power is 5×10^4 watts.
Source: Introduction to RADAR Techniques, Federal Aviation Administration, 1988

88. *Speed of a skidding car.*   Police can estimate the speed at which a car was traveling by measuring its skid marks. The function given by

$$r(L) = 2\sqrt{5L}$$

can be used, where L is the length of a skid mark, in feet, and $r(L)$ is the speed, in miles per hour. Find

the exact speed and an estimate (to the nearest tenth mile per hour) for the speed of a car that left skid marks **(a)** 20 ft long; **(b)** 70 ft long; **(c)** 90 ft long. See also Exercise 102.

89. *Wind chill temperature.*   When the temperature is T degrees Celsius and the wind speed is v meters per second, the *wind chill temperature*, T_W, is the temperature (with no wind) that it feels like. Here is a formula for finding wind chill temperature:

$$T_W = 33 - \frac{(10.45 + 10\sqrt{v} - v)(33 - T)}{22}.$$

Estimate the wind chill temperature (to the nearest tenth of a degree) for the given actual temperatures and wind speeds.

a) $T = 7°C$, $v = 8$ m/sec
b) $T = 0°C$, $v = 12$ m/sec
c) $T = -5°C$, $v = 14$ m/sec
d) $T = -23°C$, $v = 15$ m/sec

Simplify. Assume that all variables are nonnegative.

90. $\left(\sqrt{r^3t}\right)^7$ **91.** $\left(\sqrt[3]{25x^4}\right)^4$

92. $\left(\sqrt[3]{a^2b^4}\right)^5$ **93.** $\left(\sqrt{a^3b^5}\right)^7$

Draw and compare the graphs of each group of equations.

94. $f(x) = \sqrt{x^2 - 2x + 1}$,
$g(x) = x - 1$,
$h(x) = |x - 1|$

95. $f(x) = \sqrt{x^2 + 2x + 1}$,
$g(x) = x + 1$,
$h(x) = |x + 1|$

96. If $f(t) = \sqrt{t^2 - 3t - 4}$, what is the domain of f?

97. What is the domain of g, if $g(x) = \sqrt{x^2 - 6x + 8}$?

Solve.

98. $\sqrt[3]{5x^{k+1}} \; \sqrt[3]{25x^k} = 5x^7$, for k

99. $\sqrt[5]{4a^{3k+2}} \; \sqrt[5]{8a^{6-k}} = 2a^4$, for k

100. Use a graphing calculator to check your answers to Exercises 21 and 41.

101. Blair is puzzled. When he uses a graphing calculator to graph $y = \sqrt{x} \cdot \sqrt{x}$, he gets the following screen. Explain why Blair did not get the complete line $y = x$.

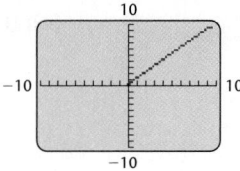

102. Does a car traveling twice as fast as another car leave a skid mark that is twice as long? (See Exercise 88.) Why or why not?

10.4 Dividing Radical Expressions

Dividing and Simplifying ■ Rationalizing Denominators or Numerators with One Term

Dividing and Simplifying

Just as the root of a product can be expressed as the product of two roots, the root of a quotient can be expressed as the quotient of two roots. For example,

$$\sqrt[3]{\frac{27}{8}} = \frac{3}{2} \quad \text{and} \quad \frac{\sqrt[3]{27}}{\sqrt[3]{8}} = \frac{3}{2}.$$

This example suggests the following.

> **The Quotient Rule for Radicals**
> For any real numbers $\sqrt[n]{a}$ and $\sqrt[n]{b}$, $b \neq 0$,
> $$\sqrt[n]{\frac{a}{b}} = \frac{\sqrt[n]{a}}{\sqrt[n]{b}}.$$

Remember that an nth root is simplified when its radicand has no factors that are perfect nth powers. Unless functions are involved, we assume that no radi-cands represent negative quantities raised to an even power.

EXAMPLE 1 Simplify by taking the roots of the numerator and the denominator.

a) $\sqrt[3]{\frac{27}{125}}$ **b)** $\sqrt{\frac{25}{y^2}}$

SOLUTION

a) $\sqrt[3]{\frac{27}{125}} = \frac{\sqrt[3]{27}}{\sqrt[3]{125}} = \frac{3}{5}$ Taking the cube roots of the numerator and the denominator

b) $\sqrt{\frac{25}{y^2}} = \frac{\sqrt{25}}{\sqrt{y^2}} = \frac{5}{y}$ Taking the square roots of the numerator and the denominator. Assume $y > 0$. **TRY EXERCISE 9**

Any radical expressions appearing in the answers should be simplified as much as possible.

EXAMPLE 2 Simplify: **(a)** $\sqrt{\frac{16x^3}{y^8}}$; **(b)** $\sqrt[3]{\frac{27y^{14}}{8x^3}}$.

SOLUTION

a) $\sqrt{\frac{16x^3}{y^8}} = \frac{\sqrt{16x^3}}{\sqrt{y^8}}$

$= \frac{\sqrt{16x^2 \cdot x}}{\sqrt{y^8}}$

$= \frac{4x\sqrt{x}}{y^4}$ Simplifying the numerator and the denominator

b) $\sqrt[3]{\dfrac{27y^{14}}{8x^3}} = \dfrac{\sqrt[3]{27y^{14}}}{\sqrt[3]{8x^3}}$

$= \dfrac{\sqrt[3]{27y^{12}y^2}}{\sqrt[3]{8x^3}}$ y^{12} is the largest perfect-cube factor of y^{14}.

$= \dfrac{\sqrt[3]{27y^{12}}\ \sqrt[3]{y^2}}{\sqrt[3]{8x^3}}$

$= \dfrac{3y^4\sqrt[3]{y^2}}{2x}$ Simplifying the numerator and the denominator

> **TRY EXERCISE** 17

If we read from right to left, the quotient rule tells us that to divide two radical expressions that have the same index, we can divide the radicands.

EXAMPLE **3**   Divide and, if possible, simplify.

a) $\dfrac{\sqrt{80}}{\sqrt{5}}$ **b)** $\dfrac{5\sqrt[3]{32}}{\sqrt[3]{2}}$

c) $\dfrac{\sqrt{72xy}}{2\sqrt{2}}$ **d)** $\dfrac{\sqrt[4]{18a^9b^5}}{\sqrt[4]{3b}}$

SOLUTION

STUDENT NOTES

When writing radical signs, pay careful attention to what is included as the radicand. Each of the following represents a *different* number:

$$\sqrt{\dfrac{5\cdot 2}{3}},\quad \dfrac{\sqrt{5\cdot 2}}{3},\quad \dfrac{\sqrt{5}\cdot 2}{3}.$$

a) $\dfrac{\sqrt{80}}{\sqrt{5}} = \sqrt{\dfrac{80}{5}} = \sqrt{16} = 4$

> Because the indices match, we can divide the radicands.

b) $\dfrac{5\sqrt[3]{32}}{\sqrt[3]{2}} = 5\sqrt[3]{\dfrac{32}{2}} = 5\sqrt[3]{16}$

$= 5\sqrt[3]{8\cdot 2}$ 8 is the largest perfect-cube factor of 16.

$= 5\sqrt[3]{8}\ \sqrt[3]{2} = 5\cdot 2\sqrt[3]{2}$

$= 10\sqrt[3]{2}$

c) $\dfrac{\sqrt{72xy}}{2\sqrt{2}} = \dfrac{1}{2}\sqrt{\dfrac{72xy}{2}}$

> Because the indices match, we can divide the radicands.

$= \dfrac{1}{2}\sqrt{36xy} = \dfrac{1}{2}\cdot 6\sqrt{xy}$

$= 3\sqrt{xy}$

d) $\dfrac{\sqrt[4]{18a^9b^5}}{\sqrt[4]{3b}} = \sqrt[4]{\dfrac{18a^9b^5}{3b}}$

$= \sqrt[4]{6a^9b^4} = \sqrt[4]{a^8b^4}\ \sqrt[4]{6a}$ Note that 8 is the largest power less than 9 that is a multiple of the index 4.

$= a^2b\sqrt[4]{6a}$ *Partial check:* $(a^2b)^4 = a^8b^4$

> **TRY EXERCISE**  2⁷

Rationalizing Denominators or Numerators with One Term*

The expressions

$$\frac{1}{\sqrt{2}} \quad \text{and} \quad \frac{\sqrt{2}}{2}$$

are equivalent, but the second expression does not have a radical expression in the denominator.[†] We can **rationalize the denominator** of a radical expression if we multiply by 1 in either of two ways.

One way is to multiply by 1 *under* the radical to make the denominator of the radicand a perfect power.

EXAMPLE 4   Rationalize each denominator.

a) $\sqrt{\dfrac{7}{3}}$ **b)** $\sqrt[3]{\dfrac{5}{16}}$

SOLUTION

a) We multiply by 1 under the radical, using $\frac{3}{3}$. We do this so that the denominator of the radicand will be a perfect square:

$$\sqrt{\frac{7}{3}} = \sqrt{\frac{7}{3} \cdot \frac{3}{3}} \qquad \text{Multiplying by 1 under the radical}$$

$$= \sqrt{\frac{21}{9}} \qquad \text{The denominator, 9, is now a perfect square.}$$

$$= \frac{\sqrt{21}}{\sqrt{9}} \qquad \text{Using the quotient rule for radicals}$$

$$= \frac{\sqrt{21}}{3}.$$

b) Note that $16 = 4^2$. Thus, to make the denominator a perfect cube, we multiply under the radical by $\frac{4}{4}$:

$$\sqrt[3]{\frac{5}{16}} = \sqrt[3]{\frac{5}{4 \cdot 4} \cdot \frac{4}{4}} \qquad \text{Since the index is 3, we need 3 identical factors in the denominator.}$$

$$= \sqrt[3]{\frac{20}{4^3}} \qquad \text{The denominator is now a perfect cube.}$$

$$= \frac{\sqrt[3]{20}}{\sqrt[3]{4^3}}$$

$$= \frac{\sqrt[3]{20}}{4}.$$

▶ **TRY EXERCISE** 41

Another way to rationalize a denominator is to multiply by 1 *outside* the radical.

EXAMPLE 5   Rationalize each denominator.

a) $\sqrt{\dfrac{4}{5b}}$ **b)** $\dfrac{\sqrt[3]{a}}{\sqrt[3]{25bc^5}}$ **c)** $\dfrac{3x}{\sqrt[5]{2x^2y^3}}$

---

*Denominators and numerators with two terms are rationalized in Section 10.5.
[†]See Exercise 73 on p. 661.

SOLUTION

a) We rewrite the expression as a quotient of two radicals. Then we simplify and multiply by 1:

$$\sqrt{\frac{4}{5b}} = \frac{\sqrt{4}}{\sqrt{5b}} = \frac{2}{\sqrt{5b}} \qquad \text{We assume } b > 0.$$

$$= \frac{2}{\sqrt{5b}} \cdot \frac{\sqrt{5b}}{\sqrt{5b}} \qquad \text{Multiplying by 1}$$

$$= \frac{2\sqrt{5b}}{\left(\sqrt{5b}\right)^2} \qquad \text{Try to do this step mentally.}$$

$$= \frac{2\sqrt{5b}}{5b}.$$

b) Note that the radicand $25bc^5$ is $5 \cdot 5 \cdot b \cdot c \cdot c \cdot c \cdot c \cdot c$. In order for this to be a cube, we need another factor of 5, two more factors of b, and one more factor of c. Thus we multiply by 1, using $\sqrt[3]{5b^2c}/\sqrt[3]{5b^2c}$:

$$\frac{\sqrt[3]{a}}{\sqrt[3]{25bc^5}} = \frac{\sqrt[3]{a}}{\sqrt[3]{25bc^5}} \cdot \frac{\sqrt[3]{5b^2c}}{\sqrt[3]{5b^2c}} \qquad \text{Multiplying by 1}$$

$$= \frac{\sqrt[3]{5ab^2c}}{\sqrt[3]{125b^3c^6}} \longleftarrow \text{This radicand is now a perfect cube.}$$

$$= \frac{\sqrt[3]{5ab^2c}}{5bc^2}.$$

c) To change the radicand $2x^2y^3$ into a perfect fifth power, we need four more factors of 2, three more factors of x, and two more factors of y. Thus we multiply by 1, using $\sqrt[5]{2^4x^3y^2}/\sqrt[5]{2^4x^3y^2}$, or $\sqrt[5]{16x^3y^2}/\sqrt[5]{16x^3y^2}$:

$$\frac{3x}{\sqrt[5]{2x^2y^3}} = \frac{3x}{\sqrt[5]{2x^2y^3}} \cdot \frac{\sqrt[5]{16x^3y^2}}{\sqrt[5]{16x^3y^2}} \qquad \text{Multiplying by 1}$$

$$= \frac{3x\sqrt[5]{16x^3y^2}}{\sqrt[5]{32x^5y^5}} \longleftarrow \text{This radicand is now a perfect fifth power.}$$

$$= \frac{3x\sqrt[5]{16x^3y^2}}{2xy} = \frac{3\sqrt[5]{16x^3y^2}}{2y}. \qquad \text{Always simplify if possible.}$$

> **TRY EXERCISE** 47

Sometimes in calculus it is necessary to rationalize a numerator. To do so, we multiply by 1 to make the radicand in the *numerator* a perfect power.

EXAMPLE 6   Rationalize the numerator: $\dfrac{\sqrt[3]{4a^2}}{\sqrt[3]{5b}}$.

SOLUTION

$$\frac{\sqrt[3]{4a^2}}{\sqrt[3]{5b}} = \frac{\sqrt[3]{4a^2}}{\sqrt[3]{5b}} \cdot \frac{\sqrt[3]{2a}}{\sqrt[3]{2a}} \qquad \text{Multiplying by 1}$$

$$= \frac{\sqrt[3]{8a^3}}{\sqrt[3]{10ba}} \longleftarrow \text{This radicand is now a perfect cube.}$$

$$= \frac{2a}{\sqrt[3]{10ab}}$$

> **TRY EXERCISE** 59

In Section 10.5, we will discuss rationalizing denominators and numerators in which two terms appear.

10.4 EXERCISE SET

👆 **Concept Reinforcement** *In each of Exercises 1–8, match the expression with an equivalent expression from the column on the right. Assume a, b > 0.*

1. ___ $\sqrt[4]{\dfrac{16a^6}{a^2}}$

2. ___ $\dfrac{\sqrt[3]{a^6}}{\sqrt[3]{b^9}}$

3. ___ $\sqrt[5]{\dfrac{a^6}{b^4}}$

4. ___ $\sqrt{\dfrac{a}{b^3}}$

5. ___ $\dfrac{\sqrt[5]{a^2}}{\sqrt[5]{b^2}}$

6. ___ $\dfrac{\sqrt{5a^4}}{\sqrt{5a^3}}$

7. ___ $\dfrac{\sqrt[5]{a^2}}{\sqrt[5]{b^3}}$

8. ___ $\sqrt[3]{\dfrac{a^2}{b^6}}$

a) $\dfrac{\sqrt[5]{a^2}\sqrt[5]{b^2}}{\sqrt[5]{b^5}}$

b) $\dfrac{a^2}{b^3}$

c) $\sqrt{\dfrac{a \cdot b}{b^3 \cdot b}}$

d) $\sqrt{a}$

e) $\dfrac{\sqrt[3]{a^2}}{b^2}$

f) $\sqrt[5]{\dfrac{a^6 b}{b^4 \cdot b}}$

g) $2a$

h) $\dfrac{\sqrt[5]{a^2 b^3}}{\sqrt[5]{b^5}}$

Simplify by taking the roots of the numerator and the denominator. Assume all variables represent positive numbers.

9. $\sqrt{\dfrac{49}{100}}$

10. $\sqrt{\dfrac{81}{25}}$

11. $\sqrt[3]{\dfrac{125}{8}}$

12. $\sqrt[3]{\dfrac{1000}{27}}$

13. $\sqrt{\dfrac{121}{t^2}}$

14. $\sqrt{\dfrac{144}{p^2}}$

15. $\sqrt{\dfrac{36y^3}{x^4}}$

16. $\sqrt{\dfrac{25a^5}{b^6}}$

17. $\sqrt[3]{\dfrac{27a^4}{8b^3}}$

18. $\sqrt[3]{\dfrac{64x^7}{216y^6}}$

19. $\sqrt[4]{\dfrac{32a^4}{2b^4 c^8}}$

20. $\sqrt[4]{\dfrac{81x^4}{y^8 z^4}}$

21. $\sqrt[4]{\dfrac{a^5 b^8}{c^{10}}}$

22. $\sqrt[4]{\dfrac{x^9 y^{12}}{z^6}}$

23. $\sqrt[5]{\dfrac{32x^6}{y^{11}}}$

24. $\sqrt[5]{\dfrac{243a^9}{b^{13}}}$

25. $\sqrt[6]{\dfrac{x^6 y^8}{z^{15}}}$

26. $\sqrt[6]{\dfrac{a^9 b^{12}}{c^{13}}}$

Divide and, if possible, simplify. Assume all variables represent positive numbers.

27. $\dfrac{\sqrt{18y}}{\sqrt{2y}}$

28. $\dfrac{\sqrt{700x}}{\sqrt{7x}}$

29. $\dfrac{\sqrt[3]{26}}{\sqrt[3]{13}}$

30. $\dfrac{\sqrt[3]{35}}{\sqrt[3]{5}}$

31. $\dfrac{\sqrt{40xy^3}}{\sqrt{8x}}$

32. $\dfrac{\sqrt{56ab^3}}{\sqrt{7a}}$

33. $\dfrac{\sqrt[3]{96a^4 b^2}}{\sqrt[3]{12a^2 b}}$

34. $\dfrac{\sqrt[3]{189x^5 y^7}}{\sqrt[3]{7x^2 y^2}}$

35. $\dfrac{\sqrt{100ab}}{5\sqrt{2}}$

36. $\dfrac{\sqrt{75ab}}{3\sqrt{3}}$

37. $\dfrac{\sqrt[4]{48x^9 y^{13}}}{\sqrt[4]{3xy^{-2}}}$

38. $\dfrac{\sqrt[5]{64a^{11}b^{28}}}{\sqrt[5]{2ab^{-2}}}$

39. $\dfrac{\sqrt[3]{x^3 - y^3}}{\sqrt[3]{x - y}}$

40. $\dfrac{\sqrt[3]{r^3 + s^3}}{\sqrt[3]{r + s}}$

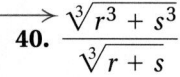

Hint: Factor and then simplify.

Rationalize each denominator. Assume all variables represent positive numbers.

41. $\sqrt{\dfrac{2}{5}}$

42. $\sqrt{\dfrac{7}{2}}$

43. $\dfrac{2\sqrt{5}}{7\sqrt{3}}$

44. $\dfrac{3\sqrt{5}}{2\sqrt{7}}$

45. $\sqrt[3]{\dfrac{5}{4}}$

46. $\sqrt[3]{\dfrac{2}{9}}$

47. $\dfrac{\sqrt[3]{3a}}{\sqrt[3]{5c}}$

48. $\dfrac{\sqrt[3]{7x}}{\sqrt[3]{3y}}$

49. $\dfrac{\sqrt[4]{5y^6}}{\sqrt[4]{9x}}$

50. $\dfrac{\sqrt[5]{3a^4}}{\sqrt[5]{2b^7}}$

51. $\sqrt[3]{\dfrac{2}{x^2 y}}$

52. $\sqrt[3]{\dfrac{5}{ab^2}}$

53. $\sqrt{\dfrac{7a}{18}}$

54. $\sqrt{\dfrac{3x}{20}}$

55. $\sqrt[5]{\dfrac{9}{32x^5 y}}$

56. $\sqrt[4]{\dfrac{7}{64a^2 b^4}}$ Aha! 57. $\sqrt{\dfrac{10ab^2}{72a^3 b}}$

58. $\sqrt{\dfrac{21x^2 y}{75xy^5}}$

Rationalize each numerator. Assume all variables represent positive numbers.

59. $\sqrt{\dfrac{5}{11}}$

60. $\sqrt{\dfrac{2}{3}}$

61. $\dfrac{2\sqrt{6}}{5\sqrt{7}}$

62. $\dfrac{3\sqrt{10}}{2\sqrt{3}}$

63. $\dfrac{\sqrt{8}}{2\sqrt{3x}}$

64. $\dfrac{\sqrt{12}}{\sqrt{5y}}$

65. $\dfrac{\sqrt[3]{7}}{\sqrt[3]{2}}$

66. $\dfrac{\sqrt[3]{5}}{\sqrt[3]{4}}$

67. $\sqrt{\dfrac{7x}{3y}}$

68. $\sqrt{\dfrac{7a}{6b}}$

69. $\sqrt[3]{\dfrac{2a^5}{5b}}$

70. $\sqrt[3]{\dfrac{2a^4}{7b}}$

71. $\sqrt{\dfrac{x^3y}{2}}$

72. $\sqrt{\dfrac{ab^5}{3}}$

73. Explain why it is easier to approximate

$$\dfrac{\sqrt{2}}{2} \quad \text{than} \quad \dfrac{1}{\sqrt{2}}$$

if no calculator is available and $\sqrt{2} \approx 1.414213562$.

74. A student *incorrectly* claims that

$$\dfrac{5 + \sqrt{2}}{\sqrt{18}} = \dfrac{5 + \sqrt{1}}{\sqrt{9}} = \dfrac{5 + 1}{3}.$$

How could you convince the student that a mistake has been made? How would you explain the correct way of rationalizing the denominator?

Skill Review

To prepare for Section 10.5, review factoring expressions and multiplying polynomials (Sections 4.6 and 5.1).

Factor. [5.1]

75. $3x - 8xy + 2xz$

76. $4a^2c + 9ac - 3a^3c$

Multiply. [4.6]

77. $(a + b)(a - b)$

78. $(a^2 - 2y)(a^2 + 2y)$

79. $(8 + 3x)(7 - 4x)$

80. $(2y - x)(3a - c)$

Synthesis

81. Is the quotient of two irrational numbers always an irrational number? Why or why not?

82. Is it possible to understand how to rationalize a denominator without knowing how to multiply rational expressions? Why or why not?

83. *Pendulums.*   The *period* of a pendulum is the time it takes to complete one cycle, swinging to and fro. For a pendulum that is L centimeters long, the period T is given by the formula

$$T = 2\pi\sqrt{\dfrac{L}{980}},$$

where T is in seconds. Find, to the nearest hundredth of a second, the period of a pendulum of length **(a)** 65 cm; **(b)** 98 cm; **(c)** 120 cm. Use a calculator's $\boxed{\pi}$ key if possible.

Perform the indicated operations.

84. $\dfrac{7\sqrt{a^2b}\,\sqrt{25xy}}{5\sqrt{a^{-4}b^{-1}}\sqrt{49x^{-1}y^{-3}}}$

85. $\dfrac{\left(\sqrt[3]{81mn^2}\right)^2}{\left(\sqrt[3]{mn}\right)^2}$

86. $\dfrac{\sqrt{44x^2y^9z}\,\sqrt{22y^9z^6}}{\left(\sqrt{11xy^8z^2}\right)^2}$

87. $\sqrt{a^2 - 3} - \dfrac{a^2}{\sqrt{a^2 - 3}}$

88. $5\sqrt{\dfrac{x}{y}} + 4\sqrt{\dfrac{y}{x}} - \dfrac{3}{\sqrt{xy}}$

89. Provide a reason for each step in the following derivation of the quotient rule:

$$\sqrt[n]{\dfrac{a}{b}} = \left(\dfrac{a}{b}\right)^{1/n} \quad \underline{}$$

$$= \dfrac{a^{1/n}}{b^{1/n}} \quad \underline{}$$

$$= \dfrac{\sqrt[n]{a}}{\sqrt[n]{b}} \quad \underline{}$$

90. Show that $\dfrac{\sqrt[n]{a}}{\sqrt[n]{b}}$ is the nth root of $\dfrac{a}{b}$ by raising it to the nth power and simplifying.

91. Let $f(x) = \sqrt{18x^3}$ and $g(x) = \sqrt{2x}$. Find $(f/g)(x)$ and specify the domain of f/g.

92. Let $f(t) = \sqrt{2t}$ and $g(t) = \sqrt{50t^3}$. Find $(f/g)(t)$ and specify the domain of f/g.

93. Let $f(x) = \sqrt{x^2 - 9}$ and $g(x) = \sqrt{x - 3}$. Find $(f/g)(x)$ and specify the domain of f/g.

10.5 Expressions Containing Several Radical Terms

Adding and Subtracting Radical Expressions ▪ Products and Quotients of Two or More Radical Terms ▪
Rationalizing Denominators or Numerators with Two Terms ▪ Terms with Differing Indices

Radical expressions like $6\sqrt{7} + 4\sqrt{7}$ or $(\sqrt{a} + \sqrt{b})(\sqrt{a} - \sqrt{b})$ contain more than one *radical term* and can sometimes be simplified.

Adding and Subtracting Radical Expressions

When two radical expressions have the same indices and radicands, they are said to be **like radicals**. Like radicals can be combined (added or subtracted) in much the same way that we combine like terms.

EXAMPLE 1 Simplify by combining like radical terms.

a) $6\sqrt{7} + 4\sqrt{7}$

b) $\sqrt[3]{2} - 7x\sqrt[3]{2} + 5\sqrt[3]{2}$

c) $6\sqrt[5]{4x} + 3\sqrt[5]{4x} - \sqrt[3]{4x}$

SOLUTION

a) $6\sqrt{7} + 4\sqrt{7} = (6 + 4)\sqrt{7}$ Using the distributive law (factoring out $\sqrt{7}$)

$\phantom{6\sqrt{7} + 4\sqrt{7}} = 10\sqrt{7}$ You can think: 6 square roots of 7 plus 4 square roots of 7 is 10 square roots of 7.

b) $\sqrt[3]{2} - 7x\sqrt[3]{2} + 5\sqrt[3]{2} = (1 - 7x + 5)\sqrt[3]{2}$ Factoring out $\sqrt[3]{2}$

$\phantom{\sqrt[3]{2} - 7x\sqrt[3]{2} + 5\sqrt[3]{2}} = (6 - 7x)\sqrt[3]{2}$ These parentheses are important!

c) $6\sqrt[5]{4x} + 3\sqrt[5]{4x} - \sqrt[3]{4x} = (6 + 3)\sqrt[5]{4x} - \sqrt[3]{4x}$ Try to do this step mentally.

$\phantom{6\sqrt[5]{4x} + 3\sqrt[5]{4x} - \sqrt[3]{4x}} = 9\sqrt[5]{4x} - \sqrt[3]{4x}$ The indices are different. We cannot combine these terms.

> TRY EXERCISE 7

Our ability to simplify radical expressions can help us to find like radicals even when, at first, it may appear that there are none.

EXAMPLE 2 Simplify by combining like radical terms, if possible.

a) $3\sqrt{8} - 5\sqrt{2}$

b) $9\sqrt{5} - 4\sqrt{3}$

c) $\sqrt[3]{2x^6y^4} + 7\sqrt[3]{2y}$

SOLUTION

a) $3\sqrt{8} - 5\sqrt{2} = 3\sqrt{4 \cdot 2} - 5\sqrt{2}$

$\phantom{3\sqrt{8} - 5\sqrt{2}} = 3\sqrt{4} \cdot \sqrt{2} - 5\sqrt{2}$ ⎱ Simplifying $\sqrt{8}$

$\phantom{3\sqrt{8} - 5\sqrt{2}} = 3 \cdot 2 \cdot \sqrt{2} - 5\sqrt{2}$ ⎰

$\phantom{3\sqrt{8} - 5\sqrt{2}} = 6\sqrt{2} - 5\sqrt{2}$

$\phantom{3\sqrt{8} - 5\sqrt{2}} = \sqrt{2}$ Combining like radicals

b) $9\sqrt{5} - 4\sqrt{3}$ cannot be simplified.   The radicands are different.

c) $\sqrt[3]{2x^6y^4} + 7\sqrt[3]{2y} = \sqrt[3]{x^6y^3 \cdot 2y} + 7\sqrt[3]{2y}$

$\qquad\qquad\qquad\quad = \sqrt[3]{x^6y^3} \cdot \sqrt[3]{2y} + 7\sqrt[3]{2y}$ Simplifying $\sqrt[3]{2x^6y^4}$

$\qquad\qquad\qquad\quad = x^2y \cdot \sqrt[3]{2y} + 7\sqrt[3]{2y}$

$\qquad\qquad\qquad\quad = (x^2y + 7)\sqrt[3]{2y}$ Factoring to combine like radical terms

> **TRY EXERCISE** 17

Products and Quotients of Two or More Radical Terms

Radical expressions often contain factors that have more than one term. Multiplying such expressions is similar to finding products of polynomials. Some products will yield like radical terms, which we can now combine.

EXAMPLE 3 Multiply.

a) $\sqrt{3}(x - \sqrt{5})$

b) $\sqrt[3]{y}\left(\sqrt[3]{y^2} + \sqrt[3]{2}\right)$

c) $(4\sqrt{3} + \sqrt{2})(\sqrt{3} - 5\sqrt{2})$

d) $(\sqrt{a} + \sqrt{b})(\sqrt{a} - \sqrt{b})$

SOLUTION

a) $\sqrt{3}(x - \sqrt{5}) = \sqrt{3} \cdot x - \sqrt{3} \cdot \sqrt{5}$ Using the distributive law

$\qquad\qquad\qquad = x\sqrt{3} - \sqrt{15}$ Multiplying radicals

b) $\sqrt[3]{y}\left(\sqrt[3]{y^2} + \sqrt[3]{2}\right) = \sqrt[3]{y} \cdot \sqrt[3]{y^2} + \sqrt[3]{y} \cdot \sqrt[3]{2}$ Using the distributive law

$\qquad\qquad\qquad\qquad = \sqrt[3]{y^3} + \sqrt[3]{2y}$ Multiplying radicals

$\qquad\qquad\qquad\qquad = y + \sqrt[3]{2y}$ Simplifying $\sqrt[3]{y^3}$

$\qquad\qquad\qquad\qquad\qquad\quad$ F $\qquad$ O $\qquad$ I $\qquad$ L

c) $(4\sqrt{3} + \sqrt{2})(\sqrt{3} - 5\sqrt{2}) = 4(\sqrt{3})^2 - 20\sqrt{3} \cdot \sqrt{2} + \sqrt{2} \cdot \sqrt{3} - 5(\sqrt{2})^2$

$\qquad\qquad\qquad\qquad\qquad = 4 \cdot 3 - 20\sqrt{6} + \sqrt{6} - 5 \cdot 2$ Multiplying radicals

$\qquad\qquad\qquad\qquad\qquad = 12 - 20\sqrt{6} + \sqrt{6} - 10$

$\qquad\qquad\qquad\qquad\qquad = 2 - 19\sqrt{6}$ Combining like terms

d) $(\sqrt{a} + \sqrt{b})(\sqrt{a} - \sqrt{b}) = (\sqrt{a})^2 - \sqrt{a}\sqrt{b} + \sqrt{a}\sqrt{b} - (\sqrt{b})^2$ Using FOIL

$\qquad\qquad\qquad\qquad\qquad = a - b$ Combining like terms

> **TRY EXERCISE** 41

In Example 3(d) above, you may have noticed that since the outer and inner products in FOIL are opposites, the result, $a - b$, is not itself a radical expression. Pairs of radical expressions like $\sqrt{a} + \sqrt{b}$ and $\sqrt{a} - \sqrt{b}$ are called **conjugates**.

Rationalizing Denominators or Numerators with Two Terms

The use of conjugates allows us to rationalize denominators or numerators that contain two terms.

EXAMPLE **4** Rationalize each denominator: **(a)** $\dfrac{4}{\sqrt{3} + x}$; **(b)** $\dfrac{4 + \sqrt{2}}{\sqrt{5} - \sqrt{2}}$.

SOLUTION

a) $\dfrac{4}{\sqrt{3} + x} = \dfrac{4}{\sqrt{3} + x} \cdot \dfrac{\sqrt{3} - x}{\sqrt{3} - x}$ Multiplying by 1, using the conjugate of $\sqrt{3} + x$, which is $\sqrt{3} - x$

$= \dfrac{4(\sqrt{3} - x)}{(\sqrt{3} + x)(\sqrt{3} - x)}$ Multiplying numerators and denominators

$= \dfrac{4(\sqrt{3} - x)}{(\sqrt{3})^2 - x^2}$ Using FOIL in the denominator

$= \dfrac{4\sqrt{3} - 4x}{3 - x^2}$ Simplifying. No radicals remain in the denominator.

b) $\dfrac{4 + \sqrt{2}}{\sqrt{5} - \sqrt{2}} = \dfrac{4 + \sqrt{2}}{\sqrt{5} - \sqrt{2}} \cdot \dfrac{\sqrt{5} + \sqrt{2}}{\sqrt{5} + \sqrt{2}}$ Multiplying by 1, using the conjugate of $\sqrt{5} - \sqrt{2}$, which is $\sqrt{5} + \sqrt{2}$

$= \dfrac{(4 + \sqrt{2})(\sqrt{5} + \sqrt{2})}{(\sqrt{5} - \sqrt{2})(\sqrt{5} + \sqrt{2})}$ Multiplying numerators and denominators

$= \dfrac{4\sqrt{5} + 4\sqrt{2} + \sqrt{2}\sqrt{5} + (\sqrt{2})^2}{(\sqrt{5})^2 - (\sqrt{2})^2}$ Using FOIL

$= \dfrac{4\sqrt{5} + 4\sqrt{2} + \sqrt{10} + 2}{5 - 2}$ Squaring in the denominator and the numerator

$= \dfrac{4\sqrt{5} + 4\sqrt{2} + \sqrt{10} + 2}{3}$ No radicals remain in the denominator.

> **TRY EXERCISE** 61

To rationalize a numerator with two terms, we use the conjugate of the numerator.

EXAMPLE **5** Rationalize the numerator: $\dfrac{4 + \sqrt{2}}{\sqrt{5} - \sqrt{2}}$.

SOLUTION

$\dfrac{4 + \sqrt{2}}{\sqrt{5} - \sqrt{2}} = \dfrac{4 + \sqrt{2}}{\sqrt{5} - \sqrt{2}} \cdot \dfrac{4 - \sqrt{2}}{4 - \sqrt{2}}$ Multiplying by 1, using the conjugate of $4 + \sqrt{2}$, which is $4 - \sqrt{2}$

$= \dfrac{16 - (\sqrt{2})^2}{4\sqrt{5} - \sqrt{5}\sqrt{2} - 4\sqrt{2} + (\sqrt{2})^2}$

$= \dfrac{14}{4\sqrt{5} - \sqrt{10} - 4\sqrt{2} + 2}$

> **TRY EXERCISE** 71

Terms with Differing Indices

To multiply or divide radical terms with identical radicands but different indices, we can convert to exponential notation, use the rules for exponents, and then convert back to radical notation.

EXAMPLE **6**

Divide and, if possible, simplify: $\dfrac{\sqrt[4]{(x+y)^3}}{\sqrt{x+y}}$.

SOLUTION

$$\frac{\sqrt[4]{(x+y)^3}}{\sqrt{x+y}} = \frac{(x+y)^{3/4}}{(x+y)^{1/2}} \qquad \text{Converting to exponential notation}$$

$$= (x+y)^{3/4-1/2} \qquad \begin{array}{l}\text{Since the bases are identical, we can}\\ \text{subtract exponents:}\\ \frac{3}{4}-\frac{1}{2}=\frac{3}{4}-\frac{2}{4}=\frac{1}{4}.\end{array}$$

$$\left.\begin{array}{l}= (x+y)^{1/4}\\ = \sqrt[4]{x+y}\end{array}\right\} \qquad \text{Converting back to radical notation}$$

TRY EXERCISE 95

STUDENT NOTES

Expressions similar to the one in Example 6 are most easily simplified by rewriting the expression using exponents in place of radicals. After simplifying, remember to write your final result in radical notation. In general, if a problem is presented in one form, it is expected that the final result be presented in the same form.

The steps used in Example 6 can be used in a variety of situations.

To Simplify Products or Quotients with Differing Indices

1. Convert all radical expressions to exponential notation.
2. When the bases are identical, subtract exponents to divide and add exponents to multiply. This may require finding a common denominator.
3. Convert back to radical notation and, if possible, simplify.

EXAMPLE **7**

Multiply and simplify: $\sqrt{x^3}\sqrt[3]{x}$.

SOLUTION

$$\begin{aligned}\sqrt{x^3}\sqrt[3]{x} &= x^{3/2} \cdot x^{1/3} &&\text{Converting to exponential notation}\\ &= x^{11/6} &&\text{Adding exponents: } \frac{3}{2}+\frac{1}{3}=\frac{9}{6}+\frac{2}{6}\\ &= \sqrt[6]{x^{11}} &&\text{Converting back to radical notation}\\ &= \sqrt[6]{x^6}\sqrt[6]{x^5}\\ &= x\sqrt[6]{x^5}\end{aligned}$$

Simplifying

TRY EXERCISE 79

EXAMPLE **8**

If $f(x) = \sqrt[3]{x^2}$ and $g(x) = \sqrt{x} + \sqrt[4]{x}$, find $(f \cdot g)(x)$.

SOLUTION   Recall from Section 7.4 that $(f \cdot g)(x) = f(x) \cdot g(x)$. Thus,

$$(f \cdot g)(x) = \sqrt[3]{x^2}\left(\sqrt{x} + \sqrt[4]{x}\right) \qquad \begin{array}{l}x \text{ is assumed to be}\\ \text{nonnegative.}\end{array}$$

$$= x^{2/3}(x^{1/2} + x^{1/4}) \qquad \begin{array}{l}\text{Converting to exponential}\\ \text{notation}\end{array}$$

$$= x^{2/3} \cdot x^{1/2} + x^{2/3} \cdot x^{1/4} \qquad \text{Using the distributive law}$$

$$= x^{2/3+1/2} + x^{2/3+1/4} \qquad \text{Adding exponents:}$$

$$= x^{7/6} + x^{11/12} \qquad \qquad \frac{2}{3}+\frac{1}{2}=\frac{4}{6}+\frac{3}{6}; \frac{2}{3}+\frac{1}{4}=\frac{8}{12}+\frac{3}{12}$$

$$= \sqrt[6]{x^7} + \sqrt[12]{x^{11}} \qquad \begin{array}{l}\text{Converting back to radical}\\ \text{notation}\end{array}$$

$$\left.\begin{array}{l}= \sqrt[6]{x^6}\sqrt[6]{x} + \sqrt[12]{x^{11}}\\ = x\sqrt[6]{x} + \sqrt[12]{x^{11}}.\end{array}\right\}$$

Simplifying

TRY EXERCISE 103

We often can write the final result as a single radical expression by finding a common denominator in the exponents.

EXAMPLE **9** Divide and, if possible, simplify: $\dfrac{\sqrt[3]{a^2b^4}}{\sqrt{ab}}$.

SOLUTION

$$\frac{\sqrt[3]{a^2b^4}}{\sqrt{ab}} = \frac{(a^2b^4)^{1/3}}{(ab)^{1/2}}$$ 　　Converting to exponential notation

$$= \frac{a^{2/3}b^{4/3}}{a^{1/2}b^{1/2}}$$ 　　Using the product and power rules

$$= a^{2/3-1/2}b^{4/3-1/2}$$ 　　Subtracting exponents

$$= a^{1/6}b^{5/6}$$

$$= \sqrt[6]{a}\,\sqrt[6]{b^5}$$ 　　Converting to radical notation

$$= \sqrt[6]{ab^5}$$ 　　Using the product rule for radicals

> **TRY EXERCISE** 91

10.5 EXERCISE SET

🖐 *Concept Reinforcement* *For each of Exercises 1–6, fill in the blanks by selecting from the following words (which may be used more than once):*

radicand(s), indices, conjugate(s), base(s), denominator(s), numerator(s).

1. To add radical expressions, the _____ and the _____ must be the same.

2. To multiply radical expressions, the _____ must be the same.

3. To find a product by adding exponents, the _____ must be the same.

4. To add rational expressions, the _____ must be the same.

5. To rationalize the _____ of $\dfrac{\sqrt{c} - \sqrt{a}}{5}$, we multiply by a form of 1, using the _____ of $\sqrt{c} - \sqrt{a}$, or $\sqrt{c} + \sqrt{a}$, to write 1.

6. To find a quotient by subtracting exponents, the _____ must be the same.

Add or subtract. Simplify by combining like radical terms, if possible. Assume that all variables and radicands represent positive real numbers.

7. $4\sqrt{3} + 7\sqrt{3}$　　　　**8.** $6\sqrt{5} + 2\sqrt{5}$

9. $7\sqrt[3]{4} - 5\sqrt[3]{4}$

10. $14\sqrt[5]{2} - 8\sqrt[5]{2}$

11. $\sqrt[3]{y} + 9\sqrt[3]{y}$

12. $4\sqrt[4]{t} - \sqrt[4]{t}$

13. $8\sqrt{2} - \sqrt{2} + 5\sqrt{2}$

14. $\sqrt{6} + 3\sqrt{6} - 8\sqrt{6}$

15. $9\sqrt[3]{7} - \sqrt{3} + 4\sqrt[3]{7} + 2\sqrt{3}$

16. $5\sqrt{7} - 8\sqrt[4]{11} + \sqrt{7} + 9\sqrt[4]{11}$

17. $4\sqrt{27} - 3\sqrt{3}$

18. $9\sqrt{50} - 4\sqrt{2}$

19. $3\sqrt{45} - 8\sqrt{20}$

20. $5\sqrt{12} + 16\sqrt{27}$

21. $3\sqrt[3]{16} + \sqrt[3]{54}$

22. $\sqrt[3]{27} - 5\sqrt[3]{8}$

23. $\sqrt{a} + 3\sqrt{16a^3}$

24. $2\sqrt{9x^3} - \sqrt{x}$

25. $\sqrt[3]{6x^4} - \sqrt[3]{48x}$

26. $\sqrt[3]{54x} - \sqrt[3]{2x^4}$

27. $\sqrt{4a - 4} + \sqrt{a - 1}$

28. $\sqrt{9y + 27} + \sqrt{y + 3}$

29. $\sqrt{x^3 - x^2} + \sqrt{9x - 9}$

30. $\sqrt{4x - 4} - \sqrt{x^3 - x^2}$

Multiply. Assume all variables represent nonnegative real numbers.

31. $\sqrt{2}(5 + \sqrt{2})$

32. $\sqrt{3}(6 - \sqrt{3})$

33. $3\sqrt{5}(\sqrt{6} - \sqrt{7})$

34. $4\sqrt{2}(\sqrt{3} + \sqrt{5})$

35. $\sqrt{2}(3\sqrt{10} - \sqrt{8})$

36. $\sqrt{3}(2\sqrt{15} - 3\sqrt{4})$

37. $\sqrt[3]{3}\left(\sqrt[3]{9} - 4\sqrt[3]{21}\right)$

38. $\sqrt[3]{2}\left(\sqrt[3]{4} - 2\sqrt[3]{32}\right)$

39. $\sqrt[3]{a}\left(\sqrt[3]{a^2} + \sqrt[3]{24a^2}\right)$

40. $\sqrt[3]{x}\left(\sqrt[3]{3x^2} - \sqrt[3]{81x^2}\right)$

41. $(2 + \sqrt{6})(5 - \sqrt{6})$

42. $(4 - \sqrt{5})(2 + \sqrt{5})$

43. $(\sqrt{2} + \sqrt{7})(\sqrt{3} - \sqrt{7})$

44. $(\sqrt{7} - \sqrt{2})(\sqrt{5} + \sqrt{2})$

45. $(2 - \sqrt{3})(2 + \sqrt{3})$

46. $(3 + \sqrt{11})(3 - \sqrt{11})$

47. $(\sqrt{10} - \sqrt{15})(\sqrt{10} + \sqrt{15})$

48. $(\sqrt{12} + \sqrt{5})(\sqrt{12} - \sqrt{5})$

49. $(3\sqrt{7} + 2\sqrt{5})(2\sqrt{7} - 4\sqrt{5})$

50. $(4\sqrt{5} - 3\sqrt{2})(2\sqrt{5} + 4\sqrt{2})$

51. $(4 + \sqrt{7})^2$ **52.** $(3 + \sqrt{10})^2$

53. $(\sqrt{3} - \sqrt{2})^2$ **54.** $(\sqrt{5} - \sqrt{3})^2$

55. $(\sqrt{2t} + \sqrt{5})^2$ **56.** $(\sqrt{3x} - \sqrt{2})^2$

57. $(3 - \sqrt{x + 5})^2$ **58.** $(4 + \sqrt{x - 3})^2$

59. $\left(2\sqrt[4]{7} - \sqrt[4]{6}\right)\left(3\sqrt[4]{9} + 2\sqrt[4]{5}\right)$

60. $\left(4\sqrt[3]{3} + \sqrt[3]{10}\right)\left(2\sqrt[3]{7} + 5\sqrt[3]{6}\right)$

Rationalize each denominator.

61. $\dfrac{6}{3 - \sqrt{2}}$ **62.** $\dfrac{5}{4 - \sqrt{5}}$

63. $\dfrac{2 + \sqrt{5}}{6 + \sqrt{3}}$ **64.** $\dfrac{1 + \sqrt{2}}{3 + \sqrt{5}}$

65. $\dfrac{\sqrt{a}}{\sqrt{a} + \sqrt{b}}$ **66.** $\dfrac{\sqrt{z}}{\sqrt{x} - \sqrt{z}}$

Aha! **67.** $\dfrac{\sqrt{7} - \sqrt{3}}{\sqrt{3} - \sqrt{7}}$ **68.** $\dfrac{\sqrt{7} + \sqrt{5}}{\sqrt{5} + \sqrt{2}}$

69. $\dfrac{3\sqrt{2} - \sqrt{7}}{4\sqrt{2} + 2\sqrt{5}}$ **70.** $\dfrac{5\sqrt{3} - \sqrt{11}}{2\sqrt{3} - 5\sqrt{2}}$

Rationalize each numerator. If possible, simplify your result.

71. $\dfrac{\sqrt{5} + 1}{4}$ **72.** $\dfrac{\sqrt{15} - 3}{6}$

73. $\dfrac{\sqrt{6} - 2}{\sqrt{3} + 7}$ **74.** $\dfrac{\sqrt{10} + 4}{\sqrt{2} - 3}$

75. $\dfrac{\sqrt{x} - \sqrt{y}}{\sqrt{x} + \sqrt{y}}$ **76.** $\dfrac{\sqrt{a} + \sqrt{b}}{\sqrt{a} - \sqrt{b}}$

77. $\dfrac{\sqrt{a + h} - \sqrt{a}}{h}$ **78.** $\dfrac{\sqrt{x - h} - \sqrt{x}}{h}$

Perform the indicated operation and simplify. Assume all variables represent positive real numbers.

79. $\sqrt[3]{a}\sqrt[6]{a}$ **80.** $\sqrt[10]{a}\sqrt[5]{a^2}$

81. $\sqrt{b^3}\sqrt[5]{b^4}$ **82.** $\sqrt[3]{b^4}\sqrt[4]{b^3}$

83. $\sqrt{xy^3}\,\sqrt[3]{x^2y}$ **84.** $\sqrt[5]{a^3b}\,\sqrt{ab}$

85. $\sqrt[4]{9ab^3}\sqrt{3a^4b}$ **86.** $\sqrt{2x^3y^3}\,\sqrt[3]{4xy^2}$

87. $\sqrt{a^4b^3c^4}\,\sqrt[3]{ab^2c}$ **88.** $\sqrt[3]{xy^2z}\sqrt{x^3yz^2}$

89. $\dfrac{\sqrt[3]{a^2}}{\sqrt[4]{a}}$ **90.** $\dfrac{\sqrt[3]{x^2}}{\sqrt[5]{x}}$

91. $\dfrac{\sqrt[4]{x^2y^3}}{\sqrt[3]{xy}}$ **92.** $\dfrac{\sqrt[5]{a^4b}}{\sqrt[3]{ab}}$

93. $\dfrac{\sqrt{ab^3}}{\sqrt[3]{a^2b^3}}$ **94.** $\dfrac{\sqrt[5]{x^3y^4}}{\sqrt{xy}}$

95. $\dfrac{\sqrt{(7 - y)^3}}{\sqrt[3]{(7 - y)^2}}$ **96.** $\dfrac{\sqrt[5]{(y - 9)^3}}{\sqrt{y - 9}}$

97. $\dfrac{\sqrt[4]{(5 + 3x)^3}}{\sqrt[3]{(5 + 3x)^2}}$ **98.** $\dfrac{\sqrt[3]{(2x + 1)^2}}{\sqrt[5]{(2x + 1)^2}}$

99. $\sqrt[3]{x^2y}\left(\sqrt{xy} - \sqrt[5]{xy^3}\right)$

100. $\sqrt[4]{a^2b}\left(\sqrt[3]{a^2b} - \sqrt[5]{a^2b^2}\right)$

101. $\left(m + \sqrt[3]{n^2}\right)\left(2m + \sqrt[4]{n}\right)$

102. $\left(r - \sqrt[4]{s^3}\right)\left(3r - \sqrt[5]{s}\right)$

In Exercises 103–106, $f(x)$ and $g(x)$ are as given. Find $(f \cdot g)(x)$. Assume all variables represent nonnegative real numbers.

103. $f(x) = \sqrt[4]{x},\ g(x) = 2\sqrt{x} - \sqrt[3]{x^2}$

104. $f(x) = \sqrt[4]{2x} + 5\sqrt{2x},\ g(x) = \sqrt[3]{2x}$

105. $f(x) = x + \sqrt{7},\ g(x) = x - \sqrt{7}$

106. $f(x) = x - \sqrt{2},\ g(x) = x + \sqrt{6}$

Let $f(x) = x^2$. Find each of the following.

107. $f(3 - \sqrt{2})$

108. $f(5 - \sqrt{3})$

109. $f(\sqrt{6} + \sqrt{21})$

110. $f(\sqrt{2} + \sqrt{10})$

111. In what way(s) is combining like radical terms similar to combining like terms that are monomials?

112. Why do we need to know how to multiply radical expressions before learning how to add them?

Skill Review

To prepare for Section 10.6, review solving equations (Sections 2.2, 5.7, and 6.6).

Solve.

113. $3x - 1 = 125$ [2.2]

114. $x + 5 - 2x = 3x + 6 - x$ [2.2]

115. $x^2 + 2x + 1 = 22 - 2x$ [5.7]

116. $9x^2 - 6x + 1 = 7 + 5x - x^2$ [5.7]

117. $\dfrac{1}{x} + \dfrac{1}{2} = \dfrac{1}{6}$ [6.6]

118. $\dfrac{x}{x - 4} + \dfrac{2}{x + 4} = \dfrac{x - 2}{x^2 - 16}$ [6.6]

Synthesis

119. Ramon *incorrectly* writes
$$\sqrt[5]{x^2} \cdot \sqrt{x^3} = x^{2/5} \cdot x^{3/2} = \sqrt[5]{x^3}.$$
What mistake do you suspect he is making?

120. After examining the expression $\sqrt[4]{25xy^3}\,\sqrt{5x^4y}$, Dyan (correctly) concludes that x and y are both nonnegative. Explain how she could reach this conclusion.

Find a simplified form for $f(x)$. Assume $x \geq 0$.

121. $f(x) = \sqrt{x^3 - x^2} + \sqrt{9x^3 - 9x^2} - \sqrt{4x^3 - 4x^2}$

122. $f(x) = \sqrt{20x^2 + 4x^3} - 3x\sqrt{45 + 9x}$
$\quad\quad + \sqrt{5x^2 + x^3}$

123. $f(x) = \sqrt[4]{x^5 - x^4} + 3\sqrt[4]{x^9 - x^8}$

124. $f(x) = \sqrt[4]{16x^4 + 16x^5} - 2\sqrt[4]{x^8 + x^9}$

Simplify.

125. $7x\sqrt{(x + y)^3} - 5xy\sqrt{x + y} - 2y\sqrt{(x + y)^3}$

126. $\sqrt{27a^5(b + 1)}\ \sqrt[3]{81a(b + 1)^4}$

127. $\sqrt{8x(y + z)^5}\ \sqrt[3]{4x^2(y + z)^2}$

128. $\frac{1}{2}\sqrt{36a^5bc^4} - \frac{1}{2}\sqrt[3]{64a^4bc^6} + \frac{1}{6}\sqrt{144a^3bc^6}$

129. $\dfrac{\dfrac{1}{\sqrt{w}} - \sqrt{w}}{\dfrac{\sqrt{w} + 1}{\sqrt{w}}}$

130. $\dfrac{1}{4 + \sqrt{3}} + \dfrac{1}{\sqrt{3}} + \dfrac{1}{\sqrt{3} - 4}$

Express each of the following as the product of two radical expressions.

131. $x - 5$ **132.** $y - 7$

133. $x - a$

Multiply.

134. $\sqrt{9 + 3\sqrt{5}}\sqrt{9 - 3\sqrt{5}}$

135. $(\sqrt{x + 2} - \sqrt{x - 2})^2$

136. Use a graphing calculator to check your answers to Exercises 25, 39, and 81.

CONNECTING | the CONCEPTS

Many radical expressions can be simplified. It is important to know under which conditions radical expressions can be multiplied and divided and radical terms can be combined.

Multiplication and division: The indices must be the same.
$$\frac{\sqrt{50t^5}}{\sqrt{2t^{11}}} = \sqrt{\frac{50t^5}{2t^{11}}} = \sqrt{\frac{25}{t^6}} = \frac{5}{t^3}; \qquad \sqrt[4]{8x^3} \cdot \sqrt[4]{2x} = \sqrt[4]{16x^4} = 2x$$

Combining like terms: The indices and the radicands must both be the same.
$$\sqrt{75x} + \sqrt{12x} - \sqrt{3x} = 5\sqrt{3x} + 2\sqrt{3x} - \sqrt{3x} = 6\sqrt{3x}$$

Radical expressions with differing indices can sometimes be simplified using rational exponents.
$$\sqrt[3]{x^2}\sqrt{x} = x^{2/3}x^{1/2} = x^{4/6}x^{3/6} = x^{7/6} = \sqrt[6]{x^7} = x\sqrt[6]{x}$$

MIXED REVIEW

Simplify. Assume that all variables represent non-negative numbers. Thus no absolute-value signs are needed in an answer.

1. $\sqrt{(t + 5)^2}$

2. $\sqrt[3]{-27a^{12}}$

3. $\sqrt{6x}\sqrt{15x}$

4. $\dfrac{\sqrt{20y}}{\sqrt{45y}}$

5. $\sqrt{15t} + 4\sqrt{15t}$

6. $\sqrt[5]{a^5b^{10}c^{11}}$

7. $\sqrt{6}(\sqrt{10} - \sqrt{33})$

8. $\dfrac{-\sqrt[4]{80a^2b}}{\sqrt[4]{5a^{-1}b^{-6}}}$

9. $\dfrac{\sqrt{t}}{\sqrt[8]{t^3}}$

10. $\sqrt[5]{\dfrac{3a^{12}}{96a^2}}$

11. $2\sqrt{3} - 5\sqrt{12}$

12. $(\sqrt{5} + 3)(\sqrt{5} - 3)$

13. $(\sqrt{15} + \sqrt{10})^2$

14. $\sqrt{25x - 25} - \sqrt{9x - 9}$

15. $\sqrt{x^3y}\sqrt[5]{xy^4}$

16. $\sqrt[3]{5000} + \sqrt[3]{625}$

17. $\sqrt{\sqrt[5]{x^2}}$

18. $\sqrt{3x^2 + 6x + 3}$

19. $\left(\sqrt[4]{a^2b^3}\right)^2$

20. $\sqrt[3]{12x^2y^5}\sqrt[3]{18x^7y}$

10.6   Solving Radical Equations

The Principle of Powers ▪ Equations with Two Radical Terms

In Sections 10.1–10.5, we learned how to manipulate radical expressions as well as expressions containing rational exponents. We performed this work to find *equivalent expressions*.

Now that we know how to work with radicals and rational exponents, we can learn how to solve a new type of equation.

The Principle of Powers

A **radical equation** is an equation in which the variable appears in a radicand. Examples are

$$\sqrt[3]{2x} + 1 = 5, \qquad \sqrt{a - 2} = 7, \qquad \text{and} \qquad 4 - \sqrt{3x + 1} = \sqrt{6 - x}.$$

To solve such equations, we need a new principle. Suppose $a = b$ is true. If we square both sides, we get another true equation: $a^2 = b^2$. This can be generalized.

> **The Principle of Powers**
> If $a = b$, then $a^n = b^n$ for any exponent n.

Note that the principle of powers is an "if–then" statement. The statement obtained by interchanging the two parts of the sentence—"if $a^n = b^n$ for some exponent n, then $a = b$"—*is not always true*. For example, "if $x = 3$, then $x^2 = 9$" is true, but the statement "if $x^2 = 9$, then $x = 3$" is *not* true when x is replaced with -3. For this reason, when both sides of an equation are raised to an even exponent, it is essential to check the answer(s) in the *original* equation.

STUDY SKILLS

Plan Your Future

As you register for next semester's courses, be careful to consider your work and family commitments. Speak to faculty and other students to estimate how demanding each course is before signing up. If in doubt, it is usually better to take one fewer course than one too many.

EXAMPLE **1** Solve: $\sqrt{x} - 3 = 4$.

SOLUTION Before using the principle of powers, we need to isolate the radical term:

$$\sqrt{x} - 3 = 4$$
$$\sqrt{x} = 7 \qquad \text{Isolating the radical by adding 3 to both sides}$$
$$(\sqrt{x})^2 = 7^2 \qquad \text{Using the principle of powers}$$
$$x = 49.$$

Check:
$$\frac{\sqrt{x} - 3 = 4}{\sqrt{49} - 3 \;\Big|\; 4}$$
$$7 - 3 \;\Big|$$
$$4 \stackrel{?}{=} 4 \quad \text{TRUE}$$

The solution is 49.

> TRY EXERCISE 7

EXAMPLE **2** Solve: $\sqrt{x} + 5 = 3$.

SOLUTION

$$\sqrt{x} + 5 = 3$$
$$\sqrt{x} = -2 \qquad \text{Isolating the radical by adding } -5 \text{ to both sides}$$

> The equation $\sqrt{x} = -2$ has no solution because the principal square root of a number is never negative. We continue as in Example 1 for comparison.

$$(\sqrt{x})^2 = (-2)^2 \qquad \text{Using the principle of powers}$$
$$x = 4$$

Check:
$$\frac{\sqrt{x} + 5 = 3}{\sqrt{4} + 5 \;\Big|\; 3}$$
$$2 + 5 \;\Big|$$
$$7 \stackrel{?}{=} 3 \quad \text{FALSE}$$

The number 4 does not check. Thus, $\sqrt{x} + 5 = 3$ has no solution.

> TRY EXERCISE 27

> *CAUTION!* Raising both sides of an equation to an even power may not produce an equivalent equation. In this case, a check is essential.

Note in Example 2 that $x = 4$ has the solution 4, but $\sqrt{x} + 5 = 3$ has *no* solution. Thus the equations $x = 4$ and $\sqrt{x} + 5 = 3$ are *not* equivalent.

> **To Solve an Equation with a Radical Term**
> 1. Isolate the radical term on one side of the equation.
> 2. Use the principle of powers and solve the resulting equation.
> 3. Check any possible solution in the original equation.

EXAMPLE **3**

Solve: $x = \sqrt{x + 7} + 5$.

SOLUTION

$$x = \sqrt{x + 7} + 5$$

$$x - 5 = \sqrt{x + 7} \qquad \text{Isolating the radical by subtracting 5 from both sides}$$

$$\left.\begin{array}{r} (x - 5)^2 = (\sqrt{x + 7})^2 \\ x^2 - 10x + 25 = x + 7 \end{array}\right\} \qquad \text{Using the principle of powers; squaring both sides}$$

$$x^2 - 11x + 18 = 0 \qquad \text{Adding } -x - 7 \text{ to both sides to write the quadratic equation in standard form}$$

$$(x - 9)(x - 2) = 0 \qquad \text{Factoring}$$

$$x = 9 \quad or \quad x = 2 \qquad \text{Using the principle of zero products}$$

The possible solutions are 9 and 2. Let's check.

Check: For 9:

$$\begin{array}{c|c} x = \sqrt{x + 7} + 5 \\ \hline 9 & \sqrt{9 + 7} + 5 \\ 9 \overset{?}{=} 9 & \text{TRUE} \end{array}$$

For 2:

$$\begin{array}{c|c} x = \sqrt{x + 7} + 5 \\ \hline 2 & \sqrt{2 + 7} + 5 \\ 2 \overset{?}{=} 8 & \text{FALSE} \end{array}$$

Since 9 checks but 2 does not, the solution is 9.

▶ **TRY EXERCISE** ▶ 39

It is important to isolate a radical term before using the principle of powers. Suppose in Example 3 that both sides of the equation were squared *before* isolating the radical. We then would have had the expression $(\sqrt{x + 7} + 5)^2$ or $x + 7 + 10\sqrt{x + 7} + 25$ on the right side, and the radical would have remained in the problem.

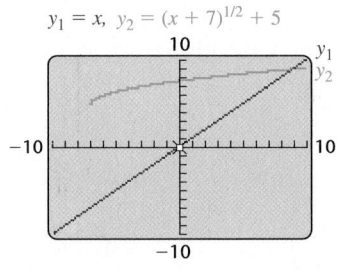
EXAMPLE **4**

Solve: $(2x + 1)^{1/3} + 5 = 0$.

SOLUTION   We can use exponential notation to solve:

$$(2x + 1)^{1/3} + 5 = 0$$

$$(2x + 1)^{1/3} = -5 \qquad \text{Subtracting 5 from both sides}$$

$$[(2x + 1)^{1/3}]^3 = (-5)^3 \qquad \text{Cubing both sides}$$

$$(2x + 1)^1 = (-5)^3 \qquad \text{Multiplying exponents. Try to do this mentally.}$$

$$2x + 1 = -125$$

$$2x = -126 \qquad \text{Subtracting 1 from both sides}$$

$$x = -63.$$

Because both sides were raised to an *odd* power, a check is not *essential*. It is wise, however, for the student to confirm that -63 checks and is the solution.

▶ **TRY EXERCISE** ▶ 25

Equations with Two Radical Terms

A strategy for solving equations with two or more radical terms is as follows.

To Solve an Equation with Two or More Radical Terms

1. Isolate one of the radical terms.
2. Use the principle of powers.
3. If a radical remains, perform steps (1) and (2) again.
4. Solve the resulting equation.
5. Check possible solutions in the original equation.

EXAMPLE **5** Solve: $\sqrt{2x - 5} = 1 + \sqrt{x - 3}$.

SOLUTION

$$\sqrt{2x - 5} = 1 + \sqrt{x - 3}$$

$$(\sqrt{2x - 5})^2 = (1 + \sqrt{x - 3})^2 \qquad \text{One radical is already isolated. We square both sides.}$$

This is like squaring a binomial. We square 1, then find twice the product of 1 and $\sqrt{x - 3}$, and finally square $\sqrt{x - 3}$. Study this carefully.

$$2x - 5 = 1 + 2\sqrt{x - 3} + (\sqrt{x - 3})^2$$

$$2x - 5 = 1 + 2\sqrt{x - 3} + (x - 3)$$

$$x - 3 = 2\sqrt{x - 3} \qquad \text{Isolating the remaining radical term}$$

$$(x - 3)^2 = (2\sqrt{x - 3})^2 \qquad \text{Squaring both sides}$$

$$x^2 - 6x + 9 = 4(x - 3) \qquad \text{Remember to square both the 2 and the } \sqrt{x - 3} \text{ on the right side.}$$

$$x^2 - 6x + 9 = 4x - 12$$

$$x^2 - 10x + 21 = 0$$

$$(x - 7)(x - 3) = 0 \qquad \text{Factoring}$$

$$x = 7 \quad or \quad x = 3 \qquad \text{Using the principle of zero products}$$

We leave it to the student to show that 7 and 3 both check and are the solutions.

▶ **TRY EXERCISE** 41

CAUTION! A common error in solving equations like

$$\sqrt{2x - 5} = 1 + \sqrt{x - 3}$$

is to obtain $1 + (x - 3)$ as the square of the right side. This is wrong because $(A + B)^2 \neq A^2 + B^2$. For example,

$$\left.\begin{array}{c} (1 + 2)^2 \neq 1^2 + 2^2 \\ 3^2 \neq 1 + 4 \\ 9 \neq 5. \end{array}\right\} \quad \text{See Example 5 for the correct expansion of } (1 + \sqrt{x - 3})^2.$$

EXAMPLE 6 Let $f(x) = \sqrt{x + 5} - \sqrt{x - 7}$. Find all x-values for which $f(x) = 2$.

SOLUTION We must have $f(x) = 2$, or

$$\sqrt{x + 5} - \sqrt{x - 7} = 2. \quad \text{Substituting for } f(x)$$

To solve, we isolate one radical term and square both sides:

Isolate a radical term.	$\sqrt{x + 5} = 2 + \sqrt{x - 7}$	Adding $\sqrt{x - 7}$ to both sides. This isolates one of the radical terms.
Raise both sides to the same power.	$(\sqrt{x + 5})^2 = (2 + \sqrt{x - 7})^2$	Using the principle of powers (squaring both sides)
	$x + 5 = 4 + 4\sqrt{x - 7} + (x - 7)$	Using $(A + B)^2 = A^2 + 2AB + B^2$
	$5 = 4\sqrt{x - 7} - 3$	Adding $-x$ to both sides and combining like terms
Isolate a radical term.	$8 = 4\sqrt{x - 7}$	Isolating the remaining radical term
	$2 = \sqrt{x - 7}$	
Raise both sides to the same power.	$2^2 = (\sqrt{x - 7})^2$	Squaring both sides
	$4 = x - 7$	
Solve.	$11 = x.$	
Check.	*Check:* $f(11) = \sqrt{11 + 5} - \sqrt{11 - 7}$	

$$= \sqrt{16} - \sqrt{4}$$
$$= 4 - 2 = 2.$$

We have $f(x) = 2$ when $x = 11$.

TRY EXERCISE 49

10.6 EXERCISE SET

For Extra Help
MyMathLab | Math XP PRACTICE | WATCH | DOWNLOAD
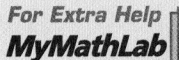

🦢 *Concept Reinforcement* *Classify each statement as either true or false.*

1. If $x^2 = 25$, then $x = 5$.

2. If $t = 7$, then $t^2 = 49$.

3. If $\sqrt{x} = 3$, then $(\sqrt{x})^2 = 3^2$.

4. If $x^2 = 36$, then $x = 6$.

5. $\sqrt{x} - 8 = 7$ is equivalent to $\sqrt{x} = 15$.

6. $\sqrt{t} + 5 = 8$ is equivalent to $\sqrt{t} = 3$.

Solve.

7. $\sqrt{5x + 1} = 4$

8. $\sqrt{7x - 3} = 5$

9. $\sqrt{3x + 1} = 5$

10. $\sqrt{2x - 1} = 2$

11. $\sqrt{y + 5} - 4 = 1$

12. $\sqrt{x - 2} - 7 = -4$

13. $\sqrt{8 - x} + 7 = 10$

14. $\sqrt{y + 4} + 6 = 7$

15. $\sqrt[3]{y + 3} = 2$

16. $\sqrt[3]{x - 2} = 3$

17. $\sqrt[4]{t - 10} = 3$

18. $\sqrt[4]{t + 5} = 2$

19. $6\sqrt{x} = x$

20. $7\sqrt{y} = y$

21. $2y^{1/2} - 13 = 7$

22. $3x^{1/2} + 12 = 9$

23. $\sqrt[5]{x} = -5$

24. $\sqrt[3]{y} = -4$

25. $z^{1/4} + 8 = 10$

26. $x^{1/4} - 2 = 1$

Aha! **27.** $\sqrt{n} = -2$

28. $\sqrt{a} = -1$

29. $\sqrt[4]{3x + 1} - 4 = -1$

30. $\sqrt[4]{2x + 3} - 5 = -2$

31. $(21x + 55)^{1/3} = 10$

32. $(5y + 31)^{1/4} = 2$

33. $\sqrt[3]{3y + 6} + 7 = 8$

34. $\sqrt[3]{6x + 9} + 5 = 2$

35. $\sqrt{3t + 4} = \sqrt{4t + 3}$

36. $\sqrt{2t - 7} = \sqrt{3t - 12}$

37. $3(4 - t)^{1/4} = 6^{1/4}$

38. $2(1 - x)^{1/3} = 4^{1/3}$

39. $3 + \sqrt{5 - x} = x$

40. $x = \sqrt{x - 1} + 3$

41. $\sqrt{4x - 3} = 2 + \sqrt{2x - 5}$

42. $3 + \sqrt{z - 6} = \sqrt{z + 9}$

43. $\sqrt{20 - x} + 8 = \sqrt{9 - x} + 11$

44. $4 + \sqrt{10 - x} = 6 + \sqrt{4 - x}$

45. $\sqrt{x + 2} + \sqrt{3x + 4} = 2$

46. $\sqrt{6x + 7} - \sqrt{3x + 3} = 1$

47. If $f(x) = \sqrt{x} + \sqrt{x - 9}$, find any x for which $f(x) = 1$.

48. If $g(x) = \sqrt{x} + \sqrt{x - 5}$, find any x for which $g(x) = 5$.

49. If $f(t) = \sqrt{t - 2} - \sqrt{4t + 1}$, find any t for which $f(t) = -3$.

50. If $g(t) = \sqrt{2t + 7} - \sqrt{t + 15}$, find any t for which $g(t) = -1$.

51. If $f(x) = \sqrt{2x - 3}$ and $g(x) = \sqrt{x + 7} - 2$, find any x for which $f(x) = g(x)$.

52. If $f(x) = 2\sqrt{3x + 6}$ and $g(x) = 5 + \sqrt{4x + 9}$, find any x for which $f(x) = g(x)$.

53. If $f(t) = 4 - \sqrt{t - 3}$ and $g(t) = (t + 5)^{1/2}$, find any t for which $f(t) = g(t)$.

54. If $f(t) = 7 + \sqrt{2t - 5}$ and $g(t) = 3(t + 1)^{1/2}$, find any t for which $f(t) = g(t)$.

55. Explain in your own words why it is important to check your answers when using the principle of powers.

56. The principle of powers is an "if–then" statement that becomes false when the sentence parts are interchanged. Give an example of another such if–then statement from everyday life (answers will vary).

Skill Review

To prepare for Section 10.7, review finding dimensions of triangles and rectangles (Sections 2.5 and 5.8).

Solve.

57. *Sign dimensions.* The largest sign in the United States is a rectangle with a perimeter of 430 ft. The length of the rectangle is 5 ft longer than thirteen times the width. Find the dimensions of the sign. [2.5]
Source: Florida Center for Instructional Technology

58. *Sign dimensions.* The base of a triangular sign is 4 in. longer than twice the height. The area of the sign is 255 in². Find the dimensions of the sign. [5.8]

59. *Photograph dimensions.* A rectangular family photo is 4 in. longer than it is wide. The area of the photo is 140 in². Find the dimensions of the photograph. [5.8]

60. *Sidewalk length.* The length of a rectangular lawn between classroom buildings is 2 yd less than twice the width of the lawn. A path that is 34 yd long stretches diagonally across the area. What are the dimensions of the lawn? [5.8]

61. The sides of a right triangle are consecutive even integers. Find the length of each side. [5.8]

62. One leg of a right triangle is 5 cm long. The hypotenuse is 1 cm longer than the other leg. Find the length of the hypotenuse. [5.8]

Synthesis

63. Describe a procedure that could be used to create radical equations that have no solution.

64. Is checking essential when the principle of powers is used with an odd power n? Why or why not?

65. *Firefighting.* The velocity of water flow, in feet per second, from a nozzle is given by

$$v(p) = 12.1\sqrt{p},$$

where p is the nozzle pressure, in pounds per square inch (psi). Find the nozzle pressure if the water flow is 100 feet per second.
Source: Houston Fire Department Continuing Education

66. *Firefighting.* The velocity of water flow, in feet per second, from a water tank that is h feet high is given by

$$v(h) = 8\sqrt{h}.$$

Find the height of a water tank that provides a water flow of 60 feet per second.
Source: Houston Fire Department Continuing Education

67. *Music.* The frequency of a violin string varies directly with the square root of the tension on the string. A violin string vibrates with a frequency of 260 Hz when the tension on the string is 28 N. What is the frequency when the tension is 32 N?

68. *Music.* The frequency of a violin string varies inversely with the square root of the density of the string. A nylon violin string with a density of 1200 kg/m^3 vibrates with a frequency of 250 Hz. What is the frequency of a silk violin string with a density of 1300 kg/m^3?
Source: www.speech.kth.se

Steel manufacturing. In the production of steel and other metals, the temperature of the molten metal is so great that conventional thermometers melt. Instead, sound is transmitted across the surface of the metal to a receiver on the far side and the speed of the sound is measured. The formula

$$S(t) = 1087.7\sqrt{\frac{9t + 2617}{2457}}$$

gives the speed of sound $S(t)$, in feet per second, at a temperature of t degrees Celsius.

69. Find the temperature of a blast furnace where sound travels 1880 ft/sec.

70. Find the temperature of a blast furnace where sound travels 1502.3 ft/sec.

71. Solve the above equation for t.

Automotive repair. For an engine with a displacement of 2.8 L, the function given by

$$d(n) = 0.75\sqrt{2.8n}$$

can be used to determine the diameter size of the carburetor's opening, in millimeters. Here n is the number of rpm's at which the engine achieves peak performance.
Source: macdizzy.com

72. If the diameter of a carburetor's opening is 81 mm, for what number of rpm's will the engine produce peak power?

73. If a carburetor's opening is 84 mm, for what number of rpm's will the engine produce peak power?

Escape velocity. A formula for the escape velocity v of a satellite is

$$v = \sqrt{2gr}\sqrt{\frac{h}{r + h}},$$

where g is the force of gravity, r is the planet or star's radius, and h is the height of the satellite above the planet or star's surface.

74. Solve for h.

75. Solve for r.

Solve.

76. $\left(\dfrac{z}{4} - 5\right)^{2/3} = \dfrac{1}{25}$

77. $\dfrac{x + \sqrt{x + 1}}{x - \sqrt{x + 1}} = \dfrac{5}{11}$

78. $\sqrt{\sqrt{y} + 49} = 7$

79. $(z^2 + 17)^{3/4} = 27$

80. $x^2 - 5x - \sqrt{x^2 - 5x - 2} = 4$
(*Hint:* Let $u = x^2 - 5x - 2$.)

81. $\sqrt{8 - b} = b\sqrt{8 - b}$

Without graphing, determine the x-intercepts of the graphs given by each of the following.

82. $f(x) = \sqrt{x - 2} - \sqrt{x + 2} + 2$

83. $g(x) = 6x^{1/2} + 6x^{-1/2} - 37$

84. $f(x) = (x^2 + 30x)^{1/2} - x - (5x)^{1/2}$

85. Use a graphing calculator to check your answers to Exercises 9, 15, and 31.

86. Saul is trying to solve Exercise 73 using a graphing calculator. Without resorting to trial and error, how can he determine a suitable viewing window for finding the solution?

87. Use a graphing calculator to check your answers to Exercises 27, 35, and 41.

CORNER

Tailgater Alert

Focus: Radical equations and problem solving

Time: 15–25 minutes

Group size: 2–3

Materials: Calculators or square-root tables

The faster a car is traveling, the more distance it needs to stop. Thus it is important for drivers to allow sufficient space between their vehicle and the vehicle in front of them. Police recommend that for each 10 mph of speed, a driver allow 1 car length. Thus a driver going 30 mph should have at least 3 car lengths between his or her vehicle and the one in front.

In Exercise Set 10.3, the function $r(L) = 2\sqrt{5L}$ was used to find the speed, in miles per hour, that a car was traveling when it left skid marks L feet long.

ACTIVITY

1. Each group member should estimate the length of a car in which he or she frequently travels. (Each should use a different length, if possible.)

2. Using a calculator as needed, each group member should complete the table below. Column 1 gives a car's speed s, and column 2

lists the minimum amount of space between cars traveling s miles per hour, as recommended by police. Column 3 is the speed that a vehicle *could* travel were it forced to stop in the distance listed in column 2, using the above function.

Column 1 s (in miles per hour)	Column 2 $L(s)$ (in feet)	Column 3 $r(L)$ (in miles per hour)
20		
30		
40		
50		
60		
70		

3. Determine whether there are any speeds at which the "1 car length per 10 mph" guideline might not suffice. On what reasoning do you base your answer? Compare tables to determine how car length affects the results. What recommendations would your group make to a new driver?

10.7   The Distance and Midpoint Formulas and Other Applications

Using the Pythagorean Theorem • Two Special Triangles • The Distance and Midpoint Formulas

Using the Pythagorean Theorem

There are many kinds of problems that involve powers and roots. Many also involve right triangles and the Pythagorean theorem, which we studied in Section 5.8 and restate here.

STUDY SKILLS

Making Sketches

One need not be an artist to make highly useful mathematical sketches. That said, it is important to make sure that your sketches are drawn accurately enough to represent the relative sizes within each shape. For example, if one side of a triangle is clearly the longest, make sure your drawing reflects this.

The Pythagorean Theorem *

In any right triangle, if a and b are the lengths of the legs and c is the length of the hypotenuse, then

$$a^2 + b^2 = c^2.$$

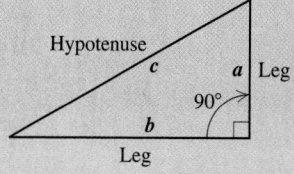

In using the Pythagorean theorem, we often make use of the following principle.

The Principle of Square Roots

For any nonnegative real number n,

If $x^2 = n$, then $x = \sqrt{n}$ or $x = -\sqrt{n}$.

For most real-world applications involving length or distance, $-\sqrt{n}$ is not needed.

EXAMPLE **1**

Baseball. A baseball diamond is actually a square 90 ft on a side. Suppose a catcher fields a ball while standing on the third-base line 10 ft from home plate, as shown in the figure. How far is the catcher's throw to first base? Give an exact answer and an approximation to three decimal places.

SOLUTION We make a drawing and let d = the distance, in feet, to first base. Note that a right triangle is formed in which the leg from home plate to first base measures 90 ft and the leg from home plate to where the catcher fields the ball measures 10 ft.

We substitute these values into the Pythagorean theorem to find d:

$$d^2 = 90^2 + 10^2$$
$$d^2 = 8100 + 100$$
$$d^2 = 8200.$$

We now use the principle of square roots: If $d^2 = 8200$, then $d = \sqrt{8200}$ or $d = -\sqrt{8200}$. Since d represents a length, it follows that d is the positive square root of 8200:

$$d = \sqrt{8200} \text{ ft} \qquad \text{This is an exact answer.}$$
$$d \approx 90.554 \text{ ft.} \qquad \text{Using a calculator for an approximation}$$

> **TRY EXERCISE** ▸ 19

---

*The converse of the Pythagorean theorem also holds. That is, if a, b, and c are the lengths of the sides of a triangle and $a^2 + b^2 = c^2$, then the triangle is a right triangle.

EXAMPLE **2**

Guy wires. The base of a 40-ft long guy wire is located 15 ft from the telephone pole that it is anchoring. How high up the pole does the guy wire reach? Give an exact answer and an approximation to three decimal places.

SOLUTION   We make a drawing and let h = the height, in feet, to which the guy wire reaches. A right triangle is formed in which one leg measures 15 ft and the hypotenuse measures 40 ft. Using the Pythagorean theorem, we have

$$h^2 + 15^2 = 40^2$$
$$h^2 + 225 = 1600$$
$$h^2 = 1375$$
$$h = \sqrt{1375}.$$

Exact answer:

$h = \sqrt{1375}$ ft   Using the positive square root

Approximation:

$h \approx 37.081$ ft   Using a calculator

40 ft h

15 ft

TRY EXERCISE 23

Two Special Triangles

When both legs of a right triangle are the same size, as shown at left, we call the triangle an *isosceles right triangle*. If one leg of an isosceles right triangle has length a, we can find a formula for the length of the hypotenuse as follows:

$45°$

a c

$45°$

a

$$c^2 = a^2 + b^2$$
$$c^2 = a^2 + a^2 \qquad \text{Because the triangle is isosceles, both legs}$$
$$\text{are the same size: } a = b.$$
$$c^2 = 2a^2. \qquad \text{Combining like terms}$$

Next, we use the principle of square roots. Because a, b, and c are lengths, there is no need to consider negative square roots or absolute values. Thus,

$$c = \sqrt{2a^2} \qquad \text{Using the principle of square roots}$$
$$c = \sqrt{a^2 \cdot 2} = a\sqrt{2}.$$

EXAMPLE **3**

One leg of an isosceles right triangle measures 7 cm. Find the length of the hypotenuse. Give an exact answer and an approximation to three decimal places.

SOLUTION   We substitute:

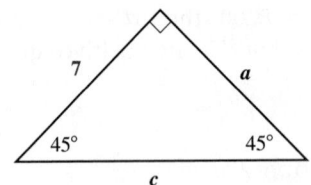

7   a

$45°$ $45°$

c

$$c = a\sqrt{2} \qquad \text{This equation is worth remembering.}$$
$$c = 7\sqrt{2}.$$

Exact answer: $c = 7\sqrt{2}$ cm

Approximation: $c \approx 9.899$ cm   Using a calculator

TRY EXERCISE 29

When the hypotenuse of an isosceles right triangle is known, the lengths of the legs can be found.

EXAMPLE 4

The hypotenuse of an isosceles right triangle is 5 ft long. Find the length of a leg. Give an exact answer and an approximation to three decimal places.

SOLUTION   We replace c with 5 and solve for a:

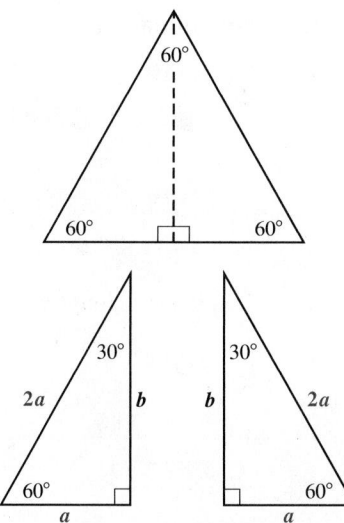

$$5 = a\sqrt{2} \qquad \text{Substituting 5 for } c \text{ in } c = a\sqrt{2}$$

$$\frac{5}{\sqrt{2}} = a \qquad \text{Dividing both sides by } \sqrt{2}$$

$$\frac{5\sqrt{2}}{2} = a. \qquad \text{Rationalize the denominator if desired.}$$

Exact answer: $a = \dfrac{5}{\sqrt{2}}$ ft, or $\dfrac{5\sqrt{2}}{2}$ ft

Approximation: $a \approx 3.536$ ft   Using a calculator

TRY EXERCISE 35

A second special triangle is known as a 30°–60°–90° triangle, so named because of the measures of its angles. Note that in an equilateral triangle, all sides have the same length and all angles are 60°. An altitude, drawn dashed in the figure, bisects, or splits in half, one angle and one side. Two 30°–60°–90° right triangles are thus formed.

If we let a represent the length of the shorter leg in a 30°–60°–90° triangle, then $2a$ represents the length of the hypotenuse. We have

$$a^2 + b^2 = (2a)^2 \qquad \text{Using the Pythagorean theorem}$$

$$a^2 + b^2 = 4a^2$$

$$b^2 = 3a^2 \qquad \text{Subtracting } a^2 \text{ from both sides}$$

$$b = \sqrt{3a^2} \qquad \text{Considering only the positive square root}$$

$$b = \sqrt{a^2 \cdot 3}$$

$$b = a\sqrt{3}.$$

EXAMPLE 5

The shorter leg of a 30°–60°–90° triangle measures 8 in. Find the lengths of the other sides. Give exact answers and, where appropriate, an approximation to three decimal places.

SOLUTION   The hypotenuse is twice as long as the shorter leg, so we have

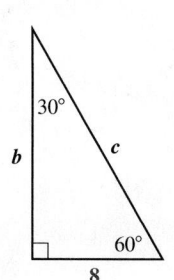

$$c = 2a \qquad \text{This relationship is worth remembering.}$$

$$= 2 \cdot 8 = 16 \text{ in.} \qquad \text{This is the length of the hypotenuse.}$$

The length of the longer leg is the length of the shorter leg times $\sqrt{3}$. This gives us

$$b = a\sqrt{3} \qquad \text{This is also worth remembering.}$$

$$= 8\sqrt{3} \text{ in.} \qquad \text{This is the length of the longer leg.}$$

Exact answer: $c = 16$ in., $b = 8\sqrt{3}$ in.

Approximation: $b \approx 13.856$ in.

TRY EXERCISE 37

EXAMPLE **6**

The length of the longer leg of a 30°–60°–90° triangle is 14 cm. Find the length of the hypotenuse. Give an exact answer and an approximation to three decimal places.

SOLUTION   The length of the hypotenuse is twice the length of the shorter leg. We first find a, the length of the shorter leg, by using the length of the longer leg:

$$14 = a\sqrt{3} \qquad \text{Substituting 14 for } b \text{ in } b = a\sqrt{3}$$

$$\frac{14}{\sqrt{3}} = a. \qquad \text{Dividing by } \sqrt{3}$$

Since the hypotenuse is twice as long as the shorter leg, we have

$$c = 2a$$

$$= 2 \cdot \frac{14}{\sqrt{3}} \qquad \text{Substituting}$$

$$= \frac{28}{\sqrt{3}} \text{ cm.}$$

Exact answer: $c = \dfrac{28}{\sqrt{3}}$ cm, or $\dfrac{28\sqrt{3}}{3}$ cm if the denominator is rationalized.

Approximation: $c \approx 16.166$ cm

TRY EXERCISE 33

STUDENT NOTES

Perhaps the easiest way to remember the important results listed in the adjacent box is to write out, on your own, the derivations shown on pp. 678 and 679.

Lengths Within Isosceles and 30°–60°–90° Right Triangles

The length of the hypotenuse in an isosceles right triangle is the length of a leg times $\sqrt{2}$.

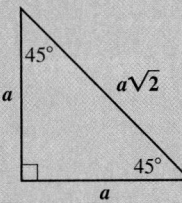

The length of the longer leg in a 30°–60°–90° right triangle is the length of the shorter leg times $\sqrt{3}$. The hypotenuse is twice as long as the shorter leg.

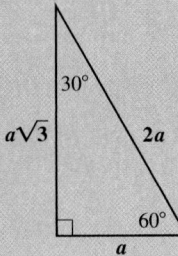

The Distance and Midpoint Formulas

We can use the Pythagorean theorem to find the distance between two points on a plane.

To find the distance between two points on the number line, we subtract. Depending on the order in which we subtract, the difference may be positive or negative. However, if we take the absolute value of the difference, we always obtain a positive value for the distance:

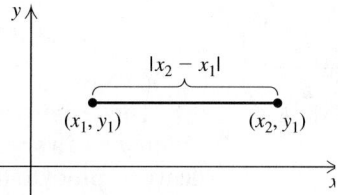

$$|4 - (-3)| = |7| = 7$$
$$|-3 - 4| = |-7| = 7$$

If two points are on a horizontal line, they have the same second coordinate. We can find the distance between them by subtracting their first coordinates and taking the absolute value of that difference.

The distance between the points (x_1, y_1) and (x_2, y_1) on a horizontal line is thus $|x_2 - x_1|$. Similarly, the distance between the points (x_2, y_1) and (x_2, y_2) on a vertical line is $|y_2 - y_1|$.

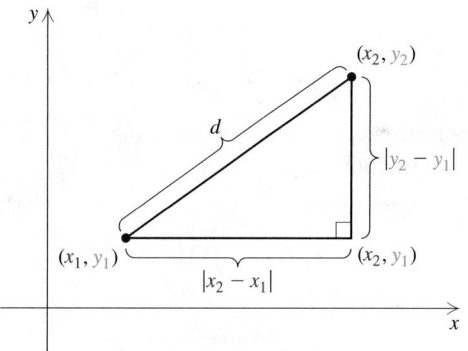

Now consider two points (x_1, y_1) and (x_2, y_2). If $x_1 \neq x_2$ and $y_1 \neq y_2$, these points, along with the point (x_2, y_1), describe a right triangle. The lengths of the legs are $|x_2 - x_1|$ and $|y_2 - y_1|$. We find d, the length of the hypotenuse, by using the Pythagorean theorem:

$$d^2 = |x_2 - x_1|^2 + |y_2 - y_1|^2.$$

Since the square of a number is the same as the square of its opposite, we can replace the absolute-value signs with parentheses:

$$d^2 = (x_2 - x_1)^2 + (y_2 - y_1)^2.$$

Taking the principal square root, we have a formula for distance.

> ### The Distance Formula
> The distance d between any two points (x_1, y_1) and (x_2, y_2) is given by
> $$d = \sqrt{(x_2 - x_1)^2 + (y_2 - y_1)^2}.$$

EXAMPLE   7   Find the distance between $(5, -1)$ and $(-4, 6)$. Find an exact answer and an approximation to three decimal places.

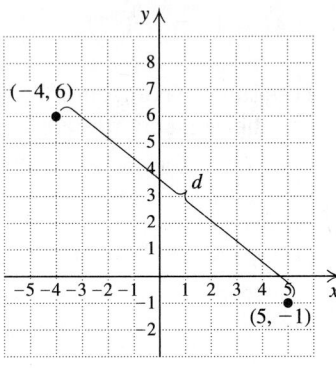

SOLUTION   We substitute into the distance formula:

$$d = \sqrt{(-4 - 5)^2 + [6 - (-1)]^2}$$ Substituting. A drawing is optional.
$$= \sqrt{(-9)^2 + 7^2}$$
$$= \sqrt{130}$$ This is exact.
$$\approx 11.402.$$ Using a calculator for an approximation

TRY EXERCISE 51

The distance formula is needed to verify a formula for the coordinates of the *midpoint* of a segment connecting two points. We state the midpoint formula and leave its proof to the exercises.

STUDENT NOTES

To help remember the formulas correctly, note that the distance formula (a variation on the Pythagorean theorem) involves both subtraction and addition, whereas the midpoint formula does not include any subtraction.

> ### The Midpoint Formula
> If the endpoints of a segment are (x_1, y_1) and (x_2, y_2), then the coordinates of the midpoint are
> $$\left(\frac{x_1 + x_2}{2}, \frac{y_1 + y_2}{2}\right).$$
>
>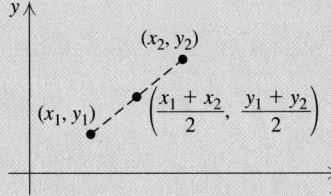
>
> (To locate the midpoint, average the x-coordinates and average the y-coordinates.)

EXAMPLE   8   Find the midpoint of the segment with endpoints $(-2, 3)$ and $(4, -6)$.

SOLUTION   Using the midpoint formula, we obtain

$$\left(\frac{-2 + 4}{2}, \frac{3 + (-6)}{2}\right), \quad \text{or} \quad \left(\frac{2}{2}, \frac{-3}{2}\right), \quad \text{or} \quad \left(1, -\frac{3}{2}\right).$$

TRY EXERCISE 65

10.7	EXERCISE SET

Concept Reinforcement *Complete each sentence with the best choice from the column on the right.*

1. In any _____ triangle, the square of the length of the hypotenuse is the sum of the squares of the lengths of the legs.

2. The shortest side of a right triangle is always one of the two _____.

3. The principle of _____ states that if $x^2 = n$, then $x = \sqrt{n}$ or $x = -\sqrt{n}$.

4. In a(n) _____ right triangle, both legs have the same length.

5. In a(n) _____ right triangle, the hypotenuse is twice as long as the shorter leg.

6. If both legs in a right triangle have measure a, then the _____ measures $a\sqrt{2}$.

a) Hypotenuse

b) Isosceles

c) Legs

d) Right

e) Square roots

f) 30°–60°–90°

In a right triangle, find the length of the side not given. Give an exact answer and, where appropriate, an approximation to three decimal places.

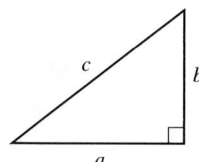

7. $a = 5, b = 3$

8. $a = 8, b = 10$

Aha! **9.** $a = 9, b = 9$

10. $a = 10, b = 10$

11. $b = 15, c = 17$

12. $a = 7, c = 25$

In Exercises 13–18, give an exact answer and, where appropriate, an approximation to three decimal places.

13. A right triangle's hypotenuse is 8 m and one leg is $4\sqrt{3}$ m. Find the length of the other leg.

14. A right triangle's hypotenuse is 6 cm and one leg is $\sqrt{5}$ cm. Find the length of the other leg.

15. The hypotenuse of a right triangle is $\sqrt{20}$ in. and one leg measures 1 in. Find the length of the other leg.

16. The hypotenuse of a right triangle is $\sqrt{15}$ ft and one leg measures 2 ft. Find the length of the other leg.

Aha! **17.** One leg in a right triangle is 1 m and the hypotenuse measures $\sqrt{2}$ m. Find the length of the other leg.

18. One leg of a right triangle is 1 yd and the hypotenuse measures 2 yd. Find the length of the other leg.

In Exercises 19–28, give an exact answer and, where appropriate, an approximation to three decimal places.

19. *Bicycling.* Clare routinely bicycles across a rectangular parking lot on her way to work. If the lot is 200 ft long and 150 ft wide, how far does Clare travel when she rides across the lot diagonally?

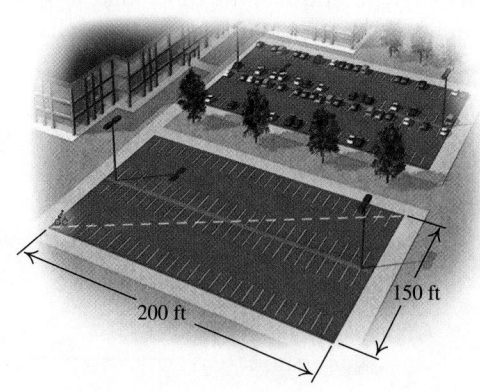

200 ft   150 ft

20. *Guy wire.* How long is a guy wire if it reaches from the top of a 15-ft pole to a point on the ground 10 ft from the pole?

21. *Softball.* A slow-pitch softball diamond is actually a square 65 ft on a side. How far is it from home plate to second base?

22. *Baseball.* Suppose the catcher in Example 1 makes a throw to second base from the same location. How far is that throw?

23. *Television sets.* What does it mean to refer to a 51-in. TV set? Such units refer to the diagonal of the screen. A 51-in. TV set has a width of 45 in. What is its height?

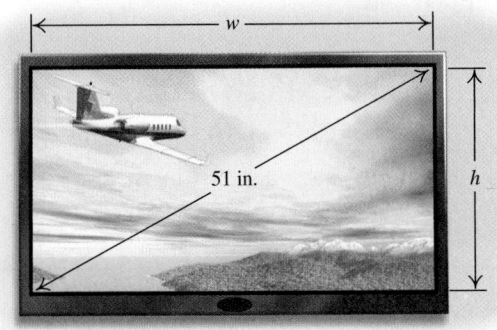

24. *Television sets.* A 53-in. TV set has a screen with a height of 28 in. What is its width? (See Exercise 23.)

25. *Speaker placement.* A stereo receiver is in a corner of a 12-ft by 14-ft room. Wire will run under a rug, diagonally, to a subwoofer in the far corner. If 4 ft of slack is required on each end, how long a piece of wire should be purchased?

26. *Distance over water.* To determine the width of a pond, a surveyor locates two stakes at either end of the pond and uses instrumentation to place a third stake so that the distance across the pond is the length of a hypotenuse. If the third stake is 90 m from one stake and 70 m from the other, what is the distance across the pond?

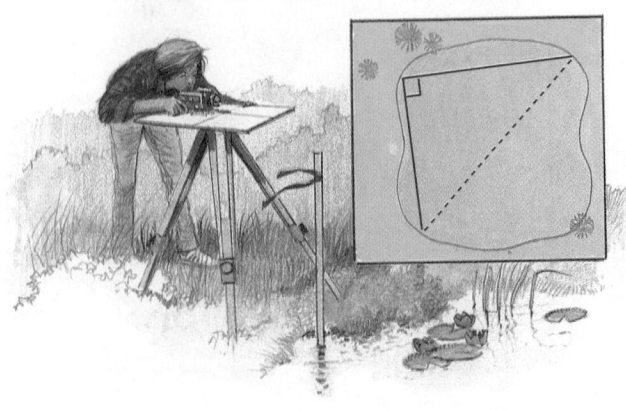

27. *Walking.* Students at Pohlman Community College have worn a path that cuts diagonally across the campus "quad." If the quad is actually a rectangle that Marissa measured to be 70 paces long and 40 paces wide, how many paces will Marissa save by using the diagonal path?

28. *Crosswalks.* The diagonal crosswalk at the intersection of State St. and Main St. is the hypotenuse of a triangle in which the crosswalks across State St. and Main St. are the legs. If State St. is 28 ft wide and Main St. is 40 ft wide, how much shorter is the distance traveled by pedestrians using the diagonal crosswalk?

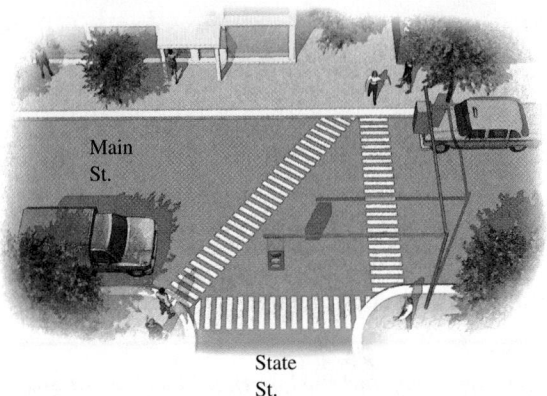

For each triangle, find the missing length(s). Give an exact answer and, where appropriate, an approximation to three decimal places.

29.

30.

31.

32.

33.

34.

35.

36.

37.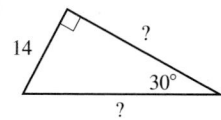

38.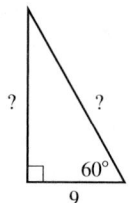

39.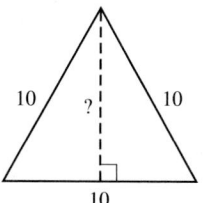

40.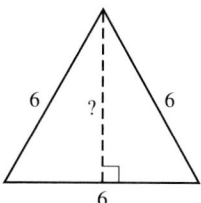

41.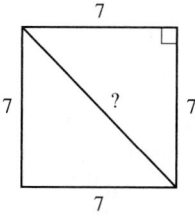

42.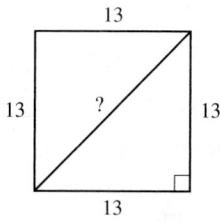

43.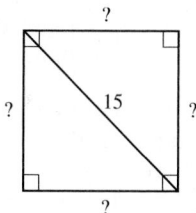

44.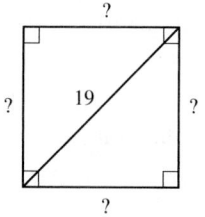

In Exercises 45–48, give an exact answer and, where appropriate, an approximation to three decimal places.

45. *Bridge expansion.* During the summer heat, a 2-mi bridge expands 2 ft in length. If we assume that the bulge occurs straight up the middle, how high is the bulge? (The answer may surprise you. Most bridges have expansion spaces to avoid such buckling.)

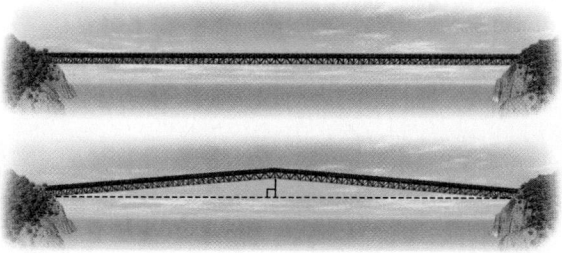

46. Triangle *ABC* has sides of lengths 25 ft, 25 ft, and 30 ft. Triangle *PQR* has sides of lengths 25 ft, 25 ft, and 40 ft. Which triangle, if either, has the greater area and by how much?

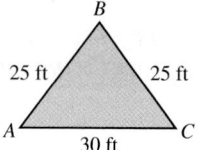
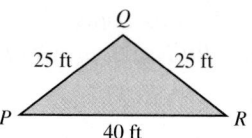

47. *Architecture.* The Rushton Triangular Lodge in Northamptonshire, England, was designed and constructed by Sir Thomas Tresham between 1593 and 1597. The building is in the shape of an equilateral triangle with walls of length 33 ft. How many square feet of land is covered by the lodge?

Source: The Internet Encyclopedia of Science

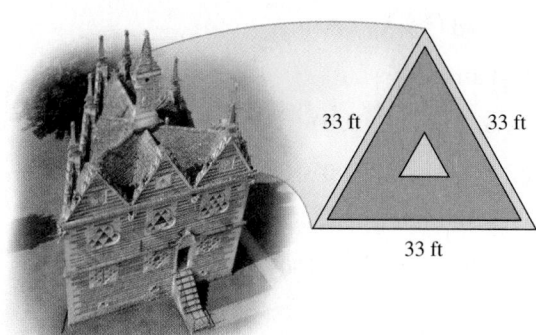

48. *Antenna length.* As part of an emergency radio communication station, Rik sets up an "Inverted-V" antenna. He stretches a copper wire from one point on the ground to a point on a tree and then back down to the ground, forming two 30°–60°–90° triangles. If the wire is fastened to the tree 34 ft above the ground, how long is the copper wire?

49. Find all points on the *y*-axis of a Cartesian coordinate system that are 5 units from the point $(3, 0)$.

50. Find all points on the *x*-axis of a Cartesian coordinate system that are 5 units from the point $(0, 4)$.

Find the distance between each pair of points. Where appropriate, find an approximation to three decimal places.

51. $(4, 5)$ and $(7, 1)$

52. $(0, 8)$ and $(6, 0)$

53. $(0, -5)$ and $(1, -2)$

54. $(-1, -4)$ and $(-3, -5)$

55. $(-4, 4)$ and $(6, -6)$

56. $(5, 21)$ and $(-3, 1)$

Aha! **57.** $(8.6, -3.4)$ and $(-9.2, -3.4)$

58. $(5.9, 2)$ and $(3.7, -7.7)$

59. $\left(\frac{1}{2}, \frac{1}{3}\right)$ and $\left(\frac{5}{6}, -\frac{1}{6}\right)$

60. $\left(\frac{5}{7}, \frac{1}{14}\right)$ and $\left(\frac{1}{7}, \frac{11}{14}\right)$

61. $(-\sqrt{6}, \sqrt{6})$ and $(0, 0)$

62. $(\sqrt{5}, -\sqrt{3})$ and $(0, 0)$

63. $(-1, -30)$ and $(-2, -40)$

64. $(0.5, 100)$ and $(1.5, -100)$

Find the midpoint of each segment with the given endpoints.

65. $(-2, 5)$ and $(8, 3)$

66. $(1, 4)$ and $(9, -6)$

67. $(2, -1)$ and $(5, 8)$

68. $(-1, 2)$ and $(1, -3)$

69. $(-8, -5)$ and $(6, -1)$

70. $(8, -2)$ and $(-3, 4)$

71. $(-3.4, 8.1)$ and $(4.8, -8.1)$

72. $(4.1, 6.9)$ and $(5.2, -8.9)$

73. $\left(\frac{1}{6}, -\frac{3}{4}\right)$ and $\left(-\frac{1}{3}, \frac{5}{6}\right)$

74. $\left(-\frac{4}{5}, -\frac{2}{3}\right)$ and $\left(\frac{1}{8}, \frac{3}{4}\right)$

75. $(\sqrt{2}, -1)$ and $(\sqrt{3}, 4)$

76. $(9, 2\sqrt{3})$ and $(-4, 5\sqrt{3})$

77. Are there any right triangles, other than those with sides measuring 3, 4, and 5, that have consecutive numbers for the lengths of the sides? Why or why not?

78. If a 30°–60°–90° triangle and an isosceles right triangle have the same perimeter, which will have the greater area? Why?

Skill Review

Review graphing (Sections 3.2, 3.3, 3.6, and 9.4).

Graph on a plane.

79. $y = 2x - 3$ [3.6]

80. $y < x$ [9.4]

81. $8x - 4y = 8$ [3.3]

82. $2y - 1 = 7$ [3.3]

83. $x \geq 1$ [9.4]

84. $x - 5 = 6 - 2y$ [3.2]

Synthesis

85. Describe a procedure that uses the distance formula to determine whether three points, (x_1, y_1), (x_2, y_2), and (x_3, y_3), are vertices of a right triangle.

86. Outline a procedure that uses the distance formula to determine whether three points, (x_1, y_1), (x_2, y_2), and (x_3, y_3), are collinear (lie on the same line).

87. The perimeter of a regular hexagon is 72 cm. Determine the area of the shaded region shown.

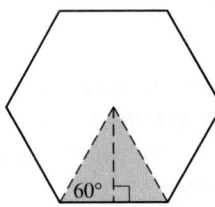

88. If the perimeter of a regular hexagon is 120 ft, what is its area? (*Hint:* See Exercise 87.)

89. Each side of a regular octagon has length *s*. Find a formula for the distance *d* between the parallel sides of the octagon.

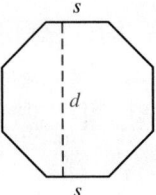

90. *Roofing.*   Kit's home, which is 24 ft wide and 32 ft long, needs a new roof. By counting clapboards that are 4 in. apart, Kit determines that the peak of the roof is 6 ft higher than the sides. A packet of shingles covers 100 ft^2. How many packets will the job require?

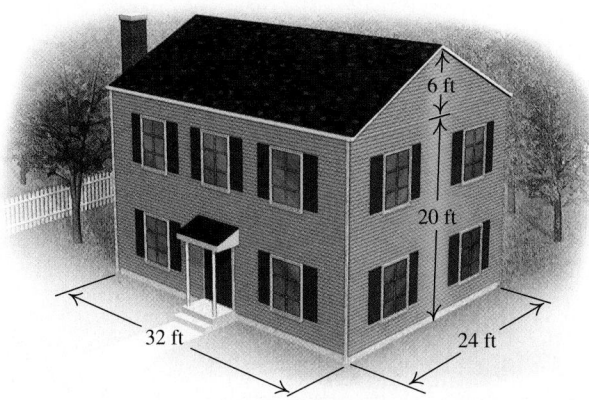

91. *Painting.*   (Refer to Exercise 90.) A gallon of Benjamin Moore® exterior acrylic paint covers 450–500 ft^2. If Kit's house has dimensions as shown above, how many gallons of paint should be bought to paint the house? What assumption(s) is made in your answer?

92. *Contracting.*   Oxford Builders has an extension cord on their generator that permits them to work, with electricity, anywhere in a circular area of 3850 ft^2. Find the dimensions of the largest square room they could work on without having to relocate the generator to reach each corner of the floor plan.

93. *Contracting.*   Cleary Construction has a hose attached to their insulation blower that permits them to reach anywhere in a circular area of 6160 ft^2. Find the dimensions of the largest square room with 12-ft ceilings in which they could reach all corners with the hose while leaving the blower centrally located. Assume that the blower sits on the floor.

94. The length and the width of a rectangle are given by consecutive integers. The area of the rectangle is 90 cm^2. Find the length of a diagonal of the rectangle.

95. A cube measures 5 cm on each side. How long is the diagonal that connects two opposite corners of the cube? Give an exact answer.

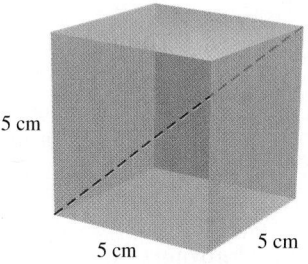

96. Prove the midpoint formula by showing that

i) the distance from (x_1, y_1) to
$$\left(\frac{x_1 + x_2}{2}, \frac{y_1 + y_2}{2}\right)$$
equals the distance from (x_2, y_2) to
$$\left(\frac{x_1 + x_2}{2}, \frac{y_1 + y_2}{2}\right);$$
and

ii) the points
$$(x_1, y_1), \left(\frac{x_1 + x_2}{2}, \frac{y_1 + y_2}{2}\right),$$
and
$$(x_2, y_2)$$
lie on the same line (see Exercise 86).

10.8 The Complex Numbers

Imaginary and Complex Numbers ▪ Addition and Subtraction ▪ Multiplication ▪ Conjugates and Division ▪ Powers of i

Imaginary and Complex Numbers

Negative numbers do not have square roots in the real-number system. However, a larger number system that contains the real-number system is designed so that negative numbers *do* have square roots. That system is called the **complex-number system**, and it will allow us to solve equations like $x^2 + 1 = 0$. The complex-number system makes use of i, a number that is, by definition, a square root of -1.

> **The Number i**
>
> i is the unique number for which $i = \sqrt{-1}$ and $i^2 = -1$.

We can now define the square root of a negative number as follows:

$$\sqrt{-p} = \sqrt{-1}\sqrt{p} = i\sqrt{p} \text{ or } \sqrt{p}i, \text{ for any positive number } p.$$

EXAMPLE 1 Express in terms of i: **(a)** $\sqrt{-7}$; **(b)** $\sqrt{-16}$; **(c)** $-\sqrt{-13}$; **(d)** $-\sqrt{-50}$.

SOLUTION

a) $\sqrt{-7} = \sqrt{-1 \cdot 7} = \sqrt{-1} \cdot \sqrt{7} = i\sqrt{7}$, or $\sqrt{7}i$ i is *not* under the radical.

b) $\sqrt{-16} = \sqrt{-1 \cdot 16} = \sqrt{-1} \cdot \sqrt{16} = i \cdot 4 = 4i$

c) $-\sqrt{-13} = -\sqrt{-1 \cdot 13} = -\sqrt{-1} \cdot \sqrt{13} = -i\sqrt{13}$, or $-\sqrt{13}i$

d) $-\sqrt{-50} = -\sqrt{-1} \cdot \sqrt{25} \cdot \sqrt{2} = -i \cdot 5 \cdot \sqrt{2} = -5i\sqrt{2}$, or $-5\sqrt{2}i$

    TRY EXERCISE 9

> **Imaginary Numbers**
>
> An *imaginary number* is a number that can be written in the form $a + bi$, where a and b are real numbers and $b \neq 0$.

Don't let the name "imaginary" fool you. Imaginary numbers appear in fields such as engineering and the physical sciences. The following are examples of imaginary numbers:

$5 + 4i$,   Here $a = 5$, $b = 4$.

$\sqrt{3} - \pi i$,   Here $a = \sqrt{3}$, $b = -\pi$.

$\sqrt{7}i$ Here $a = 0$, $b = \sqrt{7}$.

The union of the set of all imaginary numbers and the set of all real numbers is the set of all **complex numbers**.

Complex Numbers

A *complex number* is any number that can be written in the form $a + bi$, where a and b are real numbers. (Note that a and b both can be 0.)

The following are examples of complex numbers:

$7 + 3i$ (here $a \neq 0, b \neq 0$); $4i$ (here $a = 0, b \neq 0$);

8 (here $a \neq 0, b = 0$); 0 (here $a = 0, b = 0$).

Complex numbers like $17i$ or $4i$, in which $a = 0$ and $b \neq 0$, are called *pure imaginary numbers*.

For $b = 0$, we have $a + 0i = a$, so every real number is a complex number. The relationships among various real and complex numbers are shown below.

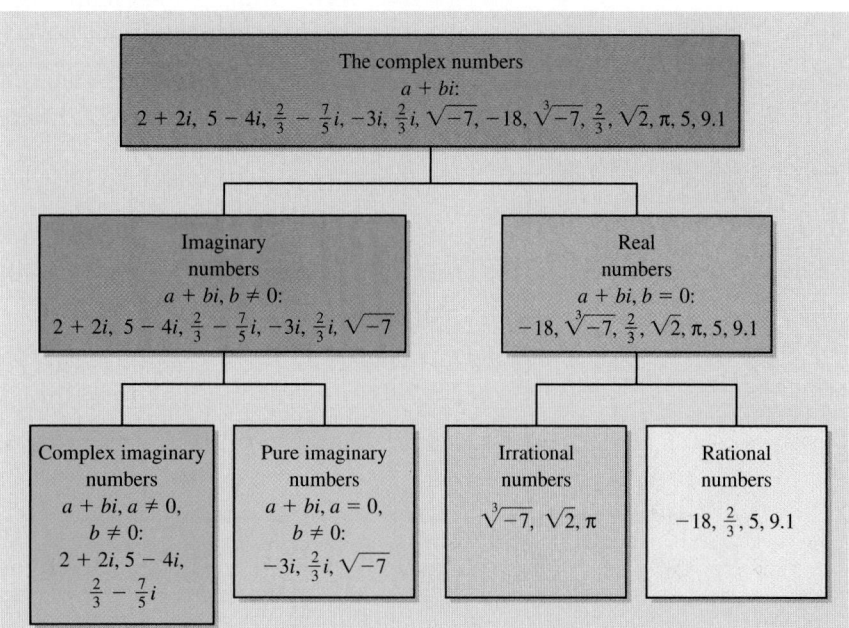

Note that although $\sqrt{-7}$ and $\sqrt[3]{-7}$ are both complex numbers, $\sqrt{-7}$ is imaginary whereas $\sqrt[3]{-7}$ is real.

Addition and Subtraction

The complex numbers obey the commutative, associative, and distributive laws. Thus we can add and subtract them as we do binomials.

EXAMPLE 2

Add or subtract and simplify.

a) $(8 + 6i) + (3 + 2i)$ **b)** $(4 + 5i) - (6 - 3i)$

SOLUTION

a) $(8 + 6i) + (3 + 2i) = (8 + 3) + (6i + 2i)$ Combining the real parts and the imaginary parts

$$= 11 + (6 + 2)i = 11 + 8i$$

b) $(4 + 5i) - (6 - 3i) = (4 - 6) + [5i - (-3i)]$ Note that the 6 and the $-3i$ are *both* being subtracted.

$$= -2 + 8i$$

TRY EXERCISE 27

STUDENT NOTES —————

The rule developed in Section 10.3, $\sqrt[n]{a} \cdot \sqrt[n]{b} = \sqrt[n]{a \cdot b}$, does *not* apply when n is 2 and either a or b is negative. Indeed this condition is stated on p. 650 when it is specified that $\sqrt[n]{a}$ and $\sqrt[n]{b}$ are both *real* numbers.

Multiplication

To multiply square roots of negative real numbers, we first express them in terms of i. For example,

$$\sqrt{-2} \cdot \sqrt{-5} = \sqrt{-1} \cdot \sqrt{2} \cdot \sqrt{-1} \cdot \sqrt{5}$$
$$= i \cdot \sqrt{2} \cdot i \cdot \sqrt{5}$$
$$= i^2 \cdot \sqrt{10}$$
$$= -1\sqrt{10} = -\sqrt{10} \text{ is correct!}$$

> *CAUTION!* With complex numbers, simply multiplying radicands is *incorrect* when both radicands are negative: $\sqrt{-2} \cdot \sqrt{-5} \neq \sqrt{10}$.

With this in mind, we can now multiply complex numbers.

EXAMPLE **3**   Multiply and simplify. When possible, write answers in the form $a + bi$.

a) $\sqrt{-4}\sqrt{-25}$ **b)** $\sqrt{-5} \cdot \sqrt{-7}$ **c)** $-3i \cdot 8i$

d) $-4i(3 - 5i)$ **e)** $(1 + 2i)(4 + 3i)$

SOLUTION

a) $\sqrt{-4}\sqrt{-25} = \sqrt{-1} \cdot \sqrt{4} \cdot \sqrt{-1} \cdot \sqrt{25}$
$$= i \cdot 2 \cdot i \cdot 5$$
$$= i^2 \cdot 10$$
$$= -1 \cdot 10 \quad\quad i^2 = -1$$
$$= -10$$

b) $\sqrt{-5} \cdot \sqrt{-7} = \sqrt{-1} \cdot \sqrt{5} \cdot \sqrt{-1} \cdot \sqrt{7}$ Try to do this step mentally.
$$= i \cdot \sqrt{5} \cdot i \cdot \sqrt{7}$$
$$= i^2 \cdot \sqrt{35}$$
$$= -1 \cdot \sqrt{35} \quad\quad i^2 = -1$$
$$= -\sqrt{35}$$

c) $-3i \cdot 8i = -24 \cdot i^2$
$$= -24 \cdot (-1) \quad\quad i^2 = -1$$
$$= 24$$

d) $-4i(3 - 5i) = -4i \cdot 3 + (-4i)(-5i)$ Using the distributive law
$$= -12i + 20i^2$$
$$= -12i - 20 \quad\quad\quad\quad\quad i^2 = -1$$
$$= -20 - 12i \quad\quad\quad\quad\quad \text{Writing in the form } a + bi$$

e) $(1 + 2i)(4 + 3i) = 4 + 3i + 8i + 6i^2$ Multiplying each term of $4 + 3i$ by each term of $1 + 2i$ (FOIL)
$$= 4 + 3i + 8i - 6 \quad\quad i^2 = -1$$
$$= -2 + 11i \quad\quad\quad\quad\quad \text{Combining like terms}$$

TRY EXERCISES 35 and 49

Conjugates and Division

Recall that the conjugate of $4 + \sqrt{2}$ is $4 - \sqrt{2}$.

Conjugates of complex numbers are defined in a similar manner.

> ### Conjugate of a Complex Number
> The *conjugate* of a complex number $a + bi$ is $a - bi$, and the *conjugate* of $a - bi$ is $a + bi$.

EXAMPLE **4**   Find the conjugate of each number.

a) $-3 - 7i$ **b)** $4i$

SOLUTION

a) $-3 - 7i$ The conjugate is $-3 + 7i$.

b) $4i$ The conjugate is $-4i$. Note that $4i = 0 + 4i$.

The product of a complex number and its conjugate is a real number.

EXAMPLE **5**   Multiply: $(5 + 7i)(5 - 7i)$.

SOLUTION

$$(5 + 7i)(5 - 7i) = 5^2 - (7i)^2 \qquad \text{Using } (A + B)(A - B) = A^2 - B^2$$
$$= 25 - 49i^2$$
$$= 25 - 49(-1) \qquad i^2 = -1$$
$$= 25 + 49 = 74$$

TRY EXERCISE ▶ 55

Conjugates are used when dividing complex numbers. The procedure is much like that used to rationalize denominators in Section 10.5.

EXAMPLE **6**   Divide and simplify to the form $a + bi$.

a) $\dfrac{-2 + 9i}{1 - 3i}$ **b)** $\dfrac{7 + 4i}{5i}$

SOLUTION

a) To divide and simplify $(-2 + 9i)/(1 - 3i)$, we multiply by 1, using the conjugate of the denominator to form 1:

$$\frac{-2 + 9i}{1 - 3i} = \frac{-2 + 9i}{1 - 3i} \cdot \frac{1 + 3i}{1 + 3i} \qquad \begin{array}{l}\text{Multiplying by 1 using the conjugate of}\\ \text{the denominator in the symbol for 1}\end{array}$$

$$= \frac{(-2 + 9i)(1 + 3i)}{(1 - 3i)(1 + 3i)} \qquad \begin{array}{l}\text{Multiplying numerators;}\\ \text{multiplying denominators}\end{array}$$

$$= \frac{-2 - 6i + 9i + 27i^2}{1^2 - 9i^2} \qquad \text{Using FOIL}$$

$$= \frac{-2 + 3i + (-27)}{1 - (-9)} \qquad i^2 = -1$$

$$= \left.\begin{array}{l}\dfrac{-29 + 3i}{10}\\[2ex] = -\dfrac{29}{10} + \dfrac{3}{10}\,i.\end{array}\right\rbrace \qquad \begin{array}{l}\text{Writing in the form } a + bi;\\ \text{note that } \dfrac{X + Y}{Z} = \dfrac{X}{Z} + \dfrac{Y}{Z}\end{array}$$

b) The conjugate of $5i$ is $-5i$, so we *could* multiply by $-5i/(-5i)$. However, when the denominator is a pure imaginary number, it is easiest if we multiply by i/i:

$$\frac{7 + 4i}{5i} = \frac{7 + 4i}{5i} \cdot \frac{i}{i}$$
Multiplying by 1 using i/i. We can also use the conjugate of $5i$ to write $-5i/(-5i)$.

$$= \frac{7i + 4i^2}{5i^2}$$
Multiplying

$$= \frac{7i + 4(-1)}{5(-1)}$$
$i^2 = -1$

$$= \frac{7i - 4}{-5} = \frac{-4}{-5} + \frac{7}{-5}i, \text{ or } \frac{4}{5} - \frac{7}{5}i.$$
Writing in the form $a + bi$

> **TRY EXERCISE** 73

Powers of i

Answers to problems involving complex numbers are generally written in the form $a + bi$. In the following discussion, we show why there is no need to use powers of i (other than 1) when writing answers.

Recall that -1 raised to an *even* power is 1, and -1 raised to an *odd* power is -1. Simplifying powers of i can then be done by using the fact that $i^2 = -1$ and expressing the given power of i in terms of i^2. Consider the following:

$$i^2 = -1,$$
$$i^3 = i^2 \cdot i = (-1)i = -i,$$
$$i^4 = (i^2)^2 = (-1)^2 = 1,$$
$$i^5 = i^4 \cdot i = (i^2)^2 \cdot i = (-1)^2 \cdot i = i,$$
$$i^6 = (i^2)^3 = (-1)^3 = -1. \longleftarrow \text{The pattern is now repeating.}$$

The powers of i cycle themselves through the values i, -1, $-i$, and 1. Even powers of i are -1 or 1 whereas odd powers of i are i or $-i$.

EXAMPLE **7** Simplify: **(a)** i^{18}; **(b)** i^{24}.

SOLUTION

a) $i^{18} = (i^2)^9$ Using the power rule

     $= (-1)^9 = -1$ Raising -1 to a power

b) $i^{24} = (i^2)^{12}$

     $= (-1)^{12} = 1$

> **TRY EXERCISE** 83

To simplify i^n when n is odd, we rewrite i^n as $i^{n-1} \cdot i$.

EXAMPLE **8** Simplify: **(a)** i^{29}; **(b)** i^{75}.

SOLUTION

a) $i^{29} = i^{28} i^1$ Using the product rule. This is a key step when i is raised to an odd power.

     $= (i^2)^{14} i$ Using the power rule

     $= (-1)^{14} i$

     $= 1 \cdot i = i$

b) $i^{75} = i^{74} i^1$ Using the product rule

     $= (i^2)^{37} i$ Using the power rule

     $= (-1)^{37} i$

     $= -1 \cdot i = -i$

> **TRY EXERCISE** 85

10.8 EXERCISE SET

For Extra Help MyMathLab | Math XL PRACTICE | WATCH | DOWNLOAD

↪ **Concept Reinforcement** *Classify each statement as either true or false.*

1. Imaginary numbers are so named because they have no real-world applications.

2. Every real number is imaginary, but not every imaginary number is real.

3. Every imaginary number is a complex number, but not every complex number is imaginary.

4. Every real number is a complex number, but not every complex number is real.

5. We add complex numbers by combining real parts and combining imaginary parts.

6. The product of a complex number and its conjugate is always a real number.

7. The square of a complex number is always a real number.

8. The quotient of two complex numbers is always a complex number.

Express in terms of i.

9. $\sqrt{-100}$

10. $\sqrt{-9}$

11. $\sqrt{-5}$

12. $\sqrt{-7}$

13. $\sqrt{-8}$

14. $\sqrt{-12}$

15. $-\sqrt{-11}$

16. $-\sqrt{-17}$

17. $-\sqrt{-49}$

18. $-\sqrt{-81}$

19. $-\sqrt{-300}$

20. $-\sqrt{-75}$

21. $6 - \sqrt{-84}$

22. $4 - \sqrt{-60}$

23. $-\sqrt{-76} + \sqrt{-125}$

24. $\sqrt{-4} + \sqrt{-12}$

25. $\sqrt{-18} - \sqrt{-64}$

26. $\sqrt{-72} - \sqrt{-25}$

Perform the indicated operation and simplify. Write each answer in the form $a + bi$.

27. $(3 + 4i) + (2 - 7i)$

28. $(5 - 6i) + (8 + 9i)$

29. $(9 + 5i) - (2 + 3i)$

30. $(8 + 7i) - (2 + 4i)$

31. $(7 - 4i) - (5 - 3i)$

32. $(5 - 3i) - (9 + 2i)$

33. $(-5 - i) - (7 + 4i)$

34. $(-2 + 6i) - (-7 + i)$

35. $5i \cdot 8i$

36. $3i \cdot 9i$

37. $(-4i)(-6i)$

38. $7i \cdot (-8i)$

39. $\sqrt{-36}\sqrt{-9}$

40. $\sqrt{-49}\sqrt{-16}$

41. $\sqrt{-3}\sqrt{-10}$

42. $\sqrt{-6}\sqrt{-7}$

43. $\sqrt{-6}\sqrt{-21}$

44. $\sqrt{-15}\sqrt{-10}$

45. $5i(2 + 6i)$

46. $2i(7 + 3i)$

47. $-7i(3 + 4i)$

48. $-4i(6 - 5i)$

49. $(1 + i)(3 + 2i)$

50. $(4 + i)(2 + 3i)$

51. $(6 - 5i)(3 + 4i)$

52. $(5 - 6i)(2 + 5i)$

53. $(7 - 2i)(2 - 6i)$

54. $(-4 + 5i)(3 - 4i)$

55. $(3 + 8i)(3 - 8i)$

56. $(1 + 2i)(1 - 2i)$

57. $(-7 + i)(-7 - i)$

58. $(-4 + 5i)(-4 - 5i)$

59. $(4 - 2i)^2$

60. $(1 - 2i)^2$

61. $(2 + 3i)^2$

62. $(3 + 2i)^2$

63. $(-2 + 3i)^2$

64. $(-5 - 2i)^2$

65. $\dfrac{10}{3 + i}$

66. $\dfrac{26}{5 + i}$

67. $\dfrac{2}{3 - 2i}$

68. $\dfrac{4}{2 - 3i}$

69. $\dfrac{2i}{5 + 3i}$

70. $\dfrac{3i}{4 + 2i}$

71. $\dfrac{5}{6i}$

72. $\dfrac{4}{7i}$

73. $\dfrac{5 - 3i}{4i}$

74. $\dfrac{2 + 7i}{5i}$

Aha! **75.** $\dfrac{7i + 14}{7i}$

76. $\dfrac{6i + 3}{3i}$

77. $\dfrac{4 + 5i}{3 - 7i}$

78. $\dfrac{5 + 3i}{7 - 4i}$

79. $\dfrac{2 + 3i}{2 + 5i}$

80. $\dfrac{3 + 2i}{4 + 3i}$

81. $\dfrac{3 - 2i}{4 + 3i}$

82. $\dfrac{5 - 2i}{3 + 6i}$

Simplify.

83. i^{32}

84. i^{19}

85. i^{15}

86. i^{38}

87. i^{42}

88. i^{64}

89. i^9

90. $(-i)^{71}$

91. $(-i)^6$

92. $(-i)^4$

93. $(5i)^3$

94. $(-3i)^5$

95. $i^2 + i^4$

96. $5i^5 + 4i^3$

97. Is the product of two imaginary numbers always an imaginary number? Why or why not?

98. In what way(s) are conjugates of complex numbers similar to the conjugates used in Section 10.5?

Skill Review

To prepare for Section 11.1, review solving quadratic equations (Section 5.7).

Solve. [5.7]

99. $x^2 - x - 6 = 0$

100. $(x - 5)^2 = 0$

101. $t^2 = 100$

102. $2t^2 - 50 = 0$

103. $15x^2 = 14x + 8$

104. $6x^2 = 5x + 6$

Synthesis

105. Is the set of real numbers a subset of the set of complex numbers? Why or why not?

106. Is the union of the set of imaginary numbers and the set of real numbers the set of complex numbers? Why or why not?

Complex numbers are often graphed on a plane. The horizontal axis is the real axis and the vertical axis is the imaginary axis. A complex number such as $5 - 2i$ *then corresponds to 5 on the real axis and* -2 *on the imaginary axis.*

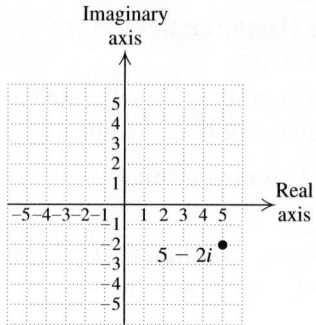

107. Graph each of the following.

   a) $3 + 2i$ **b)** $-1 + 4i$

   c) $3 - i$ **d)** $-5i$

108. Graph each of the following.

   a) $1 - 4i$ **b)** $-2 - 3i$

   c) i **d)** 4

The absolute value of a complex number $a + bi$ *is its distance from the origin. Using the distance formula, we have* $|a + bi| = \sqrt{a^2 + b^2}$. *Find the absolute value of each complex number.*

109. $|3 + 4i|$

110. $|8 - 6i|$

111. $|-1 + i|$

112. $|-3 - i|$

A function g is given by

$$g(z) = \frac{z^4 - z^2}{z - 1}.$$

113. Find $g(3i)$.

114. Find $g(1 + i)$.

115. Find $g(5i - 1)$.

116. Find $g(2 - 3i)$.

117. Evaluate

$$\frac{1}{w - w^2} \quad \text{for} \quad w = \frac{1 - i}{10}.$$

Simplify.

118. $\dfrac{i^5 + i^6 + i^7 + i^8}{(1 - i)^4}$

119. $(1 - i)^3(1 + i)^3$

120. $\dfrac{5 - \sqrt{5}i}{\sqrt{5}i}$

121. $\dfrac{6}{1 + \dfrac{3}{i}}$

122. $\left(\dfrac{1}{2} - \dfrac{1}{3}i\right)^2 - \left(\dfrac{1}{2} + \dfrac{1}{3}i\right)^2$

123. $\dfrac{i - i^{38}}{1 + i}$

Study Summary

KEY TERMS AND CONCEPTS	EXAMPLES

SECTION 10.1: RADICAL EXPRESSIONS AND FUNCTIONS

c is a **square root** of a if $c^2 = a$.

c is a **cube root** of a if $c^3 = a$.

$\sqrt{a}$ indicates the **principal** square root of a.

$\sqrt[n]{a}$ indicates the **nth root** of a.

index ⟶ $\sqrt[n]{a}$ ⟵ **radicand**

radical symbol

The square roots of 25 are -5 and 5.

The cube root of -8 is -2.

$\sqrt{25} = 5$

$\sqrt[3]{-8} = -2$

For all a,

$$\sqrt[n]{a^n} = |a| \text{ when } n \text{ is even;}$$
$$\sqrt[n]{a^n} = a \text{ when } n \text{ is odd.}$$

If a represents a nonnegative number,

$$\sqrt[n]{a^n} = a.$$

Assume that x can be any real number.
$$\sqrt{(3 + x)^2} = |3 + x|$$

Assume that x represents a nonnegative number.
$$\sqrt{(7x)^2} = 7x$$

SECTION 10.2: RATIONAL NUMBERS AS EXPONENTS

$a^{1/n}$ means $\sqrt[n]{a}$.

$a^{m/n}$ means $\left(\sqrt[n]{a}\right)^m$ or $\sqrt[n]{a^m}$.

$a^{-m/n}$ means $\dfrac{1}{a^{m/n}}$.

$64^{1/2} = \sqrt{64} = 8$

$125^{2/3} = \left(\sqrt[3]{125}\right)^2 = 5^2 = 25$

$8^{-1/3} = \dfrac{1}{8^{1/3}} = \dfrac{1}{2}$

SECTION 10.3: MULTIPLYING RADICAL EXPRESSIONS

The Product Rule for Radicals

For any real numbers $\sqrt[n]{a}$ and $\sqrt[n]{b}$,

$$\sqrt[n]{a} \cdot \sqrt[n]{b} = \sqrt[n]{a \cdot b}.$$

$$\sqrt[3]{4x} \cdot \sqrt[3]{5y} = \sqrt[3]{20xy}$$

Using the Product Rule to Simplify

For any real numbers $\sqrt[n]{a}$ and $\sqrt[n]{b}$,

$$\sqrt[n]{a \cdot b} = \sqrt[n]{a} \cdot \sqrt[n]{b}.$$

$$\sqrt{75x^8 y^{11}} = \sqrt{25 \cdot x^8 \cdot y^{10} \cdot 3 \cdot y}$$
$$= \sqrt{25} \cdot \sqrt{x^8} \cdot \sqrt{y^{10}} \cdot \sqrt{3y}$$
$$= 5x^4 y^5 \sqrt{3y} \quad \text{Assuming } y \text{ is nonnegative}$$

SECTION 10.4: DIVIDING RATIONAL EXPRESSIONS

The Quotient Rule for Radicals

For any real numbers $\sqrt[n]{a}$ and $\sqrt[n]{b}$, $b \neq 0$,

$$\sqrt[n]{\dfrac{a}{b}} = \dfrac{\sqrt[n]{a}}{\sqrt[n]{b}}.$$

$$\sqrt[3]{\dfrac{8y^4}{125}} = \dfrac{\sqrt[3]{8y^4}}{\sqrt[3]{125}} = \dfrac{2y\sqrt[3]{y}}{5}$$

$$\dfrac{\sqrt{18a^9}}{\sqrt{2a^3}} = \sqrt{\dfrac{18a^9}{2a^3}} = \sqrt{9a^6} = 3a^3 \quad \text{Assuming } a \text{ is positive}$$

We can **rationalize a denominator** by multiplying by 1.

$$\frac{\sqrt[3]{5}}{\sqrt[3]{4y}} = \frac{\sqrt[3]{5}}{\sqrt[3]{4y}} \cdot \frac{\sqrt[3]{2y^2}}{\sqrt[3]{2y^2}} = \frac{\sqrt[3]{10y^2}}{\sqrt[3]{8y^3}} = \frac{\sqrt[3]{10y^2}}{2y}$$

SECTION 10.5: EXPRESSIONS CONTAINING SEVERAL RADICAL TERMS

Like radicals have the same indices and radicands and can be combined.

$$\sqrt{12} + 5\sqrt{3} = \sqrt{4 \cdot 3} + 5\sqrt{3} = 2\sqrt{3} + 5\sqrt{3} = 7\sqrt{3}$$

Radical expressions are multiplied in much the same way that polynomials are multiplied.

$$(1 + 5\sqrt{6})(4 - \sqrt{6}) = 1 \cdot 4 - 1\sqrt{6} + 4 \cdot 5\sqrt{6} - 5\sqrt{6} \cdot \sqrt{6}$$
$$= 4 - \sqrt{6} + 20\sqrt{6} - 5 \cdot 6$$
$$= -26 + 19\sqrt{6}$$

To rationalize a denominator containing two terms, we use the **conjugate** of the denominator to write a form of 1.

$$\frac{2}{1 - \sqrt{3}} = \frac{2}{1 - \sqrt{3}} \cdot \frac{1 + \sqrt{3}}{1 + \sqrt{3}} \qquad 1 + \sqrt{3} \text{ is the}$$
$$\text{conjugate of } 1 - \sqrt{3}.$$
$$= \frac{2(1 + \sqrt{3})}{-2} = -1 - \sqrt{3}$$

When terms have different indices, we can often use rational exponents to simplify.

$$\sqrt[3]{p} \cdot \sqrt[4]{q^3} = p^{1/3} \cdot q^{3/4}$$
$$= p^{4/12} \cdot q^{9/12} \qquad \text{Finding a common denominator}$$
$$= \sqrt[12]{p^4 q^9}$$

SECTION 10.6: SOLVING RADICAL EQUATIONS

The Principle of Powers

If $a = b$, then $a^n = b^n$.

To solve a radical equation, use the principle of powers and the steps on pp. 670 and 672.

Solutions found using the principle of powers must be checked in the original equation.

$$x - 7 = \sqrt{x - 5}$$
$$(x - 7)^2 = (\sqrt{x - 5})^2$$
$$x^2 - 14x + 49 = x - 5$$
$$x^2 - 15x + 54 = 0$$
$$(x - 6)(x - 9) = 0$$
$$x = 6 \ \ or \ \ x = 9$$

Only 9 checks and is the solution.

$$2 + \sqrt{t} = \sqrt{t + 8}$$
$$(2 + \sqrt{t})^2 = (\sqrt{t + 8})^2$$
$$4 + 4\sqrt{t} + t = t + 8$$
$$4\sqrt{t} = 4$$
$$\sqrt{t} = 1$$
$$(\sqrt{t})^2 = (1)^2$$
$$t = 1$$

1 checks and is the solution.

SECTION 10.7: THE DISTANCE AND MIDPOINT FORMULAS AND OTHER APPLICATIONS

The Pythagorean Theorem

In any right triangle, if a and b are the lengths of the legs and c is the length of the hypotenuse, then

$$a^2 + b^2 = c^2.$$

Find the length of the hypotenuse of a right triangle with legs of lengths 4 and 7. Give an exact answer in radical notation, as well as a decimal approximation to three decimal places.

$$a^2 + b^2 = c^2$$
$$4^2 + 7^2 = c^2 \qquad \text{Substituting}$$
$$16 + 49 = c^2$$
$$65 = c^2$$
$$\sqrt{65} = c \qquad \text{This is exact.}$$
$$8.062 \approx c \qquad \text{This is approximate.}$$

Special Triangles

The length of the hypotenuse in an isosceles right triangle is the length of a leg times $\sqrt{2}$.

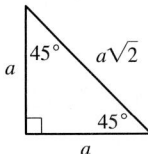

Find the missing lengths. Give an exact answer and, where appropriate, an approximation to three decimal places.

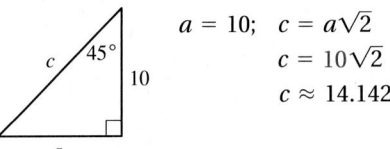

$$a = 10; \quad c = a\sqrt{2}$$
$$c = 10\sqrt{2}$$
$$c \approx 14.142$$

The length of the longer leg in a $30°$–$60°$–$90°$ triangle is the length of the shorter leg times $\sqrt{3}$. The hypotenuse is twice as long as the shorter leg.

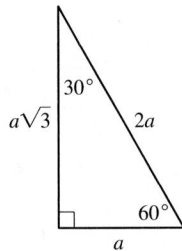

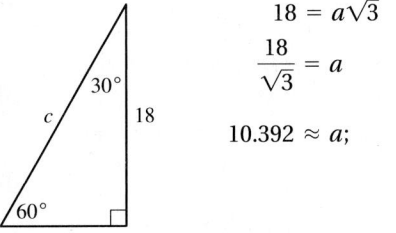

$$18 = a\sqrt{3} \qquad c = 2a$$
$$\frac{18}{\sqrt{3}} = a \qquad c = 2\left(\frac{18}{\sqrt{3}}\right)$$
$$10.392 \approx a; \qquad c = \frac{36}{\sqrt{3}}$$
$$c \approx 20.785$$

The Distance Formula

The distance d between any two points (x_1, y_1) and (x_2, y_2) is given by
$$d = \sqrt{(x_2 - x_1)^2 + (y_2 - y_1)^2}.$$

Find the distance between $(3, -5)$ and $(-1, -2)$.
$$d = \sqrt{(-1 - 3)^2 + (-2 - (-5))^2}$$
$$= \sqrt{(-4)^2 + (3)^2}$$
$$= \sqrt{16 + 9} = \sqrt{25} = 5$$

The Midpoint Formula

If the endpoints of a segment are (x_1, y_1) and (x_2, y_2), then the coordinates of the midpoint are
$$\left(\frac{x_1 + x_2}{2}, \frac{y_1 + y_2}{2}\right).$$

Find the midpoint of the segment with endpoints $(3, -5)$ and $(-1, -2)$.
$$\left(\frac{3 + (-1)}{2}, \frac{-5 + (-2)}{2}\right), \text{ or } \left(1, -\frac{7}{2}\right)$$

SECTION 10.8: THE COMPLEX NUMBERS

A **complex number** is any number that can be written in the form $a + bi$, where a and b are real numbers,
$$i = \sqrt{-1}, \quad \text{and} \quad i^2 = -1.$$

$$(3 + 2i) + (4 - 7i) = 7 - 5i$$
$$(8 + 6i) - (5 + 2i) = 3 + 4i$$
$$(2 + 3i)(4 - i) = 8 - 2i + 12i - 3i^2$$
$$= 8 + 10i - 3(-1) = 11 + 10i$$

$$\frac{1 - 4i}{3 - 2i} = \frac{1 - 4i}{3 - 2i} \cdot \frac{3 + 2i}{3 + 2i} \qquad \text{The conjugate of } 3 - 2i \text{ is } 3 + 2i.$$
$$= \frac{3 + 2i - 12i - 8i^2}{9 + 6i - 6i - 4i^2}$$
$$= \frac{3 - 10i - 8(-1)}{9 - 4(-1)} = \frac{11 - 10i}{13} = \frac{11}{13} - \frac{10}{13}i$$

Review Exercises: Chapter 10

🖐 **Concept Reinforcement** *Classify each statement as either true or false.*

1. $\sqrt{ab} = \sqrt{a} \cdot \sqrt{b}$ for any real numbers $\sqrt{a}$ and $\sqrt{b}$. [10.3]

2. $\sqrt{a + b} = \sqrt{a} + \sqrt{b}$ for any real numbers $\sqrt{a}$ and $\sqrt{b}$. [10.5]

3. $\sqrt{a^2} = a$, for any real number a. [10.1]

4. $\sqrt[3]{a^3} = a$, for any real number a. [10.1]

5. $x^{2/5}$ means $\sqrt[5]{x^2}$ and $(\sqrt[5]{x})^2$. [10.2]

6. The hypotenuse of a right triangle is never shorter than either leg. [10.7]

7. Some radical equations have no solution. [10.6]

8. If $f(x) = \sqrt{x - 5}$, then the domain of f is the set of all nonnegative real numbers. [10.1]

Simplify. [10.1]

9. $\sqrt{\dfrac{100}{121}}$

10. $-\sqrt{0.36}$

Let $f(x) = \sqrt{x + 10}$. Find the following. [10.1]

11. $f(15)$

12. The domain of f

Simplify. Assume that each variable can represent any real number. [10.1]

13. $\sqrt{64t^2}$

14. $\sqrt{(c + 7)^2}$

15. $\sqrt{4x^2 + 4x + 1}$

16. $\sqrt[5]{-32}$

17. Write an equivalent expression using exponential notation: $(\sqrt[3]{5ab})^4$. [10.2]

18. Write an equivalent expression using radical notation: $(16a^6)^{3/4}$. [10.2]

Use rational exponents to simplify. Assume $x, y \geq 0$. [10.2]

19. $\sqrt{x^6 y^{10}}$

20. $(\sqrt[6]{x^2 y})^2$

Simplify. Do not use negative exponents in the answers. [10.2]

21. $(x^{-2/3})^{3/5}$

22. $\dfrac{7^{-1/3}}{7^{-1/2}}$

23. If $f(x) = \sqrt{25(x - 6)^2}$, find a simplified form for $f(x)$. [10.3]

Simplify. Write all answers using radical notation. Assume that all variables represent nonnegative numbers.

24. $\sqrt[4]{16x^{20}y^8}$ [10.3]

25. $\sqrt{250x^3 y^2}$ [10.3]

26. $\sqrt{5a}\sqrt{7b}$ [10.3]

27. $\sqrt[3]{3x^4 b}\sqrt[3]{9xb^2}$ [10.3]

28. $\sqrt[3]{-24x^{10}y^8}\sqrt[3]{18x^7 y^4}$ [10.3]

29. $\sqrt[3]{-\dfrac{27y^{12}}{64}}$ [10.4]

30. $\dfrac{\sqrt[3]{60xy^3}}{\sqrt[3]{10x}}$ [10.4]

31. $\dfrac{\sqrt{75x}}{2\sqrt{3}}$ [10.4]

32. $\sqrt[4]{\dfrac{48a^{11}}{c^8}}$ [10.4]

33. $5\sqrt[3]{4y} + 2\sqrt[3]{4y}$ [10.5]

34. $2\sqrt{75} - 9\sqrt{3}$ [10.5]

35. $\sqrt[3]{8x^4} + \sqrt[3]{xy^6}$ [10.5]

36. $\sqrt{50} + 2\sqrt{18} + \sqrt{32}$ [10.5]

37. $(3 + \sqrt{10})(3 - \sqrt{10})$ [10.5]

38. $(\sqrt{3} - 3\sqrt{8})(\sqrt{5} + 2\sqrt{8})$ [10.5]

39. $\sqrt[4]{x}\sqrt{x}$ [10.5]

40. $\dfrac{\sqrt[3]{x^2}}{\sqrt[4]{x}}$ [10.5]

41. If $f(x) = x^2$, find $f(2 - \sqrt{a})$. [10.5]

42. Rationalize the denominator:

$$\dfrac{4\sqrt{5}}{\sqrt{2} + \sqrt{3}}.$$ [10.5]

43. Rationalize the numerator of the expression in Exercise 42. [10.5]

Solve. [10.6]

44. $\sqrt{y + 6} - 2 = 3$

45. $(x + 1)^{1/3} = -5$

46. $1 + \sqrt{x} = \sqrt{3x - 3}$

47. If $f(x) = \sqrt{x + 2} + x$, find a such that $f(a) = 4$. [10.6]

Solve. Give an exact answer and, where appropriate, an approximation to three decimal places. [10.7]

48. The diagonal of a square has length 10 cm. Find the length of a side of the square.

49. A skate-park jump has a ramp that is 6 ft long and is 2 ft high. How long is its base?

50. Find the missing lengths. Give exact answers and, where appropriate, an approximation to three decimal places.

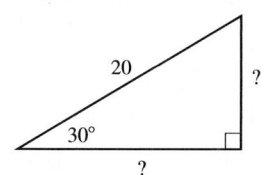

51. Find the distance between $(-6, 4)$ and $(-1, 5)$. Give an exact answer and an approximation to three decimal places. [10.7]

52. Find the midpoint of the segment with endpoints $(-7, -2)$ and $(3, -1)$. [10.7]

53. Express in terms of i and simplify: $\sqrt{-45}$. [10.8]

54. Add: $(-4 + 3i) + (2 - 12i)$. [10.8]

55. Subtract: $(9 - 7i) - (3 - 8i)$. [10.8]

Simplify. [10.8]

56. $(2 + 5i)(2 - 5i)$

57. i^{34}

58. $(6 - 3i)(2 - i)$

59. Divide. Write the answer in the form $a + bi$.

$$\frac{7 - 2i}{3 + 4i}$$ [10.8]

Synthesis

60. What makes some complex numbers real and others imaginary? [10.8]

61. Explain why $\sqrt[n]{x^n} = |x|$ when n is even, but $\sqrt[n]{x^n} = x$ when n is odd. [10.1]

62. Write a quotient of two imaginary numbers that is a real number (answers may vary). [10.8]

63. Solve:

$$\sqrt{11x + \sqrt{6 + x}} = 6.$$ [10.6]

64. Simplify:

$$\frac{2}{1 - 3i} - \frac{3}{4 + 2i}.$$ [10.8]

65. Don's Discount Shoes has two locations. The sign at the original location is shaped like an isosceles right triangle. The sign at the newer location is shaped like a 30°–60°–90° triangle. The hypotenuse of each sign measures 6 ft. Which sign has the greater area and by how much? (Round to three decimal places.) [10.7]

Test: Chapter 10

Simplify. Assume that variables can represent any real number.

1. $\sqrt{50}$

2. $\sqrt[3]{-\dfrac{8}{x^6}}$

3. $\sqrt{81a^2}$

4. $\sqrt{x^2 - 8x + 16}$

5. Write an equivalent expression using exponential notation: $\sqrt{7xy}$.

6. Write an equivalent expression using radical notation: $(4a^3b)^{5/6}$.

7. If $f(x) = \sqrt{2x - 10}$, determine the domain of f.

8. If $f(x) = x^2$, find $f(5 + \sqrt{2})$.

Simplify. Write all answers using radical notation. Assume that all variables represent positive numbers.

9. $\sqrt[5]{32x^{16}y^{10}}$

10. $\sqrt[3]{4w}\sqrt[3]{4v^2}$

11. $\sqrt{\dfrac{100a^4}{9b^6}}$

12. $\dfrac{\sqrt[5]{48x^6y^{10}}}{\sqrt[5]{16x^2y^9}}$

13. $\sqrt[4]{x^3}\sqrt{x}$

14. $\dfrac{\sqrt{y}}{\sqrt[10]{y}}$

15. $8\sqrt{2} - 2\sqrt{2}$

16. $\sqrt{x^4y} + \sqrt{9y^3}$

17. $(7 + \sqrt{x})(2 - 3\sqrt{x})$

18. Rationalize the denominator:
$$\dfrac{\sqrt{3}}{5 + \sqrt{2}}.$$

Solve.

19. $6 = \sqrt{x - 3} + 5$

20. $x = \sqrt{3x + 3} - 1$

21. $\sqrt{2x} = \sqrt{x + 1} + 1$

Solve. For Exercises 22–24, give exact answers and approximations to three decimal places.

22. A referee jogs diagonally from one corner of a 50-ft by 90-ft basketball court to the far corner. How far does she jog?

23. The hypotenuse of a 30°–60°–90° triangle is 10 cm long. Find the lengths of the legs.

24. Find the distance between the points $(3, 7)$ and $(-1, 8)$.

25. Find the midpoint of the segment with endpoints $(2, -5)$ and $(1, -7)$.

26. Express in terms of i and simplify: $\sqrt{-50}$.

27. Subtract: $(9 + 8i) - (-3 + 6i)$.

28. Multiply. Write the answer in the form $a + bi$.
$$(4 - i)^2$$

29. Divide. Write the answer in the form $a + bi$.
$$\dfrac{-2 + i}{3 - 5i}$$

30. Simplify: i^{37}.

Synthesis

31. Solve:
$$\sqrt{2x - 2} + \sqrt{7x + 4} = \sqrt{13x + 10}.$$

32. Simplify:
$$\dfrac{1 - 4i}{4i(1 + 4i)^{-1}}.$$

33. The function $D(h) = 1.2\sqrt{h}$ can be used to approximate the distance D, in miles, that a person can see to the horizon from a height h, in feet. How far above sea level must a pilot fly in order to see a horizon that is 180 mi away?

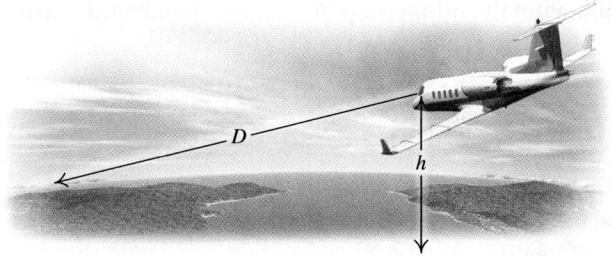

Cumulative Review: Chapters 1–10

Solve.

1. $2(x - 5) - 3 = 3(2x + 5)$ [2.2]

2. $x(x + 2) = 35$ [5.7]

3. $2y^2 = 50$ [5.7]

4. $\dfrac{1}{x} = \dfrac{2}{5}$ [6.6]

5. $\sqrt[3]{t} = -1$ [10.6]

6. $25x^2 - 10x + 1 = 0$ [5.7]

7. $|x - 2| \leq 5$ [9.3]

8. $2x + 5 > 6 \text{ or } x - 3 \leq 9$ [9.2]

9. $\dfrac{2x}{x - 1} + \dfrac{x}{x - 3} = 2$ [6.6]

10. $x = \sqrt{2x - 5} + 4$ [10.6]

11. $3x + y = 5,$
    $x - y = -5$ [8.2]

12. $2x - y + z = 1,$
    $x + 2y + z = -3,$
    $5x - y + 3z = 0$ [8.4]

Graph on a plane.

13. $3y = -6$ [3.3]

14. $y = -x + 5$ [3.6]

15. $x + y \leq 2$ [9.4]

16. $2x = y$ [3.2]

17. Determine the slope and the y-intercept of the line given by $y = -6 - x$. [3.6]

18. Find an equation for the line parallel to the line given by $y = 7x$ and passing through the point $(0, -11)$. [3.6]

Perform the indicated operations and, if possible, simplify. For radical expressions, assume that all variables represent positive numbers.

19. $18 \div 3 \cdot 2 - 6^2 \div (2 + 4)$ [1.8]

20. $(x^2y - 3x^2 - 4xy^2) - (x^2y - 3x^2 + 4xy^2)$ [4.4]

21. $(2a - 5b)^2$ [4.6]

22. $(c^2 - 3d)(c^2 + 3d)$ [4.6]

23. $\dfrac{1}{x} + \dfrac{1}{x + 1}$ [6.4]

24. $\dfrac{x + 3}{x - 2} - \dfrac{x + 5}{x + 1}$ [6.4]

25. $\dfrac{a^2 - a - 6}{a^2 - 1} \div \dfrac{a^2 - 6a + 9}{2a^2 + 3a + 1}$ [6.2]

26. $\dfrac{\dfrac{1}{x} + \dfrac{1}{x + 1}}{\dfrac{x}{x + 1}}$ [6.5]

27. $\sqrt{200} - 5\sqrt{8}$ [10.5]

28. $(1 + \sqrt{5})(4 - \sqrt{5})$ [10.5]

29. $\sqrt{10a^2b} \cdot \sqrt{15ab^3}$ [10.3]

30. $\sqrt[3]{y}\sqrt[5]{y}$ [10.5]

Factor.

31. $x^2 - 5x - 14$ [5.2]

32. $4y^8 - 4y^5$ [5.5]

33. $100c^2 - 25d^2$ [5.4]

34. $3t^2 - 5t - 8$ [5.3]

35. $3x^2 - 6x - 21$ [5.1]

36. $yt - xt - yz^2 + xz^2$ [5.1]

Find the domain of each function.

37. $f(x) = \dfrac{2x - 3}{x^2 - 6x + 9}$ [7.2]

38. $f(x) = \sqrt{2x - 11}$ [9.1]

Find each of the following, if $f(x) = \sqrt{2x - 3}$ and $g(x) = x^2$.

39. $f(14)$ [10.1]

40. $g(1 - \sqrt{5})$ [10.5]

41. $(f + g)(x)$ [7.4]

Solve.

42. *Flood rescue.*   A flood rescue team uses a boat that travels 10 mph in still water. To reach a stranded family, they travel 7 mi against the current and return 7 mi with the current in a total time of $1\frac{2}{3}$ hr. What is the speed of the current?   [6.7]

43. *Emergency shelter.*   The entrance to a tent used by a rescue team is the shape of an equilateral triangle. If the base of the tent is 4 ft wide, how tall is the tent? Give an exact answer and an approximation to three decimal places.   [10.7]

44. *Age at marriage.*   The median age at first marriage for U.S. men has grown from 25.1 in 2001 to 25.5 in 2006. Let $m(t)$ represent the median age of men at first marriage t years after 2000.   [7.3]
Source: U.S. Census Bureau

a) Find a linear function that fits the data.
b) Use the function from part (a) to predict the median age of men at first marriage in 2020.
c) In what year will the median age of men at first marriage be 28?

45. *Salary.*   Neil's annual salary is $38,849. This includes a 6% superior performance raise. What would Neil's salary have been without the performance raise?   [2.5]

46. *Food service.*   Melted Goodness mixes Swiss chocolate and whipping cream to make a dessert fondue. Swiss chocolate costs $1.20 per ounce and whipping cream costs $0.30 per ounce. How much of each does Melted Goodness use to make 65 oz of fondue at a cost of $60.00?   [8.3]

47. *Food cost.*   The average cost of a Thanksgiving dinner in the United States rose from $34.56 in 2002 to $42.26 in 2007. What was the rate of increase?   [3.4]
Sources: Purdue University; American Farm Bureau Federation

48. *Landscaping.*   A rectangular parking lot is 80 ft by 100 ft. Part of the asphalt is removed in order to install a landscaped border of uniform width around it. The area of the new parking lot is 6300 ft². How wide is the landscaped border?   [5.8]

Synthesis

49. Give an equation in standard form for the line whose x-intercept is $(-3, 0)$ and whose y-intercept is $(0, 5)$.   [3.6]

50. Solve by graphing:
$$y = x - 1,$$
$$y = x^2 - 1. \quad [8.1]$$

Solve.

51. $\dfrac{\dfrac{1}{x} + \dfrac{1}{x+1}}{\dfrac{1}{x} - 1} = 1$ [6.4], [6.5]

52. $3\sqrt{2x - 11} = 2 + \sqrt{5x - 1}$ [10.6]

Quadratic Functions and Equations

JANET FISHER
MUSIC PUBLISHER AND
RECORD LABEL OWNER
Los Angeles, California

As a music publisher and record label owner, I pay our songwriters and artists royalties based on sales and uses of their songs, both physical and digital. In the world of digital downloads, these royalties run from fractions of cents for a "streamed listen," to a set number of cents per download, depending on the site from which the purchase or stream is made. Varying advances are split between publishers and writers when a song is used in a film or TV show. When a writer has a combination of uses for his or her song, you can imagine how important math is in order to pay them properly.

AN APPLICATION

As more listeners download their music purchases, sales of compact discs are decreasing. According to Nielsen SoundScan, sales of music CDs increased from 500 million in 1997 to 700 million in 2001 and then decreased to 450 million in 2007. Find a quadratic function that fits the data, and use the function to estimate the sales of music CDs in 2009.

This problem appears as Example 3 in Section 11.8.

T he mathematical translation of a problem is often a function or an equation containing a second-degree polynomial in one variable. Such functions or equations are said to be *quadratic.* In this chapter, we examine a variety of ways to solve quadratic equations and look at graphs and applications of quadratic functions.

11.1 Quadratic Equations

The Principle of Square Roots • Completing the Square • Problem Solving

The general form of a quadratic function is

$$f(x) = ax^2 + bx + c, \quad \text{with } a \neq 0.$$

The graph of a quadratic function is a *parabola.* Such graphs open up or down and can have 0, 1, or 2 x-intercepts. We learn to graph quadratic functions later in this chapter.

ALGEBRAIC–GRAPHICAL CONNECTION

The graphs of the quadratic function $f(x) = x^2 + 6x + 8$ and the linear function $g(x) = 0$ are shown below.

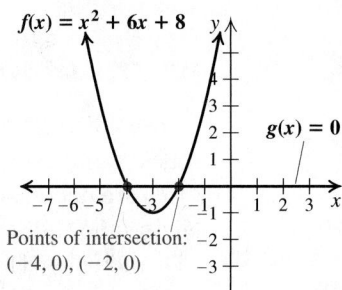

Note that $(-4, 0)$ and $(-2, 0)$ are the points of intersection of the graphs of $f(x) = x^2 + 6x + 8$ and $g(x) = 0$ (the x-axis). In Sections 11.6 and 11.7, we will develop efficient ways to graph quadratic functions. For now, the graphs help us visualize solutions.

In Chapter 5, we solved equations like $x^2 + 6x + 8 = 0$ by factoring:

$$x^2 + 6x + 8 = 0$$
$$(x + 4)(x + 2) = 0 \qquad \text{Factoring}$$
$$x + 4 = 0 \quad or \quad x + 2 = 0 \qquad \text{Using the principle of zero products}$$
$$x = -4 \quad or \qquad x = -2.$$

Note that -4 and -2 are the first coordinates of the points of intersection (or the x-intercepts) of the graph of $f(x)$ above.

In this section and the next, we develop algebraic methods for solving *any* quadratic equation, whether it is factorable or not.

EXAMPLE **1**   Solve: $x^2 = 25$.

SOLUTION   We have

$$x^2 = 25$$
$$x^2 - 25 = 0 \qquad \text{Writing in standard form}$$
$$(x - 5)(x + 5) = 0 \qquad \text{Factoring}$$
$$x - 5 = 0 \quad or \quad x + 5 = 0 \qquad \text{Using the principle of zero products}$$
$$x = 5 \quad or \qquad x = -5.$$

The solutions are 5 and −5. A graph in which $f(x) = x^2$ represents the left side of the original equation and $g(x) = 25$ represents the right side provides a check (see the figure at left). Of course, we can also check by substituting 5 and −5 into the original equation.

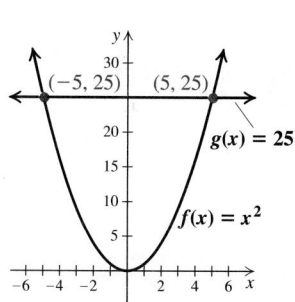

A visualization of Example 1

TRY EXERCISE ▶ 7

The Principle of Square Roots

Let's reconsider $x^2 = 25$. We know from Chapter 10 that the number 25 has two real-number square roots, 5 and −5, the solutions of the equation in Example 1. Thus we see that square roots provide quick solutions for equations of the type $x^2 = k$.

> ### The Principle of Square Roots
>
> For any real number k, if $x^2 = k$, then
> $$x = \sqrt{k} \quad or \quad x = -\sqrt{k}.$$

EXAMPLE **2**   Solve: $3x^2 = 6$. Give exact solutions and approximations to three decimal places.

SOLUTION   We have

$$3x^2 = 6$$
$$x^2 = 2 \qquad \text{Isolating } x^2$$
$$x = \sqrt{2} \quad or \quad x = -\sqrt{2}. \qquad \text{Using the principle of square roots}$$

We can use the symbol $\pm\sqrt{2}$ to represent both of the solutions.

> ***CAUTION!***   There are *two* solutions: $\sqrt{2}$ and $-\sqrt{2}$. Don't forget the second solution.

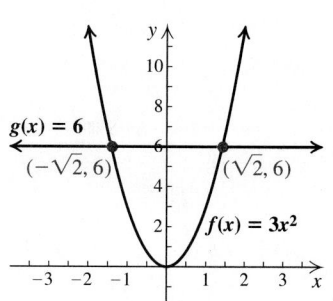

A visualization of Example 2

Check:

For $\sqrt{2}$:	For $-\sqrt{2}$:
$3x^2 = 6$	$3x^2 = 6$
$3(\sqrt{2})^2 \mid 6$	$3(-\sqrt{2})^2 \mid 6$
$3 \cdot 2 \mid$	$3 \cdot 2 \mid$
$6 \overset{?}{=} 6$ TRUE	$6 \overset{?}{=} 6$ TRUE

The solutions are $\sqrt{2}$ and $-\sqrt{2}$, or $\pm\sqrt{2}$, which round to 1.414 and −1.414.

TRY EXERCISE ▶ 11

EXAMPLE **3**

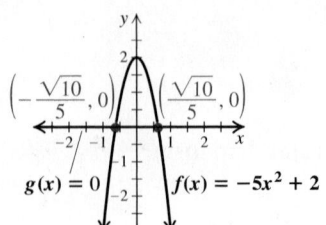

$g(x) = 0$ $f(x) = -5x^2 + 2$

A visualization of Example 3

Solve: $-5x^2 + 2 = 0$.

SOLUTION We have

$$-5x^2 + 2 = 0$$

$$x^2 = \frac{2}{5} \qquad \text{Isolating } x^2$$

$$x = \sqrt{\frac{2}{5}} \quad or \quad x = -\sqrt{\frac{2}{5}}. \qquad \text{Using the principle of square roots}$$

The solutions are $\sqrt{\frac{2}{5}}$ and $-\sqrt{\frac{2}{5}}$, or simply $\pm\sqrt{\frac{2}{5}}$. If we rationalize the denominator, the solutions are written $\pm\dfrac{\sqrt{10}}{5}$. The checks are left to the student.

TRY EXERCISE 15

Sometimes we get solutions that are imaginary numbers.

EXAMPLE **4**

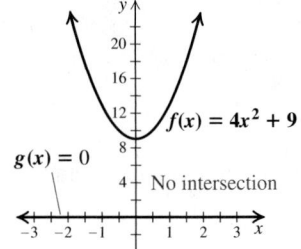

$f(x) = 4x^2 + 9$

$g(x) = 0$

No intersection

A visualization of Example 4

Solve: $4x^2 + 9 = 0$.

SOLUTION We have

$$4x^2 + 9 = 0$$

$$x^2 = -\tfrac{9}{4} \qquad \text{Isolating } x^2$$

$$x = \sqrt{-\tfrac{9}{4}} \quad or \quad x = -\sqrt{-\tfrac{9}{4}} \qquad \text{Using the principle of square roots}$$

$$x = \sqrt{\tfrac{9}{4}}\sqrt{-1} \quad or \quad x = -\sqrt{\tfrac{9}{4}}\sqrt{-1}$$

$$x = \tfrac{3}{2}i \quad\quad or \quad x = -\tfrac{3}{2}i. \qquad \text{Recall that } \sqrt{-1} = i.$$

Check: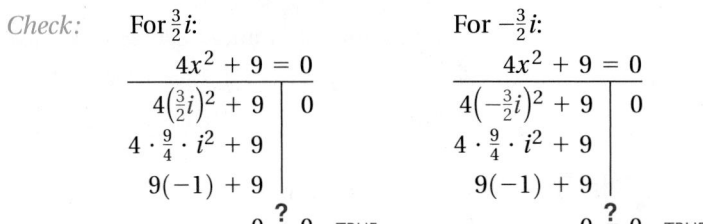

For $\frac{3}{2}i$:

$4x^2 + 9 = 0$	
$4\left(\frac{3}{2}i\right)^2 + 9$	0
$4 \cdot \frac{9}{4} \cdot i^2 + 9$	
$9(-1) + 9$	
	$0 \overset{?}{=} 0$ TRUE

For $-\frac{3}{2}i$:

$4x^2 + 9 = 0$	
$4\left(-\frac{3}{2}i\right)^2 + 9$	0
$4 \cdot \frac{9}{4} \cdot i^2 + 9$	
$9(-1) + 9$	
	$0 \overset{?}{=} 0$ TRUE

The solutions are $\frac{3}{2}i$ and $-\frac{3}{2}i$, or $\pm\frac{3}{2}i$. The graph at left confirms that there are no real-number solutions.

TRY EXERCISE 19

The principle of square roots can be restated in a more general form for any equation in which some algebraic expression squared equals a constant.

> **The Principle of Square Roots (Generalized Form)**
>
> For any real number k and any algebraic expression X:
>
> $$\text{If } X^2 = k, \quad \text{then} \quad X = \sqrt{k} \quad or \quad X = -\sqrt{k}.$$

EXAMPLE **5**

Let $f(x) = (x - 2)^2$. Find all x-values for which $f(x) = 7$.

SOLUTION We are asked to find all x-values for which

$$f(x) = 7,$$

or

$$(x - 2)^2 = 7. \qquad \text{Substituting } (x - 2)^2 \text{ for } f(x)$$

The generalized principle of square roots gives us

$$x - 2 = \sqrt{7} \qquad or \quad x - 2 = -\sqrt{7} \qquad \text{Using the principle of square roots}$$

$$x = 2 + \sqrt{7} \quad or \qquad x = 2 - \sqrt{7}.$$

Check: $f(2 + \sqrt{7}) = (2 + \sqrt{7} - 2)^2 = (\sqrt{7})^2 = 7.$

Similarly,

$$f(2 - \sqrt{7}) = (2 - \sqrt{7} - 2)^2 = (-\sqrt{7})^2 = 7.$$

The solutions are $2 + \sqrt{7}$ and $2 - \sqrt{7}$, or simply $2 \pm \sqrt{7}$.

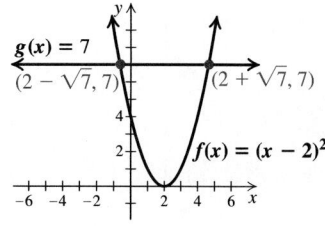

A visualization of Example 5

> TRY EXERCISE 35

Example 5 is of the form $(x - a)^2 = c$, where a and c are constants. Sometimes we must factor in order to obtain this form.

EXAMPLE 6

Solve: $x^2 + 6x + 9 = 2$.

SOLUTION We have

$$x^2 + 6x + 9 = 2 \qquad \text{The left side is the square of a binomial.}$$

$$(x + 3)^2 = 2 \qquad \text{Factoring}$$

$$x + 3 = \sqrt{2} \qquad or \quad x + 3 = -\sqrt{2} \qquad \text{Using the principle of square roots}$$

$$x = -3 + \sqrt{2} \quad or \qquad x = -3 - \sqrt{2}. \qquad \text{Adding } -3 \text{ to both sides}$$

The solutions are $-3 + \sqrt{2}$ and $-3 - \sqrt{2}$, or $-3 \pm \sqrt{2}$. The checks are left to the student.

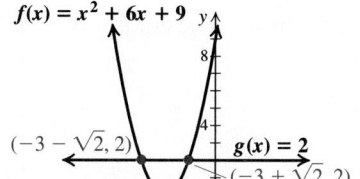

A visualization of Example 6

> TRY EXERCISE 29

Completing the Square

Not all quadratic equations are in the form $X^2 = k$. By using a method called *completing the square*, we can use the principle of square roots to solve *any* quadratic equation by writing it in this form.

Suppose we want to solve the quadratic equation

$$x^2 + 6x + 4 = 0.$$

The trinomial $x^2 + 6x + 4$ is not a perfect square. We can, however, create an equivalent equation with a perfect-square trinomial on one side:

$$x^2 + 6x + 4 = 0$$

$$x^2 + 6x \qquad = -4 \qquad \text{Only variable terms are on the left side.}$$

$$x^2 + 6x + 9 = -4 + 9 \qquad \text{Adding 9 to both sides. We explain this shortly.}$$

$$(x + 3)^2 = 5. \qquad \text{We could now use the principle of square roots to solve.}$$

We chose to add 9 to both sides because it creates a perfect-square trinomial on the left side. The 9 was determined by taking half of the coefficient of x and squaring it—that is,

$$\left(\tfrac{1}{2} \cdot 6\right)^2 = 3^2, \quad \text{or} \quad 9.$$

To understand why this procedure works, examine the following drawings.

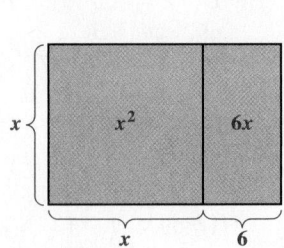 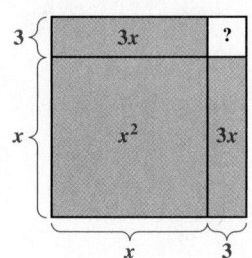

Note that the shaded areas in both figures represent the same area, $x^2 + 6x$. However, only the figure on the right, in which the $6x$ is halved, can be converted into a square with the addition of a constant term. The constant 9 is the "missing" piece that *completes* the square.

To complete the square for $x^2 + bx$, we add $(b/2)^2$.

Example 7, which follows, provides practice in finding numbers that complete the square. We will then use this skill to solve equations.

EXAMPLE **7**

Replace the blanks in each equation with constants to form a true equation.

a) $x^2 + 14x +$ ____ $= (x +$ ____$)^2$

b) $x^2 - 5x +$ ____ $= (x -$ ____$)^2$

c) $x^2 + \frac{3}{4}x +$ ____ $= (x +$ ____$)^2$

SOLUTION We take half of the coefficient of x and square it.

a) Half of 14 is 7, and $7^2 = 49$. Thus, $x^2 + 14x + 49$ is a perfect-square trinomial and is equivalent to $(x + 7)^2$. We have

$$x^2 + 14x + 49 = (x + 7)^2.$$

b) Half of -5 is $-\frac{5}{2}$, and $\left(-\frac{5}{2}\right)^2 = \frac{25}{4}$. Thus, $x^2 - 5x + \frac{25}{4}$ is a perfect-square trinomial and is equivalent to $\left(x - \frac{5}{2}\right)^2$. We have

$$x^2 - 5x + \frac{25}{4} = \left(x - \frac{5}{2}\right)^2.$$

c) Half of $\frac{3}{4}$ is $\frac{3}{8}$, and $\left(\frac{3}{8}\right)^2 = \frac{9}{64}$. Thus, $x^2 + \frac{3}{4}x + \frac{9}{64}$ is a perfect-square trinomial and is equivalent to $\left(x + \frac{3}{8}\right)^2$. We have

$$x^2 + \frac{3}{4}x + \frac{9}{64} = \left(x + \frac{3}{8}\right)^2.$$

> **TRY EXERCISE** 39

STUDENT NOTES ———————

In problems like Examples 7(b) and (c), it is best to avoid decimal notation. Most students have an easier time recognizing $\frac{9}{64}$ as $\left(\frac{3}{8}\right)^2$ than seeing 0.140625 as 0.375^2.

We can now use the method of completing the square to solve equations.

EXAMPLE **8**

Solve: $x^2 - 8x - 7 = 0$.

SOLUTION We begin by adding 7 to both sides:

$$x^2 - 8x - 7 = 0$$

$$x^2 - 8x \quad\quad = 7 \quad\quad \text{Adding 7 to both sides. We can now complete the square on the left side.}$$

$$x^2 - 8x + 16 = 7 + 16 \quad\quad \text{Adding 16 to both sides to complete the square: } \tfrac{1}{2}(-8) = -4, \text{ and } (-4)^2 = 16$$

$$(x - 4)^2 = 23 \quad\quad \text{Factoring and simplifying}$$

$$x - 4 = \pm\sqrt{23} \quad\quad \text{Using the principle of square roots}$$

$$x = 4 \pm \sqrt{23}. \quad\quad \text{Adding 4 to both sides}$$

Check: For $4 + \sqrt{23}$:

$$x^2 - 8x - 7 = 0$$

$$
\begin{array}{c|c}
(4 + \sqrt{23})^2 - 8(4 + \sqrt{23}) - 7 & 0 \\
16 + 8\sqrt{23} + 23 - 32 - 8\sqrt{23} - 7 & \\
16 + 23 - 32 - 7 + 8\sqrt{23} - 8\sqrt{23} & \\
\end{array}
$$

$$0 \overset{?}{=} 0 \quad \text{TRUE}$$

For $4 - \sqrt{23}$:

$$x^2 - 8x - 7 = 0$$

$$
\begin{array}{c|c}
(4 - \sqrt{23})^2 - 8(4 - \sqrt{23}) - 7 & 0 \\
16 - 8\sqrt{23} + 23 - 32 + 8\sqrt{23} - 7 & \\
16 + 23 - 32 - 7 - 8\sqrt{23} + 8\sqrt{23} & \\
\end{array}
$$

$$0 \overset{?}{=} 0 \quad \text{TRUE}$$

The solutions are $4 + \sqrt{23}$ and $4 - \sqrt{23}$, or $4 \pm \sqrt{23}$.   **TRY EXERCISE** 53

Recall that the value of $f(x)$ must be 0 at any x-intercept of the graph of f. If $f(a) = 0$, then $(a, 0)$ is an x-intercept of the graph.

EXAMPLE 9   Find the x-intercepts of the graph of $f(x) = x^2 + 5x - 3$.

SOLUTION   We set $f(x)$ equal to 0 and solve:

$$f(x) = 0$$

$$x^2 + 5x - 3 = 0 \qquad \text{Substituting}$$

$$x^2 + 5x = 3 \qquad \text{Adding 3 to both sides}$$

$$x^2 + 5x + \frac{25}{4} = 3 + \frac{25}{4} \qquad \begin{array}{l}\text{Completing the square:} \\ \frac{1}{2} \cdot 5 = \frac{5}{2}, \text{ and } \left(\frac{5}{2}\right)^2 = \frac{25}{4}\end{array}$$

$$\left(x + \frac{5}{2}\right)^2 = \frac{37}{4} \qquad \text{Factoring and simplifying}$$

$$x + \frac{5}{2} = \pm\frac{\sqrt{37}}{2} \qquad \begin{array}{l}\text{Using the principle of square roots} \\ \text{and the quotient rule for radicals}\end{array}$$

$$x = -\frac{5}{2} \pm \frac{\sqrt{37}}{2}, \quad \text{or} \quad \frac{-5 \pm \sqrt{37}}{2}. \qquad \text{Adding } -\frac{5}{2} \text{ to both sides}$$

The x-intercepts are

$$\left(-\frac{5}{2} - \frac{\sqrt{37}}{2}, 0\right) \quad \text{and} \quad \left(-\frac{5}{2} + \frac{\sqrt{37}}{2}, 0\right), \quad \text{or}$$

$$\left(\frac{-5 - \sqrt{37}}{2}, 0\right) \quad \text{and} \quad \left(\frac{-5 + \sqrt{37}}{2}, 0\right).$$

The checks are left to the student.   **TRY EXERCISE** 59

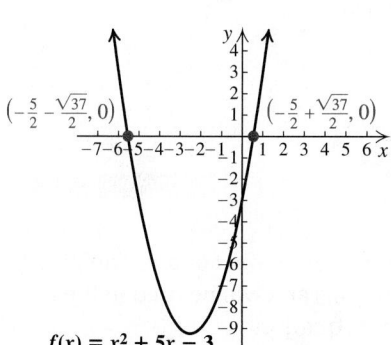

$\left(-\frac{5}{2} - \frac{\sqrt{37}}{2}, 0\right)$ $\left(-\frac{5}{2} + \frac{\sqrt{37}}{2}, 0\right)$

$f(x) = x^2 + 5x - 3$

A visualization of Example 9

Before we complete the square in a quadratic equation, the leading coefficient must be 1. When it is not 1, we divide both sides of the equation by whatever that coefficient may be.

> **To Solve a Quadratic Equation in *x* by Completing the Square**
> 1. Isolate the terms with variables on one side of the equation, and arrange them in descending order.
> 2. Divide both sides by the coefficient of x^2 if that coefficient is not 1.
> 3. Complete the square by taking half of the coefficient of *x* and adding its square to both sides.
> 4. Express the trinomial as the square of a binomial (factor the trinomial) and simplify the other side.
> 5. Use the principle of square roots (find the square roots of both sides).
> 6. Solve for *x* by adding or subtracting on both sides.

EXAMPLE 10   Solve: $3x^2 + 7x - 2 = 0$.

SOLUTION   We follow the steps listed above:

$$3x^2 + 7x - 2 = 0$$

Isolate the variable terms.

$$3x^2 + 7x = 2 \qquad \text{Adding 2 to both sides}$$

Divide both sides by the x^2-coefficient.

$$x^2 + \frac{7}{3}x = \frac{2}{3} \qquad \text{Dividing both sides by 3}$$

Complete the square.

$$x^2 + \frac{7}{3}x + \frac{49}{36} = \frac{2}{3} + \frac{49}{36} \qquad \text{Completing the square: } \left(\frac{1}{2} \cdot \frac{7}{3}\right)^2 = \frac{49}{36}$$

Factor the trinomial.

$$\left(x + \frac{7}{6}\right)^2 = \frac{73}{36} \qquad \text{Factoring and simplifying}$$

Use the principle of square roots.

$$x + \frac{7}{6} = \pm\frac{\sqrt{73}}{6} \qquad \begin{array}{l}\text{Using the principle of square roots} \\ \text{and the quotient rule for radicals}\end{array}$$

Solve for *x*.

$$x = -\frac{7}{6} \pm \frac{\sqrt{73}}{6}, \ \text{or} \ \frac{-7 \pm \sqrt{73}}{6}. \qquad \text{Adding } -\frac{7}{6} \text{ to both sides}$$

The checks are left to the student. The solutions are $-\dfrac{7}{6} \pm \dfrac{\sqrt{73}}{6}$, or $\dfrac{-7 \pm \sqrt{73}}{6}$.

This can be written as

$$-\frac{7}{6} + \frac{\sqrt{73}}{6} \ \text{and} \ -\frac{7}{6} - \frac{\sqrt{73}}{6}, \ \text{or} \ \frac{-7 + \sqrt{73}}{6} \ \text{and} \ \frac{-7 - \sqrt{73}}{6}.$$

> **TRY EXERCISE** ▶ 69

Any quadratic equation can be solved by completing the square. The procedure is also useful when graphing quadratic equations and will be used in the next section to develop a formula for solving quadratic equations.

Problem Solving

After one year, an amount of money *P*, invested at 4% per year, is worth 104% of *P*, or *P*(1.04). If that amount continues to earn 4% interest per year, after the second year the investment will be worth 104% of *P*(1.04), or $P(1.04)^2$. This is called **compounding interest** since after the first time period, interest is earned on both the initial investment *and* the interest from the first time period. Continuing the above pattern, we see that after the third year, the investment will be worth 104% of $P(1.04)^2$. Generalizing, we have the following.

> ### The Compound-Interest Formula
>
> If an amount of money P is invested at interest rate r, compounded annually, then in t years, it will grow to the amount A given by
>
> $$A = P(1 + r)^t. \qquad (r \text{ is written in decimal notation.})$$

We can use quadratic equations to solve certain interest problems.

EXAMPLE 11

Investment growth.   Katia invested \$4000 at interest rate r, compounded annually. In 2 yr, it grew to \$4410. What was the interest rate?

SOLUTION

1. **Familiarize.** We are already familiar with the compound-interest formula. If we were not, we would need to consult an outside source.

2. **Translate.** The translation consists of substituting into the formula:

   $$A = P(1 + r)^t$$
   $$4410 = 4000(1 + r)^2. \qquad \text{Substituting}$$

3. **Carry out.** We solve for r:

   $$4410 = 4000(1 + r)^2$$
   $$\tfrac{4410}{4000} = (1 + r)^2 \qquad \text{Dividing both sides by 4000}$$
   $$\tfrac{441}{400} = (1 + r)^2 \qquad \text{Simplifying}$$
   $$\pm\sqrt{\tfrac{441}{400}} = 1 + r \qquad \text{Using the principle of square roots}$$
   $$\pm\tfrac{21}{20} = 1 + r \qquad \text{Simplifying}$$
   $$-\tfrac{20}{20} \pm \tfrac{21}{20} = r \qquad \text{Adding } -1, \text{ or } -\tfrac{20}{20}, \text{ to both sides}$$
   $$\tfrac{1}{20} = r \quad \text{or} \quad -\tfrac{41}{20} = r.$$

4. **Check.** Since the interest rate cannot be negative, we need check only $\tfrac{1}{20}$, or 5%. If \$4000 were invested at 5% interest, compounded annually, then in 2 yr it would grow to $4000(1.05)^2$, or \$4410. The rate 5% checks.

5. **State.** The interest rate was 5%.   TRY EXERCISE 75

EXAMPLE 12

Free-falling objects.   The formula $s = 16t^2$ is used to approximate the distance s, in feet, that an object falls freely from rest in t seconds. The Grand Canyon Skywalk is 4000 ft above the Colorado River. How long will it take a stone to fall from the Skywalk to the river? Round to the nearest tenth of a second.

Source: www.grandcanyonskywalk.com

SOLUTION

1. **Familiarize.** We agree to disregard air resistance and use the given formula.

2. **Translate.** We substitute into the formula:

   $$s = 16t^2$$
   $$4000 = 16t^2.$$

3. **Carry out.** We solve for t:

$$4000 = 16t^2$$

$$250 = t^2$$

$$\sqrt{250} = t \qquad \text{Using the principle of square roots;}$$
rejecting the negative square root since t
cannot be negative in this problem

$$15.8 \approx t. \qquad \text{Using a calculator and rounding to}$$
the nearest tenth

4. **Check.** Since $16(15.8)^2 = 3994.24 \approx 4000$, our answer checks.

5. **State.** It takes about 15.8 sec for a stone to fall freely from the Grand Canyon Skywalk to the river.

> **TRY EXERCISE** 79

TECHNOLOGY CONNECTION

As we saw in Section 5.7, a graphing calculator can be used to find approximate solutions of any quadratic equation that has real-number solutions.

To check Example 8, we graph $y = x^2 - 8x - 7$ and use the ZERO or ROOT option of the CALC menu. When asked for a Left and Right Bound, we enter cursor positions to the left of and to the right of the root. A Guess between the bounds is entered and a value for the root then appears.

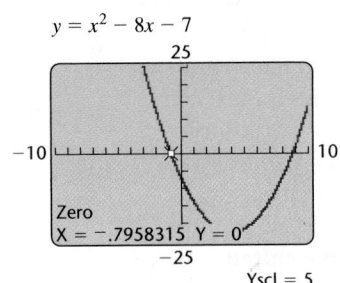

1. Use a graphing calculator to check the second solution of Example 8.
2. Use a graphing calculator to confirm the solutions in Example 9.
3. Can a graphing calculator be used to find *exact* solutions in Example 10? Why or why not?

4. Use a graphing calculator to confirm that there are no real-number solutions of $x^2 - 6x + 11 = 0$.

11.1 EXERCISE SET

Concept Reinforcement *Complete each of the following to form true statements.*

1. The principle of square roots states that if $x^2 = k$, then $x = $ ____ or $x = $ ____.

2. If $(x + 5)^2 = 49$, then $x + 5 = $ ____ or $x + 5 = $ ____.

3. If $t^2 + 6t + 9 = 17$, then (____)$^2 = 17$ and ____ $= \pm\sqrt{17}$.

4. The equations $x^2 + 8x + $ ____ $= 23$ and $x^2 + 8x = 7$ are equivalent.

5. The expressions $t^2 + 10t + $ ____ and $(t + $ ____ $)^2$ are equivalent.

6. The expressions $x^2 - 6x + $ ____ and $(x - $ ____ $)^2$ are equivalent.

Solve.

7. $x^2 = 100$

8. $t^2 = 144$

9. $p^2 - 50 = 0$

10. $c^2 - 8 = 0$

11. $5y^2 = 30$

12. $4y^2 = 12$

13. $9x^2 - 49 = 0$

14. $36a^2 - 25 = 0$

15. $6t^2 - 5 = 0$

16. $7x^2 - 5 = 0$

17. $a^2 + 1 = 0$

18. $t^2 + 4 = 0$

19. $4d^2 + 81 = 0$

20. $25y^2 + 16 = 0$

21. $(x - 3)^2 = 16$

22. $(x + 1)^2 = 100$

23. $(t + 5)^2 = 12$

24. $(y - 4)^2 = 18$

25. $(x + 1)^2 = -9$

26. $(x - 1)^2 = -49$

27. $\left(y + \frac{3}{4}\right)^2 = \frac{17}{16}$

28. $\left(t + \frac{3}{2}\right)^2 = \frac{7}{2}$

29. $x^2 - 10x + 25 = 64$ **30.** $x^2 - 6x + 9 = 100$

31. Let $f(x) = x^2$. Find x such that $f(x) = 19$.

32. Let $f(x) = x^2$. Find x such that $f(x) = 11$.

33. Let $f(x) = (x - 5)^2$. Find x such that $f(x) = 16$.

34. Let $g(x) = (x - 2)^2$. Find x such that $g(x) = 25$.

35. Let $F(t) = (t + 4)^2$. Find t such that $F(t) = 13$.

36. Let $f(t) = (t + 6)^2$. Find t such that $f(t) = 15$.

Aha! **37.** Let $g(x) = x^2 + 14x + 49$. Find x such that $g(x) = 49$.

38. Let $F(x) = x^2 + 8x + 16$. Find x such that $F(x) = 9$.

Replace the blanks in each equation with constants to complete the square and form a true equation.

39. $x^2 + 16x +$ ___ $= (x +$ ___$)^2$

40. $x^2 + 12x +$ ___ $= (x +$ ___$)^2$

41. $t^2 - 10t +$ ___ $= (t -$ ___$)^2$

42. $t^2 - 6t +$ ___ $= (t -$ ___$)^2$

43. $t^2 - 2t +$ ___ $= (t -$ ___$)^2$

44. $x^2 + 2x +$ ___ $= (x +$ ___$)^2$

45. $x^2 + 3x +$ ___ $= \left(x +$ ___$\right)^2$

46. $t^2 - 9t +$ ___ $= \left(t -$ ___$\right)^2$

47. $x^2 + \frac{2}{5}x +$ ___ $= \left(x +$ ___$\right)^2$

48. $x^2 + \frac{2}{3}x +$ ___ $= \left(x +$ ___$\right)^2$

49. $t^2 - \frac{5}{6}t +$ ___ $= \left(t -$ ___$\right)^2$

50. $t^2 - \frac{5}{3}t +$ ___ $= \left(t -$ ___$\right)^2$

Solve by completing the square. Show your work.

51. $x^2 + 6x = 7$

52. $x^2 + 8x = 9$

53. $t^2 - 10t = -23$

54. $t^2 - 4t = -1$

55. $x^2 + 12x + 32 = 0$

56. $x^2 + 16x + 15 = 0$

57. $t^2 + 8t - 3 = 0$

58. $t^2 + 6t - 5 = 0$

Complete the square to find the x-intercepts of each function given by the equation listed.

59. $f(x) = x^2 + 6x + 7$

60. $f(x) = x^2 + 10x - 2$

61. $g(x) = x^2 + 9x - 25$

62. $g(x) = x^2 + 5x + 2$

63. $f(x) = x^2 - 10x - 22$

64. $f(x) = x^2 - 8x - 10$

Solve by completing the square. Remember to first divide, as in Example 10, to make sure that the coefficient of x^2 is 1.

65. $9x^2 + 18x = -8$ **66.** $4x^2 + 8x = -3$

67. $3x^2 - 5x - 2 = 0$ **68.** $2x^2 - 5x - 3 = 0$

69. $5x^2 + 4x - 3 = 0$ **70.** $4x^2 + 3x - 5 = 0$

71. Find the x-intercepts of the function given by $f(x) = 4x^2 + 2x - 3$.

72. Find the x-intercepts of the function given by $f(x) = 3x^2 + x - 5$.

73. Find the x-intercepts of the function given by $g(x) = 2x^2 - 3x - 1$.

74. Find the x-intercepts of the function given by $g(x) = 3x^2 - 5x - 1$.

Interest. Use $A = P(1 + r)^t$ to find the interest rate in Exercises 75–78. Refer to Example 11.

75. $2000 grows to $2420 in 2 yr

76. $1000 grows to $1440 in 2 yr

77. $6250 grows to $6760 in 2 yr

78. $6250 grows to $7290 in 2 yr

Free-falling objects. Use $s = 16t^2$ for Exercises 79–82. Refer to Example 12 and neglect air resistance.

79. At a height of 290 ft, the Rainbow Bridge in Lake Powell National Monument, Utah, is the world's highest natural arch. How long would it take an object to fall freely from the bridge?
Source: *Guinness World Records* 2008

80. The Sears Tower in Chicago is 1454 ft tall. How long would it take an object to fall freely from the top?

81. At 2063 ft, the KVLY-TV tower in North Dakota is the tallest supported tower. How long would it take an object to fall freely from the top?
Source: North Dakota Tourism Division

82. El Capitan in Yosemite National Park is 3593 ft high. How long would it take a carabiner to fall freely from the top?
Source: *Guinness World Records* 2008

83. Explain in your own words a sequence of steps that can be used to solve any quadratic equation in the quickest way.

84. Write an interest-rate problem for a classmate to solve. Devise the problem so that the solution is "The loan was made at 7% interest."

Skill Review

To prepare for Section 11.2, review evaluating expressions and simplifying radical expressions (Sections 1.8, 10.3, and 10.8).

Evaluate. [1.8]

85. $b^2 - 4ac$, for $a = 3$, $b = 2$, and $c = -5$

86. $b^2 - 4ac$, for $a = 1$, $b = -1$, and $c = 4$

Simplify. [10.3], [10.8]

87. $\sqrt{200}$ **88.** $\sqrt{96}$

89. $\sqrt{-4}$ **90.** $\sqrt{-25}$

91. $\sqrt{-8}$ **92.** $\sqrt{-24}$

Synthesis

93. What would be better: to receive 3% interest every 6 months, or to receive 6% interest every 12 months? Why?

94. Write a problem involving a free-falling object for a classmate to solve (see Example 12). Devise the problem so that the solution is "The object takes about 4.5 sec to fall freely from the top of the structure."

Find b such that each trinomial is a square.

95. $x^2 + bx + 81$ **96.** $x^2 + bx + 49$

97. If $f(x) = 2x^5 - 9x^4 - 66x^3 + 45x^2 + 280x$ and $x^2 - 5$ is a factor of $f(x)$, find all a for which $f(a) = 0$.

98. If $f(x) = \left(x - \frac{1}{3}\right)(x^2 + 6)$ and $g(x) = \left(x - \frac{1}{3}\right)\left(x^2 - \frac{2}{3}\right)$, find all a for which $(f + g)(a) = 0$.

99. *Boating.*   A barge and a fishing boat leave a dock at the same time, traveling at a right angle to each other. The barge travels 7 km/h slower than the fishing boat. After 4 hr, the boats are 68 km apart. Find the speed of each boat.

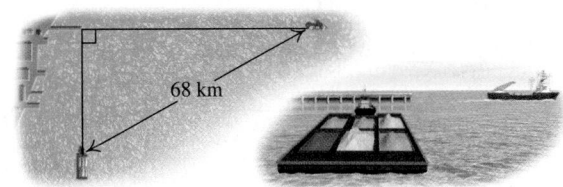

100. Find three consecutive integers such that the square of the first plus the product of the other two is 67.

101. Exercises 29, 33, and 53 can be solved on a graphing calculator without first rewriting in standard form. Simply let y_1 represent the left side of the equation and y_2 the right side. Then use a graphing calculator to determine the x-coordinate of any point of intersection. Use a graphing calculator to solve Exercises 29, 33, and 53 in this manner.

102. Use a graphing calculator to check your answers to Exercises 5, 13, 71, and 73.

103. Example 11 can be solved with a graphing calculator by graphing each side of

$$4410 = 4000(1 + r)^2.$$

How could you determine, from a reading of the problem, a suitable viewing window? What might that window be?

11.2   The Quadratic Formula

Solving Using the Quadratic Formula   ■   Approximating Solutions

We can use the process of completing the square to develop a general formula for solving quadratic equations.

Solving Using the Quadratic Formula

Each time we solve by completing the square, the procedure is the same. When a procedure is repeated many times, we can often develop a formula to speed up our work.

We begin with a quadratic equation in standard form,

$$ax^2 + bx + c = 0,$$

with $a > 0$. For $a < 0$, a slightly different derivation is needed (see Exercise 60), but the result is the same. Let's solve by completing the square. As the steps are performed, compare them with Example 10 on p. 710.

$$ax^2 + bx = -c \qquad \text{Adding to both sides}$$

$$x^2 + \frac{b}{a}x = -\frac{c}{a} \qquad \text{Dividing both sides by } a$$

Half of $\frac{b}{a}$ is $\frac{b}{2a}$ and $\left(\frac{b}{2a}\right)^2$ is $\frac{b^2}{4a^2}$. We add $\frac{b^2}{4a^2}$ to both sides:

$$x^2 + \frac{b}{a}x + \frac{b^2}{4a^2} = -\frac{c}{a} + \frac{b^2}{4a^2} \qquad \text{Adding } \frac{b^2}{4a^2} \text{ to complete the square}$$

$$\left(x + \frac{b}{2a}\right)^2 = -\frac{4ac}{4a^2} + \frac{b^2}{4a^2} \qquad \text{Factoring on the left side; finding a common denominator on the right side}$$

$$\left(x + \frac{b}{2a}\right)^2 = \frac{b^2 - 4ac}{4a^2}$$

$$x + \frac{b}{2a} = \pm\frac{\sqrt{b^2 - 4ac}}{2a} \qquad \text{Using the principle of square roots and the quotient rule for radicals. Since } a > 0, \sqrt{4a^2} = 2a.$$

$$x = \frac{-b \pm \sqrt{b^2 - 4ac}}{2a}. \qquad \text{Adding } -\frac{b}{2a} \text{ to both sides}$$

It is important to remember the quadratic formula and know how to use it.

The Quadratic Formula

The solutions of $ax^2 + bx + c = 0$, $a \neq 0$, are given by

$$x = \frac{-b \pm \sqrt{b^2 - 4ac}}{2a}.$$

EXAMPLE **1**   Solve $5x^2 + 8x = -3$ using the quadratic formula.

SOLUTION   We first find standard form and determine a, b, and c:

$5x^2 + 8x + 3 = 0$;   Adding 3 to both sides to get 0 on one side

$a = 5$, $b = 8$, $c = 3$.

Next, we use the quadratic formula:

$$x = \frac{-b \pm \sqrt{b^2 - 4ac}}{2a}$$ It is important to remember this formula.

$$x = \frac{-8 \pm \sqrt{8^2 - 4 \cdot 5 \cdot 3}}{2 \cdot 5}$$ Substituting

$$x = \frac{-8 \pm \sqrt{64 - 60}}{10}$$

Be sure to write the fraction bar all the way across.

$$x = \frac{-8 \pm \sqrt{4}}{10} = \frac{-8 \pm 2}{10}$$

$$x = \frac{-8 + 2}{10} \quad or \quad x = \frac{-8 - 2}{10}$$ The symbol $\pm$ indicates two solutions.

$$x = \frac{-6}{10} \quad or \quad x = \frac{-10}{10}$$

$$x = -\frac{3}{5} \quad or \quad x = -1.$$

The solutions are $-\frac{3}{5}$ and -1. The checks are left to the student.

> TRY EXERCISE 25

STUDY SKILLS

Know It "By Heart"

When memorizing something like the quadratic formula, try to first understand and write out the derivation. Doing this two or three times will help you remember the formula.

Because $5x^2 + 8x + 3$ can be factored, the quadratic formula may not have been the fastest way of solving Example 1. However, because the quadratic formula works for *any* quadratic equation, we need not spend too much time struggling to solve a quadratic equation by factoring.

To Solve a Quadratic Equation

1. If the equation can be easily written in the form $ax^2 = p$ or $(x + k)^2 = d$, use the principle of square roots as in Section 11.1.
2. If step (1) does not apply, write the equation in the form $ax^2 + bx + c = 0$.
3. Try factoring and using the principle of zero products.
4. If factoring seems difficult or impossible, use the quadratic formula. Completing the square can also be used.

The solutions of a quadratic equation can always be found using the quadratic formula. They cannot always be found by factoring.

Recall that a second-degree polynomial in one variable is said to be quadratic. Similarly, a second-degree polynomial function in one variable is said to be a **quadratic function**.

EXAMPLE 2

For the quadratic function given by $f(x) = 3x^2 - 6x - 4$, find all x for which $f(x) = 0$.

SOLUTION   We substitute and solve for x:

$$f(x) = 0$$
$$3x^2 - 6x - 4 = 0. \quad \text{Substituting}$$

Since $3x^2 - 6x - 4$ does not factor, we use the quadratic formula with $a = 3$, $b = -6$, and $c = -4$:

$$x = \frac{-(-6) \pm \sqrt{(-6)^2 - 4 \cdot 3 \cdot (-4)}}{2 \cdot 3}$$

$$= \frac{6 \pm \sqrt{36 + 48}}{6} \qquad (-6)^2 - 4 \cdot 3 \cdot (-4) = 36 - (-48) = 36 + 48$$

$$= \frac{6 \pm \sqrt{84}}{6} \qquad \text{Note that 4 is a perfect-square factor of 84.}$$

$$= \frac{6}{6} \pm \frac{\sqrt{84}}{6} \qquad \text{Writing as two fractions to simplify each separately}$$

$$= 1 \pm \frac{\sqrt{4}\sqrt{21}}{6} \qquad 84 = 4 \cdot 21$$

$$\left. \begin{array}{l} = 1 \pm \frac{2\sqrt{21}}{2 \cdot 3} \\[2mm] = 1 \pm \frac{\sqrt{21}}{3}. \end{array} \right\} \quad \text{Simplifying by removing a factor of 1: } \frac{2}{2} = 1$$

The solutions are $1 - \dfrac{\sqrt{21}}{3}$ and $1 + \dfrac{\sqrt{21}}{3}$. The checks are left to the student.

TRY EXERCISE 39

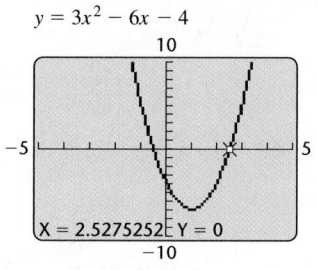
Some quadratic equations have solutions that are imaginary numbers.

EXAMPLE 3

Solve: $x(x + 5) = 2(2x - 1)$.

SOLUTION   We first find standard form:

$$x^2 + 5x = 4x - 2 \qquad \text{Multiplying}$$
$$x^2 + x + 2 = 0. \qquad \text{Subtracting } 4x \text{ and adding 2 to both sides}$$

Since we cannot factor $x^2 + x + 2$, we use the quadratic formula with $a = 1$, $b = 1$, and $c = 2$:

$$x = \frac{-1 \pm \sqrt{1^2 - 4 \cdot 1 \cdot 2}}{2 \cdot 1} \qquad \text{Substituting}$$

$$= \frac{-1 \pm \sqrt{1 - 8}}{2}$$

$$= \frac{-1 \pm \sqrt{-7}}{2}$$

$$= \frac{-1 \pm i\sqrt{7}}{2}, \text{ or } -\frac{1}{2} \pm \frac{\sqrt{7}}{2} i.$$

The solutions are $-\dfrac{1}{2} - \dfrac{\sqrt{7}}{2} i$ and $-\dfrac{1}{2} + \dfrac{\sqrt{7}}{2} i$. The checks are left to the student.

TRY EXERCISE 35

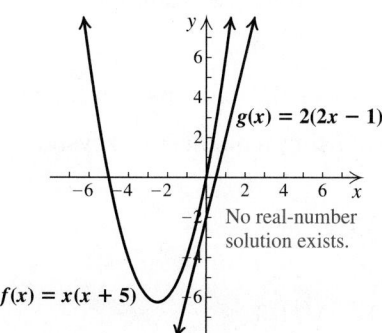

$g(x) = 2(2x - 1)$

No real-number solution exists.

$f(x) = x(x + 5)$

A visualization of Example 3

The quadratic formula can be used to solve certain rational equations.

EXAMPLE **4**

If $f(t) = 2 + \dfrac{7}{t}$ and $g(t) = \dfrac{4}{t^2}$, find all t for which $f(t) = g(t)$.

SOLUTION We set $f(t)$ equal to $g(t)$ and solve:

$$f(t) = g(t)$$

$$2 + \frac{7}{t} = \frac{4}{t^2}. \qquad \text{Substituting. Note that } t \neq 0.$$

This is a rational equation similar to those in Section 6.6. To solve, we multiply both sides by the LCD, t^2:

$$t^2\left(2 + \frac{7}{t}\right) = t^2 \cdot \frac{4}{t^2}$$

$$2t^2 + 7t = 4 \qquad \text{Simplifying}$$

$$2t^2 + 7t - 4 = 0. \qquad \text{Subtracting 4 from both sides}$$

We use the quadratic formula with $a = 2$, $b = 7$, and $c = -4$:

$$t = \frac{-7 \pm \sqrt{7^2 - 4 \cdot 2 \cdot (-4)}}{2 \cdot 2}$$

$$= \frac{-7 \pm \sqrt{49 + 32}}{4} \qquad 7^2 - 4 \cdot 2 \cdot (-4) = 49 - (-32) = 49 + 32$$

$$= \frac{-7 \pm \sqrt{81}}{4}$$

$$= \frac{-7 \pm 9}{4} \qquad \text{This means } \frac{-7 + 9}{4} \text{ or } \frac{-7 - 9}{4}.$$

$$t = \frac{2}{4} = \frac{1}{2} \quad \text{or} \quad t = \frac{-16}{4} = -4. \qquad \begin{array}{l}\text{Both answers should check}\\ \text{since } t \neq 0.\end{array}$$

You can confirm that $f\left(\frac{1}{2}\right) = g\left(\frac{1}{2}\right)$ and $f(-4) = g(-4)$. The solutions are $\frac{1}{2}$ and -4.

> TRY EXERCISE 41

Approximating Solutions

When the solution of an equation is irrational, a rational-number approximation is often useful. This is often the case in real-world applications similar to those found in Section 11.4.

EXAMPLE **5**

Use a calculator to approximate, to three decimal places, the solutions of Example 2.

SOLUTION On most calculators, one of the following sequences of keystrokes can be used to approximate $1 + \sqrt{21}/3$:

$$\boxed{1}\ \boxed{+}\ \boxed{\sqrt{}}\ \boxed{2}\ \boxed{1}\ \boxed{)}\ \boxed{\div}\ \boxed{3}\ \boxed{\text{ENTER}} \quad \text{or}$$
$$\boxed{1}\ \boxed{+}\ \boxed{2}\ \boxed{1}\ \boxed{\sqrt{}}\ \boxed{\div}\ \boxed{3}\ \boxed{=}.$$

Similar keystrokes can be used to approximate $1 - \sqrt{21}/3$.

The solutions are approximately 2.527525232 and -0.5275252317. Rounded to three decimal places, the solutions are approximately 2.528 and -0.528.

> TRY EXERCISE 45

TECHNOLOGY CONNECTION

We saw in Sections 5.7 and 11.1 how graphing calculators can solve quadratic equations. To determine whether quadratic equations are solved more quickly on a graphing calculator or by using the quadratic formula, solve Examples 2 and 4 both ways. Which method is faster? Which method is more precise? Why?

STUDENT NOTES

It is important that you understand both the rules for order of operations *and* the manner in which your calculator applies those rules.

11.2 EXERCISE SET

Concept Reinforcement *Classify each statement as either true or false.*

1. The quadratic formula can be used to solve *any* quadratic equation.

2. The steps used to derive the quadratic formula are the same as those used when solving by completing the square.

3. The quadratic formula does not work if solutions are imaginary numbers.

4. Solving by factoring is always slower than using the quadratic formula.

5. A quadratic equation can have as many as four solutions.

6. It is possible for a quadratic equation to have no real-number solutions.

Solve.

7. $2x^2 + 3x - 5 = 0$

8. $3x^2 - 7x + 2 = 0$

9. $u^2 + 2u - 4 = 0$

10. $u^2 - 2u - 2 = 0$

11. $t^2 + 3 = 6t$

12. $t^2 + 4t = 1$

13. $x^2 = 3x + 5$

14. $x^2 + 5x + 3 = 0$

15. $3t(t + 2) = 1$

16. $2t(t + 2) = 1$

17. $\dfrac{1}{x^2} - 3 = \dfrac{8}{x}$

18. $\dfrac{9}{x} - 2 = \dfrac{5}{x^2}$

19. $t^2 + 10 = 6t$

20. $t^2 + 10t + 26 = 0$

21. $p^2 - p + 1 = 0$

22. $p^2 + p + 4 = 0$

23. $x^2 + 4x + 6 = 0$

24. $x^2 + 11 = 6x$

25. $12t^2 + 17t = 40$

26. $15t^2 + 7t = 2$

27. $25x^2 - 20x + 4 = 0$

28. $36x^2 + 84x + 49 = 0$

29. $7x(x + 2) + 5 = 3x(x + 1)$

30. $5x(x - 1) - 7 = 4x(x - 2)$

31. $14(x - 4) - (x + 2) = (x + 2)(x - 4)$

32. $11(x - 2) + (x - 5) = (x + 2)(x - 6)$

33. $51p = 2p^2 + 72$

34. $72 = 3p^2 + 50p$

35. $x(x - 3) = x - 9$

36. $x(x - 1) = 2x - 7$

37. $x^3 - 8 = 0$ (*Hint:* Factor the difference of cubes. Then use the quadratic formula.)

38. $x^3 + 1 = 0$

39. Let $f(x) = 6x^2 - 7x - 20$. Find x such that $f(x) = 0$.

40. Let $g(x) = 4x^2 - 2x - 3$. Find x such that $g(x) = 0$.

41. Let
$$f(x) = \dfrac{7}{x} + \dfrac{7}{x + 4}.$$
Find all x for which $f(x) = 1$.

42. Let
$$g(x) = \dfrac{2}{x} + \dfrac{2}{x + 3}.$$
Find all x for which $g(x) = 1$.

43. Let
$$F(x) = \dfrac{3 - x}{4} \quad \text{and} \quad G(x) = \dfrac{1}{4x}.$$
Find all x for which $F(x) = G(x)$.

44. Let
$$f(x) = x + 5 \quad \text{and} \quad g(x) = \dfrac{3}{x - 5}.$$
Find all x for which $f(x) = g(x)$.

Solve using the quadratic formula. Then use a calculator to approximate, to three decimal places, the solutions as rational numbers.

45. $x^2 + 4x - 7 = 0$

46. $x^2 + 6x + 4 = 0$

Aha! 47. $x^2 - 6x + 4 = 0$

48. $x^2 - 4x + 1 = 0$

49. $2x^2 - 3x - 7 = 0$

50. $3x^2 - 3x - 2 = 0$

51. Are there any equations that can be solved by the quadratic formula but not by completing the square? Why or why not?

52. Suppose you are solving a quadratic equation with no constant term ($c = 0$). Would you use factoring or the quadratic formula to solve? Why?

Skill Review

To prepare for Section 11.3, review multiplying and simplifying radical and complex-number expressions (Sections 10.5 and 10.8).

Multiply and simplify.

53. $(x - 2i)(x + 2i)$ [10.8]

54. $(x - 6\sqrt{5})(x + 6\sqrt{5})$ [10.5]

55. $(x - (2 - \sqrt{7}))(x - (2 + \sqrt{7}))$ [10.5]

56. $(x - (-3 + 5i))(x - (-3 - 5i))$ [10.8]

Simplify.

57. $\dfrac{-6 \pm \sqrt{(-4)^2 - 4(2)(2)}}{2(2)}$ [10.3]

58. $\dfrac{-(-1) \pm \sqrt{(6)^2 - 4(3)(5)}}{2(3)}$ [10.8]

Synthesis

59. Explain how you could use the quadratic formula to help factor a quadratic polynomial.

60. If $a < 0$ and $ax^2 + bx + c = 0$, then $-a$ is positive and the equivalent equation, $-ax^2 - bx - c = 0$, can be solved using the quadratic formula.

   a) Find this solution, replacing a, b, and c in the formula with $-a$, $-b$, and $-c$ from the equation.

   b) How does the result of part (a) indicate that the quadratic formula "works" regardless of the sign of a?

For Exercises 61–63, let

$$f(x) = \frac{x^2}{x - 2} + 1 \quad and \quad g(x) = \frac{4x - 2}{x - 2} + \frac{x + 4}{2}.$$

61. Find the x-intercepts of the graph of f.

62. Find the x-intercepts of the graph of g.

63. Find all x for which $f(x) = g(x)$.

Solve. Approximate the solutions to three decimal places.

64. $x^2 - 0.75x - 0.5 = 0$

65. $z^2 + 0.84z - 0.4 = 0$

Solve.

66. $(1 + \sqrt{3})x^2 - (3 + 2\sqrt{3})x + 3 = 0$

67. $\sqrt{2}x^2 + 5x + \sqrt{2} = 0$

68. $ix^2 - 2x + 1 = 0$

69. One solution of $kx^2 + 3x - k = 0$ is -2. Find the other.

70. Use a graphing calculator to solve Exercises 9, 27, and 43.

71. Use a graphing calculator to solve Exercises 11, 33, and 41. Use the method of graphing each side of the equation.

72. Can a graphing calculator be used to solve *any* quadratic equation? Why or why not?

11.3 Studying Solutions of Quadratic Equations

The Discriminant ▪ Writing Equations from Solutions

The Discriminant

It is sometimes enough to know what *type* of number a solution will be, without actually solving the equation. Suppose we want to know if $4x^2 - 5x - 2 = 0$ has rational solutions (and thus can be solved by factoring). Using the quadratic formula, we would have

$$x = \frac{-b \pm \sqrt{b^2 - 4ac}}{2a}$$

$$= \frac{-(-5) \pm \sqrt{(-5)^2 - 4 \cdot 4(-2)}}{2 \cdot 4}.$$

Since $(-5)^2 - 4 \cdot 4 \cdot (-2) = 25 - 16(-2) = 25 + 32 = 57$ and since 57 is not a perfect square, the solutions of the equation are not rational numbers. This means that $4x^2 - 5x - 2 = 0$ *cannot* be solved by factoring. Note that the radicand, 57, determines what type of number the solutions will be.

The radicand $b^2 - 4ac$ is known as the **discriminant**. If a, b, and c are rational, then we have the following.

- When $b^2 - 4ac$ simplifies to 0, it doesn't matter if we use $+\sqrt{b^2 - 4ac}$ or $-\sqrt{b^2 - 4ac}$; we get the same solution twice. Thus, when the discriminant is 0, there is one *repeated* solution and it is rational.

   Example: $9x^2 + 6x + 1 = 0 \rightarrow b^2 - 4ac = 6^2 - 4 \cdot 9 \cdot 1 = 0$.
   Solving $9x^2 + 6x + 1 = 0$ gives the (repeated) solution $-\frac{1}{3}$.

- When $b^2 - 4ac$ is positive, there are two different real-number solutions: If $b^2 - 4ac$ is a perfect square, these solutions are rational numbers.

   Example: $6x^2 + 5x + 1 = 0 \rightarrow b^2 - 4ac = 5^2 - 4 \cdot 6 \cdot 1 = 1$.
   Solving $6x^2 + 5x + 1 = 0$ gives the solutions $-\frac{1}{3}$ and $-\frac{1}{2}$.

- When $b^2 - 4ac$ is positive but not a perfect square, there are two irrational solutions and they are conjugates of each other (see p. 663).

   Example: $x^2 + 4x + 2 = 0 \rightarrow b^2 - 4ac = 4^2 - 4 \cdot 1 \cdot 2 = 8$.
   Solving $x^2 + 4x + 2 = 0$ gives the solutions $-2 + \sqrt{2}$ and $-2 - \sqrt{2}$.

- When the discriminant is negative, there are two imaginary-number solutions and they are complex conjugates of each other.

   Example: $x^2 + 4x + 5 = 0 \rightarrow b^2 - 4ac = 4^2 - 4 \cdot 1 \cdot 5 = -4$.
   Solving $x^2 + 4x + 5 = 0$ gives the solutions $-2 + i$ and $-2 - i$.

Note that any equation for which $b^2 - 4ac$ is a perfect square can be solved by factoring.

Discriminant $b^2 - 4ac$	Nature of Solutions
0	One solution; a rational number

Positive
   Perfect square
   Not a perfect square
	Two different real-number solutions
	   Solutions are rational.
	   Solutions are irrational conjugates.

Negative	Two different imaginary-number solutions (complex conjugates)

EXAMPLE 1

For each equation, determine what type of number the solutions are and how many solutions exist.

a) $9x^2 - 12x + 4 = 0$ **b)** $x^2 + 5x + 8 = 0$ **c)** $2x^2 + 7x - 3 = 0$

SOLUTION

a) For $9x^2 - 12x + 4 = 0$, we have

$$a = 9, \quad b = -12, \quad c = 4.$$

We substitute and compute the discriminant:

$$b^2 - 4ac = (-12)^2 - 4 \cdot 9 \cdot 4$$
$$= 144 - 144 = 0.$$

There is exactly one solution, and it is rational. This indicates that $9x^2 - 12x + 4 = 0$ can be solved by factoring.

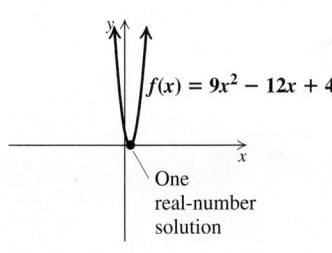

$f(x) = 9x^2 - 12x + 4$

One
real-number
solution

A visualization of part (a)

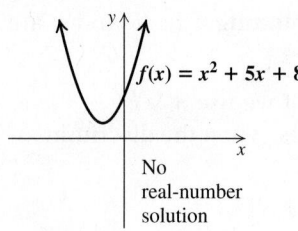

No
real-number
solution

A visualization of part (b)

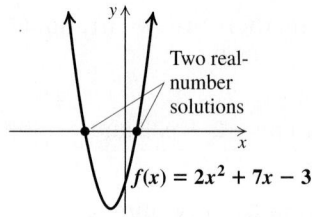

A visualization of part (c)

b) For $x^2 + 5x + 8 = 0$, we have

$$a = 1, \quad b = 5, \quad c = 8.$$

We substitute and compute the discriminant:

$$b^2 - 4ac = 5^2 - 4 \cdot 1 \cdot 8$$
$$= 25 - 32 = -7.$$

Since the discriminant is negative, there are two different imaginary-number solutions that are complex conjugates of each other.

c) For $2x^2 + 7x - 3 = 0$, we have

$$a = 2, \quad b = 7, \quad c = -3.$$

We substitute and compute the discriminant:

$$b^2 - 4ac = 7^2 - 4 \cdot 2(-3)$$
$$= 49 - (-24) = 73.$$

The discriminant is a positive number that is not a perfect square. Thus there are two different irrational solutions that are conjugates of each other.

> **TRY EXERCISE** 7

Discriminants can also be used to determine the number of x-intercepts of the graph of a quadratic function.

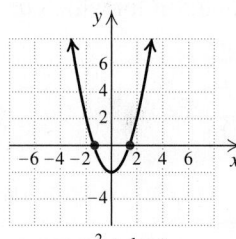

$y = ax^2 + bx + c$
$b^2 - 4ac > 0$
Two real solutions
of $ax^2 + bx + c = 0$
Two x-intercepts

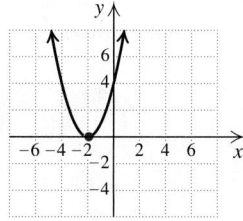

$y = ax^2 + bx + c$
$b^2 - 4ac = 0$
One real solution
of $ax^2 + bx + c = 0$
One x-intercept

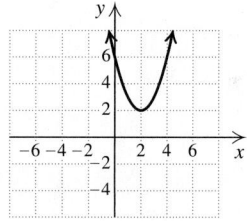

$y = ax^2 + bx + c$
$b^2 - 4ac < 0$
No real solutions
of $ax^2 + bx + c = 0$
No x-intercept

Writing Equations from Solutions

We know by the principle of zero products that $(x - 2)(x + 3) = 0$ has solutions 2 and -3. If we wish for two given numbers to be solutions of an equation, we can create such an equation, using the principle in reverse.

EXAMPLE 2   Find an equation for which the given numbers are solutions.

a) 3 and $-\frac{2}{5}$

b) $2i$ and $-2i$

c) $5\sqrt{7}$ and $-5\sqrt{7}$

d) $-4, 0$, and 1

SOLUTION

a)
$$x = 3 \quad or \quad x = -\tfrac{2}{5}$$
$$x - 3 = 0 \quad or \quad x + \tfrac{2}{5} = 0 \qquad \text{Getting 0's on one side}$$
$$(x - 3)\left(x + \tfrac{2}{5}\right) = 0 \qquad \begin{array}{l}\text{Using the principle of zero}\\ \text{products (multiplying)}\end{array}$$
$$x^2 + \tfrac{2}{5}x - 3x - 3 \cdot \tfrac{2}{5} = 0 \qquad \text{Multiplying}$$
$$x^2 - \tfrac{13}{5}x - \tfrac{6}{5} = 0 \qquad \text{Combining like terms}$$
$$5x^2 - 13x - 6 = 0 \qquad \text{Multiplying both sides by 5 to clear fractions}$$

To check Example 2(a), we can let $y_1 = 5x^2 - 13x - 6$ and verify that the x-intercepts are 3 and $-\frac{2}{5}$. One way to do this is to press (TRACE) and then enter 3. The cursor then appears on the curve at $x = 3$. Since this is an x-intercept, we know that $5x^2 - 13x - 6 = 0$ is an equation that has 3 as a solution.

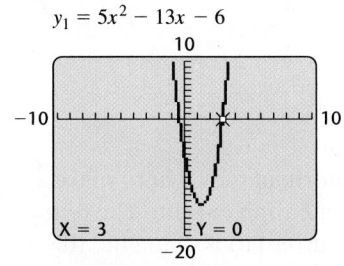

$y_1 = 5x^2 - 13x - 6$

1. Confirm that $5x^2 - 13x - 6 = 0$ also has $-\frac{2}{5}$ as a solution.
2. Check Example 2(c).
3. Check Example 2(d).

Note that multiplying both sides by the LCD, 5, clears the equation of fractions. Had we preferred, we could have multiplied $x + \frac{2}{5} = 0$ by 5, thus clearing fractions *before* using the principle of zero products.

b)
$$x = 2i \quad or \quad x = -2i$$
$$x - 2i = 0 \quad or \quad x + 2i = 0 \qquad \text{Getting 0's on one side}$$
$$(x - 2i)(x + 2i) = 0 \qquad \text{Using the principle of zero products (multiplying)}$$
$$x^2 - (2i)^2 = 0 \qquad \text{Finding the product of a sum and a difference}$$
$$x^2 - 4i^2 = 0$$
$$x^2 + 4 = 0 \qquad i^2 = -1$$

c)
$$x = 5\sqrt{7} \quad or \quad x = -5\sqrt{7}$$
$$x - 5\sqrt{7} = 0 \quad or \quad x + 5\sqrt{7} = 0 \qquad \text{Getting 0's on one side}$$
$$(x - 5\sqrt{7})(x + 5\sqrt{7}) = 0 \qquad \text{Using the principle of zero products}$$
$$x^2 - (5\sqrt{7})^2 = 0 \qquad \text{Finding the product of a sum and a difference}$$
$$x^2 - 25 \cdot 7 = 0$$
$$x^2 - 175 = 0$$

d)
$$x = -4 \quad or \quad x = 0 \quad or \quad x = 1$$
$$x + 4 = 0 \quad or \quad x = 0 \quad or \quad x - 1 = 0 \qquad \text{Getting 0's on one side}$$
$$(x + 4)x(x - 1) = 0 \qquad \text{Using the principle of zero products}$$
$$x(x^2 + 3x - 4) = 0 \qquad \text{Multiplying}$$
$$x^3 + 3x^2 - 4x = 0$$

TRY EXERCISE 29

To check any of these equations, we can simply substitute one or more of the given solutions. For example, in Example 2(d) above,

$$(-4)^3 + 3(-4)^2 - 4(-4) = -64 + 3 \cdot 16 + 16$$
$$= -64 + 48 + 16 = 0.$$

The other checks are left to the student.

11.3   EXERCISE SET

For Extra Help      PRACTICE   WATCH   DOWNLOAD

❧ *Concept Reinforcement   Match the nature of the solution(s) with each discriminant. Answers may be used more than once.*

1. ___ $b^2 - 4ac = 9$

2. ___ $b^2 - 4ac = 0$

3. ___ $b^2 - 4ac = -1$

4. ___ $b^2 - 4ac = 1$

5. ___ $b^2 - 4ac = 8$

6. ___ $b^2 - 4ac = 12$

a) One rational solution

b) Two different rational solutions

c) Two different irrational solutions

d) Two different imaginary-number solutions

For each equation, determine what type of number the solutions are and how many solutions exist.

7. $x^2 - 7x + 5 = 0$

8. $x^2 - 5x + 3 = 0$

9. $x^2 + 11 = 0$

10. $x^2 + 7 = 0$

11. $x^2 - 11 = 0$

12. $x^2 - 7 = 0$

13. $4x^2 + 8x - 5 = 0$

14. $4x^2 - 12x + 9 = 0$

15. $x^2 + 4x + 6 = 0$

16. $x^2 - 2x + 4 = 0$

17. $9t^2 - 48t + 64 = 0$

18. $10t^2 - t - 2 = 0$

Aha! **19.** $9t^2 + 3t = 0$

20. $4m^2 + 7m = 0$

21. $x^2 + 4x = 8$

22. $x^2 + 5x = 9$

23. $2a^2 - 3a = -5$

24. $3a^2 + 5 = -7a$

25. $7x^2 = 19x$

26. $5x^2 = 48x$

27. $y^2 + \frac{9}{4} = 4y$

28. $x^2 = \frac{1}{2}x - \frac{3}{5}$

Write a quadratic equation having the given numbers as solutions.

29. $-5, 4$

30. $-2, 8$

31. 3, only solution (*Hint*: It must be a repeated solution.)

32. -5, only solution

33. $-1, -3$

34. $-2, -5$

35. $5, \frac{3}{4}$

36. $4, \frac{2}{3}$

37. $-\frac{1}{4}, -\frac{1}{2}$

38. $\frac{1}{2}, \frac{1}{3}$

39. $2.4, -0.4$

40. $-0.6, 1.4$

41. $-\sqrt{3}, \sqrt{3}$

42. $-\sqrt{7}, \sqrt{7}$

43. $2\sqrt{5}, -2\sqrt{5}$

44. $3\sqrt{2}, -3\sqrt{2}$

45. $4i, -4i$

46. $3i, -3i$

47. $2 - 7i, 2 + 7i$

48. $5 - 2i, 5 + 2i$

49. $3 - \sqrt{14}, 3 + \sqrt{14}$

50. $2 - \sqrt{10}, 2 + \sqrt{10}$

51. $1 - \dfrac{\sqrt{21}}{3}, 1 + \dfrac{\sqrt{21}}{3}$

52. $\dfrac{5}{4} - \dfrac{\sqrt{33}}{4}, \dfrac{5}{4} + \dfrac{\sqrt{33}}{4}$

Write a third-degree equation having the given numbers as solutions.

53. $-2, 1, 5$

54. $-5, 0, 2$

55. $-1, 0, 3$

56. $-2, 2, 3$

57. Explain why there are not two different solutions when the discriminant is 0.

58. Describe a procedure that could be used to write an equation having the first 7 natural numbers as solutions.

Skill Review

To prepare for Section 11.4, review solving formulas and solving motion problems (Sections 6.7, 7.5, and 8.3).

Solve each formula for the specified variable. [7.5]

59. $\dfrac{c}{d} = c + d$, for c

60. $\dfrac{p}{q} = \dfrac{a + b}{b}$, for b

61. $x = \dfrac{3}{1 - y}$, for y

Solve.

62. *Boating.*   Kiara's motorboat took 4 hr to make a trip downstream with a 2-mph current. The return trip against the same current took 6 hr. Find the speed of the boat in still water. [8.3]

63. *Walking.*   Jamal walks 1.5 mph faster than Kade. In the time it takes Jamal to walk 7 mi, Kade walks 4 mi. Find the speed of each person. [6.7]

64. *Aviation.*   Taryn's Cessna travels 120 mph in still air. She flies 140 mi into the wind and 140 mi with the wind in a total of 2.4 hr. Find the wind speed. [6.7]

Synthesis

65. If we assume that a quadratic equation has integers for coefficients, will the product of the solutions always be a real number? Why or why not?

66. Can a fourth-degree equation with rational coefficients have exactly three irrational solutions? Why or why not?

67. The graph of an equation of the form

$$y = ax^2 + bx + c$$

is a curve similar to the one shown below. Determine a, b, and c from the information given.

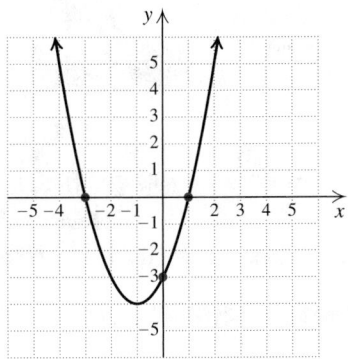

68. Show that the product of the solutions of $ax^2 + bx + c = 0$ is c/a.

For each equation under the given condition, **(a)** *find k and* **(b)** *find the other solution.*

69. $kx^2 - 2x + k = 0$; one solution is -3

70. $x^2 - kx + 2 = 0$; one solution is $1 + i$

71. $x^2 - (6 + 3i)x + k = 0$; one solution is 3

72. Show that the sum of the solutions of $ax^2 + bx + c = 0$ is $-b/a$.

73. Show that whenever there is just one solution of $ax^2 + bx + c = 0$, that solution is of the form $-b/(2a)$.

74. Find h and k, where $3x^2 - hx + 4k = 0$, the sum of the solutions is -12, and the product of the solutions is 20. (*Hint*: See Exercises 68 and 72.)

75. Suppose that $f(x) = ax^2 + bx + c$, with $f(-3) = 0$, $f\left(\frac{1}{2}\right) = 0$, and $f(0) = -12$. Find a, b, and c.

76. Find an equation for which $2 - \sqrt{3}$, $2 + \sqrt{3}$, $5 - 2i$, and $5 + 2i$ are solutions.

Aha! **77.** Write a quadratic equation with integer coefficients for which $-\sqrt{2}$ is one solution.

78. Write a quadratic equation with integer coefficients for which $10i$ is one solution.

79. Find an equation with integer coefficients for which $1 - \sqrt{5}$ and $3 + 2i$ are two of the solutions.

80. A discriminant that is a perfect square indicates that factoring can be used to solve the quadratic equation. Why?

81. While solving a quadratic equation of the form $ax^2 + bx + c = 0$ with a graphing calculator, Keisha gets the following screen. How could the sign of the discriminant help her check the graph?

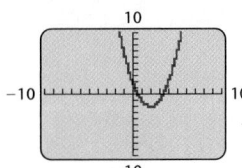

11.4   Applications Involving Quadratic Equations

Solving Problems   •   Solving Formulas

Solving Problems

As we found in Section 6.7, some problems translate to rational equations. The solution of such rational equations can involve quadratic equations.

EXAMPLE **1**   *Motorcycle travel.*   Fiona rode her motorcycle 300 mi at a certain average speed. Had she traveled 10 mph faster, the trip would have taken 1 hr less. Find Fiona's average speed.

SOLUTION

1. **Familiarize.** We make a drawing, labeling it with the information provided. As in Section 6.7, we can create a table. We let r represent the rate, in miles per hour, and t the time, in hours, for Fiona's trip.

300 miles

Time t Speed r

300 miles

Time $t - 1$ Speed $r + 10$

Distance	Speed	Time
300	r	t	$\longrightarrow r = \dfrac{300}{t}$
300	$r + 10$	$t - 1$	$\longrightarrow r + 10 = \dfrac{300}{t - 1}$

Recall that the definition of speed, $r = d/t$, relates the three quantities.

2. **Translate.** From the table, we obtain

$$r = \frac{300}{t} \quad \text{and} \quad r + 10 = \frac{300}{t - 1}.$$

3. **Carry out.** A system of equations has been formed. We substitute for r from the first equation into the second and solve the resulting equation:

$$\frac{300}{t} + 10 = \frac{300}{t - 1} \qquad \text{Substituting } 300/t \text{ for } r$$

$$t(t - 1) \cdot \left[\frac{300}{t} + 10 \right] = t(t - 1) \cdot \frac{300}{t - 1} \qquad \begin{array}{l}\text{Multiplying by the}\\\text{LCD to clear}\\\text{fractions}\end{array}$$

$$t(t - 1) \cdot \frac{300}{t} + t(t - 1) \cdot 10 = t(t - 1) \cdot \frac{300}{t - 1} \qquad \begin{array}{l}\text{Using the}\\\text{distributive law}\end{array}$$

$$\frac{\cancel{t}(t - 1)}{1} \cdot \frac{300}{\cancel{t}} + t(t - 1) \cdot 10 = \frac{t(\cancel{t - 1})}{1} \cdot \frac{300}{\cancel{t - 1}} \qquad \begin{array}{l}\text{Removing factors}\\\text{that equal 1:}\\t/t = 1 \text{ and}\\(t - 1)/(t - 1)\end{array}$$

$$\left. \begin{array}{l} 300(t - 1) + 10(t^2 - t) = 300t \\ 300t - 300 + 10t^2 - 10t = 300t \\ 10t^2 - 10t - 300 = 0 \end{array} \right\} \qquad \begin{array}{l}\text{Rewriting in}\\\text{standard form}\end{array}$$

$$t^2 - t - 30 = 0 \qquad \begin{array}{l}\text{Multiplying by } \frac{1}{10}\\\text{or dividing by 10}\end{array}$$

$$(t - 6)(t + 5) = 0 \qquad \text{Factoring}$$

$$t = 6 \quad \text{or} \quad t = -5. \qquad \begin{array}{l}\text{Principle of zero}\\\text{products}\end{array}$$

4. **Check.** Note that we have solved for t, not r as required. Since negative time has no meaning here, we disregard the -5 and use 6 hr to find r:

$$r = \frac{300 \text{ mi}}{6 \text{ hr}} = 50 \text{ mph}.$$

> **CAUTION!** Always make sure that you find the quantity asked for in the problem.

To see if 50 mph checks, we increase the speed 10 mph to 60 mph and see how long the trip would have taken at that speed:

$$t = \frac{d}{r} = \frac{300 \text{ mi}}{60 \text{ mph}} = 5 \text{ hr}.$$ Note that mi/mph $= \text{mi} \div \frac{\text{mi}}{\text{hr}} =$

$$\text{mi} \cdot \frac{\text{hr}}{\text{mi}} = \text{hr}.$$

This is 1 hr less than the trip actually took, so the answer checks.

5. **State.** Fiona traveled at an average speed of 50 mph.   ▶ **TRY EXERCISE** 1

Solving Formulas

Recall that to solve a formula for a certain letter, we use the principles for solving equations to get that letter alone on one side.

EXAMPLE 2

Period of a pendulum.   The time T required for a pendulum of length l to swing back and forth (complete one period) is given by the formula $T = 2\pi \sqrt{l/g}$, where g is the earth's gravitational constant. Solve for l.

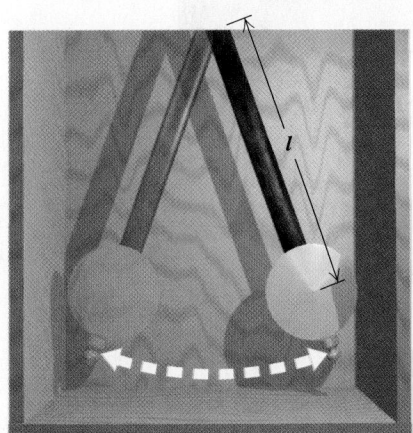

SOLUTION   We have

$$T = 2\pi \sqrt{\frac{l}{g}}$$ This is a radical equation (see Section 10.6).

$$T^2 = \left(2\pi \sqrt{\frac{l}{g}}\right)^2$$ Squaring both sides

$$T^2 = 2^2 \pi^2 \frac{l}{g}$$

$$gT^2 = 4\pi^2 l$$ Multiplying both sides by g to clear fractions

$$\frac{gT^2}{4\pi^2} = l.$$ Dividing both sides by $4\pi^2$

We now have l alone on one side and l does not appear on the other side, so the formula is solved for l.   ▶ **TRY EXERCISE** 21

In formulas for which variables represent only nonnegative numbers, there is no need for absolute-value signs when taking square roots.

EXAMPLE 3

*Hang time.** An athlete's *hang time* is the amount of time that the athlete can remain airborne when jumping. A formula relating an athlete's vertical leap V, in inches, to hang time T, in seconds, is $V = 48T^2$. Solve for T.

SOLUTION We have

$$48T^2 = V$$

$$T^2 = \frac{V}{48} \qquad \text{Dividing by 48 to isolate } T^2$$

$$T = \frac{\sqrt{V}}{\sqrt{48}} \qquad \begin{array}{l}\text{Using the principle of square roots} \\ \text{and the quotient rule for radicals.} \\ \text{We assume } V, T \geq 0.\end{array}$$

$$= \frac{\sqrt{V}}{\sqrt{16}\sqrt{3}}$$

$$= \frac{\sqrt{V}}{4\sqrt{3}}$$

$$= \frac{\sqrt{V}}{4\sqrt{3}} \cdot \frac{\sqrt{3}}{\sqrt{3}} \left.\begin{array}{l} \\ \\ \end{array}\right\} \quad \text{Rationalizing the denominator}$$

$$= \frac{\sqrt{3V}}{12}.$$

TRY EXERCISE 15

EXAMPLE 4

Falling distance. An object tossed downward with an initial speed (velocity) of v_0 will travel a distance of s meters, where $s = 4.9t^2 + v_0t$ and t is measured in seconds. Solve for t.

SOLUTION Since t is squared in one term and raised to the first power in the other term, the equation is quadratic in t.

$$4.9t^2 + v_0t = s$$

$$4.9t^2 + v_0t - s = 0 \qquad \text{Writing standard form}$$

$$a = 4.9, \quad b = v_0, \quad c = -s$$

$$t = \frac{-v_0 \pm \sqrt{(v_0)^2 - 4(4.9)(-s)}}{2(4.9)} \qquad \text{Using the quadratic formula}$$

Since the negative square root would yield a negative value for t, we use only the positive root:

$$t = \frac{-v_0 + \sqrt{(v_0)^2 + 19.6s}}{9.8}.$$

TRY EXERCISE 25

STUDENT NOTES

After identifying which numbers to use as a, b, and c, be careful to replace only the *letters* in the quadratic formula.

*This formula is taken from an article by Peter Brancazio, "The Mechanics of a Slam Dunk," *Popular Mechanics,* November 1991. Courtesy of Professor Peter Brancazio, Brooklyn College.

The following list of steps should help you when solving formulas for a given letter. Try to remember that when solving a formula, you use the same approach that you would to solve an equation.

> **To Solve a Formula for a Letter—Say, h**
> 1. Clear fractions and use the principle of powers, as needed. Perform these steps until radicals containing h are gone and h is not in any denominator.
> 2. Combine all like terms.
> 3. If the only power of h is h^1, the equation can be solved as in Sections 2.3 and 7.5. (See Example 2.)
> 4. If h^2 appears but h does not, solve for h^2 and use the principle of square roots to solve for h. (See Example 3.)
> 5. If there are terms containing both h and h^2, put the equation in standard form and use the quadratic formula. (See Example 4.)

11.4 EXERCISE SET

Solve.

1. *Car trips.* During the first part of a trip, Tara's Honda traveled 120 mi at a certain speed. Tara then drove another 100 mi at a speed that was 10 mph slower. If the total time of Tara's trip was 4 hr, what was her speed on each part of the trip?

2. *Canoeing.* During the first part of a canoe trip, Ken covered 60 km at a certain speed. He then traveled 24 km at a speed that was 4 km/h slower. If the total time for the trip was 8 hr, what was the speed on each part of the trip?

3. *Car trips.* Diane's Dodge travels 200 mi averaging a certain speed. If the car had gone 10 mph faster, the trip would have taken 1 hr less. Find Diane's average speed.

4. *Car trips.* Stuart's Subaru travels 280 mi averaging a certain speed. If the car had gone 5 mph faster, the trip would have taken 1 hr less. Find Stuart's average speed.

5. *Air travel.* A Cessna flies 600 mi at a certain speed. A Beechcraft flies 1000 mi at a speed that is 50 mph faster, but takes 1 hr longer. Find the speed of each plane.

6. *Air travel.* A turbo-jet flies 50 mph faster than a super-prop plane. If a turbo-jet goes 2000 mi in 3 hr less time than it takes the super-prop to go 2800 mi, find the speed of each plane.

7. *Bicycling.* Naoki bikes the 36 mi to Hillsboro averaging a certain speed. The return trip is made at a speed 3 mph slower. Total time for the round trip is 7 hr. Find Naoki's average speed on each part of the trip.

8. *Car speed.* On a sales trip, Mark drives the 600 mi to Richmond averaging a certain speed. The return trip is made at an average speed that is 10 mph slower. Total time for the round trip is 22 hr. Find Mark's average speed on each part of the trip.

9. *Navigation.* The Hudson River flows at a rate of 3 mph. A patrol boat travels 60 mi upriver and returns in a total time of 9 hr. What is the speed of the boat in still water?

10. *Navigation.* The current in a typical Mississippi River shipping route flows at a rate of 4 mph. In order for a barge to travel 24 mi upriver and then return in a total of 5 hr, approximately how fast must the barge be able to travel in still water?

11. *Filling a pool.* A well and a spring are filling a swimming pool. Together, they can fill the pool in 3 hr. The well, working alone, can fill the pool in 8 hr less time than the spring. How long would the spring take, working alone, to fill the pool?

12. *Filling a tank.* Two pipes are connected to the same tank. Working together, they can fill the tank in 4 hr. The larger pipe, working alone, can fill the tank in 6 hr less time than the smaller one. How long would the smaller one take, working alone, to fill the tank?

13. *Paddleboats.* Kofi paddles 1 mi upstream and 1 mi back in a total time of 1 hr. The speed of the river is 2 mph. Find the speed of Kofi's paddleboat in still water.

14. *Rowing.* Abby rows 10 km upstream and 10 km back in a total time of 3 hr. The speed of the river is 5 km/h. Find Abby's speed in still water.

Solve each formula for the indicated letter. Assume that all variables represent nonnegative numbers.

15. $A = 4\pi r^2$, for r
(Surface area of a sphere of radius r)

16. $A = 6s^2$, for s
(Surface area of a cube with sides of length s)

17. $A = 2\pi r^2 + 2\pi rh$, for r
(Surface area of a right cylindrical solid with radius r and height h)

18. $N = \dfrac{k^2 - 3k}{2}$, for k
(Number of diagonals of a polygon with k sides)

19. $F = \dfrac{Gm_1m_2}{r^2}$, for r
(Law of gravity)

20. $N = \dfrac{kQ_1Q_2}{s^2}$, for s
(Number of phone calls between two cities)

21. $c = \sqrt{gH}$, for H
(Velocity of ocean wave)

22. $V = 3.5\sqrt{h}$, for h
(Distance to horizon from a height)

23. $a^2 + b^2 = c^2$, for b
(Pythagorean formula in two dimensions)

24. $a^2 + b^2 + c^2 = d^2$, for c
(Pythagorean formula in three dimensions)

25. $s = v_0t + \dfrac{gt^2}{2}$, for t
(A motion formula)

26. $A = \pi r^2 + \pi rs$, for r
(Surface area of a cone)

27. $N = \frac{1}{2}(n^2 - n)$, for n
(Number of games if n teams play each other once)

28. $A = A_0(1 - r)^2$, for r
(A business formula)

29. $T = 2\pi\sqrt{\dfrac{l}{g}}$, for g
(A pendulum formula)

30. $W = \sqrt{\dfrac{1}{LC}}$, for L
(An electricity formula)

Aha! **31.** $at^2 + bt + c = 0$, for t
(An algebraic formula)

32. $A = P_1(1 + r)^2 + P_2(1 + r)$, for r
(Amount in an account when P_1 is invested for 2 yr and P_2 for 1 yr at interest rate r)

Solve.

33. *Falling distance.* (Use $4.9t^2 + v_0t = s$.)

  a) A bolt falls off an airplane at an altitude of 500 m. Approximately how long does it take the bolt to reach the ground?

  b) A ball is thrown downward at a speed of 30 m/sec from an altitude of 500 m. Approximately how long does it take the ball to reach the ground?

  c) Approximately how far will an object fall in 5 sec, when thrown downward at an initial velocity of 30 m/sec from a plane?

34. *Falling distance.* (Use $4.9t^2 + v_0t = s$.)

  a) A ring is dropped from a helicopter at an altitude of 75 m. Approximately how long does it take the ring to reach the ground?

  b) A coin is tossed downward with an initial velocity of 30 m/sec from an altitude of 75 m. Approximately how long does it take the coin to reach the ground?

  c) Approximately how far will an object fall in 2 sec, if thrown downward at an initial velocity of 20 m/sec from a helicopter?

35. *Bungee jumping.* Chad is tied to one end of a 40-m elasticized (bungee) cord. The other end of the cord is tied to the middle of a bridge. If Chad jumps off the bridge, for how long will he fall before the cord begins to stretch? (Use $4.9t^2 = s$.)

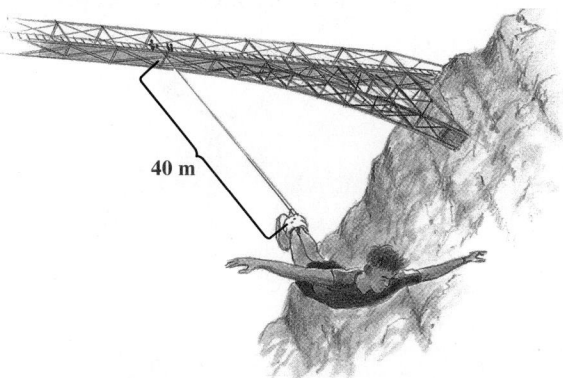

36. *Bungee jumping.* Chika is tied to a bungee cord (see Exercise 35) and falls for 2.5 sec before her cord begins to stretch. How long is the bungee cord?

37. *Hang time.* The NBA's Dwight Howard has a vertical leap of 38 in. What is his hang time? (Use $V = 48T^2$.)

Source: www.dwighthoward.com

38. *League schedules.* In a bowling league, each team plays each of the other teams once. If a total of 66 games is played, how many teams are in the league? (See Exercise 27.)

For Exercises 39 and 40, use $4.9t^2 + v_0t = s$.

39. *Downward speed.* A stone thrown downward from a 100-m cliff travels 51.6 m in 3 sec. What was the initial velocity of the object?

40. *Downward speed.* A pebble thrown downward from a 200-m cliff travels 91.2 m in 4 sec. What was the initial velocity of the object?

For Exercises 41 and 42, use $A = P_1(1 + r)^2 + P_2(1 + r)$. (See Exercise 32.)

41. *Compound interest.* A firm invests $3200 in a savings account for 2 yr. At the beginning of the second

year, an additional $1800 is invested. If a total of $5375.48 is in the account at the end of the second year, what is the annual interest rate?

42. *Compound interest.* A business invests $10,000 in a savings account for 2 yr. At the beginning of the second year, an additional $3500 is invested. If a total of $14,822.75 is in the account at the end of the second year, what is the annual interest rate?

43. Marti is tied to a bungee cord that is twice as long as the cord tied to Rafe's. Will Marti's fall take twice as long as Rafe's before their cords begin to stretch? Why or why not? (See Exercises 35 and 36.)

44. Under what circumstances would a negative value for t, time, have meaning?

Skill Review

To prepare for Section 11.5, review raising a power to a power and solving rational equations and radical equations (Sections 4.1, 6.6, and 10.6).

Simplify.

45. $(m^{-1})^2$ [4.2]

46. $(t^{1/3})^2$ [10.2]

47. $(y^{1/6})^2$ [10.2]

48. $(z^{1/4})^2$ [10.2]

Solve.

49. $t^{-1} = \dfrac{1}{2}$ [6.6], [10.6]

50. $x^{1/4} = 3$ [10.6]

Synthesis

51. Write a problem for a classmate to solve. Devise the problem so that **(a)** the solution is found after solving a rational equation and **(b)** the solution is "The express train travels 90 mph."

52. In what ways do the motion problems of this section (like Example 1) differ from the motion problems in Section 6.7?

53. *Biochemistry.* The equation

$$A = 6.5 - \frac{20.4t}{t^2 + 36}$$

is used to calculate the acid level A in a person's blood t minutes after sugar is consumed. Solve for t.

54. *Special relativity.* Einstein found that an object with initial mass m_0 and traveling velocity v has mass

$$m = \frac{m_0}{\sqrt{1 - \dfrac{v^2}{c^2}}},$$

where c is the speed of light. Solve the formula for c.

55. Find a number for which the reciprocal of 1 less than the number is the same as 1 more than the number.

56. *Purchasing.*   A discount store bought a quantity of potted plants for $250 and sold all but 15 at a profit of $3.50 per plant. With the total amount received, the manager could buy 4 more than twice as many as were bought before. Find the cost per plant.

57. *Art and aesthetics.*   For over 2000 yr, artists, sculptors, and architects have regarded the proportions of a "golden" rectangle as visually appealing. A rectangle of width w and length l is considered "golden" if

$$\frac{w}{l} = \frac{l}{w + l}.$$

Solve for l.

58. *Diagonal of a cube.*   Find a formula that expresses the length of the three-dimensional diagonal of a cube as a function of the cube's surface area.

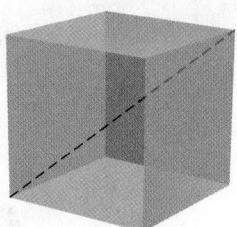

59. Solve for n:
$$mn^4 - r^2pm^3 - r^2n^2 + p = 0.$$

60. *Surface area.*   Find a formula that expresses the diameter of a right cylindrical solid as a function of its surface area and its height. (See Exercise 17.)

61. A sphere is inscribed in a cube as shown in the figure below. Express the surface area of the sphere as a function of the surface area S of the cube. (See Exercise 15.)

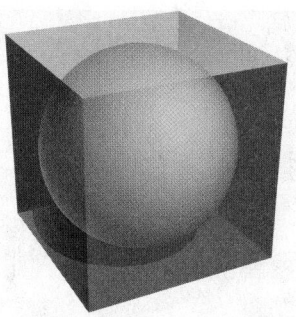

11.5   Equations Reducible to Quadratic

Recognizing Equations in Quadratic Form   ■   Radical Equations and Rational Equations

Recognizing Equations in Quadratic Form

Certain equations that are not really quadratic can be thought of in such a way that they can be solved as quadratic. For example, because the square of x^2 is x^4, the equation $x^4 - 9x^2 + 8 = 0$ is said to be "quadratic in x^2":

$$x^4 - 9x^2 + 8 = 0$$

$$(x^2)^2 - 9(x^2) + 8 = 0 \qquad \text{Thinking of } x^4 \text{ as } (x^2)^2$$

$$u^2 - 9u + 8 = 0. \qquad \text{To make this clearer, write } u \text{ instead of } x^2.$$

The equation $u^2 - 9u + 8 = 0$ can be solved for u by factoring or by the quadratic formula. Then, remembering that $u = x^2$, we can solve for x. Equations that can be solved like this are *reducible to quadratic* and are said to be *in quadratic form.*

EXAMPLE **1**

Solve: $x^4 - 9x^2 + 8 = 0$.

SOLUTION   We begin by letting $u = x^2$ and finding u^2:

$$u = x^2$$
$$u^2 = (x^2)^2 = x^4.$$

Then we solve by substituting u^2 for x^4 and u for x^2:

$$u^2 - 9u + 8 = 0$$
$$(u - 8)(u - 1) = 0 \qquad \text{Factoring}$$
$$u - 8 = 0 \quad \text{or} \quad u - 1 = 0 \qquad \text{Principle of zero products}$$
$$u = 8 \quad \text{or} \qquad u = 1.$$

STUDENT NOTES

To identify an equation in quadratic form, look for two variable expressions in the equation. The exponent in one expression is twice the exponent in the other expression.

We replace u with x^2 and solve these equations:

$$x^2 = 8 \qquad \text{or} \quad x^2 = 1$$
$$x = \pm\sqrt{8} \quad \text{or} \qquad x = \pm 1 \qquad \boxed{\text{We are solving for } x.}$$
$$x = \pm 2\sqrt{2} \quad \text{or} \qquad x = \pm 1.$$

To check, note that for both $x = 2\sqrt{2}$ and $-2\sqrt{2}$, we have $x^2 = 8$ and $x^4 = 64$. Similarly, for both $x = 1$ and -1, we have $x^2 = 1$ and $x^4 = 1$. Thus instead of making four checks, we need make only two.

Check: For $\pm 2\sqrt{2}$:

$$\begin{array}{c|c} x^4 - 9x^2 + 8 = 0 & \\ \hline (\pm 2\sqrt{2})^4 - 9(\pm 2\sqrt{2})^2 + 8 & 0 \\ 64 - 9 \cdot 8 + 8 & \\ & 0 \overset{?}{=} 0 \quad \text{TRUE} \end{array}$$

For ± 1:

$$\begin{array}{c|c} x^4 - 9x^2 + 8 = 0 & \\ \hline (\pm 1)^4 - 9(\pm 1)^2 + 8 & 0 \\ 1 - 9 + 8 & \\ & 0 \overset{?}{=} 0 \quad \text{TRUE} \end{array}$$

The solutions are $1, -1, 2\sqrt{2}$, and $-2\sqrt{2}$.

TRY EXERCISE ▶ 17

> *CAUTION!*   A common error when working on problems like Example 1 is to solve for u but forget to solve for x. Remember to solve for the *original* variable!

Equations like those in Example 1 can be solved directly by factoring:

$$x^4 - 9x^2 + 8 = 0$$
$$(x^2 - 1)(x^2 - 8) = 0$$
$$x^2 - 1 = 0 \quad \text{or} \quad x^2 - 8 = 0$$
$$x^2 = 1 \quad \text{or} \qquad x^2 = 8$$
$$x = \pm 1 \quad \text{or} \qquad x = \pm 2\sqrt{2}.$$

However, it often becomes difficult to solve the equation without first making a substitution.

To recognize an equation in quadratic form, inspect all the variable expressions in the equation. For an equation to be written in the form $au^2 + bu + c = 0$, it is necessary to identify one variable expression as u and a second variable expression as u^2.

EXAMPLE 2

Find the x-intercepts of the graph of $f(x) = (x^2 - 1)^2 - (x^2 - 1) - 2$.

SOLUTION   The x-intercepts occur where $f(x) = 0$ so we must have

$$(x^2 - 1)^2 - (x^2 - 1) - 2 = 0. \qquad \text{Setting } f(x) \text{ equal to } 0$$

If we identify $x^2 - 1$ as u, the equation can be written in quadratic form:

$$u = x^2 - 1$$
$$u^2 = (x^2 - 1)^2.$$

Substituting, we have

$$u^2 - u - 2 = 0 \qquad \begin{array}{l}\text{Substituting in}\\ (x^2 - 1)^2 - (x^2 - 1) - 2 = 0\end{array}$$

$$(u - 2)(u + 1) = 0$$

$$u = 2 \quad or \quad u = -1. \qquad \begin{array}{l}\text{Using the principle of}\\ \text{zero products}\end{array}$$

Next, we replace u with $x^2 - 1$ and solve these equations:

$$x^2 - 1 = 2 \qquad or \quad x^2 - 1 = -1$$
$$x^2 = 3 \qquad or \qquad x^2 = 0 \qquad \text{Adding 1 to both sides}$$
$$x = \pm\sqrt{3} \quad or \qquad x = 0. \qquad \begin{array}{l}\text{Using the principle of}\\ \text{square roots}\end{array}$$

The x-intercepts occur at $(-\sqrt{3}, 0)$, $(0, 0)$, and $(\sqrt{3}, 0)$.   **TRY EXERCISE** 49

Radical Equations and Rational Equations

Sometimes rational equations, radical equations, or equations containing exponents that are fractions are reducible to quadratic. It is especially important that answers to these equations be checked in the original equation.

EXAMPLE 3

Solve: $x - 3\sqrt{x} - 4 = 0$.

SOLUTION   This radical equation could be solved using the method discussed in Section 10.6. However, if we note that the square of $\sqrt{x}$ is x, we can regard the equation as "quadratic in $\sqrt{x}$."

We determine u and u^2:

$$u = \sqrt{x}$$
$$u^2 = x.$$

Substituting, we have

$$x - 3\sqrt{x} - 4 = 0$$
$$u^2 - 3u - 4 = 0$$
$$(u - 4)(u + 1) = 0$$
$$u = 4 \quad or \quad u = -1. \qquad \begin{array}{l}\text{Using the principle of}\\ \text{zero products}\end{array}$$

Next, we replace u with $\sqrt{x}$ and solve these equations:

$$\sqrt{x} = 4 \quad or \quad \sqrt{x} = -1.$$

Squaring gives us $x = 16$ or $x = 1$ and also makes checking essential.

Check: For 16:

$$x - 3\sqrt{x} - 4 = 0$$
$$\frac{16 - 3\sqrt{16} - 4}{\quad} \bigg|\; 0$$
$$16 - 3 \cdot 4 - 4 \bigg|$$
$$0 \overset{?}{=} 0 \quad \text{TRUE}$$

For 1:

$$x - 3\sqrt{x} - 4 = 0$$
$$\frac{1 - 3\sqrt{1} - 4}{\quad} \bigg|\; 0$$
$$1 - 3 \cdot 1 - 4 \bigg|$$
$$-6 \overset{?}{=} 0 \quad \text{FALSE}$$

The number 16 checks, but 1 does not. Had we noticed that $\sqrt{x} = -1$ has no solution (since principal square roots are never negative), we could have solved only the equation $\sqrt{x} = 4$. The solution is 16.   **TRY EXERCISE** ▸ 21

The following tips may prove useful.

To Solve an Equation That Is Reducible to Quadratic

1. Look for two variable expressions in the equation. One expression should be the square of the other.
2. Write down any substitutions that you are making.
3. Remember to solve for the variable that is used in the original equation.
4. Check possible answers in the original equation.

EXAMPLE **4**   Solve: $2m^{-2} + m^{-1} - 15 = 0$.

SOLUTION   Note that the square of m^{-1} is $(m^{-1})^2$, or m^{-2}. We let $u = m^{-1}$:

Determine u and u^2.

$$u = m^{-1}$$
$$u^2 = m^{-2}.$$

Substituting, we have

Substitute.

$$2u^2 + u - 15 = 0 \qquad \text{Substituting in}$$
$$\qquad\qquad\qquad\qquad\qquad 2m^{-2} + m^{-1} - 15 = 0$$

$$(2u - 5)(u + 3) = 0$$

$$2u - 5 = 0 \quad or \quad u + 3 = 0 \qquad \text{Using the principle of zero products}$$

$$2u = 5 \quad or \qquad u = -3$$

Solve for u.

$$u = \frac{5}{2} \quad or \qquad u = -3.$$

Now we replace u with m^{-1} and solve:

$$m^{-1} = \frac{5}{2} \quad or \quad m^{-1} = -3$$

$$\frac{1}{m} = \frac{5}{2} \quad or \quad \frac{1}{m} = -3 \qquad \text{Recall that } m^{-1} = \frac{1}{m}.$$

$$1 = \frac{5}{2}m \quad or \quad 1 = -3m \qquad \text{Multiplying both sides by } m$$

Solve for the original variable.

$$\frac{2}{5} = m \quad or \quad -\frac{1}{3} = m. \qquad \text{Solving for } m$$

Check.

Check:

For $\frac{2}{5}$:

$$\frac{2m^{-2} + m^{-1} - 15 = 0}{2\left(\frac{2}{5}\right)^{-2} + \left(\frac{2}{5}\right)^{-1} - 15 \,\bigg|\, 0}$$

$$2\left(\frac{5}{2}\right)^2 + \left(\frac{5}{2}\right) - 15$$

$$2\left(\frac{25}{4}\right) + \frac{5}{2} - 15$$

$$\frac{25}{2} + \frac{5}{2} - 15$$

$$\frac{30}{2} - 15$$

$$0 \overset{?}{=} 0 \;\; \text{TRUE}$$

For $-\frac{1}{3}$:

$$\frac{2m^{-2} + m^{-1} - 15 = 0}{2\left(-\frac{1}{3}\right)^{-2} + \left(-\frac{1}{3}\right)^{-1} - 15 \,\bigg|\, 0}$$

$$2\left(-\frac{3}{1}\right)^2 + \left(-\frac{3}{1}\right) - 15$$

$$2(9) + (-3) - 15$$

$$18 - 3 - 15$$

$$0 \overset{?}{=} 0 \;\; \text{TRUE}$$

Both numbers check. The solutions are $-\frac{1}{3}$ and $\frac{2}{5}$.

▶ **TRY EXERCISE** ▶ 29

Note that Example 4 can also be written

$$\frac{2}{m^2} + \frac{1}{m} - 15 = 0.$$

It can then be solved by letting $u = 1/m$ and $u^2 = 1/m^2$ or by clearing fractions as in Section 6.6.

EXAMPLE **5**

Solve: $t^{2/5} - t^{1/5} - 2 = 0$.

SOLUTION Note that the square of $t^{1/5}$ is $(t^{1/5})^2$, or $t^{2/5}$. The equation is therefore quadratic in $t^{1/5}$, so we let $u = t^{1/5}$:

$$u = t^{1/5}$$
$$u^2 = t^{2/5}.$$

Substituting, we have

$$u^2 - u - 2 = 0 \qquad \text{Substituting in } t^{2/5} - t^{1/5} - 2 = 0$$
$$(u - 2)(u + 1) = 0$$
$$u = 2 \quad or \quad u = -1. \qquad \text{Using the principle of zero products}$$

Now we replace u with $t^{1/5}$ and solve:

$$t^{1/5} = 2 \quad or \quad t^{1/5} = -1$$
$$t = 32 \quad or \quad t = -1. \qquad \text{Principle of powers; raising both sides to the 5th power}$$

Check:

For 32:

$$\frac{t^{2/5} - t^{1/5} - 2 = 0}{32^{2/5} - 32^{1/5} - 2 \,\bigg|\, 0}$$

$$(32^{1/5})^2 - 32^{1/5} - 2$$

$$2^2 - 2 - 2$$

$$0 \overset{?}{=} 0 \;\; \text{TRUE}$$

For -1:

$$\frac{t^{2/5} - t^{1/5} - 2 = 0}{(-1)^{2/5} - (-1)^{1/5} - 2 \,\bigg|\, 0}$$

$$[(-1)^{1/5}]^2 - (-1)^{1/5} - 2$$

$$(-1)^2 - (-1) - 2$$

$$0 \overset{?}{=} 0 \;\; \text{TRUE}$$

Both numbers check. The solutions are 32 and -1.

▶ **TRY EXERCISE** ▶ 33

EXAMPLE **6**   Solve: $(5 + \sqrt{r})^2 + 6(5 + \sqrt{r}) + 2 = 0$.

SOLUTION   We determine u and u^2:

$$u = 5 + \sqrt{r}$$
$$u^2 = (5 + \sqrt{r})^2.$$

Substituting, we have

$$u^2 + 6u + 2 = 0$$

$$u = \frac{-6 \pm \sqrt{6^2 - 4 \cdot 1 \cdot 2}}{2 \cdot 1} \qquad \text{Using the quadratic formula}$$

$$= \frac{-6 \pm \sqrt{28}}{2}$$

$$= \frac{-6}{2} \pm \frac{2\sqrt{7}}{2} \qquad \left.\begin{array}{l} \end{array}\right\} \text{Simplifying; } \sqrt{28} = \sqrt{4}\sqrt{7}$$

$$= -3 \pm \sqrt{7}.$$

Now we replace u with $5 + \sqrt{r}$ and solve for r:

$$5 + \sqrt{r} = -3 + \sqrt{7} \quad or \quad 5 + \sqrt{r} = -3 - \sqrt{7} \qquad u = -3 + \sqrt{7}$$
$$\sqrt{r} = -8 + \sqrt{7} \quad or \qquad \sqrt{r} = -8 - \sqrt{7}. \qquad or\ u = -3 - \sqrt{7}$$

We could now solve for r and check possible solutions, but first let's examine $-8 + \sqrt{7}$ and $-8 - \sqrt{7}$. Since $\sqrt{7} \approx 2.6$, both $-8 + \sqrt{7}$ and $-8 - \sqrt{7}$ are negative. Since the principal square root of r is never negative, both values of $\sqrt{r}$ must be rejected. Note too that in the original equation, $(5 + \sqrt{r})^2$, $6(5 + \sqrt{r})$, and 2 are all positive. Thus it is impossible for their sum to be 0.

The original equation has no solution.

TRY EXERCISE 25

11.5   **EXERCISE SET**

For Extra Help
MyMathLab   Math XL PRACTICE   WATCH   DOWNLOAD

👋 *Concept Reinforcement*   *In each of Exercises 1–8, match the equation with a substitution from the column on the right that could be used to reduce the equation to quadratic form.*

1. ____ $4x^6 - 2x^3 + 1 = 0$

2. ____ $3x^4 + 4x^2 - 7 = 0$

3. ____ $5x^8 + 2x^4 - 3 = 0$

4. ____ $2x^{2/3} - 5x^{1/3} + 4 = 0$

5. ____ $3x^{4/3} + 4x^{2/3} - 7 = 0$

6. ____ $2x^{-2/3} + x^{-1/3} + 6 = 0$

7. ____ $4x^{-4/3} - 2x^{-2/3} + 3 = 0$

8. ____ $3x^{-4} + 4x^{-2} - 2 = 0$

a) $u = x^{-1/3}$

b) $u = x^{1/3}$

c) $u = x^{-2}$

d) $u = x^2$

e) $u = x^{-2/3}$

f) $u = x^3$

g) $u = x^{2/3}$

h) $u = x^4$

Write the substitution that could be used to make each equation quadratic in u.

9. For $3p - 4\sqrt{p} + 6 = 0$, use $u =$ _____ .

10. For $x^{1/2} - x^{1/4} - 2 = 0$, use $u =$ _____ .

11. For $(x^2 + 3)^2 + (x^2 + 3) - 7 = 0$, use $u =$ _____ .

12. For $t^{-6} + 5t^{-3} - 6 = 0$, use $u =$ _____ .

13. For $(1 + t)^4 + (1 + t)^2 + 4 = 0$, use $u =$ _____ .

14. For $w^{1/3} - 3w^{1/6} + 8 = 0$, use $w =$ _____ .

Solve.

15. $x^4 - 13x^2 + 36 = 0$

16. $x^4 - 17x^2 + 16 = 0$

17. $t^4 - 7t^2 + 12 = 0$

18. $t^4 - 11t^2 + 18 = 0$

19. $4x^4 - 9x^2 + 5 = 0$

20. $9x^4 - 38x^2 + 8 = 0$

21. $w + 4\sqrt{w} - 12 = 0$

22. $s + 3\sqrt{s} - 40 = 0$

23. $(x^2 - 7)^2 - 3(x^2 - 7) + 2 = 0$

24. $(x^2 - 2)^2 - 12(x^2 - 2) + 20 = 0$

25. $r - 2\sqrt{r} - 6 = 0$

26. $s - 4\sqrt{s} - 1 = 0$

27. $(1 + \sqrt{x})^2 + 5(1 + \sqrt{x}) + 6 = 0$

28. $(3 + \sqrt{x})^2 + 3(3 + \sqrt{x}) - 10 = 0$

29. $x^{-2} - x^{-1} - 6 = 0$

30. $2x^{-2} - x^{-1} - 1 = 0$

31. $4t^{-2} - 3t^{-1} - 1 = 0$

32. $2m^{-2} + 7m^{-1} - 15 = 0$

33. $t^{2/3} + t^{1/3} - 6 = 0$

34. $w^{2/3} - 2w^{1/3} - 8 = 0$

35. $y^{1/3} - y^{1/6} - 6 = 0$

36. $t^{1/2} + 3t^{1/4} + 2 = 0$

37. $t^{1/3} + 2t^{1/6} = 3$

38. $m^{1/2} + 6 = 5m^{1/4}$

39. $(3 - \sqrt{x})^2 - 10(3 - \sqrt{x}) + 23 = 0$

40. $(5 + \sqrt{x})^2 - 12(5 + \sqrt{x}) + 33 = 0$

41. $16\left(\dfrac{x - 1}{x - 8}\right)^2 + 8\left(\dfrac{x - 1}{x - 8}\right) + 1 = 0$

42. $9\left(\dfrac{x + 2}{x + 3}\right)^2 - 6\left(\dfrac{x + 2}{x + 3}\right) + 1 = 0$

43. $x^4 + 5x^2 - 36 = 0$

44. $x^4 + 5x^2 + 4 = 0$

45. $(n^2 + 6)^2 - 7(n^2 + 6) + 10 = 0$

46. $(m^2 + 7)^2 - 6(m^2 + 7) - 16 = 0$

Find all x-intercepts of the given function f. If none exists, state this.

47. $f(x) = 5x + 13\sqrt{x} - 6$

48. $f(x) = 3x + 10\sqrt{x} - 8$

49. $f(x) = (x^2 - 3x)^2 - 10(x^2 - 3x) + 24$

50. $f(x) = (x^2 - 6x)^2 - 2(x^2 - 6x) - 35$

51. $f(x) = x^{2/5} + x^{1/5} - 6$

52. $f(x) = x^{1/2} - x^{1/4} - 6$

Aha! **53.** $f(x) = \left(\dfrac{x^2 + 2}{x}\right)^4 + 7\left(\dfrac{x^2 + 2}{x}\right)^2 + 5$

54. $f(x) = \left(\dfrac{x^2 + 1}{x}\right)^4 + 4\left(\dfrac{x^2 + 1}{x}\right)^2 + 12$

55. To solve $25x^6 - 10x^3 + 1 = 0$, Jose lets $u = 5x^3$ and Robin lets $u = x^3$. Can they both be correct? Why or why not?

56. Jenn writes that the solutions of $x^4 - 5x^2 + 6 = 0$ are 2 and 3. What mistake is she making?

Skill Review

To prepare for Section 11.6, review graphing functions (Section 7.3).

Graph. [7.3]

57. $f(x) = x$

58. $g(x) = x + 2$

59. $h(x) = x - 2$

60. $f(x) = x^2$

61. $g(x) = x^2 + 2$

62. $h(x) = x^2 - 2$

Synthesis

63. Describe a procedure that could be used to solve any equation of the form $ax^4 + bx^2 + c = 0$.

64. Describe a procedure that could be used to write an equation that is quadratic in $3x^2 - 1$. Then explain how the procedure could be adjusted to write equations that are quadratic in $3x^2 - 1$ and have no real-number solution.

Solve.

65. $3x^4 + 5x^2 - 1 = 0$

66. $5x^4 - 7x^2 + 1 = 0$

67. $(x^2 - 5x - 1)^2 - 18(x^2 - 5x - 1) + 65 = 0$

68. $(x^2 - 4x - 2)^2 - 13(x^2 - 4x - 2) + 30 = 0$

69. $\dfrac{x}{x - 1} - 6\sqrt{\dfrac{x}{x - 1}} - 40 = 0$

70. $\left(\sqrt{\dfrac{x}{x - 3}}\right)^2 - 24 = 10\sqrt{\dfrac{x}{x - 3}}$

71. $a^5(a^2 - 25) + 13a^3(25 - a^2) + 36a(a^2 - 25) = 0$

72. $a^3 - 26a^{3/2} - 27 = 0$

73. $x^6 - 28x^3 + 27 = 0$

74. $x^6 + 7x^3 - 8 = 0$

75. Use a graphing calculator to check your answers to Exercises 15, 17, 41, and 53.

76. Use a graphing calculator to solve
$$x^4 - x^3 - 13x^2 + x + 12 = 0.$$

77. While trying to solve $0.05x^4 - 0.8 = 0$ with a graphing calculator, Salam gets the screen at right. Can Salam solve this equation with a graphing calculator? Why or why not?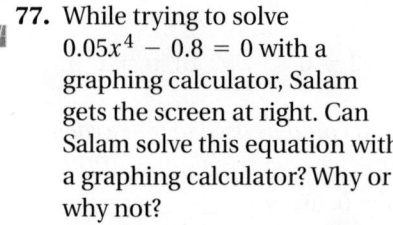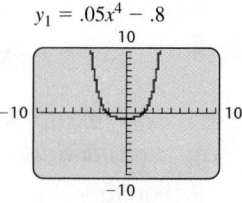

CONNECTING the CONCEPTS

We have studied four different ways of solving quadratic equations. Each method has advantages and disadvantages, as outlined below. Note that although the quadratic formula can be used to solve *any* quadratic equation, the other methods are sometimes faster and easier to use. Also note that any of these methods can be used when solving equations that are reducible to quadratic.

Method	Advantages	Disadvantages	Example
Factoring	Can be very fast.	Can be used only on certain equations. Many equations are difficult or impossible to solve by factoring.	$x^2 - x - 6 = 0$ $(x - 3)(x + 2) = 0$ $x = 3$ or $x = -2$
The principle of square roots	Fastest way to solve equations of the form $X^2 = k$. Can be used to solve *any* quadratic equation.	Can be slow when original equation is not written in the form $X^2 = k$.	$(x - 5)^2 = 2$ $x - 5 = \pm\sqrt{2}$ $x = 5 \pm \sqrt{2}$
Completing the square	Works well on equations of the form $x^2 + bx = -c$, when b is even. Can be used to solve *any* quadratic equation.	Can be complicated when $a \neq 1$ or when b is not even in $x^2 + bx = -c$.	$x^2 + 14x = -2$ $x^2 + 14x + 49 = -2 + 49$ $(x + 7)^2 = 47$ $x + 7 = \pm\sqrt{47}$ $x = -7 \pm \sqrt{47}$
The quadratic formula	Can be used to solve *any* quadratic equation.	Can be slower than factoring or the principle of square roots for certain equations.	$x^2 - 2x - 5 = 0$ $x = \dfrac{-(-2) \pm \sqrt{(-2)^2 - 4(1)(-5)}}{2 \cdot 1}$ $= \dfrac{2 \pm \sqrt{24}}{2}$ $= \dfrac{2}{2} \pm \dfrac{2\sqrt{6}}{2} = 1 \pm \sqrt{6}$

MIXED REVIEW

Solve. Examine each exercise carefully, and try to solve using the easiest method.

1. $x^2 - 3x - 10 = 0$

2. $x^2 = 121$

3. $x^2 + 6x = 10$

4. $x^2 + x - 3 = 0$

5. $(x + 1)^2 = 2$

6. $x^2 - 10x + 25 = 0$

7. $x^2 - x - 1 = 0$

8. $x^2 - 2x = 6$

9. $4t^2 = 11$

(continued)

10. $2t^2 + 1 = 3t$

11. $c^2 + c + 1 = 0$

12. $16c^2 = 7c$

13. $6y^2 - 7y - 10 = 0$

14. $y^2 - 2y + 8 = 0$

15. $x^4 - 10x^2 + 9 = 0$

16. $x^4 - 8x^2 - 9 = 0$

17. $t(t - 3) = 2t(t + 1)$

18. $(t + 4)(t - 3) = 18$

19. $(m^2 + 3)^2 - 4(m^2 + 3) - 5 = 0$

20. $m^{-4} - 5m^{-2} + 6 = 0$

11.6 Quadratic Functions and Their Graphs

The Graph of $f(x) = ax^2$ ▪ The Graph of $f(x) = a(x - h)^2$ ▪ The Graph of $f(x) = a(x - h)^2 + k$

We have seen that the graph of any linear function $f(x) = mx + b$ is a straight line. In this section and the next, we will see that the graph of any quadratic function $f(x) = ax^2 + bx + c$ is a *parabola*. We examine the shape of such graphs by first looking at quadratic functions with $b = 0$ and $c = 0$.

The Graph of $f(x) = ax^2$

The most basic quadratic function is $f(x) = x^2$.

EXAMPLE 1

Graph: $f(x) = x^2$.

SOLUTION We choose some values for x and compute $f(x)$ for each. Then we plot the ordered pairs and connect them with a smooth curve.

x	$f(x) = x^2$	$(x, f(x))$
-3	9	$(-3, 9)$
-2	4	$(-2, 4)$
-1	1	$(-1, 1)$
0	0	$(0, 0)$
1	1	$(1, 1)$
2	4	$(2, 4)$
3	9	$(3, 9)$

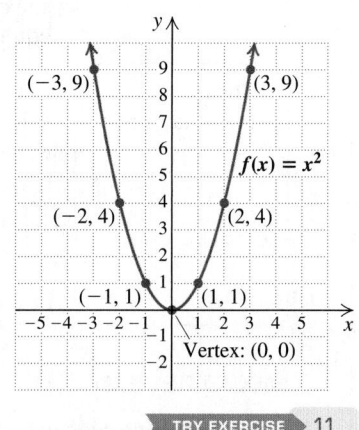

TRY EXERCISE 11

All quadratic functions have graphs similar to the one in Example 1. Such curves are called **parabolas**. They are U-shaped and can open upward, as in Example 1, or downward. The "turning point" of the graph is called the **vertex** of the parabola. The vertex of the graph in Example 1 is $(0, 0)$.

A parabola is symmetric with respect to a line that goes through the center of the parabola and the vertex. This line is known as the parabola's **axis of symmetry**. In Example 1, the y-axis (the vertical line $x = 0$) is the axis of symmetry. Were the paper folded on this line, the two halves of the curve would match.

The graph of any function of the form $y = ax^2$ has a vertex of $(0, 0)$ and an axis of symmetry $x = 0$. By plotting points, we can compare the graphs of $g(x) = \frac{1}{2}x^2$ and $h(x) = 2x^2$ with the graph of $f(x) = x^2$.

x	$g(x) = \frac{1}{2}x^2$
-3	$\frac{9}{2}$
-2	2
-1	$\frac{1}{2}$
0	0
1	$\frac{1}{2}$
2	2
3	$\frac{9}{2}$

x	$h(x) = 2x^2$
-3	18
-2	8
-1	2
0	0
1	2
2	8
3	18

Note that the graph of $g(x) = \frac{1}{2}x^2$ is "wider" than the graph of $f(x) = x^2$, and the graph of $h(x) = 2x^2$ is "narrower." The vertex and the axis of symmetry, however, remain $(0, 0)$ and the line $x = 0$, respectively.

When we consider the graph of $k(x) = -\frac{1}{2}x^2$, we see that the parabola is the same shape as the graph of $g(x) = \frac{1}{2}x^2$, but opens downward. We say that the graphs of k and g are *reflections* of each other across the x-axis.

x	$k(x) = -\frac{1}{2}x^2$
-3	$-\frac{9}{2}$
-2	-2
-1	$-\frac{1}{2}$
0	0
1	$-\frac{1}{2}$
2	-2
3	$-\frac{9}{2}$

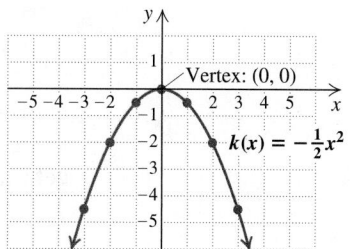

Graphing $f(x) = ax^2$

The graph of $f(x) = ax^2$ is a parabola with $x = 0$ as its axis of symmetry. Its vertex is the origin.

For $a > 0$, the parabola opens upward. For $a < 0$, the parabola opens downward.

If $|a|$ is greater than 1, the parabola is narrower than $y = x^2$.

If $|a|$ is between 0 and 1, the parabola is wider than $y = x^2$.

The width of a parabola and whether it opens upward or downward are determined by the coefficient a in $f(x) = ax^2 + bx + c$. In the remainder of this section, we graph quadratic functions that are written in a form from which the vertex can be read directly.

The Graph of $f(x) = a(x - h)^2$

EXAMPLE 2

Graph: $f(x) = (x - 3)^2$.

SOLUTION We choose some values for x and compute $f(x)$. Since $(x - 3)^2 = 1 \cdot (x - 3)^2$, $a = 1$, and the graph opens upward. It is important to note that when an input here is 3 more than an input for Example 1, the outputs match. We plot the points and draw the curve.

x	$f(x) = (x - 3)^2$
-1	16
0	9
1	4
2	1
3	0	← Vertex
4	1
5	4
6	9

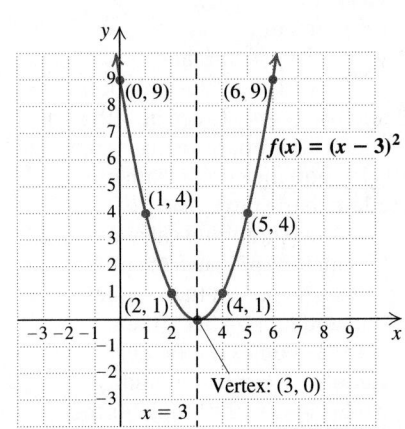

Note that $f(x)$ is smallest when $x - 3$ is 0, that is, for $x = 3$. Thus the line $x = 3$ is now the axis of symmetry and the point $(3, 0)$ is the vertex. Had we recognized earlier that $x = 3$ is the axis of symmetry, we could have computed some values on one side, such as $(4, 1)$, $(5, 4)$, and $(6, 9)$, and then used symmetry to get their mirror images $(2, 1)$, $(1, 4)$, and $(0, 9)$ without further computation.

TRY EXERCISE 19

The result of Example 2 can be generalized:

The vertex of the graph of $f(x) = a(x - h)^2$ is $(h, 0)$.

EXAMPLE 3

Graph: $g(x) = -2(x + 4)^2$.

SOLUTION We choose some values for x and compute $g(x)$. Since $a = -2$, the graph will open downward. Note that $g(x)$ is greatest when $x + 4$ is 0, that is, for $x = -4$. Thus the line given by $x = -4$ is the axis of symmetry and the point $(-4, 0)$ is the vertex. We plot some points and draw the curve.

To explore the effect of h on the graph of $f(x) = a(x - h)^2$, let $y_1 = 7x^2$ and $y_2 = 7(x - 1)^2$. Graph both y_1 and y_2 and compare y-values, beginning at $x = 1$ and increasing x by one unit at a time. The G-T or HORIZ **MODE** can be used to view a split screen showing both the graph and a table.

Next, let $y_3 = 7(x - 2)^2$ and compare its graph and y-values with those of y_1 and y_2. Then let $y_4 = 7(x + 1)^2$ and $y_5 = 7(x + 2)^2$.

1. Compare graphs and y-values and describe the effect of h on the graph of $f(x) = a(x - h)^2$.
2. If the Transfrm application is available, let $y_1 = A(x - B)^2$ and describe the effect that A and B have on each graph.

x	$g(x) = -2(x + 4)^2$
-6	-8
-5	-2
-4	0
-3	-2
-2	-8

$\longleftarrow$Vertex

$x = -4$

Vertex: $(-4, 0)$

$g(x) = -2(x + 4)^2$
or $-2(x - (-4))^2$

TRY EXERCISE 23

In Example 2, the graph of $f(x) = (x - 3)^2$ looks just like the graph of $y = x^2$, except that it is moved, or *translated*, 3 units to the right. In Example 3, the graph of $g(x) = -2(x + 4)^2$ looks like the graph of $y = -2x^2$, except that it is shifted 4 units to the left. These results are generalized as follows.

Graphing $f(x) = a(x - h)^2$

The graph of $f(x) = a(x - h)^2$ has the same shape as the graph of $y = ax^2$.

*   If h is positive, the graph of $y = ax^2$ is shifted h units to the right.
*   If h is negative, the graph of $y = ax^2$ is shifted $|h|$ units to the left.
*   The vertex is $(h, 0)$ and the axis of symmetry is $x = h$.

The Graph of $f(x) = a(x - h)^2 + k$

Given a graph of $f(x) = a(x - h)^2$, what happens if we add a constant k? Suppose that we add 2. This increases $f(x)$ by 2, so the curve is moved up. If k is negative, the curve is moved down. The axis of symmetry for the parabola remains $x = h$, but the vertex will be at (h, k), or, equivalently, $(h, f(h))$.

Because of the shape of their graphs, quadratic functions have either a *minimum* value or a *maximum* value. Many real-world applications involve finding that value. For example, a business owner is concerned with minimizing cost and maximizing profit. If a parabola opens upward ($a > 0$), the function value, or y-value, at the vertex is a least, or minimum, value. That is, it is less than the y-value at any other point on the graph. If the parabola opens downward ($a < 0$), the function value at the vertex is a greatest, or maximum, value.

Graphs of $f(x) = a(x - h)^2 + k$

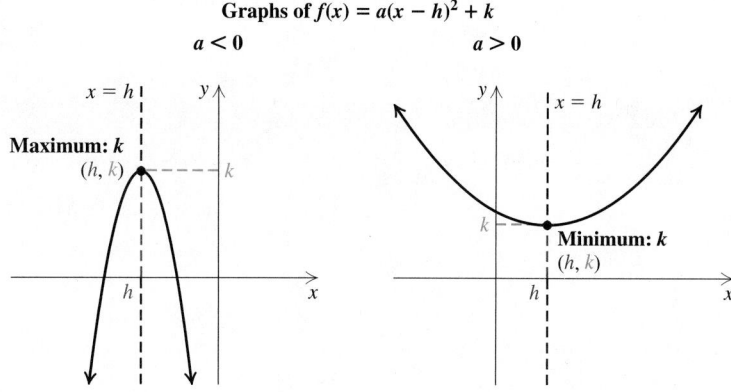

**TECHNOLOGY
CONNECTION**

To study the effect of k on the graph of
$f(x) = a(x - h)^2 + k$, let $y_1 = 7(x - 1)^2$ and
$y_2 = 7(x - 1)^2 + 2$. Graph both y_1 and y_2 in the window
$[-5, 5, -5, 5]$ and use TRACE or a TABLE to compare the
y-values for any given x-value.

   1. Let $y_3 = 7(x - 1)^2 - 4$ and compare its graph and
      y-values with those of y_1 and y_2.

2. Try other values of k, including decimals and
fractions. Describe the effect of k on the graph of
$f(x) = a(x - h)^2$.

3. If the Transfrm application is available, let
$y_1 = A(x - B)^2 + C$ and describe the effect that
A, B, and C have on each graph.

Graphing $f(x) = a(x - h)^2 + k$

    The graph of $f(x) = a(x - h)^2 + k$ has the same shape as the graph
of $y = a(x - h)^2$.

- If k is positive, the graph of $y = a(x - h)^2$ is shifted k units up.
- If k is negative, the graph of $y = a(x - h)^2$ is shifted $|k|$ units down.
- The vertex is (h, k), and the axis of symmetry is $x = h$.
- For $a > 0$, the minimum function value is k. For $a < 0$, the maximum function value is k.

EXAMPLE **4**   Graph $g(x) = (x - 3)^2 - 5$, and find the minimum function value.

SOLUTION   The graph will look like that of $f(x) = (x - 3)^2$ (see Example 2) but
shifted 5 units down. You can confirm this by plotting some points. For instance,
$g(4) = (4 - 3)^2 - 5 = -4$, whereas in Example 2, $f(4) = (4 - 3)^2 = 1$.
    The vertex is now $(3, -5)$, and the minimum function value is -5.

x	$g(x) = (x - 3)^2 - 5$	
0	4	
1	-1	
2	-4	
3	-5	← Vertex
4	-4	
5	-1	
6	4	

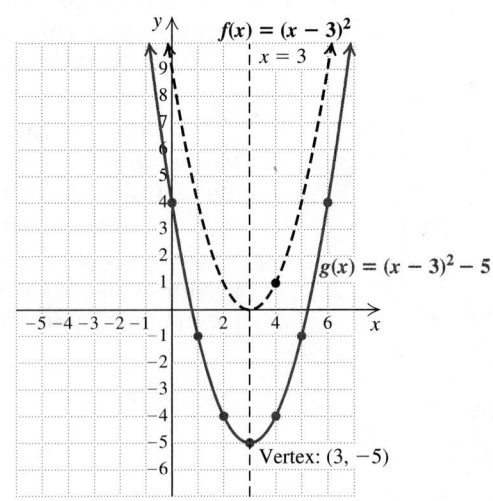

TRY EXERCISE 39

EXAMPLE 5   Graph $h(x) = \frac{1}{2}(x - 3)^2 + 6$, and find the minimum function value.

SOLUTION   The graph looks just like that of $f(x) = \frac{1}{2}x^2$ but moved 3 units to the right and 6 units up. The vertex is $(3, 6)$, and the axis of symmetry is $x = 3$. We draw $f(x) = \frac{1}{2}x^2$ and then shift the curve over and up. The minimum function value is 6. By plotting some points, we have a check.

x	$h(x) = \frac{1}{2}(x - 3)^2 + 6$
0	$10\frac{1}{2}$
1	8
3	6
5	8
6	$10\frac{1}{2}$

←——— Vertex

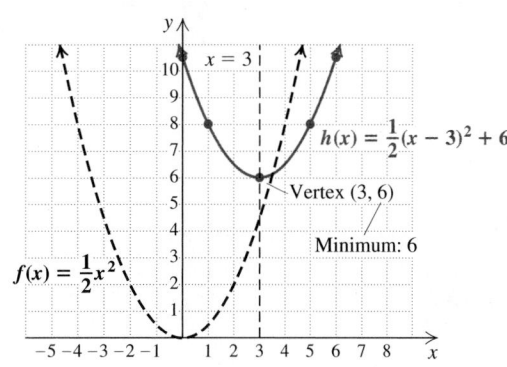

TRY EXERCISE 43

EXAMPLE 6   Graph $y = -2(x + 3)^2 + 5$. Find the vertex, the axis of symmetry, and the maximum or minimum value.

SOLUTION   We first express the equation in the equivalent form

$$y = -2[x - (-3)]^2 + 5.$$ This is in the form $y = a(x - h)^2 + k$.

The graph looks like that of $y = -2x^2$ translated 3 units to the left and 5 units up. The vertex is $(-3, 5)$, and the axis of symmetry is $x = -3$. Since -2 is negative, the graph opens downward, and we know that 5, the second coordinate of the vertex, is the maximum y-value.

We compute a few points as needed, selecting convenient x-values on either side of the vertex. The graph is shown here.

x	$y = -2(x + 3)^2 + 5$
-4	3
-3	5
-2	3

←——— Vertex

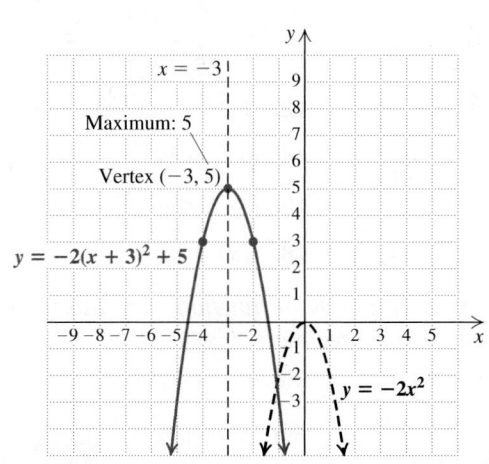

TRY EXERCISE 45

11.6   **EXERCISE SET**

Concept Reinforcement *In each of Exercises 1–8, match the equation with the corresponding graph from those shown.*

1. ___ $f(x) = 2(x - 1)^2 + 3$

2. ___ $f(x) = -2(x - 1)^2 + 3$

3. ___ $f(x) = 2(x + 1)^2 + 3$

4. ___ $f(x) = 2(x - 1)^2 - 3$

5. ___ $f(x) = -2(x + 1)^2 + 3$

6. ___ $f(x) = -2(x + 1)^2 - 3$

7. ___ $f(x) = 2(x + 1)^2 - 3$

8. ___ $f(x) = -2(x - 1)^2 - 3$

c)
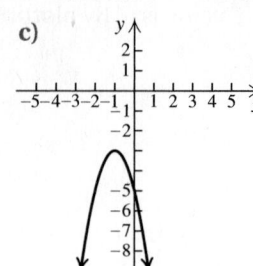

d)
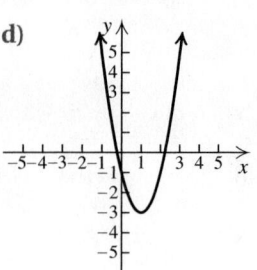

e)
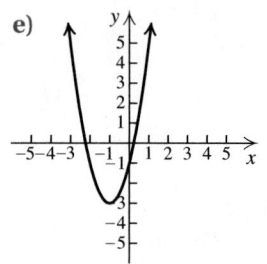

f)
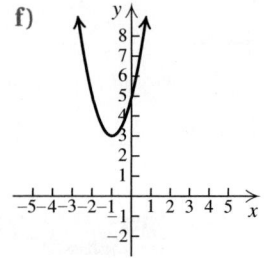

a)
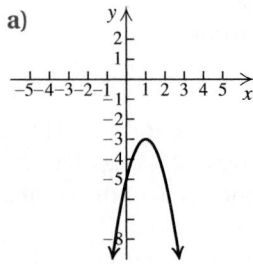

b)
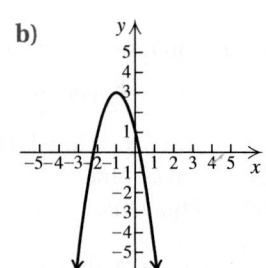

g)
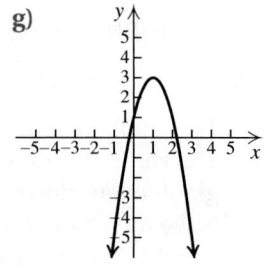

h)
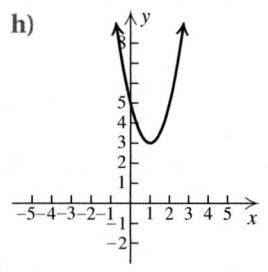

Graph.

9. $f(x) = x^2$

10. $f(x) = -x^2$

11. $f(x) = -2x^2$

12. $f(x) = -3x^2$

13. $g(x) = \frac{1}{3}x^2$

14. $g(x) = \frac{1}{4}x^2$

Aha! **15.** $h(x) = -\frac{1}{3}x^2$

16. $h(x) = -\frac{1}{4}x^2$

17. $f(x) = \frac{5}{2}x^2$

18. $f(x) = \frac{3}{2}x^2$

For each of the following, graph the function, label the vertex, and draw the axis of symmetry.

19. $g(x) = (x + 1)^2$

20. $g(x) = (x + 4)^2$

21. $f(x) = (x - 2)^2$

22. $f(x) = (x - 1)^2$

23. $f(x) = -(x + 1)^2$

24. $f(x) = -(x - 1)^2$

25. $g(x) = -(x - 2)^2$

26. $g(x) = -(x + 4)^2$

27. $f(x) = 2(x + 1)^2$

28. $f(x) = 2(x + 4)^2$

29. $g(x) = 3(x - 4)^2$

30. $g(x) = 3(x - 5)^2$

31. $h(x) = -\frac{1}{2}(x - 4)^2$

32. $h(x) = -\frac{3}{2}(x - 2)^2$

33. $f(x) = \frac{1}{2}(x - 1)^2$

34. $f(x) = \frac{1}{3}(x + 2)^2$

35. $f(x) = -2(x + 5)^2$

36. $f(x) = -3(x + 7)^2$

37. $h(x) = -3\left(x - \frac{1}{2}\right)^2$

38. $h(x) = -2\left(x + \frac{1}{2}\right)^2$

For each of the following, graph the function and find the vertex, the axis of symmetry, and the maximum value or the minimum value.

39. $f(x) = (x - 5)^2 + 2$

40. $f(x) = (x + 3)^2 - 2$

41. $f(x) = (x + 1)^2 - 3$

42. $f(x) = (x - 1)^2 + 2$

43. $g(x) = \frac{1}{2}(x + 4)^2 + 1$

44. $g(x) = -(x - 2)^2 - 4$

45. $h(x) = -2(x - 1)^2 - 3$

46. $h(x) = -2(x + 1)^2 + 4$

47. $f(x) = 2(x + 3)^2 + 1$

48. $f(x) = 2(x - 5)^2 - 3$

49. $g(x) = -\frac{3}{2}(x - 2)^2 + 4$

50. $g(x) = \frac{3}{2}(x + 2)^2 - 1$

Without graphing, find the vertex, the axis of symmetry, and the maximum value or the minimum value.

51. $f(x) = 5(x - 3)^2 + 9$

52. $f(x) = 2(x - 1)^2 - 10$

53. $f(x) = -\frac{3}{7}(x + 8)^2 + 2$

54. $f(x) = -\frac{1}{4}(x + 4)^2 - 12$

55. $f(x) = \left(x - \frac{7}{2}\right)^2 - \frac{29}{4}$

56. $f(x) = -\left(x + \frac{3}{4}\right)^2 + \frac{17}{16}$

57. $f(x) = -\sqrt{2}(x + 2.25)^2 - \pi$

58. $f(x) = 2\pi(x - 0.01)^2 + \sqrt{15}$

59. Explain, without plotting points, why the graph of $y = x^2 - 4$ looks like the graph of $y = x^2$ translated 4 units down.

60. Explain, without plotting points, why the graph of $y = (x + 2)^2$ looks like the graph of $y = x^2$ translated 2 units to the left.

Skill Review

To prepare for Section 11.7, review finding intercepts and completing the square (Sections 3.3, 5.7, and 11.1).

Find the x-intercept and the y-intercept. [3.3]

61. $8x - 6y = 24$

62. $3x + 4y = 8$

Find the x-intercepts. [5.7]

63. $y = x^2 + 8x + 15$

64. $y = 2x^2 - x - 3$

Replace the blanks with constants to form a true equation. [11.1]

65. $x^2 - 14x + \underline{\quad} = (x - \underline{\quad})^2$

66. $x^2 + 7x + \underline{\quad} = \left(x + \underline{\quad}\right)^2$

Synthesis

67. Before graphing a quadratic function, Martha always plots five points. First, she calculates and plots the coordinates of the vertex. Then she plots *four* more points after calculating *two* more ordered pairs. How is this possible?

68. If the graphs of $f(x) = a_1(x - h_1)^2 + k_1$ and $g(x) = a_2(x - h_2)^2 + k_2$ have the same shape, what, if anything, can you conclude about the a's, the h's, and the k's? Why?

Write an equation for a function having a graph with the same shape as the graph of $f(x) = \frac{3}{5}x^2$, but with the given point as the vertex.

69. $(1, 3)$

70. $(2, 8)$

71. $(4, -7)$

72. $(9, -6)$

73. $(-2, -5)$

74. $(-4, -2)$

For each of the following, write the equation of the parabola that has the shape of $f(x) = 2x^2$ or $g(x) = -2x^2$ and has a maximum value or a minimum value at the specified point.

75. Minimum: $(2, 0)$

76. Minimum: $(-4, 0)$

77. Maximum: $(0, -5)$

78. Maximum: $(3, 8)$

Use the following graph of $f(x) = a(x - h)^2 + k$ for Exercises 79–82.

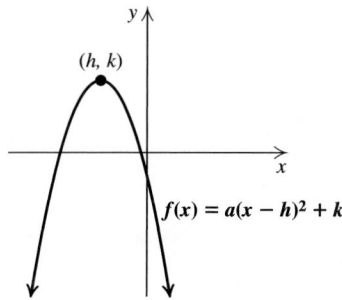

79. Describe what will happen to the graph if h is increased.

80. Describe what will happen to the graph if k is decreased.

81. Describe what will happen to the graph if a is replaced with $-a$.

82. Describe what will happen to the graph if $(x - h)$ is replaced with $(x + h)$.

Find an equation for the quadratic function F that satisfies the following conditions.

83. The graph of F is the same shape as the graph of f, where $f(x) = 3(x + 2)^2 + 7$, and $F(x)$ is a minimum at the same point that $g(x) = -2(x - 5)^2 + 1$ is a maximum.

84. The graph of F is the same shape as the graph of f, where $f(x) = -\frac{1}{3}(x - 2)^2 + 7$, and $F(x)$ is a maximum at the same point that $g(x) = 2(x + 4)^2 - 6$ is a minimum.

Functions other than parabolas can be translated. When calculating $f(x)$, if we replace x with $x - h$, where h is a constant, the graph will be moved horizontally. If we replace $f(x)$ with $f(x) + k$, the graph will be moved vertically. Use the graph below for Exercises 85–90.

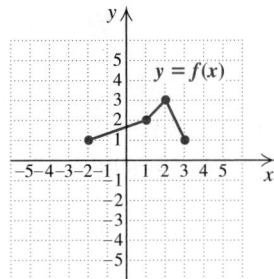

Draw a graph of each of the following.

85. $y = f(x - 1)$

86. $y = f(x + 2)$

87. $y = f(x) + 2$

88. $y = f(x) - 3$

89. $y = f(x + 3) - 2$

90. $y = f(x - 3) + 1$

91. Use the TRACE and/or TABLE features of a graphing calculator to confirm the maximum and minimum values given as answers to Exercises 51, 53, and 55. Be sure to adjust the window appropriately. On many graphing calculators, a maximum or minimum option may be available by using a CALC key.

92. Use a graphing calculator to check your graphs for Exercises 18, 28, and 48.

93. While trying to graph $y = -\frac{1}{2}x^2 + 3x + 1$, Yusef gets the following screen. How can Yusef tell at a glance that a mistake has been made?

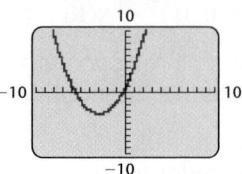

COLLABORATIVE CORNER

Match the Graph

Focus: Graphing quadratic functions

Time: 15–20 minutes

Group size: 6

Materials: Index cards

ACTIVITY

1. On each of six index cards, write one of the following equations:

   $y = \frac{1}{2}(x - 3)^2 + 1$; $y = \frac{1}{2}(x - 1)^2 + 3$;

   $y = \frac{1}{2}(x + 1)^2 - 3$; $y = \frac{1}{2}(x + 3)^2 + 1$;

   $y = \frac{1}{2}(x + 3)^2 - 1$; $y = \frac{1}{2}(x + 1)^2 + 3$.

2. Fold each index card and mix up the six cards in a hat or bag. Then, one by one, each group member should select one of the equations. Do not let anyone see your equation.

3. Each group member should carefully graph the equation selected. Make the graph large enough so that when it is finished, it can be easily viewed by the rest of the group. Be sure to scale the axes and label the vertex, but **do not label the graph with the equation used**.

4. When all group members have drawn a graph, place the graphs in a pile. The group should then match and agree on the correct equation for each graph *with no help from the person who drew the graph*. If a mistake has been made and a graph has no match, determine what its equation *should* be.

5. Compare your group's labeled graphs with those of other groups to reach consensus within the class on the correct label for each graph.

| 11.7 | More About Graphing Quadratic Functions |

Completing the Square   •   Finding Intercepts

Completing the Square

By *completing the square* (see Section 11.1), we can rewrite any polynomial $ax^2 + bx + c$ in the form $a(x - h)^2 + k$. Once that has been done, the procedures discussed in Section 11.6 will enable us to graph any quadratic function.

EXAMPLE 1

Graph: $g(x) = x^2 - 6x + 4$. Label the vertex and the axis of symmetry.

SOLUTION   We have

$$g(x) = x^2 - 6x + 4$$
$$= (x^2 - 6x) + 4.$$

STUDY SKILLS

Use What You Know

An excellent and common strategy for solving any new type of problem is to rewrite the problem in an equivalent form that we already know how to solve. Although this is not always feasible, when it is—as in most of the problems in this section—it can make a new topic much easier to learn.

To complete the square inside the parentheses, we take half the x-coefficient, $\frac{1}{2} \cdot (-6) = -3$, and square it to get $(-3)^2 = 9$. Then we add $9 - 9$ inside the parentheses:

$$g(x) = (x^2 - 6x + 9 - 9) + 4 \qquad \text{The effect is of adding 0.}$$
$$= (x^2 - 6x + 9) + (-9 + 4) \qquad \text{Using the associative law of addition to regroup}$$
$$= (x - 3)^2 - 5. \qquad \text{Factoring and simplifying}$$

This equation appeared as Example 4 of Section 11.6. The graph is that of $f(x) = x^2$ translated 3 units right and 5 units down. The vertex is $(3, -5)$, and the axis of symmetry is $x = 3$.

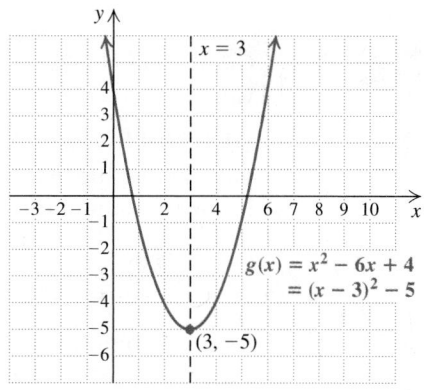

> TRY EXERCISE ▸ 19

When the leading coefficient is not 1, we factor out that number from the first two terms. Then we complete the square and use the distributive law.

EXAMPLE 2

Graph: $f(x) = 3x^2 + 12x + 13$. Label the vertex and the axis of symmetry.

SOLUTION   Since the coefficient of x^2 is not 1, we need to factor out that number—in this case, 3—from the first two terms. Remember that we want the form $f(x) = a(x - h)^2 + k$:

$$f(x) = 3x^2 + 12x + 13$$
$$= 3(x^2 + 4x) + 13.$$

Now we complete the square as before. We take half of the x-coefficient, $\frac{1}{2} \cdot 4 = 2$, and square it: $2^2 = 4$. Then we add $4 - 4$ inside the parentheses:

$$f(x) = 3(x^2 + 4x + 4 - 4) + 13.$$ Adding $4 - 4$, or 0, inside the parentheses

The distributive law allows us to separate the -4 from the perfect-square trinomial so long as it is multiplied by 3. *This step is critical*:

$$f(x) = 3(x^2 + 4x + 4) + 3(-4) + 13$$ This leaves a perfect-square trinomial inside the parentheses.

$$= 3(x + 2)^2 + 1.$$ Factoring and simplifying

The vertex is $(-2, 1)$, and the axis of symmetry is $x = -2$. The coefficient of x^2 is 3, so the graph is narrow and opens upward. We choose a few x-values on either side of the vertex, compute y-values, and then graph the parabola.

x	$f(x) = 3(x + 2)^2 + 1$
-2	1
-3	4
-1	4

⟵ Vertex

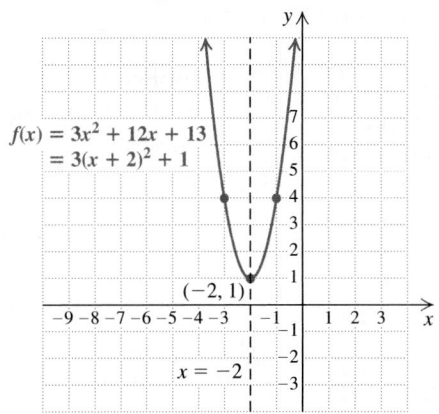

$f(x) = 3x^2 + 12x + 13$
$= 3(x + 2)^2 + 1$

$(-2, 1)$

$x = -2$

TRY EXERCISE 23

EXAMPLE **3**   Graph $f(x) = -2x^2 + 10x - 7$, and find the maximum or minimum function value.

SOLUTION   We first find the vertex by completing the square. To do so, we factor out -2 from the first two terms of the expression. This makes the coefficient of x^2 inside the parentheses 1:

$$f(x) = -2x^2 + 10x - 7$$
$$= -2(x^2 - 5x) - 7.$$

Now we complete the square as before. We take half of the x-coefficient and square it to get $\frac{25}{4}$. Then we add $\frac{25}{4} - \frac{25}{4}$ inside the parentheses:

$$f(x) = -2\left(x^2 - 5x + \frac{25}{4} - \frac{25}{4}\right) - 7$$
$$= -2\left(x^2 - 5x + \frac{25}{4}\right) + (-2)\left(-\frac{25}{4}\right) - 7$$ Multiplying by -2, using the distributive law, and regrouping

$$= -2\left(x - \frac{5}{2}\right)^2 + \frac{11}{2}.$$ Factoring and simplifying

The vertex is $\left(\frac{5}{2}, \frac{11}{2}\right)$, and the axis of symmetry is $x = \frac{5}{2}$. The coefficient of x^2, -2, is negative, so the graph opens downward and the second coordinate of the vertex, $\frac{11}{2}$, is the maximum function value.

We plot a few points on either side of the vertex, including the *y*-intercept, $f(0)$, and graph the parabola.

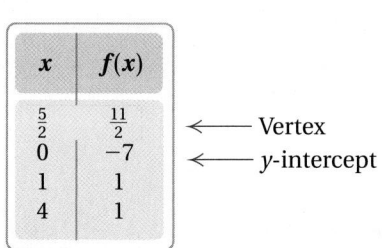

x	$f(x)$	
$\frac{5}{2}$	$\frac{11}{2}$	← Vertex
0	-7	← *y*-intercept
1	1	
4	1	

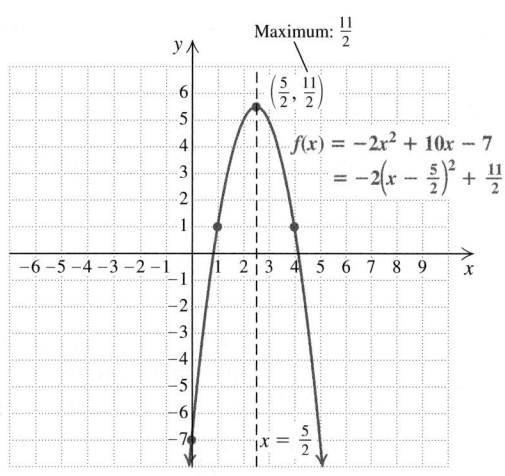

TRY EXERCISE ▶ 39

The method used in Examples 1–3 can be generalized to find a formula for locating the vertex. We complete the square as follows:

$$f(x) = ax^2 + bx + c$$

$$= a\left(x^2 + \frac{b}{a}x\right) + c. \qquad \text{Factoring } a \text{ out of the first two terms.}$$
$$\text{Check by multiplying.}$$

Half of the *x*-coefficient, $\frac{b}{a}$, is $\frac{b}{2a}$. We square it to get $\frac{b^2}{4a^2}$ and add $\frac{b^2}{4a^2} - \frac{b^2}{4a^2}$ inside the parentheses. Then we distribute the *a* and regroup terms:

$$f(x) = a\left(x^2 + \frac{b}{a}x + \frac{b^2}{4a^2} - \frac{b^2}{4a^2}\right) + c$$

$$= a\left(x^2 + \frac{b}{a}x + \frac{b^2}{4a^2}\right) + a\left(-\frac{b^2}{4a^2}\right) + c \qquad \begin{array}{l}\text{Using the}\\ \text{distributive law}\end{array}$$

$$= a\left(x + \frac{b}{2a}\right)^2 + \frac{-b^2}{4a} + \frac{4ac}{4a} \qquad \begin{array}{l}\text{Factoring and finding a}\\ \text{common denominator}\end{array}$$

$$= a\left[x - \left(-\frac{b}{2a}\right)\right]^2 + \frac{4ac - b^2}{4a}.$$

Thus we have the following.

The Vertex of a Parabola

The vertex of the parabola given by $f(x) = ax^2 + bx + c$ is

$$\left(-\frac{b}{2a}, f\left(-\frac{b}{2a}\right)\right), \quad \text{or} \quad \left(-\frac{b}{2a}, \frac{4ac - b^2}{4a}\right).$$

• The *x*-coordinate of the vertex is $-b/(2a)$.

• The axis of symmetry is $x = -b/(2a)$.

• The second coordinate of the vertex is most commonly found by computing $f\left(-\frac{b}{2a}\right)$.

Let's reexamine Example 3 to see how we could have found the vertex directly. From the formula above,

$$\text{the } x\text{-coordinate of the vertex is } -\frac{b}{2a} = -\frac{10}{2(-2)} = \frac{5}{2}.$$

Substituting $\frac{5}{2}$ into $f(x) = -2x^2 + 10x - 7$, we find the second coordinate of the vertex:

$$\begin{aligned} f\left(\tfrac{5}{2}\right) &= -2\left(\tfrac{5}{2}\right)^2 + 10\left(\tfrac{5}{2}\right) - 7 \\ &= -2\left(\tfrac{25}{4}\right) + 25 - 7 \\ &= -\tfrac{25}{2} + 18 \\ &= -\tfrac{25}{2} + \tfrac{36}{2} = \tfrac{11}{2}. \end{aligned}$$

The vertex is $\left(\frac{5}{2}, \frac{11}{2}\right)$. The axis of symmetry is $x = \frac{5}{2}$.

We have actually developed two methods for finding the vertex. One is by completing the square and the other is by using a formula. You should check to see if your instructor prefers one method over the other or wants you to use both.

Finding Intercepts

All quadratic functions have a y-intercept and 0, 1, or 2 x-intercepts. For $f(x) = ax^2 + bx + c$, the y-intercept is $(0, f(0))$, or $(0, c)$. To find x-intercepts, if any exist, we look for points where $y = 0$ or $f(x) = 0$. Thus, for $f(x) = ax^2 + bx + c$, the x-intercepts occur at those x-values for which

$$ax^2 + bx + c = 0.$$

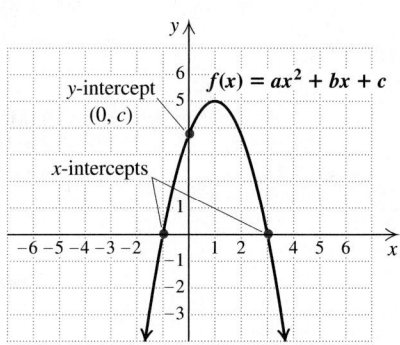

EXAMPLE   4   Find any x-intercepts and the y-intercept of the graph of $f(x) = x^2 - 2x - 2$.

SOLUTION   The y-intercept is simply $(0, f(0))$, or $(0, -2)$. To find any x-intercepts, we solve

$$0 = x^2 - 2x - 2.$$

We are unable to factor $x^2 - 2x - 2$, so we use the quadratic formula and get $x = 1 \pm \sqrt{3}$. Thus the x-intercepts are $(1 - \sqrt{3}, 0)$ and $(1 + \sqrt{3}, 0)$.

If graphing, we would approximate, to get $(-0.7, 0)$ and $(2.7, 0)$.

TRY EXERCISE ▶ 43

If the solutions of $f(x) = 0$ are imaginary, the graph of f has no x-intercepts.

Visualizing for Success

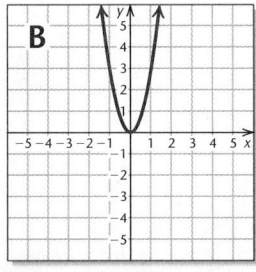

A

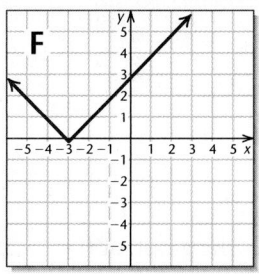

F

Match each function with its graph.

1. $f(x) = 3x^2$

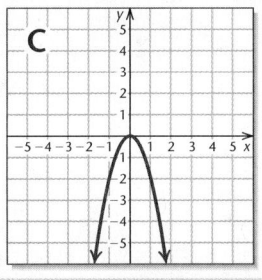

B

2. $f(x) = x^2 - 4$

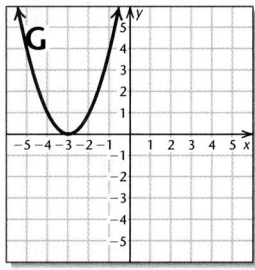

G

3. $f(x) = (x - 4)^2$

4. $f(x) = x - 4$

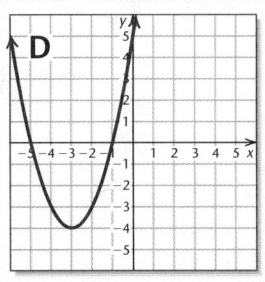

C

5. $f(x) = -2x^2$

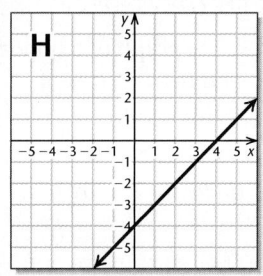

H

6. $f(x) = x + 3$

7. $f(x) = |x + 3|$

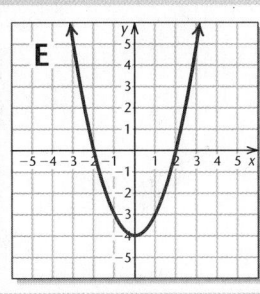

D

8. $f(x) = (x + 3)^2$

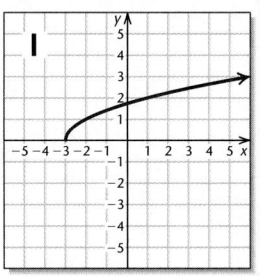

I

9. $f(x) = \sqrt{x + 3}$

10. $f(x) = (x + 3)^2 - 4$

E

Answers on page A-48.

An additional, animated version of this activity appears in MyMathLab. To use MyMathLab, you need a course ID and a student access code. Contact your instructor for more information.

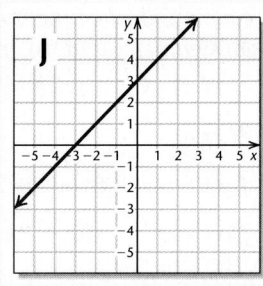

J

✎ *Concept Reinforcement*   *Classify each statement as either true or false.*

1. The graph of $f(x) = 3x^2 - x + 6$ opens upward.

2. The function given by $g(x) = -x^2 + 3x + 1$ has a minimum value.

3. The graph of $f(x) = -2(x - 3)^2 + 7$ has its vertex at $(3, 7)$.

4. The graph of $g(x) = 4(x + 6)^2 - 2$ has its vertex at $(-6, -2)$.

5. The graph of $g(x) = \frac{1}{2}\left(x - \frac{3}{2}\right)^2 + \frac{1}{4}$ has $x = \frac{1}{4}$ as its axis of symmetry.

6. The function given by $f(x) = (x - 2)^2 - 5$ has a minimum value of -5.

7. The y-intercept of the graph of $f(x) = 2x^2 - 6x + 7$ is $(7, 0)$.

8. If the graph of a quadratic function f opens upward and has a vertex of $(1, 5)$, then the graph has no x-intercepts.

Complete the square to write each function in the form $f(x) = a(x - h)^2 + k$.

9. $f(x) = x^2 - 8x + 2$

10. $f(x) = x^2 - 6x - 1$

11. $f(x) = x^2 + 3x - 5$

12. $f(x) = x^2 + 5x + 3$

13. $f(x) = 3x^2 + 6x - 2$

14. $f(x) = 2x^2 - 20x - 3$

15. $f(x) = -x^2 - 4x - 7$

16. $f(x) = -2x^2 - 8x + 4$

17. $f(x) = 2x^2 - 5x + 10$

18. $f(x) = 3x^2 + 7x - 3$

*For each quadratic function, **(a)** find the vertex and the axis of symmetry and **(b)** graph the function.*

19. $f(x) = x^2 + 4x + 5$

20. $f(x) = x^2 + 2x - 5$

21. $f(x) = x^2 + 8x + 20$

22. $f(x) = x^2 - 10x + 21$

23. $h(x) = 2x^2 - 16x + 25$

24. $h(x) = 2x^2 + 16x + 23$

25. $f(x) = -x^2 + 2x + 5$

26. $f(x) = -x^2 - 2x + 7$

27. $g(x) = x^2 + 3x - 10$

28. $g(x) = x^2 + 5x + 4$

29. $h(x) = x^2 + 7x$

30. $h(x) = x^2 - 5x$

31. $f(x) = -2x^2 - 4x - 6$

32. $f(x) = -3x^2 + 6x + 2$

*For each quadratic function, **(a)** find the vertex, the axis of symmetry, and the maximum or minimum function value and **(b)** graph the function.*

33. $g(x) = x^2 - 6x + 13$

34. $g(x) = x^2 - 4x + 5$

35. $g(x) = 2x^2 - 8x + 3$

36. $g(x) = 2x^2 + 5x - 1$

37. $f(x) = 3x^2 - 24x + 50$

38. $f(x) = 4x^2 + 16x + 13$

39. $f(x) = -3x^2 + 5x - 2$

40. $f(x) = -3x^2 - 7x + 2$

41. $h(x) = \frac{1}{2}x^2 + 4x + \frac{19}{3}$

42. $h(x) = \frac{1}{2}x^2 - 3x + 2$

Find any x-intercepts and the y-intercept. If no x-intercepts exist, state this.

43. $f(x) = x^2 - 6x + 3$

44. $f(x) = x^2 + 5x + 4$

45. $g(x) = -x^2 + 2x + 3$

46. $g(x) = x^2 - 6x + 9$

Aha! 47. $f(x) = x^2 - 9x$

48. $f(x) = x^2 - 7x$

49. $h(x) = -x^2 + 4x - 4$

50. $h(x) = -2x^2 - 20x - 50$

51. $g(x) = x^2 + x - 5$

52. $g(x) = 2x^2 + 3x - 1$

53. $f(x) = 2x^2 - 4x + 6$

54. $f(x) = x^2 - x + 2$

55. The graph of a quadratic function f opens downward and has no x-intercepts. In what quadrant(s) must the vertex lie? Explain your reasoning.

56. Is it possible for the graph of a quadratic function to have only one x-intercept if the vertex is off the x-axis? Why or why not?

Skill Review

To prepare for Section 11.8, review solving systems of three equations in three unknowns (Section 8.4).

Solve. [8.4]

57.
$x + y + z = 3,$
$x - y + z = 1,$
$-x - y + z = -1$

58.
$x - y + z = -6,$
$2x + y + z = 2,$
$3x + y + z = 0$

59.
$z = 8,$
$x + y + z = 23,$
$2x + y - z = 17$

60.
$z = -5,$
$2x - y + 3z = -27,$
$x + 2y + 7z = -26$

61.
$1.5 = c,$
$52.5 = 25a + 5b + c,$
$7.5 = 4a + 2b + c$

62.
$\frac{1}{2} = c,$
$5 = 9a + 6b + 2c,$
$29 = 81a + 9b + c$

Synthesis

63. If the graphs of two quadratic functions have the same x-intercepts, will they also have the same vertex? Why or why not?

64. Suppose that the graph of $f(x) = ax^2 + bx + c$ has $(x_1, 0)$ and $(x_2, 0)$ as x-intercepts. Explain why the graph of $g(x) = -ax^2 - bx - c$ will also have $(x_1, 0)$ and $(x_2, 0)$ as x-intercepts.

For each quadratic function, find **(a)** *the maximum or minimum value and* **(b)** *any x-intercepts and the y-intercept.*

65. $f(x) = 2.31x^2 - 3.135x - 5.89$

66. $f(x) = -18.8x^2 + 7.92x + 6.18$

67. Graph the function

$$f(x) = x^2 - x - 6.$$

Then use the graph to approximate solutions to each of the following equations.

a) $x^2 - x - 6 = 2$

b) $x^2 - x - 6 = -3$

68. Graph the function

$$f(x) = \frac{x^2}{2} + x - \frac{3}{2}.$$

Then use the graph to approximate solutions to each of the following equations.

a) $\frac{x^2}{2} + x - \frac{3}{2} = 0$

b) $\frac{x^2}{2} + x - \frac{3}{2} = 1$

c) $\frac{x^2}{2} + x - \frac{3}{2} = 2$

Find an equivalent equation of the type
$$f(x) = a(x - h)^2 + k.$$

69. $f(x) = mx^2 - nx + p$

70. $f(x) = 3x^2 + mx + m^2$

71. A quadratic function has $(-1, 0)$ as one of its intercepts and $(3, -5)$ as its vertex. Find an equation for the function.

72. A quadratic function has $(4, 0)$ as one of its intercepts and $(-1, 7)$ as its vertex. Find an equation for the function.

Graph.

73. $f(x) = |x^2 - 1|$

74. $f(x) = |x^2 - 3x - 4|$

75. $f(x) = |2(x - 3)^2 - 5|$

76. Use a graphing calculator to check your answers to Exercises 25, 41, 53, 65, and 67.

11.8 Problem Solving and Quadratic Functions

Maximum and Minimum Problems ▪ Fitting Quadratic Functions to Data

Let's look now at some of the many situations in which quadratic functions are used for problem solving.

Maximum and Minimum Problems

We have seen that for any quadratic function f, the value of $f(x)$ at the vertex is either a maximum or a minimum. Thus problems in which a quantity must be maximized or minimized can be solved by finding the coordinates of a vertex, assuming the problem can be modeled with a quadratic function.

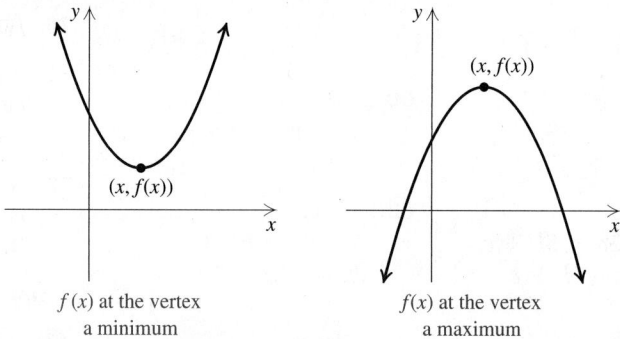

$f(x)$ at the vertex a minimum

$f(x)$ at the vertex a maximum

EXAMPLE 1

Museum attendance. After the admission fee was dropped, attendance at the Indianapolis Museum of Art began to rise after several years of decline. The number of museum admissions, in thousands, t years after 2000 can be approximated by $m(t)$, where $m(t) = 32t^2 - 320t + 975$. In what year was the museum attendance the lowest, and how many people went to the museum that year?

Source: Based on information in the *Indianapolis Star*, 9/9/07

SOLUTION

1., 2. Familiarize and **Translate.** We are given the function for museum attendance. Note that it is a quadratic function of the number of years since 2000. The coefficient of the squared term is positive, so the graph opens upward and there is a minimum value. The calculator-generated graph at left confirms this.

3. Carry out. We can either complete the square or use the formula for the vertex. Completing the square, we have

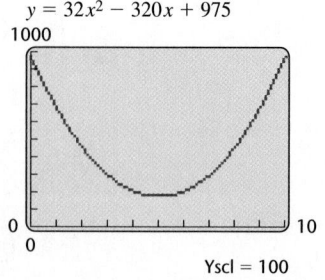

$y = 32x^2 - 320x + 975$

Yscl = 100

A visualization for Example 1

$$m(t) = 32t^2 - 320t + 975$$
$$= 32(t^2 - 10t) + 975$$
$$= 32(t^2 - 10t + 25 - 25) + 975 \qquad \text{Completing the square}$$
$$= 32(t^2 - 10t + 25) - (32)(25) + 975$$
$$= 32(t - 5)^2 + 175. \qquad \text{Factoring and simplifying}$$

There is a minimum value of 175 when $t = 5$.

4. Check. Using the formula, we have $-b/(2a) = -(-320)/64 = 5$. Then

$$m(5) = 32(5)^2 - 320(5) + 975 = 175.$$

Both approaches give the same minimum, and that minimum is also confirmed by the graph. The answer checks.

5. State. The minimum attendance was 175,000. It occurred 5 yr after 2000, or in 2005.

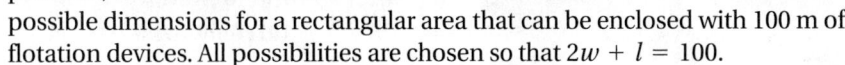 **TRY EXERCISE** 7

EXAMPLE 2

Swimming area. A lifeguard has 100 m of roped-together flotation devices with which to cordon off a rectangular swimming area at North Beach. If the shoreline forms one side of the rectangle, what dimensions will maximize the size of the area for swimming?

SOLUTION

1. Familiarize. We make a drawing and label it, letting $w =$ the width of the rectangle, in meters, and $l =$ the length of the rectangle, in meters.

Recall that Area $= l \cdot w$ and Perimeter $= 2w + 2l$. Since the beach forms one length of the rectangle, the flotation devices comprise three sides. Thus

$$2w + l = 100.$$

To get a better feel for the problem, we can look at some possible dimensions for a rectangular area that can be enclosed with 100 m of flotation devices. All possibilities are chosen so that $2w + l = 100$.

l	w	Rope Length	Area
40 m	30 m	100 m	1200 m²
30 m	35 m	100 m	1050 m²
20 m	40 m	100 m	800 m²
⋮	⋮	⋮	⋮

What choice of l and w will maximize A?

2. Translate. We have two equations: One guarantees that all 100 m of flotation devices are used; the other expresses area in terms of length and width.

$$2w + l = 100,$$
$$A = l \cdot w$$

3. Carry out. We need to express A as a function of l or w but not both. To do so, we solve for l in the first equation to obtain $l = 100 - 2w$. Substituting for l in the second equation, we get a quadratic function:

$$A = (100 - 2w)w \qquad \text{Substituting for } l$$
$$= 100w - 2w^2. \qquad \text{This represents a parabola opening downward, so a maximum exists.}$$

Factoring and completing the square, we get

$$A = -2(w^2 - 50w + 625 - 625) \qquad \text{We could also use the vertex formula.}$$

$$= -2(w - 25)^2 + 1250.$$

There is a maximum value of 1250 when $w = 25$.

4. Check. If $w = 25$ m, then $l = 100 - 2 \cdot 25 = 50$ m. These dimensions give an area of 1250 m². Note that 1250 m² is greater than any of the values for A found in the *Familiarize* step. To be more certain, we could check values other than those used in that step. For example, if $w = 26$ m, then $l = 48$ m, and $A = 26 \cdot 48 = 1248$ m². Since 1250 m² is greater than 1248 m², it appears that we have a maximum.

5. State. The largest rectangular area for swimming that can be enclosed is 25 m by 50 m.

> **TRY EXERCISE** ▶ **11**

Fitting Quadratic Functions to Data

Whenever a certain quadratic function fits a situation, that function can be determined if three inputs and their outputs are known. Each of the given ordered pairs is called a *data point*.

EXAMPLE **3**

Music CDs. As more listeners download their music purchases, sales of compact discs are decreasing. According to Nielsen SoundScan, sales of music CDs increased from 500 million in 1997 to 700 million in 2001 and then decreased to 450 million in 2007. As the graph suggests, sales of music CDs can be modeled by a quadratic function.

Years After 1997	Number of Music CDs Sold in the United States (in millions)
0	500
4	700
10	450

Source: Nielsen SoundScan

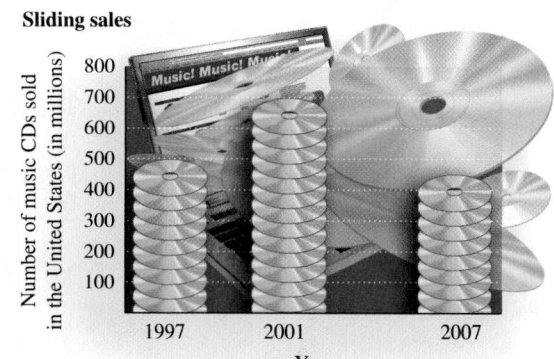

Sliding sales

(Graph: Number of music CDs sold in the United States (in millions) vs. Year, with axis values 100–800 and years 1997, 2001, 2007)

a) Let t represent the number of years since 1997 and $S(t)$ the total number of CDs sold, in millions. Use the data points (0, 500), (4, 700), and (10, 450) to find a quadratic function that fits the data.

b) Use the function from part (a) to estimate the sales of music CDs in 2009.

SOLUTION

a) We are looking for a function of the form $S(t) = at^2 + bt + c$ given that $S(0) = 500$, $S(4) = 700$, and $S(10) = 450$. Thus,

$$500 = a \cdot 0^2 + b \cdot 0 + c, \qquad \text{Using the data point (0, 500)}$$
$$700 = a \cdot 4^2 + b \cdot 4 + c, \qquad \text{Using the data point (4, 700)}$$
$$450 = a \cdot 10^2 + b \cdot 10 + c. \qquad \text{Using the data point (10, 450)}$$

After simplifying, we see that we need to solve the system

$$500 = c, \qquad\qquad\qquad \textbf{(1)}$$
$$700 = 16a + 4b + c, \qquad \textbf{(2)}$$
$$450 = 100a + 10b + c. \quad \textbf{(3)}$$

STUDENT NOTES ───────

Try to keep the "big picture" in mind on problems like Example 3. Solving a system of three equations is but one part of the solution.

To use a graphing calculator to fit a quadratic function to the data in Example 3, we first select EDIT in the **STAT** menu and enter the given data.

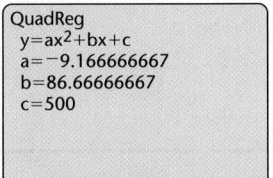

L1	L2	L3	2
0	500	------	
4	700		
10	450		

L2(4) =

To fit a quadratic function to the data, we press **STAT** **▷** **5** **VARS** **▷** **1** **1** **ENTER**. The first three keystrokes select QuadReg from the STAT CALC menu. The keystrokes **VARS** **▷** **1** **1** copy the regression equation to the equation-editor screen as y_1.

QuadReg
y=ax²+bx+c
a=⁻9.166666667
b=86.66666667
c=500

We see that the regression equation is $y = -9.166666667x^2 + 86.66666667x + 500$. We press **Y=** **△** **ENTER** to turn on the PLOT feature and **ZOOM** **9** to see the regression equation graphed with the data points.

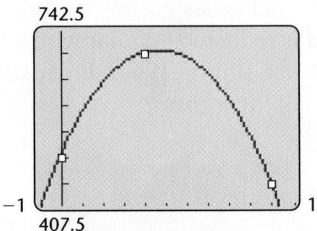

To check Example 3(b), we set Indpnt to Ask in the Table Setup and enter $X = 12$ in the table. A Y_1-value of 220 confirms our answer.

1. Use the above approach to estimate the sales of music CDs in 2005.

We know from equation (1) that $c = 500$. Substituting that value into equations (2) and (3), we have

$$700 = 16a + 4b + 500,$$
$$450 = 100a + 10b + 500.$$

Subtracting 500 from both sides of each equation, we have

$$200 = 16a + 4b, \qquad \textbf{(4)}$$
$$-50 = 100a + 10b. \qquad \textbf{(5)}$$

To solve, we multiply equation (4) by 5 and equation (5) by -2. We then add to eliminate b:

$$1000 = 80a + 20b$$
$$\underline{100 = -200a - 20b}$$
$$1100 = -120a$$
$$-\frac{1100}{120} = a, \quad \text{or} \quad a = -\frac{55}{6}. \qquad \text{Simplifying}$$

Next, we solve for b, using equation (5) above:

$$-50 = 100\left(-\frac{55}{6}\right) + 10b$$
$$-50 = -\frac{2750}{3} + 10b$$
$$\frac{2600}{3} = 10b \qquad \text{Adding } \tfrac{2750}{3} \text{ to both sides and simplifying}$$
$$\frac{2600}{30} = b, \quad \text{or} \quad b = \frac{260}{3}. \qquad \text{Dividing both sides by 10 and simplifying}$$

We can now write $S(t) = at^2 + bt + c$ as

$$S(t) = -\frac{55}{6}t^2 + \frac{260}{3}t + 500.$$

b) To find the sales of CDs in 2009, we evaluate the function. Note that 2009 is 12 yr after 1997. Thus,

$$S(12) = -\frac{55}{6} \cdot 12^2 + \frac{260}{3} \cdot 12 + 500$$
$$= 220.$$

In 2009, an estimated 220 million music CDs will be sold.

TRY EXERCISE ▶ 35

11.8 EXERCISE SET

For Extra Help
MyMathLab  MathXL
PRACTICE   WATCH   DOWNLOAD

↪ *Concept Reinforcement*   *In each of Exercises 1–6, match the description with the graph that displays that characteristic.*

1. ____ A minimum value of $f(x)$ exists.

2. ____ A maximum value of $f(x)$ exists.

3. ____ No maximum or minimum value of $f(x)$ exists.

4. ____ The data points appear to suggest a linear model for g.

5. ____ The data points appear to suggest that g is a quadratic function with a maximum.

6. ____ The data points appear to suggest that g is a quadratic function with a minimum.

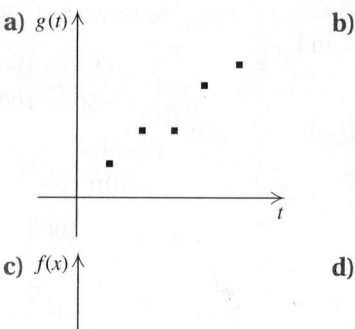

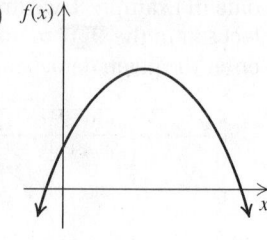

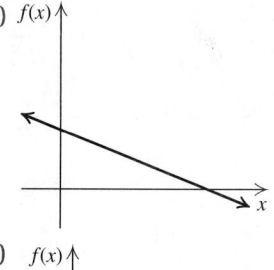

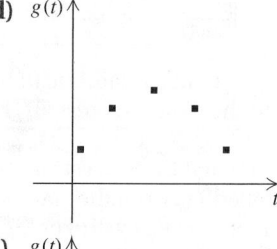

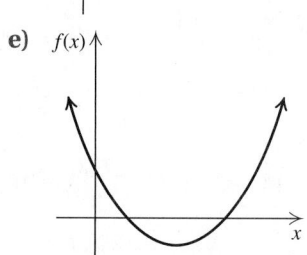

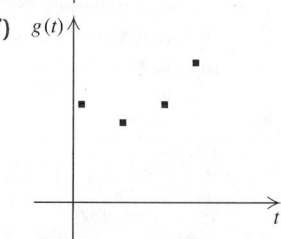

Solve.

7. *Newborn calves.*   The number of pounds of milk per day recommended for a calf that is x weeks old can be approximated by $p(x)$, where

$$p(x) = -0.2x^2 + 1.3x + 6.2.$$

When is a calf's milk consumption greatest and how much milk does it consume at that time?
Source: C. Chaloux, University of Vermont, 1998

8. *Stock prices.*   The value of a share of I. J. Solar can be represented by $V(x) = x^2 - 6x + 13$, where x is the number of months after January 2009. What is the lowest value $V(x)$ will reach, and when did that occur?

9. *Minimizing cost.*   Sweet Harmony Crafts has determined that when x hundred dulcimers are built, the average cost per dulcimer can be estimated by

$$C(x) = 0.1x^2 - 0.7x + 2.425,$$

where $C(x)$ is in hundreds of dollars. What is the minimum average cost per dulcimer and how many dulcimers should be built in order to achieve that minimum?

10. *Maximizing profit.*   Recall that total profit P is the difference between total revenue R and total cost C. Given $R(x) = 1000x - x^2$ and $C(x) = 3000 + 20x$, find the total profit, the maximum value of the total profit, and the value of x at which it occurs.

11. *Architecture.*   An architect is designing an atrium for a hotel. The atrium is to be rectangular with a perimeter of 720 ft of brass piping. What dimensions will maximize the area of the atrium?

12. *Furniture design.* A furniture builder is designing a rectangular end table with a perimeter of 128 in. What dimensions will yield the maximum area?

13. *Patio design.* A stone mason has enough stones to enclose a rectangular patio with 60 ft of perimeter, assuming that the attached house forms one side of the rectangle. What is the maximum area that the mason can enclose? What should the dimensions of the patio be in order to yield this area?

14. *Garden design.* Ginger is fencing in a rectangular garden, using the side of her house as one side of the rectangle. What is the maximum area that she can enclose with 40 ft of fence? What should the dimensions of the garden be in order to yield this area?

15. *Molding plastics.* Economite Plastics plans to produce a one-compartment vertical file by bending the long side of an 8-in. by 14-in. sheet of plastic along two lines to form a U shape. How tall should the file be in order to maximize the volume that the file can hold?

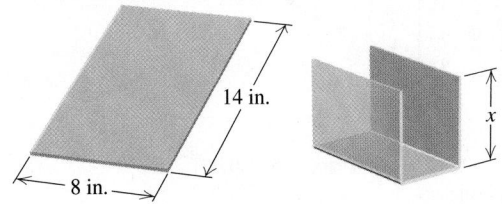

16. *Composting.* A rectangular compost container is to be formed in a corner of a fenced yard, with 8 ft of chicken wire completing the other two sides of the rectangle. If the chicken wire is 3 ft high, what dimensions of the base will maximize the container's volume?

17. What is the maximum product of two numbers that add to 18? What numbers yield this product?

18. What is the maximum product of two numbers that add to 26? What numbers yield this product?

19. What is the minimum product of two numbers that differ by 8? What are the numbers?

20. What is the minimum product of two numbers that differ by 7? What are the numbers?

Aha! **21.** What is the maximum product of two numbers that add to -10? What numbers yield this product?

22. What is the maximum product of two numbers that add to -12? What numbers yield this product?

Choosing models. *For the scatterplots and graphs in Exercises 23–34, determine which, if any, of the following functions might be used as a model for the data: Linear, with $f(x) = mx + b$; quadratic, with $f(x) = ax^2 + bx + c$, $a > 0$; quadratic, with $f(x) = ax^2 + bx + c$, $a < 0$; neither quadratic nor linear.*

23. **Sonoma Sunshine**

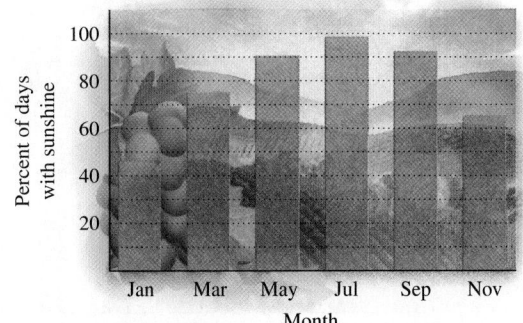

Source: www.city-data.com

24. **Sonoma Precipitation**

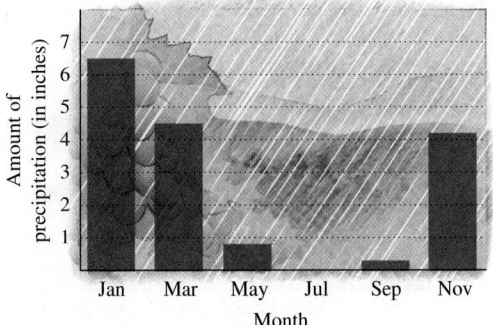

Source: www.city-data.com

25. Safe sight distance to the left

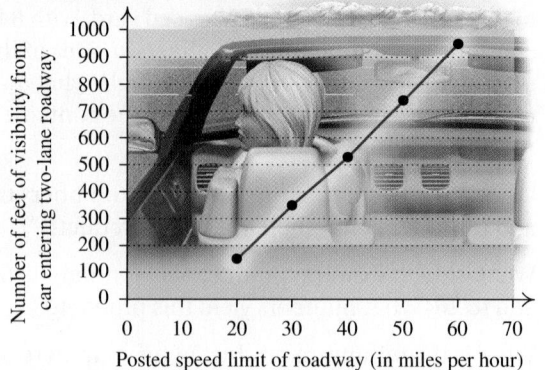

Source: Institute of Traffic Engineers

26. Safe sight distance to the right

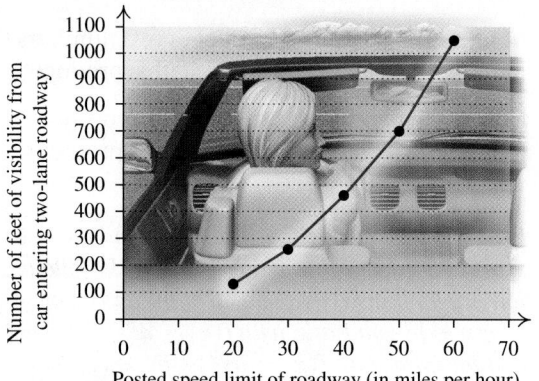

Source: Institute of Traffic Engineers

27. Winter Olympic volunteers

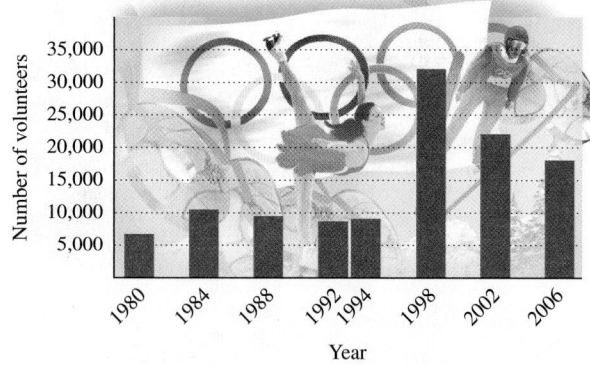

28. U.S. senior population

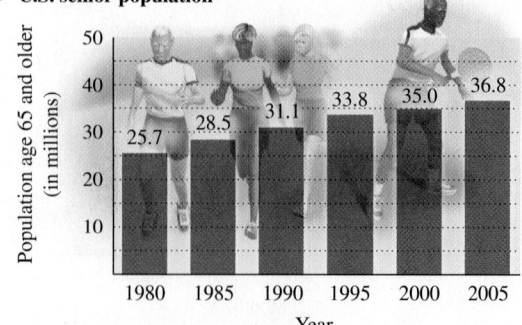

Source: U.S. Bureau of Labor Statistics

29. Changing work force

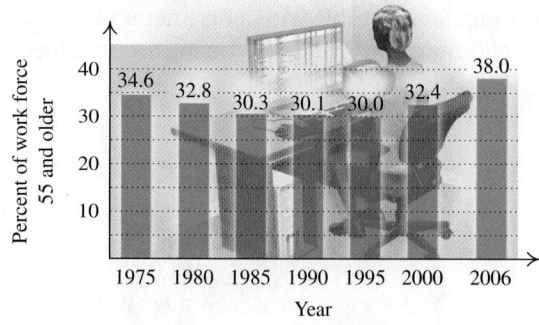

Source: U.S. Department of Labor, Bureau of Labor Statistics

30. Airline bumping rate

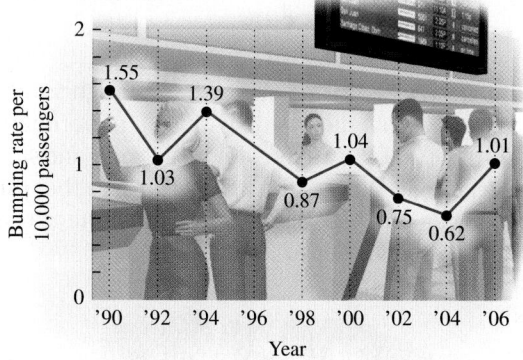

Source: U.S. Department of Transportation

31. Hybrid vehicles

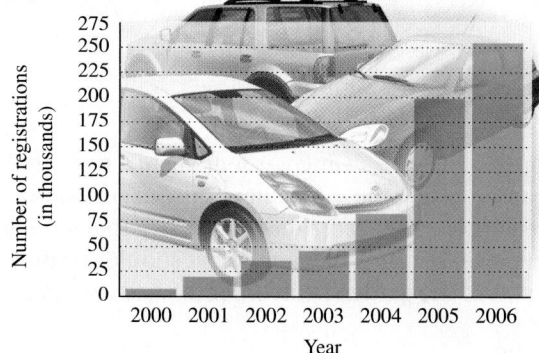

Source: R.L. Polk & Co.

32.

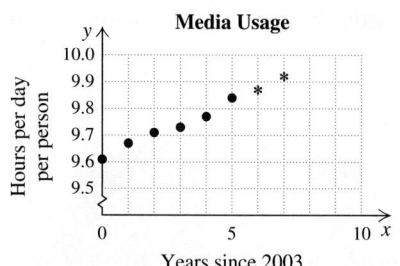

* Projected

Source: Statistical Abstract of the United States

33. **Employee contribution to health insurance premium**

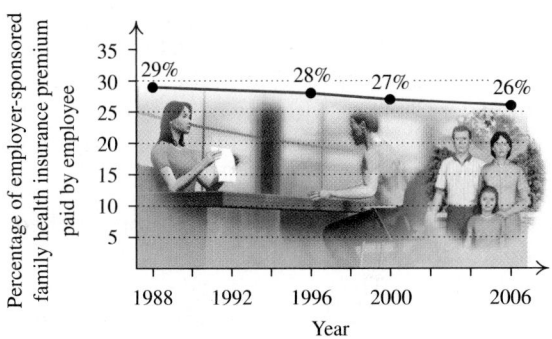

Source: Based on data from Kaiser

34. **Average number of live births per 1000 women, 2005**

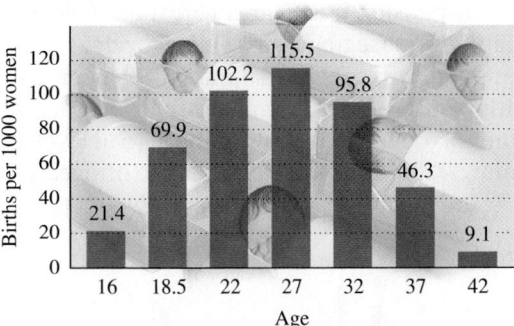

Source: U.S. Centers for Disease Control

Find a quadratic function that fits the set of data points.

35. $(1, 4), (-1, -2), (2, 13)$

36. $(1, 4), (-1, 6), (-2, 16)$

37. $(2, 0), (4, 3), (12, -5)$

38. $(-3, -30), (3, 0), (6, 6)$

39. **a)** Find a quadratic function that fits the following data.

Travel Speed (in kilometers per hour)	Number of Nighttime Accidents (for every 200 million kilometers driven)
60	400
80	250
100	250

b) Use the function to estimate the number of nighttime accidents that occur at 50 km/h.

40. **a)** Find a quadratic function that fits the following data.

Travel Speed (in kilometers per hour)	Number of Daytime Accidents (for every 200 million kilometers driven)
60	100
80	130
100	200

b) Use the function to estimate the number of daytime accidents that occur at 50 km/h.

41. *Archery.* The Olympic flame tower at the 1992 Summer Olympics was lit at a height of about 27 m by a flaming arrow that was launched about 63 m from the base of the tower. If the arrow landed about 63 m beyond the tower, find a quadratic function that expresses the height h of the arrow as a function of the distance d that it traveled horizontally.

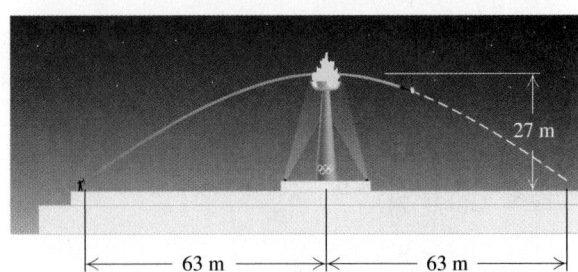

42. *Outsourcing.* The revenue, in billions of dollars, from India's outsourcing industry is shown in the following table.

Year	Outsourcing Revenue (in billions of dollars)
2001	$12
2004	21
2007	48

Source: Nasscom

a) Let t represent the number of years since 2000 and $r(t)$ the revenue, in billions of dollars. Find a quadratic function that fits the data.

b) Use the function to estimate India's outsourcing revenue in 2012.

43. Does every nonlinear function have a minimum or a maximum value? Why or why not?

44. Explain how the leading coefficient of a quadratic function can be used to determine if a maximum or a minimum function value exists.

Skill Review

To prepare for Section 11.9, review solving inequalities and rational expressions and equations (Chapters 6 and 9).

Solve.

45. $2x - 3 > 5$ [9.1]

46. $4 - x \leq 7$ [9.1]

47. $|9 - x| \geq 2$ [9.3]

48. $|4x + 1| < 11$ [9.3]

Subtract. [6.4]

49. $\dfrac{x - 3}{x + 4} - 5$

50. $\dfrac{x}{x - 1} - 1$

Solve. [6.6]

51. $\dfrac{x - 3}{x + 4} = 5$

52. $\dfrac{x}{x - 1} = 1$

53. $\dfrac{x}{(x - 3)(x + 7)} = 0$

54. $\dfrac{(x + 6)(x - 9)}{x + 5} = 0$

Synthesis

The following graphs can be used to compare the baseball statistics of pitcher Roger Clemens with the 31 other pitchers since 1968 who started at least 10 games in at least 15 seasons and pitched at least 3000 innings. Use the graphs to answer questions 55 and 56.

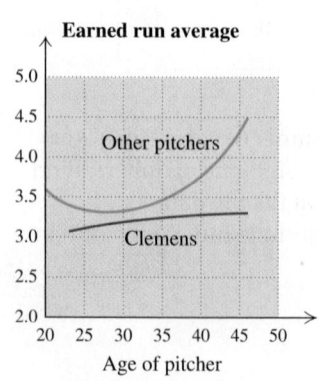

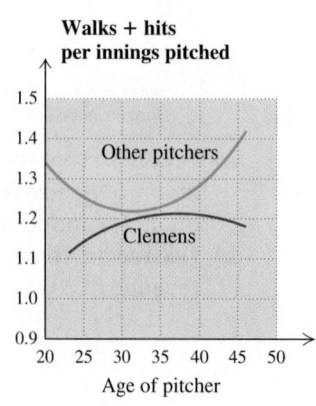

Source: *The New York Times*, February 10, 2008;
Eric Bradlow, Shane Jensen, Justin Wolfers and Adi Wyner

55. The earned run average describes how many runs a pitcher has allowed per game. The lower the earned run average, the better a pitcher. Compare, in terms of maximums or minimums, the earned run average of Roger Clemens with that of other pitchers. Is there any reason to suspect that the aging process was unusual for Clemens? Explain.

56. The statistic "Walks + hits per innings pitched" is related to how often a pitcher allows a batter to reach a base. The lower this statistic, the better. Compare, in terms of maximums or minimums, the "walks + hits" statistic of Roger Clemens with that of other pitchers.

57. *Bridge design.* The cables supporting a straight-line suspension bridge are nearly parabolic in shape. Suppose that a suspension bridge is being designed with concrete supports 160 ft apart and with vertical cables 30 ft above road level at the midpoint of the bridge and 80 ft above road level at a point 50 ft from the midpoint of the bridge. How long are the longest vertical cables?

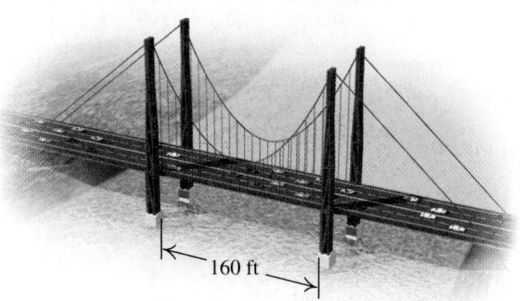

160 ft

58. *Trajectory of a launched object.* The height above the ground of a launched object is a quadratic function of the time that it is in the air. Suppose that a flare is launched from a cliff 64 ft above sea level. If 3 sec after being launched the flare is again level with the cliff, and if 2 sec after that it lands in the sea, what is the maximum height that the flare will reach?

59. *Cover charges.* When the owner of Sweet Sounds charges a $10 cover charge, an average of 80 people will attend a show. For each 25¢ increase in admission price, the average number attending decreases by 1. What should the owner charge in order to make the most money?

60. *Crop yield.* An orange grower finds that she gets an average yield of 40 bushels (bu) per tree when she plants 20 trees on an acre of ground. Each time she adds one tree per acre, the yield per tree decreases by 1 bu, due to congestion. How many trees per acre should she plant for maximum yield?

61. *Norman window.*   A *Norman window* is a rectangle with a semicircle on top. Big Sky Windows is designing a Norman window that will require 24 ft of trim. What dimensions will allow the maximum amount of light to enter a house?

62. *Minimizing area.*   A 36-in. piece of string is cut into two pieces. One piece is used to form a circle while the other is used to form a square. How should the string be cut so that the sum of the areas is a minimum?

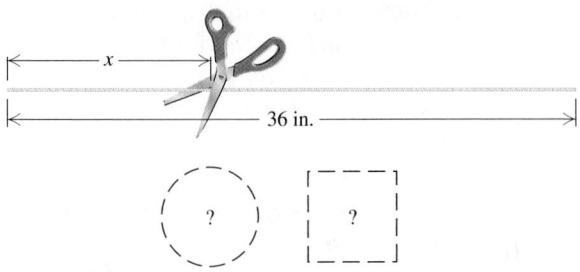

Regression can be used to find the "best"-fitting quadratic function when more than three data points are provided. In Exercises 63 and 64, six data points are given, but the approach used in the Technology Connection on p. 759 still applies.

63. *Hybrid vehicles.*   The number of hybrid vehicles in the United States during several years is shown in the table below.

Year	Number of Vehicles
2001	19,963
2002	35,934
2003	45,943
2004	83,153
2005	199,148
2006	254,545

Source: R. L. Polk & Co.

a) Use regression to find a quadratic function that can be used to estimate the number of hybrid vehicles $h(x)$ in the United States x years after 2000.

b) Use the function found in part (a) to predict the number of hybrid vehicles in the United States in 2010.

64. *Hydrology.*   The drawing below shows the cross section of a river. Typically rivers are deepest in the middle, with the depth decreasing to 0 at the edges. A hydrologist measures the depths D, in feet, of a river at distances x, in feet, from one bank. The results are listed in the table below.

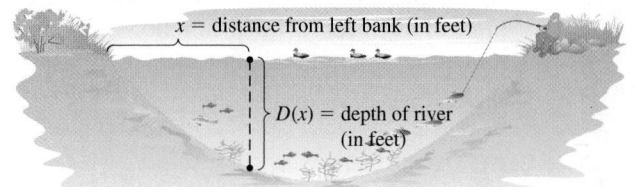

Distance x, from the Left Bank (in feet)	Depth, D, of the River (in feet)
0	0
15	10.2
25	17
50	20
90	7.2
100	0

a) Use regression to find a quadratic function that fits the data.

b) Use the function to estimate the depth of the river 70 ft from the left bank.

COLLABORATIVE CORNER

Parabolic Pizza

Focus: Modeling
Time: 20–30 minutes
Group size: 3
Materials: Graphing calculators are optional.

College Pizza on Chestnut Street in Philadelphia, PA, sells a 10-in.–diameter cheese pizza for $5.00, a 14-in. cheese pizza for $7.50, and an 18-in. cheese pizza for $11.00. Which models better the price of the pizza: a linear function or a quadratic function of the diameter?

Source: Campusfood.com

ACTIVITY

1. As a group, carefully graph the ordered pairs from the data above in the form (diameter, price). Do the data appear to be quadratic or linear?

2. Each group member should choose one of the following to fit a model to the data, where $p(x)$ is the price, in dollars, of an x-inch–diameter pizza. Then, using a different color for each graph, that member should graph the function on the same graph as the ordered pairs.

a) Linear function $p(x) = mx + b$, using the points (10, 5) and (14, 7.5)
b) Linear function $p(x) = mx + b$, using the points (10, 5) and (18, 11)
c) Quadratic function $p(x) \doteq ax^2 + bx + c$, using all three points

3. As a group, determine which function from part (1) appears to be the best fit.

4. One way to tell whether a function is a good fit is to see how well it predicts another known value. College Pizza also sells a 16-in. cheese pizza for $9.00. Each group member should use the function from part (2) to predict the price of a 16-in. cheese pizza. Which function came the closest to predicting the actual value?

5. If a graphing calculator is available, use the LINREG and QUADREG options to fit and graph linear and quadratic functions for the four data points (three pairs from part (1) and one pair from part (4)). Which function appears to give the best fit?

6. Because the area of a circle is given by $A = \pi r^2$, would you expect the price of a cheese pizza to be quadratic or linear?

11.9 Polynomial and Rational Inequalities

Quadratic and Other Polynomial Inequalities • Rational Inequalities

Quadratic and Other Polynomial Inequalities

Inequalities like the following are called *polynomial inequalities*:

$$x^3 - 5x > x^2 + 7, \quad 4x - 3 < 9, \quad 5x^2 - 3x + 2 \geq 0.$$

Second-degree polynomial inequalities in one variable are called *quadratic inequalities*. To solve polynomial inequalities, we often focus attention on where the outputs of a polynomial function are positive and where they are negative.

EXAMPLE 1 Solve: $x^2 + 3x - 10 > 0$.

SOLUTION Consider the "related" function $f(x) = x^2 + 3x - 10$. We are looking for those x-values for which $f(x) > 0$. Graphically, function values are positive when the graph is above the x-axis.

The graph of f opens upward since the leading coefficient is positive. Thus y-values are positive *outside* the interval formed by the x-intercepts. To find the intercepts, we set the polynomial equal to 0 and solve:

$$x^2 + 3x - 10 = 0$$
$$(x + 5)(x - 2) = 0$$
$$x + 5 = 0 \quad or \quad x - 2 = 0$$
$$x = -5 \quad or \quad x = 2. \qquad \text{The } x\text{-intercepts are } (-5, 0) \text{ and } (2, 0).$$

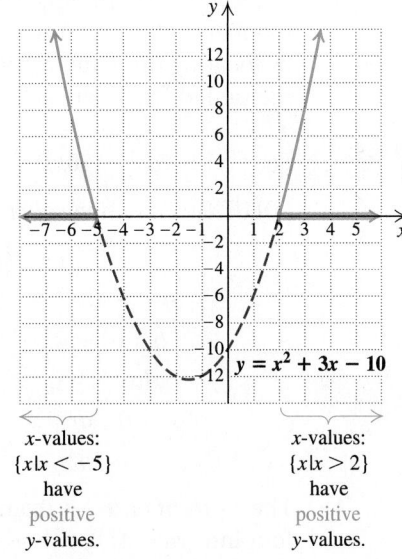

x-values: $\{x | x < -5\}$ have positive y-values.

x-values: $\{x | x > 2\}$ have positive y-values.

Thus the solution set of the inequality is

$$(-\infty, -5) \cup (2, \infty), \text{ or } \{x | x < -5 \text{ or } x > 2\}.$$

TRY EXERCISE 13

Any inequality with 0 on one side can be solved by considering a graph of the related function and finding intercepts as in Example 1. Sometimes the quadratic formula is needed to find the intercepts.

EXAMPLE **2**

Solve: $x^2 - 2x \le 2$.

SOLUTION We first write the quadratic inequality in standard form, with 0 on one side:

$$x^2 - 2x - 2 \le 0. \qquad \text{This is equivalent to the original inequality.}$$

The graph of $f(x) = x^2 - 2x - 2$ is a parabola opening upward. Values of $f(x)$ are negative for x-values between the x-intercepts. We find the x-intercepts by solving $f(x) = 0$:

$$x = \frac{-b \pm \sqrt{b^2 - 4ac}}{2a}$$
$$= \frac{-(-2) \pm \sqrt{(-2)^2 - 4 \cdot 1(-2)}}{2 \cdot 1}$$
$$= \frac{2 \pm \sqrt{12}}{2} = \frac{2}{2} \pm \frac{2\sqrt{3}}{2} = 1 \pm \sqrt{3}.$$

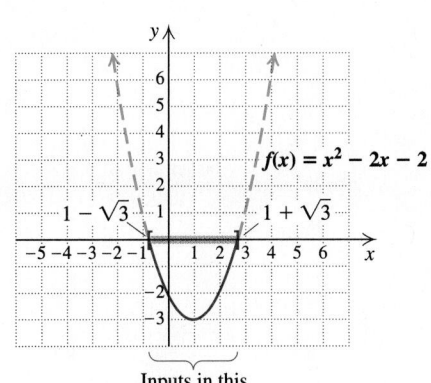

Inputs in this interval have negative or 0 outputs.

At the *x*-intercepts, $1 - \sqrt{3}$ and $1 + \sqrt{3}$, the value of $f(x)$ is 0. Since the inequality symbol is $\leq$, the solution set will include all values of *x* for which $f(x)$ is negative *or* $f(x)$ is 0. Thus the solution set of the inequality is

$$[1 - \sqrt{3}, 1 + \sqrt{3}], \quad \text{or} \quad \{x | 1 - \sqrt{3} \leq x \leq 1 + \sqrt{3}\}.$$

> **TRY EXERCISE** ▸ 21

In Example 2, it was not essential to draw the graph. The important information came from finding the *x*-intercepts and the sign of $f(x)$ on each side of those intercepts. We now solve a third-degree polynomial inequality, without graphing, by locating the *x*-intercepts, or **zeros**, of *f* and then using *test points* to determine the sign of $f(x)$ over each interval of the *x*-axis.

EXAMPLE 3

For $f(x) = 5x^3 + 10x^2 - 15x$, find all *x*-values for which $f(x) > 0$.

SOLUTION We first solve the related equation:

$$f(x) = 0$$
$$5x^3 + 10x^2 - 15x = 0 \qquad \text{Substituting}$$
$$5x(x^2 + 2x - 3) = 0$$
$$5x(x + 3)(x - 1) = 0$$
$$5x = 0 \quad or \quad x + 3 = 0 \quad or \quad x - 1 = 0$$
$$x = 0 \quad or \qquad x = -3 \quad or \qquad x = 1.$$

The zeros of *f* are -3, 0, and 1. These zeros divide the number line, or *x*-axis, into four intervals: A, B, C, and D.

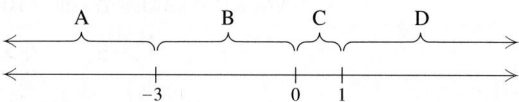

Next, selecting one convenient test value from each interval, we determine the sign of $f(x)$ for that interval. We know that, within each interval, the sign of $f(x)$ cannot change. If it did, there would need to be another zero in that interval. Using the factored form of $f(x)$ eases the computations:

$$f(x) = 5x(x + 3)(x - 1).$$

For interval A,

$$f(-4) = \underbrace{5(-4)}\underbrace{((-4) + 3)}\underbrace{((-4) - 1)}$$

$$\underbrace{\text{Negative} \cdot \text{Negative} \cdot \text{Negative}}_{\text{Negative}}$$

-4 is a convenient value in interval A.

Only the sign is important. The product of three negative numbers is negative, so $f(-4)$ is negative.

For interval B,

$$f(-1) = \underbrace{5(-1)}\underbrace{((-1) + 3)}\underbrace{((-1) - 1)}$$

$$\underbrace{\text{Negative} \cdot \text{Positive} \cdot \text{Negative}}_{\text{Positive}}$$

-1 is a convenient value in interval B.

$f(-1)$ is positive.

STUDENT NOTES ───────

When we are evaluating test values, there is often no need to do lengthy computations since all we need to determine is the sign of the result.

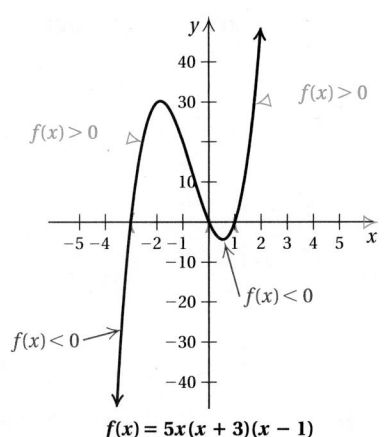

$f(x) > 0$

$f(x) > 0$

$f(x) < 0$

$f(x) < 0$

$f(x) = 5x(x + 3)(x - 1)$

A visualization of Example 3

For interval C,

$$f\left(\tfrac{1}{2}\right) = \underbrace{5 \cdot \tfrac{1}{2}}_{} \cdot \underbrace{\left(\tfrac{1}{2} + 3\right)}_{} \cdot \underbrace{\left(\tfrac{1}{2} - 1\right)}_{}. \qquad \tfrac{1}{2} \text{ is a convenient value in interval C.}$$

$$\underbrace{\text{Positive} \cdot \text{Positive} \cdot \text{Negative}}_{}$$

$$\text{Negative} \qquad f\left(\tfrac{1}{2}\right) \text{ is negative.}$$

For interval D,

$$f(2) = \underbrace{5 \cdot 2}_{} \cdot \underbrace{(2 + 3)}_{} \cdot \underbrace{(2 - 1)}_{}. \qquad 2 \text{ is a convenient value in interval D.}$$

$$\underbrace{\text{Positive} \cdot \text{Positive} \cdot \text{Positive}}_{} \qquad f(2) \text{ is positive.}$$

Recall that we are looking for all x for which $5x^3 + 10x^2 - 15x > 0$. The calculations above indicate that $f(x)$ is positive for any number in intervals B and D. The solution set of the original inequality is

$$(-3, 0) \cup (1, \infty), \quad \text{or} \quad \{x \mid -3 < x < 0 \text{ or } x > 1\}.$$

TRY EXERCISE 29

The calculations in Example 3 were made simpler by using a factored form of the polynomial and by focusing on only the *sign* of $f(x)$. By looking at how many positive or negative factors are multiplied, we are able to determine the sign of the polynomial function.

> **To Solve a Polynomial Inequality Using Factors**
> 1. Add or subtract to get 0 on one side and solve the related polynomial equation by factoring.
> 2. Use the numbers found in step (1) to divide the number line into intervals.
> 3. Using a test value from each interval, determine the sign of the function over each interval. First find the sign of each factor, and then determine the sign of the product of the factors. Remember that the product of an odd number of negative numbers is negative.
> 4. Select the interval(s) for which the inequality is satisfied and write interval notation or set-builder notation for the solution set. Include endpoints of intervals when $\leq$ or $\geq$ is used.

EXAMPLE 4   For $f(x) = 4x^3 - 4x$, find all x-values for which $f(x) \leq 0$.

SOLUTION   We first solve the related equation:

Solve $f(x) = 0$.

$$f(x) = 0$$
$$4x^3 - 4x = 0$$
$$4x(x^2 - 1) = 0$$
$$4x(x + 1)(x - 1) = 0$$
$$4x = 0 \quad or \quad x + 1 = 0 \quad or \quad x - 1 = 0$$
$$x = 0 \quad or \quad x = -1 \quad or \quad x = 1.$$

Divide the number line into intervals.

The function f has zeros at -1, 0, and 1, so we divide the number line into four intervals:

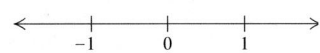

The product $4x(x + 1)(x - 1)$ is positive or negative, depending on the signs of $4x$, $x + 1$, and $x - 1$. This can be determined by making a chart.

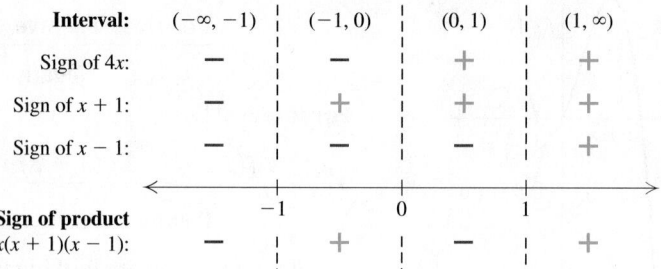

Interval:	$(-\infty, -1)$	$(-1, 0)$	$(0, 1)$	$(1, \infty)$
Sign of $4x$:	$-$	$-$	$+$	$+$
Sign of $x + 1$:	$-$	$+$	$+$	$+$
Sign of $x - 1$:	$-$	$-$	$-$	$+$
Sign of product $4x(x + 1)(x - 1)$:	$-$	$+$	$-$	$+$

Determine the sign of the function over each interval.

A product is negative when it has an odd number of negative factors. Since the $\leq$ sign allows for equality, the endpoints -1, 0, and 1 are solutions. From the chart, we see that the solution set is

Select the interval(s) for which the inequality is satisfied.

$$(-\infty, -1] \cup [0, 1], \quad \text{or} \quad \{x \mid x \leq -1 \: or \: 0 \leq x \leq 1\}.$$

TRY EXERCISE 31

TECHNOLOGY CONNECTION

To solve $2.3x^2 \leq 9.11 - 2.94x$, we write the inequality in the form $2.3x^2 + 2.94x - 9.11 \leq 0$ and graph the function $f(x) = 2.3x^2 + 2.94x - 9.11$.

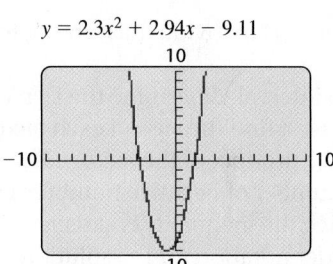

$y = 2.3x^2 + 2.94x - 9.11$

The region in which the graph lies *on or below* the x-axis begins somewhere between -3 and -2, and

continues to somewhere between 1 and 2. Using the ZERO option of CALC and rounding, we find that the endpoints are -2.73 and 1.45. The solution set is approximately $\{x \mid -2.73 \leq x \leq 1.45\}$.

Had the inequality been $2.3x^2 > 9.11 - 2.94x$, we would look for portions of the graph that lie *above* the x-axis. An approximate solution set of this inequality is $\{x \mid x < -2.73 \: or \: x > 1.45\}$.

Use a graphing calculator to solve each inequality. Round the values of the endpoints to the nearest hundredth.

1. $4.32x^2 - 3.54x - 5.34 \leq 0$
2. $7.34x^2 - 16.55x - 3.89 \geq 0$
3. $10.85x^2 + 4.28x + 4.44$
   $> 7.91x^2 + 7.43x + 13.03$
4. $5.79x^3 - 5.68x^2 + 10.68x$
   $> 2.11x^3 + 16.90x - 11.69$

Rational Inequalities

Inequalities involving rational expressions are called **rational inequalities**. Like polynomial inequalities, rational inequalities can be solved using test values. Unlike polynomials, however, rational expressions often have values for which the expression is undefined. These values must be used when dividing the number line into intervals.

EXAMPLE	5

Solve: $\dfrac{x - 3}{x + 4} \geq 2$.

SOLUTION We write the related equation by changing the $\geq$ symbol to $=$:

$$\frac{x - 3}{x + 4} = 2. \quad \text{Note that } x \neq -4.$$

Next, we solve this related equation:

$$(x + 4) \cdot \frac{x - 3}{x + 4} = (x + 4) \cdot 2 \qquad \text{Multiplying both sides by the LCD, } x + 4$$

$$x - 3 = 2x + 8$$

$$-11 = x. \qquad \text{Solving for } x$$

Since -11 is a solution of the related equation, we use -11 when dividing the number line into intervals. Since the rational expression is undefined for $x = -4$, we also use -4:

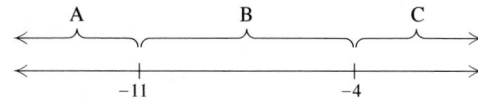

We test a number from each interval to see where the original inequality is satisfied:

$$\frac{x - 3}{x + 4} \geq 2.$$

For Interval A,

$$\text{Test } -15, \quad \frac{-15 - 3}{-15 + 4} = \frac{-18}{-11}$$

$$= \frac{18}{11} \not\geq 2. \qquad \begin{array}{l} -15 \text{ } is \text{ } not \text{ a solution, so interval A is} \\ \text{not part of the solution set.} \end{array}$$

For Interval B,

$$\text{Test } -8, \quad \frac{-8 - 3}{-8 + 4} = \frac{-11}{-4}$$

$$= \frac{11}{4} \geq 2. \qquad \begin{array}{l} -8 \text{ } is \text{ a solution, so interval B is part of} \\ \text{the solution set.} \end{array}$$

For Interval C,

$$\text{Test } 1, \quad \frac{1 - 3}{1 + 4} = \frac{-2}{5}$$

$$= -\frac{2}{5} \not\geq 2. \qquad \begin{array}{l} 1 \text{ } is \text{ } not \text{ a solution, so interval C is not} \\ \text{part of the solution set.} \end{array}$$

The solution set includes interval B. The endpoint -11 is included because the inequality symbol is $\geq$ and -11 is a solution of the related equation. The number -4 is *not* included because $(x - 3)/(x + 4)$ is undefined for $x = -4$. Thus the solution set of the original inequality is

$$[-11, -4), \quad \text{or} \quad \{x | -11 \leq x < -4\}. \qquad \boxed{\text{TRY EXERCISE} \ \blacktriangleright \ 37}$$

ALGEBRAIC–GRAPHICAL CONNECTION

To compare the algebraic solution of Example 5 with a graphical solution, we graph $f(x) = (x - 3)/(x + 4)$ and the line $y = 2$. The solutions of $(x - 3)/(x + 4) \geq 2$ are found by locating all x-values for which $f(x) \geq 2$.

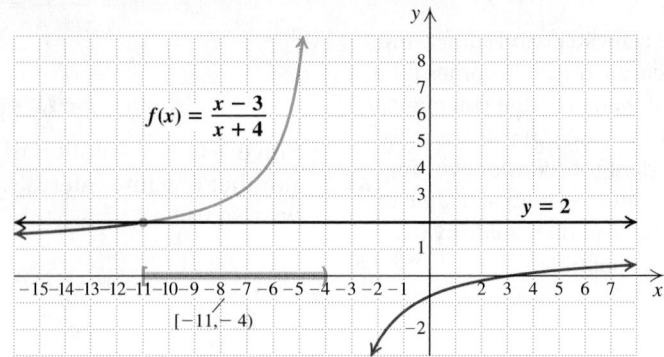

To Solve a Rational Inequality

1. Find any replacements for which any rational expression is undefined.
2. Change the inequality symbol to an equals sign and solve the related equation.
3. Use the numbers found in steps (1) and (2) to divide the number line into intervals.
4. Substitute a test value from each interval into the inequality. If the number is a solution, then the interval to which it belongs is part of the solution set.
5. Select the interval(s) and any endpoints for which the inequality is satisfied and use interval notation or set-builder notation for the solution set. If the inequality symbol is $\leq$ or $\geq$, then the solutions from step (2) are also included in the solution set. All numbers found in step (1) must be excluded from the solution set, even if they are solutions from step (2).

11.9 EXERCISE SET

🖐 *Concept Reinforcement*   *Classify each statement as either true or false.*

1. The solution of $(x - 3)(x + 2) \leq 0$ is $[-2, 3]$.

2. The solution of $(x + 5)(x - 4) \geq 0$ is $[-5, 4]$.

3. The solution of $(x - 1)(x - 6) > 0$ is $\{x \mid x < 1 \text{ or } x > 6\}$.

4. The solution of $(x + 4)(x + 2) < 0$ is $(-4, -2)$.

5. To solve $\dfrac{x + 2}{x - 3} < 0$ using intervals, we divide the number line into the intervals $(-\infty, -2)$ and $(-2, \infty)$.

6. To solve $\dfrac{x - 5}{x + 4} \geq 0$ using intervals, we divide the number line into the intervals $(-\infty, -4)$, $(-4, 5)$, and $(5, \infty)$.

Solve each inequality using the graph provided.

7. $p(x) \le 0$

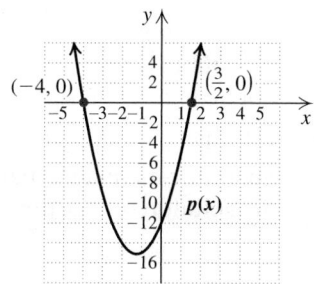

8. $p(x) < 0$

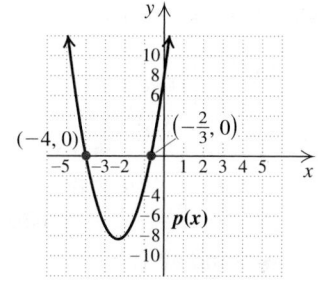

9. $x^4 + 12x > 3x^3 + 4x^2$

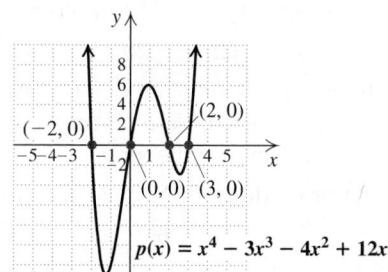

10. $x^4 + x^3 \ge 6x^2$

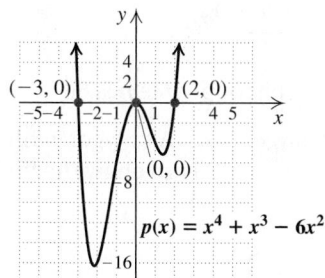

11. $\dfrac{x - 1}{x + 2} < 3$

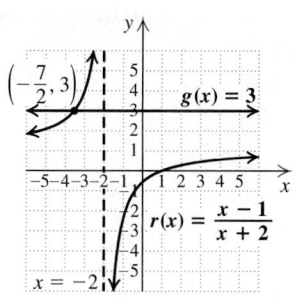

12. $\dfrac{2x - 1}{x - 5} \ge 1$

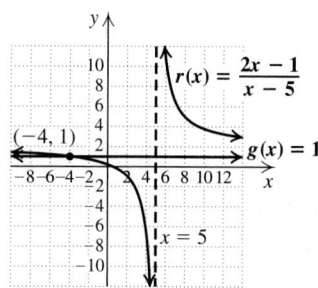

Solve.

13. $(x - 6)(x - 5) < 0$

14. $(x + 8)(x + 10) > 0$

15. $(x + 7)(x - 2) \ge 0$

16. $(x - 1)(x + 4) \le 0$

17. $x^2 - x - 2 > 0$

18. $x^2 + x - 2 < 0$

Aha! **19.** $x^2 + 4x + 4 < 0$

20. $x^2 + 6x + 9 < 0$

21. $x^2 - 4x \le 3$

22. $x^2 + 6x \ge 2$

23. $3x(x + 2)(x - 2) < 0$

24. $5x(x + 1)(x - 1) > 0$

25. $(x - 1)(x + 2)(x - 4) \ge 0$

26. $(x + 3)(x + 2)(x - 1) < 0$

27. For $f(x) = 7 - x^2$, find all x-values for which $f(x) \ge 3$.

28. For $f(x) = 14 - x^2$, find all x-values for which $f(x) > 5$.

29. For $g(x) = (x - 2)(x - 3)(x + 1)$, find all x-values for which $g(x) > 0$.

30. For $g(x) = (x + 3)(x - 2)(x + 1)$, find all x-values for which $g(x) < 0$.

31. For $F(x) = x^3 - 7x^2 + 10x$, find all x-values for which $F(x) \le 0$.

32. For $G(x) = x^3 - 8x^2 + 12x$, find all x-values for which $G(x) \ge 0$.

Solve.

33. $\dfrac{1}{x-5} < 0$

34. $\dfrac{1}{x+4} > 0$

35. $\dfrac{x+1}{x-3} \ge 0$

36. $\dfrac{x-2}{x+4} \le 0$

37. $\dfrac{x+1}{x+6} \ge 1$

38. $\dfrac{x-1}{x-2} \le 1$

39. $\dfrac{(x-2)(x+1)}{x-5} \le 0$

40. $\dfrac{(x+4)(x-1)}{x+3} \ge 0$

41. $\dfrac{x}{x+3} \ge 0$

42. $\dfrac{x-2}{x} \le 0$

43. $\dfrac{x-5}{x} < 1$

44. $\dfrac{x}{x-1} > 2$

45. $\dfrac{x-1}{(x-3)(x+4)} \le 0$

46. $\dfrac{x+2}{(x-2)(x+7)} \ge 0$

47. For $f(x) = \dfrac{5-2x}{4x+3}$, find all x-values for which $f(x) \ge 0$.

48. For $g(x) = \dfrac{2+3x}{2x-4}$, find all x-values for which $g(x) \ge 0$.

49. For $G(x) = \dfrac{1}{x-2}$, find all x-values for which $G(x) \le 1$.

50. For $F(x) = \dfrac{1}{x-3}$, find all x-values for which $F(x) \le 2$.

51. Explain how any quadratic inequality can be solved by examining a parabola.

52. Describe a method for creating a quadratic inequality for which there is no solution.

Skill Review

To prepare for Section 12.1, review function notation (Chapter 7).

Graph each function. [7.3]

53. $f(x) = x^3 - 2$

54. $g(x) = \dfrac{2}{x}$

55. If $f(x) = x + 7$, find $f\left(\dfrac{1}{a^2}\right)$. [7.1]

56. If $g(x) = x^2 - 3$, find $g(\sqrt{a-5})$. [7.1], [10.1]

57. If $g(x) = x^2 + 2$, find $g(2a+5)$. [7.1]

58. If $f(x) = \sqrt{4x+1}$, find $g(3a-5)$. [7.1]

Synthesis

59. Step (5) on p. 772 states that even when the inequality symbol is $\le$ or $\ge$, the solutions from step (2) may not be part of the solution set. Why?

60. Describe a method that could be used to create a quadratic inequality that has $(-\infty, a] \cup [b, \infty)$ as the solution set. Assume $a < b$.

Find each solution set.

61. $x^2 + 2x < 5$

62. $x^4 + 2x^2 \ge 0$

63. $x^4 + 3x^2 \le 0$

64. $\left|\dfrac{x+2}{x-1}\right| \le 3$

65. *Total profit.* Derex, Inc., determines that its total-profit function is given by
$$P(x) = -3x^2 + 630x - 6000.$$
a) Find all values of x for which Derex makes a profit.

b) Find all values of x for which Derex loses money.

66. *Height of a thrown object.* The function
$$S(t) = -16t^2 + 32t + 1920$$
gives the height S, in feet, of an object thrown from a cliff that is 1920 ft high. Here t is the time, in seconds, that the object is in the air.

a) For what times does the height exceed 1920 ft?

b) For what times is the height less than 640 ft?

67. *Number of handshakes.* There are n people in a room. The number N of possible handshakes by the people is given by the function
$$N(n) = \dfrac{n(n-1)}{2}.$$
For what number of people n is $66 \le N \le 300$?

68. *Number of diagonals.* A polygon with n sides has D diagonals, where D is given by the function
$$D(n) = \dfrac{n(n-3)}{2}.$$
Find the number of sides n if
$$27 \le D \le 230.$$

Use a graphing calculator to graph each function and find solutions of $f(x) = 0$. Then solve the inequalities $f(x) < 0$ and $f(x) > 0$.

69. $f(x) = x^3 - 2x^2 - 5x + 6$

70. $f(x) = \dfrac{1}{3}x^3 - x + \dfrac{2}{3}$

71. $f(x) = x + \dfrac{1}{x}$

72. $f(x) = x - \sqrt{x},\ x \geq 0$

73. $f(x) = \dfrac{x^3 - x^2 - 2x}{x^2 + x - 6}$

74. $f(x) = x^4 - 4x^3 - x^2 + 16x - 12$

Find the domain of each function

75. $f(x) = \sqrt{x^2 - 4x - 45}$

76. $f(x) = \sqrt{9 - x^2}$

77. $f(x) = \sqrt{x^2 + 8x}$

78. $f(x) = \sqrt{x^2 + 2x + 1}$

79. Describe a method that could be used to create a rational inequality that has $(-\infty, a] \cup (b, \infty)$ as the solution set. Assume $a < b$.

80. Use a graphing calculator to solve Exercises 43 and 49 by drawing two curves, one for each side of the inequality.

Study Summary

KEY TERMS AND CONCEPTS	EXAMPLES

SECTION 11.1: QUADRATIC EQUATIONS

A **quadratic equation in standard form** is written $ax^2 + bx + c = 0$, with a, b, and c constant and $a \neq 0$. Some quadratic equations can be solved by factoring.	$x^2 - 3x - 10 = 0$ $(x + 2)(x - 5) = 0$ $x + 2 = 0 \quad or \quad x - 5 = 0$ $\qquad x = -2 \quad or \qquad x = 5$
The Principle of Square Roots For any real number k, if $X^2 = k$, then $X = \sqrt{k} \quad or \quad X = -\sqrt{k}$.	$x^2 - 8x + 16 = 25$ $(x - 4)^2 = 25$ $x - 4 = -5 \quad or \quad x - 4 = 5$ $\quad x = -1 \quad or \qquad x = 9$
Any quadratic equation can be solved by **completing the square**.	$x^2 + 6x = 1$ $x^2 + 6x + \left(\frac{6}{2}\right)^2 = 1 + \left(\frac{6}{2}\right)^2$ $x^2 + 6x + 9 = 1 + 9$ $(x + 3)^2 = 10$ $x + 3 = \pm\sqrt{10}$ $x = -3 \pm \sqrt{10}$

SECTION 11.2: THE QUADRATIC FORMULA

The Quadratic Formula

The solutions of $ax^2 + bx + c = 0$ are given by

$$x = \frac{-b \pm \sqrt{b^2 - 4ac}}{2a}.$$

$3x^2 - 2x - 5 = 0 \qquad a = 3, b = -2, c = -5$

$$x = \frac{-(-2) \pm \sqrt{(-2)^2 - 4 \cdot 3(-5)}}{2 \cdot 3}$$

$$x = \frac{2 \pm \sqrt{4 + 60}}{6}$$

$$x = \frac{2 \pm \sqrt{64}}{6}$$

$$x = \frac{2 \pm 8}{6}$$

$$x = \frac{10}{6} = \frac{5}{3} \quad or \quad x = \frac{-6}{6} = -1$$

SECTION 11.3: STUDYING SOLUTIONS OF QUADRATIC EQUATIONS

The **discriminant** of the quadratic formula is $b^2 - 4ac$.

$b^2 - 4ac = 0 \rightarrow$ One solution; a rational number

For $4x^2 - 12x + 9 = 0$, $b^2 - 4ac = (-12)^2 - 4(4)(9)$

$\qquad = 144 - 144 = 0.$ The discriminant is zero.

Thus, $4x^2 - 12x + 9 = 0$ has one rational solution.

$b^2 - 4ac > 0 \rightarrow$ Two real solutions; both are rational if $b^2 - 4ac$ is a perfect square.

For $x^2 + 6x - 2 = 0$, $b^2 - 4ac = (6)^2 - 4(1)(-2)$

$\qquad = 36 + 8 = 44.$ The discriminant is not a perfect square.

Thus, $x^2 + 6x - 2 = 0$ has two irrational real-number solutions.

$b^2 - 4ac < 0 \rightarrow$ Two imaginary-number solutions

For $2x^2 - 3x + 5 = 0$, $b^2 - 4ac = (-3)^2 - 4(2)(5)$

$\qquad = 9 - 40 = -31.$ The discriminant is negative.

Thus, $2x^2 - 3x + 5 = 0$ has two imaginary-number solutions.

SECTION 11.4: APPLICATIONS INVOLVING QUADRATIC EQUATIONS

To solve a formula for a letter, use the same principles used to solve equations.

Solve $y = pn^2 + dn$ for n.

$pn^2 + dn - y = 0$ — Writing standard form of a quadratic equation

$$n = \frac{-d \pm \sqrt{d^2 - 4p(-y)}}{2 \cdot p}$$ — Using the quadratic formula; $a = p, b = d, c = -y$

$$n = \frac{-d \pm \sqrt{d^2 + 4py}}{2p}$$

SECTION 11.5: EQUATIONS REDUCIBLE TO QUADRATIC

Equations that are **reducible to quadratic** or in **quadratic form** can be solved by making an appropriate substitution.

$x^4 - 10x^2 + 9 = 0 \qquad$ Let $u = x^2$. Then $u^2 = x^4$.

$u^2 - 10u + 9 = 0 \qquad$ Substituting

$(u - 9)(u - 1) = 0$

$u - 9 = 0 \quad or \quad u - 1 = 0$

$u = 9 \quad or \quad u = 1 \qquad$ Solving for u

$x^2 = 9 \quad or \quad x^2 = 1$

$x = \pm 3 \quad or \quad x = \pm 1 \qquad$ Solving for x

SECTION 11.6: QUADRATIC FUNCTIONS AND THEIR GRAPHS
SECTION 11.7: MORE ABOUT GRAPHING QUADRATIC FUNCTIONS

The graph of a quadratic function

$$f(x) = ax^2 + bx + c = a(x - h)^2 + k$$

is a **parabola**. The graph opens upward for $a > 0$ and downward for $a < 0$.

The **vertex** is (h, k) and the **axis of symmetry** is $x = h$.

If $a > 0$, the function has a **minimum** value of k, and if $a < 0$, the function has a **maximum** value of k.

The vertex and the axis of symmetry occur at $x = -\dfrac{b}{2a}$.

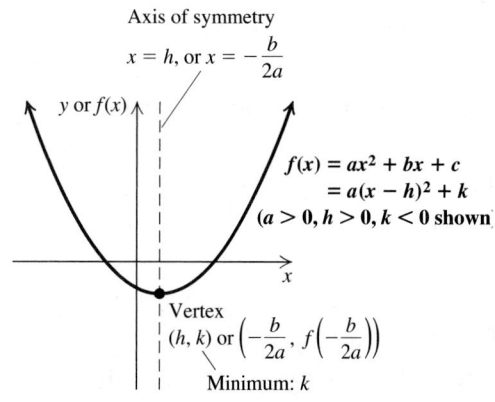

Axis of symmetry

$x = h$, or $x = -\dfrac{b}{2a}$

y or $f(x)$

$f(x) = ax^2 + bx + c$
$= a(x - h)^2 + k$
$(a > 0, h > 0, k < 0$ shown$)$

Vertex

(h, k) or $\left(-\dfrac{b}{2a}, f\left(-\dfrac{b}{2a}\right)\right)$

Minimum: k

SECTION 11.8: PROBLEM SOLVING AND QUADRATIC FUNCTIONS

Some problem situations can be **modeled** using quadratic functions. For those problems, a quantity can often be **maximized** or **minimized** by finding the coordinates of a vertex.

A lifeguard has 100 m of roped-together flotation devices with which to cordon off a rectangular swimming area at North Beach. If the shoreline forms one side of the rectangle, what dimensions will maximize the size of the area for swimming?

This problem and its solution appear as Example 2 on pp. 757–758.

SECTION 11.9: POLYNOMIAL AND RATIONAL INEQUALITIES

The x-intercepts, or **zeros**, of a function are used to divide the x-axis into intervals when solving a **polynomial inequality**. (See p. 769.)

Solve: $x^2 - 2x - 15 > 0$.

$$x^2 - 2x - 15 = 0 \qquad \text{Solving the related equation}$$
$$(x - 5)(x + 3) = 0$$
$$x = 5 \quad or \quad x = -3 \qquad \text{-3 and 5 divide the number line}$$
into three intervals.

```
      +         −         +
  <----+---------+---------->
      −3         5
```

The solutions of a rational equation and any replacements that make a denominator zero are both used to divide the x-axis into intervals when solving a **rational inequality**. (See p. 772.)

Since $f(x) = x^2 - 2x - 15 = (x - 5)(x + 3)$,

$f(x)$ is positive for $x < -3$;
$f(x)$ is negative for $-3 < x < 5$;
$f(x)$ is positive for $x > 5$.

Thus, $x^2 - 2x - 15 > 0$ for $(-\infty, -3) \cup (5, \infty)$, or $\{x | x < -3 \; or \; x > 5\}$.

Review Exercises: Chapter 11

👆 *Concept Reinforcement* *Classify each statement as either true or false.*

1. Every quadratic equation has two different solutions. [11.3]

2. Every quadratic equation has at least one solution. [11.3]

3. If an equation cannot be solved by completing the square, it cannot be solved by the quadratic formula. [11.2]

4. A negative discriminant indicates two imaginary-number solutions of a quadratic equation. [11.3]

5. The graph of $f(x) = 2(x + 3)^2 - 4$ has its vertex at $(3, -4)$. [11.6]

6. The graph of $g(x) = 5x^2$ has $x = 0$ as its axis of symmetry. [11.6]

7. The graph of $f(x) = -2x^2 + 1$ has no minimum value. [11.6]

8. The zeros of $g(x) = x^2 - 9$ are -3 and 3. [11.6]

9. If a quadratic function has two different imaginary-number zeros, the graph of the function has two x-intercepts. [11.7]

10. To solve a polynomial inequality, we often must solve a polynomial equation. [11.9]

Solve.

11. $9x^2 - 2 = 0$ [11.1]

12. $8x^2 + 6x = 0$ [11.1]

13. $x^2 - 12x + 36 = 9$ [11.1]

14. $x^2 - 4x + 8 = 0$ [11.2]

15. $x(3x + 4) = 4x(x - 1) + 15$ [11.2]

16. $x^2 + 9x = 1$ [11.2]

17. $x^2 - 5x - 2 = 0$. Use a calculator to approximate, to three decimal places, the solutions with rational numbers. [11.2]

18. Let $f(x) = 4x^2 - 3x - 1$. Find x such that $f(x) = 0$. [11.2]

Replace the blanks with constants to form a true equation. [11.1]

19. $x^2 - 18x + \underline{\quad} = (x - \underline{\quad})^2$

20. $x^2 + \frac{3}{5}x + \underline{\quad} = (x + \underline{\quad})^2$

21. Solve by completing the square. Show your work.
$$x^2 - 6x + 1 = 0 \quad [11.1]$$

22. $2500 grows to $2704 in 2 yr. Use the formula $A = P(1 + r)^t$ to find the interest rate. [11.1]

23. The U.S. Bank Tower in Los Angeles, California, is 1018 ft tall. Use $s = 16t^2$ to approximate how long it would take an object to fall from the top. [11.1]

For each equation, determine whether the solutions are real or imaginary. If they are real, specify whether they are rational or irrational. [11.3]

24. $x^2 + 3x - 6 = 0$

25. $x^2 + 2x + 5 = 0$

26. Write a quadratic equation having the solutions $3i$ and $-3i$. [11.3]

27. Write a quadratic equation having -5 as its only solution. [11.3]

Solve. [11.4]

28. Horizons has a manufacturing plant located 300 mi from company headquarters. Their corporate pilot must fly from headquarters to the plant and back in 4 hr. If there is a 20-mph headwind going and a 20-mph tailwind returning, how fast must the plane be able to travel in still air?

29. Working together, Dani and Cheri can reply to a day's worth of customer-service e-mails in 4 hr. Working alone, Dani takes 6 hr longer than Cheri. How long would it take Cheri to reply to the e-mails alone?

30. Find all x-intercepts of the graph of $f(x) = x^4 - 13x^2 + 36$. [11.5]

Solve. [11.5]

31. $15x^{-2} - 2x^{-1} - 1 = 0$

32. $(x^2 - 4)^2 - (x^2 - 4) - 6 = 0$

33. a) Graph: $f(x) = -3(x + 2)^2 + 4$. [11.6]
    b) Label the vertex.
    c) Draw the axis of symmetry.
    d) Find the maximum or the minimum value.

34. For the function given by $f(x) = 2x^2 - 12x + 23$: [11.7]

    a) find the vertex and the axis of symmetry;
    b) graph the function.

35. Find any x-intercepts and the y-intercept of the graph of
$$f(x) = x^2 - 9x + 14. \quad [11.7]$$

36. Solve $N = 3\pi\sqrt{\dfrac{1}{p}}$ for p. [11.4]

37. Solve $2A + T = 3T^2$ for T. [11.4]

State whether each graph appears to represent a quadratic function or a linear function. [11.8]

38.

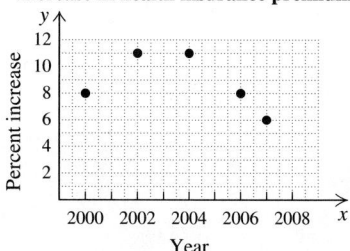

Increase in health insurance premiums
Percent increase
Year
Source: Kaiser/HRET

39.

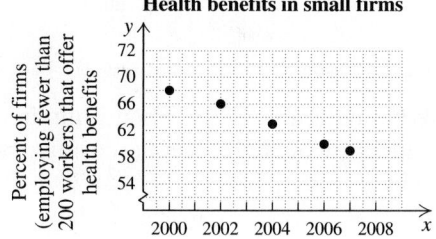

Health benefits in small firms
Percent of firms (employing fewer than 200 workers) that offer health benefits
Year
Source: Kaiser/HRET

40. Eastgate Consignments wants to build a rectangular area in a corner for children to play in while their parents shop. They have 30 ft of low fencing. What is the maximum area they can enclose? What dimensions will yield this area? [11.8]

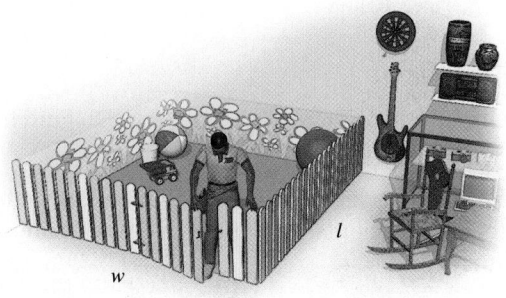

41. The following table lists the percent increase in health insurance premiums x years after 2000. (See Exercise 38.) [11.8]

Years Since 2000	Percent Increase in Health Insurance Premiums
0	8
2	11
6	8

a) Find the quadratic function that fits the data.
b) Use the function to estimate the percent increase in health insurance premiums in 2005.

Solve. [11.9]

42. $x^3 - 3x > 2x^2$

43. $\dfrac{x - 5}{x + 3} \le 0$

Synthesis

44. Explain how the x-intercepts of a quadratic function can be used to help find the maximum or minimum value of the function. [11.7], [11.8]

45. Suppose that the quadratic formula is used to solve a quadratic equation. If the discriminant is a perfect square, could factoring have been used to solve the equation? Why or why not? [11.2], [11.3]

46. What is the greatest number of solutions that an equation of the form $ax^4 + bx^2 + c = 0$ can have? Why? [11.5]

47. Discuss two ways in which completing the square was used in this chapter. [11.1], [11.2], [11.7]

48. A quadratic function has x-intercepts at -3 and 5. If the y-intercept is at -7, find an equation for the function. [11.7]

49. Find h and k if, for $3x^2 - hx + 4k = 0$, the sum of the solutions is 20 and the product of the solutions is 80. [11.3]

50. The average of two positive integers is 171. One of the numbers is the square root of the other. Find the integers. [11.5]

Test: Chapter 11

CHAPTER **Test Prep** VIDEO CD

Step-by-step test solutions are found on the video CD in the back of this book.

Solve.

1. $25x^2 - 7 = 0$

2. $4x(x - 2) - 3x(x + 1) = -18$

3. $x^2 + 2x + 3 = 0$

4. $2x + 5 = x^2$

5. $x^{-2} - x^{-1} = \frac{3}{4}$

6. $x^2 + 3x = 5$. Use a calculator to approximate, to three decimal places, the solutions with rational numbers.

7. Let $f(x) = 12x^2 - 19x - 21$. Find x such that $f(x) = 0$.

Replace the blanks with constants to form a true equation.

8. $x^2 - 20x + \underline{} = (x - \underline{})^2$

9. $x^2 + \frac{2}{7}x + \underline{} = (x + \underline{})^2$

10. Solve by completing the square. Show your work.
$$x^2 + 10x + 15 = 0$$

11. Determine the type of number that the solutions of $x^2 + 2x + 5 = 0$ will be.

12. Write a quadratic equation having solutions $\sqrt{11}$ and $-\sqrt{11}$.

Solve.

13. The Connecticut River flows at a rate of 4 km/h for the length of a popular scenic route. In order for a cruiser to travel 60 km upriver and then return in a total of 8 hr, how fast must the boat be able to travel in still water?

14. Dal and Kim can assemble a swing set in $1\frac{1}{2}$ hr. Working alone, it takes Kim 4 hr longer than Dal to assemble the swing set. How long would it take Dal, working alone, to assemble the swing set?

15. Find all x-intercepts of the graph of
$$f(x) = x^4 - 15x^2 - 16.$$

16. a) Graph: $f(x) = 4(x - 3)^2 + 5$.
b) Label the vertex.
c) Draw the axis of symmetry.
d) Find the maximum or the minimum function value.

17. For the function $f(x) = 2x^2 + 4x - 6$:
a) find the vertex and the axis of symmetry;
b) graph the function.

18. Find the x- and y-intercepts of
$$f(x) = x^2 - x - 6.$$

19. Solve $V = \frac{1}{3}\pi(R^2 + r^2)$ for r. Assume all variables are positive.

20. State whether the graph appears to represent a linear function, a quadratic function, or neither.

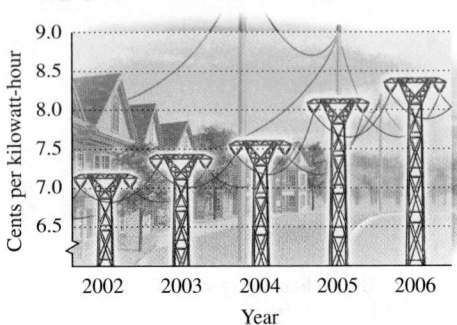

Average price of electricity

Source: Energy Information Administration, U.S. Department of Energy

21. Jay's Metals has determined that when x hundred storage cabinets are built, the average cost per cabinet is given by
$$C(x) = 0.2x^2 - 1.3x + 3.4025,$$
where $C(x)$ is in hundreds of dollars. What is the minimum cost per cabinet and how many cabinets should be built to achieve that minimum?

22. Find the quadratic function that fits the data points $(0, 0)$, $(3, 0)$, and $(5, 2)$.

Solve.

23. $x^2 + 5x < 6$

24. $x - \dfrac{1}{x} \geq 0$

Synthesis

25. One solution of $kx^2 + 3x - k = 0$ is -2. Find the other solution.

26. Find a fourth-degree polynomial equation, with integer coefficients, for which $-\sqrt{3}$ and $2i$ are solutions.

27. Solve: $x^4 - 4x^2 - 1 = 0$.

Cumulative Review: Chapters 1–11

Simplify.

1. $-3 \cdot 8 \div (-2)^3 \cdot 4 - 6(5 - 7)$ [1.8]

2. $\dfrac{18a^5bc^{10}}{24a^{-5}bc^3}$ [4.2]

3. $(5x^2y - 8xy - 6xy^2) - (2xy - 9x^2y + 3xy^2)$ [4.7]

4. $(9p^2q + 8t)(9p^2q - 8t)$ [4.7]

5. $\dfrac{t^2 - 25}{9t^2 + 24t + 16} \div \dfrac{3t^2 - 11t - 20}{t^2 + t}$ [6.2]

6. $\dfrac{1}{4 - x} + \dfrac{8}{x^2 - 16} - \dfrac{2}{x + 4}$ [6.4]

7. $\sqrt[3]{18x^4y} \cdot \sqrt[3]{6x^2y}$ [10.3]

8. $(3\sqrt{2} + i)(2\sqrt{2} - i)$ [10.8]

Factor.

9. $12x^4 - 75y^4$ [5.4]

10. $x^3 - 24x^2 + 80x$ [5.2]

11. $100m^6 - 100$ [5.5]

12. $6t^2 + 35t + 36$ [5.3]

Solve.

13. $2(5x - 3) - 8x = 4 - (3 - x)$ [2.2]

14. $2(5x - 3) - 8x < 4 - (3 - x)$ [2.6]

15. $2x - 6y = 3,$
    $-3x + 8y = -5$ [8.2]

16. $x(x - 5) = 66$ [5.7]

17. $\dfrac{2}{t} + \dfrac{1}{t - 1} = 2$ [6.6]

18. $\sqrt{x} = 1 + \sqrt{2x - 7}$ [10.6]

19. $m^2 + 10m + 25 = 2$ [11.1]

20. $3x^2 + 1 = x$ [11.2]

Graph.

21. $9x - 2y = 18$ [3.3]

22. $x < \frac{1}{2}y$ [9.4]

23. $y = 2(x - 3)^2 + 1$ [11.6]

24. $f(x) = x^2 + 4x + 3$ [11.7]

25. Find an equation in slope–intercept form whose graph has slope -5 and y-intercept $\left(0, \frac{1}{2}\right)$. [3.6]

26. Find the slope of the line containing $(8, 3)$ and $(-2, 10)$. [3.5]

27. For the function described by $f(x) = 3x^2 - 8x - 7$, find $f(-2)$. [7.1]

Find the domain of each function.

28. $f(x) = \sqrt{10 - x}$ [9.1]

29. $f(x) = \dfrac{x + 3}{x - 4}$ [9.2]

Solve each formula for the specified letter.

30. $b = \dfrac{a + c}{2a}$, for a [7.5]

31. $p = 2\sqrt{\dfrac{r}{3t}}$, for t [11.4]

Solve.

32. *Mobile ad spending.* The amount spent worldwide in advertising on mobile devices can be estimated by $f(x) = 0.4x^2 + 0.01x + 0.9$, where x is the number of years after 2005 and $f(x)$ is in billions of dollars.
    Source: Based on data from eMarketer

    a) How much was spent worldwide for mobile ads in 2008? [7.1]

    b) When will worldwide mobile ad spending reach $41 billion? [11.4]

33. *Wi-fi hotspots.* The number of Wi-fi hotspots worldwide grew from 19,000 in 2002 to 118,000 in 2005. Let $h(t)$ represent the number of hotspots, in thousands, t years after 2000. [7.1]
    Source: IDC

    a) Find a linear function that fits the data.

    b) Use the function from part (a) to predict the number of Wi-fi hotspots in 2010.

    c) In what year will there be 500,000 Wi-fi hotspots?

34. *Gold prices.* Annette is selling some of her gold jewelry. She has 4 bracelets and 1 necklace that weigh a total of 3 oz. [2.5]

a) Annette's jewelry is 58% gold. How many ounces of gold does her jewelry contain?

b) A gold dealer offers Annette $1044 for the jewelry. How much per ounce of gold was she offered?

c) The price of gold at the time of Annette's sale was $800 an ounce. What percent of the gold price was she offered for her jewelry?

35. *Education.* Sven ordered number tiles at $9 per set and alphabet tiles at $15 per set for his classroom. He ordered a total of 36 sets for $384. How many sets of each did he order? [8.3]

36. *Minimizing cost.* Dormitory Furnishings has determined that when x bunk beds are built, the average cost, in dollars, per bunk bed can be estimated by $c(x) = 0.004375x^2 - 3.5x + 825$. What is the minimum average cost per bunk bed and how many bunk beds should be built to achieve that minimum? [11.8]

37. *Volunteer work.* It takes Deanna twice as long to set up a fundraising auction as it takes Donna. Together they can set up for the auction in 4 hr. How long would it take each of them to do the job alone? [6.7]

38. *Canoeing.* Kent paddled for 2 hr with a 5-km/h current to reach a campsite. The return trip against the same current took 7 hr. Find the speed of Kent's canoe in still water. [8.3]

39. *Truck rentals.* Josh and Lindsay plan to rent a moving truck. The truck costs $70 plus 40¢ per mile. They have budgeted $90 for the truck rental. For what mileages will they not exceed their budget? [2.7]

Synthesis

Solve.

40. $\dfrac{\dfrac{1}{x}}{2 + \dfrac{1}{x-1}} = 3$ [6.5], [11.2]

41. $x^4 + 5x^2 \le 0$ [11.9]

42. The graph of the function $f(x) = mx + b$ contains the point $(2, 3)$ and is perpendicular to the line containing the points $(-1, 4)$ and $(-2, 5)$. Find the equation of the function. [3.7]

43. Find the points of intersection of the graphs of $f(x) = x^2 + 8x + 1$ and $g(x) = 10x + 6$. [11.2]

Exponential and Logarithmic Functions

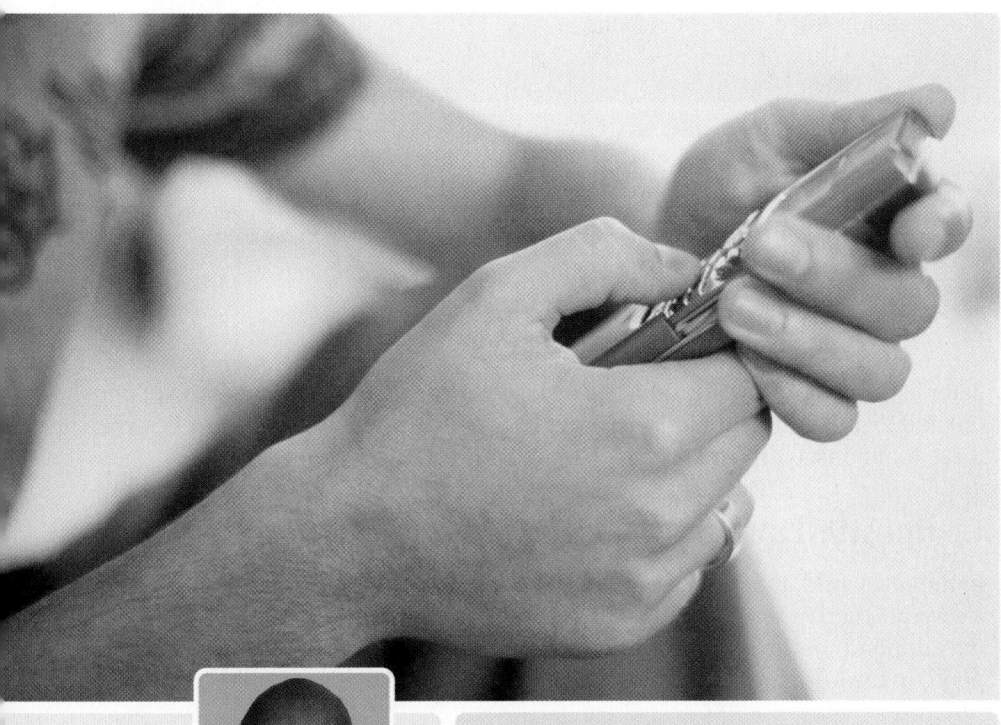

PAUL WILLIAMS
DATA SYSTEMS SPECIALIST
Indianapolis, Indiana

Because I work with computers and networking every day, everything around me is based on math. Computers and their algorithms are based in the binary number system. We use mathematical tools such as statistics and graphs to study network bottlenecks in computers and transmission pipes and to detect network saturation points.

AN APPLICATION

In 2000, there were approximately 12 million text messages sent each month in the United States. This number has increased exponentially at an average rate of 108% per year. Find the exponential growth function that models the data, and estimate the number of text messages sent each month in 2009.

Source: Based on information from www.cellulist.com

This problem appears as Example 4 in Section 12.7.

The functions that we consider in this chapter have rich applications in many fields, such as epidemiology (the study of the spread of disease), population growth, and marketing.

The theory centers on functions with variable exponents (*exponential functions*). Results follow from those functions, their properties, and properties of their closely related *inverse* functions.

12.1 Composite and Inverse Functions

Composite Functions • Inverses and One-to-One Functions • Finding Formulas for Inverses • Graphing Functions and Their Inverses • Inverse Functions and Composition

Later in this chapter, we introduce two closely related types of functions: exponential and logarithmic functions. In order to properly understand the link between these functions, we must first understand composite and inverse functions.

Composite Functions

In the real world, functions frequently occur in which some quantity depends on a variable that, in turn, depends on another variable. For instance, a firm's profits may depend on the number of items the firm produces, which may in turn depend on the number of employees hired. Functions like this are called **composite functions**.

For example, the function g that gives a correspondence between women's shoe sizes in the United States and those in Britain is given by $g(x) = x - 2$, where x is the U.S. size and $g(x)$ is the British size. Thus a U.S. size 4 corresponds to a shoe size of $g(4) = 4 - 2$, or 2, in Britain.

A second function converts women's shoe sizes in Britain to those in Italy. This particular function is given by $f(x) = 2x + 28$, where x is the British size and $f(x)$ is the corresponding Italian size. Thus a British size 2 corresponds to an Italian size $f(2) = 2 \cdot 2 + 28$, or 32.

It is correct to conclude that a U.S. size 4 corresponds to an Italian size 32 and that some function h describes this correspondence.

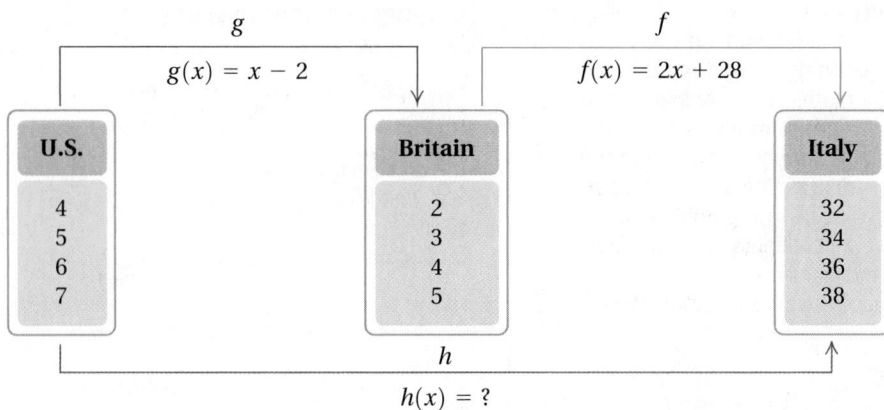

Throughout this chapter, keep in mind that equations such as $g(x) = x - 2$ and $g(t) = t - 2$ describe the same function g. Both equations tell us to find a function value by subtracting 2 from the input.

Size x shoes in the United States correspond to size $g(x)$ shoes in Britain, where

$$g(x) = x - 2.$$

Size n shoes in Britain correspond to size $f(n)$ shoes in Italy. Similarly, size $g(x)$ shoes in Britain correspond to size $f(g(x))$ shoes in Italy. Since the x in the expression $f(g(x))$ represents a U.S. shoe size, we can find the Italian shoe size that corresponds to a U.S. size x as follows:

$$f(g(x)) = f(x - 2) = 2(x - 2) + 28 \qquad \text{Using } g(x) \text{ as an input}$$
$$= 2x - 4 + 28 = 2x + 24.$$

This gives a formula for h: $h(x) = 2x + 24$. Thus U.S. size 4 corresponds to Italian size $h(4) = 2(4) + 24$, or 32. We call h the *composition* of f and g and denote it by $f \circ g$ (read "the composition of f and g," "f composed with g," or "f circle g").

> **Composition of Functions**
> The *composite function* $f \circ g$, the *composition* of f and g, is defined as
> $$(f \circ g)(x) = f(g(x)).$$

We can visualize the composition of functions as follows.

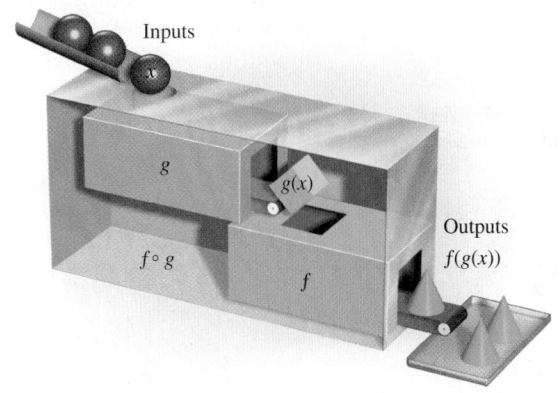

EXAMPLE **1**   Given $f(x) = 3x$ and $g(x) = 1 + x^2$:

a) Find $(f \circ g)(5)$ and $(g \circ f)(5)$.   **b)** Find $(f \circ g)(x)$ and $(g \circ f)(x)$.

SOLUTION   Consider each function separately:

$$f(x) = 3x \qquad \text{This function multiplies each input by 3.}$$

and

$$g(x) = 1 + x^2. \qquad \text{This function adds 1 to the square of each input.}$$

a) To find $(f \circ g)(5)$, we find $g(5)$ and then use that as an input for f:

$$(f \circ g)(5) = f(g(5)) = f(1 + 5^2) \qquad \text{Using } g(x) = 1 + x^2$$
$$= f(26) = 3 \cdot 26 = 78. \qquad \text{Using } f(x) = 3x$$

To find $(g \circ f)(5)$, we find $f(5)$ and then use that as an input for g:

$$(g \circ f)(5) = g(f(5)) = g(3 \cdot 5) \qquad \text{Note that } f(5) = 3 \cdot 5 = 15.$$
$$= g(15) = 1 + 15^2 = 1 + 225 = 226.$$

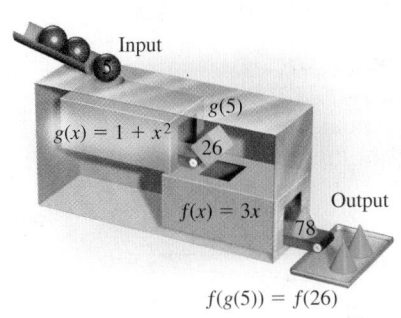

A composition machine for Example 1

b) We find $(f \circ g)(x)$ by substituting $g(x)$ for x in the equation for $f(x)$:

$$(f \circ g)(x) = f(g(x)) = f(1 + x^2) \qquad \text{Using } g(x) = 1 + x^2$$
$$= 3 \cdot (1 + x^2) = 3 + 3x^2. \qquad \text{Using } f(x) = 3x$$

To find $(g \circ f)(x)$, we substitute $f(x)$ for x in the equation for $g(x)$:

$$(g \circ f)(x) = g(f(x)) = g(3x) \qquad \text{Substituting } 3x \text{ for } f(x)$$
$$= 1 + (3x)^2 = 1 + 9x^2.$$

We can now find the function values of part (a) using the functions of part (b):

$$(f \circ g)(5) = 3 + 3(5)^2 = 3 + 3 \cdot 25 = 78;$$
$$(g \circ f)(5) = 1 + 9(5)^2 = 1 + 9 \cdot 25 = 226. \qquad \boxed{\text{TRY EXERCISE} \; 9}$$

Example 1 shows that, in general, $(f \circ g)(x) \neq (g \circ f)(x)$.

EXAMPLE **2** Given $f(x) = \sqrt{x}$ and $g(x) = x - 1$, find $(f \circ g)(x)$ and $(g \circ f)(x)$.

SOLUTION

$$(f \circ g)(x) = f(g(x)) = f(x - 1) = \sqrt{x - 1}; \qquad \text{Using } g(x) = x - 1$$
$$(g \circ f)(x) = g(f(x)) = g(\sqrt{x}) = \sqrt{x} - 1 \qquad \text{Using } f(x) = \sqrt{x}$$

$\boxed{\text{TRY EXERCISE} \; 15}$

In fields ranging from chemistry to geology and economics, one needs to recognize how a function can be regarded as the composition of two "simpler" functions. This is sometimes called *de*composition.

EXAMPLE **3** If $h(x) = (7x + 3)^2$, find f and g such that $h(x) = (f \circ g)(x)$.

SOLUTION We can think of $h(x)$ as the result of first finding $7x + 3$ and then squaring that. This suggests that $g(x) = 7x + 3$ and $f(x) = x^2$. We check by forming the composition:

$$(f \circ g)(x) = f(g(x))$$
$$= f(7x + 3) = (7x + 3)^2 = h(x), \text{ as desired.}$$

This may be the most "obvious" solution, but there are other less obvious answers. For example, if $f(x) = (x - 1)^2$ and $g(x) = 7x + 4$, then

$$(f \circ g)(x) = f(g(x)) = f(7x + 4)$$
$$= (7x + 4 - 1)^2 = (7x + 3)^2 = h(x). \qquad \boxed{\text{TRY EXERCISE} \; 21}$$

TECHNOLOGY CONNECTION

In Example 3, we see that if $g(x) = 7x + 3$ and $f(x) = x^2$, then $f(g(x)) = (7x + 3)^2$. One way to show this is to let $y_1 = 7x + 3$ and $y_2 = x^2$. If we let $y_3 = (7x + 3)^2$ and $y_4 = y_2(y_1)$, we can use graphs or a table to show that $y_3 = y_4$.

1. Check Example 2 by using the above approach.

Inverses and One-to-One Functions

Let's view the following two functions as relations, or correspondences.

Countries and Their Capitals

Domain (Set of Inputs)	Range (Set of Outputs)
Australia ⟶	Canberra
China ⟶	Beijing
Germany ⟶	Berlin
Madagascar ⟶	Antananarivo
Turkey ⟶	Ankara
United States ⟶	Washington, D.C.

Phone Keys

Domain (Set of Inputs)	Range (Set of Outputs)
a	
b	2
c	
d	
e	3
f	

Suppose we reverse the arrows. We obtain what is called the **inverse relation**. Are these inverse relations functions?

Countries and Their Capitals

Range (Set of Outputs)	Domain (Set of Inputs)
Australia ⟵	Canberra
China ⟵	Beijing
Germany ⟵	Berlin
Madagascar ⟵	Antananaviro
Turkey ⟵	Ankara
United States ⟵	Washington, D.C.

Phone Keys

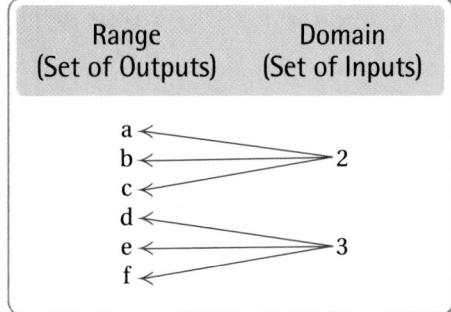

Recall that for each input, a function has exactly one output. However, it is possible for different inputs to correspond to the same output. Only when this possibility is *excluded* will the inverse be a function. For the functions listed above, this means the inverse of the "Capitals" correspondence is a function, but the inverse of the "Phone Keys" correspondence is not.

In the Capitals function, each input has its own output, so it is a **one-to-one-function**. In the Phone Keys function, a and b are both paired with 2. Thus the Phone Keys function is not a one-to-one function.

> ## One-To-One Function
>
> A function f is *one-to-one* if different inputs have different outputs. That is, if for a and b in the domain of f with $a \neq b$, we have $f(a) \neq f(b)$, then f is one-to-one. If a function is one-to-one, then its inverse correspondence is also a function.

How can we tell graphically whether a function is one-to-one?

EXAMPLE 4   At left is the graph of a function similar to those we will study in Section 12.2. Determine whether the function is one-to-one and thus has an inverse that is a function.

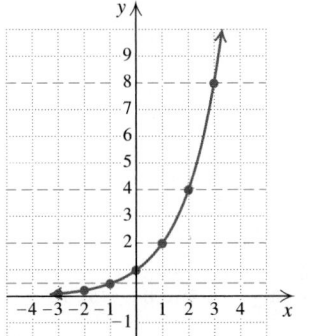

SOLUTION   A function is one-to-one if different inputs have different outputs—that is, if no two x-values have the same y-value. For this function, we cannot find two x-values that have the same y-value. Note that this means that no horizontal line can be drawn so that it crosses the graph more than once. The function is one-to-one so its inverse is a function.

> TRY EXERCISE 31

The graph of every function must pass the vertical-line test. In order for a function to have an inverse that is a function, it must pass the *horizontal-line test* as well.

> ## The Horizontal-Line Test
>
> If it is impossible to draw a horizontal line that intersects a function's graph more than once, then the function is one-to-one. For every one-to-one function, an inverse function exists.

EXAMPLE 5

Determine whether the function $f(x) = x^2$ is one-to-one and thus has an inverse that is a function.

SOLUTION   The graph of $f(x) = x^2$ is shown here. Many horizontal lines cross the graph more than once. For example, the line $y = 4$ crosses where the first coordinates are -2 and 2. Although these are different inputs, they have the same output. That is, $-2 \neq 2$, but

$$f(-2) = (-2)^2 = 4 = 2^2 = f(2).$$

Thus the function is not one-to-one and no inverse function exists.

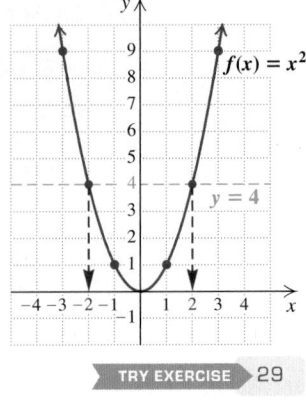

TRY EXERCISE 29

Finding Formulas for Inverses

When the inverse of f is also a function, it is denoted f^{-1} (read "f-inverse").

CAUTION!   The -1 in f^{-1} is *not* an exponent!

Suppose a function is described by a formula. If its inverse is a function, how do we find a formula for that inverse? For any equation in two variables, if we interchange the variables, we form an equation of the inverse correspondence. If it is a function, we proceed as follows to find a formula for f^{-1}.

To Find a Formula for f^{-1}

First make sure that f is one-to-one. Then:

1. Replace $f(x)$ with y.
2. Interchange x and y. (This gives the inverse function.)
3. Solve for y.
4. Replace y with $f^{-1}(x)$. (This is inverse function notation.)

EXAMPLE 6

Determine whether each function is one-to-one and if it is, find a formula for $f^{-1}(x)$.

a) $f(x) = x + 2$ **b)** $f(x) = 2x - 3$

SOLUTION

a) The graph of $f(x) = x + 2$ is shown at left. It passes the horizontal-line test, so it is one-to-one. Thus its inverse is a function.

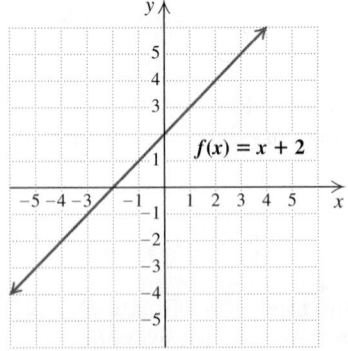

1. Replace $f(x)$ with y: $y = x + 2$.
2. Interchange x and y: $x = y + 2$.   This gives the inverse function.
3. Solve for y: $x - 2 = y$.
4. Replace y with $f^{-1}(x)$: $f^{-1}(x) = x - 2$.   We also "reversed" the equation.

In this case, the function f adds 2 to all inputs. Thus, to "undo" f, the function f^{-1} must subtract 2 from its inputs.

b) The function $f(x) = 2x - 3$ is also linear. Any linear function that is not constant will pass the horizontal-line test. Thus, f is one-to-one.

1. Replace $f(x)$ with y: $y = 2x - 3$.

2. Interchange x and y: $x = 2y - 3$.

3. Solve for y: $x + 3 = 2y$

$$\frac{x + 3}{2} = y.$$

4. Replace y with $f^{-1}(x)$: $f^{-1}(x) = \dfrac{x + 3}{2}$.

In this case, the function f doubles all inputs and then subtracts 3. Thus, to "undo" f, the function f^{-1} adds 3 to each input and then divides by 2.

> TRY EXERCISE ▸ 35

Graphing Functions and Their Inverses

How do the graphs of a function and its inverse compare?

EXAMPLE 7   Graph $f(x) = 2x - 3$ and $f^{-1}(x) = (x + 3)/2$ on the same set of axes. Then compare.

SOLUTION   The graph of each function follows. Note that the graph of f^{-1} can be drawn by reflecting the graph of f across the line $y = x$. That is, if we graph $f(x) = 2x - 3$ in wet ink and fold the paper along the line $y = x$, the graph of $f^{-1}(x) = (x + 3)/2$ will appear as the impression made by f.

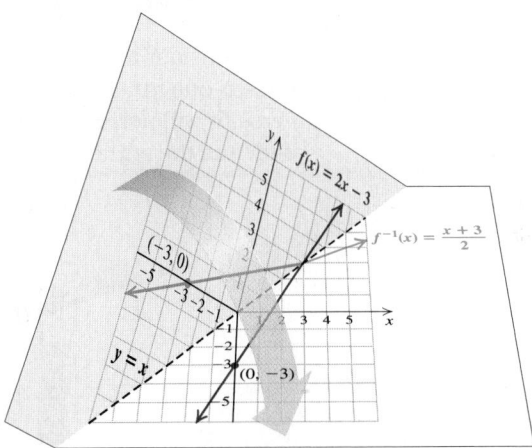

When x and y are interchanged to find a formula for the inverse, we are, in effect, reflecting or flipping the graph of $f(x) = 2x - 3$ across the line $y = x$. For example, when $(0, -3)$, the coordinates of the y-intercept of the graph of f, are reversed, we get $(-3, 0)$, the x-intercept of the graph of f^{-1}.

> TRY EXERCISE ▸ 59

Visualizing Inverses

The graph of f^{-1} is a reflection of the graph of f across the line $y = x$.

EXAMPLE 8

Consider $g(x) = x^3 + 2$.

a) Determine whether the function is one-to-one.

b) If it is one-to-one, find a formula for its inverse.

c) Graph the inverse, if it exists.

SOLUTION

a) The graph of $g(x) = x^3 + 2$ is shown at right. It passes the horizontal-line test and thus is one-to-one and has an inverse that is a function.

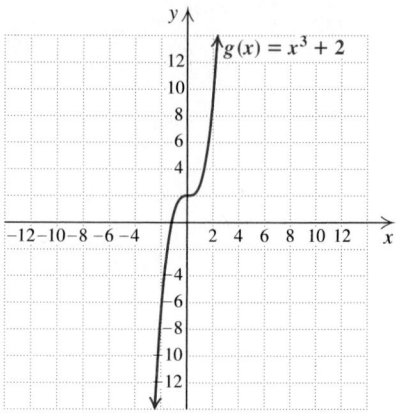

b) **1.** Replace $g(x)$ with y: $y = x^3 + 2$.   Using $g(x) = x^3 + 2$

  2. Interchange x and y: $x = y^3 + 2$.

  3. Solve for y:
  $$x - 2 = y^3$$
  $$\sqrt[3]{x - 2} = y.$$ Each real number has only one cube root, so we can solve for y.

  4. Replace y with $g^{-1}(x)$: $g^{-1}(x) = \sqrt[3]{x - 2}$.

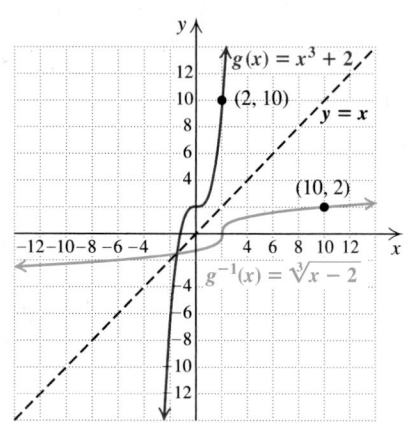

c) To graph g^{-1}, we can reflect the graph of $g(x) = x^3 + 2$ across the line $y = x$, as we did in Example 7. We also could graph $g^{-1}(x) = \sqrt[3]{x - 2}$ by plotting points. Note that $(2, 10)$ is on the graph of g, whereas $(10, 2)$ is on the graph of g^{-1}. The graphs of g and g^{-1} are shown at left.   ▶ TRY EXERCISE 61

Inverse Functions and Composition

Let's consider inverses of functions in terms of function machines. Suppose that a one-to-one function f is programmed into a machine. If the machine is run in reverse, it will perform the inverse function f^{-1}. Inputs then enter at the opposite end, and the entire process is reversed.

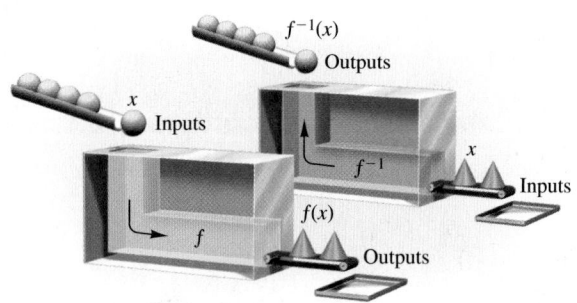

Consider $f(x) = x^3 + 2$ and $f^{-1}(x) = \sqrt[3]{x - 2}$ from Example 8. For the input 3,

$$f(3) = 3^3 + 2 = 27 + 2 = 29.$$

The output is 29. Let's now use 29 for the input in the inverse:

$$f^{-1}(29) = \sqrt[3]{29 - 2} = \sqrt[3]{27} = 3.$$

The function f takes 3 to 29. The inverse function f^{-1} takes the number 29 back to 3.

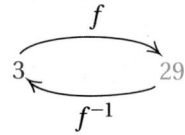

In general, if f is one-to-one, then f^{-1} takes the output $f(x)$ back to x. Similarly, f takes the output $f^{-1}(x)$ back to x.

Composition and Inverses

If a function f is one-to-one, then f^{-1} is the unique function for which

$$(f^{-1} \circ f)(x) = f^{-1}(f(x)) = x \quad \text{and} \quad (f \circ f^{-1})(x) = f(f^{-1}(x)) = x.$$

EXAMPLE 9   Let $f(x) = 2x + 1$. Show that

$$f^{-1}(x) = \frac{x - 1}{2}.$$

SOLUTION   We find $(f^{-1} \circ f)(x)$ and $(f \circ f^{-1})(x)$ and check to see that each is x.

$$(f^{-1} \circ f)(x) = f^{-1}(f(x)) = f^{-1}(2x + 1)$$

$$= \frac{(2x + 1) - 1}{2}$$

$$= \frac{2x}{2} = x \qquad \text{Thus, } (f^{-1} \circ f)(x) = x.$$

$$(f \circ f^{-1})(x) = f(f^{-1}(x)) = f\left(\frac{x - 1}{2}\right)$$

$$= 2 \cdot \frac{x - 1}{2} + 1$$

$$= x - 1 + 1 = x \qquad \text{Thus, } (f \circ f^{-1})(x) = x.$$

TRY EXERCISE 69

TECHNOLOGY CONNECTION

To determine whether $y_1 = 2x + 6$ and $y_2 = \frac{1}{2}x - 3$ are inverses of each other, we can graph both functions, along with the line $y = x$, on a "squared" set of axes. It *appears* that y_1 and y_2 are inverses of each other. A more precise check is achieved by selecting the DRAWINV option of the (DRAW) menu. The resulting graph of the inverse of y_1 should coincide with y_2.

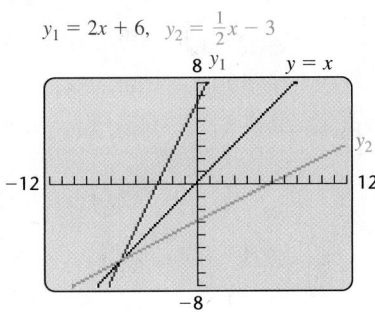

$y_1 = 2x + 6, \quad y_2 = \frac{1}{2}x - 3$

For a more dependable check, examine a TABLE in which $y_1 = 2x + 6$ and $y_2 = \frac{1}{2} \cdot y_1 - 3$. Note that y_2 "undoes" what y_1 does.

TBLSTART $= -3$ ΔTBL $= 1$ $y_2 = \frac{1}{2}y_1 - 3$

X	Y1	Y2
−3	0	−3
−2	2	−2
−1	4	−1
0	6	0
1	8	1
2	10	2
3	12	3

X = 3

1. Use a graphing calculator to check Examples 7, 8, and 9.
2. Will DRAWINV work for *any* choice of y_1? Why or why not?

12.1   EXERCISE SET

For Extra Help
MyMathLab   PRACTICE   WATCH   DOWNLOAD

🖐 **Concept Reinforcement**   *Classify each statement as either true or false.*

1. The composition of two functions f and g is written $f \circ g$.

2. The notation $(f \circ g)(x)$ means $f(g(x))$.

3. If $f(x) = x^2$ and $g(x) = x + 3$, then $(g \circ f)(x) = (x + 3)^2$.

4. For any function h, there is only one way to decompose the function as $h = f \circ g$.

5. The function f is one-to-one if $f(1) = 1$.

6. The -1 in f^{-1} is an exponent.

7. The function f is the inverse of f^{-1}.

8. If g and h are inverses of each other, then $(g \circ h)(x) = x$.

For each pair of functions, find **(a)** $(f \circ g)(1)$; **(b)** $(g \circ f)(1)$; **(c)** $(f \circ g)(x)$; **(d)** $(g \circ f)(x)$.

9. $f(x) = x^2 + 1$; $g(x) = x - 3$

10. $f(x) = x + 4$; $g(x) = x^2 - 5$

11. $f(x) = 5x + 1$; $g(x) = 2x^2 - 7$

12. $f(x) = 3x^2 + 4$; $g(x) = 4x - 1$

13. $f(x) = x + 7$; $g(x) = 1/x^2$

14. $f(x) = 1/x^2$; $g(x) = x + 2$

15. $f(x) = \sqrt{x}$; $g(x) = x + 3$

16. $f(x) = 10 - x$; $g(x) = \sqrt{x}$

17. $f(x) = \sqrt{4x}$; $g(x) = 1/x$

18. $f(x) = \sqrt{x + 3}$; $g(x) = 13/x$

19. $f(x) = x^2 + 4$; $g(x) = \sqrt{x - 1}$

20. $f(x) = x^2 + 8$; $g(x) = \sqrt{x + 17}$

Find $f(x)$ and $g(x)$ such that $h(x) = (f \circ g)(x)$. Answers may vary.

21. $h(x) = (3x - 5)^4$

22. $h(x) = (2x + 7)^3$

23. $h(x) = \sqrt{9x + 1}$

24. $h(x) = \sqrt[3]{4x - 5}$

25. $h(x) = \dfrac{6}{5x - 2}$

26. $h(x) = \dfrac{3}{x} + 4$

Determine whether each function is one-to-one.

27. $f(x) = -x$

28. $f(x) = x + 5$

Aha! **29.** $f(x) = x^2 + 3$

30. $f(x) = 3 - x^2$

31.
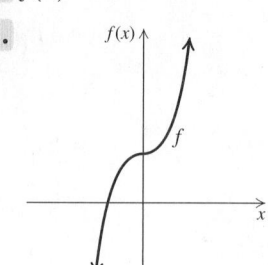

32.
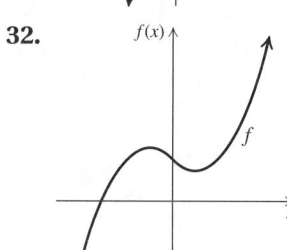

33.
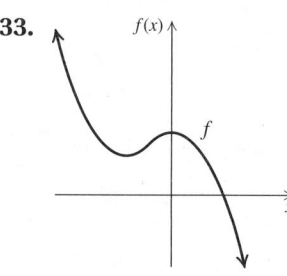

34.
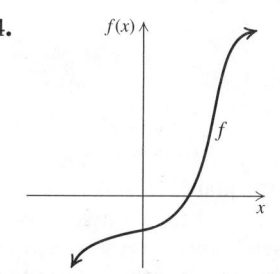

*For each function, **(a)** determine whether it is one-to-one; **(b)** if it is one-to-one, find a formula for the inverse.*

35. $f(x) = x + 3$

36. $f(x) = x + 2$

37. $f(x) = 2x$

38. $f(x) = 3x$

39. $g(x) = 3x - 1$

40. $g(x) = 2x - 3$

41. $f(x) = \dfrac{1}{2}x + 1$

42. $f(x) = \dfrac{1}{3}x + 2$

43. $g(x) = x^2 + 5$

44. $g(x) = x^2 - 4$

45. $h(x) = -10 - x$

46. $h(x) = 7 - x$

Aha! **47.** $f(x) = \dfrac{1}{x}$

48. $f(x) = \dfrac{3}{x}$

49. $g(x) = 1$

50. $h(x) = 8$

51. $f(x) = \dfrac{2x + 1}{3}$

52. $f(x) = \dfrac{3x + 2}{5}$

53. $f(x) = x^3 + 5$

54. $f(x) = x^3 - 4$

55. $g(x) = (x - 2)^3$

56. $g(x) = (x + 7)^3$

57. $f(x) = \sqrt{x}$

58. $f(x) = \sqrt{x - 1}$

Graph each function and its inverse using the same set of axes.

59. $f(x) = \dfrac{2}{3}x + 4$

60. $g(x) = \dfrac{1}{4}x + 2$

61. $f(x) = x^3 + 1$

62. $f(x) = x^3 - 1$

63. $g(x) = \dfrac{1}{2}x^3$

64. $g(x) = \dfrac{1}{3}x^3$

65. $F(x) = -\sqrt{x}$

66. $f(x) = \sqrt{x}$

67. $f(x) = -x^2, x \geq 0$

68. $f(x) = x^2 - 1, x \leq 0$

69. Let $f(x) = \sqrt[3]{x - 4}$. Show that
$$f^{-1}(x) = x^3 + 4.$$

70. Let $f(x) = 3/(x + 2)$. Show that
$$f^{-1}(x) = \dfrac{3}{x} - 2.$$

71. Let $f(x) = (1 - x)/x$. Show that
$$f^{-1}(x) = \dfrac{1}{x + 1}.$$

72. Let $f(x) = x^3 - 5$. Show that
$$f^{-1}(x) = \sqrt[3]{x + 5}.$$

73. *Dress sizes in the United States and Italy.* A size-6 dress in the United States is size 36 in Italy. A function that converts dress sizes in the United States to those in Italy is
$$f(x) = 2(x + 12).$$

a) Find the dress sizes in Italy that correspond to sizes 8, 10, 14, and 18 in the United States.

b) Determine whether f has an inverse that is a function. If so, find a formula for the inverse.

c) Use the inverse function to find dress sizes in the United States that correspond to sizes 40, 44, 52, and 60 in Italy.

74. *Dress sizes in the United States and France.* A size-6 dress in the United States is size 38 in France. A function that converts dress sizes in the United States to those in France is
$$f(x) = x + 32.$$

a) Find the dress sizes in France that correspond to sizes 8, 10, 14, and 18 in the United States.

b) Determine whether f has an inverse that is a function. If so, find a formula for the inverse.

c) Use the inverse function to find dress sizes in the United States that correspond to sizes 40, 42, 46, and 50 in France.

75. Is there a one-to-one relationship between items in a store and the price of each of those items? Why or why not?

76. Mathematicians usually try to select "logical" words when forming definitions. Does the term "one-to-one" seem logical? Why or why not?

Skill Review

To prepare for Section 12.2, review simplifying exponential expressions and graphing equations (Sections 4.2, 7.3, and 10.2).

Simplify.

77. 2^{-3} [4.2]

78. $5^{(1-3)}$ [4.2]

79. $4^{5/2}$ [10.2]

80. $3^{7/10}$ [10.2]

Graph. [7.3]

81. $y = x^3$

82. $x = y^3$

Synthesis

83. The function $V(t) = 750(1.2)^t$ is used to predict the value $V(t)$ of a certain rare stamp t years from 2008. Do not calculate $V^{-1}(t)$, but explain how V^{-1} could be used.

84. An organization determines that the cost per person $C(x)$, in dollars, of chartering a bus with x passengers is given by

$$C(x) = \frac{100 + 5x}{x}.$$

Determine $C^{-1}(x)$ and explain how this inverse function could be used.

For Exercises 85 and 86, graph the inverse of f.

85.

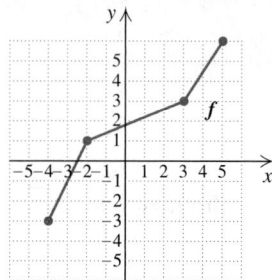

86.

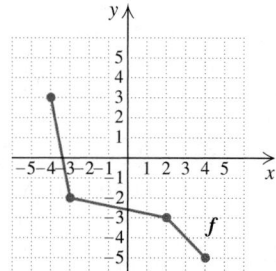

87. *Dress sizes in France and Italy.*   Use the information in Exercises 73 and 74 to find a function for the French dress size that corresponds to a size x dress in Italy.

88. *Dress sizes in Italy and France.*   Use the information in Exercises 73 and 74 to find a function for the Italian dress size that corresponds to a size x dress in France.

89. What relationship exists between the answers to Exercises 87 and 88? Explain how you determined this.

90. Show that function composition is associative by showing that $((f \circ g) \circ h)(x) = (f \circ (g \circ h))(x)$.

91. Show that if $h(x) = (f \circ g)(x)$, then $h^{-1}(x) = (g^{-1} \circ f^{-1})(x)$. (*Hint:* Use Exercise 90.)

Determine whether or not the given pairs of functions are inverses of each other.

92. $f(x) = 0.75x^2 + 2;\ g(x) = \sqrt{\dfrac{4(x-2)}{3}}$

93. $f(x) = 1.4x^3 + 3.2;\ g(x) = \sqrt[3]{\dfrac{x - 3.2}{1.4}}$

94. $f(x) = \sqrt{2.5x + 9.25};$
$g(x) = 0.4x^2 - 3.7,\ x \geq 0$

95. $f(x) = 0.8x^{1/2} + 5.23;$
$g(x) = 1.25(x^2 - 5.23),\ x \geq 0$

96. $f(x) = 2.5(x^3 - 7.1);$
$g(x) = \sqrt[3]{0.4x + 7.1}$

97. Match each function in Column A with its inverse from Column B.

Column A

(1) $y = 5x^3 + 10$

(2) $y = (5x + 10)^3$

(3) $y = 5(x + 10)^3$

(4) $y = (5x)^3 + 10$

Column B

A. $y = \dfrac{\sqrt[3]{x} - 10}{5}$

B. $y = \sqrt[3]{\dfrac{x}{5}} - 10$

C. $y = \sqrt[3]{\dfrac{x - 10}{5}}$

D. $y = \dfrac{\sqrt[3]{x - 10}}{5}$

98. Examine the following table. Is it possible that f and g are inverses of each other? Why or why not?

x	$f(x)$	$g(x)$
6	6	6
7	6.5	8
8	7	10
9	7.5	12
10	8	14
11	8.5	16
12	9	18

99. The following window appears on a graphing calculator.

X	Y₁	Y₂
0	1	−2
1	1.5	0
2	2	2
3	2.5	4
4	3	6
5	3.5	8
6	4	10

X = 0

a) What evidence is there that the functions Y₁ and Y₂ are inverses of each other?

b) Find equations for Y₁ and Y₂, assuming that both are linear functions.

c) On the basis of your answer to part (b), are Y₁ and Y₂ inverses of each other?

Graphing Exponential Functions   ▪   Equations with *x* and *y* Interchanged   ▪
Applications of Exponential Functions

In this section, we introduce a new type of function, the *exponential function*. These functions and their inverses, called *logarithmic functions*, have applications in many fields.

Consider the graph below. The rapidly rising curve approximates the graph of an *exponential function*.

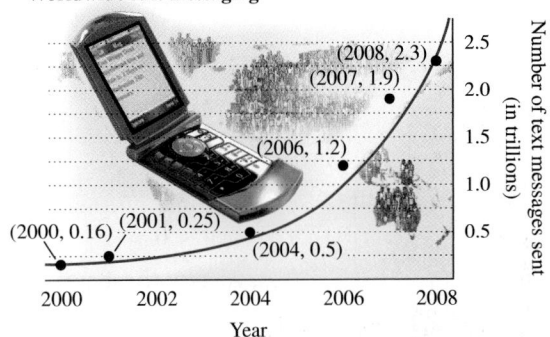

Worldwide text messaging

(2008, 2.3)
(2007, 1.9)
(2006, 1.2)
(2000, 0.16) (2001, 0.25)
(2004, 0.5)

Number of text messages sent (in trillions)

Year

Source: Mobile SMS Marketing, Gartner

Graphing Exponential Functions

In Chapter 10, we studied exponential expressions with rational-number exponents, such as

$$5^{1/4}, \quad 3^{-3/4}, \quad 7^{2.34}, \quad 5^{1.73}.$$

For example, $5^{1.73}$, or $5^{173/100}$, represents the 100th root of 5 raised to the 173rd power. What about expressions with irrational exponents, such as $5^{\sqrt{3}}$ or $7^{-\pi}$? To attach meaning to $5^{\sqrt{3}}$, consider a rational approximation, r, of $\sqrt{3}$. As r gets closer to $\sqrt{3}$, the value of 5^r gets closer to some real number p.

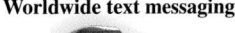

r closes in on $\sqrt{3}$.	5^r closes in on some real number p.
$1.7 < r < 1.8$	$15.426 \approx 5^{1.7} < p < 5^{1.8} \approx 18.119$
$1.73 < r < 1.74$	$16.189 \approx 5^{1.73} < p < 5^{1.74} \approx 16.452$
$1.732 < r < 1.733$	$16.241 \approx 5^{1.732} < p < 5^{1.733} \approx 16.267$

We define $5^{\sqrt{3}}$ to be the number p. To eight decimal places,

$$5^{\sqrt{3}} \approx 16.24245082.$$

Any positive irrational exponent can be interpreted in a similar way. Negative irrational exponents are then defined using reciprocals. Thus, so long as a is positive, a^x has meaning for *any* real number x. All of the laws of exponents still hold, but we will not prove that here. We can now define an *exponential function*.

> **Exponential Function**
>
> The function $f(x) = a^x$, where a is a positive constant, $a \neq 1$, and x is any real number, is called the *exponential function*, base a.

We require the base a to be positive to avoid imaginary numbers that would result from taking even roots of negative numbers. The restriction $a \neq 1$ is made to exclude the constant function $f(x) = 1^x$, or $f(x) = 1$.

The following are examples of exponential functions:

$$f(x) = 2^x, \qquad f(x) = \left(\tfrac{1}{3}\right)^x, \qquad f(x) = 5^{-3x}. \qquad \text{Note that } 5^{-3x} = (5^{-3})^x.$$

Like polynomial functions, the domain of an exponential function is the set of all real numbers. Unlike polynomial functions, exponential functions have a variable exponent. Because of this, graphs of exponential functions either rise or fall dramatically.

EXAMPLE **1**   Graph the exponential function given by $y = f(x) = 2^x$.

SOLUTION   We compute some function values, thinking of y as $f(x)$, and list the results in a table. It is a good idea to start by letting $x = 0$.

$$f(0) = 2^0 = 1; \qquad\qquad f(-1) = 2^{-1} = \frac{1}{2^1} = \frac{1}{2};$$
$$f(1) = 2^1 = 2;$$
$$f(2) = 2^2 = 4; \qquad\qquad f(-2) = 2^{-2} = \frac{1}{2^2} = \frac{1}{4};$$
$$f(3) = 2^3 = 8;$$
$$f(-3) = 2^{-3} = \frac{1}{2^3} = \frac{1}{8}$$

x	y, or $f(x)$
0	1
1	2
2	4
3	8
-1	$\frac{1}{2}$
-2	$\frac{1}{4}$
-3	$\frac{1}{8}$

Next, we plot these points and connect them with a smooth curve.

The curve comes very close to the x-axis, but does not touch or cross it.

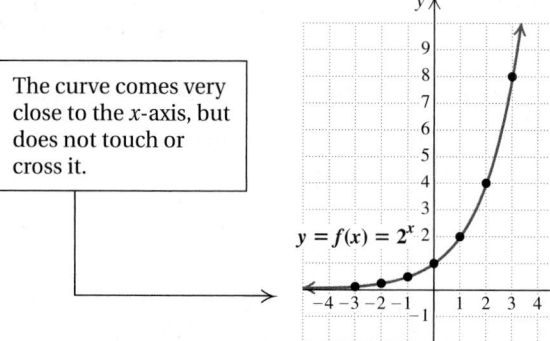

$$y = f(x) = 2^x$$

Be sure to plot enough points to determine how steeply the curve rises.

Note that as x increases, the function values increase without bound. As x decreases, the function values decrease, getting closer to 0. The x-axis, or the line $y = 0$, is a horizontal *asymptote*, meaning that the curve gets closer and closer to this line the further we move to the left.

TRY EXERCISE 7

EXAMPLE **2**   Graph: $y = f(x) = \left(\tfrac{1}{2}\right)^x$.

SOLUTION   We compute some function values, thinking of y as $f(x)$, and list the results in a table. Before we do this, note that

$$y = f(x) = \left(\tfrac{1}{2}\right)^x = (2^{-1})^x = 2^{-x}.$$

Then we have

$$f(0) = 2^{-0} = 1; \qquad\qquad f(3) = 2^{-3} = \frac{1}{2^3} = \frac{1}{8};$$
$$f(1) = 2^{-1} = \frac{1}{2^1} = \frac{1}{2}; \qquad\qquad f(-1) = 2^{-(-1)} = 2^1 = 2;$$
$$f(-2) = 2^{-(-2)} = 2^2 = 4;$$
$$f(2) = 2^{-2} = \frac{1}{2^2} = \frac{1}{4}; \qquad\qquad f(-3) = 2^{-(-3)} = 2^3 = 8.$$

x	y, or $f(x)$
0	1
1	$\frac{1}{2}$
2	$\frac{1}{4}$
3	$\frac{1}{8}$
-1	2
-2	4
-3	8

Graphing calculators are helpful when graphing equations like $y = 5000(1.075)^x$. To choose a window, we note that y-values are positive and increase rapidly. One suitable window is $[-10, 10, 0, 15000]$, with a y-scale of 1000.

$y = 5000(1.075)^x$

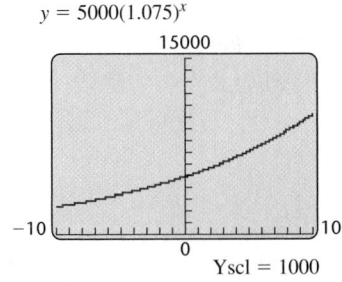

Graph each pair of functions. Select an appropriate window and scale.

1. $y_1 = \left(\frac{5}{2}\right)^x$ and $y_2 = \left(\frac{2}{5}\right)^x$
2. $y_1 = 3.2^x$ and $y_2 = 3.2^{-x}$
3. $y_1 = \left(\frac{3}{7}\right)^x$ and $y_2 = \left(\frac{7}{3}\right)^x$
4. $y_1 = 5000(1.08)^x$ and $y_2 = 5000(1.08)^{x-3}$

Next, we plot these points and connect them with a smooth curve. This curve is a mirror image, or *reflection*, of the graph of $y = 2^x$ (see Example 1) across the y-axis. The line $y = 0$ is again the horizontal asymptote.

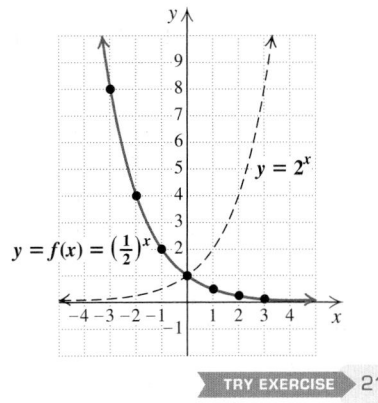

TRY EXERCISE 21

From Examples 1 and 2, we can make the following observations.

- For $a > 1$, the graph of $f(x) = a^x$ increases from left to right. The greater the value of a, the steeper the curve. (See the figure on the left below.)

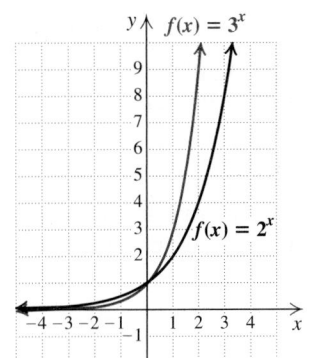 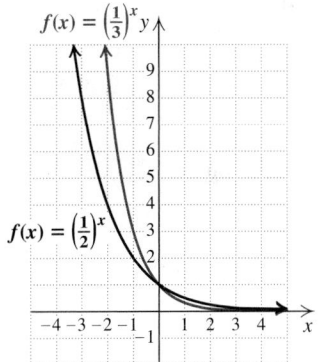

- For $0 < a < 1$, the graph of $f(x) = a^x$ decreases from left to right. For smaller values of a, the curve is steeper. (See the figure on the right above.)
- All graphs of $f(x) = a^x$ go through the y-intercept $(0, 1)$.
- All graphs of $f(x) = a^x$ have the x-axis as the horizontal asymptote.
- If $f(x) = a^x$, with $a > 0$, $a \neq 1$, the domain of f is all real numbers, and the range of f is all positive real numbers.
- For $a > 0$, $a \neq 1$, the function given by $f(x) = a^x$ is one-to-one. Its graph passes the horizontal-line test.

EXAMPLE 3

STUDENT NOTES

When using translations, make sure that you are shifting in the correct direction. When in doubt, substitute a value for x and make some calculations.

Graph: $y = f(x) = 2^{x-2}$.

SOLUTION We construct a table of values. Then we plot the points and connect them with a smooth curve. Here $x - 2$ is the *exponent*.

$f(0) = 2^{0-2} = 2^{-2} = \frac{1}{4}$;
$f(1) = 2^{1-2} = 2^{-1} = \frac{1}{2}$;
$f(2) = 2^{2-2} = 2^0 = 1$;
$f(3) = 2^{3-2} = 2^1 = 2$;
$f(4) = 2^{4-2} = 2^2 = 4$;
$f(-1) = 2^{-1-2} = 2^{-3} = \frac{1}{8}$;
$f(-2) = 2^{-2-2} = 2^{-4} = \frac{1}{16}$

x	y, or $f(x)$
0	$\frac{1}{4}$
1	$\frac{1}{2}$
2	1
3	2
4	4
-1	$\frac{1}{8}$
-2	$\frac{1}{16}$

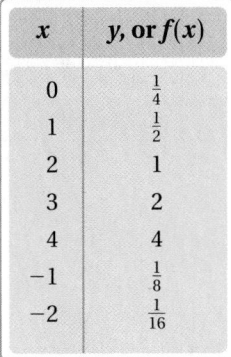

To practice graphing equations that are translations of each other, use **MODE** SIMUL and **FORMAT** EXPROFF to graph $y_1 = 2^x$, $y_2 = 2^{x+1}$, $y_3 = 2^{x-1}$, $y_4 = 2^x + 1$, and $y_5 = 2^x - 1$. Use a bold curve for y_1 and then predict which curve represents which equation. Use **TRACE** to confirm your predictions. Switching **FORMAT** to EXPRON and using **TRACE** provides a definitive check (see also Exercise 75).

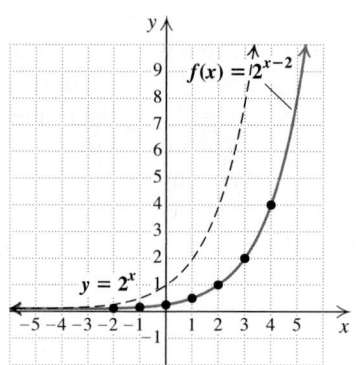

The graph looks just like the graph of $y = 2^x$, but it is translated 2 units to the right. The y-intercept of $y = 2^x$ is $(0, 1)$. The y-intercept of $y = 2^{x-2}$ is $\left(0, \frac{1}{4}\right)$. The line $y = 0$ is again the horizontal asymptote.

TRY EXERCISE 17

Equations with x and y Interchanged

It will be helpful in later work to be able to graph an equation in which the x and the y in $y = a^x$ are interchanged.

EXAMPLE 4

Graph: $x = 2^y$.

SOLUTION Note that x is alone on one side of the equation. To find ordered pairs that are solutions, we choose values for y and then compute values for x.

For $y = 0$, $x = 2^0 = 1$.

For $y = 1$, $x = 2^1 = 2$.

For $y = 2$, $x = 2^2 = 4$.

For $y = 3$, $x = 2^3 = 8$.

For $y = -1$, $x = 2^{-1} = \frac{1}{2}$.

For $y = -2$, $x = 2^{-2} = \frac{1}{4}$.

For $y = -3$, $x = 2^{-3} = \frac{1}{8}$.

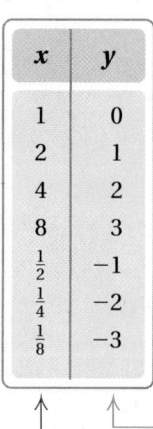

x	y
1	0
2	1
4	2
8	3
$\frac{1}{2}$	-1
$\frac{1}{4}$	-2
$\frac{1}{8}$	-3

(1) Choose values for y.

(2) Compute values for x.

We plot the points and connect them with a smooth curve.

This curve does not touch or cross the y-axis, which serves as a vertical asymptote.

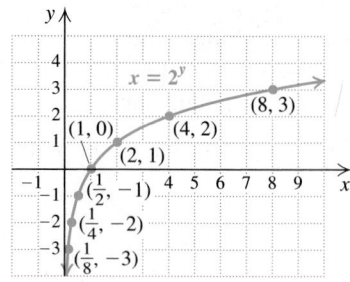

Note too that this curve looks just like the graph of $y = 2^x$, except that it is reflected across the line $y = x$, as shown here.

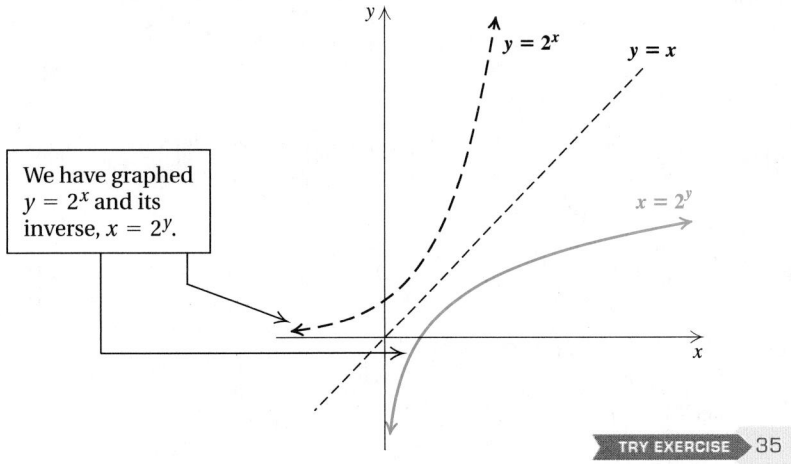

We have graphed $y = 2^x$ and its inverse, $x = 2^y$.

TRY EXERCISE 35

Applications of Exponential Functions

EXAMPLE 5

Interest compounded annually. The amount of money A that a principal P will be worth after t years at interest rate i, compounded annually, is given by the formula

$$A = P(1 + i)^t.$$ You might review Example 11 in Section 11.1.

Suppose that $100,000 is invested at 8% interest, compounded annually.

a) Find a function for the amount in the account after t years.

b) Find the amount of money in the account at $t = 0$, $t = 4$, $t = 8$, and $t = 10$.

c) Graph the function.

SOLUTION

a) If $P = 100,000$ and $i = 8\% = 0.08$, we can substitute these values and form the following function:

$$A(t) = \$100,000(1 + 0.08)^t \quad \text{Using } A = P(1 + i)^t$$
$$= \$100,000(1.08)^t.$$

b) To find the function values, a calculator with a power key is helpful.

$$A(0) = \$100,000(1.08)^0 \qquad\qquad A(8) = \$100,000(1.08)^8$$
$$= \$100,000(1) \qquad\qquad\qquad \approx \$100,000(1.85093021)$$
$$= \$100,000 \qquad\qquad\qquad\quad \approx \$185,093.02$$

$$A(4) = \$100,000(1.08)^4 \qquad\qquad A(10) = \$100,000(1.08)^{10}$$
$$= \$100,000(1.36048896) \qquad\quad \approx \$100,000(2.158924997)$$
$$\approx \$136,048.90 \qquad\qquad\qquad \approx \$215,892.50$$

TECHNOLOGY CONNECTION

Graphing calculators can quickly find many function values. Let $y_1 = 100,000(1.08)^x$. Then use the TABLE feature with INDPNT set to ASK to check Example 5(b).

c) We use the function values computed in part (b), and others if we wish, to draw the graph as follows. Note that the axes are scaled differently because of the large numbers.

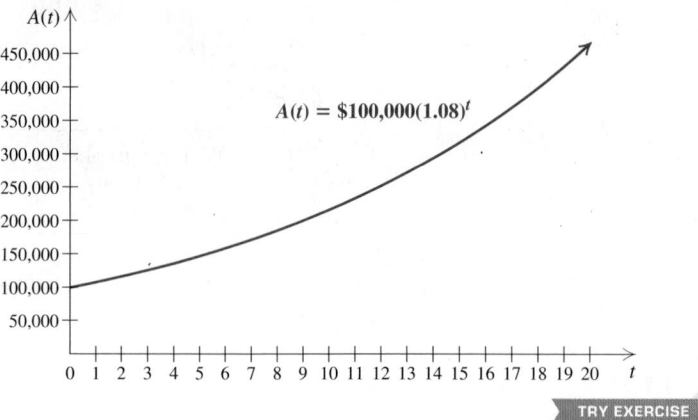

$$A(t) = \$100{,}000(1.08)^t$$

TRY EXERCISE ▶ 39

12.2 EXERCISE SET

☙ **Concept Reinforcement** *Classify each statement as either true or false.*

1. The graph of $f(x) = a^x$ always passes through the point $(0, 1)$.

2. The graph of $g(x) = \left(\frac{1}{2}\right)^x$ gets closer and closer to the x-axis as x gets larger and larger.

3. The graph of $f(x) = 2^{x-3}$ looks just like the graph of $y = 2^x$, but it is translated 3 units to the right.

4. The graph of $g(x) = 2^x - 3$ looks just like the graph of $y = 2^x$, but it is translated 3 units up.

5. The graph of $y = 3^x$ gets close to, but never touches, the y-axis.

6. The graph of $x = 3^y$ gets close to, but never touches, the y-axis.

Graph.

7. $y = f(x) = 3^x$

8. $y = f(x) = 4^x$

9. $y = 6^x$

10. $y = 5^x$

11. $y = 2^x + 1$

12. $y = 2^x + 3$

13. $y = 3^x - 2$

14. $y = 3^x - 1$

15. $y = 2^x - 5$

16. $y = 2^x - 4$

17. $y = 2^{x-3}$

18. $y = 2^{x-1}$

19. $y = 2^{x+1}$

20. $y = 2^{x+3}$

21. $y = \left(\frac{1}{4}\right)^x$

22. $y = \left(\frac{1}{5}\right)^x$

23. $y = \left(\frac{1}{3}\right)^x$

24. $y = \left(\frac{1}{10}\right)^x$

25. $y = 2^{x+1} - 3$

26. $y = 2^{x-3} - 1$

27. $x = 6^y$

28. $x = 3^y$

29. $x = 3^{-y}$

30. $x = 2^{-y}$

31. $x = 4^y$

32. $x = 5^y$

33. $x = \left(\frac{4}{3}\right)^y$

34. $x = \left(\frac{3}{2}\right)^y$

Graph each pair of equations on the same set of axes.

35. $y = 3^x, \ x = 3^y$

36. $y = 2^x, \ x = 2^y$

37. $y = \left(\frac{1}{2}\right)^x, \ x = \left(\frac{1}{2}\right)^y$

38. $y = \left(\frac{1}{4}\right)^x, \ x = \left(\frac{1}{4}\right)^y$

Solve.

39. *Music downloads.* The number $M(t)$ of single tracks downloaded, in billions, t years after 2003 can be approximated by
$$M(t) = 0.353(1.244)^t.$$
Source: International Federation of the Phonographic Industry

a) Estimate the number of single tracks downloaded in 2006, in 2008, and in 2012.

b) Graph the function.

40. *Growth of bacteria.* The bacteria *Escherichia coli* are commonly found in the human bladder. Suppose that 3000 of the bacteria are present at time $t = 0$. Then t minutes later, the number of bacteria present can be approximated by
$$N(t) = 3000(2)^{t/20}.$$

a) How many bacteria will be present after 10 min? 20 min? 30 min? 40 min? 60 min?

b) Graph the function.

41. *Smoking cessation.* The percentage of smokers P who receive telephone counseling to quit smoking and are still successful t months later can be approximated by

$$P(t) = 21.4(0.914)^t.$$

Sources: *New England Journal of Medicine;* data from California's Smokers' Hotline

a) Estimate the percentage of smokers receiving telephone counseling who are successful in quitting for 1 month, 3 months, and 1 year.

b) Graph the function.

42. *Smoking cessation.* The percentage of smokers P who, without telephone counseling, have successfully quit smoking for t months (see Exercise 41) can be approximated by

$$P(t) = 9.02(0.93)^t.$$

Sources: *New England Journal of Medicine;* data from California's Smokers' Hotline

a) Estimate the percentage of smokers not receiving telephone counseling who are successful in quitting for 1 month, 3 months, and 1 year.

b) Graph the function.

43. *Marine biology.* Due to excessive whaling prior to the mid 1970s, the humpback whale is considered an endangered species. The worldwide population of humpbacks, $P(t)$, in thousands, t years after 1900 ($t < 70$) can be approximated by*

$$P(t) = 150(0.960)^t.$$

a) How many humpback whales were alive in 1930? in 1960?

b) Graph the function.

44. *Salvage value.* A laser printer is purchased for $1200. Its value each year is about 80% of the value of the preceding year. Its value, in dollars, after t years is given by the exponential function

$$V(t) = 1200(0.8)^t.$$

a) Find the value of the printer after 0 yr, 1 yr, 2 yr, 5 yr, and 10 yr.

b) Graph the function.

45. *Marine biology.* As a result of preservation efforts in most countries in which whaling was common, the humpback whale population has grown since the 1970s. The worldwide population of hump-

backs, $P(t)$, in thousands, t years after 1982 can be approximated by*

$$P(t) = 5.5(1.047)^t.$$

a) How many humpback whales were alive in 1992? in 2004?

b) Graph the function.

46. *Recycling aluminum cans.* It is estimated that $\frac{1}{2}$ of all aluminum cans distributed will be recycled each year. A beverage company distributes 250,000 cans. The number still in use after time t, in years, is given by the exponential function

$$N(t) = 250,000\left(\tfrac{1}{2}\right)^t.$$

Source: The Aluminum Association, Inc., 2005

a) How many cans are still in use after 0 yr? 1 yr? 4 yr? 10 yr?

b) Graph the function.

47. *Spread of zebra mussels.* Beginning in 1988, infestations of zebra mussels started spreading throughout North American waters.[†] These mussels spread with such speed that water treatment facilities, power plants, and entire ecosystems can become threatened. The function

$$A(t) = 10 \cdot 34^t$$

can be used to estimate the number of square centimeters of lake bottom that will be covered with mussels t years after an infestation covering 10 cm^2 first occurs.

a) How many square centimeters of lake bottom will be covered with mussels 5 yr after an infestation covering 10 cm^2 first appears? 7 yr after the infestation first appears?

b) Graph the function.

---

*Based on information from the American Cetacean Society, 2001, and the ASK Archive, 1998.

[†]Many thanks to Dr. Gerald Mackie of the Department of Zoology at the University of Guelph in Ontario for the background information for this exercise.

48. *Cell phones.*   The number of cell phones in use in the United States is increasing exponentially. The number N, in millions, in use can be estimated by

$$N(t) = 7.12(1.3)^t,$$

where t is the number of years after 1990.
Source: Based on data from CTIA-The Wireless Association

a) Estimate the number of cell phones in use in 1995, in 2005, and in 2010.
b) Graph the function.

49. Without using a calculator, explain why 2^π must be greater than 8 but less than 16.

50. Suppose that $1000 is invested for 5 yr at 7% interest, compounded annually. In what year will the most interest be earned? Why?

Skill Review

Review factoring polynomials (Sections 5.1–5.6).

Factor.

51. $3x^2 - 48$ [5.4]

52. $x^2 - 20x + 100$ [5.4]

53. $6x^2 + x - 12$ [5.3]

54. $8x^6 - 64y^6$ [5.5]

55. $6y^2 + 36y - 240$ [5.2]

56. $5x^4 - 10x^3 - 3x^2 + 6x$ [5.1]

Synthesis

57. Examine Exercise 48. Do you believe that the equation for the number of cell phones in use in the United States will be accurate 20 yr from now? Why or why not?

58. Explain why the graph of $x = 2^y$ is the graph of $y = 2^x$ reflected across the line $y = x$.

Determine which of the two numbers is larger. Do not use a calculator.

59. $\pi^{1.3}$ or $\pi^{2.4}$

60. $\sqrt{8^3}$ or $8^{\sqrt{3}}$

Graph.

61. $f(x) = 2.5^x$

62. $f(x) = 0.5^x$

63. $y = 2^x + 2^{-x}$

64. $y = \left|\left(\frac{1}{2}\right)^x - 1\right|$

65. $y = |2^x - 2|$

66. $y = 2^{-(x-1)^2}$

67. $y = |2^{x^2} - 1|$

68. $y = 3^x + 3^{-x}$

Graph both equations using the same set of axes.

69. $y = 3^{-(x-1)}$, $x = 3^{-(y-1)}$

70. $y = 1^x$, $x = 1^y$

71. *Navigational devices.*   The number of GPS navigational devices in use in the United States has grown from 0.5 million in 2000 to 4 million in 2004 to 50 million in 2008. After pressing **STAT** and entering the data, use the ExpReg option in the STAT CALC menu to find an exponential function that models the number of navigational devices in use t years after 2000. Then use that function to predict the total number of devices in use in 2012.
Source: Telematics Research Group

72. *Keyboarding speed.*   Trey is studying keyboarding. After he has studied for t hours, Trey's speed, in words per minute, is given by the exponential function

$$S(t) = 200[1 - (0.99)^t].$$

Use a graph and/or table of values to predict Trey's speed after studying for 10 hr, 40 hr, and 80 hr.

73. The following graph shows growth in the height of ocean waves over time, assuming a steady surface wind.
Source: magicseaweed.com

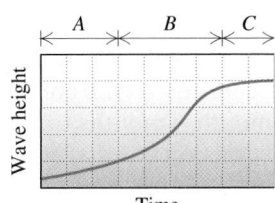

Source: magicseaweed.com

a) Consider the portions of the graph marked A, B, and C. Suppose that each portion can be labeled Exponential Growth, Linear Growth, or Saturation. How would you label each portion?
b) Small vertical movements in wind, surface roughness of water, and gravity are three forces that create waves. How might these forces be related to the shape of the wave-height graph?

74. Consider any exponential function of the form $f(x) = a^x$ with $a > 1$. Will it always follow that $f(3) - f(2) > f(2) - f(1)$, and, in general, $f(n + 2) - f(n + 1) > f(n + 1) - f(n)$? Why or why not? (*Hint*: Think graphically.)

75. On many graphing calculators, it is possible to enter and graph $y_1 = A \wedge (X - B) + C$ after first pressing **APPS** Transfrm. Use this application to graph $f(x) = 2.5^{x-3} + 2$, $g(x) = 2.5^{x+3} + 2$, $h(x) = 2.5^{x-3} - 2$, and $k(x) = 2.5^{x+3} - 2$.

CORNER

The True Cost of a New Car

Focus: Car loans and exponential functions

Time: 30 minutes

Group size: 2

Materials: Calculators with exponentiation keys

The formula

$$M = \frac{Pr}{1 - (1 + r)^{-n}}$$

is used to determine the payment size, *M*, when a loan of *P* dollars is to be repaid in *n* equally sized monthly payments. Here *r* represents the monthly interest rate. Loans repaid in this fashion are said to be *amortized* (spread out equally) over a period of *n* months.

ACTIVITY

1. Suppose one group member is selling the other a car for $2600, financed at 1% interest per month for 24 months. What should be the size of each monthly payment?

2. Suppose both group members are shopping for the same model new car. To save time, each group member visits a different dealer. One dealer offers the car for $13,000 at 10.5% interest (0.00875 monthly interest) for 60 months (no down payment). The other dealer offers the same car for $12,000, but at 12% interest (0.01 monthly interest) for 48 months (no down payment).

   a) Determine the monthly payment size for each offer. Then determine the total amount paid for the car under each offer. How much of each total is interest?

   b) Work together to find the annual interest rate for which the total cost of 60 monthly payments for the $13,000 car would equal the total amount paid for the $12,000 car (as found in part a above).

12.3	Logarithmic Functions

Graphs of Logarithmic Functions ▪ Equivalent Equations ▪ Solving Certain Logarithmic Equations

We are now ready to study inverses of exponential functions. These functions have many applications and are called *logarithm,* or *logarithmic, functions.*

Graphs of Logarithmic Functions

Consider the exponential function $f(x) = 2^x$. Like all exponential functions, f is one-to-one. Can a formula for f^{-1} be found? To answer this, we use the method of Section 12.1:

1. Replace $f(x)$ with y: $y = 2^x$.

2. Interchange x and y: $x = 2^y$.

3. Solve for y: y = the exponent to which we raise 2 to get x.

4. Replace y with $f^{-1}(x)$: $f^{-1}(x)$ = the exponent to which we raise 2 to get x.

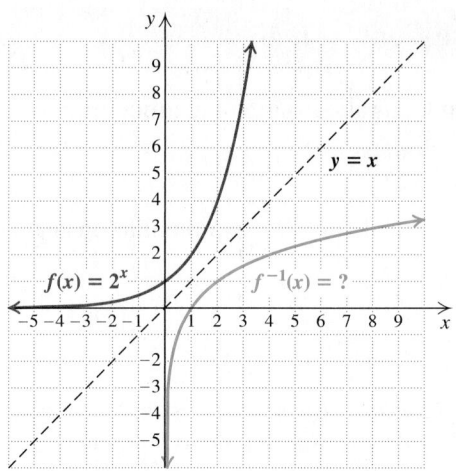

We now define a new symbol to replace the words "the exponent to which we raise 2 to get x":

> $\log_2 x$, read "the logarithm, base 2, of x," or "log, base 2, of x," means "the exponent to which we raise 2 to get x."

Thus if $f(x) = 2^x$, then $f^{-1}(x) = \log_2 x$. Note that $f^{-1}(8) = \log_2 8 = 3$, because 3 is *the exponent to which we raise* 2 *to get* 8.

EXAMPLE 1

Simplify: **(a)** $\log_2 32$; **(b)** $\log_2 1$; **(c)** $\log_2 \frac{1}{8}$.

SOLUTION

a) Think of $\log_2 32$ as the exponent to which we raise 2 to get 32. That exponent is 5. Therefore, $\log_2 32 = 5$.

b) We ask ourselves: "To what exponent do we raise 2 in order to get 1?" That exponent is 0 (recall that $2^0 = 1$). Thus, $\log_2 1 = 0$.

c) To what exponent do we raise 2 in order to get $\frac{1}{8}$? Since $2^{-3} = \frac{1}{8}$, we have $\log_2 \frac{1}{8} = -3$.

▶ **TRY EXERCISE** ▶ 9

Although numbers like $\log_2 13$ can be only approximated, we must remember that $\log_2 13$ represents *the exponent to which we raise* 2 *to get* 13. That is, $2^{\log_2 13} = 13$. A calculator indicates that $\log_2 13 \approx 3.7$ and $2^{3.7} \approx 13$.

For any exponential function $f(x) = a^x$, the inverse is called a **logarithmic function, base a**. The graph of the inverse can be drawn by reflecting the graph of $f(x) = a^x$ across the line $y = x$. It will be helpful to remember that the inverse of $f(x) = a^x$ is given by $f^{-1}(x) = \log_a x$.

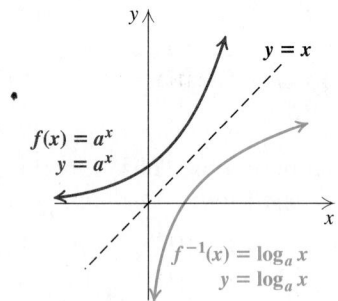

The Meaning of $\log_a x$

For $x > 0$ and a a positive constant other than 1, $\log_a x$ is the exponent to which a must be raised in order to get x. Thus,

$$\log_a x = m \quad \text{means} \quad a^m = x$$

or equivalently,

$$\log_a x \text{ is that unique exponent for which } a^{\log_a x} = x.$$

STUDENT NOTES ────────

As an aid in remembering what $\log_a x$ means, note that a is called the *base*, just as it is the base in $a^y = x$.

It is important to remember that *a logarithm is an exponent*. It might help to verbalize: "The logarithm, base *a*, of a number *x* is the exponent to which *a* must be raised in order to get *x*."

EXAMPLE 2   Simplify: $7^{\log_7 85}$.

SOLUTION   Remember that $\log_7 85$ is the exponent to which 7 is raised to get 85. Raising 7 to that exponent, we have

$$7^{\log_7 85} = 85.$$

> **TRY EXERCISE** 35

Because logarithmic and exponential functions are inverses of each other, the result in Example 2 should come as no surprise: If $f(x) = \log_7 x$, then

for $f(x) = \log_7 x$, we have $f^{-1}(x) = 7^x$

and $f^{-1}(f(x)) = f^{-1}(\log_7 x) = 7^{\log_7 x} = x.$

Thus, $f^{-1}(f(85)) = 7^{\log_7 85} = 85$.

The following is a comparison of exponential and logarithmic functions.

Exponential Function	Logarithmic Function
$y = a^x$	$x = a^y$
$f(x) = a^x$	$g(x) = \log_a x$
$a > 0, a \neq 1$	$a > 0, a \neq 1$
The domain is $\mathbb{R}$.	The range is $\mathbb{R}$.
$y > 0$ (Outputs are positive.)	$x > 0$ (Inputs are positive.)
$f^{-1}(x) = \log_a x$	$g^{-1}(x) = a^x$

EXAMPLE 3   Graph: $y = f(x) = \log_5 x$.

SOLUTION   If $y = \log_5 x$, then $5^y = x$. We can find ordered pairs that are solutions by choosing values for y and computing the x-values.

For $y = 0$, $x = 5^0 = 1$.

For $y = 1$, $x = 5^1 = 5$.

For $y = 2$, $x = 5^2 = 25$.

For $y = -1$, $x = 5^{-1} = \frac{1}{5}$.

For $y = -2$, $x = 5^{-2} = \frac{1}{25}$.

(1) Select y.

(2) Compute x.

x, or 5^y	y
1	0
5	1
25	2
$\frac{1}{5}$	-1
$\frac{1}{25}$	-2

This table shows the following:

$\log_5 1 = 0;$

$\log_5 5 = 1;$

$\log_5 25 = 2;$

$\log_5 \frac{1}{5} = -1;$

$\log_5 \frac{1}{25} = -2.$

These can all be checked using the equations above.

We plot the set of ordered pairs and connect the points with a smooth curve. The graphs of $y = 5^x$ and $y = x$ are shown only for reference.

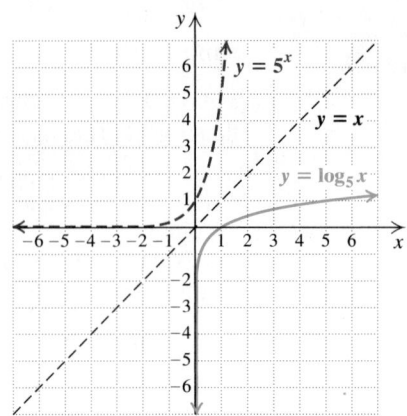

TRY EXERCISE 37

Equivalent Equations

We use the definition of logarithm to rewrite a *logarithmic equation* as an equivalent *exponential equation* or the other way around:

$$m = \log_a x \quad \text{is equivalent to} \quad a^m = x.$$

> *CAUTION!* **Do not forget this relationship!** It is probably the most important definition in the chapter. Many times this definition will be used to justify a property we are considering.

EXAMPLE 4 Rewrite each as an equivalent exponential equation: **(a)** $y = \log_3 5$; **(b)** $-2 = \log_a 7$; **(c)** $a = \log_b d$.

SOLUTION

a) $y = \log_3 5$ is equivalent to $3^y = 5$ The logarithm is the exponent.

The base remains the base.

b) $-2 = \log_a 7$ is equivalent to $a^{-2} = 7$

c) $a = \log_b d$ is equivalent to $b^a = d$

TRY EXERCISE 47

We also use the definition of logarithm to rewrite an exponential equation as an equivalent logarithmic equation.

EXAMPLE 5 Rewrite each as an equivalent logarithmic equation: **(a)** $8 = 2^x$; **(b)** $y^{-1} = 4$; **(c)** $a^b = c$.

SOLUTION

a) $8 = 2^x$ is equivalent to $x = \log_2 8$ The exponent is the logarithm.

The base remains the base.

b) $y^{-1} = 4$ is equivalent to $-1 = \log_y 4$

c) $a^b = c$ is equivalent to $b = \log_a c$

TRY EXERCISE 63

Solving Certain Logarithmic Equations

Many logarithmic equations can be solved by rewriting them as equivalent exponential equations.

EXAMPLE 6

Solve: **(a)** $\log_2 x = -3$; **(b)** $\log_x 16 = 2$.

SOLUTION

a) $\log_2 x = -3$

$\quad\quad 2^{-3} = x \quad\quad$ Rewriting as an exponential equation

$\quad\quad \frac{1}{8} = x \quad\quad$ Computing 2^{-3}

Check: $\log_2 \frac{1}{8}$ is the exponent to which 2 is raised to get $\frac{1}{8}$. Since that exponent is -3, we have a check. The solution is $\frac{1}{8}$.

b) $\log_x 16 = 2$

$\quad\quad x^2 = 16 \quad\quad\quad$ Rewriting as an exponential equation

$\quad x = 4 \;\; or \;\; x = -4 \quad$ Principle of square roots

Check: $\log_4 16 = 2$ because $4^2 = 16$. Thus, 4 is a solution of $\log_x 16 = 2$. Because all logarithmic bases must be positive, -4 cannot be a solution. Logarithmic bases must be positive because logarithms are defined using exponential functions that require positive bases. The solution is 4.

TRY EXERCISE ▸ 79

One method for solving certain logarithmic and exponential equations relies on the following property, which results from the fact that exponential functions are one-to-one.

> ### The Principle of Exponential Equality
>
> For any real number b, where $b \neq -1$, 0, or 1,
>
> $$b^{x_1} = b^{x_2} \quad \text{is equivalent to} \quad x_1 = x_2.$$
>
> (Powers of the same base are equal if and only if the exponents are equal.)

EXAMPLE 7

Solve: **(a)** $\log_{10} 1000 = x$; **(b)** $\log_4 1 = t$.

SOLUTION

a) We rewrite $\log_{10} 1000 = x$ in exponential form and solve:

$\quad\quad 10^x = 1000 \quad\quad$ Rewriting as an exponential equation

$\quad\quad 10^x = 10^3 \quad\quad$ Writing 1000 as a power of 10

$\quad\quad\quad x = 3. \quad\quad$ Equating exponents

Check: This equation can also be solved directly by determining the exponent to which we raise 10 in order to get 1000. In both cases we find that $\log_{10} 1000 = 3$, so we have a check. The solution is 3.

b) We rewrite $\log_4 1 = t$ in exponential form and solve:

$$4^t = 1 \qquad \text{Rewriting as an exponential equation}$$
$$4^t = 4^0 \qquad \text{Writing 1 as a power of 4. This can be done mentally.}$$
$$t = 0. \qquad \text{Equating exponents}$$

Check: As in part (a), this equation can be solved directly by determining the exponent to which we raise 4 in order to get 1. In both cases we find that $\log_4 1 = 0$, so we have a check. The solution is 0.

> TRY EXERCISE 81

Example 7 illustrates an important property of logarithms.

> **$\log_a 1$**
> The logarithm, base a, of 1 is always 0: $\log_a 1 = 0$.

This follows from the fact that $a^0 = 1$ is equivalent to the logarithmic equation $\log_a 1 = 0$. Thus, $\log_{10} 1 = 0$, $\log_7 1 = 0$, and so on.

Another property results from the fact that $a^1 = a$. This is equivalent to the logarithmic equation $\log_a a = 1$.

> **$\log_a a$**
> The logarithm, base a, of a is always 1: $\log_a a = 1$.

Thus, $\log_{10} 10 = 1$, $\log_8 8 = 1$, and so on.

12.3 EXERCISE SET

Concept Reinforcement *In each of Exercises 1–8, match the expression or equation with an equivalent expression or equation from the column on the right.*

1. ____ $\log_5 25$
2. ____ $2^5 = x$
3. ____ $\log_5 5$
4. ____ $\log_2 1$
5. ____ $\log_5 5^x$
6. ____ $\log_x 27 = 5$
7. ____ $5 = 2^x$
8. ____ $x^{-2} = 5$

a) 1
b) x
c) $x^5 = 27$
d) $\log_2 x = 5$
e) $\log_2 5 = x$
f) $\log_x 5 = -2$
g) 2
h) 0

Simplify.

9. $\log_{10} 1000$
10. $\log_{10} 100$
11. $\log_7 49$
12. $\log_2 8$
13. $\log_3 81$
14. $\log_3 9$
15. $\log_5 \frac{1}{25}$
16. $\log_5 \frac{1}{5}$
17. $\log_8 \frac{1}{8}$
18. $\log_8 \frac{1}{64}$
19. $\log_5 625$
20. $\log_5 125$
21. $\log_7 7$
22. $\log_9 1$
23. $\log_3 1$
24. $\log_3 3$
Aha! 25. $\log_6 6^5$
26. $\log_6 6^9$
27. $\log_{10} 0.01$
28. $\log_{10} 0.1$

29. $\log_{16} 4$

30. $\log_{100} 10$

31. $\log_9 27$

32. $\log_4 32$

33. $\log_{1000} 100$

34. $\log_{16} 8$

35. $3^{\log_3 29}$

36. $6^{\log_6 13}$

Graph.

37. $y = \log_{10} x$

38. $y = \log_2 x$

39. $y = \log_3 x$

40. $y = \log_7 x$

41. $f(x) = \log_6 x$

42. $f(x) = \log_4 x$

43. $f(x) = \log_{2.5} x$

44. $f(x) = \log_{1/2} x$

Graph both functions using the same set of axes.

45. $f(x) = 3^x$, $f^{-1}(x) = \log_3 x$

46. $f(x) = 4^x$, $f^{-1}(x) = \log_4 x$

Rewrite each of the following as an equivalent exponential equation. Do not solve.

47. $x = \log_{10} 8$

48. $y = \log_8 10$

49. $\log_9 9 = 1$

50. $\log_6 36 = 2$

51. $\log_{10} 0.1 = -1$

52. $\log_{10} 0.01 = -2$

53. $\log_{10} 7 = 0.845$

54. $\log_{10} 3 = 0.4771$

55. $\log_c m = 8$

56. $\log_b n = 23$

57. $\log_r C = t$

58. $\log_m P = a$

59. $\log_e 0.25 = -1.3863$

60. $\log_e 0.989 = -0.0111$

61. $\log_r T = -x$

62. $\log_c M = -w$

Rewrite each of the following as an equivalent logarithmic equation. Do not solve.

63. $10^2 = 100$

64. $10^4 = 10,000$

65. $5^{-3} = \frac{1}{125}$

66. $2^{-5} = \frac{1}{32}$

67. $16^{1/4} = 2$

68. $8^{1/3} = 2$

69. $10^{0.4771} = 3$

70. $10^{0.3010} = 2$

71. $z^m = 6$

72. $m^n = r$

73. $p^t = q$

74. $y^t = x$

75. $e^3 = 20.0855$

76. $e^2 = 7.3891$

77. $e^{-4} = 0.0183$

78. $e^{-2} = 0.1353$

Solve.

79. $\log_6 x = 2$

80. $\log_4 x = 3$

81. $\log_2 32 = x$

82. $\log_5 25 = x$

83. $\log_x 9 = 1$

84. $\log_x 12 = 1$

85. $\log_x 7 = \frac{1}{2}$

86. $\log_x 9 = \frac{1}{2}$

87. $\log_3 x = -2$

88. $\log_2 x = -1$

89. $\log_{32} x = \frac{2}{5}$

90. $\log_8 x = \frac{2}{3}$

91. In what way is a logarithm an exponent?

92. Is it easier to find x given $x = \log_9 \frac{1}{3}$ or given $9^x = \frac{1}{3}$? Explain your reasoning.

Skill Review

Review simplifying rational and radical expressions (Chapters 6 and 10).

Simplify.

93. $\sqrt{18a^3b}\sqrt{50ab^7}$ [10.3]

94. $(2\sqrt{3} + \sqrt{5})(2\sqrt{3} - \sqrt{10})$ [10.5]

95. $\sqrt{192x} - \sqrt{75x}$ [10.5]

96. $\sqrt[4]{\sqrt[3]{x}}$ [10.2]

97. $\dfrac{\dfrac{3}{x} - \dfrac{2}{xy}}{\dfrac{2}{x^2} + \dfrac{1}{xy}}$ [6.5]

98. $\dfrac{\dfrac{4+x}{x^2 + 2x + 1}}{\dfrac{3}{x+1} - \dfrac{2}{x+2}}$ [6.5]

Synthesis

99. Would a manufacturer be pleased or unhappy if sales of a product grew logarithmically? Why?

100. Explain why the number $\log_2 13$ must be between 3 and 4.

101. Graph both equations using the same set of axes:
$$y = \left(\tfrac{3}{2}\right)^x, \qquad y = \log_{3/2} x.$$

Graph.

102. $y = \log_2 (x - 1)$

103. $y = \log_3 |x + 1|$

Solve.

104. $|\log_3 x| = 2$

105. $\log_4 (3x - 2) = 2$

106. $\log_8 (2x + 1) = -1$

107. $\log_{10} (x^2 + 21x) = 2$

Simplify.

108. $\log_{1/4} \frac{1}{64}$

109. $\log_{1/5} 25$

110. $\log_{81} 3 \cdot \log_3 81$

111. $\log_{10} (\log_4 (\log_3 81))$

112. $\log_2 (\log_2 (\log_4 256))$

113. Show that $b^{x_1} = b^{x_2}$ is *not* equivalent to $x_1 = x_2$ for $b = 0$ or $b = 1$.

114. If $\log_b a = x$, does it follow that $\log_a b = 1/x$? Why or why not?

12.4 Properties of Logarithmic Functions

Logarithms of Products ▪ Logarithms of Powers ▪ Logarithms of Quotients ▪
Using the Properties Together

Logarithmic functions are important in many applications and in more advanced mathematics. We now establish some basic properties that are useful in manipulating expressions involving logarithms. As their proofs reveal, the properties of logarithms are related to the properties of exponents.

Logarithms of Products

The first property we discuss is related to the product rule for exponents: $a^m \cdot a^n = a^{m+n}$. Its proof appears immediately after Example 2.

> **The Product Rule for Logarithms**
> For any positive numbers M, N, and a ($a \neq 1$),
> $$\log_a (MN) = \log_a M + \log_a N.$$
> (The logarithm of a product is the sum of the logarithms of the factors.)

EXAMPLE 1 Express as an equivalent expression that is a sum of logarithms: $\log_2 (4 \cdot 16)$.

SOLUTION   We have

$$\log_2 (4 \cdot 16) = \log_2 4 + \log_2 16. \qquad \text{Using the product rule for logarithms}$$

As a check, note that

$$\log_2 (4 \cdot 16) = \log_2 64 = 6 \qquad 2^6 = 64$$

and that

$$\log_2 4 + \log_2 16 = 2 + 4 = 6. \qquad 2^2 = 4 \text{ and } 2^4 = 16$$

TRY EXERCISE ▸ 7

EXAMPLE 2 Express as an equivalent expression that is a single logarithm: $\log_b 7 + \log_b 5$.

SOLUTION   We have

$$\log_b 7 + \log_b 5 = \log_b (7 \cdot 5) \qquad \text{Using the product rule for logarithms}$$
$$= \log_b 35.$$

TRY EXERCISE ▸ 13

A Proof of the Product Rule.   Let $\log_a M = x$ and $\log_a N = y$. Converting to exponential equations, we have $a^x = M$ and $a^y = N$.

Now we multiply the left side of the first exponential equation by the left side of the second equation and similarly multiply the right sides to obtain

$$MN = a^x \cdot a^y, \quad \text{or} \quad MN = a^{x+y}.$$

Converting back to a logarithmic equation, we get

$$\log_a (MN) = x + y.$$

Recalling what x and y represent, we have

$$\log_a(MN) = \log_a M + \log_a N.$$

Logarithms of Powers

The second basic property is related to the power rule for exponents: $(a^m)^n = a^{mn}$. Its proof follows Example 3.

The Power Rule for Logarithms

For any positive numbers M and a ($a \neq 1$), and any real number p,

$$\log_a M^p = p \cdot \log_a M.$$

(The logarithm of a power of M is the exponent times the logarithm of M.)

To better understand the power rule, note that

$$\log_a M^3 = \log_a(M \cdot M \cdot M) = \log_a M + \log_a M + \log_a M = 3 \log_a M.$$

EXAMPLE **3**   Use the power rule for logarithms to write an equivalent expression that is a product: **(a)** $\log_a 9^{-5}$; **(b)** $\log_7 \sqrt[3]{x}$.

SOLUTION

a) $\log_a 9^{-5} = -5 \log_a 9$ Using the power rule for logarithms

b) $\log_7 \sqrt[3]{x} = \log_7 x^{1/3}$ Writing exponential notation

           $= \frac{1}{3} \log_7 x$ Using the power rule for logarithms

> **TRY EXERCISE** 17

A Proof of the Power Rule.   Let $x = \log_a M$. We then write the equivalent exponential equation, $a^x = M$. Raising both sides to the pth power, we get

$$(a^x)^p = M^p, \quad \text{or} \quad a^{xp} = M^p. \quad \text{Multiplying exponents}$$

Converting back to a logarithmic equation gives us

$$\log_a M^p = xp.$$

But $x = \log_a M$, so substituting, we have

$$\log_a M^p = (\log_a M)p = p \cdot \log_a M.$$

STUDENT NOTES

Without understanding and *remembering* the rules of this section, it will be extremely difficult to solve the equations of Section 12.6.

Logarithms of Quotients

The third property that we study is similar to the quotient rule for exponents: $a^m/a^n = a^{m-n}$. Its proof follows Example 5.

> ### The Quotient Rule for Logarithms
> For any positive numbers M, N, and a $(a \neq 1)$,
>
> $$\log_a \frac{M}{N} = \log_a M - \log_a N.$$
>
> (The logarithm of a quotient is the logarithm of the dividend minus the logarithm of the divisor.)

To better understand the quotient rule, note that

$$\log_2 \tfrac{8}{32} = \log_2 \tfrac{1}{4} = -2$$

and $\quad \log_2 8 - \log_2 32 = 3 - 5 = -2.$

EXAMPLE **4** Express as an equivalent expression that is a difference of logarithms: $\log_t (6/U)$.

SOLUTION

$$\log_t \frac{6}{U} = \log_t 6 - \log_t U \qquad \text{Using the quotient rule for logarithms}$$

TRY EXERCISE 23

EXAMPLE **5** Express as an equivalent expression that is a single logarithm:

$$\log_b 17 - \log_b 27.$$

SOLUTION

$$\log_b 17 - \log_b 27 = \log_b \frac{17}{27} \qquad \begin{array}{l} \text{Using the quotient rule for} \\ \text{logarithms "in reverse"} \end{array}$$

TRY EXERCISE 27

A Proof of the Quotient Rule. Our proof uses both the product rule and the power rule:

$$\begin{aligned} \log_a \frac{M}{N} &= \log_a (MN^{-1}) & \text{Rewriting } \frac{M}{N} \text{ as } MN^{-1} \\ &= \log_a M + \log_a N^{-1} & \text{Using the product rule for logarithms} \\ &= \log_a M + (-1)\log_a N & \text{Using the power rule for logarithms} \\ &= \log_a M - \log_a N. \end{aligned}$$

Using the Properties Together

EXAMPLE **6** Express as an equivalent expression, using the individual logarithms of x, y, and z.

a) $\log_b \dfrac{x^3}{yz}$ **b)** $\log_a \sqrt[4]{\dfrac{xy}{z^3}}$

SOLUTION

a) $\log_b \dfrac{x^3}{yz} = \log_b x^3 - \log_b yz \qquad \begin{array}{l}\text{Using the quotient rule for}\\ \text{logarithms}\end{array}$

$$= 3 \log_b x - \log_b yz \qquad \text{Using the power rule for logarithms}$$

$$= 3 \log_b x - (\log_b y + \log_b z) \qquad \begin{array}{l}\text{Using the product rule for}\\ \text{logarithms. Because of the}\\ \text{subtraction, parentheses are}\\ \text{essential.}\end{array}$$

$$= 3 \log_b x - \log_b y - \log_b z \qquad \text{Using the distributive law}$$

b) $\log_a \sqrt[4]{\dfrac{xy}{z^3}} = \log_a \left(\dfrac{xy}{z^3}\right)^{1/4}$ Writing exponential notation

$\qquad\qquad = \dfrac{1}{4} \cdot \log_a \dfrac{xy}{z^3}$ Using the power rule for logarithms

$\qquad\qquad = \dfrac{1}{4}\left(\log_a xy - \log_a z^3\right)$ Using the quotient rule for logarithms. Parentheses are important.

$\qquad\qquad = \dfrac{1}{4}\left(\log_a x + \log_a y - 3\log_a z\right)$ Using the product rule and the power rule for logarithms

> TRY EXERCISE   37

> *CAUTION!*   Because the product and quotient rules replace one term with two, it is often essential to apply the rules within parentheses, as in Example 6.

EXAMPLE   7   Express as an equivalent expression that is a single logarithm.

a) $\dfrac{1}{2}\log_a x - 7\log_a y + \log_a z$ **b)** $\log_a \dfrac{b}{\sqrt{x}} + \log_a \sqrt{bx}$

SOLUTION

a) $\dfrac{1}{2}\log_a x - 7\log_a y + \log_a z$

$\qquad = \log_a x^{1/2} - \log_a y^7 + \log_a z$ Using the power rule for logarithms

$\qquad = \left(\log_a \sqrt{x} - \log_a y^7\right) + \log_a z$ Using parentheses to emphasize the order of operations; $x^{1/2} = \sqrt{x}$

$\qquad = \log_a \dfrac{\sqrt{x}}{y^7} + \log_a z$ Using the quotient rule for logarithms. Note that all terms have the same base.

$\qquad = \log_a \dfrac{z\sqrt{x}}{y^7}$ Using the product rule for logarithms

b) $\log_a \dfrac{b}{\sqrt{x}} + \log_a \sqrt{bx} = \log_a \dfrac{b \cdot \sqrt{bx}}{\sqrt{x}}$ Using the product rule for logarithms

$\qquad\qquad = \log_a b\sqrt{b}$ Removing a factor equal to 1: $\dfrac{\sqrt{x}}{\sqrt{x}} = 1$

$\qquad\qquad = \log_a b^{3/2}, \text{ or } \dfrac{3}{2}\log_a b$ Since $b\sqrt{b} = b^1 \cdot b^{1/2}$

> TRY EXERCISE   49

If we know the logarithms of two different numbers (with the same base), the properties allow us to calculate other logarithms.

EXAMPLE   8   Given $\log_a 2 = 0.431$ and $\log_a 3 = 0.683$, use the properties of logarithms to calculate a value for each of the following. If this is not possible, state so.

a) $\log_a 6$ **b)** $\log_a \dfrac{2}{3}$ **c)** $\log_a 81$

d) $\log_a \dfrac{1}{3}$ **e)** $\log_a (2a)$ **f)** $\log_a 5$

SOLUTION

a) $\log_a 6 = \log_a(2 \cdot 3) = \log_a 2 + \log_a 3$ Using the product rule for logarithms

$$= 0.431 + 0.683 = 1.114$$

Check: $a^{1.114} = a^{0.431} \cdot a^{0.683} = 2 \cdot 3 = 6$

b) $\log_a \frac{2}{3} = \log_a 2 - \log_a 3$ Using the quotient rule for logarithms

$$= 0.431 - 0.683 = -0.252$$

c) $\log_a 81 = \log_a 3^4 = 4 \log_a 3$ Using the power rule for logarithms

$$= 4(0.683) = 2.732$$

d) $\log_a \frac{1}{3} = \log_a 1 - \log_a 3$ Using the quotient rule for logarithms

$$= 0 - 0.683 = -0.683$$

e) $\log_a (2a) = \log_a 2 + \log_a a$ Using the product rule for logarithms

$$= 0.431 + 1 = 1.431$$

f) $\log_a 5$ *cannot be found using these properties.* $(\log_a 5 \neq \log_a 2 + \log_a 3)$

> **TRY EXERCISE** 55

A final property follows from the product rule: Since $\log_a a^k = k \log_a a$, and $\log_a a = 1$, we have $\log_a a^k = k$.

> ## The Logarithm of the Base to an Exponent
>
> For any base a,
>
> $$\log_a a^k = k.$$
>
> (The logarithm, base a, of a to an exponent is the exponent.)

This property also follows from the definition of logarithm: k is the exponent to which you raise a in order to get a^k.

EXAMPLE **9**   Simplify: **(a)** $\log_3 3^7$; **(b)** $\log_{10} 10^{-5.2}$.

SOLUTION

a) $\log_3 3^7 = 7$ 7 is the exponent to which you raise 3 in order to get 3^7.

b) $\log_{10} 10^{-5.2} = -5.2$

> **TRY EXERCISE** 65

We summarize the properties of logarithms as follows.

> For any positive numbers M, N, and a $(a \neq 1)$:
>
> $$\log_a (MN) = \log_a M + \log_a N; \qquad \log_a M^p = p \cdot \log_a M;$$
>
> $$\log_a \frac{M}{N} = \log_a M - \log_a N; \qquad \log_a a^k = k.$$

CAUTION! Keep in mind that, in general,

$$\log_a (M + N) \neq \log_a M + \log_a N, \qquad \log_a (MN) \neq (\log_a M)(\log_a N),$$

$$\log_a (M - N) \neq \log_a M - \log_a N, \qquad \log_a \frac{M}{N} \neq \frac{\log_a M}{\log_a N}.$$

12.4 EXERCISE SET

For Extra Help

> **Concept Reinforcement** In each of Exercises 1–6, match the expression with an equivalent expression from the column on the right.

1. ____ $\log_7 20$
2. ____ $\log_7 5^4$
3. ____ $\log_7 \frac{5}{4}$
4. ____ $\log_7 7$
5. ____ $\log_7 1$
6. ____ $\log_7 5 + \log_7 6$

a) $\log_7 5 - \log_7 4$
b) 1
c) 0
d) $\log_7 30$
e) $\log_7 5 + \log_7 4$
f) $4 \log_7 5$

Express as an equivalent expression that is a sum of logarithms.

7. $\log_3 (81 \cdot 27)$
8. $\log_2 (16 \cdot 32)$
9. $\log_4 (64 \cdot 16)$
10. $\log_5 (25 \cdot 125)$
11. $\log_c (rst)$
12. $\log_t (3ab)$

Express as an equivalent expression that is a single logarithm.

13. $\log_a 2 + \log_a 10$
14. $\log_b 5 + \log_b 9$
15. $\log_c t + \log_c y$
16. $\log_t H + \log_t M$

Express as an equivalent expression that is a product.

17. $\log_a r^8$
18. $\log_b t^5$
19. $\log_2 y^{1/3}$
20. $\log_{10} y^{1/2}$
21. $\log_b C^{-3}$
22. $\log_c M^{-5}$

Express as an equivalent expression that is a difference of two logarithms.

23. $\log_2 \frac{5}{11}$
24. $\log_3 \frac{29}{13}$
25. $\log_b \frac{m}{n}$
26. $\log_a \frac{y}{x}$

Express as an equivalent expression that is a single logarithm.

27. $\log_a 19 - \log_a 2$
28. $\log_b 3 - \log_b 32$
29. $\log_b 36 - \log_b 4$
30. $\log_a 26 - \log_a 2$
31. $\log_a x - \log_a y$
32. $\log_b c - \log_b d$

Express as an equivalent expression, using the individual logarithms of w, x, y, and z.

33. $\log_a (xyz)$
34. $\log_a (wxy)$
35. $\log_a (x^3 z^4)$
36. $\log_a (x^2 y^5)$
37. $\log_a (w^2 x^{-2} y)$
38. $\log_a (xy^2 z^{-3})$
39. $\log_a \frac{x^5}{y^3 z}$
40. $\log_a \frac{x^4}{yz^2}$
41. $\log_b \frac{xy^2}{wz^3}$
42. $\log_b \frac{w^2 x}{y^3 z}$
43. $\log_a \sqrt{\frac{x^7}{y^5 z^8}}$
44. $\log_c \sqrt{\frac{x^4}{y^3 z^2}}$
45. $\log_a \sqrt[3]{\frac{x^6 y^3}{a^2 z^7}}$
46. $\log_a \sqrt[4]{\frac{x^8 y^{12}}{a^3 z^5}}$

Express as an equivalent expression that is a single logarithm and, if possible, simplify.

47. $8 \log_a x + 3 \log_a z$
48. $2 \log_b m + \frac{1}{2} \log_b n$
49. $\log_a x^2 - 2 \log_a \sqrt{x}$
50. $\log_a \frac{a}{\sqrt{x}} - \log_a \sqrt{ax}$
51. $\frac{1}{2} \log_a x + 5 \log_a y - 2 \log_a x$
52. $\log_a 2x + 3(\log_a x - \log_a y)$
53. $\log_a (x^2 - 9) - \log_a (x + 3)$
54. $\log_a (2x + 10) - \log_a (x^2 - 25)$

Given $\log_b 3 = 0.792$ *and* $\log_b 5 = 1.161$. *If possible, use the properties of logarithms to calculate values for each of the following.*

55. $\log_b 15$

56. $\log_b \frac{5}{3}$

57. $\log_b \frac{3}{5}$

58. $\log_b \frac{1}{3}$

59. $\log_b \frac{1}{5}$

60. $\log_b \sqrt{b}$

61. $\log_b \sqrt{b^3}$

62. $\log_b 3b$

63. $\log_b 8$

64. $\log_b 45$

Simplify.

Aha! **65.** $\log_t t^{10}$

66. $\log_p p^{-5}$

67. $\log_e e^m$

68. $\log_Q Q^t$

69. Explain the difference between the phrases "the logarithm of a quotient" and "a quotient of logarithms."

70. How could you convince someone that

$$\log_a c \neq \log_c a?$$

Skill Review

To prepare for Section 12.5, review graphing functions and finding domains of functions.

Graph.

71. $f(x) = \sqrt{x} - 3$ [10.1]

72. $g(x) = \sqrt[3]{x} + 1$ [10.1]

73. $g(x) = x^3 + 2$ [7.3]

74. $f(x) = 1 - x^2$ [11.7]

Find the domain of each function.

75. $f(x) = \dfrac{x - 3}{x + 7}$ [9.1]

76. $f(x) = \dfrac{x}{(x - 2)(x + 3)}$ [9.1]

77. $g(x) = \sqrt{10 - x}$ [10.1]

78. $g(x) = |x^2 - 6x + 7|$ [7.2]

Synthesis

79. A student *incorrectly* reasons that

$$\log_b \frac{1}{x} = \log_b \frac{x}{xx}$$
$$= \log_b x - \log_b x + \log_b x = \log_b x.$$

What mistake has the student made?

80. Why are properties of logarithms related to properties of exponents?

Express as an equivalent expression that is a single logarithm and, if possible, simplify.

81. $\log_a (x^8 - y^8) - \log_a (x^2 + y^2)$

82. $\log_a (x + y) + \log_a (x^2 - xy + y^2)$

Express as an equivalent expression that is a sum or a difference of logarithms and, if possible, simplify.

83. $\log_a \sqrt{1 - s^2}$

84. $\log_a \dfrac{c - d}{\sqrt{c^2 - d^2}}$

85. If $\log_a x = 2$, $\log_a y = 3$, and $\log_a z = 4$, what is

$$\log_a = \frac{\sqrt[3]{x^2 z}}{\sqrt[3]{y^2 z^{-2}}}?$$

86. If $\log_a x = 2$, what is $\log_a (1/x)$?

87. If $\log_a x = 2$, what is $\log_{1/a} x$?

Solve.

88. $\log_{10} 2000 - \log_{10} x = 3$

89. $\log_2 80 + \log_2 x = 5$

Classify each of the following as true or false. Assume a, x, P, and $Q > 0$, $a \neq 1$.

90. $\log_a \left(\dfrac{P}{Q}\right)^x = x \log_a P - \log_a Q$

91. $\log_a (Q + Q^2) = \log_a Q + \log_a (Q + 1)$

92. Use graphs to show that

$$\log x^2 \neq \log x \cdot \log x.$$

(*Note*: log means $\log_{10}$.)

12.5   Common and Natural Logarithms

Common Logarithms on a Calculator   ▪   The Base *e* and Natural Logarithms on a Calculator   ▪
Changing Logarithmic Bases   ▪   Graphs of Exponential and Logarithmic Functions, Base *e*

Any positive number other than 1 can serve as the base of a logarithmic function.
However, some numbers are easier to use than others, and there are logarithmic
bases that fit into certain applications more naturally than others.

Base-10 logarithms, called **common logarithms**, are useful because they
have the same base as our "commonly" used decimal system. Before calculators
became widely available, common logarithms helped with tedious calculations.
In fact, that is why logarithms were devised.

The logarithmic base most widely used today is an irrational number named *e*.
We will consider *e* and base *e*, or *natural*, logarithms later in this section. First we
examine common logarithms.

Common Logarithms on a Calculator

Before the advent of scientific calculators, printed tables listed common logarithms. Today we find common logarithms using calculators.

Here, and in most books, the abbreviation **log**, with no base written, is understood to mean logarithm base 10, that is, a common logarithm. Thus,

$$\log 17 \quad \text{means} \quad \log_{10} 17.$$ It is important to remember
this abbreviation.

The key for common logarithms is usually marked **LOG**. To find the common
logarithm of a number, we key in that number and press **LOG**. With most graphing calculators, we press **LOG**, the number, and then **ENTER**.

EXAMPLE   1

Use a calculator to approximate each number to four decimal places.

a) $\log 5312$

b) $\dfrac{\log 6500}{\log 0.007}$

SOLUTION

a) We enter 5312 and then press **LOG**. On most graphing calculators, we press
LOG, followed by 5312 and **ENTER**. We find that

$$\log 5312 \approx 3.7253.$$ Rounded to four decimal places

b) We enter 6500 and then press **LOG**. Next, we press ÷ , enter 0.007, and then
press **LOG** = . On most graphing calculators, we press **LOG**, key in 6500,
press) ÷ **LOG**, key in 0.007, and then press) **ENTER**. Be careful not to
round until the end:

$$\frac{\log 6500}{\log 0.007} \approx -1.7694.$$ Rounded to four decimal places

TRY EXERCISE ▸ 11

The inverse of a logarithmic function is an exponential function. Because of
this, on many calculators the **LOG** key doubles as the $\boxed{10^x}$ key after a **2ND** or
$\boxed{\text{SHIFT}}$ key is pressed. Calculators lacking a $\boxed{10^x}$ key may have a key labeled $\boxed{x^y}$,
$\boxed{a^x}$, or ⌃ . Such a key can raise any positive real number to any real-numbered
exponent.

EXAMPLE   2   Use a calculator to approximate $10^{3.417}$ to four decimal places.

SOLUTION   We enter 3.417 and then press $\boxed{10^x}$. On most graphing calculators, $\boxed{10^x}$ is pressed first, followed by 3.417 and $\boxed{\text{ENTER}}$. Rounding to four decimal places, we have

$$10^{3.417} \approx 2612.1614.$$

 TRY EXERCISE   21

The Base *e* and Natural Logarithms on a Calculator

When interest is compounded *n* times a year, the compound interest formula is

$$A = P\left(1 + \frac{r}{n}\right)^{nt},$$

where *A* is the amount that an initial investment *P* is worth after *t* years at interest rate *r*. Suppose that $1 is invested at 100% interest for 1 year (no bank would pay this). The preceding formula becomes a function *A* defined in terms of the number of compounding periods *n*:

$$A(n) = \left(1 + \frac{1}{n}\right)^n.$$

Let's find some function values. We use a calculator and round to six decimal places.

n	$A(n) = \left(1 + \dfrac{1}{n}\right)^{n}$
1 (compounded annually)	$2.00
2 (compounded semiannually)	2.25
3	2.370370
4 (compounded quarterly)	2.441406
12 (compounded monthly)	2.613035
100	2.704814
365 (compounded daily)	2.714567
8760 (compounded hourly)	2.718127

The numbers in this table approach a very important number in mathematics, called *e*. Because *e* is irrational, its decimal representation does not terminate or repeat.

> ## The Number *e*
> $$e \approx 2.7182818284\ldots$$

Logarithms base *e* are called **natural logarithms**, or **Napierian logarithms**, in honor of John Napier (1550–1617), the "inventor" of logarithms.

The abbreviation "ln" is generally used with natural logarithms. Thus,

$$\ln 53 \quad \text{means} \quad \log_e 53. \qquad \text{It is important to remember this abbreviation.}$$

On most calculators, the key for natural logarithms is marked $\boxed{\text{LN}}$.

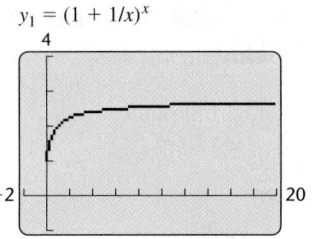

EXAMPLE 3

Use a calculator to approximate ln 4568 to four decimal places.

SOLUTION   We enter 4568 and then press **LN**. On most graphing calculators, we press **LN** first, followed by 4568 and **ENTER**. We find that

$$\ln 4568 \approx 8.4268. \qquad \text{Rounded to four decimal places}$$

TRY EXERCISE 25

On many calculators, the **LN** key doubles as the **eˣ** key after a **2ND** or SHIFT key has been pressed.

EXAMPLE 4

Use a calculator to approximate $e^{-1.524}$ to four decimal places.

SOLUTION   We enter -1.524 and then press **eˣ**. On most graphing calculators, **eˣ** is pressed first, followed by -1.524 and **ENTER**. Since $e^{-1.524}$ is irrational, our answer is approximate:

$$e^{-1.524} \approx 0.2178. \qquad \text{Rounded to four decimal places}$$

TRY EXERCISE 31

Changing Logarithmic Bases

Most calculators can find both common and natural logarithms. To find a logarithm with some other base, a conversion formula is often used.

The Change-of-Base Formula

For any logarithmic bases a and b, and any positive number M,

$$\log_b M = \frac{\log_a M}{\log_a b}.$$

(To find the log, base b, of M, we typically compute $\log M / \log b$ or $\ln M / \ln b$.)

Proof.   Let $x = \log_b M$. Then,

$$b^x = M \qquad \log_b M = x \text{ is equivalent to } b^x = M.$$
$$\log_a b^x = \log_a M \qquad \text{Taking the logarithm, base } a, \text{ on both sides}$$
$$x \log_a b = \log_a M \qquad \text{Using the power rule for logarithms}$$
$$x = \frac{\log_a M}{\log_a b}. \qquad \text{Dividing both sides by } \log_a b$$

But at the outset we stated that $x = \log_b M$. Thus, by substitution, we have

$$\log_b M = \frac{\log_a M}{\log_a b}. \qquad \text{This is the change-of-base formula.}$$

EXAMPLE **5**

Find $\log_5 8$ using the change-of-base formula.

SOLUTION We use the change-of-base formula with $a = 10$, $b = 5$, and $M = 8$:

$$\log_5 8 = \frac{\log_{10} 8}{\log_{10} 5}$$ Substituting into $\log_b M = \dfrac{\log_a M}{\log_a b}$

$$\approx \frac{0.903089987}{0.6989700043}$$ Using **LOG** twice

$$\approx 1.2920.$$ When using a calculator, it is best not to round until the end.

To check, note that $\ln 8/\ln 5 \approx 1.2920$. We can also use a calculator to verify that $5^{1.2920} \approx 8$.

TRY EXERCISE 35

EXAMPLE **6**

Find $\log_4 31$.

SOLUTION As shown in the check of Example 5, base e can also be used.

STUDENT NOTES

The choice of the logarithm base a in the change-of-base formula should be either 10 or e so that the logarithms can be found using a calculator. Either choice will yield the same end result.

$$\log_4 31 = \frac{\log_e 31}{\log_e 4}$$ Substituting into $\log_b M = \dfrac{\log_a M}{\log_a b}$

$$= \frac{\ln 31}{\ln 4} \approx \frac{3.433987204}{1.386294361}$$ Using **LN** twice

$$\approx 2.4771.$$ *Check*: $4^{2.4771} \approx 31$

TRY EXERCISE 41

Graphs of Exponential and Logarithmic Functions, Base e

EXAMPLE **7**

Graph $f(x) = e^x$ and $g(x) = e^{-x}$ and state the domain and the range of f and g.

SOLUTION We use a calculator with an (e^x) key to find approximate values of e^x and e^{-x}. Using these values, we can graph the functions.

x	e^x	e^{-x}
0	1	1
1	2.7	0.4
2	7.4	0.1
−1	0.4	2.7
−2	0.1	7.4

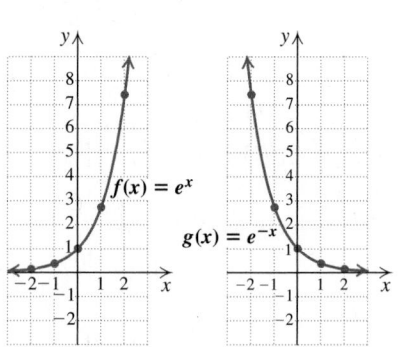

The domain of each function is $\mathbb{R}$ and the range of each function is $(0, \infty)$.

TRY EXERCISE 61

EXAMPLE **8**   Graph $f(x) = e^{-x} + 2$ and state the domain and the range of f.

SOLUTION   We find some solutions with a calculator, plot them, and then draw the graph. For example, $f(2) = e^{-2} + 2 \approx 0.1 + 2 \approx 2.1$. The graph is exactly like the graph of $g(x) = e^{-x}$, but is translated up 2 units.

x	$e^{-x} + 2$
0	3
1	2.4
2	2.1
−1	4.7
−2	9.4

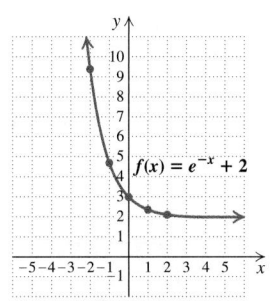

The domain of f is $\mathbb{R}$ and the range is $(2, \infty)$.

TRY EXERCISE 49

EXAMPLE **9**   Graph and state the domain and the range of each function.

a) $g(x) = \ln x$ **b)** $f(x) = \ln(x + 3)$

SOLUTION

a) We find some solutions with a calculator and then draw the graph. As expected, the graph is a reflection across the line $y = x$ of the graph of $y = e^x$.

x	$\ln x$
1	0
4	1.4
7	1.9
0.5	−0.7

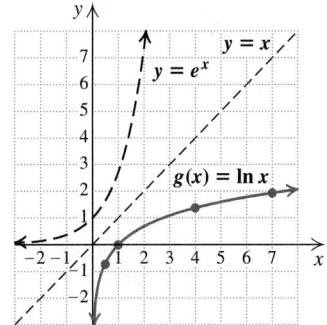

The domain of g is $(0, \infty)$ and the range is $\mathbb{R}$.

b) We find some solutions with a calculator, plot them, and draw the graph.

x	$\ln(x + 3)$
0	1.1
1	1.4
2	1.6
3	1.8
4	1.9
−1	0.7
−2	0
−2.5	−0.7

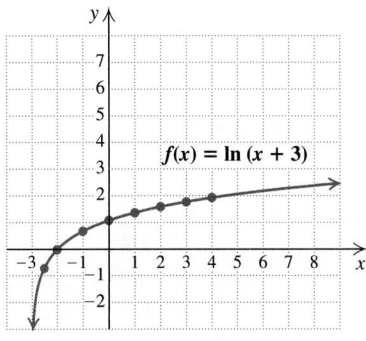

The graph of $y = \ln(x + 3)$ is the graph of $y = \ln x$ translated 3 units to the left. Since $x + 3$ must be positive, the domain is $(-3, \infty)$ and the range is $\mathbb{R}$.

TRY EXERCISE 63

Visualizing for Success

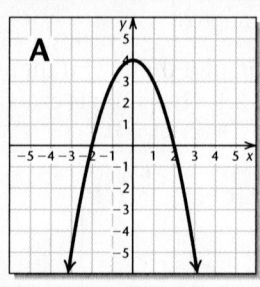

A

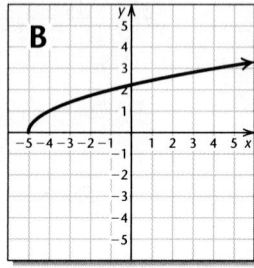

B

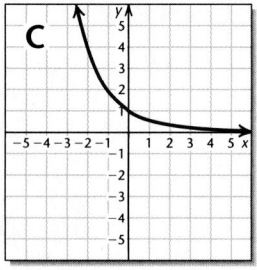

C

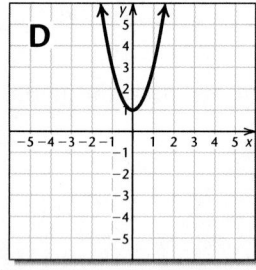

D

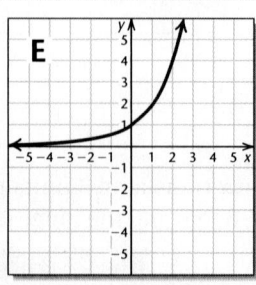

E

Match each function with its graph.

1. $f(x) = 2x - 3$

2. $f(x) = 2x^2 + 1$

3. $f(x) = \sqrt{x + 5}$

4. $f(x) = |x - 4|$

5. $f(x) = \ln x$

6. $f(x) = 2^{-x}$

7. $f(x) = -4$

8. $f(x) = \log x + 3$

9. $f(x) = 2^x$

10. $f(x) = 4 - x^2$

Answers on page A-56

An additional, animated version of this activity appears in MyMathLab. To use MyMathLab, you need a course ID and a student access code. Contact your instructor for more information.

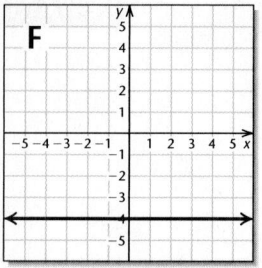

F

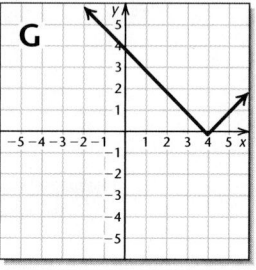

G

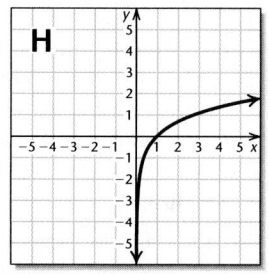

H

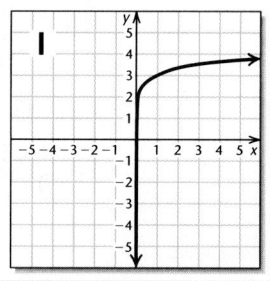

I

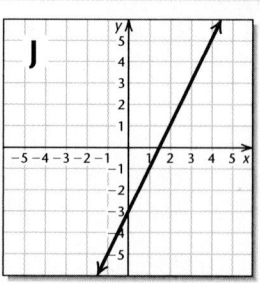

J

12.5 **EXERCISE SET**

👆 *Concept Reinforcement* *Classify each statement as either true or false.*

1. The expression log 23 means $\log_{10} 23$.

2. The expression ln 7 means $\log_e 7$.

3. The number e is approximately 2.7.

4. The expressions log 9 and log 18/log 2 are equivalent.

5. The expressions log 9 and log 18 − log 2 are equivalent.

6. The expressions $\log_2 9$ and ln 9/ln 2 are equivalent.

7. The expressions ln 81 and 2 ln 9 are equivalent.

8. The domain of the function given by $f(x) = \ln(x + 2)$ is $(-2, \infty)$.

9. The range of the function given by $g(x) = e^x$ is $(0, \infty)$.

10. The range of the function given by $f(x) = \ln x$ is $(-\infty, \infty)$.

▦ *Use a calculator to find each of the following to four decimal places.*

11. log 7

12. log 2

13. log 13.7

14. log 98.3   Aha!

15. log 1000

16. log 100

17. log 0.75

18. log 0.25

19. $\dfrac{\log 8200}{\log 2}$

20. $\dfrac{\log 5700}{\log 5}$

21. $10^{1.7}$

22. $10^{0.59}$

23. $10^{-2.9523}$

24. $10^{-3.2046}$

25. ln 9

26. ln 13

27. ln 0.0062

28. ln 0.00073

29. $\dfrac{\ln 2300}{0.08}$

30. $\dfrac{\ln 1900}{0.07}$

31. $e^{2.71}$

32. $e^{3.06}$

33. $e^{-3.49}$

34. $e^{-2.64}$

▦ *Find each of the following logarithms using the change-of-base formula. Round answers to four decimal places.*

35. $\log_3 28$

36. $\log_6 37$

37. $\log_2 100$

38. $\log_7 100$

39. $\log_4 5$

40. $\log_8 7$

41. $\log_{0.1} 2$

42. $\log_{0.25} 25$

43. $\log_2 0.1$

44. $\log_{25} 0.25$

45. $\log_\pi 10$

46. $\log_\pi 100$

▦ *Graph and state the domain and the range of each function.*

47. $f(x) = e^x$

48. $f(x) = e^{-x}$

49. $f(x) = e^x + 3$

50. $f(x) = e^x + 2$

51. $f(x) = e^x - 2$

52. $f(x) = e^x - 3$

53. $f(x) = 0.5e^x$

54. $f(x) = 2e^x$

55. $f(x) = 0.5e^{2x}$

56. $f(x) = 2e^{-0.5x}$

57. $f(x) = e^{x-3}$

58. $f(x) = e^{x-2}$

59. $f(x) = e^{x+2}$

60. $f(x) = e^{x+3}$

61. $f(x) = -e^x$

62. $f(x) = -e^{-x}$

63. $g(x) = \ln x + 1$

64. $g(x) = \ln x + 3$

65. $g(x) = \ln x - 2$

66. $g(x) = \ln x - 1$

67. $g(x) = 2 \ln x$

68. $g(x) = 3 \ln x$

69. $g(x) = -2 \ln x$

70. $g(x) = -\ln x$

71. $g(x) = \ln(x + 2)$

72. $g(x) = \ln(x + 1)$

73. $g(x) = \ln(x - 1)$

74. $g(x) = \ln(x - 3)$

75. Using a calculator, Adan gives an *incorrect* approximation for log 79 that is between 4 and 5. How could you convince him, without using a calculator, that he is mistaken?

76. Examine Exercise 75. What mistake do you believe Adan made?

Skill Review

To prepare for Section 12.6, review solving equations.

Solve.

77. $x^2 - 3x - 28 = 0$ [5.7]

78. $5x^2 - 7x = 0$ [5.7]

79. $17x - 15 = 0$ [2.2]

80. $\frac{5}{3} = 2t$ [2.2]

81. $(x - 5) \cdot 9 = 11$ [2.2]

82. $\frac{x + 3}{x - 3} = 7$ [6.6]

83. $x^{1/2} - 6x^{1/4} + 8 = 0$ [11.5]

84. $2y - 7\sqrt{y} + 3 = 0$ [11.5]

Synthesis

85. Explain how the graph of $f(x) = e^x$ could be used to graph the function given by $g(x) = 1 + \ln x$.

86. How would you explain to a classmate why $\log_2 5 = \log 5/\log 2$ *and* $\log_2 5 = \ln 5/\ln 2$?

Knowing only that $\log 2 \approx 0.301$ *and* $\log 3 \approx 0.477$, *approximate each of the following to three decimal places.*

87. $\log_6 81$

88. $\log_9 16$

89. $\log_{12} 36$

90. Find a formula for converting common logarithms to natural logarithms.

91. Find a formula for converting natural logarithms to common logarithms.

Solve for x. Give an approximation to four decimal places.

92. $\log(275x^2) = 38$

93. $\log(492x) = 5.728$

94. $\frac{3.01}{\ln x} = \frac{28}{4.31}$

95. $\log 692 + \log x = \log 3450$

For each function given below, **(a)** *determine the domain and the range,* **(b)** *set an appropriate window, and* **(c)** *draw the graph. Graphs may vary, depending on the scale used.*

96. $f(x) = 7.4e^x \ln x$

97. $f(x) = 3.4 \ln x - 0.25e^x$

98. $f(x) = x \ln(x - 2.1)$

99. $f(x) = 2x^3 \ln x$

100. Use a graphing calculator to check your answers to Exercises 49, 57, and 71.

101. Use a graphing calculator to check your answers to Exercises 48, 54, and 64.

102. In an attempt to solve $\ln x = 1.5$, Emma gets the following graph. How can Emma tell at a glance that she has made a mistake?

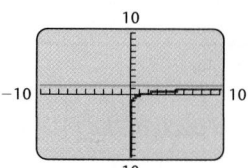

CONNECTING the CONCEPTS

It is important to distinguish between *simplifying* an exponential or logarithmic *expression* and *solving* an exponential or logarithmic *equation*. We use the following properties to simplify expressions and to rewrite equivalent logarithmic and exponential equations.

$$\log_a x = m \text{ means } x = a^m.$$

$$\log_a(MN) = \log_a M + \log_a N$$

$$\log_a \frac{M}{N} = \log_a M - \log_a N$$

$$\log_a M^p = p \cdot \log_a M$$

$$\log_b M = \frac{\log_a M}{\log_a b}$$

$$\log_a a^k = k$$

$$\log_a a = 1$$

$$\log_a 1 = 0$$

$$\log x = \log_{10} x$$

$$\ln x = \log_e x$$

MIXED REVIEW

Simplify.

1. $\log_4 16$

2. $\log_5 \frac{1}{5}$

3. $\log_{100} 10$

4. $\log_{10} 100$

5. $\log 10$

6. $\ln 1$

7. $\log 10^4$

8. $\ln e^8$

9. $e^{\ln 7}$

10. $10^{\log 3}$

Rewrite each of the following as an equivalent exponential equation.

11. $\log_x 3 = m$

12. $\log_2 1024 = 10$

Rewrite each of the following as an equivalent logarithmic equation.

13. $e^t = x$

14. $64^{2/3} = 16$

Solve.

15. $\log_x 64 = 3$

16. $\log_3 x = -1$

17. Express as an equivalent expression using $\log x$, $\log y$, and $\log z$:

$$\log \sqrt{\frac{x^2}{yz^3}}.$$

18. Express as an equivalent expression that is a single logarithm: $\log a - 2 \log b - \log c$.

Find each of the following logarithms using the change-of-base formula. Round answers to four decimal places where appropriate.

19. $\log_4 8$

20. $\log_5 100$

12.6 Solving Exponential and Logarithmic Equations

Solving Exponential Equations • Solving Logarithmic Equations

Solving Exponential Equations

Equations with variables in exponents, such as $5^x = 12$ and $2^{7x} = 64$, are called **exponential equations**. In Section 12.3, we solved certain exponential equations by using the principle of exponential equality. We restate that principle below.

> **The Principle of Exponential Equality**
>
> For any real number b, where $b \neq -1, 0$, or 1,
>
> $$b^x = b^y \quad \text{is equivalent to} \quad x = y.$$
>
> (Powers of the same base are equal if and only if the exponents are equal.)

EXAMPLE **1** Solve: $4^{3x} = 16$.

SOLUTION Note that $16 = 4^2$. Thus we can write each side as a power of the same base:

$$4^{3x} = 4^2 \qquad \text{Rewriting 16 as a power of 4}$$
$$3x = 2 \qquad \text{Since the base on each side is 4, the exponents are equal.}$$
$$x = \tfrac{2}{3}. \qquad \text{Solving for } x$$

Since $4^{3x} = 4^{3(2/3)} = 4^2 = 16$, the answer checks. The solution is $\tfrac{2}{3}$.

TRY EXERCISE 9

In Example 1, we wrote both sides of the equation as powers of 4. When it seems impossible to write both sides of an equation as powers of the same base, we use the following principle and write an equivalent logarithmic equation.

> ### The Principle of Logarithmic Equality
>
> For any logarithmic base a, and for $x, y > 0$,
>
> $$x = y \quad \text{is equivalent to} \quad \log_a x = \log_a y.$$
>
> (Two expressions are equal if and only if the logarithms of those expressions are equal.)

The principle of logarithmic equality, used together with the power rule for logarithms, allows us to solve equations in which the variable is an exponent.

EXAMPLE **2** Solve: $7^{x-2} = 60$.

SOLUTION We have

$$7^{x-2} = 60$$

Take the logarithm of both sides.

$$\log 7^{x-2} = \log 60 \qquad \begin{array}{l}\text{Using the principle of logarithmic} \\ \text{equality to take the common} \\ \text{logarithm on both sides. Natural} \\ \text{logarithms also would work.}\end{array}$$

Use the power rule for logarithms.

$$(x - 2)\log 7 = \log 60 \qquad \text{Using the power rule for logarithms}$$

$$x - 2 = \frac{\log 60}{\log 7} \qquad \leftarrow \boxed{\textit{CAUTION!} \ \text{This is } not \log 60 - \log 7.}$$

$$x = \frac{\log 60}{\log 7} + 2 \qquad \text{Adding 2 to both sides}$$

Solve for x.

$$x \approx 4.1041. \qquad \begin{array}{l}\text{Using a calculator and rounding to four} \\ \text{decimal places}\end{array}$$

Check.

Since $7^{4.1041-2} \approx 60.0027$, we have a check. We can also note that since $7^{4-2} = 49$, we expect a solution greater than 4. The solution is $\dfrac{\log 60}{\log 7} + 2$, or approximately 4.1041.

TRY EXERCISE 17

ALGEBRAIC–GRAPHICAL CONNECTION

The solution of $4^x = 16$ can be visualized by graphing $y = 4^x$ and $y = 16$ on the same set of axes. The y-values for both equations are the same where the graphs intersect. The x-value at that point is the solution of the equation. That x-value appears to be 2. Since $4^2 = 16$, we see that this is indeed the case.

Similarly, the solution of Example 2 can be visualized by graphing $y = 7^{x-2}$ and $y = 60$ and identifying the x-value at the point of intersection. As expected, this value appears to be approximately 4.1.

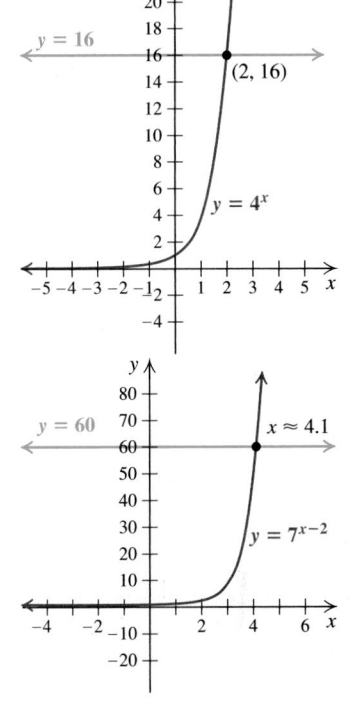

EXAMPLE 3

Solve: $e^{0.06t} = 1500$.

SOLUTION Since one side is a power of e, it is easiest to take the *natural logarithm* on both sides:

$$\ln e^{0.06t} = \ln 1500 \qquad \text{Taking the natural logarithm on both sides}$$

$$0.06t = \ln 1500 \qquad \begin{array}{l}\text{Finding the logarithm of the base to a power:}\\ \log_a a^k = k. \text{ Logarithmic and exponential}\\ \text{functions are inverses of each other.}\end{array}$$

$$t = \frac{\ln 1500}{0.06} \qquad \text{Dividing both sides by 0.06}$$

$$\approx 121.887. \qquad \begin{array}{l}\text{Using a calculator and rounding to three}\\ \text{decimal places}\end{array}$$

TRY EXERCISE 21

To Solve an Equation of the Form $a^t = b$ for t

1. Take the logarithm (either natural or common) of both sides.
2. Use the power rule for logarithms so that the variable is no longer written as an exponent.
3. Divide both sides by the coefficient of the variable to isolate the variable.
4. If appropriate, use a calculator to find an approximate solution.

Solving Logarithmic Equations

Recall from Section 12.3 that certain logarithmic equations can be solved by writing an equivalent exponential equation.

EXAMPLE **4**   Solve: **(a)** $\log_4 (8x - 6) = 3$; **(b)** $\ln (5x) = 27$.

SOLUTION

a) $\log_4 (8x - 6) = 3$

$$4^3 = 8x - 6 \qquad \text{Writing the equivalent exponential equation}$$

$$64 = 8x - 6$$

$$70 = 8x \qquad \text{Adding 6 to both sides}$$

$$x = \tfrac{70}{8}, \text{ or } \tfrac{35}{4}.$$

Check:

$$\begin{array}{c|c}
\log_4 (8x - 6) = 3 & \\
\hline
\log_4 (8 \cdot \tfrac{35}{4} - 6) & 3 \\
\log_4 (2 \cdot 35 - 6) & \\
\log_4 64 & \\
3 \stackrel{?}{=} 3 & \text{TRUE}
\end{array}$$

The solution is $\tfrac{35}{4}$.

b) $\ln (5x) = 27 \qquad$ Remember: $\ln (5x)$ means $\log_e (5x)$.

$$e^{27} = 5x \qquad \text{Writing the equivalent exponential equation}$$

$$\frac{e^{27}}{5} = x \qquad \text{This is a very large number.}$$

The solution is $\dfrac{e^{27}}{5}$. The check is left to the student.

TRY EXERCISE 45

STUDENT NOTES

It is essential that you remember the properties of logarithms from Section 12.4. Consider reviewing the properties before attempting to solve equations similar to those in Example 5.

Often the properties for logarithms are needed in order to solve a logarithmic equation. The goal is to first write an equivalent equation in which the variable appears in just one logarithmic expression. We then isolate that expression and solve as in Example 4.

EXAMPLE **5**   Solve.

a) $\log x + \log (x - 3) = 1$

b) $\log_2 (x + 7) - \log_2 (x - 7) = 3$

c) $\log_7 (x + 1) + \log_7 (x - 1) = \log_7 8$

SOLUTION

a) To increase understanding, we write in the base, 10.

Find a single logarithm.

$$\log_{10} x + \log_{10} (x - 3) = 1$$

$$\log_{10} [x(x - 3)] = 1 \qquad \text{Using the product rule for logarithms to obtain a single logarithm}$$

Write an equivalent exponential equation.

$$x(x - 3) = 10^1 \qquad \text{Writing an equivalent exponential equation}$$

$$x^2 - 3x = 10$$

$$x^2 - 3x - 10 = 0$$

$$(x + 2)(x - 5) = 0 \qquad \text{Factoring}$$

$$x + 2 = 0 \quad or \quad x - 5 = 0 \qquad \text{Using the principle of zero products}$$

Solve.

$$x = -2 \quad or \qquad x = 5$$

Check.

Check:

For -2:

$$\frac{\log x + \log (x - 3) = 1}{\log (-2) + \log (-2 - 3) \overset{?}{=} 1}$$ FALSE

For 5:

$$\frac{\log x + \log (x - 3) = 1}{\log 5 + \log (5 - 3) \,\big|\, 1}$$
$$\log 5 + \log 2$$
$$\log 10 \,\big|$$
$$1 \overset{?}{=} 1 \quad \text{TRUE}$$

The number -2 *does not check* because the logarithm of a negative number is undefined. The solution is 5.

b) We have

$$\log_2 (x + 7) - \log_2 (x - 7) = 3$$

$$\log_2 \frac{x + 7}{x - 7} = 3 \qquad \begin{array}{l}\text{Using the quotient} \\ \text{rule for logarithms} \\ \text{to obtain a single} \\ \text{logarithm}\end{array}$$

$$\frac{x + 7}{x - 7} = 2^3 \qquad \begin{array}{l}\text{Writing an equivalent} \\ \text{exponential equation}\end{array}$$

$$\frac{x + 7}{x - 7} = 8$$

$$x + 7 = 8(x - 7) \qquad \begin{array}{l}\text{Multiplying by the} \\ \text{LCD, } x - 7\end{array}$$

$$x + 7 = 8x - 56 \qquad \begin{array}{l}\text{Using the distributive} \\ \text{law}\end{array}$$

$$63 = 7x$$

$$9 = x. \qquad \text{Dividing by 7}$$

Check:

$$\frac{\log_2 (x + 7) - \log_2 (x - 7) = 3}{\log_2 (9 + 7) - \log_2 (9 - 7) \,\big|\, 3}$$
$$\log_2 16 - \log_2 2$$
$$4 - 1 \,\big|$$
$$3 \overset{?}{=} 3 \quad \text{TRUE}$$

The solution is 9.

c) We have

$$\log_7 (x + 1) + \log_7 (x - 1) = \log_7 8$$

$$\log_7 [(x + 1)(x - 1)] = \log_7 8 \qquad \begin{array}{l}\text{Using the product rule} \\ \text{for logarithms}\end{array}$$

$$\log_7 (x^2 - 1) = \log_7 8 \qquad \begin{array}{l}\text{Multiplying. Note that both} \\ \text{sides are base-7 logarithms.}\end{array}$$

$$x^2 - 1 = 8 \qquad \begin{array}{l}\text{Using the principle of} \\ \text{logarithmic equality. Study} \\ \text{this step carefully.}\end{array}$$

$$x^2 - 9 = 0$$

$$(x - 3)(x + 3) = 0 \qquad \begin{array}{l}\text{Solving the quadratic} \\ \text{equation}\end{array}$$

$$x = 3 \quad or \quad x = -3.$$

We leave it to the student to show that 3 checks but -3 does not. The solution is 3.

> **TRY EXERCISE** 55

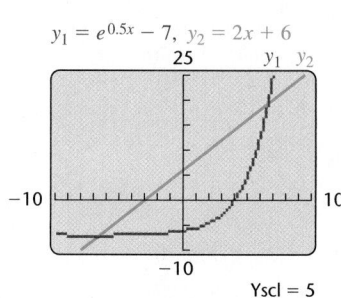

TECHNOLOGY CONNECTION

To solve exponential and logarithmic equations, we can use the INTERSECT option of the CALC menu to determine the x-coordinate at each intersection.

For example, to solve $e^{0.5x} - 7 = 2x + 6$, we graph $y_1 = e^{0.5x} - 7$ and $y_2 = 2x + 6$ as shown. Using INTERSECT twice, we find that the x-coordinates at the intersections are approximately -6.48 and 6.52.

$y_1 = e^{0.5x} - 7, \; y_2 = 2x + 6$

Use a graphing calculator to solve each equation to the nearest hundredth.

1. $e^{7x} = 14$
2. $8e^{0.5x} = 3$
3. $xe^{3x-1} = 5$
4. $4 \ln (x + 3.4) = 2.5$
5. $\ln 3x = 0.5x - 1$
6. $\ln x^2 = -x^2$

12.6 EXERCISE SET

🖐 *Concept Reinforcement* *In each of Exercises 1–8, match the equation with an equivalent equation from the column on the right that could be the next step in the solution process.*

1. ____ $5^x = 3$

2. ____ $e^{5x} = 3$

3. ____ $\ln x = 3$

4. ____ $\log_x 5 = 3$

5. ____ $\log_5 x + \log_5 (x - 2) = 3$

6. ____ $\log_5 x - \log_5 (x - 2) = 3$

7. ____ $\ln x - \ln (x - 2) = 3$

8. ____ $\log x + \log (x - 2) = 3$

a) $\ln e^{5x} = \ln 3$

b) $\log_5 (x^2 - 2x) = 3$

c) $\log (x^2 - 2x) = 3$

d) $\log_5 \dfrac{x}{x - 2} = 3$

e) $\log 5^x = \log 3$

f) $e^3 = x$

g) $\ln \dfrac{x}{x - 2} = 3$

h) $x^3 = 5$

Solve. Where appropriate, include approximations to three decimal places.

9. $3^{2x} = 81$

10. $2^{3x} = 64$

11. $4^x = 32$

12. $9^x = 27$

13. $2^x = 10$

14. $2^x = 24$

15. $2^{x+5} = 16$

16. $2^{x-1} = 8$

17. $8^{x-3} = 19$

18. $5^{x+2} = 15$

19. $e^t = 50$

20. $e^t = 20$

21. $e^{-0.02t} = 8$

22. $e^{-0.01t} = 100$

23. $4.9^x - 87 = 0$

24. $7.2^x - 65 = 0$

25. $19 = 2e^{4x}$

26. $29 = 3e^{2x}$

27. $7 + 3e^{-x} = 13$

28. $4 + 5e^{-x} = 9$

Aha! 29. $\log_3 x = 4$

30. $\log_2 x = 6$

31. $\log_4 x = -2$

32. $\log_5 x = -3$

33. $\ln x = 5$

34. $\ln x = 4$

35. $\ln (4x) = 3$

36. $\ln (3x) = 2$

37. $\log x = 1.2$

38. $\log x = 0.6$

39. $\ln (2x + 1) = 4$

40. $\ln (4x - 2) = 3$

Aha! 41. $\ln x = 1$

42. $\log x = 1$

43. $5 \ln x = -15$

44. $3 \ln x = -3$

45. $\log_2 (8 - 6x) = 5$

46. $\log_5 (7 - 2x) = 3$

47. $\log (x - 9) + \log x = 1$

48. $\log (x + 9) + \log x = 1$

49. $\log x - \log (x + 3) = 1$

50. $\log x - \log (x + 7) = -1$

Aha! 51. $\log (2x + 1) = \log 5$

52. $\log (x + 1) - \log x = 0$

53. $\log_4 (x + 3) = 2 + \log_4 (x - 5)$

54. $\log_2 (x + 3) = 4 + \log_2 (x - 3)$

55. $\log_7 (x + 1) + \log_7 (x + 2) = \log_7 6$

56. $\log_6 (x + 3) + \log_6 (x + 2) = \log_6 20$

57. $\log_5 (x + 4) + \log_5 (x - 4) = \log_5 20$

58. $\log_4 (x + 2) + \log_4 (x - 7) = \log_4 10$

59. $\ln (x + 5) + \ln (x + 1) = \ln 12$

60. $\ln (x - 6) + \ln (x + 3) = \ln 22$

61. $\log_2 (x - 3) + \log_2 (x + 3) = 4$

62. $\log_3 (x - 4) + \log_3 (x + 4) = 2$

63. $\log_{12} (x + 5) - \log_{12} (x - 4) = \log_{12} 3$

64. $\log_6 (x + 7) - \log_6 (x - 2) = \log_6 5$

65. $\log_2 (x - 2) + \log_2 x = 3$

66. $\log_4 (x + 6) - \log_4 x = 2$

67. Madison finds that the solution of $\log_3 (x + 4) = 1$ is -1, but rejects -1 as an answer. What mistake do you suspect she is making?

68. Could Example 2 have been solved by taking the natural logarithm on both sides? Why or why not?

Skill Review

To prepare for Section 12.7, review using the five-step problem-solving strategy.

Solve.

69. A rectangle is 6 ft longer than it is wide. Its perimeter is 26 ft. Find the length and the width. [2.5]

70. Under one health insurance plan offered in California, the maximum co-pay for an individual is $3000 per calendar year. The co-pay for each visit to a specialist is $40, and the co-pay for a hospitalization is $1000. With hospitalizations and specialist visits, Marguerite reached the maximum co-pay in 2008. If she was hospitalized twice, how many visits to specialists did she make? [9.1]
Source: ehealthinsurance.com

71. Joanna wants to mix Golden Days bird seed containing 25% sunflower seeds with Snowy Friends bird seed containing 40% sunflower seeds. She wants 50 lb of a mixture containing 33% sunflower seeds. How much of each type should she use? [8.3]

72. The outside edge of a picture frame measures 12 cm by 19 cm, and 144 cm² of picture shows. Find the width of the frame. [5.8]

73. Max can key in a musical score in 2 hr. Miles takes 3 hr to key in the same score. How long would it take them, working together, to key in the score? [6.7]

74. A sign is in the shape of a right triangle. The hypotenuse is 3 ft long, and the base and the height of the triangle are equal. Find the length of the base and the height. Round to the nearest tenth of a foot. [10.7]

Synthesis

75. Can the principle of logarithmic equality be expanded to include all functions? That is, is the statement "$m = n$ is equivalent to $f(m) = f(n)$" true for any function f? Why or why not?

76. Explain how Exercises 37 and 38 could be solved using the graph of $f(x) = \log x$.

Solve. If no solution exists, state this.

77. $8^x = 16^{3x+9}$

78. $27^x = 81^{2x-3}$

79. $\log_6 (\log_2 x) = 0$

80. $\log_x (\log_3 27) = 3$

81. $\log_5 \sqrt{x^2 - 9} = 1$

82. $x \log \frac{1}{8} = \log 8$

83. $2^{x^2+4x} = \frac{1}{8}$

84. $\log (\log x) = 5$

85. $\log_5 |x| = 4$

86. $\log x^2 = (\log x)^2$

87. $\log \sqrt{2x} = \sqrt{\log 2x}$

88. $1000^{2x+1} = 100^{3x}$

89. $3^{x^2} \cdot 3^{4x} = \frac{1}{27}$

90. $3^{3x} \cdot 3^{x^2} = 81$

91. $\log x^{\log x} = 25$

92. $3^{2x} - 8 \cdot 3^x + 15 = 0$

93. $(81^{x-2})(27^{x+1}) = 9^{2x-3}$

94. $3^{2x} - 3^{2x-1} = 18$

95. Given that $2^y = 16^{x-3}$ and $3^{y+2} = 27^x$, find the value of $x + y$.

96. If $x = (\log_{125} 5)^{\log_5 125}$, what is the value of $\log_3 x$?

97. Find the value of x for which the natural logarithm is the same as the common logarithm.

98. Use a graphing calculator to check your answers to Exercises 11, 31, 41, and 59.

12.7 Applications of Exponential and Logarithmic Functions

Applications of Logarithmic Functions   ■   Applications of Exponential Functions

We now consider applications of exponential and logarithmic functions.

Applications of Logarithmic Functions

EXAMPLE 1

Sound levels.   To measure the volume, or "loudness," of a sound, the *decibel* scale is used. The loudness L, in decibels (dB), of a sound is given by

$$L = 10 \cdot \log \frac{I}{I_0},$$

where I is the intensity of the sound, in watts per square meter (W/m^2), and $I_0 = 10^{-12}$ W/m^2. (I_0 is approximately the intensity of the softest sound that can be heard by the human ear.)

a) The average maximum intensity of sound in a New York subway car is about 3.2×10^{-3} W/m^2. How loud, in decibels, is the sound level?

Source: Columbia University Mailman School of Public Health

b) The Occupational Safety and Health Administration (OSHA) considers sustained sound levels of 90 dB and above unsafe. What is the intensity of such sounds?

SOLUTION

a) To find the loudness, in decibels, we use the above formula:

$$L = 10 \cdot \log \frac{I}{I_0}$$

$= 10 \cdot \log \dfrac{3.2 \times 10^{-3}}{10^{-12}}$ Substituting

$= 10 \cdot \log (3.2 \times 10^9)$ Subtracting exponents

$= 10 (\log 3.2 + \log 10^9)$ $\log MN = \log M + \log N$

$= 10 (\log 3.2 + 9)$ $\log_{10} 10^9 = 9$

$\approx 10 (0.5051 + 9)$ Approximating $\log 3.2$

$= 10 (9.5051)$ Adding within the parentheses

$\approx 95.$ Multiplying and rounding

The volume of the sound in a subway car is about 95 decibels.

b) We substitute and solve for I:

$$L = 10 \cdot \log \frac{I}{I_0}$$

$$90 = 10 \cdot \log \frac{I}{10^{-12}} \qquad \text{Substituting}$$

$$9 = \log \frac{I}{10^{-12}} \qquad \text{Dividing both sides by 10}$$

$$9 = \log I - \log 10^{-12} \qquad \text{Using the quotient rule for logarithms}$$

$$9 = \log I - (-12) \qquad \log 10^a = a$$

$$-3 = \log I \qquad \text{Adding } -12 \text{ to both sides}$$

$$10^{-3} = I. \qquad \text{Converting to an exponential equation}$$

Sustained sounds with intensities exceeding 10^{-3} W/m^2 are considered unsafe.

TRY EXERCISE ▶ 15

EXAMPLE 2

Chemistry: pH of liquids. In chemistry, the pH of a liquid is a measure of its acidity. We calculate pH as follows:

$$\text{pH} = -\log[\text{H}^+],$$

where $[\text{H}^+]$ is the hydrogen ion concentration in moles per liter.

a) The hydrogen ion concentration of human blood is normally about 3.98×10^{-8} moles per liter. Find the pH.

Source: www.merck.com

b) The average pH of seawater is about 8.2. Find the hydrogen ion concentration.

Source: www.seafriends.org.nz

SOLUTION

a) To find the pH of blood, we use the above formula:

$$\text{pH} = -\log[\text{H}^+]$$

$$= -\log[3.98 \times 10^{-8}]$$

$$\approx -(-7.400117) \qquad \text{Using a calculator}$$

$$\approx 7.4.$$

The pH of human blood is normally about 7.4.

b) We substitute and solve for $[\text{H}^+]$:

$$8.2 = -\log[\text{H}^+] \qquad \text{Using pH} = -\log[\text{H}^+]$$

$$-8.2 = \log[\text{H}^+] \qquad \text{Dividing both sides by } -1$$

$$10^{-8.2} = [\text{H}^+] \qquad \text{Converting to an exponential equation}$$

$$6.31 \times 10^{-9} \approx [\text{H}^+]. \qquad \text{Using a calculator; writing scientific notation}$$

The hydrogen ion concentration of seawater is about 6.31×10^{-9} moles per liter.

TRY EXERCISE ▶ 11

Applications of Exponential Functions

EXAMPLE 3 *Interest compounded annually.* Suppose that $25,000 is invested at 4% interest, compounded annually. In t years, it will grow to the amount A given by

$$A(t) = 25{,}000(1.04)^t.$$

(See Example 5 in Section 12.2.)

a) How long will it take to accumulate $80,000 in the account?

b) Find the amount of time it takes for the $25,000 to double itself.

SOLUTION

a) We set $A(t) = 80{,}000$ and solve for t:

$$80{,}000 = 25{,}000(1.04)^t$$

$$\frac{80{,}000}{25{,}000} = 1.04^t \qquad \text{Dividing both sides by 25,000}$$

$$3.2 = 1.04^t$$

$$\log 3.2 = \log 1.04^t \qquad \text{Taking the common logarithm on both sides}$$

$$\log 3.2 = t \log 1.04 \qquad \text{Using the power rule for logarithms}$$

$$\frac{\log 3.2}{\log 1.04} = t \qquad \text{Dividing both sides by } \log 1.04$$

$$29.7 \approx t. \qquad \text{Using a calculator}$$

Remember that when doing a calculation like this on a calculator, it is best to wait until the end to round. At an interest rate of 4% per year, it will take about 29.7 yr for $25,000 to grow to $80,000.

STUDENT NOTES

Study the different steps in the solution of Example 3(b). Note that if 50,000 and 25,000 are replaced with 6000 and 3000, the doubling time is unchanged.

b) To find the *doubling time*, we replace $A(t)$ with 50,000 and solve for t:

$$50{,}000 = 25{,}000(1.04)^t$$

$$2 = (1.04)^t \qquad \text{Dividing both sides by 25,000}$$

$$\log 2 = \log (1.04)^t \qquad \text{Taking the common logarithm on both sides}$$

$$\log 2 = t \log 1.04 \qquad \text{Using the power rule for logarithms}$$

$$t = \frac{\log 2}{\log 1.04} \approx 17.7. \qquad \text{Dividing both sides by } \log 1.04 \text{ and using a calculator}$$

At an interest rate of 4% per year, the doubling time is about 17.7 yr.

TRY EXERCISE 21

Like investments, populations often grow exponentially.

Exponential Growth

An **exponential growth model** is a function of the form

$$P(t) = P_0 e^{kt}, \quad k > 0,$$

where P_0 is the population at time 0, $P(t)$ is the population at time t, and k is the **exponential growth rate** for the situation. The **doubling time** is the amount of time necessary for the population to double in size.

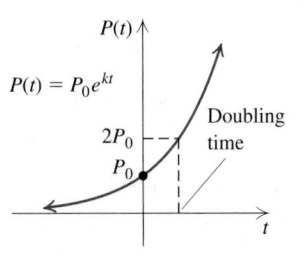

The exponential growth rate is the rate of growth of a population or other quantity at any *instant* in time. Since the population is continually growing, the percent of total growth after one year will exceed the exponential growth rate.

EXAMPLE 4   *Text messaging.*   In 2000, there were approximately 12 million text messages sent each month in the United States. This number has increased exponentially at an average rate of 108% per year.

Source: Based on information from www.cellulist.com

a) Find the exponential growth function that models the data.

b) Estimate the number of text messages sent each month in 2009.

SOLUTION

a) In 2000, at $t = 0$, the number of messages was 12 million per month. We substitute 12 for P_0 and 108%, or 1.08, for k. This gives the exponential growth function

$$P(t) = 12e^{1.08t}.$$

b) In 2009, we have $t = 9$ (since 9 yr have passed since 2000). To determine the number of messages in 2009, we compute $P(9)$:

$$P(9) = 12e^{1.08(9)} \qquad \text{Using } P(t) = 12e^{1.08t} \text{ from part (a)}$$
$$= 12e^{9.72} \approx 200{,}000. \quad \text{Using a calculator}$$

In 2009, the number of text messages sent in the United States each month will reach approximately 200,000 million, or 200 billion.

TRY EXERCISE   23

EXAMPLE 5   *Cruise ship passengers.*   In 1970, cruise lines carried approximately 500,000 passengers. This number has increased exponentially to 12.1 million in 2006.

Source: Cruise Lines International Association

a) Find the exponential growth rate and the exponential growth function.

b) Estimate the year in which cruise lines will carry 20 million passengers.

SOLUTION

a) We use $S(t) = S_0 e^{kt}$, where t is the number of years since 1970 and $S(t)$ is the number of passengers, in millions. Since 500,000 is half a million, we substitute 0.5 for S_0:

$$S(t) = 0.5e^{kt}.$$

To find the exponential growth rate k, note that after 36 yr (2006 − 1970 = 36), there were 12.1 million passengers:

$$S(36) = 0.5e^{k \cdot 36}$$
$$12.1 = 0.5e^{36k} \qquad \text{Substituting}$$
$$24.2 = e^{36k} \qquad \text{Dividing both sides by 0.5}$$
$$\ln 24.2 = \ln e^{36k} \qquad \text{Taking the natural logarithm on both sides}$$
$$\ln 24.2 = 36k \qquad \ln e^{36k} = \log_e e^{36k} = 36k$$
$$\frac{\ln 24.2}{36} = k \qquad \text{Dividing both sides by 36}$$
$$0.089 \approx k. \qquad \text{Using a calculator and rounding}$$

The exponential growth rate is 8.9% and the exponential growth function is given by $S(t) = 0.5e^{0.089t}$.

A visualization of Example 5

b) To estimate the year in which cruise lines will carry 20 million passengers, we replace $S(t)$ with 20 and solve for t:

$$20 = 0.5e^{0.089t}$$

$$40 = e^{0.089t} \qquad \text{Dividing both sides by 0.5}$$

$$\ln 40 = \ln e^{0.089t} \qquad \text{Taking the natural logarithm on both sides}$$

$$\ln 40 = 0.089t \qquad \ln e^a = a$$

$$\frac{\ln 40}{0.089} = t \qquad \text{Dividing both sides by 0.089}$$

$$41.4 \approx t. \qquad \text{Using a calculator}$$

Rounding to 41, we see that, according to this model, cruise lines will carry 20 million passengers 41 yr after 1970, or in 2011. **TRY EXERCISE ▶ 31**

EXAMPLE 6

Interest compounded continuously. When an amount of money P_0 is invested at interest rate k, compounded *continuously*, interest is computed every "instant" and added to the original amount. The balance $P(t)$, after t years, is given by the exponential growth model

$$P(t) = P_0 e^{kt}.$$

a) Suppose that $30,000 is invested and grows to $44,754.75 in 5 yr. Find the exponential growth function.

b) What is the doubling time?

SOLUTION

a) We have $P(0) = 30,000$. Thus the exponential growth function is

$$P(t) = 30,000e^{kt}, \quad \text{where } k \text{ must still be determined.}$$

Knowing that for $t = 5$ we have $P(5) = 44,754.75$, it is possible to solve for k:

$$44,754.75 = 30,000e^{k(5)}$$

$$44,754.75 = 30,000e^{5k}$$

$$\frac{44,754.75}{30,000} = e^{5k} \qquad \text{Dividing both sides by 30,000}$$

$$1.491825 = e^{5k}$$

$$\ln 1.491825 = \ln e^{5k} \qquad \text{Taking the natural logarithm on both sides}$$

$$\ln 1.491825 = 5k \qquad \ln e^a = a$$

$$\frac{\ln 1.491825}{5} = k \qquad \text{Dividing both sides by 5}$$

$$0.08 \approx k. \qquad \text{Using a calculator and rounding}$$

The interest rate is about 0.08, or 8%, compounded continuously. Because interest is being compounded continuously, the yearly interest rate is a bit more than 8%. The exponential growth function is

$$P(t) = 30,000e^{0.08t}.$$

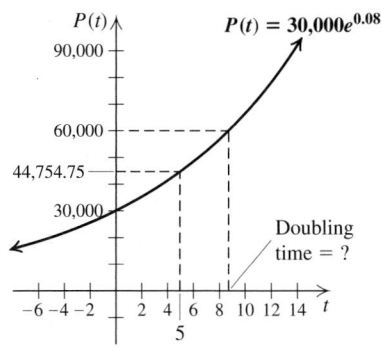

$P(t) = 30{,}000e^{0.08t}$

A visualization of Example 6

b) To find the doubling time T, we replace $P(T)$ with 60,000 and solve for T:

$$60{,}000 = 30{,}000e^{0.08T}$$

$$2 = e^{0.08T} \qquad \text{Dividing both sides by 30,000}$$

$$\ln 2 = \ln e^{0.08T} \qquad \text{Taking the natural logarithm on both sides}$$

$$\ln 2 = 0.08T \qquad \ln e^{a} = a$$

$$\frac{\ln 2}{0.08} = T \qquad \text{Dividing both sides by 0.08}$$

$$8.7 \approx T. \qquad \text{Using a calculator and rounding}$$

Thus the original investment of \$30,000 will double in about 8.7 yr.

TRY EXERCISE ▶ 41

For any specified interest rate, continuous compounding gives the highest yield and the shortest doubling time.

In some real-life situations, a quantity or population is *decreasing* or *decaying* exponentially.

Exponential Decay

An **exponential decay model** is a function of the form

$$P(t) = P_0 e^{-kt}, \quad k > 0,$$

where P_0 is the quantity present at time 0, $P(t)$ is the amount present at time t, and k is the **decay rate**. The **half-life** is the amount of time necessary for half of the quantity to decay.

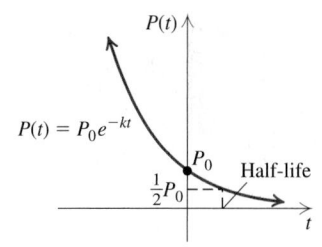

$P(t) = P_0 e^{-kt}$

EXAMPLE 7

Carbon dating. The radioactive element carbon-14 has a half-life of 5750 yr. The percentage of carbon-14 in the remains of organic matter can be used to determine the age of that material. Recently, while digging in Chaco Canyon, New Mexico, archaeologists found corn pollen that had lost 38.1% of its carbon-14. The age of this corn pollen was evidence that Indians had been cultivating crops in the Southwest centuries earlier than scientists had thought. What was the age of the pollen?

Source: *American Anthropologist*

Chaco Canyon, New Mexico

SOLUTION We first find k. To do so, we use the concept of half-life. When $t = 5750$ (the half-life), $P(t)$ is half of P_0. Then

$$0.5P_0 = P_0 e^{-k(5750)} \qquad \text{Substituting in } P(t) = P_0 e^{-kt}$$

$$0.5 = e^{-5750k} \qquad \text{Dividing both sides by } P_0$$

$$\ln 0.5 = \ln e^{-5750k} \qquad \text{Taking the natural logarithm on both sides}$$

$$\ln 0.5 = -5750k \qquad \ln e^{a} = a$$

$$\frac{\ln 0.5}{-5750} = k \qquad \text{Dividing}$$

$$0.00012 \approx k. \qquad \text{Using a calculator and rounding}$$

Now we have a function for the decay of carbon-14:

$$P(t) = P_0e^{-0.00012t}.$$ This completes the first part of our solution.

(*Note*: This equation can be used for subsequent carbon-dating problems.) If the corn pollen has lost 38.1% of its carbon-14 from an initial amount P_0, then $100\% - 38.1\%$, or 61.9%, of P_0 is still present. To find the age t of the pollen, we solve this equation for t:

$0.619P_0 = P_0e^{-0.00012t}$	We want to find t for which $P(t) = 0.619P_0$.
$0.619 = e^{-0.00012t}$	Dividing both sides by P_0
$\ln 0.619 = \ln e^{-0.00012t}$	Taking the natural logarithm on both sides
$\ln 0.619 = -0.00012t$	$\ln e^a = a$
$\dfrac{\ln 0.619}{-0.00012} = t$	Dividing both sides by -0.00012
$4000 \approx t.$	Using a calculator

The pollen is about 4000 yr old.

TRY EXERCISE 35

12.7   EXERCISE SET

Solve.

1. *Asteroids.*   The total number $A(t)$ of known asteroids t years after 1990 can be estimated by

    $$A(t) = 77(1.283)^t.$$

    Source: Based on data from NASA

    a) Determine the year in which the number of known asteroids first reached 4000.
    b) What is the doubling time for the number of known asteroids?

2. *Social networking.*   The number of unique (different) visitors per month to Facebook t months after April 2006 can be estimated by

    $$F(t) = 11.755(1.109)^t,$$

    where $F(t)$ is in millions.
    Source: Based on data from comScore World Metrix

    a) In what month will the number of Facebook visitors first reach 1 billion?
    b) What is the doubling time for the number of unique Facebook visitors per month?

3. *Health.*   The rate of number of deaths due to stroke in the United States can be estimated by

    $$S(t) = 180(0.97)^t,$$

    where $S(t)$ is the number of deaths per 100,000 people and t is the number of years since 1960.
    Source: Based on data from Centers for Disease Control and Prevention

    a) In what year was the death rate due to stroke 100 per 100,000 people?
    b) In what year will the death rate due to stroke be 25 per 100,000 people?

4. *Alternative fuels.*   The number of gallons of ethanol produced in the United States can be estimated by

    $$E(t) = 0.18(1.137)^t,$$

    where $E(t)$ is the annual production, in billions of gallons, t years after 1980.

    a) In what year did the United States produce 5 billion gal of ethanol?
    b) In what year will the United States produce 25 billion gal of ethanol?

5. *Student loan repayment.*   A college loan of $29,000 is made at 3% interest, compounded annually. After t years, the amount due, A, is given by the function

    $$A(t) = 29,000(1.03)^t.$$

    a) After what amount of time will the amount due reach $35,000?
    b) Find the doubling time.

6. *Spread of a rumor.*   The number of people who have heard a rumor increases exponentially. If all who hear a rumor repeat it to two people a day, and if 20 people start the rumor, the number of people N who have heard the rumor after t days is given by

    $$N(t) = 20(3)^t.$$

a) After what amount of time will 1000 people have heard the rumor?

b) What is the doubling time for the number of people who have heard the rumor?

7. *Health insurance.*   The percentage of workers covered by a conventional health plan is decreasing exponentially. The percentage of covered workers $W(t)$ enrolled in conventional plans t years after 1988 can be estimated by

$$W(t) = 89(0.837)^t.$$

Sources: Based on data from Kaiser and HRET

a) According to this model, in what year did the percentage of covered workers enrolled in conventional plans drop below 50%?

b) In what year will the percentage of covered workers enrolled in conventional plans drop below 1%?

8. *Smoking.*   The percentage of smokers who received telephone counseling and had successfully quit smoking for t months is given by

$$P(t) = 21.4(0.914)^t.$$

Sources: *New England Journal of Medicine:* data from California's Smoker's Hotline

a) In what month will 15% of those who quit and used telephone counseling still be smoke-free?

b) In what month will 5% of those who quit and used telephone counseling still be smoke-free?

9. *Marine biology.*   As a result of preservation efforts in countries in which whaling was once common, the humpback whale population has grown since the 1970s. The worldwide population $P(t)$, in thousands, t years after 1982 can be estimated by

$$P(t) = 5.5(1.047)^t.$$

a) In what year will the humpback whale population reach 30,000?

b) Find the doubling time.

10. *World population.*   The world population $P(t)$, in billions, t years after 2000 can be approximated by

$$P(t) = 4.553(1.014)^t.$$

Sources: Based on data from U.S. Census Bureau; International Data Base

a) In what year will the world population reach 10 billion?

b) Find the doubling time.

Use the pH formula given in Example 2 for Exercises 11–14.

11. *Chemistry.*   The hydrogen ion concentration of fresh-brewed coffee is about 1.3×10^{-5} moles per liter. Find the pH.

12. *Chemistry.*   The hydrogen ion concentration of milk is about 1.6×10^{-7} moles per liter. Find the pH.

13. *Medicine.*   When the pH of a patient's blood drops below 7.4, a condition called *acidosis* sets in. Acidosis can be deadly when the patient's pH reaches 7.0. What would the hydrogen ion concentration of the patient's blood be at that point?

14. *Medicine.*   When the pH of a patient's blood rises above 7.4, a condition called *alkalosis* sets in. Alkalosis can be deadly when the patient's pH reaches 7.8. What would the hydrogen ion concentration of the patient's blood be at that point?

Use the formula in Example 1 for Exercises 15–18.

15. *Racing.*   The intensity of sound from a race car in full throttle is about 10 W/m². How loud in decibels is this sound level?

Source: nascar.about.com

16. *Audiology.*   The intensity of sound in normal conversation is about 3.2×10^{-6} W/m². How loud in decibels is this sound level?

17. *Concerts.* The crowd at a Hearsay concert at Wembley Arena in London cheered at a sound level of 128.8 dB. What is the intensity of such a sound?
Source: www.peterborough.gov.uk

18. *City ordinances.* In Albuquerque, New Mexico, the maximum allowable sound level from a car's exhaust is 96 dB. What is the intensity of such a sound?
Source: www.cabq.gov

19. *E-mail volume.* The SenderBase® Security Network ranks e-mail volume using a logarithmic scale. The magnitude M of a network's daily e-mail volume is given by

$$M = \log \frac{v}{1.34},$$

where v is the number of e-mail messages sent each day. How many e-mail messages are sent each day by a network that has a magnitude of 7.5?
Source: forum.spamcop.net

20. *Richter scale.* The Richter scale, developed in 1935, has been used for years to measure earthquake magnitude. The Richter magnitude m of an earthquake is given by the formula

$$m = \log \frac{A}{A_0},$$

where A is the maximum amplitude of the earthquake and A_0 is a constant. What is the magnitude on the Richter scale of an earthquake with an amplitude that is a million times A_0?

Use the compound-interest formula in Example 6 for Exercises 21 and 22.

21. *Interest compounded continuously.* Suppose that P_0 is invested in a savings account where interest is compounded continuously at 2.5% per year.
a) Express $P(t)$ in terms of P_0 and 0.025.
b) Suppose that $5000 is invested. What is the balance after 1 yr? after 2 yr?
c) When will an investment of $5000 double itself?

22. *Interest compounded continuously.* Suppose that P_0 is invested in a savings account where interest is compounded continuously at 3.1% per year.
a) Express $P(t)$ in terms of P_0 and 0.031.
b) Suppose that $1000 is invested. What is the balance after 1 yr? after 2 yr?
c) When will an investment of $1000 double itself?

23. *Population growth.* In 2008, the population of the United States was 304 million and the exponential growth rate was 0.9% per year.
Source: U.S. Census Bureau
a) Find the exponential growth function.
b) Predict the U.S. population in 2012.
c) When will the U.S. population reach 325 million?

24. *World population growth.* In 2008, the world population was 6.7 billion and the exponential growth rate was 1.14% per year.
Source: U.S. Census Bureau
a) Find the exponential growth function.
b) Predict the world population in 2014.
c) When will the world population be 8.0 billion?

25. *Zebra mussels.* The number of zebra mussels in a river grows at an exponential growth rate of 340% per year. What is the doubling time for zebra mussels?

26. *Population growth.* The exponential growth rate of the population of United Arab Emirates is 4.4% per year (one of the highest in the world). What is the doubling time?
Sources: Based on data from U.S. Census Bureau; International Data Base 2007

27. *World population.* The function

$$Y(x) = 71.41 \ln \frac{x}{4.6}$$

can be used to estimate the number of years $Y(x)$ after 2000 required for the world population to reach x billion people.
Sources: Based on data from U.S. Census Bureau; International Data Base
a) In what year will the world population reach 10 billion?
b) In what year will the world population reach 12 billion?
c) Graph the function.

28. *Marine biology.* The function

$$Y(x) = 21.77 \ln \frac{x}{5.5}$$

can be used to estimate the number of years $Y(x)$ after 1982 required for the world's humpback whale population to reach x thousand whales.
a) In what year will the whale population reach 15,000?
b) In what year will the whale population reach 25,000?
c) Graph the function.

29. *Forgetting.* Students in an English class took a final exam. They took equivalent forms of the exam at monthly intervals thereafter. The average score $S(t)$, in percent, after t months was found to be given by

$$S(t) = 68 - 20 \log (t + 1), \quad t \geq 0.$$

a) What was the average score when they initially took the test, $t = 0$?
b) What was the average score after 4 months? after 24 months?
c) Graph the function.
d) After what time t was the average score 50%?

30. *Health insurance.* The amount spent each year by the U.S. government for health insurance for low-income children can be estimated by

$$h(t) = 2.6 \ln t,$$

where $h(t)$ is in billions of dollars and t is the number of years after 1998.
Source: Based on data from the Congressional Budget Office

a) How much was spent on health insurance for low-income children in 2007?
b) Graph the function.
c) In what year will $7 billion be spent on health insurance for low-income children?

31. *Wind power.* U.S. wind-power capacity has grown exponentially from about 2000 megawatts in 1990 to 17,000 megawatts in 2007.
Source: American Wind Energy Association.

a) Find the exponential growth rate k and write an equation for an exponential function that can be used to predict U.S. wind-power capacity t years after 1990.
b) Estimate the year in which wind-power capacity will reach 50,000 megawatts.

32. *Spread of a computer virus.* The number of computers infected by a virus t hours after it first appears usually increases exponentially. In 2004, the "MyDoom" worm spread from 100 computers to about 100,000 computers in 24 hr.
Source: Based on data from IDG News Service

a) Find the exponential growth rate k and write an equation for an exponential function that can be used to predict the number of computers infected t hours after the virus first appeared in 100 computers.
b) Assuming exponential growth, estimate how long it took the MyDoom worm to infect 9000 computers.

33. *Cable costs.* In 1997, the cost to construct communication cables under the ocean was approximately $8200 per gigabit per second per mile. This cost for subsea cables dropped exponentially to $500 by 2007.
Source: Based on information from TeleGeography

a) Find the exponential growth rate k, and write an equation for an exponential function that can be used to predict the cost of subsea cables t years after 1997.
b) Estimate the cost of subsea cables in 2010.
c) In what year (theoretically) will it cost only $1 per gigabit per second per mile to construct subsea cables?

34. *Decline in farmland.* The number of acres of farmland in the United States has decreased from 945 million acres in 2000 to 932 million acres in 2006. Assume the number of acres of farmland is decreasing exponentially.
Source: Statistical Abstract of the United States

a) Find the value k, and write an equation for an exponential function that can predict the number of acres of U.S. farmland t years after 2000.
b) Predict the number of acres of farmland in 2015.
c) In what year (theoretically) will there be only 800 million acres of U.S. farmland remaining?

35. *Archaeology.* A date palm seedling is growing in Kibbutz Ketura, Israel, from a seed found in King Herod's palace at Masada. The seed had lost 21% of its carbon-14. How old was the seed? (See Example 7.)
Source: Based on information from www.sfgate.com

36. *Archaeology.* Soil from beneath the Kish Church in Azerbaijan was found to have lost 12% of its carbon-14. How old was the soil? (See Example 7.)
Source: Based on information from www.azer.com

37. *Chemistry.* The exponential decay rate of iodine-131 is 9.6% per day. What is its half-life?

38. *Chemistry.* The decay rate of krypton-85 is 6.3% per year. What is its half-life?

39. *Caffeine.* The half-life of caffeine in the human body for a healthy adult is approximately 5 hr.
a) What is the exponential decay rate?
b) How long will it take 95% of the caffeine consumed to leave the body?

40. *Home construction.* The chemical urea formaldehyde was found in some insulation used in houses built during the mid to late 1960s. Unknown at the time was the fact that urea formaldehyde emitted toxic fumes as it decayed. The half-life of urea formaldehyde is 1 yr.
a) What is its decay rate?
b) How long will it take 95% of the urea formaldehyde present to decay?

41. *Value of a sports card.* Legend has it that because he objected to teenagers smoking, and because his first baseball card was issued in cigarette packs, the great shortstop Honus Wagner halted production of his card before many were produced. One of these cards was purchased in 1991 by hockey great Wayne Gretzky (and a partner) for $451,000. The same card was sold in 2007 for $2.8 million. For the following questions, assume that the card's value increases exponentially, as it has for many years.

WAGNER, PITTSBURG

a) Find the exponential growth rate k, and determine an exponential function that can be used to estimate the dollar value, $V(t)$, of the card t years after 1991.
b) Predict the value of the card in 2012.
c) What is the doubling time for the value of the card?
d) In what year will the value of the card first exceed $4,000,000?

42. *Art masterpieces.* As of April 2008, the highest auction price for a contemporary painting was $72.8 million, paid in 2007 for Rothko's *White Center* (*Yellow, Pink and Lavender on Rose*). The same painting sold for about $10,000 in 1960.

a) Find the exponential growth rate k, and determine the exponential growth function that can be used to estimate $V(t)$, the painting's value, in millions of dollars, t years after 1960.
b) Estimate the value of the painting in 2012.
c) What is the doubling time for the value of the painting?
d) How long after 1960 will the value of the painting be $1 billion?

43. Write a problem for a classmate to solve in which information is provided and the classmate is asked to find an exponential growth function. Make the problem as realistic as possible.

44. Examine the restriction on t in Exercise 29.

a) What upper limit might be placed on t?
b) In practice, would this upper limit ever be enforced? Why or why not?

Skill Review

To prepare for Section 13.1, review the distance and midpoint formulas, completing the square, and graphing parabolas (Sections 10.7, 11.1, and 11.7).

Find the distance between each pair of points. [10.7]
45. $(-3, 7)$ and $(-2, 6)$ **46.** $(1, 5)$ and $(4, 1)$

Find the coordinates of the midpoint of the segment connecting each pair of points. [10.7]
47. $(3, -8)$ and $(5, -6)$

48. $(2, -11)$ and $(-9, -8)$

Solve by completing the square. [11.1]
49. $x^2 + 8x = 1$

50. $x^2 - 10x = 15$

Graph. [11.7]
51. $y = x^2 - 5x - 6$

52. $g(x) = 2x^2 - 6x + 3$

Synthesis

53. Will the model used in Example 4 to predict the number of text messages still be realistic in 2030? Why or why not?

54. *Atmospheric pressure.* Atmospheric pressure P at an elevation of a feet above sea level is given by
$$P = P_0 e^{-0.00004a},$$
where P_0 is the pressure at sea level, which is approximately 29.9 in inches of mercury (Hg). Explain how a barometer, or some other device for measuring atmospheric pressure, can be used to find the height of a skyscraper.

55. *Sports salaries.* As of April 2008, Alex Rodriguez of the New York Yankees had the largest contract in sports history. As part of the 10-year $275-million deal, he will receive $20 million in 2016. How much money would need to be invested in 2008, at 4% interest compounded continuously, in order to have $20 million for Rodriguez in 2016? (This is much like determining what $20 million in 2016 is worth in 2008 dollars.)
Source: *The San Francisco Chronicle*

56. *Supply and demand.* The supply and demand for the sale of stereos by Sound Ideas are given by

$$S(x) = e^x \quad \text{and} \quad D(x) = 162{,}755e^{-x},$$

where $S(x)$ is the price at which the company is willing to supply x stereos and $D(x)$ is the demand price for a quantity of x stereos. Find the equilibrium point. (For reference, see Section 8.8.)

57. *Stellar magnitude.* The apparent stellar magnitude m of a star with received intensity I is given by

$$m(I) = -(19 + 2.5 \cdot \log I),$$

where I is in watts per square meter (W/m^2). The smaller the apparent stellar magnitude, the brighter the star appears.
Source: The Columbus Optical SETI Observatory

a) The intensity of light received from the sun is 1390 W/m^2. What is the apparent stellar magnitude of the sun?

b) The 5-m diameter Hale telescope on Mt. Palomar can detect a star with magnitude +23. What is the received intensity of light from such a star?

58. *Growth of bacteria.* The bacteria *Escherichia coli* (*E. coli*) are commonly found in the human bladder. Suppose that 3000 of the bacteria are present at time $t = 0$. Then t minutes later, the number of bacteria present is

$$N(t) = 3000(2)^{t/20}.$$

If 100,000,000 bacteria accumulate, a bladder infection can occur. If, at 11:00 A.M., a patient's bladder contains 25,000 *E. coli* bacteria, at what time can infection occur?

59. Show that for exponential growth at rate k, the doubling time T is given by $T = \dfrac{\ln 2}{k}$.

60. Show that for exponential decay at rate k, the half-life T is given by $T = \dfrac{\ln 2}{k}$.

61. *Generic drugs.* Largely because of budget constraints, the Food and Drug Administration (FDA) cannot keep up with the rapidly increasing number of applications for approval of generic drugs. The following table shows the number of applications and the number of approvals for generic drugs for recent years.

Year	Number of New Applications for Generic Drugs	Number of Approvals of Generic Drugs
2001	300	310
2002	361	364
2003	449	373
2004	563	413
2005	766	467
2006	810	525

Source: U.S. Food and Drug Administration

a) Graph the data for applications submitted to the FDA, and determine which would be a better fit for the data: an exponential function or a linear function. Explain your reasoning.

b) Graph the data for approvals from the FDA, and determine which would be a better fit for the data: an exponential function or a linear function. Explain your reasoning.

c) Use regression to fit a function to each set of data.

d) If the trends continue, in what year will there be only half as many approvals as applications?

CORNER

Safe Listening

Focus: Logarithmic models

Time: 30 minutes

Group size: 2–4

Materials: MP3 players (one per group) with music

The *decibel* scale is used to measure the volume, or "loudness," of a sound. Listening to music at a high volume can lead to damaged hearing, because the power required to produce louder volumes increases exponentially.

ACTIVITY

1. Group members should work together to answer the following questions.

   a) The volume *V*, in decibels (dB), of sound on an MP3 player is given by

   $$V = 10 \cdot \log \frac{P}{P_0},$$

   where *P* is the power needed to produce the sound, in milliwatts (mW), and $P_0 = 0.1$ mW is the power needed to produce the lowest volume that registers on the MP3 player. Find V_0, the volume when the lowest power setting is used $(P = P_0)$.

   b) On a typical MP3 player, each time the volume is stepped up by pressing the volume button, the volume increases by 1.5 decibels (dB). If the volume is stepped up 20 times from V_0, by how many decibels does the volume increase?

   c) If the volume is stepped up 20 times from V_0, how much power is required to produce the sound?

   d) By how much does the power level increase every time the volume is stepped up?

   e) Solve the formula for *P*. What type of function is this?

2. One group member should begin listening to a song on the MP3 player. If possible, the song should be recent and downloaded from a music site.

   a) Most MP3 songs are designed to play at 100 dB. Beyond this volume, the music begins to sound distorted. Starting at the minimum volume level, the listener should increase the volume until the music begins to sound distorted. Calculate the approximate decibel increase per step for this MP3 player by dividing 100 by the number of times the volume was stepped up. How much power would be needed to produce this volume?

   b) The Occupational Safety and Health Administration (OSHA) considers sustained sound levels of 90 dB and above unsafe. Using the decibel increase per step from part 2(a), calculate the number of steps needed to increase the volume from the minimum volume level to 90 dB.

   c) Starting again at the minimum volume level, the listener should increase the volume the calculated number of steps until the volume reaches approximately 90 dB. How much power is required to produce this volume?

   d) How many times as much power is used to produce 100 dB as to produce 90 dB?

Thanks to Greg Massey, Embedded Software Engineer, for suggesting this application.

Study Summary

KEY TERMS AND CONCEPTS	EXAMPLES

SECTION 12.1: COMPOSITE AND INVERSE FUNCTIONS

The **composition** of f and g is defined as
$$(f \circ g)(x) = f(g(x)).$$

If $f(x) = \sqrt{x}$ and $g(x) = 2x - 5$, then
$$(f \circ g)(x) = f(g(x)) = f(2x - 5)$$
$$= \sqrt{2x - 5}.$$

A function f is **one-to-one** if different inputs have different outputs. The graph of a one-to-one function passes the **horizontal-line test**.

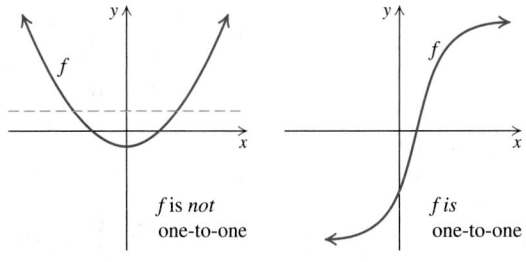

f is *not* one-to-one f is one-to-one

If f is one-to-one, it is possible to find its inverse:

1. Replace $f(x)$ with y.
2. Interchange x and y.
3. Solve for y.

4. Replace y with $f^{-1}(x)$.

If $f(x) = 2x - 3$, find $f^{-1}(x)$.

1. $y = 2x - 3$
2. $x = 2y - 3$
3. $x + 3 = 2y$
   $$\frac{x + 3}{2} = y$$

4. $\dfrac{x + 3}{2} = f^{-1}(x)$

SECTION 12.2: EXPONENTIAL FUNCTIONS
SECTION 12.3: LOGARITHMIC FUNCTIONS

For an **exponential function** f:
$f(x) = a^x$;
$a > 0,\ a \neq 1$;
Domain: $\mathbb{R}$;
$f^{-1}(x) = \log_a x$.

For a **logarithmic function** g:
$g(x) = \log_a x$;
$a > 0,\ a \neq 1$;
Domain: $(0, \infty)$;
$g^{-1}(x) = a^x$.

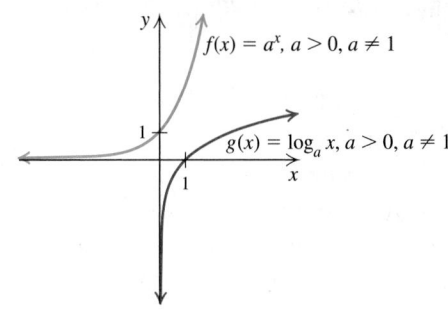

$\log_a x = m$ means $a^m = x$.

$\log_8 x = 2$
$\quad 8^2 = x \qquad$ Rewriting as an exponential equation
$\quad 64 = x$

SECTION 12.4: PROPERTIES OF LOGARITHMIC FUNCTIONS
SECTION 12.5: COMMON AND NATURAL LOGARITHMS

Properties of Logarithms

$\log_a (MN) = \log_a M + \log_a N$

$\log_a \dfrac{M}{N} = \log_a M - \log_a N$

$\log_a M^p = p \cdot \log_a M$

$\log_a 1 = 0$

$\log_a a = 1$

$\log_a a^k = k$

$\log M = \log_{10} M$

$\ln M = \log_e M$

$\log_b M = \dfrac{\log_a M}{\log_a b}$

$\log_7 10 = \log_7 5 + \log_7 2$

$\log_5 \dfrac{14}{3} = \log_5 14 - \log_5 3$

$\log_8 5^{12} = 12 \log_8 5$

$\log_9 1 = 0$

$\log_4 4 = 1$

$\log_3 3^8 = 8$

$\log 43 = \log_{10} 43$

$\ln 37 = \log_e 37$

$\log_6 31 = \dfrac{\log 31}{\log 6} = \dfrac{\ln 31}{\ln 6}$

SECTION 12.6: SOLVING EXPONENTIAL AND LOGARITHMIC EQUATIONS

The Principle of Exponential Equality

For any real number b, $b \neq -1, 0,$ or 1:

$\qquad b^x = b^y$ is equivalent to $x = y$.

$25 = 5^x$

$5^2 = 5^x$

$\ \ 2 = x$

The Principle of Logarithmic Equality

For any logarithm base a, and for $x, y > 0$:

$x = y$ is equivalent to $\log_a x = \log_a y$.

$\quad 83 = 7^x$

$\log 83 = \log 7^x$

$\log 83 = x \log 7$

$\dfrac{\log 83}{\log 7} = x$

SECTION 12.7: APPLICATIONS OF EXPONENTIAL AND LOGARITHMIC FUNCTIONS

Exponential Growth Model

$\qquad P(t) = P_0 e^{kt}, \quad k > 0$

P_0 is the population at time 0.

$P(t)$ is the population at time t.

k is the **exponential growth rate**.

The **doubling time** is the amount of time necessary for the population to double in size.

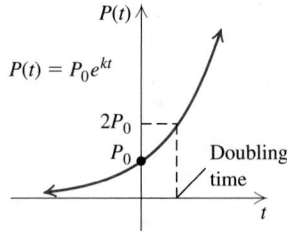

Exponential Decay Model

$\qquad P(t) = P_0 e^{-kt}, \quad k > 0$

P_0 is the quantity present at time 0.

$P(t)$ is the amount present at time t.

k is the **exponential decay rate**.

The **half-life** is the amount of time necessary for half of the quantity to decay.

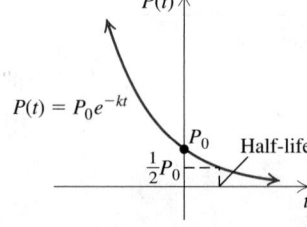

Review Exercises: Chapter 12

🖊 **Concept Reinforcement** *In each of Exercises 1–10, classify the statement as either true or false.*

1. The functions given by $f(x) = 3^x$ and $g(x) = \log_3 x$ are inverses of each other. [12.3]

2. A function's doubling time is the amount of time t for which $f(t) = 2 \cdot f(0)$. [12.7]

3. A radioactive isotope's half-life is the amount of time t for which $f(t) = \frac{1}{2} \cdot f(0)$. [12.7]

4. $\ln(ab) = \ln a - \ln b$ [12.4]

5. $\log x^a = x \ln a$ [12.4]

6. $\log_a \dfrac{m}{n} = \log_a m - \log_a n$ [12.4]

7. For $f(x) = 3^x$, the domain of f is $[0, \infty)$. [12.2]

8. For $g(x) = \log_2 x$, the domain of g is $[0, \infty)$. [12.3]

9. The function F is not one-to-one if $F(-2) = F(5)$. [12.1]

10. The function g is one-to-one if it passes the vertical-line test. [12.1]

---

11. Find $(f \circ g)(x)$ and $(g \circ f)(x)$ if $f(x) = x^2 + 1$ and $g(x) = 2x - 3$. [12.1]

12. If $h(x) = \sqrt{3 - x}$, find $f(x)$ and $g(x)$ such that $h(x) = (f \circ g)(x)$. Answers may vary. [12.1]

13. Determine whether $f(x) = 4 - x^2$ is one-to-one. [12.1]

Find a formula for the inverse of each function. [12.1]

14. $f(x) = x - 10$

15. $g(x) = \dfrac{3x + 1}{2}$

16. $f(x) = 27x^3$

Graph.

17. $f(x) = 3^x + 1$ [12.2]   18. $x = \left(\frac{1}{4}\right)^y$ [12.2]

19. $y = \log_5 x$ [12.3]

Simplify. [12.3]

20. $\log_9 81$ 21. $\log_3 \frac{1}{9}$

22. $\log_2 2^{11}$ 23. $\log_{16} 4$

Rewrite as an equivalent logarithmic equation. [12.3]

24. $2^{-3} = \frac{1}{8}$ 25. $25^{1/2} = 5$

Rewrite as an equivalent exponential equation. [12.3]

26. $\log_4 16 = x$ 27. $\log_8 1 = 0$

Express as an equivalent expression using the individual logarithms of x, y, and z. [12.4]

28. $\log_a x^4 y^2 z^3$

29. $\log_a \dfrac{x^5}{yz^2}$

30. $\log \sqrt[4]{\dfrac{z^2}{x^3 y}}$

Express as an equivalent expression that is a single logarithm and, if possible, simplify. [12.4]

31. $\log_a 5 + \log_a 8$

32. $\log_a 48 - \log_a 12$

33. $\frac{1}{2}\log a - \log b - 2\log c$

34. $\frac{1}{3}[\log_a x - 2\log_a y]$

Simplify. [12.4]

35. $\log_m m$ 36. $\log_m 1$

37. $\log_m m^{17}$

Given $\log_a 2 = 1.8301$ and $\log_a 7 = 5.0999$, find each of the following. [12.4]

38. $\log_a 14$ 39. $\log_a \frac{2}{7}$

40. $\log_a 28$ 41. $\log_a 3.5$

42. $\log_a \sqrt{7}$ 43. $\log_a \frac{1}{4}$

🖩 *Use a calculator to find each of the following to four decimal places.* [12.5]

44. $\log 75$ 45. $10^{1.789}$

46. $\ln 0.3$ 47. $e^{-0.98}$

🖩 *Find each of the following logarithms using the change-of-base formula. Round answers to four decimal places.* [12.5]

48. $\log_5 50$ 49. $\log_6 5$

Graph and state the domain and the range of each function. [12.5]

50. $f(x) = e^x - 1$ 51. $g(x) = 0.6 \ln x$

Solve. Where appropriate, include approximations to four decimal places. [12.6]

52. $5^x = 125$

53. $3^{2x} = \frac{1}{9}$

54. $\log_3 x = -4$

55. $\log_x 16 = 4$

56. $\log x = -3$

57. $6 \ln x = 18$

58. $4^{2x-5} = 19$

59. $2^x = 12$

60. $e^{-0.1t} = 0.03$

61. $2 \ln x = -6$

62. $\log (2x - 5) = 1$

63. $\log_4 x - \log_4 (x - 15) = 2$

64. $\log_3 (x - 4) = 2 - \log_3 (x + 4)$

65. In a business class, students were tested at the end of the course with a final exam. They were then tested again 6 months later. The forgetting formula was determined to be

$$S(t) = 82 - 18 \log (t + 1),$$

where $S(t)$ was the average student grade t months after taking the final exam. [12.7]

  a) Determine the average score when they first took the exam (when $t = 0$).

  b) What was the average score after 6 months?

  c) After what time was the average score 54?

66. A laptop computer is purchased for $1500. Its value each year is about 80% of its value in the preceding year. Its value in dollars after t years is given by the exponential function

$$V(t) = 1500(0.8)^t. \quad [12.7]$$

  a) After what amount of time will the computer's value be $900?

  b) After what amount of time will the computer's value be half the original value?

67. U.S. companies spent $885 million in e-mail marketing in 2005. This amount was predicted to grow exponentially to $1.1 billion in 2010. [12.7]
Source: Jupiter Research

  a) Find the exponential growth rate k, and write a function that describes the amount $A(t)$, in millions of dollars, spent on e-mail marketing t years after 2005.

  b) Estimate the amount spent on e-mail marketing in 2008.

  c) In what year will U.S. companies spend $2 billion on e-mail marketing?

  d) Find the doubling time.

68. In 2005, consumers received, on average, 3253 spam messages. The volume of spam messages per consumer is decreasing exponentially with an exponential decay rate of 13.7% per year. [12.7]

  a) Find the exponential decay function that can be used to predict the average number of spam messages, $M(t)$, t years after 2005.

  b) Predict the number of spam messages received per consumer in 2010.

  c) In what year, theoretically, will the average consumer receive 100 spam messages?

69. The value of Aret's stock market portfolio doubled in 6 yr. What was the exponential growth rate? [12.7]

70. How long will it take $7600 to double if it is invested at 4.2%, compounded continuously? [12.7]

71. How old is a skull that has lost 34% of its carbon-14? (Use $P(t) = P_0 e^{-0.00012t}$.) [12.7]

72. What is the pH of coffee if its hydrogen ion concentration is 7.9×10^{-6} moles per liter? (Use $pH = -\log [H^+]$.) [12.7]

73. The roar of a lion can reach a sound intensity of 2.5×10^{-1} W/m^2. How loud in decibels is this sound level? $\left(\text{Use } L = 10 \cdot \log \dfrac{I}{10^{-12} \text{W/m}^2}. \right)$ [12.7]
Source: en.allexperts.com

Synthesis

74. Explain why negative numbers do not have logarithms. [12.3]

75. Explain why $f(x) = e^x$ and $g(x) = \ln x$ are inverse functions. [12.5]

Solve. [12.6]

76. $\ln (\ln x) = 3$

77. $2^{x^2 + 4x} = \frac{1}{8}$

78. Solve the system:
$$5^{x+y} = 25,$$
$$2^{2x-y} = 64. \quad [12.6]$$

Test: Chapter 12

 CHAPTER Test Prep VIDEO CD · Step-by-step test solutions are found on the video CD in the back of this book.

1. Find $(f \circ g)(x)$ and $(g \circ f)(x)$ if $f(x) = x + x^2$ and $g(x) = 2x + 1$.

2. If
$$h(x) = \frac{1}{2x^2 + 1},$$
find $f(x)$ and $g(x)$ such that $h(x) = (f \circ g)(x)$. Answers may vary.

3. Determine whether $f(x) = x^2 + 3$ is one-to-one.

Find a formula for the inverse of each function.

4. $f(x) = 3x + 4$

5. $g(x) = (x + 1)^3$

Graph.

6. $f(x) = 2^x - 3$

7. $g(x) = \log_7 x$

Simplify.

8. $\log_5 125$

9. $\log_{100} 10$

10. $3^{\log_3 18}$

11. $\log_n n$

12. $\log_c 1$

13. $\log_a a^{19}$

14. Rewrite as an equivalent logarithmic equation: $5^{-4} = \frac{1}{625}$.

15. Rewrite as an equivalent exponential equation: $m = \log_2 \frac{1}{2}$.

16. Express as an equivalent expression using the individual logarithms of a, b, and c:
$$\log \frac{a^3 b^{1/2}}{c^2}.$$

17. Express as an equivalent expression that is a single logarithm:
$$\tfrac{1}{3} \log_a x + 2 \log_a z.$$

Given $\log_a 2 = 0.301$, $\log_a 6 = 0.778$, and $\log_a 7 = 0.845$, find each of the following.

18. $\log_a 14$

19. $\log_a 3$

20. $\log_a 16$

Use a calculator to find each of the following to four decimal places.

21. $\log 25$

22. $10^{-0.8}$

23. $\ln 0.4$

24. $e^{4.8}$

25. Find $\log_3 14$ using the change-of-base formula. Round to four decimal places.

Graph and state the domain and the range of each function.

26. $f(x) = e^x + 3$

27. $g(x) = \ln(x - 4)$

Solve. Where appropriate, include approximations to four decimal places.

28. $2^x = \frac{1}{32}$

29. $\log_4 x = \frac{1}{2}$

30. $\log x = -2$

31. $5^{4-3x} = 87$

32. $7^x = 1.2$

33. $\ln x = 3$

34. $\log(x - 3) + \log(x + 1) = \log 5$

35. The average walking speed R of people living in a city of population P is given by $R = 0.37 \ln P + 0.05$, where R is in feet per second and P is in thousands.
   a) The population of Tulsa, Oklahoma, is 383,000. Find the average walking speed.
   b) San Diego, California, has an average walking speed of about 3 ft/sec. Find the population.

36. The population of Nigeria was about 140 million in 2008 and the exponential growth rate was 2.4% per year.
   a) Write an exponential function describing the population of Nigeria.
   b) What will the population be in 2012? in 2016?
   c) When will the population be 200 million?
   d) What is the doubling time?

37. The average cost of a year at a private four-year college grew exponentially from $21,855 in 2001 to $27,317 in 2006.
   Source: National Center for Education Statistics
   a) Find the exponential growth rate k, and write a function that approximates the cost $C(t)$ of a year of college t years after 2001.
   b) Predict the cost of a year of college in 2012.
   c) In what year will the average cost of college be $50,000?

38. An investment with interest compounded continuously doubled itself in 16 yr. What is the interest rate?

39. How old is an animal bone that has lost 43% of its carbon-14? (Use $P(t) = P_0 e^{-0.00012t}$.)

40. Blue whales and fin whales are the loudest animals, with sound levels up to 188 dB. What is the intensity of such a sound? $\left(\text{Use } L = 10 \cdot \log \dfrac{I}{10^{-12} \text{ W/m}^2}. \right)$
   Source: Guinness World Records

41. The hydrogen ion concentration of water is 1.0×10^{-7} moles per liter. What is the pH? (Use $\text{pH} = -\log[H^+]$.)

Synthesis

42. Solve: $\log_5 |2x - 7| = 4$.

43. If $\log_a x = 2$, $\log_a y = 3$, and $\log_a z = 4$, find
$$\log_a \frac{\sqrt[3]{x^2 z}}{\sqrt[3]{y^2 z^{-1}}}.$$

Cumulative Review: Chapters 1–12

1. Evaluate $\dfrac{x^0 + y}{-z}$ for $x = 6$, $y = 9$, and $z = -5$.

[1.8], [4.1]

Simplify.

2. $(-2x^2y^{-3})^{-4}$ [4.2]

3. $(-5x^4y^{-3}z^2)(-4x^2y^2)$ [4.2]

4. $\dfrac{3x^4y^6z^{-2}}{-9x^4y^2z^3}$ [4.2]

5. $(1.5 \times 10^{-3})(4.2 \times 10^{-12})$ [4.2]

6. $3^3 + 2^2 - (32 \div 4 - 16 \div 8)$ [1.8]

Solve.

7. $3(2x - 3) = 9 - 5(2 - x)$ [2.2]

8. $4x - 3y = 15,$
$3x + 5y = 4$ [8.2]

9. $x + y - 3z = -1,$
$2x - y + z = 4,$
$-x - y + z = 1$ [8.4]

10. $x(x - 3) = 70$ [5.7]

11. $\dfrac{7}{x^2 - 5x} - \dfrac{2}{x - 5} = \dfrac{4}{x}$ [6.6]

12. $\sqrt{4 - 5x} = 2x - 1$ [10.6]

13. $\sqrt[3]{2x} = 1$ [10.6]

14. $3x^2 + 48 = 0$ [11.1]

15. $x^4 - 13x^2 + 36 = 0$ [11.5]

16. $\log_x 81 = 2$ [12.3]

17. $3^{5x} = 7$ [12.6]

18. $\ln x - \ln(x - 8) = 1$ [12.6]

19. $x^2 + 4x > 5$ [11.9]

20. If $f(x) = x^2 + 6x$, find a such that $f(a) = 11$. [11.2]

21. If $f(x) = |2x - 3|$, find all x for which $f(x) \geq 7$. [9.3]

Solve.

22. $D = \dfrac{ab}{b + a}$, for a [7.5]

23. $d = ax^2 + vx$, for x [11.4]

24. Find the domain of the function f given by
$$f(x) = \dfrac{x + 4}{3x^2 - 5x - 2}.$$ [9.2]

Perform the indicated operations and simplify.

25. $(5p^2q^3 + 6pq - p^2 + p) - (2p^2q^3 + p^2 - 5pq - 9)$ [4.7]

26. $(3x^2 - z^3)^2$ [4.7]

27. $\dfrac{1 + \dfrac{3}{x}}{x - 1 - \dfrac{12}{x}}$ [6.5]

28. $\dfrac{a^2 - a - 6}{a^3 - 27} \cdot \dfrac{a^2 + 3a + 9}{6}$ [6.2]

29. $\dfrac{3}{x + 6} - \dfrac{2}{x^2 - 36} + \dfrac{4}{x - 6}$ [6.4]

30. $\dfrac{\sqrt[3]{24xy^8}}{\sqrt[3]{3xy}}$ [10.4]

31. $\sqrt{x + 5}\ \sqrt[5]{x + 5}$ [10.5]

32. $(2 - i\sqrt{3})(6 + i\sqrt{3})$ [10.8]

33. $(x^4 - 8x^3 + 15x^2 + x - 3) \div (x - 3)$ [4.8]

Factor.

34. $27 + 64n^3$ [5.5]

35. $6x^2 + 8xy - 8y^2$ [5.3]

36. $x^4 - 4x^3 + 7x - 28$ [5.1]

37. $2m^2 + 12mn + 18n^2$ [5.4]

38. $x^4 - 16y^4$ [5.4]

39. Rationalize the denominator:
$$\dfrac{3 - \sqrt{y}}{2 - \sqrt{y}}.$$ [10.5]

40. Find the inverse of f if $f(x) = 9 - 2x$. [12.1]

41. Find a linear function with a graph that contains the points $(0, -8)$ and $(-1, 2)$. [7.3]

42. Find an equation of the line whose graph has a y-intercept of $(0, 5)$ and is perpendicular to the line given $2x + y = 6$. [3.6]

Graph.

43. $5x = 15 + 3y$ [3.3]

44. $y = \log_3 x$ [12.3]

45. $-2x - 3y \leq 12$ [9.4]

46. Graph: $f(x) = 2x^2 + 12x + 19$. [11.7]

    a) Label the vertex.
    b) Draw the axis of symmetry.
    c) Find the maximum or minimum value.

47. Graph $f(x) = 2e^x$ and determine the domain and the range. [12.5]

48. Express as a single logarithm:

$$3 \log x - \tfrac{1}{2} \log y - 2 \log z. \quad [12.4]$$

Solve.

49. *Colorado River.* The Colorado River delivers 1.5 million acre-feet of water to Mexico each year. This is only 10% of the volume of the river; the remainder is diverted at an earlier time for agricultural use. How much water is diverted each year from the Colorado River? [2.5]

Source: www.sierraclub.org

50. *Desalination.* More cities are supplying some of their fresh water through desalination, the process of removing the salt from ocean water. The worldwide desalination capacity has grown exponentially from 15 million m³ per day in 1990 to 55 million m³ per day in 2007. [12.7]

Source: Global Water Intelligence

a) Find the exponential growth rate k, and write an equation for an exponential function that can be used to predict the worldwide desalination capacity $D(t)$, in millions of cubic meters per day, t years after 1990.

b) Predict the worldwide desalination capacity in 2012.

c) In what year will the worldwide desalination capacity reach 100 million m³ per day?

51. *Gasoline consumption.* The number of barrels of gasoline consumed per day in the United States has increased from 8.5 million in 2000 to 9.3 million in 2006.

Source: U.S. Department of Energy, Energy Information Administration

a) At what rate did gasoline consumption increase from 2000 to 2006? [3.4]

b) Find a linear function g that fits the data. Let t represent the number of years since 2000. [7.3]

c) Find an exponential function G that fits the data. Let t represent the number of years since 2000. [12.7]

52. Good's Candies of Indiana makes all their chocolates by hand. It takes Anne 10 min to coat a tray of candies in chocolate. It takes Clay 12 min to coat a tray of candies. How long would it take Anne and Clay, working together, to coat the candies? [6.7]

53. Joe's Thick and Tasty salad dressing gets 45% of its calories from fat. The Light and Lean dressing gets 20% of its calories from fat. How many ounces of each should be mixed in order to get 15 oz of dressing that gets 30% of its calories from fat? [8.3]

54. A fishing boat with a trolling motor can move at a speed of 5 km/h in still water. The boat travels 42 km downriver in the same time that it takes to travel 12 km upriver. What is the speed of the river? [6.7]

55. What is the minimum product of two numbers whose difference is 14? What are the numbers that yield this product? [11.8]

56. Students in a biology class just took a final exam. A formula for predicting the average exam grade on a similar test t months later is

$$S(t) = 78 - 15 \log(t + 1).$$

a) Find the students' average score when they first took the final exam. [12.7]

b) What would the expected average score be on a retest after 4 months? [12.7]

Synthesis

Solve.

57. $\dfrac{5}{3x - 3} + \dfrac{10}{3x + 6} = \dfrac{5x}{x^2 + x - 2}$ [6.6]

58. $\log \sqrt{3x} = \sqrt{\log 3x}$ [12.6]

59. The Danville Express travels 280 mi at a certain speed. If the speed had been increased by 5 mph, the trip could have been made in 1 hr less time. Find the actual speed. [11.4]

Conic Sections

RAGHVENDRA SAHAI
ASTRONOMER
Pasadena, California

As an astronomer, I study stars and interstellar matter, that is, the gas and dust that lie in the vastness of space between the stars in our galaxy. My research includes obtaining images of the clouds of gas and dust ejected by dying sun-like stars. Calculus allows me to solve the equations governing these processes, and thus understand the physical properties of the dust and gas. Trigonometry helps me in figuring out the complex motions and velocities of the ejected matter.

AN APPLICATION

The maximum distance of the planet Mars from the sun is 2.48×10^8 mi. The minimum distance is 3.46×10^7 mi. The sun is at one focus of the elliptical orbit. Find the distance from the sun to the other focus.

This problem appears as Exercise 49 in Section 13.2.

The ellipse described on the preceding page is one example of a *conic section*, meaning that it can be regarded as a cross section of a cone. This chapter presents a variety of applications and equations with graphs that are conic sections. We have already worked with two conic sections, *lines* and *parabolas*, in Chapters 3 and 11.

13.1   Conic Sections: Parabolas and Circles

Parabolas   •   Circles

This section and the next two examine curves formed by cross sections of cones. These curves are all graphs of $Ax^2 + By^2 + Cxy + Dx + Ey + F = 0$. The constants $A, B, C, D, E,$ and F determine which of the following shapes will serve as the graph.

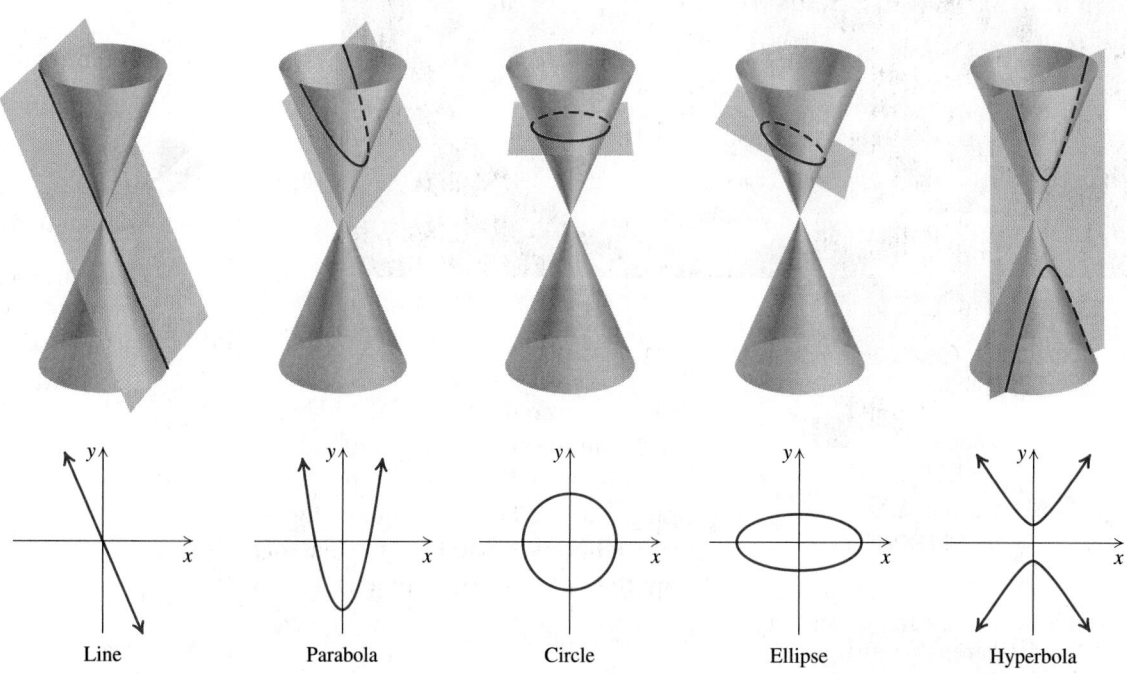

Line   Parabola   Circle   Ellipse   Hyperbola

Parabolas

When a cone is cut as shown in the second figure above, the conic section formed is a **parabola**. Parabolas have many applications in electricity, mechanics, and optics. A cross section of a contact lens or a satellite dish is a parabola, and arches that support certain bridges are parabolas.

> ## Equation of a Parabola
>
> A parabola with a vertical axis of symmetry opens upward or downward and has an equation that can be written in the form
>
> $$y = ax^2 + bx + c.$$
>
> A parabola with a horizontal axis of symmetry opens to the right or to the left and has an equation that can be written in the form
>
> $$x = ay^2 + by + c.$$

Parabolas with equations of the form $f(x) = ax^2 + bx + c$ were graphed in Chapter 11.

EXAMPLE **1**

Graph: $y = x^2 - 4x + 9$.

SOLUTION   To locate the vertex, we can use either of two approaches. One way is to complete the square:

$y = (x^2 - 4x) + 9$	Note that half of -4 is -2, and $(-2)^2 = 4$.
$\quad = (x^2 - 4x + 4 - 4) + 9$	Adding and subtracting 4
$\quad = (x^2 - 4x + 4) + (-4 + 9)$	Regrouping
$\quad = (x - 2)^2 + 5.$	Factoring and simplifying

The vertex is $(2, 5)$.

A second way to find the vertex is to recall that the x-coordinate of the vertex of the parabola given by $y = ax^2 + bx + c$ is $-b/(2a)$:

$$x = -\frac{b}{2a} = -\frac{-4}{2(1)} = 2.$$

To find the y-coordinate of the vertex, we substitute 2 for x:

$$y = x^2 - 4x + 9 = 2^2 - 4(2) + 9 = 5.$$

Either way, the vertex is $(2, 5)$. Next, we calculate and plot some points on each side of the vertex. As expected for a positive coefficient of x^2, the graph opens upward.

x	y	
2	5	←Vertex
0	9	←y-intercept
1	6	
3	6	
4	9	

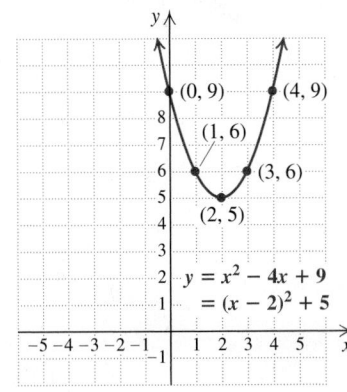

TRY EXERCISE 11

> **To Graph an Equation of the Form $y = ax^2 + bx + c$**
>
> 1. Find the vertex (h, k) either by completing the square to find an equivalent equation
>
> $$y = a(x - h)^2 + k,$$
>
>   or by using $-b/(2a)$ to find the x-coordinate and substituting to find the y-coordinate.
> 2. Choose other values for x on each side of the vertex, and compute the corresponding y-values.
> 3. The graph opens upward for $a > 0$ and downward for $a < 0$.

Any equation of the form $x = ay^2 + by + c$ represents a horizontal parabola that opens to the right for $a > 0$, opens to the left for $a < 0$, and has an axis of symmetry parallel to the x-axis.

EXAMPLE 2   Graph: $x = y^2 - 4y + 9$.

SOLUTION   This equation is like that in Example 1 but with x and y interchanged. The vertex is $(5, 2)$ instead of $(2, 5)$. To find ordered pairs, we choose values for y on each side of the vertex. Then we compute values for x. Note that the x- and y-values of the table in Example 1 are now switched. You should confirm that, by completing the square, we get $x = (y - 2)^2 + 5$.

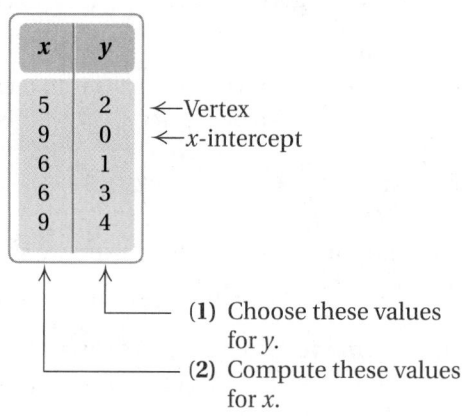

x	y	
5	2	←Vertex
9	0	←x-intercept
6	1	
6	3	
9	4	

(1) Choose these values for y.

(2) Compute these values for x.

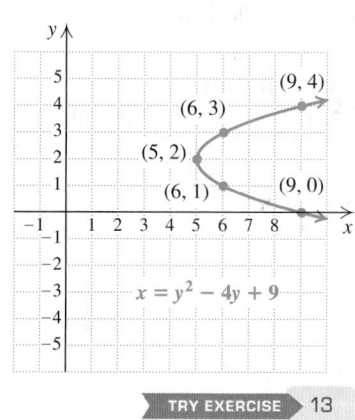

TRY EXERCISE ▶ 13

> **To Graph an Equation of the Form $x = ay^2 + by + c$**
>
> 1. Find the vertex (h, k) either by completing the square to find an equivalent equation
>
> $$x = a(y - k)^2 + h,$$
>
>   or by using $-b/(2a)$ to find the y-coordinate and substituting to find the x-coordinate.
> 2. Choose other values for y that are on either side of k and compute the corresponding x-values.
> 3. The graph opens to the right if $a > 0$ and to the left if $a < 0$.

EXAMPLE **3**   Graph: $x = -2y^2 + 10y - 7$.

SOLUTION   We find the vertex by completing the square:

$$x = -2y^2 + 10y - 7$$
$$= -2(y^2 - 5y \qquad) - 7$$
$$= -2\left(y^2 - 5y + \tfrac{25}{4}\right) - 7 - (-2)\tfrac{25}{4} \qquad \tfrac{1}{2}(-5) = \tfrac{-5}{2}; \left(\tfrac{-5}{2}\right)^2 = \tfrac{25}{4}; \text{ we add and subtract } (-2)\tfrac{25}{4}.$$
$$= -2\left(y - \tfrac{5}{2}\right)^2 + \tfrac{11}{2}. \qquad \text{Factoring and simplifying}$$

The vertex is $\left(\tfrac{11}{2}, \tfrac{5}{2}\right)$.

For practice, we also find the vertex by first computing its y-coordinate, $-b/(2a)$, and then substituting to find the x-coordinate:

$$y = -\frac{b}{2a} = -\frac{10}{2(-2)} = \frac{5}{2}$$
$$x = -2y^2 + 10y - 7 = -2\left(\tfrac{5}{2}\right)^2 + 10\left(\tfrac{5}{2}\right) - 7$$
$$= \tfrac{11}{2}.$$

To find ordered pairs, we choose values for y on each side of the vertex and then compute values for x. A table is shown below, together with the graph. The graph opens to the left because the y^2-coefficient, -2, is negative.

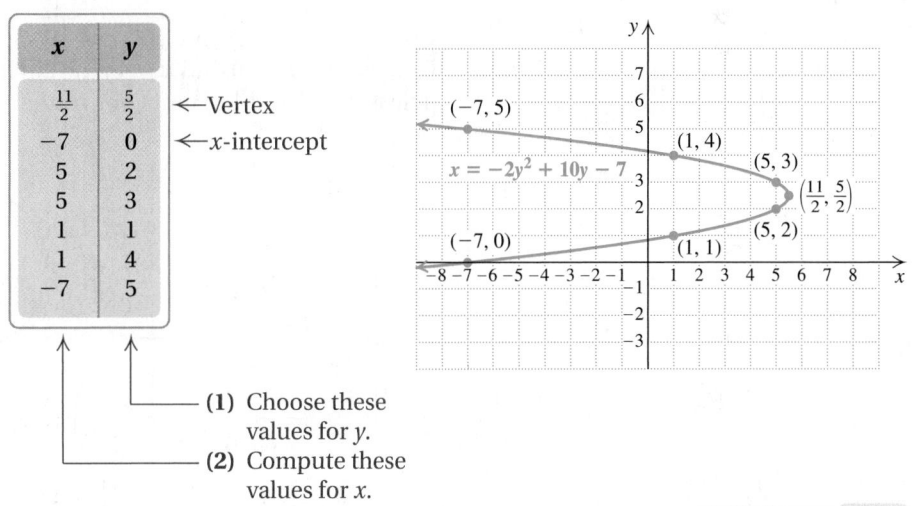

x	y	
$\tfrac{11}{2}$	$\tfrac{5}{2}$	←Vertex
-7	0	←x-intercept
5	2	
5	3	
1	1	
1	4	
-7	5	

(1) Choose these values for y.
(2) Compute these values for x.

TRY EXERCISE   19

Circles

Another conic section, the **circle**, is the set of points in a plane that are a fixed distance r, called the **radius** (plural, **radii**), from a fixed point (h, k), called the **center**. Note that the word radius can mean either any segment connecting a point on a circle to the center or the length of such a segment. Using the idea of a fixed distance r and the distance formula,

$$d = \sqrt{(x_2 - x_1)^2 + (y_2 - y_1)^2},$$

we can find the equation of a circle.

If (x, y) is on a circle of radius r, centered at (h, k), then by the definition of a circle and the distance formula, it follows that

$$r = \sqrt{(x - h)^2 + (y - k)^2}.$$

Squaring both sides gives the equation of a circle in standard form.

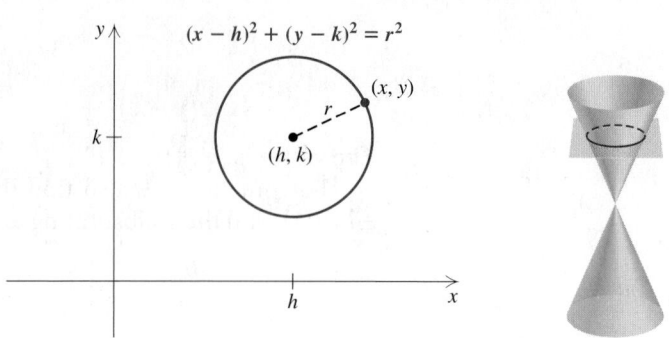

Equation of a Circle (Standard Form)

The equation of a circle, centered at (h, k), with radius r, is given by

$$(x - h)^2 + (y - k)^2 = r^2.$$

Note that for $h = 0$ and $k = 0$, the circle is centered at the origin. Otherwise, the circle is translated $|h|$ units horizontally and $|k|$ units vertically.

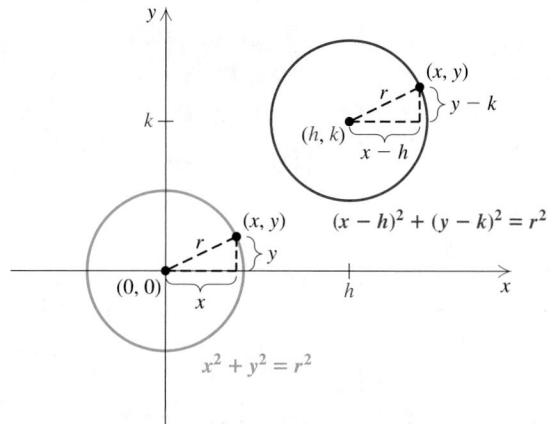

EXAMPLE   4   Find an equation of the circle centered at $(4, -5)$ with radius 6.

SOLUTION   Using the standard form, we obtain

$$(x - 4)^2 + (y - (-5))^2 = 6^2, \quad \text{Using } (x - h)^2 + (y - k)^2 = r^2$$

or

$$(x - 4)^2 + (y + 5)^2 = 36.$$

▶ TRY EXERCISE 31

EXAMPLE   5   Find the center and the radius and then graph each circle.

a) $(x - 2)^2 + (y + 3)^2 = 4^2$

b) $x^2 + y^2 + 8x - 2y + 15 = 0$

SOLUTION

a) We write standard form:

$$(x - 2)^2 + [y - (-3)]^2 = 4^2.$$

The center is $(2, -3)$ and the radius is 4. To graph, we plot the points $(2, 1)$, $(2, -7)$, $(-2, -3)$, and $(6, -3)$, which are, respectively, 4 units above, below, left, and right of $(2, -3)$. We then either sketch a circle by hand or use a compass.

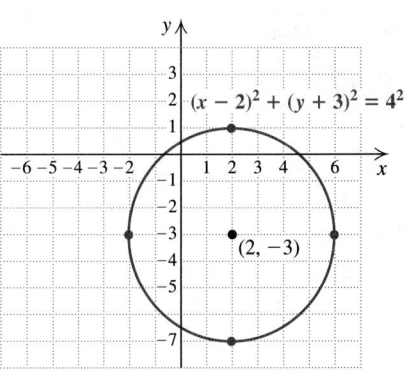

b) To write the equation $x^2 + y^2 + 8x - 2y + 15 = 0$ in standard form, we complete the square twice, once with $x^2 + 8x$ and once with $y^2 - 2y$:

$$x^2 + y^2 + 8x - 2y + 15 = 0$$
$$x^2 + 8x \qquad + y^2 - 2y \qquad = -15 \qquad \text{Grouping the } x\text{-terms and the } y\text{-terms; subtracting 15 from both sides}$$

$$x^2 + 8x + 16 + y^2 - 2y + 1 = -15 + 16 + 1 \qquad \text{Adding } \left(\tfrac{8}{2}\right)^2, \text{ or 16, and } \left(-\tfrac{2}{2}\right)^2, \text{ or 1, to both sides to get standard form}$$

$$(x + 4)^2 + (y - 1)^2 = 2 \qquad \text{Factoring}$$
$$[x - (-4)]^2 + (y - 1)^2 = (\sqrt{2})^2. \qquad \text{Writing standard form}$$

The center is $(-4, 1)$ and the radius is $\sqrt{2}$.

> **TRY EXERCISE** 51

TECHNOLOGY CONNECTION

Most graphing calculators graph only functions, so graphing the equation of a circle usually requires two steps:

1. Solve the equation for y. The result will include a $\pm$ sign in front of a radical.
2. Graph two functions, one for the $+$ sign and the other for the $-$ sign, on the same set of axes.

For example, to graph $(x - 3)^2 + (y + 1)^2 = 16$, solve for $y + 1$ and then y:

$$(y + 1)^2 = 16 - (x - 3)^2$$
$$y + 1 = \pm\sqrt{16 - (x - 3)^2}$$
$$y = -1 \pm \sqrt{16 - (x - 3)^2},$$
or
$$y_1 = -1 + \sqrt{16 - (x - 3)^2}$$
and
$$y_2 = -1 - \sqrt{16 - (x - 3)^2}.$$

When both functions are graphed (in a "squared" window to eliminate distortion), the result is as follows.

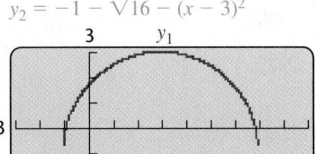

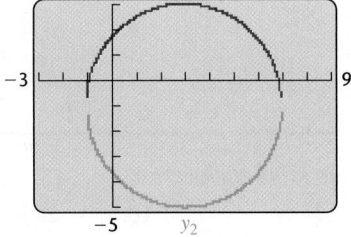

On many calculators, pressing **APPS** and selecting Conics and then Circle accesses a program in which equations in standard form can be graphed directly and then Traced.

Graph each of the following equations.

1. $x^2 + y^2 - 16 = 0$
2. $(x - 1)^2 + (y - 2)^2 = 25$
3. $(x + 3)^2 + (y - 5)^2 = 16$
4. $(x - 5)^2 + (y + 6)^2 = 49$

13.1   EXERCISE SET

For Extra Help
MyMathLab
 PRACTICE
 WATCH   DOWNLOAD

 Concept Reinforcement *In each of Exercises 1–8, match the equation with the graph of that equation from those shown.*

1. ____ $(x - 2)^2 + (y + 5)^2 = 9$

2. ____ $(x + 2)^2 + (y - 5)^2 = 9$

3. ____ $(x - 5)^2 + (y + 2)^2 = 9$

4. ____ $(x + 5)^2 + (y - 2)^2 = 9$

5. ____ $y = (x - 2)^2 - 5$

6. ____ $y = (x - 5)^2 - 2$

7. ____ $x = (y - 2)^2 - 5$

8. ____ $x = (y - 5)^2 - 2$

a)
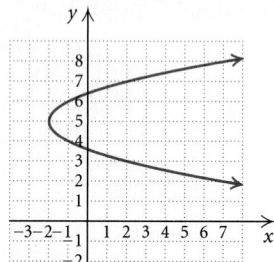

b)
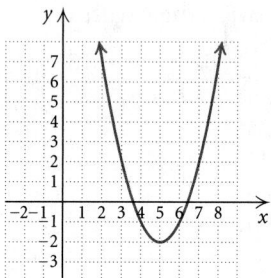

c)
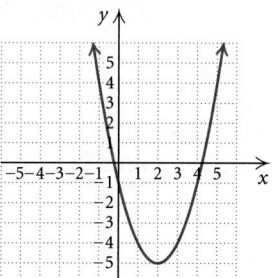

d)
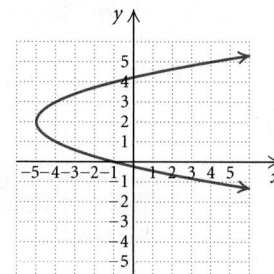

e)
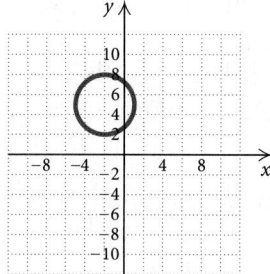

f)
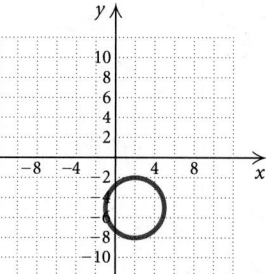

g)
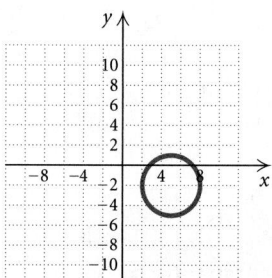

h)
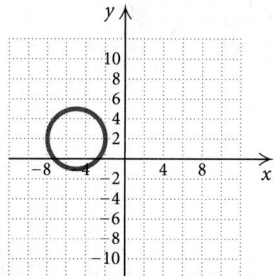

Graph. Be sure to label each vertex.

9. $y = -x^2$

10. $y = 2x^2$

11. $y = -x^2 + 4x - 5$

12. $x = 4 - 3y - y^2$

13. $x = y^2 - 4y + 2$

14. $y = x^2 + 2x + 3$

15. $x = y^2 + 3$

16. $x = -y^2$

17. $x = 2y^2$

18. $x = y^2 - 1$

19. $x = -y^2 - 4y$

20. $x = y^2 + 3y$

21. $y = x^2 - 2x + 1$

22. $y = x^2 + 2x + 1$

23. $x = -\frac{1}{2}y^2$

24. $y = -\frac{1}{2}x^2$

25. $x = -y^2 + 2y - 1$

26. $x = -y^2 - 2y + 3$

27. $x = -2y^2 - 4y + 1$

28. $x = 2y^2 + 4y - 1$

Find an equation of the circle satisfying the given conditions.

29. Center $(0, 0)$, radius 8

30. Center $(0, 0)$, radius 11

31. Center $(7, 3)$, radius $\sqrt{6}$

32. Center $(5, 6)$, radius $\sqrt{11}$

33. Center $(-4, 3)$, radius $3\sqrt{2}$

34. Center $(-2, 7)$, radius $2\sqrt{5}$

35. Center $(-5, -8)$, radius $10\sqrt{3}$

36. Center $(-7, -2)$, radius $5\sqrt{2}$

Aha! **37.** Center $(0, 0)$, passing through $(-3, 4)$

38. Center $(0, 0)$, passing through $(11, -10)$

39. Center $(-4, 1)$, passing through $(-2, 5)$

40. Center $(-1, -3)$, passing through $(-4, 2)$

Find the center and the radius of each circle. Then graph the circle.

41. $x^2 + y^2 = 1$

42. $x^2 + y^2 = 25$

43. $(x + 1)^2 + (y + 3)^2 = 49$

44. $(x - 2)^2 + (y + 3)^2 = 100$

45. $(x - 4)^2 + (y + 3)^2 = 10$

46. $(x + 5)^2 + (y - 1)^2 = 15$

47. $x^2 + y^2 = 8$

48. $x^2 + y^2 = 20$

49. $(x - 5)^2 + y^2 = \frac{1}{4}$

50. $x^2 + (y - 1)^2 = \frac{1}{25}$

51. $x^2 + y^2 + 8x - 6y - 15 = 0$

52. $x^2 + y^2 + 6x - 4y - 15 = 0$

53. $x^2 + y^2 - 8x + 2y + 13 = 0$

54. $x^2 + y^2 + 6x + 4y + 12 = 0$

55. $x^2 + y^2 + 10y - 75 = 0$

56. $x^2 + y^2 - 8x - 84 = 0$

57. $x^2 + y^2 + 7x - 3y - 10 = 0$

58. $x^2 + y^2 - 21x - 33y + 17 = 0$

59. $36x^2 + 36y^2 = 1$

60. $4x^2 + 4y^2 = 1$

61. Does the graph of an equation of a circle include the point that is the center? Why or why not?

62. Is a point a conic section? Why or why not?

Skill Review

To prepare for Section 13.2, review solving quadratic equations (Section 11.1).

Solve. [11.1]

63. $\dfrac{y^2}{16} = 1$

64. $\dfrac{x^2}{a^2} = 1$

65. $\dfrac{(x - 1)^2}{25} = 1$

66. $\dfrac{(y + 5)^2}{12} = 1$

67. $\dfrac{1}{4} + \dfrac{(y + 3)^2}{36} = 1$

68. $\dfrac{1}{9} + \dfrac{(x - 2)^2}{4} = 1$

Synthesis

69. On a piece of graph paper, draw a line and a point not on the line. Then plot several points that are the same distance from the point and from the line. What shape do the points appear to form? How is this set of points different from a circle?

70. If an equation has two terms with the same degree, can its graph be a parabola? Why or why not?

Find an equation of a circle satisfying the given conditions.

71. Center $(3, -5)$ and tangent to (touching at one point) the *y*-axis

72. Center $(-7, -4)$ and tangent to the *x*-axis

73. The endpoints of a diameter are $(7, 3)$ and $(-1, -3)$.

74. Center $(-3, 5)$ with a circumference of 8π units

75. Find the point on the *y*-axis that is equidistant from $(2, 10)$ and $(6, 2)$.

76. Find the point on the *x*-axis that is equidistant from $(-1, 3)$ and $(-8, -4)$.

77. *Wrestling.* The equation $x^2 + y^2 = \frac{81}{4}$, where *x* and *y* represent the number of meters from the center, can be used to draw the outer circle on a wrestling mat used in International, Olympic, and World Championship wrestling. The equation $x^2 + y^2 = 16$ can be used to draw the inner edge of the red zone. Find the area of the red zone.

Source: Based on data from the Government of Western Australia

78. *Snowboarding.* Each side edge of the Burton X8 155 snowboard is an arc of a circle with a "running length" of 1180 mm and a "sidecut depth" of 23 mm (see the figure below).
Source: evogear.com

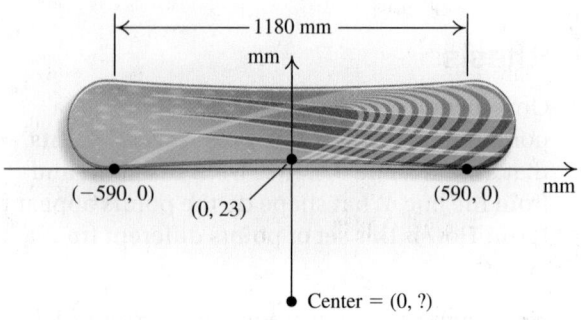

a) Using the coordinates shown, locate the center of the circle. (*Hint*: Equate distances.)
b) What radius is used for the edge of the board?

79. *Snowboarding.* The Never Summer Infinity 149 snowboard has a running length of 1160 mm and a sidecut depth of 23.5 mm (see Exercise 78). What radius is used for the edge of this snowboard?
Source: neversummer.com

80. *Skiing.* The Rossignol Blast ski, when lying flat and viewed from above, has edges that are arcs of a circle. (Actually, each edge is made of two arcs of slightly different radii. The arc for the rear half of the ski edge has a slightly larger radius.)
Source: evogear.com

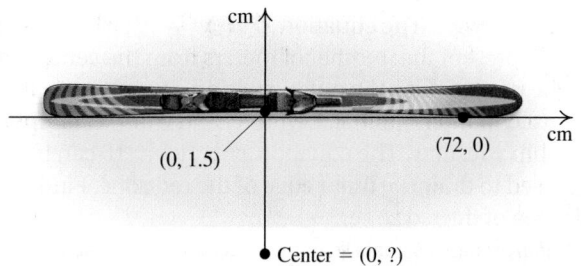

a) Using the coordinates shown, locate the center of the circle. (*Hint*: Equate distances.)
b) What radius is used for the arc passing through (0, 1.5) and (72, 0)?

81. *Doorway construction.* Engle Carpentry needs to cut an arch for the top of an entranceway. The arch needs to be 8 ft wide and 2 ft high. To draw the arch, the carpenters will use as a compass a stretched string with chalk attached at an end.

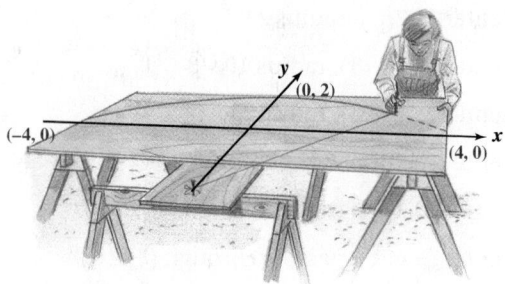

a) Using a coordinate system, locate the center of the circle.
b) What radius should the carpenters use to draw the arch?

82. *Archaeology.* During an archaeological dig, Estella finds the bowl fragment shown below. What was the original diameter of the bowl?

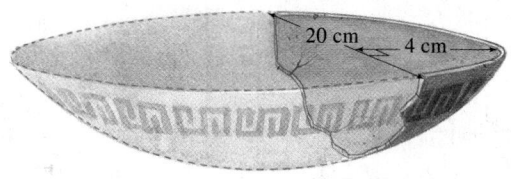

83. *Ferris wheel design.* A ferris wheel has a radius of 24.3 ft. Assuming that the center is 30.6 ft off the ground and that the origin is below the center, as in the following figure, find an equation of the circle.

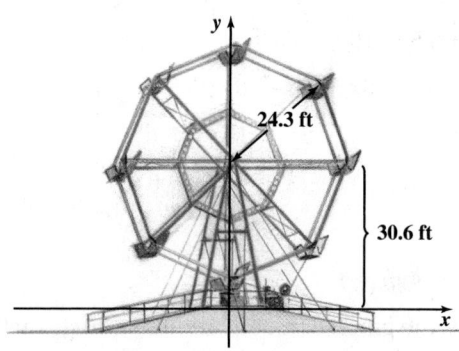

84. Use a graph of the equation $x = y^2 - y - 6$ to approximate to the nearest tenth the solutions of each of the following equations.
a) $y^2 - y - 6 = 2$
b) $y^2 - y - 6 = -3$

85. *Power of a motor.* The horsepower of a certain kind of engine is given by the formula

$$H = \frac{D^2 N}{2.5},$$

where N is the number of cylinders and D is the diameter, in inches, of each piston. Graph this equation, assuming that $N = 6$ (a six-cylinder engine). Let D run from 2.5 to 8. Then use the graph to estimate the diameter of each piston in a six-cylinder 120-horsepower engine.

 86. If the equation $x^2 + y^2 - 6x + 2y - 6 = 0$ is written as $y^2 + 2y + (x^2 - 6x - 6) = 0$, it can be regarded as quadratic in y.

a) Use the quadratic formula to solve for y.
b) Show that the graph of your answer to part (a) coincides with the graph in the Technology Connection on p. 859.

 87. How could a graphing calculator best be used to help you sketch the graph of an equation of the form $x = ay^2 + by + c$?

 88. Why should a graphing calculator's window be "squared" before graphing a circle?

13.2 Conic Sections: Ellipses

Ellipses Centered at $(0, 0)$ ▪ Ellipses Centered at (h, k)

When a cone is cut at an angle, as shown below, the conic section formed is an *ellipse*. To draw an ellipse, stick two tacks in a piece of cardboard. Then tie a loose string to the tacks, place a pencil as shown, and draw an oval by moving the pencil while stretching the string tight.

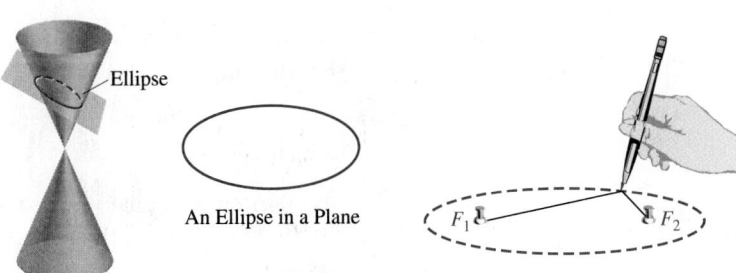

Ellipse

An Ellipse in a Plane

F_1 F_2

Ellipses Centered at $(0, 0)$

An **ellipse** is defined as the set of all points in a plane for which the sum of the distances from two fixed points F_1 and F_2 is constant. The points F_1 and F_2 are called **foci** (pronounced fō-sī), the plural of focus. In the figure above, the tacks are at the foci and the length of the string is the constant sum of the distances from the tacks to the pencil. The midpoint of the segment $F_1 F_2$ is the **center**. The equation of an ellipse follows. Its derivation is outlined in Exercise 51.

> ### Equation of an Ellipse Centered at the Origin
> The equation of an ellipse centered at the origin and symmetric with respect to both axes is
> $$\frac{x^2}{a^2} + \frac{y^2}{b^2} = 1, \quad a, b > 0. \quad \text{(Standard form)}$$

To graph an ellipse centered at the origin, it helps to first find the intercepts. If we replace x with 0, we can find the y-intercepts:

$$\frac{0^2}{a^2} + \frac{y^2}{b^2} = 1$$

$$\frac{y^2}{b^2} = 1$$

$$y^2 = b^2 \quad \text{or} \quad y = \pm b.$$

Thus the y-intercepts are $(0, b)$ and $(0, -b)$. Similarly, the x-intercepts are $(a, 0)$ and $(-a, 0)$. If $a > b$, the ellipse is said to be horizontal and $(-a, 0)$ and $(a, 0)$ are referred to as the **vertices** (singular, **vertex**). If $b > a$, the ellipse is said to be vertical and $(0, -b)$ and $(0, b)$ are then the vertices.

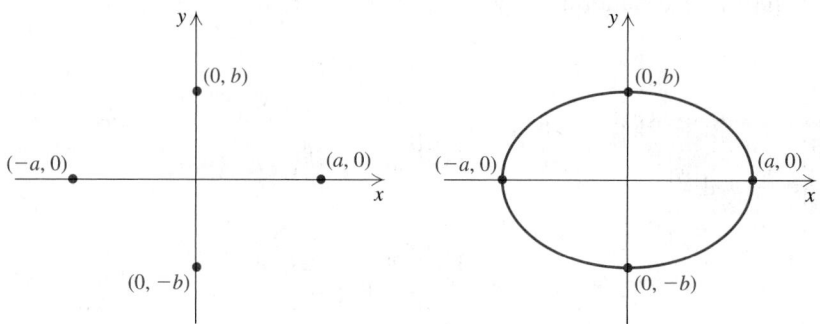

Plotting these four points and drawing an oval-shaped curve, we graph the ellipse. If a more precise graph is desired, we can plot more points.

> ## Using a and b to Graph an Ellipse
>
> For the ellipse
>
> $$\frac{x^2}{a^2} + \frac{y^2}{b^2} = 1,$$
>
> the x-intercepts are $(-a, 0)$ and $(a, 0)$. The y-intercepts are $(0, -b)$ and $(0, b)$. For $a^2 > b^2$, the ellipse is horizontal. For $b^2 > a^2$, the ellipse is vertical.

EXAMPLE **1** Graph the ellipse

$$\frac{x^2}{4} + \frac{y^2}{9} = 1.$$

SOLUTION Note that

$$\frac{x^2}{4} + \frac{y^2}{9} = \frac{x^2}{2^2} + \frac{y^2}{3^2}. \qquad \text{Identifying } a \text{ and } b. \text{ Since } b^2 > a^2, \text{ the ellipse is vertical.}$$

Since $a = 2$ and $b = 3$, the x-intercepts are $(-2, 0)$ and $(2, 0)$, and the y-intercepts are $(0, -3)$ and $(0, 3)$. We plot these points and connect them with an oval-shaped curve. To plot two other points, we let $x = 1$ and solve for y:

$$\frac{1^2}{4} + \frac{y^2}{9} = 1$$

$$36\left(\frac{1}{4} + \frac{y^2}{9}\right) = 36 \cdot 1$$

$$36 \cdot \frac{1}{4} + 36 \cdot \frac{y^2}{9} = 36$$

$$9 + 4y^2 = 36$$

$$4y^2 = 27$$

$$y^2 = \frac{27}{4}$$

$$y = \pm\sqrt{\frac{27}{4}}$$

$$y \approx \pm 2.6.$$

Thus, $(1, 2.6)$ and $(1, -2.6)$ can also be used to draw the graph. Similarly, the points $(-1, 2.6)$ and $(-1, -2.6)$ should appear on the graph.

> **TRY EXERCISE** 9

EXAMPLE **2**

Graph: $4x^2 + 25y^2 = 100$.

SOLUTION   To write the equation in standard form, we divide both sides by 100 to get 1 on the right side:

$$\frac{4x^2 + 25y^2}{100} = \frac{100}{100} \qquad \text{Dividing by 100 to get 1 on the right side}$$

$$\left.\begin{array}{c} \dfrac{4x^2}{100} + \dfrac{25y^2}{100} = 1 \\[2mm] \dfrac{x^2}{25} + \dfrac{y^2}{4} = 1 \end{array}\right\} \qquad \text{Simplifying}$$

$$\frac{x^2}{5^2} + \frac{y^2}{2^2} = 1. \qquad a = 5, b = 2$$

STUDENT NOTES

Note that any equation of the form $Ax^2 + By^2 = C$ (with $A \neq B$ and $A, B > 0$) can be rewritten as an equivalent equation in standard form. The graph is an ellipse.

The x-intercepts are $(-5, 0)$ and $(5, 0)$, and the y-intercepts are $(0, -2)$ and $(0, 2)$. We plot the intercepts and connect them with an oval-shaped curve. Other points can also be computed and plotted.

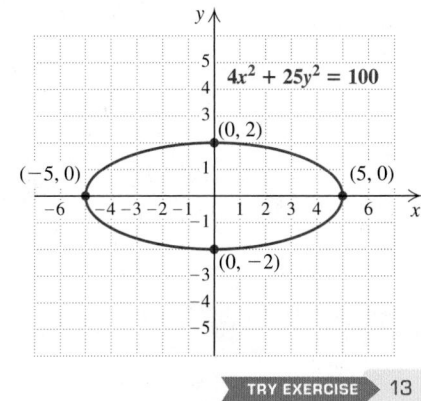

> **TRY EXERCISE** 13

Ellipses Centered at (h, k)

Horizontal and vertical translations, similar to those used in Chapter 11, can be used to graph ellipses that are not centered at the origin.

Equation of an Ellipse Centered at (h, k)

The standard form of a horizontal or vertical ellipse centered at (h, k) is

$$\frac{(x - h)^2}{a^2} + \frac{(y - k)^2}{b^2} = 1.$$

The vertices are $(h + a, k)$ and $(h - a, k)$ if horizontal; $(h, k + b)$ and $(h, k - b)$ if vertical.

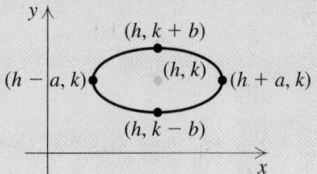

EXAMPLE 3   Graph the ellipse

$$\frac{(x - 1)^2}{4} + \frac{(y + 5)^2}{9} = 1.$$

SOLUTION   Note that

$$\frac{(x - 1)^2}{4} + \frac{(y + 5)^2}{9} = \frac{(x - 1)^2}{2^2} + \frac{(y + 5)^2}{3^2}.$$

Thus, $a = 2$ and $b = 3$. To determine the center of the ellipse, (h, k), note that

$$\frac{(x - 1)^2}{2^2} + \frac{(y + 5)^2}{3^2} = \frac{(x - 1)^2}{2^2} + \frac{(y - (-5))^2}{3^2}.$$

Thus the center is $(1, -5)$. We plot points 2 units to the left and right of center, as well as 3 units above and below center. These are the points $(3, -5)$, $(-1, -5)$, $(1, -2)$, and $(1, -8)$. The graph of the ellipse is shown at left.

Note that this ellipse is the same as the ellipse in Example 1 but translated 1 unit to the right and 5 units down.

TRY EXERCISE 27

Ellipses have many applications. Communications satellites move in elliptical orbits with the earth as a focus while the earth itself follows an elliptical path around the sun. A medical instrument, the lithotripter, uses shock waves originating at one focus to crush a kidney stone located at the other focus.

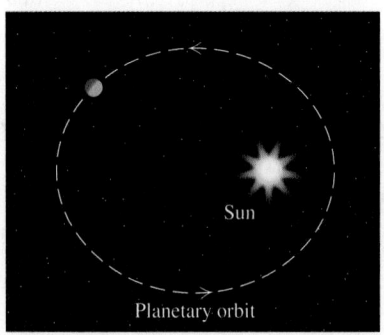

Planetary orbit

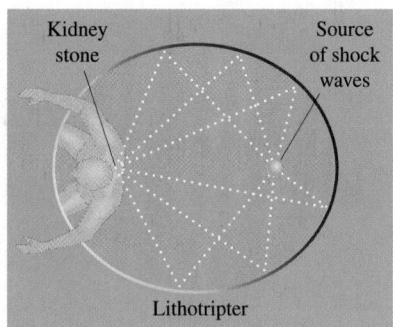

Lithotripter

In some buildings, an ellipsoidal ceiling creates a "whispering gallery" in which a person at one focus can whisper and still be heard clearly at the other focus. This happens because sound waves coming from one focus are all reflected to the other focus. Similarly, light waves bouncing off an ellipsoidal mirror are used in a dentist's or surgeon's reflector light. The light source is located at one focus while the patient's mouth or surgical field is at the other.

To graph an ellipse on a graphing calculator, we solve for y and graph two functions.

To illustrate, let's check Example 2:

$$4x^2 + 25y^2 = 100$$
$$25y^2 = 100 - 4x^2$$
$$y^2 = 4 - \frac{4}{25}x^2$$
$$y = \pm\sqrt{4 - \frac{4}{25}x^2}.$$

Using a squared window, we have our check:

$$y_1 = -\sqrt{4 - \frac{4}{25}x^2}, \quad y_2 = \sqrt{4 - \frac{4}{25}x^2}$$

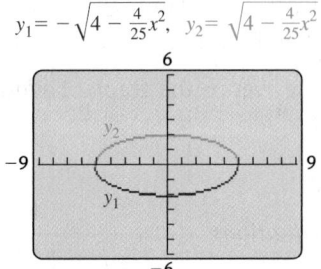

On many calculators, pressing **APPS** and selecting Conics and then Ellipse accesses a program in which equations in Standard Form can be graphed directly.

13.2   EXERCISE SET

👆 *Concept Reinforcement   Classify each statement as either true or false.*

1. The graph of $\dfrac{x^2}{25} + \dfrac{y^2}{50} = 1$ is a vertical ellipse.

2. The graph of $\dfrac{x^2}{30} + \dfrac{y^2}{20} = 1$ is a vertical ellipse.

3. The graph of $\dfrac{x^2}{25} - \dfrac{y^2}{9} = 1$ is a horizontal ellipse.

4. The graph of $\dfrac{-x^2}{20} + \dfrac{y^2}{16} = 1$ is a horizontal ellipse.

5. The graph of $\dfrac{x^2}{9} + \dfrac{y^2}{25} = 1$ includes the points $(-3, 0)$ and $(3, 0)$.

6. The graph of $\dfrac{x^2}{36} + \dfrac{y^2}{25} = 1$ includes the points $(0, -5)$ and $(0, 5)$.

7. The graph of $\dfrac{(x + 3)^2}{25} + \dfrac{(y - 2)^2}{36} = 1$ is an ellipse centered at $(-3, 2)$.

8. The graph of $\dfrac{(x - 2)^2}{49} + \dfrac{(y + 5)^2}{9} = 1$ is an ellipse centered at $(2, -5)$.

Graph each of the following equations.

9. $\dfrac{x^2}{1} + \dfrac{y^2}{4} = 1$

10. $\dfrac{x^2}{4} + \dfrac{y^2}{1} = 1$

11. $\dfrac{x^2}{25} + \dfrac{y^2}{9} = 1$

12. $\dfrac{x^2}{16} + \dfrac{y^2}{25} = 1$

13. $4x^2 + 9y^2 = 36$

14. $9x^2 + 4y^2 = 36$

15. $16x^2 + 9y^2 = 144$

16. $9x^2 + 16y^2 = 144$

17. $2x^2 + 3y^2 = 6$

18. $5x^2 + 7y^2 = 35$

Aha! 19. $5x^2 + 5y^2 = 125$

20. $8x^2 + 5y^2 = 80$

21. $3x^2 + 7y^2 - 63 = 0$

22. $3x^2 + 3y^2 - 48 = 0$

23. $16x^2 = 16 - y^2$

24. $9y^2 = 9 - x^2$

25. $16x^2 + 25y^2 = 1$

26. $9x^2 + 4y^2 = 1$

27. $\dfrac{(x - 3)^2}{9} + \dfrac{(y - 2)^2}{25} = 1$

28. $\dfrac{(x - 2)^2}{25} + \dfrac{(y - 4)^2}{9} = 1$

29. $\dfrac{(x + 4)^2}{16} + \dfrac{(y - 3)^2}{49} = 1$

30. $\dfrac{(x + 5)^2}{4} + \dfrac{(y - 2)^2}{36} = 1$

31. $12(x - 1)^2 + 3(y + 4)^2 = 48$
   (*Hint*: Divide both sides by 48.)

32. $4(x - 6)^2 + 9(y + 2)^2 = 36$

Aha! **33.** $4(x + 3)^2 + 4(y + 1)^2 - 10 = 90$

34. $9(x + 6)^2 + (y + 2)^2 - 20 = 61$

35. Explain how you can tell from the equation of an ellipse whether the graph will be horizontal or vertical.

36. Can an ellipse ever be the graph of a function? Why or why not?

Skill Review

Review solving equations.

Solve.

37. $x^2 - 5x + 3 = 0$ [11.2]   **38.** $\log_x 81 = 4$ [12.6]

39. $\dfrac{4}{x + 2} + \dfrac{3}{2x - 1} = 2$ [6.6]

40. $3 - \sqrt{2x - 1} = 1$ [10.6]

41. $x^2 = 11$ [11.1]   **42.** $x^2 + 4x = 60$ [5.7]

Synthesis

43. Explain how it is possible to recognize that the graph of $9x^2 + 18x + y^2 - 4y + 4 = 0$ is an ellipse.

44. As the foci get closer to the center of an ellipse, what shape does the graph begin to resemble? Explain why this happens.

Find an equation of an ellipse that contains the following points.

45. $(-9, 0)$, $(9, 0)$, $(0, -11)$, and $(0, 11)$

46. $(-7, 0)$, $(7, 0)$, $(0, -10)$, and $(0, 10)$

47. $(-2, -1)$, $(6, -1)$, $(2, -4)$, and $(2, 2)$

48. $(4, 3)$, $(-6, 3)$, $(-1, -1)$, and $(-1, 7)$

49. *Astronomy.* The maximum distance of the planet Mars from the sun is 2.48×10^8 mi. The minimum distance is 3.46×10^7 mi. The sun is at one focus of the elliptical orbit. Find the distance from the sun to the other focus.

50. *Theatrical lighting.* The spotlight on a violin soloist casts an ellipse of light on the floor below her that is 6 ft wide and 10 ft long. Find an equation of that ellipse if the performer is in its center, x is the distance from the performer to the side of the ellipse, and y is the distance from the performer to the top of the ellipse.

51. Let $(-c, 0)$ and $(c, 0)$ be the foci of an ellipse. Any point $P(x, y)$ is on the ellipse if the sum of the distances from the foci to P is some constant. Use $2a$ to represent this constant.

a) Show that an equation for the ellipse is given by
$$\frac{x^2}{a^2} + \frac{y^2}{a^2 - c^2} = 1.$$

b) Substitute b^2 for $a^2 - c^2$ to get standard form.

52. *President's office.* The Oval Office of the President of the United States is an ellipse 31 ft wide and 38 ft long. Show in a sketch precisely where the President and an adviser could sit to best hear each other using the room's acoustics. (*Hint*: See Exercise 51(b) and the discussion following Example 3.)

53. *Dentistry.* The light source in some dental lamps shines against a reflector that is shaped like a portion of an ellipse in which the light source is

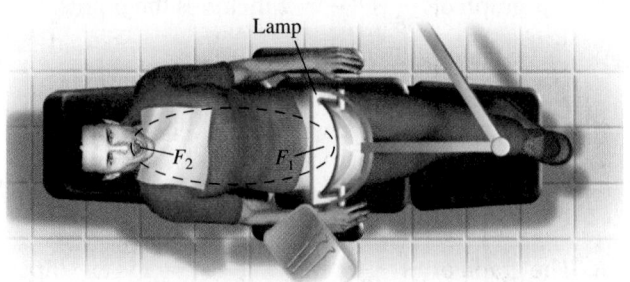

one focus of the ellipse. Reflected light enters a patient's mouth at the other focus of the ellipse. If the ellipse from which the reflector was formed is 2 ft wide and 6 ft long, how far should the patient's mouth be from the light source? (*Hint*: See Exercise 51(b).)

54. *Firefighting.* The size and shape of certain forest fires can be approximated as the union of two "half-ellipses." For the blaze modeled below, the equation of the smaller ellipse—the part of the fire moving *into* the wind—is

$$\frac{x^2}{40{,}000} + \frac{y^2}{10{,}000} = 1.$$

The equation of the other ellipse—the part moving *with* the wind—is

$$\frac{x^2}{250{,}000} + \frac{y^2}{10{,}000} = 1.$$

Determine the width and the length of the fire.

Source for figure: "Predicting Wind-Driven Wild Land Fire Size and Shape," Hal E. Anderson, Research Paper INT-305, U.S. Department of Agriculture, Forest Service, February 1983

For each of the following equations, complete the square as needed and find an equivalent equation in standard form. Then graph the ellipse.

55. $x^2 - 4x + 4y^2 + 8y - 8 = 0$

56. $4x^2 + 24x + y^2 - 2y - 63 = 0$

57. Use a graphing calculator to check your answers to Exercises 11, 25, 29, and 33.

COLLABORATIVE CORNER

A Cosmic Path

Focus: Ellipses

Time: 20–30 minutes

Group size: 2

Materials: Scientific calculators

On May 4, 2007, Comet 17P/Holmes was at the point closest to the sun in its orbit. Comet 17P is traveling in an elliptical orbit with the sun as one focus, and one orbit takes about 6.88 yr. One astronomical unit (AU) is 93,000,000 mi. One group member should do the following calculations in AU and the other in millions of miles.

Source: Harvard-Smithsonian Center for Astrophysics

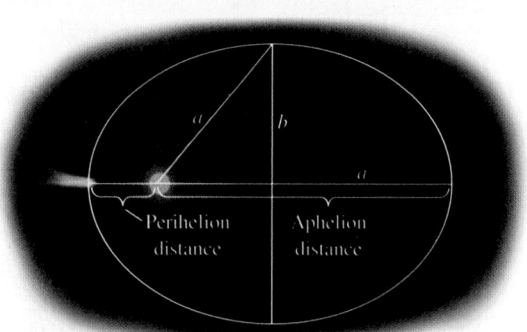

ACTIVITY

1. At its *perihelion*, a comet with an elliptical orbit is at the point in its orbit closest to the sun. At its *aphelion*, the comet is at the point farthest from the sun. The perihelion distance for Comet 17P is 2.053218 AU, and the aphelion distance is 5.183610 AU. Use these distances to find a. (See the following diagram.)

2. Using the figure above, express b^2 as a function of a. Then find b using the value found for a in part (1).

3. One formula for approximating the perimeter of an ellipse is

$$P = \pi\left(3a + 3b - \sqrt{(3a+b)(a+3b)}\right),$$

developed by the Indian mathematician S. Ramanujan in 1914. How far does Comet 17P travel in one orbit?

4. What is the speed of the comet? Find the answer in AU per year and in miles per hour.

13.3 Conic Sections: Hyperbolas

Hyperbolas • Hyperbolas (Nonstandard Form) • Classifying Graphs of Equations

Hyperbolas

A **hyperbola** looks like a pair of parabolas, but the shapes are not quite parabolic. A hyperbola has two **vertices** and the line through the vertices is known as the **axis**. The point halfway between the vertices is called the **center**. The two curves that comprise a hyperbola are called **branches**.

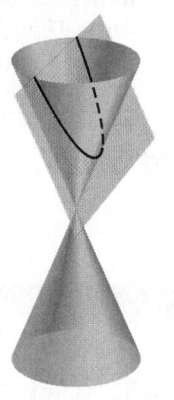

Parabola

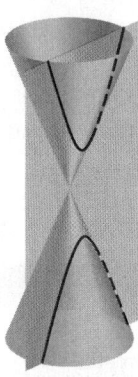

Hyperbola in three dimensions

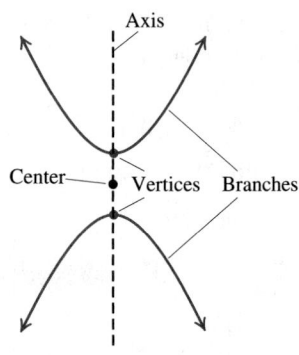

Hyperbola in a plane

Equation of a Hyperbola Centered at the Origin

A hyperbola with its center at the origin* has its equation as follows:

$$\frac{x^2}{a^2} - \frac{y^2}{b^2} = 1 \quad \text{(Horizontal axis);}$$

$$\frac{y^2}{b^2} - \frac{x^2}{a^2} = 1 \quad \text{(Vertical axis).}$$

Note that both equations have 1 on the right-hand side and subtraction between the terms. For the discussion that follows, we assume $a, b > 0$.

To graph a hyperbola, it helps to begin by graphing two lines called **asymptotes**. Although the asymptotes themselves are not part of the graph, they serve as guidelines for an accurate sketch.

As a hyperbola gets farther away from the origin, it gets closer and closer to its asymptotes. The larger $|x|$ gets, the closer the graph gets to an asymptote. The asymptotes act to "constrain" the graph of a hyperbola. Parabolas are *not* constrained by any asymptotes.

---

*Hyperbolas with horizontal or vertical axes and centers *not* at the origin are discussed in Exercises 59–64.

Asymptotes of a Hyperbola

For hyperbolas with equations as shown below, the asymptotes are the lines

$$y = \frac{b}{a}x \quad \text{and} \quad y = -\frac{b}{a}x.$$

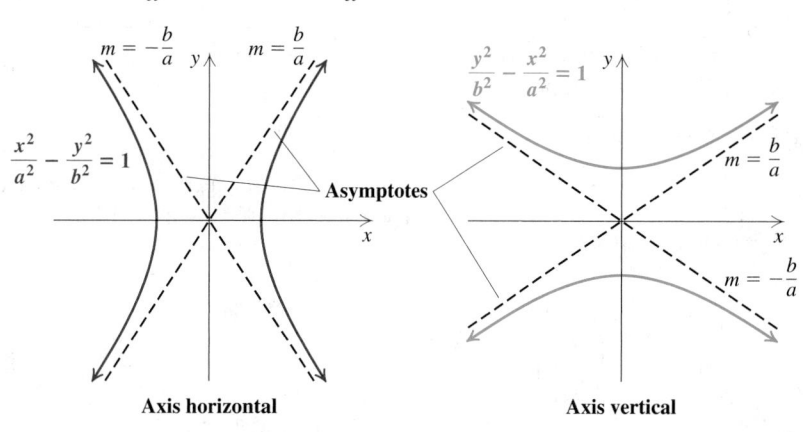

Axis horizontal   Axis vertical

In Section 13.2, we used a and b to determine the width and the length of an ellipse. For hyperbolas, a and b are used to determine the base and the height of a rectangle that can be used as an aid in sketching asymptotes and locating vertices. This is illustrated in the following example.

EXAMPLE   1   Graph: $\dfrac{x^2}{4} - \dfrac{y^2}{9} = 1$.

SOLUTION   Note that

$$\frac{x^2}{4} - \frac{y^2}{9} = \frac{x^2}{2^2} - \frac{y^2}{3^2}, \qquad \text{Identifying } a \text{ and } b$$

so $a = 2$ and $b = 3$. The asymptotes are thus

$$y = \frac{3}{2}x \quad \text{and} \quad y = -\frac{3}{2}x.$$

To help us sketch asymptotes and locate vertices, we use a and b—in this case, 2 and 3—to form the pairs $(-2, 3)$, $(2, 3)$, $(2, -3)$, and $(-2, -3)$. We plot these pairs and lightly sketch a rectangle. The asymptotes pass through the corners and, since this is a horizontal hyperbola, the vertices are where the rectangle intersects the x-axis. Finally, we draw the hyperbola, as shown below.

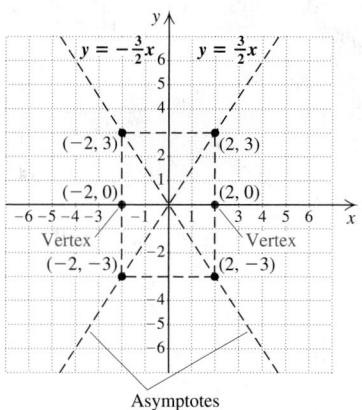

Asymptotes

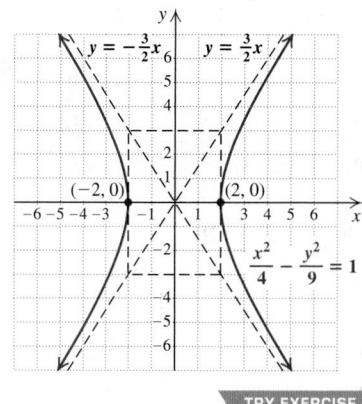

TRY EXERCISE   11

EXAMPLE **2**   Graph: $\dfrac{y^2}{36} - \dfrac{x^2}{4} = 1$.

SOLUTION   Note that

$$\dfrac{y^2}{36} - \dfrac{x^2}{4} = \dfrac{y^2}{6^2} - \dfrac{x^2}{2^2} = 1.$$

> Whether the hyperbola is horizontal or vertical is determined by which term is nonnegative. Here the y^2-term is nonnegative, so the hyperbola is vertical.

Using ± 2 as x-coordinates and ± 6 as y-coordinates, we plot $(2, 6)$, $(2, -6)$, $(-2, 6)$, and $(-2, -6)$, and lightly sketch a rectangle through them. The asymptotes pass through the corners (see the figure on the left below). Since the hyperbola is vertical, its vertices are $(0, 6)$ and $(0, -6)$. Finally, we draw curves through the vertices toward the asymptotes, as shown below.

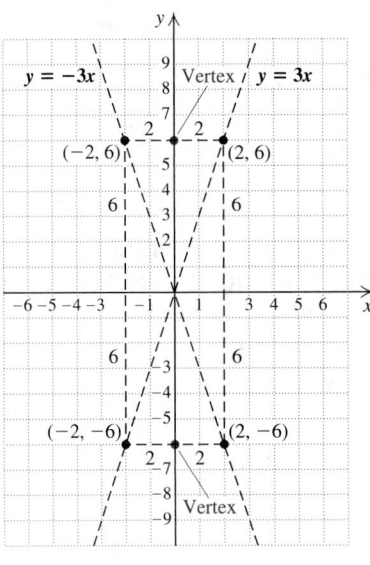
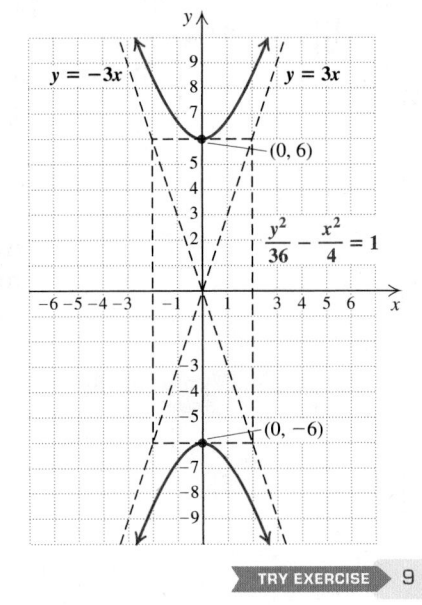

TRY EXERCISE   **9**

Hyperbolas (Nonstandard Form)

The equations for hyperbolas just examined are the standard ones, but there are other hyperbolas. We consider some of them.

> ### Equation of a Hyperbola in Nonstandard Form
> Hyperbolas having the x- and y-axes as asymptotes have equations as follows:
>
> $$xy = c, \quad \text{where } c \text{ is a nonzero constant.}$$

EXAMPLE 3   Graph: $xy = -8$.

SOLUTION   We first solve for y:

$$y = -\frac{8}{x}.$$ Dividing both sides by x. Note that $x \neq 0$.

Next, we find some solutions and form a table. Note that x cannot be 0 and that for large values of $|x|$, the value of y is close to 0. Thus the x- and y-axes serve as asymptotes. We plot the points and draw two curves.

x	y
2	-4
-2	4
4	-2
-4	2
1	-8
-1	8
8	-1
-8	1

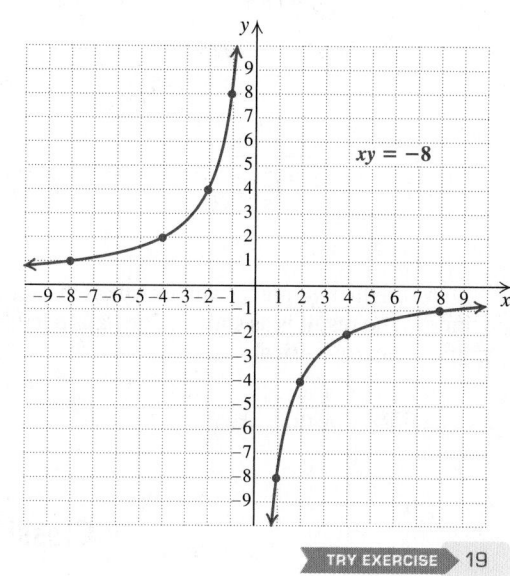

$xy = -8$

TRY EXERCISE ▶ 19

   Hyperbolas have many applications. A jet breaking the sound barrier creates a sonic boom with a wave front the shape of a cone. The intersection of the cone with the ground is one branch of a hyperbola. Some comets travel in hyperbolic orbits, and a cross section of many lenses is hyperbolic in shape.

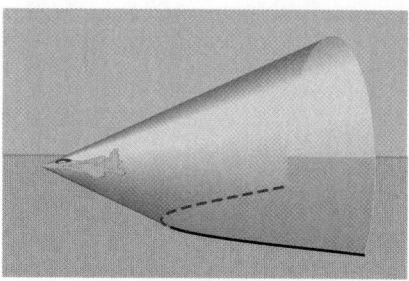

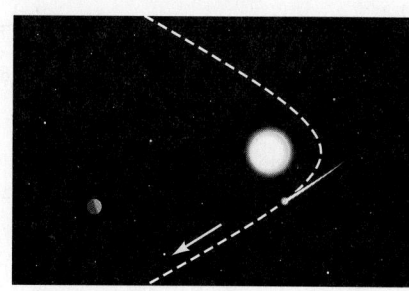

TECHNOLOGY CONNECTION

The procedure used to graph a hyperbola in standard form is similar to that used to draw a circle or an ellipse. Consider the graph of

$$\frac{x^2}{25} - \frac{y^2}{49} = 1.$$

The student should confirm that solving for y yields

$$y_1 = \frac{\sqrt{49x^2 - 1225}}{5} = \frac{7}{5}\sqrt{x^2 - 25}$$

and

$$y_2 = \frac{-\sqrt{49x^2 - 1225}}{5} = -\frac{7}{5}\sqrt{x^2 - 25},$$

or $y_2 = -y_1.$

When the two pieces are drawn on the same squared window, the result is as shown. The gaps occur where the graph is nearly vertical.

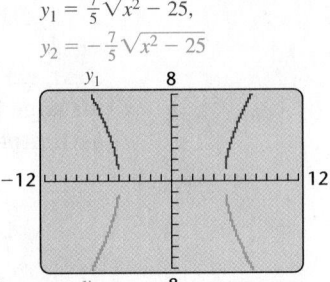

$$y_1 = \frac{7}{5}\sqrt{x^2 - 25},$$
$$y_2 = -\frac{7}{5}\sqrt{x^2 - 25}$$

On many calculators, pressing **APPS** and selecting Conics and then Hyperbola accesses a program in which hyperbolas in standard form can be graphed directly. Graph each of the following.

1. $\dfrac{x^2}{16} - \dfrac{y^2}{60} = 1$ 2. $16x^2 - 3y^2 = 64$

3. $\dfrac{y^2}{20} - \dfrac{x^2}{64} = 1$ 4. $45y^2 - 9x^2 = 441$

Classifying Graphs of Equations

By writing an equation of a conic section in a standard form, we can classify its graph as a parabola, a circle, an ellipse, or a hyperbola. Every conic section can also be represented by an equation of the form

$$Ax^2 + By^2 + Cxy + Dx + Ey + F = 0.$$

We can also classify graphs using values of A and B.

Graph	Standard Form		$Ax^2 + By^2 + Cxy + Dx + Ey + F = 0$
Parabola	$y = ax^2 + bx + c;$	Vertical parabola	Either $A = 0$ or $B = 0$, but not both.
	$x = ay^2 + by + c$	Horizontal parabola	
Circle	$x^2 + y^2 = r^2;$	Center at the origin	$A = B$
	$(x - h)^2 + (y - k)^2 = r^2$	Center at (h, k)	
Ellipse	$\dfrac{x^2}{a^2} + \dfrac{y^2}{b^2} = 1;$	Center at the origin	$A \neq B$, and A and B have the same sign.
	$\dfrac{(x - h)^2}{a^2} + \dfrac{(y - k)^2}{b^2} = 1$	Center at (h, k)	
Hyperbola	$\dfrac{x^2}{a^2} - \dfrac{y^2}{b^2} = 1;$	Horizontal hyperbola	A and B have opposite signs.
	$\dfrac{y^2}{b^2} - \dfrac{x^2}{a^2} = 1$	Vertical hyperbola	
	$xy = c$	Asymptotes are axes	Only C and F are nonzero.

Algebraic manipulations may be needed to express an equation in one of the preceding forms.

EXAMPLE 4 Classify the graph of each equation as a circle, an ellipse, a parabola, or a hyperbola. Refer to the above table as needed.

a) $5x^2 = 20 - 5y^2$
b) $x + 3 + 8y = y^2$
c) $x^2 = y^2 + 4$
d) $x^2 = 16 - 4y^2$

SOLUTION

a) We get the terms with variables on one side by adding $5y^2$ to both sides:

$$5x^2 + 5y^2 = 20.$$

Since x and y are *both* squared, we do not have a parabola. The fact that the squared terms are *added* tells us that we do not have a hyperbola. Do we have a circle? We factor the 5 out of both terms on the left and then divide by 5:

$$
\begin{aligned}
5(x^2 + y^2) &= 20 && \text{Factoring out 5} \\
x^2 + y^2 &= 4 && \text{Dividing both sides by 5} \\
x^2 + y^2 &= 2^2. && \text{This is an equation for a circle.}
\end{aligned}
$$

We see that the graph is a circle centered at the origin with radius 2.

We can also write the equation in the form

$$5x^2 + 5y^2 - 20 = 0. \quad A = 5, B = 5$$

Since $A = B$, the graph is a circle.

b) The equation $x + 3 + 8y = y^2$ has only one variable that is squared, so we solve for the other variable:

$$x = y^2 - 8y - 3. \quad \text{This is an equation for a parabola.}$$

The graph is a horizontal parabola that opens to the right.

We can also write the equation in the form

$$y^2 - x - 8y - 3 = 0. \quad A = 0, B = 1$$

Since $A = 0$ and $B \neq 0$, the graph is a parabola.

c) In $x^2 = y^2 + 4$, both variables are squared, so the graph is not a parabola. We subtract y^2 on both sides and divide by 4 to obtain

$$\frac{x^2}{2^2} - \frac{y^2}{2^2} = 1. \quad \text{This is an equation for a hyperbola.}$$

The minus sign here indicates that the graph is a hyperbola. Because it is the x^2-term that is nonnegative, the hyperbola is horizontal.

We can also write the equation in the form

$$x^2 - y^2 - 4 = 0. \quad A = 1, B = -1$$

Since A and B have opposite signs, the graph is a hyperbola.

d) In $x^2 = 16 - 4y^2$, both variables are squared, so the graph cannot be a parabola. We obtain the following equivalent equation:

$$x^2 + 4y^2 = 16. \quad \text{Adding } 4y^2 \text{ to both sides}$$

If the coefficients of the terms were the same, we would have the graph of a circle, as in part (a), but they are not. Dividing both sides by 16 yields

$$\frac{x^2}{16} + \frac{y^2}{4} = 1. \qquad \text{This is an equation for an ellipse.}$$

The graph of this equation is a horizontal ellipse.
We can also write the equation in the form

$$x^2 + 4y^2 - 16 = 0. \qquad A = 1, B = 4$$

Since $A \neq B$ and both A and B are positive, the graph is an ellipse.

▶ TRY EXERCISES ▶ 27 and 29

13.3 EXERCISE SET

✎ **Concept Reinforcement** *In each of Exercises 1–8, match the conic section with the equation in the column on the right that represents that type of conic section.*

1. ____ A hyperbola with a horizontal axis

2. ____ A hyperbola with a vertical axis

3. ____ An ellipse with its center not at the origin

4. ____ An ellipse with its center at the origin

5. ____ A circle with its center at the origin

6. ____ A circle with its center not at the origin

7. ____ A parabola opening upward or downward

8. ____ A parabola opening to the right or left

a) $\dfrac{x^2}{10} + \dfrac{y^2}{12} = 1$

b) $(x + 1)^2 + (y - 3)^2 = 30$

c) $y - x^2 = 5$

d) $\dfrac{x^2}{9} - \dfrac{y^2}{10} = 1$

e) $x - 2y^2 = 3$

f) $\dfrac{y^2}{20} - \dfrac{x^2}{35} = 1$

g) $3x^2 + 3y^2 = 75$

h) $\dfrac{(x - 1)^2}{10} + \dfrac{(y - 4)^2}{8} = 1$

Graph each hyperbola. Label all vertices and sketch all asymptotes.

9. $\dfrac{y^2}{16} - \dfrac{x^2}{16} = 1$

10. $\dfrac{x^2}{9} - \dfrac{y^2}{9} = 1$

11. $\dfrac{x^2}{4} - \dfrac{y^2}{25} = 1$

12. $\dfrac{y^2}{16} - \dfrac{x^2}{9} = 1$

13. $\dfrac{y^2}{36} - \dfrac{x^2}{9} = 1$

14. $\dfrac{x^2}{25} - \dfrac{y^2}{36} = 1$

15. $y^2 - x^2 = 25$

16. $x^2 - y^2 = 4$

17. $25x^2 - 16y^2 = 400$

18. $4y^2 - 9x^2 = 36$

Graph.

19. $xy = -6$

20. $xy = 8$

21. $xy = 4$

22. $xy = -9$

23. $xy = -2$

24. $xy = -1$

25. $xy = 1$

26. $xy = 2$

Classify each of the following as the equation of a circle, an ellipse, a parabola, or a hyperbola.

27. $x^2 + y^2 - 6x + 10y - 40 = 0$

28. $y - 4 = 2x^2$

29. $9x^2 + 4y^2 - 36 = 0$

30. $x + 3y = 2y^2 - 1$

31. $4x^2 - 9y^2 - 72 = 0$

32. $y^2 + x^2 = 8$

33. $y^2 = 20 - x^2$

34. $2y + 13 + x^2 = 8x - y^2$

35. $x - 10 = y^2 - 6y$ **36.** $y = \dfrac{5}{x}$

37. $x - \dfrac{3}{y} = 0$ **38.** $9x^2 = 9 - y^2$

39. $y + 6x = x^2 + 5$ **40.** $x^2 = 49 + y^2$

41. $25y^2 = 100 + 4x^2$

42. $3x^2 + 5y^2 + x^2 = y^2 + 49$

43. $3x^2 + y^2 - x = 2x^2 - 9x + 10y + 40$

44. $4y^2 + 20x^2 + 1 = 8y - 5x^2$

45. $16x^2 + 5y^2 - 12x^2 + 8y^2 - 3x + 4y = 568$

46. $56x^2 - 17y^2 = 234 - 13x^2 - 38y^2$

47. Explain how the equation of a hyperbola differs from the equation of an ellipse.

48. Is it possible for a hyperbola to represent the graph of a function? Why or why not?

Skill Review

To prepare for Section 13.4, review solving systems of equations and solving quadratic equations (Sections 8.2 and 11.2).

Solve.

49. $5x + 2y = -3,$
$2x + 3y = 12$ [8.2]

50. $4x - 2y = 5,$
$3x + 5y = -6$ [8.2]

51. $\frac{3}{4}x^2 + x^2 = 7$ [11.2]

52. $3x^2 + 10x - 8 = 0$ [11.2]

53. $x^2 - 3x - 1 = 0$ [11.2]

54. $x^2 + \dfrac{25}{x^2} = 26$ [11.5]

Synthesis

55. What is it in the equation of a hyperbola that controls how wide open the branches are? Explain your reasoning.

56. If, in
$$\frac{x^2}{a^2} - \frac{y^2}{b^2} = 1,$$
$a = b$, what are the asymptotes of the graph? Why?

Find an equation of a hyperbola satisfying the given conditions.

57. Having intercepts $(0, 6)$ and $(0, -6)$ and asymptotes $y = 3x$ and $y = -3x$

58. Having intercepts $(8, 0)$ and $(-8, 0)$ and asymptotes $y = 4x$ and $y = -4x$

The standard form for equations of horizontal or vertical hyperbolas centered at (h, k) are as follows:

$$\frac{(x - h)^2}{a^2} - \frac{(y - k)^2}{b^2} = 1$$

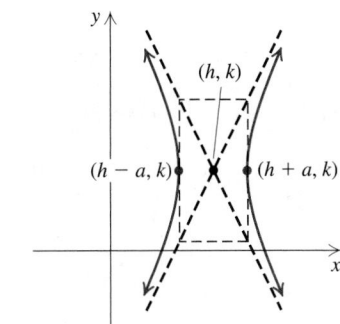

$$\frac{(y - k)^2}{b^2} - \frac{(x - h)^2}{a^2} = 1$$

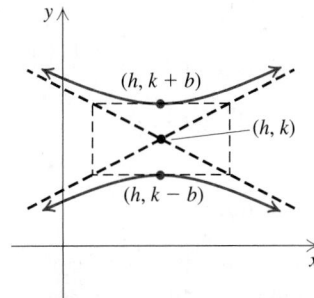

The vertices are as labeled and the asymptotes are

$$y - k = \frac{b}{a}(x - h) \quad and \quad y - k = -\frac{b}{a}(x - h).$$

For each of the following equations of hyperbolas, complete the square, if necessary, and write in standard form. Find the center, the vertices, and the asymptotes. Then graph the hyperbola.

59. $\dfrac{(x - 5)^2}{36} - \dfrac{(y - 2)^2}{25} = 1$

60. $\dfrac{(x - 2)^2}{9} - \dfrac{(y - 1)^2}{4} = 1$

61. $8(y + 3)^2 - 2(x - 4)^2 = 32$

62. $25(x - 4)^2 - 4(y + 5)^2 = 100$

63. $4x^2 - y^2 + 24x + 4y + 28 = 0$

64. $4y^2 - 25x^2 - 8y - 100x - 196 = 0$

65. Use a graphing calculator to check your answers to Exercises 13, 25, 31, and 59.

CONNECTING the CONCEPTS

When graphing equations of conic sections, it is usually helpful to first determine what type of graph the equation represents. We then find the coordinates of key points and equations of lines that determine the shape and the location of the graph.

Graph	Equation	Key Points	Equations of Lines
Parabola	$y = a(x - h)^2 + k$ $x = a(y - k)^2 + h$	Vertex: (h, k) Vertex: (h, k)	Axis of symmetry: $x = h$ Axis of symmetry: $y = k$
Circle	$(x - h)^2 + (y - k)^2 = r^2$	Center: (h, k)	
Ellipse	$\dfrac{x^2}{a^2} + \dfrac{y^2}{b^2} = 1$	x-intercepts: $(-a, 0)$, $(a, 0)$; y-intercepts: $(0, -b)$, $(0, b)$	
Hyperbola	$\dfrac{x^2}{a^2} - \dfrac{y^2}{b^2} = 1$ $\dfrac{y^2}{b^2} - \dfrac{x^2}{a^2} = 1$	Vertices: $(-a, 0)$, $(a, 0)$ Vertices: $(0, -b)$, $(0, b)$	Asymptotes (for both equations): $y = \dfrac{b}{a}x, y = -\dfrac{b}{a}x$
	$xy = c$		Asymptotes: $x = 0, y = 0$

MIXED REVIEW

1. Find the vertex and the axis of symmetry of the graph of $y = 3(x - 4)^2 + 1$.

2. Find the vertex and the axis of symmetry of the graph of $x = y^2 + 2y + 3$.

3. Find the center of the graph of $(x - 3)^2 + (y - 2)^2 = 5$.

4. Find the center of the graph of $x^2 + 6x + y^2 + 10y = 12$.

5. Find the x-intercepts and the y-intercepts of the graph of $\dfrac{x^2}{144} + \dfrac{y^2}{81} = 1$.

6. Find the vertices of the graph of $\dfrac{x^2}{9} - \dfrac{y^2}{121} = 1$.

7. Find the vertices of the graph of $4y^2 - x^2 = 4$.

8. Find the asymptotes of the graph of $\dfrac{y^2}{9} - \dfrac{x^2}{4} = 1$.

Classify each of the following as the graph of a parabola, a circle, an ellipse, or a hyperbola. Then graph.

9. $x^2 + y^2 = 36$

10. $y = x^2 - 5$

11. $\dfrac{x^2}{25} + \dfrac{y^2}{49} = 1$

12. $\dfrac{x^2}{25} - \dfrac{y^2}{49} = 1$

13. $x = (y + 3)^2 + 2$

14. $4x^2 + 9y^2 = 36$

15. $xy = -4$

16. $(x + 2)^2 + (y - 3)^2 = 1$

17. $x^2 + y^2 - 8y - 20 = 0$

18. $x = y^2 + 2y$

19. $16y^2 - x^2 = 16$

20. $x = \dfrac{9}{y}$

Systems Involving One Nonlinear Equation   ▪   Systems of Two Nonlinear Equations   ▪   Problem Solving

The equations appearing in systems of two equations have thus far in our discussion always been linear. We now consider systems of two equations in which at least one equation is nonlinear.

Systems Involving One Nonlinear Equation

Suppose that a system consists of an equation of a circle and an equation of a line. In what ways can the circle and the line intersect? The figures below represent three ways in which the situation can occur. We see that such a system will have 0, 1, or 2 real solutions.

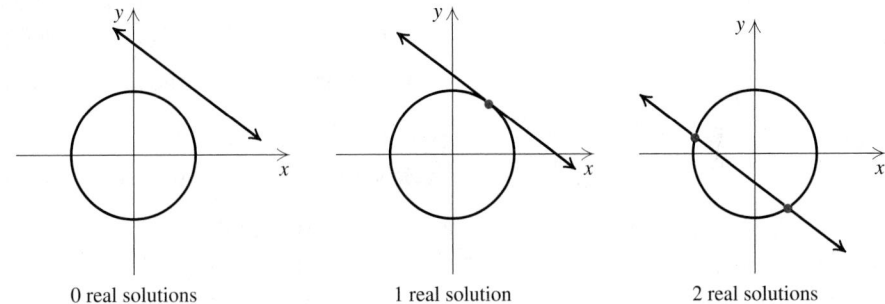

0 real solutions   1 real solution   2 real solutions

Recall that graphing, *elimination*, and *substitution* were all used to solve systems of linear equations. To solve systems in which one equation is of first degree and one is of second degree, it is preferable to use the *substitution* method.

EXAMPLE **1**   Solve the system

$$x^2 + y^2 = 25, \quad (1) \quad \text{(The graph is a circle.)}$$
$$3x - 4y = 0. \quad (2) \quad \text{(The graph is a line.)}$$

SOLUTION   First, we solve the linear equation, (2), for x:

$$x = \tfrac{4}{3}y. \quad (3) \quad \text{We could have solved for } y \text{ instead.}$$

Then we substitute $\tfrac{4}{3}y$ for x in equation (1) and solve for y:

$$\left(\tfrac{4}{3}y\right)^2 + y^2 = 25$$
$$\tfrac{16}{9}y^2 + y^2 = 25$$
$$\tfrac{25}{9}y^2 = 25$$
$$y^2 = 9 \qquad \text{Multiplying both sides by } \tfrac{9}{25}$$
$$y = \pm 3. \qquad \text{Using the principle of square roots}$$

Now we substitute these numbers for y in equation (3) and solve for x:

$$\text{for } y = 3, \quad x = \tfrac{4}{3}(3) = 4; \qquad \text{The ordered pair is } (4, 3).$$
$$\text{for } y = -3, \quad x = \tfrac{4}{3}(-3) = -4. \qquad \text{The ordered pair is } (-4, -3).$$

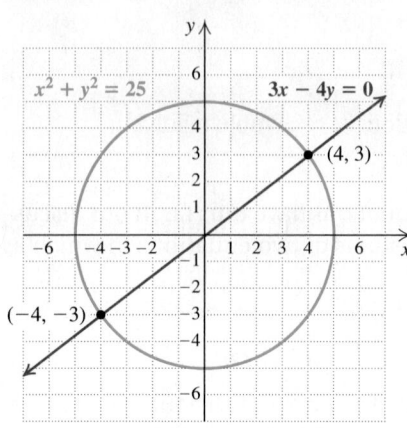

Check: For $(4, 3)$:

$$\begin{array}{c|c} x^2 + y^2 = 25 \\ \hline 4^2 + 3^2 & 25 \\ 16 + 9 & \\ 25 \overset{?}{=} 25 & \text{TRUE} \end{array} \qquad \begin{array}{c|c} 3x - 4y = 0 \\ \hline 3(4) - 4(3) & 0 \\ 12 - 12 & \\ 0 \overset{?}{=} 0 & \text{TRUE} \end{array}$$

It is left to the student to confirm that $(-4, -3)$ also checks in both equations. The pairs $(4, 3)$ and $(-4, -3)$ check, so they are solutions. The graph at left serves as a check. Intersections occur at $(4, 3)$ and $(-4, -3)$.

▶ TRY EXERCISE 7

Even if we do not know what the graph of each equation in a system looks like, the algebraic approach of Example 1 can still be used.

EXAMPLE 2   Solve the system

$$\begin{aligned} y + 3 &= 2x, & (1) & \quad \text{(A first-degree equation)} \\ x^2 + 2xy &= -1. & (2) & \quad \text{(A second-degree equation)} \end{aligned}$$

SOLUTION   First, we solve the linear equation (1) for y:

$$y = 2x - 3. \qquad (3)$$

Then we substitute $2x - 3$ for y in equation (2) and solve for x:

$$\begin{aligned} x^2 + 2x(2x - 3) &= -1 \\ x^2 + 4x^2 - 6x &= -1 \\ 5x^2 - 6x + 1 &= 0 \\ (5x - 1)(x - 1) &= 0 & \text{Factoring} \\ 5x - 1 = 0 \quad or \quad x - 1 &= 0 & \begin{array}{l}\text{Using the principle of} \\ \text{zero products}\end{array} \\ x = \tfrac{1}{5} \quad or \qquad x &= 1. \end{aligned}$$

Now we substitute these numbers for x in equation (3) and solve for y:

for $x = \frac{1}{5}$, $y = 2\left(\frac{1}{5}\right) - 3 = -\frac{13}{5}$;   The ordered pair is $\left(\frac{1}{5}, -\frac{13}{5}\right)$.

for $x = 1$, $y = 2(1) - 3 = -1$.   The ordered pair is $(1, -1)$.

You can confirm that $\left(\frac{1}{5}, -\frac{13}{5}\right)$ and $(1, -1)$ check, so they are both solutions.

▶ TRY EXERCISE 13

EXAMPLE 3   Solve the system

$$\begin{aligned} x + y &= 5, & (1) & \quad \text{(The graph is a line.)} \\ y &= 3 - x^2. & (2) & \quad \text{(The graph is a parabola.)} \end{aligned}$$

SOLUTION   We substitute $3 - x^2$ for y in the first equation:

$$\begin{aligned} x + 3 - x^2 &= 5 \\ -x^2 + x - 2 &= 0 & \text{Adding } -5 \text{ to both sides and rearranging} \\ x^2 - x + 2 &= 0. & \text{Multiplying both sides by } -1 \end{aligned}$$

Real-number solutions of systems of equations can be found using the INTERSECT option of $\boxed{\text{CALC}}$.

To solve Example 2,

$$y + 3 = 2x,$$
$$x^2 + 2xy = -1,$$

we solve each equation for y and then graph:

$$\left.\begin{array}{l} y_1 = 2x - 3, \\ y_2 = \dfrac{-1 - x^2}{2x}. \end{array}\right\} \quad \begin{array}{l} \text{Note that} \\ x, y \neq 0. \end{array}$$

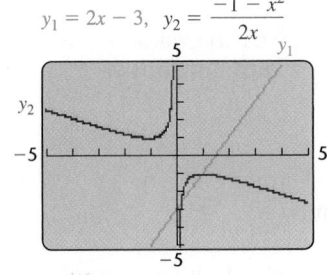

$$y_1 = 2x - 3, \quad y_2 = \dfrac{-1 - x^2}{2x}$$

Using INTERSECT, we find the solutions to be $(0.2, -2.6)$ and $(1, -1)$.

Solve each system. Round all values to two decimal places.

1. $4xy - 7 = 0,$
$\quad x - 3y - 2 = 0$
2. $x^2 + y^2 = 14,$
$\quad 16x + 7y^2 = 0$

Since $x^2 - x + 2$ does not factor, we need the quadratic formula:

$$\begin{aligned} x &= \frac{-b \pm \sqrt{b^2 - 4ac}}{2a} \\ &= \frac{-(-1) \pm \sqrt{(-1)^2 - 4 \cdot 1 \cdot 2}}{2(1)} \qquad \text{Substituting} \\ &= \frac{1 \pm \sqrt{1 - 8}}{2} = \frac{1 \pm \sqrt{-7}}{2} = \frac{1}{2} \pm \frac{\sqrt{7}}{2}i. \end{aligned}$$

Solving equation (1) for y gives us $y = 5 - x$. Substituting values for x gives

$$y = 5 - \left(\frac{1}{2} + \frac{\sqrt{7}}{2}i\right) = \frac{9}{2} - \frac{\sqrt{7}}{2}i \quad \text{and}$$

$$y = 5 - \left(\frac{1}{2} - \frac{\sqrt{7}}{2}i\right) = \frac{9}{2} + \frac{\sqrt{7}}{2}i.$$

The solutions are

$$\left(\frac{1}{2} + \frac{\sqrt{7}}{2}i, \frac{9}{2} - \frac{\sqrt{7}}{2}i\right) \quad \text{and} \quad \left(\frac{1}{2} - \frac{\sqrt{7}}{2}i, \frac{9}{2} + \frac{\sqrt{7}}{2}i\right).$$

There are no real-number solutions. Note in the figure at right that the graphs do not intersect. Getting only nonreal solutions tells us that the graphs do not intersect.

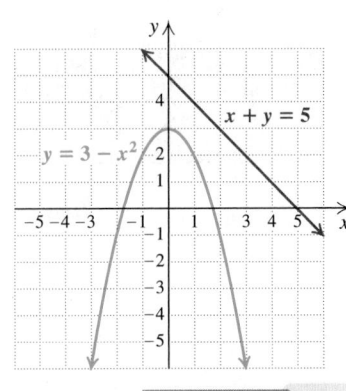

TRY EXERCISE ▶ 19

Systems of Two Nonlinear Equations

We now consider systems of two second-degree equations. Graphs of such systems can involve any two conic sections. The following figure shows some ways in which a circle and a hyperbola can intersect.

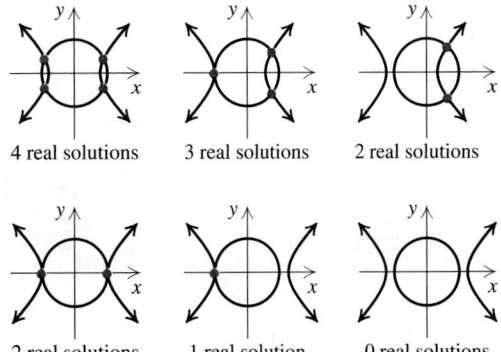

| 4 real solutions | 3 real solutions | 2 real solutions |

| 2 real solutions | 1 real solution | 0 real solutions |

To solve systems of two second-degree equations, we either substitute or eliminate. The elimination method is generally better when both equations are of the form $Ax^2 + By^2 = C$. Then we can eliminate an x^2-term or a y^2-term in a manner similar to the procedure used in Chapter 8.

Solve the system

$$2x^2 + 5y^2 = 22, \quad (1) \quad \text{(The graph is an ellipse.)}$$
$$3x^2 - y^2 = -1. \quad (2) \quad \text{(The graph is a hyperbola.)}$$

SOLUTION   Here we multiply equation (2) by 5 and then add:

$$
\begin{array}{ll}
2x^2 + 5y^2 = 22 & \\
\underline{15x^2 - 5y^2 = -5} & \text{Multiplying both sides of equation (2) by 5} \\
17x^2 = 17 & \text{Adding} \\
x^2 = 1 & \\
x = \pm 1. &
\end{array}
$$

There is no x-term, and whether x is -1 or 1, we have $x^2 = 1$. Thus we can simultaneously substitute 1 and -1 for x in equation (2):

$$
\left.
\begin{array}{l}
3 \cdot (\pm 1)^2 - y^2 = -1 \\
3 - y^2 = -1 \\
-y^2 = -4
\end{array}
\right\}
\quad
\begin{array}{l}
\text{Since } (-1)^2 = 1^2, \text{ we can evaluate for} \\
x = -1 \text{ and } x = 1 \text{ simultaneously.}
\end{array}
$$

$$y^2 = 4 \quad \text{or} \quad y = \pm 2.$$

Thus, if $x = 1$, then $y = 2$ or $y = -2$; and if $x = -1$, then $y = 2$ or $y = -2$. The four possible solutions are $(1, 2)$, $(1, -2)$, $(-1, 2)$, and $(-1, -2)$.

Check: Since $(2)^2 = (-2)^2$ and $(1)^2 = (-1)^2$, we can check all four pairs at once.

$$
\begin{array}{ll}
\underline{2x^2 + 5y^2 = 22} & \qquad \underline{3x^2 - y^2 = -1} \\
2(\pm 1)^2 + 5(\pm 2)^2 \;\big|\; 22 & \qquad 3(\pm 1)^2 - (\pm 2)^2 \;\big|\; -1 \\
2 + 20 \;\big| & \qquad 3 - 4 \;\big| \\
22 \overset{?}{=} 22 \quad \text{TRUE} & \qquad -1 \overset{?}{=} -1 \quad \text{TRUE}
\end{array}
$$

The solutions are $(1, 2)$, $(1, -2)$, $(-1, 2)$, and $(-1, -2)$.   **TRY EXERCISE** ▶ 29

When a product of variables is in one equation and the other equation is of the form $Ax^2 + By^2 = C$, we often solve for a variable in the equation with the product and then use substitution.

Solve the system

$$x^2 + 4y^2 = 20, \quad (1) \quad \text{(The graph is an ellipse.)}$$
$$xy = 4. \quad (2) \quad \text{(The graph is a hyperbola.)}$$

SOLUTION   First, we solve equation (2) for y:

$$y = \frac{4}{x}. \qquad \text{Dividing both sides by } x. \text{ Note that } x \neq 0.$$

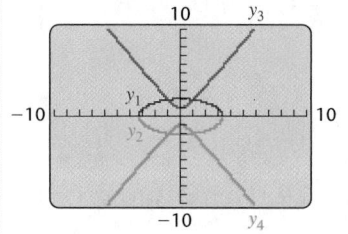

Then we substitute $4/x$ for y in equation (1) and solve for x:

$$x^2 + 4\left(\frac{4}{x}\right)^2 = 20$$

$$x^2 + \frac{64}{x^2} = 20$$

$$x^4 + 64 = 20x^2 \qquad \text{Multiplying by } x^2$$

$$x^4 - 20x^2 + 64 = 0 \qquad \begin{array}{l}\text{Obtaining standard form.} \\ \text{This equation is reducible} \\ \text{to quadratic.}\end{array}$$

$$(x^2 - 4)(x^2 - 16) = 0 \qquad \begin{array}{l}\text{Factoring. If you prefer, let} \\ u = x^2 \text{ and substitute.}\end{array}$$

$$(x - 2)(x + 2)(x - 4)(x + 4) = 0 \qquad \text{Factoring again}$$

$$x = 2 \quad or \quad x = -2 \quad or \quad x = 4 \quad or \quad x = -4. \qquad \begin{array}{l}\text{Using the principle of} \\ \text{zero products}\end{array}$$

Since $y = 4/x$, for $x = 2$, we have $y = 4/2$, or 2. Thus, $(2, 2)$ is a solution. Similarly, $(-2, -2)$, $(4, 1)$, and $(-4, -1)$ are solutions. You can show that all four pairs check.

> **TRY EXERCISE** 37

Problem Solving

We now consider applications that can be modeled by a system of equations in which at least one equation is not linear.

EXAMPLE 6 *Architecture.* For a college fitness center, an architect wants to lay out a rectangular piece of land that has a perimeter of 204 m and an area of 2565 m². Find the dimensions of the piece of land.

SOLUTION

1. **Familiarize.** We draw and label a sketch, letting $l =$ the length and $w =$ the width, both in meters.

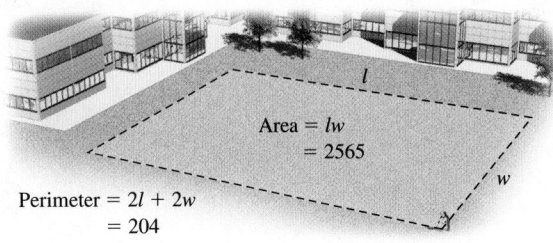

Area = lw
    = 2565

Perimeter = $2l + 2w$
    = 204

2. **Translate.** We then have the following translation:

Perimeter: $2w + 2l = 204$;

Area: $lw = 2565$.

3. **Carry out.** We solve the system

$$2w + 2l = 204,$$
$$lw = 2565.$$

Solving the second equation for l gives us $l = 2565/w$. Then we substitute $2565/w$ for l in the first equation and solve for w:

$$2w + 2\left(\frac{2565}{w}\right) = 204$$

$$2w^2 + 2(2565) = 204w \qquad \text{Multiplying both sides by } w$$

$$2w^2 - 204w + 2(2565) = 0 \qquad \text{Standard form}$$

$$w^2 - 102w + 2565 = 0 \qquad \text{Multiplying by } \tfrac{1}{2}$$

> Factoring could be used instead of the quadratic formula, but the numbers are quite large.

$$w = \frac{-(-102) \pm \sqrt{(-102)^2 - 4 \cdot 1 \cdot 2565}}{2 \cdot 1}$$

$$w = \frac{102 \pm \sqrt{144}}{2} = \frac{102 \pm 12}{2}$$

$$w = 57 \quad or \quad w = 45.$$

If $w = 57$, then $l = 2565/w = 2565/57 = 45$. If $w = 45$, then $l = 2565/w = 2565/45 = 57$. Since length is usually considered to be longer than width, we have the solution $l = 57$ and $w = 45$, or $(57, 45)$.

4. Check. If $l = 57$ and $w = 45$, the perimeter is $2 \cdot 57 + 2 \cdot 45$, or 204. The area is $57 \cdot 45$, or 2565. The numbers check.

5. State. The length is 57 m and the width is 45 m.

> **TRY EXERCISE** 47

EXAMPLE **7**   *Laptop dimensions.* The screen on Tara's new laptop has an area of 90 in^2 and a $\sqrt{200.25}$-in. diagonal. Find the width and the length of the screen.

SOLUTION

1. Familiarize. We make a drawing and label it. Note that the width, the length, and the diagonal form a right triangle. We let l = the length and w = the width, both in inches.

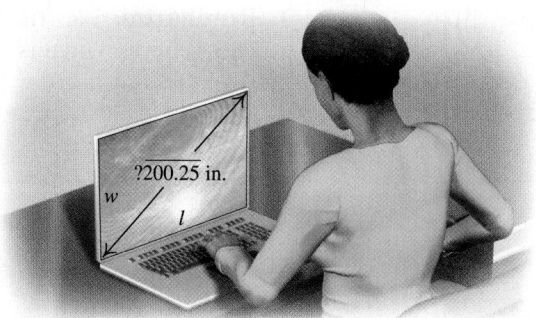

2. Translate. We translate to a system of equations:

$$l^2 + w^2 = (\sqrt{200.25})^2, \qquad \text{Using the Pythagorean theorem}$$

$$lw = 90. \qquad \text{Using the formula for the area of a rectangle}$$

3. Carry out. We solve the system

$$\left. \begin{array}{l} l^2 + w^2 = (\sqrt{200.25})^2, \\ lw = 90. \end{array} \right\} \qquad \text{You should complete the solution of this system.}$$

We get $(12, 7.5)$, $(7.5, 12)$, $(-12, -7.5)$, and $(-7.5, -12)$.

4. Check. Since measurements must be positive and length is usually greater than width, we check only $(12, 7.5)$. In the right triangle, $12^2 + 7.5^2 = 144 + 56.25 = 200.25$. The area is $12(7.5) = 90$, so our answer checks.

5. State. The length is 12 in. and the width is 7.5 in.

> **TRY EXERCISE** 51

Visualizing for Success

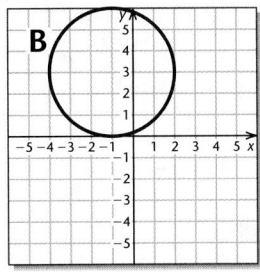

A

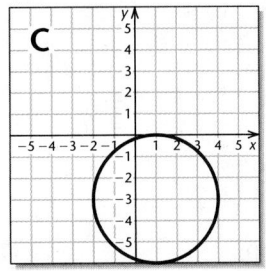

B

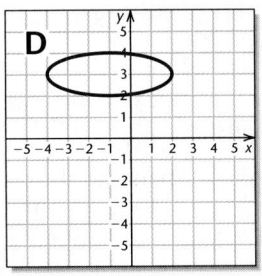

C

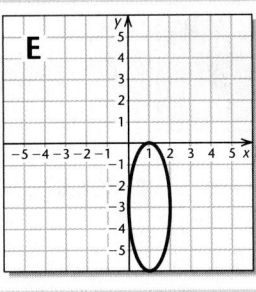

D

E

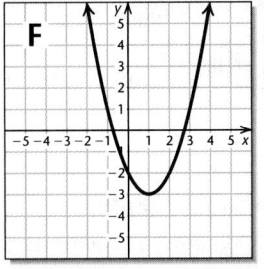

F

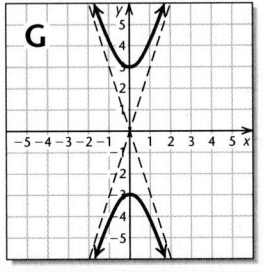

G

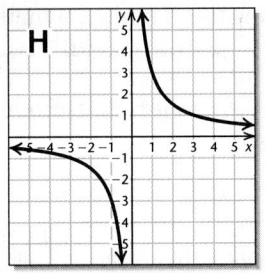

H

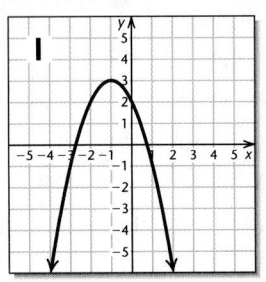

I

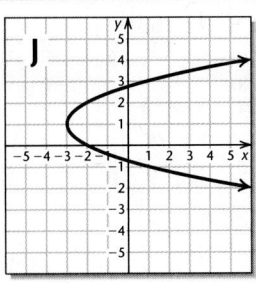

J

Match each equation with its graph

1. $(x - 1)^2 + (y + 3)^2 = 9$

2. $\dfrac{x^2}{9} - \dfrac{y^2}{1} = 1$

3. $y = (x - 1)^2 - 3$

4. $(x + 1)^2 + (y - 3)^2 = 9$

5. $x = (y - 1)^2 - 3$

6. $\dfrac{(x + 1)^2}{9} + \dfrac{(y - 3)^2}{1} = 1$

7. $xy = 3$

8. $y = -(x + 1)^2 + 3$

9. $\dfrac{y^2}{9} - \dfrac{x^2}{1} = 1$

10. $\dfrac{(x - 1)^2}{1} + \dfrac{(y + 3)^2}{9} = 1$

Answers on page A-64

An additional, animated version of this activity appears in MyMathLab. To use MyMathLab, you need a course ID and a student access code. Contact your instructor for more information.

13.4 **EXERCISE SET**

 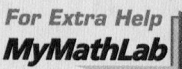
↪ *Concept Reinforcement* *Classify each statement as either true or false.*

1. A system of equations that represent a line and an ellipse can have 0, 1, or 2 solutions.

2. A system of equations that represent a parabola and a circle can have up to 4 solutions.

3. A system of equations representing a hyperbola and a circle can have no fewer than 2 solutions.

4. A system of equations representing an ellipse and a line has either 0 or 2 solutions.

5. Systems containing one first-degree equation and one second-degree equation are most easily solved using the substitution method.

6. Systems containing two second-degree equations of the form $Ax^2 + By^2 = C$ are most easily solved using the elimination method.

Solve. Remember that graphs can be used to confirm all real solutions.

7. $x^2 + y^2 = 41,$
   $y - x = 1$

8. $x^2 + y^2 = 45,$
   $y - x = 3$

9. $4x^2 + 9y^2 = 36,$
   $3y + 2x = 6$

10. $9x^2 + 4y^2 = 36,$
    $3x + 2y = 6$

11. $y^2 = x + 3,$
    $2y = x + 4$

12. $y = x^2,$
    $3x = y + 2$

13. $x^2 - xy + 3y^2 = 27,$
    $x - y = 2$

14. $2y^2 + xy + x^2 = 7,$
    $x - 2y = 5$

15. $x^2 + 4y^2 = 25,$
    $x + 2y = 7$

16. $x^2 - y^2 = 16,$
    $x - 2y = 1$

17. $x^2 - xy + 3y^2 = 5,$
    $x - y = 2$

18. $m^2 + 3n^2 = 10,$
    $m - n = 2$

19. $3x + y = 7,$
    $4x^2 + 5y = 24$

20. $2y^2 + xy = 5,$
    $4y + x = 7$

21. $a + b = 6,$
    $ab = 8$

22. $p + q = -1,$
    $pq = -12$

23. $2a + b = 1,$
    $b = 4 - a^2$

24. $4x^2 + 9y^2 = 36,$
    $x + 3y = 3$

25. $a^2 + b^2 = 89,$
    $a - b = 3$

26. $xy = 10,$
    $x + y = 7$

27. $y = x^2,$
    $x = y^2$

28. $x^2 + y^2 = 25,$
    $y^2 = x + 5$

Aha! 29. $x^2 + y^2 = 16,$
    $x^2 - y^2 = 16$

30. $y^2 - 4x^2 = 25,$
    $4x^2 + y^2 = 25$

31. $x^2 + y^2 = 25,$
    $xy = 12$

32. $x^2 - y^2 = 16,$
    $x + y^2 = 4$

33. $x^2 + y^2 = 9,$
    $25x^2 + 16y^2 = 400$

34. $x^2 + y^2 = 4,$
    $9x^2 + 16y^2 = 144$

35. $x^2 + y^2 = 14,$
    $x^2 - y^2 = 4$

36. $x^2 + y^2 = 16,$
    $y^2 - 2x^2 = 10$

37. $x^2 + y^2 = 10,$
    $xy = 3$

38. $x^2 + y^2 = 5,$
    $xy = 2$

39. $x^2 + 4y^2 = 20,$
    $xy = 4$

40. $x^2 + y^2 = 13,$
    $xy = 6$

41. $2xy + 3y^2 = 7,$
    $3xy - 2y^2 = 4$

42. $3xy + x^2 = 34,$
    $2xy - 3x^2 = 8$

43. $4a^2 - 25b^2 = 0,$
    $2a^2 - 10b^2 = 3b + 4$

44. $xy - y^2 = 2,$
    $2xy - 3y^2 = 0$

45. $ab - b^2 = -4,$
    $ab - 2b^2 = -6$

46. $x^2 - y = 5,$
    $x^2 + y^2 = 25$

Solve.

47. *Art.* Elliot is designing a rectangular stained glass miniature that has a perimeter of 28 cm and a diagonal of length 10 cm. What should the dimensions of the glass be?

48. *Geometry.* A rectangle has an area of 2 yd² and a perimeter of 6 yd. Find its dimensions.

49. *Tile design.* The Clay Works tile company wants to make a new rectangular tile that has a perimeter of 6 in. and a diagonal of length $\sqrt{5}$ in. What should the dimensions of the tile be?

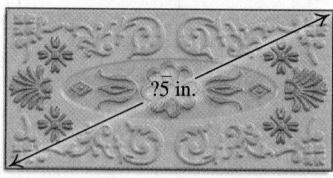

50. *Geometry.* A rectangle has an area of 20 in² and a perimeter of 18 in. Find its dimensions.

51. *Design of a van.* The cargo area of a delivery van must be 60 ft², and the length of a diagonal must accommodate a 13-ft board. Find the dimensions of the cargo area.

52. *Dimensions of a rug.* The diagonal of a Persian rug is 25 ft. The area of the rug is 300 ft². Find the length and the width of the rug.

53. The product of two numbers is 90. The sum of their squares is 261. Find the numbers.

54. *Investments.* A certain amount of money saved for 1 yr at a certain interest rate yielded $125 in simple interest. If $625 more had been invested and the rate had been 1% less, the interest would have been the same. Find the principal and the rate.

55. *Garden design.* A garden contains two square flower beds. Find the length of each bed if the sum of their areas is 832 ft² and the difference of their areas is 320 ft².

56. *TV dimensions.* The Kaplans' new LCD screen has an area of 1100 in² and has a $\sqrt{2561}$-in. diagonal. Find the width and the length of the screen.

57. The area of a rectangle is $\sqrt{3}$ m², and the length of a diagonal is 2 m. Find the dimensions.

58. The area of a rectangle is $\sqrt{2}$ m², and the length of a diagonal is $\sqrt{3}$ m. Find the dimensions.

59. How can an understanding of conic sections be helpful when a system of nonlinear equations is being solved algebraically?

60. Suppose a system of equations is comprised of one linear equation and one nonlinear equation. Is it possible for such a system to have three solutions? Why or why not?

Skill Review

To prepare for Section 14.1, review evaluating expressions (Section 1.8).

Simplify. [1.8]

61. $(-1)^9(-3)^2$

62. $(-1)^{10}(-3)^3$

Evaluate each of the following. [1.8]

63. $\dfrac{(-1)^k}{k-6}$, for $k = 7$

64. $\dfrac{(-1)^k}{k-5}$, for $k = 10$

65. $\dfrac{n}{2}(3 + n)$, for $n = 11$

66. $\dfrac{7(1-r^2)}{1-r}$; for $r = \frac{1}{2}$

Synthesis

67. Write a problem that translates to a system of two equations. Design the problem so that at least one equation is nonlinear and so that no real solution exists.

68. Write a problem for a classmate to solve. Devise the problem so that a system of two nonlinear equations with exactly one real solution is solved.

69. Find the equation of a circle that passes through $(-2, 3)$ and $(-4, 1)$ and whose center is on the line $5x + 8y = -2$.

70. Find the equation of an ellipse centered at the origin that passes through the points $(2, -3)$ and $(1, \sqrt{13})$.

Solve.

71. $p^2 + q^2 = 13$,
$\dfrac{1}{pq} = -\dfrac{1}{6}$

72. $a + b = \dfrac{5}{6}$,
$\dfrac{a}{b} + \dfrac{b}{a} = \dfrac{13}{6}$

73. *Fence design.* A roll of chain-link fencing contains 100 ft of fence. The fencing is bent at a 90° angle to enclose a rectangular work area of 2475 ft², as shown. Determine the length and the width of the rectangle.

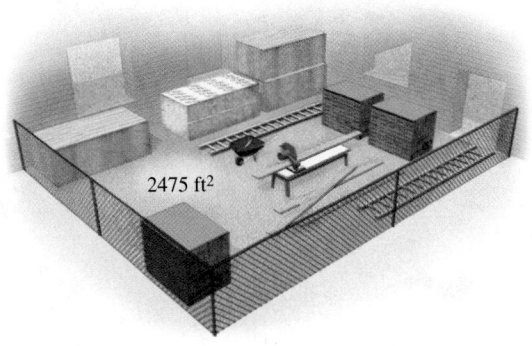

2475 ft²

74. A piece of wire 100 cm long is to be cut into two pieces and those pieces are each to be bent to make a square. The area of one square is to be 144 cm^2 greater than that of the other. How should the wire be cut?

75. *Box design.*   Four squares with sides 5 in. long are cut from the corners of a rectangular metal sheet that has an area of 340 in^2. The edges are bent up to form an open box with a volume of 350 in^3. Find the dimensions of the box.

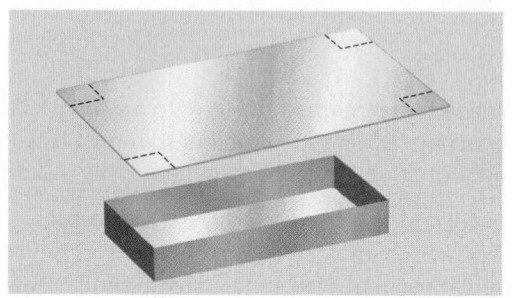

76. *Computer screens.*   The ratio of the length to the height of the screen on a computer monitor is 4 to 3. A Dell Inspiron notebook has a 15-in. diagonal screen. Find the dimensions of the screen.

77. *HDTV screens.*   The ratio of the length to the height of an HDTV screen is 16 to 9. The Sollar Lounge has an HDTV screen with a 73-in. diagonal screen. Find the dimensions of the screen.

78. *Railing sales.*   Fireside Castings finds that the total revenue R from the sale of x units of railing is given by

$$R = 100x + x^2.$$

Fireside also finds that the total cost C of producing x units of the same product is given by

$$C = 80x + 1500.$$

A break-even point is a value of x for which total revenue is the same as total cost; that is, $R = C$. How many units must be sold to break even?

79. Use a graphing calculator to check your answers to Exercises 13, 25, and 47.

Study Summary

KEY TERMS AND CONCEPTS	EXAMPLES

SECTION 13.1: CONIC SECTIONS: PARABOLAS AND CIRCLES

Parabola

$y = ax^2 + bx + c$ Opens upward $(a > 0)$ or downward $(a < 0)$

$= a(x - h)^2 + k;$ Vertex: (h, k)

$x = ay^2 + by + c$ Opens right $(a > 0)$ or left $(a < 0)$

$= a(y - k)^2 + h$ Vertex: (h, k)

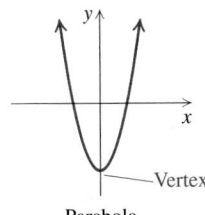

Parabola

$$x = -y^2 + 4y - 1$$
$$= -(y^2 - 4y \quad) - 1$$
$$= -(y^2 - 4y + 4) - 1 - (-1)(4)$$
$$= -(y - 2)^2 + 3 \quad a = -1; \text{parabola opens left}$$

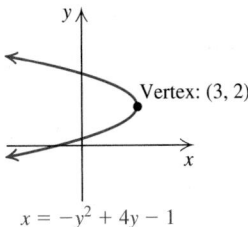

$x = -y^2 + 4y - 1$

Vertex: $(3, 2)$

Circle

$x^2 + y^2 = r^2;$ Radius: r Center: $(0, 0)$

$(x - h)^2 + (y - k)^2 = r^2$ Radius: r Center: (h, k)

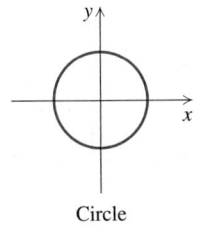

Circle

$$x^2 + y^2 + 2x - 6y + 6 = 0$$
$$x^2 + 2x + \quad y^2 - 6y \quad = -6$$
$$x^2 + 2x + 1 + y^2 - 6y + 9 = -6 + 1 + 9$$
$$(x + 1)^2 + (y - 3)^2 = 4$$
$$[x - (-1)]^2 + (y - 3)^2 = 2^2 \quad \text{Radius: } 2$$
$$\text{Center: } (-1, 3)$$

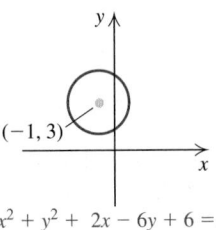

$(-1, 3)$

$x^2 + y^2 + 2x - 6y + 6 = 0$

SECTION 13.2: CONIC SECTIONS: ELLIPSES

Ellipse

$\dfrac{x^2}{a^2} + \dfrac{y^2}{b^2} = 1;$ Center: $(0, 0)$

$\dfrac{(x - h)^2}{a^2} + \dfrac{(y - k)^2}{b^2} = 1$ Center: (h, k)

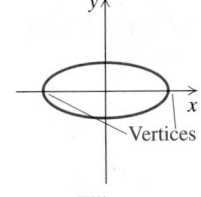

Vertices

Ellipse

$$\frac{(x - 4)^2}{4} + \frac{(y + 1)^2}{9} = 1$$
$$\frac{(x - 4)^2}{2^2} + \frac{[y - (-1)]^2}{3^2} = 1 \quad 3 > 2; \text{ellipse is vertical}$$
$$\text{Center: } (4, -1)$$

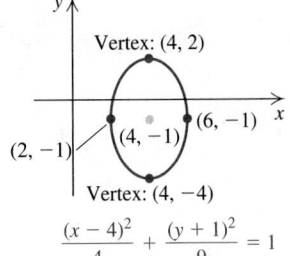

Vertex: $(4, 2)$

$(2, -1)$ $(4, -1)$ $(6, -1)$

Vertex: $(4, -4)$

$$\frac{(x - 4)^2}{4} + \frac{(y + 1)^2}{9} = 1$$

SECTION 13.3: CONIC SECTIONS: HYPERBOLAS

Hyperbola

$\dfrac{x^2}{a^2} - \dfrac{y^2}{b^2} = 1;$ Two branches opening right and left

$\dfrac{y^2}{b^2} - \dfrac{x^2}{a^2} = 1$ Two branches opening upward and downward

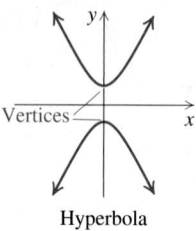

Hyperbola

$\dfrac{x^2}{4} - \dfrac{y^2}{1} = 1$

$\dfrac{x^2}{2^2} - \dfrac{y^2}{1^2} = 1$ Opens right and left

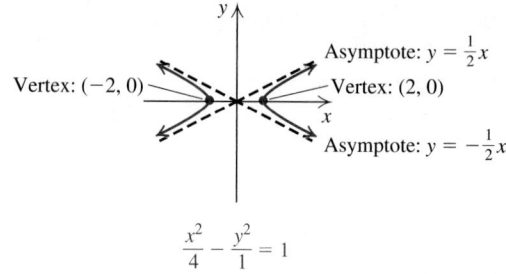

$\dfrac{x^2}{4} - \dfrac{y^2}{1} = 1$

SECTION 13.4: NONLINEAR SYSTEMS OF EQUATIONS

We can solve a system containing at least one nonlinear equation using substitution or elimination.

Solve:

$x^2 - y = -1,$ (1)   (The graph is a parabola.)
$x + 2y = 3.$ (2)   (The graph is a line.)

$x = 3 - 2y$ Solving for x in equation (2)

$(3 - 2y)^2 - y = -1$ Substituting for x in equation (1)
$9 - 12y + 4y^2 - y = -1$
$4y^2 - 13y + 10 = 0$
$(4y - 5)(y - 2) = 0$
$4y - 5 = 0$ *or* $y - 2 = 0$
$y = \tfrac{5}{4}$ *or* $y = 2$ Solving for y

If $y = \tfrac{5}{4}$, then $x = 3 - 2\left(\tfrac{5}{4}\right) = \tfrac{1}{2}$. $\left(\tfrac{1}{2}, \tfrac{5}{4}\right)$ is a solution.
If $y = 2$, then $x = 3 - 2(2) = -1$. $(-1, 2)$ is a solution.

The solutions are $\left(\tfrac{1}{2}, \tfrac{5}{4}\right)$ and $(-1, 2)$.

Review Exercises: Chapter 13

🖐 *Concept Reinforcement* *Classify each statement as either true or false.*

1. Every parabola that opens upward or downward can represent the graph of a function. [13.1]

2. The center of a circle is part of the circle itself. [13.1]

3. The foci of an ellipse are part of the ellipse itself. [13.2]

4. It is possible for a hyperbola to represent the graph of a function. [13.3]

5. If an equation of a conic section has only one term of degree 2, its graph cannot be a circle, an ellipse, or a hyperbola. [13.3]

6. Two nonlinear graphs can intersect in more than one point. [13.4]

7. Every system of nonlinear equations has at least one real solution. [13.4]

8. Both substitution and elimination can be used as methods for solving a system of nonlinear equations. [13.4]

Find the center and the radius of each circle. [13.1]

9. $(x + 3)^2 + (y - 2)^2 = 16$

10. $(x - 5)^2 + y^2 = 11$

11. $x^2 + y^2 - 6x - 2y + 1 = 0$

12. $x^2 + y^2 + 8x - 6y = 20$

13. Find an equation of the circle with center $(-4, 3)$ and radius 4. [13.1]

14. Find an equation of the circle with center $(7, -2)$ and radius $2\sqrt{5}$. [13.1]

Classify each equation as a circle, an ellipse, a parabola, or a hyperbola. Then graph.

15. $5x^2 + 5y^2 = 80$ [13.1], [13.3]

16. $9x^2 + 2y^2 = 18$ [13.2], [13.3]

17. $y = -x^2 + 2x - 3$ [13.1], [13.3]

18. $\dfrac{y^2}{9} - \dfrac{x^2}{4} = 1$ [13.3]

19. $xy = 9$ [13.3]

20. $x = y^2 + 2y - 2$ [13.1], [13.3]

21. $\dfrac{(x + 1)^2}{3} + (y - 3)^2 = 1$ [13.2], [13.3]

22. $x^2 + y^2 + 6x - 8y - 39 = 0$ [13.1], [13.3]

Solve. [13.4]

23. $x^2 - y^2 = 21,$
    $x + y = 3$

24. $x^2 - 2x + 2y^2 = 8,$
    $2x + y = 6$

25. $x^2 - y = 5,$
    $2x - y = 5$

26. $x^2 + y^2 = 25,$
    $x^2 - y^2 = 7$

27. $x^2 - y^2 = 3,$
    $y = x^2 - 3$

28. $x^2 + y^2 = 18,$
    $2x + y = 3$

29. $x^2 + y^2 = 100,$
    $2x^2 - 3y^2 = -120$

30. $x^2 + 2y^2 = 12,$
    $xy = 4$

31. A rectangular bandstand has a perimeter of 38 m and an area of 84 m². What are the dimensions of the bandstand? [13.4]

32. One type of carton used by tableproducts.com exactly fits both a rectangular plate of area 108 in² and chopsticks of length 15 in., laid diagonally on top of the plate. Find the length and the width of the carton. [13.4]

33. The perimeter of a square mirror is 12 cm more than the perimeter of another square mirror. Its area exceeds the area of the other by 39 cm². Find the perimeter of each mirror. [13.4]

34. The sum of the areas of two circles is 130π ft². The difference of the circumferences is 16π ft. Find the radius of each circle. [13.4]

Synthesis

📝 35. How does the graph of a hyperbola differ from the graph of a parabola? [13.1], [13.3]

📝 36. Explain why function notation rarely appears in this chapter, and list the graphs discussed for which function notation could be used. [13.1], [13.2], [13.3]

37. Solve: [13.4]
    $$4x^2 - x - 3y^2 = 9,$$
    $$-x^2 + x + y^2 = 2.$$

38. Find the points whose distance from $(8, 0)$ and from $(-8, 0)$ is 10. [13.1]

39. Find an equation of the circle that passes through $(-2, -4)$, $(5, -5)$, and $(6, 2)$. [13.1], [13.4]

40. Find an equation of the ellipse with the following intercepts: $(-10, 0)$, $(10, 0)$, $(0, -1)$, and $(0, 1)$. [13.2]

41. Find the point on the *x*-axis that is equidistant from $(-3, 4)$ and $(5, 6)$. [13.1]

Test: Chapter 13

Step-by-step test solutions are found on the video CD in the back of this book.

1. Find an equation of the circle with center $(3, -4)$ and radius $2\sqrt{3}$.

Find the center and the radius of each circle.

2. $(x - 4)^2 + (y + 1)^2 = 5$

3. $x^2 + y^2 + 4x - 6y + 4 = 0$

Classify the equation as a circle, an ellipse, a parabola, or a hyperbola. Then graph.

4. $y = x^2 - 4x - 1$

5. $x^2 + y^2 + 2x + 6y + 6 = 0$

6. $\dfrac{x^2}{16} - \dfrac{y^2}{9} = 1$

7. $16x^2 + 4y^2 = 64$

8. $xy = -5$

9. $x = -y^2 + 4y$

Solve.

10. $x^2 + y^2 = 36,$
$3x + 4y = 24$

11. $x^2 - y = 3,$
$2x + y = 5$

12. $x^2 - 2y^2 = 1,$
$xy = 6$

13. $x^2 + y^2 = 10,$
$x^2 = y^2 + 2$

14. A rectangular bookmark with diagonal of length $5\sqrt{5}$ has an area of 22. Find the dimensions of the bookmark.

15. Two squares are such that the sum of their areas is 8 m^2 and the difference of their areas is 2 m^2. Find the length of a side of each square.

16. A rectangular dance floor has a diagonal of length 40 ft and a perimeter of 112 ft. Find the dimensions of the dance floor.

17. Brett invested a certain amount of money for 1 yr and earned $72 in interest. Erin invested $240 more than Brett at an interest rate that was $\frac{5}{6}$ of the rate given to Brett, but she earned the same amount of interest. Find the principal and the interest rate for Brett's investment.

Synthesis

18. Find an equation of the ellipse passing through $(6, 0)$ and $(6, 6)$ with vertices at $(1, 3)$ and $(11, 3)$.

19. Find the point on the *y*-axis that is equidistant from $(-3, -5)$ and $(4, -7)$.

20. The sum of two numbers is 36, and the product is 4. Find the sum of the reciprocals of the numbers.

21. *Theatrical production.* An E.T.C. spotlight for a college's production of *Hamlet* projects an ellipse of light on a stage that is 8 ft wide and 14 ft long. Find an equation of that ellipse if an actor is in its center and *x* represents the number of feet, horizontally, from the actor to the edge of the ellipse and *y* represents the number of feet, vertically, from the actor to the edge of the ellipse.

Cumulative Review: Chapters 1–13

Simplify.

1. $(4t^2 - 5s)^2$ [4.7]

2. $\dfrac{1}{3t} + \dfrac{1}{t-3}$ [6.4]

3. $\dfrac{x - \dfrac{1}{a}}{a - \dfrac{1}{x}}$ [6.5]

4. $\sqrt{6t}\,\sqrt{15t^3w}$ [10.3]

5. $(81a^{2/3}b^{1/4})^{3/4}$ [10.2]

6. $\log_2 \dfrac{1}{16}$ [12.3]

7. $(4 + 3i)(4 - 3i)$ [10.8]

8. $\log_m 1$ [12.4]

9. -8^{-2} [4.2]

10. $\sqrt{8} - 2\sqrt{2} + \sqrt{12}$ [10.5]

Factor.

11. $100x^2 - 60xy + 9y^2$ [5.4]

12. $3m^6 - 24$ [5.5]

13. $ax + by - ay - bx$ [5.1]

14. $32x^2 - 20x - 3$ [5.3]

Solve. Where appropriate, give an approximation to four decimal places.

15. $3(x - 5) - 4x \geq 2(x + 5)$ [9.1]

16. $16x^2 - 18x = 0$ [5.7]

17. $\dfrac{2}{x} + \dfrac{1}{x-2} = 1$ [6.6]

18. $5x^2 + 5 = 0$ [11.2]

19. $\log_x 64 = 3$ [12.6]

20. $3^x = 1.5$ [12.6]

21. $x = \sqrt{2x - 5} + 4$ [10.6]

22. $x^2 + 2y^2 = 5,$
    $2x^2 + y^2 = 7$ [13.4]

Graph.

23. $3x - y = 9$ [3.3]

24. $y = \log_5 x$ [12.3]

25. $\dfrac{x^2}{25} + \dfrac{y^2}{1} = 1$ [13.2]

26. $f(x) = 2^{x-1}$ [12.2]

27. $x^2 + (y - 3)^2 = 4$ [13.1]

28. $x < 2y + 1$ [9.4]

29. Graph: $f(x) = -(x + 2)^2 + 3$. [11.7]
    a) Label the vertex.
    b) Draw the axis of symmetry.
    c) Find the maximum or minimum value.

30. Find the domain of the function given by
    $f(x) = \sqrt{5 - 3x}$. [9.1]

31. Solve $t = \dfrac{ab}{c^2}$ for c. [11.4]

32. Find the slope–intercept equation of the line containing the points $(-3, 6)$ and $(1, 2)$. [3.7]

33. Write a quadratic equation having the solutions $\sqrt{3}$ and $-\sqrt{3}$. Answers may vary. [11.3]

34. Write an equivalent exponential equation:
    $\log_t 16 = m$. [12.4]

Solve.

35. *Aviation.* BlueAir owns two types of airplanes. One type flies 60 mph faster than the other. Laura often rents a plane from BlueAir to visit her parents. The flight takes 4 hr with the faster plane and 4 hr 24 min with the slower plane. What distance does she fly? [8.3]

36. *Aviation.* It takes Greg 21 hr longer than it takes Kyle to service a Cessna 350. Together they can service the plane in 10 hr. How long would it take each of them, working alone, to service the plane? [6.7]

37. *Employment.*   The average supermarket employee worked 32.3 hr per week in 2003. This number fell to 29.5 hr per week in 2007. [7.3]
Source: *The Wall Street Journal,* 3/8/08

   a) Find a linear function that fits the data. Let t represent the number of years since 2003.
   b) Use the function from part (a) to predict the number of hours the average supermarket employee will work in 2010.
   c) Assuming the trend continues, in what year will the average supermarket employee work 20 hr per week?

38. *Picture frames.*   The outside edge of a rectangular picture frame measures 18 in. by 11 in., and 120 in^2 of picture shows. How wide is the frame? [5.8]

39. *Population.*   The population of Latvia was 2.26 million in 2007 and was decreasing exponentially at a rate of 0.95% per year. [12.7]
Source: Based on information from *CIA World Factbook*

   a) Write an exponential function describing the population of Latvia t years after 2007.
   b) Predict what the population will be in 2025.
   c) What is the half life of the population?

40. *Geometry.*   In triangle ABC, the measure of angle B is three times the measure of angle A. The measure of angle C is 105° greater than the measure of angle A. Find the angle measures. [2.5]

41. *Art.*   Elyse is designing a rectangular tray. She wants to put a row of beads around the tray, and has enough beads to make an edge that is 32 in. long. What dimensions of the tray will give it the greatest area? [11.8]

42. *Geometry.*   In right triangle ABC, the hypotenuse is 8 cm long and one leg is 3 cm long. Find the length of the other leg. Give an exact answer and an approximation to three decimal places. [10.7]

Synthesis

43. Find a and b if the graph of
$$ax - 6y = 3x + by + 2$$
is a vertical line passing through $(-1, 0)$. [3.3]

44. Solve:
$$\begin{aligned} w - \ &x + 3y - \ z = -1, \\ 2w + \ &x + \ y + 2z = 8, \\ -w - \ &2x + 5y - \ z = -1, \\ 2w + \ &3x - 4y + \ z = 0. \end{aligned}$$ [8.4]

45. If y varies inversely as the square root of x and x is multiplied by 100, what is the effect on y? [7.5], [10.1]

46. For $f(x) = x - \dfrac{1}{x^2}$, find all x-values for which $f(x) \le 0$. [11.9]

Sequences, Series, and the Binomial Theorem

**ALEXIS SPENCER-BYERS;
LEE HARPER**
COFFEE SHOP OWNERS
Jackson, Mississippi

At our coffee house, we make many different beverages. To satisfy our customers, each beverage must include the properly measured and combined ingredients to obtain the desired flavor, look, and aroma. Meanwhile, as owners, we need to know exactly how much it costs us to make each item in order to determine our prices. Behind each tasty treat is a series of calculations for determining recipes, managing cash flow, and evaluating profit margins.

AN APPLICATION

Approximately 17 billion espresso-based coffees were sold in the United States in 2007. This number is expected to grow by 4% each year. How many espresso-based coffees will be sold from 2007 through 2015?

Source: Based on data in the *Indianapolis Star,* 11/22/07

This problem appears as Exercise 65 in Exercise Set 14.3.

The first three sections of this chapter are devoted to *sequences* and *series*. A sequence is simply an ordered list. For example, when a baseball coach writes a batting order, a sequence is being formed. When the members of a sequence are numbers, they can be added. Such a sum is called a *series*. Section 14.4 presents the *binomial theorem*, which is used to expand expressions of the form $(a + b)^n$. Such an expansion is itself a series.

14.1 Sequences and Series

Sequences ▪ Finding the General Term ▪ Sums and Series ▪ Sigma Notation

Sequences

Suppose that $10,000 is invested at 5%, compounded annually. The value of the account at the start of years 1, 2, 3, 4, and so on, is

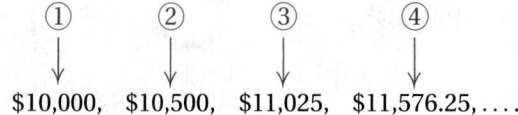

$10,000, $10,500, $11,025, $11,576.25, ...

We can regard this as a function that pairs 1 with $10,000, 2 with $10,500, 3 with $11,025, and so on. This is an example of a **sequence** (or **progression**). The domain of a sequence is a set of consecutive positive integers beginning with 1, and the range varies from sequence to sequence.

If we stop after a certain number of years, we obtain a **finite sequence**:

$10,000, $10,500, $11,025, $11,576.25.

If we continue listing the amounts in the account, we obtain an **infinite sequence**:

$10,000, $10,500, $11,025, $11,576.25, $12,155.06, ...

The three dots near the end indicate that the sequence goes on without stopping.

> ### Sequences
>
> An *infinite sequence* is a function having for its domain the set of natural numbers: $\{1, 2, 3, 4, 5, \ldots\}$.
>
> A *finite sequence* is a function having for its domain a set of natural numbers: $\{1, 2, 3, 4, 5, \ldots, n\}$, for some natural number n.

As another example, consider the sequence given by

$$a(n) = 2^n, \quad \text{or} \quad a_n = 2^n.$$

The notation a_n means $a(n)$ but is used more commonly with sequences. Some function values (also called *terms* of the sequence) follow:

$$a_1 = 2^1 = 2,$$
$$a_2 = 2^2 = 4,$$
$$a_3 = 2^3 = 8,$$
$$a_6 = 2^6 = 64.$$

The first term of the sequence is a_1, the fifth term is a_5, and the nth term, or **general term**, is a_n. This sequence can also be denoted in the following ways:

$$2, 4, 8, \ldots;$$

or $2, 4, 8, \ldots, 2^n, \ldots.$ The 2^n emphasizes that the nth term of this sequence is found by raising 2 to the nth power.

EXAMPLE **1**   Find the first four terms and the 57th term of the sequence for which the general term is given by $a_n = (-1)^n/(n + 1)$.

SOLUTION   We have

$$a_1 = \frac{(-1)^1}{1 + 1} = -\frac{1}{2}, \quad \text{Substituting in } a_n = \frac{(-1)^n}{n + 1}$$

$$a_2 = \frac{(-1)^2}{2 + 1} = \frac{1}{3},$$

$$a_3 = \frac{(-1)^3}{3 + 1} = -\frac{1}{4},$$

$$a_4 = \frac{(-1)^4}{4 + 1} = \frac{1}{5},$$

$$a_{57} = \frac{(-1)^{57}}{57 + 1} = -\frac{1}{58}.$$

Note that the expression $(-1)^n$ causes the signs of the terms to alternate between positive and negative, depending on whether n is even or odd.

TRY EXERCISE 17

Finding the General Term

By looking for a pattern, we can often write an expression for the general term of a sequence. When only a few terms are given, more than one pattern may fit.

EXAMPLE **2**   For each sequence, predict the general term.

a) $1, 4, 9, 16, 25, \ldots$

b) $2, 4, 8, \ldots$

c) $-1, 2, -4, 8, -16, \ldots$

TECHNOLOGY CONNECTION

Sequences are entered and graphed much like functions. The difference is that the SEQUENCE MODE must be selected. You can then enter U_n or V_n using n as the variable. Use this approach to check Example 1 with a table of values for the sequence.

SOLUTION

a) $1, 4, 9, 16, 25 \ldots$

These are squares of consecutive positive integers, so the general term could be n^2.

b) $2, 4, 8, \ldots$

We regard the pattern as powers of 2, in which case 16 would be the next term and 2^n the general term. The sequence could then be written with more terms as

$$2, 4, 8, 16, 32, 64, 128, \ldots.$$

c) $-1, 2, -4, 8, -16, \ldots$

These are powers of 2 with alternating signs, so the general term may be

Making sure the signs of the terms alternate
$$(-1)^n [2^{n-1}].$$
Raising 2 to a power that is 1 less than the term's position

To check, note that -4 is the third term, and $(-1)^3 [2^{3-1}] = -1 \cdot 2^2 = -4$.

> **TRY EXERCISE** 29

In part (b) above, suppose that the second term is found by adding 2, the third term by adding 4, the next term by adding 6, and so on. In this case, 14 would be the next term and the sequence would be

$$2, 4, 8, 14, 22, 32, 44, 58, \ldots.$$

This illustrates that the fewer terms we are given, the greater the uncertainty about determining the nth term.

Sums and Series

> **Series**
>
> Given the infinite sequence
>
> $$a_1, a_2, a_3, a_4, \ldots, a_n, \ldots,$$
>
> the sum of the terms
>
> $$a_1 + a_2 + a_3 + \cdots + a_n + \cdots$$
>
> is called an *infinite series* and is denoted S_∞. A *partial sum* is the sum of the first n terms:
>
> $$a_1 + a_2 + a_3 + \cdots + a_n.$$
>
> A partial sum is also called a *finite series* and is denoted S_n.

EXAMPLE 3

For the sequence $-2, 4, -6, 8, -10, 12, -14$, find: **(a)** S_2; **(b)** S_3; **(c)** S_7.

SOLUTION

a) $S_2 = -2 + 4 = 2$ This is the sum of the first 2 terms.

b) $S_3 = -2 + 4 + (-6) = -4$ This is the sum of the first 3 terms.

c) $S_7 = -2 + 4 + (-6) + 8 + (-10) + 12 + (-14) = -8$ This is the sum of the first 7 terms.

> **TRY EXERCISE** 45

Sigma Notation

When the general term of a sequence is known, the Greek letter Σ (upper-case sigma) can be used to write a series. For example, the sum of the first four terms of the sequence $3, 5, 7, 9, 11, \ldots, 2k + 1, \ldots$ can be named as follows, using *sigma notation*, or *summation notation*:

$$\sum_{k=1}^{4} (2k + 1).$$

This represents
$(2 \cdot 1 + 1) + (2 \cdot 2 + 1) + (2 \cdot 3 + 1) + (2 \cdot 4 + 1).$

This is read "the sum as k goes from 1 to 4 of $(2k + 1)$." The letter k is called the *index of summation*. The index need not always start at 1.

EXAMPLE **4**   Write out and evaluate each sum.

a) $\displaystyle\sum_{k=1}^{5} k^2$ **b)** $\displaystyle\sum_{k=4}^{6} (-1)^k (2k)$ **c)** $\displaystyle\sum_{k=0}^{3} (2^k + 5)$

STUDENT NOTES

A great deal of information is condensed into sigma notation. Be careful to pay attention to what values the index of summation will take on. Evaluate the expression following sigma, the general term, for each value and then add the results.

SOLUTION

a) $\displaystyle\sum_{k=1}^{5} k^2 = 1^2 + 2^2 + 3^2 + 4^2 + 5^2 = 1 + 4 + 9 + 16 + 25 = 55$

Evaluate k^2 for all integers from 1 through 5. Then add.

b) $\displaystyle\sum_{k=4}^{6} (-1)^k (2k) = (-1)^4 (2 \cdot 4) + (-1)^5 (2 \cdot 5) + (-1)^6 (2 \cdot 6)$

$= 8 - 10 + 12 = 10$

c) $\displaystyle\sum_{k=0}^{3} (2^k + 5) = (2^0 + 5) + (2^1 + 5) + (2^2 + 5) + (2^3 + 5)$

$= 6 + 7 + 9 + 13 = 35$

TRY EXERCISE 49

EXAMPLE **5**   Write sigma notation for each sum.

a) $1 + 4 + 9 + 16 + 25$

b) $3 + 9 + 27 + 81 + \cdots$

c) $-1 + 3 - 5 + 7$

SOLUTION

a) $1 + 4 + 9 + 16 + 25$

Note that this is a sum of squares, $1^2 + 2^2 + 3^2 + 4^2 + 5^2$, so the general term is k^2. Sigma notation is

$$\sum_{k=1}^{5} k^2.$$ The sum starts with 1^2 and ends with 5^2.

Answers can vary here. For example, another—perhaps less obvious—way of writing $1 + 4 + 9 + 16 + 25$ is

$$\sum_{k=2}^{6} (k - 1)^2.$$

b) $3 + 9 + 27 + 81 + \cdots$

This is a sum of powers of 3, and it is also an infinite series. We use the symbol ∞ for infinity and write the series using sigma notation:

$$\sum_{k=1}^{\infty} 3^k.$$

c) $-1 + 3 - 5 + 7$

Except for the alternating signs, this is the sum of the first four positive odd numbers. It is useful to remember that $2k - 1$ is a formula for the kth positive odd number. It is also important to remember that the factor $(-1)^k$ can be used to create the alternating signs. The general term is thus $(-1)^k(2k - 1)$, beginning with $k = 1$. Sigma notation is

$$\sum_{k=1}^{4} (-1)^k(2k - 1).$$

To check, we can evaluate $(-1)^k(2k - 1)$ using 1, 2, 3, and 4. Then we can write the sum of the four terms. We leave this to the student.

TRY EXERCISE ▶ 61

14.1 EXERCISE SET

🐦 **Concept Reinforcement**   *In each of Exercises 1–6, match the expression with the most appropriate expression from the column on the right.*

1. _____ $\displaystyle\sum_{k=1}^{4} k^2$

2. _____ $\displaystyle\sum_{k=3}^{6} (-1)^k$

3. _____ $5 + 10 + 15 + 20$

4. _____ $a_n = 5^n$

5. _____ $a_n = 3n + 2$

6. _____ $a_1 + a_2 + a_3$

a) $-1 + 1 + (-1) + 1$

b) $a_2 = 25$

c) $a_2 = 8$

d) $\displaystyle\sum_{k=1}^{4} 5k$

e) S_3

f) $1 + 4 + 9 + 16$

Find the indicated term of each sequence.

7. $a_n = 5n + 3;\ a_8$

8. $a_n = 3n - 4;\ a_8$

9. $a_n = (3n + 1)(2n - 5);\ a_9$

10. $a_n = (3n + 2)^2;\ a_6$

11. $a_n = (-1)^{n-1}(3.4n - 17.3);\ a_{12}$

 **12.** $a_n = (-2)^{n-2}(45.68 - 1.2n);\ a_{23}$

13. $a_n = 3n^2(9n - 100);\ a_{11}$

14. $a_n = 4n^2(2n - 39);\ a_{22}$

15. $a_n = \left(1 + \dfrac{1}{n}\right)^2;\ a_{20}$

16. $a_n = \left(1 - \dfrac{1}{n}\right)^3;\ a_{15}$

In each of the following, the nth term of a sequence is given. Find the first 4 terms; the 10th term, a_{10}; and the 15th term, a_{15}, of the sequence.

17. $a_n = 3n - 1$ **18.** $a_n = 2n + 1$

19. $a_n = n^2 + 2$ **20.** $a_n = n^2 - 2n$

21. $a_n = \dfrac{n}{n + 1}$ **22.** $a_n = \dfrac{n^2 - 1}{n^2 + 1}$

23. $a_n = \left(-\dfrac{1}{2}\right)^{n-1}$ **24.** $a_n = (-2)^{n+1}$

25. $a_n = (-1)^n/n$ **26.** $a_n = (-1)^n n^2$

27. $a_n = (-1)^n(n^3 - 1)$

28. $a_n = (-1)^{n+1}(3n - 5)$

Look for a pattern and then write an expression for the general term, or nth term, a_n, of each sequence. Answers may vary.

29. $2, 4, 6, 8, 10, \ldots$

30. $1, 3, 5, 7, \ldots$

31. $-1, 1, -1, 1, \ldots$

32. $1, -1, 1, -1, \ldots$

33. $1, -2, 3, -4, \ldots$

34. $-1, 2, -3, 4, \ldots$

35. $3, 5, 7, 9, \ldots$

36. $4, 6, 8, 10, \ldots$

37. $0, 3, 8, 15, 24, \ldots$

38. $2, 6, 12, 20, 30, \ldots$

39. $\frac{1}{2}, \frac{2}{3}, \frac{3}{4}, \frac{4}{5}, \frac{5}{6}, \ldots$

40. $1 \cdot 3, 2 \cdot 4, 3 \cdot 5, 4 \cdot 6, \ldots$

41. $0.1, 0.01, 0.001, 0.0001, \ldots$

42. $\frac{1}{2}, \frac{1}{4}, \frac{1}{8}, \frac{1}{16}, \ldots$

43. $-1, 4, -9, 16, \ldots$

44. $1, -4, 9, -16, \ldots$

Find the indicated partial sum for each sequence.

45. $-1, 2, -3, 4, -5, 6, \ldots;\ S_{10}$

46. $2, -4, 6, -8, 10, -12, \ldots;\ S_{10}$

47. $1, \frac{1}{10}, \frac{1}{100}, \frac{1}{1000}, \ldots;\ S_6$

48. $3, 6, 9, 12, 15, \ldots;\ S_6$

Write out and evaluate each sum.

49. $\displaystyle\sum_{k=1}^{5} \frac{1}{2k}$

50. $\displaystyle\sum_{k=1}^{6} \frac{1}{2k - 1}$

51. $\displaystyle\sum_{k=0}^{4} 10^k$

52. $\displaystyle\sum_{k=2}^{6} \sqrt{5k - 1}$

53. $\displaystyle\sum_{k=2}^{8} \frac{k}{k - 1}$

54. $\displaystyle\sum_{k=2}^{5} \frac{k - 1}{k + 1}$

55. $\displaystyle\sum_{k=1}^{8} (-1)^{k+1} 2^k$

56. $\displaystyle\sum_{k=1}^{7} (-1)^k 4^{k+1}$

57. $\displaystyle\sum_{k=0}^{5} (k^2 - 2k + 3)$

58. $\displaystyle\sum_{k=0}^{5} (k^2 - 3k + 4)$

59. $\displaystyle\sum_{k=3}^{5} \frac{(-1)^k}{k(k + 1)}$

60. $\displaystyle\sum_{k=3}^{7} \frac{k}{2^k}$

Rewrite each sum using sigma notation. Answers may vary.

61. $\dfrac{2}{3} + \dfrac{3}{4} + \dfrac{4}{5} + \dfrac{5}{6} + \dfrac{6}{7}$

62. $\dfrac{1}{1^2} + \dfrac{1}{2^2} + \dfrac{1}{3^2} + \dfrac{1}{4^2} + \dfrac{1}{5^2}$

63. $1 + 4 + 9 + 16 + 25 + 36$

64. $1 + \sqrt{2} + \sqrt{3} + 2 + \sqrt{5} + \sqrt{6}$

65. $4 - 9 + 16 - 25 + \cdots + (-1)^n n^2$

66. $9 - 16 + 25 + \cdots + (-1)^{n+1} n^2$

67. $6 + 12 + 18 + 24 + \cdots$

68. $11 + 22 + 33 + 44 + \cdots$

69. $\dfrac{1}{1 \cdot 2} + \dfrac{1}{2 \cdot 3} + \dfrac{1}{3 \cdot 4} + \dfrac{1}{4 \cdot 5} + \cdots$

70. $\dfrac{1}{1 \cdot 2^2} + \dfrac{1}{2 \cdot 3^2} + \dfrac{1}{3 \cdot 4^2} + \dfrac{1}{4 \cdot 5^2} + \cdots$

71. The sequence $1, 4, 9, 16, \ldots$ can be written as $f(x) = x^2$ with the domain the set of all positive integers. Explain how the graph of f would compare with the graph of $y = x^2$.

72. Consider the sums

$$\sum_{k=1}^{5} 3k^2 \quad \text{and} \quad 3\sum_{k=1}^{5} k^2.$$

a) Which is easier to evaluate and why?
b) Is it true that

$$\sum_{k=1}^{n} ca_k = c\sum_{k=1}^{n} a_k?$$

Why or why not?

Skill Review

To prepare for Section 14.2, review evaluating expressions and simplifying expressions (Section 1.8).

Evaluate. [1.8]

73. $\dfrac{7}{2}(a_1 + a_7)$, for $a_1 = 8$ and $a_7 = 20$

74. $a_1 + (n - 1)d$, for $a_1 = 3$, $n = 10$, and $d = -2$

Simplify. [1.8]

75. $(a_1 + 3d) + d$

76. $(a_1 + 5d) + (a_n - 5d)$

77. $(a_1 + a_n) + (a_1 + a_n) + (a_1 + a_n)$

78. $(a_1 + 8d) - (a_1 + 7d)$

Synthesis

79. Explain why the equation

$$\sum_{k=1}^{n} (a_k + b_k) = \sum_{k=1}^{n} a_k + \sum_{k=1}^{n} b_k$$

is true for any positive integer n. What laws are used to justify this result?

80. Can a finite series be formed from an infinite sequence? Can an infinite series be formed from a finite sequence? Why or why not?

Some sequences are given by a recursive *definition. The value of the first term, a_1, is given, and then we are told how to find any subsequent term from the term preceding it. Find the first six terms of each of the following recursively defined sequences.*

81. $a_1 = 1, a_{n+1} = 5a_n - 2$

82. $a_1 = 0, a_{n+1} = (a_n)^2 + 3$

83. *Value of a projector.* The value of an LCD projector is $2500. Its scrap value each year is 80% of its value the year before. Write a sequence listing the scrap value of the machine at the start of each year for a 10-yr period.

84. *Cell biology.* A single cell of bacterium divides into two every 15 min. Suppose that the same rate of division is maintained for 4 hr. Write a sequence listing the number of cells after successive 15-min periods.

85. Find S_{100} and S_{101} for the sequence in which $a_n = (-1)^n$.

Find the first five terms of each sequence; then find S_5.

86. $a_n = \dfrac{1}{2^n} \log 1000^n$

87. $a_n = i^n, i = \sqrt{-1}$

88. Find all values for x that solve the following:

$$\sum_{k=1}^{x} i^k = -1.$$

89. The nth term of a sequence is given by

$$a_n = n^5 - 14n^4 + 6n^3 + 416n^2 - 655n - 1050.$$

Use a graphing calculator with a TABLE feature to determine which term in the sequence is 6144.

90. To define a sequence recursively on a graphing calculator (see Exercises 81 and 82), we use the SEQ MODE. The general term U_n or V_n can often be expressed in terms of U_{n-1} or V_{n-1} by pressing **2ND** **7** or **2ND** **8**. The starting values of U_n, V_n, and n are set as one of the WINDOW variables.

Use recursion to determine how many different handshakes occur when 50 people shake hands with one another. To develop the recursion formula, begin with a group of 2 and determine how many additional handshakes occur with the arrival of each new person.

14.2   Arithmetic Sequences and Series

Arithmetic Sequences   ▪   Sum of the First n Terms of an Arithmetic Sequence   ▪   Problem Solving

In this section, we concentrate on sequences and series that are said to be arithmetic (pronounced ar-ith-MET-ik).

Arithmetic Sequences

In an **arithmetic sequence** (or **progression**), adding the same number to any term gives the next term in the sequence. For example, the sequence 2, 5, 8, 11, 14, 17, ... is arithmetic because adding 3 to any term produces the next term.

> ### Arithmetic Sequence
> A sequence is *arithmetic* if there exists a number d, called the *common difference*, such that $a_{n+1} = a_n + d$ for any integer $n \geq 1$.

EXAMPLE 1   For each arithmetic sequence, identify the first term, a_1, and the common difference, d.

a) $4, 9, 14, 19, 24, \ldots$ **b)** $27, 20, 13, 6, -1, -8, \ldots$

SOLUTION   To find a_1, we simply use the first term listed. To find d, we choose any term other than a_1 and subtract the preceding term from it.

Sequence	First Term, a_1	Common Difference, d
a) $4, 9, 14, 19, 24, \ldots$	4	$5 \leftarrow 9 - 4 = 5$
b) $27, 20, 13, 6, -1, -8, \ldots$	27	$-7 \leftarrow 20 - 27 = -7$

To find the common difference, we subtracted a_1 from a_2. Had we subtracted a_2 from a_3 or a_3 from a_4, we would have found the same values for d.

Check: As a check, note that when d is added to each term, the result is the next term in the sequence.

> TRY EXERCISE 11

To develop a formula for the general, nth, term of any arithmetic sequence, we denote the common difference by d and write out the first few terms:

$a_1,$

$a_2 = a_1 + d,$

$a_3 = a_2 + d = (a_1 + d) + d = a_1 + 2d,$ Substituting $a_1 + d$ for a_2

$a_4 = a_3 + d = (a_1 + 2d) + d = a_1 + 3d.$ Substituting $a_1 + 2d$ for a_3

Note that the coefficient of d in each case is 1 less than the subscript.

Generalizing, we obtain the following formula.

> ### To Find a_n for an Arithmetic Sequence
> The nth term of an arithmetic sequence with common difference d is
> $$a_n = a_1 + (n - 1)d, \quad \text{for any integer } n \geq 1.$$

EXAMPLE 2   Find the 14th term of the arithmetic sequence $6, 9, 12, 15, \ldots$.

SOLUTION   First we note that $a_1 = 6$, $d = 3$, and $n = 14$. Using the formula for the nth term of an arithmetic sequence, we have

$$a_n = a_1 + (n - 1)d$$
$$a_{14} = 6 + (14 - 1) \cdot 3 = 6 + 13 \cdot 3 = 6 + 39 = 45.$$

The 14th term is 45.

> TRY EXERCISE 17

EXAMPLE **3**

For the sequence in Example 2, which term is 300? That is, find n if $a_n = 300$.

SOLUTION We substitute into the formula for the nth term of an arithmetic sequence and solve for n:

$$a_n = a_1 + (n - 1)d$$
$$300 = 6 + (n - 1) \cdot 3$$
$$300 = 6 + 3n - 3$$
$$297 = 3n$$
$$99 = n.$$

The term 300 is the 99th term of the sequence.

> **TRY EXERCISE** 23

Given two terms and their places in an arithmetic sequence, we can construct the sequence.

EXAMPLE **4**

The 3rd term of an arithmetic sequence is 14, and the 16th term is 79. Find a_1 and d and construct the sequence.

SOLUTION We know that $a_3 = 14$ and $a_{16} = 79$. Thus we would have to add d a total of 13 times to get from 14 to 79. That is,

$$14 + 13d = 79. \qquad a_3 \text{ and } a_{16} \text{ are 13 terms apart; } 16 - 3 = 13$$

Solving $14 + 13d = 79$, we obtain

$$13d = 65 \qquad \text{Subtracting 14 from both sides}$$
$$d = 5. \qquad \text{Dividing both sides by 13}$$

We subtract d twice from a_3 to get to a_1. Thus,

$$a_1 = 14 - 2 \cdot 5 = 4. \qquad a_1 \text{ and } a_3 \text{ are 2 terms apart; } 3 - 1 = 2$$

The sequence is 4, 9, 14, 19, ... Note that we could have subtracted d a total of 15 times from a_{16} in order to find a_1.

> **TRY EXERCISE** 33

In general, d should be subtracted $(n - 1)$ times from a_n in order to find a_1.

Sum of the First n Terms of an Arithmetic Sequence

When the terms of an arithmetic sequence are added, an **arithmetic series** is formed. To develop a formula for computing S_n when the series is arithmetic, we list the first n terms of the sequence as follows:

This is the next-to-last term. If you add d to this term, the result is a_n.

$$a_1, (a_1 + d), (a_1 + 2d), \ldots, (a_n - 2d), \overbrace{(a_n - d)}, a_n$$

This term is two terms back from the end. If you add d to this term, you get the next-to-last term, $a_n - d$.

Thus, S_n is given by

$$S_n = a_1 + (a_1 + d) + (a_1 + 2d) + \cdots + (a_n - 2d) + (a_n - d) + a_n.$$

Using a commutative law, we have a second equation:

$$S_n = a_n + (a_n - d) + (a_n - 2d) + \cdots + (a_1 + 2d) + (a_1 + d) + a_1.$$

Adding corresponding terms on each side of the two equations above, we get

$$2S_n = [a_1 + a_n] + [(a_1 + d) + (a_n - d)] + [(a_1 + 2d) + (a_n - 2d)]$$
$$+ \cdots + [(a_n - 2d) + (a_1 + 2d)] + [(a_n - d) + (a_1 + d)] + [a_n + a_1].$$

This simplifies to

$$2S_n = [a_1 + a_n] + [a_1 + a_n] + [a_1 + a_n]$$
$$+ \cdots + [a_n + a_1] + [a_n + a_1] + [a_n + a_1].$$ There are n bracketed sums.

Since $[a_1 + a_n]$ is being added n times, it follows that

$$2S_n = n[a_1 + a_n].$$

Dividing both sides by 2 leads to the following formula.

> ## To Find S_n for an Arithmetic Sequence
> The sum of the first n terms of an arithmetic sequence is given by
>
> $$S_n = \frac{n}{2}(a_1 + a_n).$$

STUDENT NOTES

The formula for the sum of an arithmetic sequence is very useful, but remember that it does not work for sequences that are not arithmetic.

EXAMPLE 5   Find the sum of the first 100 positive even numbers.

SOLUTION   The sum is

$$2 + 4 + 6 + \cdots + 198 + 200.$$

This is the sum of the first 100 terms of the arithmetic sequence for which

$$a_1 = 2, \quad n = 100, \quad \text{and} \quad a_n = 200.$$

Substituting in the formula

$$S_n = \frac{n}{2}(a_1 + a_n),$$

we get

$$S_{100} = \frac{100}{2}(2 + 200) = 50(202) = 10,100.$$ ▶ **TRY EXERCISE** 39

The above formula is useful when we know the first and last terms, a_1 and a_n. To find S_n when a_n is unknown, but a_1, n, and d are known, we can use $a_n = a_1 + (n - 1)d$ to calculate a_n and then proceed as in Example 5.

EXAMPLE 6   Find the sum of the first 15 terms of the arithmetic sequence 4, 7, 10, 13, ...

SOLUTION   Note that

$$a_1 = 4, \quad n = 15, \quad \text{and} \quad d = 3.$$

Before using the formula for S_n, we find a_{15}:

$$a_{15} = 4 + (15 - 1)3 \quad \text{Substituting into the formula for } a_n$$
$$= 4 + 14 \cdot 3 = 46.$$

Thus, knowing that $a_{15} = 46$, we have

$$S_{15} = \tfrac{15}{2}(4 + 46) \quad \text{Using the formula for } S_n$$
$$= \tfrac{15}{2}(50) = 375.$$ ▶ **TRY EXERCISE** 37

Problem Solving

In problem-solving situations, translation may involve sequences or series. As always, there is often a variety of ways in which a problem can be solved. You should use the approach that is best or easiest for you. In this chapter, however, we will try to emphasize sequences and series and their related formulas.

EXAMPLE 7

Hourly wages. Chris accepts a job managing a music store, starting with an hourly wage of $14.60, and is promised a raise of 25¢ per hour every 2 months for 5 years. After 5 years of work, what will be Chris's hourly wage?

SOLUTION

1. **Familiarize.** It helps to write down the hourly wage for several two-month time periods.

   Beginning: 14.60,
   After two months: 14.85,
   After four months: 15.10,

   and so on.

   What appears is a sequence of numbers: 14.60, 14.85, 15.10, ... Since the same amount is added each time, the sequence is arithmetic.

   We list what we know about arithmetic sequences. The pertinent formulas are

   $$a_n = a_1 + (n - 1)d$$

   and

   $$S_n = \frac{n}{2}(a_1 + a_n).$$

   In this case, we are not looking for a sum, so it is probably the first formula that will give us our answer. We want to determine the last term in a sequence. To do so, we need to know a_1, n, and d. From our list above, we see that

   $$a_1 = 14.60 \quad \text{and} \quad d = 0.25.$$

   What is n? That is, how many terms are in the sequence? After 1 year, there have been 6 raises, since Chris gets a raise every 2 months. There are 5 years, so the total number of raises will be $5 \cdot 6$, or 30. Altogether, there will be 31 terms: the original wage and 30 increased rates.

2. **Translate.** We want to find a_n for the arithmetic sequence in which $a_1 = 14.60$, $n = 31$, and $d = 0.25$.

3. **Carry out.** Substituting in the formula for a_n gives us

   $$a_{31} = 14.60 + (31 - 1) \cdot 0.25$$
   $$= 22.10.$$

4. **Check.** We can check by redoing the calculations or we can calculate in a slightly different way for another check. For example, at the end of a year, there will be 6 raises, for a total raise of $1.50. At the end of 5 years, the total raise will be $5 \times \$1.50$, or $7.50. If we add that to the original wage of $14.60, we obtain $22.10. The answer checks.

5. **State.** After 5 years, Chris's hourly wage will be $22.10.

> TRY EXERCISE 47

EXAMPLE **8**   *Telephone pole storage.*   A stack of telephone poles has 30 poles in the bottom row. There are 29 poles in the second row, 28 in the next row, and so on. How many poles are in the stack if there are 5 poles in the top row?

SOLUTION

1. **Familiarize.**   The following figure shows the ends of the poles. There are 30 poles on the bottom and one fewer in each successive row. How many rows will there be?

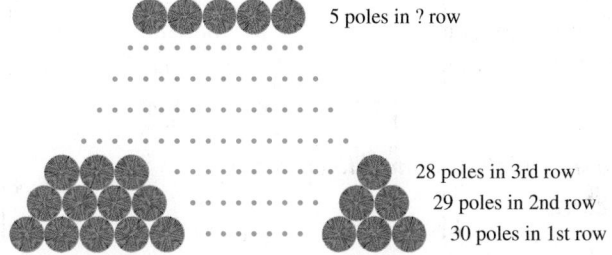

5 poles in ? row

28 poles in 3rd row
29 poles in 2nd row
30 poles in 1st row

   Note that there are $30 - 1 = 29$ poles in the 2nd row, $30 - 2 = 28$ poles in the 3rd row, $30 - 3 = 27$ poles in the 4th row, and so on. The pattern leads to $30 - 25 = 5$ poles in the 26th row.

   The situation is represented by the equation

   $$30 + 29 + 28 + \cdots + 5.$$ There are 26 terms in this series.

   Thus we have an arithmetic series. We recall the formula

   $$S_n = \frac{n}{2}(a_1 + a_n).$$

2. **Translate.**   We want to find the sum of the first 26 terms of an arithmetic sequence in which $a_1 = 30$ and $a_{26} = 5$.

3. **Carry out.**   Substituting into the above formula gives us

   $$S_{26} = \frac{26}{2}(30 + 5)$$
   $$= 13 \cdot 35 = 455.$$

4. **Check.**   In this case, we can check the calculations by doing them again. A longer, more difficult way would be to do the entire addition:

   $$30 + 29 + 28 + \cdots + 5.$$

5. **State.**   There are 455 poles in the stack.

TRY EXERCISE 49

14.2   EXERCISE SET

For Extra Help

👉 **Concept Reinforcement** *Classify each statement as either true or false.*

1. In an arithmetic sequence, the difference between any two consecutive terms is always the same.

2. In an arithmetic sequence, if $a_9 - a_8 = 4$, then $a_{13} - a_{12} = 4$ as well.

3. In an arithmetic sequence containing 17 terms, the common difference is $a_{17} - a_1$.

4. To find a_{20} in an arithmetic sequence, add the common difference to a_1 a total of 20 times.

5. The sum of the first 20 terms of an arithmetic sequence can be found by knowing just a_1 and a_{20}

6. The sum of the first 30 terms of an arithmetic sequence can be found by knowing just a_1 and d, the common difference.

7. The notation S_5 means $a_1 + a_5$.

8. For any arithmetic sequence, $S_9 = S_8 + d$, where d is the common difference.

Find the first term and the common difference.

9. $8, 13, 18, 23, \ldots$

10. $2.5, 3, 3.5, 4, \ldots$

11. $7, 3, -1, -5, \ldots$

12. $-8, -5, -2, 1, \ldots$

13. $\frac{3}{2}, \frac{9}{4}, 3, \frac{15}{4}, \ldots$

14. $\frac{3}{5}, \frac{1}{10}, -\frac{2}{5}, \ldots$

15. $\$8.16, \$8.46, \$8.76, \$9.06, \ldots$

16. $\$825, \$804, \$783, \$762, \ldots$

17. Find the 19th term of the arithmetic sequence $10, 18, 26, \ldots$.

18. Find the 23rd term of the arithmetic sequence $10, 16, 22, \ldots$.

19. Find the 18th term of the arithmetic sequence $8, 2, -4, \ldots$.

20. Find the 14th term of the arithmetic sequence $3, \frac{7}{3}, \frac{5}{3}, \ldots$.

21. Find the 13th term of the arithmetic sequence $\$1200, \$964.32, \$728.64, \ldots$.

22. Find the 10th term of the arithmetic sequence $\$2345.78, \$2967.54, \$3589.30, \ldots$.

23. In the sequence of Exercise 17, what term is 210?

24. In the sequence of Exercise 18, what term is 208?

25. In the sequence of Exercise 19, what term is -328?

26. In the sequence of Exercise 20, what term is -27?

27. Find a_{18} when $a_1 = 8$ and $d = 10$.

28. Find a_{20} when $a_1 = 12$ and $d = -5$.

29. Find a_1 when $d = 4$ and $a_8 = 33$.

30. Find a_1 when $d = 8$ and $a_{11} = 26$.

31. Find n when $a_1 = 5$, $d = -3$, and $a_n = -76$.

32. Find n when $a_1 = 25$, $d = -14$, and $a_n = -507$.

33. For an arithmetic sequence in which $a_{17} = -40$ and $a_{28} = -73$, find a_1 and d. Write the first five terms of the sequence.

34. In an arithmetic sequence, $a_{17} = \frac{25}{3}$ and $a_{32} = \frac{95}{6}$. Find a_1 and d. Write the first five terms of the sequence.

Aha!   35. Find a_1 and d if $a_{13} = 13$ and $a_{54} = 54$.

36. Find a_1 and d if $a_{12} = 24$ and $a_{25} = 50$.

37. Find the sum of the first 20 terms of the arithmetic series $1 + 5 + 9 + 13 + \cdots$.

38. Find the sum of the first 14 terms of the arithmetic series $11 + 7 + 3 + \cdots$.

39. Find the sum of the first 250 natural numbers.

40. Find the sum of the first 400 natural numbers.

41. Find the sum of the even numbers from 2 to 100, inclusive.

42. Find the sum of the odd numbers from 1 to 99, inclusive.

43. Find the sum of all multiples of 6 from 6 to 102, inclusive.

44. Find the sum of all multiples of 4 that are between 15 and 521.

45. An arithmetic series has $a_1 = 4$ and $d = 5$. Find S_{20}.

46. An arithmetic series has $a_1 = 9$ and $d = -3$. Find S_{32}.

Solve.

47. *Band formations.* The South Brighton Drum and Bugle Corps has 7 musicians in the front row, 9 in the second row, 11 in the third row, and so on, for 15 rows. How many musicians are in the last row? How many musicians are there altogether?

48. *Gardening.* A gardener is planting tulip bulbs at the entrance to a college. She puts 50 bulbs in the first row, 46 in the second row, 42 in the third row, and so on, for 13 rows. How many bulbs will be in the last row? How many bulbs will she plant altogether?

49. *Archaeology.* Many ancient Mayan pyramids were constructed over a span of several generations. Each layer of the pyramid has a stone perimeter, enclosing a layer of dirt or debris on which a structure once stood. One drawing of such a pyramid indicates that the perimeter of the bottom layer contains 36 stones, the next level up contains 32 stones, and so on, up to the top row, which contains 4 stones. How many stones are in the pyramid?

50. *Telephone pole piles.* How many poles will be in a pile of telephone poles if there are 50 in the first layer, 49 in the second, and so on, until there are 6 in the top layer?

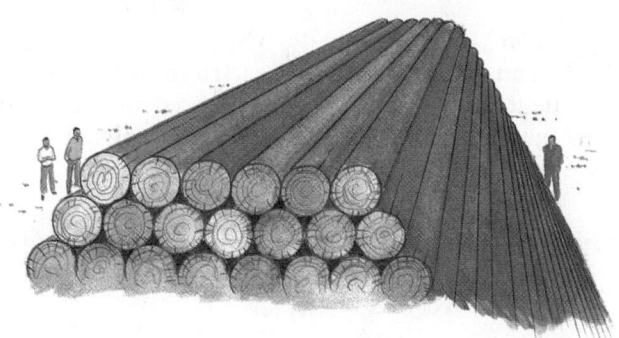

51. *Accumulated savings.* If 10¢ is saved on October 1, another 20¢ on October 2, another 30¢ on October 3, and so on, how much is saved during October? (October has 31 days.)

52. *Accumulated savings.* Carrie saves money in an arithmetic sequence: $700 for the first year, another $850 the second, and so on, for 20 yr. How much does she save in all (disregarding interest)?

53. *Auditorium design.* Theaters are often built with more seats per row as the rows move toward the back. The Community Theater has 20 seats in the first row, 22 in the second, 24 in the third, and so on, for 16 rows. How many seats are in the theater?

54. *Accumulated savings.* Shirley sets up an investment so that it yields $5000 the first year, $6125 the second year, $7250 the third year, and so on, for 25 yr. What is the total yield from the investment?

55. It is said that as a young child, the mathematician Karl F. Gauss (1777–1855) was able to compute the sum $1 + 2 + 3 + \cdots + 100$ very quickly in his head. Explain how Gauss might have done this and present a formula for the sum of the first n natural numbers. (*Hint*: $1 + 99 = 100$.)

56. Write a problem for a classmate to solve. Devise the problem so that its solution requires computing S_{17} for an arithmetic sequence.

Skill Review

Review finding equations.

Find an equation of the line satisfying the given conditions.

57. Slope $\frac{1}{3}$, y-intercept $(0, 10)$ [3.6]

58. Containing the points $(2, 3)$ and $(4, -5)$ [3.7]

59. Containing the point $(5, 0)$ and parallel to the line given by $2x + y = 8$ [3.7]

60. Containing the point $(-1, -4)$ and perpendicular to the line given by $3x - 4y = 7$ [3.7]

Find an equation of the circle satisfying the given conditions. [13.1]

61. Center $(0, 0)$, radius 4

62. Center $(-2, 1)$, radius $2\sqrt{5}$

Synthesis

63. When every term in an arithmetic sequence is an integer, S_n must also be an integer. Given that n, a_1, and a_n may each, at times, be even or odd, explain why $\frac{n}{2}(a_1 + a_n)$ is always an integer.

64. The sum of the first n terms of an arithmetic sequence is also given by

$$S_n = \frac{n}{2}\left[2a_1 + (n - 1)d\right].$$

Use the earlier formulas for a_n and S_n to explain how this equation was developed.

65. A frog is at the bottom of a 100-ft well. With each jump, the frog climbs 4 ft, but then slips back 1 ft. How many jumps does it take for the frog to reach the top of the hole?

66. Find a formula for the sum of the first n consecutive odd numbers starting with 1:

$$1 + 3 + 5 + \cdots + (2n - 1).$$

67. Prove that if p, m, and q are consecutive terms in an arithmetic sequence, then

$$m = \frac{p + q}{2}.$$

68. *Straight-line depreciation.* A company buys a color laser printer for $5200 on January 1 of a given year. The machine is expected to last for 8 yr, at the end of which time its *trade-in*, or *salvage*, *value* will be $1100. If the company figures the decline in value to be the same each year, then the trade-in values, after t years, $0 \le t \le 8$, form an arithmetic sequence given by

$$a_t = C - t\left(\frac{C - S}{N}\right),$$

where C is the original cost of the item, N the years of expected life, and S the salvage value.

a) Find the formula for a_t for the straight-line depreciation of the printer.

b) Find the trade-in value after 0 yr, 1 yr, 2 yr, 3 yr, 4 yr, 7 yr, and 8 yr.

c) Find a formula that expresses a_t recursively.

69. Use your answer to Exercise 39 to find the sum of all integers from 501 through 750.

14.3 Geometric Sequences and Series

Geometric Sequences ▪ Sum of the First n Terms of a Geometric Sequence ▪
Infinite Geometric Series ▪ Problem Solving

In an arithmetic sequence, a certain number is added to each term to get the next term. When each term in a sequence is *multiplied* by a certain fixed number to get the next term, the sequence is **geometric**. In this section, we examine both geometric sequences (or progressions) and geometric series.

Geometric Sequences

Consider the sequence

$$2, 6, 18, 54, 162, \ldots$$

If we multiply each term by 3, we obtain the next term. The multiplier is called the *common ratio* because it is found by dividing any term by the preceding term.

> ### Geometric Sequence
>
> A sequence is *geometric* if there exists a number r, called the *common ratio*, for which
>
> $$\frac{a_{n+1}}{a_n} = r, \quad \text{or} \quad a_{n+1} = a_n \cdot r \quad \text{for any integer } n \geq 1.$$

EXAMPLE **1**   For each geometric sequence, find the common ratio.

a) 4, 20, 100, 500, 2500, ...

b) 3, −6, 12, −24, 48, −96, ...

c) \$5200, \$3900, \$2925, \$2193.75, ...

SOLUTION

Sequence		*Common Ratio*	
a) 4, 20, 100, 500, 2500, ...	5	$\frac{20}{4} = 5, \frac{100}{20} = 5$, and so on	
b) 3, −6, 12, −24, 48, −96, ...	−2	$\frac{-6}{3} = -2, \frac{12}{-6} = -2$, and so on	
c) \$5200, \$3900, \$2925, \$2193.75, ...	0.75	$\frac{\$3900}{\$5200} = 0.75, \frac{\$2925}{\$3900} = 0.75$	

TRY EXERCISE 11

Note that when the signs of the terms alternate, the common ratio is negative.

To develop a formula for the general, or *n*th, term of a geometric sequence, let a_1 be the first term and let r be the common ratio. We write out the first few terms as follows:

$$a_1,$$
$$a_2 = a_1 r,$$
$$a_3 = a_2 r = (a_1 r)r = a_1 r^2, \qquad \text{Substituting } a_1 r \text{ for } a_2$$
$$a_4 = a_3 r = (a_1 r^2)r = a_1 r^3. \qquad \text{Substituting } a_1 r^2 \text{ for } a_3$$

Note that the exponent is 1 less than the subscript.

Generalizing, we obtain the following.

> ### To Find a_n for a Geometric Sequence
>
> The *n*th term of a geometric sequence with common ratio r is given by
>
> $$a_n = a_1 r^{n-1}, \quad \text{for any integer } n \geq 1.$$

EXAMPLE **2**   Find the 7th term of the geometric sequence 4, 20, 100, ...

SOLUTION   First, we note that

$$a_1 = 4 \quad \text{and} \quad n = 7.$$

To find the common ratio, we can divide any term (other than the first) by the term preceding it. Since the second term is 20 and the first is 4,

$$r = \frac{20}{4}, \quad \text{or } 5.$$

The formula

$$a_n = a_1 r^{n-1}$$

gives us

$$a_7 = 4 \cdot 5^{7-1} = 4 \cdot 5^6 = 4 \cdot 15,625 = 62,500.$$

> TRY EXERCISE 19

EXAMPLE 3   Find the 10th term of the geometric sequence

$$64, -32, 16, -8, \ldots.$$

SOLUTION   First, we note that

$$a_1 = 64, \quad n = 10, \quad \text{and} \quad r = \frac{-32}{64} = -\frac{1}{2}.$$

Then, using the formula for the nth term of a geometric sequence, we have

$$a_{10} = 64 \cdot \left(-\frac{1}{2}\right)^{10-1} = 64 \cdot \left(-\frac{1}{2}\right)^9 = 2^6 \cdot \left(-\frac{1}{2^9}\right) = -\frac{1}{2^3} = -\frac{1}{8}.$$

The 10th term is $-\frac{1}{8}$.

> TRY EXERCISE 23

Sum of the First n Terms of a Geometric Sequence

We next develop a formula for S_n when a sequence is geometric:

$$a_1, \ a_1 r, \ a_1 r^2, \ a_1 r^3, \ldots, a_1 r^{n-1}, \ldots.$$

The **geometric series** S_n is given by

$$S_n = a_1 + a_1 r + a_1 r^2 + \cdots + a_1 r^{n-2} + a_1 r^{n-1}. \qquad \textbf{(1)}$$

Multiplying both sides by r gives us

$$r S_n = a_1 r + a_1 r^2 + a_1 r^3 + \cdots + a_1 r^{n-1} + a_1 r^n. \qquad \textbf{(2)}$$

When we subtract corresponding sides of equation (2) from equation (1), the color terms drop out, leaving

$$S_n - r S_n = a_1 - a_1 r^n$$
$$S_n(1 - r) = a_1(1 - r^n), \qquad \text{Factoring}$$

or

$$S_n = \frac{a_1(1 - r^n)}{1 - r}. \qquad \text{Dividing both sides by } 1 - r$$

STUDENT NOTES —————

The three determining characteristics of a geometric sequence or series are the first term (a_1), the number of terms (n), and the common ratio (r). Be sure you understand how to use these characteristics to write out a sequence or a series.

> **To Find S_n for a Geometric Sequence**
>
> The sum of the first n terms of a geometric sequence with common ratio r is given by
>
> $$S_n = \frac{a_1(1 - r^n)}{1 - r}, \quad \text{for any } r \neq 1.$$

EXAMPLE 4   Find the sum of the first 7 terms of the geometric sequence $3, 15, 75, 375, \ldots.$

SOLUTION   First, we note that

$$a_1 = 3, \quad n = 7, \quad \text{and} \quad r = \frac{15}{3} = 5.$$

Then, substituting in the formula $S_n = \dfrac{a_1(1 - r^n)}{1 - r}$, we have

$$S_7 = \frac{3(1 - 5^7)}{1 - 5} = \frac{3(1 - 78{,}125)}{-4}$$

$$= \frac{3(-78{,}124)}{-4}$$

$$= 58{,}593.$$

▶ TRY EXERCISE 33

Infinite Geometric Series

Suppose we consider the sum of the terms of an infinite geometric sequence, such as 3, 6, 12, 24, 48, ... We get what is called an **infinite geometric series**:

$$3 + 6 + 12 + 24 + 48 + \cdots.$$

Here, as n increases, the sum of the first n terms, S_n, increases without bound. There are also infinite series that get closer and closer to some specific number. Here is an example:

$$\frac{1}{2} + \frac{1}{4} + \frac{1}{8} + \frac{1}{16} + \cdots + \frac{1}{2^n} + \cdots.$$

Let's consider S_n for the first four values of n:

$$
\begin{aligned}
S_1 &= \tfrac{1}{2} & &= \tfrac{1}{2} = 0.5,\\
S_2 &= \tfrac{1}{2} + \tfrac{1}{4} & &= \tfrac{3}{4} = 0.75,\\
S_3 &= \tfrac{1}{2} + \tfrac{1}{4} + \tfrac{1}{8} & &= \tfrac{7}{8} = 0.875,\\
S_4 &= \tfrac{1}{2} + \tfrac{1}{4} + \tfrac{1}{8} + \tfrac{1}{16} & &= \tfrac{15}{16} = 0.9375.
\end{aligned}
$$

> The denominator of each sum is 2^n, where n is the subscript of S. The numerator is $2^n - 1$.

Thus, for this particular series, we have

$$S_n = \frac{2^n - 1}{2^n} = \frac{2^n}{2^n} - \frac{1}{2^n} = 1 - \frac{1}{2^n}.$$

Note that the value of S_n is less than 1 for any value of n, but as n gets larger and larger, the value of $1/2^n$ gets closer to 0 and the value of S_n gets closer to 1. We can visualize S_n by considering a square with area 1. For S_1, we shade half the square. For S_2, we shade half the square plus half the remaining part, or $\tfrac{1}{4}$. For S_3, we shade the parts shaded in S_2 plus half the remaining part. Again we see that the values of S_n will continue to get close to 1 (shading the complete square).

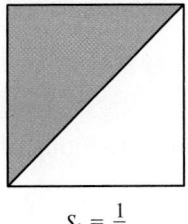
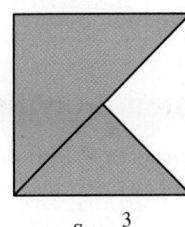
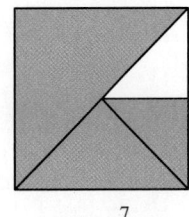
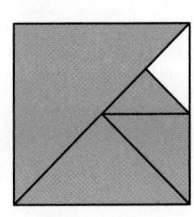

$S_1 = \dfrac{1}{2}$ $S_2 = \dfrac{3}{4}$ $S_3 = \dfrac{7}{8}$ $S_4 = \dfrac{15}{16}$

We say that 1 is the **limit** of S_n and that 1 is the sum of this infinite geometric series. An infinite geometric series is denoted S_∞. It can be shown (but we will not do so here) that the sum of the terms of an infinite geometric sequence exists if and only if $|r| < 1$ (that is, the common ratio's absolute value is less than 1).

To find a formula for the sum of an infinite geometric series, we first consider the sum of the first n terms:

$$S_n = \frac{a_1(1 - r^n)}{1 - r} = \frac{a_1 - a_1 r^n}{1 - r}. \qquad \text{Using the distributive law}$$

For $|r| < 1$, it follows that the value of r^n gets closer to 0 as n gets larger. (Check this by selecting a number between -1 and 1 and finding larger and larger powers on a calculator.) As r^n gets closer to 0, so too does $a_1 r^n$. Thus, S_n gets closer to $a_1/(1 - r)$.

The Limit of an Infinite Geometric Series

For $|r| < 1$, the limit of an infinite geometric series is given by

$$S_\infty = \frac{a_1}{1 - r}. \qquad (\text{For } |r| \geq 1, \text{ no limit exists.})$$

EXAMPLE **5**   Determine whether each series has a limit. If a limit exists, find it.

a) $1 + 3 + 9 + 27 + \cdots$ **b)** $-35 + 7 - \frac{7}{5} + \frac{7}{25} + \cdots$

SOLUTION

a) Here $r = 3$, so $|r| = |3| = 3$. Since $|r| \not< 1$, the series does *not* have a limit.

b) Here $r = -\frac{1}{5}$, so $|r| = |-\frac{1}{5}| = \frac{1}{5}$. Since $|r| < 1$, the series *does* have a limit. We find the limit by substituting into the formula for S_∞:

$$S_\infty = \frac{-35}{1 - \left(-\frac{1}{5}\right)} = \frac{-35}{\frac{6}{5}} = -35 \cdot \frac{5}{6} = \frac{-175}{6} = -29\frac{1}{6}.$$

TRY EXERCISE 41

EXAMPLE **6**   Find fraction notation for $0.63636363\ldots$.

SOLUTION   We can express this as

$$0.63 + 0.0063 + 0.000063 + \cdots.$$

This is an infinite geometric series, where $a_1 = 0.63$ and $r = 0.01$. Since $|r| < 1$, this series has a limit:

$$S_\infty = \frac{a_1}{1 - r} = \frac{0.63}{1 - 0.01} = \frac{0.63}{0.99} = \frac{63}{99}.$$

Thus fraction notation for $0.63636363\ldots$ is $\frac{63}{99}$, or $\frac{7}{11}$.

TRY EXERCISE 53

Problem Solving

For some problem-solving situations, the translation may involve geometric sequences or series.

EXAMPLE 7   *Daily wages.*   Suppose you were offered a job for the month of September (30 days) under the following conditions. You will be paid $0.01 for the first day, $0.02 for the second, $0.04 for the third, and so on, doubling your previous day's salary each day. How much would you earn? (Would you take the job? Make a guess before reading further.)

SOLUTION

1. **Familiarize.** You earn $0.01 the first day, $0.01(2) the second day, $0.01(2)(2) the third day, and so on. Since each day's wages are a constant multiple of the previous day's wages, a geometric sequence is formed.

2. **Translate.** The amount earned is the geometric series

$$\$0.01 + \$0.01(2) + \$0.01(2^2) + \$0.01(2^3) + \cdots + \$0.01(2^{29}),$$

where $a_1 = \$0.01$, $n = 30$, and $r = 2$.

3. **Carry out.** Using the formula

$$S_n = \frac{a_1(1 - r^n)}{1 - r},$$

we have

$$S_{30} = \frac{\$0.01(1 - 2^{30})}{1 - 2}$$

$$= \frac{\$0.01(-1{,}073{,}741{,}823)}{-1} \quad \text{Using a calculator}$$

$$= \$10{,}737{,}418.23.$$

4. **Check.** The calculations can be repeated as a check.

5. **State.** The pay exceeds $10.7 million for the month. Most people would probably take the job!

> **TRY EXERCISE** ▸ 69

EXAMPLE 8   *Loan repayment.*   Francine's student loan is in the amount of $6000. Interest is 9% compounded annually, and the entire amount is to be paid after 10 yr. How much is to be paid back?

SOLUTION

1. **Familiarize.** Suppose we let P represent any principal amount. At the end of one year, the amount owed will be $P + 0.09P$, or $1.09P$. That amount will be the principal for the second year. The amount owed at the end of the second year will be $1.09 \times$ New principal $= 1.09(1.09P)$, or 1.09^2P. Thus the amount owed at the beginning of successive years is as follows:

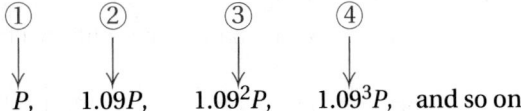

$$P, \quad 1.09P, \quad 1.09^2P, \quad 1.09^3P, \quad \text{and so on.}$$

We have a geometric sequence. The amount owed at the beginning of the 11th year will be the amount owed at the end of the 10th year.

2. **Translate.** We have a geometric sequence with $a_1 = 6000$, $r = 1.09$, and $n = 11$. The appropriate formula is

$$a_n = a_1 r^{n-1}.$$

3. Carry out. We substitute and calculate:

$$a_{11} = \$6000(1.09)^{11-1} = \$6000(1.09)^{10}$$

$$\approx \$14,204.18. \qquad \text{Using a calculator and rounding to the nearest hundredth}$$

4. Check. A check, by repeating the calculations, is left to the student.

5. State. Francine will owe $14,204.18 at the end of 10 yr.

> TRY EXERCISE 61

EXAMPLE 9

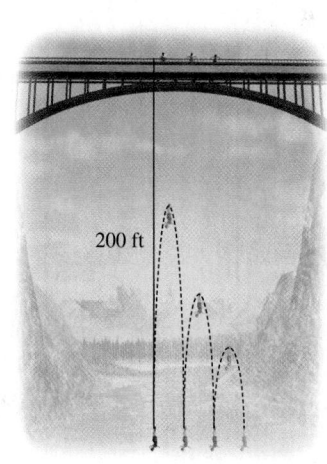

Bungee jumping. A bungee jumper rebounds 60% of the height jumped. Clyde's bungee jump is made using a cord that stretches to 200 ft.

a) After jumping and then rebounding 9 times, how far has Clyde traveled upward (the total rebound distance)?

b) Theoretically, how far will Clyde travel upward (bounce) before coming to rest?

SOLUTION

1. Familiarize. Let's do some calculations and look for a pattern.

First fall:	200 ft
First rebound:	0.6×200, or 120 ft
Second fall:	120 ft, or 0.6×200
Second rebound:	0.6×120, or $0.6(0.6 \times 200)$, which is 72 ft
Third fall:	72 ft, or $0.6(0.6 \times 200)$
Third rebound:	0.6×72, or $0.6(0.6(0.6 \times 200))$, which is 43.2 ft

The rebound distances form a geometric sequence:

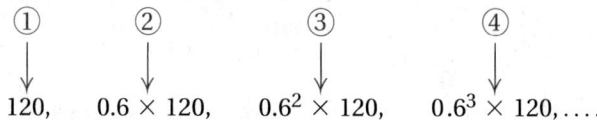

$$120, \quad 0.6 \times 120, \quad 0.6^2 \times 120, \quad 0.6^3 \times 120, \dots.$$

2. Translate.

a) The total rebound distance after 9 bounces is the sum of a geometric sequence. The first term is 120 and the common ratio is 0.6. There will be 9 terms, so we can use the formula

$$S_n = \frac{a_1(1 - r^n)}{1 - r}.$$

b) Theoretically, Clyde will never stop bouncing. Realistically, the bouncing will eventually stop. To approximate the actual distance bounced, we consider an infinite number of bounces and use the formula

$$S_\infty = \frac{a_1}{1 - r}. \qquad \text{Since } r = 0.6 \text{ and } |0.6| < 1, \text{ we know that } S_\infty \text{ exists.}$$

3. Carry out.

a) We substitute into the formula and calculate:

$$S_9 = \frac{120[1 - (0.6)^9]}{1 - 0.6} \approx 297. \qquad \text{Using a calculator}$$

b) We substitute and calculate:

$$S_\infty = \frac{120}{1 - 0.6} = 300.$$

4. **Check.** We can do the calculations again.
5. **State.**

    a) In 9 bounces, Clyde will have traveled upward a total distance of about 297 ft.

    b) Theoretically, Clyde will travel upward a total of 300 ft before coming to rest.

        TRY EXERCISE ▸ 67

14.3 EXERCISE SET

For Extra Help — MyMathLab — Math XL PRACTICE — WATCH — DOWNLOAD

🖐 **Concept Reinforcement** *Classify each of the following as an arithmetic sequence, a geometric sequence, an arithmetic series, a geometric series, or none of these.*

1. $3, 6, 12, 24, \ldots$

2. $-2, 3, 8, 13, \ldots$

3. $10, 7, 4, 1, -2, \ldots$

4. $1000, 500, 250, 125, \ldots$

5. $4 + 20 + 100 + 500 + 2500 + 12{,}500$

6. $10 + 12 + 14 + 16 + 18 + 20$

7. $3 - \frac{3}{2} + \frac{3}{4} - \frac{3}{8} + \frac{3}{16} - \cdots$

8. $1 + \frac{1}{2} + \frac{1}{3} + \frac{1}{4} + \frac{1}{5} + \frac{1}{6} + \cdots$

Find the common ratio for each geometric sequence.

9. $10, 20, 40, 80, \ldots$

10. $5, 20, 80, 320, \ldots$

11. $6, -0.6, 0.06, -0.006, \ldots$

12. $-5, -0.5, -0.05, -0.005, \ldots$

13. $\frac{1}{2}, -\frac{1}{4}, \frac{1}{8}, -\frac{1}{16}, \ldots$

14. $\frac{2}{3}, -\frac{4}{3}, \frac{8}{3}, -\frac{16}{3}, \ldots$

15. $75, 15, 3, \frac{3}{5}, \ldots$

16. $12, -4, \frac{4}{3}, -\frac{4}{9}, \ldots$

17. $\dfrac{1}{m}, \dfrac{6}{m^2}, \dfrac{36}{m^3}, \dfrac{216}{m^4}, \ldots$

18. $4, \dfrac{4m}{5}, \dfrac{4m^2}{25}, \dfrac{4m^3}{125}, \ldots$

Find the indicated term for each geometric sequence.

19. $2, 6, 18, \ldots$; the 7th term

20. $2, 8, 32, \ldots$; the 9th term

21. $\sqrt{3}, 3, 3\sqrt{3}, \ldots$; the 10th term

22. $2, 2\sqrt{2}, 4, \ldots$; the 8th term

23. $-\frac{8}{243}, \frac{8}{81}, -\frac{8}{27}, \ldots$; the 14th term

24. $\frac{7}{625}, \frac{-7}{125}, \frac{7}{25}, \ldots$; the 13th term

25. $\$1000, \$1040, \$1081.60, \ldots$; the 10th term

26. $\$1000, \$1050, \$1102.50, \ldots$; the 12th term

Find the nth, or general, term for each geometric sequence.

27. $1, 5, 25, 125, \ldots$

28. $2, 4, 8, \ldots$

29. $1, -1, 1, -1, \ldots$

30. $\frac{1}{4}, \frac{1}{16}, \frac{1}{64}, \ldots$

31. $\dfrac{1}{x}, \dfrac{1}{x^2}, \dfrac{1}{x^3}, \ldots$

32. $5, \dfrac{5m}{2}, \dfrac{5m^2}{4}, \ldots$

For Exercises 33–40, use the formula for S_n to find the indicated sum for each geometric series.

33. S_9 for $6 + 12 + 24 + \cdots$

34. S_6 for $16 - 8 + 4 - \cdots$

35. S_7 for $\frac{1}{18} - \frac{1}{6} + \frac{1}{2} - \cdots$

Aha! **36.** S_5 for $7 + 0.7 + 0.07 + \cdots$

37. S_8 for $1 + x + x^2 + x^3 + \cdots$

38. S_{10} for $1 + x^2 + x^4 + x^6 + \cdots$

39. S_{16} for $\$200 + \$200(1.06) + \$200(1.06)^2 + \cdots$

40. S_{23} for $\$1000 + \$1000(1.08) + \$1000(1.08)^2 + \cdots$

Determine whether each infinite geometric series has a limit. If a limit exists, find it.

41. $18 + 6 + 2 + \cdots$

42. $80 + 20 + 5 + \cdots$

43. $7 + 3 + \frac{9}{7} + \cdots$

44. $12 + 9 + \frac{27}{4} + \cdots$

45. $3 + 15 + 75 + \cdots$

46. $2 + 3 + \frac{9}{2} + \cdots$

47. $4 - 6 + 9 - \frac{27}{2} + \cdots$

48. $-6 + 3 - \frac{3}{2} + \frac{3}{4} - \cdots$

49. $0.43 + 0.0043 + 0.000043 + \cdots$

50. $0.37 + 0.0037 + 0.000037 + \cdots$

51. $\$500(1.02)^{-1} + \$500(1.02)^{-2} + \$500(1.02)^{-3} + \cdots$

52. $\$1000(1.08)^{-1} + \$1000(1.08)^{-2} + \$1000(1.08)^{-3} + \cdots$

Find fraction notation for each repeating decimal.

53. $0.5555\ldots$

54. $0.8888\ldots$

55. $3.4646\ldots$

56. $1.2323\ldots$

57. $0.15151515\ldots$

58. $0.12121212\ldots$

▦ *Solve. Use a calculator as needed for evaluating formulas.*

59. *Rebound distance.*   A ping-pong ball is dropped from a height of 20 ft and always rebounds one-fourth of the distance fallen. How high does it rebound the 6th time?

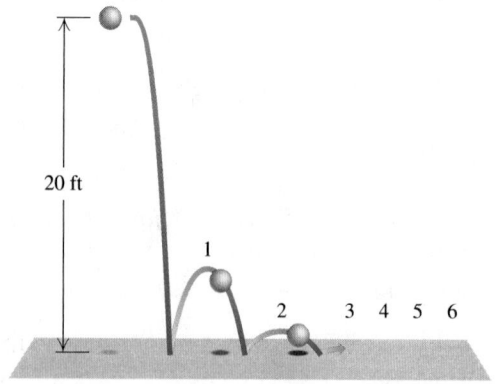

60. *Rebound distance.*   Approximate the total of the rebound heights of the ball in Exercise 59.

61. *Population growth.*   Yorktown has a current population of 100,000, and the population is increasing by 3% each year. What will the population be in 15 yr?

62. *Amount owed.*   Gilberto borrows $15,000. The loan is to be repaid in 13 yr at 5.5% interest, compounded annually. How much will be repaid at the end of 13 yr?

63. *Shrinking population.*   A population of 5000 fruit flies is dying off at a rate of 4% per minute. How many flies will be alive after 15 min?

64. *Shrinking population.*   For the population of fruit flies in Exercise 63, how long will it take for only 1800 fruit flies to remain alive? (*Hint*: Use logarithms.) Round to the nearest minute.

65. *Food service.*   Approximately 17 billion espresso-based coffees were sold in the United States in 2007. This number is expected to grow by 4% each year. How many espresso-based coffees will be sold from 2007 through 2015?
Source: Based on data in the *Indianapolis Star*, 11/22/07

66. *Text messaging.*   Approximately 160 billion text messages were sent worldwide in 2000. Since then, the number of text messages sent each year has grown by about 140% per year. How many text messages were sent worldwide from 2000 through 2010?
Source: Based on data from mobilesmsmarketing.com

67. *Rebound distance.*   A superball dropped from the top of the Washington Monument (556 ft high) rebounds three-fourths of the distance fallen. How far (up and down) will the ball have traveled when it hits the ground for the 6th time?

68. *Rebound distance.*   Approximate the total distance that the ball of Exercise 67 will have traveled when it comes to rest.

69. *Stacking paper.*   Construction paper is about 0.02 in. thick. Beginning with just one piece, a stack is doubled again and again 10 times. Find the height of the final stack.

70. *Monthly earnings.* Suppose you accepted a job for the month of February (28 days) under the following conditions. You will be paid $0.01 the first day, $0.02 the second, $0.04 the third, and so on, doubling your previous day's salary each day. How much would you earn?

Aha! **71.** Under what circumstances is it possible for the 5th term of a geometric sequence to be greater than the 4th term but less than the 7th term?

72. When *r* is negative, a series is said to be *alternating*. Why do you suppose this terminology is used?

Skill Review

To prepare for Section 14.4, review products of binomials (Section 4.5).

Multiply. [4.5]

73. $(x + y)^2$

74. $(x + y)^3$

75. $(x - y)^3$

76. $(x - y)^4$

77. $(2x + y)^3$

78. $(2x - y)^3$

Synthesis

79. Write a problem for a classmate to solve. Devise the problem so that a geometric series is involved and the solution is "The total amount in the bank is $900(1.08)^{40}$, or about $19,550."

80. The infinite series

$$S_\infty = 2 + \frac{1}{2} + \frac{1}{2 \cdot 3} + \frac{1}{2 \cdot 3 \cdot 4} + \frac{1}{2 \cdot 3 \cdot 4 \cdot 5}$$

$$+ \frac{1}{2 \cdot 3 \cdot 4 \cdot 5 \cdot 6} + \cdots$$

is not geometric, but it does have a sum. Using $S_1, S_2, S_3, S_4, S_5,$ and S_6, make a conjecture about the value of S_∞ and explain your reasoning.

Calculate each of the following sums.

81. $\displaystyle\sum_{k=1}^{\infty} 6(0.9)^k$

82. $\displaystyle\sum_{k=1}^{\infty} 5(-0.7)^k$

83. Find the sum of the first *n* terms of

$$x^2 - x^3 + x^4 - x^5 + \cdots.$$

84. Find the sum of the first *n* terms of

$$1 + x + x^2 + x^3 + \cdots.$$

85. The sides of a square are each 16 cm long. A second square is inscribed by joining the midpoints of the sides, successively. In the second square we repeat the process, inscribing a third square. If this process is continued indefinitely, what is the sum of all of the areas of all the squares? (*Hint*: Use an infinite geometric series.)

86. Show that 0.999... is 1.

87. Using Example 5 and Exercises 41–52, explain how the graph of a geometric sequence can be used to determine whether a geometric series has a limit.

88. To compare the *graphs* of an arithmetic and a geometric sequence, we plot *n* on the horizontal axis and a_n on the vertical axis. Graph Example 1(a) of Section 14.2 and Example 1(a) of Section 14.3 on the same set of axes. How do the graphs of geometric sequences differ from the graphs of arithmetic sequences?

CONNECTING the CONCEPTS

A *sequence* is simply an ordered list. A *series* is a sum of consecutive terms in a sequence. Some sequences of numbers have patterns and a formula can be found for a general term. If each pair of consecutive terms has a common difference, the sequence is *arithmetic*. If each pair of consecutive terms has a common ratio, the sequence is *geometric*. Arithmetic and geometric sequences have formulas for general terms and for sums.

Arithmetic Sequences	Geometric Sequences
$a_n = a_1 + (n - 1)d$	$a_n = a_1 r^{n-1}$
$S_n = \dfrac{n}{2}(a_1 + a_n)$	$S_n = \dfrac{a_1(1 - r^n)}{1 - r};$ $S_\infty = \dfrac{a_1}{1 - r}, \quad \lvert r \rvert < 1$

MIXED REVIEW

1. Find a_{20} if $a_n = n^2 - 5n$.

2. Write an expression for the general term a_n of the sequence $\frac{1}{2}, \frac{1}{3}, \frac{1}{4}, \frac{1}{5}, \dots$.

3. Find S_{12} for the sequence $1, 2, 3, 4, \dots$.

4. Write out and evaluate the sum

$$\sum_{k=2}^{5} k^2.$$

5. Rewrite using sigma notation:
   $1 - 2 + 3 - 4 + 5 - 6$.

6. Find the common difference for the arithmetic sequence $115, 112, 109, 106, \dots$.

7. Find the 21st term of the arithmetic sequence $10, 15, 20, 25, \dots$.

8. Which term is 22 in the arithmetic sequence $10, 10.2, 10.4, 10.6, \dots$?

9. For an arithmetic sequence, find a_{25} when $a_1 = 9$ and $d = -2$.

10. For an arithmetic sequence, find a_1 when $d = 11$ and $a_5 = 65$.

11. For an arithmetic sequence, find n when $a_1 = 5$, $d = -\frac{1}{2}$, and $a_n = 0$.

12. Find S_{30} for the arithmetic series
    $2 + 12 + 22 + 32 + \cdots$.

13. Find the common ratio for the geometric sequence $\frac{1}{3}, -\frac{1}{6}, \frac{1}{12}, -\frac{1}{24}, \dots$.

14. Find the 8th term of the geometric sequence $5, 10, 20, 40, \dots$.

15. Find the nth, or general, term for the geometric sequence $2, -2, 2, -2, \dots$.

16. Find S_{10} for the geometric series
    $\$100 + \$100(1.03) + \$100(1.03)^2 + \cdots$.

17. Determine whether the infinite geometric series $0.9 + 0.09 + 0.009 + \cdots$ has a limit. If a limit exists, find it.

18. Determine whether the infinite geometric series $0.9 + 9 + 90 + \cdots$ has a limit. If a limit exists, find it.

19. Renata earns $1 on June 1, another $2 on June 2, another $3 on June 3, another $4 on June 4, and so on. How much does she earn during the 30 days of June?

20. Dwight earns $1 on June 1, another $2 on June 2, another $4 on June 3, another $8 on June 4, and so on. How much does he earn during the 30 days of June?

COLLABORATIVE CORNER

Bargaining for a Used Car

Focus: Geometric series

Time: 30 minutes

Group size: 2

Materials: Graphing calculators are optional.

ACTIVITY *

1. One group member ("the seller") has a car for sale and is asking $3500. The second ("the buyer") offers $1500. The seller splits the difference ($3500 − $1500 = $2000, and $2000 ÷ 2 = $1000) and lowers the price to $2500. The buyer then splits the difference again ($2500 − $1500 = $1000, and $1000 ÷ 2 = $500) and counters with $2000. Continue in this manner and stop when you are able to agree on the car's selling price to the nearest penny.

2. What should the buyer's initial offer be in order to achieve a purchase price of $2000 or less? (Check several guesses to find the appropriate initial offer.)

---

*This activity is based on the article "Bargaining Theory, or Zeno's Used Cars," by James C. Kirby, *The College Mathematics Journal,* **27**(4), September 1996.

3. The seller's price in the bargaining above can be modeled recursively (see Exercises 81, 82, and 90 in Section 14.1) by the sequence

$$a_1 = 3500, \qquad a_n = a_{n-1} - \frac{d}{2^{2n-3}},$$

where d is the difference between the initial price and the first offer. Use this recursively defined sequence to solve parts (1) and (2) above either manually or by using the SEQ MODE and the TABLE feature of a graphing calculator.

4. The first four terms in the sequence in part (3) can be written as

$$a_1, \quad a_1 - \frac{d}{2}, \quad a_1 - \frac{d}{2} - \frac{d}{8},$$

$$a_1 - \frac{d}{2} - \frac{d}{8} - \frac{d}{32}.$$

Use the formula for the limit of an infinite geometric series to find a simple algebraic formula for the eventual sale price, P, when the bargaining process from above is followed. Verify the formula by using it to solve parts (1) and (2) above.

14.4 The Binomial Theorem

Binomial Expansion Using Pascal's Triangle ■ Binomial Expansion Using Factorial Notation

The expression $(x + y)^2$ may be regarded as a series: $x^2 + 2xy + y^2$. This sum of terms is the *expansion* of $(x + y)^2$. For powers greater than 2, finding the expansion of $(x + y)^n$ can be time-consuming. In this section, we look at two methods of streamlining binomial expansion.

Binomial Expansion Using Pascal's Triangle

Consider the following expanded powers of $(a + b)^n$.

$$(a + b)^0 = 1$$
$$(a + b)^1 = a + b$$
$$(a + b)^2 = a^2 + 2a^1b^1 + b^2$$
$$(a + b)^3 = a^3 + 3a^2b^1 + 3a^1b^2 + b^3$$
$$(a + b)^4 = a^4 + 4a^3b^1 + 6a^2b^2 + 4a^1b^3 + b^4$$
$$(a + b)^5 = a^5 + 5a^4b^1 + 10a^3b^2 + 10a^2b^3 + 5a^1b^4 + b^5$$

Each expansion is a polynomial. There are some patterns worth noting:

1. There is one more term than the power of the binomial, n. That is, there are $n + 1$ terms in the expansion of $(a + b)^n$.

2. In each term, the sum of the exponents is the power to which the binomial is raised.

3. The exponents of a start with n, the power of the binomial, and decrease to 0 (since $a^0 = 1$, the last term has no factor of a). The first term has no factor of b, so powers of b start with 0 and increase to n.

4. The coefficients start at 1, increase through certain values, and then decrease through these same values back to 1.

Let's study the coefficients further. Suppose we wish to expand $(a + b)^8$. The patterns we noticed above indicate 9 terms in the expansion:

$$a^8 + c_1a^7b + c_2a^6b^2 + c_3a^5b^3 + c_4a^4b^4 + c_5a^3b^5 + c_6a^2b^6 + c_7ab^7 + b^8.$$

How can we determine the values for the c's? One method seems very simple, but it has some drawbacks. It involves writing down the coefficients in a triangular array as follows. We form what is known as **Pascal's triangle**:

$$
\begin{array}{llccccccccc}
(a + b)^0: & & & & & & 1 & & & & \\
(a + b)^1: & & & & & 1 & & 1 & & & \\
(a + b)^2: & & & & 1 & & 2 & & 1 & & \\
(a + b)^3: & & & 1 & & 3 & & 3 & & 1 & \\
(a + b)^4: & & 1 & & 4 & & 6 & & 4 & & 1 \\
(a + b)^5: & 1 & & 5 & & 10 & & 10 & & 5 & & 1 \\
\end{array}
$$

There are many patterns in the triangle. Find as many as you can.

Perhaps you discovered a way to write the next row of numbers, given the numbers in the row above it. There are always 1's on the outside. Each remaining number is the sum of the two numbers above:

$$
\begin{array}{ccccccccccccc}
 & & & & & & 1 & & & & & & \\
 & & & & & 1 & & 1 & & & & & \\
 & & & & 1 & & 2 & & 1 & & & & \\
 & & & 1 & & 3 & & 3 & & 1 & & & \\
 & & 1 & & 4 & & 6 & & 4 & & 1 & & \\
 & 1 & & 5 & & 10 & & 10 & & 5 & & 1 & \\
1 & & 6 & & 15 & & 20 & & 15 & & 6 & & 1 \\
\end{array}
$$

We see that in the bottom (seventh) row

the 1st and last numbers are 1;

the 2nd number is $1 + 5$, or 6;

the 3rd number is $5 + 10$, or 15;

the 4th number is $10 + 10$, or 20;

the 5th number is $10 + 5$, or 15; and

the 6th number is $5 + 1$, or 6.

Thus the expansion of $(a + b)^6$ is

$$(a + b)^6 = 1a^6 + 6a^5b + 15a^4b^2 + 20a^3b^3 + 15a^2b^4 + 6ab^5 + 1b^6.$$

To expand $(a + b)^8$, we complete two more rows of Pascal's triangle:

$$
\begin{array}{ccccccccccccccccc}
 & & & & & & & & 1 & & & & & & & & \\
 & & & & & & & 1 & & 1 & & & & & & & \\
 & & & & & & 1 & & 2 & & 1 & & & & & & \\
 & & & & & 1 & & 3 & & 3 & & 1 & & & & & \\
 & & & & 1 & & 4 & & 6 & & 4 & & 1 & & & & \\
 & & & 1 & & 5 & & 10 & & 10 & & 5 & & 1 & & & \\
 & & 1 & & 6 & & 15 & & 20 & & 15 & & 6 & & 1 & & \\
 & 1 & & 7 & & 21 & & 35 & & 35 & & 21 & & 7 & & 1 & \\
1 & & 8 & & 28 & & 56 & & 70 & & 56 & & 28 & & 8 & & 1 \\
\end{array}
$$

The expansion of $(a + b)^8$ has coefficients found in the 9th row above:

$$(a + b)^8 = 1a^8 + 8a^7b + 28a^6b^2 + 56a^5b^3 + 70a^4b^4 + 56a^3b^5 + 28a^2b^6 + 8ab^7 + 1b^8.$$

We can generalize our results as follows:

> ### The Binomial Theorem (Form 1)
>
> For any binomial $a + b$ and any natural number n,
>
> $$(a + b)^n = c_0a^nb^0 + c_1a^{n-1}b^1 + c_2a^{n-2}b^2$$
> $$+ \cdots + c_{n-1}a^1b^{n-1} + c_na^0b^n,$$
>
> where the numbers $c_0, c_1, c_2, \ldots, c_n$ are from the $(n + 1)$st row of Pascal's triangle.

A proof of the binomial theorem is beyond the scope of this text.

EXAMPLE **1**   Expand: $(u - v)^5$.

SOLUTION   Using the binomial theorem, we have $a = u$, $b = -v$, and $n = 5$. We use the 6th row of Pascal's triangle: $1\ \ 5\ \ 10\ \ 10\ \ 5\ \ 1$. Thus,

$$
\begin{aligned}
(u - v)^5 &= [u + (-v)]^5 \qquad \text{Rewriting } u - v \text{ as a sum}\\
&= 1(u)^5 + 5(u)^4(-v)^1 + 10(u)^3(-v)^2 + 10(u)^2(-v)^3 \\
&\quad + 5(u)^1(-v)^4 + 1(-v)^5 \\
&= u^5 - 5u^4v + 10u^3v^2 - 10u^2v^3 + 5uv^4 - v^5.
\end{aligned}
$$

Note that the signs of the terms alternate between $+$ and $-$. When $-v$ is raised to an odd power, the sign is $-$; when the power is even, the sign is $+$.

EXAMPLE **2** Expand: $\left(2t + \dfrac{3}{t}\right)^6$.

SOLUTION Note that $a = 2t$, $b = 3/t$, and $n = 6$. We use the 7th row of Pascal's triangle: 1 6 15 20 15 6 1. Thus,

$$\left(2t + \frac{3}{t}\right)^6 = 1(2t)^6 + 6(2t)^5\left(\frac{3}{t}\right)^1 + 15(2t)^4\left(\frac{3}{t}\right)^2 + 20(2t)^3\left(\frac{3}{t}\right)^3$$

$$+ 15(2t)^2\left(\frac{3}{t}\right)^4 + 6(2t)^1\left(\frac{3}{t}\right)^5 + 1\left(\frac{3}{t}\right)^6$$

$$= 64t^6 + 6\left(32t^5\right)\left(\frac{3}{t}\right) + 15(16t^4)\left(\frac{9}{t^2}\right) + 20(8t^3)\left(\frac{27}{t^3}\right)$$

$$+ 15(4t^2)\left(\frac{81}{t^4}\right) + 6(2t)\left(\frac{243}{t^5}\right) + \frac{729}{t^6}$$

$$= 64t^6 + 576t^4 + 2160t^2 + 4320 + 4860t^{-2} + 2916t^{-4} + 729t^{-6}.$$

Binomial Expansion Using Factorial Notation

The drawback to using Pascal's triangle is that we must compute all the preceding rows in the table to obtain the row we need. The following method avoids this difficulty. It will also enable us to find a specific term—say, the 8th term—without computing all the other terms in the expansion. This method is useful in such courses as finite mathematics, calculus, and statistics.

To develop the method, we need some new notation. Products of successive natural numbers, such as $6 \cdot 5 \cdot 4 \cdot 3 \cdot 2 \cdot 1$ and $8 \cdot 7 \cdot 6 \cdot 5 \cdot 4 \cdot 3 \cdot 2 \cdot 1$, have a special notation. For the product $6 \cdot 5 \cdot 4 \cdot 3 \cdot 2 \cdot 1$, we write 6!, read "6 factorial."

Factorial Notation

For any natural number n,

$$n! = n(n - 1)(n - 2) \cdots (3)(2)(1).$$

Here are some examples:

$$6! = 6 \cdot 5 \cdot 4 \cdot 3 \cdot 2 \cdot 1 = 720,$$
$$5! = \qquad 5 \cdot 4 \cdot 3 \cdot 2 \cdot 1 = 120,$$
$$4! = \qquad\quad 4 \cdot 3 \cdot 2 \cdot 1 = \quad 24,$$
$$3! = \qquad\qquad 3 \cdot 2 \cdot 1 = \quad\; 6,$$
$$2! = \qquad\qquad\quad 2 \cdot 1 = \quad\; 2,$$
$$1! = \qquad\qquad\qquad\; 1 = \quad\; 1.$$

We also define 0! to be 1 for reasons explained shortly.

To simplify expressions like

$$\frac{8!}{5!3!},$$

note that

$$8! = 8 \cdot 7 \cdot 6 \cdot 5 \cdot 4 \cdot 3 \cdot 2 \cdot 1 = 8 \cdot 7! = 8 \cdot 7 \cdot 6! = 8 \cdot 7 \cdot 6 \cdot 5!$$

and so on.

CAUTION! $\dfrac{6!}{3!} \neq 2!$ To see this, note that

$$\frac{6!}{3!} = \frac{6 \cdot 5 \cdot 4 \cdot \cancel{3} \cdot \cancel{2} \cdot \cancel{1}}{\cancel{3} \cdot \cancel{2} \cdot \cancel{1}} = 6 \cdot 5 \cdot 4.$$

EXAMPLE 3   Simplify: $\dfrac{8!}{5!3!}$.

SOLUTION

$$\frac{8!}{5!3!} = \frac{8 \cdot 7 \cdot 6 \cdot 5!}{5! \cdot 3 \cdot 2 \cdot 1} = 8 \cdot 7 \qquad \text{Removing a factor equal to 1: } \frac{6 \cdot 5!}{5! \cdot 3 \cdot 2} = 1$$
$$= 56$$

TRY EXERCISE 15

STUDENT NOTES

It is important to recognize factorial notation as representing a product with descending factors. Thus, 7!, 7 · 6!, and 7 · 6 · 5! all represent the same product.

The following notation is used in our second formulation of the binomial theorem.

$\dbinom{n}{r}$ **Notation**

For n and r nonnegative integers with $n \geq r$,

$$\binom{n}{r}, \quad \text{read "}n\text{ choose }r\text{,"} \quad \text{means} \quad \frac{n!}{(n-r)!r!}.^{*}$$

EXAMPLE 4   Simplify: **(a)** $\dbinom{7}{2}$; **(b)** $\dbinom{9}{6}$; **(c)** $\dbinom{6}{6}$.

SOLUTION

a) $\dbinom{7}{2} = \dfrac{7!}{(7-2)!2!}$

$$= \frac{7!}{5!2!} = \frac{7 \cdot 6 \cdot 5!}{5! \cdot 2 \cdot 1} = \frac{7 \cdot 6}{2} \qquad \text{We can write 7! as } 7 \cdot 6 \cdot 5! \text{ to aid our simplification.}$$
$$= 7 \cdot 3$$
$$= 21$$

TECHNOLOGY CONNECTION

The PRB option of the MATH menu provides access to both factorial calculations and NCR. In both cases, a number must be entered first. To find $\dbinom{7}{2}$, we press ⑦ **MATH**, select PRB and NCR, and press ② **ENTER**.

```
7 nCr 2
                    21
```

1. Find 12!.

2. Find $\dbinom{8}{3}$ and $\dbinom{12}{5}$.

---

*In many books and for many calculators, the notation $_nC_r$ is used instead of $\dbinom{n}{r}$.

b) $\begin{pmatrix} 9 \\ 6 \end{pmatrix} = \dfrac{9!}{3!6!}$

$= \dfrac{9 \cdot 8 \cdot 7 \cdot 6!}{3 \cdot 2 \cdot 1 \cdot 6!} = \dfrac{9 \cdot 8 \cdot 7}{3 \cdot 2}$ Writing 9! as $9 \cdot 8 \cdot 7 \cdot 6!$ to help with simplification

$= 3 \cdot 4 \cdot 7$

$= 84$

c) $\begin{pmatrix} 6 \\ 6 \end{pmatrix} = \dfrac{6!}{0!6!} = \dfrac{6!}{1 \cdot 6!}$ Since $0! = 1$

$= \dfrac{6!}{6!}$

$= 1$

> TRY EXERCISE 17

Now we can restate the binomial theorem using our new notation.

The Binomial Theorem (Form 2)

For any binomial $a + b$ and any natural number n,

$$(a + b)^n = \begin{pmatrix} n \\ 0 \end{pmatrix}a^n + \begin{pmatrix} n \\ 1 \end{pmatrix}a^{n-1}b + \begin{pmatrix} n \\ 2 \end{pmatrix}a^{n-2}b^2 + \cdots + \begin{pmatrix} n \\ n \end{pmatrix}b^n.$$

EXAMPLE 5   Expand: $(3x + y)^4$.

SOLUTION   We use the binomial theorem (Form 2) with $a = 3x$, $b = y$, and $n = 4$:

$$(3x + y)^4 = \begin{pmatrix} 4 \\ 0 \end{pmatrix}(3x)^4 + \begin{pmatrix} 4 \\ 1 \end{pmatrix}(3x)^3 y + \begin{pmatrix} 4 \\ 2 \end{pmatrix}(3x)^2 y^2 + \begin{pmatrix} 4 \\ 3 \end{pmatrix}(3x)y^3 + \begin{pmatrix} 4 \\ 4 \end{pmatrix}y^4$$

$$= \frac{4!}{4!0!}3^4 x^4 + \frac{4!}{3!1!}3^3 x^3 y + \frac{4!}{2!2!}3^2 x^2 y^2 + \frac{4!}{1!3!}3xy^3 + \frac{4!}{0!4!}y^4$$

$$= 1 \cdot 81x^4 + 4 \cdot 27x^3 y + 6 \cdot 9x^2 y^2 + 4 \cdot 3xy^3 + y^4 \left.\vphantom{\begin{matrix}1\\1\end{matrix}}\right\}$$

$$= 81x^4 + 108x^3 y + 54x^2 y^2 + 12xy^3 + y^4. \qquad\qquad \text{Simplifying}$$

> TRY EXERCISE 29

EXAMPLE 6   Expand: $(x^2 - 2y)^5$.

SOLUTION   In this case, $a = x^2$, $b = -2y$, and $n = 5$:

$$(x^2 - 2y)^5 = \begin{pmatrix} 5 \\ 0 \end{pmatrix}(x^2)^5 + \begin{pmatrix} 5 \\ 1 \end{pmatrix}(x^2)^4(-2y) + \begin{pmatrix} 5 \\ 2 \end{pmatrix}(x^2)^3(-2y)^2$$

$$+ \begin{pmatrix} 5 \\ 3 \end{pmatrix}(x^2)^2(-2y)^3 + \begin{pmatrix} 5 \\ 4 \end{pmatrix}(x^2)(-2y)^4 + \begin{pmatrix} 5 \\ 5 \end{pmatrix}(-2y)^5$$

$$= \frac{5!}{5!0}x^{10} + \frac{5!}{4!1!}x^8(-2y) + \frac{5!}{3!2!}x^6(-2y)^2$$

$$+ \frac{5!}{2!3!}x^4(-2y)^3 + \frac{5!}{1!4!}x^2(-2y)^4 + \frac{5!}{0!5!}(-2y)^5$$

$$= x^{10} - 10x^8 y + 40x^6 y^2 - 80x^4 y^3 + 80x^2 y^4 - 32y^5.$$

> TRY EXERCISE 35

Note that in the binomial theorem (Form 2), $\binom{n}{0}a^n b^0$ gives us the first term, $\binom{n}{1}a^{n-1}b^1$ gives us the second term, $\binom{n}{2}a^{n-2}b^2$ gives us the third term, and so on. This can be generalized to give a method for finding a specific term without writing the entire expansion.

> ### Finding a Specific Term
>
> When $(a + b)^n$ is expanded and written in descending powers of a, the $(r + 1)$st term is
>
> $$\binom{n}{r}a^{n-r}b^r.$$

EXAMPLE   7   Find the 5th term in the expansion of $(2x - 3y)^7$.

SOLUTION   To find the 5th term, we note that $5 = 4 + 1$. Thus, $r = 4$, $a = 2x$, $b = -3y$, and $n = 7$. Using the above formula, we have

$$\binom{n}{r}a^{n-r}b^r = \binom{7}{4}(2x)^{7-4}(-3y)^4, \text{ or } \frac{7!}{3!\,4!}(2x)^3(-3y)^4, \text{ or } 22{,}680x^3y^4.$$

> **TRY EXERCISE**   45

It is because of the binomial theorem that $\binom{n}{r}$ is called a *binomial coefficient*.

We can now explain why 0! is defined to be 1. In the binomial theorem, $\binom{n}{0}$ must equal 1 when using the definition $\binom{n}{r} = \dfrac{n!}{(n-r)!\,r!}$.

Thus we must have

$$\binom{n}{0} = \frac{n!}{(n-0)!\,0!} = \frac{n!}{n!\,0!} = 1.$$

This is satisfied only if 0! is defined to be 1.

A

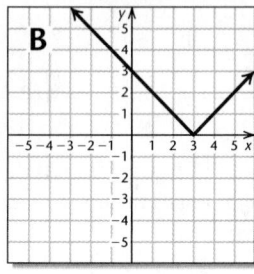

Visualizing for Success

F

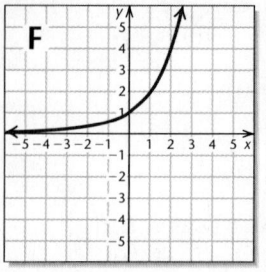

B

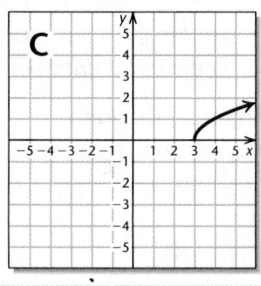

Match each equation with its graph.

1. $xy = 2$

2. $y = \log_2 x$

3. $y = x - 3$

4. $(x - 3)^2 + y^2 = 4$

5. $\dfrac{(x - 3)^2}{1} + \dfrac{y^2}{4} = 1$

6. $y = |x - 3|$

7. $y = (x - 3)^2$

8. $y = \dfrac{1}{x - 3}$

9. $y = 2^x$

10. $y = \sqrt{x - 3}$

Answers on page A-67

An additional, animated version of this activity appears in MyMathLab. To use MyMathLab, you need a course ID and a student access code. Contact your instructor for more information.

G

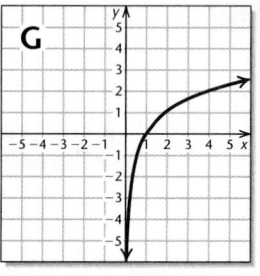

C

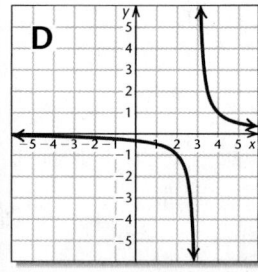

H

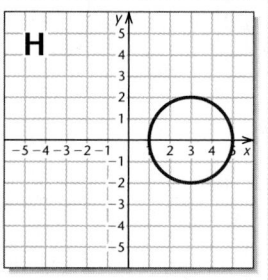

D

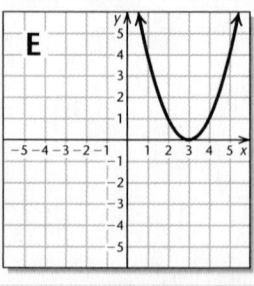

I

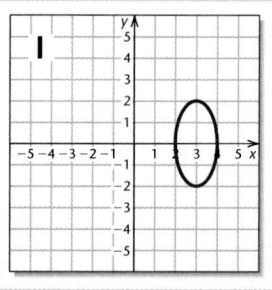

E

J

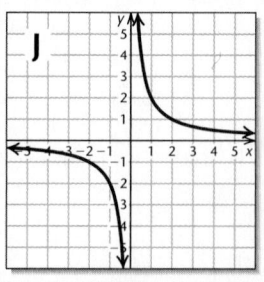

14.4 **EXERCISE SET**

↪ *Concept Reinforcement* *Complete each of the following.*

1. The last term in the expansion of $(x + 2)^5$ is _____ .

2. The expansion of $(x + y)^7$, when simplified, contains a total of _____ terms.

3. In the expansion of $(a + b)^9$, the exponents in each term add to _____ .

4. The expression _____ represents $4 \cdot 3 \cdot 2 \cdot 1$.

5. The expression _____ represents $\dfrac{8!}{3! \, 5!}$.

6. In the expansion of $(a + b)^{10}$, the coefficient of a^8b^2 is the same as the coefficient of _____ .

7. In the expansion of $(x + y)^9$, the coefficient of y^9 is _____ .

8. The notation $\dbinom{9}{5}$ is read _____ .

Simplify.

9. $4!$

10. $9!$

11. $10!$

12. $12!$

13. $\dfrac{10!}{8!}$

14. $\dfrac{12!}{10!}$

15. $\dfrac{9!}{4! \, 5!}$

16. $\dfrac{10!}{6! \, 4!}$

17. $\dbinom{10}{4}$

18. $\dbinom{8}{5}$

Aha! 19. $\dbinom{9}{9}$

20. $\dbinom{7}{7}$

21. $\dbinom{30}{2}$

22. $\dbinom{51}{49}$

23. $\dbinom{40}{38}$

24. $\dbinom{35}{2}$

Expand. Use both of the methods shown in this section.

25. $(a - b)^4$

26. $(m + n)^5$

27. $(p + q)^7$

28. $(x - y)^6$

29. $(3c - d)^7$

30. $(x^2 - 3y)^5$

31. $(t^{-2} + 2)^6$

32. $(3c - d)^6$

33. $(x - y)^5$

34. $(x - y)^3$

35. $\left(3s + \dfrac{1}{t}\right)^9$

36. $\left(x + \dfrac{2}{y}\right)^9$

37. $(x^3 - 2y)^5$

38. $(a^2 - b^3)^5$

39. $(\sqrt{5} + t)^6$

40. $(\sqrt{3} - t)^4$

41. $\left(\dfrac{1}{\sqrt{x}} - \sqrt{x}\right)^6$

42. $(x^{-2} + x^2)^4$

Find the indicated term for each binomial expression.

43. 3rd, $(a + b)^6$

44. 6th, $(x + y)^7$

45. 12th, $(a - 3)^{14}$

46. 11th, $(x - 2)^{12}$

47. 5th, $(2x^3 + \sqrt{y})^8$

48. 4th, $\left(\dfrac{1}{b^2} + c\right)^7$

49. Middle, $(2u + 3v^2)^{10}$

50. Middle two, $(\sqrt{x} + \sqrt{3})^5$

Aha! 51. 9th, $(x - y)^8$

52. 13th, $(a - \sqrt{b})^{12}$

53. Maya claims that she can calculate mentally the first two and the last two terms of the expansion of $(a + b)^n$ for any whole number n. How do you think she does this?

54. Without performing any calculations, explain why the expansions of $(x - y)^8$ and $(y - x)^8$ must be equal.

Skill Review

Review graphing equations and inequalities.

Graph.

55. $y = x^2 - 5$ [11.7]

56. $y = x - 5$ [3.6]

57. $y \geq x - 5$ [9.4]

58. $y = 5^x$ [12.2]

59. $f(x) = \log_5 x$ [12.3]

60. $x^2 + y^2 = 5$ [13.1]

Synthesis

61. Explain how someone can determine the x^2-term of the expansion of $\left(x - \dfrac{3}{x} \right)^{10}$ without calculating any other terms.

62. Devise two problems requiring the use of the binomial theorem. Design the problems so that one is solved more easily using Form 1 and the other is solved more easily using Form 2. Then explain what makes one form easier to use than the other in each case.

63. Show that there are exactly $\dbinom{5}{3}$ ways of choosing a subset of size 3 from $\{a, b, c, d, e\}$.

64. *Baseball.* During the 2007 season, Matt Holliday of the Colorado Rockies had a batting average of 0.340. In that season, if someone were to randomly select 5 of his "at-bats," the probability of Holliday getting exactly 3 hits would be the 3rd term of the binomial expansion of $(0.340 + 0.660)^5$. Find that term and use a calculator to estimate the probability.
Source: www.mlb.com

65. *Widows or divorcees.* The probability that a woman will be either widowed or divorced is 85%. If 8 women are randomly selected, the probability that exactly 5 of them will be either widowed or divorced is the 6th term of the binomial expansion of $(0.15 + 0.85)^8$. Use a calculator to estimate that probability.

66. *Baseball.* In reference to Exercise 64, the probability that Holliday will get *at most* 3 hits is found by adding the last 4 terms of the binomial expansion of $(0.340 + 0.660)^5$. Find these terms and use a calculator to estimate the probability.

67. *Widows or divorcees.* In reference to Exercise 65, the probability that *at least* 6 of the women will be widowed or divorced is found by adding the last three terms of the binomial expansion of $(0.15 + 0.85)^8$. Find these terms and use a calculator to estimate the probability.

68. Find the term of
$$\left(\frac{3x^2}{2} - \frac{1}{3x} \right)^{12}$$
that does not contain x.

69. Prove that
$$\binom{n}{r} = \binom{n}{n-r}$$
for any whole numbers n and r. Assume $r \le n$.

70. Find the middle term of $(x^2 - 6y^{3/2})^6$.

71. Find the ratio of the 4th term of
$$\left(p^2 - \frac{1}{2} p \sqrt[3]{q} \right)^5$$
to the 3rd term.

72. Find the term containing $\dfrac{1}{x^{1/6}}$ of
$$\left(\sqrt[3]{x} - \frac{1}{\sqrt{x}} \right)^7.$$

Aha! **73.** Multiply: $(x^2 + 2xy + y^2)(x^2 + 2xy + y^2)^2(x + y)$.

74. What is the degree of $(x^3 + 2)^4$?

Study Summary

KEY TERMS AND CONCEPTS	EXAMPLES

SECTION 14.1: SEQUENCES AND SERIES

An ordered list of numbers that ends is a **finite sequence**.

An ordered list of numbers that does not end is an **infinite sequence**.

A **series** is a sum of terms of a sequence.

5, 7, 8, 11, 17 is a finite sequence.

6, 9, 12, 15, … is an infinite sequence.

$6 + 9 + 12 + 15 + \cdots$ is an infinite series.

Sigma or Summation Notation

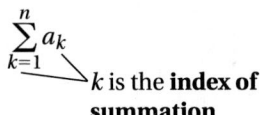

k is the **index of summation**

$$\sum_{k=3}^{5}(-1)^k(k^2) = (-1)^3(3^2) + (-1)^4(4^2) + (-1)^5(5^2)$$
$$= -1 \cdot 9 + 1 \cdot 16 + (-1) \cdot 25$$
$$= -9 + 16 - 25 = -18$$

SECTION 14.2: ARITHMETIC SEQUENCES AND SERIES

Arithmetic Sequences and Series

$a_{n+1} = a_n + d$ d is the **common difference**.

$a_n = a_1 + (n-1)d$ The nth term

$S_n = \dfrac{n}{2}(a_1 + a_n)$ The sum of the first n terms

For the arithmetic sequence $10, 7, 4, 1, \ldots$:

$d = -3;$

$a_7 = 10 + (7-1)(-3) = 10 - 18 = -8;$

$S_7 = \dfrac{7}{2}(10 + (-8)) = \dfrac{7}{2}(2) = 7.$

SECTION 14.3: GEOMETRIC SEQUENCES AND SERIES

Geometric Sequences and Series

$a_{n+1} = a_n \cdot r$ r is the **common ratio**.

$a_n = a_1 r^{n-1}$ The nth term

$S_n = \dfrac{a_1(1-r^n)}{1-r}, r \neq 1$ The sum of the first n terms

$S_\infty = \dfrac{a_1}{1-r}, |r| < 1$ Limit of an infinite geometric series

For the geometric sequence $25, -5, 1, -\frac{1}{5}, \ldots$:

$r = -\dfrac{1}{5};$

$a_7 = 25\left(-\dfrac{1}{5}\right)^{7-1} = 5^2 \cdot \dfrac{1}{5^6} = \dfrac{1}{625};$

$S_7 = \dfrac{25\left(1 - \left(-\frac{1}{5}\right)^7\right)}{1 - \left(-\frac{1}{5}\right)} = \dfrac{5^2\left(\frac{78,126}{5^7}\right)}{\frac{6}{5}} = \dfrac{13,021}{625};$

$S_\infty = \dfrac{25}{1 - \left(-\frac{1}{5}\right)} = \dfrac{125}{6}.$

SECTION 14.4: THE BINOMIAL THEOREM

Factorial Notation

$n! = n(n-1)(n-2)\cdots 3 \cdot 2 \cdot 1$

$7! = 7 \cdot 6 \cdot 5 \cdot 4 \cdot 3 \cdot 2 \cdot 1 = 5040$

Binomial Coefficient

$\dbinom{n}{r} = {}_nC_r = \dfrac{n!}{(n-r)!\,r!}$

$\dbinom{10}{3} = {}_{10}C_3 = \dfrac{10!}{7!\,3!} = \dfrac{10 \cdot 9 \cdot 8 \cdot 7!}{7! \cdot 3 \cdot 2 \cdot 1} = 120$

Binomial Theorem

$$(a + b)^n = \binom{n}{0}a^n + \binom{n}{1}a^{n-1}b + \cdots + \binom{n}{n}b^n$$

$$(r + 1)\text{st term of } (a + b)^n: \binom{n}{r}a^{n-r}b^r$$

$$(1 - 2x)^3 = \binom{3}{0}1^3 + \binom{3}{1}1^2(-2x)$$
$$+ \binom{3}{2}1(-2x)^2 + \binom{3}{3}(-2x)^3$$
$$= 1 \cdot 1 + 3 \cdot 1(-2x) + 3 \cdot 1 \cdot (4x^2) + 1 \cdot (-8x^3)$$
$$= 1 - 6x + 12x^2 - 8x^3$$

$$\text{3rd term of } (1 - 2x)^3: \binom{3}{2}(1)^1(-2x)^2 = 12x^2 \qquad r = 2$$

Review Exercises: Chapter 14

Concept Reinforcement *Classify each statement as either true or false.*

1. The next term in the arithmetic sequence $10, 15, 20, \ldots$ is 35. [14.2]

2. The next term in the geometric sequence $2, 6, 18, 54, \ldots$ is 162. [14.3]

3. $\sum_{k=1}^{3} k^2$ means $1^2 + 2^2 + 3^3$. [14.1]

4. If $a_n = 3n - 1$, then $a_{17} = 19$. [14.1]

5. A geometric sequence has a common difference. [14.3]

6. The infinite geometric series $10 - 5 + \frac{5}{2} - \cdots$ has a limit. [14.3]

7. For any natural number $n, n! = n(n - 1)$. [14.4]

8. When simplified, the expansion of $(x + y)^{17}$ has 19 terms. [14.4]

Find the first four terms; the 8th term, a_8; and the 12th term, a_{12}. [14.1]

9. $a_n = 10n - 9$

10. $a_n = \dfrac{n - 1}{n^2 + 1}$

Write an expression for the general term of each sequence. Answers may vary. [14.1]

11. $-5, -10, -15, -20, \ldots$

12. $-1, 3, -5, 7, -9, \ldots$

Write out and evaluate each sum. [14.1]

13. $\sum_{k=1}^{5} (-2)^k$

14. $\sum_{k=2}^{7} (1 - 2k)$

Rewrite using sigma notation. [14.1]

15. $7 + 14 + 21 + 28 + 35 + 42$

16. $\dfrac{-1}{2} + \dfrac{1}{4} + \dfrac{-1}{8} + \dfrac{1}{16} + \dfrac{-1}{32}$

17. Find the 14th term of the arithmetic sequence $-3, -7, -11, \ldots$. [14.2]

18. An arithmetic sequence has $a_1 = 11$ and $a_{16} = 14$. Find the common difference, d. [14.2]

19. An arithmetic sequence has $a_8 = 20$ and $a_{24} = 100$. Find the first term, a_1, and the common difference, d. [14.2]

20. Find the sum of the first 17 terms of the arithmetic series $-8 + (-11) + (-14) + \cdots$. [14.2]

21. Find the sum of all the multiples of 5 from 5 to 500, inclusive. [14.2]

22. Find the 20th term of the geometric sequence $2, 2\sqrt{2}, 4, \ldots$. [14.3]

23. Find the common ratio of the geometric sequence $40, 30, \frac{45}{2}, \ldots$. [14.3]

24. Find the nth term of the geometric sequence $-2, 2, -2, \ldots$. [14.3]

25. Find the nth term of the geometric sequence
$3, \frac{3}{4}x, \frac{3}{16}x^2, \ldots$ [14.3]

26. Find S_6 for the geometric series

$$3 + 15 + 75 + \cdots.$$

[14.3]

27. Find S_{12} for the geometric series

$$3x - 6x + 12x - \cdots.$$ [14.3]

Determine whether each infinite geometric series has a limit. If a limit exists, find it. [14.3]

28. $6 + 3 + 1.5 + 0.75 + \cdots$

29. $7 - 4 + \frac{16}{7} - \cdots$

30. $-\frac{1}{2} + \frac{1}{2} + \left(-\frac{1}{2}\right) + \frac{1}{2} + \cdots$

31. $0.04 + 0.08 + 0.16 + 0.32 + \cdots$

32. $\$2000 + \$1900 + \$1805 + \$1714.75 + \cdots$

33. Find fraction notation for $0.555555\ldots$ [14.3]

34. Find fraction notation for $1.454545\ldots$ [14.3]

35. Tyrone took a job working in a convenience store starting with an hourly wage of $11.50. He was promised a raise of 40¢ per hour every 3 mos for 8 yr. After 8 yr, what will be his hourly wage? [14.2]

36. A stack of poles has 42 poles in the bottom row. There are 41 poles in the second row, 40 poles in the third row, and so on, ending with 1 pole in the top row. How many poles are in the stack? [14.2]

37. Janine's student loan is for $12,000 at 4%, compounded annually. The total amount is to be paid off in 7 yr. How much will she then owe? [14.3]

38. Find the total rebound distance of a ball, given that it is dropped from a height of 12 m and each rebound is one-third of the preceding one. [14.3]

Simplify. [14.4]

39. $7!$

40. $\dbinom{10}{3}$

41. Find the 3rd term of $(a + b)^{20}$. [14.4]

42. Expand: $(x - 2y)^4$. [14.4]

Synthesis

43. What happens to a_n in a geometric sequence with $|r| < 1$, as n gets larger? Why? [14.3]

44. Compare the two forms of the binomial theorem given in the text. Under what circumstances would one be more useful than the other? [14.4]

45. Find the sum of the first n terms of the geometric series $1 - x + x^2 - x^3 + \cdots$. [14.3]

46. Expand: $(x^{-3} + x^3)^5$. [14.4]

Test: Chapter 14

Step-by-step test solutions are found on the video CD in the back of this book.

1. Find the first five terms and the 12th term of a sequence with general term $a_n = \dfrac{1}{n^2 + 1}$.

2. Write an expression for the general term of the sequence $\frac{4}{3}, \frac{4}{9}, \frac{4}{27}, \ldots$.

3. Write out and evaluate:

$$\sum_{k=2}^{5} (1 - 2^k).$$

4. Rewrite using sigma notation:

$$1 + (-8) + 27 + (-64) + 125.$$

5. Find the 13th term, a_{13}, of the arithmetic sequence $\frac{1}{2}, 1, \frac{3}{2}, 2, \ldots$.

6. Find the common difference d of an arithmetic sequence when $a_1 = 7$ and $a_7 = -11$.

7. Find a_1 and d of an arithmetic sequence when $a_5 = 16$ and $a_{10} = -3$.

8. Find the sum of all the multiples of 12 from 24 to 240, inclusive.

9. Find the 10th term of the geometric sequence $-3, 6, -12, \ldots$.

10. Find the common ratio of the geometric sequence $22\frac{1}{2}, 15, 10, \ldots$.

11. Find the nth term of the geometric sequence $3, 9, 27, \ldots$.

12. Find S_9 for the geometric series

$$11 + 22 + 44 + \cdots.$$

Determine whether each infinite geometric series has a limit. If a limit exists, find it.

13. $0.5 + 0.25 + 0.125 + \cdots$

14. $0.5 + 1 + 2 + 4 + \cdots$

15. $\$1000 + \$80 + \$6.40 + \cdots$

16. Find fraction notation for $0.85858585\ldots$.

17. An auditorium has 31 seats in the first row, 33 seats in the second row, 35 seats in the third row, and so on, for 18 rows. How many seats are in the 17th row?

18. Alyssa's uncle Ken gave her $100 for her first birthday, $200 for her second birthday, $300 for her third birthday, and so on, until her eighteenth birthday. How much did he give her in all?

19. Each week the price of a $10,000 boat will be reduced 5% of the previous week's price. If we assume that it is not sold, what will be the price after 10 weeks?

20. Find the total rebound distance of a ball that is dropped from a height of 18 m, with each rebound two-thirds of the preceding one.

21. Simplify: $\dbinom{12}{9}$.

22. Expand: $(x - 3y)^5$.

23. Find the 4th term in the expansion of $(a + x)^{12}$.

Synthesis

24. Find a formula for the sum of the first n even natural numbers:

$$2 + 4 + 6 + \cdots + 2n.$$

25. Find the sum of the first n terms of

$$1 + \frac{1}{x} + \frac{1}{x^2} + \frac{1}{x^3} + \cdots.$$

Simplify.

1. $\left| -\dfrac{2}{3} + \dfrac{1}{5} \right|$ [1.8]

2. $y - [3 - 4(5 - 2y) - 3y]$ [14.3]

3. $(10 \cdot 8 - 9 \cdot 7)^2 - 54 \div 9 - 3$ [1.8]

4. $(2.7 \times 10^{-24})(3.1 \times 10^9)$ [4.2]

5. Evaluate

$$\frac{ab - ac}{bc}$$

for $a = -2$, $b = 3$, and $c = -4$. [1.8]

Perform the indicated operations to create an equivalent expression. Be sure to simplify your result if possible.

6. $(5a^2 - 3ab - 7b^2) - (2a^2 + 5ab + 8b^2)$ [4.7]

7. $(2a - 1)(2a + 1)$ [4.6]

8. $(3a^2 - 5y)^2$ [4.7]

9. $\dfrac{1}{x - 2} - \dfrac{4}{x^2 - 4} + \dfrac{3}{x + 2}$ [6.4]

10. $\dfrac{x^2 - 6x + 8}{4x + 12} \cdot \dfrac{x + 3}{x^2 - 4}$ [6.2]

11. $\dfrac{3x + 3y}{5x - 5y} \div \dfrac{3x^2 + 3y^2}{5x^3 - 5y^3}$ [6.2]

12. $\dfrac{x - \dfrac{a^2}{x}}{1 + \dfrac{a}{x}}$ [6.5]

13. $\sqrt{12a}\,\sqrt{12a^3b}$ [10.3]

14. $(-9x^2y^5)(3x^8y^{-7})$ [4.2]

15. $(125x^6y^{1/2})^{2/3}$ [10.2]

16. $\dfrac{\sqrt[3]{x^2y^5}}{\sqrt[4]{xy^2}}$ [10.5]

17. $(4 + 6i)(2 - i)$, where $i = \sqrt{-1}$ [10.8]

Factor, if possible, to form an equivalent expression.

18. $4x^2 - 12x + 9$ [5.4]

19. $27a^3 - 8$ [5.5]

20. $12s^4 - 48t^2$ [5.4]

21. $15y^4 + 33y^2 - 36$ [5.3]

22. Divide:

$$(7x^4 - 5x^3 + x^2 - 4) \div (x - 2).$$ [4.8]

23. For the function described by

$$f(x) = 3x^2 - 4x,$$

find $f(-2)$. [7.1]

Find the domain of each function.

24. $f(x) = \sqrt{2x - 8}$ [10.1]

25. $g(x) = \dfrac{x - 4}{x^2 - 10x + 25}$ [9.2]

26. Write an equivalent expression by rationalizing the denominator:

$$\frac{1 - \sqrt{x}}{1 + \sqrt{x}}.$$ [10.5]

27. Find a linear equation whose graph has a y-intercept of $(0, -8)$ and is parallel to the line whose equation is $3x - y = 6$. [3.6]

28. Write a quadratic equation whose solutions are $5\sqrt{2}$ and $-5\sqrt{2}$. [11.3]

29. Find the center and the radius of the circle given by

$$x^2 + y^2 - 4x + 6y - 23 = 0.$$ [13.1]

30. Write an equivalent expression that is a single logarithm:

$$\tfrac{2}{3}\log_a x - \tfrac{1}{2}\log_a y + 5\log_a z.$$ [12.4]

31. Write an equivalent exponential equation:
$\log_a c = 5$. [12.3]

Use a calculator to find each of the following. Round to four decimal places. [12.5]

32. $\log 120$

33. $\log_5 3$

34. Find the distance between the points $(-1, -5)$ and $(2, -1)$. [10.7]

35. Find the 21st term of the arithmetic sequence $19, 12, 5, \ldots$. [14.2]

36. Find the sum of the first 25 terms of the arithmetic series $-1 + 2 + 5 + \cdots$. [14.2]

37. Write an expression for the general term of the geometric sequence $16, 4, 1, \ldots$. [14.3]

38. Find the 7th term of $(a - 2b)^{10}$. [14.4]

39. Find the sum of the first nine terms of the geometric series $4 + 6 + 9 + \cdots$. [14.3]

Solve.

40. $8(x - 1) - 3(x - 2) = 1$ [2.2]

41. $\dfrac{6}{x} + \dfrac{6}{x + 2} = \dfrac{5}{2}$ [6.6]

42. $2x + 1 > 5 \text{ or } x - 7 \le 3$ [9.2]

43. $5x + 6y = -2,$
$3x + 10y = 2$ [8.2]

44. $x + y - z = 0,$
$3x + y + z = 6,$
$x - y + 2z = 5$ [8.4]

45. $3\sqrt{x - 1} = 5 - x$ [10.6]

46. $x^4 - 29x^2 + 100 = 0$ [11.5]

47. $x^2 + y^2 = 8,$
$x^2 - y^2 = 2$ [13.4]

48. $4^x = 12$ [12.6]

49. $\log(x^2 - 25) - \log(x + 5) = 3$ [12.6]

50. $\log_5 x = -2$ [12.6]

51. $7^{2x+3} = 49$ [12.6]

52. $|2x - 1| \le 5$ [9.3]

53. $15x^2 + 45 = 0$ [11.1]

54. $x^2 + 4x = 3$ [11.2]

55. $y^2 + 3y > 10$ [11.9]

56. Let $f(x) = x^2 - 2x$. Find a such that $f(a) = 80$. [11.1]

57. If $f(x) = \sqrt{-x + 4} + 3$ and $g(x) = \sqrt{x - 2} + 3$, find a such that $f(a) = g(a)$. [10.6]

58. Solve $V = P - Prt$ for r. [2.3]

59. Solve $I = \dfrac{R}{R + r}$ for R. [7.5]

Graph.

60. $3x - y = 7$ [3.6]

61. $x^2 + y^2 = 100$ [13.1]

62. $\dfrac{x^2}{36} - \dfrac{y^2}{9} = 1$ [13.3]

63. $y = \log_2 x$ [12.3]

64. $f(x) = 2^x - 3$ [12.2]

65. $2x - 3y < -6$ [9.4]

66. Graph: $f(x) = -2(x - 3)^2 + 1$. [11.7]
a) Label the vertex.
b) Draw the axis of symmetry.
c) Find the maximum or minimum value.

Solve.

67. The Brighton recreation department plans to fence in a rectangular park next to a river. (Note that no fence will be needed along the river.) What is the area of the largest region that can be fenced in with 200 ft of fencing? [11.8]

68. The perimeter of a rectangular sign is 34 ft. The length of a diagonal is 13 ft. Find the dimensions of the sign. [13.4]

69. A movie club offers two types of membership. Limited members pay a fee of $40 a year and can rent movies for $2.45 each. Preferred members pay $60 a year and can rent movies for $1.65 each. For what numbers of annual movie rentals would it be less expensive to be a preferred member? [9.1]

70. Find three consecutive odd integers whose sum is 177. [2.5]

71. Cosmos Tastes mixes herbs that cost $2.68 an ounce with herbs that cost $4.60 an ounce to create a seasoning that costs $3.80 an ounce. How many ounces of each herb should be mixed together to make 24 oz of the seasoning? [8.3]

72. An airplane can fly 190 mi with the wind in the same time it takes to fly 160 mi against the wind. The speed of the wind is 30 mph. How fast can the plane fly in still air? [6.7]

73. Jared can tap the sugar maple trees in Southway Farm in 21 hr. Delia can tap the trees in 14 hr. How long would it take them, working together, to tap the trees? [6.7]

74. The centripetal force F of an object moving in a circle varies directly as the square of the velocity v and inversely as the radius r of the circle. If $F = 8$ when $v = 1$ and $r = 10$, what is F when $v = 2$ and $r = 16$? [7.5]

75. *Mortgages.* The loan-to-value ratio of a mortgage is the ratio of the amount owed on the loan to the value of the home. In 2002, the average homeowner owed 80% of the value of the home. By 2007, largely due to falling home prices, this amount had risen to 87%.
Source: UBS Mortgage Strategy Group

a) What was the average rate of change? [3.4]
b) Find a linear function that fits the data. Let t represent the number of years since 2000. [7.3]
c) Use the function from part (b) to predict the average loan-to-value ratio in 2010. [7.3]
d) Assuming the trend continues, in what year will the average U.S. homeowner owe 95% of the value of the home? [7.3]

76. *Mortgages.* In a reverse mortgage, the lender makes payments to the borrower and the borrower keeps control of the house. The loan is repaid when the house is sold. The number of reverse mortgages in the United States has increased exponentially from approximately 160 in 1990 to 108,000 in 2007. [12.7]
Source: U.S. Department of Housing and Urban Development

a) Find the exponential growth rate k, and write an equation for an exponential function that can be used to predict the number of reverse mortgages t years after 1990.
b) Predict the number of reverse mortgages in 2012.
c) In what year will there be 1 million reverse mortgages?

77. *Retirement.* Sarita invested $2000 in a retirement account on her 22nd birthday. If the account earns 5% interest, compounded annually, how much will this investment be worth on her 62nd birthday? [14.3]

Synthesis

Solve.

78. $\dfrac{9}{x} - \dfrac{9}{x + 12} = \dfrac{108}{x^2 + 12x}$ [6.6]

79. $\log_2 (\log_3 x) = 2$ [12.6]

80. y varies directly as the cube of x and x is multiplied by 0.5. What is the effect on y? [7.5]

81. Diaphantos, a famous mathematician, spent $\frac{1}{6}$ of his life as a child, $\frac{1}{12}$ as an adolescent, and $\frac{1}{7}$ as a bachelor. Five years after he was married, he had a son who died 4 years before his father at half his father's final age. How long did Diaphantos live? [8.5]

Elementary Algebra Review

BEN GIVENS
WIND FARM OPERATIONS
MANAGER
Trent, Texas

The operations and maintenance of a wind farm depend heavily on data. We use many complicated mathematical algorithms to analyze the performance of the turbines. Production data are used to determine royalty payments and revenue for the project on a monthly basis. Without math, there would be no way to perform these tasks.

AN APPLICATION

The number of watts of power P generated by a particular turbine at a wind speed of x miles per hour can be approximated by the polynomial

$$P = 0.0157x^3 + 0.1163x^2 - 1.3396x + 3.7063.$$

Estimate the power generated by a 10-mph wind.

Source: Based on data from *QST*, November 2006

This exercise appears as Exercise 47 in Section R.4.

This chapter is a review of the first six chapters of this text. Each section corresponds to a chapter of the text. For further explanation of the topics in this chapter, refer to the section or pages referenced in the margin.

R.1 Introduction to Algebraic Expressions

The Real Numbers ▪ Operations on Real Numbers ▪ Algebraic Expressions

The Real Numbers

Sets of real numbers (Section 1.4)

Sets of Numbers

Natural numbers: $\{1, 2, 3, \dots\}$

Whole numbers: $\{0, 1, 2, 3, \dots\}$

Integers: $\{\dots, -3, -2, -1, 0, 1, 2, 3, \dots\}$

Rational numbers: $\left\{ \dfrac{a}{b} \middle| a \text{ and } b \text{ are integers and } b \neq 0 \right\}$

Terminating decimals (p. 32)

Repeating decimals (p. 32)

Irrational numbers (p. 32)

Real numbers (p. 33)

Rational numbers can always be written as **terminating** or **repeating** decimals. **Irrational numbers,** like $\sqrt{2}$ or π, can be thought of as nonterminating and nonrepeating decimals. The set of **real numbers** consists of all rational and irrational numbers, taken together.

Real numbers can be represented by points on a number line.

Order (p. 34)

Equation (p. 6)

Inequality (p. 34)

We can compare, or **order,** real numbers by their graphs on the number line. For any two numbers, the one to the left is less than the one to the right.

Sentences like $\frac{1}{4} = 0.25$, containing an equals sign, are called **equations.** An **inequality** is a sentence containing $>$ (is greater than), $<$ (is less than), $\geq$ (is greater than or equal to), or $\leq$ (is less than or equal to). Equations and inequalities can be true or false.

EXAMPLE **1** Write true or false for each equation or inequality.

a) $-2\frac{1}{3} = -\frac{7}{3}$ **b)** $1 = -1$ **c)** $-5 < -2$

d) $-3 \geq 2$ **e)** $1.1 \leq 1.1$

SOLUTION

a) $-2\frac{1}{3} = -\frac{7}{3}$ is *true* because $-2\frac{1}{3}$ and $-\frac{7}{3}$ represent the same number.

b) $1 = -1$ is a *false* equation.

c) $-5 < -2$ is *true* because -5 is to the left of -2 on the number line.

d) $-3 \geq 2$ is *false* because neither $-3 > 2$ nor $-3 = 2$ is true.

e) $1.1 \leq 1.1$ is *true* because $1.1 = 1.1$ is true.

▶ **TRY EXERCISE** 1

Absolute value (p. 35)

The distance of a number from 0 is called the **absolute value** of the number. The notation $|-4|$ represents the absolute value of -4. The absolute value of a number is never negative.

EXAMPLE 2

Find the absolute value: **(a)** $|-4|$; **(b)** $\left|\frac{11}{3}\right|$; **(c)** $|0|$.

SOLUTION

a) $|-4| = 4$ since -4 is 4 units from 0.

b) $\left|\frac{11}{3}\right| = \frac{11}{3}$ since $\frac{11}{3}$ is $\frac{11}{3}$ units from 0.

c) $|0| = 0$ since 0 is 0 units from itself.

TRY EXERCISE 7

Operations on Real Numbers

Addition (Section 1.5)

Rules for Addition of Real Numbers

1. *Positive numbers*: Add as usual. The answer is positive.
2. *Negative numbers*: Add absolute values and make the answer negative.
3. *A positive and a negative number*: Subtract the smaller absolute value from the greater absolute value. Then:

   a) If the positive number has the greater absolute value, the answer is positive.

   b) If the negative number has the greater absolute value, the answer is negative.

   c) If the numbers have the same absolute value, the answer is 0.

4. *One number is zero*: The sum is the other number.

Opposite (p. 44)

Every real number has an **opposite**. The opposite of -6 is 6, the opposite of 3.7 is -3.7, and the opposite of 0 is itself. When opposites are added, the result is 0. Finding the opposite of a number is often called "changing its sign."

Subtraction of real numbers is defined in terms of addition and opposites.

Subtraction (Section 1.6)

Subtraction of Real Numbers

To subtract, add the opposite of the number being subtracted.

The rules for multiplication of real numbers can be stated together.

Multiplication and division (Section 1.7)

Rules for Multiplication and Division

To multiply or divide two nonzero real numbers:

1. Using the absolute values, multiply or divide, as indicated.
2. If the signs are the same, the answer is positive.
3. If the signs are different, the answer is negative.

EXAMPLE 3   Perform the indicated operations: **(a)** $-13 + (-9)$; **(b)** $-\frac{4}{5} + \frac{1}{10}$; **(c)** $-6 - (-7.3)$; **(d)** $3(-1.5)$; **(e)** $\left(-\frac{4}{9}\right) \div \left(-\frac{2}{5}\right)$.

SOLUTION

a) $-13 + (-9) = -22$

Two negatives. *Think:* Add the absolute values, 13 and 9, to get 22. Make the answer *negative*, -22.

b) $-\frac{4}{5} + \frac{1}{10} = -\frac{8}{10} + \frac{1}{10} = -\frac{7}{10}$

A negative and a positive. *Think:* The difference of absolute values is $\frac{8}{10} - \frac{1}{10}$, or $\frac{7}{10}$. The negative number has the larger absolute value, so the answer is *negative*, $-\frac{7}{10}$.

c) $-6 - (-7.3) = -6 + 7.3 = 1.3$

Change the subtraction to addition and add the opposite.

d) $3(-1.5) = -4.5$

Think: $3(1.5) = 4.5$. The signs are different, so the answer is negative.

e) $\left(-\frac{4}{9}\right) \div \left(-\frac{2}{5}\right) = \left(-\frac{4}{9}\right) \cdot \left(-\frac{5}{2}\right)$

Multiplying by the reciprocal. The answer is positive.

$$= \frac{20}{18} = \frac{10}{9} \cdot \frac{2}{2} = \frac{10}{9}$$

TRY EXERCISE 11

Division by 0 (p. 55)

Exponential notation (p. 60)

Addition, subtraction, and multiplication are defined for all real numbers, but we cannot **divide by 0**. For example, $\frac{0}{3} = 0 \div 3 = 0$, but $\frac{3}{0} = 3 \div 0$ is **undefined**.

A product like $2 \cdot 2 \cdot 2 \cdot 2$, in which the factors are the same, is called a **power**. Powers are often written using **exponential notation**:

$$2 \cdot 2 \cdot 2 \cdot 2 = 2^4. \leftarrow \text{There are 4 factors; 4 is the } \textit{exponent}.$$

$$\text{2 is the } \textit{base}.$$

A number raised to the power of 1 is the number itself; for example, $3^1 = 3$.

An expression containing a series of operations is not necessarily evaluated from left to right. Instead, we perform the operations according to the following rules.

Rules for Order of Operations

1. Calculate within the innermost grouping symbols, $(\ )$, $[\ ]$, $\{\ \}$, $|\ |$, and above or below fraction bars.
2. Simplify all exponential expressions.
3. Perform all multiplications and divisions, working from left to right.
4. Perform all additions and subtractions, working from left to right.

EXAMPLE 4   Simplify: $3 - [(4 \times 5) + 12 \div 2^3 \times 6] + 5$.

SOLUTION

$$3 - [(4 \times 5) + 12 \div 2^3 \times 6] + 5$$

$$= 3 - [20 + 12 \div 2^3 \times 6] + 5 \qquad \text{Doing the calculations in the innermost parentheses first}$$

$$= 3 - [20 + 12 \div 8 \times 6] + 5 \qquad \text{Working inside the brackets; evaluating } 2^3$$

$$= 3 - [20 + 1.5 \times 6] + 5 \qquad 12 \div 8 \text{ is the first multiplication or division working from left to right.}$$

$$= 3 - [20 + 9] + 5 \qquad \text{Multiplying}$$

$$= 3 - 29 + 5 \qquad \text{Completing the calculations within the brackets}$$

$$= -26 + 5$$
$$= -21 \qquad \text{Adding and subtracting from left to right}$$

> **TRY EXERCISE** 45

Algebraic expression (p. 3)

Constant (p. 2)

Variable (p. 2)

Evaluate (p. 3)

Substitute (p. 3)

Algebraic Expressions

In an **algebraic expression** like $2xt^3$, the number 2 is a **constant** and x and t are **variables**. Algebraic expressions containing variables can be **evaluated** by **substituting** a number for each variable in the expression and following the rules for order of operations.

EXAMPLE **5**   The perimeter P of a rectangle of length l and width w is given by the formula $P = 2l + 2w$. Find the perimeter when l is 16 in. and w is 7.5 in.

SOLUTION   We evaluate, substituting 16 in. for l and 7.5 in. for w and carrying out the operations:

$$P = 2l + 2w$$

Substitute. $= 2 \cdot 16 + 2 \cdot 7.5$

Carry out the operations. $= 32 + 15$

$= 47$ in.

> **TRY EXERCISE** 59

Expressions that represent the same number are said to be **equivalent**. The laws that follow provide methods for writing equivalent expressions.

Laws and Properties of Real Numbers

Commutative laws:	$a + b = b + a;\ ab = ba$
Associative laws:	$a + (b + c) = (a + b) + c;$ $a(bc) = (ab)c$
Distributive law:	$a(b + c) = ab + ac$
Identity property of 1:	$1 \cdot a = a \cdot 1 = a$
Identity property of 0:	$a + 0 = 0 + a = a$
Law of opposites:	$a + (-a) = 0$
Multiplicative property of 0:	$0 \cdot a = a \cdot 0 = 0$
Property of -1:	$-1 \cdot a = -a$
Opposite of a sum:	$-(a + b) = -a + (-b)$
$\dfrac{-a}{b} = \dfrac{a}{-b} = -\dfrac{a}{b},$	$\dfrac{-a}{-b} = \dfrac{a}{b}$

Factor (p. 17)

The distributive law can be used to multiply and to **factor** expressions. We factor an expression when we write an equivalent expression that is a product.

EXAMPLE **6** Write an equivalent expression as indicated.

a) Multiply: $-2(5x - 3)$. **b)** Factor: $5x + 10y + 5$.

SOLUTION

a) $-2(5x - 3) = -2(5x + (-3))$ Adding the opposite

$= -2 \cdot 5x + (-2) \cdot (-3)$ Using the distributive law

$= (-2 \cdot 5)x + 6$ Using the associative law for multiplication

$= -10x + 6$

b) $5x + 10y + 5 = 5 \cdot x + 5 \cdot 2y + 5 \cdot 1$ The common factor is 5.

$= 5(x + 2y + 1)$ Using the distributive law

Factoring can be checked by multiplying:

$5(x + 2y + 1) = 5 \cdot x + 5 \cdot 2y + 5 \cdot 1 = 5x + 10y + 5.$

TRY EXERCISES 61 and 69

Terms (p. 17)

Combine like terms (p. 41)

The **terms** of an algebraic expression are separated by plus signs. When two terms have variable factors that are exactly the same, the terms are called **like**, or **similar**, **terms**. The distributive law enables us to **combine**, or **collect**, **like terms**.

EXAMPLE **7** Combine like terms: $-5m + 3n - 4n + 10m$.

SOLUTION

$-5m + 3n - 4n + 10m$

$= -5m + 3n + (-4n) + 10m$ Rewriting as addition

$= -5m + 10m + 3n + (-4n)$ Using the commutative law of addition

$= (-5 + 10)m + (3 + (-4))n$ Using the distributive law

$= 5m + (-n)$

$= 5m - n$ Rewriting as subtraction

TRY EXERCISE 75

We can also use the distributive law to help simplify algebraic expressions containing parentheses.

EXAMPLE **8** Simplify: **(a)** $4x - (y - 2x)$; **(b)** $3(t + 2) - 6(t - 1)$.

SOLUTION

a) $4x - (y - 2x) = 4x - y + 2x$ Removing parentheses and changing the sign of every term

$= 6x - y$ Combining like terms

b) $3(t + 2) - 6(t - 1) = 3t + 6 - 6t + 6$ Multiplying each term of $t + 2$ by 3 and each term of $t - 1$ by -6

$= -3t + 12$ Combining like terms

TRY EXERCISE 81

Value (p. 3)
Solution (p. 6)

If the expressions on each side of an equation have the same **value** for a given number, then that number is a **solution** of the equation.

EXAMPLE **9**   Determine whether each number is a solution of $x - 2 = -5$.

a) 3   **b)** −3

SOLUTION

a) We have:

$$
\begin{array}{rl}
x - 2 = -5 & \text{Writing the equation} \\
\overline{3 - 2 \ \bigm| \ -5} & \text{Substituting 3 for } x \\
1 \overset{?}{=} -5 & 1 = -5 \text{ is FALSE}
\end{array}
$$

Since $3 - 2 = -5$ is false, 3 is not a solution of $x - 2 = -5$.

b) We have

$$
\begin{array}{rl}
x - 2 = -5 & \\
\overline{-3 - 2 \ \bigm| \ -5} & \\
-5 \overset{?}{=} -5 & \text{TRUE}
\end{array}
$$

Since $-3 - 2 = -5$ is true, −3 is a solution of $x - 2 = -5$.

TRY EXERCISE 89

Translating to algebraic expressions (p. 4)

Translating to equations (p. 6)

Certain word phrases can be translated to algebraic expressions. These in turn can often be used to translate problems to equations.

EXAMPLE **10**   *Energy Use.*   Translate the following problem to an equation. Do not solve.

On average, a home spa costs about \$192 a year to operate. This is 16 times as much as the average annual operating cost of a home computer. How much does it cost to operate a home computer for a year?

Source: U.S. Department of Energy

SOLUTION   We let c represent the annual cost of operating a home computer. We then reword the problem to make the translation more direct.

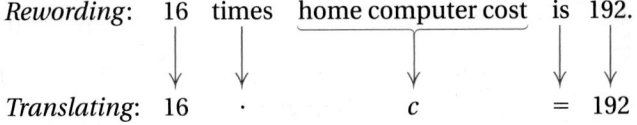

Rewording: 16   times   home computer cost   is   192.

Translating: 16   · c =   192

TRY EXERCISE 95

R.1   EXERCISE SET

Classify each equation or inequality as true or false.

1. $1.4 = 1.41$

2. $-3 \geq -3$

3. $-8 < -7$

4. $0 \leq -1$

5. $0 > -5$

6. $\frac{1}{10} = 0.1$

Find each absolute value.

7. $|22|$

8. $\left|\frac{11}{4}\right|$

9. $|-1.3|$

10. $|-105|$

Simplify.

11. $(-14) + (-11)$

12. $3 - (-2)$

13. $-\frac{1}{3} - \frac{2}{5}$

14. $\frac{3}{8} \div \frac{3}{5}$

15. $4.2 - 10.7$

16. $(-1.3)(2.8)$

17. $-9 + 0$

18. $\left(-\frac{1}{2}\right) + \frac{1}{8}$

19. $0 \div (-10)$

20. $0 - 32$

21. $\left(-\frac{3}{10}\right) + \left(-\frac{1}{5}\right)$

22. $\left(-\frac{4}{7}\right)\left(\frac{7}{4}\right)$

23. $-3.8 + 9.6$

24. $-0.01 + 1$

25. $(-12) \div 4$

26. $(-87)(0)$

27. $32 - (-7)$

28. $-100 + 35$

29. $(-10)(-17.5)$

30. $-10 - 2.68$

31. $(-68) + 36$

32. $175 \div (-25)$

33. $2 + (-3) + 7 + 10$

34. $-5 + (-15) + 13 + (-1)$

35. $3 \cdot (-2) \cdot (-1) \cdot (-1)$

36. $(-6) \cdot (-5) \cdot (-4) \cdot (-3) \cdot (-2) \cdot (-1)$

37. $(-1)^4 + 2^3$

38. $(-1)^5 + 2^4$

39. $2 \times 6 - 3 \times 5$

40. $12 \div 4 + 15 \div 3$

41. $3 - (11 + 2 \cdot 4)$

42. $3 - 11 + 2 \cdot 4$

43. $4 \cdot 5^2$

44. $7 \cdot 2^3$

45. $25 - 8 \times 3 + 1$

46. $12 - 16 \times 5 + 4$

47. $2 - (3^3 + 16 \div (-2)^3)$

48. $-7 - (8 + 10 \cdot 2^2)$

49. $|6(-3)| + |(-2)(-9)|$

50. $3 - |2 - 7 + 4|$

51. $\dfrac{7000 + (-10)^3}{10^2 \times (2 + 4)}$

52. $\dfrac{3 - 2 \times 6 - 5}{2(3 + 7)^2}$

53. $2 + 8 \div 2 \times 2$

54. $2 + 8 \div (2 \times 2)$

Evaluate.

55. $y - x$, for $x = 10$ and $y = 3$

56. $n - 2m$, for $m = 6$ and $n = 11$

57. $-3 - x^2 + 12x$, for $x = 5$

58. $14 + (y - 5)^2 - 12 \div y$, for $y = -2$

59. The area of a parallelogram with base b and height h is bh. Find the area of the parallelogram when the height is 3.5 cm and the base is 8 cm.

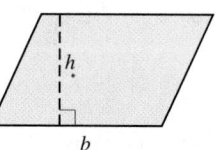

60. The area of a triangle with base b and height h is $\frac{1}{2}bh$. Find the area of the triangle when the height is 2 in. and the base is 6.2 in.

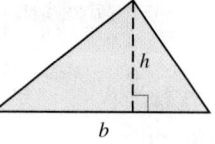

Multiply.

61. $4(2x + 7)$

62. $3(5y + 1)$

63. $-2(15 - 3x)$

64. $-7(3x - 5)$

65. $2(4a + 6b - 3c)$

66. $5(8p + q - 5r)$

67. $-3(2x - y + z)$

68. $-10(-6 - y - z)$

Factor.

69. $8x + 6y$

70. $7p + 14q$

71. $3 + 3w$

72. $4x + 4y$

73. $10x + 50y + 100$

74. $81p + 27q + 9$

Combine like terms.

75. $3p - 2p$

76. $4x + 3x$

77. $4m + 10 - 5m + 12$

78. $3a - 4b - b - 6a$

79. $-6x + 7 + x$

80. $16r + (-7r) + 3s$

Remove parentheses and simplify.

81. $2p - (7 - 4p)$

82. $4r - (3r + 5)$

83. $6x + 5y - 7(x - y)$

84. $14m - 6(2n - 3m) + n$

85. $6[2a + 4(a - 2b)]$

86. $2[2a + 1 - (3a - 6)]$

87. $3 - 2[5(x - 10y) - (3 + 2y)]$

88. $7 - 4[2(3 - 2x) - 5(4x - 3)]$

Determine whether the given number is a solution of the given equation.

89. $4; 3x - 2 = 10$

90. $12; 100 = 4x + 50$

91. $-3; 4 - x = 1$

92. $-1; 2 = 5 + 3x$

93. $4.6; \dfrac{x}{2} = 2.3$

94. $144; \dfrac{x}{9} = 16$

Translate each problem to an equation. Do not solve.

95. Three times what number is 348?

96. What number added to 256 is 113?

97. *Fast-food calories.* A McDonald's Big Mac® contains 500 calories. This is 69 more calories than a Taco Bell Beef Burrito® provides. How many calories are in a Taco Bell Beef Burrito?

98. *Coca-Cola® consumption.* The average U.S. citizen consumes 296 servings of Coca-Cola each year. This is 7.4 times the international average. What is the international average per capita consumption of Coke?

99. *Vegetable production.* It takes 42 gal of water to produce 1 lb of broccoli. This is twice the amount of water used to produce 1 lb of lettuce. How many gallons of water does it take to produce 1 lb of lettuce?

100. *Sports costs.* The average annual cost for scuba diving is $470. This is $458 more than the average annual cost to play badminton. What is the average annual cost to play badminton?

R.2 | Equations, Inequalities, and Problem Solving

Solving Equations and Formulas ▪ Solving Inequalities ▪ Problem Solving

Solving Equations and Formulas

Any replacement for the variable in an equation that makes the equation true is called a *solution* of the equation. To **solve** an equation means to find all of its solutions.

Equivalent equations (p. 79)

We use the following principles to write **equivalent equations**, or equations with the same solutions.

The addition principle (p. 79)

> ### The Addition and Multiplication Principles for Equations
>
> ***The Addition Principle***
>
> For any real numbers a, b, and c,
>
> $$a = b \quad \text{is equivalent to} \quad a + c = b + c.$$
>
> ***The Multiplication Principle***
>
> For any real numbers a, b, and c, with $c \neq 0$,
>
> $$a = b \quad \text{is equivalent to} \quad a \cdot c = b \cdot c.$$

The multiplication principle (p. 81)

To solve $x + a = b$ for x, we add $-a$ to (or subtract a from) both sides. To solve $ax = b$ for x, we multiply both sides by $\dfrac{1}{a}$ (or divide both sides by a).

To solve an equation like $-3x - 10 = 14$, we first isolate the variable term, $-3x$, using the addition principle. Then we use the multiplication principle to get the variable by itself.

EXAMPLE 1   Solve: $-3x - 10 = 14$.

SOLUTION

$$-3x - 10 = 14$$

$$-3x - 10 + 10 = 14 + 10 \qquad \text{Using the addition principle:}$$
Adding 10 to both sides

Isolate the x-term. $-3x = 24$ Simplifying

$$\frac{-3x}{-3} = \frac{24}{-3} \qquad \text{Dividing both sides by } -3$$

Isolate x. $x = -8$ Simplifying

Check:

$$\begin{array}{r|l} -3x - 10 = 14 \\ \hline -3(-8) - 10 & 14 \\ 24 - 10 & \\ 14 \overset{?}{=} 14 & \text{TRUE} \end{array}$$

The solution is -8.

TRY EXERCISE 13

Clearing fractions (p. 88)

Equations are generally easier to solve when they do not contain fractions. The easiest way to clear an equation of fractions is to multiply *every term on both sides* of the equation by the least common denominator.

EXAMPLE 2   Solve: $\frac{5}{2} - \frac{1}{6}t = \frac{2}{3}$.

SOLUTION   The number 6 is the least common denominator, so we multiply both sides by 6.

$$\frac{5}{2} - \frac{1}{6}t = \frac{2}{3}$$

$$6\left(\frac{5}{2} - \frac{1}{6}t\right) = 6 \cdot \frac{2}{3}$$ Multiplying both sides by 6

$$6 \cdot \frac{5}{2} - 6 \cdot \frac{1}{6}t = 6 \cdot \frac{2}{3}$$ Using the distributive law. Be sure to multiply every term by 6.

$$15 - t = 4$$ The fractions are cleared.

$$15 - t - 15 = 4 - 15$$ Subtracting 15 from both sides

$$-t = -11$$ $15 - t - 15 = 15 + (-t) + (-15)$
$= -t + 15 + (-15) = -t$

$$(-1)(-t) = (-1)(-11)$$ Multiplying both sides by -1 to change the sign

$$t = 11$$

Check:

$$\frac{5}{2} - \frac{1}{6}t = \frac{2}{3}$$

$$\frac{5}{2} - \frac{1}{6}(11) \quad \bigg| \quad \frac{2}{3}$$

$$\frac{5}{2} - \frac{11}{6}$$

$$\frac{15}{6} - \frac{11}{6}$$

$$\frac{2}{3} \stackrel{?}{=} \frac{2}{3} \quad \text{TRUE}$$

The solution is 11.

TRY EXERCISE 19

To solve equations that contain parentheses, we can use the distributive law to first remove the parentheses. If like terms appear in an equation, we combine them and then solve.

EXAMPLE 3   Solve: $1 - 3(4 - x) = 2(x + 5) - 3x$.

SOLUTION

$$1 - 3(4 - x) = 2(x + 5) - 3x$$

$$1 - 12 + 3x = 2x + 10 - 3x$$ Using the distributive law

$$-11 + 3x = -x + 10$$ Combining like terms;
$1 - 12 = -11$ and $2x - 3x = -x$

$$-11 + 3x + x = 10$$ Adding x to both sides to get all x-terms on one side

$$-11 + 4x = 10$$ Combining like terms

$$4x = 10 + 11$$ Adding 11 to both sides to isolate the x-term

$$4x = 21$$ Simplifying

$$x = \frac{21}{4}$$ Dividing both sides by 4

Check:

$$\frac{1 - 3(4 - x) = 2(x + 5) - 3x}{\begin{array}{c|c} 1 - 3\left(4 - \frac{21}{4}\right) & 2\left(\frac{21}{4} + 5\right) - 3\left(\frac{21}{4}\right) \\ 1 - 3\left(-\frac{5}{4}\right) & 2\left(\frac{41}{4}\right) - \frac{63}{4} \\ 1 + \frac{15}{4} & \frac{82}{4} - \frac{63}{4} \\ \frac{19}{4} & \stackrel{?}{=} \frac{19}{4} \end{array}}$$

TRUE

The solution is $\frac{21}{4}$.

TRY EXERCISE 25

Conditional equation (p. 90)
Identity (p. 90)
Contradiction (p. 90)

Formulas (Section 2.3)

The equations in Examples 1–3 are **conditional**—that is, they are true for some values of x and false for other values of x. An **identity** is an equation like $x + 1 = x + 1$ that is true for all values of x. A **contradiction** is an equation like $x + 1 = x + 2$ that is never true.

A **formula** is an equation using two or more letters that represents a relationship between two or more quantities. A formula can be solved for a specified letter using the principles for solving equations.

EXAMPLE 4

The formula

$$A = \frac{a + b + c + d}{4}$$

gives the average A of four test scores a, b, c, and d. Solve for d.

SOLUTION   We have

$$A = \frac{a + b + c + d}{4} \qquad \text{We want the letter } d \text{ alone.}$$

$$4A = a + b + c + d \qquad \text{Multiplying by 4 to clear the fraction}$$

$$4A - a - b - c = d. \qquad \begin{array}{l} \text{Subtracting } a + b + c \text{ from (or} \\ \text{adding } -a - b - c \text{ to) both sides.} \\ \text{The letter } d \text{ is now isolated.} \end{array}$$

We can also write this as $d = 4A - a - b - c$. This formula can be used to determine the test score needed to obtain a specified average if three tests have already been taken.

TRY EXERCISE 33

Solving Inequalities

Solutions of inequalities (p. 121)
Graphs of inequalities (p. 122)

A **solution of an inequality** is a replacement of the variable that makes the inequality true. The solutions of an inequality in one variable can be **graphed**, or represented by a drawing, on the number line. All points that are solutions are shaded. A parenthesis indicates an endpoint that is not a solution and a bracket indicates an endpoint that is a solution.

EXAMPLE 5

Graph each inequality: **(a)** $m \leq 2$; **(b)** $-1 \leq x < 4$.

SOLUTION

a) The solutions of $m \leq 2$ are shown on the number line by shading points to the left of 2 as well as the point at 2. The bracket at 2 indicates that 2 is a part of the graph (that is, it is a solution of $m \leq 2$).

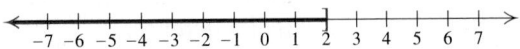

b) In order to be a solution of the inequality $-1 \leq x < 4$, a number must be a solution of both $-1 \leq x$ and $x < 4$. The solutions are shaded on the number line, with a parenthesis indicating that 4 is not a solution and a bracket indicating that -1 is a solution.

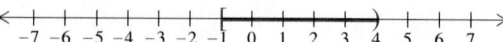

Set-builder notation (p. 123)
Interval notation (p. 123)

In Example 5, note that $m \leq 2$ and $-1 \leq x < 4$ are inequalities that describe a solution set. Since it is impossible to list all the solutions, we use **set-builder notation** or **interval notation** to write such sets.

Using set-builder notation, we write the solution set of Example 5(a) as

$$\{m \mid m \leq 2\},$$

read

"the set of all m such that m is less than or equal to 2."

Interval notation uses parentheses, ( ), and brackets, [ ], to describe a set of real numbers.

Interval Notation	Set-builder Notation	Graph
(a, b) open interval	$\{x \mid a < x < b\}$	(a, b) a b
$[a, b]$ closed interval	$\{x \mid a \leq x \leq b\}$	$[a, b]$ a b
$(a, b]$ half-open interval	$\{x \mid a < x \leq b\}$	$(a, b]$ a b
$[a, b)$ half-open interval	$\{x \mid a \leq x < b\}$	$[a, b)$ a b
(a, ∞)	$\{x \mid x > a\}$	a
$[a, \infty)$	$\{x \mid x \geq a\}$	a
$(-\infty, a)$	$\{x \mid x < a\}$	a
$(-\infty, a]$	$\{x \mid x \leq a\}$	a

Thus, for example, the solution set of Example 5(a), in interval notation, is $(-\infty, 2]$.

Equivalent inequalities (p. 124)

As with equations, our goal when solving inequalities is to isolate the variable on one side. We use principles that enable us to write **equivalent inequalities**—inequalities having the same solution set. The addition principle is similar to the addition principle for equations; the multiplication principle contains an important difference.

The addition principle for inequalities (p. 124)

The multiplication principle for inequalities (p. 126)

> ### The Addition and Multiplication Principles for Inequalities
>
> #### The Addition Principle
>
> For any real numbers a, b, and c,
>
> $$a < b \quad \text{is equivalent to} \quad a + c < b + c, \quad \text{and}$$
> $$a > b \quad \text{is equivalent to} \quad a + c > b + c.$$
>
> #### The Multiplication Principle
>
> For any real numbers a and b, and for any *positive* number c,
>
> $$a < b \quad \text{is equivalent to} \quad ac < bc, \quad \text{and}$$
> $$a > b \quad \text{is equivalent to} \quad ac > bc.$$
>
> For any real numbers a and b, and for any *negative* number c,
>
> $$a < b \quad \text{is equivalent to} \quad ac > bc, \quad \text{and}$$
> $$a > b \quad \text{is equivalent to} \quad ac < bc.$$
>
> Similar statements hold for $\leq$ and $\geq$.

Note that when we multiply both sides of an inequality by a negative number, we must reverse the direction of the inequality symbol in order to have an equivalent inequality.

EXAMPLE **6**   Solve $-2x \geq 5$ and then graph the solution.

SOLUTION   We have

$$-2x \geq 5$$

$$\frac{-2x}{-2} \leq \frac{5}{-2} \qquad \text{Multiplying by } -\frac{1}{2} \text{ or dividing by } -2$$
$$\text{The symbol must be reversed!}$$

$$x \leq -\frac{5}{2}.$$

Any number less than or equal to $-\frac{5}{2}$ is a solution. The graph is as follows:

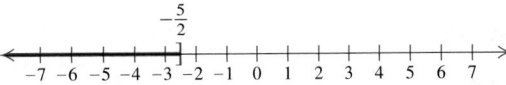

The solution set is $\left\{ x \mid x \leq -\frac{5}{2} \right\}$, or $\left(-\infty, -\frac{5}{2} \right]$.    TRY EXERCISE 47

We can use the addition and multiplication principles together to solve inequalities. We can also combine like terms, remove parentheses, and clear fractions and decimals.

EXAMPLE 7

Solve: $2 - 3(x + 5) > 4 - 6(x - 1)$.

SOLUTION We have

$$2 - 3(x + 5) > 4 - 6(x - 1)$$

$$2 - 3x - 15 > 4 - 6x + 6 \qquad \text{Using the distributive law to remove parentheses}$$

$$-3x - 13 > -6x + 10 \qquad \text{Simplifying}$$

$$-3x + 6x > 10 + 13 \qquad \text{Adding } 6x \text{ and also } 13, \text{ to get all } x\text{-terms on one side and all other terms on the other side}$$

$$3x > 23 \qquad \text{Combining like terms}$$

$$x > \frac{23}{3}. \qquad \text{Multiplying by } \frac{1}{3}. \text{ The inequality symbol stays the same because } \frac{1}{3} \text{ is positive.}$$

The solution set is $\left\{ x \mid x > \frac{23}{3} \right\}$, or $\left(\frac{23}{3}, \infty \right)$.

 **TRY EXERCISE** 53

Problem solving (Section 2.5)

Problem Solving

One of the most important uses of algebra is as a tool for problem solving. The following five steps can be used to help solve problems of many types.

> ### Five Steps for Problem Solving in Algebra
> 1. *Familiarize* yourself with the problem.
> 2. *Translate* to mathematical language. (This often means writing an equation.)
> 3. *Carry out* some mathematical manipulation. (This often means *solving* an equation.)
> 4. *Check* your possible answer in the original problem.
> 5. *State* the answer clearly, using a complete English sentence.

EXAMPLE 8

Kitchen cabinets. Cherry kitchen cabinets cost 10% more than oak cabinets. Shelby Custom Cabinets designs a kitchen using $7480 worth of cherry cabinets. How much would the same kitchen cost using oak cabinets?

SOLUTION

Familiarization step (p. 110)

Percent (Section 2.4)

1. **Familiarize.** The *Familiarize* step is often the most important of the five steps, and may require a significant amount of time. Sometimes it helps to make a drawing or a table, make a guess and check it, or look up further information. For this problem, we could review percent notation. We could also make a guess. Let's suppose that the oak cabinets cost $6500. Then the cherry cabinets would cost 10% more, or an additional $(0.10)(\$6500) = \$650$. Altogether the cherry cabinets would cost $\$6500 + \$650 = \$7150$. Since $\$7150 \neq \$7480$, our guess is incorrect, but we see that 10% of the price of the oak cabinets must be added to the price of the oak cabinets to get the price of the cherry cabinets. We let $c =$ the cost of the oak cabinets.

2. **Translate.** What we learned in the *Familiarize* step leads to the translation of the problem to an equation.

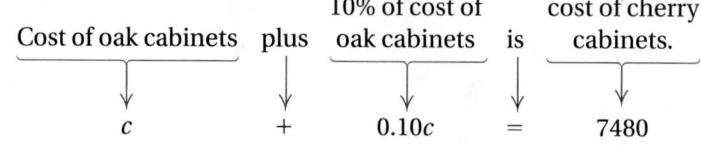

Cost of oak cabinets	plus	10% of cost of oak cabinets	is	cost of cherry cabinets.
↓	↓	↓	↓	↓
c	$+$	$0.10c$	$=$	7480

3. Carry out. We solve the equation:

$$c + 0.10c = 7480$$

$$1c + 0.10c = 7480 \qquad \text{Writing } c \text{ as } 1c \text{ before combining terms}$$

$$1.10c = 7480 \qquad \text{Combining like terms}$$

$$c = \frac{7480}{1.10} \qquad \text{Dividing by 1.10}$$

$$c = 6800.$$

4. Check. We check in the wording of the stated problem: Cherry cabinets cost 10% more, so the additional cost is

$$10\% \text{ of } \$6800 = 0.10(\$6800) = \$680.$$

The total cost of the cherry cabinets is then

$$\$6800 + \$680 = \$7480,$$

which is the amount stated in the problem.

5. State. The oak cabinets would cost $6800.   ▸ **TRY EXERCISE** 63

Sometimes the translation of a problem is an inequality.

EXAMPLE 9

Long-distance telephone usage. Elyse pays a flat rate of 6¢ per minute for long-distance telephone calls. The monthly charge for her local calls is $21.50. How many minutes can she spend calling long distance in a month and not exceed her telephone budget of $50?

SOLUTION

1. Familiarize. Suppose that Elyse spends 10 hr, or 600 min, making long-distance calls one month. Then her bill would be the local service charge plus the long-distance charges, or

$$\$21.50 + \$0.06(600) = \$57.50.$$

This exceeds $50, so we know that the number of long-distance minutes must be less than 600. We let $m =$ the number of minutes of long-distance calls in a month.

2. Translate. The *Familiarize* step helps us reword and translate.

	The local		the long-	cannot	
Rewording:	service charge	plus	distance charges	exceed	$50.
Translating:	21.50	+	0.06m	≤	50

Solving applications with inequalities (Section 2.7)

3. Carry out. We solve the inequality:

$$21.50 + 0.06m \le 50$$

$$0.06m \le 28.50 \qquad \text{Subtracting 21.50 from both sides}$$

$$m \le 475. \qquad \text{Dividing by 0.06. The inequality symbol stays the same.}$$

4. Check. As a partial check, note that the telephone bill for 475 min of long-distance charges is

$$\$21.50 + \$0.06(475) = \$50.$$

Since fewer minutes will cost even less, our answer checks. We also note that 475 is less than 600 min, as noted in the *Familiarize* step.

5. State. Elyse will not exceed her budget if she talks long distance for no more than 475 min.   ▸ **TRY EXERCISE** 71

R.2   EXERCISE SET

Solve.

1. $-6 + x = 10$

2. $y + 7 = -3$

3. $t + \frac{1}{3} = \frac{1}{4}$

4. $-\frac{2}{3} + p = \frac{1}{6}$

5. $-1.9 = x - 1.1$

6. $x + 4.6 = 1.7$

7. $-x = \frac{5}{3}$

8. $-y = -\frac{2}{5}$

9. $-\frac{2}{7}x = -12$

10. $-\frac{1}{4}x = 3$

11. $\dfrac{-t}{5} = 1$

12. $\dfrac{2}{3} = -\dfrac{z}{8}$

13. $3y + 10 = 15$

14. $12 = 5y + 18$

15. $4x + 7 = 3 - 5x$

16. $2x = 5 + 7x$

17. $2x - 7 = 5x + 1 - x$

18. $a + 7 - 2a = 14 + 7a - 10$

19. $\frac{2}{5} + \frac{1}{3}t = 5$

20. $-\frac{5}{6} + t = \frac{1}{2}$

21. $x + 0.45 = 2.6x$

22. $1.8x + 0.16 = 4.2 - 0.05x$

23. $8(3 - m) + 7 = 47$

24. $2(5 - m) = 5(6 + m)$

25. $4 - (6 + x) = 13$

26. $18 = 9 - (3 - x)$

27. $2 + 3(4 + c) = 1 - 5(6 - c)$

28. $b + (b + 5) - 2(b - 5) = 18 + b$

29. $0.1(a - 0.2) = 1.2 + 2.4a$

30. $\frac{2}{3}\left(\frac{1}{2} - x\right) + \frac{5}{6} = \frac{3}{2}\left(\frac{2}{3}x + 1\right)$

31. $A = lw$, for l

32. $A = lw$, for w

33. $I = \dfrac{P}{V}$, for P

34. $b = \dfrac{A}{h}$, for A

35. $q = \dfrac{p + r}{2}$, for p

36. $q = \dfrac{p - r}{2}$, for r

37. $A = \pi r^2 + \pi r^2 h$, for π

38. $ax + by = c$, for a

Determine whether each number is a solution of the given inequality.

39. $x \le -5$

　　a) 5

　　b) -5

　　c) 0

　　d) -10

40. $y > 0$

　　a) -1

　　b) 1

　　c) 0

　　d) 100

Solve and graph. Write each answer in set-builder notation and in interval notation.

41. $x + 3 \le 15$

42. $y + 7 < -10$

43. $m - 17 > -5$

44. $x + 9 \ge -8$

45. $2x \ge -3$

46. $-\frac{1}{2}n \le 4$

47. $-5t > 15$

48. $3x > 10$

Solve. Write each answer in set-builder notation and in interval notation.

49. $2y - 7 > 13$

50. $2 - 6y \le 18$

51. $6 - 5a \le a$

52. $4b + 7 > 2 - b$

53. $2(3 + 5x) \ge 7(10 - x)$

54. $2(x + 5) < 8 - 3x$

55. $\frac{2}{3}(6 - x) < \frac{1}{4}(x + 3)$

56. $\frac{2}{3}t + \frac{8}{9} \ge \frac{4}{6} - \frac{1}{4}t$

57. $0.7(2 + x) \ge 1.1x + 5.75$

58. $0.4x + 5.7 \le 2.6 - 3(1.2x - 7)$

Solve. Use the five-step problem-solving process.

59. Three less than the sum of 2 and some number is 6. What is the number?

60. Five times some number is 10 less than the number. What is the number?

61. The sum of two consecutive even integers is 34. Find the numbers.

62. The sum of three consecutive integers is 195. Find the numbers.

63. *Reading.* Leisa is reading a 500-page book. She has twice as many pages to read as she has already finished. How many pages has she already read?

64. *Mowing.* It takes Caleb 50 min to mow his lawn. It will take him three times as many minutes to finish as he has already spent mowing. How long has he already spent mowing?

65. *Perimeter of a rectangle.*   The perimeter of a rectangle is 28 cm. The width is 5 cm less than the length. Find the width and the length.

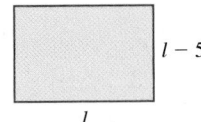

66. *Triangles.*   The second angle of a triangle is one third as large as the first. The third angle is 5° more than the first. Find the measure of the second angle.

67. *Water usage.*   Rural Water Company charges a monthly service fee of $9.70 plus a volume charge of $2.60 for every hundred cubic feet of water used. How much water was used if the monthly bill is $33.10?

68. *Telephone bills.*   Brandon pays $4.95 a month for a long-distance telephone service that offers a flat rate of 7¢ per minute. One month his total long-distance telephone bill was $10.69. How many minutes of long-distance telephone calls were made that month?

69. *Sale price.*   A can of tomatoes is on sale at 20% off for 64¢. What is the normal selling price of the tomatoes?

70. *Plywood.*   The price of a piece of plywood rose 5% to $42. What was the original price of the plywood?

71. *Practice.*   Dierdre's basketball coach requires each team member to average at least 15 min a day shooting baskets. One week Dierdre spent 10 min, 20 min, 5 min, 0 min, 25 min, and 15 min shooting baskets. How long must she practice shooting baskets on the seventh day if she is to meet the requirement?

72. *Perimeter of a garden.*   The perimeter of Garry's rectangular garden cannot exceed the 100 ft of fencing that he purchased. He wants the length to be twice the width. What widths of the garden will meet these conditions?

73. *Meeting costs.*   The Winds charges a $75 cleaning fee plus $45 an hour for the use of its meeting room. Complete Consultants has budgeted $200 to rent a room for a seminar. For how many hours can they rent the meeting room at The Winds?

74. *Meeting costs.*   Spring Haven charges a $15 setup fee, a $30 cleanup fee, and $50 an hour for the use of its meeting room. For what lengths of time will Spring Haven's room be less expensive than the room at The Winds (see Exercise 73)?

R.3   Introduction to Graphing

Points and Ordered Pairs   ▪   Graphs and Slope   ▪   Linear Equations

Points and Ordered Pairs

We can represent, or graph, pairs of numbers such as $(2, -5)$ on a plane. To do so, we use two perpendicular number lines called **axes**. The axes cross at a point called the **origin**. Arrows on the axes show the positive directions.

Graphing ordered pairs (p. 151)

The order of the **coordinates**, or numbers in a pair, is important. The **first coordinate** indicates horizontal position and the **second coordinate** indicates vertical position. Such pairs of numbers are called **ordered pairs**. Thus the ordered pairs $(1, -2)$ and $(-2, 1)$ correspond to different points, as shown in the accompanying figure.

Coordinates (p. 151)

The axes divide the plane into four regions, or **quadrants**, as indicated by Roman numerals in the figure at right. Points on the axes are not considered to be in any quadrant. The horizontal axis is often labeled the x-axis, and the vertical axis the y-axis.

Quadrants (p. 153)

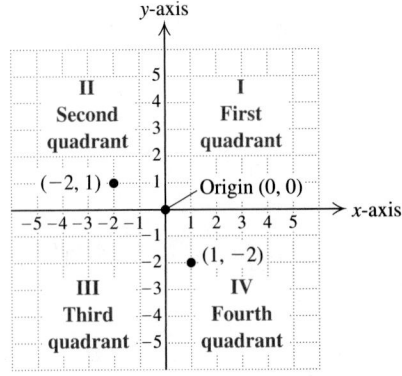

Graphs and Slope

Solutions of equations (p. 158)

When an equation contains two variables, solutions must be ordered pairs. Unless stated otherwise, the first number in each pair replaces the variable that occurs first alphabetically.

EXAMPLE **1**

Determine whether $(1, 4)$ is a solution of $y - x = 3$.

SOLUTION   We substitute 1 for x and 4 for y since x occurs first alphabetically:

$$\begin{array}{c} y - x = 3 \\ \hline 4 - 1 \mid 3 \\ 3 \overset{?}{=} 3 \quad \text{TRUE} \end{array}$$

Since $3 = 3$ is true, the pair $(1, 4)$ *is* a solution.

▶ TRY EXERCISE 9

A curve or line that represents all the solutions of an equation is called its **graph**.

EXAMPLE **2**

Graph: $y = -2x + 1$.

SOLUTION   We select a value for x, calculate the corresponding value of y, and form an ordered pair.

If $x = 0$, then $y = -2 \cdot 0 + 1 = 1$, and $(0, 1)$ is a solution. Repeating this step, we find other ordered pairs and list the results in a table. We then plot the points corresponding to the pairs. They appear to form a straight line, so we draw a line through the points.

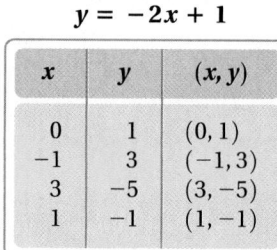

$y = -2x + 1$

x	y	(x, y)
0	1	$(0, 1)$
-1	3	$(-1, 3)$
3	-5	$(3, -5)$
1	-1	$(1, -1)$

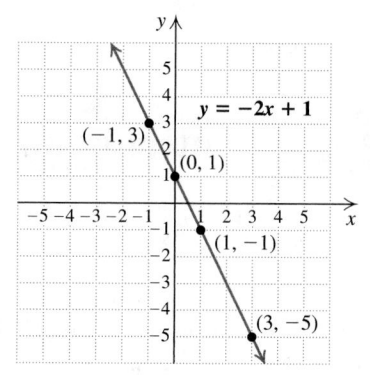

▶ TRY EXERCISE 13

The graph in Example 2 is a straight line. An equation whose graph is a straight line is a **linear equation**. The *rate of change* of y with respect to x is called the **slope** of a graph. A linear graph has constant slope. It can be found using any two points on a line.

Slope (p. 187)

Slope

The *slope* of the line containing points (x_1, y_1) and (x_2, y_2) is given by

$$m = \frac{\text{change in } y}{\text{change in } x} = \frac{\text{rise}}{\text{run}} = \frac{y_2 - y_1}{x_2 - x_1}.$$

EXAMPLE 3 Find the slope of the line containing the points $(-2, 1)$ and $(3, -4)$.

SOLUTION From $(-2, 1)$ to $(3, -4)$, the change in y, or the rise, is $-4 - 1$, or -5. The change in x, or the run, is $3 - (-2)$, or 5. Thus

$$\text{Slope} = \frac{\text{change in } y}{\text{change in } x} = \frac{\text{rise}}{\text{run}} = \frac{-4 - 1}{3 - (-2)} = \frac{-5}{5} = -1.$$

> **TRY EXERCISE** 17

The slope of a line indicates the direction and steepness of its slant. The larger the absolute value of the slope, the steeper the line. The direction of the slant is indicated by the sign of the slope, as shown in the figures below.

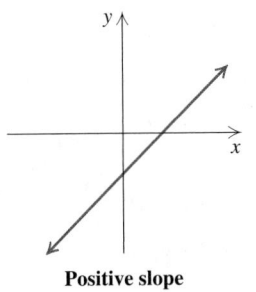

Positive slope

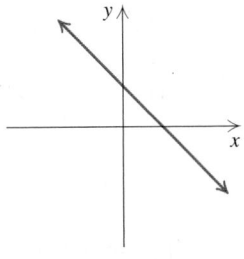

Negative slope

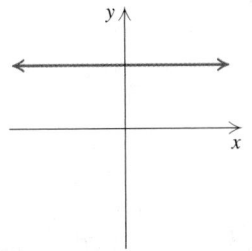

Zero slope

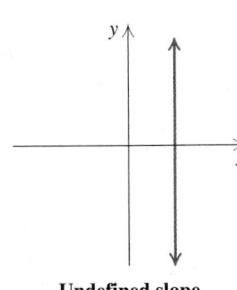

Undefined slope

***x*-intercept** (p. 168)

The ***x*-intercept** of a line, if it exists, is the point at which the graph crosses the x-axis. To find an x-intercept, we replace y with 0 and calculate x.

***y*-intercept** (p. 168)

The ***y*-intercept** of a line, if it exists, is the point at which the graph crosses the y-axis. To find a y-intercept, we replace x with 0 and calculate y.

Linear Equations

Any equation that can be written in the **standard form** $Ax + By = C$ is linear. Linear equations can also be written in other forms.

> ### Forms of Linear Equations
> Standard form: $Ax + By = C$
> Slope–intercept form: $y = mx + b$
> Point–slope form: $y - y_1 = m(x - x_1)$

The slope and y-intercept of a line can be read from the slope–intercept form of the line's equation.

> ### Slope and *y*-intercept
> For the graph of any equation $y = mx + b$,
>
> - the slope is m, and
> - the y-intercept is $(0, b)$.

EXAMPLE **4**

Find the slope and the *y*-intercept of the line given by the equation $4x - 3y = 9$.

SOLUTION   We write the equation in slope–intercept form $y = mx + b$:

$$4x - 3y = 9 \qquad \text{We must solve for } y.$$
$$-3y = -4x + 9 \qquad \text{Adding } -4x \text{ to both sides}$$
$$y = \tfrac{4}{3}x - 3. \qquad \text{Dividing both sides by } -3$$

The slope is $\tfrac{4}{3}$ and the *y*-intercept is $(0, -3)$.

> TRY EXERCISE 23

If we know an equation is a straight line, we can plot two points on the line and draw the line through those points. The intercepts are often convenient points to use.

EXAMPLE **5**

Graph $2x - 5y = 10$ using intercepts.

SOLUTION   To find the *x*-intercept, we let $y = 0$ and solve for *x*:

$$2x - 5 \cdot 0 = 10 \qquad \text{Replacing } y \text{ with } 0$$
$$2x = 10$$
$$x = 5.$$

To find the *y*-intercept, we let $x = 0$ and solve for *y*:

$$2 \cdot 0 - 5y = 10 \qquad \text{Replacing } x \text{ with } 0$$
$$-5y = 10$$
$$y = -2.$$

Thus the *x*-intercept is $(5, 0)$ and the *y*-intercept is $(0, -2)$. The graph is a line, since $2x - 5y = 10$ is in the form $Ax + By = C$. It passes through these two points.

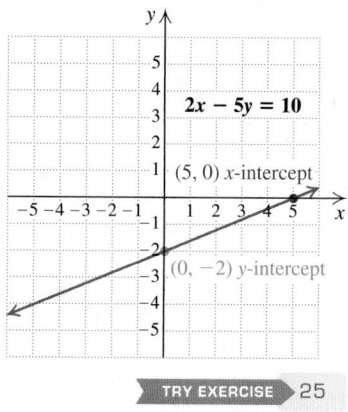

> TRY EXERCISE 25

Alternatively, if we know a point on the line and its slope, we can plot the point and "count off" its slope to locate another point on the line.

EXAMPLE **6**

Graph: $y = -\dfrac{1}{2}x + 3$.

SOLUTION   The equation is in slope–intercept form, so we can read the slope and *y*-intercept directly from the equation.

$$\text{Slope: } -\frac{1}{2}$$

$$\text{\textit{y}-intercept: } (0, 3)$$

We plot the *y*-intercept and use the slope to find another point.

Another way to write the slope is $\dfrac{-1}{2}$.

This means for a run of 2 units, there is a negative rise, or a fall, of 1 unit. Starting at $(0, 3)$, we move 2 units in the positive horizontal direction and then 1 unit down, to locate the point $(2, 2)$. Then we draw the graph. A third point can be calculated and plotted as a check.

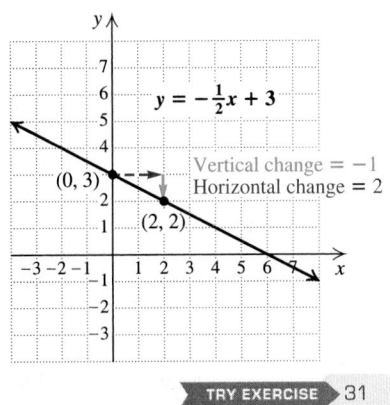

TRY EXERCISE 31

Horizontal and vertical lines intersect only one axis.

Horizontal line (p. 173)

Vertical line (p. 173)

Horizontal and Vertical Lines

Horizontal Line	*Vertical Line*
$y = b$	$x = a$
y-intercept $(0, b)$	x-intercept $(a, 0)$
Slope is 0	Undefined slope
Example: $y = -3$	Example: $x = 2$

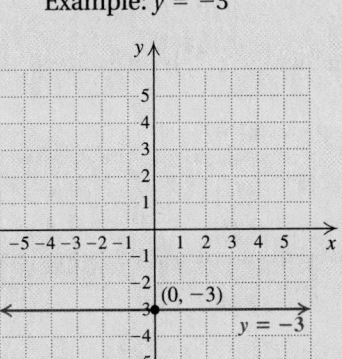

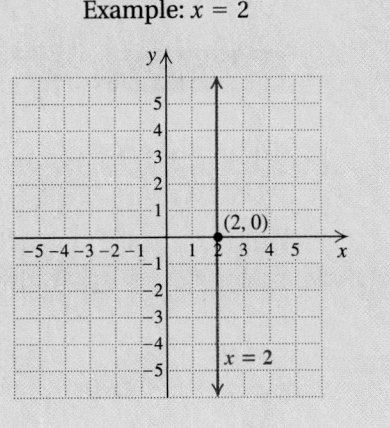

If we know the slope of a line and the coordinates of a point on the line, we can find an equation of the line, using either the slope–intercept equation $y = mx + b$, or the point–slope equation $y - y_1 = m(x - x_1)$.

EXAMPLE 7   Find the slope-intercept equation of a line given the following:

a) The slope is 2, and the y-intercept is $(0, -5)$.

b) The graph contains the points $(-2, 1)$ and $(3, -4)$.

SOLUTION

a) Since the slope and the y-intercept are given, we use the slope–intercept equation:

$$y = mx + b$$
$$y = 2x - 5. \qquad \text{Substituting 2 for } m \text{ and } -5 \text{ for } b$$

b) To use the point–slope equation, we need a point on the line and its slope. The slope can be found from the points given:

Find the slope.

$$m = \frac{1 - (-4)}{-2 - 3} = \frac{5}{-5} = -1.$$

Either point can be used for (x_1, y_1). Using $(-2, 1)$, we have

Find a point–slope equation for the line.

$$y - y_1 = m(x - x_1)$$
$$y - 1 = -1(x - (-2))\qquad \text{Substituting } -2 \text{ for } x_1, 1 \text{ for } y_1, \text{ and } -1 \text{ for } m$$
$$y - 1 = -(x + 2)$$

Find the slope–intercept equation for the line.

$$y - 1 = -x - 2$$
$$y = -x - 1.\qquad \text{This is in slope–intercept form.}$$

> **TRY EXERCISE** ▸ 39

We can tell from the slopes of two lines whether they are parallel or perpendicular.

Parallel lines (p. 201)

Perpendicular lines (p. 202)

> ## Parallel and Perpendicular Lines
> Two lines are parallel if they have the same slope.
> Two lines are perpendicular if the product of the slopes is -1.

EXAMPLE 8 Tell whether the graphs of each pair of lines are parallel, perpendicular, or neither.
a) $2x - y = 7,$
$y = 2x + 3$
b) $4x - y = 8,$
$x + 4y = 8$

SOLUTION

a) The slope of $y = 2x + 3$ is 2.
To find the slope of $2x - y = 7$, we solve for y:

$$2x - y = 7$$
$$-y = -2x + 7$$
$$y = 2x - 7.$$

The slope of $2x - y = 7$ is also 2. Since the slopes are equal, the lines are parallel.

b) We solve both equations for y in order to determine the slopes of the lines:

$$4x - y = 8$$
$$-y = -4x + 8$$
$$y = 4x - 8.$$

The slope of $4x - y = 8$ is 4.
For the second line, we have

$$x + 4y = 8$$
$$4y = -x + 8$$
$$y = -\tfrac{1}{4}x + 2.$$

The slope of $x + 4y = 8$ is $-\tfrac{1}{4}$. Since $4 \cdot \left(-\tfrac{1}{4}\right) = -1$, the lines are perpendicular.

> **TRY EXERCISE** ▸ 43

R.3 EXERCISE SET

For Extra Help

1. Plot these points.

$(2, -3), (5, 1), (0, 2), (-1, 0),$
$(0, 0), (-2, -5), (-1, 1), (1, -1)$

2. Plot these points.

$(0, -4), (-4, 0), (5, -2), (2, 5),$
$(3, 3), (-3, -1), (-1, 4), (0, 1)$

In which quadrant is each point located?

3. $(-2, 5)$

4. $(15, 27)$

5. $(3, -2.6)$

6. $(-1.7, -5.9)$

7. First coordinates are positive in quadrants ____ and ____.

8. Second coordinates are negative in quadrants ____ and ____.

Determine whether each equation has the given ordered pair as a solution.

9. $y = 2x - 5$; $(1, 3)$

10. $4x + 3y = 8$; $(-1, 4)$

11. $a - 5b = -3$; $(2, 1)$

12. $c = d + 1$; $(1, 2)$

Graph.

13. $y = \frac{1}{3}x + 3$

14. $y = -x - 2$

15. $y = -4x$

16. $y = \frac{3}{4}x + 1$

Find the slope of the line containing each given pair of points. If it is undefined, state this.

17. $(3, 6)$ and $(2, 7)$

18. $(-1, 7)$ and $(-5, 1)$

19. $\left(-2, -\frac{1}{2}\right)$ and $\left(5, -\frac{1}{2}\right)$

20. $(6.8, 7.5)$ and $(6.8, -3.2)$

Find the slope and the y-intercept of each equation.

21. $y = 2x - 5$

22. $y = 4 - x$

23. $2x + 7y = 1$

24. $x - 2y = 3$

Find the intercepts. Then graph.

25. $3 - y = 2x$

26. $2x + 5y = 10$

27. $y = 3x + 5$

28. $y = -x + 7$

29. $3x - 2y = 6$

30. $2y + 1 = x$

Determine the coordinates of the y-intercept of each equation. Then graph the equation.

31. $y = 2x - 5$

32. $y = -\frac{5}{4}x - 3$

33. $2y + 4x = 6$

34. $3y + x = 4$

Find the slope of each line, and graph.

35. $y = 4$

36. $x = -5$

37. $x = 3$

38. $y = -1$

Find the slope–intercept equation of a line given the conditions.

39. The slope is 5 and the y-intercept is $(0, 9)$.

40. The slope is $\frac{2}{3}$ and the y-intercept is $(0, -5)$.

41. The graph contains the points $(0, 3)$ and $(-1, 4)$.

42. The graph contains the points $(5, 1)$ and $(8, 0)$.

Determine whether each pair of lines is parallel, perpendicular, or neither.

43. $x + y = 5,$
$x - y = 1$

44. $2x + y = 3,$
$y = 4 - 2x$

45. $2x + 3y = 1,$
$2x - 3y = 5$

46. $y = \frac{1}{3}x - 7,$
$y + 3x = 1$

R.4 Polynomials

Exponents ▪ Polynomials ▪ Addition and Subtraction of Polynomials ▪
Multiplication of Polynomials ▪ Division of Polynomials

Exponents

We know that x^4 means $x \cdot x \cdot x \cdot x$ and that x^1 means x. Exponential notation is also defined for zero and negative exponents.

Zero and Negative Exponents

For any real number a, $a \neq 0$,

$$a^0 = 1 \quad \text{and} \quad a^{-n} = \frac{1}{a^n}.$$

EXAMPLE **1**   Simplify: **(a)** $(97)^0$; **(b)** $(-2x)^0$.

SOLUTION

The exponent zero (p. 230)

a) $(97)^0 = 1$ since any number (other than 0 itself) raised to the 0 power is 1.

b) $(-2x)^0 = 1$ for any $x \neq 0$.   **TRY EXERCISE** 1

EXAMPLE **2**   Write an equivalent expression using positive exponents.

a) x^{-2} **b)** $\dfrac{1}{x^{-2}}$ **c)** $7y^{-1}$

SOLUTION

Negative exponents (p. 236)

a) $x^{-2} = \dfrac{1}{x^2}$ x^{-2} is the reciprocal of x^2.

b) $\dfrac{1}{x^{-2}} = x^{-(-2)} = x^2$ The reciprocal of x^{-2} is $x^{-(-2)}$, or x^2.

c) $7y^{-1} = 7\left(\dfrac{1}{y^1}\right) = \dfrac{7}{y}$ y^{-1} is the reciprocal of y^1.   **TRY EXERCISE** 5

The following properties hold for any integers m and n and any real numbers a and b, provided no denominators are 0 and 0^0 is not considered.

Properties of Exponents

The Product Rule:	$a^m \cdot a^n = a^{m+n}$
The Quotient Rule:	$\dfrac{a^m}{a^n} = a^{m-n}$
The Power Rule:	$(a^m)^n = a^{mn}$
Raising a product to a power:	$(ab)^n = a^n b^n$
Raising a quotient to a power:	$\left(\dfrac{a}{b}\right)^n = \dfrac{a^n}{b^n}$

These properties are often used to simplify exponential expressions.

EXAMPLE 3   Simplify.

a) $(x^2y^{-1})(xy^{-3})$ **b)** $\dfrac{(3p)^3}{(3p)^{-2}}$ **c)** $\left(\dfrac{ab^2}{3c^3}\right)^{-4}$

SOLUTION

a) $(x^2y^{-1})(xy^{-3}) = x^2y^{-1}xy^{-3}$ Using an associative law

$= x^2x^1y^{-1}y^{-3}$ Using a commutative law; $x = x^1$

The product rule (p. 228) $= x^{2+1}y^{-1+(-3)}$ Using the product rule: Adding exponents

$= x^3y^{-4}$, or $\dfrac{x^3}{y^4}$

The quotient rule (p. 229)   **b)** $\dfrac{(3p)^3}{(3p)^{-2}} = (3p)^{3-(-2)}$ Using the quotient rule: Subtracting exponents

$= (3p)^5$

The power rule (p. 231) $= 3^5p^5$ Raising each factor to the fifth power

$= 243p^5$

Raising a product to a power (p. 232)   **c)** $\left(\dfrac{ab^2}{3c^3}\right)^{-4} = \dfrac{(ab^2)^{-4}}{(3c^3)^{-4}}$ Raising the numerator and the denominator to the -4 power

Raising a quotient to a power (p. 233) $= \dfrac{a^{-4}(b^2)^{-4}}{3^{-4}(c^3)^{-4}}$ Raising each factor to the -4 power

$= \dfrac{a^{-4}b^{-8}}{3^{-4}c^{-12}}$ Multiplying exponents

$= \dfrac{3^4c^{12}}{a^4b^8}$, or $\dfrac{81c^{12}}{a^4b^8}$ Rewriting without negative exponents

TRY EXERCISE 23

Polynomials (Section 4.3)

Polynomials

Algebraic expressions like

$$2x^3 + 3x - 5, \quad 4x, \quad -7, \quad \text{and} \quad 2a^3b^2 + ab^3$$

are all examples of **polynomials**. All variables in a polynomial are raised to whole-number powers, and there are no variables in a denominator. The **terms** of a polynomial are separated by addition signs. The **degree of a term** is the number of variable factors in that term. The **leading term** of a polynomial is the term of highest degree. The **degree of a polynomial** is the degree of the leading term. A polynomial is written in *descending order* when the leading term appears first, followed by the term of next highest degree, and so on.

Term (p. 246)
Degree of a term (p. 247)
Leading term (p. 247)
Degree of a polynomial (p. 247)

The number -2 in the term $-2y^3$ is called the **coefficient** of that term. The coefficient of the leading term is the **leading coefficient** of the polynomial. To illustrate this terminology, consider the polynomial

Coefficient (p. 247)
Leading coefficient (p. 247)

$$4y^2 - 8y^5 + y^3 - 6y + 7.$$

The *terms* are $4y^2, \quad -8y^5, \quad y^3, \quad -6y, \quad$ and $\quad 7.$

The *coefficients* are $4, \quad -8, \quad 1, \quad -6, \quad$ and $\quad 7.$

The *degree of each term* is $2, \quad 5, \quad 3, \quad 1, \quad$ and $\quad 0.$

The *leading term* is $-8y^5$ and the *leading coefficient* is -8.

The *degree of the polynomial* is 5.

Types of polynomials (p. 246)

Polynomials are classified by the number of terms and by degree.

A **monomial** has one term.	*Example*: $-2x^3y$
A **binomial** has two terms.	*Example*: $1.4x^2 - 10$
A **trinomial** has three terms.	*Example*: $x^2 - 3x - 6$
A **constant** polynomial has degree 0.	*Example*: 7
A **linear** polynomial has degree 1.	*Example*: $3x + 5$
A **quadratic** polynomial has degree 2.	*Example*: $5x^2 - x$
A **cubic** polynomial has degree 3.	*Example*: $x^3 + 2x^2 - \frac{1}{3}$
A **quartic** polynomial has degree 4.	*Example*: $-6x^4 - 2x^2 + 19$

Like, or *similar, terms* are either constant terms or terms containing the same variable(s) raised to the same power(s). Polynomials containing like terms can be simplified by *combining* those terms.

EXAMPLE 4

Combine like terms: $4x^2y + 2xy - x^2y + xy^2$.

SOLUTION   The like terms are $4x^2y$ and $-x^2y$. Thus we have

$$4x^2y + 2xy - x^2y + xy^2 = 4x^2y - x^2y + 2xy + xy^2$$
$$= 3x^2y + 2xy + xy^2.$$

▶ TRY EXERCISE 37

A polynomial can be evaluated by replacing the variable or variables with a number or numbers.

EXAMPLE 5

Evaluate $-a^2 + 2ab + 5b^2$ for $a = -1$ and $b = 3$.

SOLUTION   We replace a with -1 and b with 3 and calculate the value using the rules for order of operations:

Evaluating a polynomial (p. 248)

$$-a^2 + 2ab + 5b^2 = -(-1)^2 + 2 \cdot (-1) \cdot 3 + 5 \cdot 3^2$$
$$= -1 - 6 + 45 = 38.$$

▶ TRY EXERCISE 41

Polynomials can be added, subtracted, multiplied, and divided.

Addition and Subtraction of Polynomials

Addition of polynomials (Section 4.4)

To add two polynomials, we write a plus sign between them and combine like terms.

EXAMPLE 6

Add: $(4x^3 + 3x^2 + 2x - 7) + (-5x^2 + x - 10)$.

SOLUTION

$$(4x^3 + 3x^2 + 2x - 7) + (-5x^2 + x - 10)$$
$$= 4x^3 + (3 - 5)x^2 + (2 + 1)x + (-7 - 10)$$
$$= 4x^3 - 2x^2 + 3x - 17$$

▶ TRY EXERCISE 49

Opposite of a polynomial (p. 255)

To find the **opposite of a polynomial**, we replace each term with its opposite. This process is also called *changing the sign* of each term. For example, the opposite of

$$3y^4 - 7y^2 - \tfrac{1}{3}y + 17$$

is

$$-(3y^4 - 7y^2 - \tfrac{1}{3}y + 17) = -3y^4 + 7y^2 + \tfrac{1}{3}y - 17.$$

Subtraction of polynomials
(Section 4.4)

EXAMPLE   7

To subtract polynomials, we add the opposite of the polynomial being subtracted.

Subtract: $(3a^4 - 2a + 7) - (-a^3 + 5a - 1)$.

SOLUTION

$$(3a^4 - 2a + 7) - (-a^3 + 5a - 1)$$
$$= 3a^4 - 2a + 7 + a^3 - 5a + 1 \quad \text{Adding the opposite}$$
$$= 3a^4 + a^3 - 7a + 8 \quad \text{Combining like terms}$$

> TRY EXERCISE   51

Multiplication of polynomials
(Section 4.5)

Multiplication of Polynomials

To multiply two monomials, we multiply coefficients and then multiply variables using the product rule for exponents. To multiply a monomial and a polynomial, we multiply each term of the polynomial by the monomial, using the distributive property.

EXAMPLE   8

Multiply: $4x^3(3x^4 - 2x^3 + 7x - 5)$.

SOLUTION

$$\textit{Think: } \underbrace{4x^3 \cdot 3x^4} - \underbrace{4x^3 \cdot 2x^3} + \underbrace{4x^3 \cdot 7x} - \underbrace{4x^3 \cdot 5}$$

$$4x^3(3x^4 - 2x^3 + 7x - 5) = 12x^7 \quad - \quad 8x^6 \quad + \quad 28x^4 \quad - \quad 20x^3$$

> TRY EXERCISE   55

To multiply any two polynomials P and Q, we select one of the polynomials—say, P. We then multiply each term of P by every term of Q and combine like terms.

EXAMPLE   9

Multiply: $(2a^3 + 3a - 1)(a^2 - 4a)$.

SOLUTION   It is often helpful to use columns for a long multiplication. We multiply each term at the top by every term at the bottom, write like terms in columns, and add the results.

$$
\begin{array}{r}
2a^3 + 3a - 1 \\
a^2 - 4a \\
\hline
-8a^4 - 12a^2 + 4a \\
2a^5 + 3a^3 - a^2 \\
\hline
2a^5 - 8a^4 + 3a^3 - 13a^2 + 4a
\end{array}
$$

Multiplying the top row by $-4a$

Multiplying the top row by a^2

Combining like terms. Be sure that like terms are lined up in columns.

> TRY EXERCISE   59

We could multiply two binomials in the same manner in which we multiplied the polynomials in Example 9. However, by observing the pattern of the products formed, we can develop a method of multiplying two binomials more efficiently.

The FOIL Method

To multiply two binomials, $A + B$ and $C + D$, multiply the First terms AC, the Outer terms AD, the Inner terms BC, and then the Last terms BD. Then combine like terms, if possible.

$$(A + B)(C + D) = AC + AD + BC + BD$$

1. Multiply First terms: AC.
2. Multiply Outer terms: AD.
3. Multiply Inner terms: BC.
4. Multiply Last terms: BD.

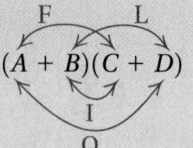

FOIL

EXAMPLE **10** Multiply: $(3x + 4)(x - 2)$.

SOLUTION

FOIL (p. 270)

$$\begin{array}{c} \text{F}\text{O}\text{I}\text{L} \\ (3x + 4)(x - 2) = 3x^2 - 6x + 4x - 8 \\ = 3x^2 - 2x - 8 \quad \text{Combining like terms} \end{array}$$

TRY EXERCISE 57

Special products occur so often that specific formulas or methods for computing them have been developed.

Special Products

The product of a sum and difference of the same two terms:

Multiplying sums and differences of two terms (p. 271)

$$(A + B)(A - B) = \underline{A^2 - B^2}$$

This is called a *difference of squares.*

Squaring binomials (p. 272)

The square of a binomial:

$$(A + B)^2 = A^2 + 2AB + B^2$$
$$(A - B)^2 = A^2 - 2AB + B^2$$

EXAMPLE **11** Multiply: **(a)** $(x + 3y)(x - 3y)$; **(b)** $(x^3 + 2)^2$.

SOLUTION

$$(A + B)(A - B) = A^2 - B^2$$

a) $(x + 3y)(x - 3y) = x^2 - (3y)^2 \qquad A = x \text{ and } B = 3y$
$$= x^2 - 9y^2$$

$$(A + B)^2 = A^2 + 2 \cdot A \cdot B + B^2$$

b) $(x^3 + 2)^2 = (x^3)^2 + 2 \cdot x^3 \cdot 2 + 2^2 \qquad A = x^3 \text{ and } B = 2$
$$= x^6 + 4x^3 + 4$$

TRY EXERCISE 61

Division of polynomials
(Section 4.8)

Division of Polynomials

Polynomial division is similar to division in arithmetic. To divide a polynomial by a monomial, we divide each term by the monomial.

EXAMPLE 12   Divide: $(3x^5 + 8x^3 - 12x) \div (4x)$.

SOLUTION   This division can be written

$$\frac{3x^5 + 8x^3 - 12x}{4x} = \frac{3x^5}{4x} + \frac{8x^3}{4x} - \frac{12x}{4x} \qquad \text{Dividing each term by } 4x$$

$$= \frac{3}{4}x^{5-1} + \frac{8}{4}x^{3-1} - \frac{12}{4}x^{1-1} \qquad \begin{array}{l}\text{Dividing coefficients} \\ \text{and subtracting} \\ \text{exponents}\end{array}$$

$$= \frac{3}{4}x^4 + 2x^2 - 3.$$

To check, we multiply the quotient by $4x$:

$$\left(\tfrac{3}{4}x^4 + 2x^2 - 3\right)4x = 3x^5 + 8x^3 - 12x. \qquad \text{The answer checks.}$$

TRY EXERCISE ▶ 69

To use long division, we write polynomials in descending order, including terms with 0 coefficients for missing terms. As shown below in Example 13, the procedure ends when the degree of the remainder is less than the degree of the divisor.

EXAMPLE 13   Divide: $(4x^3 - 7x + 1) \div (2x + 1)$.

SOLUTION   The polynomials are already written in descending order, but there is no x^2-term in the dividend. We fill in $0x^2$ for that term.

$$\begin{array}{r} 2x^2 \\ 2x + 1\overline{)4x^3 + 0x^2 - 7x + 1} \\ \underline{4x^3 + 2x^2} \\ -2x^2 \end{array}$$

Divide the first term of the dividend, $4x^3$, by the first term in the divisor, $2x$: $4x^3/(2x) = 2x^2$.

Multiply $2x^2$ by the divisor, $2x + 1$.

Subtract: $(4x^3 + 0x^2) - (4x^3 + 2x^2) = -2x^2$.

Then we bring down the next term of the dividend, $-7x$.

$$\begin{array}{r} 2x^2 - x \\ 2x + 1\overline{)4x^3 + 0x^2 - 7x + 1} \\ \underline{4x^3 + 2x^2} \\ -2x^2 - 7x \\ \underline{-2x^2 - x} \\ -6x \end{array}$$

Divide the first term of $-2x^2 - 7x$ by the first term in the divisor: $-2x^2/(2x) = -x$.

The $-7x$ has been "brought down."

Multiply $-x$ by the divisor, $2x + 1$.

Subtract: $(-2x^2 - 7x) - (-2x^2 - x) = -6x$.

Since the degree of the remainder, $-6x$, is *not* less than the degree of the divisor, we must continue dividing.

$$
\begin{array}{r}
2x^2 - x - 3 \\
2x+1 \overline{\smash{\big)}\ 4x^3 + 0x^2 - 7x + 1} \\
\underline{4x^3 + 2x^2} \\
-2x^2 - 7x \\
\underline{-2x^2 - x} \\
-6x + 1 \\
\underline{-6x - 3} \\
4
\end{array}
$$

Divide the first term of $-6x + 1$ by the first term in the divisor: $-6x/(2x) = -3$.

$\leftarrow$ The 1 has been "brought down."

$\leftarrow$ Multiply -3 by $2x + 1$.

$\leftarrow$ Subtract.

The answer is $2x^2 - x - 3$ with R4, or

Quotient $\longrightarrow 2x^2 - x - 3 + \dfrac{4}{2x+1}$. $\begin{array}{l}\leftarrow \text{Remainder} \\ \leftarrow \text{Divisor}\end{array}$

Check: To check, we can multiply by the divisor and add the remainder:

$$(2x + 1)(2x^2 - x - 3) + 4 = 4x^3 - 7x - 3 + 4$$
$$= 4x^3 - 7x + 1.$$

TRY EXERCISE 71

R.4 EXERCISE SET

For Extra Help | MyMathLab | MathXL PRACTICE | WATCH | DOWNLOAD

Solve.

1. a^0, for $a = -25$

2. y^0, for $y = 6.97$

3. $4^0 - 4^1$

4. $8^1 - 8^0$

Write an equivalent expression using positive exponents. Then, if possible, simplify.

5. 8^{-2}

6. $(-2)^{-3}$

7. $10x^{-5}$

8. $-16y^{-3}$

9. $(ab)^{-2}$

10. ab^{-2}

11. $\dfrac{1}{y^{-10}}$

12. $\dfrac{1}{x^{-t}}$

Write an equivalent expression using negative exponents.

13. $\dfrac{1}{t^4}$

14. $\dfrac{1}{a^2 b^3}$

Simplify.

15. $x^5 \cdot x^{10}$

16. $a^4 \cdot a^{-2}$

17. $\dfrac{a}{a^{-5}}$

18. $\dfrac{p^{-3}}{p^{-8}}$

19. $\dfrac{(4x)^{11}}{(4x)^2}$

20. $\dfrac{a^2 b^9}{a^9 b^2}$

21. $(7^8)^5$

22. $(x^3)^{-7}$

23. $(x^{-2} y^{-3})^{-4}$

24. $(-2a^2)^3$

25. $\left(\dfrac{y^2}{4}\right)^3$

26. $\left(\dfrac{ab^2}{c^3}\right)^4$

27. $\left(\dfrac{2p^3}{3q^4}\right)^{-2}$

28. $\left(\dfrac{2}{x}\right)^{-5}$

Identify the terms of each polynomial.

29. $8x^3 - 6x^2 + x - 7$

30. $-a^2 b + 4a^2 - 8b + 17$

Determine the coefficient and the degree of each term in each polynomial. Then find the degree of each polynomial.

31. $18x^3 + 36x^9 - 7x + 3$

32. $-8y^7 + y + 19$

33. $-x^2 y + 4y^3 - 2xy$

34. $8 - x^2 y^4 + y^7$

Determine the leading term and the leading coefficient of each polynomial.

35. $-p^2 + 5 + 8p^4 - 7p$

36. $13 + 20t - 30t^2 - t^3$

Combine like terms. Write each answer in descending order.

37. $3x^3 - x^2 + x^4 + x^2$

38. $5t - 8t^2 + 4t^2$

39. $3 - 2t^2 + 8t - 3t - 5t^2 + 7$

40. $8x^5 - \frac{1}{3} + \frac{4}{5}x + 1 - \frac{1}{2}x$

Evaluate each polynomial for the given replacements of the variables.

41. $3x^2 - 7x + 10$, for $x = -2$

42. $-y + 3y^2 + 2y^3$, for $y = 3$

43. $a^2b^3 + 2b^2 - 6a$, for $a = 2$ and $b = -1$

44. $2pq^3 - 5q^2 + 8p$, for $p = -4$ and $q = -2$

The distance s, in feet, traveled by a body falling freely from rest in t seconds is approximated by

$$s = 16t^2.$$

45. A pebble is dropped into a well and takes 3 sec to hit the water. How far down is the surface of the water?

46. An acorn falls from the top of an oak tree and takes 2 sec to hit the ground. How high is the tree?

The number of watts of power P generated by a particular turbine at a wind speed of x miles per hour can be approximated by the polynomial

$$P = 0.0157x^3 + 0.1163x^2 - 1.3396x + 3.7063.$$

Source: Based on data from QST, November 2006

47. Estimate the power generated by a 10-mph wind.

48. Estimate the power generated by a 30-mph wind.

Add or subtract, as indicated.

49. $(3x^3 + 2x^2 + 8x) + (x^3 - 5x^2 + 7)$

50. $(-6x^4 + 3x^2 - 16) + (4x^2 + 4x - 7)$

51. $(8y^2 - 2y - 3) - (9y^2 - 7y - 1)$

52. $(4t^2 + 6t - 7) - (t + 5)$

53. $(-x^2y + 2y^2 + y) - (3y^2 + 2x^2y - 7y)$

54. $(ab + x^2y^2) + (2ab - x^2y^2)$

Multiply.

55. $4x^2(3x^3 - 7x + 7)$

56. $a^2b(a^3 + b^2 - ab - 2b)$

57. $(2a + y)(4a + b)$

58. $(x + 7y)(y - 3x)$

59. $(x + 7)(x^2 - 3x + 1)$

60. $(2x - 3)(x^2 - x - 1)$

61. $(x + 7)(x - 7)$

62. $(2x + 1)^2$

63. $(x + y)^2$

64. $(xy + 1)(xy - 1)$

65. $(2x^2 + 7)(3x^2 - 2)$

66. $(1.1x^2 + 5)(0.1x^2 - 2)$

67. $(6a - 5y)(7a + 3y)$

68. $(3p^2 - q^3)^2$

Divide and check.

69. $(3t^5 + 9t^3 - 6t^2 + 15t) \div (-3t)$

70. $(4x^5 + 10x^4 - 16x^2) \div (4x^2)$

71. $(15x^2 - 16x - 15) \div (3x - 5)$

72. $(x^3 - 2x^2 - 14x + 1) \div (x - 5)$

73. $(2x^3 - x^2 + 1) \div (x + 1)$

74. $(2x^3 + 3x^2 - 50) \div (2x - 5)$

75. $(5x^3 + 3x^2 - 5x) \div (x^2 - 1)$

76. $(2x^3 + 3x^2 + 6x + 10) \div (x^2 + 3)$

R.5 Polynomials and Factoring

Common Factors and Factoring by Grouping • Factoring Trinomials • Factoring Special Forms • Solving Polynomial Equations by Factoring

Common Factors and Factoring by Grouping

Factor (p. 304)

To *factor* a polynomial is to find an equivalent expression that is a product. To factor a monomial, we find two monomials whose product is equivalent to the original monomial. For example, three factorizations of $50x^6$ are $5 \cdot 10x^6$, $5x^3 \cdot 10x^3$, and $2x \cdot 25x^5$.

Common factor (p. 305)

If all the terms in a polynomial share a common factor, that factor can be "factored out" of the polynomial. Whenever you are factoring a polynomial with two or more terms, try to first find the largest common factor of the terms, if one exists.

EXAMPLE **1**

Factor: $3x^6 + 15x^4 - 9x^3$.

SOLUTION The largest factor common to 3, 15, and -9 is 3. The largest power of x common to x^6, x^4, and x^3 is x^3. Thus the largest common factor of the terms of the polynomial is $3x^3$. We factor as follows:

$$3x^6 + 15x^4 - 9x^3 = 3x^3 \cdot x^3 + 3x^3 \cdot 5x - 3x^3 \cdot 3 \qquad \text{Factoring each term}$$

$$= 3x^3(x^3 + 5x - 3). \qquad \text{Factoring out } 3x^3$$

Factorizations can always be checked by multiplying:

$$3x^3(x^3 + 5x - 3) = 3x^6 + 15x^4 - 9x^3. \qquad \blacktriangleright\text{TRY EXERCISE} \; 1$$

A polynomial with two or more terms can be a common factor.

EXAMPLE **2**

Factor: $3x^2(x - 2) + 5(x - 2)$.

SOLUTION The binomial $x - 2$ is a factor of both $3x^2(x - 2)$ and $5(x - 2)$. Thus we have

$$3x^2(x - 2) + 5(x - 2) = (x - 2)(3x^2 + 5). \qquad \text{Factoring out the common factor, } x - 2$$

$\blacktriangleright\text{TRY EXERCISE} \; 5$

Factoring by grouping (p. 307)

If a polynomial with four terms can be split into two groups of terms, and both groups share a common binomial factor, the polynomial can be factored. This method is known as **factoring by grouping**.

EXAMPLE **3**

Factor by grouping: $2x^3 + 6x^2 - x - 3$.

SOLUTION First, we consider the polynomial as two groups of terms, $2x^3 + 6x^2$ and $-x - 3$. Then we factor each group separately:

$$2x^3 + 6x^2 - x - 3 = 2x^2(x + 3) - 1(x + 3) \qquad \text{Factoring out } 2x^2 \text{ and } -1 \text{ to give the common binomial factor, } x + 3$$

$$= (x + 3)(2x^2 - 1).$$

The check is left to the student. $\blacktriangleright\text{TRY EXERCISE} \; 23$

Prime polynomial (p. 316)

Not every polynomial with four terms is factorable by grouping. A polynomial that is not factorable is said to be **prime**.

Factoring Trinomials

Factoring trinomials of the type $x^2 + bx + c$ (Section 5.2)

Many trinomials that have no common factor can be written as the product of two binomials. We look first at trinomials of the form $x^2 + bx + c$, for which the leading coefficient is 1.

Factoring trinomials involves a trial-and-error process. In order for the product of two binomials to be $x^2 + bx + c$, the binomials must look like

$$(x + p)(x + q),$$

where p and q are constants that must be determined. We look for two numbers whose product is c and whose sum is b.

EXAMPLE 4   Factor.

a) $x^2 + 10x + 16$ **b)** $x^2 - 8x + 15$
c) $x^2 - 2x - 24$ **d)** $3t^2 - 33st + 84s^2$

SOLUTION

a) The factorization is of the form

$$(x +\quad)(x +\quad).$$

Constant term positive (p. 312)

To find the constant terms, we need a pair of factors whose product is 16 and whose sum is 10. Since 16 is positive, its factors will have the same sign as 10—that is, we need consider only positive factors of 16.

We list the possible factorizations in a table and calculate the sum of each pair of factors.

Pairs of Factors of 16	Sums of Factors
1, 16	17
2, 8	10 ←
4, 4	8

The numbers we seek are 2 and 8.

The factorization of $x^2 + 10x + 16$ is $(x + 2)(x + 8)$. To check, we multiply.

Check: $(x + 2)(x + 8) = x^2 + 8x + 2x + 16 = x^2 + 10x + 16.$

b) For $x^2 - 8x + 15$, c is positive and b is negative. Therefore, the factors of 15 will be negative. Again, we list the possible factorizations in a table.

Pairs of Factors of 15	Sums of Factors
-1, -15	-16
-3, -5	-8 ←

The numbers we need are -3 and -5.

The factorization is $(x - 3)(x - 5)$.

Check: $(x - 3)(x - 5) = x^2 - 5x - 3x + 15 = x^2 - 8x + 15.$

Constant term negative (p. 314)

c) For $x^2 - 2x - 24$, c is negative, so one factor of -24 will be negative and one will be positive. Since b is also negative, the negative factor must have the larger absolute value.

Pairs of Factors of -24	Sums of Factors
1, -24	-23
2, -12	-10
3, -8	-5
4, -6	-2 ←

The numbers we need are 4 and -6.

The factorization is $(x + 4)(x - 6)$.

Check: $(x + 4)(x - 6) = x^2 - 6x + 4x - 24 = x^2 - 2x - 24$.

d) Always look first for a common factor. There is a common factor, 3, which we factor out first:

$$3t^2 - 33st + 84s^2 = 3(t^2 - 11st + 28s^2).$$

Now we consider $t^2 - 11st + 28s^2$. Think of $28s^2$ as the "constant" term c and $-11s$ as the "coefficient" b of the middle term. We try to express $28s^2$ as the product of two factors whose sum is $-11s$. These factors are $-4s$ and $-7s$. Thus the factorization of $t^2 - 11st + 28s^2$ is

$(t - 4s)(t - 7s).$ This is not the entire factorization of $3t^2 - 33st + 84s^2$.

We now include the common factor, 3, and write

$$3t^2 - 33st + 84s^2 = 3(t - 4s)(t - 7s).$$ This is the factorization.

Check: $3(t - 4s)(t - 7s) = 3(t^2 - 11st + 28s^2) = 3t^2 - 33st + 84s^2$.

TRY EXERCISE 11

Factoring trinomials of the type $ax^2 + bx + c$ (Section 5.3)

When the leading coefficient of a trinomial is not 1, the number of trials needed to find a factorization can increase dramatically. We will consider two methods for factoring trinomials of the type $ax^2 + bx + c$: factoring with FOIL and the grouping method.

To Factor $ax^2 + bx + c$ Using FOIL

1. Make certain that all common factors have been removed. If any remain, factor out the largest common factor.
2. Find two First terms whose product is ax^2:

   $(\boxed{}x + \,)(\boxed{}x + \,) = ax^2 + bx + c.$
   ⎣————— FOIL

Factoring with FOIL (p. 320)

3. Find two Last terms whose product is c:

   $(x + \boxed{})(x + \boxed{}) = ax^2 + bx + c.$
   ⎣————— FOIL

4. Check by multiplying to see if the sum of the Outer and Inner products is bx. If necessary, repeat steps 2 and 3 until the correct combination is found.

   $(\boxed{}x + \boxed{})(\boxed{}x + \boxed{}) = ax^2 + bx + c.$
   ⎣ I ⎦
   ⎣——— O ———⎦   FOIL

If no correct combination exists, state that the polynomial is prime.

EXAMPLE **5** Factor: $20x^3 - 22x^2 - 12x$.

SOLUTION

1. First, we factor out the largest common factor, $2x$:
   $$20x^3 - 22x^2 - 12x = 2x(10x^2 - 11x - 6).$$

2. Next, in order to factor the trinomial $10x^2 - 11x - 6$, we search for two terms whose product is $10x^2$. The possibilities are
   $$(x + \quad)(10x + \quad) \quad \text{or} \quad (2x + \quad)(5x + \quad).$$

3. There are four pairs of factors of -6. Since the first terms of the binomials are different, the order of the factors is important. So there are eight possibilities for the last terms:

   $$
   \begin{array}{ccc}
   1, -6 & & -6, \ 1 \\
   -1, \ 6 & \text{and} & 6, -1 \\
   2, -3 & & -3, \ 2 \\
   -2, \ 3 & & 3, -2.
   \end{array}
   $$

4. Since each of the eight possibilities from step (3) could be used in either of the two possibilities from step (2), there are $2 \cdot 8$, or 16, possible factorizations. We check the possibilities systematically until we find one that gives the correct factorization. Let's first try factors with $(2x + \quad)(5x + \quad)$.

Pair of Factors	*Corresponding Trial*	*Product*
$1, -6$	$(2x + 1)(5x - 6)$	$10x^2 - 7x - 6$ ←Wrong middle term
$-1, \ 6$	$(2x - 1)(5x + 6)$	$10x^2 + 7x - 6$ ←Wrong middle term. Note that changing the signs in the binomials changed the sign of middle term in the product.
$2, -3$	$(2x + 2)(5x - 3)$	$10x^2 + 4x - 6$ ←Wrong middle term. We need not consider $(2x - 2)(5x + 3)$.
$-6, \ 1$	$(2x - 6)(5x + 1)$	$10x^2 - 28x - 6$←Wrong middle term. We need not consider $(2x + 6)(5x - 1)$.
$-3, \ 2$	$(2x - 3)(5x + 2)$	$10x^2 - 11x - 6$←Correct middle term

   We can stop when we find a correct factorization. Including the common factor $2x$, we now have
   $$20x^3 - 22x^2 - 12x = 2x(2x - 3)(5x + 2).$$

   This can be checked by multiplying.
   TRY EXERCISE 27

   With practice, some of the trials can be skipped or performed mentally.

The grouping method (p. 325)

The second method of factoring trinomials of the type $ax^2 + bx + c$ involves factoring by grouping.

> **To Factor $ax^2 + bx + c$, Using the Grouping Method**
> 1. Factor out the largest common factor, if one exists.
> 2. Multiply the leading coefficient a and the constant c.
> 3. Find a pair of factors of ac whose sum is b.
> 4. Rewrite the middle term, bx, as a sum or difference using the factors found in step (3).
> 5. Factor by grouping.
> 6. Include any common factor from step (1) and check by multiplying.

EXAMPLE **6**   Factor: $7x^2 + 31x + 12$.

SOLUTION

1. There is no common factor (other than 1 or -1).

2. We multiply the leading coefficient, 7, and the constant, 12:

   $7 \cdot 12 = 84.$

3. We look for a pair of factors of 84 whose sum is 31. Since both 84 and 31 are positive, we need consider only positive factors.

Pairs of Factors of 84	Sums of Factors
1, 84	85
2, 42	44
3, 28	31 ←

   $3 + 28 = 31$

4. Next, we rewrite $31x$ using the factors 3 and 28:

   $31x = 3x + 28x.$

5. We now factor by grouping:

   $7x^2 + 31x + 12 = 7x^2 + 3x + 28x + 12$ Substituting $3x + 28x$ for $31x$

   $= x(7x + 3) + 4(7x + 3)$

   $= (7x + 3)(x + 4).$ Factoring out the common factor, $7x + 3$

6. *Check:* $(7x + 3)(x + 4) = 7x^2 + 31x + 12.$ **TRY EXERCISE** 15

Factoring Special Forms

We can factor certain types of polynomials directly, without using trial and error.

> **Factoring Formulas**
>
> Perfect-square trinomial: $A^2 + 2AB + B^2 = (A + B)^2,$
> $A^2 - 2AB + B^2 = (A - B)^2$
>
> Difference of squares: $A^2 - B^2 = (A + B)(A - B)$
>
> Sum of cubes: $A^3 + B^3 = (A + B)(A^2 - AB + B^2)$
>
> Difference of cubes: $A^3 - B^3 = (A - B)(A^2 + AB + B^2)$

Before using the factoring formulas, it is important to check carefully that the expression being factored is indeed in one of the forms listed. Note that there is no factoring formula for the sum of two squares.

EXAMPLE 7   Factor: **(a)** $2x^2 - 2$; **(b)** $x^2y^2 + 20xy + 100$; **(c)** $p^3 - 64$; **(d)** $3y^2 + 27$.

SOLUTION

a) We first factor out a common factor, 2:

$$2x^2 - 2 = 2(x^2 - 1).$$

Recognizing and factoring differences of squares (pp. 331, 332)

Looking at $x^2 - 1$, we see that it is a difference of squares, with $A = x$ and $B = 1$. The factorization is thus

$$2x^2 - 2 = 2(x^2 - 1) = 2(x + 1)(x - 1).$$

$$\underbrace{A^2 - B^2}_{} \qquad \underbrace{(A + B)(A - B)}_{}$$

Recognizing and factoring perfect-square trinomials (pp. 329, 330)

b) First, we check for a common factor; there is none. The polynomial is a perfect-square trinomial, since x^2y^2 and 100 are squares; there is no minus sign before either square; and $20xy$ is $2 \cdot xy \cdot 10$, where xy and 10 are square roots of x^2y^2 and 100, respectively. The factorization is thus

$$x^2y^2 + 20xy + 100 = (xy)^2 + 2 \cdot xy \cdot 10 + 10^2 = (xy + 10)^2.$$

$$A^2 + \quad 2 \cdot A \cdot \; B + \; B^2 = (A + B)^2$$

Factoring sums or differences of cubes (Section 5.5)

c) This is a difference of cubes, with $A = p$ and $B = 4$:

$$p^3 - 64 = (p)^3 - (4)^3$$
$$= (p - 4)(p^2 + 4p + 16).$$

d) We factor out the common factor, 3:

$$3y^2 + 27 = 3(y^2 + 9).$$

Since $y^2 + 9$ is a sum of squares, no further factorization is possible.

TRY EXERCISE   7

Factoring completely (p. 333)

A polynomial is said to be *factored completely* when no factor can be factored further.

EXAMPLE 8   Factor completely: $x^4 - 1$.

SOLUTION

$$x^4 - 1 = (x^2 + 1)(x^2 - 1) \qquad \text{Factoring a difference of squares}$$
$$= (x^2 + 1)(x + 1)(x - 1) \qquad \text{The factor } x^2 - 1 \text{ is itself a difference of squares.}$$

TRY EXERCISE   21

Solving Polynomial Equations by Factoring

Polynomial equation (p. 345)
Quadratic equation (p. 345)

A **polynomial equation** is formed by setting two polynomials equal to each other. A **quadratic equation** is a polynomial equation equivalent to one of the form $ax^2 + bx + c = 0$, where $a \neq 0$. Polynomial equations that can be factored can be solved using the principle of zero products.

The principle of zero products
(p. 346)

> ### The Principle of Zero Products
>
> An equation $ab = 0$ is true if and only if $a = 0$ or $b = 0$, or both. (A product is 0 if and only if at least one factor is 0.)

If we can write an equation as a product that equals 0, we can try to use the principle of zero products to solve the equation.

EXAMPLE 9   Solve:

a) $x^2 - 11x = 12$ b) $5x^2 + 10x + 5 = 0$ c) $9x^2 = 1$

SOLUTION

a) We must have 0 on one side of the equation before using the principle of zero products:

Get 0 on one side.

Factor.

Use the principle of zero products.

Solve.

$$x^2 - 11x = 12$$
$$x^2 - 11x - 12 = 0 \qquad \text{Subtracting 12 from both sides}$$
$$(x - 12)(x + 1) = 0 \qquad \text{Factoring}$$
$$x - 12 = 0 \quad or \quad x + 1 = 0 \qquad \text{Using the principle of zero products}$$
$$x = 12 \quad or \qquad x = -1.$$

The solutions are 12 and -1. The check is left to the student.

b) We have

$$5x^2 + 10x + 5 = 0$$
$$5(x^2 + 2x + 1) = 0 \qquad \text{Factoring out a common factor}$$
$$5(x + 1)(x + 1) = 0 \qquad \text{Factoring completely}$$
$$x + 1 = 0 \quad or \quad x + 1 = 0 \qquad \text{Using the principle of zero products}$$
$$x = -1 \quad or \qquad x = -1.$$

There is only one solution, -1. The check is left to the student.

c) We have

$$9x^2 = 1$$
$$9x^2 - 1 = 0 \qquad \text{Subtracting 1 from both sides to get 0 on one side}$$
$$(3x + 1)(3x - 1) = 0 \qquad \text{Factoring a difference of squares}$$
$$3x + 1 = 0 \quad or \quad 3x - 1 = 0 \qquad \text{Using the principle of zero products}$$
$$3x = -1 \quad or \qquad 3x = 1$$
$$x = -\tfrac{1}{3} \quad or \qquad x = \tfrac{1}{3}.$$

The solutions are $\frac{1}{3}$ and $-\frac{1}{3}$. The check is left to the student.

> **TRY EXERCISE** 37

Quadratic equations can be used to solve problems. One important result that uses squared quantities is the Pythagorean theorem. It relates the lengths of the sides of a **right triangle**, that is, a triangle with a 90° angle. The side opposite the 90° angle is called the **hypotenuse**, and the other sides are called the **legs**.

The Pythagorean Theorem

The sum of the squares of the legs of a right triangle is equal to the square of the hypotenuse:

$$a^2 + b^2 = c^2.$$

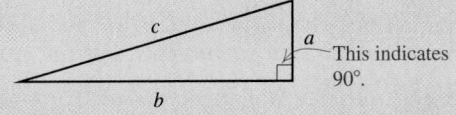

This indicates 90°.

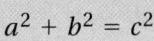 **EXAMPLE 10**

Swing sets. The length of a slide on a swing set is 5 ft. The distance from the base of the ladder to the base of the slide is 1 ft more than the height of the ladder. Find the height of the ladder.

SOLUTION

1. **Familiarize.** We first make a drawing and let x = the height of the ladder, in feet. We know then that the other leg of the triangle is $x + 1$, since it is 1 ft longer than the ladder. The hypotenuse has length 5 ft.

2. **Translate.** Applying the Pythagorean theorem gives us

$$a^2 + b^2 = c^2$$
$$x^2 + (x + 1)^2 = 5^2. \quad \text{Substituting}$$

3. **Carry out.** We solve the equation:

$$x^2 + (x + 1)^2 = 5^2$$

$x^2 + x^2 + 2x + 1 = 25$	Squaring $x + 1$; squaring 5
$2x^2 + 2x + 1 = 25$	Combining like terms
$2x^2 + 2x - 24 = 0$	Getting 0 on one side
$2(x^2 + x - 12) = 0$	Factoring out a common factor
$2(x + 4)(x - 3) = 0$	Factoring a trinomial
$x + 4 = 0 \quad or \quad x - 3 = 0$	Using the principle of zero products
$x = -4 \quad or \quad x = 3.$	

4. **Check.** We know that the integer -4 is not a solution because the height of the ladder cannot be negative. When $x = 3$, the distance from the base of the ladder to the base of the slide is $x + 1 = 4$, and $3^2 + 4^2 = 5^2$. So the solution 3 checks.

5. **State.** The ladder is 3 ft high.

TRY EXERCISE ▶ 49

R.5 EXERCISE SET

Factor completely. If a polynomial is prime, state this.

1. $18t^5 - 12t^4 + 6t^3$ **2.** $x^2y^4 - 2xy^5 + 3x^3y^6$

3. $y^2 - 6y + 9$ **4.** $4z^2 - 25$

5. $2p^3(p + 2) + (p + 2)$ **6.** $6y^2 + y - 1$

7. $x^2 + 100$ **8.** $y^3 - 1$

9. $8t^3 + 27$ **10.** $a^2b^2 + 24ab + 144$

11. $m^2 + 13m + 42$ **12.** $2x^3 - 6x^2 + x - 3$

13. $x^4 - 81$ **14.** $x^2 + x + 3$

15. $8x^2 + 22x + 15$ **16.** $4x^2 - 40x + 100$

17. $x^3 + 2x^2 - x - 2$

18. $(x + 2y)(x - 1) + (x + 2y)(x - 2)$

19. $0.001t^6 - 0.008$

20. $x^2 - 20 - x$

21. $-\frac{1}{16} + x^4$

22. $5x^8 - 5z^{16}$

23. $mn - 2m + 3n - 6$

24. $t^6 - p^6$

25. $5mn + m^2 - 150n^2$

26. $\frac{1}{27} + x^3$

27. $24x^2y - 6y - 10xy$ **28.** $-3y^2 - 12y - 12$

29. $y^2 + 121 - 22y$ **30.** $t^3 - 2t^2 - 5t + 10$

Solve.

31. $(x - 1)(x + 3) = 0$

32. $(3x - 5)(7 - 4x) = 0$

33. $8x(11 - x) = 0$

34. $(x - 3)(x + 1)(2x - 9) = 0$

35. $x^2 = 9$ **36.** $8x^2 = 2x$

37. $4x^2 - 18x = 70$ **38.** $x^2 + 6x + 9 = 0$

39. $2x^2 - 10x = 0$ **40.** $100x^2 = 9$

41. $(a + 1)(a - 5) = 7$ **42.** $d(d - 3) = 40$

43. $x^2 + 6x - 55 = 0$ **44.** $x^2 + 7x - 60 = 0$

45. $\frac{1}{2}x^2 + 5x + \frac{25}{2} = 0$ **46.** $3 + 10x^2 = 11x$

47. *Landscaping.* A triangular flower garden is 3 ft longer than it is wide. The area of the garden is 20 ft². What are the dimensions of the garden?

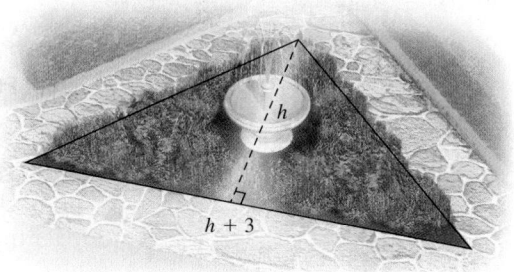

48. *Page numbers.* The product of the page numbers on two facing pages of a book is 156. Find the page numbers.

49. *Right triangles.* The hypotenuse of a right triangle is 17 ft. One leg is 1 ft shorter than twice the length of the other leg. Find the lengths of the legs.

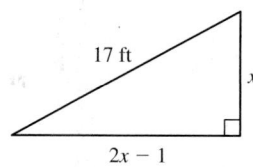

50. *Hiking.* Cheri hiked 500 ft up a steep incline. Her global positioning unit indicated that her horizontal position had changed by 100 ft more than her vertical position had changed. What was the change in altitude?

R.6 Rational Expressions and Equations

Multiplication and Division of Rational Expressions ■ Addition and Subtraction of Rational Expressions ■ Complex Rational Expressions ■ Solving Rational Equations

Rational expressions (p. 376)

A **rational expression** is a quotient of two polynomials. Because division by 0 is undefined, a rational expression is undefined for any number that will make the denominator 0.

EXAMPLE 1 Find all numbers for which the rational expression

$$\frac{2x + 5}{x^2 - 9x - 10}$$

is undefined.

SOLUTION   We set the denominator equal to 0 and solve:

$$x^2 - 9x - 10 = 0$$

$$(x - 10)(x + 1) = 0 \qquad \text{Factoring}$$

$$x - 10 = 0 \quad or \quad x + 1 = 0 \qquad \text{Using the principle of zero products}$$

$$x = 10 \quad or \qquad x = -1.$$

If x is replaced with 10 or with -1, the denominator is 0. Thus,

$$\frac{2x + 5}{x^2 - 9x - 10} \text{ is undefined for } x = 10 \text{ and } x = -1.$$

 **TRY EXERCISE   1**

Multiplication and Division of Rational Expressions

Multiplication and division of rational expressions is similar to multiplication and division with fractions.

Multiplication and division of rational expressions (Section 6.2)

> ### The Product and the Quotient of Two Rational Expressions
>
> To multiply two rational expressions, multiply numerators and multiply denominators:
>
> $$\frac{A}{B} \cdot \frac{C}{D} = \frac{AC}{BD}.$$
>
> To divide by a rational expression, multiply by its reciprocal:
>
> $$\frac{A}{B} \div \frac{C}{D} = \frac{A}{B} \cdot \frac{D}{C} = \frac{AD}{BC}.$$

EXAMPLE 2   Simplify: $\dfrac{9x^2 + 12x}{6x^2 - 3x}$.

SOLUTION   We first factor the numerator and the denominator:

$$\frac{9x^2 + 12x}{6x^2 - 3x} = \frac{3x(3x + 4)}{3x(2x - 1)}.$$

We can now write this as a product of two rational expressions using the rule for multiplying rational expressions in reverse. Then we can simplify.

$$\frac{3x(3x + 4)}{3x(2x - 1)} = \frac{3x}{3x} \cdot \frac{3x + 4}{2x - 1} \qquad \text{Rewriting as a product of two rational expressions}$$

$$= 1 \cdot \frac{3x + 4}{2x - 1} \qquad \frac{3x}{3x} = 1$$

$$= \frac{3x + 4}{2x - 1} \qquad \text{Removing the factor 1}$$

> **TRY EXERCISE** 7

Only factors can be removed. Be sure that the numerator and the denominator are factored before you attempt to remove factors equal to 1.

After multiplying or dividing rational expressions, we simplify, if possible.

EXAMPLE **3**   Perform each indicated operation and simplify.

a) $\dfrac{x^2 - x - 6}{3x} \cdot \dfrac{12x^3}{x + 2}$

b) $\dfrac{x^2 - 1}{x + 5} \div \dfrac{x^2 + 2x + 1}{2x + 10}$

SOLUTION

a) $\dfrac{x^2 - x - 6}{3x} \cdot \dfrac{12x^3}{x + 2} = \dfrac{(x^2 - x - 6)(12x^3)}{3x(x + 2)}$ Multiplying the numerators and the denominators

$$= \frac{(x - 3)(x + 2)(3x)(4x^2)}{3x(x + 2)} \qquad \begin{array}{l}\text{Factoring the numerator.} \\ \text{Try to go directly to this} \\ \text{step.}\end{array}$$

$$= \frac{(x - 3)\cancel{(x + 2)}\cancel{(3x)}(4x^2)}{\cancel{(3x)}\cancel{(x + 2)}} \qquad \begin{array}{l}\text{Removing a factor equal} \\ \text{to 1:} \dfrac{(x + 2)(3x)}{(x + 2)(3x)} = 1\end{array}$$

$$= 4x^2(x - 3)$$

b) $\dfrac{x^2 - 1}{x + 5} \div \dfrac{x^2 + 2x + 1}{2x + 10} = \dfrac{x^2 - 1}{x + 5} \cdot \dfrac{2x + 10}{x^2 + 2x + 1}$ Multiplying by the reciprocal of the divisor

$$= \frac{(x + 1)(x - 1)(2)(x + 5)}{(x + 5)(x + 1)(x + 1)} \qquad \begin{array}{l}\text{Multiplying rational} \\ \text{expressions and} \\ \text{factoring numerators} \\ \text{and denominators}\end{array}$$

$$= \frac{\cancel{(x + 1)}(x - 1)(2)\cancel{(x + 5)}}{\cancel{(x + 5)}\cancel{(x + 1)}(x + 1)} \qquad \begin{array}{l}\text{Removing a factor} \\ \text{equal to 1:} \\ \dfrac{(x + 1)(x + 5)}{(x + 1)(x + 5)} = 1\end{array}$$

$$= \frac{2(x - 1)}{x + 1} \qquad \begin{array}{l}\text{We leave the} \\ \text{numerator in} \\ \text{factored form.}\end{array}$$

> **TRY EXERCISE** 13

Addition and Subtraction of Rational Expressions

Like multiplication and division, addition and subtraction of rational expressions is similar to addition and subtraction of fractions.

Addition and subtraction of rational expressions (Sections 6.3 and 6.4)

> ### The Sum and the Difference of Two Rational Expressions
>
> To add when the denominators are the same, add the numerators and keep the same denominator:
>
> $$\frac{A}{B} + \frac{C}{B} = \frac{A + C}{B}.$$
>
> To subtract when the denominators are the same, subtract the second numerator from the first and keep the same denominator:
>
> $$\frac{A}{B} - \frac{C}{B} = \frac{A - C}{B}.$$

EXAMPLE 4   Add and simplify, if possible:

$$\frac{x - 6}{x^2 - 6x + 5} + \frac{5}{x^2 - 6x + 5}.$$

SOLUTION

$$\frac{x - 6}{x^2 - 6x + 5} + \frac{5}{x^2 - 6x + 5} = \frac{x - 6 + 5}{x^2 - 6x + 5} \qquad \text{Adding numerators}$$

$$= \frac{x - 1}{(x - 5)(x - 1)} \qquad \begin{array}{l}\text{Factoring the} \\ \text{denominator}\end{array}$$

$$= \frac{1(x - 1)}{(x - 5)(x - 1)} \qquad \begin{array}{l}\text{Removing a factor} \\ \text{equal to 1:} \\ \dfrac{x - 1}{x - 1} = 1\end{array}$$

$$= \frac{1}{x - 5}$$

TRY EXERCISE 15

Least common denominator (p. 390)
Least common multiple (p. 390)

When two rational expressions do not have a common denominator, we must rewrite them with a common denominator before we can add or subtract them. We generally rewrite them using their **least common denominator** (**LCD**), which is the **least common multiple** (**LCM**) of their denominators.

> ### To Find the Least Common Denominator (LCD)
>
> **1.** Write the prime factorization of each denominator.
> **2.** Select one of the factorizations and inspect it to see if it completely contains the other.
>
>   **a)** If it does, it represents the LCM of the denominators.
>   **b)** If it does not, multiply that factorization by any factors of the other denominator that it lacks. The final product is the LCM of the denominators.
>
> The LCD is the LCM of the denominators. It should contain each factor the greatest number of times that it occurs in any of the individual factorizations.

EXAMPLE 5

Add: $\dfrac{x-3}{x^2-1} + \dfrac{4x^2}{x^2+4x+3}$.

SOLUTION   We first find the LCD. We write the prime factorization of each denominator and construct the LCM:

$$x^2 - 1 = (x+1)(x-1);$$
$$x^2 + 4x + 3 = (x+1)(x+3).$$

The LCM must contain both factorizations. We select the factorization of $x^2 - 1$. It does not contain the factor $(x+3)$ from the factorization of $x^2 + 4x + 3$. We multiply $(x+1)(x-1)$ by $(x+3)$:

$$\text{LCM} = (x+1)(x-1)(x+3).$$

The denominator $x^2 - 1 = (x+1)(x-1)$ must be multiplied by $x+3$ in order to obtain the LCD. The denominator $x^2 + 4x + 3 = (x+1)(x+3)$ must be multiplied by $x-1$ in order to obtain the LCD. We multiply each expression by a form of 1 that is made up of these "missing" factors:

$$\frac{x-3}{x^2-1} + \frac{4x^2}{x^2+4x+3} = \frac{x-3}{(x+1)(x-1)} \cdot \frac{x+3}{x+3} + \frac{4x^2}{(x+1)(x+3)} \cdot \frac{x-1}{x-1}$$

$$= \frac{x^2-9}{(x+1)(x-1)(x+3)} + \frac{4x^3 - 4x^2}{(x+1)(x-1)(x+3)}$$

$$= \frac{4x^3 - 3x^2 - 9}{(x+1)(x-1)(x+3)}.$$

> TRY EXERCISE ▶ 25

EXAMPLE 6

Subtract: $\dfrac{x}{x+2} - \dfrac{2x-3}{3x-4}$.

SOLUTION   We have

$$\frac{x}{x+2} - \frac{2x-3}{3x-4}$$

Find the LCD.

$$= \frac{x}{x+2} \cdot \frac{3x-4}{3x-4} - \frac{2x-3}{3x-4} \cdot \frac{x+2}{x+2}$$

The LCD is $(x+2)(3x-4)$.

Rewrite each expression with the LCD.

$$= \frac{3x^2 - 4x}{(x+2)(3x-4)} - \frac{2x^2 + x - 6}{(x+2)(3x-4)}$$

Multiplying out the numerators (but not the denominators)

Subtract numerators. Keep the denominator.

$$= \frac{3x^2 - 4x - (2x^2 + x - 6)}{(x+2)(3x-4)}$$

Parentheses are important.

$$= \frac{3x^2 - 4x - 2x^2 - x + 6}{(x+2)(3x-4)}$$

Removing parentheses in the numerator; subtracting every term

$$= \frac{x^2 - 5x + 6}{(x+2)(3x-4)}$$

Simplify if possible.

$$= \frac{(x-2)(x-3)}{(x+2)(3x-4)}.$$

Factoring the numerator in hopes of simplifying. There are no common factors.

The result could be written as either of the last two expressions.

> TRY EXERCISE ▶ 33

Factors that are opposites (p. 400)   When denominators are opposites, we can find a common denominator by multiplying either rational expression by $-1/-1$.

EXAMPLE **7** Add: $\dfrac{a}{a-b} + \dfrac{5}{b-a}$.

SOLUTION

$$\dfrac{a}{a-b} + \dfrac{5}{b-a} = \dfrac{a}{a-b} + \dfrac{5}{b-a} \cdot \dfrac{-1}{-1}$$

Writing 1 as $-1/-1$ and multiplying to obtain a common denominator

$$= \dfrac{a}{a-b} + \dfrac{-5}{a-b}$$

$(b-a)(-1) = -b + a = a - b$

$$= \dfrac{a-5}{a-b}$$

> **TRY EXERCISE** 27

Complex Rational Expressions

Complex rational expressions
(Section 6.5)

A **complex rational expression** is a rational expression that has one or more rational expressions within its numerator or denominator. We will consider two methods for simplifying complex rational expressions. The first involves writing the expression as a quotient of two rational expressions.

> **To Simplify a Complex Rational Expression by Dividing**
> 1. Add or subtract, as needed, to get a single rational expression in the numerator.
> 2. Add or subtract, as needed, to get a single rational expression in the denominator.
> 3. Divide the numerator by the denominator (invert and multiply).
> 4. If possible, simplify by removing a factor equal to 1.

EXAMPLE **8** Simplify by dividing: $\dfrac{\dfrac{2}{x+1}}{\dfrac{1}{x+2} + \dfrac{1}{x}}$.

SOLUTION

1. There is already a single rational expression in the numerator.

2. We add to get a single rational expression in the denominator:

$$\dfrac{\dfrac{2}{x+1}}{\dfrac{1}{x+2} + \dfrac{1}{x}} = \dfrac{\dfrac{2}{x+1}}{\dfrac{1}{x+2} \cdot \dfrac{x}{x} + \dfrac{1}{x} \cdot \dfrac{x+2}{x+2}}$$

Multiplying by 1 to get the LCD, $x(x+2)$, for the denominator

$$= \dfrac{\dfrac{2}{x+1}}{\dfrac{x}{x(x+2)} + \dfrac{x+2}{x(x+2)}} = \dfrac{\dfrac{2}{x+1}}{\dfrac{2x+2}{x(x+2)}}.$$

Adding in the denominator

3. Next, we invert and multiply:

$$\dfrac{\dfrac{2}{x+1}}{\dfrac{2x+2}{x(x+2)}} = \dfrac{2}{x+1} \div \dfrac{2x+2}{x(x+2)} = \dfrac{2}{x+1} \cdot \dfrac{x(x+2)}{2x+2}.$$

4. Simplifying, we have:

$$\frac{2}{(x+1)} \cdot \frac{x(x+2)}{2x+2} = \frac{\cancel{2} \cdot x(x+2)}{\cancel{2}(x+1)(x+1)}$$

Factoring in the denominator and removing a factor equal to 1: $\frac{2}{2} = 1$

$$= \frac{x(x+2)}{(x+1)^2}.$$

TRY EXERCISE ▸ 35

A second method for simplifying complex rational expressions involves multiplying by the LCD.

To Simplify a Complex Rational Expression by Multiplying by the LCD

1. Find the LCD of *all* rational expressions within the complex rational expression.
2. Multiply the complex rational expression by a factor equal to 1. Write 1 as the LCD over itself (LCD/LCD).
3. Simplify. No fraction expressions should remain within the complex rational expression.
4. Factor and, if possible, simplify.

EXAMPLE **9** Simplify by multiplying by the LCD: $\dfrac{1 + \dfrac{2}{t}}{\dfrac{4}{t^2} - 1}$.

SOLUTION

1. The denominators *within* the complex rational expression are t and t^2, so the LCD is t^2.

2. We multiply by a form of 1 using t^2/t^2:

$$\frac{1 + \dfrac{2}{t}}{\dfrac{4}{t^2} - 1} = \frac{1 + \dfrac{2}{t}}{\dfrac{4}{t^2} - 1} \cdot \frac{t^2}{t^2}.$$

3. We distribute and simplify:

$$\frac{1 + \dfrac{2}{t}}{\dfrac{4}{t^2} - 1} \cdot \frac{t^2}{t^2} = \frac{1 \cdot t^2 + \dfrac{2}{t} \cdot t^2}{\dfrac{4}{t^2} \cdot t^2 - 1 \cdot t^2}$$

$$= \frac{t^2 + 2t}{4 - t^2}.$$

No rational expression remains within the numerator or denominator.

4. Finally, we simplify:

$$\frac{t^2 + 2t}{4 - t^2} = \frac{t\cancel{(t+2)}}{\cancel{(2+t)}(2-t)}$$

Factoring and simplifying; $\dfrac{t+2}{t+2} = 1$

$$= \frac{t}{2-t}.$$

TRY EXERCISE ▸ 41

Solving Rational Equations

Solving rational equations
(Section 6.6)

A **rational equation** is an equation containing one or more rational expressions, often with the variable in a denominator.

> **To Solve a Rational Equation**
> 1. List any restrictions that exist. Numbers that make a denominator equal 0 can never be solutions.
> 2. Clear the equation of fractions by multiplying both sides by the LCM of the denominators.
> 3. Solve the resulting equation using the addition principle, the multiplication principle, and the principle of zero products, as needed.
> 4. Check the possible solution(s) in the original equation.

Because a possible solution in step 3 may make a denominator 0, checking is essential when solving rational equations.

EXAMPLE 10   Solve: $x + \dfrac{10}{x} = 7$.

SOLUTION   First we note that x cannot be 0. The LCD is x, so we multiply both sides by x:

$$x + \frac{10}{x} = 7$$

$$x\left(x + \frac{10}{x}\right) = 7x \qquad \text{Don't forget the parentheses!}$$

$$x \cdot x + x \cdot \frac{10}{x} = 7x \qquad \text{Using the distributive law}$$

$$x^2 + 10 = 7x \qquad \text{We have a quadratic equation.}$$

$$x^2 - 7x + 10 = 0 \qquad \text{Getting 0 on one side}$$

$$(x - 2)(x - 5) = 0 \qquad \text{Factoring}$$

$$x - 2 = 0 \quad or \quad x - 5 = 0 \qquad \text{Using the principle of zero products}$$

$$x = 2 \quad or \qquad x = 5.$$

Check: For 2:

$$x + \frac{10}{x} = 7$$

$$\begin{array}{c|c} 2 + \dfrac{10}{2} & 7 \\ \hline 2 + 5 & \\ & 7 \overset{?}{=} 7 \quad \text{TRUE} \end{array}$$

For 5:

$$x + \frac{10}{x} = 7$$

$$\begin{array}{c|c} 5 + \dfrac{10}{5} & 7 \\ \hline 5 + 2 & \\ & 7 \overset{?}{=} 7 \quad \text{TRUE} \end{array}$$

Both numbers check, so there are two solutions, 2 and 5.   **TRY EXERCISE** 45

Work problems (p. 419)

Many problems translate to rational equations. **Work problems**, which involve the time that it takes to complete a task, can often be solved using the work principle.

> ### The Work Principle
>
> If
>
> $$a = \text{the time needed for } A \text{ to complete the work alone,}$$
> $$b = \text{the time needed for } B \text{ to complete the work alone, and}$$
> $$t = \text{the time needed for } A \text{ and } B \text{ to complete the work together,}$$
>
> then
>
> $$\frac{t}{a} + \frac{t}{b} = 1.$$
>
> The following are equivalent equations that can also be used:
>
> $$\frac{1}{a} \cdot t + \frac{1}{b} \cdot t = 1 \quad \text{and} \quad \frac{1}{a} + \frac{1}{b} = \frac{1}{t}.$$

EXAMPLE 11   *Drafting.*   It takes Kerry 30 hr to draw a set of plans for a house. It takes Jesse 45 hr to draw the same set of plans. How long would it take Kerry and Jesse, working together, to draw the set of plans?

SOLUTION

1. **Familiarize.** We could make some guesses to help us understand the problem and then list our results in a table. We could also reason that if Kerry and Jesse each drew half the plans, it would take Kerry 15 hr and Jesse $22\frac{1}{2}$ hr. So the time it takes them working together should be between 15 and $22\frac{1}{2}$ hr. We let $t = $ the time that it takes them to draw the plans, working together.

2. **Translate.** We will use the work principle to translate the problem:

   $$\frac{t}{a} + \frac{t}{b} = 1 \qquad \begin{array}{l} a \text{ is the time that it takes Kerry to draw the plans;} \\ b \text{ is the time that it takes Jesse to draw the plans.} \end{array}$$

   $$\frac{t}{30} + \frac{t}{45} = 1.$$

3. **Carry out.** We solve the equation:

   $$\frac{t}{30} + \frac{t}{45} = 1$$

   $$90\left(\frac{t}{30} + \frac{t}{45}\right) = 90 \cdot 1 \qquad \text{The LCD is } 2 \cdot 3 \cdot 3 \cdot 5, \text{ or } 90.$$

   $$90 \cdot \frac{t}{30} + 90 \cdot \frac{t}{45} = 90$$

   $$3t + 2t = 90$$

   $$5t = 90$$

   $$t = 18.$$

4. **Check.** We note that, as predicted in the *Familiarize* step, the answer is between 15 and $22\frac{1}{2}$ hr. Also, if each works 18 hr, Kerry will do $\frac{18}{30}$ of the job and Jesse will do $\frac{18}{45}$ of the job, and

   $$\frac{18}{30} + \frac{18}{45} = \frac{3}{5} + \frac{2}{5} = 1. \qquad \text{The entire job will be completed.}$$

5. **State.** Together it will take them 18 hr to draw the plans.

TRY EXERCISE 51

Motion problems (p. 422)

Motion problems deal with distance, speed (or rate), and time, and can often be translated using the distance formula $d = rt$.

EXAMPLE 12

Driving time.   Karen and Eva are each driving to a sales meeting. Because of road conditions, Karen is able to drive 15 mph faster than Eva. In the same time that it takes Karen to travel 120 mi, Eva travels only 90 mi. Find their speeds.

SOLUTION

1. **Familiarize.** We let t = the time, in hours, that is spent traveling and r = Karen's speed, in mph. Then Eva's speed = $r - 15$. We set up a table.

$$d \quad = \quad r \quad \cdot \quad t$$

	Distance	Speed	Time
Karen	120	r	t
Eva	90	$r - 15$	t

2. **Translate.** From the distance formula, we have $t = d/r$, so we can replace the times in the table with expressions involving r.

	Distance	Speed	Time
Karen	120	r	$120/r$
Eva	90	$r - 15$	$90/(r - 15)$

Since the times are the same, we have the equation

$$\frac{120}{r} = \frac{90}{r - 15}.$$

3. **Carry out.** We solve the equation:

$$\frac{120}{r} = \frac{90}{r - 15}$$

$$r(r - 15)\frac{120}{r} = r(r - 15)\frac{90}{r - 15} \qquad \text{The LCD is } r(r - 15).$$

$$120(r - 15) = 90r \qquad \text{Simplifying}$$

$$120r - 1800 = 90r \qquad \text{Removing parentheses}$$

$$-1800 = -30r \qquad \text{Subtracting } 120r$$

$$60 = r. \qquad \text{Dividing both sides by } -30$$

4. **Check.** If $r = 60$, then $r - 15 = 45$. If Karen travels 120 mi at 60 mph, she will have traveled 2 hr. If Eva travels 90 mi at 45 mph, she will also have traveled 2 hr. Since the times are the same, the speeds check.

5. **State.** Karen is traveling at 60 mph, while Eva is traveling at 45 mph.

> **TRY EXERCISE** 53

Ratio (p. 425)
Proportion (p. 425)

Another type of problem that translates to a rational equation involves proportions. A **ratio** of two quantities is their quotient. A **proportion** is an equation stating that two ratios are equal.

EXAMPLE **13** *Baking.* Rob discovers there is $2\frac{1}{2}$ cups of pancake mix left in the box. The directions on the mix indicate that $1\frac{1}{3}$ cups of milk should be added to 2 cups of mix. How much milk should Rob add to the $2\frac{1}{2}$ cups of mix?

SOLUTION Since the problem translates directly to a proportion, we will not follow all five steps of the problem-solving process. We write the ratio of mix to milk in two ways:

$$\text{Mix} \longrightarrow \frac{2}{1\frac{1}{3}} = \frac{2\frac{1}{2}}{x} \longleftarrow \text{Mix}$$
$$\text{Milk} \longrightarrow \qquad\qquad \longleftarrow \text{Milk}$$

The LCD is $x\left(1\frac{1}{3}\right)$. We solve for x:

$$x\left(1\frac{1}{3}\right)\frac{2}{1\frac{1}{3}} = x\left(1\frac{1}{3}\right)\frac{2\frac{1}{2}}{x} \qquad \text{Multiplying by the LCD}$$

$$2x = \left(1\frac{1}{3}\right)\left(2\frac{1}{2}\right) \qquad \text{Simplifying}$$

$$2x = \frac{10}{3} \qquad \text{Converting to fraction notation and multiplying}$$

$$x = \frac{5}{3}. \qquad \text{Multiplying both sides by } \frac{1}{2} \text{ and simplifying}$$

Rob needs to add $\frac{5}{3}$ or $1\frac{2}{3}$ cups of milk.

TRY EXERCISE 55

R.6 EXERCISE SET

For Extra Help  PRACTICE WATCH DOWNLOAD

List all numbers for which each rational expression is undefined.

1. $\dfrac{x-7}{3x+1}$

2. $\dfrac{10-y}{-5y}$

3. $\dfrac{p^2-1}{p^2-4}$

4. $\dfrac{10x}{x^2+9x+8}$

Simplify by removing a factor equal to 1.

5. $\dfrac{16x^2y}{18xy^2}$

6. $\dfrac{2x+10}{6x+30}$

7. $\dfrac{t^2-2t-8}{t^2-16}$

8. $\dfrac{a^3+2a^2+a}{a^2+4a+3}$

9. $\dfrac{2-x}{x^2-4}$

10. $\dfrac{n-3}{3-n}$

Perform each indicated operation. Then, if possible, simplify.

11. $\dfrac{3x}{x+y} \cdot \dfrac{2x+2y}{x^2}$

12. $\dfrac{5}{x+7} \cdot \dfrac{x+7}{10}$

13. $\dfrac{a^2+2a+1}{a} \div \dfrac{a^2}{a^2-1}$

14. $\dfrac{x}{x+3} + \dfrac{3-x}{x+3}$

15. $\dfrac{2x}{x-7} - \dfrac{x+7}{x-7}$

16. $\dfrac{x}{x+y} \div \dfrac{y}{x+y}$

17. $\dfrac{5}{x} + \dfrac{4}{x^2}$

18. $\dfrac{x^2+4x+3}{x^2+x-2} \cdot \dfrac{x^2+3x+2}{x^2+2x-3}$

19. $\dfrac{2a+b}{a-b} - \dfrac{4}{3a-3b}$

20. $(x^2-16) \div \dfrac{4x+16}{3x^2}$

21. $\dfrac{2-x}{5x^2} \div \dfrac{x^2-4}{3x}$

22. $\dfrac{2x}{x-5} + \dfrac{3}{x+4}$

23. $\dfrac{x^3+2x^2+x}{x^2-4} \cdot \dfrac{x^2-x-2}{x^4+x^3}$

24. $\dfrac{-1}{x^2+7x+10} - \dfrac{3}{x^2+8x+15}$

25. $\dfrac{2}{(x+1)^2} + \dfrac{1}{x+1}$

26. $\dfrac{2x}{x^2 - 3x} \div (x - 3)$

27. $\dfrac{x}{x - 2} + \dfrac{2}{2 - x}$

28. $\dfrac{3}{y - 1} - \dfrac{y}{1 - y}$

29. $\dfrac{t}{t^2 - 1} - \dfrac{1}{1 - t}$

30. $\dfrac{1}{5 - x} + \dfrac{x}{2x - 10}$

31. $\dfrac{x - y}{2x} \cdot \dfrac{3x^2}{y - x}$

32. $\dfrac{1}{x + y} + \dfrac{2}{x^2 + y^2}$

33. $\dfrac{x - 2}{x + 5} - \dfrac{x + 3}{x - 4}$

34. $\dfrac{z^2 + 2z + 1}{8z} \div \dfrac{z^2 - z - 2}{4z^2 - 4}$

Simplify.

35. $\dfrac{\dfrac{2}{x} - \dfrac{1}{x^2}}{\dfrac{x}{4}}$

36. $\dfrac{\dfrac{x}{3} - \dfrac{3}{x}}{\dfrac{1}{x} + \dfrac{1}{3}}$

37. $\dfrac{\dfrac{3}{x - 7}}{\dfrac{4x + 3}{x + 1}}$

38. $\dfrac{\dfrac{a}{a - b}}{\dfrac{a^2}{a^2 - b^2}}$

39. $\dfrac{x - \dfrac{3}{x - 2}}{x - \dfrac{12}{x + 1}}$

40. $\dfrac{t + \dfrac{1}{t}}{t - \dfrac{2}{t}}$

41. $\dfrac{\dfrac{1}{2} - \dfrac{1}{x}}{\dfrac{2 - x}{2}}$

42. $\dfrac{\dfrac{x}{2y^2} + \dfrac{y}{3x^2}}{\dfrac{1}{6xy} + \dfrac{2}{x^2 y}}$

Solve.

43. $\dfrac{1}{2} + \dfrac{1}{3} = \dfrac{1}{t}$

44. $\dfrac{1}{4} + \dfrac{1}{t} = \dfrac{1}{3}$

45. $x + \dfrac{1}{x} = 2$

46. $\dfrac{x - 7}{x + 1} = \dfrac{2}{3}$

47. $\dfrac{3}{y + 7} = \dfrac{1}{y - 8}$

48. $\dfrac{x + 1}{x - 2} = \dfrac{3}{x - 2}$

49. $\dfrac{1}{x - 3} - \dfrac{x - 4}{x^2 - 9} = 1$

50. $\dfrac{3}{a + 4} = \dfrac{a - 1}{4 - a}$

51. *Painting.* Quentin can paint the turret on a Queen Anne house in 40 hr. It takes Austin 50 hr to paint the same turret. How long would it take them, working together, to paint the turret?

52. *Building fences.* Lindsay can build a fence in 6 hr. Laura can do the same job in 5 hr. How long will it take them, working together, to build the fence?

53. *Snowmobiling.* Jessica can ride her snowmobile through the fields 20 km/h faster than Josh can ride his through the woods. In the time it takes Jessica to ride 18 km, Josh travels 10 km. Find the speed of each snowmobile.

54. *Bicycling.* Ani bicycles 8 mi and Lia bicycles 12 mi to meet at a park for lunch. Because Ani's trip is mostly uphill, she rides 5 mph slower than Lia. Ani and Lia leave their homes at the same time and arrive at the park at the same time. Find the speed of each bicyclist.

55. *Elk population.* To determine the size of a park's elk population, rangers tag 15 elk and set them free. Months later, 40 elk are caught, of which 12 have tags. Estimate the size of the elk population.

56. *Manufacturing pegs.* A sample of 136 wooden pegs contained 17 defective pegs. How many defective pegs would you expect in a sample of 840 pegs?

Photo Credits

1, © Reuters/Corbis   7, © Richard Chung/Reuters/Corbis   10, Copyright © Jack Stein Grove/PhotoEdit—All rights reserved.   30, travelstockphoto.com   36, National Science Federation   40, © moodboard/Corbis   51, © Peter Turnley/Corbis   58, © Reuters/Corbis   77, 93, Ryan McVay/Stone/Getty Images   97, Copyright © Park Street/PhotoEdit—All rights reserved.   107, Michelle Pemberton/*The Indianapolis Star*   111, Dennis Flaherty/Getty Images   115, AP Photo/H. Rumph, Jr.   120, © Ed Quinn/Corbis   132, © Jeremy Birch   133, © Push Pictures/Corbis   137, Brand X Pictures (Getty)   138, Chris Gardner/Associated Press   145, © David Bergman/www.DavidBergman.net/Corbis   147, AFP/Getty Images   179, © Alan Carey/Corbis/Corbis Royalty Free   180, © Earl & Nazima Kowall/Corbis   185, Photograph by John Locke/Kirtas Technologies, Inc.   195, AbbiOrca.com   210, AFP/Getty Images   215, © image100/Corbis   223, Panoramic Images/Getty Images   225, © Ted Levine/zefa/Corbis Royalty Free   227, © Richard T. Nowitz/Corbis   249, © moodboard/Corbis Royalty Free   251, Wild Pics/Getty Images   252, © Richard T. Nowitz/Corbis   298, © Vo Trung Dung/Corbis Sygma   300, © Marco Cristofori/Corbis   303, Photo by Michael Smith/Getty Images   353, Getty/Royalty Free   356, Ron Kaufmann, University of San Diego   357, Custom Medical Stock   363, Photo by Michael Smith/Getty Images   370, © Richard Klune/Corbis   375, Car Culture/Getty Images   396, Dr. Paulus Gerdes, Professor of Mathematics, Research Center for Mathematics, Culture, and Education, C.P. 915, maputo, Mozambique   397, Blend Images Photography/Veer   419, Copyright © Michael Newman/PhotoEdit—All rights reserved.   421, © Thinkstock/Corbis   427, Car Culture/Getty Images   431, Premium RF © Tim Pannell/Corbis   433, © Fritz Polking; Frank Lane Picture Agency/Corbis   434, Value RF © Richard Gross/Corbis   442, Adrian Weinbrecht/Getty Images   443, 449, © Charles Gupton/Corbis   470, Amana Productions Inc./Corbis Royalty Free   471, © Duncan Smith/Corbis Royalty Free   484, PhotoDisc/Getty Images   485, Purestock/Getty Images   487, Standard RF © moodboard/Corbis   488, Daisuke Morita/PhotoDisc/Getty Images   493, Barbara Johnson   495, © Mark Gamba/Corbis   499, © Ales Fevzer/Corbis   503, © Gail Mooney/Corbis   505, 524, © Macduff Everton/Corbis   507, PM Images/Getty Images   522, Getty Images News   526, TeaPots n Treasures, Donna Yarema Proprietress (almost) on The Circle, Indianapolis, IN 46204   533 (left), © RandyFaris/2007/Corbis Royalty Free   533 (right), © Carson Ganci/Design Pics/Corbis Royalty Free   534, James Forte/National Geographic/Getty Images   549, © Kate Mitchell/zefa/Corbis Royalty Free   550 (left), © Reuter Raymond/Corbis Sygma   550 (right), Indianapolis Symphony Orchestra   562, Sabah Arar/Stringer/AFP/Getty Images   566, Fido Products   575, © Tiziana and Gianni Baldizzone/Corbis   577, 581, Jon Feingersh/Getty Images   584 (left), Stockbyte/Getty Images   584 (right), © Keith Wood/Corbis Royalty Free   595, Altrendo Nature/Getty Images   616, © Neil C. Robinson/Corbis Royalty Free   619, © Comstock/Corbis Royalty Free   630, © Karl Weatherly/Corbis   633, 674, Manchan/RF Getty Images   655, Copyright © Tony Freeman/PhotoEdit—All rights reserved.   661, © Renaud Visage/Art Life Images   677, © Thinkstock/Corbis   685, © Skyscan/Corbis   702, Michael Nagle/Stringer/Getty Images News   703, 758, PhotoDisc/Getty Images   711, David McNew/Staff/Getty Images News   713, Jan Butchofsky-Houser/Corbis   714, Tyler Stableford/The Image Bank/Getty Images   724, © Skyscan/Corbis   728, © moodboard/Corbis   729, © Henny Ray Abrams/Reuters/Corbis   731, © Larry W. Smith/epa/Corbis   781, © John Harper/Corbis   783, 835 (top), © Markus Moellenberg/zefa/Corbis   801, Wil Meinderts/Foto Natura/Getty Images   832, Associated Press   835 (bottom), © Carl & Ann Purcell/Corbis Royalty Free   837, © Mark Karrass/Corbis Royalty Free   839 (left), © DLILLC/Corbis Royalty Free   839 (right), © Schlegelmilch/Corbis   842 (top), © Kit Kittle/Corbis   842 (bottom), AFP/Getty Images   848, Digital Vision/Getty Images   851, Stockbyte/Getty Images   853, 868 (left), Stocktrek/Corbis   868 (right), Patrick Wright, Photographic Services, Clemson University   893, © Lucidio Studio Inc./Corbis   895, 918, Heath Korvola, UpperCut Images/Getty   937, © Kevin Fleming/Corbis   991, Don Emmert/AFP/Gettty Images   994, Associated Press

Appendixes

A | Mean, Median, and Mode

Mean • Median • Mode

One way to analyze data is to look for a single representative number, called a **center point** or **measure of central tendency**. Those most often used are the **mean** (or **average**), the **median**, and the **mode**.

Mean

Let's first consider the *mean*, or *average*.

> **Mean, or Average**
> The *mean*, or *average*, of a set of numbers is the sum of the numbers divided by the number of addends.

EXAMPLE **1**   Consider the following data on revenue, in billions of dollars, for Starbucks Corporation in five recent years:

$$\$2.2, \quad \$2.6, \quad \$3.3, \quad \$4.1, \quad \$5.3.$$

What is the mean of the numbers?

SOLUTION   First, we add the numbers:

$$2.2 + 2.6 + 3.3 + 4.1 + 5.3 = 17.5.$$

Then we divide by the number of addends, 5:

$$\frac{(2.2 + 2.6 + 3.3 + 4.1 + 5.3)}{5} = \frac{17.5}{5} = 3.5.$$

The mean, or average, revenue of Starbucks for those five years is $3.5 billion.

Note that $3.5 + 3.5 + 3.5 + 3.5 + 3.5 = 17.5$. If we use this center point, 3.5, repeatedly as the addend, we get the same sum that we do when adding individual data numbers.

Median

The *median* is useful when we wish to de-emphasize extreme scores. For example, suppose five workers in a technology company manufactured the following number of computers during one day's work:

Sarah: 88
Matt: 92
Pat: 66
Jen: 94
Mark: 91

Let's first list the scores in order from smallest to largest:

66 88 91 92 94.

↑
Middle number

The middle number—in this case, 91—is the **median**.

> ### Median
>
> Once a set of data has been arranged from smallest to largest, the *median* of the set of data is the middle number if there is an odd number of data numbers. If there is an even number of data numbers, then there are two middle numbers and the median is the *average* of the two middle numbers.

EXAMPLE 2   Find the median of the following set of household incomes:

$76,000, $58,000, $87,000, $32,500, $64,800, $62,500.

SOLUTION   We first rearrange the numbers in order from smallest to largest.

$32,500, $58,000, $62,500, $64,800, $76,000, $87,000

↑
Median

There is an even number of numbers. We look for the middle two, which are $62,500 and $64,800. The median is the average of $62,500 and $64,800:

$$\frac{\$62,500 + \$64,800}{2} = \$63,650.$$

Mode

The last center point we consider is called the *mode*. A number that occurs most often in a set of data is sometimes considered a representative number or center point.

> ### Mode
>
> The *mode* of a set of data is the number or numbers that occur most often. If each number occurs the same number of times, then there is *no* mode.

EXAMPLE 3   Find the mode of the following data:

23, 24, 27, 18, 19, 27.

SOLUTION   The number that occurs most often is 27. Thus the mode is 27.

It is easier to find the mode of a set of data if the data are ordered.

EXAMPLE 4   Find the mode of the following data:

83, 84, 84, 84, 85, 86, 87, 87, 87, 88, 89, 90.

SOLUTION   There are two numbers that occur most often, 84 and 87. Thus the modes are 84 and 87.

EXAMPLE 5   Find the mode of the following data:

115, 117, 211, 213, 219.

SOLUTION   Each number occurs the same number of times. The set of data has *no* mode.

A   EXERCISE SET

For each set of numbers, find the mean (average), the median, and any modes that exist.

1. 13, 21, 18, 13, 20

2. 5, 2, 8, 10, 7, 1, 9

3. 3, 8, 20, 3, 20, 10

4. 19, 19, 8, 16, 8, 7

5. 4.7, 2.3, 4.6, 4.9, 3.8

6. 13.4, 13.4, 12.6, 42.9

7. 234, 228, 234, 228, 234, 278

8. $29.95, $28.79, $30.95, $29.95

9. *Hurricanes.* The following bar graph shows the number of hurricanes that struck the United States by month from 1851 to 2006. What is the average number for the 8 months given? the median? the mode?

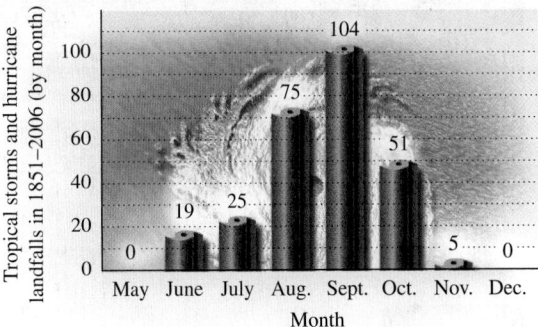

Atlantic Storms and Hurricanes

Source: Atlantic Oceanographic and Meteorological Laboratory

10. *iPod prices.* A price comparison showed the following online prices for an Apple iPod Nano:

$199, $197.97, $249.99, $179, $197.97.

What was the average price? the median price? the mode?

11. *NBA tall men.*   The following lists the heights, in inches, of the tallest men in the NBA in a recent year. Find the mean, the median, and the mode.

Zydrunas Ilgauskas	87
Yao Ming	90
Dikembe Mutombo	86
Kosta Perovic	86

Source: National Basketball Association

12. *Coffee consumption.*   The following lists the annual coffee consumption, in number of cups per person, for various countries. Find the mean, the median, and the mode.

Germany	1113
United States	610
Switzerland	1215
France	798
Italy	750

Source: Beverage Marketing Corporation

13. *PBA scores.*   Kelly Kulick rolled scores of 254, 202, 184, 269, 151, 223, 258, 222, and 202 in a recent tour trial for the Professional Bowlers Association. What was her average? her median? her mode?

Source: Professional Bowlers Association

14. *Salmon prices.*   The following prices per pound of Atlantic salmon were found at six fish markets:

$8.99, $8.49, $8.99, $9.99, $9.49, $7.99.

What was the average price per pound? the median price? the mode?

Synthesis

15. *Hank Aaron.*   Hank Aaron averaged $34\frac{7}{22}$ home runs per year over a 22-yr career. After 21 yr, Aaron had averaged $35\frac{10}{21}$ home runs per year. How many home runs did Aaron hit in his final year?

16. *Length of pregnancy.*   Marta was pregnant 270 days, 259 days, and 272 days for her first three pregnancies. In order for Marta's average length of pregnancy to equal the worldwide average of 266 days, how long must her fourth pregnancy last?
Source: David Crystal (ed.), *The Cambridge Factfinder.* Cambridge CB2 1RP: Cambridge University Press, 1993, p. 84.

17. The ordered set of data 18, 21, 24, a, 36, 37, b has a median of 30 and an average of 32. Find a and b.

18. *Male height.*   Jason's brothers are 174 cm, 180 cm, 179 cm, and 172 cm tall. The average male is 176.5 cm tall. How tall is Jason if he and his brothers have an average height of 176.5 cm?

B   Sets

Naming Sets   •   Membership   •   Subsets   •   Intersections   •   Unions

A **set** is a collection of objects. In mathematics the objects, or **elements**, of a set are generally numbers. This section provides an introduction to sets and how to combine them.

Naming Sets

To name the set of whole numbers less than 6, we can use *roster notation*, as follows:

$$\{0, 1, 2, 3, 4, 5\}.$$

The set of real numbers x for which x is less than 6 cannot be named by listing all its members because there is an infinite number of them. We name such a set using *set-builder notation*, as follows:

$$\{x \,|\, x < 6\}.$$

This is read

"The set of all x such that x is less than 6."

See Section 2.6 for more on this notation.

Membership

The symbol $\in$ means *is a member of* or *belongs to*, or *is an element of*. Thus,

$$x \in A$$

means

x is a member of A,   or x belongs to A,   or x is an element of A.

EXAMPLE **1**   Classify each of the following as true or false.

a) $1 \in \{1, 2, 3\}$
b) $1 \in \{2, 3\}$
c) $4 \in \{x \,|\, x \text{ is an even whole number}\}$
d) $5 \in \{x \,|\, x \text{ is an even whole number}\}$

SOLUTION

a) Since 1 is listed as a member of the set, $1 \in \{1, 2, 3\}$ is true.

b) Since 1 is *not* a member of $\{2, 3\}$, the statement $1 \in \{2, 3\}$ is false.

c) Since 4 is an even whole number, $4 \in \{x \,|\, x \text{ is an even whole number}\}$ is true.

d) Since 5 is *not* even, $5 \in \{x \,|\, x \text{ is an even whole number}\}$ is false.

 TRY EXERCISE   7

Set membership can be illustrated with a diagram, as shown below.

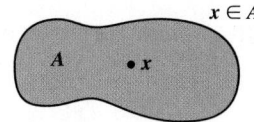

Subsets

If every element of A is also an element of B, then A is a *subset* of B. This is denoted $A \subseteq B$.

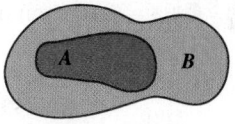

The set of whole numbers is a subset of the set of integers. The set of rational numbers is a subset of the set of real numbers.

EXAMPLE 2   Classify each of the following as true or false.

a) $\{1, 2\} \subseteq \{1, 2, 3, 4\}$

b) $\{p, q, r, w\} \subseteq \{a, p, r, z\}$

c) $\{x | x < 6\} \subseteq \{x | x \leq 11\}$

SOLUTION

a) Since every element of $\{1, 2\}$ is in the set $\{1, 2, 3, 4\}$, it follows that $\{1, 2\} \subseteq \{1, 2, 3, 4\}$ is true.

b) Since $q \in \{p, q, r, w\}$, but $q \notin \{a, p, r, z\}$, it follows that $\{p, q, r, w\} \subseteq \{a, p, r, z\}$ is false.

c) Since every number that is less than 6 is also less than 11, the statement $\{x | x < 6\} \subseteq \{x | x \leq 11\}$ is true.

> **TRY EXERCISE** 15

Intersections

The *intersection* of sets A and B, denoted $A \cap B$, is the set of members common to both sets.

EXAMPLE 3   Find each intersection.

a) $\{0, 1, 3, 5, 25\} \cap \{2, 3, 4, 5, 6, 7, 9\}$

b) $\{a, p, q, w\} \cap \{p, q, t\}$

SOLUTION

a) $\{0, 1, 3, 5, 25\} \cap \{2, 3, 4, 5, 6, 7, 9\} = \{3, 5\}$

b) $\{a, p, q, w\} \cap \{p, q, t\} = \{p, q\}$

> **TRY EXERCISE** 19

Set intersection can be illustrated with a diagram, as shown below.

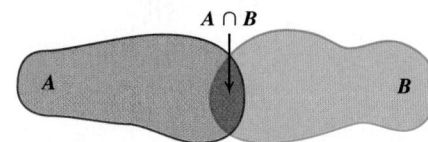

The set without members is known as the *empty set*, and is written $\varnothing$ and sometimes $\{\ \}$. Each of the following is a description of the empty set:

The set of all 12-ft–tall people;

$\{2, 3\} \cap \{5, 6, 7\};$

$\{x | x \text{ is an even natural number}\} \cap \{x | x \text{ is an odd natural number}\}.$

Unions

Two sets A and B can be combined to form a set that contains the members of both A and B. The new set is called the *union* of A and B, denoted $A \cup B$.

EXAMPLE 4   Find each union.

a) $\{0, 5, 7, 13, 27\} \cup \{0, 2, 3, 4, 5\}$

b) $\{a, c, e, g\} \cup \{b, d, f\}$

SOLUTION

a) $\{0, 5, 7, 13, 27\} \cup \{0, 2, 3, 4, 5\} = \{0, 2, 3, 4, 5, 7, 13, 27\}$

Note that the 0 and the 5 are *not* listed twice in the solution.

b) $\{a, c, e, g\} \cup \{b, d, f\} = \{a, b, c, d, e, f, g\}$

TRY EXERCISE 25

Set union can be illustrated with a diagram, as shown below.

$A \cup B$ **is shaded.**

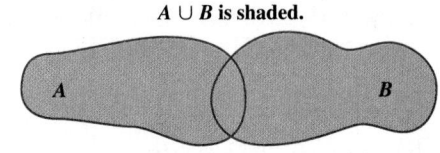

B EXERCISE SET

Name each set using the roster method.

1. The set of whole numbers 8 through 11

2. The set of whole numbers 83 through 89

3. The set of odd numbers between 40 and 50

4. The set of multiples of 5 between 10 and 40

5. $\{x \mid$ the square of x is $9\}$

6. $\{x \mid x$ is the cube of $\frac{1}{2}\}$

Classify each statement as either true or false.

7. $5 \in \{x \mid x$ is an odd number$\}$

8. $8 \in \{x \mid x$ is an odd number$\}$

9. Skiing $\in$ The set of all sports

10. Pharmacist $\in$ The set of all professions requiring a college degree

11. $3 \in \{-4, -3, 0, 1\}$

12. $0 \in \{-4, -3, 0, 1\}$

13. $\frac{2}{3} \in \{x \mid x$ is a rational number$\}$

14. $\frac{2}{3} \in \{x \mid x$ is a real number$\}$

15. $\{-1, 0, 1\} \subseteq \{-3, -2, -1, 1\, 2, 3\}$

16. The set of vowels $\subseteq$ The set of consonants

17. The set of integers $\subseteq$ The set of rational numbers

18. $\{2, 4, 6\} \subseteq \{1, 2, 3, 4, 5, 6, 7\}$

Find each intersection.

19. $\{a, b, c, d, e\} \cap \{c, d, e, f, g\}$

20. $\{a, e, i, o, u\} \cap \{q, u, i, c, k\}$

21. $\{1, 2, 3, 4, 6, 12\} \cap \{1, 2, 3, 6, 9, 18\}$

22. $\{1, 2, 3, 4, 6, 12\} \cap \{1, 5, 7, 35\}$

23. $\{2, 4, 6, 8\} \cap \{1, 3, 5, 7\}$

24. $\{a, e, i, o, u\} \cap \{m, n, f, g, h\}$

Find each union.

25. $\{a, e, i, o, u\} \cup \{q, u, i, c, k\}$

26. $\{a, b, c, d, e\} \cup \{c, d, e, f, g\}$

27. $\{1, 2, 3, 4, 6, 12\} \cup \{1, 2, 3, 6, 9, 18\}$

28. $\{1, 2, 3, 4, 6, 12\} \cup \{1, 5, 7, 35\}$

29. $\{2, 4, 6, 8\} \cup \{1, 3, 5, 7\}$

30. $\{a, e, i, o, u\} \cup \{m, n, f, g, h\}$

31. What advantage(s) does set-builder notation have over roster notation?

32. What advantage(s) does roster notation have over set-builder notation?

Synthesis

33. Find the union of the set of integers and the set of whole numbers.

34. Find the intersection of the set of odd integers and the set of even integers.

35. Find the union of the set of rational numbers and the set of irrational numbers.

36. Find the intersection of the set of even integers and the set of positive rational numbers.

37. Find the intersection of the set of rational numbers and the set of irrational numbers.

38. Find the union of the set of negative integers, the set of positive integers, and the set containing 0.

39. For a set A, find each of the following.

a) $A \cup \varnothing$
b) $A \cup A$
c) $A \cap A$
d) $A \cap \varnothing$

Classify each statement as either true or false.

40. The empty set can be written $\varnothing$, { }, or {0}.

41. For any set A, $\varnothing \subseteq A$.

42. For any set A, $A \subseteq A$.

43. For any sets A and B, $A \cap B \subseteq A$.

44. A set is *closed* under an operation if, when the operation is performed on its members, the result is in the set. For example, the set of real numbers is closed under the operation of addition since the sum of any two real numbers is a real number.

a) Is the set of even numbers closed under addition?
b) Is the set of odd numbers closed under addition?
c) Is the set {0, 1} closed under addition?
d) Is the set {0, 1} closed under multiplication?
e) Is the set of real numbers closed under multiplication?
f) Is the set of integers closed under division?

45. Experiment with sets of various types and determine whether the following distributive law for sets is true:

$$A \cap (B \cup C) = (A \cap B) \cup (A \cap C).$$

C Synthetic Division

Streamlining Long Division • The Remainder Theorem

Streamlining Long Division

To divide a polynomial by a binomial of the type $x - a$, we can streamline the usual procedure to develop a process called *synthetic division*.

Compare the following. In each stage, we attempt to write less than in the previous stage, while retaining enough essentials to solve the problem.

STAGE 1

When a polynomial is written in descending order, the coefficients provide the essential information:

$$
\begin{array}{r}
4x^2 + 5x + 11 \\
x - 2 \overline{)\, 4x^3 - 3x^2 + x + 7} \\
\underline{4x^3 - 8x^2} \\
5x^2 + x \\
\underline{5x^2 - 10x} \\
11x + 7 \\
\underline{11x - 22} \\
29
\end{array}
\qquad
\begin{array}{r}
4 + 5 + 11 \\
1 - 2 \overline{)\, 4 - 3 + 1 + 7} \\
\underline{4 - 8} \\
5 + 1 \\
\underline{5 - 10} \\
11 + 7 \\
\underline{11 - 22} \\
29
\end{array}
$$

Because the leading coefficient in $x - 2$ is 1, each time we multiply it by a term in the answer, the leading coefficient of that product duplicates a coefficient in the answer. In the next stage, rather than duplicate these numbers we focus on where -2 is used and drop the 1 from the divisor.

STAGE 2

$$
x - 2 \overline{)\begin{array}{r} 4x^2 + 5x + 11 \\ 4x^3 - 3x^2 + x + 7 \end{array}}
$$

$$
\begin{array}{r}
4x^3 - 8x^2 \\ \hline
5x^2 + x \\
5x^2 - 10x \\ \hline
11x + 7 \\
11x - 22 \\ \hline
29
\end{array}
$$

$$
-2 \overline{)\begin{array}{r} 4 + 5 + 11 \\ 4 - 3 + 1 + 7 \end{array}}
$$

$$
\begin{array}{r}
-8 \\ \hline
5 + 1 \\
-10 \\ \hline
11 + 7 \\
-22 \\ \hline
29
\end{array}
$$

— Multiply: $-2 \cdot 4 = -8$.
Subtract: $-3 - (-8) = 5$.
— Multiply: $-2 \cdot 5 = -10$.
Subtract: $1 - (-10) = 11$.
— Multiply: $-2 \cdot 11 = -22$.
Subtract: $7 - (-22) = 29$.

To simplify further, we now reverse the sign of the -2 in the divisor and, in exchange, *add* at each step in the long division.

STAGE 3

$$
x - 2 \overline{)\begin{array}{r} 4x^2 + 5x + 11 \\ 4x^3 - 3x^2 + x + 7 \end{array}}
$$

$$
\begin{array}{r}
4x^3 - 8x^2 \\ \hline
5x^2 + x \\
5x^2 - 10x \\ \hline
11x + 7 \\
11x - 22 \\ \hline
29
\end{array}
$$

$$
2 \overline{)\begin{array}{r} 4 + 5 + 11 \\ 4 - 3 + 1 + 7 \end{array}}
$$

Replace the -2 with 2.

$$
\begin{array}{r}
8 \\ \hline
5 + 1 \\
10 \\ \hline
11 + 7 \\
22 \\ \hline
29
\end{array}
$$

— Multiply: $2 \cdot 4 = 8$.
Add: $-3 + 8 = 5$.
— Multiply: $2 \cdot 5 = 10$.
Add: $1 + 10 = 11$.
— Multiply: $2 \cdot 11 = 22$.
Add: $7 + 22 = 29$.

The blue numbers can be eliminated if we look at the red numbers instead.

STAGE 4

$$
x - 2 \overline{)\begin{array}{r} 4x^2 + 5x + 11 \\ 4x^3 - 3x^2 + x + 7 \end{array}}
$$

$$
\begin{array}{r}
4x^3 - 8x^2 \\ \hline
5x^2 + x \\
5x^2 - 10x \\ \hline
11x + 7 \\
11x - 22 \\ \hline
29
\end{array}
$$

$$
2 \overline{)\begin{array}{rrrr}
4 & -3 & 1 & 7 \\
 & 8 & 10 & 22 \\ \hline
4 & 5 & 11 & 29
\end{array}}
$$

STUDENT NOTES

You will not need to write out all five stages when performing synthetic division on your own. We show the steps to help you understand the reasoning behind the method.

Note that the 5 and the 11 preceding the remainder 29 coincide with the 5 and the 11 following the 4 on the top line. By writing a 4 to the left of 5 on the bottom line, we can eliminate the top line in stage 4 and read our answer from the bottom line. This final stage is commonly called **synthetic division**.

STAGE 5

$$
2 \overline{)\begin{array}{rrrr}
4 & 5 & 11 & \\
4 & -3 & 1 & 7 \\
 & 8 & 10 & 22 \\ \hline
 & 5 & 11 & 29
\end{array}}
$$

$$
2 \,|\, \begin{array}{rrrr}
4 & -3 & 1 & 7 \\
 & 8 & 10 & 22 \\ \hline
4 & 5 & 11 & | \; 29
\end{array}
$$

— This is the remainder.
— This is the zero-degree coefficient.
— This is the first-degree coefficient.
— This is the second-degree coefficient.

The quotient is $4x^2 + 5x + 11$ with a remainder of 29.

> Remember that for this method to work, the divisor must be of the form $x - a$, that is, a variable minus a constant.

EXAMPLE 1   Use synthetic division to divide: $(x^3 + 6x^2 - x - 30) \div (x - 2)$.

SOLUTION

$$2 \underline{)}\ 1\quad 6\quad -1\quad -30 \qquad$$
$$\overline{\ 1}$$

Write the 2 of $x - 2$ and the coefficients of the dividend.

Bring down the first coefficient.

$$2 \underline{)}\ 1\quad 6\quad -1\quad -30$$
$$2$$
$$\overline{1\quad 8}$$

Multiply 1 by 2 to get 2.

Add 6 and 2.

$$2 \underline{)}\ 1\quad 6\quad -1\quad -30$$
$$2\quad 16$$
$$\overline{1\quad 8\quad 15}$$

Multiply 8 by 2.

Add -1 and 16.

$$2 \underline{)}\ 1\quad 6\quad -1\quad -30$$
$$2\quad 16\quad 30$$
$$\overline{1\quad 8\quad 15\quad 0}$$

Multiply 15 by 2 and add.

The answer is $x^2 + 8x + 15$ with R 0, or just $x^2 + 8x + 15$.

▶ **TRY EXERCISE** 7

EXAMPLE 2   Use synthetic division to divide.

a) $(2x^3 + 7x^2 - 5) \div (x + 3)$

b) $(10x^2 - 13x + 3x^3 - 20) \div (4 + x)$

SOLUTION

a) $(2x^3 + 7x^2 - 5) \div (x + 3)$

The dividend has no x-term, so we need to write 0 as the coefficient of x. Note that $x + 3 = x - (-3)$, so we write -3 inside the $\underline{)}$.

$$-3 \underline{)}\ 2\quad 7\quad 0\quad -5$$
$$-6\quad -3\quad 9$$
$$\overline{2\quad 1\quad -3\ \big|\ 4}$$

The answer is $2x^2 + x - 3$, with R 4, or $2x^2 + x - 3 + \dfrac{4}{x + 3}$.

b) We first rewrite $(10x^2 - 13x + 3x^3 - 20) \div (4 + x)$ in descending order:

$$(3x^3 + 10x^2 - 13x - 20) \div (x + 4).$$

Next, we use synthetic division. Note that $x + 4 = x - (-4)$.

$$-4 \underline{)}\ 3\quad 10\quad -13\quad -20$$
$$-12\quad 8\quad 20$$
$$\overline{3\quad -2\quad -5\ \big|\ 0}$$

The answer is $3x^2 - 2x - 5$.

▶ **TRY EXERCISE** 15

The Remainder Theorem

Because the remainder is 0, Example 1 shows that $x - 2$ is a factor of $x^3 + 6x^2 - x - 30$ and that $x^3 + 6x^2 - x - 30 = (x - 2)(x^2 + 8x + 15)$. Thus if $f(x) = x^3 + 6x^2 - x - 30$, then $f(2) = 0$ (since $x - 2$ is a factor of $f(x)$). Similarly, from Example 2(b), we know that if $g(x) = 10x^2 - 13x + 3x^3 - 20$, then $x + 4$ is a factor of $g(x)$ and $g(-4) = 0$. In both examples, the remainder from the division, 0, can serve as a function value. Remarkably, this pattern extends to nonzero remainders. For example, the remainder in Example 2(a) is 4, and if $f(x) = 2x^3 + 7x^2 - 5$, then $f(-3)$ is also 4 (you should check this). The fact that the remainder and the function value coincide is predicted by the remainder theorem.

> **The Remainder Theorem**
>
> The remainder obtained by dividing $P(x)$ by $x - r$ is $P(r)$.

A proof of this result is outlined in Exercise 31.

EXAMPLE 3   Let $f(x) = 8x^5 - 6x^3 + x - 8$. Use synthetic division to find $f(2)$.

SOLUTION   The remainder theorem tells us that $f(2)$ is the remainder when $f(x)$ is divided by $x - 2$. We use synthetic division to find that remainder:

$$
\begin{array}{r|rrrrrr}
2 & 8 & 0 & -6 & 0 & 1 & -8 \\
  &   & 16 & 32 & 52 & 104 & 210 \\
\hline
  & 8 & 16 & 26 & 52 & 105 & 202
\end{array}
$$

Although the bottom line can be used to find the quotient for the division $(8x^5 - 6x^3 + x - 8) \div (x - 2)$, what we are really interested in is the remainder. It tells us that $f(2) = 202$.

TRY EXERCISE ▸ 21

The remainder theorem is often used to check division. Thus Example 2(a) can be checked by computing $P(-3) = 2(-3)^3 + 7(-3)^2 - 5$. Since $P(-3) = 4$ and the remainder in Example 2(a) is also 4, our division was probably correct.

C   EXERCISE SET

🔖 *Concept Reinforcement   Classify each statement as either true or false.*

1. If $x - 2$ is a factor of some polynomial $P(x)$, then $P(2) = 0$.

2. If $p(3) = 0$ for some polynomial $p(x)$, then $x - 3$ is a factor of $p(x)$.

3. If $P(-5) = 39$ and $P(x) = x^3 + 7x^2 + 3x + 4$, then

$$
\begin{array}{r|rrrr}
-5 & 1 & 7 & 3 & 4 \\
   &   & -5 & -10 & 35 \\
\hline
   & 1 & 2 & -7 & 39
\end{array}
$$

4. In order for $f(x)/g(x)$ to exist, $g(x)$ must be 0.

5. In order to use synthetic division, we must be sure that the divisor is of the form $x - a$.

6. Synthetic division can be used in problems in which long division could not be used.

Use synthetic division to divide.

7. $(x^3 - 4x^2 - 2x + 5) \div (x - 1)$

8. $(x^3 - 4x^2 + 5x - 6) \div (x - 3)$

9. $(a^2 + 8a + 11) \div (a + 3)$

10. $(a^2 + 8a + 11) \div (a + 5)$

11. $(2x^3 - x^2 - 7x + 14) \div (x + 2)$

12. $(3x^3 - 10x^2 - 9x + 15) \div (x - 4)$

13. $(a^3 - 10a + 12) \div (a - 2)$

14. $(a^3 - 14a + 15) \div (a - 3)$

15. $(3y^3 - 7y^2 - 20) \div (y - 3)$

16. $(2x^3 - 3x^2 + 8) \div (x + 2)$

17. $(x^5 - 32) \div (x - 2)$

18. $(y^5 - 1) \div (y - 1)$

19. $(3x^3 + 1 - x + 7x^2) \div \left(x + \frac{1}{3}\right)$

20. $(8x^3 - 1 + 7x - 6x^2) \div \left(x - \frac{1}{2}\right)$

Use synthetic division to find the indicated function value.

21. $f(x) = 5x^4 + 12x^3 + 28x + 9;\ f(-3)$

22. $g(x) = 3x^4 - 25x^2 - 18;\ g(3)$

23. $P(x) = 2x^4 - x^3 - 7x^2 + x + 2;\ P(-3)$

24. $F(x) = 3x^4 + 8x^3 + 2x^2 - 7x - 4;\ F(-2)$

25. $f(x) = x^4 - 6x^3 + 11x^2 - 17x + 20;\ f(4)$

26. $p(x) = x^4 + 7x^3 + 11x^2 - 7x - 12;\ p(2)$

27. Why is it that we *add* when performing synthetic division, but *subtract* when performing long division?

28. Explain how synthetic division could be useful when attempting to factor a polynomial.

Synthesis

29. Let $Q(x)$ be a polynomial function with $p(x)$ a factor of $Q(x)$. If $p(3) = 0$, does it follow that $Q(3) = 0$? Why or why not? If $Q(3) = 0$, does it follow that $p(3) = 0$? Why or why not?

30. What adjustments must be made if synthetic division is to be used to divide a polynomial by a binomial of the form $ax + b$, with $a > 1$?

31. To prove the remainder theorem, note that any polynomial $P(x)$ can be rewritten as $(x - r) \cdot Q(x) + R$, where $Q(x)$ is the quotient polynomial that arises when $P(x)$ is divided by $x - r$, and R is some constant (the remainder).

　a) How do we know that R must be a constant?

　b) Show that $P(r) = R$ (this says that $P(r)$ is the remainder when $P(x)$ is divided by $x - r$).

32. Let $f(x) = 6x^3 - 13x^2 - 79x + 140$. Find $f(4)$ and then solve the equation $f(x) = 0$.

33. Let $f(x) = 4x^3 + 16x^2 - 3x - 45$. Find $f(-3)$ and then solve the equation $f(x) = 0$.

34. Use the TRACE feature on a graphing calculator to check your answer to Exercise 32.

35. Use the TRACE feature on a graphing calculator to check your answer to Exercise 33.

Nested evaluation. *One way to evaluate a polynomial function like $P(x) = 3x^4 - 5x^3 + 4x^2 - 1$ is to successively factor out x as shown:*

$$P(x) = x(x(x(3x - 5) + 4) + 0) - 1.$$

Computations are then performed using this "nested" form of $P(x)$.

36. Use nested evaluation to find $f(4)$ in Exercise 32. Note the similarities to the calculations performed with synthetic division.

37. Use nested evaluation to find $f(-3)$ in Exercise 33. Note the similarities to the calculations performed with synthetic division.

Tables

TABLE 1   Fraction and Decimal Equivalents

Fraction Notation	$\frac{1}{10}$	$\frac{1}{8}$	$\frac{1}{6}$	$\frac{1}{5}$	$\frac{1}{4}$	$\frac{3}{10}$	$\frac{1}{3}$	$\frac{3}{8}$	$\frac{2}{5}$	$\frac{1}{2}$
Decimal Notation	0.1	0.125	$0.16\overline{6}$	0.2	0.25	0.3	$0.333\overline{3}$	0.375	0.4	0.5
Percent Notation	10%	12.5%, or $12\frac{1}{2}\%$	$16.6\overline{6}\%$, or $16\frac{2}{3}\%$	20%	25%	30%	$33.3\overline{3}\%$, or $33\frac{1}{3}\%$	37.5%, or $37\frac{1}{2}\%$	40%	50%
Fraction Notation	$\frac{3}{5}$	$\frac{5}{8}$	$\frac{2}{3}$	$\frac{7}{10}$	$\frac{3}{4}$	$\frac{4}{5}$	$\frac{5}{6}$	$\frac{7}{8}$	$\frac{9}{10}$	$\frac{1}{1}$
Decimal Notation	0.6	0.625	$0.666\overline{6}$	0.7	0.75	0.8	$0.83\overline{3}$	0.875	0.9	1
Percent Notation	60%	62.5%, or $62\frac{1}{2}\%$	$66.6\overline{6}\%$, or $66\frac{2}{3}\%$	70%	75%	80%	$83.3\overline{3}\%$, or $83\frac{1}{3}\%$	87.5%, or $87\frac{1}{2}\%$	90%	100%

TABLE 2 Squares and Square Roots with Approximations to Three Decimal Places

N	$\sqrt{N}$	N^2	N	$\sqrt{N}$	N^2	N	$\sqrt{N}$	N^2	N	$\sqrt{N}$	N^2
1	1	1	26	5.099	676	51	7.141	2601	76	8.718	5776
2	1.414	4	27	5.196	729	52	7.211	2704	77	8.775	5929
3	1.732	9	28	5.292	784	53	7.280	2809	78	8.832	6084
4	2	16	29	5.385	841	54	7.348	2916	79	8.888	6241
5	2.236	25	30	5.477	900	55	7.416	3025	80	8.944	6400
6	2.449	36	31	5.568	961	56	7.483	3136	81	9	6561
7	2.646	49	32	5.657	1024	57	7.550	3249	82	9.055	6724
8	2.828	64	33	5.745	1089	58	7.616	3364	83	9.110	6889
9	3	81	34	5.831	1156	59	7.681	3481	84	9.165	7056
10	3.162	100	35	5.916	1225	60	7.746	3600	85	9.220	7225
11	3.317	121	36	6	1296	61	7.810	3721	86	9.274	7396
12	3.464	144	37	6.083	1369	62	7.874	3844	87	9.327	7569
13	3.606	169	38	6.164	1444	63	7.937	3969	88	9.381	7744
14	3.742	196	39	6.245	1521	64	8	4096	89	9.434	7921
15	3.873	225	40	6.325	1600	65	8.062	4225	90	9.487	8100
16	4	256	41	6.403	1681	66	8.124	4356	91	9.539	8281
17	4.123	289	42	6.481	1764	67	8.185	4489	92	9.592	8464
18	4.243	324	43	6.557	1849	68	8.246	4624	93	9.644	8649
19	4.359	361	44	6.633	1936	69	8.307	4761	94	9.695	8836
20	4.472	400	45	6.708	2025	70	8.367	4900	95	9.747	9025
21	4.583	441	46	6.782	2116	71	8.426	5041	96	9.798	9216
22	4.690	484	47	6.856	2209	72	8.485	5184	97	9.849	9409
23	4.796	529	48	6.928	2304	73	8.544	5329	98	9.899	9604
24	4.899	576	49	7	2401	74	8.602	5476	99	9.950	9801
25	5	625	50	7.071	2500	75	8.660	5625	100	10	10,000

Answers

The complete step-by-step solutions for the exercises listed below can be found in the *Student's Solutions Manual*, ISBN 0-321-58623-9/978-0-321-58623-0, which can be purchased online or at your bookstore.

CHAPTER 1

Technology Connection, p. 7

1. 3438   **2.** 47,531

Translating for Success, p. 9

1. H   **2.** E   **3.** K   **4.** B   **5.** O   **6.** L   **7.** M   **8.** C
9. D   **10.** F

Exercise Set 1.1, pp. 10–12

1. Expression   **2.** Equation   **3.** Equation
4. Expression   **5.** Equation   **6.** Equation
7. Expression   **8.** Equation   **9.** Equation
10. Expression   **11.** Expression   **12.** Expression
13. 45   **15.** 8   **17.** 5   **19.** 4   **21.** 5   **23.** 3
25. $24 \, \text{ft}^2$ **27.** $15 \, \text{cm}^2$ **29.** 0.345   **31.** Let r represent
Ron's age; $r + 5$, or $5 + r$ **33.** $6b$, or $b \cdot 6$ **35.** $c - 9$
37. $6 + q$, or $q + 6$ **39.** Let m represent Mai's speed;

$8m$, or $m \cdot 8$ **41.** $y - x$ **43.** $x \div w$, or $\dfrac{x}{w}$ **45.** Let l

represent the length of the box and h represent the height;
$l + h$, or $h + l$ **47.** $9 \cdot 2m$, or $2m \cdot 9$ **49.** Let y

represent "some number"; $\dfrac{1}{4}y - 13$, or $\dfrac{y}{4} - 13$ **51.** Let a

and b represent the two numbers; $5(a - b)$ **53.** Let w
represent the number of women attending; 64% of w,
or $0.64w$ **55.** Yes   **57.** No   **59.** Yes   **61.** Yes
63. Let x represent the unknown number; $73 + x = 201$
65. Let x represent the unknown number; $42x = 2352$
67. Let s represent the number of unoccupied squares;
$s + 19 = 64$ **69.** Let w represent the amount of solid
waste generated, in millions of tons; 32% of $w = 79$, or
$0.32w = 79$ **71.** (f)   **73.** (d)   **75.** (g)   **77.** (e)
79. ✑   **81.** ✑   **83.** $450   **85.** 2   **87.** 6   **89.** $w + 4$
91. $l + w + l + w$, or $2l + 2w$ **93.** $t + 8$ **95.** ✑

Exercise Set 1.2, pp. 18–20

1. Commutative   **2.** Associative   **3.** Associative
4. Commutative   **5.** Distributive   **6.** Associative
7. Associative   **8.** Commutative   **9.** Commutative
10. Distributive   **11.** $t + 11$ **13.** $8x + 4$
15. $3y + 9x$ **17.** $5(1 + a)$ **19.** $x \cdot 7$ **21.** ts

23. $5 + ba$ **25.** $(a + 1)5$ **27.** $x + (8 + y)$
29. $(u + v) + 7$ **31.** $ab + (c + d)$ **33.** $8(xy)$
35. $(2a)b$ **37.** $(3 \cdot 2)(a + b)$
39. $(s + t) + 6; (t + 6) + s$ **41.** $17(ab); b(17a)$
43. $(1 + x) + 2 = (x + 1) + 2$ Commutative law
$\qquad\qquad\quad = x + (1 + 2)$ Associative law
$\qquad\qquad\quad = x + 3$ Simplifying
45. $(m \cdot 3)7 = m(3 \cdot 7)$ Associative law
$\qquad\qquad = m \cdot 21$ Simplifying
$\qquad\qquad = 21m$ Commutative law
47. $2x + 30$ **49.** $4 + 4a$ **51.** $24 + 8y$
53. $90x + 60$ **55.** $5r + 10 + 15t$ **57.** $2a + 2b$

59. $5x + 5y + 10$ **61.** $x, xyz, 1$ **63.** $2a, \dfrac{a}{3b}, 5b$

65. x, y **67.** $4x, 4y$ **69.** $2(a + b)$ **71.** $7(1 + y)$
73. $4(8x + 1)$ **75.** $5(x + 2 + 3y)$ **77.** $7(a + 5b)$
79. $11(4x + y + 2z)$ **81.** $5, n$ **83.** $3, (x + y)$
85. $7, a, b$ **87.** $(a - b), (x - y)$ **89.** ✑   **91.** Let k

represent Kara's salary; $\dfrac{1}{2}k$, or $\dfrac{k}{2}$ **92.** $2(m + 3)$, or

$2(3 + m)$ **93.** ✑   **95.** Yes; distributive law
97. No; for example, let $m = 1$. Then $7 \div 3 \cdot 1 = \frac{7}{3}$ and
$1 \cdot 3 \div 7 = \frac{3}{7}$.   **99.** No; for example, let $x = 1$ and
$y = 2$. Then $30 \cdot 2 + 1 \cdot 15 = 60 + 15 = 75$ and
$5[2(1 + 3 \cdot 2)] = 5[2(7)] = 5 \cdot 14 = 70$.   **101.** ✑

Exercise Set 1.3, pp. 27–29

1. (b)   **2.** (c)   **3.** (d)   **4.** (a)   **5.** Composite
7. Prime   **9.** Composite   **11.** Prime   **13.** Neither
15. $1 \cdot 50; 2 \cdot 25; 5 \cdot 10; 1, 2, 5, 10, 25, 50$
17. $1 \cdot 42; 2 \cdot 21; 3 \cdot 14; 6 \cdot 7; 1, 2, 3, 6, 7, 14, 21, 42$
19. $3 \cdot 13$ **21.** $2 \cdot 3 \cdot 5$ **23.** $3 \cdot 3 \cdot 3$ **25.** $2 \cdot 3 \cdot 5 \cdot 5$
27. $2 \cdot 2 \cdot 2 \cdot 5$ **29.** Prime   **31.** $2 \cdot 3 \cdot 5 \cdot 7$ **33.** $5 \cdot 23$
35. $\frac{3}{5}$ **37.** $\frac{2}{7}$ **39.** $\frac{1}{4}$ **41.** 4   **43.** $\frac{1}{4}$ **45.** 6
47. $\frac{21}{25}$ **49.** $\frac{60}{41}$ **51.** $\frac{15}{7}$ **53.** $\frac{3}{10}$ **55.** 6   **57.** $\frac{1}{2}$

59. $\frac{7}{6}$ **61.** $\dfrac{3b}{7a}$ **63.** $\dfrac{10}{n}$ **65.** $\frac{5}{6}$ **67.** 1   **69.** $\frac{5}{18}$

71. 0   **73.** $\frac{35}{18}$ **75.** $\frac{10}{3}$ **77.** 27   **79.** 1   **81.** $\frac{6}{35}$

83. 18   **85.** ✑   **87.** $5(3 + x)$; answers may vary
88. $7 + (b + a)$, or $(a + b) + 7$ **89.** ✑
91. Row 1: 7, 2, 36, 14, 8, 8; row 2: 9, 18, 2, 10, 12, 21

93. $\frac{2}{5}$ **95.** $\dfrac{5q}{t}$ **97.** $\frac{6}{25}$ **99.** $\dfrac{5ap}{2cm}$ **101.** $\dfrac{23r}{18t}$

103. $\frac{28}{45}$ m^2 **105.** $14\frac{2}{9}$ m **107.** $27\frac{3}{5}$ cm

Technology Connection, p. 33

1. 2.236067977 **2.** 2.645751311 **3.** 3.605551275
4. 5.196152423 **5.** 6.164414003 **6.** 7.071067812

Exercise Set 1.4, pp. 35–37

1. Repeating **2.** Terminating **3.** Integer
4. Whole number **5.** Rational number
6. Irrational number **7.** Natural number
8. Absolute value **9.** $-10{,}500, 27{,}482$ **11.** $136, -4$
13. $-554, 499.19$ **15.** $650, -180$ **17.** $8, -5$
19.
21.
23. **25.** 0.875 **27.** -0.75
29. $-1.1\overline{6}$ **31.** $0.\overline{6}$ **33.** -0.5 **35.** 0.13
37. **39.**
41. $>$ **43.** $<$ **45.** $<$ **47.** $>$ **49.** $<$ **51.** $<$
53. $x < -2$ **55.** $y \geq 10$ **57.** True **59.** False
61. True **63.** 58 **65.** 12.2 **67.** $\sqrt{2}$ **69.** $\frac{9}{7}$ **71.** 0
73. 8 **75.** $-83, -4.7, 0, \frac{5}{9}, 2.\overline{16}, 62$ **77.** $-83, 0, 62$
79. $-83, -4.7, 0, \frac{5}{9}, 2.\overline{16}, \pi, \sqrt{17}, 62$ **81.** ✒ **83.** 42
84. $ba + 5$, or $5 + ab$ **85.** ✒ **87.** ✒
89. $-23, -17, 0, 4$ **91.** $-\frac{4}{3}, \frac{4}{9}, \frac{4}{8}, \frac{4}{6}, \frac{4}{3}, \frac{4}{2}$ **93.** $<$ **95.** $=$
97. $-19, 19$ **99.** $-4, -3, 3, 4$ **101.** $\frac{3}{3}$ **103.** $\frac{70}{9}$
105. $x \leq 0$ **107.** $|t| \geq 20$ **109.** ✒

Exercise Set 1.5, pp. 41–43

1. (f) **2.** (d) **3.** (e) **4.** (a) **5.** (b) **6.** (c)
7. -3 **9.** 4 **11.** -7 **13.** -8 **15.** -35 **17.** -8
19. 0 **21.** -41 **23.** 0 **25.** 9 **27.** -2 **29.** 11
31. -43 **33.** 0 **35.** 18 **37.** -45 **39.** 0 **41.** 16
43. -0.8 **45.** -9.1 **47.** $\frac{3}{5}$ **49.** $\frac{-6}{5}$ **51.** $-\frac{1}{15}$
53. $\frac{2}{9}$ **55.** -3 **57.** 0 **59.** The price rose 29¢.
61. Her new balance was \$95. **63.** The total gain was 20 yd.
65. The lake rose $\frac{3}{10}$ ft. **67.** Logan owes \$85. **69.** $17a$
71. $9x$ **73.** $25t$ **75.** $-2m$ **77.** $-10y$ **79.** $1 - 2x$
81. $12x + 17$ **83.** $7r + 8t + 16$ **85.** $18n + 16$
87. ✒ **89.** $21z + 14y + 7$ **90.** $\frac{28}{3}$ **91.** ✒
93. \$451.70 **95.** $-5y$ **97.** $-7m$ **99.** $-7t, -23$
101. 1 under par

Exercise Set 1.6, pp. 48–51

1. (d) **2.** (g) **3.** (f) **4.** (h) **5.** (a) **6.** (c)
7. (b) **8.** (e) **9.** Six minus ten
11. Two minus negative twelve **13.** Nine minus the opposite of t **15.** The opposite of x minus y
17. Negative three minus the opposite of n **19.** -51
21. $\frac{11}{3}$ **23.** 3.14 **25.** 45 **27.** $\frac{14}{3}$ **29.** -0.101

31. 37 **33.** $-\frac{2}{5}$ **35.** 1 **37.** -15 **39.** -3 **41.** -6
43. -3 **45.** -7 **47.** -6 **49.** 0 **51.** -5
53. -10 **55.** -11 **57.** 0 **59.** 0 **61.** 8 **63.** -11
65. 16 **67.** -19 **69.** 1 **71.** 17 **73.** 3 **75.** -3
77. -21 **79.** 10 **81.** -8 **83.** -60 **85.** -23
87. -7.3 **89.** 1.1 **91.** -5.5 **93.** -0.928 **95.** $-\frac{7}{11}$
97. $-\frac{4}{5}$ **99.** $\frac{5}{17}$ **101.** $3.8 - (-5.2); 9$
103. $114 - (-79); 193$ **105.** -40 **107.** 43 **109.** 32
111. -62 **113.** -139 **115.** 0 **117.** $-3y, -8x$
119. $9, -5t, -3st$ **121.** $-3x$ **123.** $-5a + 4$
125. $-n - 9$ **127.** $-3x - 6$ **129.** $-8t - 7$
131. $-12x + 3y + 9$ **133.** $8x + 66$ **135.** 214°F
137. 30,347 ft **139.** 116 m **141.** ✒ **143.** 432 ft^2
144. $2 \cdot 2 \cdot 2 \cdot 2 \cdot 2 \cdot 3 \cdot 3 \cdot 3$ **145.** ✒
147. 11:00 P.M., August 14 **149.** False. For example,
let $m = -3$ and $n = -5$. Then $-3 > -5$, but
$-3 + (-5) = -8 \not> 0$. **151.** True. For example, for
$m = 4$ and $n = -4, 4 = -(-4)$ and $4 + (-4) = 0$; for
$m = -3$ and $n = 3, -3 = -3$ and $-3 + 3 = 0$.
153. (-) 9 - (-) 7 ENTER

Exercise Set 1.7, pp. 56–58

1. 1 **2.** 0 **3.** 0 **4.** 1 **5.** 0 **6.** 1 **7.** 1 **8.** 0
9. 1 **10.** 0 **11.** -40 **13.** -56 **15.** -40 **17.** 72
19. -42 **21.** 45 **23.** 190 **25.** -132 **27.** 1200
29. -126 **31.** 11.5 **33.** 0 **35.** $-\frac{2}{7}$ **37.** $\frac{1}{12}$
39. -11.13 **41.** $-\frac{5}{12}$ **43.** 252 **45.** 0 **47.** $\frac{1}{28}$
49. 150 **51.** 0 **53.** -720 **55.** $-30{,}240$ **57.** -9
59. -4 **61.** -7 **63.** 4 **65.** -9 **67.** 5.1 **69.** $\frac{100}{11}$
71. -8 **73.** Undefined **75.** -4 **77.** 0 **79.** 0
81. $-\frac{8}{3}, \frac{8}{-3}$ **83.** $-\frac{29}{35}, \frac{-29}{35}$ **85.** $-\frac{7}{3}; \frac{7}{-3}$ **87.** $-\frac{x}{2}, \frac{x}{-2}$
89. $-\frac{5}{4}$ **91.** $-\frac{10}{51}$ **93.** $-\frac{1}{10}$ **95.** $\frac{1}{4.3}$, or $\frac{10}{43}$ **97.** $-\frac{4}{9}$
99. Does not exist **101.** $\frac{21}{20}$ **103.** -1 **105.** 1
107. $\frac{3}{11}$ **109.** $-\frac{7}{4}$ **111.** 1 **113.** $\frac{1}{10}$ **115.** $-\frac{7}{6}$
117. Undefined **119.** $-\frac{14}{15}$ **121.** ✒ **123.** $\frac{22}{39}$
124. $12x - y - 9$ **125.** ✒ **127.** $\dfrac{1}{a + b}$
129. $-(a + b)$ **131.** $x = -x$ **133.** For 2 and 3,
the reciprocal of the sum is $1/(2 + 3)$, or $1/5$. But
$1/5 \neq 1/2 + 1/3$. **135.** 5°F **137.** Positive
139. Positive **141.** Positive **143.** Distributive law;
law of opposites; multiplicative property of zero

Connecting the Concepts, p. 59

1. -10 **2.** 16 **3.** 4 **4.** -6 **5.** -120 **6.** -7
7. -23 **8.** -3 **9.** -1 **10.** -3 **11.** -0.8
12. -3.77 **13.** -7 **14.** -4.1 **15.** -12 **16.** $\frac{5}{3}$
17. 100 **18.** 77 **19.** 180 **20.** -52

Exercise Set 1.8, pp. 66–68

1. (a) Division; (b) subtraction; (c) addition;
(d) multiplication; (e) subtraction; (f) multiplication

2. (a) Multiplication; **(b)** subtraction; **(c)** addition; **(d)** subtraction; **(e)** division; **(f)** multiplication
3. x^6 **5.** $(-5)^3$ **7.** $(3t)^5$ **9.** $2n^4$ **11.** 16   **13.** 9
15. -9 **17.** 64   **19.** 625   **21.** 7   **23.** -32
25. $81t^4$ **27.** $-343x^3$ **29.** 26   **31.** 51   **33.** -6
35. 1   **37.** 298   **39.** 11   **41.** -36 **43.** 1291
45. 152   **47.** 36   **49.** 1   **51.** -44 **53.** 41   **55.** -10
57. -5 **59.** -19 **61.** -3 **63.** -75 **65.** 9   **67.** 30
69. 6   **71.** -17 **73.** $-9x-1$ **75.** $7n-8$
77. $-4a+3b-7c$ **79.** $-3x^2-5x+1$ **81.** $2x-7$
83. $-9x+6$ **85.** $21t-r$ **87.** $9y-25z$ **89.** x^2+6
91. $-t^3+4t$ **93.** $37a^2-23ab+35b^2$
95. $-22t^3-t^2+9t$ **97.** $2x-25$ **99.** ✍   **101.** Let n represent the number; $2n-9$ **102.** Let m and n represent the two numbers; $\frac{1}{2}(m+n)$ **103.** ✍
105. $-6r-5t+21$ **107.** $-2x-f$ **109.** ✍
111. True   **113.** False   **115.** 0   **117.** 17
119. 39,000   **121.** $44x^3$

Review Exercises: Chapter 1, pp. 73–74

1. True   **2.** True   **3.** False   **4.** True   **5.** False
6. False   **7.** True   **8.** False   **9.** False   **10.** True
11. 24   **12.** 4   **13.** -16 **14.** -15 **15.** $y-7$
16. $xz+10$, or $10+xz$ **17.** Let b represent Brandt's speed and w represent the wind speed; $15(b-w)$
18. No   **19.** Let d represent the number of digital prints, in billions, made in 2006; $14.1 = d + 3.2$ **20.** $t \cdot 3 + 5$
21. $2x + (y+z)$ **22.** $(4x)y, 4(yx), (4y)x$; answers may vary   **23.** $18x+30y$ **24.** $40x+24y+16$
25. $3(7x+5y)$ **26.** $11(2a+9b+1)$ **27.** $2 \cdot 2 \cdot 2 \cdot 7$
28. $\frac{5}{12}$ **29.** $\frac{9}{4}$ **30.** $\frac{19}{24}$ **31.** $\frac{3}{16}$ **32.** $\frac{3}{5}$ **33.** $\frac{27}{25}$
34. $-3600, 1350$ **35.**

$$\xleftarrow{\hspace{0.3cm}}\!\!\!\underset{-5\;-4\;-3\;-2\;-1\;\;0\;\;1\;\;2\;\;3\;\;4\;\;5}{+\!+\!+\!+\!+\!\bullet\!+\!+\!+\!+\!+}\!\!\!\xrightarrow{\hspace{0.3cm}} \quad \overset{-\frac{1}{3}}{}$$

36. $x > -3$ **37.** True   **38.** False   **39.** $-0.\overline{4}$ **40.** 1
41. -12 **42.** -10 **43.** $-\frac{7}{12}$ **44.** 0   **45.** -5 **46.** 8
47. $-\frac{7}{5}$ **48.** -7.9 **49.** 63   **50.** -9.18 **51.** $-\frac{2}{7}$
52. -140 **53.** -7 **54.** -3 **55.** $\frac{9}{4}$ **56.** 48
57. 168   **58.** $\frac{21}{8}$ **59.** 18   **60.** 53   **61.** $\frac{103}{17}$
62. $7a-b$ **63.** $-4x+5y$ **64.** 7   **65.** $-\frac{1}{7}$
66. $(2x)^4$ **67.** $-125x^3$ **68.** $-3a+9$ **69.** $11b-27$
70. $3x^4+10x$ **71.** $17n^2+m^2+20mn$ **72.** $5x+28$
73. ✍ The value of a constant never varies. A variable can represent a variety of numbers.   **74.** ✍ A term is one of the parts of an expression that is separated from the other parts by plus signs. A factor is part of a product.   **75.** ✍ The distributive law is used in factoring algebraic expressions, multiplying algebraic expressions, combining like terms, finding the opposite of a sum, and subtracting algebraic expressions.   **76.** ✍ A negative number raised to an even power is positive; a negative number raised to an odd power is negative.   **77.** 25,281   **78. (a)** $\frac{3}{11}$; **(b)** $\frac{10}{11}$
79. $-\frac{5}{8}$ **80.** -2.1 **81.** (i)   **82.** (j)   **83.** (a)
84. (h)   **85.** (k)   **86.** (b)   **87.** (c)   **88.** (e)
89. (d)   **90.** (f)   **91.** (g)

Test: Chapter 1, p. 75

1. [1.1] 4   **2.** [1.1] Let x and y represent the numbers; $xy-9$ **3.** [1.1] $240\,\text{ft}^2$ **4.** [1.2] $q+3p$
5. [1.2] $(x \cdot 4) \cdot y$ **6.** [1.1] No   **7.** [1.1] Let p represent the maximum production capability; $p-4250 = 45,950$
8. [1.2] $35+7x$ **9.** [1.7] $-5y+10$
10. [1.2] $11(1+4x)$ **11.** [1.2] $7(x+1+7y)$
12. [1.3] $2 \cdot 2 \cdot 3 \cdot 5 \cdot 5$ **13.** [1.3] $\frac{3}{7}$ **14.** [1.4] $<$
15. [1.4] $>$ **16.** [1.4] $\frac{9}{4}$ **17.** [1.4] 3.8   **18.** [1.6] $\frac{2}{3}$
19. [1.7] $-\frac{7}{4}$ **20.** [1.6] 10   **21.** [1.4] $-5 \geq x$
22. [1.6] 7.8   **23.** [1.5] -8 **24.** [1.6] -2.5 **25.** [1.6] $-\frac{7}{8}$
26. [1.7] -48 **27.** [1.7] $\frac{2}{9}$ **28.** [1.7] -6 **29.** [1.7] $\frac{3}{4}$
30. [1.7] -9.728 **31.** [1.8] -173 **32.** [1.6] 15
33. [1.8] -64 **34.** [1.8] 448   **35.** [1.6] $21a+22y$
36. [1.8] $16x^4$ **37.** [1.8] $x+7$ **38.** [1.8] $9a-12b-7$
39. [1.8] $-y-16$ **40.** [1.1] 5
41. [1.8] $9-(3-4)+5 = 15$ **42.** [1.8] 15
43. [1.8] $4a$ **44.** [1.8] False

CHAPTER 2

Exercise Set 2.1, pp. 83–85

1. (c)   **2.** (b)   **3.** (f)   **4.** (a)   **5.** (d)   **6.** (e)   **7.** (d)
8. (b)   **9.** (c)   **10.** (a)   **11.** 11   **13.** -25 **15.** -31
17. 41   **19.** 19   **21.** -6 **23.** $\frac{7}{3}$ **25.** $-\frac{1}{10}$ **27.** $\frac{41}{24}$
29. $-\frac{1}{20}$ **31.** 9.1   **33.** -5 **35.** 7   **37.** 12   **39.** -38
41. 8   **43.** -7 **45.** 8   **47.** 88   **49.** 20   **51.** -54
53. $-\frac{5}{9}$ **55.** 1   **57.** $\frac{9}{2}$ **59.** -7.6 **61.** -2.5 **63.** -15
65. -5 **67.** $-\frac{7}{6}$ **69.** -128 **71.** $-\frac{1}{2}$ **73.** -15 **75.** 9
77. 310.756   **79.** ✍   **81.** -6 **82.** 2   **83.** 1
84. -16 **85.** ✍   **87.** 11.6   **89.** 2   **91.** $-23, 23$
93. 9000   **95.** 250

Technology Connection, p. 88

1. 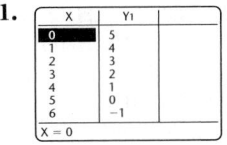   **2.**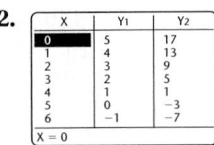

3. 4; not reliable because, depending on the choice of ΔTbl, it is easy to scroll past a solution without realizing it.

Exercise Set 2.2, pp. 91–92

1. (c)   **2.** (e)   **3.** (a)   **4.** (f)   **5.** (b)   **6.** (d)   **7.** 8
9. 7   **11.** 5   **13.** $\frac{10}{3}$ **15.** -7 **17.** -5 **19.** -4
21. 19   **23.** -2.8 **25.** 3   **27.** 15   **29.** -6 **31.** $-\frac{25}{2}$
33. All real numbers; identity   **35.** -3 **37.** -6 **39.** 2
41. 0   **43.** 6   **45.** No solution; contradiction   **47.** $-\frac{1}{2}$
49. 0   **51.** 10   **53.** 4   **55.** 0   **57.** No solution; contradiction   **59.** $\frac{5}{2}$ **61.** -8 **63.** $\frac{1}{6}$
65. All real numbers; identity   **67.** 2   **69.** $\frac{16}{3}$ **71.** $\frac{2}{5}$
73. 1   **75.** -4 **77.** $1.\overline{6}$ **79.** $-\frac{60}{37}$ **81.** 11   **83.** 8
85. $\frac{16}{15}$ **87.** $-\frac{1}{31}$ **89.** 2   **91.** ✍   **93.** -7 **94.** 15

95. -15 **96.** -28 **97.** ✒   **99.** $\frac{1136}{909}$, or $1.\overline{2497}$
101. No solution; contradiction   **103.** $\frac{2}{3}$ **105.** 0
107. $\frac{52}{45}$ **109.** All real numbers; identity

Technology Connection, p. 94

1. 800

Exercise Set 2.3, pp. 97–100

1. 309.6 m   **3.** 1423 students   **5.** 8.4734   **7.** 255 mg
9. $b = \dfrac{A}{h}$ **11.** $r = \dfrac{d}{t}$ **13.** $P = \dfrac{I}{rt}$ **15.** $m = 65 - H$
17. $l = \dfrac{P - 2w}{2}$, or $l = \dfrac{P}{2} - w$ **19.** $\pi = \dfrac{A}{r^2}$
21. $h = \dfrac{2A}{b}$ **23.** $c^2 = \dfrac{E}{m}$ **25.** $d = 2Q - c$
27. $b = 3A - a - c$ **29.** $r = wf$ **31.** $C = \frac{5}{9}(F - 32)$
33. $y = 2x - 1$ **35.** $y = -\frac{2}{5}x + 2$ **37.** $y = \frac{4}{3}x - 2$
39. $y = -\frac{9}{8}x + \frac{1}{2}$ **41.** $y = \frac{3}{5}x - \frac{8}{5}$
43. $x = \dfrac{z - 13}{2} - y$, or $x = \dfrac{z - 13 - 2y}{2}$
45. $l = 4(t - 27) + w$ **47.** $t = \dfrac{A}{a + b}$ **49.** $h = \dfrac{2A}{a + b}$
51. $L = W - \dfrac{N(R - r)}{400}$, or $L = \dfrac{400W - NR + Nr}{400}$
53. ✒   **55.** -10 **56.** -196 **57.** 0   **58.** -32
59. -13 **60.** 65   **61.** ✒   **63.** 40 yr   **65.** 27 in^3
67. $a = \dfrac{w}{c} \cdot d$ **69.** $c = \dfrac{d}{a - b}$ **71.** $a = \dfrac{c}{3 + b + d}$
73. $K = 9.632w + 19.685h - 10.54a + 102.3$

Exercise Set 2.4, pp. 104–108

1. (d)   **2.** (c)   **3.** (e)   **4.** (b)   **5.** (c)   **6.** (d)   **7.** (f)
8. (a)   **9.** (b)   **10.** (e)   **11.** 0.49   **13.** 0.01
15. 0.041   **17.** 0.2   **19.** 0.0625   **21.** 0.002   **23.** 1.75
25. 38%   **27.** 3.9%   **29.** 45%   **31.** 70%   **33.** 0.09%
35. 106%   **37.** 180%   **39.** 60%   **41.** 32%   **43.** 25%
45. 26%   **47.** $46\frac{2}{3}$, or $\frac{140}{3}$ **49.** 2.5   **51.** 10,000
53. 125%   **55.** 0.8   **57.** 50%   **59.** $33.\overline{3}$%, or $33\frac{1}{3}$%
61. 2.85 million Americans   **63.** 23.37 million Americans
65. 75 credits   **67.** 595 at bats   **69.** (a) 16%; (b) $29
71. $33.\overline{3}$%, or $33\frac{1}{3}$%; $66.\overline{6}$%, or $66\frac{2}{3}$%   **73.** $168
75. 285 women   **77.** $19.20 an hour   **79.** The actual
cost was 43.7% more than the estimate.   **81.** $45
83. $148.50   **85.** About 31.5 lb   **87.** About 2.45 billion
pieces of mail   **89.** About 165 calories   **91.** ✒
93. Let l represent the length and w represent the width;
$2l + 2w$ **94.** $0.05 \cdot 180$ **95.** Let p represent the number
of points Tino scored; $p - 5$ **96.** $15 + 1.5x$ **97.** $10\left(\frac{1}{2}a\right)$
98. Let n represent the number; $3n + 10$ **99.** Let l
represent the length and w represent the width; $w = l - 2$
100. Let x represent the first number and y represent the
second number; $x = 4y$ **101.** ✒   **103.** 18,500 people
105. About 6 ft 7 in.   **107.** About 27%   **109.** ✒

Exercise Set 2.5, pp. 116–121

1. 11   **3.** $\frac{11}{2}$ **5.** $150   **7.** $130   **9.** About 78.4 mi
11. 160 mi   **13.** 1204 and 1205   **15.** 285 and 287
17. 32, 33, 34   **19.** Man: 103 yr; woman: 101 yr
21. Non-spam: 25 billion messages; spam: 100 billion
messages   **23.** 140 and 141   **25.** Width: 100 ft; length:
160 ft; area: 16,000 ft^2 **27.** Width: 21 m; length: 25 m
29. $1\frac{3}{4}$ in. by $3\frac{1}{2}$ in.   **31.** 30°, 90°, 60°   **33.** 70°
35. Bottom: 144 ft; middle: 72 ft; top: 24 ft   **37.** 8.75 mi,
or $8\frac{3}{4}$ mi   **39.** $128\frac{1}{3}$ mi   **41.** 65°, 25°   **43.** 140°, 40°
45. Length: 27.9 cm; width: 21.6 cm   **47.** $6600
49. 830 points   **51.** $125,000   **53.** 160 chirps per
minute   **55.** ✒   **57.** <   **58.** >   **59.** >   **60.** <
61. $-4 \leq x$ **62.** $5 > x$ **63.** $y < 5$ **64.** $t \geq -10$
65. ✒   **67.** $37   **69.** 20   **71.** Half-dollars: 5;
quarters: 10; dimes: 20; nickels: 60   **73.** $95.99
75. 5 DVDs   **77.** 6 mi   **79.** ✒   **81.** Width: 23.31 cm;
length: 27.56 cm

Exercise Set 2.6, pp. 128–130

1. $\geq$ **2.** $\leq$ **3.** $<$ **4.** $>$ **5.** Equivalent
6. Equivalent   **7.** Equivalent   **8.** Not equivalent
9. (a) Yes; (b) no; (c) no   **11.** (a) Yes; (b) no; (c) yes
13. (a) Yes; (b) yes; (c) yes   **15.** (a) No; (b) yes; (c) no
17.   **19.**
21.   **23.**
25.
27. $\{y \mid y < 6\}$, $(-\infty, 6)$
29. $\{x \mid x \geq -4\}$, $[-4, \infty)$
31. $\{t \mid t > -3\}$, $(-3, \infty)$
33. $\{x \mid x \leq -7\}$, $(-\infty, -7]$
35. $\{x \mid x > -4\}$, $(-4, \infty)$ **37.** $\{x \mid x \leq 2\}$, $(-\infty, 2]$
39. $\{x \mid x < -1\}$, $(-\infty, -1)$ **41.** $\{x \mid x \geq 0\}$, $[0, \infty)$
43. $\{y \mid y > 3\}$, $(3, \infty)$
45. $\{n \mid n < 17\}$, $(-\infty, 17)$,
47. $\{x \mid x \leq -9\}$, $(-\infty, -9]$,
49. $\left\{y \mid y \leq \frac{1}{2}\right\}$, $\left(-\infty, \frac{1}{2}\right]$,
51. $\left\{t \mid t > \frac{5}{8}\right\}$, $\left(\frac{5}{8}, \infty\right)$,
53. $\{x \mid x < 0\}$, $(-\infty, 0)$,
55. $\{t \mid t < 23\}$, $(-\infty, 23)$
57. $\{y \mid y \geq 22\}$, $[22, \infty)$,
59. $\{x \mid x < 7\}$, $(-\infty, 7)$
61. $\{t \mid t < -3\}$, $(-\infty, -3)$,

63. $\{n \mid n \geq -1.5\}$, $[-1.5, \infty)$,

65. $\{y \mid y \geq -\frac{1}{10}\}$, $\left[-\frac{1}{10}, \infty\right)$

67. $\{x \mid x < -\frac{4}{5}\}$, $\left(-\infty, -\frac{4}{5}\right)$

69. $\{x \mid x < 6\}$, or $(-\infty, 6)$ **71.** $\{t \mid t \leq 7\}$, or $(-\infty, 7]$
73. $\{x \mid x > -4\}$, or $(-4, \infty)$ **75.** $\left\{y \mid y < -\frac{10}{3}\right\}$, or
$\left(-\infty, -\frac{10}{3}\right)$ **77.** $\{x \mid x > -10\}$, or $(-10, \infty)$
79. $\{y \mid y < 0\}$, or $(-\infty, 0)$ **81.** $\left\{y \mid y \geq \frac{3}{2}\right\}$, or $\left[\frac{3}{2}, \infty\right)$
83. $\{x \mid x > -4\}$, or $(-4, \infty)$ **85.** $\{t \mid t > 1\}$, or $(1, \infty)$
87. $\{x \mid x \leq -9\}$, or $(-\infty, -9]$ **89.** $\{t \mid t < 14\}$, or
$(-\infty, 14)$ **91.** $\{y \mid y \leq -4\}$, or $(-\infty, -4]$
93. $\left\{t \mid t < -\frac{5}{3}\right\}$, or $\left(-\infty, -\frac{5}{3}\right)$ **95.** $\{r \mid r > -3\}$, or $(-3, \infty)$
97. $\{x \mid x \leq 7\}$, or $(-\infty, 7]$ **99.** $\left\{x \mid x > -\frac{5}{32}\right\}$, or $\left(-\frac{5}{32}, \infty\right)$
101. ✍   **103.** $17x - 6$ **104.** $2m - 16n$
105. $7x - 8y - 46$ **106.** $-47t + 1$
107. $35a - 20b - 17$ **108.** $-21x + 32$ **109.** ✍
111. $\{x \mid x \text{ is a real number}\}$, or $(-\infty, \infty)$
113. $\left\{x \mid x \leq \frac{5}{6}\right\}$, or $\left(-\infty, \frac{5}{6}\right]$
115. $\{x \mid x \leq -4a\}$, or $(-\infty, -4a]$
117. $\left\{x \mid x > \dfrac{y - b}{a}\right\}$, or $\left(\dfrac{y - b}{a}, \infty\right)$
119. $\{x \mid x \text{ is a real number}\}$, or $(-\infty, \infty)$

Connecting the Concepts, pp. 130–131

1. 21   **2.** $\{x \mid x \leq 21\}$, or $(-\infty, 21]$ **3.** -6
4. $\{x \mid x > -6\}$, or $(-6, \infty)$ **5.** $\{x \mid x < 6\}$, or $(-\infty, 6)$
6. 3   **7.** $-\frac{1}{3}$ **8.** $\{y \mid y < 3\}$, or $(-\infty, 3)$
9. $\{t \mid t \leq -16\}$, or $(-\infty, -16]$ **10.** $\frac{11}{2}$
11. $\{a \mid a < -1\}$, or $(-\infty, -1)$ **12.** $\{x \mid x < 10.75\}$, or
$(-\infty, 10.75)$ **13.** $\{x \mid x \geq -11\}$, or $[-11, \infty)$
14. 105   **15.** -4.24 **16.** 15   **17.** $\{y \mid y \geq 3\}$, or $[3, \infty)$
18. $\frac{14}{3}$ **19.** $\left\{x \mid x > \frac{22}{5}\right\}$, or $\left(\frac{22}{5}, \infty\right)$
20. $\{a \mid a \geq 0\}$, or $[0, \infty)$

Translating for Success, p. 134

1. F   **2.** I   **3.** C   **4.** E   **5.** D   **6.** J   **7.** O   **8.** M
9. B   **10.** L

Exercise Set 2.7, pp. 135–138

1. $b \leq a$ **2.** $b < a$ **3.** $a \leq b$ **4.** $a < b$ **5.** $b \leq a$
6. $a \leq b$ **7.** $b < a$ **8.** $a < b$ **9.** Let n represent the
number; $n < 10$ **11.** Let t represent the temperature;
$t \leq -3$ **13.** Let d represent the number of years of
driving experience; $d \geq 5$ **15.** Let a represent the age of
the altar; $a > 1200$ **17.** Let h represent Tania's hourly
wage; $12 < h < 15$ **19.** Let w represent the wind speed;
$w > 50$ **21.** Let c represent the cost of a room; $c \leq 120$
23. More than 2.375 hr   **25.** At least 2.25   **27.** Scores
greater than or equal to 97   **29.** 8 credits or more
31. At least 3 plate appearances   **33.** Lengths greater than
6 cm   **35.** Depths less than 437.5 ft   **37.** Blue-book value

is greater than or equal to $10,625   **39.** Lengths less than
55 in.   **41.** Temperatures greater than 37°C
43. No more than 3 ft tall   **45.** A serving contains at least
16 g of fat   **47.** Dates after September 16   **49.** No more
than 134 text messages   **51.** Years after 2012
53. Mileages less than or equal to 225   **55.** ✍   **57.** -14
58. $-\frac{2}{3}$ **59.** -60 **60.** -11.1 **61.** 0   **62.** 5
63. -2 **64.** -1 **65.** ✍   **67.** Temperatures between
$-15°C$ and $-9\frac{4}{9}°C$ **69.** Lengths less than or equal to 8 cm
71. They contain at least 7.5 g of fat per serving.
73. At least $42   **75.** ✍

Review Exercises: Chapter 2, pp. 142–143

1. True   **2.** False   **3.** True   **4.** True   **5.** True
6. False   **7.** True   **8.** True   **9.** -25 **10.** 7
11. -65 **12.** 3   **13.** -20 **14.** 1.11   **15.** $\frac{1}{2}$ **16.** $-\frac{3}{2}$
17. -8 **18.** -4 **19.** $-\frac{1}{3}$ **20.** 4   **21.** No solution;
contradiction   **22.** 4   **23.** 16   **24.** 1   **25.** $-\frac{7}{5}$ **26.** 0
27. All real numbers; identity   **28.** $d = \dfrac{C}{\pi}$ **29.** $B = \dfrac{3V}{h}$
30. $y = \frac{5}{2}x - 5$ **31.** $x = \dfrac{b}{t - a}$ **32.** 0.012   **33.** 44%
34. 70%   **35.** 140   **36.** No   **37.** Yes   **38.** No
39.

$5x - 6 < 2x + 3$

40.

$-2 < x \leq 5$

41.

$t > 0$

42. $\left\{t \mid t \geq -\frac{1}{2}\right\}$, or $\left[-\frac{1}{2}, \infty\right)$

43. $\{x \mid x \geq 7\}$, or $[7, \infty)$ **44.** $\{y \mid y > 3\}$, or $(3, \infty)$
45. $\{y \mid y \leq -4\}$, or $(-\infty, -4]$
46. $\{x \mid x < -11\}$, or $(-\infty, -11)$ **47.** $\{y \mid y > -7\}$, or
$(-7, \infty)$ **48.** $\{x \mid x > -6\}$, or $(-6, \infty)$
49. $\left\{x \mid x > -\frac{9}{11}\right\}$, or $\left(-\frac{9}{11}, \infty\right)$ **50.** $\{t \mid t \leq -12\}$, or
$(-\infty, -12]$ **51.** $\{x \mid x \leq -8\}$, or $(-\infty, -8]$
52. About 48%   **53.** 8 ft, 10 ft   **54.** Japanese students:
41,000; Chinese students: 62,000   **55.** 57, 59
56. Width: 11 cm; length: 17 cm   **57.** $160   **58.** About 109
million subscribers   **59.** 35°, 85°, 60°   **60.** No more than
55 g of fat   **61.** 14 or fewer copies   **62.** ✍ Multiplying
both sides of an equation by *any* nonzero number results in
an equivalent equation. When multiplying on both sides of
an inequality, the sign of the number being multiplied by
must be considered. If the number is positive, the direction
of the inequality symbol remains unchanged; if the number
is negative, the direction of the inequality symbol must be
reversed to produce an equivalent inequality.   **63.** ✍
The solutions of an equation can usually each be checked.
The solutions of an inequality are normally too numerous
to check. Checking a few numbers from the solution set
found cannot guarantee that the answer is correct, although
if any number does not check, the answer found is
incorrect.   **64.** About 1 hr 36 min   **65.** Nile: 4160 mi;
Amazon: 4225 mi   **66.** $16   **67.** $-23, 23$ **68.** $-20, 20$
69. $a = \dfrac{y - 3}{2 - b}$ **70.** $F = \dfrac{0.3(12w)}{9}$, or $F = 0.4w$

Test: Chapter 2, p. 144

1. [2.1] 9   2. [2.1] 15   3. [2.1] −3   4. [2.1] 49
5. [2.1] −12   6. [2.2] 2   7. [2.1] −8   8. [2.2] $-\frac{23}{67}$
9. [2.2] 7   10. [2.2] $\frac{23}{3}$ 11. [2.2] All real numbers; identity
12. [2.6] $\{x \mid x > -5\}$, or $(-5, \infty)$
13. [2.6] $\{x \mid x > -13\}$, or $(-13, \infty)$
14. [2.6] $\{y \mid y \le -13\}$, or $(-\infty, -13]$
15. [2.6] $\left\{y \mid y \le -\frac{15}{2}\right\}$, or $\left(-\infty, -\frac{15}{2}\right]$
16. [2.6] $\{n \mid n < -5\}$, or $(-\infty, -5)$
17. [2.6] $\{x \mid x < -7\}$, or $(-\infty, -7)$
18. [2.6] $\{t \mid t \ge -1\}$, or $[-1, \infty)$
19. [2.6] $\{x \mid x \le -1\}$, or $(-\infty, -1]$
20. [2.3] $r = \dfrac{A}{2\pi h}$ 21. [2.3] $l = 2w - P$ 22. [2.4] 2.3
23. [2.4] 0.3%   24. [2.4] 14.8   25. [2.4] 44%
26. [2.6]

$y < 4$
```
<-+--+--+--+--+--+--+--+--+->
-10-8-6-4-2  0  2  4  6  8 10
```

27. [2.6]

$-2 \le x \le 2$
```
<-+--+--+--[==+==]--+--+--+->
-5-4-3-2-1  0  1  2  3  4  5
```

28. [2.5] Width: 7 cm; length: 11 cm   29. [2.5] 525 mi from Springer Mountain and 1575 mi from Mt. Katahdin
30. [2.5] 81 mm, 83 mm, 85 mm   31. [2.4] $65
32. [2.7] More than 36 one-way trips per month
33. [2.3] $d = \dfrac{a}{3}$ 34. [1.4], [2.2] −15, 15   35. [2.7] Let $h =$ the number of hours of sun each day; $4 \le h \le 6$
36. [2.5] 60 tickets

Cumulative Review: Chapters 1–2, pp. 145–146

1. −12   2. $\frac{3}{4}$ 3. −4.2   4. 10   5. 134   6. 149
7. $2x + 1$ 8. $-10t + 12$ 9. $-21n + 36$
10.
```
        -5/2
<-+--+--+●-+--+--+--+--+--+->
-5-4-3-2-1  0  1  2  3  4  5
```
11. 27
12. $6(2x + 3y + 5z)$ 13. −6   14. 16   15. 9
16. $\frac{13}{18}$ 17. 1   18. $-\frac{7}{2}$ 19. $z = \dfrac{x}{4y}$ 20. $y = \frac{4}{9}x - \frac{1}{9}$
21. $n = \dfrac{p}{a + r}$ 22. 1.83   23. 37.5%
24.
```
     -5/2      t > -5/2
<-+--+--(-+--+--+--+--+--+->
-5-4-3-2-1  0  1  2  3  4  5
```
25. $\{t \mid t \le -2\}$, or $(-\infty, -2]$
26. $\{t \mid t < 3\}$, or $(-\infty, 3)$ 27. $\{x \mid x < 30\}$, or $(-\infty, 30)$
28. $\{n \mid n \ge 2\}$, or $[2, \infty)$ 29. 48 million   30. 3.2 hr
31. $14\frac{1}{3}$ m   32. 9 ft, 15 ft   33. No more than $52
34. About 194%   35. 105°   36. For widths greater than 27 cm   37. $4t$ 38. −5, 5   39. $1025

CHAPTER 3

Exercise Set 3.1, pp. 154–157

1. (a)   2. (c)   3. (b)   4. (d)   5. 2 drinks
7. The person weighs more than 140 lb.   9. About $5156

11. $231,856,000   13. About 29.2 million tons   15. About 3.2 million tons   17. About $12 billion   19. 2001

21. 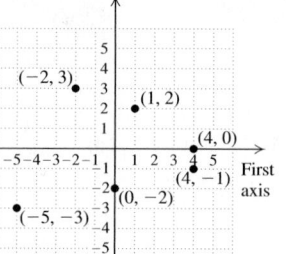   23.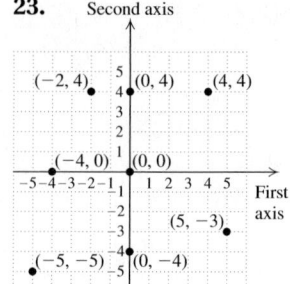

25.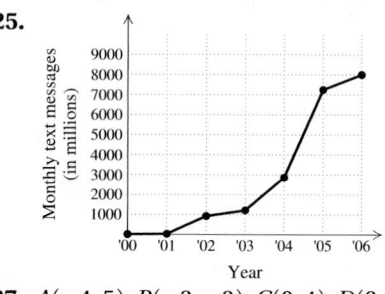

27. $A(-4, 5)$; $B(-3, -3)$; $C(0, 4)$; $D(3, 4)$; $E(3, -4)$
29. $A(4, 1)$; $B(0, -5)$; $C(-4, 0)$; $D(-3, -2)$; $E(3, 0)$

31. 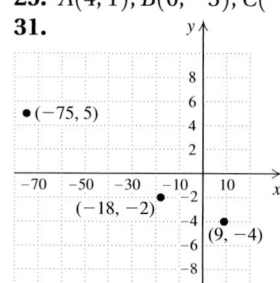   33.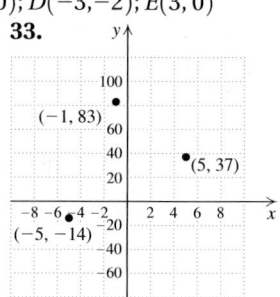

35. 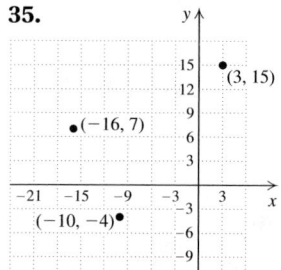   37.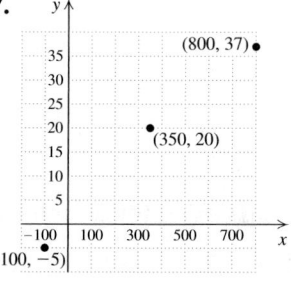

39. 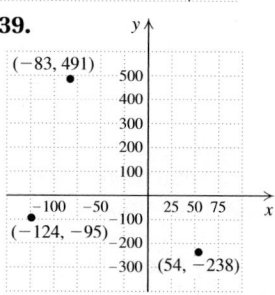   41. IV   43. III   45. I
47. II   49. I and IV
51. I and III   53. ✍
55. $y = \dfrac{2x}{5}$, or $y = \frac{2}{5}x$
56. $y = \dfrac{-3x}{2}$, or $y = -\frac{3}{2}x$
57. $y = x - 8$
58. $y = -\frac{2}{5}x + 2$
59. $y = -\frac{2}{3}x + \frac{5}{3}$
60. $y = \frac{5}{8}x - \frac{1}{8}$ 61. ✍   63. II or IV   65. $(-1, -5)$

67.

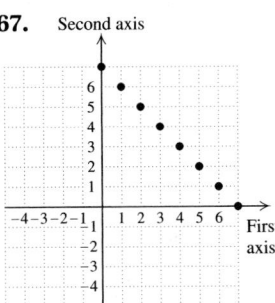

Second axis

First axis

69. $\frac{65}{2}$ sq units
71. Latitude 27° North; longitude 81° West
73.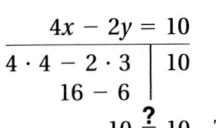

Technology Connection, p. 164

1. $y = -5x + 6.5$

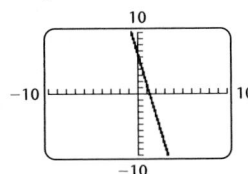

2. $y = 3x + 4.5$

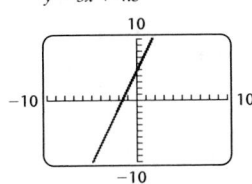

3. $7y - 4x = 22$, or $y = \frac{4}{7}x + \frac{22}{7}$

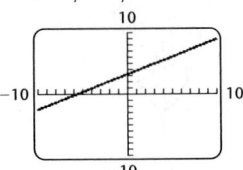

4. $5y + 11x = -20$, or $y = -\frac{11}{5}x - 4$

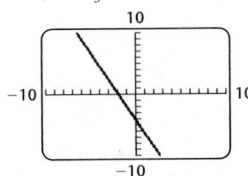

5. $2y - x^2 = 0$, or $y = 0.5x^2$

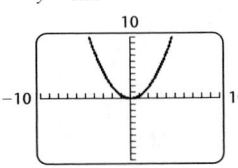

6. $y + x^2 = 8$, or $y = -x^2 + 8$

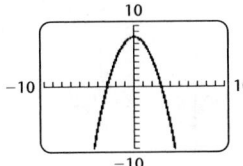

Exercise Set 3.2, pp. 165–167

1. False **2.** True **3.** True **4.** True **5.** True
6. False **7.** Yes **9.** No **11.** No

13. $\dfrac{y = x + 3}{\begin{array}{c|c} 2 & -1 + 3 \\ 2 \overset{?}{=} 2 \end{array}}$ True

$\dfrac{y = x + 3}{\begin{array}{c|c} 7 & 4 + 3 \\ 7 \overset{?}{=} 7 \end{array}}$ True

$(2, 5)$; answers may vary

15. $\dfrac{y = \frac{1}{2}x + 3}{\begin{array}{c|c} 5 & \frac{1}{2} \cdot 4 + 3 \\ & 2 + 3 \\ 5 \overset{?}{=} 5 \end{array}}$ True

$\dfrac{y = \frac{1}{2}x + 3}{\begin{array}{c|c} 2 & \frac{1}{2}(-2) + 3 \\ & -1 + 3 \\ 2 \overset{?}{=} 2 \end{array}}$ True

$(0, 3)$; answers may vary

17. $\dfrac{y + 3x = 7}{\begin{array}{c|c} 1 + 3 \cdot 2 & 7 \\ 1 + 6 & \\ 7 \overset{?}{=} 7 \end{array}}$ True

$\dfrac{y + 3x = 7}{\begin{array}{c|c} -5 + 3 \cdot 4 & 7 \\ -5 + 12 & \\ 7 \overset{?}{=} 7 \end{array}}$ True

$(1, 4)$; answers may vary

19. $\dfrac{4x - 2y = 10}{\begin{array}{c|c} 4 \cdot 0 - 2(-5) & 10 \\ 0 + 10 & \\ 10 \overset{?}{=} 10 \end{array}}$ True

$\dfrac{4x - 2y = 10}{\begin{array}{c|c} 4 \cdot 4 - 2 \cdot 3 & 10 \\ 16 - 6 & \\ 10 \overset{?}{=} 10 \end{array}}$ True

$(2, -1)$; answers may vary

21.

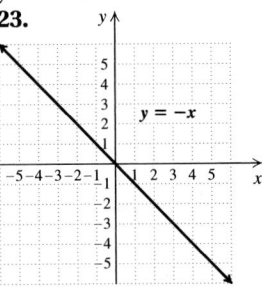

$y = x + 1$

23.

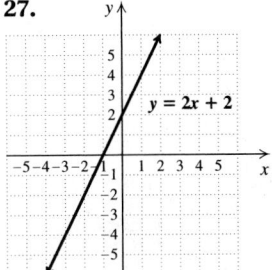

$y = -x$

25.

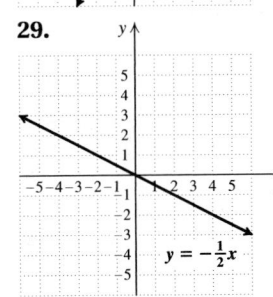

$y = 2x$

27.

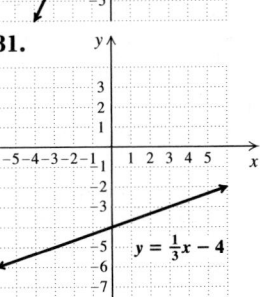

$y = 2x + 2$

29.

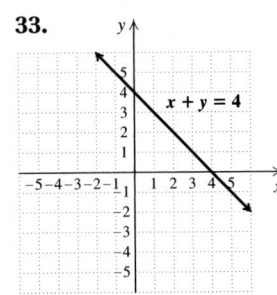

$y = -\frac{1}{2}x$

31.

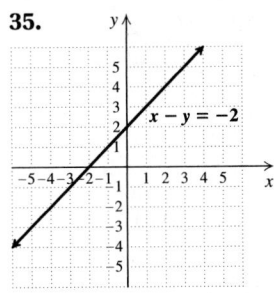

$y = \frac{1}{3}x - 4$

33.

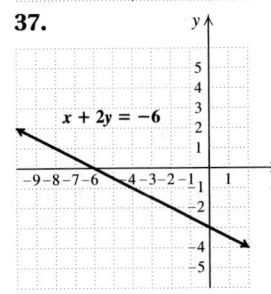

$x + y = 4$

35.

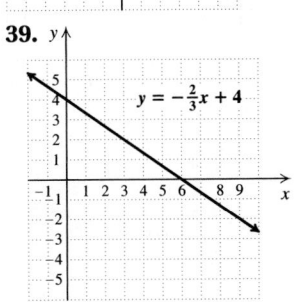

$x - y = -2$

37.

$x + 2y = -6$

39.

$y = -\frac{2}{3}x + 4$

41.
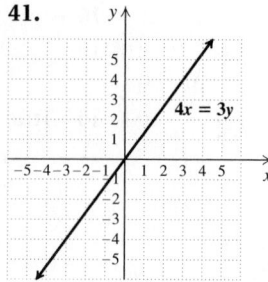
$4x = 3y$

43.
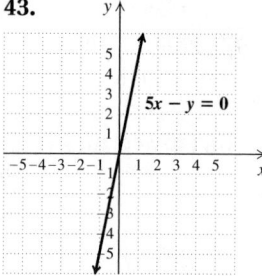
$5x - y = 0$

45.
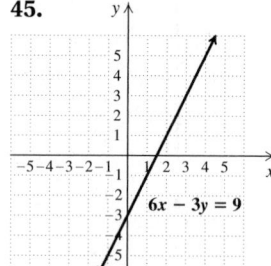
$6x - 3y = 9$

47.
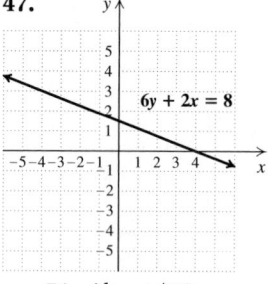
$6y + 2x = 8$

49. About $3800   **51.** About $49

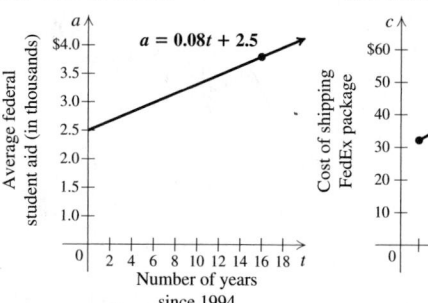
$a = 0.08t + 2.5$

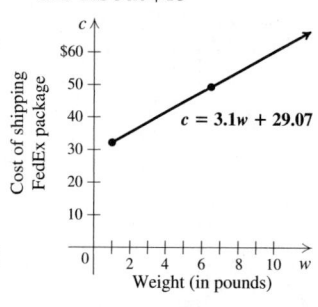
$c = 3.1w + 29.07$

53. About $96   **55.** About 33 gal

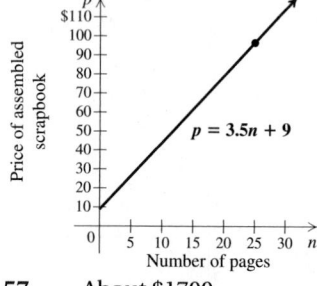
$p = 3.5n + 9$

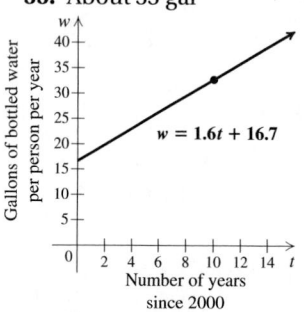
$w = 1.6t + 16.7$

57. About $1700
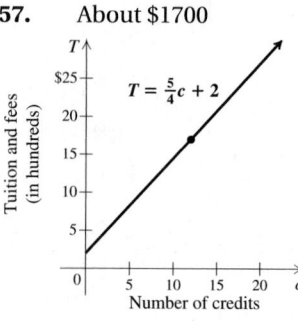
$T = \frac{5}{4}c + 2$

59. 📝   **61.** $\frac{12}{5}$ **62.** $-\frac{3}{4}$
63. 3   **64.** $-\frac{21}{5}$
65. $Q = 2A - T$
66. $p = \dfrac{w}{q + 1}$
67. $y = \dfrac{C - Ax}{B}$
68. $y = m(x - h) + k$
69. 📝

71.
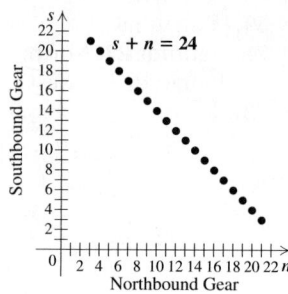
$s + n = 24$

73. $x + y = 5$, or
$y = -x + 5$
75. $y = x + 2$

77.
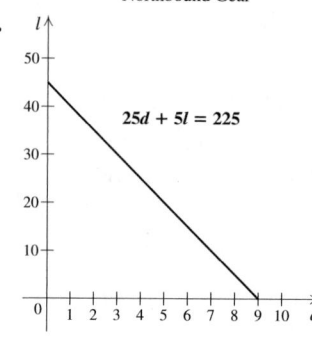
$25d + 5l = 225$

Answers may vary.
1 dinner, 40 lunches;
5 dinners, 20 lunches;
8 dinners, 5 lunches

79.
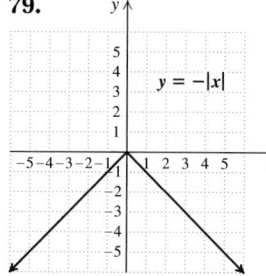
$y = -|x|$

81.
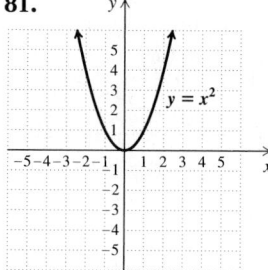
$y = x^2$

83. $y = -2.8x + 3.5$
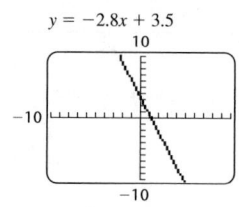

85. $y = 2.8x - 3.5$
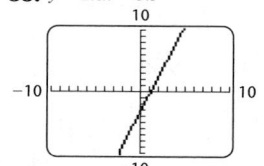

87. $y = x^2 + 4x + 1$
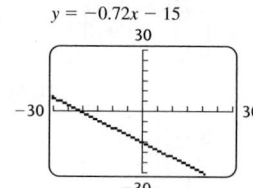

89. $56.62; 16.2 gal

Technology Connection, p. 171

1. $y = -0.72x - 15$
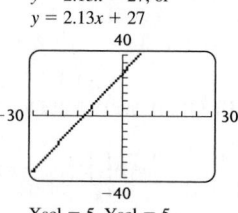
Xscl = 5, Yscl = 5

2. $y - 2.13x = 27$, or
$y = 2.13x + 27$

Xscl = 5, Yscl = 5

3. $5x + 6y = 84$, or
$y = -\frac{5}{6}x + 14$

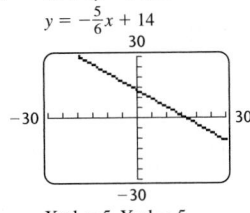

Xscl = 5, Yscl = 5

4. $2x - 7y = 150$, or
$y = \frac{2}{7}x - \frac{150}{7}$

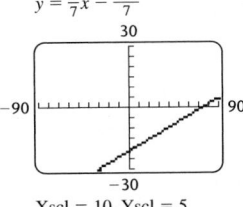

Xscl = 10, Yscl = 5

5. $19x - 17y = 200$, or
$y = \frac{19}{17}x - \frac{200}{17}$

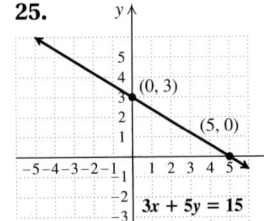

6. $6x + 5y = 159$, or
$y = -\frac{6}{5}x + \frac{159}{5}$

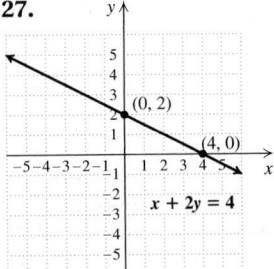

Xscl = 5, Yscl = 5

Exercise Set 3.3, pp. 173–175

1. (f)   **2.** (e)   **3.** (d)   **4.** (c)   **5.** (b)   **6.** (a)
7. (a) $(0, 5)$; **(b)** $(2, 0)$ **9. (a)** $(0, -4)$; **(b)** $(3, 0)$
11. (a) $(0, -2)$; **(b)** $(-3, 0), (3, 0)$ **13. (a)** $(0, 0)$;
(b) $(-2, 0), (0, 0), (5, 0)$ **15. (a)** $(0, 3)$; **(b)** $(5, 0)$
17. (a) $(0, -18)$; **(b)** $(4, 0)$ **19. (a)** $(0, 16)$; **(b)** $(-20, 0)$
21. (a) None; **(b)** $(12, 0)$ **23. (a)** $(0, -9)$; **(b)** none

25.

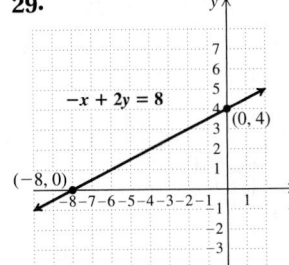

27.

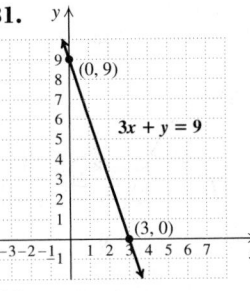

29.

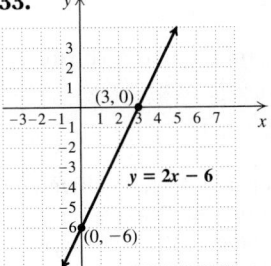

31.

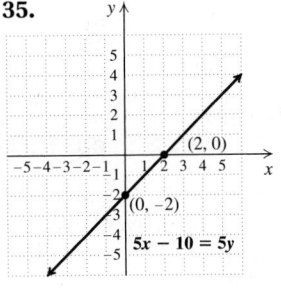

33.

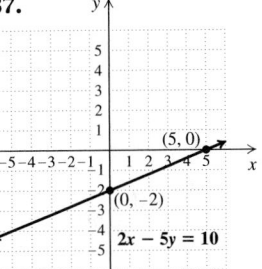

35.

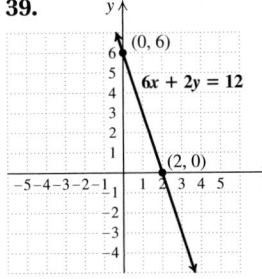

37.

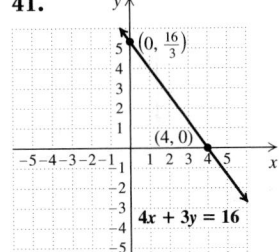

39.

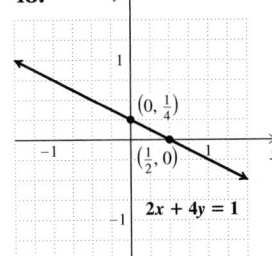

41.

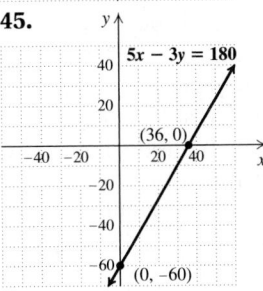

43.

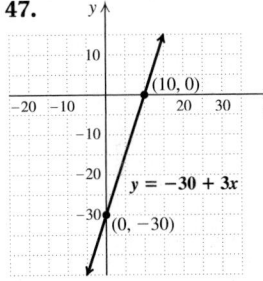

45.

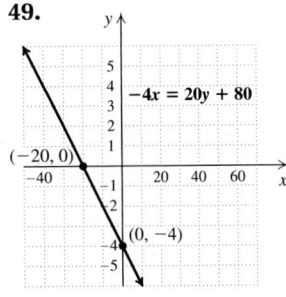

47.

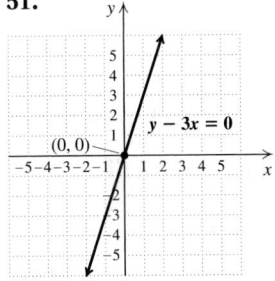

49.

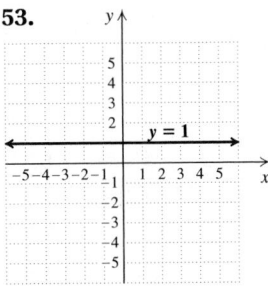

51.

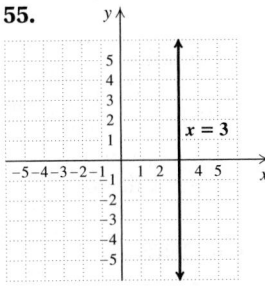

53.

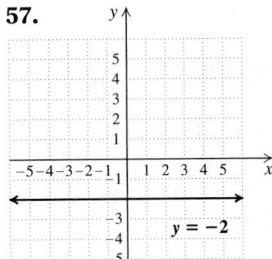

55.

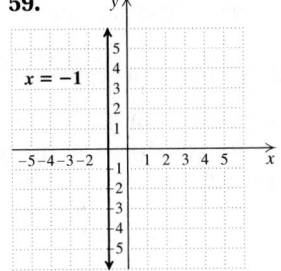

57.

59.

61.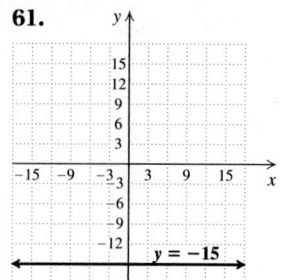

63.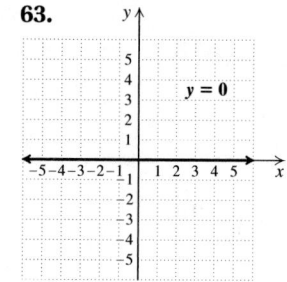

65.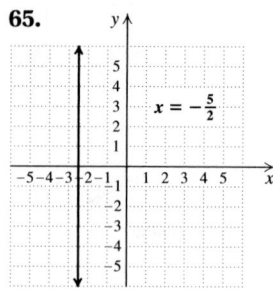

67.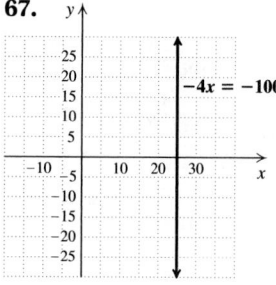

69.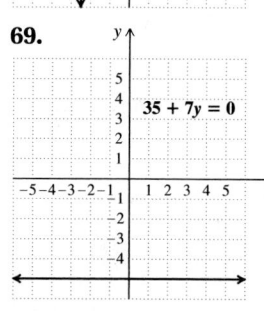

71. $y = -1$ **73.** $x = 4$
75. $x = 0$ **77.** 🖋
79. $d - 7$
80. $w + 5$, or $5 + w$
81. Let n represent the number; $7 + 4n$
82. Let n represent the number; $3n$
83. Let x and y represent the numbers; $2(x + y)$
84. Let a and b represent the numbers; $\frac{1}{2}(a + b)$
85. 🖋   **87.** $y = 0$ **89.** $x = -2$ **91.** $(-3, 4)$
93. $-5x + 3y = 15$, or $y = \frac{5}{3}x + 5$ **95.** -24
97. $\left(\dfrac{C - D}{A}, 0\right)$ **99.** $\left(0, -\frac{80}{7}\right)$, or $(0, -11.\overline{428571})$; $(40, 0)$
101. $(0, -9)$; $(45, 0)$ **103.** $\left(0, \frac{1}{25}\right)$, or $(0, 0.04)$; $\left(\frac{1}{50}, 0\right)$, or $(0.02, 0)$

Exercise Set 3.4, pp. 179–184

1. Miles per hour, or $\dfrac{\text{miles}}{\text{hour}}$

2. Hours per chapter, or $\dfrac{\text{hours}}{\text{chapter}}$

3. Dollars per mile, or $\dfrac{\text{dollars}}{\text{mile}}$

4. Petunias per foot, or $\dfrac{\text{petunias}}{\text{foot}}$

5. Minutes per errand, or $\dfrac{\text{minutes}}{\text{errand}}$

6. Cups of flour per cake, or $\dfrac{\text{cups of flour}}{\text{cake}}$

7. (a) 30 mpg; **(b)** \$39.33/day; **(c)** 130 mi/day; **(d)** 30¢/mi
9. (a) 7 mph; **(b)** \$7.50/hr; **(c)** \$1.07/mi
11. (a) \$22/hr; **(b)** 20.6 pages/hr; **(c)** \$1.07/page
13. \$568.4 billion/yr **15. (a)** 14.5 floors/min;
(b) 4.14 sec/floor   **17. (a)** 23.42 ft/min; **(b)** 0.04 min/ft

19.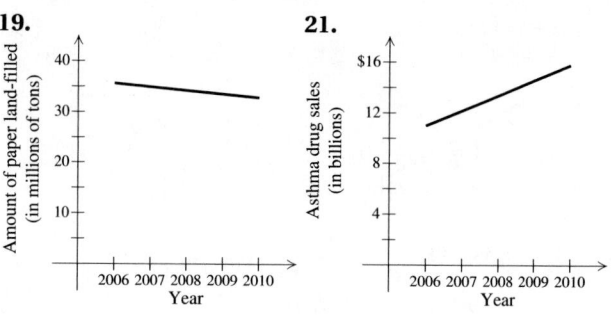

21.

23.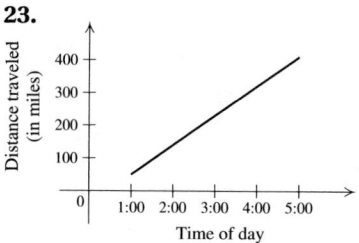

25.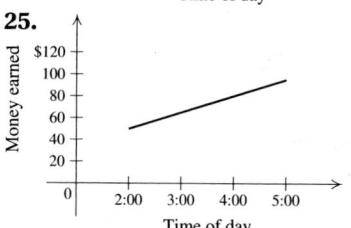

27.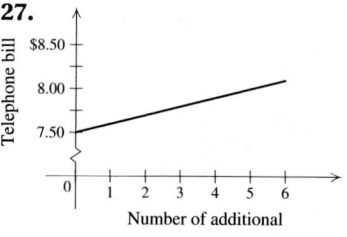

29. 20 calls/hr   **31.** 75 mi/hr   **33.** 12¢/min
35. -2000 people/yr   **37.** 0.02 gal/mi   **39.** (e)   **41.** (d)
43. (b)   **45.** 🖋   **47.** 5   **48.** -6 **49.** -1 **50.** $-\frac{4}{3}$
51. $-\frac{4}{3}$ **52.** 1   **53.** 0   **54.** Undefined   **55.** 🖋
57.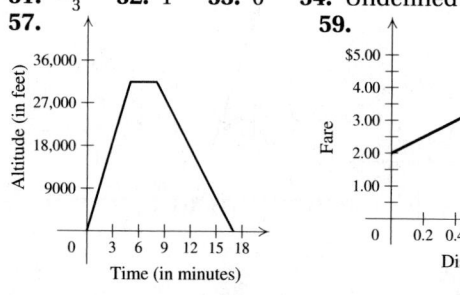

59.

61. 0.45 min/mi   **63.** About 41.6 min   **65.** 3.6 bu/hr

Exercise Set 3.5, pp. 190–196

1. Positive   **2.** Negative   **3.** Negative   **4.** Positive
5. Positive   **6.** Negative   **7.** Zero   **8.** Positive
9. Negative   **10.** Zero   **11.** \$60/blog   **13.** $-\$6$/month
15. 1 point/\$1000 income   **17.** About $-2.1°$/min   **19.** $\frac{4}{3}$
21. $\frac{3}{2}$ **23.** 2   **25.** -1 **27.** 0   **29.** $-\frac{1}{3}$

31. Undefined   **33.** $-\frac{3}{4}$ **35.** $\frac{1}{4}$ **37.** 0   **39.** $\frac{5}{4}$
41. $-\frac{4}{5}$ **43.** $\frac{2}{3}$ **45.** -1 **47.** $-\frac{1}{2}$ **49.** 0
51. 1   **53.** Undefined   **55.** 0   **57.** Undefined
59. Undefined   **61.** 0   **63.** 15%   **65.** 35%
67. $\frac{29}{98}$, or about 30%   **69.** About 5.1%; yes   **71.**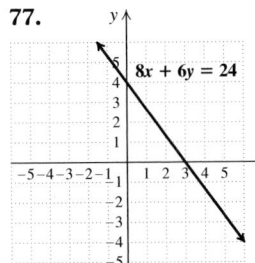
73. $y = \dfrac{c - ax}{b}$ **74.** $r = \dfrac{p + mn}{x}$ **75.** $y = \dfrac{ax - c}{b}$

76. $t = \dfrac{q - rs}{n}$

77. 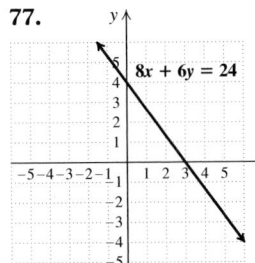   **78.**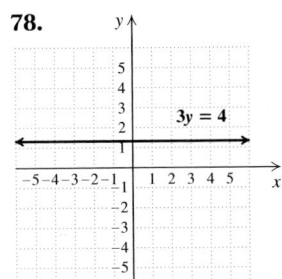

79. 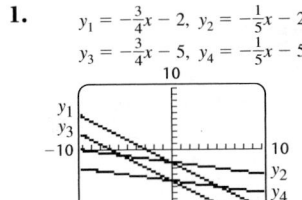   **81.** 0.364, or 36.4%   **83.** $\{m \mid m \geq \frac{5}{2}\}$ **85.** $\frac{1}{2}$

Technology Connection, p. 201

1. $y_1 = -\frac{3}{4}x - 2,\ y_2 = -\frac{1}{5}x - 2,$
$y_3 = -\frac{3}{4}x - 5,\ y_4 = -\frac{1}{5}x - 5$

Exercise Set 3.6, pp. 203–205

1. (f)   **2.** (b)   **3.** (d)   **4.** (c)   **5.** (e)   **6.** (a)
7. 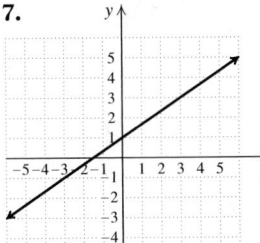   **9.**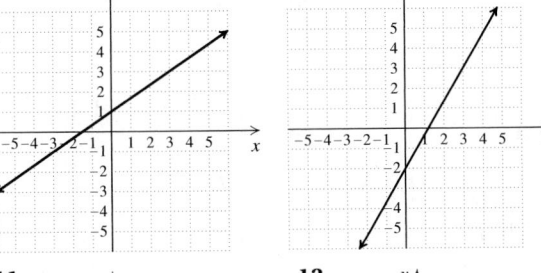

11. 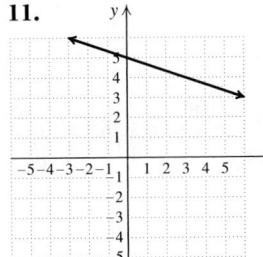   **13.**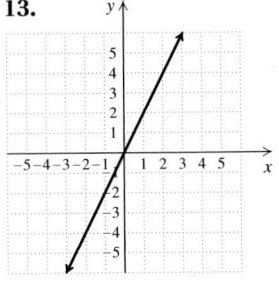

15. 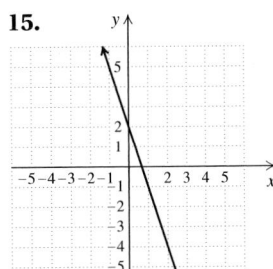   **17.**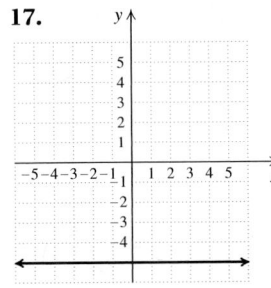

19. $-\frac{2}{7}$; $(0, 5)$ **21.** $\frac{1}{3}$; $(0, 7)$ **23.** $\frac{9}{5}$; $(0, -4)$ **25.** 3; $(0, 7)$
27. -2; $(0, 4)$ **29.** 0; $(0, 3)$ **31.** $\frac{2}{5}$; $\left(0, \frac{8}{5}\right)$ **33.** $\frac{9}{8}$; $(0, 0)$
35. $y = 5x + 7$ **37.** $y = \frac{7}{8}x - 1$ **39.** $y = -\frac{5}{3}x - 8$
41. $y = \frac{1}{3}$ **43.** $y = \frac{3}{2}x + 17$, where y is the number of gallons per person and x is the number of years since 2000   **45.** $y = \frac{2}{5}x + 15$, where y is the number of jobs, in millions, and x is the number of years since 2000
47. 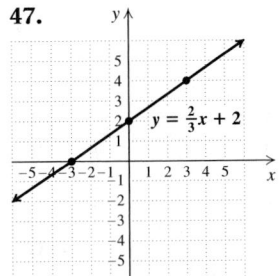   **49.**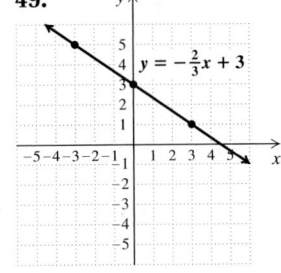

51. 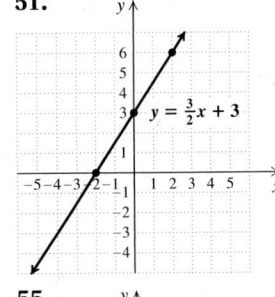   **53.**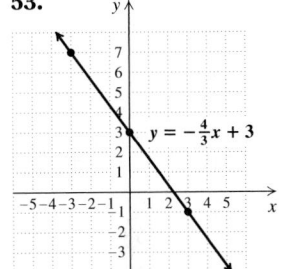

55. 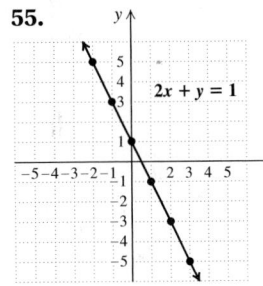   **57.**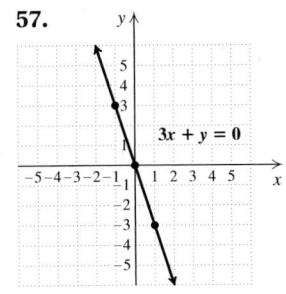

59. 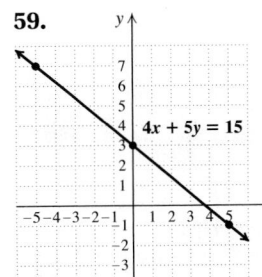   **61.**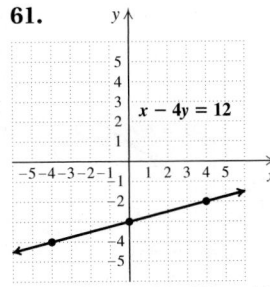

63. Yes   **65.** No   **67.** Yes   **69.** Yes   **71.** No   **73.** Yes

75. $y = 5x + 11$ **77.** $y = \frac{1}{2}x$ **79.** $y = x + 3$
81. $y = x - 4$ **83.** 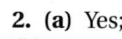   **85.** $y = m(x - h) + k$
86. $y = -2(x + 4) + 9$ **87.** -7 **88.** 13 **89.** -9
90. -11 **91.** ☞   **93.** When $x = 0, y = b$, so $(0, b)$ is
on the line. When $x = 1, y = m + b$, so $(1, m + b)$ is on the
line. Then

$$\text{slope} = \frac{(m + b) - b}{1 - 0} = m.$$

95. $y = \frac{1}{3}x + 3$ **97.** $y = -\frac{5}{3}x + 3$ **99.** $y = -\frac{2}{3}x$

Connecting the Concepts, p. 206

1. (a) Yes;
(b)

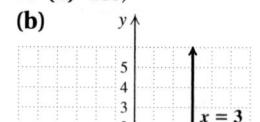

2. (a) Yes;
(b)

3. (a) Yes;
(b)

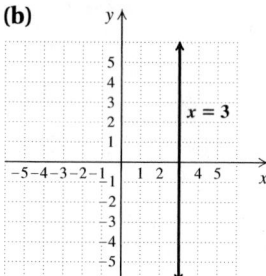

4. (a) Yes;
(b)

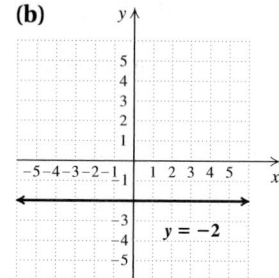

5. (a) Yes;
(b)

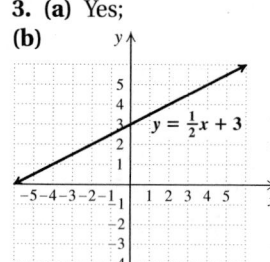

6. (a) Yes;
(b)

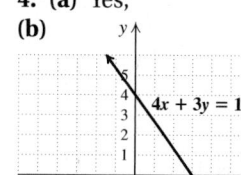

7. (a) No
8. (a) Yes;
(b)

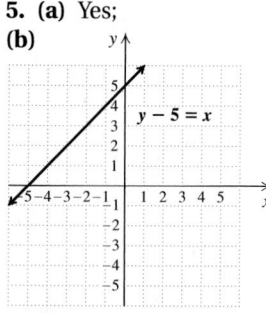

9. (a) Yes;
(b)

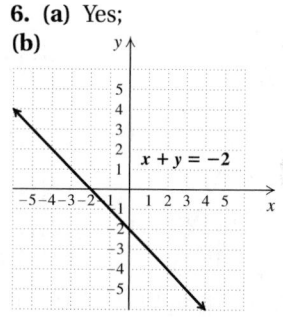

10. (a) No

11. (a) Yes;
(b)

12. (a) Yes;
(b)

13. (a) Yes;
(b)

14. (a) Yes;
(b)

15. (a) Yes; **(b)**   **16. (a)** No

17. (a) Yes;
(b)

18. (a) Yes;
(b)

19. (a) Yes;
(b)

20. (a) Yes;
(b)

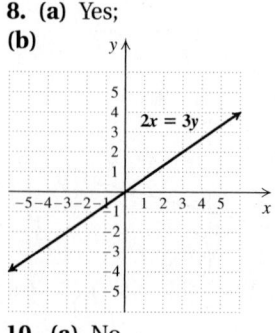

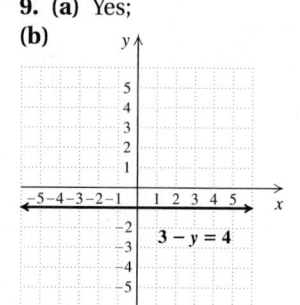

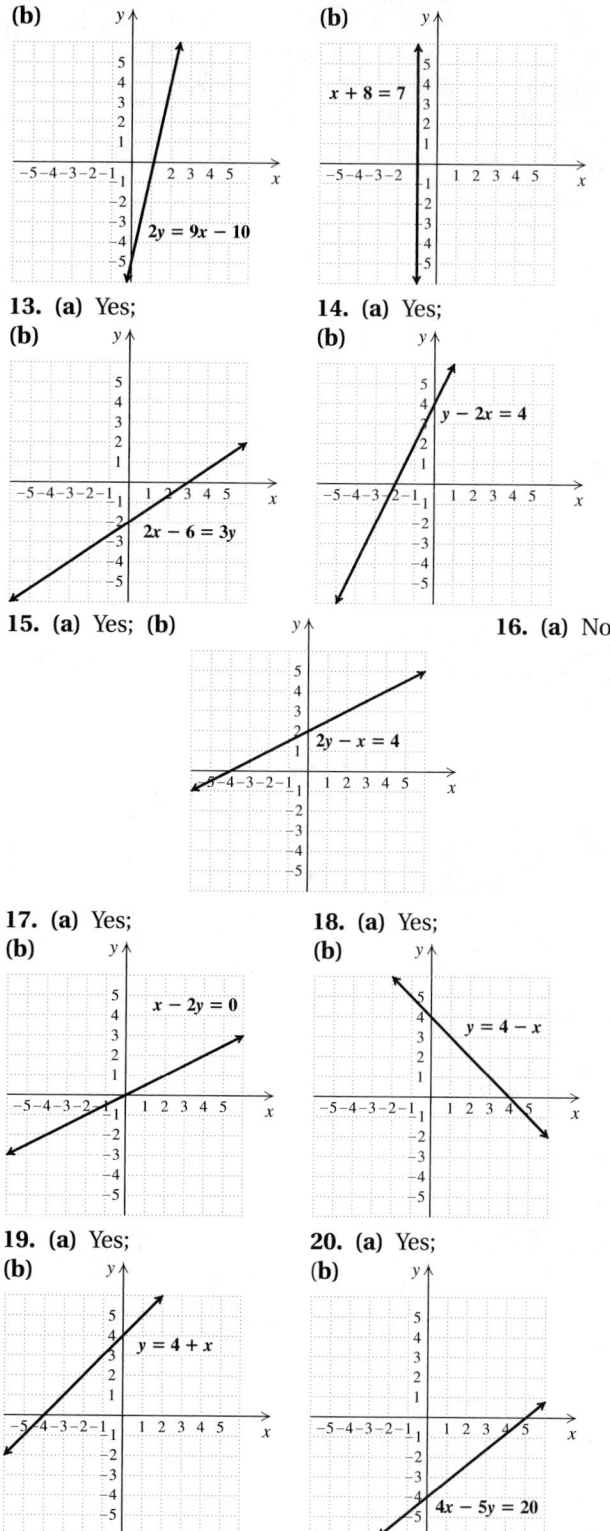

Technology Connection, p. 208

1. 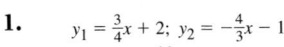 $y_1 = \frac{3}{4}x + 2;\ y_2 = -\frac{4}{3}x - 1$ **2.** $y_1 = -\frac{2}{5}x - 4;\ y_2 = \frac{5}{2}x + 3$

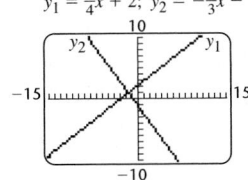 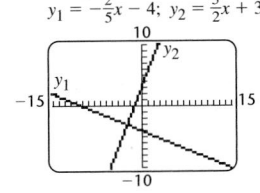

3. $y_1 = \frac{31}{40}x + 2;\ y_2 = -\frac{40}{30}x - 1$ No: $-\frac{40}{30} \neq -\frac{1}{\frac{31}{40}}$

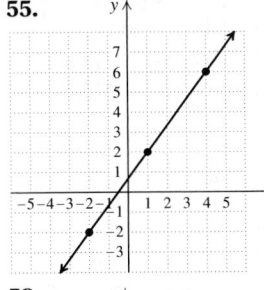

Although the lines appear to be perpendicular, they are not, because the product of their slopes is not -1:

$$\frac{31}{40}\left(-\frac{40}{30}\right) = -\frac{1240}{1200} \neq -1.$$

Visualizing for Success, p. 212

1. C **2.** G **3.** F **4.** B **5.** D **6.** A **7.** I
8. H **9.** J **10.** E

Exercise Set 3.7, pp. 213–216

1. (g) **2.** (b) **3.** (d) **4.** (h) **5.** (e) **6.** (a)
7. (f) **8.** (c) **9.** (c) **10.** (b) **11.** (d) **12.** (a)
13. $y - 6 = 3(x - 1)$ **15.** $y - 8 = \frac{3}{5}(x - 2)$
17. $y - 1 = -4(x - 3)$ **19.** $y - (-4) = \frac{3}{2}(x - 5)$
21. $y - 6 = -\frac{5}{4}(x - (-2))$
23. $y - (-1) = -2(x - (-4))$
25. $y - 8 = 1(x - (-2))$ **27.** $y = 4x - 7$
29. $y = \frac{7}{4}x - 9$ **31.** $y = -2x + 1$ **33.** $y = -4x - 9$
35. $y = \frac{2}{3}x + \frac{8}{3}$ **37.** $y = -\frac{5}{6}x + 4$ **39.** $y = \frac{1}{2}x + 4$
41. $y = 4x - 5$ **43.** $y = -\frac{2}{3}x - \frac{13}{3}$ **45.** $x = 5$
47. $y = -\frac{3}{2}x + \frac{11}{2}$ **49.** $y = x + 6$ **51.** $y = -\frac{1}{2}x + 6$
53. $x = -3$
55.   **57.**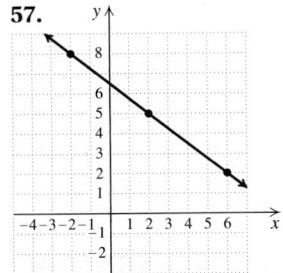

59. 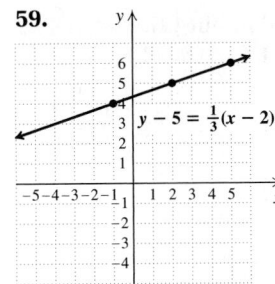 **61.**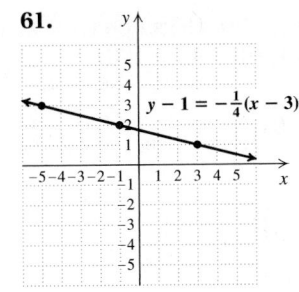

63. 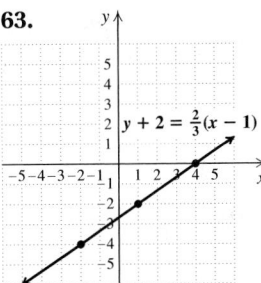 **65.**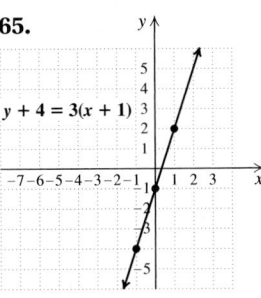

67. 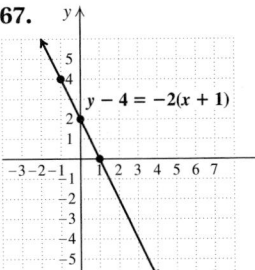 **69.**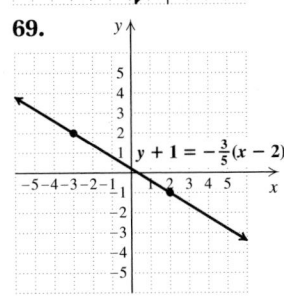

71. (a) 51.6 births per 1000 females; (b) 39.8 births per 1000 females **73.** (a) About 13.1%; (b) 4%
75. (a) 16.47 million students; (b) 18.95 million students
77. (a) 33.65 million residents; (b) about 38.6 million residents
79. $y = -x + 5$ **81.** $y = \frac{2}{3}x + 3$ **83.** $y = \frac{2}{5}x - 2$
85. $y = 2x + 7$ **87.** ✍ **89.** -125 **90.** 64 **91.** -64
92. 8 **93.** 28 **94.** -4 **95.** ✍
97.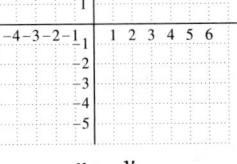

99. $y = 2x - 9$
101. $y = -\frac{4}{3}x + \frac{23}{3}$
103. $y = -4x + 7$
105. $y = \frac{10}{3}x + \frac{25}{3}$
107. $(2, 0), (0, 5)$

109. $-\frac{x}{4} + \frac{y}{3} = 1;\ (-4, 0), (0, 3)$ **111.** ✍

Connecting the Concepts, pp. 216–217

1. Slope–intercept form **2.** Standard form
3. None of these **4.** Standard form
5. Point–slope form **6.** None of these
7. $2x - 5y = 10$ **8.** $x - y = 2$
9. $-2x + y = 7$, or $2x - y = -7$ **10.** $\frac{1}{2}x + y = 3$
11. $3x - y = -23$, or $-3x + y = 23$ **12.** $x + 0y = 18$
13. $y = \frac{2}{7}x - \frac{8}{7}$ **14.** $y = -x - 8$ **15.** $y = 8x - 3$
16. $y = -\frac{3}{5}x + 3$ **17.** $y = -\frac{8}{9}x + \frac{5}{9}$ **18.** $y = -x$
19. $y = -\frac{1}{2}$ **20.** $y = \frac{1}{2}x$

Review Exercises: Chapter 3, pp. 220–222

1. True **2.** True **3.** False **4.** False **5.** True
6. True **7.** True **8.** False **9.** True **10.** True
11. About 1.3 billion searches **12.** About 1137 searches

13.–15.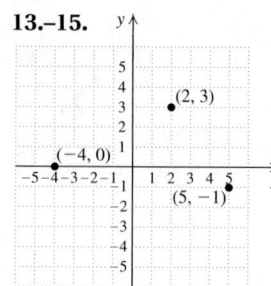

16. III   **17.** IV   **18.** II
19. $(-5, -1)$ **20.** $(-2, 5)$
21. $(3, 0)$

22.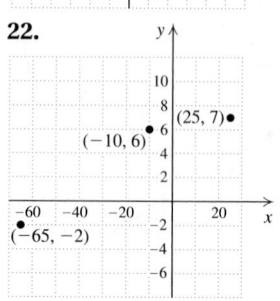

23. (a) No; **(b)** yes

24.

$$\frac{2x - y = 3}{2 \cdot 0 - (-3) \mid 3}$$
$$0 + 3$$
$$3 \overset{?}{=} 3 \quad \text{True}$$

$$\frac{2x - y = 3}{2 \cdot 2 - 1 \mid 3}$$
$$4 - 1$$
$$3 \overset{?}{=} 3 \quad \text{True}$$

$(-1, -5)$; answers may vary

25.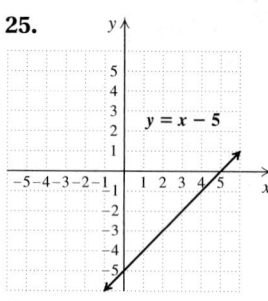

26.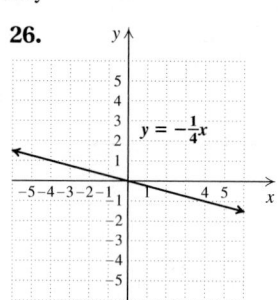

27.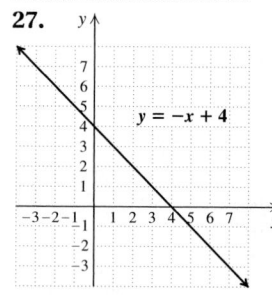

28.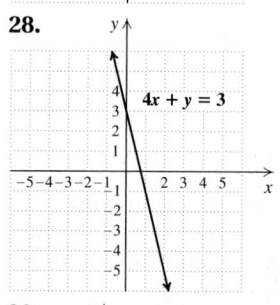

29.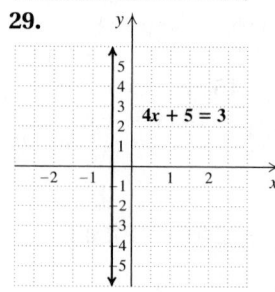

30.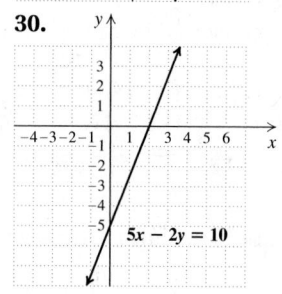

31. About 7 million viewers

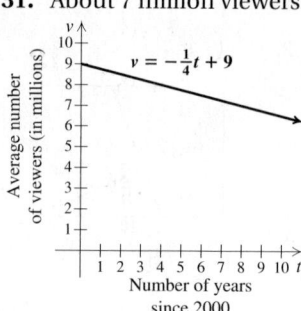

32. (a) $\frac{2}{15}$ mi/min;
(b) 7.5 min/mi
33. 12 mpg   **34.** 0
35. $\frac{7}{3}$ **36.** $-\frac{3}{7}$ **37.** $-\frac{6}{5}$
38. 0   **39.** Undefined
40. 2   **41.** $8.\overline{3}\%$
42. $(16, 0), (0, -10)$
43. $-\frac{3}{5}; (0, 9)$
44. Perpendicular
45. Parallel
46. $y = \frac{3}{8}x + 7$ **47.** $y - 9 = -\frac{1}{3}(x - (-2))$
48. (a) \$1725; **(b)** \$2701   **49.** $y = 5x - 25$
50. $y = -\frac{5}{3}x - \frac{5}{3}$

51.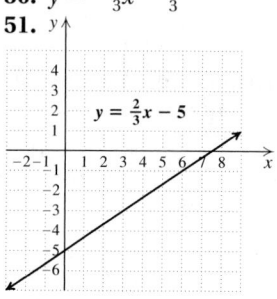

52.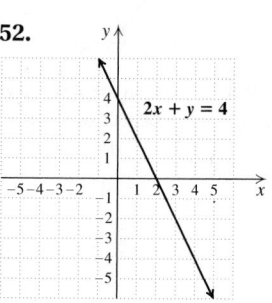

53.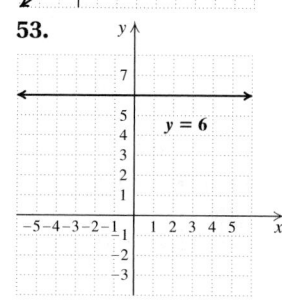

54.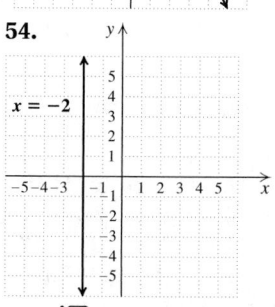

55.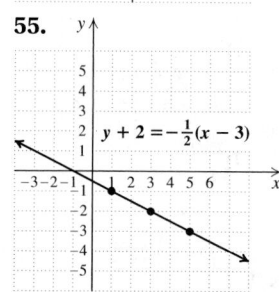

56. ✍ Two perpendicular lines share the same y-intercept if their point of intersection is on the y-axis.
57. ✍ The graph of a vertical line has only an x-intercept. The graph of a horizontal line has only a y-intercept. The graph of a nonvertical, non-horizontal line will have only one intercept if it passes through the origin: $(0, 0)$ is both the x-intercept and the y-intercept.   **58.** -1
59. 19   **60.** Area: 45 sq units; perimeter: 28 units
61. $(0, 4), (1, 3), (-1, 3)$; answers may vary

Test: Chapter 3, pp. 222–223

1. [3.1] About 95 students   **2.** [3.1] About 137 students
3. [3.1] III   **4.** [3.1] II   **5.** [3.1] $(3, 4)$ **6.** [3.1] $(0, -4)$
7. [3.1] $(-5, 2)$

8. [3.2]

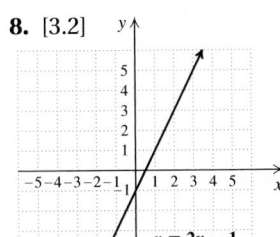

$y = 2x - 1$

9. [3.3]

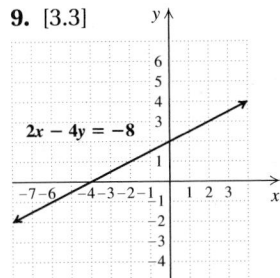

$2x - 4y = -8$

10. [3.3]

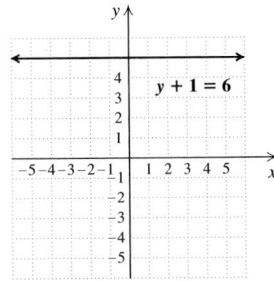

$y + 1 = 6$

11. [3.2]

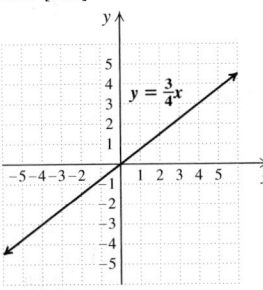

$y = \frac{3}{4}x$

12. [3.2]

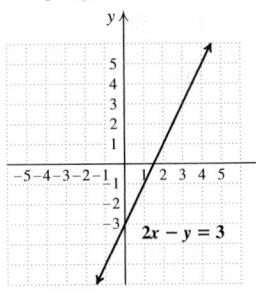

$2x - y = 3$

13. [3.3]

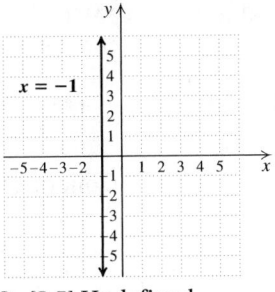

$x = -1$

14. [3.5] 5   **15.** [3.5] $-\frac{9}{4}$ **16.** [3.5] Undefined
17. [3.4] $\frac{1}{3}$ km/min   **18.** [3.5] 31.5%
19. [3.3] $(6, 0)$, $(0, -30)$ **20.** [3.6] 8; $(0, 10)$
21. [3.6] $y = -\frac{1}{3}x - 11$ **22.** [3.6] Parallel
23. [3.6] Perpendicular   **24.** [3.7] $y = -\frac{5}{2}x - \frac{11}{2}$
25. [3.7] **(a)**

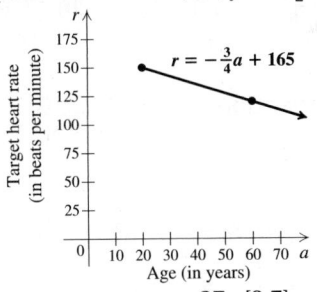

$r = -\frac{3}{4}a + 165$

(b) 138 beats per minute

26. [3.6]

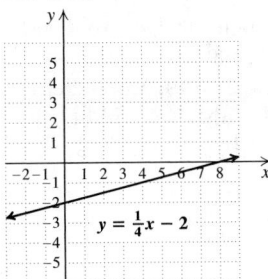

$y = \frac{1}{4}x - 2$

27. [3.7]

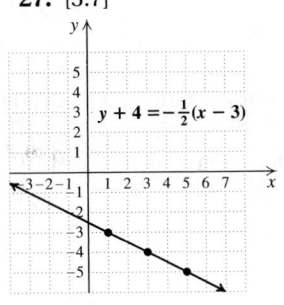

$y + 4 = -\frac{1}{2}(x - 3)$

28. [3.6] $y = \frac{2}{5}x + 9$ **29.** [3.1] Area: 25 sq units; perimeter: 20 units   **30.** [3.2], [3.7] $(0, 12)$, $(-3, 15)$, $(5, 7)$

Cumulative Review: Chapters 1–3, pp. 224–225

1. 7   **2.** $12a - 6b + 18$ **3.** $4(2x - y + 1)$ **4.** $2 \cdot 3^3$
5. -0.15 **6.** 37   **7.** $\frac{1}{10}$ **8.** -10 **9.** 0.367   **10.** $\frac{11}{60}$
11. 2.6   **12.** 7.28   **13.** $-\frac{5}{12}$ **14.** -3 **15.** 27
16. $-2y - 7$ **17.** $5x + 11$ **18.** -2.6 **19.** -27
20. 16   **21.** -6 **22.** 2   **23.** $\frac{7}{9}$ **24.** -17 **25.** 2
26. $\{x | x < 16\}$, or $(-\infty, 16)$ **27.** $\left\{x | x \le -\frac{11}{8}\right\}$, or
$\left(-\infty, -\frac{11}{8}\right]$ **28.** $h = \dfrac{A - \pi r^2}{2\pi r}$ **29.** IV

30.

$-1 < x \le 2$

31.

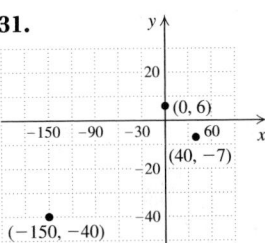

$(0, 6)$
$(40, -7)$
$(-150, -40)$

32.

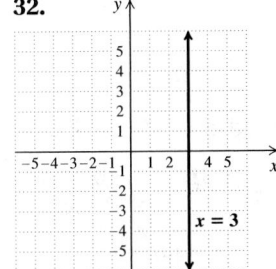

$x = 3$

33.

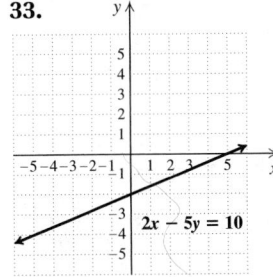

$2x - 5y = 10$

34.

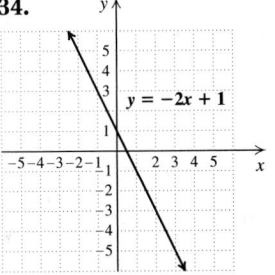

$y = -2x + 1$

35.

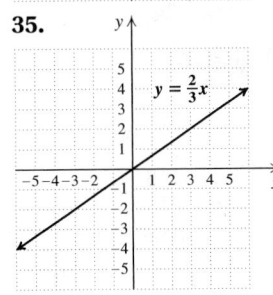

$y = \frac{2}{3}x$

36.

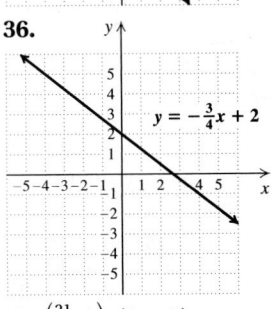

$y = -\frac{3}{4}x + 2$

37.

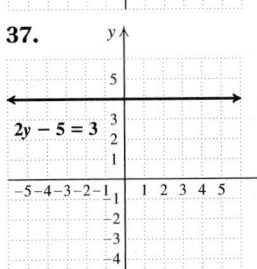

$2y - 5 = 3$

38. $\left(\frac{21}{2}, 0\right)$, $(0, -3)$
39. $\left(-\frac{5}{4}, 0\right)$, $(0, 5)$
40. 3; $(0, -2)$ **41.** $-\frac{1}{3}$
42. $y = \frac{2}{7}x - 4$
43. $y - 4 = -\frac{3}{8}(x - (-6))$
44. $y = -\frac{3}{8}x + \frac{7}{4}$
45. $y = 2x + 1$
46. 0.5 million bicycles per year

47. (a)

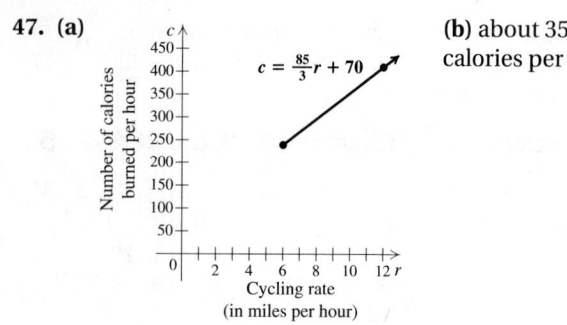

$c = \frac{85}{3}r + 70$

Number of calories burned per hour (vertical axis, 0 to 450 by 50)
Cycling rate (in miles per hour) (horizontal axis, 0 to 12 r)

(b) about 353 calories per hour

48. \$210 billion   **49.** \$54,533   **50.** 21 million Americans
51. \$120   **52.** 50 m, 53 m, 40 m   **53.** 4 hr
54. \$25,000   **55.** $-4, 4$ **56.** 2   **57.** -5 **58.** 3
59. No solution   **60.** $Q = \dfrac{2 - pm}{p}$
61. $y = -\frac{7}{3}x + 7; y = -\frac{7}{3}x - 7; y = \frac{7}{3}x - 7; y = \frac{7}{3}x + 7$

CHAPTER 4

Exercise Set 4.1, pp. 234–235

1. (e)   **2.** (f)   **3.** (b)   **4.** (h)   **5.** (g)   **6.** (a)
7. (c)   **8.** (d)   **9.** Base: $2x$; exponent: 5
11. Base: x; exponent: 3   **13.** Base: $\dfrac{4}{y}$; exponent: 7
15. d^{13} **17.** a^7 **19.** 6^{15} **21.** $(3y)^{12}$ **23.** $(8n)^{10}$
25. a^5b^9 **27.** $(x+3)^{13}$ **29.** r^{10} **31.** m^4n^9 **33.** 7^3
35. t^7 **37.** $5a$ **39.** 1   **41.** $(r+s)^8$ **43.** $\frac{4}{5}d^7$
45. $4a^7b^6$ **47.** $x^{12}y^7$ **49.** 1   **51.** 5   **53.** 2   **55.** -4
57. x^{33} **59.** 5^{32} **61.** t^{80} **63.** $100x^2$ **65.** $-8a^3$
67. $25n^{14}$ **69.** $a^{14}b^7$ **71.** $r^{17}t^{11}$ **73.** $24x^{19}$
75. $\dfrac{x^3}{125}$ **77.** $\dfrac{49}{36n^2}$ **79.** $\dfrac{a^{18}}{b^{48}}$ **81.** $\dfrac{x^8y^4}{z^{12}}$ **83.** $\dfrac{a^{12}}{16b^{20}}$
85. $-\dfrac{125x^{21}y^3}{8z^{12}}$ **87.** 1   **89.** ✎   **91.** -24 **92.** -15
93. -11 **94.** 16   **95.** -14 **96.** 4   **97.** ✎
99. ✎   **101.** Let $x = 1$; then $3x^2 = 3$, but $(3x)^2 = 9$.
103. Let $t = -1$; then $\dfrac{t^6}{t^2} = 1$, but $t^3 = -1$.   **105.** y^{6x}
107. x^t **109.** 13   **111.** $<$ **113.** $<$ **115.** $>$
117. 4,000,000; 4,194,304; 194,304   **119.** 2,000,000,000;
2,147,483,648; 147,483,648   **121.** 1,536,000 bytes, or
approximately 1,500,000 bytes

Technology Connection, p. 241

1. 1.71×10^{17} **2.** $5.\overline{370} \times 10^{-15}$ **3.** 3.68×10^{16}

Exercise Set 4.2, pp. 242–244

1. (c)   **2.** (d)   **3.** (a)   **4.** (b)   **5.** $\dfrac{1}{2^3} = \dfrac{1}{8}$
7. $\dfrac{1}{(-2)^6} = \dfrac{1}{64}$ **9.** $\dfrac{1}{t^9}$ **11.** $\dfrac{x}{y^2}$ **13.** $\dfrac{t}{r^5}$ **15.** a^8

17. $\dfrac{1}{7}$ **19.** $\left(\dfrac{5}{3}\right)^2 = \dfrac{25}{9}$ **21.** $\left(\dfrac{2}{x}\right)^5 = \dfrac{32}{x^5}$ **23.** $\left(\dfrac{t}{s}\right)^7 = \dfrac{t^7}{s^7}$
25. 9^{-2} **27.** y^{-3} **29.** 5^{-1} **31.** t^{-1} **33.** 2^3, or 8
35. $\dfrac{1}{x^{12}}$ **37.** $\dfrac{1}{t^2}$ **39.** $\dfrac{1}{n^{15}}$ **41.** t^{18} **43.** $\dfrac{1}{t^{12}}$
45. $\dfrac{1}{m^7n^7}$ **47.** $\dfrac{9}{x^8}$ **49.** $\dfrac{25t^6}{r^8}$ **51.** t^{14} **53.** $\dfrac{1}{y^4}$
55. $5y^3$ **57.** $2x^5$ **59.** $\dfrac{-3b^9}{2a^7}$ **61.** 1   **63.** $\dfrac{8y^7z}{x^3}$
65. $3s^2t^4u^4$ **67.** $\dfrac{1}{x^{12}y^{15}}$ **69.** $\dfrac{m^{10}n^6}{9}$ **71.** $\dfrac{b^5c^4}{a^8}$
73. $\dfrac{9}{a^8}$ **75.** $\dfrac{n^{12}}{m^3}$ **77.** $\dfrac{27b^{12}}{8a^6}$ **79.** 1   **81.** $\dfrac{2b^3}{a^4}$
83. $\dfrac{5y^4z^{10}}{4x^{11}}$ **85.** 4920   **87.** 0.00892   **89.** 904,000,000
91. 0.000003497   **93.** 3.6×10^7 **95.** 5.83×10^{-3}
97. 7.8×10^{10} **99.** 5.27×10^{-7} **101.** 1.032×10^{-6}
103. 6×10^{13} **105.** 2.47×10^8 **107.** 3.915×10^{-16}
109. 2.5×10^{13} **111.** 5.0×10^{-6} **113.** 3×10^{-21}
115. ✎   **117.** $8x$ **118.** $-3a - 6b$ **119.** $-2x - 7$
120. $-4t - r - 5$ **121.** 1004   **122.** 9   **123.** ✎
125. 8×10^5 **127.** 2^{-12} **129.** 5   **131.** 5^6
133. $\frac{1}{3} + \frac{1}{4} = \frac{7}{12}$ **135.** 9   **137.** $6.304347826 \times 10^{25}$
139. $1.19140625 \times 10^{-15}$ **141.** 3×10^8 mi
143. \$2.31 \times 10^8$ **145.** 2.277×10^{10} min

Connecting the Concepts, p. 245

1. x^{14} **2.** $\dfrac{1}{x^{14}}$ **3.** $\dfrac{1}{x^{14}}$ **4.** x^{14} **5.** x^{40} **6.** x^{40}
7. c^8 **8.** $\dfrac{1}{c^8}$ **9.** $16x^{12}y^4$ **10.** $\dfrac{1}{16x^{12}y^4}$ **11.** 1
12. a^5 **13.** $\dfrac{a^{15}}{b^{20}}$ **14.** $\dfrac{b^{20}}{a^{15}}$ **15.** $\dfrac{5x^3}{2y^4}$ **16.** $\dfrac{6a^2}{7b^5}$
17. $\dfrac{7t^6}{xp^5}$ **18.** $\dfrac{16a^2}{9b^6}$ **19.** $18p^4q^{14}$ **20.** $\dfrac{9x^3}{2y^5}$

Technology Connection, pp. 249–250

1. 9.8

Exercise Set 4.3, pp. 250–254

1. (b)   **2.** (f)   **3.** (h)   **4.** (d)   **5.** (g)   **6.** (e)
7. (a)   **8.** (c)   **9.** $8x^3, -11x^2, 6x, 1$
11. $-t^6, -3t^3, 9t, -4$ **13.** Coefficients: 8, 2; degrees: 4, 1
15. Coefficients: 9, -3, 4; degrees: 2, 1, 0
17. Coefficients: 6, 9, 1; degrees: 5, 1, 3   **19.** Coefficients:
1, -1, 4, -3; degrees: 4, 3, 1, 0   **21. (a)** 1, 3, 4; **(b)** $8t^4$, 8;
(c) 4   **23. (a)** 2, 0, 4; **(b)** $2a^4$, 2; **(c)** 4
25. (a) 0, 2, 1, 5; **(b)** $-x^5$, -1; **(c)** 5

27.

Term	Coefficient	Degree of the Term	Degree of the Polynomial
$8x^5$	8	5	
$-\frac{1}{2}x^4$	$-\frac{1}{2}$	4	
$-4x^3$	-4	3	5
$7x^2$	7	2	
6	6	0	

29. Trinomial   **31.** Polynomial with no special name
33. Binomial   **35.** Monomial   **37.** $11n^2 + n$ **39.** $4a^4$
41. $7x^3 + x^2 - 6x$ **43.** $11b^3 + b^2 - b$ **45.** $-x^4 - x^3$
47. $\frac{1}{15}x^4 + 10$ **49.** $-1.1a^2 + 5.3a - 7.5$ **51.** $-3; 21$
53. $16; 34$ **55.** $-38; 148$ **57.** $159; 165$ **59.** $-39; 21$
61. \$1.93 billion   **63.** 1112 ft   **65.** 62.8 cm
67. 153.86 m^2 **69.** About 75 donations
71. About 9 words   **73.** About 6   **75.** About 16;
about 19   **77.** ⬓   **79.** $x - 3$ **80.** $-2x - 6$
81. $6a - 3$ **82.** $-t - 1$ **83.** $-t^4 + 17t$
84. $0.4a^2 - a + 11$ **85.** ⬓
87. $2x^5 + 4x^4 + 6x^3 + 8$; answers may vary   **89.** \$2510
91. $3x^6$ **93.** 3 and 8   **95.** 85.0
97.

t	$-t^2 + 10t - 18$
3	3
4	6
5	7
6	6
7	3

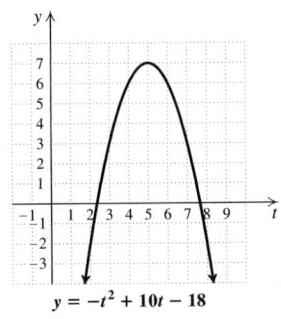

$y = -t^2 + 10t - 18$

Technology Connection, p. 258

1. In each case, let $y_1 =$ the expression before the addition or subtraction has been performed, $y_2 =$ the simplified sum or difference, and $y_3 = y_2 - y_1$; and note that the graph of y_3 coincides with the x-axis. That is, $y_3 = 0$.

Exercise Set 4.4, pp. 259–261

1. x^2 **2.** -6 **3.** $-$ **4.** $+$ **5.** $4x + 9$ **7.** $-6t + 8$
9. $-3x^2 + 6x - 2$ **11.** $9t^2 + t + 3$
13. $8m^3 - 3m - 7$ **15.** $7 + 13a - a^2 + 7a^3$
17. $2x^6 + 9x^4 + 2x^3 - 4x^2 + 5x$
19. $x^4 + \frac{1}{4}x^3 - \frac{3}{4}x^2 - \frac{5}{6}x + 3$
21. $4.2t^3 + 3.5t^2 - 6.4t - 1.8$ **23.** $-4x^3 + 4x^2 + 6x$
25. $1.3x^4 + 0.35x^3 + 9.53x^2 + 2x + 0.96$
27. $-(-3t^3 + 4t^2 - 7); 3t^3 - 4t^2 + 7$
29. $-(x^4 - 8x^3 + 6x); -x^4 + 8x^3 - 6x$
31. $-9x + 10$ **33.** $-3a^4 + 5a^2 - 1.2$

35. $4x^4 - 6x^2 - \frac{3}{4}x + 8$ **37.** $-2x - 7$
39. $-t^2 - 12t + 13$ **41.** $8a^3 + 8a^2 + a - 10$
43. $4.6x^3 + 9.2x^2 - 3.8x - 23$ **45.** 0
47. $1 + a + 12a^2 - 3a^3$ **49.** $\frac{9}{8}x^3 - \frac{1}{2}x$
51. $0.05t^3 - 0.07t^2 + 0.01t + 1$ **53.** $2x^2 + 6$
55. $-3x^4 - 8x^3 - 7x^2$ **57.** **(a)** $5x^2 + 4x$; **(b)** $145; 273$
59. $16y + 26$ **61.** $(r + 11)(r + 9); 9r + 99 + 11r + r^2$
63. $(x + 3)^2; x^2 + 3x + 9 + 3x$ **65.** $m^2 - 40$
67. $\pi r^2 - 49$ **69.** $(x^2 - 12)$ ft^2 **71.** $(z^2 - 36\pi)$ ft^2
73. $\left(144 - \frac{d^2}{4}\pi\right)$ m^2 **75.** ⬓   **77.** $2x^2 - 2x + 6$
78. $-15x^2 + 10x + 35$ **79.** x^8 **80.** y^7 **81.** $2n^3$
82. $-6n^{12}$ **83.** ⬓   **85.** $9t^2 - 20t + 11$
87. $-6x + 14$ **89.** $250.591x^3 + 2.812x$ **91.** $20w + 42$
93. $2x^2 + 20x$ **95.** $8x + 24$ **97.** ⬓

Technology Connection, p. 266

1. Let $y_1 = (5x^4 - 2x^2 + 3x)(x^2 + 2x)$ and $y_2 = 5x^6 + 10x^5 - 2x^4 - x^3 + 6x^2$. With the table set in AUTO mode, note that the values in the Y_1- and Y_2-columns match, regardless of how far we scroll up or down.
2. Use TRACE, a table, or a boldly drawn graph to confirm that y_3 is always 0.

Exercise Set 4.5, pp. 267–269

1. (c)   **2.** (d)   **3.** (d)   **4.** (a)   **5.** (c)   **6.** (b)
7. $21x^5$ **9.** $-x^7$ **11.** x^8 **13.** $36t^4$ **15.** $-0.12x^9$
17. $-\frac{1}{20}x^{12}$ **19.** $5n^3$ **21.** $-44x^{10}$ **23.** $72y^{10}$
25. $20x^2 + 5x$ **27.** $3a^2 - 27a$ **29.** $x^5 + x^2$
31. $-6n^3 + 24n^2 - 3n$ **33.** $-15t^3 - 30t^2$
35. $4a^9 - 8a^7 - \frac{5}{12}a^4$ **37.** $x^2 + 7x + 12$
39. $t^2 + 4t - 21$ **41.** $a^2 - 1.3a + 0.42$ **43.** $x^2 - 9$
45. $28 - 15x + 2x^2$ **47.** $t^2 + \frac{17}{6}t + 2$
49. $\frac{3}{16}a^2 + \frac{5}{4}a - 2$
51. 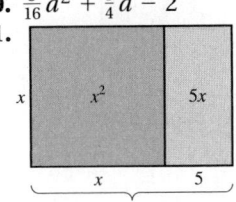   **53.**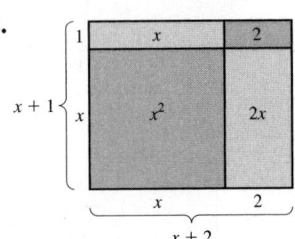

55. 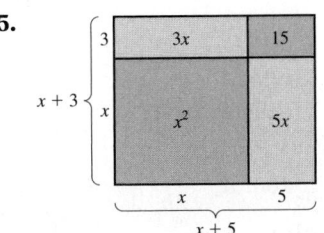   **57.** $x^3 + 2x + 3$

59. $2a^3 - a^2 - 11a + 10$
61. $3y^6 - 21y^4 + y^3 + 2y^2 - 7y - 14$
63. $33x^2 + 25x + 2$ **65.** $x^4 + 4x^3 - 3x^2 + 16x - 3$
67. $10t^4 + 3t^3 - 20t^2 + \frac{9}{2}t - 2$
69. $x^4 + 8x^3 + 12x^2 + 9x + 4$ **71.** ⬓   **73.** 0   **74.** 0
75. 8   **76.** 7   **77.** 32   **78.** 50   **79.** ⬓

81. $75y^2 - 45y$ **83.** 5   **85.** $V = (4x^3 - 48x^2 + 144x)$ in³;
$S = (-4x^2 + 144)$ in²   **87.** $(x^3 - 5x^2 + 8x - 4)$ cm³
89. 16 ft by 8 ft   **91.** 0   **93.** $x^3 + x^2 - 22x - 40$ **95.** 0

Visualizing for Success, p. 275

1. E, F   **2.** B, O   **3.** S, K   **4.** R, G   **5.** D, M   **6.** J, P
7. C, L   **8.** N, Q   **9.** A, H   **10.** I, T

Exercise Set 4.6, pp. 276–278

1. True   **2.** False   **3.** False   **4.** True
5. $x^3 + 3x^2 + 2x + 6$ **7.** $t^5 + 7t^4 - 2t - 14$
9. $y^2 - y - 6$ **11.** $9x^2 + 21x + 10$
13. $5x^2 + 17x - 12$ **15.** $15 - 13t + 2t^2$
17. $x^4 - 4x^2 - 21$ **19.** $p^2 - \frac{1}{16}$ **21.** $x^2 - 0.6x + 0.09$
23. $-3n^2 - 19n + 14$ **25.** $x^2 + 20x + 100$
27. $1 - 3t + 5t^2 - 15t^3$ **29.** $x^5 + 3x^3 - x^2 - 3$
31. $3x^6 - 2x^4 - 6x^2 + 4$ **33.** $4t^6 + 20t^3 + 25$
35. $8x^5 + 16x^3 + 5x^2 + 10$ **37.** $100x^4 - 9$
39. $x^2 - 64$ **41.** $4x^2 - 1$ **43.** $25m^4 - 16$
45. $81a^6 - 1$ **47.** $x^8 - 0.01$ **49.** $t^2 - \frac{9}{16}$
51. $x^2 + 6x + 9$ **53.** $49x^6 - 14x^3 + 1$
55. $a^2 - \frac{4}{5}a + \frac{4}{25}$ **57.** $t^8 + 6t^4 + 9$
59. $4 - 12x^4 + 9x^8$ **61.** $25 + 60t^2 + 36t^4$
63. $49x^2 - 4.2x + 0.09$ **65.** $14n^5 - 7n^3$
67. $a^3 - a^2 - 10a + 12$ **69.** $49 - 42x^4 + 9x^8$
71. $5x^3 + 30x^2 - 10x$ **73.** $q^{10} - 1$
75. $15t^5 - 3t^4 + 3t^3$ **77.** $36x^8 - 36x^5 + 9x^2$
79. $18a^4 + 0.8a^3 + 4.5a + 0.2$ **81.** $\frac{1}{25} - 36x^8$
83. $a^3 + 1$ **85.** $x^2 + 6x + 9$ **87.** $t^2 + 7t + 12$
89. $a^2 + 10a + 25$ **91.** $x^2 + 10x + 21$
93. $a^2 + 8a + 7$ **95.** $25t^2 + 20t + 4$
97. 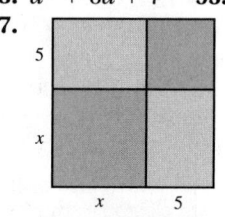   **99.**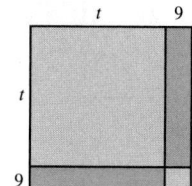
101. 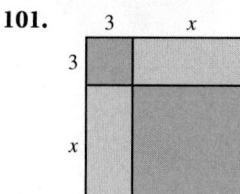   **103.** ✍
105. Washing machine: 9 kWh/mo; refrigerator:
189 kWh/mo; freezer: 99 kWh/mo   **106.** About
\$5.8 billion   **107.** $y = \dfrac{8}{5x}$ **108.** $a = \dfrac{c}{3b}$
109. $x = \dfrac{by + c}{a}$ **110.** $y = \dfrac{ax - c}{b}$ **111.** ✍
113. $16x^4 - 81$ **115.** $81t^4 - 72t^2 + 16$
117. $t^{24} - 4t^{18} + 6t^{12} - 4t^6 + 1$ **119.** 396   **121.** -7
123. $17F + 7(F - 17)$, $F^2 - (F - 17)(F - 7)$; other
equivalent expressions are possible.   **125.** $(y + 1)(y - 1)$,
$y(y + 1) - y - 1$; other equivalent expressions are possible.
127. $y^2 - 4y + 4$ **129.** ◿

Connecting the Concepts, p. 279

1. Addition; $3x^2 + 3x + 3$ **2.** Subtraction; $6x + 13$
3. Multiplication; $48x^5 - 42x^3$ **4.** Multiplication;
$6x^2 + x - 2$ **5.** Subtraction; $9x^3 - 5x^2 - 7x + 13$
6. Multiplication; $2x^3 + 3x^2 - 5x - 3$ **7.** Multiplication;
$81x^2 - 1$ **8.** Addition; $9x^4 + 2x^3 - 5x$
9. $-6x^2 + 2x - 12$ **10.** $9x^2 + 45x + 56$
11. $40x^9 - 48x^8 + 16x^5$ **12.** $t^9 + 5t^7$
13. $4m^2 - 4m + 1$ **14.** $x^3 - 1$ **15.** $5x^3 + 3$
16. $c^2 - 9$ **17.** $16y^6 + 56y^3 + 49$
18. $3a^4 - 13a^3 - 13a^2 - 4$ **19.** $16t^4 - 25$
20. $a^8 - 5a^4 - 24$

Technology Connection, p. 283

1. 36.22   **2.** 22,312

Exercise Set 4.7, pp. 284–287

1. (a)   **2.** (b)   **3.** (b)   **4.** (a)   **5.** (c)   **6.** (c)
7. (a)   **8.** (a)   **9.** -13 **11.** -68 **13.** 3.51 L
15. 1889 calories   **17.** 73.005 in²   **19.** 66.4 m
21. Coefficients: 3, -5, 2, -11; degrees: 3, 2, 2, 0; 3
23. Coefficients: 7, -1, 1, 9; degrees: 0, 3, 3, 3; 3
25. $2r - 6s$ **27.** $5xy^2 - 2x^2y + x + 3x^2$
29. $9u^2v - 11uv^2 + 11u^2$ **31.** $6a^2c - 7ab^2 + a^2b$
33. $11x^2 - 10xy - y^2$ **35.** $-6a^4 - 8ab + 7ab^2$
37. $-6r^2 - 5rt - t^2$ **39.** $3x^3 - x^2y + xy^2 - 3y^3$
41. $-2y^4x^3 - 3y^3x$ **43.** $-8x + 8y$
45. $12c^2 + 5cd - 2d^2$ **47.** $x^2y^2 + 4xy - 5$
49. $4a^2 - b^2$ **51.** $20r^2t^2 - 23rt + 6$
53. $m^6n^2 + 2m^3n - 48$ **55.** $30x^2 - 28xy + 6y^2$
57. $0.01 - p^2q^2$ **59.** $x^2 + 2xh + h^2$
61. $16a^2 - 40ab + 25b^2$ **63.** $a^2b^2 - c^2d^4$
65. $x^3y^2 + x^2y^3 + 2x^2y^2 + 2xy^3 + 3xy + 3y^2$
67. $a^2 + 2ab + b^2 - c^2$ **69.** $a^2 - b^2 - 2bc - c^2$
71. $x^2 + 2xy + y^2$ **73.** $\frac{1}{2}a^2b^2 - 2$
75. $a^2 + c^2 + ab + 2ac + ad + bc + bd + cd$
77. $m^2 - n^2$
79. We draw a rectangle with   **81.**
dimensions $r + s$ by $u + v$.
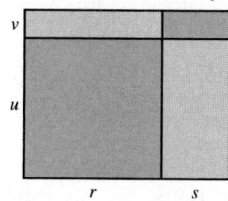
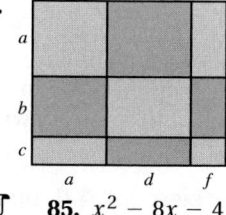
83. ✍   **85.** $x^2 - 8x - 4$
86. $2x^3 - x^2 - x + 4$
87. $-2x + 5$ **88.** $5x^2 + x$ **89.** $13x^2 + 1$
90. $-x - 3$ **91.** ✍   **93.** $4xy - 4y^2$
95. $2\pi ab - \pi b^2$ **97.** $x^3 + 2y^3 + x^2y + xy^2$
99. $2x^2 - 2\pi r^2 + 4xh + 2\pi rh$ **101.** ✍   **103.** 40
105. $P + 2Pr + Pr^2$ **107.** \$15,638.03

Exercise Set 4.8, pp. 292–293

1. $8x^6 - 5x^3$ **3.** $1 - 2u + u^6$ **5.** $6t^2 - 8t + 2$
7. $7x^3 - 6x + \frac{3}{2}$ **9.** $-4t^2 - 2t + 1$ **11.** $4x - 5 + \dfrac{1}{2x}$

13. $x + 2x^3y + 3$ **15.** $-3rs - r + 2s$ **17.** $x - 6$

19. $t - 5 + \dfrac{-45}{t - 5}$ **21.** $2x - 1 + \dfrac{1}{x + 6}$

23. $t^2 - 3t + 9$ **25.** $a + 5 + \dfrac{4}{a - 5}$

27. $x - 3 - \dfrac{3}{5x - 1}$ **29.** $3a + 1 + \dfrac{3}{2a + 5}$

31. $t^2 - 3t + 1$ **33.** $x^2 + 1$ **35.** $t^2 - 1 + \dfrac{3t - 1}{t^2 + 5}$

37. $3x^2 - 3 + \dfrac{x - 1}{2x^2 + 1}$ **39.** ✍

41.

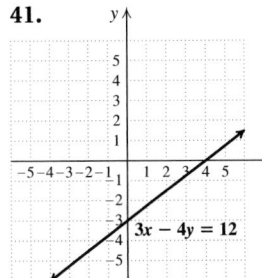

42.

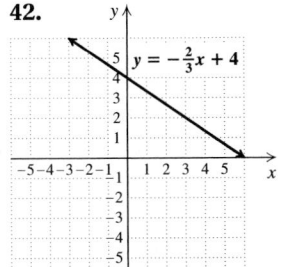

43.

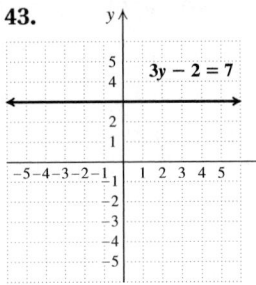

44.

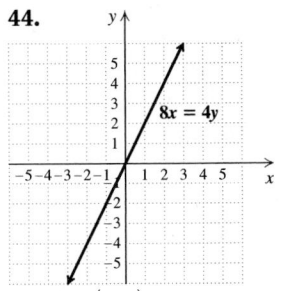

45. $-\dfrac{3}{10}$ **46.** Slope: 4; y-intercept: $\left(0, \dfrac{7}{2}\right)$

47. $y = -5x - 10$ **48.** $y = \dfrac{5}{4}x - \dfrac{9}{2}$ **49.** ✍

51. $5x^{6k} - 16x^{3k} + 14$ **53.** $3t^{2h} + 2t^h - 5$

55. $a + 3 + \dfrac{5}{5a^2 - 7a - 2}$ **57.** $2x^2 + x - 3$

59. 3   **61.** -1

Review Exercises: Chapter 4, pp. 297–298

1. True   **2.** True   **3.** True   **4.** False   **5.** False
6. False   **7.** True   **8.** True   **9.** n^{12} **10.** $(7x)^{10}$
11. t^6 **12.** 4^3, or 64   **13.** 1   **14.** $-9c^4d^2$
15. $-8x^3y^6$ **16.** $18x^5$ **17.** a^7b^6 **18.** $\dfrac{4t^{10}}{9s^8}$

19. $\dfrac{1}{8^6}$ **20.** a^{-9} **21.** $\dfrac{1}{4^2}$, or $\dfrac{1}{16}$ **22.** $\dfrac{2b^9}{a^{13}}$ **23.** $\dfrac{1}{w^{15}}$

24. $\dfrac{x^6}{4y^2}$ **25.** $\dfrac{y^3}{8x^3}$ **26.** 470,000,000   **27.** 1.09×10^{-5}
28. 2.09×10^4 **29.** 5.12×10^{-5}
30. $8x^2, -x, \dfrac{2}{3}$ **31.** $-4y^5, 7y^2, -3y, -2$ **32.** 9, -1, 7
33. 7, $-\dfrac{5}{6}$, -4, 10   **34.** (a) 2, 0, 5; (b) $15t^5$, 15; (c) 5
35. (a) 5, 0, 2, 1; (b) $-2x^5, -2$; (c) 5   **36.** Trinomial
37. Polynomial with no special name   **38.** Monomial
39. $-x^2 + 7x$ **40.** $-\dfrac{1}{4}x^3 + 4x^2 + 7$ **41.** $t - 1$
42. $14a^5 - 2a^2 - a - \dfrac{2}{3}$ **43.** -24 **44.** 16
45. $x^5 + 8x^4 + 6x^3 - 2x - 9$ **46.** $6a^5 - a^3 - 12a^2$
47. $-3y^2 + 8y + 3$ **48.** $x^5 - 3x^3 - 2x^2 + 8$

49. $\dfrac{3}{4}x^4 + \dfrac{1}{4}x^3 - \dfrac{1}{3}x^2 - \dfrac{7}{4}x + \dfrac{3}{8}$
50. $-x^5 + x^4 - 5x^3 - 2x^2 + 2x$ **51.** (a) $4w + 6$;
(b) $w^2 + 3w$ **52.** $-30x^5$ **53.** $49x^2 + 14x + 1$
54. $a^2 - 3a - 28$ **55.** $d^2 - 64$
56. $12x^3 - 23x^2 + 13x - 2$ **57.** $x^2 - 16x + 64$
58. $15t^5 - 6t^4 + 12t^3$ **59.** $4a^2 - 81$
60. $x^2 - 1.3x + 0.4$
61. $x^7 + x^5 - 3x^4 + 3x^3 - 2x^2 + 5x - 3$
62. $16y^6 - 40y^3 + 25$ **63.** $2t^4 - 11t^2 - 21$
64. $a^2 + \dfrac{1}{6}a - \dfrac{1}{3}$ **65.** $-49 + 4n^2$ **66.** 49
67. Coefficients: 1, -7, 9, -8; degrees: 6, 2, 2, 0; 6
68. Coefficients: 1, -1, 1; degrees: 13, 22, 15; 22
69. $-4u + 4v - 7$ **70.** $6m^3 + 4m^2n - mn^2$
71. $2a^2 - 16ab$ **72.** $11x^3y^2 - 8x^2y - 6x^2 - 6x + 6$
73. $2x^2 - xy - 15y^2$ **74.** $25a^2b^2 - 10abcd^2 + c^2d^4$
75. $\dfrac{1}{2}x^2 - \dfrac{1}{2}y^2$ **76.** $y^4 - \dfrac{1}{3}y + 4$

77. $3x^2 - 7x + 4 + \dfrac{1}{2x + 3}$ **78.** $t^3 + 2t - 3$

79. ✍ In the expression $5x^3$, the exponent refers only to the x. In the expression $(5x)^3$, the entire expression $5x$ is the base.   **80.** ✍ It is possible to determine two possibilities for the binomial that was squared by using the equation $(A - B)^2 = A^2 - 2AB + B^2$ in reverse. Since, in $x^2 - 6x + 9, A^2 = x^2$ and $B^2 = 9$, or 3^2, the binomial that was squared was $A - B$, or $x - 3$. If the polynomial is written $9 - 6x + x^2$, then $A^2 = 9$ and $B^2 = x^2$, so the binomial that was squared was $3 - x$. We cannot determine without further information whether the binomial squared was $x - 3$ or $3 - x$.   **81.** (a) 9; (b) 28   **82.** $64x^{16}$
83. $8x^4 + 4x^3 + 5x - 2$ **84.** $-16x^6 + x^2 - 10x + 25$
85. $\dfrac{94}{13}$ **86.** 2.28×10^{11} platelets

Test: Chapter 4, p. 299

1. [4.1] x^{13} **2.** [4.1] 3   **3.** [4.1] 1   **4.** [4.1] t^{45}
5. [4.1] $-40x^{19}y^4$ **6.** [4.1] $\dfrac{6}{5}a^5b^3$ **7.** [4.2] $\dfrac{1}{y^7}$

8. [4.2] 5^{-6} **9.** [4.2] $\dfrac{1}{t^9}$ **10.** [4.2] $\dfrac{3y^5}{x^5}$ **11.** [4.2] $\dfrac{b^4}{16a^{12}}$

12. [4.2] $\dfrac{c^3}{a^3b^3}$ **13.** [4.2] 3.06×10^9 **14.** [4.2] 0.00000005
15. [4.2] 1.75×10^{17} **16.** [4.2] 1.296×10^{22}
17. [4.3] Binomial   **18.** [4.3] 3, -1, $\dfrac{1}{9}$ **19.** [4.3] Degrees
of terms: 3, 1, 5, 0; leading term: $7t^5$; leading coefficient: 7;
degree of polynomial: 5   **20.** [4.3] -7 **21.** [4.3] $5a^2 - 6$
22. [4.3] $\dfrac{7}{4}y^2 - 4y$ **23.** [4.3] $4x^3 + 4x^2 + 3$
24. [4.4] $4x^5 + x^4 + 5x^3 - 8x^2 + 2x - 7$
25. [4.4] $5x^4 + 5x^2 + x + 5$
26. [4.4] $-2a^4 + 3a^3 - a - 7$
27. [4.4] $-t^4 + 2.5t^3 - 0.6t^2 - 9$
28. [4.5] $-6x^4 + 6x^3 + 10x^2$ **29.** [4.6] $x^2 - \dfrac{2}{3}x + \dfrac{1}{9}$
30. [4.6] $25t^2 - 49$ **31.** [4.6] $6b^2 + 7b - 5$
32. [4.6] $x^{14} - 4x^8 + 4x^6 - 16$
33. [4.6] $48 + 34y - 5y^2$ **34.** [4.5] $6x^3 - 7x^2 - 11x - 3$
35. [4.6] $64a^6 + 48a^3 + 9$ **36.** [4.7] 24
37. [4.7] $-4x^3y - x^2y^2 + xy^3 - y^3 + 19$
38. [4.7] $8a^2b^2 + 6ab + 6ab^2 + ab^3 - 4b^3$
39. [4.7] $9x^{10} - y^2$ **40.** [4.8] $4x^2 + 3x - 5$

41. [4.8] $2x^2 - 4x - 2 + \dfrac{17}{3x + 2}$ **42.** [4.5], [4.6]

$V = l(l - 2)(l - 1) = l^3 - 3l^2 + 2l$ **43.** [2.2], [4.6] $\dfrac{100}{21}$

44. [4.2] $\frac{1}{2} - \frac{1}{4} = \frac{1}{4}$ **45.** [4.2] About 1.4×10^7 hr

Cumulative Review: Chapters 1–4, pp. 300–301

1. 6   **2.** -8 **3.** $-\frac{7}{45}$ **4.** 6   **5.** $y + 10$ **6.** t^{12}
7. $-4x^5y^3$ **8.** $50a^4b^7$ **9.** $2(5a - 3b + 6)$ **10.** $\frac{11}{16}$
11. 6   **12.** $\frac{1}{4}t^3 - 5t^2 - 0.35$ **13.** II
14.

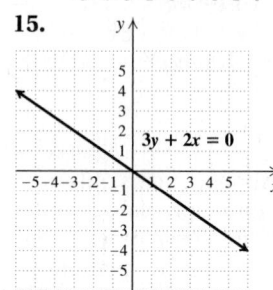

15.

16.

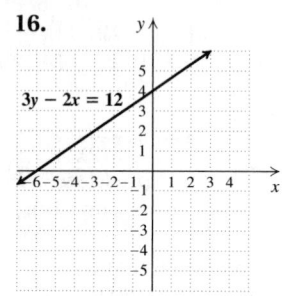

17.

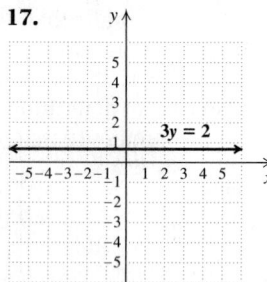

18.

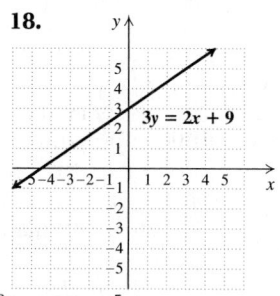

19. Slope: $\frac{1}{10}$; y-intercept: $(0, \frac{3}{8})$ **20.** $-\frac{5}{8}$
21. $y = -\frac{2}{3}x - 10$ **22.** x-intercept: $(4, 0)$; y-intercept:
$(0, \frac{8}{5})$ **23.** -4 **24.** $\frac{3}{5}$ **25.** $\frac{8}{3}$ **26.** 22   **27.** $\frac{5}{3}$
28. -7 **29.** $\{t | t \geq -8\}$, or $[-8, \infty)$ **30.** $\{x | x < \frac{1}{2}\}$, or
$\left(-\infty, \frac{1}{2}\right)$ **31.** $t = \dfrac{5pq}{2c}$ **32.** $7u^2v - 2uv^2 + uv + 3u^2$
33. $3x^3 - 3x^2 + 12x - 7$ **34.** $x^5 - 2x$
35. $20a - 30b + 70$ **36.** $-16x^5 - 48x^4 + 56x^3$
37. $6a^2 + 11a - 7$ **38.** $x^3 - x^2 - 7x + 10$
39. $16t^4 + 24t^2 + 9$ **40.** $\frac{1}{4}x^2 - 1$
41. $6r^4 - 5r^2s - 4s^2$ **42.** $x + \dfrac{3}{x - 1}$ **43.** $\dfrac{1}{7^{10}}$
44. $\dfrac{1}{x^7}$ **45.** $\dfrac{9}{16s^2t^{10}}$ **46.** $\dfrac{x^7y^2}{3}$ **47.** 48 thousand
megawatts   **48.** About 1298 megawatt hours
49. Washer: 20 kWh; dryer: 50 kWh
50. (a)   **(b)** 375 kWh

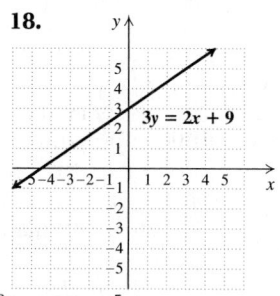

51. No more than 6 hr   **52.** $160 billion   **53.** About 3.6
54. $\$\frac{12}{7}$ billion per year   **55.** 0   **56.** No solution
57. $\frac{22}{9}$ **58.** $y = \frac{1}{2}x + 4$ **59.** $\frac{1}{7} + 1 = \frac{8}{7}$ **60.** $15x^{12}$

CHAPTER 5

Technology Connection, p. 310

1. Correct   **2.** Correct   **3.** Not correct
4. Not correct   **5.** Not correct   **6.** Correct
7. Not correct   **8.** Correct

Exercise Set 5.1, pp. 310–311

1. (h)   **2.** (f)   **3.** (b)   **4.** (e)   **5.** (c)   **6.** (g)
7. (d)   **8.** (a)   **9.** Answers may vary. $(14x)(x^2)$,
$(7x^2)(2x)$, $(-2)(-7x^3)$ **11.** Answers may vary. $(-15)(a^4)$,
$(-5a)(3a^3)$, $(-3a^2)(5a^2)$ **13.** Answers may vary.
$(5t^2)(5t^3)$, $(25t)(t^4)$, $(-5t)(-5t^4)$
15. $8(x + 3)$ **17.** $6(x - 5)$ **19.** $2(x^2 + x - 4)$
21. $t(3t + 1)$ **23.** $-5y(y + 2)$ **25.** $x^2(x + 6)$
27. $8a^2(2a^2 - 3)$ **29.** $-t^2(6t^4 - 9t^2 + 4)$
31. $6x^2(x^6 + 2x^4 - 4x^2 + 5)$
33. $x^2y^2(x^3y^3 + x^2y + xy - 1)$
35. $-5a^2b^2(7ab^2 - 2b + 3a)$ **37.** $(n - 6)(n + 3)$
39. $(x + 3)(x^2 - 7)$ **41.** $(2y - 9)(y^2 + 1)$
43. $(x + 2)(x^2 + 5)$ **45.** $(a + 3)(5a^2 + 2)$
47. $(3n - 2)(3n^2 + 1)$ **49.** $(t - 5)(4t^2 + 3)$
51. $(7x + 5)(x^2 - 3)$ **53.** $(6a + 7)(a^2 + 1)$
55. $(x + 6)(2x^2 - 5)$ **57.** Not factorable by grouping
59. $(y + 8)(y^2 - 2)$ **61.** $(x - 4)(2x^2 - 9)$ **63.** ✍
65. $x^2 + 9x + 14$ **66.** $x^2 - 9x + 14$ **67.** $x^2 - 5x - 14$
68. $x^2 + 5x - 14$ **69.** $a^2 - 4a + 3$ **70.** $t^2 + 8t + 15$
71. $t^2 + 5t - 50$ **72.** $a^2 - 2a - 24$ **73.** ✍
75. $(2x^3 + 3)(2x^2 + 3)$ **77.** $2x(x + 1)(x^2 - 2)$
79. $(x - 1)(5x^4 + x^2 + 3)$ **81.** Answers may vary.
$8x^4y^3 - 24x^2y^4 + 16x^3y^4$

Exercise Set 5.2, pp. 317–318

1. Positive; positive   **2.** Negative; negative
3. Negative; positive   **4.** Positive; positive
5. Positive   **6.** Negative   **7.** $(x + 4)(x + 4)$
9. $(x + 1)(x + 10)$ **11.** $(x + 3)(x + 7)$
13. $(t - 2)(t - 7)$ **15.** $(b - 4)(b - 1)$
17. $(a - 3)(a - 4)$ **19.** $(d - 2)(d - 5)$
21. $(x - 5)(x + 3)$ **23.** $(x + 5)(x - 3)$
25. $2(x + 2)(x - 9)$ **27.** $-x(x + 2)(x - 8)$
29. $(y - 5)(y + 9)$ **31.** $(x - 6)(x + 12)$
33. $-5(b - 3)(b + 10)$ **35.** $x^3(x - 2)(x + 1)$
37. Prime   **39.** $(t + 4)(t + 8)$ **41.** $(x + 9)(x + 11)$
43. $3x(x - 25)(x + 4)$ **45.** $-2(x - 24)(x + 3)$
47. $(y - 12)(y - 8)$ **49.** $-a^4(a - 6)(a + 15)$
51. $\left(t + \frac{1}{3}\right)^2$ **53.** Prime   **55.** $(p - 5q)(p - 2q)$
57. Prime   **59.** $(s - 6t)(s + 2t)$ **61.** $6a^8(a - 2)(a + 7)$
63. ✍   **65.** $6x^2 + 17x + 12$ **66.** $6x^2 + x - 12$
67. $6x^2 - x - 12$ **68.** $6x^2 - 17x + 12$

69. $5x^2 - 36x + 7$ **70.** $3x^2 + 13x - 30$
71. ✍   **73.** $-5, 5, -23, 23, -49, 49$
75. $(y + 0.2)(y - 0.4)$ **77.** $-\frac{1}{3}a(a - 3)(a + 2)$
79. $(x^m + 4)(x^m + 7)$ **81.** $(a + 1)(x + 2)(x + 1)$
83. $(x + 3)^3$, or $(x^3 + 9x^2 + 27x + 27)$ cubic meters
85. $x^2\left(\frac{3}{4}\pi + 2\right)$, or $\frac{1}{4}x^2(3\pi + 8)$ **87.** $x^2\left(9 - \frac{1}{2}\pi\right)$
89. $(x + 4)(x + 5)$

Exercise Set 5.3, pp. 327–328

1. (c)   **2.** (a)   **3.** (d)   **4.** (b)   **5.** $(2x - 1)(x + 4)$
7. $(3x + 1)(x - 6)$ **9.** $(2t + 1)(2t + 5)$
11. $(5a - 3)(3a - 1)$ **13.** $(3x + 4)(2x + 3)$
15. $2(3x + 1)(x - 2)$ **17.** $t(7t + 1)(t + 2)$
19. $(4x - 5)(3x - 2)$ **21.** $-1(7x + 4)(5x + 2)$, or
$-(7x + 4)(5x + 2)$ **23.** Prime   **25.** $(5x + 4)^2$
27. $(20y - 1)(y + 3)$ **29.** $(7x + 5)(2x + 9)$
31. $-1(x - 3)(2x + 5)$, or $-(x - 3)(2x + 5)$
33. $-3(2x + 1)(x + 5)$ **35.** $2(a + 1)(5a - 9)$
37. $4(3x - 1)(x + 6)$ **39.** $(3x + 1)(x + 1)$
41. $(x + 3)(x - 2)$ **43.** $(4t - 3)(2t - 7)$
45. $(3x + 2)(2x + 5)$ **47.** $(y + 4)(2y - 1)$
49. $(3a - 4)(2a - 1)$ **51.** $(16t + 7)(t + 1)$
53. $-1(3x + 1)(3x + 5)$, or $-(3x + 1)(3x + 5)$
55. $10(x^2 + 3x - 7)$ **57.** $3x(3x - 1)(2x + 3)$
59. $(x + 1)(25x + 64)$ **61.** $3x(7x + 1)(8x + 1)$
63. $-t^2(2t - 3)(7t + 1)$ **65.** $2(2y + 9)(8y - 5)$
67. $(2a - b)(a - 2b)$ **69.** $2(s + t)(4s + 7t)$
71. $3(3x - 4y)^2$ **73.** $-2(3a - 2b)(4a - 3b)$
75. $x^2(2x + 3)(7x - 1)$ **77.** $a^6(3a + 4)(3a + 2)$
79. ✍   **81.** $x^2 - 4x + 4$ **82.** $x^2 + 4x + 4$
83. $x^2 - 4$ **84.** $25t^2 - 30t + 9$ **85.** $16a^2 + 8a + 1$
86. $4n^2 - 49$ **87.** $9c^2 - 60c + 100$
88. $1 - 10a + 25a^2$ **89.** $64n^2 - 9$ **90.** $81 - y^2$
91. ✍   **93.** $(3xy + 2)(6xy - 5)$ **95.** Prime
97. $(4t^5 - 1)^2$ **99.** $-1(5x^m - 2)(3x^m - 4)$, or
$-(5x^m - 2)(3x^m - 4)$ **101.** $(3a^{3n} + 1)(a^{3n} - 1)$
103. $[7(t - 3)^n - 2][(t - 3)^n + 1]$

Connecting the Concepts, pp. 328–329

1. $6x^2(x^3 - 3)$ **2.** $(x + 2)(x + 8)$ **3.** $(x + 7)(2x - 1)$
4. $(x + 3)(x^2 + 2)$ **5.** $5(x - 2)(x + 10)$ **6.** Prime
7. $7y(x - 4)(x + 1)$ **8.** $3a^2(5a^2 - 9b^2 + 7b)$
9. $(b - 7)^2$ **10.** $(3x - 1)(4x + 1)$
11. $(c + 1)(c + 2)(c - 2)$ **12.** $2(x - 5)(x + 20)$
13. Prime   **14.** $15(d^2 - 2d + 5)$
15. $(3p + 2q)(5p + 2q)$ **16.** $-2t(t + 2)(t + 3)$
17. $(x + 11)(x - 7)$ **18.** $10(c + 1)^2$
19. $-1(2x - 5)(x + 1)$ **20.** $2n(m - 5)(m^2 - 3)$

Exercise Set 5.4, pp. 334–335

1. Prime polynomial   **2.** Difference of squares
3. Difference of squares   **4.** None of these
5. Perfect-square trinomial   **6.** Perfect-square trinomial
7. None of these   **8.** Prime polynomial
9. Difference of squares   **10.** Perfect-square trinomial

11. Yes   **13.** No   **15.** No   **17.** Yes   **19.** $(x + 8)^2$
21. $(x - 5)^2$ **23.** $5(p + 2)^2$ **25.** $(1 - t)^2$, or $(t - 1)^2$
27. $2(3x + 1)^2$ **29.** $(7 - 4y)^2$, or $(4y - 7)^2$
31. $-x^3(x - 9)^2$ **33.** $2n(n + 10)^2$ **35.** $5(2x + 5)^2$
37. $(7 - 3x)^2$, or $(3x - 7)^2$ **39.** $(4x + 3)^2$
41. $2(1 + 5x)^2$, or $2(5x + 1)^2$ **43.** $(3p + 2q)^2$
45. Prime   **47.** $-1(8m + n)^2$, or $-(8m + n)^2$
49. $-2(4s - 5t)^2$ **51.** Yes   **53.** No   **55.** Yes
57. $(x + 5)(x - 5)$ **59.** $(p + 3)(p - 3)$
61. $(7 + t)(-7 + t)$, or $(t + 7)(t - 7)$
63. $6(a + 2)(a - 2)$ **65.** $(7x - 1)^2$
67. $2(10 + t)(10 - t)$ **69.** $-5(4a + 3)(4a - 3)$
71. $5(t + 4)(t - 4)$ **73.** $2(2x + 9)(2x - 9)$
75. $x(6 + 7x)(6 - 7x)$ **77.** Prime
79. $(t^2 + 1)(t + 1)(t - 1)$ **81.** $-3x(x - 4)^2$
83. $3t(5t + 3)(5t - 3)$ **85.** $a^6(a - 1)^2$
87. $10(a + b)(a - b)$ **89.** $(4x^2 + y^2)(2x + y)(2x - y)$
91. $2(3t + 2s)(3t - 2s)$ **93.** ✍   **95.** $8x^6y^{12}$
96. $-125x^6y^3$ **97.** $x^3 + 3x^2 + 3x + 1$
98. $x^3 - 3x^2 + 3x - 1$ **99.** $p^3 + 3p^2q + 3pq^2 + q^3$
100. $p^3 - 3p^2q + 3pq^2 - q^3$ **101.** ✍
103. $(x^4 + 2^4)(x^2 + 2^2)(x + 2)(x - 2)$, or
$(x^4 + 16)(x^2 + 4)(x + 2)(x - 2)$
105. $2x\left(3x - \frac{2}{5}\right)\left(3x + \frac{2}{5}\right)$
107. $[(y - 5)^2 + z^4][(y - 5) + z^2][(y - 5) - z^2]$, or
$(y^2 - 10y + 25 + z^4)(y - 5 + z^2)(y - 5 - z^2)$
109. $-1(x^2 + 1)(x + 3)(x - 3)$, or $-(x^2 + 1)(x + 3)(x - 3)$
111. $(y + 4)^2$ **113.** $(3p + 5)(3p - 5)^2$
115. $(9 + b^{2k})(3 + b^k)(3 - b^k)$ **117.** $2x^3 - x^2 - 1$
119. $(y + x + 7)(y - x - 1)$ **121.** 16
123. $(x + 1)^2 - x^2 = [(x + 1) + x][(x + 1) - x] = $
$2x + 1 = (x + 1) + x$

Exercise Set 5.5, p. 339

1. Difference of cubes   **2.** Sum of cubes   **3.** Difference
of squares   **4.** None of these   **5.** Sum of cubes
6. Difference of cubes   **7.** None of these   **8.** Difference
of squares   **9.** Difference of cubes   **10.** None of these
11. $(x - 4)(x^2 + 4x + 16)$ **13.** $(z + 1)(z^2 - z + 1)$
15. $(t - 10)(t^2 + 10t + 100)$
17. $(3x + 1)(9x^2 - 3x + 1)$
19. $(4 - 5x)(16 + 20x + 25x^2)$
21. $(x - y)(x^2 + xy + y^2)$ **23.** $\left(a + \frac{1}{2}\right)\left(a^2 - \frac{1}{2}a + \frac{1}{4}\right)$
25. $8(t - 1)(t^2 + t + 1)$ **27.** $2(3x + 1)(9x^2 - 3x + 1)$
29. $rs(s + 4)(s^2 - 4s + 16)$
31. $5(x - 2z)(x^2 + 2xz + 4z^2)$
33. $\left(y - \frac{1}{10}\right)\left(y^2 + \frac{1}{10}y + \frac{1}{100}\right)$
35. $(x + 0.1)(x^2 - 0.1x + 0.01)$
37. $8(2x^2 - t^2)(4x^4 + 2x^2t^2 + t^4)$
39. $2y(3y - 4)(9y^2 + 12y + 16)$
41. $(z + 1)(z^2 - z + 1)(z - 1)(z^2 + z + 1)$
43. $(t^2 + 4y^2)(t^4 - 4t^2y^2 + 16y^4)$
45. $(x^4 - yz^4)(x^8 + x^4yz^4 + y^2z^8)$ **47.** ✍
49. $-\frac{1}{5}$ **50.** $\frac{1}{4}$

51.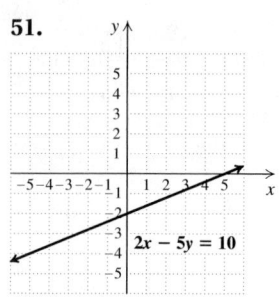

52.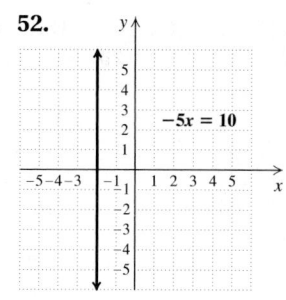

53.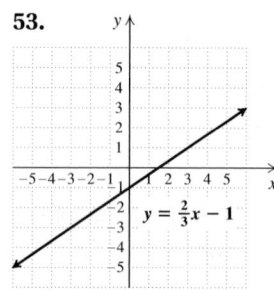

54.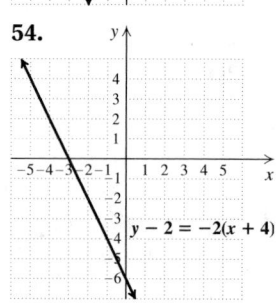

55. ✑   **57.** $(x^{2a} - y^b)(x^{4a} + x^{2a}y^b + y^{2b})$
59. $2x(x^2 + 75)$ **61.** $5(xy^2 - \frac{1}{2})(x^2y^4 + \frac{1}{2}xy^2 + \frac{1}{4})$
63. $-(3x^{4a} + 3x^{2a} + 1)$ **65.** $(t - 8)(t - 1)(t^2 + t + 1)$

Exercise Set 5.6, pp. 343–344

1. Common factor   **2.** Perfect-square trinomial
3. Grouping   **4.** Multiplying   **5.** $5(a + 5)(a - 5)$
7. $(y - 7)^2$ **9.** $(3t + 7)(t + 3)$ **11.** $x(x + 9)^2$
13. $(x - 5)^2(x + 5)$ **15.** $3t(3t + 1)(3t - 1)$
17. $3x(3x - 5)(x + 3)$ **19.** Prime   **21.** $6(y - 5)(y + 8)$
23. $-2a^4(a - 2)^2$ **25.** $5x(x^2 + 4)(x + 2)(x - 2)$
27. $(t^2 + 3)(t^2 - 3)$ **29.** $-x^4(x^2 - 2x + 7)$
31. $(x - y)(x^2 + xy + y^2)$ **33.** $a(x^2 + y^2)$
35. $2\pi r(h + r)$ **37.** $(a + b)(5a + 3b)$
39. $(x + 1)(x + y)$ **41.** $(a - 2)(a - y)$
43. $(3x - 2y)(x + 5y)$ **45.** $8mn(m^2 - 4mn + 3)$
47. $(a - 2b)^2$ **49.** $(4x + 3y)^2$ **51.** Prime
53. $(a^2b^2 + 4)(ab + 2)(ab - 2)$ **55.** $4c(4d - c)(5d - c)$
57. $(2t + 1)(4t^2 - 2t + 1)(2t - 1)(4t^2 + 2t + 1)$
59. $-1(xy + 2)(xy + 6)$, or $-(xy + 2)(xy + 6)$
61. $5(pq + 6)(pq - 1)$
63. $2a(3a + 2b)(9a^2 - 6ab + 4b^2)$
65. $x^4(x + 2y)(x - y)$ **67.** $(6a - \frac{5}{4})^2$ **69.** $(\frac{1}{9}x - \frac{4}{3})^2$
71. $(1 + 4x^6y^6)(1 + 2x^3y^3)(1 - 2x^3y^3)$ **73.** $(2ab + 3)^2$
75. $z(z + 6)(z^2 - 6)$ **77.** ✑   **79.** $\frac{9}{8}$ **80.** $-\frac{5}{3}$
81. $-\frac{7}{2}$ **82.** $\frac{1}{4}$ **83.** 3   **84.** 11   **85.** -1 **86.** 3
87. ✑   **89.** $-x(x^2 + 9)(x^2 - 2)$
91. $-1(x^2 + 2)(x + 3)(x - 3)$, or
$-(x^2 + 2)(x + 3)(x - 3)$ **93.** $(y + 1)(y - 7)(y + 3)$
95. $(y + 4 + x)^2$ **97.** $(a + 3)^2(2a + b + 4)(a - b + 5)$
99. $(7x^2 + 1 + 5x^3)(7x^2 + 1 - 5x^3)$

Technology Connection, p. 351

1. $-4.65, 0.65$ **2.** $-0.37, 5.37$ **3.** $-8.98, -4.56$
4. No solution   **5.** $0, 2.76$

Exercise Set 5.7, pp. 351–353

1. (c)   **2.** (a)   **3.** (d)   **4.** (b)   **5.** $-9, -2$ **7.** $-6, \frac{3}{2}$
9. $\frac{1}{7}, \frac{3}{10}$ **11.** $0, 7$ **13.** $\frac{1}{21}, \frac{18}{11}$ **15.** $-\frac{8}{3}, 0$ **17.** $50, 70$
19. $-5, \frac{2}{3}, 1$ **21.** $1, 6$ **23.** $-7, 3$ **25.** $-9, -2$ **27.** $0, 10$
29. $-6, 0$ **31.** $-6, 6$ **33.** $-\frac{7}{2}, \frac{7}{2}$ **35.** -5 **37.** 8
39. $0, 2$ **41.** $-\frac{5}{4}, 3$ **43.** 3   **45.** $0, \frac{4}{3}$ **47.** $-\frac{7}{6}, \frac{7}{6}$
49. $-4, -\frac{2}{3}$ **51.** $-3, 1$ **53.** $-1, 0, \frac{3}{2}$ **55.** $-7, -\frac{8}{3}, \frac{5}{2}$
57. $-1, 4$ **59.** $-3, 2$ **61.** $(-2, 0), (3, 0)$ **63.** $(-4, 0)$,
$(2, 0)$ **65.** $(-3, 0), (\frac{3}{2}, 0)$ **67.** ✑   **69.** Let m and n
represent the numbers; $(m + n)^2$ **70.** Let m and n represent the numbers; $m^2 + n^2$ **71.** Let x represent the first
integer; then $x + 1$ represents the second integer; $x(x + 1)$
72. Mother's Day: \$13.8 billion; Father's Day: \$9 billion
73. $140°, 35°, 5°$ **74.** Length: 64 in.; width: 32 in.
75. ✑   **77.** (a) $x^2 - x - 20 = 0$; (b) $x^2 - 6x - 7 = 0$;
(c) $4x^2 - 13x + 3 = 0$; (d) $6x^2 - 5x + 1 = 0$;
(e) $12x^2 - 17x + 6 = 0$; (f) $x^3 - 4x^2 + x + 6 = 0$
79. $-5, 4$ **81.** $-\frac{3}{5}, \frac{3}{5}$ **83.** $-4, 2$
85. (a) $2x^2 + 20x - 4 = 0$; (b) $x^2 - 3x - 18 = 0$;
(c) $(x + 1)(5x - 5) = 0$; (d) $(2x + 8)(2x - 5) = 0$;
(e) $4x^2 + 8x + 36 = 0$; (f) $9x^2 - 12x + 24 = 0$
87. ✑   **89.** $2.33, 6.77$ **91.** $-9.15, -4.59$ **93.** $-3.76, 0$

Connecting the Concepts, p. 354

1. Expression   **2.** Equation   **3.** Equation
4. Expression   **5.** Expression   **6.** Equation
7. $2x^3 + x^2 - 8x$ **8.** $-2x^2 - 11$ **9.** $-10, 10$
10. $6a^2 - 19a + 10$ **11.** $(n - 1)(n - 9)$ **12.** $2, 8$
13. $-\frac{5}{2}$ **14.** $7x^3 + 2x - 7$ **15.** $(4x + 9)(4x - 9)$
16. $-3, 8$ **17.** $-4a^2 - a - 11$
18. $2x^2(9x^2 - 12x + 10)$ **19.** $-1, -\frac{2}{3}$
20. $8x^5 - 20x^4 + 12x^2$

Translating for Success, p. 361

1. O   **2.** M   **3.** K   **4.** I   **5.** G   **6.** E   **7.** C
8. A   **9.** H   **10.** B

Exercise Set 5.8, pp. 362–366

1. $-2, 3$ **3.** 6 m, 8 m, 10 m   **5.** 11, 12
7. -14 and -12; 12 and 14   **9.** Length: 30 ft; width: 6 ft
11. Length: 6 cm; width: 4 cm   **13.** Base: 12 in.; height: 9 in.
15. Foot: 7 ft; height: 12 ft   **17.** 1 min, 3 min   **19.** In 2008
21. 16 teams   **23.** 66 handshakes   **25.** 12 players
27. 9 ft   **29.** 32 ft   **31.** 300 ft by 400 ft by 500 ft
33. Dining room: 12 ft by 12 ft; kitchen: 12 ft by 10 ft
35. 20 ft   **37.** 1 sec, 2 sec   **39.** ✑   **41.** $-\frac{12}{35}$ **42.** $-\frac{21}{20}$
43. -1 **44.** $-\frac{7}{4}$ **45.** $\frac{1}{4}$ **46.** $\frac{4}{5}$ **47.** $\frac{53}{168}$ **48.** $\frac{19}{18}$
49. ✑   **51.** \$180   **53.** 39 cm   **55.** 4 in., 6 in.
57. 35 ft   **59.** 2 hr, 4.2 hr   **61.** 3 hr

Review Exercises: Chapter 5, pp. 369–371

1. False   **2.** True   **3.** True   **4.** False   **5.** False
6. True   **7.** True   **8.** False   **9.** Answers may vary.
$(4x)(5x^2), (-2x^2)(-10x), (x^3)(20)$

10. Answers may vary. $(-3x^2)(6x^3)$, $(2x^4)(-9x)$, $(-18x)(x^4)$ **11.** $6x^3(2x - 3)$ **12.** $4a(2a - 3)$
13. $(10t + 1)(10t - 1)$ **14.** $(x + 4)(x - 3)$
15. $(x + 7)^2$ **16.** $3x(2x + 1)^2$ **17.** $(2x + 3)(3x^2 + 1)$
18. $(6a - 5)(a + 1)$ **19.** $(5t - 3)^2$
20. $2(24t^2 - 14t + 3)$ **21.** $(9a^2 + 1)(3a + 1)(3a - 1)$
22. $3x(3x - 5)(x + 3)$ **23.** $2(x - 5)(x^2 + 5x + 25)$
24. $(x + 4)(x^3 - 2)$ **25.** $(ab^2 + 8)(ab^2 - 8)$
26. $-4x^4(2x^2 - 8x + 1)$ **27.** $3(2x - 5)^2$
28. Prime   **29.** $-t(t + 6)(t - 7)$ **30.** $(2x + 5)(2x - 5)$
31. $(n + 6)(n - 10)$ **32.** $5(z^2 - 6z + 2)$
33. $(4t + 5)(t + 2)$ **34.** $(2t + 1)(t - 4)$
35. $7x(x + 1)(x + 4)$ **36.** $(2y + 3x^2)(4y^2 - 6x^2y + 9x^4)$
37. $5(2x - 1)^2$ **38.** $-6x(x + 5)(x - 5)$
39. $(5 - x)(3 - x)$ **40.** Prime   **41.** $(xy + 8)(xy - 2)$
42. $3(2a + 7b)^2$ **43.** $(m + 5)(m + t)$
44. $32(x^2 + 2y^2z^2)(x^2 - 2y^2z^2)$ **45.** $(2m + n)(3m + n)$
46. $(3r + 5s)(2r - 3s)$ **47.** $-11, 9$ **48.** $-7, 5$
49. $-\frac{3}{4}, \frac{3}{4}$ **50.** $\frac{2}{3}, 1$ **51.** $-\frac{5}{2}, 6$ **52.** $-2, 3$ **53.** $0, \frac{3}{5}$
54. 1 **55.** $-3, 4$ **56.** 10 teams   **57.** $(-1, 0), \left(\frac{5}{2}, 0\right)$
58. Height: 14 ft; base: 14 ft   **59.** 10 holes
60. ✒ Answers may vary. Because Celia did not first factor out the largest common factor, 4, her factorization will not be "complete" until she removes a common factor of 2 from each binomial. The answer should be $4(x - 5)(x + 5)$. Awarding 3 to 7 points would seem reasonable.   **61.** ✒ The equations solved in this chapter have an x^2-term (are quadratic), whereas those solved previously have no x^2-term (are linear). The principle of zero products is used to solve quadratic equations and is not used to solve linear equations.   **62.** 2.5 cm   **63.** $0, 2$ **64.** Length: 12 cm; width: 6 cm   **65.** 100 cm^2, 225 cm^2 **66.** $-3, 2, \frac{5}{2}$
67. No real solution

Test: Chapter 5, pp. 371–372

1. [5.1] Answers may vary. $(3x^2)(4x^2)$, $(-2x)(-6x^3)$, $(12x^3)(x)$ **2.** [5.2] $(x - 4)(x - 9)$ **3.** [5.4] $(x - 5)^2$
4. [5.1] $2y^2(2y^2 - 4y + 3)$ **5.** [5.1] $(x + 1)(x^2 + 2)$
6. [5.1] $t^5(t^2 - 3)$ **7.** [5.2] $a(a + 4)(a - 1)$
8. [5.3] $2(5x - 6)(x + 4)$ **9.** [5.4] $(2t + 5)(2t - 5)$
10. [5.2] $(x + 2)(x - 3)$ **11.** [5.3] $-3m(2m + 1)(m + 1)$
12. [5.5] $3(r - 1)(r^2 + r + 1)$ **13.** [5.4] $5(3r + 2)^2$
14. [5.4] $3(x^2 + 4)(x + 2)(x - 2)$ **15.** [5.4] $(7t + 6)^2$
16. [5.1] $(x + 2)(x^3 - 3)$ **17.** [5.2] Prime
18. [5.3] $(2x + 3)(2x - 5)$ **19.** [5.3] $3t(2t + 5)(t - 1)$
20. [5.3] $3(m - 5n)(m + 2n)$ **21.** [5.7] $1, 5$
22. [5.7] $-\frac{3}{2}, 5$ **23.** [5.7] $0, \frac{2}{5}$ **24.** [5.7] $-\frac{1}{5}, \frac{1}{5}$
25. [5.7] $-4, 5$ **26.** [5.7] $(-1, 0), \left(\frac{8}{3}, 0\right)$
27. [5.8] Length: 10 m; width: 4 m   **28.** [5.8] 10 people
29. [5.8] 5 ft   **30.** [5.8] 15 cm by 30 cm
31. [5.2] $(a - 4)(a + 8)$ **32.** [5.7] $-\frac{8}{3}, 0, \frac{2}{5}$

Cumulative Review: Chapters 1–5, pp. 372–373

1. $\frac{1}{2}$ **2.** $\frac{9}{32}$ **3.** $\frac{9}{8}$ **4.** 29   **5.** $\frac{1}{9x^4y^6}$ **6.** t^{10}
7. $3x^4 + 5x^3 + x - 10$ **8.** $-3a^2b - ab^2 + 2b^3$

9. $\dfrac{t^8}{4s^2}$ **10.** $-\dfrac{8x^6y^3}{27z^{12}}$ **11.** 8   **12.** -8 **13.** $-x^4$
14. $2x^3 - 5x^2 + \frac{1}{2}x - 1$ **15.** $-4t^{11} + 8t^9 + 20t^8$
16. $9x^2 - 30x + 25$ **17.** $100x^{10} - y^2$
18. $x^3 - 2x^2 + 1$ **19.** $(c + 1)(c - 1)$
20. $5(x + y)(1 + 2x)$ **21.** $(2r - t)^2$
22. $(2x - 5)(3x - 2)$ **23.** $10(y^2 + 4)$
24. $y(x - 1)(x - 2)$ **25.** $(3x - 2y)(4x + y)$
26. $(5a + 4b)(25a^2 - 20ab + 16b^2)$ **27.** $\frac{1}{12}$
28. 1   **29.** $-\frac{9}{4}$ **30.** $\{x | x \geq 1\}$, or $[1, \infty)$ **31.** $-3, 1$
32. $-4, 3$ **33.** $-2, 2$ **34.** $0, 4$ **35.** $c = \dfrac{a}{b + d}$ **36.** 0
37. $-2; (0, 5)$ **38.** $y = 5x - \frac{1}{3}$ **39.** $y = 5x + \frac{5}{3}$

40. 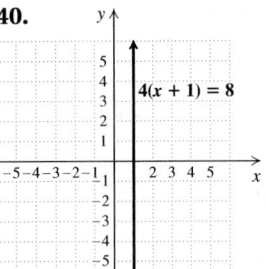   **41.**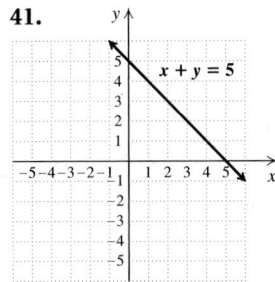

42. 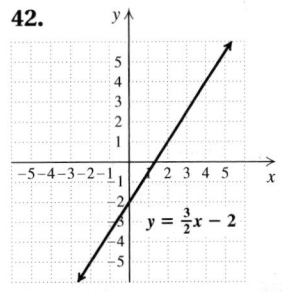   **43.**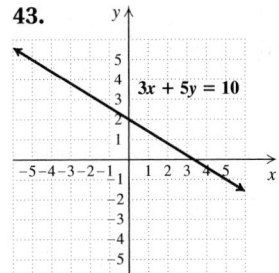

44. 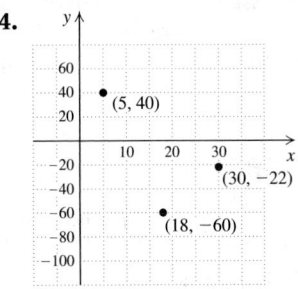   **45.** 372 min
46. 11,616,000 subscribers per year   **47.** 60,000,000 users   **48.** 101,180 people
49. Bottom of ladder to building: 5 ft; top of ladder to ground: 12 ft
50. Length: 12 ft; width: 2 ft
51. Scores that are 9 and higher

52. (a) 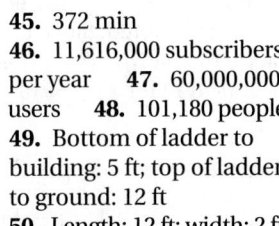   **(b)** 950 mg per serving
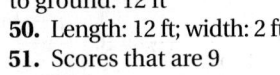

53. $b = \dfrac{2}{a + 1}$ **54.** $y = -8$
55. (a) $9y^2 + 12y + 4 - x^2$; **(b)** $(3y + 2 + x)(3y + 2 - x)$
56. $-1, 0, \frac{1}{3}$

CHAPTER 6

Technology Connection, p. 379

1. Correct   **2.** Correct   **3.** Not correct   **4.** Not correct

Exercise Set 6.1, pp. 381–382

1. (e)   **2.** (a)   **3.** (d)   **4.** (b)   **5.** (c)   **6.** (f)

7. 0   **9.** -5 **11.** 5   **13.** $-4, 7$ **15.** $-6, \frac{1}{2}$ **17.** $\frac{5a}{4b^2}$

19. $\frac{t+2}{t-3}$ **21.** $\frac{7}{8}$ **23.** $\frac{a-3}{a+1}$ **25.** $-\frac{2x^3}{3}$ **27.** $\frac{y-3}{4y}$

29. $\frac{t-4}{t-5}$ **31.** $\frac{a+4}{2(a-4)}$ **33.** $\frac{x-4}{x+4}$ **35.** $\frac{1}{n^2+2n+4}$

37. $t-1$ **39.** $\frac{y^2+4}{y+2}$ **41.** $\frac{1}{2}$ **43.** $\frac{y}{2y+1}$

45. $\frac{2x-3}{5x+2}$ **47.** -1 **49.** -7 **51.** $-\frac{1}{4}$ **53.** $-\frac{3}{2}$

55. -1 **57.** 📙   **59.** $-\frac{4}{21}$ **60.** $-\frac{5}{3}$ **61.** $-\frac{15}{4}$

62. $-\frac{21}{16}$ **63.** $\frac{13}{63}$ **64.** $\frac{5}{48}$ **65.** 📙   **67.** $-(2y+x)$

69. $\frac{x^3+4}{(x^3+2)(x^2+2)}$ **71.** $\frac{(t-1)(t-9)^2}{(t+1)(t^2+9)}$

73. $\frac{(x-y)^3}{(x+y)^2(x-5y)}$ **75.** 📙

Technology Connection, pp. 385–386

1. Let $y_1 = ((x^2 + 3x + 2)/(x^2 + 4))/(5x^2 + 10x)$ and $y_2 = (x + 1)/((x^2 + 4)(5x))$. With the tables set in AUTO mode, note that the values in the $Y1$- and $Y2$-columns match except for $x = -2$.   **2.** ERROR messages occur when division by 0 is attempted. Since the simplified expression has no factor of $x + 5$ or $x + 1$ in a denominator, NO ERROR message occurs in Y2 for $x = -5$ or -1.

Exercise Set 6.2, pp. 386–388

1. $\frac{3x(x+2)}{8(5x-1)}$ **3.** $\frac{(a-4)(a+2)}{(a+6)^2}$ **5.** $\frac{(2x+3)(x+1)}{4(x-5)}$

7. $\frac{(n-4)(n+4)}{(n^2+4)(n^2-4)}$ **9.** $\frac{(y+6)(y-3)}{(1+y)(y+3)}$ **11.** $\frac{6t}{5}$

13. $\frac{4}{c^2d}$ **15.** $\frac{(x-5)(x+2)}{x-2}$ **17.** $\frac{(n-5)(n-1)(n-6)}{(n+6)(n^2+36)}$

19. $\frac{7(a+3)}{a(a+4)}$ **21.** $\frac{3v}{v-2}$ **23.** $\frac{t-5}{t+5}$ **25.** $\frac{9(y-5)}{10(y-1)}$

27. 1   **29.** $\frac{(t-2)(t+5)}{(t-5)(t-5)}$ **31.** $(2x-1)^2$ **33.** $c(c-2)$

35. $\frac{9}{2x}$ **37.** $\frac{1}{a^4+3a}$ **39.** $\frac{20}{27}$ **41.** $\frac{x^2}{20}$ **43.** $\frac{a^3}{b^3}$

45. $\frac{4(t-3)}{3(t+1)}$ **47.** $4(y-2)$ **49.** $-\frac{a}{b}$

51. $\frac{(n+3)(n+3)}{n-2}$ **53.** $\frac{15}{16}$ **55.** $\frac{-x^2-4}{x+2}$

57. $\frac{a-5}{3(a-1)}$ **59.** $\frac{(2x-1)(2x+1)}{x-5}$

61. $\frac{(w-7)(w-8)}{(2w-7)(3w+1)}$ **63.** $\frac{1}{(c-5)(5c-3)}$

65. $\frac{(x-4y)(x-y)}{(x+y)^3}$ **67.** $\frac{x^2+4x+16}{(x+4)^2}$ **69.** $\frac{(2a+b)^2}{2(a+b)}$

71. 📙   **73.** $\frac{19}{12}$ **74.** $\frac{41}{24}$ **75.** $\frac{1}{18}$ **76.** $-\frac{1}{6}$ **77.** $x^2 + 3$

78. $-2x^2 - 4x + 1$ **79.** 📙   **81.** $\frac{3}{7x}$ **83.** 1

85. $\frac{1}{(x+y)^3(3x+y)}$ **87.** $\frac{a^2-2b}{a^2+3b}$ **89.** $\frac{(z+4)^3}{3(z-4)^2}$

91. $\frac{a-3b}{c}$ **93.** 📈

Exercise Set 6.3, pp. 395–397

1. Numerators; denominator   **2.** Term   **3.** Least common denominator; LCD   **4.** Factorizations; denominators   **5.** $\frac{8}{t}$ **7.** $\frac{3x+5}{12}$ **9.** $\frac{9}{a+3}$

11. $\frac{8}{4x-7}$ **13.** $\frac{2y+7}{2y}$ **15.** 6   **17.** $\frac{4(x-1)}{x+3}$

19. $a + 5$ **21.** $y - 7$ **23.** 0   **25.** $\frac{1}{x+2}$ **27.** $\frac{t-4}{t+3}$

29. $\frac{y+2}{y-4}$ **31.** $-\frac{5}{x-4}$, or $\frac{5}{4-x}$ **33.** $-\frac{1}{x-1}$, or $\frac{1}{1-x}$

35. 180   **37.** 72   **39.** 60   **41.** $18t^5$ **43.** $30a^4b^8$

45. $6(y-3)$ **47.** $(x-5)(x+3)(x-3)$

49. $t(t-4)(t+2)^2$ **51.** $120x^2y^3z^2$

53. $(a+1)(a-1)^2$ **55.** $(2n-1)(n+1)(n+2)$

57. $12x^3(x-5)(x-3)(x-1)$

59. $2(x+1)(x-1)(x^2+x+1)$ **61.** $\frac{15}{18t^4}, \frac{st^2}{18t^4}$

63. $\frac{21y}{9x^4y^3}, \frac{4x^3}{9x^4y^3}$ **65.** $\frac{2x(x+3)}{(x-2)(x+2)(x+3)}, \frac{4x(x-2)}{(x-2)(x+2)(x+3)}$ **67.** 📙   **69.** $\frac{-5}{8}, \frac{5}{-8}$

70. $\frac{-4}{11}, -\frac{4}{11}$ **71.** $-x + y$, or $y - x$ **72.** $-3 + a$, or $a - 3$ **73.** $-2x + 7$, or $7 - 2x$ **74.** $-a + b$, or $b - a$ **75.** 📙

77. $\frac{18x+5}{x-1}$ **79.** $\frac{x}{3x+1}$ **81.** 30 strands

83. 60 strands   **85.** $(2x+5)(2x-5)(3x+4)^4$

87. 30 sec   **89.** 7:55 A.M.   **91.** 📙

Exercise Set 6.4, pp. 402–404

1. LCD   **2.** Missing; denominator   **3.** Numerators; LCD

4. Simplify   **5.** $\frac{3+5x}{x^2}$ **7.** $-\frac{5}{24r}$ **9.** $\frac{3u^2+4v}{u^3v^2}$

11. $\frac{-2(xy+9)}{3x^2y^3}$ **13.** $\frac{7x+1}{24}$ **15.** $\frac{-x-4}{6}$

17. $\dfrac{a^2 + 13a - 5}{15a^2}$ **19.** $\dfrac{7z - 12}{12z}$ **21.** $\dfrac{(3c - d)(c + d)}{c^2d^2}$

23. $\dfrac{4x^2 - 13xt + 9t^2}{3x^2t^2}$ **25.** $\dfrac{6x}{(x + 2)(x - 2)}$

27. $\dfrac{(t - 3)(t + 1)}{(t - 1)(t + 3)}$ **29.** $\dfrac{11x + 2}{3x(x + 1)}$ **31.** $\dfrac{-5t + 3}{2t(t - 1)}$

33. $\dfrac{a^2}{(a - 3)(a + 3)}$ **35.** $\dfrac{16}{3(z + 4)}$ **37.** $\dfrac{5q - 3}{(q - 1)^2}$

39. $\dfrac{1}{t^2 + t + 1}$ **41.** $\dfrac{9a}{4(a - 5)}$ **43.** 0

45. $\dfrac{10}{(a - 3)(a + 2)}$ **47.** $\dfrac{x - 5}{(x + 5)(x + 3)}$

49. $\dfrac{3z^2 + 19z - 20}{(z - 2)^2(z + 3)}$ **51.** $\dfrac{-7}{x^2 + 25x + 24}$ **53.** $\dfrac{3x - 1}{2}$

55. $y + 3$ **57.** 0   **59.** $\dfrac{p^2 + 7p + 1}{(p - 5)(p + 5)}$

61. $\dfrac{(x + 1)(x + 3)}{(x - 4)(x + 4)}$ **63.** $\dfrac{-a - 2}{(a + 1)(a - 1)}$, or

$\dfrac{a + 2}{(1 + a)(1 - a)}$ **65.** $\dfrac{2(5x + 3y)}{(x - y)(x + y)}$ **67.** $\dfrac{2x - 3}{2 - x}$

69. 3   **71.** 0   **73.** ✍   **75.** $-\frac{3}{22}$ **76.** $\frac{7}{9}$ **77.** $\frac{9}{10}$

78. $\frac{16}{27}$ **79.** $\frac{2}{3}$ **80.** $\dfrac{(x - 3)(x + 2)}{(x - 2)(x + 3)}$ **81.** ✍

83. Perimeter: $\dfrac{2(5x - 7)}{(x - 5)(x + 4)}$; area: $\dfrac{6}{(x - 5)(x + 4)}$

85. $\dfrac{x}{3x + 1}$ **87.** $\dfrac{x^4 + 4x^3 - 5x^2 - 126x - 441}{(x + 2)^2(x + 7)^2}$

89. $\dfrac{5(a^2 + 2ab - b^2)}{(a - b)(3a + b)(3a - b)}$ **91.** $\dfrac{a}{a - b} + \dfrac{3b}{b - a}$;

answers may vary.   **93.** ✍, 〰

Connecting the Concepts, pp. 404–405

1. Addition; $\dfrac{3x + 10}{5x^2}$ **2.** Multiplication; $\dfrac{6}{5x^3}$

3. Division; $\dfrac{3x}{10}$ **4.** Subtraction; $\dfrac{3x - 10}{5x^2}$

5. Multiplication; $\dfrac{x - 3}{15(x - 2)}$ **6.** Multiplication;

$\dfrac{6}{(x + 3)(x + 4)}$ **7.** Division; $\frac{1}{3}$ **8.** Subtraction;

$\dfrac{x^2 - 2x - 2}{(x - 1)(x + 2)}$ **9.** Addition; $\dfrac{5x + 17}{(x + 3)(x + 4)}$

10. Addition; -5 **11.** Subtraction; $\dfrac{5}{x - 4}$ **12.** Division;

$\dfrac{(2x + 3)(x + 3)}{(x + 1)^2}$ **13.** Subtraction; $\frac{1}{6}$ **14.** Multiplication;

$\dfrac{x(x + 4)}{(x - 1)^2}$ **15.** Addition; $\dfrac{x + 7}{(x - 5)(x + 1)}$ **16.** Division;

$\dfrac{9(u - 1)}{16}$ **17.** Addition; $\dfrac{7t + 8}{30}$ **18.** Multiplication;

$(t + 5)^2$ **19.** Division; $\dfrac{a - 1}{(a + 2)(a - 2)^2}$

20. Subtraction; $\dfrac{-3x^2 + 9x - 14}{2x}$

Technology Connection, p. 409

1. $(1 - 1/x)/(1 - 1/x^2)$ **2.** Parentheses are needed to group separate terms into factors. When a fraction bar is replaced with a division sign, we need parentheses to preserve the groupings that had been created by the fraction bar. This holds for denominators and numerators alike.

Exercise Set 6.5, pp. 409–411

1. (d)   **2.** (a)   **3.** (b)   **4.** (c)   **5.** $\frac{5}{11}$ **7.** 10

9. $\dfrac{5x^2}{4(x^2 + 4)}$ **11.** $\dfrac{-10t}{5t - 2}$ **13.** $\dfrac{2(2a - 5)}{a - 7}$

15. $\dfrac{x^2 - 18}{2(x + 3)}$ **17.** $-\frac{1}{5}$ **19.** $\dfrac{1 + t^2}{t(1 - t)}$ **21.** $\dfrac{x}{x - y}$

23. $\dfrac{c(4c + 7)}{3(2c^2 - 1)}$ **25.** $\dfrac{15(4 - a^3)}{14a^2(9 + 2a)}$ **27.** 1

29. $\dfrac{3a^2 + 4b^3}{b^3(5 - 3a^2)}$ **31.** $\dfrac{(t - 3)(t + 3)}{t^2 + 4}$ **33.** $\dfrac{a^2b^2}{b^2 - ab + a^2}$

35. $\dfrac{a(3ab^3 + 4)}{b^2(3 + a)}$ **37.** $\dfrac{t^2 + 5t + 3}{(t + 1)^2}$ **39.** $\dfrac{x^2 - 2x - 1}{x^2 - 5x - 4}$

41. ✍   **43.** -4 **44.** -4 **45.** $\frac{19}{3}$ **46.** $-\frac{14}{27}$ **47.** 3, 4

48. $-15, 2$ **49.** ✍   **51.** 6, 7, 8   **53.** $-3, -\frac{4}{5}, 3$

55. $\dfrac{A}{B} \div \dfrac{C}{D} = \dfrac{\frac{A}{B}}{\frac{C}{D}} = \dfrac{\frac{A}{B}}{\frac{C}{D}} \cdot \dfrac{BD}{BD} = \dfrac{AD}{BC} = \dfrac{A}{B} \cdot \dfrac{D}{C}$

57. $\dfrac{x^2 + 5x + 15}{-x^2 + 10}$ **59.** 0   **61.** $\dfrac{2z(5z - 2)}{(z + 2)(13z - 6)}$

63. ✍

Exercise Set 6.6, pp. 417–418

1. False   **2.** True   **3.** True   **4.** True   **5.** $-\frac{2}{5}$ **7.** $\frac{6}{7}$

9. $\frac{24}{5}$ **11.** $-4, -1$ **13.** $-6, 6$ **15.** 12   **17.** $\frac{14}{3}$

19. -10 **21.** $-4, -3$ **23.** $\frac{23}{2}$ **25.** $\frac{5}{2}$ **27.** -1

29. No solution   **31.** -10 **33.** $-\frac{7}{3}$ **35.** $-2, \frac{7}{3}$

37. $-3, 13$ **39.** 2   **41.** -8 **43.** No solution

45. $-6, 6$ **47.** ✍   **49.** 137, 139   **50.** 14 yd

51. Base: 9 cm; height: 12 cm   **52.** $-8, -6; 6, 8$

53. 0.06 cm per day   **54.** 0.28 in. per day   **55.** ✍

57. -2 **59.** 3   **61.** 4   **63.** 4   **65.** -2 **67.** 〰

Connecting the Concepts, pp. 418–419

1. Expression; $\dfrac{x - 2}{x + 1}$ **2.** Expression; $\dfrac{19n - 2}{5n(2n - 1)}$

3. Equation; 8   **4.** Expression; $\dfrac{(z + 1)^2(z - 1)}{z}$

5. Equation; $-\frac{1}{2}$ **6.** Expression; $\dfrac{8(t + 1)}{(t - 1)(2t - 1)}$

7. Expression; $\dfrac{2a}{a-1}$ **8.** Equation; $-5, 3$

9. Expression; $\dfrac{3x}{10}$ **10.** Equation; $-10, 10$

Translating for Success, p. 429

1. K   **2.** E   **3.** C   **4.** N   **5.** D   **6.** O   **7.** F   **8.** H
9. B   **10.** A

Exercise Set 6.7, pp. 430–435

1. $\frac{1}{2}$ cake per hour   **2.** $\frac{1}{3}$ cake per hour   **3.** $\frac{5}{6}$ cake per hour
4. 1 lawn per hour   **5.** $\frac{1}{3}$ lawn per hour   **6.** $\frac{2}{3}$ lawn per hour
7. $3\frac{3}{7}$ hr   **9.** $\frac{42}{13}$ min, or $3\frac{3}{13}$ min   **11.** $19\frac{1}{11}$ min
13. H470: 135 min; K5400: $67\frac{1}{2}$ min   **15.** Austin: 15 min;
Airgle: 30 min   **17.** 300 min, or 5 hr
19.

	Distance (in km)	Speed (in km/h)	Time (in hours)
B & M	330	$r - 14$	$\dfrac{330}{r-14}$
AMTRAK	400	r	$\dfrac{400}{r}$

AMTRAK: 80 km/h; B & M: 66 km/h
21. 7 mph   **23.** 5.2 ft/sec   **25.** 3 hr   **27.** 9 km/h
29. 2 km/h   **31.** 20 mph   **33.** 10.5   **35.** $\frac{8}{3}$ **37.** $3\frac{3}{4}$ in.
39. 20 ft   **41.** 15 ft   **43.** 12.6   **45.** 1440 messages
47. About 231,000 people   **49.** 18 gal   **51.** $26\frac{2}{3}$ cm
53. 52 bulbs   **55.** $7\frac{1}{2}$ oz   **57.** 90 whales
59. (a) 1.92 T; **(b)** 28.8 lb   **61.**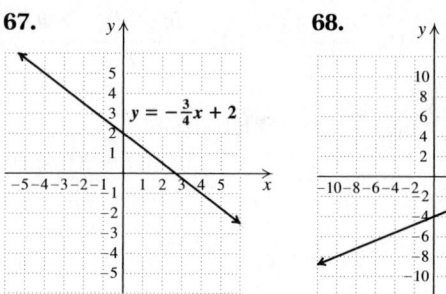
63.
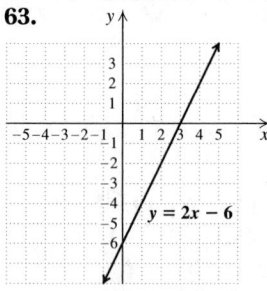
64.
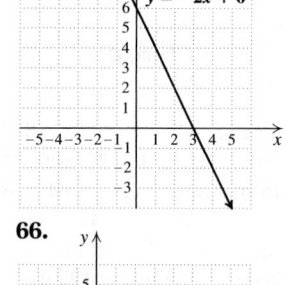
65.
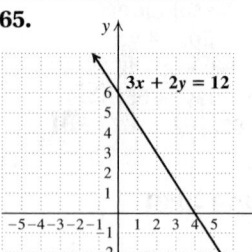
66.
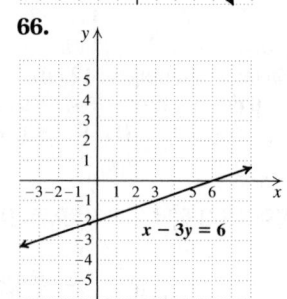

67.

68.

69. ⌖   **71.** $49\frac{1}{2}$ hr   **73.** 2250 people per hour
75. $14\frac{7}{8}$ mi   **77.** About 57%   **79.** Page 278
81. $\dfrac{B}{A} = \dfrac{D}{C}; \dfrac{A}{C} = \dfrac{B}{D}; \dfrac{C}{A} = \dfrac{D}{B}$ **83.** $51\frac{3}{7}$ mph   **85.** ⌖

Review Exercises: Chapter 6, pp. 439–440

1. False   **2.** True   **3.** False   **4.** True   **5.** True
6. False   **7.** False   **8.** False   **9.** 0   **10.** -5
11. $-6, 6$ **12.** $-6, 5$ **13.** -2 **14.** $\dfrac{x-3}{x+5}$ **15.** $\dfrac{7x+3}{x-3}$
16. $\dfrac{3(y-3)}{2(y+3)}$ **17.** $-5(x+2y)$ **18.** $\dfrac{a-6}{5}$
19. $\dfrac{3(y-2)^2}{4(2y-1)(y-1)}$ **20.** $-32t$ **21.** $\dfrac{2x(x-1)}{x+1}$
22. $\dfrac{(x^2+1)(2x+1)}{(x-2)(x+1)}$ **23.** $\dfrac{(t+4)^2}{t+1}$ **24.** $60a^5b^8$
25. $x^4(x-1)(x+1)$ **26.** $(y-2)(y+2)(y+1)$
27. $\dfrac{15-3x}{x+3}$ **28.** $\dfrac{4}{x-4}$ **29.** $\dfrac{x+5}{2x}$
30. $\dfrac{2a^2-21ab+15b^2}{5a^2b^2}$ **31.** $y-4$ **32.** $\dfrac{t(t-2)}{(t-1)(t+1)}$
33. $d+2$ **34.** $\dfrac{-x^2+x+26}{(x+1)(x-5)(x+5)}$ **35.** $\dfrac{2(x-2)}{x+2}$
36. $\dfrac{3(7t+2)}{4t(3t+2)}$ **37.** $\dfrac{z}{1-z}$ **38.** $\dfrac{10x}{3x^2+16}$ **39.** $c-d$
40. 4   **41.** $\frac{7}{2}$ **42.** $-6, -1$ **43.** $5\frac{1}{7}$ hr   **44.** Core 2
Duo: 45 sec; Core 2 Quad: 30 sec   **45.** Car: 105 km/h;
train: 90 km/h   **46.** 24 mph   **47.** 55 seals
48. 6   **49.** ⌖ The LCM of denominators is used to clear
fractions when simplifying a complex rational expression
using the method of multiplying by the LCD, and when
solving rational equations.   **50.** ⌖ Although multiplying
the denominators of the expressions being added results in a
common denominator, it is often not the *least* common
denominator. Using a common denominator other than the
LCD makes the expressions more complicated, requires
additional simplifying after the addition has been performed,
and leaves more room for error.   **51.** $\dfrac{5(a+3)^2}{a}$
52. $\dfrac{10a}{(a-b)(b-c)}$ **53.** 0

Test: Chapter 6, pp. 440–441

1. [6.1] 0   **2.** [6.1] −8   **3.** [6.1] −1, 1   **4.** [6.1] 1, 2
5. [6.1] $\dfrac{3x + 7}{x + 3}$ **6.** [6.2] $\dfrac{2t(t + 3)}{3(t - 1)}$
7. [6.2] $\dfrac{(5y + 1)(y + 1)}{3y(y + 2)}$ **8.** [6.2] $\dfrac{(2a + 1)(4a^2 + 1)}{4a^2(2a - 1)}$
9. [6.2] $(x + 3)(x - 3)$ **10.** [6.3] $(y - 3)(y + 3)(y + 7)$
11. [6.3] $\dfrac{-3x + 9}{x^3}$ **12.** [6.3] $\dfrac{-2t + 8}{t^2 + 1}$ **13.** [6.4] 1
14. [6.4] $\dfrac{3x - 5}{x - 3}$ **15.** [6.4] $\dfrac{11t - 8}{t(t - 2)}$
16. [6.4] $\dfrac{y^2 + 3}{(y - 1)(y + 3)^2}$ **17.** [6.4] $\dfrac{x^2 + 2x - 7}{(x + 1)(x - 1)^2}$
18. [6.5] $\dfrac{3y + 1}{y}$ **19.** [6.5] $x - 8$ **20.** [6.6] $\frac{8}{3}$
21. [6.6] −3, 5   **22.** [6.7] 12 min   **23.** [6.7] $1\frac{1}{4}$ mi
24. [6.7] Ryan: 65 km/h; Alicia: 45 km/h   **25.** [6.7] Rema:
4 hr; Pe'rez: 10 hr   **26.** [6.5] a **27.** [6.7] −1

Cumulative Review: Chapters 1–6, pp. 441–442

1. $a + cb$ **2.** −25   **3.** 25   **4.** $-3x + 33$ **5.** 10
6. −7, 7   **7.** $-\frac{5}{3}$ **8.** $\frac{8}{5}$ **9.** 4   **10.** −10, −1   **11.** −8
12. 1, 4   **13.** $\left\{y \,|\, y \le -\frac{2}{3}\right\}$ **14.** −17   **15.** $-4, \frac{1}{2}$
16. $\{x \,|\, x > 43\}$ **17.** 5   **18.** $-\frac{7}{2}, 5$ **19.** −13
20. $b = 3a - c + 9$ **21.** $y = \dfrac{4z - 3x}{6}$

22.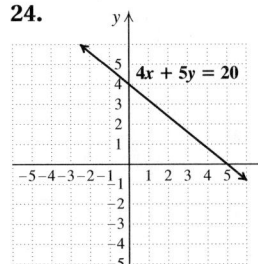
23.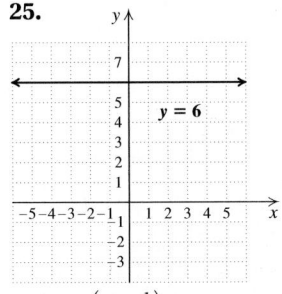

24.
25.

26. −2   **27.** Slope: $\frac{1}{2}$; y-intercept: $\left(0, -\frac{1}{4}\right)$
28. $y = -\frac{5}{8}x - 4$ **29.** $\dfrac{1}{x^2}$ **30.** y^{-7}, or $\dfrac{1}{y^7}$
31. $-4a^4b^{14}$ **32.** $-y^3 - 2y^2 - 2y + 7$
33. $-15a + 10b - 5c$ **34.** $2x^5 + x^3 - 6x^2 - x + 3$
35. $36x^2 - 60xy + 25y^2$ **36.** $3n^2 - 13n - 10$
37. $4x^6 - 1$ **38.** $2x(3 - x - 12x^3)$
39. $(4x + 9)(4x - 9)$ **40.** $10(t + 1)(t^2 - t + 1)$
41. $(4x + 3)(2x + 1)$ **42.** $2(3x - 2)(x - 4)$

43. $(5t + 4)^2$ **44.** $(xy - 5)(xy + 4)$
45. $(x + 2)(x^3 - 3)$ **46.** $\dfrac{4}{t + 4}$ **47.** 1
48. $\dfrac{a^2 + 7ab + b^2}{(a + b)(a - b)}$ **49.** $\dfrac{2x + 5}{4 - x}$ **50.** $\dfrac{x}{x - 2}$
51. $\dfrac{y(3y^2 + 2)}{y^3 - 3}$ **52.** $6x^2 - 5x + 2 + \dfrac{4}{x} + \dfrac{1}{x^2}$
53. $15x^3 - 57x^2 + 177x - 529 + \dfrac{1605}{x + 3}$ **54.** 15 books
55. About 340.8 billion admissions   **56.** 33,339,507 and
33,339,508   **57.** 12 ft   **58.** $\frac{75}{4}$ min, or $18\frac{3}{4}$ min
59. 90 antelope   **60.** 150 calories   **61.** −144, 144
62. $16y^6 - y^4 + 6y^2 - 9$ **63.** −7, 4, 12   **64.** 18

CHAPTER 7

Exercise Set 7.1, pp. 450–455

1. Correspondence   **2.** Exactly   **3.** Domain
4. Range   **5.** Horizontal   **6.** Vertical   **7.** "f of 3," "f at 3,"
or "the value of f at 3"   **8.** Vertical   **9.** Yes   **11.** Yes
13. No   **15.** Yes   **17.** Function   **19.** Function
21. (a) −1; (b) −3   **23.** (a) 3; (b) 3   **25.** (a) 3; (b) 0
27. (a) 3; (b) −3   **29.** (a) 1; (b) 3   **31.** (a) 4; (b) −1, 3
33. (a) 2; (b) $\{x \,|\, 0 < x \le 2\}$, or $(0, 2]$ **35.** Yes   **37.** Yes
39. No   **41.** (a) 5; (b) −3; (c) −9; (d) 21;
(e) $2a + 9$; (f) $2a + 7$ **43.** (a) 0; (b) 1; (c) 57;
(d) $5t^2 + 4t$; (e) $20a^2 + 8a$; (f) 48   **45.** (a) $\frac{3}{5}$; (b) $\frac{1}{3}$; (c) $\frac{4}{7}$;
(d) 0; (e) $\dfrac{x - 1}{2x - 1}$; (f) $\dfrac{a + h - 3}{2a + 2h - 5}$
47. $4\sqrt{3}$ cm$^2 \approx 6.93$ cm^2 **49.** 36π in$^2 \approx 113.10$ in^2
51. 164.98 cm   **53.** 23°F   **55.** 75 heart attacks per
10,000 men   **57.** 500 movies
59. 19 watts; 30 watts

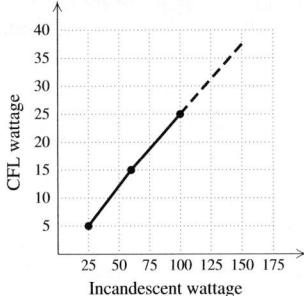

61. 3.5 drinks; 6 drinks

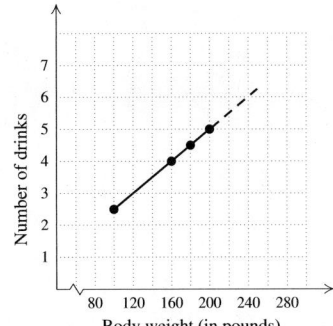

63. $257,000; $306,000

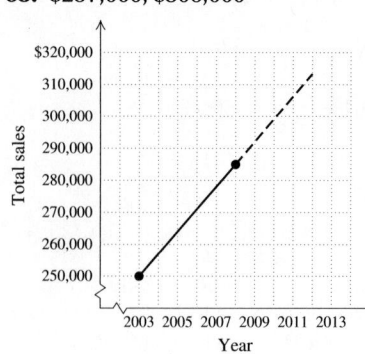

85.

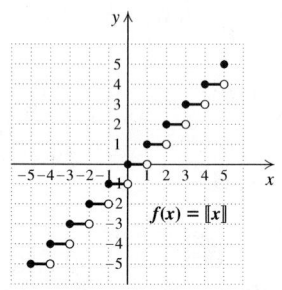

65. 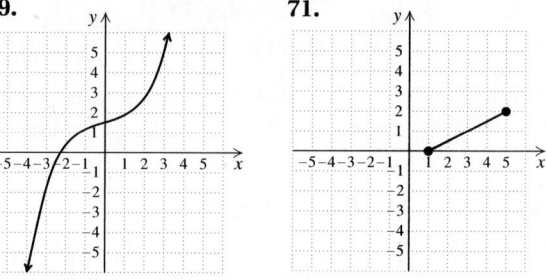 **67.** 6
68. $-6, 6$ **69.** $-\frac{1}{2}$
70. $-1, 1$
71. $-4, 3$ **72.** $\frac{9}{32}$
73. $\frac{1}{7}$ **74.** $-4, 8$
75. **77.** 26; 99
79. Worm
81. About 2 min 50 sec
83. 1 every 3 min

63. Slope: $\frac{2}{3}$; y-intercept: $(0, -4)$
64. Slope: $-\frac{1}{4}$; y-intercept: $(0, 6)$
65. Slope: $\frac{4}{3}$; y-intercept: $(0, 0)$
66. Slope: -5; y-intercept: $(0, 0)$ **67.**
69. **71.**

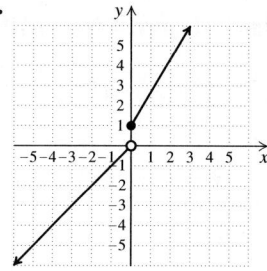 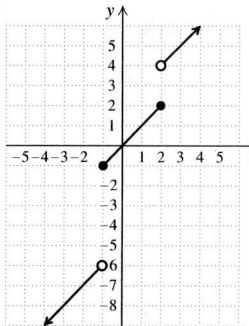

73. Domain: $\{x \mid x$ is a real number $and\ x \neq 0\}$;
range: $\{y \mid y$ is a real number $and\ y \neq 0\}$
75. Domain: $\{x \mid x < -2\ or\ x > 0\}$; range:
$\{y \mid y < -2\ or\ y > 3\}$ **77.** Domain: $\mathbb{R}$; range: $\{y \mid y \geq 0\}$,
or $[0, \infty)$ **79.** Domain: $\{x \mid x$ is a real number $and\ x \neq 2\}$;
range: $\{y \mid y$ is a real number $and\ y \neq 0\}$
81. $\{h \mid 0 \leq h \leq 144\}$, or $[0, 144]$
83. **85.**

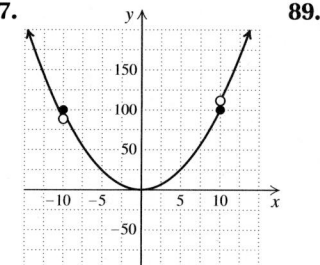

87. **89.**

Exercise Set 7.2, pp. 461–463

1. (c) **2.** (d) **3.** (d) **4.** (b) **5.** (c) **6.** (b)
7. Domain: $\{2, 9, -2, -4\}$; range: $\{8, 3, 10, 4\}$
9. Domain: $\{0, 4, -5, -1\}$; range: $\{0, -2\}$
11. Domain: $\{-4, -2, 0, 2, 4\}$; range: $\{-2, -1, 0, 1, 2\}$
13. Domain: $\{-5, -3, -1, 0, 2, 4\}$; range: $\{-1, 1\}$
15. Domain: $\{x \mid -4 \leq x \leq 3\}$, or $[-4, 3]$; range:
$\{y \mid -3 \leq y \leq 4\}$, or $[-3, 4]$ **17.** Domain:
$\{x \mid -4 \leq x \leq 5\}$, or $[-4, 5]$; range: $\{y \mid -2 \leq y \leq 4\}$,
or $[-2, 4]$ **19.** Domain: $\{x \mid -4 \leq x \leq 4\}$, or $[-4, 4]$;
range: $\{-3, -1, 1\}$ **21.** Domain: $\mathbb{R}$; range: $\mathbb{R}$
23. Domain: $\mathbb{R}$; range: $\{4\}$ **25.** Domain: $\mathbb{R}$; range:
$\{y \mid y \geq 1\}$, or $[1, \infty)$ **27.** Domain: $\{x \mid x$ is a real number
$and\ x \neq -2\}$; range: $\{y \mid y$ is a real number $and\ y \neq -4\}$
29. Domain: $\{x \mid x \geq 0\}$, or $[0, \infty)$; range: $\{y \mid y \geq 0\}$,
or $[0, \infty)$ **31.** $\{x \mid x$ is a real number $and\ x \neq 3\}$
33. $\{x \mid x$ is a real number $and\ x \neq \frac{1}{2}\}$ **35.** $\mathbb{R}$ **37.** $\mathbb{R}$
39. $\{x \mid x$ is a real number $and\ x \neq 3\ and\ x \neq -3\}$ **41.** $\mathbb{R}$
43. $\{x \mid x$ is a real number $and\ x \neq -1\ and\ x \neq -7\}$
45. $\{t \mid 0 \leq t < 624\}$, or $[0, 624)$
47. $\{p \mid \$0 \leq p \leq \$10.60\}$, or $[0, 10.60]$ **49.** $\{d \mid d \geq 0\}$,
or $[0, \infty)$ **51.** $\{t \mid 0 \leq t \leq 5\}$, or $[0, 5]$ **53.** (a) -5;
(b) 1; (c) 21 **55.** (a) 0; (b) 2; (c) 7 **57.** (a) 100;
(b) 100; (c) 131 **59.**
61. **62.**

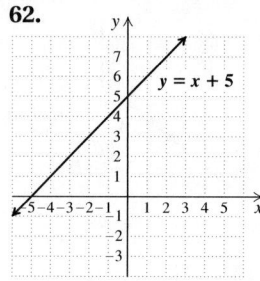

Visualizing for Success, p. 469

1. D **2.** J **3.** A **4.** E **5.** B **6.** C **7.** I **8.** F
9. G **10.** H

Exercise Set 7.3, pp. 470–473

1. False **2.** True **3.** False **4.** True
5. True **6.** True

7.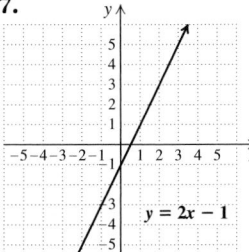

$y = 2x - 1$

9.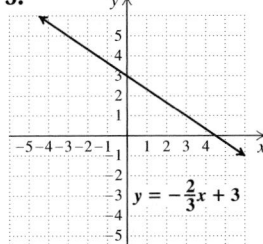

$y = -\frac{2}{3}x + 3$

11.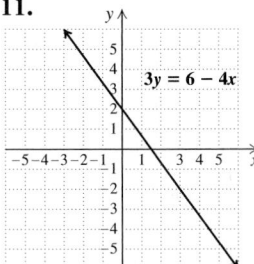

$3y = 6 - 4x$

13.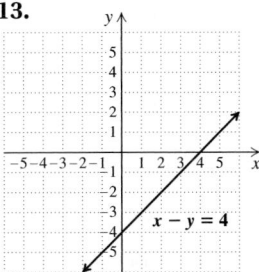

$x - y = 4$

15.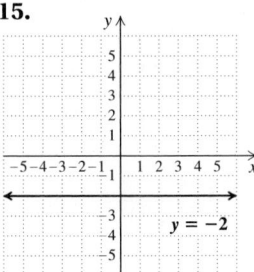

$y = -2$

17.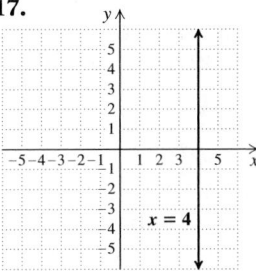

$x = 4$

19.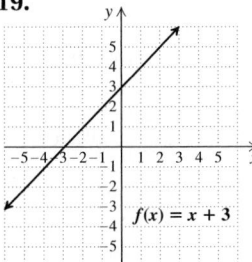

$f(x) = x + 3$

21.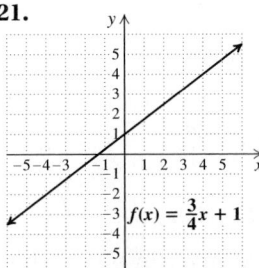

$f(x) = \frac{3}{4}x + 1$

23.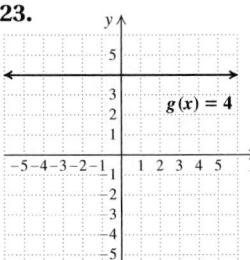

$g(x) = 4$

25. $C(d) = 0.75d + 30$; 60 miles
27. $L(t) = \frac{1}{2}t + 5$; 20 months after the haircut
29. $A(t) = 849t + 5960$; 5 years after 2006, or 2011
31. (a) $a(t) = 0.9t + 12.7$; (b) 24.4 million cars; (c) about 2014
33. (a) $E(t) = 0.14t + 78.44$; (b) 81.52 yr
35. (a) $A(t) = 22.525t + 236.95$; (b) \$462.2 million
37. (a) $N(t) = 9.4t + 16$; (b) 110 million Americans; (c) 2015   **39.** (a) $A(t) = \frac{41}{110}t + \frac{1615}{22}$; (b) about 80.9 million acres   **41.** Linear function; $\mathbb{R}$
43. Quadratic function; $\mathbb{R}$ **45.** Rational function; $\{t \mid t \text{ is a real number } and\ t \neq -\frac{4}{3}\}$ **47.** Polynomial function; $\mathbb{R}$
49. Rational function; $\{x \mid x \text{ is a real number } and\ x \neq \frac{5}{2}\}$
51. Rational function; $\{n \mid n \text{ is a real number } and\ n \neq -1\ and\ n \neq -2\}$ **53.** Linear function; $\mathbb{R}$ **55.** $\{y \mid y \geq 0\}$, or $[0, \infty)$ **57.** $\mathbb{R}$ **59.** $\{y \mid y \leq 0\}$, or $(-\infty, 0]$

61.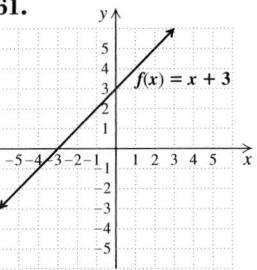

$f(x) = x + 3$

63.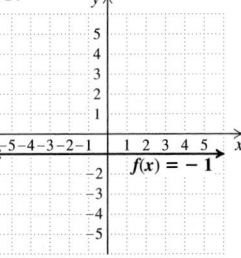

$f(x) = -1$

Domain: $\mathbb{R}$; range: $\mathbb{R}$ Domain: $\mathbb{R}$; range: $\{-1\}$

65.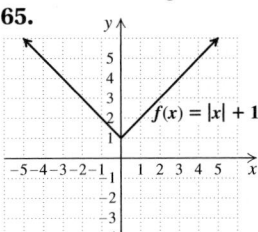

$f(x) = |x| + 1$

67.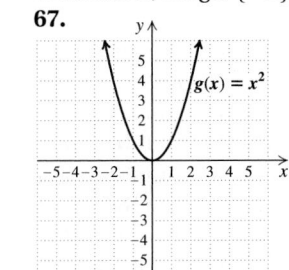

$g(x) = x^2$

Domain: $\mathbb{R}$;
range: $\{y \mid y \geq 1\}$, or $[1, \infty)$
Domain: $\mathbb{R}$;
range: $\{y \mid y \geq 0\}$, or $[0, \infty)$
69. ✍   **71.** $4x^2 + 2x - 1$ **72.** $2x^3 - x^2 - x + 7$
73. $2x^2 - 13x - 7$ **74.** $x^2 + x - 12$
75. $x^3 - 2x^2 - 3x + 5$ **76.** $x^3 + 2x^2 + x + 4$
77. ✍   **79.** False   **81.** False   **83.** 21.1°C   **85.** \$60
87. (a) $g(x) = x - 8$; (b) -10; (c) 83

Connecting the Concepts, p. 474

1. $\{-1, 0, 3, 4\}$ **2.** $\{-2, 0, 6, 8\}$ **3.** -2
4.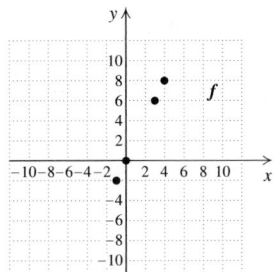

f

5. $\mathbb{R}$

6. $\{x \mid x \text{ is a real number } and\ x \neq 0\}$ **7.** $\frac{1}{5}$ **8.** 2   **9.** 3
10. $\{x \mid x \text{ is a real number } and\ x \neq 0\ and\ x \neq 1\}$
11. Linear function   **12.** Rational function
13. Range: $\mathbb{R}$

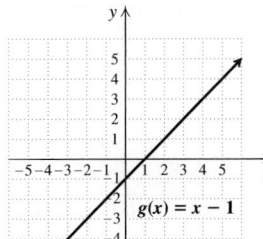

$g(x) = x - 1$

14. Yes   **15.** -4
16. $\{x \mid x \geq -3\}$, or $[-3, \infty)$
17. $\{y \mid y \geq -4\}$, or $[-4, \infty)$
18. 4   **19.** 10   **20.** 13

Exercise Set 7.4, pp. 480–483

1. Sum   **2.** Subtract   **3.** Evaluate   **4.** Domains
5. Excluding   **6.** Sum   **7.** 1   **9.** 5   **11.** -7 **13.** 1
15. -5 **17.** $x^2 - 2x - 2$ **19.** $x^2 - x + 3$ **21.** 5

23. $-a^3 + 5a^2 + 2a - 10$ **25.** $\dfrac{x^2 - 2}{5 - x}, x \neq 5$ **27.** $\frac{7}{2}$

29. -2 **31.** $1.2 + 2.9 = 4.1$ million   **33.** 4%
35. About 95 million; the number of tons of municipal solid waste that was composted or recycled in 2005   **37.** About 215 million; the number of tons of municipal solid waste in 1996   **39.** About 230 million; the number of tons of municipal solid waste that was not composted in 2004
41. $\mathbb{R}$ **43.** $\{x \mid x$ is a real number $and\ x \neq -5\}$
45. $\{x \mid x$ is a real number $and\ x \neq 0\}$
47. $\{x \mid x$ is a real number $and\ x \neq 1\}$
49. $\{x \mid x$ is a real number $and\ x \neq -\frac{9}{2}$ $and\ x \neq 1\}$
51. $\{x \mid x$ is a real number $and\ x \neq 3\}$
53. $\{x \mid x$ is a real number $and\ x \neq -4\}$
55. $\{x \mid x$ is a real number $and\ x \neq 4$ $and\ x \neq 5\}$
57. $\{x \mid x$ is a real number $and\ x \neq -1$ $and\ x \neq -\frac{5}{2}\}$
59. 4; 3   **61.** 5; -1 **63.** $\{x \mid 0 \leq x \leq 9\}$;
$\{x \mid 3 \leq x \leq 10\}$; $\{x \mid 3 \leq x \leq 9\}$; $\{x \mid 3 \leq x \leq 9\}$

65.

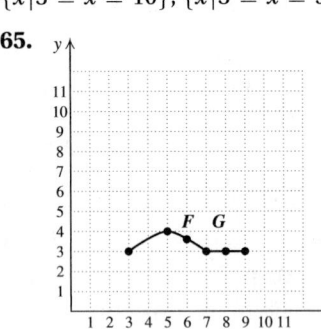

67. 🖾   **69.** $c = \dfrac{b}{a}$

70. $w = \dfrac{x - y}{z}$

71. $q = \dfrac{st}{p - r}$

72. $b = \dfrac{d}{a + c}$

73. $b = \dfrac{cd + d}{a - 3}$

74. $d = \dfrac{ab - 3b}{c + 1}$

75. 🖾   **77.** $\{x \mid x$ is a real number $and\ x \neq 4$ $and\ x \neq 3$ $and\ x \neq 2$ $and\ x \neq -2\}$
79. Answers may vary.

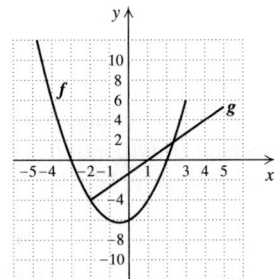

81. $\{x \mid x$ is a real number $and\ -1 < x < 5$ $and\ x \neq \frac{3}{2}\}$

83. Answers may vary. $f(x) = \dfrac{1}{x + 2}, g(x) = \dfrac{1}{x - 5}$

85. 〰

Exercise Set 7.5, pp. 490–495

1. (d)   **2.** (f)   **3.** (e)   **4.** (b)   **5.** (a)   **6.** (c)
7. Inverse   **8.** Direct   **9.** Direct   **10.** Inverse

11. Inverse   **12.** Direct   **13.** $d = \dfrac{L}{f}$

15. $v_1 = \dfrac{2s}{t} - v_2$, or $\dfrac{2s - tv_2}{t}$ **17.** $b = \dfrac{at}{a - t}$

19. $g = \dfrac{Rs}{s - R}$ **21.** $n = \dfrac{IR}{E - Ir}$ **23.** $q = \dfrac{pf}{p - f}$

25. $t_1 = \dfrac{H}{Sm} + t_2$, or $\dfrac{H + Smt_2}{Sm}$ **27.** $r = \dfrac{Re}{E - e}$

29. $r = 1 - \dfrac{a}{S}$, or $\dfrac{S - a}{S}$ **31.** $a + b = \dfrac{f}{c^2}$

33. $r = \dfrac{A}{P} - 1$, or $\dfrac{A - P}{P}$

35. $t_1 = t_2 - \dfrac{d_2 - d_1}{v}$, or $\dfrac{vt_2 - d_2 + d_1}{v}$ **37.** $t = \dfrac{ab}{b + a}$

39. $Q = \dfrac{2Tt - 2AT}{A - q}$ **41.** $w = \dfrac{4.15c - 98.42}{p + 0.082}$

43. $k = 6; y = 6x$ **45.** $k = 1.7; y = 1.7x$

47. $k = 10; y = 10x$ **49.** $k = 100; y = \dfrac{100}{x}$

51. $k = 44; y = \dfrac{44}{x}$ **53.** $k = 9; y = \dfrac{9}{x}$ **55.** $33\frac{1}{3}$ cm

57. 3.5 hr   **59.** 56 in.   **61.** 32 kg   **63.** 286 Hz
65. About 21 min   **67.** About 110,000,000 tons

69. $y = \frac{1}{2}x^2$ **71.** $y = \dfrac{5000}{x^2}$ **73.** $y = 1.5xz$

75. $y = \dfrac{4wx^2}{z}$ **77.** 61.3 ft   **79.** 308 cm^3

81. About 57 mph   **83.** 🖾   **85.** $y = \frac{1}{6}x - \frac{1}{2}$
86. $y = \frac{3}{8}x - \frac{5}{8}$ **87.** $y = -\frac{5}{2}x - \frac{3}{2}$ **88.** $y = -\frac{1}{8}x + \frac{1}{2}$
89. Let n represent the number; $2n + 5 = 49$ **90.** Let x represent the number; $\frac{1}{2}x - 3 = 57$ **91.** Let x represent the number; $x + (x + 1) = 145$ **92.** Let n represent the number; $n - (-n) = 20$ **93.** 🖾   **95.** 567 mi

97. Ratio is $\dfrac{a + 12}{a + 6}$; percent increase is $\dfrac{6}{a + 6} \cdot 100\%$, or

$\dfrac{600}{a + 6}\%$ **99.** $t_1 = t_2 + \dfrac{(d_2 - d_1)(t_4 - t_3)}{a(t_4 - t_2)(t_4 - t_3) + d_3 - d_4}$
101. The intensity is halved.   **103.** About 1.7 m

105. $d(s) = \dfrac{28}{s}$; 70 yd

Review Exercises: Chapter 7, pp. 498–500

1. True   **2.** True   **3.** False   **4.** True   **5.** False
6. True   **7.** True   **8.** True   **9.** True   **10.** False
11. (a) 3; **(b)** $\{x \mid -2 \leq x \leq 4\}$, or $[-2, 4]$; **(c)** -1;
(d) $\{y \mid 1 \leq y \leq 5\}$, or $[1, 5]$ **12.** $\frac{3}{2}$ **13.** $4a^2 + 4a - 3$
14. 10.53 yr   **15.** About $9.30;

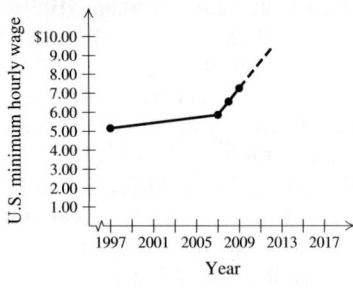

16. (a) Yes;   **(b)** Domain: $\mathbb{R}$; range: $\{y \mid y \geq 0\}$, or $[0, \infty)$

17. **(a)** No   **18.** **(a)** No   **19.** **(a)** Yes; **(b)** Domain: $\mathbb{R}$;
range: $\{-2\}$ **20.** $\mathbb{R}$
21. $\{x \mid x \text{ is a real number } and \ x \neq 1\}$
22. $\{t \mid t \text{ is a real number } and \ t \neq -1 \ and \ t \neq -4\}$
23. $\{t \mid 0 \leq t \leq 60\}$, or $[0, 60]$ **24.** **(a)** 5; **(b)** 4;
(c) 16; **(d)** 35   **25.** $C(t) = 30t + 90$; 7 months
26. **(a)** $R(t) = -\frac{43}{2400}t + \frac{15{,}843}{800}$; **(b)** about 19.21 sec; about
19.09 sec   **27.** Absolute-value function   **28.** Polynomial
function   **29.** Quadratic function   **30.** Linear function
31. Rational function
32.

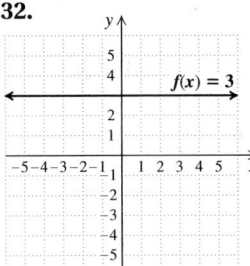

33.

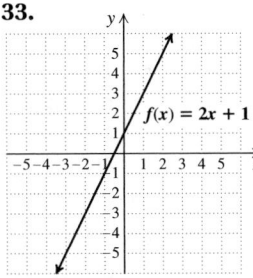

Domain: $\mathbb{R}$; range: $\{3\}$ Domain: $\mathbb{R}$; range: $\mathbb{R}$
34.

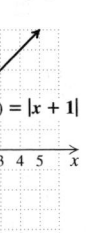

35. 102   **36.** -17
37. $-\frac{9}{2}$ **38.** $\mathbb{R}$

Domain: $\mathbb{R}$; range: $\{y \mid y \geq 0\}$,
or $[0, \infty)$

39. $\{x \mid x \text{ is a real number } and \ x \neq 2\}$ **40.** $r = \dfrac{2V - IR}{2I}$,
or $\dfrac{V}{I} - \dfrac{R}{2}$ **41.** $m = \dfrac{H}{S(t_1 - t_2)}$ **42.** $c = \dfrac{b + 3a}{2}$
43. $t_1 = \dfrac{-A}{vT} + t_2$, or $\dfrac{-A + vTt_2}{vT}$ **44.** $y = \frac{15}{2}x$
45. $y = \dfrac{\frac{3}{4}}{x}$ **46.** $y = \dfrac{1}{2}\dfrac{xw^2}{z}$ **47.** 15 min
48. 22.5 lb   **49.** About 2.9 sec   **50.** ✍ Two functions
that have the same domain and range are not necessarily
identical. For example, the functions f: $\{(-2, 1), (-3, 2)\}$
and g: $\{(-2, 2), (-3, 1)\}$ have the same domain and range
but are different functions.   **51.** ✍ Jenna is not correct.
Any value of the variable that makes a denominator 0
is not in the domain; 0 itself may or may not make a
denominator 0.   **52.** $f(x) = 10.94x + 20$ **53.** Domain:
$\{x \mid x \geq -4 \ and \ x \neq 2\}$; range: $\{y \mid y \geq 0 \ and \ y \neq 3\}$

Test: Chapter 7, pp. 501–502

1. [7.1], [7.2] **(a)** 1; **(b)** $\{x \mid -3 \leq x \leq 4\}$, or $[-3, 4]$; **(c)** 3;
(d) $\{y \mid -1 \leq y \leq 2\}$, or $[-1, 2]$ **2.** [7.1] About 49 million
international visitors   **3.** **(a)** [7.1] Yes; **(b)** [7.2] domain: $\mathbb{R}$;
range: $\mathbb{R}$ **4.** **(a)** [7.1] Yes; **(b)** [7.2] domain: $\mathbb{R}$;
range: $\{y \mid y \geq 1\}$, or $[1, \infty)$ **5.** **(a)** [7.1] No
6. [7.2] $\{t \mid 0 \leq t \leq 4\}$, or $[0, 4]$ **7.** [7.2] **(a)** -5; **(b)** 10
8. [7.3] $C(t) = 55t + 180$; 12 months

9. [7.3] **(a)** $C(m) = 0.3m + 25$; **(b)** \$175
10. [7.3] Linear function; $\mathbb{R}$ **11.** [7.3] Rational function;
$\{x \mid x \text{ is a real number } and \ x \neq -4 \ and \ x \neq 4\}$
12. [7.3] Quadratic function; $\mathbb{R}$
13. [7.3]

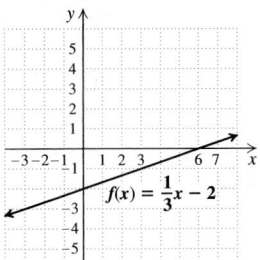

14. [7.3]

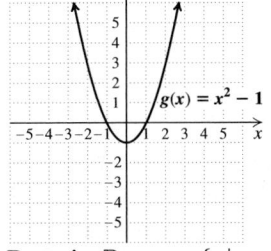

Domain: $\mathbb{R}$; range: $\mathbb{R}$ Domain: $\mathbb{R}$; range: $\{y \mid y \geq -1\}$,
or $[-1, \infty)$

15. [7.3]

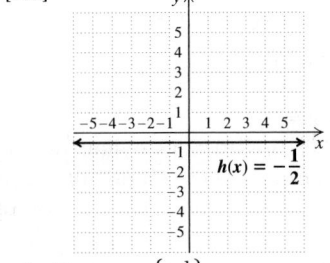

16. [7.1] -1
17. [7.1] $10a + 1$
18. [7.4] $\frac{1}{x} + 2x + 1$

Domain: $\mathbb{R}$; range: $\left\{-\frac{1}{2}\right\}$
19. [7.1] $\{x \mid x \text{ is a real number } and \ x \neq 0\}$
20. [7.4] $\{x \mid x \text{ is a real number } and \ x \neq 0\}$
21. [7.4] $\left\{x \mid x \text{ is a real number } and \ x \neq 0 \ and \ x \neq -\frac{1}{2}\right\}$
22. [7.5] $s = \dfrac{Rg}{g - R}$ **23.** [7.5] $y = \frac{1}{2}x$ **24.** [7.5] 30 workers
25. [7.5] 637 in^2 **26.** [7.3] **(a)** 30 mi; **(b)** 15 mph
27. [7.4] $h(x) = 7x - 2$

Cumulative Review: Chapters 1–7, pp. 502–504

1. 10   **2.** 3.91×10^8 **3.** Slope: $\frac{7}{4}$; y-intercept: $(0, -3)$
4. $y = -2x + 5$ **5.** **(a)** 0; **(b)** $\{x \mid x \text{ is a real number } and$
$x \neq 5 \ and \ x \neq 6\}$

6.

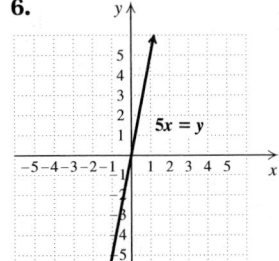

7.

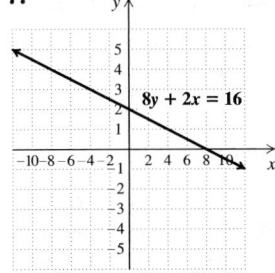

8.

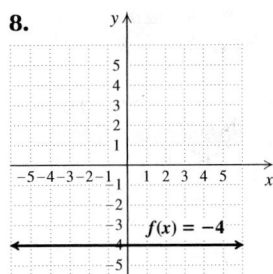

9.

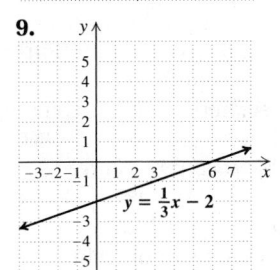

10. $-24x^4y^4$ **11.** $15x^4 - x^3 - 9x^2 + 5x - 2$

12. $9x^4 + 6x^2y + y^2$ **13.** $4x^4 - 81$

14. $-m^3n^2 - m^2n^2 - 5mn^3$ **15.** $\dfrac{y-6}{2}$ **16.** $x - 1$

17. $\dfrac{a^2 + 7ab + b^2}{(a-b)(a+b)}$ **18.** $\dfrac{-m^2 + 5m - 6}{(m+1)(m-5)}$ **19.** $\dfrac{3y^2 - 2}{3y}$

20. $\dfrac{y-x}{xy(x+y)}$ **21.** $4x(x^2 + 100)$ **22.** $(x-6)(x+14)$

23. $(4y-5)(4y+5)$ **24.** $8(2x+1)(4x^2 - 2x + 1)$

25. $(t-8)^2$ **26.** $x^2(x-1)(x+1)(x^2+1)$

27. $\left(\frac{1}{2}b - c\right)\left(\frac{1}{4}b^2 + \frac{1}{2}bc + c^2\right)$ **28.** $(3t-4)(t+7)$

29. $(x^2 - y)(x^3 + y)$ **30.** $\frac{1}{4}$ **31.** $-12, 12$

32. $\{x \mid x \geq -1\}$, or $[-1, \infty)$ **33.** $\frac{3}{5}$ **34.** -1

35. No solution   **36.** $a = \dfrac{Pb}{4-P}$ **37.** -5 **38.** 8

39. 0 **40.** $y = -x + 3$ **41.** 99 **42.** $\dfrac{x^2 - 1}{x+5}$

43. $\{x \mid x \text{ is a real number } and \ x \neq -6\}$

44. Domain: $\{x \mid -4 \leq x \leq 3\}$; or $[-4, 3]$;
range: $\{y \mid -3 \leq y \leq 0 \ or \ y = 2\}$

45. (a) $r(t) = 45.95t + 20$; (b) \$1076.85 million; (c) in
2016–2017   **46.** 34 performances per month, or 408 performances per year   **47.** 72 in.   **48.** IQAir HealthPro:
10 min; Austin Healthmate: 15 min   **49.** 12 billion hr

50. $22\frac{1}{2}$ min   **51.** \$1.09 billion

52. $x^3 - 12x^2 + 48x - 64$ **53.** $-3, 3, -5, 5$

54. All real numbers except 9 and -5 **55.** $-\frac{1}{4}, 0, \frac{1}{4}$

56. $C(x) = \begin{cases} 12, & \text{if } x \leq 20, \\ 12 + 0.75(x - 20), & \text{if } x > 20 \end{cases}$

CHAPTER 8

Technology Connection, p. 510

1. $(1.53, 2.58)$ **2.** $(-0.26, 57.06)$ **3.** $(2.23, 1.14)$
4. $(0.87, -0.32)$

Visualizing for Success, p. 512

1. C   **2.** H   **3.** J   **4.** G   **5.** D   **6.** I   **7.** A   **8.** F
9. E   **10.** B

Exercise Set 8.1, pp. 513–515

1. False   **2.** True   **3.** True   **4.** True   **5.** True
6. False   **7.** False   **8.** True   **9.** Yes   **11.** No
13. Yes   **15.** Yes   **17.** $(3, 2)$ **19.** $(2, -1)$ **21.** $(1, 4)$
23. $(-3, -2)$ **25.** $(3, -1)$ **27.** $(3, -7)$ **29.** $(7, 2)$
31. $(4, 0)$ **33.** No solution   **35.** $\{(x, y) \mid y = 3 - x\}$
37. All except Exercise 33   **39.** Exercise 35
41. Let x represent the first number and y the second number; $x + y = 10, x = \frac{2}{3}y$ **43.** Let p represent the number
of personal e-mails and b the number of business e-mails;
$p + b = 578, b = p + 30$ **45.** Let x and y represent the
angles; $x + y = 180, x = 2y - 3$ **47.** Let x represent the
number of two-point shots and y the number of foul shots;
$x + y = 64, 2x + y = 100$ **49.** Let x represent the

number of hats sold and y the number of tee shirts sold;
$x + y = 45, 14.50x + 19.50y = 697.50$ **51.** Let h represent the number of vials of Humalog sold and n the number
of vials of Lantus; $h + n = 50, 83.29h + 76.76n = 3981.66$
53. Let l represent the length, in yards, and w the width, in
yards; $2l + 2w = 340; l = w + 50$ **55.** ✒   **57.** $\frac{8}{13}$
58. -1 **59.** $-\frac{1}{10}$ **60.** 11   **61.** $y = 3x - 4$
62. $x = \frac{5}{2}y - \frac{7}{2}$ **63.** ✒   **65.** Answers may vary.
(a) $x + y = 6, x - y = 4$; (b) $x + y = 1, 2x + 2y = 3$;
(c) $x + y = 1, 2x + 2y = 2$ **67.** $A = -\frac{17}{4}, B = -\frac{12}{5}$
69. Let x and y represent the number of years that Dell
and Juanita have taught at the university, respectively;
$x + y = 46, x - 2 = 2.5(y - 2)$ **71.** Let s and v represent
the number of ounces of baking soda and vinegar needed,
respectively; $s = 4v, s + v = 16$ **73.** Mineral oil: 12 oz;
vinegar: 4 oz   **75.** $(0, 0), (1, 1)$ **77.** $(0.07, -7.95)$
79. $(0.00, 1.25)$

Exercise Set 8.2, pp. 521–523

1. (d)   **2.** (e)   **3.** (a)   **4.** (f)   **5.** (c)   **6.** (b)
7. $(2, -1)$ **9.** $(-4, 3)$ **11.** $(2, -2)$
13. $\{(x, y) \mid 2x - 3 = y\}$ **15.** $(-2, 1)$ **17.** $\left(\frac{1}{2}, \frac{1}{2}\right)$
19. $(2, 0)$ **21.** No solution   **23.** $(1, 2)$ **25.** $(7, -2)$
27. $(-1, 2)$ **29.** $\left(\frac{49}{11}, -\frac{12}{11}\right)$ **31.** $(6, 2)$ **33.** No solution
35. $(20, 0)$ **37.** $(3, -1)$ **39.** $\{(x, y) \mid -4x + 2y = 5\}$
41. $\left(2, -\frac{3}{2}\right)$ **43.** $(-2, -9)$ **45.** $(30, 6)$
47. $\{(x, y) \mid 4x - 2y = 2\}$ **49.** No solution
51. $(140, 60)$ **53.** $\left(\frac{1}{3}, -\frac{2}{3}\right)$ **55.** ✒   **57.** Toaster oven:
3 kWh; convection oven: 12 kWh   **58.** 90   **59.** \$105,000
60. 290 mi   **61.** First: 30 in.; second: 60 in.; third: 6 in.
62. 165 min   **63.** ✒   **65.** $m = -\frac{1}{2}, b = \frac{5}{2}$
67. $a = 5, b = 2$ **69.** $\left(-\frac{32}{17}, \frac{38}{17}\right)$ **71.** $\left(-\frac{1}{5}, \frac{1}{10}\right)$ **73.** ✒

Connecting the Concepts, pp. 523–524

1. $(1, 1)$ **2.** $(9, 1)$ **3.** $(4, 3)$ **4.** $(5, 7)$ **5.** $(5, 10)$
6. $\left(2, \frac{2}{5}\right)$ **7.** No solution   **8.** $\{(x, y) \mid x = 2 - y\}$
9. $(1, 1)$ **10.** $(0, 0)$ **11.** $(6, -1)$ **12.** No solution
13. $(3, 1)$ **14.** $\left(\frac{95}{71}, -\frac{1}{142}\right)$ **15.** $(1, 1)$ **16.** $(11, -3)$
17. $\{(x, y) \mid x - 2y = 5\}$ **18.** $\left(1, -\frac{1}{19}\right)$ **19.** $\left(\frac{201}{23}, -\frac{18}{23}\right)$
20. $\left(\frac{40}{9}, \frac{10}{3}\right)$

Exercise Set 8.3, pp. 532–536

1. $4, 6$ **3.** Personal e-mails: 274; business e-mails: 304
5. $119°, 61°$ **7.** Two-point shots: 36; foul shots: 28
9. Hats: 36; tee shirts: 9   **11.** Humalog vials: 22; Lantus
vials: 28   **13.** Length: 110 yd; width: 60 yd   **15.** Regular
paper: 32 reams; recycled paper: 84 reams   **17.** 13-watt
bulbs: 60; 18-watt bulbs: 140   **19.** HP C7115A cartridges:
180; M3908GA cartridges: 270   **21.** Mexican: 14 lb;
Peruvian: 14 lb   **23.** Custom-printed M&Ms: 64 oz; bulk
M&Ms: 256 oz   **25.** 50%-chocolate: 7.5 lb; 10%-chocolate:
12.5 lb   **27.** Deep Thought: 12 lb; Oat Dream: 8 lb
29. \$7500 at 6.5%; \$4500 at 7.2%   **31.** Steady State: 12.5 L;
Even Flow: 7.5 L   **33.** 87-octane: 2.5 gal; 95-octane: 7.5 gal
35. Whole milk: $169\frac{3}{13}$ lb; cream: $30\frac{10}{13}$ lb   **37.** 375 km

39. 14 km/h   **41.** About 1489 mi   **43.** Length: 265 ft; width: 165 ft   **45.** Wii game machines: 3.63 million; PlayStation 3 consoles: 1.21 million   **47.** $8.99 plans: 182; $4.99 plans: 68   **49.** Quarters: 17; fifty-cent pieces: 13
51. 🔊   **53.** 1   **54.** $\frac{1}{2}$ **55.** -13 **56.** 17   **57.** 7
58. $\frac{13}{4}$ **59.** 🔊   **61.** 0%: 20 reams; 30%: 40 reams
63. $10\frac{2}{3}$ oz   **65.** 12 sets   **67.** Brown: 0.8 gal; neutral: 0.2 gal   **69.** City: 261 mi; highway: 204 mi
71. $P(x) = \dfrac{0.1 + x}{1.5}$ (This expresses the percent as a decimal quantity.)

Exercise Set 8.4, pp. 543-544

1. True   **2.** False   **3.** False   **4.** True   **5.** True
6. False   **7.** Yes   **9.** $(3, 1, 2)$ **11.** $(1, -2, 2)$
13. $(2, -5, -6)$ **15.** No solution   **17.** $(-2, 0, 5)$
19. $(21, -14, -2)$ **21.** The equations are dependent.
23. $\left(3, \frac{1}{2}, -4\right)$ **25.** $\left(\frac{1}{2}, \frac{1}{3}, \frac{1}{6}\right)$ **27.** $\left(\frac{1}{2}, \frac{2}{3}, -\frac{5}{6}\right)$
29. $(15, 33, 9)$ **31.** $(3, 4, -1)$ **33.** $(10, 23, 50)$
35. No solution   **37.** The equations are dependent.
39. 🔊   **41.** Let x and y represent the numbers: $x = \frac{1}{2}y$
42. Let x and y represent the numbers; $x - y = 2x$
43. Let x represent the first number; $x + (x + 1) + (x + 2) = 100$
44. Let x, y, and z represent the numbers; $x + y + z = 100$
45. Let x, y, and z represent the numbers; $xy = 5z$
46. Let x and y represent the numbers; $xy = 2(x + y)$
47. 🔊   **49.** $(1, -1, 2)$ **51.** $(1, -2, 4, -1)$
53. $\left(-1, \frac{1}{5}, -\frac{1}{2}\right)$ **55.** 14   **57.** $z = 8 - 2x - 4y$ **59.** 🔊

Exercise Set 8.5, pp. 548-551

1. 8, 15, 62   **3.** 8, 21, -3 **5.** $32°, 96°, 52°$
7. Reading: 502; mathematics: 515; writing: 494
9. Bran muffin: 1.5 g; banana: 3 g; 1 cup of Wheaties: 3 g
11. Basic price: $30,610; tow package: $205; camera: $750
13. 12-oz cups: 17; 16-oz cups: 25; 20-oz cups: 13
15. Bank loan: $15,000; small-business loan: $35,000; mortgage: $70,000
17. Gold: $30/g; silver: $3/g; copper: $0.02/g
19. Roast beef: 2 servings; baked potato: 1 serving; broccoli: 2 servings
21. First mezzanine: 8 tickets; main floor: 12 tickets; second mezzanine: 20 tickets
23. Asia: 5.5 billion; Africa: 2.0 billion; rest of the world: 1.9 billion   **25.** 🔊   **27.** $-4x + 6y$ **28.** $-x + 6y$
29. $7y$ **30.** $11a$ **31.** $-2a + b + 6c$
32. $-50a - 30b + 10c$ **33.** $-12x + 5y - 8z$
34. $23x - 13z$ **35.** 🔊   **37.** Applicant: $87; spouse: $87; first child: $47; second child: $42   **39.** 20 yr   **41.** 35 tickets

Exercise Set 8.6, pp. 555-556

1. Matrix   **2.** Horizontal; columns   **3.** Entry
4. Matrices   **5.** Rows   **6.** First   **7.** $(3, 4)$ **9.** $(-2, 5)$
11. $\left(\frac{3}{2}, \frac{5}{2}\right)$ **13.** $\left(2, \frac{1}{2}, -2\right)$ **15.** $(2, -2, 1)$ **17.** $\left(4, \frac{1}{2}, -\frac{1}{2}\right)$
19. $(1, -3, -2, -1)$ **21.** Dimes: 18; nickels: 24
23. Dried fruit: 9 lb; macadamia nuts: 6 lb

25. $400 at 7%; $500 at 8%; $1600 at 9%   **27.** 🔊
29. 17   **30.** -19 **31.** 37   **32.** 422   **33.** 🔊   **35.** 1324

Exercise Set 8.7, pp. 560-561

1. True   **2.** True   **3.** True   **4.** False   **5.** False
6. False   **7.** 4   **9.** -50 **11.** 27   **13.** -3 **15.** -5
17. $(-3, 2)$ **19.** $\left(\frac{9}{19}, \frac{51}{38}\right)$ **21.** $\left(-1, -\frac{6}{7}, \frac{11}{7}\right)$
23. $(2, -1, 4)$ **25.** $(1, 2, 3)$ **27.** 🔊   **29.** 9700
30. $70x - 2500$ **31.** -1800 **32.** 4500   **33.** $\frac{250}{7}$
34. $\frac{250}{7}$ **35.** 🔊   **37.** 12   **39.** 10

Exercise Set 8.8, pp. 565-567

1. (b)   **2.** (f)   **3.** (h)   **4.** (a)   **5.** (e)   **6.** (d)
7. (c)   **8.** (g)
9. (a) $P(x) = 20x - 200{,}000$; (b) (10,000 units, $550,000)
11. (a) $P(x) = 25x - 3100$; (b) (124 units, $4960)
13. (a) $P(x) = 45x - 22{,}500$; (b) (500 units, $42,500)
15. (a) $P(x) = 16x - 50{,}000$; (b) (3125 units, $125,000)
17. (a) $P(x) = 50x - 100{,}000$; (b) (2000 units, $250,000)
19. ($60, 1100)   **21.** ($22, 474)   **23.** ($50, 6250)
25. ($10, 1070)   **27.** (a) $C(x) = 45{,}000 + 40x$;
(b) $R(x) = 130x$;   (c) $P(x) = 90x - 45{,}000$;
(d) $225,000 profit, $9000 loss   (e) (500 phones, $65,000)
29. (a) $C(x) = 10{,}000 + 30x$; (b) $R(x) = 80x$;
(c) $P(x) = 50x - 10{,}000$; (d) $90,000 profit, $7500 loss;
(e) (200 seats, $16,000)   **31.** 🔊   **33.** 6   **34.** -2
35. $-\frac{11}{9}$ **36.** $\frac{3}{4}$ **37.** -6 **38.** -5 **39.** 🔊
41. ($5, 300 yo-yo's)   **43.** (a) $8.74; (b) 24,509 units

Review Exercises: Chapter 8, pp. 571-572

1. Substitution   **2.** Elimination   **3.** Graphical
4. Dependent   **5.** Inconsistent   **6.** Contradiction
7. Parallel   **8.** Square   **9.** Determinant   **10.** Zero
11. $(4, 1)$ **12.** $(3, -2)$ **13.** $\left(\frac{8}{3}, \frac{14}{3}\right)$ **14.** No solution
15. $\left(-\frac{4}{5}, \frac{2}{5}\right)$ **16.** $\left(\frac{9}{4}, \frac{7}{10}\right)$ **17.** $\left(\frac{76}{17}, -\frac{2}{119}\right)$ **18.** $(-2, -3)$
19. $\{(x, y)\,|\,3x + 4y = 6\}$
20. Melon: $2.49; pineapple: $3.98   **21.** 4 hr
22. 8% juice: 10 L; 15% juice: 4 L   **23.** $(4, -8, 10)$
24. The equations are dependent.   **25.** $(2, 0, 4)$
26. $\left(\frac{8}{9}, -\frac{2}{3}, \frac{10}{9}\right)$ **27.** $A: 90°; B: 67.5°; C: 22.5°$
28. Oil: $21\frac{1}{3}$ oz; lemon juice: $10\frac{2}{3}$ oz
29. Man: 1.4; woman: 5.3; one-year-old child: 50
30. $\left(55, -\frac{89}{2}\right)$ **31.** $(-1, 1, 3)$ **32.** -5 **33.** 9
34. $(6, -2)$ **35.** $(-3, 0, 4)$ **36.** ($3, 81)
37. (a) $C(x) = 4.75x + 54{,}000$; (b) $R(x) = 9.25x$;
(c) $P(x) = 4.5x - 54{,}000$; (d) $31,500 loss, $13,500 profit;
(e) (12,000 pints of honey, $111,000)
38. 🔊 To solve a problem involving four variables, go through the *Familiarize* and *Translate* steps as usual. The resulting system of equations can be solved using the elimination method just as for three variables but likely with more steps.   **39.** 🔊 A system of equations can be both dependent and inconsistent if it is equivalent to a system with fewer equations that has no solution. An example is a system of three equations in three unknowns in which two

of the equations represent the same plane, and the third represents a parallel plane.   **40.** 20,000 pints
41. Round Stic: 15 packs; Matic Grip: 9 packs
42. $(0, 2), (1, 3)$ **43.** $a = -\frac{2}{3}, b = -\frac{4}{3}, c = 3$; $f(x) = -\frac{2}{3}x^2 - \frac{4}{3}x + 3$

Test: Chapter 8, pp. 572–573

1. [8.1] $(2, 4)$ **2.** [8.2] $\left(3, -\frac{11}{3}\right)$
3. [8.2] $\{(x, y)\,|\,x = 2y - 3\}$ **4.** [8.2] $(2, -1)$
5. [8.2] No solution   **6.** [8.2] $\left(-\frac{3}{2}, -\frac{3}{2}\right)$
7. [8.3] Length: 94 ft; width: 50 ft
8. [8.3] Pepperidge Farm Goldfish: 120 g;
Rold Gold Pretzels: 500 g   **9.** [8.3] 5.5 hr   **10.** [8.4] The equations are dependent.   **11.** [8.4] $\left(2, -\frac{1}{2}, -1\right)$
12. [8.4] No solution   **13.** [8.4] $(0, 1, 0)$
14. [8.6] $\left(\frac{22}{5}, -\frac{28}{5}\right)$ **15.** [8.6] $(3, 1, -2)$ **16.** [8.7] -14
17. [8.7] -59 **18.** [8.7] $\left(\frac{7}{13}, -\frac{17}{26}\right)$ **19.** [8.5] Electrician: 3.5 hr; carpenter: 8 hr; plumber: 10 hr   **20.** [8.8] ($3, 55$)
21. [8.8] **(a)** $C(x) = 25x + 44,000$; **(b)** $R(x) = 80x$; **(c)** $P(x) = 55x - 44,000$; **(d)** $27,500 loss, $5500 profit; **(e)** $(800$ hammocks, $64,000$)
22. [7.3], [8.3] $m = 7, b = 10$ **23.** [8.3] $\frac{120}{7}$ lb

Cumulative Review: Chapters 1–8, pp. 574–575

1. 24   **2.** $\frac{1}{10}$ **3.** -10 **4.** $13x^2 - y$ **5.** -24
6. 2   **7.** x^{11} **8.** $-\frac{9x^4}{2y^6}$ **9.** $-\frac{2a^{11}}{5b^{33}}$ **10.** $\frac{81x^{36}}{256y^8}$
11. 1.87×10^6 **12.** 4×10^6 **13.** $\frac{9}{7}$ **14.** $-3, -2$
15. $2, 3$ **16.** $\frac{1}{9}$ **17.** $\left\{y\,|\,y \geq -\frac{2}{3}\right\}$, or $\left[-\frac{2}{3}, \infty\right)$
18. $-10, 10$ **19.** $(1, 2)$ **20.** -11 **21.** $-\frac{1}{2}, \frac{2}{3}$
22. $(7, 3)$ **23.** $(-3, 2, -4)$ **24.** $(0, -1, 2)$
25. $p = \frac{3t}{q}$ **26.** $s = 2A - r$ **27.** $3a^3 + 8a^2 - 14$
28. $\frac{m(m + 2n)}{(m - n)(2m + n)}$ **29.** $-3a^3 + 8a^2 - 12a$
30. $\frac{2x + 4}{x - 2}$ **31.** $25a^4 - b^2$
32. $-15x^5 + 18x^4 + 6x^3 - 3x^2$ **33.** $4n^2 + 20n + 25$
34. $8t^5 + 5t^3 + 32t^2 + 20$ **35.** $\frac{(x - 1)(x + 2)}{(x - 2)^2}$
36. $2x + 1$ **37.** $\frac{x}{2(x + 1)(x + 2)}$ **38.** $(x + 1)(x^2 + 2)$
39. $(m^2 + 1)(m + 1)(m - 1)$ **40.** $2x(x + 4)(x + 5)$
41. $(x^2 - 3)^2$ **42.** $2(2x^2 - x - 5)$
43. $(2x - 5)(5x - 2)$
44.   **45.**

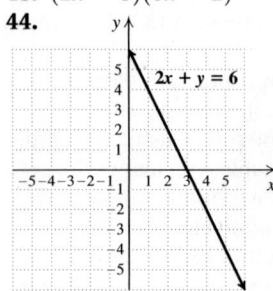

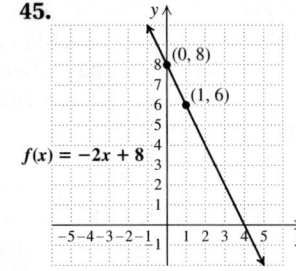

46.   **47.**

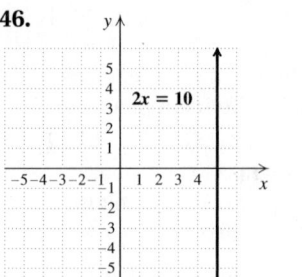

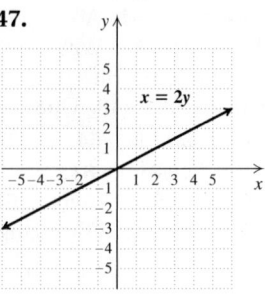

48. x-intercept: $(6, 0)$; y-intercept: $(0, -4)$
49. $y = -2x + \frac{4}{7}$ **50.** $y = -\frac{1}{10}x + \frac{12}{5}$ **51.** Parallel
52. $y = -2x + 5$ **53.** $\{-5, -3, -1, 1, 3\}$; $\{-3, -2, 1, 4, 5\}$; -2; 3   **54.** $\{x\,|\,x$ is a real number and $x \neq -10\}$ **55.** -31 **56.** 3   **57.** $2a^2 + 4a - 4$
58. 14   **59.** $1260   **60.** Dog sled: 12 mph; snowmobile: 52 mph   **61.** At least 35 patients   **62.** 11 and 13
63. 100 min   **64.** $32.5°, 65°, 82.5°$ **65.** 42 hr
66. $x = \dfrac{t}{p - q}$ **67.** $y = 2x + 7$ **68.** $\dfrac{3t^2(2t^2 - t + 1)}{4(t - 1)}$
69. $m = -\frac{5}{9}, b = -\frac{2}{9}$ **70.** Length: 6 in.; width: $4\frac{1}{4}$ in

CHAPTER 9

Exercise Set 9.1, pp. 583–586

1. Equivalent equations   **2.** Equivalent expressions
3. Equivalent inequalities   **4.** Not equivalent
5. Not equivalent   **6.** Equivalent equations
7. Equivalent expressions   **8.** Not equivalent
9. Not equivalent   **10.** Equivalent inequalities
11. $\{x\,|\,x < 2\}$, or $(-\infty, 2)$ **13.** $\{x\,|\,x \leq -9\}$, or $(-\infty, -9]$
15. $\{x\,|\,x < -26\}$, or $(-\infty, -26)$
17. $\left\{t\,|\,t \geq -\frac{13}{3}\right\}$, or $\left[-\frac{13}{3}, \infty\right)$ **19.** $\left\{y\,|\,y \leq -\frac{3}{2}\right\}$, or $\left(-\infty, -\frac{3}{2}\right]$
21. $\left\{t\,|\,t < \frac{29}{5}\right\}$, or $\left(-\infty, \frac{29}{5}\right)$ **23.** $\left\{m\,|\,m > \frac{7}{3}\right\}$, or $\left(\frac{7}{3}, \infty\right)$
25. $\{x\,|\,x \geq 2\}$, or $[2, \infty)$ **27.** $\{x\,|\,x > 0\}$, or $(0, \infty)$
29. $\{x\,|\,x < 7\}$, or $(-\infty, 7)$ **31.** $\{x\,|\,x \geq 2\}$, or $[2, \infty)$
33. $\{x\,|\,x < 8\}$, or $(-\infty, 8)$ **35.** $\{x\,|\,x \leq 2\}$, or $(-\infty, 2]$
37. $\{x\,|\,x \geq 2\}$, or $[2, \infty)$ **39.** $\left\{x\,|\,x > \frac{2}{3}\right\}$, or $\left(\frac{2}{3}, \infty\right)$
41. $\{x\,|\,x \geq 10\}$, or $[10, \infty)$ **43.** $\{x\,|\,x \leq 3\}$, or $(-\infty, 3]$
45. $\left\{x\,|\,x \geq -\frac{7}{2}\right\}$, or $\left[-\frac{7}{2}, \infty\right)$ **47.** $\{x\,|\,x \leq 4\}$, or $(-\infty, 4]$
49. $\{x\,|\,x \geq -15\}$, or $[-15, \infty)$ **51.** $\{x\,|\,x \geq 5\}$, or $[5, \infty)$
53. Lengths of time less than $7\frac{1}{2}$ hr   **55.** At least 56 questions correct   **57.** Years after 2010   **59.** Gross sales greater than $7000   **61.** For more than $6000
63. **(a)** Body densities less than $\frac{99}{95}$ kg/L, or about 1.04 kg/L; **(b)** body densities less than $\frac{495}{482}$ kg/L, or about 1.03 kg/L
65. **(a)** $\left\{x\,|\,x < 3913\frac{1}{23}\right\}$, or $\{x\,|\,x \leq 3913\}$;
(b) $\left\{x\,|\,x > 3913\frac{1}{23}\right\}$, or $\{x\,|\,x \geq 3914\}$ **67.** ✑
69. $\{x\,|\,x$ is a real number and $x \neq 0\}$
70. $\{x\,|\,x$ is a real number and $x \neq 6\}$
71. $\left\{x\,|\,x$ is a real number and $x \neq -\frac{1}{2}\right\}$
72. $\left\{x\,|\,x$ is a real number and $x \neq \frac{7}{5}\right\}$
73. $\mathbb{R}$ **74.** $\{x\,|\,x$ is a real number and $x \neq 0\}$ **75.** ✑

77. $\left\{x \mid x \le \dfrac{2}{a-1}\right\}$ **79.** $\left\{y \mid y \ge \dfrac{2a+5b}{b(a-2)}\right\}$

81. $\left\{x \mid x > \dfrac{4m-2c}{d-(5c+2m)}\right\}$ **83.** False; $2 < 3$ and $4 < 5$, but $2 - 4 = 3 - 5$. **85.** ✎

87. $\mathbb{R}$ ⟵―――|―――⟶
           0

89. $\{x \mid x$ is a real number $and\ x \ne 0\}$

⟵―――✕―――⟶
      0

91. $\{x \mid x \le 6\}$, or $(-\infty, 6]$ **93. (a)** $\{x \mid x < 4\}$, or $(-\infty, 4)$; **(b)** $\{x \mid x \ge 2\}$, or $[2, \infty)$; **(c)** $\{x \mid x \ge 3.2\}$, or $[3.2, \infty)$

95. ⬚

Exercise Set 9.2, pp. 593-596

1. (h) **2.** (j) **3.** (f) **4.** (a) **5.** (e) **6.** (d)
7. (b) **8.** (g) **9.** (c) **10.** (i) **11.** $\{4, 16\}$
13. $\{0, 5, 10, 15, 20\}$ **15.** $\{b, d, f\}$ **17.** $\{u, v, x, y, z\}$
19. $\varnothing$ **21.** $\{1, 3, 5\}$

23. ⟵―――(――)―――⟶ $(1, 3)$
      $-2\ -1\ 0\ 1\ 2\ 3\ 4\ 5\ 6\ 7\ 8$

25. ⟵―[――――]――⟶ $[-6, 0]$
      $-8\ -7\ -6\ -5\ -4\ -3\ -2\ -1\ 0\ 1\ 2$

27. ⟵―――)―――(―――⟶ $(-\infty, -1) \cup (4, \infty)$
      $-3\ -2\ -1\ 0\ 1\ 2\ 3\ 4\ 5\ 6\ 7$

29. ⟵――]―――(――――⟶ $(-\infty, -2] \cup (1, \infty)$
      $-5\ -4\ -3\ -2\ -1\ 0\ 1\ 2\ 3\ 4\ 5$

31. ⟵――(――――]―――⟶ $(-2, 4]$
      $-5\ -4\ -3\ -2\ -1\ 0\ 1\ 2\ 3\ 4\ 5$

33. ⟵――(――――)―――⟶ $(-2, 4)$
      $-5\ -4\ -3\ -2\ -1\ 0\ 1\ 2\ 3\ 4\ 5$

35. ⟵――――)―――(―⟶ $(-\infty, 5) \cup (7, \infty)$
      $-1\ 0\ 1\ 2\ 3\ 4\ 5\ 6\ 7\ 8\ 9$

37. ⟵―]―――[―――――⟶ $(-\infty, -4] \cup [5, \infty)$
      $-5\ -4\ -3\ -2\ -1\ 0\ 1\ 2\ 3\ 4\ 5$

39. ⟵―[―――――)――⟶ $[-3, 7)$
      $-3\ -2\ -1\ 0\ 1\ 2\ 3\ 4\ 5\ 6\ 7$

41. ⟵――(―――――]――⟶ $(-7, 0]$
      $-8\ -7\ -6\ -5\ -4\ -3\ -2\ -1\ 0\ 1\ 2$

43. ⟵―――――――)―⟶ $(-\infty, 5)$
      $-4\ -3\ -2\ -1\ 0\ 1\ 2\ 3\ 4\ 5\ 6$

45. $\{x \mid -5 \le x < 7\}$, or $[-5, 7)$ ⟵―[――――)―⟶
$-5\quad 0\quad 7$

47. $\{t \mid 4 < t \le 8\}$, or $(4, 8]$ ⟵――(――]―⟶
$0\quad 4\quad 8$

49. $\{a \mid -2 \le a < 2\}$, or $[-2, 2)$ ⟵――[――)――⟶
$-2\quad 0\quad 2$

51. $\mathbb{R}$, or $(-\infty, \infty)$ ⟵―――――――⟶
0

53. $\{x \mid -3 \le x \le 2\}$, or $[-3, 2]$ ⟵―[―――]――⟶
$-3\quad 0\quad 2$

55. $\{x \mid 7 < x < 23\}$, or $(7, 23)$ ⟵――(――――)―⟶
$0\quad 7\quad 23$

57. $\{x \mid -32 \le x \le 8\}$, or $[-32, 8]$
⟵―[―――――]―⟶
$-32\quad 0\quad 8$

59. $\{x \mid 1 \le x \le 3\}$, or $[1, 3]$ ⟵―――[――]――⟶
$0\ 1\quad 3$

61. $\{x \mid -\frac{7}{2} < x \le 7\}$, or $\left(-\frac{7}{2}, 7\right]$ ⟵――(―――]―⟶
$-\frac{7}{2}\quad 0\quad 7$

63. $\{t \mid t < 0\ or\ t > 1\}$, or $(-\infty, 0) \cup (1, \infty)$
⟵――)――(――⟶
$0\quad 1$

65. $\{a \mid a < \frac{7}{2}\}$, or $\left(-\infty, \frac{7}{2}\right)$ ⟵――――――)―⟶
$0\quad\quad \frac{7}{2}$

67. $\{a \mid a < -5\}$, or $(-\infty, -5)$ ⟵―――)―――――⟶
$-5\qquad 0$

69. $\varnothing$

71. $\{t \mid t \le 6\}$, or $(-\infty, 6]$ ⟵――――――]―⟶
$0\qquad\quad 6$

73. $(-\infty, -6) \cup (-6, \infty)$ **75.** $(-\infty, 0) \cup (0, \infty)$
77. $(-\infty, 4) \cup (4, \infty)$ **79.** ✎

81. 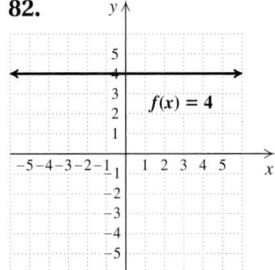 **82.**

83. **84.**

85. -1 **86.** -4 **87.** ✎ **89.** $(-1, 6)$
91. Between 2003 and 2009 **93.** Sizes between 6 and 13
95. Densities between 1.03 kg/L and 1.04 kg/L
97. More than 12 and fewer than 125 trips
99. $\left\{m \mid m < \frac{6}{5}\right\}$, or $\left(-\infty, \frac{6}{5}\right)$ ⟵――――――|――⟶
$0\quad \frac{6}{5}$

101. $\left\{x \mid -\frac{1}{8} < x < \frac{1}{2}\right\}$, or $\left(-\frac{1}{8}, \frac{1}{2}\right)$ ⟵――(――)――⟶
$-\frac{1}{8}\ 0\qquad \frac{1}{2}$

103. False **105.** True
107. $(-\infty, -7) \cup \left(-7, \frac{3}{4}\right]$ **109.** ⬚ **111.** ⬚

Technology Connection, p. 602

1. The x-values on the graph of $y_1 = |4x + 2|$ that are *below* the line $y = 6$ solve the inequality $|4x + 2| < 6$.
2. The x-values on the graph of $y_1 = |3x - 2|$ that are below the line $y = 4$ are in the interval $\left(-\frac{2}{3}, 2\right)$.
3. The graphs of $y_1 = \text{abs}(4x + 2)$ and $y_2 = -6$ do not intersect.

Exercise Set 9.3, pp. 603-605

1. True **2.** False **3.** True **4.** True **5.** True
6. True **7.** False **8.** False **9.** (g) **10.** (h) **11.** (d)
12. (a) **13.** (a) **14.** (b) **15.** $\{-10, 10\}$ **17.** $\varnothing$
19. $\{0\}$ **21.** $\left\{-\frac{1}{2}, \frac{7}{2}\right\}$ **23.** $\varnothing$ **25.** $\{-4, 8\}$
27. $\{6, 8\}$ **29.** $\{-5.5, 5.5\}$ **31.** $\{-8, 8\}$ **33.** $\{-1, 1\}$
35. $\left\{-\frac{11}{2}, \frac{13}{2}\right\}$ **37.** $\{-2, 12\}$ **39.** $\left\{-\frac{1}{3}, 3\right\}$ **41.** $\{-7, 1\}$
43. $\{-8.7, 8.7\}$ **45.** $\left\{-\frac{9}{2}, \frac{11}{2}\right\}$ **47.** $\{-8, 2\}$ **49.** $\left\{-\frac{1}{2}\right\}$
51. $\left\{-\frac{3}{5}, 5\right\}$ **53.** $\mathbb{R}$ **55.** $\left\{\frac{1}{4}\right\}$

57. $\{a \,|\, -3 \le a \le 3\}$, or $[-3, 3]$

59. $\{t \,|\, t < 0 \text{ or } t > 0\}$, or $(-\infty, 0) \cup (0, \infty)$

61. $\{x \,|\, -3 < x < 5\}$, or $(-3, 5)$

63. $\{n \,|\, -8 \le n \le 4\}$, or $[-8, 4]$

65. $\{x \,|\, x < -2 \text{ or } x > 8\}$, or $(-\infty, -2) \cup (8, \infty)$

67. $\mathbb{R}$, or $(-\infty, \infty)$

69. $\left\{a \,\middle|\, a \le -\frac{10}{3} \text{ or } a \ge \frac{2}{3}\right\}$, or $\left(-\infty, -\frac{10}{3}\right] \cup \left[\frac{2}{3}, \infty\right)$

71. $\{y \,|\, -9 < y < 15\}$, or $(-9, 15)$

73. $\{x \,|\, x \le -8 \text{ or } x \ge 0\}$, or $(-\infty, -8] \cup [0, \infty)$

75. $\left\{x \,\middle|\, x < -\frac{1}{2} \text{ or } x > \frac{7}{2}\right\}$, or $\left(-\infty, -\frac{1}{2}\right) \cup \left(\frac{7}{2}, \infty\right)$

77. $\varnothing$

79. $\left\{x \,\middle|\, x < -\frac{43}{24} \text{ or } x > \frac{9}{8}\right\}$, or $\left(-\infty, -\frac{43}{24}\right) \cup \left(\frac{9}{8}, \infty\right)$

81. $\{m \,|\, -9 \le m \le 3\}$, or $[-9, 3]$

83. $\{a \,|\, -6 < a < 0\}$, or $(-6, 0)$

85. $\left\{x \,\middle|\, -\frac{1}{2} \le x \le \frac{7}{2}\right\}$, or $\left[-\frac{1}{2}, \frac{7}{2}\right]$

87. $\left\{x \,\middle|\, x \le -\frac{7}{3} \text{ or } x \ge 5\right\}$, or $\left(-\infty, -\frac{7}{3}\right] \cup [5, \infty)$

89. $\{x \,|\, -4 < x < 5\}$, or $(-4, 5)$

91. ✍

93.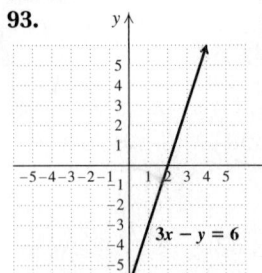

94.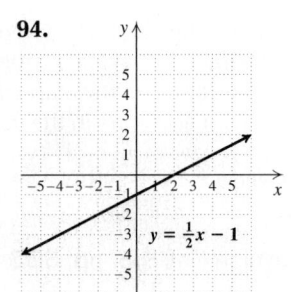

95.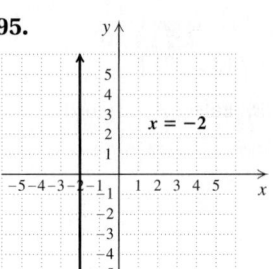

96.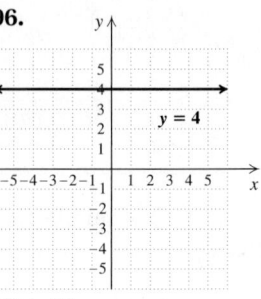

97. $\left(4, -\frac{4}{3}\right)$ **98.** $(5, 1)$ **99.** $\left(\frac{5}{7}, -\frac{18}{7}\right)$ **100.** $(-2, -5)$

101. ✍   **103.** $\left\{t \,\middle|\, t \ge \frac{5}{3}\right\}$, or $\left[\frac{5}{3}, \infty\right)$

105. $\mathbb{R}$, or $(-\infty, \infty)$ **107.** $\left\{-\frac{1}{7}, \frac{7}{3}\right\}$ **109.** $|x| < 3$

111. $|x| \ge 6$ **113.** $|x + 3| > 5$

115. $|x - 7| < 2$, or $|7 - x| < 2$ **117.** $|x - 3| \le 4$

119. $|x + 4| < 3$ **121.** Between 80 ft and 100 ft

123. $\{x \,|\, 1 \le x \le 5\}$, or $[1, 5]$ **125.** ✍, 📈

Connecting the Concepts, pp. 605–606

1. 2   **2.** $\{x \,|\, x > 3\}$, or $(3, \infty)$ **3.** $\frac{9}{2}$ **4.** $\{-4, 3\}$

5. $\{x \,|\, x \ge -6\}$, or $[-6, \infty)$ **6.** $\{t \,|\, -4 < t < 4\}$, or $(-4, 4)$ **7.** $-\frac{11}{8}$ **8.** $\frac{62}{3}$ **9.** $\left\{-\frac{8}{3}, \frac{8}{3}\right\}$

10. $\{x \,|\, -7 \le x \le 13\}$, or $[-7, 13]$ **11.** 31   **12.** $\varnothing$

13. $\left\{x \,\middle|\, x \le -\frac{17}{2} \text{ or } x \ge \frac{7}{2}\right\}$, or $\left(-\infty, -\frac{17}{2}\right] \cup \left[\frac{7}{2}, \infty\right)$

14. $\varnothing$ **15.** $\{m \,|\, -24 < m < 12\}$, or $(-24, 12)$

16. $\{-42, 38\}$ **17.** $\{t \,|\, t \le 4 \text{ or } t \ge 10\}$, or $(-\infty, 4] \cup [10, \infty)$ **18.** $\frac{7}{2}$ **19.** $\{a \,|\, -7 < a < -5\}$, or $(-7, -5)$ **20.** $\mathbb{R}$, or $(-\infty, \infty)$

Technology Connection, p. 610

1.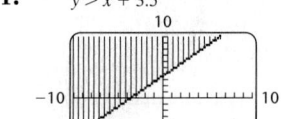

2.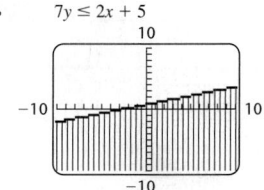

3.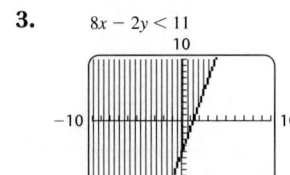

4.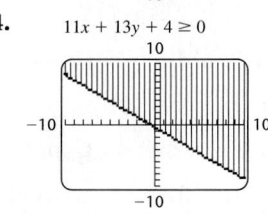

Technology Connection, p. 612

1.

Visualizing for Success, p. 613

1. B   **2.** F   **3.** J   **4.** A   **5.** E   **6.** G   **7.** C   **8.** D
9. I   **10.** H

Exercise Set 9.4, pp. 614–616

1. (e)   **2.** (c)   **3.** (d)   **4.** (a)   **5.** (b)   **6.** (f)
7. No   **9.** Yes

11.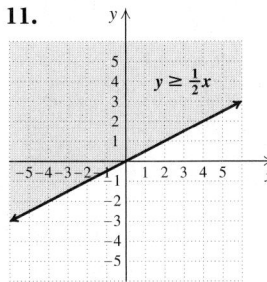

13.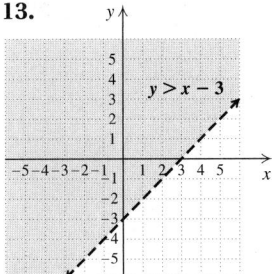

15.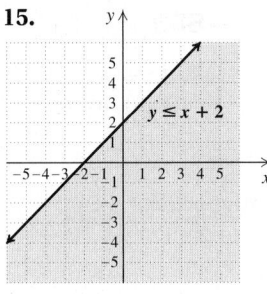

17.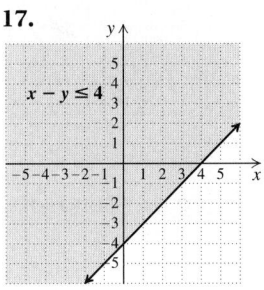

19.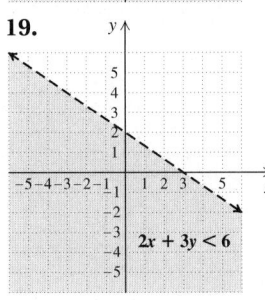

21.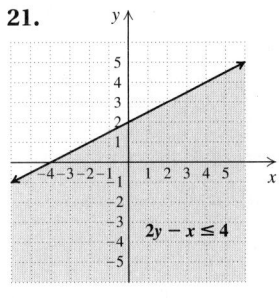

23.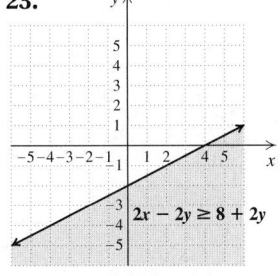

25.

27.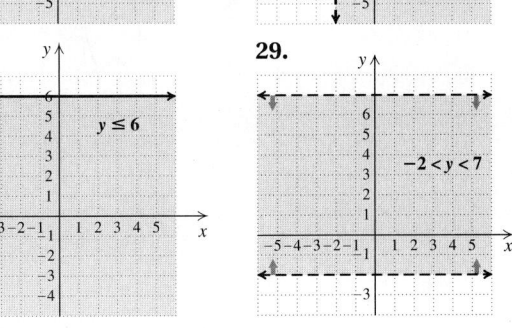

29.

31.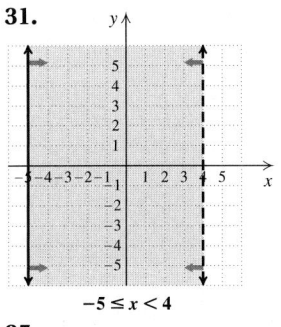
$-5 \le x < 4$

33.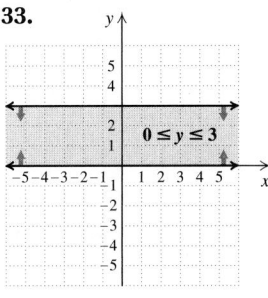
$0 \le y \le 3$

35.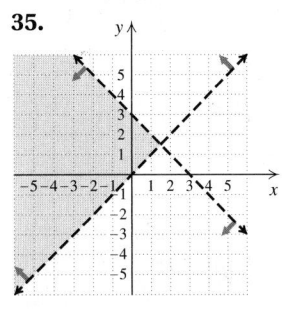

37.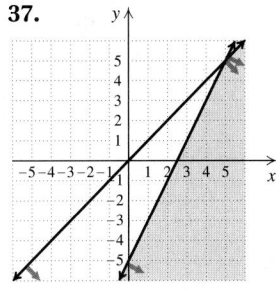

39.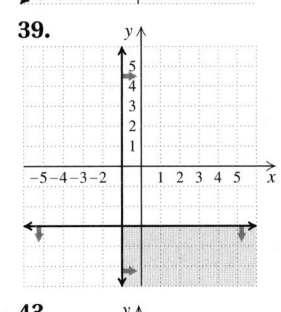

41.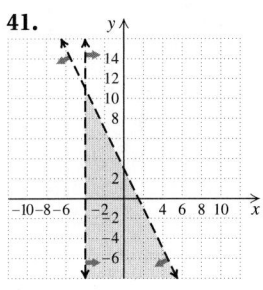

43.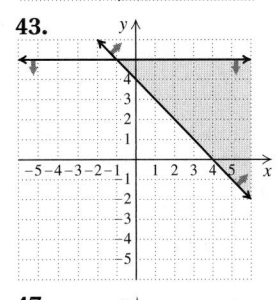

45.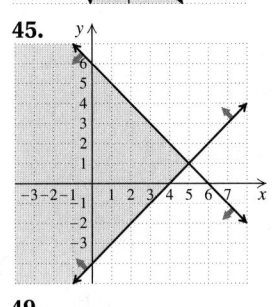

47.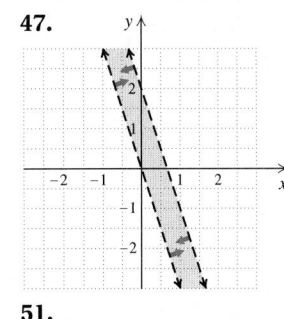

49.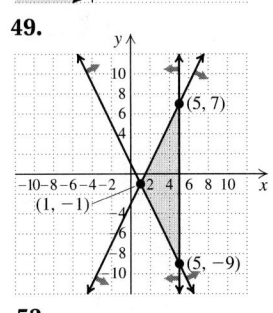

51.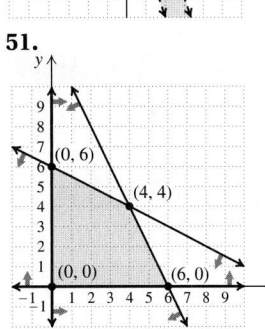

53.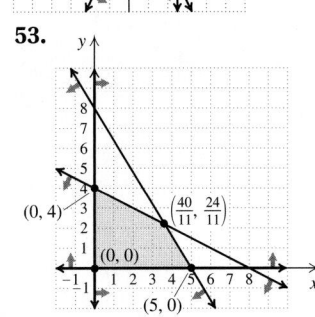

55.

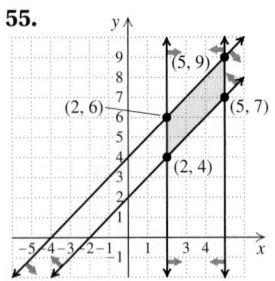

57. 🔽 **59.** 3.25% **60.** $228

61. 3%: $3600; 5%: $6400 **62.** Carrots: 12 lb;
broccoli: 8 lb **63.** Student tickets: 62; adult tickets: 108
64. Corn: 240 acres; soybeans: 160 acres **65.** 🔽

67.

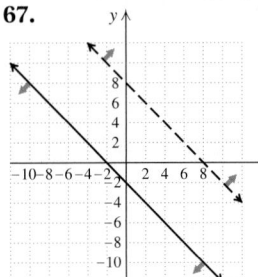

69.

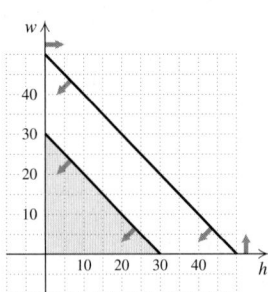

71.
$$w > 0,$$
$$h > 0,$$
$$w + h + 30 \le 62, \text{ or}$$
$$w + h \le 32,$$
$$2w + 2h + 30 \le 130, \text{ or}$$
$$w + h \le 50$$

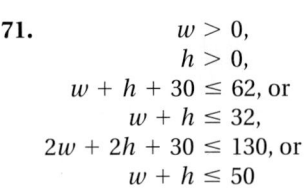

73. $q + v \ge 1150,$
$q \ge 700,$
$q \le 800,$
$v \ge 400,$
$v \le 800$

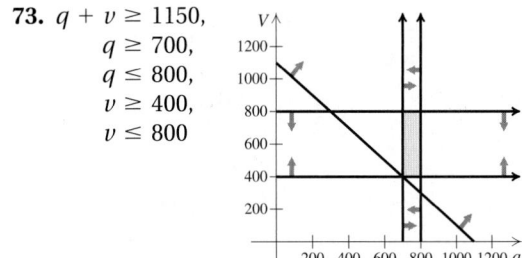

75. $35c + 75a > 1000,$
$c \ge 0,$
$a \ge 0$

77. $h < 2w,$
$w \le 1.5h,$
$h \le 3200,$
$h \ge 0,$
$w \ge 0$

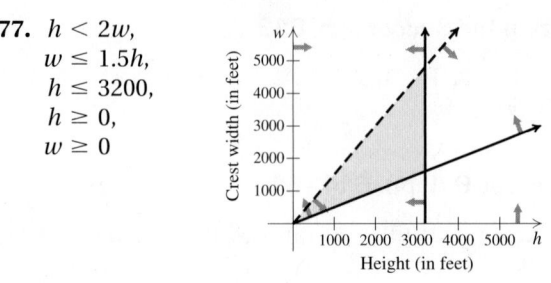

79. (a) $3x + 6y > 2$ **(b)** $x - 5y \le 10$

(c) $13x - 25y + 10 \le 0$ **(d)** $2x + 5y > 0$

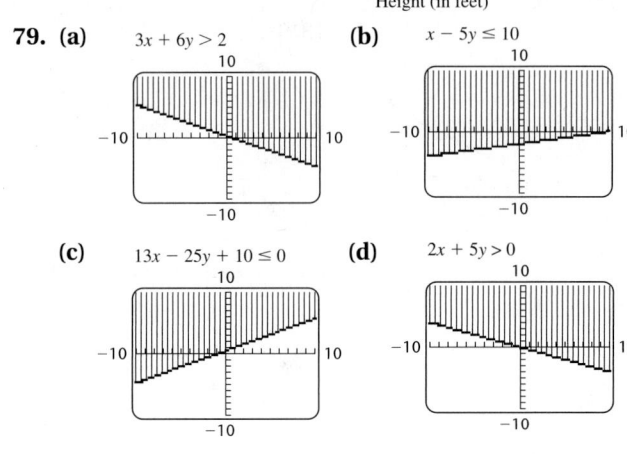

Connecting the Concepts, pp. 617–618

1. ◄———————●———►
        0   5

2. ◄—————————(—►
        0   5

3. ◄———————|———►
        0   5

4. ◄—————————●———►
        0   13

5. ◄———]|——►
     -1 0

6. ◄—————————————►
    -2   0

7.

8.

9.

10.

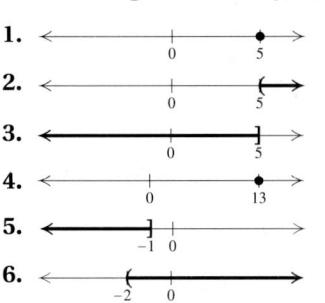

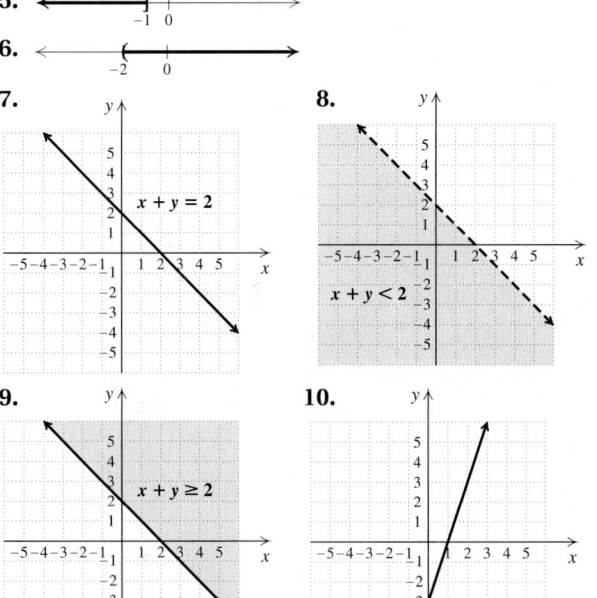

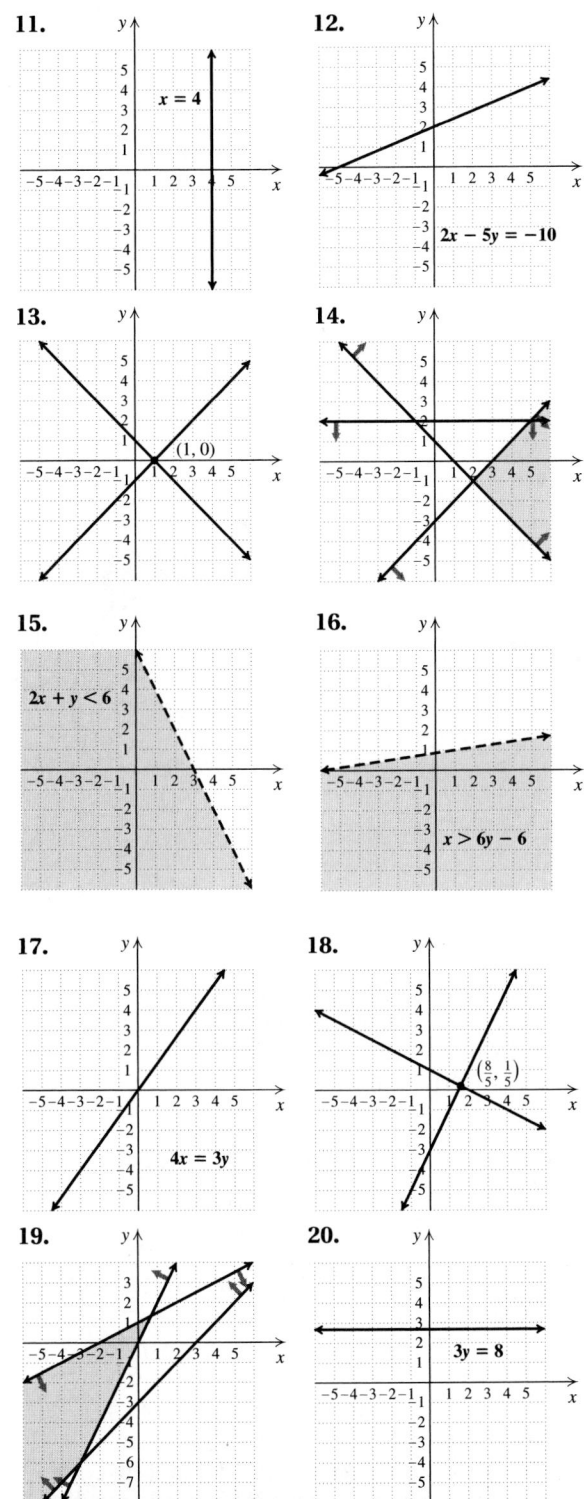

11.

12.

$x = 4$

$2x - 5y = -10$

13.

$(1, 0)$

14.

15.

$2x + y < 6$

16.

$x > 6y - 6$

17.

$4x = 3y$

18.

$\left(\frac{8}{5}, \frac{1}{5}\right)$

19.

20.

$3y = 8$

Exercise Set 9.5, pp. 622–624

1. Objective **2.** Constraints **3.** Corner
4. Feasible **5.** Vertices **6.** Vertex **7.** Maximum
84 when $x = 0, y = 6$; minimum 0 when $x = 0, y = 0$

9. Maximum 76 when $x = 7, y = 0$; minimum 16 when
$x = 0, y = 4$ **11.** Maximum 5 when $x = 3, y = 7$;
minimum -15 when $x = 3, y = -3$ **13.** Gumbo: 40
orders; sandwiches: 50 orders **15.** 4-photo pages: 5;
6-photo pages: 15; 110 photos **17.** Corporate bonds:
$22,000; municipal bonds: $18,000; maximum: $3110
19. Short-answer questions: 12; essay questions: 4
21. Merlot: 80 acres; Cabernet: 160 acres
23. 2.5 servings of each **25.** ✍ **27.** $4a - 7 + h$
28. $4a + 4h - 7$ **29.** $\left\{x \mid x \text{ is a real number } and \; x \neq -\frac{1}{2}\right\}$,
or $\left(-\infty, -\frac{1}{2}\right) \cup \left(-\frac{1}{2}, \infty\right)$ **30.** $\mathbb{R}$
31. $\{x \mid x \geq -4\}$, or $[-4, \infty)$
32. $\{x \mid x \text{ is a real number } and \; x \neq -1 \text{ and } x \neq 1\}$, or
$(-\infty, -1) \cup (-1, 1) \cup (1, \infty)$ **33.** ✍
35. T3's: 30; S5's: 10 **37.** Chairs: 25; sofas: 9

Review Exercises: Chapter 9, pp. 627–628

1. True **2.** False **3.** True **4.** False **5.** True
6. True **7.** True **8.** False **9.** False **10.** False
11. $\left\{x \mid x > -\frac{3}{2}\right\}$, or $\left(-\frac{3}{2}, \infty\right)$ **12.** $\{x \mid x < -3\}$, or
$(-\infty, -3)$ **13.** $\left\{y \mid y > -\frac{220}{23}\right\}$, or $\left(-\frac{220}{23}, \infty\right)$
14. $\left\{x \mid x \leq -\frac{5}{2}\right\}$, or $\left(-\infty, -\frac{5}{2}\right]$ **15.** $\{x \mid x < 5\}$, or $(-\infty, 5)$
16. $\{x \mid x < 1\}$, or $(-\infty, 1)$ **17.** $\{x \mid x \geq -4\}$, or $[-4, \infty)$
18. $\{x \mid x \geq 6\}$, or $[6, \infty)$ **19.** $\{x \mid x \leq 2\}$, or $(-\infty, 2]$
20. More than 125 hr **21.** $3000 **22.** $\{a, c\}$
23. $\{a, b, c, d, e, f, g\}$
24. ⟵(| |]⟶ $(-3, 2]$
          -3 0 2
25. ⟵ | ⟶ $(-\infty, \infty)$
          0
26. $\{x \mid -8 < x \leq 0\}$, or $(-8, 0]$ ⟵(| |]⟶
                                                    -8 0
27. $\{x \mid -\frac{5}{4} < x < \frac{5}{2}\}$, or $\left(-\frac{5}{4}, \frac{5}{2}\right)$ ⟵(|)⟶
                                                          $-\frac{5}{4}$ 0 $\frac{5}{2}$
28. $\{x \mid x < -3 \text{ or } x > 1\}$, or $(-\infty, -3) \cup (1, \infty)$
          ⟵) | (⟶
          -3 0 1
29. $\{x \mid x < -11 \text{ or } x \geq -6\}$, or $(-\infty, -11) \cup [-6, \infty)$
          ⟵) [| ⟶
          -11 -6 0
30. $\{x \mid x \leq -6 \text{ or } x \geq 8\}$, or $(-\infty, -6] \cup [8, \infty)$
          ⟵] | [⟶
          -6 0 8
31. $\{x \mid x < -\frac{2}{5} \text{ or } x > \frac{8}{5}\}$, or $\left(-\infty, -\frac{2}{5}\right) \cup \left(\frac{8}{5}, \infty\right)$
          ⟵) | (⟶
          $-\frac{2}{5}$ 0 $\frac{8}{5}$
32. $(-\infty, -3) \cup (-3, \infty)$ **33.** $[2, \infty)$ **34.** $\left(-\infty, \frac{1}{4}\right]$
35. $\{-11, 11\}$ **36.** $\{t \mid t \leq -21 \text{ or } t \geq 21\}$, or
$(-\infty, -21] \cup [21, \infty)$ **37.** $\{5, 11\}$
38. $\left\{a \mid -\frac{7}{2} < a < 2\right\}$, or $\left(-\frac{7}{2}, 2\right)$
39. $\left\{x \mid x \leq -\frac{11}{3} \text{ or } x \geq \frac{19}{3}\right\}$, or $\left(-\infty, -\frac{11}{3}\right] \cup \left[\frac{19}{3}, \infty\right)$
40. $\left\{-14, \frac{4}{3}\right\}$ **41.** $\varnothing$ **42.** $\{x \mid -16 \leq x \leq 8\}$, or $[-16, 8]$
43. $\{x \mid x < 0 \text{ or } x > 10\}$, or $(-\infty, 0) \cup (10, \infty)$
44. $\{x \mid -6 \leq x \geq 4\}$, or $[-6, 4]$ **45.** $\varnothing$

46.
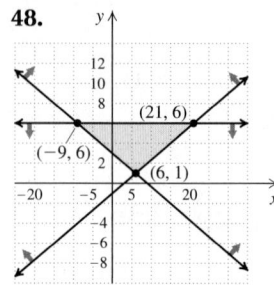
$x - 2y \geq 6$

47.
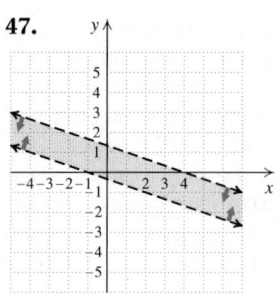

48.
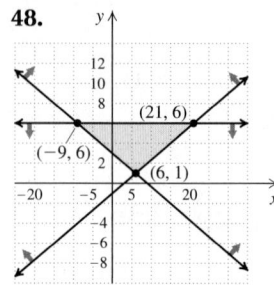
(21, 6)
(−9, 6)
(6, 1)

49. Maximum 40 when $x = 7$, $y = 15$; minimum 10 when $x = 1$, $y = 3$ **50.** Ohio plant: 120 computers; Oregon plant: 40 computers   **51.** The equation $|X| = p$ has two solutions when p is positive because X can be either p or $-p$. The same equation has no solution when p is negative because no number has a negative absolute value.
52. The solution set of a system of inequalities is all ordered pairs that make *all* the individual inequalities true. This consists of ordered pairs that are common to all the individual solution sets, or the intersection of the graphs.
53. $\left\{x \mid -\frac{8}{3} \leq x \leq -2\right\}$, or $\left[-\frac{8}{3}, -2\right]$ **54.** False: $-4 < 3$ is true, but $(-4)^2 < 9$ is false.   **55.** $|d - 2.5| \leq 0.003$

Test: Chapter 9, p. 629

1. [9.1] $\{y \mid y \leq -2\}$, or $(-\infty, -2]$ **2.** [9.1] $\left\{x \mid x > \frac{16}{5}\right\}$, or $\left(\frac{16}{5}, \infty\right)$ **3.** [9.1] $\left\{x \mid x \leq \frac{9}{16}\right\}$, or $\left(-\infty, \frac{9}{16}\right]$
4. [9.1] $\{x \mid x > 1\}$, or $(1, \infty)$ **5.** [9.1] $\{x \mid x \geq 4\}$, or $[4, \infty)$
6. [9.1] $\{x \mid x > 1\}$, or $(1, \infty)$ **7.** [9.1] More than $187\frac{1}{2}$ mi
8. [9.1] Less than or equal to 2.5 hr   **9.** [9.2] $\{a, e\}$
10. [9.2] $\{a, b, c, d, e, i, o, u\}$ **11.** [9.1] $(-\infty, 2]$
12. [9.2] $(-\infty, 7) \cup (7, \infty)$
13. [9.2] $\left\{x \mid -\frac{3}{2} < x \leq \frac{1}{2}\right\}$, or $\left(-\frac{3}{2}, \frac{1}{2}\right]$
14. [9.2] $\{x \mid x < 3 \text{ or } x > 6\}$, or $(-\infty, 3) \cup (6, \infty)$
15. [9.2] $\left\{x \mid x < -4 \text{ or } x \geq -\frac{5}{2}\right\}$, or $(-\infty, -4) \cup \left[-\frac{5}{2}, \infty\right)$
16. [9.2] $\{x \mid -3 \leq x \leq 1\}$, or $[-3, 1]$
17. [9.3] $\{-15, 15\}$
18. [9.3] $\{a \mid a < -5 \text{ or } a > 5\}$, or $(-\infty, -5) \cup (5, \infty)$

19. [9.3] $\left\{x \mid -2 < x < \frac{8}{3}\right\}$, or $\left(-2, \frac{8}{3}\right)$
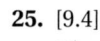
20. [9.3] $\left\{t \mid t \leq -\frac{13}{5} \text{ or } t \geq \frac{7}{5}\right\}$, or $\left(-\infty, -\frac{13}{5}\right] \cup \left[\frac{7}{5}, \infty\right)$
21. [9.3] $\varnothing$
22. [9.2] $\left\{x \mid x < \frac{1}{2} \text{ or } x > \frac{7}{2}\right\}$, or $\left(-\infty, \frac{1}{2}\right) \cup \left(\frac{7}{2}, \infty\right)$
23. [9.3] $\left\{-\frac{3}{2}\right\}$
24. [9.4]   **25.** [9.4]
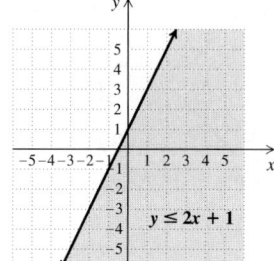
$y \leq 2x + 1$
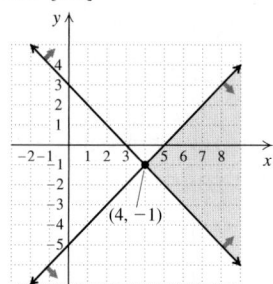
(4, −1)
26. [9.4]
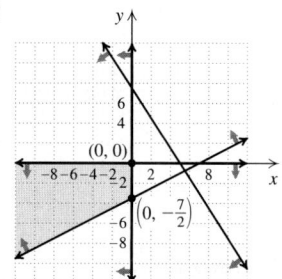
(0, 0)
$\left(0, -\frac{7}{2}\right)$

27. [9.5] Maximum 57 when $x = 6$, $y = 9$; minimum 5 when $x = 1$, $y = 0$ **28.** [9.5] Manicures: 35; haircuts: 15; maximum: $690   **29.** [9.3] $[-1, 0] \cup [4, 6]$
30. [9.2] $\left(\frac{1}{5}, \frac{4}{5}\right)$ **31.** [9.3] $|x + 3| \leq 5$

Cumulative Review: Chapters 1–9, pp. 630–631

1. 22   **2.** $c - 6$ **3.** $-\frac{1}{100}$ **4.** $-\frac{6x^4}{y^3}$ **5.** $\frac{9a^6}{4b^4}$
6. $\frac{2}{x^2 + 5x + 25}$ **7.** $x^2 - 25$ **8.** $15n^2 + 11n - 14$
9. $\frac{-10}{(x + 5)(x - 5)}$ **10.** $\frac{(x - 3)^2}{x^2(2x - 3)}$ **11.** $25(x - 1)^2$
12. $(4m + 7)(2n - 3)$ **13.** $8(t^2 + 100)$ **14.** 5
15. $\mathbb{R}$ **16.** $-2, 4$ **17.** $-4, 3$ **18.** No solution
19. $(5, 1)$ **20.** $\left\{-\frac{7}{2}, \frac{9}{2}\right\}$ **21.** $\left\{x \mid x < \frac{13}{2}\right\}$, or $\left(-\infty, \frac{13}{2}\right)$
22. $\{t \mid t < -3 \text{ or } t > 3\}$, or $(-\infty, -3) \cup (3, \infty)$
23. $\left\{x \mid -2 \leq x \leq \frac{10}{3}\right\}$, or $\left[-2, \frac{10}{3}\right]$

24.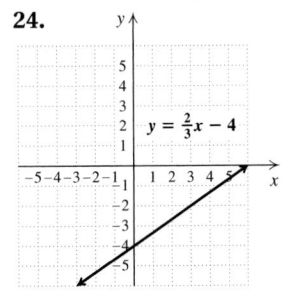

25.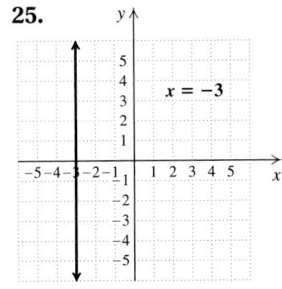

26.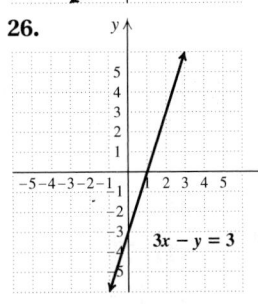

27.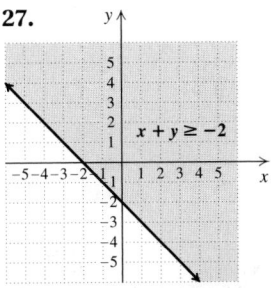

28.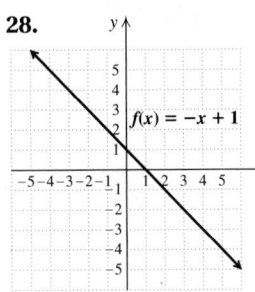

29.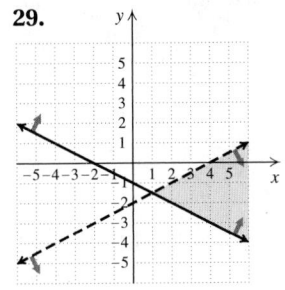

30. Slope: $\frac{4}{9}$; y-intercept: $(0, -2)$ **31.** $y = -7x - 25$
32. $y = \frac{2}{3}x + 4$ **33.** Domain: $\mathbb{R}$; range: $\{y \mid y \geq -2\}$,
or $[-2, \infty)$ **34.** $\left\{x \mid x \text{ is a real number } and\, x \neq -\frac{5}{2}\right\}$,
or $\left(-\infty, -\frac{5}{2}\right) \cup \left(-\frac{5}{2}, \infty\right)$ **35.** 22   **36.** $x^2 + 6x - 9$
37. ←———[———+————→
              -1 0   4

38. $\left\{x \mid x \text{ is a real number } and\, x \neq 0 \, and\, x \neq \frac{1}{3}\right\}$

39. $t = \dfrac{c}{a - d}$ **40.** 183 cannons   **41.** Beef: 9300 gal;
wheat: 2300 gal   **42.** (a) $b(t) = \frac{34}{15}t + 34.6$; (b) \$55 billion;
(c) 2008   **43.** (a) More than \$40; (b) costs greater than \$30
44. Length: 10 cm; width: 6 cm   **45.** \$640
46. \$50 billion   **47.** Vegetarian dinners: 12; steak
dinners: 16   **48.** $m = \frac{1}{3}$, $b = \frac{16}{3}$ **49.** $\frac{64}{3}$
50. $[-4, 0) \cup (0, \infty)$ **51.** 2^{11a-16}

CHAPTER 10

Technology Connection, p. 636

1. False   **2.** True   **3.** False

Visualizing for Success, p. 640

1. B   **2.** H   **3.** C   **4.** I   **5.** D   **6.** A   **7.** F   **8.** J
9. G   **10.** E

Exercise Set 10.1, pp. 641-642

1. Two   **2.** Negative   **3.** Positive   **4.** Negative
5. Irrational   **6.** Real   **7.** Nonnegative   **8.** Negative
9. 8, -8 **11.** 10, -10 **13.** 20, -20 **15.** 25, -25
17. 7   **19.** -4 **21.** $\frac{6}{7}$ **23.** -13 **25.** $-\frac{4}{9}$ **27.** 0.2
29. 0.09   **31.** $p^2 + 4$; 2   **33.** $\dfrac{x}{y+4}$; 5   **35.** $\sqrt{5}$; 0;
does not exist; does not exist   **37.** -7; does not exist; -1;
does not exist   **39.** 1; $\sqrt{2}$; $\sqrt{101}$ **41.** $10|x|$
43. $|8 - t|$ **45.** $|y + 8|$ **47.** $|2x + 7|$ **49.** -4
51. -1 **53.** $\frac{2}{3}$ **55.** $|x|$ **57.** t **59.** $6|a|$ **61.** 6
63. $|a + b|$ **65.** $|a^{11}|$ **67.** Cannot be simplified
69. $4x$ **71.** $-3t$ **73.** $5b$ **75.** $a + 1$ **77.** 3   **79.** $2x$
81. $x - 1$ **83.** $5y$ **85.** t^9 **87.** $(x - 2)^4$
89. 2; 3; -2; -4 **91.** 2; does not exist; does not exist; 3
93. $\{x \mid x \geq 6\}$, or $[6, \infty)$ **95.** $\{t \mid t \geq -8\}$, or $[-8, \infty)$
97. $\{x \mid x \leq 5\}$, or $(-\infty, 5]$ **99.** $\mathbb{R}$ **101.** $\left\{z \mid z \geq -\frac{2}{5}\right\}$, or
$\left[-\frac{2}{5}, \infty\right)$ **103.** $\mathbb{R}$ **105.** ✍   **107.** a^6b^2 **108.** $15x^3y^9$
109. $\dfrac{125x^6}{y^9}$ **110.** $\dfrac{a^3}{8b^6c^3}$ **111.** $\dfrac{x^3}{2y^6}$ **112.** $\dfrac{y^4z^8}{16x^4}$
113. ✍   **115.** About 1404 species
117. $\{x \mid x \geq -5\}$, or $[-5, \infty)$
119. $\{x \mid x \geq 0\}$, or $[0, \infty)$

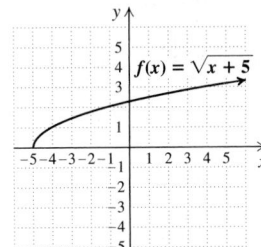   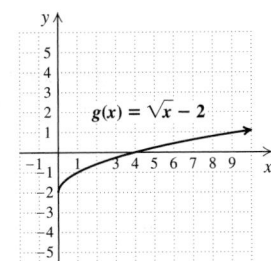

121. $\{x \mid -3 \leq x < 2\}$, or $[-3, 2)$
123. $\{x \mid x < -1 \, or\, x > 6\}$, or $(-\infty, -1) \cup (6, \infty)$
125. 📈

Technology Connection, p. 644

1. Without parentheses, the expression entered would be $\dfrac{7^2}{3}$.

2. For $x = 0$ or $x = 1$, $y_1 = y_2 = y_3$; on $(0, 1)$, $y_1 > y_2 > y_3$;
on $(1, \infty)$, $y_1 < y_2 < y_3$.

Technology Connection, p. 646

1. Many graphing calculators do not have keys for radicals
of index 3 or higher. On those graphing calculators that offer
$\sqrt[x]{\ }$ in a MATH menu, rational exponents still require fewer
keystrokes.

Exercise Set 10.2, pp. 647-649

1. (g)   **2.** (c)   **3.** (e)   **4.** (h)   **5.** (a)   **6.** (d)   **7.** (b)
8. (f)   **9.** $\sqrt[3]{y}$ **11.** 6   **13.** 2   **15.** 8   **17.** $\sqrt{xyz}$
19. $\sqrt[5]{a^2b^2}$ **21.** $\sqrt[6]{t^5}$ **23.** 8   **25.** 625   **27.** $27\sqrt[4]{x^3}$
29. $125x^6$ **31.** $18^{1/3}$ **33.** $30^{1/2}$ **35.** $x^{7/2}$ **37.** $m^{2/5}$

39. $(pq)^{1/4}$ **41.** $(xy^2z)^{1/5}$ **43.** $(3mn)^{3/2}$
45. $(8x^2y)^{5/7}$ **47.** $\dfrac{2x}{z^{2/3}}$ **49.** $\dfrac{1}{a^{1/4}}$ **51.** $\dfrac{1}{(2rs)^{3/4}}$
53. 8   **55.** $8a^{3/5}c$ **57.** $\dfrac{2a^{3/4}c^{2/3}}{b^{1/2}}$ **59.** $\dfrac{a^3}{3^{5/2}b^{7/3}}$
61. $\left(\dfrac{3c}{2ab}\right)^{5/6}$ **63.** $\dfrac{6a}{b^{1/4}}$ **65.** $11^{5/6}$ **67.** $3^{3/4}$
69. $4.3^{1/2}$ **71.** $10^{6/25}$ **73.** $a^{23/12}$ **75.** 64   **77.** $\dfrac{m^{1/3}}{n^{1/8}}$
79. $\sqrt[3]{x}$ **81.** y^5 **83.** $\sqrt{a}$ **85.** x^2y^2 **87.** $\sqrt{7a}$
89. $\sqrt[4]{8x^3}$ **91.** $\sqrt[10]{m}$ **93.** x^3y^3 **95.** a^6b^{12}
97. $\sqrt[12]{xy}$ **99.** ✍   **101.** $x^2 - 25$ **102.** $x^3 - 8$
103. $(2x + 5)^2$ **104.** $(3a - 4)^2$ **105.** $5(t - 1)^2$
106. $3(n + 2)^2$ **107.** ✍   **109.** $\sqrt[6]{x^5}$
111. $\sqrt[7]{c - d}, c \geq d$ **113.** $2^{7/12} \approx 1.498 \approx 1.5$
115. (a) 1.8 m; (b) 3.1 m; (c) 1.5 m; (d) 5.3 m
117. 338 cubic feet   **119.** 〰

Technology Connection, p. 651

1. The graphs differ in appearance because the domain of y_1 is the intersection of $[-3, \infty)$ and $[3, \infty)$, or $[3, \infty)$. The domain of y_2 is $(-\infty, -3] \cup [3, \infty)$.

Exercise Set 10.3, pp. 654–655

1. True   **2.** False   **3.** False   **4.** False   **5.** True
6. True   **7.** $\sqrt{30}$ **9.** $\sqrt[3]{35}$ **11.** $\sqrt[4]{54}$ **13.** $\sqrt{26xy}$
15. $\sqrt[5]{80y^4}$ **17.** $\sqrt{y^2 - b^2}$ **19.** $\sqrt[3]{0.21y^2}$
21. $\sqrt[5]{(x - 2)^3}$ **23.** $\sqrt{\dfrac{6s}{11t}}$ **25.** $\sqrt[7]{\dfrac{5x - 15}{4x + 8}}$ **27.** $2\sqrt{3}$
29. $3\sqrt{5}$ **31.** $2x^4\sqrt{2x}$ **33.** $2\sqrt{30}$ **35.** $6a^2\sqrt{b}$
37. $2x\sqrt[3]{y^2}$ **39.** $-2x^2\sqrt[3]{2}$ **41.** $f(x) = 2x^2\sqrt[3]{5}$
43. $f(x) = |7(x - 3)|$, or $7|x - 3|$
45. $f(x) = |x - 1|\sqrt{5}$ **47.** $a^5b^5\sqrt{b}$ **49.** $xy^2z^3\sqrt[3]{x^2z}$
51. $2xy^2\sqrt[4]{xy^3}$ **53.** $x^2yz^3\sqrt[3]{x^3y^3z^2}$ **55.** $-2a^4\sqrt[3]{10a^2}$
57. $5\sqrt{2}$ **59.** $3\sqrt{22}$ **61.** 3   **63.** $24y^5$ **65.** $a\sqrt[3]{10}$
67. $2x^3\sqrt{5x}$ **69.** $s^2t^3\sqrt[3]{t}$ **71.** $(x - y)^4$
73. $2ab^3\sqrt[4]{5a}$ **75.** $x(y + z)^2\sqrt[5]{x}$ **77.** ✍   **79.** $9abx^2$
80. $\dfrac{(x - 1)^2}{(x - 2)^2}$ **81.** $\dfrac{x + 1}{2(x + 5)}$ **82.** $\dfrac{3(4x^2 + 5y^3)}{50xy^2}$
83. $\dfrac{b + a}{a^2b^2}$ **84.** $\dfrac{-x - 2}{4x + 3}$ **85.** ✍   **87.** 175.6 mi
89. (a) $-3.3°C$; (b) $-16.6°C$; (c) $-25.5°C$; (d) $-54.0°C$
91. $25x^5\sqrt[3]{25x}$ **93.** $a^{10}b^{17}\sqrt{ab}$
95.

$f(x) = h(x); f(x) \neq g(x)$

97. $\{x \mid x \leq 2 \text{ or } x \geq 4\}$, or $(-\infty, 2] \cup [4, \infty)$ **99.** 6
101. 〰, ✍

Exercise Set 10.4, pp. 660–661

1. (g)   **2.** (b)   **3.** (f)   **4.** (c)   **5.** (h)   **6.** (d)   **7.** (a)
8. (e)   **9.** $\frac{7}{10}$ **11.** $\frac{5}{2}$ **13.** $\dfrac{11}{t}$ **15.** $\dfrac{6y\sqrt{y}}{x^2}$ **17.** $\dfrac{3a\sqrt[3]{a}}{2b}$
19. $\dfrac{2a}{bc^2}$ **21.** $\dfrac{ab^2}{c^2}\sqrt[4]{\dfrac{a}{c^2}}$ **23.** $\dfrac{2x}{y^2}\sqrt[5]{\dfrac{x}{y}}$ **25.** $\dfrac{xy}{z^2}\sqrt[6]{\dfrac{y^2}{z^3}}$
27. 3   **29.** $\sqrt[3]{2}$ **31.** $y\sqrt{5y}$ **33.** $2\sqrt[3]{a^2b}$ **35.** $\sqrt{2ab}$
37. $2x^2y^3\sqrt[4]{y^3}$ **39.** $\sqrt[3]{x^2 + xy + y^2}$ **41.** $\dfrac{\sqrt{10}}{5}$
43. $\dfrac{2\sqrt{15}}{21}$ **45.** $\dfrac{\sqrt[3]{10}}{2}$ **47.** $\dfrac{\sqrt[3]{75ac^2}}{5c}$ **49.** $\dfrac{y\sqrt[4]{45y^2x^3}}{3x}$
51. $\dfrac{\sqrt[3]{2xy^2}}{xy}$ **53.** $\dfrac{\sqrt{14a}}{6}$ **55.** $\dfrac{\sqrt[5]{9y^4}}{2xy}$ **57.** $\dfrac{\sqrt{5b}}{6a}$
59. $\dfrac{5}{\sqrt{55}}$ **61.** $\dfrac{12}{5\sqrt{42}}$ **63.** $\dfrac{2}{\sqrt[3]{6x}}$ **65.** $\dfrac{7}{\sqrt[3]{98}}$
67. $\dfrac{7x}{\sqrt{21xy}}$ **69.** $\dfrac{2a^2}{\sqrt[3]{20ab}}$ **71.** $\dfrac{x^2y}{\sqrt{2xy}}$ **73.** ✍
75. $x(3 - 8y + 2z)$ **76.** $ac(4a + 9 - 3a^2)$
77. $a^2 - b^2$ **78.** $a^4 - 4y^2$ **79.** $56 - 11x - 12x^2$
80. $6ay - 2cy - 3ax + cx$ **81.** ✍   **83.** (a) 1.62 sec;
(b) 1.99 sec; **(c)** 2.20 sec   **85.** $9\sqrt[3]{9n^2}$ **87.** $\dfrac{-3\sqrt{a^2 - 3}}{a^2 - 3}$,
or $\dfrac{-3}{\sqrt{a^2 - 3}}$ **89.** Step 1: $\sqrt[n]{a} = a^{1/n}$, by definition;
Step 2: $\left(\dfrac{a}{b}\right)^n = \dfrac{a^n}{b^n}$, raising a quotient to a power;
Step 3: $a^{1/n} = \sqrt[n]{a}$, by definition   **91.** $(f/g)(x) = 3x$,
where x is a real number and $x > 0$
93. $(f/g)(x) = \sqrt{x + 3}$, where x is a real number and $x > 3$

Exercise Set 10.5, pp. 666–668

1. Radicands; indices   **2.** Indices   **3.** Bases
4. Denominators   **5.** Numerator; conjugate   **6.** Bases
7. $11\sqrt{3}$ **9.** $2\sqrt[3]{4}$ **11.** $10\sqrt[3]{y}$ **13.** $12\sqrt{2}$
15. $13\sqrt[3]{7} + \sqrt{3}$ **17.** $9\sqrt{3}$ **19.** $-7\sqrt{5}$ **21.** $9\sqrt[3]{2}$
23. $(1 + 12a)\sqrt{a}$ **25.** $(x - 2)\sqrt[3]{6x}$ **27.** $3\sqrt{a - 1}$
29. $(x + 3)\sqrt{x - 1}$ **31.** $5\sqrt{2} + 2$ **33.** $3\sqrt{30} - 3\sqrt{35}$
35. $6\sqrt{5} - 4$ **37.** $3 - 4\sqrt[3]{63}$ **39.** $a + 2a\sqrt[3]{3}$
41. $4 + 3\sqrt{6}$ **43.** $\sqrt{6} - \sqrt{14} + \sqrt{21} - 7$ **45.** 1
47. -5 **49.** $2 - 8\sqrt{35}$ **51.** $23 + 8\sqrt{7}$ **53.** $5 - 2\sqrt{6}$
55. $2t + 5 + 2\sqrt{10t}$ **57.** $14 + x - 6\sqrt{x + 5}$
59. $6\sqrt[4]{63} + 4\sqrt[4]{35} - 3\sqrt[4]{54} - 2\sqrt[4]{30}$ **61.** $\dfrac{18 + 6\sqrt{2}}{7}$
63. $\dfrac{12 - 2\sqrt{3} + 6\sqrt{5} - \sqrt{15}}{33}$ **65.** $\dfrac{a - \sqrt{ab}}{a - b}$ **67.** -1
69. $\dfrac{12 - 3\sqrt{10} - 2\sqrt{14} + \sqrt{35}}{6}$ **71.** $\dfrac{1}{\sqrt{5} - 1}$
73. $\dfrac{2}{14 + 2\sqrt{3} + 3\sqrt{2} + 7\sqrt{6}}$ **75.** $\dfrac{x - y}{x + 2\sqrt{xy} + y}$
77. $\dfrac{1}{\sqrt{a + h} + \sqrt{a}}$ **79.** $\sqrt{a}$ **81.** $b^2\sqrt[10]{b^3}$
83. $xy\sqrt[6]{xy^5}$ **85.** $3a^2b\sqrt[4]{ab}$ **87.** $a^2b^2c^2\sqrt[6]{a^2bc^2}$
89. $\sqrt[12]{a^5}$ **91.** $\sqrt[12]{x^2y^5}$ **93.** $\sqrt[10]{ab^9}$ **95.** $\sqrt[6]{(7 - y)^5}$
97. $\sqrt[12]{5} + 3x$ **99.** $x\sqrt[6]{xy^5} - \sqrt[15]{x^{13}y^{14}}$
101. $2m^2 + m\sqrt[4]{n} + 2m\sqrt[3]{n^2} + \sqrt[12]{n^{11}}$

103. $2\sqrt[4]{x^3} - \sqrt[12]{x^{11}}$ **105.** $x^2 - 7$ **107.** $11 - 6\sqrt{2}$
109. $27 + 6\sqrt{14}$ **111.** ✍   **113.** 42   **114.** $-\frac{1}{3}$
115. $-7, 3$ **116.** $-\frac{2}{5}, \frac{3}{2}$ **117.** -3 **118.** $-6, 1$
119. ✍   **121.** $f(x) = 2x\sqrt{x-1}$
123. $f(x) = (x + 3x^2)\sqrt[4]{x-1}$ **125.** $(7x^2 - 2y^2)\sqrt{x+y}$
127. $4x(y+z)^3\sqrt[9]{2x(y+z)}$ **129.** $1 - \sqrt{w}$
131. $(\sqrt{x} + \sqrt{5})(\sqrt{x} - \sqrt{5})$
133. $(\sqrt{x} + \sqrt{a})(\sqrt{x} - \sqrt{a})$ **135.** $2x - 2\sqrt{x^2 - 4}$

Connecting the Concepts, pp. 668–669

1. $t + 5$ **2.** $-3a^4$ **3.** $3x\sqrt{10}$ **4.** $\frac{2}{3}$ **5.** $5\sqrt{15t}$
6. $ab^2c^2\sqrt[5]{c}$ **7.** $2\sqrt{15} - 3\sqrt{22}$ **8.** $-2b\sqrt[4]{a^3b^3}$
9. $\sqrt[8]{t}$ **10.** $\frac{a^2}{2}$ **11.** $-8\sqrt{3}$ **12.** -4 **13.** $25 + 10\sqrt{6}$
14. $2\sqrt{x-1}$ **15.** $xy\sqrt[10]{x^7y^3}$ **16.** $15\sqrt[3]{5}$ **17.** $\sqrt[5]{x}$
18. $(x + 1)\sqrt{3}$ **19.** $ab\sqrt{b}$ **20.** $6x^3y^2$

Technology Connection, p. 671

1. The *x*-coordinates of the points of intersection should approximate the solutions of the examples.

Exercise Set 10.6, pp. 673–675

1. False   **2.** True   **3.** True   **4.** False   **5.** True
6. True   **7.** 3   **9.** $\frac{16}{3}$ **11.** 20   **13.** -1 **15.** 5
17. 91   **19.** $0, 36$ **21.** 100   **23.** -125 **25.** 16
27. No solution   **29.** $\frac{80}{3}$ **31.** 45   **33.** $-\frac{5}{3}$ **35.** 1
37. $\frac{106}{27}$ **39.** 4   **41.** $3, 7$ **43.** $\frac{80}{9}$ **45.** -1
47. No solution   **49.** $2, 6$ **51.** 2   **53.** 4   **55.** ✍
57. Length: 200 ft; width: 15 ft   **58.** Base: 34 in.; height: 15 in.
59. Length: 14 in.; width: 10 in.
60. Length: 30 yd; width: 16 yd   **61.** $6, 8, 10$ **62.** 13 cm
63. ✍   **65.** About 68 psi   **67.** About 278 Hz
69. $524.8°C$ **71.** $t = \frac{1}{9}\left(\frac{S^2 \cdot 2457}{1087.7^2} - 2617\right)$

73. 4480 rpm   **75.** $r = \frac{v^2h}{2gh - v^2}$ **77.** $-\frac{8}{9}$ **79.** $-8, 8$
81. $1, 8$ **83.** $\left(\frac{1}{36}, 0\right), (36, 0)$ **85.** ▱   **87.** ▱

Exercise Set 10.7, pp. 683–687

1. (d)   **2.** (c)   **3.** (e)   **4.** (b)   **5.** (f)   **6.** (a)
7. $\sqrt{34}$; 5.831   **9.** $9\sqrt{2}$; 12.728   **11.** 8   **13.** 4 m
15. $\sqrt{19}$ in.; 4.359 in.   **17.** 1 m   **19.** 250 ft
21. $\sqrt{8450}$, or $65\sqrt{2}$ ft; 91.924 ft   **23.** 24 in.
25. $(\sqrt{340} + 8)$ ft; 26.439 ft
27. $(110 - \sqrt{6500})$ paces; 29.377 paces
29. Leg $= 5$; hypotenuse $= 5\sqrt{2} \approx 7.071$
31. Shorter leg $= 7$; longer leg $= 7\sqrt{3} \approx 12.124$
33. Leg $= 5\sqrt{3} \approx 8.660$; hypotenuse $= 10\sqrt{3} \approx 17.321$
35. Both legs $= \frac{13\sqrt{2}}{2} \approx 9.192$
37. Leg $= 14\sqrt{3} \approx 24.249$; hypotenuse $= 28$
39. $5\sqrt{3} \approx 8.660$ **41.** $7\sqrt{2} \approx 9.899$

43. $\frac{15\sqrt{2}}{2} \approx 10.607$ **45.** $\sqrt{10,561}$ ft ≈ 102.767 ft
47. $\frac{1089}{4}\sqrt{3}$ ft$^2 \approx 471.551$ ft^2 **49.** $(0, -4), (0, 4)$
51. 5   **53.** $\sqrt{10} \approx 3.162$ **55.** $\sqrt{200} \approx 14.142$
57. 17.8   **59.** $\frac{\sqrt{13}}{6} \approx 0.601$ **61.** $\sqrt{12} \approx 3.464$
63. $\sqrt{101} \approx 10.050$ **65.** $(3, 4)$ **67.** $\left(\frac{7}{2}, \frac{7}{2}\right)$
69. $(-1, -3)$ **71.** $(0.7, 0)$ **73.** $\left(-\frac{1}{12}, \frac{1}{24}\right)$
75. $\left(\frac{\sqrt{2} + \sqrt{3}}{2}, \frac{3}{2}\right)$ **77.** ✍

79.
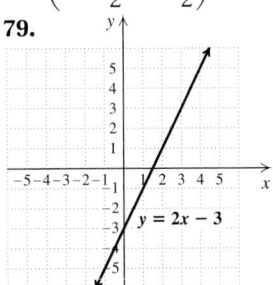
80.
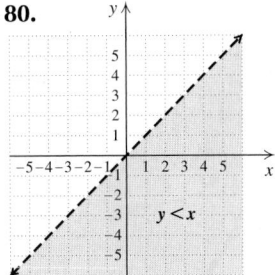

81.
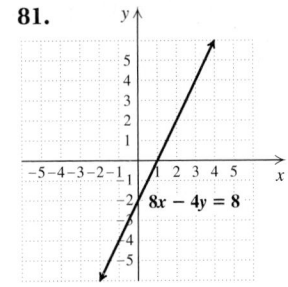
82.
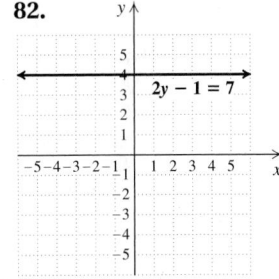

83.
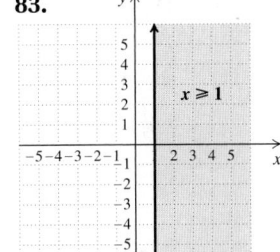
84.
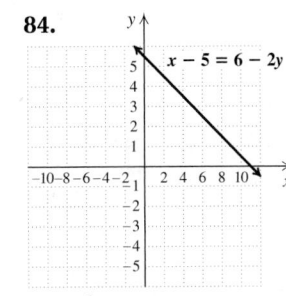

85. ✍   **87.** $36\sqrt{3}$ cm^2; 62.354 cm^2 **89.** $d = s + s\sqrt{2}$
91. 5 gal. The total area of the doors and windows is 134 ft^2 or more.   **93.** 60.28 ft by 60.28 ft   **95.** $\sqrt{75}$ cm

Exercise Set 10.8, pp. 693–694

1. False   **2.** False   **3.** True   **4.** True   **5.** True
6. True   **7.** False   **8.** True   **9.** $10i$ **11.** $i\sqrt{5}$, or $\sqrt{5}i$
13. $2i\sqrt{2}$, or $2\sqrt{2}i$ **15.** $-i\sqrt{11}$, or $-\sqrt{11}i$ **17.** $-7i$
19. $-10i\sqrt{3}$, or $-10\sqrt{3}i$ **21.** $6 - 2i\sqrt{21}$, or $6 - 2\sqrt{21}i$
23. $(-2\sqrt{19} + 5\sqrt{5})i$ **25.** $(3\sqrt{2} - 8)i$ **27.** $5 - 3i$
29. $7 + 2i$ **31.** $2 - i$ **33.** $-12 - 5i$ **35.** -40
37. -24 **39.** -18 **41.** $-\sqrt{30}$ **43.** $-3\sqrt{14}$
45. $-30 + 10i$ **47.** $28 - 21i$ **49.** $1 + 5i$ **51.** $38 + 9i$
53. $2 - 46i$ **55.** 73   **57.** 50   **59.** $12 - 16i$
61. $-5 + 12i$ **63.** $-5 - 12i$ **65.** $3 - i$ **67.** $\frac{6}{13} + \frac{4}{13}i$
69. $\frac{3}{17} + \frac{5}{17}i$ **71.** $-\frac{5}{6}i$ **73.** $-\frac{3}{4} - \frac{5}{4}i$ **75.** $1 - 2i$
77. $-\frac{23}{58} + \frac{43}{58}i$ **79.** $\frac{19}{29} - \frac{4}{29}i$ **81.** $\frac{6}{25} - \frac{17}{25}i$ **83.** 1

85. $-i$ **87.** -1 **89.** i **91.** -1 **93.** $-125i$
95. 0 **97.** 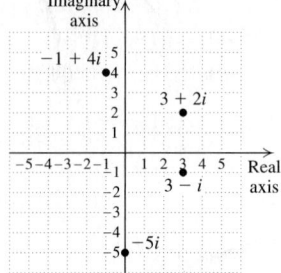 **99.** $-2, 3$ **100.** 5 **101.** $-10, 10$
102. $-5, 5$ **103.** $-\frac{2}{5}, \frac{4}{3}$ **104.** $-\frac{2}{3}, \frac{3}{2}$ **105.**
107.

113. $-9 - 27i$ **115.** $50 - 120i$ **117.** $\frac{250}{41} + \frac{200}{41}i$
119. 8 **121.** $\frac{3}{5} + \frac{9}{5}i$ **123.** 1

Review Exercises: Chapter 10, pp. 698–699

1. True **2.** False **3.** False **4.** True **5.** True
6. True **7.** True **8.** False **9.** $\frac{10}{11}$ **10.** -0.6
11. 5 **12.** $\{x | x \geq -10\}$, or $[-10, \infty)$ **13.** $8|t|$
14. $|c + 7|$ **15.** $|2x + 1|$ **16.** -2 **17.** $(5ab)^{4/3}$
18. $8a^4\sqrt{a}$ **19.** x^3y^5 **20.** $\sqrt[3]{x^2y}$ **21.** $\dfrac{1}{x^{2/5}}$
22. $7^{1/6}$ **23.** $f(x) = 5|x - 6|$ **24.** $2x^5y^2$
25. $5xy\sqrt{10x}$ **26.** $\sqrt{35ab}$ **27.** $3xb\sqrt[3]{x^2}$
28. $-6x^5y^4\sqrt[3]{2x^2}$ **29.** $-\dfrac{3y^4}{4}$ **30.** $y\sqrt[3]{6}$ **31.** $\dfrac{5\sqrt{x}}{2}$
32. $\dfrac{2a^2\sqrt[4]{3a^3}}{c^2}$ **33.** $7\sqrt[3]{4y}$ **34.** $\sqrt{3}$ **35.** $(2x + y^2)\sqrt[3]{x}$
36. $15\sqrt{2}$ **37.** -1 **38.** $\sqrt{15} + 4\sqrt{6} - 6\sqrt{10} - 48$
39. $\sqrt[4]{x^3}$ **40.** $\sqrt[12]{x^5}$ **41.** $4 - 4\sqrt{a} + a$
42. $-4\sqrt{10} + 4\sqrt{15}$ **43.** $\dfrac{20}{\sqrt{10} + \sqrt{15}}$ **44.** 19
45. -126 **46.** 4 **47.** 2 **48.** $5\sqrt{2}$ cm; 7.071 cm
49. $\sqrt{32}$ ft; 5.657 ft
50. Short leg $= 10$; long leg $= 10\sqrt{3} \approx 17.321$
51. $\sqrt{26} \approx 5.099$ **52.** $\left(-2, -\frac{3}{2}\right)$ **53.** $3i\sqrt{5}$, or $3\sqrt{5}i$
54. $-2 - 9i$ **55.** $6 + i$ **56.** 29 **57.** -1 **58.** $9 - 12i$
59. $\frac{13}{25} - \frac{34}{25}i$ **60.** A complex number $a + bi$ is real
when $b = 0$. It is imaginary when $b \neq 0$. **61.** An
absolute-value sign must be used to simplify $\sqrt[n]{x^n}$ when n
is even, since x may be negative. If x is negative while n is
even, the radical expression cannot be simplified to x, since
$\sqrt[n]{x^n}$ represents the principal, or nonnegative, root. When n
is odd, there is only one root, and it will be positive or negative depending on the sign of x. Thus there is no absolute-
value sign when n is odd. **62.** $\dfrac{2i}{3i}$; answers may vary

63. 3 **64.** $-\frac{2}{5} + \frac{9}{10}i$ **65.** The isosceles right triangle is
larger by about 1.206 ft^2.

Test: Chapter 10, p. 700

1. [10.3] $5\sqrt{2}$ **2.** [10.4] $-\dfrac{2}{x^2}$ **3.** [10.1] $9|a|$
4. [10.1] $|x - 4|$ **5.** [10.2] $(7xy)^{1/2}$ **6.** [10.2] $\sqrt[6]{(4a^3b)^5}$

7. [10.1] $\{x | x \geq 5\}$, or $[5, \infty)$ **8.** [10.5] $27 + 10\sqrt{2}$
9. [10.3] $2x^3y^2\sqrt[5]{x}$ **10.** [10.3] $2\sqrt[3]{2wv^2}$ **11.** [10.4] $\dfrac{10a^2}{3b^3}$
12. [10.4] $\sqrt[5]{3x^4y}$ **13.** [10.5] $x\sqrt[4]{x}$ **14.** [10.5] $\sqrt[5]{y^2}$
15. [10.5] $6\sqrt{2}$ **16.** [10.5] $(x^2 + 3y)\sqrt{y}$
17. [10.5] $14 - 19\sqrt{x} - 3x$ **18.** [10.5] $\dfrac{5\sqrt{3} - \sqrt{6}}{23}$
19. [10.6] 4 **20.** [10.6] $-1, 2$ **21.** [10.6] 8
22. [10.7] $\sqrt{10,600}$ ft ≈ 102.956 ft **23.** [10.7] 5 cm;
$5\sqrt{3}$ cm ≈ 8.660 cm **24.** [10.7] $\sqrt{17} \approx 4.123$
25. [10.7] $\left(\frac{3}{2}, -6\right)$ **26.** [10.8] $5i\sqrt{2}$, or $5\sqrt{2}i$
27. [10.8] $12 + 2i$ **28.** [10.8] $15 - 8i$ **29.** [10.8] $-\frac{11}{34} - \frac{7}{34}i$
30. [10.8] i **31.** [10.6] 3 **32.** [10.8] $-\frac{17}{4}i$ **33.** [10.6] $22,500$ ft

Cumulative Review: Chapters 1–10, pp. 701–702

1. -7 **2.** $-7, 5$ **3.** $-5, 5$ **4.** $\frac{5}{2}$ **5.** -1 **6.** $\frac{1}{5}$
7. $\{x | -3 \leq x \leq 7\}$, or $[-3, 7]$ **8.** $\mathbb{R}$, or $(-\infty, \infty)$
9. $-3, 2$ **10.** 7 **11.** $(0, 5)$ **12.** $(1, -1, -2)$
13.

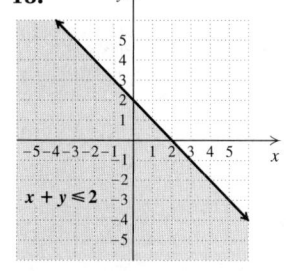

14.

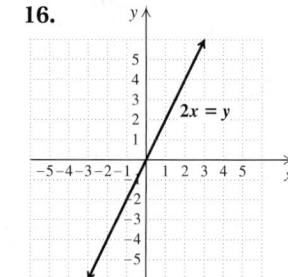

15.

16.

17. Slope: -1; y-intercept: $(0, -6)$
18. $y = 7x - 11$ **19.** 6 **20.** $-8xy^2$
21. $4a^2 - 20ab + 25b^2$ **22.** $c^4 - 9d^2$ **23.** $\dfrac{2x + 1}{x(x + 1)}$
24. $\dfrac{x + 13}{(x - 2)(x + 1)}$ **25.** $\dfrac{(a + 2)(2a + 1)}{(a - 3)(a - 1)}$ **26.** $\dfrac{2x + 1}{x^2}$
27. 0 **28.** $-1 + 3\sqrt{5}$ **29.** $5ab^2\sqrt{6a}$ **30.** $\sqrt[15]{y^8}$
31. $(x - 7)(x + 2)$ **32.** $4y^5(y - 1)(y^2 + y + 1)$
33. $25(2c + d)(2c - d)$ **34.** $(3t - 8)(t + 1)$
35. $3(x^2 - 2x - 7)$ **36.** $(y - x)(t - z^2)$
37. $\{x | x \text{ is a real number } and \ x \neq 3\}$, or $(-\infty, 3) \cup (3, \infty)$
38. $\left\{x | x \geq \frac{11}{2}\right\}$, or $\left[\frac{11}{2}, \infty\right)$ **39.** 5 **40.** $6 - 2\sqrt{5}$
41. $(f + g)(x) = x^2 + \sqrt{2x - 3}$ **42.** 4 mph
43. $2\sqrt{3}$ ft ≈ 3.464 ft **44.** (a) $m(t) = 0.08t + 25.02$;
(b) 26.62; (c) 2037 **45.** $\$36,650$ **46.** Swiss chocolate:
45 oz; whipping cream: 20 oz **47.** 4×10^5 programs
48. $\$1.54$ per year **49.** 5 ft **50.** $5x - 3y = -15$
51. $(0, -1), (1, 0)$ **52.** -2 **53.** 10

CHAPTER 11

Technology Connection, p. 712

1. The right-hand x-intercept should be an approximation of $4 + \sqrt{23}$.   **2.** x-intercepts should be approximations of $(-5 + \sqrt{37})/2$ and $(-5 - \sqrt{37})/2$.
3. Most graphing calculators can give only rational-number approximations of the two irrational solutions. An *exact* solution cannot be found with a graphing calculator.
4. The graph of $y = x^2 - 6x + 11$ has no x-intercepts.

Exercise Set 11.1, pp. 712–714

1. $\sqrt{k}$; $-\sqrt{k}$ **2.** 7; -7 **3.** $t + 3$; $t + 3$ **4.** 16
5. 25; 5 **6.** 9; 3 **7.** ± 10 **9.** $\pm 5\sqrt{2}$ **11.** $\pm\sqrt{6}$
13. $\pm\frac{7}{3}$ **15.** $\pm\sqrt{\frac{5}{6}}$, or $\pm\frac{\sqrt{30}}{6}$ **17.** $\pm i$ **19.** $\pm\frac{9}{2}i$
21. $-1, 7$ **23.** $-5 \pm 2\sqrt{3}$ **25.** $-1 \pm 3i$
27. $-\frac{3}{4} \pm \frac{\sqrt{17}}{4}$, or $\frac{-3 \pm \sqrt{17}}{4}$ **29.** $-3, 13$ **31.** $\pm\sqrt{19}$
33. $1, 9$ **35.** $-4 \pm \sqrt{13}$ **37.** $-14, 0$
39. $x^2 + 16x + 64 = (x + 8)^2$
41. $t^2 - 10t + 25 = (t - 5)^2$
43. $t^2 - 2t + 1 = (t - 1)^2$ **45.** $x^2 + 3x + \frac{9}{4} = \left(x + \frac{3}{2}\right)^2$
47. $x^2 + \frac{2}{5}x + \frac{1}{25} = \left(x + \frac{1}{5}\right)^2$
49. $t^2 - \frac{5}{6}t + \frac{25}{144} = \left(t - \frac{5}{12}\right)^2$ **51.** $-7, 1$ **53.** $5 \pm \sqrt{2}$
55. $-8, -4$ **57.** $-4 \pm \sqrt{19}$
59. $(-3 - \sqrt{2}, 0), (-3 + \sqrt{2}, 0)$
61. $\left(-\frac{9}{2} - \frac{\sqrt{181}}{2}, 0\right), \left(-\frac{9}{2} + \frac{\sqrt{181}}{2}, 0\right)$, or
$\left(\frac{-9 - \sqrt{181}}{2}, 0\right), \left(\frac{-9 + \sqrt{181}}{2}, 0\right)$
63. $(5 - \sqrt{47}, 0), (5 + \sqrt{47}, 0)$ **65.** $-\frac{4}{3}, -\frac{2}{3}$
67. $-\frac{1}{3}, 2$ **69.** $-\frac{2}{5} \pm \frac{\sqrt{19}}{5}$, or $\frac{-2 \pm \sqrt{19}}{5}$
71. $\left(-\frac{1}{4} - \frac{\sqrt{13}}{4}, 0\right), \left(-\frac{1}{4} + \frac{\sqrt{13}}{4}, 0\right)$, or $\left(\frac{-1 - \sqrt{13}}{4}, 0\right)$,
$\left(\frac{-1 + \sqrt{13}}{4}, 0\right)$ **73.** $\left(\frac{3}{4} - \frac{\sqrt{17}}{4}, 0\right), \left(\frac{3}{4} + \frac{\sqrt{17}}{4}, 0\right)$,
or $\left(\frac{3 - \sqrt{17}}{4}, 0\right), \left(\frac{3 + \sqrt{17}}{4}, 0\right)$ **75.** 10% **77.** 4%
79. About 4.3 sec   **81.** About 11.4 sec   **83.** ◫
85. 64 **86.** -15 **87.** $10\sqrt{2}$ **88.** $4\sqrt{6}$ **89.** $2i$
90. $5i$ **91.** $2i\sqrt{2}$, or $2\sqrt{2}i$ **92.** $2i\sqrt{6}$ or $2\sqrt{6}i$
93. ◫   **95.** ± 18 **97.** $-\frac{7}{2}, -\sqrt{5}, 0, \sqrt{5}, 8$
99. Barge: 8 km/h; fishing boat: 15 km/h   **101.** ◹
103. ◫, ◹

Exercise Set 11.2, pp. 719–720

1. True   **2.** True   **3.** False   **4.** False   **5.** False
6. True   **7.** $-\frac{5}{2}, 1$ **9.** $-1 \pm \sqrt{5}$ **11.** $3 \pm \sqrt{6}$
13. $\frac{3}{2} \pm \frac{\sqrt{29}}{2}$ **15.** $-1 \pm \frac{2\sqrt{3}}{3}$ **17.** $-\frac{4}{3} \pm \frac{\sqrt{19}}{3}$
19. $3 \pm i$ **21.** $\frac{1}{2} \pm \frac{\sqrt{3}}{2}i$ **23.** $-2 \pm \sqrt{2}i$ **25.** $-\frac{8}{3}, \frac{5}{4}$

27. $\frac{2}{5}$ **29.** $-\frac{11}{8} \pm \frac{\sqrt{41}}{8}$ **31.** $5, 10$ **33.** $\frac{3}{2}, 24$
35. $2 \pm \sqrt{5}i$ **37.** $2, -1 \pm \sqrt{3}i$
39. $-\frac{4}{3}, \frac{5}{2}$ **41.** $5 \pm \sqrt{53}$ **43.** $\frac{3}{2} \pm \frac{\sqrt{5}}{2}$
45. $-5.317, 1.317$ **47.** $0.764, 5.236$ **49.** $-1.266, 2.766$
51. ◫   **53.** $x^2 + 4$ **54.** $x^2 - 180$ **55.** $x^2 - 4x - 3$
56. $x^2 + 6x + 34$ **57.** $-\frac{3}{2}$ **58.** $\frac{1}{6} \pm \frac{\sqrt{6}}{3}i$ **59.** ◫
61. $(-2, 0), (1, 0)$ **63.** $4 - 2\sqrt{2}, 4 + 2\sqrt{2}$
65. $-1.179, 0.339$ **67.** $\frac{-5\sqrt{2} \pm \sqrt{34}}{4}$ **69.** $\frac{1}{2}$ **71.** ◹

Technology Connection, p. 723

1. $(-0.4, 0)$ is the other x-intercept of $y = 5x^2 - 13x - 6$.
2. The x-intercepts of $y = x^2 - 175$ are $(-13.22875656, 0)$ and $(13.22875656, 0)$, or $(-5\sqrt{7}, 0)$ and $(5\sqrt{7}, 0)$.
3. The x-intercepts of $y = x^3 + 3x^2 - 4x$ are $(-4, 0), (0, 0)$, and $(1, 0)$.

Exercise Set 11.3, pp. 723–725

1. (b)   **2.** (a)   **3.** (d)   **4.** (b)   **5.** (c)   **6.** (c)
7. Two irrational   **9.** Two imaginary   **11.** Two irrational
13. Two rational   **15.** Two imaginary   **17.** One rational
19. Two rational   **21.** Two irrational
23. Two imaginary   **25.** Two rational
27. Two irrational   **29.** $x^2 + x - 20 = 0$
31. $x^2 - 6x + 9 = 0$ **33.** $x^2 + 4x + 3 = 0$
35. $4x^2 - 23x + 15 = 0$ **37.** $8x^2 + 6x + 1 = 0$
39. $x^2 - 2x - 0.96 = 0$ **41.** $x^2 - 3 = 0$
43. $x^2 - 20 = 0$ **45.** $x^2 + 16 = 0$
47. $x^2 - 4x + 53 = 0$ **49.** $x^2 - 6x - 5 = 0$
51. $3x^2 - 6x - 4 = 0$ **53.** $x^3 - 4x^2 - 7x + 10 = 0$
55. $x^3 - 2x^2 - 3x = 0$ **57.** ◫   **59.** $c = \frac{d^2}{1 - d}$
60. $b = \frac{aq}{p - q}$ **61.** $y = \frac{x - 3}{x}$, or $1 - \frac{3}{x}$
62. 10 mph   **63.** Jamal: 3.5 mph; Kade: 2 mph
64. 20 mph   **65.** ◫   **67.** $a = 1, b = 2, c = -3$
69. (a) $-\frac{3}{5}$; (b) $-\frac{1}{3}$ **71.** (a) $9 + 9i$; (b) $3 + 3i$
73. The solutions of $ax^2 + bx + c = 0$ are
$x = \frac{-b \pm \sqrt{b^2 - 4ac}}{2a}$. When there is just one solution,
$b^2 - 4ac$ must be 0, so $x = \frac{-b \pm 0}{2a} = \frac{-b}{2a}$.
75. $a = 8, b = 20, c = -12$ **77.** $x^2 - 2 = 0$
79. $x^4 - 8x^3 + 21x^2 - 2x - 52 = 0$ **81.** ◫, ◹

Exercise Set 11.4, pp. 729–732

1. First part: 60 mph; second part: 50 mph   **3.** 40 mph
5. Cessna: 150 mph, Beechcraft: 200 mph; or Cessna: 200 mph, Beechcraft: 250 mph   **7.** To Hillsboro: 12 mph; return trip: 9 mph   **9.** About 14 mph   **11.** 12 hr
13. About 3.24 mph   **15.** $r = \frac{1}{2}\sqrt{\frac{A}{\pi}}$

17. $r = \dfrac{-\pi h + \sqrt{\pi^2 h^2 + 2\pi A}}{2\pi}$ **19.** $r = \dfrac{\sqrt{Gm_1 m_2}}{F}$

21. $H = \dfrac{c^2}{g}$ **23.** $b = \sqrt{c^2 - a^2}$

25. $t = \dfrac{-v_0 + \sqrt{(v_0)^2 + 2gs}}{g}$ **27.** $n = \dfrac{1 + \sqrt{1 + 8N}}{2}$

29. $g = \dfrac{4\pi^2 l}{T^2}$ **31.** $t = \dfrac{-b \pm \sqrt{b^2 - 4ac}}{2a}$

33. (a) 10.1 sec; **(b)** 7.49 sec; **(c)** 272.5 m   **35.** 2.9 sec
37. 0.890 sec   **39.** 2.5 m/sec   **41.** 4.5%   **43.**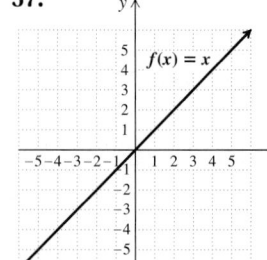

45. m^{-2}, or $\dfrac{1}{m^2}$ **46.** $t^{2/3}$ **47.** $y^{1/3}$ **48.** $z^{1/2}$ **49.** 2

50. 81   **51.**

53. $t = \dfrac{-10.2 + 6\sqrt{-A^2 + 13A - 39.36}}{A - 6.5}$ **55.** $\pm\sqrt{2}$

57. $l = \dfrac{w + w\sqrt{5}}{2}$

59. $n = \pm\sqrt{\dfrac{r^2 \pm \sqrt{r^4 + 4m^4 r^2 p - 4mp}}{2m}}$

61. $A(S) = \dfrac{\pi S}{6}$

Exercise Set 11.5, pp. 737–738

1. (f)   **2.** (d)   **3.** (h)   **4.** (b)   **5.** (g)   **6.** (a)
7. (e)   **8.** (c)   **9.** $\sqrt{p}$ **10.** $x^{1/4}$ **11.** $x^2 + 3$ **12.** t^{-3}
13. $(1 + t)^2$ **14.** $w^{1/6}$ **15.** $\pm 2, \pm 3$ **17.** $\pm\sqrt{3}, \pm 2$
19. $\pm\dfrac{\sqrt{5}}{2}, \pm 1$ **21.** 4   **23.** $\pm 2\sqrt{2}, \pm 3$
25. $8 + 2\sqrt{7}$ **27.** No solution   **29.** $-\frac{1}{2}, \frac{1}{3}$ **31.** $-4, 1$
33. $-27, 8$ **35.** 729   **37.** 1   **39.** No solution   **41.** $\frac{12}{5}$
43. $\pm 2, \pm 3i$ **45.** $\pm i, \pm 2i$ **47.** $\left(\dfrac{4}{25}, 0\right)$
49. $\left(\dfrac{3}{2} + \dfrac{\sqrt{33}}{2}, 0\right), \left(\dfrac{3}{2} - \dfrac{\sqrt{33}}{2}, 0\right), (4, 0), (-1, 0)$
51. $(-243, 0), (32, 0)$ **53.** No x-intercepts   **55.**
57.

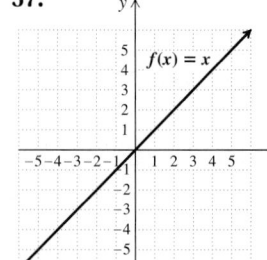

58.

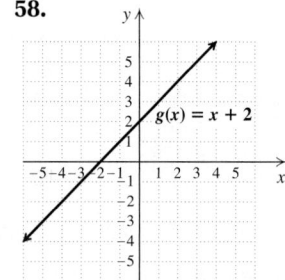

59.

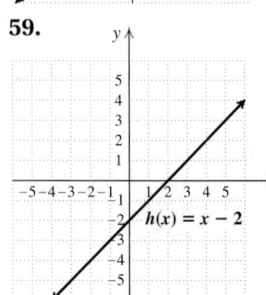

60.

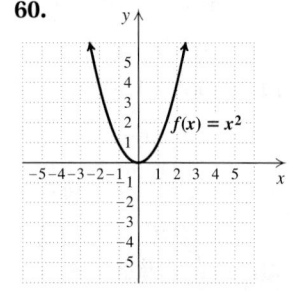

61.

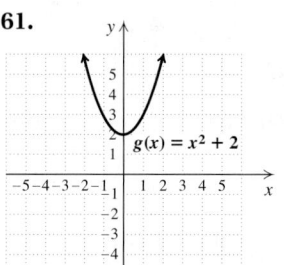

62.

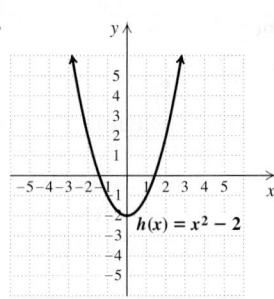

63.   **65.** $\pm\sqrt{\dfrac{-5 \pm \sqrt{37}}{6}}$ **67.** $-2, -1, 6, 7$

69. $\dfrac{100}{99}$ **71.** $-5, -3, -2, 0, 2, 3, 5$ **73.** $1, 3, -\dfrac{1}{2} + \dfrac{\sqrt{3}}{2} i,$

$-\dfrac{1}{2} - \dfrac{\sqrt{3}}{2} i, -\dfrac{3}{2} + \dfrac{3\sqrt{3}}{2} i, -\dfrac{3}{2} - \dfrac{3\sqrt{3}}{2} i$ **75.**

77. ,

Connecting the Concepts, pp. 739–740

1. $-2, 5$ **2.** ± 11 **3.** $-3 \pm \sqrt{19}$ **4.** $-\dfrac{1}{2} \pm \dfrac{\sqrt{13}}{2}$

5. $-1 \pm \sqrt{2}$ **6.** 5   **7.** $\dfrac{1}{2} \pm \dfrac{\sqrt{5}}{2}$ **8.** $1 \pm \sqrt{7}$

9. $\pm\dfrac{\sqrt{11}}{2}$ **10.** $\frac{1}{2}, 1$ **11.** $-\dfrac{1}{2} \pm \dfrac{\sqrt{3}}{2} i$ **12.** $0, \frac{7}{16}$

13. $-\frac{5}{6}, 2$ **14.** $1 \pm \sqrt{7}i$ **15.** $\pm 1, \pm 3$ **16.** $\pm 3, \pm i$
17. $-5, 0$ **18.** $-6, 5$ **19.** $\pm\sqrt{2}, \pm 2i$

20. $\pm\dfrac{\sqrt{3}}{3}, \pm\dfrac{\sqrt{2}}{2}$

Technology Connection, p. 741

1. The graphs of y_1, y_2, and y_3 open upward. The graphs of y_4, y_5, and y_6 open downward. The graph of y_1 is wider than the graph of y_2. The graph of y_3 is narrower than the graph of y_2. Similarly, the graph of y_4 is wider than the graph of y_5, and the graph of y_6 is narrower than the graph of y_5.
2. If A is positive, the graph opens upward. If A is negative, the graph opens downward. Compared with the graph of $y = x^2$, the graph of $y = Ax^2$ is wider if $|A| < 1$ and narrower if $|A| > 1$.

Technology Connection, p. 743

1. Compared with the graph of $y = ax^2$, the graph of $y = a(x - h)^2$ is shifted left or right. It is shifted left if h is negative and right if h is positive.   **2.** The value of A makes the graph wider or narrower, and makes the graph open downward if A is negative. The value of B shifts the graph left or right.

Technology Connection, p. 744

1. The graph of y_2 looks like the graph of y_1 shifted up 2 units, and the graph of y_3 looks like the graph of y_1 shifted down 4 units.   **2.** Compared with the graph of $y = a(x - h)^2$, the graph of $y = a(x - h)^2 + k$ is shifted up or down. It is shifted down if k is negative and up if k is positive.

3. The value of A makes the graph wider or narrower, and makes the graph open downward if A is negative. The value of B shifts the graph left or right. The value of C shifts the graph up or down.

Exercise Set 11.6, pp. 746–748

1. (h)   **2.** (g)   **3.** (f)   **4.** (d)   **5.** (b)   **6.** (c)
7. (e)   **8.** (a)

9.
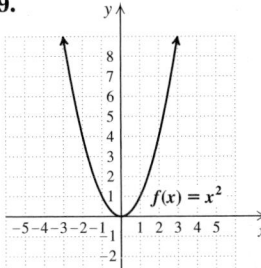

11.
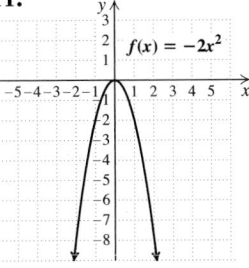

13.
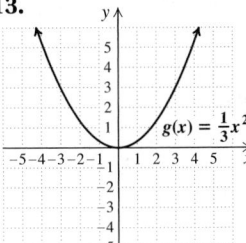

15.
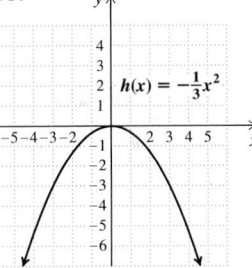

17.
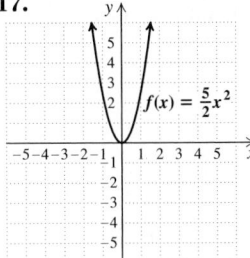

19. Vertex: $(-1, 0)$;
axis of symmetry: $x = -1$
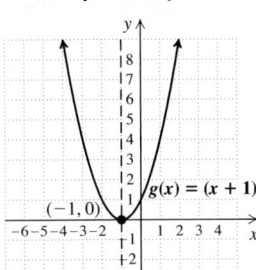

21. Vertex $(2, 0)$;
axis of symmetry: $x = 2$
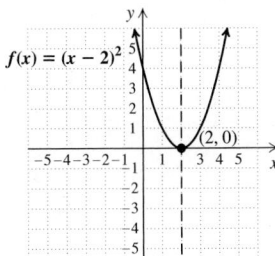

23. Vertex: $(-1, 0)$;
axis of symmetry: $x = -1$
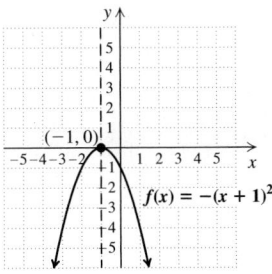

25. Vertex: $(2, 0)$;
axis of symmetry: $x = 2$
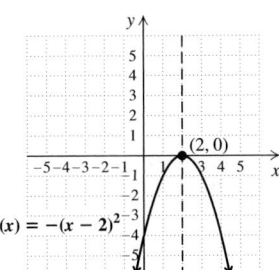

27. Vertex: $(-1, 0)$;
axis of symmetry: $x = -1$
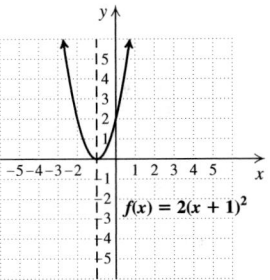

29. Vertex: $(4, 0)$;
axis of symmetry: $x = 4$
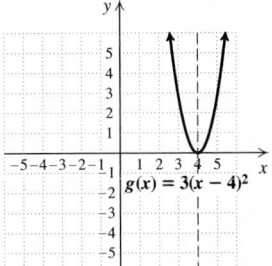

31. Vertex: $(4, 0)$;
axis of symmetry: $x = 4$
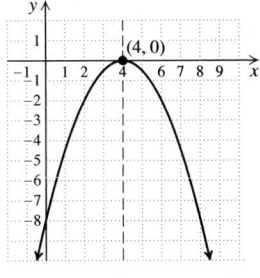

33. Vertex: $(1, 0)$;
axis of symmetry: $x = 1$
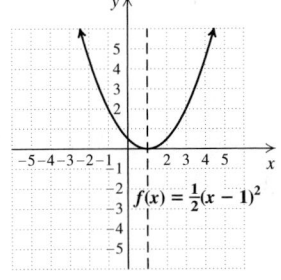

35. Vertex: $(-5, 0)$;
axis of symmetry: $x = -5$
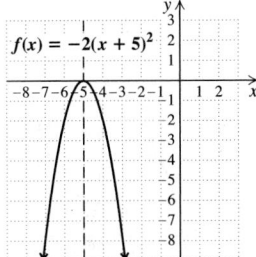

37. Vertex: $\left(\frac{1}{2}, 0\right)$;
axis of symmetry: $x = \frac{1}{2}$
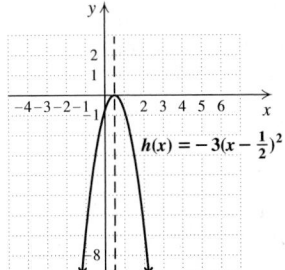

39. Vertex: $(5, 2)$;
axis of symmetry: $x = 5$;
minimum: 2
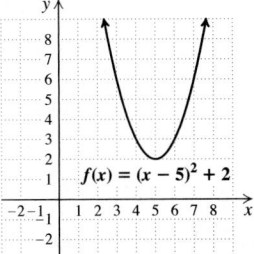

41. Vertex: $(-1, -3)$;
axis of symmetry: $x = -1$;
minimum: -3

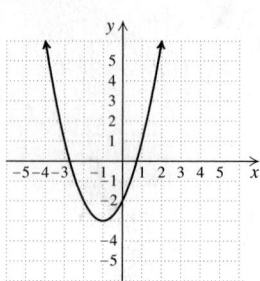

$$f(x) = (x + 1)^2 - 3$$

43. Vertex: $(-4, 1)$;
axis of symmetry: $x = -4$;
minimum: 1

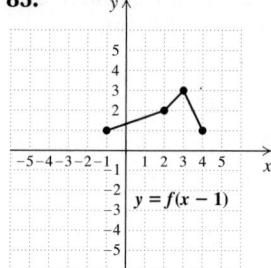

$$g(x) = \tfrac{1}{2}(x + 4)^2 + 1$$

45. Vertex: $(1, -3)$;
axis of symmetry: $x = 1$;
maximum: -3

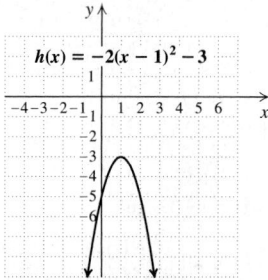

$$h(x) = -2(x - 1)^2 - 3$$

47. Vertex: $(-3, 1)$;
axis of symmetry: $x = -3$
minimum: 1

$$f(x) = 2(x + 3)^2 + 1$$

49. Vertex: $(2, 4)$;
axis of symmetry: $x = 2$;
maximum: 4

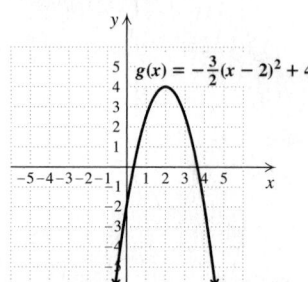

$$g(x) = -\tfrac{3}{2}(x - 2)^2 + 4$$

51. Vertex: $(3, 9)$; axis of symmetry: $x = 3$; minimum: 9
53. Vertex: $(-8, 2)$; axis of symmetry: $x = -8$; maximum: 2
55. Vertex: $\left(\tfrac{7}{2}, -\tfrac{29}{4}\right)$; axis of symmetry: $x = \tfrac{7}{2}$;
minimum: $-\tfrac{29}{4}$ **57.** Vertex: $(-2.25, -\pi)$; axis of
symmetry: $x = -2.25$; maximum: $-\pi$ **59.** ✍
61. x-intercept: $(3, 0)$; y-intercept: $(0, -4)$
62. x-intercept: $\left(\tfrac{8}{3}, 0\right)$; y-intercept: $(0, 2)$
63. $(-5, 0), (-3, 0)$ **64.** $(-1, 0), \left(\tfrac{3}{2}, 0\right)$
65. $x^2 - 14x + 49 = (x - 7)^2$
66. $x^2 + 7x + \tfrac{49}{4} = \left(x + \tfrac{7}{2}\right)^2$ **67.** ✍
69. $f(x) = \tfrac{3}{5}(x - 1)^2 + 3$ **71.** $f(x) = \tfrac{3}{5}(x - 4)^2 - 7$
73. $f(x) = \tfrac{3}{5}(x + 2)^2 - 5$ **75.** $f(x) = 2(x - 2)^2$

77. $g(x) = -2x^2 - 5$ **79.** The graph will move to the
right.   **81.** The graph will be reflected across the x-axis.
83. $F(x) = 3(x - 5)^2 + 1$
85.

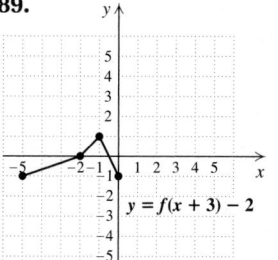

87.

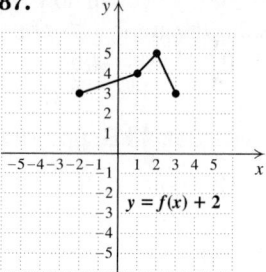

89.

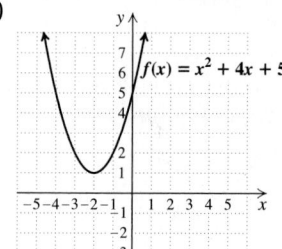

91. 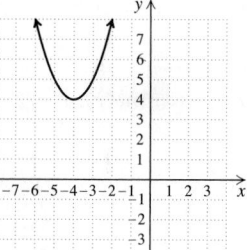   **93.** ✍,

Visualizing for Success, p. 753

1. B   **2.** E   **3.** A   **4.** H   **5.** C   **6.** J   **7.** F
8. G   **9.** I   **10.** D

Exercise Set 11.7, pp. 754–755

1. True   **2.** False   **3.** True   **4.** True   **5.** False
6. True   **7.** False   **8.** True
9. $f(x) = (x - 4)^2 + (-14)$
11. $f(x) = \left(x - \left(-\tfrac{3}{2}\right)\right)^2 + \left(-\tfrac{29}{4}\right)$
13. $f(x) = 3(x - (-1))^2 + (-5)$
15. $f(x) = -(x - (-2))^2 + (-3)$
17. $f(x) = 2\left(x - \tfrac{5}{4}\right)^2 + \tfrac{55}{8}$
19. (a) Vertex: $(-2, 1)$;   **21. (a)** Vertex: $(-4, 4)$;
axis of symmetry: $x = -2$; axis of symmetry: $x = -4$;
(b)   **(b)**

$$f(x) = x^2 + 4x + 5$$

$$f(x) = x^2 + 8x + 20$$

23. (a) Vertex: $(4, -7)$; axis of symmetry: $x = 4$;
(b)

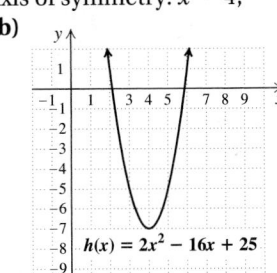

$h(x) = 2x^2 - 16x + 25$

25. (a) Vertex: $(1, 6)$; axis of symmetry: $x = 1$;
(b)

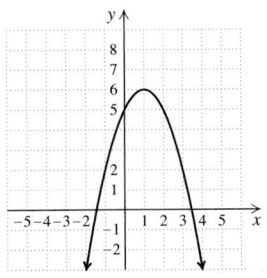

$f(x) = -x^2 + 2x + 5$

27. (a) Vertex: $\left(-\frac{3}{2}, -\frac{49}{4}\right)$; axis of symmetry: $x = -\frac{3}{2}$;
(b)

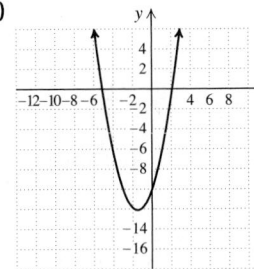

$g(x) = x^2 + 3x - 10$

29. (a) Vertex: $\left(-\frac{7}{2}, -\frac{49}{4}\right)$; axis of symmetry: $x = -\frac{7}{2}$;
(b)

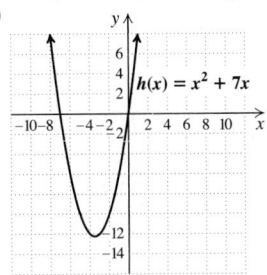

$h(x) = x^2 + 7x$

31. (a) Vertex: $(-1, -4)$; axis of symmetry: $x = -1$;
(b)

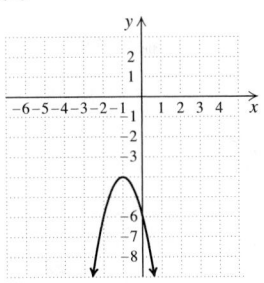

$f(x) = -2x^2 - 4x - 6$

33. (a) Vertex: $(3, 4)$; axis of symmetry: $x = 3$; minimum: 4;
(b)

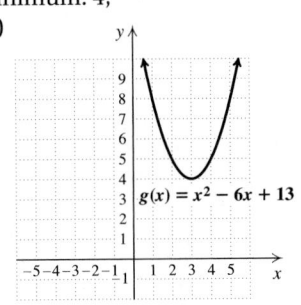

$g(x) = x^2 - 6x + 13$

35. (a) Vertex: $(2, -5)$; axis of symmetry: $x = 2$; minimum: -5;
(b)

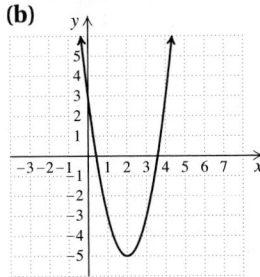

$g(x) = 2x^2 - 8x + 3$

37. (a) Vertex: $(4, 2)$; axis of symmetry: $x = 4$; minimum: 2;
(b)

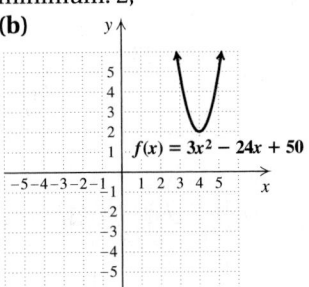

$f(x) = 3x^2 - 24x + 50$

39. (a) Vertex: $\left(\frac{5}{6}, \frac{1}{12}\right)$; axis of symmetry: $x = \frac{5}{6}$; maximum: $\frac{1}{12}$;
(b)

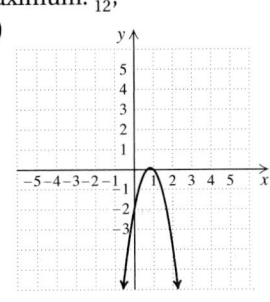

$f(x) = -3x^2 + 5x - 2$

41. (a) Vertex: $\left(-4, -\frac{5}{3}\right)$; axis of symmetry: $x = -4$; minimum: $-\frac{5}{3}$;
(b)

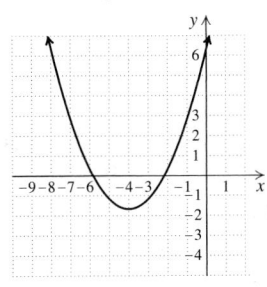

$h(x) = \frac{1}{2}x^2 + 4x + \frac{19}{3}$

43. $(3 - \sqrt{6}, 0), (3 + \sqrt{6}, 0)$; $(0, 3)$ **45.** $(-1, 0), (3, 0)$; $(0, 3)$ **47.** $(0, 0), (9, 0)$; $(0,0)$ **49.** $(2, 0)$; $(0, -4)$
51. $\left(-\frac{1}{2} - \frac{\sqrt{21}}{2}, 0\right), \left(-\frac{1}{2} + \frac{\sqrt{21}}{2}, 0\right)$; $(0, -5)]$
53. No x-intercept; $(0, 6)$ **55.** 🖲 **57.** $(1, 1, 1)$
58. $(-2, 5, 1)$ **59.** $(10, 5, 8)$ **60.** $(-3, 6, -5)$
61. $(2.4, -1.8, 1.5)$ **62.** $\left(\frac{1}{3}, \frac{1}{6}, \frac{1}{2}\right)$ **63.** 🖲
65. (a) Minimum: -6.953660714; **(b)** $(-1.056433682, 0)$, $(2.413576539, 0)$; $(0, -5.89)$ **67. (a)** $-2.4, 3.4$;
(b) $-1.3, 2.3$ **69.** $f(x) = m\left(x - \frac{n}{2m}\right)^2 + \frac{4mp - n^2}{4m}$
71. $f(x) = \frac{5}{16}x^2 - \frac{15}{8}x - \frac{35}{16}$, or $f(x) = \frac{5}{16}(x - 3)^2 - 5$
73.

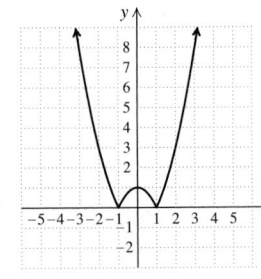

$f(x) = |x^2 - 1|$

75.

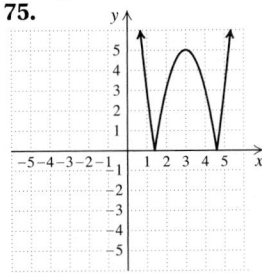

$f(x) = |2(x - 3)^2 - 5|$

Technology Connection, p. 759

1. About 607 million CDs

Exercise Set 11.8, pp. 760–765

1. (e) **2. (b)** **3. (c)** **4. (a)** **5. (d)** **6. (f)**
7. $3\frac{1}{4}$ weeks; 8.3 lb of milk per day **9.** \$120/dulcimer; 350 dulcimers **11.** 180 ft by 180 ft **13.** 450 ft²; 15 ft by 30 ft (The house serves as a 30-ft side.) **15.** 3.5 in.
17. 81; 9 and 9 **19.** -16; 4 and -4 **21.** 25; -5 and -5
23. $f(x) = ax^2 + bx + c, a < 0$ **25.** $f(x) = mx + b$
27. Neither quadratic nor linear
29. $f(x) = ax^2 + bx + c, a > 0$
31. $f(x) = ax^2 + bx + c, a > 0$ **33.** $f(x) = mx + b$
35. $f(x) = 2x^2 + 3x - 1$ **37.** $f(x) = -\frac{1}{4}x^2 + 3x - 5$

39. (a) $A(s) = \frac{3}{16}s^2 - \frac{135}{4}s + 1750$; **(b)** about 531 accidents
41. $h(d) = -0.0068d^2 + 0.8571d$ **43.**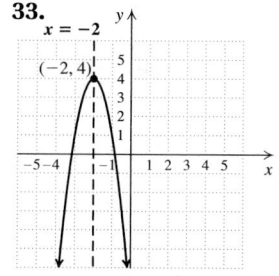
45. $\{x \mid x > 4\}$, or $(4, \infty)$ **46.** $\{x \mid x \geq -3\}$, or $[-3, \infty)$
47. $\{x \mid x \leq 7 \ or \ x \geq 11\}$, or $(-\infty, 7] \cup [11, \infty)$
48. $\left\{x \mid -3 < x < \frac{5}{2}\right\}$, or $\left(-3, \frac{5}{2}\right)$
49. $\dfrac{-4x - 23}{x + 4}$ **50.** $\dfrac{1}{x - 1}$ **51.** $-\frac{23}{4}$ **52.** No solution
53. 0   **54.** $-6, 9$ **55.** 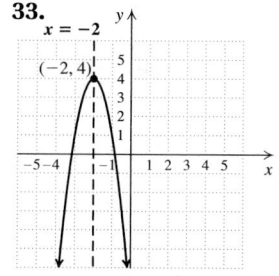   **57.** 158 ft   **59.** $15
61. The radius of the circular portion of the window and the height of the rectangular portion should each be $\dfrac{24}{\pi + 4}$ ft.
63. (a) $h(x) = 11,090.60714x^2 - 29,069.62143x + 39,983.8$; **(b)** 858,348 vehicles

Technology Connection, p. 770

1. $\{x \mid -0.78 \leq x \leq 1.59\}$, or $[-0.78, 1.59]$
2. $\{x \mid x \leq -0.21 \ or \ x \geq 2.47\}$, or $(-\infty, -0.21] \cup [2.47, \infty)$
3. $\{x \mid x < -1.26 \ or \ x > 2.33\}$, or $(-\infty, -1.26) \cup (2.33, \infty)$
4. $\{x \mid x > -1.37\}$, or $(-1.37, \infty)$

Exercise Set 11.9, pp. 772–775

1. True   **2.** False   **3.** True   **4.** True   **5.** False
6. True   **7.** $\left[-4, \frac{3}{2}\right]$, or $\left\{x \mid -4 \leq x \leq \frac{3}{2}\right\}$
9. $(-\infty, -2) \cup (0, 2) \cup (3, \infty)$, or $\{x \mid x < -2 \ or \ 0 < x < 2 \ or \ x > 3\}$
11. $\left(-\infty, -\frac{7}{2}\right) \cup (-2, \infty)$, or $\left\{x \mid x < -\frac{7}{2} \ or \ x > -2\right\}$
13. $(5, 6)$, or $\{x \mid 5 < x < 6\}$
15. $(-\infty, -7] \cup [2, \infty)$, or $\{x \mid x \leq -7 \ or \ x \geq 2\}$
17. $(-\infty, -1) \cup (2, \infty)$, or $\{x \mid x < -1 \ or \ x > 2\}$
19. $\varnothing$ **21.** $[2 - \sqrt{7}, 2 + \sqrt{7}]$, or $\{x \mid 2 - \sqrt{7} \leq x \leq 2 + \sqrt{7}\}$ **23.** $(-\infty, -2) \cup (0, 2)$, or $\{x \mid x < -2 \ or \ 0 < x < 2\}$ **25.** $[-2, 1] \cup [4, \infty)$, or $\{x \mid -2 \leq x \leq 1 \ or \ x \geq 4\}$ **27.** $[-2, 2]$, or $\{x \mid -2 \leq x \leq 2\}$ **29.** $(-1, 2) \cup (3, \infty)$, or $\{x \mid -1 < x < 2 \ or \ x > 3\}$ **31.** $(-\infty, 0] \cup [2, 5]$, or $\{x \mid x \leq 0 \ or \ 2 \leq x \leq 5\}$ **33.** $(-\infty, 5)$, or $\{x \mid x < 5\}$
35. $(-\infty, -1] \cup (3, \infty)$, or $\{x \mid x \leq -1 \ or \ x > 3\}$
37. $(-\infty, -6)$, or $\{x \mid x < -6\}$ **39.** $(-\infty, -1] \cup [2, 5)$, or $\{x \mid x \leq -1 \ or \ 2 \leq x < 5\}$ **41.** $(-\infty, -3) \cup [0, \infty)$, or $\{x \mid x < -3 \ or \ x \geq 0\}$ **43.** $(0, \infty)$, or $\{x \mid x > 0\}$
45. $(-\infty, -4) \cup [1, 3)$, or $\{x \mid x < -4 \ or \ 1 \leq x < 3\}$
47. $\left(-\frac{3}{4}, \frac{5}{2}\right]$, or $\left\{x \mid -\frac{3}{4} < x \leq \frac{5}{2}\right\}$ **49.** $(-\infty, 2) \cup [3, \infty)$, or $\{x \mid x < 2 \ or \ x \geq 3\}$ **51.**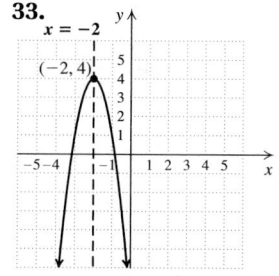
53.

[graph: $f(x) = x^3 - 2$]

54.

[graph: $g(x) = \dfrac{2}{x}$]

55. $\dfrac{1}{a^2} + 7$ **56.** $a - 8$ **57.** $4a^2 + 20a + 27$

58. $\sqrt{12a - 19}$ **59.** 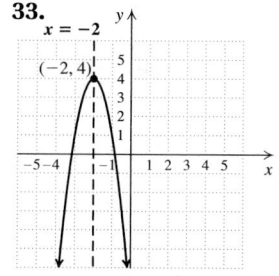   **61.** $(-1 - \sqrt{6}, -1 + \sqrt{6})$, or $\{x \mid -1 - \sqrt{6} < x < -1 + \sqrt{6}\}$ **63.** $\{0\}$
65. (a) $(10, 200)$, or $\{x \mid 10 < x < 200\}$; **(b)** $[0, 10) \cup (200, \infty)$, or $\{x \mid 0 \leq x < 10 \ or \ x > 200\}$
67. $\{n \mid n \text{ is an integer } and \ 12 \leq n \leq 25\}$
69. $f(x) = 0$ for $x = -2, 1, 3$; $f(x) < 0$ for $(-\infty, -2) \cup (1, 3)$, or $\{x \mid x < -2 \ or \ 1 < x < 3\}$; $f(x) > 0$ for $(-2, 1) \cup (3, \infty)$, or $\{x \mid -2 < x < 1 \ or \ x > 3\}$
71. $f(x)$ has no zeros; $f(x) < 0$ for $(-\infty, 0)$, or $\{x \mid x < 0\}$; $f(x) > 0$ for $(0, \infty)$, or $\{x \mid x > 0\}$ **73.** $f(x) = 0$ for $x = -1, 0$; $f(x) < 0$ for $(-\infty, -3) \cup (-1, 0)$, or $\{x \mid x < -3 \ or -1 < x < 0\}$; $f(x) > 0$ for $(-3, -1) \cup (0, 2) \cup (2, \infty)$, or $\{x \mid -3 < x < -1 \ or \ 0 < x < 2 \ or \ x > 2\}$
75. $(-\infty, -5] \cup [9, \infty)$, or $\{x \mid x \leq -5 \ or \ x \geq 9\}$
77. $(-\infty, -8] \cup [0, \infty)$, or $\{x \mid x \leq -8 \ or \ x \geq 0\}$
79.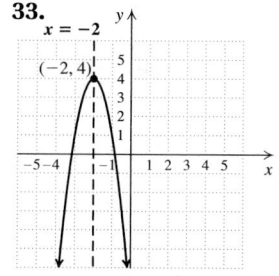

Review Exercises: Chapter 11, pp. 778–779

1. False   **2.** True   **3.** True   **4.** True   **5.** False
6. True   **7.** True   **8.** True   **9.** False   **10.** True
11. $\pm\dfrac{\sqrt{2}}{3}$ **12.** $0, -\frac{3}{4}$ **13.** $3, 9$ **14.** $2 \pm 2i$ **15.** $3, 5$
16. $-\dfrac{9}{2} \pm \dfrac{\sqrt{85}}{2}$ **17.** $-0.372, 5.372$ **18.** $-\frac{1}{4}, 1$
19. $x^2 - 18x + 81 = (x - 9)^2$
20. $x^2 + \frac{3}{5}x + \frac{9}{100} = \left(x + \frac{3}{10}\right)^2$ **21.** $3 \pm 2\sqrt{2}$
22. 4%   **23.** 8.0 sec   **24.** Two irrational real numbers
25. Two imaginary numbers   **26.** $x^2 + 9 = 0$
27. $x^2 + 10x + 25 = 0$ **28.** About 153 mph   **29.** 6 hr
30. $(-3, 0), (-2, 0), (2, 0), (3, 0)$ **31.** $-5, 3$
32. $\pm\sqrt{2}, \pm\sqrt{7}$
33.

[graph: $f(x) = -3(x + 2)^2 + 4$, $x = -2$, $(-2, 4)$, Maximum: 4]

$f(x) = -3(x + 2)^2 + 4$
Maximum: 4

34. (a) Vertex: $(3, 5)$; axis of symmetry: $x = 3$;
(b)

[graph: $f(x) = 2x^2 - 12x + 23$]

35. $(2, 0), (7, 0); (0, 14)$ **36.** $p = \dfrac{9\pi^2}{N^2}$

37. $T = \dfrac{1 \pm \sqrt{1 + 24A}}{6}$ **38.** Quadratic   **39.** Linear
40. 225 ft²; 15 ft by 15 ft   **41. (a)** $f(x) = -\frac{3}{8}x^2 + \frac{9}{4}x + 8$;
(b) about 10%   **42.** $(-1, 0) \cup (3, \infty)$, or
$\{x | -1 < x < 0 \ or \ x > 3\}$ **43.** $(-3, 5]$, or
$\{x | -3 < x \le 5\}$ **44.** ✍ The x-coordinate of the maximum or minimum point lies halfway between the
x-coordinates of the x-intercepts.   **45.** ✍ Yes; if the discriminant is a perfect square, then the solutions are rational
numbers, p/q and r/s. (Note that if the discriminant is 0,
then $p/q = r/s$.) Then the equation can be written in
factored form, $(qx - p)(sx - r) = 0$.   **46.** ✍ Four; let
$u = x^2$. Then $au^2 + bu + c = 0$ has at most two solutions,
$u = m$ and $u = n$. Now substitute x^2 for u and obtain
$x^2 = m$ or $x^2 = n$. These equations yield the solutions
$x = \pm\sqrt{m}$ and $x = \pm\sqrt{n}$. When $m \ne n$, the maximum
number of solutions, four, occurs.   **47.** ✍ Completing
the square was used to solve quadratic equations and to
graph quadratic functions by rewriting the function in the
form $f(x) = a(x - h)^2 + k$.   **48.** $f(x) = \frac{7}{15}x^2 - \frac{14}{15}x - 7$
49. $h = 60, k = 60$ **50.** $18, 324$

Test: Chapter 11, p. 780

1. [11.1] $\pm\dfrac{\sqrt{7}}{5}$ **2.** [11.1] $2, 9$ **3.** [11.2] $-1 \pm \sqrt{2}i$
4. [11.2] $1 \pm \sqrt{6}$ **5.** [11.5] $-2, \frac{2}{3}$ **6.** [11.2] $-4.193, 1.193$
7. [11.2] $-\frac{3}{4}, \frac{7}{3}$ **8.** [11.1] $x^2 - 20x + 100 = (x - 10)^2$
9. [11.1] $x^2 + \frac{2}{7}x + \frac{1}{49} = \left(x + \frac{1}{7}\right)^2$ **10.** [11.1] $-5 \pm \sqrt{10}$
11. [11.3] Two imaginary numbers   **12.** [11.3] $x^2 - 11 = 0$
13. [11.4] 16 km/h   **14.** [11.4] 2 hr   **15.** [11.5] $(-4, 0), (4, 0)$
16. [11.6]   **17.** [11.7] **(a)** $(-1, -8), x = -1$;
(b)

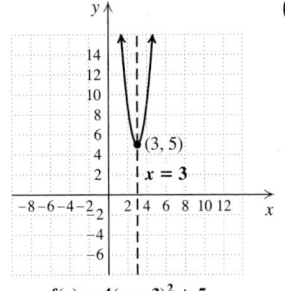

$f(x) = 4(x - 3)^2 + 5$
Minimum: 5

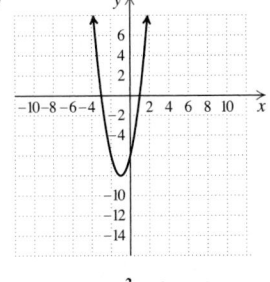

$f(x) = 2x^2 + 4x - 6$

18. [11.7] $(-2, 0), (3, 0); (0, -6)$ **19.** [11.4] $r = \sqrt{\dfrac{3V}{\pi} - R^2}$

20. [11.8] Quadratic   **21.** [11.8] Minimum: \$129/cabinet
when 325 cabinets are built   **22.** [11.8] $f(x) = \frac{1}{5}x^2 - \frac{3}{5}x$
23. [11.9] $(-6, 1)$, or $\{x | -6 < x < 1\}$
24. [11.9] $[-1, 0) \cup [1, \infty)$, or $\{x | -1 \le x < 0 \ or \ x \ge 1\}$
25. [11.3] $\frac{1}{2}$ **26.** [11.3] $x^4 + x^2 - 12 = 0$; answers may vary
27. [11.5] $\pm\sqrt{\sqrt{5} + 2}, \pm\sqrt{\sqrt{5} - 2}i$

Cumulative Review: Chapters 1–11, pp. 781–782

1. 24   **2.** $\dfrac{3a^{10}c^7}{4}$ **3.** $14x^2y - 10xy - 9xy^2$

4. $81p^4q^2 - 64t^2$ **5.** $\dfrac{t(t + 1)(t + 5)}{(3t + 4)^3}$ **6.** $-\dfrac{3}{x + 4}$

7. $3x^2\sqrt[3]{4y^2}$ **8.** $13 - \sqrt{2}i$
9. $3(2x^2 + 5y^2)(2x^2 - 5y^2)$ **10.** $x(x - 20)(x - 4)$
11. $100(m + 1)(m^2 - m + 1)(m - 1)(m^2 + m + 1)$
12. $(2t + 9)(3t + 4)$ **13.** 7   **14.** $\{x | x < 7\}$, or $(-\infty, 7)$
15. $\left(3, \frac{1}{2}\right)$ **16.** $-6, 11$ **17.** $\frac{1}{2}, 2$ **18.** 4   **19.** $-5 \pm \sqrt{2}$
20. $\dfrac{1}{6} \pm \dfrac{\sqrt{11}}{6}i$

21. 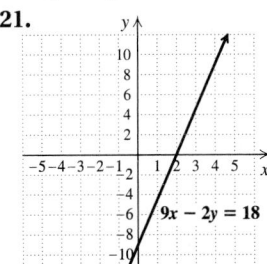   **22.**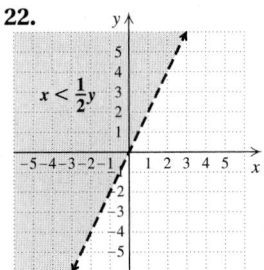

23. 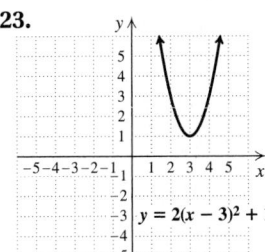   **24.**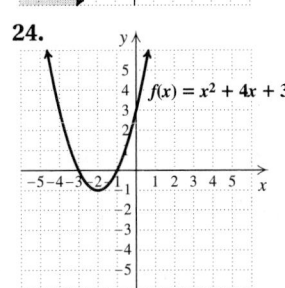

25. $y = -5x + \frac{1}{2}$ **26.** $-\frac{7}{10}$ **27.** 21
28. $(-\infty, 10]$, or $\{x | x \le 10\}$
29. $\{x | x \text{ is a real number } and \ x \ne 4\}$, or $(-\infty, 4) \cup (4, \infty)$

30. $a = \dfrac{c}{2b - 1}$ **31.** $t = \dfrac{4r}{3p^2}$ **32. (a)** \$4.53 billion;
(b) 2015   **33. (a)** $h(t) = 33t - 47$; **(b)** 283,000 hotspots;
(c) about 2017   **34. (a)** 1.74 oz; **(b)** \$600 per ounce;
(c) 75%   **35.** Number tiles: 26 sets; alphabet tiles: 10 sets
36. \$125/bunk bed; 400 bunk beds
37. Deanna: 12 hr; Donna: 6 hr
38. 9 km/h   **39.** Mileages no greater than 50 mi
40. $\dfrac{1}{3} \pm \dfrac{\sqrt{2}}{6}i$ **41.** $\{0\}$ **42.** $f(x) = x + 1$
43. $(1 - \sqrt{6}, 16 - 10\sqrt{6}), (1 + \sqrt{6}, 16 + 10\sqrt{6})$

CHAPTER 12

Technology Connection, p. 786

1. To check $(f \circ g)(x)$, we let $y_1 = \sqrt{x}, y_2 = x - 1$,
$y_3 = \sqrt{x - 1}$, and $y_4 = y_1(y_2)$. A table shows that we have
$y_3 = y_4$. The check for $(g \circ f)(x)$ is similar. A graph can also
be used.

Technology Connection, p. 791

1. Graph each pair of functions in a square window along
with the line $y = x$ and determine whether the first two
functions are reflections of each other across $y = x$. For further verification, examine a table of values for each pair of
functions.   **2.** Yes; most graphing calculators do not
require that the inverse relation be a function.

Exercise Set 12.1, pp. 792–794

1. True   **2.** True   **3.** False   **4.** False   **5.** False
6. False   **7.** True   **8.** True   **9. (a)** $(f \circ g)(1) = 5$;
(b) $(g \circ f)(1) = -1$; **(c)** $(f \circ g)(x) = x^2 - 6x + 10$;
(d) $(g \circ f)(x) = x^2 - 2$ **11. (a)** $(f \circ g)(1) = -24$;
(b) $(g \circ f)(1) = 65$; **(c)** $(f \circ g)(x) = 10x^2 - 34$;
(d) $(g \circ f)(x) = 50x^2 + 20x - 5$
13. (a) $(f \circ g)(1) = 8$; **(b)** $(g \circ f)(1) = \frac{1}{64}$;
(c) $(f \circ g)(x) = \dfrac{1}{x^2} + 7$; **(d)** $(g \circ f)(x) = \dfrac{1}{(x + 7)^2}$
15. (a) $(f \circ g)(1) = 2$; **(b)** $(g \circ f)(1) = 4$;
(c) $(f \circ g)(x) = \sqrt{x + 3}$; **(d)** $(g \circ f)(x) = \sqrt{x} + 3$
17. (a) $(f \circ g)(1) = 2$; **(b)** $(g \circ f)(1) = \frac{1}{2}$;
(c) $(f \circ g)(x) = \sqrt{\dfrac{4}{x}}$; **(d)** $(g \circ f)(x) = \dfrac{1}{\sqrt{4x}}$
19. (a) $(f \circ g)(1) = 4$; **(b)** $(g \circ f)(1) = 2$;
(c) $(f \circ g)(x) = x + 3$; **(d)** $(g \circ f)(x) = \sqrt{x^2 + 3}$
21. $f(x) = x^4; g(x) = 3x - 5$ **23.** $f(x) = \sqrt{x}$;
$g(x) = 9x + 1$ **25.** $f(x) = \dfrac{6}{x}; g(x) = 5x - 2$ **27.** Yes

29. No   **31.** Yes   **33.** No   **35. (a)** Yes;
(b) $f^{-1}(x) = x - 3$ **37. (a)** Yes; **(b)** $f^{-1}(x) = \dfrac{x}{2}$

39. (a) Yes; **(b)** $g^{-1}(x) = \dfrac{x + 1}{3}$ **41. (a)** Yes;
(b) $f^{-1}(x) = 2x - 2$ **43. (a)** No   **45. (a)** Yes;
(b) $h^{-1}(x) = 10 - x$ **47. (a)** Yes; **(b)** $f^{-1}(x) = \dfrac{1}{x}$

49. (a) No   **51. (a)** Yes; **(b)** $f^{-1}(x) = \dfrac{3x - 1}{2}$
53. (a) Yes; **(b)** $f^{-1}(x) = \sqrt[3]{x - 5}$ **55. (a)** Yes;
(b) $g^{-1}(x) = \sqrt[3]{x} + 2$ **57. (a)** Yes;
(b) $f^{-1}(x) = x^2, x \geq 0$
59.

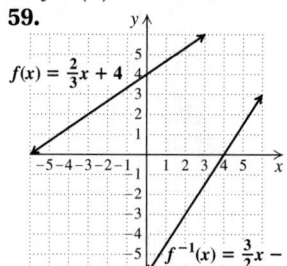

61.

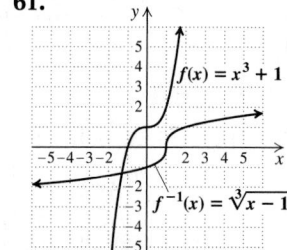

63.

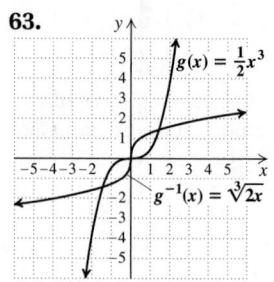

65.

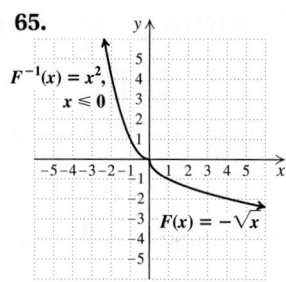

67.

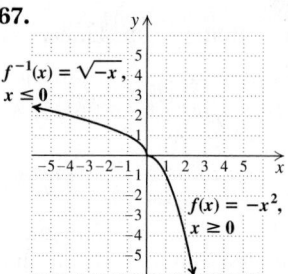

69. (1) $(f^{-1} \circ f)(x) = f^{-1}(f(x))$
$$= f^{-1}(\sqrt[3]{x - 4}) = (\sqrt[3]{x - 4})^3 + 4$$
$$= x - 4 + 4 = x;$$
(2) $(f \circ f^{-1})(x) = f(f^{-1}(x))$
$$= f(x^3 + 4) = \sqrt[3]{x^3 + 4 - 4}$$
$$= \sqrt[3]{x^3} = x$$

71. (1) $(f^{-1} \circ f)(x) = f^{-1}(f(x)) = f^{-1}\left(\dfrac{1 - x}{x}\right)$

$$= \dfrac{1}{\left(\dfrac{1 - x}{x}\right) + 1}$$

$$= \dfrac{1}{\dfrac{1 - x + x}{x}}$$

$$= x;$$

(2) $(f \circ f^{-1})(x) = f(f^{-1}(x)) = f\left(\dfrac{1}{x + 1}\right)$

$$= \dfrac{1 - \left(\dfrac{1}{x + 1}\right)}{\left(\dfrac{1}{x + 1}\right)}$$

$$= \dfrac{\dfrac{x + 1 - 1}{x + 1}}{\dfrac{1}{x + 1}} = x$$

73. (a) 40, 44, 52, 60; **(b)** $f^{-1}(x) = (x - 24)/2$, or $\dfrac{x}{2} - 12$

(c) 8, 10, 14, 18   **75.** 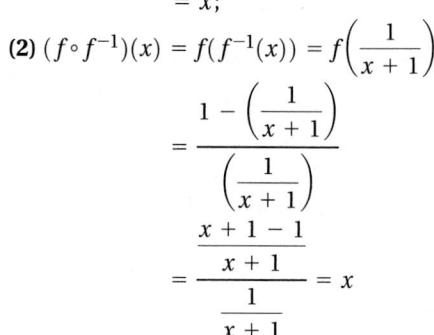   **77.** $\frac{1}{8}$ **78.** $\frac{1}{25}$ **79.** 32
80. Approximately 2.1577
81.

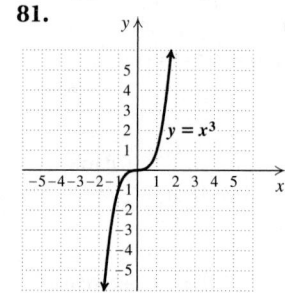

82.

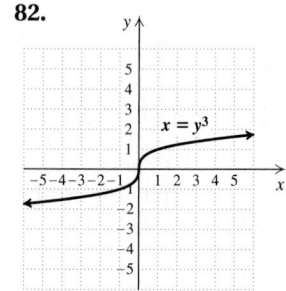

83. 📝

85.

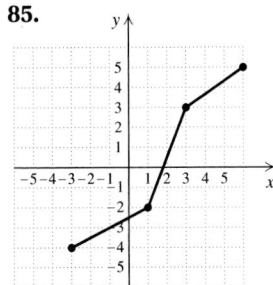

87. $g(x) = \dfrac{x}{2} + 20$ **89.** ✍

91. Suppose that $h(x) = (f \circ g)(x)$. First, note that for $I(x) = x, (f \circ I)(x) = f(I(x)) = f(x)$ for any function f.

(i) $((g^{-1} \circ f^{-1}) \circ h)(x) = ((g^{-1} \circ f^{-1}) \circ (f \circ g))(x)$
$= ((g^{-1} \circ (f^{-1} \circ f)) \circ g)(x)$
$= ((g^{-1} \circ I) \circ g)(x)$
$= (g^{-1} \circ g)(x) = x$

(ii) $(h \circ (g^{-1} \circ f^{-1}))(x) = ((f \circ g) \circ (g^{-1} \circ f^{-1}))(x)$
$= ((f \circ (g \circ g^{-1})) \circ f^{-1})(x)$
$= ((f \circ I) \circ f^{-1})(x)$
$= (f \circ f^{-1})(x) = x.$

Therefore, $(g^{-1} \circ f^{-1})(x) = h^{-1}(x)$.

93. Yes   **95.** No   **97. (1)** C; **(2)** A; **(3)** B; **(4)** D
99. ✍

Technology Connection, p. 797

1. $y_1 = \left(\dfrac{5}{2}\right)^x;\ y_2 = \left(\dfrac{2}{5}\right)^x$ **2.** $y_1 = 3.2^x;\ y_2 = 3.2^{-x}$

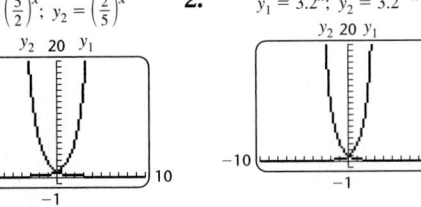

3. $y_1 = \left(\dfrac{3}{7}\right)^x;\ y_2 = \left(\dfrac{7}{3}\right)^x$ **4.** $y_1 = 5000(1.08)^x;\ y_2 = 5000(1.08)^{x-3}$

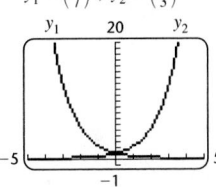 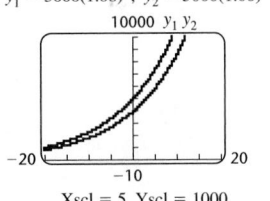

Xscl = 5, Yscl = 1000

Exercise Set 12.2, pp. 800–802

1. True   **2.** True   **3.** True   **4.** False   **5.** False
6. True
7.

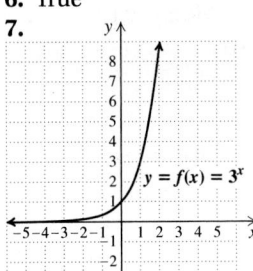

9.

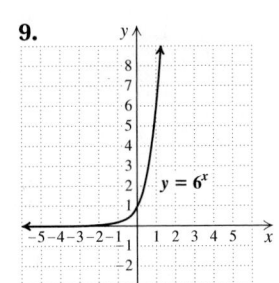

11.

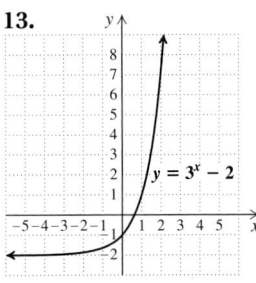

$y = 2^x + 1$

13.

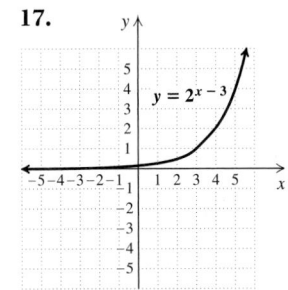

$y = 3^x - 2$

15.

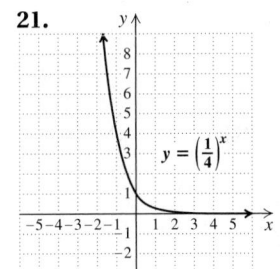

$y = 2^x - 5$

17.

$y = 2^{x-3}$

19.

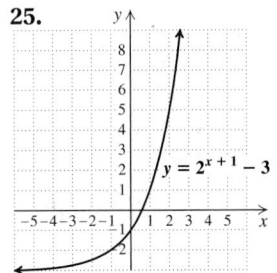

$y = 2^x + 1$

21.

$y = \left(\dfrac{1}{4}\right)^x$

23.

$y = \left(\dfrac{1}{3}\right)^x$

25.

$y = 2^{x+1} - 3$

27.

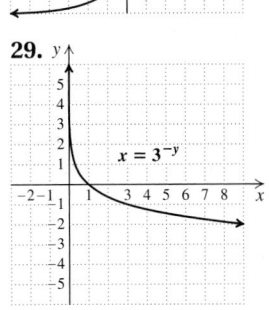

$x = 6^y$

29.

$x = 3^{-y}$

31.

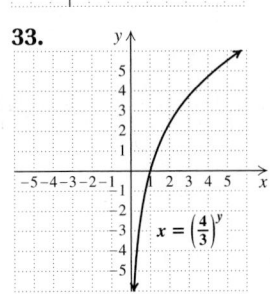

$x = 4^y$

33.

$x = \left(\dfrac{4}{3}\right)^y$

35.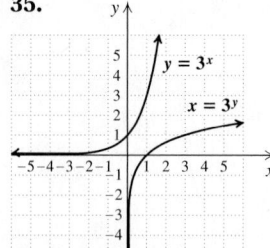

37.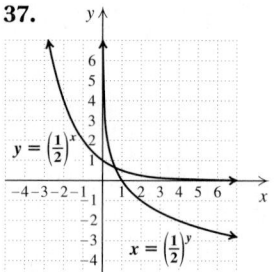

39. (a) About 0.68 billion tracks; about 1.052 billion tracks; about 2.519 billion tracks;

(b)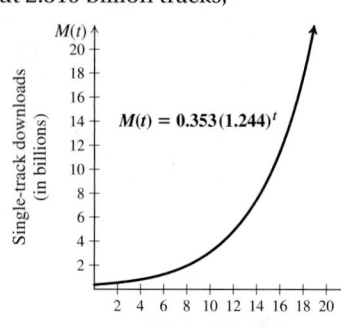

41. (a) 19.6%; 16.3%; 7.3%

(b)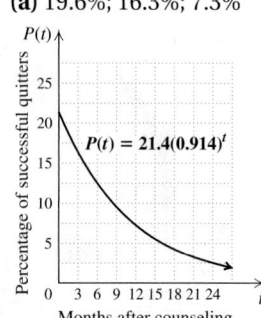

43. (a) About 44,079 whales; about 12,953 whales;

(b)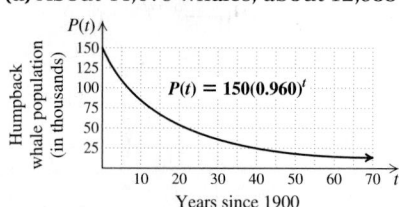

45. (a) About 8706 whales; about 15,107 whales;

(b)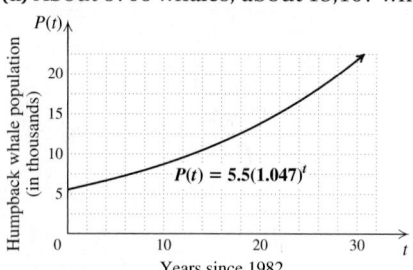

47. (a) 454,354,240 cm^2; 525,233,501,400 cm^2;

(b)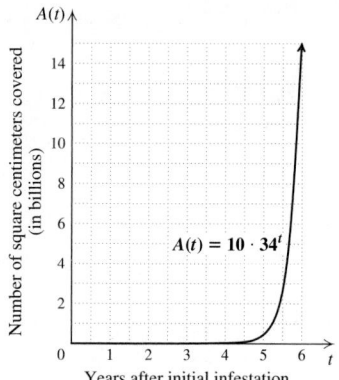

49. 🖉   **51.** $3(x + 4)(x - 4)$ **52.** $(x - 10)^2$
53. $(2x + 3)(3x - 4)$
54. $8(x^2 - 2y^2)(x^4 + 2x^2y^2 + 4y^4)$
55. $6(y - 4)(y + 10)$ **56.** $x(x - 2)(5x^2 - 3)$
57. 🖉   **59.** $\pi^{2.4}$

61. 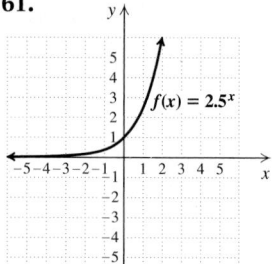   **63.**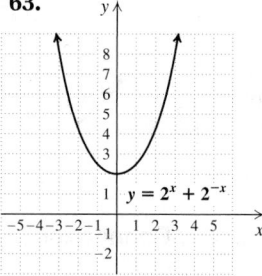

65. 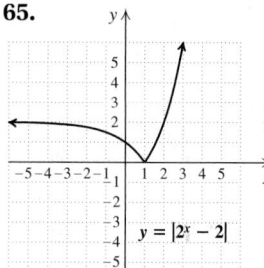   **67.**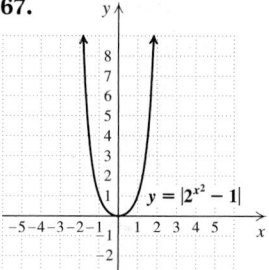

69.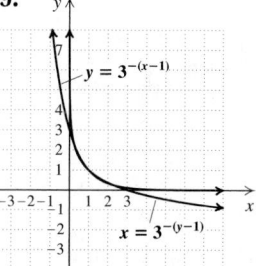

71. $N(t) = 0.464(1.778)^t$; about 464 million devices
73. 🖉   **75.**

Exercise Set 12.3, pp. 808–809

1. (g) **2.** (d) **3.** (a) **4.** (h) **5.** (b) **6.** (c)
7. (e) **8.** (f) **9.** 3 **11.** 2 **13.** 4 **15.** -2
17. -1 **19.** 4 **21.** 1 **23.** 0 **25.** 5 **27.** -2
29. $\frac{1}{2}$ **31.** $\frac{3}{2}$ **33.** $\frac{2}{3}$ **35.** 29
37.

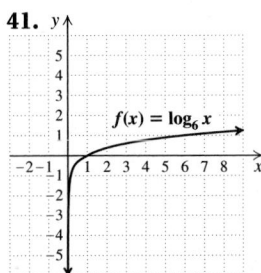

39.

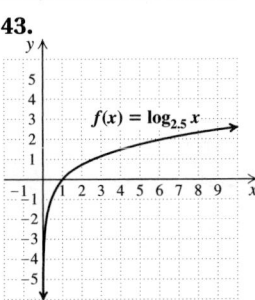

41.

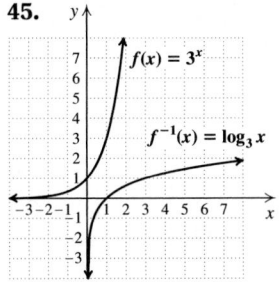

43.

45.

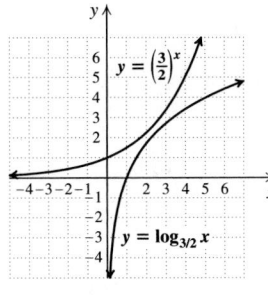

47. $10^x = 8$ **49.** $9^1 = 9$ **51.** $10^{-1} = 0.1$
53. $10^{0.845} = 7$ **55.** $c^8 = m$ **57.** $r^t = C$
59. $e^{-1.3863} = 0.25$ **61.** $r^{-x} = T$ **63.** $2 = \log_{10} 100$
65. $-3 = \log_5 \frac{1}{125}$ **67.** $\frac{1}{4} = \log_{16} 2$
69. $0.4771 = \log_{10} 3$ **71.** $m = \log_z 6$ **73.** $t = \log_p q$
75. $3 = \log_e 20.0855$ **77.** $-4 = \log_e 0.0183$ **79.** 36
81. 5 **83.** 9 **85.** 49 **87.** $\frac{1}{9}$ **89.** 4 **91.** ✍
93. $30a^2b^4$ **94.** $12 - 2\sqrt{30} + 2\sqrt{15} - 5\sqrt{2}$

95. $3\sqrt{3x}$ **96.** $\sqrt[12]{x}$ **97.** $\dfrac{x(3y-2)}{2y+x}$ **98.** $\dfrac{x+2}{x+1}$

99. ✍
101.

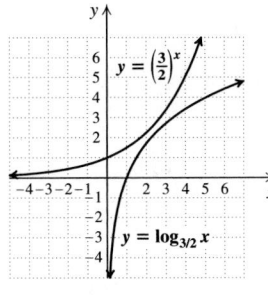

103.

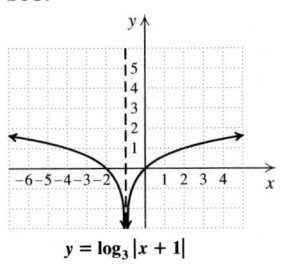

105. 6 **107.** $-25, 4$ **109.** -2 **111.** 0 **113.** Let $b = 0$, and suppose that $x_1 = 1$ and $x_2 = 2$. Then $0^1 = 0^2$, but $1 \neq 2$. Then let $b = 1$, and suppose that $x_1 = 1$ and $x_2 = 2$. Then $1^1 = 1^2$, but $1 \neq 2$.

Exercise Set 12.4, pp. 815–816

1. (e) **2.** (f) **3.** (a) **4.** (b) **5.** (c) **6.** (d)
7. $\log_3 81 + \log_3 27$ **9.** $\log_4 64 + \log_4 16$
11. $\log_c r + \log_c s + \log_c t$ **13.** $\log_a (2 \cdot 10)$, or $\log_a 20$
15. $\log_c (t \cdot y)$ **17.** $8 \log_a r$ **19.** $\frac{1}{3} \log_2 y$
21. $-3 \log_b C$ **23.** $\log_2 5 - \log_2 11$
25. $\log_b m - \log_b n$ **27.** $\log_a \frac{19}{2}$ **29.** $\log_b \frac{36}{4}$, or $\log_b 9$
31. $\log_a \dfrac{x}{y}$ **33.** $\log_a x + \log_a y + \log_a z$
35. $3 \log_a x + 4 \log_a z$ **37.** $2 \log_a w - 2 \log_a x + \log_a y$
39. $5 \log_a x - 3 \log_a y - \log_a z$
41. $\log_b x + 2 \log_b y - \log_b w - 3 \log_b z$
43. $\frac{1}{2}(7 \log_a x - 5 \log_a y - 8 \log_a z)$
45. $\frac{1}{3}(6 \log_a x + 3 \log_a y - 2 - 7 \log_a z)$ **47.** $\log_a (x^8 z^3)$
49. $\log_a x$ **51.** $\log_a \dfrac{y^5}{x^{3/2}}$ **53.** $\log_a (x-3)$ **55.** 1.953
57. -0.369 **59.** -1.161 **61.** $\frac{3}{2}$ **63.** Cannot be found
65. 10 **67.** m **69.** ✍
71.

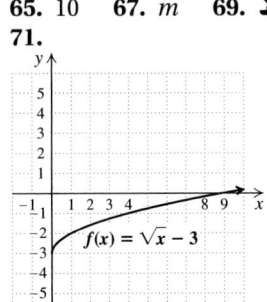

72.

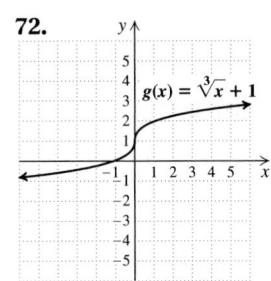

73.

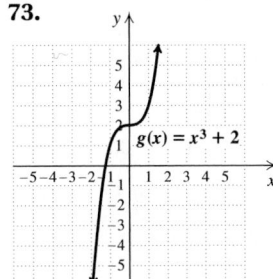

74.

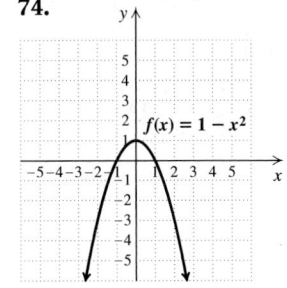

75. $(-\infty, -7) \cup (-7, \infty)$, or $\{x \,|\, x \text{ is a real number } and \, x \neq -7\}$
76. $(-\infty, -3) \cup (-3, 2) \cup (2, \infty)$, or $\{x \,|\, x \text{ is a real number } and \, x \neq -3 \, and \, x \neq 2\}$
77. $(-\infty, 10]$, or $\{x \,|\, x \leq 10\}$ **78.** $(-\infty, \infty)$, or $\mathbb{R}$
79. ✍ **81.** $\log_a (x^6 - x^4 y^2 + x^2 y^4 - y^6)$
83. $\frac{1}{2} \log_a (1 - s) + \frac{1}{2} \log_a (1 + s)$ **85.** $\frac{10}{3}$ **87.** -2
89. $\frac{2}{5}$ **91.** True

Technology Connection, p. 817

1. ⬭LOG⬭ ⬭7⬭ ⬭)⬭ ⬭÷⬭ ⬭LOG⬭ ⬭3⬭ ⬭)⬭ ⬭ENTER⬭

Technology Connection, p. 818

1. As x gets larger, the value of y_1 approaches 2.7182818284.... **2.** For large values of x, the graphs of y_1 and y_2 will be very close or appear to be the same curve, depending on the window chosen. **3.** Using (TRACE), no y-value is given for $x = 0$. Using a table, an error message appears for y_1 when $x = 0$. The domain does not include 0 because division by 0 is undefined.

Technology Connection, p. 821

1. $y = \log x/\log 7$
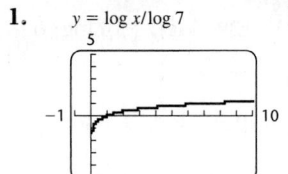

2. $y = \log (x+2)/\log 5$
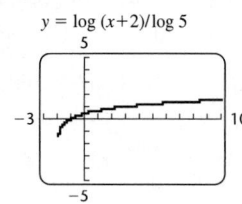

3. $y = \log x/\log 7 + 2$
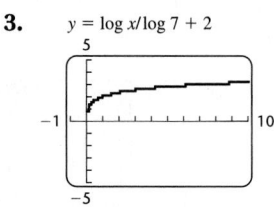

Visualizing for Success, p. 822

1. J **2.** D **3.** B **4.** G **5.** H **6.** C **7.** F **8.** I
9. E **10.** A

Exercise Set 12.5, pp. 823–824

1. True **2.** True **3.** True **4.** False **5.** True
6. True **7.** True **8.** True **9.** True **10.** True
11. 0.8451 **13.** 1.1367 **15.** 3 **17.** −0.1249
19. 13.0014 **21.** 50.1187 **23.** 0.0011 **25.** 2.1972
27. −5.0832 **29.** 96.7583 **31.** 15.0293 **33.** 0.0305
35. 3.0331 **37.** 6.6439 **39.** 1.1610 **41.** −0.3010
43. −3.3219 **45.** 2.0115
47. Domain: $\mathbb{R}$;
range: $(0, \infty)$
49. Domain: $\mathbb{R}$;
range: $(3, \infty)$

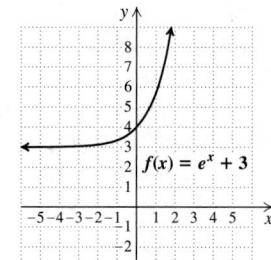
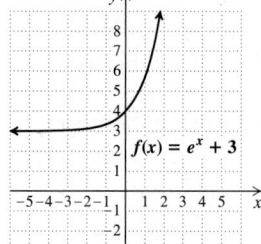

51. Domain: $\mathbb{R}$;
range: $(-2, \infty)$
53. Domain: $\mathbb{R}$;
range: $(0, \infty)$

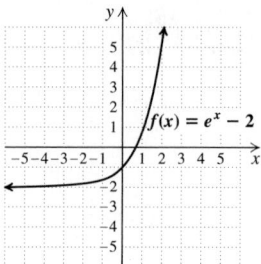
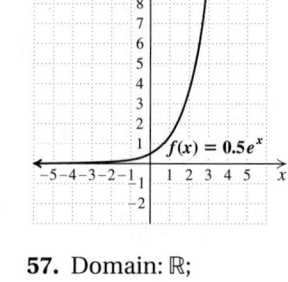

55. Domain: $\mathbb{R}$;
range: $(0, \infty)$
57. Domain: $\mathbb{R}$;
range: $(0, \infty)$

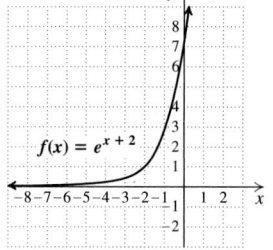

59. Domain: $\mathbb{R}$;
range: $(0, \infty)$
61. Domain: $\mathbb{R}$;
range: $(-\infty, 0)$

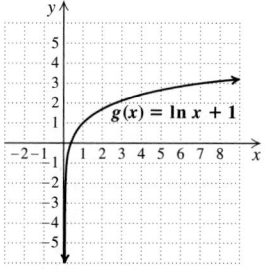

63. Domain: $(0, \infty)$;
range: $\mathbb{R}$
65. Domain: $(0, \infty)$;
range: $\mathbb{R}$

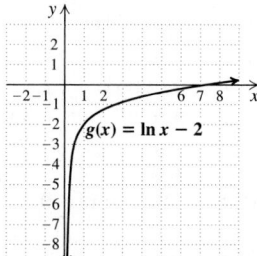

67. Domain: $(0, \infty)$;
range: $\mathbb{R}$
69. Domain: $(0, \infty)$;
range: $\mathbb{R}$

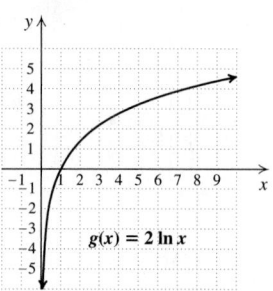
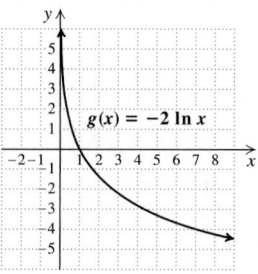

71. Domain: $(-2, \infty)$; range: $\mathbb{R}$

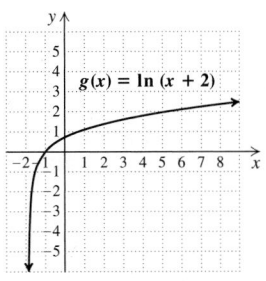

$g(x) = \ln(x + 2)$

73. Domain: $(1, \infty)$; range: $\mathbb{R}$

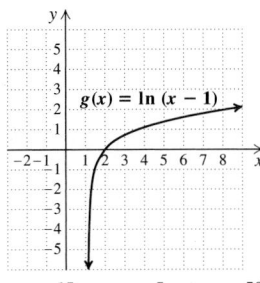

$g(x) = \ln(x - 1)$

75. ⌨ **77.** $-4, 7$ **78.** $0, \frac{7}{5}$ **79.** $\frac{15}{17}$ **80.** $\frac{5}{6}$ **81.** $\frac{56}{9}$
82. 4 **83.** $16, 256$ **84.** $\frac{1}{4}, 9$ **85.** ⌨ **87.** 2.452
89. 1.442 **91.** $\log M = \dfrac{\ln M}{\ln 10}$ **93.** 1086.5129
95. 4.9855 **97. (a)** Domain: $\{x | x > 0\}$, or $(0, \infty)$; range: $\{y | y < 0.5135\}$, or $(-\infty, 0.5135)$; **(b)** $[-1, 5, -10, 5]$; **(c)** $y = 3.4 \ln x - 0.25e^x$

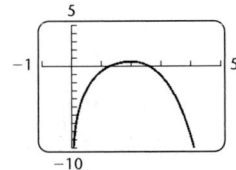

99. (a) Domain: $\{x | x > 0\}$, or $(0, \infty)$; range: $\{y | y > -0.2453\}$, or $(-0.2453, \infty)$; **(b)** $[-1, 5, -1, 10]$; **(c)** $y = 2x^3 \ln x$

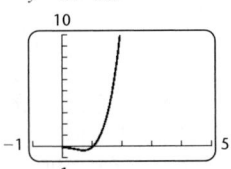

101. ◿

Connecting the Concepts, pp. 824–825

1. 2 **2.** -1 **3.** $\frac{1}{2}$ **4.** 2 **5.** 1 **6.** 0 **7.** 4 **8.** 8
9. 7 **10.** 3 **11.** $x^m = 3$ **12.** $2^{10} = 1024$
13. $t = \ln x$ **14.** $\frac{2}{3} = \log_{64} 16$ **15.** 4 **16.** $\frac{1}{3}$
17. $\log x - \frac{1}{2}\log y - \frac{3}{2}\log z$ **18.** $\log \dfrac{a}{b^2 c}$ **19.** 1.5
20. 2.8614

Technology Connection, p. 829

1. 0.38 **2.** -1.96 **3.** 0.90 **4.** -1.53 **5.** $0.13, 8.47$
6. $-0.75, 0.75$

Exercise Set 12.6, pp. 830–831

1. (e) **2.** (a) **3.** (f) **4.** (h) **5.** (b) **6.** (d)
7. (g) **8.** (c) **9.** 2 **11.** $\frac{5}{2}$ **13.** $\dfrac{\log 10}{\log 2} \approx 3.322$
15. -1 **17.** $\dfrac{\log 19}{\log 8} + 3 \approx 4.416$ **19.** $\ln 50 \approx 3.912$

21. $\dfrac{\ln 8}{-0.02} \approx -103.972$ **23.** $\dfrac{\log 87}{\log 4.9} \approx 2.810$
25. $\dfrac{\ln\left(\frac{19}{2}\right)}{4} \approx 0.563$ **27.** $\dfrac{\ln 2}{-1} \approx -0.693$ **29.** 81
31. $\frac{1}{16}$ **33.** $e^5 \approx 148.413$ **35.** $\dfrac{e^3}{4} \approx 5.021$
37. $10^{1.2} \approx 15.849$ **39.** $\dfrac{e^4 - 1}{2} \approx 26.799$
41. $e \approx 2.718$ **43.** $e^{-3} \approx 0.050$ **45.** -4 **47.** 10
49. No solution **51.** 2 **53.** $\frac{83}{15}$ **55.** 1 **57.** 6
59. 1 **61.** 5 **63.** $\frac{17}{2}$ **65.** 4 **67.** ⌨
69. Length: 9.5 ft; width: 3.5 ft **70.** 25 visits or more
71. Golden Days; $23\frac{1}{3}$ lb; Snowy Friends: $26\frac{2}{3}$ lb
72. 1.5 cm **73.** $1\frac{1}{5}$ hr **74.** Approximately 2.1 ft
75. ⌨ **77.** -4 **79.** 2 **81.** $\pm\sqrt{34}$ **83.** $-3, -1$
85. $-625, 625$ **87.** $\frac{1}{2}, 5000$ **89.** $-3, -1$
91. $\frac{1}{100,000}, 100,000$ **93.** $-\frac{1}{3}$ **95.** 38 **97.** 1

Exercise Set 12.7, pp. 838–843

1. (a) Approximately 2006; **(b)** 2.8 yr
3. (a) Approximately 1979; **(b)** approximately 2025
5. (a) 6.4 yr; **(b)** 23.4 yr **7. (a)** 1991; **(b)** 2013
9. (a) 2018; **(b)** 15.1 yr **11.** 4.9
13. 10^{-7} moles per liter **15.** 130 dB **17.** 7.6 W/m²
19. Approximately 42.4 million messages per day
21. (a) $P(t) = P_0 e^{0.025t}$; **(b)** \$5126.58; \$5256.36; **(c)** 27.7 yr
23. (a) $P(t) = 304 e^{0.009t}$; **(b)** 315 million; **(c)** about 2015
25. 0.2 yr **27. (a)** About 2055; **(b)** about 2068;
(c)

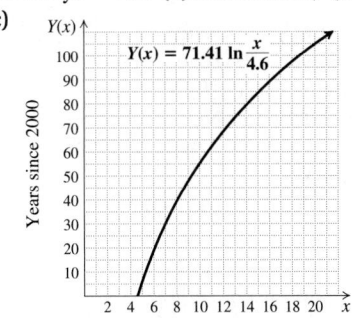

29. (a) 68%; **(b)** 54%, 40%
(c) **(d)** 6.9 months

$S(t) = 68 - 20 \log(t + 1), t \geq 0$

31. (a) $k \approx 0.126$; $P(t) = 2000 e^{0.126t}$; **(b)** 2015

33. (a) $k \approx 0.280$; $P(t) = 8200e^{-0.280t}$; **(b)** \$215 per gigabit per second per mile; **(c)** 2029   **35.** About 1964 yr
37. 7.2 days   **39. (a)** 13.9% per hour; **(b)** 21.6 hr
41. (a) $k \approx 0.114$; $V(t) = 451{,}000e^{0.114t}$; **(b)** \$4.9 million;
(c) 6.1 yr; **(d)** 2010   **43.** 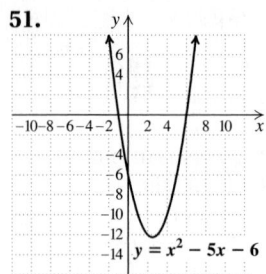   **45.** $\sqrt{2}$ **46.** 5
47. $(4, -7)$ **48.** $\left(-\frac{7}{2}, -\frac{19}{2}\right)$ **49.** $-4 \pm \sqrt{17}$
50. $5 \pm 2\sqrt{10}$

51.

52.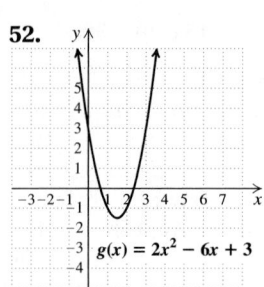

53.   **55.** \$14.5 million   **57. (a)** -26.9;
(b) 1.58×10^{-17} W/m^2 **59.** Consider an exponential growth function $P(t) = P_0 e^{kt}$. At time T, $P(T) = 2P_0$. Solve for T:
$$2P_0 = P_0 e^{kt}$$
$$2 = e^{kt}$$
$$\ln 2 = kT$$
$$\frac{\ln 2}{k} = T.$$

61.

Review Exercises: Chapter 12, pp. 847–848

1. True   **2.** True   **3.** True   **4.** False   **5.** False
6. True   **7.** False   **8.** False   **9.** True   **10.** False
11. $(f \circ g)(x) = 4x^2 - 12x + 10$; $(g \circ f)(x) = 2x^2 - 1$
12. $f(x) = \sqrt{x}$; $g(x) = 3 - x$ **13.** No
14. $f^{-1}(x) = x + 10$ **15.** $g^{-1}(x) = \dfrac{2x - 1}{3}$
16. $f^{-1}(x) = \dfrac{\sqrt[3]{x}}{3}$ **17.**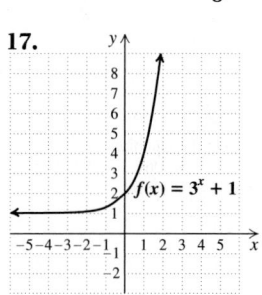

18. 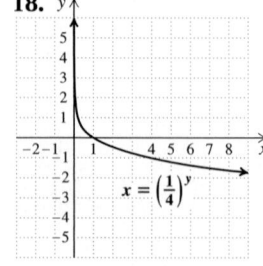   **19.**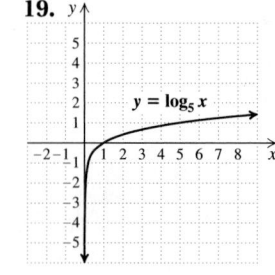

20. 2   **21.** -2 **22.** 11   **23.** $\frac{1}{2}$ **24.** $\log_2 \frac{1}{8} = -3$
25. $\log_{25} 5 = \frac{1}{2}$ **26.** $16 = 4^x$ **27.** $1 = 8^0$

28. $4 \log_a x + 2 \log_a y + 3 \log_a z$
29. $5 \log_a x - (\log_a y + 2 \log_a z)$, or
$5 \log_a x - \log_a y - 2 \log_a z$
30. $\frac{1}{4}(2 \log z - 3 \log x - \log y)$ **31.** $\log_a (5 \cdot 8)$, or $\log_a 40$
32. $\log_a \frac{48}{12}$, or $\log_a 4$ **33.** $\log \dfrac{a^{1/2}}{bc^2}$ **34.** $\log_a \sqrt[3]{\dfrac{x}{y^2}}$
35. 1   **36.** 0   **37.** 17   **38.** 6.93   **39.** -3.2698
40. 8.7601   **41.** 3.2698   **42.** 2.54995   **43.** -3.6602
44. 1.8751   **45.** 61.5177   **46.** -1.2040 **47.** 0.3753
48. 2.4307   **49.** 0.8982
50. Domain: $\mathbb{R}$; range: $(-1, \infty)$

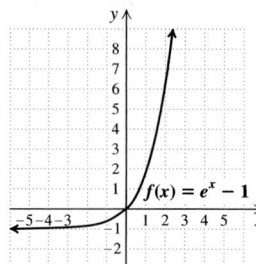

51. Domain: $(0, \infty)$; range: $\mathbb{R}$

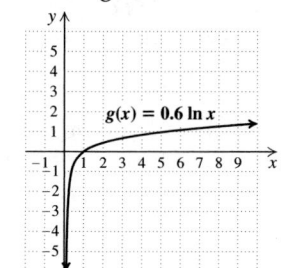

52. 3   **53.** -1 **54.** $\frac{1}{81}$ **55.** 2   **56.** $\frac{1}{1000}$
57. $e^3 \approx 20.0855$ **58.** $\frac{1}{2}\left(\dfrac{\log 19}{\log 4} + 5\right) \approx 3.5620$
59. $\dfrac{\log 12}{\log 2} \approx 3.5850$ **60.** $\dfrac{\ln 0.03}{-0.1} \approx 35.0656$
61. $e^{-3} \approx 0.0498$ **62.** $\frac{15}{2}$ **63.** 16   **64.** 5
65. (a) 82; **(b)** 66.8; **(c)** 35 months   **66. (a)** 2.3 yr;
(b) 3.1 yr   **67. (a)** $k \approx 0.043$; $A(t) = 885e^{0.043t}$;
(b) \$1.0 billion; **(c)** 2023; **(d)** 16.1 yr
68. (a) $M(t) = 3253e^{-0.137t}$; **(b)** 1640 spam messages per consumer; **(c)** 2030   **69.** 11.553% per year   **70.** 16.5 yr
71. 3463 yr   **72.** 5.1   **73.** About 114 dB
74. Negative numbers do not have logarithms because logarithm bases are positive, and there is no exponent to which a positive number can be raised to yield a negative number.   **75.** If $f(x) = e^x$, then to find the inverse function, we let $y = e^x$ and interchange x and y: $x = e^y$. If $x = e^y$, then $\log_e x = y$ by the definition of logarithms. Since $\log_e x = \ln x$, we have $y = \ln x$ or $f^{-1}(x) = \ln x$. Thus, $g(x) = \ln x$ is the inverse of $f(x) = e^x$. Another approach is to find $(f \circ g)(x)$ and $(g \circ f)(x)$:
$$(f \circ g)(x) = e^{\ln x} = x, \text{ and}$$
$$(g \circ f)(x) = \ln e^x = x.$$
Thus, g and f are inverse functions.
76. e^{e^3} **77.** $-3, -1$ **78.** $\left(\frac{8}{3}, -\frac{2}{3}\right)$

Test: Chapter 12, p. 849

1. [12.1] $(f \circ g)(x) = 2 + 6x + 4x^2$; $(g \circ f)(x) = 2x^2 + 2x + 1$
2. [12.1] $f(x) = \dfrac{1}{x}$; $g(x) = 2x^2 + 1$ **3.** [12.1] No
4. [12.1] $f^{-1}(x) = \dfrac{x - 4}{3}$ **5.** [12.1] $g^{-1}(x) = \sqrt[3]{x} - 1$

6. [12.2]

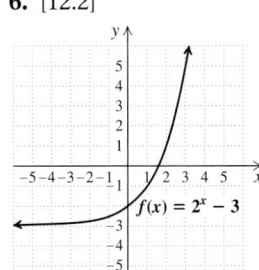

7. [12.3]

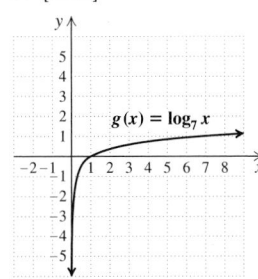

8. [12.3] 3   **9.** [12.3] $\frac{1}{2}$ **10.** [12.3] 18   **11.** [12.4] 1
12. [12.4] 0   **13.** [12.4] 19   **14.** [12.3] $\log_5 \frac{1}{625} = -4$
15. [12.3] $2^m = \frac{1}{2}$ **16.** [12.4] $3\log a + \frac{1}{2}\log b - 2\log c$
17. [12.4] $\log_a\left(z^2 \sqrt[3]{x}\right)$ **18.** [12.4] 1.146   **19.** [12.4] 0.477
20. [12.4] 1.204   **21.** [12.5] 1.3979   **22.** [12.5] 0.1585
23. [12.5] -0.9163 **24.** [12.5] 121.5104   **25.** [12.5] 2.4022
26. [12.5]

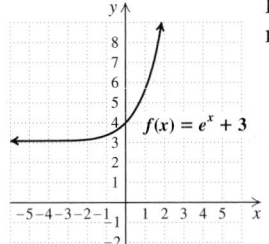

Domain: $\mathbb{R}$;
range: $(3, \infty)$

27. [12.5]

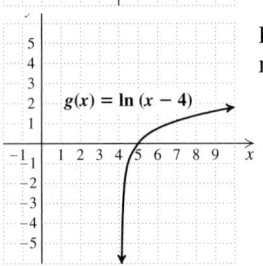

Domain: $(4, \infty)$;
range: $\mathbb{R}$

28. [12.6] -5 **29.** [12.6] 2   **30.** [12.6] $\frac{1}{100}$

31. [12.6] $-\frac{1}{3}\left(\dfrac{\log 87}{\log 5} - 4\right) \approx 0.4084$

32. [12.6] $\dfrac{\log 1.2}{\log 7} \approx 0.0937$ **33.** [12.6] $e^3 \approx 20.0855$

34. [12.6] 4   **35.** [12.7] **(a)** 2.25 ft/sec; **(b)** 2,901,000
36. [12.7] **(a)** $P(t) = 140e^{0.024t}$, where t is the number of
years after 2008 and $P(t)$ is in millions; **(b)** 154 million;
170 million; **(c)** 2023; **(d)** 28.9 yr
37. [12.7] **(a)** $k \approx 0.045; C(t) = 21{,}855e^{0.045t}$; **(b)** \$35,853;
(c) 2019   **38.** [12.7] 4.3%   **39.** [12.7] 4684 yr
40. [12.7] 6.3×10^6 W/m^2 **41.** [12.7] 7.0
42. [12.6] $-309{,}316$ **43.** [12.4] 2

Cumulative Review: Chapters 1–12, pp. 850–851

1. 2   **2.** $\dfrac{y^{12}}{16x^8}$ **3.** $\dfrac{20x^6 z^2}{y}$ **4.** $-\dfrac{y^4}{3z^5}$ **5.** 6.3×10^{-15}
6. 25   **7.** 8   **8.** $(3, -1)$ **9.** $(1, -2, 0)$ **10.** $-7, 10$
11. $\frac{9}{2}$ **12.** $\frac{3}{4}$ **13.** $\frac{1}{2}$ **14.** $\pm 4i$ **15.** $\pm 2, \pm 3$ **16.** 9
17. $\dfrac{\log 7}{5 \log 3} \approx 0.3542$ **18.** $\dfrac{8e}{e - 1} \approx 12.6558$

19. $(-\infty, -5) \cup (1, \infty)$, or $\{x \mid x < -5 \text{ or } x > 1\}$
20. $-3 \pm 2\sqrt{5}$ **21.** $\{x \mid x \le -2 \text{ or } x \ge 5\}$,
or $(-\infty, -2] \cup [5, \infty)$ **22.** $a = \dfrac{Db}{b - D}$

23. $x = \dfrac{-v \pm \sqrt{v^2 + 4ad}}{2a}$

24. $\left\{x \mid x \text{ is a real number } and \; x \ne -\frac{1}{3} \; and \; x \ne 2\right\}$, or
$\left(-\infty, -\frac{1}{3}\right) \cup \left(-\frac{1}{3}, 2\right) \cup (2, \infty)$
25. $3p^2 q^3 + 11pq - 2p^2 + p + 9$ **26.** $9x^4 - 6x^2 z^3 + z^6$
27. $\dfrac{1}{x - 4}$ **28.** $\dfrac{a + 2}{6}$ **29.** $\dfrac{7x + 4}{(x + 6)(x - 6)}$
30. $2y^2 \sqrt[3]{y}$ **31.** $\sqrt[10]{(x + 5)^7}$ **32.** $15 - 4\sqrt{3}i$
33. $x^3 - 5x^2 + 1$ **34.** $(3 + 4n)(9 - 12n + 16n^2)$
35. $2(3x - 2y)(x + 2y)$ **36.** $(x - 4)(x^3 + 7)$
37. $2(m + 3n)^2$ **38.** $(x - 2y)(x + 2y)(x^2 + 4y^2)$
39. $\dfrac{6 + \sqrt{y} - y}{4 - y}$ **40.** $f^{-1}(x) = \dfrac{x - 9}{-2}$, or $f^{-1}(x) = \dfrac{9 - x}{2}$
41. $f(x) = -10x - 8$ **42.** $y = \frac{1}{2}x + 5$
43.

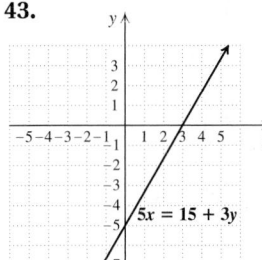

44.

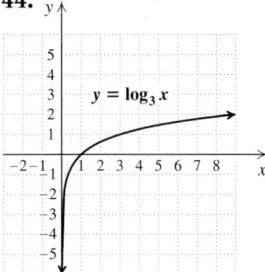

45.

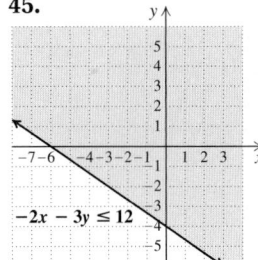

46.

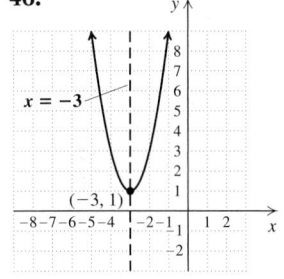

$f(x) = 2x^2 + 12x + 19$
Minimum: 1

47.

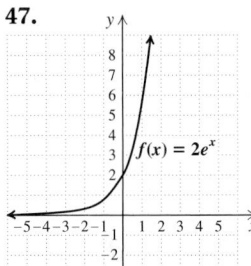

Domain: $\mathbb{R}$;
range: $(0, \infty)$

48. $\log\left(\dfrac{x^3}{y^{1/2} z^2}\right)$ **49.** 13.5 million acre-feet

50. (a) $k \approx 0.076; D(t) = 15e^{0.076t}$; **(b)** 79.8 million cubic
meters per day; **(c)** 2015   **51. (a)** $\frac{2}{15}$ million barrels per
day per year; **(b)** $g(t) = \frac{2}{15}t + 8.5$; **(c)** $G(t) = 8.5e^{0.015t}$
52. $5\frac{5}{11}$ min   **53.** Thick and Tasty: 6 oz; Light and Lean:
9 oz   **54.** $2\frac{7}{9}$ km/h   **55.** -49; -7 and 7

56. (a) 78; **(b)** 67.5   **57.** All real numbers except 1 and -2 **58.** $\frac{1}{3}, \frac{10,000}{3}$ **59.** 35 mph

CHAPTER 13

Technology Connection, p. 859

1. $x^2 + y^2 - 16 = 0$
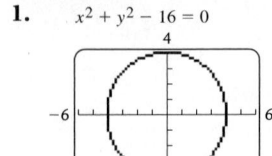

2. $(x - 1)^2 + (y - 2)^2 = 25$
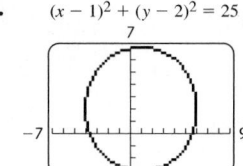

3. $(x + 3)^2 + (y - 5)^2 = 16$
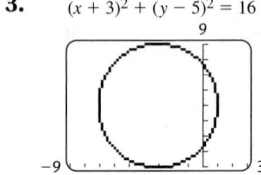

4. $(x - 5)^2 + (y + 6)^2 = 49$
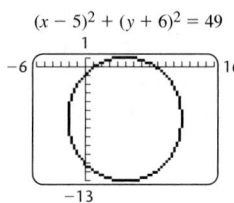

Exercise Set 13.1, pp. 860–863

1. (f)   **2.** (e)   **3.** (g)   **4.** (h)   **5.** (c)   **6.** (b)
7. (d)   **8.** (a)

9.
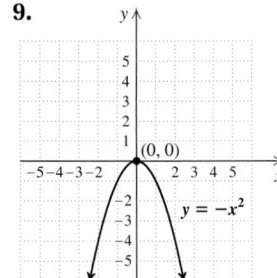

11.
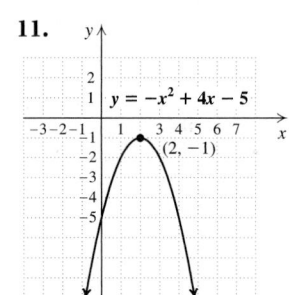

13.
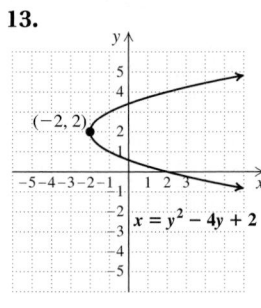

15.
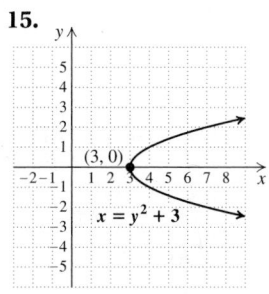

17.
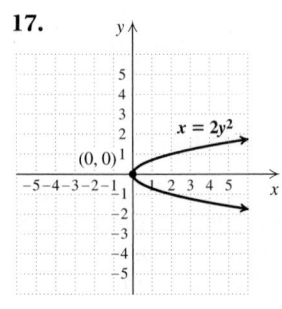

19.
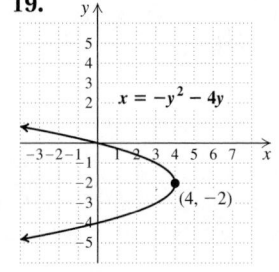

21.
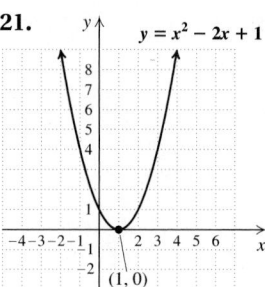

23.
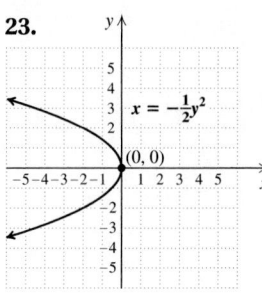

25.
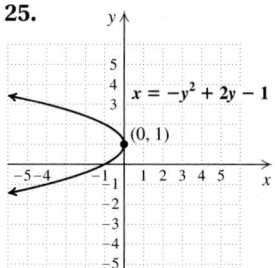

27.
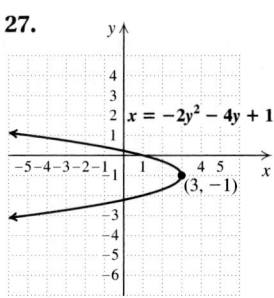

29. $x^2 + y^2 = 64$ **31.** $(x - 7)^2 + (y - 3)^2 = 6$
33. $(x + 4)^2 + (y - 3)^2 = 18$
35. $(x + 5)^2 + (y + 8)^2 = 300$
37. $x^2 + y^2 = 25$ **39.** $(x + 4)^2 + (y - 1)^2 = 20$
41. $(0, 0); 1$ **43.** $(-1, -3); 7$

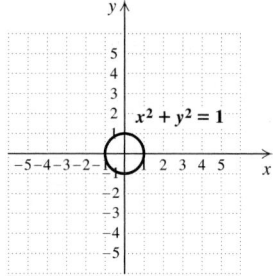

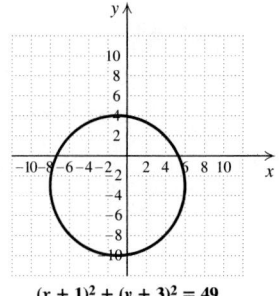
$(x + 1)^2 + (y + 3)^2 = 49$

45. $(4, -3); \sqrt{10}$ **47.** $(0, 0); 2\sqrt{2}$

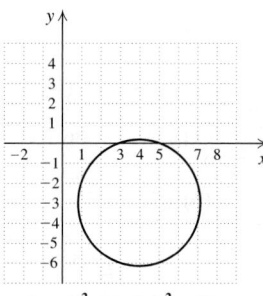
$(x - 4)^2 + (y + 3)^2 = 10$

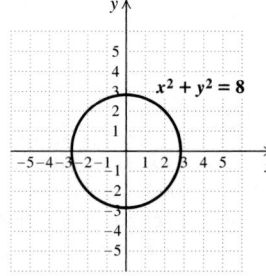

49. $(5, 0); \frac{1}{2}$ **51.** $(-4, 3); \sqrt{40}$, or $2\sqrt{10}$

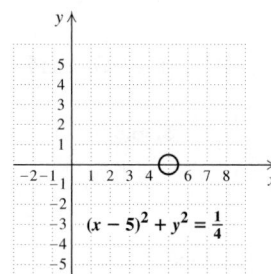

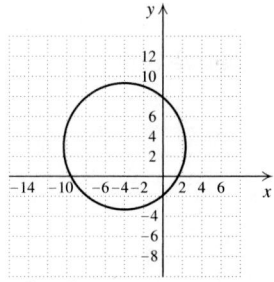
$x^2 + y^2 + 8x - 6y - 15 = 0$

53. $(4, -1); 2$ **55.** $(0, -5); 10$

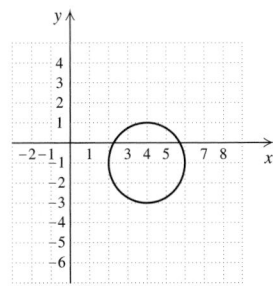

$x^2 + y^2 - 8x + 2y + 13 = 0$

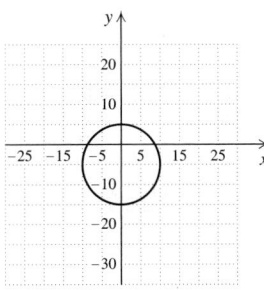

$x^2 + y^2 + 10y - 75 = 0$

57. $\left(-\dfrac{7}{2}, \dfrac{3}{2}\right); \sqrt{\dfrac{98}{4}}$, or $\dfrac{7\sqrt{2}}{2}$ **59.** $(0, 0); \dfrac{1}{6}$

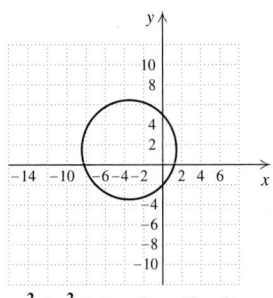

$x^2 + y^2 + 7x - 3y - 10 = 0$

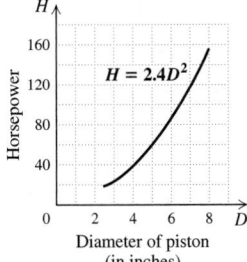

61. 📺   **63.** ± 4 **64.** $\pm a$ **65.** $-4, 6$
66. $-5 \pm 2\sqrt{3}$ **67.** $-3 \pm 3\sqrt{3}$ **68.** $2 \pm \dfrac{4\sqrt{2}}{3}$
69. 📺   **71.** $(x - 3)^2 + (y + 5)^2 = 9$
73. $(x - 3)^2 + y^2 = 25$ **75.** $(0, 4)$ **77.** $\dfrac{17}{4}\pi$ m^2, or
approximately 13.4 m^2 **79.** 7169 mm
81. (a) $(0, -3)$; (b) 5 ft   **83.** $x^2 + (y - 30.6)^2 = 590.49$
85.   7 in.   **87.** 📺,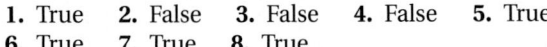

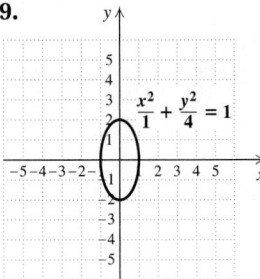

Horsepower
$H = 2.4D^2$
Diameter of piston (in inches)

Exercise Set 13.2, pp. 867–869

1. True   **2.** False   **3.** False   **4.** False   **5.** True
6. True   **7.** True   **8.** True
9.

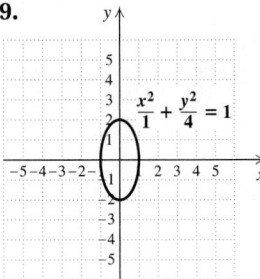

$\dfrac{x^2}{1} + \dfrac{y^2}{4} = 1$

11.

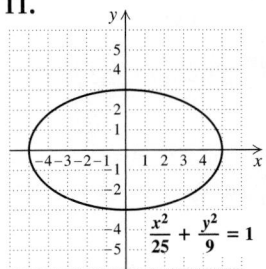

$\dfrac{x^2}{25} + \dfrac{y^2}{9} = 1$

13.

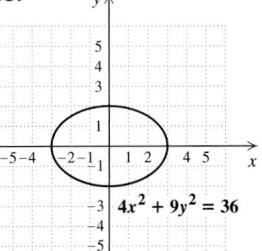

$4x^2 + 9y^2 = 36$

15.

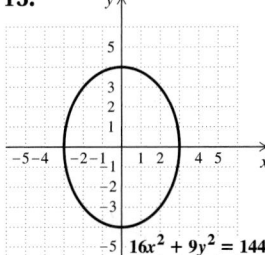

$16x^2 + 9y^2 = 144$

17.

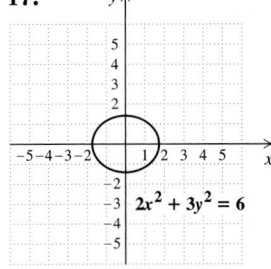

$2x^2 + 3y^2 = 6$

19.

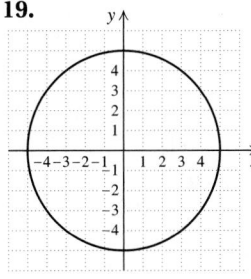

$5x^2 + 5y^2 = 125$

21.

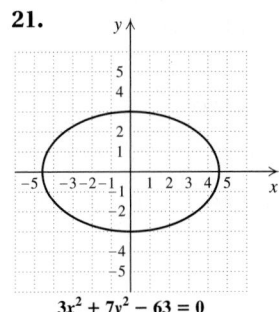

$3x^2 + 7y^2 - 63 = 0$

23.

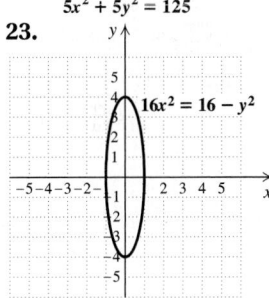

$16x^2 = 16 - y^2$

25.

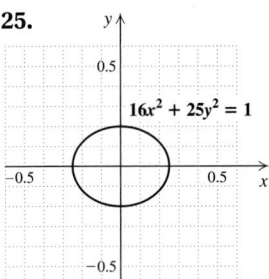

$16x^2 + 25y^2 = 1$

27.

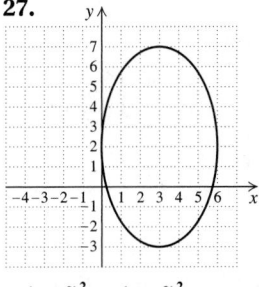

$\dfrac{(x - 3)^2}{9} + \dfrac{(y - 2)^2}{25} = 1$

29.

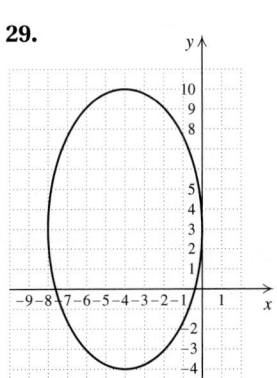

$\dfrac{(x + 4)^2}{16} + \dfrac{(y - 3)^2}{49} = 1$

31.

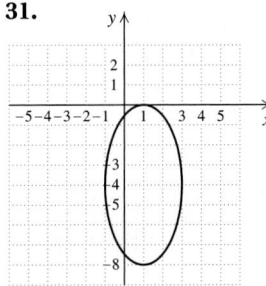

$12(x - 1)^2 + 3(y + 4)^2 = 48$

33.

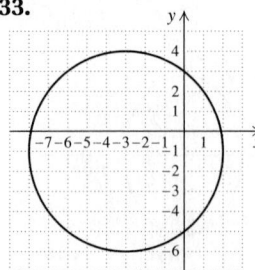

$$4(x + 3)^2 + 4(y + 1)^2 - 10 = 90$$

35.   **37.** $\dfrac{5}{2} \pm \dfrac{\sqrt{13}}{2}$ **38.** 3   **39.** $-\dfrac{3}{4}, 2$ **40.** $\dfrac{5}{2}$

41. $-\sqrt{11}, \sqrt{11}$ **42.** $-10, 6$ **43.**

45. $\dfrac{x^2}{81} + \dfrac{y^2}{121} = 1$ **47.** $\dfrac{(x - 2)^2}{16} + \dfrac{(y + 1)^2}{9} = 1$

49. $2.134 \times 10^8 \, \text{mi}$ **51.** **(a)** Let $F_1 = (-c, 0)$ and $F_2 = (c, 0)$. Then the sum of the distances from the foci to P is $2a$. By the distance formula,

$$\sqrt{(x + c)^2 + y^2} + \sqrt{(x - c)^2 + y^2} = 2a, \text{ or}$$
$$\sqrt{(x + c)^2 + y^2} = 2a - \sqrt{(x - c)^2 + y^2}.$$

Squaring, we get

$$(x + c)^2 + y^2 = 4a^2 - 4a\sqrt{(x - c)^2 + y^2} + (x - c)^2 + y^2,$$

or

$$x^2 + 2cx + c^2 + y^2 = 4a^2 - 4a\sqrt{(x - c)^2 + y^2} + x^2 - 2cx + c^2 + y^2.$$

Thus,

$$-4a^2 + 4cx = -4a\sqrt{(x - c)^2 + y^2}$$
$$a^2 - cx = a\sqrt{(x - c)^2 + y^2}.$$

Squaring again, we get

$$a^4 - 2a^2cx + c^2x^2 = a^2(x^2 - 2cx + c^2 + y^2)$$
$$a^4 - 2a^2cx + c^2x^2 = a^2x^2 - 2a^2cx + a^2c^2 + a^2y^2,$$

or

$$x^2(a^2 - c^2) + a^2y^2 = a^2(a^2 - c^2)$$
$$\dfrac{x^2}{a^2} + \dfrac{y^2}{a^2 - c^2} = 1.$$

(b) When P is at $(0, b)$, it follows that $b^2 = a^2 - c^2$. Substituting, we have

$$\dfrac{x^2}{a^2} + \dfrac{y^2}{b^2} = 1.$$

53. 5.66 ft   **55.**

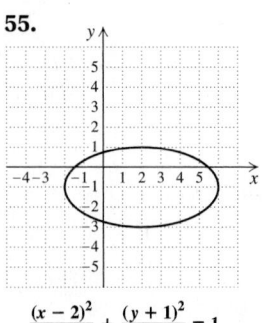

$$\dfrac{(x - 2)^2}{16} + \dfrac{(y + 1)^2}{4} = 1$$

57.

Technology Connection, p. 874

1.

$$y_1 = \dfrac{\sqrt{15x^2 - 240}}{2};$$
$$y_2 = -\dfrac{\sqrt{15x^2 - 240}}{2}$$

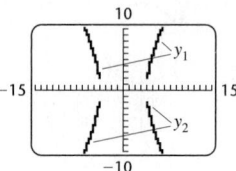

2.

$$y_1 = \sqrt{\dfrac{16x^2 - 64}{3}};$$
$$y_2 = -\sqrt{\dfrac{16x^2 - 64}{3}}$$

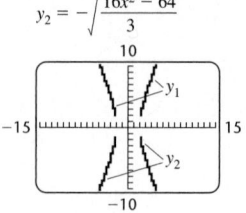

3.

$$y_1 = \dfrac{\sqrt{5x^2 + 320}}{4};$$
$$y_2 = -\dfrac{\sqrt{5x^2 + 320}}{4}$$

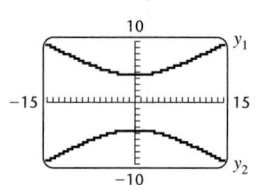

4.

$$y_1 = \sqrt{\dfrac{9x^2 + 441}{45}};$$
$$y_2 = -\sqrt{\dfrac{9x^2 + 441}{45}}$$

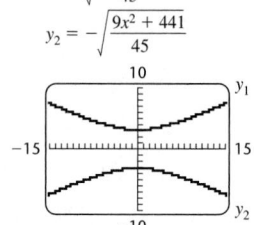

Exercise Set 13.3, pp. 876–877

1. (d)   **2.** (f)   **3.** (h)   **4.** (a)   **5.** (g)   **6.** (b)
7. (c)   **8.** (e)

9.

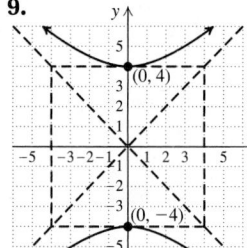

$$\dfrac{y^2}{16} - \dfrac{x^2}{16} = 1$$

11.

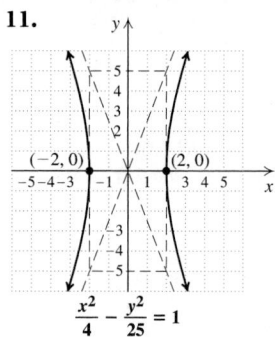

$$\dfrac{x^2}{4} - \dfrac{y^2}{25} = 1$$

13.

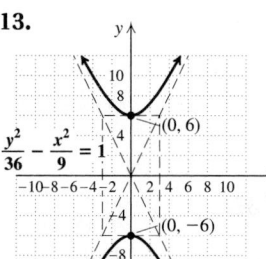

$$\dfrac{y^2}{36} - \dfrac{x^2}{9} = 1$$

15.

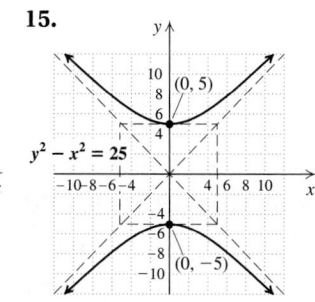

$$y^2 - x^2 = 25$$

17.

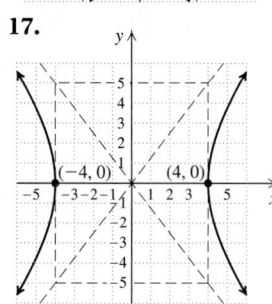

$$25x^2 - 16y^2 = 400$$

19.

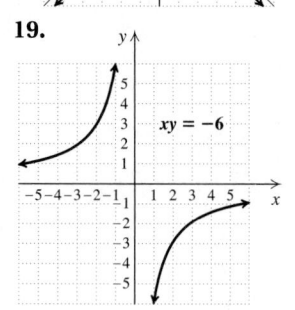

$$xy = -6$$

21.

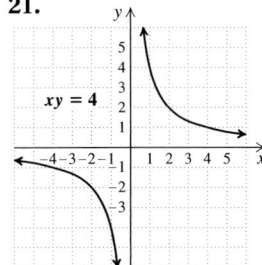

$xy = 4$

23.

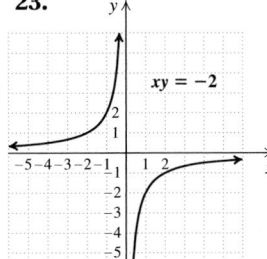

$xy = -2$

25.

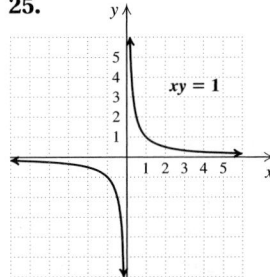

$xy = 1$

27. Circle   **29.** Ellipse   **31.** Hyperbola   **33.** Circle
35. Parabola   **37.** Hyperbola   **39.** Parabola
41. Hyperbola   **43.** Circle   **45.** Ellipse   **47.**
49. $(-3, 6)$ **50.** $\left(\frac{1}{2}, -\frac{3}{2}\right)$ **51.** $-2, 2$ **52.** $-4, \frac{2}{3}$
53. $\frac{3}{2} \pm \frac{\sqrt{13}}{2}$ **54.** $\pm 1, \pm 5$ **55.**   **57.** $\frac{y^2}{36} - \frac{x^2}{4} = 1$
59. C: $(5, 2)$; V: $(-1, 2), (11, 2)$; asymptotes:
$y - 2 = \frac{5}{6}(x - 5), y - 2 = -\frac{5}{6}(x - 5)$

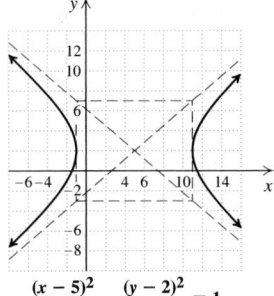

$\frac{(x - 5)^2}{36} - \frac{(y - 2)^2}{25} = 1$

61. $\frac{(y + 3)^2}{4} - \frac{(x - 4)^2}{16} = 1$; C: $(4, -3)$; V: $(4, -5), (4, -1)$;
asymptotes: $y + 3 = \frac{1}{2}(x - 4), y + 3 = -\frac{1}{2}(x - 4)$

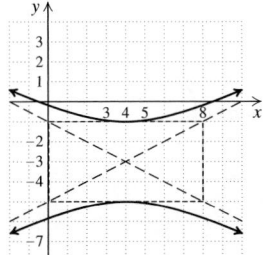

$8(y + 3)^2 - 2(x - 4)^2 = 32$

63. $\frac{(x + 3)^2}{1} - \frac{(y - 2)^2}{4} = 1$; C: $(-3, 2)$; V: $(-4, 2), (-2, 2)$;
asymptotes: $y - 2 = 2(x + 3), y - 2 = -2(x + 3)$

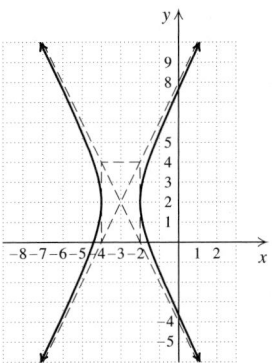

$4x^2 - y^2 + 24x + 4y + 28 = 0$

65.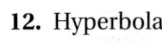

Connecting the Concepts, p. 878

1. $(4, 1)$; $x = 4$ **2.** $(2, -1)$; $y = -1$ **3.** $(3, 2)$
4. $(-3, -5)$ **5.** $(-12, 0), (12, 0), (0, -9), (0, 9)$
6. $(-3, 0), (3, 0)$ **7.** $(0, -1), (0, 1)$ **8.** $y = \frac{3}{2}x, y = -\frac{3}{2}x$
9. Circle   **10.** Parabola

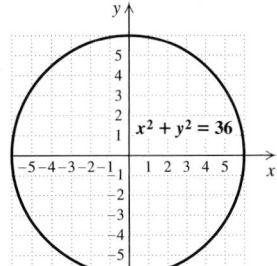

$x^2 + y^2 = 36$

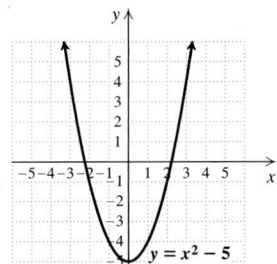

$y = x^2 - 5$

11. Ellipse   **12.** Hyperbola

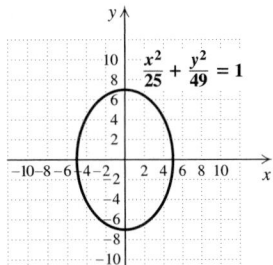

$\frac{x^2}{25} + \frac{y^2}{49} = 1$

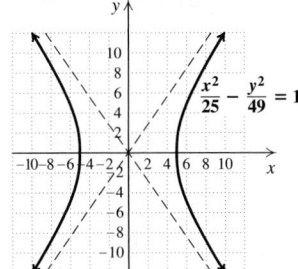

$\frac{x^2}{25} - \frac{y^2}{49} = 1$

13. Parabola   **14.** Ellipse

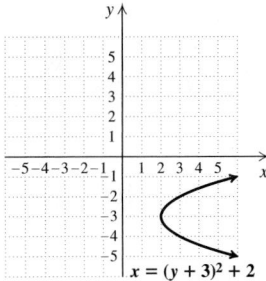

$x = (y + 3)^2 + 2$

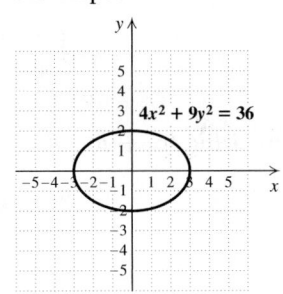

$4x^2 + 9y^2 = 36$

15. Hyperbola

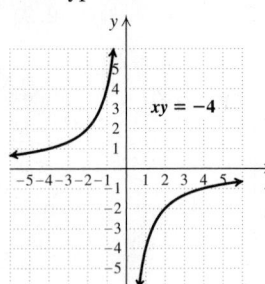

$xy = -4$

16. Circle

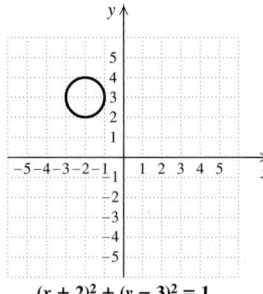

$(x + 2)^2 + (y - 3)^2 = 1$

17. Circle

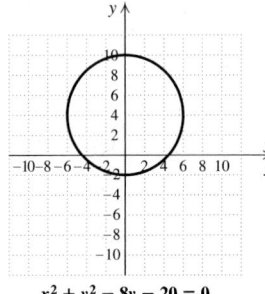

$x^2 + y^2 - 8y - 20 = 0$

18. Parabola

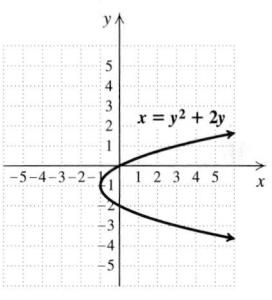

$x = y^2 + 2y$

19. Hyperbola

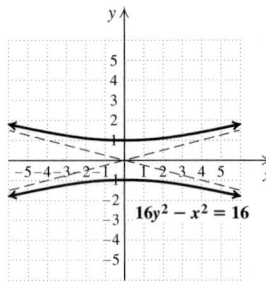

$16y^2 - x^2 = 16$

20. Hyperbola

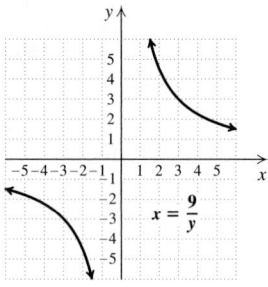

$x = \dfrac{9}{y}$

Technology Connection, p. 881

1. $(-1.50, -1.17); (3.50, 0.50)$
2. $(-2.77, 2.52); (-2.77, -2.52)$

Technology Connection, p. 882

1. $y_1 = \sqrt{(20 - x^2)/4}; \; y_2 = -\sqrt{(20 - x^2)/4}; \; y_3 = 4/x$

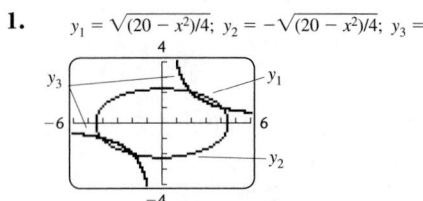

Visualizing for Success, p. 885

1. C   **2.** A   **3.** F   **4.** B   **5.** J   **6.** D   **7.** H   **8.** I
9. G   **10.** E

Exercise Set 13.4, pp. 886–888

1. True   **2.** True   **3.** False   **4.** False   **5.** True

6. True   **7.** $(-5, -4), (4, 5)$ **9.** $(0, 2), (3, 0)$
11. $(-2, 1)$
13. $\left(\dfrac{5 + \sqrt{70}}{3}, \dfrac{-1 + \sqrt{70}}{3}\right), \left(\dfrac{5 - \sqrt{70}}{3}, \dfrac{-1 - \sqrt{70}}{30}\right)$
15. $\left(4, \dfrac{3}{2}\right), (3, 2)$ **17.** $\left(\dfrac{7}{3}, \dfrac{1}{3}\right), (1, -1)$ **19.** $\left(\dfrac{11}{4}, -\dfrac{5}{4}\right), (1, 4)$
21. $(2, 4), (4, 2)$ **23.** $(3, -5), (-1, 3)$
25. $(-5, -8), (8, 5)$ **27.** $(0, 0), (1, 1),$
$\left(-\dfrac{1}{2} + \dfrac{\sqrt{3}}{2}i, -\dfrac{1}{2} - \dfrac{\sqrt{3}}{2}i\right), \left(-\dfrac{1}{2} - \dfrac{\sqrt{3}}{2}i, -\dfrac{1}{2} + \dfrac{\sqrt{3}}{2}i\right)$
29. $(-4, 0), (4, 0)$ **31.** $(-4, -3), (-3, -4), (3, 4), (4, 3)$
33. $\left(\dfrac{16}{3}, \dfrac{5\sqrt{7}}{3}i\right), \left(\dfrac{16}{3}, -\dfrac{5\sqrt{7}}{3}i\right), \left(-\dfrac{16}{3}, \dfrac{5\sqrt{7}}{3}i\right),$
$\left(-\dfrac{16}{3}, -\dfrac{5\sqrt{7}}{3}i\right)$ **35.** $(-3, -\sqrt{5}), (-3, \sqrt{5}), (3, -\sqrt{5}),$
$(3, \sqrt{5})$ **37.** $(-3, -1), (-1, -3), (1, 3), (3, 1)$
39. $(4, 1), (-4, -1), (2, 2), (-2, -2)$ **41.** $(2, 1), (-2, -1)$
43. $\left(2, -\dfrac{4}{5}\right), \left(-2, -\dfrac{4}{5}\right), (5, 2), (-5, 2)$ **45.** $(-\sqrt{2}, \sqrt{2}),$
$(\sqrt{2}, -\sqrt{2})$ **47.** Length: 8 cm; width: 6 cm
49. Length: 2 in.; width: 1 in.   **51.** Length: 12 ft; width: 5 ft
53. 6 and 15; -6 and -15 **55.** 24 ft, 16 ft   **57.** Length:
$\sqrt{3}$ m; width: 1 m   **59.** 📈   **61.** -9 **62.** -27
63. -1 **64.** $\dfrac{1}{5}$ **65.** 77   **66.** $\dfrac{21}{2}$ **67.** 📈
69. $(x + 2)^2 + (y - 1)^2 = 4$ **71.** $(-2, 3), (2, -3),$
$(-3, 2), (3, -2)$ **73.** Length: 55 ft; width: 45 ft
75. 10 in. by 7 in. by 5 in.   **77.** Length: 63.6 in.;
height: 35.8 in.   **79.** 📈

Review Exercises: Chapter 13, pp. 891–892

1. True   **2.** False   **3.** False   **4.** True   **5.** True
6. True   **7.** False   **8.** True   **9.** $(-3, 2), 4$
10. $(5, 0), \sqrt{11}$ **11.** $(3, 1), 3$ **12.** $(-4, 3), 3\sqrt{5}$
13. $(x + 4)^2 + (y - 3)^2 = 16$
14. $(x - 7)^2 + (y + 2)^2 = 20$
15. Circle

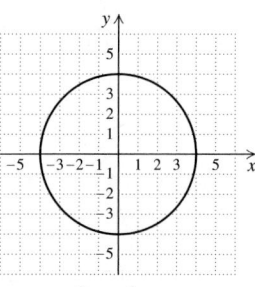

$5x^2 + 5y^2 = 80$

16. Ellipse

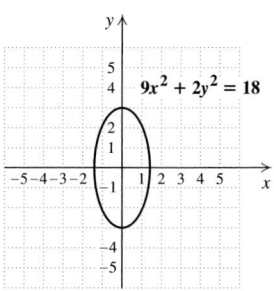

$9x^2 + 2y^2 = 18$

17. Parabola

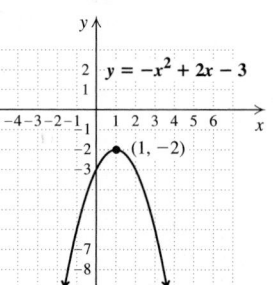

$y = -x^2 + 2x - 3$
$(1, -2)$

18. Hyperbola

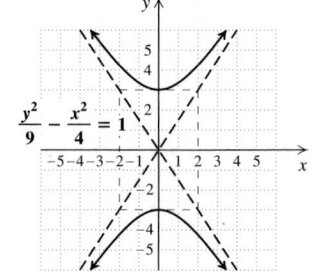

$\dfrac{y^2}{9} - \dfrac{x^2}{4} = 1$

19. Hyperbola

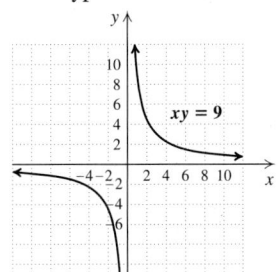

$xy = 9$

20. Parabola

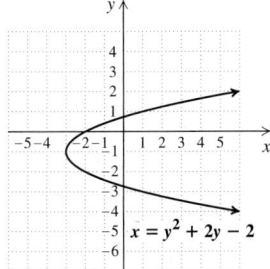

$x = y^2 + 2y - 2$

21. Ellipse

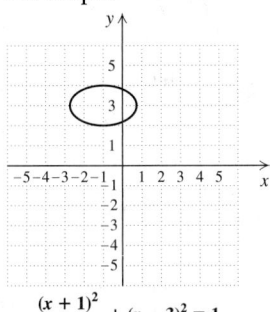

$\dfrac{(x+1)^2}{3} + (y - 3)^2 = 1$

22. Circle

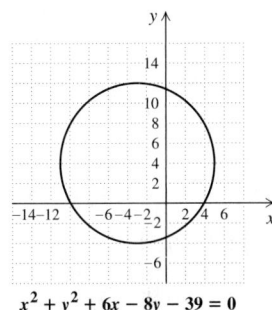

$x^2 + y^2 + 6x - 8y - 39 = 0$

23. $(5, -2)$　**24.** $(2, 2), \left(\frac{32}{9}, -\frac{10}{9}\right)$　**25.** $(0, -5), (2, -1)$
26. $(4, 3), (4, -3), (-4, 3), (-4, -3)$　**27.** $(2, 1), (\sqrt{3}, 0),$
$(-2, 1), (-\sqrt{3}, 0)$　**28.** $(3, -3), \left(-\frac{3}{5}, \frac{21}{5}\right)$　**29.** $(6, 8),$
$(6, -8), (-6, 8), (-6, -8)$　**30.** $(2, 2), (-2, -2),$
$(2\sqrt{2}, \sqrt{2}), (-2\sqrt{2}, -\sqrt{2})$　**31.** Length: 12 m; width: 7 m
32. Length: 12 in.; width: 9 in.　**33.** 32 cm, 20 cm
34. 3 ft, 11 ft　**35.** ✍ The graph of a parabola has one
branch whereas the graph of a hyperbola has two branches.
A hyperbola has asymptotes, but a parabola does not.
36. ✍ Function notation rarely appears in this chapter
because many of the relations are not functions. Function
notation could be used for vertical parabolas and for hyperbolas that have the axes as asymptotes.
37. $(-5, -4\sqrt{2}), (-5, 4\sqrt{2}), (3, -2\sqrt{2}), (3, 2\sqrt{2})$
38. $(0, 6), (0, -6)$　**39.** $(x - 2)^2 + (y + 1)^2 = 25$
40. $\dfrac{x^2}{100} + \dfrac{y^2}{1} = 1$　**41.** $\left(\frac{9}{4}, 0\right)$

Test: Chapter 13, p. 892

1. [13.1] $(x - 3)^2 + (y + 4)^2 = 12$　**2.** [13.1] $(4, -1), \sqrt{5}$
3. [13.1] $(-2, 3), 3$
4. [13.1], [13.3] Parabola　**5.** [13.1], [13.3] Circle

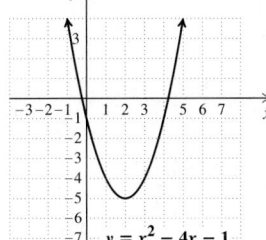

$y = x^2 - 4x - 1$

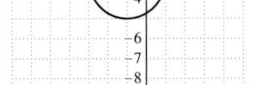

$x^2 + y^2 + 2x + 6y + 6 = 0$

6. [13.3] Hyperbola

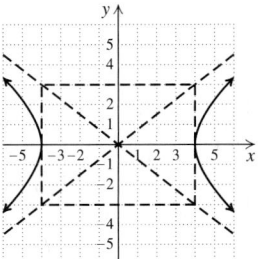

$\dfrac{x^2}{16} - \dfrac{y^2}{9} = 1$

7. [13.2], [13.3] Ellipse

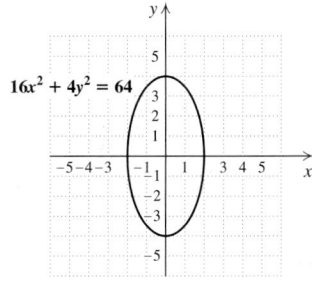

$16x^2 + 4y^2 = 64$

8. [13.3] Hyperbola

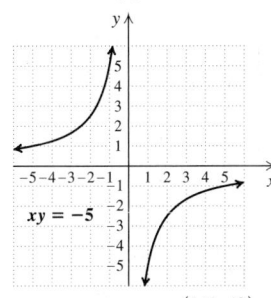

$xy = -5$

9. [13.1], [13.3] Parabola

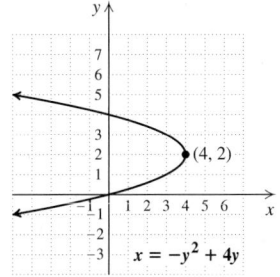

$(4, 2)$

$x = -y^2 + 4y$

10. [13.4] $(0, 6), \left(\frac{144}{25}, \frac{42}{25}\right)$　**11.** [13.4] $(-4, 13), (2, 1)$
12. [13.4] $(3, 2), (-3, -2), \left(-2\sqrt{2}i, \frac{3\sqrt{2}}{2}i\right), \left(2\sqrt{2}i, -\frac{3\sqrt{2}}{2}i\right)$
13. [13.4] $(\sqrt{6}, 2), (\sqrt{6}, -2), (-\sqrt{6}, 2), (-\sqrt{6}, -2)$
14. [13.4] 2 by 11　**15.** [13.4] $\sqrt{5}$ m, $\sqrt{3}$ m
16. [13.4] Length: 32 ft; width: 24 ft　**17.** [13.4] $1200, 6\%$
18. [13.2] $\dfrac{(x - 6)^2}{25} + \dfrac{(y - 3)^2}{9} = 1$　**19.** [13.1] $\left(0, -\frac{31}{4}\right)$
20. [13.4] 9　**21.** [13.2] $\dfrac{x^2}{16} + \dfrac{y^2}{49} = 1$

Cumulative Review: Chapters 1–13, pp. 893–894

1. $16t^4 - 40t^2 s + 25s^2$　**2.** $\dfrac{4t - 3}{3t(t - 3)}$　**3.** $\dfrac{x}{a}$
4. $3t^2\sqrt{10w}$　**5.** $27a^{1/2}b^{3/16}$　**6.** -4　**7.** 25　**8.** 0
9. $-\frac{1}{64}$　**10.** $2\sqrt{3}$　**11.** $(10x - 3y)^2$
12. $3(m^2 - 2)(m^4 + 2m^2 + 4)$　**13.** $(x - y)(a - b)$
14. $(4x - 3)(8x + 1)$　**15.** $\left(-\infty, -\frac{25}{3}\right]$, or $\left\{x \mid x \le -\frac{25}{3}\right\}$
16. $0, \frac{9}{8}$　**17.** 1, 4　**18.** $\pm i$　**19.** 4
20. $\dfrac{\log 1.5}{\log 3} \approx 0.3691$　**21.** 7
22. $(-\sqrt{3}, -1), (-\sqrt{3}, 1), (\sqrt{3}, -1), (\sqrt{3}, 1)$
23.

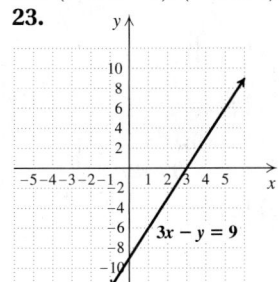

$3x - y = 9$

24.

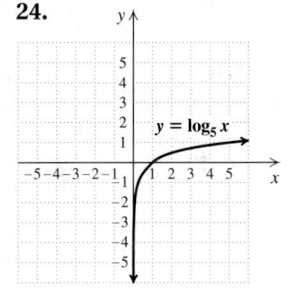

$y = \log_5 x$

25.

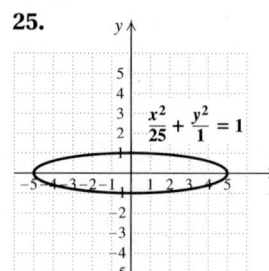

$$\frac{x^2}{25} + \frac{y^2}{1} = 1$$

26.

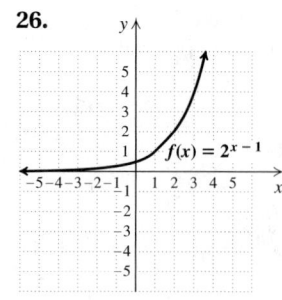

$f(x) = 2^{x-1}$

27.

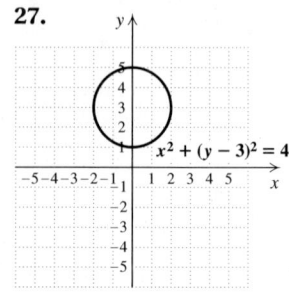

$x^2 + (y - 3)^2 = 4$

28.

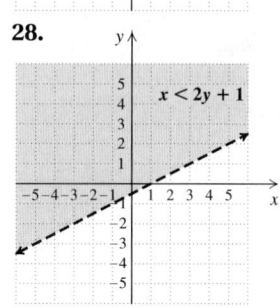

$x < 2y + 1$

29.

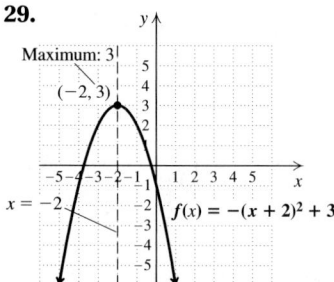

Maximum: 3

$(-2, 3)$

$x = -2$ $f(x) = -(x + 2)^2 + 3$

30. $\left(-\infty, \frac{5}{3}\right]$, or $\left\{x \mid x \le \frac{5}{3}\right\}$ **31.** $c = \pm\sqrt{\dfrac{ab}{t}}$

32. $y = -x + 3$ **33.** $x^2 - 3 = 0$ **34.** $t^m = 16$

35. 2640 mi   **36.** Greg: 35 hr; Kyle: 14 hr

37. (a) $h(t) = -0.7t + 32.3$; **(b)** 27.4 hr per week;

(c) approximately 2021   **38.** 1.5 in.

39. (a) $P(t) = 2.26e^{-0.0095t}$; **(b)** 1.90 million;

(c) 73 yr   **40.** A: 15°; B: 45°; C: 120°   **41.** 8 in. by 8 in.

42. $\sqrt{55}$ cm $\approx$ 7.416 cm   **43.** $a = 1, b = -6$

44. $(1, -2, 0, 4)$ **45.** y is divided by 10.

46. $(-\infty, 0) \cup (0, 1]$, or $\{x \mid x < 0 \, or \, 0 < x \le 1\}$

CHAPTER 14

Exercise Set 14.1, pp. 900–902

1. (f)   **2.** (a)   **3.** (d)   **4.** (b)   **5.** (c)   **6.** (e)   **7.** 43

9. 364   **11.** -23.5 **13.** -363 **15.** $\frac{441}{400}$

17. 2, 5, 8, 11; 29; 44   **19.** 3, 6, 11, 18; 102; 227

21. $\frac{1}{2}, \frac{2}{3}, \frac{3}{4}, \frac{4}{5}; \frac{10}{11}; \frac{15}{16}$ **23.** $1, -\frac{1}{2}, \frac{1}{4}, -\frac{1}{8}; \frac{1}{512}; \frac{1}{16,384}$

25. $-1, \frac{1}{2}, -\frac{1}{3}, \frac{1}{4}; \frac{1}{10}; -\frac{1}{15}$ **27.** $0, 7, -26, 63; 999; -3374$

29. $2n$ **31.** $(-1)^n$ **33.** $(-1)^{n+1} \cdot n$ **35.** $2n + 1$

37. $n^2 - 1$, or $(n + 1)(n - 1)$ **39.** $\dfrac{n}{n + 1}$

41. $(0.1)^n$, or 10^{-n} **43.** $(-1)^n \cdot n^2$ **45.** 5

47. 1.11111, or $1\frac{11,111}{100,000}$ **49.** $\dfrac{1}{2} + \dfrac{1}{4} + \dfrac{1}{6} + \dfrac{1}{8} + \dfrac{1}{10} = \dfrac{137}{120}$

51. $10^0 + 10^1 + 10^2 + 10^3 + 10^4 = 11{,}111$

53. $2 + \dfrac{3}{2} + \dfrac{4}{3} + \dfrac{5}{4} + \dfrac{6}{5} + \dfrac{7}{6} + \dfrac{8}{7} = \dfrac{1343}{140}$

55. $(-1)^2 2^1 + (-1)^3 2^2 + (-1)^4 2^3 + (-1)^5 2^4 +$
$(-1)^6 2^5 + (-1)^7 2^6 + (-1)^8 2^7 + (-1)^9 2^8 = -170$

57. $(0^2 - 2 \cdot 0 + 3) + (1^2 - 2 \cdot 1 + 3) +$
$(2^2 - 2 \cdot 2 + 3) + (3^2 - 2 \cdot 3 + 3) +$
$(4^2 - 2 \cdot 4 + 3) + (5^2 - 2 \cdot 5 + 3) = 43$

59. $\dfrac{(-1)^3}{3 \cdot 4} + \dfrac{(-1)^4}{4 \cdot 5} + \dfrac{(-1)^5}{5 \cdot 6} = -\dfrac{1}{15}$ **61.** $\displaystyle\sum_{k=1}^{5} \dfrac{k + 1}{k + 2}$

63. $\displaystyle\sum_{k=1}^{6} k^2$ **65.** $\displaystyle\sum_{k=2}^{n} (-1)^k k^2$ **67.** $\displaystyle\sum_{k=1}^{\infty} 6k$

69. $\displaystyle\sum_{k=1}^{\infty} \dfrac{1}{k(k + 1)}$ **71.** ✒   **73.** 98   **74.** -15

75. $a_1 + 4d$ **76.** $a_1 + a_n$ **77.** $3(a_1 + a_n)$, or $3a_1 + 3a_n$

78. d **79.** ✒   **81.** 1, 3, 13, 63, 313, 1563   **83.** \$2500,
\$2000, \$1600, \$1280, \$1024, \$819.20, \$655.36, \$524.29,
\$419.43, \$335.54   **85.** $S_{100} = 0; S_{101} = -1$

87. $i, -1, -i, 1, i; i$ **89.** 11th term

Exercise Set 14.2, pp. 908–910

1. True   **2.** True   **3.** False   **4.** False   **5.** True

6. True   **7.** False   **8.** False   **9.** $a_1 = 8, d = 5$

11. $a_1 = 7, d = -4$ **13.** $a_1 = \frac{3}{2}, d = \frac{3}{4}$

15. $a_1 = \$8.16, d = \$0.30$ **17.** 154   **19.** -94

21. $-\$1628.16$ **23.** 26th   **25.** 57th   **27.** 178

29. 5   **31.** 28   **33.** $a_1 = 8; d = -3; 8, 5, 2, -1, -4$

35. $a_1 = 1; d = 1$ **37.** 780   **39.** 31,375   **41.** 2550

43. 918   **45.** 1030   **47.** 35 musicians; 315 musicians

49. 180 stones   **51.** \$49.60   **53.** 560 seats   **55.** ✒

57. $y = \frac{1}{3}x + 10$ **58.** $y = -4x + 11$ **59.** $y = -2x + 10$

60. $y = -\frac{4}{3}x - \frac{16}{3}$ **61.** $x^2 + y^2 = 16$

62. $(x + 2)^2 + (y - 1)^2 = 20$ **63.** ✒   **65.** 33 jumps

67. Let $d =$ the common difference. Since $p, m,$ and q form
an arithmetic sequence, $m = p + d$ and $q = p + 2d$.

Then $\dfrac{p + q}{2} = \dfrac{p + (p + 2d)}{2} = p + d = m$.   **69.** 156,375

Exercise Set 14.3, pp. 917–919

1. Geometric sequence   **2.** Arithmetic sequence

3. Arithmetic sequence   **4.** Geometric sequence

5. Geometric series   **6.** Arithmetic series

7. Geometric series   **8.** None of these   **9.** 2   **11.** -0.1

13. $-\frac{1}{2}$ **15.** $\frac{1}{5}$ **17.** $\dfrac{6}{m}$ **19.** 1458   **21.** 243

23. 52,488   **25.** \$1423.31   **27.** $a_n = 5^{n-1}$

29. $a_n = (-1)^{n-1}$, or $a_n = (-1)^{n+1}$

31. $a_n = \dfrac{1}{x^n}$, or $a_n = x^{-n}$ **33.** 3066   **35.** $\frac{547}{18}$

37. $\dfrac{1 - x^8}{1 - x}$, or $(1 + x)(1 + x^2)(1 + x^4)$ **39.** \$5134.51

41. 27   **43.** $\frac{49}{4}$ **45.** No   **47.** No   **49.** $\frac{43}{99}$

51. \$25,000   **53.** $\frac{5}{9}$ **55.** $\frac{343}{99}$ **57.** $\frac{5}{33}$ **59.** $\frac{5}{1024}$ ft

61. 155,797   **63.** 2710 flies   **65.** Approximately 179.9 billion coffees   **67.** 3100.35 ft   **69.** 20.48 in.   **71.** 📜
73. $x^2 + 2xy + y^2$ **74.** $x^3 + 3x^2y + 3xy^2 + y^3$
75. $x^3 - 3x^2y + 3xy^2 - y^3$
76. $x^4 - 4x^3y + 6x^2y^2 - 4xy^3 + y^4$
77. $8x^3 + 12x^2y + 6xy^2 + y^3$
78. $8x^3 - 12x^2y + 6xy^2 - y^3$ **79.** 📜   **81.** 54
83. $\dfrac{x^2[1 - (-x)^n]}{1 + x}$ **85.** 512 cm² **87.** 📜, 〰

Connecting the Concepts, p. 920

1. 300   **2.** $\dfrac{1}{n + 1}$ **3.** 78   **4.** $2^2 + 3^2 + 4^2 + 5^2 = 54$

5. $\displaystyle\sum_{k=1}^{6} (-1)^{k+1} \cdot k$ **6.** -3 **7.** 110   **8.** 61st   **9.** -39
10. 21   **11.** 11   **12.** 4410   **13.** $-\frac{1}{2}$ **14.** 640
15. $2(-1)^{n+1}$ **16.** \$1146.39   **17.** 1   **18.** No
19. \$465   **20.** \$1,073,741,823

Technology Connection, p. 925

1. 479,001,600   **2.** 56; 792

Visualizing for Success, p. 928

1. J   **2.** G   **3.** A   **4.** H   **5.** I   **6.** B   **7.** E   **8.** D
9. F   **10.** C

Exercise Set 14.4, pp. 929-930

1. 2^5, or 32   **2.** 8   **3.** 9   **4.** 4!   **5.** $\binom{8}{5}$, or $\binom{8}{3}$

6. a^2b^8 **7.** 1   **8.** 9 choose 5   **9.** 24   **11.** 3,628,800
13. 90   **15.** 126   **17.** 210   **19.** 1   **21.** 435   **23.** 780
25. $a^4 - 4a^3b + 6a^2b^2 - 4ab^3 + b^4$
27. $p^7 + 7p^6q + 21p^5q^2 + 35p^4q^3 + 35p^3q^4 + 21p^2q^5 + 7pq^6 + q^7$
29. $2187c^7 - 5103c^6d + 5103c^5d^2 - 2835c^4d^3 + 945c^3d^4 - 189c^2d^5 + 21cd^6 - d^7$
31. $t^{-12} + 12t^{-10} + 60t^{-8} + 160t^{-6} + 240t^{-4} + 192t^{-2} + 64$
33. $x^5 - 5x^4y + 10x^3y^2 - 10x^2y^3 + 5xy^4 - y^5$
35. $19{,}683s^9 + \dfrac{59{,}049s^8}{t} + \dfrac{78{,}732s^7}{t^2} + \dfrac{61{,}236s^6}{t^3} + \dfrac{30{,}618s^5}{t^4} + \dfrac{10{,}206s^4}{t^5} + \dfrac{2268s^3}{t^6} + \dfrac{324s^2}{t^7} + \dfrac{27s}{t^8} + \dfrac{1}{t^9}$
37. $x^{15} - 10x^{12}y + 40x^9y^2 - 80x^6y^3 + 80x^3y^4 - 32y^5$
39. $125 + 150\sqrt{5}t + 375t^2 + 100\sqrt{5}t^3 + 75t^4 + 6\sqrt{5}t^5 + t^6$
41. $x^{-3} - 6x^{-2} + 15x^{-1} - 20 + 15x - 6x^2 + x^3$
43. $15a^4b^2$ **45.** $-64{,}481{,}508a^3$ **47.** $1120x^{12}y^2$
49. $1{,}959{,}552u^5v^{10}$ **51.** y^8 **53.** 📜

55.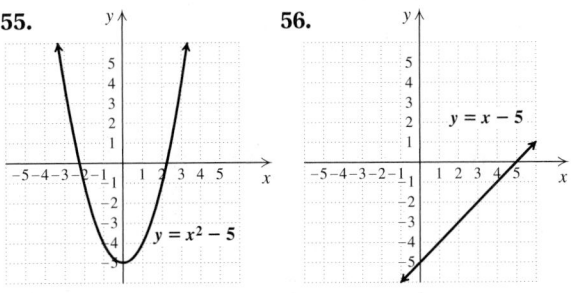

56.

57.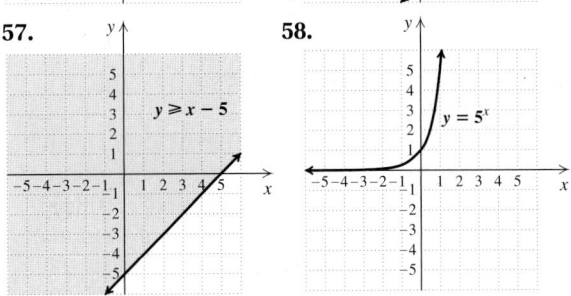

58.

59.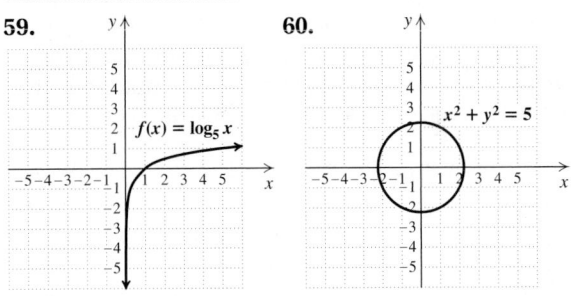

60.

61. 📜   **63.** List all the subsets of size 3: $\{a, b, c\}$, $\{a, b, d\}$, $\{a, b, e\}$, $\{a, c, d\}$, $\{a, c, e\}$, $\{a, d, e\}$, $\{b, c, d\}$, $\{b, c, e\}$, $\{b, d, e\}$, $\{c, d, e\}$. There are exactly 10 subsets of size 3 and $\binom{5}{3} = 10$, so there are exactly $\binom{5}{3}$ ways of forming a subset of size 3 from $\{a, b, c, d, e\}$.

65. $\binom{8}{5}(0.15)^3(0.85)^5 \approx 0.084$

67. $\binom{8}{6}(0.15)^2(0.85)^6 + \binom{8}{7}(0.15)(0.85)^7 + \binom{8}{8}(0.85)^8 \approx 0.89$

69. $\binom{n}{n - r} = \dfrac{n!}{[n - (n - r)]!(n - r)!}$
$= \dfrac{n!}{r!(n - r)!} = \binom{n}{r}$

71. $\dfrac{-\sqrt[3]{q}}{2p}$ **73.** $x^7 + 7x^6y + 21x^5y^2 + 35x^4y^3 + 35x^3y^4 + 21x^2y^5 + 7xy^6 + y^7$

Review Exercises: Chapter 14, pp. 932-933

1. False   **2.** True   **3.** True   **4.** False   **5.** False
6. True   **7.** False   **8.** False   **9.** 1, 11, 21, 31; 71; 111

10. $0, \frac{1}{5}, \frac{1}{5}, \frac{3}{17}, \frac{7}{65}; \frac{11}{145}$ **11.** $a_n = -5n$
12. $a_n = (-1)^n(2n - 1)$
13. $-2 + 4 + (-8) + 16 + (-32) = -22$
14. $-3 + (-5) + (-7) + (-9) + (-11) + (-13) = -48$
15. $\sum_{k=1}^{6} 7k$ **16.** $\sum_{k=1}^{5} \frac{1}{(-2)^k}$ **17.** -55 **18.** $\frac{1}{5}$
19. $a_1 = -15, d = 5$ **20.** -544 **21.** $25{,}250$
22. $1024\sqrt{2}$ **23.** $\frac{3}{4}$ **24.** $a_n = 2(-1)^n$
25. $a_n = 3\left(\frac{x}{4}\right)^{n-1}$ **26.** $11{,}718$ **27.** $-4095x$ **28.** 12
29. $\frac{49}{11}$ **30.** No   **31.** No   **32.** $\$40{,}000$ **33.** $\frac{5}{9}$ **34.** $\frac{16}{11}$
35. $\$24.30$ **36.** 903 poles   **37.** $\$15{,}791.18$ **38.** 6 m
39. 5040 **40.** 120 **41.** $190a^{18}b^2$
42. $x^4 - 8x^3y + 24x^2y^2 - 32xy^3 + 16y^4$
43. 🖎 For a geometric sequence with $|r| < 1$, as n gets larger, the absolute value of the terms gets smaller, since $|r^n|$ gets smaller.   **44.** 🖎 The first form of the binomial theorem draws the coefficients from Pascal's triangle; the second form uses factorial notation. The second form avoids the need to compute all preceding rows of Pascal's triangle, and is generally easier to use when only one term of an expansion is needed. When several terms of an expansion are needed and n is not large (say, $n \le 8$), it is often easier to use Pascal's triangle.   **45.** $\dfrac{1 - (-x)^n}{x + 1}$
46. $x^{-15} + 5x^{-9} + 10x^{-3} + 10x^3 + 5x^9 + x^{15}$

Test: Chapter 14, p. 934

1. [14.1] $\frac{1}{2}, \frac{1}{5}, \frac{1}{10}, \frac{1}{17}, \frac{1}{26}; \frac{1}{145}$ **2.** [14.1] $a_n = 4\left(\frac{1}{3}\right)^n$
3. [14.1] $-3 + (-7) + (-15) + (-31) = -56$
4. [14.1] $\sum_{k=1}^{5} (-1)^{k+1}k^3$ **5.** [14.2] $\frac{13}{2}$ **6.** [14.2] -3
7. [14.2] $a_1 = 31.2; d = -3.8$ **8.** [14.2] 2508
9. [14.3] 1536 **10.** [14.3] $\frac{2}{3}$ **11.** [14.3] 3^n
12. [14.3] 5621 **13.** [14.3] 1 **14.** [14.3] No
15. [14.3] $\frac{\$25{,}000}{23} \approx \$1086.96$ **16.** [14.3] $\frac{85}{99}$
17. [14.2] 63 seats   **18.** [14.2] $\$17{,}100$
19. [14.3] $\$5987.37$ **20.** [14.3] 36 m   **21.** [14.4] 220
22. [14.4] $x^5 - 15x^4y + 90x^3y^2 - 270x^2y^3 + 405xy^4 - 243y^5$ **23.** [14.4] $220a^9x^3$ **24.** [14.2] $n(n + 1)$
25. [14.3] $\dfrac{1 - \left(\frac{1}{x}\right)^n}{1 - \frac{1}{x}}$, or $\dfrac{x^n - 1}{x^{n-1}(x - 1)}$

Cumulative Review/Final Exam: Chapters 1–14, pp. 935–937

1. $\frac{7}{15}$ **2.** $-4y + 17$ **3.** 280 **4.** 8.4×10^{-15}
5. $\frac{7}{6}$ **6.** $3a^2 - 8ab - 15b^2$ **7.** $4a^2 - 1$
8. $9a^4 - 30a^2y + 25y^2$ **9.** $\dfrac{4}{x + 2}$ **10.** $\dfrac{x - 4}{4(x + 2)}$
11. $\dfrac{(x + y)(x^2 + xy + y^2)}{x^2 + y^2}$ **12.** $x - a$ **13.** $12a^2\sqrt{b}$

14. $-27x^{10}y^{-2}$, or $-\dfrac{27x^{10}}{y^2}$ **15.** $25x^4y^{1/3}$
16. $y\sqrt[12]{x^5y^2}, y \ge 0$ **17.** $14 + 8i$
18. $(2x - 3)^2$ **19.** $(3a - 2)(9a^2 + 6a + 4)$
20. $12(s^2 + 2t)(s^2 - 2t)$ **21.** $3(y^2 + 3)(5y^2 - 4)$
22. $7x^3 + 9x^2 + 19x + 38 + \dfrac{72}{x - 2}$ **23.** 20
24. $[4, \infty)$, or $\{x | x \ge 4\}$
25. $(-\infty, 5) \cup (5, \infty)$, or $\{x | x < 5 \text{ or } x > 5\}$
26. $\dfrac{1 - 2\sqrt{x} + x}{1 - x}$ **27.** $y = 3x - 8$ **28.** $x^2 - 50 = 0$
29. $(2, -3); 6$ **30.** $\log_a \dfrac{\sqrt[3]{x^2} \cdot z^5}{\sqrt{y}}$ **31.** $a^5 = c$
32. 2.0792 **33.** 0.6826 **34.** 5 **35.** -121 **36.** 875
37. $16\left(\frac{1}{4}\right)^{n-1}$ **38.** $13{,}440a^4b^6$ **39.** $\frac{19{,}171}{64}$, or 299.546875
40. $\frac{3}{5}$ **41.** $-\frac{6}{5}, 4$ **42.** $\mathbb{R}$, or $(-\infty, \infty)$ **43.** $\left(-1, \frac{1}{2}\right)$
44. $(2, -1, 1)$ **45.** 2 **46.** $\pm 2, \pm 5$
47. $(\sqrt{5}, \sqrt{3}), (\sqrt{5}, -\sqrt{3}), (-\sqrt{5}, \sqrt{3}), (-\sqrt{5}, -\sqrt{3})$
48. 1.7925 **49.** 1005 **50.** $\frac{1}{25}$ **51.** $-\frac{1}{2}$
52. $\{x | -2 \le x \le 3\}$, or $[-2, 3]$ **53.** $\pm i\sqrt{3}$
54. $-2 \pm \sqrt{7}$ **55.** $\{y | y < -5 \text{ or } y > 2\}$, or $(-\infty, -5) \cup (2, \infty)$ **56.** $-8, 10$ **57.** 3
58. $r = \dfrac{V - P}{-Pt}$, or $\dfrac{P - V}{Pt}$ **59.** $R = \dfrac{lr}{1 - I}$

60.

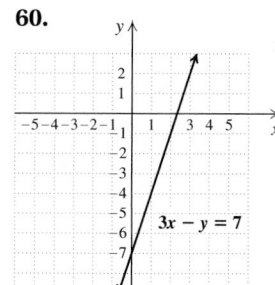

$3x - y = 7$

61.

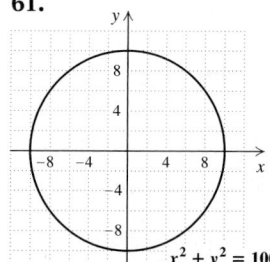

$x^2 + y^2 = 100$

62.

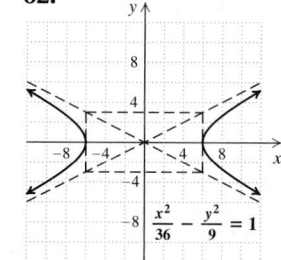

$\dfrac{x^2}{36} - \dfrac{y^2}{9} = 1$

63.

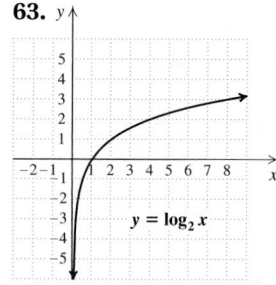

$y = \log_2 x$

64.

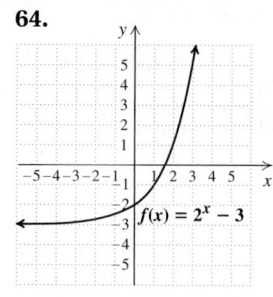

$f(x) = 2^x - 3$

65.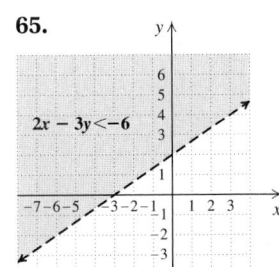

66.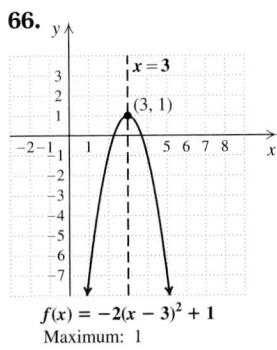

$f(x) = -2(x - 3)^2 + 1$
Maximum: 1

67. 5000 ft^2 **68.** 5 ft by 12 ft   **69.** More than 25 rentals
70. 57, 59, 61   **71.** $2.68 herb: 10 oz; $4.60 herb: 14 oz
72. 350 mph   **73.** $8\frac{2}{5}$ hr, or 8 hr 24 min   **74.** 20
75. **(a)** The loan-to-value ratio increased 1.4% per year;
(b) $f(t) = 1.4t + 77.2$, where $f(t)$ is the loan-to-value ratio,
in percent; **(c)** 91.2%; **(d)** about 2013
76. **(a)** $k \approx 0.383$; $P(t) = 160e^{0.383t}$; **(b)** 730,273 reverse
mortgages; **(c)** about 2013   **77.** $14,079.98
78. All real numbers except 0 and -12 **79.** 81
80. y gets divided by 8   **81.** 84 yr

CHAPTER R

Exercise Set R.1, pp. 945–946

1. False   **3.** True   **5.** True   **7.** 22   **9.** 1.3   **11.** -25
13. $-\frac{11}{15}$ **15.** -6.5 **17.** -9 **19.** 0   **21.** $-\frac{1}{2}$ **23.** 5.8
25. -3 **27.** 39   **29.** 175   **31.** -32 **33.** 16   **35.** -6
37. 9   **39.** -3 **41.** -16 **43.** 100   **45.** 2   **47.** -23
49. 36   **51.** 10   **53.** 10   **55.** -7 **57.** 32
59. 28 cm^2 **61.** $8x + 28$ **63.** $-30 + 6x$
65. $8a + 12b - 6c$ **67.** $-6x + 3y - 3z$
69. $2(4x + 3y)$ **71.** $3(1 + w)$ **73.** $10(x + 5y + 10)$
75. p **77.** $-m + 22$ **79.** $-5x + 7$
81. $6p - 7$ **83.** $-x + 12y$ **85.** $36a - 48b$
87. $-10x + 104y + 9$ **89.** Yes   **91.** No   **93.** Yes
95. Let n represent the number; $3n = 348$
97. Let c represent the number of calories in a Taco Bell
Beef Burrito; $c + 69 = 500$ **99.** Let l represent the
amount of water used to produce 1 lb of lettuce; $42 = 2l$

Exercise Set R.2, pp. 954–955

1. 16   **3.** $-\frac{1}{12}$ **5.** -0.8 **7.** $-\frac{5}{3}$ **9.** 42   **11.** -5
13. $\frac{5}{3}$ **15.** $-\frac{4}{9}$ **17.** -4 **19.** $\frac{69}{5}$ **21.** $\frac{9}{32}$ **23.** -2
25. -15 **27.** $\frac{43}{2}$ **29.** $-\frac{61}{115}$ **31.** $l = \dfrac{A}{w}$ **33.** $P = IV$

35. $p = 2q - r$ **37.** $\pi = \dfrac{A}{r^2 + r^2 h}$

39. **(a)** No; **(b)** yes; **(c)** no; **(d)** yes
41. $\{x \mid x \le 12\}$, or $(-\infty, 12]$

43. $\{m \mid m > 12\}$, or $(12, \infty)$

45. $\left\{x \mid x \ge -\frac{3}{2}\right\}$, or $\left[-\frac{3}{2}, \infty\right)$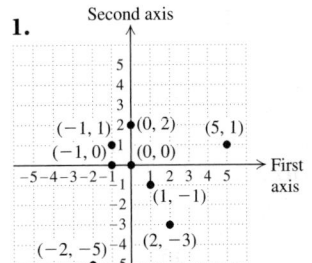

47. $\{t \mid t < -3\}$, or $(-\infty, -3)$

49. $\{y \mid y > 10\}$, or $(10, \infty)$
51. $\{a \mid a \ge 1\}$, or $[1, \infty)$
53. $\left\{x \mid x \ge \frac{64}{17}\right\}$, or $\left[\frac{64}{17}, \infty\right)$
55. $\left\{x \mid x > \frac{39}{11}\right\}$, or $\left(\frac{39}{11}, \infty\right)$
57. $\{x \mid x \le -10.875\}$, or $(-\infty, -10.875]$
59. 7   **61.** 16, 18   **63.** $166\frac{2}{3}$ pages   **65.** 4.5 cm, 9.5 cm
67. 900 cubic feet   **69.** 80¢   **71.** 30 min or more
73. For $2\frac{7}{9}$ hr or less

Exercise Set R.3, pp. 961

1.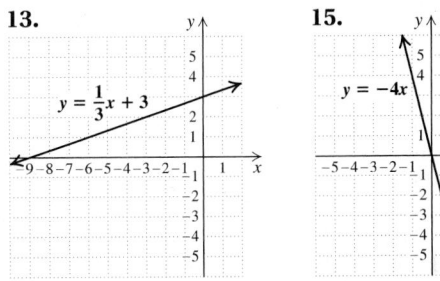

3. II   **5.** IV   **7.** I, IV
9. No   **11.** Yes

13. 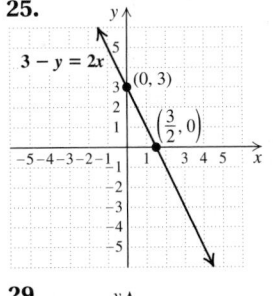   **15.**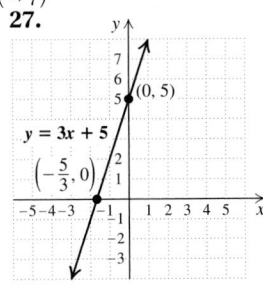

17. -1 **19.** 0   **21.** Slope: 2; y-intercept: $(0, -5)$
23. Slope: $-\frac{2}{7}$; y-intercept: $\left(0, \frac{1}{7}\right)$
25. 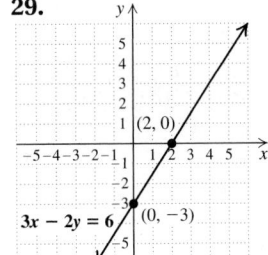   **27.**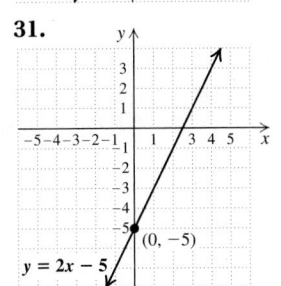

29. 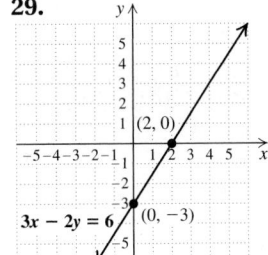   **31.**

33.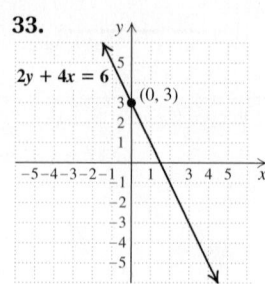

35. 0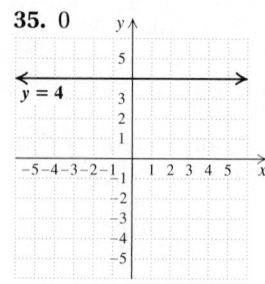

37. Undefined

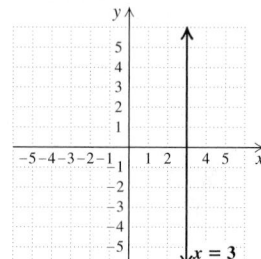

39. $y = 5x + 9$
41. $y = -x + 3$
43. Perpendicular
45. Neither

Exercise Set R.4, pp. 968–969

1. 1   **3.** -3 **5.** $\dfrac{1}{8^2} = \dfrac{1}{64}$ **7.** $\dfrac{10}{x^5}$ **9.** $\dfrac{1}{(ab)^2}$
11. y^{10} **13.** t^{-4} **15.** x^{15} **17.** a^6 **19.** $(4x)^9$
21. 7^{40} **23.** $x^8 y^{12}$ **25.** $\dfrac{y^6}{64}$ **27.** $\dfrac{9q^8}{4p^6}$
29. $8x^3, -6x^2, x, -7$ **31.** 18, 36, -7, 3; 3, 9, 1, 0; 9
33. $-1, 4, -2; 3, 3, 2; 3$ **35.** $8p^4; 8$ **37.** $x^4 + 3x^3$
39. $-7t^2 + 5t + 10$ **41.** 36   **43.** -14 **45.** 144 ft
47. About 17.6 watts   **49.** $4x^3 - 3x^2 + 8x + 7$
51. $-y^2 + 5y - 2$ **53.** $-3x^2y - y^2 + 8y$
55. $12x^5 - 28x^3 + 28x^2$ **57.** $8a^2 + 2ab + 4ay + by$
59. $x^3 + 4x^2 - 20x + 7$ **61.** $x^2 - 49$
63. $x^2 + 2xy + y^2$ **65.** $6x^4 + 17x^2 - 14$
67. $42a^2 - 17ay - 15y^2$ **69.** $-t^4 - 3t^2 + 2t - 5$
71. $5x + 3$ **73.** $2x^2 - 3x + 3 + \dfrac{-2}{x + 1}$
75. $5x + 3 + \dfrac{3}{x^2 - 1}$

Exercise Set R.5, p. 978

1. $6t^3(3t^2 - 2t + 1)$ **3.** $(y - 3)^2$ **5.** $(p + 2)(2p^3 + 1)$
7. Prime   **9.** $(2t + 3)(4t^2 - 6t + 9)$
11. $(m + 6)(m + 7)$ **13.** $(x^2 + 9)(x + 3)(x - 3)$
15. $(2x + 3)(4x + 5)$ **17.** $(x + 2)(x + 1)(x - 1)$
19. $(0.1t^2 - 0.2)(0.01t^4 + 0.02t^2 + 0.04)$
21. $\left(x^2 + \tfrac{1}{4}\right)\left(x + \tfrac{1}{2}\right)\left(x - \tfrac{1}{2}\right)$ **23.** $(n - 2)(m + 3)$
25. $(m + 15n)(m - 10n)$ **27.** $2y(3x + 1)(4x - 3)$
29. $(y - 11)^2$ **31.** $-3, 1$ **33.** $0, 11$ **35.** $-3, 3$
37. $-\tfrac{5}{2}, 7$ **39.** 0, 5   **41.** $-2, 6$ **43.** $-11, 5$
45. -5 **47.** Base: 8 ft; height: 5 ft   **49.** 8 ft, 15 ft

Exercise Set R.6, pp. 988–989

1. $-\tfrac{1}{3}$ **3.** $-2, 2$ **5.** $\dfrac{8x}{9y}$ **7.** $\dfrac{t + 2}{t + 4}$ **9.** $\dfrac{-1}{x + 2}$

11. $\dfrac{6}{x}$ **13.** $\dfrac{(a + 1)^3(a - 1)}{a^3}$ **15.** 1   **17.** $\dfrac{5x + 4}{x^2}$
19. $\dfrac{6a + 3b - 4}{3a - 3b}$ **21.** $\dfrac{-3}{5x(x + 2)}$ **23.** $\dfrac{(x + 1)^2}{x^2(x + 2)}$
25. $\dfrac{x + 3}{(x + 1)^2}$ **27.** 1   **29.** $\dfrac{2t + 1}{(t + 1)(t - 1)}$
31. $\dfrac{-3x}{2}$ **33.** $\dfrac{-14x - 7}{(x + 5)(x - 4)}$ **35.** $\dfrac{8x - 4}{x^3}$
37. $\dfrac{3(x + 1)}{(x - 7)(4x + 3)}$ **39.** $\dfrac{(x + 1)^2}{(x - 2)(x + 4)}$ **41.** $\dfrac{-1}{x}$
43. $\dfrac{6}{5}$ **45.** 1   **47.** $\dfrac{31}{2}$ **49.** $-4, 4$ **51.** $22\tfrac{2}{9}$ hr
53. Jessica: 45 km/h; Josh: 25 km/h   **55.** 50

APPENDIXES

Exercise Set A, pp. 993–994

1. Mean: 17; median: 18; mode: 13   **3.** Mean: $10.\overline{6}$;
median: 9; mode: 3, 20   **5.** Mean: 4.06; median: 4.6;
mode: none   **7.** Mean: $239.\overline{3}$; median: 234; mode: 234
9. Average: 34.875; median: 22; mode: 0   **11.** Mean: 87.25;
median: 86.5; mode: 86   **13.** Average: $218.\overline{3}$; median: 222;
mode: 202   **15.** 10 home runs   **17.** $a = 30, b = 58$

Exercise Set B, pp. 997–998

1. $\{8, 9, 10, 11\}$ **3.** $\{41, 43, 45, 47, 49\}$ **5.** $\{-3, 3\}$
7. True   **9.** True   **11.** False   **13.** True   **15.** False
17. True   **19.** $\{c, d, e\}$ **21.** $\{1, 2, 3, 6\}$ **23.** $\varnothing$
25. $\{a, e, i, o, u, q, c, k\}$ **27.** $\{1, 2, 3, 4, 6, 9, 12, 18\}$
29. $\{1, 2, 3, 4, 5, 6, 7, 8\}$ **31.** ✍   **33.** The set of integers
35. The set of real numbers   **37.** $\varnothing$ **39.** **(a)** A; **(b)** A;
(c) A; **(d)** $\varnothing$ **41.** True   **43.** True   **45.** True

Exercise Set C, pp. 1001–1002

1. True   **2.** True   **3.** True   **4.** False   **5.** True
6. False   **7.** $x^2 - 3x - 5$ **9.** $a + 5 + \dfrac{-4}{a + 3}$
11. $2x^2 - 5x + 3 + \dfrac{8}{x + 2}$ **13.** $a^2 + 2a - 6$
15. $3y^2 + 2y + 6 + \dfrac{-2}{y - 3}$
17. $x^4 + 2x^3 + 4x^2 + 8x + 16$
19. $3x^2 + 6x - 3 + \dfrac{2}{x + \tfrac{1}{3}}$ **21.** 6   **23.** 125   **25.** 0
27. ✍   **29.** ✍
31. **(a)** The degree of R must be less than 1, the degree of
$x - r$;   **(b)** Let $x = r$. Then
$$\begin{aligned} P(r) &= (r - r) \cdot Q(r) + R \\ &= 0 \cdot Q(r) + R \\ &= R. \end{aligned}$$
33. $0; -3, -\tfrac{5}{2}, \tfrac{3}{2}$ **35.** 〰   **37.** 0

Glossary

A

Absolute value [1.4]   The distance that a number is from 0 on the number line.

Additive inverse [1.6]   A number's opposite. Two numbers are additive inverses of each other if their sum is zero.

Algebraic expression [1.1]   A collection of numerals and/or variables on which the operations $+$, $-$, $\cdot$, $\div$, $(\;)^n$, or $\sqrt[n]{\;(\;)}$ are performed.

Arithmetic sequence [14.2]   A sequence in which the difference between any two successive terms is constant.

Arithmetic series [14.2]   A series for which the associated sequence is arithmetic.

Ascending order   A polynomial written with the terms arranged according to degree of one variable, from least to greatest.

Associative law of addition [1.2]   The statement that when three numbers are added, regrouping the addends gives the same sum.

Associative law of multiplication [1.2]   The statement that when three numbers are multiplied, regrouping the factors gives the same product.

Asymptote [12.2], [13.3]   A line that a graph approaches more and more closely as x increases or as x decreases.

Average [2.7]   Most commonly, the mean of a set of numbers found by adding the numbers and dividing by the number of addends.

Axes (singular, axis) [3.1]   Two perpendicular number lines used to identify points in a plane.

Axis of symmetry [11.6]   A line that can be drawn through a graph such that the part of the graph on one side of the line is an exact reflection of the part on the opposite side.

B

Bar graph [3.1]   A graphic display of data using bars proportional in length to the numbers represented.

Base [1.8]   In exponential notation, the number or expression being raised to a power. In expressions of the form a^n, a is the base.

Binomial [4.2]   A polynomial composed of two terms.

Branches [13.3]   The two curves that comprise a hyperbola.

Break-even point [8.8]   In business, the point of intersection of the revenue function and the cost function.

C

Circle [13.1]   A set of points in a plane that are a fixed distance r, called the radius, from a fixed point (h, k), called the center.

Circle graph [3.1]   A graphic display of data using sectors of a circle to represent percents.

Circumference [2.3]   The distance around a circle.

Closed interval $[a, b]$ [2.6]   The set of all numbers x for which $a \le x \le b$. Thus, $[a, b] = \{x | a \le x \le b\}$.

Coefficient [2.1]   The numerical multiplier of a variable or variables.

Combined variation [7.5]   A mathematical relationship in which a variable varies directly and/or inversely, at the same time, with more than one other variable.

Common logarithm [12.5]   A logarithm with base 10.

Commutative law of addition [1.2]   The statement that when two numbers are added, changing the order in which the numbers are added does not affect the sum.

Commutative law of multiplication [1.2]   The statement that when two numbers are multiplied, changing the order in which the numbers are multiplied does not affect the product.

Complementary angles [2.5]   A pair of angles, the sum of whose measures is 90°.

Completing the square [11.1]   The procedure in which one adds a particular constant to an expression so that the resulting sum is a perfect square.

Complex number [10.8]   Any number that can be written as $a + bi$, where a and b are real numbers and $i = \sqrt{-1}$.

Complex rational expression [6.5]   A rational expression that has one or more rational expressions within its numerator and/or denominator.

Complex-number system [10.8]   A number system that contains the real-number system and is designed so that negative numbers have defined square roots.

Composite function [12.1]   A function in which a quantity depends on a variable that, in turn, depends on another variable.

Composite number [1.3]   A natural number, other than 1, that is not prime.

Compound inequality [9.2]   A statement in which two or more inequalities are combined using the word *and* or the word *or.*

Compound interest [11.1]   Interest computed on the sum of an original principal and the interest previously accrued by that principal.

Conditional equation [2.2]   An equation that is true for some replacements of a variable(s) and false for others.

Conic section [13.1]   A curve formed by the intersection of a plane and a cone.

Conjugates [10.5], [10.8]   Pairs of radical expressions, like $a\sqrt{b} + c\sqrt{d}$ and $a\sqrt{b} - c\sqrt{d}$, for which the product does not have a radical term or pairs of imaginary numbers, like $a + bi$ and $a - bi$, for which the product is real.

Conjunction [9.2]   A sentence in which two statements are joined by the word *and*.

Consecutive numbers [2.5]   Integers that are one unit apart.

Consistent system of equations [8.1], [8.4]   A system of equations that has at least one solution.

Constant [1.1]   A known number.

Constant function [7.3]   A function given by an equation of the form $f(x) = b$, where b is a real number.

Constant of proportionality [7.5]   The constant in an equation of variation.

Constraint [9.5]   A requirement imposed on a problem.

Contradiction [2.2]   An equation that is never true.

Coordinates [3.1]   The numbers in an ordered pair.

Cube root [10.1]   The number c is called the cube root of a if $c^3 = a$.

D

Data point [11.8]   A given ordered pair of a function, usually found experimentally.

Degree of a polynomial [4.3]   The degree of the term of highest degree in a polynomial.

Degree of a term [4.3]   The number of variable factors in a term.

Demand function [8.8]   A function modeling the relationship between the price of a good and the quantity of that good demanded.

Denominator [1.3], [6.1]   The number below the fraction bar in a fraction or the expression below the fraction bar in a rational expression.

Dependent equations [8.1]   Equations in a system from which one equation can be removed without changing the solution set.

Descending order [4.3]   A polynomial written with the terms arranged according to degree of one variable, from greatest to least.

Determinant [8.7]   A descriptor of a square matrix. The determinant of a two-by-two matrix $\begin{bmatrix} a & c \\ b & d \end{bmatrix}$ is denoted $\begin{vmatrix} a & c \\ b & d \end{vmatrix}$ and represents $ad - bc$.

Difference of squares [4.6], [5.4]   An expression that can be written in the form $A^2 - B^2$.

Direct variation [7.5]   A situation that translates to an equation of the form $y = kx$, with k a nonzero constant.

Discriminant [11.3]   The expression $b^2 - 4ac$ from the quadratic formula.

Disjunction [9.2]   A sentence in which two statements are joined by the word *or*.

Distributive law [1.2]   The statement that multiplying a factor by the sum of two addends gives the same result as multiplying the factor by each of the two addends and then adding.

Domain [7.2]   The set of all first coordinates of the ordered pairs in a function.

Doubling time [12.7]   The time necessary for a population to double in size.

E

e [12.5]   An irrational number that is approximately 2.7182818284, which is used in many applications.

Element [8.6]   An entry in a matrix.

Elimination method [8.2]   An algebraic method that uses the addition principle to solve a system of equations.

Ellipse [13.2]   The set of all points in a plane for which the sum of the distances from two fixed points F_1 and F_2, called foci, is constant.

Equation [1.1]   A number sentence with the verb $=$.

Equation of variation [7.5]   An equation used to represent direct, inverse, or combined variation.

Equilibrium point [8.8]   The point of intersection between the demand function and the supply function.

Equivalent equations [2.1]   Equations with the same solutions.

Equivalent expressions [1.2]   Expressions that have the same value for all allowable replacements.

Equivalent inequalities [2.6]   Inequalities that have the same solution set.

Evaluate [1.1]   To substitute a value for each occurrence of a variable in an expression and calculate the result.

Exponent [1.8]   The power to which a base is raised. In expressions of the form a^n, the number n is an exponent. For n a natural number, a^n represents n factors of a.

Exponential decay [12.7]   A decrease in quantity over time that can be modeled by an exponential equation of the form $P(t) = P_0 e^{-kt}$, $k > 0$.

Exponential equation [12.6]   An equation in which a variable appears in an exponent.

Exponential function [12.2]   A function that can be described by an exponential equation.

Exponential growth [12.7]   An increase in quantity over time that can be modeled by an exponential function of the form $P(t) = P_0 e^{kt}$, $k > 0$.

Exponential notation [1.8]   A representation of a number using a base raised to a power.

Extrapolation [3.7]   The process of predicting a future value on the basis of given data.

F

Factor [1.2]   *Verb*: to write an equivalent expression that is a product. *Noun*: a multiplier.

Finite sequence [14.1]   A function having for its domain a set of natural numbers: $\{1, 2, 3, 4, 5, \ldots, n\}$, for some natural number n.

Fixed costs [8.8]   In business, costs that are incurred regardless of how many items are produced.

Focus (plural, foci) [13.2]   One of two fixed points that determine the points of an ellipse or a hyperbola.

FOIL [4.5]   An acronym for a procedure for multiplying two binomials by multiplying the First terms, the Outer terms, the Inner terms, and the Last terms, and then adding the results.

Formula [2.3]   An equation that uses numbers and/or letters to represent a relationship between two or more quantities.

Fraction notation [1.3]   A number written using a numerator and a denominator.

Function [7.1]   A correspondence that assigns to each member of a set called the domain exactly one member of a set called the range.

G

General term of a sequence [14.1]   The nth term, denoted a_n.

Geometric sequence [14.3]   A sequence in which the ratio of every pair of successive terms is constant.

Geometric series [14.3]   A series for which the associated sequence is geometric.

Grade [3.5]   The ratio of the vertical distance a road rises over the horizontal distance it runs, expressed as a percent.

Graph [3.1]   A picture or diagram of the data in a table, or a line, a curve, a plane, or collection of points, etc., that represents all the solutions of an equation or an inequality.

H

Half-life [12.7]   The amount of time necessary for half of a quantity to decay.

Half-open interval [2.6]   An interval that includes exactly one endpoint.

Horizontal-line test [12.1]   The statement that if it is impossible to draw a horizontal line that intersects the graph of a function more than once, then that function is one-to-one.

Hyperbola [13.3]   The set of all points P in the plane such that the difference of the distance from P to two fixed points is constant.

Hypotenuse [5.8]   In a right triangle, the side opposite the right angle.

I

i [10.8]   The square root of -1. That is, $i = \sqrt{-1}$ and $i^2 = -1$.

Identity [2.2]   An equation that is always true.

Identity property of 0 [1.5]   The statement that the sum of a number and 0 is always the original number.

Identity property of 1 [1.3]   The statement that the product of a number and 1 is always the original number.

Imaginary number [10.8]   A number that can be written in the form $a + bi$, where a and b are real numbers and $b \neq 0$ and $i = \sqrt{-1}$.

Inconsistent system of equations [8.1]   A system of equations for which there is no solution.

Independent equations [8.1]   Equations that are not dependent.

Index (plural, indices) [10.1]   In the radical $\sqrt[n]{a}$, the number n is called the index.

Inequality [1.4]   A mathematical sentence using $<, >, \leq, \geq,$ or $\neq$.

Infinite geometric series [14.3]   The sum of the terms of an infinite geometric sequence.

Infinite sequence [14.1]   A function having for its domain the set of natural numbers: $\{1, 2, 3, 4, 5, \ldots\}$.

Input [7.1]   A member of the domain of a function.

Integers [1.4]   The whole numbers and their opposites.

Interpolation [3.7]   The process of estimating a value between given values.

Intersection of two sets [9.2]   The set of all elements that are common to both sets.

Interval notation [2.6]   The use of a pair of numbers inside parentheses and/or brackets to represent the set of all numbers between those two numbers. *See also* Closed, Open, and Half-open intervals.

Inverse relation [12.1]   The relation formed by interchanging the members of the domain and the range of a relation.

Inverse variation [7.5]   A situation that translates to an equation of the form $y = k/x$, with k a nonzero constant.

Irrational number [1.4]   A real number that cannot be named as a ratio of two integers. In decimal notation, irrational numbers neither terminate nor repeat.

Isosceles right triangle [10.7]   A right triangle in which both legs have the same length.

J

Joint variation [7.5]   A situation that translates to an equation of the form $y = kxz$, with k a nonzero constant.

L

Largest common factor [5.1]   The common factor of the terms of a polynomial with the largest possible coefficient and the largest possible exponent(s).

Leading coefficient [4.3]   The coefficient of the term of highest degree in a polynomial.

Leading term [4.3]   The term of highest degree in a polynomial.

Least common denominator (LCD) [6.3]   The least common multiple of the denominators of two or more rational expressions.

Least common multiple (LCM) [6.3]   The multiple of all expressions under consideration that has the smallest positive coefficient and the least possible degree.

Legs [5.8]   In a right triangle, the two sides that form the right angle.

Like radicals [10.5]   Radical expressions that have identical indices and radicands.

Like terms [1.5], [4.3]   Terms that have exactly the same variable factors.

Line graph [3.1]   A graph in which quantities are represented as points connected by straight-line segments.

Linear equation [3.2], [8.4]   In two variables, any equation that can be written in the form $y = mx + b$, or $Ax + By = C$, where x and y are variables and m, b, A, B, and C are constants and A and B are not both 0. In three variables, an equation that is equivalent to one of the form $Ax + By + Cz = D$, where x, y, and z are variables, and A, B, and C are constants that are not all 0.

Linear function [7.3]   A function that can be described by an equation of the form $f(x) = mx + b$, where m and b are constants.

Linear inequality [9.4]   An inequality whose related equation is a linear equation.

Linear programming [9.5]   A branch of mathematics involving graphs of inequalities and their constraints.

Logarithmic equation [12.6]   An equation containing a logarithmic expression.

Logarithmic function, base *a* [12.3]   The inverse of an exponential function with base a.

M

Matrix (plural, matrices) [8.6]   A rectangular array of numbers.

Maximum value [11.6]   The greatest function value (output) achieved by a function.

Mean [2.7]   The sum of a set of numbers divided by the number of addends.

Minimum value [11.6]   The least function value (output) achieved by a function.

Monomial [4.3]   A constant, a variable, or a product of a constant and one or more variables.

Motion problem [6.7]   A problem that deals with distance, speed, and time.

Multiplicative inverses [1.3]   Reciprocals; two numbers whose product is 1.

Multiplicative property of zero [1.7]   The statement that the product of 0 and any real number is 0.

N

Natural logarithm [12.5]   A logarithm with base e.

Natural numbers [1.3]   The counting numbers: 1, 2, 3, 4, 5, ...

Nonlinear function [7.3]   A function whose graph is not a straight line.

Numerator [1.3], [6.1]   The number above the fraction bar in a fraction or the expression above the fraction bar in a rational expression.

O

Objective function [9.5]   In linear programming, the function in which the expression being maximized or minimized appears.

One-to-one function [12.1]   A function for which all different inputs have different outputs.

Open interval (*a*, *b*) [2.6]   The set of all numbers x for which $a < x < b$. Thus, $(a, b) = \{x | a < x < b\}$.

Opposite [1.6]   The additive inverse of a number. Opposites are the same distance from 0 on the number line but on different sides of 0.

Ordered pair [3.1]   A pair of numbers of the form (h, k) for which the order in which the numbers are listed is important.

Origin [3.1]   The point on a graph in a coordinate plane where the two axes intersect.

Output [7.1]   A member of the range of a function.

P

Parabola [11.6], [13.1]   A graph of a second degree polynomial equation in one variable.

Parallel lines [3.6]   Lines that extend indefinitely without intersecting.

Pascal's triangle [14.4]   A triangular array of coefficients of the expansion $(a + b)^n$ for $n = 0, 1, 2, ...$

Perfect square [10.1]   A rational number for which there exists a number a for which $a^2 = p$.

Perfect-square trinomial [4.6], [5.4]   A trinomial that is the square of a binomial.

Perpendicular lines [3.6]   Lines that intersect at a right angle.

Piecewise function [7.2]   A function that is defined by different equations for various parts of its domain.

Point–slope equation [3.7]   An equation of the type $y - y_1 = m(x - x_1)$, where x and y are variables, and m is the slope of the line and (x_1, y_1) is a point on the line.

Polynomial [4.2]   A monomial or a sum of monomials.

Price [8.8]   The amount a purchaser pays for an item.

Polynomial equation [5.7]   An equation in which two polynomials are set equal to each other.

Polynomial inequality [11.9]   An inequality that is equivalent to an inequality with a polynomial as one side and 0 as the other.

Prime factorization [1.3]   The factorization of a natural number into a product of its prime factors.

Prime number [1.3]   A natural number that has exactly two different natural number factors: the number itself and 1.

Principal square root [10.1]   The nonnegative square root of a number.

Proportion [6.7]   An equation stating that two ratios are equal.

Pure imaginary number [10.8]   A complex number of the form $a + bi$, with $a = 0$ and $b \neq 0$.

Pythagorean theorem [5.8]   The theorem that states that in any right triangle, if a and b are the lengths of the legs and c is the length of the hypotenuse, then $a^2 + b^2 = c^2$.

Q

Quadrants [3.1]   The four regions into which the axes divide a plane.

Quadratic equation [5.7]   An equation equivalent to one of the form $ax^2 + bx + c = 0$, where $a \neq 0$.

Quadratic formula [11.2] $x = \dfrac{-b \pm \sqrt{b^2 - 4ac}}{2a}$, which gives the solutions of $ax^2 + bx + c = 0$, where $a \neq 0$.

Quadratic function [11.1]   A second-degree polynomial function in one variable.

Quadratic inequality [11.9]   A second-degree polynomial inequality in one variable.

R

Radical equation [10.6]   An equation in which a variable appears in a radicand.

Radical expression [10.1]   An algebraic expression in which a radical sign appears.

Radical sign [10.1]   The symbol $\sqrt{\ }$ or $\sqrt[n]{\ }$, where $n > 2$.

Radical term [10.5]   A term in which a radical sign appears.

Radicand [10.1]   The expression under a radical sign.

Radius (plural, radii) [13.1]   The distance from the center of a circle to a point on the circle. Also, a segment connecting the center to a point on the circle.

Range [7.2]   The set of all second coordinates of the ordered pairs in a function.

Rate [3.4]   A ratio that indicates how two quantities change with respect to each other.

Ratio [6.7]   The quotient of two quantities. The ratio of a to b is a/b, also written $a:b$.

Rational equation [6.6]   An equation containing one or more rational expressions.

Rational expression [6.1]   A quotient of two polynomials.

Rational inequality [11.9]   An inequality containing a rational expression.

Rational number [1.4]   A number that can be written in the form $\dfrac{a}{b}$, where a and b are integers and $b \neq 0$. In decimal notation, a rational number repeats or terminates.

Rationalizing the denominator [10.4], [10.5]   A procedure for finding an equivalent expression without a radical in its denominator.

Rationalizing the numerator [10.4], [10.5]   A procedure for finding an equivalent expression without a radical in its numerator.

Real number [1.4]   Any number that is either rational or irrational.

Reciprocal [1.3]   A multiplicative inverse. Two numbers are reciprocals if their product is 1.

Reflection [11.6]   The mirror image of a graph.

Relation [7.1]   A correspondence between two sets, called the domain and the range, such that each member of the domain corresponds to at least one member of the range.

Repeating decimal [1.4]   A decimal in which a block of digits repeats indefinitely.

Right triangle [5.8]   A triangle that includes a right angle.

Row-equivalent operations [8.6]   Operations used to produce equivalent systems of equations.

S

Scientific notation [4.2]   A number written in the form $N \times 10^m$, where m is an integer, $1 \leq N < 10$, and N is expressed in decimal notation.

Sequence [14.1]   A function for which the domain is a set of consecutive natural numbers beginning with 1.

Series [14.1]   The sum of specified terms in a sequence.

Set [1.4]   A collection of objects.

Set-builder notation [2.6]   The naming of a set by describing basic characteristics of the elements in the set.

Sigma notation [14.1]   The naming of a sum using the Greek letter Σ (sigma) as part of an abbreviated form.

Similar triangles [10.7]   Triangles in which corresponding sides are proportional and corresponding angles have the same measure.

Simplify   To rewrite an expression in an equivalent, abbreviated, form.

Slope [3.5]   The ratio of the rise to the run for any two points on a line.

Slope–intercept equation [3.6]   An equation of the form $y = mx + b$, where x and y are variables, m is the slope of the line, and $(0, b)$ is its y-intercept.

Solution [1.1], [2.1], [2.6], [8.1]   A replacement or substitution that makes an equation, an inequality, or a system of equations or inequalities true.

Solution set [2.1], [2.6], [8.1]   The set of all solutions of an equation, an inequality, or a system of equations or inequalities.

Solve [2.1], [2.6], [8.1]   To find all solutions of an equation, an inequality, or a system of equations or inequalities; to find the solution(s) of a problem.

Speed [6.7]   The ratio of distance traveled to the time required to travel that distance.

Square matrix [8.7]   A matrix with the same number of rows and columns.

Square root [10.1]   The number c is a square root of a if $c^2 = a$.

Substitute [1.1]   To replace a variable with a number or an expression that represents a number.

Substitution method [8.2]   An algebraic method for solving systems of equations.

Supplementary angles [2.5]   A pair of angles, the sum of whose measure is 180°.

Supply function [8.8]   A function modeling the relationship between the price of a good and the quantity of that good supplied.

System of equations [8.1]   A set of two or more equations that are to be solved simultaneously.

T

Term [1.2], [4.3]   A number, a variable, or a product or a quotient of numbers and/or variables.

Terminating decimal [1.4]   A number in decimal notation that can be written using a finite number of decimal places.

Total cost [8.8]   The amount spent to produce a product.

Total profit [8.8]   The amount taken in less the amount spent, or total revenue minus total cost.

Total revenue [8.8]   The amount taken in from the sale of a product.

Trinomial [4.3]   A polynomial that is composed of three terms.

U

Union of *A* and *B* [9.2]   The set of all elements belonging to either *A* or *B* or both.

Undefined [1.7]   An expression that has no meaning attached to it.

V

Value [1.1]   The numerical result after a number has been substituted into an expression.

Variable [1.1]   A letter that represents an unknown number.

Variable costs [8.8]   In business, costs that vary according to the amount produced.

Variable expression [1.1]   An expression containing a variable.

Vertex (plural, vertices) [11.6], [13.1], [13.2], [13.3]   The point at which the graph of a parabola, an ellipse, or a hyperbola crosses its axis of symmetry.

Vertical-line test [7.1]   The statement that a graph represents a function if it is impossible to draw a vertical line that intersects the graph more than once.

W

Whole numbers [1.3]   The natural numbers and 0: 0, 1, 2, 3, . . .

X

***x*-intercept** [3.3]   A point at which a graph crosses the x-axis.

Y

***y*-intercept** [3.3]   A point at which a graph crosses the y-axis.

Z

Zeros [11.9]   The x-values for which $f(x)$ is 0, for any function f.

Index

Index of Applications

Geometric Formulas

Plane Geometry

Rectangle
Area: $A = lw$
Perimeter: $P = 2l + 2w$
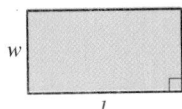

Square
Area: $A = s^2$
Perimeter: $P = 4s$
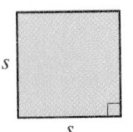

Triangle
Area: $A = \frac{1}{2}bh$
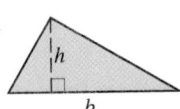

Triangle
Sum of Angle Measures:
$A + B + C = 180°$
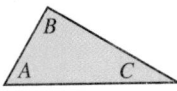

Right Triangle
Pythagorean Theorem
(Equation):
$a^2 + b^2 = c^2$
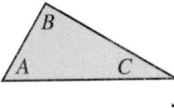

Parallelogram
Area: $A = bh$
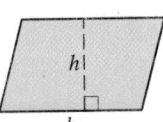

Trapezoid
Area: $A = \frac{1}{2}h(b_1 + b_2)$
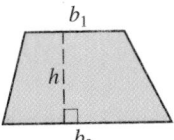

Circle
Area: $A = \pi r^2$
Circumference:
$C = \pi d = 2\pi r$
$\left(\frac{22}{7}\right.$ and 3.14 are different
approximations for $\pi\left.\right)$
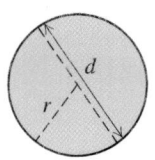

Solid Geometry

Rectangular Solid
Volume: $V = lwh$
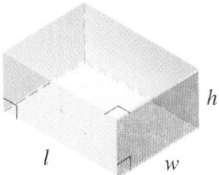

Cube
Volume: $V = s^3$
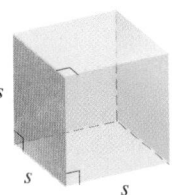

Right Circular Cylinder
Volume: $V = \pi r^2 h$
Total Surface Area:
$S = 2\pi rh + 2\pi r^2$
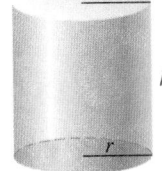

Right Circular Cone
Volume: $V = \frac{1}{3}\pi r^2 h$
Total Surface Area:
$S = \pi r^2 + \pi rs$
Slant Height:
$s = \sqrt{r^2 + h^2}$
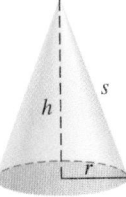

Sphere
Volume: $V = \frac{4}{3}\pi r^3$
Surface Area: $S = 4\pi r^2$
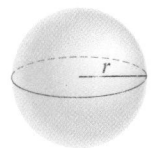

Selected Keys of the
Scientific Calculator

This secondary function takes the square root of number displayed.

Squares number displayed.

Activates secondary functions printed above certain keys. Also denoted INV or 2nd.

Used when entering numbers in scientific notation. Also denoted EXP.

Finds reciprocal of number displayed.

Used to raise any base to a power. Also denoted y^x, a^x, or ⌃.

Stores number displayed in memory. Also denoted MIN or M.

Recalls number stored in memory. Also denoted MR.

This secondary function raises 10 to any power entered.

Clears all preceding numbers and operations. Also used to turn calculator on.

Used as an approximation for pi.

Used to perform indicated operation.

Used to control order in which certain operations are performed.

Clears last number displayed but not preceding operations.

Used when entering decimal notation.

Used to change sign of number displayed.

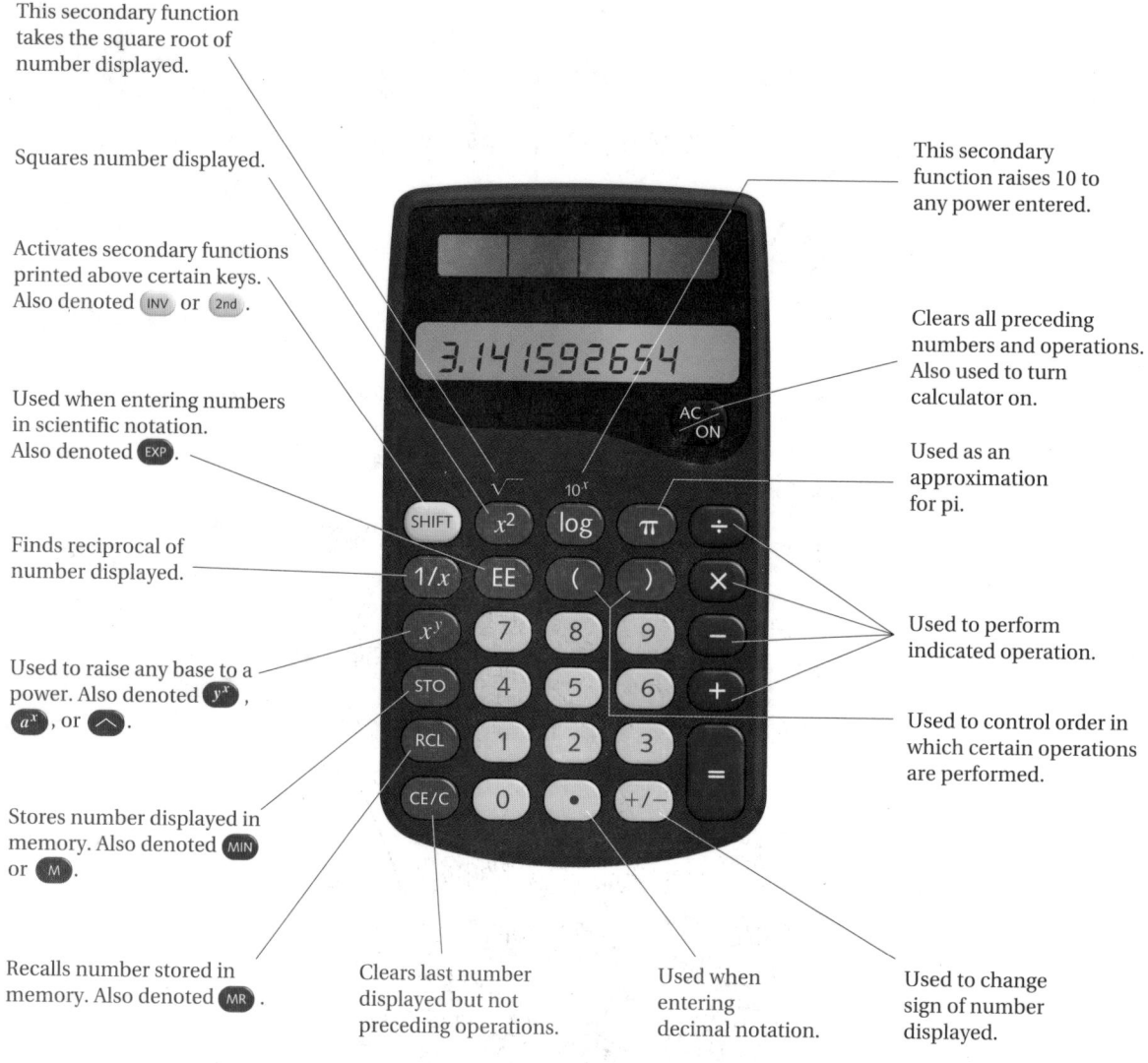